WATER CHLORINATION
Chemistry,
Environmental Impact
and Health Effects
Volume 5

WATER CHLORINATION
Chemistry, Environmental Impact and Health Effects
Volume 5

Edited by

Robert L. Jolley
Richard J. Bull
William P. Davis
Sidney Katz
Morris H. Roberts, Jr.
Vivian A. Jacobs

Proceedings of the Fifth Conference on
Water Chlorination: Environmental Impact and Health Effects
Williamsburg, Virginia
June 3–8, 1984

Sponsored by
Electric Power Research Institute, National Cancer Institute,
Oak Ridge National Laboratory, Tennessee Valley Authority,
U.S. Department of Energy, U.S. Environmental Protection Agency,
Virginia Institute of Marine Science – School of Marine Science, and
College of William and Mary

Library of Congress Cataloging-in-Publication Data
Main entry under title:

Water chlorination. Volume 5.

   Papers presented at the fifth Conference on Water
Chlorination: Environmental Impact and Health Effects,
held at the College of William and Mary, Williamsburg,
Virginia, June 3–8, 1984.
   Includes bibliographies and index.
   1. Water—Purification—Chlorination—Environmental
aspects. 2. Water—Purification—Chlorination—Hygienic
aspects. 3. Water chemistry. I. Jolley, Robert L.
II. Conference on Water Chlorination: Environmental
Impact and Health Effects (5th : 1984 : College of
William and Mary)

TD462.W38   1985   363.6′1   85-18122
ISBN 0-87371-005-3

COPYRIGHT © 1985 by LEWIS PUBLISHERS, INC.
ALL RIGHTS RESERVED

Neither this book nor any part may be reproduced or transmitted in
any form or by any means, electronic or mechanical, including
photocopying, microfilming, and recording, or by any information
storage and retrieval system, without permission in writing from the
publisher.

LEWIS PUBLISHERS, INC.
121 South Main Street, P.O. Drawer 519, Chelsea, Michigan 48118

PRINTED IN THE UNITED STATES OF AMERICA

PREFACE

"No man is an island . . ."

John Donne, 1573-1631

Our mystical relationship with water is born of practical necessity. For the human species, both individually and collectively, the quality of life parallels the availability and quality of water. Thus, high-quality water continues to be of international concern, and the effects of chlorine and other oxidants as used in water treatment continue to be the subject of active research and study. Much of the research and development that occurred within the last several years was documented at the "Fifth Conference on Water Chlorination: Environmental Impact and Health Effects," held at the College of William and Mary, Williamsburg, Virginia, June 3-8, 1984. This volume represents the permanent record of that conference.

The technical presentations and associated discussions at the Williamsburg conference were enhanced by the venerable university and the historical surroundings. However, the excellent technical quality and success of the conference are a tribute to the enthusiastic participants, both attendees and presenters. We are most grateful for their contributions and are pleased to have been associated with them in this continuing saga.

The Williamsburg conference would not have been possible without the sponsorship of the Electric Power Research Institute, National Cancer Institute, Oak Ridge National Laboratory, Tennessee Valley Authority, U.S. Department of Energy, U.S. Environmental Protection Agency, School of Marine Science of the Virginia Institute of Marine Science, and The College of William and Mary.

We wish to thank the Conference Committee for their guidance, for chairing the sessions, and for efficiently organizing the peer review process. Committee members were Nathaniel Barr, U.S. Department of Energy; Roger Bean, Battelle Pacific Northwest Laboratories; Herbert Brass, Joseph Cotruvo, John Couch, and Lee McCabe, U.S. Environmental Protection Agency; William Cooper, Florida International University; Robert Cumming, Oak Ridge National Laboratory; James Fava, EA Engineering Science and Technology, Inc.; George Helz, University of Maryland; Joseph Hunter, Rutgers University; J. Donald Johnson, University of North Carolina; Herman Kraybill, National Cancer Institute; John Lehr, U.S. Nuclear Regulatory Commission; Jack Mattice, Electric Power Research Institute; Roger Minear, University of Illinois; Vincent Olivieri, The Johns Hopkins University; Rip Rice, Rip G.

Rice, Inc.; Jim Ruane, Tennessee Valley Authority; John Veenstra, Oklahoma State University; and Rodney Zika, University of Miami.

Details and logistics of conference management were capably handled by the staff of the Oak Ridge National Laboratory, which included Bonnie Reesor, Conference Coordinator; Martha Dawson, Conference Secretary; Debbie Brown, Housing Arrangements; and Janice Shannon, Registration Assistant. Members of the Conference Publications Office, Raleigh H. Powell, Jr., and George C. Battle, Jr., were instrumental in editing these proceedings; in addition, the cooperation of the Technical Publication Department is most appreciated.

The assistance of Dean Kenneth E. Smith, Betty Kelly, and Mary McDevitt of The College of William and Mary enhanced the success of the conference.

Several students from the Johns Hopkins University and the Virginia Institute of Marine Science assisted with the audiovisual equipment.

Working with Ed and Jon Lewis of Lewis Publishers, Inc., has been most enjoyable, and we value their assistance in publishing this volume.

Each paper in this proceedings was critically reviewed by at least two peers. We wish to express appreciation to the many reviewers who, although they must remain anonymous, contributed so much of their time and effort to this arduous task. This extensive review to ensure high quality and logical validity resulted in improvement of most of the papers. However, the ultimate value of each paper must stand on its own merit. Because of the peer review process, not all presented papers were accepted for publication. Although many of the papers in this proceedings describe research funded or partially funded by U.S. federal agencies, the individual papers do not necessarily reflect the views of these agencies, and no endorsement should be inferred.

To all who contributed by giving so generously of their time, we voice a sincere thank you.

> Robert L. Jolley
> Richard J. Bull
> William P. Davis
> Sidney Katz
> Morris H. Roberts, Jr.
> Vivian A. Jacobs

Robert L. Jolley is the Manager of the Water Quality Program for the Engineering Development Section, Chemical Technology Division, of the Oak Ridge National Laboratory. Dr. Jolley did his undergraduate studies at Friends University, Wichita, Kansas, and his graduate studies at the University of Chicago and the University of Tennessee. His doctoral thesis was the landmark pioneering research on formation of chloroorganics in chlorinated wastewaters. His current research areas include environmental control technology, chemical characterization and toxicology of wastewater effluents, and the chemistry and environmental effects of low-level radioactive waste.

Dr. Jolley has organized and chaired each of the five conferences on Water Chlorination: Environmental Impact and Health Effects, and has been the principal editor for each proceedings. He has been past chairman and secretary of the Environmental Chemistry Division of the American Chemical Society, and he is currently assuming the position of Program Chairman for that ACS division. He received the 1979 Distinguished Alumni Award from Friends University for outstanding contributions to the field of science.

Richard J. Bull obtained his Bachelor of Science degree in Pharmacy from the University of Washington in Seattle in 1964. At that time, he joined the U.S. Public Health Service and entered the area of environmental research. He obtained his PhD at the University of California, San Francisco Medical Center, where he majored in pharmacology. Dr. Bull was detailed to the Environmental Protection Agency in 1970, where he held positions of Research Pharmacologist, Chief of the Toxicological Assessment Branch, and Director of the Toxicology and Microbiology Division. In 1985, he accepted a position in the College of Pharmacy at Washington State University. Dr. Bull is very active in investigating the toxicological properties of drinking water disinfectants and their by-products, and this interest continues into his academic career.

William P. Davis is currently a Research Aquatic Ecologist at the Gulf Breeze Environmental Research Laboratory, U.S. Environmental Protection Agency, Pensacola Beach, Florida. He was previously Laboratory Chief of the Bears Bluff Field Station, Wadmalaw Island, South Carolina. He received a PhD in Biological Oceanography, an MS degree in Marine Biological Sciences from the Rosenstiel School of Marine and Atmospheric Sciences, University of Miami, and a BS degree from Cornell University. His professional employment has included directing contract research projects in design and operation of marine fish culture systems, as well as management of a marine specimen sorting/identification center under the auspices of the Smithsonian Institution in Tunisia, prior to joining EPA. Dr. Davis' research has included systematics and ecology of fishes, reproductive and developmental effects of toxicants on fishes, the ecological effects and assessment methodologies of oil and hazardous substance spills in aquatic ecosystems, and the effects of pesticides and biocides on adjacent wetland and marsh habitats. He is currently evaluating developmental and teratogenic expressions among offspring of fishes exposed to known and suspected toxic agents.

Sidney Katz is a consultant with the Oak Ridge National Laboratory, having retired from ORNL in 1981. Dr. Katz received his BA degree from Kalamazoo College, Kalamazoo, Michigan, and a PhD in Analytical Chemistry from Michigan State University. His researches have been in the field of inorganic chemistry, particularly with radioactive species and with gas-solid and gas-gas reactions, in analytical instrumentation for continuous control of productions processes, and in the application of high-resolution liquid chromatography to the analysis of body fluids and water. Dr. Katz has helped organize each of the water chlorination conferences.

Morris H. Roberts, Jr., is a Senior Marine Scientist in the Department of Estuarine and Coastal Ecology, Virginia Institute of Marine Science, and an Associate Professor, School of Marine Science, College of William and Mary, Gloucester Point, Virginia. Dr. Roberts received both an MA and a PhD degree in Marine Science from the College of William and Mary and a BA degree in Zoology from Kenyon College, Gambier, Ohio. Since completing doctoral studies on decapod larval development, he has principally studied the effects of toxicants on invertebrates and fishes. Principal research interests include effects of chlorine and chlorinated sewage on invertebrate and fish species; bioconcentration of pesticides, polycyclic aromatic hydrocarbons, other organic compounds, and heavy metals; and culture of invertebrates.

Dr. Roberts has published over 50 technical papers describing his research. He is a member of the World Mariculture Society, National Shellfish Association, Society of Environmental Toxicology and Chemistry, American Society of Zoology, American Society of Limnology and Oceanography, and American Society for Testing and Materials.

Vivian A. Jacobs is coordinator of the Conference Publications Office in the Information Processing Section of the Information Resources Organization at Oak Ridge National Laboratory. She has extensive experience as a technical editor and has assisted in editing the last three volumes in this series of symposia proceedings, *Water Chlorination: Environmental Impact and Health Effects*. Several of the publications she has worked on have won local and international awards. Ms. Jacobs received a BS degree from the University of Illinois. She is currently serving as secretary for the Society for Technical Communication, the world's largest professional organization devoted to technical communication.

This volume is dedicated to
J. Carrell Morris,
whose rigorous experimental studies
elucidated the aqueous chemistry of chlorine;
and Herman F. Kraybill,
who early focused attention on
the health effects of environmental pollutants in water.

CONTENTS

SECTION I
WATER CHLORINATION: BASIC ISSUES

1. Water Chlorination: Crossroad of Uncertainties and Decisions, *William P. Davis and Morris H. Roberts, Jr.* 3
2. Human Pathogens, Disinfection, and Chlorine, *Vincent P. Olivieri* 5
3. Basic Issues in Water Chlorination: A Chemical Perspective, *Robert L. Jolley* 19
4. Chlorination of Power Plant Cooling Waters, *Jack S. Mattice* 39
5. Environmental Impacts of Chlorine Discharges: A Utility Industry Perspective, *Peter M. Cumbie, Thomas A. Miskimen, and James K. Rice* 63
6. Conceptual Approach to Evaluate Alternatives to Chlorination for Biofouling Control, *James A. Fava, William J. Rue, Paul Chrostowski, Joseph S. Ferris, and Hans Plugge* 73
7. The Chlorine Dilemma: An Environmental Organization's Perspective, *David S. Bailey* 85
8. Regulatory Aspects of Disinfection, *Joseph A. Cotruvo and Craig D. Vogt* 91

SECTION II
RISK: THE BOTTOM LINE

9. Microbiological Risks Associated with Changes in Drinking Water Disinfection Practices, *Elmer W. Akin and John C. Hoff* 99
10. Risk Assessment Issues in Evaluating the Health Effects of Alternate Means of Drinking Water Disinfection, *R. J. Bull and L. J. McCabe* 111

SECTION III
EPIDEMIOLOGICAL CONSIDERATIONS

11. Epidemiologic Considerations for Evaluating Associations Between the Disinfection of Drinking Water and Cancer in Humans, *Gunther F. Craun* 133
12. Drinking Water Source and Risk of Bladder Cancer: A Case-Control Study, *Kenneth P. Cantor, Robert Hoover, Patricia Hartge, Thomas J. Mason, Debra T. Silverman, and Lynn I. Levin* 145
13. A Case-Control Study of Colon Cancer and Water Chlorination in North Carolina, *Donna L. Cragle, Carl M. Shy, Robert J. Struba, and Edward J. Siff* 153
14. Reactions of Chlorine in Drinking Water, with Humic Acids and In Vivo, *F. C. Kopfler, H. P. Ringhand, W. E. Coleman, and J. R. Meier* 161
15. Reactions of Hypochlorite and Organic N-Chloramines in Stomach Fluid, *Frank E. Scully, Kathryn E. Mazina, Daniel E. Sonenshine, and F. B. Daniel* 175

SECTION IV
CARCINOGENIC AND MUTAGENIC EFFECTS

16. Mutagenic and Carcinogenic Properties of Drinking Water, *H. J. Kool, C. F. van Kreijl, and J. Hrubec* 187
17. Mutagenic Properties of Drinking Water Disinfectants and By-Products, *John R. Meier and Richard J. Bull* 207
18. Carcinogenic Activity of Haloacetonitrile and Haloacetone Derivatives in the Mouse Skin and Lung, *Richard J. Bull and Merrell Robinson* 221
19. Relationship Between Metabolism and Haloacetonitriles and Chloroform and Their Carcinogenic Activity, *Michael A. Pereira, F. Bernard Daniel, and Edith L. C. Lin* 229
20. Mutagenicity Produced by Aqueous Chlorination of Tyrosine, *W. Howard Rapson, Bonnie Isacovics, and C. Ian Johnson* .. 237
21. Identification of Carcinogens by Measurement of Cell-Mediated Immunity vs. Antitumor Immunity in Rats to Halogen-Containing Organic Compounds, *Reggie H. Stevens, Dean A. Cole, Paul A. Lindholm, Paul T. Liu, Margaret L. Gourlay, and H. F. Cheng* 251
22. Formation of Genotoxic Compounds by Chlorination of Residues from Oil Refinery Effluents, *Christopher D. Metcalfe and Ronald A. Sonstegard* 265

SECTION V
TOXICOLOGY OF DISINFECTANTS AND THEIR BY-PRODUCTS

23. Pharmacokinetics of Chlorine Obtained from Chlorine Dioxide, Chlorine, Chloramine, and Chloride, *Mohamed S. Abdel-Rahman* 281
24. Reproductive Effects of Alternate Disinfectants and Their By-Products, *Betsy D. Carlton and M. Kate Smith* 295
25. Toxicity of 2-Chlorophenol, 2,4-Dichlorophenol, and 2,4,6-Trichlorophenol, *Jerry H. Exon and Loren D. Koller* 307
26. Toxicological Evaluation of Selected Chlorinated Phenols, *Joseph F. Borzelleca, Lyman W. Condie, and Johnnie R. Hayes* 331
27. Target Organ Effects of Disinfectants and Their By-Products, *Lyman W. Condie and J. Peter Bercz* 345
28. Effects of Chlorine Dioxide on Neurobehavioral Development of Rats, *Douglas H. Taylor and Ronald J. Pfohl* 355
29. Effects of Chlorinated Drinking Water on Myocardial Structure and Functions in Pigeons and Rabbits, *N. W. Revis, T. R. Osborne, G. Holdsworth, and P. McCauley* 365

SECTION VI
AQUATIC MODELS AND TUMOR INDUCTION

30. Aquatic Models and Tumor Induction, *Herman F. Kraybill* .. 375
31. Attempts to Abbreviate Time to Endpoint in Fish Hepatocarcinogenesis Assays, *John A. Couch and Lee A. Courtney* 377
32. Occurrence of Hepatic Neoplasms and Other Lesions in Bottom-Dwelling Fish and Relationship to Pollution in Puget Sound, Washington, *Donald C. Malins, Bruce B. McCain, Margaret M. Krahn, Mark S. Myers, John E. Stein, William T. Roubal, Donald W. Brown, Usha Varanasi, Harold O. Hodgins, and Sin-Lam Chan* 399
33. Carcinogenic Effects of River Sediment Extracts in Fish and Mice, *John Black, Helen Fox, Penny Black, and Fred Bock* .. 415
34. Tumor Induction in Several Small Fish Species by Classical Carcinogens and Related Compounds, *William E. Hawkins, Robin M. Overstreet, William W. Walker, and C. Steve Manning* 429
35. Japanese Medaka Liver Tumor Model: Review of Literature and New Findings, *David E. Hinton, James A. Hampton, and Patricia A. McCuskey* 439

36. Black Bullhead: An Indicator of the Presence of Chemical Carcinogens, *John M. Grizzle* 451

SECTION VII
ENVIRONMENTAL EFFECTS

37. Interactions of Chlorine-Produced Oxidants, Salinity, and a Protistan Parasite in Affecting Lethal and Sublethal Physiological Effects in the Eastern or American Oyster, *Geoffrey I. Scott, Edward O. Oswald, Tommy I. Sammons, Douglas S. Baughman, and Douglas P. Middaugh* 463
38. Response of Sheep River, Alberta, Macroinvertebrate Communities to Discharge of Chlorinated Municipal Sewage Effluent, *Lewis L. Osborne* 481
39. Comparison of Acute Toxicity and Avoidance Responses of Atlantic Silverside and White Perch to Chlorinated Estuarine Waters, *James A. Fava and John W. Meldrim* 493
40. Depression of Larval Growth and Metamorphosis of Oysters Exposed to Chlorinated Sewage, *Morris H. Roberts, Jr., and Beverly B. Casey* 509
41. Effect of Selected Chlorine-Produced Oxidants on Oyster Larvae, *Mary Elizabeth Stewart and Walter Blogoslawski* 521
42. Delayed Effects of Chlorine on Early Life Stages of the Mayfly, *Sylvia A. Murray, Kenneth J. Tennessen, and Susan M. Laborde* 533
43. Inhibition of Phytoplankton Photosynthesis by Chlorinated Sewage in the James River, *Soon Lin Ho and Morris H. Roberts, Jr.* 541

SECTION VIII
DISINFECTION

44. Inability of Laboratory Models to Accurately Predict Field Performance of Disinfectants, *Roy L. Wolfe and Betty H. Olson* 555
45. Aspects of the Mode of Action of Monochloramine, *J. G. Jacangelo and V. P. Olivieri* 575
46. Disinfection of *E. coli* in the Presence of N-Organic Compounds, *Neil M. Ram, James P. Malley, Jr., Cynthia A. Parks, and Brian Dudley* 587

47. Disinfection Resistance of *Legionella pneumophila* and *Escherichia coli* Grown in Continuous and Batch Culture, *James D. Berg, John C. Hoff, Paul V. Roberts, and Abdul Matin* 603
48. Response of Chemostat-Grown Enteric Bacteria to Chlorine Dioxide, *M. S. Harakeh, J. C. Hoff, and A. Matin* 615
49. Mode of Action of Chlorine Dioxide on Selected Viruses, *V. P. Olivieri, F. S. Hauchman, C. I. Noss, and R. Vasl* 619
50. Relative Disinfection Potentials of Chlorine and Chlorine Dioxide in Drinking Water, *M. M. Varma, G. Torrence, R. C. Chawla, and H. Okrend* 635
51. Recurrent Coliforms in Water Distribution Systems in the Presence of Free Residual Chlorine, *Vincent P. Olivieri, Alexander E. Bakalian, Keith W. Bossung, and Ernest D. Lowther* 651
52. Sensitivity of Vegetative Protozoa to Free and Combined Chlorine, *Charles N. Haas, Kamel M. Khater, and Allen T. Wojtas* 667
53. Factors Influencing Chlorine Disinfection of Wastewater Effluent Contaminated by Rotaviruses, Enteroviruses, and Bacteriophages, *M. S. Harakeh* 681
54. Ozonation as a Stage in Upgrading Secondary Wastewater Effluents for Reuse, *Yehuda Kott* 691

SECTION IX
REACTION DYNAMICS IN WATER CHLORINATION

55. Reaction Dynamics in Water Chlorination, *J. Carrell Morris* . 701
56. Chlorine Decay Chemistry in Natural Waters, *Douglas Dotson and George R. Helz* 713
57. A Kinetic Model of Chlorination of Natural Water: The Roles of Organic Nitrogen and Humic Substances, *Robert G. Qualls and J. Donald Johnson* 723
58. Seawater Chlorination: Influence of Ammonia Concentration, *Jean-Marie Fiquet* 737
59. Chlorination Kinetics of Surface and Deep Tropical Seawater, *Francis J. Sansone and Terrence J. Kearney* 755
60. Decomposition of Bromamines in Aqueous Solutions: Preliminary Report of the Decomposition Kinetics and Disproportionation of NH_2Br, *S. Pasquini Christina, M. T. Azure, H. J. Workman, and E. T. Gray, Jr.* 763
61. Influence of Sodium, Potassium, and Lithium on Hypochlorite Solution Equilibria, *Charles N. Haas and Delores M. Brncich* 775

62. The Chemistry of Oxo-Chlorine Compounds Relevant to Chlorine Dioxide Generation, *E. Marco Aieta and Paul V. Roberts* 783

SECTION X
CHLORINE DEMAND REACTIONS: PROTEINS AND OTHER ORGANICS

63. Chlorination of the Peptide Nitrogen, *R. C. Ayotte and E. T. Gray, Jr.* 797
64. Contribution of Proteins to Formation of Trihalomethanes on Chlorination of Natural Waters, *Frank E. Scully, Jr., Robert Kravitz, C. Dean Howell, Mark A. Speed, and Richard P. Arber* 807
65. Evolution of Amino Acids in Water Treatment Plants and the Effect of Chlorination on Amino Acids, *C. LeCloirec and G. Martin* 821
66. Characterization of the Products from the Reaction of Hydroxybenzoic and Hydroxycinnamic Acids with Aqueous Solutions of Chlorine, Chlorine Dioxide, and Chloramine, *Robert M. Carlson and Sechoing Lin* 835
67. Formation of Aryl-Chlorinated Aromatic Acids and Precursors for Chloroform in Chlorination of Humic Acid, *Ed W. B. deLeer, Jaap S. Sinninghe Damste, and Leo de Galan* 843
68. Formation of Acidic Trace Organic By-Products from Chlorination of Humic Acids, *Dennis R. Seeger, Leown A. Moore, and Alan A. Stevens* 859
69. Nonpurgeable Organohalide Formation on Chlorination of Algal Extracellular Material, *Jan K. Wachter and Julian B. Andelman* 875
70. Novel Precursor of Trihalomethanes, *William J. Cooper and Delia M. Kaganowicz* 895
71. Factors Affecting Incorporation of Bromide into Brominated Trihalomethanes During Chlorination, *Gary L. Amy, Paul A. Chadik, Zaid K. Chowdhury, Paul H. King, and William J. Cooper* 907
72. Formation of Iodinated Trihalomethanes, *Joseph P. Gould, Maurizio Giabbai, and Jong-Soo Kim* 923

SECTION XI
CHEMISTRY OF CHLORAMINATION

73. Characterization of the Reaction Between Monochloramine and Isolated Aquatic Fulvic Acid, *James N. Jensen, Jessica J. St. Aubin, Russell F. Christman, and J. Donald Johnson* 939
74. Significant Findings Related to Formation of Chlorinated Organics in the Presence of Chloramines, *Richard Arber, Mark A. Speed, and Frank Scully* 951
75. Analysis and Formation Mechanisms of Mixed N-Halogenated Methylamines, *Terrence J. Kearney and Francis J. Sansone* .. 965
76. Disappearance of Monochloramine in the Presence of Nitrite, *Richard L. Valentine* 975
77. Subbreakpoint Modeling of the HOBr-NH_3-Org-N Reactions, *Russell A. Isaac, Johannes Edmund Wajon, and J. Carrell Morris* 985
78. Reversibility in the Reactions of Chloramines with Bromide: Dimethylchloramine Reaction, *Werner R. Haag* 999

SECTION XII
PHOTOCHEMISTRY OF OXIDANTS

79. Degradation of Compounds in Water by Singlet Oxygen, *Werner R. Haag and Jurg Hoigne* 1011
80. Utilization of Molecular Oxygen and Sunlight in the Oxidative Purification of Water, *William Cherry and Brian Jessen* 1021
81. Hydrogen Peroxide in Estuarine Waters: A Minor But Significant Contributor to Chlorine Demand, *George R. Helz and Robert J. Kieber* 1033
82. Sunlight-Induced Photodecomposition of Chlorine Dioxide, *Rod G. Zika, Cynthia A. Moore, Louis T. Gidel, and William J. Cooper* 1041
83. Photodegradation of Water Pollutants in Chlorinated Water, *Lisa H. Nowell and Donald G. Crosby* 1055

SECTION XIII
CHEMICAL METHODS

84. Analytical Determination of Inorganic Chlorination Products in Water Treatment by an HPLC Method, *M. Dreux, M. Lafosse, M. Gibert, and A. Blaison* 1065

85. Instrument for Total Chlorine Amperometric Back Titration Using Coulometric Iodine Generation, *Daniel H. Raab and Calvin O. Huber* 1073
86. Rapid Oxidant Demand: Methods for Study, *Donald A. Jaworske and George R. Helz* 1081
87. Anodic Voltammetric Determination of Monochloramine in Water, *Debra A. Davies and Calvin O. Huber* 1091
88. Broad-Spectrum Analysis of Organics in Drinking Water Using Macroreticular Resins — A Quality Assurance Evaluation, *J. Gibbs, B. Najar, and I. H. Suffet* 1099
89. Monitoring Trichloroacetic Acid in Municipal Drinking Water, *Daniel L. Norwood, Gavin P. Thompson, J. Donald Johnson, and Russell F. Christman* 1115
90. Monitoring for Volatile Organohalides Using Purgeable and Total Organic Halide as Surrogates, *R. K. Sorrell, E. Daly, L. Boyer, and H. J. Brass* 1123

SECTION XIV
DRINKING WATER TREATMENT

91. Influence of Water Treatment Processes on Formation of Organic Halogens and Mutagenic Activity by Postchlorination, *J. C. Kruithof, A. Noordsij, L. M. Puijker, and M. A. van der Gaag* 1137
92. Characterization of Total Halogenated Compounds During Various Water Treatment Processes, *A. Bruchet, Y. Tsutsumi, J. Duguet, and J. Mallevialle* 1165
93. Trihalomethane Formation and Control Through a Direct Filtration Water Treatment System, *John N. Veenstra and Parweiz A. Khan* 1185
94. Chloropicrin in Potable Water: Conditions of Formation and Production During Treatment Processes, *J. P. Duguet, Y. Tsutsumi, A. Bruchet, and J. Mallevialle* 1201
95. Preozonation in Drinking Water Treatment: Nondisinfection Applications of Ozone, *Rip G. Rice* 1215
96. Mechanisms of Organic Halide Formation During Fulvic Acid Chlorination and Implications with Respect to Preozonation, *David A. Reckhow and Phillip C. Singer* 1229
97. Effect of Cyanuric Acid, a Chlorine Stabilizer, on Trihalomethane Formation, *Caren M. Feldstein, Janet Rickabaugh, and Richard J. Miltner* 1259
98. Potential New Water Disinfectants, *S. D. Worley, D. E. Williams, H. D. Burkett, S. B. Barnela, and L. J. Swango* ... 1269

99. Properties of Ferrate(VI) in Aqueous Solution: An Alternate Oxidant in Wastewater Treatment, *James D. Carr, Paul B. Kelter, Alireza Tabatabai, David Splichal, John Erickson, and C. William McLaughlin* 1285
100. Effect of a Spill Event on an Ozone Granular-Activated-Carbon Treatment Plant, *Howard M. Neukrug, Matthew G. Smith, Stephen W. Maloney, and Irwin H. Suffet* 1299
101. Activated Carbon: An Oxidant Producing Hydroxylated PCBs, *Evangelos A. Voudrias, Richard A. Larson, Vernon L. Snoeyink, and A. S.-C. Chen* 1313
102. Mutagenic Residues Recovered from Granular-Activated Carbon After Use in Drinking Water Treatment, *John C. Loper, M. Wilson Tabor, and Laura Rosenblum* 1329
103. Effect of Dechlorinating Agents on the Mutagenic Activity of Chlorinated Water Samples, *Philip Wilcox and Susan Denny* . 1341

SECTION XV
COOLING WATER TREATMENT

104. Analysis of Sediment Matter for Halogenated Products from Chlorination of Power Plant Cooling Water, *Roger M. Bean, Bertha L. Thomas, and Duane A. Neitzel* 1357
105. Halogenated Compounds Discharged from a Coastal Power Plant, *Robert S. Grove, Edward J. Faeder, Jean Ospital, and Roger M. Bean* 1371
106. Results of Analyzing Simulated Cooling Tower Blowdown for Organic Priority Pollutants, *James Rios* 1381
107. Chlorination of Coal Slurry Transport Waters, *John W. Davis, M. Carrington Reid, Roger A. Minear, and Gary S. Sayler* 1399
108. Effects of Chlorination on the Levels of Mutagens in Contaminated Estuarine Sediments, *Carol B. Daniels, Sandra M. Baksi, Allen D. Uhler, and Jay C. Means* 1411
109. Chlorine Minimization in Macrofouling Control in the Netherlands, *Henk A. Jenner* 1425
110. A Predictive Model for Destruction of Biofilms with Chlorine, *Meletios Platon and Thomas D. Waite* 1435
111. Targeted Chlorination: Design and Field Tests, *Robert D. Moss, Stephen P. Gautney, and Patrick A. March* 1447
112. Concentrations of Chlorine Around Marine Cooling Water Outfalls: Validation of a Model, *Jack Coughlan and Martin H. Davis* 1459
113. Predicting Chlorine Compounds in Power Plant Cooling Tower Systems, *Vito L. Punzi and Rutton D. Patel* 1469

114. Prediction of Total Residual Chlorine in Power Plant Discharges and Receiving Waters: Application to Effluent Limitations and Water Quality Standards, *John P. Lawler, Thomas B. Vanderbeek, and Peter M. Cumbie* 1489

SECTION XVI
WASTEWATER TREATMENT

115. Discharge of Halogenated Octylphenol Polyethoxylate Residues in a Chlorinated Secondary Effluent, *Harold A. Ball and Martin Reinhard* 1505
116. Effect of Ozonation and Chlorination on Organic Environmental Protection Agency Priority Pollutants, *Yun-Shen Lee and Joseph V. Hunter* 1515

Epilogue 1527

List of Authors 1529

Index 1543

SECTION I

Water Chlorination: Basic Issues

"The objectives of this conference, to see what is known, what is being done, and what should be done concerning the chlorination of various waters, are very important, because the real endpoint of this is to establish a base of data, models, and understanding that will permit those who have the awesome responsibility of setting standards and regulations to do so based on facts."

Herman Postma, Director
Oak Ridge National Laboratory
First Water Chlorination Conference, 1975

Reason, of course, is weak when measured against its never-ending task. Weak, indeed, compared with the follies and passions of mankind, which, we must admit, almost entirely control our human destinies, in great things and small.

Albert Einstein, 1870–1955
Isaac Newton, 1942

CHAPTER 1

Water Chlorination: Crossroad of Uncertainties and Decisions

William P. Davis and Morris H. Roberts, Jr.

The chlorination of water is at the crossroads of the issues of health, quality of water resources, scientific uncertainties, and decisions to regulate or revise our theories and practices of application. In our current rush to decrease regulation per se, and to increase common consensus for mutual benefit, water chlorination stands out clearly as the proverbial horns of a dilemma. It is not an issue that we seek to minimize disease; healthy societies are those where potable water sources abound. It is not an issue that efficiency of power generation is equated with minimizing interfering surface growth and deposits in heat exchangers. Few public spirited citizens argue against achieving the objective of swimmable, drinkable, or fishable water in our ponds, streams, lakes, rivers, and bays.

The objectives of water chlorination are repeated often enough to become a litany of tenets in the dogma surrounding public health and welfare and the richness of our society. Chlorination is the application of our commonest microbial pesticide to achieve those uncontested objectives. Dogmas, however, can be dangerously simplistic in the realm of reality where we are faced with expecting the unexpected. Thus, chlorination also is a source of demonstrable mutagenic and carcinogenic compounds. These compounds arise from reactions with precursors from natural products of forests and streams as well as from our treated wastes or industrial and agricultural by-products. The water we use for cooling electric power generators totals more than all other uses (irrigation and drinking combined). Are we chlorinating all this water? Because water is man's single most important resource, our responsibilities in the use and treatment of water rank among the most strategic decisions we regularly make.

We expect, as a society, to enjoy good health and receive basically disease- and toxicant-free water and food. Public discussion of any possible loss of those "deliverables" results in panic and outrage. Uncertainty regarding risk on the part of a research scientist becomes evidence for public reprimand, if not cause for dismissal or banishment. The issue, therefore, is not chlorination, but truly the ready supply of "safe" water to every citizen. The citizen is not amused by computer games that estimate how much time has elapsed since the last use of his or her water. This citizen doesn't want to know there is risk of a toxicant in the water, nor does he or she want to catch a fish with signs of abnormalities, tumors, or disease. The citizen feels that taxes, elected officials, and public supported technocrats are part of a universal guarantee of health and clean water.

It is popular to say that we excel in science, interpretation and evaluation of risk, and application of knowledge to policy decisions. When we examine the scientific knowledge relative to chlorination, by-products, and environmental effects together with current policies, the statement of one researcher comes to mind: "It is clear that chlorination is the disinfectant of choice because man is the resistant species!"

In these proceedings of the Fifth Water Chlorination Conference, we present technical reports on all aspects of the subject, which most people would rather not know about personally. The avoidance of the subject is apparently spreading to many managers and decision makers as evidenced by their absence from the conference, which may have unpleasant portents for the future. If technical data are to be interpreted and evaluated effectively, there simply must be realistic support for the effort from environmental managers and government agencies. Risk assessments are not reliable without well-designed research and careful challenge and interpretation. As we approach this crossroads of issues, we encounter a number of flashing yellow warning lights:

1. The need to better detect, analyze, and track carcinogenic and mutagenic by-products.

2. The need to be able to moderate if not minimize chlorination to rates that are realistic for high-use waters and to be able to calculate and analyze what realistic means.

3. The need to assess and to anticipate the potential to produce chlorine-resistant pathogens.

4. The need to examine disinfectant alternatives, especially for municipal and industrial treatment, to minimize or prevent formation of toxic by-products in the chlorination-reactors of treatment plants.

Progress can be achieved through open dialogue and development of conceptual approaches and decisions. In the case of water chlorination, there is a very real dilemma. The presence of cancer-inducing compounds is established, but both chemical and biological investigations to follow up on these observations and to define critical parameters have not been adequate. The mutualism necessary to pursue aggressive approaches to problem solution is absent. Many decision makers and managers appear to "wish away" any acknowledgment that this problem exists, while researchers find and present, even on television, new evidence for concern. How long can we continue to say, "The research results are inconclusive?" This absence of dialogue may be the most crucial gap in our continued well being, much less our very survival. At the very least, it is a time bomb yet to be deactivated.

CHAPTER 2

Human Pathogens, Disinfection, and Chlorine

Vincent P. Olivieri

Disease may be transmitted by the consumption of contaminated water and shellfish or close-contact recreation. The human pathogens of major concern in drinking water and shellfish consumption follow the anal-oral route of transmission. In both marine and fresh recreational waters, water-washed diseases (skin, eye, ear, nose, and throat infections) are an additional concern. The transmission of infectious disease requires that a susceptible host be exposed to a sufficient number of pathogenic microorganisms. The pathogen must enter the host through the appropriate portal, overcome the host defenses, and proliferate to a sufficient degree to cause symptoms of the disease, and the key words are exposure and number. To prevent the transmission of disease by water, a complex sanitary fabric has been woven to reduce the number of pathogens that the human population is exposed to in water.

The disinfection of drinking water before consumption and the terminal disinfection of wastewater prior to discharge to the aquatic environment have evolved as important threads in this sanitary fabric. Disinfection, as the word implies, is the removal of infectious agents. Disinfection should not be confused with sterilization, the complete inactivation of living material. The intention of the disinfection of water and wastewater was never sterilization. The primary disinfection process has been the intentional addition of biocidal agents to inactivate microorganisms. In the United States, chlorine has been the most widely used disinfectant.

HUMAN PATHOGENS

The clinical and epidemiological features of human pathogens responsible for the majority of disease associated with water are shown in Table I.[1] The enteric bacteria are responsible for a wide variety of diseases that range from mild gastrointestinal upset to typhoid fever. Respiratory illness (Legionellosis and tuberculosis) has occasionally been associated with water. In general, high numbers of enteric pathogens are shed in feces by a small percentage of the population. The shedding of pathogens typically lasts several weeks, but a carrier state has been observed for members of the genus *Salmonella* and *Shigella*. The infectious dose of enteric bacteria for healthy individuals generally exceeds 10,000. However, much lower levels of *Shigella* and *Salmonella typhi* have been observed to cause infection.

Table I. Summary of Taxonomic, Clinical, and Epidemiological Features of Potential Drinking Water Pathogens[1]

Organism or group	No. Types	Major disease	Major reservoirs, primary sources	Concentration in primary source (per g)	Infect. dose	Prevalence (Av % excretion)	Duration of shedding	Carrier state
Bacteria								
Salmonella typhi	1	Typhoid fever	Human feces	10^6	Low		Typically	
Salmonella paratyphi	1	Paratyphoid fever	Human feces	10^6	High	1–3.9	4 weeks	+
Other salmonellae	1000	Salmonellosis	Human/animal feces	10^6	High		occ. 1 year	
Shigella	4	Bacillary dysentary	Human feces	10^6	Medium		1 week	+ (?)
Vibrio cholerae	?	Cholera	Human feces	10^6	High		?	
Enteropathogenic E. coli	?	Gastroenteritis	Human feces	10^8	High	1.2–15.5	?	
Yersinia enterocolitica	?	Gastroenteritis	Human/animal feces	?	High		?	
Campylobacter fetus (sub sp. jejuni)	1	Gastroenteritis	Human/animal (?) feces	?	?		?	
Legionella pneumophila and related bacteria	4	Acute respiratory illness (legionellosis)	Thermally enriched waters	?	High?		?	
Mycobacterium tuberculosis	1	Tuberculosis	Human respiratory exudates	?	?		?	
Other (atypical) mycobacteria	2	Pulmonary illness	Soil and water	?	?		?	
Opportunistic bacteria	?	Variable	Natural waters	?	?		?	
Enteric Viruses								
Enteroviruses								
Polioviruses	3	Poliomyelitis	Human feces					
Coxsackieviruses A	23	Aseptic meningitis	Human feces	10^6				
Coxsackieviruses B	6	Aseptic meningitis	Human feces		1–3 weeks			
Echovirus	31	Aseptic meningitis	Human feces					
Other enteroviruses	4	AHC; encephalitis	Human feces					
Reoviruses	1–3	Mild UR and GI illness	Human/animal feces	10^9 Parts	Low		1–2 weeks	—
Rotaviruses	2	Gastroenteritis	Human feces				1 week	—
Adenoviruses	37	UR and GI illness	Human feces	10^6			? week	?
Hepatitis A virus	1	Infectious hepatitis	Human feces				3 weeks	—
Norwalk and related gastrointestinal viruses	3	Gastroenteritis	Human feces				1 week?	—

Table I, continued

Organism or group	No. Types	Major disease	Major reservoirs, primary sources	Concentration in primary source (per g)	Infect. dose	Prevalence (Av % excretion)	Duration of shedding	Carrier state
Protozoans								
Acanthamoeba castellani	1	Amoebic meningoence	Soil and water	?	?			
Balantidium coli	1	Balantidiasis (dysentery)	Human feces	?	?		None	+
Entamoeba histolytica	1	Amoebic dysentery	Human feces	10^5	Low	10		
Giardia lamblia	1	Giardiasis (gastroenteritis)	Human/animal feces	10^5	Low	1.5–2.0	6–7 weeks	+ (months to years)
Naegleria fowleri	1	Primary amoebic	Soil and water	?	?			
Helminths								
Nematodes (roundworms)								
Ascaris lumbricoides	1	Ascariasis	Human/animal feces					
Trichuris trichiura	1	Trichuriasis	Human feces					
Enterobius vermicularis	1	Pinworms	Human feces					
Hookworms								
Ancylostoma duodenale	1	Hookworm disease	Human feces					
Necator americanus	1	Hookworm disease	Human feces	$10–10^2$				
Strongyloides stercoralis		Threadworm disease	Human/animal feces					
Cestodes (tapeworms)								
Taenia saginata and T. Solium	2	Taeniasis	Human feces					
Hymenolepsis nana	1	Gastroenteritis	Human/rodent feces					
Algae								
Cyanobacteria Microcystis aeruginosa Anabaena flos-aqua Aphanizomenon flos-aqua	8	Gastroenteritis	Natural waters	?	N/A[a]	N/A	N/A	

[a]Not applicable.

Goldfield[2] reviewed the epidemiological evidence for the transmission of viral diseases by the water route. He concluded, similar to Mosely,[3] that the demonstrated health hazard of viruses in water has been limited to an occasional outbreak of infectious hepatitis associated with the direct consumption of contaminated water and raw shellfish, a rare occurrence of poliomyelitis, and adenovirus infection associated with swimming pools. More recently, evidence has been reported to suggest that viruses (Norwalk agent and rotovirus) may be implicated as the etiological agents for acute gastroenteritis.[4-8] Astrovirus, calicivirus, and coronavirus have been associated with gastrointestinal disease and have been found in feces.[9] The role of water in the transmission of these viruses is not clear.

Enteric viruses are shed in feces at high levels for several weeks. Levels of 10^9 per gram for reovirus and 10^{10} per gram for rotavirus[10] have been reported. A carrier state similar to that observed for some members of the enterobacteria has generally not been found. The exact quantity of enteroviruses that must be ingested to produce injurious infections has received considerable attention. The authors of several reviews of the problem of viral minimal infectious dose (MID)[10-12] generally arrived at the conclusion that one tissue culture infective dose correlates well with one MID for a broad spectrum of viruses. This principle applies to both water and airborne infections[10] and is based not only on work with experimental animals but administration of viruses to humans as well. It is particularly germane to this discussion that these observations on MID included human viruses such as poliovirus 1 and 3,[11,13] coxsackievirus A21,[14] coxsackievirus B4, rhinovirus, and adenovirus.[10] Admittedly, the viruses achieved their high degree of efficiency after careful instillation of the inoculum under optimal conditions with minimum interference from environmental factors and host resistance factors. Nevertheless, the potential for establishing the infection warrants concern. We should recognize that infection does not always lead to overt disease.

Waterborne diseases caused by protozoans have been primarily associated with *Entamoeba histolytica*, and more recently, *Giardia lamblia*. Since 1974, waterborne giardiasis has been consistently reported to the Centers for Disease Control.[15-17] From 1978 through 1981, *Giardia lamblia* was the most frequently identified pathogen associated with waterborne disease. Waterborne amoebic dysentery has not been reported in the United States for some time.

The cysts of *Giardia lamblia* and *Entamoeba histolytica* have been estimated to be found at 10^5 per gram of feces. Akin et al.[18] reported levels of *Giardia* up to 2.2×10^6 per gram in infected children and as high as 9.7×10^7 per gram in asymptomatic carriers. *Giardia* is shed for 6 to 7 weeks, and 1 to 2% of the population appears to be shedding the cysts. In a study of *Giardia lamblia* infections in man, Rendtorff[19] administered varying numbers of cysts to volunteers. Five individuals receiving only 1 cyst demonstrated no cysts or trophozoites in their stools, while five receiving 100 cysts became infected. At no time were clinical complaints observed in any of the volunteers, even in those receiving up to 1 million cysts. In a similar study, Rendtorff[20] gave 1,

100, and 10,000 *Entamoeba coli* cysts to volunteers, with a range of infection rates from 12 to 25%. Diagnosis was made by stool examination, and in only one case was clinical illness observed.

In the United States, transmission of helminth diseases by the water route appears to be rare. No recent reports can be found. A number of additional requirements must be met for parasitic helminths to cause infection and disease in man. The parasites must be viable and, in many cases, must have the opportunity to develop to the infective stage of their life cycles. After release from the host, parasitic helminths may require simple or complex periods of incubation in the environment before being infective to a new host. With few exceptions, one or more intermediate hosts are required in the life cycle of the tapeworms of man.

INACTIVATION OF MICROORGANISMS WITH CHLORINE

Water and wastewater disinfectants belong to a class of general protoplasmic poisons and are potent biocides. The factors that influence the inactivation of microorganisms are well known, and an excellent review of the chemistry and microbiology of disinfection can be found in Volumes 2 and 3 of *Drinking Water and Health*.[21,22] Table II from this reference summarizes the factors that affect microbial inactivation and compares the efficacy of the commonly used water and wastewater disinfectants. The information presented in Table II is for clean laboratory systems with little or no disinfectant demand. The presence of demand will dramatically alter the concentration and species of disinfectant. The microorganisms used in the studies, from which Table II was compiled, were highly purified and washed laboratory cultures and, in general, were more sensitive than the microorganisms found in the real world. Despite the shortcomings of these studies, the assemblage of data does provide information to evaluate the efficacy of disinfectants and the parameters that influence the inactivation of microorganisms in water.

The disinfectant concentration and contact time (CT) are the prime factors in the inactivation of microorganisms. The product of the concentration and contact time for a specified level of kill, in this case 99% kill, provides a useful term to compare disinfectants and factors affecting microbial inactivation. The CT values for each of the disinfectants are variable. The free species of chlorine are potent disinfectants and rapidly inactivate bacteria, viruses, and protozoan cysts at low concentrations. At comparable pH and temperature values, HOCl and OCl$^-$ have CT values of 0.04 and 0.92 for *Escherichia coli* and 2.0 and 10.5 for poliovirus 1 respectively. Cysts of *Entamoeba histolytica* are more resistant and have a CT value of 20 at pH 7 and 30°C. Monochloramine, a less potent biocide, has CT values of 64 and 900 for *E. coli* and poliovirus 1, respectively, at pH 9.0 and 5°C. The pH of the water controls the species of the halogen disinfectants and the rate of reaction with the demand.

Table II. Comparative Efficacy of Disinfectants in the Production of 99% Inactivation of Microorganisms in Demand-Free Systems[21,22]

Disinfection Agent	E. coli pH[a]	E. coli Temperature (°C)[a]	E. coli c·t[b]	Poliovirus I pH[a]	Poliovirus I Temperature (°C)[a]	Poliovirus I c·t[b]	Entamoeba histolytica cysts pH[a]	Entamoeba histolytica cysts Temperature (°C)[a]	Entamoeba histolytica cysts c·t[b]
Hypochlorous acid	6.0	5	0.04	6.0	0	1.0	7	30	20
				6.0	5	2.0			
				7.0	0	1.0			
Hypochlorite ion	10.0	5	0.92	10.5	5	10.5		NDR[c]	
Ozone	6.0	11	0.031	7.0	20	0.005	7.5–8.0	19	1.5[d]
	7.0	12	0.002	7.0	25	0.42			
Chlorine dioxide	6.5	20	0.18	7.0	15	1.32		NDR[c]	
	6.5	15	0.38	7.0	25	1.90			
	7.0	25	0.28						
Iodine	6.5	20–25	0.38	7.0	26	30	7.0	30	80
	7.5	20–25	0.40						
Bromine		NDR[c]		7.0	20	0.06	7.0	30	18
Chloramines									
Monochloramine	9.0	15	64	9.0	15	900		NDR[c]	
	9.0	25	40	9.0	25	320			
Dichloramine	4.5	15	5.5	4.5	15	5000		NDR[c]	

[a]Conditions closest to pH 7.0 and 20°C were selected from studies discussed in the text. Values for other conditions and agents appear in the text along with discussions of the cited studies.
[b]Concentration of disinfectant (mg/L) times contact time (min).
[c]Either no data reported or only available data were not free from confounding factors, thus rendering them not amenable to comparison with other data.
[d]This value was derived primarily from experiments that were conducted with tap water; however, some parallel studies with distilled water showed essentially no differences in inactivation rates.

For chlorine, the microbial inactivation decreases as the pH increases. The CT values for HOCl (pH 6) and OCl⁻ (pH 10) for *E. coli* at 5°C were 0.04 and 0.92 respectively. Temperature influences the rate of microbial inactivation and the rate of disinfectant-consuming side reactions. As temperature increases, the rates of both reactions increase.

The extent and rate of disinfection are influenced by the type and physiological state of the microorganism. In general, bacteria are more susceptible than viruses, which are more susceptible than protozoan cysts. The CT values of HOCl for *E coli*, poliovirus 1, and *Entamoeba histolytica* cysts are 0.04, 1.0, and 20.0 respectively. The previous culture history and physiological condition of the microorganism also play an important but less understood role in disinfection. Laboratory strains are generally more susceptible than microorganisms found in nature.

RECENT OUTBREAKS OF WATERBORNE DISEASE

Drinking Water

The Centers for Disease Control in cooperation with the Environmental Protection Agency and state and local health agencies provide a surveillance system to keep track of waterborne disease. Table III shows data from the three most recent annual reports, 1980, 1981, and 1982,[15-17] for outbreaks of disease from drinking water. In 1980, 50 outbreaks were reported with a total

Table III. Etiology of Waterborne Disease Outbreaks Related to Drinking Water from 1980 to 1982[33-35]

	1980		1981		1982	
Agent	Outbreaks	Cases	Outbreaks	Cases	Outbreaks	Cases
AGI[a]	28	13,220	14	1,893	16	1,836
Giardia	7	1,724	9	297	12	561
Chemical	7	2,298	5	128	2	18
Shigella	1	4	1	253	2	172
Campylobacter	1	800	1	81	nr	nr
V. cholera	nr[b]	nr	1	17	nr	nr
Rotavirus	nr	nr	1	1,761	nr	nr
Norwalk	5	1,914	nr	nr	4	750
Hepatits A	1	48	nr	nr	3	103
Yersinia	nr	nr	nr	nr	1	16
Total	50	20,008	32	4,430	40	3,456

[a]Acute gastrointestinal illness.
[b]None reported.

of 20,008 cases. Acute gastrointestinal illness (AGI) of undefined etiology accounted for 28 outbreaks and 13,220 cases. The 1980 report was dominated by the Georgetown, Texas, epidemic with almost 8000 cases of AGI and 36 cases of hepatitis A. Hepatitis A antigen was detected in three of five sewage concentrates and in one well-water concentrate. While bacteriological data prior to the outbreak were limited, high densities of fecal coliforms were observed in central city wells after the second peak of AGI. The water supply for Georgetown comes from seven wells, four of them located in the central city. The water from the central city wells was chlorinated prior to storage in a 1 ML/d tank with a calculated residence time of 20 to 30 min. A constant chlorine dose from a single-cylinder chlorinator was used. No backup chlorinator was available. The disinfection practices appear questionable, with a situation where lapses in chlorination may occur and inadequate chlorination would be likely if the water became contaminated and the chlorine demand increased.[23]

In 1981, fewer outbreaks (32) were reported with 4430 cases. Acute gastrointestinal illness was still the predominant disease, followed by giardiasis. A cholera outbreak (17 reported cases) was the largest of its kind in the United States in this century. Waterborne outbreaks caused by Norwalk agent were not reported in 1981. The failure to find this etiological agent may be the result of the short supply of diagnostic reagents. *Shigella*, *Campylobacter*, and *Rotavirus* were responsible for one outbreak each.

Forty outbreaks consisting of 3456 cases related to drinking water were reported in 1982. Acute gastrointestinal illness was again the predominant disease, with 16 outbreaks and 1836 cases. *Giardia lamblia* was responsible for the largest number of outbreaks where the etiological agent was identified. There were 12 outbreaks of giardiasis with 561 cases. For the past 5 years, *G. lamblia* was the most frequently identified waterborne pathogen. Norwalk agent and hepatitis A accounted for four and three incidents, respectively. The former involved 750 cases, whereas the latter had only 103 cases. Two outbreaks were caused by *Shigella* and one outbreak was attributed to *Yersinia*. The number of cases involving Norwalk agent and hepatitis A were 172 and 16, respectively. While waterborne diseases attributable to *Campylobacter* were reported in 1980 and 1981, none were reported in 1982. There were two suspected instances associated with *Campylobacter* in Washington state. These outbreaks, however, were not proven to be waterborne. No waterborne *Cholera* or *Rotavirus* outbreaks were observed in 1982.

The largest number of outbreaks in recent years has been associated with noncommunity water supplies. The number of cases in each of these outbreaks was generally low. The number of outbreaks associated with community systems, while fewer in number, usually affected a much higher number of people. The dramatic potential for disease in these systems can be seen in the recent Georgetown, Texas, outbreak where almost 8000 cases of AGI were observed.

Recreational Waters

In the United States, both fresh and saline waters are used extensively for contact recreation. Early studies in the United Kingdom[24,25] suggested that the risk to health by bathing in sewage-contaminated waters was negligible. The diseases followed in these studies were poliomyelitis and salmonellosis. Poliomyelitis, after almost 50 years of epidemiological study, does not appear to be waterborne[2,3] and thus would not be expected to be transmitted by contact recreation. Salmonellosis has been transmitted by water; however, the number of *Salmonella* that must be ingested to cause disease are greater than 10 thousand.[26] The quantity of recreational water that would contain this number of *Salmonella* would not be expected to be consumed by bathers. Thus, it was not surprising that the early studies reported that enteric diseases were not associated with swimming. Stevenson,[27] in the United States, reported the results of a series of field studies to determine the relationship between health and bathing water quality. The studies demonstrated that swimmers had a higher overall incidence of disease compared to nonswimmers, regardless of water quality. However, more recent studies by Cabelli et al.[28-30] showed that enteric disease was transmitted by contact recreation and that the level of disease was related to the level of contamination, as indicated by the density of enterococci.

Recent outbreaks demonstrated that other enteric diseases are transmitted by contact recreation. A shigellosis outbreak in 1974 in Dubuque, Iowa, was related to swimming in the Mississippi River.[31] The mean level of fecal coliforms (FC) in the water was 17,500 FC per 100 mL, with counts as high as 5 million per 100 mL below the sewage treatment plant and 400,000 per 100 mL at the beach. *Shigella sonnei* isolated from the water 1 month after the outbreak had the same antibiogram, colicin type and phage type as that isolated from six swimmers. Although other sources of contamination were possible, the Dubuque sewage treatment plant outfall 5 miles above the beach could have been the source of *Shigella*. The disinfection practices at the sewage treatment plant were marginally effective at best. Effluent data for June through September showed high levels of fecal coliforms (3,960 to 3,230,000 per 100 mL) in the effluent and chlorine residuals of 0 to 13.4 mg/L. The chlorinator was not functioning during a 12-h period in August.

Waterborne disease outbreaks related to recreational water in 1980, 1981, and 1982 are shown in Table IV.[15-17] Skin diseases dominate the recreational water picture. In 1980, however, four outbreaks of shigellosis and one of AGI were reported. In 1981, no enteric outbreaks were reported. Infectious hepatitis and salmonellosis have previously been reported to be transmitted by recreational contact. In 1982, 27 outbreaks of disease involving 784 cases related to recreational waters were reported. This statistic represents the highest number of reported incidents since surveillance began in 1971. Dermatitis was responsible for 25 of the 27 outbreaks, and *Pseudomonas aeruginosa* caused 24 of these. The majority of outbreaks were associated with whirlpools and hot tubs in public facilities. A second outbreak of legionellosis was reported in Michi-

Table IV. Etiology of Waterborne Disease Outbreaks Related to Recreational Water from 1980 to 1982[33-35]

Agent	1980		1981		1982	
	Outbreaks	Cases	Outbreaks	Cases	Outbreaks	Cases
Pseudomonas	4	78	7	642	24	697
Shigella	4	325	nr[a]	nr	nr	nr
Adenovirus	1	15	nr	nr	1	66
AGI[b]	2	83	nr	nr	nr	nr
Legionella	nr	nr	1	34	1	14
Cercarial	nr	nr	nr	nr	1	7
Total	11	511	8	676	27	784

[a]None reported.
[b]Acute gastrointestinal illness.

gan. *L. pneumophila* serogroup 6 was isolated from a whirlpool bath. One outbreak of "swimmers itch" (seven cases) caused by schistosomal cercariae was reported, and one outbreak of fever and pharyngitis from a community swimming pool occurred. Adenovirus 7A was recovered from throat cultures of 7 of the 66 cases in the latter outbreak.

Shellfish

The protection of shellfish harvesting areas is an important function of sewage disinfection. The reductions of enteric pathogens, bacteria, viruses, and parasites coupled with a no-harvest buffer zone below the plant and modern shellfish sanitation programs have resulted in an enviable record. However, a dramatic increase in reported illnesses associated with shellfish has renewed concern and prompted a report by the United States General Accounting Office (GAO).[32] Appended to this document is a list of reported shellfish illnesses in the United States and Canada from 1900 through 1983. Similar to the observations in drinking water in recent years, the etiology has shifted dramatically. Few recent outbreaks were caused by bacteria, and no typhoid has been associated with shellfish since 1954. Infectious hepatitis outbreaks have been regularly reported since 1961, and AGI where no agent was identified has predominated in the most recent outbreaks. The few incidents of illness since 1970 associated with bacteria were caused by *Vibrio cholera* (4), *Vibrio parahaemolyticus* (3), *Shigella* (3), *Staphylococcus* (1), and *Plesiomonas shigelloides* (2). Multiple pathogenic bacteria were isolated in one outbreak in Florida. The *Vibrio* and *Plesiomonas* proliferate in the marine environment.

Table V is a summary of reported illnesses associated with shellfish. The summary was prepared from the GAO report for 1980 to 1983. The overwhelming problem associated with shellfish was AGI, with 38 episodes and 4171 cases during this period. Most notable were the more than 1900 cases of

Table V. Etiology of Reported Illnesses Attributable to the Consumption of Shellfish in the United States from 1980 to 1983

Agent	Outbreaks	Cases
AGI[a]	38	4171
Infectious hepatitis	1	1
Vibrio cholera	3	5
Vibrio parahaemolyticus	1	4
Plesiomonas shigelloides	2	11
Norwalk agent	1	6
Mixed	1	18
Total	47	4216

[a] Acute gastrointestinal illness.

AGI reported in the New York–New Jersey area in 1983. While most shellfish are taken from approved beds, the GAO report suggests that illegal harvesting of shellfish may have contributed to the outbreaks. Besides infectious hepatitis, other viral agents have recently been associated with disease caused by the consumption of shellfish. Norwalk agent was reported responsible for a small outbreak in Florida[32] and has been identified as the causative agent in an outbreak in Australia.[33] Shellfish episodes have also been reported for parvovirus[34] and unidentified small round viruses.[35] The latter outbreak was associated with oysters that had been depurated for 72 h. Gastrointestinal disease has also been associated with depurated oysters in a feeding study conducted by Grohman and co-workers.[36] The previous two reports and an earlier study conducted with bacterial virus[37] suggest that shellfish cleansed sufficiently to eliminate indicator bacteria may still contain sufficient quantities of virus to cause disease.

A previous report to Congress by the Comptroller General of the United States,[38] although critical of sewage disinfection with chlorine, recognized the value of terminal disinfection of sewage effluents for the protection of shellfish-growing water.

Other Water-Associated Outbreaks

Other outbreaks occurred in 1982[17] that resulted from exposure to contaminated water but were not directly related to drinking water or recreational waters in the conventional sense. Three outbreaks resulted from swimming in surface waters. *Entamoeba* and *Giardia* were isolated from the stools of professional divers in the Hudson river who were employed by the police and fire departments. The waterborne route of transmission was suspected in two additional outbreaks. *Giardia* was found in the stools of children who used a community swimming pool in Washington state, and an outbreak of gastroenteritis caused by *Shigella sonnei* showed an increased risk of infection associated with swimming in a reservoir.

SUMMARY

Despite the lack of attention, waterborne disease is alive and well in the United States. Epidemics associated with drinking water, recreational water, and shellfish occur with alarming frequency. Large common-source outbreaks associated with water are still observed (Georgetown, Texas; Rome, New York; and New York City), and thousands of individuals are involved. The outbreaks of disease can be attributed to situations where the aquatic environment was contaminated with the wastes of man.

The intentional addition of a biocide in the water and wastewater treatment systems is still necessary. Wastewater that does not receive terminal disinfection will load the streams, lakes, and coastal waters and place an additional stress on water-treatment facilities, bathing beaches, and shellfish harvesting waters. The disinfection of water prior to consumption and the maintainance of a disinfectant residual throughout the distribution system provide an effective barrier against the transmission of infectious disease.

Chlorine is an effective biocide that inactivates pathogenic bacteria, viruses, and protozoan cysts in properly designed and operated water and wastewater systems. Chlorination effectively reduces the numbers of disease-causing microorganisms in drinking water, such that the transmission of disease is below the detection of currently available epidemiological methods. Wastewater chlorination effectively reduces the numbers of disease-causing microorganisms in the effluents and minimizes the dissemination of these agents in the receiving waters. Terminal chlorination minimizes the stress (increased microbial loads) on downstream water-treatment plants, recreational areas, and shellfish beds. The rationale for the chlorination of wastewater is as valid today as it was in the early part of the twentieth century.

REFERENCES

1. Sobsey, M., and B. Olson. "Agents of Waterborne Disease," in Assessment of Microbiology and Turbidity Standards for Drinking Water, (U.S. Environmental Protection Agency Workshop, 1983).
2. Goldfield, M. "Epidemiological Indicator For Transmission of Viruses by Water," in *Viruses in Water*, G. Berg, H. L. Bodily, E. H. Lennette, J. L. Melnick, and T. G. Metcalf, Eds. (Washington, DC: American Public Health Association, 1976).
3. Mosely, J. W. "Transmission of Viral Diseases by Drinking Water," in *Transmission of Viruses by the Water Route*, G. Berg, Ed. (New York: Interscience, 1967), pp. 5–23.
4. Kaplan, J. E., R. Feldman, D. Campbell, C. Lookabaugh, and G. W. Gary. "The Frequency of Norwalk-like Pattern of Illness in Outbreaks of Acute Gastroenteritis," *Am. J. Public Health* 72(12):1329–1332 (1982).
5. Blacklow, N. R., and G. Cukor. "Viral Gastroenteritis," *N. Eng. J. Med.* 304:397–406 (1981).

6. Taylor, J. W., G. W. Gay, and H. B. Greenberg. "Norwalk-related Viral Gastroenteritis Due to Contaminated Drinking Water," *Am. J. Epidemiol.* 114(4):584–592 (1981)
7. Baron, R. C., F. D. Murphy, and H. B. Greenberg. "Norwalk Gastrointestinal Illness: An Outbreak Associated with Swimming in a Recreational Lake and Secondary Person-to-Person Transmission," *Am. J. Epidemiol.* 115:163–172 (1982).
8. Wilson, R., L. J. Anderson, R. C. Holman, G. W. Gary, and H. B. Greenberg. "Waterborne Gastroenteritis Due to the Norwalk Agent: Clinical and Epidemiologic Investigation," *Am. J. Public Health* 72(1):72–74 (1982).
9. Banatvala, J. E. "Viruses in Feces," *Proceedings of Viruses and Wastewater Treatment*, M. Goddard and M. Butler, Eds. (Oxford, England: Pergamon Press, 1981).
10. Westwood, J. C. M., and S. A. Sattar. "The Minimal Infective Dose," in *Viruses in Water*, G. Berg, H. L. Bodily, E. H. Lennette, J. L. Melnick, and T. G. Metcalf, Eds. (Washington, DC: American Public Health Association, 1976).
11. Plotkin, S. A., and M. Katz. "Minimal Infective Doses for Man by the Oral Route," in *Transmission of Viruses by the Water Route*, G. Berg, Ed. (New York: Interscience, 1967).
12. Taylor, F. B. "Viruses—What Is Their Significance in Water Supply?" *J. Am. Water Works Assoc.* 66:306 (1974).
13. Koprowski, H. J. "Immunization Against Poliomyelitis with Living Attenuated Virus," *Am. J. Trop. Med. Hyg.* 5:440–452 (1956).
14. Couch, R. B., T. R. Cote, P. J. Gerone, W. F. Fleet, D. J. Lang, W. R. Griffith, and V. Knight. "Production of Illness With a Small Particle Aerosol of Coxsackie A21," *J. Clin. Invest.* 44:535–542 (1965).
15. Centers for Disease Control. "Water-Related Disease Outbreaks," Annual Summary, 1980 (Atlanta, GA: 1982).
16. Centers for Disease Control. "Water-Related Disease Outbreaks," Annual Summary, 1981 (Atlanta, GA: 1982).
17. Centers for Disease Control. "Water-Related Disease Outbreaks," Annual Summary, 1982 (Atlanta, GA: 1983).
18. Akin, E. W., et al. "Health Hazards Associated with Wastewater Effluents and Sludge: Microbiological Considerations," in *Proceedings of the Conference on Risk Assessment and Health Effects of Land Application of Municipal Wastewater and Sludges*," B. P. Sagik and C. A. Sorber, Eds. (San Antonio, TX: University of Texas, 1978).
19. Rendtorff, R. "The Experimental Transmission of Human Intestinal Protozoan Parasites. II. *Giardia lamblia* Cysts given in Capsules," *Am. J. Hyg.* 59:209–220 (1954).
20. Rendtorff, R. "The Experimental Transmission of Human Intestinal Protozoan Parasites. I. *Entamoeba coli* Cysts Given in Capsules," *Am. J. Hyg.* 9:196–208 (1954).
21. *Drinking Water and Health, Vol. 2* (Washington, DC: National Academy Press, 1980).
22. *Drinking Water and Health, Vol. 3* (Washington, DC: National Academy Press, 1980).
23. Hejkal, T. W., B. Keswick, R. L. LaBelle, C. P. Gerba, Y. Sanchez, G. Dreesman, B. Hafkin, and J. L. Melnick. "Viruses in a Community Water Supply Associated with an Outbreak of Gastroenteritis and Infectious Hepatitis," *J. Am. Water Works Assoc.* 318–321 (1982).

24. Moore, B. "A Survey of Beach Pollution at a Seaside Resort," *J. Hyg. Camb.* 52:71 (1954).
25. Moore, B. "Sewage Contamination of Coastal Bathing Waters," *Bull. Hyg.* 29:689 (1954).
26. Bryan, F. L. "Diseases Transmitted by Foods Contaminated by Wastewater," in *Proceedings, U.S. Department of Health, Education, and Welfare, Public Health Service* (1974).
27. Stevenson, A. H. "Studies of Bathing Water Quality and Health," *Am. J. Public Health* 43:529–538.
28. Cabelli, V. J., A. P. Dufour, M. A. Levin, and L. J. McCabe. "The Development of Criteria for Recreational Waters," in *Discharge of Sewage from Sea Outfalls*, A. L. H. Gameson, Ed. (New York: Pergamon Press, 1975), pp. 63–73.
29. Cabelli, V. J., A. P. Dufour, M. A. Levin, and P. W. Habermann. "The Impact of Pollution on Marine Bathing Beaches: An Epidemiological Study," *Limnol. Oceanogr.* 2:424–432 (1976).
30. Cabelli, V. J. "Indicators of Recreational Water Quality," in *Bacterial Indicators/ Health Hazards Associated with Water*, A. W. Hoadley and B. J. Dutka, Eds. *Tech. Publ.* (Philadelphia: American Society for Testing and Materials) 635 pp. 222–238.
31. Rosenberg, M. S., et al. "Transmission of Shigellosis by Swimming in a Contaminated River," EPA-75-18-2 (Atlanta, GA: Centers for Disease Control, 1975).
32. U.S. General Accounting Office. "Problems in Protecting Consumers from Illegally Harvested Shellfish (Clams, Mussels, and Oysters)," GAO/HRD-84-36, report to the Honorable Thomas J. Downey, House of Representatives, June 14, 1984.
33. Murphy, A. M., G. S. Grohman, P. J. Christopher, W. A. Lopey, G. R. Davey, and R. H. Millson. "An Australia-wide Outbreak of Gastroenteritis from Oysters Caused by Norwalk Virus," *Med. J. Aust.* 2:329 (1979).
34. Anonymous. "Viral Infection Spread by Shellfish," *Environ. Health* (October 1981), p. 26.
35. Gill, O. N., W. D. Cubitt, D. A. McSwiggan, B. M. Watney, and C. L. R. Barlett. "Epidemic of Gastroenteritis Caused by Oysters Contaminated with Small Round Structured Viruses," *Br. Med. J.* 287:1532 (1983).
36. Grohman, G. S., A. M. Murphy, P. J. Christopher, E. Auty, and H. B. Greenberg. "Norwalk Virus Gastroenteritis in Volunteers Consuming Depurated Oysters," *Aust. J. Exp. Biol. Med. Sci.* 59:219 (1981).
37. Canzonier, W. J. "Accumulation and Elimination of Coliphage S-13 by the Hard Clam Mercenaria mercenaria," *Appl. Microbiol.* 21:1024 (1971).
38. Comptroller General of the United States. "Unnecessary and Harmful Levels of Domestic Sewage Chlorination Should be Stopped," CED-77-108, report to the Congress (Aug. 30, 1977).

CHAPTER 3

Basic Issues in Water Chlorination: A Chemical Perspective

Robert L. Jolley

Chemistry as a science deals with the composition and nature of materials or substances and the interactions and transformations that these substances undergo. Thus, the chemistry of water chlorination includes the reactions of chlorine and chlorine-produced oxidants with constituents, both living and nonliving, in the water and the various interactions and transformations of the oxidants and their reaction products. Any discussion, from a chemical perspective, of basic issues in water chlorination must consider what of fundamental significance is known, and what is thought to be unknown, concerning the chemistry of water chlorination. Such a discussion must be selective in order to remain succinct. Furthermore, ultimate significance must be defined in ecological rather than purely chemical terms.

The yardstick for measurement of significance to humans must be an ecological one, embracing not only physical phenomena but also biological (including human and social) interactions. Of what significance to the human species is the chemistry of water chlorination, or any technological accomplishment for that matter, if the ecological niche of the human species is so altered as to greatly deteriorate the quality of life? Balanced against this is the necessity of pathogen-free drinking water for the preservation of the quality of human life itself and clean waters for the rearing of edible aquatic organisms such as oysters, crabs, and fish.[1,2] For the past eight decades in the United States, chlorine has been the principal disinfectant for drinking water and has achieved much success in reduction of water-diseases. Only within the last decade have possible adverse ancillary effects of water chlorination been noted. Historically, this observation was not possible until the proper combination of scientific methodology (modern instrumental analysis) and scientific investigators was achieved.[3,4] Because of the subtle ecological endpoints, the ultimate significance of water chlorination practices may not be known for many decades to come. The human situation is never simple.

The chemistry of water chlorination is thought of, rather narrowly, in terms of chemical reactions, kinetics, equilibria, yields, and products. This restriction or specialization permits the use of classical scientific methodology for precise quantitation and careful control of reaction conditions and variables in order to prove or disprove hypotheses and permit corroborative experiments. Such fundamental information is acutely important for understanding and predicting the effects of water chlorination. But in the broader ecological sense, the world may be viewed as a macrocosm in which chemical experiments have been historically, and are currently, being conducted without the benefit of full knowledge concerning chemical effects. The world, because of its relatively vast size in an anthropomorphic sense, has been thought of as an enormous chemical buffer capable of absorbing an astronomically large number of small insults without change of environmental quality. Hopefully, this viewpoint is changing. For example, we are becoming increasingly aware of the possible significance of cumulative small insults, such as the production of low concentrations of low-molecular-weight halogenated organic constituents during water chlorination.

We are at the interface of the relatively easily verifiable chemistry of reactions and products and the less easily verifiable chemistry of subtle ecological effects. What endpoints do we examine to study this lesser understood chemistry? Do we collectively have enough available information to postulate the necessary hypotheses and to define the scientific experiments to objectively prove or disprove the hypotheses? Such questions form the reason for this conference series.

SCOPE OF REVIEW: WATER CHLORINATION AS A CHEMICAL PROCESS

Most of the principal scientists in the area of water chlorination have participated in this conference series. Consequently, much of the research and information developed within the last decade has been summarized in the previous conference proceedings.[5-9] The aqueous chemistry of chlorine as it pertains to water chlorination was specifically reviewed in previous conference proceedings.[10-17] The purpose of this summary is not to repeat the past reviews but to update them and to emphasize the basic issues, including areas of needed research and possible information gaps. Inclusive bibliographies may be found in the other reviews.[3,18-25]

From the standpoint of treatment, water chlorination may be conveniently grouped into four areas: drinking water treatment, wastewater treatment, cooling water treatment, and industrial processes. The latter includes a wide diversity of industrial processes (e.g., food treatment and paper pulp bleaching). For purposes of this presentation, a review of the production of organic chemicals is not included. The unifying principle for these several process waters is that they are governed by the same chemical laws although each water may have an array of similar or dissimilar constituents.

From the standpoint of the unit chemical process, water chlorination may be divided into the following categories: reactions of chlorine with water; formation of chloramines and other chlorine-produced oxidants (CPO); transformation and decomposition reactions of chlorine and CPO; reactions of chlorine and CPO with constituents in the water (chlorine demand) and other media (e.g., food materials); and interaction of chlorine, CPO, and their reaction products with biota, including the human species.

It can be readily seen that the extent of our knowledge is much greater in some areas than in others. For example, much information exists concerning the reactions of chlorine with water and the subsequent decomposition of chlorine and CPO oxidants, whereas relatively less information exists concerning the reaction mechanisms that govern the toxicity of chlorine, CPO, and associated reaction products to living organisms.

CHEMICAL PERSPECTIVE

Reactions of Chlorine with Water

Much information is available concerning the dissolution and hydrolysis of chlorine and the dissociation of hypochlorous acid in aqueous systems. The values of currently accepted kinetic and thermodynamic constants are adequate for the development of models and predictive relationships. Further refinements may be anticipated with respect to precision of values and temperature range.

Formation of Chloramines and Other Chlorine-Produced Oxidants

The formation and decomposition of chloramines and bromamines have been thoroughly studied by several groups. Over a four-decade period, Morris and co-workers have elucidated the chemistry of the chlorine-ammonia-water system.[13,17,26] During the last two decades, Johnson and co-workers have provided much insight into the bromine-ammonia-water system.[18,27] Approximately 5 years ago, two young investigators, W. R. Haag and T. W. Trofe, independently postulated the existence of mixed halamines and catalyzed research in that area.[28,29] Much progress has also been made in the determination of rates of halamine formation and equilibrium constants.[17,29-31]

Analytical Considerations

Process control needs and federal and state regulations require continuing attention to statistically reliable, as well as practicable, analytical methods for chlorine and CPO species, especially at the parts-per-billion or less level. The subject was reviewed by Jolley and Carpenter[32,33] and much progress was reported at the Asilomar Conference.[16]

Sixteen laboratories participated in a comparison of several analytical methods for free available chlorine. Cooper et al. concluded from the resulting series of 192 analyses that FACTS (free available chlorine test with syringaldazine) is equivalent to both the DPD (N,N-diethyl-*p*-phenylenediamine) and amperometric titration methods for free chlorine.[34]

Cooper et al. made a thorough comparative study of several instrumental methods for measuring chlorine residuals in water.[35] Amperometric titration was used as the referee method in this study. They presented ample statistical data and concluded the following:

> The total chlorine analyzer was shown to be analyst independent, providing online continuous measurement. This instrument would be the method of choice when continuous or many total chlorine analyses are required. Although operation of the potentiometric electrode is easy, the results can vary from analyst to analyst and would require additional quality assurance for in-plant multianalyst operation. No conclusion can be made regarding the amperometric membrane electrodes because of the variability in the results.

Wong investigated factors affecting the amperometric titration of trace quantities of chlorine (<0.1 mg/L) in seawater. He concluded that titration at pH 2 gave true concentrations of total residual concentration after correction for the concentration of naturally occurring iodate. The interference of nitrite, significant at <0.1 mg/L chlorine concentrations, can be removed by sulfamic acid addition.[36]

Hatch and Yang concluded that concentrations of chlorine as low as 0.25 mg/L can be determined accurately by the starch-iodine titrimetric method as long as the sample temperature is $<20°C$.[37]

Transformation and Decomposition Reactions of Chlorine and CPO

Our understanding of the transformation and decomposition reactions of aqueous chlorine solutions and CPO formed in natural and process waters has increased in the last several years. This is due principally to the development of highly sophisticated computer techniques for examining the end result of the integration of the many rate equations dealing with the formation and decomposition of chlorine and CPO species.[38-40] It is also due to new insight into the decomposition of chlorine in waters containing bromide (e.g., seawater). Haag observed, with regard to chlorination practices using seawater, that (1) halate formation is minimal in the dark; (2) in the sunlight, halate formation is one of the major causes for halogen disappearance at all salinities; (3) bromate is the major product of the decomposition of hypobromite (formed by the oxidation of the bromide in seawater by hypochlorite) in sunlight-exposed seawater, and that chlorate is not a product; and (4) remaining oxidant loss is probably related to organic demand and oxygen-evolving reactions.[41]

Under conditions of chlorination of "once-through" cooling water (freshwater), Qualls and Johnson concluded that (1) the most important reactions consuming free residual chlorine were formation of NH_2Cl and consumption by humic substances, and (2) most reduction in total residual chlorine was caused by humic substances (i.e., organic demand).[42] Evans studied the decomposition of combined chlorine (chloramines) over the pH range of 3 to 12. Typical first-order rate constants were 2.5×10^{-2}, 1.3×10^{-2}, and $1.9 \times$

10^{-2} min^{-1} at pH 5.0, 7.5, and 11.0, respectively, and 25°C. A rate minimum of 1.1 × 10^{-2} min^{-1} occurred at pH 9.8, where monochloramine should have been the only reactive species.[43] Similar studies should be conducted in selected natural waters.

The speciation of CPO in freshwater and seawater as a function of time has been described by several investigators.[38-40] Morris and Isaac[17] have estimated rate constants for the formation of NH$_2$Cl and NHCl$_2$ at some variance with the values selected by Haag and Lietzke.[39] Furthermore, 22 of the 33 rate constants used in the Haag and Lietzke model were estimated. Although Lietzke concluded that the effect of this estimation is minimal,[44] a more precise measurement of these rate constants would provide greater confidence in the significance of the model. Three of the estimated rate constants relate to organic demand reactions of HOCl, HOBr, and bromamines; four to the formation of RNHBr, RNBr$_2$, RNBrCl (from RNHCl), and RNBrCl (from RNHBr); and three to the hydrolysis of RNHBr, RNBr$_2$, and RNBrCl.[39] These rate constants must remain estimates until the organic reactions involved in demand reactions are better defined and more information concerning the nature of significant organic halamines is obtained. The other estimated rate constants are: (1) formation of NHBr$_2$, NBr$_3$, NHBrCl (from NH$_2$Cl and HOBr), and NHBrCl (from NH$_2$Cl and Br$^-$); hydrolysis of NBr$_3$, NHBrCl, and NH$_2$Br; decomposition of HOBr to O$_2$ and Br$^-$; disproportionation of NH$_2$Br and of NHBr$_2$; reaction of NH$_3$ and NHBr$_2$; and reaction of NH$_2$Br and NBr$_3$.[39]

The effect on speciation models of the use of measured and better estimated values for the above rate constants should be determined. It would be useful to have more rate constant values determined for temperatures other than 25°C. In addition, an effort to unify or integrate the bromamine decomposition chemistry might result in an improved understanding of the decomposition mechanism(s).

Reactions of Chlorine and CPO With Constituents in Water (Chlorine Demand) and Other Reaction Media

Much research in this area in the last several years dealt with trihalomethane (THM) formation and the chlorination of humic materials, wastewaters, cooling waters, and paper industry effluents. However, additional insight has also been given in aqueous chlorine reactions with ozone, nitrite, and uracil. Continued research efforts are needed to more completely identify the products of water chlorination practices and to more fully understand the chemical reactions, rates, and mechanisms that are involved.

General

Wong and Oatts observed that the dissipation of chlorine in seawater occurs in two phases and that the consumption rate of chlorine in the second (slower) phase was $\sim 1 \times 10^{-4}$ mg L^{-1} min^{-1} for all seawater samples studied. They concluded that the decrease of chlorine demand with increasing depth (e.g., 0.9 mg/L at 25 m, 0.5 mg/L at 250 m, and 0.4 mg/L at 2000 m) is caused principally by a decrease in the organic chlorine demand.[45]

The high chlorine demand of fully nitrified wastewaters (low ammonia and high nitrite concentrations) indicates that nitrite is apparently oxidized by free chlorine but not monochloramine.[46]

Gould et al. studied the chlorination of uracil over a pH range of 5 to 9 and at a chlorine-to-uracil mol ratio of 0.5 to 5. The second-order reaction rate constant increased from 0.3 M^{-1} s^{-1} at pH 5 to 5 M^{-1} s^{-1} above pH 7. At low chlorine-to-uracil ratios the chlorination product was exclusively 5-chlorouracil, while at higher ratios a dichlorouracil was formed and, ultimately, ring breakage occurred.[47]

Haag and Hoigne studied the reaction of ozone with aqueous chlorine and chloramines. Ozone was found to react with free chlorine present as OCl$^-$ with a second-order rate constant of 120 M^{-1} s^{-1} at 20°C forming Cl$^-$ (77% yield) and ClO$_3^-$ (23% yield). The reaction of ozone with monochloramine is slower (second-order rate constant, 26 M^{-1} s^{-1} at 20°C) forming Cl$^-$ and NO$_3^-$. The direct reaction of ozone and chlorine accounts for a significant amount of the chlorine and ozone demand found when the two oxidants are used in combination for water treatment.[48]

Trihalomethane Formation

One of the major precursors of THM in the chlorination process is believed to be humic substances.[49,50] Other organic materials such as algae and algal extracellular metabolic products may also be significant THM precursors.[51-53] The THM formation potential for fulvic acid molecular weight fractions was investigated by Lynn.[54] The molecular weight fraction between 5000–10,000 appeared to contribute most to THM formation while the <1000 fraction contributed least.

A predictive model for chloroform formation from humic acid was developed by Engerholm and Amy.[55]

Urano et al. derived an empirical rate equation for THM formation from humic materials and applied it to THM formation in several river and lake waters.[56]

De Laat et al. studied the formation of THM from several model compounds. They reported reaction rate constants (k) of 1.5×10^3, 1.5, and 0.5 M^{-1} s^{-1} for resorcinol, phloroglucinol, and 3,5-dichlorophenol, respectively, where v = k[organic][Cl$_2$]. They concluded that significant amounts of THM can be formed with highly reactive precursors (e.g., metapolyhydroxybenzenes) in the presence of ammonia before reaching breakpoint. Even with a highly reactive compound such as resorcinol, the formation of THM was accompanied by the formation of trichloracetic acid ($\sim$10% yield) and monochloromaleic acid ($\sim$60% yield).[57]

Boyce and Hornig studied the chlorination of 1,3-dihydroxyaromatic compounds and concluded that the formation of THM was generally consistent with a haloform-type reaction mechanism, although several competitive pathways might also be involved. At high chlorine concentrations a variety of chlorine-substituted carboxylic acids were formed in addition to THMs, and at low chlorine concentrations several polyhalogenated intermediates were identified.[58,59]

Trihalomethane Control

A principal method for THM control or for improvement of drinking water quality, in general, should be development of the highest quality source water. For example, THM concentrations can be reduced by observation and control of the natural phytoplankton communities in the water sources for domestic water supplies.[51,53]

One method for controlling THM during water treatment is to remove precursor material. Significant reductions in THM formation potential (THMFP) were achieved by treating the raw water with coagulants before chlorination.[60,61] The effectiveness of the coagulants was dependent upon the coagulant type and dosage, pH, and water characteristics. Glaze and Wallace state that granular activated carbon (GAC) adsorption appears to be the most effective treatment for reduction of THMFP in waters containing high THM precursor levels. They suggested that two mechanisms accounted for the removal of natural organic materials: physical adsorption during the early stages of GAC column operation, and microbiological degradation of the organics adsorbed on the columns. Preozonation of the waters at 2- to 3.5-mg/L ozone dosage before adsorption on the GAC columns did not have significant effect on the removal of THM precursors.[62]

Reed, using a pilot-plant train that provided alum coagulation, sedimentation, filtration, and carbon adsorption, concluded that prechlorination with chloramines and prechlorination with free chlorine were equally effective in removing and controlling THM precursors. Control of bacterial population (as determined by standard plate count) was more consistent with chloramine treatment than with free chlorine, using approximately equivalent chlorine residual levels.[63]

Humic Acid Chlorination

Considerable progress has been made in studying the complex aqueous chlorination reactions of humic substances.[25,42,50,54,56,64-66] However, continuing efforts are necessary to more fully understand the nature of the nonvolatile halogenated organic products of water chlorination. For example, Rook has shown that the nonvolatile, organic-bound chlorine exceeds by a factor of 3 to 5 the volatile fraction produced from water chlorination.[14]

Oliver observed the formation of dihaloacetonitriles (DHAN) on chlorinating aquatic fulvic acids and several aquatic algae. In a survey of 10 chlorinated drinking water samples, he found the average molar DHAN concentration was ~10% of the average molar THM concentration.[64]

Kringstad et al. observed that the strong, direct-acting mutagens 1,3-dichloroacetone and 2-chloropropenal are produced at low concentrations during the chlorination of humic acids. They concluded that these compounds might be responsible for some of the mutagenic activity found in chlorinated drinking waters.[66]

Miller and Uden studied the effects of reaction time, chlorine-to-carbon ratio, pH, and humic acid source on the chlorination products of humic substances. Seventeen major chlorinated products were produced. The principal products were chloroform, dichloroacetic acid, trichloroacetic acid, and chloral hydrate.[65] The principal chlorination products of aquatic humic and fulvic acid samples identified by Christman et al. were, in order of decreasing yield, trichloroacetic acid, chloroform, dichloroacetic acid, and dichlorosuccinic acid.[25,50] The four products accounted for ~50% of the total organic halogen (TOX) produced in the reactions.

McCreary et al. chlorinated fulvic acid samples that were adsorbed on activated carbon using chlorination conditions typical of water processing plants. Phthalates, methyl esters of fatty acids, and mutagenic compounds (Ames test) were isolated from the influent portion of the activated carbon filters.[67] Previously, McCreary et al. had observed that activated carbon promoted the reaction of free chlorine at pH 6 with phenolic acids to produce a variety of products including polyphenols and quinones that were apparently formed by decarboxylation, hydroxylation, oxidation, and chlorine substitution reactions.[68]

Drinking Waters

There is an ongoing national and international effort to identify the products of chlorination and other oxidant disinfection treatments and to determine the toxicological implications of such products.

Many investigators have concluded that nonvolatile organohalide compounds are usually formed in much greater quantities than THMs during water chlorination.[14,69] According to Fleischacker and Randtke, organohalide formation can be minimized by avoiding prechlorination, avoiding higher concentrations of chlorine than are necessary for good disinfection, using combined rather than free chlorine, maintaining a high pH, using a source water with low concentrations of precursors, and maintaining a cool temperature.[70]

Uden and Miller detected dichloroacetic acid and trichloroacetic acid in the concentration range of 30 to 160 μg/L (comparable to chloroform) in tap water samples from two water treatment plants. Chloral concentrations were lower by a factor of ten. Qualitative studies indicated that all the chlorinated compounds found in the tap water samples were also formed in the chlorination of soil fulvic acids.[71]

Because only ~25% of the organic halides in chlorinated drinking water are detectable and measurable by chromatographic means, the analysis of organic halides as a group parameter may be of value as a unique indicator of potentially harmful organic halides. A method and instrument have been developed that measure the organic halide content of a water sample by carbon adsorption and pyrolysis.[72]

Wastewaters

Progress is continuing in identification of chlorination and disinfection byproducts of wastewaters. However, national effort in this area is considerably diminished. The basic question—"Is wastewater disinfection necessary?"—remains unresolved.

Thorough studies on the products of chlorination of wastewaters and superchlorination of wastewater sludges have been continued by Glaze and coworkers.[73-75] Among the products they have identified are monochloro- and dichlorotyrosine. By analogy with brominated tyrosine, it is postulated that a possible toxic effect of these chlorinated tyrosines is an inhibition of thyroid hormone synthesis.[73] Other identified compounds are trichloropropionitrile, polychlorinated acetone, and chlorinated alkylbenzenes.[74] The latter type of compounds may result from decarboxylation of chlorinated aromatic acids.[76] However, Hofler et al. indicate that a possible source of chlorinated benzenes is the direct reaction of aqueous chlorine with benzene, even in the absence of light.[77]

Grady et al. concluded that a major effect of chlorination of wastewaters was a drastic reduction in molecular size of constituents.[78] This is consistent with the findings of Jolley that the major reaction occurring during chlorination of wastewaters is oxidation rather than chlorine substitution.[79]

Breakpoint chlorination of wastewater effluents (activated sludge treatment) resulted in chlorine incorporation into the organic matter of 0.019- to 0.067-mg chlorine per mg carbon.[80] In effluents with high dissolved oxygen (DO) (e.g., 7 mg/L), the chlorine incorporation in organic matter was significantly less than that in effluents with low DO (e.g., 1 mg/L). This fact implies that oxidation of organic constituents or sites susceptible to chlorine substitution occurs at higher DO levels. The bulk of the chlorine incorporated was associated with organics of <1000 mol wt.[48,81]

Haas and Karra describe a two-phase chlorine demand model for the kinetics of wastewater chlorination. The model is derived from the assumption of parallel first-order decay of two principal components of total chlorine residual. Approximately 10 to 40% of the applied chlorine decays with a rate constant of ~1.0 min^{-1}.[82] A possible basis for this rapid loss may be reactions with organic material.[83-85] The values for the rate constant of the second, slow rate of decomposition cluster around 0.003 min^{-1}, a rate consistent with the decomposition of pure inorganic and organic chloramines.[82]

Cooling Waters

Power plants are the largest water users in the United States, exceeding the demand for both potable and irrigation water. About one-third of the cooling waters are supplied by estuarine and marine waters and about two-thirds are supplied by freshwaters. There is considerable potential for ecological effects from chlorination of cooling waters because of the large amounts of water used.[85] During the last several years, research in this area has been severely curtailed. The apparent rationale is that no earth-shaking chemical and biological effects have been observed to date and that minimization programs will mitigate known effects. However, there are still many information gaps. A basic question is, "Are we asking the appropriate questions and examining the correct endpoints?"

Bean[24] and Hollod and Wilde[86] corroborated the previous reports of Jolley et al.[12] and Smith et al.[87] that THMs are formed at the parts-per-billion concentration level during the chlorination of cooling waters. However, although Hollod and Wilde assumed insignificance,[86] they did not report sufficient data to permit calculation of THM production and loss to the atmosphere. Jolley et al. conservatively estimated that the production of THMs from chlorination of cooling waters at electric power plants in the United States was in the range of 100 to 200 tons per year.[24] The ecological significance of this THM release to the atmosphere has not been established. It should be noted in this regard that THMs and other halogenated organics are produced naturally by some algae.[88] For example, bromoform was detected at the 1- to 100-μg/L range at all depths of the Arctic Ocean.[89]

Halogenated phenols have been detected in the sediments in chlorinated discharges from nuclear power plants and in river sediments in the Netherlands.[24,90] In the latter study, the highest concentrations of halogenated phenols (up to 6.3 μg/kg sediment on a dry basis) were found in sediment samples in rivers draining highly industrialized areas.[90] Bean reported formation of THMs at the parts-per-billion level in the chlorinated cooling waters of seven nuclear power plants throughout the United States. A complex mixture of halogenated phenols was also found in the discharges. Cooling towers concentrated the halogenated phenols to the parts-per-billion level. No significant concentrations of lipophilic base-neutral compounds were identified in the chlorinated discharges.[24]

Sanders observed that chlorination of estuarine cooling waters with >5 ppt salinity resulted in an average organic constituent oxidation of 35 μmol/L. In every case, scission of macromolecules was apparent because of the increase of organic carbon associated with the <1000 and <10,000 mol wt fractions. He observed that coprecipitation of metal oxyhydroxides may be enhanced by chlorination.[91]

Helz et al. found that ~90% of the applied dosage (~1.5 mg/L chlorine) to estuarine cooling water reacted very rapidly and did not exit the major electric power plant. The remainder of the oxidant residual decayed by a first-order rate law ($T_{1/2}$ = 0.6 to 4.6 h). Although colloidal bromocarbons were detected in the discharge plume >6 km from the plant, only traces of bromoform were observed in the effluent. An increase in ammonia concentration and a decrease in soluble manganese concentration appeared to be associated with the chlorination of the cooling waters.[85]

Paper Industry Effluents

A significant portion of the nation's chlorine production is used for bleaching paper pulp. Much active research is occurring in the area of identification of chlorination by-products from the paper industry.

Aqueous chlorination of dihydroconiferyl alcohol, a phenol present in pulping liquors, produced several products. The two major products were chlorinated epoxides.[32]

Talka examined several chlorinated effluents from pulp bleaching processes. He detected a higher proportion of chlorocatechols in the chlorination stage and a higher proportion of chloroguaiacols in the alkali extraction stage. The bleaching effluents of hardwood pulp contained chlorosyringols and chlorinated syringaldehydes.[93] Chlorinated guaiacols at concentrations of 4 µg/L to 1.6 mg/L were found in the extraction stage spent bleach liquor from a pine kraft pulp mill.[94]

The following major chlorinated compounds were found by Voss in biologically treated effluents from bleached kraft mills: chloroform, dichlorosulfone, trichlorosulfone, and tetrachlorosulfone. He observed that the chlorinated sulfones were not highly toxic.[95]

Kringstad et al. observed that a minor part of the mutagenic activity of the spent chlorination liquor from the bleaching of softwood kraft pulp was found in highly lipophilic compounds. The lipophilic compounds appeared to be quite stable under conditions similar to that of receiving waters. Consequently, they recommended bioaccumulation studies with appropriate species.[96]

Food Industry

Only limited research has occurred in this area of considerable importance. This is an area of such potential significance for human health that a greater national effort should be made towards identifying and quantifying possible chlorination and disinfection by-products, and toward determining possible toxicological effects. Kirk, Ghanbari, and their co-workers are currently the principal national investigators.[97] Some of the potential areas of concern are by-products from chlorine washing of fish, poultry, and meats; and by-products from bleaching flour.

Interaction of Chlorine, CPO, and Reaction Products with Biota

During the last two decades, the toxicities of chlorine, CPO, and some chlorination reaction products (e.g., chlorinated phenols) to microorganisms and larger biological species have been better defined.[9,18,98,99] However, fundamental studies are lacking on the chemical mechanisms involved in the biotoxicity of chlorine, CPO, and chlorination reaction products. Progress in this area with respect to microorganisms was summarized by Olivieri and Fujioka and their co-workers at the Colorado Springs (1979) and the Asilomar (1981) Conferences.[100-103]

Fundamental questions, such as the following, remain unanswered or only partially answered:

- Is bioaccumulation strictly a function of lipophilicity? Are chlorine, CPO, and reaction products incorporated into biological/cellular components? For example, Cumming demonstrated that mice drinking water containing 5-chlorouracil incorporated the halogenated pyrimidine into the DNA of the mice.[104] Are there genetic or other subtle endpoints that should be examined? Crathorne and co-workers detected both 5-chlorouracil and 5-chlorouridine in drinking waters from several sources.[105] Thus, there exists an unknown but possible human health risk from the presence of these chlorinated pyrimidine and nucleoside compounds in drinking waters. Because of the potential implications of these findings, corroborative research is needed.
- What individual behavioral or species behavioral changes are attributable to water chlorination practices? For example, are pheromonic communications of anadromous fish altered by cooling water chlorination? Further research in this area would be desirable.
- What chemical reactions are involved in toxicity? Oxidation? Chloramination? Chlorine-substitution? Additional understanding of the chemical reactions responsible for toxicity will facilitate optimization of disinfection processes and minimization of deleterious by-product effects.
- What chlorination products and by-products are responsible for observed toxicity? Efforts regarding clarification of mechanisms of toxicity should be targeted at those chemical species which are most abundant and toxic.
- Are oxidants transferred through cell walls? Through cells into the blood serum? How are chlorine, CPO, and/or reaction products passed through the cell membranes and walls? Is the chlorine reactivity and oxidant potential retained within the cell? Within the blood serum? Further information in this area may permit examination of additional, and perhaps heretofore unknown, toxicological endpoints.

SUMMARY

Although much progress has been made toward developing predictive models for chlorination discharges, additional data are needed for reaction rates and mechanisms over a wider temperature range, especially for transformation and decomposition reactions of chlorine and chlorine-produced oxidants. Significant advances have also been made toward understanding the formation of THMs, chlorination of humic materials, and identification of chlorination products in cooling waters, paper industry effluents, and wastewaters; but, because of the vast number of chlorination products, much remains unknown in this area. Only very limited information is available concerning products from water chlorination treatments of food materials; much research is needed.

In summary, since the Asilomar Conference (1981) significant progress has occurred in the classical areas of the chemistry of water chlorination. There has been much less progress toward understanding the chemistry underlying the toxicity of chlorine, CPO, and reaction products. It is in the latter area that ultimate ecological (including mankind) significance may lie.

ACKNOWLEDGMENTS

Appreciation is expressed to Dr. Nathaniel F. Barr, Manager, Health and Environmental Risk Analysis Program, Office of Health and Environmental Research, United States Department of Energy, for making this review possible. Appreciation is also expressed to C. E. Lamb and O. K. Tallent for critical review, to D. R. Reichle for editing assistance, and to M. M. Dawson for typing this manuscript.

Research sponsored by the Health and Environmental Risk Analysis Program, Office of Health and Environmental Research, United States Department of Energy under contract DE-AC05-84OR21400 with Martin Marietta Energy Systems, Inc.

REFERENCES

1. Orihuela, L. A., R. C. Ballance, and R. Novick. "Worldwide Aspects of Water Chlorination," in *Water Chlorination: Environmental Impact and Health Effects, Vol. 2*, R. L. Jolley, H. Gorchev, and D. H. Hamilton, Eds. (Ann Arbor, MI: Ann Arbor Science Publishers, Inc., 1978), pp. 493–507.
2. Lippy, E. C., and S. C. Waltrip. "Waterborne Disease Outbreaks—1946–1980: A Thirty-Five Year Perspective," *J. Am. Water Works Assoc.* 76(2):60–67 (1984).
3. Jolley, R. L. *Chlorination Effects on Organic Constituents in Effluents from Domestic Sanitary Sewage Treatment Plants*, ORNL/TM-4290 (Oak Ridge, TN: Oak Ridge National Laboratory, 1973).
4. Rook, J. J. "Formation of Haloforms During Chlorination of Natural Waters," *Water Treat. Exam.* 23:234–243 (1974).
5. Jolley, R. L. *Water Chlorination: Environmental Impact and Health Effects, Vol. 1*, (Ann Arbor, MI: Ann Arbor Science Publishers, Inc., 1978).
6. Jolley, R. L., H. Gorchev, and D. H. Hamilton. *Water Chlorination: Environmental Impact and Health Effects, Vol. 2* (Ann Arbor, MI: Ann Arbor Science Publishers, Inc., 1978).

7. Jolley, R. L., W. A. Brungs and R. B. Cumming. *Water Chlorination: Environmental Impact and Health Effects, Vol. 3* (Ann Arbor, MI: Ann Arbor Science Publishers, Inc., 1980).
8. Jolley, R. L., W. A. Brungs, J. A. Cotruvo, R. B. Cumming, J. S. Mattice,and V. A. Jacobs. *Water Chlorination: Environmental Impact and Health Effects, Vol. 4, Book 1, Chemistry and Water Treatment* (Ann Arbor, MI: Ann Arbor Science Publishers, Inc., 1983).
9. Jolley, R. L., W. A. Brungs, J. A. Cotruvo, R. B. Cumming, J. S. Mattice, and V. A. Jacobs. *Water Chlorination: Environmental Impact and Health Effects, Vol. 4, Book 2, Environment, Health, and Risk* (Ann Arbor, MI: Ann Arbor Science Publishers, Inc., 1983).
10. Carlson, R. M., and R. Caple. "Organochemical Implications of Water Chlorination," in *Water Chlorination: Environmental Impact and Health Effects, Vol. 1*, R. L. Jolley, Ed. (Ann Arbor, MI: Ann Arbor Science Publishers, Inc., 1978), pp. 65-75
11. Jolley, R. L., G. Jones, W. W. Pitt, and J. E. Thompson. "Chlorination of Organics in Cooling Waters and Process Effluents," in *Water Chlorination: Environmental Impact and Health Effects, Vol. 1*, R. L. Jolley, Ed. (Ann Arbor, MI: Ann Arbor Science Publishers, Inc., 1978), pp. 105-138.
12. Jolley, R. L., W. W. Pitt, F. G.Taylor, S. J. Hartmann, G. Jones, and J. E. Thompson. "An Experimental Assessment of Halogenated Organics in Waters from Cooling Towers and Once-Through Systems," in *Water Chlorination: Environmental Impact and Health Effect, Vol. 2*, R. L. Jolley, H. Gorchev, and D. H. Hamilton, Eds. (Ann Arbor, MI: Ann Arbor Science Publishers, Inc., 1978), pp. 695-705.
13. Morris J. C. "The Chemistry of Aqueous Chlorine in Relation to Water Chlorination," in *Water Chlorination: Environmental Impact and Health Effects, Vol. 1*, R. L. Jolley, Ed. (Ann Arbor, MI: Ann Arbor Science Publishers, Inc., 1978), pp. 21-35.
14. Rook, J. J. "Possible Pathways for the Formation of Chlorinated Degradation Products During Chlorination of Humic Acids and Resorcinol," in *Water Chlorination: Environmental Impact and Health Effects, Vol. 3*, R. L. Jolley, W. A. Brungs, and R. B. Cumming, Eds. (Ann Arbor, MI: Ann Arbor Science Publishers, Inc., 1980), pp. 85-98.
15. Bean, R. M. "Recent Progress in the Organic Chemistry of Water Chlorination," in *Water Chlorination: Environmental Impact and Health Effects, Vol. 4*, R. L. Jolley, W. A. Brungs, J. A. Cotruvo, R. B. Cumming, J. S. Mattice, and V. A. Jacobs, Eds. (Ann Arbor, MI: Ann Arbor Science Publishers, Inc., 1983), pp. 843-850.
16. Jolley, R. L., and J. H. Carpenter. "A Review of the Chemistry and Environmental Fate of Reactive Oxidant Species in Chlorinated Water," in *Water Chlorination: Environmental Impact and Health Effects, Vol. 4*, R. L. Jolley, W. A. Brungs, J. A. Cotruvo, R. B. Cumming, J. S. Mattice and V. A. Jacobs, Eds. (Ann Arbor, MI: Ann Arbor Science Publishers, Inc., 1983), pp. 3-47.
17. Morris, J. C., and R. A. Isaac. "A Critical Review of Kinetic and Thermodynamic Constants for the Aqueous Chlorine-Ammonia System," in *Water Chlorination: Environmental Impact and Health Effects, Vol. 4*, R. L. Jolley, W. A. Brungs, J. A. Cotruvo, R. B.Cumming, J. S. Mattice, and V. A. Jacobs, Eds. (Ann Arbor, MI: Ann Arbor Science Publishers, Inc., 1983), pp. 49-62.

18. Johnson, J. D. *Disinfection, Water and Wastewater* (Ann Arbor, MI: Ann Arbor Science Publishers, Inc., 1975).
19. Morris, J. C. *Formation of Halogenated Organics by Chlorination of Water Supplies (A Review)*, EPA-600/1-75-002 (Washington, DC: United States Environmental Protection Agency, 1975).
20. Carlson, R. M., and R. Caple. *Chemical Implications of Using Chlorine and Ozone for Disinfection*, EPA-600/3-77-066 (Duluth, MN: United States Environmental Protection Agency, 1977).
21. Pierce, R. C. *The Aqueous Chlorination of Organic Compounds: Chemical Reactivity and Effects on Environmental Quality*, NRCC 16450 (Ottawa, Ontario: National Research Council, 1978).
22. Morris, J. C., R. F. Christman, W. H. Glaze, R. C. Hoehn, and R. L. Jolley. "The Chemistry of Disinfectants in Water: Reactions and Products," in *Drinking Water and Health, Vol. 2,* Safe Drinking Water Comittee (Washington, DC: National Academy Press, 1980), pp. 139–249.
23. Hall, L. W., G. R. Helz, and D. T. Burton. *Power Plant Chlorination: A Biological and Chemical Assessment* (Ann Arbor, MI: Ann Arbor Science Publishers, Inc., 1981).
24. Bean, R. M. *Organohalogen Products from the Chlorination of Cooling Water at Nuclear Power Stations*, NUREG/CR-3408(PNL-4788), (Washington, DC: United States Nuclear Regulatory Commission, 1983).
25. Christman, R. F., J. D. Johnson, D. S. Millington, and A. A. Stevens. *Chemical Reactions of Aquatic Humic Materials with Selected Oxidants*, EPA-600/D-83-117 (Cincinnati: United States Environmental Protection Agency, 1983).
26. Isaac, R. A., and J. C. Morris. "Modeling of Reactions Between Aqueous Chlorine and Nitrogenous Compounds," in *Water Chlorination: Environmental Impact and Health Effects, Vol. 4,* R. L. Jolley, W. A. Brungs, J. A. Cotruvo, R. B. Cumming, J. S. Mattice, and V. A. Jacobs, Eds., (Ann Arbor, MI: Ann Arbor Science Publishers, Inc., 1983), pp. 63–75.
27. Johnson, J. D., and R. Overby. "Bromine and Bromamine Disinfection Chemistry," *J. Sanit. Eng. Div., Amer. Soc. Civ. Eng.* 97:617–627. (1971).
28. Haag, W. R. "Formation of N-Bromo-N-Chloramines in Chlorinated Saline Waters," in *Water Chlorination: Environmental Impact and Health Effects, Vol. 3,* R. L. Jolley, W. A. Brungs, and R. B. Cumming, Eds., (Ann Arbor, MI: Ann Arbor Science Publishers, Inc., 1980), pp. 193–201.
29. Trofe, T. W., G. W. Inman, and J. D. Johnson. "The Kinetics of Monochloramine Decomposition in the Presence of Bromide," *Environ. Sci. Technol.* 14(5):544–549 (1980).
30. Cromer, J. L., G. W. Inman, and J. D. Johnson. "Dibromamine Decomposition Kinetics," in *Chemistry of Wastewater Technology*, A. J. Rubin, Ed. (Ann Arbor, MI: Ann Arbor Science Publishers, Inc., 1978), pp. 213–225.
31. Wajon, J. E., and J. C. Morris. "Bromamination Chemistry: Rates of Formation of NH_2Br and Some N-Bromoamino Acids," in *Water Chlorination: Environmental Impact and Health Effects, Vol. 3,* R. L. Jolley, W. A. Brungs, and R. B. Cumming, Eds. (Ann Arbor, MI: Ann Arbor Science Publishers, Inc., 1980), pp. 171–181.
32. Jolley, R. L., and J. H. Carpenter. *Aqueous Chemistry of Chlorine: Chemistry, Analysis, and Environmental Fate of Reactive Oxidant Species*, ORNL/TM-7788 (Oak Ridge, TN: Oak Ridge National Laboratory, 1982).

33. Jolley, R. L., and J. H. Carpenter. "Review of Analytical Methods for Reactive Oxidant Species in Chlorinated Water," in *Water Chlorination: Environmental Impact and Health Effects, Vol. 4,* R. L. Jolley, W. A. Brungs, J. A. Cotruvo, R. B. Cumming, J. S. Mattice, and V. A. Jacobs, Eds. (Ann Arbor, MI: Ann Arbor Science Publishers, Inc., 1983), pp. 611–652.
34. Cooper, W. J., P. H. Gibbs, E. M. Ott, and P. Patel. "Equivalency Testing of Procedures for Measuring Free Available Chlorine: Amperometric Titration, DPD, and FACTS," *J. Am. Water Works. Assoc.* 75(12):625–629 (1983).
35. Cooper, W. J., M. F. Mehran, R. A. Slifker, D. A. Smith, J. T. Villate, and P. H. Gibbs. "Comparison of Several Instrumental Methods for Determining Chlorine Residuals in Drinking Water," *J. Am. Water Works Assoc.* 74(10):546–552 (1982).
36. Wong, G. T. F. "Factors Affecting the Amperometric Determination of Trace Quantities of Total Residual Chlorine in Seawater," *Environ. Sci. Technol.* 16:785–790 (1982).
37. Hatch, G. L., and V. Yang. "Determining Residual Chlorine: Effect of Temperature on the Titrametric Starch-Iodine End Point, *J. Am. Water Works Assoc.* 75(3):154–156 (1983).
38. Lietzke, M. H. "A Kinetic Model for Predicting the Composition of Chlorinated Water Discharged from Power Plant Cooling Systems," in *Water Chlorination: Environmental Impact and Health Effects, Vol. 2,* R. L. Jolley, H. Gorchev, and D. H. Hamilton, Eds. (Ann Arbor, MI: Ann Arbor Science Publishers, Inc., 1978), pp. 707–716.
39. Haag, W. R., and M. H. Lietzke. "A Kinetic Model for Predicting the Concentrations of Active Halogen Species in Chlorinated Saline Cooling Waters," in *Water Chlorination: Environmental Impact and Health Effects, Vol. 3,* R. L. Jolley, W. A. Brungs, and R. B. Cumming, Eds. (Ann Arbor, MI: Ann Arbor Science Publishers, Inc., 1980), pp. 425–426.
40. Haag, W. R., and M. H. Lietzke. *A Kinetic Model for Predicting the Concentrations of Active Halogen Species in Chlorinated Saline Cooling Waters,* ORNL/TM-7942 (Oak Ridge, TN: Oak Ridge National Laboratory, 1981).
41. Haag, W. R. "On the Disappearance of Chlorine in Seawater," *Water Res.* 15:937–940 (1981).
42. Qualls, R. G., and J. D. Johnson. "Kinetics of the Short-Term Consumption of Chlorine by Fulvic Acid," *Environ. Sci. Technol.* 17(11):692–698 (1983).
43. Evans, O. M. "Voltammetric Determination of the Decomposition Rates of Combined Chlorine in Aqueous Solution," *Anal. Chem.* 54:1579–1582 (1982).
44. Lietzke, M. H. Personal communication (Knoxville: University of Tennessee, 1984).
45. Wong, G. T. F., and T. J. Oatts. "Chlorine Demand in Deep Seawater," *Water Res.* 17(11):1533–1535 (1983).
46. Dhaliwal, B. S., and R. A. Baker. "Role of Ammonia-N in Secondary Effluent Chlorination," *J. Water Pollut. Control Fed.* 55(5):454–456 (1983).
47. Gould, J. P., J. T. Richards, and M. G. Miles. "The Kinetics and Primary Products of Uracil Chlorination," *Water Res.* 18(2):205–212 (1984).
48. Haag, W. R., and J. Hoigne. "Ozonation of Water Containing Chlorine or Chloramines," *Water Res.* 17(10):1397–1402 (1983).
49. Stevens, A. A., C. J. Slocum, D. R. Seeger, and G. G. Roebeck. "Chlorination of Organics in Drinking Water," in *Water Chlorination: Environmental Impact*

and Health Effects, Vol. 1, R. L. Jolley, Ed. (Ann Arbor, MI: Ann Arbor Science Publishers, Inc., 1978), pp. 77–104.
50. Christman, R. F., D. L. Norwood, D. S. Millington, J. D. Johnson, and A. A. Stevens. "Identity and Yields of Major Halogenated Products of Aquatic Fulvic Acid Chlorination," *Environ. Sci. Technol.* 17:625–628 (1983).
51. Hoehn, R. C., C. W. Randall, R. P. Goode, and P. T. B. Shaffer. "Chlorination and Water Treatment for Minimizing Trihalomethanes in Drinking Water," in *Water Chlorination: Environmental Impact and Health Effects, Vol. 2,* R. L. Jolley, H. Gorchev, and D. H. Hamilton, Eds. (Ann Arbor, MI: Ann Arbor Science Publishers, Inc., 1978), pp. 519–535.
52. Morris, J. C., and B. Baum. "Precursors and Mechanisms of Haloform Formation in the Chlorination of Water Supplies," in *Water Chlorination: Environmental Impact and Health Effects, Vol. 2,* R. L. Jolley, H. Gorchev, and D. H. Hamilton, Eds. (Ann Arbor, MI: Ann Arbor Science Publishers, Inc., 1978), pp. 29–48.
53. Briley, K. F., R. F. Williams, and C. A. Sorber. *Alternative Water Disinfection Schemes for Reduced Trihalomethane Formation, Volume II, Algae as Precursors for Trihalomethanes in Chlorinated Drinking Water,* EPA-600/S2-84-005 (Cincinnati: United States Environmental Protection Agency, 1984).
54. Lynn, S. W. *An Analytical Survey of Chloroform Formed from the Chlorination of Humic Substances,* Ph.D. Dissertation (Amherst: University of Massachusetts, 1982).
55. Engerholm, B. A., and G. L. Amy. "A Predictive Model for Chloroform Formation from Humic Acid," *J. Am. Water Works Assoc.* 75(8):418–423 (1983).
56. Urano, K., H. Wada, and T. Takemasa. "Empirical Rate Formation for Trihalomethane Formation with Chlorination of Humic Substances in Water," *Water Res.* 17(12):1797–1802 (1983).
57. De Laat, J., N. Merlet, and M. Dore. "Chloration de Composes Organiques: Demande en Chlore et Reactivite vis-a-vis de la Formation des Trihalomethanes," *Water Res.* 16:1437–1450 (1982).
58. Boyce, S. D., and J. F. Hornig. "Reaction Pathways of Trihalomethane Formation from the Halogenation of Dihydroxyaromatic Model Compounds for Humic Acid," *Environ. Sci. Technol.* 17:202–211 (1983).
59. Boyce, S. D., and J. F. Hornig. "Formation of Trihalomethanes from the Halogenation of 1,3-Dihydroxybenzenes in Dilute Aqueous Solution," in *Water Chlorination: Environmental Impact and Health Effects, Vol. 4,* R. L. Jolley, W. A. Brungs, J. A. Cotruvo, R. B. Cumming, J. S. Mattice, and V. A. Jacobs, Eds. (Ann Arbor, MI: Ann Arbor Science Publishers, Inc., 1983), pp. 277–267.
60. Amy, G. L., and P. A. Chadik. "Cationic Polyelectrolytes as Primary Coagulants for Removing Trihalomethane Precursors," *J. Am. Water Works Assoc.* 75(10):527–531 (1983).
61. Chadik, P. A., and G. L. Amy. "Removing Trihalomethane Precursors from Various Natural Waters by Metal Coagulants," *J. Am. Water Works Assoc.* 75(10):532–536 (1983).
62. Glaze, W. H., and J. L. Wallace. "Control of Trihalomethane Precursors in Drinking Water: Granular Activated Carbon With and Without Preozonation," *J. Am. Water Works Assoc.* 76(2):68–75 (1984).
63. Reed, G. D. "Effects of Prechlorination on THM Formation and Microbial Growth in Pilot-Plant Units," *J. Am. Water Works Assoc.* 75(8):426–439 (1983).

64. Oliver, B. G. "Dihaloacetonitriles in Drinking Water: Algae and Fulvic Acid as Precursors," *Environ. Sci. Technol.* 17:80–83 (1983).
65. Miller, J. W., and P. C. Uden. "Characterization of Nonvolatile Aqueous Chlorination Products of Humic Substances," *Environ. Sci. Technol.* 17:150–157 (1983).
66. Kringstad, K. P., P. O. Ljungquist, F. de Sousa, and L. M. Stromberg. "On the Formation of Mutagens in the Chlorination of Humic Acid," *Environ. Sci. Technol.* 17:553–555 (1983).
67. McCreary, J. J., K. R. Batsel, and J. R. Rivera. *Reactions of Chlorine Disinfectant with Organics Adsorbed on Granular Activated Carbon*, WRRC-PUB-70 (Gainesville: Florida Water Resources Research Center, 1983).
68. McCreary, J. J., V. L. Snoeyink, and R. A. Larson. "Comparison of the Reaction of Aqueous Free Chlorine with Phenolic Acids in Solution and Adsorbed on Granular Activated Carbon," *Environ. Sci. Technol.* 16:339–344 (1982).
69. Johnson, D. E., and S. J. Randtke. "Removing Nonvolatile Organic Chlorine and Its Precursors by Coagulation and Softening," *J. Am. Water Works Assoc.* 75(5):249–253 (1983).
70. Fleischacker, S. J., and S. J. Randtke. "Formation of Organic Chlorine in Public Water Supplies," *J. Am. Water Works Assoc.* 75(3):132–138 (1983).
71. Uden, P. C., and J. W. Miller. "Chlorinated Acids and Chloral in Drinking Water," *J. Am. Water Works Assoc.* 75(10):524–527 (1983).
72. Dressman, R. C., and A. A. Stevens. "The Analysis of Organohalides in Water—An Evaluation Update," *J. Am. Water Works Assoc.* 75(8):431–434 (1983).
73. Burleson, J. L., G. R. Peyton, and W. H. Glaze. "Gas Chromatographic/Mass Spectrometric Analysis of Derivatized Amino Acids in Municipal Wastewater Products," *Environ. Sci. Technol.* 14:1354–1359 (1980).
74. Henderson, J. E., and W. H. Glaze. "GC/MS Analysis of XAD-2 Extracts of Superchlorinated Sludges," *Water Res.* 16:211–218 (1982).
75. Henderson, J. E. *GC/MS Analysis of Chlorinated Organic Compounds in Municipal Wastewater After Chlorination*, Ph.D. Dissertation, (Denton, TX: North Texas State University, 1982).
76. Rockwell, A. L., and R. A. Larson. "Aqueous Chlorination of Some Phenolic Acids," in *Water Chlorination: Environmental Impact and Health Effects, Vol. 2*, R. L. Jolley, H. Gorchev, and D. H. Hamilton, Eds. (Ann Arbor, MI: Ann Arbor Science Publishers, Inc., 1978), pp. 67–74.
77. Hofler, M., E. S. Lahaniatis, D. Bieniek, and F. Korte. "Reaktionsverhalten von Benzol—in ppm Konzentrationen bei der Chlorierung mit NaOCl in Wasseriger Losung," *Chemosphere* 12(2):217–224 (1983).
78. Grady, C. P. L., E. J. Kirsch, M. K. Koczwara, B. Trgovich, and R. D. Watt. "Molecular Weight Distributions in Activated Sludge Effluents," *Water Res.* 18(2):239–246 (1984).
79. Jolley, R. L. "Chlorine-Containing Organic Constituents in Chlorinated Effluents," *J. Water Pollut. Control Fed.* 47(3):601–618 (1975).
80. Koczwara, M. K., E. J. Kirsch, and C. P. L. Grady. "Formation of Organic Chlorine in Activated Sludge Effluents," *Water Res.* 17(2):1863–1869 (1983).
81. Trgovich, B., E. J. Kirsch, and C. P. L. Grady. "Characteristics of Activated Sludge Effluents Before and After Breakpoint Chlorination," *J. Water Pollut. Control Fed.* 55(7)966–976 (1983).
82. Haas, C. N., and S. B. Karra. "Kinetics of Wastewater Chlorine Demand Exertion," *J. Water Pollut. Control Fed.* 56(2):170–173 (1984).

83. Helz, G. R., R. Sugam, and R. Y. Hsu. "Chlorine Degradation and Halocarbon Production in Estuarine Waters," in *Water Chlorination: Environmental Impact and Health Effects, Vol. 2,* R. L. Jolley, H. Gorchev, and D. H. Hamilton, Eds. (Ann Arbor, MI: Ann Arbor Science Publishers, Inc., 1978), pp. 209–222.
84. Helz, G. R., D. A. Dotson, and A. C. Sigleo. "Chlorine Demand: Studies Concerning its Chemical Basis," in *Water Chlorination: Environmental Impact and Health Effects, Vol. 4,* R. L. Jolley, W. A. Brungs, J. A. Cotruvo, R. B. Cumming, J. S. Mattice, and V. A. Jacobs, Eds. (Ann Arbor, MI: Ann Arbor Science Publishers, Inc., 1983), pp. 181–190.
85. Helz, G. R., R. Sugam, and A. C. Sigleo. "Chemical Modifications of Estuarine Water by a Power Plant Using Continuous Chlorination," *Environ. Sci. Technol.* 18(3):192–199 (1984).
86. Hollod, G. J., and E. W. Wilde. "Trihalomethanes in Chlorinated Cooling Waters of Nuclear Reactors," *Bull. Environ. Contam. Toxicol.* 28:404–408 (1983).
87. Smith, J. H., J. C. Harper, and B. C. DaRoss. "Atmospheric Emissions from Electric Power Plant Cooling Systems," in *Water Chlorination: Environmental Impact and Health Effects, Vol. 4,* R. L. Jolley, W. A. Brungs, J. A. Cotruvo, R. B. Cumming, J. S. Mattice, and V. A. Jacobs, Eds. (Ann Arbor, MI: Ann Arbor Science Publishers, Inc., 1983), pp. 391–404.
88. Siuda, J. "1980. Natural Production of Organohalogens," in *Water Chlorination: Environmental Impact and Health Effects, Vol. 3,* R. L. Jolley, W. A. Brungs, and R. B. Cumming, Eds. (Ann Arbor, MI: Ann Arbor Science Publishers, Inc., 1980), pp. 63–72.
89. Dyrssen, D., and E. Fogelqvist. "Bromoform Concentrations of the Arctic Ocean in the Svalbard Area," *Oceanol. Acta.* 4:313 (1981).
90. Wegman, R. C. C., and H. H. van den Broek. "Chlorophenols in River Sediment in the Netherlands," *Water Res.* 17:227–230 (1983).
91. Sanders, J. G. "Chlorination of Estuarine Water: The Occurrence and Magnitude of Carbon Oxidation and Its Impact on Trace Metal Transport," *Environ. Sci. Technol.* 16:791–796 (1982).
92. McKague, A. B., S. J. Rettig, J. Trotter, G. R. Douglas, and E. Nestmann. "The Chlorination of Dihydroconiferyl Alcohol," *Can. J. Chem.* 61:545–549 (1983).
93. Talka, E. "Determination of Low Molecular Weight Chlorinated Compounds in Pulp Mill Bleaching Effluents by Mass Spectrometry," *Internat. J. Mass Spectrom. Ion Phys.* 48:295–298 (1983).
94. Knuutinen, J. "Analysis of Chlorinated Guaiacols in Spent Bleach Liquor from a Pulp Mill," *J. Chromatogr.* 248:289–295 (1982).
95. Voss, R. H. "Chlorinated Neutral Organics in Biologically Treated Kraft Mill Effluents," *Environ. Sci. Technol.* 17:530–537 (1983).
96. Kringstad, K. P., F. de Sousa, and L. M. Stromberg. "Evaluation of Lipophilic Properties of Mutagens Present in the Spent Chlorination Liquor from Pulp Bleaching," *Environ. Sci. Technol.* 18(3):200–203 (1984).
97. Ghanbari, H. A., W. B. Wheeler, and J. R. Kirk. "Reactions of Chlorine and Chlorine Dioxide with Free Fatty Acids, Fatty Acid Esters, and Triglycerides," in *Water Chlorination: Environmental Impact and Health Effects, Vol. 4,* R. L. Jolley, W. A. Brungs, J. A. Cotruvo, R. B. Cumming, J. S. Mattice, and V. A. Jacobs, Eds. (Ann Arbor, MI: Ann Arbor Science Publishers, Inc., 1983), pp. 167–177.

98. Davis, W. P., and D. P. Middaugh. "A Revised Review of the Impact of Chlorination Processes Upon Marine Ecosystems: Update 1977," in *Water Chlorination: Environmental Impact and Heatlh Effects, Vol. 1,* R. L. Jolley, Ed. (Ann Arbor, MI: Ann Arbor Science Publishers, Inc., 1978), pp. 283–210.
99. Brooks, A. S., and G. L. Seegert. "The Toxicity of Chlorine to Freshwater Organisms Under Varying Environmental Conditions," in *Water Chlorination: Environmental Impact and Health Effects, Vol. 1,* R. L. Jolley, Ed. (Ann Arbor, MI: Ann Arbor Science Publishers, Inc., 1978), pp. 261–282.
100. Olivieri, V. P., W. H. Dennis, M. C. Snead, D. T. Richfield, and C. W. Kruse. "Reaction of Chlorine and Chloramines with Nucleic Acids Under Disinfection Conditions," in *Water Chlorination: Environmental Impact and Health Effects, Vol. 3,* R. L. Jolley, W. A. Brungs, and R. B. Cumming, Eds. (Ann Arbor, MI: Ann Arbor Science Publishers, Inc., 1980), pp. 651-663.
101. Tenno, K. M., R. S. Fujioka, and P. C. Loh. "The Mechanisms of Poliovirus Inactivation by Hypochlorous Acid," in *Water Chlorination: Environmental Impact and Health Effects, Vol. 3,* R. L. Jolley, W. A. Brungs, and R. B. Cumming, Eds. (Ann Arbor, MI: Ann Arbor Science Publishers, Inc., 1980), pp. 665–675.
102. Fujioka, R. S., K. M. Tenno, and P. C. Loh. "Mechanism of Chloramine Inactivation of Poliovirus: A Concern for Regulators?" in *Water Chlorination: Environmental Impact and Health Effects, Vol. 4,* R. L. Jolley, W. A. Brungs, J. A. Cotruvo, R. B. Cumming, J. S. Mattice, and V. A. Jacobs, Eds. (Ann Arbor, MI: Ann Arbor Science Publishers, Inc., 1983), pp. 1067–1076.
103. Noss, C. I., W. H. Dennis, and V. P. Olivieri. "Reactivity of Chlorine Dioxide with Nucleic Acids and Proteins," in *Water Chlorination: Environmental Impact and Health Effects, Vol. 4,* R. L. Jolley, W. A. Brungs, J. A. Cotruvo, R. B. Cumming, J. S. Mattice, and V. A. Jacobs, Eds. (Ann Arbor, MI: Ann Arbor Science Publishers, Inc., 1983), pp. 1077–1086.
104. Cumming R. B. "The Potential for Increased Mutagenic Risk to the Human Population Due to the Products of Water Chlorination," in *Water Chlorination: Environmental Impact and Health Effects, Vol. 1,* R. L. Jolley, Ed. (Ann Arbor, MI: Ann Arbor Science Publishers, Inc., 1978), pp. 229–241.
105. Crathorne, B., C. D. Watts, and M. Fielding. "The Analysis of Nonvolatile Organic Compounds in Water by High-Performance Liquid Chromatography," *J. Chromatogr.* 185(1):671–690 (1979).

CHAPTER 4

Chlorination of Power Plant Cooling Waters

Jack S. Mattice

The sequence of water chlorination conferences is an indication of the need perceived by both industry and government for balancing the known benefits and potential costs of chlorination. Chlorination is important in many industrial processes, but utility chlorine use has been a major focus of attention for more than a decade. Questions were initially raised concerning the potential environmental effects of residual chlorine compounds released in power plant cooling waters. The scope of the questions broadened with the discovery of chlorinated organics in cooling waters[1] and preliminary evidence that some of these compounds could have health and/or environmental consequences. Government regulatory and energy agencies responded by increasing funding for chlorine research and by increasing efforts directed toward development of regulations on chlorinated effluents at power plants. The utility industry quickly recognized the implications of these efforts for industry practices and economics and increased its own sponsorship of research on chlorination. Industry-sponsored research has been a consequential share of the total effort. For example, the Electric Power Research Institute (EPRI), which is voluntarily sponsored by most utilities nationwide, has funded about $6 million of chlorine-related research since 1977.[2] An additional $6 million has been funded by other industry groups, such as the Edison Electric Institute, and by individual utilities. The figure for these other groups is underestimated, because reporting is not coordinated industrywide, and minimization studies, most of which have been voluntary, are not included. Industrial and governmental research programs on chlorine have covered the same general areas with respect to power plants. In fact, although it was not as clear in the early to mid-1970s, both groups are seeking the same goal: determination of reasonable limits on chlorine and chlorination-produced compounds in power plant effluents.

The existing national regulation dealing with chlorine in power plant effluents appears simple but is not. Residual chlorine (oxidants) is limited depending on plant size and design (Table I).[3] In addition, seven polychlorinated biphenyls cannot be released in power plant effluents. Other chlorinated or brominated compounds on the priority pollutant list are excluded from regulation. However, one clause of the regulation states that NPDES permit-issuing authorities may "require limitation of pollutants not covered by this regulation (or require more stringent limits on covered pollutants)." This clause is in use.

Table I. Summary of the National Residual Chlorine Limits for Steam-Electric Power Plants.[3] The term oxidant can be substituted for chlorine in estuarine or marine environments where other halogens may be prevalent.

Plant type	Units treated simultaneously	Chlorine type	Time	Concentration (mg/L) Max	Av
Once-through cooling <25 MW(e)	1	Free residual	≤2 h[a]	0.5[b]	0.2[b]
Cooling tower	1	Free residual	≤2 h[a]	0.5[b]	0.2[b]
Once-through cooling >25 MW(e)	≥1	Total residual	≤2 h[c,d]	0.2	

[a] Limit applies to free and total residual chlorine for each unit at a plant.
[b] Unless the utility can demonstrate to the administrator (or state if the state has NPDES issuing authority) that units cannot operate at or below this level.
[c] Limit applies to total residual chlorine for each unit at a plant.
[d] Unless the discharger demonstrates to the permitting authority that discharge for more than 2 h is required for macroinvertebrate control.

For example, one utility has had to dechlorinate in some instances to meet a regional limit of 0.0 mg/L total residual chlorine (TRC). There is thus no assurance that other compounds will not be subject to future regulation or that other limits will not be enforced.

Site-specific limits per se are not a focus of disagreement between industry and regulatory agencies. In fact, representatives on both sides have supported their use.[4,5] Any disagreement is based on the reality of the ecological representation, the same argument that has surfaced in comment on technology-based limits such as the current national regulation.[3] These disagreements and their economic implications indicate the necessity for determining both the need for chlorination by the utility industry and the levels that can be released without deleteriously affecting the receiving water body.

In this chapter I will present (1) a format for evaluating utility chlorine use, (2) a summary of selected information relating to this evaluation, and (3) a proposal for research that is needed to complete this evaluation.

OPTIMUM POWER PLANT CHLORINATION

Considering what is involved in defining optimum chlorination practices will help provide a format for examining what we know (or think we know) and what we need to know about power plant chlorination (Figure 1). The first criterion for optimum chlorination is maintenance of efficient energy production. Chlorine is applied to condenser-cooling and service water systems at many power plants to control fouling. In both systems, fouling control is required for maintaining water-flow and heat-exchange efficiency. Costs of decreased efficiencies under certain assumptions of fouling severity have been hypothesized to reach $100 million per year in the United States[6] and £20 million per year in the United Kingdom.[7] A single incident of fouling by Asiatic clams at a freshwater power plant resulted in a multi-million dollar loss.[8] With the potential for costs such as these, chlorine application must be maintained at a level to accomplish fouling control. On the other hand, applications over and above those required to control fouling are economically costly and also increase the potential for environmental impact. For example, one utility conducted a minimization study and decreased chlorine use by about 30% without affecting system operation.[9] Determining the necessary and sufficient chlorine application schedule and concentration to control fouling is thus not an inconsequential problem.

The second criterion for optimum chlorination (Figure 1) is maintenance of environmental quality. As with the first criterion, incorrect decisions about whether a chlorinated effluent is environmentally benign or will degrade the environment can have economic implications. We have reached a period in which competition for environmental and economic resources is the rule rather than the exception. We as citizens will pay if the environment is degraded; we will also pay if environmental regulations are overstrict and require unnecessary technological fixes. Thus, there is an economic as well as an environmental payoff in determining what levels of chlorinated products under what release regimes will provide for environmental protection.

Each of these criteria for optimum power plant chlorination can be subdivided into two categories of information needs (Figure 1). The subdivisions are somewhat arbitrary. For each of the criteria, we are concerned about what the exposures are and how the organisms (target or nontarget) respond to them.

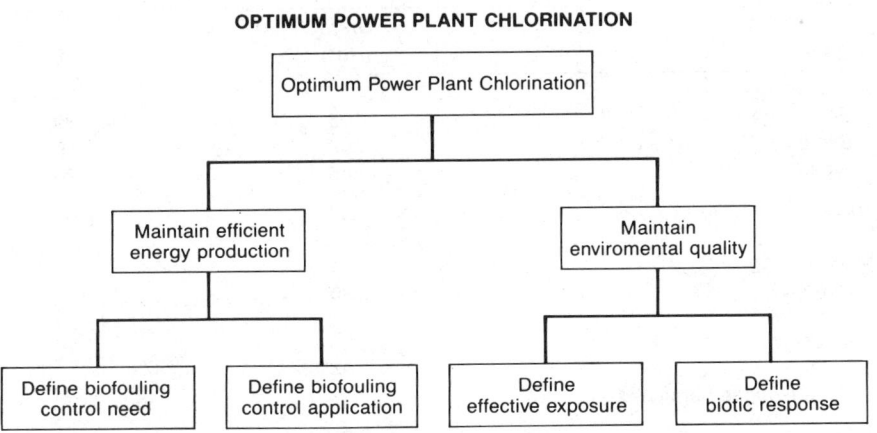

Figure 1. Conceptual format for evaluating progress and remaining needs for information to optimize chlorination procedures at power plants.

However, for convenience of discussion here, the biofouling control need is defined by the questions, Is biofouling a problem at all? and When (how often) is biofouling control required? whereas biofouling control application is defined by the question, How much chlorine is required to control fouling? Similarly, effective exposure is defined by the question, What are the fates of the toxic products of chlorine (in both temporal and spatial terms) in the environment? whereas biotic response is defined by the questions, To what concentration of toxic chlorine-produced compounds are organisms exposed? and Are populations and communities of organisms deleteriously affected by these exposures?

DEFINITION OF BIOFOULING CONTROL NEEDED

Biofouling is a problem at most power plants, but biofouling problems differ from plant to plant. General categorization of fouling organisms (Table II) conceals a multitude of problem species; however, biofouling problems can be pragmatically classified as microfouling and macrofouling. Because these two groups differ in susceptibility to control by chlorine, they are best treated separately.

MICROFOULING NEED

Chlorination has long been the preferred method for controlling microfouling organisms. Microfouling organisms include bacteria with their associated extracellular secretions (commonly called slimes) and algae (Table II).

Table II. Fouling Organisms Reported at Operating Power Plants[a]

Organism	Units (No.)	%
Microfouling		
Slime/algae	105	30
Slime	32	9
Algae	85	24
Total	222	63
Macrofouling		
Hydroids	25	7
Barnacles	11	3
Mussels	56	16
Clams[b]	34	10
Oysters/algae/barnacles	7	2
Total	133	37

[a]Modified from Reference 10.
[b]All reports from freshwater sites are assumed to be the Asiatic clam, *Corbicula fluminea*.

Although these organisms can form on conduits and piping throughout a power plant, they represent a real problem only at condensers, where they settle on heat-exchange surfaces and form an insulation layer, and in cooling towers, where they affect evaporation rates. The decreased heat exchange that results can cause changes in plant operating temperatures and decrease the efficiency of energy generation. Not all power plants are vulnerable to microfouling. Where it has not been a problem, scouring by suspended sediments has usually been presumed to provide natural control, but the variability of the microfouling control need at plants located on a single water body[6] suggests that the reasons may be more complex. Where microfouling has been a problem, intermittent chlorination has proven effective as long as interchlorination periods are not extensive.[11] Because intermittent chlorination has generally not proven effective for macrofouling control, the small number of units chlorinated continuously and the large number of units that are chlorinated for 2 h/d or less (Figure 2) indicate that chlorine application at power plants is primarily focused on microfouling problems. Applications as short as 3 min several times a day have proven effective at some sites, but longer and/or more frequent application is required at other sites. In addition, at many plants, winter chlorination is either not required or can be reduced from summer levels;[11] for other plants, summer and winter needs are the same.[12] It is clear that the need for microfouling control varies substantially, depending at least in part on attachment and growth rates.

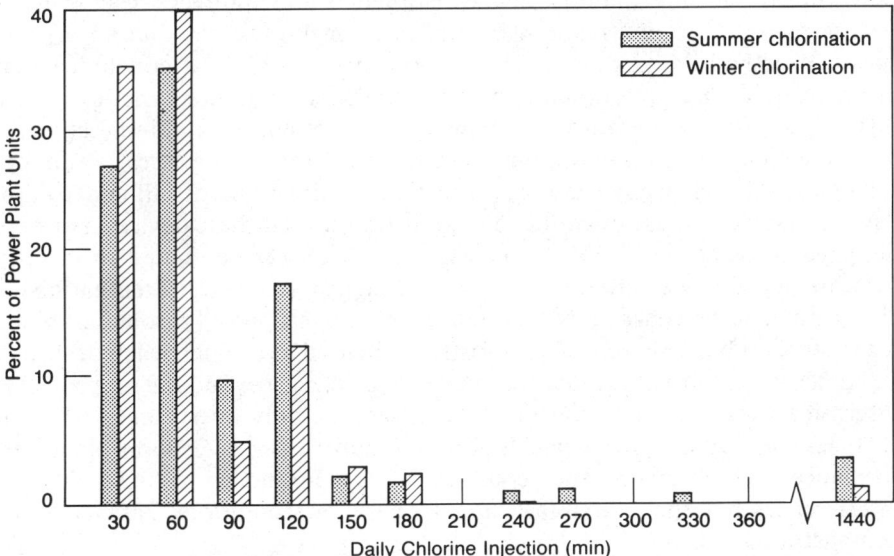

Figure 2. Distribution of power plants injecting chlorine into cooling water for up to a given time each day. Data are for a subsample of existing power plant units, but reasonably represent the industry with respect to cooling water type and plant design (modified from Reference 10).

We have made significant progress in the last few years in describing the process of microfouling and in identifying parameters that affect fouling rate and biofilm thickness.[7,13,14] Microfouling appears to involve a number of phases, but any division into phases is somewhat arbitrary. Perhaps more important, physical, chemical, and biological parameters are all essential in determining rates of fouling. These include flow velocity and characteristics; surface charge and production of extracellular polysaccharides; and water quality parameters such as pH, temperature, nutrients, and dissolved oxygen. However, the relationships of these parameters to fouling are not simple. For example, the microfouling rate is maximum at an intermediate temperature (i.e., fouling is less at temperatures above and below an optimum). In addition, surface irregularities and fluid velocity can increase or decrease the fouling rate, depending on other conditions and/or the phase of fouling. Also, the effect of nutrients on fouling rate is controversial.[13,15] Finally, biofilm formation is more rapid following chlorination than it is on a virgin surface. It is apparent that substantial work will be required to develop capabilities for predicting microfouling rates.

In the absence of capability for predicting microfouling, optimization of chlorine application requires that we be able to monitor microfouling via effects on plant operation. Several authors have examined such methods. In 1979, Rice[12] concluded that there was no generally accepted method to relate condenser performance to biofilm formation. Sharpe[7] summarized information on a number of potential methods for measuring microfouling, including both direct (biofilm quantity and constituents) and indirect (heat transfer resistance, fluid friction resistance, condenser back pressure, and absorption spectra) methods. Sharpe criticized existing techniques and concluded that direct methods had questionable on-line applicability, indirect methods were difficult to interpret because of variability, and both direct and indirect methods were relatively insensitive, particularly during the early phases of fouling. Hillman et al.[16] critically examined direct and indirect microfouling detection devices that were in use or could be used by utilities. Of these devices, ten were designed to detect heat transfer coefficient, three were designed to directly measure biofilm weight, and two were designed to detect attenuation or change in acoustic characteristics of ultrasonic signals passed through or along a condenser tube. Thirteen characteristics were used to evaluate each device. A full description of their evaluations and conclusions is beyond the scope of this discussion; however, the acoustic devices were the only ones judged to be of sufficient sensitivity to indicate fouling buildup before effects on plant performance were observed. The acoustic devices were judged to be sensitive to interference, but their reliability under operating conditions has yet to be demonstrated.

We are not yet able to predict the rate of microfouling development or to measure the rate accurately enough to preclude effects on plant performance. The complexity of the biofouling process and the number of parameters affecting rate of formation suggest that the development of accurate, sensitive measurement devices offers a more pragmatic approach to the development of optimum chlorine application regimes. The EPRI is currently investigating the use of such devices.

MACROFOULING NEED

Mechanisms involved in macrofouling are not very well understood.[8,17] If we ignore those organisms that present power plant operators with trash or impingement problems, we find that all major macrofouling organisms (Table I) have a dispersal phase in their life cycle, and all have attachment organs. During the dispersal phase the animals are carried to the site of active attachment within the plant. They then grow to a size that can obstruct flows through conduits or clog condenser tubes. These organisms do not grow within condenser tubes; after they have reached fouling size, they are carried into the tubes, either alive (mussels, clams) or dead (oysters, barnacles, mussels, clams), by water currents from the original attachment site.

Marine macrofouling occurs in stages: biofilm formation precedes settlement of macrofouling organisms, apparently providing a necessary surface conditioning; this is followed by settlement of barnacles, bryozoans, and hydroids, and somewhat later by coelenterates and bivalves.[17] Freshwater macrofouling is primarily caused by *Corbicula fluminea*, the Asiatic clam.[8] Behavioral studies (see discussion and references in Newell[18]) have indicated that many fouling organisms have some control over when and where they settle out of the plankton; the attachment process has been described for some species. However, the rate of fouling (numbers of organisms that attach and the rates at which they grow) is highly variable from site to site, season to season, and year to year.

Some of the variability in fouling rate is undoubtedly the result of the risky existence of organisms with planktonic life stages, the variation in reproduction, and the changes in environmental conditions; however, without an understanding of at least some of these factors, the prediction of macrofouling rates will be impossible. Plant design, such as condenser tube diameter,[19] can also be important in determining susceptibility to macrofouling. It seems clear that information is needed for a quantitative description of factors that affect (1) reproductive timing and production, (2) settling, (3) attachment and detachment, and (4) growth of attached organisms before it will be possible to predict a macrofouling control need.

As with microfouling, until capabilities to predict a macrofouling control need are developed, an existing method of monitoring plant fouling seems to be an attractive alternative. Jenner[20] described a technique for monitoring mussel settlement and control that uses a trap on a bypass line from the intake water flow to the condenser. This technique has proven successful in determining when control is necessary at a single site, and offers possibilities for monitoring other macrofouling organisms and other sites. Its sensitivity and accuracy have yet to be documented.

DEFINITION OF BIOFOULING CONTROL APPLICATION

A number of studies have been designed to answer the question, How much chlorine is required to control fouling? These studies include standard laboratory toxicity testing, observations of responses of fouling communities to the chlorination of surrogate condenser systems in the laboratory or at field sites, and direct observations of condenser systems themselves.

All studies present both advantages and disadvantages. Laboratory toxicity studies are relatively simple, and conditions can be controlled to allow cause/effect conclusions to be drawn; however, such studies can raise questions regarding their applicability to conditions that actually exist in power plant cooling water systems. Field studies using operating power plant condenser systems raise no questions as to applicability; however, they are more costly and generally must be relied on to supply site-specific information because of the numerous parameters that cannot be controlled. Model condensers offer an intermediate between the two extremes but leave all questions regarding applicability unanswered unless the system is validated. Using all studies in combination appears to be the only approach that will eventually provide the necessary generic answers to allow development of predictive capabilities under a given set of water-quality, plant-design, and operating conditions. This approach, however, has yet to be taken. The following summaries describing what we have learned about chlorine requirements for microfouling and macrofouling should thus be considered tentative and preliminary. (We may know much less than we think.)

MICROFOULING CONTROL

We know from both minimization studies[11,21,22] and from industry surveys[23,24] that chlorine application levels required to control microfouling vary widely. Some of these differences undoubtedly result from the effects of water quality on the portion of chlorine applied that actually reaches the biofilms in active form. For example, Moss et al.[10] found that the feed rate required to control microfouling was directly related to chlorine demand and water temperature; Schumacher and Lingle[22] found that the relationship between biofilm removal and free residual chlorine concentration was much better than that between biofilm removal and total residual chlorine concentration; and Characklis et al.[13] found that free residual chlorine was much more effective at destroying biofilm at high pH. These results strongly suggest that hypochlorite is the most bioactive of the residual chlorine forms. Thus, water quality appears to affect the required dosage via the effects on proportionation between the different forms of residual chlorine.

We can be sure that other factors also affect biofilm destruction by chlorine and chlorine-produced oxidants; however, information on the few that have been considered is inferential or controversial. For example, Rippon[23] reported that required chlorine application levels, which differed at power plant condensers, also differed in microfouling community structure. By inference, this suggests that requirements differed because of the different types of bacteria, although this requires direct confirmation. Condenser tube material may also affect chlorine requirements, but results are apparently controversial. Rippon[23] and Yasui et al.[25] found different microbial populations on tubing of different materials, although no studies were conducted of chlorine requirements for destruction. On the other hand, Battaglia et al.[24] concluded from a selected survey of power plants in the United States that the tube material does not affect chlorine requirements. Further studies of these and other potential influences are needed.

The most comprehensive studies of microfilm control using chlorine were those conducted by Characklis and co-workers with a model condenser (see Bryers et al.[26] and references therein). From results of these studies, they concluded that disinfection and biofilm destruction proceeded via different mechanims, because hypochlorous acid is the most effective for disinfection, whereas the hypochlorite ion is most effective for biofilm destruction. They proposed that hypochlorite ions interacted with the polysaccharide matrix of the biofilm, causing release of organic matter, which weakened the matrix and rendered it vulnerable to removal by fluid shear. Characklis and his colleagues have been careful to qualify their conclusions to the conditions tested, but if they can be generalized, these results imply that future work must be conducted on real or simulated condensers to ensure applicability of results; batch tests, commonly used in disinfection studies, would not be useful.

Unfortunately, it is not yet possible to unequivocally evaluate the conclusions of Characklis and colleagues. Sugam et al.[27] and Guerra et al.[28] found that the model condenser system they used accurately reflected response in nearby power plants; however, the tube materials were identical. The tubes studied by Characklis et al.[13] were fabricated with glass, which may have affected results.[23,25] However, the conclusion by Characklis et al. that shorter, higher-concentration doses of chlorine would be more effective in control of fouling than longer, lower-concentration doses with equivalent chlorine was supported by studies of Moss et al.[10] at an operating power plant. More direct comparisons of model and power plant results will be required to validate the model.

The development of a microfouling monitor (see Microfouling Need) may be the best solution for defining microfouling control application. With such a device, it should be possible to determine economically the application schedule and level on a plant-specific basis, thereby bypassing the extensive research that will be required to develop reliable predictive relationships. The EPRI is currently funding an evaluation of biofilm monitoring devices.

MACROFOULING CONTROL

Chlorine is not routinely used for macrofouling control in the United States. Macrofouling organisms (Table II) are generally refractory to chlorine once they have settled, requiring continuous or nearly continuous[29-31] applications of chlorine during reproductive or settling periods for successful control. Despite this demonstrated effectiveness, few plants are chlorinated continuously (Figure 2). One can only conclude that utilities do not apply for, or that authorities issuing permits do not normally grant, variances to the 2-h limit in the national effluent limitations guidelines (Figure 1). However, chlorine is applied continuously at only a few U.S. power plants.

Numerous studies have been conducted to define levels of chlorine or chlorine-produced oxidants (CPO) required for macrofouling control. Standard toxicity tests, using available water, have been conducted to identify toxic chlorine or CPO levels for various species of hydroids, barnacles, sponges, and bivalves.[8,31-33] However, the applicability of these results is questionable. For example, Tilly[34] reported that 65% of the Asiatic clams exposed to 10 to 40 mg/L of chlorine (species unknown) in pump wells were still alive after 7 d, although Gooch et al.[35] found that exposures of 0.3 to 0.4 mg/L total residual chlorine would kill all exposed larvae of the same species in less than 100 h in laboratory studies. Tilly[34] attributed survival to the fact that the clams were partially or completely buried in mud, which shielded them from exposure. However, unless off-stream or laboratory studies are designed to replicate plant conditions, the applicability of the test results must be questioned. Field studies using a power plant as the experimental system have been conducted at a large number of facilities, but the range of concentrations required for control[20,25,30] indicates that macrofouling control is a complex problem. Based on studies at three power plants in France, Khalanski and Bordet[36] developed a relationship for predicting exposures that were required to control mussels; however, this relationship has not been tested for generic applicability.

It does appear that minimization studies at a specific site can be used to identify levels of chlorine required to control macrofouling, but that prediction of such levels is not possible. Because it does not appear that variances from the 2-h limit on chlorination will be commonplace, the research effort required to define such levels does not seem warranted, and efforts would seem better spent in examining chlorine alternatives. The current EPRI efforts in macrofouling control are directed toward an evaluation of alternative control technologies.

DEFINITION OF EFFECTIVE EXPOSURE

From the perspective of chemistry, a definition of effective exposure is dependent on identification and quantification, through dimensions of both time and space, of the toxic reaction products of chlorination of cooling waters. Inclusion of the word toxic indicates the necessity for interaction between chemists and toxicologists in arriving at this definition. The high reactivity of chlorine and the large number of potential reactants in natural waters makes the problem particularly complex and resistant to simple solution. Since the First Conference on Water Chlorination, great strides have been made toward identifying and quantifying chlorine reaction products found in power plant cooling waters. For the sake of organization, I have arbitrarily separated these products into four categories: volatile haloorganics, nonvolatile lipophilic haloorganics, nonvolatile hydrophilic haloorganics, and residual oxidants. Most chlorinated organics identified in power plant effluents (e.g., chlorophenols) fall into the hydrophilic category. The lipophilic component can be obtained (almost certainly overestimated) from the difference between the hydrophilic component and total organic halogen. This order is not in accordance with the potential for environmental effects (see Definition of Biotic Response). The halo and oxidant terms are used to attempt to combine work conducted in both saline and fresh waters. Reactions in saline waters are more complex because of the presence of other reduced halogen ions (Br^-, I^-), but the similarity of the generic problems associated with power plants using both types of waters seems to warrant combination.

The predominant volatile haloorganics produced at power plants appear to be the bromine- and/or chlorine-containing trihalomethanes.[37-39] The formation of the trihalomethanes (THMs) occurs at both closed-cycle and once-through plants, but losses to volatilization and drift at closed-cycle plants keep discharge concentrations low. Two groups[38,40] have reported higher concentrations of THMs in chlorinated power plant intake waters than those found in discharge waters, indicating the need for caution in extrapolations from laboratory to the field. However, information currently available supports the conclusion that volatile haloorganics released in the discharges of power plants are below 100 ppb, with the relative proportion of chlorinated or brominated forms being the result of the presence of the bromide ion in the water being chlorinated.[37-42]

The nonvolatile lipophilic haloorganics have been the least studied of all the haloorganics. Onsite studies by Bean et al.[43] suggest that these compounds are present only at low parts per billion levels or less. In addition, surveys by the EPA for priority pollutants[3] indicated very low or undetectable concentrations. Obviously, more extensive surveys to identify and quantify these compounds in power plant effluents would be appropriate.

In the mid-1970s, Jolley et al.[1] reported the formation of nonvolatile hydrophilic chlorinated organics following chlorination at two power plants. They concluded from ^{36}Cl tracer studies that more than 50 such compounds were formed, of which 17 were tentatively identified. Others have subsequently been identified under various conditions of chlorination of natural waters.[39,40,43] However, in all cases, these compounds represented < 4% of the total chlorine applied, and individual compounds were found at low parts per billion concentration ranges.

Considerations thus far have been limited to chlorine reaction products in water. Bean et al. (this volume) examined sediments near a power plant for chlorinated organics. They found that concentrations were only marginally higher in sediments exposed to the chlorinated power plant effluent than in reference sediments. Thus, unless new information is discovered to contradict these findings, the emphasis on water concentration appears justified.

We know more about the reactions of residual oxidants than about the products previously discussed; however, critical gaps in our knowledge remain. Some of the reactions of chlorine with other halogens, ammonia, and inorganic amines have been studied[44] and quantitatively described with certain accuracy.[45] Substantial effort will be required to provide the information that we need for a quantitative prediction of the inorganic residual halogen species (absent chlorine demand), especially in saline waters; however, it seems likely that most of the primary reactions and reaction products have been identified. Conversely, this does not appear to be true for reactions with organic amines. Reaction products and the decay of organic haloamines are dependent on the amines present in the water being chlorinated.[46,47] These products and reactions have not been identified for a single site; therefore, generalization is impossible.

The extraction of organic haloamines from power plant effluents is not a simple process because they are thermally labile.[48] However, the fact remains that we do not have the information at present to evaluate the significance of their (halomines) formation at power plants.[49] Perhaps it will be possible to use the relationship between the octanol/water partition coefficient and the toxicity of short-term exposures[50] to organic haloamines to substantially reduce the workload required, but the relationship needs to be confirmed for longer exposures and other biota. Nevertheless, the toxicity of the residual halogen species, including some organic chloramines, indicates the critical need for further studies of the rates of formation and decay.

The numerous potential products, the difficulty in measuring the products in real time, the dependence of the products on water quality, and the dependence of biotic response on concentration and time of exposure to toxic residual halogens have led a number of investigators to try to model the fate of chlorine in natural waters.[51] These models have ranged from empirical "black box" types to those incorporating individual reactions and rate constants (or estimates of them). Despite the differences in the models, they all have two common flaws: estimating chlorine demand and predicting organic haloamine formation. Chlorine demand encompasses a plethora of competing reactions and results in a myriad of products. Even if it is assumed that the component products of chlorine demand at power plants are nontoxic (this appears likely, except for perhaps the nonvolatile lipophilic haloorganics), their function being only to reduce concentrations of the toxic residual halogen species, an accurate estimation of the chlorine demand rate is critical for determining biotic response. The complexity of the estimating problem is highlighted by the inclusion of whole sessions on chlorine demand at the last two Water Chlorination Conferences. Problems in identifying and quantifying the organic haloamines were discussed previously. Despite their complexity, the problems discussed here are the most critical ones facing chemists. Until they are solved it will not be possible to provide toxicologists with the necessary information to accurately estimate effective exposure and provide a logical realistic framework for toxicology-based effluent limits at power plants.

DEFINITION OF BIOTIC RESPONSE

At the Third Conference on Water Chlorination, Maki[52] proposed a tiered format for conducting hazard assessments of toxicants. The format includes consideration of effective exposure levels and biotic response, including bioaccumulation. An examination of power plant chlorination using this format will help to focus attention on the areas of research most likely to provide relevant information for establishing realistic regulations for power plant chlorination. For ease of comparison, this section is organized to coincide with the previous section.

Volatile haloorganic compounds released from power plants, primarily THMs, do not appear to pose a significant environmental hazard. THMs are found in power plant effluents at concentrations below 100 ppb. In comparison, toxicity tests conducted with several species of freshwater and marine organisms all indicated that acute LC_{50} values (≥ 96 h) were at least 7 mg/L or higher.[53] Sublethal effects were not observed in life-cycle exposures of *Daphnia* to bromoform at concentrations below 10 mg/L.[54] The latter concentration is somewhat suspect because the studies were conducted in static systems with intermittent toxicant addition; however, intermittent exposure is most likely to be the real case at power plants. Animals at the next lower nominal concentration of 1 mg/L (in these studies) had significantly longer life spans and higher reproductive rates than controls. In addition to this difference between effluent and effects concentrations, THMs do not appear likely to persist in the aquatic environment. Half-lives of THMs in effluent streams are difficult to predict, but it is likely that they are relatively short.[53] Existing evidence[55-57] indicates that bioaccumulation of volatile haloorganics is low. Uptake is rapid, but only to one to ten times the water concentration, and depuration is also rapid. Further studies of volatile haloorganics and their application to power plants seem unwarranted.

An evaluation of the environmental hazard resulting from the release of nonvolatile lipophilic halogenated organics is not currently possible. We think that they are present in low concentrations (≤ 100 ppb) in power plant effluents. Some are likely to be organic chloramines[58] and probably are included in the measure of TRC. Thus, in our categorization, a fraction of these compounds is probably already considered with other residual chlorine species. An analogy with other chlorinated organics, including chlorinated amines, would suggest that the nonvolatile lipophilic haloorganics would not be toxic at the concentrations and times that they could be released at power plants, but this expectation remains to be confirmed. Direct information concerning bioaccumulation of these lipophilic compounds does not exist. Lee et al.[59] did not find conclusive evidence that the clam, *Corbicula*, bioaccumulated any chlorinated organics, but how much these results can be generalized is unknown. However, given the low concentrations in and intermittence of power plant releases, bioaccumulation factors would have to be quite high to expect a major problem. If further investigations of the nonvolatile lipophilic haloorganics are conducted, they should be focused on chemical identification and characterization (e.g., octanol/water partition coefficient).

Continued study of the nonvolatile hydrophilic haloorganics appears to be analogous to the proverbial search for the needle in the haystack. Only a fraction of these compounds have been identified,[37] but those that have been both identified and quantified in power plant effluents are released at concentrations two to four orders of magnitude below those required to produce sublethal effects, even after long-term exposure.[54,60] Even the longer half-lives (than THMs) in the environment would not modify the conclusion that they do not cause environmental problems. Bioaccumulation factors also are not expected to be high because of the water solubility of these compounds. Thus, it appears that continuing to search for a nonvolatile hydrophilic haloorganic that is orders of magnitude more toxic than those already studied is likely to produce only negative findings.

It is time to confine the focus of chlorine research as it affects the utility industry. The intense concern spawned by identification of chlorinated organics in power plant effluents was directed toward a potential that we now can admit with reasonable risk was not realized. We do not know all the answers; we never will. However, at the levels they are formed and released at power plants, chlorinated organics (with the possible exception of the nonvolatile lipophilic haloorganics) do not appear to present a significant problem. Other industries and their regulatory agencies should bear the brunt of research efforts on these compounds. The utility industry should share the costs of that part of the chlorine problem which, potentially, it most affects and which most affects it, that is, the release of the residual chlorine species.

The results of research on the toxicity of residual halogens to aquatic organisms serve notice that the development of a reasonable regulation governing halogens in power plant effluents will be a difficult task. Early workers had a strong tendency to treat chlorine as a conservative chemical and to conduct static tests. Nominal concentrations were generally reported and were often estimated from the weight or volume of chlorine added. Other toxicologists followed the lead of the disinfection researchers who found that free residual chlorine (FRC) effectiveness was far better than that of combined residual chlorine (CRC), and thus they only reported FRC. As toxicological testing and chlorine analytical techniques have become more sophisticated, it has become obvious that organisms respond very differently to the different components of TRC.[50,61,62] Similar conclusions can be expected with regard to other residual halogen species, but studies have yet to be conducted. Mattice and Tsai[50] reported a 1600-fold difference in LC_{50} between inorganic dichloramine and N-chloroglycine for the same exposure time; organic chloramines also varied 1000-fold in toxicity, with several being as toxic as FRC.

These foregoing differences appear important for assessment because, for various reasons,[46,63] organic haloamines can form a significant fraction of the combined halogen in freshwater and marine power plant effluents. The problems that have thwarted identification and quantification of the organic haloamines were mentioned previously, but it is clear that overcoming these problems is a major hindrance in toxicity assessment. The structure-activity relationship found by Tsai et al.,[64] if confirmed, may provide a way around these problems. In addition to determining the toxicity of individual residual halogen species to aquatic organisms, it is also important to determine responses to mixtures, if the toxicity of a whole effluent is to be predicted. What little work has been done[50,61,64] suggests that toxicities are additive, but further confirmatory studies are needed.

The fact that time (Figure 3), concentration, and the halogen species all affect toxicity is particularly important for assessing the toxicity of power plant effluents. At most power plant units, chlorine is applied intermittently (Figure 2), but regulations permit the chlorination of each unit for up to 2 h; therefore, the time of occurrence of residual halogens in the plant effluent is partially dependent on the number of units. (Fish do recover during periods between chlorine exposures,[62] but it is not yet possible to quantify that recovery for specific exposures to an effluent.) Exposure time is also dependent on the decline of chlorine or chlorine-produced oxidants once the effluent enters the receiving water body. Organisms in a water body cannot all live at the entry point of the discharge. In fact, at plants with multiport jet diffusers, water velocities and turbulence may be sufficient to create an exclusion area in which motile organisms will rarely be found, and then only briefly. The decline of residual halogens is primarily due to dilution and halogen demand reactions during mixing. Both will vary from site to site, depending on plant design and water quality. Thus, the concentrations of halogens and exposure times experienced by resident organisms, as well as the volume of water containing residual halogens, will also vary from site to site. The time of exposure will also vary with the location of organisms within the discharge plume. There is not a relevant model of fish behavior that would be helpful, but we do know, from studies sponsored primarily by utilities or utility groups,[65-67] that under many conditions fish will avoid acutely toxic oxidant levels. Many questions remain, but field observations have generally validated laboratory results, and it appears that avoidance should receive consideration in determining the site-specific regulation of residual halogens in power plant discharges.

The temperature and life stage of the organisms also affect toxicity.[68] In particular, for short exposures, increase in temperature increases toxicity. The effect of life stage is somewhat less clear, but the larvae, at least of fish, tend to be more sensitive than eggs, juveniles, or adults.

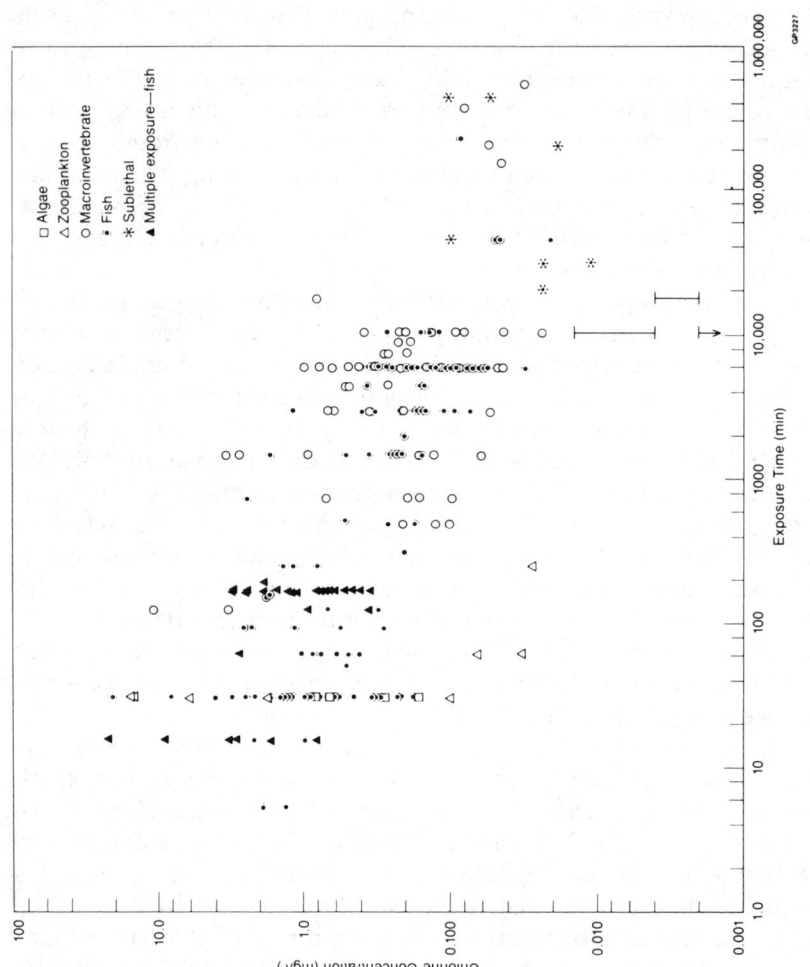

Figure 3. Chlorine toxicity data for freshwater organisms that survived critical scrutiny, using criteria reported by Fava and Seegert.[71] Unless noted, data points represent the exposure (concentration, time) that resulted in the death of 50% of the organisms (LC_{50}). Concentrations are reported as total residual chlorine regardless of the chlorine species composition.

Clearly, the toxicity of residual chlorine or chlorine-produced oxidants is difficult to define. This problem is exacerbated when the focus is shifted to ecosystem protection. Different species respond differently even to the same oxidant exposure. Thus, even though for some organisms in some instances we can reasonably and accurately estimate response to a given exposure, it is important to develop a method or methods to determine safe levels of oxidants that may be released at power plants. Two such methods have been proposed.[32,68,69] Both have been criticized with alacrity, but more importantly with reason. One reason common to both methods is the lack of consideration of the factors (aside from time and concentration) that are important in determining toxicity.[70]

Each of the foregoing methods has proven useful and has focused interest on the problem of toxicity prediction; however, it is time to make a new attempt to develop a more realistic model. Most power plants are sited on fresh waters, and the data base, even after selecting only results that meet stringent criteria,[71] for freshwater organisms is large (Figure 3). These data are certainly not complete; no studies of the organic haloamines are included. However, since organic haloamine data will be sparse for some time to come, it does not appear logical to wait. A method, even if incomplete, will provide for a biologically based regulation and help focus research efforts where they will be most effective and useful to society.

Table III. Research Needs for Evaluating Optimum Chlorination at Power Plants

Biofouling control need and application
  Develop technology to monitor microfouling, especially during early stages

Effective exposure
  Identify, quantify, and characterize (octanol/water partition coefficient) nonvolatile lipophilic haloorganics in power plant cooling waters
  Identify, quantify, and characterize (octanol/water partition coefficient) organic chloramines in power plant cooling waters

Biotic response
  Evaluate toxicity of representative (range of octanol/water coefficients) nonvolatile lipophilic haloorganics
  Evaluate octanol/water partition coefficient as a predictor for toxicity of organic chloramines
  Develop a model to summarize toxicity data for predicting effects of total residual oxidants on mortality of aquatic communities. Model should be capable of including exposure, total residual oxidant species, and temperature, as each can be quantified regarding effects on toxicity

RESEARCH NEEDS—CHLORINATION AT POWER PLANTS

Lacking a method that is generally accepted, a list of priority areas of research must be based on these discussions. The developed list (Table III) of such areas is biased toward determining optimum chlorination practices at power plants and is not intended to suffice for other industries or municipal treatment facilities where different problems may be of higher priority. However, since many of the items on the list deal with problems faced by more than just the utility industry and its regulatory agencies, responsibility for developing the relevant information should be shared among industries and government funding agencies.

SUMMARY

Criteria for defining optimum chlorination practices provide a format for considering questions about power plant chlorination and directing attention to future research needs. The criteria are (1) maintenance of efficient energy production and (2) maintenance of environmental quality. The first of these criteria can be subdivided into a biofouling control need and a biofouling control application. Current industry practice suggests that obtaining a variance to the 2-h chlorination limit is difficult. Research on a macrofouling need or control application is therefore best directed toward development of alternatives to chlorine. Research on a microfouling control need should be directed toward development of devices to measure microfouling levels rather than to prediction based on water quality. With such devices, control applications are probably best determined on a site-by-site basis. The second criterion can also be subdivided into effective exposure and biotic response, but risk assessment involves iteration between the two areas. An evaluation of effluent concentrations, environmental persistence, direct toxicity levels, and bioaccumulation suggests that volatile haloorganics and nonvolatile hydrophilic haloorganics released from power plants will not result in environmental degradation. Nonvolatile lipophilic haloorganics appear to be released at very low levels, but isolation, identification, and examination of structure/activity relationships will be required for further evaluation.

Research on power plant chlorination should be focused on residual halogens. Efforts should be directed toward an examination of the occurrence and toxicity of organic haloamines and toward the development of a toxicity relationship that incorporates residual halogen species, exposures, and temperatures to predict safe halogen concentrations in power plant effluents.

REFERENCES

1. Jolley, R. L., C. W. Gehrs, and W. W. Pitt. "Chlorination of Cooling Water: A Source of Chlorine-Containing Organic Compounds With Possible Environmental Significance," in *Radioecology and Energy Resources*, C. E. Cushing, Ed. (Stroudsberg, PA: Dowden, Hutchinson & Ross, Inc., 1976), pp. 21–28.
2. Laliberte, M. "EPRI Condenser-Related Research Projects," CS-3196-SR (Palo Alto, CA: Electric Power Research Institute, 1983).
3. U.S. Environmental Protection Agency. "Steam Electric Power Generating Point Source Category; Effluent Limitations Guidelines, Pretreatment Standards and New Source Performance Standards," *Fed. Regist.* 47:52290–52309 (1982).
4. U.S. Environmental Protection Agency. "Policy for the Development of Water Quality-Based Effluent Limitations for Toxic Pollutants," (Washington, DC: USEPA, 1983).
5. Hunton and Williams. "Comments on Policy for the Development of Water Quality-Based Permit Limitations for Toxic Pollutants," (Richmond, VA: Utility Water Act Group, Edison Electric Institute, and National Rural Electric Cooperative, 1983).
6. Chow, W., and R. K. Kawaratani. "Biofouling Assessment and Control: An Electric Power Research Institute Overview," in *Water Chlorination: Environmental Impact and Health Effects, Vol. 4.*, R. L. Jolley, W. A. Brungs, J. A. Cotruvo, R. B. Cumming, J. S. Mattice, and V. A. Jacobs, Eds. (Ann Arbor, MI: Ann Arbor Science Publishers, Inc., 1983), pp. 887–900.
7. Sharpe, V. J. "Biofilm Formation and Control—A Review," 82-252-K (Toronto: Ontario Hydro Research Division, 1982).
8. Mattice, J. S. "Freshwater Macrofouling and Control with Emphasis on *Corbicula*," in *Symposium on Condenser Macrofouling and Control Technologies: The State of the Art*, CS-3343, I. A. Diaz-Tous, M. J. Miller, and Y. G. Mussalli, Eds. (Palo Alto, CA: Electric Power Research Institute, 1983), Chap. 4.
9. Truchan, J. G. "Power Plant Chlorination: Regulatory Considerations," SR-38 (Palo Alto, CA: Electric Power Research Institute, 1976), pp. 229–234.
10. Graham, J. "Biofouling Control Assessment—A Preliminary Data Base Analysis," CS-2469 (Palo Alto, CA: Electric Power Research Institute, 1982).
11. Moss, R., H. B. Flora, R. A. Hiltunen, and N. D. Moore. "Chlorine Minimization/Optimization at a TVA Steam Plant," in *Condenser Biofouling Control: Symposium Proceedings*, J. F. Garey, R. M. Jordan, A. H. Aitken, D. T. Burton, and R. H. Gray, Eds. (Ann Arbor, MI: Ann Arbor Science Publishers, Inc., 1980), pp. 325–337.
12. Rice, J. K. "Chlorine Minimization—An Overview," in *Condenser Biofouling Control: Symposium Proceedings*, J. F. Garey, R. M. Jordan, A. H. Aitken, D. T. Burton, and R. H. Gray, Eds. (Ann Arbor, MI: Ann Arbor Science Publishers, Inc., 1980), pp. 295–299.
13. Characklis, W. G., J. D. Bryers, M. G. Trulear, and N. Zelver. "Biofouling Film Development and Its Effects on Energy Losses: A Laboratory Study," in *Condenser Biofouling Control: Symposium Proceedings*, J. F. Garey, R. M. Jordan, A. H. Aitken, D. T. Burton, and R. H. Gray, Eds. (Ann Arbor, MI: Ann Arbor Science Publishers, Inc., 1980), pp. 49–76.

14. Tarbuck, L., and C. H. E. Wyborn. "Microbial Growth and Activity in Condenser Tubes," in *Condenser Biofouling Control: Symposium Proceedings*, J. F. Garey, R. M. Jordan, A. H. Aitken, D. T. Burton, and R. H. Gray, Eds. (Ann Arbor, MI: Ann Arbor Science Publishers, Inc., 1980), pp. 85-103.
15. Sakaguchi, I. "Studies on the Biofouling and Corrosion of Condensers in Steam Power Plants in Japan," in *Condenser Biofouling Control: Symposium Proceedings*, J. F. Garey, R. M. Jordan, A. H. Aitken, D. T. Burton, and R. H. Gray, Eds. (Ann Arbor, MI: Ann Arbor Science Publishers, Inc., 1980), pp. 301-324.
16. Hillman, R. E., D. Anson, J. M. Corliss, B. W. Vigor, and R. H. Gray. "Biofouling Detection Monitoring Devices," Status Assessment, Draft Final Report RP2300-1 (Palo Alto, CA: Electric Power Research Institute, 1984).
17. Garey, J. F., D. T. Burton, and E. P. Taft. "Marine Macrofouling," in *Symposium on Condenser Macrofouling Control Techologies: The State of The Art*, CS-3343, I. A. Diaz-Tous, M. J. Miller, and Y. G. Mussalli, Eds. (Palo Alto, CA: Electric Power Research Institute, 1983), Chap. 3.
18. Newell, R. C. "Biology of Intertidal Animals," (New York: American Elsevier Publishing Company, Inc., 1970).
19. McMahon, R. F. "Shell Size-Frequency Distributions of *Corbicula manilensis* Philippi from a Clam-fouled Steam Condenser," *Nautilus* 71:51-59 (1977).
20. Jenner, H. A. "A Microcosm Monitoring Mussel Fouling," in *Symposium on Condenser Macrofouling Control Technologies: The State of the Art*, CS-3343, I. A. Diaz-Tous, M. J. Miller, and Y. G. Mussalli, Eds. (Palo Alto, CA: Electric Power Research Institute, 1983), Chap. 6.
21. Gaulke, A. E., T. E. Webb, and F. L. Stokes. "A Chlorine Minimization Program for an Electric Generating Facility Located on the Ohio River," in *Condenser Biofouling Control: Symposium Proceedings*, J. F. Garey, R. M. Jordan, A. H. Aitken, D. T. Burton, and R. H. Gray, Eds. (Ann Arbor, MI: Ann Arbor Science Publishers, Inc., 1980), pp. 339-354.
22. Schumacher, P. D., and J. W. Lingle. "Chlorine Minimization Studies at the Wisconsin Electric Power Co. Valley and Oak Creek Power Plants," in *Condenser Biofouling Control: Symposium Proceedings*, J. F. Garey, R. M. Jordan, A. H. Aitken, D. T. Burton, and R. H. Gray, Eds. (Ann Arbor, MI: Ann Arbor Science Publishers, Inc., 1980), pp. 301-324.
23. Rippon, J. E. "United Kingdom Biofouling Control Practices," in *Condenser Biofouling Control: Symposium Proceedings*, J. F. Garey, R. M. Jordan, A. H. Aitken, D. T. Burton, and R. H. Gray, Eds. (Ann Arbor, MI: Ann Arbor Science Publishers, Inc., 1980), pp. 279-294.
24. Battaglia, P. J., D. P. Bour, and R. M. Burd. "Biofouling Control Practice and Assessment," CS-1796 (Palo Alto, CA: Electric Power Research Institute, 1981).
25. Yasui, K., A. Kawabe, I. Sakaguchi, K. Aoki, and K. Fukuhara. "Velocity vs. Settling Success of Macroinvertebrates," in *Symposium on Condenser Macrofouling Control Technologies: The State of the Art*, CS-3343, I. A. Diaz-Tous, M. J. Miller, and Y. G. Mussalli, Eds. (Palo Alto, CA: Electric Power Research Institute, 1983), Chap. 22.
26. Bryers, J. D., W. G. Characklis, and G. R. Brown. "Measurement of Primary Biofilm Formation," in *Condenser Biofouling Control: Symposium Proceedings*, J. F. Garey, R. M. Jordan, A. H. Aitken, D. T. Burton, and R. H. Gray, Eds. (Ann Arbor, MI: Ann Arbor Science Publishers, Inc., 1980), pp. 169-183.

27. Sugam, R., C. R. Guerra, J. L. Del Monaco, J. H. Singletary, and W. A. Sandvik. "Biofouling Control With Ozone at the Bergen Generating Station," CS-1629 (Palo Alto, CA: Electric Power Research Institute, 1980).
28. Guerra, C. R., L. J. Wyzalek, and A. C. Ciallella. "Use of Pilot Scale Condensers for Biofouling Measurement and Control," in *Condenser Biofouling Control: Symposium Proceedings*, J. F. Garey, R. M. Jordan, A. H. Aitken, D. T. Burton, and R. H. Gray, Eds. (Ann Arbor, MI: Ann Arbor Science Publishers, Inc., 1980), pp. 205–237.
29. Beauchamp, R. S. A. "The Use of Chlorine in the Cooling Water System of a Coastal Power Station," *Chesapeake Sci.* 10:280 (1969).
30. Whitehouse, J. W., M. Khalanski, and M. Saroglia. "Marine Macrofouling Control Experience in the U.K. with an Overview of European Practices," in *Symposium on Condenser Macrofouling Control Technologies: The State of the Art*, CS-3343, I. A. Diaz-Tous, M. J. Miller, and Y. G. Mussalli, Eds. (Palo Alto, CA: Electric Power Research Institute, 1983), Chap. 17.
31. Morris, D. W., J. H. Tackett, J. F. Garey, and J. W. Egan. "Minimizing Chlorine Application Consistent with Effective Macrofouling Control: A Pilot Study of Continuous Low Level Chlorination," in *Symposium on Condenser Macrofouling Control Technologies: The State of the Art*, CS-3343, I. A. Diaz-Tous, M. J. Miller, and Y. G. Mussalli, Eds. (Palo Alto, CA: Electric Power Research Institute, 1983), Chap. 7.
32. Mattice, J. S., and H. E. Zittel. "Site-Specific Evaluation of Power Plant Chlorination," *J. Water Pollut. Control Fed.* 48(10):2284–2308 (1976).
33. Rodgers, E. B., and J. T. Johnson. "The Control by Chlorination of Fresh-Water Sponge Fouling Found in the Vicinity of a Nuclear Power Plant," in *Symposium on Condenser Macrofouling Control Technologies: The State of the Art*, CS-3343, I. A. Diaz-Tous, M. J. Miller, and Y. G. Mussalli, Eds. (Palo Alto, CA: Electric Power Research Institute, 1983), Chap. 8.
34. Tilly, L. J. "Clam Survival in Chlorinated Water," DP-1398 (Aiken, SC: Savannah River Laboratory, 1976).
35. Gooch, C., B. G. Isom, J. Moses, and L. Neill. "*Corbicula* Chlorine Bioassay—*Corbicula* Control Project," Final Report I-WQ-67-10 (Muscle Shoals, AL: Tennessee Valley Authority, 1978), pp. 1–36.
36. Khalanski, M., and F. Bordet. "Effects of Chlorination on Marine Mussels," in *Water Chlorination: Environmental Impact and Health Effects, Vol. 3*, R. L. Jolley, W. A. Brungs, and R. B. Cumming, Eds. (Ann Arbor, MI: Ann Arbor Science Publishers, Inc., 1980), pp. 557–567.
37. Jolley, R. L., W. W. Pitt, F. G. Taylor, S. J. Hartmann, G. Jones, and J. E. Thompson. "An Experimental Assessment of Halogenated Organics in Waters from Cooling Towers and Once-through Systems," in *Water Chlorination: Environmental Impact and Health Effects, Vol. 2*, R. L. Jolley, H. Gorchev, and D. H. Hamilton, Jr., Eds. (Ann Arbor, MI: Ann Arbor Science Publishers, Inc., 1978), pp. 695–705.
38. Bean, R. M. "Recent Progress in the Organic Chemistry of Water Chlorination," in *Water Chlorination: Environmental Impact and Health Effects, Vol. 4*, R. L. Jolley, W. A. Brungs, J. A. Cotruvo, R. B. Cumming, J. S. Mattice, and V. A. Jacobs, Eds. (Ann Arbor, MI: Ann Arbor Science Publishers, Inc., 1983), pp. 843–850.

39. Seaman, C. V., L. O. Hill, B. W. Vignon, T. B. Stanford, and M. D. Hunter, "Halogenated Organic Study at Selected Tennessee Valley Authority Fossil-Fueled Power Plants," in *Water Chlorination: Environmental Impact and Health Effects, Vol. 4,* R. L. Jolley, W. A. Brungs, J. A. Cotruvo, R. B. Cumming, J. S. Mattice, and V. A. Jacobs, Eds. (Ann Arbor, MI: Ann Arbor Science Publishers, Inc., 1983), pp. 373-382.
40. Helz, G. R., R. Sugam, and R. Y. Hsu. "Chlorine Degradation and Halocarbon Production in Estuarine Waters," in *Water Chlorination: Environmental Impact and Health Effects, Vol. 2,* R. L. Jolley, H. Gorchev, and D. H. Hamilton, Jr., Eds. (Ann Arbor, MI: Ann Arbor Science Publishers, Inc., 1978), pp. 209-222.
41. Fogelquist, E., B. Josefsson, and C. Roas. "Halocarbons as Tracer Substances in Studies of the Distribution Patterns of Chlorinated Waters in Coastal Areas," *Environ. Sci. Technol.* 16:479-482 (1982).
42. Hollod, G. J., and E. W. Wilde. "Trihalomethanes in Chlorinated Cooling Waters of Nuclear Reactors," *Bull. Environ. Contam. Toxicol.* 28:404-408 (1982).
43. Bean, R. M., D. C. Mann, and D. A. Neitzel. "Organohalogens in Chlorinated Cooling Waters Discharged from Nuclear Power Stations," in *Water Chlorination: Environmental Impact and Health Effects, Vol. 4,* R. L. Jolley, W. A. Brungs, J. A. Cotruvo, R. B. Cumming, J. S. Mattice, and V. A. Jacobs, Eds. (Ann Arbor, MI: Ann Arbor Science Publishers, Inc., 1983), pp. 383-389.
44. Jolley, R. L., and J. H. Carpenter. "Chemistry of Nitrogenous and Other Compounds," in *Water Chorination: Environmental Impact and Health Effects, Vol. 4,* R. L. Jolley, W. A. Brungs, J. A. Cotruvo, R. B. Cumming, J. S. Mattice, and V. A. Jacobs, Eds. (Ann Arbor, MI: Ann Arbor Science Publishers, Inc., 1983), pp. 3-47.
45. Morris, J. C., and R. A. Isaac. "A Critical Review of Kinetic and Thermodynamic Constants for the Aqueous Chlorine-Ammonia System," in *Water Chlorination: Environmental Impact and Health Effects, Vol. 4,* R. L. Jolley, W. A. Brungs, J. A. Cotruvo, R. B. Cumming, J. S. Mattice, and V. A. Jacobs, Eds. (Ann Arbor, MI: Ann Arbor Science Publishers, Inc., 1983), pp. 49-62.
46. Isaac, R. A., and J. C. Morris, "Rates of Transfer of Active Chlorine Between Nitrogenous Substrates," in *Water Chlorination: Environmental Impact and Health Effects, Vol. 3,* R. L. Jolley, W. A. Brungs, and R. B. Cumming, Eds. (Ann Arbor, MI: Ann Arbor Science Publishers, Inc., 1980), pp. 183-191.
47. Isaac, R. A., and J. C. Morris. "Modeling of Reactions Between Aqueous Chlorine and Nitrogenous Compounds," in *Water Chlorination: Environmental Impact and Health Effects, Vol. 4,* R. L. Jolley, W. A. Brungs, J. A. Cotruvo, R. B. Cumming, J. S. Mattice, and V. A. Jacobs, Eds. (Ann Arbor, MI: Ann Arbor Science Publishers, Inc., 1983), pp. 63-75.
48. Scully, F. E., J. P. Yang, M. A. Bempong, and F. B. Daniel. "Analysis of Organic N-Chloramines," in *Water Chlorination: Environmental Impact and Health Effects, Vol. 4,* R. L. Jolley, W. A. Brungs, J. A. Cotruvo, R. B. Cumming, J. S. Mattice, and V. A. Jacobs, Eds. (Ann Arbor, MI: Ann Arbor Science Publishers, Inc., 1983), pp. 555-563.
49. Carpenter, J. H., R. G. Zika, and C. A. Moore. "Sensitive Simple Determination of Oxidants in Chlorinated Waters," in *Water Chlorination: Environmental Impact and Health Effects, Vol. 4,* R. L. Jolley, W. A. Brungs, J. A. Cotruvo, R. B. Cumming, J. S. Mattice, and V. A. Jacobs, Eds. (Ann Arbor,MI: Ann Arbor Science Publishers, Inc., 1983), pp. 681-697.

50. Mattice, J. S., and S. C. Tsai. "Total Residual Chlorine as a Regulatory Tool," in *Water Chlorination: Environmental Impact and Health Effects, Vol. 4,* R. L. Jolley, W. A. Brungs, J. A. Cotruvo, R. B. Cumming, J. S. Mattice, and V. A. Jacobs, Eds. (Ann Arbor, MI: Ann Arbor Science Publishers, Inc., 1983), pp. 901–912.
51. Mattice, J. S. "Current Status of Models in Determining Optimum Chlorination Practices," *Environ. Internat.* 4:175–187 (1980).
52. Maki, A. W. "Design and Conduct of Hazard Evaluation Programs for the Aquatic Environment," in *Water Chlorination: Environmental Impact and Health Effects, Vol. 3,* R. L. Jolley, W. A. Brungs, and R. B. Cumming, Eds. (Ann Arbor, MI: Ann Arbor Science Publishers, Inc., 1980), pp. 949–959.
53. Mattice, J. S., S. C. Tsai, and M. B. Burch. "Toxicity of Trihalomethanes to Common Carp Embryos," *Trans. Am. Fish. Soc.* 110:261–269 (1981).
54. Trabalka, J. R., S. C. Tsai, J. S. Mattice, and M. B. Burch. "Effect on Carp Embryos (*Cyprinus carpio*) and *Daphnia pulex* of Chlorinated Organic Compounds Producing During Control of Fouling Organisms," in *Water Chlorination: Environmental Impact and Health Effects, Vol. 3,* R. J. Jolley, W. A. Brungs, and R. B. Cumming, Eds. (Ann Arbor, MI: Ann Arbor Science Publishers, Inc., 1980), pp. 599–606.
55. Gibson, C. I., F. C. Tone, R. E. Schirmer, and J. W. Blaylock. "Bioaccumulation and Depuration of Bromoform in Five Marine Species," in *Water Chlorination: Environmental Impact and Health Effects, Vol. 3,* R. L. Jolley, W. A. Brungs, and R. B. Cumming, Eds. (Ann Arbor, MI: Ann Arbor Science Publishers, Inc., 1980), pp. 517–533.
56. Scott, G. I., D. P. Middaugh, A. M. Crane, N. P. McGlothin, and N. Watabe. "Physiological Effects of Chlorine-Produced Oxidants and Uptake of Chlorination By-Products in the American Oyster, *Crassostrea virginica* (Gemlin)," in *Water Chlorination: Environmenal Impact and Health Effects, Vol. 3,* R. L. Jolley, W. A. Brungs, and R. B. Cumming, Eds. (Ann Arbor, MI: Ann Arbor Science Publishers, Inc., 1980), pp. 501–516.
57. Scott, G. I., D. P. Middaugh, and S. Klingensmith. "Bioconcentration of Bromoform by American Oysters, *Crassostrea virginica* (G), Exposed to Chlorinated and Dechlorinated Seawater, with Notes on Survival and Feeding," in *Water Chlorination: Environmental Impact and Health Effects, Vol. 4,* R. L. Jolley, W. A. Brungs, J. A. Cotruvo, R. B. Cumming, J. S. Mattice, and V. A. Jacobs, Eds. (Ann Arbor, MI: Ann Arbor Science Publishers, Inc., 1983), pp. 1029–1037.
58. Carpenter, J. H., C. A. Smith, and R. G. Zika. "Reaction Products Formed in the Chlorination of Seawater," in *Water Chlorination: Environmental Impact and Health Effects, Vol. 3,* R. L. Jolley, W. A. Brungs, and R. B. Cumming, Eds. (Ann Arbor, MI: Ann Arbor Science Publishers, Inc., 1980), pp. 379–385.
59. Lee, N. R., W. R. Haag, and R. L. Jolley. "Cooling Water Pollutants: Bioaccumulation by *Corbicula*," in *Water Chlorination: Environmental Impact and Health Effects, Vol. 4,* R. L. Jolley, W. A. Brungs, J. A. Cortruvo, and R. B. Cumming, J. S. Mattice, and V. A. Jacobs, Eds. (Ann Arbor, MI: Ann Arbor Science Publishers, Inc., 1983), pp. 851–870.
60. Trabalka, J. R., and M. B. Burch. "Investigation of the Effects of Halogenated Organic Compounds Produced in Cooling Systems and Process Effluents on Aquatic Organisms," in *Water Chlorination: Environmental Impact and Health Effects, Vol. 2,* R. L. Jolley, H. Gorchev, and D. H. Hamilton, Jr., Eds. (Ann Arbor, MI: Ann Arbor Science Publishers, Inc., 1978), pp. 163–173.

61. Mattice, J. S., S. C. Tsai, and M. B. Burch. "Comparative Toxicity of Hypochlorous Acid and Hypochlorite Ions to Mosquitofish," *Trans. Am. Fish. Soc.* 110:519–529 (1981).
62. Brooks, A. S., J. M. Bartos, and P. T. Danos. "The Effects of Chlorine on Freshwater Fish Under Various Time and Chemical Conditions," EA-2481 (Palo Alto, CA: Electric Power Research Institute, 1982).
63. Wajon, J. E., and J. C. Morris. "Bromamination Chemistry: Rates of Formation of NH_2 BR and Some N-Bromamino Acids," in *Water Chlorination: Environmental Impact and Health Effects, Vol. 3,* R. L. Jolley, W. A. Brungs, and R. B. Cumming, Eds. (Ann Arbor, MI: Ann Arbor Science Publishers, Inc., 1980), pp. 171–181.
64. Tsai, S. C., J. S. Mattice, K. B. Packard, and W. K. Roy. "Partition Coefficient and Fish Toxicity of Residual Halogen Species in Chlorinated Waters," unpublished manuscript.
65. Cherry, D. S., S. R. Larrick, J. D. Giattina, and J. Cairns. "Continuation of Laboratory and Field Determined Avoidance Responses of Ohio River, New River, and Lake Michigan Fish to Thermal and Chlorinated Discharges," Biology Department and Center for Environmental Studies (Blacksburg, VA: Virginia Polytechnic Institute and State University, 1980).
66. Hose, J. E., T. D. King, K. E. Zerba, R. J. Stoffel, J. S. Stephens, and J. A. Dickinson. "Does Avoidance of Chlorinated Seawater Protect Fish Against Toxicity? Laboratory and Field Observations," in *Water Chlorination: Environmental Impact and Health Effects, Vol. 4,* R. L. Jolley, W. A. Brungs, J. A. Cotruvo, R. B. Cumming, J. S. Mattice, and V. A. Jacobs, Eds. (Ann Arbor, MI: Ann Arbor Science Publishers, Inc., 1983), pp. 967–982.
67. Hall, L. W., S. L. Margrey, W. C. Graves, and D. T. Burton. "Avoidance Responses of Juvenile Atlantic Menhaden *Brevoortia tyrannus*, Subjected to Simultaneous Chlorine and $\triangle T$ Conditions," in *Water Chlorination: Environmental Impact and Health Effects, Vol. 4,* R. L. Jolley, W. A. Brungs, J. A.Cotruvo, R. B. Cumming, J. S. Mattice, and V. A. Jacobs, Eds. (Ann Arbor, MI: Ann Arbor Science Publishers, Inc., 1983), pp. 983–991.
68. Mattice, J. S. "Power Plant Discharges: Toward More Reasonable Effluent Limits on Chlorine," *Nucl. Saf.* 18(6):802–819 (1977).
69. Turner, A., and T. A. Thayer. "Chlorine Toxicity in Aquatic Ecosystems," in *Water Chlorination: Environmental Impact and Health Effects, Vol. 3,* R. L. Jolley, W. A. Brungs, and R. B. Cumming, Eds. (Ann Arbor, MI: Ann Arbor Science Publishers, Inc., 1980), pp. 607–630.
70. Seegert, G. L., and R. B. Bogardus. "Ecological and Environmental Factors to be Considered in Developing Chlorine Criteria," in *Water Chlorination: Environmental Impact and Health Effects, Vol. 3,* R. L. Jolley, W. A. Brungs, and R. B.Cumming, Eds. (Ann Arbor, MI: Ann Arbor Science Publishers, Inc., 1980), pp. 961–971.
71. Fava, J. A., and G. L. Seegert. "Factors in the Design of Chlorine Toxicological Research," in *Water Chlorination: Environmental Impact and Health Effects, Vol. 4,* R. L. Jolley, W. A. Brungs, J. A. Cotruvo, R. B. Cumming, J. S. Mattice, and V. A. Jacobs, Eds. (Ann Arbor, MI: Ann Arbor Science Publishers, Inc., 1983), pp. 913–925.

CHAPTER 5

Environmental Impacts of Chlorine Discharges: A Utility Industry Perspective

Peter M. Cumbie, Thomas A. Miskimen, and James K. Rice

Biofouling control through the use of chlorine is an operating necessity for most steam-electric power plants to maintain acceptably efficient condenser heat exchange and to protect cooling tower structures. Most chlorination of fresh or saline cooling water is used for slime control in the main condenser. Since the cost associated with even a small loss of efficiency in these facilities is enormous, it is necessary to minimize both the environmental impacts associated with the discharge of chlorinated cooling water and the costs associated with condenser fouling. As a practical matter, at most power plants, chlorination is the most effective means of accomplishing both objectives. The various available alternatives are either less effective or entail environmental risks of their own.[1]

Three types of environmental hazard may be associated with power plant discharges of residual chlorine (excluded from consideration in this discussion are any environmental hazards that may be associated with accidents occurring during rail or truck transport of chlorine gas or liquid bleach, since these hazards are not specific to power plant chlorine use). The first is toxicity of residual chlorine to humans through downstream intakes of drinking water. Because the residuals maintained at power plant condenser outfalls (less than 0.2 mg/L before dilution) are typically only about 2% of the 10-mg/L total chlorine residual that can be tolerated in finished drinking water,[2] this hazard appears to be minimal and will not be discussed further.

A second type of hazard is the potential effect of discharges of chlorinated organics formed during the chlorination of water containing organic substances. As will be shown later, power plant chlorination is a very minor contributor to this risk, based on current data.

A third type of hazard is the toxic effect of chlorine residuals on aquatic organisms in the receiving waters. Impacts on aquatic organisms are of concern in both fresh- and saltwater discharges.

This chapter presents a discussion of the latter two types of hazard as they relate specifically to power plant cooling water discharges. Also described are the chlorination practices of the steam-electric utility industry and the suggested manner in which the interactions of chlorinated cooling water discharges with receiving waters should be analyzed.

UTILITY INDUSTRY CHLORINATION PRACTICES

Steam-electric utility chlorination practices have been described in the U.S. Environmental Protection Agency's (EPA) 1982 BAT Effluent Guidelines Development Document.[3] Once-through condenser cooling water is typically treated by intermittent application of chlorine to the raw intake water at a rate sufficient to achieve a free chlorine residual at the condenser outlet. A free residual (as distinguished from a total residual) of 0.1 ppm is generally necessary at this point, because in most instances it is the free residual chlorine that achieves control of biological growth on the condenser tube surfaces. Apparently, at most plants, disinfection of condenser tube surfaces and disintegration of the biofilm cannot be accomplished by monochloramine, but instead requires free available chlorine. Under EPA's 1982 BAT regulations,[4] chlorination at a plant with once-through cooling is limited to 2 h/d per unit, and plant discharges of total residual chlorine (TRC) are limited to a level 0.2 of mg/L. However, a continuous discharge can be allowed where chlorination is used to control fouling by macroinvertebrates (e.g., barnacles, both in marine intake structures and in marine and freshwater auxiliary cooling systems). A report by the Electric Power Research Institute,[5] based on a survey of power industry chlorination practices, shows that, of those facilities that chlorinate, 95% discharge TRC for 2 h or less per day and 70% discharge TRC for 1 h or less per day. Cooling tower chlorination also involves intermittent treatment of the recirculating cooling water to obtain a chlorine residual; residual chlorine discharges are controlled by simply blocking off the sidestream discharge (blowdown) from the tower for a period that will allow sufficient chlorine decay to comply with the 2-h/d limit on the discharge of any TRC.

To eliminate any misconceptions regarding the extent of chlorinated discharges from U.S. steam-electric power plants, we note that only about 42% of the installed capacity of all power plants discharge chlorinated once-through cooling water as described above.[6] Moreover, only about 36% of the installed capacity of all plants discharge chlorinated cooling tower blowdown. We also note that the cooling tower blowdown flow rate is only about 0.5% of the flow rate of once-through cooling water discharges per megawatt; therefore, these plants contribute very little of the total cooling water discharged. The remaining plants do not chlorinate cooling water because of intake water characteristics (i.e., sand or silt, which can act as scouring agents), or because they use alternative biocides or other means to control biofouling (i.e., mechanical brush systems that are designed into the condenser cooling system when it is installed).

POWER PLANT CHLORINATION AS A SOURCE OF CHLORINATED ORGANICS

The discharge from power plants of some 100 to 200 tons per year of chloroform (as a measure of total chlorinated hydrocarbons) has been estimated by Jolley et al.[7] We believe that this estimate overstates the magnitude of such discharges.

The Utility Water Act Group (UWAG)* has estimated that plants producing approximately 277 GW of U.S. capacity use once-through cooling, whereas plants producing 197 GW of U.S. capacity use cooling towers.[6] Of plants with once-through cooling, 73% chlorinate (42% of total installed capacity), and of plants with cooling towers, 85% chlorinate (36% of total installed capacity). Assuming that (1) the average flow for once-through cooling is 750,000 gpm/GW, (2) 13,000 gpm/GW is the average makeup rate for cooling tower units, and (3) 1 h/d is the general industry chlorination practice,[5] the amount of water that was chlorinated in 1982 can be calculated based on that year's generation of electricity for steam-electric power plants in the United States of 1.9 million GWh.[8]

The Tennessee Valley Authority showed that the average net increase (condenser discharge prior to dilution by the receiving stream, less intake) of chloroform in a chlorinated once-through cooling system was 4 ppb.[9] This value compares favorably with the increase noted in a UWAG study at Pacific Gas and Electric Company's Potrero once-through seawater-cooled plant,[10] where 12 measurements each of unchlorinated inlet and chlorinated outlet water over a period of 10 d showed an average increase in the chloroform concentration of 3.9 ppb, $\pm$ 9.6 ppb, with four of the differences being negative. Based on 4 ppb chloroform added to the foregoing chlorinated volume calculated above, the total amount of chloroform released to the aquatic environment from the U.S. steam-electric power industry is approximately 32 tons for the year 1982, rather than the 100 to 200 tons per year given by Jolley et al.[7]

To put this 32 tons per year estimate into some perspective, in 1982, publicly owned treatment works (POTWs) in the United States reportedly discharged an average of approximately 50,000 mgd of treated sewage.[11] Assume that all of such discharge is from secondary treatment, and assume, based on data from Cooper et al.[12] (for 30-min contact time and maximum monochloramine concentration prior to breakpoint), that 24 ppb is the average chloroform increase resulting from chlorination of treated secondary sewage effluents. Since substantially less than 100% of sewage treatment plants use secondary treatment,[11] the estimate of 24 ppb is likely to be conservative. Assume also that most plants chlorinate continuously most of the year. Based on these assumptions, the annual release of chloroform from POTW sewage treatment operations in the United States is approximately 1800 tons per year, or 50 times more than that estimated to be discharged from power plants.

In a report by Bean,[13] chloroform accounts for some 12 to 40% of the total organic halogens (TOX) in once-through cooling water effluents from two

---

*The Utility Water Act Group is a consortium of 73 electric power generating companies representing more than 50% of the electric power generating capacity of the United States. The Edison Electric Institute, the American Public Power Association, and the National Rural Electric Cooperative Association are also members.

nuclear power plants. Thus, the amount of TOX in U.S. power plant discharges is probably only 100 to 300 tons per year.

Even allowing for the many approximations involved in these estimates of chloroform release, and of TOX in relation to chloroform, the foregoing discussion demonstrates that chlorinated hydrocarbon production by power plants is a relatively minor problem when viewed in relation to other known sources of these compounds. These statistics certainly give little support for expending additional effort (beyond achieving 1984 BAT limitations) on treatment or modification of power plant discharges for the purpose of reducing chlorinated hydrocarbon production.

INTERACTIONS OF POWER PLANT RESIDUAL CHLORINE DISCHARGES WITH RECEIVING WATERS

Much effort has been expended to date in gathering data for analyses of the environmental impacts of chlorine residuals in cooling water discharges. However, most analyses of such impacts fall considerably short of a realistic appraisal because of incomplete consideration concerning the fate of the residual chlorine that is discharged. Environmental impact analyses should include consideration of the following processes:

1. the demand of the intake cooling water stream for both free and combined residual chlorine,
2. the simple dilution of the discharge from the plant by the receiving waterbody,
3. the decay of the residual chlorine by numerous reactions following discharge from the plant, and
4. the rapid (30-s) demand of the receiving water for free and combined residual chlorine in the discharge.

Several papers presented at the Fifth Water Chlorination Conference, including one by Qualles and Johnson entitled "A Kinetic Model of the Chlorination of Natural Waters,"[14] have discussed the fast and slow reactions of chlorination that would take place in the original stream of cooling water. Beyond these processes, the dilution of the cooling water discharge stream by the receiving water is frequently the only fate process that is considered in projecting impacts of cooling water discharges. Dilution is a well understood hydraulic process, amenable to detailed mathematical analysis and modeling.[15,16] Dilution effects make a major contribution to the reduction of chlorine concentrations in both near- and far-field analyses of chlorine discharge plumes.

In addition to these fast and slow reactions in the original cooling water stream, and in addition to simple dilution of the discharge, consideration must also be given to several other processes that reduce the concentration of the residual oxidant in the receiving water. These include fast reactions with receiving water demand, losses of volatile forms, and photochemically induced reactions. The reduction of residual oxidant concentration thus proceeds under either plug-flow or turbulent, mixed conditions.

Chlorine decay, following an initial rapid reaction when chlorine is added to water, is generally considered to be a first-order reaction, with the rate dependent on the residual chlorine concentration. Such first-order chlorine decay is a factor in the far-field analysis of discharge plumes, particularly under plug-flow or poorly mixed conditions because of the relatively slow rates of the reactions involved. The first-order rate constant for TRC in river water is typically 20 to 30/d.[17] This slow decay of chlorine residuals is not always considered in discharge plume analysis, although the phenomenon is well understood and applicable data are readily collected.

The Utility Water Act Group recently commissioned a review and analysis of dilution models recommended by EPA for projecting chlorine discharge plume impacts.[17] As a part of this project, a first-order decay subroutine was incorporated into the computer program for the dilution models. Lawler et al. reported on this project at the Fifth Water Chlorination Conference.[18]

However, an additional major factor affecting the chlorine residuals in cooling water discharge plumes has not been generally recognized. This factor is the rapid (30-s) demand of the unchlorinated receiving water for the combined chlorine residuals (CRC) in the cooling water discharge stream.

During initial mixing in the power plant, the free chlorine demand of the cooling water is generally exceeded for the time of travel (10 s to 10 min) from the point of application to the condenser outlet to produce the disinfecting action of free available chlorine (FAC) in the condenser tubes. However, the discharged cooling water is then generally mixed with the receiving water; thus, a new source of demand for FAC and CRC is encountered. Then, because FAC residuals in the discharge stream react rapidly with ammonia and other demands, any FAC is rapidly consumed, and that which reacted with ammonia or organic amines generates additional CRC. Further mixing with more of the receiving water causes the CRC to react with a portion of the demand of the subsequently added unchlorinated water. If the dilution process is conceptualized as a stepwise series of dilutions of the cooling water by the receiving water, we see that additional chlorine demand will be exerted on each dilution step.

Chlorine decay rate constants, as noted above, are generally based on the more slowly reacting components of chlorine demand in a single volume of water to which chlorine has been added. Therefore, a consideration of decay based on this slow reaction rate data does not adequately treat the effect of instantaneous free and combined residual chlorine demands. While the collection of FAC and CRC rapid demand data for the receiving water requires more effort, these data can be readily generated for any cooling or receiving water.

Utility Water Act Group trial runs with one of the EPA-recommended dilution models showed that, under a realistic combination of discharge and receiving water flows, first-order chlorine decay had relatively little effect on the plume residual prediction beyond that resulting from dilution alone in the stream reach of interest, largely because of the relatively rapid downstream travel of the plume (1–3 ft/s). When an instantaneous 30-s combined residual chlorine demand term was included in the stepwise dilution model routine, plume front prediction changed significantly.

Analyses of the Ohio River near Cincinnati have shown 30-s CRC demands of 0.02 to 0.08 mg/L.[19] Therefore, if a plant effluent TRC of 0.5 mg/L, a receiving water 30-s (CRC) demand of 0.05 mg/L, and a dilution ratio of 10:1 are assumed, a prediction of 0.0 mg/L TRC in the diluted effluent results rather than the 0.05 mg/L that would be predicted from a 10:1 dilution alone. This model of the reaction of chlorine in discharge plumes appears to be much more realistic than a simple dilution plus first-order decay approach and should be used in the analysis of such plumes and their impacts on receiving water biota. It is gratifying to note that Coughlan and Davis, in their presentation at this conference, made similar observations and corrections to the simple decay and dilution models.[20]

IMPACT ON AQUATIC ORGANISMS

Analyses of environmental impacts of power plant cooling water discharges should include evaluation of the discharge regime that is used. Once-through condenser chlorination is usually intermittent in nature. As noted earlier, approximately 70% of those who chlorinate discharge chlorine residuals for 1 h/d or less, and 95% discharge chlorine residuals for 2 h/d or less. Therefore, there is generally a well-defined limitation to the duration of exposure that aquatic life in the vicinity of the discharge could experience. Cooling tower blowdown, while normally continuous, is also chlorinated intermittently and may be held up during and immediately following a chlorination episode so that the chlorine residual might be allowed to decay prior to resumption of the blowdown discharge.

The general application of water quality standards to intermittent power plant discharges, based on continuous exposure to chlorine residuals, is clearly inappropriate. Both once-through condenser cooling water discharges and cooling tower blowdown discharges containing chlorine are usually intermittent. Moreover, the steam-electric guidelines expressly limit discharges of chlorine residuals in once-through cooling waters and cooling tower blowdown to 2 h/d per unit. Permit writers and industry alike can legitimately expect EPA to develop clear recommendations for dealing with the regulation of intermittent chlorine discharges, apart from application of the continuous exposure criteria as such.

Research supports the foregoing conclusion. Toxicity testing under a variety of conditions has demonstrated unequivocally that organisms differ in their toxic responses to continuous and intermittent chlorine exposure. The 50 data points on the lethality of intermittent exposure to chlorine contained in Table

11-6 (based on seven research reports) of EPA's Draft Multi-Media Chlorine Criteria Document[2] indicate that organisms are generally much more tolerant of intermittent chlorine exposures than of continuous exposures. Turner and Thayer[21] used the available LC_{50} data on intermittent chlorine exposure to construct a model for predicting protective water quality criteria in such situations.

The EPA's current proposed chlorine criteria are based solely on continuous exposure data.[22] While such criteria may prove useful in other contexts, a realistic regulation of power plant effluent environmental impacts should explicitly recognize the reduced toxic responses of aquatic organisms to intermittent chlorine exposures.

As vehicles for accomplishing this regulatory consideration, models are now available that will predict acceptable chlorine exposure time-concentration combinations.[21] We note that the states of California and West Virginia have used such models to develop intermittent exposure standards for residual chlorine.[23,24] Since the EPA has apparently approved California's use of intermittent chlorine criteria, the Agency should proceed to ultimately develop nationally applicable intermittent chlorine exposure criteria. In the interim, the EPA should issue guidance for permit writers for developing intermittent exposure standards from the existing criteria for continuous exposure.

Although behavioral responses such as avoidance by aquatic organisms of intermittent chlorine residuals have not been discussed here, they can be very important.[25] Regulators should include a consideration of such responses in their analyses of the impact of chlorinated cooling water discharges on the aquatic environment. They should also give guidance to permit writers.

CONCLUSIONS

Biofouling control through the addition of chlorine to condenser cooling water is an operating necessity for most steam-electric power plants. On a national basis, it is conservatively estimated that chlorination of cooling water results in the discharge of only 1/50 of the amount of chloroform produced during chlorination of POTW discharges, and thus appears to be a relatively minor contributor of chlorinated organics to the environment. Furthermore, the impact of chlorine in power plant discharges on aquatic life is probably much less than supposed, since dissipation of TRC is rapid and most discharges are intermittent and limited by BAT guidelines to a total duration of only 2 h/d per unit.

Analyses of the environmental impacts of cooling water chlorine discharges should evaluate all of the major interactions of such discharges with receiving waters, including rapid demand and decay as well as simple dilution.

Since the EPA's proposed water quality criteria for chlorine are based solely upon continuous exposure data, and since it is well known that intermittent exposure to TRC is less toxic than continuous exposure, it is essential that the

EPA develop criteria for intermittent exposure that can be properly applied to existing intermittent discharges.

Because the chlorine demand reactions in the discharge dilution plumes have not been adequately studied, it is important that further research be conducted on such problems (e.g., reactions of monochloramine with demands created by dilution in the receiving waters).

ACKNOWLEDGMENTS

The authors wish to express their appreciation to Ms. Kristy Niehaus, Esq., and the staff of Hunton & Williams for their assistance in the preparation of this report on behalf of the Utility Water Act Group.

REFERENCES

1. Helz, G. R., and L. Kosack-Channing. "Dechlorination of Wastewater and Cooling Water," *Environ. Sci. Technol.* 18:48A–55A (1984).
2. "Ambient Water Quality Criteria for Chlorine," Draft (Washington, DC: U.S. Environmental Protection Agency, 1981).
3. *Development Document for Effluent Limitations Guidelines and Standards and Pretreatment Standards for the Steam Electric Point Source Category,* EPA 400/1-82/029 (Washington, DC: U.S. Environmental Protection Agency, 1982).
4. U.S. Environmental Protection Agency. "Steam Electric Power Generating Point Source Category," U.S. Code of Federal Regulations, Vol. 40, Part 423.13 (1982).
5. *Biofouling Control Assessment—A Preliminary Data Base Assessment,* EPRI CS-2469 (Palo Alto, CA: Electric Power Research Institute, 1982).
6. Utility Water Act Group. "Economic and Financial Impacts of EPA's October 14, 1980 Proposed Regulations," in *Comments of the Utility Water Act Group on EPA's Proposed Effluent Limitations for the Steam Electric Generating Point Source Category,* Sect. XI, (Hunton & Williams, January 1981).
7. Jolley, R. L., W. W. Pitt, Jr., F. G. Taylor, Jr., S. J. Hartmann, G. Jones, Jr., and J. E. Thompson. "An Experimental Assessment of Halogenated Organics in Waters from Cooling Towers and Once-Through Systems," in *Water Chlorination: Environmental Impact and Health Effects, Vol. 2,* R. L. Jolley, H. Gorchev, and D. H. Hamilton, Jr., Eds. (Ann Arbor, MI: Ann Arbor Science Publishers, Inc., 1978), pp. 695–706.
8. Energy Information Administration. "1982 Annual," *Electric Power Annual,* DOE/EIA-0348(82) (Washington, DC: U.S. Department of Energy, 1983).
9. Tennessee Valley Authority. *Chlorine Minimization/Optimization for Condenser Biofouling Control: Final Report,* EPA-600170-80-143 (Washington, DC: U.S. Environmental Protection Agency, 1980).
10. Utility Water Act Group. "The Proposed Limitations and Standards for Once-Through Cooling Water and Cooling Tower Blowdown," *Comments of the Utility Water Act Group On EPA's Proposed Effluent Limitations for the Steam Electric Generating Point Source Category,* Sect. III, (Hunton & Williams, Jan. 19, 1981).

BASIC ISSUES 71

11. Personal communication, J. K. Rice. Consulting Engineer, Olney, MD, and Wen Huang, U.S. Environmental Protection Agency, Water Programs Office (May 18, 1984).
12. Cooper, W. J., J. T. Villate, E. M. Ott, R. A. Slifker, F. Z. Parsons, and G. A. Graves. "Formation of Organohalogen Compounds in Chlorinated Secondary Wastewater Effluent," in *Water Chlorination: Environmental Impact and Health Effects, Vol. 4,* R. L. Jolley, W. A. Brungs, J. A. Cotruvo, R. B. Cumming, J. S. Mattice, and V. A. Jacobs, Eds. (Ann Arbor, MI: Ann Arbor Science Publishers, Inc., 1983), pp. 483–497.
13. *Organohalogen Products from Chlorination of Cooling Water at Nuclear Power Stations*, NUREG/CR 3408, PNL-4708 (Washington, DC: U.S. Nuclear Regulatory Commission, 1983).
14. Qualles, R. G., and J. R. Johnson. "A Kinetic Model of Chlorination of Natural Water: The Roles of Organic Nitrogen and Humic Substances," Chapter 57, this volume.
15. Jirka, G. H., G. Abraham, and D. R. F. Harlemonn. "An Assessment of Techniques for Hydrothermal Prediction," Ralph M. Parsons Laboratory for Water Resources and Hydrodynamics, Report 203 (Cambridge, MA: Massachusetts Institute of Technology, 1975).
16. "Technical Guidance Manual for the Regulation Promulgated Pursuant to Section 301(g) of the Clean Water Act of 1977, 40 CFR. 125 (Subpart F)," Draft (Washington, DC: U.S. Environmental Protection Agency, 1982).
17. Utility Water Act Group. "Chlorine Plume Modeling Study," (Lawler, Matusky, and Skelly Engineers, 1983).
18. Lawler, J. P., T. B. Vanderbeek, and P. M. Cumbie. "Prediction of Total Residual Chlorine in Power Plant Discharges and Receiving Waters: Application to Effluent Limitations and Water Quality Standards," Chapter 114, this volume.
19. Dames and Moore. "Chlorine Dispersion Modeling for the Cincinnati Gas & Electric Company Miami Fort Generating Station," (Cincinnati: Cincinnati Gas & Electric Co., 1984).
20. Coughlan, J., and M. H. Davis. "Concentration of Chlorine Around Marine Cooling Water Outfalls: Validation of a Model," Chapter 112, this volume.
21. Turner, A., and T. A. Thayer. "Chlorine Toxicity in Aquatic Ecosystems," in *Water Chlorination: Environmental Impact and Health Effects, Vol. 3,* R. L. Jolley, W. A. Brungs, and R. B. Cummings, Eds. (Ann Arbor, MI: Ann Arbor Science Publishers, Inc., 1980), pp. 607–630.
22. "Ambient Aquatic Life Water Quality Criteria for Chlorine" (Washington, DC: U.S. Environmental Protection Agency, 1983).
23. "Water Quality Control Plan for Ocean Water of California" (Sacramento, CA: California State Water Resources Control Board, 1983), pp. 6–7.
24. State of West Virginia Code, Ch.20-5, 20-5a, Sect. 8 (1983).
25. Giattina, J. D., D. S. Cherry, J. Cairns, Jr., and S. R. Larrick. "Comparison of Laboratory and Field Avoidance Behavior of Fish in Heated Chlorinated Water," *Trans. Am. Fish. Soc.* 110:526–535 (1981).

CHAPTER **6**

Conceptual Approach to Evaluate Alternatives to Chlorination for Biofouling Control

James A. Fava, William J. Rue, Paul Chrostowski, Joseph S. Ferris, and Hans Plugge

Chlorine is the most commonly used chemical to control biofouling at power plants and to disinfect wastewater from sewage treatment plants. Chlorination is probably one of the most extensively studied chemical processes in the field of environmental chemistry. Potential negative impacts to both human health and aquatic communities have been identified or suggested, but the widespread use of chlorination continues.

This chapter has two objectives: (1) to conceptually evaluate the appropriateness of existing data for making management decisions; and (2) to identify elements of a conceptual framework, using existing and new data, which would allow for technically based environmental decisions on the use of chlorination or alternatives for biofouling control.

APPROPRIATENESS OF EXISTING DATA

To assess the appropriateness of existing data for making managerial decisions, the proceedings of the previous Water Chlorination Conferences[1-4] were reviewed to characterize the papers and information into four basic categories: Chemistry, Effects, Exposure (i.e., expected environmental concentrations), and Policy and Management. Chapters in the Chemistry category were concerned with the formation of chlorination organic by-products, decay kinetics, and approaches for measuring chlorine. Effects chapters dealt with the responses of mammalian (including man) or aquatic organisms to chlorination. Exposure chapters either measured or predicted (via models) chlorine concentrations the organisms would encounter in the environment. The Policy and Management papers presented criteria for decision-making or information on the decision-making process. Chapters on risk assessment were included in this last category. After these categories were defined, each chapter within the four preceding Water Chlorination Conferences, plus the abstracts for the 1984 Conference, were reviewed and placed into one of the four categories (Table I). We further subdivided chapters within each category as to whether they related primarily to ecological concerns or human health. Since the categorization was based on the study design and the discussion of results, the 1984

Conference proceedings were not subdivided because only the abstracts were available.

The percentage of chapters making up each of the four categories within each proceedings document was plotted (Figure 1). While this assessment is limited to the Water Chlorination Conferences, we believe it is generally representative of the emphasis in chlorination research to date.

Several observations can be made from Table I and Figure 1. First, papers on chemistry and biological effects have dominated water chlorination research. During the five Water Chlorination Conferences, a change in relative emphasis among the various categories has occurred. Increasing emphasis (from a percentage standpoint) has occurred in the Chemistry and Effects categories, whereas emphasis on Exposure and Policy and Management has declined (from a relative standpoint). While the absolute numbers for Exposure have remained basically the same (Table I), the percentage of Exposure papers has decreased significantly (Figure 1). At the same time, the number of Effects papers has increased significantly from 5 (in 1975) to 41 (in 1984); concomitant with that is the large increase in the percentage of these papers (Figure 1). Although it is recognized that some data that might provide estimates of the exposure concentrations in the receiving waters may exist in other places (e.g., individual plant records and monitoring reports), many of these data are not readily available for use in making management decisions.

A potential problem with this research emphasis on biological effects and chemistry is that management decisions concerning the use of chlorine may be made without adequate estimates of the concentrations the organisms are expected to encounter in the environment. Without obtaining adequate information on the estimated environmental exposure concentrations, we have only generated half of the information needed to make proper management decisions. As a result, a manager might conclude that chlorine is a major problem

Table I. Emphasis on Topics Presented at Previous Conferences on Water Chlorination

Category	Total Number of Papers Each Year				
	1975	1977	1979	1981	1984
Effects, total	5	18	29	26	41
Ecological	3	14	13	17	
Human health	2	4	16	9	
Exposure, total	5	5	7	9	7
Ecological	4	4	4	6	
Human health	1	1	3	3	
Chemistry, total	8	21	42	55	66
Ecological	3	6	14	19	
Human health	1	4	8	12	
Other	4	11	20	24	
Policy and management, total	4	9	14	11	8
Ecological	2	3	3	4	
Human health	2	6	11	7	

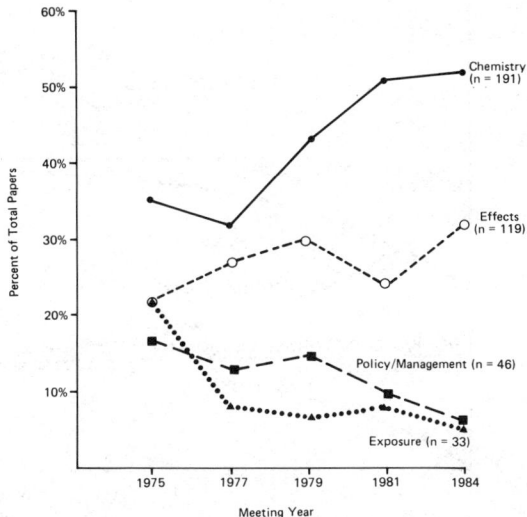

Figure 1. Comparison of topics presented at previous water chlorination conferences.

because biological effects are predicted to occur based on laboratory tests. However, until we can compare the biological no-effect levels with real-world environmental exposure data, this may be a premature conclusion.

To illustrate further, Figure 2 conceptually presents the hazard-assessment concept that relates toxicological effects to exposure concentrations. The key to the assessment is that if the estimated environmental exposure concentration is less than the no-observed-effect levels, then an impact is unlikely to occur. Clearly, both biological effects data and exposure data are needed to make decisions concerning the potential impact of chlorine. Although Maki[5] urged the incorporation of a hazard-assessment concept into the water-chlorination field, this approach has had only limited applications. Additional research should be directed toward the exposure element of the assessment so that more effective management decisions can be made. This exposure research should be published and made more available for review and use in assessing impacts.

PROPOSED CONCEPTUAL FRAMEWORK

As a result of several scientific efforts,[6-8] we have developed a cost-effective assessment methodology that has been used for evaluating the relative risk to man and the environment from waste disposal alternatives. In the context of this discussion, a cost-effective assessment methodology is defined as a technique for constructing a matrix whereby the cost of an engineering option, its effectiveness in a given application, and the risks generated by application of

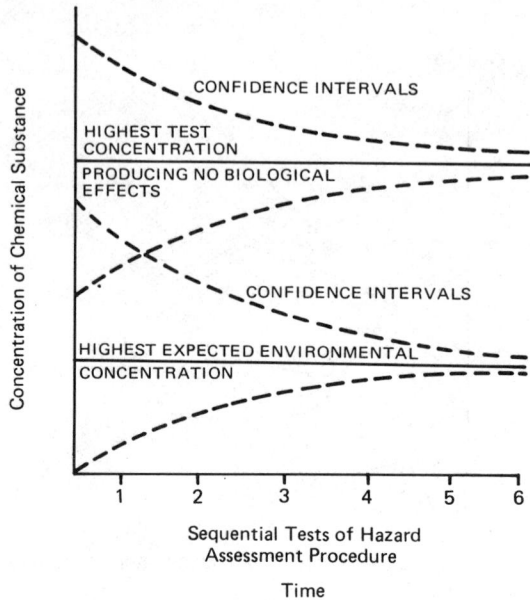

Figure 2. Hazard assessment concept.[5]

the option may be evaluated. This methodology incorporates the relationship between no-observed-effect levels and estimated environmental exposure concentrations.

We are currently applying cost-effective methodology in evaluating alternatives to chlorination for biofouling control from the Electric Power Research Institute. This approach contains quantitative assessments of human health, costs, and environmental impact. The methodology can also be focused on real-world engineering and environmental and management expectations. Cost-effective methodology is capable of yielding a holistic assessment of all major aspects of chlorination management, including engineering and technology, human health risks, environmental impact, social perception and acceptability, and costs. A schematic of the methodology's application to biofouling control is shown in Figure 3. This section contains a more detailed description of the conceptual basis for the methodology.

Biofouling Control Options and Engineering Considerations

As the first step in cost-effective analysis, each major biofouling control option should be assessed on the efficacy of the process or agent to control biofouling target organisms. The process selected will be site specific and based on the actual cooling system configuration in use. Many chemicals may control biofouling if used in sufficient quantities; however, numerous other

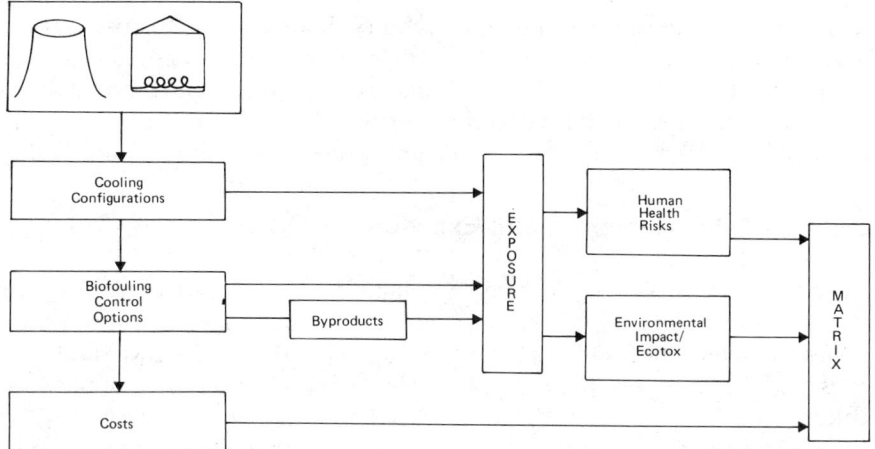

Figure 3. Conceptual method of approach.

technical aspects such as actual operational experience and quality vs quantity of appropriate data should be considered before alternatives are selected.

Once options that control biofouling efficiently are identified, the next phase of the analysis (engineering design and operating requirements) involves determining the facility changes that would be needed to implement each alternative (e.g., ozone generators, evaporators for bromine chloride, feed systems, retrofitting of dechlorination systems). Required consumables (e.g., chemicals, maintenance equipment) should also be considered as part of the design evaluation, as should estimates of fuel and power cost and their availability.

Following design evaluation, the reliability of each option should be assessed using actual data for plants having specific on-line experience (to the extent possible) in preference to manufacturers specifications or theoretical expectations. Factors to be considered in this phase of analysis include the expected percentage of time on-line and corrosion resistance (as they relate to equipment replacement). Should the treatment under consideration be highly efficient for target organisms control, but have a history of mechanical difficulties or high frequency of repairs or replacement, the actual efficiency would be lower and operating costs would be higher.

The final engineering consideration is cost. Each option should be evaluated for required initial investment (e.g., retrofitting or dechlorination equipment) and operating and maintenance costs (equipment, chemicals, power), based on actual operating conditions where possible. If actual operating data are not available for the target process, they may be extrapolated from a similar process (e.g., from potable water disinfection to biofouling control). Costs must be analyzed in a manner consistent with the practices of the industry. For utilities, the present value of the revenue requirement approach is often used.[9]

It is anticipated that for each site where biofouling control alternatives are considered, this series of engineering assessments should be made for each alternative. Therefore, by using appropriate plant-specific data, the final matrix becomes site specific in its applicability. For example, corrosion resistance (a reliability factor) is known to be different for freshwater and saltwater scenarios. Again, although the final matrix and overall methodology are generic, the analyses and data required to fill the matrix can and must be site specific if this information is to be maximally useful to the individual utility.

Chemical Fate, Transport, and Exposure Analysis

The engineering options deemed acceptable for site-specific biofouling control are selected, and initial exposure concentrations needed to control target species are determined. Once a biocide is applied to the water, it will undergo a series of chemical reactions resulting in the formation of chemical by-products.

Subsequent human health and environmental assessments, however, require knowledge of the concentrations of both biocides and by-products to which nontarget species will be exposed. Such a site-specific determination requires assessments of the chemical fate, transport, and dilution of the introduced biofoulant compounds. Critical components of this analysis should consider:

1. chemical reactions/transformations that occur within specific waterbodies (e.g., chlorine reaction with ammonia to form chloramines);
2. dilution with distance from the point of discharge by advection and dispersion;
3. volatilization from the water to the atmosphere (e.g., chloroform volatilization);
4. partitioning between the water column, suspended particulates, and bottom sediments (e.g., sorption of protonated 2,4,6-trichlorophenol); and
5. bioaccumulation/biotransformation by microscopic and macroscopic species.

Since it is clearly beyond the state-of-the-art to evaluate the human health and environmental effects of all measurable biofoulant reaction by-products, it will be necessary to select carefully those chemical species determined to be most important. Criteria for selecting chemicals to evaluate must necessarily account for expected concentrations within the water column and the extent of chemical by-product formation, persistence of the chemical species (prolonged or transient), known or expected toxicity, ability of the chemical to bioaccumulate or become biomagnified, and ability of a chemical to act as a surrogate or indicator of other similar chemicals.

For each biofouling control option, the fate, transport, and transformation subsection must identify major transformation products of any chemical additive (e.g., sodium thiosulfate for dechlorination, chlorine, chlorine dioxide, ozone) and estimates of the concentrations expected to reach nontarget organisms in the receiving water.

Human-Health Risk

Potential human-health impacts associated with a specific biofouling control option could affect two populations: occupationally exposed individuals and the population surrounding the facility. In all likelihood, both groups will be exposed to more than one chemical from each control option. For example, possible chemical species to which individuals could be exposed from the chlorination alternative include chlorine gas, chloramines, and trihalomethanes. The proposed risk-assessment methodology allows quantification of both chemical-specific risk and total risk for exposure to all chemicals of concern previously identified.

The chemical-specific risk estimates should be obtained as the product of three quantifiable parameters: exposed population, average exposure (per time period), and probability of health effect.

A typical risk assessment flow chart used in this context is shown in Figure 4. Identification of the population exposed and the average exposure (both occupational and surrounding community) can be site specific. The probability of a health effect will differ with the route of exposure (e.g., inhalation vs drinking water) and is derived from studies for each chemical including the toxicodynamics, systemic effects, oncogenicity, mutagenicity, and teratogenicity. Linear interpolation with consideration of a threshold (nononcogenic effects) or an appropriate carcinogenic dose-response model for oncogenic effects provides actual quantitative risk estimates for each chemical.

Quantification of the total human-health risk for chemicals identified in the exposure analysis (for each control option) must incorporate a normalization factor to allow for the summation of chemical-specific risks. One normalization approach that can be used involves health-effect perception factors.[7] These or other factors can then be multiplied by the chemical-specific risk to provide estimates of relative risk in common units. These risk estimates may

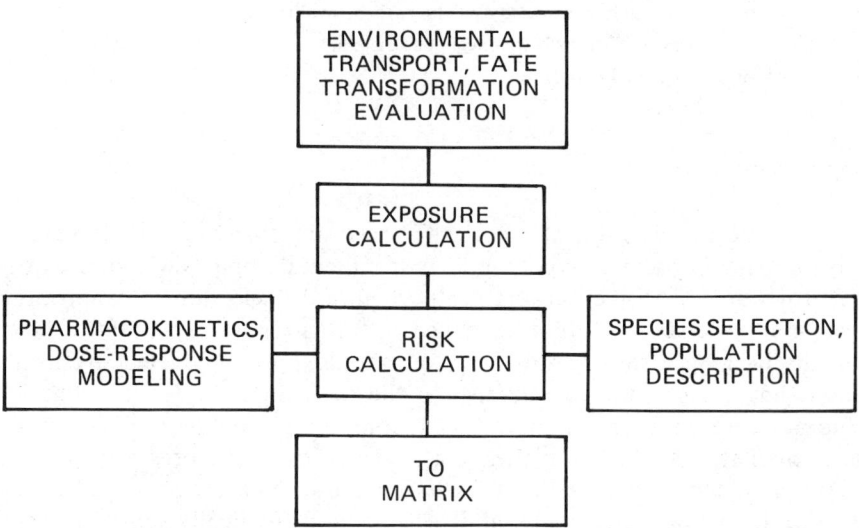

Figure 4. Conceptual risk assessment portion of the methodology.

then be summed for all chemicals in all media yielding a total quantitative health risk estimate.

The health-effect perception factors developed by Ecological Analysts, Inc. and SEAMOcean, Inc. are unique and have potential for this methodology. They are derived from a large post-stratified random-sample survey of nearly 800 urban and rural residents, in which respondents were asked to rank the severity of 20 different illnesses that can result from chemical exposures.

The resulting total human-health risk value, in common units, also allows comparison of risk for several waste-disposal or pollution-control treatment options. Thus, the total health risk for both occupationally exposed individuals and the surrounding community associated with each biofouling control option may be compared in the final cost-effective assessment matrix.

Environmental Impact Assessment

As its starting point, the environmental impact evaluation requires the chemical-specific concentration (exposure) data developed previously in the chemistry and fate assessment. Once developed, these compound-specific exposure concentrations can be compared to existing toxicological data bases to help predict anticipated impacts on the aquatic environment.

The types of environmental information required for the cost-effective assessment matrix include site-specific assessments of expected acute effects, chronic effects, and bioaccumulation. The determination of each of these toxicological endpoints relies on the following basic assessment parameters:

- selection of appropriate sensitive species for assessing toxicological effects,
- selection of specific chemicals and by-products to be evaluated,
- exposure concentrations of each chemical species, and
- duration of the expected exposure.

Species Selection

The selection of appropriate species for assessing toxicological effects is an important step in predicting potential environmental impacts. Guidance documents for water quality criteria typically require the selection of "appropriate sensitive aquatic species." In accordance with this regulatory intent, a power plant discharging into a freshwater river should optimally select freshwater species that are considered sensitive to the toxicant(s) under investigation, including representatives of important taxonomic groups (e.g., fish, invertebrates, plants) or ecological niches (e.g., carnivores, herbivores, primary producers). Similarly, power facilities discharging into estuarine or marine ecosystems should evaluate the potential effects on toxicologically sensitive species that are expected to inhabit those water bodies. Alternatively, a species of local economic importance may be selected.

Selection of Chemicals and By-products

When evaluating the environmental effects of power plant biofouling control options, it is essential that the appropriate chemical by-products are investigated and incorporated into the exposure assessment. For example, a variety of chlorination by-products result from the reaction of chlorine with natural waters (e.g., hypochlorous acid, inorganic chloramines, organic chloramines, trihalomethanes, halophenols). The prediction of which by-products might result depends on the site-specific water quality characteristics (pH, salinity, dissolved organic matter, and ammonia). Other biofouling control options (e.g., ozone, bromine chloride, and dechlorination) also lead to the formation of toxicologically active reaction products that are affected by water quality.

Concentrations of Chemical Species

It is essential to determine the chemical fate and transport of introduced biofoulants. This chemical fate assessment and the quantitative prediction of exposure concentrations for all important by-products are as critical for environmental impact assessment as they are for human-health assessment.

Exposure Duration

Predicting toxicological effects also requires the knowledge of whether the biocide will be administered continuously or intermittently. If the engineering specifications allow for intermittent chemical additions, the frequency and duration of these treatments must also be determined. This exposure information can be related to the toxicological effects data base to assess whether effects would be expected to occur.

Environmental Assessment Summarization

Thus, once the chemicals to which various populations would be exposed are identified and their concentrations computed, the determination of anticipated aquatic effects involves comparing these exposure data with existing toxicological data bases. Concentrations could be compared to a variety of acute, chronic, and sublethal endpoints used routinely in aquatic toxicology including LC_{50}, ED_{50}, MATC values, and avoidance responses. Comparisons could also be made with state and federal water quality criteria. Several techniques[10,11] that involve quantification of environmental effects are being assessed for inclusion in this stage of the process.

Expected Outcomes

Integrating risk, hazard, and cost values from individual analyses into an overall cost-effective assessment matrix will allow for an objective technical comparison of biofouling control alternatives. The cost value for each biofouling control alternative is derived from a comprehensive analysis of the efficacy, design, and reliability of the biofouling control operation (e.g., chlorination, ozonation, bromine chloride, dehalogenation). The fate-and-transport analysis yields the estimated exposure concentrations for aquatic biota and human populations of both the parent compound (e.g., chlorine, bromine, sodium thiosulfate, ozone) and associated transformation products. Based on the expected concentrations and chemical species, the environmental impact and human-health risks may be assessed. Inserting the site-specific results from each analysis into a cost-effective assessment matrix allows for the evaluation of the costs and the environmental and human-health impacts of biofouling control on a site-specific basis.

Each alternative is then ranked according to its estimated value in each of the four assessment parameters. A hypothetical matrix derived by this process is shown in Figure 5. The matrix may then be used in a management context through a process of optimization (i.e., selecting the alternative with the highest effectiveness and lowest cost, environmental impact, and human-health risk). In constructing the matrix, the highest values are assigned to the most favorable case. Thus, most effective, least costly, and least toxic will be ranked

	Biofouling Control Alternative			
	I	II	III	IV
Effectiveness	4	4	3	1
Environmental Impact	2	4	3	1
Human Health Risk	3	4	2	1
Cost	3	2	1	4
Overall	12	14	9	7

Figure 5. Conceptual matrix for selection of biofouling control alternatives. Numbers are relative ranking except "overall," which is the sum. High numbers represent best case, whereas low numbers represent worst case.

highest. Those alternatives with the lowest overall values are considered to be the least desirable. In the hypothetical example, applying the cost-effective matrix, option II would be the preferred biofouling control alternative.

CONCLUSIONS

Based on information acquired while developing the approach discussed here, a number of conclusions can be made:

1. Although considerable understanding has been developed on the subject of chlorination, many gaps still remain in our knowledge. We need to ask more appropriate questions concerning the use of chlorination to fill these gaps. The area of research that appears to require the most effort is exposure.
2. Holistic, multidisciplinary approaches are required to more effectively provide the types of data and information necessary for making decisions.
3. A conceptual framework, with well-developed decision criteria, should result in a formalized rational approach for cost-effective environmental management decisions.

We must manage the environment while recognizing that man is an integral part of the system. A worthwhile management goal is to develop a more effective use of the environment while minimizing the impacts to man and the environment in a cost-effective manner.

REFERENCES

1. *Water Chlorination*: *Environmental Impact and Health Effects, Vol. 1,* R. L. Jolley, Ed. (Ann Arbor, MI: Ann Arbor Science Publishers, Inc., 1978).
2. *Water Chlorination*: *Environmental Impact and Health Effects, Vol. 2,* R. L. Jolley, H. Gorchev, and D. H. Hamilton, Jr., Eds. (Ann Arbor, MI: Ann Arbor Science Publishers, Inc., 1978).
3. *Water Chlorination*: *Environmental Impact and Health Effects, Vol. 3,* R. L. Jolley, W. A. Brungs, and R. B. Cumming, Eds., (Ann Arbor, MI: Ann Arbor Science Publishers, Inc., 1983).
4. *Water Chlorination*: *Environmental Impact and Health Effects, Vol. 4,* R. L. Jolley, W. A. Brungs, J. A. Cotruvo, R. B. Cumming, J. S. Mattice, and V. A. Jacobs, Eds. (Ann Arbor, MI: Ann Arbor Science Publishers, Inc., 1983).
5. Maki, A. "Design and Conduct of Hazard and Evaluation Programs for the Aquatic Environment," in *Water Chlorination*: *Environmental Impact and Health Effects, Vol. 3,* R. L. Jolley, W. A. Brungs, R. B. Cumming, and V. A. Jacobs, Eds., (Ann Arbor, MI: Ann Arbor Science Publishers, Inc., 1980), pp. 949-960.
6. Plugge, H. "Issues in Human Health Risk Assessment," in *Water Chlorination*: *Environmental Impact and Health Effects, Vol. 4,* R. L. Jolley, W. A. Brungs, J. A. Cotruvo, R. B. Cumming, J. S. Mattice, and V. A. Jacobs, Eds. (Ann Arbor, MI: Ann Arbor Science Publishers, Inc., 1983), pp. 1457-1468.

7. Ecological Analysts, Inc., and SEAMOcean, Inc. "A Special Permit Application for the Disposal of Sewage Sludge from Twelve New York City Water Pollution Control Plants at the 12-Mile Site" (New York: Department of Environmental Protection, 1983).
8. Gift, J. J., H. Plugge, W. J. Rue, B. L. Rubin, J. A. Fava, S. E. Storms, D. A. Segar, E. Stamman. "Comparative Multi-Media Risk Management of Sewage Sludge Disposal Incineration Versus Ocean Disposal, A Case Study for the City of New York," I. W. Duedall et al., Eds. (New York: in press).
9. *Technical Assessment Guide*, EPRI P-2410-SR, (Palo Alto, CA: Electric Power Research Institute, 1982).
10. Buikema, A. L. "Proposed Approach for Estimating Aquatic Toxicological Effects for Program Integration Studies," Report prepared by Life Systems, Inc. (1982).
11. Hakanson, L. "Aquatic Contamination and Ecological Risk," *Water Res*. 18(9):1107–1118 (1984).

CHAPTER 7

The Chlorine Dilemma: An Environmental Organization's Perspective

David S. Bailey

The traditional roles for environmental groups, particularly established organizations, are those of advocates and catalysts for environmentally responsible action. Most frequently, there is a need for a voice to counter special interests that may paralyze or unduly sway the government decision process. At other times, environmentalists seek to force government action from a system stifled by bureaucratic timidity or lethargy. Sometimes governmental inaction results when competing government agencies are unable to resolve an issue requiring compromise and decision in a coordinated fashion. It is this last situation that the Environmental Defense Fund (EDF) faces in Virginia with respect to the control of chlorine discharges to the aquatic environment.

The EDF was founded in 1967. It is a national, not-for-profit environmental organization with offices in Washington, D.C., New York, Colorado, California, and Virginia. The primary objectives of the organization are to reduce the exposure of humans and their environment to toxic chemicals and to promote and encourage wise use of natural resources. EDF's national programs include toxic wastes, acid rain, endangered species, water conservation, energy development, wetlands protection, and many other issues. The organization has a national membership of 50,000 and gains additional support from foundation and corporation grants. The Virginia Office, which concentrates almost exclusively on Virginia issues, receives its support in large part from the Virginia Environmental Endowment and nearly 2000 Virginia EDF members.

In Virginia, the chlorine issue has all the classic elements for controversy and governmental inaction. Chlorine is a widely used chemical typically discharged from municipal sewage treatment plants, which, as a group, have enjoyed widespread discretion in enforcement of water pollution provisions. Chlorine has proven to be of immense benefit to public health for wastewater disinfection, but its adverse toxic effects have only recently become known. Reliable, economical analytical equipment to measure low concentrations of chlorine is still not commonly available. Furthermore, in Virginia, responsibility for regulation of chlorine discharges is divided between the State Water Control Board, which regulates water quality, and the State Health Department, which dictates disinfection requirements.

As a result of these factors, Virginia has been slow to regulate chlorine discharges, even though significant environmental impacts from chlorine discharges have been known as early as 1971. The Virginia State Water Control Board found chlorine to be primarily responsible for massive fish kills in the James River in 1971 and ordered reductions in chlorine discharges from area sewage treatment plants.[1]

Existing regulations assigned dischargers arbitrary chlorine ranges, often without maximum limitations. These chlorine limitations were usually determined by what the treatment facility could achieve, rather than what the receiving stream required to protect aquatic life.

In 1977, an interagency task force appointed by the Governor recognized that chlorine discharges could be toxic to aquatic life and recommended that chlorine residuals not exceed 0.02 mg/L outside established mixing zones in receiving streams.[2] The recommendations received only minimal application by the respective regulatory agencies.

The Governor's Task Force failed, however, to resolve the single greatest factor impeding movement toward chlorine regulation: the separation of responsibility for public health and water quality between two state agencies. The Health Department, comfortable with the disinfection capability of chlorine, was reluctant to reduce chlorine discharges or shift to newer untested alternatives; the Water Control Board balked at large-scale dechlorination because it believed the Health Department's chlorination demands were excessive, and the Water Control Board would not force dischargers into dechlorination costs without more consideration from the Health Department. The State Health Department never directly opposed dechlorination but did insist that a 30-min contact with 2.0 mg/L be maintained at the sewage treatment facility. This would necessitate the Water Control Board uniformly requiring construction of dechlorination facilities at most treatment plants. The Water Control Board wanted to reduce chlorination to minimum disinfection levels and only require dechlorination facilities when such reductions failed to meet water quality requirements. The Water Control Board felt the 2.0 mg/L chlorine minimum was arbitrary; the Health Department was reluctant to depart from the established chlorination practice that had been shown to be effective in pathogen control.[3]

Thus, as late as 1982, chlorine regulation in Virginia was making little or no progress, and most National Pollutant Discharge Elimination System (NPDES) permits gave wide latitude to the discharge of chlorine, whereas effluent limits bore little or no relationship to water quality requirements.

Obviously, this was a situation ripe for outside intervention by an environmental group. EDF has intervened in similar controversies before in an effort to spur resolution of complex issues. EDF has been an active participant in the Denver Water Round Table, helping to resolve highly controversial water issues in the West. In Virginia, EDF played a major role in developing a hazardous waste siting law through the hazardous waste siting committee and is actively involved in the resolution of toxic pollution issues through the efforts of the toxic roundtable. In the summer of 1983, EDF, joined by several other environmental, fisherman, and oystermen groups, petitioned the Water Control Board to promulgate a water quality standard for chlorine;[4] as a result of that petition, EDF accepted a membership role in an Agency Task Force to consider chlorine regulation.

In this kind of situation, the role of an outside nongovernment agency environmental group is far more than just a stimulus to action. To satisfy all concerns, the resolution of the chlorine issue must involve not only a determination of water quality requirements but also a reevaluation of the whole disinfection process. Broad-scale dechlorination, without proper consideration of other issues, is not the answer to the chlorine discharge problem. In terms of public health, little or no improvement is achieved by simple dechlorination; in fact, some loss of residual disinfection may occur in the older treatment systems that depended on contact time in the discharge pipe itself to complete the disinfection process. Nor does dechlorination address concerns about the creation of chlorinated organic compounds. Furthermore, dechlorination adds another chemical feed process to systems that often did not manage chlorine feed well in the first instance.[5] Economically, dechlorination can interfere with other alternatives. Once dischargers add the expense of dechlorination to already existing chlorination facilities and at the same time satisfy immediate environmental concerns, there is little interest or incentive to evaluate alternative forms of disinfection.

The dilemma, as EDF sees it, is to maintain a broad perspective in solving chlorination problems in the face of regulatory agencies and affected parties with generally narrow authority and interests. Chlorine toxicity must be considered along with disinfection requirements. If one evaluates all the costs and benefits of chlorination vs alternative means of disinfection, the alternatives can be very attractive indeed; however, if the chlorine issue is reduced to one of chlorine vs aquatic life, the costs of dechlorination are valued separately and independently of the larger problem.

There remains another alternative to chlorine discharge that must be carefully evaluated: no disinfection by any chemical means. Except for a small number of discharges into receiving streams with great dilution capability, this option was not considered acceptable in the United States until recently, although it is widely practiced in foreign jurisdictions. Nondisinfected waste is not really the proper term when applied to modern advanced treatment plants, particularly those with long-term detention as part of the treatment process. In these facilities, some pathogen removal is effected by chemical coagulation and filtration, with long-term detention providing for die-off of most remaining pathogenic organisms. Even without long-term detention, seasonal disinfection may be a possible alternative in certain circumstances.

Using a nondisinfection approach to eliminate chlorine toxicity raises several problems. Many public health officials will never feel comfortable about nondisinfected waste. Although they may admit and agree that retention times are sufficient to eliminate many pathogens, they also agree that bacteria which yield positive results on the fecal coliform test media often remain, and disinfection is therefore required. Seasonal disinfection may involve issues of health warnings or stream-use restrictions that are incompatible with state and federal law.

In some respects, however, chlorine disinfection is little more than an illusion presenting a false sense of security about public health. In facilities with long-term detention, such as lagoons, chlorination may disinfect little more than common bacterial organisms that we are continually exposed to in soil and water, because fecal coliform organisms rarely survive in the environment for more than a few days to a few weeks. Chlorine is not often effective on viruses, yet the illusion of disinfection bars consideration of alternatives that might further reduce virus survival.

The Virginia State Department of Health acknowledged that fecal coliform survival was significantly reduced in retention ponds, polishing ponds, and stabilization ponds, depending on the detention time. Notwithstanding this fact, however, health officials were reluctant to eliminate chlorination because the fecal coliform bacteria indicator test still yielded positive results. It was assumed that these positive results were from either wildlife or other nonpathogenic bacteria capable of producing positive coliform test results. The State Health Department compromised this position by agreeing to consider detention time as a factor in determining whether chlorination would be required.[6] Other states have recognized pathogenic die-off in such ponds and have not required chlorination: Wisconsin and Minnesota (no chlorination required for stabilization ponds); Texas (no chlorination required for 21-d retention).

In Virginia, EDF has sought to resolve the chlorine problem by forcing the consideration of chlorine toxicity to aquatic life in evaluating individual discharge permit requirements while, to the extent possible, keeping the emphasis on an evaluation of the need for chlorination in general. Thus, EDF has pursued a state water quality standard for chlorine to address aquatic toxicity while also supporting a broad implementation policy that encourages a thorough reevaluation of disinfection options at each site.

However, just encouraging an evaluation of alternatives may not make it happen. In fact, when pressed hard, many dischargers and regulatory agencies simply take the path of least resistance. Since dechlorination can often be applied quickly and relatively inexpensively, serious consideration of alternatives is often foregone. This is especially true if federal or state funding is available. Environmental organizations can help force alternative evaluations by scrutinizing the application of dechlorination very carefully and endorsing it only when it is clearly appropriate following the consideration of other available alternatives. Environmentalists should consider the costs, disinfection ability, treatment reliability, operational requirements, and residual products of each alternative to chlorination or dechlorination before accepting dechlorination as the solution to chlorine toxicity.

In some cases, dechlorination is the only immediately feasible alternative available. Semipermanent dechlorination facilities can usually be constructed quite inexpensively and can be used seasonally or year-round to mitigate immediate serious effects of chlorine discharges. Although some regulatory agencies are inclined to do so, one cannot wait for another round of federal grant funding to remove the toxic effects of chlorine discharges from municipal facilities. We see no reason why special funding cannot or should not be made available for this purpose now. It has always seemed amazing to us that an agency could authorize millions of dollars to build a sewage treatment plant that can guarantee dissolved oxygen levels satisfactory for aquatic life but then balks at spending a few thousand more to prevent eradicating the very aquatic life it sought to protect.

Although painfully slow, progress is being made in Virginia. State funding for dechlorination has been included in the authorization for Virginia's part of the Chesapeake Bay Program. New treatment facilities are getting stricter chlorine requirements or are using alternative disinfection methods such as ozone and ultraviolet light.

As an interim response to the EDF petition, state officials began issuing discharge permits with maximum chlorine limitations as well as restricting the operating range of chlorination. Some new treatment facilities, which are in a better position to consider disinfection alternatives at the planning stage, are installing ozone (e.g., Henrico County Sewage Treatment Plant, design flow, 12 Mgd); others are considering ultraviolet light.

An agency task force has spent the last year exploring the chlorine issue and will make its recommendations to the Virginia State Water Control Board in 1984.[6] Although the task force is still divided on the most effective approach to control chlorine, the Board will likely authorize public hearings in 1984 to consider a chlorine standard.

Proper control of chlorine discharges in Virginia has been long in coming and still is not in effect. Environmentalists have led the principal effort to abate chlorine toxicity to aquatic life as well as seek a review of all disinfection procedures. Efforts in this area still proceed in direct proportion to the political and legal pressures exerted by environmental groups. Now, however, engineers, planners, and even municipalities, as well as environmental groups and natural resource organizations, recognize that alternatives to chlorination have solid economic and social benefits. In the future, there will be changes in chlorine regulation in Virginia.

After this chapter had been prepared, the Virginia State Water Control Board proposed a water quality standard for chlorine (0.14 mg/L for fresh water, and 0.13 mg/L for salt water) and endorsed a series of task force recommendations relating to disinfection alternatives.[7]

REFERENCES

1. Bellanca, M. A., and D. S. Bailey. "Effects of Chlorinated Effluents on Aquatic Ecosystems in the Lower James River," *J. Water Pollut. Control Fed.* 49:639 (1977).
2. Douglas, J. E., Jr. "Summary Report of the Select Inter-Agency Task Force on Chlorine," (Richmond, VA: Virginia Marine Resources Commission, Commonwealth of Virginia, 1977).
3. State Water Control Board and Health Department personnel, Commonwealth of Virginia.
4. Environmental Defense Fund, Inc., et al. "Petition for Rule-making", before the Virginia State Water Control Board, July 22, 1982.
5. A recent draft Environmental Protection Agency report found that 11% of municipal sewage treatment plants in the United States are violating the Clean Water Act because of deficient operation and maintenance procedures. *Environmental Reporter* (BNA), 15(24), Oct. 12, 1984.
6. Virginia State Water Control Board, "Final Report of the Disinfection Task Force, 1984."
7. Meeting of the Virginia State Water Control Board, Sept. 24–25, 1984.

CHAPTER 8

Regulatory Aspects of Disinfection

Joseph A. Cotruvo and Craig D. Vogt

The Environmental Protection Agency (EPA) is currently engaged in the most detailed and comprehensive assessment of drinking water quality specifications ever attempted. Under the Safe Drinking Water Act (SDWA),[1] Revised National Primary Drinking Water Regulations must be developed for any substances that may have any adverse effect on health. Recommended maximum contaminant levels (RMCLs) must be determined at the concentrations in drinking water that would result in no known or anticipated adverse effect on health, including an adequate margin of safety. These RMCLs are not enforceable standards, but they are goals to strive for. Subsequent to the RMCLs, the enforceable maximum contaminant levels (MCLs) will be developed. These must be as close to the RMCLs as is feasible, taking costs and other factors into consideration.

The revised regulations will span all classes of drinking water contaminants, including biological contaminants, organic and inorganic chemicals, and radionuclides. The complete process will require 4 years of formal regulatory action, and it has been preceded by several years of detailed assessment. For each substance being examined, supporting assessments are produced, including environmental occurrence, human exposure, toxicology, analytical methods, treatment technology, unit costs, implementation forecasts, costs to communities and consumers, and the national economic impact assessments, along with several other analyses required by Executive Orders and/or statutory imperatives.

Management of this immense and continuing task requires separation of the mass of candidates for regulation into four large groupings or phases.

- Phase I — Volatile Synthetic Organic Chemicals
- Phase II — Synthetic Organic Chemicals, Inorganic Chemicals, and Microbiological Contaminants
- Phase III — Radionuclides
- Phase IV — Disinfection By-products, Including Trihalomethanes

In general the approach for all four phases will be similar.

Initially, an advance notice of proposed rule making (ANPRM) will be published, followed by a comment period and a public meeting. Public technical workshops will also be held. The workshops provide an opportunity for EPA to present the issues that must be addressed in development of the regulations and to receive information on scientific and technical matters, as well as to receive comments on regulatory approaches. RMCLs will then be proposed, followed by a public comment period and a public hearing(s).

RMCLs will then be promulgated, and proposals will be published for MCLs, monitoring and reporting, and other requirements, followed by a public comment period and a public hearing(s). Technologies will be identified that were used as the basis of determining the MCLs; in addition, generally available treatment technologies (GATs) will be identified for use in the issuance of variances. The MCLs, monitoring and reporting, and other requirements, including GATs, will then be promulgated.

The legal and public policy issues associated with these regulations are described in detail in three documents published since 1982.[2-4] Subjects of interest include Phase IV—Disinfection; the Microbiology portion of Phase II; and some limited contribution from Phase I—Volatile Synthetic Organic Chemicals, most of which are chlorinated solvents.

PHASE I—VOLATILE SYNTHETIC ORGANIC CHEMICALS

RMCLs have been proposed for nine chemicals, eight of which are chlorinated. They include trichloroethylene, tetrachloroethylene, carbon tetrachloride, 1,2-dichloroethane, vinyl chloride, 1,1-dichloroethylene, benzene, 1,1,1-trichloroethane and 1,4-dichlorobenzene. These are among the most frequently detected synthetic organic contaminants in groundwater. According to the legislative history of the SDWA,[5] RMCLs should be set at the zero level if there is no "safe" threshold for a contaminant. Proposed RMCLs for 1,1,1-trichloroethane and 1,4-dichlorobenzene were 200 and 750 ppb, respectively.[3] The principal issue to be resolved in the proposal is the criterion for determining that there is a sufficient evidentiary basis for treating a substance as a potential human carcinogen for regulatory purposes.

PHASE II—MICROBIOLOGY

The control of biological pathogens is still the most significant drinking water quality goal. Of course, disinfection using chlorine and other oxidizing species is one of the principal methods of water treatment to ensure the biological safety of water.

Despite improvements in disinfection and other types of water treatment, outbreaks of waterborne disease still occur, particularly in smaller communities. From 1971–80, there were 315 reported outbreaks of waterborne disease involving almost 78,000 cases; 50 outbreaks and 20,000 cases occurred in 1980 alone.[4] At least two deaths were involved. Major causes of outbreaks in community water systems were contamination of the distribution system and treatment deficiencies such as inadequate filtration and interruption of disinfection. Specific causes of other outbreaks could not be determined. In noncommunity water systems, contamination of groundwater used without treatment or with treatment deficiencies (usually interruption of or inadequate disinfection) was responsible for most outbreaks and cases.

Many outbreaks, probably the great majority, are not reported to the Centers for Disease Control (CDC), which keeps records on the occurrence of reportable diseases, because few waterborne diseases are required to be reported, and because of difficulties in identifying the etiology of these occurrences. In Colorado, a current pilot effort to improve the outbreak reporting system indicated that perhaps only about one-fifth of the actual outbreaks were being recognized and reported. As recognition of waterborne illness has improved, the trend in the reported (although not necessarily the actual) number of outbreaks and cases has increased.

Proposals are being developed to revise the existing regulations for coliforms and turbidity and to develop new regulations for giardia and pathogenic viruses. Monitoring by public water systems for giardia and virus contamination would be much more complicated and costly than monitoring for the traditional contaminants. In such cases, the SDWA requires that treatment techniques be identified such that public water systems would be using the best generally available technologies to ensure that the drinking water is as close as possible to the quality goals. Thus EPA would propose filtration and disinfection (conventional treatment) for surface supplies and disinfection for groundwater supplies as part of the revised primary drinking water regulations. Under this type of provision, each supply would be required to demonstrate via sanitary surveys, source protection, and measurement that those technologies would not be necessary to ensure the safety of the finished water from contamination by giardia, viruses, and other pathogenic organisms. Those that could not adequately do so would be required to install the indicated treatment.

PHASE IV—DISINFECTION (OXIDATION) CHEMISTRY AND TOXICOLOGY

The objective of disinfection treatment is attainment of the maximum control of biological contaminants while introducing the minimum possible amounts of potentially toxic chemical by-products. Oxidation and other disinfection processes must be integrated with physical, biological, or chemical removal processes to ensure that this objective is achieved.

Oxidation and other disinfection processes are today the major sources of anthropogenic organic chemicals and of some inorganic chemicals in finished drinking waters. This fact has been known since 1975, when major studies began to expand upon the initial identifications of trihalomethanes (THMs) as by-products of chlorination and to identify numerous other products. The following products are now under consideration as candidates for comprehensive evaluation of possible MCLs for inclusion in Phase IV of the Revised Primary Drinking Water Regulations:

- Trihalomethanes
- Dihaloacetonitriles
- Haloacid derivatives
- Halophenols
- Chloramines
- Chlorine dioxide and by-product ions
- Residual chlorine

Following are several brief summaries of concerns associated with by-products of various disinfection agents.

Principal Disinfectants

Chlorine-containing species including hypochlorous acid, hypochlorite, and mono- and dichloramines are the primary agents used in drinking water disinfection that would be likely residues in water in the United States. Chlorine dioxide and its by-products chlorite and chlorate would also be frequently found. Stable by-products of ozone (other than oxygen and hydroxide) are not known. Other oxidizing species include bromine, bromine chloride, permanganate, ferrate, hydrogen peroxide, and iodine; however, only permanganate and iodine are used in public drinking water supplies in the United States, whereas the others may be used in wastewater treatment.

Chlorine

The most immediate organic oxidation processes of chlorine in water result in fragmentation of macromolecules that may be present, with a commensurate lowering of average molecular weight. Various oxygen-containing functional groups are generated, including alcohols, aldehydes, and acids. Halogenation products that have been detected include trihalomethanes (containing chlorine, bromine, and iodine), mixed haloacetonitriles, halophenols, haloalcohols, haloaldehydes, haloacids, and N-haloamines.

Inorganic ions are often oxidized, for example, nitrite to nitrate and cyanide to cyanate; undoubtedly halogens like bromine and iodine are converted into forms that readily halogenate organic molecules. Ammonia can be halogenated to mono-, di-, or trichloramine. Reduced metallic ions such as chromium (III) or ferrous can be oxidized to higher valence states.

In vivo chemistry of chlorine includes generation of chloroform and chloroacetic acids in gut and plasma, as well as dihalonitriles in gut contents. Recent studies where low concentrations of chlorine in drinking water were consumed by rats, pigeons, and rabbits indicated susceptibility to elevated cholesterol levels when the animals were maintained on high-fat and calcium-deficient diets. If these experiments are borne out by further studies, this finding could be of considerable significance, given the commonality of high fat and low calcium consumption in the typical U.S. diet. This is an area of accelerated experimentation by EPA.

Chloramines

Preformed chloramines are chemically less active than free chlorine; thus, very limited organic by-product formation has been noted to date from drinking water studies. Organic chloramines may form by transfer reactions. Mono- and dichloramines are currently under test in the National Toxicology Program Bioassay to determine carcinogenicity potential.

Chlorine Dioxide

Only limited organic by-product formation has been noted in studies under drinking water treatment conditions, with aldehydes having been occasionally found. The residues of chlorine dioxide, chlorite, and chlorate seem to be of greatest toxicological interest. EPA has recommended that the total residues of these products not exceed 1 mg/L in finished drinking water because of concerns about hematopoietic and possibly thyroid effects.

Ozone

Ozone exhibits a substantial degree of oxidation activity, producing aldehydes, ketones, and metastable peroxy species, among undoubtedly many others. Because of the instability or hydrophylic nature of the products, low concentrations are especially difficult to detect by commonly used analytical techniques. One might speculate that much of the oxidation chemistry of ozone is similar to the oxidation chemistry of chlorine and chlorine dioxide.

THMs AND BEYOND

Drinking water standards for trihalomethanes were promulgated by EPA in 1979 after these were determined to be common contaminants in chlorinated drinking water. Among the reasons given at the time were that: THMs were the principal identified by-products of chlorination; there was virtually universal exposure of very large populations; these exposures were at relatively high concentrations compared with those of other synthetic organic drinking water contaminants that had been detected; and, perhaps most importantly, THMs are indicative of the presence of other chlorination by-products yet to be identified that are much more difficult to analyze.

Water disinfection is clearly the major source of synthetic organic chemicals in public drinking water supplies. Phase IV of the revised regulations is intended to take a major step toward examining most of the principal chemical contaminants in drinking water resulting from the disinfection process, including a reassessment of THMs. As chemists continue the exploration of these by-products that are being produced, toxicologists will continue to raise questions about certain of the substances that are being detected, and treatment engineers will be asked to find ways of maintaining the biological safety of public drinking water supplies while minimizing the chemical by-product formation.

These concerns were anticipated when the THM standard was written in 1978 and 1979. Alternative disinfectants and processes were discussed at the Second Water Chlorination Conference along with the fact that the best long-term answer is the obvious one: the design of complete water treatment systems should maximize the removal of organic precursors, so that the amount of disinfectant needed to produce biologically safe water will be minimized along with the amounts of residual organic by-products.[6]

In practice, now as then, this means using the best available physical and chemical treatment technologies to produce water of the highest chemical quality prior to application of the disinfectant. At a minimum, this involves filtration of surface waters as well as more advanced physical and chemical control technologies when needed.

Since 1977, much of the suggested research work on the chemistry and toxicology of disinfectants and their by-products has been done, and more is on the way. The time is fast approaching when that information will be utilized in the construction of drinking water standards designed to assure the public that their drinking water is both clean and safe regardless of its source, because those contaminants that were in the source have been removed by treatments that do not leave undesirable residues.

REFERENCES

1. The Safe Drinking Water Act, 42 U.S.C. 300f et seq.
2. Advance Notice of Proposed Rulemaking, Volatile Synthetic Organic Chemicals in Drinking Water, *Fed. Regist.* 47(43):9350–9358 (Mar. 4, 1982).
3. National Primary Drinking Water Regulations Proposed Rulemaking, *Fed. Regist.* 49(114):24330–24355 (June 12, 1984).
4. Advance Notice of Proposed Rulemaking, National Revised primary Drinking Water Regulations; *Fed. Regist.* 48(194):455020–455021 (1983).
5. Committee on Interstate and Foreign Commerce, Report No. 93-1185 (July 10, 1974), p. 20.
6. Cotruvo, J. A. "Regulatory Aspects of Potable Water Disinfection," in *Water Chlorination: Environmental Impact and Health Effects, Vol. 2*, R. L. Jolley, H. Gorchev, and D. H. Hamilton, Eds. (Ann Arbor, MI: Ann Arbor Science Publishers, Inc., 1978), pp. 817–821.

SECTION II

Risk:
The Bottom Line

"I see nobody on the road," said Alice.
"I only wish I had such eyes," the king remarked in a fretful tone. "To be able to see Nobody! And at that distance too! Why, it's as much as I can do to see real people, by this light!"

Lewis Carroll, 1832–1898
Through the Looking Glass

All risk estimates have inherent uncertainties.

Hans Plugge
Issues in Risk Assessment, 1981

Probable impossibilities are to be preferred to improbable possibilities.

Aristotle, 384–322 B.C.

CHAPTER 9

Microbiological Risks Associated with Changes in Drinking Water Disinfection Practices

Elmer W. Akin and John C. Hoff

Since its first application in Jersey City in 1908, the chlorination of drinking water to control microbiological contamination has become common practice in the United States. Although not all community water supplies in this country are disinfected with chlorine or any other agent or process, it is a widely held expert opinion that they should be disinfected to reduce the risk of waterborne infectious diseases. In fact, the disinfection committee of the American Water Works Association (AWWA) has stated that *all* public water supplies *must* be disinfected.[1] The Office of Drinking Water of the Environmental Protection Agency (EPA) is now considering proposing a legal requirement for the application of this water treatment process.

Disinfection of water has been practically synonymous with chlorination for the 77 years it has been practiced in this country. Even with renewed interest in alternative disinfectants because of the chlorinated-organics health issue, there is strong economic and technical pressure to avoid the displacement of chlorine as the primary disinfectant in resolving this health concern. For this reason and because most of the microbiological disinfection data have been obtained with chlorine, this report will focus on chlorination.

The decision to apply a "poison" or at best a "forced medication" to contaminated water to make it potable was not supported by all in the early years of chlorination. As chlorination became more widely used for water treatment, public opinion often considered bad-tasting water synonymous with chlorinated water. Therefore, good public health practice was considered to be the addition of the absolute minimum amount of chlorine that would kill indicator bacteria. This resulted in a common practice of adding about 0.25 ppm total residual chlorine, which typically yielded less than 0.1-ppm *o*-tolidine-detectable residual after a 15-min contact time.[2]

The public health significance of the chlorination of drinking water was demonstrated most clearly with the precipitous drop in typhoid fever deaths in the first third of this century. At the turn of the century, about 25,000 deaths and an untold number of cases of typhoid fever occurred annually in the United States. Waterborne transmission of the bacterium was believed to be responsible for perhaps 40% of the typhoid fever cases. By 1935, a 10-fold drop in typhoid fever deaths had occurred, and improved water sanitation was given much of the credit. It was fortunate indeed that the chlorine level that was fairly acceptable from a taste and odor standpoint and reduced the *Bacillus* (now *Escherichia*) *coli* fecal indicator was also effective against *Salmonella typhosa*, the causative agent of typhoid fever.

Of course *S. typhosa* was not the only bacterial pathogen in sewage-contaminated drinking water that was a major public health hazard. An observation was made in 1903 that when a community's water supply was changed from "poor" to "excellent," every person saved from a typhoid fever death was accompanied by three others saved from causes not obviously related to water contamination.[2] Therefore, the typhoid fever death rate became somewhat of a gauge for the sanitary condition of a community.

By midcentury the chemistry of water chlorination was more clearly understood. Also, the existence of a large number of fecally shed viruses had been detected through the application of cell-culture isolation procedures. Studies on virus inactivation by chlorine species which soon followed these developments showed enteric viruses to be significantly more resistant to chlorine than bacterial indicators and enteric pathogens. These observations led to considerable concern over the drinking water transmission of viruses and the need for a more rigorous disinfection practice.

At about the same time, disinfection studies with the agent of amoebic dysentery, *E. histolytica*, revealed that the cysts of this waterborne pathogen were highly resistant to chlorine. Since the cysts were large enough to be removed by filtration, this method of control was considered to be the more practical approach.

This in turn led to the recommended chlorination practice, based on virus inactivation data, of treating water with 0.4 to 1.0 mg/L free residual chlorine for at least a 30-min contact time.[3,4] However, in common practice, prechlorination was incorporated into most conventional treatment schemes. That extended the chlorine contact time to 2–4 h, usually at significantly elevated chlorine levels.

Whereas control of waterborne virus transmission had been viewed as the most significant problem facing disinfection in the 1970s, control of waterborne giardiasis became the problem of the 1980s. Since its first recognition as an important waterborne pathogen in 1974, *Giardia* has become the most frequently identified etiologial agent of waterborne outbreaks in the United States. Like *E. histolytica*, the environmentally stable cyst form of this parasite is also more resistant to chlorine inactivation than enteric bacteria and viruses.

COMPARATIVE MICROBIAL INACTIVATION

Efforts have been made to provide guidance on disinfection practice using linear rate-reduction data. Different types of enteric microorganisms were used in studies with different chlorine species. The products of the chlorine concentration (C) and contact time (t) for a given level of microbial inactivation (e.g., 99%) were obtained. Factors such as chlorine demand, pH, and temperature that were known to affect the reaction were stabilized. The product of C times t ($C \cdot t$) then can be considered as a constant that reflects the relative disinfection sensitivity of each organism tested. The results of studies with some microorganisms are shown in Table I (see References 5-10).

Full conventional treatment (prechlorination, flocculation, sedimentation, rapid filtration, and postdisinfection), as it had evolved prior to the chlorinated-organics health concern, was believed to be completely adequate for producing drinking water safe from the spread of infectious diseases. This empirical knowledge was accepted even though a large percentage of the U.S. population was consuming treated water derived from sewage-polluted surface-water sources. There is no clinical or epidemiological evidence to refute this claim when water is fully treated to meet the coliform and turbidity standards. However, incomplete treatment or deficiencies in unit processes, as well as subsequent recontamination of treated water in distribution systems, has resulted in documented episodes of illness.

WATERBORNE OUTBREAKS

The formal collection and reporting of waterborne disease outbreak data have been ongoing in this country since 1920. Although reporting seems to have improved over the years, there is no claim that the data base is complete, even for recent years. Investigation and reporting of infectious disease outbreaks is primarily a function of state and local health authorities and is strongly dependent on interest, funding, and the availability of properly skilled personnel. Nonetheless, these data provide considerable insight into the factors important in waterborne disease occurrence subsequent to the availability of enlightened water treatment technology.

Table I. Comparative Inactivation of Selected Enteric Organisms by Different Species of Chlorine

| Microorganism | Water | | Chlorine | | | | |
	pH	Temp (°C)	Species	Conc (mg/L)	Time[a] (min)	$C \cdot t$[b]	Reference
E. coli	6.0	5	HOCl	0.1	0.4	.04	5
	10.0	5	OCl$^-$	1.0	0.9	0.9	5
	9.0	5	NH$_2$Cl	1.0	175	175	6
	9.0	15	NH$_2$Cl	1.0	64	64	6
Poliovirus 1	6.0	5	HOCl	0.5	2.0	1.0	7
	10.0	5	OCl$^-$	0.5	21	10.5	7
	9.0	15	NH$_2$Cl	10.0	90	900	6
E. histolytica	6.0	5	HOCl	5.0	18	90	8
G. lamblia (cysts)	6.0	5	HOCl	2.0	40	80	9
	7.5	3	NH$_2$Cl	2.4	220	528	10

[a]Contact time to yield 99% inactivation.
[b]Product of concentration (C) times contact time (t).

The annual number of reported waterborne infectious disease outbreaks has shown an upward trend since 1966 to a modern-day high in 1980 of 43 with 17,710 cases of illness.[11,12] Lippy and Waltrip[11] have reviewed the data on 623 outbreaks that were reported for the 35-year period 1946–80. These outbreaks generally occurred in small surface-water and groundwater systems where full conventional treatment was not applied. In fact, a majority of the outbreaks occurred in systems that failed to apply or maintain any treatment (Table II).[11] One may conclude from these data that the risk of waterborne infectious disease is increased in the absence of full conventional treatment. The data also show the health significance of recontaminated water in the distribution system. Of considerable importance in understanding and controlling waterborne illness is the fact that the causative agent was not identified in over 50% of the outbreaks.

It is of interest to note some general features of water treatment plant facilities and operations relative to the occurrence of waterborne outbreaks. Fuhs[13] pointed out a basic consideration regarding the nature of source waters and the potential for waterborne transmission of disease. He distinguished between source waters receiving consistent pollution from large populations, such as the Mississippi and lower Hudson rivers, and source waters from smaller streams or impoundments in watershed areas virtually devoid of human populations. In the first case, pathogens are constantly present in the source water, and any major deficiency in water treatment would likely result in a detectable outbreak of waterborne disease. In the second case, pathogens will be present only intermittently, and operational deficiencies will result in waterborne outbreaks only when a particular set of circumstances occurs coincidentally with the defective treatment. In this situation, water-treatment personnel may become lax in operational matters because they observe that whether the water is treated apparently makes no difference. This circumstance, coupled with the fact that systems with relatively uncontaminated source waters are often those of small communities with limited facilities and less well-trained operators, is likely a major factor in the predominance of waterborne outbreaks in these types of supplies.

Table II. Deficiencies Associated with Waterborne Disease Outbreaks in Public Water System for the Period 1946 – 1980[11]

Deficiency	Outbreaks (%)
Contaminated untreated groundwater	35.3
Inadequate or interrupted treatment	27.2
Distribution network problems	20.8
Contaminated untreated surface water	8.3
Miscellaneous	8.3

RECENT CONCERNS REGARDING ESTABLISHED PRACTICE

In practice, it has been accepted that full conventional treatment is not required for all water supplies to protect the public health. Some drinking water sources, particularly groundwaters, are thought to be free of pathogenic microorganisms and require no treatment. At best, marginal disinfection may be applied to comply with regulations or satisfy a view that chance contamination of the source water would be controlled. The outbreak record indicates that such treatment can be inadequate.

If water treatment is implemented for a given water supply, disinfection invariably will be included as the only or the final step. This practice is consistent with the widely held view, with considerable research support, that disinfection is the true barrier to waterborne disease transmission and that prior treatment steps only enhance the capability for effective disinfection. If sanitary surveys do not reveal a source of fecal contamination and the source water is relatively clear, disinfection, perhaps marginal, has been viewed as the only logical treatment needed. Many supplies have operated this way for decades without the known occurrence of waterborne disease.

Of the many waterborne outbreaks that have occurred in the past few years, two stand out that especially shake confidence in the ability to predict water health risks and in the capability of limited treatment to protect against major episodes of waterborne disease. In June 1980, an outbreak of gastroenteritis (about 8000 cases) followed by an outbreak of about 36 cases of hepatitis A occurred in a small Texas community that utilized chlorinated, deep groundwater as its drinking water.[14] The yet unexplained contamination of wells in the central-city area is believed to be the source of the unidentified etiological agents. Chlorination of the water supply was believed to be adequate with an assumed 30-min contact time to protect against unexpected contamination of the source water. After performing an engineering evaluation of the water system, Lippy[15] found that the contact time was considerably less and that chlorine was not applied proportional to water volume delivered. Treatment that had been considered sufficient for decades was determined after the occurrence of a major outbreak to be inadequate.

Similar confidence for many years has been placed in disinfection-only treatment of several impounded surface water sources in eastern Pennsylvania. In recent months, at least 400 confirmed cases of giardiasis were believed to have been caused by contaminated drinking water from two chlorinated supplies. Across the state, a filtered and chlorinated river supply produced by an outdated plant was believed to be responsible for an additional 300 cases. During a period in late December to early January an abnormally high demand for water in this community had left insufficient volume to backwash the filters. For 6 d during this period, turbidity rose from a typical value of <1.0 turbidity unit to >4.0 units. Chlorine values during the 4 weeks immediately prior to the outbreak remained practically unchanged at averages of 1.2 and

1.5 mg/L free and total residuals, respectively. Coliform tests on 10-mL volumes of the finished water samples during and immediately prior to the outbreak did not exceed two positive tubes of the 35 tubes inoculated each week. These data indicate that a slight deterioration in water quality occurred, but typically would not be interpreted to indicate a health hazard. Drinking water samples collected from the three supplies subsequent to the outbreaks have been analyzed for *Giardia* cysts. Investigators from EPA analyzed 12 samples and detected cysts in 5 of these. During this period, over 350,000 people on treated water supplies in Pennsylvania were advised by health authorities to boil their water to assure protection from waterborne giardiasis.

OCCURRENCE OF VIRUS IN DRINKING WATER

In addition to the outbreak data, the occurrence of infectious viruses in treated water has been demonstrated on a number of occasions in the absence of known waterborne disease (Table III).[16,17] The most notable of these virus isolations have been those from studies in Canada and Mexico. Payment et al.[18] found 11 of 155 (7%) 1000-L samples of fully treated drinking water from the Montreal area to be positive for infectious viruses. Viruses were recovered from 121 of 153 (79%) of the 100-L samples of the surface source water. In a more recent study of a Mexican community supply, Gerba and Keswick[19] found that a significant percentage of finished water samples that were positive for enteric viruses had acceptable free chlorine residuals and met the U.S. standard for coliforms and turbidity.

By using improved recovery methods, viruses have been detected in treated drinking waters in sample volumes of a few hundred liters. These findings are at odds with prediction models based on typical virus levels found in raw

Table III. Reported Isolations of Human Viruses from Treated Drinking Water During the Past Decade

Location	Date	Water treatment	Samples No. pos.	Samples No. taken	Virus type(s) isolated
Canada	1983	Complete[a]	11	155	Polio, coxsackie
France	1982	NR[b]	2	219	Polio, coxsackie
Texas	1980	Disinfection	3	6	Coxsackie
Mexico	1978	Complete	11	11	Coxsackie, rota
Israel	1978	Disinfection	12	18	Polio, echo
India	1978	Disinfection	19	74	NR
Romania	1972–77	Complete	8	220	Polio, coxsackie
Virginia	1975	Complete	4	12	Polio
Florida	1975	Disinfection	1	10	Echo

[a]Sedimentaiton, filtration, and disinfection.
[b]Not reported.

source waters and reductions achieved with in vitro treatment systems using seeded viruses. Sproul[20] predicted that a water treatment scheme that incorporates coagulation, filtration, and chlorination (0.5 mg/L free residual for 30-min contact) could reduce the virus level by 7 logs. Therefore, source water having a high contamination level of about 1 viral unit per liter would be treated to contain 1 viral unit per 10^7 L in the finished product. The finding of viruses in much smaller volumes of treated water than the model predicted suggests that naturally occurring viruses may not be removed or inactivated in actual plant operation as effectively as are laboratory-grown seeded viruses. Indeed, this possibility is supported by studies showing reduced disinfectant inactivation of clumped viruses and viruses adsorbed on cell debris.[21,22]

Even considering the unexplained positive virus findings in fully treated drinking water and the occurrence of waterborne disease outbreaks associated with chlorinated supplies, confidence in chlorination as an effective disinfection process remains high. No microbial pathogens are known to be innately resistant to chlorine in an absolute sense. Therefore, a rigorous chlorination procedure that provided adequate contact time, continuous application, and appropriate residuals during treatment and distribution to a clarified water was believed to be adequate to produce microbiologically safe drinking water.

ALTERNATIVE DISINFECTANTS AND PROCEDURES

The previously assumed safety of relatively high concentrations of chlorine in drinking water to control microorganisms has been tempered by the more recent concern over health effects of chlorinated organic chemicals produced during the chlorination process when free chlorine residuals occur. This concern and the resultant regulatory limits on trihalomethane (THM) levels in drinking water[23] have stimulated considerable interest in alternative disinfection processes. In a recent survey of 24 community water supplies that had modified their treatment process to lower THM levels, 9 had changed the point of chlorination, 10 had changed to chloramination, and 1 had changed to chlorine dioxide.[24]

Change in chlorine application in the treatment chain from a presedimentation to a postsedimentation point reduces the level of THM-precursor chemicals available for chlorine reactivity. Chlorine application rates often are reduced in this process because chlorine-demand materials are removed. This relatively simple change has been successful in lowering THMs to an acceptable level. However, the chlorine contact time for microbial destruction prior to distribution of the water may be significantly reduced. One comprehensive study has reported on the occurrence of enteric viruses in a water treatment plant that modified the point of chlorination.[25] The supply was found to have high virus levels in the surface source and high THM concentrations (average of five samples was 338 μg/L) in the finished water. Treatment prior to the study consisted of adding 7 to 10 mg/L chlorine just prior to the rapid mixing phase of the sedimentation process (prechlorination). During the study, chlorination was moved to a point immediately following the sand filtration step. Water samples were collected three times per month for 13 months and analyzed for THM levels, viruses, and coliform indicators.

Key results of this study are summarized in Table IV. The change in chlorination practice did not appear to compromise the microbiological quality of the finished water during the study period, as indicated by negative virus and coliform findings. It should be noted, however, that even under the modified chlorination procedure, chlorine residual and contact time at this plant far exceeded the recommended values. The filtered water, having a pH below 8.0 and turbidity below 1.0 Nephelometric Turbidity Units, had been dosed with 5-7 mg/L chlorine and stored in a reservoir for at least 16 h prior to sample collection. In this system an acceptable THM level (i.e., ≤ 100 µg/L) was not achieved with the treatment modification.

The addition of ammonia prior to or just after chlorine addition in a ratio of 3:1 has been used to produce combined chlorine residuals that do not react further to produce THMs. This procedure has been in use for many years to prevent sporadic taste and odor problems and, because of the stability of the combined form, to ensure chloramine residual throughout a distribution system. Chloramination has been discouraged by many in public health because several studies have shown the combined chlorine residuals to be considerably less microbicidal than the free residuals for the same contact period.[26] However, in practice, chloramination has produced finished water that meets the microbiological standards, and no known waterborne outbreaks have been attributed to an inadequacy of this disinfectant.

Table IV. Water Quality Parameters Associated with a Michigan Water Treatment Plant That Applied Chlorine Just Prior to Sand Filtration[25]

Date	Source Water		Finished Water			
	Viruses[a]	Coliforms[b]	Chlorine[c]	Viruses[d]	Coliforms[b]	THM[e]
5/81	3	1,574	3.8	<1	<1	189
6/81	14	11,210	3.0	<1	<1	221
7/81	23	48,080	3.1	<1	<1	242
8/81	3	8,303	4.8	<1	<1	277
9/81	90	12,439	3.0	<1	<1	262
10/81	54	6,810	3.9	<1	<1	187
11/81	13	2,908	5.0	<1	<1	176
12/81	64	1,974	2.6	<1	<1	124
1/82	48	1,346	4.6	<1	<1	107
2/82	293	7,540	3.1	<1	<1	114
3/82	71	1,760	2.6	<1	<1	113
4/82	5	3,818	2.8	<1	<1	131
5/82	0	2,155	2.5	<1	<1	149

[a]Monthly PFU averages/380 L (100 gal).
[b]Monthly geometric mean CFU/100 mL.
[c]Monthly mg/L averages (total residuals).
[d]PFU/1900 L.
[e]µg/L.

In addition to the one water supply in the survey that converted from chlorine, about 85 other treatment plants in the United States use chlorine dioxide for disinfection.[27] Chlorine dioxide is a stronger oxidant than chlorine and has been shown to be an effective microbicide. Although only a limited number of studies have been conducted with waterborne pathogens, it is likely to be superior to chlorine as a disinfectant, especially in alkaline waters.

Although ozone disinfection was not selected by any of the plants in the survey, it is a viable alternative to chlorine that does not yield THMs. About 52 supplies in the United States are now or soon will be using ozone for water disinfection.[27] It can be as effective as chlorine in primary disinfection of drinking water but has the disadvantage when used alone of not providing a residual to protect the distribution system against subsequent contamination. Other alternative disinfectants to those mentioned above do not now appear practical due to cost, technical, or other disadvantages.

RISK OF WATERBORNE DISEASE

The basic expectation concerning drinking water quality in this country is that no infectious agents should be present. Disinfection has been relied upon to achieve this expectation. Animal feeding studies conducted in the 1950s and 1960s[28,29] had indicated that chlorine posed no health hazard at concentrations up to 200 mg/L of drinking water. Therefore, high levels were freely applied to drinking water derived from contaminated sources. Application rates of 10 mg/L as a prechlorination step have not been uncommon. It is doubtful that very many of the supplies derived from river and stream sources limited their chlorination practice to the recommended 0.5 to 1.0 mg/L of free residual chlorine for a 30-min contact period. Apparently, the full-conventional treatment process that evolved for sewage-contaminated source waters has been effective. No outbreaks have been attributed to an inadequacy of this process. However, the occurrence of sporadic cases of infectious disease, not obviously associated with water, has been suggested.[30]

Gastroenteritis is a common malady in the U.S. population. Studies of persons of ages <1 through 50 years indicate an average incidence of 1.2 episodes of enteric illness each year.[31] A number of potentially waterborne enteric microorganisms can cause gastroenteritis and, indeed, this disease is the one most commonly associated with waterborne outbreaks. The significance, if any, of the water route in the transmission of this disease in other than outbreak occurrences has not been determined. However, the isolation of enteric virus from large-volume samples of fully treated water meeting current standards is cause for concern. Human feeding studies have determined that many of the enteric pathogens have relatively low infective dose numbers for healthy volunteers (Table V).[32,33] More susceptible segments of the population would likely be infected with significantly lower numbers, perhaps at levels undetectable in water with current methodology.

Table V. Dose Response of Human Subjects to Oral Exposure to Relatively Low Levels of Enteric Microorganisms

Organism	Dose	Positive (No.)	Dosed (No.)	%
Shigella flexneri	180	9	36	22
Shigella dysenteriae	10	1	10	10
Giardia lamblia	10	2	2	100
Poliovirus 1	50	3	6	50
Echovirus 12	330	15	50	30

Even though no firm evidence exists for the occurrence of sporadic waterborne infections, reductions in disinfection capability are of concern. The effectiveness of 0.5 mg/L chlorine for only 30 min of contact before water is distributed to the consumer has not been thoroughly evaluated under worst-case conditions. Reductions in disinfection potency resulting from a switch to combined chlorine as the primary disinfectant may allow low-level penetration of the more resistant pathogens. The switch to potent oxidizing agents such as chlorine dioxide and ozone may also increase the likelihood of pathogen penetration. This could occur from failure to consistently maintain a sufficient disinfection level in the water because of personnel inexperience with the process or technological failure of unproven delivery systems. The existence of increased risk of infectious disease from the use of alternative disinfection agents and practices is currently unproven. However, very little study has addressed this question. Increased surveillance for deterioration in microbiological water quality and the occurrence of waterborne disease in the consuming population is certainly warranted.

ACKNOWLEDGMENTS

The authors thank Jon Capacasa, John Stoecker, John Ashton, and Judith Sauch for information concerning the Pennsylvania *Giardia* outbreaks, and Dixie White for clerical assistance in preparing the manuscript.

REFERENCES

1. White, G. C., R. J. Baker, N .J. Davoust, R. C. Hoehn, J. O. Johnson, K. E. Longley, J. C. Morris, J. J. Morrow, R. H. Moser, A. T. Palin, J. M. Symons, and R. S. Woodhull. "Disinfection—Committee Report," *J. Am. Water Works Assoc.* 70(4):219–222 (1978).
2. White, G. C. *Handbook of Chlorination* (New York: Van Nostrand Reinhold Co., 1972).
3. "Manual for Evaluating Public Drinking Water Supplies," USPHS Publ. 182 (1969). [Reprinted as EPA-430/9-75-011 (Washington, DC: U.S. Environmental Protection Agency, 1971, 1974, and 1975).

4. Akin, E. W., G. Berg, N. A. Clarke, R. Culp, R. S. Engelbrecht, E. H. Lennette, T. Metcalf, J. W. Mosley, H. E. Pearson, R. Sullivan, and H. W. Wolf. "Viruses in Drinking Water—Committee Report," *J. Am. Water Works Assoc.* 71(8):441-444 (1979).
5. Scarpino, P. V., M. Lucas, D. R. Dahling, G. Berg, and S. L. Chang. "Effectiveness of Hypochlorous Acid and Hypochlorite Ion in Destruction of Viruses and Bacteria," in *Chemistry of Water Supply Treatment and Distribution*, A. J. Rubin, Ed. (Ann Arbor, MI: Ann Arbor Science Publishers, Inc., 1974), pp. 359-368.
6. Siders, D. L., P. V. Scarpino, M. Lucas, G. Berg, and S. L. Chang. "Destruction of Viruses and Bacteria in Water by Monochloramine," *Abst. Annu. Meet. Am. Soc. Microbiol.*, Abst. E27 (1973).
7. Engelbrecht, R. S., M. J. Weber, B. L. Salter, and C. A. Schmidt. "Comparative Inactivation of Viruses by Chlorine," *Appl. Env. Microbiol.* 40:249-256 (1980).
8. Snow, W. B. "Recommended Residuals for Military Water Supplies," *J. Am. Water Works Assoc.* 48:1510-1515 (1956).
9. Jarroll, E. L., A. K. Bingham, and E. A. Meyer. "Effect of Chlorine on *Giardia lamblia* Viability," *Appl. Env. Microbiol.* 41:483-487 (1981).
10. E. H. Meyer and A. K. Bingham. "*Giardia* Cyst Inactivation by Chloramine," (Submitted for publication).
11. Lippy, E. C., and S. C. Waltrip. "Waterborne Disease Outbreaks—1946-1980: A Thirty-Five-Year Perspective," *J. Am. Water Works Assoc.* 76(2):60-67 (1984).
12. "Water-Related Disease Outbreaks—Annual Summary 1980," HHS Publ. (CDC) 82-8385 (Atlanta: U.S. Department of Health and Human Services, PHS, CDC, 1982).
13. Fuhs, G. W. "Alternative Means of Measuring Compliance with Bacteriologic Standards of Public Water Supplies," in *Evaluation of the Microbiology Standards for Drinking Water*, C. W. Hendricks, Ed., Report EPA-570/9-78-00C (Washington, DC: U.S. Environmental Protection Agency, 1978), pp. 219-227.
14. Hejkal, T. W., B. Keswick, R. L. LaBelle, C. P. Gerba, Y. Sanchez, G. Dreesman, B. Hafkin, and J. L. Melnick. "Viruses in a Community Water Supply Associated with an Outbreak of Gastroenteritis and Infectious Hepatitis," *J. Am. Water Works Assoc.* 74(6):318-321 (1982).
15. Lippy, E. C. "Contaminated Source Water and Inadequate Chlorination Led to Hepatitis Outbreak, Letters to the Editor," *J. Am. Water Works Assoc.* 74(10):14 (1982).
16. Akin, E. W., and W. O. K. Grabow, Eds. "Newsletter on Water Virology," No. 4, January 1984, (London: Int. Water Pollut. Res. Control 1984).
17. Melnick, J. L., and C. P. Gerba. "Viruses in Surface and Drinking Waters," *Environ. Int.* 7(1):3-7 (1982).
18. Payment, P., M. Trudel, and R. Plante. "Removal of Viruses During Drinking Water Treatment Processes," *Abstracts of the Annual Meeting of the American Society for Microbiology*, Abst. Q66 (1984).
19. Gerba, C. P., and B. H. Keswick. "Virus Removal During Conventional Drinking Water Treatments—Final Report," U.S. Environmental Protection Agency Cooperative Agreement 809331, Cincinnati (in preparation).
20. Sproul, O. J. "Removal of Viruses by Treatment Processes," in *Viruses in Water*, G. Berg, Ed. (Washington, DC: American Public Health Association, 1976), pp. 167-179.
21. Sharp, D. G., R. Floyd, and J. J. Johnson. "Nature of the Surviving Plaque-Forming Unit of Reovirus in Water Containing Bromine," *Appl. Microbiol.* 29(1):94-101 (1975).

22. Hoff, J. C. "The Relationship of Turbidity to Disinfection of Potable Water," in *Evaluation of the Microbiology Standards for Drinking Water*, C. H. Hendricks, Ed., EPA-570/9-78-00C (Washington, DC: U.S. Environmental Protection Agency, 1978), pp. 103-118.
23. "National Interim Primary Drinking Water Regulations: Control of Trihalomethanes in Drinking Water; Final Rule," *Fed. Regist.* 6864 (Nov. 29, 1979).
24. "Evaluation of Treatment Effectiveness for Reducing Trihalomethanes in Drinking Water," Final Report, Contract 68-01-6292 (Cincinnati: U.S. Environmental Protection Agency, TSD, ODW, 1983).
25. Stetler, R. E., R. L. Ward, and S. C. Waltrip. "Enteric Virus and Indicator Bacteria Levels in a Water Treatment System Modified to Reduce Trihalomethane Production," *Appl. Env. Microbiol.* 47(2):319-324 (1984).
26. Hoff, J. C., and E. E. Geldreich. "Comparison of the Biocidal Efficiency of Alternative Disinfectants," *J. Am. Water Works Assoc.* 73(1):40-44 (1981).
27. Anderson, A. C., R. S. Reimers, and P. DeKernion. "A Brief Review of the Current Status of Alternatives to Chlorine Disinfection of Water," *Am. J. Public Health* 72(11):1290-1293 (1982).
28. Blabaum, C. J., and M. S. Nichols. "Effects of Highly Chlorinated Drinking Water on White Mice," *J. Am. Water Works Assoc.* 48:1503-1506 (1956).
29. Druckery, H. "Chlorinated Drinking Water Toxicity Studies in Seven Generations of Rats," *Food Cosmet. Toxicol.* (England) 6:147-152 (1968).
30. Berg, G. "Introduction," in *Transmission of Viruses by the Water Route*, G. Berg, Ed. (New York: Interscience Publishers, 1967), p. 1.
31. Monto, A. S., and J. S. Koopman. "The Tecumseh Study XI. Occurrence of Acute Enteric Illness in the Community," *Am. J. Epid.* 112:323-333 (1980).
32. Akin, E. W. "A Review of Infective Dose Data for Entroviruses and Other Enteric Microorganisms in Human Subjects," in *Microbial Health Considerations of Soil Disposal of Domestic Wasterwater: Proceedings*, L. W. Canter, E. W. Akin, J. F. Kreissl, and J. F. McNabb, Eds. EPA-600/9-83-017 (Cincinnati: U.S. Environmental Protection Agency, 1983), pp. 304-322.
33. Schiff, G. M., G. M. Stefanovic, E. C. Young, D. S. Sander, J. K. Pennekamp, and R. L. Ward. "Studies of Echovirus 12 in Volunteers; Determination of Minimum Infective Dose and Effect of Previous Infection on Infectious Dose" *J. Infect. Dis.* 150:858-866 (1984).

CHAPTER 10

Risk Assessment Issues in Evaluating the Health Effects of Alternate Means of Drinking Water Disinfection

R. J. Bull and L. J. McCabe

The first realization that there may be some hazards to health arising from the use of drinking water disinfectants came from the observation that chloroform and other trihalomethanes (THMs) were produced as by-products of chlorination.[1,2] The immediate response to the fact that chlorination was producing a carcinogenic by-product was to seek methods of disinfection that did not give rise to THMs. It soon became apparent that chlorite, a by-product in the use of chlorine dioxide as a disinfectant, was capable of producing hemolytic anemia in experimental animals.[3,4] The formation of mutagenic by-products as a result of the use of chlorine in drinking water and in other analogous situations has been documented.[5-9] Similar results have been reported for ozonated, recycled wastewater.[10] These findings forced the realization that all methods of drinking water disinfection in common use involved the use of reactive chemicals. As such, the use of these chemicals will change the nature of background material present in the source of drinking water.

Recently, it has become apparent that consideration of reaction products of disinfectants must also include products that are produced by residual levels of disinfectant within the gastrointestinal tract.[11,12] Although studied much less extensively, it is now becoming clear that the formation of potentially toxic by-products in vivo is not limited to chlorine but may be responsible for some of the effects seen with chlorine dioxide[13] and chloramine[14] as well.

Although considerable research has been done in the past decade, the ability to deal effectively with this information in a regulatory way is still quite primitive. It is clear that the research information must begin to appear in a form that can be examined and understood by the public. The organization of this information is assuming the aspects of what has been referred to recently as risk assessment. Examples of different types of regulatory actions that might be taken are provided in Table I.

Where risk assessment techniques have been applied in recent years, very simplistic decision-making processes have almost universally been involved, usually dealing with one kind or source of hazard that is considered potentially dangerous by everyone. Essentially, the risk assessment procedures simply decide the extent of the hazard. In the case of drinking water disinfection there are benefits as well as hazards; in addition, a variety of health hazards associated with each option are poorly characterized.

Table I. Regulatory Options for Disinfectants

1. Specify degree of disinfection required and need for disinfectant residual in the distribution system to control microorganisms.
2. Ban use of particular disinfectants because of ineffectiveness or toxicological effects.
3. Define circumstances under which use of a particular disinfectant might be counterindicated.
4. Set Maximum Contaminant Level (MCL) for maximum dose of disinfectant.
5. Set MCL for maximum residual.
6. Set MCLs for individual by-products.
7. Define use of disinfectant that minimizes risk.
8. Specify what disinfectant should be used under specified circumstances.

Because of the unknowns, it is clear that any risk assessments that attempt to be comprehensive today may well have to be modified tomorrow. Therefore, we must ask what risks can be identified and quantitated at this time? Then, to what extent can we act on these data without forcing a decision that will have to be reversed later? When do we know that our information base is such that we can make that final decision (i.e., the safest means of producing public drinking water that gives the best protection against the spread of waterborne infectious disease in each likely circumstance)? An ability to estimate the magnitude of the unknowns involved in a complex circumstance is implied; therefore, the unknowns can be dismissed as negligible when compared to the quantified factors.

This chapter cannot begin to address these problems in detail. However, drawing from the information that is available we will attempt to (1) indicate those areas in which hazards have been identified; (2) provide some feel for the likely dimensions of these hazards; and (3) identify those areas for which there is so little information that making informed judgments, much less quantitative assessments of risk, is virtually impossible.

RANGE OF TOXICOLOGICAL EFFECTS IDENTIFIED WITH DISINFECTANTS OR DISINFECTANT BY-PRODUCTS

Table II summarizes the types of toxicological effects that have been associated directly with a disinfectant, the chemicals that have been more commonly shown to be by-products (an abbreviated list), and a brief description of the health effects identified with each by-product. In reviewing this listing, it should be recognized that much of the existing information involves data from

Table II. Summary of Health Effects Identified with Alternate Drinking Water Disinfectants or Their By-products

Disinfectant	Direct Effects	By-products	Effects of By-products
Chlorine	Increased spermhead abnormalities	Chloroform	Hepatotoxic, renal toxicity, carcinogenic
	Increased serum cholesterol	Dichlorobromomethane	Hepatotoxic, renal toxicity
		Dibromochloromethane	Hepatotoxic
		Bromoform	Hepatotoxic
	Myocardial hypertrophy	Dichloroacetic acid	Decreased serum glucose and lactate, neurotoxic, aspermatogenesis, ocular lesions
	Mutagenic in bacteria		Induction of peroxisomes
	Hyperplasia in skin		
		Trichloroacetic acid	Clastogenic, carcinogenic
		Chloroacetonitrile	Mutagenic, clastogenic
		Dichloroacetonitrile	Clastogenic
		Trichloroacetonitrile	Mutagenic, clastogenic, carcinogenic
		Bromochloroacetonitrile	Clastogenic, carcinogenic
		Dibromoacetonitrile	Fetotoxic, tumor promoter
		2-Chlorophenol	
		2,4-Dichlorophenol	Fetotoxic, tumor promoter
		2,4,6-Trichlorophenol	Carcinogenic
		Halogenated ketones	Mutagenic
		Halogenated aldehydes	Mutagenic
		Organic N-chloramines	Model compounds genotoxic
Monochloramine	Hemolytic anemia	Organic by-products poorly studied	
	Abnormal mitotic figures in liver	Miscellaneous direct-acting mutagens	
	Mutagenic in bacteria		

Table II, Continued

Disinfectant	Direct Effects	By-products	Effects of By-products
Chlorine dioxide	Antithyroid effects	Chlorite	Hemolytic anemia, spermhead abnormalities, decreased sperm motility
	Prenatal exposure Depressed cerebellar weight Decreased cell number in cerebellum Delayed in neurobehavioral effects Postnatal exposure Depressed cerebral weight Decreased cell number in cerebrum Delayed neurobehavioral effects Hyperplasia in skin	Organic by-products poorly studied	
Ozone		Mutagenic by-products poorly studied	

what are essentially screening tests. However, these data are suggestive of adverse health effects and are not to be dismissed lightly. On the other hand, the data should not be interpreted as indicating health hazards of such magnitude as to cause immediate concern. The large gaps in information that bear directly on the overall problem should indicate the need to keep these data in perspective, because to do otherwise might tend to push decisions to final resolution prematurely. For example, the ability to identify a large number of mutagenic chemicals with chlorination is at least partially due to the fact that reaction products of chlorine have been much better investigated than those of any of the alternatives.

DISINFECTANTS

Chlorine

In contrast to what we know about by-products of disinfectants, surprisingly little toxicological information exists concerning the toxicology of chlorine itself. Chlorine has been reported to produce a weak mutagenic effect in bacteria[15] and to preferentially kill DNA-deficient strains of bacteria.[16]

Recent data indicate that sodium hypochlorite, but not hypochlorous acid, is capable of inducing spermhead abnormalities in mice when administered by gavage.[17] From a risk assessment point of view this effect poses a problem. The maximum response is observed at 4 mg/kg body weight, and higher doses provide no further increase in response. The degree of effect observed, while repeatable, is small and would be difficult to associate with adverse reproductive or teratogenic effects experimentally. In fact, Druckrey[18] failed to note any adverse effects of chlorine administered to rats over seven generations. However, the data do suggest the in vivo formation of a very potent mutagen, a situation that would necessarily carry some probability of producing an adverse health effect, whether it is a reproductive or carcinogenic effect. Since we do not have a quantitative relationship between the percent abnormal sperm and the degree of carcinogenic or reproductive hazard, we cannot estimate the actual hazard to man. It is distinctly possible that the formation of a product in the gastrointestinal tract could pose a greater hazardous health effect than a known by-product such as chloroform. Obviously, more research will be needed to develop the information necessary to estimate the hazards associated with this effect. However, the task will be very difficult because the chemical responsible has not been identified.

Another problem in risk assessment can be illustrated by the fact that solutions of chlorine (100 mg/L) applied to the skin result in hyperplastic response in the mouse skin. Chlorine was more effective in producing this effect at pH 6.5 than at 8.5, thus indicating that hypochlorous acid is more active in this regard. Such activity is associated with chemicals that act as tumor promoters but is apparently not specific to promoters.[19] The ability of chlorine solutions to promote skin tumors has not been well researched. Hayatsu et al.[20] reported that sodium hypochlorite does increase skin tumor yields when applied in conjunction with a sub-threshold dose of 4-nitro-quinoline-1-oxide. On the other hand, Pfeiffer[21] reported an absence of such activity and a tendency for a reduction in benzo(*a*)pyrene-induced cancer with chlorine. If further work confirms the tumor-promoting activity of chlorine, the risk assessment questions that must be answered are (1) how can risks be assessed that depend on the presence of another chemical, and (2) what models are appropriate for assessing the risk of tumor promoters?

Very recent information indicates that chlorine residuals within an order of magnitude of those found in drinking water interact with a diet marginal in calcium to increase serum cholesterol[22] and produce signs of myocardial hypertrophy and arteriosclerosis in rabbits and pigeons.[23] Restoration of calcium intake to a normal level considerably modifies the response, thus increasing the concentrations of chlorine required to produce the effect by at least a factor of 3. In addition to the usual questions of across-species extrapolation, in this case the degree of hazard to the human population that consumes chlorinated water obviously depends on the extent to which that population is marginal in its calcium intake. There are data that suggest certain segments of the population may well be deficient to the levels used in these experiments. For example, the median calcium intake by blacks routinely falls below the recommended daily requirement at age 18 and above, and whites fall below that level by age 35.[24]

The relationship between cardiovascular disease and chlorine has been studied and observed only in rabbits and pigeons and may not be a factor in humans. However, as will be discussed later, there has been a long-standing association of water of low hardness and cardiovascular disease mortality in the human population.[25] A systematic investigation of a mineral basis for this effect in pigeons indicated some interactions between essential trace metals and toxic trace metals;[26] however, the magnitude of these effects was considerably below that observed in the chlorine and calcium interaction.

Chloramine

As with chlorine, chloramine has been shown to be mutagenic in bacteria.[27] Although tested under similar conditions for its ability to produce spermhead abnormalities in mice as did chlorine, it has not been possible to obtain consistent results with chloramine.[17] Consequently, if the same by-product is formed with chloramine, the reaction must be attenuated relative to the reaction with hypochlorite. On the other hand, subchronic studies preliminary to lifetime bioassays that were conducted by the National Toxicology Program indicated the appearance of abnormal mitotic figures in the liver of B6CF1 mice, which suggests that chloramine could possess some of the properties that have been associated with chemical carcinogens.[14] Again we are faced with the fact that there is little information on which to base an estimate of the health hazard that chloramine poses at the levels ordinarily encountered in drinking water, although there is reason for some concern. Hopefully, the studies that EPA has undertaken in cooperation with the National Toxicology Program will clarify this issue.

It is of interest that chloramine has been associated with the development of hemolytic anemia in dialysis patients who have been dialyzed against fluids containing water that had been disinfected with chloramine. Experimental studies in animals have shown that whatever risk does exist with the use of chloramine, it is considerably less than that which would be associated with the use of chlorine dioxide.[28]

Chlorine Dioxide

Of the disinfectants commonly proposed for drinking water, chlorine dioxide is unique in that it is associated with chlorite, a toxicologically important inorganic by-product. Chlorite is capable of producing methemoglobinemia both in vivo and in vitro.[3,4] However, this effect is only seen in vivo with fairly large bolus doses and has been uniformly absent from studies that involve more gradual exposures via drinking water.[3,29] Drinking water exposure does, however, result in the production of a hemolytic anemia, signs of which can be observed at chlorite concentrations as low as 50 mg/L. At this level of exposure, the gross picture of hemolytic anemia is compensated for in that red cell and hemoglobin concentrations return to normal levels with continued exposure. The hematological effects of chlorite are also observed with chlorine dioxide, probably as a result of its conversion to chlorite in the gastrointestinal tract.[29]

Chlorite has been examined in reproductive studies and does not appear to present a measurable reproductive or teratogenic hazard in rats.[30] However, chronic exposure at levels of 100 mg/L in drinking water does increase the number of abnormal sperm and impair certain measures of sperm motility. The significance of the latter observations in the absence of clear-cut reproductive effects is uncertain. Unlike results that were discussed earlier with chlorine, the abnormalities observed with chronic treatment cannot be associated with mutagenic response. However, it may be indicative of a reproductive effect that could be expressed in individuals having minimal reproductive capacity.

More recently, chlorine dioxide has been shown to be capable of affecting thyroid metabolism in primates.[13] This effect of chlorine dioxide cannot be attributed to its conversion to chlorite, since equivalent doses of chlorite fail to produce the effect. Because of the reactive nature of chlorine dioxide, it does not seem likely that it could be absorbed intact. Consequently, the possibility that this effect is secondary to the formation of a reaction product in the gastrointestinal tract must be considered.

The administration of chlorine dioxide to adult rats at 100 mg/L fails to produce significant decreases in serum thyroxine concentrations. On the other hand, administration of the same concentration to pregnant and lactating females does substantially reduce the serum thyroxine levels in rat pups at 21 days of postnatal age.[31] Similar depressions of serum thyroxine levels are observed if rat pups are directly exposed to aqueous solutions of chlorine dioxide at doses of 14 mg kg^{-1} d^{-1} (a dose equivalent to that taken in by a lactating female consuming drinking water containing chlorine dioxide at concentrations of 100 mg/L). Although the depressions of serum thyroxine associated with these exposures are rather small (i.e., depressed 20% relative to controls), these exposures do significantly delay brain development. Exposure via the dam, which includes prenatal exposure, results in decreased cell numbers and weight of the cerebellum.[32] The exclusive postnatal exposure of rat pups by gavage primarily affects the same parameters in the cerebral cortex. Both types of exposure result in behavioral effects that are still apparent 40 d after exposure has ceased. The extent to which biochemical measures of brain development are depressed has not been documented at lower doses or at later times. However, depressed exploratory behavior is observed at doses as low as 2 mg/L drinking water at 60 d of age in the offspring of female rats that were exposed during pregnancy and lactation. Obviously, these effects of chlorine dioxide must be included in any assessment of health risks associated with alternative disinfectants.

In addition, recent data indicate that like chlorine, chlorine dioxide produces hyperplasia in the mouse skin. The meaning of the result with respect to chlorine dioxide's ability to act as a tumor promoter in this tissue remains to be explored.

What is notable about research results with chlorine dioxide is that the effects are noncarcinogenic with the exception of hyperplasia in the skin. We have yet to identify any mutagenic activity with chlorine dioxide by-products in our laboratory, which suggests that deciding between chlorine dioxide and chlorine will involve the trading of carcinogenic and noncarcinogenic effects. Under present day conventions, these types of effects are handled by different models. Appropriate decision making may depend heavily on the accuracy of the assumptions that underlie these two methodologies when considering them as alternatives for drinking water disinfection.

Ozone

The use of ozone in drinking water does not ordinarily result in residual concentrations of disinfectant at the consumer's tap. Consequently, little hazard would be anticipated to result directly from human exposure to ozone in drinking water. If there are hazards to the consumer's health in this case, they would have to be associated with the production of by-products in the finished drinking water.

DISINFECTANT BY-PRODUCTS

Despite the considerable efforts of a number of laboratories, the characterization of disinfection by-products is far from complete. The limited information in this area is illustrated by the fact that we are presently unable to chemically identify more than 90% of the mutagenic activity that is generated through the reaction of chlorine with humic acids.[33] To make matters worse, chlorination by-products have been more thoroughly investigated than the reaction products of the alternate disinfectants. With the exception of chloramine, we have not observed substantial increases in mutagenic activity with other disinfectants. Because of our ignorance of the chemical and toxicological characteristics of by-products, assessment of health hazards cannot be comprehensively accomplished.

We must also recognize that the absence of mutagen generation by a disinfectant offers little assurance that toxicologically important by-products do not result. This is perhaps best illustrated by the clearly important effects of different disinfectants on the cardiovascular system and thyroid functions— effects that are not necessarily dependent on properties of mutagens.

Nevertheless, the ability of different disinfectants to increase mutagenic activity is illustrative of the general problem with by-products. Although they do not establish the existence of an unacceptable risk, they do demonstrate that we have much to accomplish before we can account for these biologically important chemicals. The question that must be asked is how can the health risks associated with these by-products be estimated? Until that and similar questions can be answered, quantitative risk assessments are of questionable relevance. The remainder of this chapter will consider those chemicals that have been clearly identified as products of disinfectant reactions in water. However, it should continually be kept in mind that a large number of by-products with biological properties of concern still remain to be identified.

By-products of Chlorination

Trihalomethanes (THMs) are the most familiar by-products of drinking water chlorination. Most of the concern about THMs arises from the demonstrated carcinogenic properties of chloroform in rats and mice.[34] Although other THMs share many of the toxicological properties of chloroform, the results of carcinogenesis bioassays have not been reported.

The recently completed study of chloroform carcinogenicity, sponsored by EPA at Stanford Research Institute International, will necessitate a reexamination of previous estimations of the carcinogenic risk associated with chloroform.[35] This study[36] used exposures to chloroform via drinking water, whereas the previous National Cancer Institute (NCI) bioassay used corn oil gavage.[34] As reported by Jorgensen et al.,[37] the results obtained in the Osborne-Mendel rat closely approximated the results obtained in the previous NCI bioassay.

However, despite the use of comparable daily doses in drinking water, the study failed to confirm the induction of hepatocellular carcinomas in B6C3F1 mice seen in the earlier study. These data suggest that the tumorigenic response in the mouse was in some way secondary to the use of corn oil gavage and not simply attributable to chloroform alone. Since Withey et al.[38] found that corn oil gavage actually decreases the rate of chloroform absorption, it would appear that this result is not due to a changed pharmacokinetic pattern. A better explanation may be related to the observation that incorporation of corn oil into the diet increased the yield of liver tumors in aflatoxin B_1-initiated rats.[39] In any case, this result points up the general difficulty involved in the extrapolation of data obtained in experimental animals under one set of conditions to man, who is exposed under an entirely different set of circumstances.

The haloacetonitriles are another group of by-products and products formed by the chlorination of drinking water[40,41] and in vivo, respectively.[12] In addition to being to being mutagenic in *Salmonella* and capable of inducing sister chromatid exchange in mammalian cells (in vitro), these chemicals have been shown to induce tumors of the skin[42] and lungs in mice.[43] Those who estimate the risks from such by-products must consider not only the amount ingested from drinking water but also the extent of their formation in vivo. The question then must be asked: Is a microgram formed in vivo equivalent to a microgram in a glass of water? Since these chemicals are reasonably reactive, the extent to which a microgram ingested in water could impact a target cell may be considerably less than a microgram formed in the gastrointestinal tract, simply because of the different time, distance, and competing reaction sites available before its target cell. Similar arguments can be made for other direct-acting mutagens such as the halogenated acetones and aldehydes. However, the carcinogenicity of these products remains to be confirmed.

One must not conclude that the haloacetonitriles are the only toxicologically important compounds produced on reaction between chlorine and amino acids. Sussmuth[44] reported that very potent mutagens were formed in the reaction of methionine, tyrosine, phenylalanine, and glycine with chlorine. Glycine did not form dihaloacetonitriles (DHAN) with chlorine,[40] whereas tryptophan was a better precursor for DHAN than tyrosine. Tyrosine served as a precursor of mutagenic activity under conditions in which tryptophan did not.[44] One possible group of by-products of chlorine reaction with amino acids that includes members with genotoxic properties is known as organic N-chloramines.[45] However, such products have yet to receive serious toxicological study.

Haloacid derivatives make up another class of compounds formed in the chlorination of drinking water and on ingestion of aqueous solutions of chlorine. The most prominent of these are dichloroacetic acid and trichloroacetic acid. Various salts of dichloroacetate have been used therapeutically to control blood glucose levels in diabetics.[46] However, its use in humans was curtailed because of the development of polyneuropathy[47] in a patient at doses of about 50 mg kg^{-1} d^{-1}. The disease slowly reversed on termination of treatment. Katz et al.[48] confirmed the toxic properties of dichloroacetate in 90-d studies conducted in rats and dogs. They found ocular lesions, gall bladder anomalies (dog), brain lesions (both species), loss of testicular germinal epithelium, and aspermatogenesis (rats). Serum glucose and lactic acid levels were significantly depressed at the lowest doses tested (i.e., 50 mg/kg in dogs and 125 mg/kg in rats). Consequently, it is not possible to estimate the dose at which no effect would be observed.

Halogenated phenols form another class of chemicals that have been identified as by-products of chlorination. Within this class, the chemical that has been best characterized toxicologically is 2,4,6-trichlorophenol. The principal finding is that it is capable of inducing hepatocellular carcinomas in male and female mice (B6C3F1) and increasing the incidence of lymphomas and leukemia in male rats.[49]

The chemicals 2-chlorophenol and 2,4-dichlorophenol are apparently weakly fetotoxic and seem to have the capability of acting as cocarcinogens.[50] These two chemicals also appear capable of promoting the development of dimethylbenzanthracene (DMBA)-initiated skin tumors in mice.[51] Although these lower-substituted chlorinated phenols have not been capable of inducing cancer on their own, they must be kept in mind when estimating the overall carcinogenic risk that may be associated with the use of chlorinated drinking water. They again bring up the general problem of how one should formally work with a phenomenon such as tumor promotion in a regulatory arena.

By-products of Chloramination

Unlike chlorination, chloramination has received very little attention as a process that gives rise to toxicologically important by-products. Again this seems to stem from the general theory that the use of chloramination suppresses the formation of the trihalomethanes. Obviously, the data presented in this conference[52] indicate that chloramine does give rise to direct-acting mutagens in much the same manner as chlorine. To date no products of this reaction have been identified or studied toxicologically. Since the two processes differ substantially in the level of trihalomethanes (THMs) produced, it is possible that the nature of the products formed may be substantially different. Therefore, to conclude from the Ames assay results that the reduced level of mutagenic activity seen with chloramine actually represents a reduced carcinogenic hazard would be premature.

Reaction Products of Chlorine Dioxide

There has been little characterization of the toxicology of by-products associated with the use of chlorine dioxide as a disinfectant in drinking water. As pointed out earlier, the emphasis with this disinfectant has been on the hematological effects of its inorganic by-product, chlorite. Another problem is that since the yield of chlorinated compounds is much less with chlorine dioxide, there is no convenient way of tagging or identifying by-products. This has been further complicated by the lack of a convenient biological indicator of by-product toxicity for chlorine dioxide. However, the observed antithyroid activity of chlorine dioxide discussed previously points out the need for much better characterization of the organic by-products of chlorine dioxide.

By-products of Ozonation

Although the general types of products formed in reactions of ozone with organic chemicals are well understood, no particular effort has been made to identify the products formed when ozone is used in the treatment of drinking water. Consequently, the toxicological characterization of ozonation by-products is in the most primitive state of all. This may be partially attributable to the inability to detect any significant generation of mutagenic activity when ozone disinfection of drinking water has been studied using the Ames test.[7] On the other hand, increased mutagenic activity has been observed with ozonation of wastewater.[10] Of course, mutagenicity testing can only be taken as a qualitative indication of the presence of putative carcinogenic or mutagenic hazard and not as a comprehensive means of assessing health hazards. Consequently, it cannot be said that the by-products of ozonation have been studied sufficiently to allow characterization of health hazards relative to other means of drinking water disinfection.

MAGNITUDE OF HEALTH RISKS THAT MIGHT BE ATTRIBUTABLE TO DISINFECTANT USE

Relatively little reliable human data are applicable to health risks caused by chemical exposures via drinking water. In particular, data that could be directly relevant to a practice as widespread as chlorination are not generally available. Essentially two areas have been extensively investigated with some consistency with respect to the chemical qualities of water during the years observed: the negative relationship between cardiovascular disease mortality and the hardness of drinking water,[25] and the positive associations that have been noted between cancer mortality and chlorination of drinking water.[53] Neither association has been universally accepted, partially because confounding variables have not been satisfactorily controlled, and because there is little basic data to provide a causal explanation of the mortality rates attributed to these variables. For example, Crump and Guess[53] pointed out that the risk attributable to chlorination in epidemiological studies actually exceeds that which can be attributed to the THMs, assuming they all have a carcinogenic potency similar to that of chloroform.

Table III illustrates the problem with the relationship between cardiovascular disease mortality and water quality. It was observed rather early[54] that a series of statistically significant correlations exists between the mineral content of drinking water and a variety of causes of death. In many of these cases, it is difficult to rationalize any causal connection between this gross parameter of water quality and the cause of death. For example, the positive relationship between death by accident and the dissolved solids content of drinking water is just as strong as the negative association with diseases of the heart. Obviously, an investigation of the relationship between diseases of the heart and water quality is intuitively more attractive to pursue than the relationship with accidents. On the other hand, Sharrett and Feinleib[25] and Comstock[54] all pointed out that the relationship of water quality with cardiovascular disease mortality is reasonably and consistently observed across a wide variety of studies. However, there has been little progress in identifying a specific constituent as being responsible for this correlation.

To what extent may these data be useful in placing into perspective the maximum contribution that chlorine could be making to the two most common causes of death in the United States?

The relationship between cancer mortality and chlorination has been studied most directly through epidemiology. Since the risk associated with chlorination appears to be larger in human populations than can be accounted for by the

Table III. Significant ($p < 0.01$) Correlations Between Twenty Leading Causes of Death and the Mineralization of Drinking Water[a]

Disease	Mineral Content	
Male (white)		
Diseases of heart	−0.731	Decrease as solids increase
Malignant neoplasms	−0.720	
Cirrhosis of liver	−0.665	
Ulcer of stomach and duodenum	−0.676	
Vascular lesions, CNS[b]	+0.742	Increase as solids increase
Accidents	+0.725	
Diseases of infancy	+0.868	
Congenital malformations	+0.604	
Gastritis	+0.815	
All other than 20 leading	+0.566	
Female (white)		
Diseases of heart	−0.747	Decrease
Cirrhosis of liver	−0.725	
Vascular lesions, CNS[b]	+0.626	Increase
Accidents	+0.736	
Diseases of infancy	+0.698	
Gastritis	+0.769	
Complications of pregnancy	+0.637	
All other	+0.808	

[a]Based on data in Reference 55.
[b]Central nervous system.

trihalomethanes,[54] we must ask the question of whether this higher risk can be accounted for by new experimental data. Although this question cannot be answered in quantitative terms, it can be stated that other by-products and biological effects that are consistent with this idea have been associated with chlorine. The principal data are:

1. the carcinogenic properties associated with the haloacetonitriles;
2. evidence that a wide variety of mutagenic by-products are formed with chlorination;
3. the production of spermhead abnormalities by direct administration of chlorine to mice, indicating the in vivo formation of mutagenic and potentially carcinogenic by-products; and
4. evidence that chlorine is capable of producing many of the same carcinogenic by-products in the gastrointestinal tract that have been identified in finished drinking water.

Those interested in the epidemiological study of the associations between chlorination and cancer or in assessing these risks must, however, pay particular attention to the conditions under which these effects are observed. The production of mutagenic activity with chlorine is very much favored at acid pH. In parallel with this observation is that the formation of total organic chlorine is also favored at acid pH.[9] Conversely, the production of trihalomethanes is favored at alkaline pH.[41] In addition, spermhead abnormalities were produced only when mice were intubated with chlorine at pH 8.5 and not when intubated at pH 6.5. These data indicate that the products associated with chlorine are going to be very dependent on other water conditions. Therefore, consistency in epidemiological associations in different geographical locations should not necessarily be expected.

In future studies it would also be appropriate to concentrate on documenting some secondary water parameters such as pH and total organic carbon as surrogates of the types and extent of chlorine reaction that might be expected at a given site rather than to focus so closely on a few specific chemicals as has been done in the past. To concentrate on individual compounds when the majority of by-products are yet to be identified chemically may well be counterproductive. Obviously, an estimation of the cancer risk from animal experimentation must also consider these variables. In general, selection of a single chemical or class of chemicals as a comprehensive indicator of exposure to disinfectant by-products is not appropriate, considering our present understanding of the conditions that will modify the form of chlorination reaction products.

At present, there are no human studies that support the idea that chlorine contributes to cardiovascular disease mortality. The data in experimental animals suggest that any effect that chlorine has on the cardiovascular system would be very dependent on the calcium status of the individual.

This observation would indicate a reexamination of some data used in an earlier study of cardiovascular mortality and hardness of water[55] and how this relationship might be confounded by chlorination. Data in Table IV provide evidence that surface water supplies that receive chlorine as their only treatment present a greater risk for contributing to arteriosclerotic heart disease than both groundwater supplies (also chlorinated with one exception) and surface waters that receive treatment in addition to chlorination. These results do not appear to be confounded by an urban–rural effect, since essentially the same relationships are observed if the data are stratified by population size. Clearly, these data follow the usual inverse relationship between water hardness and cardiovascular mortality, since the mean calcium content of the water varies inversely with the mortality ratios. However, in view of recent work in experimental animals, two other possibilities have to be considered:

1. The higher Ca^{2+} content of groundwater and surface waters receiving other treatment ameliorated the cardiovascular effects of chlorination;
2. A substrate with which chlorine reacts to produce a toxic intermediate is not present in groundwater and is perhaps removed by other treatment in surface waters.

However, these data do not provide a clear-cut relationship between chlorination and cardiovascular disease mortality. They simply illustrate the difficulties that will be involved in assessing and establishing the extent to which any additional cardiovascular disease risks are attributable to chlorination. The only data that we are aware of in the literature that bear directly on this issue is the recent paper of Wilkins and Comstock[56] that identifies a near-significant relationship between chlorinated water supplies and arteriosclerotic heart disease in Washington County, Maryland. As pointed out by the authors, this study is complicated by the fact that chlorinated water was obtained from surface water, whereas the control population got its water from deep groundwater sources. This results in an urban–rural complication common to many of the studies of water quality and cardiovascular disease. Nevertheless, the relative risk ratio observed was 1.13, a level consistent with the projected maximum impact of a water factor in cardiovascular disease mortality. Interestingly, associations between water hardness and cardiovascular disease mortality gave conflicting results in a previous epidemiological study of this population.[54]

Table IV. Average Mortality Ratios for Arteriosclerotic Heart Disease from Chlorinated Water Supplies (1949 – 1951)

Water Sources	Calcium Concentration (mg/L)	N[a]	Male	Female
Groundwater	53.9	24	1.02 ± 0.04[b]	0.89 ± 0.04
Surface water, disinfection only	17.7	16	1.12 ± 0.03	1.05 ± 0.06
Surface water, disinfection and other treatment	27.7	60	1.04 ± 0.02	0.91 ± 0.02

[a]Number of individuals tested.
[b]Relative risk ratio ± standard error of the mean.

SUMMARY AND CONCLUSIONS

There are epidemiological associations between drinking water quality and the two leading causes of death in the United States. In the case of cancer, the problem has been studied directly; in the case of cardiovascular mortality, one has to rely primarily on supposition. However, we must recognize that data accumulated on animals in the past few years are consistent with the role of chlorine in the development of both diseases in the general population.

There will be some risk associated with any alternative means to chlorination for the disinfection of drinking water. The risks projected for a contribution to these diseases by chlorinated drinking water are sufficiently high to warrant a continued research effort to map out a course of least risk. Conversely, the risks associated with chlorine are still small enough so that there is no reason to institute changes in a disinfection practice (except where there is clear violation of a particular drinking water standard) that involves the use of alternatives for which little information exists concerning the health hazards that may be associated with their use. We need to take the time to accumulate the necessary health effects information before regulatory decisions are made that are not easy to reverse.

Retrospective epidemiological studies are largely restricted to classified disease categories. We should recognize that many of the effects being seen in experimental studies attributed to chlorine or other disinfectants do not clearly fit into such categories, particularly if the effects are relatively mild. Good examples of problems that would be difficult to solve using retrospective approaches would be any adverse reproductive outcomes that might be associated with the use of chlorine, and the ability of chlorine dioxide to delay development of the cerebellum and cerebral cortex. These are not overt lesions that are likely to be noticed by physicians in the first place and would probably not be associated with drinking water disinfection if they were observed in an individual patient. These effects are actually observed at lower doses in animal experiments than any other effects of the disinfectants or their by-products. In this sense, there is less of a margin of safety for these effects than for cancer. It is only the way in which data demonstrating carcinogenic responses are used in risk assessment (based on our current theoretical understanding of the processes involved in chemical carcinogenesis) that places such emphasis on cancer. It appears likely that decisions involving choices between chlorine and chlorine dioxide will necessarily focus on trade-offs between carcinogenic and noncarcinogenic effects, respectively. Consequently, regulatory decision makers in the disinfectant area must avoid the trap of concentrating so heavily on carcinogenesis as the pivotal issue.

REFERENCES

1. Bellar, T. A., J. J. Lichtenberg, and R. C. Kroner. "The Occurrence of Organohalides in Chlorinated Drinking Waters," *J. Am. Water Works Assoc.* 66:703-706 (1974).
2. Rook, J. J. "Formation of Haloforms During Chlorination of Natural Waters," *Water Treat. Exam.* 23:234-243 (1974).
3. Heffernan, W. P., C. Guion, and R. J. Bull. "Oxidative Damage to the Erythrocyte Induced by Sodium Chlorite In Vivo," *J. Environ. Pathol. Toxicol.* 2:1487-1499 (1979).
4. Heffernan, W. P., C. Guion, and R. J. Bull. "Oxidative Damage to the Erythrocyte Induced by Sodium Chlorite In Vitro," *J. Environ. Pathol. Toxicol.* 2:1501-1510 (1979).
5. Cheh, A. M., J. Skochdopole, P. Koski, and L. Cole. "Nonvolatile Mutagens in Drinking Water: Production by Chlorination and Destruction by Sulfite," *Science* 207:90-92 (1980).
6. Zoeteman, B. C. J., J. Hrubec, E. de Greef, and H. J. Kool. "Mutagenic Activity Associated with By-products of Drinking Water Disinfection by Chlorine, Chlorine Dioxide, Ozone and UV-Irradiation," *Environ. Health Perspect.* 46:197-205.
7. Kool, H. J., C. F. van Kreijl, E. de Greef, and H. J. van Kranan. "Presence, Introduction and Removal of Mutagenic Activity During the Preparation of Drinking Water in the Netherlands," *Environ. Health Perspect.* 46:207-214 (1982).
8. Bull, R. J., M. Robinson, J. R. Meier, and J. Stober. "Use of Biological Assay Systems to Assess the Relative Carcinogenic Hazards of Disinfection By-products," *Environ. Health Perspect.* 46:215-227 (1982).
9. Meier, J. R., R. D. Lingg, and R. J. Bull. "Formation of Mutagens Following Chlorination of Humic Acid: A Model for Mutagen Formation During Drinking Water Treatment," *Mutat. Res.* 118:25-41 (1983).
10. Gruener, N. "Mutagenicity of Ozonated, Recycled Water," *Bull. Environ. Contam. Toxicol.* 20:522-526 (1978).
11. Vogt, C. R., J. C. Liao, G. Y. Sun, and A. V. Sun. Proc. 13th Annu. Conf. Trace Subst. Environ. Health, (Columbia, MO: University of Missouri, 1979), pp. 453-460.
12. Mink, F. L., W. E. Coleman, J. W. Munch, W. H. Kaylor, and H. P. Ringhand. "In Vivo Formation of Halogenated Reaction Products Following Peroral Sodium Hypochlorite," *Bull. Environ. Contam. Toxicol.* 30:394-399 (1983).
13. Bercz, J. P., L. Jones, L. Garner, D. Murray, D. A. Ludwid, and J. Boston. "Subchronic Toxicity of Chlorine Dioxide and Related Compounds in Drinking Water in the Nonhuman Primate," *Environ. Health Perspect.* 46:47-55 (1982).
14. National Toxicology Program. "A Subchronic Study of Chloramine Generated In Situ in the Drinking Water of F344 Rats and B6C3F1 Mice," report from Gulf South Research Institute to Tracor Jitco, Inc. (Managers of Chloramine Bioassay in the National Toxicology Program, 1981).
15. Wlodkowski, T. J., and H. S. Rosenkranz. "Mutagenicity of Sodium Hypochlorite for *Salmonella* Typhimurium," *Mutat. Res.* 31:39-42 (1975).
16. Rosenkranz, H. S. "Sodium Hypochlorite and Sodium Perborate: Preferential Inhibitors of DNA Polymerase-Deficient Bacteria," *Mutat. Res.* 21:171-174 (1973).

17. Meier, J. R., R. J. Bull, J. A. Stober, and M. C. Cimino. "Evaluation of Chemicals used for Drinking Water Disinfection for Production of Chromosomal Damage and Spermhead Abnormalities in Mice," *Environ. Mutagen.* (in press).
18. Druckrey, H. "Chlorine Drinking Water, Toxicity Tests Involving Seven Generations of Rats," *Food Cosmet. Toxicol.* 6:147–154 (1968).
19. Slaga, T. J., G. T. Bowden, and R. K. Boutwell. "Acetic Acid, A Potent Stimulator of Mouse Epidermal Macromolecular Synthesis and Hyperplasia but with Weak Tumor-Promoting Ability," *J. Nat. Cancer Inst.* 55:983–987 (1975).
20. Hayatsu, H., H. Hoshino, and Y. Kawazoe. "Potential Cocarcinogenicity of Sodium Hypochlorite," *Nature* 233:495 (1971).
21. Pfeiffer, E. H. "Health Aspects of Water Chlorination with Special Consideration to the Carcinogenicity of Chlorine. II. Communication on the Cocarcinogenicity," *Zentralbl. Bakteriol. Parasitenk. Infectionskr. Hyg. Abt. 1 Orig. Reihe B.* 166:185–211 (1978).
22. Revis, N. W. "The Cardiovascular System as a Target Organ for Chlorine and Chlorination By-products," Chapter 29, this volume.
23. Revis, N. W., B. H. Douglas, P. T. McCauley, H. P. Witschi, and R. J. Bull. "The Relationship of Drinking Water Chlorine to Coronary Atherosclerosis," *Pharmacologist* 25:732 (1983).
24. Dresser, C. M., M. D. Carroll, and S. Abraham. "Vital and Health Statistics," Series 11, National Health and Nutrition Examination Surveys, National Center for Statistics, Department of Health and Human Services, Hyattsville, MD (1984).
25. Sharrett, A. R., and M. Feinleib. "Water Constituents and Trace Elements in Relation to Cardiovascular Diseases," *Prev. Med.* 4:20–36 (1975).
26. Revis, N. W., A. R. Zinsmeister, and R. J. Bull. "Atherosclerosis and Hypertension Induction by Lead and Cadmium Ions: An Effect Prevented by Calcium Ion." *Proc. Nat. Acad. Sci. USA* 78:6494–6498 (1981).
27. Shih, K. L., and J. Lederberg. "Chloramine Mutagenesis in *Bacillus Subtilis*," *Science* 192:1141–1143 (1976).
28. Bull, R. J. "Health Effects of Alternate Disinfectants and Their Reaction Products," *J. Am. Water Works Assoc.* 72:299–303 (1980).
29. Abdel-Rahman, M. S., D. Couri, and R. J. Bull. "Kinetics of ClO_2 and Effect of ClO_2, ClO_2^- and ClO_2^- in Drinking Water on Blood Glutahione and Hemolysis in Rat and Chicken," *J. Environ. Pathol. Toxicol.* 3:431–449 (1980).
30. Carlton, B. D., and M. K. Smith. "Reproductive Effects of Alternate Disinfectants and Their By-products," Chapter 24, this volume.
31. Orme, J., D. H. Taylor, R. D. Laurie, and R. J. Bull. "Effects of Chlorine Dioxide on Thyroid Function in Neonatal Rats," *J. Toxicol. Environ. Health*, (in press).
32. Taylor, D. H., and R. J. Pfol. "Effects of Chlorine Dioxide on Neurobehavioral Development of Rats," Chapter 28, this volume.
33. Coleman, W. E., J. W. Munch, W. H. Kaylor, R. P. Streicher, H. P. Ringhand, and J. R. Meier. "GC/MS Analysis of Mutagenic Extracts of Aqueous Chlorinated Humic Acid—A Comparison of the By-products to Drinking Water Contaminants," *Environ. Sci. Technol.* 18:674–681 (1984).
34. "National Cancer Institute (U.S.) Carcinogenesis Bioassay of Chloroform," NCI, NITS NO. PB264018/AS (1976).
35. "U.S. Environmental Protection Agency Water Quality Criteria Documents; Availability, *Fed. Regist.* 45, No. 231, 79317-79 (1980).

36. Jorgensen, T. A., C. F. Rushbrook, and O. C. L. Jones. "Dose-Response Study of Chloroform Carcinogenesis in the Mouse and Rat: Status Report," *Environ. Health Perspect.* 46:141-149 (1982).
37. Jorgensen, T., E. Meierhenry, C. F. Rushbrook, R. J. Bull, M. Robinson, and C. E. Whitmire. "Carcinogenicity of Chloroform in Drinking Water to Male Osborne-Mendel Rats and Female B6C3F1 Mice," *Fundam. Appl. Toxicol.* (in press).
38. Withey, J. R., B. T. Collins, and P. G. Collins. "Effect of Vehicle on the Pharmacokinetics and Uptake of Four Halogenated Hydrocarbons from the Gastrointestinal Tract of the Rat," *J. Appl. Toxicol.* 3:249-253 (1983).
39. Newberne, P. M., J. Weigert, and N. Kula. "Effects of Dietary Fat on Hepatic Mixed Function Oxidases and Hepatocellular Carcinoma Induced by Aflatoxin B in Rats," *Cancer Res.* 39:3986-3991 (1979).
40. Trehy, M. L., and T. I. Bieber. "Certain Amino Acids as Probable Precursors of Dihaloacetonitriles in Chlorinated Natural Waters," in *Advances in the Identification and Analysis of Organic Pollutants in Water II*, L. H. Keith, Ed. (Ann Arbor, MI: Ann Arbor Science Publishers, Inc., 1981), pp. 941-975.
41. Oliver, B. G. "Effect of Temperature, pH and Bromide Concentration on the Trihalomethane Reaction of Chlorine with Aquatic Humic Material," in *Water Chlorination: Environmental Impact and Health Effects, Vol. 3,* R. L. Jolley, W. A. Brungs, and R. B. Cumming, Eds. (Ann Arbor, MI: Ann Arbor Science Publishers, Inc., 1979), pp. 141-149.
42. Bull, R. J., J. R. Meier, M. Robinson, H. P. Ringhand, R. D. Laurie, and J. Stober. "Evaluation of the Mutagenic and Carcinogenic Properties of Brominated and Chlorinated Acetonitriles: By-products of Chlorination," *Fundam. Appl. Toxicol.* (in press).
43. Bull, R. J., and M. Robinson. "Carcinogenic Activity of Haloacetonitriles and Haloacetone Derivatives in the Mouse Skin and Lung," Chapter 18, this volume.
44. Sussmuth, R. "Genetic Effects of Amino Acids after Chlorination," *Mutat. Res.* 105:23-28.
45. Scully, F. E., Jr., and M. A. Bempong. "Organic N-chloramines: Chemistry and Toxicology," *Environ. Health Perspect.* 46:111-116 (1982).
46. Stacpoole, P. W., G. W. Moore, and D. M. Kornhauser. "Metabolic Effects of Dichloroacetate in Patients with Diabetes Mellitus and Hyperlepoproteinemia," *New England J. Med.* 298:526-530 (1978).
47. Moore, G. W., L. L. Swift, D. Robinowitz, O. B. Crofford, J. A. Oates, and P. W. Stacpoole. "Reduction of Serum Cholesterol in Two Patients with Homozygous Familial Hypercholesterolemia by Dichloroacetate," *Atherosclerosis* 33:285-293 (1979).
48. Katz, R., C. N. Tai, R. M. Diener, R. F. McConnell, and D. E. Semonick. "Dichloroacetate, Sodium: 3-month Oral Toxicity Studies in Rats and Dogs," *Toxicol. Appl. Pharmacol.* 57:273-287 (1981).
49. "National Cancer Institute Bioassay of 2,4,6-Trichlorophenol for Possible Carcinogenicity," DHEW Publ. No. 79-1711, National Institutes of Health, NCI (1979).
50. Exon, J., and L. Koller. "Co-carcinogenic and Reproductive Effects of Chlorinated Phenols," Chapter 25, this volume.
51. Boutwell, R. K., and D. K. Bosch. "The Tumor-Promoting Action of Phenol and Related Compounds for Mouse Skin," *Cancer Res.* 19:413-424 (1959).
52. Meier, J. R., and R. J. Bull. "Mutagenic Properties of Drinking Water Disinfectants and By-products," Chapter 17, this volume.

53. Crump, K. S., and H. A. Guess. "Drinking Water and Cancer: Review of Recent Epidemiological Findings and Assessment of Risks," *Annu. Rev. Public Health* 3:339–357 (1982).
54. Comstock, G. W. "The Epidemiological Perspective: Water Hardness and Cardiovascular Disease," *J. Environ. Pathol. Toxicol.* 4:925 (1980).
55. Winton, E. F., and L. J. McCabe. "Studies Relating to Water Mineralization and Health," *J. Am. Water Works Assoc.* 62:26–30 (1970).
56. Wilkins, J. R., and G. W. Comstock. "Source of Drinking Water at Home and Site-Specific Cancer Incidence in Washington County, Maryland," *Am. J. Epidemiol.* 114:178–190 (1981).

SECTION III

Epidemiological Considerations

We should keep in mind that epidemiologic studies are comparative studies. An initial task of the epidemiologist is to define populations for comparison, either on the basis of differential mortality or morbidity rates, or of gradients in exposure to known or suspected toxic substances.

Kenneth P. Cantor
*The Epidemiologic Approach
to the Evaluation
of Water-Borne Carcinogens*, 1975

The task of determining whether a particular drinking water is a cancer hazard epidemiologically is a formidable undertaking . . .

Michael Alavanja, Inge Goldstein, and **Mervyn Susser**
*A Case Control Study of Gastrointestinal
and Urinary Tract Cancer Mortality
and Drinking Water Chlorination*, 1977

Dripping water hollows out a stone.
Ovid, 43 B.C. – 17 A.D.

SECTION III

Epidemiological Considerations

CHAPTER **11**

Epidemiologic Considerations for Evaluating Associations Between the Disinfection of Drinking Water and Cancer in Humans

Gunther F. Craun

An epidemiologist is primarily interested in the causes of disease. While experiments can be conducted in human populations to seek cures or preventives for disease, the epidemiologist, in seeking causes of disease, must rely on data obtained from nonexperimental settings, because it is clearly unethical to knowingly cause illness or harm in studies of human populations. Thus, an epidemiologic study is nothing more than an attempt to describe nature, and the essential concern is with comparisons. Illness in an exposed group of individuals is compared with illness in an unexposed group, or the proportion of exposed individuals in a diseased group is compared with the proportion of exposed individuals in a group without the disease. Because of the largely unknown factors that lead to development of disease, an overriding consideration is comparability of the study groups. There is basically a lack of comparability in groups of individuals with and without specific characteristics, and it is essential in epidemiologic studies to collect and analyze data in a comparable manner. In experimental studies, the investigator has the ability to assign exposure at random, and this tends to assure comparability over the long term.

The epidemiologist seeks to make a quantitative statement about the association between exposure and disease, and rates of similar types can be compared. The basic measures are the rate ratio or relative risk and rate difference or attributable risk, and they are generally expressed as a point estimate (a single value) or as an interval estimate (a range of possible values consistent with the data). The appropriateness or accuracy of this measure is dependent on components of study design, data collection, and analysis of epidemiologic data. These must be evaluated for each study, and the primary considerations are (1) precision or lack of random error, and (2) validity or lack of systematic error. Precision is influenced primarily by the size of the study population and the efficiency of information obtained for each individual. An interval estimate is preferable to a point estimate, because it accounts for random variability in the data. A point estimate is usually guaranteed to be wrong, because it is only one point on an infinite scale of values. A point estimate does serve as an anchor point for a given confidence interval and is useful in that context.

Assessing the validity of epidemiologic associations requires a search for potential sources of systematic bias that might have influenced the observed results. Both internal and external validity are important. Internal validity involves the making of a valid inference about the association between exposure and disease in a particular study and must be assured before considering external validity. External validity concerns extending the results of several studies to a target population and is often referred to as scientific generalization or interpretation. The results of epidemiologic studies must be interpreted in the context of other information, and no single study is likely to provide a definitive answer. The interpretation of epidemiologic data must be cautious, as today's facts may become tomorrow's fallacies.[1] The interpretation of epidemiologic data requires an awareness by both epidemiologists and nonepidemiologists of the potential shortcomings of such data, and an assessment of the internal validity of each study must be undertaken. Internal validity has three basic components that should be considered, and each of these will be discussed in more detail later:

1. validity of selection of subjects (selection bias);
2. validity of information (information or observation bias);
3. validity of comparison (confounding bias).

Conscientious investigators will discuss in their publications methods used to prevent selection bias, to minimize observation bias, and to assess, prevent, and/or control confounding bias, and they will discuss how these are likely to influence interpretation of the data.

TYPES OF EPIDEMIOLOGIC STUDIES

To better understand the sources of potential bias that affect internal validity, it is necessary to have a basic understanding of the various types of epidemiologic studies. These are briefly reviewed according to a classification scheme proposed by Monson[1] (Table I). In an experimental study, the investigator randomly assigns exposure to a study participant, whereas in a nonex-

Table I. Types of Epidemiologic Studies[a]

    I. Experimental
   II. Nonexperimental
      A. Descriptive
      B. Analytic
         1. Longitudinal
            a. Cohort or Follow-up
               (1) Prospective
               (2) Retrospective
            b. Case-comparison
         2. Cross-sectional

[a]Reprinted with permission from Monson, R. R., *Occupational Epidemiology*. Copyright CRC Press, Inc., Boca Raton, FL (1980).

perimental study, the investigator has no control over exposure. In a descriptive study, information is available on exposure and disease for groups of persons only, or information is available only on exposure or on disease. In an analytic study, information on exposure *and* disease is available for each individual, and a measure of the association can be obtained. In a longitudinal study, the time sequence between the exposure and disease can be inferred, but in a cross-sectional study, the data on exposure and disease relate to the same point in time. In a case-comparison study, individuals enter the study on the basis of disease and nondisease status, and various exposures are determined. In a cohort study, individuals enter the study on the basis of exposure status. In a prospective cohort study, the disease has not occurred at the time the exposed and nonexposed groups are defined. In a retrospective cohort study, the disease has occurred at the time the exposed and nonexposed groups are defined.[1]

I would classify as descriptive studies all of the early epidemiologic studies that report associations between cancer mortality and surface or chlorinated water, because information was available on disease and exposure in groups of people rather than individuals. These studies have been extensively reviewed in the scientific literature[2-5] and at previous chlorination conferences.[6,7] More recently, variations of traditional case-comparison study designs have been conducted using mortality data from death certificates. These studies were reviewed by Crump and Guess[5] and also by Cantor[7] at the 1981 conference; and only those studies published in the peer-reviewed literature since these reviews will be considered here. These include studies reported by Lawrence et al.,[8] Wilkins and Comstock,[9] and Gottlieb et al.[10]

VALIDITY OF SELECTION OF SUBJECTS

If the criteria used to enroll subjects in a study are not comparable, the data cannot be used to measure an association between exposure and disease because of selection bias. For selection bias to occur, the disease must have taken place prior to enrollment of the subjects in the study; thus, it is important to consider selection bias in all case-comparison and retrospective cohort studies. In a case-comparison study, selection bias results from selective admission of exposed persons into the diseased group, unexposed persons into the diseased group, exposed persons into the comparison group, or unexposed persons into the comparison group. In a retrospective cohort study, selection bias occurs when there is the selective admission of diseased persons into the exposed group or selective admission of nondiseased persons into the unexposed group. Selection bias can be prevented in case-comparison studies if knowledge of exposure is masked in selecting cases and comparison subjects; it can be prevented in retrospective cohort studies if knowledge of disease is masked in selecting cohorts and determining exposure. Selection bias must be prevented; it cannot be controlled.[1]

There must be a difference in selection criteria between the two groups in order for selection bias to result. An inaccurate definition of disease or of exposure that applies equally to the two groups results in random misclassification, and this can only alter the results of a study toward no association between exposure and disease.[1]

VALIDITY OF INFORMATION

If data are collected on two groups using methods that are not comparable, the data contain incorrect information on the association between exposure and disease, because of observation bias. In cohort studies, observation bias results when information on disease outcome is obtained differently for exposed and nonexposed groups. In case-comparison studies, observation bias results when information on exposure is obtained differently for cases and comparison subjects. A way to prevent observation bias in a cohort study is not to know the exposure status of study participants when information on disease is obtained; any errors in measurement should be made equally in members of the exposed and nonexposed groups. In a case-comparison study, no observation bias is possible if neither the patient nor the data collector knows the diagnosis when information on exposure is collected. However, this blindness is not always possible in a case-comparison study. To minimize observation bias in situations where the interviewer or patient knows case status, objectivity is sought in obtaining information. Questions are asked that require objective answers (closed-ended) rather than subjective answers (open-ended). This does not prevent observation bias but tends to minimize it.[1]

VALIDITY OF COMPARISON

If a characteristic exists that is a cause of the disease and is also associated with the exposure and disease, the data relating exposure to disease may convey an appearance of association because of confounding bias. Although negative confounding can also occur, the primary concern is that confounding has led to the erroneous observation of an association. Confounding bias is potentially present in all epidemiologic data, and it must always be considered as the possible explanation for any association seen. Confounding bias does not result from any error of the investigator. It is a basic characteristic of any epidemiologic study.[1]

It is never possible to know all of the effects of confounding bias, but information can be collected on known or suspected confounding characteristics to prevent or control any bias introduced by those particular characteristics. If a characteristic can be made or demonstrated to have no association with exposure or with disease, that characteristic cannot be confounding. Matching is a technique generally employed in the study design to prevent confounding, and stratification or multivariate techniques are employed to assess and control confounding at the time of data analysis.

MEASURE OF ASSOCIATION

As previously noted, it is important to measure the association between exposure and disease, and the usual measure is the rate ratio or relative risk. A rate ratio of unity (1.0) indicates no association; any other ratio indicates some association, either positive or negative. Based on Monson's experience,[1] ranges of the rate ratio may be used to judge the strength of the association (Table II). Any rate ratio between 0.9 and 1.2 indicates essentially no association. This is not to say that the true association is 1.0, but that associations in this range are generally considered too weak to be detected by epidemiologic methods. This means that if the exposure raises the rate of disease by 20% among the exposed, nonexperimental epidemiologic methods are unlikely to detect this association.

In situations where a true association is very weak, the range of variability in any data collected is large relative to the true rate ratio, and, even with large numbers, random variation may lead to an observed rate ratio of 1.0. Similarly, if the true rate ratio is 1.0, an observed rate ratio of 1.2 would not be unexpected because of random variation. Confounding bias is much more likely to influence the interpretation of small rate ratios. There may be one or more confounding factors that lead to a weak association between exposure and disease, and an epidemiologist is limited in his ability to identify and measure such weak confounding factors.

Variability and confounding bias are much less likely to influence the interpretation of a large rate ratio when the association is based on reasonably large numbers. It is generally accepted that if some confounding factor exists that accounts totally for such a strong association, its detection should be relatively simple. The size of a rate ratio, however, has little to do with the possibility that an association could be due to selection bias or observation bias. Either of these forms of bias can lead to a total misrepresentation of the underlying association between exposure and disease.

Table II. A Guide to Strength of Association[a]

Rate Ratio		Strength of Association
0.9 – 1.0	1.0 – 1.2	None
0.7 – 0.9	1.2 – 1.5	Weak
0.4 – 0.7	1.5 – 3.0	Moderate
0.1 – 0.4	3.0 – 10.0	Strong
<0.1	>10.0	Infinite

[a]Reprinted with permission from Monson, R. R., *Occupational Epidemiology*. Copyright CRC Press, Inc., Boca Raton, FL (1980).

It should be noted that the ranges presented as a guide to the strength of an association are clearly arbitrary but are based on experience. However, two general points that Monson[1] suggests to keep in mind are accepted by most epidemiologists:

> "If the true association between exposure and disease is very weak, the detection of such an association by nonexperimental methods is unlikely.
>
> If the true association between exposure and disease is very strong, no epidemiologic expertise is necessary to detect it."

INTERPRETATION OF EPIDEMIOLOGIC DATA SHOWING NO ASSOCIATION BETWEEN DRINKING WATER EXPOSURE AND CANCER

The study by Lawrence et al.[8] reported no association between trihalomethanes (THM) in surface water and cancer mortality. To interpret data reporting no association, it is important to determine whether random misclassification could have resulted in not observing an association when one was actually present. As previously noted, random misclassification results when there is imprecision in classifying study participants as diseased or nondiseased or as exposed or unexposed and tends to alter a rate ratio toward unity (1.0). A rate ratio greater than 1.0 is lessened, while one less than 1.0 is raised; with complete random misclassification the rate ratio is 1.0. Selection, observation, and confounding bias and random variability should also be considered, but random misclassification is always the important consideration when no association is reported in studies that are judged to have sufficient information and to be free of systematic bias.

The relation of THMs to colorectal cancer was evaluated by Lawrence et al.[8] A total of 395 colorectal cancer deaths among white female teachers in New York State was compared with an equal number of deaths of teachers from noncancerous causes. Cumulative chloroform ($CHCl_3$) exposure was estimated by the application of a statistical model to operational records from the individual water treatment facilities that served the home and work addresses of each study subject during the 20 years prior to death. The odds of exposure to a surface source containing THMs was no greater for cases than for controls (odds ratio = 1.07; the 90% confidence interval = 0.79, 1.43). The distribution of $CHCl_3$ exposure was not significantly different between cases and controls as rated by the Wilcoxon signed rank statistic.

No effect of cumulative $CHCl_3$ exposure on outcome was seen in a logistic analysis controlling for average source type, population density, marital status, age, and year of death. Misclassification of outcome was minimized by use of predominantly histologically confirmed cases of colorectal cancer and by searching the state tumor registry for possible cancer cases in the comparison group. Exposure misclassification was minimized by the collection of extensive records on home and work addresses, water treatment service areas, and estimates of exposure to THMs over the 20-year period prior to death; however, it is possible that exposures greater than 20 years may be important, and this was not assessed. There appeared to be no selection or observation bias that might have been responsible for observing no association. To influence the results toward no association, the direction of bias would have to be opposite to the true association, and more complete information would have to have been collected for the comparison group. This seems unlikely.

Potential confounding factors were assessed and controlled either in the study design by matching or in the analysis of the data. As with the other forms of bias, the direction of confounding must be opposite to the true association; for this to bias the results, negative confounding must have occurred. The search for negative confounding is usually of little value.[1] The authors thoroughly discussed possible sources of bias, methods for control, and possible effects on the interpretation of results. Since all measures of association are subject to random variation, it is not possible to completely rule out random variation as a reason for no observed association. The study population was of reasonable size and the confidence interval for the odds ratio is narrow; however, a more convincing case could be made for concluding that THMs in chlorinated surface water are not associated with colorectal cancer if there were two or more studies where a rate ratio of 1.0 was found.

The selection of comparison subjects is one of the most important aspects of a case-comparison study. Comparison subjects should be representative of the population from which cases are derived to avoid the selective admission of persons with specific diseases or exposures. The Lawrence et al. study population was basically homogenous, but most of the comparison group died of cardiovascular disease.[8]

Recent evidence from animal studies suggests that cardiovascular disease may be associated with exposure to chlorine in drinking water where populations are deficient in calcium intake; if this evidence applies to humans, the use of cardiovascular deaths as a comparison would tend to obscure a true association between THMs and cancer mortality. Currently, this possibility should not be ruled out as a possible reason for no observed association.

Evaluation of precision or random variation is the primary consideration in interpreting results of the study conducted by Wilkins and Comstock[9] in Washington County, Maryland, to investigate the postulated relationship between organic chemical by-products of water chlorination and risk of human cancer. Vital records and nonofficial census data available for each of nearly 31,000 study subjects were used to compute selected sex- and site-specific cancer incidence rates in a well-defined county population. Age, socioeconomic status, smoking history, source of drinking water at home, and other individual characteristics of the study population were examined in relation to the cancer rates. Three historical cohorts, each distinguished by a different degree of exposure to chloroform and other chlorination by-products, were studied. Incidence rates for cancer of the bladder (relative risk = 1.80, 95% confidence interval = 0.80, 4.75) among men and for cancer of the liver (relative risk = 1.80, 95% confidence interval = 0.64, 6.79) among women were reported to be nearly twofold higher in the drinking water cohort that had been supplied chlorinated surface water at home when compared with the cohort with a history of consumption of unchlorinated groundwater.

The results of this study cannot be interpreted as a true association, because the reported rates are unstable and subject to random variation. There were few cancer deaths in the cohort, and the confidence intervals reported for the rate ratios are wide. Because the confidence intervals include 1.0, neither can the null hypothesis of no association be rejected. Random misclassification is not likely to have influenced the results toward no association; however, because of the instability of the rates, these results should be interpreted as inconclusive.

INTERPRETATION OF EPIDEMIOLOGIC DATA SHOWING A POSITIVE ASSOCIATION BETWEEN DRINKING WATER EXPOSURE AND CANCER

Selection, observation, and confounding bias and random variability should be considered as the reason(s) for the observed associations in the studies conducted by Gottlieb et al.[10] Case-comparison mortality studies of 17 cancer sites (bladder, colon, kidney, liver, lymphoma, rectum, stomach, Hodgkin's lymphoma, leukemia, lung, malignant melanoma, multiple myeloma, prostate, breast, brain, esophagus, and pancreas) were conducted in 13 parishes (counties) of southern Louisiana. Parishes were grouped for similarities in industrialization and approximately equal exposure to surfacewater and groundwater. Noncancer deaths were randomly selected and matched to cancer deaths on age, gender, year of death, and parish where death occurred.

Associations with exposure to chlorinated surface water (Mississippi River) compared with exposure to groundwater were observed for cancer of the rectum in males (odds ratio = 3.18, 95% confidence interval = 1.96, 5.19) but not in females (odds ratio = 1.73, 95% confidence interval = 0.97, 3.10) when lifetime water use was considered. Lifetime water use was defined as type of water used by cases and comparison subjects at birth and at death, and the analysis was confined to those subjects where surface water was used at both birth and death or groundwater was used at both birth and death. Associations that were not observed for cancer sites such as bladder, colon, kidney, liver, and stomach must be interpreted in light of the previous discussion of random misclassification of exposures, since water exposures for study subjects could not be traced for more than 10 years and lifetime exposures were estimated by classification of water source only at birth and death.

The association seen for cancer of the rectum in males must be interpreted in light of a possible bias in selection of cases and comparison subjects, observation bias, or confounding bias. Because data were available on both disease and exposure when subjects were selected, selection and observation bias were possible. The effect of these possible sources of bias could not be adequately assessed, because control measures were inadequately discussed in the publication.

Confounding bias was controlled to the extent possible for factors that could be readily obtained from death certificates. (Gender, race, age, year of death were controlled by matching, and data were obtained on occupation and ancestry). However, it was not possible to accurately assess individual occupational exposures, and occupation may be a possible confounding bias which contributed to the association observed for cancer of the rectum in males and not observed in females. There is also the possibility that this association was observed because of random variability due to a possible multiple comparison problem. Gottlieb et al.[10] suggested that the multiple comparison problem was minimized, because each case-comparison series was selected independently within the study population, but the degree to which these may be independent observations can be questioned. The studies were conducted within the same small geographic area, and the multiple comparison problem still exists within each series of observations.

SUMMARY

It must be recognized that all epidemiologic data are imperfect, no matter what their source or how they were collected. "There are few absolutes. Few sets of epidemiologic data are worthless, and few convey absolute truth. The task is to decide how to judge the validity of the data."[1] Any judgment must consider not only the potential for bias but also how bias might have contributed to the observed association (would the true association likely be altered toward no observed association, or how likely is the observed association a true association). "The source of the data may not be ideal, but it may be the only source available. The data may not have been collected under absolutely correct circumstances, but the error introduced may be small. The analysis of the data may have been superficial, but further analysis may add little to the understanding of the meaning of the association seen. The interpretation of the meaning of the association must take into account the source of the data, methods of collection of information, and methods of analysis."[1] This is largely a scientific process, but decisions as to the utility of the data are generally based on an independent process that is primarily political.

The large number of epidemiologic studies conducted to date suggest that there may be an association between exposure to surface waters that have been chlorinated and mortality due to cancer at one or more sites. However, these studies have done little to clarify the specific association(s) and the possible magnitude of the association.

In a case-comparison study of 2982 incident cases of bladder cancer and 5782 population-based comparison subjects from ten areas of the United

States, a relative risk of 3.1 (confidence interval = 1.5, 6.5) was observed by Cantor[11] in nonsmokers who had used chlorinated surface water for 60 or more years compared with nonsmokers who had never used surface or chlorinated water.

In a case-comparison study of 200 incident cases of colon cancer and 407 hospital-comparison subjects among North Carolina white residents, Cragle[12] observed odds ratios of 1.38, 2.15, and 3.36 for home consumption of chlorinated water for 16 or more years and colon cancer in 60, 70, and 80 year olds. These studies are suggestive of an association between bladder and colon cancer and consumption of chlorinated water over a long period in nonsmokers and over a moderate period in an elderly population, but additional case-comparison studies are required before conclusions can be extended to other populations. Retrospective cohort studies should also be conducted where suitable cohorts can be assembled and followed. Both of these study designs, however, require an accurate assessment of past exposure to contaminants of interest for a sufficient length of time (which may be greater than 20 years) to minimize random misclassification.

ACKNOWLEDGMENT

This paper was based largely on concepts and principles discussed by Monson either in his lectures or in his book,[1] and no claim is made by the author for the originality of these principles and concepts. In several instances the statements are paraphrased or taken directly from Monson's work, as I know of no better way to present or state these principles. The reviews of the published works[8-10] are, however, solely the opinions of the author.

REFERENCES

1. Monson, R. R. *Occupational Epidemiology* (Boca Raton, FL: CRC Press, Inc., 1980).
2. Wilkins, J. R., III, N. A. Reiches, and C. W. Kruse. "Organic Chemicals in Drinking Water and Cancer," *Am. J. Epidemiol.* 110:420–448 (1979).
3. National Academy of Sciences. *Drinking Water and Health*, Vol. 3 (Washington, DC: National Academy of Sciences, 1980), pp. 5–21.
4. Hoel, D. G., and K. S. Crump. "Waterborne Carcinogens: A Scientist's View," in *The Scientific Basis of Health and Safety Regulation*, R. W. Crandall and L. B. Love, Eds. (Washington, DC: Brookings Inst., 1981), pp. 1973–1995.
5. Crump, K. S., and H. A. Guess. "Drinking Water and Cancer: Review of Recent Epidemiological Finding and Assessment of Risks," *Ann. Rev. Public Health* 3:339–357 (1982).
6. Shy, C. M., and R. J. Struba. "Epidemiologic Evidence for Human Cancer Risk Associated with Organics in Drinking Water," in *Water Chlorination: Environmental Impact and Health Effects, Vol. 3*, R. L. Jolley, W. A. Brungs, and R. B.

Cumming, Eds. (Ann Arbor, MI: Ann Arbor Science Publishers, Inc., 1980), pp. 1029–1042.
7. Cantor, K. P. "Epidemiologic Studies of Chlorination By-Products in Drinking Water: An Overview," in *Water Chlorination: Environmental Impact and Health Effects, Vol. 4*, R. L. Jolley, W. A. Brungs, J. A. Cotruvo, R. B. Cumming, J. S. Mattice, and V. A. Jacobs, Eds. (Ann Arbor, MI: Ann Arbor Science Publishers, Inc., 1983), pp. 1381–1397.
8. Lawrence, C. E., P. R. Taylor, B. J. Trock, and A. A. Reilly. "Trihalomethanes in Drinking Water and Human Colorectal Cancer," *J. Nat. Cancer Inst.* 72:563–568 (1984).
9. Wilkins, J. R., and G. W. Comstock. "Source of Drinking Water at Home and Site-Specific Cancer Incidence in Washington County, Maryland." *Am. J. Epidemiol.* 114:178–190 (1981).
10. Gottlieb, M. A., J. K. Carr, and J. R. Clarkson. "Drinking Water and Cancer in Louisiana," *Am. J. Epidemiol.* 116:652–667 (1982).
11. Cantor, K. P., et al. "Drinking Water Source and Risk of Bladder Cancer," Chapter 12, this volume.
12. Cragle, D. L., C. M. Shy, R. J. Struba, and E. J. Siff. "A Case-Control Study of Colon Cancer and Water Chlorination in North Carolina," Chapter 13, this volume.

CHAPTER 12

Drinking Water Source and Risk of Bladder Cancer: A Case-Control Study

Kenneth P. Cantor, Robert Hoover, Patricia Hartge,
Thomas J. Mason, Debra T. Silverman, and Lynn I. Levin

Chlorine reacts with naturally occurring organic compounds in water treatment plants to produce halogenated by-products, including chloroform and other trihalomethanes. Among these compounds, chloroform is a carcinogen,[1] others are mutagenic in bacterial tester strains,[2] and some concentrated mixtures of higher-molecular-weight organic fractions transform mammalian cells in tissue culture[3] and induce skin-painted tumors in rodents.[4] This raises the possibility that the time-tested benefits of chlorine to control infectious disease may be, in part, offset by increased cancer risk in continuously exposed populations.

Some drinking water supplies are also contaminated by other potentially toxic agents from industrial or municipal outfalls, agricultural and municipal runoff, or toxic waste dumps. While organic chemicals such as trichloroethylene, benzene, or perchloroethylene, and cations such as cadmium or arsenic, pose important public health threats to some exposed populations, affected water supplies are usually limited geographically and temporally. This does not minimize the threat posed by such contaminants in some local water supplies nor the threat of future problems in many others. In contrast, chlorination by-products have been widespread, and their historical distribution patterns can be deduced from knowledge of water sources and disinfection practices used in the past.

Several epidemiologic studies have evaluated the possibility of a link between drinking water contaminants and cancer in human populations. The first studies were ecologic in design. More recent work used a case-control approach based on death certificates. Many of these studies linked bladder cancer mortality with exposure indicators—water source type and treatment—that were used as surrogates of contaminant levels.[5-7] Among these surrogate measures are surface (as contrasted with ground) source, chlorinated (as compared with nonchlorinated) source, and recent measures of chloroform concentration. To further pursue these observations, we incorporated a water source component in a large population-based case-control interview study of bladder cancer designed at the National Cancer Institute (NCI) in 1977.

Eligible cases included all persons between the ages of 21 and 84 diagnosed with cancer of the urinary bladder in 1978 and residing in ten areas of the United States, including Connecticut, Iowa, New Jersey, New Mexico, and Utah, and the metropolitan areas of Atlanta, Detroit, New Orleans, San Francisco, and Seattle. Slightly less than a third of the cases were from New Jersey. Connecticut, Iowa, Detroit, and San Francisco each accounted for more than 10%. Bladder cancer is primarily a disease of older men. There were three times as many men as women cases, and the median age was 67. A total of 2982 cases, 73% of the eligible pool, were interviewed. Controls were randomly selected from the population of each area, frequency matching on sex, 5-year age group, and study area. Controls between 21 and 64 years of age were selected by a random-digit dialing method, and controls between 65 and 84 years of age were randomly selected from a roster provided by the Health Care Financing Agency; 5782 population-based controls were interviewed. Details of the study design and methods have been published.[8]

Cases and controls were interviewed at home by trained interviewers. Items on the questionnaire included demographic background; a smoking, occupational, and medical history; artificial sweetener use; and other factors possibly linked to bladder cancer, including hair dyes, coffee and tea consumption, and fluid ingestion. Each respondent was also asked to name each city or town in which he or she had lived for a year or more, the years moved into and out of that place, and whether the primary source of drinking water at each place was a private well, the community water supply, bottled water, or another source. We coded geographic areas by a standard coding scheme.[9]

In collaboration with the Cincinnati Health Effects Research Laboratory of the Environmental Protection Agency, we independently surveyed all community water supplies that served more than 1000 persons in the ten study areas. We collected historical information on water source, treatment, and geographic distribution since 1900. Water sources were classified as surface or ground, and further details on source characteristics and potential contamination were recorded. Treatment information, especially chlorination, was also gathered. Although details on amounts of added chlorine were often lacking, we were able to ascertain the years in which chlorination disinfection had been used. The towns and cities historically served by each water source were listed and coded with the same geocoding scheme used for residential histories.

A year-by-year record of water source and treatment was created for each study respondent. For each year that a respondent lived in one of the ten study areas and used a community supply, we looked up water source and treatment information in the water supply data file. We were not able to describe water source for years when respondents used community sources outside of the study areas or when they lived in very small communities with supplies not covered by our survey.

Of the 587,565 person-years lived by all respondents since 1900, 444,735 (76%) were at a known water source. This ranged from 63% in New Mexico to 83% in Iowa. The year-by-year profile of water source and treatment information for each person provided us with a flexible tool to look at patterns of water use in the study population as well as to define individual exposures.

We have estimated relative risk according to several different measures derived from drinking water histories. Here we report on risk as related to the number of years that a respondent lived at a residence served by a chlorinated surface source. Most chlorinated surface sources have much higher levels of chlorination by-products than most chlorinated or nonchlorinated groundwater sources;[10] therefore, duration of exposure is a crude index of dose.

We used the odds ratio to estimate the relative risk. Logistic regression for unmatched data was used to obtain a maximum likelihood point and 95% confidence interval estimates of the odds ratio, and also to control for the potential confounding effects of selected variables.[11,12] Among the potential confounders in most calculations were geographic area (ten levels), six levels of cigarette smoking intensity, three age groups, a 1/0 variable for usual employment as a farmer, and race and sex when the analyses were not race- or sex-specific.

Table I shows the overall relative risk among whites by the number of years at a chlorinated surface water source. All risks are relative to those who lived at places never served by such sources. Compared to this base-line measure are respondents with less than 20, 20 to 39, 40 to 59, and 60 or more years at places with chlorinated surface sources. Relative risks are not elevated in the exposed groups, and there is no suggestion of a duration-response relationship.

Eligibility for inclusion in the analyses reported here was restricted in two major ways: (1) Preliminary analyses suggested a potential for confounding of drinking water associations by employment in a high-risk occupation for bladder cancer (as identified by D. Silverman, L. Levin, and R. Hoover at NCI). As a control, we included only persons who never held a high-risk job. (2) Among some persons in low-exposure categories (i.e., those with few years known to be served by a chlorinated surface source), there was uncertainty as to exposures during years they were not known to be served by a chlorinated surface source. Some of these years were classified as "municipal, not otherwise specified," or "unknown," and, in such cases, assignment to a more precise exposure category was not possible. We wished to remove from the analysis those persons whose exposures were least certain. In addition, we desired to maximize the number of nonexposed years spent at nonchlorinated ground sources (low exposure) while still including enough subjects to maintain adequate statistical power. To these ends, we further reduced the analysis population to those whose years at a chlorinated surface source plus years at a nonchlorinated ground source summed up to at least half of their lifetimes.

Interesting differences in relative risk for duration of exposure to chlorinated surface waters are observed within geographic regions (Table II). Elevated relative risks are seen in New Mexico, Utah, and Iowa, with the number of years resident at a place served by a chlorinated surface source. The number of respondents in New Mexico and Utah is small, but the result is statistically significant. The pattern in Iowa is also interesting and is apparent among

Table I. Relative Risks (RR) and 95% Confidence Intervals (CI) for Bladder Cancer According to Number of Years at a Residence Served by a Chlorinated Surface Drinking Water Source[a]

Years	RR	95% CI	Number Cases	Number Controls
0	1.0		231	570
1–19	1.1	0.8–1.4	141	285
20–39	1.0	0.8–1.3	324	650
40–59	1.0	0.8–1.3	437	849
60+	1.1	0.8–1.5	111	196

[a]Whites, from logistic regression adjusted for study area (10 strata), sex, age (3 strata), smoking level (6 strata), and usual employment as a farmer (2 strata).

Table II. Relative Risks (RR) and 95% Confidence Intervals (CI) for Bladder Cancer According to Number of Years at a Residence Served by a Chlorinated Surface Drinking Water Source[a]

Study area	Years	RR	95% CI	Number Cases	Number Control
San Francisco, Seattle	0	1.0		20	26
	1–19	0.7	0.3–2.0	16	37
	20+	1.05	0.4–2.7	163	332
New Mexico, Utah	0	1.0		15	75
	1–19	4.5	1.03–19.5	6	8
	20+	11.8	2.5–55.1	8	5
New Orleans, Atlanta	0	1.0		5	8
	1–19	0.2	0.03–1.1	5	23
	20+	0.4	0.1–1.9	70	156
Iowa	0	1.0		111	323
	1–19	1.04	0.6–1.8	26	64
	20+	1.6	0.94–2.7	35	55
Detroit	0	1.0		6	9
	1–19	1.4	0.4–5.4	12	15
	20+	1.2	0.4–3.6	197	278
Connecticut	0	1.0		17	53
	1–19	1.7	0.8–3.4	36	64
	20+	1.17	0.7–2.1	158	378
New Jersey	0	1.0		57	76
	1–19	0.7	0.4–1.3	40	76
	20+	0.7	0.5–1.1	241	491

[a]Whites, from logistic regression adjusted for sex, age (3 strata), smoking level (6 strata), and usual employment as a farmer (2 strata).

Table III. Relative Risks (RR) and 95% Confidence Intervals (CI) for Bladder Cancer According to Number of Years at a Residence Served by a Chlorinated Surface Drinking Water Source (by cigarette smoking status)[a]

Cigarette smoking status	Years	RR	95% CI	Number Cases	Controls
Never smoked	0	1.0		61	268
	1–19	1.3	0.7–2.2	29	110
	20–39	1.5	0.9–2.4	73	236
	40–59	1.4	0.9–2.3	108	348
	60+	2.3	1.3–4.2	46	77
Past smokers	0	1.0		83	193
	1–19	1.0	0.7–1.7	49	104
	20–39	1.1	0.7–1.8	115	228
	40–59	1.2	0.8–1.8	163	290
	60+	0.8	0.5–1.5	38	82
Current smokers	0	1.0		87	109
	1–19	0.9	0.6–1.5	63	71
	20–39	0.7	0.4–1.1	136	186
	40–59	0.7	0.5–1.2	166	211
	60+	0.6	0.3–1.2	27	37

[a]Whites, from logistic regression adjusted for study area (10 strata); sex; age (3 strata); usual employment as a farmer (2 strata); 2 smoking levels for past smokers, and 3 smoking levels for current smokers.

smokers and nonsmokers. It may be important that these three areas are the most intensely agricultural of the study areas. In New Jersey, the overall relative risk was less than 1.0, and risk decreased with the number of years that a chlorinated surface source was used. We cannot explain this observation by confounding with other risk factors for bladder cancer that we have evaluated. Associations of bladder cancer with contaminated groundwater in New Jersey is one issue that deserves exploration.

Table III shows results of analysis by major smoking category, that is, nonsmokers, former smokers, and current smokers. Among nonsmokers, the relative risk generally increases with the number of years at a residence with a chlorinated surface source. The relative risk among those with the longest exposure, 60 or more years, is 2.3, and the increase is statistically significant. Former smokers show an uneven risk pattern, and the risk among current smokers appears to vary inversely with the number of years exposed. The unusual and unexpected inverse risk pattern among current smokers could not be explained by confounding by smoking level within broad smoking category or by several other factors that we investigated, including age. Among nonsmokers, there are similar risk patterns for each sex, with the relative risk increasing with the number of years at a surface source, rising to 2.2 in men and 2.5 in women in the longest exposure category. Former smokers show divergent patterns for the sexes. Current smokers of both sexes show inverse risk patterns with the number of years exposed.

When evaluating these results, some limitations must be considered. The study had its origins in the issue of saccharin as a human bladder carcinogen, and, therefore, study areas were not selected with the water source hypothesis in mind.[13] Some of the places, notably the five metropolitan areas, are predominantly served by one water source. This limits intraregional variability of exposure in some areas and dampens the statistical power of our large numbers. The exposure measure used here simplifies a complex world by dichotomizing water sources into chlorinated surface and nonchlorinated ground. While based on extensive environmental information, this classification ignores other differences among sources and may completely misclassify exposure in places with contaminated groundwater.

The detection of relatively small risk differences expected in environmental epidemiologic studies is a challenging task, and great care must be taken to minimize bias and account for risk factors that may confound the result.[14] Given the stringent study design and its careful execution, it is unlikely that bias from case or control selection, or from differentially conducted interviews of cases and controls, has influenced our findings.[8] Information on factors that remain unknown or that are now thought to possibly influence bladder cancer risk were not available and could have confounded the results. Although unlikely, confounding could occur if drinking water source is correlated with ingestion of beta-carotene, retinol, or other micronutrients that may behave as tumor promotors or anticarcinogens.

Our finding is that there is no overall elevation in bladder cancer risk among persons who have lived at places with chlorinated surface water as compared with those who have lived at places with nonchlorinated groundwater; and no dose response is observed. Among the 10 study areas, respondents from three places with agricultural land use show elevated risk for bladder cancer with the number of years at a surface source. This is in contrast with decreased risk in the largest study area, New Jersey. Cigarette smoking is a well-known bladder cancer risk factor.[15] Among smokers, there is an unexplained negative association with the number of years at a chlorinated surface source that is also consistent in the sexes. The pattern among former smokers is variable. Among nonsmokers who never were employed in a high-risk occupation, a group otherwise at low risk for bladder cancer, the risk is elevated among those served by chlorinated surface sources; there is a duration-response relationship, and the pattern is similar in men and women.

Although the overall result is reassuring, these findings raise questions warrant further elaboration. Although smokers within each water exposure category are at higher risk for bladder cancer than nonsmokers, the risk of bladder cancer among smokers with lifetime exposure to nonchlorinated groundwater appears to be higher than that among smokers with chlorinated surface drinking water. Whether this is due to confounding by unmeasured risk factors, such as chance or biological interaction, is not currently understood. Geographical differences in bladder cancer risk patterns suggest the possibility of water contaminants in agricultural areas or groundwater contaminants in New Jersey. An increasing risk of bladder cancer with duration of exposure to chlorinated surface water is observed among persons otherwise at lowest risk and is consistent across the sexes. An analytical study with limited statistical power has also noted positive associations of bladder cancer risk with the use of chlorinated surface water sources.[16] A causal interpretation of these findings would be strengthened by similar observations in another setting.

REFERENCES

1. Page, N. P., and U. Saffiotti. *Report on Carcinogenesis Bioassay of Chloroform*, (Bethesda, MD: National Cancer Institute, Division of Cancer Cause and Prevention, 1976).
2. Loper, J. C. "Mutagenic Effects of Organic Compounds in Drinking Water," *Mutat. Res.*; *Rev. Genet. Toxicol.* 76:241-267 (1980).
3. Land, D. R., H. Kurzepa, M. S. Cole, J. C. Loper. "Malignant Transformation of BALB/3T3 Cells by Residue Organic Mixtures from Drinking Water," *J. Environ. Pathol. Toxicol.* 4:41-54 (1980).
4. Robinson, M., J. W. Glass, D. Cmehil, R. J. Bull, and J. G. Orthoefer. "Initiating and Promoting Activity of Chemicals Isolated from Drinking Waters in the SENCAR Mouse—a Five City Survey," in *Short-Term Bioassays in the Analysis of Complex Environmental Mixtures, II (Environmental Science Research, Vol. 22)*, M. D. Waters et al., Eds. (New York: Plenum Press, 1980), pp. 177-188.
5. National Academy of Sciences-National Research Council Assembly of Life Sciences. *Drinking Water and Health, Vol. 3* (Washington, DC: National Academy of Sciences, 1980).
6. Wilkins, J. R., III, N. A. Reiches, and C. W. Kruse. "Organic Chemicals in Drinking Water and Cancer," *Am. J. Epidemiol.* 110:420-448 (1979).
7. Crump, K. S., and H. A. Guess. "Drinking Water and Cancer: Review of Recent Epidemiological Findings and Assessment of Risks," *Ann. Rev. Public Health* 3:339-57 (1982).
8. Hartge, P., J. I. Cahill, D. West, M. Hauck, D. Austin, D. Silverman, and R. Hoover. "Design and Methods in a Multi-Center Case-Control Interview Study," *Am. J. Public Health* 74:52-56 (1984).
9. U.S. General Service Administration, Office of Finance. *Worldwide Geographic Location Codes* (Washington, DC: U.S. Government Printing Office, 1981).
10. Symons, J. M. et al. "National Organics Reconnaissance Survey for Halogenated Organics," *J. Am. Water Works Assoc.* 667:634-647 (1975).
11. Breslow, N. E., and N. E. Day. *Statistical Methods in Cancer Research: Vol. I. The Analysis of Case-Control Studies* (Lyon: IARC, Sci. Publ. Issue 32, 1980).
12. SAS Institute. *SAS User's Guide: Statistics* (Cary, NC: SAS Institute, 1982), pp. 257-286.

13. Hoover, R. N. et al. "Artificial Sweeteners and Human Bladder Cancer," *Lancet* 1:837–840 (1980).
14. Cantor, K. P. "Epidemiologic Studies to Estimate Effects of Low-level Exposures," in *Methods of Estimating Risk of Chemical Injury: Human and Non-human Biota and Ecosystems*, V. B. Vouk, et al. Eds. (Geneva: Wiley & Sons, 1984), pp. 303–325.
15. Matanoski, G. M., and E. A. Elliott. "Bladder Cancer Epidemiology," *Epidemiol. Rev.* 3:203–229 (1981).
16. Wilkins, J. R., and G. W. Comstock. "Source of Drinking Water at Home and Site-Specific Cancer Incidence in Washington County, Maryland," *Am. J. Epidemiol.* 114:178–90 (1981).

CHAPTER 13

A Case-Control Study of Colon Cancer and Water Chlorination in North Carolina

Donna L. Cragle, Carl M. Shy, Robert J. Struba, and Edward J. Siff

Ecologic studies of water chlorination and cancer occurrence in a number of different populations have shown elevated rates of colon or rectal cancer mortality associated with surface or chlorinated water supplies.[1-3] These consistent observations strengthen the probability that a real association exists; however, they fall short of establishing a causal relationship due to the inherent drawbacks of the ecological approach. Case-control studies[4-5] have produced small but significant odds ratios for the association between colon cancer and water chlorination. These case-control studies used the place of last residence, which was listed on the death certificate, to assign the exposure variable; therefore, water type used at the home at the time of death was the only measure of exposure available.

The purpose of this study was to investigate the relationship between colon cancer and water chlorination using incident cases of colon cancer and measuring exposure through 25-year residence histories obtained during interviews with the study subjects.

METHODS

All incident colon cancer cases (International Classification of Diseases, Eighth Revision, Codes 153.0–154.0) between September 1, 1978, and May 31, 1980, were ascertained from seven hospitals in North Carolina. One or both of the following sources were used at each hospital to generate a list of potential cases: (1) hospital diagnostic index and (2) tumor registry.

Medical records were reviewed for each potential case to ensure that the person had been diagnosed as a new case of colon cancer between the dates of the study. Also, the medical record was reviewed for evidence of a recent move to North Carolina. Since water chlorination was one of the major exposure variables to be measured, and also a complex variable to measure, it was decided that greater accuracy would result from assessing water chlorination thoroughly for the state of North Carolina, and for no other state. Therefore, the case had to be a resident of North Carolina for at least 10 years to be considered for the study. The medical record was reviewed to locate a pathology report that confirmed the diagnosis as a primary cancer of the colon. If the colon cancer case met the time-of-diagnosis and residence requirements, and if the cancer represented a primary tumor, the case was entered into the study.

Two controls were selected for each case and three controls were selected for every eighth case. The variable matching scheme was used to overcome an anticipated higher participation refusal rate in the controls. The controls were chosen to be the persons having the closest admission date to the date of diagnosis of the case (backward or forward in time), who matched on (1) age, (2) race, (3) sex, (4) vital status (living vs dead), and (5) hospital. Since the information for this study was collected largely by interview, no controls were chosen who had a discharge diagnosis of mental disorder. Additionally, the control had to fulfill the 10-year residence requirement, never have had any cancer, and not have had any history of familial polyposis, ulcerative colitis, adenomatous polyposis, or any other major chronic intestinal disorder.

The study subjects were allowed to choose whether they preferred to be interviewed by telephone or to complete a mailed questionnaire. The telephone interviews were conducted by a trained interviewer, and the quality of the interview was judged by the interviewer immediately following completion. Questionable interviews were reviewed to determine whether the information that had been collected was complete enough to include the person in the study. All mailed, self-completed questionnaires were reviewed upon receipt, and questionable responses were clarified by telephone contact with the study subject.

Water exposure was trichotomized on the basis of the following classification scheme: (1) groundwater, no chlorination; (2) groundwater, chlorination; and (3) surface water, chlorination. The residence history obtained in the questionnaire was used to establish which water type a person consumed. All persons who had private wells were in category 1. Persons reporting municipal water supplies at any of their past residences were assigned to a category after query of the local water treatment plant to establish water source and treatment procedures during the time period in question. Also, the company was asked to verify service to the address in question. If the company was not able to verify service to the residence, they were asked to supply the name of the company that might serve the residence, and that company was similarly contacted.

The exposure of every study subject was measured as the number of years during 1953–1978 that the person was exposed to each water type at his or her residence.

Descriptive statistics, correlation coefficients, and stratified analyses were used to investigate the nature of the data to be analyzed. This was followed by a logistic regression using PROC LOGIST, a Statistical Analysis System[6] procedure that fits the logistic multiple regression model to a single binary dependent variable. The regression was performed for the entire group, using cancer as the dependent variable and chlorination as the exposure variable, and the best model was formed.

RESULTS

Less than 7% of the cases and controls were found to have consumed chlorinated wellwater at their residence during the study period. For the purposes of analysis, these study subjects were included in the chlorinated surface water group and the exposure variable was changed to dichotomous (chlorinated vs nonchlorinated water).

Table I displays characteristics of the cases and controls by sex. The average years of schooling was less than a high school education, and cases and controls were similar in years of education.

The average number of years of living in specific regions of the state was calculated only for the study period from 1953 to 1978. Male and female cases and controls showed a similar distribution. However, the females tended to

Table I. Study Population Characteristics[a]

	White Females		White Males	
Variable	Cases N = 107	Controls N = 225	Cases N = 93	Controls N = 182
Average years of schooling	10.9	10.3	10.7	10.1
Average years living in				
Mountain	5.1	4.9	2.8	3.5
Piedmont	13.4	13.2	13.0	12.4
Coast	5.3	5.5	8.2	8.3
Place of birth				
North Carolina	94	177	73	149
	(87.9)[a]	(78.7)	(78.5)	(81.9)
Outside North Carolina	12	46	20	29
	(11.2)	(20.4)	(22.5)	(15.9)
Don't know	1	2	0	4
	(0.09)	(0.09)	(0.0)	(2.2)
Number dead at interview	37	50	35	45
	(34.6)	(22.2)	(37.6)	(24.7)
Type of residence most of study				
Urban	57	101	52	93
	(53.5)	(44.9)	(55.9)	(51.1)
Rural	46	112	32	83
	(43.0)	(50.0)	(34.4)	(45.6)
Mixture	4	12	9	6
	(3.7)	(5.3)	(9.7)	(3.3)
Average age at index admission	64.7	63.3	66.4	64.0
Age range	21–91	15–97	23–95	22–88
Average age at death	66.2	73.9	68.4	67.4
Age range at death	21–91	43–99	31–89	43–84

[a]Numbers in parentheses are percentages.

have lived longer in the mountains and for less time at the coast than the males. The majority of the cases and controls in each group were born in the state of North Carolina.

Both sex groups began with equal proportions of living and dead study subjects, yet a higher proportion of cases were dead at the time of the interview than controls. This is consistent with a higher mortality rate in the persons diagnosed with colon cancer.

Type of residence during most of the study was derived by determining whether the study subject had lived 16 or more years in a rural or urban setting. If they lived equally in the two types of environment, they were placed in the "mixture" category. The cases lived more frequently in the city than the controls.

The average age at index admission was similar for cases and controls. Age at death was a matching variable initially, but deaths also occurred during the identification of the study sample and prior to the actual interview. Therefore, it is not expected that the average age at death would be comparable between the cases and the controls.

A tabulation of discharge diagnoses for the controls by sex revealed that diseases of the circulatory system were the one category with the largest percentage of the controls. Nearly 27% of the female controls and 39% of the male controls were in this group.

The stratified analysis, correlation coefficients, and literature review indicated that the following variables should be considered in the logistic regression analysis: alcohol consumption, genetic risk, diet, region of North Carolina, urban residence, education, and number of pregnancies. The coding of the variables is detailed in Table II. Every model that was considered included the variables of sex and age, because these were matching variables in the original study design. The final model is presented in Table III.

Variables positively associated with colon cancer are genetic risk, a product term between alcohol consumption and high-fat diet, and an interaction term between age and chlorination. Smoking and number of pregnancies are negatively associated with colon cancer. This negative association indicates that the probability of developing colon cancer decreases with increasing smoking measured in packs per day. Also, for females, the probability of developing colon cancer decreases with increasing number of pregnancies.

The two most significant variables in the model are genetic risk and the pregnancy term. Next in significance is the interaction term between age and chlorination, followed by a product term between alcoholic drinks per day and high-fat diet (regardless of fiber consumption). The least significant variable in the model is smoking, with a χ^2 of 5.33. The significance of the single terms for chlorination, age, high-fat diet, and drinks per day cannot be evaluated from the χ^2 value because of the inclusion of these variables in product or interaction terms. The likelihood ratio test was used to compare the model illustrated in Table III with a model that did not contain the chlorination term or the age/chlorination interaction term. The result of this test was a χ^2 of 13.61 with 2 degrees of freedom ($p = 0.0001$).

Table II. Variables Considered in Logistic Regression Analysis

Exposure variable	
Chlorination	0 = No chlorinated water at residence between 1953 and 1978
	1 = 1 to 15 years of chlorinated water
	2 = 16 to 25 years of chlorinated water
Possible confounders	
Sex	0 = Male
	1 = Female
Age	Age on 9/1/78
Genetic risk	Score derived from the number of first-degree relatives with cancer
Diet (entered as a group)	
Low fiber, high fat	0,1
High fiber, high fat	0,1
High fiber, low fat	0,1
Regions (entered as a group)	
Mountain	1 if lived in this region more than 15 years between 1953 and 1978
Piedmont	
Coast	
Urban residence	1 if lived in an urban residence more than 15 years between 1953 and 1978
Smoking	Packs per day
Alcohol consumption	0 = No drinks
	1 = Social drinking
	2 = 1 or more drinks per day
Education	Years of schooling (for women this is years of schooling of spouse)
Number of pregnancies	Number of pregnancies reported

Table III. Logistic Regression Model for Colon Cancer[a]

Variable	Coefficient	Standard error	χ^2
Intercept	−0.0254	0.7015	0.00
Chlorination	−1.1705	0.5365	4.76[b]
Confounders			
Sex	0.2670	0.2540	1.11
Age	−0.0100	0.0105	0.92
High fat diet	−0.4008	0.2890	1.92
Genetic risk	0.2407	0.0715	11.32[c]
Alcohol	−0.0937	0.1842	0.26
Smoking	−0.3284	0.1423	5.33[b]
Number of pregnancies	−0.1947	0.0583	11.17[c]
Alcohol-high fat diet	0.7634	0.3049	6.27[b]
Interaction			
Age-chlorination	0.0222	0.0081	7.41[d]

[a]Model χ^2 = 54.62 with 10 degrees of freedom (p = 0.0001).
[b]p < 0.05.
[c]p < 0.001.
[d]p < 0.01.

Table IV. Chlorination Odds Ratio by Age, Derived from the Logistic Regression Model (Table III)

Age	Length of Exposure (years)	OR	95% CI
20	<15	0.23	(0.11, 0.49)
	>15	0.48	(0.23, 1.01)
30	<15	0.36	(0.20, 0.66)
	>15	0.60	(0.33, 1.09)
40	<15	0.57	(0.36, 0.88)
	>15	0.75	(0.48, 1.18)
50	<15	0.89	(0.83, 1.21)
	>15	0.94	(0.69, 1.29)
60	<15	1.18	(0.94, 1.47)
	>15	1.38	(1.10, 1.72)
70	<15	1.47	(1.16, 1.84)
	>15	2.15	(1.70, 2.69)
80	<15	1.83	(1.32, 2.53)
	>15	3.36	(2.41, 4.61)

Table IV displays the odds ratios and confidence intervals for different ages, computed from the logistic model. It is necessary to specify an age because of the interaction between the chlorination variable and age. The odds ratios increase with age. Within each age group, the odds ratios increase with increasing duration of exposure.

SUMMARY

Consumption of chlorinated water is strongly associated with cancer of the colon in this study. The relationship is highly dependent on age. There is no relationship between chlorination and colon cancer below the age of 60. Above age 60, there is a statistically significant relationship between chlorination and colon cancer, controlling for possible confounders. The odds ratios for persons who drank chlorinated water at their home for 16 years or more are consistently higher than those for 15 years or less, estimated for every age group; however, the 95% confidence intervals overlap in all cases.

ACKNOWLEDGMENT

This research was supported by Public Health Service Grant 5-RO1ES-2232 from the National Cancer Institute.

REFERENCES

1. De Rouen, T. A., and J. E. Diem. "Relationships Between Cancer Mortality in Louisiana, Drinking Water Source and Other Possible Causitive Agents," in *Origins of Human Cancer, Book A.*, H. H. Hiatt, et al., Eds. (Cold Spring Harbor, ME: Cold Spring Harbor Laboratory, 1977), pp. 331-345.
2. Page, T., R. H. Harris, and S. S. Epstein. "Drinking Water and Cancer Mortality in Louisiana," *Science* 193:55-57 (1976).
3. Hogan, M. D., et al. "Association Between Chloroform Levels in Finished Drinking Water Supplies and Various Site-Specific Cancer Mortality Rates," *J. Environ. Pathol. Toxicol.* 2:873-887 (1979).
4. Struba, R. J. *Cancer and Drinking Water Quality*, Doctoral dissertation (Chapel Hill, NC: University of North Carolina, 1979).
5. Alavanja, M., I. Goldstein, and M. Susser. "A Case-Control Study of Gastrointestinal and Urinary Tract Cancer Mortality and Drinking Water Chlorination," in *Water Chlorination, Environmental Impact and Health Effects, Vol. 2*, R. L. Jolley, H. Gorchev, and D. H. Hamilton, Jr., Eds. (Ann Arbor, MI: Ann Arbor Science Publishers, Inc., 1978), pp. 395-409.
6. Harrell, F. "The LOGIST Procedure," in *SAS Supplemental Library User's Guide*, P. S. Reinhardt, Ed. (Cary, NC: SAS Institute, Inc., 1980).

CHAPTER **14**

Reactions of Chlorine in Drinking Water, with Humic Acids and In Vivo

F. C. Kopfler, H. P. Ringhand, W. E. Coleman, and J. R. Meier

Disinfection of drinking water with chlorine is known to produce a variety of chlorinated derivatives by reaction of the chlorine with the organic substances found in the source water.[1-4] Substantial evidence has been accumulated that humic substances, which constitute approximately 30 to 50% of the dissolved organic carbon in water,[5] are the principal precursors of trihalomethanes (THMs) and numerous other chlorinated organic species that are present in drinking water as a result of water chlorination.[6-9] Recent studies from our laboratory have demonstrated that the chlorination of both fulvic and humic acids results in the formation of direct-acting mutagenic chemicals.[10-11] The similarity of mutagenic compounds identified in chlorinated drinking water and in these chlorinated solutions of humic materials suggests that the reaction of chlorine with natural aquatic humic material is a likely source of mutagen formation in drinking water.[9] This chapter describes our work to date on the characterization and identification of reaction products resulting from the chlorination of humic material.

Additional studies were conducted to determine the potential for in vivo formation of halogenated or oxidized derivatives of endogenous organic material resulting from the ingestion of residual chlorine in potable water. Gas chromatography/mass spectrometry (GC/MS) analyses of extracts of stomach contents and plasma, taken from fasted and nonfasted Sprague-Dawley rats dosed with NaOCl by gavage, demonstrated the in vivo formation of di- and trichloroacetic acids. Chloroform was identified in the stomach contents of the fasted and nonfasted animals dosed with NaOCl but was not detected in the control groups.

The potential for in vivo production of mutagenic products formed by the reaction of ingested residual chlorine with endogenous organic material is another area of major concern related to drinking water disinfection. The fact that chlorine is known to react with amino acids to form dihaloacetonitriles[12] that exhibit mutagenic activity in the Ames *Salmonella* bioassay and tumor-initiating activity in the Sencar mouse skin bioassay[13] supports this concern.

EXPERIMENTAL

Preparation of Chlorinated Humic Material

Humic material was purchased from Fluka AG (Switzerland). The preparation and chlorination of the humic material was performed according to Meier et al.[11] at a concentration of 1 g/L total organic carbon (TOC), using a 1:1 ratio of chlorine equivalents per mol of carbon. The pH 7 soluble portion of the Fluka humic material before chlorination was found to contain a humic acid:fulvic acid ratio of approximately 10:1, using pH selective precipitation.[14]

Molecular Weight Fractionation

Chlorinated solutions of the humic material were fractionated by diafiltration using an Amicon stirrer/ultrafiltration cell (model 52) equipped with Nucleopore® ultrafiltration membranes having nominal molecular weight cutoffs of 500, 5,000, and 50,000.

A 30-mL aliquot of the chlorinated humic solution was placed in the filtration cell containing a 500-mol wt cutoff filter; a pressure of 70 psi was then applied using helium gas. The cell volume was held constant by the continual addition of sterile distilled water from the reservoir at a rate equal to the diafiltration output. Following collection of 210 mL of filtrate (mol wt < 500), the retentate (mol wt > 500) was transferred to another unit, and the process was repeated for the 5,000- and 50,000-mol wt cutoff filters.

Liquid-Liquid Extraction

Aliquots of chlorinated humic material at pH 2.5 were extracted with two portions of diethyl ether at a solvent:sample ratio of 1:2. Following centrifugation to separate the layers, the ether extracts were combined, dried with sodium sulfate, and concentrated 200× by Kuderna-Danish evaporation. Portions of the aqueous layer and concentrated ether extract were saved for mutagenesis testing along with the original samples. GC/MS analyses and quantification were also performed on a separate portion of the ether concentrate containing internal standards.[9]

Closed-Loop Stripping Analysis

Closed-loop stripping (CLS) analyses were performed on the aqueous chlorinated humic acid solution to provide qualitative information on compounds that were present below the detection limits of the liquid-liquid extraction procedure. The CLS system and conditions used for the analyses were reported by Coleman et al.[15]

Steam Distillation

Three-hundred milliliters of an aqueous solution of the chlorinated Fluka humic material was placed in a distillation apparatus[16] equipped with a special condensate collection chamber designed to concentrate volatile, polar, low-molecular-weight organics. The flask temperature was controlled to provide a distillate at a rate of approximately 1 mL/90 min.

GC/MS Analyses

GC/MS analyses were performed using a Carlo-Erba Fractovap 4160 gas chromatograph and a Finnigan 3300 quadrupole mass spectrometer equipped with an Incos 2300 data system. The gas chromatograph used a 60 m by 0.25 mm ID wall-coated open-tubular (WCOT) fused silica SE-30 column for separating compounds in the steam distillate and CLS samples. Samples were injected in the splitless mode and the column vent (20:1 ratio) was opened 30 s after injection. The helium carrier gas flow rate was approximately 3 mL/min at 25°C at a linear velocity of 25 cm/s. The injector temperature was maintained at 260°C. Samples were injected at an oven temperature of 20°C, held for 6 min, then programmed to 250°C at a rate of 4°C/min. Mass spectra were acquired at the rate of 1 spectrum per 2 s over a mass range of 14 to 450 amu at 70 eV. A 30 m by 0.2 mm ID DB 1701 WCOT fused silica capillary column was used for the GC analysis of liquid-liquid extracts. The oven temperature was programmed at a rate of 6°C/min and mass spectra were acquired at a rate of 1 spectrum/s. All other GC conditions were as listed above.

Mutagenicity Testing

Solutions of chlorinated humic substances, or organic solvent extracts, were tested for mutagenicity using the *Salmonella* histidine reverse mutation assay described by Ames et al.[17] The assays were performed with *S. typhimurium* his⁻ strains TA98 and TA100. Tests were made with and without rat liver homogenate fraction (S9).

In Vivo Studies

GC/MS analyses were performed on the plasma and stomach contents of rats dosed orally with a single administration of 7 mL of an 8 mg/mL solution of NaOCl at pH 7.9. The methods for sample preparation and analyses have been described by Mink et al.[18] Attempts at compound identification were also performed on urine samples collected from rats dosed with daily administrations of 8 and 16 mg (NaOCl) $kg^{-1}d^{-1}$ for 8 d. Following the final dose the rats were placed in metabolism cages, and urine was collected in water-cooled vials. The urine samples from each dose group (10 animals per group) were pooled. The extraction and concentration of the urine samples were performed using the procedure of Yamasaki and Ames.[19] The acetone eluates were concentrated and subjected to GC/MS analyses.

RESULTS AND DISCUSSION

Since it has been demonstrated previously in this laboratory[10,11] and by Kringstad et al.[20] that the chlorination products of humic materials exhibit mutagenic activity in the Ames test, various fractionation and isolation procedures were used in an attempt to characterize and identify the mutagenic compounds. The molecular weight fractionation of chlorinated humic material by diafiltration suggested that the majority of the mutagenic activity was in the <500 mol wt fraction (Table I). The finding of a significant portion of the mutagenic activity in the greater than 50,000 mol wt fraction for TA98, however, was unexpected. An explanation may be the binding or association of mutagens possessing hydrophobic properties to dissolved humic material, similar to that discussed by Carter and Suffet[21] for the pesticide DDT. The inability to account for a higher percentage of the mutagenic activity in the whole sample following fractionation was probably attributable to adsorption losses on the ultrafiltration membranes and to the detection limits of the assay.

Using the liquid-liquid extraction procedure, diethyl ether recovered nearly all the mutagenic activity present in the chlorinated solution of humic material.[9] The ability of diethyl ether to recover approximately 100% of the activity, as compared to only 40% with methylene chloride, strongly suggests the presence of oxygen-containing functional groups in the mutagenic compounds. GC/MS analyses of the same methylene chloride and ether extracts (but without solvent exchange), which were tested for mutagenicity, revealed the presence of a number of halogenated organic compounds with carbonyl functional groups (Table II). Confirmed compound identifications were made for chloroform, dichloroacetonitrile, trichloroacetonitrile, dichloroacetic acid, trichloroacetic acid, 1,1-dichloropropanone, and 1,1,1,-trichloropropanone. Confirmations were based on mass spectra and relative retention times of authentic standards run under identical GC/MS conditions as the sample. Tentative identifications were made for the other compounds given in Table II for which authentic standards were not available. These were based on a manual interpretation of the data. Table II also indicates compounds that have been previously identified in drinking water. The similarity of identified substances, a number with known mutagenic activity, suggests that the reaction of chlorine with humic material is a likely source of mutagen formation in drinking water.

Two other isolation techniques were used in an attempt to identify chlorinated reaction products that might be present. CLS was selected because it is an

Table I. Molecular Weight Distribution of Chlorinated Humic Material After Diafiltration

Fraction	TOX		TA98 Mutagenesis[a]		TA100 Mutagenesis[a]	
	mg/L	% Total	Net Revertants/mL	% Total	Net Revertants/mL	% Total
Whole sample	425	100	242 ± 17	100	2,467 ± 48	100
<500	200	47	86 ± 12	36	1,465 ± 33	59
>500	138	33	44 ± 17	18		<18
>500, <5,000	61	14		<17		<18
>5,000	59	14		<17		<18
>5,000, <50,000	12	3		<17		<18
>50,000	38	9	63 ± 13	26		<18

[a]Negative response is based on a less than twofold above background increase in the number of revertants per plate at highest dose tested. "Less than" calculations are based on background values of 16 for TA98, 180 for TA100, and maximum dose levels of 0.4-mL sample.

Table II. Halogenated Compounds Identified in Chlorinated Humic Acid Solutions by GC/MS Analysis Following Liquid-Liquid Extraction, Closed-Loop Stripping, or Steam Distillation

Compound	Method identified			Ames Test Mutagenicity
	Solvent Extracts	Closed-loop Stripping	Steam Distillate	
Trihalomethanes				
•Chloroform[a]	+	+	+	−
Nitriles				
Chloroacetonitrile[b]	−	−	+	+
•Dichloroacetonitrile[b]	+	+	+	+
•Trichloroacetonitrile[b]	+	+	+	?
Dichloropropanenitrile	−	−	+	?
Dichloropropenenitrile	+	+	+	?
•Trichloropropenenitrile	+	+	+	?
Acids				
•Dichloroacetic acid[b]	+	−	−	−
•Trichloroacetic acid[b]	+	−	−	−
Aldehydes				
•Dichloroacetaldehyde	−	−	+	?
•Trichloroacetaldehyde	−	−	+	?
Dichloropropanal	−	−	+	+
Trichloropropanal	−	−	+	+
2-Chloropropenal[b]	−	−	+	+
2,3-Dichloropropenal	−	+	+	+
3,3-Dichloropropenal[b]	−	−	+	+
2,3,3-Trichloropropenal[b]	−	−	+	+
Trichlorobutanal	−	−	+	?
Dichlorobutenal	−	−	+	?

Table II, continued

Compound				
Ketones				
Chloropropanone			+	−
•1,1-Dichloropropanone[b]	+	+	+	+
1,3-Dichloropropanone[b]	+	+	+	+
•1,1,1-Trichloropropanone[b]	+	+	+	+
1,1,3-Trichloropropanone[b]	+	+	+	+
1,1,1,3-Tetrachloropropanone	+	+	+	−
1,1,3,3-Tetrachloropropanone[b]	+	−		
Pentachloropropanone[b]	−	+	+	
3-Chloro-2-butanone	−	+	−	
•1,1-Dichloro-2-butanone	−	−	+	
1,3-Dichloro-2-butanone	+	+	+	
3,3-Dichloro-2-butanone	−	+	−	
•1,1,1-Trichloro-2-butanone	−	+	+	
1,1,3-Trichloro-2-butanone	−	+	+	
1-Chloro-3-buten-2-one	−	+	+	
3-Chloro-3-buten-2-one	+	+	+	
Trichlorocyclopentenedione	+	+	+	
Tetrachloro-3-buten-2-one	−	+	+	
Pentachloro-3-buten-2-one		+		
Miscellaneous				
•Hexachloroethane[b]	−	−	+	−
Pentachloropropene	+	−	+	+
Tetrachlorocyclopropene	−	−	−	+
•Hexachlorocyclopentadiene[b]	+	−	−	+
Tetrachlorothiophene	−	−	+	−
2,4,6-Trichlorophenol	+	−	−	−

[a] •Indicates compound identified with drinking water.
[b] Compounds for which quantifications were made by Coleman et al.[9]

ultrasensitve method for detecting purgeable organics. This technique resulted in the tentative identification of several nonpolar chlorinated reaction products (Table II, Miscellaneous). Use of a distillation technique was made to determine if low-molecular-weight moderately polar compounds that azeotrope with water were responsible for the mutagenic activity. Some success was attained in that additional compounds, including several known mutagens (primarily aldehydes) were identified in the condensate (Table II). These compounds had not been previously observed in the CLS or liquid-liquid extracts. However, the recovery of mutagenic activity in the distillate following fractionation by steam distillation was less than 15% of the total sample activity. Synergistic or antagonistic interactions of the components did not appear to be a factor, because the recombined pot and distillate fractions gave an additive mutagenic response (Table III). The loss of mutagenic activity is probably not due to thermal degradation during distillation because autoclaving of similar samples in this laboratory resulted in only minimal losses of activity.

An attempt was made to estimate the contribution of the mutagenic activity of the identified known mutagens, for which quantitations were made,[9] to the overall mutagenicity level of the humic acid sample that had been chlorinated at neutral pH. This estimation was done by multiplying the specific mutagenicities (net revertants per milligram) of the compounds by their concentration in the original sample, and then dividing by the mutagenicity level (net revertants per milliliter) of the sample. Based on this calculation, the identified mutagens in Table II accounted for only 5 to 10% of the total mutagenicity present.

Because of the lack of standards for the compounds tentatively identified in Table II, quantification as well as the determination of the specific mutagenicity on a compound-by-compound basis was not possible. Consequently, the importance of these compounds in terms of their contribution to the overall

Table III. Fractionation of Mutagenic Activity in Chlorinated Humic Material By Steam Distillation

		TA100 Mutagenicity	
Sample	Fraction	Net revertants per mL[a]	Recovery (%)
Nonchlorinated	Original	ND[b]	
Nonchlorinated	Pot	ND	
Nonchlorinated	Distillate	ND	
Chlorinated	Original	2380	100
Chlorinated	Pot	149	6.3
Chlorinated	Distillate	318	13.4
Chlorinated	Pot and distillate combined	443	18.6
Chlorinated	Original, refluxed for 2 h	1140	47.9

[a]Results represent combined data from three experiments. Assays were conducted using four dose levels with triplicate plates per dose. Mutagenicity values are calculated from initial slopes.
[b]ND, not detected.

mutagenicity of the samples is not known. It is interesting to note the identification of several chlorinated propenals (i.e., chloroacroleins), because these compounds have been reported by Rosen et al.[22] to be very potent mutagens. The presence of 2-chloropropenal following humic acid chlorination has also been reported by Kringstad et al.[20]

An attempt was made to determine whether the compounds responsible for the mutagenic activity were amenable to the GC/MS procedure described. Preliminary heat stability tests were performed on the residue of ether extracts of chlorinated humic acid solutions. Ames testing indicated that approximately 50% of the mutagenic activity was associated with compounds that were degraded when heated to 220°C in an evacuated flask for 1 min (Figure 1). The mutagenic activity also appeared to be associated with relatively nonvolatile compound(s), based on recovery values of about 90% of the mutagenic activity following lyophilization of chlorinated solutions of humic material, as previously reported.[11]

Subsequent trapping studies of the gas stream exiting the GC system (using ether traps) agreed with the preliminary findings, since less than 10% of the mutagenic activity was recovered (Figure 2). An on-column injection technique that avoided the high injector block temperature was also tried with essentially the same results. These findings strongly suggest that the majority of the mutagenic activity is not associated with compounds quantified in Table II.

Because of the similarity of mutagenic compounds common to solutions of lignin and humic acid, the separation procedure for strong acids recently reported by Holmbom et al.[23] for the fractionation of chlorinated Kraft pulp was attempted. Approximately 70% of the mutagenic activity was retained in the strong acid fraction, supporting the possibility that a chlorinated hydroxy furanone such as the 3-chloro-4-(dichloromethyl)-5-hydroxy-2(5H)-furanone, tentatively identified by Holmbom et al.,[23] may be responsible for the activity in chlorinated humic acid solutions, as well as for chlorinated pulp effluents. Also supporting this possibility is that prior work in this laboratory[11] demonstrated the loss of mutagenic activity with increases in pH and temperature similar to that shown for the chlorinated hydroxy furanone. Attempts to analyze mucochloric acid, a closely related compound, by direct GC/MS resulted in a nonquantitative decarboxylation of the acid to 2,3-dichloropropenal. Similarly, decarboxylation of the proposed highly mutagenic chlorinated hydroxy furanone should produce a trichloroisobutenal. Dichlorobutenals have been tentatively identified by GC/MS.

The essential issue of the single dose rate study was to establish the presence of chlorinated reaction products in vivo. GC/MS analyses by Mink et al.[18] in this laboratory demonstrated the in vivo formation of dichloroacetonitrile, dichloroacetic acid (DCA), trichloroacetic acid (TCA), and chloroform. Chloroform was detected in the stomach contents, whereas DCA and TCA were detected in both the stomach contents and plasma of the fasted and nonfasted dosed animals. Dichloroacetonitrile, a known chlorination by-product of several amino acids,[12] was found only in the stomach contents of nonfasted rats.

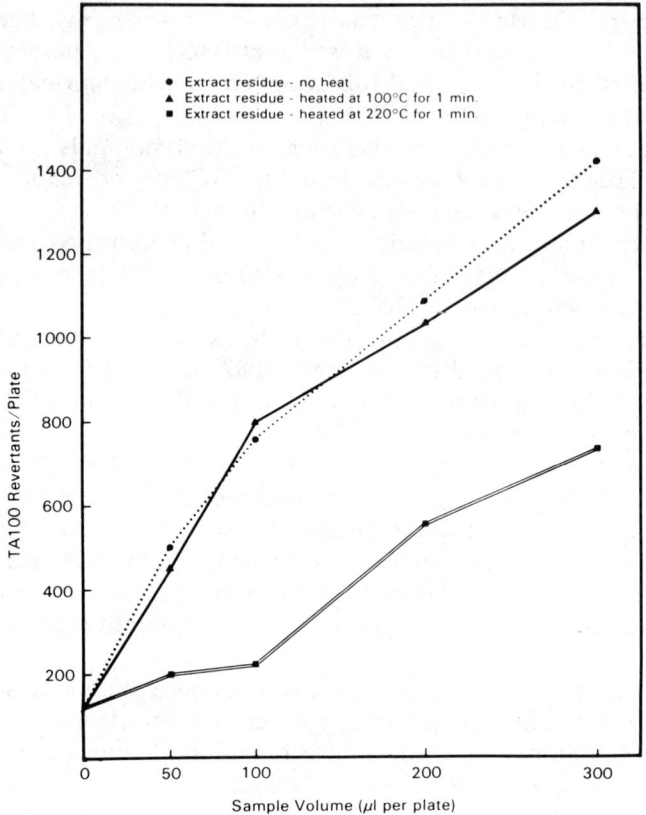

Figure 1. Effect of temperature on the mutagenic activity of chlorinated humic acid extracts.

The multiple dose study used a sodium hypochlorite concentration that was more consistent with drinking water intake and had produced cardiovascular effects in pigeons and rabbits[24] and spermhead abnormalities in the mouse.[25] GC/MS analyses of the urine extracts failed to demonstrate the presence of any chlorinated reaction products.

SUMMARY

The by-products of the chlorination of humic material were similar to those found in drinking water (Table II).

The majority of the mutagenic activity of chlorination by-products was associated with compounds that were heat labile at 220°C, were relatively

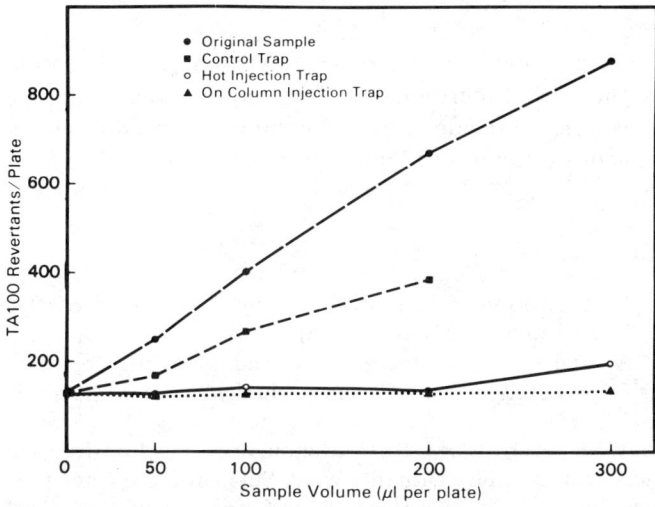

Figure 2. Comparison of gas chromatographic injection methods for recovery of mutagenic activity from chlorinated humic acid extracts.

nonvolatile (could be lyophilized), and had a molecular weight of less than 500 amu.

Only 5 to 10% of the mutagenic activity can be accounted for by specific compounds identified in Table II.

Less than 10% of mutagenic activity was recovered in diethyl ether traps using the GC procedure described.

Approximately 70% of the mutagenic activity was associated with compounds that are strongly acidic. This finding coupled with heat lability data, instability at high pH, low molecular weight, the likelihood of carbonyl or hydroxyl functional groups (extraction data), the inability to recover the activity following direct GC injection, and the fact that lignin structural units are present in humic material of terrestrial origin suggests that chlorinated hydroxy furanones may also be responsible for an appreciable portion of the activity in chlorinated humic material as well as in chlorinated kraft pulp effluent.

Further research is required to determine whether chlorinated hydroxy furanones are responsible for an appreciable portion of the mutagenic activity. Synthesis of appropriate standards will be required.

Future in vivo studies require the development of techniques with increased sensitivity to detect microgram and nanogram quantities of chlorinated reaction products.

DISCLAIMER

The research described in this paper has been peer and administratively reviewed by the U.S. Environmental Protection Agency and approved for publication. Mention of trade names or commercial products does not constitute endorsement or recommendation for use.

REFERENCES

1. Rook, J. J. "Formation of Haloforms During Chlorination of Natural Water," *J. Water Treat. Exam.* 23:234-243 (1974).
2. Bellar, T. A., and J. J. Lichtenberg. "Determining Volatile Organics at the Microgram per Liter Levels by Gas Chromatography," *J. Am. Water Works Assoc.* 66:739-744 (1974).
3. Coleman, W. E., R. D. Lingg, R. G. Melton, and F. C. Kopfler. "The Occurrence of Volatile Organics in Five Drinking Water Supplies Using Gas Chromatography/Mass Spectrometry," in *Identification and Analysis of Organic Pollutants in Water*, L. H. Keith, Ed. (Ann Arbor, MI: Ann Arbor Science Publishers, Inc., 1976), pp. 305-327.
4. Giger, W., M. Reinhard, C. Schaffner, and F. Zürcher. "Analyses of Organic Constituents in Water by High-Resolution Gas Chromatography in Combination with Specific Detection and Computer-Assisted Mass Spectrometry," in *Identification and Analysis of Organic Pollutants in Water*, L. H. Keith, Ed. (Ann Arbor, MI: Ann Arbor Science Publishers, Inc., 1976), pp. 433-452.
5. Thurman, E. M., and R. L. Malcolm. "Preparative Isolation of Aquatic Humic Substances," *Environ. Sci. Technol.* 15:463-466 (1981).
6. Rook, J. J. "Chlorination Reactions of Fulvic Acids in Natural Waters," *Environ. Sci. Technol.* 11:478-482 (1977).
7. Symons, J. M., T. A. Bellar, J. Carswell, J. DeMarco, K. L. Kropp, G. C. Robeck, D. R. Seeger, C. J. Slocum, B. C. Smith, and A. A. Stevens. "National Organics Reconnaissance Survey for Halogenated Organics," *J. Am. Water Works Assoc.* 67:634-647 (1975).
8. Christman, R. F., J. D. Johnson, F. K. Pfaender, D. L. Norwood, and M. R. Webb. "Chemical Identification of Aquatic Humic Chlorination Products," in *Water Chlorination: Environmental Impact and Health Effects, Vol. 3*, R. L. Jolley, W. A. Brungs, and R. B. Cumming, Eds. (Ann Arbor, MI: Ann Arbor Science Publishers, Inc., 1978), pp. 75-83.
9. Coleman, W. E., J. W. Munch, W. H. Kaylor, H. P. Ringhand, and J. R. Meier. "GC/MS Analysis of Mutagenic Extracts of Aqueous Chlorinated Humic Acid—A Comparison of the By-Products to Drinking Water Contaminants," *Environ. Sci. Technol.* 18:674-681(1984).
10. Bull, R. J., M. Robinson, J. R. Meier, and J. Stober. "The Use of Biological Assay Systems to Assess the Relative Carcinogenic Hazards of Disinfection By-products," *Environ. Health Perspect.* 46:215-227 (1982).
11. Meier, J. R., R. D. Lingg, and R. J. Bull. "Formation of Mutagens Following Chlorination of Humic Acid—A Model for Mutagen Formation During Drinking Water Treatment," *Mutat. Res.* 118:25-41 (1983).

12. Trehy, M. L., and T. I. Bieber. "Detection, Identification and Quantitative Analysis of Dihaloacetonitriles in Chlorinated Natural Waters," in *Advances in the Identification and Analysis of Organic Pollutants in Water, Vol. 2*, L. H. Keith, Ed. (Ann Arbor, MI: Ann Arbor Science Publishers, Inc., 1981), pp. 433-452.
13. Bull, R. J., and M. Robinson. "Carcinogenic Activity of Haloacetonitriles and Haloacetone Derivatives in the Mouse Skin and Lung," Chapter 18, this volume.
14. Christman, R. F., W. T. Liao, D. S. Millington, and J. D. Johnson. "Oxidative Degradation of Aquatic Humic Material," in *Advances in the Identification and Analysis of Organic Pollutants in Water, Vol. 2*, L. H. Keith, Ed. (Ann Arbor, MI: Ann Arbor Science Publishers, Inc., 1981), pp. 979-999.
15. Coleman, W. E., J. W. Munch, R. W. Slater, R. G. Melton, and F. C. Kopfler. "Optimization of Purging Efficiency and Quantification of Organic Contaminants from Water Using a 1-L Closed-Loop-Stripping Apparatus and Computerized Capillary Column GC/MS," *Environ. Sci. Technol.* 17:571-576 (1983).
16. Peters, T. L. "Steam Distillation Apparatus for Concentration of Trace Water Soluble Organics," *Anal. Chem.* 52:211-213 (1979).
17. Ames, B. N., J. McCann, and E. Yamasaki. "Methods for Detecting Carcinogens and Mutagens with the Salmonella/Mammalian-Microsome Mutagenicity Test," *Mutat. Res.* 31:347-363 (1975).
18. Mink, F. L, W. E. Coleman, J. W. Munch, W. H. Kaylor, and H. P. Ringhand. "*In Vivo* Formation of Halogenated Reaction Products Following Peroral Sodium Hypochlorite" *Bull. Environ. Contam. Toxicol.* 30:394-399 (1983).
19. Yamasaki, E., and B. N. Ames. "Concentration of Mutagens from Urine by Adsorption with the Nonpolar Resin XAD 2: Cigarette Smokers Have Mutagenic Urine," *Proc. Nat. Acad. Sci.* 74:3555-3559 (1977).
20. Kringstad, K. P., P. O. Ljungquist, F. de Sousa, and L. M. Strömberg. "On the Formation of Mutagens in the Chlorination of Humic Acid," *Environ. Sci. Technol.* 17:553-555 (1983).
21. Carter, C. W., and I. H. Suffet. "Binding of DDT to Dissolved Humic Materials," *Environ. Sci. Technol.* 16:735-740 (1982).
22. Rosen, J. D., Y. Segall, and J. E. Casida. "Mutagenic Potency of Haloacroleins and Related Compounds," *Mutat. Res.* 78:113-119 (1980).
23. Holmbom, B., R. H. Voss, R. D. Mortimer, and A. Wong. "Fractionation, Isolation, and Characterization of Ames Mutagenic Compounds in Kraft Chlorination Effluents," *Environ. Sci. Technol.* 18:333-337 (1984).
24. Revis, N. W., T. R. Osborne, P. T. McCauley, and G. Holsworth. "The Effects of Chlorinated Drinking Water on Myocardial Structure and Function in Pigeons and Rabbits," Chapter 29, this volume.
25. Meier, J. R., and R. J. Bull. "Mutagenic Properties of Drinking Water Disinfectants and By-Products," Chapter 17, this volume.

CHAPTER 15

Reactions of Hypochlorite and Organic N-Chloramines in Stomach Fluid

Frank E. Scully, Jr., Kathryn E. Mazina, Daniel E. Sonenshine, and F. B. Daniel

Over the past 10 years, it has been recognized that chlorine, used to disinfect drinking water, reacts with trace organic compounds dissolved in natural waters to produce by-products that may have adverse health effects in man. However, little attention has been given to the possible reactions of hypochlorous acid that may take place in the organic-rich medium of the stomach on ingestion of chlorinated drinking water. The average person's daily diet includes a minimum of 30 to 45 g of protein. Through the action of digestive enzymes in the stomach, proteins are broken down into peptones, large polypeptides, and about 15% amino acids.[1] Therefore, a variety of amino nitrogen compounds are available substrates for reaction with aqueous hypochlorous acid in the stomach. The objective of this study was to examine the possibility that organic N-chloramines could be formed on ingestion of hypochlorous acid, and to determine whether these compounds could be absorbed into the bloodstream for circulation to other parts of the body. Two chloramines were selected to probe the reactions of hypochlorite and the stabilities of chloramines in the stomach: N-chloroglycine and N-chloropiperidine. The first is a relatively stable chloramino acid[2] that is likely to be present in the stomach because of ingestion or as a product of proteolytic activity. The second is a model of more stable chloramines. The parent amine, piperidine, has been identified in drinking water[3] and its N-chloro derivative, N-chloropiperidine, has been found to be mutagenic.[4]

Recently, a method for the derivatization and analysis of organic N-chloramines in dilute aqueous solution was described.[5] In the method, solutions containing N-chloramines were reacted with 5-dimethylamino-naphthalene-1-sulfinic acid (DANSO$_2$H) to produce highly fluorescent sulfonamide derivatives (dansyl derivatives) that could be analyzed by high-performance liquid chromatography (HPLC). In the present study, this derivatization method is used to detect the formation of N-chloroglycine in stomach fluid, but results are corroborated by chromatography of an underivatized chloramine (N-chloropiperidine) and its radiolabeled counterpart.

EXPERIMENTAL

The method used for the derivatization of organic N-chloramines and the liquid chromatographic equipment used in these studies have been described elsewhere.[5] ^{14}C-Glycine was purchased from New England Nuclear and Amersham Corporation. ^{3}H-Piperidine hydrochloride was purchased from Amersham Corporation. Radiochemicals were greater than 98% pure. Solutions of sodium hypochlorite were prepared and standardized as described in EPA Method 510.1. Total Kjeldahl nitrogen (TKN) determinations were performed according to EPA Method 351.3, using titration with standard sulfuric acid. Protein concentrations were determined by the buret method.[6] A standard curve was constructed using bovine serum albumin (BSA) as the standard.

Synthesis of Tritiated N-Chloropiperidine

^{3}H-Piperidine (200 μL of a 1 μCi/μL solution) was added dropwise with stirring to 3.3 ml of an ice-chilled aqueous solution of sodium hypochlorite (2000 mg/L as Cl_2). Unlabeled piperidine (200 μL of a 0.2 M solution) was added with stirring to produce the solution used in animal studies or in exchange experiments.

Animals

Adult Sprague-Dawley rats weighing between 250 and 400 g were fasted for 48 h before each experiment. Prior to administration of a solution by gavage, the animals were anesthetized with ether. All solutions were administered by gavage using a syringe fitted with a 3-in. curved intubation needle.

Dosing and Recovery of Stomach Fluid

The contents of the stomach were recovered within 3 to 4 min after administration of a test solution by removal of the whole stomach. The fluid was chilled immediately on ice. Centrifugation was occasionally necessary to remove particulate matter before analysis was carried out.

To characterize the concentration of nitrogen compounds in the stomachs of test animals, 4 mL of deionized water was routinely administered. To determine whether organic N-chloramines could be formed in stomach fluid, solutions of sodium hypochlorite (4 mL) with concentrations of either 200 or 1000 mg/L were administered by gavage. Within 30 min of recovery, the stomach fluid was derivatized with $DANSO_2H$.

When the formation of either N-chloropiperidine or N-chloroglycine in stomach fluid was examined, dosing of animals was carried out in two stages. Initially 0.5 or 1.0 mL of an amine solution (0.2 M piperidine or 0.5 M glycine) in water was administered. This was followed within 2 to 3 min by a sufficient volume of standardized hypochlorite (200, 220, or 1000 mg/L) to give a total administered volume of 4.0 mL. Stomach fluid containing N-chloroglycine was passed through a Waters Assoc. C_{18} SEP-PAK octadecylsilica cartridge, and the chloramine was recovered by washing the cartridge with 2.0 mL of a solution of 20% acetonitrile/80% water (1% acetic acid). The wash was immediately derivatized with $DANSO_2H$. The procedure for handling stomach fluid containing N-chloropiperidine was similar, except that the chloramine was recovered using a 1.0-mL wash of acetonitrile. This latter chloramine was analyzed without derivatization.

When the appearance of N-chloropiperidine in plasma was examined, 1.2 mL of a 0.017 M N-chloropiperidine solution containing tritiated N-chloropiperidine (100 μCi) was administered by gavage. Animals were sacrificed at 30, 60, and 120 min to obtain blood. The blood was centrifuged to remove red blood cells, and proteins were removed from the plasma by alcohol precipitation at $-80°C$. Samples were then concentrated and analyzed directly by liquid chromatography.

Chlorine Demand of Stomach Fluid

After recovery of stomach fluid from animals administered 4 mL of deionized water, 1.5 mL of the fluid was diluted to 100 mL with deionized water, and 20-mL aliquots were chlorinated to 5, 10, 15, 20, and 30 mg Cl_2 per liter. Samples were incubated in the dark at room temperature for 1 h before being analyzed for residual chlorine by the DPD titrimetric method. A breakpoint curve was constructed, and the chlorine demand in milligrams per liter as nitrogen was determined by dividing the amount of chlorine required to reach breakpoint by 10, since this is the Cl_2:N weight ratio at which glycine reaches a breakpoint.

Chromatographic Conditions

Chromatographic conditions for the analysis of dansylglycine have been described.[5] The gradient program used to analyze samples containing radiolabeled N-chloropiperidine included isocratic elution at 10% acetonitrile/90% H_2O (1% acetic acid) for the first 5 min. During this time the solvent flow was programmed from 0.5 to 2.0 mL/min. After 5 min, the solvent was programmed by a linear gradient over the next 10 min up to 30% acetonitrile (constant flow). For the remaining 5 min of the chromatography, the solvent was programmed from 30 to 50% acetonitrile (constant flow).

RESULTS AND DISCUSSION

Amino Nitrogen Compounds in Stomach Fluid

Typical protein and TKN concentrations of the stomach fluid in laboratory rats have been measured, and results are listed in Table I. It was necessary to acquire these data so that experiments involving various dosage levels of hypochlorite could be designed with a clear understanding of the effect these levels would have on the stability of organic amino nitrogen compounds in the stomach. Since N,N-dichloro derivatives of primary amines are considerably less stable than their monochloro counterparts, administration of too high a concentration of hypochlorite would cause decomposition of the N-chloro compounds that later experiments were designed to detect.

The protein concentrations as determined by the biuret method are recorded in Table I in mg/L as BSA. If the percentage nitrogen in the stomach proteins is assumed to be identical to the percentage nitrogen in the BSA (15.7%) used to construct the standard curve, then the protein contribution to the TKN value can be estimated. The results of this calculation suggest that 70 to 99% of the TKN nitrogen is proteinaceous. In addition, protein nitrogen concentrations in the stomach fluid of rats (as calculated from the biuret results) range from 100 to 400 mg/L. These values have not been normalized for the weight of the animal for which they were determined. However, variations among animals are not believed to be due to weight differences, but rather to the amount of food remaining in the stomach at the time the fluid was acquired, as well as to metabolic differences of the animals.

The values measured for the chlorine demand of rat stomach fluid are listed in the last column of Table I in mg/L as N. They were obtained from breakpoint curves constructed from the chlorine residuals measured 1 h after chlorinating diluted samples of stomach fluid. The curves showed a definite maximum followed by a relatively large "irreducible minimum," beyond which free residual chlorine was measured. The shape of the curve looked much like that obtained with model solutions containing proteins and amino acids. Chlorine demand in stomach fluid can be exerted by many different species: sulfhydryl residues, amino nitrogen end groups and side chains on proteins, reducing sugars, trace metals, and other reducing species. If all of the demand is assumed to be exerted by amino nitrogen moieties, a maximum value for the contribution of amino nitrogen to the TKN measurement can be calculated. Although these values vary considerably, they are generally less than one-half of the TKN values. This also suggests that a large portion of the nitrogen is found within proteins and peptides. Taking an average of the values for the chlorine demand (in mg/L as N) gives 74 mg/L. Since a 1:1 ratio of chlorine to nitrogen would be achieved at a chlorine concentration of about 370 mg/L, chlorine concentrations higher than this would result in decomposition of N-chloro derivatives of simple primary amines and amino acids.

Table I. Protein and Nitrogen Concentrations in Rat Stomach Fluid

Animal No.	Protein[a] (mg/l as BSA)	Protein-N[b] (mg/l as N)	TKN (mg/l as N)	Chlorine Demand (mg/l as N)
1	1340	201	208	125
2[c]	2745	412	415	
3[c]	1780	267	357	
4	2112	317	307.5	
5	800	120	171.7	42
6	740	116		45
7	2430	382		83
8	2732	429		60
9	2940	462		41
10	2318	364		73
11	1828	287		73

[a]By biuret method.
[b]0.157 × Protein as BSA.
[c]Composite sample.

In Vivo Formation of Stomach N-Chloramines

Laboratory rats were administered 4 mL sodium hypochlorite (1000 mg/L), and the stomach fluid was recovered shortly thereafter and derivatized with $DANSO_2H$. HPLC analysis of the resulting solution yielded the chromatogram shown in Figure 1. There are several major peaks in the chromatogram. The peak with a retention time of about 3 min is due to unreacted $DANSO_2H$, and the peak with a retention time of 21 min is an artifact due to oxidative degradation of the derivatizing reagent at high concentrations of hypochlorite or chloramines.[5] The remaining peaks in the chromatogram are believed to be derivatives of chloramines formed by derivatization of N-chloramines produced in the stomach when chlorinated water was administered. The compounds that appear in the chromatogram probably have relatively low molecular weights, since large proteins and polypeptides are retained on the SEP-PAK octadecylsilica cartridge used in the cleanup step. Consequently, there are likely to be other N-chloramines formed on the side chains of proteins that are not detected here. There is conspicuous absence of any significant amount of derivative of NH_2Cl (retention time = 10.5 min). This suggests that ammonia does not contribute significantly to the breakpoint curve of stomach fluid discussed above.

Because of the complex nature of stomach fluid, speculation about the identities of the compounds in the chromatogram by correlation of their retention times with those of known derivatives is not satisfactory. Therefore, the identity of an N-chloro derivative of a known amine was examined.

180   WATER CHLORINATION

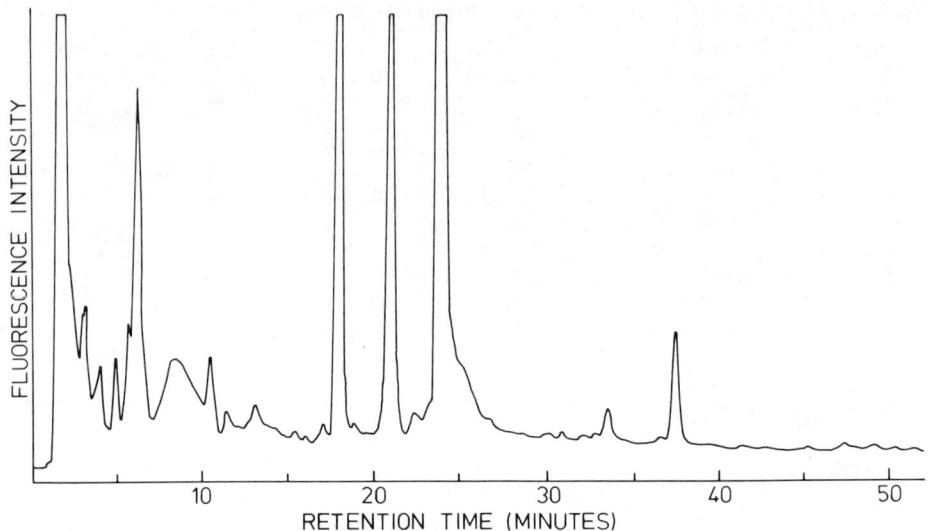

Figure 1. Chromatogram of stomach fluid from a rat that had been given 4 mL of a solution of hypochlorite (1000 mg/L) 5 min before removal of the stomach contents. The fluid was derivatized with $DANSO_2H$ before chromatography. Injection size = 20 µL. The 21-min peak is an artifact.

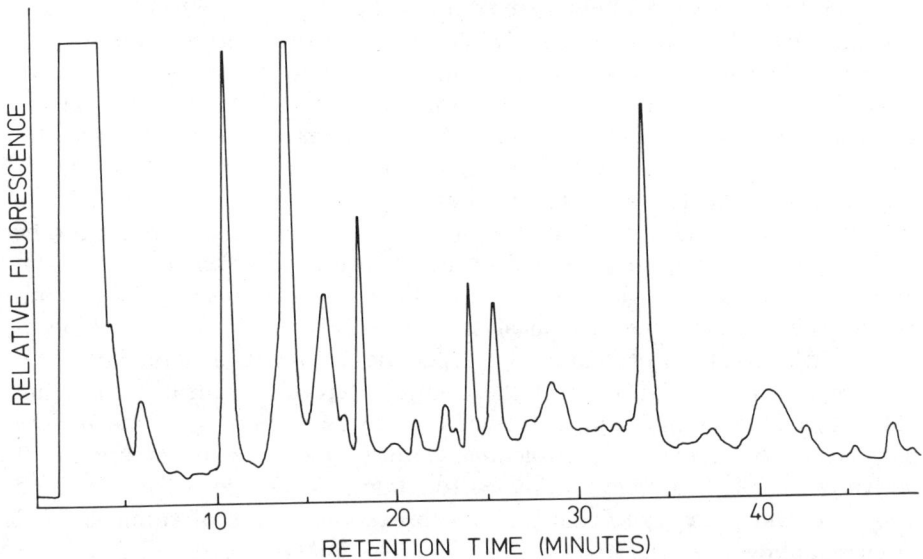

Figure 2. Chromatogram of stomach fluid from a rat that had been administered 0.5 mL of a 0.5 M solution of glycine followed by 3.5 mL of a solution of hypochlorite (1000 mg/L). The fluid was derivatized with $DANSO_2H$ before chromatography. Injection size = 37 µL. Dansylglycine has a retention time of 14 min.

In Vivo Formation of N-Chloroglycine

Administration of a solution of glycine (0.5 M) followed by a solution of hypochlorite (either 200 or 1000 mg/L) resulted in formation of N-chloroglycine. The N-chloroglycine was detected by derivatization with DANSO$_2$H and HPLC analysis (Figure 2, dansylglycine retention time, 14 min). When the 1000-mg/L solution of hypochlorite was used, the amount of dansylglycine formed was found to be proportionally higher than the amount of dansylglycine formed when the 200-mg/L solution was used.

Additional confirmation of N-chloroglycine formation was obtained using ^{14}C-glycine. Rats were dosed with 10 to 12 μCi of the radiolabeled glycine in a total volume of 1 mL and then with hypochlorite. Approximately 40 to 50% of the radiolabel was routinely recovered from the stomach. The derivatized extract was separated by HPLC, and fractions were collected and analyzed by liquid scintillation counting. When 1000 mg/L hypochlorite was used, approximately 4.5% of the radiolabel eluted with the retention time of dansylglycine (Figure 3). When 200 mg/L hypochlorite was used, only 0.44% of the recovered radiolabel was present in the fraction that eluted with the retention time of dansylglycine. The low recovery is probably due to a combination of factors: absorption of N-chloroglycine or glycine into stomach tissues, and decompsition of the N-chloroglycine in vivo and during workup.

In Vivo Chlorination of Piperidine

The derivatization technique used to detect the formation of N-chloroglycine in stomach fluid as described above is subject to the criticism that the derivatization is carried out at high pH (pH 9–10) and may not reflect what happens in the stomach at lower pH. Since N-chloropiperidine can be separated from its unchlorinated amine by liquid chromatography, the analysis of this chloramine is possible without derivatization. Elution of N-chloropiperidine from the chromatography column can be detected by its UV absorbance at 254 nm. For low concentrations of N-chloropiperidine, radiolabeled N-chloropiperidine can be used.

In general, the in vivo chlorination of piperidine was carried out in a manner similar to the chlorination of glycine. An aqueous solution of piperidine (0.5 mL of a 0.2 M solution) containing 35 μCi of ^3H-piperidine was administered to the test animals shortly before dosing with hypochlorite. However, when the stomach fluid was isolated, only 30% of the radiolabel could be recovered. The material that was not recovered was believed to be absorbed into tissue. Approximately 30 to 40% of the recovered label was lost in the cleanup of the fluid with a SEP-PAK octadecylsilica cartridge. The radiochromatogram of the labeled material recovered when a 1000-mg/L solution was used showed that 8% was N-chloropiperidine. When 220 mg/L hypochlorite was used (Figure 4), 3% of the recovered material was N-chloropiperidine. Using the aver-

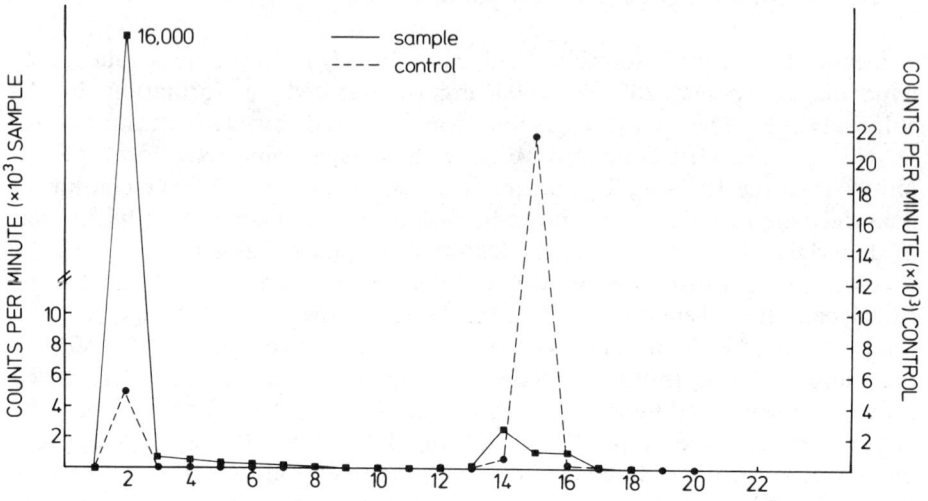

Figure 3. Radiochromatogram of stomach fluid from a rat that had been given 2 mL H_2O containing 12 μCi ^{14}C-glycine by gavage followed by 3 ml of a solution of hypochlorite (1000 mg/l). The fluid was derivatized with $DANSO_2H$ before chromatography. Fraction size = 2 mL. The radiochromatogram of ^{14}C-glycine derivatized with dansyl chloride (● – – – – – ●) was used to check the purity of the labeled glycine. Dansylglycine elutes in fractions 14 to 17.

age chlorine demand of the stomach fluid estimated above (74 mg/L) and adding it to the demand of the piperidine administered, an expected yield of N-chloropiperidine can be calculated. Such a calculation suggests that 19% of the piperidine should be chlorinated when 3 mL of 1000 mg/L hypochlorite is administered and 4% of the piperidine should be chlorinated when 3 mL of 220 mg/L hypochlorite is administered. The yields actually recovered (8 and 3%) may be low due to some decomposition of the chloramine in the stomach, but they may only appear low due to inadequate recovery of the chloramine during workup. In any case, the actual recovered yields suggest not only that the formation of N-chloropiperidine is an important reaction of the hypochlorite in the stomach in this experiment, but also that the formation of other N-chloramino nitrogen compounds in the stomach probably takes place as well.

Absorption of N-Chloropiperidine into Blood Plasma

Since organic N-chloramines were shown to be formed in the stomach, an experiment was designed to determine whether they could be absorbed into the bloodstream and possibly distributed to other parts of the body. Animals were administered 100 μCi of ^{3}H-N-chloropiperidine. Of the 100 μCi administered,

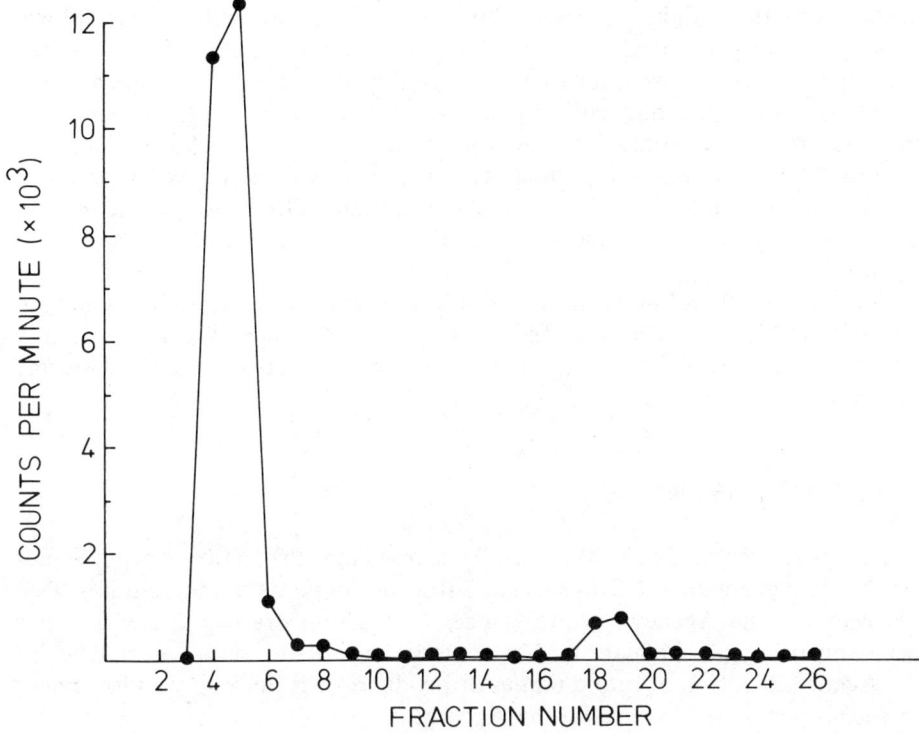

Figure 4. Radiochromatogram of stomach fluid from a rat that had been given 0.5 mL of a 0.2 M solution of piperidine (containing 35 μCi of ^{3}H-piperidine) followed by 3.5 mL of hypochlorite (220 mg/L). The injection size was 200 μL. N-Chloropiperidine elutes in fractions 18 and 19.

only a small amount (adjusted for the entire volume of the plasma) appeared in the plasma (1 to 3 μCi) at these intervals. Of the radiolabeled material found in the plasma, the major portion was unchlorinated ^{3}H-piperidine. However, 5.5, 16, and 1% of the labeled material recovered at 30, 60, and 120 min, respectively, was due to the presence of tritiated N-chloropiperidine.

CONCLUSIONS

This study has demonstrated that (1) organic N-chloramines can be formed in stomach fluid when water containing high concentrations of hypochlorite is ingested, and (2) these chloramines can be absorbed into the bloodstream. However, several facts must be considered before the implications of these results on the health effects of drinking chlorinated water can be made. For instance, the high dosage levels used could be overwhelming any reducing mechanisms that might otherwise function to deactivate chlorine or chloramines at lower concentrations in the stomach. The low yields of chloramines recovered in blood plasma may be evidence that such mechanisms are surprisingly effective even at high chlorine dosages. On the other hand, absorption of the chloramine into stomach tissue or exchange of the active chlorine between the test chloramine and unchlorinated amines in the stomach may also account for the low concentration of the model chloramine in the plasma. The mechanisms by which active chlorine compounds are decomposed in the stomach will require further study.

Finally, a single administration of chlorine or a test chloramine is not an adequate model of the daily ingestion of chlorinated water. Work is currently under way to evaluate the effect of multiple administration of active chlorine compounds.

ACKNOWLEDGMENT

This work was funded by the U.S. Environmental Protection Agency under assistance agreement CR-810459. The information contained herein has been subjected to the Agency's required peer and administrative review and has been approved for publication. The contents reflect the views and policies of the Agency. Mention of brand names or products does not imply endorsement or recommendation for use by the Agency.

REFERENCES

1. Guyton, A. C. *Textbook of Medical Physiology*, Fifth Ed. (Philadelphia: W. B. Saunders Co., 1976), pp. 885, 970.
2. Isaac, R. A. and J. C. Morris. "Modeling of Reactions Between Aqueous Chlorine and Nitrogenous Compounds," in *Water Chlorination: Environmental Impact and Health Effects, Vol. 4,* R. L. Jolley, W. A. Brungs, J. A. Cotruvo, R. B. Cumming, J. S. Mattice, V. A. Jacobs, eds. (Ann Arbor, MI: Ann Arbor Science Publishers, Inc., 1983), pp. 63-75.
3. U.S. Environmental Protection Agency, Health Effects Research Laboratory. *Organic Compounds Identified in U.S. Drinking Water* (Cincinnati: 1978).
4. Bempong, M. A., and F. E. Scully, Jr. "Mutagenic Activity of N-Chloropiperidine," *J. Environ. Path. Toxicol.* 4(2,3):345-354 (1983).
5. Scully, F. E., Jr., J. P. Yang, K. Mazina, and F. B. Daniel. "Derivatization of Organic and Inorganic N-Chloramines for HPLC Analysis of Chlorinated Water," *Environ. Sci. Technol.* (in press).
6. Clark, J. M., Jr., and R. L. Switzer. *Experimental Biochemistry*, Second Ed. (San Francisco: W. H. Freeman and Co., 1977), p. 12.

SECTION IV

Carcinogenic and Mutagenic Effects

To many critics of cost-benefit analysis, a central problem is the difficulty in placing a value on life; to an epidemiologist or toxicologist, a central problem is establishing a causal link between chemicals in drinking water and increased cancer rates.

Talbot Page and **Robert Harris**
A Cost-Benefit Approach to Drinking Water and Cancer, 1981

Although the medical consequences of genetic disease may be altogether as devastating as the consequences of cancer, identifying the environmental factors that lead to the damage is very much more difficult. The expression of genetic disease is usually far removed in time from the induction of the mutations which caused it, and thus, from the toxic agent which led to the mutations. The consequences of genetic damage would not be expected to fall upon the individual in which the mutations were induced but upon his progeny or descendants—perhaps several or many generations later.

Robert B. Cumming
The Potential for Increased Mutagenic Risk to the Human Population Due to the Products of Water Chlorination, 1975

CHAPTER 16

Mutagenic and Carcinogenic Properties of Drinking Water

H. J. Kool, C. F. van Kreijl, and J. Hrubec

In the Netherlands, about 60% of the drinking water is produced from groundwater and does not receive a chlorine or other oxidative treatment. The bulk of the use of a chlorination or other oxidation step is related to surface water treatment. Mainly in drinking water prepared from surface water, to date hundreds of organic constituents, including chlorinated hydrocarbons, have been identified. Among these compounds, several mutagenic and (suspect) carcinogenic organic compounds could also be identified in this type of drinking water used in several cities in The Netherlands.[1] Similar findings have also been reported in other countries.[2-5] Some data on these compounds identified in drinking water in The Netherlands are shown in Table I.[6-9] Limitations of scientific information have not permitted an in-depth evaluation of many compounds recently found in drinking water; therefore, relatively little is known about their toxic effects including their carcinogenic potential. In addition to this, it is recognized that constituents of the nonpurgeable fraction that comprises 90 to 95% of the total organics in water have not been identified.[2,10] The reason for this is that this fraction cannot be readily volatized for separation and identification by gas chromatography/mass spectrometry (GC/MS) analysis. It is therefore apparent that there is an advantage in combining analytical procedures and biological testing to establish whether water treatment processes change biologically inactive fractions into biologically active fractions and to permit us to isolate the bioactive subfractions. That this approach may finally lead to the identification of the bioactive compounds was demonstrated by Tabor and Loper.[11] They identified the structure of a highly mutagenic compound in Cincinnati tap water.[12]

In this chapter results of oxidation treatments with chlorine, ozone, chlorine dioxide, and ultraviolet (UV), with respect to their effects on activity (Ames test) in drinking water supplies are reviewed. In addition, we present the preliminary results of a pilot plant study on the effects of chlorine and chlorine dioxide on mutagenicity. Furthermore, results of several carcinogenicity studies performed with organic drinking water concentrates are discussed in relation to the results of a Dutch carcinogenicity study with mutagenic drinking water concentrates.

Table I. Review of Some Suspect Carcinogenic and Mutagenic Compounds Detected in the Rhine and Meuse Rivers and in Drinking Water Prepared from These Rivers in 1973–77.

Compound	Maximum Concentration (µg/l)		Carcinogenicity		Mutagenic[e]		In Vivo
					In Vitro		
	In Surface Water	In Drinking Water	In Humans	In Lower Animals	Prokaryotic	Eukaryotic	Eukaryotic
Benzene	0.03	0.005	•[a]				
3,4-Benzo(a)pyrene	1	0.005		•[a]		•	•
Benzo(b)fluoranthene	0.21	0.045		•[a]		•	•
Indeno(1,2,3-c,d)pyrene	1.4	0.0075		•[a]			
Bis(2-chloroethyl) ether	0.01	0.03		•[c]	•		
Bis(2-chloroisopropyl) ether[d]	25	3					
Bromodichloromethane[d]	2	55		•[a]	•		
Chloroform	30	100		•[a]	•		•
Tetrachloromethane	4.2	0.7		•[a]	•		
1,2-Dichloroethane	20	0.06		•[b]	•		
Heptachlor	0.04	0.01		•[b]		•	
Lindane	0.03	0.01		•[b]		•	
DDE	0.2	0.2		•[c]		•	

[a]Sufficient evidence for carcinogenicity (see Reference 6).
[b]Limited evidence for carcinogenicity (see References 7 and 8).
[c]Suspect carcinogenic compounds that do not fulfill requirements a and b.
[d]Compounds currently tested at the National Cancer Institute, Bethesda, Md.
[e]Mutagenicity data obtained from NIH-EPA, CIS, NIOSH, and RTECS and from Reference 9.

MATERIAL AND METHODS

XAD Resins

Amberlite XAD-4 and -8 resins were obtained from Serva GmbH, Heidelberg, F.R.G. Purification by repeated Soxhlet extraction, control by GC analysis, and storage of the resins in methanol have been described previously.[13]

XAD Concentration Procedure

Water samples were taken before and after treatment in the pilot plant. To obtain about 7000-fold concentration, 150 L of the filtered water was passed over columns containing 20-cm^3 XAD at a flow rate of maximal 4 bed vol/min and at a constant temperature of 15°C. Elution of the adsorbed neutral fraction of the organic constituents was carried out with the appropriate volume ($\geq$ 1 bed volume) of either dimethylsulfoxide (DMSO) or acetone. The XAD filtrate was adjusted to pH 2 with HCl and readsorbed on XAD-4/8. Subsequent elution of this acid fraction was carried out with DMSO or acetone. For lower or higher concentration factors, correspondingly smaller or larger volumes of water were passed through the XAD column until the desired water/eluate ratio (v/v) was obtained.

Bacterial Strains

The *Salmonella typhimurium* strains TA98 and TA100 were used.[14] They were stored frozen at -80°C in nutrient broth containing 10% DMSO.

Ames *Salmonella*/Microsome Assay

Mutagenicity testing of the organic concentrates of drinking water was carried out according to the plate incorporation assay.[14] The induction of microsomal enzymes with Aroclor 1254 and the preparation of the rat liver homogenates (S9) have also been described previously.[14] In the S9 mix, 0.075 mL of liver homogenate was added per milliliter of mix. All water concentrates were tested in 3-5 replicates, and the results were considered significant when a 2-fold increase above the background and dose-response effects were observed. The deviation of the mean was usually below 20%. Routine controls to check for the presence of factors affecting bacterial growth were incorporated as described previously.[13]

Chemical Analyses

The analysis of water quality parameters in the unconcentrated water samples was carried out according to routine procedures, described previously. Extractable organic chlorine (EOCl) was analyzed by microcoulometry.[15] Adsorbable organic chlorine (AOCl) was determined with the aid of activated carbon.[16] Total organic carbon (TOC) was analyzed with a Beckman TOC analyzer (Tocomaster model 915 B). Trihalomethanes (THM) were analyzed with a Carlo Erba 2900 analyzer containing a capillary column G.C. OV/225, diam 0.5 mm, length 50 mm, and equipped with an automatic headspace sampler, model 250. Volatile organic halogens (VOCl) were determined by microcoulometry.[17]

Pilot Plant

The schematic flow diagram, the description of the pilot plant, and the application of chlorine and chlorine dioxide have been described previously.[18]

Chemical Analysis in the Pilot Plant Study

Residual concentrations of chlorine and chlorine dioxide in water were measured by the DPD-FAS technique.[19] Chlorine and chlorine dioxide in stock solution were determined by the iodometric method described by Berndt[20] and Valenta and Gabler.[21]

MUTAGENIC EFFECTS

Chlorination

The Rhine and Meuse rivers serve as important sources of drinking water supply; the two rivers provide the potable water for about 5 million people in The Netherlands. Both rivers contain mutagenic activity (Ames test)[13,22-26] that was predominantly observed with strain TA98 and was most pronounced with metabolic activation.[13,23,24]

In chlorine-treated water in The Netherlands, in which treatment is generally applied in the form of breakpoint chlorination, transport chlorination, and postchlorination, the mutagenic activity is most pronounced without metabolic activation and is observed in both strain TA98 and strain TA100. The latter suggests that during drinking water treatment a change in mutagenic activity occurred. Several drinking water studies have shown that a chlorine treatment is able to increase mutagenic activity in water. Cheh et al.[27] showed an increase in the direct mutagenic activity with strain TA100 by treating water with chlorine in a pilot plant study. Similar results were obtained after chlorination in other pilot plant studies.[18,28-29] Zoeteman et al.[18] showed that the effect of chlorination on the mutagenic activity depended strongly on the type of organics present in the water. It was also shown that in some experiments a reduction of promutagenic activity in chlorinated Rhine water was observed. Several studies with respect to a chlorine treatment have also been carried out in some waterworks.[26,30-34] These studies support the results obtained in the pilot plant studies and clearly show that a chlorine treatment (transport, post-, and breakpoint chlorination) is able to increase or even to generate mutagenic activity. Only in exceptional cases could a decrease[31] or no effect of such treatment[33] be observed. The type and level of mutagenic activity appeared to be dependent on the organic composition of the treated water. A brief summary of chlorination studies carried out in drinking water treatment with respect to mutagenic activity is shown in Table II.[18,26-29,31-33,35]

Ozonation

Ozone, which is a powerful disinfectant, may be an alternative for chlorination in drinking water treatment and is nowadays more frequently used than in the past in the purification of drinking water. Therefore this agent must be evaluated in order to determine whether it will also cause the production of organic mutagens and carcinogens in water supplies. Studying the effect of an ozone treatment on the mutagenicity of recycled water of a municipal waste water plant, Gruener[36] showed that ozone induced direct and promutagenic activity with strains TA98 and TA100. Cotruvo et al.[37] studied the effect of ozone on 28 selected organic compounds, including polycyclic aromatic hydrocarbons (PAHs), polychlorinated biphenyls (PCBs), and humic acid, and found that this oxidation step did not result in substantial mutagenic activities. A similar study, however, with 36 selected organic compounds that were mutagens, carcinogens, or both was carried out by Burleson and Chambers.[38] They showed that ozonation decreased the mutagenicity of PAHs and aromatic amines, did not change the mutagenicity of alkylating agents, nitro aromatics, and nitroso compounds, and increased the mutagenicity of hydrazines.

Investigations by Dolara et al.[31] with ozone in drinking water treatment revealed that in 1 out of 4 samples an increase of mutagenic activity with strain TA100 was observed, while in the other samples the mutagenic activity drastically decreased after this treatment. The effect of an ozone treatment on stored Rhine and Meuse river water was investigated by Zoeteman et al.[18] in a pilot plant study. They found almost a complete reduction of the mutagenic activity when this water was treated with ozone. Other studies with ozone in drinking water treatment[18,26,35,39,40] showed more or less similar results: ozone decreased or had hardly any influence on the mutagenic activity in the neutral fraction. The mutagenic activity with strain TA100 may in some cases show a slight increase after this treatment in the acid fraction,[26,39] although the increase of mutagenicity due to a chlorine treatment in general is much higher than that

Table II. Summary of Chlorination Studies with Respect to Mutagenic Activity in Drinking Water Supply

Type of Water	Chlorine (mg/L)	Contact Time[a]	pH[a]	TA1538	TA1538 +S9	TA98	TA98 +S9	TA100	TA100 +S9	Water Equivalent Tested per Plate (L)	Reference
Stored Meuse water	0 5 15	24 min	6.2	NT	NT	68 151 156	12 60 57	15 37 70	– – –	1.5	28
Lake water	0 5			NT NT	NT NT	+ –	– –	– 694	– –	1.2	
River water	0 5	20 h	7.1	NT NT	NT NT	65 +	434 876	43 258	102 162	1.2	29
Groundwater	0 5					– –	– –	– –	– –	0.6	
Drinking water	0 4.2	300 min	7.7	NT	NT	* *	* *	100 980	– *	0.5	27
Arno river water March	0 2.5–7.5			84 20	68 10	NT	NT	0 70	0 40	10	31
July	0 2.5–7.5	*	*	17 68	7 58	NT	NT	20 30	2 53	10	
Treated Rhine water	0 0.2[b]		7–8	NT NT	NT NT	* *	*	0 228	40 122	2	26
Stored Meuse water	0 1.85		7–8	NT NT	NT NT	31 105	47 57	38 203	23 151		
Stored Rhine water	0 5	24 min	8–9	NT	NT	10 34	45 51	32 54	– –	1.5	32,33
Stored Meuse water	0 1 5	24 min	7–8	NT	NT	29 51 82	50 66 91	– – –	– – –	1.5	18,35

[a] *, Data not presented; –, no effect detectable; +, fraction toxic for bacteria; NT, not tested.
[b] After contact time of 20 min.

obtained with ozone. A brief summary of representative results of ozone studies carried out in drinking water supply with respect to mutagenic activity is shown in Table III.[18,26,31,32,35,39]

Chlorine Dioxide and Ultraviolet

Both chlorine dioxide and UV treatments are under investigation to see whether these agents will change or introduce mutagenic activity in the preparation of drinking water. Very few data are, however, available about UV irradiation with respect to mutagenicity. Some results have shown that UV irradiation (120 mW s cm^{-2}, detention time 26 s) did not affect the characteristic mutagenic activity of stored Rhine water.[18] Additional experiments carried out later confirmed these results.[35] A study of UV applied in wastewater treatment showed similar results; a slight increase of mutagenic activity in only one case was reported by Jolley et al.,[41] while several mutagenic constituents were destroyed by this treatment.

Considering chlorine dioxide for drinking water treatment, De Greef et al.[28] showed that relatively high doses of chlorine dioxide (5 and 15 mg/L) increased the direct mutagenic activity significantly only with strain TA98. In a similar experiment, Zoeteman et al.[18] partly confirmed these results; a significant increase of direct-acting mutagens with strain TA98 but a reduction of promutagenic activity with this strain was found in the case of a chlorine dioxide (5 mg/L) application to stored Rhine water. Additional experiments with stored river water showed that varying results, increase as well as decrease of mutagenic activity, are obtained by this treatment (Figure 1). In general, however, the increase of mutagenic activity is significantly lower than that obtained with a chlorine treatment (Figure 2).

Pilot Plant Study on Chlorine and Chlorine Dioxide

Results

In experiments recently carried out in a pilot plant study in The Netherlands, effects of chlorine and chlorine dioxide (both doses 1 mg/L) on mutagenic activity in the drinking water of six cities that prepare their drinking water mainly from groundwater are under investigation. Besides the mutagenic activity, the levels of volatile and nonvolatile halogenated hydrocarbons before and after both treatments were determined. Preliminary results of this study are shown in Figure 3 and Table IV.

As shown in Table IV, after a chlorine treatment, a drastic increase of all halogenated hydrocarbons was observed, with some exceptions in the case of EOCl. When a chlorine dioxide treatment was applied, the EOCl levels decreased while an increase of AOCl was observed in two cases (cities 3 and 6). Figure 3 shows that a chlorine dose of 1 mg/L generally increased the direct mutagenic activity with strains TA98 and TA100 (top 2 graphs in Figure 3) in

Table III. Summary of Ozone Studies with Respect to Mutagenic Activity in Drinking Water Supply

Type of Water	Ozone (mg/L)	Contact Time[a] (min)	pH[a]	Number of Induced Revertants per Plate[a]						Water Equivalent Tested per Plate (L)	Reference
				TA1538	TA1538 +S9	TA98	TA98 +S9	TA100	TA100 +S9		
Arno River water											
June 1980	0[b]	8	*	0	30	NT	NT	—	58	10	31
	0.4			18	28				130		
July 1980	0			68	58			30	53		
	0.4			8	5			5	55		
Stored Rhine water											
October 1980	0	30	7.5–8.5	NT	NT	14	44	—	—	1	18,35
	5					0	0	—	—		
June 1980	0					59	21	—	—		
	10					0	0	—	—		
Treated Meuse water											
January 1981	0	10	7.5–8.5	NT	NT	58	80	—	—	3	32
	2					58	40	—	—		
April 1981	0					20	41	—	—	1.5	
	2					11	7	—	—		
Treated Meuse water	0[b]	35	8	NT	NT	241	135	454	85		39
	2					—	—	131	50		
	4					—	—	—	47		
	8					—	—	—	—		
Treated Rhine water	0	20	*	NT	NT	*	*	57	319	2	26
	1.5							38	32		
	0	50	*	NT	NT	*	*	0	40		
	4							40	0		

[a]NT, not tested; *, data not given; —, no effect detectable.
[b]Water treated with chlorine.

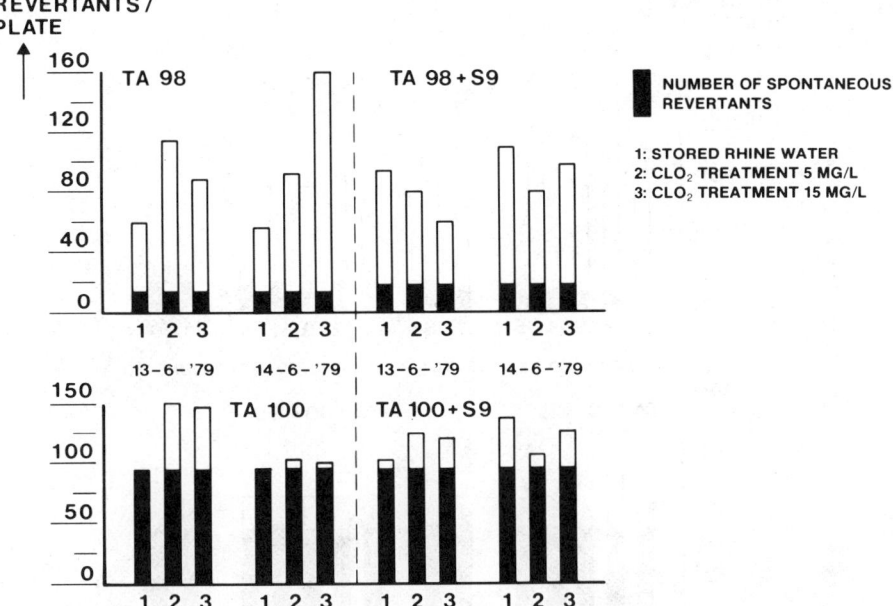

Figure 1. Effect of a chlorine dioxide treatment on mutagenic activity. The sampling: 1800-fold concentration of stored Rhine water on XAD-4/8, elution with DMSO, and testing the DMSO concentrate in the Ames test. The results correspond to 0.75 L of water per plate.

all cities where no toxic effects were observed. Chlorine dioxide (1 mg/L), however, showed varying results (no effect to a slight increase), indicating that the composition of the treated drinking water may play a significant role in this matter. It appeared that the acid fraction often showed toxic properties toward the bacteria, and therefore it is rather difficult to interpret these results. The promutagenic activity with strain TA98 + S9 (next-to-bottom graph in Figure 3) showed after the chlorine treatment an increase in activity with the exception of city 6, while application of chlorine dioxide increased the activity significantly only in city 4. The promutagenic activity with strain TA100 + S9 (bottom graph in Figure 3) showed for both treatments almost similar results: hardly any or no effect in cities 1, 2, and 5, an increase in city 4, and an apparent decrease in activity in the acid fraction in cities 3 and 6.

CARCINOGENIC EFFECTS

The possible toxicity of drinking water and organic concentrates prepared from drinking water has been studied both in vivo and in vitro. Within the scope of this chapter, discussion of these studies will be restricted mainly to carcinogenic studies of organic drinking water concentrates.

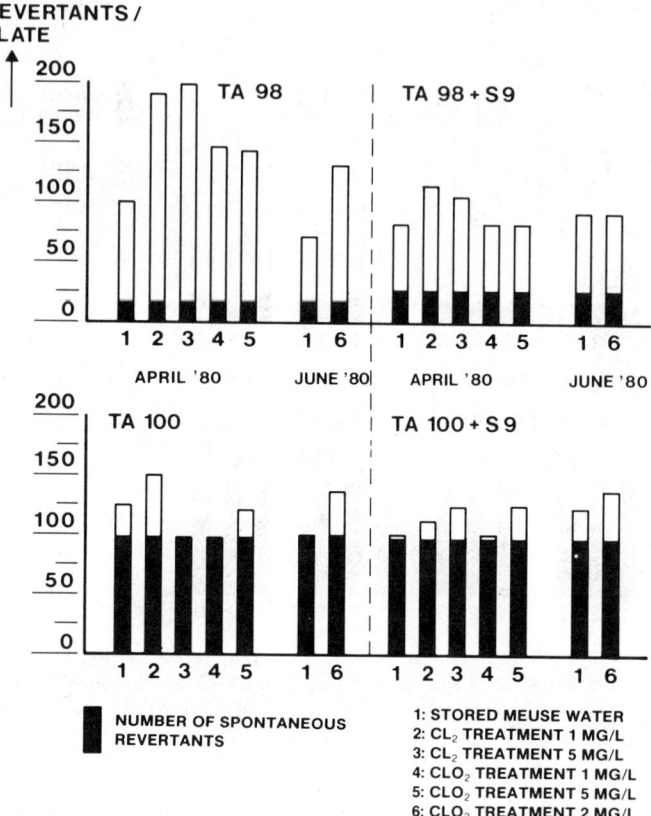

Figure 2. Effect of a chlorine and a chlorine dioxide treatment on mutagenic activity. The sampling: 4000-fold concentration of stored Meuse water on XAD-4/8, elution with DMSO, and testing the DMSO concentrate in the Ames test. The results correspond to 2 L of water per plate.

Several studies regarding the carcinogenicity of drinking water organics have been carried out in which the mouse (skin) was exposed to drinking water concentrates. In somewhat older studies in the United States, chloroform–activated carbon extracts showed both negative[42] and positive[43] results. In the experiment with a positive result, a total dose of 56 mg organic material was administered subcutaneously. Using the same procedure, drinking water concentrates from two areas with respectively a high and low incidence of bladder cancer did not induce carcinogenic effects after subcutaneous injection in mice.[44] This negative result may, however, be due to the relatively low dose to which the mice were exposed (total dose 5 mg organic material). Applying another chloroform extraction procedure,[45] organics in French drinking water indeed showed tumor-promoting activity in a mouse skin test.[46]

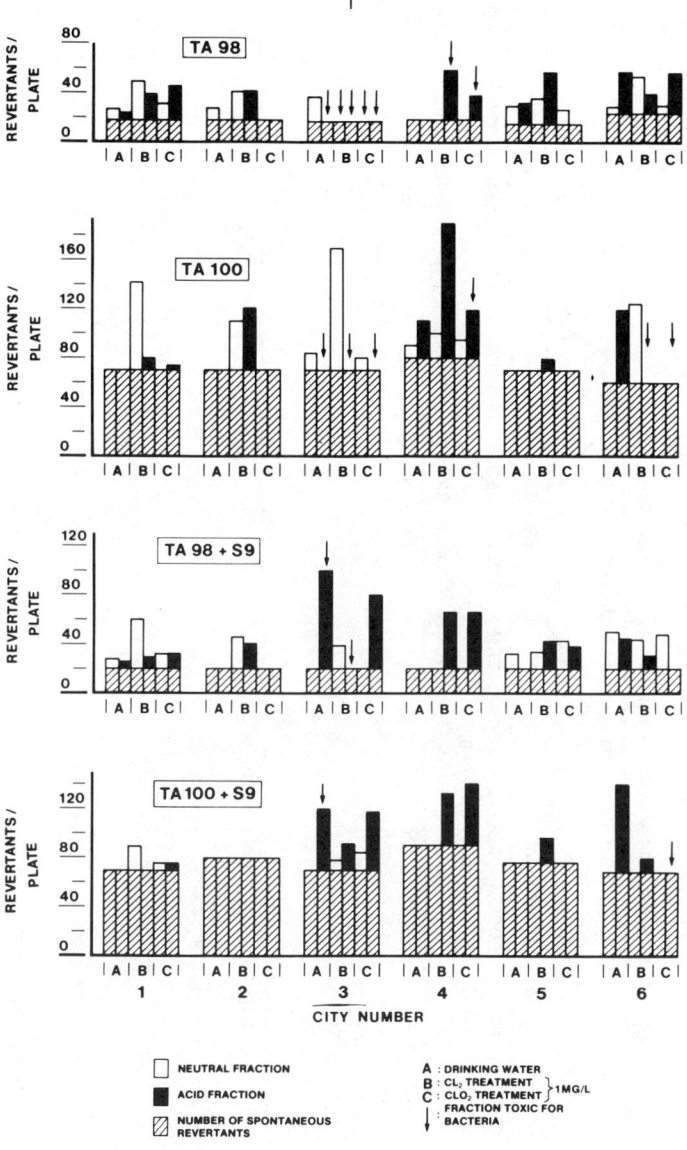

Figure 3. Effect of a chlorine and a chlorine dioxide treatment on mutagenic activity in drinking water. The sampling: 7000-fold concentration of drinking water on XAD-4/8, elution with DMSO, and testing the DMSO concentrate in the Ames test. The city numbers refer to the six cities depicted in Table IV, and the results correspond to 3.5 L of water per plate.

WATER CHLORINATION

Table IV. Effect of ClO_2 and Cl_2 Treatment on the Level of Halogenated Hydrocarbons in Drinking Water

		Halogenated Hydrocarbons in Drinking Water[b]												
		Before Treatment				After Treatment							
						With ClO_2				With Cl_2			
City No.	Source[a] of Drinking Water	THM (µg/L)	VOCl (nmol/L)	EOCl (nmol/L)	AOCl (µmol/L)	TOC (mg C/L)	THM (µg/L)	VOCl (nmol/L)	EOCl (nmol/L)	AOCl (µmol/L)	THM (µg/L)	VOCl (nmol/L)	EOCl (nmol/L)	AOCl (µmol/L)
1	SW	ND	ND	35	1.2	2.5	ND	9	30	1.0	75	1380	55	2.9
2	GW	ND	ND	10	0.6	1.8	<0.1	<6	10	0.6	29.4	480	40	2.7
3	GW	ND	ND	70	0.2	2.0	0.9	24	50	2.3	12.7	655	175	5.3
4	GW	ND	ND	50	1.0	8	0.05	6	35	1.3	5.2	18	45	2.8
5	GW	0.5	<6	5	<0.1	<0.1	1.2	6	3	<0.1	0.65	27	13	0.4
6	GW	ND	ND	5	<0.1	4.0	ND	<6	0.9	0.9	11.7	415	15	4

[a]GW, groundwater (not chlorinated); SW, surface water (not chlorinated).
[b]ND, not detectable.

In another French drinking water study, with concentrates of Paris drinking water (neutral chloroform extract), Truhaut et al.[47] showed that an increase in tumor induction was observed in rats and mice exposed to 100 and 200 times the human dose, expressed in milligrams per kilogram of body weight; a dose-response effect was seen only in the rats. A list of specific tumors found in the experiment, however, was not presented in the study. The positive results in 3 studies[43,46,47] out of 5 suggest that a chloroform extraction procedure seems effective to concentrate those organics that may give a carcinogenic response. However, in 2 of these 3 studies[43,47] a matter of concern is that the control groups did not receive a "control" chloroform extract.[48] Additionally, information as to whether these drinking water concentrates showed mutagenic activity is unfortunately not available. Mutagenic drinking water concentrates[49,50] obtained by reverse osmosis (RO) in combination with XAD resin also gave positive results in a mouse skin initiation/promotion test.[51,52] Concentrates of drinking water of 2 cities out of 5 (one RO and one XAD concentrate) increased the number of papillomas in the presence of the promoter phorbol myristate acetate, proving initiating properties. The tumor-promoting potential of all 10 samples (5 XAD and 5 RO concentrates) was also tested and found to be negative. All samples failed to be complete carcinogens at a total dose up to 30 mg organic material per mouse. These results are questionable, however, since the duration of the experiment (38 weeks) is certainly not long enough to test for complete carcinogenicity.

Bull[53] investigated the health effects of disinfection with chlorine, chloramine, chlorine dioxide, and ozone and their reaction products in coagulated and sand-filtrated Ohio River water, also in the mouse skin test. The dose of disinfectants was adjusted to demand (chlorine or ClO_2, 2 to 2.5 mg/L; chloramine, 3 mg/L; or ozone, 1 to 1.15 mg/L). It was found that the nondisinfected water, after 100-fold concentration of the organics, was inactive both in the absence and in the presence of phorbol myristate acetate, while increased skin tumor incidences were observed with water treated with chlorine, ozone, and chloramine (106-, 186-, and 142-fold concentration). Chlorine dioxide-treated water (160-fold concentration) did not increase the tumor incidence. In a similar experiment about 20 months later, the same water was examined before and after disinfection; in that experiment, the nondisinfected water also showed an increase in tumor yield. In another study, Bull et al.[52] found an increase in carcinogenic activity by disinfection of drinking water with chlorine, ozone, chloramine, and chlorine dioxide. Subsequent experiments, however, failed to confirm this pattern; it appeared here too that the nondisinfected water also showed this kind of activity. These results suggest that the composition as well as certain precursors present in the untreated water will greatly influence the toxicity of the water.

In The Netherlands, a long-term carcinogenicity study with a mixture of chlorinated hydrocarbons frequently detected in drinking water supplies has been recently carried out.[54] The mixture consisted of 11 compounds: trichloroethene, tetrachloroethane, tetrachloromethane, chloroform, dichlorobromomethane, dichlorobenzenes (*o, m,* and *p*), and trichlorobenzenes (1,2,3; 1,2,4;

and 1,2,5) and was based on equal weights. The organic mixture was administered in the drinking water to 4 groups of rats, which received, daily, 0, 0.22, 2.2, and 22 mg of the mixture per kilogram of body weight during a period of 27 months. After this period, no increase of tumor incidence was observed.[15] A long-term carcinogenicity study of organic drinking water concentrates that contained mutagenic activity and that were obtained by the XAD-4/8 procedure has been finished recently.[55] In this study, rats were administered, in their drinking water, organic concentrates prepared from chlorinated drinking water (source: river Rhine) for a period of 24 months. This study showed that mutagenic organic drinking water concentrates up to 68 times the calculated human exposure level (on a per body weight basis) from water consumption (2 L/day, body weight 70 kg) did not increase the tumor incidence in rats.[55]

What is the significance of this negative carcinogenicity study with mutagenic drinking water concentrates? One of the problems in this respect is the fact that the organic mutagens concentrated by the XAD-4/8 procedure only represent a small part, of the order of 1%, of the total organic material present in drinking water. Therefore, it cannot be excluded that the organic constituents left in drinking water may influence the results obtained in this present study. However, it is not unrealistic to assume that the mutagenic drinking water concentrates (XAD concentrates) contain most organic mutagens/carcinogens, since organic mutagens and carcinogens tend to be nonpolar[56] and the XAD procedure is well known to concentrate these less polar compounds.[57] If it is further assumed that only the organics present in the mutagenic drinking water concentrate will contribute to carcinogenicity and an extrapolation of data from lower animals to man is allowed, then the negative results in this carcinogenicity study, in which the highest group received 68 times the expected human exposure level, may be used to estimate a risk for people who consume this drinking water. Statistically, the absence of cancer in a group of 100 test animals means only that at a 99% confidence level the true incidence is less than 5%.[58] At 99% confidence level the true incidence in this experiment with a "natural" incidence of malignant tumors of 10% is less than 0.22 for females.[58] In this study no malignant tumors were induced at a dose of 68 times the expected human exposure level. By linear extrapolation,[59] one can estimate that the risk of cancer at the expected human exposure level is less than $(1/68) \times 0.22$. For the population of this Dutch city (110,000), this would imply $(1/68) \times 0.22 \times 110,000 = 356$. Therefore, the negative results in this experiment tell us that fewer than 356 people might be at risk. On this basis the contribution of drinking water, if there is a contribution at all, should be relatively small, less than 1.1% of the expected tumor incidence of 33,000 in a population of 110,000. The latter value is based on an average tumor incidence of 30 per 100 due to background processes.[60]

To improve, however, the reliability of the estimation of the risk to people from consumption of drinking water, organics of drinking water excluded in this carcinogenicity study by the XAD procedure should, in combination with a dose of the XAD concentrate higher than that used in this study, be examined for carcinogenic activity.

Besides the examination for complete carcinogenicity, these organic concentrates should be tested for promoting and/or initiating activity. Identification of the organics responsible for these effects should also be carried out.

CONCLUSIONS

A chlorine treatment generally increases mutagenic activity significantly. An ozone treatment generally decreases mutagenic activity with strain TA98 but is able to increase the mutagenic activity slightly with strain TA100. Chlorine dioxide is able to increase mutagenic activity in some cases. The increase, however, is much less than that with chlorine treatment. Chlorine dioxide seems, therefore, a good alternative for chlorine treatment particularly when less than 1 mg/L ClO_2 is applied. Mutagenic organic drinking water concentrates obtained by RO and/or XAD concentration procedures, when administered in drinking water, do not induce carcinogenic effects at the dose levels tested.

REFERENCES

1. Zoeteman, B. C. J. *Sensory Assessment and Chemical Composition of Drinking Water*, Thesis, (Utrecht, The Netherlands: State University of Utrecht, 1978).
2. National Academy of Sciences. "Drinking water and Health," Vols. 1 and 2 (Washington DC: National Academy of Sciences, 1980).
3. Packham, R. F., S. A. Beresford, and M. Fielding. "Health Related Studies of Organic Compounds in Relation to Reuse in the U.K.," in: *Water Supply and Health*, H. van Lelyveld and B. C. J. Zoeteman, Eds. (New York: Elsevier Scientific, 1981), pp. 167-186.
4. Eklund, G., B. Josefsson, and C. Ross. Trace Analyses of Volatile Organic Substances in Göteborg Municipal Drinking Water," *Vatten* 34:195-206 (1978).
5. Benoit, F. M., G. L. Lebel, and D. T. Williams. "The Determination of Polycyclic Aromatic Hydrocarbons at the ng/L Level in Ottawa Tap Water." *Int. J. Environ. Anal. Chem.* 6:277-287 (1979).
6. "IARC Annual Report," (Lyon, France: World Health Organization, International Agency for Research on Cancer, 1979).
7. *IARC Monographs on the Evaluation of the Carcinogenic Risk of Chemicals to Humans, Vol. 20,* "Some Halogenated Hydrocarbons" (Lyon, France, World Health Organization, International Agency for Research on Cancer, 1980).
8. "Report of an IARC Working Group. An Evaluation of Chemicals and Industrial Processes Associated with Cancer in Humans Based on Human and Animal Data: IARC Monographs Volumes 1 to 20," *Cancer Res.* 40:1-12 (1980).
9. Simmon, V. F., K. Kauhanen, and R. C. Tardiff. Mutagenic Activity of Chemicals Identified in Drinking Water," in *Progress in Genetic Toxicology*, D. Scott, B. A. Bridges, and F. H. Sobels, Eds. (New York: Elsevier/North Holland, Biomedical Press, 1977), pp. 249-258.
10. National Academy of Sciences. *Drinking Water and Health, Vols. 3 and 4* (Washington DC: National Academy of Sciences, 1980, 1981).

11. Tabor, M. W., and J. C. Loper. "Separation of Mutagens from Drinking Water Concentrates Using Coupled Bioassay/Analytical Fractionation," *Int. J. Environ. Anal. Chem.* 8:197-215 (1980).
12. Tabor, M. W. "Structure Elucidation of 3-(2-chloroethoxy)-1,2-dichloro propene, a New Promutagen from an Old Drinking Water Residue," paper presented at the 12th Annual Symposium on the Analytical Chemistry of Pollutants, Amsterdam, April 14-16, 1982.
13. Kool, H. J., C. F., van Kreijl, H. J. van Kranen, and E. de Greef. "The Use of XAD Resins for the Detection of Mutagenic Activity in Water. I. Studies with Surface Water," *Chemosphere* 10:85-89 (1981).
14. Ames, B. N., J. McCann, and E. Yamasaki. "Methods for Detecting Carcinogens and Mutagens with the *Salmonella*/Mammalian-Microsome Mutagenicity Test," *Mutat. Res.* 31:347-364 (1975).
15. Wegman, R. C. C., and P. A. Greve. "The Microcoulometric Determination of Extractable Organic Halogen in Surface Water. Application to Surface Waters of The Netherlands," *Sci. Total Environ.* 7:235-245 (1977).
16. Sander, R. "Verbesserung des Pyrohydrolyse Verfahrens," *Veroeff. Ber. Lehrstuhls Wasserchem.* 15:128-162 (1980).
17. Wegman, R. C. C., and A. W. M. Hofstee. "The Microcoulometric Determination of Volatile Organic Halogen in Water Samples," paper presented at the European Symposium on Analysis of Organic Micropollutants in Water, Berlin, December 11-13, 1979.
18. Zoeteman, B. C. J., J. Hrubec, E. de Greef, and H. J. Kool. "Mutagenic Activity Associated with By-Products of Drinking Water Disinfection by Chlorine, Chlorine Dioxide, Ozone and U.V. Irradiation," *Environ. Health Perspect.* 46:197-205 (1982).
19. Paling, A. T. "Analytical Control of Water Disinfection with Special References to Differential DPD Methods for Chlorine, Chlorine Dioxide, Bromine, Iodine and Ozone," *J. Inst. Water Eng.* 28:139-154 (1974).
20. Berndt, H. "Untersuchungsmethoden zur Bestimmung von ClO_2, $NaClO_2$ und Cl_2," *Staedthygiene* 11:224-228 (1960).
21. Valenta, J., and W. Gabler. "Chlordioxid Anlage," *Gas, Wasser, Abwasser* 55:566-569 (1975).
22. Prein, A. E., G. M. Thie, G. M. Alink, J. H. Koeman, and C. L. M. Poels. "Cytogenic Changes in Fish Exposed to Water of the River Rhine," *Sci. Total Environ.* 9:287-291 (1978).
23. Van Kreijl, C. F., H. J. Kool, M. de Vries, H. J. van Kranen, and E. de Greef. "Mutagenic Activity in the Rivers Rhine and Meuse in The Netherlands," *Sci. Total Environ.* 15:137-147 (1980).
24. Slooff, W., and C. F. van Kreijl. "Monitoring the Rivers Rhine and Meuse in The Netherlands for Mutagenic Activity Using the Ames Test in Combination with Rat or Fish Liver Homogenates," *Aquat. Toxicol.* 2:89-98 (1982).
25. C. L. Poels. "Untersuchung uber die Mutagene Eigenschaften des Rheinwassers" (in "8er IAWR Tagung", 107-119, IAWR, Amsterdam), paper presented at the 8th IAWR Conference, Amsterdam, May 18-20, 1981.
26. Van der Gaag, M. A., A. Noordsij, and J. P. Oranje, "Presence of Mutagens in Dutch Surface Water and Effects of Water Treatment Processes for Drinking Water Preparation," in *Progress in Clinical and Biological Research, Vol. 109, Mutagens in Our Environment* M. Sorse and H. Vaino, Eds. (New York: Alan R. Liss, Inc., 1982), pp. 277-286.

27. Cheh, A. M., J. Skochdopole, P. Koski, and L. Cole. "Nonvolatile Mutagens in Drinking Water: Production by Chlorination and Destruction by Sulfite," *Science* 207:90-92 (1980).
28. De Greef, E., J. C. Morris, C. F. van Kreijl, and C. H. F. Morra. "Health Effects in the Chemical Oxidation of Polluted Waters," in *Water Chlorination: Environmental Impact and Health Effects, Vol. 3,* R. L. Jolley, W. A. Brungs, and R. B. Cumming, Eds. (Ann Arbor, MI: Ann Arbor Science Publishers, Inc., 1980), pp. 913-924.
29. Maruoka, S., and S. Yamanaka. "Mutagenic Potential of Laboratory Chlorinated Water," *Sci. Total Environ.* 29:143-154 (1983).
30. Maruoka, S., and S. Yamanaka. "Production of Mutagenic Substances by Chlorination of Waters," *Mutat. Res.* 79:381-386 (1980).
31. Dolara, P., V. Ricci, D. Burrini, and O. Griffini. "Effect of Ozonation and Chlorination on the Mutagenic Potential of Drinking Water," *Bull. Environ. Contam. Toxicol.* 27:1-6 (1981).
32. Kool, H. J., C. F. van Kreijl, E. de Greef, and H. J. van Kranen. "Presence, Introduction and Removal of Mutagenic Activity During the Preparation of Drinking Water in The Netherlands," *Environ. Health Perspect.* 46:207-214 91982).
33. Kool, H. J., and C. F. van Kreijl. "Formation and Removal of Mutagenic Activity During Drinking Water Preparation, *Water Res.* 18:1011-1016 (1984).
34. Flanagan, E. P., and H. E. Allen. "Effect of Water Treatment on Mutagenic Potential," *Bull. Environ. Contam. Toxicol.* 27:765-772 (1981).
35. Hrubec, J., and H. J. Kool. Unpublished data (1982).
36. Gruener, N. "Mutagenicity of Ozonated Recycled Water," *Bull. Environ. Contam. Toxicol.* 20:522-526 (1978).
37. Cotruvo, J. A., V. E. Simmon, and R. J. Spanggord. "Investigations of Mutagenic Effects of Products of Ozonation Reactions in Water," *Ann. N.Y. Acad. Sci.* 298:124-140 (1978).
38. Burleson, G. R., and T. M. Chambers. "Effect of Ozonation on the Mutagenicity of Carcinogens in Aqueous Solution," *Environ. Mutat.* 4:469-476 (1982).
39. Van Hoof, F. "Influence of Ozonation on Direct Acting Mutagens Formed During Drinking Water Chlorination," in *Water Chlorination: Environmental Impact and Health Effects, Vol. 4,* R. L. Jolley, W. A. Brungs, J. A. Cotruvo, R. B. Cumming, J. S. Mattice, and V. A. Jacobs, Eds. (Ann Arbor, MI: Ann Arbor Science Publishers, Inc., 1983), pp. 1211-1220.
40. Bourbigot, M. M., L. H. Pottenger, J. L. Paquin, M. F. Bleck, and P. Hartman. "Study of Mutagenic Activity in Water in a Progressive Ozonation Unit," *Aqua* 3:99-102 (1983).
41. Jolley, R. L., R. B. Cumming, N. E. Lee, L. R. Lewis, J. E. Thompson, and C. I. Mashni. "Nonvolatile Organics in Disinfected Wastewater Effluents: Chemical Characterization," in *Water Chlorination: Environmental Impact and Health Effects, Vol. 4,,* R. L. Jolley, W. A. Brungs, J. A. Cotruvo, R. B. Cumming, J. S. Mattice, and V. A. Jacobs, Eds. (Ann Arbor, MI: Ann Arbor Science Publishers, Inc., 1983), pp. 449-523.
42. Hueper, W. C., and C. C. Ruchhoft, "Carcinogenic Studies on Adsorbates of Industrially Polluted Raw and Finished Water Supplies," *Arch. Ind. Hyg.* 9:488-495 (1954).
43. Hueper, W. C., and W. W. Payne. "Carcinogenic Effects of Adsorbates of Raw and Finished Water Supplies," *Am. J. Clin. Pathol.* 39:475-481 (1963).

44. Dunham, L. J., R. W. O'Hara, and F. B. Taylor. "Studies on Pollutants from Processed Water: Collection from Three Stations and Biologic Testing for Toxicity and Carcinogenesis," *Am. J. Public. Health* 57:2178–2185 (1967).
45. Cabridenc, R., and A. Sdika. "Quelques Aspects de l'Extraction et de l'Identification des Micropollutants des Eaux," *Tech. Sci. Munic.* 70:285–388 (1975).
46. Hemon, D., P. Lazare, R. Cabridenc, A. Sdika, B. Festy, C. Gérin-Roze, and I. Chouroulinkov. "Micropollution Organique des Eaux Destinées à la Consommation Humaine," *Rev. Epidem. Santé Publ.* 26:441–450 (1978).
47. Truhaut, R., J. C. Gak, and C. Graillot. "Recherches sur les Risques Pouvant Resulter de la Pollution Chimique des Eaux d'Alimentation," *Water Res.* 13:689–697 (1979).
48. Kool, H. J., C. F. van Kreijl, and B. C. J. Zoeteman. "Toxicology Assessment of Organic Compounds in Drinking Water," *CRC Environ. Control* 12:307–358 (1982).
49. Loper, J. C., D. R. Lang, R. S. Schoeny, R. B. Richmond, P. M. Gallagher, and C. C. Smith. "Residue Organic Mixtures from Drinking Water Show in Vitro Mutagenic and Transformation Activity," *J. Toxicol. Environ. Health* 4:919–938 (1978).
50. Kopfler, F. C., W. E. Coleman, R. G. Melton, R. G. Tardiff, S. C. Lynch, and J. H. F. Smith. "Extraction and Identification of Organic Micropollutants: Reverse Osmosis Method," *Ann. N.Y. Acad. Sci.* 298:20–30 (1977).
51. Robinson, M., J. Glass, D. Chmehil, R. J. Bull, and J. Orthoefer. "The Initiating and Promoting Activity of Chemicals Isolated from Drinking Waters in the Sencar Mouse: A Five City Survey," in *Short-Term Bioassays in the Analysis of Complex Environmental Mixtures, Vol. II*, M. D. Waters, S. S. Sandhu, J. L. Huisingh, L. Claxton, and S. Nesnow, Eds. (New York: Plenum Press, 1981).
52. Bull, R. J., M. Robinson, J. R. Meier, and J. Stober. "Use of Biological Assay Systems to Assess the Relatively Carcinogenic Hazards of Disinfection By-Products," *Environ. Health Perspect.* 46:215–227 (1982).
53. Bull, R. J. "Health Effects of Alternative Disinfectants and Their Reaction Products," *J. Am. Water Works Assoc.* 72:299–303 (1980).
54. Van der Heijden, C. A., and G. J. Van Esch. "Onderzoek naar de Carcinogeniteit bij de Rat van Gechloreerde Koolwaterstoffen die in Grondwater Voorkomen," (Abstract), Microsymposium, Section on Genetic Toxicology and Chemical Carcinogenesis, Rijksinstituut voor Drinkwatervoorziening, Leidschendam, The Netherlands, 1980.
55. Kool, H. J., F. Kuper, H. van Haeringen, and J. H. Koeman. "A Carcinogenicity Study with Mutagenic Organic Drinking Water Concentrates in The Netherlands," *Food Chem. Toxicol.* 23:79–85 (1985).
56. Yamasaki, E., and B. N. Ames. "Concentration of Mutagens from Urine by Adsorption with the Non-Polar Resin XAD-2: Cigarette Smokers Have Mutagenic Urine," *Proc. Natl. Acad. Sci. USA* 74:3555–3559 (1979).
57. Webb, R. G. "Isolating Organic Water Pollutants, XAD Resins, Urethane Foam, Solvent extraction," EPA-660/4-75-003, (Corvallis, OR: U.S. Environmental Protection Agency, 1975).
58. Hoel, D., D. W. Gaylor, R. K. Kirschstein, U. Saffiotti, and M. A. Schneiderman, "Estimation of Risks of Irreversible, Delayed Toxicity," *J. Toxicol. Environ. Health* 1:133–151 (1975).

59. Weinhouse, S. "Problems in the Assessment of Human Risk of Carcinogenesis from Chemicals," in *Origins of Human Cancer, Vol. 4*, H. H. Hiatt, J. D. Watson, and J. A. Winsten, Eds. (Cold Spring Harbor, NY: Cold Spring Harbor Laboratory, 1977).
60. Gezondheidsraad. "Advies Inzake de Beoordeling van Carcinogeniteit van Chemische Stoffen," Rapport No. 1978/19. Rijswijk, The Netherlands: Uitgebracht door een Commissie van de Gezondheidsraad, 1978.

CHAPTER 17

Mutagenic Properties of Drinking Water Disinfectants and By-products

John R. Meier and Richard J. Bull

The identification of a number of mutagenic and carcinogenic chemicals in our public water supplies has raised concern over potential genetic and carcinogenic hazards to the human population.[1-3] Growing evidence indicates that these chemicals are produced during water chlorination,[4,5] and, consequently, alternative strategies for water disinfection are being considered. Unfortunately, it is not known to what extent the mutagenic activity in chlorinated drinking water and its associated potential health risks are accounted for by chemicals identified thus far. Our laboratory has been exploring the use of humic acid for studying the mutagenic and carcinogenic properties of disinfection by-products.[6-8] This chapter extends our findings on the mutagenic properties of chlorinated humic acids in the Ames test; it includes results from studies on the ability of chlorinated and nonchlorinated humic acids to induce sister chromatid exchange (SCE) in vitro and to produce spermhead abnormalities and micronuclei in bone marrow in mice in vivo.

Since disinfectant chemicals are generally added at levels sufficient to produce disinfectant residuals during distribution, the concern over potential health risks arising form the use of disinfectants may extend to the disinfectants themselves or to by-products formed in vivo. This idea is supported by results in bacterial assays which suggest that chlorine and monochloramine are capable of inducing DNA damage and causing mutation.[9-11] In addition, halogenated organic compounds with known mutagenic and carcinogenic properties have been shown to be formed in vivo following oral dosing of rats with sodium hypochlorite.[12] Because of these findings, tests have been conducted on the mutagenic potential of various disinfectants in vivo by examining chromosomal damage in bone marrow and spermhead abnormalities in mice. The results of these tests are summarized here.

Studies are also being conducted on the toxicological properties of drinking water samples prepared using alternative techniques for disinfection and post-disinfection treatment. Preliminary results on the mutagenic activities of these samples are discussed.

MATERIALS

Humic acid solutions were prepared from Fluka humic acid (Fluka AG, Buchs, Switzerland) at a concentration of 1-g total organic carbon (TOC) per liter distilled water. Chlorination was performed at pH 7 or 11.5 without buffer, and at pH 7 with the addition of 0.25 M (final concentration) sodium phosphate buffer. At all pH reaction conditions, a 1:1 molar ratio of chlorine (as Cl) to carbon (as TOC) was used. Further details of the methods for preparation and chlorination of the humic acids have been described previously.[7]

All other chemicals were purchased as reagent grade quality from commercial suppliers.

METHODS

In Vitro Bioassays

Mutagenicity assays were conducted with *Salmonella typhimurium* strains TA98 and TA100 according to the methods described by Maron and Ames.[13] Testing was performed at three or four dose levels using duplicate or triplicate plates per dose. Liver homogenate fraction (S9) was prepared from Aroclor 1254 treated male Sprague-Dawley rats and added to the S9 mix at a 5% (v/v) level. Assessment of a positive mutagenic response was based on a dose-related increase in the number of histidine revertant colonies that exceeded twice the background number of revertants for at least one dose level.

The methods used for evaluating the ability of test samples to induce SCEs in vitro using Chinese hamster ovary (CHO) cells have been described in detail elsewhere.[14] A one-tailed Student's t-test at the 0.01 level of significance was used to test for significant increases in the number of SCEs per cell over the concurrent control levels.

In Vivo Bioassays

The procedures for conducting the mouse bone-marrow micronucleus and chromosomal aberration assays and the mouse sperm morphology assay are described elsewhere.[15] Strain CD-1 mice (Charles River Breeding Laboratories) were used for the micronucleus and chromosomal aberration assays, and B6C3F1 hybrid mice (Harlan Industries) were used for the sperm morphology assay. The same subchronic dosing regimen was used for studies involving both the disinfectant chemicals and the humic acid solutions. This involved administration of the dosing solutions by oral gavage at three dose levels (undiluted, 1:1 dilution, and 1:4 dilution) in five 1-mL doses on consecutive days.

Preparation of Water Samples Treated with Alternative Disinfectants

Mississippi River water that had been clarified and sand filtered was subjected to disinfection by treatment with either ozone, chlorine, chlorine dioxide, or monochloramine. The treatment contact time was 2 h at initial disinfectant concentrations of 2 to 5 mg/L for chlorine, monochloramine, and chlorine dioxide, and 0.2 to 5 mg/L for ozone. Granular activated carbon (GAC) treatment of the chlorine-treated water, with or without post-GAC rechlorination, was an additional treatment option. Organic concentrates were prepared by passing the water samples (after acidification to pH 2) over XAD-8 and XAD-2 resins in sequence, eluting with acetone, and reducing the volume of the combined acetone eluates by rotary evaporation. Following removal of the acetone, the volume was adjusted with emulphor and distilled water to obtain a 4000-fold concentrate in 2% emulphor. The samples were tested for mutagenicity in the *Salmonella*/microsome mutagenicity assay.[13]

RESULTS AND DISCUSSION

Studies on Mutagenic Potential of Disinfectants

Chlorine, chlorine dioxide, monochloramine, chlorite, and chlorate were tested for their ability to produce chromosomal damage in a somatic cell line (as evaluated with the mouse bone marrow micronucleus and chromosomal aberration assays) and for their mutagenic potential to a germ cell line (as evaluated in the mouse sperm morphology assay). Since the species of chlorine that predominates during water treatment is pH dependent,[16] chlorine was prepared and tested at both pH 6.5 (where HOCl predominates) and at pH 8.5 (where the OCl$^-$ predominates). Sodium chlorite and sodium chlorate were included in the testing of the disinfectants to determine whether any activity associated with chlorine dioxide might be due to the presence of chlorite or chlorate, which are major inorganic by-products formed upon the addition of chlorine dioxide to water.[17] The results are summarized in Table I. The detailed data from these evaluations are discussed elsewhere.[18]

Chlorine at pH 8.5 elicited a positive response in the mouse sperm morphology assay. No other effects with any of the disinfectants were noted. The effect with chlorine, which was seen at the spermatocyte stage of spermatogenesis, was weak (about twofold above background at the maximum response) but reproducible and statistically significant for at least two dose levels in both the initial and repeat trials. Failure to observe an effect with chlorine as OCl$^-$ in the bone-marrow micronucleus or chromosomal aberration assays may reflect differences in assay sensitivities or differences in tissue distribution of OCl$^-$ or its by-products. The work of Bruce and Heddle[18] seems to support the former explanation, since results from their studies on 61 compounds indicated that the sperm assay has a greater success rate than the micronucleus assay for detecting agents that are active in vivo. Induction of spermhead abnormalities in mice is not considered as definitive evidence of the mutagenicity or carcinogenicity of a chemical.[19] Nevertheless, the positive results obtained with chlorine raise concern over potential genetic and carcinogenic risks to the human population.

Because it seemed likely that a reaction product formed in vivo (rather than OCl^- itself) might be responsible for the effect on sperm morphology, we decided to examine for the presence of mutagenic activity in the urine from animals treated with chlorine at pH 8.5. A similar dosing regimen to that used in the sperm morphology assay was followed. In these experiments, rats were used to obtain sufficient volumes for adequate testing. *Salmonella*/microsome mutagenicity assays were conducted on both unconcentrated urine and 50-fold XAD-2 concentrates. Assays were performed both in the presence and absence of β-glucuronidase for detecting mutagens as their glucuronide conjugates. The results from these assays were negative and therefore provided no evidence for the presence of mutagens in the urine.

Genotoxic Properties of Humic Acid Chlorination By-Products

In Vitro Studies

Previous work[6,7] demonstrated that chlorination of humic acid resulted in the formation of direct-acting mutagens in the Ames test. The level of mutagenicity formed was dependent on the chlorination pH, although this observation appeared to be largely attributable to the instability of the mutagenic activity at alkaline pH. We were interested in knowing if the genotoxic potential and the pH dependency of activity of chlorinated humic acid would be

Table I. Evaluation of Mutagenic and Carcinogenic Potential of Disinfectants in Three In Vivo Bioassays

Disinfectant	Mouse micronucleus assay	Mouse bone marrow cytogenetics assay	Mouse spermhead abnormality assay
Chlorine, pH 6.5 (HOCl)	−	−	−
Chlorine, pH 8.5 (OCl^-)	−	−	+[a]
Monochloramine	−	−	−
Chlorine dioxide	−	−	−
Chlorite	−	−	−
Chlorate	−	−	−

[a]Based on reproducible, dose-related, and statistically significant elevations in percent spermhead abnormalities over concurrent control values at the 3-week posttreatment kill time.

observed in other assay systems. Therefore, the same humic acid samples (before and after chlorination) were tested for mutagenic activity in the Ames test and SCE induction in CHO cells in vitro. The samples were also tested for micronuclei induction in mouse bone marrow and for alteration of mouse spermhead morphology in vivo.

Ames test results demonstrate the presence of mutagenic activity after, but not before, chlorination (Table II). It is also clear that the pH during chlorination has a substantial effect on the level of mutagenicity observed. The results from testing these samples for SCE induction in CHO cells are also summarized in Table II. Each of the humic acid samples produced linear dose-dependent increases in the number of SCEs per cell. For all of the samples, at least the three highest responses were significantly greater than the concurrent control values at the 0.01 level of statistical significance. The slopes of the least-squares regression lines were used in determining the potencies of the samples for SCE induction.

Comparison of the results in the Ames test with those in the SCE assay reveals clear differences in the two assay systems for detecting activity of the samples. When expressed as a percentage of the activities of low pH (sample A), the levels of SCE induction at the higher pHs during chlorination (73% for sample B, 37% for sample C) are substantially higher than the corresponding mutagenic activities in the Ames test (18 to 22% for sample A, 0 to 29% for sample B). Furthermore, there is clear evidence of SCE induction by non-chlorinated humic acids (sample D), whereas humic acids prior to chlorination show no evidence of mutagenicity. SCE induction is generally considered to be evidence of primary DNA damage rather than mutation per se, although there

Table II. Dependence (pH) of Mutagenic Activity in the Ames Test and SCE Induction in CHO Cells Upon Treatment of Humic Acid with Chlorine (HOCl/OCl)

Humic acid sample	Mutagenic activity[a]		SCE Induction[b]
	TA98	TA100	
A-Chlorinated (pH7.0 → 2.5)	339 ± 29 (100)	1696 ± 148 (100)	2.96 (100)
B-Chlorinated[c] (pH7.0 → 6.5)	62 ± 10 (18)	367 ± 34 (22)	2.15 (73)
C-Chlorinated (pH11.5 → 6.5)	NS[d]	490 ± 33 (29)	1.10 (37)
D-Nonchlorinated[e]	NS[d]	NS[d]	0.62 (21)

[a]Net revertants per milliliter of sample, calculated from the linear portion of the dose-response curve from assays without S9 added. Parentheses indicate the percent of activity in sample A.
[b]SCEs per cell per percent sample incorporated into medium, calculated from the linear portion of the dose-response curve from assays without S9 added. Parentheses indicate the percent of activity in sample A.
[c]0.25 M Sodium phosphate buffer was added during chlorination to stabilize the pH.
[d]NS = Not significant (i.e., less than twofold above background at any dose tested).
[e]The nonchlorinated humic acid was prepared at pH 7; the pH was then lowered to 2.5 with HCl.

is generally a good correlation between the two end points with regard to the types of compounds detected.[20] The positive results showing SCE induction in a mammalian cell line provide confirmation of the genotoxic properties of chlorinated humic acids at pHs relevant to drinking water chlorination pH. Test results also demonstrate for the first time the genotoxic potential of nonchlorinated humic acid.

Figures 1 through 4 illustrate some of the problems that might occur when evaluating the mutagenic activity of complex mixtures using in vitro methods. In these experiments, known mutagens were added to the assay plates in the Ames test along with either chlorinated or nonchlorinated humic acids to determine whether the expected additivity of response would be obtained. Figures 1 and 2 show that the mutagenic activities of 2-aminoanthracene (2-AA) and benzo(*a*)pyrene [B(*a*)P] were greatly suppressed by both chlorinated and nonchlorinated humic acids. Since both 2-AA and B(*a*)P require metabolic activation to become mutagenic, it occurred to us that the humic materials might interfere with one or more components of the S9 activation mixture. However, inhibition of the activities of two direct-acting mutagens, sodium azide and 2-nitrofluorene, was also observed (Figures 3 and 4). To better compare the extent of inhibition among the four known mutagens, the observed vs expected activities were determined, and the percent inhibition was calculated. The data in Table III indicate that the extent of inhibition is compound dependent and is dependent on the nature of the humic material (chlorinated vs nonchlorinated). Overall, the activities of the two promutagens (those requiring S9 activation) were apparently inhibited to a greater degree than those of the two direct-acting compounds.

Our results suggest that components in both chlorinated and unfinished water might interfere with the metabolic activation capacity of the S9 mixture, and this interference may mask the presence of promutagens. Humic acids are known to chelate metal ions,[21] including Mg^{2+}, which is a co-factor in the S9 mix needed for NADPH generation. Humic materials are also known to bind other organics,[21] and such binding could serve to make direct-acting mutagens or mutagenic metabolites inaccessible to DNA binding. In the case of chlorinated humic acid samples, which are toxic (i.e., lethal) as well as mutagenic, the toxicity may also contribute to the inhibition of activity. Regardless of the reasons for the inhibition, the present experiments clearly demonstrate antagonism by chlorinated and nonchlorinated humic acids toward the in vitro mutagenic activity of known mutagens. In view of these results, the potential for antagonism among constituents of raw and finished water seems a distinct possibility. Since such antagonism could have a substantial impact on mutagenicity assay results and their interpretation, we believe this finding warrants further attention in mutagenicity studies of water samples.

Previous results from testing humic materials obtained from two natural sources as well as a commercial humic source indicated that the mutagenic activity formed following chlorination was decreased to varying extents by the addition of S9 mix to the assay.[6,7] This observation is also common in muta-

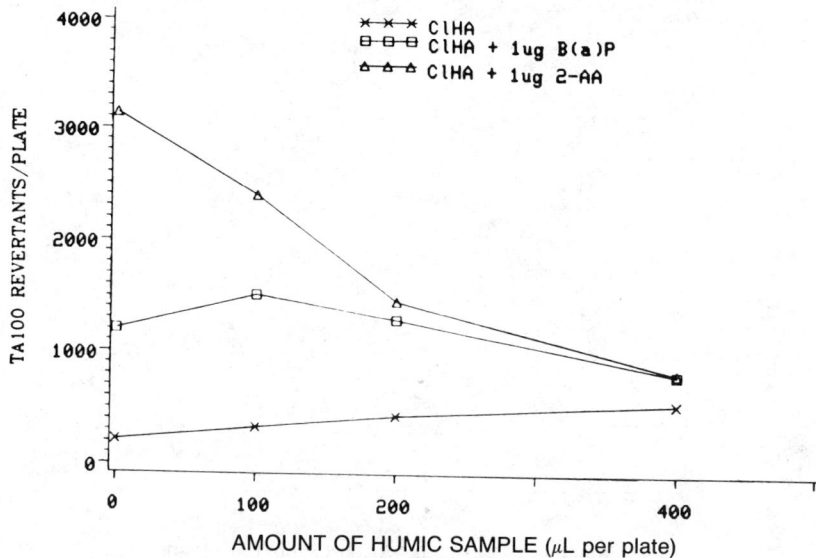

Figure 1. Inhibition of the mutagenicity of 2-aminoanthracene and benzo(a)pyrene for strain TA100 by chlorinated humic acid.

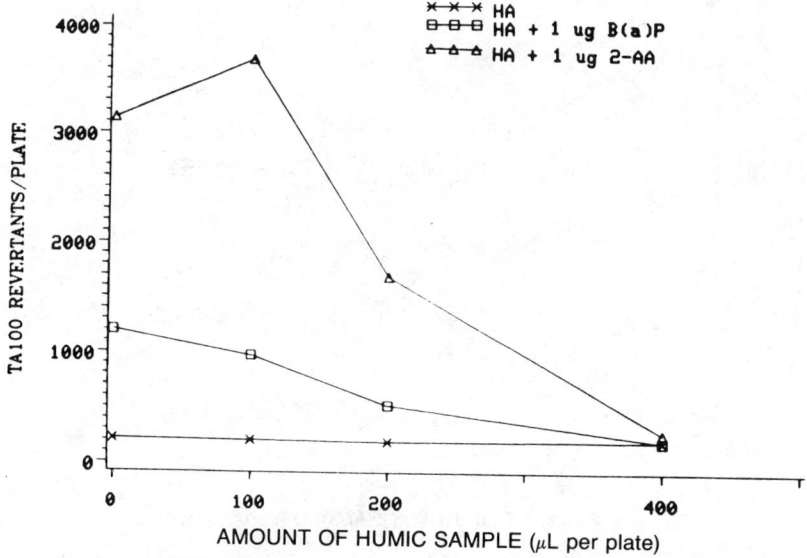

Figure 2. Inhibition of the mutagenicity of 2-aminoanthracene and benzo(a)pyrene for strain TA100 by nonchlorinated humic acid.

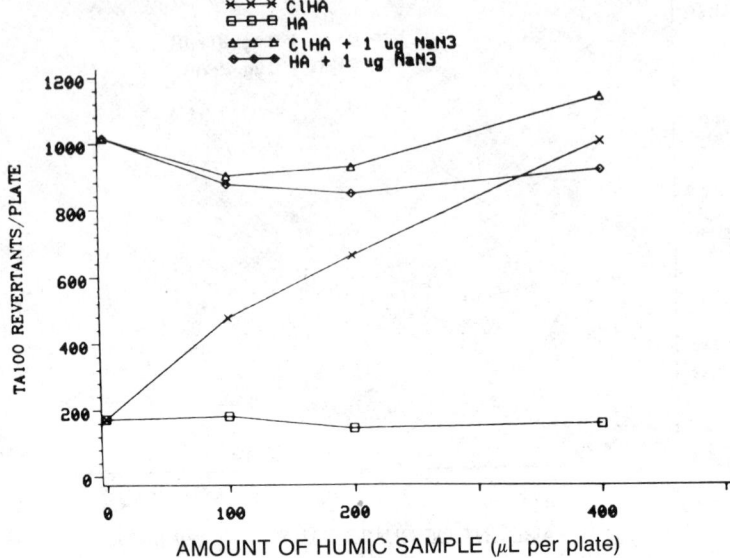

Figure 3. Effect of chlorinated and nonchlorinated humic acids on the direct-acting mutagenicity of sodium azide for strain TA100.

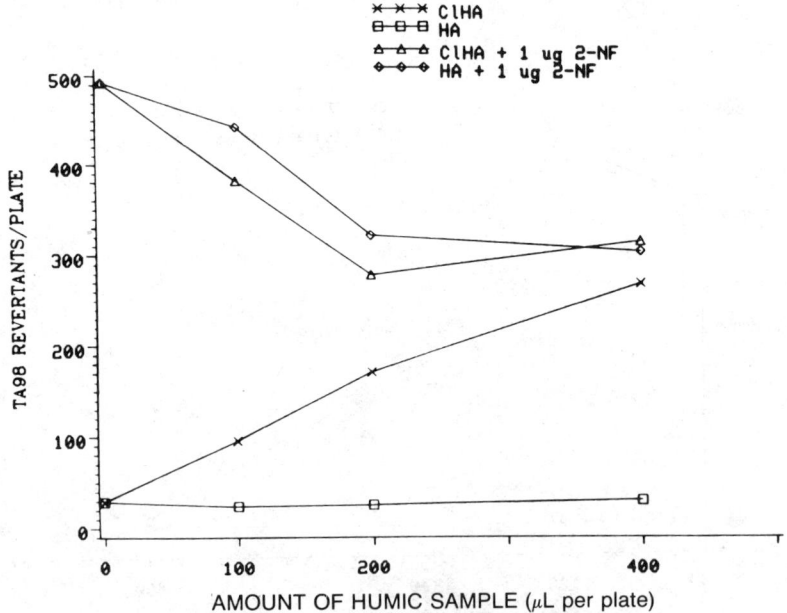

Figure 4. Effect of chlorinated and nonchlorinated humic acids on the direct-acting mutagenicity of 2-nitrofluorene for strain TA98.

Table III. Effect of Chlorinated and Nonchlorinated Humic Acids on Activity of Known Mutagens in the Ames Test

Humic acid sample	Amount sample added (μL/plate)	Inhibition of Mutagenicity (%)					
		TA98[a]			TA100[a]		
		2-NF	2-AA	B(a)P	NaN$_3$	2-AA	B(a)P
Nonchlorinated	100	10	7	72	18	−19	23
	200	37	40	99	18	49	69
	400	41	79	100	10	98	100
Chlorinated	100	34	28	39	37	28	−17
	200	59	57	55	44	60	11
	400	59	98	65	42	81	55

[a]2-NF = 2-Nitrofluorene; 2-AA = 2-aminoanthracene; B(a)P = benzo(a) pyrene; NaN$_3$ = sodium azide.

genicity studies conducted on drinking water concentrates.[3] Eder et al.[22] have suggested that the glutathione (GSH) in liver S9 preparations might be responsible for the reduction of mutagenicity of direct-acting allylic compounds by S9. Because chlorallylic compounds have been identified in chlorinated humic acid solutions,[23] we decided to see whether GSH might be responsible for the S9 inactivation of mutagens in chlorinated humic acids. The results in Figure 5 show that a substantial portion of the mutagenic activity is quenched by GSH. The biphasic shape of the inhibition curve without S9 indicates the presence of mutagenic compounds that are highly sensitive to GSH, as well as compounds that are relatively insensitive.

The fact that much less reduction in activity by GSH occurs when S9 is present is probably attributable to the fact that the activity of GSH-sensitive compounds has already been quenched by the S9. Based on levels of 1.3 to 1.7 mM GSH in S9 mix (i.e., approximately 230 μg per plate),[22] it appeared that the degree of inactivation seen with S9 alone (Figure 5) could be largely accounted for by the GSH present. To confirm this possibility, GSH concentrations were measured in our S9 using the method of Hissin and Hilf[24] and were found to be 1.27 mM in the S9 fraction, or about 5 mM in the whole liver. This value agrees well with reported values for rat liver, which are typically 2 to 10 mM.[25]

Although the levels of S9 fraction in S9 mix were not indicated by Eder et al.,[22] the levels generally used are 4 to 10% (v/v). This would mean that the GSH levels in S9 mix reported by Eder et al.[22] are 10 to 25 times higher than ours. The reason for this discrepancy is unknown; however, it may reflect an error in terminology by Eder et al.[22] (i.e., S9 mix vs S9 fraction). In any case, the level of GSH in our assays with S9 amounts to only about 10 μg per plate, which would account for only a slight reduction in mutagenicity of chlorinated humic acid (Figure 5). Thus, GSH appears to be of only minor importance in the mutagen inactivation by S9.

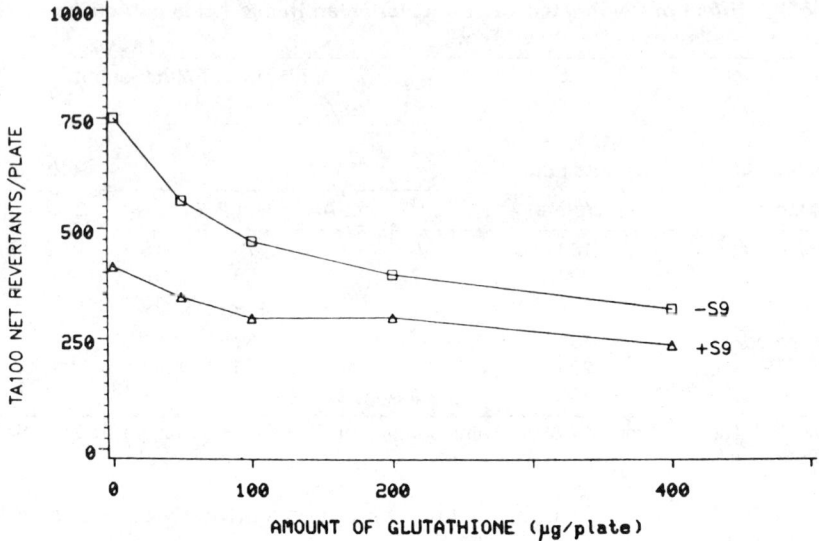

Figure 5. Effect of glutathione on the mutagenicity of chlorinated humic acid. (Note: Spontaneous revertants have been subtracted.)

In Vivo Studies

The results obtained from testing the chlorinated and nonchlorinated humic acid samples in the mouse micronucleus assay are shown in Table IV. There were no significant increases in the percent of micronucleated cells above control values for any of the samples tested, thus providing no evidence for clastogenic effects on mouse bone marrow in vivo. Similarly, the results from testing the samples in the mouse sperm morphology assay were also negative (Table V), thus providing no evidence for the mutagenic potential of these samples in vivo.

The reason for the discrepancies between the in vivo and in vitro assay results is not known, but several explanations are possible. The recommendation made by the respective Gene-Tox groups for the micronucleus assay[26] and the sperm morphology assay[19] is that testing be conducted with doses approaching LD_{50} for the test to be considered adequate. The dose levels administered in the present studies were limited by the solubility of humic acid and the maximum volumes that could be administered. The highest dose levels for any of the samples produced no overt signs of toxicity or lethality. Thus, the negative results obtained under the conditions of our assays should probably be regarded as inconclusive. The same logic would apply to the disinfectant studies in vivo where no lethality of dosed animals was observed.

Table IV. Activity of Chlorinated and Nonchlorinated Humic Acids in the Mouse Micronucleus Assay

Humic acid sample	Micronucleated Cells[a] (%) Dose[b]			
	0	1:4	1:1	Undiluted
Nonchlorinated	0.25 ± 0.06	0.23 ± 0.03	0.40 ± 0.09	0.24 ± 0.06
Chlorinated (pH 7.0 → 2.5)	0.43 ± 0.12	0.24 ± 0.06	0.39 ± 0.12	0.24 ± 0.09
Chlorinated (pH 7.0 → 6.5)	0.47 ± 0.08	0.41 ± 0.17	0.40 ± 0.14	0.33 ± 0.08
Chlorinated (pH 11.5 → 6.5)	0.74 ± 0.09	0.41 ± 0.05	0.38 ± 0.07	0.33 ± 0.07

[a]Mean per animal ± standard error of the mean based on 10 animals per dose group (5 males + 5 females).
[b]Animals received five consecutive daily administrations of the humic acid solutions at the indicated concentratons in 1-mL volumes by oral gavage.

Table V. Activity of Chlorinated and Nonchlorinated Humic Acids in the Mouse Spermhead Abnormality Assay

Humic acid sample	Sacrifice time (weeks)	Spermhead Abnormalities[a] (%) Dose[b]			
		0	1:4	1:1	Undiluted
Nonchlorinated	3	1.49 ± 0.14	1.05 ± 0.12	0.83 ± 0.15	1.00 ± 0.11
	5	1.03 ± 0.11	0.75 ± 0.07	0.82 ± 0.12	0.90 ± 0.10
Chlorinated (pH 7.0 → 2.5)	3	0.75 ± 0.11	0.87 ± 0.10	0.92 ± 0.11	0.99 ± 0.13
	5	0.89 ± 0.09	0.86 ± 0.08	0.76 ± 0.10	1.27 ± 0.20
Chlorinated[c] (pH 7.0 → 6.5)	3	0.72 ± 0.10	0.74 ± 0.13	1.12 ± 0.14	1.02 ± 0.12
	5	0.96 ± 0.09	1.12 ± 0.12	0.88 ± 0.07	0.96 ± 0.09
Chlorinated (pH 11.5 → 6.5)	3	0.81 ± 0.12	0.75 ± 0.10	0.88 ± 0.15	1.00 ± 0.11
	5	0.71 ± 0.10	0.84 ± 0.07	0.69 ± 0.05	1.20 ± 0.12

[a]Mean per animal ± standard error of the mean based on 10 animals per dose group.
[b]The dose schedule was the same as that used for the mouse micronucleus assay.

The other explanation for the negative results in vivo might be that the direct-acting mutagens may be effectively detoxified during in vivo metabolism. As previously indicated, liver S9 preparations have been found to decrease the mutagenicity of chlorinated humic acids in vitro. It seems reasonable to expect that similar inactivation might also occur in vivo. Although GSH does not appear to be important for inactivation of mutagens by S9 in vitro, we note that the observations of significant GSH inactivation at levels above 100 μg per plate (0.1 mM in the top agar) in vitro may have substantial importance for detoxification in vivo where liver and kidney GSH levels are 20

Table VI. Mutagenic Activity of Organic Concentrates from Raw and Treated Mississippi River Water in *Salmonella typhimurium* Strains TA98 and TA100 in the Absence of Metabolic Activation

Sample	Induced Revertants/Control Revertants[a]			
	TA98		TA100	
	Trial 1	Trial 2	Trial 1	Trial 2
Influent	0.55	0.08	0.20	−0.06
Ozone	0.15	−0.18	−0.07	−0.33
Chlorine dioxide	1.10	0.29	0.37	0.12
Chloramine	2.10	0.69	1.82	1.09
Chlorine	3.00	2.33	3.26	2.77
Chlorine and GAC	0.20	0.12	0.24	0.03
Chlorine, GAC, and rechlorination	1.15	0.49	1.45	1.21

[a]Average of the net revertants/control revertants observed in duplicate plates treated with 100 μ of a 4000-fold concentrate produced by combined XAD-2 and XAD-8 resins. Trials indicate independent experiments.

to 100 times higher.[25] Thus, failure to observe any effects in the two in vivo assays may be accounted for by both the in vivo detoxification of the mutagens and the failure to administer high enough doses to produce an effect. In addition, the two assays used will only detect effects on bone marrow and testis. There may be specificity in the tissue distribution of the active compounds, which might make these tissues unlikely targets.

Studies on Alternative Disinfectant By-Products

To assess the relative safety of chlorine vs other chemicals used for disinfection, studies are being conducted to examine the mutagenic activity of water before and after treatment with various disinfectants. Preliminary results are presented in Table VI. Based on a twofold above-background criterion for assessing a positive response (i.e., induced/control revertants $\geq$ 1.0), it is clear that both chlorine- and monochloramine-treated water exhibit reproducible mutagenicities for strains TA98 and TA100. The dose-response curves in both cases were essentially linear over the dose ranges tested. Treatment with GAC that had been in use for about 7 months removed the mutagenic activity of chlorine-treated water; however, rechlorination after GAC resulted in the reformation of about one-third of the activity. This latter result is in contrast to the findings of Loper et al.,[27] who found that both the mutagenicity and the mutagen-forming potential of chlorinated Ohio River water were completely removed by GAC that had been in use for a similar period of time.

To our knowledge, the mutagenicity of chloramine-treated water has not been previously reported. This finding should be taken into consideration in the overall safety assessment of disinfectant treatment alternatives.

ACKNOWLEDGMENTS AND DISCLAIMER

Portions of this work were performed under contract 68-03-2977 with Litton Bionetics, Inc., Kensington, Maryland. We wish to thank Carolyn Smallwood and Dana Laurie for the measurement of liver GSH levels, and Pat Underwood for her excellent typing.

This article has been reviewed by the Health Effects Research Laboratory, U.S. Environmental Protection Agency, and approved for publication. Mention of trade names or commercial products does not constitute endorsements or recommendation for use.

REFERENCES

1. Simmon, V. F., K. Kauhanen, and R. G. Tardiff. "Mutagenic Activity of Chemicals Identified in Drinking Water," in *Progress in Genetic Toxicology*, D. Scott, B. A. Bridges, and F. H. Sobels, Eds. (Amsterdam: Elsevier, North-Holland, 1970), pp. 249-258.
2. Bull, R. J. "Experimental Methods for Evaluating the Health Risks Associated with Organic Chemicals in Drinking Water," *Toxicol. Environ. Chem.* 6:1-17 (1982).
3. Loper, J. C. "Mutagenic Effects of Organic Compounds in Drinking Water," *Mutat. Res.* 76:241-268 (1980).
4. Maruoka, S., and S. Yamanaka. "Production of Mutagenic Substances by Chlorination of Waters," *Mutat. Res.* 79:381-386 (1980).
5. Cheh, A. M., J. Sknochdopole, P. Koski, and L. Cole. "Nonvolatile Mutagens in Drinking Water: Production by Chlorination and Destruction by Sulfite," *Science* 207:90-92 (1980).
6. Bull, R. J., M. Robinson, J. R. Meier, and J. Stober. "Use of Biological Assay Systems to Assess the Relative Carcinogenic Hazards of Disinfection By-products," *Environ. Health Perspect.* 46:215-227 (1982).
7. Meier, J. R., R. D. Lingg, and R. J. Bull. "Formation of Mutagens Following Chlorination of Humic Acid: A Model for Mutagen Formation During Drinking Water Treatment," *Mutat. Res.* 118:25-41 (1983).
8. Coleman, W. E., J. W. Munch, W. H. Kaylor, R. P. Streicher, H. P. Ringhand, and J. R. Meier. "GC/MS Analysis of Mutagenic Extracts of Aqueous Chlorinated Humic Acid—A Comparison of the By-products to Drinking Water Contaminants," *Environ. Sci. Technol.* 18:674-681 (1984).
9. Shih, K. L., and J. Lederberg. "Chloramine Mutagenesis in Bacillus Subtilis," *Science* 192:1141-1143 (1976).
10. Rosenkranz, H. S. "Sodium Hypochlorite and Sodium Perborate: Preferential Inhibitors of DNA Polymerase-Deficient Bacteria," *Mutat. Res.* 31:39-42 (1973).
11. Wlodkowski, T. J., and H. S. Rosenkranz. "Mutagenicity of Sodium Hypochlorite for *Salmonella typhimurium*," *Mutat. Res.* 31:39-42 (1975).
12. Mink, F. L., W. E. Coleman, J. W. Munch, W. H. Kaylor, and H. P. Ringhand. "In Vivo Formation of Halogenated Reaction Products Following Peroral Sodium Hypochlorite," *Bull. Environ. Contam. Toxicol.* 30:394-399 (1983).

13. Maron, D. M., and B. N. Ames. "Revised Methods for the *Salmonella* Mutagenicity Test," *Mutat. Res.* 113:173–215 (1983).
14. Bull, R. J., J. R. Meier, M. Robinson, H. P. Ringhand, R. D. Laurie, and J. Stober. "Mutagenic and Carcinogenic Properties of Brominated and Chlorinated Acetonitriles: By-Products of Chlorination." (in press).
15. Meier, J. R., R. J. Bull, J. A. Stober, and M. C. Cimino. "Evaluation of Chemicals Used for Drinking Water Disinfection for Production of Chromosomal Damage and Spermhead Abnormalities in Mice," *Environ. Mutag.* (in press).
16. Morris, J. C. "The Chemistry of Aqueous Chlorine in Relation to Water Chlorination," in *Water Chlorination: Environmental Impact and Health Effects, Vol. 1*, R. L. Jolley, Ed. (Ann Arbor, MI: Ann Arbor Science Publishers, Inc., 1978), pp. 21–35.
17. Stevens, A. A. "Reaction Products of Chlorine Dioxide," *Environ. Health Perspect.* 46:101–110 (1982).
18. Bruce, W. R., and J. A. Heddle. "The Mutagenic Activity of 61 Agents as Determined by the Micronucleus, *Salmonella*, and Sperm Abnormality Assay," *Can. J. Genet. Cytol.* 22:319–334 (1979).
19. Wyrobek, A. J., A. Gordon, J. G. Burkhart, M. W. Francis, R. W. Kopp, G. Letz, H. V. Malling, J. C. Topham, and M. D. Whorton. "An Evaluation of the Mouse Sperm Morphology Test and Other Sperm Tests in Nonhuman Animals. A Report of the U.S. Environmental Protection Agency Gene-Tox Program," *Mutat. Res.* 115:1–72 (1983).
20. Latt, S. A. "Sister Chromatid Exchange Formation," *Annu. Rev. Genet.* 15:11–55 (1981).
21. Steelink, C. "Humates and Other Natural Organic Substances in the Aquatic Environment," *J. Chem. Ed.* 54(10):599–603 (1977).
22. Eder, E., D. Henschler, and T. Neudecker. "Mutagenic Properties of Allylic and α, β-Unsaturated Compounds: Considerations of Alkylating Mechanisms," *Xenobiotica* 12(12):831–848 (1982).
23. Kopfler, F. C., H. P. Ringhand, W. E. Coleman, and J. R. Meier. "Reactions of Chlorine in Drinking Water, with Humic Acids and In Vivo," Chapter 14, this volume.
24. Hissin, P. J., and R. Hilf. "A Fluorometric Method for Determination of Oxidized and Reduced Glutathione in Tissues," *Anal. Biochem.* 74:214–226 (1975).
25. Glatt, H., M. Protic-Sabljic, and F. Oesch. "Mutagenicity of Glutathione and Cysteine in the Ames Test," *Science* 220:961–963 (1983).
26. Heddle, J. A., M. Hite, B. Kirkhart, K. Mavourinin, J. T. MacGregor, G. W. Newell, and M. F. Salamone. "The Induction of Micronuclei as a Measure of Genotoxicity," *Mutat. Res.* 123:61–118 (1983).
27. Loper, J. C., L. Rosenblum, M. W. Tabor, and J. DeMarco. "Mutagenic Residues Recovered from Granular Activated Carbon After its Use in Drinking Water Treatment," Chapter 102, this volume.

CHAPTER 18

Carcinogenic Activity of Haloacetonitrile and Haloacetone Derivatives in the Mouse Skin and Lung

Richard J. Bull and Merrell Robinson

Chlorinated and brominated acetonitrile and acetone derivatives have been shown to be produced as a result of the chlorination of drinking water[1,2] or humic acids.[3] Haloacetonitriles have also been observed in the stomach contents of rats treated with aqueous solutions of chlorine[4] and result from reactions of chlorine with certain amino acids.[5] Because there have been a number of studies that associate the incidence of certain cancers with the chlorination of drinking water, it is important to assess the carcinogenic properties of products of chlorination.

The intent of this chapter is to review the evidence that haloacetonitrile and haloacetone derivatives that have been identified as by-products of chlorination possess biological properties that are associated with chemical carcinogens.

METHODS

The methodology utilized in these experiments has been published in detail elsewhere.[6] The mutagenic activity of these compounds was evaluated using the standard plate assay of Ames et al.,[7] with and without a fraction of rat liver homogenate; by examining the induction of sister chromatid exchange in Chinese hamster ovary (CHO) cells in vitro, also with and without metabolic activation; by enumerating the number of micronuclei in polychromatic erythrocytes obtained from the bone marrow of CD-1 mice treated in vivo; and by determining the percentage of spermhead abnormalities at 1, 3, and 5 weeks following in vivo exposure of B6C3F1 mice. The chemicals were tested for carcinogenic activity as tumor initiators in the skin of female Sencar mice or by their ability to increase the yield of lung tumors in strain A/J mice.

RESULTS

Table I summarizes the activity of the haloacetonitriles and haloacetone derivatives in *Salmonella typhimurium* strain TA100.[6,8-10] In general, this is the strain that gives the greatest response to these chemicals. Of five haloacetonitriles tested, two are positive in the Ames test, dichloroacetonitrile (DCAN) and bromochloroacetonitrile (BCAN). Both chemicals gave evidence of activity in the absence of rat liver homogenate 9000 × g supernatant (S9) in strain TA1535, whereas such direct action was evident only for DCAN in strains TA98 and TA100. The absence of direct activity with BCAN in TA100 is probably more of a technicality since there was a clear indication of increased activity in the absence of S9 fraction, but the response failed to reach the twofold over background criterion used to designate a positive response. This appeared to result from the cytotoxic properties of BCAN. In point of fact, the initial slope of this response indicates slightly greater specific activity than that noted for BCAN with S9 in TA100.[6] Six chlorinated acetones have been tested for mutagenic activity in *Salmonella typhimurium*, 1,1-dichloropropanone (1,1-DCP), 1,3-dichloropropanone (1,3-DCP), 1,1,1-trichloropropanone (1,1,1-TCP), 1,1,3-trichloropropanone (1,1,3-TCP), 1,1,3,3-tetrachloropropanone (1,1,3,3-TCP), and pentachloropropanone (PCP). The aldehydes 2-chloropropenal and 2,3,3-trichloropropenal have also been identified as by-products of the chlorination of humic acids[8,11] and are included in Table I for perspective. These chemicals are the most potent mutagens in *Salmonella* (Table I) yet identified as chlorination by-products. Of the six chlorinated

Table I. Mutagenic Activity of Haloacetonitriles and Chlorinated Acetone and Aldehyde Derivatives in *Salmonella typhimurium* Strain TA100

Chemical	Net Revertants/μmol[a]
Chloroacetonitrile (CAN)[b]	0
Dichloroacetonitrile (DCAN)[b]	7.1×10^1
Trichloroacetonitrile (TCAN)[b]	0
Bromochloroacetonitrile (BCAN)[b]	(8.6×10^2)
Dibromoacetonitrile (DBAN)[b]	0
1,1-Dichloropropanone (1,1-DCP)[c]	4.8×10^0
1,3-Dichloropropanone (1,3-DCP)[c]	1.4×10^4
1,1,1-Trichloropropanone (1,1,1,-TCP)[c]	9.3×10^2
1,1,3-Trichloropropanone (1,1,3-TCP)[d]	4.0×10^3
1,1,3,3-Tetrachloropropanone (1,1,3,3-TCP)[c]	1.5×10^3
Pentachloropropanone (PCP)[c]	8.5×10^2
2-Chloropropenal (2-CP)[e]	1.1×10^5
2,3,3-Trichloropropenal (2,3,3-TCP)[e]	2.2×10^5

[a]The net revertants/μmol were calculated from the linear portion of the dose response curves. Numbers in parentheses indicate result in the presence of a 9000 × g supernatant of a rat liver homogenate.
[b]Based on data of Bull et al.[6]
[c]Based on data of Meier et al.[8]
[d]Based on data of Douglas et al.[9]
[e]Based on data of Rosen et al.[10]

Table II. Induction of Sister Chromatid Exchange in Chinese Hamster Ovary Cells in Vitro by Haloacetonitriles

Chemical	Induced SCEs
Chloroacetonitrile	50 ± 11[a]
Dichloroacetonitrile	103 ± 26
Trichloroacetonitrile	230 ± 26
Bromochloroacetonitrile	396 ± 69
Dibromoacetonitrile	469 ± 17

[a] Based on data of Bull et al.[6]

acetone derivatives tested, 1,3-DCP was the most potent, followed by 1,1,3-TCP, 1,1,3,3-TCP, PCP, 1,1,1-TCP and 1,1-DCP, respectively. In all cases, the activity of these compounds was expressed in the absence of metabolic activation.

The haloacetonitriles have also been evaluated for their ability to induce sister chromatid exchange (SCE) in CHO cells (Table II). As in the case with the Ames test, the activity of the haloacetonitriles was clearly evident in the absence of S9 fraction, suggesting that these compounds do not require metabolic activation. The ability to induce SCE in CHO cells was seen to follow an orderly progression with the degree of chlorine substitution and was further enhanced by the substitution of bromine for chlorine.

Two attempts have been made to determine whether the mutagenic activity of the haloacetonitriles could be demonstrated in vivo: induction of micronuclei in polychromatic erythrocytes, and spermhead abnormalities. None of the haloacetonitriles were found capable of inducing micronuclei in polychromatic erythrocytes isolated from the bone marrow of mice treated orally with five consecutive daily doses of 0, 12.5, 25, and 50 mg/kg.[6] The same dosing schedule failed to induce spermhead abnormalities in B6C3F1 mice sacrificed at 1, 3, or 5 weeks following the last dose with all five haloacetonitriles.[12]

Three of the five haloacetonitriles tested were clearly capable of acting as tumor initiators in the skin of Sencar mice (Table III). In this case, the chemicals were applied topically in doses of 200, 400, and 800 mg/kg 3 times weekly for 2 weeks. Two weeks following the last dose of haloacetonitrile, the animals were subjected to a promotion schedule of topical applications of 12-O-tetradecanoyl-phorbol-13-acetate (TPA) for 20 weeks. DBAN, BCAN, and CAN were all found capable of acting as tumor initiators under these conditions. DBAN was found to be the most potent, followed by BCAN and then CAN. The yield of squamous cell carcinomas at 1 year was also increased significantly with these three compounds; it was roughly proportional to the benign tumor yield. TCAN significantly increased tumor yields at a total dose of 2400 mg/kg; however, 4800 mg/kg failed to produce a consistent response. DCAN tended to increase tumor yield in a dose-related manner, but the magnitude of the response must be considered marginal.

Table IV provides data on the tumor-initiating activity of haloacetone derivatives. 1,1-DCP and 1,1,1-TCP are the only members of this class that have been tested to date. Doses were single oral doses of 200 mg/kg 1,1-DCP, or 50 mg/kg 1,1,1-TCP, or topical doses of 400 mg/kg for both compounds. Twenty-four weeks following the start of the promotion schedule with TPA outlined in the preceding paragraph, there was no evidence of an increased yield of skin tumors as a result of these treatments.

The results of a preliminary evaluation of the capability of the haloacetonitriles to increase the lung tumor yield in strain A/J mice are provided in Table V. In these experiments, a single oral dose level, 10 mg/kg administered 3 times weekly for 8 weeks, was utilized. Although all of the haloacetonitriles increased the yield of lung tumors, only in the case of CAN, TCAN, and BCAN were the increases statistically significant ($P < 0.05$). Due to the relatively large variation in the background levels of lung tumors in this strain,

Table III. Tumor-Initiating Activity of Haloacetonitriles in the Skin of the Sencar Mouse

Chemical	ED_{35}[b]	Percentage of Animals with Squamous Cell Carcinomas at 2.4 g/kg
Chloroacetonitrile	2.9	10.8[c]
Dichloroacetonitrile		8.6
Trichloroacetonitrile		9.7
Bromochloroacetonitrile	2.2	18.9[c]
Dibromoacetonitrile	1.6	23.5[c]
Control		4.9

[a]Based on data of Bull et al.[6]
[b]Total dose in g/kg body weight necessary to initiate papillomas in 35% of the animals.
[c]Overall response (including higher doses) is significantly different from control incidence ($P < 0.05$ using the Fisher exact test.

Table IV. Tumor-Initiating Activity of Chlorinated Acetone Derivatives in Mouse Skin[a]

Chemical	Route	Dose	N[b]	S[c]	Cumulative Papillomas[d]
1,1-Dichloropropanone	Oral	200 mg/kg	40	32	1
	Topical	400 mg/kg	40	40	1
1,1,1-Trichloropropanone	Oral	50 mg/kg	40	37	1
	Topical	400 mg/kg	40	38	1
DMSO	Oral	0.2 mL	40	38	1
Ethanol	Topical	0.2 mL	40	38	2

[a]Chemicals were administered in a single application of 0.2 mL of the appropriate vehicle by the indicated route. Two weeks later, a promotion schedule was begun with the application of 1 μg 12-O-tetradecanoyl-phorbol-13-acetate (TPA) 3 times weekly for 20 weeks.
[b]Original animals started on treatment.
[c]Animals surviving to 24 weeks.
[d]Cumulative numbers of papillomas observed 24 weeks after the beginning of TPA applications.

Table V. Effects of Orally Administered Haloacetonitriles on the Development of Lung Adenomas in A/J Mice

Chemical	Dose[a] (× 24)	Animals Necropsied (No.)	Animals with Tumors (%)	Tumors/Animals
Chloroacetonitrile	10 mg/kg	28	32	0.43
Dichloroacetonitrile	10 mg/kg	30	23	0.23
Trichloroacetonitrile	10 mg/kg	32	28	0.38
Bromochloroacetonitrile	10 mg/kg	32	31	0.34
Dibromoacetonitrile	10 mg/kg	31	16	0.19
Vehicle (emulphor 10%)	0.2 mL/mouse	31	10	0.10
Ethyl carbamate	42 mg/kg	29	100	9.00

[a]Forty female strain A/J mice were administered the indicated doses of each chemical 3 times weekly for a period of 8 weeks. Treatment was begun at 10 weeks of age. Animals were sacrificed at 9 months of age.

even these results should be interpreted with caution. However, it should be pointed out that 10 mg/kg is considerably below the maximally tolerated dose for these chemicals. They are currently being tested at higher doses.

DISCUSSION

It is clear from the above information that consideration of the carcinogenic hazards that may be associated with the chlorination of drinking water must go beyond formation of the trihalomethanes. At least two additional classes of chemicals must be considered: halogenated acetone and acetonitrile derivatives.

In the case of the haloacetone derivatives, testing has been limited to mutagenicity testing for most of the class. Carcinogenicity testing has been limited to two of the derivatives that are relatively weak mutagens, 1,1-DCP and 1,1,1-TCP. It is relatively clear that substitution of chlorine on each of the two available carbon atoms considerably enhances the mutagenic potency. Consequently, even though the 1,3-substituted compounds are usually reported at lower concentrations, they are 2 to 4 orders of magnitude more potent than the corresponding 1,1-derivatives; therefore, they should be subjected to a more thorough evaluation of their carcinogenic properties. For similar reasons, the extremely potent by-product 2-chloro-propenal deserves further attention as well. It also should be remembered that the biological properties beyond bacterial mutagenesis assays of the brominated derivatives of these two classes of by-products have yet to be considered.

The haloacetonitriles clearly possess carcinogenic as well as mutagenic properties. Although there does not appear to be a consistent relationship in the potency of these compounds in the various test systems to which they have been subjected, it appears that the brominated derivatives are the most potent carcinogens in the mouse skin. However, CAN and TCAN appear capable of inducing lung tumors in A/J mice, while CAN will also clearly act as a tumor initiator in the skin of the Sencar mouse. Only DCAN has failed to produce a carcinogenic response in tests conducted to date. The most critical issue that remains with the haloacetonitriles is an evaluation of their ability to produce cancer by a systemic route of administration. Although suggestive, the results that have been obtained to date on lung tumor induction in strain A/J mice are quite marginal. Preliminary information would suggest, however, that these compounds are capable of interacting with DNA in vivo.[13] Nevertheless, the only means of assessing the carcinogenic hazard these chemicals might pose for consumers of drinking water in quantitative terms are long-term bioassays in experimental animals.

REFERENCES

1. Trehy, M. L., and T. I. Bieber. "Detection, Identification, and Quantitative Analysis of Dihaloacetonitriles in Chlorinated Natural Water," in *Advances in the Identification and Analysis of Organic Pollutants in Water, Vol. 2*, L. H. Keith, Ed. (Ann Arbor, MI: Ann Arbor Science Publishers, Inc., 1981), pp. 941–975.
2. Oliver, B. G. "Dihaloacetonitriles in Drinking Water: Algae and Fulvic Acid as Precursors," *Environ. Sci. Technol.* 17:80–83 (1983).
3. Coleman, W. E., J. W. Munch, W. H. Kaylor, R. P. Streicher, H. P. Ringhand, and J. R. Meier. "GC/MS Analysis of Mutagenic Extracts of Aqueous Chlorinated Humic Acid. A Comparison of the By-products to Drinking Water Contaminants," *Environ. Sci. Technol.* 18:674–681 (1984).
4. Mink, F. L., W. E. Coleman, J. W. March, W. H. Kaylor, and H. P. Ringhand. "In vivo Formation of Halogenated Reaction Products Following Peroral Sodium Hypochlorite," *Bull. Environ. Contam. Toxicol.* 30:394–399 (1983).
5. Bieber, T. I., and M. L. Trehy. "Dihaloacetonitriles in Chlorinated Natural Waters," in *Water Chlorination: Environmental Impact and Health Effects, Vol. 4*, R. L. Jolley, W. A. Brungs, J. A. Cotruvo, R. B. Cumming, J. S. Mattice, and V. A. Jacobs, Eds. (Ann Arbor, MI: Ann Arbor Science Publishers, Inc., 1983), pp. 85–96.
6. Bull, R. J., J. R. Meier, M. Robinson, H. P. Ringhand, R. D. Laurie, and J. Stober. "Mutagenic and Carcinogenic Properties of Brominated and Chlorinated Acetonitriles: By-products of Chlorination," *Fund. Appl. Toxicol.* (in press).
7. Ames, B. N., J. McCann, and E. Yaniasaki. "Methods for Detecting Carcinogens and Mutagens with the *Salmonella*/Mammalian Microsome Mutagenicity Test," *Mutat. Res.* 31:347–364 (1975).
8. Meier, J. R., H. P. Ringhand, W. E. Coleman, J. W. Munch, R. P. Streicher, W. H. Kaylor, and K. M. Schenck. "Identification of Mutagenic Constituents Formed During Chlorination of Humic Acid," *Mutat. Res.* (in press).

9. Douglas, G. R., E. R. Nestmann, A. B. McKague, R. H. C. San, E. G.-H. Lee, V. W. Lie-Lee, and D. J. Kowbel. "Determination of Potential Hazard from Pulp and Paper Mills: Mutagenicity and Chemical Analysis," in *Carcinogens and Mutagens in the Environment. Vol. 4, The Workplace*, H. F. Stich, Ed. (Cleveland: CRC Press Inc., in press).
10. Rosen, J. D., Y. Segall, and J. E. Casida. "Mutagenic Potency of Haloacroleins and Related Compounds," *Mutat. Res.* 78:113–119 (1980).
11. Kringstad, K. P., P. O. Ljungquist, F. de Sousa, and L. M. Stromberg. "On the Formation of Mutagens in the Chlorination of Humic Acid," *Environ. Sci. Technol.* 17:553–555 (1983).
12. Meier, J. R., R. J. Bull, J. Stober, and M. Cimino. "Evaluation of Chemicals Used for Drinking Water Disinfection for Production of Chromosomal Damage and Spermhead Abnormalities," *Environ. Mutagenesis* (in press).
13. Pereira, M. A., Health Effects Research Laboratory, U.S. Environmental Protection Agency, Cincinnati, personal communication.

CHAPTER **19**

Relationship Between Metabolism of Haloacetonitriles and Chloroform and Their Carcinogenic Activity

Michael A. Pereira, F. Bernard Daniel, and Edith L. C. Lin

The chlorination of drinking water results in the formation of low-molecular-weight by-products including halogenated acetonitriles and trihalomethanes. Dihaloacetonitriles have been found in chlorinated drinking water at a concentration of 0.3 to 8.1 ppb.[1,2] Haloacetonitriles are direct-acting alkylating agents that bind the dye, *p*-nitrobenzyl-pyridine,[3] and are direct-acting mutagens in *Salmonella typhimurium*.[4] Several haloacetonitriles induced tumors when applied to the skin of Sencar mice.[4] Thus, haloacetonitriles would appear to possess carcinogenic activity. Chloroform, a trihalomethane found in chlorinated drinking water,[5-7] has been shown (in a long-term bioassay), to cause liver tumors in B6C3F1 mice and kidney tumors in Osborne-Mendel rats.[8] In this chapter, we summarize results on the possible mechanisms of the apparent carcinogenicity of haloacetonitriles and chloroform, especially as related to their metabolism.

HALOACETONITRILES

DNA Alteration or Damage

The initiating event of chemical carcinogenesis is believed to be the interaction of the carcinogen with DNA. In the absence of a metabolic activation system, we determined the ability of the haloacetonitriles to bind to various polynucleotides. Dichloroacetonitrile (DCAN)-^{14}C bound polyadenylic acid to produce an adduct that, after acid hydrolysis, was isolated by HPLC.[9] No adducts were eluted from the HPLC column when a hydrolysate of polyguanylic acid previously incubated with DCAN-^{14}C was analyzed. When calf thymus DNA was incubated with DCAN-^{14}C and the resultant product hydrolyzed, the HPLC elution profile contained a peak of radioactivity identical to that obtained with polyadenosine (Figure 1).

We then determined the ability of dibromoacetonitrile (DBAN) to covalently bind to liver and kidney DNA in rats. When administered orally to rats, DBAN-^{14}C appeared to bind covalently to liver and kidney DNA (Table I). However, we are still in the process of attempting to demonstrate that the

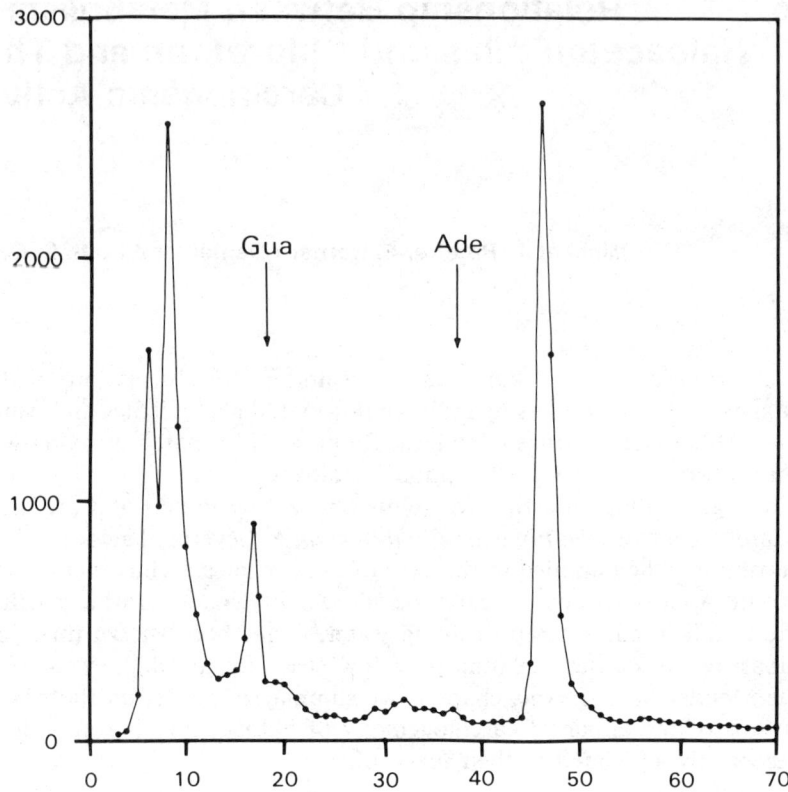

Figure 1. Elution profile of the products of DCAN binding to DNA. DCAN-^{14}C was incubated with calf thymus DNA at 37°C for 22 h. After isolation of the DNA and extraction of unbound DCAN, the DNA was hydrolyzed by treatment with 0.1 N HCl at 70°C for 30 min. The hydrolysate was analyzed by HPLC using a Whatman Magnum 9 SCX column eluted with 0.025 M ammonium phosphate (pH 4.0 for 10 min), followed by a linear gradient to 0.2 M ammonium phosphate (pH 4.0 in 10% methanol for 15 min) at a flow rate of 1.0 mL/min. The adenine (Ade) and guanine (Gua) in the hydrolyzate were followed by absorbance at 260 nm.

radioactivity present in the isolated DNA results from actual DNA adducts and not from contaminating protein adducts or from DNA synthesis. The covalent binding of haloacetonitriles to DNA would indicate potential carcinogenic activity.

The ability of a carcinogen to interact and to alter DNA can be demonstrated by an induction of DNA strand breaks. Haloacetonitriles induced

Table I. Binding of Dibromoacetonitrile to Liver and Kidney DNA in Rats[a]

Organ (h)	DNA Binding (pmol/mg DNA)
Liver	
4	43.8 ± 12.9[b]
24	434 ± 153
Kidney	
4	10.8 ± 1.0
24	62.1

[a]The rats were administered 0.52 mmol/kg DBAN-^{14}C (8.6 mCi/mmol) and were sacrificed either 4 or 24 h later. The DNA was isolated by phenol extraction and the amount of radioactivity and DNA were measured (see Reference 10).
[b]Results are means ± standard error for 2 to 4 rats each, except for the 24-h kidney in which the kidneys from 2 rats were pooled.

Table II. DNA Strand Breaks Produced by 1-h Exposure to Haloacetonitriles in Human CCRF-CEM cells[a]

Haloacetonitrile	DNA Strand Break (Breaks cell^{-1} μmol^{-1} ± SD[b])
TCAN	74.5 ± 27.6
BCAN	12.8 ± 3.7
DBAN	6.9 ± 0.4
DCAN	4.3 ± 1.1
CAN	2.0 ± 1.4
Controls	
DMS[c]	158 ± 16
MMS[c]	33 ± 2.5

[a]The human CCRF-CEM cells were incubated in a serum-free medium with different dose levels of the haloacetonitriles for 1 h. The number of breaks per cell was determined by the previously described alkaline unwinding procedure using hydroxylapatite to separate single and double-stranded DNA.[11] The number of breaks per cell per μmol was calculated from the slope of the dose-response curve.
[b]Significantly greater than solvent control (P ≤ 0.01) by the Student t test.
[c]DMS, dimethysulfate; MMS, methylmethane sulfonate.

DNA strand breakage in cultured human CCRF-CEM cells (Table II).[11] Trichloroacetonitrile (TCAN) was a very potent DNA strand breaker, whereas DCAN and chloroacetonitrile (CAN) were marginally active. DBAN and bromochloroacetonitrile (BCAN) were of intermediate activity. Therefore, the interaction of haloacetonitriles with DNA results in alterations that can cause DNA strand breakage.

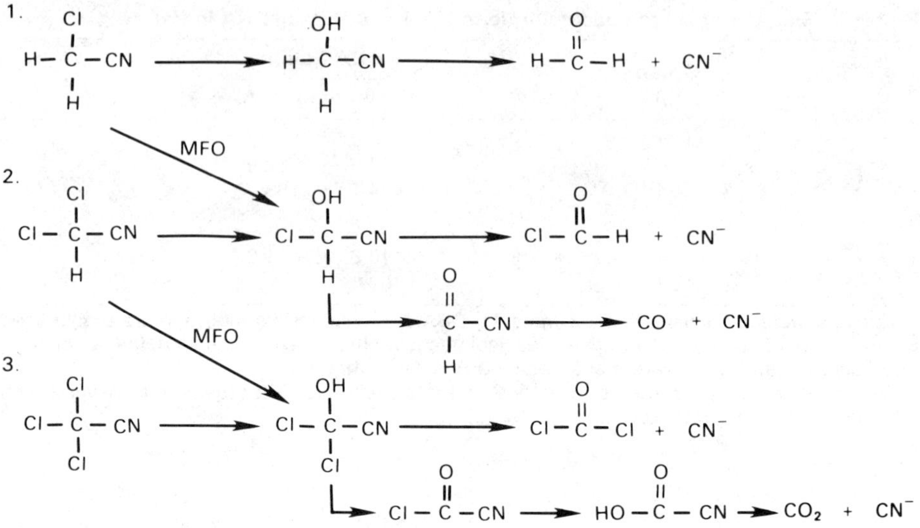

Figure 2. Proposed metabolic pathways for the metabolism of chloroacetonitrile, dichloroacetonitrile, and trichloroacetonitrile.

Metabolism

We have proposed a scheme for the metabolism of haloacetonitriles that is presented in Figure 2. Haloacetonitriles can be metabolized with the release of cyanide by two different mechanisms: nonenzymatic hydrolysis and oxidation via the mixed function oxidase (MFO) of the endoplasmic reticulum. According to this metabolic scheme, CAN will yield formaldehyde by nonenzymatic hydrolysis, and formyl cyanide and formyl chloride by MFO metabolism; DCAN will yield formyl cyanide and formyl chloride by nonenzymatic hydrolysis, and phosgene and cyanoformyl chloride by MFO metabolism; and TCAN will yield phosgene and cyanoformyl chloride by nonenzymatic hydrolysis. That these reactions occur in vivo was evident by the excretion of thiocyanates in the urine of rats administered the haloacetonitriles (Table III). Thiocyanate is produced by the further metabolism of the released cyanide by rhodanese (EC 2.8.1.1). The extent of urinary excretion in 24 h of thiocyanates as the percentage of the administered dose was greater for CAN and BCAN than for DBAN and DCAN, which, in turn, were much greater than for TCAN. Thus, the urinary excretion of thiocyanates would indicate that the haloacetonitriles are absorbed systemically. In conclusion, the haloacetonitriles appear capable, after systemic absorption, of binding or damaging

Table III. Urinary Excretion of Haloacetonitriles as Thiocyanates[a]

Haloacetonitriles	Urinary Thiocyanates [Dose (%)]
CAN	14.2 ± 3.8[b]
BCAN	12.8 ± 1.5
DCAN	9.28 ± 1.09
DBAN	7.67 ± 2.13
TCAN	2.25 ± 0.82

[a]The haloacetonitriles (0.75 mmol/kg) were administered by gavage in tricaprylin to male rats; a 24-h urine sample was obtained.[3] For each animal, the thiocyanates content in the urine collected during the 24 h prior to he administration of the haloacetonitriles was subtracted from the value obtained for the 24 h after its administration.
[b]Results are mean ± standard error for 5 to 6 rats.

DNA. Since DNA alteration or damage is believed to be an important initiating event in chemical carcinogenesis, the haloacetonitriles warrant further study to determine whether their oral administration can cause cancer or some other disease having a genotoxic basis (mutation).

CHLOROFORM

In a U.S. National Cancer Institute (NCI)-sponsored study, chloroform administered by stomach gavage in corn oil was demonstrated to induce liver tumors in mice.[8] However, the lack of genotoxic activity of chloroform has resulted in the hypothesis that the mechanism of chloroform heptatocarcinogenicity is tumor promotion.[12,13] As a result of these observations, we tested chloroform for tumor-promoting activity in mouse liver.

In summary, our study consisted of administering to 15-d-old outbred Swiss mice either 0, 5, or 20 mg/kg ethylnitrosourea (ENU) dissolved in saline. At weaning, some of the mice from each ENU treatment group started to receive 1800 ppm chloroform or 500 ppm sodium phenobarbital (positive control for mouse liver tumor promotion) in their drinking water. The animals were sacrificed at 51 weeks of age (after 46 weeks of chloroform exposure), and a complete histopathological evaluation was performed on the liver, lung, kidneys, and all gross lesions. Ethylnitrosourea at 5 and 20 mg/kg induced a dose-dependent increase in liver adenomas and in hepatocellular carcinomas (Table IV). The subsequent administration of chloroform inhibited the occurrence of liver tumors both in animals that did not receive ENU and in those that did. Therefore, in Swiss mice, chloroform administered in the drinking water acts as an inhibitor of hepatocarcinogenesis.

The inhibitory action on hepatocarcinogenesis of chloroform in our study is in contrast to the NCI-sponsored study where chloroform demonstrated hepatocarcinogenicity.[8] In the NCI study, the chloroform was administered to

Table IV. Effect of Subsequent Treatment with Chloroform on the Incidence of Liver Tumors in Male Mice Initiated by Ethylnitrosourea (ENU) at 15 d of Age[a]

Group	Treatment ENU dosage (mg/kg)	Additive [dosage (ppm)]	Animals (No.)	Animals with Adenomas (No.)	Adenomas per Animal	Animals with Carcinomas (No.)	Carcinomas per Animal
1	20		30	22 (73)[b]	3.1	10 (33)[b]	0.83
2	5		39	8 (21)	0.51	2 (5)	0.10
3	0		37	2 (5)	0.19	2 (5)	0.08
4	20	Chloroform (1800)	29	12 (41)	1.0	5 (17)	0.21
5	5	Chloroform (1800)	25	1 (5)	0.05	0 (0)	0
6	0	Chloroform (1800)	23	0 (0)	0	0 (0)	0
7	20	Phenobarbital (500)	25	22 (88)	2.2	17 (68)	2.5
8	5	Phenobarbital (500)	36	14 (39)	0.56	10 (28)	0.44
9	0	Phenobarbital (500)	30	6 (20)	0.27	6 (20)	0.23

[a] Fifteen-day-old mice were administered ENU by stomach gavage; at weaning the animals started to receive either 1800-ppm chloroform or 500-ppm phenobarbital (positive control for liver tumor promoter). The animals were sacrificed at 51 weeks of age.
[b] Numbers in parentheses are the percentage of animals with the lesion.

B6C3F1 mice in corn oil by stomach gavage.[8] In our study, the strain of mice was CD-1 (Swiss) and the route of administration was drinking water. Preliminary results (gross observation) of an ongoing study in B6C3F1 indicate that chloroform administered in the drinking water to this strain of mice also inhibits both spontaneous and chemically (ENU and diethylnitrosamine) induced hepatocarcinogenesis. The 1800-ppm dose of chloroform administered in the drinking water is very similar to the amount administered by stomach gavage in the long-term bioassay. However, the pharmacokinetic consequences of a bolus vs continuous dose might be the reason for these differing effects of chloroform.

An alternative possibility for the differing results in the two studies is that a synergistic interaction between the chloroform and the corn oil results in the hepatocarcinogenic activity. In fact, increasing fat consumption of mice, especially polyunsaturated fat, has been shown to result in an increased incidence of spontaneous liver tumors.[14-16] In rats, the concurrent feeding of a lipotrope-deficient high polyunsaturated fat diet with diverse carcinogens such as aflatoxin B_1, dibutylnitrosamine, diethylnitrosamine, or N-2-fluorenylacetamide enhanced their hepatocarcinogenicity.[17,18] The mechanism of the enhancement by polyunsaturated fat might result (1) from the induction of liver-mixed function oxidase activity by polyunsaturated fat as compared to saturated fat[19-20] or (2) from the hepatic toxicity of a lipotrope-deficient high polyunsaturated fat diet. These metabolic alterations in the liver produced by high levels of polyunsaturated fat might change the hepatic activity of chloroform from an inhibitor (administered in drinking water) to an inducer (administered in corn oil) of hepatocarcinogenesis. Further studies, especially in the area of metabolism, are required to determine the mechanisms by which the chloroform can, under different circumstances, act either as an inhibitor or as an inducer of liver cancer.

REFERENCES

1. Oliver, B. G. "Dihaloacetonitriles in Drinking Water: Algae and Fulvic Acid as Precursors," *Environ. Sci. Technol.* 17:80–83 (1983).
2. Trehy, M. L., and T. I. Bieber. "Detection, Identification and Quantitative Analysis of Dihaloacetonitriles in Chlorinated Natural Waters," in *Advances in Identification and Analysis of Organic Pollutants in Water, Vol. 2*, L. H. Keith, Ed. (Ann Arbor, MI: Ann Arbor Science Publishers, Inc., 1981), pp. 941–975.
3. Pereira, M. A., L.-H. C. Lin, and J. K. Mattox. "Haloacetonitriles Excretion as Thiocyanate and Inhibition of Dimethylnitrosamine Demethylase: A Proposed Metabolic Scheme," *J. Toxicol. Environ. Health* 13:633–641 (1984).
4. Bull, R. J., and M. Robinson. "Carcinogenic Activity of Haloacetonitriles and Haloacetone Derivatives in the Mouse Skin and Lung," Chapter 18, this volume.
5. Rook, J. J. "Formation of Haloforms during Chlorination of Natural Waters," *Water Treat. Exam.* 23:234–243 (1974).

6. Bellar, T. A., J. J. Lichtenberg, and R. C. Kroner. "The Occurrence of Organohalides in Chlorinated Drinking Water," *J. Am. Water Works Assoc.* 66:703–706 (1974).
7. Symons, J. M., T. A. Bellar, J. K. Carswell, J. Demarco, K. L. Kropp, G. G. Robeck, D. R. Seeger, C. J. Slocum, B. L. Smith, and A. A. Stevens. "National Organics Reconnaissance Survey for Halogenated Organics" *J. Am. Water Works Assoc.* 67:634–647 (1975).
8. National Cancer Institute. "*Carcinogenesis Bioassay of Chloroform*," National Technical Information Form, Service No. PB 264018/AS (Washington, DC: National Cancer Institute, 1976).
9. Daniel, F. B., M. A. Pereira, L.-H. C. Lin, D. L. Haas, J. K. Mattox, and K. Schenck. "Genotoxicity of Haloacetonitriles," *Toxicologist* 3:37 (1983).
10. Pereira, M. A., L.-H. C. Lin, and L. W. Chang. "Dose-Dependency of 2-Acetylaminofluorene Binding to Liver DNA and Hemoglobin in Mice and Rats," *Toxicol. Appl. Pharmacol.* 60:472–478 (1981).
11. Daniel, F. B., D. H. Haas, and S. M. Pyle. "Quantitation of Chemically Induced DNA Strand Breaks in Human Cells Via an Alkaline Unwinding Assay," submitted for publication.
12. Pereira, M. A., L.-H. C. Lin, H. M. Lippitt, and S. L. Herren. "Trihalomethanes as Initiators and Promoters of Carcinogenesis," *Environ. Health Perspect.* 46:151–156 (1982).
13. Reitz, R. H., T. R. Fox, and J. F. Quast. "Mechanistic Considerations for Carcinogenic Risk Estimation: Chloroform," *Environ. Health Perspect.* 46:163–168 (1982).
14. Tannenbaum, A. "The Genesis and Growth of Tumors. III. Effects of a High-Fat Diet," *Cancer Res.* 2:468–475 (1942).
15. Silverstone, H., and A. Tannenbaum. "The Influence of Dietary Fat and Riboflavin on the Formation of Spontaneous Hepatomas in the Mouse," *Cancer Res.* 11:200–206 (1951).
16. Gellatly, J. B. M. "The Natural History of Hepatic Parenchymal Nodule Formation in a Colony of C57BL Mice with Reference to the Effect of Diet," in *Mouse Hepatic Neoplasia*, W. H. Butler and P. M. Newberne, Eds. (New York: Elsevier Press, 1975), pp. 77–110.
17. Rogers, A. E. "Influence of Dietary Content of Lipids and Lipotropic Nutrients on Chemical Carcinogenesis in Rats," *Cancer Res.* 43:2477–2484 (1983).
18. Rogers, A. E., G. Lenhart, and G. Morrison. "Influence of Dietary Lipotropes and Lipid Content on Aflatoxin B_1, N-2-Fluorenylacetamide and 1,2-Dimethylhydrazine Carcinogenesis in Rats," *Cancer Res.* 40:2802–2807 (1980).
19. Norred, W. P., and A. E. Wade. "Dietary Fatty Acid-Induced Alteration of Hepatic Microsomal Drug Metabolism," *Biochem. Pharm.* 21:2887–2892 (1977).
20. Newberne, P. M., J. Weigert, and N. Kula. "Effects of Dietary Fat on Hepatic Mixed Function Oxidases and Hepatocellular Carcinoma Induced by Aflatoxin B_1 in Rats," *Cancer Res.* 39:3986–3991 (1979).

CHAPTER 20

Mutagenicity Produced by Aqueous Chlorination of Tyrosine

W. Howard Rapson, Bonnie Isacovics, and C. Ian Johnson

In 1978, Rapson et al.[1] reported that all phenols among 40 compounds tested produced mutagenicity by the Ames test when treated with chlorine in dilute aqueous solution. That work was carried out after it was reported that the filtrate from the chlorination of wood pulp in the first stage of bleaching is mutagenic by the Ames test.[2,3] It was shown that the lignin component of wood is the source of such mutagenicity.[1,4] Chlorophenols, chlorocatechols, and chloroguaiacols were shown to be toxic but not mutagenic.[1]

It was also found that if wood pulp is treated with chlorine dioxide instead of chlorine in the first stage of bleaching, the filtrate is not mutagenic.[1,3,4] As chlorine dioxide is substituted for equivalent chlorine, the mutagenicity decreases almost linearly to zero.

Pure substituted phenols, catechols, and guaiacols, typical of structural units in the lignin macromolecule, were treated with chlorine and chlorine dioxide in water under conditions typical of those used in the first stage of pulp bleaching. The mutagenicity produced by treatment of the first such model compound tried (acetovanillone) almost duplicated that obtained by substituting chlorine dioxide for equivalent chlorine in the treatment of unbleached kraft wood pulp.[1] Studies with such model compounds were continued, because the filtrate from the chlorination of wood pulp contains an extremely large number of products, and it is very difficult to isolate and identify the mutagens. With pure, relatively simple compounds the number of possible products is much smaller; therefore, analysis and identification of products should be easier.

Among the products of chlorination of wood pulp, lignin, and lignin model compounds, the following substances have been found to be mutagenic:

bromodichloromethane[5,6]	o-quinone[1,*]
dibromomethane[5,6]	dibromochloromethane[5,6]
tetrachloroethylene[5-7]	trichloroethylene[5-7]
pentachloropropene[8,9]	tetrachloropropene[8,9]
dichloro-p-cymene[10]	bromo-p-cymene[10]
chloroacetaldehyde[5,12]	trichloro-1,2,3-trihydroxybenzene[11]
3,5,5-trichloropent-4-ene-2-one	2-chloropropenal[5,13]
1,1,3,3-tetrachloroacetone[1,5]	1,3-dichloroacetone[1,5]
3,6-dichloro-2-hydroxybenzaldehyde[1,*]	hexachloroacetone[1,5]
3-chloro-4-dichloromethyl-5-hydroxy-2(5H)-furanone[14,15]	mucochloric acid[1,*]

The compounds marked with an asterisk (*) have not yet been found in the solutions produced by the chlorination of lignin or model compounds, but it is highly probable that they are formed. O-Quinone, for example, is highly mutagenic but unstable. Tetrachloroquinone is found in pulp chlorination filtrates, but it is not mutagenic.

3,5,5-Trichloropent-4-ene-2-one was identified in our laboratory by GC/MS analysis of the solution resulting from the chlorination of acetovanillone. It was predicted that it would be mutagenic because it contains the α-chlorocarbonyl and α-chloroethylene groups. It was synthesized[16] and found to give a positive response in the Ames test. It is probable that many compounds containing these functional groups (like most of those listed above) and produced by the aqueous chlorination of phenols will be found to be mutagenic.

Many municipal potable water supplies disinfected with chlorine contain mutagens.[17-20] It may be assumed that any phenols, lignin or lignin fragments, or humic acids contained in the water being chlorinated will produce mutagens.

It occurred to us that any amino acids or peptides or proteins that contain phenolic groups are also likely to produce mutagens on chlorination in dilute aqueous solution. Payne and Rahimtula[17] reported in 1981 that aqueous chlorination of tyrosine did not produce a positive response in the Ames test for mutagenicity. In 1982, we found that tyrosine produced substantially higher and more stable mutagenicity by the Ames test on chlorination at pH 2 than the same molar concentration of acetovanillone. We also found that tyrosine chlorinated at pH 7 produced some stable mutagenicity, whereas kraft pulp and lignin model compounds produced little or no mutagenicity on chlorination at pH 7. This means that tyrosine in water or in sewage plant effluent would produce more mutagens than an equivalent concentration of lignin or lignin-derived compounds.

3,5-Dibromotyrosine, 3,5-diiodotyrosine, and thyroxine, which are related to tyrosine, produced mutagens on chlorination. Tryptophane and phenylalanine also produced mutagens, but much less per mol of amino acid than tyrosine. Chlorination of glycine and glutamic acid produced no mutagens detectable by the Ames TA100 test.

EXPERIMENTAL

To a 4 mM solution of L-tyrosine (Aldrich Chemical Co., 99+% pure) in 0.01 N HCl (pH 2), an increasing ratio of chlorine or chlorine dioxide (from 0 to 16 oxidizing equiv/mol) in water solution was added at room temperature (about 22°C). In each experiment, sufficient 0.01 N HCl was added to bring all solutions to the same total volume so that the results of the Ames test would be comparable. After either 30 min or 24 h at room temperature, any residual chlorine or chlorine dioxide was removed by bubbling nitrogen through the solution.

Each solution was submitted to the Ames test[21,22] using the TA100 strain of *Salmonella* without liver enzyme to measure the direct-acting mutagens.

Ames *Salmonella* strain TA100 cultures were grown from master plates prepared from a slant. Diagnostic tests used to check for genetic markers included histidine requirement, sensitivity to ultraviolet light, crystal violet and ampicillin, and response to the standard mutagen methylmethane-sulphonate (MMS). In a test tube, 0.2 mL of the solution to be tested was mixed rapidly into 2 mL of warm overlay agar, 0.025 mM in histidine and in biotin, and 0.1 mL of the bacterial culture was added and mixed rapidly. The contents of the tube were poured onto the Petri plate. The plates were incubated at 37°C for 48 h and the colonies were counted.

If the test solution contains substances toxic to the *Salmonella* strain TA100, some of the bacteria will die. This decreases the concentration of bacteria available for mutation, which therefore lowers the mutated colony count. A rough quantitative estimate of the fraction of bacteria surviving may be made by looking at the "background lawn," the large number of microcolonies of unmutated bacteria on the test plate under a microscope. If no difference is observed between a plate with no added solution and the test plate, it is assumed there is no toxicity and the test plate is arbitrarily given the number 10. This indicates 100% survival of the bacteria, and the mutated colony count is assumed to be a reliable measure of the mutagenicity of the test solution.

If the background lawn on the test plate is entirely absent, it is assigned the number 0, indicating death of all bacteria and therefore no mutated colonies. This indicates severe toxicity.

Between these two limits, the observer of the background lawn subjectively estimates the degree of survival with an integer between 0 and 10. Thus, a rough scale of increasing toxicity is created as the bacterial survival decreases from 10 to 0.

The number of revertant colonies for 0.2 mL of each solution per plate is shown in the figures, along with the number of spontaneous colonies produced without adding any chemical (called background count). By using the same scale for the number of revertants per plate in all figures, the variations in mutagenicity produced by variations in the conditions of treatment of tyrosine may be compared. The number subjectively assigned to each plate representing bacterial survival is recorded on each experimental point as an aid in interpreting the effect of toxicity on the results.

RESULTS AND DISCUSSION

Figure 1 shows that for 4 mM tyrosine, as the chlorine-to-tyrosine ratio increased the number of revertant colonies increased to a maximum and then decreased sharply to background. The survival numbers on the points indicate that the toxicity of the solution toward the *Salmonella* bacteria increased with the increasing chlorine-to-tyrosine ratio. The sharp decrease in revertant count

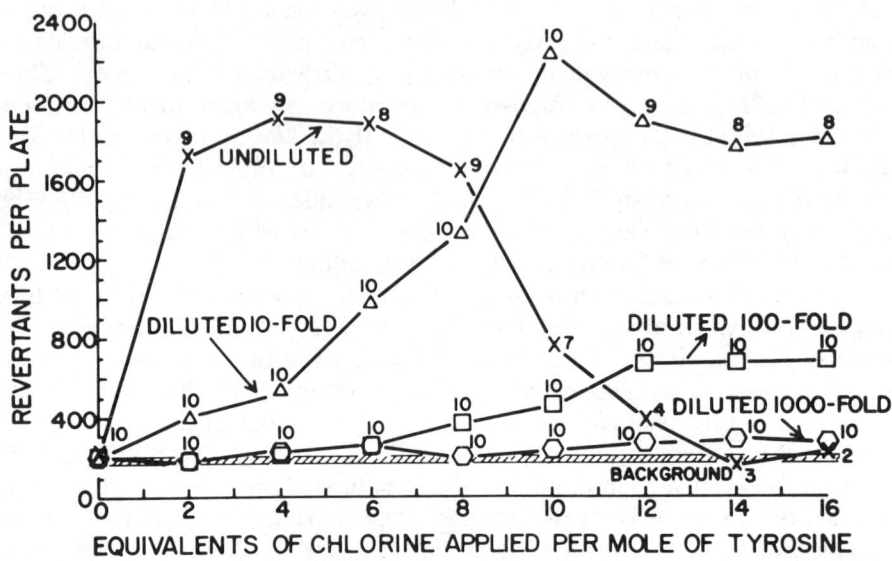

Figure 1. Ames mutagenicity produced by chlorination of 4 mM aqueous tyrosine diluted after chlorination. Conditions: 4.000 mM tyrosine; 0 to 16 equiv chlorine applied per mol of tyrosine; pH 2; room temperature, 22°C; 30 min; residual chlorine removed with nitrogen; tested undiluted, diluted 10-, 100-, and 1000-fold; 0.2 mL/plate; numbers on points are estimated bacterial survival, 10 representing no toxicity, 0 representing death of all bacteria.

was accompanied by a sharp increase in toxicity. Presumably, the toxicity could be due to excessive mutagenicity or to nonmutagenic toxins.

DILUTION OF THE CHLORINATED TYROSINE SOLUTION

Each solution of chlorinated 4 mM tyrosine was diluted 10-, 100-, and 1000-fold with 0.01 N HCl and tested again (Figure 1). The pH 2 solution used for dilution maintains sterility. The buffering capacity of the medium applied to the plate is sufficient to accept the small amount of acid in the 0.2-mL sample without significant change in pH.

At tenfold dilution, up to 8 equiv chlorine per mol of tyrosine, the mutagenicity was lower than that of the undiluted solution, as would be expected.

At a higher chlorine-to-tyrosine ratio, the diluted solution gave a much higher revertant count than that of the undiluted solution. At tenfold dilution, toxicity only became recognizable at about 12 equiv chlorine/mol, where the mutagenicity decreased. This is good evidence that toxicity decreases the apparent mutagenicity by killing some of the bacteria. Dilution decreases the toxicity, and the apparent mutagenicity increases above that of the highest value with the undiluted solution.

At 100-fold lower concentration, the mutagenicity decreased substantially, but no toxicity was observed up to the limit of 16 equiv chlorine/mol tyrosine applied. At 1000-fold dilution, the revertant count was close to background, with no observable toxicity.

Although with the higher chlorine-to-tyrosine ratios the toxicity decreased on dilution and the mutagenicity increased substantially, this does not prove that the toxicity is the result of excessive mutagenicity. It may be caused by nonmutagenic products of chlorination such as chlorotyrosine or other chlorinated phenolic products, which are known to be produced on aqueous chlorination of tyrosine.[23,24] Their toxicity may be decreased on dilution, thus allowing the revertant counts to increase.

It has been established that dichloroacetonitrile is one of the products of aqueous chlorination of tyrosine,[25] which has been reported to be mutagenic.[6] It is highly probable that many other mutagens such as those found on chlorination of wood pulp, lignin, and simple phenols contribute to the mutagenicity of chlorinated tyrosine.[26]

DILUTION OF THE TYROSINE SOLUTION BEFORE CHLORINATION

For comparison, instead of diluting the chlorinated tyrosine solution, the concentration of tyrosine in the solution being chlorinated was decreased 10- and 100-fold before applying 0 to 16 equiv chlorine/mol tyrosine. This means that both the tyrosine and the chlorine applied were diluted 10- and 100-fold before reacting rather than after, thus simulating concentrations of tyrosine that may possibly be encountered in the chlorination of potable water supplies and sewage plant effluents.

The results in Figure 2 with 4 mM tyrosine are those in Figure 1. With 0.4 and 0.04 mM tyrosine, the absolute revertant counts were substantially lower than when the chlorination was carried out at a lower concentration, compared with chlorinating 4 mM tyrosine and diluting 10- and 100-fold (shown in Figure 1), as might be expected from kinetic considerations.

This means that the mutagenicity decreased as the concentrations of both tyrosine and chlorine decreased, even though the same ratio of chlorine to tyrosine was applied. The pattern of mutagenicity increase with increasing chlorine-to-tyrosine ratio was similar; however, there was no decrease in density of bacteria in the background lawn, indicating lack of toxicity toward the

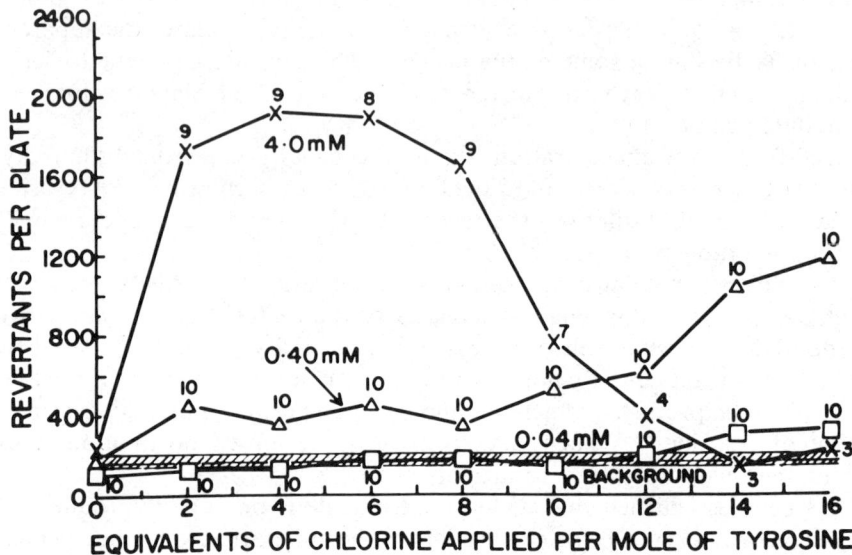

Figure 2. Ames mutagenicity produced by chlorination of diluted aqueous tyrosine. Conditions: 4.00, 0.400, and 0.0400 m*M* aqueous tyrosine; 0 to 16 equiv chlorine applied per mol of tyrosine; pH 2; room temperature; 30 min; residual chlorine removed with nitrogen; 0.2 mL/plate; numbers on points estimated toxicity as in Figure 1.

Salmonella bacteria on chlorination of tyrosine in the diluted solution, as previously discussed.

The significant mutagenicity produced by the higher chlorine-to-tyrosine ratio, even in 0.04 *M* tyrosine solution (6.7 ppm by weight), indicates that tyrosine may be one of many sources of the mutagenicity found after chlorination of many municipal water supplies and sewage plant effluents.

EFFECT OF pH ON MUTAGENICITY

Mutagenicity produced by chlorination of lignin model compounds at pH 2 decreases with time when the pH is raised. It was shown by Nazar and Rapson[27] that the decrease in mutagenicity is accompanied by the formation of hydrochloric acid. This means that hydrolysis of chlorine in the mutagens decreases the mutagenicity.

Figure 3 shows the rate of decrease in mutagenicity of chlorinated tyrosine and chlorinated acetovanillone solutions with time, both when the pH was

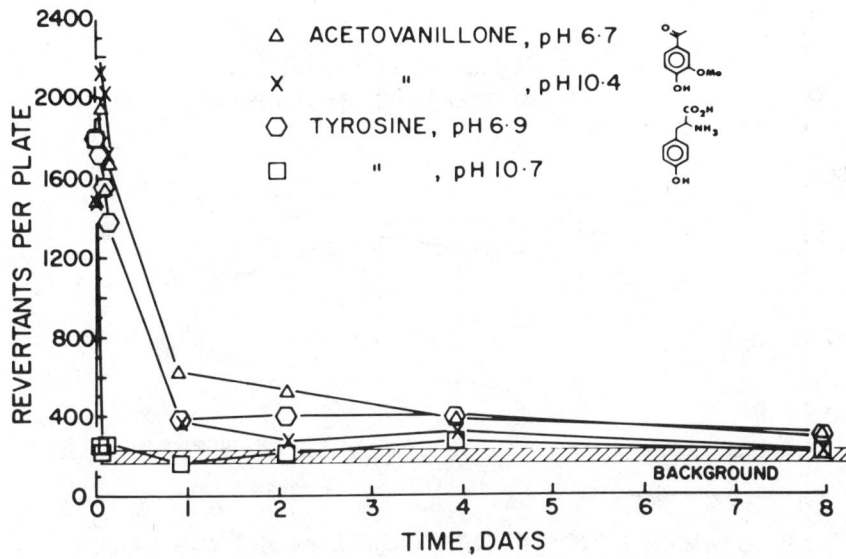

Figure 3. Decrease of mutagenicity with time after raising pH of 0.400 mM tyrosine and 4.00 mM acetovanillone chlorinated at pH 2. Conditions: 0.400 mM tyrosine, 4.000 mM acetovanillone; each treated with 16 equiv chlorine/mol; pH 2; room temperature; 24 h; residual chlorine removed with nitrogen; the chlorinated acetovanillone solution was diluted with an equal volume of 0.01 N NCl to eliminate toxicity; phosphate buffer or sodium carbonate was added to aliquots to raise pH to the values shown; after 1, 2, and 3 h, and daily after that, all aliquots were reacidified to pH 2; 0.2 mL/plate in all cases; no toxicity was observed, since all background lawns were estimated at 10; revertants per plate for the pH 2 solution are shown on the 0 time line.

raised to near neutrality by adding phosphate buffer and again when the pH was raised above 10 by adding sodium carbonate buffer. The chlorinated tyrosine decreased in mutagenicity even faster than chlorinated acetonvanillone did, indicating that the mutagens were even more sensitive to hydrolysis. The fact that at neutral pH the revertant count dropped relatively rapidly to a low but still mutagenic level indicates that some of the mutagens are easily hydrolyzed, others are relatively stable at pH 7, but unstable at pH 10.7. At the high pH, mutagenicity was decreased to background in less than 1 h, the time at which the first sample was tested.

Figure 4 shows the results of an experiment in which tyrosine was chlorinated at pH 2, stripped with nitrogen to remove any unreacted chlorine, then increased in pH by adding increasing amounts of sodium hydroxide and letting

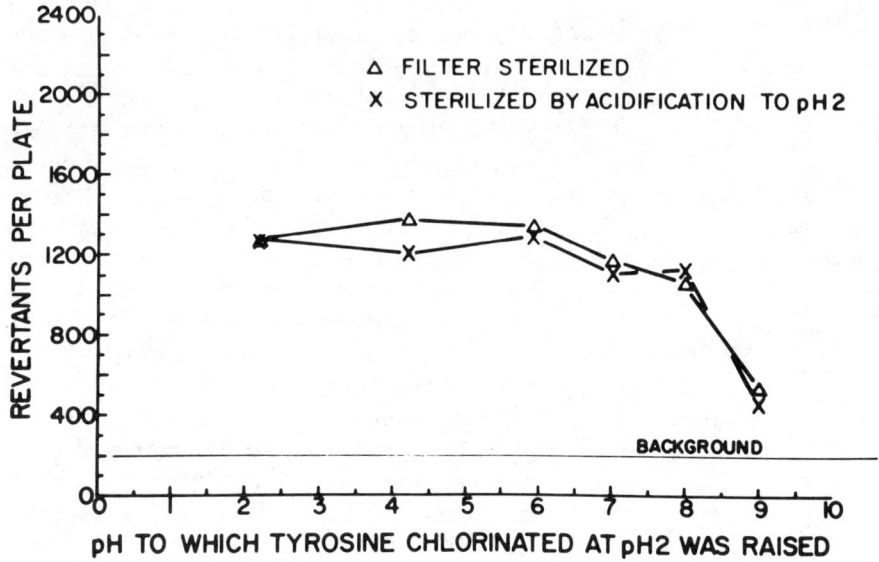

Figure 4. Effect of raising pH of tyrosine chlorinated at pH 2. Conditions: 0.400 mM tyrosine; chlorinated with 16 equiv chlorine/mol; pH 2; room temperature; after 30 min, pH was raised to values shown by adding sodium hydroxide; let stand for 3 h; one series was filter sterilized for plating; another series was sterilized by decreasing pH to 2 and then plated; 0.2 mL/plate in all cases.

the solution stand for 3 h. One set of samples was sterilized by ultrafiltration, and an identical set was acidified to pH 2 for sterilization. The reacidification did not restore mutagenicity lost at the higher pH. The number of revertants decreased a little in the neutral range, but the major decrease occurred between pH 8 and 9.

CHLORINATION OF TYROSINE AT pH 7

Because chlorination of potable water and sewage plant effluent is normally carried out in the neutral range, a set of experiments was carried out with increasing oxidation equivalents of sodium hypochlorite/mol tyrosine at pH 7, as shown in Figure 5. Substantial mutagenicity was produced, compared with chlorination of acetovanillone at pH 7. At up to 4 equiv hypochlorite/mol, the mutagenicity was not significantly higher than the background count; however, the increase was substantial as the number of equivalents applied per mol increased to 12.

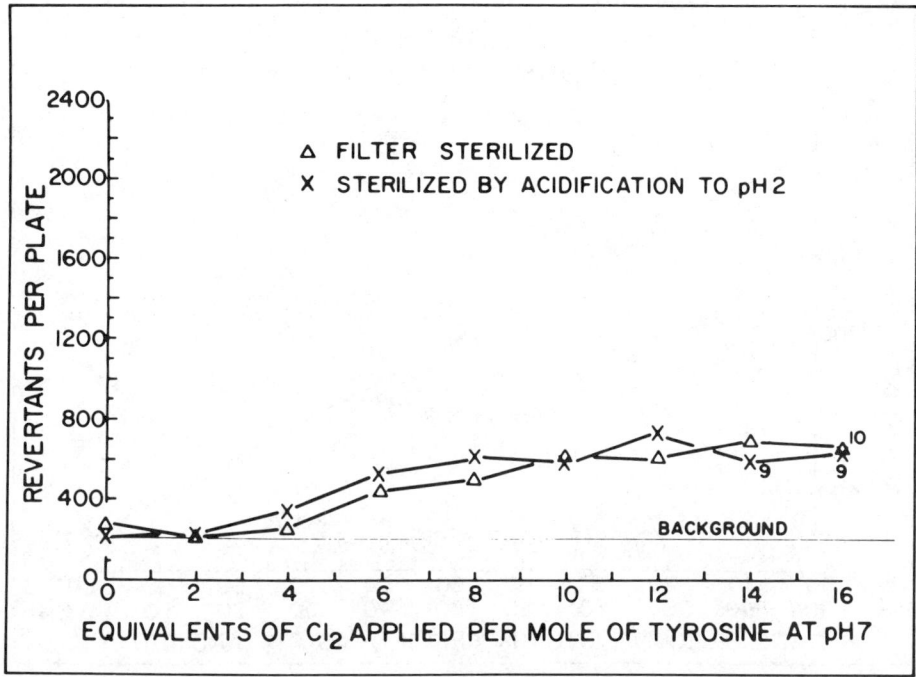

Figure 5. Chlorination of tyrosine at pH 7. Conditions: 4.000 mM tyrosine; chlorinated with 0 to 16 equiv/mol chlorine as sodium hypochlorite; pH 7 maintained with phosphate buffer; 30 min; room temperature; one set filter sterilized before plating; another set sterilized by acidification of pH 2 with hydrochloric acid; 0.2 mL/plate.

The method of sterilization had a small effect on the revertant count. Acidification gave higher revertant counts but produced some toxins above 12 equiv/mol of tyrosine.

This experiment shows that tyrosine is likely to be a significant contributor to the mutagenicity resulting from the disinfection of water or wastewater by chlorination.

TREATMENT OF TYROSINE WITH CHLORINE DIOXIDE

The use of chlorine dioxide for disinfection of water and sewage plant effluent is increasing. It has been found that in the treatment of unbleached pulp, pure lignin, and pure lignin model compounds, as chlorine dioxide is substituted for equivalent chlorine (1 wt unit chlorine dioxide for 2.63 wt units chlorine), the mutagenicity decreases almost linearly down to zero with pure chlorine dioxide.[1,3,4]

Figure 6 shows that the same kind of decrease of mutagenicity occurs as chlorine dioxide replaces equivalent chlorine in the treatment of tyrosine.

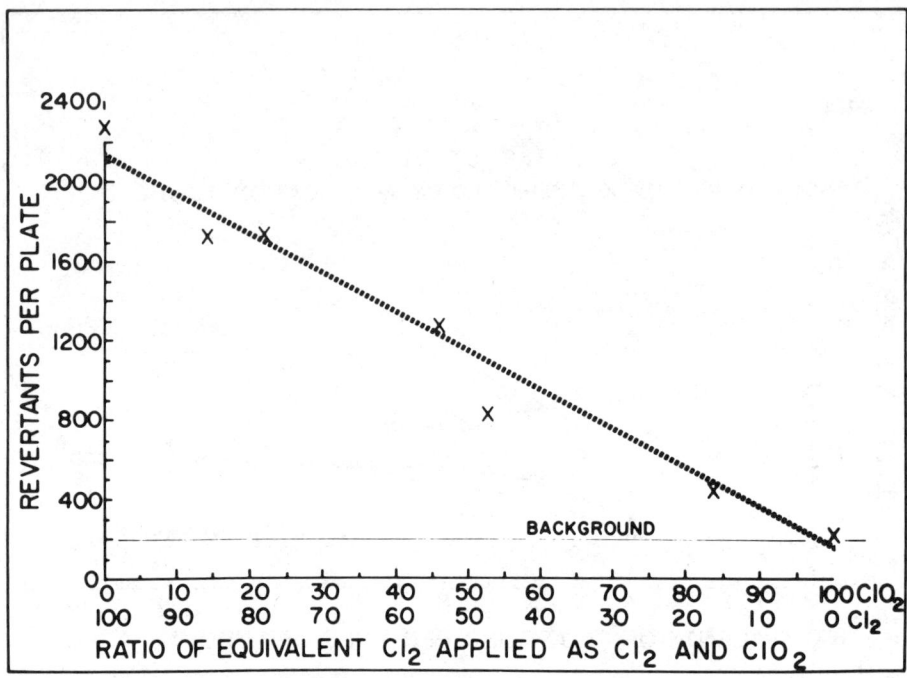

Figure 6. Treatment of 4 mM tyrosine with chlorine and chlorine dioxide. Conditions: 4.000 mM tyrosine; 4 oxidizing equiv applied per mol tyrosine, chosen to avoid toxicity; ratio varied from 100% chlorine to 100% chlorine dioxide (2.63 g chlorine is equivalent to 1 g chlorine dioxide in oxidizing power); pH 2; room temperature; 30 min; 0.2 mL/plate.

Within the limits of experimental error, the decrease in revertant counts is linear with respect to the degree of substitution of chlorine dioxide for equivalent chlorine. With tyrosine, as with lignin and lignin model compounds, mutagenicity is virtually eliminated with 100% chlorine dioxide.

CONCLUSIONS

On exposure to chlorine in dilute water solution at pH 2, tyrosine produced mutagens giving a strong positive response in the Ames test using *Salmonella typhimurium* tester strain TA100. As the ratio of chlorine to tyrosine increased, the number of revertants per plate increased. Simultaneously, the toxicity of the solution toward the bacteria increased as the chlorine-to-tyrosine ratio increased. With 4 mM tyrosine, the toxicity decreased the revertant count, giving a false apparent decrease in mutagens.

When the solution was diluted many times, the toxicity decreased and the revertant counts increased, with an increasing chlorine-to-tyrosine ratio up to the 16 equiv chlorine/mol tyrosine applied. Dilution of the tyrosine solution before applying the same ratio of chlorine to tyrosine decreased the revertant count substantially, but the same trend is apparent.

Raising the pH of the tyrosine solution chlorinated at pH 2 to neutrality decreased the mutagenicity substantially, but it is apparent that some mutagens are stable for a long time in the normal neutral range for potable water. Raising the pH above 10 decreased the revertant counts very rapidly to background. The decrease in mutagenicity is caused by hydrolysis of chlorine from the mutagens to produce the chloride ion.

Chlorination of tyrosine at pH 7 produced significant mutagenicity, substantially less than chlorination at pH 2, but substantially more than is produced by the chlorination of lignin model compounds such as acetovanillone at pH 7. Tyrosine can therefore be considered a potential source, among many, of the mutagenicity produced by disinfection of water with chlorine.

Treatment with chlorine dioxide instead of chlorine produced no mutagens. As chlorine dioxide replaced equivalent chlorine, the mutagenicity decreased almost linearly from about 2100 revertant counts above background to 0 with 100% chlorine dioxide. Because chlorine dioxide has the same effect in the first stage of pulp bleaching and in the treatment of humic acids and lignin model compounds, it can be assumed that substitution of chlorine dioxide for chlorine in the disinfection of potable water, sewage plant effluent, power plant cooling water, or any other water supply requiring disinfection, will alleviate or perhaps eliminate the problem of mutagen production on disinfection with chlorine.

REFERENCES

1. Rapson, W. H., M. A. Nazar, and V. V. Butsky. "Mutagenicity Produced by Aqueous Chlorination of Organic Compounds," *Bull. Environ. Contam. Toxicol.* 24:590–596 (1980).
2. Ander, P., K.-E. Eriksson, M.-C. Kolar, K. Kringstad, U. Rannug, and C. Ramel. "Studies on the Mutagenic Properties of Bleaching Effluents," *Svensk. Papper.* 80(14):454–459 (1977).
3. Eriksson, K.-E., M-C. Kolar, and K. Kringstad. "Studies on the Mutagenic Properties of Bleaching Effluents. II," *Svensk Papper.* 82(4):95–104 (1979).
4. Nazar, M. A., and W. H. Rapson. "Elimination of the Mutagenicity of Bleach Plant Effluents," *Pulp Paper Can.* 81(8):T191–196 (1980).
5. Kringstad, K. P., P. O. Ljungquist, F. de Sousa, and L. M. Stromberg. "Identification and Mutagenic Properties of Some Chlorinated Aliphatic Compounds in the Spent Liquor from Kraft Pulp Chlorination," *Environ. Sci. Technol.* 15:562–566 (1981).
6. Simmon, V. F., K. Kauhanen, and R. G. Tardiff. "Mutagenic Activity of Chemicals Identified in Drinking Water," *Prog. Gen. Toxicol.* 2:249–258 (1977).

7. Cerna, M., and H. Kytenova. "Mutagenic Activity of Chloroethylenes Analysed by Screening System Tests," *Mutat. Res.* 46:214 (1977).
8. Nestmann, E. R., E. G.-H. Lee, T. Matula, G. R. Douglas, and J. C. Meuller. "Mutagenicity of Constituents Identified in Pulp and Paper Mill Effluents Using the *Salmonella* Mammalian Microsome Assay," *Mutat. Res.* 79:203–212 (1980).
9. *Biological Characteristics of Pulp Mill Effluents (Part 1)*, CPAR Project Report 6-78-1, (Ottawa, Ont., Canada: Department of Environment, 1978).
10. Bjorseth, A., G. E. Carlberg, and M. Moller. "Determination of Halogenated Organic Compounds and Mutagenic Testing of Spent Bleach Liquors," *Sci. Total Environ.* 11:197–211 (1979).
11. Carlberg, G. E., N. Gjos, K. O. Gustavsen, G. Tuelan, and L. Renberg. "Chemical Characterization and Mutagenicity Testing of Chlorinated Trihydroxybenzenes Identified in Spent Bleaching Liquors from a Sulphite Plant," *Sci. Total Environ.* 15:3–15 (1980).
12. McCann, H., V. Simmon, D. Streitweiser, and B. N. Ames. "Mutagenicity of Chloroacetaldehyde, A Possible Metabolic Product of 1,2-Dichloroethane (Ethylene Dichloride), Chloroethanol (Ethylene Chlorohydrin), Vinyl Chloride and Cyclophosphamide," *Proc. Natl. Acad. Sci.* 72:3190–3193 (1975).
13. Rosen, J. D., Y. Segall, and J. E. Casida. "Mutagenic Potency of Haloacroleins and Related Compounds," *Mutat. Res.* 78:113–119 (1980).
14. Holmbom, B. R., R. H. Voss, R. D. Mortimer, and A. Wong. "Isolation and Identification of an Ames Mutagenic Compound Present in Kraft Chlorination Effluents," *Tappi* 64(3):172–174 (1981).
15. Holmbom, B., R. H. Voss, R. D. Mortimer, and A. Wong. "Fractionation, Isolation and Characterization of Ames Mutagenic Compounds in Kraft Chlorination Effluents," *Environ. Sci. Technol.* 18(5):333–337 (1980).
16. Kiehlmann, E., P.-W. Loo, B. C. Menon, and N. McGillivray. "Synthesis and Novel Rearrangement of 1,1,1-Trichloro-2-alkene-4-ones," *Can. J. Chem.* 49:2964–2976 (1971).
17. Payne, J. F., and A. Rahimtula. "Water Chlorination as a Source of Aquatic Environmental Mutagens," in *Toxicology of Halogenated Hydrocarbons: Health and Ecological Effects*, M. A. Q. Khan and R. H. Stanton, Eds. (New York: Pergamon Press, 1981), pp. 209–221.
18. Bedding, N. D., A. E. McIntyre, R. Perry, and J. N. Lester. "Organic Contaminants in the Aquatic Environment I. Sources and Occurrence," *Sci. Total Environ.* 25:143–167 (1982).
19. Bedding, N. D., A. E. McIntyre, R. Perry, and J. N. Lester. "Organic Contaminants in the Aquatic Environment II. Behaviour and Fate in the Hydrological Cycle," *Sci. Total Environ.* 26:255–312 (1983).
20. Cheh, A. M., R. E. Carlson, J. R. Hildebrandt, C. Woodward, and M. A. Pereira. "Contamination of Purified Water by Mutagenic Electrophiles," in *Water Chlorination: Environmental Impact and Health Effects, Vol. 4*, R. L. Jolley, W. A. Brungs, J. A. Cotruvo, R. B. Cumming, J. S. Mattice, V. A. Jacobs, Eds. (Ann Arbor, MI: Ann Arbor Science Publishers, 1983), pp. 1221–1235.
21. Ames, B. N., J. McCann, and E. Yamasaki. "Methods for Detecting Carcinogens and Mutagens with the *Salmonella* Mammalian-Microsome Mutagenicity Test," *Mutat. Res.* 31:347–364 (1975).
22. Maron, D. M., and B. N. Ames. "Revised Methods for the *Salmonella* Mutagenicity Test," *Mutat. Res.* 113:173–215 (1983).

23. Burleson, J. L., G. R. Peyton, and W. H. Glaze. "Chlorinated Tyrosine in Municipal Waste Treatment Plant Products after Superchlorination," *Bull. Environ. Contam. Toxicol.* 19:724–728 (1978).
24. Burleson, J. L., G. R. Peyton, and W. H. Glaze. "Gas-Chromatographic Mass Spectrographic Analysis of Derivatized Amino Acids in Municipal Wastewater Products," *Environ. Sci. Technol.* 14(11):1354–1359 (1980).
25. Trehy, M. L., and T. I. Bieber. "Detection, Identification and Quantitative Analysis of Dihaloacetonitriles in Chlorinated Natural Waters," in *Advances in the Identification and Analysis of Organic Pollutants in Water, Vol. 2*, L. M. Keith, Ed. (Ann Arbor, MI: Ann Arbor Science Publishers, Inc., 1981), pp. 941–975.
26. Nazar, M. A., W. H. Rapson, M. A. Brook, S. May, and J. Tarhanen. "Mutagenic Reaction Products of Aqueous Chlorination of Catechol," *Mutat. Res.* 89:45–55 (1981).
27. Nazar, M. A., and W. H. Rapson. "pH Stability of Some Mutagens Produced by Aqueous Chlorination of Organic Compounds," *Environ. Mutagen.* 4:435–444 (1982).

CHAPTER 21

Identification of Carcinogens by Measurement of Cell-Mediated Immunity v. Antitumor Immunity in Rats to Halogen-Containing Organic Compounds

Reggie H. Stevens, Dean A. Cole, Paul A. Lindholm, Paul T. Liu, Margaret L. Gourlay, and H. F. Cheng

Much consideration is being given to the possible influence of environmental pollutants on cancer incidence rates occurring in developed countries. During the past two decades, estimates have been published suggesting that as many as 70 to 90% of all human cancers may be the result of exposure to agents in the environment.[1-3] Because of such a possible relationship, interest is being expressed in the identification of those environmental carcinogens that could be contributing to the rates of this disease, the thought being that if such substances can be recognized, then exposure to them can be reduced and possibly even eliminated. One important source for such cancer-causing agents that requires careful consideration, because of possible widespread exposures, is public drinking water. Experimentally, studies have now indicated a possible link between the consumption of polluted water and the occurrence of certain forms of cancer. For example, Loper and co-workers[4] applied the Ames *Salmonella typhimurium* assay to evaluate water supplies obtained from several American cities and obtained positive indication of mutagenicity for all of them. Analogous studies have been performed on various other drinking water supplies with similar indications of the presence of mutagens.[5-8]

What substances may actually be responsible for the apparent mutagenicity remains to be revealed. Recent findings point to the possibility that the very chemicals being used to disinfect the water may be the responsible agents, with chlorine being a substance of primary suspicion. In the mid 1970s, the U.S. Environmental Protection Agency published results of a national survey that indicated chloroform, a product resulting from chlorine, was a ubiquitous substance in drinking water. In 1976, National Cancer Institute studies indicated that chloroform was an animal carcinogen.[9] Epidemiological data have since indicated an apparent association between chlorination of drinking water and the development of colon, rectal, and bladder cancer, with a possible risk ratio ranging from 1.13 to 1.93 being reported.[10-12]

While the pollution of drinking water may contribute to the incidence of cancer, before any reasonable step can be taken to reduce such a risk, usable tests must be available to identify the particular carcinogenic contaminants in question. Characteristics for such an ideal test are that it should be rapid, easily performed, inexpensive, and most importantly, capable of distinguishing not only carcinogens that act via genetics but also those expressing their effects through epigenetic mechanistic steps. An ideal test should also be capable of detecting cocarcinogens as well as cancer-promoting agents.

There are a number of physical techniques quite capable of identifying known carcinogens existing at very low concentrations in drinking water; however, such analyses are incapable of presenting information regarding the possible interactions within a mixture that could alter the carcinogenic potential of a particular contaminant. This is certainly a major limitation, since the true effective carcinogenicity potential of a substance would be expected to depend on both the concentration and its relative carcinogenicity. Consequently, there has been a recent concentrated effort to develop biological tests that can be used for the identification of such agents. These tests can be essentially grouped in a tier scheme according to their biological complexity, as suggested by Bridges.[13] The least biologically complex tests involve the submammalian and in vitro tests such as that exemplified by the Ames assay.[14] The second level of testing involves using the whole animal as a test system. The generally accepted procedure to establish experimentally that a substance is a carcinogen in this level of testing is to administer the test substance over nearly the entire lifetime of the animal and then to evaluate for the presence of cancer.[15] However, such a procedure has both very great physical and financial requirements that are associated with the handling of large numbers of animals for the long periods necessary for them to ultimately develop detectable cancer. For example, Clayson[16] in 1980 estimated the cost required for testing a single substance for carcinogenicity was about $500,000. Thus, there is an interest in the development of short-term bioassays that can be applied for the identification of carcinogens that could act as a complement to the expensive long-term or chronic whole-animal testing procedure.

Our approach in the development of such an inexpensive short-term whole-animal bioassay has been to evaluate antitumor immune responses that are induced following the exposure to a carcinogenic substance.[17] The concept has been based on the idea that such a substance will result in the formation of cells that have unique properties that will be recognized as being foreign to the host's immune system. Therefore, following such an exposure, measurable antitumor cell-mediated immune responses will shortly develop (Figure 1). By evaluating such induced immune responses, we believe that it should be possible to identify, in this indirect fashion, those substances that are capable of inducing foreign-like cells (i.e., carcinogens). In this study, we have evaluated several halogen-containing organic compounds, and the results suggest that

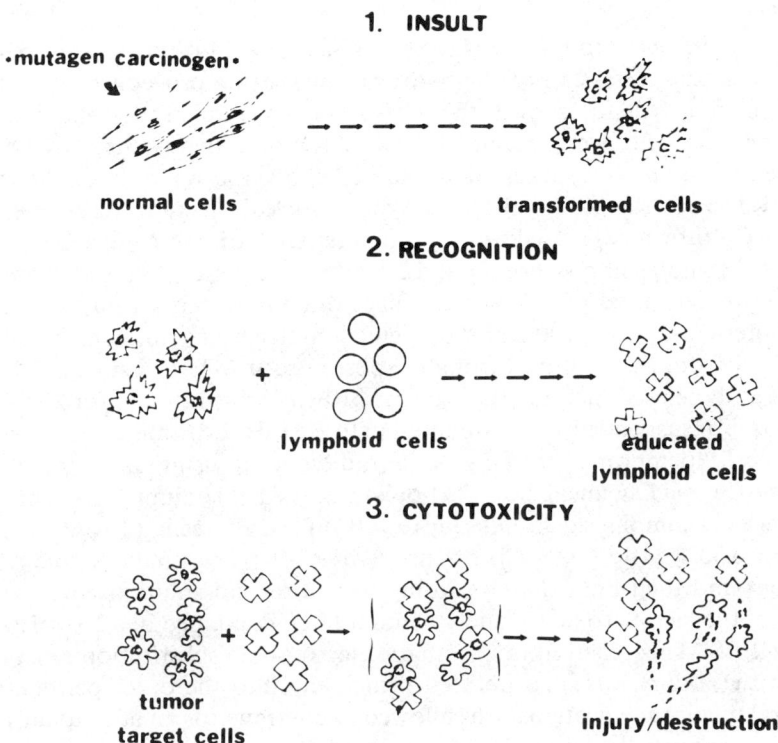

Figure 1. Generalized scheme for development of cell-mediated immune (CMI) response after exposure to an environmental insult.

such measurements may serve as a useful basis for developing a biological analysis for carcinogens existing in drinking water.

MATERIALS AND METHODS

Organohalogen Exposures

Inbred Fischer F344 (Microbiological Associates, Walkerville, Maryland) rats were maintained on a 12-h light cycle with food (Purina Laboratory Chow) and water provided ad libitum. Certified pesticide-grade halogenated organic compounds, obtained from Fischer Scientific Company (Pittsburgh), were dissolved in spectroanalyzed-grade dimethyl sulfoxide (DMSO), also obtained from Fischer, and administered by intraperitoneal injection. Controls consisted of rats that were administered only the DMSO solution.

Cell-Mediated Immune Measurements

Lymphoid-cell cytotoxicity was determined at a 14-d postexposure interval following administration of the hydrocarbons, using procedures that we outlined previously.[18] In essence, the measurements entailed the following steps: tumor target cells were obtained from a tumor cell line that we originally derived from an X-ray induced, small-bowel adenocarcinoma of the rat (IA-XrSBR Cells: CRL 1677, American Type Culture Collection, Rockville, Maryland). Culture media used for the maintenance of these cells consisted of Ham's F-10 nutrient mixture (Grand Island Biological Co., Grand Island, New York) supplemented with 10% heat-inactivated fetal calf serum, 0.25 mg/mL amphotericin B, and 100 units of penicillin–streptomycin. Single cells were obtained from this culture through exposure to 0.25% trypsin for 10 min at 37°C, and then sedimenting them by centrifugation at $100 \times g$ for 20 min. The cells were resuspended in phosphate-buffered (pH 7.4) saline (PBS), and the outer cell membrane proteins were radiolabeled using a lactoperoxidase enzyme (Sigma Chemical Co., St. Louis) catalysis of sodium ^{125}I.[19] This procedure was accomplished by adding to 1.0 mL of the cells (2 to 3×10^6) an aliquot of 3.7 MBq (100 µCi) of carrier-free ^{125}I (New England Nuclear Co., Boston) and then immediately adding to it a second solution containing 6.2 nmol of the lactoperoxidase and 100 nmol H_2O_2 contained in a 1.0 mL volume of the PBS. This reaction (an aromatic electrophilic substitution reaction), in which the iodine was incorporated principally into the outer peripheral and integral membrane proteins, was allowed to continue for an additional 10 min. The reaction was then quenched by adding 5.0 mL of the culture medium. The cells were washed six times in this medium to remove any unincorporated ^{125}I. The cells were placed in a multiwell test plate at a concentration of 3 to 8×10^3 cells per well, and then allowed to adhere to the surface of the plate by incubating overnight in a 95 to 5% air–CO_2 atmosphere at 37°C.

On the following day, day 14 of the exposure interval, peripheral blood was collected from the test and appropriate control rats from a freely bleeding tail vein. The peripheral blood lymphoid cells (PBLC) were partially purified[20] by their sedimentation through a Ficoll-Hypaque gradient (Gallard-Schlessinger, Long Island). The interfacial cells were harvested from this gradient, washed with culture medium, and then suspended at a concentration so that on addition of 0.5 mL of of these cells, there was a working concentration of 30:1 PBLC-to-tumor target cells.

The test and control PBLC were arranged in a 2×2 randomized block pattern, as suggested by Brown and co-workers, for the best statistical evaluation of such microcytotoxicity tests.[21] The cell mixture was then incubated in the 95 to 5% air–CO_2 atmosphere for an 18-h interval, and the amount of iodinated membrane proteins released from the injured and destroyed tumor target cells was determined using a Beckman 4000 gamma spectrometer (Beckman Instruments, Irvine, California). The CMI status of the test animal was established by relating the increased protein release, as compared with the appropriate control rat that had been exposed to only the DMSO. This increase release was determined as a percent release according to the following formula:

$$CMI = [(Rest - Control)/(Control)] \times 100$$

Test = equals the release of the radioiodinated tumor target cell membrane proteins in the presence of the PBLC obtained from the rats exposed to the halogenated organic compounds, and Control equals the release occurring from the targets in the presence of the control PBLC.

Use of the values for the control in these determinations thus allows an accounting for possible background noise in the system such as might occur from natural cytotoxicity being exhibited towards the tumor cells (NK effects) and spontaneous release of the label during the incubation periods. Statistical evaluation for significance was established using the Student's two-tailed t-test.[22]

RESULTS

During the past decade, we have focused much of our effort on determining possible changes in normal immune reactivity that might occur following exposure to environmental carcinogens. To ascertain whether detectable alterations might occur following exposure to halogen-containing organic compounds, we chose to evaluate those which have been reported to be carcinogens, and to first determine whether any measurable effects might exist following exposure to what would be considered as very high concentrations of these substances. The initial series of findings, in which the test rats were exposed to millimolar concentrations of dichloromethane, dichloroethane, chloroform, and carbon tetrachloride, indicated that identifiable responses were being elicited (Table I). Attempts to evaluate bromoform and carbon tetrabromide at these relatively high concentrations were unsuccessful because of their extreme toxicity. The positive findings of the chlorinated substances were, however, encouraging, because such results tended to support the predication that it might indeed be possible to use the measurement of altered immune responses as a means of bioassaying the carcinogenic properties of environmentally occurring cancer-inducing substances.

It would not be expected that such very high concentrations (millimol) of the chlorine-containing hydrocarbons would environmentally occur in drinking water; therefore, a second series of studies was set up to determine the dose-response relationships in an attempt to estimate what might be the threshold detection levels of these chemicals in this particular type of an analysis. A clear-cut dose-response relationship was not observed, however, for these compounds, which were studied over a 6-log concentration range of 10^{-5} to 10^{-10}

Table I. Cell-Mediated Immunity (CMI) Induced in the Rat Following Exposure to mmol/kg Body Weight Concentrations of Chlorine-Containing Hydrocarbons

Chemical	CMI[a] (%)
Dichloromethane	3.2 ± 0.3 (2)
Dichloroethane	8.1 ± 0.5 (3)
Chloroform	5.8 ± 2.2 (3)
Carbon tetrachloride	0.9 ± 0.7 (3)
Chlorobenzene	0.0 (2)
Positive Control[b]	18.8 ± 6.9 (3)

[a]Values represent the mean ± SE of the number ( ) of rats evaluated two times, with the results being derived from triplicate samples obtained from 12 wells of a test plate for each animal undergoing testing.
[b]Positive control for quality assurance consisted of rats with 7,12-dimethylbenz[a]anthracene implanted in the "head" of the pancreas (Reference 23).

mol/kg body weight exposure (Table II). Consequently, it would appear that a lack of such a relationship might preclude the use of this methodology for quantification purposes. While this may indeed be true, the information at present is still too preliminary to completely dismiss the procedure as being nonquantitative.

The reason for not dismissing the procedure is that we have evaluated only three animals per group over a very wide concentration range in this particular study. In addition, it would appear that (1) we failed to actually determine what might be the threshold range for dichloromethane and dichloroethane, since the range appears to be occurring at considerably less than the concentrations studied; and (2) the immunoreactivities may not be a simple direct response to the insult by the chlorinated hydrocarbon such as that seemingly illustrated by chloroform. This particular chemical appeared to be both an inhibitor at its high concentrations and a stimulator of the responses at the lesser doses, thus resulting in a biphasic type of response being observed. The other halogenated compounds might be acting in a similar but less apparent biphasic fashion.

To emphasize that the findings do indeed merit further investigation, the CMI values determined for the 70 rats that were evaluated during the five individual studies, which were completed in the attempt to establish the dose-response relationship, were summarized (Table III). Each distinct study also was accompanied with PBLC that were obtained from rats that had cancer of the pancreas induced through implantation of 7,12-dimethylbenz[a]-anthracene[23] into the "head" of this organ. These cancer-bearing animals were selected to act as positive controls for this study. The findings summarized in this table imply that a similar cytotoxicity is being expressed by the lymphoid cells obtained from the rats exposed to the halogenated hydrocarbons, a cytotoxicity such as that observed for the immune cells obtained from those animals having pancreas cancer.

Table II. Cell-Mediated Immunity (CMI) Expressed by Rats Exposed to Varying Doses of Halogenated Hydrocarbons

Chemical	Dose (mol/kg)[a]					
	1×10^{-5}	1×10^{-6}	1×10^{-7}	1×10^{-8}	1×10^{-9}	1×10^{-10}
Bromoform	1.6 ± 1.3	0.0	0.0	0.0	0.0	0.0
Dichloromethane	16.4 ± 6.7	27.5 ± 9.1	18.5 ± 7.7		4.5 ± 3.7	11.8 ± 3.8
Dichloroethane	20.9 ± 2.3	14.9 ± 4.9	14.4 ± 0.9	19.5 ± 2.6	15.2 ± 1.3	14.9 ± 0.8
Chloroform	3.8 ± 0.7	6.1 ± 3.2	11.0 ± 1.4	9.9 ± 4.8	21.6 ± 4.3	2.8 ± 2.3
Carbon tetrachloride	22.3 ± 3.1	7.1 ± 1.5	14.6 ± 5.6	3.4 ± 1.4	6.0 ± 2.9	3.4 ± 2.8
Chlorobenzene	0.0	0.0				

[a]CMI values (%) comprised the mean ± SE for a group of 3 animals for each exposure with triplicate samples gathered from 12 wells of a test plate (36 determinations) for each test animal. Positve control for each assay consisted of rats having 7,12-dimethylbenz[a]anthracene-induced pancreas cancer (Reference 23).

Table III. Average Cell-Mediated Immunity (CMI) Expressed by Rats Exposed to the Organohalogens (100 pmol to 10 μmol)

Chemical	Percent CMI[a]
Bromoform	1.6 ± 1.3 (3)
Dichloromethane	22.9 ± 5.5 (18)
Dichloroethane	16.4 ± 1.3 (14)
Chloroform	10.6 ± 2.1 (17)
Carbon tetrachloride	9.4 ± 2.0 (18)
Positive control (DMBA)[b]	16.6 ± 8.5 (5)

[a] Values for number ( ) of animals represent the mean ± SE of triplicate samples obtained from 12 wells of a test plate for each rat assayed.
[b] Positive control to evaluate quality assurance for each study consisted of the rat pancreas model: 7,12-dimethylbenz[a]anthracene implanted in the "head" of the pancreas (Reference 23).

Table IV. Lowest Concentration of Halogenated Hydrocarbon Evaluated for Induction of Antitumor Cell-Mediated Immunity (CMI)

Chemical	Test Dose (mol/kg)	Percent CMI
Bromoform	1×10^{-5}	1.6 ± 1.3
Dichloromethane	1×10^{-10}	4.1 ± 2.1
Dichloroethane	1×10^{-10}	14.9 ± 0.7
Chloroform	1×10^{-10}	2.8 ± 2.3
Carbon tetrachloride	1×10^{-10}	3.7 ± 2.7
Chlorobenzene	1×10^{-3}	0.0
Positive control[a]	$\sim 4.8 \times 10^{-5}$	18.5 ± 7.1

[a] Positive control for quality assurance represented by pancreas cancer model: rats with 7,12-dimethylbenz[a]anthracene implanted in the "head" of the pancreas (Reference 23).

The findings at this stage of our investigations continued to support the concept of using alterations in the immune responses as indicators of exposure to carcinogenic compounds; consequently, we tried to estimate the possible threshold detection level for this particular methodology. Except for bromoform, exposure to all the chlorinated hydrocarbons evaluated could be readily detected at concentration levels of nanomol per kilogram exposures (Table IV).

One of the important shortcomings experienced by the nonbiological tests currently being conducted to determine the carcinogenicity of drinking water sources is the inability to provide an understanding regarding possible interactions resulting from complex mixtures. Without an assessment for the possible additive, synergistic, and inhibitory effects resulting from the combination of the various substances present in a complex mixture, the true effects exerted upon the cancer incidence by a particular carcinogen, such as certain of the halogenated organic compounds, cannot truly be established. We have recently completed a study in which urine that contained carcinogens served as an example of a complex mixture[24] so that we could gage the usefulness of this methodology for evaluating such solutions. The investigation involved administering the colon carcinogen 1,2-dimethylhydrazine (DMH) to rats and collecting their urine for the subsequent 24 h. The DMH is a relatively specific carcinogen for the rodent colon,[25] and it is known that a certain percentage will be eliminated in the animals' urine. Urine obtained from the exposed animals was representative of a complex mixture, and unexposed rat urine was used for the control; both were administered to rats as the test agents. CMI was detectable over a dose response amounting to an initial exposure ranging from 0.1 to 20 mg/kg. The threshold level of detection for this carcinogen was in the range of 100×10^{-6} µg/kg body weight (CMI = $8.8\% \pm 2.2\%$), which represented an exposure of 1.7×10^{-6} mol DMH/kg, or 100 ppb. This amounted to an exposure of less than 1×10^{-4} times the LD_{50} (1 g/kg) and to just 1×10^{-3} the amount of the chemical that was found to be necessary to induce tumors following a 20-month latency period.[25]

Since a test to evaluate the potential carcinogenicity of drinking water ideally should involve administering the water orally, we initiated an investigation to determine whether it might be possible to detect the induction of CMI following the gastric intubation of the DMH. The study entailed the dosing of the rats with this carcinogen and then evaluating the rats for cytotoxic lymphoid cells at both a 7- and 14-d postexposure sampling interval. The levels of CMI responses (Table V) induced via this gastric intubation route were found to be similar to those previously observed on administering the chemical subcutaneously.[17] Apparently, the immune assay may be suitable for monitoring exposures occurring by way of oral ingestion.

A major criterion for any potential assay to determine carcinogenicity is that the test possess specificity. We have yet to complete a large battery of tests to determine the possible false positive and negative substances associated with

Table V. Cell-Mediated Immunity (CMI) in the Rat Following Subcutaneous or Oral Administration of 1,2-Dimethylhydrazine

Dose (mg/kg)	Percent CMI[a]			
	Subcutaneous		Oral[b]	
	Day 7	Day 14	Day 7	Day 14
10	3.2 ± 0.4 (3)	5.2 ± 0.3 (3)	2.3 ± 1.8 (6)	9.2 ± 2.3 (7)
1	4.7 ± 0.5 (3)	5.1 ± 0.4 (3)	3.4 ± 2.8 (3)	
0.1	1.2 ± 0.7 (3)	3.0 ± 0.6 (3)	3.9 ± 1.8 (6)	7.7 ± 2.1 (6)
0.01	0.0 (3)	0.0		

[a] Values for number ( ) of animals is represented by the mean ± SE of triplicate samples from 12 wells of a test plate for each assay. Number of animals determined at 7 and 14 d postexposure.
[b] Oral exposure was accomplished by gastric intubation of the DMH.

Table VI. Cell-Mediated Immunity (CMI) Expressed by Apparently Noncarcinogenic Insults

Insult	Dose	Percent CMI[a]
Chemical		
Theophylline	10 mg	0.0 (2)
Insulin	10 mg	0.0 (4)
Sucrose	10 mg	0.0 (3)
Ascorbic acid	10 mg	0.0 (3)
Saline	0.9%	0.0 (5)
Distilled water		0.0 (3)
X rays		
Right hind limb	2000 rad	0.0 (4)
Thoracic cavity	2000 rad	0.0 (4)
Surgery		
Midline peritoneal laporatomy		0.0 (5)
Small intestinal transection		0.0 (3)
Small intestinal resection		0.0 (4)
Beeswax implant pancreas	3 mg/kg	0.0 (10)

[a]Number of rats that were evaluated listed in parentheses. At the time all of these studies were being completed, rats having small bowel, colon, and pancreas cancers exhibited measurable CMI (References 17,18).

the CMI determinations; however, the limited observations gathered to date suggest that the methodology is capable of differentiating between carcinogen and noncarcinogen substances (Table VI). Included in this list is a study in which 6-week-old rats were maintained on distilled water for 2 months and then evaluated for the possible induction of CMI. No cytotoxic responses were found to be associated with these animals' lymphoid cells, thus indicating that for an evaluation of drinking water supplies for carcinogenicity, it should be possible to maintain the animals involved in the study on distilled (known carcinogen-free) water.

DISCUSSION

A major problem that exists today in establishing the carcinogenicity of drinking water entails the development of sensitive and specific assays that can be applied to determine both the direct and indirect induction of cancer. To be usable in a practical fashion, such a test must also have the characteristics of being rapid, economical, and relevant. The concept of relevance implies that the test substance is biologically processed in such a manner that the toxicological endpoints of the bioassay are homologous to those observed in man.[26]

A theoretical approach for possibly devising such an assay was brought forth by Clayson,[27] who noted that if cancer possesses some unique mechanism in its genesis, it should be possible to devise a test for experimentally determining a similar mechanism that could be used for identifying the tumor-inducing substance. One successful application of this concept has been to adopt the supposition that cancer-causing agents interact with DNA. This singular idea led to the development of a variety of bioassays for cancer-causing agents such as the widely applied microbial and mutagenesis tests[14,28] and the DNA repair assays.[29] There are several advantages to the use of these tests in that they fulfill the requirements of being rapid, inexpensive, and easy to perform; however, they do have two major shortcomings.

First, these tests require that the test agent be in an activated form before mutagenicity can be expressed. It has been observed that most carcinogens are not active as such, but require metabolic activation. Cells in general, and more specifically, cells from various tissues and organs, are known to vary widely in their ability to effect the required metabolism for carcinogens to become activated. Consequently, using selected enzyme sources, such as the liver S-9 mixture commonly applied for the activation, could possibly fail to provide the conditions necessary for a substance to express mutagenicity.

Second, the test results have to be extrapolated from the in vitro stage to that of the actual human exposure condition. In regard to the latter shortcoming, it has been previously pointed out that it is simply illogical to ignore the vast differences that exist in the various detoxification and activation mechanisms occurring in the intact animal, that is, in contrast to those in the microbial systems or even in the exogenous-added mammalian enzymes.[30]

Even the selection of the animal providing the source of the metabolizing enzymes may lead to misinterpretation. For example, the animal carcinogen dimethylnitrosamine was found to be mutagenic in the *Salmonella* plate incorporation assay in the presence of hamster liver S-9; however, it was not mutagenic when the S-9 was derived from either mouse or rat.[31] Of course, a similar argument can be raised about an assay based on the immune reactivity of the rat, since the rat itself, in certain studies, has been found to be poorly predictive of the metabolism occurring in humans.

Yet, despite differences in test results, it is obvious that the extent of extrapolation of the results is less from one mammal to another, such as from inducing cancer in the rat to doing the same in the human. In addition, the use of the whole animal as a test system allows the occurrence of possible multistep processes in which "new cell populations represent stages in the cellular evolution from normal, through initiated, preneoplastic and premalignant cells to the highly malignant neoplasia."[32]

For the reasons described, it is highly desirable to have short-term, inexpensive, in vivo bioassays that can be readily applied to aid in the ultimate determination for carcinogenicity. Such tests can then be coupled with information gathered from in vitro studies to form a test battery with which to arrive at a decision regarding the carcinogenic risk of an agent.[13] Bioassays currently being used[33] include (1) skin tumor induction in mice, (2) pulmonary tumor induction in mice, (3) breast cancer induction in female Sprague-Dawley rats,

(4) altered foci induction in rodent liver, and (5) assays for tumor promoters. A positive result for a substance in the in vitro studies, coupled with a definitive positive measure in such a bioassay, would be recognized as making the agent highly suspect as a human carcinogen.

This study represents an outgrowth of our continuing efforts to develop a practical bioassay for identifying tumor-inducing agents based on the measurement of antitumor immunity brought about through exposure to a carcinogen. The preliminary information gathered to date for the halogenated hydrocarbons suggests that the measurement of CMI responses to such agents may form the basis for such an analysis. One possible shortcoming seemingly evident in this particular study that has not been previously observed is the apparent lack of clear-cut dose-response relationships. This would preclude the application of the CMI measurements in a quantitative manner for establishing carcinogenic potential. However, based on our previous experiences with other carcinogenic insults, we believe that a more extensive investigation using larger groups of animals over a smaller exposure range will result in the desirable dose relationship.

CONCLUSIONS

Measurable antitumor cell-mediated immune (CMI) responses can be detected in inbred rats exposed to halogenated hydrocarbons known to be cancer-inducing agents. In the limited studies accomplished to date, the procedure appears to be sensitive, specific, rapid, and economical to perform. If it is found that the procedure continues to possess these desirable characteristics, then the evaluation of CMI responses may assume the role as a short-term bioassay for evaluating the carcinogenicity of environmental substances such as drinking water.

ACKNOWLEDGMENT

This investigation was supported by PHS Grant CA 30967. Drs. Lindholm and Liu received support from training grant T32-CA 09125.

REFERENCES

1. Higginson, J. "Present Trends in Cancer Epidemiology," in *Proceedings Eighth Canadian Cancer Conference* (Honey Harbor, Ontario: 1969), pp. 40–75.
2. Doll, R. "Prevention of Cancer-Pointers for Epidemiology," Rock Carling Fellowship, Nuffield Provincial Hospitals Trust (London: Witefriars Press, Ltd., 1967).
3. Boyland, E. "A Chemist's View of Cancer Prevention," *Proc. R. Soc. Med.* 60:93–99 (1969).

4. Loper, J. C., D. R. Lang, and C. C. Smith. "Mutagenicity of Complex Mixtures from Drinking Water," in *Water Chlorination: Environmental Impact and Health Effects, Vol. 2*, R. L. Jolley, H. Gorchev, and D. H. Hamilton, Eds. (Ann Arbor, MI: Ann Arbor Science Publishers, Inc.), pp. 443–450.
5. Bull, R. J., M. A. Pereira, and K. L. Blackburn. "Bioassay Techniques for Evaluating Possible Carcinogenicity of Absorber Effluents" presented at the Adsorption Technologies Conference, (Washington, DC: 1979).
6. *Drinking Water and Cancer: Findings and Assessment of Risks* (Washington, DC: Council on Environmental Quality, 1980).
7. Harris, R. H. "Implications of Cancer Causing Substances in Mississippi River Water" (Washington, DC: Environmental Defense Fund, 1974).
8. Page, T., R. H. Harris, and S. S. Epstein. "Drinking Water and Cancer Mortality in Louisiana," *Science* 193:55(1976).
9. *Report on Carcinogenesis Bioassay of Chloroform* (Washington, DC: National Cancer Institute, 1976).
10. Alavanja, M., I. Goldstein, and M. Susser. "A Case Control Study of Gastrointestinal and Urinary Tract Cancer Mortality and Drinking Water Chlorination," in *Water Chlorination: Environmental Impact and Health Effects, Vol. 2* R. L. Jolley, H. Gorchev, and D. H. Hamilton, Eds. (Ann Arbor, MI: Ann Arbor Science Publishers, Inc., 1978), pp. 395–409.
11. Council on Environmental Quality. *Drinking Water and Cancer* (Washington, DC: 1980).
12. Maugh II, T., "New Study Links Chlorination and Cancer," *Science* 211:694 (1981).
13. Bridges, B. A. "The Three-Tier Approach to Mutagenic Screening and the Concept of Radiation-Equivalent Dose," *Mutat. Res.* 26:335–340 (1974).
14. Ames, B. N., F. D. Leed, and W. E. Durston. "An Improved Bacterial Test System for the Detection and Classification of Carcinogens," *Proc. Nat. Acad. Sci.* 70:782–786 (1973).
15. Sontag, J. M., N. P. Page, and U. Saffiotti. *Guidelines for Carcinogen Bioassay in Small Rodents*. NCI-CG-R-1, National Cancer Institute Technical Report, Series No. 1 (Washington, DC: Department of Health, Education, and Welfare, 1976).
16. Clayson, D. B. "Comparison Between in vitro and in vivo Tests for Carcinogenicity. An Overview," *Mutat. Res.* 75:205–213 (1980).
17. Stevens, R. H., D. A. Cole, R. S. Rana, and J. M. Graves. "Identification of Carcinogens by Measurement of Cell-Mediated Immunity I. Immunity Induced by Rat Pancreas and Colon Carcinogens," *Environ. Res.* 21:143–149 (1980).
18. Stevens, R. H., G. P. Brooks, J. W. Osborne, C. W. Englund, and D. W. White. "Lymphocyte Cytotoxicity in X-Irradiation Induced Adenocarcinoma of the Rat Small Bowel I. Measurement of Target Cell Destruction by Release of Radioiodinated Membrane Proteins," *J. Nat. Cancer Inst.* 59:1315–1319 (1977).
19. Bayse, G. S., A. W. Michaels, and M. Morrison. "Lactoperoxidase-Catalyzed Iodination of Tyrosine Peptides," *Biochem. Biophys. Acta.* 284:30–33 (1972).
20. Boyum, A. "Separation of Blood Leukocytes, Granulocytes, and Lymphocytes," *Antigens* 4:269–274 (1974).
21. Brown, J., P. G. VanBelle, and I. Hellstrom. "Design of Experiments Using the Microcytotoxicity Assay," *Int. J. Cancer* 18:230–235 (1976).
22. Pollard, J. H. *A Handbook of Numerical Statistical Techniques* (London: Cambridge Press, 1977), pp. 90–99.

23. Stevens, R. H., J. M. Graves, E. S. Meek, J. W. Osborne, H. F. Cheng, and D. P. Loven. "Cyclic Nucleotide Concentrations in 7,12-Dimethylbenz[a]anthracene-Induced Pancreatic Cancer in Rats," *J. Nat. Cancer Inst.* 61:1281-1284 (1978).
24. Stevens, R. H., and D. A. Cole. "Detection of Carcinogenic Exposures by Urinalysis: Induction of Cell Mediated Immunity," *Toxicol. Lett.* 11:299-303 (1982).
25. Hawks, H., R. M. Hicks, J. M. Holsman, and P. N. Magee. "Morphological and Biological Effects of 1,2-Dimethylhydrazine and 1-Methylhydrazine in Rats and Mice," *Br. J. Cancer* 30:429 (1974).
26. Wolff, G. L. "Some Genetic Considerations for the Design of Better Mammalian Assay Systems for the Detection of Chemical Mutagens and Carcinogens," *J. Environ. Toxicol.* 1:79 90 (1977).
27. Clayson, D. B. "Relationships Between Laboratory and Human Studies," *J. Environ. Pathol. Toxicol.* 1:31-44 (1977).
28. Ames, B. N., W. E. Durston, E. Yamasaki, and F. D. Leed. "Carcinogens Are Mutagens — A Simple Test Combining Liver Homogenates for Activation and Bacterial Detection," *Proc. Nat. Acad. Sci.* 70:2281-2285 (1973).
29. Huberman, E., and L. Sachs. "Cell-Radiation Mutagenesis of Mammalian Cell with Chemical Carcinogens," *Int. J. Cancer.* 13:326-33 (1974).
30. Stich, J. F., R. H. C. San, J. A. Miller, and E. C. Miller. "Various Levels of DNA Synthesis in Xeroderma Pigmentosum Cells Exposed to the Carcingoens N-Hydroxy and N-Acetoxy-2-acetyaminofluorene," *Nature New Biol.* 232:9-10 (1972).
31. Prival, M. J., and V. D. Mitchel. "Influence of Microsomal and Cytosolic Fractions from Mouse, Rat, and Hamster Liver on the Mutagenicity of Dimethylnitrosamine in *Salmonella* Plate Incorporation Assay," *Cancer Res.* 41:4361-4367 (1981).
32. Farber, E., and R. Cameron. "The Sequential Analysis of Cancer Development," *Adv. Cancer Res.* 31:125-226 (1980).
33. Weisburger, J. H., and G. M. Williams. "Carcinogen Testing: Current Problems and New Approaches," *Science* 214:401-407 (1980).

CHAPTER 22

Formation of Genotoxic Compounds by Chlorination of Residues from Oil Refinery Effluents

Christopher D. Metcalfe and Ronald A. Sonstegard

Disinfection by chlorination has been shown to dramatically increase the mutagenic activity of drinking water.[1-4] However, the mechanisms by which mutagens are formed are largely unknown. Carcinogenic alkyl halides can be formed by chlorination of the phenolic compounds consisting of humic, tannic, and fulvic acids,[5-7] as well as by chlorination of phenols and anilines.[8] Nonvolatile genotoxic compounds are also formed by chlorination of humic acids.[2,9]

There is potential for the formation of genotoxic and carcinogenic compounds by the reaction of chlorine with environmental contaminants in drinking water. One of the major mechanisms by which our aquatic resources become contaminated is through the discharge of industrial effluents. The petroleum and petrochemical industry is ranked third in its worldwide potential for industrial pollution of aquatic resources.[10] If significant quantities of refinery residues circulate into public water treatment facilities, chlorination of the water may produce hazardous compounds.

In this study, we used in vitro assays (Ames mutagenicity test, sister chromatid exchange assay) to investigate the genotoxic activity associated with the chlorination of water containing various amounts of oil refinery effluents. Comparisons were made between the relative contribution of refinery effluents and natural organic acids to genotoxicity following chlorination.

MATERIALS AND METHODS

Sample Collection and Preparation

Effluent samples were collected from the final holding pond of a refinery discharging into Lake Ontario, Canada. Grab samples (4 L) were collected in glass solvent bottles and stored at 4°C under nitrogen until used in chlorination tests.

Effluents were analyzed for pH, total suspended matter, concentrations of oil and grease, phenol, and ammonia-nitrogen. Ammonia was determined by the sodium salicylate method of Verdouw et al.[11] The other parameters were assayed by the methods outlined in Reference 12.

Effluents were diluted with volumes of organic-free water prepared by passing double-distilled water through 0.45-μm membrane filters and Norganic resin cartridges (Millipore). A stock chlorine solution was prepared by bubbling high-purity chlorine gas (Matheson Corporation) through organic-free water. The chlorine content of this solution was determined by the DPD colorimetric method.[12]

Effluent solutions were chlorinated by aliquoting an appropriate volume of chlorine stock (2–8 mL) into 2 L of effluent solution in stoppered amber-glass bottles. Solutions were incubated at 20°C, with gentle vortex stirring at chlorine concentrations of 2, 4, 8, or 16 mg/L for 1, 2, or 4 h. The pH of the solution was periodically adjusted with NaOH or HCl solutions (0.1 M) to 4, 6, or 8.

In a separate experiment, amounts of tannic acid, humic acid, and refinery effluents equivalent to 10 mg organic carbon were added to 2 L organic-free water and chlorinated at 8 mg/L for 1 h at pH 6.0. The organic carbon content of organic acid residues and effluents was determined by the modified Walkley and Black dichromate oxidation method.[13]

Organic compounds were extracted from chlorinated solutions by passing a 2-L volume through an Amberlite XAD-2 resin column (8 g dry wt.; 8- by 2.3-cm ID) at a flow rate of 25 mL/min. Compounds retained on the column were eluted with 100 mL of diethyl ether following equilibration of the solvent for 15 min. Extracts were evaporated to approximately 15 mL on a rotary evaporator, passed through sodium sulfite (15 g) to remove residual chlorine, and then through sodium sulfate (20 g) to dehydrate the sample. The sample was rotary evaporated to 2 mL and then to dryness under a stream of nitrogen. Extracts were made up to volume in acetone for genotoxicity testing.[14]

In a separate series of experiments, diluted effluent (2.5%), composed of large volumes (20 L) of surface water from the Hamilton Harbor area of Lake Ontario, was chlorinated at 8 mg/L for 1 h (pH 6.0, 20°C) and extracted by the XAD-2 method, essentially as described previously. However, the chlorinated sample was passed through the resin column at a faster flow rate of 120 mL/min. Column elution, residual chlorine removal, sample drying, and evaporation of solvent were completed as described previously.

Extracts were subfractionated into seven fractions of increasing polarity by silica gel column chromatography, according to the method of Coleman et al.[15] Extract residue with a dry weight of between 20 and 25 mg was dissolved in 0.5 mL of hexane and applied to the top of a 14-cm by 15-mm ID microchromatographic column packed with 0.2 g of 5% deactivated silica gel. Organics were partitioned into seven subfractions by eluting with the following solvents: (1) 0.5 mL hexane, (2) 1 mL hexane, (3) 4 mL hexane, (4) 4 mL hexane/benzene, (5) 4 mL benzene (6) 4 mL methylene chloride, and (7) 4 mL methanol.

Genotoxicity Assays

When testing nonvolatile extracts for mutagenicity, the *Salmonella*/ mammalian microsome assay was conducted using strains TA100 and TA98, as described by Ames et al.,[16] for the plate-incorporation assay. Rat-liver microsomes (S-9) were prepared from Aroclor 1254-induced rats, also according to the procedures outlined by Ames et al.[16] In all Ames assays, the number of revertant colonies for each treatment was calculated from the mean of three replicate plates. All samples were tested for mutagenicity at four to six concentrations within the range of a linear dose-response curve.

For the sister-chromatid exchange assay, CHO cells were cultured in αMEM medium with 10% fetal calf serum. Cells were inoculated into 25 mL of medium in 250-mL tissue culture flasks and incubated for 12 h. Test extracts (0.1 mL) were then added to the culture flasks and, after 1 h, the medium was changed. The flasks were then incubated with $10^{-5}M$ 5-bromodeoxyuridine (5-BrdU) for 22 h, after which 2 µg/mL colchicine was added. After a 2-h exposure to colchicine, the cells in metaphase were harvested by shaking the culture flasks. When metabolic activation was required, S-9 was added to the culture flasks according to the method of Latt et al.[17] Fresh medium and 10% S-9 mix (prepared as in the Ames test) were added with extract to the culture flask. After incubation from 1 to 2.5 h, the medium was changed and 5-BrdU was added.

Harvested cells were treated for 7 min with 10 mL of hypotonic solution (0.075 M KCl) at 37°C. They were then hardened by the addition of 2 mL of 3:1 methanol-acetic acid fixative. After 15 min of hardening, cells were centrifuged and fixed overnight at 4°C in methanol-acetic acid. Metaphases were spread on slides, and after 1 week of drying, they were stained with 5 mg per 100 mL of Hoechst 33258 fluorochrome. Slides were flooded with Sorensen's buffer (pH 8.0) and exposed to black light for 3 min at 50°C. They were then stained with Geimsa, dried, and mounted. Sister chromatid exchanges were scored for 25 metaphases per treatment. All samples were tested for a dose response.

RESULTS

Chlorination experiments with 2-L volumes of organic-free water as diluent were conducted using a June 1982 refinery effluent sample. Tests with 20-L volumes of surface water as diluent were conducted with an August 1982 effluent sample. The oil refinery from which samples were taken refines approximately 70,000 bbl/d crude into primarily fuel products using low-temperature ($\simeq 500°C$) catalytic-cracking technologies. Effluent discharges of approximately 4×10^6 L/d are treated by API separator, flocculation with aluminum sulfate, primary clarification, activated sludge treatment, and secondary clarification before discharge into Lake Ontario. The effluent consists of storm runoff and treated process water used in crude desalting, catalytic cracking, and steam injection processes.

Table I summarizes the water quality data for the two effluents used in chlorination experiments. These parameters fall within the range of values reported for other Canadian refineries of this type.[18] Extracts were made of the particulate and dissolved nonvolatile components of effluent samples before exposure to chlorine. The results of Ames assays of these extracts (Table II) indicate that the particulate and dissolved components of the effluents were not mutagenic before chlorination.

Figure 1 illustrates the results of Ames assays for a nonvolatile extract from a dilute effluent solution (10%) chlorinated at 8 mg/L for 1 h. The concentrations of extract in the Ames plates are presented in terms of the equivalent

Table I. Water Quality Parameters for June and August 1982 Refinery Effluent Samples Used in Chlorination Tests

Parameter	Value	
	June	August
pH	8.0	7.5
$NH_3 - N$, ppm	17.4	3.9
Phenol, ppb	32.0	42.0
Suspended solids, mg/L	34.0	29.0
Oil and grease, mg/L	6.4	4.0

Table II. Results of Ames Tests with and without S-9 Activation for Extracts of Particulate and Dissolved Compounds from Refinery Effluent Samples Collected in June and August 1982. Results are Presented for the Test Strain Giving the Maximum Response to the Extract

Extract	Strain	mL Equiv/ Plate[a]	Revertants/Plate[b]			
			With S-9		Without S-9	
			June	August	June	August
Particulates	TA100	320	107(102)	122(100)	115(112)	112(115)
Dissolved	TA98	80	72(46)	50(42)	47(49)	50(46)

[a]The dose of extract applied to each plate is calculated in terms of the volume of original effluent concentrated to the 0.1-mL plating volume.
[b]Spontaneous revertant values are presented in parentheses.

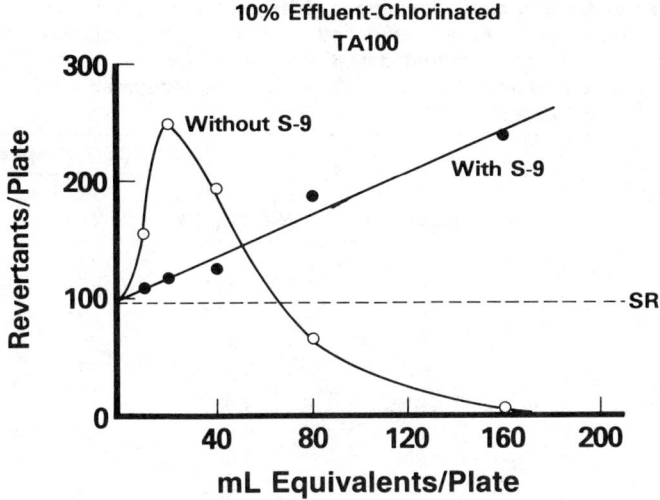

Figure 1. Mutagenicity of extract from chlorinated (8 mg/L chlorine for 1 h at pH 6.0) 10% effluent tested over a range of concentrations using strain TA100 with and without S-9.

volume of original chlorinated solution extracted and then concentrated to the 0.1-mL plating volume (milliliter equivalent per plate). Thus, 80 mL of a chlorinated solution concentrated to 0.1 mL is a dose of 80-mL equivalents. If the solution consisted of 10% effluent, this dose represents 8 mL of original effluent in the chlorinated solution. Extracts tested without S-9 were toxic at high concentrations, but showed considerable mutagenic activity at low sample doses (< 40 mL equivalents). Extracts tested with S-9 were much less toxic, but showed less mutagenic activity at low doses. Test strain TA100 was more sensitive to extracts from chlorinated solution than TA98.

The different dose-response curves for tests conducted with and without S-9 may represent the activity of two classes of mutagenic agents: one direct-acting and the other requiring S-9 activation. The extracts gave a positive response in the SCE assay without microsomal activation (Table III); however, there was no response in assays with S-9.

When extracts from chlorinated 10% effluent solutions were subfractioned by silica-gel column chromatography, mutagenicity was confined to fractions 3 and 5 in Ames tests without S-9 (Figure 2a). In tests with S-9, activity was lower but mutagenicity was detected in fractions 5 and 7 (Figure 2b). The two patterns of activity in tests with and without S-9 again indicate the presence of two classes of genotoxic agents in the chlorinated sample.

Chlorinated solutions containing a range of effluent concentrations (2.5 to 20%), as well as whole effluents, were mutagenic in tests conducted without

Table III. Mean Sister Chromatid Exchanges (SCEs) Induced by Extracts from a Control Solution (Organic-Free Water) and a 10% Effluent Solution (Organic-Free Water as Diluent) Chlorinated at 8 mg/L for 1 h (pH 6). Assays Run with and without Microsomal Activation (S-9). Extracts Incubated with S-9 in Assay Flasks for 1 or 2.5 h

Sample	mL Equiv/Flask	Mean SCEs/Metaphase		
		Without S-9	With S-9 (1 h)	With S-9 (2.5 h)
Solvent control		4.7	4.9	5.2
10% Effluent solution	320	Toxic	5.7	Toxic
	160	11.6[a]	4.6	5.9
	80	8.3[a]	4.7	5.3
	40	6.9[a]	5.2	5.0
Control solution	320	4.9		4.8
	160	4.6		4.7
	80	4.6		5.0
	40	4.8		4.8

[a]Mean SCEs significantly different from solvent controls ($t_{0.05}$).

S-9 (Figure 3). Linear dose-response curves were generated for all chlorination tests, but only mutagenicity at the highest doses of the curve (20-mL equivalent) was shown. Extracts from chlorinated control solutions (organic-free water) gave a slight response in the Ames assay (Figure 3). The XAD-2 resin used for extractions contains naphthalenic preservatives[19] that may have reacted with residual chlorine in the water passing through the column to yield mutagenic compounds. High-performance liquid chromatographic analysis of the organic-free water indicated that it contained only low levels of organic contaminants.

In many studies to determine the genotoxic activity of chlorinated solutions, sodium sulfite was added to stop the chlorination process before extraction. Since Cheh et al.[1] reported that sodium sulfite reduced the mutagenic yield of chlorinated drinking water, this treatment was eliminated in our tests. However, to reduce toxicity in the in vitro assays, it was necessary to pass ether extracts through a bed of sodium sulfite after elution from the XAD-2 column. This did not appear to reduce the mutagenicity of the samples. Moreover, when a 15-mL sample of ether containing 100 μg each of chloronaphthalene, hexachlorophenol, and tetrachlorobiphenyl was passed through sodium sulfite, gas chromatographic analysis revealed that the three compounds were recovered with efficiencies of 72, 87, and 89%, respectively.

Tests were conducted to determine the effect of various chlorination conditions on the mutagenic activity of nonvolatile extracts. The mutagenicity of a 10% effluent solution increased with the chlorine concentration over a range of 2 to 8 mg/L chlorine (Figure 4a). The mutagenicity of chlorinated effluent solutions also increased with contact time, with the greatest increment in muta-

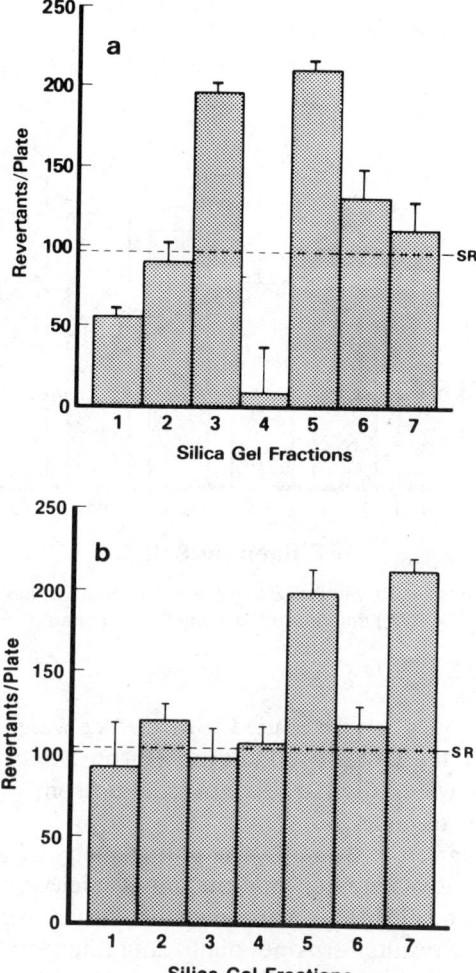

Figure 2. Mutagenicity (TA100) of silica-gel subfractions from extracts of chlorinated (8 mg/L) 10% effluent; (a) results of Ames tests without S-9 at a dose of 160-mL equiv/plate; (b) results with S-9 at a dose of 320-mL equiv/plate.

genic activity occurring in the first hour of contact with chlorine (Figure 4b). There was little difference in the mutagenicity of extracts from solutions chlorinated at pH 4 and 6; however, mutagenicity was reduced in tests at pH 4 (Table IV).

Although effluents diluted with organic-free water were mutagenic when chlorinated, it was not clear whether effluents diluted with surface water would give similar results. Results are summarized in Table V for a chlorina-

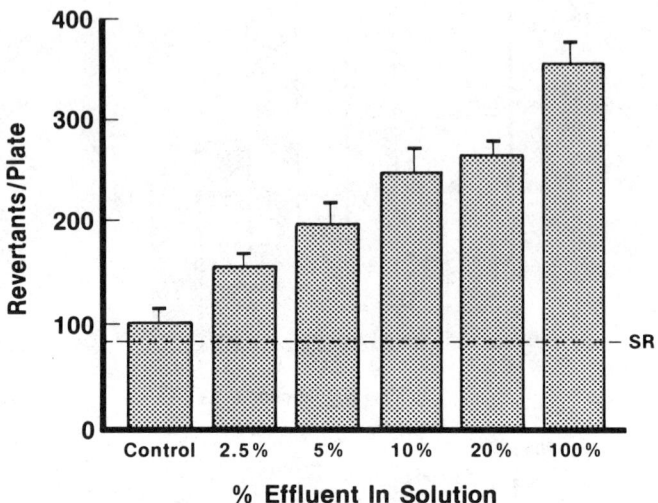

Figure 3. Mutagenicity (TA100 without S-9) of extracts from whole effluents and diluted effluents (2.5 – 20%) chlorinated at 8 mg/L and plated at a dose of 20-mL equiv/plate.

tion test in which effluents were diluted with surface water (20 L) taken from an embayment of Lake Ontario (Hamilton Harbor) that receives industrial and domestic wastewater discharges. For comparison, effluents were also diluted with 20 L of organic-free water.

Following chlorination at 8 mg/L, the mutagenicity of surface water containing 2.5% effluent was greater than the mutagenicity of a sample containing surface water alone (Table V). Dose-response curves were generated for all samples, but only the results for Ames plates containing 400-mL equivalent of chlorinated solution are shown in Table V. When the effluent was diluted with 20 L of organic-free water, the extract was slightly more mutagenic (Table V), probably because of the absence of other organic materials that react competitively with chlorine. The XAD-2 extraction procedure, in which 2-L volumes of sample were passed through the column, recovers most compounds with >70% efficiency, but the recovery efficiency using 20-L samples was not determined. While the recovery of mutagens in tests using 20-L solutions may not be quantitatively accurate, comparisons among treatments are valid.

Mutagen-directed subfractionation of extracts prepared from effluents diluted with surface water indicates that mutagenicity in Ames tests without S-9 (TA100) is confined to fractions 3 and 5 (Figure 5). These are the same fractions that yielded activity in extracts from 2-L chlorinated solutions (Figure 2). Chlorinated surface water alone also contained some mutagenic activity in fraction 3 (Figure 5).

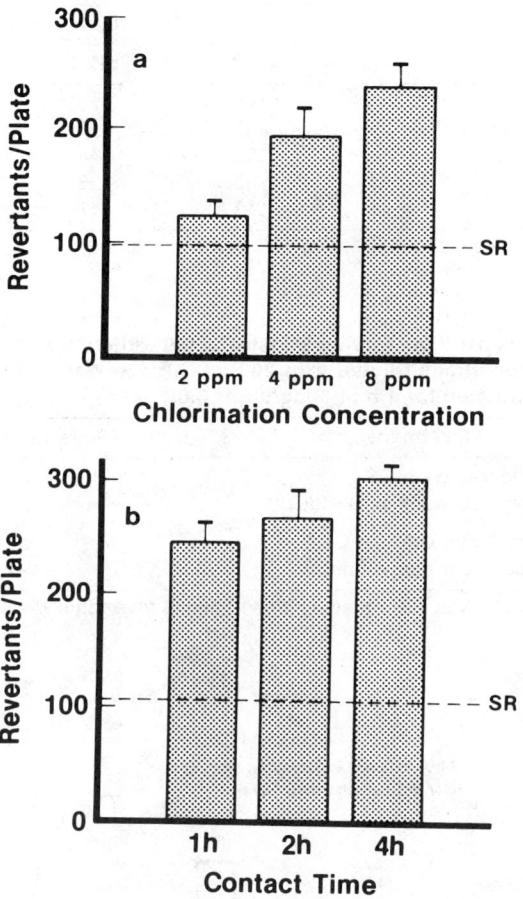

Figure 4. Mutagenicity (TA100 without S-9) of extracts (20 mL equiv/plate) from 10% effluent chlorinated at (a) 2, 4, and 8 mg/L chlorine concentrations; (b) 8 mg/L chlorine concentration for a 1-, 2-, or 4-h contact time.

Naturally occurring organic compounds such as humic and tannic acids have been shown to contribute to the formation of nonvolatile genotoxic compounds after aqueous chlorination.[2,9] The relative mutagenic potential of chlorinated refinery effluents, tannic acid, and humic acid were determined by chlorinating samples of organic-free water containing these substances at concentrations of 4-mg/L organic carbon. The extracts of chlorinated solutions were toxic in the Ames tests (TA100 without S-9) at doses greater than 40 mL

Table IV. Revertants per Plate for Ames Tests (TA100, without S-9) of Extracts from a 10% Effluent Solution Chlorinated for 1 h at 8 mg/L Chlorine and at pH 4, 6, and 8[a]

| Dose/Plate | Revertants/Plate | | |
(mL equiv)	pH 4	pH 6	pH 8
80	146	173	124
40	232	223	162
20	196	244	145
10	190	153	108
5	122	117	106

[a]Spontaneous revertants without S-9 were 102.

Table V. Revertants per Plate for Ames Tests (TA100 without S-9) of Extracts from 2.5% Effluent Solutions Diluted with 20 L of Surface Water or Organic-Free Water and Chlorinated for 1 h at 8 mg/L (pH 6.0)[a]

Treatment	Revertants/Plate
Organic-free water	82
Organic-free water plus effluent	227
Surface water	97
Surface water plus effluent	191

[a]Sample dose per plate was 400 ml equiv; spontaneous revertants without S-9 were 96.

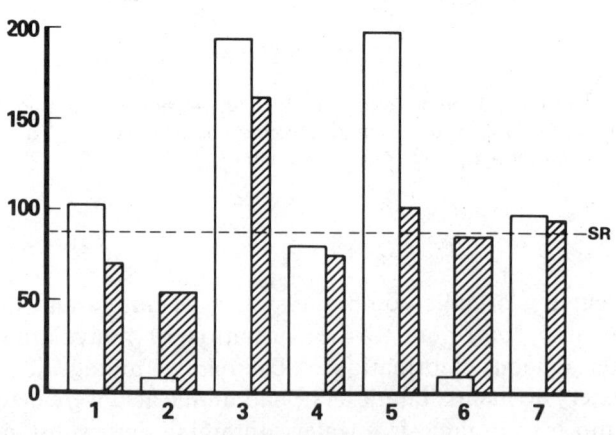

Figure 5. Mutagenicity (TA100 without S-9) of silica-gel subfractions (20-mL equiv/plate) from extracts of chlorinated (8 mg/L) surface water (Hamilton Harbor water) and chlorinated surface water plus 2.5% effluent.

Table VI. Mutagenicity of Diluted Refinery Effluents, Tannic Acid, and Humic Acid (4 mg/L organic carbon) Chlorinated at 8 mg/L for 1 h (pH 6.0)[a]

Dose/Plate (mL equiv)	Revertants/Plate		
	Effluent	Tannic Acid	Humic Acid
160	Toxic	66	102
80	61	147	146
40	185	155	162
20	244	148	145
10	153	100	140
5	126	98	111

[a]Spontaneous revertants for strain TA100 without S-9 were 104.

equivalents per plate; however, at lower doses, the relative mutagenicity of the chlorinated effluent solutions was greater than solutions containing chlorinated humic and tannic acids (Table VI).

DISCUSSION

Chlorination of water containing various concentrations of refinery effluent increased the mutagenic activity of the samples. The mutagenicity of chlorinated solutions increased with effluent concentration, contact time with chlorine, and chlorine concentration. Most surface waters used for drinking water are disinfected at chlorine concentrations >6 mg/L.[20] Therefore, the chlorine concentrations used in these tests (2–8 mg/L) represent the lower range of concentrations used in drinking water treatment.

Chlorination at pH 4 and 6 produced high mutagenic activity, but mutagenicity was reduced in tests at pH 8. The mechanisms of substitution, oxidation, and amination in aqueous chlorination are highly pH dependent.[21] Oyler et al.[22] found that aqueous chlorination of polynuclear aromatic hydrocarbons at pH >6 resulted in the production of oxygenated compounds, whereas reactions at pH <6 yielded some oxygenated (quinones) and chlorinated compounds.

The difference in the dose-response curves for chlorinated extracts tested in the Ames assay with and without S-9 (Figure 1) indicates that there may be two classes of mutagens (one direct acting and one requiring metabolic activation) in chlorinated samples. This indication is further supported by the fractionation of mutagens into subfractions 3 and 5 in Ames tests without S-9, and into subfractions 5 and 7 in tests with S-9 (Figure 2). The direct-acting component is the most mutagenic in the Ames assay, whereas elevated sister chromatid exchanges were only found in assays conducted without S-9 activation (Table III). The clastogenic activity of the direct-acting agent(s) must have been destroyed by metabolic activity or by adsorption onto S-9 proteins.

The biological implications of these results are difficult to assess without studies using pilot-scale waste treatment systems (i.e., charcoal and sand filtration, coagulation, flocculation). Contaminants originating from an oil refinery may be eliminated by water treatment systems before the application of chlorine. The contribution of naturally occurring contaminants (e.g., humic acids) to the yield of genotoxic compounds formed by chlorination may exceed the contribution by effluents. Refinery effluents may come into contact with chlorine by other mechanisms. Effluents at some refineries are not treated onsite, but are discharged into municipal sewage systems.[23] Also, municipal drinking water supplies may be used as process water in some refineries.[24] In both of these instances, there is potential for the formation of genotoxic compounds by the reaction of effluents with chlorine.

ACKNOWLEDGMENTS

This work was supported by a grant in aid for research from the Petroleum Association for Conservation of the Canadian Environment (PACE). It was also supported by a grant to R. A. Sonstegard from the National Science and Engineering Research Council (Canada). We thank Pat Henry for her technical assistance.

REFERENCES

1. Cheh, A., J. Skochdopole, P. Koski, and L. Cole. "Nonvolatile Mutagens in Drinking Water: Production by Chlorination and Destruction by Sulfite," *Science* 207:90–92 (1979).
2. Fallon, R., and C. Fliermans. "Formation of Nonvolatile Mutagens by Water Chlorination: Persistence and Relationship to Molecule Weight of Organic Material in Water," *Chemosphere* 9:385–391 (1980).
3. Dolara, P., V. Ricci, D. Burrini, and O. Griffini. "Effects of Ozonation and Chlorination on the Mutagenic Potential of Drinking Water," *Bull. Environ. Contam. Toxicol.* 27:1–6 (1981).
4. Zoeteman, B., J. Hrubec, E. de Greef, and H. Kool. "Mutagenic Activity Associated with Byproducts of Drinking Water Disinfection by Chlorine, Chlorine Dioxide, Ozone and UV-Irradiation," *Environ. Health Perspect.* 46:197–205 (1982).
5. Youssefi, M., S. Zenchelsky, and S. Faust. "Chlorination of Naturally Occurring Organic Compounds in Water," *J. Environ. Sci. Health* A13:629–637 (1978).
6. Rook, J. "Chlorination Reactions of Fulvic Acids in Natural Waters," *Environ. Sci. Technol.* 11:478–484 (1977).
7. Dowty, B., L. Green, and J. Laster. "Automated Gas Chromatographic Procedure to Analyze Volatile Organics in Water and Biological Fluids," *Anal. Chem.* 48:946–953 (1976).
8. Hirose, H., and T. Okitsu. "Formation of Trihalomethanes by Reaction of Halogenated Phenols or Halogenated Anilines with Sodium Hypochlorite," *Chemosphere* 11:81–87 (1982).

9. Watts, C., B. Crathorne, M. Fielding, and S. Killops. "Nonvolatile Organic Compounds in Treated Waters," *Environ. Health Perspect.* 46:87-99 (1982).
10. Gattelier, C. R. "Les Outères Analytiques de la Pollution des Eaux par les Produites Pétroliers," *Rev. Assoc. Fr. Tech. Pét.* 206:59-60 (1971).
11. Verdouw, H., C. Van Echteld, and E. Dekkers. "Ammonia Determinations Based on Indophenol Formation with Sodium Salicylate," *Water Res.* 23a:399-404 (1977).
12. *Standard Methods for the Examination of Water and Wastewater*, 14th edition. M. Rand, A. Svenberg, M. Taras, Eds. (Washington, DC: Am. Public Health Assoc., 1975), pp. 332-334.
13. Bremner, J., and D. Jenkinson. "Determination of Organic Carbon in Soil I. Oxidation by Dichromate of Organic Matter in Soil and Plant Materials," *J. Soil Sci.* 11:394-402 (1960).
14. Pagel, J., and R. Smillie. *Development of Concentration Techniques for Mutagenic Substances in Environmental Samples. Part I: Concentration of Volatile Organics*, OTC Report 8001 (Toronto: Ontario Ministry of Environment, 1980).
15. Coleman, W., R. Melton, F. Kopfler, K. Barone, T. Aurand, and J. Jellison. "Identification of Organic Compounds in a Mutagenic Extract of a Surface Drinking Water by a Computerized Gas Chromatography/Mass Spectrometry System," *Environ. Sci. Technol.* 14:576-588 (1980).
16. Ames, B., J. McCann, and E. Yamasaki. "Models for Detecting Carcinogens and Mutagens with the *Salmonella*/Mammalian Microsome Mutagenicity Test," *Mutat. Res.* 31:347-364 (1975).
17. Latt, S., J. Allen, S. Bloom, A. Carrans, E. Falke, D. Krom, E. Schneider, R. Schreck, R. Tice, B. Whitfield, and S. Wolff. "Sister-Chromatid Exchanges: A Report of the Gene-Tox Program," *Mutat. Res.* 87:17-62 (1981).
18. *Survey of Trace Substances in Canadian Petroleum Industry Effluents*, PACE Report No. 81-4, (Toronto: Petroleum Association for Conservation of the Canadian Environment, 1981).
19. Junk, G., J. Richard, M. Griser, D. Witiak, J. Witiak, M. Arguello, R. Vick, H. Avec, J. Fritz, and G. Calder. "Use of Macroreticular Resins in the Analysis of Water for Trace Organic Contaminants," *J. Chromatogr.* 99:745-762 (1974).
20. Cantor, K. "Epidemiological Evidence of Carcinogenicity of Chlorinated Organics in Drinking Water," *Environ. Health Perspect.* 46:187-195 (1982).
21. Pierce, R. C. *The Aqueous Chlorination of Organic Compounds: Chemical Reactivity and Effects on Environmental Quality*, NRCC No. 16460 (Ottawa: National Research Council of Canada, 1978).
22. Oyler, A., R. Liukkonen, M. Lukasewycz, D. Cox, D. Peake, and R. Carlson. "Implications of Treating Water Containing Polynuclear Aromatic Hydrocarbons with Chlorine: A Gas Chromatographic-Mass Spectrometric Study," *Environ. Health Perspect.* 46:73-86 (1982).
23. Glaze, W. H., and J. E. Henderson. "Formation of Organochlorine Compounds from the Chlorination of a Municipal Secondary Effluent," *J. Water Pollut. Control Fed.* 47:2511-2515 (1975).
24. Coté, R. P. *The Effects of Petroleum Refinery Liquid Wastes on Aquatic Life, with Special Emphasis on the Canadian Environment*, NRCC No. 15021 (Ottawa; National Research Council of Canada, 1976).

SECTION V

Toxicology of Disinfectants and Their By-Products

The discovery . . . that the origin of certain chlorinated organics was the action of chlorine on natural substances rather than the careless handling of the corresponding "man-sponsored" organics themselves was a brutal reminder of the frequently obscure fragility of our ecosystem, and the full implications of this discovery are as yet unknown.

William H. Glaze and **Gary R. Peyton**
Soluble Organic Constituents of Natural Waters and Wastewaters Before and After Chlorination, 1977

In considering the toxicity of organic compounds present in finished drinking water, there should be less concern about acute toxic effects because the concentrations of acutely toxic organic compounds in drinking water are likely to be too low to be of concern. Consequently, there is more concern about the chronic toxicity of low levels of toxic organic compounds in drinking water, that is, repeated exposure to low levels of toxic compounds which may lead to chronic effects such as cancer.

Robert A. Neal
Known and Projected Toxicology of Chlorination By-Products, 1979

CHAPTER 23

Pharmacokinetics of Chlorine Obtained from Chlorine Dioxide, Chlorine, Chloramine, and Chloride

Mohamed S. Abdel-Rahman

The treatment of drinking water has been the most successful public health measure since the turn of the century. As early as 1827, it was suggested that chlorite (ClO_2^-) be used for disinfection of water supply. However, widespread use of chlorine (HOCl) for water treatment began only in 1908 when the United States introduced it as a water disinfectant.[1] Since then, chlorination has remained the most widely used method for disinfection of public water supplies. There exists a variety of convincing incidences that support the clear significance of the role of chlorination. For example, Craun and Gunn[2] reported a total of 59 waterborne outbreaks, with almost 16,000 cases of disease in 1975 and 1976. All of these epidemics were associated with the use of untreated surface or groundwater. In recent years, however, it has become clear that treatment of drinking water with chlorine results in the formation of trihalomethanes (THMs).[3] Also, Bellar et al.[4] documented that the halogenated organic chemicals most commonly found in drinking water were produced by chlorination.

In the past, chemical hazards in drinking water had been viewed simply as one of industrial and agricultural contamination of the source water. To assess health hazards in drinking water, we must now evaluate the qualitative changes in chemical composition that are produced from water disinfection. The precursors for THMs include many classes of organics: (1) compounds containing carboxylic groups such as humic, fulvic, hydroxybenzoic, citric, β-ketoglutaric, fumaric, malic, and pyruvic acids; (2) ketones and aldehydes such as acetone, acetylacetone, acetaldehydes, and alcohols.[5-8] Among these precursors, humic and fulvic acids deserve special attention since they are major sources for the formation of THMs. Concern over the possible adverse effects of THMs has increased considerably.[9] One of the THMs, chloroform ($CHCl_3$), has been shown to be carcinogenic in mice and rats.[10]

Epidemiological studies have demonstrated statistical associations between increased cancer mortality and the practice of chlorination of drinking water.

Multivariant regression analysis indicates a statistically significant relation between cancer mortality rates in Louisiana and drinking water obtained from the Mississippi River.[11] Also, cancer mortality rates were higher (analysis of covariance) in the surface waters of Ohio counties.[12] These observations stimulated a search for a means to minimize the formation of these products.

One of the most attractive alternatives is to substitute disinfection methods that do not promote the formation of chlorinated by-products. A number of alternatives such as chlorine dioxide (ClO_2), monochloramine (NH_2Cl), and ozone (O_3) have been suggested. Particularly, ClO_2 has received the most attention because of its better biocidal property and because it does not react with phenol to produce the same taste and odor problems that result from chlorine treatment.[13,14] Moreover, ClO_2 has been used as a water disinfectant in Europe as well as in Canada for more than three decades.

Recent studies at our laboratory demonstrated that rat blood chloroform levels were significantly decreased after 1 year of treatment with ClO_2 in drinking water,[15] whereas no significant changes occurred after NH_2Cl treatment, as compared with the control.[16] Metabolism studies revealed that chloride ($^{36}Cl^-$) was the major metabolite after administering $^{36}ClO_2$, $NH_2^{36}Cl$, or $HO^{36}Cl$ to rats.[17,18]

The studies described in this chapter were conducted to provide information on the pharmacokinetics of chlorine obtained from chlorine dioxide, chlorine, chloramine, and chloride (Cl compounds). Also, the mechanism of $CHCl_3$ formation by HOCl and its inhibition by ClO_2 in the presence of HOCl was studied.

METHODS

Synthesis of ^{36}Cl-Labelled Disinfectants and Na^{36}Cl

The radioisotopes of these disinfectants are not commercially available; therefore, the commercially available ^{36}Cl-labelled HCl with radionuclidic purity >99% (New England Nuclear) was used as the source of ^{36}Cl for the synthesis of $^{36}ClO_2$, $NH_2^{36}Cl$, $HO^{36}Cl$, and $^{36}Cl^-$. Potassium chlorate ($K^{36}ClO_3$) was synthesized as described by Abdel-Rahman et al.[17] Fast neutron activation analysis indicated that $K^{36}ClO_3$ purity was 95%. The generation of $^{36}ClO_2$ from $K^{36}ClO_3$ was obtained by the following reaction:

$$2K^{36}ClO_3 + (COOH)_2 + 2H_2SO_4 \rightarrow 2^{36}ClO_2 + 2KHSO_4 + 2CO_2 + 2H_2O$$

$HO^{36}Cl$ ($^{36}Cl_2$ in H_2O) was formed by the following reaction:

$$KMnO_4 + 16H^{36}Cl \rightarrow 2K^{36}Cl + Mn^{36}Cl_2 + 5^{36}Cl_2 + 8H_2O$$

$NH_2{}^{36}Cl$ was synthesized from $HO^{36}Cl$, sodium bicarbonate buffer pH 9.0 to 9.1, and ammonium hydroxide as described previously.[18] ^{36}Cl-Labelled NaCl was prepared by the addition of $H^{36}Cl$ to NaOH to yield 0.171-mmol $Na^{36}Cl$. The pH was adjusted to 5.5 by 0.1 N NaOH, and the amount of sodium was quantitated by a Corning flame photometer 430.

Absorption and Elimination Studies

Four groups of four male Sprague-Dawley rats (250–350 g) each, which had been fasted overnight, were used in these experiments. ^{36}Cl-compounds were administered by gavage as follows:

Body wt (mg/kg)	Compound (mg/L)	μCi
1.5	100 $^{36}ClO_2$	0.7
3.3	370 $NH_2{}^{36}Cl$	1.66
1.8	200 $HO^{36}Cl$	0.6
1.8	200 $Na^{36}Cl$	0.42

Heparinized blood samples were collected by cardiac puncture at 10, 20, 30, and 60 min, and at 2, 4, 8, 16, 24, 48, 72, 96, and 120 h. The blood was centrifuged at 1000 g for 15 min to separate the red blood cells from the plasma. Radioactivity was then counted in a liquid scintillation counter (Beckman LS 7500), and the rate constant and half-life ($T^{1/2}$) of absorption and elimination from plasma were calculated after the administration of $^{36}ClO_2$, $NH_2{}^{36}Cl$, $HO^{36}Cl$, and $Na^{36}Cl$.

Excretion, Metabolism, and Distribution Studies

In another identical group, fasted rats were administered orally the same concentrations of ^{36}Cl compounds. Animals were housed in modified Roth all-glass metabolism chambers for the collection of expired air (using 1 N NaOH to trap ^{36}Cl) and fecal and urine samples. The radioactivity of the total ^{36}Cl compounds was measured. The analysis of radioactive metabolites such as chloride, chlorite, and chlorate (Cl^-, ClO_2^-, ClO_3^-, respectively) was performed and represented as a percentage of the initial dose.

Rats were sacrificed after 120 h from the administration of $NH_2{}^{36}Cl$ or $Na^{36}Cl$. The $HO^{36}Cl$ and $^{36}ClO_2$ groups were sacrificed after 96 and 72 h, respectively. Tissue specimens were prepared to determine the ^{36}Cl distribution as detailed in a previous report.[19]

Chloroform Formation from Organic Substances and HOCl and Inhibition by ClO_2

The reaction was started by the addition of 5, 10, and 20 mg/L HOCl or ClO_2 to the reaction mixture containing 0.1 mL of various concentrations of citrate or β-ketoglutaric acid (β-KGA) and 0.9 mL of 20 mM phosphate buffer (pH 7.0). The vials were incubated at 37°C for 30 min. Pentane (1 mL) was then added to the reaction mixture using a two-way stopcock and syringe assembly to prevent any loss of the products. $CHCl_3$ and any other by-products were extracted. The products in the organic layer were quantitated using a gas chromatograph (GC) equipped with a ^{63}Ni electron capture detector (ECD). A Hewlett-Packard 5995A gas chromatograph/mass spectrometer (GC/MS) was used to identify chloroacetones and dimethyl malonate.

RESULTS

Chlorine Dioxide, Monochloramine, Hypochlorous Acid, and Chloride Absorption and Elimination from Rat Plasma

The rate constant and the $T^{1/2}$ for absorption and elimination from plasma of these disinfectants and NaCl are described in Table I. The rate constant for absorption was highest for $^{36}ClO_2$ (3.77 ± 0.24/h), followed by $HO^{36}Cl$, $NH_2^{36}Cl$, and lowest for $Na^{36}Cl$ (0.04 ± 0.006/h). The $T^{1/2}$ for absorption, which is inversely proportional to the rate constant, was longest for $Na^{36}Cl$, (19.2 ± 2.96 h), followed by $NH_2^{36}Cl$ and $HO^{36}Cl$, and was shortest for $^{36}ClO_2$ (0.18 ± 0.01 h). Although the range of rate constant for the elimination of these compounds was close (0.013–0.018/h), the values for $^{36}ClO_2$ and $HO^{36}Cl$ were the same (0.016 ± 0.001/h). The $T^{1/2}$ during the elimination phase was longest for $Na^{36}Cl$ (51.9 ± 4.0 h) and shortest for $NH_2^{36}Cl$ (38.8 ± 3.7 h), whereas the values for $^{36}ClO_2$ and $HO^{36}Cl$ were almost the same (43.9 vs 44.1 h, respectively).

Table I. Kinetic Constants of ClO_2, NH_2Cl, HOCl, and NaCl in the Rat[a]

Treatment (mg/kg)	Absorption, h^{-1}		Elimination, h^{-1}	
	Rate Constant	$T^{1/2}$[b]	Rate Constant	$T^{1/2}$[b]
$^{36}ClO_2$ (1.5)	3.77 ± 0.240	0.18 ± 0.01	0.016 ± 0.001	43.9 ± 2.3
$NH_2^{36}Cl$ (3.3)	0.28 ± 0.041	2.5 ± 0.31	0.018 ± 0.002	38.8 ± 3.7
$HO^{36}Cl$ (1.8)	0.32 ± 0.095	2.2 ± 0.49	0.016 ± 0.001	44.1 ± 1.1
$Na^{36}Cl$ (1.8)	0.04 ± 0.006	19.2 ± 2.96	0.013 ± 0.001	51.9 ± 4.0

[a]Values were calculated from the time course of the elimination from rat plasma; four rats per treatment.
[b]Half-life.

Excretion, Metabolism, and Distribution of ^{36}Cl Compounds

The total ^{36}Cl compound recovered in the urine and feces 120 h after the administration of $NH_2^{36}Cl$ or $Na^{36}Cl$, or 96 and 72 h after the administration of $HO^{36}Cl$ and $^{36}ClO_2$, respectively, was determined for each compound and calculated as the percentage of the initial dose. In the case of $^{36}ClO_2$, 75% of the recovered dose was found in the urine and 25% was found in the feces. For both $NH_2^{36}Cl$ and $Na^{36}Cl$, approximately 95 and 5% were found in urine and feces, respectively; for $HO^{36}Cl$, 71 and 29% were found in urine and feces, respectively. ^{36}Cl was not detected in the expired air throughout the time studied (Table II).

After $^{36}ClO_2$ administration, ^{36}Cl was recovered in the urine in the form of $^{36}Cl^-$, ClO_2^-, and ClO_3^-; however, $^{36}Cl^-$ was the major metabolite, but after the administration of $NH_2^{36}Cl$, HO^3Cl, and $Na^{36}Cl$, chloride was the only metabolite detected (Table III).

The distribution of ^{36}Cl compounds 120 h from the time of oral administration of $NH_2^{36}Cl$ or $Na^{36}Cl$, or after 72 and 96 h from the administration of $^{36}ClO_2$ and $HO^{36}Cl$, respectively, is shown in Table IV. The distribution of $^{36}ClO_2$ 72 h after the administration was approximately the same in the kidney,

Table II. Excretion of ClO_2, NH_2Cl, $HOCl$, and $NaCl$ in the Rat

Treatment (mg/kg)	Time (h)	Percentage of Initial Dose[a]	
		Urine	Feces
$^{36}ClO_2$ (1.5)	72	30.81 ± 0.80	10.1 ± 1.7
$NH_2^{36}Cl$ (3.3)	120	25.15 ± 13.32	1.98 ± 0.29
$HO^{36}Cl$ (1.8)	96	36.43 ± 5.67	14.80 ± 3.70
$Na^{36}Cl$ (1.8)	120	57.20 ± 10.60	3.0 ± 0.91

[a]Values represent the mean ± SE as percentage of the initial dose from four rats per treatment throughout the indicated time; ^{36}Cl was not detected in expired air.

Table III. Metabolism of ClO_2, NH_2Cl, $HOCl$, and $NaCl$ in Rat Urine

Treatment (mg/kg)	Time (h)	Percentage of Initial Dose[a]		
		Cl^-	ClO_2^-	ClO_3^-
$^{36}ClO_2$ (1.5)	72	26.93 ± 1.6	3.46 ± 1.0	0.73 ± 0.73
$NH_2^{36}Cl$ (3.3)	120	25.15 ± 13.32	[b]	
$HO^{36}Cl$ (1.8)	96	36.43 ± 5.67		
$Na^{36}Cl$ (1.8)	120	57.2 ± 10.6		

[a]Values represent the mean ± SE as percentage of the initial dose from four rats per treatment throughout the indicated time.
[b]None detected.

Table IV. ClO_2, NH_2Cl, $HOCl$, and $NaCl$ Distribution in the Rat

Tissue	Percentage of Initial Dose[a]			
	$^{36}ClO_2$ (1.5)	$NH_2^{36}Cl$ (3.3)	$NO^{36}Cl$ (1.8)	$Na^{36}Cl$ (1.8)
Whole blood	ND[b]	0.24 ± 0.03	0.27 ± 0.01	0.20 ± 0.01
Plasma	0.72 ± 0.02	0.28 ± 0.04	0.32 ± 0.01	0.23 ± 0.01
Packed cells		0.17 ± 0.01	0.17 ± 0.01	0.18 ± 0.03
Kidney	0.81 ± 0.15	0.15 ± 0.02	0.19 ± 0.01	0.13 ± 0.01
Lung	0.74 ± 0.15	0.14 ± 0.02	0.17 ± 0.01	0.12 ± 0.00
Stomach	0.70 ± 0.15	0.14 ± 0.02	0.12 ± 0.01	0.09 ± 0.01
Duodenum	0.29 ± 0.07	0.11 ± 0.68	0.12 ± 0.01	0.08 ± 0.01
Ileum	0.48 ± 0.09	0.05 ± 0.01	0.04 ± 0.01	0.03 ± 0.01
Liver	0.38 ± 0.09	0.07 ± 0.01	0.09 ± 0.01	0.05 ± 0.00
Spleen	0.25 ± 0.04	0.10 ± 0.02	0.11 ± 0.01	0.07 ± 0.01
Bone marrow	0.16 ± 0.03	0.14 ± 0.02	0.26 ± 0.01	0.12 ± 0.02
Testes		0.19 ± 0.02	0.21 ± 0.02	0.18 ± 0.02
Thyroid		0.12 ± 0.02	0.11 ± 0.02	
Carcass		0.07 ± 0.01	0.07 ± 0.00	0.05 ± 0.03

[a]Values represent the mean ± SE as percentage of the initial dose from four rats per treatment.
[b]Not determined.

lung, plasma, and stomach. At the same time, these tissues represent the highest activity of ^{36}Cl metabolites, followed by the ileum, liver, duodenum, spleen, and bone marrow. The distribution of $NH_2^{36}Cl$ 120 h after administration was highest in plasma, followed by whole blood, testes, packed cells, kidney, lung, stomach, and bone marrow. In the case of $HO^{36}Cl$ 96 h from the administration, plasma was highest followed by whole blood, bone marrow, testes, kidney, packed cells, and lung. After 120 h from the administration of $Na^{36}Cl$, the distribution was highest in plasma, followed by whole blood, packed cells, testes, kidney, lung, and bone marrow.

Formation of $CHCl_3$ by HOCl and Its Inhibition by ClO_2 in vitro

Chloroform production from citrate and HOCl at pH 7 indicated a linear relationship between the formation of $CHCl_3$ and the concentration of HOCl (Table V). However, $CHCl_3$ formation was inhibited by 5-mg/L ClO_2 in the reaction mixture containing HOCl and citrate, or β-KGA. When sodium citrate was used as an organic substance, 71, 40, or 17% inhibition of $CHCl_3$ formation was observed with 5 mg/L ClO_2 in the presence of 5, 10, or 20 mg/L HOCl, respectively. In the case of β-KGA, 5 mg/L ClO_2 in combination with 5, 10, or 20 mg/L HOCl also decreased the formation of $CHCl_3$ by about 82, 57, or 39%, respectively (Table V).

Table V. Effect of ClO_2 in Combination with HOCl on $CHCl_3$ Formation

HOCl Conc. (mg/L)	Citrate		β-Ketoglutaric Acid	
	HOCl	HOCl + 5 mg/L ClO_2	HOCl	HOCl + 5 mg/L ClO_2
5	60.9 ± 1.8[a]	17.4 ± 0.6 (71.4)[b]	425.9 ± 32.2	77.1 ± 4.8 (81.9)
10	113.7 ± 2.8	69.6 ± 8.5 (39.6)	753.3 ± 8.3	327.3 ± 19.0 (56.6)
20	215.0 ± 4.2	168.6 ± 10.2 (16.8)	747.8 ± 7.3	460.0 ± 2.5 (38.5)

[a] Values represent the mean ± SE from four samples and are expressed as μg/L $CHCl_3$, using 1 mM citrate or 10 μM β-ketoglutaric acid as an organic substance.
[b] Values in parentheses indicate the percentage of inhibition in the presence of 5 mg/L ClO_2.

Table VI. Chloroform Formation from β-Ketoglutaric Acid and HOCl

HOCl (mg/L)	β-Ketoglutaric Acid (μM)			
	1	10	100	1000
5	64.3 ± 0.2[a]	425.9 ± 32.2	42.1 ± 4.8	10.9 ± 1.5
10	61.7 ± 2.3	753.3 ± 8.3	136.6 ± 15.3	27.1 ± 5.4
20	71.9 ± 2.5	747.8 ± 7.3	466.0 ± 63.0	73.5 ± 2.9

[a]Values represent the mean ± SE from four samples and are expressed as $\mu g/L$ $CHCl_3$.

$CHCl_3$ production from various concentrations of β-KGA and HOCl indicated that the maximum yield of $CHCl_3$ formation was obtained when 10 μM β-KGA was used with 5, 10, or 20 mg/L HOCl (Table VI). When the concentration of β-KGA was increased to 100 and 1000 μM, the formation of $CHCl_3$ continued to decrease at every concentration of HOCl used. At a very low β-KGA concentration (1 μM), the formation of $CHCl_3$ was not dependent on the concentration of HOCl (Table VI).

High-performance liquid chromatographic (HPLC) analysis of the reaction mixture containing sodium citrate and HOCl revealed the formation of β-KGA as an intermediate. When sodium citrate or β-KGA was used as an organic substance, three distinct peaks representing mono-, di-, and trichloroacetones, besides a $CHCl_3$ peak, were observed by GC (Figure 1). GC/MS analysis revealed the presence of a base peak at m/e 43 in all three compounds. The key fragment ions were recorded at m/e 49, 83, and 117 for mono-, di-, and trichloroacetones, respectively.

The formation of monochloroacetone from β-KGA and HOCl is summarized in Table VII. The amount of monochloroacetone produced at 100 or 1000 μM was higher than at 10 μM β-KGA and increased with increasing HOCl concentration.

When citrate was incubated with different concentrations of ClO_2 (up to 150 mg/L), the concentration of citrate did not change for 30 min. On the other hand, within 5 min after ClO_2 addition to β-KGA, 63% of the β-KGA concentration was decreased. When HOCl was used instead of ClO_2, 80% of the β-KGA concentration disappeared from the reaction mixture within 1 min. The concentration of β-KGA without ClO_2 or HOCl in the mixture at the same conditions was not changed throughout the time studied (Figure 2). This indicated that no spontaneous degradation of β-KGA alone occurred under these experimental conditions. Malonic acid as a product of the reaction between β-KGA and ClO_2 was identified by HPLC and confirmed by GC/MS following methylation to yield dimethyl malonate. Mass spectra of dimethyl malonate showed that the highest fragment was at m/e 59, represented as the CH_3COO^+ fragment (Figure 3). Three abundant fragments at m/e 42, 74, and 101 corresponded to CH_3CO^+, $CH_3COOCH_3^+$, and $CH_3COOCH_2CO^+$, respectively. When the mass spectra of the compound were with those of the authentic compound, the identical group of fragments were observed (Figure 3). No malonic acid was produced when citrate or β-KGA was incubated with HOCl.

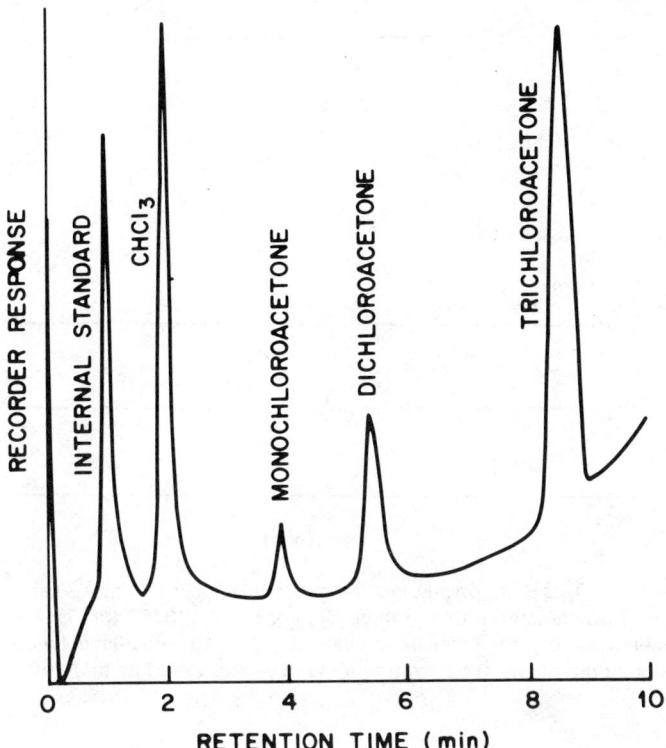

Figure 1. GC chromatogram of reaction mixture containing β-ketoglutaric acid and HOCl. A solution of 0.1 mM β-ketoglutaric acid was incubated with 10 mg/L HOCl in phosphate buffer (pH 7) at 37°C for 30 min. After extraction of the mixture with pentane, the organic layer was injected into GC equipped with ECD.

Table VII. Formation of Monochloroacetone from β-Ketoglutaric Acid and HOCl

HOCl (mg/L)	β-Ketoglutaric Acid (μM)			
	1	10	100	1000
5	23.5 ± 0.6[a]	26.0 ± 1.2	78.6 ± 3.7	169.0 ± 20.4
10	30.4 ± 1.8	65.0 ± 2.0	109.6 ± 5.2	386.3 ± 21.2
20	40.8 ± 2.4	72.3 ± 3.7	134.3 ± 20.5	730.5 ± 23.0

[a]Values represent the mean ± SE from four samples and are expressed as μg/L monochloroacetone.

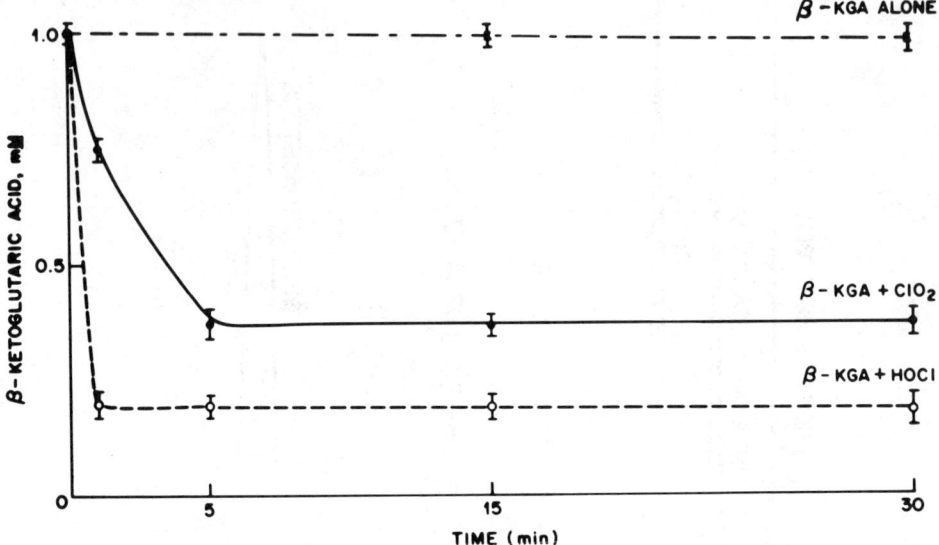

Figure 2. Effect of ClO_2 or HOCl on β-ketoglutaric acid. A solution of 1 mM β-ketoglutaric acid (β-KGA) was reacted with 150 mg/L ClO_2 or HOCl at 37°C (pH 7). The remaining concentration of β-KGA was determined at 0, 1, 5, 15, and 30 min by HPLC. Values represent the mean ± SE from four samples expressed as mM β-KGA.

DISCUSSION

The pharmacokinetic parameters obtained in this study revealed a considerable similarity between distribution and metabolism of $^{36}ClO_2$, $NH_2{}^{36}Cl$, $HO^{36}Cl$, and $Na^{36}Cl$. However, $^{36}ClO_2$ was more rapidly absorbed into the bloodstream than the other compounds. The present results revealed that after oral administration, $^{36}Cl^-$ was slowly absorbed from the gastrointestinal tract in contrast to the rapid absorption of $^{36}ClO_2$ (19.2 vs 0.18 h, respectively). This difference may be explained by the charge of the chloride ion. On the other hand, the elimination half-lives from the plasma were not significantly different in all the treatment groups. This similarity may be related to their existence mainly in the chloride form, as described in Table III.

The distribution experiment revealed that plasma and whole blood contain the highest activity. However, it is important to indicate that at the end of the experiment (3-5 d), high amounts of radioactivity were still found in the stomach, duodenum, and ileum in all the treatments.

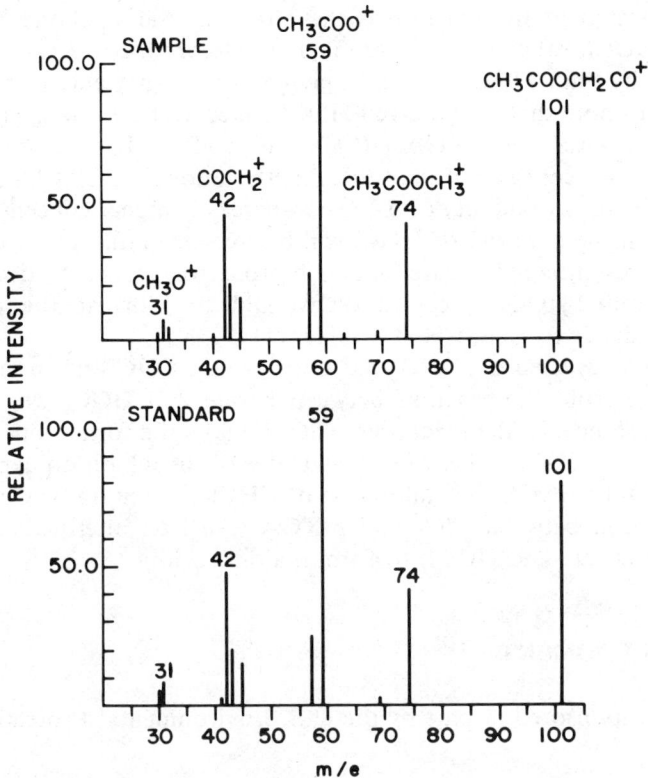

Figure 3. Mass spectra of dimethyl malonate. A reaction mixture containing β-ketoglutaric acid and ClO_2 was incubated at 37°C for 30 min (pH 7). Following the extraction and methylation, the sample was analyzed by GC/MS.

Effect of ClO_2 on the Formation of $CHCl_3$

A previous study from this laboratory revealed that rat blood $CHCl_3$ levels were significantly decreased in vivo following the administration of ClO_2 in drinking water daily for 1 year, compared to a control group.[15] This phenomenon may be explained by the direct inhibition of $CHCl_3$ formation by ClO_2. The hypothesis was tested using sodium citrate and β-KGA as organic substances. To understand the role of ClO_2 on the inhibition of $CHCl_3$ formation, it is worth discussing the mechanism of $CHCl_3$ formation from citrate and HOCl.

The reaction pathway for the formation of $CHCl_3$ from citrate and HOCl is as follows:

Citrate→β-Ketoglutaric acid→Monochloroacetone→ Dichloroacetone→Trichloroacetone→Chloroform

The first step of the reaction is the attack of the alcoholic hydrogen of citrate by HOCl, followed by decarboxylation to produce β-KGA. The β-KGA reacts rapidly with HOCl at pH 7, giving very high yields of $CHCl_3$. It is important to note that, for a given HOCl concentration, there exists an optimum concentration of β-KGA (10 μM) to produce that maximum yield of $CHCl_3$. At a low concentration of β-KGA, the amount of $CHCl_3$ produced will be limited by the availability of β-KGA, whereas at higher concentrations (100 or 1000 μM), the majority of HOCl will be consumed during the chlorination of β-KGA, resulting in increased monochloroacetone concentrations. The outcome is a limited HOCl concentration available to chlorinate the remainder of the intermediates to yield $CHCl_3$ (Table VI).

This study revealed that ClO_2 did not produce $CHCl_3$ when citrate or the intermediate from the reaction between citrate and HOCl was used as an organic substance. In the meantime, ClO_2 inhibits the formation of $CHCl_3$ in the presence of HOCl and citrate. The degree of the inhibition depends on the ratio of ClO_2 to HOCl. This inhibition of $CHCl_3$ formation by ClO_2 is related to the reaction between ClO_2 and β-KGA (the first intermediate from the reaction of citrate and HOCl) to form malonic acid.

ACKNOWLEDGMENT

Research sponsored in part by the U.S. Environmental Protection Agency.

REFERENCES

1. Kogan, B. *Health, Man in a Changing Environment*, 2nd ed., (New York: Harcourt Brace Jovanovich, Inc., 1974).
2. Craun, G. F., and R. A. Gunn. "Outbreaks of Waterborne Disease in the United States: 1975-1976," *J. Am. Water Works Assoc.* 71:422 (1979).
3. Rook, J. J. "Formation of Haloforms during Chlorination of Natural Waters," *J. Water Treat. Exam.* 23:234-236 (1974).
4. Bellar, T. A., J. J. Lichtenberg, and R. D. Kroner. "The Occurrence of Organohalides in Chlorinated Drinking Water," *J. Am. Water Works Assoc.* 66:703-706 (1974).
5. Morris, J. C. *Formation of Halogenated Organics by Chlorination of Water Supplies*, EPA-600/1-74-002 (Washington, DC: U.S. Environmental Protection Agency, 1975).
6. Rook, J. J. "Possible Pathways for the Formation of Chlorinated Degradation Products During Chlorination of Humic Acids and Resorcinol," in *Water Chlorination: Environmental Impact and Health Effects, Vol. 3*, R. L. Jolley, W. A. Brungs, and R. B. Cumming, Eds. (Ann Arbor, MI: Ann Arbor Science Publishers, Inc., 1980), pp. 95-98.
7. Larson, R. A., and A. L. Rockwell. "Chloroform and Chlorophenol Production by Decarboxylation of Natural Acids During Aqueous Chlorination," *Environ. Sci. Technol.* 13(3):325-329 (1979).

8. Dragunov, S. I. *Soil Organic Matter*, M. M. Kononova, Ed. (New York: Pergamon Press, 1961), p. 65.
9. Jolley, R. L., H. Gorchev, and D. H. Hamilton, Jr., Eds. *Water Chlorination: Environmental Impact and Health Effects*, Vol. 2 (Ann Arbor, MI: Ann Arbor Science Publishers, Inc., 1978).
10. National Cancer Institute. *Report on the Carcinogenesis Bioassay of Chloroform*, PB-264018 (Springfield, VA: National Technical Information Service).
11. Page, T., R. H. Harris, and S. S. Epstein. "Drinking Water and Cancer Mortality in Louisiana," *Science* 193:55-57 (1976).
12. Kuzma, R. J., D. M. Kuzma, and C. R. Buncher. "Ohio Drinking Water Source and Cancer Rates," *Am. J. Public Health* 67(8):725-729 (1977).
13. Akin, E. W., J. C. Hoff, and E. C. Lippy. "Waterborne Outbreak Control: Which Disinfectant?" *Environ. Health Perspect.* 46:7-12 (1982).
14. Enger, M. "Treatment of Water with Chlorine Dioxide to Improve the Taste," *Gass. Wasser Fach.* 14:330-336 (1960).
15. Suh, D. H., M. S. Abdel-Rahman, and R. J. Bull. "Biochemical Interactions of Chlorine Dioxide and Its Metabolites in Rats," *Arch. Environ. Contam. Toxicol.* 13:163-169 (1984).
16. Abdel-Rahman, M. S., D. H. Suh, and R. J. Bull. "Toxicity of Monochloramine in Rat: An Alternative Drinking Water Disinfectant," *J. Toxicol. Environ. Health* 13:825-834 (1984).
17. Abdel-Rahman, M. S., D. Couri, and J. D. Jones. "Chlorine Dioxide Metabolism in Rat," *J. Environ. Pathol. Toxicol.* 3:421-430 (1980).
18. Abdel-Rahman, M. S., D. M. Waldron, and R. J. Bull. "A Comparative Kinetics Study of Monochloramine and Hypochlorous Acid in Rat," *J. Appl. Toxicol.* 3,4:175-179 (1983).
19. Abdel-Rahman, M. S., D. Couri, and R. J. Bull. "Kinetics of ClO_2 and Effects of ClO_2, ClO_2^- and ClO_3^- in Drinking Water on Blood Glutathione and Hemolysis in Rat and Chicken," *J. Environ. Pathol. Toxicol.* 3:431-449 (1980).

CHAPTER 24

Reproductive Effects of Alternate Disinfectants and Their By-Products

Betsy D. Carlton and M. Kate Smith

There is evidence that the fertility of the U.S. population is declining. Infertility will be experienced by approximately 15% of the U.S. population of childbearing age.[1] If we consider three studies of selected fertile male populations conducted between 1938 and 1951,[2-4] we find that the estimated median sperm concentration value ranged from 85 to 120 million/mL. In comparison, the median sperm concentration values for three studies reported in the 1970s[5-7] ranged from 38 to 65 million/mL. Although 44% of MacLeod and Gold's 1951 study[4] of 1000 fertile men had sperm concentrations in excess of 100 million/mL, only 21% of Zukerman's, 1977 study[5] of 4122 prevasectomy patients had sperm counts in this upper range. We note that the median counts observed in the more recent studies approach the value (40 million) that MacLeod suggested was the threshold of infertility in 1951.

It is against this background that we must consider the potential reproductive effects of life-time exposure to chlorine or the proposed alternative water disinfectants and their by-products. The populations potentially at risk include not only male and female populations of childbearing age, but also those perinatally exposed. The effects on the perinatal population may be immediate, resulting in birth defects or increased perinatal mortality or morbidity. The reproductive sequelae may not be immediately apparent in that the germ cells may be affected, thus altering the fertility of the F_1 or succeeding generations (or both). The perinatal population may be at increased risk, not only because of the immaturity of fetal or newborn enzyme systems, but also because infants consume approximately three times more liquid per pound body weight than do adults.[8] This increased fluid intake relative to body weight would maximally expose the perinatal group to the products of water disinfection.

Despite the widespread and long-term use of chlorine as a disinfectant for drinking water, there are limited data in the literature with regard to its reproductive or other health effects. With increasing awareness of the potential health effects of chlorination by-products such as the trihalomethanes and haloacetonitriles, alternative water disinfectants have been proposed. Chlorine dioxide, ozone, and chloramine are among the alternative disinfectants under consideration. The reproductive effects of these chemicals and their degradation products such as chlorate, chlorite, and hypochlorite are also largely undetermined.

A review of the reproductive and perinatal effects of chlorine, chlorine dioxide (ClO_2), chlorite (ClO_2^-), chlorate (ClO_3^-), and monochloramine (NH_2Cl) will be presented. In addition, the results of a study from our laboratory on the reproductive effects of the administration of sodium chlorite in the drinking water of Long-Evans rats will be discussed. The effects of chlorite on sperm parameters have not been reported previously.

HUMAN STUDIES

The available data on the health effects of water disinfection in man are extremely limited. Michael et al.[9] reported no significant adverse health effects in a human population exposed to 5-ppm chlorine dioxide (ClO_2) in drinking water for 3 months. Decreases in hemoglobin, hematocrit, and red blood cell count were seen in one case of a glucose-6-phosphate (G-6-P)-deficient diabetic individual who was considered high risk for oxidant stress. A causal relationship between ClO_2 exposure and hematologic changes in this individual could not be established; however, the potential for hematologic changes resulting from ClO_2 disinfection has implications for the newborn population, which is also at increased risk for oxidant stress.

The possibility of perinatal susceptibility was examined by Tuthill et al.[10] and reported in 1982. Tuthill's epidemiologic historic record study examined the mortality and morbidity experience of newborns in two contiguous communities, Holyoke and Chicopee, Massachusetts. Holyoke used chlorine and Chicopee used high levels of ClO_2 in their water supplies during 1945, the year under study. Tuthill reported statistically significant positive association between ClO_2 exposure and a diagnosis of prematurity and increased weight loss after birth. No differences in birth defects or mortality were observed. Significant questions must be raised that cannot be answered by the Tuthill study. Was the observed change in perinatal morbidity a result of ClO_2-based water disinfection, or was the increased morbidity a response to the high levels of contamination in the water supply that necessitated the use of high levels of ClO_2 in lieu of chlorine?

ANIMAL STUDIES

In 1982, preliminary data on the effects of chlorine and monochloramine in the developing rat fetus were reported.[11] These teratologic evaluations revealed little or no effect of either chemical at levels up to 100 ppm. A slight increase in skeletal variants was noted following chlorine exposure, but the number of dams treated and fetuses examined was insufficient for both chemicals studied.

The teratogenicity of ClO_2 and its degradation products was studied by Suh, Abdel-Rahman, and Bull.[12] As in the previously described study, the number of litters per dose group was small (6–9 per group). No significant differences were observed in maternal toxicity, litter size, fetal weight, or gross, visceral, or skeletal anomalies following exposure to ClO_2, ClO_2^-, and ClO_3^-.

Couri et al.[13] examined the teratogenic potential of sodium chlorite exposure by gavage, intraperitoneal (IP) injection, or in the drinking water of Sprague-Dawley rats. Again, the number of litters per group was small (4–13 dams per group). Decreases in maternal food and water consumption and body weight were observed in all treatment groups except the dose group receiving 0.1% chlorite in the drinking water. Intraperitoneal injection with 20 or 50 mg/kg chlorite daily and gavaging with 200 mg/kg resulted in vaginal and urethral bleeding and, ultimately, death. Drinking water containing 2% sodium chlorite, or injection of 10 or 20 mg/kg, resulted in increased stillbirths and resorption sites, probably reflecting maternal toxicity. No fetal malformations were found with any treatment regimen.

The effect of chlorite administration during pregnancy was examined in two strains of mice, A/J and C57BL/J.[14] The primary focus of this study was on G-6-P dehydrogenase activity and hematologic parameters. Conception rate and litters were examined for A/J mice. A statistically significant ($p < 0.05$) decrease was found in the average weight of pups at weaning and in the birth-to-weaning growth rate. When compared with controls, birth weight, litter size, perinatal survival, gestation length, and water consumption were not different for those mice receiving 100-ppm chlorite.

Reproductive parameters, other than conception rate and perinatal survival, growth, and malformation rates, have not been previously reported in the literature. In their study of the toxicologic effects of ClO_2, ClO_2^-, and ClO_3^-, Couri et al.[15] included a study of tritiated-thymidine incorporation into the nuclei of liver, kidney, intestinal mucosa, and testes. Their studies suggested the possibility of increased DNA turnover in the intestinal mucosa and decreased DNA synthesis in several organs. Decreased tritiated-thymidine incorporation was observed in the testes following treatment with 10- or 100-ppm ClO_2, ClO_2^- or ClO_3^- in the drinking water for 3 months. The biological significance is difficult to interpret, but the suggestion of decreased DNA synthesis in the testes is of interest with regard to potential effects on the reproductive system.

CURRENT STUDIES

We have examined the effects of the administration of 0-, 1-, 10-, 100-, or 500-ppm sodium chlorite in the drinking water of Long-Evans rats in a series of three experiments.

Methods

Long-Evans rats, 4 to 6 weeks of age, were obtained from Charles River Breeding Laboratories (Portage, Michigan) and were housed (two per cage) in polycarbonate cages with Absorb-Dri® bedding. The animal room was maintained at 40 to 60% humidity and 69 to 75°F, and with room lights on from 6 a.m. to 6 p.m. Animals were given ad libitum access to Purina Certified Rodent Chow 5002 and deionized water or deionized water containing sodium chlorite as drinking water. In the first study, the effect of chlorite on reproductive outcome was evaluated. Twelve male rats per dose group received 0-, 1-, 10-, or 100-ppm chlorite for 56 d prior to breeding and throughout the 10-d breeding period. Twenty-four female rats per dose group received the same dose levels of chlorite for 14 d prior to breeding and throughout breeding, gestation, and lactation until the pups were weaned on day 21. Water consumption was recorded three times per week. During the 10-d breeding period, animals were housed one male with two females. Females were checked for the presence of sperm in a lavage smear prior to 9:30 a.m. each day. Females observed to be sperm positive (gestation day zero) were individually housed. The day of parturition was designated lactation day zero.

Following the breeding period, males were bled for a complete blood count, killed by pentobarbital overdose, and given a complete gross necropsy. The reproductive tract was removed, weighed, and, except for the right cauda epididymis, preserved for histopathologic evaluation. The right cauda epididymis was weighed and finely minced in 10 mL of phosphate-buffered saline (PBS) with 0.1% glucose (pH 7.4) maintained at 37°C. The sperm suspension and all slides, coverslips, solutions, and pipettes were maintained at 37°C. One to two drops of sperm suspension were placed on a microscope slide, coverslipped, and observed using phase-contrast microscopy. A modification of Katz and Overstreet's[16] videomicrography method was used to videotape and evaluate a minimum of ten 10-s fields at 160X magnification and 10-s fields at 400X magnification. An aliquot of the sperm suspension was heat killed and used for determining sperm counts. A 0.5 mL aliquot was stained with aqueous eosin Y and used to prepare 4 to 6 slides per animal for sperm morphology evaluation according to the method of Wyrobek and Bruce.[17] The videotapes were later evaluated for percent sperm motility (10 fields) and sperm drive range (μm/s; 5 sperm per field; 10 fields).

Dams were observed for conception rate, length of gestation, and bodyweight gain. At necropsy on lactation day 21, dams were bled for complete blood count, killed by pentobarbital overdose, and given a complete gross necropsy. Reproductive tracts were taken, weighed, and preserved for histopathologic evaluation.

Litters were evaluated for size, perinatal survival, day of eye opening, day of vaginal patency, body-weight gain, and gross external abnormalities. At necropsy on lactation day 21, pups were bled for complete blood counts, and reproductive tracts were weighed and preserved.

In experiment 2, groups of 12 male rats were given 0-, 100-, or 500-ppm sodium chlorite in deionized drinking water for 72 to 76 d. At necropsy, animals were bled for complete blood counts and methemoglobin determinations. Sperm suspensions were prepared, and sperm count, percent motility, drive range, and morphology abnormalities were determined as described previously.

The third study replicated the second experiment; however, groups of 12 male rats received 0-, 10-, or 100-ppm sodium chlorite in deionized drinking water.

Results

The body weights for adult animals in the three studies were unaffected by chlorite exposure. Water consumption was decreased by 28% among 500-ppm dose group animals.

No effects of chlorite exposure were noted when general reproductive ability was evaluated (Table I). Conception rates varied between 67 and 96% but did not differ in a dose-dependent manner. Perinatal survival was comparable for all groups, as were litter size and pup body weight. The median day of eye opening (day 16), indicative of general pup maturity, and the day of vaginal patency (day 32), indicative of sexual maturation, were the same for all groups.

Clinical pathology, including methemoglobin levels and histopathologic evaluation of reproductive tracts, revealed no chlorite-induced abnormalities. Organ weights and organ-to-body weight ratios were also unchanged in chlorite-exposed rats.

Table I. Reproductive Performance of Long-Evans Rats Treated with 0-, 1-, 10-, or 100-ppm Sodium Chlorite

Indicator	Dose (ppm)			
	0	1	10	100
Females bred	24	24	24	24
Positive sperm	23	24	22	22
Litters, D 0[a]	23	16	22	21
Litters, D 7[b]	23	16	22	21
Litters, D 21[c]	23	16	22	21
Litter size, $\bar{X}$	13	13	11	13
Median day of eye opening	16	16	16	16
Median day of vagina opening	32			32

[a]Lactation day zero.
[b]Lactation day 7.
[c]Weaning day 21.

When sperm counts, morphology, and videomicrographic records were analyzed, comparable results were observed for the three study replicates. Sperm count was unaltered by chlorite exposure (Figure 1). The percent of motile sperm, when counted as motile if any movement was detected, was not statistically different between groups, but a slight trend toward decreasing motility was observed among rats exposed to 10-ppm chlorite or greater (Figure 2). Standard deviations were relatively large, with coefficients of variation ranging from 34 to 58%. For animals in the 100- and 500-ppm dose groups, mean sperm drive range was significantly reduced ($p < 0.01$) (Figure 3). Sperm drive range dropped from 20.3 μm/s for control animals to 13.4 μm/s and 11.0 μm/s for 100- and 500-ppm dose group animals, respectively. Decreasing velocity (17.5 μm/s) was also observed for 10-ppm dose group males, but the difference from control values was not statistically significant.

Sperm morphology was determined for 500 sperm per rat from eosin Y-stained morphology slides. Approximately 0.8% of sperm from control animals exhibited abnormal morphology. Rats from the 1- and 10-ppm dose

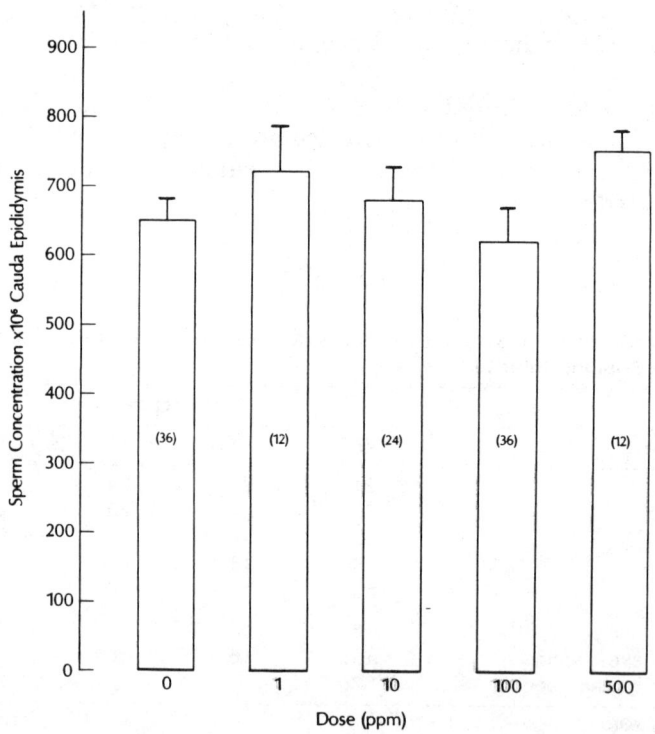

Figure 1. Sperm concentration; parentheses indicate number of animals. Results are pooled data from three studies ($\bar{x} \pm$ SEM).

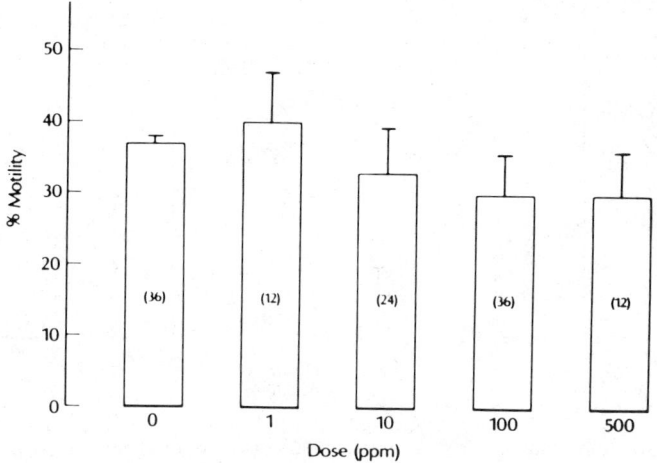

Figure 2. Sperm mortality (10 independent fields per study); parentheses indicate number of animals. Results are pooled data from three studies ($\bar{x} \pm$ SEM).

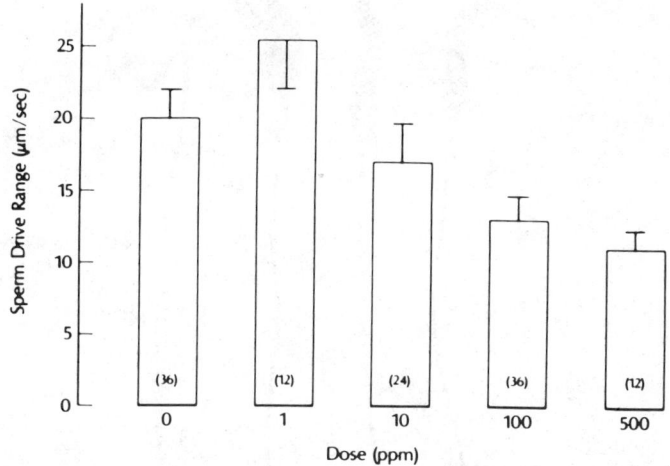

Figure 3. Sperm drive range; 5 sperm per field for each of 10 fields per study; parentheses indicate number of animals. Results are pooled data from three studies ($\bar{x} \pm$ SEM).

groups had comparable rates of abnormal sperm morphology, 0.6 and 0.9%, respectively. A significant ($p < 0.001$) increase in abnormal sperm forms was observed among 100- and 500-ppm dose group rats. In the 100- and 500-ppm dose groups, 2.9% (36 rats) and 1.8% (12 rats), respectively, exhibited abnormal sperm morphology (Figure 4). The lower percent of abnormal sperm

302   WATER CHLORINATION

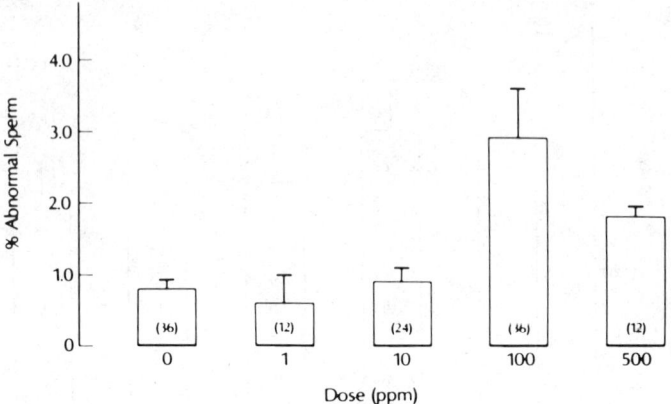

Figure 4. Abnormal sperm morphology; 500 sperm counted per animal; parentheses indicate number of animals. Results are pooled data from three studies ($\bar{x} \pm SD$).

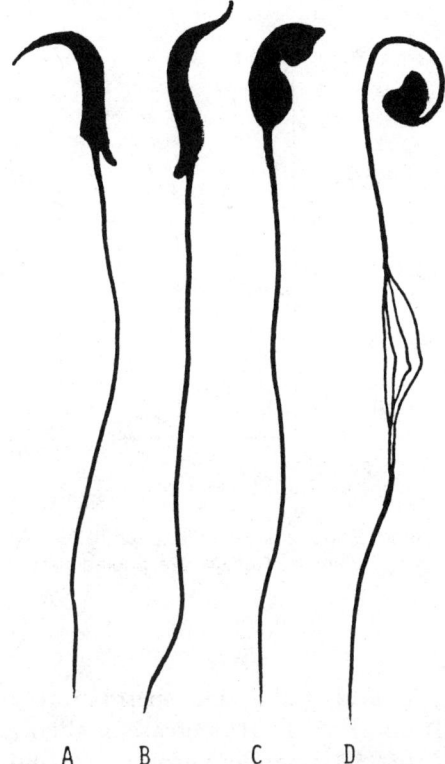

Figure 5. Commonly observed sperm morphology; Normal (A); Open hook (B); Amorphous (C); and Amorphous with frayed tail (D).

observed for the 500-ppm dose group animals relative to those in the 100-ppm dose group may be influenced by the fact that only one-third as many rats were evaluated in the former group as in the latter. The most commonly observed morphologic abnormalities were frayed tails, open hooks, and amorphous spermheads (Figure 5).

Discussion

While the magnitude of changes in sperm parameters reported here is comparatively small, the rats in this study were exposed for the duration of one spermatogenic cycle only. The human population would be exposed to chlorite in the water supply for a lifetime. Daily human sperm production is generally below that of other mammals, primarily because of a lower number of germ cells within the seminiferous tubules.[18] Although the incidence of abnormal sperm forms for the control group rats was 0.8%, as many as 20 to 40% of sperm forms may be abnormal in man.[19]

Meistrich and Brown[20] mathematically estimated the increased relative risk of human infertility from alterations in semen characteristics. They found that increased risk following exposure to industrial and environmental pollutants was dependent on and inversely related to the sperm count of the exposed population.

Sperm morphology appears to be a useful indicator of chemical toxicity in rodents[21] and man.[22] Increased incidence of abnormal sperm forms and decreased count and motility have been associated with increased incidence of abortion, abnormal pregnancies, and infertility.[23-25] Further studies are required to provide information to assess the potential impact of aqueous chlorite exposure on human reproduction performance, given man's possible declining fertility and the potential for lifetime exposure.

ACKNOWLEDGMENT

This work was supported by the U.S. Environmental Protection Agency Project No. CR-810301, but does not represent the policy or opinion of this agency.

REFERENCES

1. Menning, B. E. "The Emotional Needs of Infertile Couples," *Fertil. Steril.* 34:313–319 (1980).
2. Falk, H. G., and J. C. Kaufman. "What Constitutes Normal Semen?," *Fertil. Steril.* 1: 489–493 (1950).

3. Hotchkiss, R. S., E. K. Brunner, and P. G. Grenley. "Semen Analysis of Two Hundred Fertile Men," *Am. J. Med. Sci.* 196:362–368 (1938).
4. MacLeod, J., and R. Z. Gold. "The Male Factor in Fertility and Infertility II. Spermatozoan Counts in 1000 Men of Known Fertility and in 1000 Cases of Infertile Marriages," *J. Urol.* 66(3):436–449 (1951).
5. Zukerman, Z., et al. "Frequency Distribution of Sperm Counts in Fertile and Infertile Males," *Fertil. Steril.* 28(12):1310–1313 (1977).
6. Nelson, C. M. K., and R. G. Bunge. "Semen Analysis: Evidence for Changing Parameters of Male Fertility Potential," *Fertil. Steril.* 25(6):503–507 (1974).
7. Rehan, N. E., et al. "The Semen of Fertile Men: Statistical Analysis of 1300 men," *Fertil. Steril.* 26(6):492–502 (1975).
8. Hansen, H. E., and M. J. Bennett. In: *Textbook of Pediatrics*, W. E. Nelson, Ed. (Philadelphia: W. B. Saunders Co., 1964), p. 109.
9. Michael, G. E., et al. "Chlorine Dioxide Water Disinfection: A Prospective Epidemiology Study," *Arch. Environ. Health* 36:20–27 (1981).
10. Tuthill, R. W., et al. "Health Effects Among Newborns After Prenatal Exposure to ClO_2-Disinfected Drinking Water," *Environ. Health Perspect.* 46:39–45 (1982).
11. Abdel-Rahman, M. S., M. R. Berardi, and R. J. Bull. "Effect of Chlorite and Monochloramine in Drinking Water on the Developing Rat Fetus," *J. Appl. Toxicol.* 2(3):156–159 (1982).
12. Suh, D. H., M. S. Abdel-Rahman, and R. J. Bull. "The Effect of Chlorine Dioxide and Its Metabolites in Drinking Water on Fetal Development in the Rat," *J. Appl. Toxicol.* 3(2):75–79 (1983).
13. Couri, D., et al. "Assessment of Maternal Toxicity, Embryotoxicity, and Teratogenic Potential of Sodium Chlorite in Sprague-Dawley Rats," *Environ. Health Perspect.* 46:25–29 (1982).
14. Moore, G. S., and E. J. Calabrese. "Toxicological Effects of Chlorite in the Mouse," *Environ. Health Perspect.* 46:31–37 (1982).
15. Couri, D., M. S. Abdel-Rahman, and R. J. Bull, "Toxicological Effects of Chlorine Dioxide, Chlorite and Chlorate," *Environ. Health Perspect.* 46:13–17 (1982).
16. Katz, D. F., and J. W. Overstreet. "Sperm Motility Assessment by Videomicrography," *Fertil. Steril.* 35:188–193 (1981).
17. Wyrobek, A. J., and W. R. Bruce. "Chemical Induction of Sperm Abnormalities in Mice," *Proc. Nat. Acad. Sci. USA* 72(11):4425–4429 (1975).
18. Amann, R. P., and S. S. Howards. "Daily Spermatozoa Production and Epididymal Spermatozoal Reserves of the Human Male," *J. Urol.* 124:211–215 (1980).
19. Dixon, R. L. "Toxic Responses of the Reproductive System," in *The Basic Science of Poisons*, 2nd ed., J. Doull. C. D. Klaasen, M. O. Amdur, Eds. (New York: MacMillan Publishing Co., 1980), p. 345.
20. Meistrich, M. L., and C. C. Brown. "Estimation of the Increased Risk of Human Infertility From Alterations in Semen Characteristics," *Fertil. Steril.* 40:220–230 (1983).
21. Wyrobek, A. J., et al. "An Evaluation of the Mouse Sperm Morphology Test and Other Sperm Tests in Nonhuman Mammals," *Mutat. Res.* 115:1–72 (1983).
22. Wyrobek, A. J., et al. "An Evaluation of Human Sperm as Indicators of Chemically Induced Alterations in Spermatogenic Function," *Mutat. Res.* 115:73–148 (1983).
23. Furuhjelm, M., B. Jonsen, and C. G. Lagergren. "The Quality of Human Semen in Spontaneous Abortion," *Int. J. Fertil.* 7:17–21 (1962).

24. Furuhjelm, M., B. Jonsen, and C. G. Lagergren. "The Quality of Human Semen in Relation to Perinatal Mortality," *Acta Obstet. Gynecol. Scand.* 39:499–505 (1960).
25. Damarajan, M. "Effect on the Embryo of Staleness of the Sperm at the Time of Fertilization of the Domestic Hen," *Nature* 165:398 (1950).

CHAPTER 25

Toxicity of 2-Chlorophenol, 2,4-Dichlorophenol, and 2,4,6-Trichlorophenol

Jerry H. Exon and Loren D. Koller

The use of chlorinated phenols such as 2-chlorophenol (2-CP), 2,4-dichlorophenol (2,4-DCP), 2,4,6-trichlorophenol (2,4,6-TCP), and pentachlorophenol (PCP) in industry and agriculture has resulted in extensive contamination of the ecosystem. Residues of these compounds have been detected in otherwise relatively uncontaminated aquatic, terrestrial, and atmospheric environments in the world, thus attesting to their ubiquitous occurrence.[1] These chemicals have been detected in the urine of occupationally and nonoccupationally exposed individuals, in human and animal blood and adipose tissue, and in various food and water supplies consumed by humans and animals.[2,3] The recently discovered inadvertent spontaneous formation of chlorinated phenols, especially 2-CP, 2,4-DCP, and 2,4,6-TCP, as by-products of water chlorination for disinfection or deodorization was recognized as a significant source of low-level exposure of large populations to these chemicals.[4-7]

Until recently, it was believed that purified forms of chlorinated phenols were not very toxic to humans and other mammals, and that massive doses of these chemicals would be required to produce significant health effects. However, the toxic effects of chronic exposure to relatively low levels of chlorinated phenols, except for PCP, and their metabolites are virtually unknown. Preliminary data suggest these compounds may act indirectly by rather subtle insidious means by enhancing the tumorigenicity of known carcinogens and adversely altering immune responsiveness. Altered immunocompetence could result in decreased resistance to infectious diseases and oncogenesis or increased incidence of autoimmune or immunodeficiency disorders. The outcome of these subtle effects could range from death to a general decrease in the quality of life. The purpose of this study was to determine (1) if 2-CP and 2,4-DCP were carcinogenic to laboratory rats chronically exposed to these chemicals by the oral route; (2) if 2-CP and 2,4-DCP act as promotors or cocarcinogens by enhancing the tumorigenic effects of a known carcinogen, ethylnitrosourea (ENU); and (3) the toxic effects produced in laboratory rats following chronic oral exposure to low levels of 2-CP, 2,4-DCP, or 2,4,6-TCP as monitored by body weight, hematologic parameters, histopathologic changes in major organs, and effects on reproduction and immunocompetence.

MATERIALS AND METHODS

Chemicals

The chlorophenols used in this study were 2-CP (Aldrich Chemical Co., Milwaukee, Wisconsin, 98%, No. 18-577-9), 2,4-DCP (Aldrich, 99%, No. 576-24-9), 2,4,6-TCP (Aldrich, 98%, No. 75530-1), ethylurea (Aldrich, 97%, No. ES100-7), and sodium nitrite (Aldrich, 97%, No. 20783-7).

Experimental Design

Male and female Sprague-Dawley (S-D) outbred rats obtained from Washington State University, Department of Laboratory Animal Resources, were used in all experiments. The animals were housed four per cage in racks of stainless steel hanging wire cages, except for pregnant female rats and their preweaning age pups, which were housed in polycarbonate cages containing heat-treated hardwood shavings. The animal rooms were maintained on a 12-h on/off automatic light cycle, and temperature was controlled within a range of 20–23°C (68–73°F). Water bottles, cages, and feed trays were sanitized at least twice weekly, and each animal room was washed down weekly. Feed and water were available ad libitum throughout the experiments. A commercial rodent chow was purchased and used as the basic diet in all experiments. The drinking water was deionized and provided in clear-glass bottles (16 oz.) with rubber stoppers and stainless steel controlled-flow sipper tubes. Appropriate concentrations of chlorophenol were added to the drinking water, as indicated in Table I. Fresh batches of treated drinking water were prepared weekly.

Reproduction Parameters

Weanling female S-D rats were placed on chlorophenol treatment from 3 weeks of age through breeding and parturition. The females were bred to untreated males at 90 d of age. The parameters of reproduction recorded included percent conception, litter size, number of stillborn, birth and weaning weight, and survival to weaning.

Table I. Chlorophenol Concentrations in Water and Duration of Exposure

Chemicals	Concentrations of chemicals[a,b,c]		
	Low dose (*ppm*)	Medium dose (*ppm*)	High dose (*ppm*)
2-Chlorophenol	5	50	500
2,4-Dichlorophenol	3	30	300
2,4,6-Trichlorophenol	3	30	300

[a]Prenatal exposure. The chlorinated phenol was given to dams from 3 weeks through parturition. The dams were bred at 90 d.
[b]Postnatal exposure. Progeny of untreated dams were weaned at 3 weeks and placed on chlorophenol treatment until tumor development or termination of the experiment at approximately 24 months.
[c]Pre- and postnatal exposure. The chlorinated phenols were given to dams from 3 weeks through parturition (bred at 90 d) and lactation. The progeny were weaned at 3 weeks and continued on treatment until tumors developed or termination of the experiment at approximately 24 months.

Immunologic Assays

Progeny of female rats exposed from weaning age through breeding at 90 d and parturition and lactation were obtained at 3 weeks of age and continued on chlorophenol treatment for an additional 12 to 15 weeks. Immunologic competence was assessed in each animal by measuring the ability to elicit three major types of immune responses, namely, humoral immunity, cell-mediated immunity (CMI), and macrophage function. Humoral immune responses were assessed by using an enzyme-linked immunosorbent assay (ELISA) to quantitate specific serum IgG antibody levels to the T cell-dependent protein antigens bovine serum albumin (BSA) or keyhole limpet hemocyanin (KLH). CMI responses were measured by a delayed-type hypersensitivity (DTH) response to either oxazolone, applied to the ears, or heat-aggregated BSA, injected into the footpads. Macrophage function was assessed by the ability of peritoneal cavity-derived, elicited, glass-adherent cells to phagocytize sheep red blood cells (SRBC) in vitro. Body and organ (liver, spleen, and thymus) weights were recorded from all rats for which immunologic competence was assessed. The methodology for immunoassays used in the 2-CP,[8] 2,4-DCP, and 2,4,6-TCP[9] experiments has been described previously in detail.

Carcinogen and Cocarcinogen Bioassay

These experiments were designed to determine if the chlorinated phenols, 2-CP and 2,4-DCP, were carcinogenic in S-D rats or if these chemicals acted as promotors or cocarcinogens when given in conjunction with the known carcinogen, ENU. A design of the treatment protocols for these experiments is presented in Tables I and II to supplement this section. 2-CP was given at doses of 5, 50, or 500 ppm. 2,4-DCP and 2,4,6-TCP were given in doses of 3, 30, or 300 ppm (Table I). ENU was given as the precursors; ethylurea (EU) was mixed into the feed, and nitrite (NO_2) as sodium nitrite was added to the drinking water. Concentrations of 0.316% EU and 1-ppm NO_2 were used in the 2-CP experiment. Concentrations of 0.150% EU and 1-ppm NO_2 were used in the 2,4-DCP experiment.

The chlorophenol treatment regimens included rats that were exposed to these chemicals either prenatally, postnatally, or both pre- and postnatally (Table I). Prenatal treatment consisted of exposing 3-week-old weanling female rats to chlorophenols continuously through breeding (90 d of age) until parturition. The progeny of these females were weaned at 3 weeks and maintained on control feed for the remainder of the experiment.

Postnatal treatments consisted of exposing offspring of nontreated female rats to chlorophenols continuously from weaning age at 3 weeks until death or 24 months. Combined pre- and postnatal treatments consisted of exposing 3-week-old weanling female rats to chlorophenols continuously through breeding (90 d of age), gestation, parturition, and lactation. The progeny of this latter group were weaned at 3 weeks and continued on chlorophenol treatment until death or 24 months. EU and NO_2 were administered only on days 14 to 21 of gestation to pregnant female rats selected from each chlorophenol treatment regimen. Each treatment group consisted of 12 to 22 breeding females and 48 to 52 offspring composed of approximately equal numbers of each sex (Table II).

Body weights of the progeny placed on experiment were recorded monthly. Blood samples were collected via the tail veins of five randomly selected male and female rats in each treatment group every 2 months for hematologic analysis. Blood values, determined using a Coulter counter Model ZB_1, included red and white cell (RBC and WBC) counts, hemoglobin (Hgb) concentration, mean corpuscular volume (MCV), and packed-cell volume (PCV). All rats were observed daily for gross signs of morbidity, including rough appearance of the hair coat, diarrhea, abnormal nasal and eye discharges, unusual motor activity, alertness, and gross tumor development. Moribund or tumor-bearing rats were terminated by carbon dioxide asphyxiation, and complete necropsies were performed. A gross and microscopic examination was made of all major organs including lung, heart, liver, spleen, kidney, adrenal, intestine, stomach, urinary bladder, and brain. Sections of spinal cord, muscle, and tumor tissue (if present) were also collected. All tissues were fixed in 10% buffered formalin, processed by standard microtechniques, and stained with Harris' hematoxylin and eosin prior to examination by light microscopy. Tumor incidence, latency, and type were recorded for each treatment group.

Table II. Experimental Design for Chlorophenol (CP) Carcinogenesis Experiments in Conjunction with Ethylnitrosourea (ENU)

Treatment regimen[a]	Exposure to CP[b]		ENU Exposure	Females (No.)	Progeny on Experiment (No.)	
	Prenatal	Postnatal			Male	Female
Negative control	−	−	−	12−13	32	28
Chlorophenol only						
Low CP only	+	+	−	12−13	24	24
Med CP only	+	+	−	12−13	24	24
High CP only	+	+	−	13−14	28	28
Ethylnitrosourea only						
ENU only	−	−	+	13−14	28	24
Prenatal chlorophenol + ENU						
Low CP + ENU	+	−	+	13−14	24	24
Med CP + ENU	+	−	+	13−14	24	24
High CP + ENU	+	−	+	13−14	28	28
Postnatal chlorophenol + ENU						
Low CP + ENU	−	+	+	13−14	24	24
Med CP + ENU	−	+	+	13−14	24	24
High CP + ENU	−	+	+	13−16	28	24
Pre- and postnatal chlorophenol + ENU						
Low CP + ENU	+	+	+	13−14	24	24
Med CP + ENU	+	+	+	13−14	24	24
High CP + ENU	+	+	+	13−22	32	28

[a]See Table I for concentrations of chlorophenols.
[b]See Table I for duration of exposure to chlorophenols.

RESULTS

Reproduction

Litter sizes of female rats treated with 500-ppm 2-CP (Table III), 300-ppm 2,4-DCP (Table IV), or 300-ppm 2,4,6-TCP (Table V) were significantly smaller than the respective controls. The percent stillborn pups tended to be increased in all groups that received 2,4-DCP (Table IV) or 2,4,6-TCP (Table V). The percent of stillborn pups was significantly increased in groups treated with 500-ppm 2-CP, as compared with controls (Table III). Survival to weaning was significantly less in the group of rats given 30-ppm 2,4-DCP (Table IV).

Immune Responses

Serum antibody levels to BSA in 2-CP-treated rats averaged consistently less than controls, but in no instance were these decreases statistically significant (Table VI). DTH responses, phagocytic activity of macrophages, and numbers of peritoneal exudate cells (PEC) in 2-CP-exposed groups of rats were not significantly different from the corresponding controls (Table VI).

Antibody levels to KLH in the serum of rats exposed to 2,4-DCP were consistently greater than controls (Table VII). These increases in anti-KLH antibody levels appeared to be dose related to 2,4-DCP exposure and were significantly different than controls at the 300-ppm dose. DTH responses in

Table III. Effects of Prenatal Exposure to 2-Chlorophenol (2-CP) on Reproduction of Sprague-Dawley Rats[a]

Treatment (ppm)	Conception (%)	Litter size (mean ± SE)[b]	Stillborn (%)	Birth weight (mean ± SE)[b]	Survival to weaning (%)[c]
Controls	67	11.4 ± 1.1	0	2.2 ± 0.4	100
2-CP					
5	75	11.6 ± 1.0	2	2.3 ± 0.4	100
50	75	10.1 ± 1.0	0	2.5 ± 0.4	100
500	86	9.1[d] ± 0.9	5[d]	2.4 ± 0.4	99

[a] 2-Chlorophenol was given in the drinking water prenatally by exposing 12 to 14 dams from 3 weeks through parturition (bred at 90 d).
[b] Inclusive of stillborn pups.
[c] Noninclusive of stillborn pups.
[d] $P \leq 0.10$ compared with controls by analysis of variance and least-square means (litter size, birth weight) or chi-square analysis (percent stillborn, survival to weaning).

Table IV. Effects of Prenatal Exposure to 2,4-Dichlorophenol (2,4-DCP) on Reproduction of Sprague-Dawley Rats[a]

Treatment	Conception (%)	Litter size (mean ± SE)[b]	Stillborn (%)	Birth weight (Mean ± SE)[b]	Survival to weaning (%)[c]
Controls	80	9.8 ± 1.3	0	2.7 ± 0.3	98
2,4-DCP (ppm)					
3	80	8.6 ± 1.2	1	2.5 ± 0.3	98
30	70	8.7 ± 1.4	2	2.5 ± 0.3	87[d]
300	80	6.3[d] ± 1.6	2	2.2 ± 0.2	96

[a]2,4-DCP was administered in the drinking water to groups of 13 dams each from 3 weeks of age through parturition (bred at 90 d).
[b]Inclusive of stillborn pups.
[c]Noninclusive of stillborn pups.
[d]$P \leq 0.10$ compared with controls by analysis of variance and least-square means (litter size) or chi-square analysis (% survival).

Table V. Effects of Prenatal Exposure to 2,4,6-Trichlorophenol (2,4,6-TCP) on Reproduction of Sprague-Dawley Rats[a]

Treatment	Conception (%)	Litter size (mean ± SE)[b]	Stillborn (%)	Birth weight (Mean ± SE)[b]	Survival to weaning (%)[c]
Controls	77	12.1 ± 1.1	0	2.9 ± 0.4	100
2,4,6-TCP (ppm)					
3	69	11.3 ± 1.1	2	2.7 ± 0.3	99
30	85	11.2 ± 1.0	3	2.8 ± 0.5	99
300	77	9.1[d] ± 0.9	3	2.6 ± 0.5	97

[a]2,4,6-Trichlorophenol was given in the drinking water prenatally by exposure to 12 to 14 dams from 3 weeks through parturition (bred at 90 d).
[b]Inclusive of stillborn pups.
[c]Noninclusive of stillborn pups.
[d]$P \leq 0.10$ compared with controls by analysis of variance and least-square means.

Table VI. Effect of Pre- and Postnatal Exposure to 2-Chlorophenol (2-CP) on Immune Parameters of Sprague-Dawley Rats[a]

Treatment (ppm)	Antibody production[b]	DTH response[c]	Phagocytic activity[d]	PEC[d]/Rat (No.)
	Mean (±SE) absorbance at 405 nm	Mean (±SE) DPM treated: DPM control	Mean (±SE) % PEC ingesting SRBC	Mean (±SE) × 10^7
Controls	0.96 ±0.16	1.55 ±0.20	24.0 ±1.5	2.6 ±0.5
2-CP				
5	0.69 ±0.16	1.57 ±0.12	23.0 ±1.7	2.2 ±0.4
50	0.84 ±0.16	1.60 ±0.17	26.1 ±1.7	2.6 ±0.4
500	0.66 ±0.16	1.35 ±0.06	24.1 ±1.7	2.8 ±0.4

[a]Dams were exposed to 2-CP in the drinking water continuously from 3 weeks through parturition (bred at 90 d) and lactation. Eight randomly selected pups from each group were weaned at 3 weeks and continued on 2-CP treatment for 15 weeks.

[b]Serum antibodies to bovine serum albumin (BSA) were generated by injecting 1-mg BSA/rat SC at the base of the tail 26 and 6 d prior to serum collection by cardiac puncture. Antibody levels were measued by an indirect ELISA at a 1:1000 serum dilution.

[c]The DTH response was measured by sensitizing the ears of each rat with oxazolone 3 d apart. Ten days after the second sensitization, each rat was injected IP with 1-μCi ^{3}H-thymidine/g body weight. Twenty-four hours later, the left ear was challenged with oxazolone in olive oil and the right ear (control) was treated with olive oil only. All rats were terminated the following day and radioactive content in each ear was determined. The values in the table are the mean ratio of DPM of left to right ear.

[d]Phagocytic activity of peritoneal exudate cells (PEC) was measured in vitro as the percent of adherent PEC ingesting opsonized SRBC. PEC were induced by injecting 10 mL of thioglycollate medium IP 4 d before collection by peritoneal lavage.

2,4-DCP-exposed rats were significantly suppressed when compared with controls in a dose-related manner (Table VII). Phagocytic activity of macrophages from 2,4-DCP-exposed animals was similar to controls (Table VII). The numbers of PEC harvested from animals treated with 2,4-DCP increased with dosage, but not significantly compared with controls (Table VII).

No significant effects of 2,4,6-TCP treatment on immune responses were observed (Table VIII). However, antibody levels, DTH reactions, and macrophage numbers were consistently greater in 2,4,6-TCP-exposed animals compared with controls.

Body, liver, spleen, and thymus weights were recorded from rats from which immunologic competence was assessed. Body weights of rats exposed pre- and postnatally to 2-CP (Table IX), 2,4-DCP (Table X), or 2,4,6-TCP (Table XI) were not significantly different from the respective controls. Liver weights were significantly increased compared with control rats exposed to 300-ppm

Table VII. Effect of Pre- and Postnatal Exposure to 2,4-Dichlorophenol (2,4-DCP) on Immune Parameters of Sprague-Dawley Rats[a]

Treatment (ppm)	Antibody production[b] Mean (±SE) absorbance at 405 nm	DTH response[c] Mean (±SE) footpad swelling (mm)	Phagocytic activity[d] Mean (±SE) cpm/100 μg protein	PEC[d]/Rat (No.) Mean (±SE) × 10[7]
Controls	1.24 ±0.10	1.10 ±0.13	40201 ±4720	3.7 ±0.8
2,4-DCP				
3	1.30 ±0.10	0.85 ±0.11	33316 ±4376	3.5 ±0.8
30	1.39 ±0.10	0.67[e] ±0.11	39782 ±7565	4.5 ±0.8
300	1.68[e] ±0.08	0.63[e] ±0.11	40873 ±3379	5.0 ±0.9

[a] Dams were exposed to 2,4-DCP in the water continuously from 3 weeks of age through parturition (bred at 90 d) and lactation. Ten randomly selected pups from each group were weaned at 3 weeks of age and continued on 2,4-DCP treatment for 15 weeks.
[b] Serum antibodies were generated to keyhole limpet hemocyanin (KLH) by injecting 1 mg KLH per rat SC at base of tail 14 and 6 d prior to serum collection by cardiac puncture. Antibody was measured by an indirect enzyme-linked immunosorbent assay (ELISA) at a 1:1000 serum dilution.
[c] DTH response was elicited by sensitizing with 100 μg of bovine serum albumin (BSA) emulsified 1:1 in Freund's complete adjuvant (FCA) and given SC at the base of the tail. A challenge injection of 75 μL of heat-aggregated BSA was given in the left rear footpad 7 d later, and the right footpad was sham-injected with saline. Footpad swelling was determined 24 h later by the difference in swelling between the left and right footpads.
[d] Peritoneal exudate cells (PEC) were elicited by IP injection of 10 mL of 5% sodium caseinate 4 d prior to harvesting by peritoneal lavage. The adherent cells were isolated, and phagocytic activity to ^{51}Cr-labeled sheep red blood cells was tested in vitro and standardized to amount of protein in each culture.
[d] $P \leq 0.05$ compared with controls by analysis of variance and least-square means.

2,4-DCP (Table X) or 30- or 300-ppm 2,4,6-TCP (Table XI). Spleen weights of rats exposed to 300-ppm 2,4-DCP or 2,4,6-TCP were significantly greater than controls. Spleen and liver weights of 2-CP-treated rats (Table IX) were also generally elevated compared with controls, but not significantly. Thymus weights were not significantly altered in any of the chlorophenol-treated groups of rats. However, thymus weights of 2,4-DCP-treated groups (Table X) were consistently less than controls, whereas those of 2-CP-treated groups (Table IX) were usually greater.

Table VIII. Effect of Pre- and Postnatal Exposure to 2,4,6-Trichlorophenol (2,4,6-TCP) on Immune Parameters of Sprague-Dawley Rats[a]

Treatment (ppm)	Antibody Production[b] Mean (±SE) absorbance at 405 nm	DTH Response[c] Mean (±SE) footpad swelling (mm)	Phagocytic Activity[d] Mean (±SE) cpm/100 μg protein	PEC[d]/Rat (No.) Mean (±SE) × 10^7
Controls	1.37 ±0.11	1.59 ±0.20	34.1 ±1.5	3.3 ±1.7
2,4,6-TCP				
3	1.38 ±0.11	1.64 ±0.12	36.5 ±1.7	4.2 ±1.8
30	1.50 ±0.11	1.69 ±0.1	38.8 ±1.7	3.9 ±1.6
300	1.51 ±0.10	1.74 ±0.20	35.4 ±1.7	4.0 ±1.9

[a]Dams were exposed to 2,4,6-TCP in the drinking water continuously from 3 weeks through parturition (bred at 90 d) and lactation. Ten randomly selected pups from each group were weaned at 3 weeks and continued on 2,4,6-TCP treatment for 12 weeks.
[b]See Table VII, footnote b.
[c]See Table VII, footnote c.
[d]See Table VII, footnote d.

Table IX. Effects on Body and Organ Weights of Pre- and Postnatal Exposure of Rats to 2-Chlorophenol (2-CP)

Treatment[a]	Mean weight (g ± SE)			
	Body	Thymus	Spleen	Liver
Controls	290 ±6	0.25 ±.04	0.88 ±.05	15.3 ±0.5
2-CP, ppm				
5	273 ±8	0.25 ±.06	0.93 ±.06	15.4 ±0.3
50	291 ±7	0.27 ±.05	1.05 ±.06	14.6 ±0.4
500	292 ±9	0.31 ±.07	1.02 ±.08	16.3 ±0.5

[a]2-Chlorophenol was given to dams in the drinking water from 3 weeks through parturition (bred at 90 d) and lactation. Eight randomly selected pups were weaned at 3 weeks and continued on 2-CP treatment for approximately 15 weeks.

Table X. Effects on Body and Organ Weights of Pre- and Postnatal Exposure of Rats to 2,4-Dichlorophenol (2,4-DCP)

Treatment[a]	Mean weight (g ± SE)			
	Body	Thymus	Spleen	Liver
Controls	287 ±5	0.30 ±.03	0.95 ±.05	13.0 ±0.4
2,4-DCP, ppm				
3	290 ±6	0.29 ±.01	0.88 ±.04	13.3 ±0.6
30	283 ±6	0.29 ±.02	0.91 ±.04	14.3 ±0.6
300	299 ±8	0.27 ±.02	1.86[b] ±.05	15.5[b] ±0.9

[a]Dams were exposed to 2,4-DCP in the water continuously from 3 weeks through parturition (bred at 90 d) and lactation. Ten randomly selected pups from each group were weaned at 3 weeks and continued on 2,4-DCP treatment for approximately 15 weeks.
[b]$P \le 0.05$ compared with controls by analysis of variance and least-square means.

Table XI. Effects on Body and Organ Weights of Pre- and Postnatal Exposure of Rats to 2,4,6-Trichlorophenol (2,4,6-TCP)

Treatment[a]	Mean weight (g ± SE)			
	Body	Thymus	Spleen	Liver
Controls	271 ±4	0.38 ±.08	0.93 ±.09	10.9 ±0.4
2,4,6-TCP, ppm				
3	282 ±5	0.36 ±.10	0.95 ±.04	11.9 ±0.3
30	256 ±8	0.32 ±.08	0.89 ±.03	12.5[b] ±0.5
300	262 ±4	0.40 ±.08	1.07[b] ±.07	14.1[b] ±0.6

[a]2,4,6-Trichlorophenol was given to dams in the drinking water from 3 weeks through parturition (bred at 90 d) and lactation. Ten randomly selected pups were weaned at 3 weeks and continued on 2,4,6-TCP treatment for approximately 12 weeks.
[b]$P \le 0.05$ compared with controls by analysis of variance and least-square means.

Carcinogenicity

Tumor incidence, latency, or type in male and female rats exposed to 2-CP (Table XII) or 2,4-DCP (Table XIII) was not significantly different from that in their respective untreated controls. The carcinogenicity of 2,4,6-TCP was not tested in these studies.

Table XII. Tumor Incidence and Latency in Rats Exposed Pre- and Postnatally to 2-Chlorophenol (2-CP)

Treatment	Rats (No.)		Tumor incidence, % (Total)		Tumor latency (d) (mean ± SE)	
	Male	Female	Male	Female	Male	Female
Controls	32	28	13	5	350 ± 61	543 ± 65
2-CP,[a] ppm						
5	24	24	17	0	496 ± 61	
50	24	24	8	13	473 ± 75	483 ± 65
500	28	28	18	18	503 ± 75	507 ± 46
ENU only[b]						
0.316%	28	24	68[c]	71[c]	330[c] ± 24	337[c] ± 26

[a]Dams were exposed to 2-CP in the drinking water from 3 weeks through parturition (bred at 90 d) and lactation. The progeny were weaned at 3 weeks and continued on 2-CP treatment until tumors developed or the experiment was terminated at 24 months.
[b]ENU was given as 0.316% ethylurea in the feed and 1-ppm nitrite in the water to pregnant females during days 14 to 21 of gestation. The progeny were weaned at 3 weeks and maintained on control feed for the remainder of the experiment.
[c]$P \leq 0.10$ compared with controls by analysis of variance and least-square means (latency) or chi-square analysis (incidence).

Tumor incidence and latency in groups of rats exposed to chlorinated phenols in conjunction with ENU were compared with the positive control group treated only with ENU. Because of the unexpected high incidence of tumors that developed in ENU-treated controls used in the 2-CP experiment, tumor incidence and latency comparisons were made at three different times during the experiments, corresponding to approximately 25, 50, and 75% combined tumor incidence in male and female ENU-only treated groups (Tables XIV and XV). The level of ENU exposure was reduced in the 2,4-DCP experiment; consequently, comparisons of tumor incidence and latency among groups were made only at the termination of the study (Table XVI).

Tumor incidence was increased in all groups of male rats treated with ENU and exposed pre- and postnatally to 2-CP compared with male rats that were given only ENU (Table XIV). This effect was particularly evident when compared at an approximate tumor incidence of 50% (i.e., 46%) in the ENU-only group. Tumor latency also decreased in all groups of male rats that were treated with ENU and exposed to 2-CP pre- and postnatally, as compared with the ENU-only controls (Table XIV). This effect was most evident if comparison was made at a time when tumor incidence in the ENU-only treated group

Table XIII. Tumor Incidence and Latency in Rats Exposed Pre- and Postnatally to 2,4-Dichlorophenol (2,4-DCP)

Treatment	Rats (No.) Male	Rats (No.) Female	Tumor incidence, % (Total) Male	Tumor incidence, % (Total) Female	Tumor latency (d) Male	Tumor latency (d) Female
Controls	28	29	11	10	499 ± 15	549 ± 17
2,4-DCP,[a] ppm						
3	23	22	4	23	484 ± 0	423 ± 21
30	23	25	9	8	421 ± 125	442 ± 122
300	28	29	14	24	394 ± 73	454 ± 40
ENU only[b]						
0.150%	28	25	11	8	434 ± 72	512 ± 40

[a]Dams were exposed to 2,4-DCP in the drinking water from 3 weeks through parturition (bred at 90 d) and lactation. The progeny were weaned at 3 weeks and continued on 2,4-DCP treatment until tumors developed or the experiment was terminated at 24 months.
[b]ENU was given as 0.150% ethylurea in the feed and 1-ppm nitrite as sodium nitrite in the drinking water to pregnant females during days 14 to 21 of gestation. The progeny were weaned at 3 weeks and maintained on control feed for the remainder of the experiment.

was approximately 75% and tumor latency was 330 d. A similar but less marked decrease in tumor latency also occurred in groups of male rats given ENU and treated only prenatally with 2-CP (Table XIV). In fact, all groups of male rats that received combined exposure to ENU and 2-CP had reduced mean tumor latencies when compared with male rats that received only ENU and were compared at the approximate 25 and 75% tumor-incidence intervals (Table XIV). ENU-treated male rats given 500-ppm 2-CP postnatally also developed fewer tumors than the ENU-only controls.

The tumor incidence in female rats treated prenatally or pre- and postnatally with 2-CP, and also exposed to ENU, was not significantly different than females that received only ENU (Table XV). Female rats treated only postnatally with 500-ppm 2-CP and exposed to ENU developed significantly fewer tumors than ENU-only treated females compared at the 50 and 75% (i.e., 54 and 71%) tumor-incidence intervals. Tumor latency was significantly less in ENU-treated females exposed postnatally or pre- and postnatally to 5-ppm 2-CP (Table XV). No other consistent effects of 2-CP on tumor incidence or latency in ENU-treated female rats were evident.

Tumor incidence in male and female rats treated with ENU and exposed to 2,4-DCP was not different than ENU-only treated animals (Table XVI). The results of the 2,4-DCP-ENU study were confounded by the lack of tumor

Table XIV. Ethylnitrosourea (ENU)-Induced Tumor Incidence and Latency in Male Rats Treated with 2-Chlorophenol (2-CP) and/or Ethylurea and Nitrite

Treatment	Rats (No.)	Tumor incidence (%)[a]			Tumor latency (d)[a]		
		25%	50%	75%	25%	50%	75%
ENU only[b]	28	29	46	68	221 ± 12	266 ± 22	330 ± 24
Prenatal[b] + ENU							
5 ppm 2-CP	24	29	67	83	217 ± 13	265 ± 19	308 ± 24
50 ppm 2-CP	24	29	42	50	189[c] ± 13	244 ± 25	278 ± 30
500 ppm 2-CP	28	36	68	79	189[c] ± 11	249 ± 18	279 ± 23
Postnatal[b] + ENU							
5 ppm 2-CP	24	17	67	88	204 ± 17	291 ± 19	332 ± 23
50 ppm 2-CP	24	21	58	63	213 ± 15	291 ± 21	302 ± 27
500 ppm 2-CP	24	21	36	43[c]	189[c] ± 14	251 ± 25	286 ± 31
Pre- and postnatal[b] + ENU							
5 ppm 2-CP	24	58[c]	88[c]	96[c]	192[c] ± 9	235 ± 17	259[c] ± 22
50 ppm 2-CP	24	46	75	83	196 ± 10	239 ± 18	268[c] ± 24
500 ppm 2-CP	32	50	77[c]	80	199 ± 9	245 ± 16	258[c] ± 22

[a] Tumor incidence and latency in all groups was calculated at three time intervals corresponding to 25, 50, and 75% of combined tumor incidence in males and females exposed to ENU only. Latency is mean ± SE.
[b] ENU was administered as the precursors, ethylurea (0.316 %) and nitrite (1 ppm) to pregnant females during days 14 to 21 of gestation.
[c] $P \leq 0.10$ compared with the ENU-only group by analysis of variance and least-square means (latency) or chi-square analysis (incidence).

development in the ENU-only group. Because of the unusually high incidence of tumors resulting from ENU treatment in the 2-CP experiment, the dose of ENU was decreased from 0.316% to 0.150% in the 2,4-DCP experiment. The resulting tumor incidence in the ENU-only treated group was subsequently not significantly different than the untreated controls (Table XVI). It is possible that all or most of the tumors that developed in animals used in the 2,4-DCP experiment were spontaneous.

Table XV. Ethylnitrosourea (ENU)-Induced Tumor Incidence and Latency in Male Rats Treated with 2-Chlorophenol (2-CP) and/or Ethylurea and Nitrite

Treatment	Rats (No.)	Tumor incidence (%)[a]			Tumor latency (d)[a]		
		25%	50%	75%	25%	50%	75%
ENU only[b]	24	21	54	71	184 ± 15	296 ± 22	337 ± 28
Prenatal[b] + ENU							
5 ppm 2-CP	24	25	58	79	189 ± 14	278 ± 21	338 ± 26
50 ppm 2-CP	24	38	71	88	194 ± 12	265 ± 19	311 ± 25
500 ppm 2-CP	28	11	36	50	199 ± 12	285 ± 25	349 ± 30
Postnatal[b] + ENU							
5 ppm 2-CP	24	46	71	75	209 ± 10	256 ± 19	268[c] ± 27
50 ppm 2-CP	24	4	29	58	154 ± 35	310 ± 29	391 ± 30
500 ppm 2-CP	24	4	21[c]	25[c]	179 ± 35	307 ± 35	332 ± 46
Pre- and postnatal[b] + ENU							
5 ppm 2-CP	24	38	75	83	183 ± 12	243[c] ± 18[c]	266[c] ± 25
50 ppm 2-CP	24	21	42	54	218 ± 15	256 ± 25	308 ± 32
500 ppm 2-CP	32	23	57	83	204 ± 13	269 ± 19	343 ± 23

[a] Tumor incidence and latency in all groups was calculated at three time intervals corresponding to 25, 50, and 75% of combined tumor incidence in males and females exposed to ENU only. Latency is mean ± SE.
[b] Refer to Table XIV for dose and duration of exposure to ENU.
[c] $P \leq 0.10$ compared with ENU-only group by analysis of variance and least-square means (latency) or chi-square analysis (incidence).

Hematology

Some hematologic parameters were altered in rats exposed pre- and postnatally to chlorinated phenols (Table XVII). Numbers of RBC and PCV and hemoglobin levels were generally increased in all groups of rats treated with 500-ppm 2-CP or 300-ppm 2,4-DCP compared with the control group. This effect was most evident following 14 months of exposure. No significant differences were noted between hematologic parameters of male and female rats; therefore, the data for both sexes were combined.

Table XVI. Ethylnitrosourea(ENU)-Induced Tumor Incidence and Latency in Rats Treated with 2,4-Dichlorophenol (2,4-DCP) and/or Ethylurea and Nitrite

Treatment	Rats (No.)		Tumor incidence (%)		Tumor latency (d)[a]	
	Male	Female	Male	Female	Male	Female
Controls	28	29	11	10	499 ± 15	549 ± 17
ENU only[b]			11	8	434 ± 72	512 ± 40
Prenatal[c] + ENU						
3 ppm 2,4-DCP	24	25	13	8	448 ± 85	540 ± 18
30 ppm 2,4-DCP	23	24	22	4	491 ± 37	488 ± 0
300 ppm 2,4-DCP	26	30	15	10	492 ± 44	439 ± 85
Postnatal[c] + ENU						
3 ppm 2,4-DCP	22	23	5	9	304 ± 0	523 ± 69
30 ppm 2,4-DCP	22	24	14	17	486 ± 83	504 ± 42
300 ppm 2,4-DCP	28	23	11	22	499 ± 80	496 ± 29
Pre- and postnatal[c] + ENU						
3 ppm 2,4-DCP	22	21	9	13	485 ± 75	448 ± 18
30 ppm 2,4-DCP	19	22	21	18	459 ± 40	513 ± 29
300 ppm 2,4-DCP	25	34	16	18	428 ± 51	485 ± 43

[a] Mean ± SE.
[b] ENU was given as the precursors, ethylurea (0.150%) and nitrite (1 ppm) to pregnant females during days 14 to 21 of gestation.
[c] Refer to Table II for duration of exposure.

DISCUSSION

The chlorinated phenols, 2-CP and 2,4-DCP, used in these experiments were not innately carcinogenic in S-D outbred rats as determined by a 2-year bioassay. However, three subtle but major toxic effects were observed that could be attributed to one or the other of these chemicals. First, these data suggest that 2-CP may enhance the tumorigenicity of ENU. Second, the immune system appears to be a sensitive target to toxic insult induced by chronic exposure to

Table XVII. Effect of Chronic Exposure to Chlorophenols (CP) on Hematologic Parameters of Sprague-Dawley Rats (14 months exposure)

Treatment	Parameters[a]		
	RBC (10^4/mm^3)	PCV (%)	Hgb (g/dL)
	2-Chlorophenol (2-CP)[b]		
Controls	773 ± 57	37 ± 3	16.1 ± 1.3
500 ppm CP	914 ± 61[c]	44 ± 4[c]	18.7 ± 2.2[c]
	2,4-Dichlorophenol (2,4-DCP)[b]		
Controls	745 ± 65	38 ± 3	15.1 ± 1.0
300 ppm CP	815 ± 11[c]	39 ± 1	17.5 ± 0.4[c]

[a]Hematologic parameters were determined using a Coulter counter; RBC = red blood cells; PCV = packed-cell volume; Hgb = hemoglobin.
[b]2-CP and 2,4-DCP were given in the drinking water to dams from 3 weeks through parturition (bred at 90 d) and lactation. Progeny were weaned at 3 weeks and continued on CP treatment. Five males and five females were selected at random from each group, and blood samples were collected from the tail veins for analysis.
[c]$P \leq 0.05$ compared with respective controls by analysis of variance and least square means.

2,4-DCP and 2,4,6-TCP. Third, 2-CP, 2,4-DCP, and 2,4,6-TCP may be embryotoxic when administered transplacentally. These results suggest that the chlorinated phenols may produce toxic effects relevant to human and animal health (following chronic exposure) that were previously unrecognized and may be useful in establishing guidelines for acceptable residues of these chemicals in the environment.

The tumor incidence and latency in groups of rats treated only with 2-CP or 2,4-DCP were not significantly different than untreated controls. However, the incidence of ENU-induced tumors was increased and tumor latency decreased in several groups of male rats exposed to 2-CP when compared with rats treated only with ENU. This effect was most pronounced in ENU-treated rats exposed to 2-CP by combined pre- and postnatal exposure. These results suggest that 2-CP and 2,4-DCP are not complete carcinogens in S-D rats; however, 2-CP may act as a cocarcinogen or promotor of carcinogenesis. A previous experiment has been reported which suggests that certain chlorinated phenols may act as promotors or cocarcinogens.[10] Also, phenols in general are believed to act as promotors of carcinogens in cigarette smoke.[11]

The methods used in this study were not conventional in terms of available classic models for testing initiation and promotion. However, it was possible to detect promoting or cocarcinogenic properties of the chlorinated phenols, because these chemicals were administered both simultaneously with (e.g., prenatally) and subsequent to (e.g., postnatally) the initiating dose of ENU.

The mechanisms by which chlorophenols alter the carcinogenicity of ENU is not known at this time. The proposed action of tumor promotors, or cocarcinogens, has been the subject of several recent reviews.[12-14] Some mechanisms of action that have been proposed include (1) alteration of metabolic activation, detoxification, or excretion of the active carcinogenic moieties; (2) inhibition of DNA repair mechanisms; (3) perturbation of the production or action of growth factors or hormones; and (4) altered immunosurveillance. The effects of chlorophenols on the carcinogenicity of ENU could occur by any one or several of these mechanisms, based on the known effects of these chemicals or phenols in general.

Chlorophenols may alter the carcinogenicity of ENU by affecting its formation in the stomach. ENU was administered orally to rats in these studies as the precursors EU and NO_2. The formation of ENU from these precursors has been shown to occur spontaneously at a low pH, such as is present in the stomach.[15] Cotreatment with a chlorophenol could either directly affect the reaction of EU with NO_2 or indirectly alter ENU formation by causing a change in the pH of gastric contents, thus facilitating formation of the carcinogen. The formation of ENU in vivo has been shown to be altered by ascorbates and antioxidants such as vitamin E.[16,17] Administration of preformed ENU to rats would help to clarify this effect if it were the mechanism by which chlorophenols were acting to alter tumorigenesis.

Chlorinated phenols have been reported to affect mixed-function oxidases (MFO) and other enzymes in the liver.[18-21] Perturbation of this MFO drug-metabolizing system has been shown many times to alter the normal metabolism of chemicals and thus their toxicity. In fact, PCP has been shown to alter the metabolism and subsequent toxicity of hexachlorobenzene in rats.[19] This PCP-induced effect was correlated to altered MFO activity. Also, PCP has been shown to alter the toxicity of the known carcinogen, N-hydroxy-2-acetylaminofluorene (2-AAF), by inhibiting the conjugation of 2-AAF metabolites by sulfatransferases.[22] Conjugation of toxic metabolites to more polar forms aids in their excretion. Phenols and chlorophenols are known to be conjugated to sulfur groups or glucuronide during metabolism.[23,24] Depletion of these conjugate substrates could result in prolonged retention and increased action of electrophilic carcinogenic species, such as the genotoxic ethyldiazonium ion produced by enzymatic cleavage of ENU.

Chlorophenols have been shown to be mutagenic in some assays[25] and cause chromosomal aberrations.[26] These effects could result in increased DNA damage and altered DNA repair, which could enhance the effects of a genotoxic carcinogen such as ENU.

Apparently, hormonal influences can affect ENU tumorigenicity, since male rats in this study were more severely affected than females. The incidence of ENU-induced central nervous system tumors in mice has been shown to be altered by the production of nerve growth factors.[27] Chlorophenol-induced changes in neuroendocrine function/products could affect ENU carcinogenesis.

The effects of chlorophenols on immune responsiveness could alter the effects of ENU. Immunosurveillance has been shown to be an important defense to neoplasia.[28-31] Also, immunodeficient nude mice are more susceptible to ENU-induced tumorigenesis than their immunocompetent littermates.[32] Several investigators have reported that the immune system is a sensitive target for chlorophenol-induced toxicity.[33-35]

It is apparent that further research is required to clarify the mechanisms by which chlorophenols may act as cocarcinogens or promotors. But, regardless of the mechanisms, it is important to note that the effects of 2-CP on ENU-induced tumorigenesis occurred at very low doses (e.g., 5 ppm). This would indicate that this chemical may produce subtle toxic effects at doses below those previously suspected and may require a reevaluation of the potential hazards posed to human and animal health.

Immune responses in rats were significantly altered following treatment with 2,4-DCP. Immune responsiveness in rats treated similarly with 2-CP or 2,4,6-TCP were not different than their corresponding controls. Rats exposed to 2,4-DCP had significantly reduced DTH responses, whereas antibody production was significantly enhanced. Macrophage function of 2,4-DCP-treated rats, as measured by their ability to phagocytize SRBC, was not different than the controls. Significant effects on the immune function of rats exposed to 2,4-DCP occurred at doses as low as 30 ppm. These results suggest that the immune system is sensitive to chlorophenol-induced perturbation at relatively low doses. We have previously reported that immune functions of rats are altered following exposure to 5 ppm PCP.[9] Impairment of immune functions by chlorophenols could have serious health implications. A normal functioning immune system is essential for protecting the host against infectious diseases and oncogenic agents, as well as to spontaneously occurring neoplasia.

The mechanism by which chlorinated phenols alter immune functions is not known at this time. Chlorophenols are uncouplers of oxidative phosphorylation and result in decreased cellular energy sources.[36] Impairment of cellular energy production in immunocompetent cells during periods of acute energy demand, such as following antigenic challenge, could result in a less efficient manifestation of their ability to respond optimally.

The effect of PCP and 2,4-DCP on immune functions could also result from a direct toxic effect of these chemicals on subpopulations of cells involved in the production or regulation of immune responses. For instance, rats exposed to 2,4-DCP tended to have decreased thymus weights and increased spleen weights that were accompanied by depressed CMI responses (i.e., DTH) and enhanced humoral immunity (i.e., antibody production). A selective toxic effect of 2,4-DCP on subpopulations of T lymphocytes would be consistent with decreased thymus weight and impaired CMI responses. A reduction in the T-cell suppressor population that controls antibody production could result in enhanced humoral immunity and splenomegaly. It has been well documented that certain chemicals, drugs, or metals can be selectively toxic to certain subpopulations of lymphocytes. Cyclophosphamide, administered at appropriate doses, results in a selective decrease in T-suppressor cells.[37] Dialkyltin is selectively toxic to T lymphocytes without any apparent direct effects on B cells.[38]

Although the mechanism of chlorophenol-induced immune dysfunction is not known, it is clear that these chemicals are toxic to the immune system, as the data presented in this study and by others demonstrate.[8,20,33] The fact that these chemicals are ubiquitous within our environment, and that impaired immune function has serious health implications, indicates that further research dealing with the immunotoxicity of these compounds is warranted.

Transplacental exposure to chlorinated phenols may adversely affect the fetus. Female rats in these experiments that were treated during gestation with high doses of 2-CP, 2,4-DCP, or 2,4,6-TCP had significantly fewer pups than nontreated controls. Also, the numbers of stillborn pups were significantly greater in litters of females exposed to 500 ppm 2-CP. The number of pups that survived to weaning was significantly decreased in groups that received 30 ppm 2,4-DCP. PCP has been previously reported to adversely affect rat and hamster embryonal and fetal development,[39-41] but experiments of this type using 2-CP and 2,4-DCP have not been reported. Schwetz et al.[42] reported that PCP was most toxic to the developing embryo when administered during early organogenesis (days 8–11). They also reported that the LD_{50} of PCP in 3- to 4-day-old neonates was 65 mg/kg compared with 150 mg/kg in adult rats exposed to the same batch of this chemical. This difference in LD_{50} values suggests that chlorophenols may be particularly toxic to the embryo or fetus. Apparently, chlorinated phenols, or some metabolite of these chemicals, are able to cross the placenta and are toxic to the developing fetus. The toxic effects of these compounds to the fetus could be related to their effects on oxidative phosphorylation at a time during organogenesis of fetal tissues when maximal production of cellular energy is required. This type of effect may even be more pronounced in regard to human embryonic development, since the uncoupling of oxidative phosphorylation of chlorophenols in preparations of human mitochondria has been shown to be ten times greater than in rat mitochondria.[43]

Increased liver weights observed in animals exposed to high levels of 2,4-DCP or 2,4,6-TCP in this study are consistent with the reported toxic effects of chlorophenols. These chemicals are mild hepatotoxins, and several investigators have reported increased liver weights in rats chronically treated with PCP.[41,42,44,45] Liver weight changes, however, have not been consistently associated with microscopic changes of hepatocytes. Apparently, the liver weight changes are due to an effect at the subcellular level or to a general hyperplasia. Increased proliferation of SER in hepatocytes is a common change in chlorophenol-treated animals.[19] This effect is probably due to a stimulation of hepatic enzyme activity. Regardless of the effect, these compounds do not appear to cause serious damage to the liver even at high doses. However, chlorophenol-induced effects on liver enzymes such as MFO have been reported and could alter the metabolism and, thus, the toxic effects of other chemicals. Chlorophenols have been reported to alter the metabolism of other xenobiotics (e.g., HCB and 2-AAF).[19,22]

Hematologic parameters have been reported to be altered in rodents chronically exposed to PCP.[41,42] These effects seem to manifest as increased RBC, PCV, and Hgb concentration. RBC counts and Hgb levels were significantly increased in rats exposed to high levels of 2-CP or 2,4-DCP following 14 months of treatment. It could be postulated that effects on RBC could be secondary to effects on liver enzymes involved in the breakdown of RBC and Hgb, or effects on heme synthesis. A direct effect of chlorophenols on hematopoietic stem cells could also occur. The effect of chlorophenols on RBC has not been extensively investigated, but it does not appear to pose a significant health problem in animals exposed to these chemicals.

The fact that chlorophenols are ubiquitous environmental contaminants to which a large segment of the human and animal population are exposed is reason to warrant further research pertaining to the chronic toxicity of these chemicals. This is especially apparent in view of the potential embryolethal, immunotoxic, and tumor promoting or cocarcinogenic effects of these compounds, which have only recently been elucidated. It is evident that a more complete understanding of the potential subtle, insidious, toxic effects of low-level chronic exposure to chlorinated phenols is required to establish intelligent regulatory guidelines and accurately assess the potential hazards to human health.

REFERENCES

1. Ahlborg, U. G., and T. M. Thunberg. "Chlorinated Phenols: Occurrence, Toxicity, Metabolism and Environmental Impact," *CRC Crit. Rev. Toxicol.* 7:1–35 (1980).
2. Dougherty, R. C. "Human Exposure to Pentachlorophenol," in *Pentachlorophenol: Chemistry, Pharmacology and Environmental Toxicology*, K. R. Rao, Ed. (New York: Plenum Press, 1978).
3. Morgrade, C., A. Garquet, and C. D. Pfaffenberger. "Determination of Polyhalogenated Phenolic Compounds in Drinking Water, Human Blood Serum and Adipose Tissue," *Bull. Environ. Contam. Toxicol.* 24:257–264 (1980).
4. *Ambient Water Quality Criteria for 2,4-Dichlorophenol*, EPA 44/5-80-042, Office of Water Regulations and Standards Division, (Washington, DC: U.S. Environmental Protection Agency, 1980).
5. *Ambient Water Quality Criteria for 2-Chlorophenol*, EPA 44/5-80-038, Water Regulation and Standards Division, (Washington, DC: U.S. Environmental Protection Agency, 1980).
6. Barnhart, E. L., and G. R. Campbell. *The Effect of Chlorination on Selected Organic Chemicals*, Water Pollut. Control Research Ser. 12020, Exg 03/72, (Washington, DC: U.S.Environmental Protection Agency, 1972).
7. Dietz, F., and J. Troud. "Geruchs- und Geschmacks-Schwellen-Konzentrationen von Phenolkorpera," *Gas-Wasserfach. Wasser-Abwasser* 119:318–325 (1978).
8. Exon, J. H., and L. D. Koller. "Effects of Chlorinated Phenols on Immunity in Rats," *Int. J. Immunopharmacol.* 5:131–136 (1983).

9. Exon, J. H., L. D. Koller, G. M. Henningsen, and C. A. Osborne. "Multiple Immunoassay in a Single Animal: A Practical Approach to Immunotoxicologic Testing," *Fund. Appl. Toxicol.* 4:278-283 (1984).
10. Boutwell, R. K., and K. K. Bosch. "The Tumor-Promoting Action of Phenol and Related Compounds for Mouse Skin," *Cancer Res.* 19:413-424 (1959).
11. Van Duuren, B. L., and B. M. Goldschmidt. "Cocarcinogenic and Tumor-Promoting Agents in Tobacco Carcinogenesis," *J. Nat. Cancer Inst.* 56:1237-1242 (1976).
12. Trosko, J. E., C. Chang, and A. Medcalf. "Mechanisms of Tumor Promotion: Potential Role of Intercellular Communication," *Cancer Invest.* 1:511-526 (1983).
13. Potter, V. R. "Alternative Hypotheses for the Role of Promotion in Chemical Carcinogenesis," *Environ. Health Perspect.* 50:139-148 (1983).
14. Slaga, T. J. "Overview of Tumor Promotion in Animals," *Environ. Health Perspect.* 50:3-14 (1983).
15. Mirvish, S. S., and C. Chu. "Chemical Determination of Methylnitrosourea and Ethylnitrosourea in Stomach Contents of Rats after Intubation of Alkylureas plus Sodium Nitrite," *J. Nat. Cancer Inst.* 50:745-750 (1973).
16. Rustin, M. "Inhibitory Effect of Sodium Ascorbate on Ethylurea and Sodium Nitrite Carcinogenesis and Negative Findings in Progeny After Intestinal Inoculation of Precursors into Pregnant Hamsters," *J. Nat. Cancer Inst.* 55:1389-1393 (1975).
17. Mirvish, S. S., S. Salmasi, S. M. Cohen, K. Patil, and E. Mahboubi. "Liver and Forestomach Tumors and Other Forestomach Lesions in Rats Treated with Morpholine and Sodium Nitrite, With and Without Sodium Ascorbate," *J. Nat. Cancer Inst.* 1:81-85 (1983).
18. Arrhenius, E., L. Renberg, and L. Johansson. "Subcellular Distribution, a Factor in Risk Evaluation of Pentachlorophenol," *Chem. Biol. Interact.*, 18:23-24 (1977).
19. Debets, F. M. H., J. J. T. W. A. Strik, and K. Olie. "Effects of Pentachlorophenol on Rat Liver Changes Induced by Hexachlorobenzene, with Special Reference to Porphyria, and Alterations of Mixed Function Oxidases," *Toxicology* 15:181-195 (1980).
20. McConnell, E. E., J. A. Moore, B. N. Gupta, A. H. Rabes, M. I. Luster, J. A. Goldstein, J. K. Haseman, and C. E. Parker. "The Chronic Toxicity of Technical and Analytical Pentachlorophenol in Cattle. I. Clinicopathology," *Toxicol. Appl. Pharmacol.* 52:468-490 (1980).
21. Stockdale, M., and M. J. Selwyn. "Effects of Ring Substitutes on the Activity of Phenols as Inhibitors and Uncouplers of Mitochondrial Respiration," *Eur. J. Biochem.* 21:565-574 (1971).
22. Meerman, J. H. N. and G. J. Mulder. "Prevention of the Hepatotoxic Action of N-Hydroxy-2-acetylaminofluorene in the Rat by Inhibition of N-O-Sulfation by Pentachlorophenol," *Life Sci.* 28:2361-2365 (1981).
23. Babich, H., and D. L. Davis. "Phenol: A Review of Environmental and Health Risks," *Regul. Toxicol. Pharmacol.*, 1:90-109 (1981).
24. Braun, W. H., G. E. Blau, and M. B. Chenoweth. "The Metabolism/Pharmacokinetics of Pentachlorophenol in Man, and a Comparison with Rat and the Monkey Model," *Toxicol. Appl. Pharmacol.* 45:278-289 (1978).
25. Fahrig, R., C. A. Nilsson, and C. Rappe. "Genetic Activity of Chlorophenols and Chlorophenol Impurities," in *Pentachlorophenol: Chemistry, Pharmacology and*

Environmental Toxicology K. R. Rao, Ed. (New York: Plenum Press, 1978), pp. 325-338.
26. Bauchinger, M., J. Dresp, E. Schmid, and R. Hauf. "Chromosomal Changes in Lymphocytes After Occupational Exposure to Pentachlorophenol (PCP)," *Mutat. Res.* 102:83-88 (1982).
27. Vinores, S. A., and A. Koestner. "Reduction of Ethylnitrosourea-Induced Neoplastic Proliferation in Rat Trigeminal Nerves by Nerve Growth Factor," *Cancer Res.*, 42:1038-1040 (1982).
28. Trainin, Z., and M. Essex. "Immune Response to Tumor Cells in Domestic Animals," *J. Am. Vet. Med. Assoc.* 181:125-1133 (1982).
29. Wheelock, E. F., and M. K. Robinson. "Biology of Disease: Endogenous Control of the Neoplastic Process," *Lab. Invest.* 48:120-139 (1983).
30. Keller, R. "Host Defense Mechanisms Against Tumors as the Principal Targets of Tumor Promotors," *J. Cancer Res. Clin. Oncol.*, 105:203-211 (1983).
31. Albright, A. L., T. J. Gill, and S. J. Geyer. "Immunogenetic Control of Brain Tumor Growth in Rats," *Cancer Res.,* 37:2512-2521 (1977).
32. Anderson, L. M., K. L. Barney, and J. M. Budinger. "Sensitivity to Carcinogenesis in Nude Mice: Skin Tumors Caused by Transplacental Exposure to Ethylnitrosourea," *Science* 218:682-684 (1982).
33. Kerkvliet, N. I., L. Baecher-Steppan, A. T. Claycomb, A. M. Craig, and G. G. Sheggeby. "Immunotoxicity of Technical Pentachlorophenol (PCP-T): Depressed Humoral Immune Responses to T-Dependent and T-Independent Antigen Stimulation in PCP-T Exposed Mice," *Fund. Appl. Toxicol.* 2:90-99 (1982).
34. LaVia, M. F., and D. S. LaVia. "Phenol Derivatives are Immunosuppressive in Mice," *Drug Chem. Toxicol.* 2:90-99 (1982).
35. Prescott, C. A., B. N. Wilkie, B. Hunter, and R. J. Julian. "Influence of a Purified Grade of Pentachlorophenol on the Immune Response of Chickens," *Am. J. Vet. Res.* 43:481-487 (1982).
36. Weinbach, E. D., and J. Garbus. "Mechanism of Action of Reagents That Uncouple Oxidative Phosphorylation," *Nature (London)* 221:1016-1018 (1969).
37. Fast, P. E., C. A. Hatfield, C. L. Franz, E. G. Adams, N. J. Licht, and M. V. Merritt. "Effects of Treatment with Immunomodulatory Drugs on Thymus and Spleen Lymphocyte Subpopulations and Serum Corticosterone Levels," *Immunopharmacology*, 5:135-155 (1982).
38. Seinen, W. "Immunotoxicology of Alkyltin Compounds." In *Immunologic Considerations in Toxicology, Vol. II* R. R. Sharma, Ed. (Boca Raton, FL: CRC Press Inc., 1981), pp. 103-120.
39. Courtney, K. D. "The Effect of Pentachloronitrobenzene on Fetal Kidneys," *Toxicol. Appl. Pharmacol.* 25:455-461 (1983).
40. Hinkle, D. K. "Fetotoxic Effects of Pentachlorophenol in the Golden Syrian Hamster," *Toxicol. Appl. Pharmacol.* 25:455-461 (1973).
41. Schwetz, B. A., J. F. Quast, P. A. Keeler, C. G. Humiston, and R. J. Kociba. "Results of Two-Year Toxicity and Reproduction Studies on Pentachlorophenol in Rats," in *Pentachlorophenol: Chemistry, Pharmacology and Environmental Toxicology*, K. R. Rao, Ed. (New York: Plenum Press, 1978), pp. 301-309.
42. Schwetz, B. A., P. A. Keeler, and P. J. Gehring. "The Effect of Purified and Commercial Grade Pentachlorophenol on Rat Embryonal and Fetal Development," *Toxicol. Appl. Pharmacol.* 28:151-161 (1974).

43. Mitsuda, H., K. Murakami, and F. Kawai. "Effect of Chlorophenol Analogues on the Oxidative Phosphorylation in Rat Liver Mitochondria," *Agric. Biol. Chem.* 27:366–372 (1963).
44. Kobayashi, S., S. Toida, H. Kawamura, H. S. Chang, T. Fukuda, and K. Kawaguchi. "Chronic Toxicity of 2,4-Dichlorophenol in Mice," *J. Med. Soc. Toho Univ. Japan*, 19:356–362 (1972).
45. Kociba, R. J. "Toxicological Evaluation of Rats Maintained on Diets Containing Pentachlorophenol Sample XD-8108 OOL for 90 Days," Mar. 2, 1973, Chem. Biol. Section, (Midland, MI: Dow Chemical Co., 1973).

CHAPTER 26

Toxicological Evaluation of Selected Chlorinated Phenols

Joseph F. Borzelleca, Lyman W. Condie, and Johnnie R. Hayes

Chlorinated phenols (CPs) have been used as chemical intermediates in the manufacture of many products. Some have been used as antiseptics since 1893, whereas others are used as fungicides and preservatives in nonfood items.

The U.S. Environmental Protection Agency (EPA) and other regulatory bodies and organizations involved with the safety of drinking water are interested in these compounds because they are sometimes found in finished water and because there is a relative paucity of toxicological data concerning them. They do not appear to be a severe problem in drinking water for a number of reasons: (1) organoleptic properties, which affect the taste of water and intake, would be self-limiting; (2) the limited solubility of some of these compounds in water; and (3) available toxicological data, which suggest that they generally are not highly toxic.

The CPs are formed from the breakdown of chlorobenzenes and from the chlorination of water containing phenol. The presence of CPs in water is often the cause of its medicinal odor and taste.

In conjunction with the EPA, selected literature was reviewed and a series of investigations were conducted to evaluate the acute and subchronic toxicity of a selected number of CPs. The physical and chemical characteristics of a series of CPs and those investigated in our study are listed in Table I.

EXPERIMENTAL DESIGN

The basic experimental design consisted of an evaluation of the effects of orally administered CPs in mice and rats. These species were selected because of their sensitivity to xenobiotics and their size, cost, and ease of handling. The agents tested were selected by the EPA. Exposures were either acute, repeated, or subchronic, and by either gavage or in drinking water. Details of the major procedures used are summarized in Tables II[1,2] through IV and in Figure 1.

Table I. Physical and Chemical Properties of Phenol and Chlorinated Phenols

	Formula	Mol Wt	Physical State	Specific Gravity	Melting Point (°C)	Boiling Point (°C)	Solubility[a]			pKa
							Water (ppm)	Alcohol	Ether	
Phenol	C_6H_5OH	94.11	Crystals	1.072	41	182	8.2 at 15°	S	VS	9.89
2-CP	ClC_6H_4OH	128.56	Liquid	1.241 (18/15°C)	α −7, β −0, γ −4.1	175−176	2.85	S	S-alk	8.65
3-CP	ClC_6H_4OH	128.56	Needles	1.268 (25°C)	32−33	214	2.6	S	S	8−12
4-CP	ClC_6H_4OH	128.56	Needles	1.306 (20/4°C)	41−43	217	2.7	VS	VS	9.37
2,3-DCP	$Cl_2C_6H_3OH$	163.01	Needles	1.383 (60/25°C)	45	209−210	0.45	VS	VS	
2,4-DCP	$Cl_2C_6H_3OH$	163.01	Hexagonal needles		45	206−208	SS	VS	VS	7.85
2,5-DCP	$Cl_2C_6H_3OH$	163.01	Prisms		59	211	SS	VS	VS	7.50
2,6-DCP	$Cl_2C_6H_3OH$	163.01	Needles		68−69	219−220		VS	VS	6.91
3,4-DCP	$Cl_2C_6H_3OH$	163.01	Needles		68	253.5				8.58
3,5-DCP	$Cl_2C_6H_3OH$	163.01	Prisms		68	233	SS	VS		
2,3,5-TCP	$Cl_3C_6H_2OH$	197.46	Needles		55	249−250	SS	S	S	
2,4,5-TCP	$Cl_3C_6H_2OH$	197.46	Needles		61−63	252	I	S	S	7.07
2,3,6-TCP	$Cl_3C_6H_2OH$	197.46	Needles		58		S	V	V	5.98
2,4,6-TCP	$Cl_3C_6H_2OH$	197.46	Needles		68−69	246	0.09	VS	VS	6.62
3,4,5-TCP	$Cl_3C_6H_2OH$	197.46	Needles		101	271−277		S		7.83
2,3,4,5-TCP	Cl_4C_6HOH	231.90	Needles	1.6 (60/4°C)	69−70	164	SS	VS	VS	
2,3,4,6-TCP	Cl_4C_6HOH	231.90	Needles		70	150				5.46
2,3,5,6-TCP	Cl_4C_6HOH	231.90	Leaflets		115					
2,3,4,5,6-PCP	Cl_5C_6OH	266.35	Solid	1.85 (22°C)	188−189	310				5.00

[a]S = soluble; SS = slightly soluble; VS = very soluble; I = insoluble.

Table II. Experimental Design: Acute Oral Toxicity Study

Animals
  Adult male and female CD-1 ICR mice; healthy, acclimated, randomized
  Adult male and female CD rats; healthy, acclimated, randomized
  Overnight (18 h) fast (water, no food)
  10 males, 10 females per dosage level; 5 levels
  Housing: One per plastic shoebox cage with hardwood sawdust bedding
  Feed: Purina Rodent Chow No. 5001; deionized water (in bottles)
  Environmental conditions
    Rooms: 22° ± 2°C; relative humidity, 40 to 60%
    Light cycle: 12 light/12 dark (7 a.m. – 7 p.m.)

Test Material
  Identity confirmed
  Prepared day of administration
  Administered by gavage (stomach tube)
  Volume: 10 mL/kg body weight

Specific Conditions
  Food withheld for 2 h after dosing
  Continuous observations for 1 h after dosing, hourly for next 4 h, and twice daily for the next 14 d
  Cageside observations include changes in behavior, respiration, circulation, skin and fur, eyes, mucous membranes; evidence of tremors or convulsions, lethargy, sleep, coma, salivation, diarrhea; time of onset of changes and duration; time of death
  All animals that died during the observation period and survivors necropsied

Calculations of LD_{50}
  Log probit analysis of Finney[1]
  Litchfield-Wilcoxon[2]

RESULTS

The acute oral toxicity data, summarized in Table V, indicate that pentachlorophenol is the most toxic, and the dichlorophenols are the least toxic.

2-Chlorophenol was evaluated following 14 d of exposure at doses of 35, 69, and 175 mg $kg^{-1}d^{-1}$ administered by stomach tube. The data are summarized in Table VI. 2,4-Dichlorophenol was evaluated following 14 and 90 d of exposure. The exposure levels for the repeated dosing, 14-d study were 64, 128, and 638 mg $kg^{-1}d^{-1}$, and for the subchronic study they were 50, 150, and 500 mg $kg^{-1}d^{-1}$. The test compounds were administered by gavage for the 14-d study and in drinking water for the subchronic study. The data are summarized in Tables VII and VIII.

At the termination of the 2,4-DCP subchronic exposure study, 10 males and 10 females per group were randomly selected for an in vivo fertility study. Dosing was continued throughout mating and gestation. Eighteen days after mating all females were sacrificed, and the following parameters were measured or calculated: total implants, total resorptions, total number live pups, weight of individual pups, and fertility index. The data are summarized in

Table III. Experimental Design: Short-Term (14-d) Repeated Oral Dosing Study

Animals
  Adult male and female CD-1 ICR mice; healthy, acclimated, randomized; earpunched for identification
  Adult male and female CD rats; healthy, acclimated, randomized; 10 males, 10 females per dosage level; 5 levels

Housing: One per plastic shoebox cage with hardwood sawdust bedding
Feed: Purina Rodent Chow No. 5001; deionized water (in bottles)
Environmental conditions
    Rooms: 22° ± 2°C; relative humidity, 40 to 60%
    Light cycle: 12 light/12 dark (7 a.m. – 7 p.m.)

Test Material
  Identity confirmed
  Prepared day of administration (gavage) or twice weekly (drinking water)
  Vehicles: water, corn oil, emulphor
  Gavage: 10 mL/kg body wt; drinking water: chemical, toxicological, or palatability limit

Specific Conditions
  Groups: controls; naive, vehicle, positive (occasionally); 3 treated groups, at least 20 males and 20 females per group
  Cageside observations: twice daily; behavior, morbidity, mortality; exceptions noted
  Body weights: onset, days 8 and 15 and at necropsy
  Fluid consumption: twice weekly (unless gavaged)
  Hematology: blood collected at necropsy; wbc (total, differential); rbc, platelets, HCrt, Hb, coagulation
  Serum chemistries: LDH, SGOT, SGPT, SAP, BUN, bilirubin, protein, glucose, cholesterol, ALB/GLOB, P, Ca, Na, Cl, K
  Liver microsomal activities: cytochrome P_{450}, cytochrome b_5, microsomal protein, aminopyrine demethylase, aniline hydroxylase, arylhydrocarbon hydroxylase
  Immune response
  Genetic toxicology: testicular DNA synthesis and sperm morphology; SCE (testes, bone marrow); mitotic index (bone marrow)
  Reproductive toxicology: in vitro penetration, fertilization, blastula formation
  Activity and behavioral measurements
  Necropsy: gross observations, organ weights
  Statistical analysis: appropriate tests including Bartlett's test for homogeneity, parametric ANOVA, Dunnett's multirange test, Wilcoxon nonparametric test; significance, $p \leq 0.05$

Table IV. Experimental Design: Subchronic Oral Toxicity Study, 90-d

Animals
  Adult male and female CD-1 ICR mice; healthy, acclimated, randomized; earpunched for identification; 10 males, 10 females per dosage level; 5 levels
  Housing: One per plastic shoebox cage with hardwood sawdust bedding
  Feed: Purina Rodent Chow No. 5001; deionized water (in bottles)
  Environmental conditions
    Rooms: 22° ± 2°C; relative humidity, 40 to 60%
    Light cycle: 12 light/12 dark (7 a.m. – 7 p.m.)

Test Material
  2,4-Dichlorophenol
  Identity confirmed
  Prepared twice weekly (drinking water)
  Vehicles: emulphor (10% aqueous)

Specific Conditions
  Groups: controls; naive, vehicle, positive (cyclophosphamide); 3 treated groups, at least 20 males and 20 females per group
  Cageside observations: twice daily; behavior, morbidity, mortality; exceptions noted
  Body weights: onset, once weekly, and at necropsy
  Fluid consumption: twice weekly
  Hematology: blood collected at necropsy; wbc (total, differential); rbc, platelets, HCrt, Hb, coagulation
  Serum chemistries: LDH, SGPT, SGOT, SAP, BUN, bilirubin, protein, glucose, cholesterol, ALB/GLOB, P, Ca, Na, Cl, K
  Liver microsomal activities: cytochrome P_{450}, cytochrome b_5, microsomal protein, aminopyrine demethylase, aniline hydroxylase, arylhydrocarbon hydroxylase, ethoxycoumarin-O-deethylase, testosterone hydroxylation, cytochrome c reductase
  Genetic toxicology: testicular DNA synthesis and sperm morphology; SCE (testes, bone marrow); mitotic index (bone marrow)
  Reproductive toxicology: in vitro penetration, fertilization, blastula formation
  Activity and behavioral measurements
  Necropsy: gross observations, organ weights
  Statistical analysis: appropriate tests including Bartlett's test for homogeneity, parametric ANOVA, Dunnett's multirange test, Wilcoxon nonparametric test; significance, $p \leq 0.05$

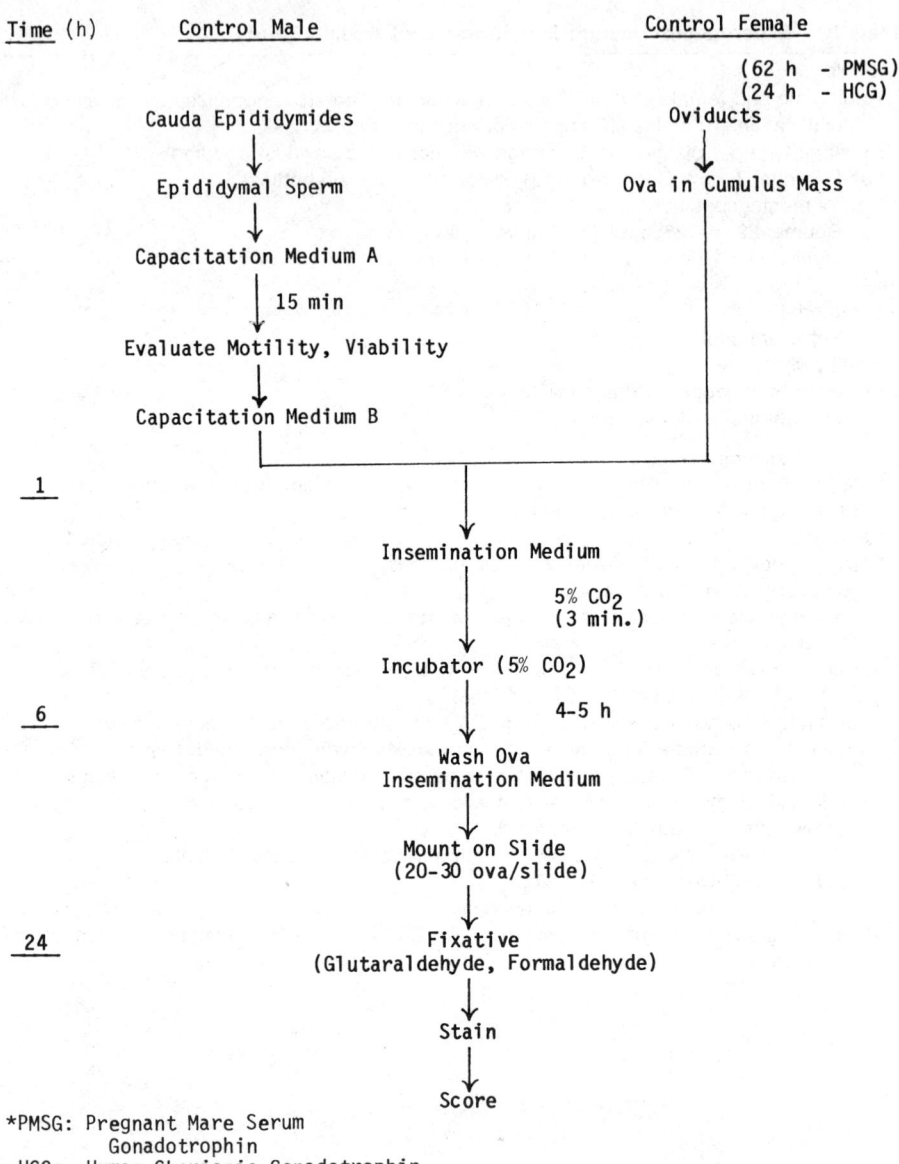

Figure 1. In vitro penetration and fertilization scheme.

Table V. Acute Oral Toxicity of Chlorophenols in CD-1 Mice

Compound	Mol wt	Purity (%)	Vehicle/Solvent	Acute Oral LD_{50} (mg/kg) Male	Acute Oral LD_{50} (mg/kg) Female
2-Chlorophenol	128.56	98+	Deionized water	347 (239 – 393)[a]	345 (321 – 381)[a]
3-Chlorophenol	128.56	99	Deionized water	521 (463 – 586)	530 (468 – 601)
4-Chlorophenol	128.56	99+	Corn oil	1373 (1191 – 1583)	1422 (1333 – 1518)
2,3-Dichlorophenol	163.00	98	Corn oil	2585 (2046 – 3266)	2376 (2186 – 2585)
2,4-Dichlorophenol	163.00	99	Corn oil	1276 (982 – 1569)	1352 (1094 – 1670)
2,5-Dichlorophenol	163.00	98	Corn oil	1600 (1233 – 2075)	946 (623 – 1438)
2,6-Dichlorophenol	163.00	99	Corn oil	2198 (1727 – 2797)	2120 (1799 – 2498)
3,4-Dichlorophenol	163.00	99	Corn oil	1685 (1504 – 1887)	2046 (1472 – 2846)
3,5-Dichlorophenol	163.00	99	Corn oil	2643 (2269 – 3078)	2389 (1829 – 3120)
Pentachlorophenol	226.34	99+	10% Emulphor	177 (125 – 252)	117 (65 – 212)

[a]Confidence limits.

Table VI. Short-Term Repeated Dosing (14 d) 2-CP, Mice

Animals: 12 males, 12 females per group
Exposure levels: naive control; vehicle control (corn oil); 35, 69, 175 mg kg^{-1} d^{-1}; positive control (cyclophosphamide, 25 mg/kg)
Body weights: days 1, 8, 15; lower at 69 mg/kg
Mortality: all died at 175 mg/kg
Organ weights and ratios: females only—↓ brain, liver, spleen
Hematology: NCRE[a]
Clinical chemistry: NCRE
Hepatic microsomal MFO activity: NCRE
Immune response, cell mediated and humoral: NCRE
Behavioral parameters: hyperactivity at both doses
Sister chromatid exchange: NCRE
Gross pathological findings: None

[a]NCRE—no biologically or statistically significant compound-related adverse effects.

Table VII. Short-Term Repeated Dosing (14 d) 2,4-DCP, Mice

Animals: 12 males, 12 females per group
Exposure levels: naive control; vehicle control (corn oil); 64, 128, 638 mg kg^{-1} d^{-1}; positive control (cyclophosphamide, 25 mg/kg)
Body weights: days 1, 8, 15—NCRE[a]
Mortality: 1 male at the 638-mg/kg level—NCRE
Organ weights and ratios: NCRE
Hematology: NCRE (↑ in platelets at 638 mg/kg)
Clinical chemistry: ↓ bilirubin among all treated females only; slight ↑ in SGOT, SGPT in male mice at mid dose
Hepatic microsomal MFO activity: ↑ glutathione, ↑ microsomal protein, ↑ cytochrome b$_5$ among females and males at 638-mg/kg dose
Behavioral parameters: NCRE
Sister chromatid exchange: NCRE
Gross pathological findings: None

[a]NCRE—no biologically or statistically significant compound-related adverse effects.

Table VIII. Subchronic Oral Toxicity (90 d) 2,4-DCP, Mice

Animals: 20 males, 20 females per group
Exposure levels: naive control (deionized water); vehicle control (10% emulphor); 0.2-, 0.6-, 2.0-mg/mL drinking solution (equivalent to 50-, 150-, 500-mg/kg body wt per d).
Fluid consumption: measured twice weekly;—naive controls (M and F) had highest fluid intake; treated females consumed more than vehicle controls
Body weights: weekly—NCRE[a]
Mortality: 1 at 150 mg/kg—NCRE
Organ weights and ratios: NCRE
Hematology: NCRE
Clinical chemistry: NCRE
Hepatic microsomal MFO activity: NCRE
Sister chromatid exchange: NCRE
Behavioral parameters: NCRE
Gross pathological findings: None
In vitro penetration: NCRE
Reproductive effects: NCRE

[a]NCRE—no biologically or statistically significant compound-related adverse effects.

Table IX. The increased resorption rate at the mid dose is not statistically different from the vehicle control. There were no consistent adverse compound-related effects. In vitro reproductive data are summarized in Table X. The effects of exposure to selected CPs on sister chromatid exchange (SCE) are summarized in Table XI.

Table IX. Reproductive Effects of 90-d Treatment with 2,4-Dichlorophenol

Treatment	FI[a]	Total Implants	Total Resorptions	Total live pups	Mean litter size	Mean pup wt (g)
Control (naive)	90	198	5	193	11	1.08
10% Emulphor	75	166	12	154	10	1.09
0.2 mg/mL	80	181	1	180	11	1.01
0.6 mg/mL	90	198	16	182	10	1.13
2.0 mg/mL	65	160	1	159	12	1.11

[a] Fertility index $= \dfrac{\text{number of pregnant females}}{\text{number of females mated}} \times 100$.

DISCUSSION

Acute Toxicity

Available acute oral toxicity data of phenol and CPs in mice and rats are summarized in Table XII.[3-11] These compounds are not highly toxic, with the exception of pentachlorophenol. The order of toxicity (most to least) is PCP > tetrachlorophenols > monochlorophenols > trichlorophenols > dichlorophenols. The order is similar in both mice and rats and suggests similar species sensitivity. The effects produced by these compounds are also similar in both species and in both sexes, suggesting similar mechanisms of action. Increased respiration, motor weakness, tremors, CNS depression, convulsions, dyspnea, coma, and then death were observed. Body temperature was decreased with the mono- and dichlorophenols but was increased with the other chlorophenols.

Short-Term Repeated Dosing

2-CP at a dose of 175 mg kg^{-1}d^{-1} was lethal. At doses of 35 and 69 mg kg^{-1}d^{-1}, slight toxic effects were noted among the females only. The 2,4-DCP data essentially confirm and extend the findings of Kobayashi et al.,[11] whose mice received 667-mg 2,4-DCP/kg body wt for 10 d without evidence of any adverse effect. The liver appeared to be a target organ in both studies.

Subchronic Toxicity

The data presented confirm and extend the earlier work of Kobayashi et al.[11] They exposed mice for 6 months to doses of 45, 100, or 230 mg kg^{-1}d^{-1}.

Table X. Percent Penetration (Mean ± S.E.) of Mouse Ova Following Exposure to Various Chlorophenols at Three Concentrations

Treatment	Compound								
	2-CP	3-CP[a]	2,3-DCP	2,4-DCP	2,5-DCP	2,6-DCP	3,4-DCP	3,5-DCP	CdCl$_2$
Control	65 ± 6 (90)[b]	58 (41)	68 ± 5 (158)	76 ± 5 (128)	75 ± 3 (135)	78 ± 4 (190)	78 ± 4 (218)	62 ± 0 (181)	66 ± 6 (393)
0.1 mM	65 ± 1 (57)	82 (40)	74 ± 1 (131)	74 ± 10 (100)	82 ± 5 (90)	83 ± 5 (132)	71 ± 14 (140)	66 ± 2 (115)	58 ± 7 (427)
0.3 mM	65 ± 5 (94)	50 (28)	82 ± 6 (95)	57 ± 6 (112)	79 ± 5 (155)	78 ± 3 (145)	64 ± 9 (163)	53 ± 3 (121)	26 ± 7[c] (219)
1.0 mM	65 ± 1 (91)	68 (28)	78 ± 3 (104)	60 ± 8 (124)	58 ± 3[c] (131)	67 ± 5 (150)	8 ± 7[c] (138)	7 ± 3[c] (109)	4 ± 2[c,d] (156)
F-value	1.41		1.36	1.22	5.00[e]	1.85	8.24[e]	18.68[e]	11.01[e]

[a]One experiment.
[b]Total ova observed in 4 to 5 experiments in parentheses.
[c]Significantly different from control.
[d]Concentration of CdCl$_2$ is 0.6 mM in this experiment.
[e]This F-value is significant at $p \leq 0.05$.

Table XI. Sister Chromatid Exchange (SCE) Response Following Exposure to Chlorinated Phenols[a]

Compound	Administration Route	Duration (h)	Response
2,3-DCP	IP	24	b
2,4-DCP	IP	24	b
2,4-DCP	Drinking water	24	b
2,5-DCP	IP	24	±[c]
2,6-DCP	IP	24	b
3,4-DCP	IP	24	b
PCP	IP	24	b

[a]Intraperitoneal (IP) injections are routinely used as a standard predrinking water exposure screen. This may become important when pharmacokinetic considerations are brought to bear on the problem. To date, a discrepancy between the data obtained by either route of exposure has not been found.
[b]The compound was negative in the SCE assay.
[c]To be considered a positive compound in the SCE assay, the results must exhibit a dose-response relationship and be 2 times greater than background. With 2,5-DCP, a dose response was established, but we were unable to achieve a twofold increase in the SCE level because of toxicity.

Table XII. Acute Oral Toxicity (LD_{50}) of Phenol and Chlorinated Phenols

	Mouse (mg/kg)		Rat (mg/kg)	
	Male	Female	Male	Female
Phenol	300 (3)[a]		530 (4)	
2-CP	347 (5), 670 (6)	345 (5)	670 (7)	
3-CP	521 (5)	530 (5)	570 (7)	
4-CP	1373 (5)	1422 (5)	670 (7), 261 (8)	
2,3-DCP	2585 (5)	2376 (5)		
2,4-DCP	1276 (5), 1630 (9)	1352 (5)	580 (7), 2830 (9), 4000 (11)	
2,5-DCP	1600 (5)	946 (5)		
2,6-DCP	2198 (5)	2120 (5)	2940 (8)	
3,4-DCP	1685 (5)	2046 (5)		
3,5-DCP	2643 (5)	2389 (5)		
2,3,5-TCP				
2,4,5-TCP			820 (7), 1620 (8)	
2,3,6-TCP				
2,4,6-TCP			820 (8)	
3,4,5-TCP				
2,3,4,5-TCP	400 (8)		140 (7)	
2,3,4,6-TCP				
2,3,5,6-TCP	109 (8)			
PCP	177 (5)	117 (5)	80 (7), 78 (9), 146 (10)	175 (10)

[a]References in parentheses.

The only effect reported was nonspecific microscopic liver changes. In our studies, the highest concentration that animals would accept was 2.0 mg/mL using 10% emulphor as the vehicle (resulting in a dose of 500 mg kg^{-1}d^{-1}). No significant biological effects were observed.

In Vitro Reproductive Effects

Eight chlorophenols were evaluated. Penetration was depressed by the 2,5-DCP, 3,4-DCP, and 3,5-DCP at the highest concentration tested (1.0 mM). 2,4-DCP was inactive in vitro. Sperm and ova were removed and evaluated from the animals exposed subchronically. There were no effects on penetration. The in vivo findings confirm the in vitro effects.

Sister Chromatid Exchange

Sister chromatid exchange rates of mice exposed either acutely or for 14 or 90 d were not above background rates.

SUMMARY AND CONCLUSIONS

Chlorophenols (except pentachlorophenol) demonstrate a relatively low order of toxicity. The order of toxicity in mice and rats (most to least) is PCP > tetra CPs > mono CPs > tri CPs > di CPs.

Short-term (14-d) repeated exposure to 2-CP at doses (gavage) of 35, 69, or 175 mg kg^{-1}d^{-1} (approximately 0.1, 0.2, 0.5 the acute oral LD$_{50}$) resulted in 100% lethality at the highest dose and no biologically significant compound-related effects at the lower doses. Short-term (14-d) repeated exposure to 2,4-DCP at doses (gavage) of 64, 128, or 638 mg kg^{-1}d^{-1} (approximately 0.05, 0.1, 0.5 the acute oral LD$_{50}$) failed to induce significant compound-related toxicity. The liver was identified as a possible target organ.

Subchronic exposure to 2,4-DCP at drinking water concentrations of 0.2, 0.6, and 2.0 mg/mL (limit of solubility and acceptability was 2.0 mg/mL) for 90 d (approximate doses of 50, 150, and 500 mg kg^{-1}d^{-1}) failed to induce significant compound-related toxicity. The data presented and a review of the available literature support the relatively low order of toxicity of most CPs ingested orally (except PCP).

REFERENCES

1. Finney, D. G. *Probit Analysis*, 3rd ed. (London: Cambridge University Press, 1971).
2. Litchfield, J. T., Jr., and F. Wilcoxon, *J. Pharmacol. Exper. Therap. 96*:99–113 (1949).
3. von Oettingen, W. F., and N. E. Sharpless. *J. Pharmacol. Exper. Therap. 88*:400 (1946).
4. Diechmann, W. B., and S. Witherup. *J. Pharmacol. Exper. Therap. 80*:233 (1944).
5. Borzelleca, J. F., J. R. Hayes, L. W. Condie, and J. L. Egle, Jr. *Toxicol. Appl. Pharmacol.* (in press).
6. Bubnov, V. D. *Tr. Vses. Nauchno-Issled. Inst. Vet. Savil. 33*:258, (1969), cited in *Ambient Water Quality Criteria Document for 2-Chlorophenol*, (Washington, DC: U.S. Environmental Protection Agency, 1980).
7. Deichmann, W. B.: *Fed. Proc. 2*:76 (1943).
8. Lewis, R. J., Jr. and R. L. Tatkew. *Registry of Toxic Effects of Chemical Substances,* (1979).
9. Vernot, E. H., J. D. MacEwen, C. C. Haun, and E. R. Kikhead. *Toxicol. Appl. Pharmacol. 42*:417 (1977).
10. Gaines, T. D. *Toxicol. Appl. Pharmacol. 14*:515 (1969).
11. Kobayashi, S., S. Toida, H. Kawamura, H. W. Chang, T. Fubuda, and K. Kawaguchi. *J. Med. Soc. Toho, Japan, 19*:356, (1972), cited in *Ambient Water Quality Criteria Document for 2,4-Dichlorophenol*, (Washington, DC: U.S. Environmental Protection Agency, 1980).

CHAPTER 27

Target Organ Effects of Disinfectants and Their By-products

Lyman W. Condie and J. Peter Bercz

Toxicological information relevant to drinking water disinfection practices has proliferated during the past decade. The subject of two recent conferences has been the complex issues involved in evaluating the health effects of drinking water disinfectants and their reaction by-products.[1,2] Although chlorine has been the primary disinfectant used in the United States, its widespread use has been questioned following the observation that chlorination resulted in the formation of mutagenic compounds such as trihalomethanes (THM).[3,4] Alternatives such as chlorine dioxide (ClO_2) and monochloramine were subsequently used at numerous water treatment facilities to prevent the unwanted generation of THMs. Within the last few years, research has been conducted to evaluate the possible health hazards arising from the use of alternative disinfectants.

Research on the adverse health effects of ClO_2 focused on hematological effects reported to be associated with ClO_2 and its inorganic by-products, chlorite (ClO_2^-) and chlorate (ClO_3^-).[5] While it was demonstrated that ClO_2^- could produce methemoglobinemia, additional research indicated that ClO_2^- produced hemolytic anemia in experimental animals at exposures lower than those required to produce significant increases in methemoglobinemia. Chlorine dioxide has also been demonstrated to exert an antithyroid effect in African green monkeys.[6] The mechanism of action of the antithyroid effect of ClO_2 is currently being investigated by us.

In addition to the potential health hazards from human exposure to residual disinfectants, risks from exposure to by-products from the reactions of disinfectants with organic material present in source water must be evaluated. A myriad of substances other than THMs are thought to arise from the oxidation and chlorination of the heterogeneous organic material present in source water (primarily humic and fulvic acids). Because the majority of these contaminants have not been identified, the health risks resulting from the ingestion of these compounds remain unknown.

Concern about the potential adverse health effects of disinfectant by-products has been increased by the recent scientific recognition of mutagenic activity exhibited by the uncharacterized organic compounds.[7] Direct support for the hypothesis that mutagenic chlorination by-products result from the reaction of chlorine with humic materials present in source water has been documented recently.[8,9] However, little is known about nonmutagenic target organ toxicities of chlorinated humic substances. This chapter presents data that summarize recent findings on subchronic toxicology studies with chlorinated and nonchlorinated humic acids.[10]

MATERIALS AND METHODS

In nonhuman primate studies, a healthy colony of *Cercopitheous aethiops* (African green monkey) consisting of seven females and four males were used. Conditions of animal husbandry, details of protocol for exposure to ClO_2, and clinical testing of in vivo systemic response were detailed by Bercz et al.[6]

Radioiodine uptake by the thyroid gland was determined as follows: After mild sedation the animal was chaired, and a bolus dose of ~ 0.5 to $1.0~\mu Ci$ of $Na^{131}I$ in pyrogen-free saline was injected via the saphenous vein. Organ counts were measured using a 2-cm γ-scintillation crystal at 2π geometry at a fixed and reproducible distance from the laryngeal region. Efficiency of the scintillation counts was determined by using a neck phantom. Organ counts were taken at fixed intervals (5 min to 24 h). Radioiodine uptake values were computed by dividing the corrected organ counts by the total administered dose.

Male Sprague-Dawley rats (250 g) were used in the rodent experiments. The conditions of husbandry and protocols for isolated organ iodination, for in vivo iodine transport, and for metabolism of radioiodine-treated feed were published by Harrington et al.[11] Covalently bound metabolites of radioiodinated feed were separated from rat urine by ion exchange chromatography, and the amount of organically bound radioactivity was expressed as percent of the total administered dose.

Solutions of commercially available humic acid and chlorinated humic solutions were prepared on a monthly basis by a previously described method.[9] Conditions of husbandry, exposure, and testing protocols have been described by Condie et al.[10] for the subchronic studies. Eighty rats were randomly assigned to five equal-exposure groups during the different experiments—distilled water, nonchlorinated (non-Cl) humic acid (1.0 g/L total organic carbon, TOC), and three chlorinated (Cl) humic groups (0.1, 0.5, and 1.0 g/L TOC). Experiments were conducted for 30 and 90 d.

Experimental observations were statistically evaluated by analysis of variance. When the analysis of variance indicated a dose-related effect, F-distribution was used to detect differences between treatment groups.

RESULTS

Thyroid Studies

Table I summarizes the decrease in serum thyroxine of the monkeys after a 6-week exposure to the various chlorine oxides and to chlorine. Table II provides the results of radioiodine uptake studies at the start, immediately after, and 6 months after cessation of exposure to 100-ppm ClO_2.

Chlorine dioxide treatment of isolated rat stomachs produces an increase in binding of ^{131}I to gastric tissue (Table III). Leaching each organ in saline for an additional 24 h did not liberate further radioactivity, indicating that the residual radioactivity bound to the tissue was not dialyzable.

Table I. Effect of Chlorine Dioxide, Chlorine, and Monochloramine in Drinking Water on Serum Thyroxine Levels of Nonhuman Primates[a]

Chemical	Drinking Water Conc (ppm)	Equivalent Dose (mg kg^{-1} d^{-1})	Average Shift from Control Values, % (n = 12)[b]	
			7 d	42 d
ClO_2	30	3		+ 4
ClO_2	100	9	− 6	− 26[c]
NH_2Cl	100	10		+ 9
Cl_2	125	10	+ 3	− 1
$NaClO_2$	400	58		− 11
$NaClO_3$	400	54		+ 8

[a] African green monkey (*Cercopithecus aethiops*).
[b] Range of control group means of serum thyroxine levels was 4.7 to 5.3 µg/dL; n = number of animals tested.
[c] Significant decrease, $p < 0.05$.

Table II. ^{131}I Thyroid Gland Uptake (Over 24-h Period) in Female Monkeys (n = 11)[a]

Condition	Mean Uptake (% ± S.E.)
Preexposure	
Over 1 year with tap water	9.15 ± 1.14
90 d	
100-ppm ClO_2 in drinking water	15.48 ± 1.32[b]
90-d Recovery period	10.23 ± 0.57

[a] n = Number of animals tested.
[b] Significant increase, $p < 0.01$.

Administration of radioiodine, after it is bound to feed nutrients by prior treatment with ClO_2 in vitro, resulted in more pronounced differences in gastrointestinal retention of ^{131}I (Table IV). After a 24-h digestion period, the gut retained significantly larger portions of the radioisotope in the animals receiving ClO_2-treated chow. Fecal elimination of the radioiodine was found to be significantly increased by ClO_2 treatment. Commensurate with the high fecal clearance, the amount of ^{131}I excreted in the urine decreased and the organically bound urinary iodine was tenfold greater.

Studies with Humic Acids

All animals survived the 30-d exposure period and appeared to tolerate the drinking of chlorinated and nonchlorinated humic substances. Food and fluid consumption, weight gain, organ weights, biochemical measurements, and most clinical chemistry and hematologic parameters were similar in control and treated animals (Table V). The most striking result of this experiment was the occurrence of hematuria in treated animals. Therefore, a 90-d subchronic

Table III. Effect of ClO_2 on Iodine Binding to Whole Isolated Stomach[a]

Dose of ^{131}I Bound to Stomach (%/g)	
Controls (n = 5)[b]	ClO_2 treated (n = 5)[b]
0.32 ± 0.04	2.10 ± 0.22[c]

[a]Final concentration of ClO_2 in stomachs is about 15 ppm.
[b]n = number of animals tested.
[c]Data are expressed as mean values ± S.E. and are analyzed by Student's t-test; significant increase, $p < 0.01$.

Table IV. Gastrointestinal Distribution and Elimination Pattern of ^{131}I in Monkeys Fed High-Protein Chow Exposed to ClO_2 In Vitro[a]

	Feed	
	With ^{131}I	With ^{131}I and ClO_2
Stomach and contents	1.6 ± 0.3	1.1 ± 0.2
Intestine and contents	0.76 ± 0.11	1.2 ± 0.2[b]
Colon and contents	0.69 ± 0.09	2.3 ± 0.5[b]
Thyroid gland	14.0 ± 2.3	10.5 ± 1.6
Fecal collection, 24h	1.9 ± 0.6	17.4 ± 1.1[b]
Urine collection, 24 h	44.4 ± 4.1	34.8 ± 2.7
Amount organically bound ^{131}I in urine	0.24 ± 0.5	2.2 ± 0.4[b]

[a]Each value represents the mean ± S.E. of the percent of total administered ^{131}I retained per organ or excreted in 24 h; 11 animals tested.
[b]Significant increase, $p < 0.001$.

Table V. Summary of Negative Observations Following 30-d Exposure to Chlorinated and Nonchlorinated Humic Substances

Body weights	Clinical chemistry
Organ weights	Creatinine
Organ/body weights	Albumin
	Protein
	SGPT
	BUN
	LDH
	SGOT
Food consumption	Biochemical measurements
	Hepatic microsomal cytochrome P-450
Fluid intake	Renal slice uptake of P-aminohippurate
Hematology	Urinalysis
WBC	PH
MCV	Glucose
HCT	Ketones
HB	Bilirubin
	Urobilinogen

study was conducted to further evaluate the relationship between chlorinated humic substance exposure and production of hematuria in Sprague-Dawley rats.

Significant results of the 90-d study are summarized as follows: The average weight gain was significantly decreased in the high-chlorinated humic acid group when compared with controls. The only organ weight that was significantly larger than the corresponding control value was the kidney weight of the 1.0-Cl humic group. There were no significant differences between groups in regard to hematological parameters. Although there were statistically significant differences in some of the clinical chemistry values, the changes were slight and were within the normal distribution range. The microscopic observations and chemical analyses of the urine collected at approximately 30 and 90 d of the experiment were insignificant except for the occurrence of hematuria and the demonstration of renal pelvic crystals in many of the hematuric rats. Hematuria was most pronounced in the 1.0-Cl humic group at both 30 and 90 d. The incidence and severity of hematuria was greater at 90 d when compared with the 30-d period.

A thorough histopathological evaluation was performed on the kidneys and urinary bladders from the rats killed following the 30- and 90-d studies, as well as on the entire urinary tract following the second 90-d subchronic exposure. A few changes, such as chronic inflammation and epithelial hyperplasia, were observed in the urinary bladders of these rats; however, these changes were not considered unusual for Sprague-Dawley rats and were interpreted as spontaneous incidental lesions. Renal changes including nephropathy, interstitial inflammation, and pyelonephritis were observed in each group; however, these changes did not appear to be treatment related. The cause of the hematuria could not be explained by the histopathological examination, but may be related to renal crystal formation.

DISCUSSION

A statistical analysis of each monkey's serum thyroxine levels following exposure to various disinfectants indicated that only ClO_2 affected thyroid hormone homeostasis. The lack of thyroid effects of the other chlorine oxides, including that of perchlorate (a known thyroid suppressant), indicates that chlorine dioxide is a unique inhibitor of thryoid metabolism.

Thyroid suppression by perchlorate salts is known to occur via competitive inhibition of iodine uptake into the thyroid gland,[12,13] which requires a certain level of saturation by perchlorate. Since we have demonstrated that ClO_2 is rapidly destroyed in contact with saliva and with stomach contents (over 98% reduced to chloride within seconds),[6] one may conclude that the effect of ClO_2 on the thyroid gland is due neither to its absorption nor to the absorption of reduction products that could subsequently inhibit iodine uptake in a manner analogous to perchlorate.

The near doubling of radioiodine uptake at the end of subchronic exposure implies that the monkeys developed a degree of iodine deficiency during the subchronic ingestion of ClO_2. The effect, however, shows clear signs of reverting back to normal after cessation of exposure. Based on the data presented here and previously,[11] it is possible to postulate a plausible mechanism by which iodine deficiency could develop.

Chlorine dioxide is a very potent oxidizing species possessing a free radical structure

$$Cl \overset{\displaystyle \nearrow O}{\searrow_O} \quad \bullet$$

its redox potential (1.9 V) being the highest among the chlorine oxides, which is about 1 V above that of hypochlorite. Because of its exaggerated chemical potential, this disinfectant reacts in the alimentary canal with inorganic iodide of nutritional origin. The elemental iodine in turn reacts with organic molecules present in saliva and with the stomach contents. A few plausible organic substances that could covalently add iodine across reactive double bonds or in complex polyalcohol matrices are:

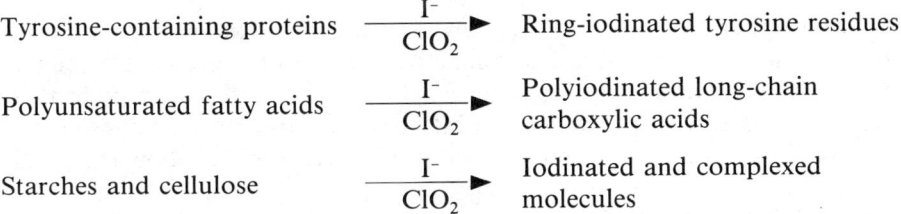

Since iodine (I⁻) is the only form of iodine that can be used for thyroxine synthesis,[14] a steady-state chronic shift of the inorganic iodide pool into covalently bound forms could establish a bona fide iodine deficiency, even though many of the iodinated molecules may be absorbed.

In fact, our experiments with isolated rat stomachs (Table III), with in vivo administration of radioiodine and ClO_2,[11] and in vivo metabolism of in vitro radioiodinated nutrients (Table IV) provide support for this hypothesis. These studies have demonstrated that when chlorine dioxide and radioiodine are instilled into the stomach in a single acute dose, the radiolabel becomes partly attached to the gastrointestinal wall and contents. When ClO_2, radioiodine, and feed are premixed and then administered to rats, a very sizable portion of the radiolabel is lost via the feces unabsorbed. Furthermore, a significant portion (about 2%) of the urinary iodine appears as organically bound.

In the evaluation of the health impact of ClO_2, the potential for in vivo formation of iodinated organics should be a focus of concern. This could be analogous to the in vivo formation of chlorinated organics during oral ingestion of hypochlorite disinfected water.[15] Our present hypothesis is that some of these iodinated compounds may be thyroid suppressors acting by enzymatic or feedback inhibition on T_4 synthesis. Thus, the endocrinological effect of ClO_2 may be the result of a combination of several factors. Since the other disinfectants are themselves oxidizing agents, it is possible that they also give rise to similar iodination reactions, albeit to a much lesser extent. Currently, the question of in vivo formation of iodinated organics, their chemical and toxicological characterization, as well as measurement of iodide deficiency are under detailed investigation in our laboratory.

It is not surprising that drinking water disinfectants can react with humic material to produce a diverse group of by-products. A key issue the Health Effects Research Laboratory (HERL) is currently addressing is whether chemicals produced through disinfection of drinking water have significant biological activity. Our laboratory has published research findings that the chlorination of humic substance yields mutagenic by-products.[8,9] Because human health risk assessment cannot be solely estimated by in vitro test results, subchronic exposures were conducted with chlorinated and nonchlorinated humic acid solutions.

The most significant finding of these studies was the increased incidence and severity of hematuria in the 1.0-Cl humic group. Increased kidney weight in the 1.0-Cl humic group may represent a response to a subchronic insult, possibly from irritation from renal crystals. Microscopic examination of freshly removed kidneys from the 1.0-Cl humic rats revealed the presence of numerous cyrstalline bodies in the renal pelvis. Unfortunately, the quantity of these materials isolated was insufficient for infrared spectroscopy. Perhaps the renal crystals from chlorinated humic acids are accelerating idiopathic renal changes.

This study was designed to determine if chlorination of humic acid material, which is found in finished drinking water, presents a health problem. The highest no-observable-effect level in this study was 500 mg/L TOC, which represents a safety factor to the human population of least 100-fold, since 95% of the municipal drinking water systems tested in a national survey have TOC levels less than 5 mg/L. Other studies are under way at HERL to better understand the relative hazards of drinking water chlorination and its various alternatives (ie., chlorine dioxide, chloramines, ozone).

ACKNOWLEDGMENTS

We wish to thank Dana Laurie, Robert Harrington, John Glass, Carolyn Smallwood, and Lillian Jones for their excellent assistance in this research, and Patricia Underwood for accurately typing the manuscript.

REFERENCES

1. "International Symposium on Health Effects of Drinking Water Disinfectants and Disinfectant By-products," *Environ. Health Perspect.* 46:1-241 (1982).
2. Jolley, R. L., W. A. Brungs, J. A. Cotruvo, R. B. Cumming, J. S. Mattice, and V. A. Jacobs, Eds. *Water Chlorination: Environment Impact and Health Effects, Vol. 4* (Ann Arbor, MI: Ann Arbor Science Publishers, 1983), pp. 1067-1491.
3. Bellar, T. A., J. J. Lichtenberg, and A. D. Kroner. "The Occurrence of Organohalides in Chlorinated Drinking Water," *J. Am. Water Works Assoc.* 66:703-706 (1974).
4. Rook, J. J. "Formation of Haloforms During Chlorination of Natural Waters," *J. Water Treat. Exam.* 22:234-243 (1974).
5. Couri, D., M. S. Abdel-Rahman, and R. J. Bull. "Toxicological Effects of Chlorine Dioxide, Chlorite and Chlorate," *Environ. Health Perspect.* 46:13-17 (1982).
6. Bercz, J. P., L. Jones, L. Garner, D. Murray, D. A. Ludwig, and J. Boston. "Subchronic Toxicity of Chlorine Dioxide and Related Compounds in Drinking Water in the Nonhuman Primate," *Environ. Health Perspect.* 46:47-55 (1982).
7. Loper J. C. "Mutagenic Effects of Organic Compounds in Drinking Water," *Mutation Res.* 76:241-268 (1980).
8. Bull, R. J., M. Robinson, J. R. Meier, and J. Stober. "The Use of Biological Assay Systems to Assess the Relative Carcinogenic Hazards of Disinfection By-Products," *Environ. Health Perspect.* 46:215-227 (1982).

9. Meier, J. R., R. D. Lingg, and R. J. Bull. "Formation of Mutagens Following Chlorination of Humic Acid: A Model for Mutagen Formation During Drinking Water Treatment," *Mutation Res.* 118:25-41 (1983).
10. Condie, L. W., R. D. Laurie, and J. P. Bercz. "Subchronic Toxicology of Humic Acid Following Chlorination in the Rat," (Submitted for publication).
11. Harrington, R., H. Shertzer, and J. P. Bercz. "Effects of ClO_2 on the Absorption and Distribution of Dietary Iodide in the Rat," *Fundamen. Appl. Toxicol.* (in press).
12. Bobek, S., and S. Kahl. "Effect of Perchlorate and Thiocyanate Anions on Protein Bound Iodine and Binding of Exogeneous Thyroxine by Blood Plasma Proteins in Rats in In Vivo and In Vitro Investigations," *Endokrynol. Pol.* 24:21-31 (1973).
13. Broadhead, G., I. Pearson, and G. Wilson. "The Effect of Prolonged Feeding of Goitrogens on Thyroid Function in the Rat," *J. Endocrinol.* 32:341-345 (1965).
14. Brown, G. K. "Extrathyroidal Iodine Concentrating Mechanisms," *Physiol. Review* 41:189-213 (1961).
15. Mink, F. L., W. E. Coleman, J. W. Munch, W. H. Kaylor, and H. P. Ringhand. "In Vivo Formation of Halogenated Reaction Products Following Peroral Sodium Hypochlorite," *Bull. Environ. Contam. Toxicol.* 30:394-399 (1983).

CHAPTER 28

Effects of Chlorine Dioxide on Neurobehavioral Development of Rats

Douglas H. Taylor and Ronald J. Pfohl

The production of carcinogenic trihalomethanes[1,2] in drinking water disinfected with chlorine is cause for concern and has led to a search for an alternative drinking water disinfectant. One such alternative is chlorine dioxide (ClO_2). The potential toxicity of ClO_2 and its metabolites (chlorate and chlorite ions) is therefore of interest from a human health perspective. Recent work indicates that exposure to relatively low levels of ClO_2^- (50 mg/L in drinking water) significantly depletes glutathione and increases 2,3-diphosphoglycerate in the red blood cells of rats.[3] Abdel-Rahman et al.[4] found similar results with ClO_2 and ClO_3^-. Suh et al.[5] found that ClO_2 exposure (100 mg/L in drinking water) resulted in depressed DNA synthesis in the testis and intestinal epithelium of rats. African green monkeys (*Cercopitheous aethiops*) exhibited reduced levels of thyroid metabolism as a result of exposure to 100-ppm ClO_2 administered in their drinking water (9 mg kg^{-1} d^{-1}).[6]

The potential role of ClO_2 as an antithyroid agent has recently been tested by Orme et al.[7] Exposure to ClO_2 either indirectly, via the dam's drinking water (100 ppm) from 14 d prior to breeding until the pups reached 21 d of age, or directly, by oral gavage from 5 through 20 d of age (14 mg kg^{-1} d^{-1}), resulted in significantly depressed levels of serum thyroxine (T_4) in 21-d-old pups.

Since ClO_2 appears to exert an antithyroid effect,[7] and since thyroid function is intimately involved with the control of neurobehavioral development,[8-10] the current study was designed to evaluate the effects of ClO_2 exposure on selected behavioral and brain growth processes. In particular, we report here the effects of ClO_2 exposure on the development of locomotor activity and exploratory behavior, and on the cell numbers and densities in the cerebellum and forebrain of rats.

MATERIALS AND METHODS

Exposure Paradigms

Rat pups were derived from Sprague-Dawley dams. At birth all litters were culled to eight pups (all males if possible). Only the males were used for experimental purposes.

Indirect Exposure

Rat pups were derived from dams that received 100-ppm ClO_2 in their drinking water from 14 d prior to breeding until the pups were weaned at 21 d of age. Fresh ClO_2 solutions were prepared[6] at 2-d intervals to maintain relatively constant concentrations. Exposure of pups to ClO_2 was terminated at weaning (21 d of age).

Direct Exposure

Rat pups derived from unexposed dams were given a daily ClO_2 dose of 14 mg/kg body wt in water from age 5 to 20 d postpartum by oral gavage. Control pups received an appropriate amount of distilled water.

Dam food and water consumption, as well as dam and pup body weights, were monitored weekly for both the indirect- and direct-exposure groups.

Behavioral Measures

Locomotor Activity

Two different measures were used: (1) A home cage apparatus first described by Croften et al.[11] was used by Orme et al.[7] to evaluate the development of locomotor activity in young rat pups that were exposed to ClO_2 indirectly (via their dams) or directly (by oral gavage) from 5 through 20 d of age. Individual dams and their litters were placed in cages to which the dam was restricted. The activity of the litter of pups was measured when the pups crossed into a small compartment via small connecting holes and interrupted the path of an infrared photobeam that transected the smaller compartment. The unit of measure in this system was the activity of the litter, which was monitored continuously by a computer. The data from the activity cages were analyzed by analysis of variance procedures.

(2) A residential cage system[12] was used to monitor feeding, drinking, and running-wheel activity of individual rats from 50 through 60 d of age. Thus, rats were tested in this apparatus beginning at 29 d after cessation of ClO_2 dosing in both exposure paradigms. The data were collected continuously for 10 d and were analyzed by repeated measures analyses of variance.

Exploratory Behavior

The behavior of 28- and 60-d-old rats was examined in an open-field apparatus during a 15-min sampling period. Specifically, exploratory behavior was evaluated by monitoring the activity of each rat in a small alcove provided with small openings that served to stimulate investigatory behavior. Infrared photobeams traversed each opening, and a computer monitored the duration and frequency with which the beams were broken. One-way analyses of variance and Kruskal-Wallace tests were used to analyze the data.

Biochemical Measures

At either 11 or 21 d postpartum, rats from each exposure paradigm were decapitated. The cerebellum (minus the paraflocculi) and the forebrain (minus olfactory lobes) were quickly removed, weighed on a Mettler PC 180 balance, and frozen at $-80°C$.

Individual cerebella and forebrains were homogenized in deionized water (Milli-Q Reagent Water System). For the DNA analyses, duplicate 1-mL samples of the homogenates were precipitated with 1-mL cold trichloroacetic acid (TCA). DNA was extracted from the precipitates by hydrolysis in 5% TCA for 15 min at 90°C.[13] Following centrifugation, supernatant fluids were assayed for deoxypentose content with the diphenylamine reagent of Burton.[14] DNA concentrations were estimated by reference to standards prepared from salmon testis DNA (Type III, sigma). A value of 6.2-pg DNA per diploid rat nucleus was used for calculating cell numbers.[15]

RESULTS

Indirect Exposure Group

Behavior

There were no significant differences in body weight of the dams exposed to 100-ppm ClO_2 in their drinking water nor in the body weights of pups born to these dams, as compared with appropriate control animals.

We found that the locomotor activity of the developing pups (10 through 20 d of age) exposed to 100-ppm ClO_2 via their dams was consistently lower than control pups when tested in the home cage.[7] These differences, however, were not statistically different from the controls ($p = 0.08$; N = 13 control litters; N = 16 litters of ClO_2-exposed rats). In terms of exploratory behavior, these animals were not significantly different from controls when tested at 28 d of age. However, at 60 d of age they exhibited significantly depressed levels of exploration, as compared with controls ($p < 0.05$ by Anova; N = 12 for controls; N = 11 for the ClO_2 group).

Body and Brain Weights
(21-day-old pups)

No significant differences were detected in the body weights of exposed vs control animals. Exposure to ClO_2, however, resulted in a significant decrease in the mean whole-brain weight ($p < 0.05$; ClO_2 group = 1.39 g, N = 14; controls = 1.43 g, N = 15). This decrease is primarily accounted for by a reduction in cerebellar weight ($p = 0.0001$; ClO_2 group = 0.170 g, N = 14; controls = 0.186 g, N = 15). Forebrain weights did not differ.

DNA Content (21-d-old pups)

The total DNA content of the cerebella was significantly lower in ClO_2-exposed animals (938 µg, N = 12) than in controls (1008 µg, N = 12; $p < 0.005$). The difference of 70 µg represents a deficit of 11 to 12 × 10^6 in the total number of cells (7.4% reduction; Figure 1). However, the density of cells in the cerebella of ClO_2-exposed animals, based on DNA per gram of tissue, did not differ from controls. The total number of cells and cells per gram of forebrain tissue were the same in exposed (N = 10) and control animals (N = 10; Figure 1).

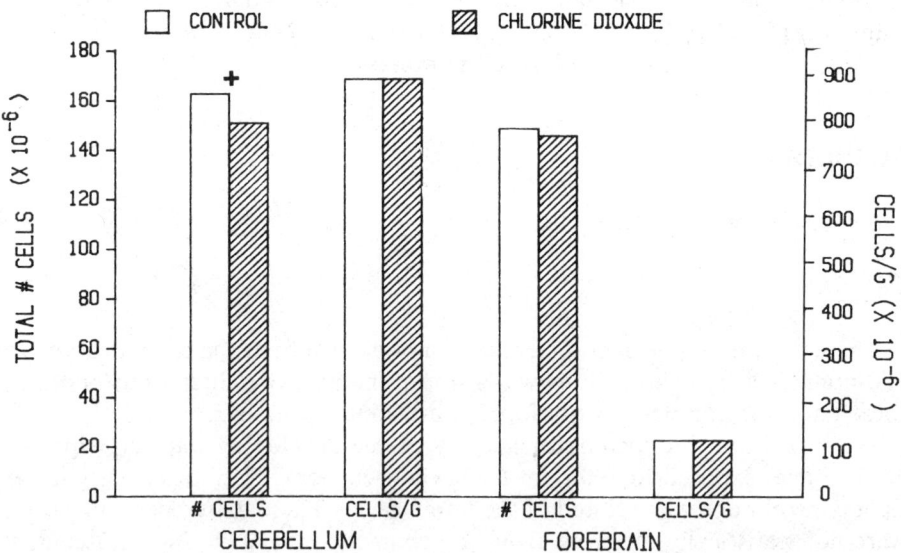

Figure 1. Cell number and density in cerebellum and forebrain of 21-d-old rats indirectly exposed to ClO_2 prenatally and postnatally. Pairs of means denoted with + are significantly different from each other.

Direct Exposure Group

Behavior

Orme et al.[7] found that at 14 d of age, the pups directly exposed to ClO_2 by gavage weighed significantly less than control pups (20 vs 24 g respectively). By 21 d of age, the difference between the ClO_2 and control animals was even more marked (31 g for ClO_2 and 46 g for control animals) and represented a significant effect ($p < 0.05$).

The pups directly exposed to ClO_2 showed a significant depression in total activity for days 18 and 19 postpartum in the home cage apparatus ($p < 0.05$, N = 18 litters for the ClO_2-exposed group and 15 for controls).[7] Animals randomly selected from these two groups were placed in residential cages for 10 d at 50 d of age. The ClO_2-treated rats exhibited significantly depressed wheel running activity during the night (Figure 2; $p < 0.05$; N = 7 for ClO_2-treated rats, N = 8 for control rats). Significant treatment effects on drinking

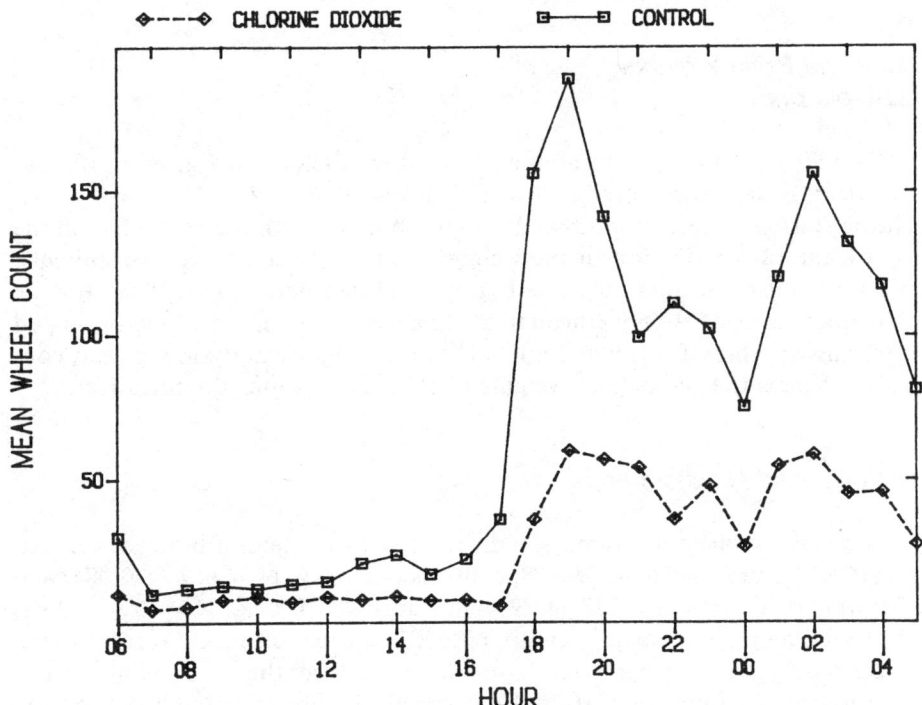

Figure 2. Mean locomotor activity (wheel running) of rats from 50 through 60 d of age exposed to ClO_2 by oral gavage from 5 through 20 d of age. Means for all days collapsed to a single 24-h period.

behavior were also detected between the ClO_2 and control rats ($p < 0.05$). Although the ClO_2-treated rats were more active at the water bottle than the controls, there were no significant differences in water consumption. No treatment differences were detected in feeding behavior or food consumption.

At 28 d of age, the ClO_2-treated animals exhibited significantly depressed exploratory behavior compared with controls ($p < 0.05$; N = 15 for both groups). Littermates of these animals exhibited no significant differences in exploratory behavior when tested at 60 d of age.

Body and Brain Weights (11-d-old pups)

The body weights of animals gavaged with ClO_2 tended to be lower than those of controls; the difference in mean body weights approached significance ($p = 0.08$; ClO_2 group = 21.8 g, N = 8; controls = 24.3 g, N = 9). In general, the whole brain and brain part weights also tended to be lower in the exposed animals than in the control animals; in the whole brain and cerebellum, the differences approached significance ($p = 0.06$).

Body and Brain Weights (21-d-old pups)

The mean body weight of animals treated with ClO_2 (38.2 g, N = 10) was significantly depressed compared with controls (48.6 g, N = 11; $p = 0.0001$). The deleterious effect of intubated ClO_2 on body growth is also evident in the significant 12% reduction in the weight of brain tissue of exposed animals. When the mean weights of whole brains and brain parts are adjusted for the differences in body weight (through covariance analysis), the whole-brain and forebrain weights of intubated animals are still significantly lower than controls, whereas the cerebellum weights of these two groups do not differ.

DNA Content (11-d-old pups)

Compared with control animals, the total DNA content of both the cerebellum (ClO_2 group = 476 μg, N = 8; controls = 578 μg, N = 8; $p < 0.004$) and forebrain (ClO_2 group = 728 μg, N = 6; controls = 800 μg, N = 9; $p = 0.01$) of treated animals was significantly reduced. Intubation of ClO_2 from day 5 through day 11 of postnatal development thus reduces the total cell number in the cerebella by more than 16×10^6 cells (17.6% reduction) and in the forebrain by 12×10^6 cells (9.3% reduction; Figure 3). The number of cells per gram of tissue was not affected.

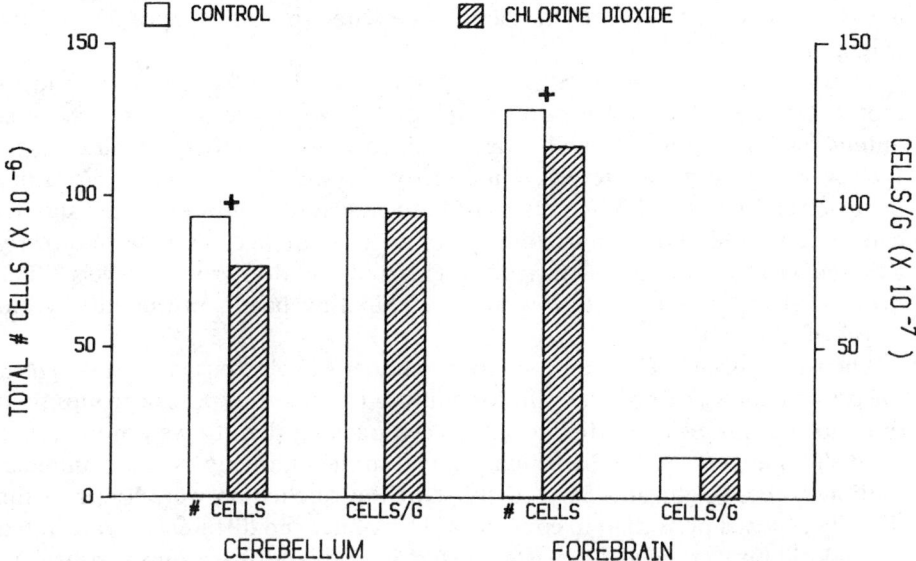

Figure 3. Cell number and density in cerebellum and forebrain of 11-d-old rats exposed to ClO_2 by oral gavage beginning at 5 d of age. Pairs of means denoted with + are significantly different from each other.

DNA Content (21-d-old pups)

No significant treatment effect on the DNA content of the cerebellum was found. The total DNA in forebrains of treated animals (848 μg, N = 10) was significantly lower than in control animals (937 μg, N = 10; p = 0.0001). The forebrains of treated animals therefore contained 14 × 10⁶ fewer cells than forebrains of controls for a 9.3% reduction. No differences, however, were noted in the number of cells per gram of tissue.

DISCUSSION

These data indicate that rat pups exposed to ClO_2, both prenatally and postnatally, until 21 d of age, and those exposed postnatally from 5 through 20 d of age, exhibited behavioral deficits and depressed brain growth consistent with effects produced by depressed thyroid function. Neonatal hypothyroidism results in depressed thyroid hormone levels and is characterized by a delay in the age of eye opening, decreased body, whole brain, and cerebellum weights, and decreased activity.[8,16-19] Both exposure groups in the current study exhibited delays in the development of locomotor activity, as measured in the home-cage apparatus. In the case of the rats exposed to ClO_2 by oral gavage, this depressed locomotor activity persisted through at least 60 d of age (Figure 3).

To date, only directly exposed rats have been evaluated for long-term locomotor activity effects; therefore, it is not known if the indirectly exposed animals will continue to exhibit depressed locomotor activity at older ages. However, rats exposed to ClO_2 indirectly do show depressed exploratory behavior at 60 d of age. Littermates of both exposure groups exhibited significantly depressed serum thyroxine levels at 21 d of age, and the indirectly exposed group exhibited significantly elevated triiodothyronine levels.[7] The behavioral data are consistent with an explanation based on the antithyroid effect of ClO_2.

The whole-brain weight and the total number of cells, based on total DNA measurements, was lower in brains of animals from both exposure groups than in controls; however, no differences in cell packing density were noted. The cerebellum appears to be the primary target in the indirectly exposed animals. Although the weight and DNA content of the cerebellum were lower in the directly exposed pups than in controls at 11 d of age, no differences were noted in 21-d-old pups; the most persistent effect in this exposure group appeared to be on the forebrain. The reasons for this differential effect of ClO_2 remain unresolved.

A decreased rate of cell acquisition and a prolonged period of cell proliferation are observed in the hypothyroid cerebellum.[8] Thus, the time of onset of neuronal differentiation (marked by termination of cell proliferation) may be delayed. A reduction in the number of functional cerebellar neurons and an increase in the number of glial cells may occur as a consequence.[20] Lauder[8] concludes that the thyroxine plays an important role in the timing, rate, and quantity of cell proliferation. Although opposing views are held relative to the influence of thyroxine on brain development during the fetal period,[9,21] there is general agreement that deprivation of thyroid hormone during a critical period of postnatal development can lead to defects in the development of some central nervous system components, particularly the cerebellum. Depressed levels of thyroxine during early development could therefore have profound and lasting effects on the animal's behavior.

In addition to the lower brain cell numbers and the behavioral deficits in animals from both ClO_2 exposure regimes, the animals exposed only postnatally also had significantly lower body weights.

Our data support the view that ClO_2 exposure may adversely affect thyroid function and that the time of exposure (i.e., prenatal/postnatal vs postnatal) may result in differential effects. The results clearly indicate that ClO_2 exposure affects neurobehavioral development in the rat and that the effects are complex. Cross-fostering studies are in progress to evaluate the effects of prenatal and postnatal exposure to ClO_2 on serum thyroxine levels, brain development, and long-term behavioral processes in rats.

ACKNOWLEDGMENTS

This research was supported by a Cooperative Agreement (CR809618-01-1) with the Health Effects Research Laboratory, U.S. Environmental Protection Agency, Cincinnati. We wish to thank K. LaGory, D. Wines, and P. Wuest for their skilled assistance in conducting this research.

REFERENCES

1. Safe Drinking Water Committee. *Drinking Water and Health, Vol. 1* (Washington, DC: National Academy of Science, 1977), pp. 715-716.
2. Symons, J. M., J. K. Carswell, R. M. Clark, P. Dorsey, E. E. Geldreich, W. P. Heffernan, J. C. Hoff, O. T. Love, Jr., L. J. McCabe, and A. A. Stevens. "Ozone, Chlorine Dioxide, and Chloramines as Alternatives to Cl_2 for Disinfection of Drinking Water: State of the Art," in *Water Chlorination: Environmental Impact and Health Effects, Vol. 2*, R. L. Jolley, H. Gorchev, and D. H. Hamilton, Jr., Eds. (Ann Arbor, MI: Ann Arbor Science Publishers, Inc., 1978), pp. 555-560.
3. Heffernan, W. P., G. Guion, and R. J. Bull. "Oxidative Damage to the Erythrocyte Induced by Sodium Chloride, in Vivo," *J. Environ. Pathol. Toxicol.* 2:1487-1499 (1979).
4. Abdel-Rahman, M. S., D. Couri, and R. J. Bull. "Kinetics of ClO_2 and Effects of ClO_2, ClO_2^-, and ClO_3^- in Drinking Water on Blood Glutathione and Hemolysis in Rat and Chicken," *J. Environ. Pathol. Toxicol.* 3:431-449 (1980).
5. Suh, D. H., M. S. Abdel-Rahman, and R. J. Bull. "Biochemical Interactions of Chlorine Dioxide and Its Metabolites in Rats," *Arch. Environ. Contam. Toxicol.* 13:163-169 (1984).
6. Bercz, J. P., L. Jones, L. Garner, D. Murray, A. Ludwig, and J. Boston. "Subchronic Toxicity of Chlorine Dioxide and Related Compounds in Drinking Water in the Nonhuman Primate," *Environ. Health Perspect.* 46:47-55 (1982).
7. Orme, J., D. H. Taylor, R. D. Laurie, and R. J. Bull. "Effects of ClO_2 on the Neurobehavioral Development of Rats," *J. Toxicol. Environ. Health* (in press).
8. Lauder, J. M. "Effects of Thyroid State on Development of Rat Cerebellar Cortex," in *Thyroid Hormones and Brain Development*, G. D. Grave, Ed. (New York: Raven Press, 1977), pp. 235-254.
9. Hamburgh, M. "An Analysis of Thyroid Hormone on Development Based on in Vivo and in Vitro Studies," *Gen. Comp. Endocrinol.* 10:198-213 (1968).
10. Sokoloff, L. "Biochemical Mechanisms of the Action of Thyroid Hormones: Relationship to Their Role in Brain," in *Thyroid Hormones and Brain Development*, G. D. Grave, Ed. (New York: Raven Press, 1977), pp. 73-91.
11. Crofton, K. M., D. H. Taylor, R. J. Bull, D. J. Sivulka, and S. D. Lutkenhoff. "Developmental Delays in Exploration and Locomotor Activity in Male Rats Exposed to Low-Level Lead," *Life Sci.* 26:823-831 (1980)
12. Doran, S. M. "Modification of Behavioral Activity in Laboratory Rats Exposed to Trimethyltin, Dimethyltin, and Trichloroethylene," Master's Thesis (Oxford, OH: Miami University, 1983).

13. Schneider, W. C. "Phosphorous Compounds in Animal Tissues. I. Extraction and Estimation of Desoxypentose Nucleic Acid and of Pentose Nucleic Acid," *J. Biol. Chem.* 161:293-303 (1945).
14. Burton, K. "A Study of the Conditions and Mechanisms of the Diphenylamine Reaction for the Colorimetric Estimation of Deoxyribonucleic Acid," *Biochem. J.* 62:315-322 (1956).
15. Enesco, M., and C. P. Leblond. "Increase in Cell Number as a Factor in the Growth of the Organs of the Young Male Rat," *J. Embryol. Exp. Morphol.* 10:530-562 (1962).
16. Eayers, J. "Thyroid and Developing Brain: Anatomical and Behavioral Effects," in *Hormones in Development*, M. Hamburgh and E. Barrington, Eds. (New York: Appleton-Century-Crofts, 1971), pp. 345-355.
17. Shapiro, S. "Hormonal and Environmental Influences on Rat Brain Development and Behavior," in *Brain Development and Behavior*, M. Sterman, D. McGinty and A. Adinolfi, Eds. (New York: Academic Press, Inc., 1971), pp. 307-333.
18. Hamburgh, M., L. Mendoza, I. Bennett, P. Krupa, Y. Kim, R. Kahn, K. Hogreff, and H. Frankfort. "Some Unresolved Questions in the Brain Thyroid Relationship," in *Thyroid Hormones and Brain Development*, G. Grave, Ed. (New York: Raven Press, 1977), pp. 49-72.
19. Schalock, R. L., W. J. Brown, and R. L. Smith. "Long-Term Effects of Propylthiouracil-Induced Neonatal Hypothyroidism," *Dev. Psychobiol.* 12(3):187-199 (1979).
20. Nicholson, J. L., and J. Altman. "The Effects of Early Hypo- and Hyperthyroidism on the Development of Rat Cerebellar Cortex. I. Cell Proliferation and Differentiation," *Brain Res.* 44:13-23 (1972).
21. Oklund, S., and P. S. Timiras. "Influences of Thyroid Levels in Brain Ontogenesis in Vivo and in Vitro," in *Thyroid Hormones and Brain Development*, G. Grave, Ed. (New York: Raven Press, 1977), pp. 33-45.

CHAPTER **29**

Effect of Chlorinated Drinking Water on Myocardial Structure and Functions in Pigeons and Rabbits

N. W. Revis, T. R. Osborne, G. Holdsworth, and P. McCauley

Chlorination of foods and drinking water has been associated with the formation of a variety of toxic compounds,[1-3] which may include chlorinated nucleic acids, amino acids, lipids, and other organic compounds. Several chlorinated nucleic acids are mutagens, whereas compounds such as trihalomethanes are hepatotoxins.[4-7] Chlorine reacts with the double bond in fatty acids and cholesterol producing chlorinated derivatives. These derivatives are absorbed from the gastrointestinal tract and metabolized by a variety of organs.[8]

Significant increases in heart weight occur in rats given chlorinated fatty acids. Furthermore, we observed hypercholesterolemia in pigeons and rabbits exposed to chlorinated water. Collectively these results suggest that chlorine or chlorination by-products may adversely affect the cardiovascular system through the well-known association of plasma cholesterol levels with the onset of atherosclerosis and of cardiac hypertrophy with hypertension.

The present studies were performed to further explore this suggestion. Specific research objectives were to determine the effect of chlorinated water on blood pressure and on the contractility and morphology of the heart.

METHODS

Male white Carneau pigeons and New Zealand white rabbits, 3 months of age, were observed for infections for 20 d prior to treatment. The effect of chlorinated drinking water on the cardiovascular system was determined by exposing groups of five pigeons to drinking water containing 0, 0.1, 10, 15 or 30 mg/L (ppm) chlorine for 9 months (or rabbits for 3 months). Animals were given a calcium-deficient diet (80% minimum daily requirement) and the drinking water ad libitum. Chlorine was added as sodium hypochlorite to deionized water at the indicated concentrations, and the drinking water used for exposure was prepared three times per week.

After 9 months of exposure, 5 pigeons from each group were restrained and a cannula (PE 50) was introduced into the brachial artery. After recording blood pressure, the cannula was advanced into the left ventricle and left ventricular pressure and pressure-time (dp/dt) measurements were made. Pressure was measured through the cannula by a strain-g pressure transducer (Stratham, P23ID). Body weight was recorded after pigeons were killed by injecting air. The hearts were removed, flushed with 0.9% saline solution, blotted, and weighed. Connective tissue and fat were removed before the hearts were weighed. Sections were made (approximately 0.25-cm thick) through the base, middle, and upper half of the heart, and these sections were fixed overnight in 10% buffered formalin. Eight sections from these areas were stained with Masson's trichrome stain and examined with a light microscope for morphological changes in the myocardium and coronary arteries.

After removing sections for morphological studies, the remainder of the left ventricle was dissected out, cut into small pieces, blotted, weighed, and lypophilized to constant weight. The dried tissue was hydrolysed in 6 N HCl for 24 h in sealed test tubes at 105°C. Humin was removed by filtration, and the resulting suspension was dried with a rotary evaporator and diluted to known volume. Hydroxyproline was determined by the modified Stegman method[9] and, also, with a Beckman amino acid Auto-Analyzer.[10]

Serum thyroxine (T_4) and the thyroxine metabolite, triiodothyroxine (T_3), were determined in the pigeon using a radioimmunoassay kit. The antibody in this kit was prepared for measuring human serum T_3 and T_4.

Results were analyzed for statistical significance by the Student's t-test of difference between means and by the Student's t-test of paired observations.[11] The expressions $p < 0.05$, $p < 0.01$, and $p < 0.001$ are used to indicate significance at the 5, 1, and 0.1% levels respectively.

RESULTS

Systolic and diastolic pressure increased in pigeons exposed to 10- to 30-ppm chlorine for 9 months (Table I). The range of increase for systolic pressure was 8 to 13 mmHg and for diastolic pressure, 9 to 17 mmHg. However, the observed change in blood pressure was not statistically significant. In contrast, a significant decrease in dp/dt was observed in pigeons given 15- or 30-ppm chlorine. The dp/dt measurement describes the rate of rise of ventricular pressure.[12] A decrease in this rate (i.e., dp/dt) is frequently observed in chronic heart failure secondary to cardiac hypertrophy.[12]

Data in Table I suggest cardiac hypertrophy and chronic heart failure. This suggestion is partly supported by the results shown in Table II. Significant increases in heart weight were observed in all treatment groups. The increase in heart weight ranged from 22 to 49% in pigeons treated with 0.1- to 30-ppm chlorine respectively. The body weight in the treated groups did not change relative to the control group. Thus the increase in ratio of heart weight to body weight reflects the absolute increase in heart weight.

Changes in the morphology of the heart were associated with increases in fibrous tissue. The severity of fibrosis, determined qualitatively, appeared to be associated with the level of chlorine in the drinking water. For example, endocardial and myocardial fibrosis appeared to be more intense in pigeons treated with 30-ppm chlorine than in animals given 0.1-ppm chlorine or deionized water (controls). We also observed qualitatively, in the coronary arteries, more atherosclerotic plaques in 30-ppm-treated animals than in the control pigeons or pigeons given 0.1-ppm chlorine.

Table I. Effect of Chlorinated Drinking Water on Blood Pressure and Contractile Properties of the Heart (i.e., dp/dt) in the White Carneau Pigeon[a]

Treatment Group	Blood Pressure		dp/dt $(mmHg\ s^{-1}\ cm^{-1})$
	Systolic (mmHg)	Diastolic (mmHg)	
Control	198 ± 8	143 ± 10	3816 ± 417
Chlorine, ppm			
0.1	192 ± 12	154 ± 18	3710 ± 954
10	206 ± 10	156 ± 13	3011 ± 766
15	211 ± 15	160 ± 20	2740 ± 832[b]
30	210 ± 12	158 ± 16	2384 ± 398[c]

[a]Pigeons were fed a calcium-deficient diet and exposed to deionized drinking water (the controls) or water containing chlorine at the indicated concentration for 9 months. Blood pressure was measured from the brachial artery, after which the cannula was advanced to the left ventricle and the contractile properties (i.e., dp/dt) determined. Five pigeons were used in each experimental group. The data are presented as the mean ± standard error of the mean.
[b]$p < 0.05$
[c]$p < 0.01$

Table II. Body and Heart Weight in the White Carneau Pigeon Following Exposure to Chlorinated Drinking Water[a]

Treatment Group	Body Weight		Heart wt (g)	Heart wt / Body wt
	Initial (g)	Final (g)		
Control	450 ± 19	530 ± 29	4.5 ± 0.3	0.009 ± 0.0002
Chlorine, ppm				
0.1	480 ± 35	550 ± 18	5.5 ± 0.5	0.010 ± 0.0002
10	435 ± 40	514 ± 33	6.3 ± 0.3[c]	0.012 ± 0.0003[b]
15	490 ± 22	552 ± 46	6.7 ± 0.6[c]	0.012 ± 0.0003[b]
30	459 ± 38	534 ± 41	6.7 ± 0.7[c]	0.013 ± 0.0001[c]

[a]Pigeons (5 per experimental group) were exposed to chlorinated drinking water or deionized water (the controls) for 9 months. After exposure, animals were killed and weighed, and the heart was removed, blotted, and weighed. Data are expressed as mean ± standard error of the mean.
[b]$p < 0.05$
[c]$p < 0.01$

Table III. Effect of Chlorinated Drinking Water on the Level of Hydroxyproline in the Heart of Pigeons and Rabbits[a]

Treatment Group	Rabbits (New Zealand White)	Pigeons (White Carneau)
	μg hydroxyproline/mg dry wt heart	
Control	3.01 ± 0.29 (5)	2.22 ± 0.17 (6)
Chlorine, ppm		
0.1	3.8 ± 0.16 (5)	3.09 ± 0.21 (6)[b]
10		5.07 ± 0.91 (5)[c]
15	5.4 ± 0.63 (4)[c]	5.20 ± 0.52 (6)[c]
30		4.99 ± 0.66 (6)[c]

[a]Pigeons were fed a calcium-deficient diet and drinking water containing chlorine at the indicated concentrations for 9 months. Rabbits were fed a calcium-deficient diet with 10% lard added and chlorinated drinking water for 3 months. The controls were fed the respective diets and deionized drinking water. The number in parenthesis equals the number of animals per experimental group. Results are expressed as mean ± standard error of the mean.
[b]$p < 0.05$
[c]$p < 0.01$

Data in Table III support the suggestion of increases in fibrous tissue in the hearts of chlorine-treated pigeons. Hydroxyproline is an amino acid specifically associated with collagen, and collagen is specifically associated with fibrous tissue. Thus, changes in hydroxyproline should reflect changes in fibrous tissue. As shown in Table III, hydroxyproline levels were more than double in pigeons treated with 10-, 15- or 30-ppm chlorine. Even in pigeons treated with 0.1-ppm chlorine, hydroxyproline levels were significantly increased by 40%. In other studies, rabbits exposed to 0.1- and 15-ppm chlorine for 3 months showed hydroxyproline levels increased by 26 and 70% respectively (Table III). However, the increase in hydroxyproline in the heart of rabbits was not as great as that observed in pigeons, although this observation may be related to the exposure period (i.e., 3 vs 9 months).

Plasma T_3 and T_4 levels were measured in pigeons exposed to the chlorinated water. There was a significant decrease, ranging from 25 to 45%, in T_4 in pigeons given 10-, 15-, and 30-ppm chlorine. Studies to determine the effect of chlorinated water on plasma T_3 and T_4 in pigeons are presently being expanded using different methods for measuring the plasma levels of thyroglobulin metabolites (i.e., T_3, T_4, monoiodothyroxine, and diiodothyroxine).

DISCUSSION

Several disease conditions in humans and experimental animals are associated with enlargement of the heart. These conditions include hypothyroidism, hyperthyroidism, coarctation of the aorta, hypertension, coronary artery disease in association with myocardial infarction, and several genetic disorders.[13-15] Of these conditions, two may be possible candidates in explaining the observed increase in heart weight in pigeons treated with chlorinated water. These are hypothyroidism and myocardial infarction secondary to coronary artery disease. Since blood pressure was not significantly increased and coarctations of the aorta were not observed, we suggest that these conditions are not associated with the observed changes in the myocardium. Hyperthyroidism is associated with cardiac hypertrophy; however, hypercholesterolemia is not observed in this condition.[13] We thus suggest an association between the observed increases in heart weight, fibrous tissue in the heart, and plasma cholesterol with hypothyroidism. Furthermore, we suggest that the observed qualitative increase in atherosclerosis in the coronary arteries is secondary to the increase in plasma cholesterol.[16]

Although hypothyroidism may have developed following treatment with chlorinated drinking water, other factors may have contributed to the observed change in the heart. For example, the formation of chlorine derivatives (i.e., chlorinated fatty acids or cholesterol) may have directly affected the metabolism of the heart, leading to these morphological and physiological changes. In previous studies we were unable to detect cholesterol chlorohydrin in the plasma of pigeons treated with chlorinated water for 9 months. Cunningham used in vivo studies to compare the catabolic rate of ^{3}H-oleic acid to ^{36}Cl-oleic acid in the heart of rats and observed a faster rate for ^{36}Cl-oleic acid.[17] These results for the chlorinated lipids do not provide support to the suggestion that they may affect the metabolism of the heart. However, since other derivatives may be formed following chlorination, additional studies may be necessary to determine the effect of chlorinated products on myocardial function.

SUMMARY

Results from the present studies suggest a relationship between drinking water chlorination and hypothyroidism. Since chlorine is very reactive, we assume the formation of a chlorinated product(s). If this assumption is correct, then the observed hypothyroidism may be related to product(s) of drinking water chlorination. Studies are presently in progress to determine the effect of various chlorinated derivatives on myocardial function and morphology.

The present studies were performed to determine the effect of chlorinated drinking water on myocardial weight, function, and morphology because of previous studies that showed significant increases in heart weight in rats treated with chlorinated foods. Our results showed significant increases in heart weight and myocardial fibrosis and decreased myocardial function (i.e., dp/dt). Previous investigations in both human and animal studies have observed these myocardial changes and/or hypercholesterolemia following the prolonged decrease in the plasma level of T_3 and T_4. Therefore, we suggest a relationship between the myocardial changes observed in the present studies and hypothyroidism. The mechanism associated with the induction of hypothyroidism following the exposure to chlorinated drinking water remains to be determined.

ACKNOWLEDGMENTS

This research was supported jointly by the Oak Ridge Research Institute and the Environmental Protection Agency under contract 810053-01 with the Oak Ridge Research Institute.

REFERENCES

1. Daniels, N. W. R., D. L. Frape, P. W. Russell Eggitt, and J. B. M. Coppock. "Studies on the Lipids of Flour. II. Chemical and Toxicological Studies on the Lipid of Chlorine-Treated Cake Flour," *J. Sci. Food Agric.* 14:883–893 (1963).
2. Bellar, T. A., J. J. Lichtenberg, and R. C. Kroner. "The Occurrence of Organohalides in Chlorinated Drinking Water," *J. Am. Water Works Assoc.* 66:703 (1974).
3. Klaassen, C. D., and G. L. Plaa. "Relative Effects of Various Chlorinated Hydrocarbons on Liver and Kidney Function in Mice," *Toxicol. Appl. Pharmacol.* 9:139–153 (1966).
4. Walton, M. F., and R. B. Cumming. "5-Halogenated Uracil Base Analog Mutagenesis," *Mutat. Res.* 38:371 (1976).
5. National Cancer Institute. *Carcinogenesis Bioassy of Chloroform*, Nat. Tech. Inf. Service No. PB 264018/AS (1976).
6. Munson, A. E., U. M. Sanders, B. M. Kauffman, K. L. White, Jr., and J. F. Borzelleca. "Toxicology of Trihalomethanes in Mice," in *Health Effects of Drinking Water Disinfectants and Disinfectant By-products*, EPA International Symposium, April 1981.
7. Reitz, R. H., J. F. Quast, W. T. Scott, P. G. Watanabe, and P. J. Gehring. "Pharmacokinetics and Macromolecular Effects of Chloroform in Rats and Mice: Implication for Carcinogenic Risk Estimation," in *Water Chlorination: Environmental Impact and Health Effects, Vol. 3*, R. L. Jolley, W. A. Brungs, and R. B. Cumming, Eds. (Ann Arbor, MI: Ann Arbor science Publishers, Inc., 1980), pp. 983–994.
8. Cunningham, H. M., and G. A. Lawrence. "Absorption and Metabolism of Chlorinated Fatty Acids and Triglycerides in Rats," *Food Cosmet. Toxicol.* 15:101–103 (1977).
9. Woessner, J. F. "The Determination of Hydroxyproline in Tissues and Protein Samples Containing Small Proportions of this Amino Acid," *Arch. Biochem. Biophy.* 93:440–447 (1961).
10. Benson, J. V., and J. A. Patterson. "Accelerated Chromatographic Analysis of Amino Acids Commonly Found in Physiological Fluids on a Spherical Resin of Specific Design," *Anal. Chem.* 13:265–280 (1965).
11. Snedecor, G. W., and W. G. Cochran. *Statistical Methods*. (Ames: Iowa State University, 1967).

12. Spann, J. F. Jr., D. T. Mason, and R. F. Zelis. "The Altered Performance of the Hypertrophied and Failing Heart," *Am. J. Med. Sci.* 258(5):291-303 (1969).
13. Means, J. H., L. H. DeGroot, and J. B. Stanbury. In *The Thyroid and Its Diseases*, 3rd ed. (New York: McGraw-Hill Book Company, 1963).
14. Meerson, F. Z. "The Myocardium in Hyperfunction, Hypertrophy and Heart Failure," *Cir. Res.* 25(Suppl.2):1-163 (1969).
15. Meerson, F. Z. "A Mechanism of Hypertrophy and Wear of the Myocardium," *Am. J. Cardiol.* 15:755-760 (1967).
16. Kannel, W. B., W. P. Castelli, and T. Gordon. "Cholesterol in the Prediction of Atherosclerotic Disease. New Perspectives on the Framingham Study," *Ann. Intern. Med.* 90:85-91 (1979).
17. Cunningham, H. M. "Toxicology of Compounds Resulting from the Use of Chlorine in Food Processing," in *Water Chlorination: Environmental Impact and Health Effects, Vol. 3*, R. L. Jolley, W. A. Brungs, and R. B. Cumming, Eds. (Ann Arbor, MI: Ann Arbor Science Publishers, Inc., 1980), pp. 995-1005.

SECTION VI

Aquatic Models and Tumor Induction

Water contaminants, in the form of inorganic and organic carcinogens, add to the total carcinogenic environmental load encountered from all stresses.

Herman F. Kraybill
Origin, Classification and Distribution of Chemicals in Drinking Water with an Assessment of Their Carcinogenic Potential, 1975

Of what is significant in one's own existence one is hardly aware . . . What does a fish know about the water in which he swims all his life?

Albert Einstein, 1879–1955
Self-Portrait, 1936

CHAPTER **30**

Aquatic Models and Tumor Induction

Herman F. Kraybill

Contamination of the aquatic environment has produced a wide spectrum of biological effects on fauna, especially aquatic animals or organisms. While such adverse biological effects have been recognized for some time, as exemplified in several episodes of fish kills in our rivers and waterways, it is only in more recent years that attention has focused on certain disease states such as an epizootic of fish tumors. Perhaps an increase in irresponsible pollution over the years has reached, in some cases, the saturation point wherein aquatic animals exist in an environment where they become moribund.

National and international recognition and concerns are quite evident relevant to tumorigenesis in fish and neoplastic-like lesions in shellfish. These concerns are reflected in enhanced research in this area and federal support for such research, since these adverse effects, demonstrated in feral populations of fish in streams, rivers, lakes, and embayments, now have implications for public health. Invariably, these incidences of fish neoplasia have, associated with these registered effects, a problem of aquatic pollution. With such investigations come the inevitable question as to the etiologic agent(s), which can be delineated if possible, so that remedial and regulatory measures can be taken. Because these aquatic animals are food sources or are in the food chain, the ultimate question develops as to what the epidemiological consequences may be from continuous human consumption of seafoods that are reflective of inorganic-organic contaminants that evoked neoplasia in these marine animals. Some human population groups such as sports fishermen, who may receive continuous insults from consumption of such fish, may be at risk. Some generalizations have been made regarding the need for epidemiological studies in this area to search for any association with human neoplastic disease.

The extensive experimental studies on fish tumorigenesis have demonstrated that fish, shellfish, fish cultures, and fish embryos may now be suitable models or techniques for carcinogenesis bioassays. As difficult as it may be for the traditional exponents for rodent bioassays, there is unequivocal evidence that aquatic animal models may take their place alongside mammalian models in the testing and evaluation of environmental chemicals for their carcinogenic properties and/or potency.

CHAPTER 31

Attempts to Abbreviate Time to Endpoint in Fish Hepatocarcinogenesis Assays

John A. Couch and Lee A. Courtney

In recent years, the use of freshwater and marine fishes in carcinogen research and in environmental carcinogen monitoring has grown substantially.[1-4] Several advances must be made with selected species to make fishes advantageous and more practical as assay subjects. Some of these advances should be (1) precise characterization of neoplastic endpoints and progression in experimentally exposed fishes, (2) abbreviation of length in time needed for risk evaluation of carcinogens or suspect agents in fishes, and (3) correlation of endpoints for carcinogen effects in fishes with those in other more-routine test species such as rodents (mammals).

Because we believe that fishes, as a phyletic group, have much to teach us about neoplasia and environmental carcinogenesis, we are studying the experimental induction, progression, and fate of neoplasms in the liver of a marine coastal fish, the sheepshead minnow (*Cyprinodon variegatus*). The agent used to induce liver lesions in these studies was N-nitrosodiethylamine (DEN). The sheepshead minnow has been used for several years as a toxicological and carcinogen assay subject in our laboratories.[5,6] The objectives of this study were to (1) characterize liver neoplastic development and (2) reduce or abbreviate times to endpoints in liver carcinogen assays using the sheepshead minnow with histological, ultrastructural, and enzyme histochemical endpoints.

MATERIALS AND METHODS

A breeding and experimental stock of adult *C. variegatus* was collected in late April 1983 from the Range Point and Big Sabine areas of Santa Rosa Sound in Escambia County, Florida. An experimental group of approximately 400 fry were cultured from this stock and maintained under flow-through conditions until the start of the experimental exposure in mid-October 1983, at which time the fish were approximately 2-months old. Also, two beginning groups of 30 fish from the April 1983 collection, 15 male and 15 female, were used for control and exposed adult experimental groups.

Exposures were conducted in two glove-box systems, one for the adults and one for the juveniles, similar to those systems described by Courtney and Couch.[6] The experimental design and chronology are diagrammed in Figure 1.

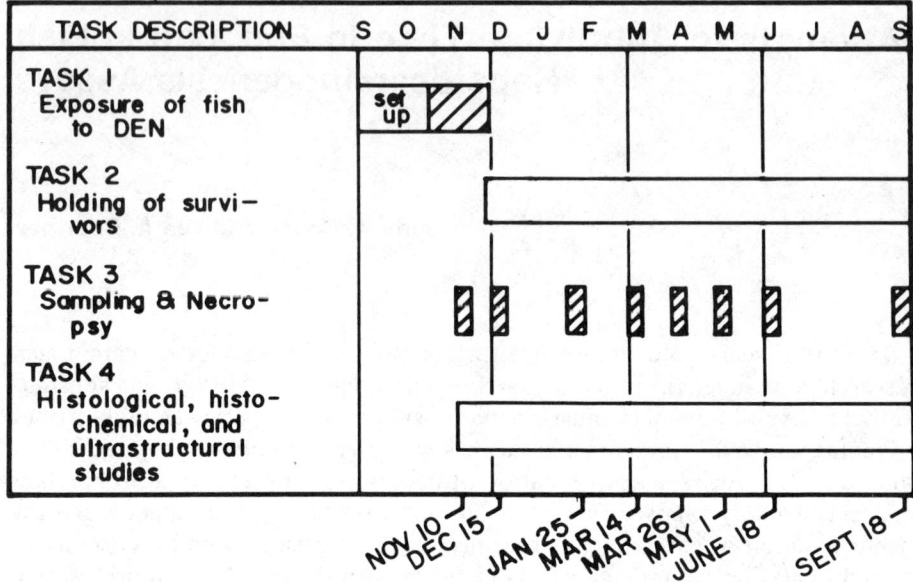

Figure 1. Chart showing experimental schedule. Dates are those of actual samples taken.

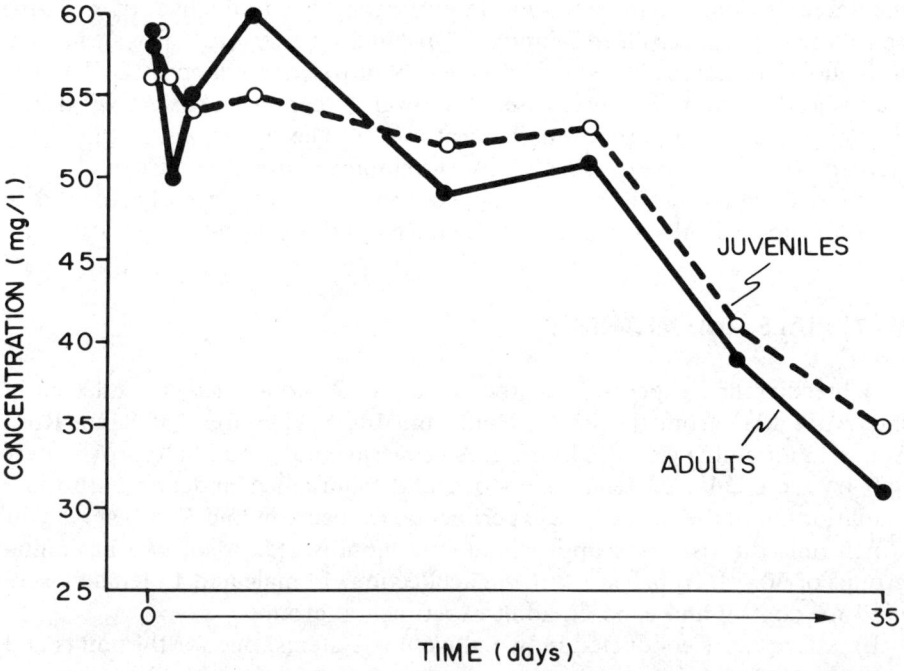

Figure 2. Concentration of N-nitrosodiethylamine (DEN) in water column of static test systems (adult and juvenile) during exposure phase of tumor characterization study.

The DEN, Sigma Chemical Company, was selected as a test compound because of its known carcinogenic properties. DEN was dissolved in seawater, which was then injected into the exposure tanks. The resulting nominal concentration of DEN in each exposure system was 66.7 mg/L. DEN concentrations in each test system were monitored using gas chromatography of seawater samples. Initial measured concentrations of DEN in the water column were 55 to 59 mg/L, which gradually decreased during exposure to approximately 33 mg/L at 5 weeks (Figure 2.).

Exposure of the juvenile fish to the DEN lasted 5 weeks, at which time the fish were removed from the system, rinsed in several changes of clean seawater, and transferred to flow-through holding tanks for growth and observation. Adult fish were exposed for a total of 6 weeks and then also transferred to flow-through holding tanks.

During the course of the exposure and subsequent holding, periodic samples of fish were taken from each experimental group (exposed and control) for histological, enzyme histochemical, and ultrastructural evaluation. Scheduled samples were taken at 3, 8, 14, 23, and 28 weeks (Figure 1). In addition, a single juvenile, found in a stressed and disoriented condition at 21 weeks, was sampled for light microscopy only. Each fish sampled was anesthetized with methane sulfonate (100 mg/L), then dissected and examined grossly.

The most critical phase of processing with respect to time was the enzyme histochemical procedure; rapid freezing was required for optimum preservation of enzyme activity. Juvenile samples were frozen whole in 2-methylbutane cooled to liquid nitrogen (LN_2) temperature (or directly in LN_2 if 2-methylbutane was unavailable) immediately following anesthesia. Livers from adult fish were separated from the visceral mass and bisected, and one segment of liver was frozen in LN_2-cooled 2-methylbutane. Following freezing, each sample was placed in an LN_2-cooled Nunc vial, labeled, and stored in an ultralow-temperature freezer at −80°C until shipment to D. E. Hinton at the Medical Center School of Medicine in Morgantown, West Virginia, for enzyme histochemical analysis and evaluation. In Morgantown, samples were stored in LN_2 until histochemical processing could begin.

Prior to processing for glucose-6-phosphate dehydrogenase (G6PD) activity, tissue samples were allowed to equilibrate in a cryostat at −15°C for 10 min. Fresh frozen cryostat sections (10μ) were then reacted for G6PD using a formazan method.[7] Additional methods are being investigated for determining glucose-6-phosphatase (GGP), adenosine triphosphatase (ATPase), and gammaglutamyl transpeptidase (GGTP) activities. These will be reported at a later date.

For electron microscopy processing, a small piece of tissue from the liver of either juvenile (with liver still in visceral cavity) or adult fish (unfrozen half of bisected liver) was placed in a 3% solution of gluteraldehyde (GTA) in 0.1 M Millonig's phosphate buffer (pH 7.4) and minced into smaller (~ 1 mm³) pieces. Following approximately 1-h GTA fixation, the tissue was postfixed in a 1% solution of osmium tetroxide (OsO_4) for 4 to 5 h,[8] rinsed overnight in phosphate buffer, then dehydrated in a series of acetone solutions and embedded in Epon 812. Sections were cut at 500 to 600 Å, stained with uranyl acetate and lead citrate,[8] and examined for ultrastructural pathology.

Samples for histology were fixed in Davidson's fixative for 24 to 48 h. Adult livers were processed as discrete samples. Visceral masses from each adult fish were also processed and embedded. Juvenile fish were processed whole. Serial sections were cut at 5 μ and slides were stained with a standard Harris' hematoxylin and eosin procedure.[9]

RESULTS

The incidence of microscopic toxic, preneoplastic, neoplastic, and neoplastic-like lesions is given for both exposed adult and juvenile *C. variegatus* in Table I. A spectrum of toxicity-related lesions and at least three probable neoplasms were found in fish samples between 3 and 28 weeks following onset of exposure to DEN (Table I). Lesions are described in the chronology of their appearance in the experimental fish.

Responses in Livers of Fish Sampled at 3 and 8 Weeks

Control fish examined during the study all had normal livers, based on gross and microscopic examinations. For comparison with lesions described later, a brief description of the normal liver of *C. variegatus*, typical of most teleost fish, is provided.

The liver parenchyma (Figure 3) is composed of hepatic cells of fairly uniform size and shape. Architecturally, the hepatocytes are arranged in cords, generally 2-cells thick, along blood sinusoids with bile canuliculi and ductiles leading to bile ducts between the cell layers. Nuclei are homogeneous in size, generally round, and fairly regular in shape, with a single, centrally located nucleolus.

Experimental juvenile and adult livers in 3- and 8-week samples (Table I) demonstrated the most severely toxic cellular and histologic responses. Microscopic examination revealed that hepatocytes were slightly to dramatically enlarged, rounded, and without visible muralia or cord relationship to one another (Figure 4). Throughout liver sections there were foci and bands of lymphocytes and fibroblasts, and granulomata were found where cell death and necrosis were evident (Figure 5). In the livers of some adults, small islets of dying hepatocytes were obvious, and early indicators of cirrhotic change (fibrous tissue being deposited) were evident. Presumed ceroid deposits were abnormally abundant in the 8-week adult livers.

With light microscopy, some possible preneoplastic foci were found in at least two adult fish from the 8-week sample (Table I). These foci consisted of

Table I. Incidence of Microscopic Toxic, Preneoplastic, Neoplastic, and Neoplastic-Like Lesions in Livers of Sheepshead Minnows (*Cyprinodon variegatus*) Experimentally Exposed to 50 mg/L N-Nitrosodiethylamine for 5 (juvenile) and 6 (adult) weeks, Followed by Holding in Flow-Through Systems

Observation Period[a] (weeks)	Cytotoxicity and Inflammation		Clear Cell		Atypical Nodular or Trabecular Foci		Trabecular[b] HC[c]		Nodular[b] HC[c]		Cholangiolar Carcinoma		Hemangiosarcoma	
	J[d]	A[e]	J	A	J	A	J	A	J	A	J	A	J	A
3	9/9[f]	4/4	1/9	0/0	0	0	0	0	0	0	0	0	0	0
8	2/2	4/4	2/2	3/4	0	2/4	0	0	0	0	0	0	0	0
14	3/3	3/3	2/3	2/3	2/3	2/3	1/3	0	0	0	1/3	0	0	0
21[g]	0		0		0		1/1		1/1		1/1		0	
23	4/4	4/5	3/4	4/5	3/4	4/5	0	3/5	0	1/5	0	3/5	0	0
28	3/3	2/3	3/3	2/3	1/3	1/3	2/3	2/3	0	1/3	2/3	2/3	0	2/3

[a] From onset of exposure.
[b] Some of these cases are mixed-cell carcinomas.
[c] Hepatocellular carcinoma.
[d] Juvenile.
[e] Adult.
[f] Number of fishes with lesion sampled.
[g] Single juvenile fish in distressed condition when sampled.

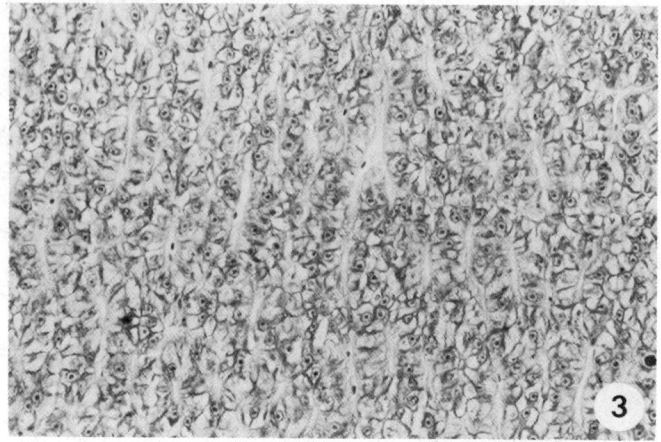

Figure 3. A control liver from an adult *C. variegatus* detailing hepatocyte-sinusoid cord pattern. The cells are generally cuboidal; nuclei are relatively consistent in size, containing a single centrally located nucleolus (1365X).

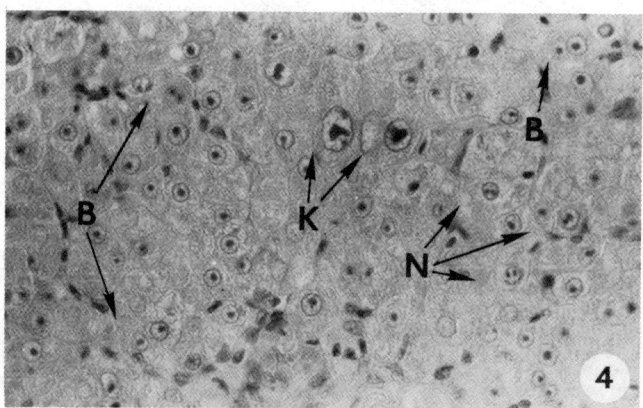

Figure 4. Hepatocytes in exposed adult liver sampled at 8 weeks and displaying severe toxic response. Cords are indistinguishable and cytomegaly and karyomegaly (K) evident; binucleate cells (B) and nuclei with multiple nucleoli (N) are present (1820X).

two apparent cell types: larger, possibly altered hepatocytes arranged as disorganized nodules, and smaller spindle-to-oval cells of uncertain origin (Figure 6). Both cell types were pleomorphic, displaying cytomegaly and karyomegaly, and showed evidence of mitotic activity. In both 3- and 8-week livers, occa-

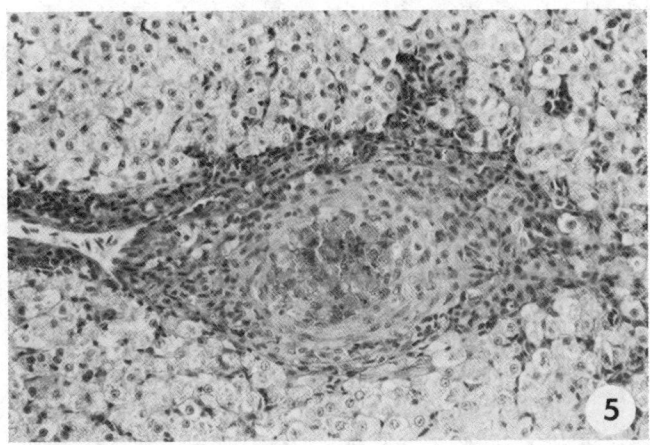

Figure 5. Granuloma in liver of 3-week juvenile fish. Foci of necrotic cell material resulting from cytotoxic effects is surrounded and encapsulated by basophilic cells consisting of lymphocytes, macrophages, and fibroblast-like cells (1365X).

sional eosinophilic foci or widespread eosinophilia (of less distinct limits) were noted. Clear-cell areas (see Figure 7) were found in focal and trabecular patterns in some adult and juvenile liver sections (Table I). The toxic effects of the DEN exposure appeared similar in the 3- and 8-week samples, but were more severe in the adult fish than in the juveniles.

Most electron microscopic (EM) findings in the 3- and 8-week samples probably reflected the intoxicated state of hepatocytes rather than distinct preneoplastic changes, although a few changes in the exposed fish hepatocytes may relate to early neoplastic alteration. The better EM results were obtained from the 8-week adults. Sections from control livers revealed expected, normal hepatocyte profiles. Plasmalemma were distinct, nuclei were round to oblong, and endoplasmic reticulum (ER) was distributed in a perinuclear configuration with large lakes of glycogen occupying most of the cell.

The exposed liver sections possessed alterations as follow: Nuclei were slightly irregular in profile in comparison with control specimens. Some nuclei were severely lobulated with intrusions of cytoplasm appearing as bay regions within nuclei or as pseudo-inclusions (membrane-bound nuclear inclusions). Organelles, in general, were diffusely spread throughout the cell. Mitochondria were enlarged, and the ER was diffuse and lacked organization. The glycogen content was moderate to light, and lipid content was minimal. Tonofibrillar material was abnormally abundant. Bile canuliculi appeared normal.

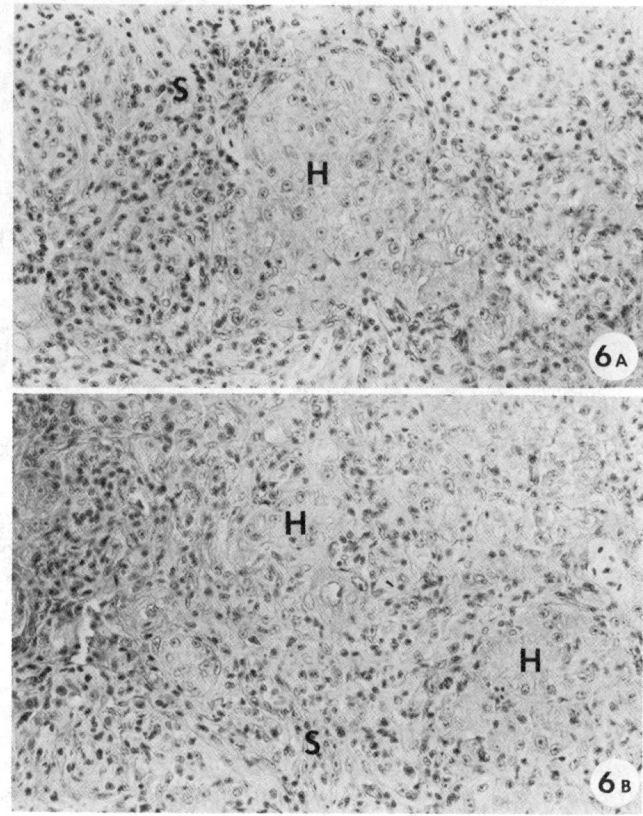

Figure 6. Liver sections of 8-week adult *C. variegatus* showing possible preneoplastic nodule formations. Nodules are composed of altered hepatic cells (H), highly pleomorphic in nature and surrounded by smaller oval cells (S), also displaying pleomorphism. The nuclei of the hepatic cells vary greatly in size and shape (karyomegaly) and are generally larger than those of the second cell type(S). Note also the loss of normal liver cord pattern (A and B, 1365X).

To date, enzyme histochemical analyses of samples taken during the exposure and holding phases of the study are under way but incomplete. Early indications show promise for the G6PD reaction. Control livers of both juvenile and adult *Cyprinodon* displayed a diffuse, moderate, blue reaction over hepatocyte cytoplasm throughout all regions of the sections. By contrast, areas of toxic response in liver sections exposed to DEN reacted with less intensity than that of control liver. By 8 weeks, rounded foci of intense reactivity characterized liver sections from some individuals. The enzyme alterations

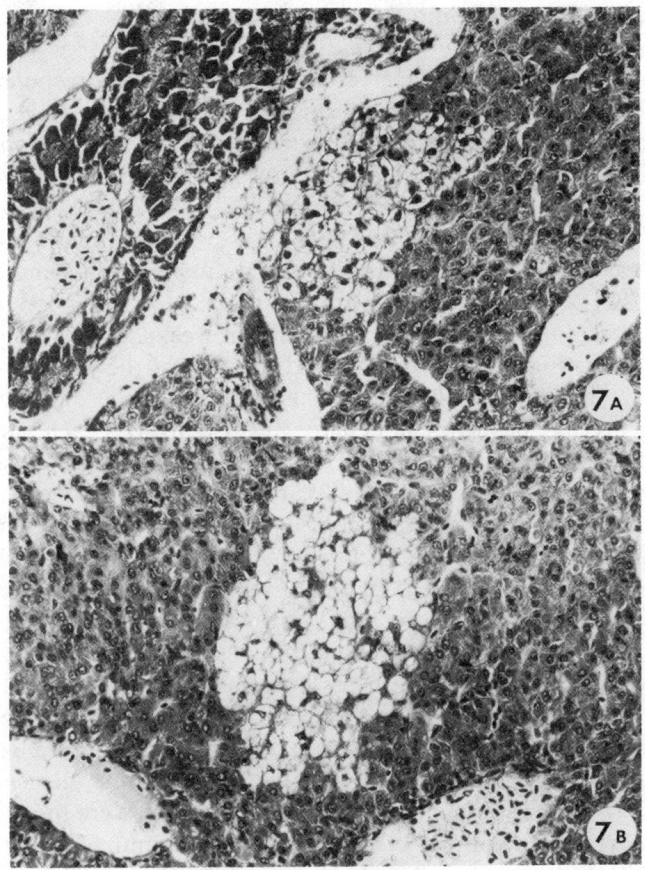

Figure 7. Examples of clear-cell areas found in specimens from most samples. A: Clear-cell foci consisting of cells with clear-to-lacy cytoplasm and centrally located nuclei. Similar clear cells were also noted in more diffuse trabecular patterns. B: Vacuolated clear-cell area in which large lipid droplets displace the nuclei to the periphery of the cell and give the cell a clear appearance (A and B 1365X).

evidenced by this reaction and that of three other enzymes commonly used in mammalian liver carcinogenesis studies suggest a progressive pattern of enzyme activity alteration in the development of liver lesions in *Cyprinodon*. This sequence is under investigation to determine its relationship to overt liver tumors. It may be possible to use early enzyme-altered foci to determine the role of specific populations of cells in tumor development, and to possibly decrease the time between exposure and endpoint (early altered foci).

Responses in Livers of Fish Samples at 14 Weeks

Microscopic evidence of toxic effects and inflammatory responses (Table I) persisted in juvenile and adult livers sampled at 14 weeks. However, the general view indicated that livers in both groups were regenerating and resorbing necrotic cell material while moderating the inflammatory response described for the 3- and 8-week samples (less massive infiltrates or foci of lymphocytes, fibroblasts, or macrophages). There were, however, very abundant and extensive abnormal deposits of ceroid-like and hemosiderin-like materials in all livers from this sample. This was probably related to the prolonged intoxication of the liver during the exposure of several weeks.

Clear-cell foci (Figure 7) were observed in most of the juvenile- and adult-exposed fish. These clear-cell foci consisted of hepatocytes in a focal or quasi-trabecular pattern. The clear-cell hepatocytes (Figure 7A) each had a clear-to-lacy cytoplasm with a distinct centrally located nucleus, closely resembling the clear-cell foci described by Frith and Ward[10] for livers of mice exposed to a variety of chemical hepatocarcinogens. Additionally, some clear-cell areas were noted consisting of cells with large vacuoles and nuclei displaced to the periphery of the cell (Figure 7B). Similar areas, referred to as vacuolated cell foci, were also noted by Frith and Ward[10] in rats undergoing hepatocarcinogenesis.

Generally, among most of the livers examined from this sample, the most striking histopathological characteristic was the considerable pleomorphism expressed among hepatocytes. Cytomegaly, karyomegaly (cells and nuclei three to five times enlarged), and nuclear and cytoplasmic inclusions (eosinophilic and basophilic) were common features. Mitoses were more common in these hepatocytes than in those of the 3- and 8-week samples or in those of any of the control fish livers examined. Many of the pleomorphic enlarged cells were arranged in trabecular or nodular foci and may represent centers of early neoplastic hepatocytes.

One of three juveniles examined from the 14-week samples had advanced trabecular hepatocellular carcinoma (Figure 8). This neoplasm occupied most of the liver and was characterized by cellular pleomorphism, loss of any normal liver cord architecture, cytomegaly and karyomegaly of hepatocytes, some multinucleate cells, and very high mitotic activity with abundant, abnormal (tripolar, polyploid) mitoses (Figure 8A). Additionally, cholangiolar transformation and proliferation had occurred, producing mixed, hepatocholangiolar carcinoma.

Electron microscopy of these samples revealed the most striking differences between control and exposed hepatocytes to date. Control hepatocyte (Figure 9) nuclei were normal in profile, appearing round to oblong. The cytoplasm was also normal in appearance, with small-to-large lakes of glycogen and a perinuclear-to-diffuse distribution of organelles. Most hepatocytes had extensive rough ER, visible Golgi apparatus, and normal mitochondria (round-to-elongate tubular forms with shelf-like cristae). Only modest lipid inclusions

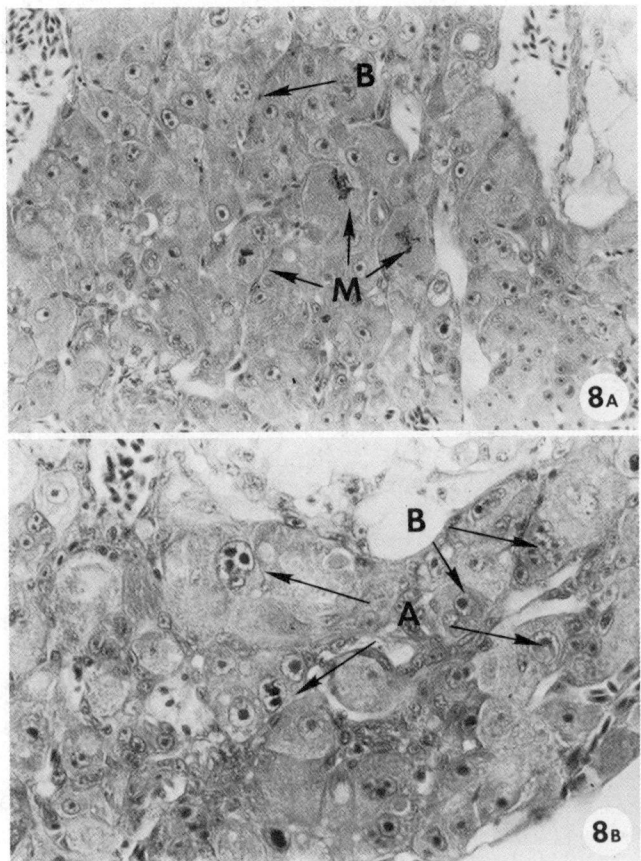

Figure 8. Advanced hepatocellular carcinoma found in juvenile *C. variegatus* sampled at 14 weeks. Note extreme pleomorphism of neoplastic hepatocytes, cytomegaly, and karyomegaly. Binucleate cells (B), abnormal nuclei (A), containing multiple nucleoli and inclusions, and abnormal division figures (M) were found throughout the liver (A, 1365X; B, 1820X).

were noted in the hepatocytes, and the plasmalemma appeared normal. Intercellular relationships, sinusoids, endothelium, and bile canuliculi were normal.

Exposed hepatocytes (Figure 10) generally had nuclei that were more irregular in profile. Nucleolar organization was altered (more condensed), and bays or invaginations of cytoplasm into the nucleus were observed. The cytoplasm was without normal lakes of glycogen in most instances. Rough ER was less abundant and considerably more dilated; smooth ER was present. Phagolysosomes were abundant in the cytoplasm. Most mitochondria were abnormally

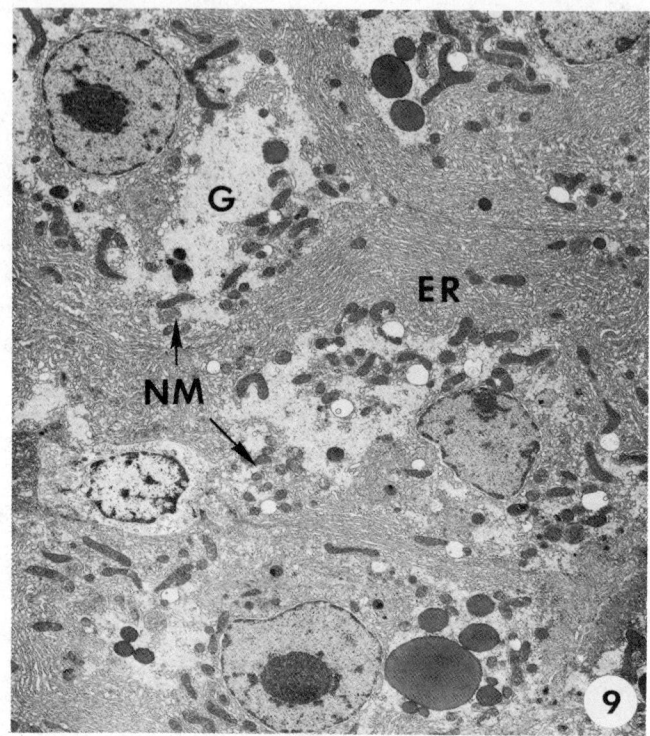

Figure 9. Electron micrograph of control adult liver showing typical ultrastructure of hepatocytes. Note lakes of glycogen (G), extensive, highly organized, rough endoplasmic reticulum (ER), and appearance of normal mitochondria (NM)(3640X).

swollen and were round-to-globose in appearance. Only a few lipid inclusions were found. Plasmalemmae were difficult to distinguish for many hepatocytes, and bizarre folding and interdigitation of the plasmalemma was obvious. Little intercellular space was visible, and cells appeared to be swollen or enlarged. Significantly increased amounts of tonofibrillar material were noted within cells (Figure 10). Recent reports describe similar increased tonofilaments in human carcinomas of the gall bladder.[11] Bile canuliculi were difficult to discern (occluded).

Pathology of Liver of Single Moribund Fish Examined at 21 Weeks

A single exposed juvenile fish became disoriented and moribund, but was taken alive, fixed, and processed for histopathology at 21 weeks from the

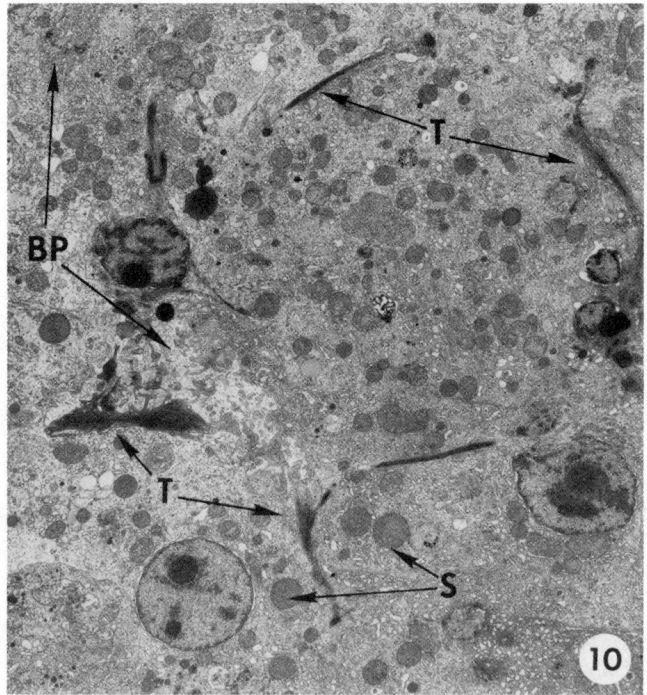

Figure 10. Electron micrograph of experimental adult liver (14 weeks) showing alterations in ultrastructure relating to DEN exposure. Note irregular lobulation of some nuclei, almost total lack of glycogen, greatly reduced rough ER and considerable increase in dilated smooth ER (lacking organization seen in controls), and swollen mitochondria(S). Unusual folding and interdigitation of plasmalemma (BP) and abnormally large amounts of tonofibrillar material (T) were noted in exposed liver sections (3640X).

onset of exposure (Table I). The condition of the liver was so remarkable that a more detailed description is in order.

Grossly, the liver appeared very abnormal. It was pale white, with apparent small, firm nodules beneath its capsule. Microscopically, all liver sections showed advanced neoplasia. Two major neoplastic cell types were evident. First, transformed hepatocytes made up nodules and trabeculae that occupied 50 to 60% of the liver (Figure 11A). These cells and nodules probably represent advanced hepatocellular carcinoma, in which all normal liver cord architecture was lost. The second transformed cell type appeared to be ductile in origin and formed disorganized or degenerate ductile and tubular elements in the remaining 40 to 50% of the liver sections (Figure 11B). These cells were oval to oblong, and only one-fourth to one-half the size of the neoplastic hepatocytes. The cells probably constituted cholangiolar carcinoma. Both cell types had extensive mitotic activity, indicating continued proliferation.

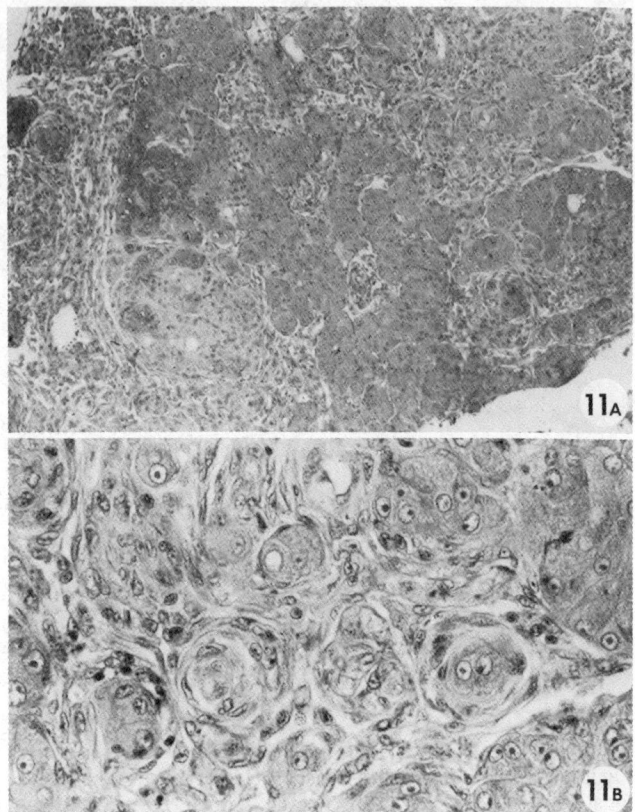

Figure 11. A: Liver of individual sampled at 21 weeks showing advanced trabecular and nodular hepatocellular carcinoma occupying approximately 50 to 60% of the hepatic parenchyma. B: Discrete hepatocellular and cholangiolar elements are present throughout this specimen; however, areas of mixed-cell neoplasia (hepatocholangiolar carcinoma) are also evident (A, 455X; B, 1820X).

Most likely, this fish was dying because of liver failure resulting from almost complete loss of normal liver structure and function.

Responses in Livers of Fish Samples at 23 Weeks

Grossly, the livers of adults from this sample appeared muddy brown and mottled with white. The livers of the juveniles appeared red to beige. Both groups had abundant pigmented macrophage centers beneath the capsule. These foci appeared as black splotches or spots and were absent in control fish livers.

Microscopically, the majority of the liver parenchyma had recovered from the toxic effects in both adults and juveniles. There were, however, certain residual indicators of cytotoxic reaction in the sections. Lesions resembling spongiosis hepatis (large, cystlike spaces) were evident in most livers (Figure 12). These lesions may result from breakdown and lysis of necrotic cells that had been injured by the direct toxic effects of the DEN. To our knowledge, this lesion has been suggested only once previously in fishes;[12] however, it is similar to lesions described in mammalian species exposed to carcinogens.[13] Occasionally, granulomata were found in the liver sections, also indicating residual repair and cleanup of necrotic foci by encapsulating and phagocytic cells.

Adult livers revealed overtly developing hepatocellular and cholangiolar carcinomas. The hepatocellular neoplasms were both nodular (Figure 13) and trabecular (Figure 14). The most striking findings in these sections were several neoplastic foci of mixed cells, both hepatocytes and ductile cells. The cholangiolar cells were arranged in sinuous strands projecting from central foci into surrounding areas of neoplastic and more normal hepatocytes (Figure 14). Livers of juveniles sampled at this time revealed suspect atypical nodular and trabecular foci but did not have advanced neoplasia (Table I).

Responses in Livers of Fish Samples at 28 Weeks

The majority of livers from exposed juvenile and adult fish taken at 28 weeks had either incipient or advanced (or both) neoplasms. At least three

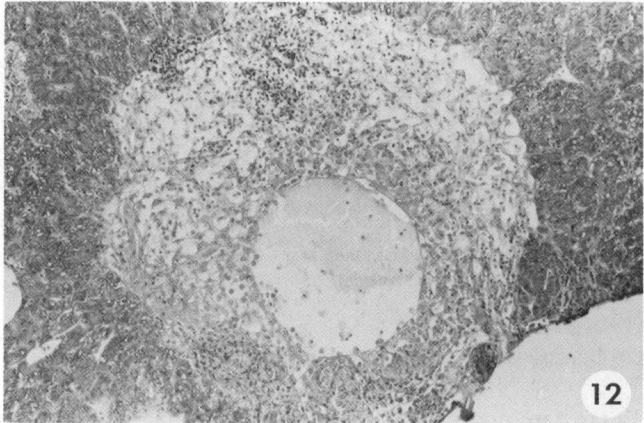

Figure 12. Spongiosis hepatis lesion from adult exposed fish sampled at 23 weeks. Lesions of similar nature were found in other specimens from the 14-, 23-, and 28-week samples (455X).

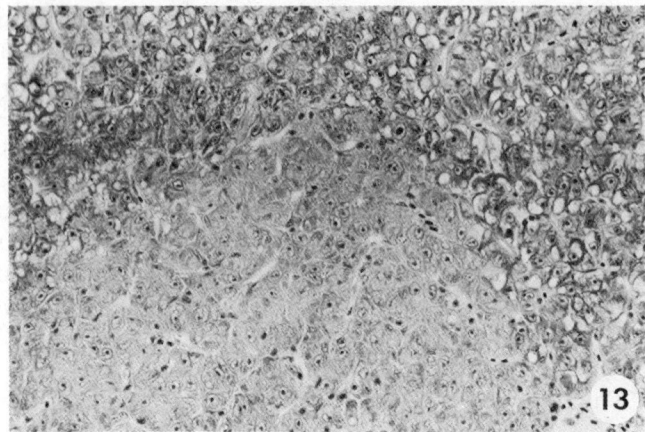

Figure 13. Nodular hepatocellular carcinoma in adult taken in 23-week sample; normal cells above, vacuolated. (1365X).

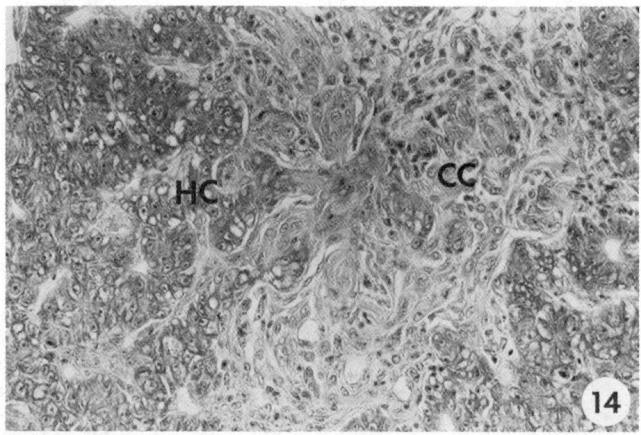

Figure 14. Mixed-cell neoplasia (at 23 weeks) consisting of cholangiolar (CC) and trabecular hepatocellular (HC) elements (1365X).

kinds of neoplasms were tentatively identified in the livers of these fish (Table I).

Hepatocellular carcinoma, cholangiolar carcinoma, and hemangiosarcoma were found in two of three adult livers. Juveniles suffered from hepatocellular carcinoma and cholangiolar carcinoma. The occurrence of presumptive differentiated hemangiosarcoma in livers of exposed fish was first noted in this sample (Table I). Both trabecular and nodular hepatocellular carcinoma and cholangiolar carcinoma were observed in sections of adult livers, and trabecular hepatocellular and cholangiolar carcinoma were found in juveniles.

The presumed hemangiosarcomas observed in adult livers will be described in some detail. At lower magnification (125X), invasive anastomosing tracts of probable blood vessel endothelial cells were apparent in areas of the liver (Figure 15). These tracts had their origins in the walls of large blood vessels

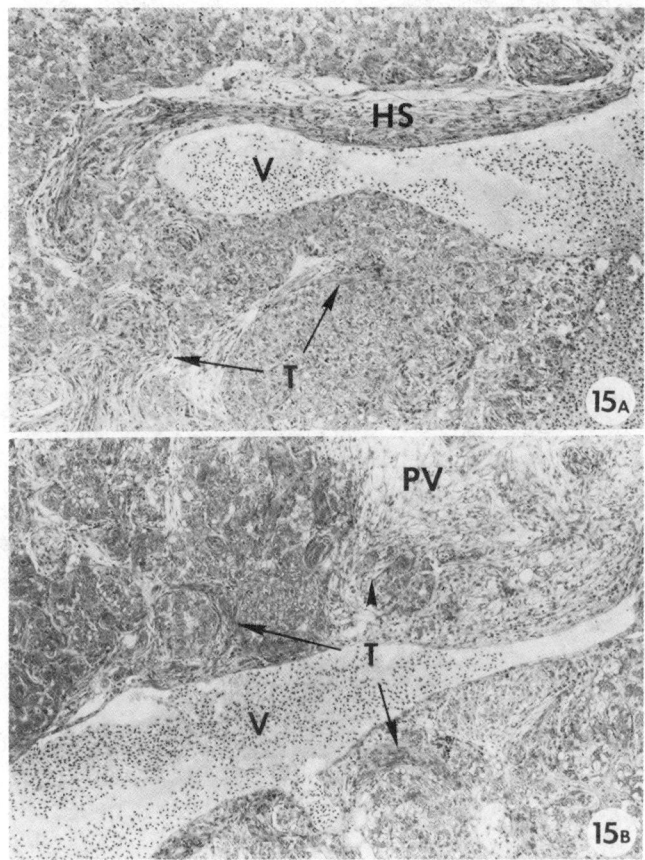

Figure 15. Presumed hemangiosarcoma (HS) observed in adult liver from 28-week sample. Invasive anastomosing tracts (T) run through liver parenchyma, which is composed of elements of hepatocellular and cholangiolar carcinoma and normal hepatocytes. Origin of neoplasm appears to be from blood vessel (V) endothelium (see also Figure 16A). Note pseudo-vascular region (PV) formed by proliferating cells assuming a squamous form (A and B, 455X).

394 WATER CHLORINATION

(Figure 16A). The cells that composed these tracts were oval-to-elongate-to-spindle shaped and were actively proliferating (Figure 16A). These cells were similar in size and form and continuous with endothelial cells (Figure 17) in the hepatic blood vessels from which they apparently arose.

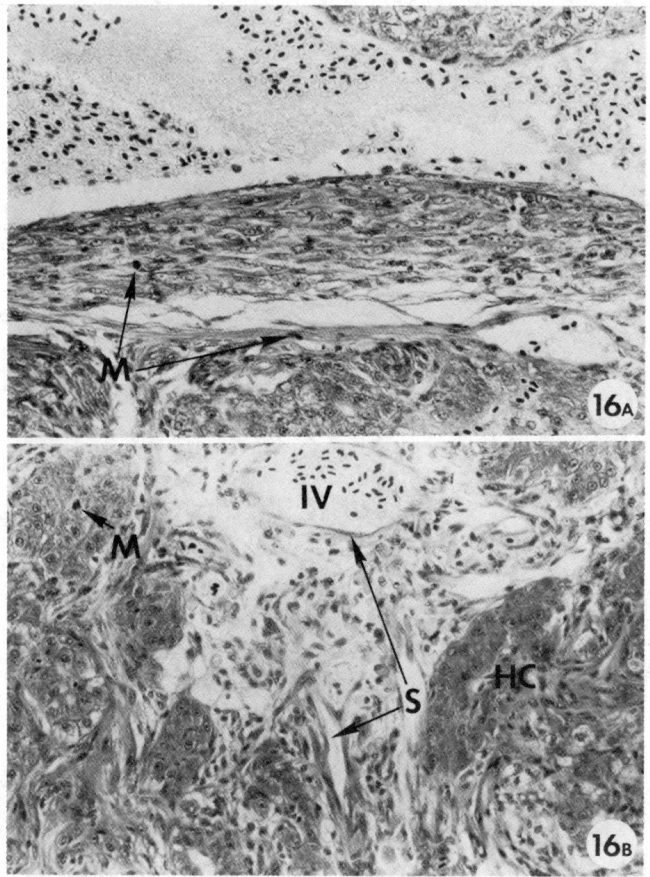

Figure 16. A: Proliferating endothelial lining of hepatic blood vessels from 28-week exposed adult fish. High mitotic activity is indicated by division figures (M). Cells are oval to elongate to spindle-shaped, with nuclei similar to normal blood vessel endothelial cells (see Fig. 17). B: Pseudo-vascular region of hemangiosarcoma from exposed adult sampled at 28 weeks. Note squamous form of the cells (S) forming abnormal, incomplete blood vessels (IV). Pseudo-vascular region depicted here is surrounded by actively proliferating (M: mitotic figures) elements of hepatocellular carcinoma (HC) (A and B, 1365X).

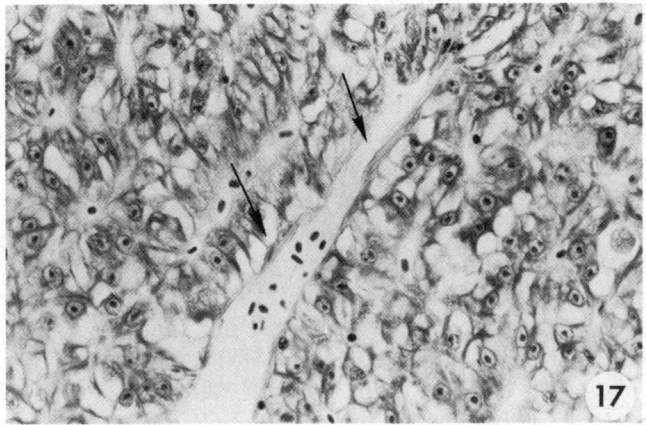

Figure 17. Endothelial cell lining of blood vessels and sinusoids in the liver of control *C. variegatus*. Note the normal 1- to 2-cell thickness of vessel wall (1820X).

In some cases the invasive tracts formed by these cells gave rise to large pseudo-vascular regions with anastomosing lumina (Figures 15B and 16B). The cells making up these pseudo-vascular-like regions often assumed a squamous form and formed incomplete endothelia of the blood vessels (Figure 16B) surrounding the red blood cells.

Examination of other tissues revealed no evidence that any of the tumor types described had metastasized.

DISCUSSION

The fate of these liver lesions is being monitored in the remaining fish in this experiment and will be reported later. The observations and data gathered to date, however, permit us to suggest possible trends in the progression of the lesions through the first 28 weeks of their development. Comparisons have been made of the development of lesions in *C. variegatus* with the occurrence of chemically induced liver lesions in mice,[10] in rats,[14-16] in rainbow trout,[17] and in the medaka fish.[18] Further comparisons of the sheepshead minnow lesion development are made with the cellular lineage model for experimental hepatocellular carcinomas in rats as summarized and described by Sell and Leffert.[19]

The earliest detectable, probable preneoplastic lesions occurred between 8 and 14 weeks following onset of exposure. These initial indicator lesions were clear-cell foci and atypical eosinophilic, and, more rarely, basophilic nodules or trabeculae (Figures 6 and 7). The first overtly microscopic neoplasms were trabecular and nodular hepatocellular carcinomas (14-week sample and 21-week individual, Table I) and cholangiolar carcinomas (14-week sample and 21-week individual, Table I). Indeed, when examined in detail, it is quite possible that these neoplasms were actually mixed hepatocholangiolar carcinomas similar to those induced in rat livers with DEN.[14]

It is interesting to compare the hemangiosarcomas induced in rat liver[14] with DEN with those presumptive ones previously described that were induced with DEN in *C. variegatus* liver. These lesions in rat liver and sheepshead minnow consist partially of spindle-shaped to ovoid cells arranged in anastomosing networks or tracts (Figure 15) that form vascular spaces.

Enzyme histochemistry conducted as part of this study is far from complete; however, as previously discussed, this research, particularly with G6PD, offers promise for the possible detection of early altered, transformed foci of induced liver neoplasms in *C. variegatus*. This work will continue and will be reported more completely at a later time.

As stated earlier, not all of the electron microscopy results have been analyzed; in fact, the samples from the livers with more advanced tumors are still being evaluated. However, distinct ultrastructural differences between exposed and control tissues have been noted in earlier samples. Most of the differences are possibly related to the cytotoxic effect of DEN exposure (i.e., reduced glycogen reserves, swollen mitochondria, and ER alterations). Some changes, for example, alterations of nuclear morphology (lobulation of nucleus, alteration of chromatin material, and nuclear inclusions), plasmalemmal alterations, and particularly the drastic increase in tonofibrillar elements, may be related to preneoplastic and neoplastic changes. A further investigation into the significance of the noted ultrastructural changes, along with an examination of unfinished samples, is in progress.

Though the timing of the appearance of specific microscopic liver lesions in sheepshead minnows differs somewhat from that in rodents, the sequence of lesion types, ranging from toxic to preneoplastic to neoplastic, is generally similar to that described by Sell and Leffert[19] for the rat. The sheepshead minnow sequence resulting from a relatively high-dose exposure (50–60 mg/L) to DEN is:

General cytotoxic response (3–8 weeks) Figures 4, 5, and 12 → Clear-cell foci, nodular foci, and trabeculae (8–14 weeks) Figures 6 and 7 → Hepatocellular and cholangiolar carcinoma, mixed-cell hepatocholangiolar carcinoma, and hemangiosarcoma (presumed) (14–28 weeks) Figures 8, 11, 13, 14, 15, and 16

The sequence of lesion development described here for *C. variegatus* as induced by DEN may be representative of that which occurs in other fish exposed to a variety of chemical agents. Some liver tumors induced in the Japanese killifish, the medaka, were morphologically similar and within a similar latency period, following exposure to DEN,[18,20] to those induced in the sheepshead minnow. Indeed, the sequence of events for chemically induced liver lesions leading to neoplasia in rainbow trout[17] exposed to the more potent hepatocarcinogen, Aflatoxin B_1, is more discrete and perhaps less complex than that described here for the sheepshead minnow, even though the time required for lesion development is longer in the cold-water rainbow trout.

The response of the sheepshead minnow to the known carcinogen, N-nitrosodiethylamine, relegates this species to the useful category of small fish species available for carcinogen monitoring and assays.

ACKNOWLEDGMENTS

This research was partially supported by an interagency agreement with the U.S. Army.

We thank Dr. David Hinton, Medical Center School of Medicine, Morgantown, West Virginia, for his work on enzyme histochemistry. We also wish to thank Steven Foss, Susan Martin, and David Bartee for their assistance during the research and in the preparation of the manuscript. We extend our appreciation to Jim Moore and Emile Lores for analyzing water samples taken during the exposure phase of the study.

REFERENCES

1. Dawe, C. J., and J. A. Couch. "Debate: Mouse Versus Minnow: The Future of Fish in Carcinogenicity Testing," in *Use of Small Fish Species in Carcinogenicity Testing*, Monograph 65, K. L. Hoover, Ed. (Washington, DC: National Cancer Institute, 1984), pp. 223–235.
2. Hendricks, J. D. "The Use of Rainbow Trout (*Salmo gairdneri*) in Carcinogen Bioassay, with Special Emphasis on Embryonic Exposure," in *Phyletic Approaches to Cancer*, C. J. Dawe, J. C. Harshbarger, S. Kondo, T. Sugimura, and S. Takayama, Eds. (Tokyo: Japan Scientific Societies Press, 1981), pp. 227–240.
3. Black, J. J. "Epidermal Hyperplasia and Neoplasia in Brown Bullheads (*Ictalurus nebulosus*) in Response to Repeated Applications of PAH-containing Extract of Polluted River Sediment," in *Proceedings of the Seventh International Symposium on Polynuclear Aromatic Hydrocarbons*, M. Cooke and A. J. Dennis, Eds. (Washington, DC: U.S. Environmental Protection Agency, 1983), pp. 99–111.
4. Hoover, K. L. *Use of Small Fish Species in Carcinogenicity Testing*, Monograph 65, (Washington, DC: National Cancer Institute, 1984).
5. Couch, J. A., L. A. Courtney, and S. S. Foss. "Laboratory Evaluation of Marine Fishes as Carcinogen Assay Subjects," in *Phyletic Approaches to Cancer*, C. J.

Dawe, J. C. Harshbarger, S. Kondo, T. Sugimura, and S. Takayama, Eds. (Tokyo: Japan Scientific Societies Press, 1981), pp. 125–139.
6. Courtney, L. A., and J. A. Couch. "Usefulness of *Cyprinodon variegatus* and *Fundulus grandis* in Carcinogenicity Testing: Advantages and Special Problems," in *Use of Small Fish Species in Carcinogenicity Testing*, Monograph 65, K. L. Hoover, Ed. (Washington, DC: National Cancer Institute, 1984), pp. 83–96.
7. Negi, D. S., and R. J. Stephens. "An Improved Method for the Histochemical Localization of Glucose-6-Phosphate Dehydrogenase in Animal and Plant Tissue," *J. Histochem. Cytochem.* 25:149–154 (1977).
8. Hayat, M. A. *Principles and Techniques of Electron Microscopy: Biological Applications, Vol. 1* (New York: Van Nostrand Reinhold Company, 1970), pp. 5–105, and pp. 241–318.
9. Luna, L. G. *Manual of Histological Staining Methods of the Armed Forces Institute of Pathology*, 3rd ed. (New York: McGraw-Hill Book Company, 1968), pp. 32–39.
10. Frith, C. H., and J. M. Ward. "A Morphologic Classification of Proliferative and Neoplastic Hepatic Lesions in Mice," *J. Environ. Pathol. Toxicol.* 3:329–351 (1980).
11. Alpers, C. E., and E. A. Smuckler. "Pleomorphic Carcinoma of The Gallbladder: Case Report and Ultrastructural Study," *Ultrastruct. Pathol.* 6(1):29–38 (1984).
12. Hinton, D. E., R. C. Lantz, and J. A. Hampton. "Effects of Age and Exposure to a Carcinogen on the Structure of the Medaka Liver: A Morphometric Study," in *Use of Small Fish Species in Carcinogenicity Testing*, Monograph 65, K. L. Hoover, Ed. (Washington, DC: National Cancer Institute 1984), pp. 239–249.
13. Bannasch P., M. Bloch, and H. Zerban. "Spongiosis Hepatis: Specific Changes of the Perisinusoidal Liver Cells Induced in Rats by N-nitroso-morpholine," *Lab. Invest.* 44:252–264 (1981).
14. Schauer, A., and E. Kunze. "Tumors of the Liver," in *Pathology of Tumors in Laboratory Animals, Vol. 1, Tumors of the Rat, Part 2*, V. S. Turusov, Ed. (Lyon: International Agency for Research on Cancer, 1976), pp. 41–72.
15. Squire, R. A., and M. H. Levitt. "Report of a Workshop on Classification of Specific Hepatocellular Lesions in Rats." *Cancer Res.* 35:3214–3223 (1975).
16. Stewart, H. L., G. Williams, C. H. Keysser, L. S. Lombard, and R. J. Montali. "Histologic Typing of Liver Tumors of the Rat," *J. Nat. Cancer Inst.* 64(1):179–206 (1980).
17. Hendricks, J. D. "Chemical Carcinogenesis," in *Aquatic Toxicology, Vol. 1*, L. J. Weber, Ed. (New York: Raven Press, 1982), pp. 149–211.
18. Ishikawa, T., T. Shimamine, and S. Takayama. "Histologic and Electron Microscopic Observations on Diethylnitrosoamine-Induced Hepatomas in Small Aquarium Fish (*Oryzias latipes*)," *J. Nat. Cancer Inst.* 55:909–911 (1975).
19. Sell, S., and H. Leffert. "An Evaluation of Cellular Lineages in the Pathogenesis of Experimental Hepatocellular Carcinoma," *Hepatology*, 2(1):77–86 (1982).
20. Takayama, S., and T. Ishikawa. "Comparability of Histological Alterations During Carcinogenesis in Animals and Man with Special Reference to Hepatocarcinogenesis in Fish," in *Air Pollution and Cancer in Man*, Proceedings of the Second Hanover International Carcinogenesis Meeting, (Lyon, France: International Agency for Research on Cancer, 1977), pp. 271–286.

CHAPTER 32

Occurrence of Hepatic Neoplasms and Other Lesions in Bottom-Dwelling Fish and Relationship to Pollution in Puget Sound, Washington

Donald C. Malins, Bruce B. McCain, Margaret M. Krahn, Mark S. Myers, John E. Stein, William T. Roubal, Donald W. Brown, Usha Varanasi, Harold O. Hodgins, and Sin-Lam Chan

Several recent reports have described high prevalences of neoplasms in bottom-dwelling fish found in polluted waters of the United States. For example, particularly high prevalences of hepatic neoplasms have been reported in freshwater fish from Torch Lake, Michigan;[1] the Niagara River, New York;[2] and in demersal saltwater fish from the urban areas of Puget Sound, Washington.[3] In each instance, anthropogenic chemicals have been implicated in the pathogenesis of the observed neoplastic disorders, but as yet there is only limited evidence concerning their etiologies. In this chapter we review our findings about the nature and prevalence of liver neoplasms in certain species of bottom-dwelling marine fishes, and discuss existing information concerning possible causative factors.

In our Puget Sound pollution research,[3-5] we focused on three common species of demersal fish: English sole (*Parophrys vetulus*), rock sole (*Lepidopsetta bilineata*), and Pacific staghorn sculpin (*Leptocottus armatus*). The organs of these fish containing the greatest number of lesions of all types were the liver, kidney, and gills. Some lesions were associated with parasites or microorganisms, but other lesions had no such apparent association and were thus classified as idiopathic.

Idiopathic lesions (four major histopathological classifications) were found most frequently in the liver where neoplasms constituted one of the four major types (Figures 1–5). The neoplasms were further classified as hepatocellular (minimum deviation basophilic nodule, liver cell adenoma, hepatocellular carcinoma), cholangiocellular (cholangioma, cholangiocellular carcinoma), and mesenchymal (hemangioma, fibroma).

The other three major types of hepatic lesions were (1) hyperplasia/foci of cellular alteration, a category which included several potentially preneoplastic lesions, such as eosinophilic foci, hyperbasophilic foci, clear-cell foci, and hyperplastic regenerative foci; (2) megalocytic hepatosis; and (3) steatosis/hemosiderosis.

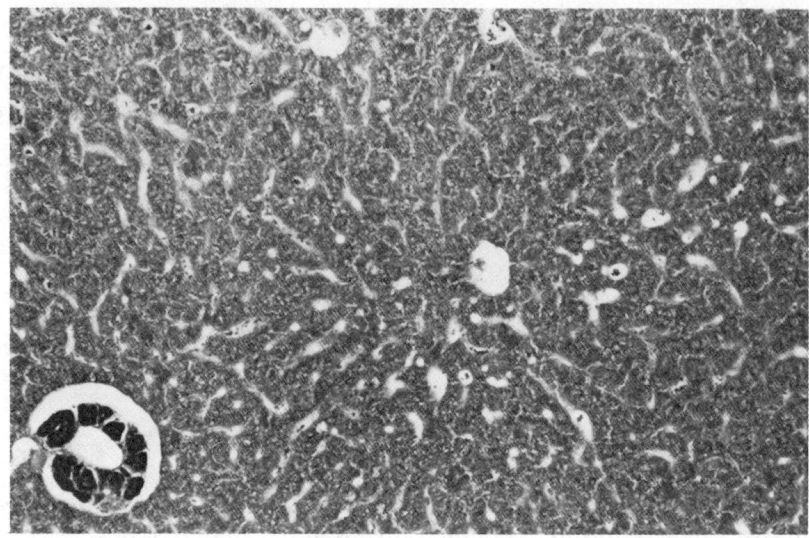

Figure 1. Normal English sole liver tissue. Acinar tissue at lower left is exocrine pancreas. Hematoxylin and eosin (160X).

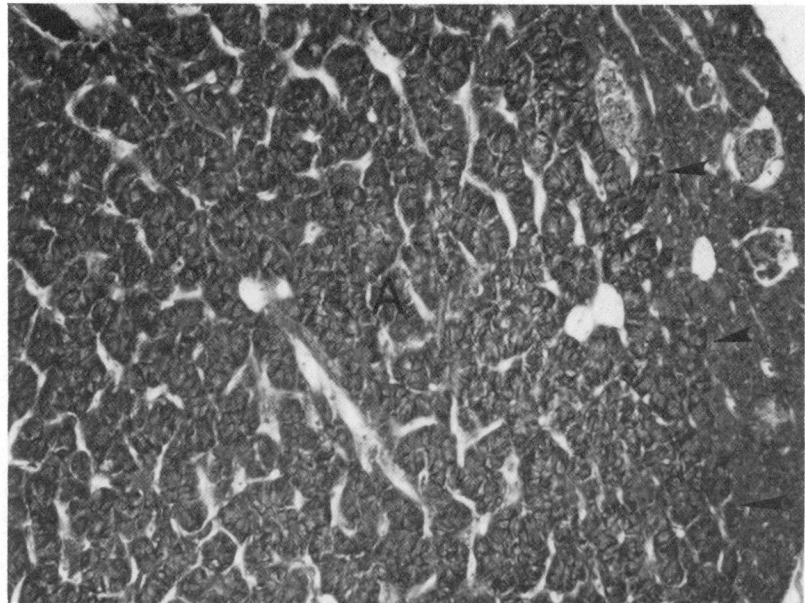

Figure 2. English sole liver with highly basophilic, well-organized liver cords that are part of a liver cell adenoma (A). This neoplasm compresses the adjacent normal parenchyma (arrows delineate the margin of neoplastic tissue); hematoxylin and eosin (160X).

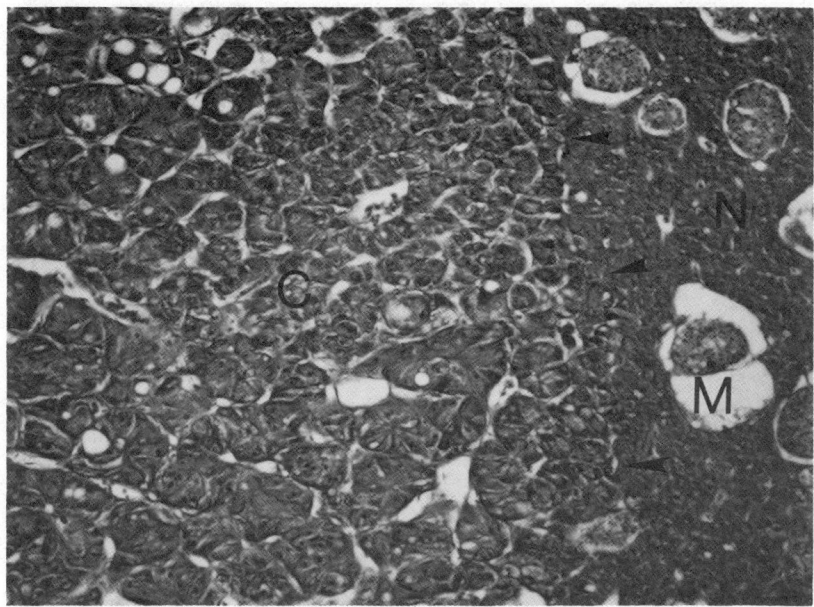

Figure 3. English sole liver with basophilic, poorly organized liver cords composed of enlarged pleomorphic, anaplastic hepatocytes. This hepatocellular carcinoma (C) is bordered (see arrows) by normal parenchyma (N) containing melanomacrophage centers (M); hematoxylin and eosin (160X).

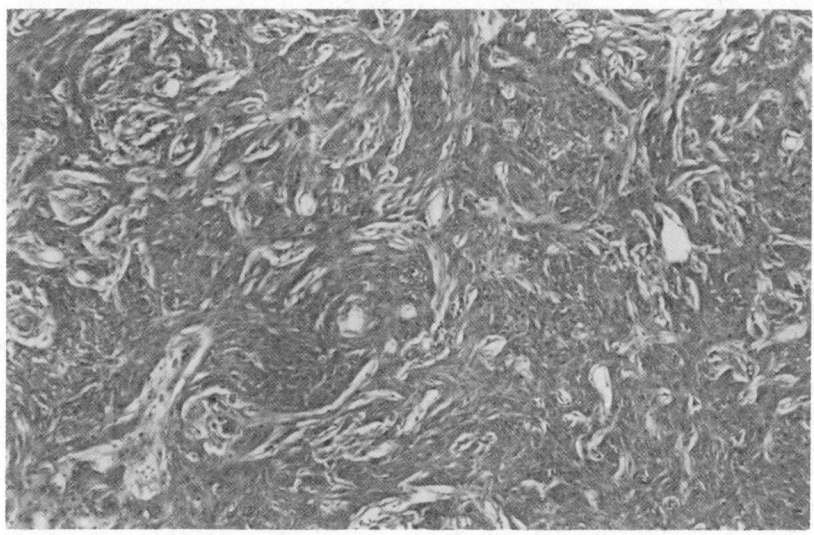

Figure 4. Cholangiocellular carcinoma in the liver of an English sole showing a poorly differentiated variant of this type of neoplasm. Metastatic foci were found in the heart, spleen, and kidney of this fish; hematoxylin and eosin (160X).

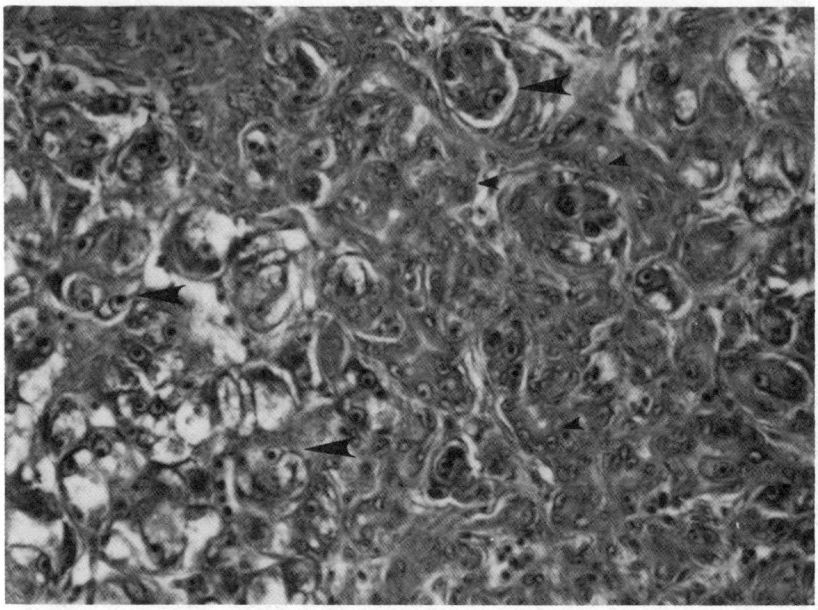

Figure 5. A mixed cholangio/hepatocellular carcinoma in an English sole showing clearly separated components. The larger cells with prominent nuclei are neoplastic hepatocytes (large arrows) that form the hepatocellular component, whereas the smaller cells are arranged in a tubular pattern (small arrows) forming the cholangiocellular component; hematoxylin and eosin (400X).

With few exceptions, hepatic neoplasms were observed in fish from urban-associated waters and not in fish from nonurban waters of Puget Sound (Figure 6). For example, high prevalences of hepatic neoplasms (see Figure 7) were found in English sole from the Duwamish Waterway in Seattle (8.2%, n = 537), from Everett Harbor (12%, n = 66), from the Hylebos Waterway in Tacoma (3.4%, n = 297), and from near Mukilteo (7.5%, n = 66).

Many diverse chemicals were detected in Puget Sound sediments, and the highest concentrations were in urban-associated sediments (Figure 8).[3,5] For example, the mean concentration of polychlorinated biphenyls (PCBs) in Seattle's Duwamish Waterway sediment was more than 100 times greater than in sediment from areas of Port Madison and Case Inlet (reference areas). Sediments with the highest concentrations of aromatic hydrocarbons (AHs) also had the highest concentrations of certain polynuclear aromatic hydrocarbon carcinogens including benz(a)anthracene (BaA, 7600 ppb) and benzo(a)pyrene (BaP, 2400 ppb). In addition, numerous methylated AHs such as dimethylbenz(a)anthracene and methylchrysene were detected in these contaminated sediments. Over 900 individual organic compounds were detected in Com-

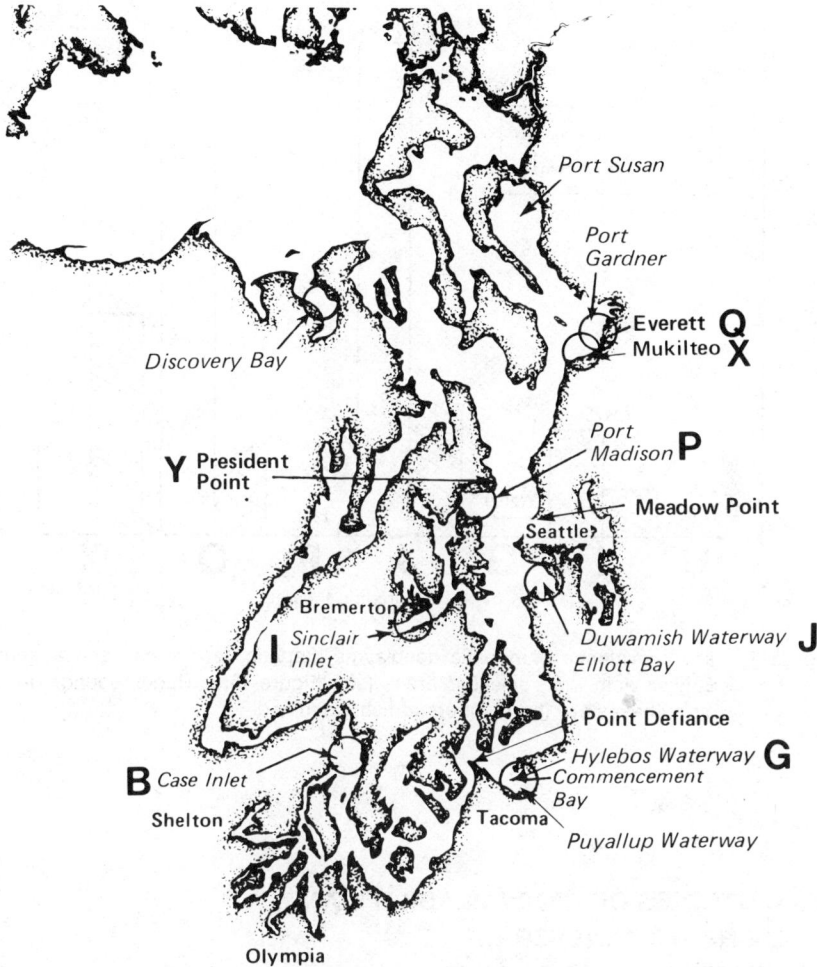

Figure 6. Map of Puget Sound showing sampling areas (B, G, J, P, Q, X, and Y) that were among those areas where sediment and bottom fish were collected.

mencement Bay (Tacoma) sediments. These included more than 500 AHs, hundreds of chlorinated hydrocarbons, and various bromine-, sulfur-, nitrogen-, and oxygen-containing compounds. The numbers and identities of these compounds have not been fully determined because of the complexity of the chemical mixtures and the lack of chemical standards. Generally, mean concentrations of sediment-associated metals were higher in the urban areas (Sinclair Inlet, Commencement Bay, and Elliott Bay) than in nonurban areas (e.g., arsenic in Figure 8). The mean concentration of cadmium, however, was similar in urban and nonurban sediments.

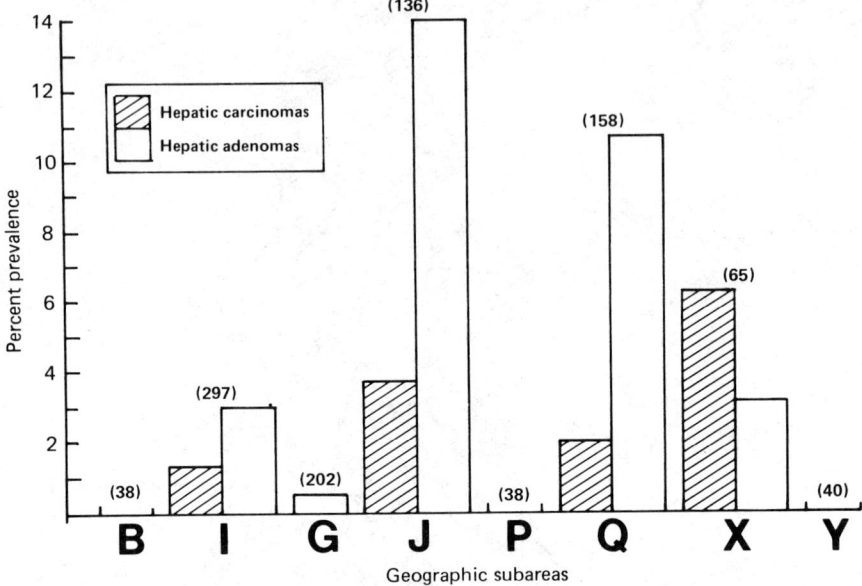

Figure 7. The prevalence of hepatic neoplasms (hepatic carcinomas and adenomas) in English sole from selected areas (see Figure 6) in Puget Sound. Numbers in parentheses are the numbers of fish examined.

FIELD STUDIES OF BIOAVAILABILITY AND FOOD-CHAIN TRANSFER OF SEDIMENT-ASSOCIATED CHEMICALS

The bioavailabilities of sediment-associated chemicals and their potential for food-chain transfer are important considerations in understanding relationships between the chemicals and hepatic lesions. In this regard, evidence has been obtained from studies of Puget Sound that relates to both the transport of chemicals from sediments to organisms and to factors that govern their subsequent transfer through food chains.[5]

Aromatic hydrocarbons were not detected or were present in only trace amounts in livers of English sole captured from polluted areas, even when the sediments contained high concentrations. However, stomach contents (consisting mainly of Annelida) of English sole captured from Mukilteo had concentrations of AHs, such as BaA and BaP (1000 and 570 ppb, respectively), similar to those in the sediment (830 and 360 ppb, respectively) from which the fish were taken.[6] The reason for the low accumulation of AHs in fish liver can

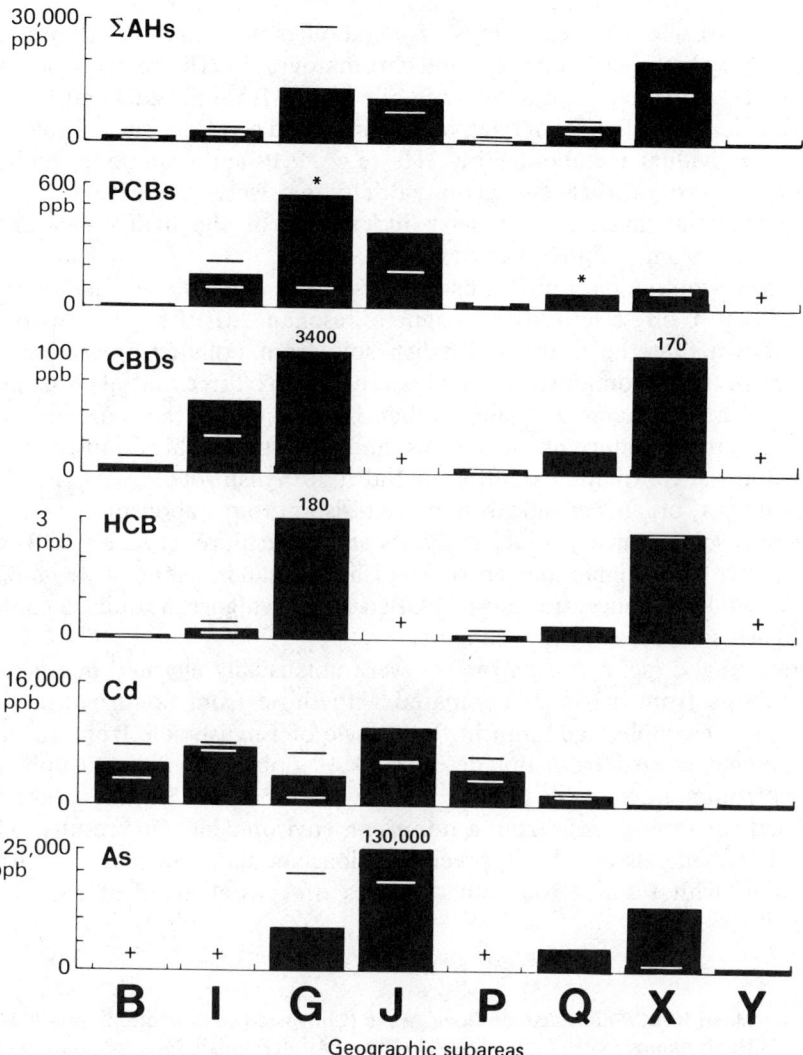

Figure 8. Mean concentrations of chemicals in sediments from selected areas (see Figure 6) of Puget Sound. Note: A large number of organic chemicals in extracts of Everett Harbor sediments could not be resolved and identified; * denotes that PCBs were not quantified in sediments from certain stations in subareas G & Q because of interference from other compounds; + denotes chemicals not detected.

be explained by the fact that AHs are extensively converted to metabolites in this organ.[7-13] It is noteworthy that routine analytical methods do not determine these metabolites in biota.

Recent studies have shown that AHs are bioavailable to English sole living in polluted sediments, and that these compounds form metabolites that are excreted into bile. Concentrations of metabolites with BaP-like fluorescence (as shown by high-performance liquid chromatography/fluorescence detection techniques) were many times higher in bile of fish from the Duwamish Waterway than in bile of fish from relatively nonpolluted reference sites (Table I).[6,14] Several individual metabolites of AHs (e.g., 9,10-anthraquinone, hydroxyfluorene, and anthracenecarboxyaldehyde) were identified by gas chromatography/mass spectrometry in extracts of the hydrolyzed bile of English sole from polluted waterways.

Nitrogen-containing compounds have also been shown to be metabolized by marine fish. Using electron paramagnetic resonance (EPR) spectroscopy, we have shown that the livers of English sole from polluted areas (e.g., the Duwamish River) contained carbazole-derived N-oxyl free radicals.[15] Comparative studies with mammals[16] suggest that the oxidation of the benzene rings of carbazoles may produce phenols, diols, and other metabolites similar to those formed in the enzymatic oxidation of BaP by English sole.[17]

In contrast, organic chemicals that are resistant to metabolism accumulated in English sole. Concentrations of PCBs and hexachlorobenzene (HCB) were consistently much higher in livers of English sole than in sediments from which they were taken. Concentrations of PCBs were also higher in stomach contents of English sole than in the sediments.

Interestingly, metal concentrations were not usually elevated in tissues of English sole from urban as compared with those from nonurban environments. For example, cadmium in the muscle of English sole from an urban environment ranged from not detected (<10 ppb wet wt) to 20 ppb; lead ranged from not detectable (<20 ppb wet wt) to 250 ppb. Similar values were obtained for English sole from a nonurban environment. Our results, which suggest that metals are not appreciably bioaccumulated in marine fish, are consistent with results from other studies of coastal areas of the United States.[18,19]

Table I. Mean HPLC/Fluorescence Peak Areas [Converted to Benzo(a)pyrene (BaP) or Naphthalene (NPH) Equivalents] in Bile of Adult English Sole Captured in Puget Sound[a]

	Equivalents (ng/g wet wt)	
	NPH	BaP
Duwamish Waterway[b]	150,000 ± 130,000	1,800 ± 2,600
Mukilteo[c]	27,000 ± 16,000	440 ± 340
Meadow Point[b]	17,000 ± 15,000	97 ± 110
President Point[c]	8,600 ± 9,700	83 ± 52

[a]Fluorescence was monitored at wavelengths for BaP and naphthalene metabolites; excitation/emission, 380/430 nm and 290/335 nm, respectively. Integrated peak areas were converted to AH equivalents, the amount of the particular AH that would be present if the integrated area were attributed to only that compound.
[b]Adapted from Reference 14.
[c]Adapted from Reference 6.

LABORATORY STUDIES ON BIOACCUMULATION AND METABOLISM OF SEDIMENT-ASSOCIATED CHEMICALS

Our studies[12,13,20] have shown that sediment-associated chemicals (e.g., AHs and PCBs) are readily bioavailable to English sole. Moreover, we have evidence that the ability of fish to metabolize xenobiotics has a significant influence on bioconcentration. When English sole were exposed simultaneously to sediment-associated naphthalene (NPH) and BaP, the bioconcentration factor (pmol of AH equivalents per gram tissue divided by pmol AH per gram sediment) for NPH was considerably higher than that for BaP;[12] it was shown that even though both AHs were metabolized by English sole, BaP was metabolized to a much greater extent. However, in addition to the effect of metabolism on the potential for bioconcentration of an organic compound, the ability of an organism to excrete the parent compounds also affects bioconcentration. It has been shown that low-molecular-weight AHs, such as NPH, are readily excreted via the gills or skin.[21,22]

Unlike AHs, PCBs are neither extensively metabolized nor excreted via the gills.[23] This fact may, in part, account for the high concentrations of PCBs detected in fish from polluted areas.[5] Stein et al.[20] demonstrated that when English sole were exposed simultaneously to sediment containing ^{3}H-BaP and Aroclor 1254 (^{14}C-PCBs), both BaP and PCBs were taken up; however, the BaP was extensively metabolized compared with the PCBs. Furthermore, the results suggest that the rate of excretion of BaP metabolites was similar to the rate of uptake of BaP, whereas the rate of excretion of PCBs or their metabolites was considerably less than the rate of uptake of PCBs. Thus, with time, the PCB body burdens in English sole increased significantly, whereas body burdens of BaP remained virtually unchanged (Figure 9). The bile of sole contained substantially higher concentrations of BaP metabolites compared with PCB metabolites, which suggests that while the concentrations of BaP in sole from contaminated areas may be low, BaP is continually being taken up and metabolized at a high rate.

STUDIES OF RELATIONSHIPS BETWEEN XENOBIOTICS AND HEPATIC LESIONS

Mathematical and statistical analyses were used to compare chemicals in Puget Sound sediments with prevalences of hepatic lesions in English sole. A mathematical method for grouping chemicals whose concentrations in sediments correlated positively with each other (factor analysis) was applied to sediment chemistry data and yielded four major groupings: AHs, metals, PCBs, and other chlorinated hydrocarbons. Using Spearman's rank correlation coefficient procedure, the prevalences of hepatic neoplasms and various

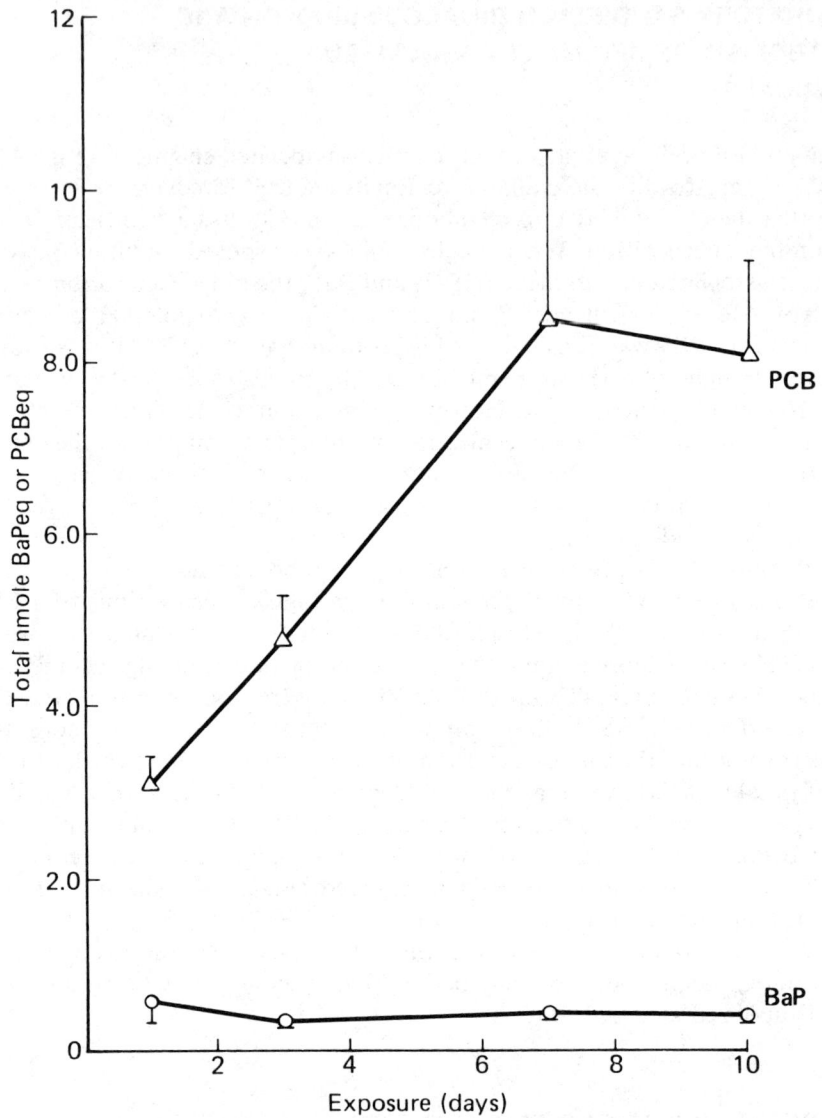

Figure 9. Accumulation of sediment-associated radiolabeled benzo(a)pyrene (–O–O–) and polychlorinated biphenyls (–△–△–) in English sole; values include parent xenobiotic and metabolites (adapted Reference 20).

nonneoplastic liver disorders were positively correlated ($p < 0.004$) with sediment concentrations of AHs and metals (Table II).[3]

Also, in studies concerned with metabolites of AHs, English sole with liver lesions (including neoplasia) had significantly higher bile concentrations of metabolites with BaP-like fluorescence than did fish without liver lesions.[14] In

Table II. Spearman's Rank Correlation Coefficients (r_s) and Significance Levels for Prevalences of Hepatic Lesions in English Sole and Chemical Concentrations in Bottom Sediment

Lesion Type	Chemical Group[a]	r_s	Significance Level
Neoplasms	AHs	0.48	0.003
Hyperplasia/FCA	AHs	0.47	0.004
Megalocytic hepatosis	AHs	0.54	0.001
Steatosis/hemosiderosis	AHs	0.49	0.002
One or more hepatic lesions	AHs	0.58	0.001
	Metals	0.54	0.001

[a]Chemical groups selected by factor analysis included aromatic hydrocarbons, metals, selected metals plus PCBs, and chlorinated compounds (Reference 3).

Table III. Covalent Binding Index (CBI) for Benzo(a)pyrene/DNA[a,b]

Compound	Species	Exposure (h)	Organ	CBI
Benzo(a)pyrene	Mouse	16	Liver	7
		4	Skin	24
	English sole	16	Liver	41
		24	Liver	45

[a]CBI represents polynuclear aromatic hydrocarbon (PAH) molecules bound per 10^6 nucleotides after administration of a theoretical dose of 1 mmol PAH/kg body weight.
[b]Varanasi et al.[26]

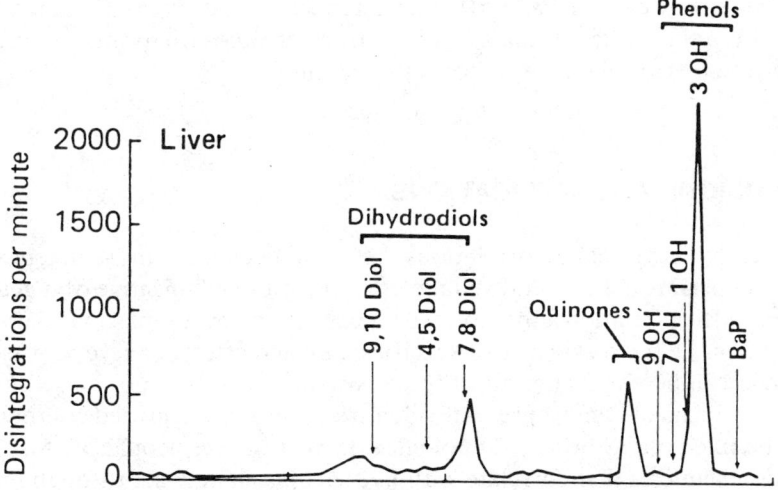

Figure 10. High-performance liquid chromatogram of organic solvent-soluble metabolites released after β-glucuronidase and arylsulfatase treatment of conjugates present in liver of English sole exposed to benzo(a)pyrene (adapted from Reference 25).

addition, using electron paramagnetic resonance (EPR) spectroscopy, as previously indicated, we have shown that the concentrations of free radicals (N-oxyl derivatives of carbazoles) were significantly higher ($p < 0.05$) in microsomes from sole livers with lesions (including hepatic neoplasms) than in microsomes from livers without lesions.[24] Subsequent studies have identified a variety of nitrogen-containing compounds in polluted Puget Sound sediments, including carbazoles.[5] Whether the bile analysis and EPR findings relate directly to the etiology of the observed hepatic neoplasms is, of course, not yet clear because of the unknown, presumably long latent period between carcinogen exposure and overt neoplasia.

In an attempt to understand relationships between hepatic neoplasms in English sole and AHs, we elected to study BaP as a model carcinogenic hydrocarbon. In one study, radiolabeled BaP was orally administered to English sole. A relatively high percentage of 7,8-dihydro-7,8-dihydroxy BaP (the penultimate carcinogen of BaP) was found in the liver (Figure 10).[17,25] Moreover, a higher percentage of this metabolite and a greater degree of covalent binding to DNA (Table III) occurred in English sole liver than in mouse liver.[26] Also, recent findings suggest that DNA excision repair is distinctly low in some fish cell cultures when compared with mammalian cell lines;[27] however, no direct evidence has yet been presented to indicate whether comparably low excision repair occurs with hepatocytes (e.g., from English sole). The findings that a high degree of DNA binding persisted for a period of several days in the liver of English sole exposed to a single dose of BaP[26] suggest that excision repair of modified DNA may be slow in English sole. Of particular interest is the fact that BaP has thus far not been clearly implicated as a liver carcinogen in mammals, although neoplastic lesions are produced at other sites in rodents.[28] We do not know whether BaP can produce liver neoplasms in English sole; however, it has now been established that liver carcinomas are produced in rainbow trout (*Salmo gairdneri*) after exposure to this carcinogen in either the diet or via intraperitoneal injection.[29]

CONCLUSIONS AND IMPLICATIONS

Statistically significant correlations between chemicals in sediment and hepatic neoplasms in bottom-dwelling fish appear to be indicative of a general cause-and-effect relationship; however, such an association should not be interpreted as de facto evidence of specific cause and effect for several reasons. One obvious inherent limitation of these correlations is that not all sediment-associated chemicals can be presently identified; thus, unidentified compounds cannot be ruled out as principal etiological factors in liver neoplasia. Also, the classes of chemicals that correlate with liver neoplasia may act through highly complex and difficult-to-define synergistic or antagonistic interactions, thus making the clarification of cause-and-effect relationships extremely difficult.

Some sediment-bound chemicals appear to be readily taken up by bottom-dwelling fish through a variety of routes, including the diet and direct uptake from sediment.[12,20]

Another gap in our understanding of associations between bottom-dwelling fish and chemicals in sediment concerns the mobility of the fish. Although migration and movement patterns of the three fish species we studied in Puget Sound are largely unknown, English sole, for example, are largely territorial, forming localized discrete subpopulations.[30] Although adult sole do make limited seasonal migrations for spawning (in winter), they apparently return to their home territory. Accordingly, it appears that this species of fish resides in polluted areas for relatively long, though presently undefined, periods of time.

It is clear that organic xenobiotics in sediments are readily transferred to bottom-dwelling fish. Little is known, however, about the transfer of many xenobiotics from contaminated bottom fish to predators, including the human consumer. Substantial metabolism of AHs in fish obviously reduces the potential for food-chain transfer of the parent AHs, although metabolites may be transferred. In contrast, highly lipophilic refractory compounds (e.g., PCBs and HCB) appear to have a relatively high potential for being transferred through food chains to birds and mammals.

The occurrences of liver neoplasia in bottom-dwelling fish from Puget Sound are not isolated examples. As indicated previously, high prevalences of neoplastic lesions have been found in other areas where there are chemically contaminated sediments. Even though there are numerous clues and indicators, there is little firm evidence concerning the specific causes of the observed neoplasms. Further insight into the etiology of these neoplasms will require considerable study in the field and especially in the laboratory. Moreover, as we stated in a previous publication, "It also seems important to stress that polluted environments which induce neoplasia in fish must surely also be responsible for a host of other serious changes at both the organismal and ecosystem level that have thus far not been recognized."[31]

ACKNOWLEDGMENTS

We thank Dr. Edmundo Casillas for reviewing this manuscript; Marylyn L. West for typing; James Peacock and Carol Hastings for preparation of the illustrations; William D. Gronlund and John R. Hughes for assistance with the field studies; Paul D. Plesha for taxonomic identification of food organisms; Leslie K. Moore, Victor D. Henry, Andrew J. Friedman, Ronald W. Pearce, Orlando Maynes, Richard G. Bogar, Catherine A. Wigren, Karen L. Grams, and Douglas G. Burrows for assistance in chemical analyses; Linda D. Rhodes for assistance in histopathological examinations and performance of statistical analyses of field data; and Marc Nishimoto and Tom Hom for assistance in biochemical studies. This study was supported, in part, by the Ocean Assessments Division, National Ocean Services, National Oceanic and Atmospheric Administration.

REFERENCES

1. Black, J. J., L. D. Evans, J. C. Harshbarger, and R. F. Zeigel. "Epizootic Neoplasms in Fishes from a Lake Polluted by Copper Mining Wastes," *J. Nat. Cancer Inst.* 69:915 (1982).
2. Black, J. J. "Field and Laboratory Studies of Environmental Carcinogenesis in Niagara River Fish," *J. Great Lakes Res.* 9:326 (1983).
3. Malins, D. C., B. B. McCain, D. W. Brown, S-L. Chan, M. S. Myers, J. T. Landahl, P. G. Prohaska, A. J. Friedman, L. D. Rhodes, D. G. Burrows, W. D. Gronlund, and H. O. Hodgins. "Chemical Pollutants in Sediments and Diseases of Bottom-Dwelling Fish in Puget Sound, Washington," *Environ. Sci. Technol.* 18:705 (1984).
4. Malins, D. C., B. B. McCain, D. W. Brown, A. K. Sparks, and H. O. Hodgins. *Chemical Contaminants and Biological Abnormalities in Central and Southern Puget Sound*, NOAA Tech. Memo OMPA-2 (Washington, DC: National Oceanic and Atmospheric Administration, 1980).
5. Malins, D. C., B. B. McCain, D. W. Brown, A. K. Sparks, H. O. Hodgins, and S-L. Chan. *Chemical Contaminants and Abnormalities in Fish and Invertebrates from Puget Sound*, NOAA Tech. Memo OMPA-19 (Washington, DC: National Oceanic and Atmospheric Administration, 1982).
6. Malins, D. C., M. M. Krahn, D. W. Brown, L. D. Rhodes, M. S. Myers, B. B. McCain, and S-L. Chan. "Toxic Chemicals in Marine Sediment and Biota from Mukilteo, Washington: Relationships with Hepatic Neoplasms and Other Hepatic Lesions in English Sole (*Parophrys vetulus*)," *J. Nat. Cancer Inst.* 74:487 (1985).
7. Roubal, W. T., T. K. Collier, and D. C. Malins. "Accumulation and Metabolism of Carbon-14 Labeled Benzene, Naphthalene, and Anthracene by Coho Salmon (*Oncorhynchus kisutch*)," *Arch. Environ. Contam. Toxicol.* 5:513 (1977).
8. Varanasi, U., and D. C. Malins. "Metabolism of Petroleum Hydrocarbons: Accumulation and Biotransformations in Marine Organisms," in *Effects of Petroleum on Arctic and Subarctic Marine Environments and Organisms, Vol. II*, D. C. Malins, Ed. (New York: Academic Press, 1977), pp. 175–270.
9. Varanasi, U., D. J. Gmur, and P. A. Treseler. "Influence of Time and Mode of Exposure on Biotransformation of Naphthalene by Juvenile Starry Flounder (*Platichthys stellatus*) and Rock Sole (*Lepidopsetta bilineata*)," *Arch. Environ. Contam. Toxicol.* 8:673 (1979).
10. Gruger, E. H., J. V. Schnell, P. S. Fraser, D. W. Brown, and D. C. Malins. "Metabolism of 2,6-Dimethylnaphthalene in Starry Flounder (*Platichthys stellatus*) Exposed to Naphthalene and p-Cresol," *Aquat. Toxicol.* 1:37 (1981).
11. Malins, D. C., and H. O. Hodgins. "Petroleum and Marine Fishes: A Review of Uptake, Disposition and Effects," *Environ. Sci. Technol.* 15:1272 (1981).
12. Varanasi, U., and D. J. Gmur. "Hydrocarbons and Metabolites in English Sole (*Parophrys vetulus*) Exposed Simultaneously to [^{3}H]Benzo[a]pryene and [^{14}C]Naphthalene in Oil-Contaminated Sediment," *Aquat. Toxicol.* 1:49 (1981).

13. Varanasi, U., and D. J. Gmur. "In vivo Metabolism of Naphthalene and Benzo[a]pyrene by Flatfish," in *Chemical Analysis and Biological Fate: Polynuclear Aromatic Hydrocarbons*, M. W. Cooke and A. J. Dennis, Eds. (Columbus, OH: Battelle Press, 1981), pp. 367-376.
14. Krahn, M. M., M. S. Myers, D. G. Burrows, and D. C. Malins. "Determination of Metabolites of Xenobiotics in the Bile of Fish from Polluted Waterways," *Xenobiotica* 14(8):633 (1984).
15. Malins, D. C., M. S. Myers, W. D. MacLeod, and W. T. Roubal. Unpublished results, presented at the Second International Symposium on Responses of Marine Organisms to Pollutants, Woods Hole Oceanographic Institution, Woods Hole, MA (1983).
16. Perin, F., M. Dufour, J. Mispelter, B. Ekert, C. Kunneke, F. Oesch, and F. Zajdela. "Heterocyclic Polycyclic Aromatic Hydrocarbon Carcinogenesis: ^{7}H-Dibenzo[c.g]carbazole Metabolism by Microsomal Enzymes from Mouse and Rat Liver," *Chem.-Biol. Interact.* 35:267 (1981).
17. Gmur, D. J., and U. Varanasi. "Characterization of Benzo[a]pyrene Metabolites Isolated from Muscle, Liver and Bile of a Juvenile Flatfish," *Carcinogenesis* 3:1397 (1982).
18. Young, D. R. "A Comparative Study of Trace Metal Contamination in Southern California and New York Bight," in *Ecological Stress and the New York Bight*, G. F. Mayer, Ed. (Columbia, SC: Science and Management, Estuarine Research Federation, 1982), p. 249.
19. Sherwood, M. J. "Fin Erosion, Liver Condition, and Trace Contaminant Exposure in Fishes from Three Coastal Zones," in *Ecological Stress and the New York Bight*, G. F. Mayer, Ed. (Columbia, SC: Science and Management, Estuarine Research Foundation, 1982), p. 359.
20. Stein, J. E., T. Hom, and U. Varanasi. "Simultaneous Exposure of English Sole (*Parophrys vetulus*) to Sediment-Associated Xenobiotics: Part 1. Uptake and Disposition of ^{14}C-Polychlorinated Biphenyls and ^{3}H-Benzo[a]pyrene," *Mar. Environ. Res.* 13:97 (1984).
21. Varanasi, U., M. Uhler, and S. I. Stranahan. "Uptake and Release of Naphthalene and Its Metabolites in Skin and Epidermal Mucus of Salmonids," *Toxicol. Appl. Pharamcol.* 44:277 (1978).
22. Thomas, R. E., and S. D. Rice. "Excretion of Aromatic Hydrocarbons and Their Metabolites by Freshwater and Saltwater Dolly Varden Char," in *Biological Monitoring of Marine Pollutants*, F. J. Vernberg, F. P. Thurberg, A. Calabrese, W. B. Vernberg, Eds. (New York: Academic Press, 1972), pp. 161-176.
23. McKim, J. M., and E. M. Heath. "Dose Determinations for Waterborne 2,5,2',5'-[^{14}C]Tetrachlorobiphenyl and Related Pharmacokinetics in Two Species of Trout (*Salmo gairdneri* and *Salvelinus fontinalis*): A Mass-Balance Approach," *Toxicol. App. Pharmacol.* 68:177 (1983).
24. Malins, D. C., M. S. Myers, and W. T. Roubal. "Organic Free Radicals Associated with Idiopathic Liver Lesions of English sole (*Parophrys vetulus*) from Polluted Marine Environments," *Environ. Sci. Technol.* 17:679 (1983).
25. Varanasi, U., J. E. Stein, M. Nishimoto, and T. Hom. "Benzo[a]pyrene Metabolites in Liver, Muscle, Gonads and Bile of Adult English sole (*Parophrys vetulus*)," reprinted from *Polynuclear Aromatic Hydrocarbons: Seventh International Symposium on Formation, Metabolism, and Measurement*, M. W. Cooke and A. J. Dennis, Eds. (Columbus, OH: Battelle Press, 1982), pp. 1221-1234.

26. Varanasi, U., J. E. Stein, and T. Hom. "Covalent Binding of Benzo[a]pyrene to DNA in Fish Liver," *Biochem. Biophys. Res. Commun.* 103(2):780 (1981).
27. Regan, J. D., R. D. Synder, A. A. Francis, and B. L. Olla. "Excision Repair of Ultraviolet- and Chemically-Induced Damage in the DNA of Fibroblasts Derived from Two Closely Related Species of Marine Fish," *Aquat. Toxicol.* 4:181 (1983).
28. Zedeck, M. S. "Polycyclic Aromatic Hydrocarbons, a Review," *J. Environ. Pathol. Toxicol.* 3:537 (1980).
29. Hendricks, J. D., T. R. Meyers, D. W. Shelton, J. L. Casteel, and G. S. Bailey. "The Hepatocarcinogenesis of Benzo[a]pyrene to Rainbow Trout by Dietary Exposure and Intraperitoneal Ingestion," *J. Nat. Cancer Inst.* (in press).
30. Day, D. E. "Homing Behavior and Population Stratification in Central Puget Sound English Sole (*Parophrys vetulus*)," *J. Fish. Res. Board Can.* 33:278 (1976).
31. Malins, D. C. "The Occurrence and Etiology of Liver Neoplasia in Bottom-Dwelling Fish from Polluted Areas of Puget Sound, Washington," in *Carcinogens and Mutagens in the Environment, Vol. VI*, H. Stich, Ed. (Boca Raton, FL: CRC Press, Inc., in press).

CHAPTER 33

Carcinogenic Effects of River Sediment Extracts in Fish and Mice

John Black, Helen Fox, Penny Black, and Fred Bock

An association between water pollution and neoplastic disease in native fish populations has been implied by a number of published reports in which a high incidence(s) of neoplastic disease has been discovered in a specific aquatic environment suspected of harboring chemical pollutants that are carcinogenic to fish.[1-5] Although chemical induction of cancer was amply demonstrated in the laboratory prior to 1940,[6,7] the realization that pollutants may be involved in the etiology of fish neoplasms has been slow to develop. In 1941, Lucke and Schlumberger[8] noted epithelial and oral neoplasms occurring among brown bullheads (*Ictalurus nebulosus*) from the Schuylkill and Delaware rivers in the vicinity of Philadelphia; however, there is no indication that these authors suspected that such lesions might be caused by pollutants. The studies by Dawe et al.[9] of hepatic neoplasms occurring in bottom-feeding species in Deep Creek Lake, Maryland, are a focal point, for they appear to be the first instance in which chemical pollutants were suspected as the cause of neoplasia in a wild fish population. Dawe suggested the practical use of bottom-feeding fish species as "useful indicators of environmental carcinogens."

Polluted river sediments are recognized sources of sequestered chemical pollutants,[10] and contaminated sediments are suspected of playing a role in the etiology of epizootics of neoplasia occurring in some wild fish populations.[11-13] Several types of neoplasms have been noted in an examination of brown bullheads collected from heavily polluted aquatic areas in eastern Lake Erie, the Buffalo River, and the upper Niagara River, Erie County, New York. The neoplasms observed included both hepatic (hepatocellular and cholangiocellular) and epidermal (papillomas and carcinomas) lesions.[12] Similar neoplasms have been noted in bullheads collected from other polluted waterways, especially from the Black River, Lorain County, Ohio.[13]

Feral bullheads from these polluted waterways show a spectrum of skin lesions ranging from simple hyperplasia to invasive skin cancers consistent with the diagnosis of melanomas and squamous cell carcinomas. Frequently, the skin lesions are located in anatomic sites (e.g., ventral surface, lips, and head), suggesting that contact with carcinogenic chemicals encountered by feeding or resting in and on polluted sediments may be a causal factor. A postulated route of exposure involved in the induction of liver neoplasms in this fish species is less obviously construed; however, in polluted aquatic systems, fish are exposed to waterborne chemicals via three routes: aqueous exposure with uptake via skin and gills, direct contact with pollutants in sediments, and the dietary route through ingestion of contaminated food-chain organisms and/or detritus.

That carcinogenic chemicals are the cause of epizootic neoplasia in some wild fish populations is also supported by laboratory experiments that have induced neoplasms in fish in response to chemicals classified as carcinogens on the basis of their activity in laboratory animals.[14] Thus, although some field studies have shown associations between pollution and epizootic fish neoplasia, there has been no direct proof of the pollutant hypothesis. Experiments using single-compound exposure in laboratory animals and in a variety of aquarium and hatchery-reared fish species are of significant value in assessing a chemical's carcinogenic potential and/or the suitability of various species as surrogates for human risk assessment. However, these types of experiments will probably not identify specific neoplasms in free-living fish populations as resulting from exposure to chemical pollutants. As a first step toward demonstrating a cause-and-effect relationship between pollutant exposure and feral fish neoplasia, we have tested complex river sediment extracts for oncogenic potential in laboratory-adapted brown bullheads and Swiss-strain laboratory mice. A previous report described the results of skin painting in bullheads.[15] Additionally, the relative carcinogenicity of the extracts was compared with a carcinogenic polycyclic aromatic hydrocarbon (PAH), 3,4-benzo(*a*)pyrene (BaP), in mouse skin paintings.

METHODS AND MATERIALS

High-Performance Liquid Chromatography (HPLC) Analysis of the Sediments and Extracts

Details of the methods and apparatus used for HPLC have been documented.[16] The methods used for isolation and prefractionation of PAH from river sediments and extracts were similar to those described by Dunn.[17] Samples of sediments and/or residues derived from 12-h Soxhlet extractions of dried river sediments were solubilized in refluxing ethanol/potassium hydroxide. Following solubilization, hydrocarbons were extracted in cyclohexane by liquid/liquid partitioning and a PAH-containing fraction isolated by chromatography on partially deactivated Florisil (60–100 mesh). Additional aliphatic hydrocarbons were removed by partitioning and back extractions between dimethyl sulfoxide (DMSO) and hexane. The PAH were concentrated into an injection volume of DMSO and analyzed by reverse-phase HPLC.

The PAH content of resulting analytical fractions was evaluated by gradient HPLC using acetonitrile–water as the eluting solvent system. The column (Vydac 201TP, 54.6) was eluted at 1.5 mL/min using a linear solvent gradient (to 100% acetonitrile) of 40 min in length with an initial solvent concentration of 40% acetonitrile:water. Compounds separated by HPLC were detected sequentially by UV absorbance at 254 nm and by fluorescence emission at 305 nm (exciting wavelength, 295 nm). Using this dual detection system, low-molecular-weight PAHs were detected by their absorbance, and most high-molecular-weight PAHs were detected by their fluorescence signals. Compounds were identified on the basis of retentions relative to chrysene and phenanthrene, by cochromatography with PAH standards, and by their fluorescence and absorbance spectra. Data were quantitated on the basis of peak areas and individual PAH response factors. Quantitation of most PAH was based on NBS standard reference materials 1647.

Preparation of River Sediment Extracts

Sediments obtained from the industrialized estuarine portions of the Buffalo and Black rivers were air dried at room temperature. After drying, a stock complex organic fraction was prepared by 12-h Soxhlet extractions with hexane–acetone (1:1 volume) as the extracting solvent. Following extraction, the organic soluble materials were pooled and stored under nitrogen in brown glass containers. Organic residue concentration was determined by air evaporating 1.0 mL of extract to dryness in a preweighed aluminum weighing dish. Concentrations of residue content were adjusted by concentrating or diluting the stock solution. In the case of mouse skin painting preparations, the hexane–acetone solvent was exchanged for acetone.

Carcinogenesis Experiments

Mice

Groups of 2- to 3-month-old Swiss IcR$_{Ha}$ female mice were randomized in cages containing five animals each. Hair was shaved from the backs of the mice, and they were treated five times a week with 0.1 mL of an acetone solution of carcinogen. In initiation-promotion assays, 0.25 mL of a single-dose exposure of the carcinogen was given 3 weeks before treatment with the promoting solution. The promoter was also delivered in a 0.1-mL acetone solution and applied five times per week. Tumor development was recorded at weekly intervals.

Fish Skin Painting

Four groups of 16 to 18 bullheads (total 69 fish, average size 6 in.) obtained from a nonpolluted pond were held in large polyethylene tanks equipped with individual recirculation pumps and biological filtration. A continuous trickle of tap water was added to maintain water levels and quality. For treatment, fish were removed from the holding tanks, and the area between the upper lip and the occiput was blotted dry with a gauze pad. Buffalo River extract (BuRSE) at 5.0% was applied to the area with a cotton-tipped swab. A brief period was allowed for absorption of the test material, and the fish were then placed in a recovery tank prior to their return to the holding tanks. This procedure was used to minimize the transfer of RSE to the holding tanks. A single group of 16 control fish were handled in the same manner, except that they were treated with the extracting solvent (hexane–acetone) only. All fish were treated once a week except during occasional outbreaks of a bacterial disease (*Aeromonas sp.*) during which paintings were temporarily suspended.

Fish Feeding

Twelve bullheads were fed a commercial trout diet (Ziegler Brothers, 1/8-in. pellets) adulterated to 5.0 parts per thousand (ppt) by adding corn oil (5.0 w/v%) containing the BuRSE. The diet was prepared by adding 2.5 g of extractable residues to 500 g of trout diet. After addition of the contaminated corn oil, the diet was thoroughly mixed by slowly hand rotating the materials in a 2-L polyethylene bottle.

Six fish were sacrificed after 4 months exposure, and the remaining six were sacrificed after 7 months on the diet. Control fish were fed unadulterated trout pellets.

RESULTS

PAH Analyses

Sediments used to prepare the organic extract for carcinogenesis testing were grossly similar in that both Buffalo and Black River materials consisted of fine silts that exhibited strong chemical or petroleum-like odors and released significant amounts of oily surface films when disturbed in the water. Dried aliquots of Buffalo and Black River sediments contained up to 7.0% ignitable organic matter and yielded from 0.3 to 1.0% Soxhlet extractable residues.

Table I compares the concentrations of 18 PAH compounds estimated by HPLC. Data are not corrected for recovery rates, which vary slightly for each PAH compound. Overall recovery rates average between 70 to 90%, as determined by trace enrichment studies on paired duplicates. Data from the Buffalo

AQUATIC MODELS AND TUMOR INDUCTION

Table I. PAH Content of Sediment Extracts[a]

PAH Compound	BIRSE[b] (ng/0.1 mL)	BuRSE[c] (ng/0.1 mL)	BIS[d] (ng/g)	BIS[e] (ng/g)	BuS[f] (ng/g)
Fluorene	1,888	144	2,270		640
Phenanthrene	11,044	1,326	17,271	390,000	5,950
Anthracene	3,620	216	3,848		1,950
Fluoranthene	31,768	2,806	28,584	220,000	7,000
MePhenanthrene	1,132	316	2,377		960
Pyrene	12,748	1,826	14,810	140,000	13,750
MeAnthracene	758	182	1,109		700
Benzofluorene	2,634	1,046	5,246		6,350
Benzanthracene	5,692	622	5,817	51,000	1,445
Chrysene	5,180	331	4,115	51,000	365
Benzo(e)pyrene	6,674	194	9,486	28,000	4,500
Perylene	9,234	1,282	12,338	12,000	9,400
Benzo(b)fluoranthene	4,900	712	5,607		5,500
Benzo(k)fluoranthene	2,308	342	2,625		780
Benzo(a)pyrene	4,484	588	5,411	43,000	1,060
Dibenz(a,h)anthracene	1,038	148	1,071	9,400[g]	3,220
Benzo(g,h,i)perylene	3,382	454	3,448	24,000	1,155
Indeno(1,2,3-c,d)pyrene	3,954	490	3,640	26,000	1,400

[a]Sediment extracts used for skin painting of mice (concentration for bullheads was 2.5 times greater).
[b]Black River sediment extract.
[c]Buffalo River sediment extract.
[d]Black River sediment.
[e]Data taken from Baumann et al.[13]
[f]Buffalo River sediment.
[g]Dibenz(a,h) + (A,C) anthracene.

River sediment are estimates based on analyses of samples from two sites where sediments were collected to prepare the extracts.

Bullhead Skin Painting

About 3 months after beginning treatments with BuRSE, bullheads underwent a series of slow progressive skin changes. Initially, this consisted of a slight blanching of the normal coloration accompanied by an apparent coarsening of the skin. Additionally, a slight hyperemia developed and appeared to be more prominent about the lips. These symptoms persisted for approximately 6 months past their initial development. These mild alterations were superficial and tended to regress rapidly on cessation of the paintings. Microscopic examinations of histology sections taken from the skin of several animals were insignificant; no obvious differences were noted between experimental and control animals.

As paintings continued, the mild, coarsened appearance subsided and the skin returned to a more or less normal color and texture. Between 10 and 14 months, the skin began to slowly darken, and some of the animals appeared irregularly pitted in the BuRSE-treated area. The pitted appearance was due to irregular hyperplasia of the epidermis (Figure 1).

Although the histologic appearance of skin sections exhibiting this type of change was variable, in general, there was increased basophilia, an irregular increase in cell numbers (including increased thickness of the basal cell layer), and a loss of the normal skin architecture (i.e., loss of dermal pegs and decreased numbers of alarm substance cells). Between 14 and 18 months, 5 of 22 surviving fish from the experimental group were grossly hyperplastic, and several fish developed papillomas. Two fish developed coalescing groups of multiple papillomas superimposed on the hyperplasia in the area exposed to the BuRSE (Figure 2). Following termination of the experiments, 8 of 22 surviving fish had developed papillomas. Controls appeared normal throughout the experiment.

Bullhead Dietary Exposure

Among 5 bullheads receiving the 5.0 ppt dietary BuRSE for 4 months, one fish liver exhibited grossly visible nodules. When the liver was examined microscopically, several types of lesions were observed. These included a small basophilic hepatocellular focus/nodule, a small focal proliferation of bile ductules, and a large lesion of mixed fibrosis and duct-like formations of hepatocytes. The first two lesions were considered incipient neoplasms (i.e., well-differentiated early-stage hepatocellular carcinoma and early-stage cholangioma). The third lesion was considered to be a nonneoplastic toxic event. Among 6 fish receiving the contaminated diet for 7 months, numerous

AQUATIC MODELS AND TUMOR INDUCTION 421

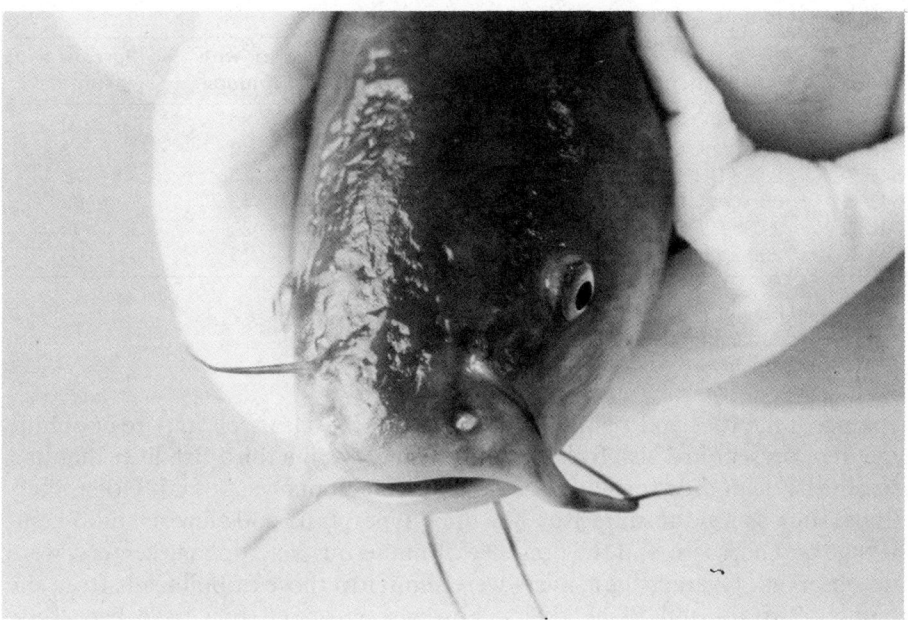

Figure 1. Gross aspect of epidermal hyperplasia induced in a laboratory-adapted brown bullhead.

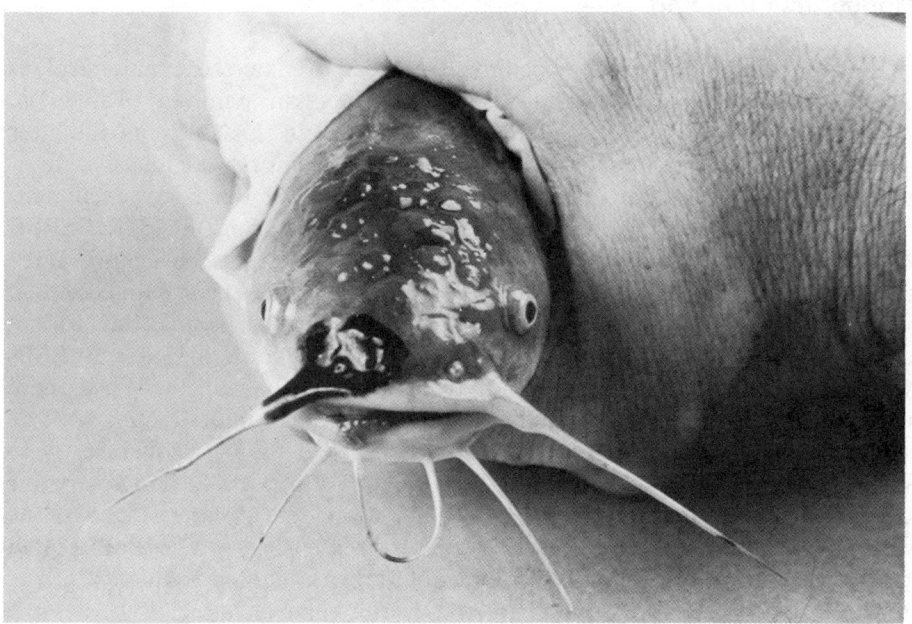

Figure 2. Multiple epidermal papillomas induced in a laboratory bullhead after approximately 22 months of exposure to 5.0% Buffalo River sediment extract (BuRSE) applied once per week.

Table II. Complete Carcinogen Assay

Group	Treatment[a]	No. per Group	No. with Tumors[b]	Percent with Tumors[b]
154B 5	Acetone	50	0	0
154B 6	66 µg/mL BaP	50	6	12
154B 7	2.0% BRSE[c]	50	15	30
154B 8	66 µg/mL BaP + 2.0% BRSE	50	43	86
154C 1	2.0% BlRSE[d]	25	20	80
154C 2	200 µg/mL BaP	25	16	64

[a]With 0.1 mL.
[b]Mice with skin papillomas at 30 weeks.
[c]Buffalo River sediment extract.
[d]Black River sediment extract.

hepatocellular foci and neoplastic nodules of the clear-cell and eosinophilic type were present in 2 fish livers (Figures 3 and 4). In a third fish liver that had grossly visible nodule formation, a large cholangioma was found. Other alterations, such as mild-to-moderate bile duct hyperplasia and fibrosis, mild hemorrhage and necrosis, and the presence of numerous macrophage centers, were also observed. Overall, these livers were similar to those of bullheads from the polluted Buffalo and Black rivers; however, clear-cell foci have been only rarely observed in the tumorous livers of feral bullheads.

Mouse Skin Painting

Both Buffalo and Black river sediments yielded organic extractable residues that were carcinogenic when tested by mouse skin painting (Table II). Although all tumors were tabulated as papillomas in this study, in the later stages of exposure, an undetermined percentage of tumors were invasive carcinomas (squamous cell). In order of carcinogencity, 2.0% Black River sediment extract (BlRSE) > 200 µg/mL BaP > 2.0% BuRSE > 66 µg/mL BaP. BuRSE applied in combination with BaP (i.e., 2.0% BuRSE plus 66 µg/mL BaP) appeared to enhance tumor formation when results were tabulated at 30 weeks. Specifically, mice treated with 0.1 mL daily (66 µg/mL BaP alone) exhibited a tumor frequency of 12 ± 4.6%, and 0.1 mL of 2.0% BuRSE, when applied alone, yielded a tumor frequency of 30 ± 6.5%. When these treatments were applied in combination, at 30 weeks a tumor frequency of 85 ± 4.9% resulted. This was greater than expected if tumor frequencies were simply additive: 86 vs 42% (30 + 12%, $Z = 4.58$, $p > 0.05$). However, this cocarcinogenic enhancement was not nearly as evident at later stages of the experiments because of increases in tumor frequencies in the BaP-treated group between weeks 33 and 34 (data not shown). Tumor frequency in the 154B Group 6 increased from 18 to 34% at this time.

Only BuRSE was assessed for initiating and promoting activity (Table III). A single initiating dose of 0.25 mL of BuRSE (154B, Group 4) followed by daily administration of 0.1 mL of 1.5 µg/mL tetradecanoyl phorbol acetate

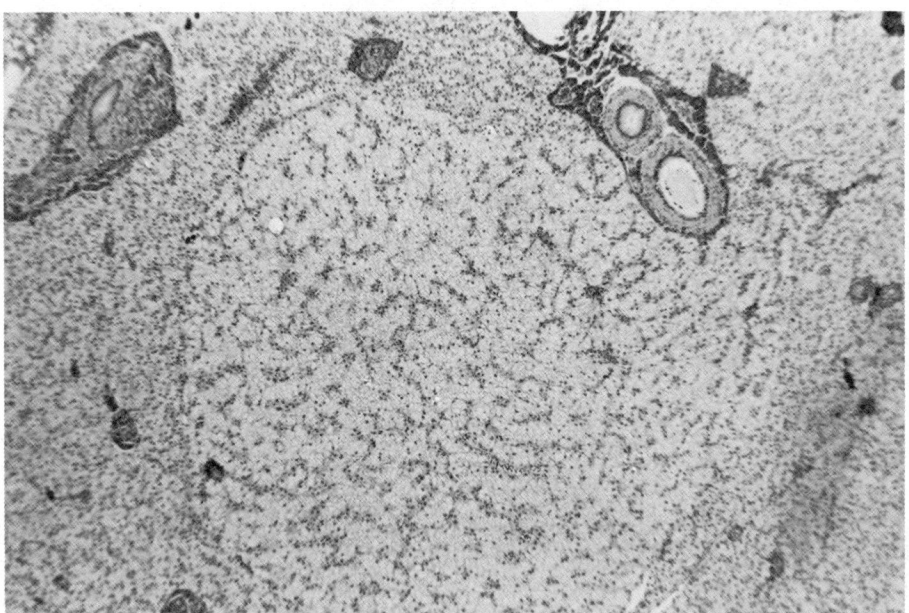

Figure 3. Hepatocellular nodule of the clear-cell type in a bullhead receiving dietary Buffalo River sediment extract (BuRSE) at 5 parts per thousand for 7 months.

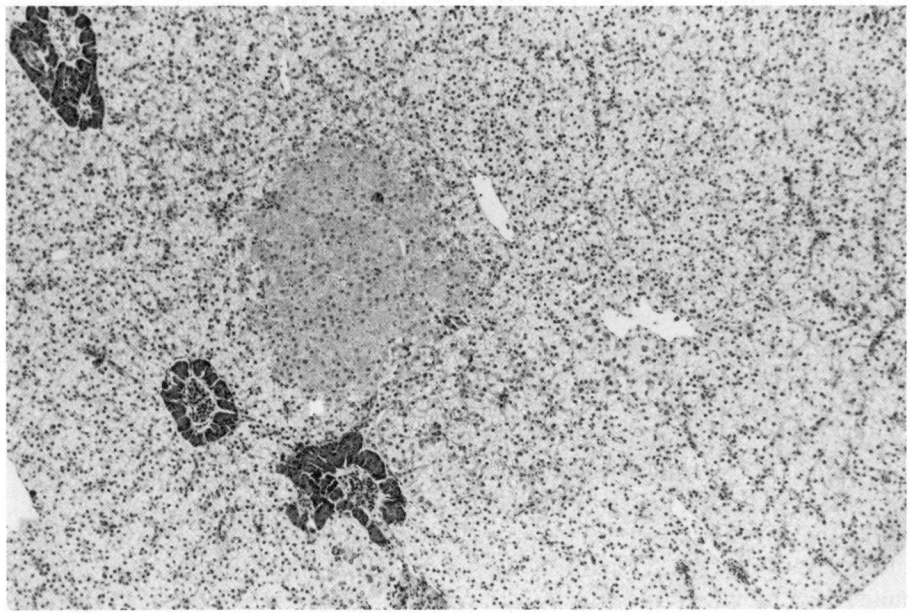

Figure 4. Eosinophilic focus of hepatocytes in a bullhead receiving dietary Buffalo River sediment extract (BuRSE) for 7 months.

Table III. Initiation-Promotion Assay

Group	Initiator[a]	Promotor[a]	No. per Group	No. with Tumors[b]	Percent with Tumors[b]
154B 1	0.05% DMBA	Acetone	50	0	0
154B 4b	Acetone	1.5 µg/mL TPA[c]	50	0	0
154B 4	2.0% BRSE[d]	1.5 µg/mL TPA	50	4	8
154B 2	0.05% DMBA	1.5 µg/mL TPA	50	20	40
154B 3	0.05% DMBA	2.0% BRSE	50	26	52

[a]Initiating dose (0.25 mL) given as a single application followed in 3 weeks by daily applications of promoter (0.1 mL).
[b]Mice exhibiting skin papillomas at 30 weeks.
[c]Tetradecanoyl phorbol acetate.
[d]Buffalo River sediment extract.

(TPA) resulted in a low frequency of tumors (8%) at 30 weeks. BuRSE at 2.0% was also used as a promoter following initiation with a single 0.25 mL dose of 0.5% dimethylbenz(a)anthracene (DMBA), 154B, Group 3. This resulted in 52% tumor frequency, which was significantly in excess ($Z = 2.24$, $p > 0.05$, 22% excess) of that observed for 2.0% BuRSE applied as a complete carcinogen (see Table II, 154B, Group 7) and which resulted in only a 30% tumor incidence. The 22% excess is interpreted as a cocarcinogenic effect/promoting activity present in the river sediment extract.

DISCUSSION

The brown bullhead is a widely distributed catfish (*Ictaluridae*) that exhibits several types of spontaneous neoplasms, including hepatic (hepatocellular and cholangiocellular) and epidermal (oral and epidermal papillomas, melanocytomas, and carcinomas).[18] The frequent occurrence of tumorous specimens in polluted aquatic environments (e.g., the Buffalo and Black rivers) has raised concern about the impacts of pollution on the health and safety of aquatic environments.[19]

Although the experiments described here do not prove conclusively that tumors in feral bullhead populations are universally caused by exposure to anthropogenic pollutants, nor do they reveal the identity of the causal agents of liver and/or skin neoplasms in either the Buffalo or Black rivers, they do strengthen the data base. In general, this suggests that some neoplasms in populations of bottom-feeding fish species are indicative of carcinogenic pollutant exposures. From this perspective, epidermal and hepatic neoplasms in bullhead populations specifically appear to be consistent with this hypothesis.

The limited data package presented in this study prevents extensive speculation about the role of PAHs relative to the carcinogenic response in either bullheads or mice. Based on the results of skin painting experiments with mice, the amount of BaP in the sediments alone is unlikely to account for the degree of carcinogenicity observed. In addition, the data indicate that river sediment extracts exhibited a low level of promoting activity when applied following a subcarcinogenic dose of DMBA. Sediments from the Black River yielded higher estimated concentrations of all PAH compounds analyzed except benzofluorene and dibenz(*a,h*)anthracene. Previous analytical studies of Black River sediments by Baumann et al.[13] indicated much higher concentrations of PAHs (i.e., approximately one log unit higher). However, these differences probably reflect variation with respect to the proximity of sediment collections to pollutant sources.

When Buffalo and Black river sediment extracts prepared for bioassay work were compared on an equal basis (i.e., both extracts were 2.0 w/v% residue), Black River extracts again yielded higher PAH estimates. Because of the very similar organic residue contents of the two sediments, it is likely that PAHs comprise a relatively greater percentage of the pollutant burden of Black River sediments. Although the rigorous extraction and concentration procedures may have favored the predominance of PAH in these extracts, because other likely environmental pollutants such as aliphatic hydrocarbons and aromatic amines were not analyzed, the contribution of other contaminants to the observed carcinogenicity cannot be assessed. The difference in the two PAH burdens appears to be roughly consistent with the differences in carcinogenicity, both in the experimental work described here and in the differences in tumor prevalence among the feral bullhead populations of the two rivers. For example, in the Black River, approximately 30% of the adult bullheads exhibited visible liver tumors and 25% exhibited either skin or lip tumors (N = 275).[13] In the Buffalo River, only approximately 17% of the large adult bullheads exhibited grossly visible skin or liver tumors (N = 28).[20]

ACKNOWLEDGMENTS

The authors wish to thank H. Meyers, B. Held, T. Hart, Jr., R. Gross, D. Paulk, and E. Weyand for their technical assistance. The authors also thank George Fox for photographic assistance, Ruth Weaver for the preparation of tissue specimens, and Dr. John Harshbarger, Registry for Tumors in Lower Animals, for helpful suggestions.

This work was supported in part by a grant from the National Science Foundation, Grant PFR78-22625, and by an EPA grant, C164851. Although the research described in this chapter has been funded in part by the U.S. Environmental Protection Agency, it has not been subjected to the Agency's required peer and policy review; therefore, it does not necessarily reflect the views of the Agency, and no official endorsement should be inferred.

REFERENCES

1. Brown, E. R., J. J. Hazdra, L. Keith, I. Greenspan, J. B. G. Kwapinski, and P. Beamer. "Frequency of Fish Tumors Found in a Polluted Watershed as Compared to Nonpolluted Canadian Waters," *Cancer Res.* 33:189-198 (1973).
2. Falkmer, S., S. Marklund, P. E. Mattsson, and C. Rappe. "Hepatomas and other Neoplasms in the Atlantic Hagfish (*Myxine glutinosa*), A Histopathological and Chemical Study," *Ann. N.Y. Acad. Sci.* 298:342-355 (1977).
3. Smith, C. E., T. H. Peck, R. H. Klauda, and J. B. McLaren. "Hepatomas in Atlantic Tomcod (*Microgadus tomcod*) Collected in the Hudson River Estuary, New York," *J. Fish Dis.* 2:313-319 (1979).
4. Grizzle, J., T. E. Schwedler, and A. L. Scott. "Papillomas of Black Bullheads, *Ictalurus melas* (Rafinesque), Living in a Chlorinated Sewage Pond," *J. Fish Dis.* 4:345-351 (1981).
5. Pierce, K. V., B. B. McCain, and S. R. Wellings. "Pathology of Hepatomas and Other Liver Abnormalities in English Sole (*Parophrys vetulus*) from The Duwamish River Estuary, Seattle, Washington," *J. Nat. Cancer Inst.* 60:1445-1453 (1978).
6. Yamagiwa, K., and K. Ichakawa. "Experimentelle Studie uber die Pathologenese der Epithelialgeschwulste," *Mitteilungen Med. Facultat Kaiserl. Univ. Tokoyo* 15:295-344 (1915).
7. Sasaki, T., and T. Yoshida. "Experimentelle Erzeugung des Lebercarcinoma durch Futterung mit O-Amidoazotoluol," *Virchows Arch. A: Pathol. Anat.* 295:175 (1935).
8. Lucke, B., and H. Schlumberger. "Transplantable Epitheliomas of the Lip and Mouth of Catfish," *J. Exp. Med.* 74:397-408 (1941).
9. Dawe, C. J., M. F. Stanton, and F. J. Schwartz. "Hepatic Neoplasms in Bottom-Feeding Fish of Deep Creek Lake, Maryland," *Cancer Res.* 24:1194-1201 (1964).
10. Lopez-Avila, V., and R. A. Hites. "Organic Compounds in an Industrial Wastewater. Their Transport into Sediments," *Environ. Sci. Technol.* 14:1382-1390 (1980).
11. Malins, D. C., B. B. McCain, D. W. Brown, A. K. Sparks, and H. O. Hodgins. "Chemical Contaminants and Biological Abnormalities in Central and Southern Puget Sound," NOAA Tech. Memorandum OMPA-2 (Boulder, CO: National Oceanic & Atmospheric Administration, 1980).
12. Black, J. J. "Field and Laboratory Studies of Environmental Carcinogenesis in Niagara River Fish," *J. Great Lakes Res.* 9:326-334 (1983).
13. Baumann, P. C., W. D. Smith, and M. Ribick. "Hepatic Tumor Levels and Polynuclear Aromatic Hydrocarbon Levels in Two Populations of Brown Bullheads (*Ictalurus nebulosus*)," in *Polynuclear Aromatic Hydrocarbons: Physical and Biological Chemistry*, M. Cooke, A. J. Dennis, and G. L. Fisher, Eds. (Columbus, OH: Battelle Press, 1982), pp. 93-102.
14. Hendricks, J. D. "Chemical Carcinogenesis in Fish," in *Aquatic Toxicology*, L. J. Weber, Ed. (New York: Raven Press, 1982), pp. 149-211.
15. Black, J. J. "Epidermal Hyperplasia and Neoplasia in Brown Bullheads (*Ictalurus nebulosus*) in Response to Repeated Applications of a PAH Containing Extract of Polluted River Sediment," in *Polynuclear Aromatic Hydrocarbons: Formation, Metabolism, and Measurement*, M. W. Cooke and A. J. Dennis, Eds. (Columbus, OH: Battelle Press, 1982), pp. 99-111.
16. Black, J. J., T. F. Hart, Jr., and E. Evans. "HPLC Studies of PAH Pollution in a Michigan Trout Stream," in *Polynuclear Aromatic Hydrocarbons: Chemical Anal-*

ysis and Biological Fate, M. Cooke and A. J. Dennis, Eds. (Columbus, OH: Battelle Press, 1981), pp. 343-355.
17. Dunn, B. P. "Techniques for Determination of Benzo(a)pyrene in Marine Organisms and Sediments" *Environ. Sci. Technol.* 10:1018-1021 (1976).
18. Harshbarger, J. C. "Activities Report: Registry of Tumors in Lower Animals," (Washington, D.C.: Random house, Smithsonian Institution press, 1965-1973, plus annual supplements 1974-1983).
19. Harshbarger, J. C. "Testimony before U.S. House of Representatives Subcommittee of Fisheries and Wildlife Conservation and the Environment, Sept. 21, 1983." *Federal Register* (in press).
20. Baumann, P. C., J. J. Black, and J. C. Harshbarger. Unpublished data (1983).

CHAPTER 34

Tumor Induction in Several Small Fish Species by Classical Carcinogens and Related Compounds

William E. Hawkins, Robin M. Overstreet, William W. Walker, and C. Steve Manning

The use of small fish species for broad-scale carcinogenicity testing has been encouraged[1] but not yet attempted. Small fish species appear to be favorable bioassay models for testing waterborne mixtures of compounds such as contaminants of drinking water.[2] Some of these contaminants, primarily halogenated organics, have been identified as carcinogens and mutagens.[3] Before such studies are conducted, however, several preliminary investigations must determine the appropriate species and exposure systems. Species considered for testing should meet several criteria. They should be susceptible to tumor formation when exposed to proven carcinogens, and they should be easily reared and maintained under laboratory conditions. Exposure systems must be reliable, must be safe to workers and the environment, and must provide constant levels of toxicants to test organisms.

In this chapter we report preliminary findings on the tumorigenic response of several small fish species to methylazoxymethanol acetate (MAM-Ac) and N-methyl-N'-nitro-N-nitrosoguanidine (MNNG). Both of these carcinogens are direct acting; that is, they do not require metabolic activation to exert their carcinogenic effect. Both have also been shown to produce neoplasms in livers of various fish species.[4]

We also report on the development of exposure systems for testing individual and mixtures of drinking water contaminants for carcinogenicity using small fish species as bioassay models.

MATERIALS AND METHODS

Japanese medaka (*Oryzias latipes*), Gulf killifish (*Fundulus grandis*), sheepshead minnow (*Cyprinodon variegatus*), guppy (*Poecilia reticulata*), fathead minnow (*Pimephales promelas*), inland silverside (*Menidia beryllina*), and rivulus (*Rivulus marmoratus*) were tested in single-pulse dose, intermittent-pulse dose, and flow-through evaluations.

Single-Pulse-Dose Tests

Specimens 6 to 10 d old were dosed with MAM-Ac or MNNG for 2 h in 1-L beakers in a vented glove box, then placed in 38-L grow-out aquaria containing uncontaminated water. Nominal exposure concentrations of up to 100 ppm (mg/L) MAM-Ac or 50-ppm MNNG were used. Specimens were removed from the aquaria for histological examination at monthly intervals. Moribund specimens were also examined.

Intermittent-Pulse-Dose Tests

Specimens were exposed to trichloroethylene (TCE), vinylidene chloride, or a 6-compound halomethane mixture in 38-L glass aquaria for 24-h periods at weekly intervals for 4 weeks. Following four exposures, fish were placed in grow-out aquaria and examined histopathologically at regular intervals.

Flow-Through Tests

Specimens were continuously exposed to TCE or the 6-compound halomethane mixture for periods up to 28 d in a specially constructed vented exposure chamber under conditions of intermittent flow. The exposure system is schematically represented in Figure 1. Unamended well water was fed by gravity from a reservoir to a seven-compartment water partitioner. Float switches in the water partitioner activated a programmable laboratory controller which, in turn, activated a series of syringe injectors that injected varying amounts of TCE or mixed halomethanes into the system. Injected toxicant was combined with clean diluent water in a mixing chamber before being carried by a self-starting siphon to a splitter box and by standpipes to four replicate exposure aquaria. Test fish were housed within each exposure aquarium in glass petri dishes fitted with Nytex sleeves. Mixing chambers, splitter boxes, and exposure aquaria were enclosed within a fiberglass and wood exposure chamber maintained under a slight negative pressure. Treatment aquaria were maintained at $27 \pm 1°C$ in a circulating water bath with a regime ratio of 16 h (light) : 8h (dark). As each exposure aquarium filled to a depth of approximately 10 cm, toxicant-laden water exited through a self-starting siphon, passed through an activated carbon filter, and was pumped into a clay-lined evaporation pond.

RESULTS

Pulse exposures with MAM-Ac produced hepatic neoplasms in some specimens of each species tested. These results are summarized in Table I. Neoplastic lesions and lesions known to accompany neoplasia first occurred 1 month

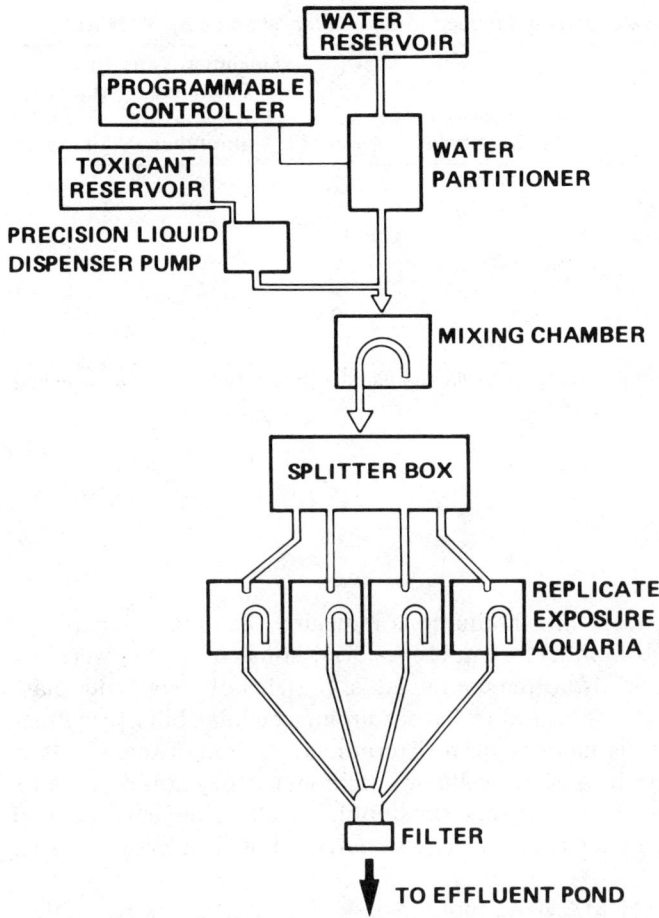

Figure 1. Schematic diagram of the flow-through exposure system.

postexposure in medaka, guppy, and inland silverside; 2 months in sheepshead minnow and rivulus; 3 months in Gulf killifish; and 7 months in fathead minnow. Lesions were most pronounced in medaka and guppy and usually occurred sooner and more frequently in specimens exposed to 50- or 100-ppm MAM-Ac than in those exposed to 10-ppm MAM-Ac. These experiments were not designed to study histogenesis of MAM-Ac-induced liver lesions. Nevertheless, some stages of a histologic progression were observed. By 1 month postexposure, livers of medaka and guppy often contained hypertrophied cells, usually together with biliary ductule hyperplasia (Figure 2). Preliminary

Table I. Hepatic Neoplasms Induced in Small Fish Species by MAM-Ac

Species	No. Neoplasms/ No. Examined[a]	Percent	Maximum Time Examined Postexposure (months)	Latent Period (months)	Exposure Conc (ppm)
Guppy	67/143	47	9	1	50
Japanese medaka	141/285	49	9	1	50
Sheepshead minnow	23/50	46	9	2	50
Fathead minnow	5/90	6	10	7	100
Gulf killifish	7/38	18	6	3	50
Inland silverside	17/47	36	5	1	50
Rivulus	3/21	14	6	2	50

[a]Data are from monthly samples beginning about 1 month postexposure, including moribund specimens.

electron microscopical examinations indicated that some hypertrophied cells contained large numbers of mitochondria, some of which were irregular in form. These cells also often contained large fields of rough endoplasmic reticulum. Other cells appeared to harbor an intracellular biliary channel, but the fate of these cells has not been determined. Although these cells might not develop directly into hepatocellular carcinomas, they could prove to be early indicators of neoplastic change. Eosinophilic cells, similarly rich in mitochondria, have been shown to give rise to carcinomas in livers of rats exposed to methapyrilene.[5]

Other types of MAM-Ac-induced lesions in fish livers resembled those in rats treated with potent liver carcinogens.[6] Preliminary analyses indicated that each type of lesion did not occur in each species and that some species had a tendency to develop specific kinds of lesions more rapidly and frequently than did other species. Hepatocellular carcinomas contained hyperbasophilic cells, often arranged in cords that were discontinuous with those of adjacent normal liver parenchyma (Figure 3). These lesions occurred often in both the guppy and medaka, but more frequently in medaka. In medaka, liver lesions that appeared to be in advanced stages of development consisted of undifferentiated or spindle-shaped cells with numerous mitotic figures. Neoplastic lesions in livers were often accompanied by proliferated bile ducts and by large multilocular cysts.

Some fish exposed to MAM-Ac also developed grossly visible soft tissue neoplasms. We tentatively classified these lesions as undifferentiated sarcomas, fibrosarcomas, rhabdomyosarcomas, and leiomyosarcomas. The lesions were unencapsulated, invasive, and contained a wide variety of cell shapes and sizes with numerous mitotic figures.

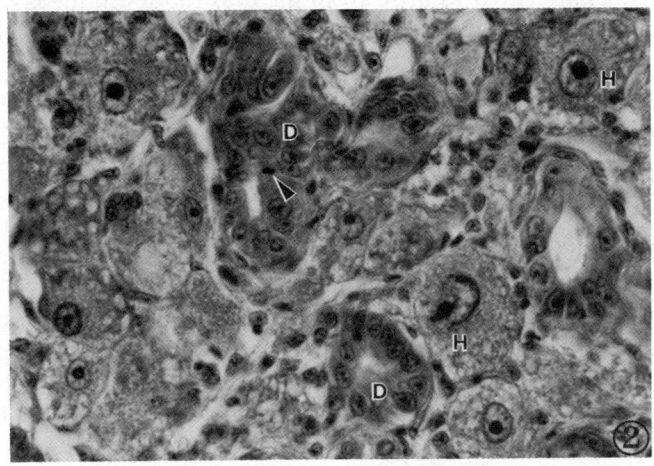

Figure 2. Medaka liver 1 month after exposure to 50-ppm MAM-Ac. Note numerous biliary ductules (D), hypertrophied hepatocytes (H), and mitotic figure (arrowhead). Paraffin section stained with hematoxylin and eosin (365X).

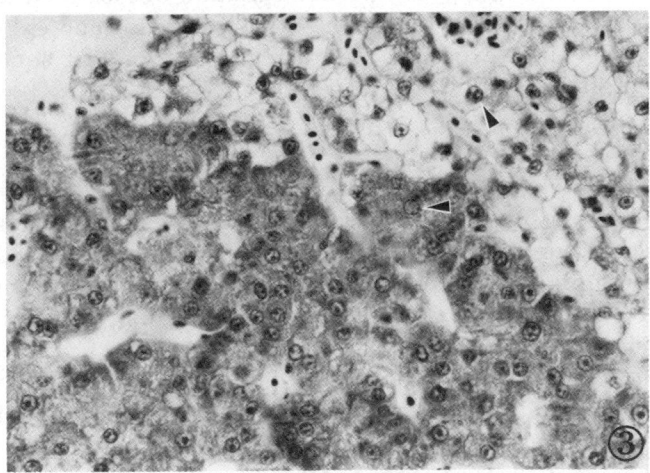

Figure 3. Guppy liver 5 months after exposure to 100-ppm MAM-Ac. This illustrates the interface between normal liver tissue (upper right) and a portion of a densely stained hepatocellular carcinoma. Hepatocyte nuclei (arrowheads). Paraffin section stained with hematoxylin and eosin (440X).

Other lesions appeared to arise from nervous tissues in medaka exposed to MAM-Ac. The tumor illustrated in Figure 4 developed near the tail of a medaka 5 weeks postexposure to 50-ppm MAM-Ac and was characterized by numerous rosettes of neuroepithelial cells with mitotic figures and areas of hemorrhage. This tumor may have originated from lateral line epithelium. Dysplastic and neoplastic lesions occurred in the eyes of MAM-Ac-exposed medaka. Ocular neoplasms included nonpigmented, pigmented, and mixed types of medulloepitheliomas. Nonpigmented types resembled retinoblastomas, whereas the pigmented types appeared to arise from melanin-containing cells of the retinal pigment epithelium or the choroid (Figure 5). These tumors appeared to represent a histologic progression that included a disorganization of the retinal layers and the formation of neuroepithelial tubes and rosettes followed by the development of neoplasms.

In contrast with MAM-Ac, MNNG induced no confirmed neoplastic lesions in any of the species tested and examined to date. Lesions that might become neoplastic, however, were seen in the gill of medaka and in the retina of medaka and fathead minnow. The retinal lesions resembled dysplastic ones seen in MAM-Ac-treated medaka.

Histological examinations of specimens exposed to trichloroethylene (TCE), vinylidene chloride, or a 6-compound halomethane mixture in intermittent pulse exposures have revealed no neoplasms to date.

Medaka and guppy were continuously exposed to TCE and the 6-compound halomethane mixture for 28 d under conditions of intermittent flow. Five toxicant concentrations were used in these exposures, the highest of which was at or near the 28-d LC_{50} for the species. Variation of measured toxicant concentrations about the mean during these exposures was approximately 10%. Specimens from these exposures are currently in grow-out aquaria and are being periodically sampled for histopathological examination.

DISCUSSION

Criteria for choosing small aquarium fishes for carcinogen testing should include ease of rearing and maintenance, resistance to disease, and propensity to develop tumors when exposed to known carcinogens. Although tumors have been reported from tissues of nearly every organ in fishes,[7] most experimental carcinogenesis studies with aquarium fishes have produced only liver tumors.[8] However, MAM-Ac-exposed medaka in our study developed several different neoplasms. Neoplasms of the eye, the peripheral nervous system, and soft tissues were diagnosed, in addition to the expected hepatic neoplasms. Induction of tumors derived from all three germ layers in medaka exposed to MAM-Ac appears to significantly expand the potential for small fish species, particularly medaka, in carcinogen testing, because this fish and probably some others are likely to respond to carcinogens other than hepatocarcinogens.

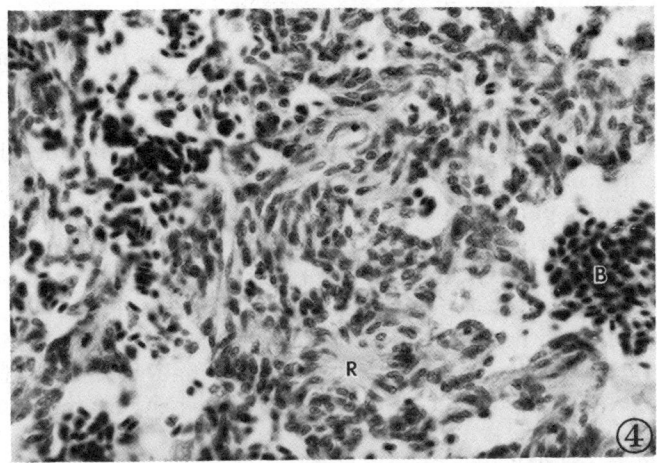

Figure 4. Section of a tumor near the tail of a medaka 5 weeks after exposure to 50-ppm MAM-Ac. The presence of neuroepithelial rosettes (R) suggests that the tumor originated from nervous tissue. Blood cells (B). Paraffin section stained with hematoxylin and eosin (440X).

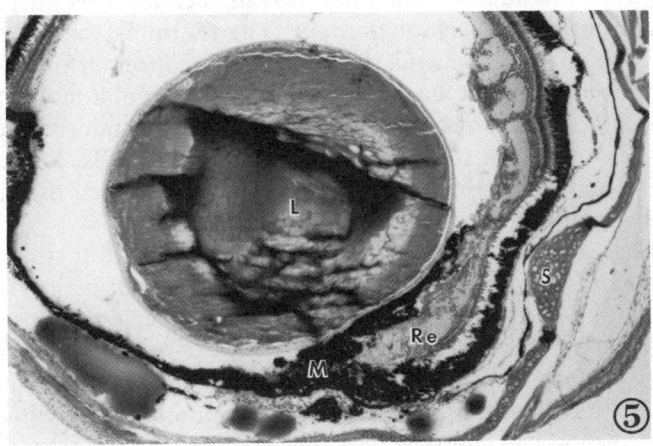

Figure 5. Section through the eye of a medaka 35 weeks after exposure to 50-ppm MAM-Ac. Note melanotic tissue (M) invading the retina (Re). Lens (L); sclera (S). Paraffin section stained with hematoxylin and eosin (65X).

The age of fish at the time they are exposed to carcinogens appears to influence the rate and sites of tumor development. In platyfish/swordtail hybrids, 14 types of neoplasms were encountered after various genotypes were treated with intermittent pulse doses of N-methyl-N-nitrosourea or X rays.[9] Tumors occurred most frequently in backcross hybrids and developed much more frequently in fish treated at 6 weeks of age than those treated at 6 months of age. This may partly explain why Aoki and Matsudaira[10] reported only hepatic tumors in medaka exposed to MAM-Ac at 1 year of age compared with tumors in multiple sites in medaka exposed when they were 6- to 10-d old. Thorough examination of older medaka and additional species exposed to various carcinogens, however, should reveal carcinogenic effects in other tissues and organs. A major advantage of using small aquarium fishes is that most major organs of one or more whole fish can be examined on a single microscope slide.

Exposure to MNNG produced no confirmed neoplasms. In contrast, Kimura et al.[11] reported gastric, kidney, air bladder, and gonad tumors in rainbow trout given 10-ppm MNNG for 24 h as embryos. This longer exposure period compared with only 2 h in our studies might account for the difference in tumor production. Differences may also result from age, species, or strain of fish.

The occurrence of retinal lesions in exposed medaka was an unexpected finding. The lesions included dysplasias and medulloepitheliomas that, in some respects, resembled retinoblastomas and ocular (choroidal) malignant melanomas in humans. Retinoblastomas are the most common ocular neoplasms in children,[12] whereas malignant melanomas are the most common ocular neoplasms in adults.[13] Retinoblastomas include hereditary, nonhereditary, and chromosomal deletion forms.[14] The retinoblastoma gene has been studied as a model for genetic mechanisms of cancers. An environmental etiology, on the other hand, has been linked to some ocular malignant melanomas.[13] The pathogenesis of human ocular neoplasms is poorly understood, in part, because of the lack of adequate experimental models. Models based on the MAM-Ac-induced retinal lesions in medaka should help fill this gap plus identify other compounds that induce retinal neoplasms.

Our results indicate that substantial differences exist in the tumorigenic response among small fish species exposed to direct-acting carcinogens. Whereas medaka and guppy showed strong tumorigenic responses to MAM-Ac, the fathead minnow, widely used in freshwater toxicity bioassays, showed virtually no response and thus appears to be a poor experimental model for bioassays that include carcinogenicity. It is encouraging that saltwater species such as the sheepshead minnow and the inland silverside responded well to MAM-Ac. Possibly, the fact that saltwater species drink water can be exploited in future carcinogen tests, particularly those tests involving long-term exposure to waterborne carcinogens. The sheepshead minnow has been recognized as the species of choice for saltwater toxicity bioassays and appears to be a good model for carcinogen testing as well.[15] The other two tested saltwater species, the inland silverside and rivulus, are less desirable models, in part, because of problems in rearing and maintaining them. The inland silverside is difficult to maintain for long periods in some testing systems. Although rivulus develops hepatic tumors after being treated with nitrosamines[16] as well as MAM-Ac, and is a homozygous species, which makes it attractive for genetic studies, maintaining these fish in adequate numbers is difficult. Each fish must be kept in an individual chamber because of its tendency toward cannibalism. We expect to concentrate primarily on the guppy and medaka for long-term carcinogenicity studies because of their strong tumorigenic responses, resistance to diseases, ability to tolerate saltwater, and ease of obtaining and rearing large numbers of specimens.

ACKNOWLEDGMENTS

This research is supported by National Cancer Institute Contract NO1-CP-26008. We thank Susan Fink, Joan Durfee, Beryl Story, Robert Allen, David Barnes, David Burke, and Alex Schesny for their assistance.

REFERENCES

1. Dawe, C. J., and J. A. Couch. "Debate: Mouse Versus Minnow: The Future of Fish in Carcinogenicity Testing," *Nat. Cancer Inst. Monogr.* 65:223-235 (1984).
2. Kraybill, H. F. "Animal Models and Systems for Risk Evaluation of Low-Level Carcinogenic Contaminants in Water," in *Water Chlorination: Environmental Impact and Health Effects, Vol. 3*, R. L. Jolley, W. A. Brungs, and R. B. Cumming, Eds. (Ann Arbor, MI: Ann Arbor Science Publishers, Inc., 1980), pp. 973-982.
3. Kraybill, H. F., C. T. Helmes, and C. C. Sigman. "Biomedical Aspects of Biorefractories in Water," in *Aquatic Pollutants, Transformation and Biological Effects*, O. Hutzinger, L. H. Van Lelyveld, and B. C. J. Zoetemann, Eds. (Oxford: Pergamon Press, 1978), pp. 419-461.
4. Hendricks, J. D. "Chemical Carcinogenesis in Fish," in *Aquatic Toxicology*, L. J. Weber, Ed. (New York: Raven Press, 1982), pp. 149-211.
5. Reznik-Schuller, H. M., and M. Gregg. "Sequential Morphologic Changes During Methapyrilene-Induced Hepatocellular Carcinogenesis in Rats," *J. Nat. Cancer Inst.* 71:1021-1031 (1983).
6. Stewart, H. L., G. Williams, C. H. Keysser, L. S. Lombard, and R. J. Montali. "Histologic Typing of Liver Tumors of the Rat," *J. Nat. Cancer Inst.* 64:179-206 (1980).
7. Harshbarger, J. C. "RTLA Supplements: Registry of Tumors in Lower Animals," (Washington, DC: Smithsonian Institution, 1965-1981).
8. Matsushima, T., and T. Sugimura. "Experimental Carcinogenesis in Small Aquarium Fishes," *Prog. Exp. Tumor Res.* 20:367-379 (1976).

9. Schwab, M., J. Haas, S. Abdo, M. R. Ahuja, G. Kollinger, A. Anders, and F. Anders. "Genetic Basis of Susceptibility for Development of Neoplasms Following Treatment with N-Methyl-N-Nitrosourea (MNU) or X-Rays in the Platyfish/Swordtail System," *Experientia* 34:780-782 (1978).
10. Aoki, K., and H. Matsudaira. "Induction of Hepatic Tumors in a Teleost (*Oryzias latipes*) After Treatment with Methylazoxymethanol Acetate: Brief Communication," *J. Nat. Cancer Inst.* 59:1747-1749 (1977).
11. Kimura, I., H. Kitaori, K. Yoshizaki, K. Tayama, M. Ito, and S. Yamada. "Development of Tumors in Rainbow Trout Following Embryonic Exposure to N-Nitroso Compounds," in *Phyletic Approaches to Cancer*, C. J. Dawe et al., Eds. (Tokyo: Japan Scientific Society Press, 1981), pp. 241-252.
12. Ackerman, L. V., and J. A. del Regato. "Cancer of the Eye," in *Cancer: Diagnosis, Treatment, and Prognosis*, 4th ed. (St. Louis: C. V. Mosby Company, 1970), pp. 956-977.
13. Albert, D. M., C. A. Puliafito, A. B. Fulton, N. L. Robinson, Z. N. Zakov, T. P. Dryja, A. B. Smith, E. Egan, and S. S. Leffingwell. "Increased Incidence of Chloroidal Malignant Melanoma Occurring in a Single Population of Chemical Workers," *Amer. J. Opthal.* 89:323-337 (1980).
14. Murphree, A. L., and W. F. Benedict. "Retinoblastoma: Clues to Human Oncogenesis," *Science* 223:1028-1033 (1984).
15. Couch, J. A., L. A. Courtney, and S. S. Foss. "Laboratory Evaluation of Marine Fishes as Carcinogen Assay Subjects," in *Phyletic Approaches to Cancer*, C. J. Dawe et al., Eds. (Tokyo: Japan Scientific Society Press, 1981), pp. 125-140.
16. Koenig, C. C., and M. P. Chasar. "Usefulness of the Hermaphroditic Marine Fish *Rivulus marmoratus* as a Test Animal for Carcinogenicity Studies," *Nat. Cancer Inst. Monogr.* 65:15-33 (1984).

CHAPTER **35**

Japanese Medaka Liver Tumor Model: Review of Literature and New Findings

David E. Hinton, James A. Hampton, and Patricia A. McCuskey

Tumors of the liver appear in Japanese medaka after exposure to solutions of carcinogens [methylazoxymethanol acetate (MAM),[1-5] diethylnitrosamine (DEN),[5-13] or ortho-aminoazotoluene[5]] in aquarium water or after dietary exposure with carcinogens of low water solubility (aflatoxins B_2 and G_2 and sterigmatocystin[5]). Table I lists the compounds and exposure conditions which, to date, have led to tumor production in the liver of Japanese medaka. When compared with tumor induction studies using rodent species, where latency periods of greater than 1 year are common,[14] liver tumors develop rapidly in Japanese medaka (Table I).

Additional advantages of studies with medaka include[15] (1) ease of handling—large numbers of fish may be exposed simultaneously in exposure aquarium; (2) assurance of thorough histopathologic sampling of liver and potential sites of metastasis, entire horizontal-longitudinal or sagittally sectioned fish can be mounted in paraffin[5] or glycolmethacrylate blocks, and serial or near-serial sections through the entire visceral mass can be analyzed on relatively few slides; (3) reduced expense, not only from shortened latency period but from reduced quantities of laboratory space, food, and contaminated water required for exposure of large numbers of individual fish; and (4) the ability to study the effects of temperature on the carcinogenic process.[8-11]

MEDAKA TUMOR MODEL

Protocol for Medaka Liver Tumorigenesis

The basic scheme for producing liver tumors in young adult medaka is shown in Figure 1. Two carcinogens, DEN and MAM, have received most attention with this model. Since only one study[5] included a screen of several different compounds, major discussion of tumor production in this review will involve these two carcinogens. Carcinogens (dissolved in aquarium water to the desired concentration) and young adult fish (approximately 1 year old) are

Table I. Carcinogen Exposure Regimes Known to Produce Liver Tumors in Japanese Medaka (*Oryzias latipes*)

Reference	Carcinogen	Concentration (ppm)	Exposure Duration	Latency Period Postexposure (weeks)
Ishikawa et al.[6]	DEN[a]	(15–135)	8 weeks	1–5
Ishikawa and Takayama[7]	DEN[a]	(15–45)	8 weeks	5–8
Kyono[8]	DEN[a]	(100)	8 weeks	1–4
Kyono and Egami[9]	DEN[a]	(50–100)	8 weeks	4
Kyono et al.[10]	DEN[a]	(100)	6 weeks	5–7
Kyono-Hamaguchi[11]	DEN[a]	(100)	6–8 weeks	3–5
Egami et al.[12]	DEN[a]	(50–100)	8 weeks	4
Klaunig et al.[13]	DEN[a]	(25–100)	10 d	12–24[b]
Aoki and Matsudaira[1-3]	MAM[c]	(0.1–3)	1 h – 120 d	9–13[b]
	MAM[c]	(0.1–10)	1 h – 120 d	9–13[b]
	MAM[c]	(0.1–10)	1 h – 120 d	9–13[b]
Hinton et al.[4]	MAM[c]	(0.5)	10 d	9–17[b]
Hatanaka et al.[5]	MAM[c]	(0.3)	9 weeks	9
	DEN[a]	(100)	2–4 weeks	15–24
	OAAT[d]	(600)	12 weeks	12
	AFB$_1$[e]	(2.5–5)	6–24 weeks	0–19
	AFG$_1$[f]	(2.5)	12–24 weeks	0–12
	SMC[g]	(1–5)	4–24 weeks	0–19

[a]Diethylnitrosamine dissolved in aquarium water.
[b]Due to experiment design, numbers may not reflect minimal latency period.
[c]Methylazoxymethanol acetate.
[d]Orthoaminoazotoluene.
[e]Aflatoxin B$_1$, a dietary supplement.
[f]Aflatoxin G$_1$, a dietary supplement.
[g]Sterigmatocystin, a dietary supplement.

placed in the exposure tank. Depending on the concentration of MAM, exposure periods have been as short as 1 h and as long as several weeks[1-3] (Table I). DEN exposures have commonly been from 6 to 8 weeks duration[6-12] (Table I).

After exposure, fish are placed in water containing no carcinogen and observed for a varying number of weeks. At the end of the observation period, fish are anesthetized, fixed, embedded in paraffin or methacrylate, and mounted on slides for histologic analysis. Since both of the tumor production schemes involve a discrete period of carcinogen exposure (i.e., initiation) followed by a tumor development phase (observational period) in water containing no carcinogen, the model is quite readily applied to the study of cocarcinogens[16] and promoting substances (Figure 1). Caffeine was administered after MAM exposure, and a slight promoting effect was noted.[2] Partial hepatectomy (PH) has become a standard procedure in mammalian liver carcinogenesis bioassay;[16] the PH may be performed at the time of initiation or after

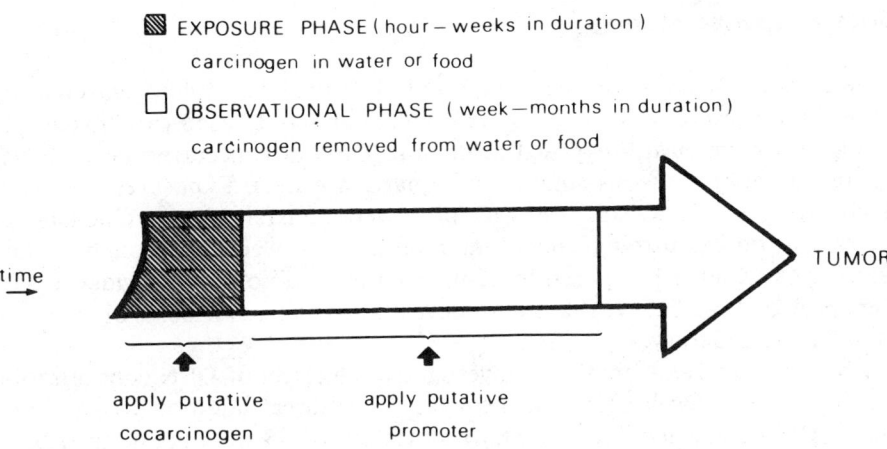

Figure 1. Basic protocol for Japanese medaka (*Oryzias latipes*) liver tumor production. Carcinogen (see Table I) is administered either dissolved in aquarium water or in food during exposure phase (striped area on large arrow). After a discrete time interval, carcinogen is removed and fish are placed in clean aquarium water and fed a diet containing no carcinogen. The latter phase, termed observational, is usually from 1 week to 2 months duration. To study effects of cocarcinogens and promoters on the tumor-forming process, they should be applied during exposure-phase (initiation) or observational-phase (tumorigenesis) intervals.[16]

administration of the chemical under test. In the former usage, PH has been termed a cocarcinogen (i. e., any process increasing the carcinogenic response of an animal to a carcinogen by a procedure or second chemical performed at or very close to the time of carcinogen administration and not itself shown to be carcinogenic), whereas in the latter usage, PH is considered a promoter.[16]

Kyono-Hamaguchi[11] performed PH on medaka 3 d prior to exposure to DEN (100 or 200 ppm for 7 or 14 d). When compared with DEN-treated but sham-operated controls, at 6, 11, and 15 weeks after DEN treatment, PH- and DEN-treated medakas showed advanced changes in stages of tumor development. This illustrates a cocarcinogenic effect of PH in DEN-induced medaka liver tumorigenesis and is similar to that reported by Scherer and Emmelot[17] in rat liver studies with the same compound. Further use of the medaka model to study possible promoting effects of halogenated methanes and other compounds associated with chlorination of water seem readily applicable.

Dose-Response Studies

Since concentration of the compound is critical in carcinogen-screening assays, dose-response data on the two most commonly used carcinogens are included. Egami et al.[12] exposed medaka to DEN at concentrations of 0, 25, 50, or 100 ppm (aqueous solution in aquarium water). Exposures were conducted from 1 to 13 weeks, after which fish were transferred to carcinogen-free water and studied histologically at intervals up to 13 weeks. Although no data were reported on fish exposed to 25-ppm DEN, and no tumor-incidence data were provided, we believe that tumor development after exposures of 50 or 100 ppm was satisfactory.

Ishikawa and Takayama[7] also investigated the effect of DEN concentration on liver tumor development in medaka. After direct addition to aquarium water, DEN concentrations (in ppm) of 15, 30, or 45 were maintained for 8 weeks. Beginning 2 weeks after termination of exposure, 3 fish were sampled biweekly for up to 8 weeks. In the 15-ppm group, only 1 fish of a total of 12 sampled showed a liver tumor, and this was seen 8 weeks after termination of exposure. At 30 ppm, 5 of 12 fish showed liver tumors (2 of 3 at 6 weeks and 3 of 3 at 8 weeks postexposure, respectively). At the highest concentration (45 ppm), considering only those biweekly intervals for which data exist for lower concentrations, results identical to those seen with 30 ppm were reported. Since these workers sampled the highest concentration group weekly, they were able to show that the initial indication of DEN-induced liver tumor appeared at 5 weeks after cessation of 8-weeks exposure, and that from 7 weeks on, all or nearly all survivors of a 45-ppm DEN exposure should show tumors. Earlier, Ishikawa et al.[6] exposed medaka to a higher DEN concentration (135 ppm); however, toxicity was extensive (only 3 fish survived the 8-weeks exposure).

Aoki and Matsudaira[2,3] designed their studies with MAM to yield dose-response data. MAM concentrations ranged from 0 to 10 ppm. The incidence of liver tumors in medaka increased linearly with the dose of MAM. In addition, increasing the concentration of MAM decreased the latency period of tumor production. For a given yield of tumor, the product of MAM (ppm) times the exposure (days) became smaller with higher levels of carcinogen.

Histogenesis of Tumor

Tissue alterations appear to follow a pattern of progression that has only been partially clarified. Egami et al.[12] studied sequential development of hepatomas produced in medaka after exposure to 100-ppm DEN. Ishikawa and Takayama[7] serially assayed medakas during and after 8-weeks exposure to 45 ppm DEN. In both studies,[7,12] exposed fish were removed for analysis at weekly intervals for 8 weeks, after which the remaining fish were transferred to tap water and studied following 1, 2, 3, 4, or 5 additional weeks. In fish exposed to DEN for 1 to 2 weeks, degenerative and necrotic changes prevailed. These are described as single-cell[7] or small[12] (cellular aggregates?) necroses. At this time hepatocytes showed increased cytoplasmic basophilia.[7] Phagocytes filled with brown-yellow ceroid pigment increased in number and were the prominent features of early degenerative lesions.[7]

Later liver lesions, seen after 3-weeks exposure, include increased basophilia and enhanced mitoses.[7,12] This change is followed by the appearance of multicentric clusters of pleomorphic and basophilic cells termed foci.[7,12] To date, no enzyme histochemical studies characterizing phenotypes of cells in foci vs control hepatocytes have been reported. Routinely used in rat[16,17] and mouse[18,19] liver bioassays, histochemical characteristics of foci in medaka liver need to be established and related to subsequent stages in the neoplastic progression.[20] The coupling of altered enzyme histochemistry to labeling and mitotic indices[2,3,7,10] of foci would constitute strong evidence for the role of enzyme-altered foci in tumorigenesis.

After 8-weeks exposure and an additional 5 weeks in tap water, tumor nodules are seen in which hyperbasophilic cells with polymorphic nuclei are common.[7,12] Some nodules are trabecular hepatomas and others are believed to be liver-cell carcinomas. The basis for differentiation between the two is not clear from the descriptions given.[7,12] Aoki and Matsudaira[2,3] stated that MAM-induced liver tumors in medaka follow a similar progression. Further support for this progression of lesions has been obtained with medaka embryos. When Klaunig et al.[13] exposed medaka embryos to DEN (from 4 h until 10 d after fertilization) and performed histopathologic analyses 1, 3, and 6 months later, similar sequential alterations were observed. Despite the fact that both MAM[2] and DEN[7] exposure are associated with its development, little attention has been given to the sequence of alterations leading from normal medaka liver morphology to cholangiolar carcinoma.

Hinton et al.[4] performed morphometric analyses of medaka liver at 66 and 116 d following a 10-d exposure to 0.5-ppm MAM. At 66 d, the volume compartment of the liver occupied by bile ducts in MAM-treated medaka was 17-fold that of controls and remained higher at 116 d. By electron microscopy, a major finding at 116 d after a 10-day (0.5-ppm) MAM exposure was the presence of large tracts of cells dissecting between nests of hepatocytes (Figure 2). On the basis of their well-developed desmosomes and association with a basal lamina (Figure 3), we considered these tracts as part of a process of bile ductular and ductal proliferation. In our experience,[4] the extensive nature of this lesion warrants further consideration in medaka liver tumorigenesis studies.

Labeling and Mitotic Indices During Tumorigenesis

The medaka liver responds initially to the toxic, necrotizing effect of the carcinogen, and then shows a period of 3 to 4 weeks in which enhanced mitotic

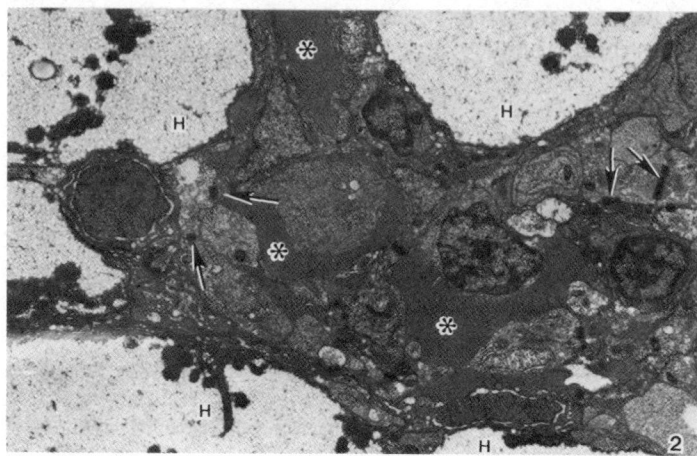

Figure 2. Transmission electron micrograph of liver of Japanese medaka 116 d after a 10-d exposure to an aqueous solution (0.5 ppm) of MAM. Numerous desmosomes (arrows) are shared by epithelial cells. Tracts of these cells and associated extracellular material of medium electron density (asterisks) separate hepatocytes (H). Fixation was by immersion in a modified Karnovsky's solution. Uranyl acetate and lead citrate stain (4300X).

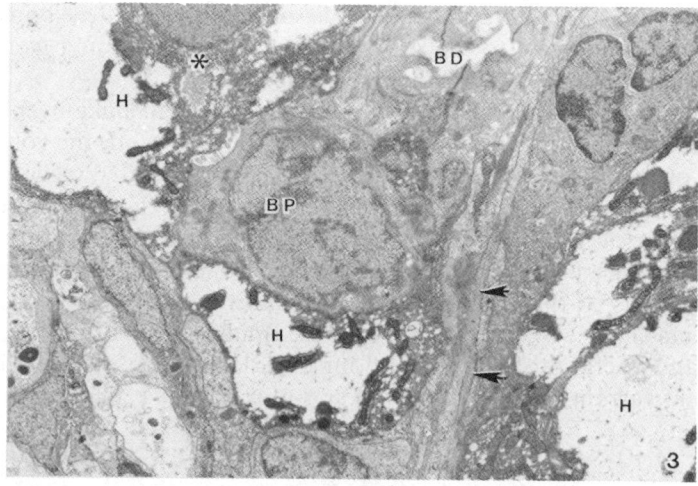

Figure 3. Transmission electron micrograph showing hepatocytes (H) separated by bile preductular (BP) and bile ductular cells. Lumen of bile ductule (BD) is shown. Arrows point to basement membrane material associated with bile ductules. Hepatocyte (upper left) shows intracellular canaliculus (asterisk) and also contributes microvilli to a bile preductule. Fish exposure conditions and techniques for microscopy are identical to those for Figure 2. Uranyl acetate and lead citrate stain (5000X).

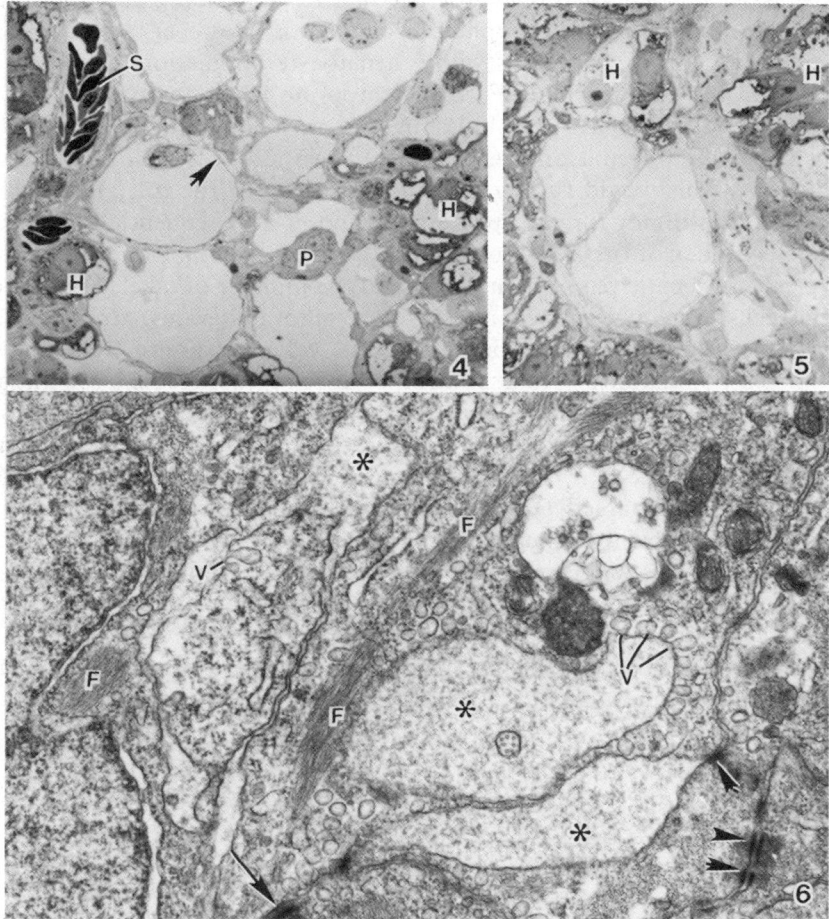

Figure 4. High-resolution-light micrograph of liver from medaka treated with MAM (at 116 d after 10-d exposure, 0.5-ppm solution). Multilocular cyst-like spaces predominate in this field. Comparison of densities of cyst-like lumina with that of sinusoidal lumen (S) shows that the former contain a material of low staining intensity. Hepatocytes (H) are separated by the cyst-like spaces. Perisinusoidal cells (P) send out fine diameter extensions (arrow) that line cyst-like spaces. Toluidine blue stain (620X).

Figure 5. Exposed medaka liver shows two adjacent cyst-like spaces containing material of low staining intensity that separate adjacent hepatocytes (H). Toluidine blue stain (660X).

Figure 6. Transmission electron micrograph of portions of three adjacent fibroblast-like perisinusoidal cells from medaka liver exposed to MAM (at 116 d after 10-d exposure, 0.5-ppm solution). Cytofilaments (F) are numerous. Cytoplasmic vesicles (V) are numerous and communicate with regions of extracellular material of low electron density (asterisks). Note numerous junctional complexes. Uranyl acetate and lead citrate stain (20,000X).

activity is seen. Some of this mitotic activity, by autoradiography, has been localized to foci of bizarre basophilic hepatocytes.[7] However, not all mitotic figures or label incorporation is within hepatocyte populations. Both parenchymal and nonparenchymal[11,12] cells incorporate label. In future studies it may be possible to use high-resolution light microscopy to prepare autoradiograms that would permit differentiation of specific cell types of labeled cell populations. This would further our understanding of the role(s) of cellular "actors" in the tumor-forming process. Kyono et al.[10] sequentially analyzed the labeling index of two cell populations (the number of labeled parenchymal cells per 1000 hepatic parenchymal cells and the number of labeled nonparenchymal cells were determined on standard autoradiograms) during DEN-induced medaka liver tumorigenesis. After 1-week exposure to DEN, indices of both cell populations were at control level. A rapid increase (seen initially after 2 weeks) continued throughout DEN exposure in both cell populations. After the fish were returned to carcinogen-free aquarium water, labeling indices dropped to the control level except in the tumor nodules, where indices remained high.[10] Coupled DNA Feulgen cytofluorometry studies of tumor-bearing liver revealed some hepatocytes with six times the DNA content of one set of chromosomes (6 C) or more DNA vs the control diploid value of 2 C.[10] The correlation of increased mitotic and labeling indices within cells of enzyme alteration foci of rat liver[21] has demonstrated the proliferative nature of these populations of transformed hepatocytes. Such coupled studies are needed in the medaka liver tumorigenesis model.

Temperature Effects on Liver Tumor Formation

Kyono[8] was the first to study in detail the effects of temperature on liver tumorigenesis in medaka. The design of that study included the typical 8-week exposure to DEN. Aquarium water for the exposure phase was maintained at either 22 or 5 to 6°C. Following exposure, the fish population from the high-temperature aquarium was split into two populations. The first was maintained at 22°C, and a second group was maintained at 5 to 6°C for the entire tumor development phase. In a similar way, medakas exposed to DEN (5 to 6°C for 8 weeks) were divided into a high- and low-temperature group for the development phase. The fish were histologically analyzed at 8 or 12 weeks following cessation of exposure. Elevated temperature during the exposure (initiation) period is of most importance. All fish treated with DEN at 22°C developed tumors whether their development phase temperature was 5 to 6 or 22°C. However, temperature elevation during both the initiation and developmental phases caused larger and a higher number of tumors. In fish exposed to DEN at 5 to 6°C, no tumors developed regardless of the temperature at the subsequent developmental phase. Aoki and Matsudaira[2,3] routinely use and recommend a temperature of 25°C during MAM exposures.

Ishikawa and Takayama[7] used two different temperatures (18 and 25°C) during exposure of medaka to DEN. They also found enhanced tumor production at the higher temperature. In an attempt to further analyze the effect of temperature on liver tumor production in DEN-exposed medaka, Kyono-Hamaguchi[11] followed the effects of exposure temperatures at 5 to 8°C and at 22 to 25°C on labeling and mitotic indices. A close positive relationship between DNA synthesis, cell divisions, and DEN treatment at a high temperature was established.

Biology of Liver Tumor

To date, major emphasis with the medaka liver tumor model has been on the production of tumor, whereas studies to characterize its biological behavior are few. Kyono-Hamaguchi[11] transplanted liver grafts from control and tumor-bearing fish into the anterior eye chamber of medaka. Although no control or tumor-bearing liver graft growth was observed, grafts from DEN-treated and DEN tumor-bearing liver maintained their original sizes longer. The inherent difficulty of performing such a procedure with these small fish is readily apparent. Interestingly, Kyono-Hamaguchi[11] cites work indicating that cell lines from DEN-induced liver tumor in medaka have been established. In vitro characterization of these lines may yield important biological information concerning the medaka liver tumor.

Aoki and Matsudaira[3] cite work indicating the establishment of inbred strains of medaka. The development of such strains should intensify further transplantation studies designed to yield more information on the biology of medaka liver tumor. In addition to chromosome studies[11] and biological characterization studies, biochemical and histochemical properties, as well as cell composition and structure at the light and electron microscopy level, remain fruitful areas for future research. Ishikawa and Takayama[7] reported invasion of the adjacent intestinal wall by DEN-induced medaka liver tumor cells. The important question of the malignant nature of liver neoplasms in medaka may be more readily addressed by long-term follow-up after embryo larval carcinogen exposure. Sufficient time for full expression of tumor biologic potential may be achieved in tumors produced by egg-embryo-larval exposures[13] and studied over the life span of individual fish.

Spongiosis Hepatis

Large cyst-like spaces were seen in livers of medaka at 66 and 116 d after a 10-d exposure to 0.5-ppm MAM.[4] Impressed by the similarity to spongiosis hepatis,[22] we made an additional high-resolution light and transmission-electron microscopic analysis of the specimens. Both spongiosis hepatis in the rat[22] and similar lesions in medaka involve perisinusoidal liver cells.[4] Further, a comparison with ultrastructural descriptions in various teleost species[23-25] shows a common feature of abundant cytofilaments, causing us to conclude that some cells rich in filaments and found in medaka liver following MAM exposure are empty, fat-storing cells.[23]

A typical field from a toluidine-blue-stained liver section shows the appearance of this lesion (Figure 4). Multilocular cyst-like spaces are numerous. These are lined by attenuated processes from cells that are not hepatocytic or endothelial. With toluidine-blue-stained preparations (Figures 4 and 5), only slight staining of material in cyst-like spaces is seen. However, in suitably reacted specimens from rat liver, the spaces contain acid mucopolysaccharide and may reflect a carcinogen-induced metabolic alteration in perisinusoidal (fat-storing) cells.[21] With transmission electron microscopy, we confirmed and extended our previous light microscopy findings. A typical electron micrograph from a MAM-exposed medaka liver is shown in Figure 6. Perisinusoidal (fat-storing) cells contained abundant filaments and numerous cytoplasmic vesicles that appeared to be associated with the production (or uptake?) of a flocculent material of low electron density (Figure 6).

SUMMARY

The Japanese medaka liver tumor model provides a rapid, in vivo screen for laboratory testing the carcinogenicity of aquatic pollutants. The high yield of tumors and associated brief induction period are strong arguments for using this model as an initial indicator in a multitiered testing program. Future studies to biologically characterize the medaka liver tumor(s) are needed. In addition, screening of a variety of compounds including proven mammalian carcinogens, promoters, cocarcinogens, weakly positive carcinogens, compounds considered negative, and compounds that may afford protection against carcinogenesis should be conducted. For other features of the medaka model and for a more complete discussion concerning fish as test organisms in carcinogenicity studies, the consensus report[26] should be consulted.

ACKNOWLEDGMENTS

We wish to thank Stephen Fidler for his technical assistance. This work was supported by Project No. CR-811017-01-0 of the U.S. Environmental Protection Agency and by the Water Research Institute, West Virginia University, with funds allocated under the Water Resources Act of 1964 (PL88-379), administered by the Office of Water Research and Technology, U.S. Department of the Interior, as part of Project A-037-WVa.

REFERENCES

1. Aoki, K., and H. Matsudaira. "Induction of Hepatic Tumors in a Teleost (*Oryzias latipes*) After Treatment with Methylazoxymethanol Acetate: Brief Communication," *J. Nat. Cancer Inst.* 59:1747-1749 (1977).
2. Aoki, K., and H. Matsudaira. "Factors Influencing Tumorigenesis in the Liver After Treatment with Methylazoxymethanol Acetate in a Teleost, *Oryzias latipes*," in *Phyletic Approaches to Cancer*, C. J. Dawe, J. C. Harshbarger, S. Kondo, T. Sugimura, and S. Takayama, Eds. (Tokyo: Japan Sci. Soc. Press, 1981), pp. 250-216.
3. Aoki, K., and H. Matsudaira. "Factors Influencing Methylazoxymethanol Acetate Initiation of Liver Tumors in *Oryzias latipes*: Carcinogen Dosage and Time of Exposure," Nat. Cancer Institute Monograph 65 (Bethesda, Md: U.S. Dept. of Health and Human Services, 1984), pp. 345-351.
4. Hinton, D., C. Lantz, and J. Hampton. "Effect of Age and Exposure to a Carcinogen on the Structure of the Medaka Liver: A Morphometric Study," Nat. Cancer Institute Monograph 65 (Bethesda, Md: U.S. Dept. of Health and Human Services, 1984), pp. 239-249.
5. Hatanaka, J., N. Doke, T. Harada, T. Aikawa, and M. Enomoto. "Usefulness and Rapidity of Screening for the Toxicity and Carcinogenicity of Chemicals in Medaka *Oryzias latipes*," *Japan. J. Exper. Med.* 52(5):243-253 (1982).
6. Ishikawa, T., T. Shimamine, and S. Takayama. "Histologic and Electron Microscopy Observations on Diethylnitrosamine-Induced Hepatomas in Small Aquarium Fish (*Oryzias latipes*)," *J. Nat. Cancer Inst.* 55(4):909-916 (1975).
7. Ishikawa, T., and S. Takayama. "Importance of Hepatic Neoplasms in Lower Vertebrate Animals as a Tool in Cancer Research," *J. Toxicol. Environ. Health* 5:537-550 (1979).
8. Kyono, Y. "Temperature Effects During and After the Diethylnitrosamine Treatment of Liver Tumorigenesis in the Fish, *Oryzias latipes*," *Eur. J. Cancer* 14:1089-1097 (1978).
9. Kyono, Y., and N. Egami. "The Effect of Temperature During the Diethylnitrosamine Treatment on Liver Tumorigenesis in the Fish, *Oryzias latipes*," *Eur. J. Cancer* 13:1191-1194 (1977).
10. Kyono, Y., A. Shima, and N. Egami. "Changes in the Labeling Index and DNA Content of Liver Cells During Diethylnitrosamine-Induced Liver Tumorigenesis in *Oryzias latipes*," *J. Nat. Cancer Inst.* 63(1):71-74 (1979).
11. Kyono-Hamaguchi, Y. "Effects of Temperature and Partial Hepatectomy on the Induction of Liver Tumors in *Oryzias latipes*," Nat. Cancer Institute Monograph 65 (Bethesda, Md: U.S. Dept. of Health and Human Services, 1984), pp. 337-344.
12. Egami, N., Y. Kyono-Hamaguchi, H. Mitani, and A. Shima. "Characteristics of Hepatoma Produced by Treatment with Diethylnitrosamine in the Fish, *Oryzias latipes*," in *Phyletic Approaches to Cancer*, C. J. Dawe, J. C. Harshbarger, S. Kondo, T. Sugimura, and S. Takayama, Eds. (Tokyo: Japan Sci. Soc. Press, 1981), pp. 205-216.
13. Klaunig, J., B. Barut, and P. Goldblatt. "Preliminary Studies on the Usefulness of Medaka, *Oryzias latipes*, Embryos in Carcinogenicity Testing," Nat. Cancer Institute Monograph 65 (Bethesda, Md: U.S. Dept. of Health and Human Services, 1984), pp. 151-161.
14. Lipsky, M., D. Hinton, J. Klaunig, and B. Trump. "Biology of Hepatocellular

Neoplasia in the Mouse. I. Histogenesis of Safrole-Induced Hepatocellular Carcinoma," *J. Nat. Cancer Inst.* 67(2):365–376 (1981).
15. Matsushima, T., and T. Sugimura. "Experimental Carcinogenesis in Small Aquarium Fishes," *Prog. Exp. Tumor Res.* 20:367–379 (1976).
16. Pereira, M. "Rat Liver Foci Bioassay," *J. Am. Coll. Toxicol.* 1:101–117 (1982).
17. Scherer, E., and P. Emmelot. "Kinetics of Induction and Growth of Enzyme-Deficient Islands Involved in Hepatocarcinogenesis," *Cancer Res.* 36:2544–54 (1976).
18. Ohmori, T., J. Rice, and G. Williams. "Histochemical Characteristics of Spontaneous and Chemically Induced Hepatocellular Neoplasms in Mice and the Development of Neoplasms with Gamma-Glutamyl Transpeptidase Activity During Phenobarbital Exposure," *Histochem. J.* 13:85–99 (1981).
19. Lipsky, M., D. Hinton, J. Klaunig, P. Goldblatt, and B. Trump. "Biology of Hepatocellular Neoplasia in the Mouse. II. Sequential Enzyme Histochemical Analysis of BALB/c Mouse Liver During Safrole-Induced Carcinogenesis," *J. Nat. Cancer Inst.* 62(2):377–392 (1981).
20. Couch, J. A., and L. Courtney. "Attempts to Abbreviate Time to Endpoint in Fish Hepatocarcinogenesis," Chapter 31, this volume.
21. Ogawa, K., D. Solt, and E. Farber. "Phenotypic Diversity as an Early Property of Putative Preneoplastic Hepatocyte Populations in Liver Carcinogenesis," *Cancer Res.* 40:725–33 (1980).
22. Bannasch, P., M. Bloch, and H. Zerban. "Spongiosis Hepatis," *Lab. Invest.* 44:252–264 (1981).
23. Takahashi, Y., H. Tsubouchi, and K. Kobayashi. "Effects of Vitamin A Administration upon Ito's Fat-storing Cells of the Liver of Carp," *Arch. Histol. Jpn.* 41:339–349 (1978).
24. Tanuma, Y., and Y. Ito. "Electron Microscopic Study on the Sinusoidal Wall of the Liver of the Crucian, *Carassius carassius*, with Special Remarks on the Fat-Storing Cell (FSC)," *Arch. Histol. Jpn.* 43:241–263 (1980).
25. Fujita, H., T. Tamaru, and J. Miyagawa. "Fine Structural Characteristics of the Hepatic Sinusoidal Walls of the Goldfish (*Carassius auratus*)," *Arch. Histol. Jpn.* 43:265–273 (1980).
26. Couch, J., C. Dawe, H. Dupree, J. Gratzek, A. Guarino, J. Harshbarger, J. Hendricks, K. Hoover, T. Ishikawa, I. Kimura, A. Kligerman, J. Lech, and R. Schultz. "Summary and Recommendations: A Consensus Report," Nat. Cancer Institute Monograph 65 (Bethesda, Md: U.S. Dept. of Health and Human Services, 1984), pp. 397–404.

CHAPTER 36

Black Bullhead: An Indicator of the Presence of Chemical Carcinogens

John M. Grizzle

Fish are useful indicators of pollutants in the aquatic environment because they accumulate and respond to toxicants. The ability of fish to develop neoplasms after exposure to carcinogens, their relatively short tumor-induction time, and the lower cost of carcinogenicity studies with fish than with mammals indicate that fish have potential for routine use as laboratory animals in carcinogenesis studies.[1,2] Several types of fish tumors can be induced by chemical carcinogens,[3] and the occurrence of tumors in wild fish has been used as evidence for the presence of carcinogenic substances in aquatic environments.[4,5] However, in some cases, viral or genetic causes cannot be eliminated because of uncontrolled conditions.

The presence of a chemical carcinogen in the chlorinated effluent entering the final oxidation pond of the south Tuskegee, Alabama, wastewater treatment plant was first suspected because black bullheads, *Ictalurus melas*, in this pond had a 73% prevalence of oral papillomas and indications of mixed function oxidase (MFO) induction.[6,7] Although papillomas are a common type of fish neoplasm,[8] those in this case were distinctive because of the high prevalence and because most were located in the oral fornices rather than in variable locations. The enzyme induction and high prevalence of papillomas in black bullheads from the final oxidation pond receiving chlorinated wastewater (no tumors or enzyme induction were found in fish from unpolluted ponds nearby) suggested that the tumors were related to the effluent entering the pond. However, analysis of water and sediment from the pond did not indicate the presence of toxicants considered capable of causing the papillomas.[6,9] These findings suggested three possibilities. First, that the papillomas were caused by a chemical carcinogen not detected by water analysis; second, that the papillomas were related to a tumor virus; or third, that the papillomas were genetically transmitted.

Because fish in this pond were confined and water was from a single source, the final oxidation pond offered an opportunity for experiments involving fish as indicators of the presence of chemical carcinogens. Black bullheads are suited for chronic exposure to chlorinated effluents because of their relatively high tolerance to chlorine. The 96-h LC_{50} for black bullheads is 1.41 mg/L compared with 0.156 mg/L for channel catfish, *Ictalurus punctatus*, and 0.172 mg/L for rainbow trout, *Salmo gairdneri*.[10]

Several approaches, in addition to chemical analysis of the water, were used to study the cause of the papillomas on black bullheads in the wastewater pond. Studies were conducted with both caged fish in controlled experiments and fish free in the pond. This chapter summarizes the oncology at this wastewater treatment facility and illustrates the usefulness of field studies with black bullheads and other fish species for detecting carcinogens.

STUDY SITE

The south Tuskegee wastewater treatment facility, located 7 km southwest of Tuskegee, Alabama, began operating during January 1977. This plant receives only domestic wastewater; there is no known discharge from industries or other sources suspected of releasing carcinogens. However, surface runoff from surrounding agricultural and residential areas probably enters the sewage system after heavy rainfall.

The 0.8-ha final oxidation pond was designed for postsecondary wastewater treatment. Before entering the final oxidation pond, wastewater undergoes activated sludge treatment, sedimentation, and disinfection by chlorination. Four mechanical aerators in the pond prevent stratification and maintain dissolved oxygen near 100% saturation. The pond has an earthen bottom with a concrete apron at the water-surface line and is surrounded by a dike that prevents runoff drainage from entering. At the average flow rate of 4×10^6 L/d, water retention time is approximately 5 d.

During January to November 1979, monthly averages of residual chlorine concentrations ranged from 1.3 to 3.1 mg/L in the effluent leaving the chlorine contact chamber and entering the final oxidation pond. Since November 1979, the chlorine concentration has ranged from 0.25 to 1.2 mg/L. The residual chlorine concentration of water leaving the pond was usually 0.1 mg/L, approximately the same as before November 1979.

Organic extracts of the pond water, tested with the Ames test, were mutagenic to bacterial tester strains TA98 and TA100, but only after metabolic activation by Aroclor-induced rat liver microsomes.[9] The acidic fraction of the wastewater was more mutagenic than the basic fraction, and the highest mutagenic activity occurred during the summer.

The only species of fish present in the final oxidation pond before May 1980 were black bullhead, green sunfish (*Lepomis cyanellus*), and golden shiner (*Notemigonus crysoleucas*). During May and July 1980, a total of 2000 silver carp (*Hypophthalmichthys molitrix*) was released in the pond.

METHODS

The prevalence of tumors in wild fish from the final oxidation pond was determined in samples of 20 to 106 specimens from ten collections between November 1979 and March 1983. Gill nets were used for capturing most fish, but on one occasion, moribund fish were collected with a dip net. Most fish were examined grossly for external lesions. Selected fish were necropsied, and representative lesions were examined by light and electron microscopy. Fish were returned to the pond if they were not necropsied or used in experiments.

Black bullheads, brown bullheads (*Ictalurus nebulosus*), yellow bullheads (*Ictalurus natalis*), and channel catfish were placed in cages in the final oxidation pond and in a control pond at the Alabama Agricultural Experiment Station. The 1-m^3 cages were constructed of plastic netting supported at the top and bottom by steel hoops. Mesh size was 6 or 19 mm, depending on the size of the fish. Cages were allowed to rest on the pond bottom (sinking cages) or suspended off the bottom by floats (floating cages). Cages were placed at three locations in the final oxidation pond: near the inlet (inlet A), approximately 60 m from the inlet (inlet B), and near the pond outlet (outlet). Most cages were stocked with 50 fish of a single species, with total lengths of individual fish between 5 and 10 cm. All caged fish were fed a commercial catfish ration. Average time between examinations of caged fish for oral lesions was 74 d.

RESULTS AND DISCUSSION

Wild Fish

The prevalence of papillomas in wild, adult black bullheads in the final oxidation pond decreased from 73% in 1980 to 23% in March 1983. The size of papillomas also decreased with time, and the large papillomas seen protruding from the mouth of black bullheads in 1979–80[6] were no longer present. A relationship between chlorination and the black bullhead papillomas was indicated by the reduced prevalence of tumors after the chlorination rate was reduced. Tumors were not found in any species other than black bullhead.

Examination of black bullhead papillomas with transmission electron microscopy failed to reveal virus-like structures. Attempts to demonstrate a viral cause by injecting black bullheads with cell-free homogenized papillomas or by cell-culture techniques did not provide evidence that viruses were associated with the papillomas.[6,9] The inclusion bodies and electron-dense 35-nm particles in epithelial cells of some tumors examined ultrastructurally[6] probably do not indicate a viral cause of these papillomas, because cells with these inclusions and particles were not consistently present, and the 35-nm particles do not resemble any known tumor virus. If viruses are associated with the black bullhead papillomas, they are unlike those previously reported in fish neoplasms.[11]

Caged Fish

The most common type of lesion that developed on fish confined to cages in the final oxidation pond was mucosal hyperplasia (sometimes accompanied by stomatitis) in the oral fornices (Table I). The location of these lesions was the same as the usual location of papillomas on wild black bullheads. Characteristics of these hyperplastic lesions that distinguished them from papillomas were their smooth surface, mucosal thickness less than 1 mm, and lack of intermixing of the mucosa and submucosa. Histologically, these lesions were similar to the epidermal hyperplasia that developed on brown bullheads painted with an extract of Buffalo River sediment; papillomas later developed on these brown bullheads.[12,13]

In addition to histological appearance, the nonneoplastic nature of hyperplastic lesions on caged fish was indicated by their relatively quick disappearance from the fish (Figure 1). In black bullheads stocked during October 1980 and February 1981, hyperplastic lesions developed during the first 2 to 6 months after the fish were placed into the pond, then most or all of the lesions regressed during the next 1 to 2 months. The development and regression of hyperplastic lesions in the first two groups of black bullheads stocked occurred two to three times in each group. Other species and black bullheads stocked later developed lower prevalences of hyperplastic lesions, which usually reached a maximum prevalence after most of the fish stocked in a cage had died (Table I). Therefore, the pattern of lesion development and regression could not be determined because of the low number of fish with lesions.

Table I. Maximum Prevalence of Oral Mucosa Hyperplasia and Papillomas on Caged Fish in the Final Oxidation Pond of the Tuskegee, Alabama, Wastewater Treatment Facility

Species	Stocking Date	Hyperplasia		Papillomas		Control[a]
		%	N[b]	%	N	%
Black bullhead	21 Oct 80	95	22	80	5	4
	23 Feb 81	50	12	0		
	29 Oct 81	33	6	0		0
	9 Apr 82	25	8	0		
	8 Jul 82	17	12	8	12	9
Yellow bullhead	12 May 81	0		0		14
Brown bullhead	29 Oct 81	0		0		0
Channel catfish	16 Jan 81	30	10	0		8
	29 Oct 81	4	28	0		4
	8 Jul 82	0		0		5

[a]Control is the maximum prevalence of hyperplastic oral lesions in caged fish kept in a control pond. Neoplasms did not occur in control fish.
[b]Number of samples on the date when the maximum prevalence occurred.

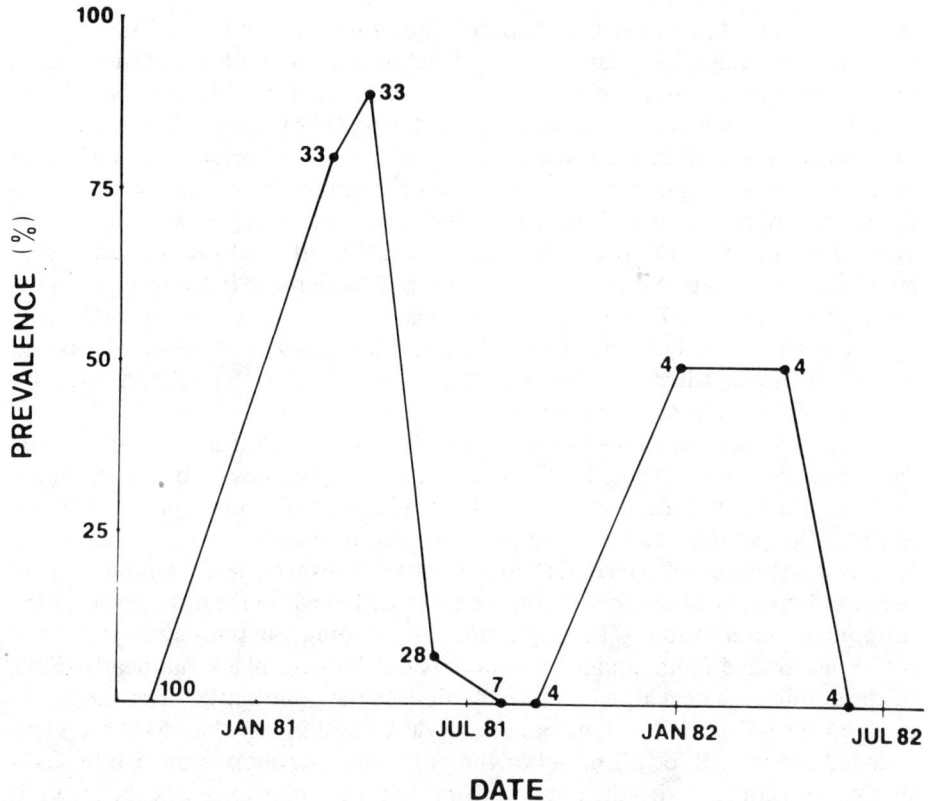

Figure 1. Prevalence of oral mucosa hyperplasia in black bullheads confined to floating cages stocked on October 21, 1980, near the outlet of the final oxidation pond of the Tuskegee, Alabama, wastewater treatment plant. The occurrence of lesions in this group of fish is representative of the pattern of lesion development and regression on most groups of black bullheads stocked on October 21, 1980, and February 23, 1981. The number at each data point indicates number of samples.

Attempts to correlate the prevalence of hyperplastic lesions with the season of the year or periods of high mutagenic activity in organic extracts of the wastewater were not successful. The lack of correlation between these lesions and mutagenicity may indicate a variable induction period, perhaps because of varied toxicant concentration or environmental conditions, or because different chemicals were involved with the lesions and mutagenicity. More frequent observation of the fish would have been helpful in better defining the length of time required for lesion development and regression, and might have made the influence of other factors (e.g., temperature, rain, short-term variation in chlorine concentration, or variation in the quantity of wastewater being treated) more evident.

The maximum prevalence of hyperplastic oral lesions on caged fish during exposures of comparable lengths in the final oxidation pond varied, depending on the fish species and the time that exposure began (Table I). The highest prevalence was in black bullheads stocked during October 1980, and maximum prevalence was less in each successive stocking. The decrease in maximum prevalence of hyperplastic lesions indicates a continuing decrease in the toxicity of the pond environment for a few years after the chlorination rate decreased. Of the four species tested, black bullheads had the highest prevalence of oral lesions. Similar lesions occurred on control fish but at a prevalence of 9% or less. There was no significant difference in the prevalence of hyperplastic lesions between floating and sinking cages or between inlet B and outlet locations. Most fish died within a few days at the inlet A location because of the high chlorine levels.

Papillomas developed on five of the black bullheads confined to cages in the final oxidation pond (Table I). The induction of papillomas on black bullheads from a different population indicates that the tumors are not genetically transmitted. The papillomas were in the same mouth location as the hyperplastic lesions developing on caged fish in this pond; however, lesions intermediate between hyperplasia and neoplasia were not observed, perhaps because of the infrequent observation. The papillomas developing on the caged fish were indistinguishable from smaller papillomas on the wild black bullheads. Four of the papillomas developed on caged fish near the pond outlet after they were exposed for 467 to 537 d. One papilloma developed in less than 57 d on a fish caged at the inlet B location. All of the papillomas developed on fish in cages that were in contact with the pond bottom, but the importance of cage location and contact with the sediment is uncertain because of the small number of fish that developed papillomas.

The reason for the uniform location of papillomas in black bullheads and hyperplastic lesions in caged ictalurids may be related to the presence of the oral valve in these species. The oral valve is a thin fold of mucosa located posterior to the upper and lower lips. The lateral ends of the valve are near the oral fornices, which were the most common locations of lesions. The oral valve is the only morphological distinction of the area where most lesions occur, but the reason for the sensitivity of the mucosa in this area has not been determined.

The usefulness of exposing caged fish to polluted water to detect chemical carcinogens was demonstrated by the development of neoplasms on black bullheads. Chemical analysis in this case failed to indicate that mutagenic or carcinogenic chemicals were present, perhaps because of infrequent sampling. Laboratory exposures of black bullheads to sediment extracts (unpublished data) and brown bullhead embryos to organic extracts of the Tuskegee wastewater[14] failed to cause tumors or other lesions that occurred on fish exposed in cages. Additional refinement of techniques may improve the usefulness of laboratory exposures; however, field exposure of caged fish is currently most satisfactory for the conditions and the carcinogen(s) present in the Tuskegee wastewater.

In other situations, laboratory exposures have been important in demonstrating the association between fish tumors and chemical carcinogens associated with pollution. Brown bullhead skin exposed to an organic extract from Buffalo River sediment developed epidermal papillomas, one of the types of neoplasms occurring in wild brown bullheads from the Niagara River and Lake Erie, which receive water from the Buffalo River.[12,13] The induction of papillomas during controlled exposures supports the hypothesis that tumors on wild fish from polluted environments can result from the pollutants, and that chemical carcinogens can cause fish papillomas.

The caged fish in the Tuskegee final oxidation pond were also useful for determining enzyme induction. Strength[15] determined that glucuronosyltransferase (UDP-GT) activity in caged channel catfish from the pond was higher than in controls. UDP-GT in other caged species, including black bullheads, and sulfotransferase activity in all caged species were not consistently different than controls. The induction of UDP-GT in channel catfish is similar to induction that occurs in rats and fish given oral or intraperitoneal doses of carcinogens[16-18] and is probably important in the excretion of carcinogens.

CONCLUSIONS

Field exposure of fish to effluents appears to be a useful method for detecting carcinogens. During field exposures, fish can be examined for lesions, enzyme induction, or other effects of carcinogens. Direct field exposure is advantageous because extraction of the carcinogen from the effluent is not necessary, and exposure levels correspond to those actually present in the environment. The duplication of some field conditions, such as complex mixtures of chemicals at variable concentrations, could not be accomplished in laboratory experiments.

Chemical analysis and laboratory exposures of fish to carcinogens extracted from water have the disadvantage of, in most cases, requiring collection of water samples that may miss the intermittent presence of toxicants. Field exposures ensure that fish are exposed to carcinogens with intermittent presence. Exposed fish can accumulate some carcinogens, and those accumulated are evidence of exposure. Furthermore, carcinogens extracted from fish can be used to identify those that have been present in the water.

Tumor induction in fish is more conclusive evidence than in vitro tests for the presence of a carcinogen that has human-health implications, because carcinogenesis in vertebrate test animals is more directly related to human cancer. However, in vitro tests have the advantages of lower cost, quicker results, and, in some cases, of indicating something about the carcinogen, such as the type of mutation resulting from exposure. The Ames test was useful in demonstrating the presence of a mutagen in organic extracts of the Tuskegee wastewater, but the development of tumors on experimentally exposed fish was more convincing evidence for the presence of a carcinogen.

The use of cages for containment of test fish permits frequent observation as well as confinement to the location of interest. Observation or sampling of test fish is necessary to detect lesions, enzyme induction, or other changes, and to ensure that indirect results of effluent exposure, such as infectious diseases, are detected. The confinement of fish to a specific location prevents them from avoiding the effluent and permits a better determination of the extent of exposure. Confinement is especially important when the fish could easily leave the area of interest.

The disadvantages of field exposures include more difficult observation of lesion development and less control over exposure concentration and environmental conditions than in laboratory experiments. In many polluted environments, the chances that test fish have for surviving chronic exposures are low because of periods of lethal conditions. Field exposures may also subject the fish to toxicants other than those of interest; therefore, the results are more difficult to interpret. Although the response time for tumor induction is long compared to chemical analysis of water or in vitro tests, the use of preneoplastic lesions or metabolic changes may be useful in reducing the time required for obtaining results from fish exposures.

In polluted environments that are not suitable for chronic fish exposures, field exposures may be useful for determining acute effects but would not be useful for demonstrating carcinogenic effects. In these cases, the extraction of suspected carcinogens and subsequent laboratory exposure would be more appropriate than field studies. It is probable that the Tuskegee pond would not have been suitable for field exposures if the pond had not been aerated as part of the treatment process. In some environments that cannot support fish life, artificial aeration or other changes might allow fish to live, but possible changes in the toxicants of interest must be considered.

The most desirable species for field exposures depends on the situation being studied. The selection of a test species must consider the environmental conditions, the nature of the toxicant, and the objectives of the experiment. When there is no compelling reason for selecting a particular species, choosing one for which there is basic information, such as culture requirements and sensitivity to toxicants, will improve the chances of a successful experiment. The survival time of exposed fish must be long enough for the desired response to occur. In many polluted aquatic environments, factors such as dissolved oxygen, pH, or toxicant concentrations are too adverse for chronic exposures of some or all fish species.

In the case of the Tuskegee wastewater pond, the black bullhead was an obvious choice for experimental exposures because of the occurrence of papillomas on wild black bullheads. This species would probably be desirable for other experimental situations because of its tolerance to pollution, wide geographical occurrence,[19] and ease of culture. Its relatively high tolerance to chlorine is an advantage for chronic exposure to chlorinated effluents that exceed the tolerance of other species. Although the reasons for the differences are not yet known, black bullheads seem more susceptible to the carcinogenic effects of the chemical carcinogen(s) in the Tuskegee wastewater pond than other exposed species.

ACKNOWLEDGMENTS

This study was supported by cooperative agreements CR807844010 and CR809336010 in the NCI/EPA Collaborative Program, Project No. 3, "Effects of Carcinogens, Mutagens and Teratogens in Non-human Species (Aquatic Animals)," administered by the Gulf Breeze Environmental Research Laboratory.

REFERENCES

1. Matsushima, T., and T. Sugimura. "Experimental Carcinogenesis in Small Aquarium Fishes," *Prog. Exp. Tumor Res.* 20:367-379 (1976).
2. Sinnhuber, R. O., J. D. Hendricks, J. H. Wales, and G. B. Putnam. "Neoplasms in Rainbow Trout, a Sensitive Animal Model for Environmental Carcinogenesis," *Ann. N.Y. Acad. Sci.* 298:389-408 (1977).
3. Meyers, T. ., and J. D. Hendricks. "A Summary of Tissue Lesions in Aquatic Animals Induced by Controlled Exposures to Environmental Contaminants, Chemotherapeutic Agents, and Potential Carcinogens," *Mar. Fish. Rev.* 44(12):1-17 (1982).
4. Sonstegard, R. A. "Feral Aquatic Organisms as Indicators of Waterborne Environmental Carcinogens," in *Aquatic Pollutants: Transformation and Biological Effects*, O. Hutzinger, I. H. Van Lelyveld, and B. C. J. H. Zoeteman, Eds. (New York: Pergamon Press, 1978), pp. 349-358.
5. Black, J. J., E. D. Evans, J. C. Harshbarger, and R. F. Zeigel. "Epizootic Neoplasms in Fishes From a Lake Polluted by Copper Mining Wastes," *J. Nat. Cancer Inst.* 69(4):915-926 (1982).
6. Grizzle, J. M., T. E. Schwedler, and A. L. Scott. "Papillomas of Black Bullheads, *Ictalurus melas* (Rafinesque), Living in a Chlorinated Sewage pond," *J. Fish Dis.* 4:345-351 (1981).
7. Tan, B., P. Melius, and J. M. Grizzle. "Hepatic Enzymes and Tumor Histopathology of Black Bullheds with Papillomas," in *Chemical Analysis and Biological Fate: Polynuclear Aromatic Hydrocarbons. Fifth International Symposium*, M. Cooke and A. Dennis, Ed. (Columbus, OH: Battelle Press, 1981), pp. 377-386.
8. Harshbarger, J. C. "Role of the Registry of Tumors in Lower Animals in the Study of Environmental Carcinogenesis in Aquatic Animals," *Ann. N.Y. Acad. Sci.* 298:280-289 (1977).
9. Grizzle, J. M., and P. Melius. *Causes of Papillomas on Fish Exposed to Chlorinated Sewage Effluent*, CR809336010 (Gulf Breeze, FL: U.S. Environmental Protection Agency, 1983).
10. Marking, L. L., and T. D. Bills. "Chlorine: Its Toxicity to Fish and Detoxification and Antimycin," *U.S. Fish Wildl. Ser. Invest. Fish Control* 74:1-5 (1977).

11. Pilcher, K. S., and J. L. Fryer. "The Viral Diseases of Fish: A Review Through 1978. Part 2: Diseases in Which a Viral Etiology is Suspected but Unproven," *CRC Critical Rev. Microbiol.* 8:1-24 (1980).
12. Black, J. J. "Epidermal Hyperplasia and Neoplasia in Brown Bullheads (*Ictalurus nebulosus*) in Response to Repeated Applications of a PAH Containing Extract of Polluted River Sediment," in *Polynuclear Aromatic Hydrocarbons: Seventh International Symposium on Formation, Metabolism and Measurement*, M. W. Cooke and A. J. Dennis, Eds. (Columbus, OH: Battelle Press, 1982), pp. 99-111.
13. Black, J. J. "Field and Laboratory Studies of Environmental Carcinogenesis in Niagara River Fish." *Great Lakes Res.* 9(2):326-334 (1983).
14. Biba, D. M. *Effects of Aflatoxin on the Brown Bullhead Ictalurus nebulosus*, M.S. Thesis (Auburn, AL: Auburn University, 1983).
15. Strength, D. R. Auburn University. Personal communication.
16. Bock, K. W., D. Josting, W. Lilienblum, and H. Pfeil."Purification of Rat-Liver Microsomal UDP-Glucuronyltransferase," *Eur. J. Biochem.* 98:19-26 (1979).
17. Strength, D. R., H. H. Daron, J. L. Aull, and J. F. Wilson. "The Induction of Glucuronide and Sulfate Transferases by Phenobarbital and Polycyclic Aromatic Hydrocarbons," *Fed. Proc.* 39:1694 (1980).
18. Strength, D. R., D. V. Saradambal, S.-L. Wang, and H. H. Daron. "Glucuronosyl- and Sulfo-Transferases in Fish Exposed to Environmental Carcinogens." *Fed. Proc.* 41:1147 (1982).
19. Scott, W. B., and E. J. Crossman. *Freshwater Fishes of Canada* (Ottawa, Canada: Fisheries Research Board of Canada, 1973), pp. 592-593.

SECTION VII

Environmental Effects

Since the discovery by man that water could be used to dilute and to transport unwanted materials from the location at which they were generated to some distant place where their adverse effects would not be so unpleasantly evident, the aquatic environment has been the unfortunate recipient of much of his waste products.

Joseph V. Hunter
Origin of Organics from Artificial Contamination, 1971

A thing is right when it tends to preserve the integrity, stability, and beauty of the biotic community. It is wrong when it tends otherwise.

Aldo Leopold
A Sand County Almanac, 1949

"Even God cannot change the past."

Agathon 446–401 B.C.

CHAPTER 37

Interactions of Chlorine-Produced Oxidants, Salinity, and A Protistan Parasite in Affecting Lethal and Sublethal Physiological Effects in the Eastern or American Oyster

Geoffrey I. Scott, Edward O. Oswald, Tommy I. Sammons, Douglas S. Baughman, and Douglas P. Middaugh

The discharge of chlorine into estuarine habitats may seriously affect organisms residing there.[1,2] Several studies[3-6] have shown that the ability of oysters to survive chlorine and bromoform exposure is clearly dependent on the physiological condition of the organism. In comparisons of spring and summer chronic exposures at identical chlorine concentrations, salinities, and temperatures, oyster survival rates were three to four times greater during the spring (prespawning period) than in the summer (postspawning period).[4] Additionally, summer respiration rates were significantly higher than spring values in chlorine-exposed oysters.[6] Both spring and summer groups were infected with the protistan parasite, *Perkinsus marinus*.

Studies on the epizootiology of *P. marinus* have indicated that this parasite is basically a summer disease causing peak mortalities in the eastern or American oyster, *Crassostrea virginica*, of greater than 50% in August and September when water temperatures and salinity values are maximal.[7,8] Salinity has been shown to be an important factor affecting oyster mortalities; *P. marinus* infection intensities and associated mortalities are greatest in high-salinity waters.[9,10] Low-salinity exposure tends to depress infection intensities and significantly enhance oyster survival.[11-14]

As earlier discussions[4-6] indicated, oysters are much more susceptible to chlorination exposure during the summer months. This increased summer susceptibility to chlorine exposure may be caused by lessened resistance in the oyster, resulting from *P. marinus* infections. Several authors[15,16] have described the interactions between increased levels of pollution and increased incidences of parasitism and disease in marine organisms.

We investigated the interactions of chlorination exposure and salinity in affecting lethal and sublethal responses in oysters, *C. virginica*, infected with the protistan parasite, *P. marinus*.

MATERIALS AND METHODS

Collection and Acclimation

Adult oysters (6–12 cm in height) were collected from Leadenwah Creek (latitude, N 32°36'; longitude, W 80°15'), a large tidal tributary of the North Edisto River Estuary in South Carolina on June 21, 1982. Following collection, oysters were immediately transported to the laboratory where they were scrubbed and all fouling organisms were removed. After numbering and measuring the height of individual oysters, 120 animals were placed in each test aquarium (circular tanks: 152.4-cm radius × 30-cm depth with a working volume on 310 L) and were acclimated for 39 d in unfiltered seawater pumped in daily from the Stono River at high tide. Tanks were cleaned daily during acclimation.

Following acclimation, oysters in each tank were exposed for 60 d to one of several treatments described in Table I. Tests were conducted from August through October 1982.

Exposure Conditions

Low-salinity water (treatment E) was prepared by diluting incoming ambient estuarine water with groundwater from a shallow well. Normal ambient salinity at high tide ranged from 21 to 25 parts per thousand (ppt). Low-salinity water of 8 to 10 ppt was desired, which required approximately a 40:60% dilution of estuarine water:groundwater.

Preparations of chlorinated seawater were made at both low (8–10 ppt) and high (21–25 ppt) salinities by daily addition of NaOCl (Fisher Reagent Grade, 5%) to each tank at appropriate concentrations and dilutions. Nominal concentrations were 1.00, 0.56, and 0.10 mg/L in both high-salinity (treatments B, C, and D, respectively) and low-salinity (treatments F, G, and H, respectively) waters. Oysters were also exposed to ambient chlorine-free high-salinity and low-salinity estuarine water (treatments A and E, respectively) to serve as indicators of natural or control conditions.

A total of 120 oysters were placed in each test aquarium and were exposed to each treatment for 60 d. Oysters fed on phytoplankton in the seawater and were exposed to the ambient light:dark cycle (14:10 h) and seawater temperatures. Water changes were made and test aquaria were cleaned daily throughout the experiment.

Test Parameters Measured

During each experiment, the following measurements were used to assess the physiological effects of each treatment on adult oysters:

Table I. Description of Treatment Codes and Physical and Chemical Parameters Measured in Each of the Various Salinity and Chlorination Treatments used in This Study. Statistical measurements included calculations of mean ($\bar{x}$) and standard error (S.E.). Single (*) or double (**) asterisks indicate measured CPO concentrations in each low-salinity CPO treatment were significantly ($^b p < 0.05$) or very significantly ($p < 0.01$) different from similar high-salinity CPO treatments.

Treatment Code	Treatment Description	Measured CPO Concentration (mg/L)[a]		Exposure Salinity (ppt)[b]		Exposure Temperature (°C)[c]	
		$\bar{x}$	S.E.	$\bar{x}$	S.E.	$\bar{x}$	S.E.
A	High-salinity control	0.00	0.000	22.6	0.42	24.2	0.40
B	High salinity 1.00-mg/L NaOCl[d] concentration	0.06	0.007	22.9	0.21	24.2	0.40
C	High salinity 0.56-mg/L NaOCl concentration	0.03	0.005	22.7	0.04	24.0	0.39
D	High salinity 0.10-mg/L NaOCl concentration	0.01	0.003	22.9	0.21	24.0	0.39
E	Low-salinity control	0.00	0.000	9.2	0.18	23.9	0.38
F	Low salinity 1.00-mg/L NaOCl concentration	0.11**	0.008	9.3	0.18	23.6	0.39
G	Low salinity 0.56-mg/L NaOCl concentration	0.06**	0.007	9.6	0.59	23.6	0.39
H	Low salinity 0.10-mg/L NaOCl concentration	0.02*	0.004	9.0	0.17	23.6	0.39

[a]Sample size (n) ranged from 36 to 38.
[b]Sample size ranged from 57 to 58.
[c]Sample size ranged from 51 to 52.
[d]NaOCl refers to nominal NaOCl concentrations added.

1. Survival and mortality percentages were recorded daily, using three replicates per treatment.

2. Chlorine-produced oxidant (CPO) concentrations (free plus combined chlorine, bromine, and other residual oxidants) were measured with a Fisher Titrator (Model 397). Phenylarsine oxide and pH 4 buffer were standard Fisher reagents. The 5% potassium iodide (KI) solution and chlorine (Fisher NaOCl analytical reagent, minimum 5% NaOCl) stock solutions were prepared with deionized water. Stock solutions were kept in lightproof containers to prevent photodecomposition. Measurements of CPO concentration were derived according to procedures outlined in *Standard Methods*.[17] All water samples were taken directly from treatment aquaria at least every 2 to 3 d during each experiment. Agitation of samples was avoided to reduce the potential for error by photodecomposition or flashing-off of CPO.[18] The mixing order of Crecelius et al.[19] (i.e., the addition of 2 mL of pH 4 buffer and 2 mL of 5% KI solution to the sampling cup, followed by the addition of the saline water sample) was used in each titration. Values are reported in milligrams per liter.

3. Ambient water temperature (°C) and salinity (ppt) were measured daily in the incoming seawater from the Stono River in each experiment. A mercury thermometer and a refractometer (American Optics) were used to measure temperature and salinity, respectively. In addition, periodic measurements of dissolved oxygen (Yellow Springs Instrument Company) and pH (Corning pH meter) were made. Dissolved oxygen measurements indicated that oxygen concentrations were always at saturation. The pH ranged from 7.40 to 8.20 during the study.

4. Fecal production or biodeposition rate measurements were made every 2 to 3 d in all treatments by observing the number of oysters producing feces.

5. Respiration was measured on excised gill tissues removed from adult oysters (7–10 cm height) that had been exposed to the various treatments, using methods described in earlier reports.[4,5]

6. *P. marinus* densities in oyster tissues were determined using previously described methods.[20-24] Fifteen adult oysters were removed from each test aquaria on days 0, 15, 30, and 60 of exposure and examined for the *P. marinus* densities. A total of 490 oysters were examined.

The incidence of infection and infection density were determined by counting the number of prezoosporangia in either a 5-mm (50X) or 2-mm (100X) field of view. Three replicate counts of the most heavily infected area of each tissue were made to provide some statistical quantification of infection density.

Infection density code numbers were averaged for each group of oysters to produce the weighted incidence of infection, which is a value reflecting both the proportion of the population infected and the intensity of infection.[8]

Statistical Treatment

Statistical treatment of each experimental measurement in oysters from each treatment included determination of mean, standard deviation, and standard error. T-tests (parametric) and Wilcoxon T-tests (nonparametric) were used to determine if measured differences among various treatments were statistically significant. The 95% confidence level ($p \leq 0.05$) was the minimum statistical significance accepted.

RESULTS AND INTERPRETATION

Salinity Effects

Results of tests comparing the interactive effects of high- and low-salinity exposure on *P. marinus* infection interactions in adult oysters are listed in Tables I through III and depicted in Figures 1A and 1B.

Physical parameters measured in this experiment are listed in Table I. Dilution of seawater with groundwater resulted in low-salinity seawater (E), which had a mean salinity of 9.2 ppt and was well within the desired range of 8 to 10 ppt. Salinity in the ambient high-salinity seawater (A) averaged 22.6 ppt.

Dilution of seawater with groundwater did not significantly lower water temperature, because the mean exposure temperatures in both the high- and low-salinity controls were very similar, 23.9°C in the low-salinity controls as compared with 24.6°C in the high-salinity controls.

Oyster mortalities (Figure 1A) were virtually the same in high and low salinity during the first 33 d of exposure (16 vs 18% mortality, respectively); however, after 60 d of exposure, oyster mortalities were significantly ($p < 0.05$) reduced in the low-salinity treatment by an average of 25% ($\bar{x} = 35\%$ mortality in low salinity compared with 46.7% in high salinity).

Measurements of fecal production or biodeposition rates are listed in Table II. Fecal production was significantly ($p \leq 0.01$) lower in low-salinity controls ($\bar{x} = 64.7\%$) when compared with high-salinity controls ($\bar{x} = 83.4\%$). The lower rate of fecal production in the low-salinity oysters was probably related to the lower phytoplankton density (food availability) in the low-salinity treatment.

Mean gill tissue respiration rates declined throughout the study in both high- ($\bar{x} = 1280.6$ to 1201.5 μL O_2 g^{-1} h^{-1}) and low-salinity ($\bar{x} = 1280.6$ to 1201.3 μL O_2 g^{-1} h^{-1}) controls in response to declining seawater temperatures (28–20°C) (Figure 1B). There were no significant differences in respiration rates measured among oysters in comparing high- and low-salinity controls.

Prezoosporangia densities of *P. marinus* in rectal and labial palp tissues of

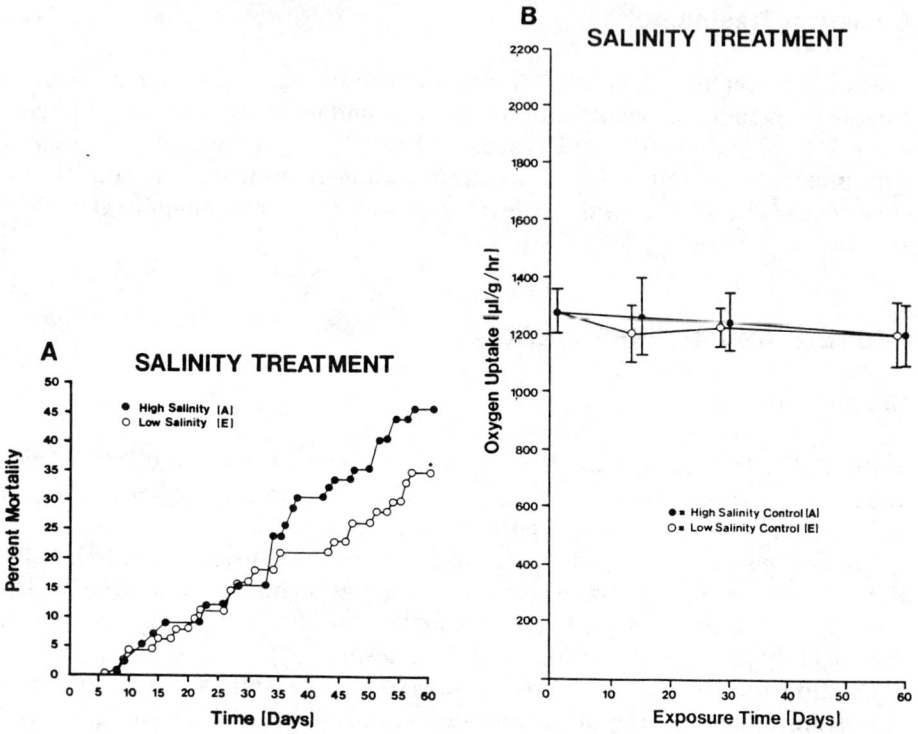

Figure 1. A: Mortality data for oysters exposed to high- (A) and low- (E) salinity treatments. Note that mortality differences between the two treatments were only significant after 30 d of exposure when temperatures fell below 25°C. Asterisk (*) indicates that 60-d mortality results were significantly ($p \leq 0.05$) different; B: Mean gill tissue respiration rates ($\mu L\ O_2\ g^{-1}\ h^{-1}$) measured in oysters exposed to high- (A) and low- (E) salinity treatments. Note the similarity in respiration rates among high- and low-salinity oysters throughout the entire 60 d of exposure. Bars represent sample standard error.

surviving oysters are listed in Table III. The incidence of infection was 100% for both high- and low-salinity-exposed oysters. Prezoosporangia densities in rectal tissues increased throughout the 60 d of exposure in both high- ($\bar{x}$ = 66.23 to 670.18 prezoosporangia per 5-mm field) and low-salinity controls ($\bar{x}$ = 66.23 to 631.14 prezoosporangia per 5-mm field). Prezoosporangia densities in labial palp tissues showed a similar increase in both high- ($\bar{x}$ = 75.8 to 752.9 prezoosporangia per 5-mm field) and low-salinity controls ($\bar{x}$ = 75.8 to 1333.3 prezoosporangia per 5-mm field) throughout the 60 d of exposure. Statistical analysis indicated that prezoosporangia densities were not significantly different in comparisons of rectal and labial palp tissues in oysters exposed to either high or low salinity. Additionally, prezoosporangia densities in rectal tissues were not significantly different in comparisons of high- and

Table II. Summary of Fecal Production Values Measured in Oysters Exposed to Various Salinity/Chlorination Conditions. Note the significant ($p \leq 0.01$) reduction in fecal production in low-salinity and low-salinity CPO-exposed oysters. Statistical parameters include calculations of mean ($\bar{x}$), standard error (S.E.), sample size (n), and Students T-value. Asterisks (**) indicate sample means were significantly ($p \leq 0.01$) different.

Treatment	Nominal NaOCl Concentration (mg/L)	Range	Fecal Production (%)			T-Value
			$\bar{x}$	S.E.	n	
A	0.00	60–100	83.4	1.54	41	
B	1.00	20–100	79.2	1.56	40	
C	0.56	48–100	80.3	2.08	40	
D	0.10	64–100	82.5	1.65	39	
E	0.00	4–100	64.7**	3.61	38	−4.76[a]
F	1.00	16–92	65.5**	2.61	38	−4.51[b]
G	0.56	12–92	72.5**	2.28	39	−2.53[c]
H	0.10	12–92	66.9**	3.19	39	4.34[d]

[a]Calculated from a comparison of treatments A and E.
[b]Calculated from a comparison of treatments B and F.
[c]Calculated from a comparison of treatments C and G.
[d]Calculated from a comparison of treatments D and H.

Table III. Mean Rectal and Labial Palp Hypnospore Densities in Oysters Exposed to Various Chlorination and Salinity Conditions. Hypnospore densities were not significantly different in comparison of high- and low-salinity CPO-exposed oysters or in comparisons with controls. Statistical parameters include calculations of mean ($\bar{x}$), standard error (S.E.), sample size (n), and Student's T-value.

Treatment	Exposure Time (d)	Exposure Temperature (°C)	Rectum Hypnospore Density[a]					Labial Palp Hypnospore Density[a]				
			$\bar{x}$	S.E.	n	T-Value[b]	T-Value[c]	$\bar{x}$	S.E.	n	T-Value[b]	T-Value[c]
A	0	28	77.4	66.23	15			75.8	64.89	15		
A	15	27	61.1	34.74	15			37.5	25.56	15		
B	15	27	2,021.4	1,065.53	15		+1.84	2,007.7	1,068.30	15		+1.84
C	15	27	1,585.2	939.67	15		+1.62	1,495.1	916.13	15		+1.59
D	15	27	81.6	70.77	15		+0.26	62.5	53.88	15		+0.42
E	15	27	3.3	1.41	15	−1.66		2.9	1.24	15	−1.35	
F	15	27	1,058.4	559.58	15	+0.80	+1.89	1,225.2	668.95	15	−0.62	+1.83
G	15	27	2,122.9	1,131.79	15	+0.37	+1.87	1,729.9	998.62	15	+0.17	+1.73
H	15	27	39.5	22.55	15	−0.57	+1.59	32.9	23.04	15	−0.51	+1.30
A	30	24	703.3	692.84	15			753.9	691.70	15		
B	30	24	27.7	7.41	15		−0.98	29.1	15.75	15		−1.05
C	30	24	712.8	697.73	15		+0.01	747.8	697.41	15		−0.01
D	30	24	771.0	713.80	15	+0.70		142.2	85.31	15		
E	30	24	14.1	9.00	15		−0.99	10.5	6.07	15		−1.07
F	30	24	1,009.5	726.23	15	+1.35	+0.23	997.7	726.04	15	+1.33	+1.17
G	30	24	795.6	759.27	15	+0.08	+0.02	799.9	76.85	15	+0.05	+0.84
H	30	24	155.3	90.69	15	+1.55	−0.86	130.4	64.92	15	+1.84	−0.11
A	60	20	765.0	670.18	15			752.9	658.44	15		
B	60	20	1,578.1	961.99	15		+0.69	1,508.2	952.28	15		+0.65
C	60	20	546.6	538.36	15		−0.25	696.5	684.71	15		−0.06
D	60	20	800.1	698.15	15		+0.04	809.1	714.65	15		+0.06
E	60	20	702.2	631.14	15	−0.07		1,333.3	902.11	15	−0.52	
F	60	20	2,141.2	1,100.10	15	+0.39	+1.13	1,792.7	945.77	15	+0.21	+0.35
G	60	20	802.9	706.70	15	+0.29	+0.11	659.2	538.94	15	−0.04	−0.64
H	60	20	746.4	724.33	15	−0.05	+0.05	756.1	742.73	15	−0.05	−0.49

[a] Number per 5-mm field.
[b] Comparisons of high-salinity CPO treatments with similar low-salinity CPO treatments (B vs F, C vs G, and D vs H), and comparisons of high- and low-salinity controls (A vs E).
[c] Comparisons of high-salinity controls (A) with high-salinity CPO-exposed oysters (B, C, and D), and comparisons of low-salinity controls (E) with low-salinity exposed oysters (F, G, and H).

low-salinity treatments. Prezoosporangia densities in labial palp tissues were not significantly different in comparisons of high- and low-salinity controls.

Weighted indices of infection-intensity calculations for *P. marinus* in rectal and labial palp tissues are listed in Figure 2. Initially, low-salinity exposure significantly ($p \leq 0.01$) lowered infection intensities in both rectal and labial palp tissues as prezoosporangia densities were substantially reduced during the

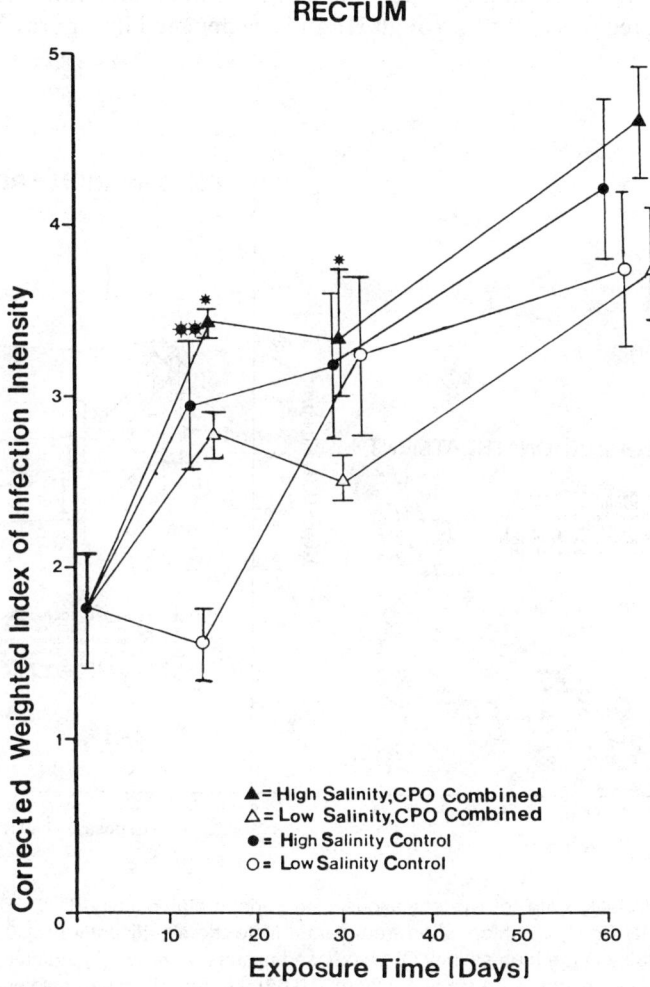

Figure 2. Mean weighted index of infection intensity measured in high- and low-salinity chlorination-exposed oysters. High- and low-salinity control oysters are also included. Values are pooled for surviving oysters and autopsied dead oysters. Asterisks (*) indicate that values were significantly ($p \leq 0.05$) different in comparisons of high- and low-salinity CPO-exposed oysters. Double asterisks (**) indicated values between high- and low-salinity controls were very significantly ($p \leq 0.01$) different.

first 15 d of exposure. However, after 30 d of exposure and continuing through 60 d, infection intensities were practically the same for low- and high-salinity controls in both rectal and labial palp tissues.

Chlorination Treatment

The results of tests comparing the effects of various concentrations of CPO at both low- and high-salinity exposures on *P. marinus* infection rates in adult oysters are listed in Tables I through III and are depicted in Figures 2, 3A, and 3B.

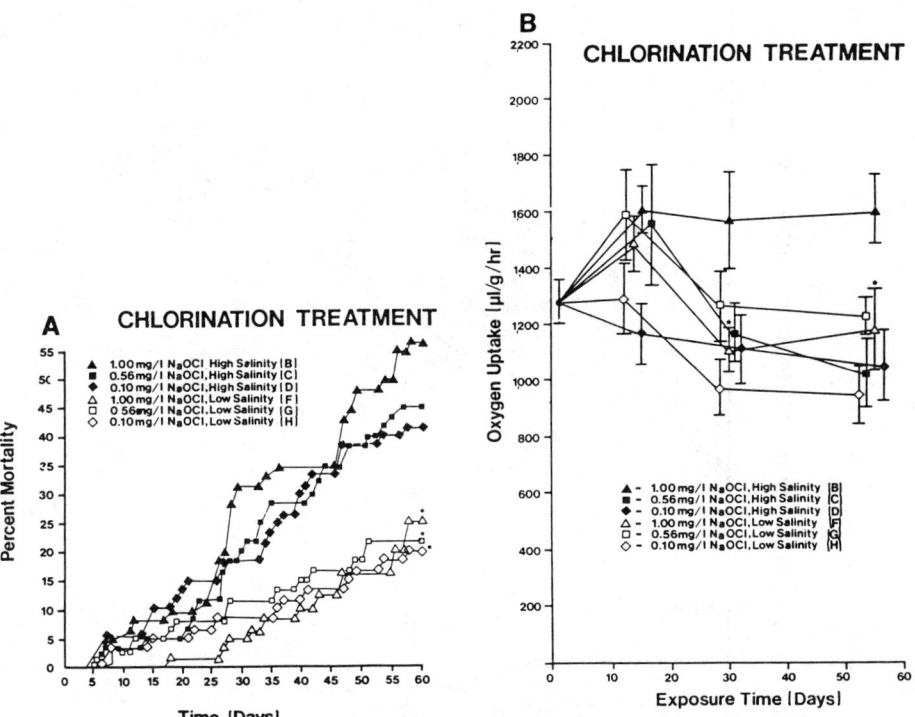

Figure 3. A: Mortality data for oysters exposed to various high-salinity (B, C, D) and low-salinity (F, G, H) chlorination treatments. Note the significantly ($p \leq 0.05$) higher mortality in the high-salinity CPO-exposed oysters. Asterisk (*) indicates that 60-d mortality results were significantly ($p \leq 0.05$) different; B: mean gill tissue respiration rates ($\mu L\ O_2\ g^{-1}\ h^{-1}$) measured in adult oysters exposed to various high-salinity (B, C, D) and low-salinity (F, G, H) chlorination treatments for 60 d. Gill tissue respiration rates generally declined throughout the 60 d of exposure because of declining seawater temperatures. Note the similarity in gill tissue respiration rates between high- and low-salinity CPO-exposed oysters, with the exception of treatment B at exposure days 30 and 60. Asterisk (*) indicates that the 30- and 60-d gill tissue respiration rates in treatment B were significantly ($p \leq 0.01$) higher than in treatment F. Bars represent sample standard error.

Dilution of seawater with groundwater resulted in low-salinity chlorinated seawater with a mean salinity of 9.3 ppt, well within the desired range of 8 to 10 ppt (Table I). Salinity in the ambient, high-salinity chlorinated seawater averaged 22.8 ppt. Mean exposure temperatures averaged 23.6°C in low- and 24.1°C in high-salinity CPO-exposed oyster treatments.

Measured CPO concentrations for each selected nominal NaOCl concentration varied significantly in both high- and low-salinity chlorinated seawater combinations (Table I). Further analysis of these results indicated that measured CPO concentrations for each selected nominal NaOCl concentration were significantly ($p \leq 0.01$–0.05) higher (45–50%) in low-salinity chlorinated seawater exposures (F, G, H) when compared with high-salinity chlorinated seawater exposures (B, C, D).

Oyster mortalities were significantly ($p \leq 0.05$) lower in each low-salinity chlorination treatment when compared with equivalent high-salinity chlorination treatments (Figure 3A). Mean oyster mortalities averaged 22.2%, ranging from 20 to 25% in the various low-salinity chlorination exposures (F, G, H), compared with an average mortality rate of 51.1%, ranging from 41.7 to 56.7% in the various high-salinity chlorination treatments (B, C, D).

Comparisons of mortality rates in the various high-salinity chlorination exposures (B, C, D) with high- and low-salinity controls (A and E, respectively) indicated the mean mortality rate in the nominal 1.00-mg/L high-salinity treatment (B = 56.7%) was significantly ($p \leq 0.05$) higher than the mean mortality rates recorded in high- and low-salinity control oysters (A = 46.7% and E = 35%, respectively). The mean mortality rates in the nominal 0.56-mg/L and 0.10-mg/L high-salinity treatments (C = 45% and D = 41.7%, respectively) were slightly lower but not significantly different from the mortality rate for high-salinity control oysters (A = 46.7%). The mean mortality rates in the nominal 0.56-mg/L and 0.10-mg/L high-salinity treatments (C = 45% and D = 41.7%, respectively) were significantly ($p \leq 0.05$) higher than the mortality rate for low-salinity control oysters (E = 35%).

Comparisons of mortality rates in the various low-salinity chlorination exposures (F, G, H) with high- and low-salinity controls (A and E, respectively) indicated that the mean mortality rates were significantly ($p \leq 0.05$) lower in each of the low-salinity chlorination treatments (F = 25%, G = 21.7%, and H = 20%) when compared to either the high-salinity (A = 46.7%) or low-salinity (E = 35%) controls.

Mortality was lowest (averaging 20%) in the 0.10-mg/L nominal NaOCl concentration, low-salinity treatment (H). This represented a 57% reduction in the mortality rate measured in high-salinity (A) control oysters and a 43% reduction in the mortality rate for low-salinity (E) control oysters.

These findings indicated that the combination of low-salinity (< 10 ppt) and low-chlorine concentrations (0.10–1.00mg/L as nominal NaOCl concentrations) significantly enhanced oyster survival above survival rates measured in both high- and low-salinity regimes, approximating a full salinity range of estuarine conditions.

Fecal production rates were significantly ($p \leq 0.01$) lower in each low-salinity chlorination exposure when compared with equivalent high-salinity chlorination exposures (Table II). Mean fecal production rates averaged 68.3% (ranging from 65.6 to 72.5%) in the various low-salinity chlorination exposures (F, G,H) compared with an average rate of 80.7% (ranging from 79.2 to 82.5%) in the different high-salinity chlorination exposures (B, C, D).

These results indicated that oyster fecal production in the chlorinated high-salinity seawater was significantly higher than in chlorinated low-salinity seawater. Further comparisons with high- and low-salinity control oysters indicated that reduced salinity (or dilution) rather than chlorination had a more profound effect on fecal production, probably related to the lower phytoplankton densities (food availability) in the low-salinity chlorination exposures.

Gill-tissue respiration rates in both high- and low-salinity chlorination exposures initially increased from days 0 to 15, then gradually declined from days 15 to 60 (Figure 3B). Gill tissue respiration rates in high-salinity CPO exposed oysters (B, C, D) generally were not significantly different from values measured in the low-salinity CPO exposures (F, G, H). The only exception to this trend occurred at days 30 and 60 in the 1.00-mg/L nominal NaOCl concentration, high-salinity-treatment values (B = 1568.7 and 1598.3 μL O_2 g^{-1} h^{-1}, respectively), which were significantly ($p \leq 0.01$) higher than corresponding low-salinity chlorination-treatment values (F = 1110.5 and 1179.2 μL O_2 g^{-1} h^{-1}, respectively).

Comparisons of high- (A) and low-salinity (E) controls with the different high-salinity chlorination exposures (B, C, D) indicated that gill-tissue respiration rates generally (in 78% of the samples) were not significantly different between controls and high-salinity CPO exposures. Similar comparisons of different low-salinity CPO-exposed oysters (F, G, H) with high- (A) and low-salinity (E) controls indicated that, generally (in 67% of the samples), gill tissue respiration rates were not significantly different between controls and low-salinity CPO exposures.

The incidence of *P. marinus* infection was 100% for both high- (B, C, D) and low-salinity (F, G, H) CPO-exposed oysters, based on examination of rectal and labial palp tissues (Table III). Prezoosporangia densities were highly variable in all CPO and salinity exposures; however, these densities generally increased in both rectum and labial palp tissues throughout the 60 d of exposure in both high- and low-salinity CPO-exposed oysters. Statistical analysis for high- (B, C, D) and low-salinity (F, G, H) CPO-exposed oysters indicated that there were no significant differences in comparisons with high- and low-salinity chlorination treatments. Comparisons within each salinity and chlorination treatment indicated that rectal prezoosporangia densities were generally greater than labial palp tissue densities.

Additional comparisons of high- (B, C, D) and low-salinity (F, G, H) CPO-exposed oysters with high- (A) and low-salinity (E) controls indicated no significant differences between controls and the various chlorination treatments in prezoosporangia densities of rectal and labial palp tissues.

Weighted index of infection-intensity calculations for *P. marinus* in rectal tissues are shown in Figure 2. Infection intensity generally increased throughout the 60 d of exposure in both rectal and labial palp tissues of high- (B, C, D) and low-salinity (F, G, H) CPO-exposed oysters. This same trend was observed in high- and low-salinity control oysters.

Statistical analysis of high- (B, C, D) and low-salinity (F, G, H) CPO-exposed oysters indicated there were significant ($p \leq 0.05$) differences in comparison of the weighted index of infection intensities among rectal tissues at days 15 and 30 of exposure. Comparisons of high- (B, C, D) and low-salinity (F, G, H) CPO-exposed oysters with high- (A) and low-salinity (E) controls indicated there were generally no significant differences in rectal infection intensities among controls and the various salinity and chlorination treatments. In similar comparisons of labial palp tissue, there were generally no significant differences among controls and the various salinity- and chlorination-treated oysters.

DISCUSSION

Results from this study have generally indicated the significant interactive effects of salinity and CPO exposure in affecting oyster mortalities resulting from *P. marinus* infection.

Comparisons of control oysters in high- and low-salinity exposures generally indicated that low-salinity conditions resulted in exposure of oysters to water which was reduced in both salt content (major ions) and phytoplankton density (as food availability was reduced by 66%). Biodeposition rates (or feeding rates) were significantly reduced in low-salinity-exposed oysters, probably related to less food availability. Galtsoff[25] reported similar reduced feeding rates in oysters exposed to low salinity. Exposure to low-salinity conditions and accompanying reduced fecal production rates did not significantly affect oyster physiology, because measurements of condition index, gonadal index, and gill tissue respiration rates were not significantly different (and were generally identical) throughout the 60 d of exposure.[26]

Analysis of mortality data from high- and low-salinity control conditions indicated that low-salinity exposure significantly increased oyster survivability of *P. marinus* infections by 25%. This significant increase in survival occurred in low-salinity control oysters even though there were generally no statistically significant differences in the incidence of infection, prezoosporangia density, and weighted index of infection intensity in comparisons of high- and low-salinity control oysters. Analysis of the weighted index of infection-intensity data suggested that low-salinity exposure initially decreased infection intensity (at day 15); however, by days 30 and 60, infection intensities were equal for both high and low salinities. Earlier studies[9,12] have indicated similar delays in prezoosporangia development from low-salinity exposure.

Although infection-intensity differences in comparisons of high- and low-salinity control oysters were only slight, actual differences may have been obscured by the higher mortality rate in high-salinity control oysters, because the most heavily infected oysters were presumably dying off first (this can be presumed since measurements were not made of gapers but rather of living specimens). The death of the most diseased individuals would result in the selection of more-resistant or less-infected oysters, which would tend to bias infection intensities downward to much lower levels than if all heavily diseased or dead individuals were included.

The mode of action for low-salinity exposure in significantly reducing oyster mortalities has been reviewed by several authors,[9,11] who generally concluded that low-salinity exposure must enhance survival because of the possible reductions in the number of waterborne prezoosporangia that would tend to limit and reduce both the incidence and intensity of infection. However, the fact that all oysters in this study were infected with *P. marinus* at day 0, since the infection spreads only from the initial site of infection, suggests that differences in survival were not related to differences in prezoosporangia density in seawater but rather to physiological differences in oysters exposed to different salinity.

Results from CPO exposure treatment indicated that the combination of low salinity and CPO exposure significantly ($p \leq 0.05$) reduced oyster mortalities by 47 to 57% below high-salinity controls at all concentrations tested (0.10–1.00 mg/L nominal NaOCl concentration; measured CPO concentrations of 0.02–0.11 mg/L). The combination of high salinity and CPO exposure increased mortality by 21% above high-salinity control levels at high chlorination doses (1.00 mg/L nominal NaOCl concentration; measured CPO concentration of 0.06 mg/L), whereas in lower concentrations (0.56- and 0.10-mg/L nominal NaOCl concentrations; measured CPO concentrations of 0.03 and 0.01 mg/L, respectively) mortality was equal to or slightly lower than high-salinity control mortality rates. Earlier studies[4,5] reported similar increased mortality 14% above control values in oysters exposed to a 1.00-mg/L nominal NaOCl concentration in high-salinity waters during the same time period of the year.

Additional comparisons indicated that the combination of low salinity and CPO exposure significantly reduced oyster mortality below low- and high-salinity control values.

Fecal production measurements indicated increased feeding rates in high-salinity CPO-exposed oysters (79.2–82.5%) when compared to low-salinity CPO-exposed oysters (65.5–72.5%). Fecal production rates in high- and low-salinity CPO-exposed oysters were not significantly different from rates for each respective high- and low-salinity control. Decreased fecal production was considered attributable to low-salinity effects rather than CPO exposure.

These results differ significantly from earlier[4-6] studies that reported CPO exposure caused significant reductions in fecal production rates at low chlorine concentrations (nominal 1.00 mg/L NaOCl). These earlier studies were continuous-exposure flow-through experiments as compared with the intermittent (8–12 h/d CPO exposure) conditions used in this study.

Comparisons of gill tissue respiration rates in high- and low-salinity CPO-exposed oysters generally indicated that respiration rates were slightly lower in low-salinity CPO-exposed oysters. Gill respiration rates were only significantly ($p \leq 0.01$) different in comparison with 1.00-mg/L nominal NaOCl concentration, high-salinity-treatment oysters and high-salinity control oysters, which indicated increased oyster respiration rates in the 1.00-mg/L nominal NaOCl concentration, high-salinity exposure treatment. Earlier studies[4-6] have reported similar increased gill tissue respiration rates in oysters chronically (60 d) exposed to 1.00-mg/L nominal NaOCl concentrations under similar exposure conditions, which resulted from increased anaerobic metabolism by oysters to survive CPO exposure.

Results from this study suggested that CPO exposure at similar chlorine concentrations (1.00-mg/L nominal NaOCl) increased gill tissue respiration rates, but only under high-salinity conditions. Higher gill tissue respiration rates resulted from increased reliance on gonadal glycogen reserves to survive CPO exposure under high-salinity conditions. Absence of these effects in lower CPO concentrations under high-salinity exposure and in low-salinity CPO exposures implied that CPO exposure must interact with factors representative of normal high-salinity exposure conditions per se, such as increased *P. marinus* infection intensity, to cause increased stress in these oysters. Such interaction would result in a more rapid depletion of stored gonadal glycogen reserves because of the increased metabolic utilization of glycogen by the oyster to survive CPO exposure and by the parasite as the infection intensifies and spreads. There was a rapid decline of gonadal indices in high-salinity CPO-exposed oysters in this study, which may be indicative of this effect.[26] The rapid depletion of gonadal glycogen stores resulting from this combined host-parasite effect would probably result in earlier utilization of protein stores (amino acids) as glycogen levels declined. This would signify utilization of different metabolic pathways (nitrogen), both by the host (oysters) and parasite.

Consideration must be given to physiological and biochemical differences between high- and low-salinity control oysters. Oysters are euryhaline organisms surviving wide ranges of salinity as low as 5 to 7 ppt.[25,27] Additionally, oysters are considered osmoconformers that prefer salinities of 11 to 30 ppt.[25,28] Osmoregulation in low-salinity seawater appears to result from volume regulation, the result of intracellular solute loss.[29] The source of these intracellular solutes has been reported to be the intracellular, free-amino-acid pool. Decreased levels of taurine, alanine, glycine, and proline have been reported in intertidal bivalve species, whereas reduced levels of taurine, alanine, and glycine have been reported in subtidal bivalve species.[29] Under low-salinity exposure, free-amino-acid levels may decline by as much as 84%, particularly ninhydrin positive (i.e., glycine and alanine) substances.[30,31] Low-salinity exposure would significantly reduce the amount of intracellular, free amino acids in oysters during this study.

This reduction of free-amino-acid levels in low-salinity control oysters may, perhaps, explain the higher survival observed under low-salinity exposure. Low-salinity oysters may have increased survival because of the reduced availability of protein or amino acids as food substrate for *P. marinus*, particularly if the infection intensifies to heavy intensities.

Additional consideration must be given to results of studies by Soniat and Koenig.[32] They found that free-amino-acid levels in oysters declined with increasing levels of *P. marinus* infection. This suggests increased use of free amino acids by both the host (oyster) and parasite during periods of infection. Soniat and Koenig[32] also reported increased levels of taurine and aspartic acid with increased infection intensity, although total free-amino-acid levels declined with increasing infection intensity. Lange[33] similarly reported increased levels of free amino acids with increased salinity in *Mytilus edulis*, with taurine having the most dramatic increase with increased salinity. Giles[34] reported a 55% decrease in taurine and an 82% decline in aspartic acid levels in *M. edulis* exposed to low-salinity seawater. These studies suggest that low-salinity exposure may reduce free-amino-acid levels as the oyster osmoregulates in response to salinity stress. This response may reduce the use of the free amino acids by the parasite, thus limiting the extent of infection and perhaps enhancing oyster survival.

Exposure of oysters to low salinity and low-salinity CPO conditions probably resulted in oysters that had lowered levels of both glycogen and free amino acids. This would initially tend to retard secondary spread and intensification of the disease; however, the fact that infection intensities were generally equal among all treatments after 60 d of exposure, and that mortality was generally higher in high-salinity control oysters and high-salinity CPO-exposed oysters, may suggest that the lowered levels of free amino acids in low-salinity control oysters and low-salinity CPO-exposed oysters offered some degree of protection to the oyster by preventing the disease from becoming toxic. Further research in this area is needed to better understand the exact mechanism of host-parasite interactions to chlorine as well as to other environmental pollutants.

REFERENCES

1. Davis, W. P., and D. P. Middaugh. "A Revised Review of the Impact of Chlorination Processes Upon Marine Ecosystems: Update 1977," in *Water Chlorination: Environmental Impact and Health Effects, Vol. 1*, R. L. Jolley, Ed. (Ann Arbor Science Publishers, Inc., Ann Arbor, MI, 1978), pp. 283–310.
2. Khlanski, M., and F. Bordet. "Effects of Chlorination on Marine Mussels," in *Water Chlorination: Environmental Impact and Health Effects, Vol. 3*, R. L. Jolley, W. A. Brungs, and R. B. Cumming, Eds. (Ann Arbor, MI: Ann Arbor Science Publishers, Inc., 1980), pp. 557–567.
3. Scott, G. I., and D. P. Middaugh. "Seasonal Chronic Toxicity of Chlorination to the American Oyster, *Crassostrea virginica*," in *Water Chlorination: Environmen-*

tal Impact and Health Effects, Vol. 2, R. L. Jolley, H. Gorchev, and D. H. Hamilton, Jr., Eds. (Ann Arbor, MI: Ann Arbor Science Publishers, Inc., 1978), pp. 311–328.
4. Scott, G. I. "The Effects of Seasonal Chronic Chlorination on the Growth, Survival, and Physiology of the American Oysters, *Crassostrea virginia* (Gmelin)," Ph.D. Dissertation (Columbia: University of South Carolina, 1979).
5. Scott, G. I., and W. B. Vernberg. "Seasonal Effects of Chlorine Produced Oxidants on the Growth, Survival, and Physiology of the American Oyster, *Crassostrea virginica* (Gmelin)," in *Marine Pollution: Functional Responses*, W. B. Vernberg, F. P. Thurberg, A. Calabrese, and F. J. Vernberg, Eds. (New York: Academic Press, 1979).
6. Scott, G. I., D. P. Middaugh, A. M. Crane, N. H. McGlothlin, and N. Watabe. "Physiological Effects of Chlorine-Produced Oxidants and Uptake of Chlorination By-products in the American Oyster, *Crassostrea virginica* (Gmelin)," in *Water Chlorination: Environmental Impact and Health Effects*, Vol. 3, R. L. Jolley, W. A. Brungs, and R. B. Cumming, Eds. (Ann Arbor, MI: Ann Arbor Science Publishers, Inc., 1980), pp. 501–516.
7. Mackin, J. G. "Oyster Disease Caused by *Dermocystidum marinum* and Other Microorganisms in Louisiana," *Publ. Inst. Mar. Sci.*, 7:132–229 (1962).
8. Quick, J. A., and J. G. Mackin. "Oyster Parasitism by *Labyrinthomyxa marina* in Florida," Prof. Pap. Ser. 13, (Tallahassee: Florida Dept. Nat. Res. Mar. Lab., 1971).
9. Mackin, J. G., and D. A. Wray. "Report on the Second Study of Mortality of Oysters in Barataria Bay, Louisiana, and Adjacent Areas": Res. Proj. No. 9, (College Station, TX: Texas A&M University, 1952).
10. Ray, S. M. "Biological Studies of *Dermocystidium marinum*, a Fungus Parasite of Oysters," in *Monographs in Biology*, Special Issue Pamphlet, (Houston: Rice Institute, 1954).
11. Andrews, J. D., and W. G. Hewatt. "Temperature Control Experiments on the Fungus Disease, *Dermocystidium marinum*, of Oysters," *Virginia Fisheries Laboratory* 62:129–133 (1957).
12. Mackin, J. G. "*Dermocystidium marinum* and Salinity," *Proc. Nat. Shellfish Assoc.* 46:116–128 (1956).
13. Ray, S. M. "Studies on the Occurrence of *Dermocystidium marinum* in Young Oysters," *Proc. Nat. Shellfish Assoc.* 1953:80–88.
14. Ray, S. M., J. G. Mackin, and J. L. Boswell. "Quantitative Measurement of the Effect on Oysters of the Disease Caused by *Dermocystidium marinum*," *Bull. Mar. Sci. Gulf Carib.* 3:6–33 (1953).
15. Couch, J. A. "Diseases, Parasites, and Toxic Responses of Commercial Penaeid Shrimps of the Gulf of Mexico and South Atlantic Coasts of North America," *Fish. Bull.* 76(1):1–44 (1978).
16. Sinderman, C. J. "Pollution-Associated Disease and Abnormalities of Fish and Shellfish," *Fish. Bull.* 76(4):717–749 (1978).
17. *Standard Methods for Examination of Water and Wastewater*, 14th ed., (Washington, DC: American Public Health Association, 1971).
18. *Handbook of Analytical Control in Water and Wastewater Laboratories*, (Cincinnati: U.S. Environmental Protection Agency, 1974).
19. Crecelius, E. A., G. Roesijadi, and T. O. Thatcher. "Correspondence," *Environ. Sci. Technol.* 12:1988 (1978).

20. Ray, S. M. "A Culture Technique for the Diagnosis of Infections with *Dermocystidium marinum* (Mackin, Owen, and Collier) in Oysters," *Science* 116(3014):360–361 (1952).
21. Ray, S. M., and A. C. Chandler. "*Dermocystidium marinum* a Parasite of Oysters," *Exp. Parisitol.* 4:172–200 (1955).
22. Ray, S. M. "A Review of the Culture Method for Detecting *Dermocystidium marinum*, with Suggested Modifications and Precautions," *Proc. Nat. Shellfish Assoc.* 54:55–69 (1966).
23. Mackin, J. G. "Status of Researches on Oyster Disease in North America," *Proc. Gulf. Carib. Fish. Inst.*, 13th Annual Session, 1960, pp. 98–109.
24. Farley, C. A., "Acid-Fast Staining of Haplosporidian Spores in Relation to Oyster Pathology," *J. Invert. Pathol.* 7(2):144–147 (1965).
25. Galtsoff, P. S. "The American Oyster, *Crassostrea virginica* (Gmelin)," *Fish Bull.* 64:1–80 (1964).
26. Scott, G. I., Sammons, T. I., and J. Hagan. "An Examination of Factors Affecting the Survival of Oysters Infected With the Fungus, *Labyrinthoxyma marina* and Application of These Factors to Improve Pond Aquaculture in the American Oyster, *Crassostrea virginica* (Gmelin)," N.S.F. Report (Washington, DC: National Science Foundation, 1983).
27. Vernberg, F. J., and W. B. Vernberg. "Environmental Physiology of Marine Animals": (New York: Springer-Verlag Press, 1972).
28. Dupuy, J. L., N. T. Windsor, and C. E. Sutton. "Manual For the Design and Operation of an Oyster Seed Hatchery For the American Oyster, *Crassostrea virginica*," Mar. Sci. Ocean Eng., VIMS Spec. Rep. No. 142, (Gloucester Point, VA: Virginia Institute of Marine Sciences, 1977), pp. 104.
29. Pierce, Jr., S. K. "A Source of Volume Regulation in Marine Mussels": *Comp. Biochem. Physiol.* 38A:619–635 (1971).
30. Lange, R. "Isosmotic Intracellular Regulation in Marine Bivalves": *J. Exp. Mar. Biol. Ecol.* 5:170–179 (1970).
31. Vikar, R. A., and K. L. Webb. "Amino Acids in the Clam, *Mya*," *Comp. Biochem. Physiol.* 32:775–783 (1970).
32. Soniat, T. M., and M. L. Koenig. "The Effects of Parasitism by *Perkinsus marinus* on the Free Amino Acid Composition of *Crassostrea virginica* Mantle Tissue," *J. Shellfish Res.*, 2(1):25–28 (1982).
33. Lange, R. "The Osmotic Function of Free Amino Acids and Taurine in the Mussel, *Mytilus edulis*," *Comp. Biochem. Physiol.* 10:173–179 (1963).
34. Giles, R. "Osmoregulation in Three Molluscs: *Acanthochitona discrepans* (Brown), *Glycymeris glycymeris* (L), and *Mytilus edulis* (L)," *Biol. Bull.* 142:25–35 (1972).

CHAPTER **38**

Response of Sheep River, Alberta, Macroinvertebrate Communities to Discharge of Chlorinated Municipal Sewage Effluent

Lewis L. Osborne

Since Tsai's publication,[1] most research efforts on the environmental effects of chlorinated effluents have been directed towards determining impacts on fish and fish populations.[2-7] Despite this emphasis, a relatively large body of evidence now exists which indicates that different invertebrate species exhibit varying degrees of sensitivity to chlorinated effluents,[3,8-10] and that temperature profoundly affects the acute and chronic toxicity of chlorine.[11-14]

Our knowledge of invertebrate community responses to the combined effects of chlorinated effluents and temperature is based primarily on laboratory investigations using either single- or multiple-species tests, supplemented by a limited number of field investigations.[10,14-16] Most field studies have been in either marine or lentic systems. This study was designed to assess the combined effects of chlorinated municipal sewage and thermal discharges on the structure and composition of the aquatic macroinvertebrate communities of the Sheep River. Previous studies on this same stream[2,14,17,18] indicated that the chlorinated municipal discharge (1) increased fish mortality and altered fish behavior; (2) increased the mutagenic activity within fish, water, and invertebrates; (3) increased the numbers of sessile and planktonic bacteria below the sewage outfall; and (4) did not alter the standing crop of epilithic periphyton.

DESCRIPTION OF STUDY AREA

This study was conducted in the Sheep River near Turner Valley, Alberta, 64 km southwest of the city of Calgary. The Turner Valley municipal sewage treatment plant serves 1100 residents and consists of secondary treatment followed by chlorination (dissolved calcium hypochlorite; 20 min retention) before the effluent is released into the river. Annually, the average daily municipal sewage discharge is 0.10 to 0.50% of the average daily river discharge. During low-flow events (late summer and autumn), the municipal effluent is roughly 5% of the total river discharge. Once discharged, the

482   WATER CHLORINATION

municipal effluent is restricted to the left side of the river for approximately 500 m until it is completely mixed (Figure 1).

The only other point source input upstream of Turner Valley is a thermal discharge from a sour-gas plant approximately 300 m upstream of the sewage outfall (Figure 1). The gas plant withdraws water from the river, cycles it through a cooling channel, and releases the water back into the stream. This process increases the temperature of the river water from 1 to 10°C.

Location of Sampling Sites

Six station transects were established for chemical and macroinvertebrate collections. Station 1 was located 1.5 km upstream of the gas plant and used as a control site; station 2 was located 200 m downstream of the thermal outfall (not chlorinated) and 100 m upstream of the municipal sewage outfall; station 3 was located 5 m downstream of the sewage effluent discharge; stations 4, 5, and 6 were located 100 m, 500 m, and 1.5 km, respectively, downstream of station 3. Preliminary studies indicated complete mixing of the sewage effluent with the river water occurred just below station 5 (Figure 1) because of a right-angle change in channel direction.

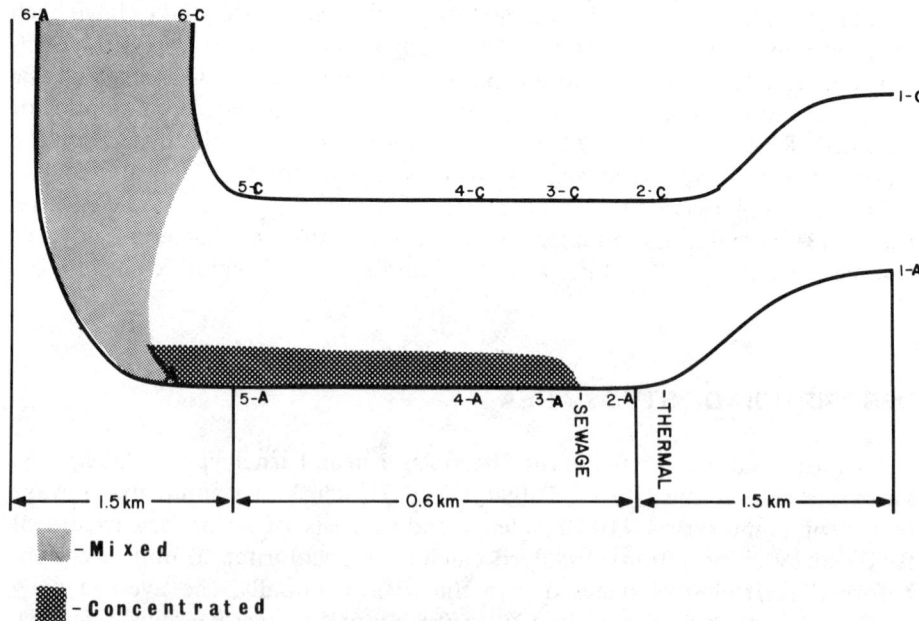

Figure 1. Sheep River study area, location of sampling substations, and mixing pattern of the Turner Valley municipal effluent within the Sheep River with respect to sampling stations.

Each station transect was longitudinally divided into three equal areas across the stream because of the restricted mixing pattern of the effluent plume. This provided both a latitudinal and longitudinal means of data comparison between the 18 sampling substations within the six station transects. The substations were denoted as A, for the left bank; B, for the middle channel; and C, for the right bank. Physical-chemical data were collected at all three substations during the study, and biological data were collected from only the lateral substations A and C.

METHODS

Physical-chemical samples were collected in October 1977, and then monthly from January to December 1978 at each substation. Water temperature, pH, dissolved oxygen, chloride, and total residual chlorine (TRC) concentrations were determined in the field. TRC was determined using a Fisher Chlorine Amperometric Titrimeter (Model 393) powered with an ac/dc converter and battery. Water samples were collected from each substation in 500-mL sterilized glass reagent bottles, returned to the laboratory on ice, and analyzed within 5 h of collection for total alkalinity, total hardness, nitrate-nitrogen, phosphate-phosphorous, sulfate, turbidity, and specific conductance.

The dominant substrate materials within the study area were coarse gravel and rubble, which are typical of most montaine systems. An earlier study[17] found no significant differences in the composition of substrate particle sizes between substations, although small amounts of silt were noted in the vicinity of substation 3-A during low-flow events. The typical phi scale for the study area ranged from −5 to −8.

Benthic samples were collected with a modified Niel sampler (area 0.1 m^2; net 250 μm; penetration 5.0 cm). Five replicates each were collected from substations A and C, providing a statistical estimate within 20% of the actual mean number of individuals and less than 5% of the mean number of taxa.[19] At substations 3-A, 4-A, and 5-A, care was taken to ensure that all samples were collected from within the sewage effluent plume. Samples were preserved, returned to the laboratory, and sorted (5X magnification). With the exception of the Oligochaeta, attempts were made to identify all taxa to the species level. Where species keys did not exist, individuals were separated on the basis of distinct morphological characteristics.

The five replicate benthic samples were combined, and a mean number of organisms per 0.1 m^2 was calculated for each substation and month. A Shannon-Weaver diversity index (H'; $\log_2$) and an evenness value (J)[20] were calculated for each substation.

Intersite and temporal differences in the community statistics were analyzed using a Kruskal-Wallis one-way ANOVA and the associated post hoc multiple confidence interval comparison procedure.[21] Differences in the physical-chemical variables were determined using a parametric one-way ANOVA and the SNK a posteriori test. The relationship between biological and physical-chemical parameters was examined using the Kendall rank nonparametric correlation test. For all statistical analyses, an alpha of 0.05 was used.

RESULTS

During the study, significant intersubstation variations were found in TRC, sulfate (SO_4), phosphate-phosphorous (P–PO_4), and Cl^- concentrations, and in temperature. TRC was detected at only 3 of the 18 sampling locations (i.e., 3-A, 4-A, and 5-A, Table I). P–PO_4 and SO_4 concentrations were significantly higher at the left-bank substations, immediately downstream of the municipal sewage outfall (Table I), and temperature was highest immediately downstream of the thermal effluent discharge (substations 2-A and 2-B). A maximum increase over stream ambient of 10°C was recorded at substation 2-A, whereas the mean temperature increase was generally 5 to 8°C.[17]

Significant temporal variation occurred in all physical-chemical parameters measured with the exception of TRC, pH, nitrite-nitrogen, and dissolved oxygen, which was always at least 90% saturated.[17] Dissolved substance concentrations were inversely related to stream discharge.

A total of 309 aquatic invertebrate taxa were collected during the study. Prior to chlorination, 81.8% of the total number of invertebrates at substation 3-A were oligochaetes, and 6.1% belonged to the family Chironomidae. Oligochaeta and chironomid species were also numerically dominant at substation 4-A; however, they comprised only 28.2 and 27.6% of the total number. Diptera (15.5%) and Ephemeroptera (16.6%) also comprised a substantial

Table I. Mean Recorded and (Minimum-Maximum) Total Residual Chlorine (TRC), Phosphate-Phosphorus (PO_4P), Sulfate Concentrations, and Temperature at each Sheep River Substation Throughout the Study (All Values Include October 1977 Prechlorination Data.)[a]

Substation	TRC (mg/L)	PO_4P (mg/L)	Sulfate (mg/L)	Temp (°C)
1-A	0.0	0.10 (0 – 0.40)	32.7 (11 – 92)	3.6 (0 – 10)
1-C	0.0	0.09 (0 – 0.30)	33.2 (13 – 82)	3.5 (0 – 10)
2-A	0.0	0.16 (0 – 0.80)	36.6 (16 – 59)	10.7 (1 – 17)
2-C	0.0	0.19 (0 – 0.90)	35.2 (18 – 62)	4.5 (0 – 11)
3-A	2.04 (0.0 – 6.8)	1.67 (0.08 – 4.60)	62.0 (14 – 135)	9.2 (0 – 15)
3-C	0.0	0.19 (0 – 0.60)	35.4 (10 – 60)	5.1 (0 – 11)
4-A	0.42 (0.0 – 1.85)	0.43 (0 – 1.70)	42.2 (12 – 76)	9.0 (4 – 14)
4-C	0.0	0.14 (0 – 0.50)	39.9 (27 – 63)	4.3 (0 – 11)
5-A	0.16 (0.0 – 0.50)	0.22 (0 – 0.70)	40.0 (20 – 62)	7.9 (3 – 13)
5-C	0.0	0.13 (0 – 0.50)	39.2 (27 – 63)	4.9 (0 – 11)
6-A	0.0	0.13 (0 – 0.60)	36.6 (19 – 53)	5.8 (1 – 11)
6-C	0.0	0.16 (0 – 0.60)	38.9 (25 – 53)	5.6 (1 – 12)

[a]First values are means; parenthetical values are minimum and maximum, respectively.

portion of the total number of individuals at substation 4-A. Most of the individuals at substation 5-A were members of the Chironomidae (27.9%) and Ephemeroptera (33.3%). The composition of the benthic communities at substations 3-A, 4-A, and 5-A changed dramatically following chlorination (Table II).

Significant temporal and intersite variations occurred in the total number of individuals and taxa and in the calculated H' and J indices. The H' values and seasonal trends for substations 1-A and 1-C were very similar to substations 6-A and 6-C (Figure 2), and those for substations 2-A and 2-C were virtually identical to substations 3-A and 3-C, respectively (Figure 2). Significantly fewer taxa occurred at substations 2-A, 3-A, and 4-A throughout the study, whereas significantly more individuals were collected at substations 2-A, 3-A, 4-A, 5-A, 6-A, and 6-C. The lowest H' diversity and evenness values occurred at stations 2-A, 3-A, 4-A, and 5-A.

Species richness (number of taxa) was negatively correlated with $P-PO_4$, SO_4, and TRC concentrations (Table III). A significant positive correlation was found among the total number of individuals and all environmental

DISCUSSION

By necessity, the experimental design of this study included only one month of prechlorination data (i.e., October 1977) because of a local water shortage during the winter of 1977-78, which prompted provincial authorities to require chlorination of the municipal effluent. Thus, interpretations were based on intersite comparisons and an examination of the October 1977 prechlorination data with appropriate postchlorination data.

The significant increase in TRC, $P-PO_4$, and SO_4 concentrations at substations 3-A, 4-A, and 5-A can be attributed to the discharge of the municipal sewage effluent (Table I). Previous studies[17,18] have demonstrated that the discharge of the chlorinated effluent was responsible for an increase in the concentration of sessile bacteria, but did not affect the standing crop of epilithic algae within these same areas. The variable concentrations of TRC at substations 3-A, 4-A, and 5-A were the result of the constant-drip effluent chlorination process,[17] which does not compensate for variability in the quantity of effluent.

Following chlorination, a significant difference in the magnitude of the macroinvertebrate H' values was apparent between the left-(A) and right-(C) bank substations between stations 2 and 5 (Figure 2), although a similar annual pattern existed. The similarity in seasonal patterns appears to be associated with the response of benthic communities to macroenvironmental factors (e.g., spring spates and temperature). The substantially lower left-bank H' values (Figure 3) at some stations suggests that (1) impacts do exist within the

Table II. Mean (%) Abundance of Each Major Taxon (in parenthesis) and the Average Number of Taxa Within Each Major Taxon Collected During the Post-Chlorination Period

Taxon	1-A		1-C		2-A		2-C		3-A		3-C	
Coleoptera	1	(0.4)	1	(0.3)	1	(<0.1)	1	(0.2)	1	(<0.1)	1	(0.1)
Diptera[a]	4	(7.0)	4	(5.8)	3	(1.6)	4	(6.4)	4	(1.8)	4	(7.3)
Chironomidae	13	(19.9)	15	(40.1)	5	(19.8)	14	(60.9)	10	(8.5)	16	(70.9)
Ephemeroptera	9	(52.0)	9	(33.9)	1	(1.0)	6	(20.5)	1	(0.2)	4	(4.4)
Gastrapoda	0	(0)	0	(0)	1	(1.9)	0	(0)	1	(<0.1)	0	(0)
Oligochaeta	1	(5.7)	1	(6.1)	1	(76.8)	1	(7.3)	1	(89.3)	1	(13.1)
Plecoptera	7	(10.5)	10	(7.9)	1	(<0.1)	4	(4.2)	0	(0)	2	(3.5)
Trichoptera	6	(6.5)	7	(5.9)	1	(<0.1)	3	(0.7)	0	(0)	1	(1.5)
Miscellaneous	0	(0)	0	(0)	1	(<0.1)	1	(<0.1)	0	(0)	1	(<0.1)
	4-A		4-C		5-A		5-C		6-A		6-C	
Coleoptera	1	(0.1)	1	(0.2)	1	(0.3)	1	(0.2)	1	(0.3)	3	(0.3)
Diptera[a]	5	(1.1)	4	(9.3)	5	(6.2)	5	(14.1)	5	(6.3)	7	(7.6)
Chironomidae	10	(5.6)	13	(53.0)	12	(16.4)	16	(52.9)	15	(37.0)	21	(49.0)
Ephemeroptera	1	(0.5)	6	(8.2)	2	(1.0)	6	(10.2)	8	(38.8)	7	(21.5)
Gastrapoda	2	(0.2)	1	(0.1)	2	(0.5)	1	(0.2)	1	(0.1)	1	(<0.1)
Oligochaeta	1	(92.0)	1	(25.5)	1	(75.4)	1	(17.9)	1	(9.6)	1	(18.4)
Plecoptera	1	(0.1)	3	(2.7)	1	(0.1)	4	(2.0)	6	(5.3)	7	(1.6)
Trichoptera	1	(0.1)	2	(1.1)	1	(0.1)	4	(2.6)	5	(3.2)	6	(1.5)
Miscellaneous	0	(0)	0	(0)	0	(0)	1	(0)	0	(0)	0	(0)

[a]Chironomidae not included in these values.

Table III. Results of Kendall Rank Correlation Analyses for Evenness, Number of Individuals, Species Richness, and H' Diversity with Various Physiochemical and Biological Parameters. (* = $p \leq 0.05$; ** = $p \leq 0.01$; *** = $p \leq 0.001$; NS = not significant; (+) = positive relationship; (−) = negative relationship)

Parameters	Species Richness	Evenness	H' Diversity	Individuals (No.)
Temperature	(−)*	(−)*	(+)**	NS
Phosphate-Phosphorus	(−)***	(−)***	(−)***	(+)***
Sulfate	(−)***	(−)***	(−)***	(+)***
TRC[a]	(−)***	(−)***	(−)***	(+)***

[a]Total residual chlorine.

parameters except temperature (Table III). Shannon-Weaver diversity was positively correlated with temperature and current velocity and negatively correlated with P–PO$_4$, SO$_4$, and TRC. Evenness was negatively correlated with all of the selected environmental variables (Table II).

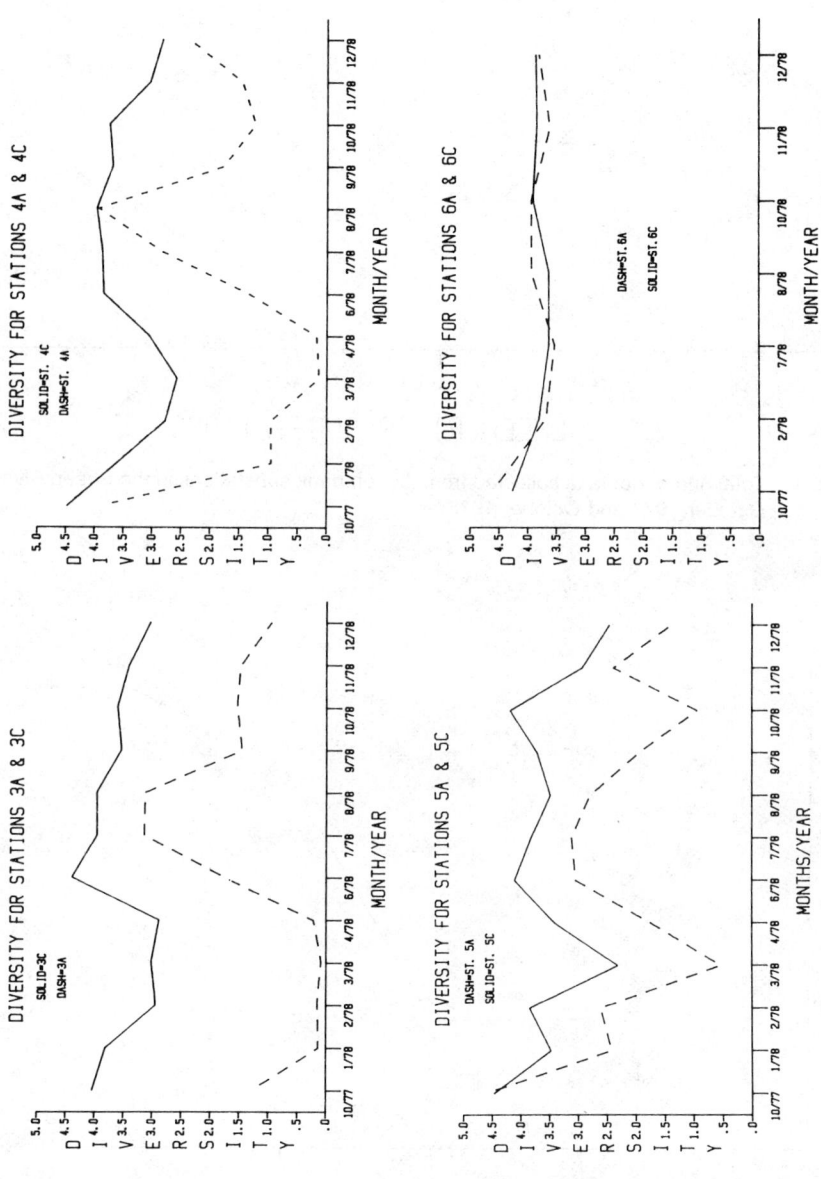

Figure 2. Shannon-Weaver diversity values with respect to substations. Trends for substations 2-A and 2-C are similar to substations 3-A and 3-C, respectively. Trends for substations 1-A and 1-C were similar to substations 6-A and 6-C, respectively.

488   **WATER CHLORINATION**

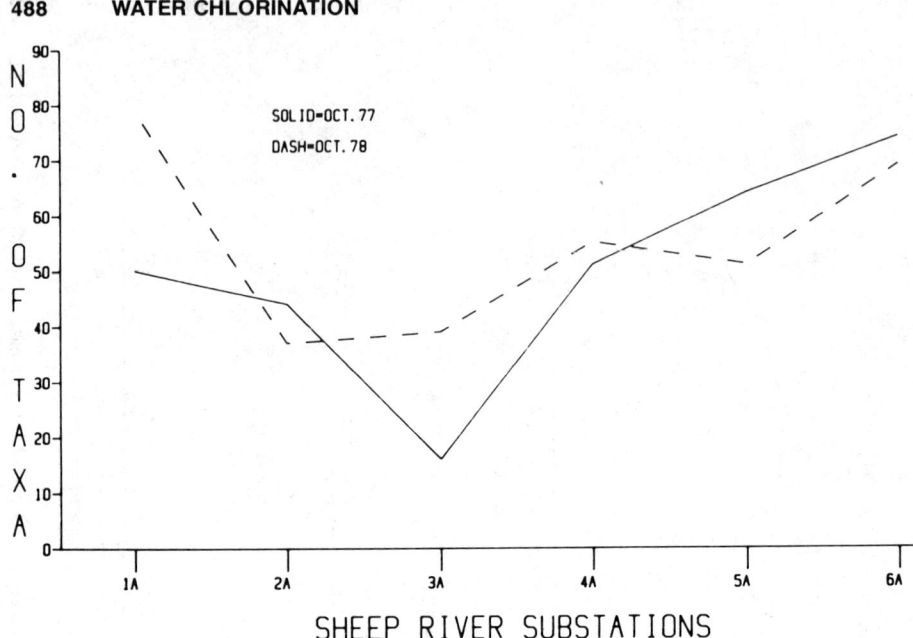

Figure 3. Total number of taxa collected from the left-bank substations in the Sheep River in October 1977 and October 1978.

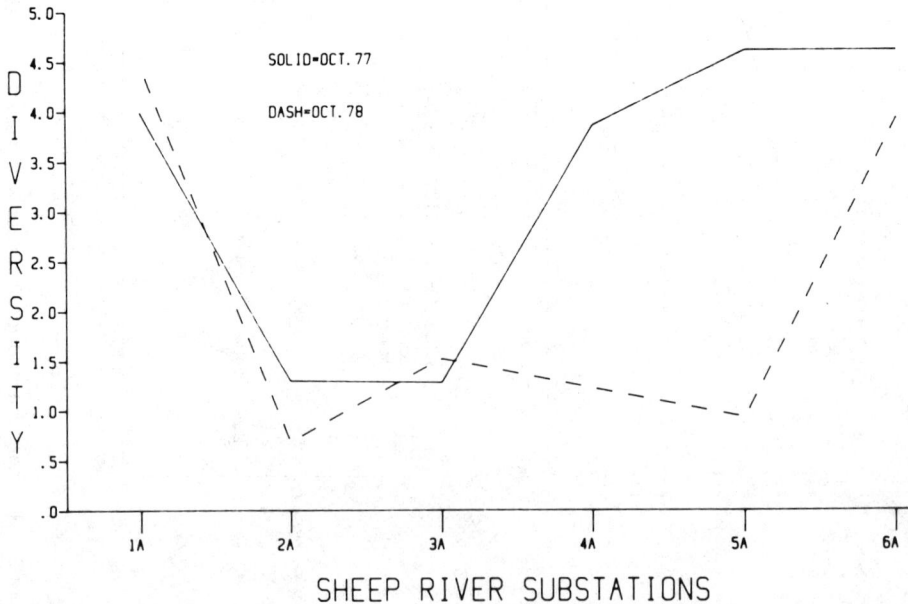

Figure 4. Calculated Shannon-Weaver (H′) macroinvertebrate diversity for the left-bank Sheep River substations for October 1977 and October 1978.

study area, (2) impacts are restricted to the left bank of the river, and (3) impacts are confined to the benthic communities located above station 6.

The restricted distribution of TRC within the stream (Table I) and the lower H' diversities suggest that the chlorinated municipal effluent is partially responsible for the depressed community structure at substations 3-A, 4-A, and 5-A. The significant negative relationships between TRC concentration and macroinvertebrate species richness, J, and H' diversities at substations 2-A and 3-A, however, suggest that the chlorinated municipal effluent is not the only factor responsible for the depressed macroinvertebrate community structure at substations 3-A, 4-A, and 5-A.

Extensive physiochemical analysis on the thermal effluent by Alberta Environment and others[17] failed to detect any aberrant concentration of natural or anthropogenic generated substances within the effluent. The gas-plant cooling water discharge was generally 6 to 8°C above ambient stream temperature (Table I), which resulted in extensive growths of mucilagenous strands of a bacteria-fungi-algae matrix on the substrate for roughly 100 m downstream of the thermal discharge.[17] This slime matrix apparently increased available microhabitat space for *Thienannimyia* sp. 2, *Cricotopus bicinctus* and Oligochaeta, which were the most numerically dominant taxa at substation 2-A. The numerical dominance of these taxa significantly depressed the macroinvertebrate diversity within the vicinity of substation 2-A, although other less abundant taxa were collected (Table II). Prior to chlorination, any significant adverse effects of temperature and unchlorinated sewage on benthic community structure appear to have been moderated in the vicinity of substation 4-A (Figure 3).

A total of 41 macroinvertebrate taxa from substation 2-A and 11 taxa from substation 3-A were collected in October 1977. Both of these sites displayed a depressed H' diversity (Figure 4). The substantially lower species richness appears to have been a function of the unchlorinated municipal sewage input. The reduced species richness and diversity at substation 3-A and the fact that 82% of the invertebrates collected were oligochaetes (Table III) is reflective of benthic communities below municipal sewage outfalls.

In October 1977, macroinvertebrate diversity and species richness at substations 4-A and 5-A were similar to right-bank substations and to upstream control substations (Figures 3 and 4). This suggests that the benthic communities below substation 3-A were not significantly affected by the municipal sewage and thermal discharges. Following chlorination of the municipal effluent, both substations 2-A and 3-A continued to display depressed H' (Figure 2) and species richness values (Table II). An extended area of downstream impact was also apparent following chlorination. This is reflected in the decreased H' values at substations 4-A and 5-A (Figure 4) and in the magnitude of the H' difference between the A and C substations at stations 4 and 5 (Figure 2).

A change in the composition and structure of the benthic communities at substations 4-A and 5-A was also apparent following chlorination (Table II). Prior to chlorination, mayflies constituted a substantial portion of the fauna at substations 4-A (16.6%) and 5-A (33.3%). Following chlorination, Ephemeroptera constituted less than 1.0% of the total number of individuals at these sites (Table II). Similar reductions in the numerical importance of chironomids, caddisflies, and stoneflies were also apparent. The fact that the area of reduced community structure corresponded with the occurrence of TRC (Table I) suggests that chlorination of the municipal sewage significantly reduced macroinvertebrate community structure and extended the area of stream impact beyond that which existed prior to chlorination.

A significant increase in the number of oligochaetes also occurred at substations 4-A and 5-A following chlorination. In October 1977, oligochaetes comprised 28.2% of the total at substation 4-A and 12.5% at substation 5-A. In October 1978, oligochaetes comprised over 85% of the total number of individuals, suggesting that certain oligochaete taxa are much less susceptible to the toxic effects of TRC than are most aquatic insect taxa.

The absence of detectable levels of TRC and any noticeable adverse effects on community structure and composition below station 5 can be attributed to the rapid and abrupt mixing of the effluent with the river water below that station. These results suggest that the area of impact might be reduced if the chlorinated effluent were rapidly mixed with river water, thereby diluting TRC concentrations within the stream.

SUMMARY

The results of this study demonstrate that chlorination of the Turner Valley municipal effluent significantly reduced the structure of downstream macroinvertebrate communities and increased the extent of the stream impacted by the discharge of unchlorinated sewage. No significant effects of the chlorinated municipal wastes were detected 1.5-km downstream after the effluent was completely mixed and diluted with river water. It is suggested that small effluents that constitute less than 0.5% of the total daily river discharge be mixed as rapidly as possible to reduce the area of stream impact.

REFERENCES

1. Tsai, C. F., "Effects of Chlorinated Sewage Effluents on Fish in Upper Patuxent River, Maryland." *Chesapeake Sci.* 9:83–93 (1969).
2. Osborne, L. L., D. R. Iredale, F. J. Wrona, and R. W. Davies. "The Effects of Chlorinated Sewage Effluents on Fish in the Sheep River, Alberta." *Trans. Am. Fish. Soc.* 110:536–540 (1981).
3. Brooks, A. S., and G. L. Seegert. *The Effects of Intermittent Chlorination on the Biota of Lake Michigan*, Center for Great Lake Studies, Special Report No. 31, (Milwaukee: University of Wisconsin-Milwaukee, 1977).
4. Larson, G. L., and D. A. Schlesinger. "Toward an Understanding of the Toxicity of Intermittent Exposures of Total Residual Chlorination to Freshwater Fish," in

Water Chlorination: Environmental Impacts and Health Effects, Vol. 2, R. L. Jolley, H. Gorchev, and D. H. Hamilton, Eds. (Ann Arbor, MI: Ann Arbor Science Publishers, Inc., 1978), pp. 111-122.
5. Heath, A. G. "Influence of Chlorine Form and Ambient Temperature on the Toxicity of Intermittent Chlorination to Freshwater Fish," in *Water Chlorination: Environmental Impact and Health Effects, Vol. 2*, R. L. Jolley, H. Gorchev, and D. H. Hamilton, Eds. (Ann Arbor, MI: Ann Arbor Science Publishers, Inc. 1978), pp. 123-134.
6. Thomas, P., J. M. Bartos, and A. S. Brooks. "Comparison of the Toxicities of Monochloramine and Dichloramine to Rainbow Trout, *Salmo gairdneri*, Under Various Time Conditions," in *Water Chlorination: Environmental Impact and Health Effects, Vol. 3*, R. L. Jolley, W. A. Brungs, and R. B. Cumming, Eds. (Ann Arbor, MI: Ann Arbor Science Publishers, Inc. 1980), pp. 581-588.
7. Trabalka, J. R., S. C. Tsai, J. S. Mattice, and M. B. Burch. "Effects on Carp Embryos (*Cyprinus carpio*) and *Daphnia pulex* of Chlorinated Organic Compounds Produced During Control of Fouling Organisms," in *Water Chlorination: Environmental Impact and Health Effects, Vol. 3*, R. L. Jolley, W. A. Brungs, and R. B. Cumming, Eds. (Ann Arbor, MI: Ann Arbor Science Publishers, Inc. 1980), pp. 599-606.
8. Bender, M. E., M. H. Roberts, R. J. Diaz, and R. J. Huggert. "Effects of Residual Chlorine on Estuarine Organisms," in *Biofouling Control Procedures Technology and Ecological Effects*, L. D. Jensen, Ed. (New York: Marcel Dekker, Inc., 1977), pp. 101-108.
9. Turner, A., and T. A. Thayer. "Chlorine Toxicity in Aquatic Ecosystems," in *Water Chlorination: Environmental Impact and Health Effects, Vol. 3*, R. L. Jolley, W. A. Brungs, and R. B. Cumming, Eds. (Ann Arbor, MI: Ann Arbor Science Publishers, Inc., 1980), pp. 607-630.
10. Arthur, J. W., R. W. Andrew, V. R. Mattson, D. T. Olson, G. E. Glass, B. J. Halligan, and C. T. Walbridge. *Comparative Toxicity of Sewage-Effluent Disinfection to Freshwater Aquatic Life*, EPA-600/3-75-012 (Washington, DC: U.S. Environmental Protection Agency, 1975).
11. Capuzzo, M. J. "The Effects of Free Chlorine and Chloramine on Growth and Respiration Rates of Larval Lobsters (*Homarus americanus*)," *Water Res.* 11:1021-1024 (1977).
12. Capuzzo, M. J. "The Effects of Halogen Toxicants on Survival, Feeding, and Egg Production of the Rotifer *Brachionus plicatilis*," *Estuarine Coastal Mar. Sci.* 8:307-316 (1979).
13. Osborne, L. L., R. W. Davies, and J. B. Rasmussen. "The Effects of Total Residual Chlorine on the Respiration Rate of Two Species of Freshwater Leech (Hirudinoidea)," *Comp. Biochem. Physiol.* 67C:203-207 (1980).
14. Moore, R. L., L. L. Osborne, and R. W. Davies. "The Mutagenic Activity in a Section of the Sheep River, Alberta, Receiving a Chlorinated Sewage Effluent," *Water Res.* 14:917-920 (1980).
15. Khalanski, M., and F. Bordet. "Effects of Chlorination on Marine Mussels," in *Water Chlorination: Environmental Impact and Health Effects, Vol. 3*, R. L. Jolley, W. A. Brungs, and R. B. Cumming, Eds. (Ann Arbor, MI: Ann Arbor Science Publishers, Inc., 1980), pp. 557-579.
16. Goldman, J. C., J. M. Capuzzo, and G. T. F. Wong. "Biological and Chemical Effects of Chlorination at Coastal Power Plants," in *Water Chlorination: Environ-*

mental Impact and Health Effects, Vol. 2, R. L. Jolley, H. Gorchev, and D. H. Hamilton, Eds. (Ann Arbor, MI: Ann Arbor Science Publishers, Inc. 1978), pp. 291–305.
17. Osborne, L. L. *The Effects of Chlorine on the Benthic Communities of the Sheep River*, Ph.D. Dissertation, (Calgary, Alberta: University of Calgary, 1981).
18. Osborne, L. L., R. W. Davies, T. I. Ladd, R. M. Ventullo, and J. W. Costerton. "The Effects of Chlorinated Municipal Sewage on the Abundance of Bacteria in the Sheep River, Alberta." *Can. J. Microbiol.* 29:261–270 (1983).
19. Elliott, J. M. *Some Methods for the Statistical Analysis of Samples of Benthic Invertebrates*. (England: Freshwater Biological Association, 1971).
20. Pielou, E. C. *Ecological Diversity*, (New York: Wiley Interscience, Inc., 1975).
21. Marascuilo, L. A., and M. McSweeney. *Nonparametric and Distribution-Free Methods for the Social Sciences* (Belmont, CA: Wadsworth Publication Co., 1977).

CHAPTER 39

Comparison of Acute Toxicity and Avoidance Responses of Atlantic Silverside and White Perch to Chlorinated Estuarine Waters

James A. Fava and John W. Meldrim

Many power plants and other industrial water users eliminate or reduce biofouling by using chlorine as a biocide. With increased use of estuarine waters for cooling, the environmental impact of chlorinated estuarine water has caused growing concern. The impact is not simply a function of the amount of chlorine used. Estuaries are dynamic environments in which variables such as salinity, oxygen, pH, and light interact with chlorinated estuarine waters. Also, the use of estuaries by organisms varies from one season to another.

Literature describing the environmental impact associated with the use and release of chlorinated waters has been reviewed by Brungs,[1,2] Mattice and Zittel,[3] and Hall et al.[4] Previous studies have examined the toxicity of chlorinated estuarine waters in relation to organisms[5-8] and have addressed the avoidance responses of estuarine organisms.[9-13] Few studies actually relate concentrations that would be toxic to those that organisms may avoid. Furthermore, a critical discussion is needed on the potential use of avoidance responses to assess receiving water impacts of complex chlorinated effluents.

This research had four objectives: (1) to further investigate the effects of chemical form on avoidance responses of Atlantic silverside and white perch; (2) to investigate the effects of chemical form and various environmental factors on the toxicity of Atlantic silverside and white perch; (3) to compare the avoidance and toxicity responses of the two species; and (4) to present a critical discussion on the potential use of avoidance responses as part of an impact assessment associated with chlorinated estuarine waters released by industrial facilities.

METHODS AND MATERIALS

General

Atlantic silverside and white perch were collected by seine from the estuarine portions of the Delaware River and Chesapeake Bay. Only fish <200 mm in total length were used. Organisms were transported to the laboratory where they were held for at least 18 to 24 h before testing at temperatures and salinities close to those at capture.

The temperature at which fish were collected was considered to be that to which they had been acclimated. Holding facilities were adjusted to simulate field temperatures, and studies were performed at temperatures between 4 and 28°C. The light duration of the holding facilities was maintained at the naturally occurring photoperiod at the time of testing. Light intensity was 645 1x at the water surface and was provided by Duro Test Vita-Lites fluorescent bulbs, which have a spectral distribution similar to sunlight.

The water used in testing was pumped at high tide from the Appoquinimink River, a tributary of the Delaware River, into two 1250-gal polyethylene storage tanks and settled prior to use.

Concentrations of chlorine-produced oxidants (CPO) were measured with a Fischer and Porter Model 17T1010 amperometric titrator. Phenylarsine oxide, potassium iodide solution, and pH buffers used were standard Fischer and Porter reagents. Free chlorine and total chlorine were measured following the Fischer and Porter procedures. Because traditional chlorination nomenclature is inappropriate for seawater, the following terms modified from Block et al.[14] will be used:

1. *CPO* refers to chlorine-produced oxidant measured at pH 4; traditionally, it has been called total chlorine.
2. *pH 7 oxidant* refers to chlorine-produced oxidant measured at pH 7; traditionally, it has been called free chlorine.

Avoidance Response

The apparatus, water and test solution delivery systems, water quality monitoring, and test procedures followed methods described by Meldrim and Gift[15] and Meldrim and Fava.[10] Briefly, the behavioral responses were determined using a modified Shelford-Allee apparatus. In this system, water flowed via gravity from the storage tanks into a temperature-controlled water bath. Water was conditioned in the bath to the acclimation temperature, and was then pumped into dose boxes. One box received chlorinated estuarine water from a constant head reservoir. The chlorine dose was regulated by a flowmeter. The other box provided control conditions since it did not receive chlorinated estuarine water. The chlorinated and control water flowed at 3 L/min from the dose boxes to each of two parallel experimental tanks. Each experimental tank received chlorinated water at one end, control water at the other, and was drained through a central pipe. Opposite ends of the parallel experimental tanks received chlorinated water to account for positional effects on fish response.

The general experimental procedure follows:

1. Establish conditions in both troughs so that water in diagonally opposite halves of the two experimental troughs would have the same oxygen content and chemical composition.
2. Place equal numbers of fish (generally four) into each trough.
3. After a 30-min orientation, record the time (during 5-min test period) that the fish spent in each half of the trough.
4. Analyze the fish-time distribution by t-test to determine if a significant response has taken place.

If no significant avoidance response occurred, the concentration of the chlorinated estuarine water was increased in stepwise fashion until a response did occur.

Meldrim and Fava[10] reported preliminary results regarding the avoidance of these two species to chlorinated estuarine waters and suggested that various environmental factors influenced the avoidance concentration. Subsequently, additional avoidance tests were conducted with both species under a variety of environmental conditions. Preliminary data analysis indicated that although environmental factors did influence the avoidance response, the chemical form may also be important. To further examine the influence of chemical form, avoidance concentrations were grouped into four class intervals: 0.01 to 0.10, 0.11 to 0.20, 0.21 to 0.30, and >0.31 mg/L CPO. The number of avoidance responses where the pH 7 oxidant made up $\geq 50\%$ of the CPO was determined for each interval. This number was used to assess the influence of the chemical form on the avoidance response.

Toxicity Tests

A Mount-Brungs diluter system[16] was used to deliver 500 mL each of five chlorine test concentrations and one chlorine-free control every 3 min. Water flowed by gravity from the storage tanks to the diluter system. The chlorine stock solution was made by adding calcium hypochlorite to distilled water to obtain a concentration (as CPO) of ~ 200 mg/L.

The experimental apparatus consisted of 12 tanks, each containing ~ 25 L of water. Tanks were made of 3/4-in. exterior grade plywood and coated with an epoxy resin. The experimental procedure was first to establish and stabilize five concentrations. Concentrations tested ranged from 0.05 to 1.5 mg/L CPO. After stabilization, an equal number of fish were placed randomly in each of the tanks. Death was defined by lack of response to general probing with a blunt rod. Time to death was recorded for periods up to 96 h. The results of water quality analyses in both the toxicity and avoidance studies were identical. Studies were conducted throughout the year following annual cycles of temperature and salinity.

Twelve experiments each were conducted with Atlantic silverside and white perch. Two separate calculations were made using the results of toxicity tests. Median survival times (MST) were calculated for each test concentration and subjected to stepwise (backward) multiple regression analysis (using a Hewlett-Packard 9830 programmable calculator) to determine the influence of the environmental factors on MST. The generalized regression equation is

$$Y = a + b_i X_i + b_j X_j + \ldots b_N X_N + e$$

where Y = median survival time (MST), h
 a = constant
 b_i, b_j, and b_N = regression coefficients
 X_i, X_j, and X_N = independent variables
 e = *the error term.*

Total fish length, temperature, salinity, pH, and percent pH 7 oxidant were incorporated as independent variables in the multiple regression analysis. Percent pH 7 oxidant was calculated by dividing the concentration measured as pH 7 oxidant by CPO and multiplying by 100 to estimate the relative amount of pH 7 oxidant to CPO present in each test. The MST was used as the dependent variable.

All variables were entered into the equation, and a partial F-statistic was used to determine the insignificance of each value. The "best" equation was defined as that which explained the maximum amount of variation in the response with the minimum number of variables. Additionally, the variable should be significant at the 1% level or, if not significant at this level, it should explain at least an additional 10% of the variation. Variables not meeting these criteria were removed from the equation.

In addition LC_{50} values at 4 and 96-h were estimated using graphical interpretation. Mean LC_{50} values and 95% confidence limits were then calculated. Because the avoidance response test results were generally 4 h or less duration, the 4-h LC_{50} values were used for comparison.

RESULTS

Avoidance Studies

A total of 144 tests were conducted with Atlantic silverside and 148 with white perch. Table I lists the conditions during avoidance tests. Test results are presented in Table II. Both species avoided concentrations as low as 0.02 mg/L

Table I. Means ($\bar{X}$), Minimum and Maximum Values for Independent Variables Tested for Atlantic Silverside and White Perch Avoidance Studies

Variables	Atlantic Silverside $\bar{X}$ (min-max)	White Perch $\bar{X}$ (min-max)
Temp, °C	18.4 (5.0 – 27.0)	13.8 (4.0 – 28.0)
Salinity, ppt	4.5 (2.5 – 7.0)	3.0 (0.0 – 7.0)
Dissolved oxygen, % saturation	90.0 (52.0 – 100.0)	87.0 (47.0 – 100.0)
Light, lx	430 or 1076	430 or 1076
pH	7.48 (7.2 – 7.8)	7.58 (7.2 – 8.0)
Total fish length, mm	69.8 (29.0 – 104.0)	116.0 (74.0 – 160.0)

Table II. Results from Avoidance Response Studies with Atlantic Silverside and White Perch (in mg/L CPO)

Species	Tests (No.)	Avoidance Concentration (mean ± S.D.)	Range (min-max)	95% Confidence Interval
Atlantic silverside	144	0.15 ± 0.12	0.02 – 0.64	0.13 – 0.17
White perch	148	0.22 ± 0.18	0.02 – 0.80	0.19 – 0.25

CPO. The mean avoidance concentrations of CPO were 0.15 mg/L for Atlantic silverside and 0.22 mg/L for white perch. Because the 95% confidence intervals for each species did not overlap (Table II), the two species avoided significantly different levels of chlorinated estuarine waters. Atlantic silverside avoided lower levels than white perch.

Table III lists the number of instances in which the pH 7 oxidant made up 50% or more of the measured avoidance concentration. Several observations can be made. First, the percentage of instances in which the pH 7 oxidant was ≥50% of the measured avoidance concentration increased as the avoidance concentration decreased. This suggests that both species avoided chlorinated estuarine waters at a lower concentration when the pH 7 oxidant was ≥ 50% of the measured CPO. Second, 75% of the avoidance concentrations were 0.20 mg/L CPO or lower for Atlantic silverside; with white perch, only 55% of the concentrations were 0.20 mg/L CPO or lower. This observation further supports the conclusion that Atlantic silverside will avoid chlorinated estuarine water at a lower level than white perch.

Table III. Instances at Various Ranges of Avoidance Concentrations when Measured pH 7 Oxidant was ≥50% of CPO for Atlantic Silverside and White Perch

	Avoidance Concentrations CPO, mg/L			
	0.01–0.10	0.11–0.20	0.21–0.30 mg/L	>0.31 mg/L
Atlantic silverside				
Number/total	66/69	34/38	12/18	4/19
Total within each interval, %	96	89	67	21
White perch				
Number/total	50/55	17/27	6/29	6/37
Total within each interval, %	91	63	21	16

Table IV. Mean ($\bar{X}$), Minimum, and Maximum Values for Variables Used to Evaluate Toxicity of Atlantic Silverside and White Perch to CPO

Variables	Atlantic Silverside $\bar{X}$ (min – max)	White Perch $\bar{X}$ (min – max)
Temp, °C	21.1 (10.0 – 28.0)	17.0 (10.0 – 27.0)
Salinity, ppt	4.64 (3.0 – 7.0)	2.74 (0.5 – 8.0)
pH	7.48 (7.1 – 7.8)	7.61 (7.2 – 8.0)
Total fish length, mm	63.5 (26.2 – 100.9)	128.3 (60.0 – 169.0)

Toxicity Studies

Table IV lists the conditions during chlorine toxicity tests with Atlantic silverside and white perch. A total of 42 MSTs were calculated for Atlantic silverside. Analysis indicated that the equation which best estimates MST contained only CPO (Figure 1). The remaining independent variables did not significantly influence the MST. This equation accounted for 77% of the variability associated with MST.

A total of 40 MSTs were calculated for white perch. Analysis indicated that the equation which best estimates MST contained CPO and pH (Figure 1). These two variables accounted for 65% of the variation in MST. An increase in CPO caused a decrease in MST, whereas an increase in pH caused an increase in MST. CPO was more toxic to white perch at lower pH values under the test conditions shown in Table IV.

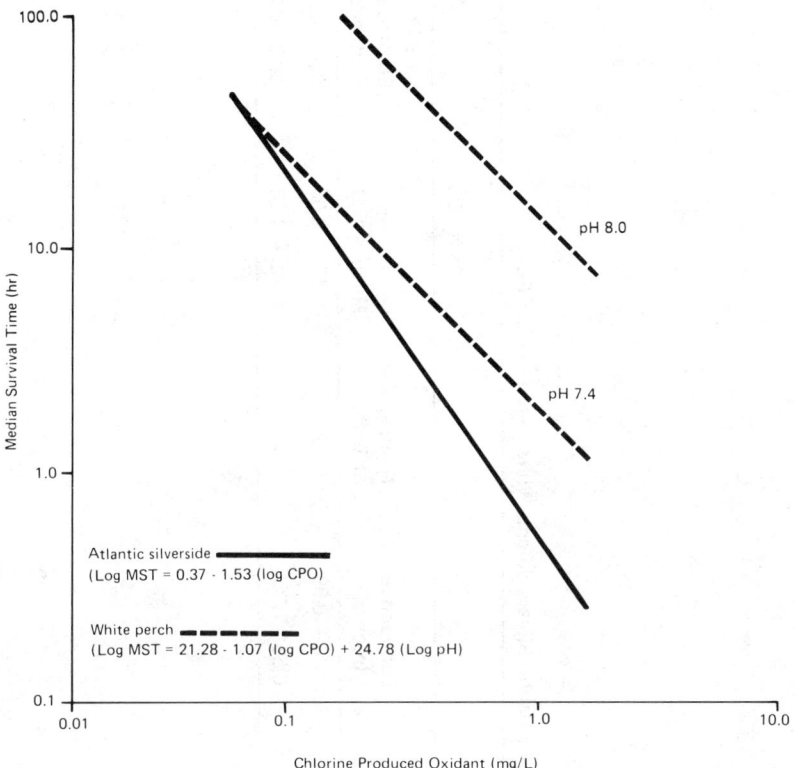

Figure 1. Relationship between median survival time (MST) and chlorine-produced oxidant (CPO) for Atlantic silverside and white perch exposed to chlorinated estuarine water.

The mean LC_{50} values calculated at 4 and 96 h are presented in Table V. The 4-h LC_{50} values for both species were higher than the 96-h value. After 96-h exposure, Atlantic silverside were slightly more sensitive than white perch to chlorinated estuarine waters. At 4-h, however, Atlantic silverside were significantly (95% confidence intervals do not overlap) more sensitive than white perch.

Avoidance vs Toxicity

Concentrations of CPO that elicited avoidance were compared with CPO concentrations that were lethal. Because the duration of the avoidance response test was generally 4 h or less, the 4-h LC_{50} values are more appropriate for comparison because of the influence of exposure duration on the

Table V. Acute Toxicity Results of 4- and 96-h LC_{50} Values for Atlantic Silverside and White Perch (mg/L CPO)

Species	4-h LC_{50}			96-h LC_{50}		
	Number of Tests	Mean ± S.D.	95% Confidence Interval	Number of Tests	Mean ± S.D.	95% Confidence Interval
Atlantic Silverside	12	0.29 ± 0.15	0.20 – 0.38	12	0.15 ± 0.06	0.11 – 0.19
White Perch	10	1.04 ± 0.69	0.55 – 1.53	12	0.22 ± 0.08	0.17 – 0.27

organism's response. For each species, the mean avoidance concentration and the mean 4-h and 96-h LC_{50} values were compared (Figure 2). The upper 95% confidence bound for the avoidance concentration did not overlap with the lower 95% confidence bound for the 4-h LC_{50} values; therefore, the two values were significantly different. The avoidance concentrations were lower than the 4-h LC_{50} values. A comparison of the 96-h LC_{50} values with avoidance concentrations indicated that the two endpoints resulted in similar values of chlorinated estuarine waters.

DISCUSSION

Relationship of Avoidance Responses to Acute Toxicity Levels

The Public Service Electric and Gas Company (PSE&G),[17] using methods identical to those described in this chapter, presented the results of avoidance tests on 12 additional species (Figure 3). The test results[17] for spot and Atlantic menhaden are similar to those published by Middaugh et al.[8] for spot and Hall et al.[12] for Atlantic menhaden. PSE&G reported a range of 0.03 to 0.40 mg/L CPO for spot, whereas Middaugh et al. reported that the avoidance response was 0.05 or 0.18 mg/L CPO, depending on temperature. Hall et al.[12] reported slight avoidance of 0.10 to 0.15 mg/L CPO, and PSE&G reported avoidance concentrations of 0.04 to 0.17 mg/L CPO.

Figure 3 also shows the acute toxicity results for 48-h or 96-h LC_{50} values determined by PSE&G for 8 estuarine species. Although the number of data points for some species are small, the observation can again be generally made that the concentrations avoided by the estuarine fish and invertebrates tested were less than or similar to the reported LC_{50} values. The avoidance concentrations and 96-h LC_{50} values in this study were similar for Atlantic silverside and white perch.

Hall et al.[13] reported that striped bass (in a test period < 1 h) did not avoid a concentration that resulted in 15% mortality during a 96-h exposure period. They further suggested that the lack of avoidance could have been caused by insufficient exposure time. In another study,[8] the avoidance concentration for spot was 0.05 mg/L CPO when tested at 14 and 20°C. This concentration compared to an incipient LC_{50} at 15°C (0.06 mg/L). At 10°C the avoidance concentration was 0.18 mg/L, whereas the incipient LC_{50} was 0.12 mg/L CPO.

A comparison of avoidance concentrations (short duration) with concentrations lethal to organisms after a 96-h exposure shows that, in many cases, the two values are similar. This finding was observed in our research, and by PSE&G,[17] Hall et al.,[13] and Middaugh et al.[8] An implication with the similarity between avoidance concentrations (based on generally < 4-h exposure) and lethal concentrations (based on 96-h exposure) is that the fish will not avoid

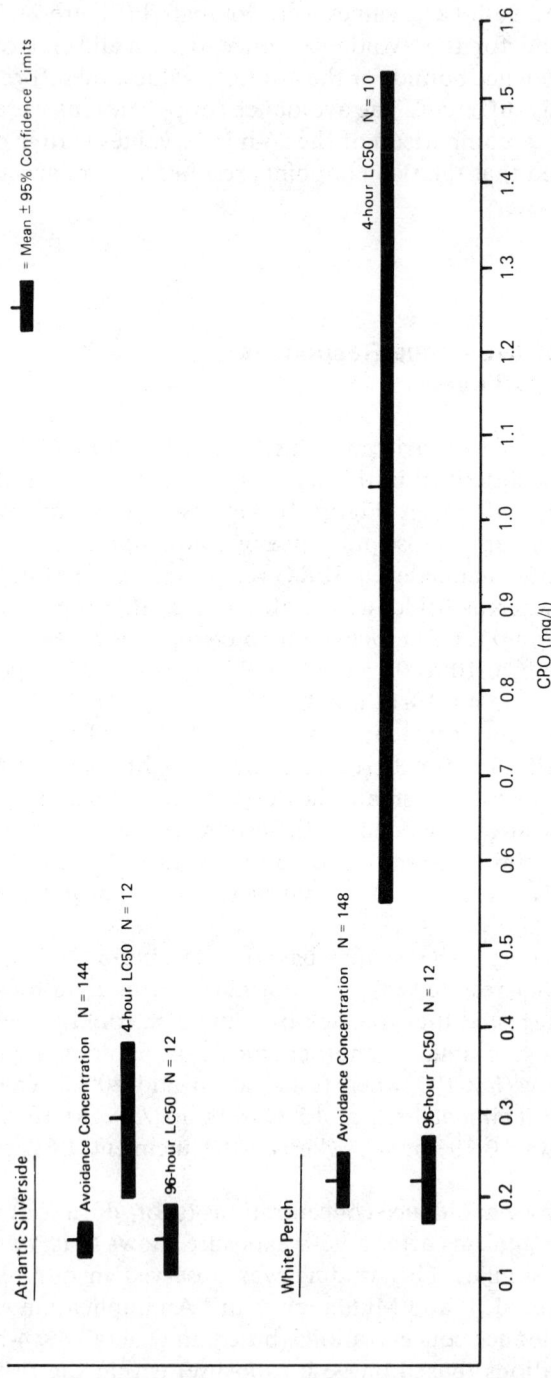

Figure 2. Relationship between avoidance concentrations and 4- and 96-h LC_{50} values for white perch and Atlantic silverside.

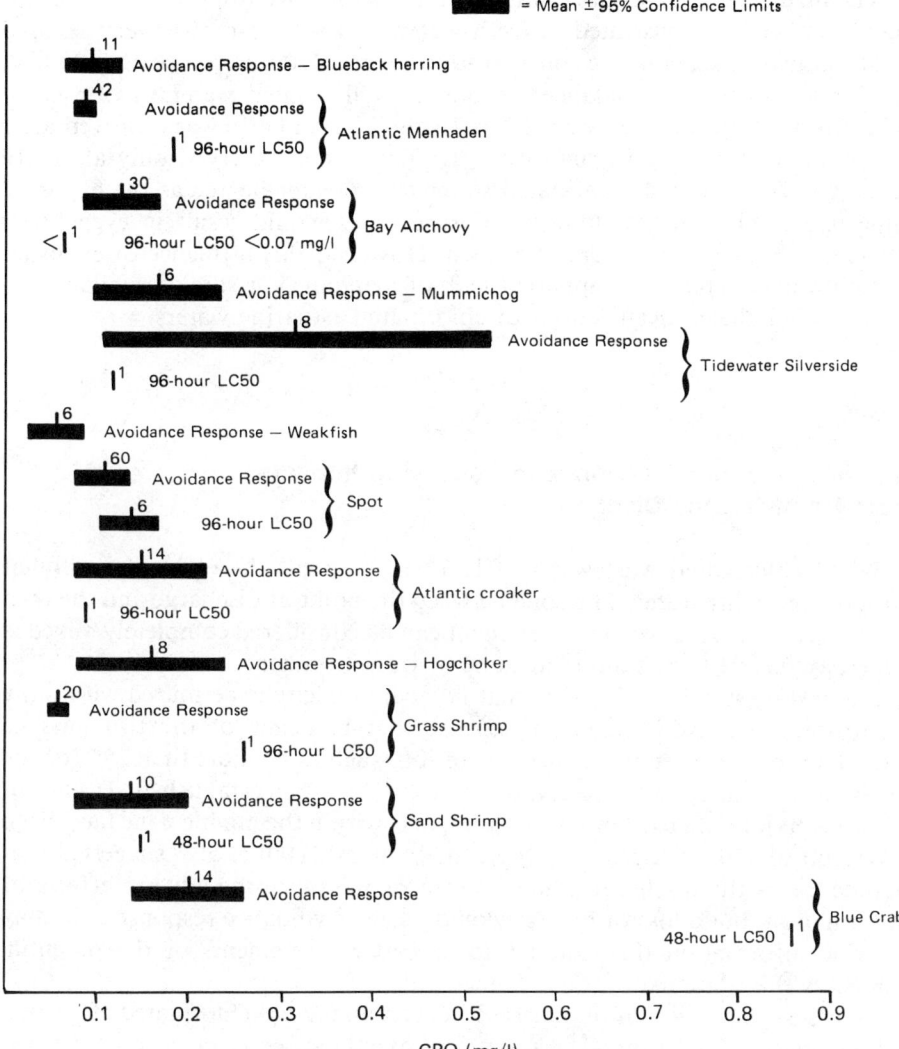

Figure 3. Relationship between avoidance concentrations and 48- or 96-h LC_{50} values for various fishes and invertebrates. The number of data points are identified. Source: Reference 17.

the lethal concentrations. The question that must be answered here is whether a longer exposure time to similar or lower lethal concentrations (based on 96-h LC_{50} results) would result in an avoidance concentration less than 96-h LC_{50} values. Except for the work by Cripe,[18] we are not aware of any studies in which avoidance tests were conducted longer than a 4-h test duration. In that study, 96-h mortality data were not generated for comparison.

The influence of exposure duration on the avoidance response of blacknose dace has been demonstrated in fresh water.[19] In that test, fish were exposed continuously to set chlorine concentrations over a 220-min test period. At 0.47 mg/L, a significant avoidance response to chlorinated water was observed after 30-min exposure, whereas for 0.21 mg/L, avoidance was observed after 120 min; for 0.07 mg/L, significant avoidance was observed only after 210 min. The fish avoided chlorinated water to a greater degree as the exposure time increased. Whether longer exposure times would result in even lower avoidance concentrations is not known. However, this influence of exposure duration may explain the apparent lack of avoidance or similar avoidance to potentially lethal concentrations of chlorinated estuarine waters.

Use of Avoidance Response in Assessing Impacts from Point-Source Discharges

When chlorinated wastewater effluent is released, the effluent is diluted with the receiving water. The zone between the point of discharge and the area in the receiving water where the effluent can be considered completely mixed is generally referred to as a mixing zone or transition zone.

Many state standards indicate that no acute toxicity is permitted within the transition zone. Additionally, to ensure that blockage of the fish passage would not occur, a geometric size dimension (such as no more than 25% of the cross-sectional area of the receiving waterbody) is often established. Traditionally, the major data used in assessing impacts within the mixing zone have been the result of acute toxicity tests (i.e., LC_{50} values). Hall et al.[20] suggested that avoidance of the discharge could prevent or minimize the adverse effects of chlorination, but a loss of habitat would occur. Avoidance response data may provide information that could lead to better assessments of the potential impact within the mixing zone.

To assess the ability of organisms to detect and avoid chlorinated effluents, information is needed on (1) estimates of avoidance concentrations for organisms found in the receiving water and (2) estimates of environmental concentrations that organisms would encounter in nature. In theory, if the concentration that would cause an avoidance response is greater than the estimated environmental concentration, then one might predict that a loss of habitat would occur. Although this overall concept has some usefulness, it simplifies complex analyses. This may result in inaccurate estimates.

The accuracy of estimates of avoidance concentrations and environmental concentrations is dependent on site-specific environmental, biological, and physical factors (Figure 4), which may influence the estimates individually or in combination. The potential influence of several of these factors can be illustrated in the following examples.

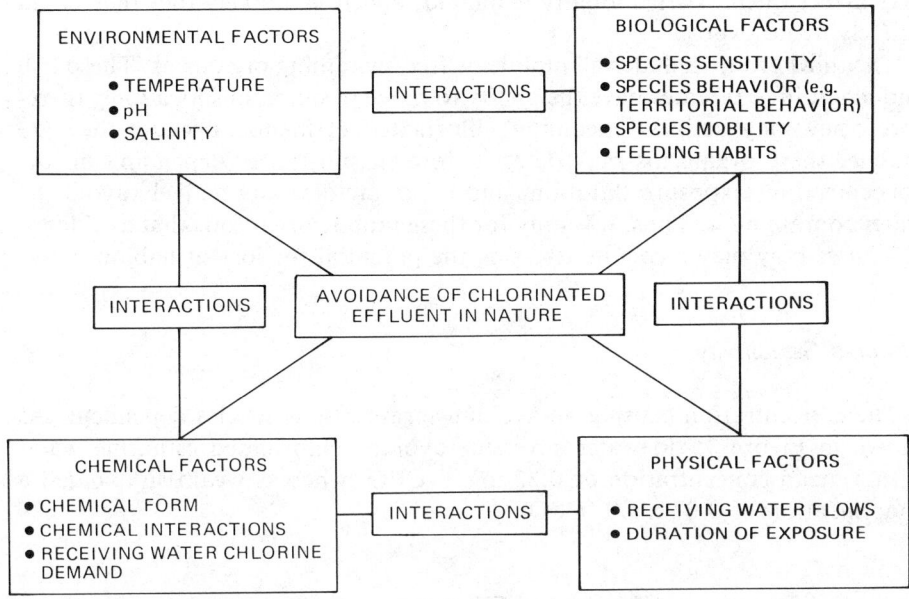

Figure 4. Factors affecting the potential use of avoidance response of mobile organisms to assess impacts of chlorinated discharges.

Species Mobility

For organisms to actively avoid a chlorinated effluent, they must be mobile enough to encounter the effluent and swim to clean water. Generally, organisms or the life stages of organisms can be classified into one of three groups: (1) nonmobile or sessile organisms such as bivalve mollusks, barnacles, and many benthic organisms; (2) planktonic organisms such as copepods, fish eggs, and early larval stages, and (3) mobile organisms such as fishes and macroinvertebrates.

Nonmobile or sessile organisms may encounter chlorinated effluents in nature only if the substance in the water reaches them. The magnitude of the effect, if any, depends on many physical, chemical, and biological factors. Sessile organisms generally do not have the ability to avoid exposure, although bivalve mollusks will close up for short periods. Planktonic organisms generally exist in a floating state or, if they are mobile, have only limited mobility in the environment. The potential exposure for these organisms may be similar to that of sessile organisms. However, because of the influence of tidal or other water movement, these organisms may be transported in and out of chlorinated waters which, depending on the concentrations and exposure duration, may affect them. Their mobility is limited, and it is unlikely that they could actively avoid exposure.

The third group includes all mobile or free-swimming organisms. These fish and macroinvertebrates have the ability to move to numerous locations; therefore, they may actively encounter chlorinated effluents. Because they are mobile, these organisms may detect a foreign substance (depending on the concentration, exposure duration, and other factors) and actively avoid the water containing it. Thus, it is only for these mobile organisms that avoidance responses may play a role in assessing the potential for loss of habitat.

Species Sensitivity

The concentration causing an avoidance response is species dependent. As shown in Figure 3, tidewater silverside avoided chlorinated estuarine water with a mean concentration of 0.32 mg/L CPO, whereas weakfish avoided a concentration of 0.07 mg/L CPO.

Nature of Release of Chlorinated Effluents

Exposure concentration and duration will be affected by whether the chlorinated effluent is released continuously, intermittently, or as a single slug. This is an important consideration in light of the test results. For power plants that chlorinate (in most cases) for 2 h/d or less, the observation that fish will avoid concentrations lower than lethal levels, at similar exposure durations, suggests that avoidance tests may be useful in evaluating intermittent chlorinated discharges (recognizing the influence of interaction with other factors, as discussed below). If the effluent is released intermittently or on a one-time basis, the organisms may return to the area when the chemical is removed from the water. In these instances, the loss of habitat may be only temporary. In situations where the effluent is continuously chlorinated, the loss of habitat could be permanent.

Interactions

When mobile organisms are exposed simultaneously to more than one stress, as would occur in a complex effluent, this interaction may create conditions that would increase or decrease their ability to avoid chlorinated effluents. For example, Hall et al.[13] and Meldrim and Fava[10] observed that, at certain acclimation temperatures, Δ T's of 2 to 6°C in conjunction with chlorinated water would override the avoidance to only chlorinated water. Similarly, Hose and Stoffel[21] and Janssen and Giesy[22] observed that the availability of food in chlorinated seawater or thermal effluents may modify the normal avoidance response.

CONCLUSION

This discussion describes examples that illustrate the complexity of using avoidance responses of mobile organisms to assess impacts of chlorinated effluents. Without adequately considering real-world interactions, univariate laboratory avoidance responses may not always provide accurate estimates of the ability of mobile organisms to detect and avoid chlorinated discharges in nature. Giattina et al.,[23] on the other hand, reported that chlorine avoidance concentrations determined in the laboratory could be used to accurately predict those that would be avoided by fish species in the field.

Conceptually, for avoidance responses or any other toxicological endpoint to be used realistically in assessing impacts, the actual exposure conditions (e.g., duration, chemical form, interactions) that organisms encounter in nature must be known and appropriately simulated in the laboratory or addressed in an assessment. This incorporation of actual exposure conditions in the assessment allows us to put into perspective laboratory biological effects data. To date, our laboratory tests have only mimicked a portion of these conditions. While it is not possible to address all combinations, it has become evident that the use of avoidance responses to assess impacts in the receiving waters is more complex than originally conceived. This observation does not preclude the continued examination of the use of avoidance responses; however, it does suggest the need to clearly understand, from a holistic standpoint, the dynamics of chlorinated effluents discharged into complex ecosystems.

REFERENCES

1. Brungs, W. A. "Effects of Residual Chlorine on Aquatic Life," *J. Water Pollut. Control Fed.* 45(10):2180–2193 (1973).
2. Brungs, W. A. *Effects of Wastewater and Cooling Water Chlorination on Aquatic Life*, EPA-600/3-76-098. (Duluth, MN: U.S. Environmental Protection Agency, 1976).
3. Mattice, J. S., and H. E. Zittel. "Site-Specific Evaluation of Power Plant Chlorination," *J. Water Pollut. Control Fed.* 48(10):2284–2308 (1976).
4. Hall, L. W., G. R. Helz, and D. T. Burton. *Power Plant Chlorination. A. Biological and Chemical Assessment*, (Ann Arbor, MI: Ann Arbor Science Publishers, Inc., 1981).
5. Stober, Q. L., and C. H. Hanson. "Toxicity of Chlorine and Heat to Pink (*Oncorhynchus gorbuscha*) and Chinook Salmon (*O. tshawytscha*)," *Trans. Am. Fish. Soc.* 103(3):569–576 (1974).
6. Roberts, M. H., R. J. Diaz, M. E. Bender, and R. J. Huggett. "Acute Toxicity of Chlorine to Selected Estuarine Species," *J. Fish. Res. Bd. Can.* 32(12):2525–2528 (1975).
7. Morgan, R. P., and R. D. Prince. "Chlorine Toxicity to Eggs and Larvae of Five Chesapeake Bay Fishes," *Trans. Am. Fish Soc.* 106(4):380–385 (1977).
8. Middaugh, D. P., A. M. Crane, and J. A. Couch. "Toxicity of Chlorine of Juvenile Spot, *Leiostomus xanthurus*," *Water Res.* 11:1089–1096 (1977).

9. Meldrim, J. W., J. J. Gift, and B. R. Petrosky. "The Effect of Temperature and Chemical Pollutants on the Behavior of Several Estuarine Organisms," Bulletin 11, (Middletown, DE: Ichthyological Associates, Inc., 1974).
10. Meldrim, J. W., and J. A. Fava. "Behavioral Avoidance Responses of Estuarine Fishes to Chlorine," *Chesapeake Sci.* 18(1):154–157 (1977).
11. Middaugh, D. P., J. A. Couch, and A. M. Crane. "Responses of Early Life History Stages of the Striped Bass, *Morone saxatilis*, to Chlorination," *Chesapeake Sci.* 18(1):141–153 (1977).
12. Hall, L. W., S. L. Margrey, W. C. Graves, and D. T. Burton. "Avoidance Responses of Juvenile Atlantic Menhaden, *Brevoortia tyrannus*, Subjected to Simultaneous Chlorine and Delta T Conditions," in *Water Chlorination: Environmental Impact and Health Effects., Vol. 4*, R. L. Jolley, W. A. Brungs, J. A. Cotruvo, R. B. Cumming, J. S. Mattice, and V. A. Jacobs, Eds. (Ann Arbor Science Publishers, Inc., 1983), pp. 983–991.
13. Hall, L. W., S. L. Margrey, D. T. Burton, W. C. Graves. "Avoidance Behavior of Juvenile Striped Bass, *Morone saxatilis*, Subjected to Simultaneous Chlorine and Elevated Temperature Conditions," *Arch. Environ. Contam. Toxicol.* 12:715–720 (1983).
14. Block, R. M., G. R. Helz, and W. P. Davis. "The Fate and Effects of Chlorine in Coastal Waters: Summary and Recommendations," *Chesapeake Sci.* 18(1):97–101 (1977).
15. Meldrim, J. W., and J. J. Gift. "An Experimental Study of the Behavior of Estuarine Fishes to a Proposed Thermal Effluent," Proc. 4th Mid-Atlantic Indust. Water Conf. (Newark, DE: University of Delaware, 1971), pp. 65–74.
16. Mount, D., and W. A. Brungs. "A Simplified Dosing Apparatus for Fish Toxicity Studies," *Water Res.* 1:21–29 (1967).
17. *Annual Environmental Operating Report (Non-Radiological) Salem Nuclear Generating Station—Unit No. 1.* Docket No. 50-272; Operating License No. DPR-70. Vol. 3 of 3 (Newark, NJ: Public Service Electric and Gas Company, 1978).
18. Cripe, C. R. "An Automated Device (AGARS) for Studying Avoidance of Pollutant Gradients by Aquatic Organisms," *J. Fish. Res. Bd. Can.* 36:11–16 (1979).
19. Fava, J., and C. Tsai. "Delayed Behavioral Responses of the Blacknose Dace (*Rhinichthys atratulus*) to Chloramines and Free Chlorine," *Comp. Biochem. Physiol.* 60C:123–128 (1978).
20. Hall, L. W., D. T. Burton, and L. H. Liden. "Power Plant Chlorination Effects on Estuarine and Marine Organisms," *CRC Crit. Rev. Toxicol.* 10(1):27–47 (1982).
21. Hose, J. E., and R. J. Stoffel. "Avoidance Response of Juvenile *Chromis punctipinnis* to Chlorinated Seawater," *Bull. Environ. Contam. Toxicol.* 25:929–935 (1980).
22. Janssen, J., and J. P. Giesy. "A Thermal Effluent as a Sporadic Cornucopia: Effects on Fish and Zooplankton," *Environ. Biol. Fishes* 11(3):191–203 (1984).
23. Giattina, J. D., D. S. Cherry, J. Cairns, Jr., and S. R. Larrick. "Comparison of Laboratory and Field Avoidance Behavior of Fish in Heated Chlorinated Water," *Trans. Am. Fish. Soc.* 110:526 (1981).

CHAPTER 40

Depression of Larval Growth and Metamorphosis of Oysters Exposed to Chlorinated Sewage

Morris H. Roberts, Jr., and Beverly B. Casey

Chlorine is widely used to disinfect sewage plant effluents and to control fouling organisms in power plants.[1,2] Sewage effluents in Virginia are generally chlorinated to a 2 mg/L total residual chlorine (TRC) after 30 min contact time. Oxidant residues in excess of this concentration result in a receiving water that is potentially toxic to fishes, oyster larvae, and other aquatic organisms.[3-7] At least one fish kill has been linked to elevated concentrations of chlorine-produced oxidants (CPO).[8]

Based on oyster embryo survival from fertilization to the straight-hinge stage, the 48-h median lethal concentration (LC_{50}) for CPO in estuarine water has been estimated to be 0.023 mg/L.[6] Only certain larval fishes have a lower LC_{50}. Roosenburg et al.[9] reported a 96-h LC_{50} for post straight-hinge larvae of 0.3 mg/L TRC and suggested that the LC_{50} for recently set spat was greater than 0.5 mg/L TRC.

There are four secondary treatment plants operated by the Hampton Roads Sanitation District Commission that discharge chlorinated effluent into the James River in close proximity to the major oyster seed beds of Virginia (Figure 1). Together, these plants discharge about 65 million gal/d treated effluent. Only one plant (the Nansemond Plant) dechlorinates the final chlorinated effluent before discharge.

Dramatic declines in spatfall on the oyster seed beds in the James River have been documented since the early 1960s.[10] Various explanations have been proposed for the decline in oyster spatfall, including a decline in brood stocks resulting from the MSX epidemic (an Apicomplexan disease of oysters) beginning in the late 1950s. It is reasonable, however, to ask whether chlorinated sewage has affected the rate of spatfall. The reported lethal concentrations for early larval stages cited above are unlikely to occur except in close proximity to an outfall chlorinating at 2 mg/L. However, this does not preclude possible inhibition of settlement or disruption of larval metamorphosis at CPO concentrations well below the LC_{50} for early-stage larvae.

The objectives of the present study were to determine (1) whether chlorinated sewage diluted with estuarine water would affect spatfall of oyster larvae

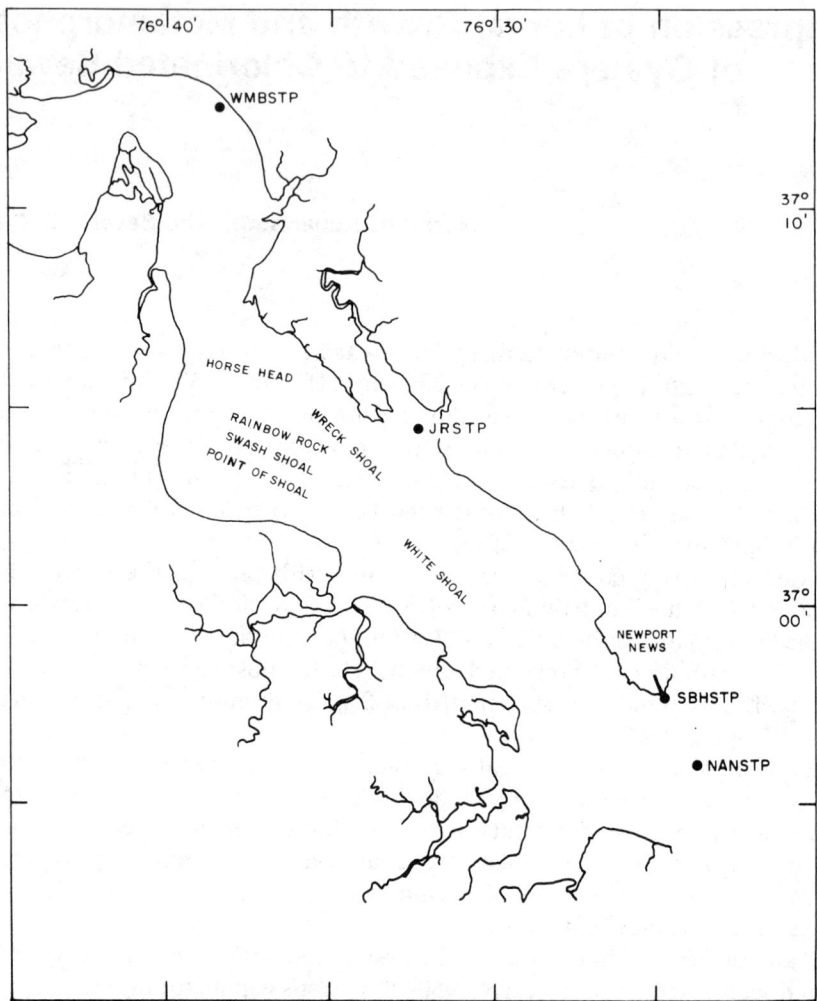

Figure 1. Lower James River, Virginia, with four major sewage outfalls: Williamsburg (WMSTP), James River (JRSTP), Small Boat Harbor (SBHSTP), and Nansemond (NANSTP) Sewage Treatment Plants. Several major shoal areas that serve as oyster rocks, providing seed oysters to Virginia's oyster industry, are also indicated.

exposed just prior to setting, and (2) whether dechlorination of sewage prior to dilution with estuarine water would change the observed toxicity.

MATERIALS AND METHODS

Each test consisted of exposing oyster larvae in a flow-through system to 1:20 dilutions of sewage, sewage chlorinated to one of four 30-min residuals,

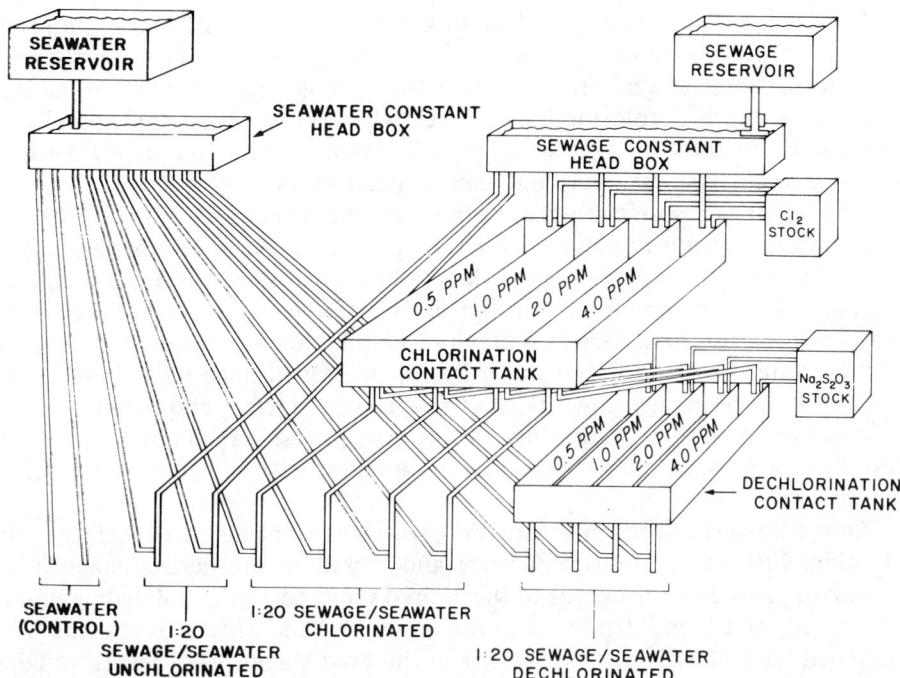

Figure 2. Schematic representation of the diluter system used to prepare the unchlorinated, chlorinated, and dechlorinated sewage/estuarine water mixtures to which oyster larvae were exposed.

dechlorinated sewage, and estuarine water alone. Chlorine residuals tested were selected to bracket the 2.0 mg/L TRC stipulated for effluents from sewage plants in Virginia.

The diluter system consisted of a sewage head tank, an estuarine water head tank, a 4-channel chlorination tank, and a 4-channel dechlorination tank (Figure 2). Flow rates of sewage and seawater were controlled by calibrated glass pipet tips. The chlorination and dechlorination tanks were built to scale so that the transit time for assumed plug flow was 30 min for each channel. Chlorinated or dechlorinated sewage solutions were mixed with 1 μm filtered York River, Virginia, estuarine water in a 1:20 ratio. All tanks in the diluter were fabricated from acrylic plastic, and all delivery tubes were glass.

Sewage used for this study was secondary clarifier effluent collected with a submersible pump from the discharge of the final clarifier at the James River Sewage Treatment Plant and stored in 55-gal polyethylene drums for 3 or 4 d prior to the expected start date for the experiment. A sample of each collection was analyzed for ammonia nitrogen and nitrite-nitrate nitrogen by standard procedures[11] using a Scientific Instruments Corporation autoanalyzer. To determine the appropriate stock concentrations of chlorine that would achieve the desired test concentrations, a 30-min residual/dose curve was produced.

Sewage was chlorinated by injecting a small volume of the appropriate calcium hypochlorite stock solution at the head of each chlorination channel. Dechlorination was similarly accomplished by injecting sodium thiosulfate at the head of each dechlorination channel. The sodium thiosulfate stock solutions were prepared at concentrations calculated to react completely with the prescribed chlorine residuals and leave a slight excess.

The larval exposure chambers were 90-cm-diameter glass cylinders 205 cm long and fitted with fine mesh screen at one end. Baskets were prepared with one of three mesh sizes: 74 μm (nylon), 104 μm (stainless steel), and 200 μm (stainless steel) to accommodate larvae of different sizes. The baskets were suspended in 8-L glass jars to a depth of about 140 cm, yielding a volume of 0.9 L. Water from the diluter was delivered directly into the larval baskets. To initiate an experiment, flows of all solutions were started, and the system was allowed to come to a steady state with respect to residual chlorine concentrations in the 30-min contact tanks and in the larval baskets before introduction of larvae.

Oyster larvae used for this study were spawned in the hatchery facility at the Virginia Institute of Marine Science and reared to the desired stage using standard procedures.[12] Larvae of the desired stage were then isolated, counted, measured, and then introduced to the larval baskets. Most experiments were initiated with 5000 larvae per basket in the eyed stage (mean length ranging from 274 to 308 μm, depending on the experiment). Larvae at this stage will usually set within 24 to 48 h. Two experiments were initiated with larvae in the umbo stage (mean length ranging from 80 to 120 μm, depending on the experiment). For experiments with umbo-stage larvae and some experiments with eyed-stage larvae, a 50:50 food mixture of the microflagellates *Pavlova lutheri* and *Isochrysis galbana* was injected into the diluent water head tank to yield a concentration of about 10^3 cells/mL.

Artificial shell strings were suspended within the larval baskets for the pediveliger larvae to set on. Each shell string consisted of five 5-by-5-cm sheets of flex board strung onto parallel strands of plastic-coated wire. The flex board was presoaked in dilute acid followed by seawater prior to use.[13] Spatfall plate strings were removed after 24 h and disassembled. Spat were counted on both the top and bottom of each plate using a dissecting microscope. Freshly prepared strings were introduced into each larval basket. Spat strings were counted daily until no further spatfall was observed. At that time, all remaining unattached larvae in the baskets were counted.

In tests with umbo larvae, the larvae were washed out of the baskets onto a sieve after 2 or 3 d, resuspended in 200 mL of water, and a 1-mL sample was counted on a Sedgwick-Rafter cell. Live and dead larvae were counted separately. Larval shell length was measured with a Filar micrometer eyepiece on a compound microscope. The remainder of the culture was returned to the appropriate larval basket, which was cleaned by immersion in a chlorine solution and then in thiosulfate to remove bacterial slime.[14] Counts and measurements were repeated at 2- or 3-d intervals.

Total residual chlorine in sewage effluent or CPO in seawater were determined by amperometric titration with phenylarsine oxide at pH 4 using a Fischer-Porter amperometric titrator.[11,15] The only exception was the first test with pediveliger larvae, in which the DPD titration method[11] was used. Stock chlorine solutions were analyzed by iodometric titration with sodium thiosulfate and a starch indicator.[11] Chlorine residuals in the 30-min contact tanks were determined thrice daily. The CPOs in the exposure tanks were determined at random times.

RESULTS AND DISCUSSION

Five tests were performed with pediveliger larvae. The temperature averaged 27.2 to 27.9°C in all experiments except the first, with a mean temperature of 25.5°C. The salinity was about 15 ppt. Dissolved oxygen and pH were, in all cases, within normal ranges (Table I).

The sewage effluent used in the five experiments differed in ammonia nitrogen and nitrite-nitrate nitrogen concentrations, but were always within normal operating ranges for the James River Treatment Plant. Ammonia concentrations ranged from 4.7 mg/L in the third experiment to 13.0 mg/L in the second experiment. Nitrite-nitrate concentrations ranged from 0.8 to 10.7 mg/L. Only in the first experiment was the nitrite concentration determined separately. In this case the nitrite-nitrate concentration was 4.1 mg/L, with 3.3 mg/L nitrite nitrogen. High nitrite levels might be expected in experiments 3 and 5 when the nitrite-nitrate concentrations were even higher than in experiment 1 and ammonia concentrations were relatively low (Table I).

The measured TRCs in the contact tanks varied somewhat among experiments (Table I). The variability was greater in the low nominal concentrations than in the high. In those cases in which the concentration of CPO in the larval baskets or the test tanks was measured after the 1:20 dilution with estuarine water, the exposure concentration in the tanks receiving chlorinated sewage was 5% of that in the corresponding contact tank. In no case was TRC ever detected in the dechlorination-contact or exposure tanks.

Pediveliger larvae used in these tests ranged in mean shell length from 274 to 308 μm. In every case, over 70% of the larvae were in the eyed condition and deemed ready to set. In the first two experiments with no food added and 70 to 80% eyed larvae, the percent set of control larvae was low (1.6-2.5%), whereas in the last two experiments with food added and 70% eyed, the percent set of control larvae was improved (7.2-9.6%). In experiment 3, with 100% of the larvae eyed at the initiation of the experiment and no food added, 29% of the larvae set.

While these set rates are low in absolute terms, the usual set rate for unfed oyster larvae under static hatchery conditions, when provided with frosted mylar plastic sheets as a setting substrate, varies from 10 to 20% for what are deemed good cultures. Windsor[16] observed spat settlement rates ranging from

Table I. Summary of Experimental Conditions for Exposure of Pediveliger and Umbo-Stage Oyster Larve

Experiment	Temp. (°C)	Salinity (ppt)	Dissolved Oxygen (mg/L)	pH	NH_4 (mg/L)	NO_3 (mg/L)	Residual Chlorine Concn (mg/L)				Mean Larval Size (μm)	Percent Eyed	Food
							1	2	3	4			
Pediveliger tests													
1	12.5	15.2	7.6	7.9	7.3	4.1	0.29	0.71	1.87	3.54	274	70	No
2	NR[a]	NR	NR	NR	13.0	0.8	0.44	1.23	2.87	4.45	306	70	No
3	27.9	15.0	6.8	7.5	4.7	10.7	0.69	1.58	3.13	4.32	308	100	No
4	27.2	14.9	8.3	7.9	10.3	2.3	0.65	1.13	2.16	3.72	282	70	Yes
5	27.4	15.4	7.9	7.8	6.7	6.5	0.60	0.96	1.95	3.76	288	80	Yes
Umbo-stage larval tests													
1	27.4	15.4	7.0	7.5	9.5	1.9	0.98	1.04	2.44	3.92	97	NA[b]	Yes
2	28.0	15.8	7.4	7.4	16.3	0.02	0.36	0.77	1.87	3.86	117	NA	Yes

[a]Not recorded.
[b]No applicable.

<1 to a maximum of 18% for larvae cultured on a variety of algal diets prior to being provided mylar sheets under standard hatchery conditions. Even with the best larval diet, setting varied from 6 to 13%.

Pediveligers exposed to unchlorinated sewage exhibited different setting rates among experiments, presumably reflecting variations in the sewage effluent. In experiments 1 and 5, setting in excess of that in the diluent control treatment was observed (146 and 120% of controls, respectively), whereas in all other cases, sewage effluent appeared somewhat toxic, with setting rates less than those of the diluent controls (38 to 82% of the diluent controls, Figure 3).

In experiments 1 and 4, chlorination at every concentration tested depressed spatfall in comparison with control treatment, and dechlorination did little to modify this response. One might discount the observations in experiment 1 because of the very low spatfall rate in the control (2%); however, in experiment 4, the control spatfall percentage was at a more reasonable level of 7%. In contrast, the results for experiments 2, 3, and 5 indicate moderate to poor spatfall at all doses above 1 mg/L TRC (in sewage or 0.05 mg/L CPO in the test tanks). In every case, dechlorination produced marked enhancement in spatfall, usually to a level greater than that in the nonchlorinated sewage control, and often to levels greater than that in the diluent controls.

Only in experiments 2 and 3 was there evidence of a dose-response relationship between spatfall percentage and chlorination dose. In both cases there was 60 to 70% set at the lowest chlorination dose, dropping to less than 20% at all higher doses. Thus, in these experiments, the EC_{50} based on spatfall percentage would seem to fall between 0.03 and 0.06 mg/L CPO (calculated concentration in the test baskets), whereas in the other three experiments, the EC_{50} was apparently less than the lowest dose tested (0.01 to 0.03 mg/L CPO in the test baskets).

In experiments 2 and 3, dechlorination at the lowest TRC dose had little or no beneficial effect. At the higher chlorination levels, dechlorination caused a dramatic improvement in spatfall, but again with no clear dose-response relationship. There was gradual improvement in spatfall with increasing chlorination concentration in experiment 3, but the reverse was true in experiment 2. Experiment 5 yielded similar results to experiment 2 in all regards except that there was virtually no spatfall in any chlorination treatment.

Roosenburg et al.[9] reported on the survival of pediveligers allowed to attach to a glass substrate and then exposed to chlorinated seawater (no sewage present). Survival was greater than 50% even at their highest test concentration of 0.3 mg/L, thus obviating determination of an EC_{50}. The great difference between the observations of Roosenburg et al.[9] and those reported here are attributable in part to the absence of sewage in the experiment of Roosenburg et al. In addition, the biological process being evaluated was different in the two studies, involving metamorphosis and postmetamorphic survival in the study of Roosenburg et al.,[9] whereas in the present case, initial attachment was also considered during the exposure period.

516 WATER CHLORINATION

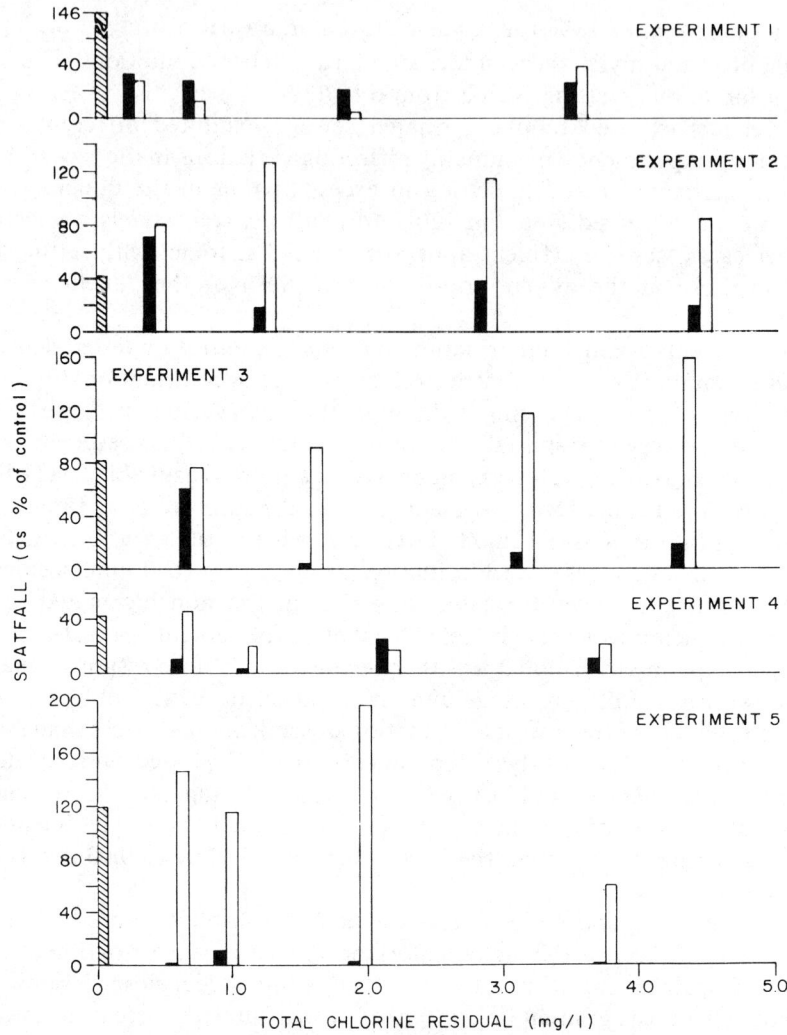

Figure 3. Spatfall under various exposures to chlorinated (solid bars) and dechlorinated (open bars) sewage/estuarine water mixtures. The response to unchlorinated sewage/estuarine water mixtures (hatched bars) is shown for comparison.

The concentration range within which the EC_{50} for spatfall probably lies, ranging from <0.01 to no more than 0.06 mg/L CPO (in the test tanks), is comparable to the 96-h LC_{50} for prestraight hinge larvae exposed to chlorinated seawater[6] and the 96-h EC_{50} for adult oysters exposed to chlorinated sewage.[17]

Two experiments were conducted starting with early umbo larvae (Table I). The temperature in these experiments averaged 27.4 and 28.0°C, respectively. Salinity averaged 15.4 and 15.8 ppt. In both experiments, the sewage effluent

Table II. Survivorship Data for Umbo-Stage Oyster Larvae Exposed to Chlorinated, Dechlorinated, and Unchlorinated Sewage Mixed with Estuarine Water.

	Experiment 1 with Umbo Larvae	
	72 h	
Treatment	Survivors (No.)	Survivors (%)
4.0 Dechlor	ND[a]	
2.0 Dechlor	800	15.4
1.0 Dechlor	600	11.5
0.5 Dechlor	100[a]	1.9
4.0 Chlor	300	5.8
2.0 Chlor	100	1.9
1.0 Chlor	0	0
0.5 Chlor	0	0
Sewage a	ND	
Sewage b	5700	110
Control a	8300[b]	
Control b	3700[a]	71

[a]No data; washout observed; no chlorinated treatment experienced washout (water level in basket never rose).
[b]Unexplained high count.

	Experiment 2 with Umbo Larvae			
	48 h		96 h	
Treatment	Survivors No.	Survivors (%)	Survivors No.	Survivors (%)
4.0 Dechlor	3400	33.0	400[a]	3.9
2.0 Dechlor	3100	30.0	1700	16.5
1.0 Dechlor	1700	16.5	1500	14.6
0.5 Dechlor	2300	22.3	200[a]	1.9
4.0 Chlor	100	1.0	200	1.9
2.0 Chlor	500	4.9	200	1.9
1.0 Chlor	300	2.9	100	1.0
0.5 Chlor	200	1.9	0	0.0
Sewage a	3700	35.9	800	7.8
Sewage b	3800	37.9	100[a]	1.0
Control a	7500	72.8	1600	15.5
Control b	7800	75.7	2500	24.3

[a]Overflow was observed in these tanks, resulting from a high suspended solids level in the sewage effluent added to the reservoir at about 80 h.

had a relatively high ammonia concentration, whereas the nitrite-nitrate nitrogen concentration was exceedingly low. In the first experiment, the two lowest TRC doses had virtually the same average TRC concentration. In the second experiment, the TRC concentrations ranged from 0.36 to 3.86 mg/L in the desired sequence.

In the first experiment with umbo-stage larvae, no reliable survival data were obtained because of the overflow from some larval chambers. It was clear, however, that all chlorination treatments in which no overflow was observed caused nearly total mortality. In the chlorinated treatments, only dead shells were recovered. Survival of the larvae exposed to chlorinated sewage was equally poor in the second experiment (Table II). Thus, in both experiments, the 96-h LC_{50} was apparently <0.02 mg/L CPO in the test tanks. Survival in the dechlorinated and sewage control treatments was approximately half that in the diluent controls after 48 h and about equal to that in the diluent controls after 96 h.

The growth of larval shell length was determined for both experiments. In the first experiment, the diluent and sewage control larvae grew from 80 μm to 95 to 97 μm in 3 d. This compares favorably with the growth of larvae retained in the hatchery that grew to 97.4 μm. Larvae exposed to chlorinated sewage exhibited no growth, whereas larvae exposed to dechlorinated sewage exhibited some growth, reaching 86 to 90 μm in the four treatments. Survival, although not unquestionable, also seemed to improve at higher chlorination/dechlorination levels.

In the second experiment, the diluent and sewage control larvae grew from 117 μm to 126 to 128 μm after 2 d of exposure, and 152 to 166 μm after 4 d. This was somewhat less than the growth observed in the hatchery where larvae reached 150 and 226 μm, respectively, for the same intervals. Larvae exposed to chlorinated sewage reached 106 to 118 μm in 2 d and 130 to 140 μm in 4 d. There was no clear relationship to the chlorine dosage. Larvae exposed to dechlorinated sewage reached 126 to 136 μm in 2 d and 154 to 157 μm in 4 d, which was virtually the same as either set of controls.

The low EC_{50} suggested by the present study for umbo-stage larvae is comparable to the results of Roberts and Gleeson,[6] but it differs dramatically from the value suggested by the data of Roosenburg et al.[9] One might argue that this difference reflects the presence of sewage in the test medium of the present study, thus changing the active chlorine species. However, the study of Roberts and Gleeson[6] was also performed without sewage. An alternative explanation would be that the estuarine waters at the two laboratories differed in a way that led to very different chlorine speciation.

It is clear that chlorinated sewage diluted 1:20 with estuarine water has a severe deleterious effect on larval development and spatfall, even at 30-min contact concentrations, which are far below those normally used in treatment plants in Virginia. The 1:20 dilution rate is that expected in the immediate vicinity of the discharge outfall at the James River Treatment Plant. Under the conditions of the present experiment, the test tank concentration produced by the lowest nominal dose was below that reliably measurable by the procedure used, and yet, at least in some experiments, it was lethal to virtually all larvae exposed. In terms of chlorine concentration, although not effluent concentration, this treatment would be comparable to a field location at some distance from an outfall. Thus, we suggest that there is potential for present chlorination practice to be a contributing factor in the continued low spatfall rate observed in the James River.

Bender et al.[18] reported to the Virginia State Legislature their findings on the effect of chlorination at the James River Treatment Plant on oyster setting in the Warwick River (Figure 1). During the primary setting period of June to October 1978, no chlorine-produced oxidants were detected (minimum detectable concentration = 0.01 mg/L CPO) even at a station immediately adjacent to the discharge from the treatment plant. They reported no effect on oyster spatfall in this tributary of the James River, but added the caveat that "the set was extremely low all over the James in 1978." This suggests that the supply of larvae produced in the river was not large enough to produce a significant set, or that the effect of chlorination on spatfall is very widespread in the James River. The latter view does not seem reasonable despite the presence of several operating treatment plants discharging into the river.

Dechlorination, under experimental conditions, had either no measurable effect on spatfall success or caused a marked improvement in spatfall, even at chlorination doses well above those used in sewage treatment plants in Virginia. If these results can validly be applied to the field situation, dechlorination would seem potentially beneficial, although the benefit may not be realized because of low larval production. While it would be prudent to verify the utility of dechlorination for enhancement of spatfall in a field study, there are plans to upgrade treatment plants in Virginia by addition of dechlorination as part of the Bay cleanup activities. Such action may result in increased spat production, but only if there is an increase in larval production by natural brood stocks.

ACKNOWLEDGMENTS

Larvae for these experiments were produced by D. Hepworth, who also measured all larvae. Algae were grown with the assistance of D. Abernathy. Nitrogen analyses were graciously performed by S. Sturm of the Hampton Roads Sanitation District (HRSD), North Shore Laboratory. D. Francis and J. Williams of the HRSD North Shore Laboratory provided the DPD reagent. D. Wheeler and N. Leblanc of HRSD were most cooperative and supportive throughout this study. The role of M. Bellanca of the Virginia State Water Control Board, whose questions regarding the impact of chlorinated sewage specifically on oyster spatfall sparked our initial interest, and his continued support are most gratefully acknowledged. This project was funded in part by the Virignia Environmental Endowment under Grant No. 82-11. This chapter is Contribution No. 1237 from the School of Marine Science, Virginia Institute of Marine Science, College of William and Mary, Gloucester Point, Virginia.

REFERENCES

1. White, G. C. *Handbook of Chlorination*, (New York: Van Nostrand Reinhold Co., 1972).

2. Hall, L. W., Jr., G. R. Helz, and D. T. Burton. *Power Plant Chlorination; A Biological and Chemical Assessment* (Ann Arbor, MI: Ann Arbor Science Publishers, Inc., 1981).
3. Mattice, J. S., and H. E. Zittel. "Site-specific Evaluation of Power Plant Chlorination," *J. Water Pollut. Control Fed.* 48:2284-2308 (1976).
4. Roberts, M. H., Jr., R. J. Diaz, M. E. Bender, and R. J. Huggett. "Acute Toxicity of Chlorine on Selected Estuarine Species," *J. Fish. Res. Bd. Canada* 32:2525-2528 (1975).
5. Bender,M. E., M. H. Roberts, Jr., R. J. Diaz, and R. J. Huggett. "Effects of Residual Chlorine on Estuarine Organisms," in *Biofouling Control Procedures, Technology and Ecological Effects*, L. D. Jensen, Ed. (New York: Marcel Dekker, Inc., 1977), pp. 101-108.
6. Roberts, M. H., Jr., and R. A. Gleeson. "Acute Toxicity of Bromochlorinated Seawater to Selected Estuarine Species with a Comparison to Chlorinated Seawater Toxicity," *Mar. Environ. Res.* 1:19-30 (1978).
7. Roberts, M. H., Jr. "Acute Toxicity Potential of Chlorination in Estuarine Waters," in *Chlorine-Bane or Benefit?*, CRC Publ. No. 104 (Shady Side, MD: Chesapeake Research Consortium, 1982), pp. 28-35.
8. Bellanca, M. A., and D. S. Bailey. "Effects of Chlorinated Effluents on the Aquatic Ecosystems of the Lower James River," *J. Water Pollut. Control Fed.* 49:639-645 (1977).
9. Roosenburg, W. H., J. C. Rhoderick, R. M. Block, V. S. Kennedy, S. R. Gullans, S. M. Vreenegoor, A. Rosenkranz, and C. Collette. "Effects of Chlorine-Produced Oxidants on Survival of Larvae of the Oyster *Crassostrea virginica*," *Mar. Ecol. Prog. Ser.* 3:93-96 (1980).
10. Haven, D. S., W. J. Hargis, Jr., and P. C. Kendall. *The Oyster Industry in Virginia: Its Status, Problems, and Promise*, Special Papers in Marine Science, No. 4 (Gloucester Point, VA: Virginia Institute of Marine Science, 1978).
11. *Standard Methods for the Examination of Water and Wastewater*, 15th Ed. (Washington, DC: American Public Health Association, 1980).
12. Dupuy, J. L., N. T. Windsor, and C. E. Sutton. "Manual for Design and Operation of an Oyster Seed Hatchery," Spec. Rep. Appl. Mar. Sci. Ocean Engin., No. 142 (Gloucester Point, VA: Virginia Institute of Marine Science, 1977).
13. Shaw, W. N. "Seasonal Fouling and Oyster Setting on Asbestos Plates in Broad Creek, Talbot County, Maryland, 1963-65," *Chesapeake Sci.* 8:228-236 (1967).
14. Roberts, M. H. Jr. "Flow-Through Toxicity Testing System for Molluscan Larvae as Applied to Halogen Toxicity in Estuarine Water," in *Aquatic Invertebrate Bioassays, ASTM STP 715*, A. L. Buikema, Jr., and J. Cairnes, Jr., Eds. (Philadelphia: American Society for Testing Materials, 1980), pp. 131-139.
15. *Instruction Bulletin for Series 17T2000 Amperometric Titrator, Instruction Bulletin 17T2000, Revision 1* (Warminster, PA: Fisher & Porter, Co., 1982).
16. Windsor, N. T. "Effect of Various Algal Diets and Larval Density in the Larviculture of the American oyster, *Crassostrea virginica* (Gmelin), M.S. Thesis (Williamsburg, VA: College of William and Mary, 1977).
17. Roberts, M. H., Jr. "Detoxification of Chlorinated Sewage Effluent by Dechlorination in Estuarine Waters," *Estuaries* 3:184-191 (1980).
18. Bender, M. E., D. S. Haven, and H. D. Slone. "Report on the Effect of Chlorine on Oysters in the Warwick River," submitted to the Virginia Legislature in Response to House Joint Resolution Number 162, (Gloucester Point, VA: Virginia Institute of Marine Science, 1978).

CHAPTER 41

Effect of Selected Chlorine-Produced Oxidants on Oyster Larvae

Mary Elizabeth Stewart and Walter Blogoslawski

The American eastern oyster, *Crassostrea virginica* (Gmelin), has been the most intensely studied and commercially important species of shellfish in the United States.[1] The harbor of New Haven, Connecticut was the center of the U.S. oyster industry from the early nineteenth century until World War I, supporting over 20 oyster companies with an average annual yield of 10 million pounds of oyster meats. Since 1919, development of the New Haven harbor into a major port has subjected the waters to problems attendant to heavily populated shipping centers (e.g., spoils from maintenance dredging, chronic minor oil spills, and municipal and industrial wastes). The oyster industry declined accordingly until approximately 10 years ago when efforts to reduce pollution in the harbor began.

The East Shore Water Pollution Abatement Facility, opened in 1979, is viewed as a step toward further improvement of water quality in the harbor.[2] This plant is designed to run at a maximum flow of 100 mgd and eventually is expected to treat all sewage from New Haven. The installation uses sodium hypochlorite in a secondary treatment process, discharging chlorine-treated effluent into the harbor.

Many data indicate that exposure of the larvae of *C. virginica* to chlorinated seawater produces an increase in larval mortality.[3-6] Early toxicity studies assumed that chlorine itself was responsible for the mortality observed. It is now recognized that adverse larval effects may also be ascribed to by-products resulting from the interactions of chlorine and the medium to which it is applied.[7-11]

The waters from the East Shore Facility are rich in organic amines, amino acids, and ammonia. Screening of harbor water indicates that monochloramine and dichloramine are the dominant chlorine-produced products to be found in effluent from that wastewater treatment plant.

In June 1980, a field and laboratory study that examined the concentration of free chlorine and the chloramines in New Haven Harbor was combined with a study to determine the effect of these oxidants on the larvae of *C. virginica*. This work, continued in 1981 and 1982, is presented in this chapter.

METHODS AND MATERIALS

Field Observations

Six sampling stations were established in New Haven Harbor (Figure 1). All fieldwork was conducted with samples taken from these stations.

Presence of Free Chlorine, Monochloramine, and Dichloramine

A 500-mL sample of seawater was gathered throughout the spawning season at each station each week. An all-glass container was submerged to mid-depth, filled, and capped. Upon retrieval, the sample was iced and stored until analysis on shore could be carried out. The DPD ferrous titrimetric and colorimetric method was used to determine the concentrations of free chlorine, monochloramine, and dichloramine.[12] The amperometric titration method, using a Fischer and Porter Model No. 17T2012, served as a check on the colorimetric method.[12] It should be noted that the values assigned as measurements of chlorine compounds are relative values and not to be regarded as absolute due to the time required for transporting the water to the laboratory.

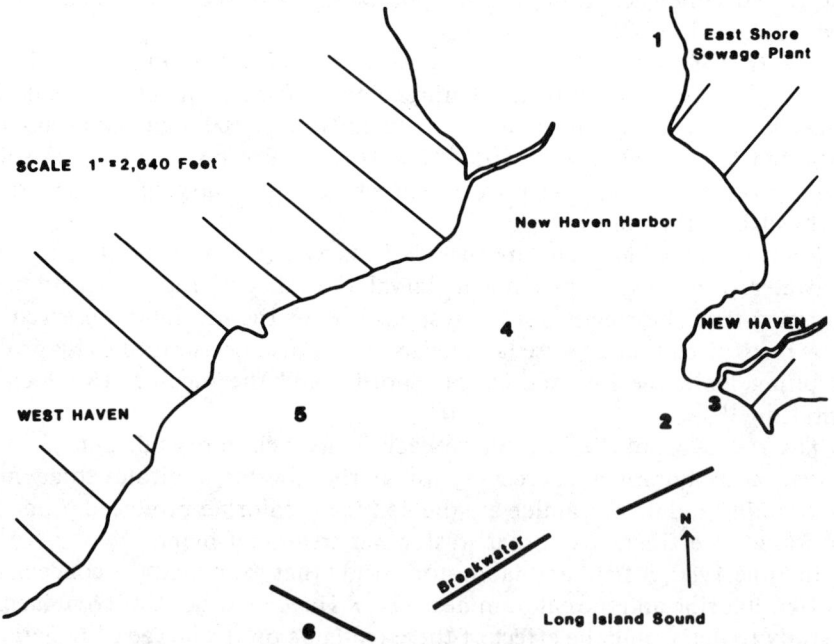

Figure 1. Sampling station locations in New Haven Harbor.

Plankton Counts

A 1.9-m^3 (500-gal) sample of seawater was taken from mid-depth and filtered through a plankton net (#20) to collect larvae of *C. virginica*. The content of the net was rinsed thoroughly into a sample bottle that contained a fixative. On shore, the sample was agitated to ensure even distribution and five 1-mL samples were examined per station. The samples were gathered and counted twice a week throughout the spawning season (approximately 15 weeks) yielding values for approximately 30 counts per station per year.

Set Bag Counts

Twice a week throughout the spawning season, identical spat sampling devices were deployed at each station. These devices consisted of 20 clean oyster shells enclosed in a PVC mesh bag attached to four limestone weights. One bag was lowered to the bottom at each station and marked with a flotation buoy. The bags were collected approximately twice a week and the shells examined microscopically for oyster spatfall.

Laboratory Observations

Survival of Larvae in Seawater from the Stations

Three liters of seawater were taken from mid-depth at each station once a week throughout the spawning season and filtered through a plankton net (#20) to remove any larvae of *C. virginica* from the water. The filtered water was brought to the laboratory where it was used to fill two 1-L beakers per station. Approximately 15,000 newly spawned larvae of *C. virginica* were added to each beaker. After 48 h, the contents of each beaker were sampled and the unfixed larvae were classified as alive, dead, or abnormal.[13]

Survival of Larvae in Conditioned Seawater

Once a week, 1-L glass beakers were filled with aged, filtered seawater (0.45-μm filter) and placed in a temperature-controlled water bath (26°C). Stock solutions of free chlorine (sodium hypochlorite) and chloramine (equimolar quantities of ammonium hydroxide and sodium hypochlorite) were used to dose duplicate beakers with relative concentrations of 50-ppb free chlorine, 300-ppb monochloramine, or 50-ppb dichloramine. Concentrations of free chlorine, monochloramine, and dichloramine were measured with the DPD ferrous titrimetric and colorimetric methods and amperometric titration method.[12] Two beakers contained only undosed seawater and served as controls.

Approximately 15,000 larvae of *C. virginica* were added to each beaker. After 48 h, the larvae were sampled and classified as alive, dead, or abnormal.[13]

DATA AND RESULTS

There is a high correlation between the presence of free chlorine, monochloramine, and dichloramine and all of the stations (Tables I to III). Chlorine-seawater reaction products are present in highest concentrations in the discharge effluent of the East Shore Water Pollution Abatement Facility (Station 1) and at Station 5 which receives chlorine-treated effluent from a small wastewater treatment plant on the west shore of the harbor (Table IV; Figure 2). Monochloramine is the predominant chlorine-produced compound detected at every station.

While analyses of plankton samples indicate that spawn of *C. virginica* is fairly uniform throughout New Haven Harbor, set bag counts reveal greatly reduced set at the discharge sites of wastewater treatment plant effluents, Stations 1 and 5 (Table V; Figure 3).

Table I. Presence of Free Chlorine vs Monochloramine at Sampling Stations in New Haven Harbor

Station	Correlation Coefficient	T value[a] Total	Pairs Used (No.)
1	0.886288	10.3054	31
2	0.852362	8.7773	31
3	0.65098	4.6181	31
4	0.730645	5.7628	31
5	0.801612	7.2206	31
6	0.753224	6.1667	31

[a]Significant at the 0.001 level.

Table II. Presence of Free Chlorine vs Dichloramine at Sampling Stations in New Haven Harbor

Station	Correlation Coefficient	T value[a] Total	Pairs Used (No.)
1	0.788857	6.9122	31
2	0.858219	9.0040	31
3	0.759669	6.2907	31
4	0.790667	6.9568	31
5	0.903834	11.3753	31
6	0.505614	3.15594	31

[a]Significant at the 0.05 level.

Table III. Presence of Monochloramine vs Dichloramine at Sampling Stations in New Haven Harbor

Station	Correlation Coefficient	T value[a] Total	Pairs Used (No.)
1	0.749759	6.1017	31
2	0.815307	7.5826	31
3	0.772672	6.5547	31
4	0.712692	5.4712	31
5	0.73244	5.7933	31
6	0.545489	3.5049	31

[a]Significant at the 0.05 level.

Table IV. Presence of Free Chlorine, Monochloramine, and Dichloramine in New Haven Harbor. Each number is reported in ppb and represents 62 observations.

	Station					
	1	2	3	4	5	6
Free chlorine	47.8	41.5	44.1	38.4	45.0	8.2
Monochloramine	232.19	120.5	101.48	110.0	185.1	25.3
Dichloramine	50.2	28.5	34.58	19.2	42.2	2.4

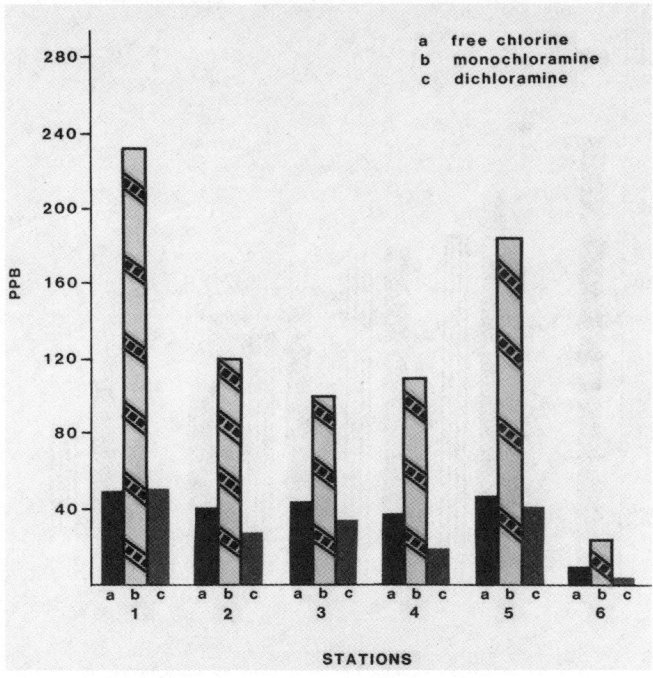

Figure 2. Free chlorine, monochloramine, and dichloramine at sampling stations in New Haven Harbor.

Table V. Spawn and Set of *Crassostrea virginica* at the Sampling Stations

	Mean Number of Oysters per Station[a]	
	Spawn	Set
Station 1	5.96	0.42
2	5.25	4.61
3	4.46	4.30
4	3.72	3.30
5	2.60	0.68
6	2.72	2.07

[a]Data reduction for 57 observations June 1980 – August 1982.

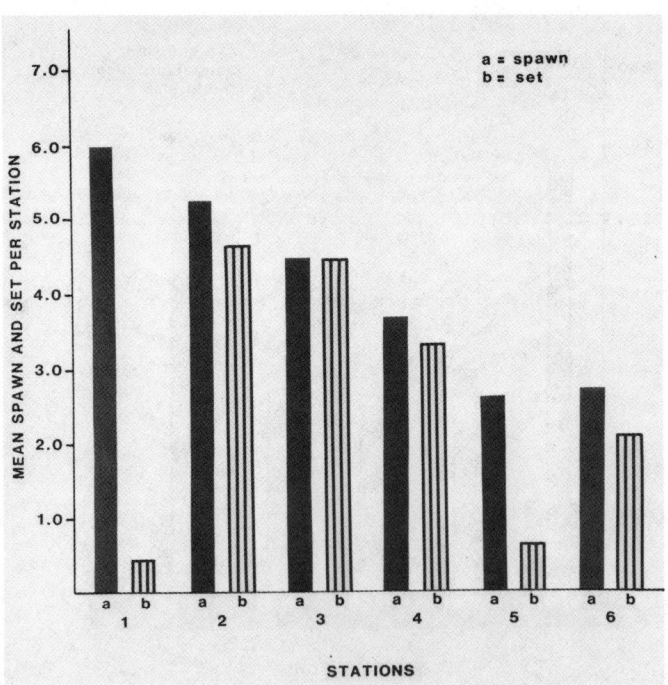

Figure 3. *Crassostrea virginica* spawn and set at the sampling stations.

Most of the laboratory studies show that increased concentrations of chlorine compounds can cause progressively higher mortality of larvae of *C. virginica*. The survival of larvae in filtered seawater differed significantly with respect to the station (Table VI). Larvae exposed to the effluent taken from the treatment plant (Station 1) experienced higher mortality than larvae exposed to water taken from other areas of the harbor. It is notable that larvae exposed to water from Station 5 showed a marked increase in mortality with respect to the larvae exposed to water from Stations 2 to 4 and 6 (Figure 4).

The survival of larvae exposed to seawater dosed with 50-ppb free chlorine, 300-ppb monochloramine, 50-ppb dichloramine, or untreated seawater differed significantly with respect to water treatment (Table VII). Monochloramine proved to be the most toxic compound tested, followed by dichloramine and free chlorine, respectively (Figure 5).

Some laboratory bioassays indicate that occasionally larvae of *C. virginica* are able to survive well when exposed to chlorine-produced compounds at levels that normally cause high mortality of the larvae (Figures 6 and 7).

CONCLUSIONS

Chemical analyses reveal elevated levels of chlorine-seawater reaction products, monochloramine in particular, in areas adjacent to the East Shore Water Pollution Abatement Facility, and in waters adjacent to a smaller wastewater plant that uses chlorine treatment as opposed to other areas of New Haven Harbor. Decreased levels of oyster set are found in these same areas, whereas plankton studies reveal roughly comparable levels of spawning throughout the harbor.

Many laboratory experiments confirm the field observations as survival of most larvae of *C. virginica* decreased upon exposure to water taken from points closest to the sewage plants. Similar results were recorded for larvae in clean filtered seawater dosed with free chlorine, monochloramine, or dichloramine at the highest levels found in the harbor. Increased doses of these selected chlorine species, particularly monochloramine, caused increased mortality to larvae of *C. virginica* in New Haven Harbor and often in the laboratory.

Some laboratory studies indicate that occasionally larvae of *C. virginica* are able to survive well when exposed to chlorine-seawater compounds.

In view of the poor survival of larvae of *C. virginica* in the field and generally in the laboratory when exposed to chlorine species, it would appear worthwhile to investigate further whether certain "groups" of larvae that show higher survival relative to normal field or laboratory mortality patterns are representative of an oyster stock that differs subtly from that found commonly in the harbor or used most frequently in the laboratory bioassays. The potential for use of such a resistant stock, if indeed this attribute is real, should not be overlooked in either theoretical evolutionary studies, where interesting

Table VI. Analysis of Variance Table for Survival of Larvae of *Crassostrea virginica* in Filtered Seawater from the Six Sampling Stations in New Haven Harbor (square-root transformation).

Source	D.F.	S.S.	M.S.	F Ratio
Treatments[a]	5	176.1	35.2	54.6
Error	282	181.8	0.6	
Total	287	5.7		

[a]Six sampling stations.

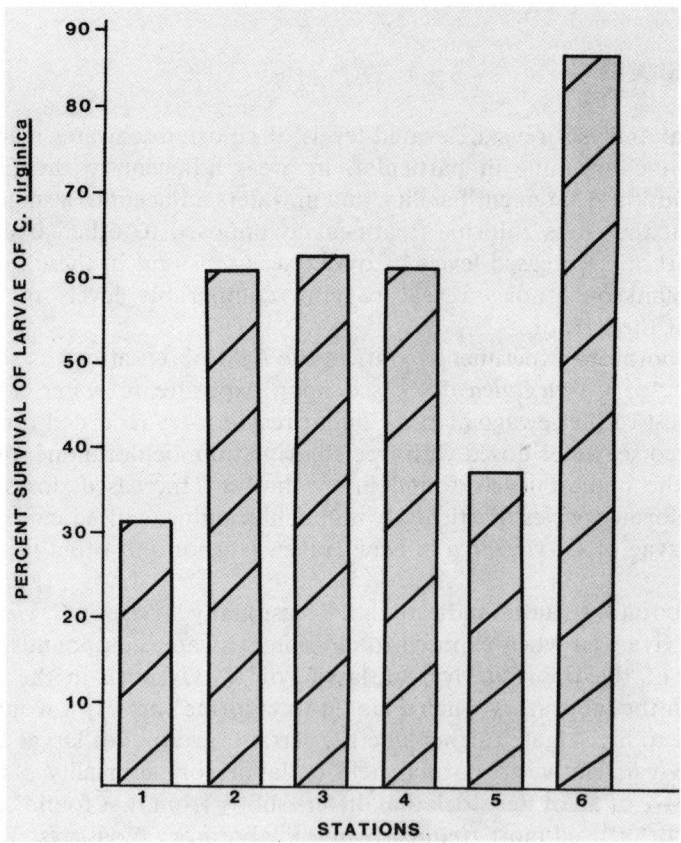

Figure 4. Survival of larvae of *Crassostrea virginica* in water collected at sampling stations.

Table VII. Analysis of Variance Table for Survival of *Crassostrea virginica* Larvae in Conditioned Seawater (square-root transformation).

Source	D.F.	S.S.	M.S.	F Ratio
Treatments[a]	3	96.3	32.1	58.2
Error	188	103.8	0.6	
Total	191	7.4		

[a]Free chlorine (50 ppb), monochloramine (300 ppb), dichloramine (50 ppb) and seawater control.

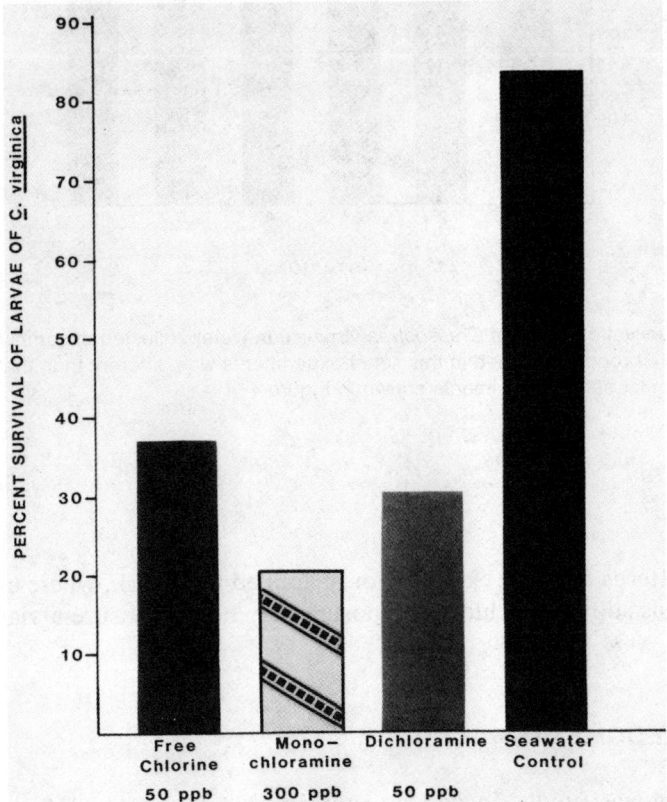

Figure 5. Survival of larvae of *Crassostrea virginica* in filtered and oxidant-dosed seawater.

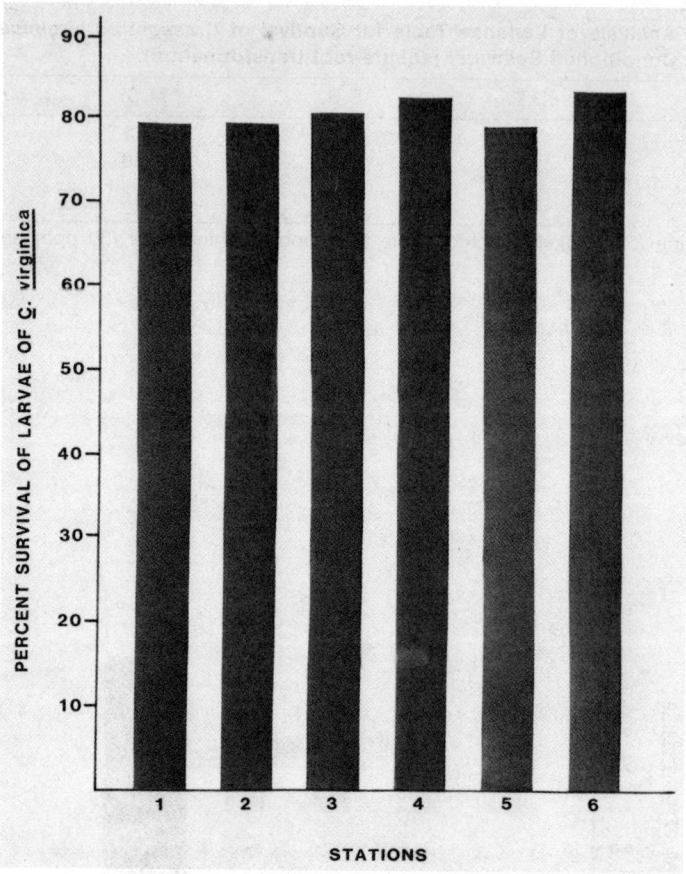

Figure 6. Survival of larvae of *Crassostrea virginica* in water collected at sampling stations. The brood stock used in this set of experiments was different than that used for a similar set of experiments shown in Figure 4.

adaptive patterns could be explored, or in applied fieldwork, where use of such stock in areas subject to chlorine exposure may help to ensure a viable oyster industry.

ACKNOWLEDGMENTS

This study was funded in part by contracts NA-80-FA-C-0026 and NA-81-FA-C-0036 from the National Marine Fisheries Service.

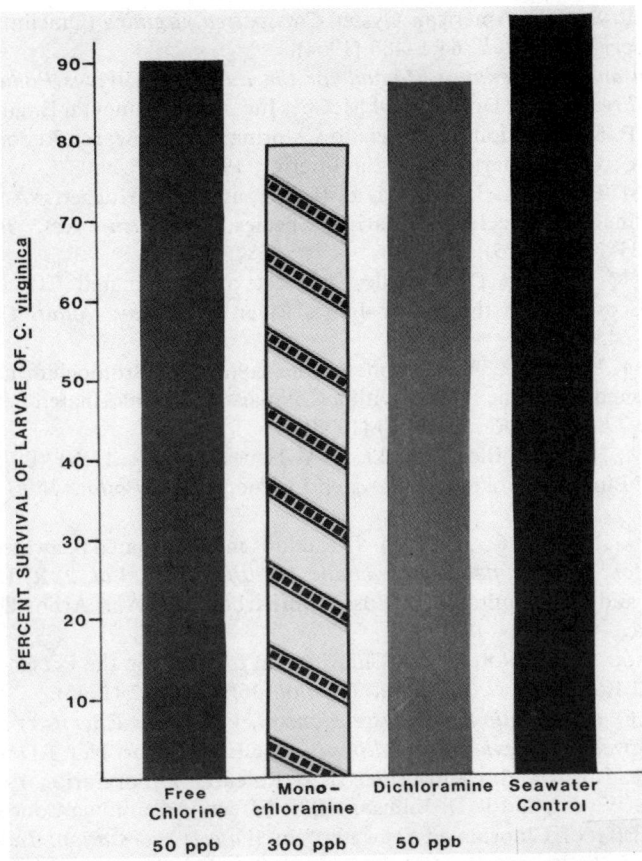

Figure 7. Survival of larvae of *Crassostrea virginica* in filtered and oxidant-dosed seawater. The brood stock used for this set of experiments was different from that used for a similar set of experiments shown in Figure 5.

The work would not have been possible without the cooperation of Long Island Oyster Farms, Inc., particularly J. R. Nelson. The assistance of Charles Johnson of the same firm is deeply appreciated. Additional boat time was provided by Harry Anastasio and Robert Anastasio. John Volk provided valuable advice with regard to oyster beds in New Haven; George Helz, University of Maryland, kindly furnished information with regard to chlorine-seawater chemistry.

REFERENCES

1. Galtsoff, P. S. "The American Oyster *Crassostrea virginica* (Gmelin)," *U.S. Fish Wildlife Serv. Fish. Bull.* 64:1-480 (1964).
2. *Operation and Maintenance Manual for the East Shore Water Pollution Abatement Facility*. (Camp, Dresser, and McGee, Inc., Environmental Engineers, 1980).
3. Galtsoff, P. S. "Reaction of Oysters to Chlorination," Research Report 11 (Washington, DC: U.S. Department of the Interior, 1946).
4. Roberts, M. H., Jr., R. J. Diaz, M. E. Bender, and R. J. Huggett. "Acute Toxicity of Chlorine to Selected Estuarine Species," *J. Fish. Res. Board Can.* 32(12):2525-2528 (1975).
5. Bellanca, M. A., and D. S. Bailey. "Effect of Chlorinated Effluents on the Aquatic Ecosystem in the Lower James River," *J. Water Pollut. Control Fed.* 49(4):639-645 (1977).
6. Roberts, M. H., and R. A. Gleeson. "Acute Toxicity of Bromochlorinated Seawater to Selected Estuarine Species with a Comparison to Chlorinated Seawater Toxicity," *Mar. Environ. Res.* 1:19-30 (1978).
7. Stewart, M. E., W. J. Blogoslawski, R. Y. Hsu, and G. R. Helz. "By-Products of Oxidative Biocides: Toxicity to Oyster Larvae," *Mar. Pollut. Bull.* 10:166-169 (1979).
8. Carpenter, J. H., and C. A. Smith. "Reactions in Chlorinated Seawater," in *Water Chlorination: Environmental Impact and Health Effects, Vol. 2*, R. L. Jolley, H. Gorchev, and D. Hamilton, Jr., Eds. (Ann Arbor, MI: Ann Arbor Science Publishers, Inc., 1979), pp. 195-207.
9. Johannesson, J. K. "Note on the Chlorination of Water in the Presence of Traces of Natural Bromide," *N. Z. J. Sci. Technol.* 36B:600-602 (1955).
10. Lewis, B. G. *Chlorination and Mussel Control, Vol. I. The Chemistry of Chlorination in Seawater: A Review of the Literature*, CERL Report NO. RD/L/N 106/66 (Leatherhead, England: Central Electricity Research Laboratories, 1966).
11. Inman, G. W., Jr., and J. D. Johnson. "The Effect of Ammonia Concentration on the Chemistry of Chlorinated Seawater," in *Water Chlorination: Environmental Impact and Health Effects, Vol. 2*, R. L. Jolley, H. Gorchev, and D. Hamilton, Jr., Eds. (Ann Arbor, MI: Ann Arbor Science Publishers, Inc., 1979), pp. 235-252.
12. *Standard Methods for the Examination of Water and Wastewater*, 14th ed. (Washington, DC: American Public Health Association, 1976).
13. Loosanoff, V. L., H. Davis, and P. Chanley. "Dimensions and Shapes of Larvae of Some Marine Bivalve Mollusks," *Malacologia* 4(2):351-435 (1966).

CHAPTER 42

Delayed Effects of Chlorine on Early Life Stages of the Mayfly

Sylvia A. Murray, Kenneth J. Tennessen, and Susan M. Laborde

Chlorination is one method used to control biofouling in power plants. The introduction of chlorine into receiving waters, however, may be harmful to some organisms.[1,2] Mayflies, an important fish food, are one of the affected organisms, although chlorine tolerances seem to vary among the different species, their life stage, and the size of the organisms.[2] In addition, chlorine is a suspected causal agent for altering tracheal gills in some aquatic insect larvae.[3] Since many power plants do not chlorinate the entire year, it is important to know if delayed effects of chlorination do exist biologically and if predisposal to chlorine (i.e., prior exposure in the embryo stage) changes the nymphs' chlorine tolerance on subsequent exposure to chlorination. No information is available on this subject.

The specific objectives in this study were to determine the (1) impact of low chlorination concentration on hatching success of mayfly eggs, (2) delayed effect of chlorine on nymph survival, (3) effect of chlorine predisposal and subsequent chlorination on nymph survival, and (4) effects of chlorination and chlorine predisposal on nymph growth and gill development.

MATERIALS AND METHODS

Adult females of *Hexagenia bilineata* Say were collected between 9:00 and 11:00 p.m. on July 12, 1981, as they were attracted to a lighted pier along Shoal Creek Embayment near Killen, Alabama. Females released eggs into large glass dishes filled with dechlorinated water. Eggs were pooled, stirred, and pipetted into 5-mL plastic petri dishes, about 1000 per dish. The first chlorination study began 2 d later when the eggs adhered to the bottom of the dishes. Chlorine was administered using a sodium hypochlorite solution. Preliminary work showed that resident time for chlorine in the containers used in these studies was no more than 180 min. Free available chlorine (FAC) was determined by the diethyl-*p*-phenylenediamine method.[4] All experiments were conducted outdoors with ambient photoperiod (14 light:10 dark) and temperature fluctuations.

For the hatching study, there were 33 dishes for each of the chlorine treatments, namely, 0, 0.21 ± 0.02, and 2.39 ± 0.14 mg/L FAC. Solutions were changed daily for 7 d. Temperatures ranged from 24.1 ± 0.4 to 29.4 ± 0.5°C. Hatching success was enumerated on the tenth day. The data were analyzed by a one-way analysis of variance.

For the survival studies, approximately 30 newly hatched nymphs from unchlorinated eggs and eggs chlorinated at 0.2 mg/L FAC were placed in small petri dishes containing 5 mL of one of the treatment solutions: 0, 0.5 ± 0.006, and 0.10 ± 0.004 mg/L FAC. Solutions were changed daily during the 96-h study. Nymphs were not fed. Temperatures ranged from 22.0 ± 0.8 to 27.6 ± 0.8°C. Mortality was determined by no response after nymphs were gently probed for several minutes. Data were analyzed by a two-way balanced analysis of variance with 15 replicates for each treatment combination.

For the growth studies, newly hatched nymphs from unchlorinated eggs and eggs chlorinated at 0.2 mg/L FAC were pipetted into 90- by 50-mm dishes (about 50 per dish) containing 3-mm deep, settled, autoclaved substrate[5] in 150 mL of dechlorinated water. Nymphs were fed weekly.[5] Solutions were changed daily. The nymphs were allowed to acclimate to substrate for 22 d. Half the nymphs were then treated with 0.15 ± 0.01 mg/L FAC daily for 18 d. Samples were preserved in 5% formalin. The dead nymphs were too decomposed for reliable enumeration. In addition, live and dead nymphs could have been siphoned out when the solutions were changed. Because it was not possible to account for all the nymphs, mortality rates could not be determined. Head width on the recovered live nymphs was measured with an ocular micrometer. The fourth left gill (abdominal) was removed, mounted on a slide, and measured with a filar micrometer. Supplemental information for the control group was obtained from preserved nymphs from a previous study[6] cultured at approximately the same temperatures as the ones in this study. However, only two nymphs were recovered from the group of nymphs not predisposed to chlorine, but were subsequently exposed to 0.1 mg/L FAC. Twenty nymphs were examined for each of the other treatment combinations. Data were analyzed by a two-way unbalanced analysis of variance.[7]

RESULTS AND DISCUSSION

Nearly 50% of the nymphs had hatched on the tenth day from both the 0.2 and 2.4 mg/L FAC concentrations. Rather than lose the nymphs in the solution changes at 0.2 mg/L FAC, these nymphs were used for the survival and growth studies, thereby terminating the hatching study. Although the hatching peak would probably have occurred on the 14th day,[6] the data indicate no effect of intermittent low-level chlorination on hatching success of mayfly eggs (Table I).

Nymph survival was inversely proportional to chlorine concentration less than 0.14 mg/L FAC, the current maximum proposed federal chlorine stan-

Table I. Effect of Intermittent Low-Level Chlorination (9 d) on Hatching Success of Mayfly Eggs. (Means and standard errors are indicated.)

Initial chlorine concentration (mg/L FAC)[a]	Hatch (%)
0	48.54 ± 3.10
0.21 ± 0.02	45.03 ± 3.10
2.39 ± 0.14	48.40 ± 3.10
Effect of chlorine	NS[b]

[a]Free available chlorine.
[b]Not significant.

dard.[8] Nymphs were extremely sensitive to 0.10 mg/L FAC, with only 34% surviving (Table II). Predisposal of eggs to 0.2 mg/L FAC had no apparent effect on nymph survival in the absence of further exposure to chlorination (0 mg/L FAC in Figure 1). This information indicates an absence of delayed chlorine effect on nymph mortality. However, nymph survival decreased by about 18% (significance >99.9%) with subsequent exposures to 0.10 mg/L FAC (see P1,P2, Table II). Tolerance to chlorine deteriorated with chlorine predisposition by 14% in the control group, by 50% for exposures of 0.05 mg/L FC, and by 78% for exposures of 0.10 mg/L FAC (Table II, CP interaction term). In addition, possible cannibalism was observed more in the predisposed groups where active nymphs attached their mouths to weak, moribund, or dead nymphs. This behavior was observed for nymphs exposed to only 0.05 mg/L FAC. These studies indicate that freshly hatched nymphs are very sensitive to intermittent chlorine treatment. However, it is not known what effect, if any, this may have on fish populations near a chlorinating power plant.

Since mortality rates of the older nymphs could not be determined, it is not known whether older nymphs (40-d old) were more resistant to chlorination than the 96-h nymphs. As for the survivors, chlorination had no apparent effect on the growth of nymphs, as measured by head width and gill development (Table II and Figure 2). Gills were not stunted as might be expected.[3] Even gills from treated nymphs were feathery and well developed. However, gill growth was significantly greater (by 51%) in nymphs predisposed to chlorine. Since predisposal had no effect on head growth, a larger than normal gill-length:head-width ratio resulted in the predisposed groups. Gills of mayflies are known to be important respiratory organs,[9,10] although in *Hexagenia* they also function as water-circulating pumps to bring food and oxygenated water into the burrow. The data indicate irreversible morphological change in the gills. Chlorine predisposition and subsequent chlorine exposure resulted in

Table II. Effect of Predisposal to Chlorine and Subsequent Chlorination on Survival (96 h) and Growth (40 d) of Mayfly Nymphs

Treatment	Percent survival (96 h)	Head width (mm) (40 d)	Gill length (mm) (40 d)	Gill length/ head width
Initial chlorine concentration, C				
0 mg/L FAC,[1] C1	88.43	0.46	0.77	1.66
0.05 mg/L FAC, C2	62.86			
0.10 mg/L FAC, C3	34.36			
0.15 mg/L FAC, C4		0.54	0.94	1.78
Significance[2] of C	3	NS[4]	NS	NS
95% L.S.D. for C	6.62	NS	NS	NS
Predisposition, P				
Not Predisposed, P1	70.84	0.43	0.63	1.42
Predisposed, P2	52.93	0.52	0.95	1.87
Significance of P	3	NS	3	3
95% L.S.D. for P	5.40	NS	0.16	0.19
C × P Interaction				
C1 P1	90.26	0.42	0.62	1.43
C1 P2	86.60	0.50	0.95	1.94
C2 P1	75.84			
C2 P2	49.87			
C3 P1	46.41			
C3 P2	22.31			
C4 P1		0.58	0.75	1.37
C4 P2		0.54	0.95	1.81
Significance of CP	3	NS	NS	NS
95% L.S.D. for CP	9.36	NS	NS	NS

[1]Free available chlorine.
[2]Significance determined by analysis of variance.
[3]PR > F = 0.01.
[4]Not significant.

longer gills, a phenomenon apparently triggered during embryonic development only in eggs exposed to chlorine. Irreversible morphological change has also been observed in salt-uptake studies, where the length of anal papillae was inversely related to chlorine ion concentration in *Culex* mosquito larvae.[9] Apparently, these irreversible morphological changes are associated with environmental stresses, causing biochemical and physiological changes during embryogenesis. Further studies should investigate morphological-physiological relationships of more mature predisposed nymphs and adults with regard to variable power plant chlorination schedules.

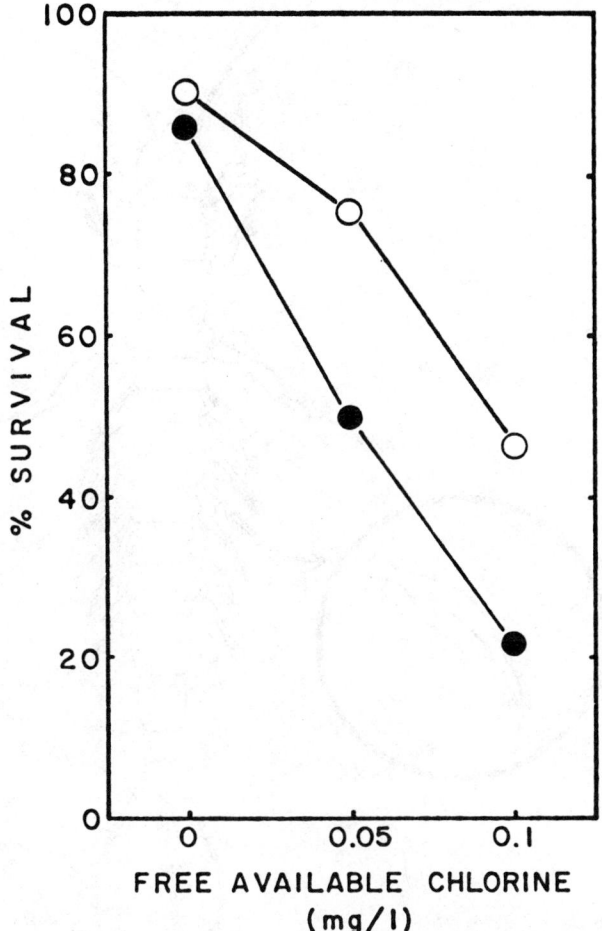

Figure 1. Effect of chlorine and predisposal to chlorine in the egg stage on nymph survival. Open circles, not predisposed; closed circles, predisposed.

CONCLUSIONS

Low-level intermittent chlorination does not affect hatching success of mayfly eggs or head growth and gill growth development in mayfly nymphs. Delayed mortality effects were not observed in newly hatched nymphs, but chlorine tolerance deteriorated in nymphs predisposed to chlorine in the egg stage and further exposed to chlorine after hatching. Gills were not affected by chlorine exposure, but gill development was enhanced in nymphs predisposed to chlorine in the egg stage and subsequently exposed to chlorine 3 weeks after hatching.

538 WATER CHLORINATION

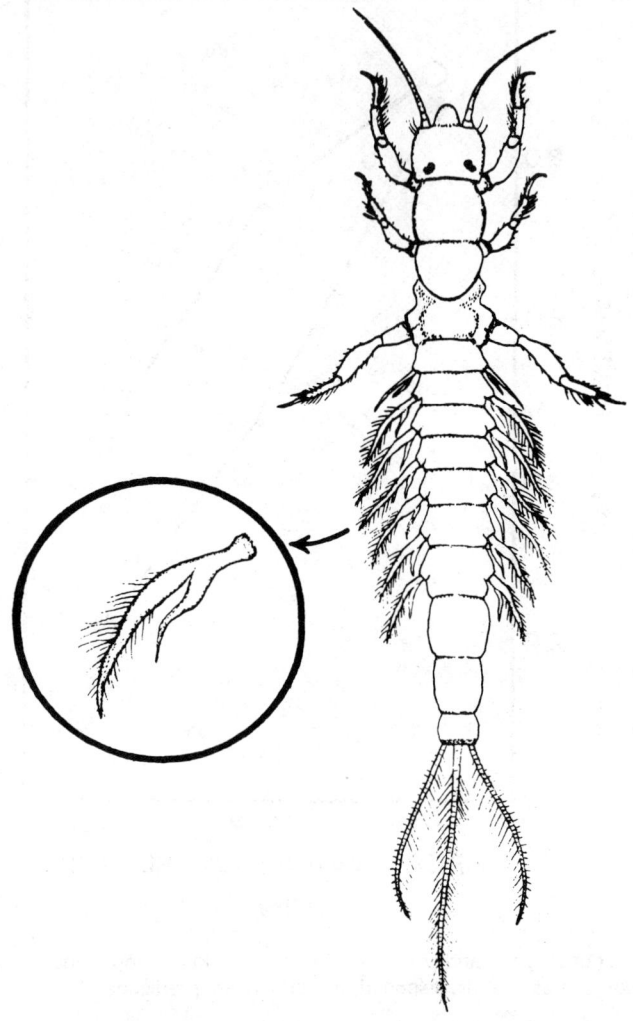

Figure 2. Nymph of *Hexagenia bilineata* Say, approximately eighth or ninth instar, with enlargement of fourth abdominal gill. Length excluding antennae and caudal filaments = 3.4 mm.

ACKNOWLEDGMENTS

This work was conducted as part of the Federal Interagency Energy/Environmental Research and Development Program with funds administered through the Environmental Protection Agency (EPA Contract No. 79-D-

X0511/TV-50447A). The authors gladly acknowledge Billy G. Isom and R. J. Ruane for making this study possible, and A. H. Rhodes for technical assistance.

REFERENCES

1. Mattice, J. S., and H. E. Zittel. "Site Specific Evaluation of Power Plant Chlorination," *J. Water Pollut. Control Fed.* 48:2284–2308 (1976).
2. Gregg, B. C. *The Effects of Chlorine and Heat on Selected Stream Invertebrates*, Ph.D. Thesis (Blacksburg: Virginia Polytechnic Inst. and State University, 1974).
3. Simpson, K .W. "Abnormalities in Tracheal Gills of Aquatic Insects Collected from Streams Receiving Chlorinated or Crude Oil Wastes," *Freshwater Biol.* 10:581–583 (1980).
4. *Standard Methods for Exmaination of Water and Wastewater*, 13th ed. (Washington, DC: American Public Health Association, 1971).
5. Wright, L. L., and J. S. Mattice. "Effects of Temperature on Adult Size and Emergence Success of *Hexagenia bilineata* Under Laboratory Conditions," *J. Freshwater Ecol.* 1:27–39 (1981).
6. Tennessen, K. J. Unpublished data, (Muscle Shoals, AL: Tennessee Valley Authority, 1980).
7. Steel, R. G. D., and J. H. Torrie. *Principles and Procedures of Statistics*, (New York: McGraw-Hill Book Co., Inc., 1960) Chap. 13.
8. Costle, D. M., R. B. Schaffer, J. Lunn, and T. Wright. *Development Document for Effluent Limitation Guidelines and Standards for Steam Electric Point Source Category*, EPA 440/1-80-029-B (Washington, DC: U.S. Environmental Protection Agency, 1980).
9. Wigglesworth, V. B. *The Principles of Insect Physiology* (London: Methuen & Co., Ltd., 1965).
10. McCaferty, W. P., and A. V. Provonsha. *Aquatic Entomology* (Boston: Science Books International, 1981).

CHAPTER 43

Inhibition of Phytoplankton Photosynthesis by Chlorinated Sewage in the James River

Soon Lin Ho and Morris H. Roberts, Jr.

Chlorine is used in water and wastewater treatment for (1) disinfection of wastewaters for the protection of public health and (2) control of biofouling in cooling water systems.[1] Chlorine has been and remains the principal disinfectant for drinking water and wastewater. This is evidenced by the increasing use of chlorination. Indeed, the use of chlorination for water and wastewater treatment has increased 20 to 40% since 1975.[2]

Morgan and Stross[3] were the first researchers to report an adverse effect of power plant chlorination on estuarine phytoplankton. Since that time, extensive evidence has been developed that both freshwater and estuarine phytoplankton or communities are extremely sensitive to chlorine, responding adversely to concentrations as low as 0.05 mg/L total residual chlorine (TRC).[4-9] All these studies have examined the effect of chlorinated water without sewage effluent.

Chlorine was the alleged cause of a fish kill in the lower James River, Virginia, during 1973.[10] In response to this event, the impact of chlorine on natural mixed populations of phytoplankton was investigated.[6,11] Thus far, however, no one has examined the effect of chlorinated sewage on resident phytoplankton communities in the James River. The research described in this chapter was designed to fill that gap. The objective was to evaluate the toxicity of chlorinated sewage discharged from the James River Sewage Treatment Plant on the photosynthetic activity of indigenous phytoplankton communities in the James River.

MATERIALS AND METHODS

Natural phytoplankton communities were exposed to five concentrations of chlorinated sewage, namely, 1.25, 2.5, 5.0, 10.0, and 20.0% (1:80, 1:40, 1:20, 1:10, and 1:5 dilutions), and 20% solutions (1:5 dilutions) of unchlorinated sewage and dechlorinated sewage. The percent concentration refers to the final concentration of the sewage after dilution with the phytoplankton samples.

For dilutions less than 1:5 (20% sewage), distilled water was added so that the volume of freshwater added to the mixture was constant. For the control, a 20% distilled water mixture was used. Thus, the salinity was held constant in every treatment.

Radiolabeled bicarbonate was used to estimate the amount of carbon uptake.[12,13] After addition of the sewage, samples were inoculated with 0.1 to 0.2 μCi of ^{14}C-labeled bicarbonate (New England Nuclear, Boston) per 5 mL, and then incubated in a light incubator at the ambient temperature of the site from which the phytoplankton community was collected. Triplicate samples were prepared for each treatment with two samples incubated for 2 h at a light intensity of 190 $\pm$ 10 μEm^{-2}s^{-1} and one in the dark. Incubation was terminated by adding a mixture of 1 mL of 95:5 (v/v) methanol:glacial acetic acid. After termination of the incubation the samples were transported to the laboratory. Chlorophyll concentration was determined by extraction of separate samples with a dimethylsulfoxide:acetone:water mixture;[14] fluorescence was measured with a Turner III fluorometer, which had been calibrated spectrophotometrically.[15] The total inorganic carbon of the samples was also determined.[16]

The samples were dried at 75°C in the laboratory. After drying, sample residues were dissolved in 2 mL of distilled water and solubilized with 10 mL of scintillation fluid (Aquasol 2). The samples were then counted in a Beckman LS150 scintillation spectrometer.

Surface samples (depth, 1 m) of phytoplankton communities were collected at two stations in the James River, designated as J11 and HRS (Figure 1). Samples were collected from a stratified water column on August 6, 1983, and from a destratified water column on August 12, 1983. Station J11 (76° 33.66'W, 37° 3.14'N) is located along the channel 4000 m from the outfall of the James River Sewage Treatment Plant at Menchville, Virginia, whereas station HRS (76° 32.20', 37° 4.20'N) is within 100 m of the effluent diffuser. The temperature and salinity of the samples were determined at the time of collection with standard equipment.

Grab samples of chlorinated and unchlorinated sewage effluent were collected from the James River Sewage Treatment Plant in the morning 1 or 2 d prior to each experiment. Unchlorinated sewage was collected at the discharge of the secondary clarifier, whereas chlorinated sewage was collected from the discharge of the 30-min contact tanks. Samples were placed in 4-L amber bottles, transported to the laboratory in an icebox, and stored in a refrigerator until used in an experiment.

The chlorinated and unchlorinated sewage effluents collected as stock sewage were used without any pretreatment. In addition, a sample of chlorinated sewage was dechlorinated by adding a suitable amount of sodium thiosulfate solution to react quantitatively with 2 mg/L TRC, which is the concentration after 30-min contact time at the James River Sewage Treatment Plant.[17]

All sewage effluents collected were analyzed for total chlorine residuals and the presence of inorganic nutrients. Total chlorine residuals in the sewage effluent were determined by amperometric titration with phenylarsine oxide at

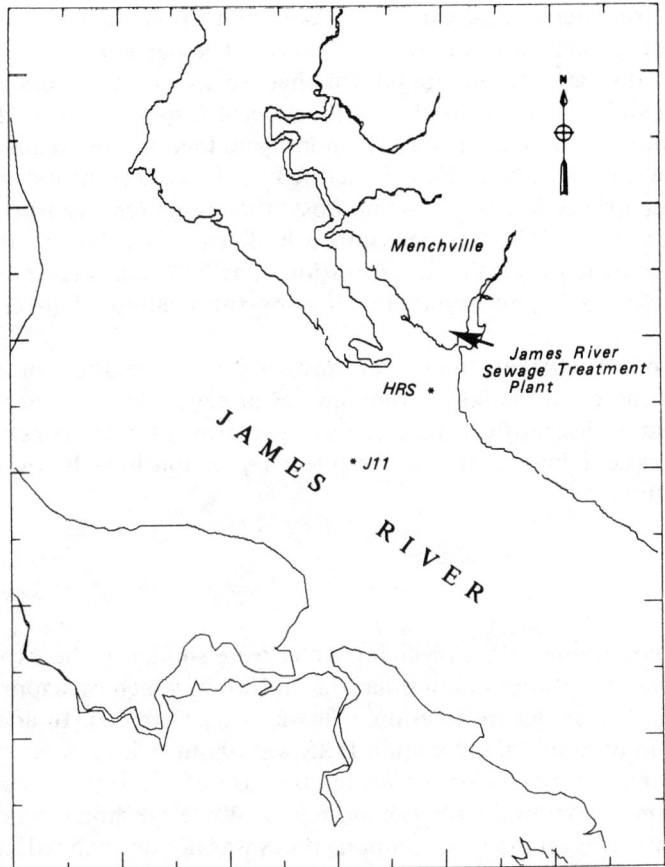

Figure 1. Location of stations J11 and HRS on James River where phytoplankton samples were collected for exposure to sewage effluent.

pH 4 using a Fischer-Porter amperometric titrator.[18] Inorganic nitrogen was determined using an autoanalyzer in the North Shore Laboratory of the Hampton Roads Sanitation District Commission located in Menchville, Virginia.

Nutrients in the estuarine water (nitrate, nitrite, and ammonia) were determined using autoanalyzer techniques. Procedures for nitrate and nitrite were similar to standard colorimetric cadmium reduction–diazotization methods,[19] except that the sample-to-reagent ratio was increased to improve sensitivity. Analytical determination of the ammonium ion was based on a modification of the phenolhypochlorite method.[20]

For each treatment, total carbon uptake due to photosynthesis was calculated according to the formula of Strickland.[12] The percent response for each treatment is the ratio of the uptake for that treatment to the uptake for the control. An arcsine transformation of the percent response was used to linearize the relationship between percent response and logarithmic toxicant concentration. The photosynthetic EC_{50} is defined as the concentration of toxicant that causes a 50% reduction in the photosynthetic rate relative to the control. The photosynthetic EC_{50} was determined by linear regression of the percent response against log toxicant concentration. The 95% confidence limits were determined for each photosynthetic EC_{50} by the method of inverse prediction.[21]

To determine whether there was any difference in the photosynthetic response of the phytoplankton communities at each station to each toxicant before and after destratification, a test was performed for heterogeneity of the slopes[22] and the Y-intercepts[23] of the linear regression lines before and after destratification.

RESULTS

Nutrient conditions in the receiving water were similar at the two sampling times with the exception of ammonia concentration, which was approximately twice as high before destratification as it was after (Table I). In addition, the concentration of ammonia at station HRS was about twice that at station J11 at both sampling times. This difference because of location may reflect the input of ammonia from the sewage discharge. While the ammonia concentration of the effluent on the field sampling dates was not determined, the ammonia concentration of the effluent 1 to 2 d prior to the field sampling dates was high (Table II), and presumably remained high throughout the first half of August.

The chlorophyll concentration of the community at station HRS was larger than that at station J11. The primary productivity potential, expressed as mg $C\ L^{-1}\ h^{-1}$, was similar at the two stations, but the assimilation rate, expressed as mg C/mg Chl a per h, was reduced by nearly half at station HRS (Table III).

The chlorinated and unchlorinated sewage effluent samples were quite similar at each sampling time, with ammonia representing the principal form of inorganic nitrogen. Nitrite-nitrate concentrations were negligible in both types of effluents collected for the first test and below detection limits during the second test. The chlorinated effluent contained the prescribed 2 mg/L TRC at the end of the contact tank (Table II).

Phytoplankton communities from both stations in the James River were equally sensitive to chlorinated sewage. The EC_{50} estimated from the phytoplankton communities from station J11 before destratification was 4.12% sewage; after destratification at the same station it was 2.72% sewage (Table IV; Figure 2). There was no significant difference in the dose-response lines

Table I. Nutrient Concentrations in Receiving Water and Sewage Effluent Samples; Nutrient Level (μM) in the Phytoplankton Samples Collected in the James River in 1983

Station	Destratification	Date	NH_4^+	NO_2^-	NO_3^-
J11	Before	August 6	3.01	0.93	3.12
	After	August 12	1.96	1.67	3.19
HRS	Before	August 6	7.6	1.02	3.31
	After	August 12	3.34	1.05	2.20

Table II. Nutrient Concentrations in Receiving Water and Sewage Effluent Samples; Total Residual Chlorine (TRC) and Inorganic Nitrogen Levels in Sewage Effluents Collected at the James River Sewage Treatment Plant

	Date of Collection	
Effluent	August 4	August 11
Chlorinated sewage		
TRC, mg/L	2.00	1.92
$NH_3 - N$, μM	181.7	248.2
$(NO_2 - NO_3) - N$, μM	9.7	<2.8
Unchlorinated sewage		
TRC, mg/L	0.0	0.0
$NH_3 - N$, μM	168.6	227.5
$(NO_2 - NO_3) - N$, μM	10.2	<2.8

Table III. Chlorophyll Concentration and Primary Production Potential of the Control Samples of Phytoplankton Communities Exposed to Chlorinated Sewage

Sample	Before or after destratification	Chlorophyll concentration (μg Chl a/L)	Primary production	
			(mg C $L^{-1} h^{-1}$)	(mg C/mg Chl a per h)
J11	Before	9.25	0.217	23.48
	After	8.70	0.110	12.59
HRS	Before	11.28	0.144	12.79
	After	20.64	0.178	8.64

before and after destratification for samples collected at this station. Similarly, at station HRS, the EC_{50} was a 4.27% sewage before destratification and 3.02% sewage after destratification (Table IV; Figure 3). Again there was no significant difference in the dose-response lines for the phytoplankton communities at HRS before and after destratification.

Table IV. Photosynthetic EC_{50} and Confidence Limits for Phytoplankton Communities from James River Exposed to Chlorinated Sewage

Sample	Destratification	EC_{50} (% sewage)	Lower interval[a] (% sewage)	Upper interval[a] (% sewage)
J11	Before	4.12	2.76	6.10
	After	2.72	0.98	5.36
HRS	Before	4.27	2.36	7.56
	After	3.02	1.30	5.55

[a]95% Confidence level.

Unchlorinated sewage was less toxic to phytoplankton at both stations than all tested dilutions of chlorinated sewage. Unchlorinated sewage at 1:5 dilution reduced photosynthesis by 25 to 35% in communities collected at stations J11 and HRS before and after destratification. At the same dilution, chlorinated sewage decreased photosynthesis at both stations by 92 to 95% (Figures 2 and 3).

The toxicity of chlorinated sewage to phytoplankton at both stations was reduced after dechlorination with thiosulfate. Nearly 50% inhibition of photosynthesis was observed in communities collected at station J11 before and after destratification when exposed to a 1:5 dilution of dechlorinated sewage (Figure 2). Photosynthesis of samples collected at station HRS was inhibited 35 to 45% by a 1:5 dilution of dechlorinated sewage before and after destratification (Figure 3). There was no significant difference between the response to the dechlorinated and to the unchlorinated sewage at either station. Further, the responses at both stations to either toxicant were not significantly different. However, both unchlorinated and dechlorinated sewage were significantly less toxic than a 1:5 dilution of chlorinated sewage.

DISCUSSION

This study showed a definite toxicity of chlorinated sewage effluent to natural phytoplankton communities. Chlorinated sewage effluent toxicity to fish is already well established.[1,10,24,25] Chlorinated estuarine water without sewage is also known to be toxic to single species and mixed communities of phytoplankton.[5,6,11]

The average EC_{50} for the phytoplankton communities at both stations was 3.5% sewage. Expressed in terms of chlorine concentration, the EC_{50} was 0.07 mg/L TRC. At the James River Sewage Treatment Plant, the near-field dilution for sewage effluent at the point of discharge into the river is 1:20.[17,26] At this dilution of an effluent chlorinated to 2 mg/L TRC, the residual chlorine

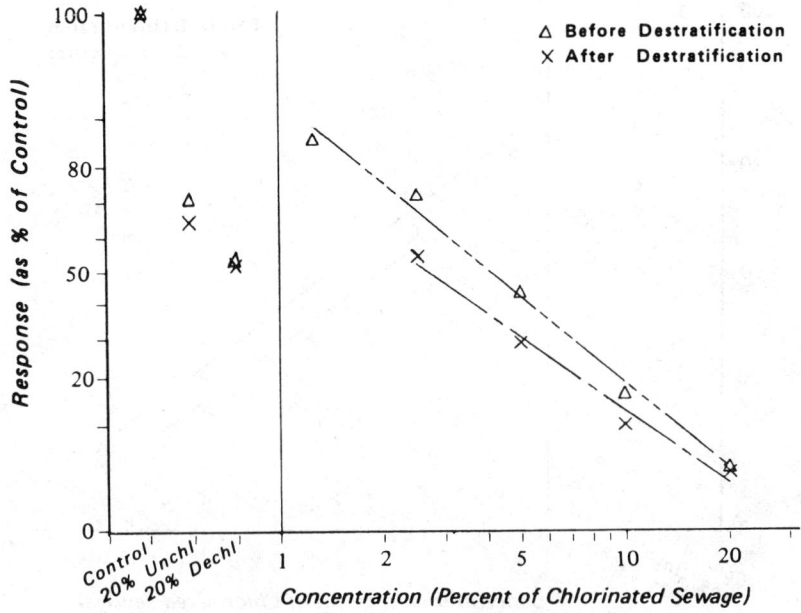

Figure 2. Concentration and response curve of James River phytoplankton at station J11 to sewage effluent before and after destratification (Unchl, unchlorinated sewage; Dechl, dechlorinated sewage).

concentration at the point of discharge would be 0.1 mg/L TRC. Thus, one would expect some inhibition of photosynthesis for the phytoplankton communities in the James River, particularly in the vicinity of station HRS. At locations more remote from the discharge point, the chlorine level would be progressively diluted and result in decreased toxicity. A comparison of the primary productivity potential at station J11 vs HRS (Table II) seems to support this expectation, at least during the period of this study.

The EC_{50} for chlorinated sewage effluent, expressed as 0.07 mg/L TRC (as previously calculated), is comparable to that previously estimated for estuarine phytoplankton species exposed to chlorinated seawater. Over a range of three salinities and three temperatures, four nanoplankton species in unialgal culture exhibited a 4-h EC_{50} from 0.01 to 0.47 mg/L.[6] Mixed phytoplankton communities have exhibited 2- or 3-h EC_{50}s to chlorinated water of 0.09 to 0.10 mg/L.[9,27]

The presence of chlorine and its associated residuals does not account for the total toxicity of the chlorinated sewage, because unchlorinated sewage samples were also found to be toxic to the phytoplankton communities at a 1:5

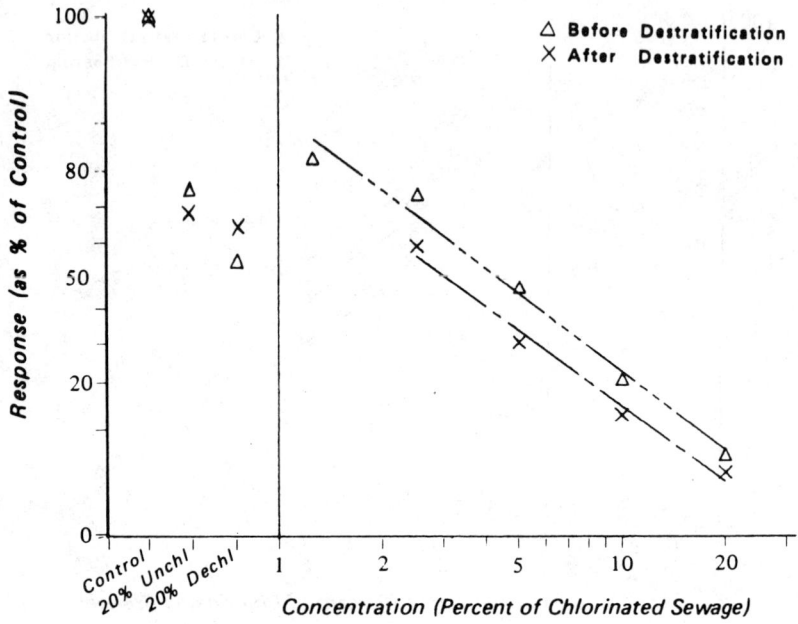

Figure 3. Concentration and response curve of James River phytoplankton at station HRS to sewage effluent before and after destratification (Unchl, unchlorinated sewage; Dechl, dechlorinated sewage).

dilution. However, there are no data to assess the impact of sewage alone at the typical 1:20 dilution calculated to occur at the diffuser of the James River Plant. Since the EC_{50}s observed (which approximate the typical 1:20 dilution found at the diffuser), when expressed in terms of chlorine content, were essentially the same as those previously determined for chlorinated estuarine water, one might suggest that there would be no effect of a 1:20 dilution of sewage on the phytoplankton community. It would be appropriate to verify this suggestion by direct testing.

Dechlorination of the chlorinated sewage removed almost all the toxicity attributable to chlorination; however, the combination of chlorination-dechlorination did not produce a decrease in toxicity compared with unchlorinated sewage. This observation conflicts with the results of Esvelt et al.,[24] who found that chlorination-dechlorination of sewage effluent with sodium bisulfite reduced the toxicity to fish below that of unchlorinated sewage. The difference most probably stems from the different sources of sewage effluent in the two studies, although one cannot rule out an effect of the reducing agent.

It has been established that the James River, in the vicinity of the stations compared in this study, is stratified during neap tide, with a small but recognizable halocline. Over a 3-d period during the spring tide, the water column becomes destratified, a condition that lasts only 3 to 5 d.[28] This turnover and homogenization of the water column affects the distribution of all nutrients, dissolved oxygen concentration, and phytoplankton distribution, as well as salinity. Therefore, one might expect that the surface phytoplankton communities before and after destratification might differ in physiological state to the extent that they would respond differently to a pollutant such as chlorinated sewage. The data presented here give no statistically significant evidence of such an effect.

ACKNOWLEDGMENTS

We wish to express our appreciation to Dr. L. W. Haas for allowing the senior author to work aboard the R/V Ridgely-Warfield during cruises supported by the National Science Foundation grant OCE81-20842 to Dr. Haas. The College of William and Mary provided a minor research grant to the senior author to conduct the study. We also thank N. LeBlanc of the Hampton Roads Sanitation District (HSRD) Commission for permitting us to collect sewage effluent from the James River Sewage Treatment Plant and for copies of plant records regarding sewage effluents. Susan Sturm of HRSD kindly performed nutrient analyses of the sewage effluents. Dr. K. L. Webb performed the nutrient analyses of the phytoplankton samples used in the studies. The senior author also wishes to thank the Rotary Foundation of Rotary International, Evanston, Illinois, for awarding him a Rotary Graduate Fellowship for International Understanding, which allowed him to study in the United States. This chapter is Contribution No. 1238 from the School of Marine Science, Virginia Institute of Marine Science, College of William and Mary, Gloucester Point, Virginia.

REFERENCES

1. Brungs, W. A. "Effects of Residual Chlorine on Aquatic Life," *J. Water Pollut. Control Fed.* 45:2180-2193 (1974).
2. Davis, W. P., and J. A. Fava. "Interaction of Aquatic Ecosystem Components with Chlorination: An Overview," in *Water Chlorination: Environmental Impact and Health Effects, Vol. 4*, R. L. Jolley, W. A. Brungs, J. A. Cotruvo, R. B. Cumming, J. S. Mattice, and V. A. Jacobs, Eds. (Ann Arbor, MI: Ann Arbor Science Publishers, Inc., 1983), pp. 791-796.
3. Morgan, R. P., and R. F. Stross. "Destruction of Phytoplankton in the Cooling Water Supply of a Steam Electric Station," *Chesapeake Sci.* 10:165-171 (1969).
4. Hamilton, D. H., D. A. Flemer, C. V. Keefe, and J. A. Mihursky. "Power Plants: Effects of Chlorination on Estuary Primary Productivity," *Science* 169:197-198 (1970).

5. Hirayama, K., and R. Hirano. "Influence of High Temperature and Residual Chlorine on Marine Phytoplankton," *Mar. Biol.* 7:205-213 (1970).
6. Bender, M. E., M. H. Roberts, Jr., R. J. Diaz, and R. J. Huggett. "Effects of Residual Chlorine on Estuarine Organisms," in *Biofouling Control Procedures, Technology and Ecological Effects*, L. Jensen, Ed. (New York: Marcel Dekker, Inc., 1977), pp. 101-108.
7. Brook, A. J., and A. L. Baker. "Chlorination at Power Plants: Impact on Phytoplankton Productivity," *Science* 176:1414-1415 (1972).
8. Brooks, A. S., and G. L. Seegert. "The Toxicity of Chlorine to Freshwater Organisms under Varying Environmental Conditions," in *Water Chlorination: Environmental Impact and Health Effects, Vol. 1*, R. L. Jolley, Ed. (Ann Arbor, MI: Ann Arbor Science Publishers, Inc., 1978), pp. 261-282.
9. Davis, M. H., and J. Coughlan. "Response of Entrained Plankton to Low Level Chlorination at a Coastal Power Station," in *Water Chlorination: Environmental Impact and Health Effects, Vol. 2*, R. L. Jolley, H. Gorchev, and D. H. Hamilton, Jr., Eds. (Ann Arbor, MI: Ann Arbor Science Publishers, Inc., 1978), pp. 369-376.
10. Bellanca, M. A., and D. S. Bailey. "Effect of Chlorinated Effluents on the Aquatic Ecosystems of the Lower James River," *J. Water Pollut. Control Fed.* 49:639-645 (1977).
11. Roberts, M. H., Jr. "Bioassay Procedures for Marine Phytoplankton with Special Reference to Chlorine," *Chesapeake Sci.* 18:137-139 (1977).
12. Strickland, J. D. H. "Measuring the Production of Marine Phytoplankton," *Bull. Fish. Res. Bd. Can.* 122:1-172 (1960).
13. Goldman, J. C., C. D. Taylor, and P. M. Gilbert. "Nonlinear Time-Course Uptake of Carbon and Ammonium by Marine Phytoplankton," *Mar. Ecol. Prog. Ser.* 6:137-148 (1981).
14. Burnison, B. K. "Modified Dimethyl Sulfoxide (DMSO) Extraction for Chlorophyll Analysis of Phytoplankton," *Can. J. Fish. Aquat. Sci.* 37:729-733 (1980).
15. Jeffrey, S. W., and G. F. Humphrey. "New Spectrometric Equations for Determining Chlorophylls a, b, c1 and c2 in Higher Plants, Algae and Natural Phytoplankton," *Biochem. Physiol. Pflanz.* 167:191-194 (1975).
16. Strickland, J. D. H., and T. R. Parsons. "A Practical Handbook of Seawater," *Bull. Fish. Res. Bd. Can.,* 167:1-310 (1972).
17. LeBlanc, N. E., M. H. Roberts, Jr., and D. R. Wheeler. *Disinfection Efficiency and Relative Toxicity of Chlorine and Bromine Chloride, a Pilot Plant Study in an Estuarine Environment*, Special Report in Applied Marine Science and Ocean Engineering No. 206, (Gloucester Point, VA: Virginia Institute of Marine Science, 1978).
18. *Standard Methods for the Examination of Water and Wastewater,* 15th ed. (Washington, DC: American Public Health Association, 1980).
19. *Method for Chemical Analysis of Water and Wastes.* EPA-600/4-79-020. (Cincinnati: U.S. Environmental Protection Agency, 1979).
20. Helder, W., and R. T. P. DeVries. "An Automatic Phenol-Hypochlorite Method for the Determination of Ammonia in Sea- and Brackish Waters," *Netherlands J. Sea Res.* 13:154-160 (1979).
21. Sokal, R. R., and F. J. Rolhf. *Biometry*, 2nd ed. (San Francisco: W. H. Freeman and Company, 1981).
22. Steele, R. G. D., and J. H. Torrie. *Principles and Procedures of Statistics* (New York: McGraw-Hill Book Company, Inc., 1960).

23. Snedecor, G. W., and W. G. Cochran. *Statistical Methods, 7th ed.* (Ames, IA: Iowa State University Press, 1980).
24. Esvelt, L. A., W. J. Kaufman, and R. E. Selleck. "Toxicity Assessment of Treated Municipal Wastewaters," *J. Water Pollut. Control Fed.* 45:1558–1572 (1973).
25. Roberts, M. H.,Jr. "Survival of Juvenile Spot (*Leiostomus xanthurus*) Exposed to Bromochlorinated and Chlorinated Sewage in Estuarine Waters," *Mar. Environ. Res.* 3:63–80 (1980).
26. Roberts, M. H., Jr., N. E. LeBlanc, D. R. Wheeler, N. E. Lee, J. E. Thompson, and R. L. Jolley. *Production of Halogenated Organics During Wastewater Disinfection,* Special Report in Applied Marine Science and Ocean Engineering No. 239, (Gloucester Point, VA: Virginia Institute of Marine Science, 1980).
27. Roberts, M. H., Jr., and J. P. Illowsky. Unpublished data (Gloucester Point, VA: Virginia Institute of Marine Science).
28. Haas, L. Personal communication (Gloucester Point, VA: Virginia Institute of Marine Science).

SECTION VIII

Disinfection

The determination of what we want to achieve and why we want to achieve it will require far more serious concentration, and it is none too soon to begin the study of this problem.

Herbert N. Woodward
The Human Dilemma, 1971

That use of chlorination provides benefits to man is not an issue—the questions are how much to use and what risks and costs are involved.

William P. Davis and **James A. Fava**
Interaction of Aquatic Ecosystem Components with Chlorination, 1981

SECTION VIII

Disinfection

CHAPTER **44**

Inability of Laboratory Models to Accurately Predict Field Performance of Disinfectants

Roy L. Wolfe and Betty H. Olson

Traditionally, disinfection experiments have been conducted with pure cultures of batch-grown bacteria suspended in particle-free, chlorine demand-free test solutions.[1] Data from these studies have indicated that chlorine is a potent bactericidal and virucidal agent at concentrations of less than 0.1 mg/L.[2] However, in spite of the demonstrated effectiveness of chlorine, field investigations have indicated that bacteria, including indicator organisms and human pathogens, may persist in municipal water systems, even with free chlorine levels exceeding 1.0 mg/L.[3,4] In the case of chloramines, however, the opposite is observed, since laboratory results have greatly underestimated the apparent effectiveness in field application.[5] Most pure culture studies have shown that chloramines require much higher concentrations and considerably longer contact times to achieve levels of inactivation comparable to those attained by free chlorine.[6,7] Field studies have revealed that chloramines are as effective as chlorine in reducing and maintaining low coliform and total bacterial counts when sufficient contact times are provided.[8-10] The inability of traditional laboratory experiments, using culturally grown organisms tested under demand-free conditions, to reliably predict the effectiveness of chlorine and chloramines in drinking water systems suggests the need for alternative test procedures for disinfectant evaluation. Research in our laboratory with the total count group has provided valuable information concerning discrepancies in disinfection performance. This chapter discusses the reasons for these discrepancies and identifies factors that should be taken into consideration in future disinfection studies.

MATERIALS AND METHODS

Sampling Regime

Water samples were obtained during summer and winter months from a fully treated drinking water reservoir located in Orange County, California.

The water in the reservoir has a pH that consistently ranges from 7.9 to 8.2 and carries no detectable disinfectant residual.

Disinfectant Test Solutions

Samples were exposed to free chlorine, prereacted chloramines, and chloramines formed by preammoniation and concurrent addition techniques. Solutions were prepared with chlorine demand-free deionized-distilled water as previously described.[7] All chloramine solutions were prepared at a $Cl_2:N$ ratio (by weight) of 3:1. Concentrations were measured by the ferrous ammonium sulfate diethyl-*p*-phenylenediamine (FAS-DPD) and amperometric titration procedures as outlined in *Standard Methods*.[11]

Pure Culture Bacteria

For pure culture studies, bacteria were statically grown for 20 to 22 h in brain heart infusion broth (Difco) at 35°C. Chlorine-interfering compounds were removed from the cultures by centrifuging (3000 rpm for 10 min) and by washing three times with 10-mM pH 7 phosphate buffer.

Inactivation Assay

Inactivation studies were initiated with the addition of the disinfectant solution to a reaction vessel containing either the water sample or the pure culture test organism. At preselected contact times, aliquots were removed and neutralized with sodium thiosulfate and peptone. Portions of the neutralized samples were filtered through 0.45-μm cellulose acetate membranes (Gelman GN-6), washed with a sterile rinse solution, and placed on membrane standard plate count (mSPC) agar. Total count bacteria were enumerated after a 7-h incubation at 23°C, whereas pure culture bacteria were counted after a 48-h incubation at 35°C. The detailed protocol for this assay has been described elsewhere.[12]

Identification of Total Count Bacteria

Gram-negative nonfermentative bacilli were identified according to the taxonomic scheme outlined by Justice et al.[13] Gram-negative fermentative bacilli were identified with the API 20 E diagnostic system (Analytical Products), whereas gram-positive cocci and bacilli were identified according to *Bergey's Manual of Determinative Bacteriology*.[14]

RESULTS AND DISCUSSION

Ultimately, the goal of laboratory studies is to simulate the performance of a disinfectant under field conditions. Although many of the laboratory studies performed to date have provided valuable information concerning disinfection in controlled settings, the investigations appear to have overestimated the microbicidal potency of chlorine and underestimated the disinfectant ability of chloramines in drinking water. This is not altogether surprising, considering the physiological differences in the bacterial populations used in laboratory and field studies. Moreover, numerous physicochemical factors exist in the natural environment that may selectively alter the potency of a disinfectant or provide organisms protection from disinfection.

Our studies have identified four distinct factors in population dynamics that should be taken into consideration in disinfection studies. These include (1) differences in tolerance between naturally occurring and subcultured organisms; (2) differences in inactivation kinetics, reflecting the inactivation of mixed bacterial communities vs pure cultures; (3) differences in the variability and reproducibility of inactivation rates; and (4) differences in the techniques used to recover bacteria exposed to disinfectants. Moreover, the presence of interfering compounds can alter the microbicidal activity of the disinfectant. In the case of chloramines, data are presented to demonstrate that the presence of nitrogenous organic compounds interferes with inorganic chloramine formation and reduces the bactericidal activity of this disinfectant.

Influence of Subculturing

The subculturing of resistant bacteria can lead to increased sensitivity to chlorine, as illustrated in Figure 1. Curve A represents the survival of bacteria from the natural water sample exposed to 0.75 mg/L free chlorine (pH 8). A red-pigmented bacterium, identified as *Flavobacterium* spp., was the only organism recovered from the sample after 1-min exposure to chlorine, and less than 10% of this population was inactivated after 60 min of contact time. Curve B represents the survival response of the *Flavobacterium* organism exposed to chlorine following a single subculture. A dramatic reduction in chlorine resistance of the subcultured *Flavobacterium* was observed as ~99.9% of this organism was inactivated in <90 s.

A reduction in resistance to disinfection following subculturing has also been observed by others. Data presented by Carson et al.[15] revealed that cells of a strain of *Pseudomonas aeruginosa* adapted to a distilled water environment were significantly more resistant to chlorine and chlorine dioxide than cells that had been subcultured only once. Similar findings have been observed by Schaffer et al.[16] with poliovirus 1, by Carson and Favero[17] with nontuberculer *Mycobacteria*, and by Kutcha[18] with *Legionella pneumophila*.

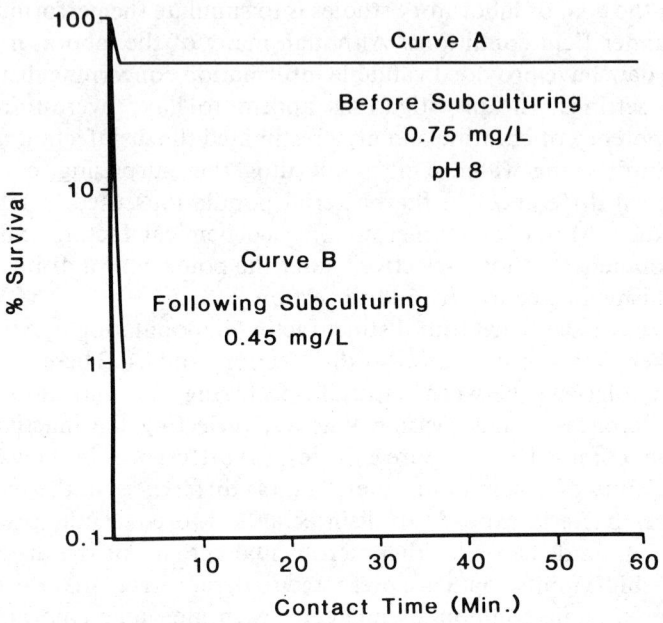

Figure 1. Sensitivity of *Flavobacterium* spp. to free chlorine (pH 8) before and after subculturing. Each datum point represents the mean of two observations.

Interestingly, the physiological basis for the increased resistance observed in naturally occurring bacteria is not well understood. Some of the proposed mechanisms of resistance include (1) modification of cell surfaces that may lead to increased aggregation or clumping of cells under in situ conditions,[2] (2) extrusion of protective extracellular capsular or slime layers,[19] and (3) development of cell membranes with decreased permeability to disinfectants.[20]

To determine the mechanisms of increased resistance associated with naturally occurring strains, Berg et al.[20] and Leyval et al.[21] have attempted to develop resistant strains by altering laboratory cultivation techniques. Berg et al.[20] found that *E. coli* cells cultured in a chemostat at slow growth rates ($D = 0.06$ h^{-1}) and low temperature ($T = 15°C$) were significantly more resistant to chlorine dioxide than cells grown at higher rates ($D = 0.40$ h^{-1}) and higher temperatures ($T = 37°C$). Additional research indicated that the structural integrity of the outer membrane, as determined by the retention of calcium and magnesium ions, was important in conferring increased resistance to cells cultured at the slow growth rates and low temperatures.

Leyval et al.[21] recently described a laboratory technique to select resistant *E. coli* cells. Their data indicated that 55 times more chlorine was required to inactivate resistant *E. coli* cells than control cells (i.e., 55 mg/L chlorine as compared to 0.01 mg/L).

Kinetics and Variability of Total Count Inactivation Rates

Typical curves for the inactivation of total counts by chlorine and chloramines are shown in Figure 2. The inactivation process is biphasic in that the rate of kill is rapid at the shorter contact times and slower with longer contact times. This shape most likely results from the inactivation of mixed bacterial communities possessing innate differences in sensitivity to disinfection. Disinfectant-sensitive populations are quickly inactivated and account for the initial rapid kill rate, whereas the more tolerant groups account for the gradual kill portion of the curve. This nonlinear inactivation curve is seldom observed in traditional pure culture studies. Figure 3 shows that the die-off curve of pure culture *E. coli* (exposed to prereacted chloramines) conforms to a linear process that can be used to accurately predict the survival of cells at a given length of exposure. Clearly, linear models of inactivation derived from pure culture studies are inappropriate for predicting the survival response of mixed communities, since linear models do not account for populations with widely varying sensitivities to disinfection.

Data for the overall inactivation of total count bacteria from samples obtained during summer and winter months by prereacted chloramines and free chlorine are presented in Table I. With prereacted chloramines (1.6 mg/L, pH 8), counts were frequently reduced by ~90% after 15 min of contact time and to >99% after 60 min. Inactivation by free chlorine (0.25-1.30 mg/L, pH 8) was characterized by an initial rapid decrease (99%) in counts in <1 min of exposure, followed by a gradual decrease in counts to 30 min.

Data in Table I also indicate the considerable sample-to-sample variability in the amount of total count inactivation. For example, in samples exposed to prereacted chloramines for 5, 15, and 30 min, the ranges in percent inactivation were 0.0 to 92.3, 35 to 99.22, and 88.4 to 100, respectively. In addition, the standard deviation of percent inactivation was greater than the mean for most of the contact times. In samples exposed to free chlorine for 0.5, 1, and 5 min, the ranges were 77.8 to 100, 82.2 to 100, and 96.1 to 100, respectively. Again, the standard deviation of percent inactivation was frequently greater than the mean survival.

The large variations in the efficiency of inactivation suggest that the effectiveness of a disinfectant in the field cannot be rigorously predicted from individual die-off curves generated from pure culture studies. This is because the amount of inactivation is predominantly controlled by changes in the relative abundance in the types of bacterial populations. In the aquatic envi-

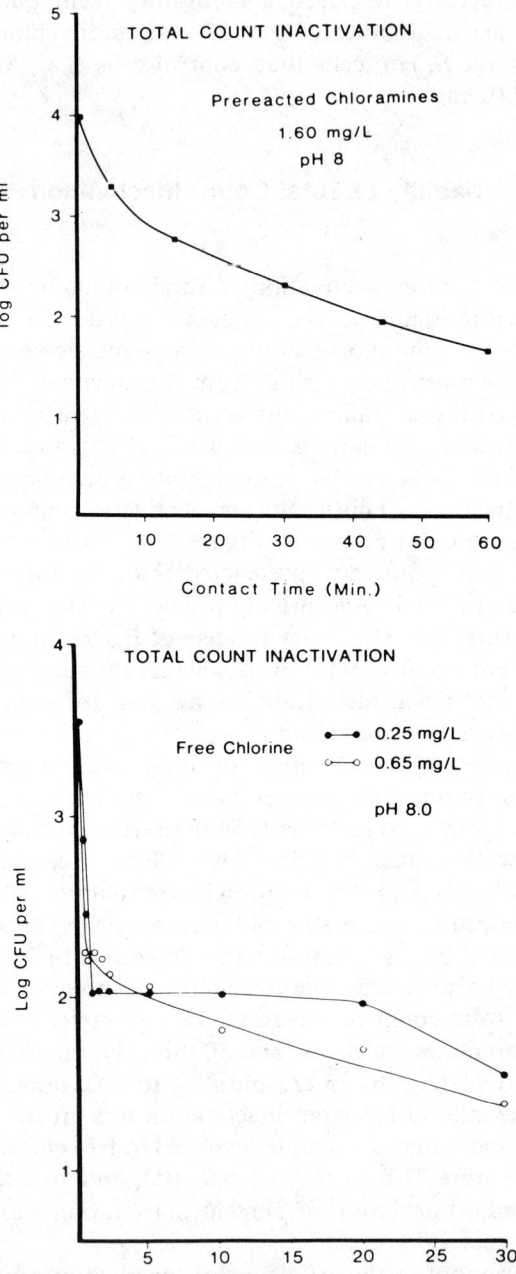

Figure 2. Typical inactivation curves for inactivation of total count bacteria from the San Joaquin Reservoir by free chlorine (0.25 and 0.45 mg/L Cl_2; pH 8) and by prereacted chloramines (1.60 mg/L; 3:1 Cl_2:N by weight; pH 8). Each datum point represents the mean of two observations.

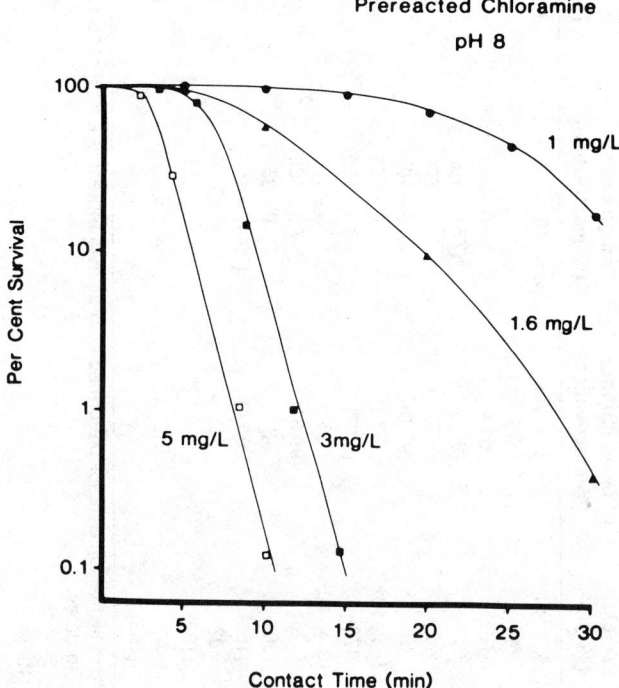

Figure 3. Typical curve for die-off of pure culture *E. coli* by prereacted chloramines at pH 8. Each datum point represents the mean of two observations.

ronment, a bacterial community is composed of populations that are sensitive and resistant to disinfection. Depending on the ratio of these populations in the sample, a disinfectant may appear highly effective or only weakly microbicidal. Thus, when the levels of disinfectant-tolerant bacteria outnumber the sensitive populations, the observed rates of total count inactivation greatly diminish. Conversely, when the disinfectant-sensitive populations outnumber the more tolerant ones, an initial rapid die-off is observed, followed by a more gradual die-off.

The rationale behind this generalization becomes clear when organisms surviving exposure to disinfection are identified. Figures 4 and 5 illustrate the impact of the survival of individual populations on the overall inactivation of total counts by chloramines and free chlorine. In Figure 4, the shape of the die-off curve from a sample exposed to 1.6 mg/L prereacted chloramines was

Table I. Summary of Inactivation Data for Total Count Bacteria from the San Joaquin Reservoir by Preformed Chloramines and Free Chlorine[a]

Contact Time min	Preformed chloramines[b]			Free chlorine[c]		
	Grand[d] mean cfu/mL (% inactivation)	Range cfu/mL (% inactivation)	SD (cfu) (% SD)	Mean cfu/mL (% inactivation)	Grand[d] range cfu/mL (% inactivation)	SD (% SD)
0	8.2×10^3 (0)	$2.2 \times 10^3 - 1.5 \times 10^4$	3.4×10^3 (41)	7.4×10^3 (0)	$2.2 \times 10^3 - 1.3 \times 10^4$	3×10^3 (40)
0.5	ND[e]	ND	ND	510 (92.3)	$0 - 2 \times 10^3$ (77.8–100)	680 (133)
1.0	ND	ND	ND	155 (97.0)	0 – 800 (82.2–100)	246 (159)
5.0	2.5×10^3 (71)	$220 - 5.5 \times 10^3$ (0–92.3)	1.9×10^3 (76)	67 (98.7)	0 – 155 (96.1–100)	64 (96)
15.0	914 (89)	$43 - 3.4 \times 10^3$ (35–99.22)	1.15×10^3 (126)	35 (98.7)	0 – 100 (97.5–100)	39 (111)
30.0	195 (97.5)	$0 - 1.1 \times 10^3$ (88.4–100)	301 (154)	8 (99.31)	0 – 35 (99.12–100)	12 (150)
45.0	43 (99.26)	0 – 260 (95–100)	82 (190)	ND (99.84)	ND	ND
60.0	26 (99.68)	0 – 130 (98.6–100)	41 (158)	ND	ND	ND

[a]Summarized from raw data collected during summer and winter months. Inactivation data for preformed chloramines represent a total of 15 sampling dates; inactivation data for free chlorine represent a total of 10 sampling dates.
[b]1.60 mg/L at pH 8.
[c]0.45 to 1.30 mg/L at pH 8; data do not include sample obtained on 7/16/82.
[d]Grand mean cfu/mL (survivors) = $\dfrac{\text{mean cfu/mL for all samples at a contact time}}{\text{total number of samples}}$.
[e]Not done.

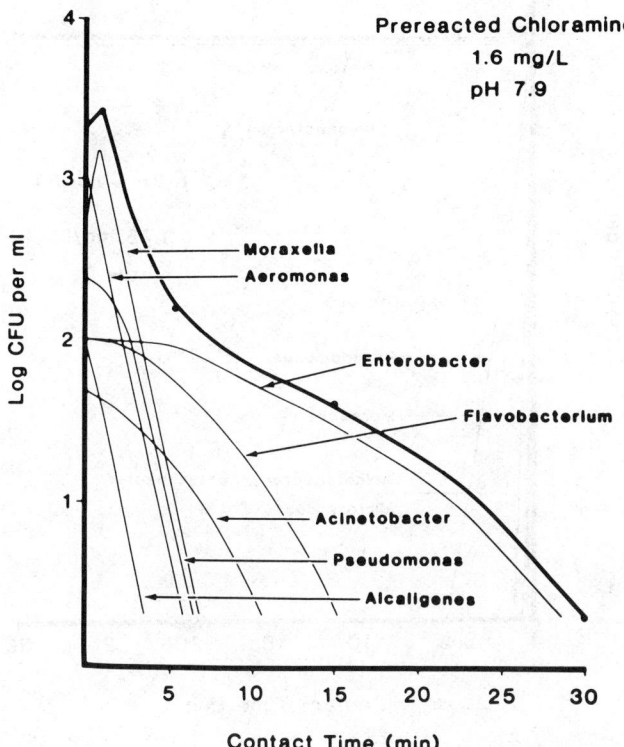

Figure 4. Composite die-off of total counts and calculated inactivation curves for genera identified from a San Joaquin Reservoir sample exposed to prereacted chloramines (1.6 mg/L; 3:1 Cl_2:N by weight; pH 7.9).

predominantly influenced by the presence of *Enterobacter* spp. Members of this genus were the only bacteria recovered after 15 min exposure to chloramines. Clearly, if these organisms had not been present in the sample, the inactivation rate would appear to be considerably more rapid. Other genera (*Moraxella, Pseudomonas, Aeromonas,* and *Alcaligenes*) appear to be highly sensitive to chloramines and were responsible for the rapid initial die-off. The influence of disinfectant-sensitive and disinfectant-tolerant populations on the apparent effectiveness to free chlorine was also observed (Figure 5). In this particular sample, the presence of the highly chlorine-tolerant, red-pigmented

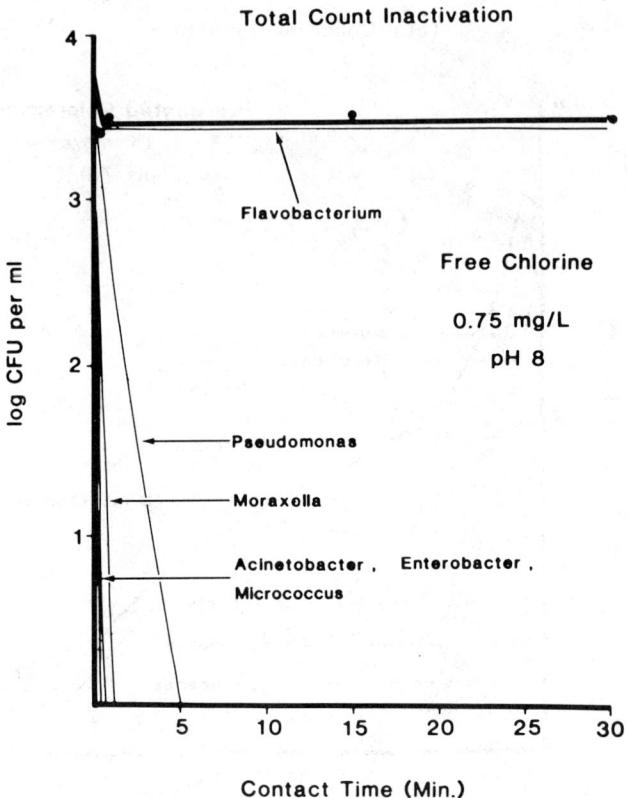

Figure 5. Composite die-off of total counts and calculated inactivation curves for genera identified from a San Joaquin Reservoir sample exposed to 0.75 mg/L free chlorine at pH 8.

Flavobacterium spp. greatly reduced the observed amount of inactivation by free chlorine. Once again, if this population had not been present, the overall die-off rate of the bacterial community would have been reflected by the other genera (*Moraxella, Enterobacter, Acinetobacter, Alcaligenes*), since they were inactivated within 5 min of exposure to 0.75 mg/L.

The large variation in the amount of inactivation observed for the total count community also suggests that changes in the numbers of total count bacteria do not necessarily provide an accurate indication of the microbial water quality of overall disinfection effectiveness. In support, Goshko et al.[22] found new significant correlations between standard plate count levels and five water quality parameters, including coliform occurrence, turbidity, and free chlorine, in seven small distribution systems. Haas et al.[23] also found little relationship between the levels of free chlorine and standard plate count densities in two New Hampshire distribution systems. One explanation for these observations is that changes in the numbers of bacteria recovered from disinfected distribution water simply denote a shift in the relative numbers of disinfectant-tolerant and disinfectant-sensitive populations within the total count community. This shift may be important, however, if there is an influx of disinfectant-tolerant bacteria into the public water distribution system.

Differences in Recovery Techniques

Field investigations are more likely to be influenced by incubation time, recovery method, incubation temperatures, and recovery media than laboratory studies. The changes in these factors make it difficult to compare the microbial quality among different water sources. Figures 6 and 7 illustrate the influence of altering selected recovery techniques on the observed inactivation of total count bacteria. Figure 6 depicts the survival of bacteria in a sample

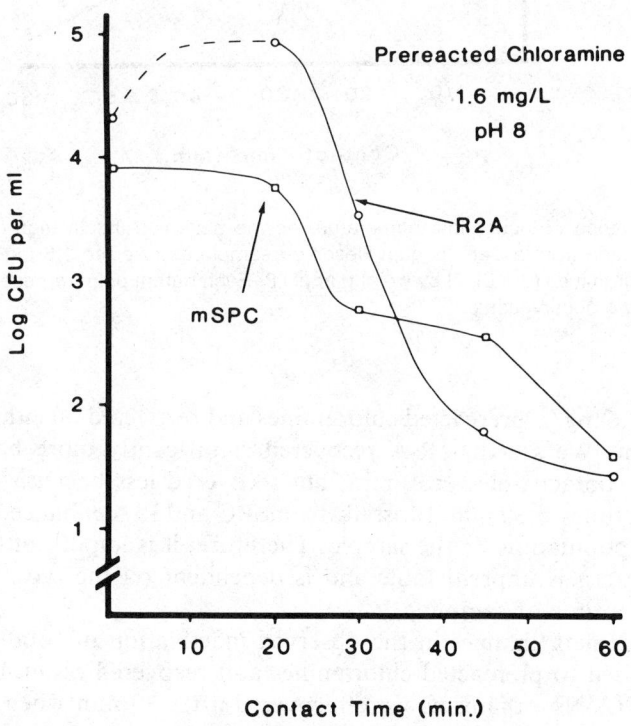

Figure 6. Influence of recovery media on the observed inactivation of total count bacteria from a San Joaquin Reservoir sample exposed to 1.6 mg/L prereacted chloramines (3:1 Cl_2:N by weight) at pH 8. Each datum point represents the mean of two observations.

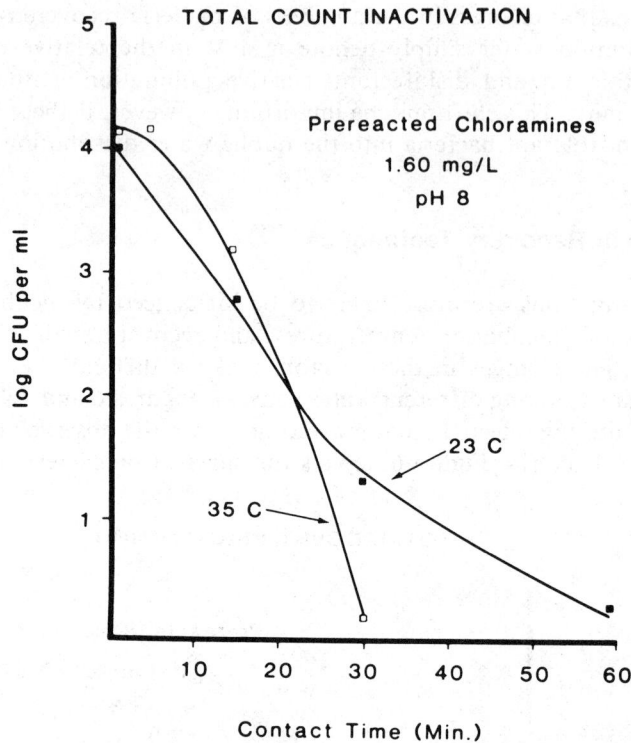

Figure 7. Influence of incubation temperature on the observed inactivation of total count bacteria from a San Joaquin Reservoir sample exposed to 1.6 mg/L prereacted chloramines (3:1 Cl_2:N by weight) at pH 8. Each datum point represents the mean of two observations.

exposed to 1.6 mg/L prereacted chloramines and recovered on either mSPC or R_2A medium. We see that R_2A recovered significantly more bacteria than mSPC with contact times <30 min, but recovered less bacteria than mSPC with contact times >30 min. Most likely, mSPC and R_2A enhanced the growth of different populations in the sample. Therefore, it is important to recognize that this pattern is unpredictable and is dependent on the types of bacteria present at the time of sampling.

Figure 7 depicts changes in the observed inactivation of total counts in a sample exposed to prereacted chloramines and recovered on mSPC at either 24°C or 35°C. No organisms were recovered after 30 min when plates were incubated at 35°C, whereas organisms could still be recovered after 60 min of exposure when incubated at 23°C. Once again, this pattern is probably not generalizable at this time and will not be until seasonal and temporal changes in community dynamics are better understood.

Interfering Nitrogenous Compounds

The presence of selected nitrogenous organic compounds can reduce the microbicidal effectiveness of chloramines by interfering with the formation and stability of this disinfectant. Chemical studies have indicated that certain nitrogenous organic compounds bind more rapidly to chlorine than does ammonia,[24,25] and that chlorine can be transferred from NH_2Cl to various forms of organic nitrogen.[26] Significantly, many organic N-chloramines have weak or nonbactericidal[27] activities and cannot be distinguished from inorganic chloramines by conventional residual measurement techniques.[28]

Our research findings using preammoniation and the concurrent addition technique suggested that during certain times of the year nitrogenous compounds were interfering with the formation of inorganic chloramines. In these instances, we observed that prereacted chloramines inactivated total count bacteria more rapidly than the preammoniation and concurrent addition techniques. A typical example of the interference is presented in Figure 8. Nitrogenous organic compounds in the reservoir were implicated as the interfering agents because the residual chloramine concentrations in the three solutions were identical (as measured by FAS-DPP and amperometric titration procedures). To test this premise indirectly, an experiment was conducted in which the amino acid glycine was supplemented into the test samples at levels of 0.1 to 0.55 mg/L (as nitrogen) to artificially increase the concentration of organic nitrogen. This amino acid was selected because it binds to chlorine 22 times more rapidly than ammonia, and because it is commonly found in aquatic environments. The concentration of ammonia-nitrogen in each of the solutions was 0.55 mg/L. These data indicated that the bactericidal activity with preammoniation was significantly reduced when a concentration of 0.1 mg/L organic nitrogen (as glycine) was tested (Figure 9). After 60 min, the counts in the prereacted and preammoniated solutions that did not receive glycine supplementation were reduced from 7000 cfu/mL to approximately 60 cfu/mL, whereas the levels of bacteria in the glycine-supplemented solutions after 60 min of exposure to preammoniation were 500, 2000, and 4000 cfu/mL (0.1, 0.25, and 0.55 mg/L organic nitrogen, respectively). Furthermore, the amperometric and FAS-DPD titrimetric procedures were unable to distinguish between inorganic and organic chloramine residuals (Table II). The impact of relatively small concentrations of amino acids on inorganic chloramine formation is shown by the theoretically calculated concentrations of inorganic chloramines. For example, the presence of 0.1 and 0.25 mg/L glycine (as nitrogen) reduces the active inorganic chloramines by 13 and 51%, respectively.

Friend[24] and Morris[25] have studied the rate of reaction of chlorine with amino acids and other nitrogenous compounds. Some of Morris' data are presented in Table III. We see that the reaction of chlorine with certain nitrogenous compounds (alanine, glycine, methylamine, dimethylamine, and diethylamine) is more rapid than with others (ethylaminoacetate, glycylglycylglycine). From these data the type of organic amine compounds in the water

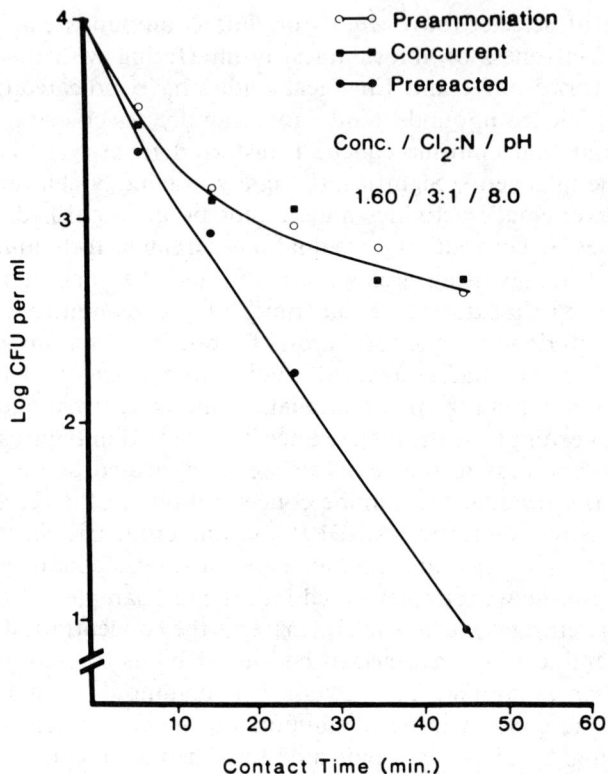

Figure 8. Evidence of interference in the inactivation of total bacterial populations from a natural water sample using prereacted chloramines, preammoniation, and concurrent application techniques. Residual chloramine concentrations of each solution was 1.6 mg/L (at pH 8 and 3:1 Cl_2:N by weight). Each datum point represents the mean of two observations.

will be more important in terms of their relative rates of N-chlorination and interference with inorganic chloramine formation than merely the amount of organic-N present.

In addition, recent studies by Margarum et al.[29] and Isaac and Morris[26] have shown that chlorine (Cl^+) can be transferred from inorganic chloramines to amino acids and peptides via direct transfer and by hydrolysis of monochloramine. The rate and extent to which this reaction proceeds is largely dependent on the type and concentration of the N-substrate. These researchers observed that with equimolar concentrations of NH_2Cl (pH 7, 25°C) and several organic nitrogen compounds at 10^{-5} M, 50% of the NH_2Cl was converted to organic chloramine in <2 d. For some compounds, the conversion occurred in <3 h. With concentrations of the nitrogenous compounds and

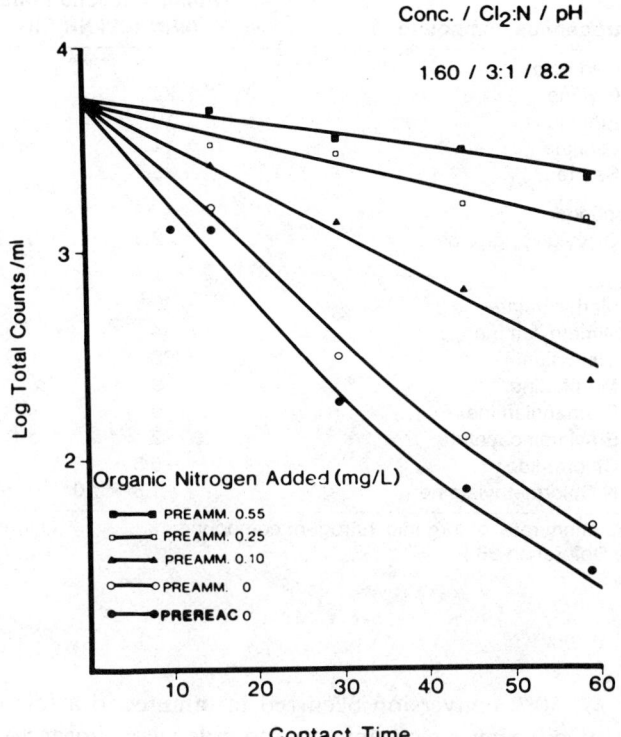

Figure 9. Inactivation of total count bacteria in a San Joaquin Reservoir sample using preammoniation and prereacted application techniques. Glycine was supplemented into the samples at levels of 0.1, 0.25, and 0.55 mg/L (as nitrogen) prior to preammoniation treatment. Each datum point represents the mean of two observations.

Table II. Combined Chlorine Determinations Using the Amperometric and FAS-DPD Procedures for San Joaquin Reservoir Samples Supplemented with Glycine

Application Technique	Glycine (mg/L N)	Ammonia (mg/L N)	NH_2Cl measured[a] Amperometric	FAS-DPD	Theoretical NH_2Cl Concentration[b]
Prereacted	0	0.55	1.54	1.63	1.60
Preammoniation	0	0.55	1.58	1.62	1.60
Preammoniation	0.1	0.55	1.50	1.60	1.39
Preammoniation	0.25	0.55	1.50	1.63	0.79
Preammoniation	0.55	0.55	1.52	1.63	0.32

[a] Residual monochloramine concentration (inorganic + organic chloramine).
[b] Theoretical inorganic chloramine based on (1) organic monochloramine formation (glycine derived) rate 22 times faster than inorganic monochloramine formation (ammonia derived), and (2) that nitrogen binds a maximum of five times the weight of chlorine (i.e., 5:1 Cl_2:N by weight).

Table III. Chlorination Rates for Organic Nitrogen Compounds Relative To Ammonia

Nitrogenous compound	Relative reaction rate (k_{RNHCl}/k_{NH_2Cl})[a]
Amino acids	
Glycine	22
Alanine	19
Leucine	14
Serine	6.7
Peptides	
Glycylglycylglycine	2.3
Others	
Methylamine	60
Dimethylamine	54
Diethylamine	23
Morpholine	9
Diethanolamine	9
Ethylaminoacetate	2
Chloramide	5.5×10^{-5}
N-Chlormethylamine	1.8×10^{-4}

[a] k_{RNHCl} = chlorination rate of organic nitrogen compound; k_{NH_2Cl} = chlorination rate of ammonia. (See Reference 26.)

NH_2Cl at 10^{-4} M, 50% conversion occurred in minutes to a few hours. The concentrations of nitrogenous organic compounds were chosen to reflect the levels in clean-water and wastewater systems.

Although the impact of chlorine transfer reactions on disinfection efficiency has not yet been tested, these reactions suggest that the disinfectant activity of inorganic chloramines could decrease over time in the presence of organic amines. Hence, long retention times in distribution systems using inorganic chloramines for disinfection may result in the formation of the weakly microbicidal organic chloramines.

Unfortunately, knowledge of the types and concentrations of nitrogenous organic compounds in surface waters is limited.[30] Amino acids and other primary amines are likely to be among the most environmentally significant forms of organic nitrogen because of their ubiquity and interference with inorganic chloramine formation. In general, the concentrations of free amino acids in freshwater range from a few to several hundred micrograms per liter.[31] By comparison, the concentration of free amino nitrogen in sewage effluent can be expected to be an order of magnitude greater than the concentrations found in freshwater. Isaac and Morris[32] have estimated that free amino nitrogen ranges from 0.2 to 0.25 mg/L in domestic sewage effluent. Current research in our laboratory is directed toward determining the types and concentrations of organic amine compounds in drinking water and their impact on disinfection with inorganic chloramines.

SUMMARY

Our research with the total count group has indicated that disinfection results obtained in traditional laboratory studies are inappropriate for predicting performance under field conditions. Factors that limit the usefulness of laboratory-derived findings include:

1. The use of subcultured bacteria that are considerably less resistant to disinfection than are naturally occurring counterparts. Although the mechanisms for increased resistance found in naturally occurring organisms are unknown, the presence of an extracellular polysaccharide layer or alteration of cell membrane permeability is thought to be important.
2. Differences in inactivation kinetics among pure cultures and mixed populations. Laboratory cultures produce constant-rate die-off curves, whereas this type of kinetics is seldom observed with mixed bacterial communities.
3. Differences in the variability of inactivation data. Laboratory models can be controlled to greatly reduce sample-to-sample variability of results, whereas field investigations indicate that variability fluctuates by orders of magnitude on a daily or weekly basis. This is caused by rapid changes in the composition of bacterial communities in the water system under investigation. Consequently, the observed effectiveness of chlorine and chloramines is greatly dependent on the types of bacteria present at the time of sampling.
4. Differences in recovery techniques. Field investigations are greatly influenced by the recovery methodology because different techniques may support the growth of different bacterial populations. Thus, the apparent effectiveness of a disinfectant is affected by the recovery of more tolerant or sensitive populations.
5. The presence of interfering compounds that alter the potency of a disinfectant. In the case of chloramines, evidence was presented that selected nitrogenous organic compounds can significantly reduce the disinfection efficiency of chloramines by interfering with the formation of this disinfectant. Based on chlorine binding rates, the types of nitrogenous organic compounds that are most likely to impair chloramine disinfection are amino acids and other primary amines.

To increase the predictiveness of laboratory-derived findings, future disinfection research should evaluate the die-off of organisms in their natural environments.

ACKNOWLEDGMENTS

Support for this research was provided by a grant from the Metropolitan Water District of Southern California. We would like to thank Alan Kelly of the Medical Microbiology Department at the University of California, Irvine, for his excellent photographic service.

REFERENCES

1. Hoff, J. C., and E. E. Geldreich. "Comparison of the Biocidal Efficiency of Alternative Disinfectants," *J. Am. Water Works Assoc.* 73(1):40-44 (1981).
2. National Research Council. *Drinking Water and Health, Vol. 2*, (Washington DC: National Academy Press, 1980).
3. Le Chevallier, M. W., R. J. Seidler, and T. M. Evans. "Enumeration and Characterization of Standard Plate Count Bacteria in Chlorinated and Raw Water Supplies," *Appl. Environ. Microbiol.* 44:922-930 (1981).
4. Ridgway, H. F.,and B. H. Olson. "Chlorine Resistance Patterns of Bacteria from Two Drinking Water Distribution Systems," *Appl. Environ. Microbiol.* 44:972-987 (1982).
5. Wolfe, R. L, N. R. Ward, and B. H. Olson. "Inorganic Chloramines as Drinking Water Disinfectants: A Review," *J. Am. Water Works Assoc.* 76(5):74-88 (1984).
6. Butterfield, C. T. "Bactericidal Properties of Chloramines and Free Chlorine in Water," *Public Health Rep.* 63:934-940 (1948).
7. Ward, N. R., R. L. Wolfe, and B. H. Olson. "Disinfectant Activity of Inorganic Chloramines with Pure Culture Bacteria: Effect of pH, Application Technique and Chlorine to Nitrogen Ratio," *Appl. Environ. Microbiol.* 48(3):508-514 (1984).
8. Norman, T. S., L. L. Harms, and R. W. Looyenga. "The Use of Chloramines to Prevent Trihalomethane Formation," *J. Am. Water Works Assoc.* 72(3):176-180 (1980).
9. Brodtmann, N. V., Jr., and P. J. Russo. "The Use of Chloramines for Reduction of Trihalomethanes and Disinfection of Drinking Water," *J. Am. Water Works Assoc.* 71(1):40-42 (1979).
10. Mitcham, R. P., M. W. Shelley, and C. M. Wheadon. "Free Chlorine Versus Ammonia-Chlorine: Disinfection, Trihalomethane Formation, and Zooplankton Removal," *J. Am. Water Works Assoc.* 75(4)196-200 (1983).
11. *Standard Methods For the Examination of Water and Wastewater*, 15th ed. (New York: American Public Health Association, 1980).
12. Wolfe, R. L., N. R. Ward, and B. H. Olson. "The Inactivation of Naturally Occurring Bacteria by Chloramines and Free Chlorine," submitted for publication.
13. Justice, C. A., N. R. Ward, R. L. Wolfe, and B. H. Olson. "A Simple and Reliable Scheme for Identifying Gram Negative, Nonfermentative Isolates from Freshwater," Annu. Meet. Abstr. Q112. (Washington, DC: American Society of Microbiology, 1983), p 279.
14. Buchanan, R. E., and N. E. Gibbons, Eds. *Bergey's Manual of Determinative Bacteriology*, 8th ed. (Baltimore: The Williams and Wilkins Co., 1974).
15. Carson, L. A., M. S. Favero, W. W. Bond, and N. J. Peterson. "Factors Affecting Comparative Resistance of Naturally Occurring and Subcultured *Pseudomonas aeruginosa* to Disinfectants," *Appl. Microbiol.* 23:863-867 (1972).
16. Schaffer, P. T. B., T. G. Metcalf, and O. J Sproul. "Chlorine Resistance of Poliovirus Isolants Recovered from Drinking Water," *Appl. Environ. Microbiol.* 40:1115-1121 (1980).
17. Carson, L. A., and M. S. Favero. "Comparative Resistance of Nontuberculosis Mycobacteria to Iodophor Disinfectants," Annu. Meet. Abstr. Q 101 (Washington DC: American Society of Microbiology, 1983), p. 221.
18. Kutcha, J. M. "Enhanced Chlorine Resistance of Tap Water Grown *Legionella pneumophila* as Compared Passaged Strains," Annu. Meet. Abstr. Q2 (Washington DC: American Society of Microbiology, 1984), p. 204.

19. Seyfried, P. L., and D. J. Fraser. "Persistance of *Pseudomonas aeruginosa* in Chlorinated Swimming Pools," *Can. J. Microbiol.* 26:350–355 (1980).
20. Berg, J. D., A. Matin, and P. V. Roberts. "Growth of Disinfection Resistant Bacteria and Simulation of Natural Aquatic Environments in the Chemostat," in *Water Chlorination: Environmental Impact and Health Effects, Vol. 4*, R. L. Jolley, W. A. Brungs, J. A. Cotruvo, R. B. Cumming, J. S. Mattice, and V. A. Jacobs, Eds. (Ann Arbor, MI: Ann Arbor Science Publishers, Inc., 1983), pp. 1137–1147.
21. Leyval, C., C. Arz, J. C. Block, and M. Rizet. "*Escherichia coli* Resistance to Chlorine After Successive Chlorinations," (submitted for publication).
22. Goshko, M. A., H. A. Minnigh, W. O. Pipes, and R. R. Christian. "Relationships Between Standard Plate Counts and Other Parameters in Water Distribution Systems," *J. Am. Water Works Assoc.* 75:568–671 (1983).
23. Haas, C. N., M. A. Meyer, and M. S. Paller. "Microbial Alterations in Water Distribution Systems and Their Relationships to Physical-Chemical Characteristics," *J. Am. Water Works Assoc.* 75:475–481 (1983).
24. Friend, A. G. "Rates of N-Chlorination of Amino Acids," Ph.D. Thesis, (Cambridge, MA: Harvard University, 1956).
25. Morris, J. C. "Kinetics of Reactions Between Aqueous Chlorine and Nitrogen Compounds," in *Principles and Application of Water Chemistry*, S. D. Faust and J. V. Hunter, Eds. (New York: John Wiley and Sons, 1967), pp. 23–53.
26. Isaac, R. A., and J. C. Morris. "Transfer of Active Chlorine from Chloramine to Nitrogenous Organic Compounds," *Environ. Sci. Technol.* 17:738–742 (1983).
27. Feng, T. H. "Behavior of Organic Chloramines in Disinfection," *J. Water Pollut. Control Fed.* 38:614–628 (1966).
28. White, G. C. *Disinfection of Wastewater and Water for Reuse* (New York: Van Nostrand Reinhold, 1978).
29. Margarum, D. W., E. T. Gray, and R. P. Huffman. "Chlorination and the Formation of N-Chloro Compounds in Water Treatment," in *Organo-metals and Organometalloids: Occurrence and Fate in the Environment*, R. F. Brinkman and J. M. Bellama, Eds. (Washington, DC: American Chemical Society, 1978), pp. 278–290.
30. Morris, J. C., N. Ram, B. Baum, and E. Wajon. *Formation and Significance of N-Chloro Compounds in Water Supplies*, EPA–600/2–80–031 (Washington, DC: U.S. Environmental Protection Agency, 1980).
31. Zygmuntowa, J. "Occurrence of Free Amino Acids in Pond Water." *Acta Hydrobiol.* 14:317–325 (1972).
32. Isaac, R. A., and J. C. Morris. "Rates of Transfer of Active Chlorine Between Nitrogenous Substrates," in *Water Chlorination: Environmental Impact and Health Effects, Vol. 3*, R. L. Jolley, W. A. Brungs and R. B. Cumming, Eds. (Ann Arbor, MI: Ann Arbor Science Publishers, Inc., 1980), pp. 183–191.

CHAPTER 45

Aspects of the Mode of Action of Monochloramine

J. G. Jacangelo and V. P. Olivieri

Combined chlorine plays an important role in both water and wastewater disinfection. In recent years, there has been an increasing interest in a return to the intentional application of combined chlorine for both primary and residual disinfection during water treatment. Using monochloramine (NH_2Cl) as an alternative to free chlorine has been one solution practiced by the water industry to meet the recent trihalomethane regulations. In wastewater disinfection, free chlorine combines with the ammonia component of the effluent to form monochloramine, which thus acts as a disinfectant. Although there has been considerable experience with monochloramine in the past, reactions of the disinfectant with certain biological macromolecules and their constituents are not well defined. A better understanding of these reactions is necessary so that we can begin to define its mode of action in inactivating microorganisms; this is particularly important because mutagens may be produced as a result of chlorination. A greater understanding of these reactions will also be helpful in assessing the impact of monochloramine on the aquatic environment. The purpose of our study was to provide some fundamental information on the chemical reactivity of monochloramine with nucleic acids, with their structural components, and with amino acids.

METHODS AND MATERIALS

Monochloramine solutions were prepared by the method of Chapin[1] with some modifications. Ammonium chloride (NH_4Cl) was reacted with previously prepared aqueous chlorine at a 3:1 molar ratio (NH_4Cl to OCl^-) at pH 10. Three-times distilled water was used in all preparations, and all glassware was chlorine demand free. Concentrations of monochloramine were measured by spectrophotometry,[2] by the syringaldazine colorimetric method,[3] or by amperometric titration.[4] The latter two methods were also used to ensure the absence of free chlorine in monochloramine stock solutions. All monochloramine concentrations were reported as mol per liter monochloramine.

Spectral changes of nucleic acids and their constituents were carried out using a Heath-Schlumberger Model EU–707–11 double-beam spectrophotometer. Changes in their respective maximum absorption bands were observed over time. Similar studies with free chlorine were used for comparison because these reactions are documented in the literature.[5] Consumption of monochloramine by various biochemical substrates obtained from the Sigma Chemical Company and the Aldrich Chemical Company were measured kinetically by the above methods for determining disinfectant concentrations.

The reactivity of nuclcotidcs was also evaluated by high-performance liquid chromatography (HPLC). Changes in chromatographic characteristics were determined isocratically using a Partisil-10 SCX ion exchange column. Ammonium dihydrogen phosphate, 0.035 M (pH 3.5) served as the mobile phase and detection was by ultraviolet absorption at 254 nm.

RESULTS

Consumption of Monochloramine by Nucleic Acids

The consumption of monochloramine by nucleic acids and their constituents at pH 7.0 and a 5:1 molar ratio is shown in Figures 1 and 2. The concentration of NH_2Cl is expressed as the log fraction of the initial concentration and is

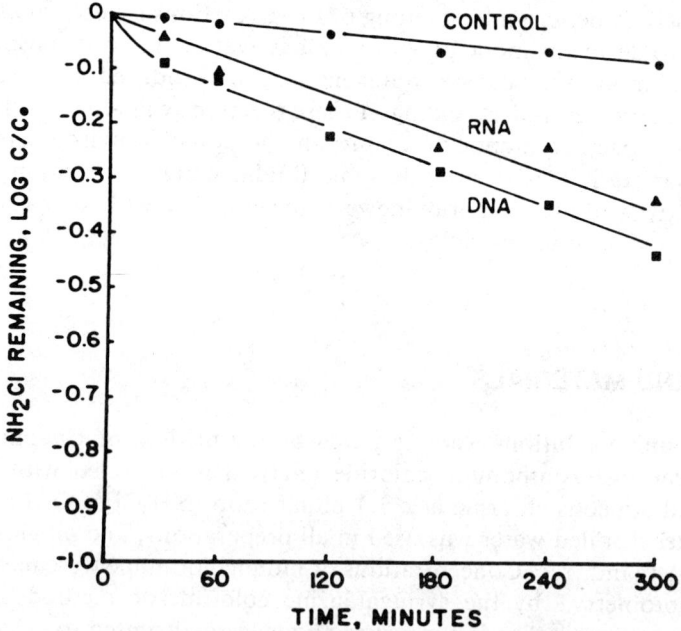

Figure 1. Comparative consumption of monochloramine by yeast RNA (– ▲ –) and yeast DNA (–■–) at 23°C and pH 7.0. The molar ratios of nucleic acids-to-monochloramine were 5:1. Initial monochloramine concentration was 1.9 × $10^{-4}M$.

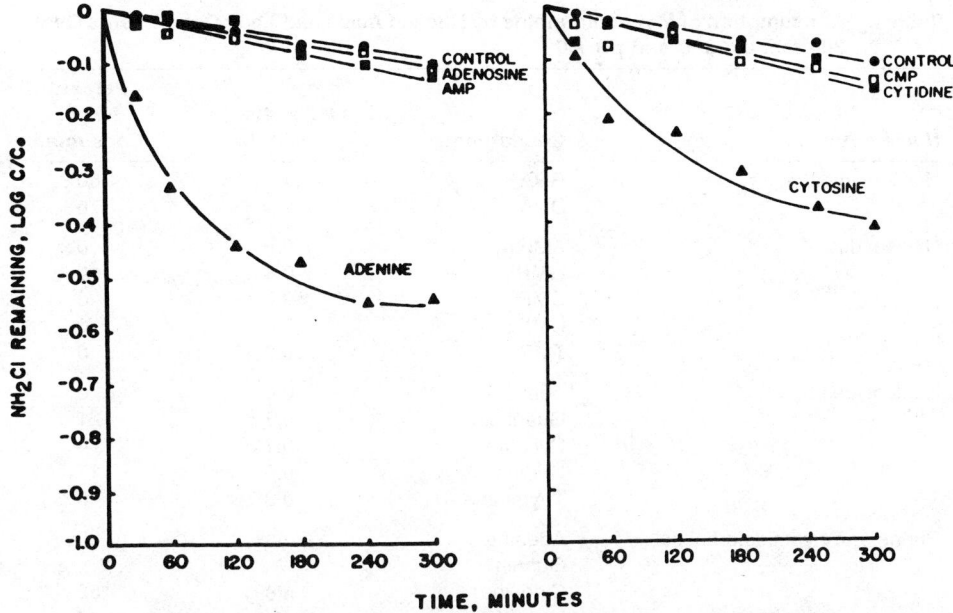

Figure 2. Comparative consumption of monochloramine by adenine and cytosine nucleotides (-■-), nucleosides (-□-), and free bases (-▲-) at 23°C and pH 7.0. Molar ratios of nucleotide, nucleoside, and free-base concentrations to monochloramine were 5:1. Initial monochloramine concentration was $1.9 \times 10^{-4} M$.

plotted against time. Controls consisted of monochloramine in pH 7.0 phosphate buffer. Figure 1 shows the consumption of monochloramine by yeast ribonucleic acid (RNA) and yeast deoxyribonucleic acid (DNA). RNA and DNA consumed 60 and 76% of the disinfectant, respectively, after 300 min. The approximate linear plot indicates pseudo-first-order kinetics. From these data, the pseudo-first-order reaction rate with RNA was 1.7×10^{-3} min^{-1}, and with DNA it was 2.4×10^{-3} min^{-1}. The calculated half-lives for monochloramine were 420 and 290 min with RNA and DNA, respectively.

Figure 2 shows monochloramine consumption patterns typical of purine and pyrimidine bases and their associated nucleosides and nucleotides. The nucleosides, adenosine and cytidine, and the nucleotides, adenosine monophosphate (AMP) and cytidine monophosphate (CMP), consumed little disinfectant when compared with controls. The free bases (adenine and cytosine in Figure 2) consumed between 22 and 60% of monochloramine after 300 min. The consumption patterns of the purines and pyrimidines tested, however, did not appear to follow first-order kinetics. Data for monochloramine consumption studies by nucleic acid constituents are shown in Table I. First-order rate constants were calculated where possible, and percent consumption after 300 min are presented. The sugars (ribose and deoxyribose) and salt (phosphate),

Table I. Consumption of Monochloramine by Nucleic Acids and Their Constituents Over 300 min at 23°C and pH 7.0

Nucleic Acids	Constituents	$k\ (\times 10^{-3})$ min^{-1}	Percent NH$_2$Cl Consumed
Nucleic acids	RNA	1.7	60
	DNA	2.4	76
Nucleotides	AMP	0.2	6
	GMP	0.4	10
	CMP	0.2	6
	UMP	0.0	0
	TMP	0.0	0
Nucleosides	Adenosine	0.0	0
	Guanosine	0.8	21
	Cytidine	0.1	4
	Uridine	0.6	17
	Thymidine	0.3	8
Purine and pyrimidine bases	Adenine	nfo[a]	60
	Guanine	[b]	[b]
	Cytosine	nfo	52
	Uracil	nfo	47
	Thymine	nfo	22
Sugars and salt	Ribose	0.0	0
	Deoxyribose	0.0	0
	Phosphate	0.0	0

[a]Non-first order.
[b]No data.

which are components of nucleic acids, did not react with monochloramine. Of the free bases tested, adenine was the most reactive, consuming 60% of the disinfectant after 300 min. Over the same reaction time, guanosine monophosphate consumed the most monochloramine of all the nucleotides (10%); guanosine was the most reactive nucleoside, consuming 21% of the disinfectant.

HPLC Analysis of Nucleotides

Monochloramine-treated nucleotides were further analyzed by high-performance liquid chromatography to detect changes that may not be solely associated with ring structures of the substrates. Table II presents the percent loss of nucleotide concentration for all nucleotides after exposure to a 5-M excess of monochloramine. For comparative purposes, similar trials were conducted with free chlorine. Of the five nucleotides, only guanosine monophosphate (GMP) and uridine monophosphate (UMP) appeared to react with

Table II. Change in Nucleotide Concentration Caused by Reaction with Monochloramine and Free Chlorine After 180 min Reaction Time at 23°C and pH 7.0 (analysis by HPLC)

Nucleotide[a]	Percent loss of nucleotide concentration	
	Treatment with monochloramine	Treatment with free chlorine
Adenosine	0	92
Guanosine	4	91
Cytidine	0	49
Uridine	2	2
Thymidine	0	7

[a]All monophosphates.

monochloramine over 180 min (4 and 2% loss of nucelotide concentration, respectively). In accord with the literature,[5-7] AMP, GMP, and CMP reacted readily with free chlorine at this pH, whereas UMP did not. The lack of reactivity of the latter nucleotide may be the result of steric hindrances presented by the associated phosphate and ribose constituents, since uracil has been shown to react readily with free chlorine.[5,8]

Spectrophotometric Analysis of Nucleic Acids

To determine whether monochloramine affected the ring structure of nucleic acids, a 10-M excess of the disinfectant was reacted with RNA and its constituents. The respective wavelengths of maximum absorption were scanned periodically. The data in Figures 3 to 5 are presented as the fraction of the initial absorption of the nucleic acid or constituent plotted against time. Controls consisted of the biochemical substrate in pH 7.0 phosphate buffer. Figure 3 shows that there were few spectral changes associated with the 257-nm absorption band of RNA when exposed to NH_2Cl for 60 min. Even after 1440 min, little difference from the control was seen. A similar experiment conducted with free chlorine showed substantial degradation of the RNA maximum absorption band. Figures 4 and 5 present the same experiments with the nucleotides and the purine and pyrimidine bases composing RNA. Similar patterns at the lower levels of nucleic acid organization were observed. Spectral changes of the nucleotides and free bases were not detected when reacted with monochloramine; however, free chlorine showed characteristic ring degradation.

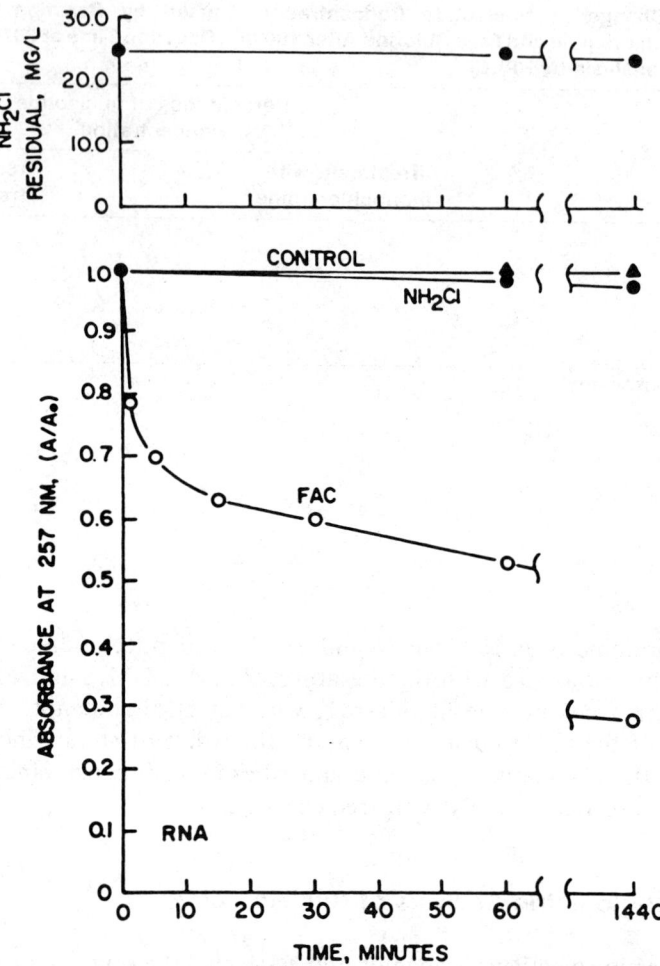

Figure 3. Comparative disappearance of the 257-nm absorption band of yeast RNA resulting from the reactions with monochloramine and free chlorine (FAC) at 23°C and pH 7.0. The molar ratios of monochloramine and free chlorine to RNA were 5:1. Initial RNA concentrations were 8.1 × $10^{-5}M$ and 1.6 × $10^{-5}M$ for monochloramine and free chlorine treatments, respectively.

Consumption of Monochloramine by Amino Acids

Table III lists the percent consumption of monochloramine by amino acids at pH 7.0 and 23°C after 2- and 180-min reaction times. Disinfection consumption was measured by the syringaldazine or amperometric methods; a 5:1 molar ratio of biochemical substrate to disinfectant was used. Nucleic acids

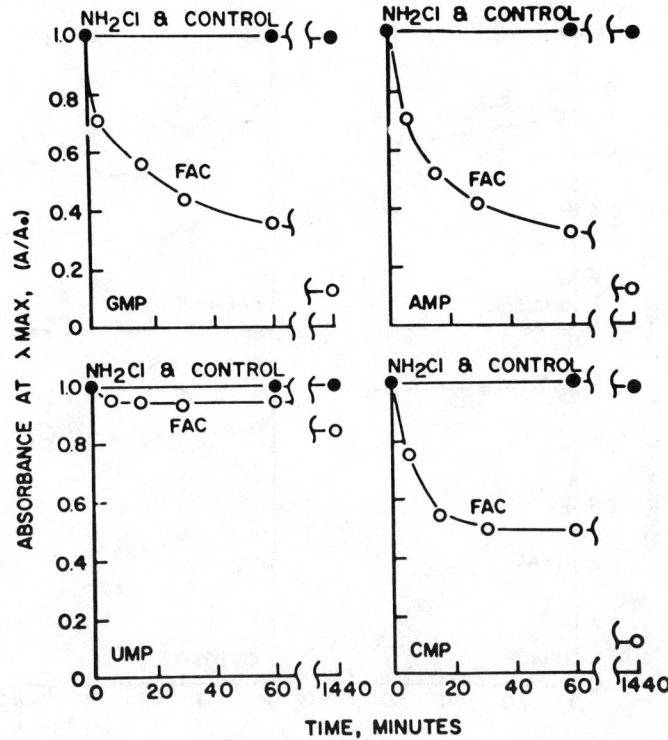

Figure 4. Comparative disappearance of the maximum absorption bands of RNA nucleotides resulting from the reactions with monochloramine and free chlorine (FAC) at 23°C and pH 7.0. Molar ratios of monochloramine and free chlorine to nucleotides were approximately 10:1. Initial nucleotide concentrations were between $7.0 \times 10^{-5} M$ and $8.0 \times 10^{-5} M$.

are also presented for comparison purposes. The most reactive amino acids after 2 min were those containing sulfur residues (cysteine, cystine, and methionine) and a ring-containing amino acid, tryptophan. All consumed 100% monochloramine in less than 2 min. Tyrosine, threonine, histidine, and leucine were the only other amino acids to consume at least 10% of the disinfectant over the same reaction time. After 180 min, all of the amino acids except glycine consumed more disinfectant than the nucleic acids.

Cysteine Oxidation by Monochloramine

A stoichiometric analysis of the reaction of the sulfhydryl group (-SH) of cysteine with monochloramine is shown in Table IV. Various molar ratios of cysteine-to-monochloramine were reacted at 23°C and pH 7.0. Cysteine was

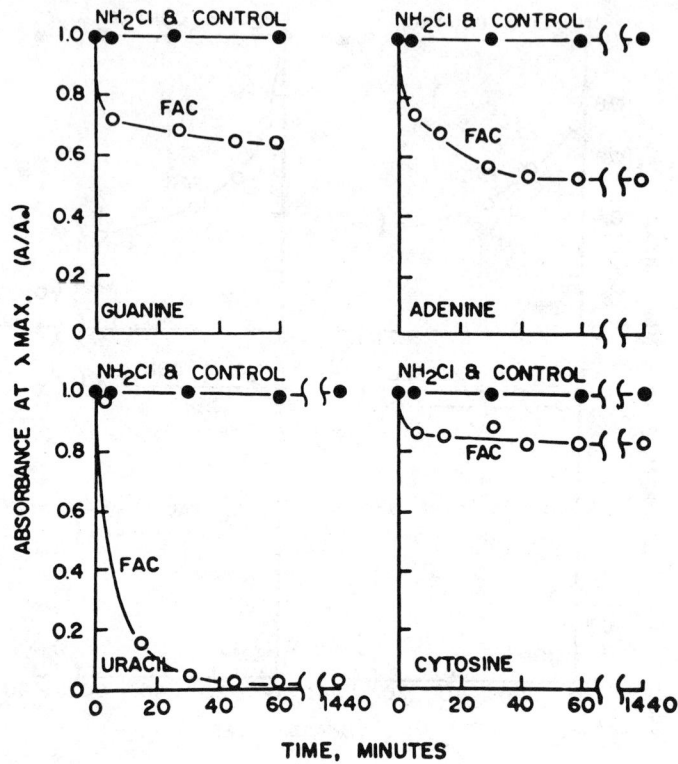

Figure 5. Comparative disappearance of the maximum absorption bands of RNA purine and pyrimidine bases resulting from the reactions with monochloramine and free chlorine (FAC) at 23°C and pH 7.0. Molar ratios of monochloramine and free chlorine to free bases were ~10:1. Initial base concentrations were between $2.7 \times 10^{-5}M$ and $6.7 \times 10^{-5}M$.

always in excess. The number of unreacted -SH groups was then measured colorimetrically with 5,5'-dithiobis(2-nitrobenzoic acid). The data show that the ratio of the number of mols of thiol groups oxidized to the number of mols of monochloramine reacted was approximately 2 (mean, 1.98; standard deviation, 0.13). The correlation of this 2:1 ratio was 0.9913.

Under similar conditions, an estimate was made of the stoichiometry of the cystine formed from the oxidation of cysteine by monochloramine. The data are presented in Table V. Cystine disulfide residues (-S-S-) were measured by the change in absorbance at 250 nm after the -SH groups were exposed to monochloramine. The ratio of the number of mols of NH_2Cl reacting with the sulfhydryl to the number of mols of the cystine disulfide formed from the reaction was ~1 (mean, 1.16; standard deviation, 0.26).

Table III. Percent Consumption of Monochloramine by Amino Acids and Nucleic Acids After 2- and 180-min Reaction Times at 23°C and pH 7.0. (Initial NH_2Cl Concentrations Between 1.7×10^{-4} M and 2.1×10^{-4} M)

	Constituents	Percent NH_2Cl Consumed	
		2 min	180 min
Amino acids	Alanine	0	86
	Cystine	100	100
	Phenylalanine	2	96
	Tyrosine	10	100
	Cysteine	100	100
	Isoleucine	3	99
	Serine	6	98
	Glycine	0	6
	Threonine	10	82
	Histidine	11	100
	Aspartic acid	8	100
	Glutamic acid	0	99
	Tryptophan	100	100
	Valine	2	93
	Lysine	3	100
	Glutamine	3	97
	Arginine	8	98
	Asparagine	2	100
	Proline	2	90
	Methionine	100	100
	Leucine	13	95
Nucleic acids	RNA	0	57
	DNA	1	43

Table IV. Oxidation of Cysteine Residues by Monochloramine at 23°C and pH 7.0 NH_2Cl is the Limiting Reactant; SH Groups are Measured with Dithiobisnitrobenzoic Acid

NH_2Cl Dose (μmol)	Ratio Cysteine:NH_2Cl	−SH Oxidized (μmol)	Ratio −SH Oxidized:NH_2Cl[a]
464	4.1:1	911	2.15
155	12.3:1	305	1.95
348	5.5:1	682	1.96
93	20.5:1	162	1.74
116	16.5:1	233	1.97
309	6.2:1	600	1.94
371	5.2:1	800	2.15
232	8.2:1	452	1.94

[a]Mean = 1.98; SD = 0.13; r^2 = 0.9913.

Table V. Oxidation of Cysteine to Cystine by Monochloramine at 23°C and pH 7.0 NH_2Cl is the Limiting Reactant; Cystine was Measured by Absorbance at 250 nm Before and After Reaction

Cysteine (μmol)	NH_2Cl (μmol)	$-S-S-$ Formed (μmol)	Ratio[a] $NH_2Cl: -S-S-$
20.5	5.0	5.0	1.00
10.1	5.0	5.1	0.98
19.2	5.8	4.8	1.21
10.7	4.4	3.0	1.50
13.1	4.4	4.1	1.07
20.5	2.5	3.3	0.75
11.8	4.4	3.5	1.23
38.4	5.8	3.7	1.57

[a]Mean = 1.16; SD = 0.26.

DISCUSSION

Our data demonstrate some aspects of the chemical reactivity of monochloramine with nucleic acids at various levels of biomolecular organization. Through disinfectant consumption studies, reactivity was observed only with the entire nucleic acid and with the free purine and pyrimidine bases. The absence of reactivity of these bases within the nucleotides and nucleosides may be caused by steric hindrances presented by the appropriate sugar (ribose or deoxyribose) or phosphate moieties associated with these structures. Such findings are consistent with those found by Dennis[5] with free chlorine and UMP.

Spectrophotometric studies of the monochloramine-treated nucleic acids showed little or no alteration in the respective maximum absorption bands at any level of organization. The data indicate that the chemical reactivity observed in the consumption studies is not associated with purine or pyrimidine ring degradation in the nucleic acid. Snead[9] demonstrated that the nucleic acid of f2 bacteriophage was inactivated by monochloramine, although the rate of inactivation was slower than that for the whole virus. If a chemical reaction is responsible for the nucleic acid inactivation or mutagenic alterations, the probable mode of action involves a cut or scission in the entire nucleic acid.[10-11] Additional studies are being conducted in our laboratory to evaluate this hypothesis.

As demonstrated through disinfectant consumption studies, amino acids were more reactive than nucleic acids; the sulfur-containing amino acids and tryptophan consumed the disinfectant most readily. Because of their reactivity, it is probable that these residues are specific sites of microbial inactivation when they are essential to the functional integrity of the microorganism and are available for reaction. They may be particularly important when the rate of microbial inactivation by monochloramine is rapid (as with bacteria).

The sulfhydryl group of cysteine plays an important role in the functional and macromolecular integrity of enzymes and proteins. The cysteine-monochloramine reaction has been proposed by Ingols et al.[12] and Boyle[13] as the lethal event paramount to inactivation. This study showed that when the sulfhydryl groups of cysteine are in excess, 1 mol of monochloramine reacts with 2 mol of cysteine to form 1 mol of the cystine disulfide. This stoichiometric relationship supports the reaction involving a two-electron transfer

$$-2 === -2e == >-1-1$$

$$NH_2Cl + 2\ -SH <---------> -S-S- + NH_3 + HCl$$

$$+1 ========= +2e ========= <-1$$

We note, however, that when the monochloramine is in excess, the reaction probably proceeds beyond the disulfide state. When exposed to a reducing agent such as glutathione, the disulfide can be reduced back to the sulfhydryl.[14] This reversibility may explain several observations in the literature on the reactivation of enzymatic activity[15] or the recovery of stressed microorganisms[16-17] after disinfection.

CONCLUSIONS

Monochloramine reacts more readily with the whole nucleic acid and with free purine and pyrimidine bases than with nucleotides or nucleosides. The consumption of monochloramine with RNA and DNA appears to follow pseudo-first-order kinetics. Monochloramine reacts more readily with the amino acids (except glycine) than with nucleic acids.

Cysteine, cystine, methionine, and tryptophan are the most reactive amino acids with monochloramine. The mode of action of monochloramine in inactivating microorganisms probably involves an amino acid reaction; however, a nucleic acid reaction cannot be ruled out when the reactive amino acids are unessential or buried.

ACKNOWLEDGMENTS

The authors would like to thank the American Water Works Association Research Foundation for their support of this ongoing research.

REFERENCES

1. Chapin, R. "Dichloramine," *J. Am. Chem. Soc.* 51:2112-2117 (1929).
2. Granstrom, M. L. "The Disproportionation of Monochloramine," Ph.D. Thesis, (Cambridge, MA: Harvard University, 1954).
3. Roscher, N. M., R. Lieberman, W. J. Cooper, and E. P. Meier. "Development of FACTS Procedures for Combined Chlorine and Ozone in Aqueous Solutions," (Fort Detrick, MD: U.S. Army Medical Research Development Command, 1978).
4. *Standard Methods for the Examination of Water and Wastewater*, 15th ed. (Washington, D.C.: American Public Health Association, 1981), pp. 286-289.
5. Dennis, W. "The Mode of Action of Chlorine on f2 Virus," Ph.D. Thesis, (Baltimore: The Johns Hopkins University, 1977).
6. Olivieri, V. P., W. H. Dennis, M. C. Snead, D. T. Richfield, and C. W. Kruse. "Reaction of Chlorine and Chloramines with Nucleic Acids Under Disinfection Conditions," in *Water Chlorination: Environmental Impact and Health Effects, Vol. 3*, R. L. Jolley, W. A. Brungs, and R. B. Cumming, Eds. (Ann Arbor: Ann Arbor Science Publishers, Inc., 1980), pp. 651-663.
7. Dennis, W. H. "The Reaction of Nucleotides with Aqueous Hypochlorous Acid," *Water Res.* 13:357-362 (1979).
8. Dennis, W. H., V. P. Olivieri, and C. W. Kruse. "Reaction of Uracil with Hypochlorous Acid," *Biochem. Biophys. Res. Comm.* 83:168-171 (1978).
9. Snead, M. "Inactivation of f2 Bacterial Virus by Monochloramine," Sc.M. Thesis, (Baltimore: The Johns Hopkins University, 1976).
10. Shih, K. L., and J. Lederberg. "Effects of Chloramine on *Bacillus subtilus* Deoxyribonucleic Acid," *J. Bacteriol.* 125:934-945 (1976).
11. Ingols, R. S. "The Effect of Monochloramine and Chromate on Bacterial Chromosomes," *Public Works*. Dec.: 105-106 (1958).
12. Ingols, R. S., H. A. Wycoff, T. W. Kethley, H. W. Hodgden, E. L. Fincher, J. C. Hildebrand, and J. E. Mandel. "Bacterial Studies of Chlorine," *Ind. Eng. Chem.* 45:996-1000 (1953).
13. Boyle, W. C. "Studies on the Biochemistry of Disinfection by Monochloramine," Ph.D. Thesis, (Pasadena: California Institute of Technology, 1963).
14. Barron, E. S. "Thiol Groups of Biological Importance," in *Advances in Enzymology*, F. F. Nord, Ed. (New York: Interscience Publishers, Inc., 1951), pp. 201-266.
15. Venkobachar, C., L. Iyengar, and A. V. Rao. "Mechanism of Disinfection: Effect of Chlorine on Cell Membrane Functions," *Water Res.* 11:727-729 (1977).
16. McFeters, G. A., A. K. Camper. "Enumeration of Indicator Bacteria Exposed to Chlorine," *Adv. Appl. Microbiol.* 29:177-193 (1983).
17. Verville, K. M. Personal communication (Newark: DL: University of Delaware) 1984.

CHAPTER 46

Disinfection of *E. coli* in the Presence of N-Organic Compounds

Neil M. Ram, James P. Malley, Jr., Cynthia A. Parks, and Brian Dudley

The U.S. Environmental Protection Agency currently permits the substitution of free-available-chlorine (FAC) residual monitoring for as much as 75% of the bacteriological samples required for public water supplies.[1] This substitution is permitted as long as no less than 0.2 mg/L FAC is measured at representative points in the distribution system. FAC monitoring is presumably stipulated because of the marked differences in the biocidal activity of FAC and chloramines.[2-7] A secure hygienic quality in water supplies is therefore dependent on the maintenance of the FAC residual. Chlorine additionally reacts with nitrogenous organic compounds to form N-chloroorganic substances, such as N-chloramino acids, which are essentially nongermicidal.[8] The N-chloroorganic forms, however, retain an oxidizing capacity and tend to react similarly with many analytical reagents for active chlorine. When these N-chloro compounds are formed, tests for free chlorine may be falsely positive and may indicate a nonexistent germicidal behavior.

Numerous investigations of the reactions of free aqueous chlorine with ammonia are epitomized by the work of Palin,[9] Wei and Morris,[10] and Saunier and Selleck.[11] Studies such as these have provided reasonably accurate information about the conditions of formation and the properties of ammonia chloramines, such that their behavior and germicidal effectiveness relative to free aqueous chlorine can be considered satisfactorily known. False positives have been observed in solutions containing chloramines using the SNORT, Syringaldazine, and DPD analytical methods of chlorine determination,[12-16] with the level of false positives increasing in proportion to higher chloramine concentrations. False positives have also been observed by Snead et al.[17] in the DPD, FACTS, amperometric titration, and membrane electrode methods, which use a biological reference procedure. A modification of the DPD procedure (DPD Steadifac) reduced the frequency and magnitude of false positives. The FACTS procedure was the most specific for FAC. Other investigators have shown the FACTS procedure to be relatively free of interferences from NH_2Cl and $NHCl_2$,[12,18-9] although interference was observed with NCl_3.[14]

Several analytical reagents have been tested for their response to N-chloro compounds. Palin[9] found that DPD could distinguish among free chlorine and several types of chlorinated amino acids. The OTA procedure was able to distinguish between free and combined oxidant when seawater containing several amino acids (e.g., diphenylamine and uracil) was chlorinated.[20]

Wajon and Morris[21] found that none of the methods for free chlorine could distinguish between free and combined chlorine in chlorinated cyanurate solutions. DPD was able to differentiate among free chlorine and some chlorinated amino acids, whereas discrimination with others (creatine and sarcosine) was not possible. In the presence of N-organic heterocyclic bases, all of the analytical methods were subject to false-positive readings for free chlorine. The implication of the Wajon and Morris study is that, in drinking water containing nitrogenous organic compounds, especially N-organic heterocycles, N-chloroorganic compounds will form and may result in an overestimate of the disinfection potential of the measured residual chlorine.

This study examined the true level of disinfection, as measured by the percent kill of *E. coli* in N-organic solutions buffered to pH 7, 8, or 9. *E. coli* were enumerated in solutions containing 1 mg/L model nitrogenous organic compounds, following 15 or 5 min of chlorine contact and at a final FAC level of 0.2 or 0.05 mg/L as Cl_2, to determine the validity of FAC monitoring as a substitute for microbiological examination of water supplies.

EXPERIMENTAL

Materials

Chlorine Demand-Free Water (CDFW)

Chlorine demand-free water was prepared by dosing distilled water with 4 mg/L free chlorine (as Cl_2) and then stirring for 6 to 8 h. The remaining chlorine was destroyed by UV irradiation and constant mixing for 24 h. Water having a 30-min demand of greater than 0.02 mg/L was discarded.

Buffer Solutions

A buffer solution (pH 7) was prepared by sterilizing a 1×10^{-3} M $NaHCO_3$ solution in CDFW at 121°C and 15 psi for 15 min and then adjusting with CO_2. Additional buffers (pH 8 and 9) were prepared by adding 50 mL of 0.025 M sodium borate to 1530 mL of CDFW, which had been autoclaved for 15 min at 121°C and 15 psi. The solutions were then adjusted to either pH 8 with 0.1 M HCl or pH 9 with 0.1 M NaOH. The pH 9 buffer was equilibrated to 10°C prior to the addition of sodium hydroxide.

Model Compounds

Stock nitrogenous organic compounds used in the study were of the highest quality commercially available. The compounds, with the exceptions of pyrrole and gelatin which are heat labile, were initially dried at 50°C for 1 h. Stock solutions of 100 mg/L were prepared in CDFW. Guanine solutions were prepared by direct addition because of their limited solubility.

Bacterial Inocula

These were prepared by transferring six loops full of bacterial cell mass grown on BBL nutrient agar slants into 250 mL of sterile pH 7, 8, or 9 buffer. After vigorous shaking, followed by 10 min of settling, 100 mL was transferred into 400 mL of N-organic test solutions, and 20 mL was transferred into 80 mL of buffer for initial microbial density determination.

Methods

Disinfection Schemes

Three disinfection schemes were used during the study (Figure 1). In all three schemes, solutions of 1-mg/L N-organic compounds were reacted ini-

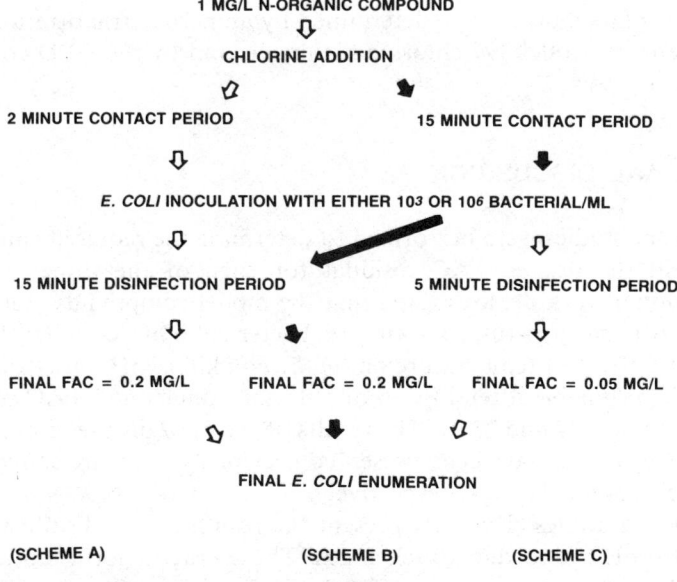

Figure 1. Disinfection schemes.

tially with chlorine and then inoculated with *E. coli*. The reactions were terminated with 2 mL of $4 \times 10^{-3} \, N \, Na_2S_2O_3$. Final *E. coli* and FAC levels were then determined. The initial chlorine contact was included to permit chlorine demand to be exerted prior to the bacterial inoculation. The longer initial chlorine contact period in schemes B and C was used to obtain lower FAC levels at the time of bacterial inoculation. Schemes A and B represent types of chlorine application that might be encountered in water supply treatment, where higher chlorine residuals prevail at the point of application so that 0.2 mg/L FAC residuals can be maintained within the distribution system. Scheme C was used to test the disinfection efficiency under conditions of lower disinfectant potency. The pH 7 and 8 studies were performed at 25°C. The influence of temperature on *E. coli* kill has been previously shown not to be marked at pH 7.0 except at FAC residuals of less than 0.02 ppm. However, the effects of temperature become more pronounced above pH 8.5.[2] Later studies at pH 9 were therefore performed at 10°C.

Bacterial Enumeration

E. coli were enumerated using either the membrane filtration or standard plate count technique after appropriate serial dilution.[21]

Chlorine Determination

Free and total chlorine were determined by amperometric titration, using a Fisher Scientific Model 397 chlorine titrimeter, and by the DPD colorimetric method of analysis.[22]

RESULTS AND DISCUSSION

Preliminary studies were performed to determine the required chlorine dosage to yield the desired FAC residual for each of the three disinfection schemes. Initial work demonstrated that the model compounds alone did not exhibit significant growth-promoting or bactericidal effects after 15 min of contact. The effect of temperature on the *E. coli* kill at pH 8 was additionally shown to be negligible after 15 min of chlorine contact at a FAC residual of 0.2 mg/L as Cl_2 at 10 and 25°C. The results of the pH 7 disinfection studies for schemes A, B, and C have been presented previously[23] and are summarized in Figures 2 and 3 and Table I, respectively.

Figure 4 and Tables II and III present the results of the disinfection study performed at pH 8 according to scheme A. The average chlorine dosage (at pH 8) required to produce an apparent FAC of 0.2 mg/L as Cl_2 after an initial 2-min chlorine-compound reaction period, followed by a 15 min disinfection

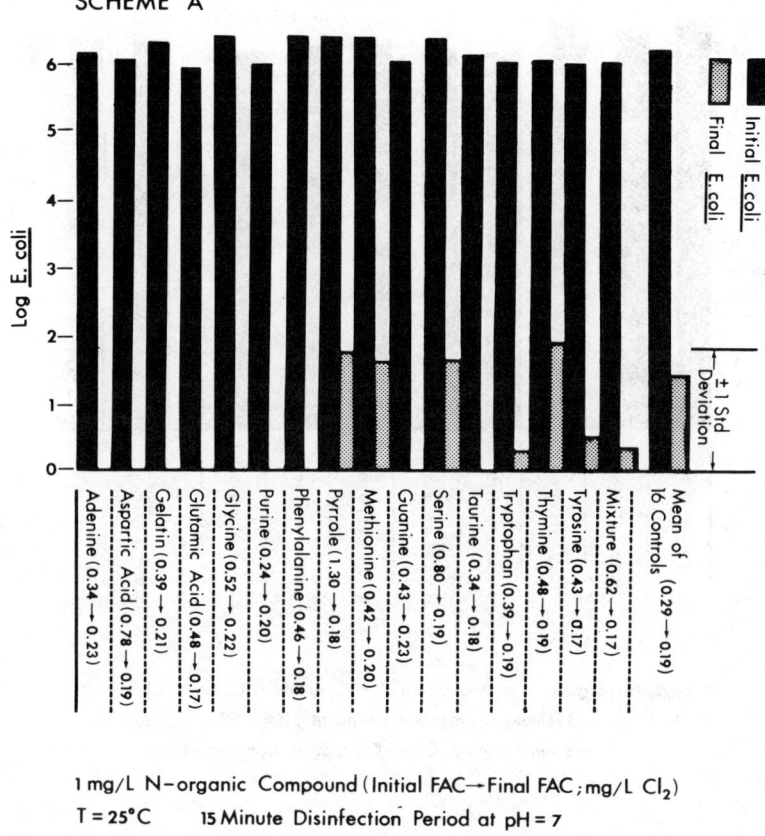

Figure 2. Disinfection of *E. coli* at pH 7 (according to scheme A).

period for the 16 N-organic solutions, was 1.08 mg/L Cl_2 [standard deviation (S.D.), 0.77]. Pyrrole exhibited the greatest chlorine requirement (3.43 mg/L) and purine and guanine the smallest (0.33 and 0.31 mg/L, respectively). The controls exhibited an average Cl_2 requirement of 0.32 mg/L (S.D., 0.04) to produce a 0.2 mg/L FAC after the total 17-min chlorine-contact period. The chlorine demand of the control, equal to 0.12 mg/L Cl_2, consisted of 0.05 mg/L Cl_2 buffer demand and 0.07-mg inoculum demand. The chlorine concentration at the time of the *E. coli* inoculation (2-min DPD free residual) ranged from 0.96 mg/L (pyrrole) to 0.23 mg/L (gelatin), with an average of 0.43 mg/L (S.D., 0.19). The bacteria were, therefore, exposed to levels above

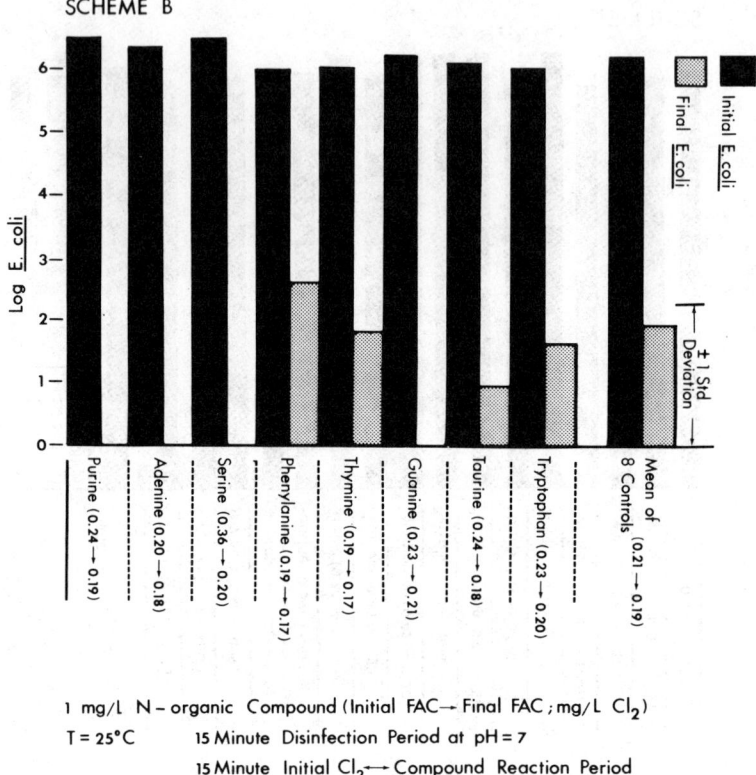

Figure 3. Disinfection of E. coli at pH 7 (according to scheme B).

0.2 mg/L Cl₂ FAC at the beginning of the 15-min disinfection period. The large percentage of *E. coli* kill (average, 99.9999%) was therefore not surprising.

No significant decrease in bactericidal effectiveness attributable to the possible presence of falsely positive tests for FAC were observed in any of the test solutions at pH 7 or 8. The number of surviving bacteria in each of the test (with N-organic compound) solutions was within one standard deviation of the mean of control replicates. Therefore, the differences in bacterial kills were attributed to chance variation and not to an assignable cause, such as falsely positive tests for FAC. The survival of some coliform in both test and control solutions, even after 15 min of contact at a FAC residual of 0.2 mg/L Cl₂, might be explained by bacterial repair mechanisms occurring when surviving cells were transferred to enrichment media for enumeration. Other explana-

Table I. Testing of N-Chloroorganics According to Scheme C; pH 7[a] (all Cl$_2$ units are mg/L as Cl$_2$)

Compound	Chlorine added	Mol Cl$_2$ Mol Compound	Mol Cl$_2$ Mol Nitrogen	15-min DPD Free residual	15-min DPD Total residual	20-min DPD Free residual	20-min Titrimeter Free residual	20-min Titrimeter Total residual	20-min Titrimeter Combined residual	Kill (%)
Taurine	0.68	1.20	1.20	0.08	0.67	0.07	0.07	0.67	0.60	100.0
Control	0.09			0.05	0.09	0.05	0.05	0.08	0.03	100.0
Purine	0.15	0.25	0.06	0.12	0.15	0.06	0.07	0.15	0.08	100.0
Control	0.09			0.07	0.09	0.05	0.06	0.09	0.03	100.0
Phenylalanine	0.76	1.77	1.77	0.05	0.50	0.05	0.05	0.46	0.41	100.0
Control	0.09			0.05	0.10	0.07	0.05	0.09	0.04	100.0
Thymine	0.12	0.21	0.11	0.05	0.12	0.05	0.05	0.07	0.02	100.0
Control	0.10			0.07	0.09	0.05	0.05	0.09	0.04	100.0
Adenine	0.40	0.76	0.15	0.08	0.38	0.05	0.05	0.34	0.29	100.0
Control	0.09			0.07	0.09	0.05	0.05	0.09	0.04	100.0
Tryptophan	1.85	5.32	5.32	0.07	0.45	0.05	0.05	0.43	0.38	100.0
Control	0.09			0.05	0.09	0.03	0.05	0.09	0.04	100.0
Serine	0.98	1.45	1.45	0.05	0.68	0.03	0.05	0.66	0.61	100.0
Control	0.10			0.05	0.09	0.04	0.05	0.09	0.04	100.0
Guanine	0.75	1.60	0.32	0.07	0.40	0.03	0.05	0.38	0.33	100.0
Control	0.09			0.05	0.10	0.05	0.05	0.10	0.05	100.0
Mean of 8 samples	0.71	1.57	1.30	0.07	0.42	0.05	0.05	0.40	0.34	100.0
Standard deviation of 8 samples	0.55	1.63	1.76	0.02	0.21	0.01	0.01	0.21	0.21	0.0
Mean of 8 controls	0.09			0.06	0.09	0.05	0.05	0.09	0.04	100.0
Standard deviation of 8 controls	0.005			0.01	0.005	0.01	0.004	0.005	0.01	0.0

[a]Scheme C is a modification of scheme B; the *E. coli* inoculum of about 4000/mL was added after 15 min to solutions containing a 0.05 mg/L FAC level and enumerated after a 5-min disinfection period (t = 20 min).

594 WATER CHLORINATION

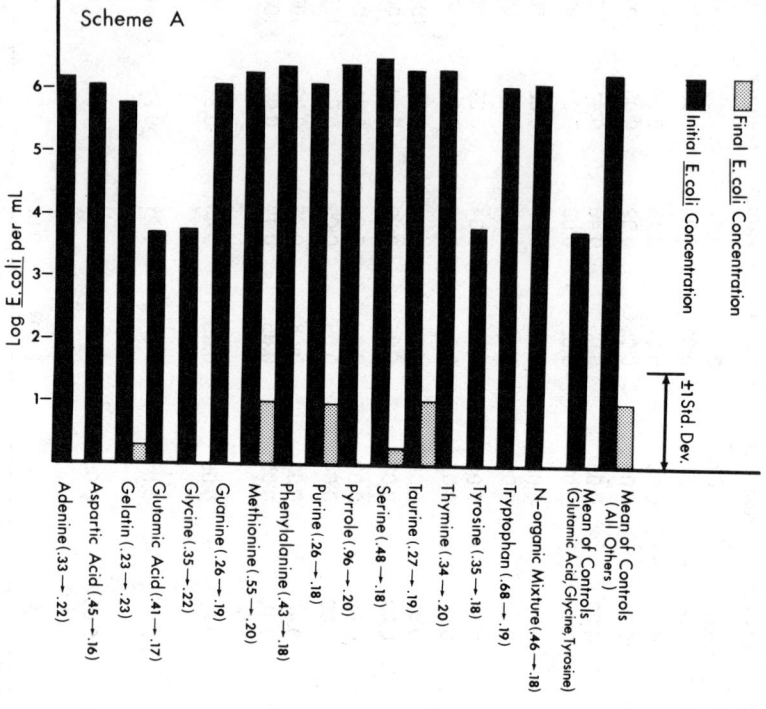

Figure 4. Disinfection of *E. coli* at pH 8 (according to scheme A).

tions for decreases in death rate with time include poor distribution of disinfectant, localized concentrations of organisms, declining disinfectant concentration, and other interfering factors.[24,25]

The results of chlorinating *E. coli* in 1-mg/L taurine solutions (at pH 9 and 10°C) in accordance with scheme C are shown in Figure 5. An average 20% *E. coli* kill resulting from 5 min of exposure to 10°C without added chlorine is also shown. The experimental conditions of higher pH, colder temperature, lower FAC residual (0.05 mg/L as Cl_2), and shorter disinfection period (5 min) were chosen to obtain conditions of weaker disinfectant potency. The results clearly indicate decreased disinfection efficiency under these conditions, in the presence of 1 mg/L taurine, over controls containing no added nitrogen compound. An average of 550 *E. coli* survived a 5-min exposure to an apparent

Table II. Disinfection Effectiveness of Chlorine on *E. coli* in the Presence of N-Organic Compounds According to Scheme A (pH 8)[a]

Compound	Initial *E. coli* Concentration (bacteria/mL × 10³)			Final *E. coli* Concentration (bacteria/mL)			Kill (%)
	MF	Standard plate count	Mean	MF	Standard plate count	Mean	
Adenine	1,710	1,630	1,670	0	0	0	100.000
Control	1,710	1,630	1,670	0	0	0	100.000
Aspartic acid	1,470	1,060	1,270	0	0	0	100.000
Control	1,470	1,060	1,270	0	5	3	99.9998
Gelatin	970	210	590	0	4	2	99.9997
Control	970	210	590	0	1	1	99.9998
Glutamic acid	6,300	5,700	6,000	0	0	0	100.000
Control	6,300	5,700	6,000	0	0	0	100.000
Glycine	6,000	6,900	6,500	0	0	0	100.000
Control	6,000	6,900	6,500	0	0	0	100.000
Guanine	1,050	1,460	1,260	0	0	0	100.000
Control	1,050	1,460	1,260	0	0	0	100.000
Methionine	1,590	2,090	1,840	2	15	9	99.9995
Control	1,590	2,090	1,840	18	27	23	99.9988
Phenylalanine	2,610	2,610	2,610	0	0	0	100.000
Control	2,610	2,610	2,610	0	0	0	100.000
Purine	1,610	1,330	1,470	6	11	9	99.9994
Control	1,610	1,330	1,470	5	9	7	99.9995
Pyrrole	2,530	2,730	2,630	0	0	0	100.000
Control	2,530	2,730	2,630	44	85	65	99.9975
Serine	2,870	3,430	3,150	1	5	3	99.9999
Control	2,870	3,430	3,150	0	0	0	100.000
Taurine	2,300	2,310	2,310	6	13	10	99.9996
Control	2,300	2,310	2,310	6	10	8	99.9997

Table II, continued

Compound	Initial E. coli Concentration (bacteria/mL × 10³)			Final E. coli Concentration (bacteria/mL)			Kill (%)
	MF	Standard plate count	Mean	MF	Standard plate count	Mean	
Thymine	2,100	2,070	2,090	0	0	0	100.000
Control	2,100	2,070	2,090	0	0	0	100.000
Tryptophan	1,270	1,160	1,220	0	0	0	100.000
Control	1,270	1,160	1,220	0	0	0	100.000
Tyrosine	6,800	6,000	6,400	0	0	0	100.000
Control	6,800	6,000	6,400	0	0	0	100.000
N-mixture[b]	1,400	1,470	1,440	0	0	0	100.000
Control	1,400	1,470	1,440	0	0	0	100.000
Mean of 12 controls[c]	1,865	1,838	1,843	6	11	8.9	99.9996
Standard deviation of 12 controls[c]	626	876	738	13	24	18.8	0.0007

[a] N-Organic compounds present at 1-mg/L concentrations; study conducted according to scheme A (at pH 8) after a 15-min disinfection period with a final FAC residual of 0.2 mg/L Cl_2; MF, membrane filtration enumeration technique.
[b] Mixture consisted of compounds listed, excluding guanine, prepared to contain a 1-mg/L nitrogen concentration.
[c] Excluding glutamic acid, tyrosine, glycine, and N-organic mixture.

DISINFECTION   597

Table III. Testing of N-Organic Compounds According to Scheme A (pH 8; all units are in mg/L as Cl_2[a])

Compound	Chlorine added	Mol Cl_2 / Mol compound	Mol Cl_2 / Mol Nitrogen	2-min DPD Free residual	17-min DPD Free residual	17-min Titrimeter Free residual	17-min Titrimeter Total residual	17-min Titrimeter Combined residual	Kill (%)
Adenine	0.43	0.82	0.16	0.33	0.22	0.20	0.42	0.22	100.000
Control	0.35				0.20	0.18	0.34	0.16	100.000
Aspartic acid	1.015	1.91	1.91	0.45	0.16	0.15	0.38	0.23	100.000
Control	0.33				0.20	0.14	0.25	0.11	99.9998
Gelatin	0.61	1.11	0.79	0.23	0.23	0.05	0.53	0.48	99.9997
Control	0.33				0.28	0.08	0.24	0.16	99.9998
Glutamic acid[b]	0.87	1.80	1.80	0.41	0.17	0.18	0.55	0.37	100.000
Control	0.25				0.20	0.20	0.25	0.05	100.000
Glycine[b]	1.20	1.27	1.27	0.35	0.22	0.25	1.21	0.96	100.000
Control	0.25				0.20	0.20	0.25	0.05	100.000
Guanine	0.31	0.66	0.13	0.26	0.19	0.15	0.24	0.09	100.000
Control	0.31				0.20	0.16	0.25	0.09	100.000
Methionine	1.53	3.22	3.22	0.55	0.20	0.18	0.77	0.59	99.9995
Control	0.35				0.17	0.18	0.32	0.14	99.9988
Phenylalanine	0.89	2.07	2.07	0.43	0.18	0.20	0.63	0.43	100.000
Control	0.35				0.18	0.19	0.32	0.13	100.000
Purine	0.34	0.58	0.14	0.26	0.18	0.18	0.26	0.08	99.9994
Control	0.33				0.18	0.17	0.30	0.13	99.9995
Pyrrole	3.43	3.25	3.25	0.96	0.20	0.17	0.77	0.60	100.00
Control	0.35				0.20	0.20	0.35	0.15	99.9975
Serine	1.11	1.64	1.64	0.48	0.18	0.17	0.88	0.71	99.9999
Control	0.35				0.20	0.21	0.34	0.13	100.0000
Taurine	0.89	1.57	1.57	0.27	0.19	0.18	0.82	0.64	99.9996
Control	0.35				0.20	0.20	0.32	0.12	99.9997

Table III, continued

Compound	Chlorine added	Mol Cl₂ / Mol compound	Mol Cl₂ / Mol Nitrogen	2-min DPD Free residual	17-min DPD Free residual	17-min Titrimeter Free residual	17-min Titrimeter Total residual	17-min Titrimeter Combined residual	Kill (%)
Thymine	0.60	1.07	0.54	0.34	0.20	0.20	0.33	0.13	100.000
Control	0.35				0.18	0.20	0.30	0.10	100.000
Tryptophan	2.02	5.81	5.81	0.68	0.19	0.20	0.26	0.06	100.000
Control	0.31				0.19	0.20	0.38	0.18	100.000
Tyrosine[b]	0.86	2.20	2.20	0.35	0.18	0.17	0.50	0.33	100.000
Control	0.25				0.20	0.22	0.25	0.03	100.000
N-Mixture[c]	1.28		0.25	0.46	0.18	0.17	0.48	0.31	100.000
control	0.31				0.18	0.16	0.25	0.09	100.000
Mean of 16 samples	1.08	1.93	1.67	0.43	0.19	0.18	0.56	0.39	99.9999
Standard deviation of 16 samples	0.77	1.34	1.50	0.19	0.02	0.04	0.27	0.26	0.0002
Mean of 16 controls	0.32				0.20	0.18	0.29	0.11	99.9997
Standard deviation of 16 samples				0.02	0.03	0.05	0.04	0.26	0.0006

[a]N-Organic compounds present at 1 mg/L concentrations; study conducted according to scheme A (pH 8) after a 15-min disinfection period (final FAC, 0.2 mg/L Cl₂).
[b]E. coli inocula for glutamic acid, tyrosine, and glycine was equal to 6000 bacteria per mL; all other inocula equal to 2 × 10⁶ bacteria per mL.
[c]Mixture consisted of compounds listed, exlcuding guanine, prepared to contain a 1 mg/L nitrogen concentration.

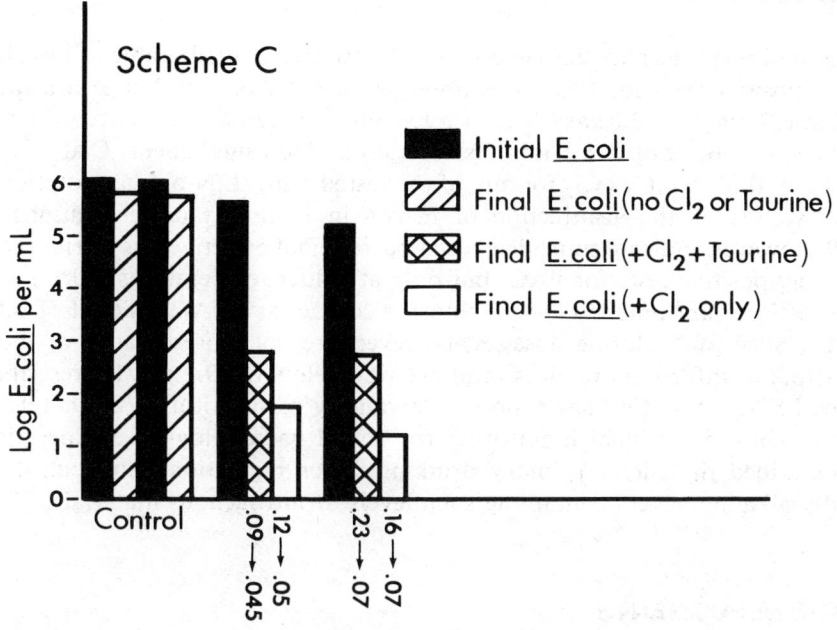

Figure 5. Disinfection of *E. coli* in 1 mg/L taurine at pH 9 and 10°C (according to scheme C).

final FAC residual of about 0.05 mg/L as Cl_2 as compared with an average of 33 *E. coli* in controls without taurine at the same measured FAC residual. The FAC and total chlorine levels in the taurine solutions at the end of the disinfection period were 0.045 and 0.68 mg/L as Cl_2, respectively, for replicate one, and 0.07 and 0.77 mg/L as Cl_2, respectively, for replicate two. These values compared with mean FAC and total chlorine concentrations in the controls of 0.06 and 0.16 mg/L, respectively. The higher total residual chlorine levels in the taurine solutions were therefore most likely comprised of nongermicidal or very weakly germicidal forms. The greater *E. coli* survival in the taurine solutions of almost two orders of magnitude over that in controls at the same apparent FAC can be attributed to an overestimation of the bactericidal potency in the chlorinated taurine solutions resulting from a measured falsely positive level of FAC.

CONCLUSIONS

No significant decrease in bactericidal effectiveness, attributable to possible falsely positive tests for FAC, was observed at pH 7 or 8 and at a measured FAC of 0.2 mg/L $\pm$ 0.02 as Cl_2 after a 15-min disinfection period according to schemes A or B, or after a 5-min disinfection period using scheme C at a FAC of 0.05 $\pm$ 0.01 mg/L as Cl_2 for any of the tested 1 mg/L N-organic solutions. Initial results on the disinfection of *E. coli* in 1 mg/L taurine solutions at pH 9, however, indicate some decreased bactericidal effectiveness attributable to falsely positive tests for FAC, but only at colder temperatures (10°C) and lower disinfectant potency (5-min chlorine contact at a FAC residual of 0.05 mg/L). Such low chlorine dosages, however, are not representative of most municipal disinfection practices, and are well below the 0.2 mg/L prescribed by the EPA. These findings support the validity of substituting 0.2 mg/L as Cl_2-free-chlorine residual monitoring for actual bacteriological enumeration (as described in federal primary drinking water regulations) in neutral or slightly alkaline waters containing such levels of nitrogenous materials.

ACKNOWLEDGMENTS

Research was supported by the U.S. Environmental Protection Agency, project No. R809767. Thanks are extended to Kevin Sheehan for assisting in the analytical determinations, and to Dorothy Pascoe for typing the final manuscript.

REFERENCES

1. "Environmental Protection Agency, Water Programs, National Interim Primary Drinking Water Regulations," *Fed. Reg.*, 40(248):59566 (Dec. 24, 1975).
2. Butterfield, C. T., et al. "Influence of pH and Temperature on the Survival of Coliform and Enteric Pathogens when Exposed to Free Chlorine," *Public Health Rep.* 58(51):1837 (1943).
3. Butterfield, C. T., and E. Wattie. "Influence of pH and Temperature on the Survival of Coliform and Enteric Pathogens When Exposed to Chloramine," *Public Health Rep.* 61(6):157 (1946).
4. Kelly, S., and W. W. Sanderson. "The Effect of Chlorine in Water on Enteric Viruses," *Am. J. Public Health*, 48(10):1328 (1958).
5. Kelly, S., and W. W. Sanderson. "The Effect of Chlorine in Water on Enteric Viruses. II. The Effect of Combined Chlorine on Poliomyelitis and Coxsackie Viruses," *Am. J. Public Health*, 50(1):14 (1960).
6. Kruse, C. W., et al. "Halogen Action on Bacteria, Viruses and Protozoa," in *Proceedings of the National Specialty Conference on Disinfection*, (New York: American Society of Civil Engineers, 1970), p. 113.

7. Olivieri, V. P. et al. "Inactivation of Virus in Sewage," in *Proceedings of the National Specialty Conference on Disinfection*, (New York: American Society of Civil Engineers, 1970), p. 365.
8. Feng, T. H. "Behavior of Organic Chloramines in Disinfection," *J. Water Pollut. Control Fed.* 38(4):614 (1966).
9. Palin, A. "Study of Chloroderivatives of Ammonia and Related Compounds With Specific Reference to Their Formation in Chlorination of Natural and Polluted Waters," *Water Water Eng.* 54(10):151; 54(11):189; 54(12)248 (1950).
10. Wei, I. W., and J. C. Morris. "Dynamics of Breakpoint Chlorination," in *Chemistry of Water Supply, Treatment and Distribution*, A. Rubin, Ed. (Ann Arbor, MI: Ann Arbor Science Publishers, Inc., 1974).
11. Saunier, B. M., and R. E. Selleck. "The Kinetics of Breakpoint Chlorination in Continuous Flow Systems," *J. Am. Water Works Assoc.* 71(3):164 (1979).
12. Cooper, W. J., C. A. Sorber, and E. P. Meier. "A Rapid Specific Free Available Chlorine Test with Syringaldazine (FACTS)," *J. Am. Water Works Assoc.* 67(1):34 (1975).
13. Strupler, N. "A Study of Interferences in the Measurement of Free and Combined Chlorine in Water by the DPD and Syringaldazine Methods," in *Proceedings AWWA Water Quality Technology Conference*, Louisville, KY, (Denver: American Water Works Association, 1978).
14. Cooper, W. J., N. M. Roscher, and R. A. Slifker. "Determining Free Available Chlorine by DPD-Colorimetric, DPD-Steadifac, and FACTS Procedures," *J. Am. Water Works Assoc.* 74(7):362 (1982).
15. Guter, K. J., and W. J. Cooper. *The Evaluation of Existing Field Test Kits for Determining Free Chlorine Residuals in Aqueous Solutions*, USAMEERU Report 73-03, (Aberdeen Proving Ground, MD: US Army Medical Envir. Engrg. Res. Unit, 1972).
16. Guter, K. J., W. J. Cooper, and C. A. Sorber. "Evaluation of Existing Field Test Kits for Determining Free Chlorine Residuals in Aqueous Solutions," *J. Am. Water Works Assoc.* 66(1):38 (1974).
17. Snead, M. C., V. P. Olivieri, and W. H. Dennis. "Biological Evaluation of Methods for the Determination of Free Available Chlorine," in *Chemistry in Water Reuse, Vol. 1*, W. J. Cooper, Ed. (Ann Arbor, MI: Ann Arbor Science Publishers, Inc., 1981).
18. Meier, E. P., and C. A. Sorber. *Development of a Rapid Specific Free Available Chlorine Test with Syringaldazine [FACTS]*, Tech. Rept. 7405, (Aberdeen Proving Ground, MD: US Army Medical Bioengrg. Res. and Devel. Lab., 1974).
19. Sorber, C. A., W. J. Cooper, and E. P. Meier. "Selection of a Field Method for Free Available Chlorine," in *Disinfection Water and Wastewater*, J. D. Johnson, Ed. (Ann Arbor, MI: Ann Arbor Science Publishers, Inc., 1975).
20. Duursma, E. K. and P. Parsi. "Persistence of Total and Combined Chlorine in Sea Water," *Netherlands J. Sea Res.* 10(2):192 (1976).
21. Wajon, J. E. and J. C. Morris. "The Analysis of Free Chlorine in the Presence of Nitrogenous Organic Compounds," *Environ. Int.* 3(1):41 (1980).
22. *Standard Methods for the Examination of Water and Wastewater*, 15th Ed., (Washington, DC: American Public Health Association, 1980).
23. Ram, N. M., and J. P. Malley. "Validity of Chlorine Residual Monitoring in the Presence of N-Organic Compounds," *J. Am. Water Works Assoc.* 76(9):74 (1984).

24. Weber, W. *Physicochemical Processes for Water Quality Control*, (New York: Wiley Interscience, 1972).
25. Fair, G. M., J. C. Geyer, and D. A. Okun. *Elements of Water Supply and Wastewater Disposal*, (New York: John Wiley and Sons, Inc., 1971).

CHAPTER 47

Disinfection Resistance of *Legionella pneumophila* and *Escherichia coli* Grown in Continuous and Batch Culture

James D. Berg, John C. Hoff, Paul V. Roberts
and Abdul Matin

Several *Legionella* species isolated from a diverse set of aquatic environments[1-8] have been reported to exhibit differing sensitivities to disinfectants.[9,10] Laboratory studies have shown that batch-grown *Legionella pneumophila* is readily inactivated by chlorine as the hypochlorite ion.[11] However, we demonstrated previously that batch-grown cultures of bacteria are more sensitive to disinfectants; therefore, they are not representative of their naturally occurring counterparts.[12-14] No studies have been reported that evaluate the efficacy of chlorine dioxide (ClO_2) against *Legionella*.

The objectives of this study were (1) to assess the sensitivity of *L. pneumophila*, grown under nutrient limitation in a chemostat, to ClO_2; (2) to compare the sensitivity of *L. pneumophila* and *Escherichia coli*, predominant constituents of the widely accepted coliform indicator group; and (3) to contrast the efficacy against *L. pneumophila* of ClO_2 with that of chlorine.

MATERIALS AND METHODS

Cultures of *L. pneumophila* and *E. coli* were grown in batch and continuous culture as described previously.[13-16] Simulation of nutrient-limited growth at submaximal rates was accomplished in a chemostat. Variables were growth temperature and growth rate, while substrate concentration, pH, dissolved oxygen, and agitation rate were held constant.

A special disinfection reactor was built to contain the aerosol-transmitted *Legionella* (Figure 1). The vessel was designed to be filled, inoculated, dosed, drained, and rinsed aseptically, and to be sampled rapidly and under negative pressure into a biohazard cabinet. Suspensions of cultures at a concentration of about 10^6 cfu/mL were dosed with ClO_2 or chlorine to achieve an initial concentration equal to 0.75 mg/L as ClO_2. Survivors were enumerated on CYE agar[11] (*L. pneumophila*) or *m*-Endo agar[12] (*E. coli*). Each set of growth conditions was evaluated in triplicate experiments.[12-16]

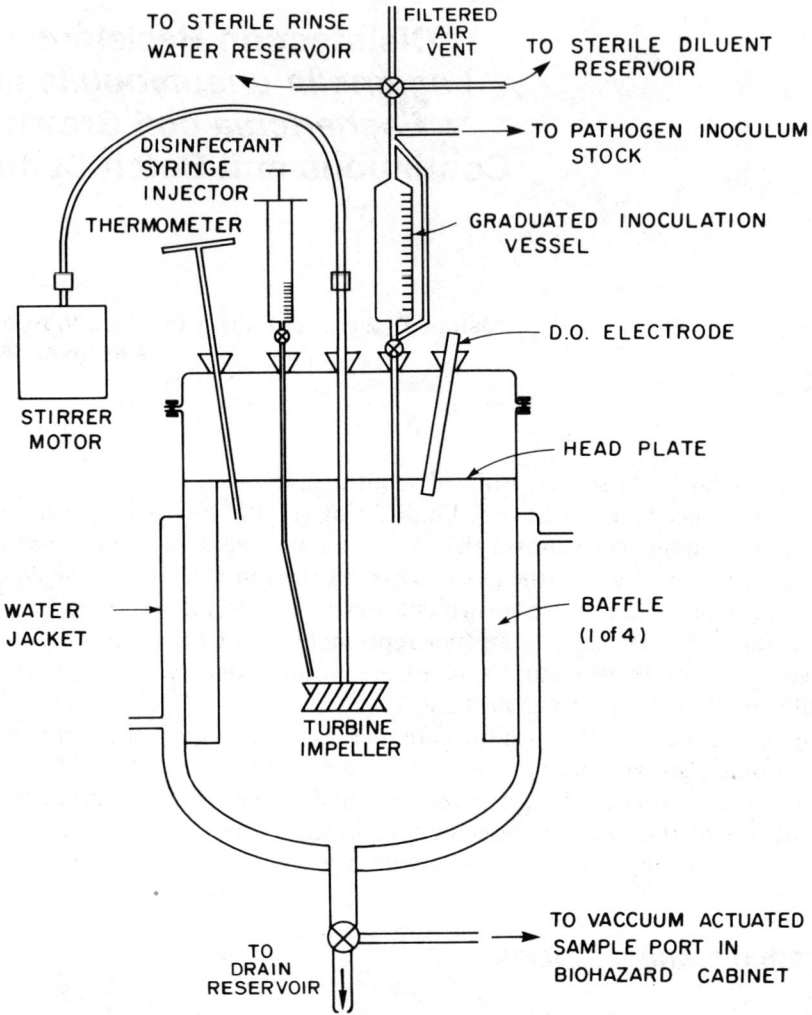

Figure 1. Schematic of 1-L disinfection reactor for *Legionella pneumophila*.

A biphasic survivor curve with rapid initial inactivation followed by a constant plateau was consistently observed. The plateau represents a resistant surviving subpopulation, the size of which is determined by the growth environment. This phenomenon is discussed in detail elsewhere.[12,16,17] The ratio of the size of the surviving fraction at 15 min (i.e., log N_{15}/N_0) was the basis for comparing the effects of the different environmental variables.

RESULTS

Effect of Growth Rate and Growth Temperature on Sensitivity of L. pneumophila

The results of experiments to assess the effects of growth rate and growth temperature on the sensitivity of *L. pneumophila* to ClO_2 are shown in Table I. An approximate fivefold or greater difference in dilution rate was required before a difference in sensitivity was observed at a growth temperature equal to 37°C. For example, there was a significant difference between the survival ratios at $D = 0.03$ h^{-1} and $D = 0.15$ h^{-1}. However, at a growth temperature equal to 44°C, a significant difference in survival ratios was observed over a twofold difference in growth rate, that is, $D = 0.03$ to $D = 0.06$ h^{-1}. Otherwise, smaller changes in D do not reveal a substantial difference in sensitivity at either growth temperature.

The basis for comparing the effects of growth conditions on sensitivity was the t-test using alpha = 0.05 with 16 degrees of freedom (Table II).[18] There was no consistent relationship between growth temperature and sensitivity, although only two temperatures, 37 and 44°C, were studied. Statistically significant differences in sensitivity were found for differences in growth rate (D) of 0.03 vs 0.06 and 0.06 vs 0.20 at a growth temperature equal to 44°C. These particular comparisons were chosen to demonstrate that large differences in growth rate are typically required to achieve significant differences in sensitivity.

Comparison of L. pneumophila with E. coli

The rates of initial inactivation and size of the recalcitrant fraction (i.e., log N_{15}/N_0) of *L. pneumophila* and of *E. coli* are compared in Figures 2 and 3. A large difference in sensitivity between the two species is revealed when either is

Table I. Effect of Growth Temperature and Growth Rate on Sensitivity of *L. pneumophila* to 0.75 mg/L ClO_2

Dilution Rate D (h^{-1})	Growth Temp (°C)	Surviving Fraction (log(N_t/N_0) at $t = 15$ min)
0.03	44	−2.99 ± 0.46
0.06	44	−4.29 ± 0.37
	37	−4.13 ± 0.36
0.15	44	−4.33 ± 0.32
	37	−5.06 ± 0.40
0.20	44	−4.86 ± 0.49
	37	−4.75 ± 0.45
0.30	37	−5.09 ± 0.30

Table II. Comparison of Selected Values from Table I Using t-Test[18] to Determine Differences in Sensitivity Resulting from Different Growth Conditions

Conditions Compared (Temp, °C, or D[a])	$\overline{Y}_2$	S_2^2	$\overline{Y}_1$	S_1^2	t[b]	$t_{[0.05,16]}$	Means Significantly Different
37° vs 44° at D = 0.06	4.13	0.13	4.29	0.14	0.92	2.12	no
37° vs 44° at D = 0.15	4.33	0.10	5.06	0.16	5.72	2.12	yes
0.03 vs 0.06 at T = 44°	2.99	0.21	4.29	0.14	8.79	2.12	yes
0.06 vs 0.20 at T = 44°	4.29	0.14	4.86	0.24	2.77	2.21	yes

[a]Dilution rate.

[b]$t = \dfrac{(\overline{Y}_1 - \overline{Y}_2) - (u_1 - u_2)}{\sqrt{\dfrac{1}{n}(S_1^2 + S_2^2)}}$

where $u_1 = u_2$, and $df = n_1 + n_2 - 2 = 16$; $\overline{Y}_1$ and $\overline{Y}_2$ are observed population means; S_2 and S_1 are observed population standard deviations; u_1 and u_2 are the population means, n_1 and n_2 are numbers of observations, and df is the degrees of freedom.

grown in batch culture (Figure 2). The difference is observed in the initial rate of inactivation as well as in the size of the recalcitrant fraction. When chemostat-grown cultures were used, however, there was a relationship between dilution rate and the size of the recalcitrant fraction that held for both species (Figure 3). Similar complex media were used in the experiments chosen for comparison. *L. pneumophila* was grown in yeast extract and *E. coli* was grown in nutrient broth at the same growth temperature, 37°C. Within the range of dilution rates studied, *E. coli* and *L. pneumophila* do not exhibit significantly different sensitivities to ClO_2.

Comparison of ClO_2 and Chlorine

One experiment was conducted to compare the efficacy of treatments with ClO_2 or chlorine. The results indicate that ClO_2 was more effective than chlorine (Figure 4) in the inactivation of batch-grown *L. pneumophila*. Experiments were not conducted with chemostat-grown cultures.

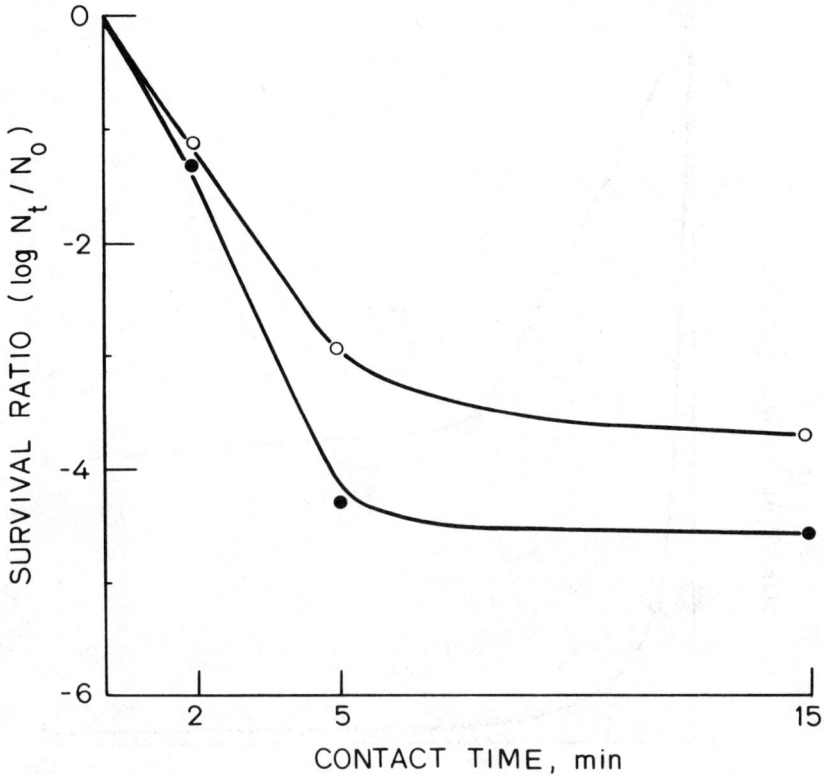

Figure 2. Comparison of inactivation of batch-grown *E. Coli* (o) and *L. pneumophila* (•); cultures were grown at 37°C and harvested at late log phase; ClO_2 dose was 0.75 mg/L.

DISCUSSION

Effect of Growth Rate and Growth Temperature on Sensitivity of *L. pneumophila*

The same trend that had been observed for *E. coli* grown in a complex medium in the chemostat[12] was also observed for *L. pneumophila*; that is, increasing growth rate yielded a more sensitive population as characterized by the decreased size of the resistant fraction. The effect of varying the growth rate was less pronounced for *L. pneumophila* than for *E. coli*. This may be explained in either of two ways: (1) the incremental changes in the growth rate did not affect the physiological target for inactivation, or (2) the precision of the experiments was such that more subtle changes in sensitivity could not be

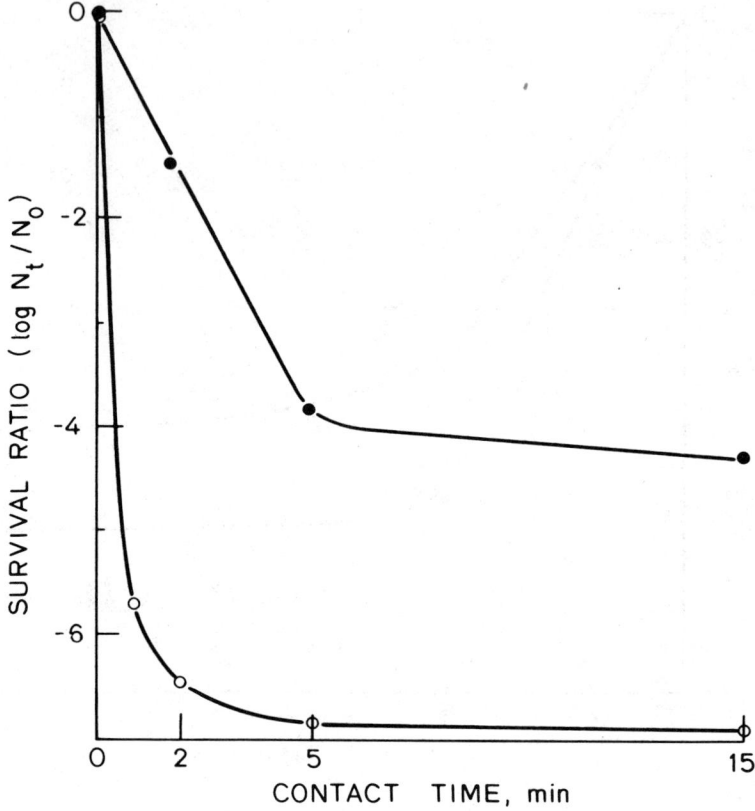

Figure 3. Effect of dilution rate on sensitivity to a dose of 0.75 mg/L ClO_2; *L. pneumophila* (•); *E. Coli* (○).

distinguished. The most notable difference in sensitivity is found at the two extremes of dilution rates (D = 0.03 h^{-1} and D = 0.30 h^{-1}) studied. If it is assumed that these two conditions more closely approximate the fast growth rate obtained at log phase in batch culture (D = 0.30 h^{-1}) and the submaximal rates achieved in natural systems (D = 0.03 h^{-1}), then the results support the major thesis of the study: conditions more closely resembling nutrient-limited, suboptimal growth produce bacterial populations that are more resistant to chemical disinfection.

Based on previous experiments with *E. coli*,[12] a relationship between growth temperature and sensitivity for *L. pneumophila* was plausible, although none

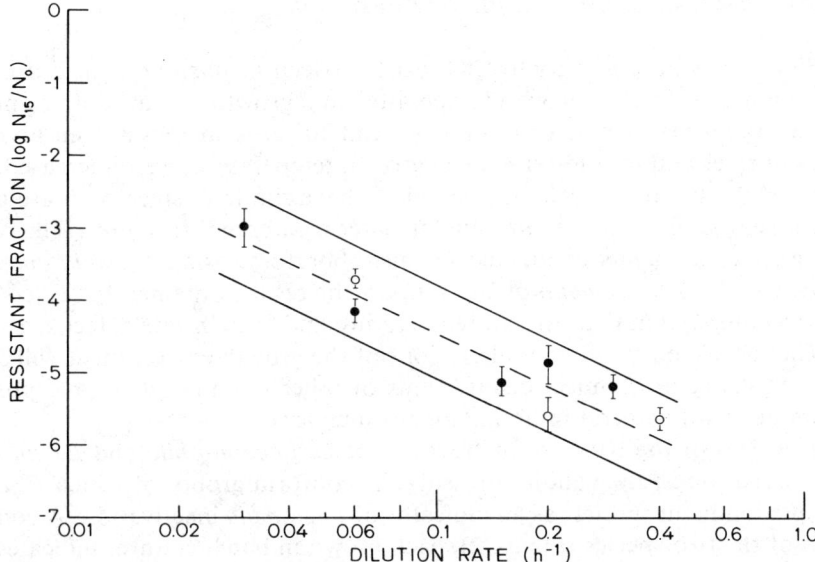

Figure 4. Comparison of ClO_2 and chlorine against batch-grown *L. pneumophila*; dose was 0.75 mg/L as ClO_2 (•) or Cl_2 (○).

was observed. This may reflect the capability of the organism to grow over a wide range of temperatures, with an optimum within the range bracketed by the temperatures that were studied (37° to 44°C). It is reasonable to expect that an organism with a broad range of growth temperatures would not experience physiological changes as a result of a small change in temperature sufficient to effect sensitivity.

Attempts to grow cultures in the laboratory at higher (50°C) and lower (25°C) temperatures, which were desirable from the perspective of modeling environmental conditions, were of limited success. Growth of a steady-state culture at 25°C was not attained at $D \leq 0.03$ h^{-1} (i.e., washout occurred). The culture remained viable at a temperature of 50°C, but a steady state was not attained. It was determined from the washout rate[19] that the U_{max} at 50°C was $\cong D = 0.02$ h^{-1}. It has been observed that changes in growth temperature affect the lipid structure and outer membrane permeability.[20] Further studies could verify whether changes in a wider range of growth temperatures affect the sensitivity to chemical disinfectants. *L. pneumophila* would be an ideal organism because of its ability to grow over a wide temperature range in nature.[3]

Comparison of L. pneumophila with E. coli

Similarities in sensitivity to ClO_2 exist between *L. pneumophila* and *E. coli* regarding the effect of growth temperature and growth rate in complex media. Both organisms exhibited a general trend towards increasing sensitivity as growth rate, and to a lesser extent, growth temperature, were increased. It is reasonable that both organisms should behave similarly since both are gram-negative organisms possessing similar outer membrane structures. The results of other experiments conducted in this laboratory, using *Yersinia enterocolitica* and *Klebsiella pneumoniae*, support the results obtained for *E. coli* and *L. pneumophila* in that growth temperature and growth rate affect sensitivity to ClO_2.[21] We postulate that alterations of the growth environment will affect the sensitivity to chemical disinfectants of other gram-negative (and possibly gram-positive) bacteria in an analogous manner.

The sizes of the recalcitrant fractions of *L. pneumophila* and *E. coli* were compared to test the validity of using the coliform group, of which *E. coli* is the predominant species, as an indicator of *Legionella* inactivation. A comparison of the two species (Figure 2), each grown in batch culture, indicates that *L. pneumophila* was more resistant to ClO_2 with respect to initial rate and extent of inactivation. The results of this conventional batch-culture procedure lead us to conclude that *E. coli* is not a conservative indicator of the inactivation of *L. pneumophila*. However, a comparison of chemostat-grown populations (Figure 3) leads to a different conclusion. The apparent linear relationship of the data strongly suggests that (1) the two species behave similarly over a range of submaximum growth rates, and (2) a positive correlation exists between increasing sensitivity to ClO_2 and increasing dilution rate. It can be inferred from these data that *E. coli* and *L. pneumophila* exhibit similar sensitivity to ClO_2 under nutrient-limited conditions likely to be found in nature. Therefore, *E. coli* should be an adequate indicator for the effectiveness of *L. pneumophila* inactivation by ClO_2. Clearly, further testing would be required along with in situ verification to extrapolate these conclusions to real situations.

The fact that such different conclusions can be drawn from either the use of batch-grown or chemostat-grown cultures is supported by comparing the results of Kuchta et al.[22] with this study. Kuchta et al.[22] have shown batch-grown *L. pneumophila* to be significantly more resistant to chlorine than *E. coli*, whereas we show that the two species respond similarly to exposure to ClO_2 following growth in the chemostat.

Comparison of Chlorine Dioxide and Chlorine

The results of these experiments indicate that ClO_2 is superior to chlorine when compared at equal mass dose. Similar results have been observed by other investigators using other species.[23-28] Since the molecular weights are nearly identical, a comparison on an equimolar basis would yield essentially the same results. Although the comparison on a mass-dose or molar basis is practical from an economic perspective, other factors must be considered before concluding which of the two disinfectants is superior. If we were to compare the disinfectants on a residual basis rather than a dose basis, the composition of the aqueous solution would be a critical factor. For example, water containing a high concentration of ammonia nitrogen would produce predominantly monochloramine, an inferior bactericide, when treated with chlorine, whereas the ClO_2 would be largely unaffected by the presence of ammonia. In that case, comparison on a total residual basis would yield drastically different information than a comparison on a dose basis because of significant extraneous side reactions with one disinfectant.

CONCLUSIONS

Chlorine dioxide was effective in inactivating *L. pneumophila*, using as a criterion the size of the resistant fraction, when compared with chlorine on an equal-dose basis. On this basis ClO_2 can be considered to be a superior alternative to chlorine for the control of *L. pneumophila* in water treatment.

The validity of *E. coli* as an indicator for the control of *L. pneumophila* was not supported by experiments using batch-grown cultures. The opposite conclusion could be drawn from the results of the chemostat studies. If the chemostat as used in this study is accepted as a laboratory model of natural aquatic environments, then *E. coli* would be an acceptable indicator organism. The results, of course, represent only a limited number of laboratory experiments and would need to be verified by further experiments.

The results cast a new light on the customary methods used to evaluate microbial contaminants of concern to public health. The use of the chemostat is an innovative methodology for the evaluation of antimicrobial agents in the laboratory and may be preferable to the use of conventional, batch-culture techniques. The simulation in the laboratory of the environmental system of concern can provide populations that more closely approximate the resistance to treatment that can be expected in a field evaluation.

ACKNOWLEDGMENTS

This work represents a portion of the Ph.D. dissertation research completed by James D. Berg in the Civil Engineering Department at Stanford University in 1984. Support by Cooperative Agreement (CR 808986010) from the U.S. Environmental Protection Agency is gratefully acknowledged. This chapter has not been subjected to the Agency's required administrative and peer review and therefore does not necessarily reflect the views of the Agency, and no official endorsement should be inferred. The results of the study were presented, in part, at the Second International Symposium on *Legionella* in Atlanta, June 19-23, 1983.

REFERENCES

1. Morris, G. K., C. M. Patten, J. C. Feeby, S. E. Johnson, G. Gorman, and W. T. Martin. "Isolation of the Legionnaires' Disease Bacterium from Environmental Samples," *Ann. Int. Med.* 90:664-666 (1979).
2. Fliermans, C. B., W. B. Cherry, L. H. Orrison, S. J. Smith, D. L. Tison, and D. H. Pope. "Ecological Distribution of *Legionella pneumophila*," *Appl. Environ. Microbiol.* 41:9-16 (1981).
3. Fliermans, C. B., W. B. Cherry, L. H. Orrison, and L. Thacker. "Isolation of *Legionella pneumophila* from Nonepidemic-Related Aquatic Habitats," *Appl. Environ. Microbiol.* 37:1239-1243 (1979).
4. Orrison, L. H., W. B. Cherry, and D. Milan. "Isolation of *Legionella pneumophila* from Cooling Tower Water by Filtration," *Appl. Environ. Microbiol.* 41:1202-1205 (1981).
5. Orrison, L. H., W. B. Cherry, R. L. Tyndall, C. B. Fliermans, S. B. Gough, M. A. Lamber, L. K. McDougal, W. F. Bibb, and D. J. Brenner, "*Legionella oakridgensis*: Unusual New Species Isolated from Cooling Tower Water," *Appl. Environ. Microbiol.* 45:536-545 (1983).
6. Orrison, L. H., W. B. Cherry, C. B. Fliermans, S. B. Dees, L. K. McDougal, and D. J. Dodd, "Characteristics of Environmental Isolates of *Legionella pneumophila*," *Appl. Environ. Microbiol.* 42:109-115 (1981).
7. Tison, D. L., D. H. Pope, W. B. Cherry, and C. B. Fliermans. "Growth of *Legionella pneumophila* in Association with Blue-Green Algae (Cyanobacteria)," *Appl. Environ. Microbiol.* 39:456-459 (1980).
8. Wadowsky, R. M., and R. B. Yee, "Glycine-Containing Selective Medium for Isolation of Legionellaceae from Environmental Specimens," *Appl. Environ. Microbiol.* 42:768-772 (1981).
9. Edelstein, P. H., R. E. Whittaker, R. L. Kreiling, and C. L. Howell, "Efficacy of Ozone in Eradication of *Legionella pneumophila* from Hospital Plumbing Fixtures," *Appl. Environ. Microbiol.* 44:1330-1334 (1982).
10. Tison, D. L., and R. J. Seidler, "*Legionella* Incidence and Density in Potable Drinking Water Supplies," *Appl. Environ. Microbiol.,* 45:337-339 (1983).
11. Wang, W. L. L., M. J. Blaser, J. Cravens, and M. A. Johnson. "Growth, Survival, and Resistance of the Legionnaires' Disease Bacterium," *Ann. Int. Med.* 90:614-619 (1979).
12. Berg, J. D., A. Matin, and P. V. Roberts. "Effect of Antecedent Growth Conditions on Sensitivity of *E. coli* to ClO_2," *Appl. Environ. Microbiol.* 44(4):814-819 (1982).

13. Berg, J. D., J. C. Hoff, P. V. Roberts, and A. Matin. "Growth of *Legionella pneumophila* in Continuous Culture and its Sensitivity to Inactivation by Chlorine Dioxide," in *Proceedings of the Second International Symposium on Legionella*, C. Thornsberry, A. Balows, J. C. Feeley, and W. Jakubowski, Eds., (Washington, DC: American Society for Microbiology, 1984), pp. 68-70.
14. Berg, J. D., A. Matin, and P. V. Roberts. "Growth of Disinfectant-Resistant Bacteria and Simulation of Natural Aquatic Environments in the Chemostat," in *Water Chlorination: Environmental Impact and Health Effects, Vol. 4*, R. L. Jolley, W. A. Brungs, J. A. Cotruvo, R. B. Cumming, J. S. Mattice, and V. A. Jacobs, Eds. (Ann Arbor, MI: Ann Arbor Science Publishers, Inc., 1983), pp. 1137-1147.
15. Berg, J. D., J. C. Hoff, A. Matin, and P. V. Roberts. 'Growth of *Legionella pneumophila* in Continuous Culture," (Submitted for publication).
16. Berg, J. D. "Effect of the Antecedent Growth Environment on the Sensitivity of *Legionella pneumophila* and *Escherichia coli* to Chlorine Dioxide," Ph.D. Thesis, (Stanford, CA: Stanford University, 1984).
17. Berg, J. D., J. C. Hoff, A. Matin, and P. V. Roberts. "Significance of the Bacterial Subpopulation Resistant to Disinfection," (Submitted for publication).
18. Sokal, R. R., and F. J. Rohlf. *Biometry*. (San Francisco: W. H. Freeman and Co., 1969).
19. Jannasch, H. W. "Estimation of Bacterial Growth Rates in Natural Waters," *J. Bacteriol.* 99:156-160 (1969).
20. Singh, U. N., "Adaptation in Microorganisms: Variation in Macromolecular Composition With Growth Rate," *J. Theoret. Biol.* 59:107-126 (1976).
21. Harakeh, M. S. "Response of Chemostat-Growth Enteric Bacteria to Chlorine Dioxide," Chapter 48, this volume.
22. Kuchta, J. M., S. J. States, A. M. McNamara, W. M. Wadowsky, and R. B. Yee. "Susceptibility of *Legionella pneumophila* to Chlorine in Tap Water," *Appl. Environ. Microbiol.* 46(5):1134-1139 (1983).
23. E. M. Aieta, J. D. Berg, P. V. Roberts, and R. C. Cooper. "Comparison of Chlorine Dioxide and Chlorine in the Disinfection of Wastewater," *J. Water Pollut. Control Fed., 52*(4):810-822 (1980).
24. Scarpino, P. V., et al. *Effect of Particulates on Disinfection of Enteroviruses in Water by Chlorine Dioxide*, EPA-600/2-79-054, (Cincinnati: U.S. Environmental Protection Agency, 1979).
25. Benarde, M. A., B. M. Israel, V. P. Olivieri, and M. L. Granstrom. "Efficiency of Chlorine Dioxide as a Bactericide," *App. Microbiol.* 13:776-780 (1965).
26. Benarde, M. A., W. B. Snow, V. P. Olivieri, and B. Davidson. "Kinetics and Mechanism of Bacterial Disinfection by Chlorine Dioxide," *Appl. Microbiol.* 15(2):257-265 (1967).
27. Ridenour, G. M., and E. H. Armbruster. "Bactericidal Effect of Chlorine Dioxide," *J. Am. Water Works Assoc.* 41:537-550 (1949).
28. Ridenour, G. M., and R. S. Ingols, "Bactericidal Properties of Chlorine Dioxide," *J. Am. Water Works Assoc.* 39:561-567 (1947).

CHAPTER 48

Response of Chemostat-Grown Enteric Bacteria to Chlorine Dioxide

M. S. Harakeh, J. C. Hoff, and A. Matin

We have previously demonstrated that growth of *Escherichia coli* at submaximal rates caused by nutrient limitation and relatively low temperatures greatly increases its resistance to chlorine dioxide (ClO_2) and phenylphenol.[1,2] We were interested in investigating whether other organisms behave similarly to *E. coli*. For this reason, we decided to extend this study to include two pathogens of considerable concern to public health, *Klebsiella pneumoniae* and *Yersinia enterocolitica*.

K. pneumoniae and *Y. enterocolitica* were grown in a glucose mineral salt medium. Populations were grown either in a chemostat (Bio Flo, New Brunswick Scientific) or in batch culture. Dose-response experiments indicated that a dose of 0.25 mg/L ClO_2 resulted in significant kills while providing measurable concentrations of residual disinfectant (0.12 mg/L) and surviving organisms. Disinfection experiments were conducted in a 1-L well-mixed pyrex flask held at 23°C $\pm$ 1°. Samples were withdrawn from the reactor and mixed with $Na_2S_2O_3$ to neutralize any residual disinfectant. Samples were suitably diluted and filtered in triplicate through Gelman-GN-6 membrane filters. The filters were placed on m-endo agar plates (Difco) and incubated at 25°C for 72 h in the case of *Y. enterocolitica*, and at 37°C for 24 h in the case of *K. pneumoniae*.

Resistance was presented in terms of survival ratio (N_t/N_0) after t min of contact with chlorine dioxide.

The effect of dilution rate (D) on sensitivity was tested at two temperatures (15 and 29°C) in the case of *Y. enterocolitica*, and at three temperatures (15, 25, and 37°C) in the case of *K. pneumoniae*; *Y. enterocolitica* did not grow at 37°C in the medium used. The plots show the resistant fraction (N_{15}/N_0) as a function of D or growth temperature. At a given temperature, the culture resistance to ClO_2 increased with decreasing D.

At a given dilution rate, temperature determined the extent of the sensitivity of *K. pneumoniae*, and an increase in resistance occurred at low temperatures in the case of *K. pneumoniae* (Figure 1). Temperature had only a marginal effect on the sensitivity of *Y. enterocolitica* to ClO_2.

We have shown previously[2] that culture cell density influences the sensitivity of *E. coli* to ClO_2; that is, a higher culture cell density during growth is accompanied with greater resistance. Studying the effects of cell density on

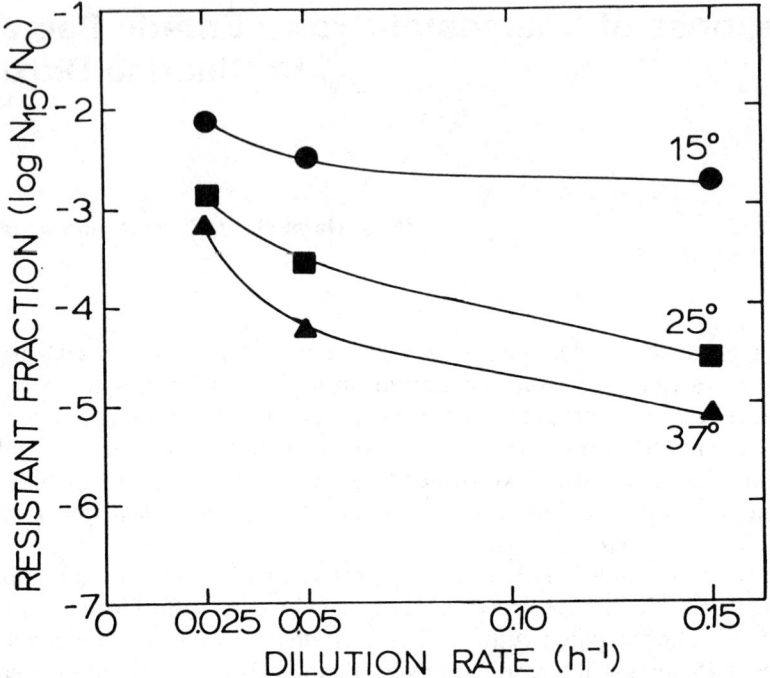

Figure 1. Effect of dilution rate on sensitivity of *Klebsiella pneumoniae* to chlorine dioxide grown under glucose limitation at three temperatures. The concentration of glucose in the inflow medium was 0.4% (●, 15°C growth temperature; ■, 25°C growth temperature; ▲, 37°C growth temperature).

sensitivity is important, because organisms in most aquatic environments grow at low cell densities compared with laboratory batch cultures. *Y. enterocolitica* cultivated at a fixed D in a chemostat at two different glucose concentrations in the inflow medium (which determines the steady-state culture density) did exhibit differences in sensitivity, the lower-density population being more resistant. *K. pneumoniae*, on the other hand, exhibited the same sensitivity after growth at two different culture densities.

The results clearly demonstrate that differences in sensitivity of microorganisms to antimicrobial agents could be attributed to their different antecedent growth conditions, and that slowly growing populations of *Y. enterocolitica* and *K. pneumoniae* are more resistant to disinfection by ClO_2 than populations grown at more rapid rates.

A comparison of the sensitivity patterns of the two microorganisms indicates that *Y. enterocolitica* was more sensitive to ClO_2 than *K. pneumoniae*.

For instance, at 15°C and $D = 0.25$ h^{-1}, treatment of *Y. enterocolitica* with ClO$_2$ produced a 5-log reduction, whereas under the same experimental conditions, *K. pneumoniae* showed a 2-log reduction.

We have concluded that populations grown under conditions that more closely approximate the natural environment are more resistant than those grown under conventional batch-culture conditions; therefore, the sensitivity of batch-culture-grown cells to antimicrobials of at least these organisms is not a suitable index of the sensitivity of natural populations.

REFERENCES

1. Abou-Schleib, H., J. D. Berg, and A. Matin. "Effect of Antecedent Growth Conditions on Sensitivity of *E. coli* to Phenylphenol," *FEMS Lett.* 19:183 (1983).
2. Berg, J. D., A. Matin, and P. V. Roberts. "Effect of Antecedent Growth Conditions on Sensitivity of *E. coli* to ClO$_2$," *Appl. Environ. Microbiol.* 44:814 (1982).

CHAPTER 49

Mode of Action of Chlorine Dioxide on Selected Viruses

V. P. Olivieri, F. S. Hauchman, C. I. Noss, and R. Vasl

Relatively few investigations have attempted to determine the mechanism of virus inactivation by chlorine dioxide. Alvarez and O'Brien[1] reported that chlorine dioxide (ClO_2) changed the isoelectric point of poliovirus 1 from pH 7.0 to 5.8, indicating that the capsid protein was altered. Treatment of the virus with 1.0 mg/L ClO_2 at pH 10.0 resulted in the inactivation of 99.9% of the infectious virions and a release of the nucleic acid, which was further evidence of a major alteration in the viral capsid. However, virus treated with the same dose of ClO_2 at pH 6.0 remained intact after 90% of the virions were inactivated. The protein-specific attachment function appeared unaltered, as demonstrated by the ability of virus to attach to host cells after treatment with ClO_2 at an unspecified pH. Sedimentation analysis of the attached virions revealed modified particles having sedimentation coefficients lower than that of intact virus, suggesting that partial uncoating of the virion had occurred. Alvarez and O'Brien[1] also examined the ability of untreated and treated poliovirus to incorporate ^{14}C-uridine into new viral nucleic acid in host cells. Untreated virus was capable of incorporating nearly four times the amount of uridine into TCA-precipitable material. The investigators suggested a mechanism of inactivation directed against the viral nucleic acid such that the replication process in host cells was prevented.

In contrast, Brigano et al.[2] used a thermodynamic analysis of poliovirus inactivation data to suggest that the fundamental reaction responsible for virus inactivation with ClO_2 was the disruption of virus protein. Thermodynamic values calculated from inactivation data were similar to those associated with protein denaturation.

Limited information has been available on the reactivity of ClO_2 with the components of nucleic acids and proteins, and little attention has been directed toward relating observed reactions to microbial inactivation. Schirle[3] reported that cysteine moieties incorporated in the proteins of wool were oxidized to cysteic acid when ClO_2 was applied as a bleaching agent. Schmidt and Braunsdorf[4] reported on the reactions of ClO_2 with tryptophan, tyrosine, histidine, and cystine, but did not characterize the reaction products of these reactive amino acids. They stated that the amino acids alanine, phenylalanine, valine, leucine, asparagine, aspartic acid, glutamine, and serine did not react with

ClO_2. Fujii and Ukita[5] reported that tryptophan was degraded to a mixture containing isatin, indoxyl, and indigo red upon exposure to ClO_2.

METHODS

Chemical Methods

Preparation and Determination of Chlorine Dioxide

Chlorine dioxide was prepared according to the method described by Granstrom and Lee.[6] Chlorine dioxide concentrations of stock solutions were determined by direct spectrophotometric measurement at 357 nm using a molar absorption coefficient of 1242 L/(mol · cm). During virus inactivation trials, the chlorine dioxide concentrations were measured by the leuco crystal violet method reported by Roller et al.[7] All ClO_2 concentrations were reported as milligrams per liter ClO_2 as available chlorine.

Microbiological Methods

Preparation, Purification, and Assay of Virus

The bacterial virus, f2, was prepared according to the Loeb and Zinder[8] method, with *Escherichia coli* k13 as host. It was purified by polyethylene glycol precipitation, followed by density gradient centrifugation in cesium chloride and exhaustive dialysis with phosphate buffer pH 7.2. The density of f2 was determined by the agar overlay procedure reported by Adams,[9] using *E. coli* k13 as host.

Poliovirus 1 was propagated in Buffalo green monkey (BGM) kidney cells grown on Eagle's minimal essential medium (MEM). The virus was released from the cells by three cycles of freezing and thawing and then concentrated and washed by filtration with a YM30 Amicon filter. The virus was then submitted to density gradient centrifugation in cesium chloride (0.47 g/mL) and exhaustively dialyzed against phosphate buffer. Poliovirus 1 density was determined by the plaque assay technique described by Dahling et al.,[10] using BGM cells as host.

The RNA of f2 virus was labeled with ^{32}P by the method described by Vinuela et al.[11] The protein component was labeled with tritium-labeled leucine similar to the procedure for ^{32}P, except that the growth medium was leucine deficient. The virus was purified as previously described.

The RNA of poliovirus 1 was labeled with ^{32}P incorporated into low-phosphate Eagle's MEM added to the BGM cell monolayers just after infection. The virus was prepared and purified as above.

Infectious RNA and Virus Attachment

The infectivity of f2 RNA was determined by the method of Hofschneider and Delius.[12] Poliovirus RNA infectivity was determined according to Parag.[13]

The attachment of the f2 virus to host *E. coli* was determined by the method described by Brinton and Beer[14] as modified by Olivieri et al.[15] A similar method was used for poliovirus 1, but the host was BGM cells.

Structural Integrity

The sedimentation characteristics of f2 and poliovirus were determined by rate-zonal centrifugation in 5 to 30% (w/w) sucrose gradients at 150,000 × G in a Beckman SW50.1 rotor at 20°C for 55 min. The virus sedimentation coefficient was calculated according to the method of Griffith.[16] Similar rate-zonal centrifugation methods were used for poliovirus, except that the centrifugation time was 30 min.

Viral proteins were analyzed by SDS-polyacrylamide gel electrophoresis according to the method of Laemmli.[17]

Cellular Functions After Infection with Poliovirus 1

The repression of cellular protein synthesis by poliovirus 1 and the synthesis of poliovirus RNA in BGM cell monolayers were evaluated by the procedure described by Young.[18]

Protein and Amino Acid Analysis

The f2 protein was purified by the method of Frankel-Conrat[19] and then freeze dried and hydrolyzed with hydrochloric acid at 110°C for 24 h. Hydrolysis for tryptophan analysis was performed by the methanesulfonic acid procedure of Fishbein et al.[20]

The protein hydrolysates were analyzed in a Durrum D500 amino acid analyzer.

Experimental Methods

Virus inactivation studies were conducted in a reaction system described by Olivieri et al.[21] All solutions were prepared in 0.002 M phosphate solution and adjusted to the desired pH prior to each trial. Chlorine dioxide was rapidly added at time zero, and samples were collected with time for determining virus survival, virus structural and biochemical characteristics, ClO_2 residual, and pH. The ClO_2 in the samples for subsequent biological analysis was neutralized at the time of sample collection with sodium thiosulfate. The disinfection reaction temperature was 5°C.

RESULTS AND DISCUSSION

Virus Inactivation and RNA Infectivity

The inactivation of the bacterial virus, f2, and poliovirus 1 by ClO_2 was compared to the loss of infectivity of RNA extracted from treated virus (RNA') and naked RNA (RNA). Figure 1 shows the results of the disinfection trials carried out at pH 7.2. Poliovirus was noticeably more resistant to ClO_2.

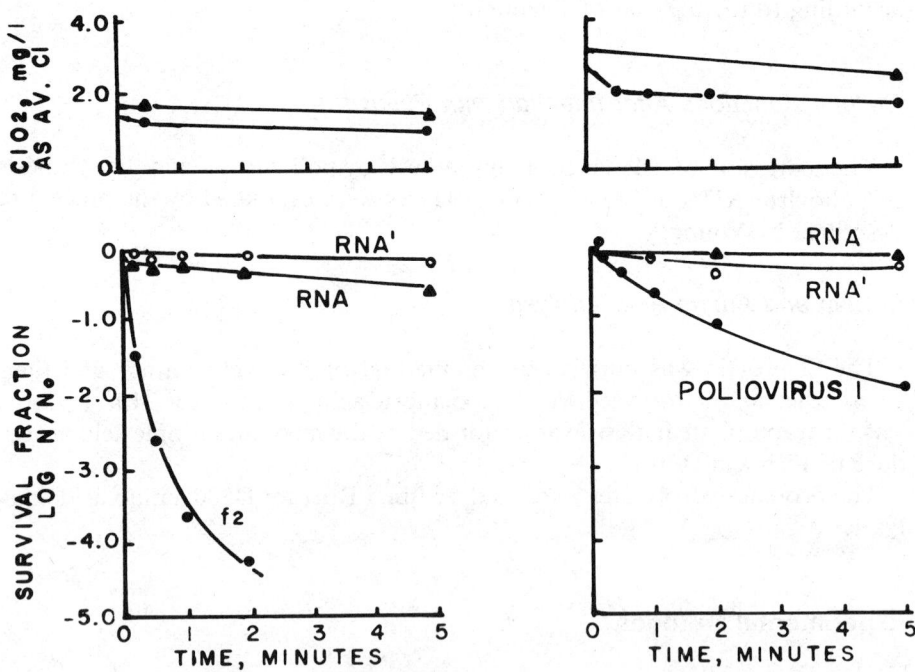

Figure 1. Inactivation of ClO_2-treated f2 virus and poliovirus, naked viral RNA, and RNA extracted from ClO_2-treated virus (RNA'). Chlorine dioxide dosage and residual are indicated in the upper panels. The inactivation conditions were pH 7.2 in 0.002 M phosphate at 5°C. Sensitivity limits for RNA and RNA' assays were three orders of magnitude for both f2 virus and poliovirus.

Greater than four orders of magnitude of kill were observed after 2-min contact for f2 virus, whereas only slightly more than one order of magnitude was observed for poliovirus. The loss of infectivity of RNA extracted from the treated viruses (RNA') lagged far behind the inactivation of virus for both f2 and poliovirus. While the magnitude of inactivation was less at pH 5.0 and greater at pH 8.7, similar results were observed relative to the loss of infectivity of the RNA. Figure 1 also shows the inactivation of naked f2 poliovirus RNA after treatment with ClO_2. Loss of infectivity of the nucleic acid was also observed, but the rate of inactivation was slow. Untreated control viral RNA remained stable during the course of each trial.

The sensitivity limit for all the infectious nucleic acid studies was ~3 log units. Although the efficiency of recovery of RNA was inherently low, all the data points in the experimental trials were well above the sensitivity limit for the assay.

Analysis of Virus Attachment Function

Figure 2 shows the loss of attachment of the f2 virus and poliovirus to host cells after treatment with ClO_2. The attachment data are presented as the log of the fraction remaining (cpm at time t/cpm at t_0) compared with the log of the virus survival fraction for the time course of disinfection. For f2 (upper panel) at pH 5.0 and 7.2, virus inactivation proceeded at a slightly faster rate than the loss of the ability of the virus attachment. At pH 8.7, no difference was detected between the rates of inactivation and inhibition of attachment. The inhibition of the attachment of f2 virus to host cells appears to closely parallel the inactivation of this bacterial virus. For poliovirus 1 (lower panel) at pH 5.0, 7.2, and 8.7, little inhibition of attachment was observed. Despite ~2.5 log units of inactivation at pH 5.0, 7.2, and 8.7 within contact times of 10.0, 5.0, and 0.5 min, respectively, poliovirus exhibited little loss in its ability to attach to host cells.

Virion Structural Integrity

Density Gradient Analysis

The effect of ClO_2 treatment on the sedimentation of f2 and poliovirus 1 in 5 to 30%(w/w) sucrose gradients is shown in Figure 3. The data for sedimentation rate are presented as the number, N, of cpm at time t divided by the number, N_0, of cpm at time zero in the 85S (f2) or 160S (poliovirus) fraction of the sucrose gradient. Changes in the sedimentation rate were compared with changes in the survival fraction of the virus, both presented as log N/N_0. At pH 7.2, there was little change in the sedimentation rate for f2 virus, even though 3 log units of inactivation were observed in 1 min. Both untreated f2

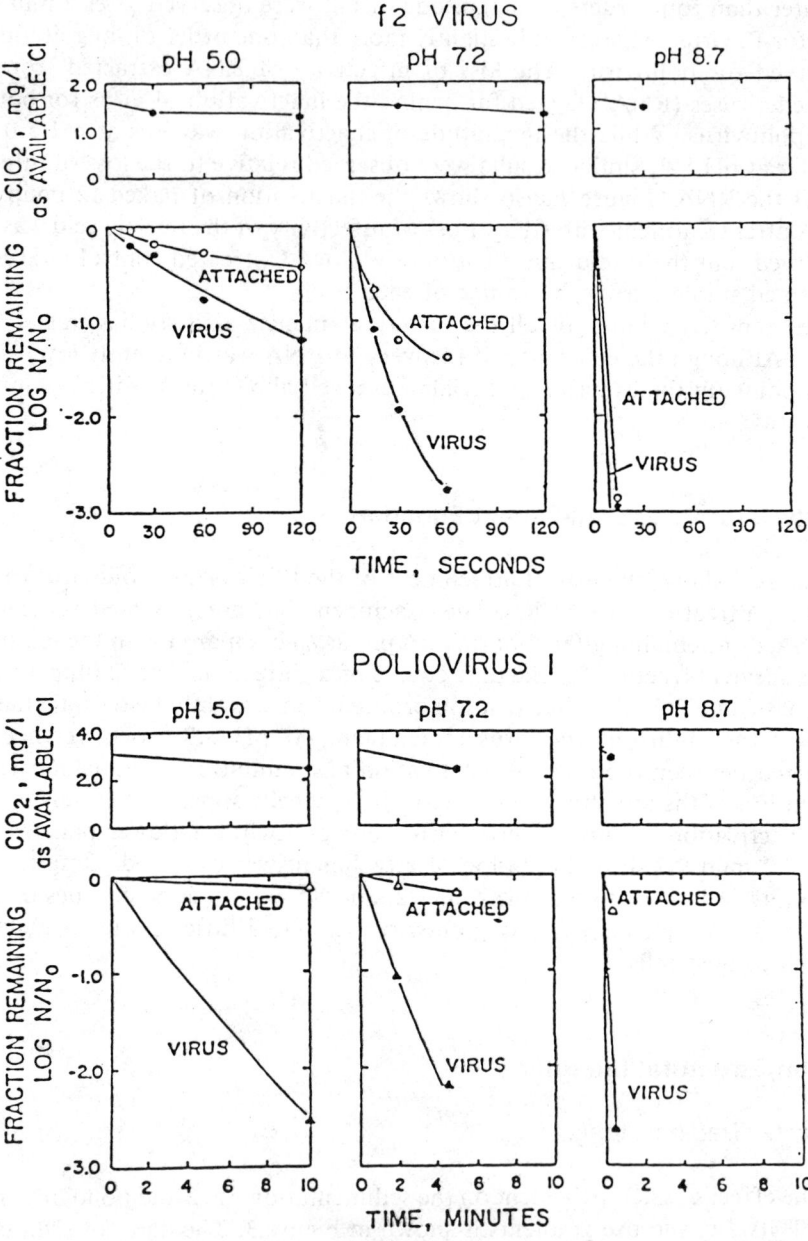

Figure 2. Inactivation of f2 virus and poliovirus with ClO_2 (as available chlorine) compared to the specific attachment of treated virus to host cells (*E. coli* k13 for f2 and BGM cells for poliovirus 1). Chlorine dioxide dosage and residual are indicated in the upper panels. Inactivation experiments were performed at pH 5.0, 7.2, and 8.7 in 0.002 *M* phosphate at 5°C. Sensitivity limit for attachment assays was three orders of magnitude. Each point represents the average of two trials.

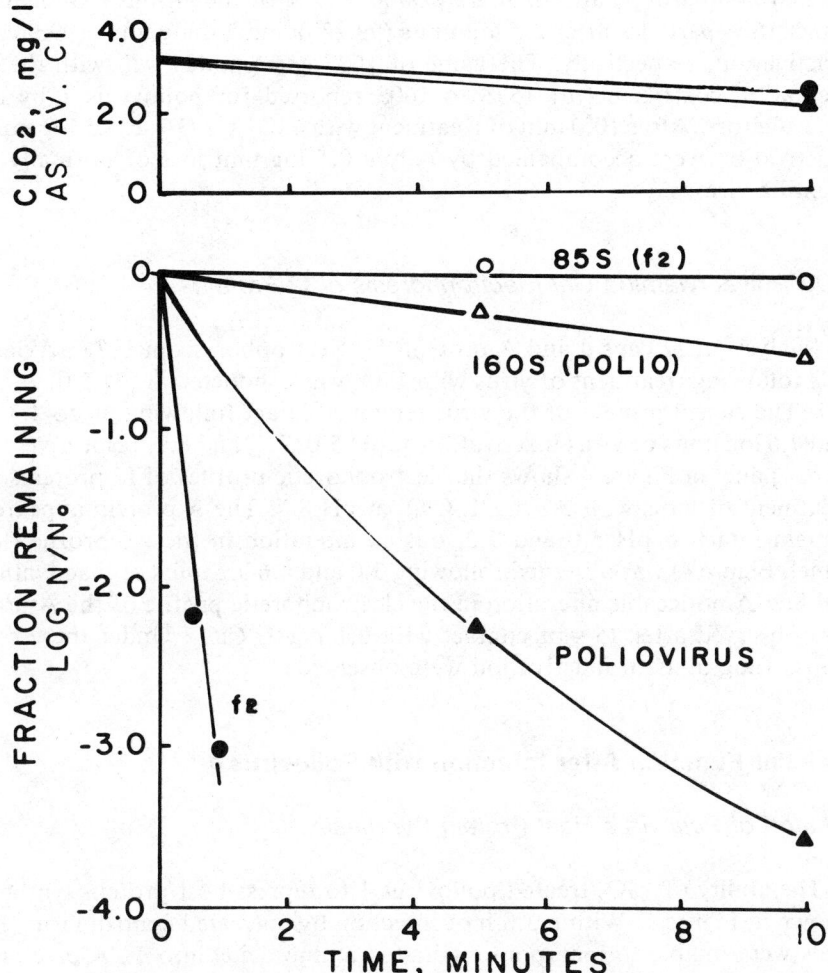

Figure 3. Inactivation of f2 and poliovirus with ClO_2 compared to the change in the sedimentation coefficient determined in 5 to 30% sucrose. Chlorine dioxide dosage and residual are indicated in the upper panels. Inactivation conditions were pH 7.2 and 5°C. Effect of ClO_2 on f2 virus and poliovirus infectivity and sedimentation in 5 to 30% sucrose gradients following treatment of virus with 2.7 to 3.5 mg/L ClO_2 (as available chlorine) at pH 7.2 and 5°C.

marker virus containing protein labeled with tritiated leucine and test virus containing ^{32}P-labeled RNA displayed a sedimentation coefficient of ~85S. This value is slightly higher than that reported for R17, MS2 and for all small RNA viruses related to f2 with sedimentation coefficients of 77S to 81S. Trials with f2 virus treated with chlorine at pH 5.0 and 8.7 also showed little change in the sedimentation rate despite 3.2 and 5.0 log units of inactivation. For ^{32}P-labeled poliovirus 1, approximately 50 and 25% of the virions sedimented as intact 160S particles after 2.5 log units (99.7%) and 3.4 log units (99.96%) of inactivation, respectively. The value of 160S corresponds well with the sedimentation coefficients of 153S to 160S reported for poliovirus 1 by other investigators. After 10.0 min of treatment with ClO_2 at pH 7.2, 3.5 log units of inactivation were accompanied by only a 0.5 log unit loss of normally sedimenting virions.

SDS-Polyacrylamide Gel Electrophoresis of f2 Proteins

Analysis of f2 capsid and A proteins by electrophoresis in 10% acrylamide gels following treatment of virus with ClO_2 was conducted at pH 5.0, 7.2, and 8.7. The capsid protein of the virus remained intact following up to 2.3, 3.2, and 4.6 log units of virus inactivation at pH 5.0, 7.2, and 8.7, respectively. The upper panel in Figure 4 shows the electrophoretic profiles of f2 proteins after treatment of virus with 2.1 mg/L ClO_2 at pH 8.7. The A protein appeared to remain intact at pH 5.0 and 7.2, but an alteration in the gel profile (lower panel, Figure 4) was observed following 3.0 and 4.6 log units of inactivation at pH 8.7. A noticeable alteration in the electrophoretic profile of the A protein was observed after 15 s of contact with 2.1 mg/L ClO_2. Under these conditions, 3 log units of inactivation were observed.

Cellular Function After Infection with Poliovirus 1

Shutoff of Poliovirus Host Protein Synthesis

The ability of ClO_2-treated poliovirus 1 to repress host protein synthesis is shown in Figure 5. Within 3.0 h of infection by untreated control virus, BGM cells were unable to incorporate tritiated amino acids into TCA-precipitable proteins. The level of uptake of amino acids in host cells infected with inactivated virus approached the normal levels of uptake exhibited by mock infected cells. Virus treated with ClO_2 was unable to shut off host protein synthesis.

Synthesis of Intracellular Poliovirus RNA

An evaluation of the ability of poliovirus 1 to synthesize RNA in host cells after treatment of virus with ClO_2 is shown in Figure 6. Untreated control virus was able to infect BGM cells and synthesize new viral RNA, as evidenced by the incorporation of increasing amounts of tritiated uridine into TCA-precipitable nucleic acid. Levels of TCA-precipitable radioactivity in lysates of

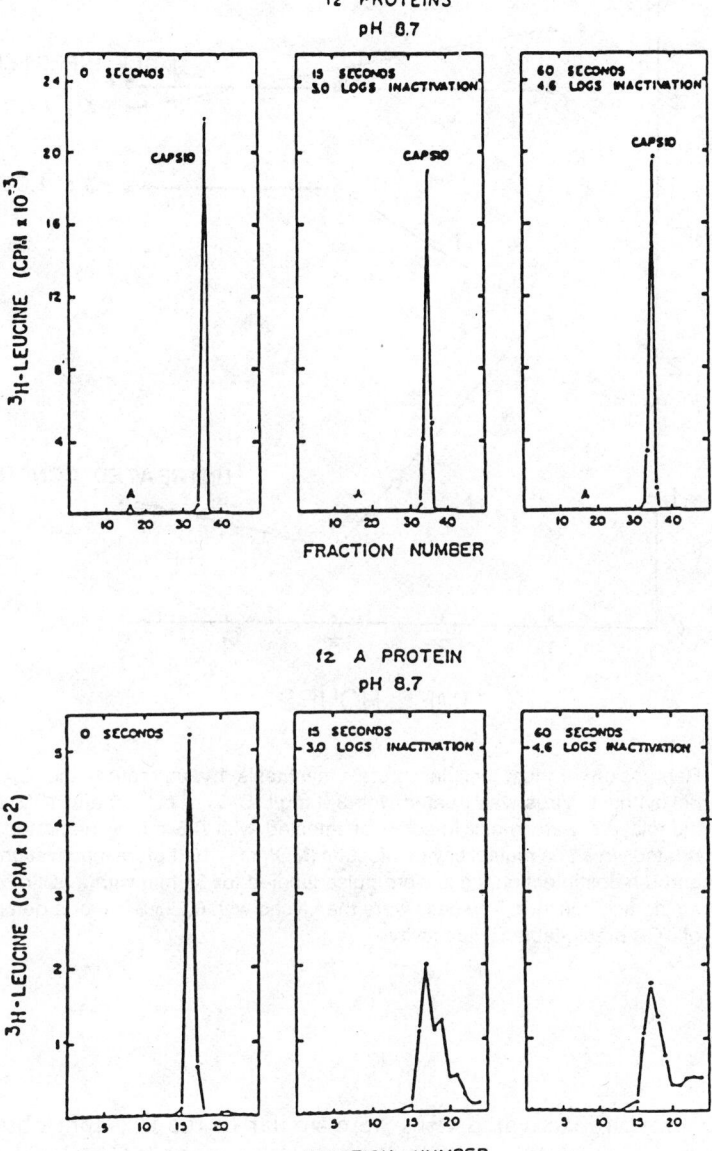

Figure 4. SDS polyacrylamide gel electrophoresis of f2 A protein following treatment of virus with 2.1 mg/L ClO_2 (as available chlorine) at pH 8.7 and 5°C. Upper panel shows capsid and lower panel shows A protein.

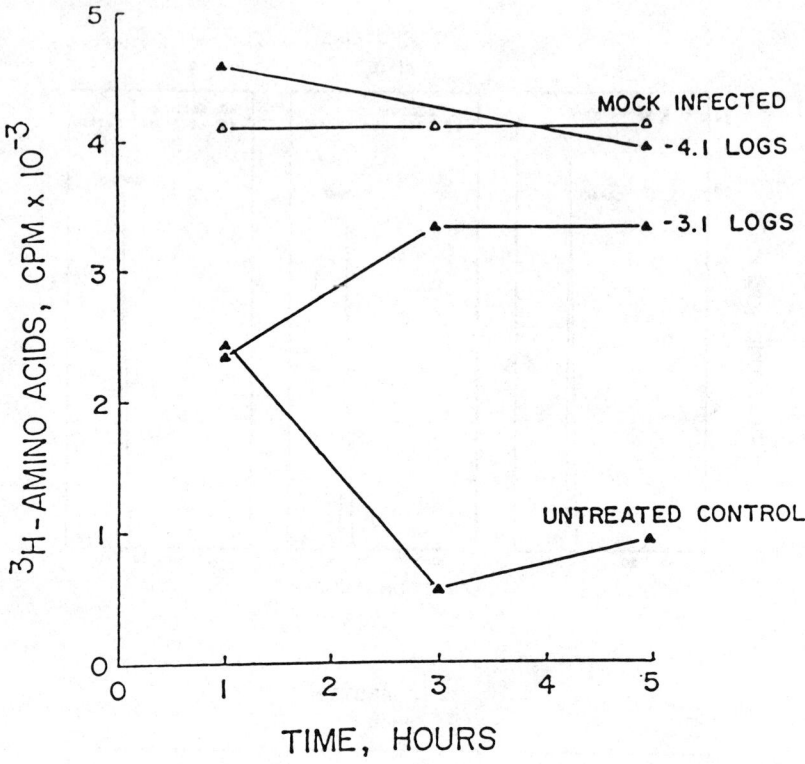

Figure 5. Repression of host cellular protein synthesis by untreated and ClO_2-treated poliovirus 1. Virus was treated with 3.2 mg/L ClO_2 at pH 7.2 and 5°C. BGM cell monolayers were mock infected or infected with 0.5 mL of untreated or ClO_2-treated virus at a multiplicity of infection (MOI) of ~10 (before inactivation). At 1, 3, and 5 h postinfection, cells were pulse labeled for 30 min with 2 μCi of a tritiated amino acid mixture. The cells were then lysed with 0.2% SDS for a determination of TCA-precipitable radioactivity.

cells infected with inactivated virus were similar to the low cpm observed in mock infected control cells, indicating that genome-length viral RNA was not synthesized.

The mechanism of inactivation proposed by Alvarez and O'Brien[1] was based on the assumption that complete uncoating and penetration of the viral genome had occurred. The modified particles observed by these investigators, designated M (modified) and C (chemotrypsin modified), were transient intermediates in the uncoating process formed as a result of an interaction of virus with host-cell membrane and membrane-associated stabilizing factors.[22] The

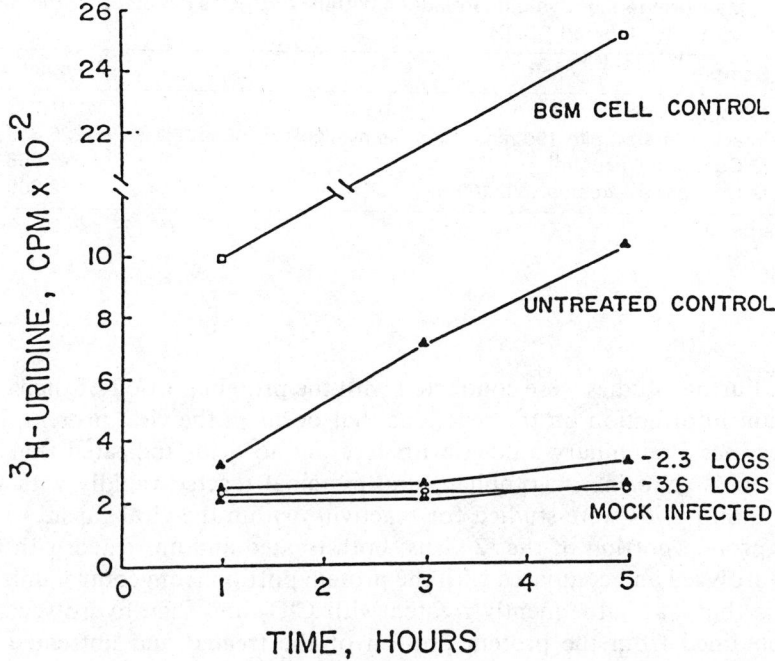

Figure 6. Synthesis of intracellular RNA by untreated and ClO_2-treated poliovirus 1. Virus was treated with 2.4 mg/L ClO_2 at pH 7.2 and 5°C. BGM cell monolayers were mock infected or infected with 0.5 mL of untreated or ClO_2-treated virus at an MOI of approximately 10 (before inactivation). At 30-min postinfection, 20 μg of actinomycin D and 2 μCi of tritiated uridine were added to each well. Cells were lysed with 0.2% SDS at 1, 3, and 5 h postinfection for a determination of TCA-precipitable radioactivity. A set of uninfected cells designated BGM cell controls received tritiated uridine but no actinomycin D.

appearance of these particles was not evidence for complete uncoating or penetration of the genome as Alvarez and O'Brien proposed. Their conclusion that an inactivated genome was responsible for virus inactivation may be incorrect if an infectious nucleic acid was never present in the host cell.

Protein Reactions

The reaction rate of ClO_2 with the RNA portion of f2 virus and poliovirus does not appear to be consistent with the inactivation rate of the virus. The mode of action appeared to be associated with the protein portion of the

Table I. Measurement of Cysteine Residues Within f2 Virus as Indicated by the Reaction with ^{203}Hg-labeled PCMB

Sample	cpm[a]
Intact f2 control	231,796
Intact f2 treated with 190 mg/L ClO$_2$ (as available chlorine)	225,858
f2 Coat protein control	176,968
f2 Coat protein treated with 100 mg/L ClO$_2$	32,300

[a]Counts per minute.

virion. Further studies were conducted with the protein portion of the f2 virus to obtain information on the reactions that occur in the viral protein during disinfection. Preliminary studies with free amino acids indicated that three amino acids (tyrosine, tryptophan, and cysteine) reacted rapidly with ClO$_2$. These amino acids were studied for reactivity within the viral capsid.

The protein portion of the f2 virus, both treated and untreated with ClO$_2$, was hydrolyzed and compared with the protein portion from control untreated f2 virus that was subsequently treated with ClO$_2$ and then hydrolyzed. The data obtained from the protein portion of the treated and untreated virus provide information on the interaction of ClO$_2$ with the protein components as they exist in the intact virion, whereas the data obtained with the treated and untreated f2 protein provide information on the interaction of ClO$_2$ with the protein but not as it exists in the virus. The levels of tyrosine and tryptophan were followed on an amino acid analyzer, whereas the cysteine residues were followed by reaction with ^{203}Hg-labeled p-chloromercuribenzoate (PCMB).

Table I demonstrates the loss of cysteine sulfhydryl groups within f2 virus upon treatment with 190 mg/L ClO$_2$ at pH 7.0 and 23°C. Even after 9.4 logs of f2 were inactivated, the number of cpm decreased only from 231,796 to 225,858 and suggested little alteration in the cysteine residues in the intact virus. Denatured f2 protein lost its ability to bind PCMB after treatment with ClO$_2$ under conditions similar to the treatment of the intact virus. The number of cpm dropped from 176,968 for control denatured-coat protein to 32,300 for ClO$_2$-treated coat protein. These data indicate that cysteine residues were buried within the viral capsid and were not available for reaction with ClO$_2$ in a time frame that could account for rapid f2 inactivation.

Figure 7 shows the chromatogram of the f2 proteins; the peak for the amino acid, tyrosine, is indicated by the label. A 30% change in the tyrosine peak was observed after treatment of the virus with 95 mg/L ClO$_2$ for 2.0 min, with a corresponding inactivation of 12 log units (high virus titers were necessary to provide sufficient material for amino acid analysis). The amino acid determinations suggest that either tyrosine reactions with ClO$_2$ were not given enough time to go to completion, or that tyrosine residues were buried within the viral capsid and were not available for reaction. Little change was seen in the other

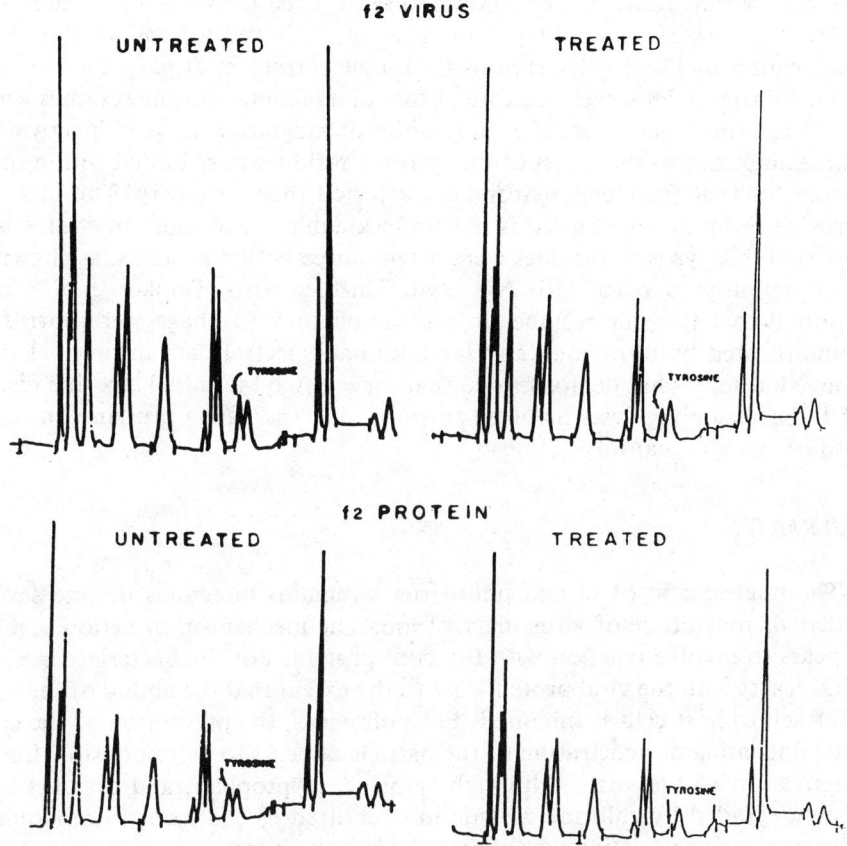

Figure 7. Untreated f2 virus protein (left panel) and f2 virus protein treated with 95 mg/L ClO_2 (as available chlorine) for 2 min at pH 7.0 and 4°C (right panel) were freeze dried and hydrolyzed with HCl. Amino acids were separated and identified with a Durrum amino acid analyzer.

amino acids in this portion of the chromatogram. The chromatograms prepared from untreated (left) and ClO_2-treated (right) f2 protein are shown in the lower panel. The tyrosine peak was not observed after exposure of the protein to chlorine dioxide. It appears that most tyrosine residues were buried within the viral capsid. When the capsid protein was disrupted and allowed to react with ClO_2, essentially all the tyrosine reacted. Little or no change in other amino acid peaks in this portion of the chromatogram was observed. Similar chromatograms were prepared for tryptophan. Little or no change was observed in the tryptophan peak in the hydrolyzed protein when the intact virus was treated with ClO_2. However, no tryptophan was found after the virus was disrupted and the protein was treated with ClO_2. The tryptophan in the coat protein of the intact virus appears to be buried and unavailable for reaction with ClO_2.

Only tyrosine residues were shown to react when the virus was treated with ClO_2. The loss of available tyrosine residues followed a trend similar to the inactivation of f2 virus by various ClO_2 concentrations at pH values ranging from 5.0 to 9.0. However, the rate of loss of available tyrosine residues within the viral capsid was not of the same order of magnitude as viral inactivation. These data suggest that most of the tyrosine residues were buried within the f2 capsid, or that the local environment affected their reactivity. The idea that tyrosine residues were buried is not unreasonable, since Riordan et al.[23] have reported that tyrosine residues were often buried within proteins, as shown by their inability to react with N-acetylimidazole. Also, Dunker et al.[24] have reported that tyrosine residues in the filamentous fd phage were buried, as demonstrated by ultraviolet and laser Raman spectral data during pH titration. Matthews[25] has demonstrated that the carboxyl-terminal tyrosine residue of f2 bacteriophage was involved in stabilizing the native protein conformation of the viral capsid.

SUMMARY

The nucleic acid of f2 and poliovirus 1 remains infectious despite several orders of magnitude of virus inactivation. The mechanism of action of ClO_2 appears to involve reaction with the viral protein. For the bacterial virus, f2, ClO_2 reacts with the viral protein coat to the extent that the ability of the virus to attach to host cells is inhibited. For poliovirus, the prevention of the complete uncoating or penetration of the particle appears to be responsible for the inactivation of the virus. Although tyrosine, tryptophan, and cysteine were easily degraded by chlorine dioxide in denatured f2 coat protein monomers, only tyrosine degradation was observed when the intact virus was treated with ClO_2. The reaction of ClO_2 with tyrosine appears to be the mechanism of action of ClO_2 on f2 virus.

REFERENCES

1. (a) Alvarez, M. E., and R. T. O'Brien. "Mechanisms of Inactivation of Poliovirus by Chlorine Dioxide and Iodine," *Appl. Environ. Microbiol.* 44:1064–1071 (1982); (b) DeSena, J., and B. Torian. "Studies on the In Vitro Uncoating of Poliovirus. III. Roles of Membrane-Modifying and -Stabilizing Factors in the Generation of Subviral Particles," *Virology* 104:149–163 (1980).
2. Brigano, F. A. O., P. V. Scarpino, S. Cronier, and M. L. Zink. "Effect of Particulates on Inactivation of Enteroviruses in Water by Chlorine Dioxide." In *Progress in Wastewater Disinfection Technology*, A. F. Venosa, Ed. EPA-600/9-79-018. (Cincinnati: U.S. Environmental Protection Agency, 1979).
3. Schirle, C. "Contribution a l'Etude de l'Action du Chlorite et du Bioxyde de Chlore sur la Laine," *Bull. Inst. Textile, Fr.*, 41:21 (1953).

4. Schmidt, E., and K. Braunsdorf. "Verhalten von Chlordioxyd Gegenuber Organischen Erbindungen," *Ber.* 55:1529 (1922).
5. Fujii, M., and M. Ukita. *Nippon Nogei Kagaku Kaishi* 31:101 (1957).
6. Granstrom, M. L., and G. F. Lee. "Generation and Use of Chlorine Dioxide in Water Treatment," *J. Am. Water Works Assoc.* 38:1301 (1958).
7. Roller, S., V. P. Olivieri, and K. Kawata. "Mode of Bacterial Inactivaiton by Chlorine Dioxide," *Water Res.* 14:635-641 (1980).
8. Loeb, T., and N. D. Zinder. "A Bacteriophage Specific for the f+ and Hfr Mating Types of *E. coli* K12," *Science* 131:932 (1961).
9. Adams, M. H. *Bacteriophages*, (New York: Interscience, 1959).
10. Dahling, D. R., G. Berg, and D. Berman. "BGM, a Continuous Cell Line More Sensitive Than Primary Rhesus and African Green Kidney Cells for the Recovery of Viruses From Water," *Health Lab. Sci.* 11:275-282 (1974).
11. Vinuela, E., I. D. Algranati, and S. Ochoa. "Synthesis of Virus-Specific Proteins in *Escherichia coli* Infected with the RNA Bacteriophage MS2," *Eur. J. Biochem.* 1:3-11 (1967).
12. Hofschneider, P. H., and H. Delius. "Assay of M12 Phage RNA Infectivity in Spheroplasts," in *Methods in Enzymology*, Vol. 12, L. Grossman and K. Moldave, Eds. (New York: Academic Press, 1968).
13. Parag, G., Personal communication. (Madison, WI: University of Wisconsin, 1982).
14. Brinton, C. C., and H. Beer. "The Interaction of Male-Specific Bacteriophages with f Pili," in *The Molecular Biology of Viruses*, J. S. Coulter and W. Paranchych, Eds. (New York: Academic Press, 1967).
15. Olivieri, V. P., C. W. Kruse, Y. C. Hsu, A. C. Griffiths, and K. Kawata. "The Comparative Mode of Action of Chlorine, Bromine, and Iodine on f2 Bacterial Virus," in *Disinfection, Water and Wastewater*, J. D. Johnson, Ed. (Ann Arbor, MI: Ann Arbor Science Publishers, Inc., 1975).
16. Griffith, O. M. Identifying Bands by Sedimentation Coefficients," in *Techniques of Preparative, Zonal and Continuous Flow Ultracentrifugation,* (Fullerton, CA: Beckman Instruments, Inc., 1979), p. 16.
17. Laemmli, V. J. "Cleavage of Structural Proteins During the Assembly of the Head of Bacteriophage T4," *Nature* 227:680-685 (1970).
18. Young, D. C. *The Inactivation of Picornaviruses in Water by Chlorine,* Ph.D. Thesis (Chapel Hill, NC: University of North Carolina, 1981).
19. Frankel-Conrat, H. "Degradation of Tobacco Mosaic Virus with Acetic Acid," *Virology* 4:1 (1957).
20. Fishbein, J. C., A. R. Place, I. J. Ropson, D. A. Powers, and W. Sofer. "Thin-Layer Peptide Mapping: Quantitative Analysis and Sequencing at Nanomole Level," *Anal. Biochem.* 108:193 (1980).
21. Olivieri, V. P., T. K. Donovan, and K. Kawata. "Inactivaiton of Viruses in Sewage," *J. San. Eng. Div., Am. Soc. Civil Eng.* 97 (SA5):661-673 (1971).
22. DeSena, J., and B. Torian. "Studies on the in vitro Uncoating of Poliovirus. III. Roles of Membrane-Modifying and -Stabilizing Factors in the Generation of Subviral Particles," *Virology* 104:149-163 (1980).
23. Riordan, J. F., W. E. C. Wacker, and B. L. Vallee. N-Acetylimidazole: A reagent for Determination of "Free" Tyrosyl Residues of Proteins," *Biochemistry* 4:1758-1765 (1965).

24. Dunker, A. K., R. W. Williams, and W. L. Peticolas. "Ultraviolet and Laser Raman Investigation of the Buried Tyrosines in fd Phage," *J. Biol. Chem.* 254:6444 (1979).
25. Matthews, C. K. *Bacteriophage Biochemistry*, A. C. S. Monograph, (New York: Van Nostrand Reinhold Co., 1971).

CHAPTER 50

Relative Disinfection Potentials of Chlorine and Chlorine Dioxide in Drinking Water

M. M. Varma, G. Torrence, R. C. Chawla, and H. Okrend

The formation of harmful by-products such as trihalomethanes (THMs), which have resulted from the use of chlorine, has prompted increased interest in further research for alternative means of water disinfection. Because of adverse health effects caused by THMs, the U.S. Environmental Protection Agency (EPA) has recently announced that the arithmetic sum of all species of THMs should not exceed 100 μg/L.[1] Other contributing factors are the rising cost of chlorine and the biomagnification of chlorinated hydrocarbons in aquatic biota, including the concomitant effects on the food web.

Chlorine dioxide (ClO_2) has been found to be five times more soluble than chlorine, has 2.5 times the oxidizing capability of chlorine, and generates no such by-products as THMs or chloramines, which have been identified with conventional chlorination. A persistent residual throughout a distribution system is apparently also more easily achieved with ClO_2.

The purpose of this research was to evaluate the disinfection potentials of ClO_2 and chlorine and to examine the practical use of such disinfectants.

METHODOLOGY

Disinfection tests of chlorine and ClO_2 were performed on pure cultures of *Escherichia coli* in batch flasks. The *E. coli* cultures were suspended in sterile phosphate buffer, chlorine-free deionized water. The pH of the test run was controlled by the pH of the phosphate buffer, which was adjusted prior to autoclaving at one of three tested pH levels, 6.0, 7.0, and 8.0. Initial dosages of chlorine varied from 0.50 to 4.00 mg/L at each pH value. Reaction times varied from close to 0 to 400 s. Bacterial counts were determined by micropipettes on nutrient agar[2] before and after chlorine and ClO_2 were added to the batch flasks. Colonies were counted 24 h after incubation at 37°C. At a predetermined time, two 10-mL samples were removed after reaction with one of the disinfectants. In one sample, the residual disinfectant was neutralized with sodium thiosulfite, after which viable plate counts were made. The other sample was tested for free residual available chlorine by the DPD ferrous titrimetric method.[3] Chlorine dioxide was measured by the DPD method.

Disinfection tests of chlorine and ClO_2 were also run on unchlorinated water obtained from the Potomac Water Treatment Plant. Bacterial counts were determined by the micropipette method before and after disinfection. Total solids were determined before and after disinfection by the gravimetric method. Other physicochemical parameters (chemical oxygen demand, ammonia, nitrite, and nitrate) were measured as prescribed in *Standard Methods*.[3]

RESULTS AND DISCUSSION

Potomac Water

Unchlorinated water samples were obtained from the Potomac Water Treatment Plant. The major physicochemical parameters of the samples tested for bacterial inactivation by chlorine and ClO_2 were in close agreement. Since particulate matter in water can act as a protective bacterial coating, its existence could account for, at least in part, the disinfectant demand. Hence, this parameter was strictly controlled and maintained in the range of 180 to 190 mg/L total dissolved solids. Samples under parallel conditions were spiked with predetermined concentrations of chlorine and ClO_2. The chlorine dose was ~4 mg/L, and the ClO_2 dose varied between 4.2 and 5.3 mg/L. Bacterial counts for all tests (unchlorinated water) were in about 10^3 cfu/mL.

Figure 1 shows residual chlorine in potable water at pH 6, 7, 8. The initial concentration was 4 mg/L; at the end of 300 s, the residual was about 2.1 mg/L.

Figure 2 is a graphical representation of bacterial inactivation by aqueous chlorine (pH values of 6, 7, and 8) at various time intervals. At 300 s at pH 6 and 7, approximately 50% of the bacteria were inactivated. At pH 6, from the beginning of the reaction to about 60 s, the rate of inactivation was high, after which it was dramatically reduced.

Figure 3 represents the decay of ClO_2 in potable water with respect to time at various pH levels. The initial concentration varied from 4.2 to 5.3 mg/L. At the end of the 300 s, the residual varied from 1.5 to 2.5 mg/L.

Figure 4 shows the plot of bacterial decay by ClO_2 at various pH levels. At the end of the test, the inactivation was 74, 73, and 75% at pH 6, 7, and 8, respectively.

Apparently, ClO_2 showed better potential as a disinfectant than chlorine. The bactericidal decay rate with ClO_2 was independent of pH in the ranges used in this research, which means ClO_2 disinfection may not require stringent pH control.

Pure Cultures

Tests were conducted with ClO_2 at pH 7 and 8 to study the decay of overnight-grown pure cultures of *E. coli*. The chlorine doses used were approximately 0.5, 1.0, and 4.0 mg/L.

Figures 5 and 6 represent bacterial decay plots as a function of time. The bacterial inactivation rate in the first few seconds was very high, and it gradu-

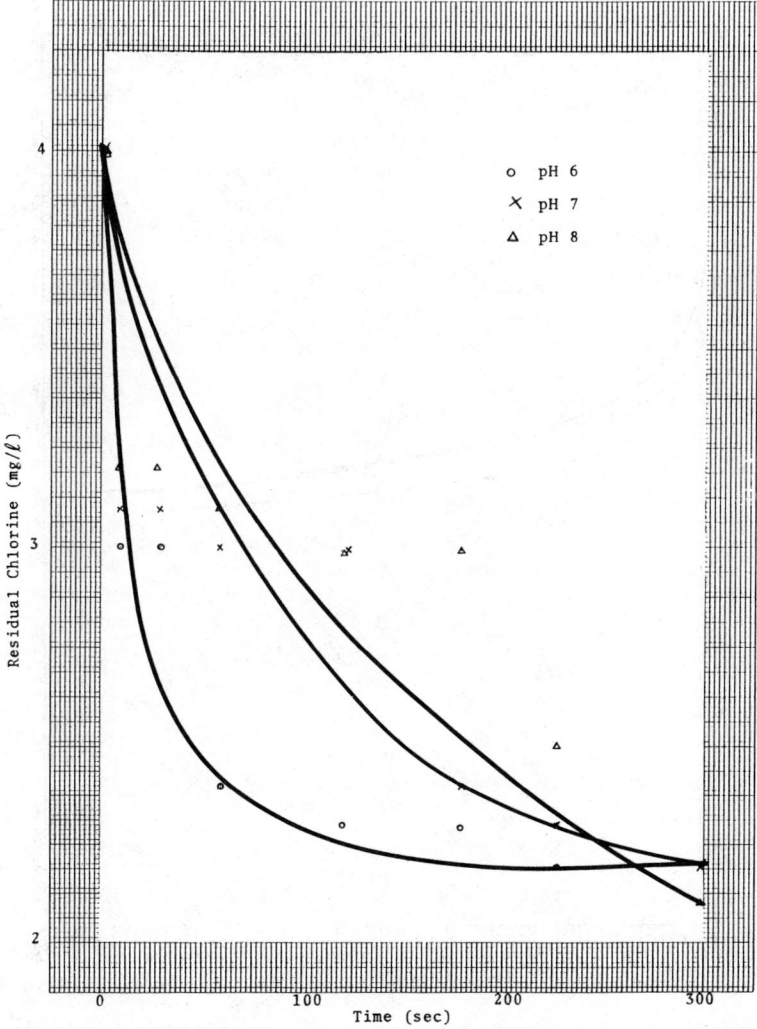

Figure 1. Residual chlorine as a function of time (at various pH values) in Potomac water.

638 WATER CHLORINATION

ally diminished as the residual concentration of the disinfectant decreased. At pH 7, maximum kill was achieved in < 120 s for an initial ClO_2 concentration of 4.0 mg/L and a bacterial count of 1.5×10^7 cfu/mL. At pH 8, maximum kill was obtained in about half the time (60 s), even though the initial disinfectant concentration was a little lower (3.5 mg/L) and the bacterial count was a little higher (18×10^6 cfu/mL). Figures 7 and 8 depict the decay of ClO_2 in the reaction mixture. The corresponding disinfectant consumptions were 2.7 and

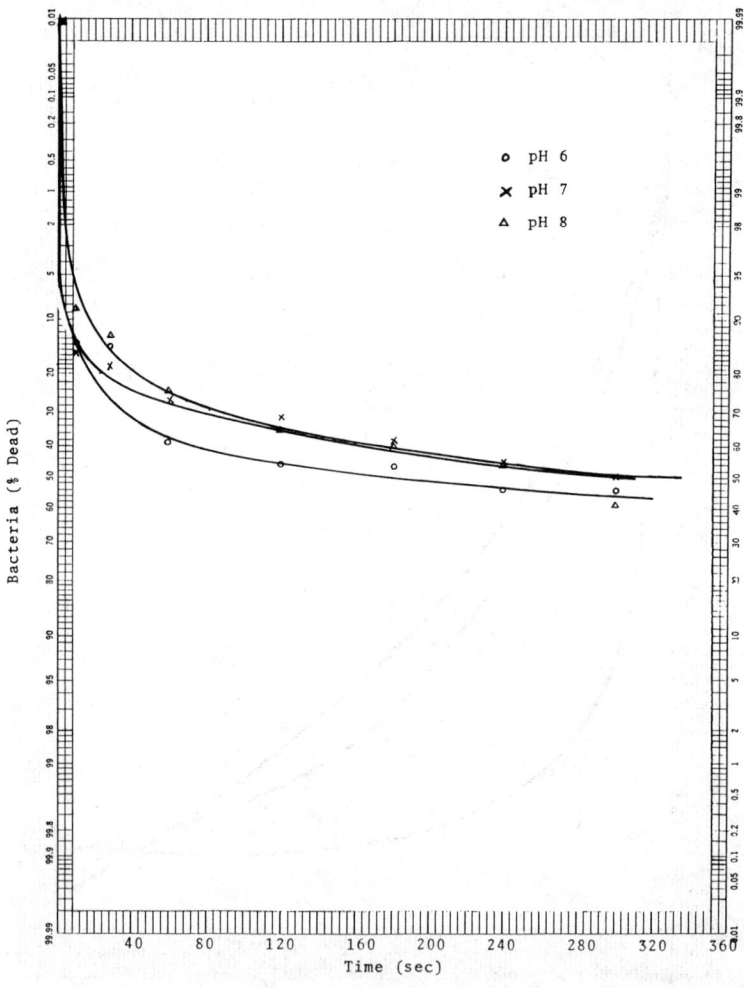

Figure 2. Bacterial inactivation with chlorine (4 mg/L) as a function of time (at various pH levels) in Potomac water.

2.4 mg/L, respectively. The results indicated that ClO_2 exhibited better efficiency on all counts at pH 8. However, this pH dependency was not observed when unchlorinated water samples with very low bacterial counts ($\sim 10^3$ cfu/mL) were tested, as discussed earlier.

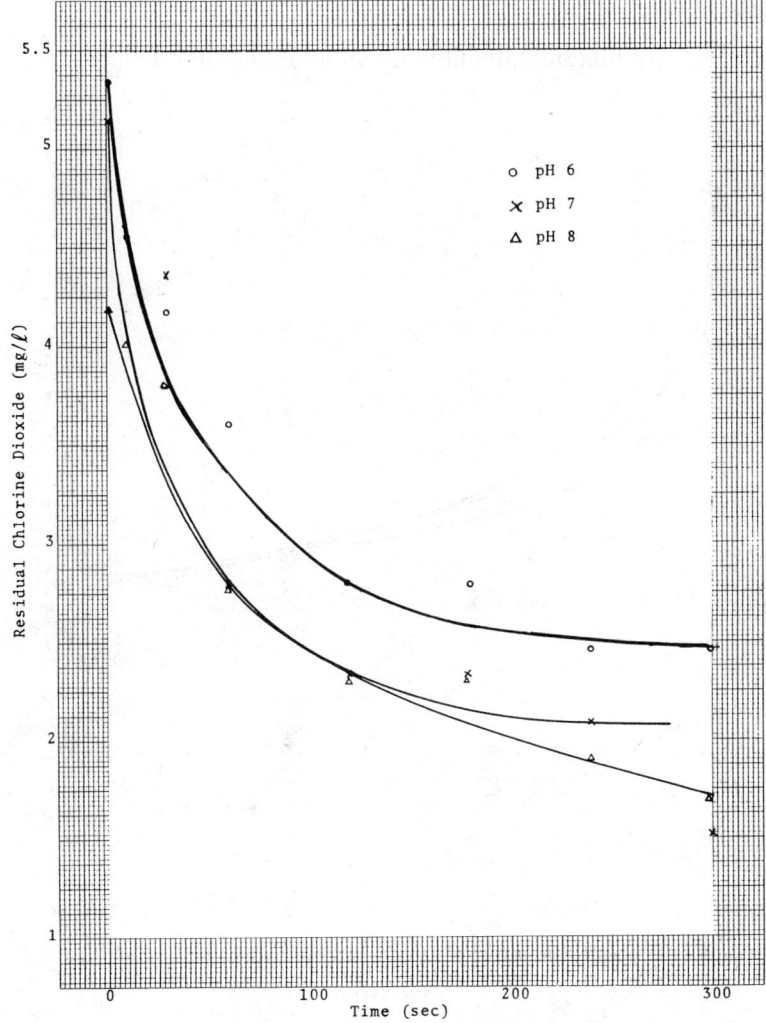

Figure 3. Residual ClO_2 as a function of time (at various pH levels) in Potomac water.

Small disinfectant doses (about 0.6 mg/L) were used in similar tests for initial bacterial counts of about 13×10^7 cfu/mL at pH 7 and 8, but no appreciable differences were observed between the two pH values. Bacterial kill plateaued at about 40% in about 150 to 200 s for both cases. At this time ClO_2 was completely exhausted (Figures 7 and 8).

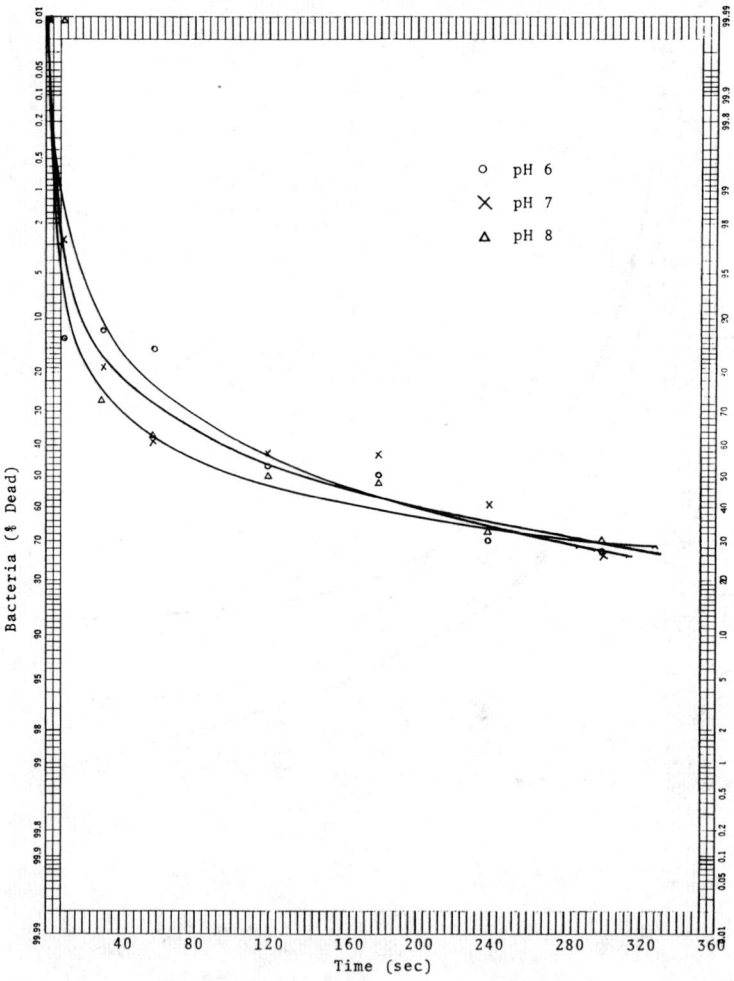

Figure 4. Bacterial inactivation with ClO_2 as a function of time (at various pH levels) in Potomac water.

Bacterial concentrations were increased 5 to 7 times and exposed to low ClO_2 dosage (about 1 mg/L) in another series of tests. Again pH 8 proved to be better (20% kill) than pH 7 (10% kill) at the point of total disinfectant consumption. The process was very fast, requiring only about 50 s to reach the plateau.

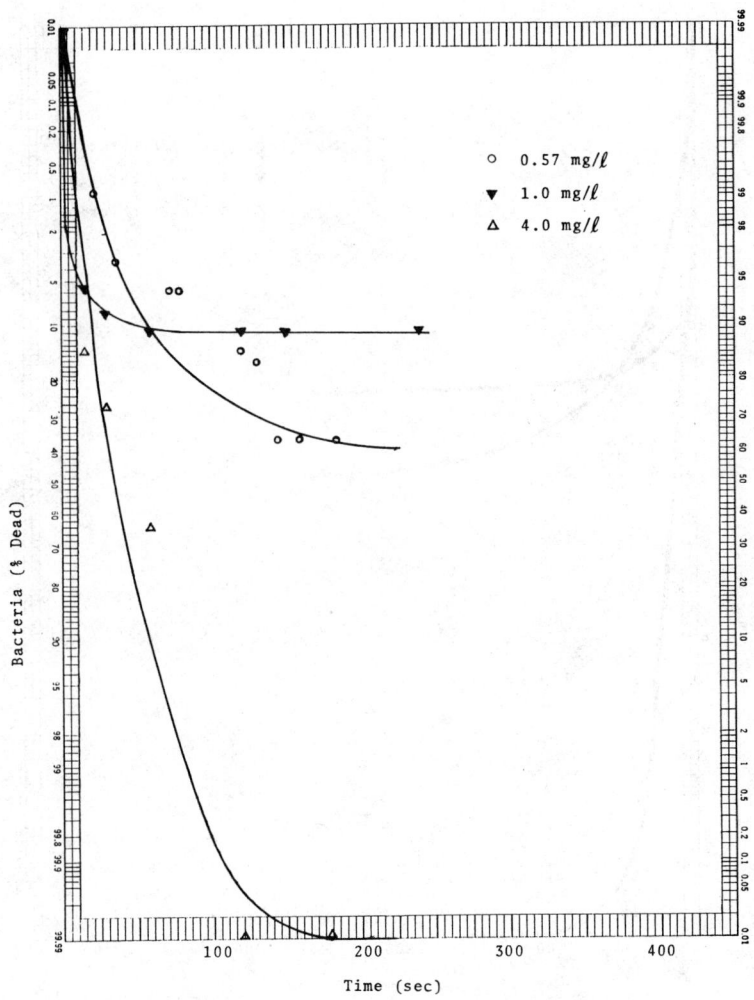

Figure 5. *E. coli* inactivation vs time at pH 7 for various ClO_2 doses.

642 **WATER CHLORINATION**

In parallel experiments, chlorine was used as a disinfectant in pure cultures. Figures 9 and 10 show bacterial inactivation as a function of time for three different initial doses of chlorine (about 0.5, 1.0, and 4.0 mg/L) at pH values of 7 and 8. The general inactivation behavior of the *E. coli* population was similar to that obtained for the ClO_2 system. At pH 7, maximum kill was achieved in <120 s for an initial chlorine concentration of 4.0 mg/L and a bacterial count of 35×10^6 cfu/mL. At the end of the experiment, the residual

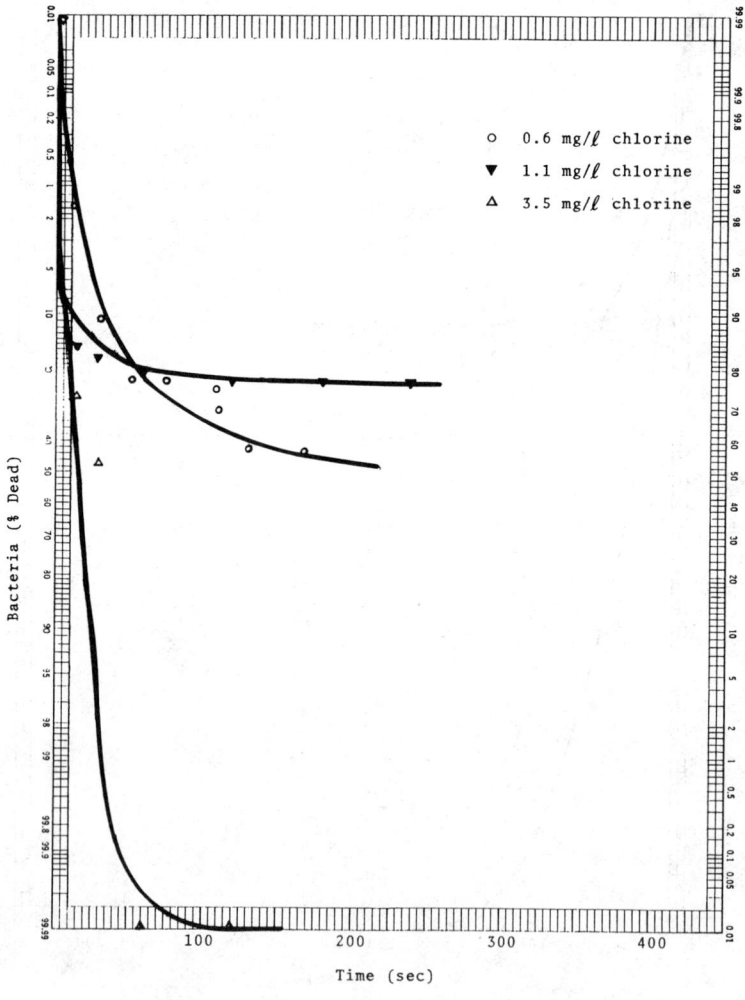

Figure 6. *E. coli* inactivation vs time at pH 8 for various ClO_2 doses.

chlorine was 2.5 mg/L (Figure 11). At pH 8, it took more than twice the time (≧240 s) for much less bacterial count (20% of the bacteria at pH 7, 7×10^6/mL) for maximum kill. At the end of the test all chlorine was exhausted. The chemistry of aqueous chlorine suggests that HOCl is a more potent disinfectant than OCl⁻ (HOCl is about 80% more effective than OCl⁻ in killing *E. coli*). At about pH 7.5, the concentrations of HOCl and OCl⁻ are

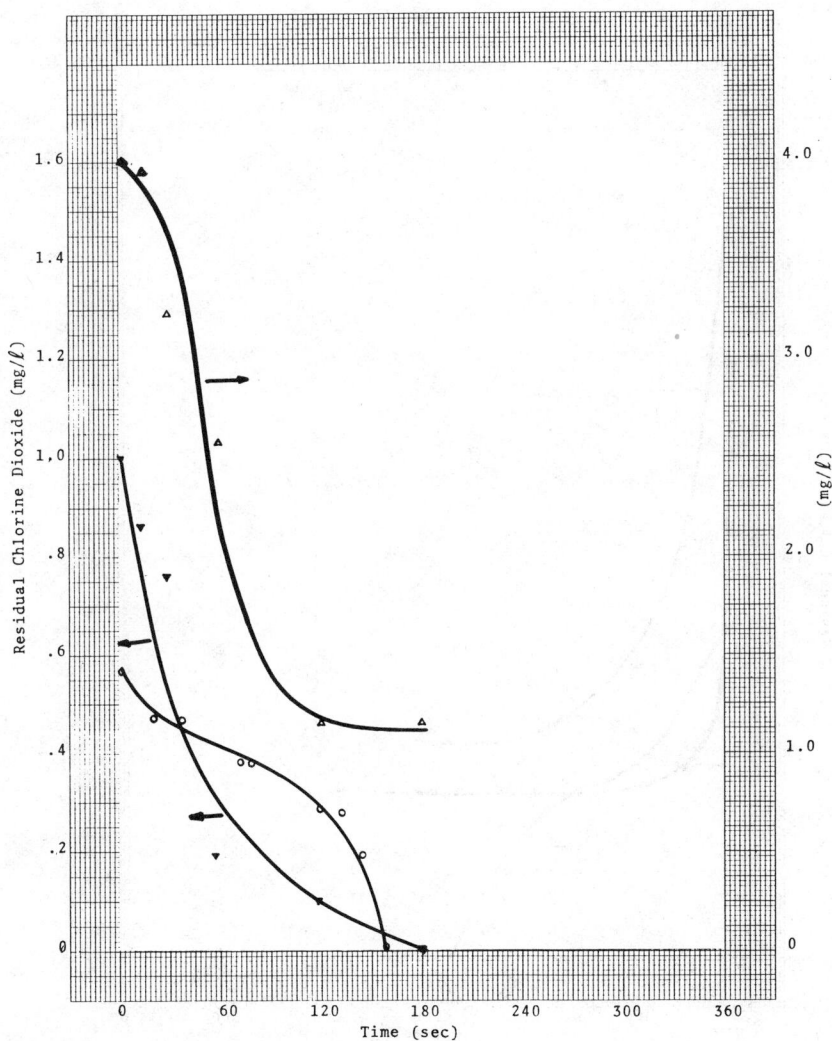

Figure 7. Residual ClO_2 vs time at pH 7 for pure culture disinfection.

644 WATER CHLORINATION

approximately equal (~50%). Above this pH, OCl⁻ predominates; below this pH, HOCl is dominant. This explains the better efficiency of chlorine at pH 7 for smaller doses (1.0 mg/L at pH 7 and 1.5 mg/L at pH 8). Also, pH 7 gave better disinfection efficiencies (90% kill) than pH 8 (60% kill) at total chlorine consumption. The data for much lower concentrations of chlorine (0.5 mg/L

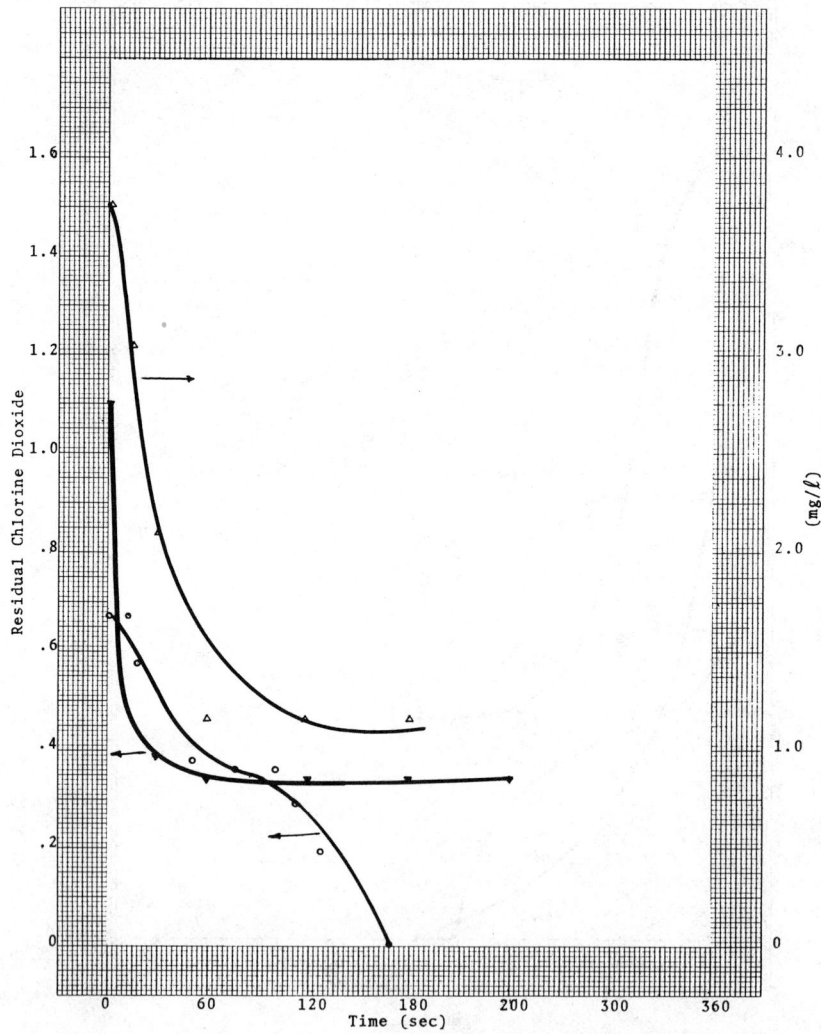

Figure 8. Residual ClO_2 vs time at pH 8 for pure culture disinfection.

at pH 7 and 0.75 mg/L at pH 8) did not show the pH effect to any significant degree. For both ClO_2 and chlorine systems, the data at very low disinfectant concentrations (~0.5 mg/L) may have larger errors, because the depletion of total available chlorine is very rapid and the probability of survivors is much greater.

In selecting an alternative disinfectant to chlorine, several parameters must be considered. Some of these are biocidal activity, toxicity of by-products, and economics. Humic substances primarily composed of humic and fulvic acids

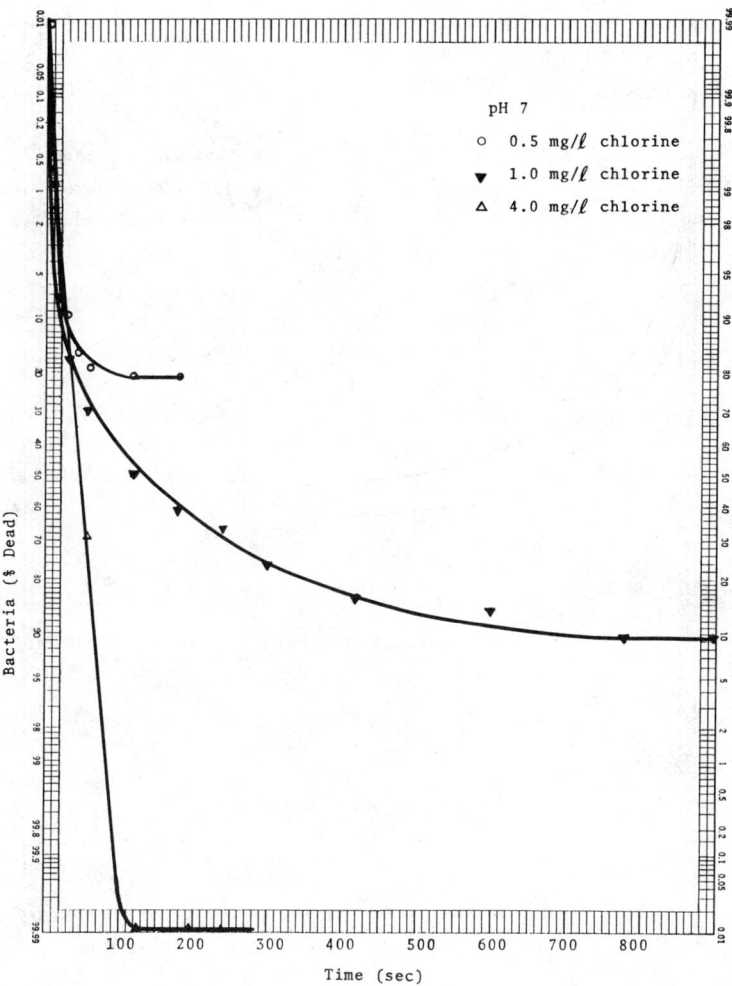

Figure 9. *E. coli* inactivation vs time at pH 7 for various chlorine doses.

are ubiquitous in surface water. The humic substances react with aqueous chlorine to form halogenated compounds. Prominent among these are THMs, which are classified as carcinogens.[4-6]

The ubiquitous presence of THM precursors necessitates the use of alternative disinfection. Chlorine dioxide does not react with humic acid to produce THMs. Studies[7,8] indicate that ClO_2 in aqueous solution will produce quinones, aldehydes, ketones, and epoxides, depending on the nature and dose

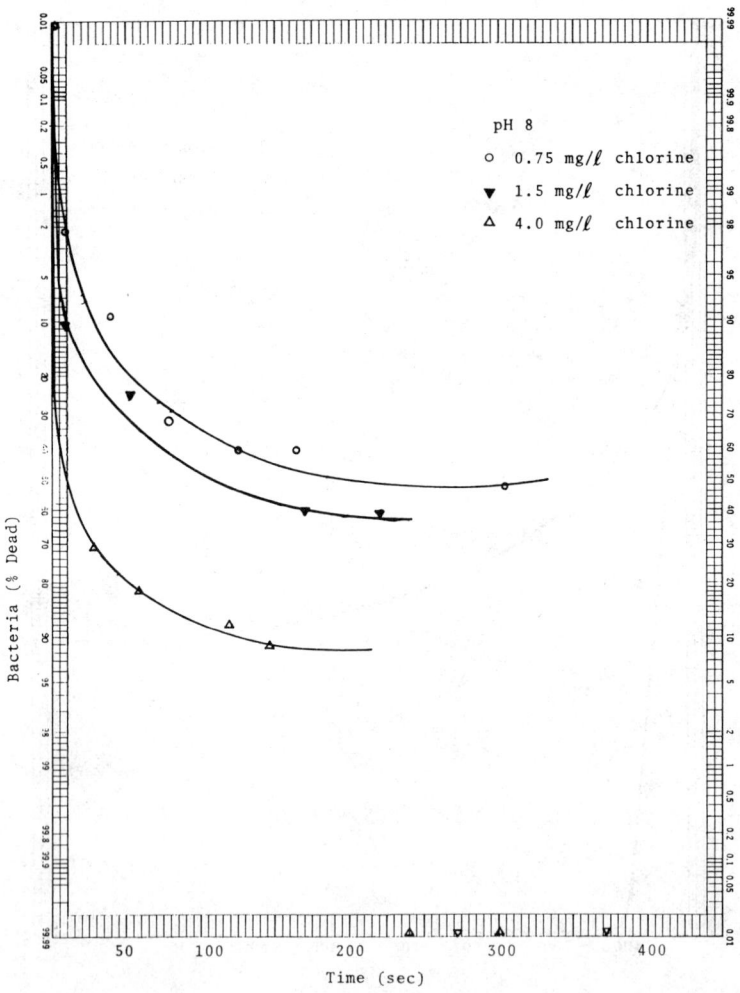

Figure 10. *E. coli* inactivation vs time at pH 8 for various chlorine doses.

of the organic matter present. Much more research has to be completed before any definitive health effects of these by-products are quantitatively evaluated. Until then, ClO_2 may be used as an alternative disinfectant.

Before deciding whether chlorine or ClO_2 should be used for water or wastewater disinfection, the evaluation should be made on the cost effectiveness of the disinfectant. This can be measured by the concentration of the disinfectant

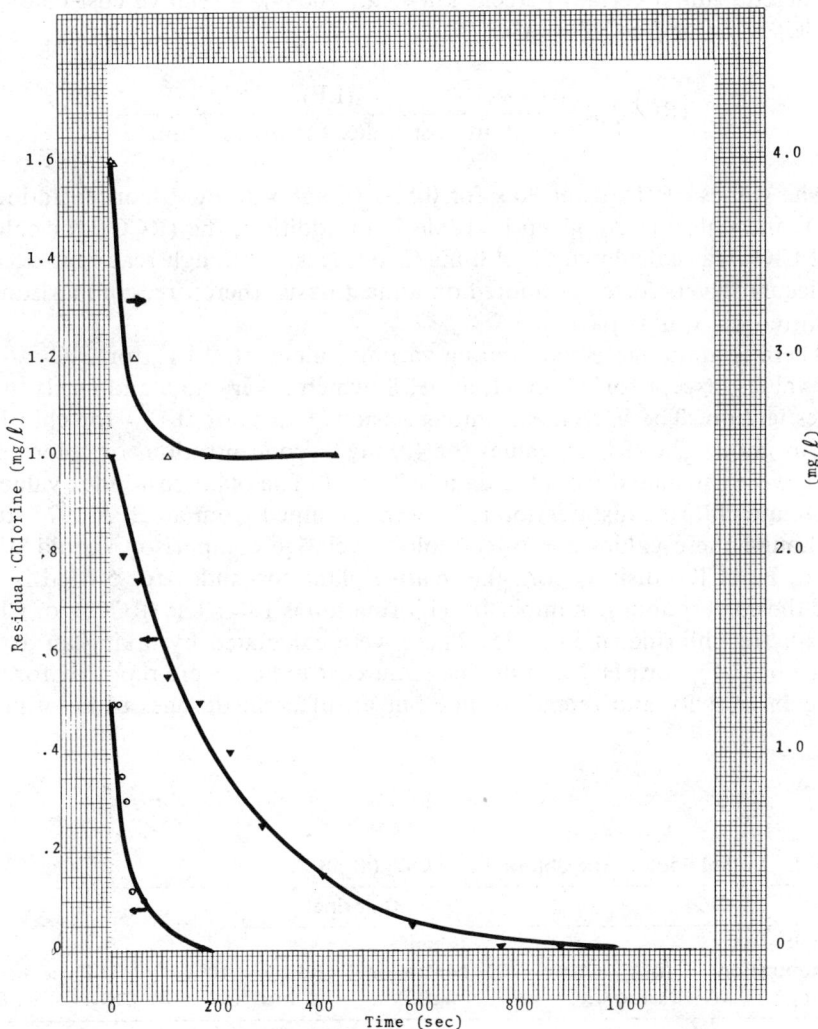

Figure 11. Residual chlorine vs time at pH 7 for pure culture disinfection.

used to achieve a given level of inactivation for a fixed contact time. We have formulated the following efficiency parameter, called lethal factor (LF):

$$(LF)_\theta = \frac{\text{fraction killed}}{(\text{time}, \theta)(\text{disinfectant consumed}, \triangle C)}$$

$(LF)_\theta$ has units of [L/(mg disinfectant) (s)].

At maximum kill, the terminal LF is denoted by $(LF)_t$. To design a treatment plant requiring a certain percent kill (e.g., 100%), a relative cost coefficient (RCC) is defined as follows:

$$(RCC)_{t,i} = \frac{(LF)_t}{\text{plant cost index for disinfectant}, i}$$

The values of $(LF)_{30}$ at 30 s for 0.5, 1.0, and 4.0 mg/L concentrations of ClO_2 and chlorine are given in Table I. In addition, the $(RCC)_t$ for chlorine and ClO_2 was calculated for 4.0 mg/L dosages. Although reactions occur at molecular level, costs are quoted on a mass basis; therefore, comparisons are reported on a mass basis.

There is good agreement among various values of $(LF)_{30}$ for ClO_2 at both pH values, except for 0.7 mg/L at pH 8, which is very high and needs further investigation. The agreement among various values of $(LF)_{30}$ for chlorine is not as good. The $(RCC)_t$ values for 4.0 mg/L concentrations of chlorine and ClO_2 were calculated from the data in Table I. The plant cost index values for ClO_2 and chlorine disinfection units were obtained from an EPA 1977 study.[9] Although these values are 7 years old, a relative comparison should still be valid. For ClO_2 disinfection, the relative plant cost index for 5 Mgd is 1.90, and the corresponding number for chlorination is 1.44. The $(RCC)_t$ for ClO_2 is 26.40; for chlorine, it is 25.35. These were calculated by taking an average value of $(LF)_t$ for pH 7 and 8. The plant cost indices were reported for some biocidal activity and retention time but at different dosages of chlorine and

Table I. Lethal Factors for Chlorine and ClO_2 (at pH 7)

ClO_2		Chlorine			
Initial Concentration (mg/L)	$(LF)_{30} \times 10^4$ L mg^{-1}s^{-1}	Initial Concentration (mg/L)	$(LF)_{30} \times 10^4$ L mg^{-1}s^{-1}	$(LF)_t \times 10^4$ ClO_2	Cl_2
0.6	8.5	0.5	120		
1.0	8.2	1.0	278		
4.0	14.1	4.0	232	30.86	55.56

ClO_2; better numbers for these indices may be calculated by incorporating more recent plant data. On the basis of LF values, chlorine is a better disinfectant (about ten times faster) for short retention times. However, the 100% kill, average $(LF)_t$ values for ClO_2 (pH 7 and 8) are 37% larger than for chlorine, indicating the superiority of ClO_2 over chlorine as a disinfectant. For rural water supplies, where contact times for disinfection are very small, chlorine, because of its higher initial biocidal rates, may be a better disinfectant than ClO_2.

CONCLUSIONS

Disinfection studies with potable water at pH 7 and 8, containing a low concentration of bacteria (10^3 cfu/mL), indicated that ClO_2 was a better disinfectant than aqueous chlorine; total solids were between 180 and 190 mg/L. Test results using pure cultures of *E. coli* showed that:

1. At pH 7, maximum kill was obtained with ClO_2 (4.0 mg/L) in 120 s; similar disinfection was obtained in 60 s at pH 8.
2. No appreciable difference in bacterial decay between pH 7 and 8 was observed at lower doses of ClO_2 (0.6 mg/L).
3. Chlorine proved to be a better disinfectant at pH 7 than at pH 8 in pure cultures. The lethal factor

$$(LF)_\theta = \frac{\text{fraction killed}}{(\text{time}, \theta)(\text{disinfectant consumed}, \triangle C)}$$

was formulated for comparing the results. $(LF)_{30}$ values at 30 s for ClO_2 were in close agreement at the concentrations tested. Corresponding $(LF)_{30}$ values for chlorine were not close to each other.
4. The relative cost coefficient

$$(RCC)_{t,i} = \frac{(LF)_t}{\text{plant cost index for disinfectant}, i}$$

was formulated. The value of $(RCC)_{t,i}$ for ClO_2 was 26.40, and the corresponding value for chlorine was 25.35 indicating insignificant cost advantage of ClO_2 over chlorine, based only on bacterial inactivation efficiencies. This does not include by-product removal costs for public health safety.

ACKNOWLEDGMENT

This study was supported in part by a U.S. Environmental Protection Agency research grant, R-806659.

REFERENCES

1. U.S. Environmental Protection Agency, "Control of Organic Chemical Contaminants in Drinking Water," *Fed. Reg.* 43(28):5755 (Feb. 9, 1978).
2. Chappelle, E. W., et al. *Laboratory Procedure Manual for the Firefly Luciferase Assay for Adenosine Triphosphate (ATP),* NASA/GSFC Document X-726-75-1, (Washington, DC: National Aeronautics and Space Administration, 1975).
3. *Standard Methods for the Examination of Water and Wastewater*, 15th ed. (Washington, DC: American Public Health Association, 1980).
4. Rook, J. J. "Formation of Haloforms During Chlorination of Natural Waters," *Water Treat. Exam.* 231 (1974).
5. Bellar, T. A., et al. "The Occurrence of Organohalides in Finished Water," *J. Am. Water Works Assoc.* 66:703 (1974).
6. Varma, M. M., et al. "Analysis of Trihalomethanes in Aqueous Solution: A Comparative Study," *J. Am. Water Works Assoc.* 71:389 (1979).
7. Stevens, A. A., D. R. Seeger, and C. J. Slocum. "Products of Chlorine Dioxide Treatment of Organic Materials in Water," in *Ozone/Chlorine Dioxide Oxidation Products of Organic Materials*, R. G. Rice and J. A. Cotruvo, Eds. (Cleveland: Ozone Press International, 1978), pp.383-399.
8. Dowling, L. T. "Chlorine Dioxide in Potable Water Treatment," *Water Treat. Exam.* 22:190 (1973).
9. *Ozone, Chlorine Dioxide, and Chloramine as Alternatives to Chlorine for Disinfection of Drinking Water*, EPA 600/D-77-003 (Cincinnati: U.S. Environmental Protection Agency, 1977).

CHAPTER 51

Recurrent Coliforms in Water Distribution Systems in the Presence of Free Residual Chlorine

Vincent P. Olivieri, Alexander E. Bakalian,
Keith W. Bossung, and Ernest D. Lowther

Several communities across the United States have experienced the persistent low-level occurrence of coliform bacteria in their water distribution systems.[1-4] In many cases, breakpoint chlorination was practiced at the treatment plant, and free chlorine residuals were maintained throughout the water distribution system. Selected samples were submitted to further microbiological analysis to obtain information to evaluate the possible source of contamination. These samples were positive for coliforms or contained high levels of background bacteria, and data presented for this particular set of samples are intended to provide information on the comparative frequency of isolation of the various microbial indicators. This presentation reports the occurrence of coliforms at one particular water system during 1983, possible sources of the coliforms, and the results of preliminary experiments to evaluate the resistance of random coliform isolates recovered from the samples collected from this system.

The water system serves a small community of 16,500 people and delivers an average of 1.2 Mgd (4.5 ML/d). Raw water is obtained from a small river and receives coagulation, flocculation, and sedimentation followed by rapid sand filtration. Breakpoint chlorination is practiced, and a free chlorine residual is carried through the distribution system. Portions of the pipe network are located in areas of a high water table and are sometimes submerged. No episodes of negative pressure were recorded at a gage located in the distribution system.

MATERIALS AND METHODS

Samples from various locations in the distribution system and the treatment plant were taken according to methods recommended by the U.S. Environmental Protection Agency.

Total and fecal coliforms were determined by the membrane filter technique.[5] Occasionally, the multiple-tube dilution technique was used.[5]

Coliform-positive samples were stored at 4°C, then assayed for the presence of *Clostridium perfringens* using the British multiple-tube dilution procedure described by Cabelli.[6]

The chlorine resistance of the recovered coliforms was studied as follows: Two isolates were picked up from positive plates and were grown on standard plate count agar (DIFCO) for 2 d at 24°C. The surface growth was washed off with 10 mL of 2-mM phosphate buffer saline (pH 7.9), and the suspension was centrifuged and washed three times (at 5,000, 7,500, and 10,000 × G, respectively). The final suspension was diluted and used as the working inoculum. The reaction vessel consisted of a baffled beaker connected to an automatic syringe adapted to dispense the appropriate volume for chemical and biological analysis during the time course of disinfection. The biological samples were plated on standard plate-count agar and incubated at 24 ± 1°C for up to 72 h. Coliforms recovered from these plates were regrown and reexposed to chlorine under the same conditions.

Parallel experiments were conducted with the same coliform isolates (grown for 5 d at room temperature) on microscope slides suspended in 10% plate count broth (DIFCO). The slides were then exposed to chlorine solutions (2.0 and 1.8 mg/L). In each case, one slide was removed at zero time and at 60 and 600 s. The slides were then dipped in 10% sodium thiosulfate solution to neutralize the chlorine. The biofilm was removed by scraping and resuspended in 5 mL of phosphate-buffered saline. The suspensions were then plated as previously described.

Chlorine residuals were measured at the time of sample collection by the N,N-diethyl-*p*-phenylene-diamine (DPD) method.[5] For chlorine resistance experiments, stock solutions were prepared by bubbling chlorine gas into triple-distilled water. The concentration of chlorine was measured by the FACTS (free available chlorine test with syringaldazine) and amperometric titration methods[5] and reported as milligrams-per-liter free available chlorine (FAC).

RESULTS

Occurrence

Routine data collected by the water utility are shown in Figure 1. The frequency of positive coliform samples collected each day in 1983 in the water distribution network is shown in the upper panel and can be compared with the temperature (middle panel) and residual chlorine (lower panel). Coliforms were not recovered during the early part of the year; however, they began to appear about 120 d into 1983 at about the beginning of May. Prior to their appearance, the temperature was slowly rising from a minimum in late January and early February to about 60°F (15°C). The residual chlorine measured at the plant was slightly more than 2.0 mg/L FAC, which was quickly increased when coliforms were observed (after exhausting lab error, sampler

DISINFECTION 653

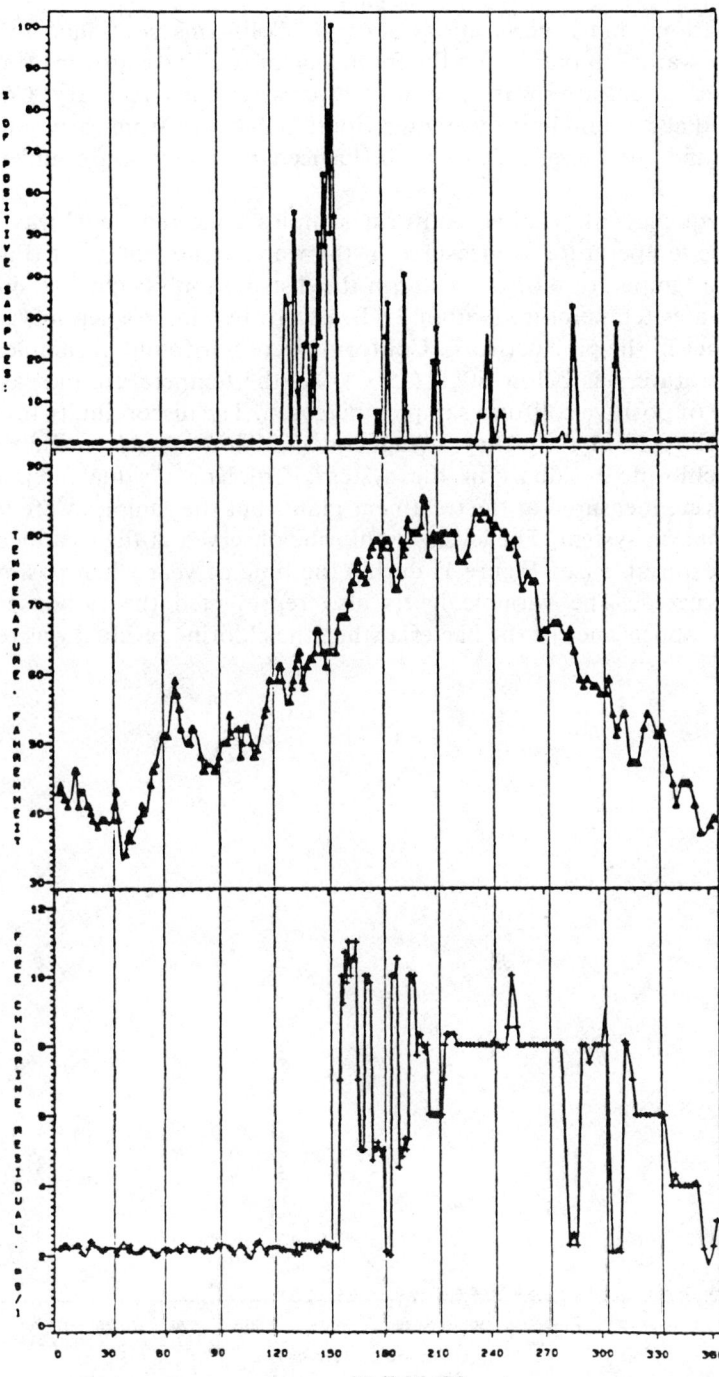

Figure 1. Residual chlorine and temperature at the water treatment plant, and daily frequency of positive coliform samples in the distribution system throughout 1983.

654 **WATER CHLORINATION**

contamination, and a connections survey). Coliforms continued to appear during the warmer months, and the frequency of positive coliforms fluctuated as the residual chlorine was altered in response to their recovery. Coliforms were found at random in the water distribution network from early summer to late fall, and they appeared to be influenced by temperature and residual chlorine.

The frequency of positive coliform samples collected daily was plotted against the temperature as measured at the water treatment plant (Figure 2). The actual temperature of the water in the distribution system was similar to the raw water temperature (within 2°C), except in winter when it was about 3°C warmer in the pipe network. Coliforms were not found in samples where the temperature was below 60°F (15°C). As the temperature increased, the frequency of positive coliform samples increased. The discontinuity in the data and the decrease in frequency at higher temperatures may be the result of elevated chlorine residuals in the system. Chlorine residuals reported in Figure 3 were measured at the treatment plant, but the samples were taken in the distribution system. The levels of chlorine observed at the treatment plant were quite variable (see Figure 1) during the time of year when positive coliforms occurred. The variable levels also represented the response to the repeated reappearance of the bacteria when the chlorine residual was reduced.

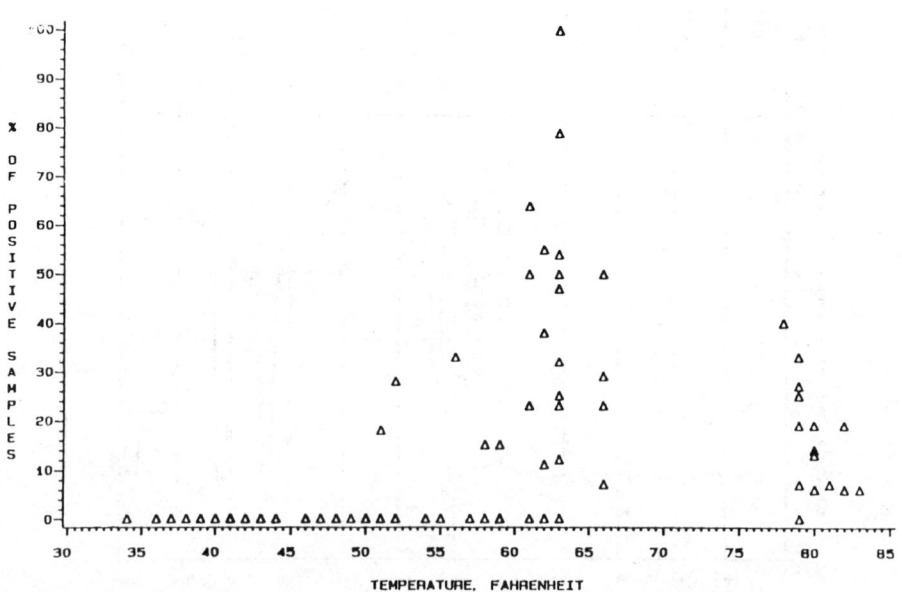

Figure 2. Frequency of positive coliform samples with respect to temperature (days where the chlorine level was >3 mg/L were omitted).

The level of chlorine at the treatment plant may have been considerably different than the level in the distribution system because of the attempt to return to normal chlorine dose and travel time. To correct for this difference, the values of chlorine residual in the distribution system for a given day were extrapolated from the chlorine residual leaving the plant the day before (offset by 1 d). The extrapolated residual levels, used for comparison to the frequency of positive daily samples, were within 1 mg/L of the observed chlorine residuals at each station. Each point represents the frequency of positive coliforms for 10 to 30 samples collected in the water distribution system on a given day. The frequency of detection of positive coliforms decreased as the chlorine residual increased. The levels of residual chlorine were not continuous; however, they did reflect the stepwise increase in chlorine dose in response to the occurrence of positive coliforms. Free chlorine at 3 mg/L appears to be a reasonable level to segregate the data for this system.

Figure 4 is a scatter diagram of 201 samples showing the number of total coliforms per 100 mL compared with the free chlorine residual observed for samples collected in the distribution network. A considerable free chlorine

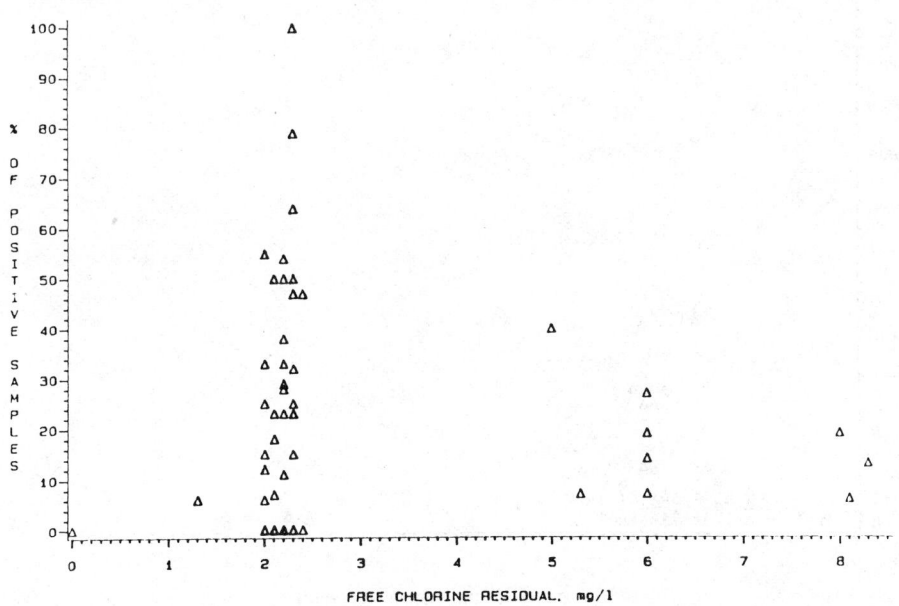

Figure 3. Daily percentage of positive coliform samples with respect to residual chlorine in the distribution system. The values for residual chlorine in the distribution system were extrapolated from the residual chlorine leaving the plant, offset by 1 d.

residual was maintained in the system throughout 1983 and was generally about 2 mg/L. Few sample stations had less than 1 mg/L. During the warmer months, the free chlorine residual was increased to as high as 10 mg/L to combat the recovery of positive coliforms. No coliforms were recovered at a residual of 10 mg/L, but they were found at levels as high as 8 mg/L. With two exceptions, the levels of coliforms were low at residual chlorine levels greater than 3 mg/L.

The distribution of samples collected in the water distribution system with respect to the levels of chlorine residual is shown in Table I. Since Figure 3 shows a dramatic break in coliform levels when the residual was greater than 3 mg/L, this residual level was used to separate the samples. Samples with coliform levels of <1, 1 to 10, 11 to 100, and >100 per 100 mL were recorded for residual chlorine levels of ≤ 3 mg/L. A total of 2515 samples were collected during 1983. Almost twice as many samples were collected at residual chlorine levels about 3 mg/L. The increased residual chlorine and sampling frequency were a part of the response of the utility to positive coliform samples; the number of samples was increased and residual chlorine levels were raised. The overwhelming majority of samples at both chlorine levels had less

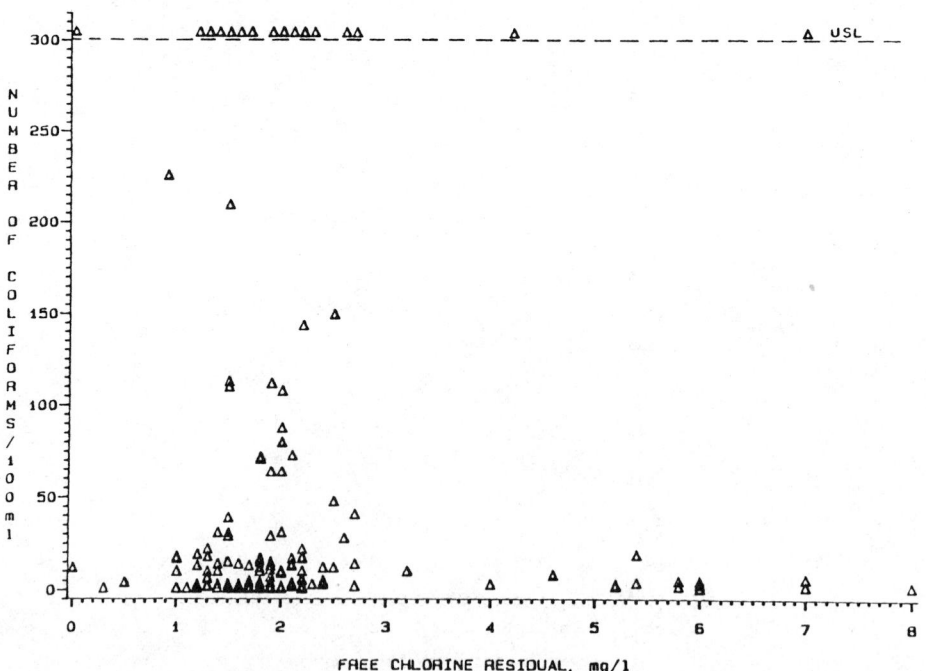

Figure 4. The number of coliforms per 100 mL vs free residual chlorine for all positive samples collected from the distribution system during 1983 (N = 201); USL, upper sensitivity limit.

Table I. Distribution of Samples from the Water Distribution System with Respect to Levels of Coliforms and Level of Free Residual Chlorine During 1983

Free Residual Chlorine (mg/L)		Coliforms per 100 mL							
		<1		1 to 10		11 to 100		>100	
	No.	No.	%	No.	%	No.	%	No.	%
<3	846	668	79.0	96	11.3	50	5.9	32	3.8
>3	1669	1646	98.6	19	1.1	1	<0.1	3	0.2
Total	2515	2314	92.0	115	4.6	51	2.0	35	1.4

than 1 coliform per 100 mL: 668 of 846 (or 79.0%) at residuals ≤ 3 mg/L and 1646 of 1669 (or 98.6%) at residuals > 3mg/L. The distribution of coliform levels in positive coliform samples differed for each level of chlorine. For 3 mg/L or less, 96 of 846 (11.3%), 50 of 846 (5.9%), and 32 of 846 (3.9%) of the samples had coliform levels of from 1 to 10, 11 to 100, and > 100 per 100 mL. For > 3 mg/L, only 19 of 1669 (1.1%), 1 of 1669 (< 0.1%), and 3 of 1669 (0.2%) of the samples were found in similar categories of coliform levels. At higher chlorine levels, fewer positive coliforms were found, and the densities observed were lower.

The quality of the raw water entering the treatment plant was variable and was heavily influenced by runoff. Figure 5 shows the final turbidity of the treatment water before it enters the distribution system compared with raw water turbidity in nephelometric turbidity units (NTU). Turbidities greater

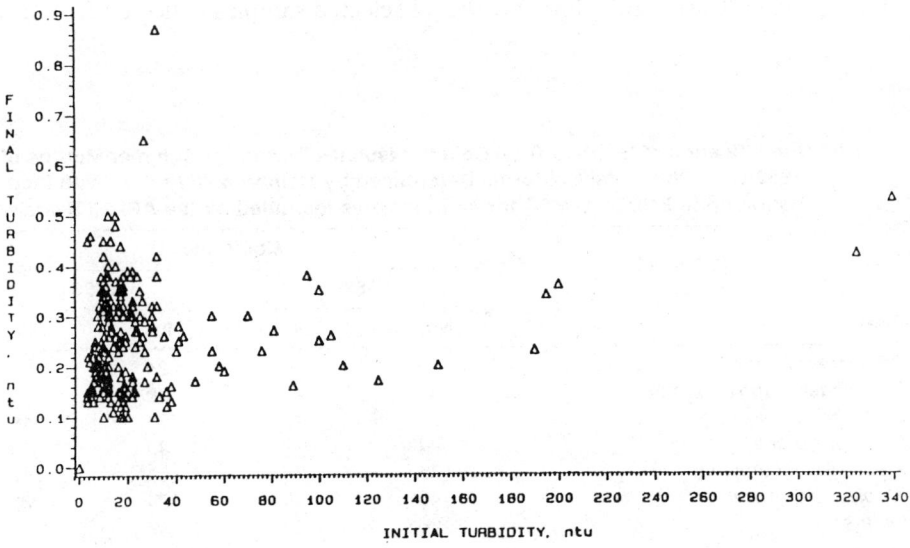

Figure 5. Comparison between initial and final turbidity at the treatment plant for 1983.

than 300 NTU were observed for the raw water. Regardless of raw water turbidity, the levels of turbidity of the finished water leaving the plant never exceeded 1 NTU.

Table II identifies total coliform isolates obtained from membrane filter cultures on m-ENDO medium during the summer months of 1982 and 1983. A total of 550 isolates were submitted to further identification by the API 20 method at the water utility laboratory. Only 2% (4 of 219) in 1982 and 14% (45 of 331) in 1983 were identified as *E. coli*. The predominant isolates in 1982 were *Klebsiella pneumoniae* (34%), *Enterobacter agglomerans* (22%), and *Klebsiella oxytoca* (18%). In 1983, the predominant isolates were *K. oxytoca* (38%), *Enterobacter cloacae* (19%), and *K. pneumoniae* (16%). Each of these microorganisms has been found in other episodes of positive coliform samples.

Source of Coliforms

Selected samples exhibiting high background bacterial growth were further assayed for levels of perfringens to provide information to evaluate the source of the coliforms. *C. perfringens* is an anaerobic spore-forming bacteria found in high numbers in feces and soils and not commonly found in water distribution systems. Table III summarizes the microbiological data for the special samples collected in the water system. No coliforms or *C. perfringens* were recovered from 18 samples of finished water leaving the water treatment plant. Routine daily samples and special 8-h and hourly sampling programs conducted by the water company consistently yielded negative coliforms for the finished water during this time. Of the 44 selected samples collected from the

Table II. Identification of Positive Total Coliform Isolates During the Summer Months of 1982 and 1983. Total Coliforms Determined by Membrane Filtration with Incubation on m-ENDO at 35°C for 48 h. Isolates Identified by the API 20 System

	Coliforms			
	1982		1983	
Agent	No.	%	No.	%
Citrobacter fruendii	0	0	24	7
Enterobacter agglomerans	48	22	18	5
Enterobacter cloacae	27	12	64	19
Escherichia coli	4	2	45	14
Klebsiella pneumoniae	74	34	54	16
Klebsiella oxytoca	40	18	124	38
Others	26	12	2	1
Total	219	100	331	100

pipe network, 16 (36%) were positive for total coliforms, and 12 (27%) were positive for fecal coliforms. The levels varied from 2 to >200 for both total and fecal coliforms. No *C. perfringens* were recovered from any of the suspect selected samples collected in the pipe network. Two storage tanks were in the water distribution system. Six (67%) and 2 (22%) of the 9 suspect samples collected from the tanks were positive for total and fecal coliforms, respectively. As in the finished water and the pipe network, no *C. perfringens* were recovered from samples taken from the tanks. Total coliforms were recovered from 18 of 22 (or 82%) well-point samples from the water table. The four negative samples had standard plate counts far in excess of the 500 cfu/mL, which suggested that the presence of coliforms may have been masked. *C. perfringens* were found in every one of the well-point samples. The levels recovered ranged from 2 to 9200 per 100 mL. Note that these data were based on selected samples that were positive for total coliforms or high background levels of other bacteria. The percentages reported are not intended to reflect the water quality in the pipe network, but rather the comparative frequency of recovery of microbial indicators in the samples assayed in our laboratory during this study.

Coliform Resistance

Several coliform isolates recovered from the water distribution system were selected at random for further study to evaluate their resistance to chlorine. Isolates 108 and 157 were submitted to further biochemical tests (API 20E) and tentatively identified as *K. pneumoniae*. Isolate 147 recovered from a well-point sample was not identified.

Each isolate was grown overnight in broth culture, washed in buffer and diluted so that the reaction density was about 100,000 cfu/mL, and exposed to

Table III. Recovery Frequency of Microbial Indicators for Special Samples in the Water System During Late Summer and Fall 1983

		Positive Samples						
		Total Coliform		Fecal Coliform		*Clostridium perfringens*		
Sample	No.	No.	%	No.	%	No.	%	
Finished water	18	0	0	0	0	0	0	
Pipe network	44	16	36	12	27	0	0	
Storage tanks	9	6	67	2	22	0	0	
Well points	22	18	82[a]	ND[b]		22	100	

[a]Negative samples had standard plate counts >500 cfu/mL.
[b]Not determined.

chlorine levels found in the water distribution system. Figure 6 shows the inactivation curves for coliform isolates and a laboratory strain of *Escherichia coli* B. The upper panel shows the level of free chlorine (as milligrams-per-liter available chlorine) during the course of the disinfection trial. The free chlorine dose was 1.2 mg/L, and the free residual chlorine after 300 s was between 0.7 and 1.0 mg/L. The log of the survival fraction N/N_0 was plotted against the contact time in seconds. *E. coli* B, isolates 108, 157 (pipe network), and 146

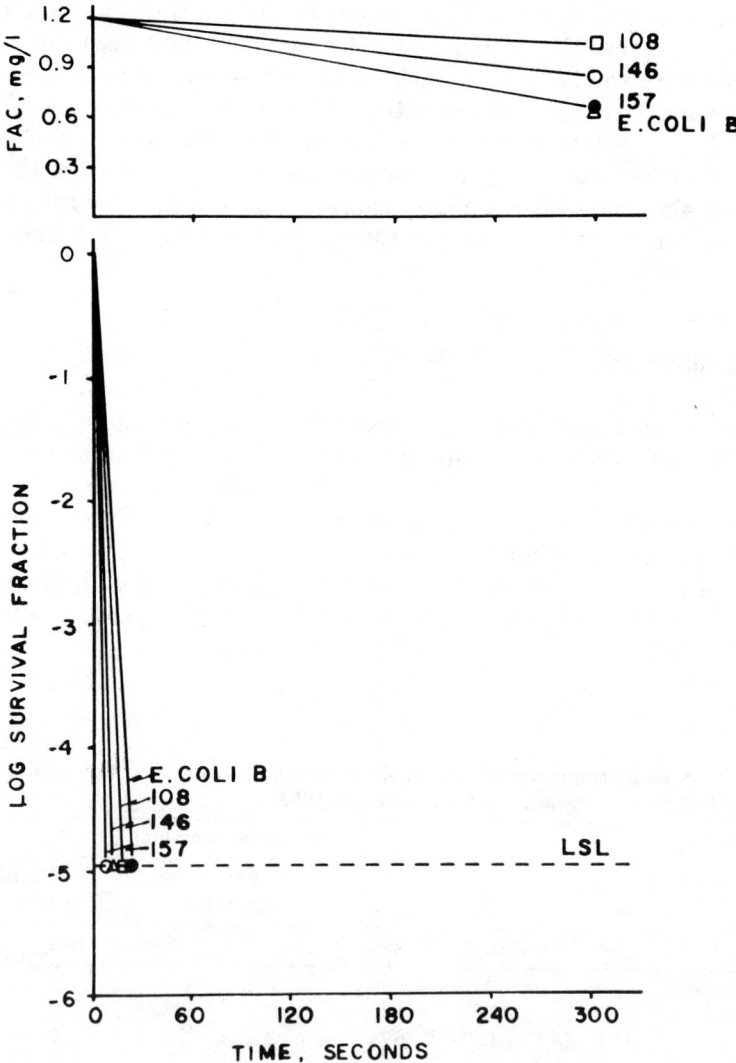

Figure 6. Typical inactivation curves for coliforms recovered from the water system and a laboratory strain of *E. coli* B (LSL, lower sensitivity limit).

(well point) were rapidly inactivated in the first few seconds of exposure to free chlorine. The levels of these microorganisms were reduced to below the sensitivity limit of the enumeration procedure well within 30 s. No microorganisms were recovered in the trials after this short contact with free chlorine. Similar studies are shown in Figure 7 and include trials with *C. perfringens* isolated from well-point samples. The free chlorine dose was slightly lower (1.0 mg/L), and the free residual chlorine after 300 s ranged between 0.5 and 0.9 mg/L. *E. coli* B and the coliform isolates were again rapidly inactivated; however, in this trial the densities of the survivors were not reduced to the sensitivity limit, and microorganisms were recovered even after a 300-s contact with free chlorine. Greater than three orders of magnitude of inactivation (99.9% kill) were found. No significant difference between the laboratory culture *E. coli* B and the coliform isolates was observed. The *C. perfringens* isolate was noticeably resistant. Just under one order of magnitude of inactivation was observed even after 300 s.

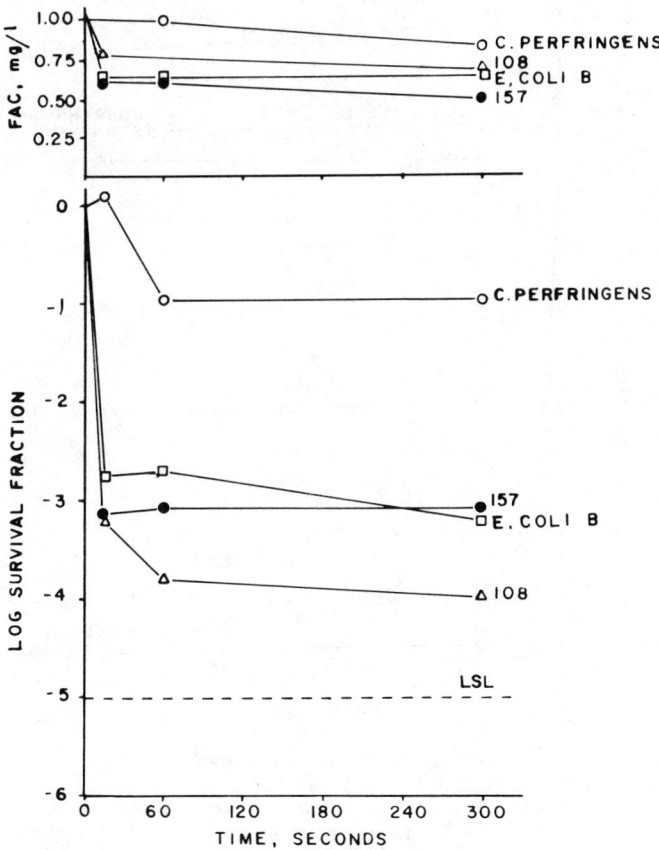

Figure 7. Comparison of relative resistance of *C. perfringens*, *E. coli* B., and two coliform isolates (108 and 157); LSL, lower sensitivity limit.

Since no particular resistance was observed for the coliform isolates, a series of disinfection trials was conducted to provide further information on the resistance of the coliform isolates from the water distribution system. Cultures were prepared from the survivors of the disinfection studies, and biofilms of isolates 108 and 157 were allowed to develop on reduced-strength media on glass slides. The intact biofilm on the slide was then exposed to free chlorine and compared with replicate cultures of original isolates and survivors grown on agar plates and resuspended in buffer prior to exposure to free chlorine. The results are shown in Figures 8 and 9 for isolates 108 and 157, respectively. The upper panel, as in the previous figures, shows the free available chlorine residual during the course of disinfection. The chlorine dose for the slide biofilm trial was 2.0 mg/L, and the 600-s free available residual chlorine was

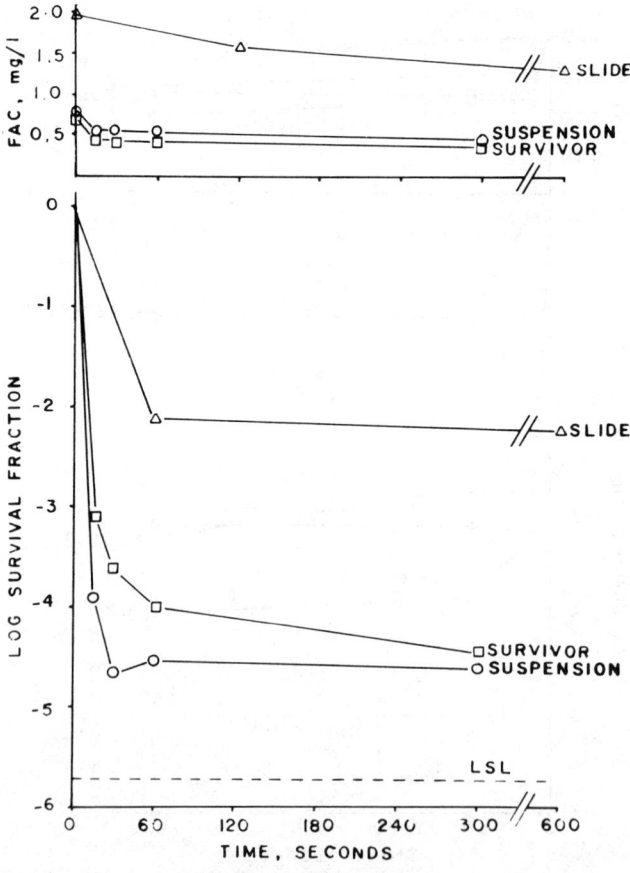

Figure 8. Relative resistance to chlorine of isolate 108 exposed on slide, in suspension, and survivor in suspension (so-called survivor is isolate 108 previously exposed to chlorine in our laboratory); LSL, lower sensitivity limit.

about 1.3 mg/L. The chlorine dose for the suspension trials was about 0.8 mg/L, and the free available chlorine residual after 300 s was in the neighborhood of 0.5 mg/L. Little difference was found between the suspension preparations of original isolate and survivor 108. Both showed no particular resistance to free chlorine. Four orders of magnitude inactivation were observed in 60 s. Survivors appeared to escape kill by chance rather than by development of resistance; however, the biofilm preparation on the glass slide was dramatically more resistant to higher levels of free chlorine. Only 2 logs inactivation were observed in 60 s, and little or no further kill was found after 600 s. The preparation and mode of exposure of the microorganisms had a greater effect on inactivation. Similar results were obtained for isolate 157 shown in Figure 9. While the magnitude of inactivation differs slightly for different isolates, the trends were the same.

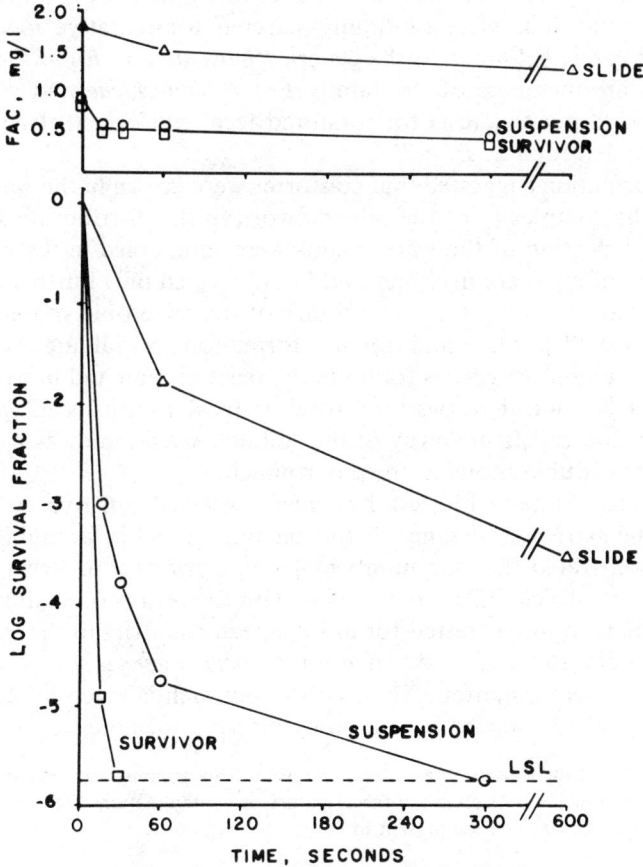

Figure 9. Relative resistance to chlorine of isolate 157 exposed on slide, in suspension, and survivor in suspension (so-called survivor is isolate 157 previously exposed to chlorine in our laboratory); LSL, lower sensitivity limit.

DISCUSSION

The recovery of coliforms from the water distribution system appeared to be seasonal and heavily influenced by temperature. No positive coliform samples were found during the winter months. The frequency of occurrence increased dramatically in May and continued through November. The routine data collected by the water utility indicated that below 60°F (15°C), the daily frequency of recovery of coliforms was low. Once they were found, the water utility's response was to increase the residual chlorine. Further correlations with temperature were difficult to establish because of the stress provided by the elevated chlorine levels. Coliforms were consistently recovered when the water system was operated at normal free chlorine levels. Prior to the recovery of coliforms, finished water typically was dosed to a free residual chlorine of about 2.0 mg/L, and a free residual chlorine in the neighborhood of 1.0 mg/L was carried throughout most of the distribution system. Coliforms recovered under these conditions were randomly selected for tentative identification, and, in most cases, belonged to the genera *Klebsiella* and *Enterobacter*. Both these genera are members of the family *Enterobacteriacae* and will ferment lactose and yield positive tests for total and fecal coliform. Infrequently, the isolates were identified as *E. coli*.

Initial information suggested that coliforms were grown in the water system. Because of the complexity of the pipe network in the distribution system and the fact that a portion of the water mains were submerged in the water table, the possibility of cross connections could not be ruled out. Further analysis of the positive water samples for the presence of the anaerobic spore former, *C. perfringens*, would provide additional information to evaluate the source of the coliforms. *C. perfringens* is found in the feces of man and most animals at lower frequencies and densities than total or fecal coliforms. *C. perfringens* forms spores and resists die-away in the soil and water, and, as a result, this microorganism is ubiquitous in the environment.

In the United States, this test has never received positive consideration because of the extreme resistance of this bacterium and its abundance in soil. Levine[7] demonstrated that the number of *C. perfringens* in streams was too constant, and the levels failed to correlate with the results of sanitary surveys. This organism is routinely tested for in European countries in the examination of potable waters. In the United Kingdom, *C. perfringens* is sometimes used as a supplemental determination. The British approach is given in Reference 8; Report 71, as this document is commonly referred to, notes:

> "The chief value of the *Cl. perfringens* test, therefore, is in demonstrating remote or intermittent pollution or in confirming the faecal nature of contamination when only coliform organisms other than *E. coli* are present in the water."

The test for the spores of *C. perfringens* was used to provide information to evaluate the source of coliforms in the distribution system. Since *C. perfringens* is a fecal indicator and is commonly found in soils and decaying organic matter, any external source of coliforms would also contain this bacterium. Once in the distribution system, the spores of *C. perfringens* would be more resistant to chlorine than coliforms. Thus, if the source of the coliforms was external, *C. perfringens* would be found with the coliforms.

The spores of *C. perfringens* were recovered from all well-point samples of the water table but were not recovered from any of the samples collected from the finished water at the treatment plant or in the distribution system. The data strongly suggest that the source of the coliforms was not external. The negative coliform results for extensive routine and special samples of the finished water suggest that coliforms found in the distribution system are not gaining entrance through the treatment plant. While neither of the above two possible sources of coliforms can be absolutely ruled out, the information supports the contention that the coliforms were a result of regrowth in the distribution system.

The regrowth of members of the coliform group in the absence of chlorine residuals was not unexpected. However, the regrowth in the presence of relatively high levels of free chlorine was somewhat unique. Coliforms were recovered in the presence of free residual chlorine as high as 5 to 8 mg/L. Preliminary studies were conducted to evaluate the resistance of coliforms isolated from the water system and to provide information to explain their occurrence. The coliform isolates, when exposed to chlorine levels in the neighborhood of 1 mg/L as a suspension, were no more resistant than a laboratory strain of *E. coli* B. Subsequent retesting of survivors of chlorination trials did not yield a culture of increased resistance. Similar inactivation curves were obtained for suspensions of the original isolates and the survivors. Little difference was observed for the inactivation rates of these isolates and other vegetative bacteria. No particular resistance of these coliforms to free chlorine could be documented.

However, when the same cultures were prepared and exposed to free chlorine as biofilms on glass slides, the coliforms were noticeably resistant and persisted in the presence of free chlorine. Though free chlorine was measured in the bulk solution, the chlorine apparently does not reach the microorganisms in the biofilm. Whether the chlorine is consumed or does not penetrate the biofilm remains to be determined.

CONCLUSIONS

Coliforms in the distribution system appeared to be a result of regrowth rather than passage through the treatment process of cross connections. Those isolated from the pipe network were not particularly resistant to free chlorine when exposed as suspensions. Biofilms of the same microorganisms were resistant to levels of free chlorine commonly found in water distribution systems. Tests for the spores of *C. perfringens* were useful and provided information to evaluate intrusions into the water system.

ACKNOWLEDGMENTS

The authors wish to acknowledge the support of the American Water Works Service Company for this project and the National Council for Scientific Research of Lebanon for the support given A. E. Bakalian. The technical assistance of G. Toles and M. Sarai is gratefully acknowledged.

REFERENCES

1. Tracy, H. W., V. M. Camarena, and F. Wing. "Coliform Persistence in Highly Chlorinated Waters," *J. Am. Water Works Assoc.* 58:1151–1159 (1966).
2. Hudson, L. D., J. W. Hankins, and M. Battaglia. "Coliforms in a Water Distribution System: A Remedial Approach," *J. Am. Water Works Assoc.* 75(4):564–568 (1983).
3. Goshko, M. A., W. O. Pipes, and R. R. Christian. "Coliform Occurrence and Chlorine Residual in Small Water Distribution Systems," *J. Am. Water Works Assoc.* 75:371–374 (1983).
4. Martin, R. S., W. H. Gates, R. S. Tobin, D. Grantham, R. Sumarah, P. Wolfe, and P. Forestall. "Factors Affecting Coliform Bacteria Growth in Distribution Systems," *J. Am. Water Works Assoc.* 74:34–37 (1982).
5. *Standard Methods for the Examination of Water and Wastewater*, 15th ed. (Washington, DC: American Public Health Association, 1980).
6. Cabelli, V. J. "*Clostridium perfringens* as a Water Quality Indicator," in *Bacterial Indicators—Health Hazards Associated with Water*, A. W. Hoadley and B. J. Dutka, Eds. (Philadelphia: American Society for Testing and Materials, 1977).
7. Levine, M. "Bacteria Fermenting Lactose and Their Significance in Water Analysis," Bulletin 62, Official Publication 20, Vol. 31 (Ames, IA: Iowa State College of Agriculture and Mechanical Arts, 1921).
8. Barrow, G. I., "Bacterial Indicators and Standards of Water Quality in Britain," in *Bacterial Indicators—Health Hazards Associated with Water*. A. W. Hoadley and B. J. Dutka, Eds. (Philadelphia: American Society for Testing and Materials, 1977), p. 289.

CHAPTER 52

Sensitivity of Vegetative Protozoa to Free and Combined Chlorine

Charles N. Haas, Kamel M. Khater, and Allen T. Wojtas

As early as 1936, it was observed[1] that bacteria surviving wastewater chlorination could regrow in a receiving stream. This observation has been replicated by numerous other workers.[2-4] Kinney et al.[5] have demonstrated that, at least in certain circumstances, the coliform count in a receiving water may be greater if the wastewater effluent is chlorinated.

The regrowth phenomenon, and in particular the finding that wastewater chlorination may adversely influence the downstream coliform count, has been attributed to predation by indigenous protozoa,[6-7] which have been suggested as the dominant means whereby coliforms decay in natural waters.[8] It has been shown[7] theoretically that if protozoa are even slightly less sensitive to chlorine residuals than coliforms, as long as sufficient nutrients are present in the receiving environment, the phenomenon of anomalous regrowth will occur.

However, to our knowledge, no attempts to quantify the sensitivity of vegetative protozoa to chlorine residuals have been made previously. The objective of this work was to perform such quantification and, subsequently, to assess whether the predation argument for anomalous regrowth is plausible.

MATERIALS AND METHODS

Tetrahymena pyriformis strain W was used as a model organism, because it is a vegetative, free-swimming ciliate that can consume bacteria, tissues of higher organisms, or soluble nutrients as food sources.[9] The organism was grown in proteose peptone broth test tubes at 25°C for 48 h for subsequent use as an inoculum for a flask culture. Following 72 h incubation at 25°C, this suspension, which was found to be in late log-phase growth, was washed and resuspended three times in pH 7 chlorine demand-free water (CDFW). The washed, resuspended culture was used in the chlorination experiments.

Experimental procedures (preparation of CDFW, use of chlorine, and cell controls) were identical to those of Engelbrecht et al.[10] Monochloramine solutions were prepared at a 5:1 weight ratio (Cl_2:N) following the procedure of Butterfield and Wattie.[11] Chlorine residuals were measured using forward amperometric titration.[12]

Cell viability was assessed microscopically under the assumption that all motile cells were viable. The nonmotile cells were counted following dechlorination in a Petroff-Hauser chamber (Hausser Scientific, Blue Bell, PA). A second aliquot of dechlorinated suspension was treated with 0.5 vols of 5% buffered formalin as a fixative,[13] and the total count was determined microscopically. The difference between total and nonmotile counts represented the motile population, which was taken to be the viable population. Generally, nonchlorinated controls contained less than 2% nonmotile cells.

Two sets of experiments were conducted. In the first series of experiments, preformed monochloramine was used as a disinfectant at pH 7. In the second series of experiments, the cell suspensions were dosed with free chlorine at pH 7.

INACTIVATION BY MONOCHLORAMINE

Since microbial viability was assessed microscopically in this study, it was necessary to use densities of protozoa of 50,000 to 100,000/mL. It was ascertained that doses of monochloramine below 0.6 mg/L resulted in minimal inactivation, whereas doses above 1 mg/L resulted in a more rapid rate of inactivation than could readily be measured. In this range of doses, it was found that the microorganisms themselves, despite triple washing in CDFW, still exhibited substantial chlorine demand. Hence, it was necessary to measure the decay of monochloramine residual during the time course of inactivation so that a quantitative comparison could be made with the rate of inactivation of other microorganisms.

Figure 1 presents a typical monochloramine survival curve. Three replicate inactivation experiments were conducted at each of three initial monochloramine concentrations (1.0, 0.85, and 0.65 mg/L). Also indicated on this figure is the loss of monochloramine during the experiment. The bulk of monochloramine demand was exerted prior to the first sampling time; however, a slow, continual demand was evidenced. In separate experiments, it was determined that, at the chlorine doses used, no titratable chlorine residual other than monochloramine was present.

The nine monochloramine inactivation experiments were qualitatively similar to that of Figure 1. At the beginning of the experiment, a lag phase or "shoulder" was apparent. Following this, a second stage of linear inactivation occurred. It was evident that the duration of the lag phase and the slope of the linear region were dependent on the monochloramine dose used.

The shape of the inactivation curve is similar to those observed by a number of other authors. Wei and Chang[14] observed shoulders in inactivation curves of a variety of microorganisms. They attributed this phenomenon to clumping or to the existence of multiple targets. McGrath and Johnson[15] attributed shoulders in spore inactivation curves to diffusional lags. In the present instance, since cells were observed microscopically and no visible clumping

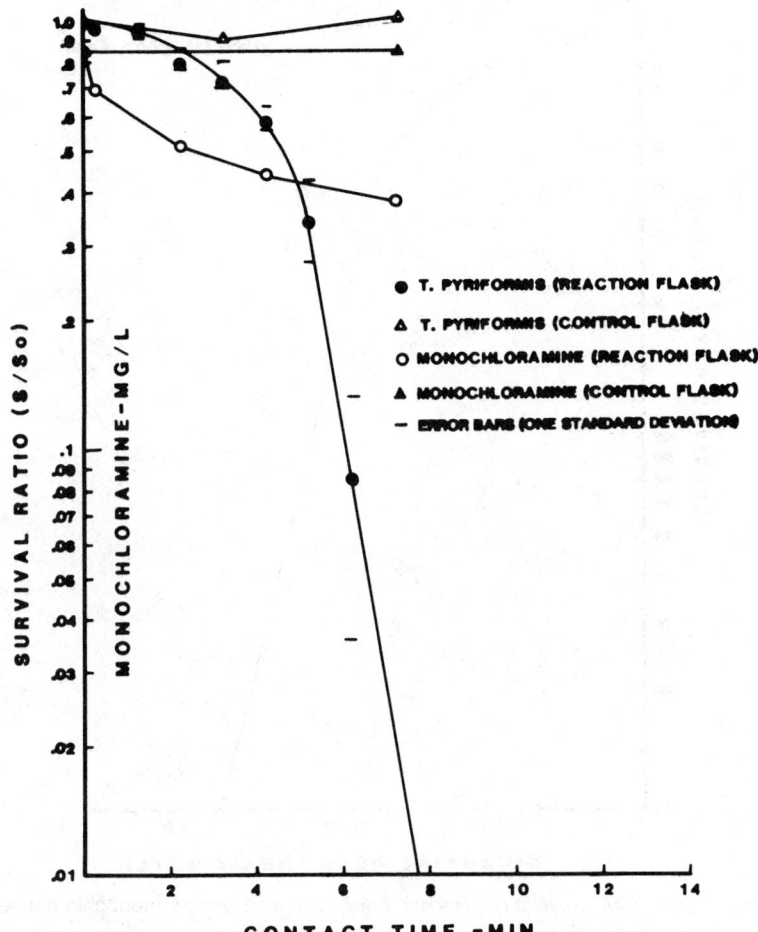

Figure 1. *T. pyriformis* survival ratio and monochloramine residuals vs contact time for 0.85 mg/L monochloramine applied at pH 7 and 25°C.

was detected, this rationale may be excluded. However, multiple internal targets or the presence of diffusional or chemical[16] lags may explain the shape of the observed curves.

The fact that monochloramine concentrations diminished with time during an experiment made the usual application of the Chick-Watson[17,18] equation invalid, because the integrated semilogarithmic form implicitly assumes constant disinfectant residual. Instead, as a means of comparison, the microbial dose concept of Hall[19] and Morris[20] was used. From the basic Chick-Watson relationship, the following equation can be written:

$$\mathrm{Ln}(N/N_0) = -\Lambda \int_0^t [C(t)]^n \, dt \tag{1}$$

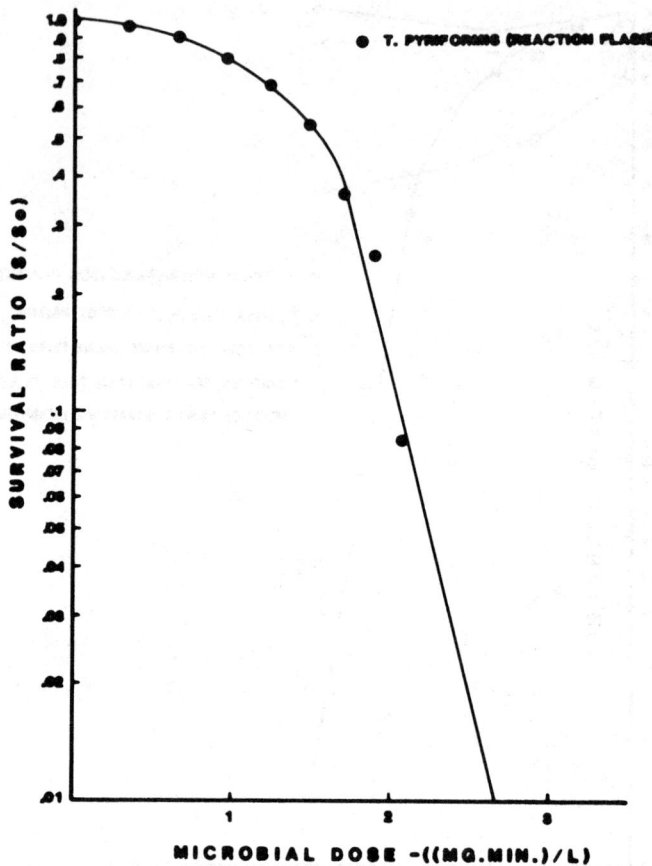

Figure 2. T. pyriformis survival ratio vs microbial dose for 1.0 mg/L monochloramine applied at pH 7 and 25°C.

If the coefficient of dilution (n) is assumed equal to unity, which appears to be approximately true for a number of organisms and monochloramine,[21] then the integral in Equation (1) can be evaluated numerically, and the inactivation curves vs the value of this integral can be plotted. This is termed the microbial dose.

As anticipated, the curves of survival vs microbial dose (Figures 2 through 4) also have a lag phase. Furthermore, the inactivation curves for the three different initial doses were not coincident, indicating that Equation (1) does not account for all of the concentration dependency of inactivation of T. pyriformis by monochloramine.

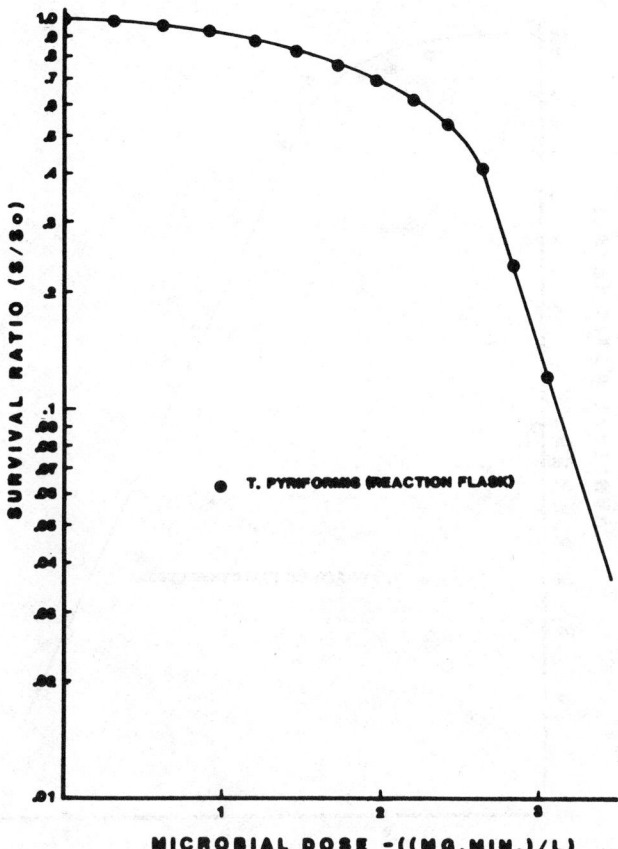

Figure 3. *T. pyriformis* survival ratio vs microbial dose for 0.85 mg/L monochloramine applied at pH 7 and 25°C.

Using the microbial dose concept, the value for the coefficient Λ, termed the specific lethality coefficient, was calculated for each of the nine experiments (Table I).

To compare the observed sensitivity of *T. pyriformis* with that of other microorganisms, it was necessary to analyze reported kinetic data in a similar manner. The investigations of Butterfield and Wattie.[11] on bacterial inactivation and Kelley and Sanderson.[22] on viral inactivation by monochloramine formed such suitable data sets. Table II summarizes specific lethality coefficients for these two sets.

A comparison of specific lethality coefficients for *T. pyriformis* (Table I) with those of other organisms (Table II) indicates that the protozoan is up to six times more sensitive to monochloramine than either *Escherichia coli* or *Salmonella (Eberthella) typhosa*. The two poliovirus strains exhibited up to 80 times greater resistance to monochloramine than the protozoan species. These

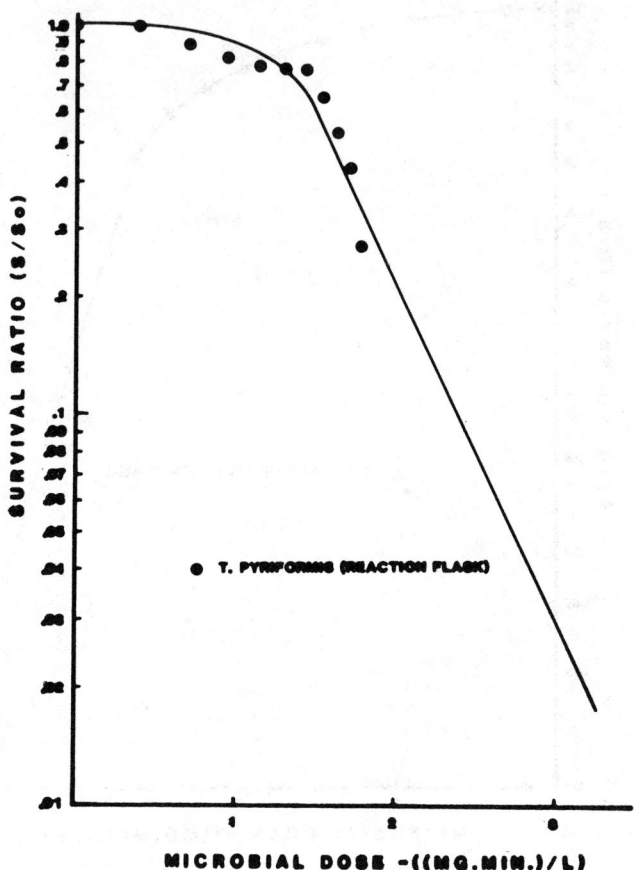

Figure 4. *T. pyriformis* survival ratio vs microbial dose for 0.65 mg/L monochloramine applied at pH 7 and 25°C.

Table I. Specific Lethality Coefficients for Monochloramine[a]

Initial Concentration (mg/L)	Specific Lethality Coefficient (L mg^{-1}min^{-1})
0.65	0.85 ± 0.026
0.85	1.15 ± 0.089
1.00	1.63 ± 0.029

[a]Reults are mean ± standard deviation of mean for three separate experiments at each concentration.

Table II. Specific Lethality Coefficients for Data from Butterfield and Wattie[11] and Kelley and Sanderson[22] for pH 7 at 20 to 25°C

Monochloramine (mg/L)	Specific Lethality Coefficient (L mg^{-1}min^{-1})			
	E. coli[a]	Eber. typhosa[a]	Poliovirus[b]	Poliovirus[c]
1.2	0.300	0.315		
1.0			0.024	0.020
0.9	0.196	0.210		
0.6	0.156	0.167		

[a]For data from Reference 11.
[b]Type I (Mahoney) data from Reference 22.
[c]Type I (Mk 500) data from Reference 22.

data thus demonstrate that, at least at pH 7 and ambient temperature, the vegetative protozoan *T. pyriformis* is more sensitive to preformed monochloramine than *E. coli*.

INACTIVATION BY FREE CHLORINE

The experiments using free chlorine were conducted in triplicate at each of three disinfectant doses, 0.7, 0.85, and 1.0 mg/L. Under these conditions, it was found that almost immediately (within 30 s) all the titratable free chlorine had disappeared and only a lesser amount of monochloramine could be measured. Thus, survival and monochloramine residual were recorded as functions of time. Figure 5 presents a typical survival curve.

The shape of the survival curves was similar to that obtained using preformed monochloramine. Because of the quantitative decay in chlorine residual, as well as the rapid conversion of free chlorine to combined residual, it was not possible to quantitatively determine the relative sensitivity of the protozoan to free chlorine. However, by comparing the observed inactivation in experiments in which free chlorine was added to the inactivation, which would have occurred if only the measured monochloramine was an active biocide, a qualitative indication of the efficiency of the initial, transient free chlorine may be obtained.

Let $I(M)$ be the fraction of microorganisms inactivated if only monochloramine was present; $I(F)$, the fraction of microorganisms that would be inactivated if only free chlorine was present; and $I(T)$, the inactivation observed when free chlorine is dosed and when monochloramine also exists. If it is assumed that the inactivation by monochloramine is not influenced by the presence of free chlorine, nor vice versa, then $I(T)$ is determined by the present series of experiments, $I(M)$ may be determined from the prior series of experi-

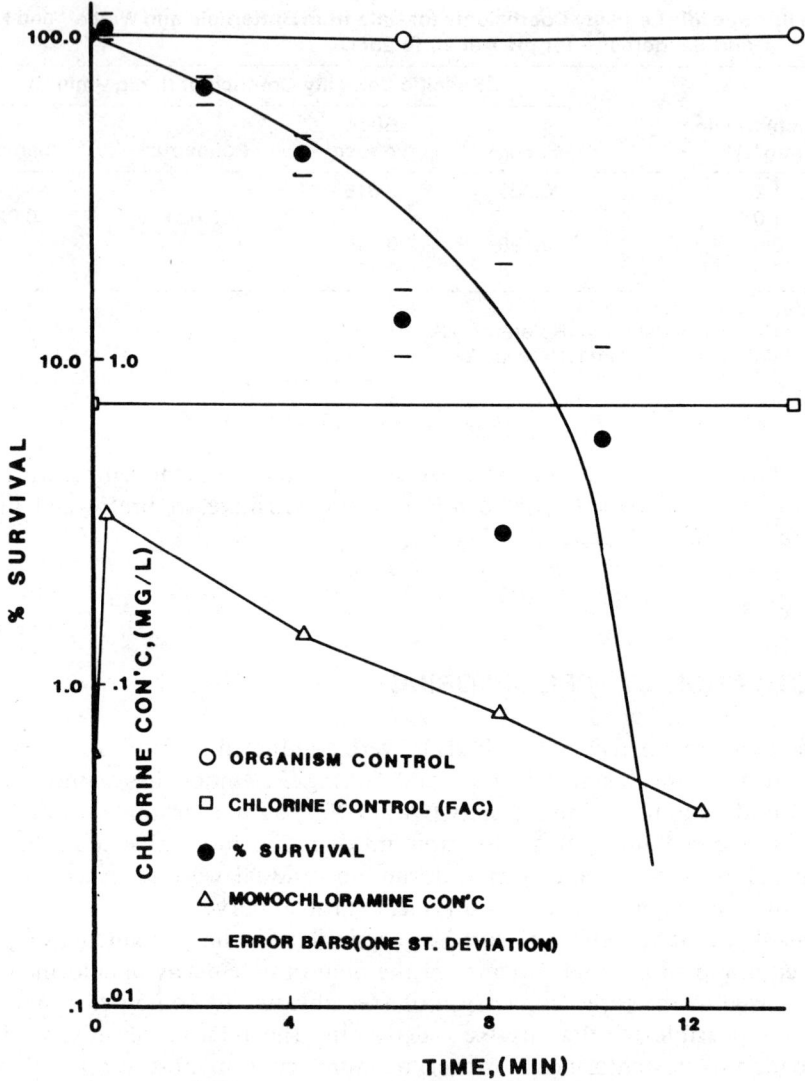

Figure 5. *T. pyriformis* survival ratio vs contact time for 0.72 mg/L free chlorine at pH 7 and 25°C.

ments where monochloramine alone was present, and $I(F)$ can be calculated by use of Equation (2).

$$[1 - I(T)] = [1 - I(M)][1 - I(F)] \tag{2}$$

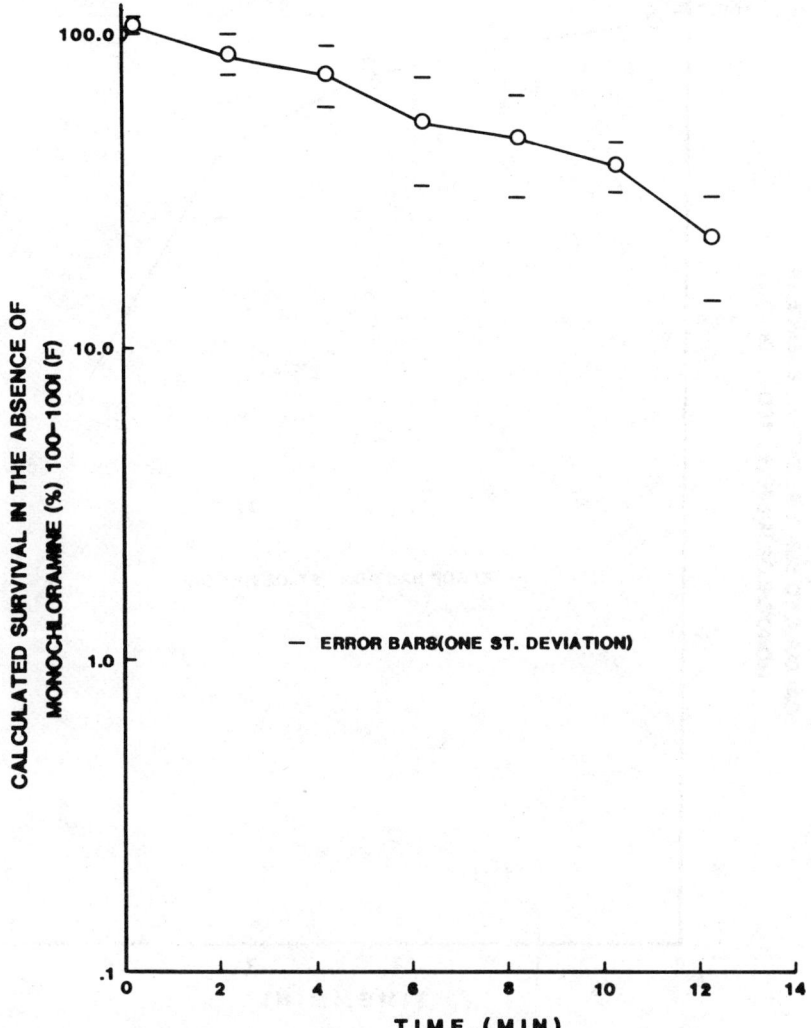

Figure 6. Calculated survival of *T. pyriformis* in the absence of monochloramine vs time (from the experimental data) at 0.70 mg/L free chlorine.

which may be rearranged to yield

$$I(F) = [(I(T) - I(M)]/[1 - I(M)] \tag{3}$$

At each free available chlorine dose, the microbial dose was calculated using the measured monochloramine vs time data. Using the information in Table I from experiments most closely corresponding to the initial monochloramine concentration, the quantity $I(M)$ was estimated. Since $I(T)$ was also known,

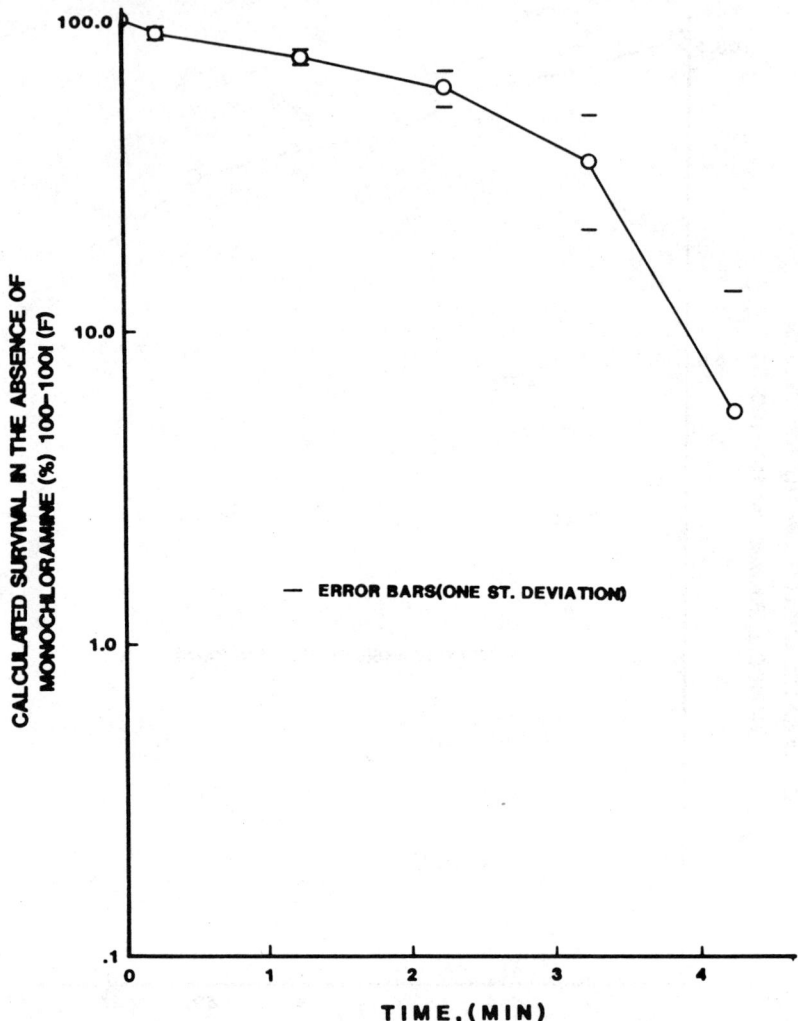

Figure 7. Calculated survival of *T. pyriformis* in the absence of monochloramine vs time (from the experimental data) at 0.85 mg/L free chlorine.

the value for $I(F)$ could be calculated from Equation (3). This term is denoted as "excess inactivation attributed to free chlorine," and $1 - I(F)$ is denoted as "calculated survival in the absence of monochloramine." Figures 6 through 8 present this latter quantity as a function of time for each of the three initial free chlorine doses used. (Each figure represents the average of three replicate experiments.)

Examination of these figures indicates that substantial excess inactivation can be attributed to the free chlorine that occurs, which is of particular interest

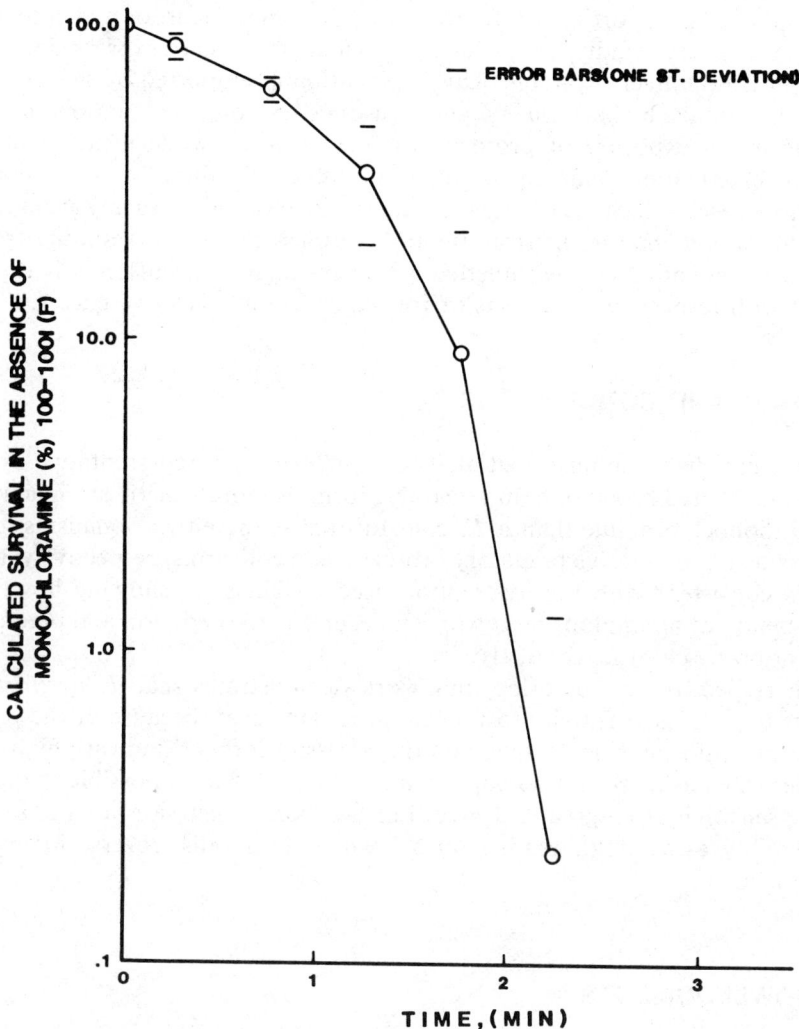

Figure 8. Calculated survival of *T. pyriformis* in the absence of monochloramine vs time (from the experimental data) at 1.0 mg/L free chlorine.

because there is a very rapid loss of titratable free chlorine from solution. In all cases, as time increases the inactivation increases, and the extent of this inactivation increases with initial dose.

Although no quantitative analysis of the relative sensitivity of *T. pyriformis* to free chlorine (either vis-a-vis monochloramine or vis-a-vis other organisms) can be made, the finding of a time- and dose-dependent excess inactivation does indicate that free chlorine also possesses substantial activity against the vegetative protozoan. There are two possible explanations for the time dependency of the excess inactivation, despite the inability to measure free chlorine.

The first is that a portion of free chlorine becomes incorporated into the microbial cell very rapidly, and, while its biocidal effect is not expressed immediately, it is no longer titratable. This explanation is supported by the finding of chlorine uptake by bacteria [23-25] and viruses.[26] A second explanation may be that the initial exposure of protozoa to free chlorine, while insufficient to result in inactivation, may cause sufficient sublethal injury so as to render them more susceptible to the longer-lasting monochloramine. In any event, the monochloramine may result from the lysis of microorganisms resulting from injury or inactivation.[27] The sublethal injury explanation is sufficiently documented with respect to the effects of free chlorine on vegetative bacteria.[28]

SUMMARY AND CONCLUSIONS

This study has demonstrated that *T. pyriformis*, a representative free-swimming ciliated protozoan, in vegetative form, is more sensitive to inactivation by monochloramine than is *E. coli*. Insofar as these two organisms may be taken as representative predatory protozoa and coliforms, respectively, this work is consistent with the assumption used by Haas[7] in showing how the phenomenon of anamolous regrowth can occur due to predation and predator inactivation by chlorine residuals.

With respect to free chlorine, this work demonstrates that *T. pyriformis* appears to possess sensitivity to free chlorine. However, because of the presence of intrinsic chlorine demand and the necessity for working with high cell densities when using the microscopic count method, it was impossible to quantify the sensitivity to a greater degree. Further work which uses a more sensitive viability assay (e.g., plating on a lawn of host cells, reverse MPN) is recommended.

ACKNOWLEDGMENTS

Charles N. Haas is Associate Professor of Environmental Engineering at the Illinois Institute of Technology (IIT). At the time this work was conducted, Kamel M. Khater and Allen T. Wojtas were graduate students at IIT. This work represents a portion of the M.S. theses research of Khater and Wojtas in the School of Advanced Studies at IIT.

REFERENCES

1. Rudolfs, W., and H. Gehm. "Sewage Chlorination Studies," Bull. 601 (New Brunswick, NJ: New Jersey Agricultural Experiment Station, 1936).
2. Heukelekian, H. "Disinfection of Sewage with Chlorine. IV. Aftergrowth of Coliform Organisms in Streams Receiving Chlorinated Sewage," *Sew. Ind. Wastes* 23:273-77 (1951).

3. Eliassen, R. "Coliforms Aftergrowths in Chlorinated Storm Overflows," *Am. Soc. Civil. Eng. J. San. Eng. Div.* 94(2):371 (1968).
4. Shuval, H., J. Cohen, and R. Kolodney. "Regrowth of Coliforms and Fecal Coliforms in Chlorinated Wastewater Effluent," *Water Res.* 7:537-46 (1973).
5. Kinney, E. C., D. W. Drummond, and N. B. Hanes. "Effects of Chlorination on Differentiated Coliform Groups." *J. Water Pollut. Control Fed.* 50:2307-12 (1978).
6. Haas, C. N. "Discussion: Effects of Chlorination on Differentiated Coliform Groups," *J. Water Pollut. Control Fed.* 51:2961-62 (1979).
7. Haas, C. N. "Application of Predator-Prey Models to Disinfection," *J. Water Pollut. Control Fed.* 53:378-86 (1981).
8. Enzinger, R., and R. Cooper. "Role of Bacteria and Protozoa in the Removal of *Escherichia coli* from Estuarine Waters," *Appl. Environ. Microbiol.* 31:758-63 (1976).
9. Elliot, A. M. "Life Cycle and Distribution of *Tetrahymena*," in *Biology of Tetrahymena* (Stroudsberg, PA: Dowden, Hutchinson and Ross, 1973).
10. Engelbrecht, R. S., B. F. Severin, M. T. Masarik, S. Farooq, S. H. Lee, C. N. Haas, and A. Lalchandani. *New Microbial Indicators of Disinfection Efficiency*, EPA 600/2-77-052 (Cincinnati: U.S. Environmental Protection Agency, 1977).
11. Butterfield, C. T. and E. Wattie. "Influence of pH and Temperature on the Survival of Coliforms and Enteric Pathogens When Exposed to Chloramines." *Public Health Rep.* 61:157-93 (1946).
12. *Standard Methods for the Examination of Water and Wastewater*, 15th ed. (Washington, DC: American Public Health Association, 1980).
13. Everhart, L. P. "Methods with *Tetrahymena*," in *Methods in Cell Physiology* 5:219-88 (1972).
14. Wei, J. H., and S. L. Chang. "A Multi-Poisson Distribution Model for Treating Disinfection Data." in *Disinfection of Water and Wastewater*, J. D. Johnson, Ed. (Ann Arbor, MI: Ann Arbor Science Publishers, 1975), pp. 11-47.
15. McGrath, S. T., and J. D. Johnson. "Microbial Dose as a Measure of Disinfection," in *Water Chlorination: Environmental Impact and Health Effects, Vol. 3*, R. L. Jolley, W. A. Brungs, and R. B. Cumming, Eds. (Ann Arbor, MI: Ann Arbor Science Publishers, Inc., 1980), pp. 687-695.
16. Haas, C. N. "Rational Approaches in the Analysis of Chemical Disinfection Kinetics," in *Chemistry in Water Reuse*, Vol. 1, W. J. Cooper, Ed. (Ann Arbor, MI: Ann Arbor Science Publishers, 1981).
17. Chick, H. "An Investigation of the Laws of Disinfection," *J. Hyg.* 8:92-158 (1908).
18. Watson, H. E. "A Note on the Variation of the Rate of Disinfection with Change in the Concentration of the Disinfectant," *J. Hyg.* 8:536 (1908).
19. Hall, E. L. "Quantitative Assessment of Disinfection Interferences," *Water Treat. Exam.* 22:153 (1973).
20. Morris, J. C. "Aspects of the Quantitative Assessment of Germicidal Efficiency," in *Disinfection of Water and Wastewater*, J. D. Johnson, Ed. (Ann Arbor, MI: Ann Arbor Science Publishers, 1975), pp. 1-10.
21. Haas, C. N., and S. B. Karra. "Kinetics of Microbial Inactivation by Chlorine. Part I. Demand-Free Systems," *Water Res.* 18:1443-9 (1984).
22. Kelley, S. M., and W. W. Sanderson. "The Effect of Chlorine in Water on Enteric Viruses. II. The Effect of Combined Chlorine on Poliomyelitis and Coxsackie Viruses." *Am. J. Public Health*, 50:14-19 (1960).

23. Haas, C. N., and R. S. Engelbrecht. "Chlorine Dynamics During Inactivation of Coliforms, Acid-Fast Bacteria Yeasts," *Water Res.* 14:1749-57 (1980).
24. Friberg, L. "Quantitative Studies on the Reaction of Chlorine with Bacteria in Water Disinfection," *Acta Pathol. Microbiol. Scand.* 38:135-44 (1956).
25. Friberg, L. "Further Quantitative Studies on the Reaction of Chlorine with Bacteria in Water Disinfection," *Acta Pathol. Microbiol. Scand.* 40:67-80 (1957).
26. Dennis, W. H., V. P. Olivieri, and C. W. Kruse. "Mechanism of Disinfection: Incorporation of Cl-36 Into f2 Virus," *Water Res.* 13:363-69 (1979).
27. Haas, C. N. and R. S. Engelbrecht. "Physiological Alterations of Vegetative Microorganisms Resulting from Chlorination," *J. Water Pollut. Control Fed.* 52(7):1976-89 (1980).
28. Camper, A. K. and G. A. McFeters. "Chlorine Injury and the Enumeration of Waterborne Coliform Bacteria," *Appl. Environ. Microbiol.* 37:633 (1979).

CHAPTER 53

Factors Influencing Chlorine Disinfection of Wastewater Effluent Contaminated by Rotaviruses, Enteroviruses, and Bacteriophages

M. S. Harakeh

The literature indicates a considerable degree of ambiguity and contradiction in the efficiency of chlorine as a wastewater disinfectant. This largely reflects on the absence of standardization in these studies and the use of only a single virus rather than different viruses belonging to different virus groups. There are also some contradictory findings reported on whether hypochlorous acid or the hypochlorite ion is a more effective virucidal agent. For instance, Scarpino et al.[1,2] found that the hypochlorite ion was a better virucide in a borate-potassium chloride-sodium hydroxide buffer than hypochlorous acid, and they attributed this unusual finding to the buffer used. However, Weidenkopf,[3] Clark et al.,[4] and Engelbrecht et al.[5-6] reported that hypochlorous acid was a better virucide than the hypochlorite ion. This study was undertaken to examine the effect of virus type, pH, temperature, and the presence of organic matter on the inactivation of viruses by chlorine in a municipal sewage effluent.

MATERIALS AND METHODS

Viruses

Stock cultures of poliovirus type 1 (LSc 2ab), coxsackievirus B5, and echovirus 1 were prepared in BGM cell cultures where viral infectivity was assayed by the microtiter method. The simian rotavirus, SA11, was cultivated in MA 104 cells in a serum-free medium containing trypsin, and infectivity was assayed by the plaque test in the same cells supported by a medium containing DEAE dextran and pancreatin.[7] The human rotavirus (from a stool specimen) was assayed in MA 104 cells by detecting cell-associated viral antigens by indirect immunofluorescence.[8] Bacteriophage f_2 was grown in *Escherichia coli* K_{12} Hfr and assayed by the soft-agar lawn-plate method.[9]

Disinfectant

Chlorine was obtained by bubbling chlorine gas into chilled distilled water. The concentration of stock solution was determined by the iodometric method.[10]

Inactivation Experiments

Experiments were conducted in 500-mL Pyrex beakers mounted in a water bath; each beaker was provided with a stirrer motivated by an overhead drive. Effluent (100 mL) was added to each beaker; virus was then added to provide approximately 10^5 to 10^6 infectious units per mL, and the effluent was dosed with different concentrations of the disinfectant. After the samples had been collected, the disinfectant was immediately neutralized with sodium thiosulfate, and aliquots were stored at $-20°C$ until required for viral assay.

Effluents

All experiments were conducted using good-quality activated sludge effluent from batches that were stored at $-12°C$ until required. A slight alkaline shift in pH (from 7.2 to 7.8) was detected after the samples were thawed; however, this was not considered important because pH values are commonly adjusted before each experiment. Other physical and chemical characteristics were unaltered (suspended solids, 12.5 mg/L; ammonia, 1.55 mg/L; biological oxygen demand, 10.5 mg/L; and chemical oxygen demand, 37.22 mg/L).

Data Presentation

Each data point represents the average of triplicate experiments conducted on the same day under exactly the same conditions. The t-test analysis of the data indicated that variations among triplicate trials were not significant at 0.05 level. Resistance has been quantifed in terms of percent survival after t-minutes of contact with the disinfectant chlorine.

RESULTS

The inactivation of viruses by chlorine at 15°C and pH 7.2 showed a characteristic biphasic mode. Among the six viruses tested, as shown in Figure 1(A-F), coxsackievirus B5 was the most resistant, with 99.99% inactivation achieved at a dose of 18 ppm (5 min contact time), whereas simian rotavirus was most sensitive, with a dose of only 5 ppm needed to achieve 99.99%

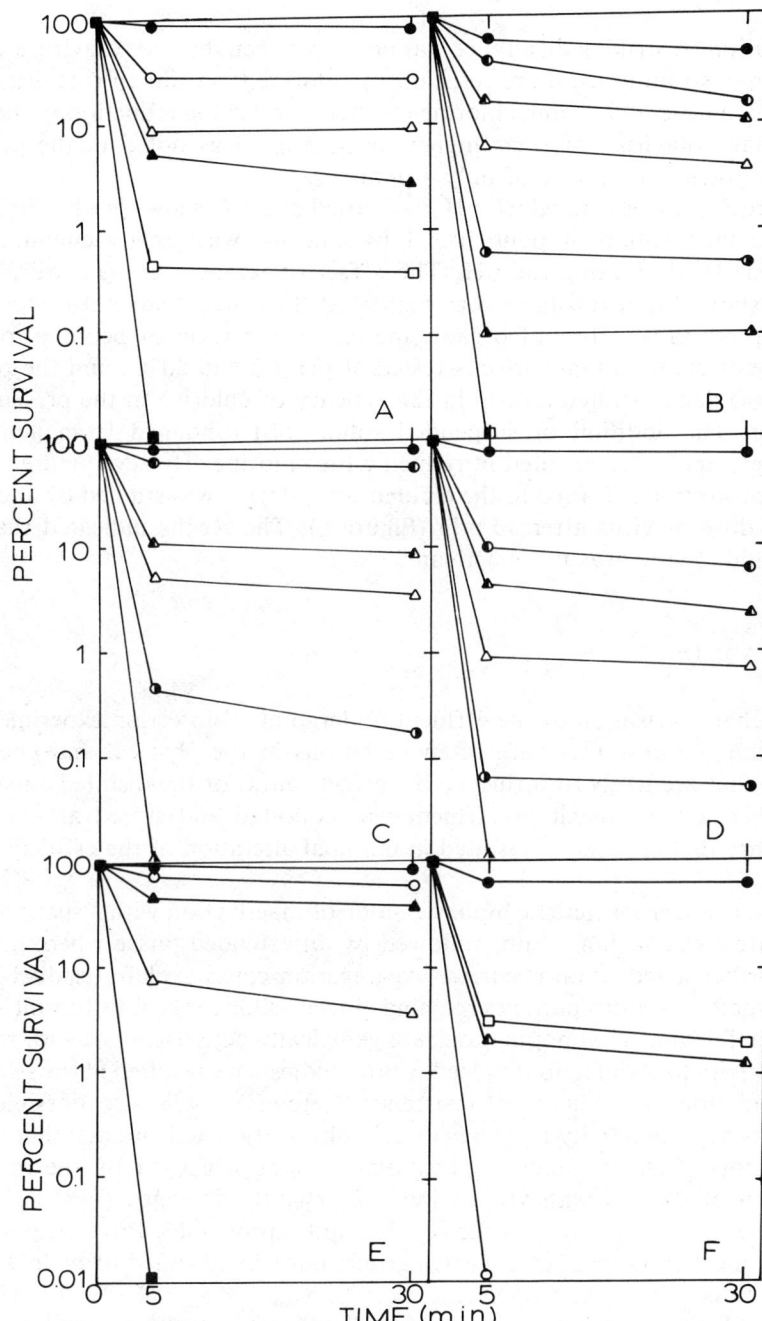

Figure 1. Inactivation of viruses by different doses (ppm) of chlorine in effluent (pH 7.2, 15°C). A: bacteriophage f_2 (symbols: ●, 0.0; ○, 8.0; △, 10.0; ▲, 12.5; □, 15.0; ■, 17.5); B: coxsackievirus B5; C: echovirus 1; D: poliovirus 1 (symbols: ●, 0.0; ◐, 5.0; ▲, 7.5; △, 9.5; ◑, 14.0; ▲, 16.0; ◨, 18.0); E: human rotavirus (symbols: ●, 0.0; ○, 4.5; ▲, 8.0; △, 9.5; ■, 14.0); F: simian rotavirus (SA11) (symbols: ●, 0.0; □, 3.5; ▲, 4.0; ○, 5.0).

inactivation. A striking difference was noted between the two rotaviruses, with the human strain much more resistant to chlorine than simian rotavirus. For example, a dose of 9.5-ppm chlorine resulted in a 1.5 log reduction in the case of human rotavirus, whereas higher inactivation was noted in the case of simian rotavirus at a dose of only 4 ppm.

The results of tests in which pH was varied at 20°C show clearly (Figure 2) that the inactivation of poliovirus 1 by chlorine was greatly enhanced, as expected, by decreasing the pH. The effect of temperature (Figure 3) also clearly showed that resistance was highest at the lowest temperature tested.

The predictable effect of organic matter (in the form of peptone) on the efficacy of chlorination was only tested at pH 7.2 and 20°C, and the results (Figure 4) demonstrated a drop in the efficacy of chlorine in the presence of peptone. The addition of suspended solids (SS) (obtained from activated sludge effluent) also resulted in reducing the chlorine efficacy (Figure 5).

The protection afforded to the effluent by chlorine was studied by adding a second dose of virus after 30 min (Figure 6). The results indicated that the newly added virus was not inactivated.

DISCUSSION

The characterization of the effluent undergoing disinfection experiments is important, because there are often variations in the physical and chemical quality that are likely to influence the effectiveness of the disinfectants.[11] To limit this, a large batch of effluent was collected and stored at −12°C, a procedure that apparently resulted in minimal alteration of the effluent quality.[12]

There was a characteristic biphasic mode of inactivation with a sharp loss of viral infectivity within 5 min, followed by an extended phase where little, if any, further inactivation occurred. Aggregation could explain residual infectivity where infectious particles remained inaccessible to the disinfectant.[13] It is also possible that viral populations are genetically heterogeneous with respect to sensitivity to disinfection; indeed, some studies have resulted in the selection of populations with increased resistance.[14] However, it is also possible that some residual infectivity is the result of multiplicity reactivation,[15] that is, the restoration of the complete replication mechanism caused by the multiple infection of the cell with virions but only slightly damaged genomes. Such doubts reemphasize the need for reliable and reproducible infectivity assays, and, in most situations, each virus particle must be assumed to be infectious and its infectivity determinable.

It is well established that for adequate disinfection of effluents it is essential to use sufficient amounts of chlorine to satisfy nitrogenous demand, the so-called "break point" phenomenon. The virucidal efficacy of the products of hydrolysis of chlorine and its combined forms has been widely studied, and the free species HOCl and OCl⁻ are superior to such combined forms as the chloramines.[5,6] Of the combined forms, Kelly and Sanderson[16] reported that

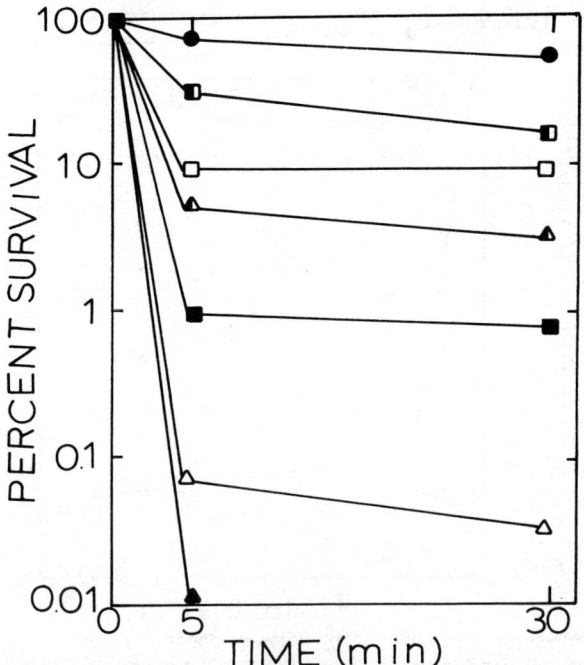

Figure 2. Inactivation of poliovirus 1 by different doses of chlorine (ppm) in effluent at various pH values and 20°C. Legend: ●, 0.0, pH 7.2; ◧, 5.0, pH 9.0; □, 5.0, pH 7.2; ▲, 7.5, pH 9.0; ■, 5.0, pH 4.0; △, 7.5, pH 7.2; ▲, 7.5, pH 4.0.

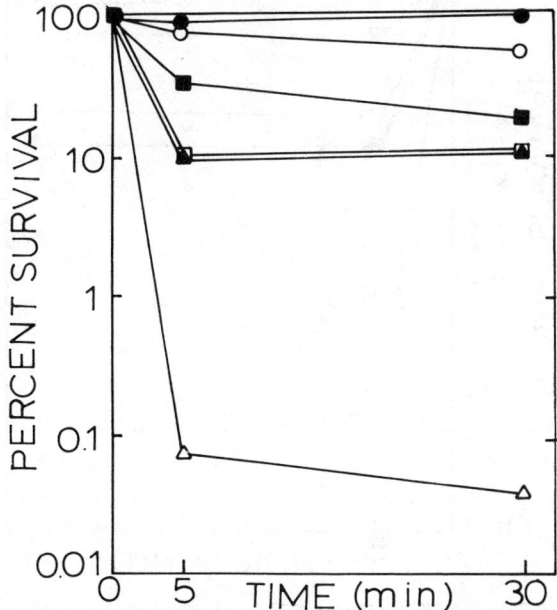

Figure 3. Inactivation of poliovirus 1 by different doses of chlorine (ppm) in effluent at various temperatures and pH 7.2. Legend: ●, 0.0, 4°C; ○, 0.0, 20°C; ■, 5.0, 4°C; □, 7.5, 4°C; ▲, 5.0, 20°C; △, 7.5, 20°C.

686   WATER CHLORINATION

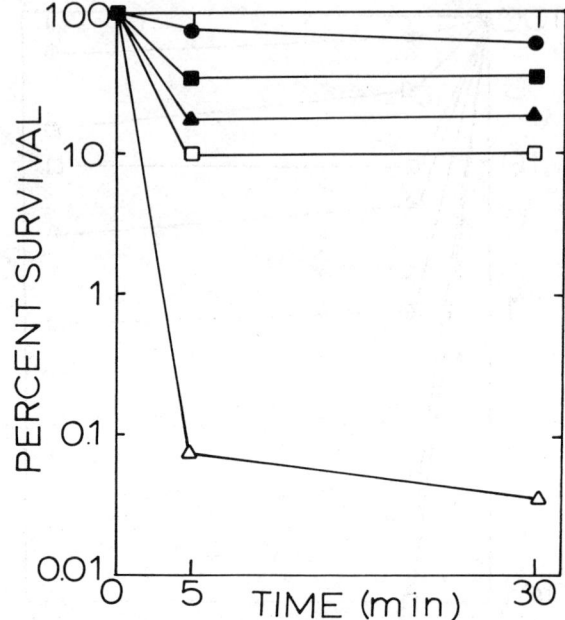

Figure 4. Inactivation of poliovirus 1 by different doses of chlorine (ppm) in effluent in the presence of added peptone (pH 7.2, 20°C). Legend: ●, 0.0, no peptone; ▲, 5.0, no peptone; ■, 5.0, 50-ppm peptone; △, 7.5, no peptone; □, 7.5, 50-ppm peptone.

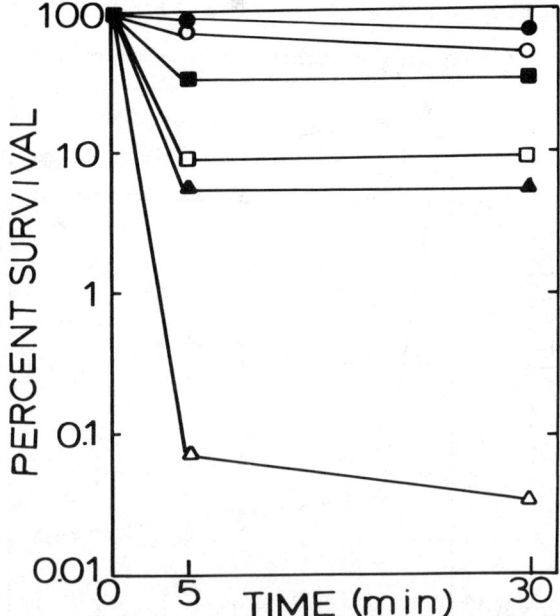

Figure 5. Inactivation of poliovirus 1 by different doses of chlorine (ppm) in effluent in the presence of added suspended solids, SS (pH 7.2, 20°C). Legend: ●, 0.0, 50 ppm SS; ○, 0.0, no SS; □, 5.0, no SS; ■, 5.0, 50 ppm SS; △, 7.5, no SS; ▲, 7.5, 50-ppm SS.

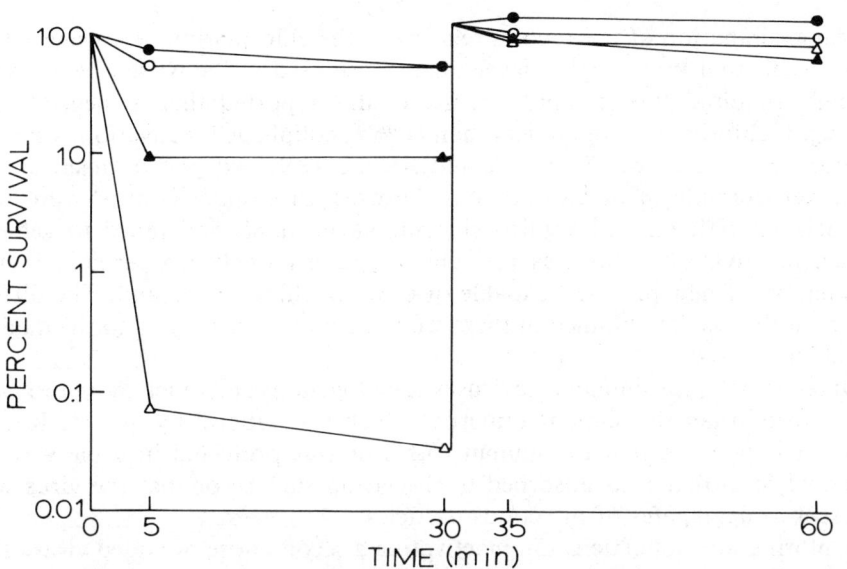

Figure 6. Chlorination of two successive doses of poliovirus 1 in effluent (pH 7.2, 15°C). Legend: ●, 0.0; ○, 3.0; ▲, 5.0; △, 14.0.

the dichloramines were superior to the monochloramines. In general, the HOCl, which is predominant at low pH, is the most active species,[17,18] but it is worth noting that in a comparative study on the chlorination of poliovirus 1 and *E. coli*,[1,2] OCl⁻ was a better virucide than HOCl. This finding is contradictory to most of the published data and draws attention to the fact that certain other factors may also play a role in the efficacy of chlorine disinfection. For instance, Hajenian and Butler[12] reported that poliovirus 1 required less chlorine for 99.99% inactivation at pH 4 and 7.7 than at pH 6.8, and they suggested that the virus may be more stable at a pH close to one of its suggested isoelectric points, pH 7.0.[19,20]

The influence of temperature, although measurable, requires a less complex explanation, which is probably simply due to slower reaction kinetics of chlorine as HOCl and OCl⁻.[17] Earlier, White[18] had reported that for adequate chlorination in the winter season, as much as a fivefold increase in contact time was required to equal that achieved by the standard dose used in the summer.

The presence of suspended solids (SS) of both organic and inorganic nature, characteristically present in colloidal suspensions, is well known to adversely affect chlorination. SS are usually present in complex ill-defined forms and contain a significant fraction of protein that will react actively with chlorine. The removal of such suspended matter from the effluent, therefore, is essential for optimal disinfection. The SS not only exert an increased demand on chlorine but may also protect the virus.[21,22] However, Boardman and Sproul[13] concluded that viral adsorption to some particulate surfaces provided negligible, if any, protection from disinfectant.

An examination of the reports relating to the chlorination of sewage effluent reveals that much higher levels than those used in the present work were usually required. For example, Kruse et al.[23] reported that at neutral pH, 30 mg/L chlorine resulted in less than 80% f_2 coliphage inactivation, whereas in our work, a dose of 17.5 ppm could effect a 99.99% degree of inactivation. Another example of inconsistency is the work of Cramer et al.,[24] who used autoclaved effluent and applied chlorine at 30 mg/L and failed to achieve much inactivation. This was probably because autoclaving resulted in the formation of new products capable of creating chlorine demand. The differences in the results obtained in these studies could be attributed to the quality of effluent used.

Interestingly, the human rotavirus was endogenous and much more resistant to chlorine than the simian rotavirus, which was laboratory grown. In this case, it is possible that the human rotavirus was protected in some way by residual fecal materials adsorbed to the virion surface or that the virus was present as aggregates of infectious particles.

Chlorine was not efficient in inactivating a second dose of added virus. The implications are that residual chlorine was absent. This was also noted earlier by Hajenian and Butler.[25]

The mechanism by which chlorine inactivates viruses is still not clear. Kruse et al.[23] postulated that halogenation causes the oxidation of the sulfhydryl (-SH) groups on the protein coat of the virus, thus denaturing the protein and preventing virus adsorption to the host. However, the halogenation of -SH groups is, apparently, not necessarily the prime inactivation process in all viruses. O'Brien and Newman[26] reported that chlorination of poliovirus resulted in the loss of viral ribonucleic acid, converting the viruses from 156s to 80s particles, although it was, in fact, observed that virus inactivation occurred before the nucleic acid was released. In a more recent study, Alvarez and O'Brien[27] reported that chlorination of poliovirus did lead to the release of the RNA, suggesting that the loss of nucleic acid was a critical event during disinfection. Tenno et al.[28] concluded that HOCl inactivated poliovirus by acting on the protein component of the virus and that the inactivation reaction did not result in any detectable change in the structure of the virus, nor did it affect the ineffectivity of the viral RNA.

The conclusion from this study is that there is considerable variation in the resistance of various viruses to chlorine; consequently, it should be emphasized that the choice of any one virus as a model is inappropriate.

ACKNOWLEDGMENTS

I wish to thank Professor A. Matin for the excellent secretarial assistance of Mrs. K. Redman and the Lebanese National Research Council for financing this study. The work was conducted in Dr. M. Butler's Laboratory, Department of Microbiology, Surrey University, United Kingdom.

REFERENCES

1. Scarpino, P. V., G. Berg, S. L. Chang, D. Dahling, and M. Lucas. "A Comparative Study of the Inactivation of Virus in Water by Chlorine," *Water Res.* 6:959-965 (1972).
2. Scarpino, P. V., M. Lucas, D. R. Dahling, G. Berg, and S. L. Chang. "Effectiveness of Hypochlorous Acid and Hypochlorite Ion in Destruction of Viruses and Bacteria," in *Chemistry of Water Supply, Treatment and Distribution*. A. J. Rubin, Ed. (Ann Arbor, MI: Ann Arbor Science Publishers, Inc., 1974), pp. 359-368.
3. Weidenkopf, S. "Inactivation of Type 1 Poliomyelitis Virus with Chlorine," *Virology* 5:56-57 (1958).
4. Clark, N. A., R. E. Stevenson, and P. W. Kabler. "The Inactivation of Purified Type 3 Adenovirus in Water by Chlorine," *Am. J. Hyg.* 64:314-319 (1956).
5. Engelbrecht, R. S., M. J. Weber, C. A. Schmidt, and B. L. Salter. "Virus Sensitivity to Chlorine Disinfection of Water Supplies," EPA 600/2-78-123, (Cincinnati: 1978).
6. Engelbrecht, R. S., M. J. Weber, B. L. Salter, and C. A. Schmidt. "Comparative Inactivation of Viruses by Chlorine," *Appl. Environ. Microbiol.* 40:249-256 (1980).
7. Smith, E. M., M. K. Estes, T. E. Larson, and J. D. Johnson. "A Plaque Assay for the Simian Rotavirus SA11," *J. Gen. Virol.* 43:513-519 (1975).
8. Banatvala, J. E., B. Totterdell, I. L. Chrystie, and G. N. Woode. "In Vitro Detection of Human Rotaviruses," *Lancet* 1975:821.
9. Balluz, S. A., M. Butler, and H. H. Jones. "The Behavior of f_2 Coliphage in Inactivated Sludge Treatment," *J. Hyg.* 80:237-242 (1978).
10. *Standard Methods for the Examination of Water and Wastewater*, 14th ed. (New York: American Public Health Association, 1976).
11. Tonelli, F. A. "General Considerations in Wastewater Disinfection," *Water Pollut. Control* 114:23-24, 28, 30-32, 46-47 (1976).
12. Hajenian, H., and M. Butler. "Inactivation of Viruses in Municipal Effluent by Chlorine," *J. Hyg.* 84:63-69 (1980).
13. Boardman, G. D., and O. J. Sproul. "Protection of Viruses During Disinfection by Adsorption to Particulate Matter," *J. Water Pollut. Control Fed.* 49:1857-1861 (1977).
14. Bates, R. C., P. T. B. Shaffer, and S. M. Sutherland. "Development of Poliovirus Having Increased Resistance to Chlorine Inactivation," *Appl. Environ. Microbiol.* 34:849-853 (1977).
15. Young, D. C., and D. G. Sharp. "Partial Inactivation of Chlorine Treated Echovirus," *Appl. Environ. Microbiol.* 37:766-773 (1979).
16. Kelly, S. M., and W. W. Sanderson. "The Effect of Chlorine in Water on Enteric Viruses II. The Effect of Combined Chlorine on Poliomyelitis and Coxsackieviruses," *Am. J. Public Health* 50:14-20 (1960).
17. White, G. C. *Handbook of Chlorination* (New York: Van Nostrand Reinhold Co., 1972).
18. White, G. C., Ed. *Disinfection of Wastewater and Water for Reuse*. (New York: Van Nostrand Reinhold Co., 1978).
19. Mandel, B. "Characterization of Type 1 Poliovirus by Electrophoretic Analysis," *Virology* 44:554-568 (1971).

20. Taylor, G. R. *The Disinfection of Viruses in Water*, Doctoral Thesis, (Guildford, Surrey, England: Surrey University, 1980).
21. Hejkal, T. W., F. M. Wellings, P. A. Larock, and A. L. Lewis. "Survival of Poliovirus with Organic Solids During Chlorination," *Appl. Environ. Microbiol.* 38:114–118 (1976).
22. Stagg, C. H., C. Wallis, and C. H. Ward. "Inactivation of Clay-Associated Bacteriophage MS-2 by Chlorine," *Appl. Environ. Microbiol.* 33:385–391 (1977).
23. Kruse, C. W., V. P. Olivieri, and K. Kawata. "The Enhancement of Viral Inactivation by Halogens," *Water Sewage Works* 118:187–193 (1971).
24. Cramer, M. W., K. Kawata, and C. W. Kruse. "Chlorination and Iodination of Poliovirus and F_2," *J. Water Pollut. Control Fed.* 48:61–76 (1976).
25. Hajenian, H., and M. Butler. "Inactivation of f_2 Coliphage in Municipal Effluent by the Use of Various Disinfectants," *J. Hyg.* 84:247–255 (1980).
26. O'Brien, R. T., and J. Newman. "Structural and Compositional Changes Associated with Chlorine Inactivation of Polioviruses, *Appl. Environ. Microbiol.* 38:1034–1039 (1979).
27. Alvarez, M. E., and R. I. O'Brien. "Effect of Chlorine Concentration on the Structure of Poliovirus," *Appl. Environ. Microbiol.* 43:237–239 (1982).
28. Tenno, K. M., R. S. Fujioko, and P. C. Loh. "The Mechanisms for Poliovirus Inactivation by Hypochlorous Acid," in *Water Chlorination: Environmental Impact and Health Effects, Vol. 3.* R. L. Jolley, W. A. Brungs, and R. B. Cumming, Eds. (Ann Arbor, MI: Ann Arbor Science Publishers, Inc., 1980), pp. 665–675.

CHAPTER 54

Ozonation as a Stage in Upgrading Secondary Wastewater Effluents for Reuse

Yehuda Kott

A shortage of water for extended agricultural use in Israel stimulated research on possible utilization of wastewater for this purpose. Upgrading secondary wastewater was also investigated by several research groups from the health viewpoint.[1-4] Recently published guidelines for use of waters produced from wastewater sources prescribe limits of effluent quality for irrigation.[4] The aim of this study was to determine the yield of bacterial kill and virus inactivation for the various stages of wastewater treatment at the pilot plant operated by the Environmental and Water Resources Engineering Laboratories. Special emphasis was placed on following the dieaway of the bacteria and viruses at each stage and after ozonation at different concentrations.

MATERIAL AND METHODS

Pilot Plant Operation

About 1 m^3/h raw wastewater was pumped from the main sewer of nearby residential quarters and treated in an extended aeration unit. Secondary effluents were lime treated, recarbonated, passed through secondary settling, pressure filtered through quartz sand, and ozonated. The ozonated effluents were subsequently passed through either a slow-sand filter or an activated-carbon filter.[5]

Bacterial Counts

Coliform, fecal coliform, fecal streptococcus, and total bacterial counts were performed by the membrane filtration technique or the most-probable-number method, according to *Standard Methods*.[6] *Salmonella typhimurium*, whenever applied, was filtered through membrane filters (50 mm diam, 0.45-μm pore size), placed on plastic petri dishes containing *Salmonella shigella* agar, and incubated at 37°C for 24 h; only typical colonies were counted. Confirmation was obtained by specific antiserum.

Viral Counts

Poliovirus type 1 (Sabin-attenuated strain) was applied at concentrations of about 10^3 plaque-forming units (PFU)/count to various points at the pilot plant. Isolation and counts were done as described in Reference 7.

Phage Assay

Coliphage, either naturally found or introduced, were assayed by enrichment on *Escherichia coli* B bacteria, using the most-probable-number method, and f_2 phage were assayed by enrichment on *E. coli* K_{12} Hfr strain, using the most-probable-number method.[9]

RESULTS

Microbiological results for the pilot-plant effluents (Table I) show typical counts—coliform and fecal coliform about 10^6 to 10^5/100 mL, fecal streptococcus about 10^4/100 mL. Enteric viruses ranged at a few hundred per liter, which is about two orders of magnitude lower than the average sampled at the Haifa municipal wastewater treatment plant.[10] Table II shows the counts of indicator bacteria and poliovirus after application of various concentrations of ozone to the secondary effluents. It can be seen that 14 to 20 mg/L applied ozone reduced the number of fecal coliform less than two orders of magnitude at the highest dose. Poliovirus 1 added to the effluents proved relatively sensitive to the high ozone dose. In another series, where ozone was applied at dosages of 10 to 15 mg/L, inactivation of the virus became very significant. For example, 18.7 mg/L ozone yielded a reduction from 1.0×10^5 to 1.2×10^1 of virus PFU/L. A 29.3 mg/L ozone dose reduced the number of viruses to 4 PFU/L, whereas 51 mg/L ozone yielded 2 PFU/L.

The results achieved to date strongly indicate that to gain significant ozone activity, further treatment of the secondary effluents is needed.

Table I. Indicator Bacteria and Enteric Virus Counts in Technion Pilot Plant Effluents

Sample No.	Total Coliform (Bacteria/ 100 mL)	Fecal Coliform (Bacteria/ 100 mL)	Fecal Streptococcus (Bacteria/ 100 mL $\times 10^4$)	Enteric Viruses (PFU/L $\times 10^1$)
1	5.4×10^6	6.0×10^5	3.6	4.34
2	1.7×10^7	1.5×10^6	3.0	3.99
3	9.5×10^6	2.4×10^6	8.6	4.44
4	1.8×10^6	2.6×10^5	2.3	5.60
5	1.2×10^5	1.8×10^4	6.0	1.80

Table III shows that the addition of lime significantly reduced the number of indicator bacteria and enteric viruses. The decrease in coliforms and fecal coliforms was about three orders of magnitude. At the same time, the decrease in fecal streptococci bacteria was less at about two orders of magnitude, while poliovirus seemed to be more sensitive to lime treatment. *Salmonella typhimurium* applied at about $10^3/100$ mL to the secondary effluents was not recovered after the lime treatment. To confirm the activity of the low-sand filter, total coliforms, fecal coliforms, and fecal streptococcus bacteria were examined before and after filtration.

The resulting absorption figures were 99.54% for coliform bacteria, about 73.35% for fecal coliforms, and 85% for fecal streptococci. Table IV shows effluent quality data after lime treatment, recarbonation, pressure filtration, and ozonation at two ozone doses. Coliform and fecal coliform bacteria that survived prior treatment had disappeared. Fecal streptococci, *S. typhimurium*, and poliovirus 1 did not stand 9.93 mg/L ozone. Coliphage apparently had higher resistance, but at 21.84 mg/L ozone, the effluents retained neither native nor applied microorganisms.

Activated carbon filters are commonly used for upgrading the final water quality before distribution, hence the interest in finding the ozone dose needed for full disinfection of the final effluents when passed through activated carbon columns. Table V shows typical results obtained at relatively low ozone concentrations. It is seen that 2.2 mg/L ozone sufficed to reduce the bacterial counts such that no bacteria were detected in a 1-L sample. Poliovirus was reduced by three orders of magnitude. Results for *Salmonella kanton* were the same as for *S. typhimurium*. Complete inactivation of poliovirus was recorded in most of these experiments.

These results, representing the treatment yield of different parts of the pilot plant, refer to naturally found bacteria, to indicator and pathogenic bacteria, and to poliovirus. It is known, however, that water supplies, as well as wastewater, may contain various other bacteria, and it should be noted that in some countries bacterial water quality is measured in terms of bacteria count.[11]

Table II. Survival of Indicator Bacteria and Poliovirus in Pilot Plant Effluents.

Ozone Application (mg/L)	Total Coliform (Bacteria/ 100 mL)	Fecal Coliform (Bacteria/ 100 mL)	Fecal Streptococcus (Bacteria/ 100 mL)	Poliovirus[a] (PFU/L)
Control	2.0×10^6	3.5×10^4	5.4×10^4	3.6×10^3
14.2	1.9×10^5	3.2×10^4	5.0×10^4	8.2×10^2
Control	9.0×10^5	5.0×10^4	1.7×10^4	4.8×10^3
16.2	1.0×10^5	4.5×10^3	1.6×10^4	1.1×10^3
Control	2.8×10^6	2.0×10^4	6.0×10^4	6.0×10^3
20.16	2.7×10^5	7.0×10^2	2.0×10^2	6.0×10^1

[a]Poliovirus 1 attenuated strain was applied to the effluents.

Table III. Number of Indicator Bacteria and Poliovirus 1 After Lime Treatment of Pilot Plant Effluents

Lime Treatment	Total Coliform (Bacteria/100 mL)	Fecal Coliform (Bacteria/100 mL)	Fecal Streptococcus (Bacteria/100 mL)	Coliphage (Phage/100 mL)	f_2 Phage (Phage/100 mL)	Poliovirus 1 (PFU/L)
Control	1.2×10^5	1.8×10^4	6.0×10^4	9.2×10^3	ND[a]	48
Lime	3.0×10^2	6.0×10^1	1.3×10^3	1.1×10^2	ND[a]	0
Control	1.2×10^6	6.2×10^5	1.2×10^5	7.0×10^3	ND[a]	4.4×10^4
Lime	3.0×10^1	1.0×10^{-1}	5.0×10^3	2.2×10^1	ND[a]	2.0×10^1
Control	5.5×10^6	1.3×10^6	9.7×10^4	3.3×10^4	2.4×10^5	3.0×10^4
Lime	2.0×10^{-1}	9.0×10^{-1}	1.1×10^2	7.8×10^{-1}	<2	3.0×10^1
Control	2.3×10^6	7.0×10^5	1.1×10^5	2.8×10^3	2.2×10^4	ND[a]
Lime	4.2×10^2	1.0×10^2	1.1×10^2	1.4×10^1	6.1×10^{-1}	ND[a]
Control	5.6×10^4	3.0×10^4	9.2×10^2	ND[a]	2.3×10^2	ND[a]
Lime	0	0	4.0×10^1	ND[a]	0	ND[a]

[a]ND, Not done.

Table VI, a summary of the results for the various sampling points, shows that addition of lime to the secondary effluents caused a decrease in the bacterial count by three orders of magnitude. This was followed by regrowth, but ozone treatment again reduced the count. A slight decrease was seen in effluents passed through the activated carbon columns.

DISCUSSION

It is commonly agreed that secondary wastewater effluents are unsuitable for unrestricted agricultural use. It is very often claimed, however, that further treatment would make it possible to extend their acceptability to a wider group of vegetables and fruits. The present study followed the various stages of pilot plant treatment at the Environmental and Water Resources Engineering laboratories. Neither extended aeration nor ozonation alone yielded a satisfactory quality for the purpose in question, but addition of lime at pH 10 to 11.5 caused a very pronounced decrease in all bacteria examined and in poliovirus 1 (attenuated strain), so that subsequent ozonation at about 10 and 20 mg/L sufficed to kill all bacteria and the poliovirus. Passage of the lime-recarbonated effluents through activated carbon columns reduced the amount of ozone needed for final destruction of indicator and pathogenic bacteria and for poliovirus. Thus it is assumed that ozonation prior to the activated carbon

Table IV. Survival of Bacteria and Viruses After Ozonation of High-Quality Effluents. The bacteria and viruses were applied before the ozonation columns.

		Ozone Applied	
	Control	9.93 mg/L	21.84 mg/L
Total Coliforms, bacteria/100 mL	6.8×10^2 6.0×10^3	0 1.4×10^1	0 0
Fecal Coliforms, bacteria/100 mL	4.0×10^2 3.5×10^3	0 6.0×10^1	0 0
Fecal Streptococcus, bacteria/100 mL	1.2×10^2 1.8×10^6 6.0×10^5 1.2×10^6	0 0 0 0	0 0 0 0
Salmonella typhimurium, bacteria/100 mL	1.7×10^5 1.5×10^5 2.7×10^3	0 0 0	0 0 0
Coliphage, phage/100 mL	9.2×10^2 3.5×10^5 3.0×10^6	0 7.0×10^1 3.3×10^1	0 0 0
f_2 Phage, phage/100 mL	7.0×10^4 1.7×10^5	0 0	0 0
Poliovirus 1, PFU/L	1.32×10^4 3.6×10^3 1.02×10^3 2.8×10^3	0 1 0 0	0 0 0 0

Table V. Typical Results Obtained for 1 of 10 Different Experiments on Bacteria and Virus Survival After Ozonation of Activated Carbon Effluents. The bacteria, phage, and poliovirus were added to activated carbon effluents.

Ozone Applied (mg/L)	Escherichia Coli (Bacteria/100 mL)	Fecal Streptococcus (Bacteria/100 mL)	Salmonella typhimurium (Bacteria/100 mL)	f_2 Phage (Phage/100 mL)	Poliovirus 1 (PFU/L)
0	1.0×10^5	1.6×10^6	1.3×10^6	3.5×10^8	3.8×10^4
0.7	0	3.7	0.2	2.0×10^{-1}	7.0×10^1
1.5	0	2	0	0	0.42×10^0
2.2	0	0	0	0	1.02×10^1
3.2	0	0	0	0	0.36×10^1

Table VI. Bacteria Count of Effluents at Various Sampling Points of the Pilot Plant

Sampling Point	Bacteria Count (No./mL)		
	Lowest	Arithmetic Mean	Highest
Secondary	4.3×10^4	1.3×10^6	1.6×10^7
Lime	2.7×10^0	1.2×10^3	1.7×10^4
Sand filter	3.0×10^0	2.1×10^5	1.0×10^6
9.93 mg/L Ozone	6.5×10^1	1.2×10^5	4.2×10^5
21.84 mg/L Ozone	2.9×10^1	1.3×10^5	1.2×10^6
Activated carbon[a]	0	1.1×10^4	1.6×10^5
Activated carbon[b]	0	1.6×10^3	2.8×10^4

[a]Activated carbon receiving effluents dosed at 9.93 mg/L ozone.
[b]Activated carbon receiving effluents dosed at 21.84 mg/L ozone.

columns might enhance the ability of nonpathogenic bacteria to utilize ozone-destructed organic matter and permit their growth. The total bacterial counts support this theory. However, it should be borne in mind that these saprophytic bacteria are irrelevant to water quality in terms of agricultural use. Use of ozone in upgrading wastewater in its final stage would constitute an important safeguard in any future reuse of such water in agriculture.

ACKNOWLEDGMENTS

This study was supported in part by joint German and Israeli Grant No. WA-071-WT 2. The author wishes to express his thanks to the Environmental and Water Resources Engineering staff for their helpful and cooperative work.

REFERENCES

1. Arthur, J. P. *Notes on the Design and Operation of Waste Stabilization Ponds in Warm Climates of Developing Countries*, (Washington, DC: The World Bank, 1983), p. 90.
2. Idelovitch, E. R., R. Terkeltoub and M. Michail. "The Role of Groundwater Recharge in Wastewater Reuse: Israel's Dan Region Project," *J. Am. Water Works Assoc.* 72(7):391-400 (1980).
3. Yannai, S. "Safety Evaluation of Water Reclaimed from Wastewater by Various Techniques," in *Proceedings of the Symposium of the Israel European Community*, (Herzlia, Israel: 1981).
4. *Criteria for Quality of Treated Wastewater Effluent to be Used for Irrigation* (Israel Ministry of Health, 1979).
5. Rom, D., A. M. Wachs, and M. Rotel. "Pilot Plant Studies of Water Renovation In a System Combining Ozonation with Activated Carbon Treatment," in *Sorption Process*, Proceedings of the Research Symposium, 53rd Annual Conference, Las Vegas, NV, (Washington, DC: Water Pollut. Control Fed., 1980), pp. 1-37.

6. *Standard Methods for the Examination of Water and Wastewater*, 15th ed. (Washington, DC: American Public Health Association, 1980).
7. Kott, Y., L. Vinokur, and H. Ben Ari. "Combined Effects of Disinfectants on Bacteria and Viruses," in *Water Chlorination: Environmental Impact and Health Effects, Vol. 3*, R. L. Jolley, W. A. Brungs, and R. B. Cumming, Eds. (Ann Arbor, MI: Ann Arbor Science Publishers, Inc., 1980), pp. 677-686.
8. Loeb, T., and N. D. Zinder. "A Bacteriophage Containing RNA," *Nat. Acad. Sci.* 47:282-289 (1961).
9. Kott, Y. "Estimation of Low Number of *Escherichia coli* Bacteriophage by Use of the Most Probable Number," *Ap. Microbiol.* 14(2):141-144 (1966).
10. Kott, Y., H. Ben Ari, and L. Vinokur. "Coliphages Survival as Viral Indicator in Various Wastewater Quality Effluents," *Prog. Wat. Tech.* 10:337-346 (1978).
11. Suess, M. J. *Examination of Water for Pollution Control—A Reference Handbook*. (Oxford: Pergamon Pres, 1982).

SECTION IX

Reaction Dynamics in Water Chlorination

It is well known to almost everyone, I presume, that the term aqueous chlorine is a misnomer when it is applied to the usual conditions of water and wastewater treatment.

J. Carrell Morris
The Chemistry of Aqueous Chlorine in Relation to Water Chlorination, 1975

We generally describe as fast anything that takes place quickly compared to the rate of resolution of our sense perceptions. However, since our perceptions are in turn based on chemical processes . . . these processes must necessarily be fast—indeed extremely fast.

Manfred Eigen
Nobel Lecture, 1967

SECTION IX

Reaction Dynamics in Water Chlorination

CHAPTER 55

Reaction Dynamics in Water Chlorination

J. Carrell Morris

Reaction dynamics may be defined as the study of matter in the process of chemical change. A major ultimate objective of such studies is the elucidation of the nature and sequence of elementary component reactions in the overall chemical process.

Three principal stages may be distinguished in investigations of reaction dynamics: (1) description of the kinetics of the overall process, including reaction orders and effects of environmental factors; (2) development of a hypothetical reaction mechanism (reaction model) composed of a set or sequence of elementary chemical steps whose composite functioning is consistent both with the empirical overall kinetics and with the result of specialized studies on individual segments of the reaction scheme; and (3) appreciation of the energetics of the elementary steps and overall process, so that basic thermodynamic principles are not violated.

Dynamic considerations are important in the investigation of water chlorination reactions, because the principal concern is with processes, that is, with changes that occur within a limited time or as a function of time. Stoichiometric or equilibrium relations are not enough. For applied scientists and engineers, limitations on reaction completion or conversion efficiencies as a result of time restrictions are very important.

In chemical research, the first interest, generally speaking, is in the nature of the changes that occur and the conditions under which they occur. The next concern has to do with how fully and how fast the overall reactions proceed as a function of experimental conditions. The final concern is to understand how the changes occur, the intermediates and pathways of the overall reaction, and to develop and verify reaction mechanisms.

This pattern of investigation is evident in the past decade of research into the reactions of aqueous chlorine with organic matter in natural water. During this period, much has been learned about the reaction products and about the conditions affecting a number of reactions. Recently, reports on the speed of some of these reactions and on their kinetic relations have begun to appear. As yet, however, little firm evidence exists on the specific pathways that yield most of the products.

The main part of this chapter will consider two specific systems of reactions in water chlorination in the light of some of the general principles of reaction dynamics. This approach should allow us to see where we stand with regard to understanding the kinetics and mechanisms in these systems. It should also serve to indicate the directions in which future research would be most helpful. The two reaction systems to be dealt with are the reaction of natural organic matter with aqueous chlorine, and the breakpoint reaction, the oxidation of ammonia by aqueous chlorine.

REACTION BETWEEN AQUEOUS CHLORINE AND FULVATE

Great difficulties must be overcome to make detailed dynamic studies of the reaction between aqueous chlorine and fulvate or humate, because the composition and structure of the organic reactant is known only in a proximate way and, moreover, is not constant from sample to sample. There is no known stoichiometry and the factor of particular interest, the formation of chlorinated products, accounts for only a minor part (about 20%) of the overall chlorine reaction.

Until now kinetic investigations have been largely confined to measurements of the reduction in oxidizing chlorine or the formation of chlorinated organic products as a function of time. These measurements have led to the formulation of some empirical correlation equations, but such measurements are usually not very helpful in clarifying the dynamics of complex systems. They give little information about the true kinetic order of the reaction, nor do they yield other significant dynamic insights.

As Laidler[1] and other kineticists have stressed, true kinetic relations can be obtained only through measurements of changes in the initial rates of reaction when the initial concentrations of reactants are varied. A start in this direction has been made by Qualls and Johnson[2] and by Gurol et al.,[3] but studies that use a wider range of variation in initial concentrations, and particularly with greater excesses of aqueous chlorine, are needed. Stoichiometric relations between the initial reduction of aqueous chlorine and the formation of specific chlorinated products would also be very helpful.

Some insight into the kind of relationships to be expected can be obtained from a reexamination of the mechanism of the classical alkaline haloform reaction for acetone, as depicted in Figure 1. The complex reaction sequence is initiated by the ionization of the ketone to produce a nucleophilic carbanion, which then reacts readily with electrophilic hypochlorous acid to form a chlorinated derivative. Successively more rapid ionizations and hypochlorinations occur until a methyl group is fully chlorinated; then, base-catalyzed hydrolysis produces the chloroform end product.[4]

Under the usual laboratory conditions for the classical haloform reaction (pH ~ 13 and 0.01 M concentrations of reagents), the initial ionization reaction is the slowest reaction in the sequence, the subsequent steps being relatively rapid. For this situation, the overall reaction is first order with respect to concentration of organic compound and zero order with respect to chlorine or other halogen concentration. Moreover, once steady state has been estab-

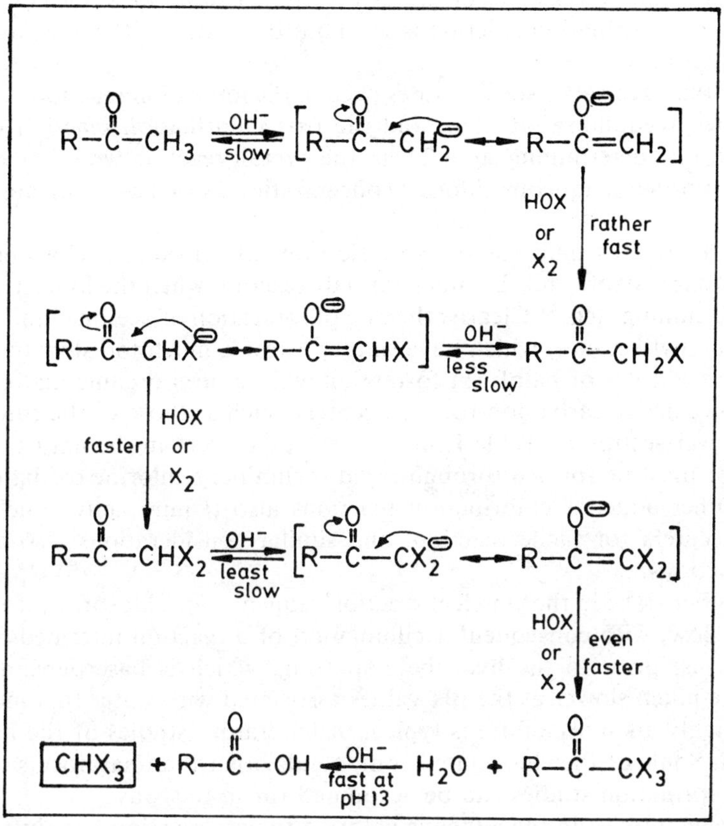

Figure 1. Detailed reaction mechanism for the haloform reaction.

lished, the rate of formation of product chloroform is the same as the rate of the initial ionization reaction.

A useful generalization for reaction chains, such as that shown in Figure 1, is that the first step in the chain, excluding preequilibration reactions, is likely to be the slow step in the chain. When a later step is the slow step, the reaction tends to come to a halt just prior to the slow step, and the intermediate that has been formed at this point accumulates in identifiable or isolable quantities. It is then natural and logical to separate such a reaction sequence into two segments to be studied individually.

This classical haloform pattern may not be wholly valid for the reaction of aqueous chlorine with fulvate in water chlorination, however, because rates of carbanion formation vary greatly with organic structure. The active centers for fulvate chlorination may ionize so much more rapidly than those of acetone that the subsequent chlorination becomes the slow step. Such an effect is enhanced at the lower concentrations ($\sim 10^{-4}$ M) used in water chlorination because the chlorination reaction is second order, whereas the ionization reaction is first order.

When this reversal of specific rates occurs, the ionization reaction becomes a reversible preequilibration step, and the first chlorination reaction becomes the slow, rate-determining step. Then, the overall reaction would be expected to be first order in aqueous chlorine concentration as well as in fulvate concentration.

Table I presents data for the specific rates of ionization of a number of haloform precursors and the times for half reaction when the ionization is the rate-determining step.[5,6] Clearly, the rate of ionization of acetone (and of other simple methyl ketones, which have similar rates) is much too slow to account for observed rates of haloform formation with natural organic matter. Thus, some more-active carbanion-forming centers, such as those of the compounds in the lower section of Table I, must either be present in the natural organic matter or must be formed through rapid preliminary chlorine oxidation reactions. Other aqueous chlorination reactions also require active nucleophilic carbon centers for facile reaction, and similar considerations also apply to such reactions.

One other stage in the classical reaction sequence for chloroform formation may be slow, with consequent accumulation of a reaction intermediate. This stage is just prior to the hydrolytic splitting, which is base dependent and therefore much slower at the pH values associated with water treatment than in the highly alkaline solutions typical of laboratory studies of the haloform reaction. Some of the chlorinated compounds other than haloforms found in water chlorination studies can be accounted for in this way.

Recently, interest in the aqueous chlorine-fulvate reaction has shifted from the formation of trihalomethanes to the formation of other chlorinated materials, grouped as TOX. It is not always clear from the reports of investigations, however, how much the measured nonhaloform TOX values represent intermediate substances that will become haloforms upon further reaction and how much they represent wholly by-product material from alternate reaction sequences. The results of some studies, such as that of Reckhow and Singer,[7]

Table I. Rates of Ionization of Some Ketones

Substance	k_1 (s^{-1})	k_{OH} (M^{-1} s^{-1})	t_{50},[a] pH 7	t_{50},[a] pH 8.5
Acetone	4.7×10^{-10}	0.25	320 d	16 d
Chloroacetone	5.5×10^{-8}	93	21 h	1.0 h
as-Dichloroacetone	7.3×10^{-7}	4.5×10^2	3.7 h	0.2 h
Pyruvic acid	4.7×10^{-7}			
Ethyl pyruvate	4.7×10^{-7}			
Malonic ester	1.0×10^{-5}	10^4	12 m	0.4 m
Acetylacetone	1.7×10^{-2}	2×10^6	3.5 s	0.1 s

[a]Calculated by the formula $t_{50} = 0.693 [k_1 + k_2(OH^-)]^{-1}$.

suggest that the TOX is largely by-product material. Other experiences, such as the increased trihalomethane (THM) values observed when previously chlorinated water samples are heated or the high THM values obtained with direct-injection gas chromatography, indicate that much TOX represents chlorinated compounds intermediate to eventual haloform production. The full technical and hygienic significance of TOX cannot be assessed until this situation is resolved. Clearly, many more refined and specific investigations are needed to construct a detailed picture of the dynamics of the interactions between aqueous chlorine and natural organic matter.

BREAKPOINT REACTION

The second reaction system to be considered, the oxidation of ammonia by aqueous chlorine, seems much less complex than the first one. The reactants, ammonia and hypochlorous acid, are simple and well characterized; the reaction products have been established; the stoichiometry, even if not simple, is known; and several of the steps involved have been investigated kinetically. Nonetheless, a satisfactory formulation of the pathway of the reaction remains elusive.

A review of the current understanding of the dynamics of the breakpoint reaction is warranted in part because of the recent appearance of a paper by Hand and Margerum.[8] This paper is a most important and provocative contribution to our knowledge of the dynamics of the ammonia–chlorine system. However, some of the deduced conclusions and proposed mechanisms, in contrast to the experimental findings, can be confusing and misleading.

It is important, as this reaction system is reviewed, to recognize certain fundamental principles that must be followed to assemble properly a reaction mechanism composed of elementary reaction steps.[9] First, because of the improbability of simultaneous contact of more than two particles, elementary reaction steps for dilute solution reactions normally should involve no more than two reactant particles, other than molecules of the solvent. In addition, because an elementary reaction surface must vary smoothly in either direction to accord with the principle of microsocopic reversibility, the same restriction applies to the reverse reaction.

Second, because the binding energies of chemical bonds are typically >50 kJ and because reaction probability decreases very rapidly with required energy, elementary reactions will not have a net formation or breakage of more than one chemical bond. Many elementary reactions have net formation or breakage equal to zero, of course.

These considerations limit elementary reactions in dilute aqueous solution almost wholly to one of three types.

1. Two-center reactions, that is, association or dissociation reactions of the type

$$A - B \rightleftarrows A + B$$

2. Three-center reactions, that is, substitution or exchange reactions of the type
$$A + B - C \rightleftarrows A - B + C$$
3. Four-center reactions, that is, metathetical or double-exchange reactions of the type
$$A - B + C - D \rightleftarrows A - C + B - D$$

Reaction mechanisms containing steps that cannot be viewed as one of these types should be regarded rather sceptically.

One of the very ingenious techniques used by Hand and Margerum[8] was the preparation of solutions of dichloramine free from other chlorinated species or excess ammonia by ion-exchange methodology. They were thus able to study the decomposition of $NHCl_2$ by itself and to show that extraneous HOCl, NCl_3, or NH_3 had enormous effects on the decomposition rate of $NHCl_2$. The findings are highly significant even though the direct role attributed to NCl_3 seems quite implausible, in part because of the difficulty of finding an elementary reaction that would explain the action of the NCl_3.

Some years ago, Weil et al. conducted some similar studies that were presented to the International Union of Pure and Applied Chemistry in 1952 but were never published.[10] Dichloramine solutions were prepared at pH 5 with a ratio of aqueous chlorine-to-ammonia slightly greater than 2. After several minutes for completion of reaction, the solutions were extracted three or four times with CCl_4 to remove NCl_3. Stock solutions (10^{-3} M) prepared in this way were found to be stable enough at pH 5 to be used for several days. Some monochloramine (<10%) was present, however.

Portions of the stock solutions were then dispensed into aqueous media buffered at pH values between 7 and 10. The $NHCl_2$ was found to decompose by a first-order reaction, in accord with Hand and Margerum,[8] and the specific rate was found to vary with pH as shown in Figure 2. Accelerating effects of HOCl and NCl_3 were recognized but could not be made quantitative at that time.

Some studies of the rate of the breakpoint reaction itself—the reduction of chlorine in mixed aqueous chlorine–ammonia solutions—were conducted at pH values 7.5 to 10. Under these pH conditions, the slow, rate-determining step was not the decomposition of $NHCl_2$, but rather the formation of this compound. A comparison of the measured breakpoint rates with rates of formation of dichloramine computed from the equation of Morris and Isaac[11] is shown in Figure 3. The agreement strongly confirms the hypothesis.

A diagram to indicate the expected overall behavior of the breakpoint reaction as a function of pH on the basis of these findings is shown in Figure 4. The expected overall rate is governed by the rate of the slow process at each pH value, that is, the rate of the HOCl-catalyzed decomposition of $NHCl_2$ at pH < 7.5 and by the rate of formation of $NHCl_2$ at pH >7.5. The indicated absolute rate near the pH 7.5 maximum is in approximate accord with that found by Palin[12] at pH 7.3 for a temperature 12 to 13°C lower.

One other finding of Weil et al. is shown in Figure 5.[10] Rates of decomposition of $NHCl_2$ at pH 9 for solutions with and without extra monochloramine were compared. It was found that neither the amount nor the rate of active

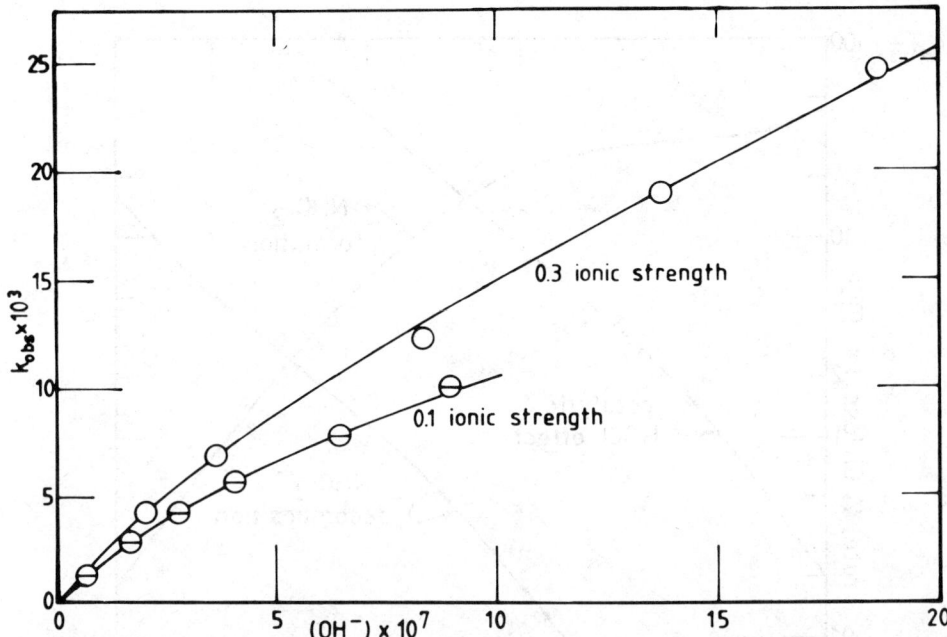

Figure 2. First-order rate constants for the decomposition of preformed dichloramine in alkaline solutions at 25°C.

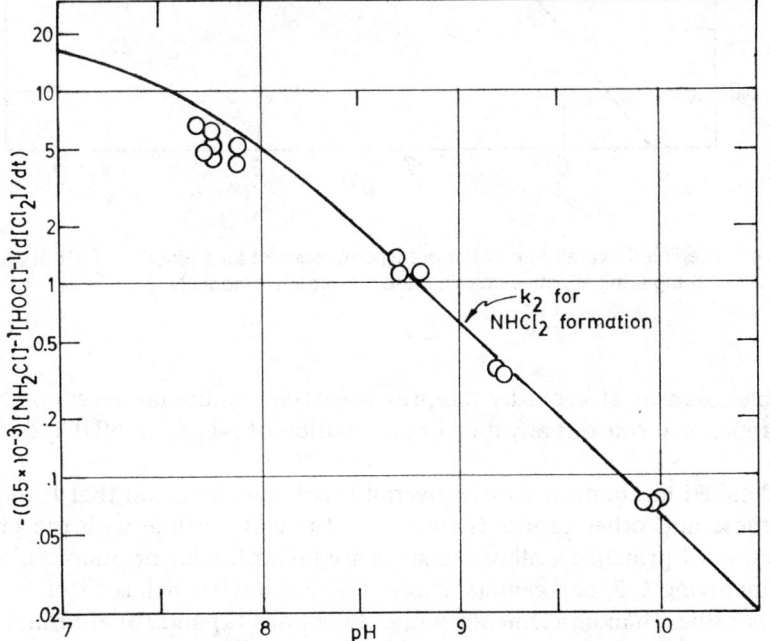

Figure 3. Comparison of the rate of the breakpoint reaction in alkaline solution with the rate of formation of dichloramine.

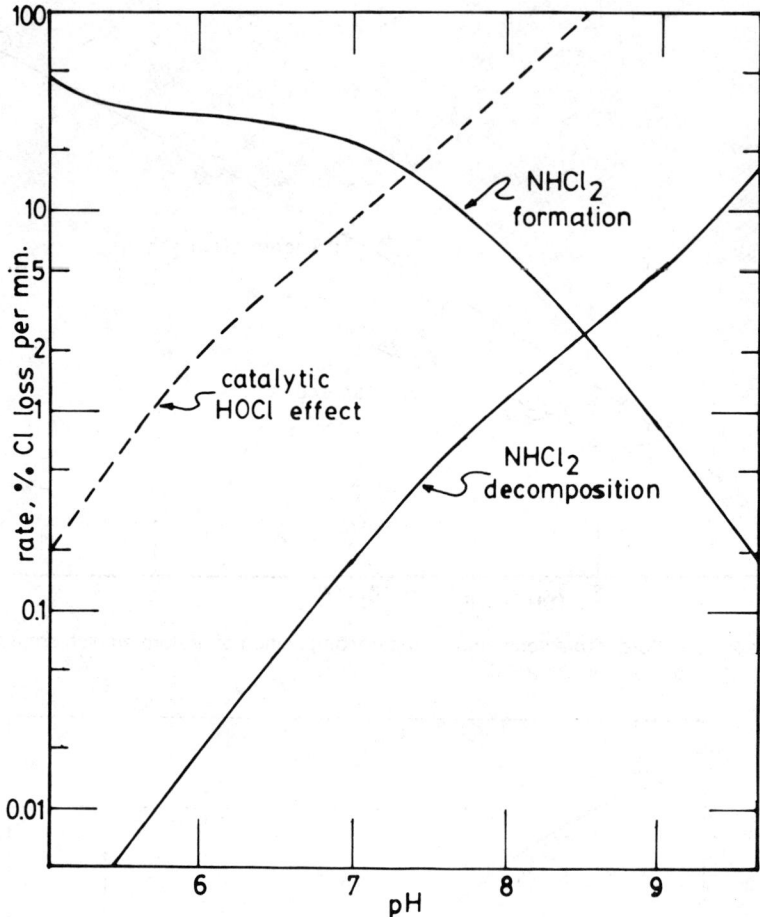

Figure 4. Predicted overall rate of the breakpoint reaction as a function of pH at 25°C with 2.0 mg/L initial active chlorine and 0.2 mg/L ammonia-N.

chlorine loss was affected by the presence of an equimolar excess of NH_2Cl. This appears to rule out any direct participation of NH_2Cl in $NHCl_2$ decomposition.

A detailed mechanism for the overall breakpoint reaction that is in accord with these and other results is shown in Table II. In line with the previous discussion of principles, all of the steps are unimolecular or bimolecular reactions involving 2, 3, or 4 centers, except for reaction 9 which is a stoichiometric process rather than an elementary one. Reactions (a) and (b) are rapid equilibrations that occur prior to the main sequence. Consequently, they enter into the kinetic equation only in terms of their equilibrium relations. Except in very

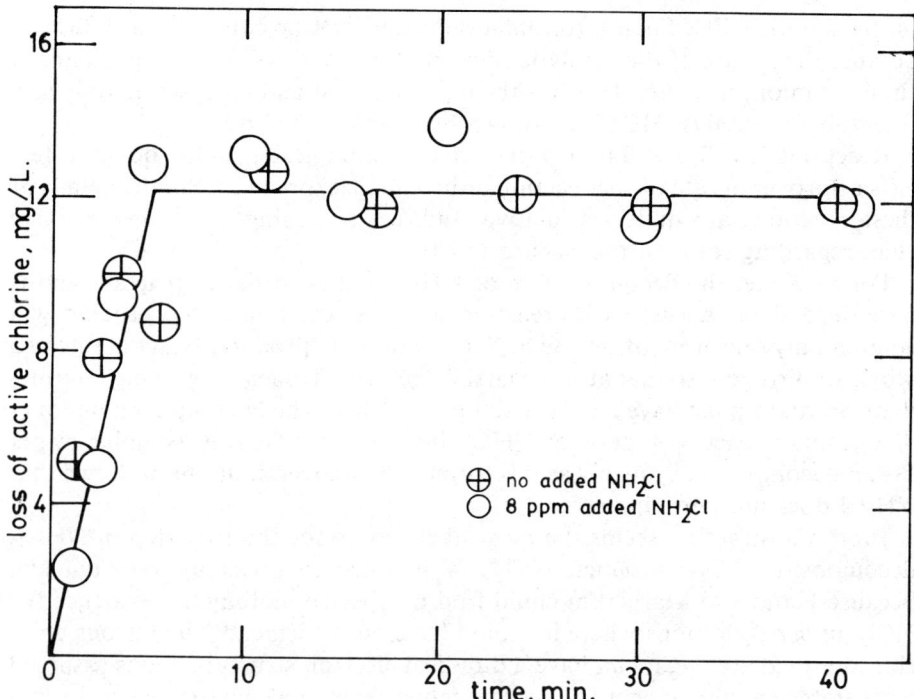

Figure 5. Effect of added NH_2Cl on loss of active chlorine from aqueous solutions of $NHCl_2$ at pH 9 and 25°C.

Table II. Prospective Breakpoint Reaction Mechansim

a. $HOCl \rightleftarrows H^+ + OCl^-$
b. $NH_4^+ \rightleftarrows NH_3 + H^+$
 1. $HOCl + NH_3 \rightarrow NH_2Cl + H_2O$
 2. $HOCl + NH_2Cl \rightarrow NHCl_2 + H_2O$
 3. $HOCl + NHCl_2 \rightarrow NCl_3 + H_2O$
 4. $NCl_3 + H_2O \rightarrow NHCl_2 + HOCl$
 5. $NHCl_2 + OH^- \rightarrow NCl_2^- + H_2O$
 6. $NCl_2^- \rightarrow NCl + Cl^-$
 7. $2NCl \rightarrow N_2 + Cl_2$
 8. $Cl_2 + H_2O \rightarrow HOCl + H^+ + Cl^-$
 9. $NCl + 2HOCl + H_2O \rightarrow HNO_3 + 3H^+ + 3Cl^-$

acidic solution, reaction 1 is very rapid compared with the later rate-determining steps and may in most instances be assumed to be complete before the remainder of the sequence begins. It affects the kinetic equation only by determining initial reactant concentrations for reaction 2.

Reaction 2 is the rate-determining step in alkaline solution, whereas reaction 5 is rate-determining in acid solutions. A complete empirical kinetic equation will, therefore, contain contributions from both of these reactions, depending on the solution pH. Such a formulation cannot yet be constructed, however, because the nature of the catalytic roles of HOCl and NCl_3 is not understood. In my opinion, however, HOCl is the direct catalyst and NCl_3 serves only as a reservoir for catalyst HOCl, as shown by reactions 3 and 4.

Reactions 6, 7, 8, and 9 follow the rate-determining step and so do not affect either the overall rate of the reaction or its kinetic expression. So, the details of these reactions are quite speculative. Indirect reasoning can, however, give clues regarding some of their characteristics.

For example, the decomposition of $NHCl_2$ is first order in reactant and is base dependent, as shown by reaction 6. Thus, the immediate product will contain only one atom of nitrogen. Yet the ultimate product, N_2, contains two atoms of nitrogen, so that at some later stage two nitrogen-containing intermediate products must have a chance to link together. The N-containing intermediate cannot react with another $NHCl_2$ because no $NHCl_2$ is available at pH >8; it decomposes as quickly as it is formed. Moreover, it has been shown that NH_2Cl does not participate.[10]

Proton abstraction seems the most likely route for the first step in $NHCl_2$ decomposition. The product, NCl_2^-, is assumed to break up very quickly, because Hand and Margerum could find no spectrophotometric evidence for NCl_2^- under conditions where it should have been formed.[8] The nitrous chloride intermediate, NCl, can have a diradical electron structure and is assumed to be stable enough to wait for another molecule with which to react according to reaction 7.

Comparison of Palin's data[12] regarding nitrate formation in the breakpoint reaction with the nearly exact stoichiometric chlorine-to-ammonia ratio (1.5) found by Pressley et al.[13] for reactant concentrations 20 times as great indicates that the extent of the side reactions leading to nitrate is strongly dependent on total concentration. This would be the case if the nitrogenous intermediate product from reaction 6 or 7 reacts in a second-order process to give the major N_2 product but in a first-order process to give NO_3^-.

Clearly, some progress is being made in elucidating the elementary reaction pattern for the breakpoint reaction, but even in this instance, much remains to be done to complete our understanding of the dynamic behavior of the chlorine-ammonia system.

CONCLUSION

Investigations of reaction kinetics provide the basis for understanding the pathways of chemical processes and deducing the reaction mechanisms. In the development of a set of elementary reactions that constitutes a complex reaction mechanism, there must be adherence to the fundamental principles for elementary reactions if the mechanism is to be reasonable.

Development of reaction mechanisms is not just a satisfying intellectual exercise. Knowledge of reaction mechanisms furnishes guidance for predicting the effects of changing treatment or environmental conditions on the overall chemical process and provides a basis for developing techniques to control or modify the overall reaction.

The aqueous chlorination of humic matter and the breakpoint reaction are two complex reaction systems in water chlorination where additional studies of reaction dynamics are likely to yield valuable insights into the nature of the processes that are occurring.

REFERENCES

1. Laidler, K. W. *Chemical Kinetics*, 2nd ed. (New York: McGraw-Hill Book Co., 1965), p. 18.
2. Qualls, R. G., and J. D. Johnson. "Kinetics of the Short-Term Consumption of Chlorine by Fulvic Acid," *Environ. Sci. Technol.* 17(11):692-698 (1983).
3. Gurol, M. D., A. Wowk, S. Myers, and I. H. Suffet. "Kinetics and Mechanisms of Haloform Formation: Chloroform Formation from Trichloroacetone" in *Water Chlorination: Environmental Impact and Health Effects, Vol. 4*, R. L. Jolley, W. A. Brungs, J. A. Cotruvo, R. B. Cumming, J. S. Mattice, and V. A. Jacobs, Eds. (Ann Arbor, MI: Ann Arbor Science Publishers, Inc., 1983), pp. 269-284.
4. Sykes, P. *A Guidebook to Mechanism in Organic Chemistry* (New York: John Wiley and Sons, Inc., 1961), pp. 204-207.
5. Pearson, R. G., and R. L. Dillon. "Ionization of Carbon Acids in Water," *J. Am. Chem. Soc.* 75:2439-2445 (1953).
6. Bell, R. P. *The Proton in Chemistry* (Ithaca, NY: Cornell University Press, 1959), pp. 161-164.
7. Reckhow, D. A., and P. C. Singer. "The Removal of Organic Halide Precursors by Preozonation and Alum Coagulation," *J. Am. Water Works Assoc.* 76:151-155 (1984).
8. Hand, V. C., and D. W. Margerum. "Kinetics and Mechanisms of the Decomposition of Dichloramine in Aqueous Solution," *Inorg. Chem.* 22(10):1449-1456 (1983).
9. Jackson, R. A. *Mechanism, An Introduction to the Study of Organic Reactions* (London: Oxford University Press, 1972), pp. 4-11.
10. Weil, I., J. C. Morris, and R. H. Culver. "Kinetic Studies on the Breakpoint with Ammonia and Glycine," presented at the International Union for Pure and Applied Chemistry, New York, Sept. 20, 1952.
11. Morris, J. C., and R. A. Isaac. "A Critical Review of Kinetic and Thermodynamic Constants for the Aqueous Chlorine-Ammonia System," in *Water Chlorination: Environmental Impact and Health Effects, Vol. 4*, R. L. Jolley, W. A. Brungs, J. A. Cotruvo, R. B. Cumming, J. S. Mattice,and V. A. Jacobs, Eds. (Ann Arbor, MI: Ann Arbor Science Publishers, Inc., 1983), pp. 49-61.
12. Palin, A. T. "A Study of the Chloro Derivatives of Ammonia and Related Compounds," *Water Water Eng.* 54(10):151-200; 54(11):189-200; 54(12):248-258 (1950).
13. Pressley, T. A., D. F. Bishop, and S. G. Roan. "Ammonia-Nitrogen Removal by Breakpoint Chlorination," *Environ. Sci. Technol.* 6(7):622-628 (1972).

CHAPTER **56**

Chlorine Decay Chemistry in Natural Waters

Douglas Dotson and George R. Helz

There are two major reasons for studying the chemical mechanisms responsible for the disappearance of residual chlorine: to predict new by-products of possible health or environmental significance, and to refine decay models and thus improve estimates of exposure time for aquatic organisms in waters receiving chlorinated effluents.

Chlorine decay is controlled mainly by oxidation rather than by substitution reactions.[1-3] These operate over a wide range of time scales. In much less than a second, significant amounts of oxidant can be consumed by very fast, highly selective reactions involving organic sulfur and certain aromatic and nitrogen heterocyclic rings.[4] In seconds to hours, oxidation of inorganic and organic amines is important.[5-10] Over minutes to days, relatively nonselective oxidations of organic matter can be important and, in photolyzed systems, products such as bromate can be produced.[11]

This chapter concerns the nature of the processes controlling chlorine decay on the minutes-to-days time scale in nonphotolyzed water dosed with chlorine gas to 140 μM (10 ppm). This dose was in excess of the ammonia breakpoint in all cases. We have collected a large number of river and estuarine water samples, characterized them with respect to total organic carbon, inorganic nitrogen species, iron and manganese, and then measured the decay curves of chlorine and the growth curves of carbon dioxide (CO_2). From this work, it is possible to show (1) that organic carbon is the only component sufficiently abundant to account for most of the decay when doses of about 140 μM are used; (2) that a wide variety of waters yield very similar decay curves when treated under standardized conditions in the laboratory; (3) that after satisfying initial fast demand, 0.2 to 0.4 mol CO_2 is generated per mol of chlorine lost; (4) that this ratio is in some cases constant over a period of days; and (5) that humic acid in distilled water yields a good model for the behavior observed in natural waters.

EXPERIMENTAL

Various types of natural waters from various locations were used. Freshwater samples were taken from the Delaware River in Trenton, New Jersey, and from the Little Patuxent River at Fort Meade, Maryland. More-saline samples were taken further downstream in the Patuxent River near the towns of Benedict and Solomons.

The analytical procedure for CO_2 and experimental apparatus have been described previously.[3] The analytical methods for ammonia, nitrite, primary amines, trace metals, and total organic carbon (TOC) are standard procedures; these, as well as the sample locations and sampling methods, have also been presented elsewhere.[12]

RESULTS

Natural Waters

The ranges of the initial concentrations of potentially chlorine-reactive species and the amount of CO_2 produced per mol of organic carbon are given in Table I. These ranges apply to Delaware River water samples taken in February, March, April, May, June, and October, and to saline Patuxent Estuary water samples taken at Solomons in March, April, and May. In most cases, organic matter was the only constituent that was sufficiently abundant to account for consumption of more than 10% of a 140-μM chlorine dose. The trace metals and nitrogen-containing species were usually present in low enough concentrations so that their demand could not account for consumption of a major part of the dose. The most notable exceptions to this were the ammonia levels in the February and October Delaware River samples in which the respective concentrations were 26.4 and 22.7 μM. Saunier and Selleck[6] chlorinated a solution containing 20 mg/L ammonia nitrogen and found that about 10% of this nitrogen was oxidized to nitrate. They assumed that the

Table I. Concentrations of Reactants and Products in Delaware and Patuxent River Water Samples

Species[a]	Delaware River (6 samples)	Patuxent Estuary[b] (3 samples)
Dissolved iron[c]	2 – 14	<0.1 – 11
Dissolved manganese[c]	0.4 – 2.0	0.5 – 2.1
Ammonia	2 – 26	<0.7 – 5.9
Nitrite	1 – 7	0.4 – 1.2
Primary amines	0.5 – 1.7	0.8 – 2.1
Total organic carbon	270 – 650	318 – 346
Carbon dioxide produced per mol organic carbon	0.051 – 0.10	0.10 – 0.11

[a]All concentrations are in micromolar units.
[b]Salinity about 10 g/kg.
[c]Concentrations of iron and manganese that passed a 0.45-μm filter; these data establish merely upper limits on the amounts of these two metals in the reduced, divalent oxidation state.

remaining ammonia was converted to nitrogen. If the same proportions of these products formed in the present experiments, the quantities of chlorine that would have decayed by reaction with ammonia would be 46 and 40 μM in the February and October samples, respectively. This means that an upper limit of 38 and 29%, respectively, of the applied chlorine doses reacted with ammonia.

The chlorine decay and CO_2 production curves for Delaware River water samples taken in March, April, and October, and the Patuxent River water sample taken in April 1983 are shown in Figure 1. These results are representative of a larger body of data obtained for the Delaware and for freshwater and saline water sites on the Patuxent. They illustrate that qualitatively similar Cl_2 decay and CO_2 production curves are observed uniformly throughout the year in both freshwater and estuarine water. We have found CO_2 production to be a universal phenomenon in natural waters dosed with 140 μM Cl_2.

Of the Delaware River water samples shown, chlorine decay is most rapid in the March sample. In the October sample, decay is initially very rapid but becomes much slower after about 3 h. In the April sample, the initial decay is not nearly as rapid as in any of the other samples, but residual chlorine is not as persistent as in the October sample.

The quantities of CO_2 produced per mol organic carbon are given in Table I. The range in this ratio is probably due to variations in the nature of the organic matter. The smallest amount of CO_2 per mol organic carbon was produced in March, when chlorine decay was fairly rapid. The initial chlorine decay in the April Delaware sample, on the other hand, is not as quick as in any of the other samples, and the greatest quantity of CO_2 is produced.

Stoichiometric Considerations

An experiment was performed to determine the relationship between chlorine dose and CO_2 production. Water used in this experiment came from Benedict, Maryland (salinity ~5 g/kg). The samples were chlorinated to 56, 140, and 282 μM with aqueous sodium hypochlorite. The chlorine decay and CO_2 production curves are shown in Figure 2. In Figure 3, CO_2 production is plotted as a function of chlorine decay. In the most highly dosed sample, CO_2 production ceased before the chlorine had entirely disappeared. The amount of chlorine remaining when CO_2 production ceased was therefore subtracted from the dose to determine the amount of chlorine that decayed. The relationship is found to be linear with a nonzero intercept (Figure 3).

Extrapolation of the line to the X axis indicates that at zero CO_2 production there is a chlorine decay of 30 μM. If the amount of CO_2 produced in each sample is divided by the quantity obtained when 30 μM is subtracted from the amount of chlorine that decayed, a ratio is obtained (0.27) that is constant for the three samples. In other words, 0.27 mol CO_2 is procured per mol of chlorine decayed once an initial non-CO_2-producing demand of 30 μM is satis-

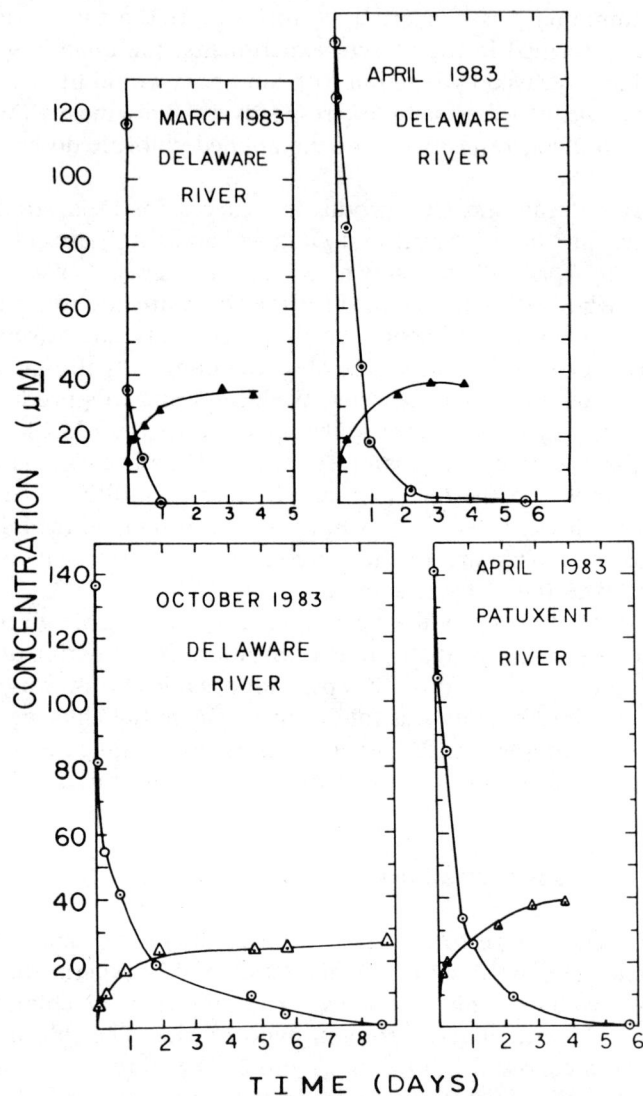

Figure 1. Chlorine decay (circles) and CO_2 production (triangles) in selected Delaware and Patuxent river water samples. The salinity of the Patuxent sample was about 10 g/kg; the other samples were freshwater.

fied. When the curves were broken into 12-h increments and the CO_2 release: chlorine decay ratios were computed, they were found to be approximately constant (0.27) throughout the course of the experiments, as seen in Figure 4. In each case, 30 μM was subtracted from the quantity of chlorine that decayed in the first 12 h.

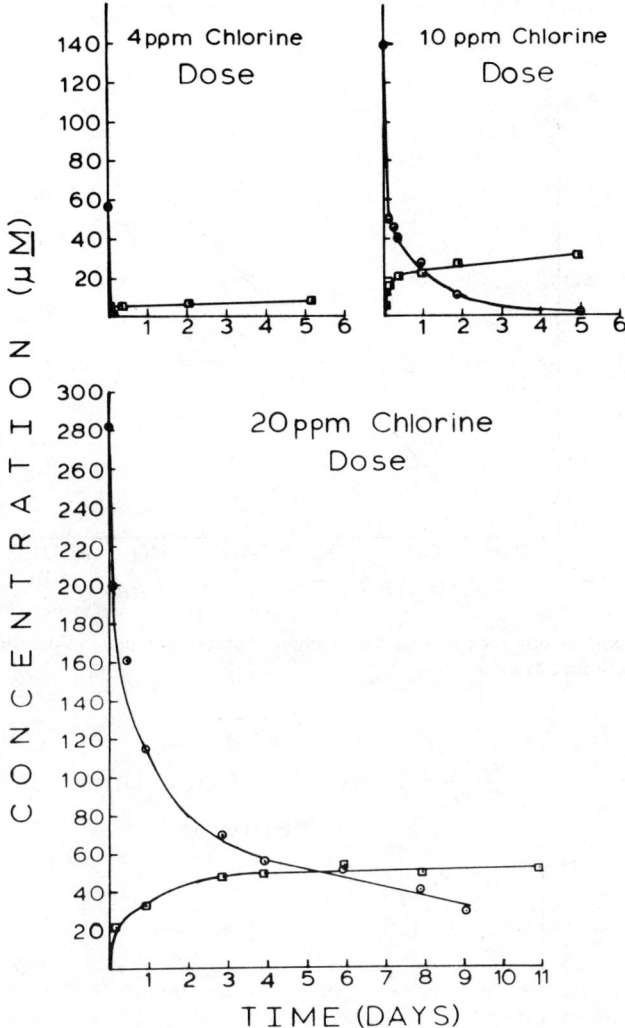

Figure 2. Chlorine decay (circles) and CO_2 production (squares) in Patuxent River water samples dosed to 56, 140, and 282 μM chlorine.

The apparent constancy in the ratio of chlorine decayed to CO_2 produced (Figure 4) is quite important. It implies that the CO_2-producing process is probably the dominant chlorine decay pathway in this sample after the initial fast demand has been satisfied. If another significant decay pathway (not leading to CO_2) exists, its rate must be very similar to that of the CO_2-producing pathway to maintain the constancy observed in Figure 4.

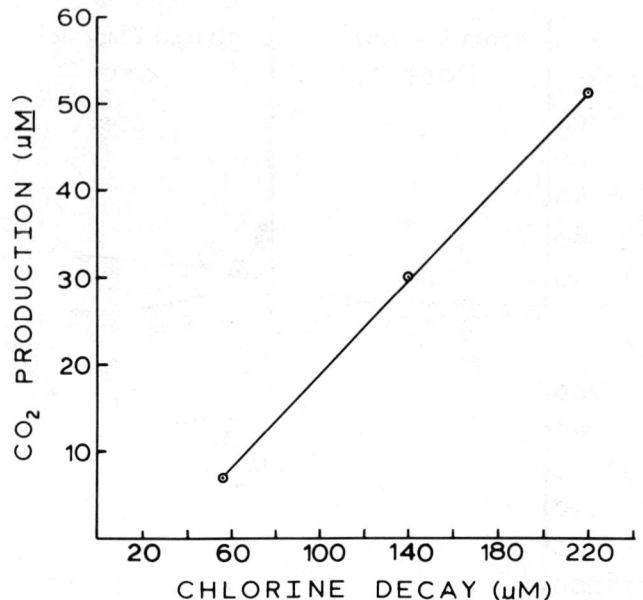

Figure 3. Carbon dioxide production as a function of chlorine decay in Patuxent River water. Data from Figure 2.

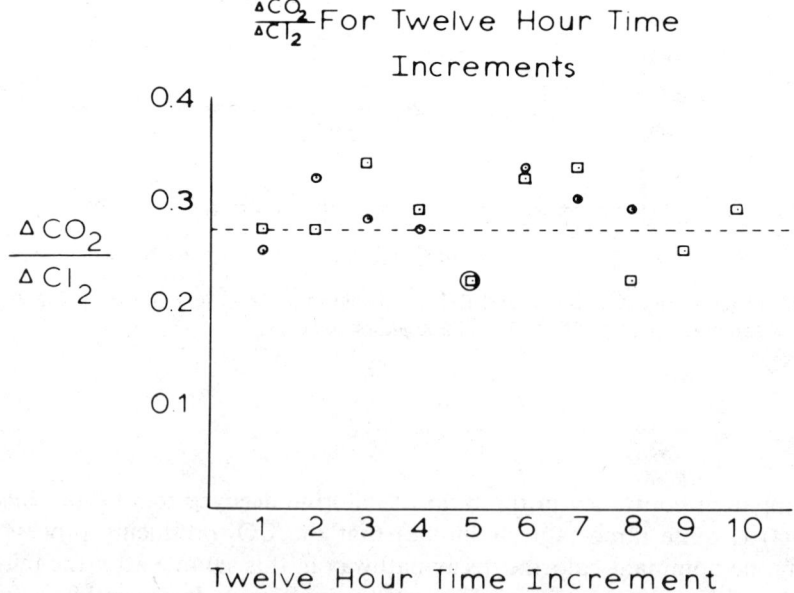

Figure 4. Carbon dioxide released per mole of chlorine decayed over 12-h time increments based on the data in Figure 2.

Model Compounds

An attempt was made to determine the type of organic matter responsible for the observed CO_2 yields and chlorine decay patterns. Natural waters contain many types of organic matter, including proteins, carbohydrates, lipids, humic materials, and a large variety of low-molecular-weight compounds. The lipids and low-molecular-weight materials are usually present in minute quantities. Furthermore, lipids are not very reactive to chlorine when both reactants are present in low concentrations. Consequently, the model compounds chosen were macromolecular materials that represent the three remaining classes of organic matter.

Bovine serum albumin (BSA) was used as a model for the proteins, and indicator starch was chosen to represent the carbohydrates. The third model compound used was humic acid. According to Wershaw et al.,[13] humic and fulvic acids are the most abundant organic compounds present in natural waters. The material used was isolated from river sediment that came from the Potomac River in the vicinity of Point of Rocks, Maryland. Point of Rocks is located 33 miles northwest of the Washington metropolitan area. There are no large population centers upstream in the Potomac watershed. The extraction procedure has been described elsewhere.[14] The elemental analysis of the humic acid is as follows: 54.7% C, 6.3% H, 5.8% N, and 33.2% O + S. The H/C molar ratio is 1.4. The humic acid composition as determined by CPMAS ^{13}C NMR spectroscopy is as follows: 29% paraffinic C, 29% aromatic C, and 18% phenolic + carboxylic C.[15]

Of the three compounds tested, humic acid was the only one that provided a good model of the natural waters (Figure 5). Starch reacted much too slowly, and bovine serum albumin did not produce sufficient CO_2. For humic acid, both CO_2 production and chlorine decay were complete within the first 24 h of reaction and 31 μM CO_2 was produced.

DISCUSSION

One mechanism that could account for a portion of the CO_2 produced in these experiments was proposed by Rockwell and Larson.[16] The reaction is a two-step process in which electrophilic substitution of chlorine to an aromatic acid occurs. In this mechanism, chlorine attacks the carbon atom attached to the carboxyl carbon. For high yields, the ring must be activated by electron-donating substituents such as $-OH$ and $-OCH_3$. Another possibility is suggested by the work of Boyce and Hornig[17] and Rook,[18] who chlorinated dihydroxyaromatic compounds. They concluded that decarboxylation can occur following oxidative ring opening. Because humic acid is thought to contain aromatic rings having $-OH$, $-OCH_3$, and $-COOH$ substituents, and because a humic acid solution was found to be a good model for natural

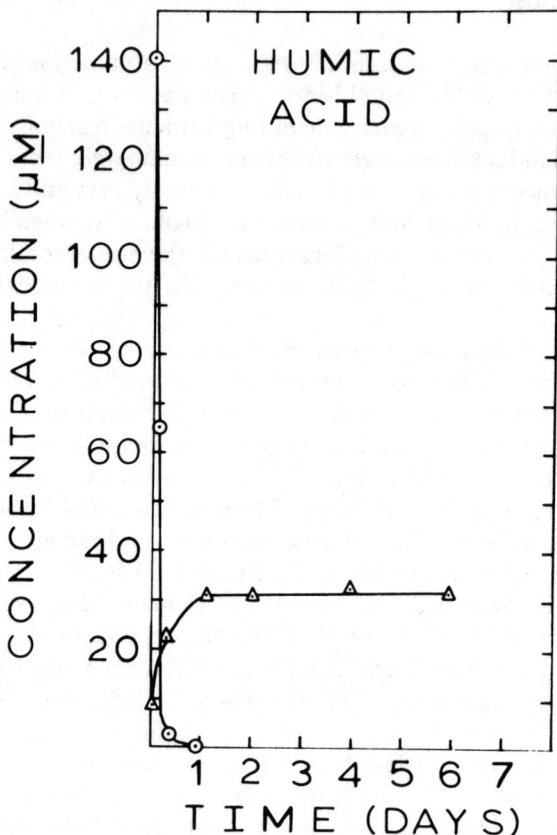

Figure 5. Chlorine decay and CO_2 production in a solution of 10 mg/L humic acid at neutral pH.

waters, it is plausible that the above mechanisms contribute to the CO_2 produced by chlorination. It is not likely that they account for the majority of the CO_2, however, because these mechanisms involve chlorine substitution in amounts equivalent to or greater than the CO_2 yield. Such high yields of halocarbons have never been observed. For example, Jolley[1] found that less than 1% of a 6-mg/L chlorine dose added to wastewater treatment plant effluent resulted in chlorine substitution products. Thus, one important requirement for a mechanism to account satisfactorily for CO_2 production in natural waters is that it produce small halocarbon yields relative to the CO_2 yield.

One way of producing CO_2 without large yields of halocarbons is by N-chlorination of amino acids;[8] in this process, the carboxyl group is lost and the alpha carbon is oxidized. The labile N-Cl bond decomposes, releasing chloride and leaving no halocarbon residue. However, the difficulty with this as a mechanism for accounting for a major fraction of the CO_2 produced by chlorination is that free amino acid concentrations are far too low (note primary organic amine concentrations in Table I).

This research has raised two important questions that have yet to be answered: (1) what are the primary mechanisms responsible for producing CO_2 without comparable amounts of halocarbons, and (2) how are the oxidations that produce CO_2 and those that produce other organic oxidation products coupled such that a fairly constant ratio of chlorine consumption to CO_2 production, as in Figure 4, is observed? If a detailed knowledge existed of the structural character of macromolecular material in natural waters, answering these questions would probably not be difficult, but unfortunately such knowledge is not at hand.

ACKNOWLEDGMENTS

The authors wish to express their appreciation to Richard Sugam and William Sandvick for furnishing water samples and to Gregory Diachenko for supplying humic acid. Funding for this research was provided by the Maryland Power Plant Siting Program, Maryland Sea Grant, and Public Service Electric and Gas. Support for presenting this paper was provided by the Office of the Dean for Graduate Studies and Research of the University of Maryland.

REFERENCES

1. Jolley, R. L. "Determination of Chlorine Containing Organics in Chlorinated Sewage Effluents by Coupled ^{36}Cl Tracer-High-Resolution Chromatography," *Environ. Lett.* 7(4):321–340 (1974).
2. Sigleo, A. C., G. R. Helz, and W. H. Zoller. "Organic-Rich Colloidal Material in Estuaries and Its Alteration by Chlorination," *Environ. Sci. Technol.* 14:673–679 (1980).
3. Helz, G. R., D. A. Dotson, and A. C. Sigleo. "Chlorine Demand: Studies Concerning its Chemical Basis," in *Water Chlorination: Environmental Impact and Health Effects, Vol. 4*, R. L. Jolley, W. A. Brungs, J. A. Cotruvo, R. B. Cumming, J. S. Mattice, V. A. Jacobs, Eds. (Ann Arbor, MI: Ann Arbor Science Publishers, Inc., 1983), pp. 181–190.
4. Jaworske, D. A. *The Kinetics of Rapid Oxidant Demand: Measurements with a Rotating Ring Disc Electrode*, PhD Thesis, (College Park, MD: University of Maryland, 1983).
5. Wei, I. W., and J. C. Morris. "Dynamics of Breakpoint Chlorination," in *Chemistry of Water Supply, Treatment and Distribution,* A. J. Rubin, Ed. (Ann Arbor, MI: Ann Arbor Science Publishers, Inc., 1974), pp. 297–332.

6. Saunier, B. M., and R. E. Selleck. "The Kinetics of Breakpoint Chlorination in Continuous Flow Systems," *J. Am. Water Works Assoc.* 71:164-172 (1979).
7. Hand, V. C., and D. W. Margerum. "Kinetics and Mechanisms of the Decomposition of Dichloramine in Aqueous Solution," *Inorg. Chem.* 22:1449-1456 (1983).
8. Stanbro, W. D., and W. D. Smith. "Kinetics and Mechanism of the Decomposition of N-Chloroalanine in Aqueous Solution," *Environ. Sci. Technol.* 13:446-451 (1979).
9. Katz, B. M. "Chlorine Dissipation and Toxicity Presence of Nitrogenous Compounds," *J. Water Pollut. Control Fed.* 49:1627-1635 (1977).
10. Inman, G. W., and J. D. Johnson. "The Effect of Ammonia Concentration on the Chemistry of Chlorinated Sea Water," in *Water Chlorination: Environmental Impact and Health Effects, Vol. 2,* R. L. Jolley, H. Gorchev, and D. H. Hamilton, Jr., Eds. (Ann Arbor, MI: Ann Arbor Science Publishers, Inc., 1978), pp. 235-252.
11. Macalady, D. L., Carpenter, J. H., and Moore, C. A. "Sunlight Induced Bromate Formation in Chlorinated Seawater," *Science* 195:1335-1337 (1977).
12. Dotson, D. A. *Chemical Modification of Natural Waters by Treatment with Chlorine,* PhD Thesis, (College Park, MD: University of Maryland, 1984).
13. Wershaw, R. L., M. A. Mikita, and C. Steelink. "Direct ^{13}C NMR Evidence for Carbohydrate Moieties in Fulvic Acids," *Environ. Sci. Technol.* 15(12):1461-1463 (1981).
14. Diachenko, G. W. *Sorptive Interactions of Selected Volatile Halocarbons with Humic Acids from Different Environments.* PhD Thesis, (College Park, MD: University of Maryland, 1981).
15. Stevenson, F. J. *Humus Chemistry* (New York: John Wiley and Sons, Inc., 1982), pp. 258, 259.
16. Rockwell, A. L., and R.. A. Larson. "Aqueous Chlorination of Some Phenolic Acids," in *Water Chlorination: Environmental Impact and Health Effects, Vol. 2,* R. L. Jolley, H. Gorchev, and D. H. Hamilton, Jr., Eds. (Ann Arbor, MI: Ann Arbor Science Publishers, Inc., 1978), pp. 67-74.
17. Boyce, S. D., and J. F. Hornig. "Reaction Pathways of Trihalomethane Formation from the Halogenation of Dihydroxyaromatic Model Compounds for Humic Acid," *Environ. Sci. Technol.* 17(4):202-211 (1983).
18. Rook, J. J. "Possible Pathways for the Formation of Chlorinated Degradation Products During Chlorination of Humic Acids and Resorcinol," in *Water Chlorination: Environmental Impact and Health Effects, Vol. 4,* R. L. Jolley, W. A. Brungs, J. A. Cotruvo, R. B. Cumming, J. S. Mattice, V. A. Jacobs, Eds. (Ann Arbor, MI: Ann Arbor Science Publishers, Inc., 1983), pp. 85-98.

CHAPTER 57

A Kinetic Model of Chlorination of Natural Water: The Roles of Organic Nitrogen and Humic Substances

Robert G. Qualls and J. Donald Johnson

Chlorination is used in the cooling systems of most power plants to prevent fouling of the condensers. To optimize dose, the concentration of free residual chlorine (FRC) at the condensers is needed to judge biofouling control, and the concentration and speciation of total residual chlorine (TRC) at the point of discharge is needed to judge environmental toxicity.[1]

Modeling the chlorination process is particularly useful because (1) chlorine consumption is difficult to predict[2] and (2) there are analytical problems in distinguishing the chemical forms of chlorine and in determining the rapidly changing concentrations of these different residuals.[3,4]

Modeling of chlorine demand in natural waters may be based on three contributions: ammonia breakpoint reactions, consumption by organic carbon reactions, and consumption by organic amines. Models for reactions of chlorine and ammonia are well established,[5,6] and we have recently developed expressions for the short-term consumption of chlorine by humic substances in natural waters.[7] Previous efforts to model the kinetics of chlorination of cooling waters[2,8,9] have met with limited success in predicting FRC concentrations because of the lack of data on the kinetics of chlorine consumption by humic substances and appropriate model compounds for the organic nitrogen fractions in natural waters.

Except in heavily polluted water, the concentrations of organic nitrogen are generally greater than those of free inorganic ammonia, frequently by several fold.[10,11] Because of the diversity of organic nitrogen compounds in natural waters, no study has even broadly characterized all of the organic nitrogen. Proteinaceous material is usually the largest single fraction, accounting for about 17 to 50% of the organic nitrogen, of which only a few percent consists of free amino acids.[10,12] The peptide bond itself is relatively resistant to chlorination.[13-15] The fastest initial reactions with chlorine probably occur with the nucleophilic terminal amino groups of peptides rather than with the peptide nitrogen, with reactive side groups of certain amino acids, and with the relatively small concentrations of free amino acids.[14,15] Based on its reactivity and the stability of the chlorinated product, as well as the reported nature of organic nitrogen, the dipeptide glycylglycine was chosen as a more reasonable model for organic nitrogen than the methylamine used earlier.[2,8,9,14] It is also interesting that the peptide terminal amino nitrogen competes so effectively with ammonia for HOCl even when ammonia is higher in concentration.[15] Taken together, these studies demonstrate that neither ammonia alone, nor total organic nitrogen content provides a good measure of nitrogenous chlorine consumption because of the wide range in the rates of reaction of the different types of nitrogen compounds.

The objective of this study was to develop kinetic expressions for chlorine consumption, including organic compounds, for use in a model for chlorination of nonsaline water. In this study, we were primarily concerned with "once-through" chlorination, and thus with the reactions occurring in the first few minutes of chlorination. The model should not require direct field measurement of chlorine consumption, but only a minimal amount of water quality data such as temperature, pH, dissolved organic carbon (DOC), dissolved organic nitrogen (DON), and NH_3. Our approach was to treat the chlorine consuming substances as either NH_3, chloramine-forming organic nitrogen, or humic substances. The experimental objectives were (1) to evaluate kinetic equations for the decomposition of the model compound, glycylglycine; (2) to compare simulations with this model, including organic nitrogen, ammonia, and humic or fulvic acid demand, to experiments using chlorinated samples of natural water and of natural water in which these components were removed or isolated and individually tested; and (3) to test the overall kinetic expressions using the model compounds for simulating short-term chlorination kinetics in natural water samples.

THE MODEL

The basic "breakpoint" reaction model of Morris and Wei for chlorine-ammonia reactions was used for Equations 1–8 for our overall kinetic model.[5,15,16] Glycylglycine was used as a model compound to represent the chloramine-forming organic nitrogen. We used the rate constants found by Margerum et al.[17] for the formation of N-chloroglycylglycine (ClHNR) and N,N-dichloroglycylglycine (Cl_2NR) (Table I). An equation (No. 11) for the

Table I. Rates of Reactions for Chlorine with Glycylglycine (Equations 9 – 11) and fulvic acid (Equations 12 – 13). All constants are for 25°C.

Equation No.	Rate Equations	Rate Constants, k
9[a]	$k_9[HOCl][H_2NR]$	$5.3 \times 10^6\ M^{-1}\ s^{-1}$
10[a]	$k_{10}[HOCl][ClHNR]$	$8.7\ M^{-1}\ s^{-1}$
11	$k_{11}[Cl_2NR]$	$5.6 \times 10^{-5}\ M\ s^{-1}$
12	$k_{12}[Cl][F_1]$ where: $[F_1] = 0.020\ M$ sites/M [C]	$6.3 \times 10^3\ M^{-1}\ s^{-1}$
13	$k_{13}[Cl][F_2]$ where: $[F_2] = 0.078\ M$ sites/M [C]	$42\ M^{-1}\ s^{-1}$

[a]See Reference 17.

decay of N,N-dichloroglycylglycine was included, based on the results described later. The concentration of the reactive form of glycylglycine (H_2NR), the unprotonated amine species, was calculated from equilibria cited by Margerum et al.[17] The amount of glycylglycine necessary to represent the chloramine-forming organic nitrogen was estimated from experiments with natural water samples.

The consumption of FRC by fulvic acids (F) was represented by the sum of two second-order equations.[7] The concentration of each of the two types of fulvic acid reactive sites, F_1 and F_2, was related to the total carbon concentration of the fulvic acid by the proportion found in the earlier work[7] as shown in Table I. The proportion of the DOC that was fulvic acid in the natural water samples was estimated from the UV absorbance of the isolated fulvic acid fraction at the same pH and ionic strength. A small proportion of the chlorine consumption by the isolated fulvic acid fraction resulted in formation of organic chloramines from nitrogen associated with the fulvic acid. This consumption was excluded from the equations representing the fulvic acid demand and was included in the organic nitrogen demand (Equation 9, Table I).

Simulations were performed using the Continuous Systems Modeling program (IBM) for solving a set of simultaneous differential equations by numerical techniques. A special algorithm for "stiff" differential equations was used.

METHODS

Analytical Procedures

FRC and TRC were determined in most experiments by amperometric titration[4] using equipment described elsewhere.[18] Small aliquots of a standardized 5 mM NaOCl stock solution were added to rapidly stirred water samples. Samples were covered with parafilm over minimal headspace and incubated at 25 ($\pm$0.2)°C. Titrations after short reaction times were done rapidly using an automatic pipet to bring the titration very quickly near the end point, which had been estimated from a preliminary run. The titration was then completed in <15 s. Ammonia, DON, and DOC were determined by methods detailed elsewhere.[7]

Experiments with Natural Water Samples

Three types of experiments were conducted using natural water samples:

1. Concentration of FRC and combined residual chlorine (CRC) measured over time after an initial dose of chlorine that yielded about 3 mg/L chlorine after 2 min.
2. "Breakpoint curves" (or chlorine residual as a function of chlorine added) at 2, 10, or 120 min, to reveal information on stoichiometry and decay reactions.

3. Concentration-time curves on FRC consumption rates for samples from which ammonia or fulvic materials had been removed to partially isolate components of the chlorine demand.

Samples were collected from three sources (Table II), filtered with a 0.45-μm Millipore filter and stored at 3°C. To strip subsamples of ammonia, the pH was raised to 10.2 to 10.4, and they were then lypholyzed and redissolved twice. In one case, the ammonia was stripped by bubbling with washed nitrogen gas. Samples were readjusted to the original pH with sulfuric acid. To partially remove hydrophobic substances including humic substances, a subsample of Cape Fear River water was brought to pH 2 and passed through an XAD-8 resin column. The effluent from the column, adjusted to the original pH, was used in chlorination experiments. Controls consisted of chlorine-demand-free water to which chlorine was added, and they were incubated the same as samples in each set of experiments. Results indicated no significant loss of chlorine as a consequence of the procedure.

RESULTS

Decay of N,N-dichloroglycylglycine

The decay rate of N,N-dichloroglycylglycine was not significant in the short time span of several minutes (Figure 1). The slope of the line for the 3:1 Cl:glycylglycine mol ratio indicated a first-order decomposition rate constant of about 5.6×10^{-5} s^{-1} at pH 7 and 25°C. The slope of the line for the 2.5:1 mol ratio was similar. This constant can only be regarded as an approximation because the slow formation of N,N-dichloroglycylglycine made complete resolution of the decay reaction difficult. For decay of N-chloroglycylglycine, Isaac and Morris[15] quote a rate constant of 1.2×10^{-5} s^{-1} at pH 6.2 and 22°C, decreasing to 8.3×10^{-6} s^{-1} at pH 8.5, using a 10:1 mol ratio of glycylglycine to chlorine. Thus, their slower rate probably represents decomposition of monochloroglycylglycine. Because the decay rate of the dichloroglycylglycine we measured was so slow, it is not important over the length of time in which we were interested, and we did not attempt to confirm either the validity of the

Table II. Description of Water Samples

	Holston River 7-20-81	Tennessee River 7-20-81	Cape Fear River 12-20-81
pH in Laboratory	7.5	7.5	6.9
NH$_3$-N, mg/L	0.126	0.011	0.269
NH$_3$-N, μM	9.0	0.8	19.2
Org-N, mg/L	0.30	0.18	0.42
Org-N, μM	21	13	30
DOC, mg/L	3.2	1.9	10.0

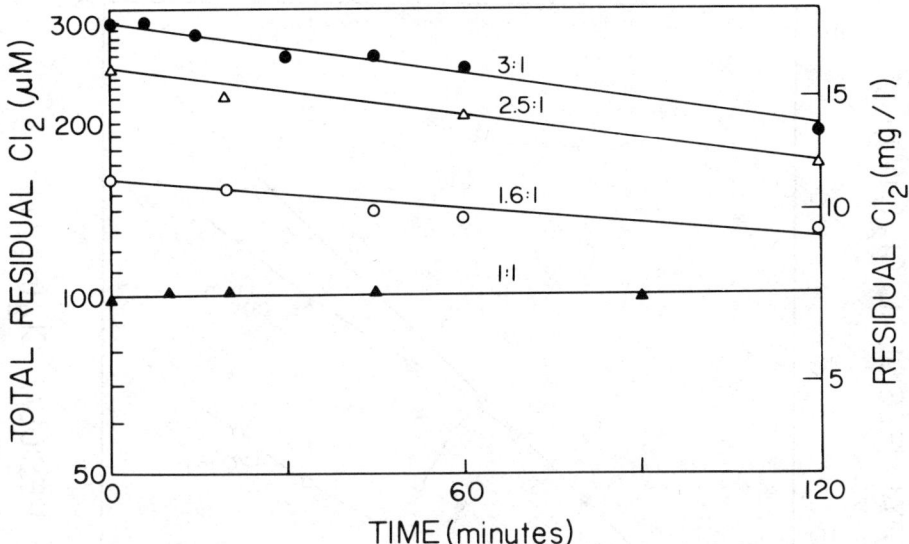

Figure 1. Decay of chlorinated glycylglycine (pH 7, 25°C). Decline in total residual chlorine (TRC) added to 100 μM glycylglycine is shown. The molar ratio of chlorine added to glycylglycine is indicated.

first-order decay expression over a range of concentrations or the reaction stoichiometry.

A simulation of a breakpoint curve for the chlorination of a natural water was performed for a hypothetical mixture of 9 μM glycylglycine (as the model of organic nitrogen) and 9 μM ammonia. The simulation generated the breakpoint curve shown as solid lines for 10 and 120 min in Figure 2. The curve for TRC remaining at 2 h is a classic ammonia breakpoint curve superimposed on a relatively stable plateau of N-chloroglycylglycine. The persistent presence of 9 μM CRC throughout the curve reflects the slow formation of N,N-dichloroglycylglycine and its slow decay rate. The nature of the competition between ammonia and glycylglycine for available chlorine was indicated by the concentrations of each component.[16] The rate of ammonia chlorination is lower than for many organic amines such as glycylglycine.[15] Glycylglycine outcompeted ammonia for the available HOCl until an excess of HOCl was added beyond that required to form N-chloroglycylglycine.

This breakpoint curve is somewhat similar to the generalized breakpoint curve for water containing equal amounts of ammonia and organic nitrogen presented by White.[19] Curves of this type having a so-called irreducible minimum are often found in natural waters in contrast to the complete loss of residual at the breakpoint found with ammonia above pH 7.

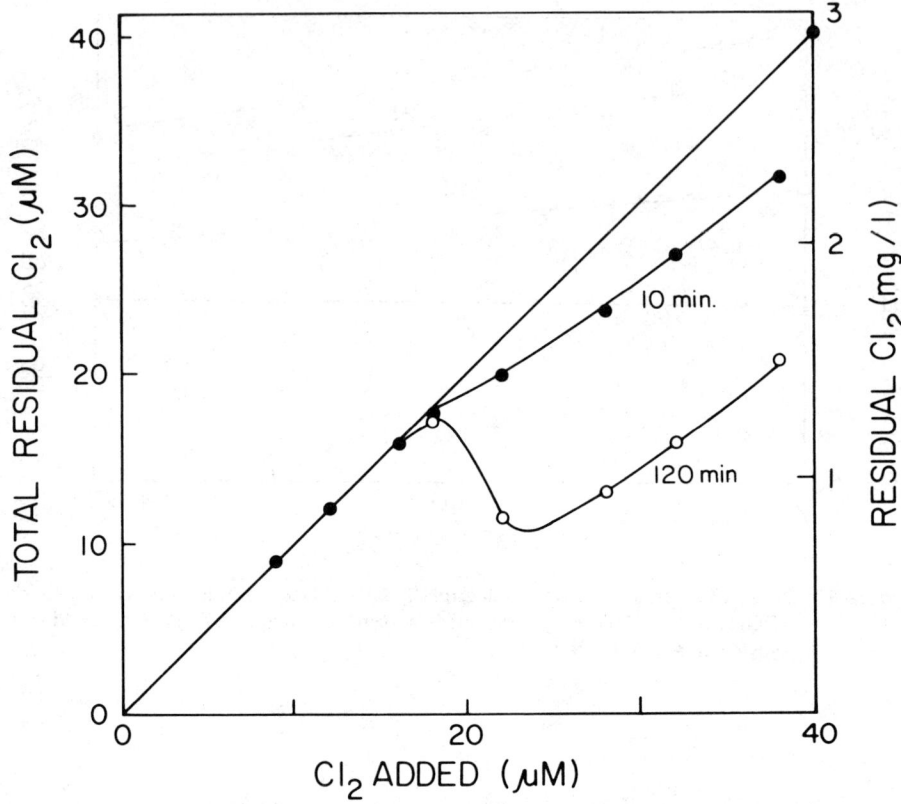

Figure 2. Simulations of reactions of chlorine added to 9 μM glycylglycine with 9 μM NH_3 (pH 7, 25°C).

Experiments with Natural Water Samples

The water quality of the samples (Table II) ranged from the very low nitrogen and DOC concentrations found in the Tennessee River, to that of the Cape Fear River samples, rich in DON and DOC. Concentrations of ammonia and DON are also given in micromolar units to show the Cl:N stoichiometry clearly.

The amount of chloramine-forming organic nitrogen was estimated from the maximum amount of CRC formed in the ammonia-stripped samples. This was done by plotting the amount of TRC and CRC remaining at either 2 or 10 min vs the amount of oxidizing chlorine initially added.[16] The time period was chosen because the initial monochlorination step should be complete in seconds whereas the formation of dichloroorganic amines should have been minimal, judging from the rate constants of the reactions for the formation of dichloroglycylglycine.[15] The chloramine-forming organic nitrogen was estimated to be a relatively small percentage of the total DON content of the three samples, ranging from 7 to 15%. For purposes of the simulation, we assumed that chloramine-forming organic nitrogen was 10% of the DON content. Despite the fact that there was more DON than ammonia in all three river water samples, a higher concentration of NH_2Cl was formed when enough chlorine was added to completely chlorinate the reactive nitrogen, because only about 10% of the DON was capable of reacting with chlorine.

We found it unnecessary to add as much chlorine as would be needed to exceed the long-term breakpoint. Sufficient FRC to prevent fouling[3] (about 3 μM, or 0.2 mg/L, Cl_2) was present during the 2- to 10-min period, when the amounts of chlorine added were much less than that necessary to exceed the 2-h breakpoint. An example for the Holston River sample is shown in Figure 3. Adding 22 μM Cl_2 was sufficient to provide 3 μM FRC after 10 min, whereas it was necessary to add over 40 μM chlorine to provide FRC after 2 h, when a classic breakpoint curve had formed (not shown).

The chlorine consumption rate in the Holston River, Tennessee River, and Cape Fear River samples showed that the FRC concentration declined rapidly in the first few minutes (Figures 4 and 5) and, in the case of the Holston and Cape Fear River samples, had disappeared by 30 min. The rate of decline demonstrates that chlorine should be injected as closely as possible to the condensers. For example, in the Holston River sample, over twice as much FRC was present at 1 min as at 10 min.

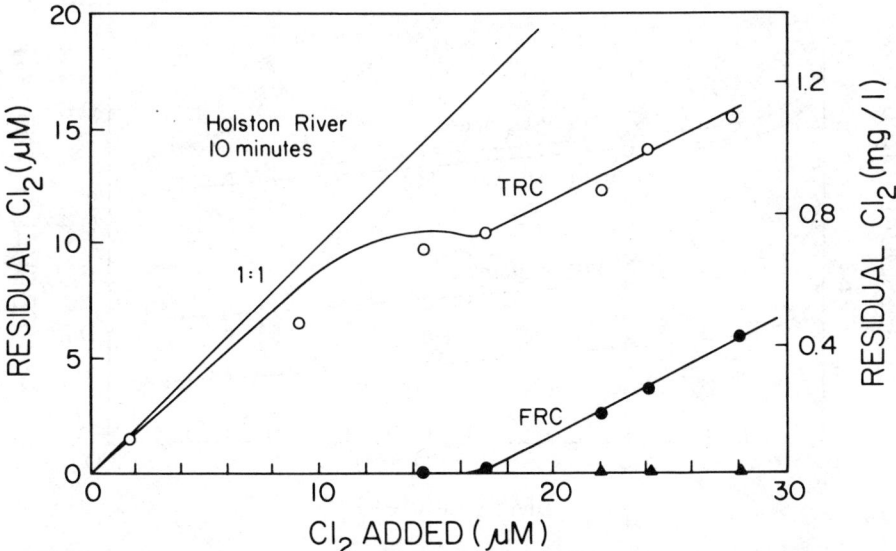

Figure 3. Simulated (lines) vs measured (data points) breakpoint curves for TRC and FRC after 10 min for the Holston River samples. Triangles indicate FRC at 120 min.

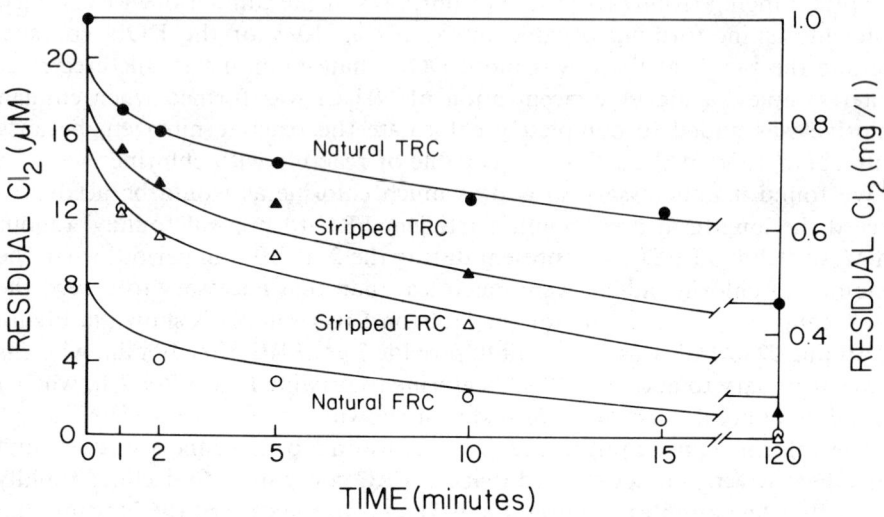

Figure 4. Chlorine consumption rates of Holston River samples with 22 μM Cl_2 added. Decline in total residual chlorine (TRC, solid symbols) and free residual chlorine (FRC, hollow symbols) is shown for natural (circles) and ammonia stripped (triangles) samples. Curves are the results of simulations.

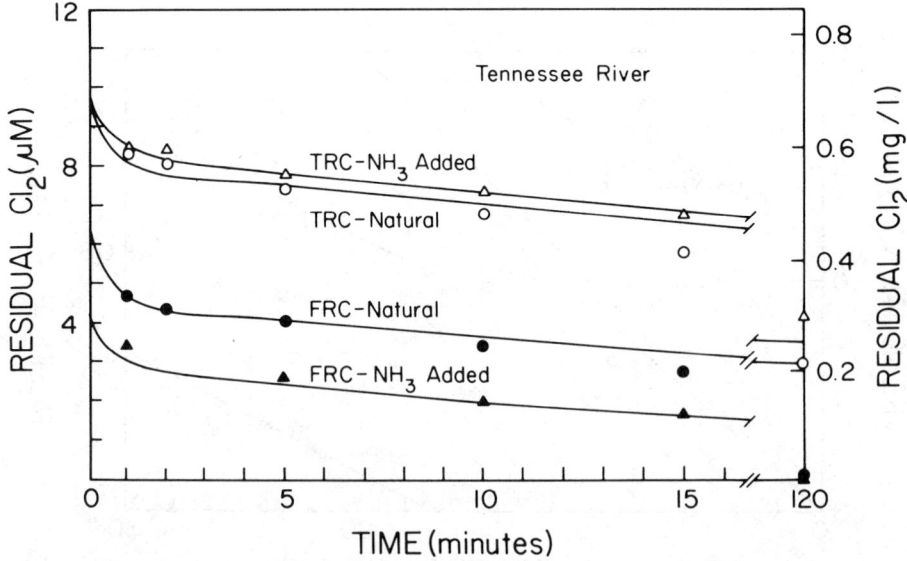

Figure 5. Simulated (lines) vs measured (data points) residual Cl_2 as a function of time after 10 μM chlorine was added to the Tennessee River water samples, for both the natural sample and a sample to which 3 μM NH_4 had been added (pH 7.5, 25°C).

The simulations, which fit the data reasonably well, allowed us to examine the reactions contributing to various phases of the chlorine consumption. The formation of monochloramine and organic N-chloramines in these natural water samples was complete within a few seconds, and the "fast" consumption reaction of fulvic acid was complete within about 30 s. The consumption of free chlorine by these reactions represents a practical minimum FRC consumption in the sense that they are almost immediate. In the case of the Holston River sample, these fast reactions consumed 16 μM FRC within 30 s, according to the assumptions of the model. An estimate of this minimum FRC consumption can be made from the following equation, according to the assumptions of the model:

$$\text{Fast Cl}_2 \text{ consumption} = [NH_3] + 0.1[DON] + 0.02[DOC]$$

where all units are micromolar. Because these reactions are fast and complete, this fast component of the consumption is relatively independent of pH and temperature in natural water. The actual Cl_2 consumption will usually be greater than this minimum because a variable amount of the "slower" reactions occur during this initial period.

The second phase of FRC consumption was dominant between 1 and 10 min. According to the simulation, the most important competing reactions during these phases were the slow consumption by fulvic acid and the formation of dichloramine.

A comparison of demand rates of the natural and the ammonia stripped samples from the Holston River (Figure 4) shows an interesting effect of the competition between the demand of the fulvic acid and ammonia. As might be expected, the FRC declined less rapidly in the ammonia stripped sample. By contrast, the TRC declines more slowly in the natural sample. The simulation suggested that the ammonia outcompetes the slower organic demand for FRC. The formation of monochloramines has the effect of stabilizing the TRC. The FRC would otherwise have reacted with other demand sources such as fulvic acid.

A portion of the chlorine consumption by fulvic acid is, however, fast enough to compete with the ammonia and organic nitrogen for limited amounts of added chlorine. Simulations suggest this competition was largely responsible for deviation of the TRC in the first part of the breakpoint curve (Figure 3) from the line indicating the 1:1 (chlorine added to chloramine formed) stoichiometry. This is shown for example in Figure 3 for 10 min.

When larger amounts of chlorine were added, the FRC consumption by fulvic acid was even more important in competing with the slow dichloramine formation reaction. Thus, when limited amounts of chlorine are added, the effect of the fulvic acid demand can be to inhibit the formation and subsequent decay of dichloramine. In a sense, the monochloramine is stablized under those circumstances. These effects are important only when chlorine is added at levels below the long-term breakpoint in the short times (several minutes) important in once-through cooling water chlorination.

The effects of fulvic substances on the rates of chlorine consumption can be seen by comparing Figures 6 and 7, in which removal of the humic substances decreased the FRC demand but allowed a sufficient excess of FRC to form dichloramine, which subsequently decayed.

For the Cape Fear River water experiment (Figure 6) that the model simulated, the relative rates of the FRC-consuming reactions were (from fastest to slowest) (1) N-chloroglycylglycine formation, (2) NH_2Cl formation, (3) consumption by the fast fulvic acid reaction, (4) $NHCl_2$ formation, and (5) the slower fulvic acid reaction.

At conditions sufficient to provide a few micromolar FRC concentration at 2 min, the total amount of FRC consumption was mainly the result of the following reactions (largest to smallest): (1) NH_2Cl formation, (2) the fast consumption by fulvic acid, (3) organic chloramine formation, (4) the slow consumption by fulvic acid, and (5) $NHCl_2$ formation. Altogether, the reduction in FRC was the result mainly, and about equally, of NH_2Cl formation and fulvic acid. The reduction in TRC resulted primarily from fulvic acid. Although the quantities were different for the other samples, the same general pattern was true.

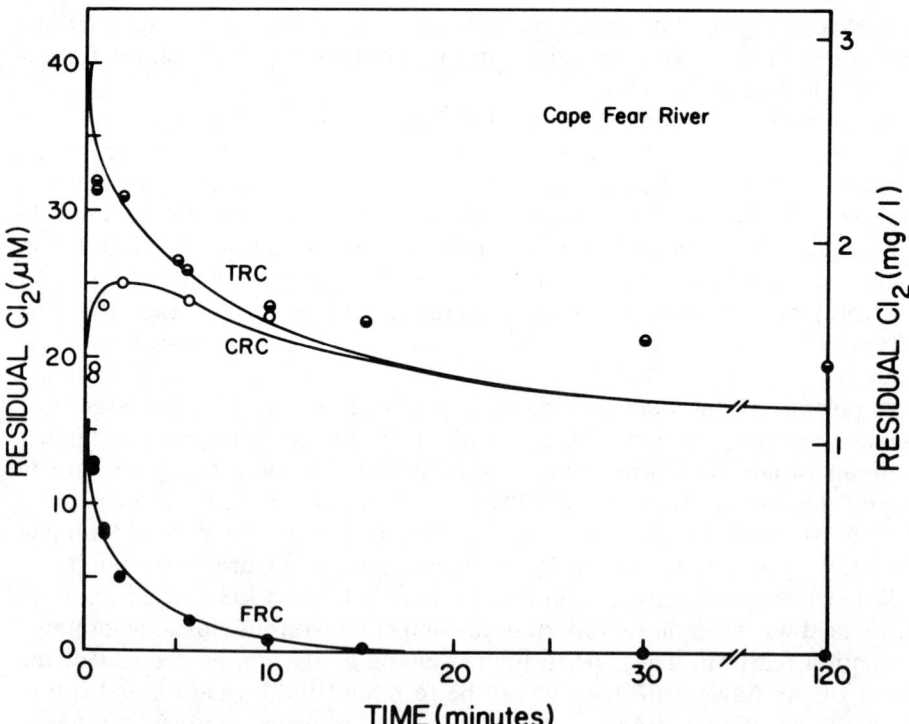

Figure 6. Simulated (lines) vs measured (data points) residual Cl_2 as a function of time after 50 μM Cl_2 was added to the Cape Fear River water sample (pH 6.9, 25°C).

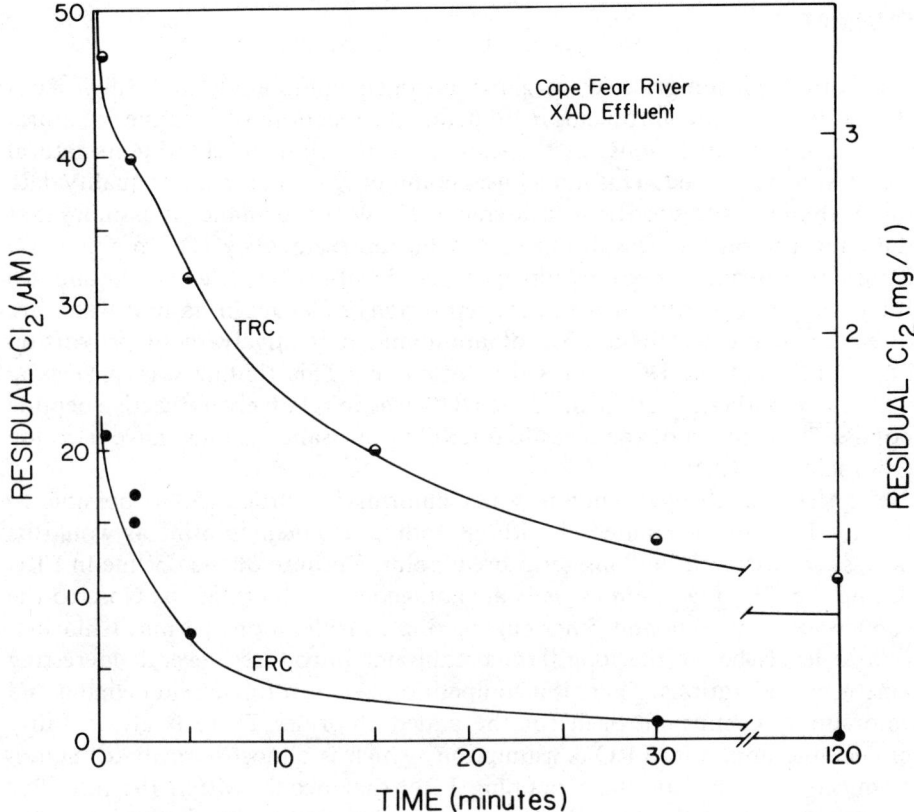

Figure 7. Simulated (lines) vs measured (data points) residual Cl_2 as a function of time after 50 μM Cl_2 was added to the Cape Fear River water subsample from which most fulvic substances have been removed by XAD-8 chromatography (pH 6.9, 25°C).

Application of this model requires the following water chemistry data: (1) temperature, (2) pH, (3) ammonia concentration, (4) chloramine-forming organic N, and (5) concentration of fulvic acid substances. Although it may be true in many natural waters that most of the DOC is humic and fulvic substances, it would be more accurate to use some surrogate measure to estimate the proportion of the DOC that is fulvic acid. Either UV absorbance or color might be the most useful measurement. Temperature coefficient data for the reactions of fulvic acid and chlorine would extend the application of the model to other temperatures. From our results we would suggest that the organic chloramine-forming nitrogen could be estimated as about 10% of the DON.

The reasonably good fit of the model to the data, at least for short time periods, indicates that the depiction of the fast reactions of the ammonia and organics is fairly accurate. Simulation of this type of experiment, in which components are added or removed, provides a stringent test of the model and highlights the competition and interaction among the components.

SUMMARY

Chlorine consumption by organic nitrogen compounds and fulvic substances has presented problems in modeling the reactions of chlorine in natural water. Kinetic expressions for the short-term reactions of chlorine in natural water were developed. This model uses commonly available water quality data rather than the site-specific measurement of chlorine demand. It assumes that the chlorine-consuming substances can be represented by (1) ammonia, (2) chloramine-forming organic nitrogen (modeled by glycylglycine), and (3) fulvic substances. Although the concentrations of DON in natural water are often much greater than those of ammonia, a relatively small proportion (about 10%) of the DON formed chloramines. This finding was in keeping with the hypothesis that much of the DON was in relatively unreactive peptide bonds. Thus, much of the measured CRC in our samples must have been the fish toxin NH_2Cl.[1]

For the once-through cooling water chlorination process, it is possible to provide FRC in the condenser without adding as much chlorine as would be necessary to exceed the long-term breakpoint. Because of the decline in FRC during the first few minutes, it is advantageous to chlorinate as close to the condensers as mixing and other engineering considerations permit. Chlorination at levels below the long-term breakpoint introduces several interesting kinetic considerations. There is a competition between fulvic acid demand and chloramine-forming nitrogen for the added chlorine. There is also a fairly predictable minimum FRC consumption, which is almost immediate. Subsequent competition for the remaining FRC that occurs within the next few minutes between fulvic acid consumption and dichloramine formation determines the amount of persistent combined residual chlorine.

ACKNOWLEDGMENTS

We appreciate the contributions of Guy Inman, Grace Brashear, Barbara Wustenhagen, Theodore Walters, and R. D. Moss. This research was supported by the U.S. Nuclear Regulatory Commission under Contract NRC-04-77-119, Dr. Phillip Reed, contract officer.

REFERENCES

1. Brungs, W. A. "Effects of Residual Chlorine on Aquatic Life," *J. Water Pollut. Control Fed.* 45(10):2180-2193 (1973).
2. Lietzke, M. H. "A Kinetic Model for Predicting the Composition of Chlorinated Water Discharged from Power Plant Cooling Systems," in *Water Chlorination: Environmental Impact and Health Effects, Vol. 2,* R. L. Jolley, H. Gorchev, and D. H. Hamilton, Eds. (Ann Arbor, MI: Ann Arbor Science Publishers, Inc., 1978), pp. 707-716.

3. Moss, R. D., H. B. Flora II, R. A. Hiltunen, and C. V. Seaman. *Chlorine Minimization/Optimization for Condenser Biofouling Control: Final Report*, EPA-600/7-80-143 (Washington, DC: U.S. Environmental Protection Agency, 1980).
4. *Standard Methods for the Examination of Water and Wastewater*, 14th ed. (Washington, DC: American Public Health Association, 1976).
5. Morris, J. C., and R. A. Isaac. "A Critical Review of Kinetic and Thermodynamic Constants for the Aqueous Chlorine-Ammonia System," in *Water Chlorination: Environmental Impact and Health Effects, Vol. 4,* R. L. Jolley, W. A. Brungs, J. A. Cotruvo, R. B. Cumming, J. S. Mattice, and V. A. Jacobs, Eds. (Ann Arbor, MI: Ann Arbor Science Publishers, Inc., 1983), pp. 49-62.
6. Palin, A. T. "A Study of the Chloro Derivatives of Ammonia and Related Compounds, with Special Reference to Their Formation in the Chlorination of Natural and Polluted Waters," *Water Water Eng.* 5(656):151-200, 248-256 (1950).
7. Qualls, R. G., and J. D. Johnson. "Kinetics of the Short-term Consumption of Chlorine by Fulvic Acid," *Environ. Sci. Technol.* 17(11):692-698 (1983).
8. Haag, W. R., and M. H. Lietzke. *A Kinetic Model for Predicting the Concentration of Active Halogen Species in Chlorinated Saline Cooling Waters: A Final Report*, ORNL/TM-7942 (Oak Ridge, TN: Oak Ridge National Laboratory, 1981).
9. Zielke, R. L., H. B. Flora II, and S. K. Macey. "Validation of a Kinetic Model to Predict Total Residual Chlorine in Freshwater," in *Water Chlorination: Environmental Impact and Health Effects, Vol. 3,* R. L. Jolley, W. A. Brungs, and R. B. Cumming, Eds. (Ann Arbor, MI: Ann Arbor Science Publishers, Inc., 1980), pp. 445-451.
10. Wetzel, R. G. *Limnology.* (Philadelphia: W. B. Saunders Co., 1975).
11. Weiss, C. M., and E. J. Kuenzler. "The Trophic State of North Carolina Lakes," Report #119 (Raleigh, NC: Water Resources Research Institute, University of North Carolina), p. 215.
12. Tuschall, J. R., and P. L. Brezonik. "Characterization of Organic Nitrogen in Natural Waters: Its Molecular Size, Protein Content and Interactions with Heavy Metals," *Limnol. Oceanogr.* 25:495 (1980).
13. Pereira, W. W., Y. Hoyano, R. E. Summons, V. A. Bacon, and A. M. Duff. "Chlorination Studies II. The Reaction of Aqueous Hypochlorous Acid with Amino Acids and Dipeptides," *Biochem. Biophys. Acta* 313:170-180 (1973).
14. Helz, G. R., D. A. Dotson, and A. C. Sigleo. "Chlorine Demand: Studies Concerning Its Chemical Basis," in *Water Chlorination: Environmental Impact and Health Effects, Vol. 4,* R. L. Jolley, W. A. Brungs, J. A. Cotruvo, R. B. Cumming, J. S. Mattice, and V. A. Jacobs, Eds. (Ann Arbor, MI: Ann Arbor Science Publishers, Inc., 1982), pp. 181-200.
15. Isaac, R. A., and J. C. Morris. "Modeling of Reactions Between Aqueous Chlorine and Nitrogenous Compounds," in *Water Chlorination: Environmental Impact and Health Effects, Vol. 4,* R. L. Jolley, W. A. Brungs, J. A. Cotruvo, R. B. Cumming, J. S. Mattice, and V. A. Jacobs, Eds. (Ann Arbor, MI: Ann Arbor Science Publishers, Inc., 1982), pp. 63-75.
16. Johnson, J. D., and R. G. Qualls. "A Kinetic Model for the Chlorination of Power Plant Cooling Waters," NUREG/CR-2806 (Washington, DC: U.S. Nuclear Regulatory Commission, 1983).
17. Margerum, D. W., E. T. Gray, and R. P. Huffman. "Chlorination and the Formation of N-chloro Compounds in Water Treatment," in *Organometals and Organo-*

metalloids. Occurrence and Fate in the Environment, American Chemical Society Symposium Series Vol. 82, F. E. Brickman and J. M. Bellama, Eds. (Washington, DC: American Chemical Society, 1979), pp. 278-291.
18. Johnson, J. D., and G. W. Inman, Jr. "Cooling Water Chlorination: The Kinetics of Chlorine, Bromine, and Ammonia in Sea Water," NUREG/CR-1522 (Washington, DC: U.S. Nuclear Regulatory Commission, 1982).
19. White, G. C. *Handbook of Chlorination* (New York: Van Nostrand Reinhold Company, 1972).

CHAPTER 58

Seawater Chlorination: Influence of Ammonia Concentration

Jean Marie Fiquet

For biofouling prevention, especially where mussels are involved, in the condenser cooling systems of French power-generating plants, continuous chlorination is used at chlorine concentrations between 0.5 and 1 mg/L when seawater temperature is above 10°C. In seawater containing very low levels of ammonia it is now well known that chlorine reacts with bromides to give hypobromous acid and hypobromite.[1-6] In nonpolluted seawater, the ammonia concentration is between 2.5 and 50 µgL (0.15 to 3 × 10^{-6} M).[7] High ammonia concentrations can occur, for example, in harbor waters. Khalanski found 0.2 to 3.5 mg/L ammonia at Dunkirk power plant,[8] and our own measurements in Le Havre power plant gave concentrations to 1.5 mg/L. It is important to know the reaction products formed during cooling water chlorination because monochloramine is more persistent than combined bromine and may be the major component in the effluents.[4-9] Furthermore, the toxicities of free bromine, combined bromine, and combined chlorine are different.[10-14]

Inman and Johnson[15] and Courtot and Péron[16] have already studied these reactions, although at higher levels than those used in practice, and their results are rather qualitative. Computer models have also been elaborated by Sugam and Helz[3,17] and Haag and Lietzke.[18,19] The latter proposed the possible formation of bromochloramine, a mixed haloamine also considered by Trofe et al.[20]

The object of this study is to quantify the residual oxidant concentrations known as free bromine, combined bromine and combined chlorine, using N,N-diethyl-p-phenylenediamine (DPD), and to compare seawater chlorination and seawater bromination.

EXPERIMENTAL

Reagents

Low-Chlorine-Demand Seawater

Seawater collected at the Paluel Nuclear Plant was filtered through 0.2-µm Millipore filters. It was chlorinated with 5 mg/L Cl_2 for at least 24 h and dechlorinated with coconut charcoal. It was then stored in a brown bottle protected against ammonia contamination by a sulfuric scrubber. The pH was 8.00 ± 0.05 and the salinity 31.6 ± 0.2 ‰.

Hypochlorite Solution

A chlorine stock solution (~250 mg/L as Cl_2) was prepared by diluting pure sodium hypochlorite (10°) with demand-free water. It was used freshly prepared after determining its titer by iodometric titration.

Hypobromite Solution

The bromine stock solution (~250 mg/L as Cl_2) was prepared by diluting a concentrated sodium hypobromite solution obtained by reaction of bromine with soda[21] at a temperature below 5°C to avoid bromate formation. The stock solution was also used freshly prepared and its titer determined by iodometric titration in the presence of sulfuric acid.

Ammonium Chloride Solution

A stock ammonium chloride solution (100 mg/L as NH_4^+, 5.5 μM) was prepared by dissolving 0.297 g of ammonium chloride (ultra pure grade) in demand-free water and diluting to 1 L.

Reagent for DPD Titration

The usual reagents for Palin's method were used.[22] However, no mercuric chloride was added to the buffer solution.

Method

Fiquet[23] showed that iodide at normal concentrations in seawater has a catalytic effect on the interference of monochloramine in free chlorine measurements by the diethyl-*p*-phenylenediamine (DPD) colorimetric titration method. Thus color development is recorded vs time. Oxidant concentrations were calculated using the following mathematical expression:

$$C = C_o + a\, b^{ct} \tag{1}$$

At t = 0, the expression gives the oxidant concentration (i.e., the free and combined bromine); at t = ∞, the total oxidant concentration (i.e., free oxidant and combined chlorine) is calculated.

Procedure

Avoiding contact of seawater with ambient air, 100 mL of seawater was sampled at room temperature by weighing to the nearest centigram. The volume of ammonium chloride solution required to obtain the desired concentration of ammonia and the volume of sodium hypochlorite or hypobromite required to obtain a final concentration of halogen of 1 mg/L as Cl_2 are added sequentially to the 100-mL seawater sample.

Taking $t = o$ for the beginning of halogen injection, the reaction mixture was gently stirred for 30 s; then the solution was allowed to settle for a total contact time of 1 min. The same test was repeated for 10- and 20-min contact times. After the appropriate contact time, the entire reaction mixture volume was added to the DPD reagents.

The absorption of the colored reaction mixture was measured at 510 nm using 10-mm cells and a Beckman UV 24 recording spectrophotometer. Zero time of the recording was assumed to be the beginning of sample and reagents mixing.

CHLORINATION OF SEAWATER WITH AMMONIA CONCENTRATION BELOW 0.15 mg/L

The breakpoint reaction

$$3Br_2 + 2NH_3 \rightarrow 6Br^- + 6H^+ + N_2 \quad (2)$$

involves a molar bromine-to-ammonia ratio of 1.5. Therefore, for ammonia concentrations below 0.15 mg/L, a chlorine dosage of 1 mg/L Cl_2 corresponds to an excess of oxidant.

Reaction Involved

Figure 1 shows that residual bromine concentration (no combined chlorine was found) linearly decreases when the ammonia level increases when seawater containing ≤0.15 mg/L ammonia is chlorinated with a chlorine dosage of 1 mg/L as Cl_2.

Regression equations were determined for 1-, 10-, and 20-min reaction times. For 1-min reaction time and $[NH_4^+] \leq 0.08$ mg/L,

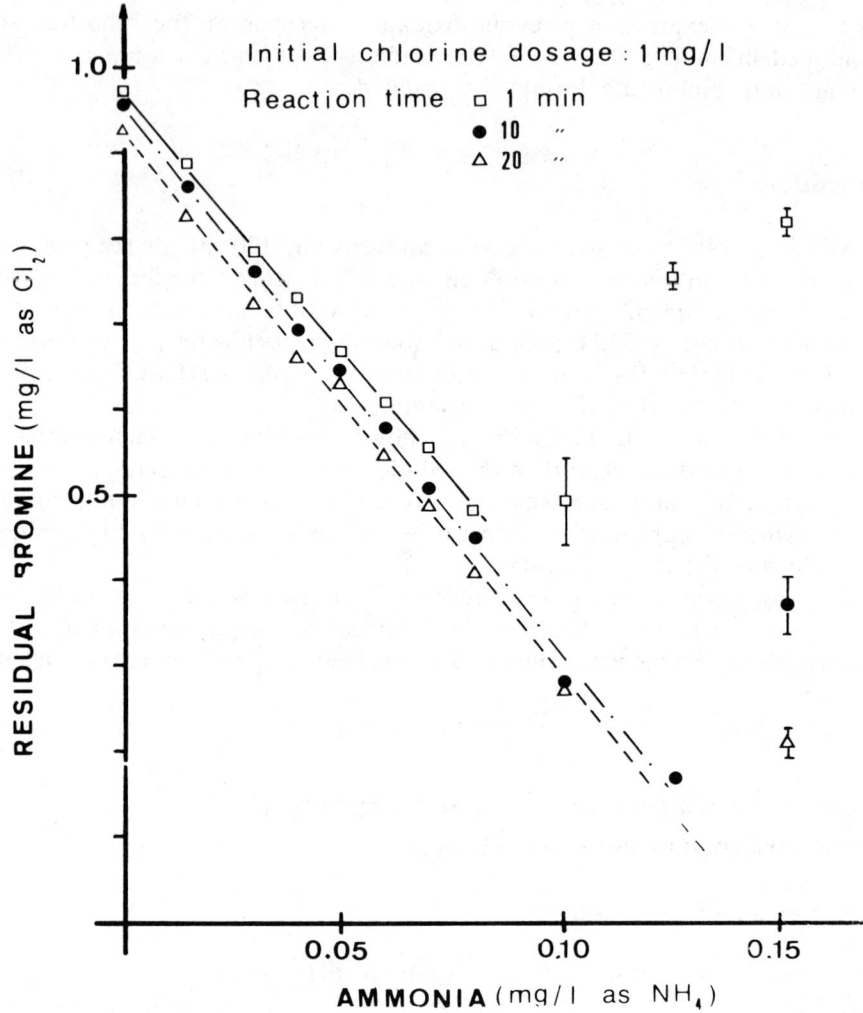

Figure 1. The residual bromine concentration when seawater is chlorinated at 1 mg/L level (as Cl_2) is shown as a function of initial ammonia concentration for different reaction times.

$$[Br_2]_1 = 0.975 - 6.08\ [NH_4^+]\ (r^2 = 0.986 \text{ for } 30\ DF) \tag{3}$$

where r = correlation coefficient and DF = degrees of freedom. For 10 min and $[NH_4^+] \leq 0.126$ mg/L,

$$[Br_2]_{10} = 0.958 - 6.44\ [NH_4^+]\ (r^2 = 0.990 \text{ for } 33\ DF) \tag{4}$$

For 20 min and $[NH_4^+] \leq 0.126$ mg/L,

$$[Br_2]_{20} = 0.926 - 6.35\,[NH_4^+] \quad (r^2 = 0.976 \text{ for } 40\ DF) \quad (5)$$

The slope of the straight lines permits the calculation of the molar ratio (R) for the reaction between bromine and ammonia. Table I shows that the ratios are close to stoichiometric (i.e., 1.5). Thus we can say that there is a total destruction of the ammonia and that residual bromine could be present as free bromine. Inman and Johnson[15] did not find these linear relations with either free or combined bromine (tribromamine). As we did not study the NBr_3 and DPD reaction we cannot make conclusions about the presence of combined bromine.

Comparison with Seawater Bromination

Figure 2 shows similar results when seawater is brominated with a bromine dosage of 1 mg/L (as Cl_2). Regression equations were determined for 1-, 10-, and 20-min reaction times. For 1-min reaction time and $[NH_4^+] \leq 0.080$ mg/L,

$$[Br_2]_1 = 0.969 - 5.79\,[NH_4^+] \quad (r^2 = 0.977 \text{ for } 28\ DF) \quad (6)$$

For 10 min and $[NH_4^+] \leq 0.126$ mg/L,

$$[Br_2]_{10} = 0.952 - 5.96\,[NH_4^+] \quad (r^2 = 0.995 \text{ for } 36\ DF) \quad (7)$$

For 20 min and $[NH_4^+] \leq 0.126$ mg/L,

$$[Br_2]_{20} = 0.922 - 5.96\,[NH_4^+] \quad (r^2 = 0.988 \text{ for } 36\ DF) \quad (8)$$

The calculated Br/N molar ratios for the ammonia and bromine reactions are given in Table II. These values are equal to the stoichiometric ratio 1.5 and are quite similar to those obtained with chlorination. Therefore, we concluded that seawater chlorination does not lead to combined chlorine formation when the molar ratio Cl/N is greater than 2.

Table I. Molar Ratio (R) of Cl/N Calculated for Different Reaction Times Between Chlorine and Ammonia When Seawater Is Chlorinated

Reaction time (min)	R
1	1.54 ± 0.06
10	1.64 ± 0.06
20	1.61 ± 0.08

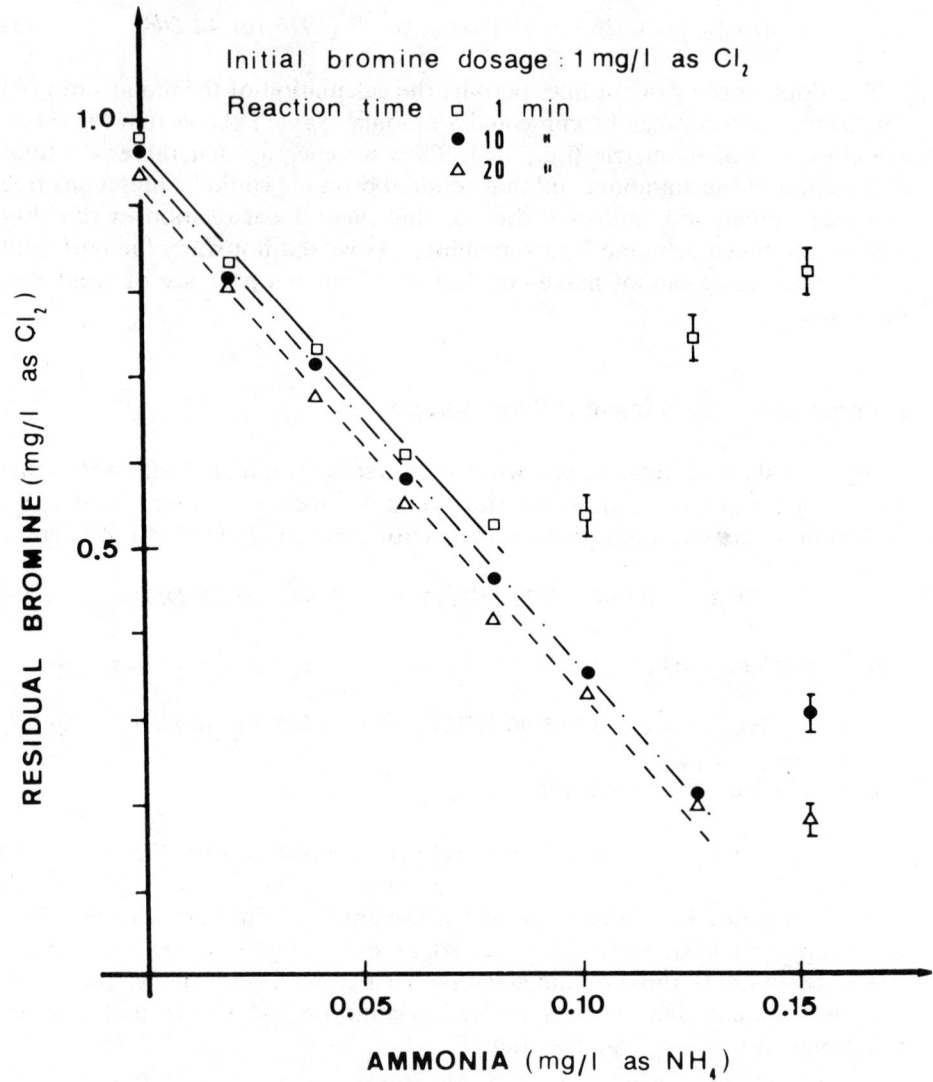

Figure 2. The residual bromine concentration when seawater is brominated at 1 mg/L (as Cl_2) is shown as a function of initial ammonia concentration for different reaction times.

Table II. Molar Ratio (R) of Br/N Calculated for Different Reaction Times Between Chlorine and Ammonia When Seawater Is Brominated

Reaction time (min)	R
1	1.47 ± 0.09
10	1.52 ± 0.04
20	1.52 ± 0.06

Kinetics of the Reaction

The reaction between ammonia and bromine is very fast when the molar ratio is greater than 3. This is not true when this ratio is near the stoichiometric value. In the latter case there is probably coexistence of free and combined bromine, the decomposition of which occurs slowly. Thus, a zero residual bromine concentration could never be obtained.

CHLORINATION OF SEAWATER WITH AMMONIA CONCENTRATION EQUAL TO OR GREATER THAN 0.15 mg/L

Composition of the Residual Oxidant Concentration

In contrast to the above results, when seawater containing ≥ 0.15 mg/L ammonia is chlorinated with 1 mg/L chlorine, combined chlorine, or something reacting with DPD like combined chlorine, is found. Figures 3 through 5 give the composition of the residual oxidant concentration as a function of initial ammonia concentration.

Combined Chlorine

The combined chlorine concentration increases with the ammonia concentration and is always lower than the bromine concentration. It is only at an ammonia concentration of 0.5 mg/L (Cl/N molar ratio = 0.5) that there is the same concentration of combined chlorine and bromine.

Combined Bromine

Since there is a competition between bromide and ammonia to react with chlorine, the combined bromine concentration shows a maximum value. This maximum shifts to higher levels when the reaction time increases.

Total Oxidant

For 1-min reaction time, the total oxidant concentration is close to the theoretical value of 1 mg/L as the Cl/N molar ratio drops below 1.5; but for 10- and 20-min reaction times, the total oxidant concentration differs significantly from that value, proving the instability of the components. The data also show that a high ammonia concentration tends to stabilize the combined oxidants and thus to increase the total residual oxidant concentration. This observation agrees with the conclusions of Inman and Johnson.[15]

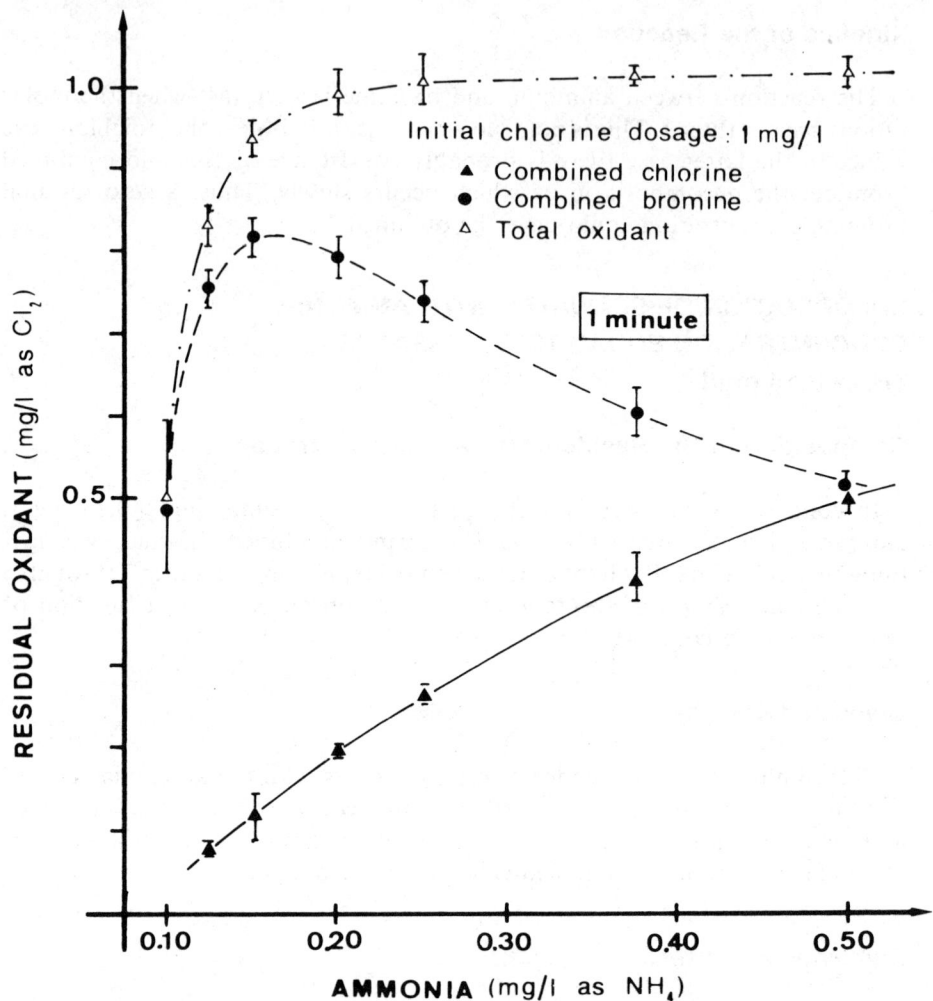

Figure 3. The composition of the residual oxidant concentration at 1 min is shown as a function of initial ammonia concentration.

Kinetics

Combined Chlorine

Combined chlorine formed during seawater chlorination is not stable. From the data it can not be concluded whether the reaction is zero, first, or second order (Table III). However, by considering the possible rate equations, appropriate constants and ammonia concentrations can be calculated. The general equation for a first-order reaction is the following:

$$\text{Ln } [Cl_2] = \text{Ln } [Cl_2]_0 + \beta t \tag{9}$$

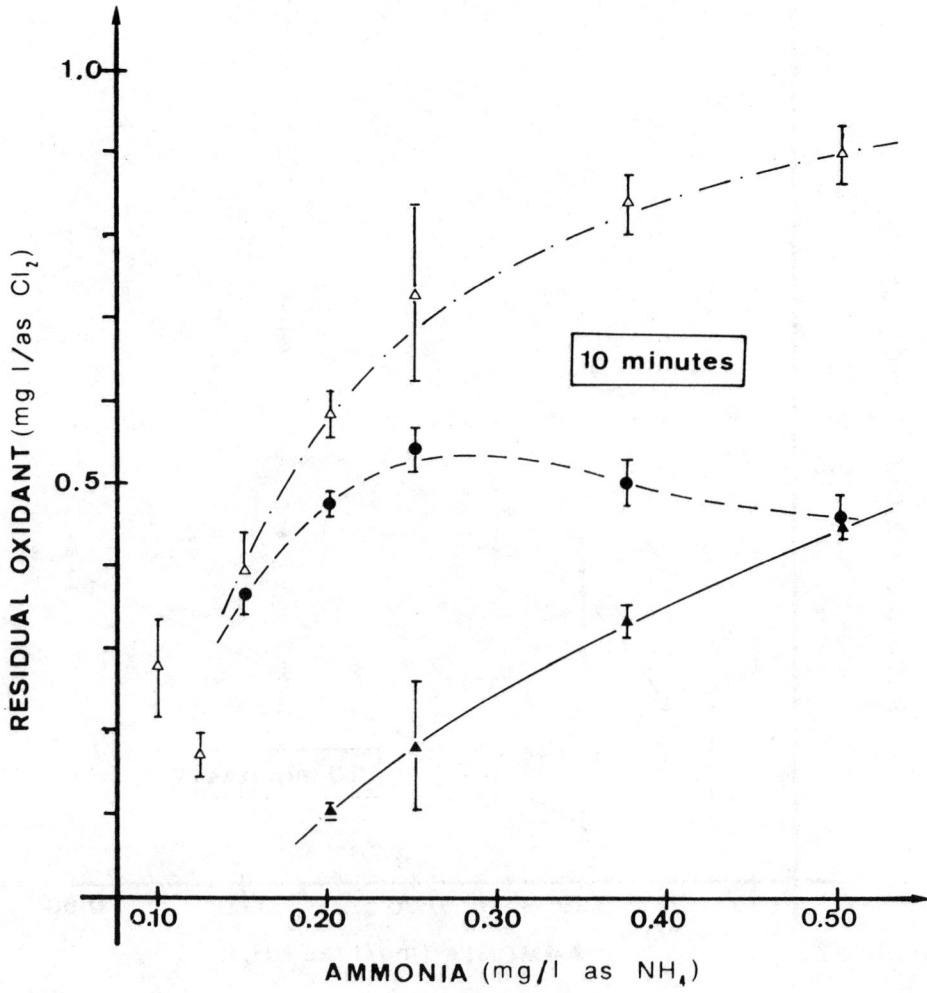

Figure 4. The composition of the residual oxidant concentration at 10 min is shown as a function of initial ammonia concentration.

If the formation of combined chlorine is a first-order reaction, then

$$\text{Ln}[Cl_2]_0 = -0.348\,[NH_4^+]^{-0.948} \quad (r^2 = 0.999 \text{ for } 2\ DF) \tag{10}$$

$$\text{and } \beta = -0.0050\,[NH_4^+]^{-1.54} \quad (r^2 = 0.988) \tag{11}$$

The general equation for a second-order reaction is the following:

$$\frac{1}{[Cl_2]} = \frac{1}{[Cl_2]_0} + \beta t \tag{12}$$

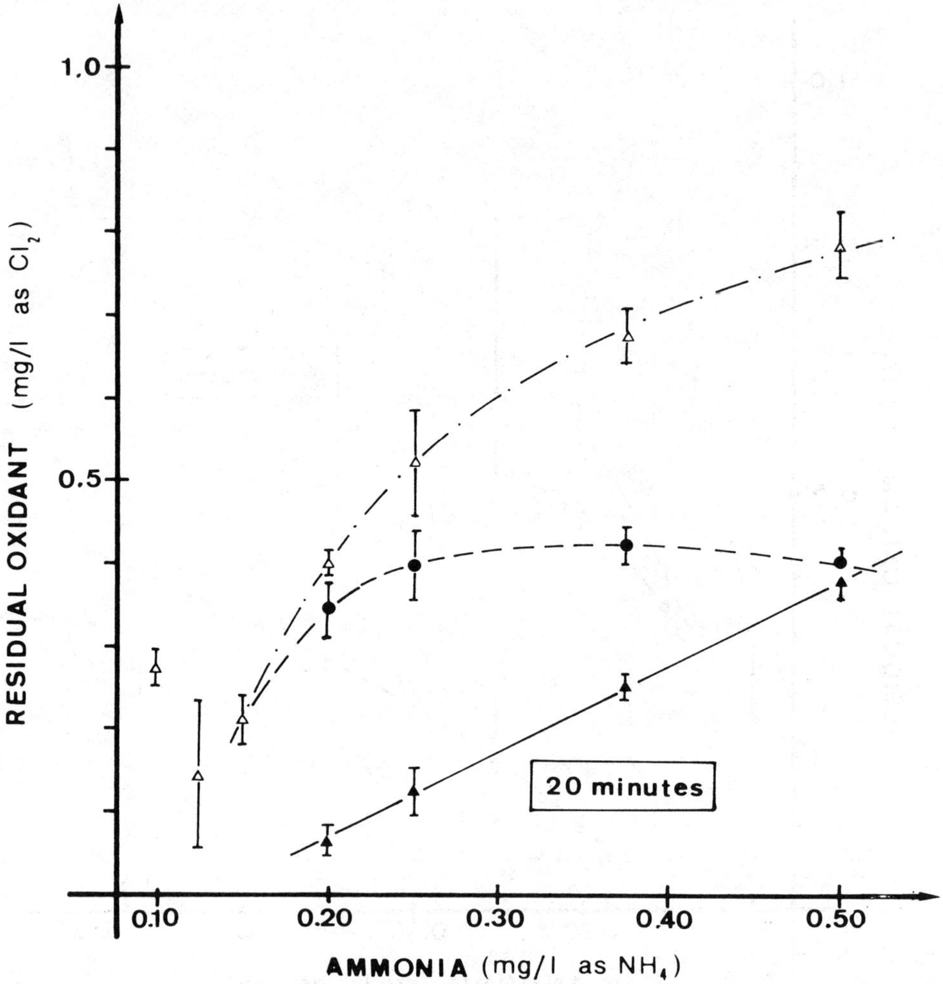

Figure 5. The composition of the residual oxidant concentration at 20 min is shown as a function of initial ammonia concentration.

If the formation of combined chlorine is a second-order reaction, then

$$\frac{1}{[Cl_2]_0} = \frac{1}{0.731 + 0.316 \, Ln[NH_4^+]} \quad (r^2 = 0.997 \text{ for } 2 \, DF) \qquad (13)$$

$$\beta = 0.0038 \, [NH_4^+]^{-3.05} \quad (r^2 = 0.994) \qquad (14)$$

Table III. Combined Chlorine Decomposition: Fitting of Data Points to Zero-, First-, and Second-Order Rate Laws (Seawater Chlorination)

Ammonia concentration (mg/L)	Zero-order reaction rate		First-order reaction rate		Second-order reaction rate		Degrees of freedom
	$[Cl_2]$	r^2	$Ln[Cl_2]$	r^2	$\frac{1}{[Cl_2]}$	r^2	
0.202	0.27 − 0.0075 t	0.931	−1.60 − 0.059 t	0.971	4.36 + 0.55 t	0.961	9
0.252			−1.28 − 0.040 t	0.935	3.46 + 0.22 t	0.921	9
0.376	0.41 − 0.0080 t	0.964	−0.865 − 0.025 t	0.961	2.32 + 0.08 t	0.946	11
0.501	0.51 − 0.0060 t	0.935	−0.677 − 0.014 t	0.933	1.96 + 0.03 t	0.926	13

Table IV. Persistence of Monochloramine in Seawater

Reaction time (min)	0.25 mg/L NH$_2$Cl (as Cl$_2$)			0.50 mg/L NH$_2$Cl (as Cl$_2$)		
	m[a]	s[b]	n[c]	m[a]	s[b]	n[c]
1	0.25	0.009	4	0.48	0.024	5
10	0.24	0.015	5	0.46	0.021	5
20	0.23	0.019	6	0.47	0.014	5

[a]m, Mean value.
[b]s, Standard deviation.
[c]n, Number of measurements.

This empirical model permits the determination of the combined chlorine concentration when seawater containing ammonia concentrations between 0.2 and 0.5 mg/L is chlorinated with 1 mg/L Cl$_2$ for reaction times below 20 min. What kind of combined chlorine is formed under these experimental conditions? Monochloramine, the most probable one, is generally considered as being stable in seawater because the reaction with bromide is slow at such low concentrations.[20] We checked this assertion with two concentrations of NH$_2$Cl: 0.25 and 0.50 mg/L as Cl$_2$. Table IV shows that the monochloramine concentration is independent of contact time.

Why does the combined chlorine found during seawater chlorination not have the same stability? There are several possible hypotheses:

- The monochloramine used in the experiment was prepared with a large excess of ammonium ions that could have a stabilizing effect. Ammonia excess is much lower during seawater chlorination.
- Monochloramine may react with combined bromine to give bromochloramine that may not react with DPD. This is not probable, because bromochloramine should have oxidant properties similar to that of combined chlorine and combined bromine and thus react rapidly with DPD.
- The possible reaction between monochloramine and combined bromine leads to the disappearance of both compounds according to the following reaction:

$$NHBr_2 + NH_2Cl \rightarrow N_2 + 2HBr + HCl \tag{15}$$

These different possibilities have to be studied to know the true chemistry of seawater chlorination.

Combined Bromine

As with combined chlorine, the statistical studies do not permit conclusions to be made about the order of the decomposition reaction of combined bromine (i.e., dibromamine, according to Imman and Johnson).[15] Table V shows that, in some cases, both first- and second-order reactions can be considered.

Table V. Combined Bromine Decomposition: Fitting of Data Points to First- and Second-Order Rate Law (Seawater Chlorination)

Ammonia concentration (mg/L)	First-order reaction rate		Second-order reaction rate		Degrees of freedom
	Ln [Br²]	r^2	$\dfrac{1}{[Br_2]}$	r^2	
0.152			0.97 + 0.185 t	0.971	10
0.202			1.19 + 0.086 t	0.978	9
0.252	−0.269 − 0.033 t	0.990	1.27 + 0.061 t	0.985	9
0.376	−0.489 − 0.019 t	0.952	1.62 + 0.037 t	0.950	11
0.502	−0.644 − 0.013 t	0.903	1.90 + 0.029 t	0.903	13

Table VI. Combined Bromine Decomposition: Fitting of Data Points to a Second-Order Rate Law (Seawater Bromination)

Ammonia concentration (mg/L)	Equation $\dfrac{1}{[Br_2]}$	r^2	Degrees of freedom
0.152	0.99 + 0.234 t	0.965	17
0.202	1.05 + 0.116 t	0.978	13
0.252	1.04 + 0.085 t	0.980	13
0.376	1.03 + 0.058 t	0.974	13
0.501	1.04 + 0.046 t	0.961	14

If the formation of combined bromine is a second-order reaction, the general rate equation is the following:

$$\frac{1}{[Br_2]} = \frac{1}{[Br_2]_0} + kt \qquad (16)$$

The rate constants are correlated with ammonia concentration according to:

$$\frac{1}{[Br_2]_0} = 2.38 + 0.76 \, Ln[NH_4^+] \quad (r^2 = 0.987 \text{ for 3 } DF) \qquad (17)$$

and

$$\frac{1}{k} = 51.4 + 24.75 \, Ln[NH_4^+] \quad (r^2 = 0.997 \text{ for 3 } DP) \qquad (18)$$

This leads to the following empirical model for combined bromine during seawater chlorination at 1 mg/L Cl_2 for ammonia concentration between 0.15 and 0.50 mg/L and for contact time less than 20 min:

$$\frac{1}{[Br_2]} = 2.38 + 0.76 \, Ln[NH_4^+] + \frac{t}{51.4 + 24.75 \, Ln[NH_4^+]} \qquad (19)$$

Comparison with Seawater Bromination

In the case of seawater bromination only combined bromine is formed. Figure 6 shows that residual bromine increases with ammonia concentration showing the same tendency as with chlorination. However, with chlorination the total oxidant level is always higher than during bromination.

Combined bromine (probably dibromamine) decomposes with time according to the second-order rate law. Table VI gives the equations developed to fit the data.

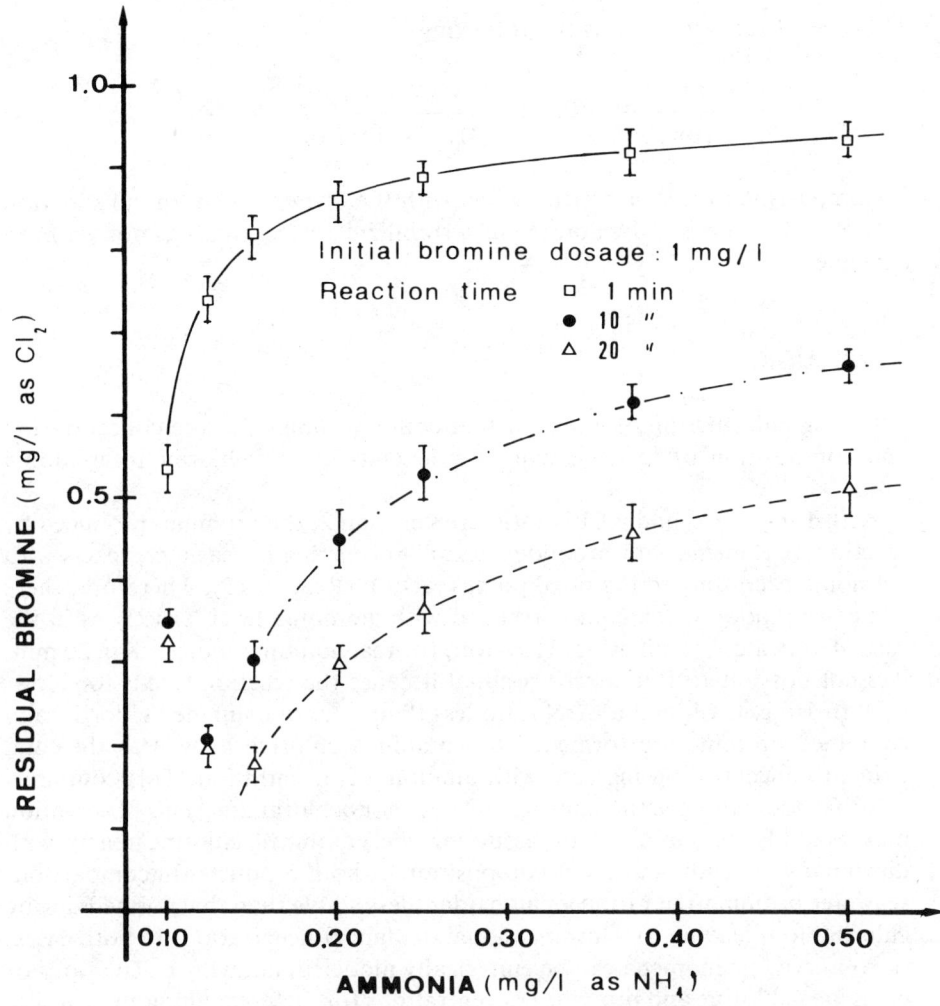

Figure 6. the residual bromine concentration when sea water is brominated at 1 mg/L (as Cl_2) is shown as a function of initial ammonia concentration for different reaction times.

In the general second-order relationship given in Equation (16) k is correlated with ammonia levels, and $1/[Br_2]_0$ is considered constant and equal to 1.03. Thus, the following relationship was determined for k:

$$k^{-1} = 31.8 + 14.6 \text{ Ln } [NH_4^+] \; (r^2 = 0.999 \text{ for 3 } DF) \qquad (20)$$

The overall rate equation is the following:

$$\frac{1}{[Br_2]} = 1.03 + \frac{t}{31.8 + 14.6 \, Ln \, [NH_4^+]} \tag{21}$$

Comparison of this empirical model with the one obtained for chlorination indicates that combined chlorine has a stabilizing effect that has not yet been explained.

CONCLUSION

During chlorination of seawater that contains ammonia, the concentration and composition of residual oxidants depends on the chlorine-to-ammonia ratio.

With 1 mg/L Cl_2 and a Cl:N ratio greater than 2, the bromine, produced by reaction of chlorine with bromide naturally occurring in seawater, reacts with ammonia according to the breakpoint reaction (Reaction 2). Therefore, there is a consumption of oxidant correlated with ammonia level, exactly as in the case of seawater bromination. However, for reaction time shorter than 20 min, it is not possible to find a zero residual because the reaction rate is too slow.

With 1 mg/L Cl_2 and a Cl:N ratio less than 2, both combined chlorine and combined bromine are formed. The combined chlorine as well as the total oxidant concentration increase with ammonia concentration. This combined chlorine does not have the same stability as monochloramine. This observation may possibly be explained by assuming the combined chlorine reacts with dibromamine, leading to the decomposition of both products. In comparison, seawater bromination produces an oxidant less stable than that formed during chlorination, leading to a lower residual oxidant concentration. In both cases, the observed phenomena can be empirically modeled, allowing calculations of combined chlorine and bromine concentrations for different ammonia concentrations and contact time shorter than 20 min. However, this model does not take into account any chlorine demand for species other than ammonia.

REFERENCES

1. Lewis, B. G. *Chlorination and Mussel Control. The Chemistry of Seawater. A Review of the Literature,* CERL Report No. RD/L/N 106/66. (Leatherhead, England: Central Electricity Research Laboratories, 1966).
2. Dove, R. A. *Reaction of Small Dosage of Chlorine in Seawater,* Research Report 42/70. File No. 0.307 0/ID, Job No. 10665 CEGB. (Southampton, England: University of Southampton, 1970).
3. Sugam, R., and G. R. Helz. "Speciation of Chlorine Produced Oxidants in Marine Waters. Theoretical Aspects," *Chesapeake Sci.* 18:113 (1977).

4. Johnson, J. D. "Analytical Problems in Chlorination of Saline Waters," *Chesapeake Sci.* 18:116 (1977).
5. Fiquet, J. M. "Contribution à l'Étude du Dosage du Chlore dans l'Eau de Mer," *TSM L'eau*. 239-245 (April 1978).
6. Peron, A., and J. Courtot-Perez. "Dosage du Brome par le Rouge de Phénol. Application au Dosage du Brome dans l'Eau de Mer," *Analysis* 6(9):389-394 (1978).
7. Riley, J. P. "Analytical Chemistry of Seawater," in *Chemical Oceanography*, J. P. Riley, and G. Skirrow, Ed. (London: Academic Press, 1975), p. 409.
8. Khalanski, M. "Structure et Production du Phytoplancton du Port de Dunkerque. Incidence du Fonctionnement de la Centrale Thermique EDF," in *2èmes Journées de la Thermoécologie*, (Paris: Electricité de France, Direction de l'Equipement, 1981), pp. 621-649.
9. Mills, J. F. "Bromine Chloride Can Treat Once Through Cooling Water," *Power* 127-129 (May 1980).
10. Morgan, R. P., and E. J. Carpenter. "Biocides," in *Power Plant Entrainment: A Biological Assessment*, J. R. Schubel and B. C. Marcy, Eds. (New York: Academic Press, Inc., 1978).
11. Capuzzo, J. M. "The Effect of Temperature on the Toxicity of Chlorinated Cooling Waters to Marine Animals. A Preliminary Review," *Mar. Pollut. Bull.* 10(2):45-47 (1979).
12. Capuzzo, J. M. "The Effects of Free Chlorine and Chloramine on Growth and Respiration Rates of Larval Lobsters," *Water Res.* 11:1021-1024 (1977).
13. Capuzzo, J. M., J. C. Goldman, J. A. Davidson, and S. A. Lawrence. "Chlorinated Cooling Waters in the Marine Environment: Development of Effluent Guidelines," *Mar. Pollut. Bull* 8(7):161-163 (1977).
14. Mattice, J. S. "Power Plant Discharges: Toward More Reasonable Effluent Limits on Chlorine," *Nucl. Safety* 18(16):802-819 (1977).
15. Inman, G. W., and J. D. Johnson. "The Effect of Ammonia Concentration on the Chemistry of Chlorinated Seawater," in *Water Chlorination: Environmental Impact and Health Effects, Vol. 2*, R. L. Jolley, H. Gorchev, and D. H. Hamilton, Eds. (Ann Arbor, MI: Ann Arbor Science Publishers, Inc., 1978), pp 235-252.
16. Courtot, J., and A. Péron. *Etude Physicochimique de la Chloration de l'Eau de Mer en Présence d'Azote Ammoniacal: Nature et Évolution des Haloamines Formées*, Rapport EDF-DER HE/33 79.11, (Paris: Electricité de France, 1979).
17. Sugam, R., and G. R. Helz. "Seawater Chlorination: A Description of Chemical Speciation," in *Water Chlorination: Environmental Impact and Health Effects, Vol. 3*, R. L. Jolley, W. A. Brungs, and R. B. Cumming, Eds. (Ann Arbor, MI: Ann Arbor Science Publishers, Inc., 1980), pp. 427-433.
18. Haag, W. R., and M. H. Lietzke. "A Kinetic Model Predicting the Concentrations of Active Halogen Species in Chlorinated Saline Cooling Waters," in *Water Chlorination: Environmental Impact and Health Effects, Vol. 3*, R. L. Jolley, W. A. Brungs, and R. B. Cumming, Eds. (Ann Arbor, MI: Ann Arbor Science Publishers, Inc., 1980), pp. 415-426.
19. Haag, W. R., and M. H. Lietzke. *A Kinetic Model for Predicting the Concentrations of Active Halogen Species in Chlorinated Saline Cooling Waters. A Final Report*, ORNL/TM-7942 (Oak Ridge, TN: Oak Ridge National Laboratory, 1981).
20. Trofe, T. W., J. D. Johnson, and G. W. Inman. *The Kinetics of Monochloramine*

Decomposition in the Presence of Bromides, NUREG/CR 1116 (Washington, DC: U.S. Nuclear Regulatory Commission, 1980).
21. Charlot, G. *Les Methodes de la Chimie Analytique. Analyse Quantitative Minérale* (Paris: Masson et Cie., Ed., 1966), p. 467.
22. Palin, A. T. "Analytical Control of Water Disinfection with Special Reference to Differential D.P.D. Methods for Chlorine, Chlorine Dioxide, Bromine, Iodine and Ozone," *J. Inst. Water Engr.* 20:139–154 (1974).
23. Fiquet, J. M. "Dosage de la Monochloramine à la N.N. Diethyl-p-phénylène Diamine: Influence de Traces d'Iodures," *TSM L'eau* 243–249 (May 1982).

CHAPTER 59

Chlorination Kinetics of Surface and Deep Tropical Seawater

Francis J. Sansone and Terrence J. Kearney

Chlorination is well established as an effective and convenient means of controlling biofouling by macrofauna in power plant heat exchangers that use marine or estuarine cooling water.[1,2] Chlorination has also been proposed for control of microbial film accumulation in heat exchangers used in ocean thermal energy conversion (OTEC) plants.[3,4] These plants, which are designed to use the steep surface-temperature gradient in the tropical ocean as their energy source, will require very large, highly efficient heat exchangers to be used with warm (surface) and cold (deep) seawater. Successful OTEC operation will require heat exchanger surfaces to be essentially bare metal (i.e., free of bacterial attachment).[3]

Because very large volumes of seawater will be processed by such plants [about 900 m^3s^{-1} for a 100-MW(e) plant],[5] it is desirable to predict the reaction kinetics and fate of the added chlorine. Previous studies on seawater chlorination kinetics, however, have dealt only with temperate or semitropical coastal seawater;[1,6-9] to our knowledge, no data are available concerning the chlorination kinetics of oligotrophic tropical seawater.

We present here measurements of the kinetics of the loss of free and combined chlorine-produced oxidants (CPOs) and the production of halocarbons in surface and deep tropical seawater; these reactions are about a hundredfold slower than those previously reported for temperate and semitropical coastal seawater. Our results imply environmental effects much different from those predicted from nontropical studies and illustrate the need for further work on tropical seawater chemistry to enable accurate assessment of the impact of OTEC chlorination on tropical marine environments.

MATERIALS AND METHODS

Seawater for these experiments was obtained from the surface (8 m depth) and deep (600 m) seawater pumping systems at the Natural Energy Laboratory of Hawaii (NELH), Keahole Point, Island of Hawaii;[10] in Table I, the characteristics of these oligotrophic tropical waters are summarized and compared with typical values for temperate coastal seawater. Keahole Point seawater is typical of open-ocean tropical seawaters: surface and deep waters both contain

Table I. Characteristics of NELH Oligotrophic Tropical Seawater and Temperate Coastal Surface Seawater[a]

Seawater Type	Depth (m)	Temperature (°C)	$NO_2 + NO_3$ (μM)	NH_4 (μM)	DON[b] (μM, as N)	TDP[c] (μM, as P)	TOC[d] (μM, as C)
NELH Surface	8	24–28	0.05–0.47	0.27–1.3	3.5–4.9	0.19–0.50	51–106
NELH Deep	600	5.5–6.5	38–42	0.06–0.76	0.75–1.5	2.9–3.2	19–61
Temperate Coastal Surface	0–30	2–28	0.1–8	0.2–2.3	8–40	0.3–10	50–480

[a] Data compiled from the following sources: NELH temperature, Reference 23; other data, T. Walsh, personal communication, 1983; temperate coastal $NO_2 + NO_3$ and NH_4, References 24, 25; DON, Reference 26; TDP, Reference 27; TOC, Reference 28.
[b] DON, Dissolved organic nitrogen.
[c] TDP, Total dissolved phosphorus.
[d] TOC, Total organic carbon.

very low levels of dissolved organic carbon (DOC) and ammonia; surface water, although depleted of inorganic nutrients, is enriched with dissolved organic nitrogen (DON); deep water is depleted of DON but is enriched with inorganic nutrients, except ammonia. These differences appear to explain the major variations in chlorination kinetics in different seawaters.

Chlorination kinetics were studied by adding 10 μL of dilute hypochlorite solution to 40-mL samples of NELH seawaters in glass containers with Teflon-lined caps, storing at ~22°C in the dark, and then monitoring changes in dissolved chemical constituents. The initial hypochlorite concentration was 3.4 μM (250 μg/L). Samples were carefully shielded from UV radiation, sunlight, or fluorescent lights; the success of these measures was indicated by the fact that the UV-oxidation product, bromate,[2,11,12] was not observed during these experiments (data not shown). Bromate was measured directly by differential pulse polarography.[11,12]

CPOs were determined by using differential pulse polarography to measure the loss of excess phenylarsine oxide added to terminate individual time-course experiments: "free available oxidants" (hypohalous acids, hypohalites) were measured at pH 6.5 to 7.5 without KI addition; "total residual oxidants" (halamines, etc.) were measured at pH 3.5 to 4.5 after addition of 50-mg KI per 10-mL sample.[13]

Inorganic nutrient analyses were made with a Technicon Autoanalyzer II using standard spectrophotometric techniques;[14] DON was measured by oxidizing samples with UV light and measuring the increase in dissolved nitrate and nitrite. DOC was measured using an Oceanography International Model 0524B Total Carbon System.[15]

Similar time-course experiments were conducted to measure the production rates of halocarbons (one- and two-carbon halogenated hydrocarbons) in chlorinated NELH seawaters. Halocarbons were measured by pure and trap gas chromatography (GC) (Varian 3700) using a halogen-selective Hall 700A electrolytic conductivity detector (Tracor) and a Tekmar LSC-2 liquid sample concentrator (U.S. EPA Method 601).[16] GC peak assignments were confirmed with using a Finnigan OWA GC/Mass Spectrometer (GC/MS).

RESULTS AND DISCUSSION

Kinetics of CPO Reduction

Data from the CPO time-course experiments are shown in Figure 1; the most striking feature is the very slow kinetics measured. This is particularly true for free available oxidants, which in the few previously reported studies (all using coastal temperate seawater) decreased to a few percent of the added chlorine level within an hour or less.[7,8] In contrast, total residual oxidants, which is the class of chlorination-induced oxidants most often reported in the literature, persist about as long as has been reported for other natural

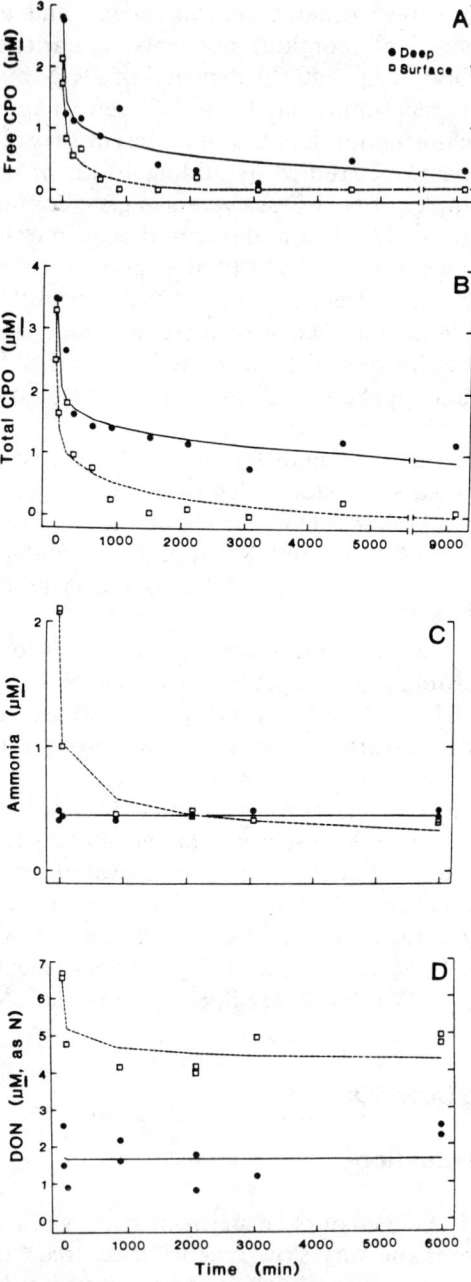

Figure 1. Time variation of (A) free available oxidants, (B) combined residual oxidants, (C) ammonia, and (D) dissolved organic nitrogen in chlorinated surface and deep NELH seawater. Standard deviation of the mean of quadruplicate samples in (A) and (B) are smaller than the plotted symbols; data in (C) and (D) are means of duplicate samples.

waters.[2,7,9,11,12,17] Our residual oxidant results are also consistent with the 5-min residual oxidant demand values previously reported for Keahole Point waters.[4]

During these experiments, the sum of dissolved nitrite and nitrate remained constant for both surface and deep water (data not shown) despite large differences in initial concentration. Dissolved ammonia, however, decreased from 2 to 0.45 μM in the surface water but remained at 0.45 μM in the deep water (Figure 1). This result is not surprising, because monochloramine production from ammonia and hypochlorous acid is a second-order reaction, first-order in each reactant.[18] DON showed a similar effect (Figure 1): DON declined steadily in the surface water but did not vary consistently from its lower initial values in the deep water.

From the above data, it appears likely that nitrogen–halogen reactions are important in controlling the kinetics and extent of halogen loss in oligotrophic tropical seawater. Indeed, on a molar basis, ammonia and DON removal in the chlorinated surface water could account for nearly all of the observed oxidant disappearance. It is also possible that the slower kinetics observed in the deep water is a reflection of the lower ammonia and DON levels originally present. The importance of nitrogen–halogen reactions is underscored by the extremely low production of halocarbons from organic carbon.

Halocarbon Production Kinetics

Only the seawater halocarbon end product bromoform[7,19] displayed net accumulation during our experiments (Figure 2); its slow rate of production provided another contrast with the much faster halogenation reactions

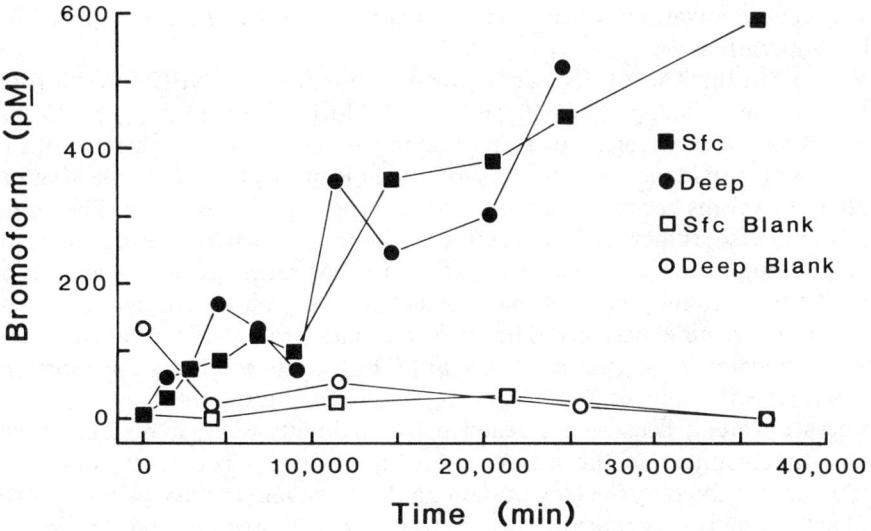

Figure 2. Time variation of bromoform in chlorinated and nonchlorinated (Blank) surface (Sfc) and deep NELH seawater.

reported for other chlorinated marine waters. For example, Crane et al.[20] found that total halocarbon production in estuarine water was complete within 30 min after addition of 5 mg/L chlorine; in contrast, after several thousand minutes halocarbon production was not complete in chlorinated NELH waters (Figure 2).

Additions of 3.4 μM hypochlorite resulted in the production of 600 and 500 pM of bromoform in NELH surface and deep seawater, respectively, during our experiments (Figure 2). These values are much lower than the halocarbon formation reported previously for other chlorinated natural waters in which halocarbon levels corresponded to 0.05 to 5% of the added chlorine.[4,6,7] We attribute these results to the very low organic carbon content of these waters (Table I) and the resultant lack of organic halocarbon precursors.

Implications for OTEC Biofouling Control

The above experiments are useful in explaining the unusual findings of ongoing OTEC biofouling tests at NELH using surface and deep tropical seawater. These latter studies have demonstrated that microbial film production from surface seawater can be eliminated on heat exchanger surfaces by chlorinating for 1 h/d at levels less than 70 ppb.[21] This contrasts with the experience at coastal temperate powerplants where daily treatments of one to several parts per million chlorine are required to control macrofaunal fouling. Even higher dosage rates would presumably be required in these latter systems to prevent colonization of heat exchanger surfaces by bacteria, as required for OTEC operations.[3,21]

We suggest that the low levels of chlorination required at NELH for biofouling control are a direct result of the lack of chlorine demand (e.g., oxidizable organic carbon or nitrogen, or ammonia) in tropical seawater, which results in very slow free oxidant loss; the chlorine added is more effective at inactivating fouling organisms because there are fewer competing side reactions. The implications of these studies are important in estimating the environmental effect of OTEC plants on tropical marine systems, in that much less chlorine will be needed for biofouling control than predicted from coastal temperate powerplant chlorination experience. However, the data presented here also suggest that the chlorine levels chosen for use in OTEC plants will have a proportionally greater environmental effect on tropical receiving waters than on coastal temperate waters, because the reactive free oxidants (e.g., hypohalites) will persist much longer in the former. Assessment of environmental impact is further complicated by the lack of data on the acute and chronic effects of free oxidants on marine organisms, although there are limited data on the toxicity of chlorine-produced residuals.[5,8,22]

ACKNOWLEDGMENTS

We thank T. H. Daniel and B. Lee of NELH for their support and assistance, and S. Kansako and R. Goo of the Sand Island (Honolulu) Wastewater Treatment Plant for access to their GC/MS. This study was supported by U.S. Department of Energy and State of Hawaii funds administered through the Hawaii Natural Energy Institute (DOE grant DE-FG03-81ER10250). Hawaii Institute of Geophysics Contribution No. 1615.

REFERENCES

1. White, G. "Current Chlorination and Dechlorination Practices in the Treatment of Potable Water, Wastewater and Cooling Water," in *Water Chlorination: Environmental Impact and Health Effects, Vol. 1,* R. Jolley, Ed. (Ann Arbor, MI: Ann Arbor Science Publishers, 1975), pp. 1-18.
2. Carpenter, J., C. Smith, and R. Zika. *Reaction Products from the Chlorination of Seawater,* PB81-172280 (Springfield, VA: National Technical Information Service, 1981), pp. 1, 11-16.
3. Yuen, P. *Ocean Thermal Energy Conversion: A Review* (Honolulu: Hawaii Natural Energy Institute, University of Hawaii, 1981), pp. 76-77.
4. Hartwig, E., and R. Valentine. "Bromoform Production in Tropical Open-Ocean Waters: Ocean Thermal Energy Conversion Chlorination," in *Water Chlorination: Environmental Impact and Health Effects, Vol. 4,* R. L. Jolley, W. A. Brungs, J. A. Cotruvo, R. B. Cumming, J. S. Mattice, and V. A. Jacobs, Eds. (Ann Arbor, MI: Ann Arbor Science Publishers, 1983), pp. 311-330.
5. Walsh, J. *The Potential Environmental Consequences of Ocean Thermal Energy Conversion (OTEC) Plants* (Washington, DC: U.S. Department of Energy, Office of Energy Research, 1981), pp. 4, 7.
6. Bean, R. "Recent Progress in the Organic Chemistry of Water Chlorination," in *Water Chlorination: Environmental Impact and Health Effects, Vol. 4,* R. L. Jolley, W. A. Brungs, J. A. Cotruvo, R. B. Cumming, J. S. Mattice, and V. A. Jacobs, Eds. (Ann Arbor, MI: Ann Arbor Science Publishers, 1983), pp. 843-870.
7. Helz, G., R. Sugam, and R. Hsu. "Chlorine Degradation and Halocarbon Production in Estuarine Waters," in *Water Chlorination: Environmental Impact and Health Effects, Vol. 2,* R. Jolley, H. Gorchev, and D. H. Hamilton, Jr., Eds. (Ann Arbor, MI: Ann Arbor Science Publishers, 1978), pp. 209-222.
8. Eppley, R., E. Renger, and P. Williams. "Chlorine Reactions with Seawater Constituents and Inhibition of Photosynthesis of Natural Marine Phytoplankton," *Estuarine Coastal Mar. Sci.* 4:147 (1976).
9. Wong, G., and J. Davidson. "The Fate of Chlorine in Sea-Water," *Water Res.* 11:971 (1977).
10. "Hawaii Deep Seawater Pipeline," *Eos* 63:141 (1982).
11. Macalady, D., J. Carpenter, and C. Moore. "Sunlight-Induced Bromate Formation in Chlorinated Seawater," *Science,* 195:1335 (1977).
12. Wong, G. "The Effects of Light on the Dissipation of Chlorine in Seawater," *Water Res.* 14:1263 (1980).

13. Smart, R., J. Lowry, and K. Mancy. "The Analysis for Ozone and Residual Chlorine by Differential Pulse Polarography of Phenylarsine Oxide," *Environ. Sci. Tech.* 13:89 (1979).
14. Strickland, J., and T. Parsons. *A Practical Handbook of Seawater Analysis*, 2nd ed. (Ottawa: Supply and Services Canada, 1977), pp. 121-134.
15. Smith, S., W. Kimmerer, E. Laws, R. Brock, and T. Walsh. "Kaneohe Bay Sewage Diversion Experiment: Perspectives on Ecosystem Responses to Nutritional Perturbation," *Pac. Sci.* 35:387 (1981).
16. Longbottom, J., and J. Lichtenberg. "Purgeable Halocarbons—Methods 601," in *Methods for Organic Chemical Analysis of Municipal and Industrial Wastewater* (Cincinnati: U.S. Environmental Protection Agency, 1982), pp. 601-1-601-10.
17. Duursma, E., and P. Parsi. "Persistence of Total and Combined Chlorine in Sea Water," *Neth. J. Sea Res.* 10:192 (1976).
18. Morris, J., and R. Isaac. "A Critical Review of Kinetic and Thermodynamic Constants for the Aqueous Chlorine-Ammonia System," in *Water Chlorination: Environmental Impact and Health Effects, Vol. 4,* R. L. Jolley, W. A. Brungs, J. A. Cotruvo, R. B. Cumming, J. S. Mattice, and V. A. Jacobs, Eds. (Ann Arbor, MI: Ann Arbor Science Publishers, 1983), p. 50.
19. Helz, G. R., and R. Y. Hsu. "Volatile Chloro- and Bromocarbons in Coastal Waters," *Limnol. Oceanogr.* 23:858 (1978).
20. Crane, A., S. Erickson, and C. Hawkins. "Contribution of Marine Algae to Trihalomethane Production in Chlorinated Estuarine Water," *Estuarine Coastal Mar. Sci.* 11:239 (1980).
21. Larsen-Basse, J. "Effect of Biofouling and Countermeasures on Heat Transfer in Surface and Deep Ocean Hawaiian Water—Early Results from the Seacoast Test Facility," in *ASME-JSME Thermal Engineering Joint Conference Proceedings*, Y. Mori and W. Yang, Eds., ASME Book I00158-B (New York: American Society of Mechanical Engineers, 1983), pp. 285-289.
22. Davis, W., and J. Fava. "Interaction of Aquatic Ecosystem Components with Chlorination: An Overview," in *Water Chlorination: Environmental Impact and Health Effects, Vol. 4,* R. L. Jolley, W. A. Brungs, J. A. Cotruvo, R. B. Cumming, J. S. Mattice, and V. A. Jacobs, Eds. (Ann Arbor, MI: Ann Arbor Science Publishers, 1983), pp. 791-796.
23. Noda, E. *OTEC Environmental Benchmark Survey Off Keahole Point, Hawaii,* Rept. 80-1 (Honolulu: Look Laboratory, University of Hawaii, 1980), pp. 3-26.
24. Vaccaro, R. "Inorganic Nitrogen in Sea Water," in *Chemical Oceanography, Vol. 1,* J. Riley and G. Skirrow, Eds . (London: Academic Press, 1965), pp. 383-390.
25. Riley, J., and R. Chester. *Introduction to Marine Chemistry* (London: Academic Press, 1971), pp. 161-164.
26. Duursma, E. "Dissolved Organic Carbon, Nitrogen, and Phosphorus in the Sea," *Neth. J. Sea Res.* 1:1 (1961).
27. Armstrong, F. "Phosphorus," in *Chemical Oceanography, Vol. 1,* J. Riley and G. Skirrow, Eds. (London: Academic Press, 1965), pp. 323-337.
28. Mackinnon, M. "The Measurement of Organic Carbon in Sea Water," in *Marine Organic Chemistry*, E. Duursma and R. Dawson, Eds. (Amsterdam: Elsevier Scientific, 1981), p. 416.

CHAPTER 60

Decomposition of Bromamines in Aqueous Solution: Preliminary Report of the Decomposition Kinetics and Disproportionation of NH_2Br

S. Pasquini Cristina, M. T. Azure, H. J. Workman, and E. T. Gray, Jr.

During the last 20 years, valuable information and insight have been added to the literature of the aqueous solution chemistry of bromamine systems, especially that of the ammoniacal bromamines.[1-4] Because of the complex and overlapping mechanisms involved in the formation, interconversion, and decomposition of aqueous bromamines, the best of these studies have restricted their scope in an effort to be as complete as possible within defined boundaries. Even so, the mechanisms deduced are typically capable of predicting bromamine behavior over little more than the first half-life. Furthermore, most of the rate constants for the interconversions of various bromamines are estimates.

This chapter marks the beginning of a concerted effort to attack the problem of bromamine kinetics through a broader approach. The use of excess amine and rapid mixing to produce bromamines has given encouraging results. We present here an initial experimental overview of the decomposition of ammoniacal bromamines as well as the kinetics and proposed mechanisms for the disproportionation of monobromamine. The data collected for the disproportionation reaction implies that general and specific acid-catalyzed hydrolysis of the monobromamine is the major pathway to dibromamine formation. An estimate of the uncatalyzed hydrolysis rate constant for monobromamine, coupled with the data of Wajon and Morris,[4] gives an estimate for the equilibrium constant for the formation of monobromamine from NH_3 and HOBr, and the oxidation potential for monobromamine. These results offer an alternative to the hypobromite pathway in the proposed mechanism for the formation of monobromamine.[4]

EXPERIMENTAL

Reagents

All solutions of bromine were generated as the hypobromite ion by the reaction of sodium bromide (Baker reagent grade) with sodium hypochlorite at pH 11. Sodium hypochlorite (Baker reagent grade) was standardized spectrophotometrically[5] at 292 nm using a molar absorptivity of 350 $M^{-1}\text{cm}^{-1}$. Solutions of sodium hydroxide were prepared from electrolytic pellets (Baker reagent grade, low carbonate) and standardized against KHP. Sodium perchlorate was prepared by neutralization of sodium bicarbonate with $HClO_4$ (Baker reagent grade). Ammonia, sodium iodide, sodium chloride, sodium hydrogen phosphate, sodium acetate, and sodium nitrate (Baker reagent grade) were used without further purification.

Apparatus

All spectrophotometric measurements and slower kinetic measurements were made using a Cary 219 interfaced to a Z-80 microprocessor (Tandy). Kinetics experiments on the Cary 219 were initiated using a hand-driven rapid-mixing unit and a glass Y-mixer mounted ahead of the 1.00-cm flow cell. All pH measurements were made using an Orion model 701 meter, a Beckman glass electrode (pH range 0–14), and a commercial calomel reference electrode. The reference electrode was bridged to the test solution from a saturated KCl solution through an agar-agar salt bridge (3 M $NaNO_3$). The meter was standardized using standard pH 6.86 and pH 4.01 buffers (Fisher), followed by $HClO_4$/NaOH titrations held at an ionic strength of 0.1 M $NaClO_4$. The standardization titrations resulted in the relationship $-\log[H^+] = 1.02$ pH $- 0.04$.

The stopped-flow instrument used in these experiments was constructed in this laboratory.[6] All measurements were carried out at 25.0 $\pm$ 0.1°C, and 0.1 M ionic strength was maintained with $NaClO_4$.

Solution Preparation

All bromamine solutions were prepared through a two-jet or a double two-jet mixer to ensure rapid attainment of homogeneity.

RESULTS AND DISCUSSION

Decomposition of Bromamines in Neutral and Basic Solution

Cromer et al.,[3] using initial rate methods, observed that the decomposition of ammoniacal bromamines at pH 7 and 8 was second order with respect to total bromamine but changed to 2.5 order at pH 6. To analyze as much of the reaction as possible, computer programs used to analyze first-order kinetics in our laboratory were modified to accomplish linear and nonlinear regressions for second-order-equal kinetics. The modified programs were used to

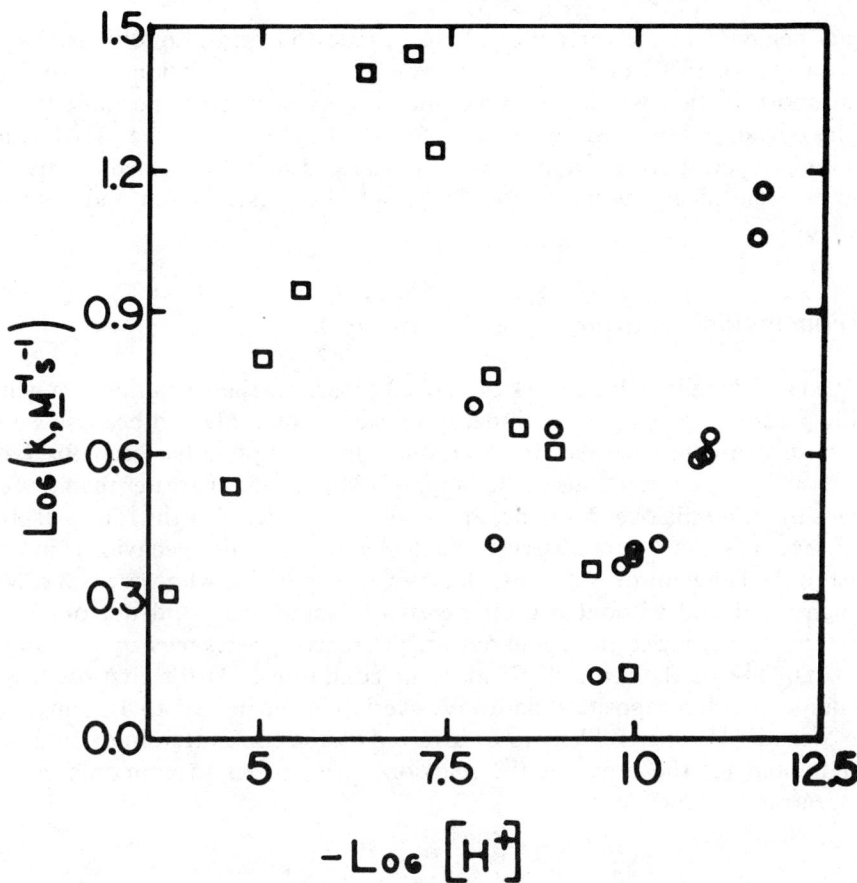

Figure 1. Plot of the log of the observed second-order rate constant vs $-\log[H^+]$ for the decomposition of bromamines (25°C, ionic strength = 0.10 M NaClO$_4$). Circles—[NH$_3$] = 0.0100 M, [phosphate buffer] = 0.0050 M; [bromine]$_{total}$ = 4.11 × 10^{-4} M. Squares—[NH$_3$] = 0.0050 M, [phosphate buffer] = 0.0050 M, [acetate buffer] = 0.0050 M, [bromine]$_{total}$ = 2.33 × 10^{-4} M. (Acetate buffer was used below $-\log[H^+]$ = 5.6, phosphate buffer was used in all other solutions.)

analyze the decomposition of bromamines over the pH range shown in Figure 1. Under these conditions, clean second-order-equal kinetics were observed for the last 90% of the reaction.

In the range of pH 9.3 to 9.5, the decomposition reaction followed second-order-equal kinetics over the entire reaction, but at higher or lower pH, an irregularity was observed at the beginning of the reaction. Therefore, the rate constants used for the pH profile were obtained by analyzing only the last 90% of the observed absorbance change of the reaction. It should be noted that the major bromamine species in solutions above pH 9 is NH$_2$Br, but below pH 7 is NHBr$_2$. The disproportionation reaction of monobromamine to dibromamine was initially believed responsible for the variation from simple second-order-equal behavior in the early part of the reaction in acidic conditions, but as shown below, this can only account for part of the variation, because the disproportionation is acid-catalyzed and is too fast to be responsible for the entire deviation observed below pH 6.5. The interference above pH 10 is not yet understood, because monobromamine should be fully formed at this pH, and no significant amount of $NHBr_2$ should be present, even under kinetic control.

Decomposition of Bromamines Below pH 3.5

We have done very little work at pH <3.5 because these solutions, in which NBr_3 predominates, are slow to decay in excess ammonia and because we do not wish to use initial rates. However, one important point has been observed: the decomposition reactions in the range of pH 1 to 2.5, in greater than tenfold excess of ammonia over bromine, are purely first order over the four half-lives that these reactions were observed. Such clean, first-order behavior is in contrast to the behavior of NBr_3 observed by Cromer et al.,[3] who were working at a higher pH and without a great excess of amine. They did not observe a unique reaction order and could not predict reaction rates over more than the first half-life of the reaction. Under our conditions, (1) the first-order and second-order decomposition pathways overlap in the pH 2.5 to 3.5 range, (2) the overlap pH range is likely to be affected by the concentration of ammonia $[NH_3]$, and (3) the order of the reaction with respect to ammonia can be concentration dependent.

Dependence of [NH$_3$] upon the Decomposition of Monobromamine in Base

Figure 2a shows the dependence of $[NH_3]$ upon the observed second-order rate constant for the decomposition of monobromamine at pH 12.7. Figure 2b shows a plot of the data which, if linear, would be consistent with a mechanism involving one term in inverse $[NH_3]$:

$$k_{obsd} = k'/[NH_3] \tag{1}$$

The curvature in this plot indicates that some higher inverse order in $[NH_3]$ must be invoked. A successful fit is presented in Figure 2c, which represents an expression such as

$$k_{obsd} = \frac{k'}{[NH_3]} + \frac{k''}{[NH_3]^2} \tag{2}$$

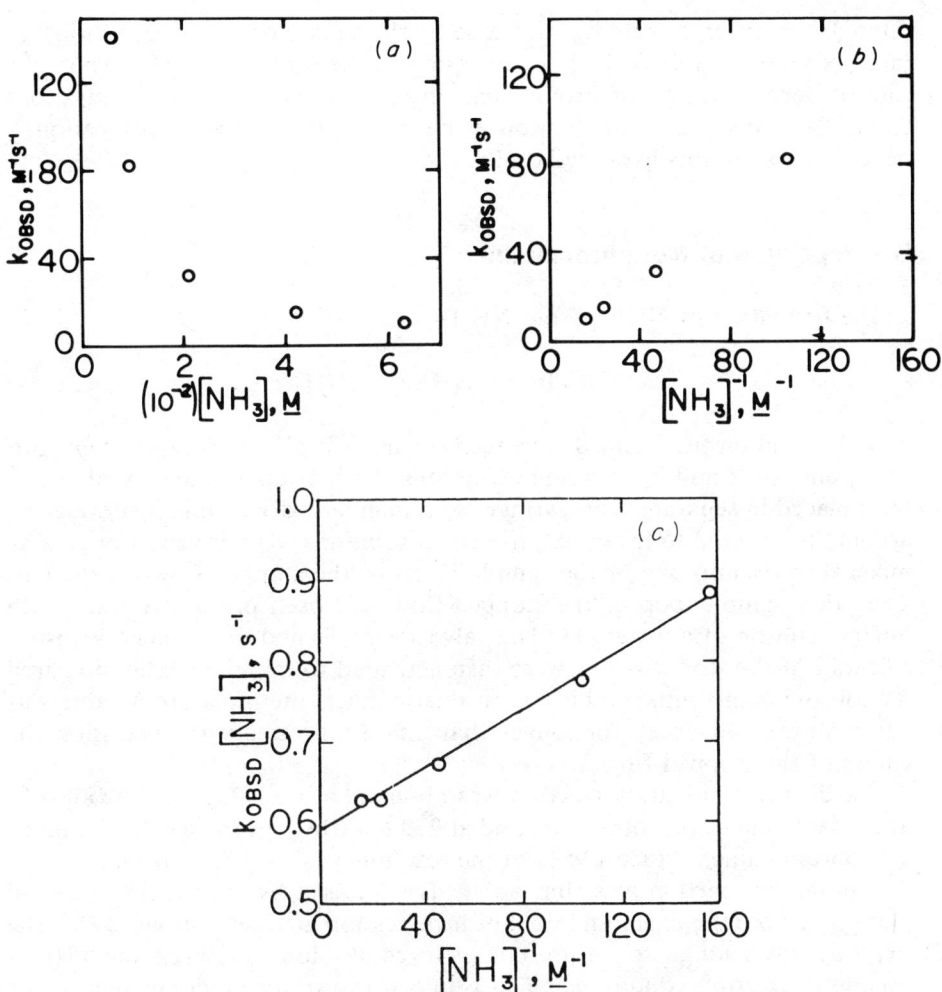

Figure 2. Plots to show the nature of the dependence of [NH$_3$] on the observed second-order rate constant for the decomposition of monobromamine ($-\log[H^+]$ = 12.7, ionic strength = 0.10 M NaClO$_4$, [bromine]$_{total}$ = 2.28 × 10^{-4} M, [Br$^-$] = 1.00 × 10^{-2} M, wavelength = 279 nm, 25.0°C).

or, rearranging

$$k_{obsd}[NH_3] = k' + k''/[NH_3] \tag{3}$$

This mixed-order relationship results in some interesting kinetic predictions. As the [NH$_3$] decreases, the decomposition rate will decrease as the square of the [NH$_3$]. However, if a solution of monobromamine and ammonia is simply diluted, the observed rate will decrease by the square of the concentration of monobromamine [NH_2Br] while increasing by the square of [NH_3]. Thus, the rate of decomposition of bromamines in strong base at low concentrations should be independent of dilution at constant pH. Further investigation is necessary to test this hypothesis fully.

Disproportion of Monobromamine

The formation of $NHBr_2$ from NH_2Br,

$$2NH_2Br \rightarrow NHBr_2 + NH_3 \qquad (4)$$

was observed on the stopped-flow instrument by a pH-jump experiment. Initially, ammonia and hypobromite solutions, both at approximately pH 10.5, were placed in separate drive syringes of a manually driven mixing device. To accomplish a jump to lower pH, these two solutions were driven through a Y-mixer directly into one of the sample loops of the stopped-flow instrument. The other sample loop of the stopped-flow had been previously filled with buffer solution of a lower pH. The valve solenoids and the pneumatic piston solenoid of the stopped-flow were then activated to mix the freshly prepared monobromamine with the buffer. In this manner, the monobromamine was never allowed to decay for longer than the 3 s necessary to reposition the valves of the stopped-flow.

The disproportionation reaction was observed at both 232 nm (the absorbance maximum of dibromamine) and at 279 nm (the absorbance maximum of monobromamine). At least 95% of the reaction was used for kinetic analysis. The observed reaction was first order, as witnessed by the random residual pattern of both the linear and nonlinear regression analyses, provided that the pH was low enough to ensure complete conversion to dibromamine (i.e., pseudo-irreversible conditions). This is in contrast to the predominance of the second-order reaction in the disproportionation of monochloramine at high concentrations.[7] It is also in contrast to the results recently presented by Inman and Johnson,[8] who determined that the reaction was second order in total bromamine at pH 7.00 using initial rate methods. When our reaction data are subject to analysis using the integrated form of a second-order-equal mechanism, a nonrandom pattern of residuals results, clearly showing that the reactions we are observing are not second order in total bromine concentration. The major difference between our experiments and those of Inman and Johnson is that they observed only the initial rate directly after mixing ammonia and bromine solutions, both of which are initially at pH 7.00. Regardless, it is not possible to rationalize the difference in the observed order with respect to bromine (or, in our case, bromamine) at this time. We also observed that the disproportionation reaction was independent of [NH_3]. The dependence on the concentration of the buffers will be discussed below.

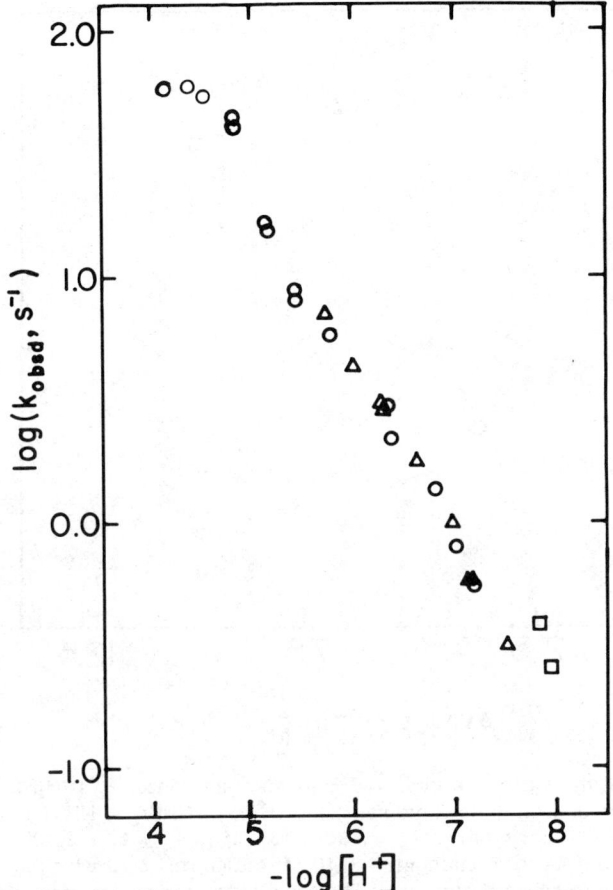

Figure 3. A plot of the log of the observed first-order rate constant vs log[H$^+$] for the disproportionation of monobromamine, $2NH_2Br \rightarrow NHBr_2 + NH_3$ (25.0°C, wavelength = 232 nm, [Br$^-$] = 2.5 × 10^{-3} M, ionic strength = 0.10 M NaClO$_4$). Circles—[bromine]$_{total}$ = 2.00 × 10^{-4} M, [NH$_3$] = 2.41 × 10^{-3} M, [phosphate buffer] = [acetate buffer] = 5.0 × 10^{-3} M. Triangles—[bromine]$_{total}$ = 2.00 × 10^{-4} M, [NH$_3$] = 2.41 × 10^{-3} M, [phosphate buffer] = 5.0 × 10^{-3} M. Squares—[bromine]$_{total}$ = 2.62 × 10^{-4} M, [NH$_3$] = 4.55 × 10^{-3} M, [phosphate buffer] = 5.0 × 10^{-3} M.

A pH profile of the data for the disproportionation reaction is shown in Figure 3. The reason for the "roller-coaster" shape of this plot becomes clearer when the dependence of [phosphate buffer] is investigated, as shown in Figure 4. This dependence on the [buffer] strongly implies general acid catalysis by the buffer. Figure 3 is, then, the result of two overlapping "S" curves, because the data in Figure 3 were obtained with both phosphate and acetate buffers. Although the dependence on phosphate is small, it is definitely present, supporting the findings of Inman and Johnson.[8]

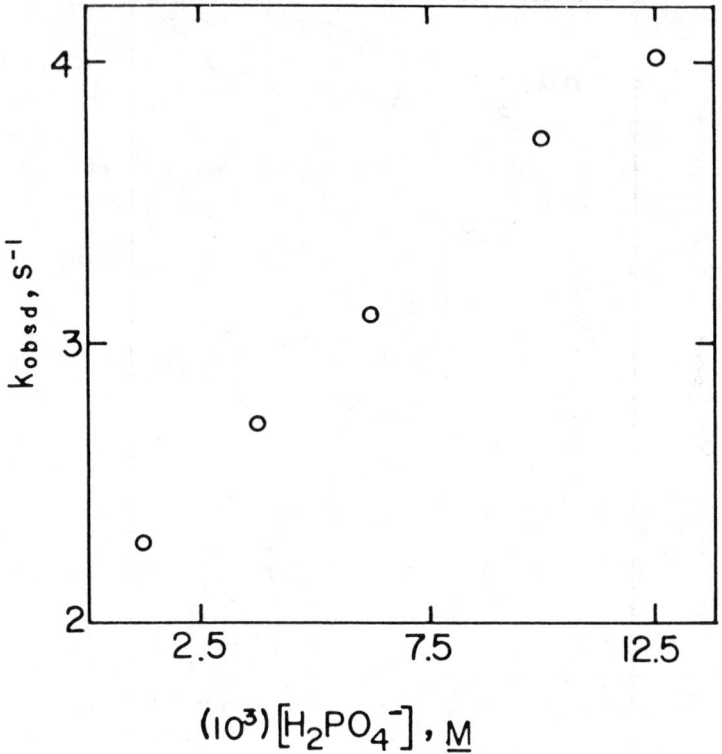

Figure 4. A plot of the log of the observed first-order-rate constant vs [$H_2PO_4^-$] for the disproportionation of monobromamine, $2NH_2Br \rightarrow NHBr_2 + NH_3$ (25.0°C, wavelength = 232 nm, [bromine]$_{total}$ = 2.00 × 10^{-4} M, [NH_3] = 1.21 × 10^{-3} M, [Br^-] = 2.5 × 10^{-3} M, ionic strength = 0.10 M NaClO$_4$, pH 6.70 which is the pK_a of $H_2PO_4^-$ under these conditions). The slope at low phosphate concentration is k_1 and the intercept is $k_5 + k_4K_3[H^+]$. The curvature at high concentrations of general acid is caused by the changing of the rate-determining step from protonation of NH_2Br to hydrolysis of NH_3Br^+.

Because our observed reactions of the formation of dibromamine are first order, the pH profile and the phosphate dependence are consistent with a rate-determining step composed of (1) general-acid catalysis of the hydrolysis of monobromamine, followed by a fast step in which the uncombined bromine can react with monobromamine to form dibromamine; (2) specific-acid catalysis in which the hydrolysis of NH_3Br^+ is probably rate determining, and (3) direct hydrolysis of NH_2Br. Thus, the intercept of Figure 4 would represent the phosphate-independent segment of the reaction, which is the hydrolysis of unprotonated NH_2Br and specific acid catalysis leading to the hydrolysis of NH_3Br^+. The slope at low [buffer] represents the general acid catalysis by $H_2PO_4^-$, and the curvature at higher [buffer] represents the shifting of the rate-determining step from protonation of NH_2Br to hydrolysis of NH_3Br^+. This implies the rate law

$$d[NHBr_2]/dt = (k_5 + k_4K_3[H^+] + k_1[H_2PO_4^-])[NH_2Br] \quad (5)$$

When acetic acid is present, a second general acid term must be added, $k_2[HOAc][NH_2Br]$. The general shape of the pH profile is indicative of the standard "S" curve for the region of each buffer. This rate law (Equation 5) is taken from the following scheme showing the mechanism, rate law, and calculated constants for the disproportionation of monobromamine as deduced from the data shown in Figure 3.

$$H_2PO_4^- + NH_2Br \rightleftarrows NH_3Br^+ + HPO_4^- \qquad K_1 = k_1/k_{-1} \quad (6)$$

$$HOAc + NH_2Br \rightleftarrows NH_3Br^+ + acetate^- \qquad K_2 = k_2/k_{-2} \quad (7)$$

$$H^+ + NH_2Br \rightleftarrows NH_3Br^+ \qquad K_3 \quad (8)$$

$$H_2O + NH_3Br \rightarrow HOBr + NH_4^+ \qquad k_4 \quad (9)$$

$$H_2O + NH_2Br \rightarrow HOBr + NH_3 \qquad k_5 \sim 0.3 \text{ s}^{-1} \quad (10)$$

$$\text{Rate} = \{k_5 + k_4K_3[H^+] + k_1[H_2PO_4^-] + k_2[HOAc]\}[NH_2Br] \quad (11)$$

when $k_{-1}[HPO_4^-]$ and $k_{-2}[acetate^-]$ are both much less than k_4, thus making k_1 and k_2 the rate determining steps of their pathways, respectively.

Although we are still in the process of carefully resolving these rate constants, a few points can be made at this juncture. The linearity of the data in Figure 4 indicate that the proton transfer from the $H_2PO_4^-$ general acid to monobromamine is rate determining, indicating that monobromamine must be a much stronger acid than NH_3Br^+. The "S" curve in the pH region of the acetic acid buffer implies that the same argument could be made, implying that the pK_a of NH_3Br^+ is well below the pK_a of acetic acid. The value of k_5, the uncatalyzed hydrolysis of NH_2Br, can be estimated to be ~ 0.3 s^{-1} from observed rate constant approached at high pH in Figure 3, since the concentration of the general acid is very small, as is the concentration of the specific acid ($[H^+]$).

The equilibrium constant for the formation of monobromamine from NH_3 and HOBr can be estimated using the hydrolysis constant determined in this work and the formation rate constant of Wajon and Morris.[4] From a calculated value for the formation equilibrium constant and the electrochemical potential of HOBr, electrochemical potentials for NH_2Br under various conditions can be estimated. Table I is a summary of these values for the bromamine and the comparable chloramine systems.

Table I. Summary of Rate Constants, Equilibrium Constants, and Standard Reduction Potentials for Selected Chlorine and Bromine Systems (25°C)[a]

System		X	
		Cl[b]	Br[c]
1. $NH_3 + HOX \rightleftarrows NH_2X + H_2O$	k_f	$2.9 \times 10^6 \ M^{-1}s^{-1}$	$7.4 \times 10^7 \ M^{-1}s^{-1}$ [d]
	k_r	$1.9 \times 10^{-5} s^{-1}$	$0.3 \ s^{-1}$
	K	$1.5 \times 10^{11} \ M^{-1}$	$3 \times 10^8 M^{-1}$
2. $HOX + H^+ + 2e^- \rightarrow H_2O + X^-$ [e]	$E°$	1.49 v	1.3 v
3. $NH_2X + H_2O + 2e^- \rightarrow X^- + NH_3 + OH^-$ [f]	$E°$	0.75 v	0.7 v
4. $NH_2X + 2H^+ + 2e^- \rightarrow X^- + NH_4^+$ [g]	$E°$	1.44 v	1.4 v

[a] K_f for System 1 was evaluated at a temperature other than 25°C.
[b] Reference 10; ionic strength = 0.5 M $NaClO_4$ (except for System 2).
[c] This work except for K_f evaluation in System 1 and for System 2.
[d] Reference 4.
[e] Reference 11.
[f] 1 M NH_3 and 1 M OH^-.
[g] 1 M NH_4^+ and 1 M H^+.

Mechanism of the Formation Reaction of NH_2Br

Having estimated a value for the hydrolysis rate constant of NH_2Br, it is possible to offer a different mechanistic interpretation to the experimental data of Wajon and Morris,[4] concerning the formation reaction of NH_2Br. They proposed that both HOBr and OBr^- can lose bromine to ammonia to form NH_2Br. The OBr^- path was invoked to explain observed rate constant values that were too large to be reconciled by the HOBr pathway alone for pH >11. However, NH_2Br decomposes via reduction of the bromine to bromide quite rapidly above pH 11. Under conditions of low $[NH_3]$, as shown above, decomposition is even more rapid. Therefore, initial rate data were used by Wajon and Morris[4] to determine these formation rate constants. Consequently, it was not possible to show if monobromamine was fully formed in the experiment with pH >11. If the reaction is not pseudo-irreversible, the observed rate constant for Equation 12 is that shown in Equation 13:

$$NH_3 + HOBr \underset{k_r}{\overset{k_f}{\rightleftarrows}} NH_2Br + H_2O \qquad (12)$$

$$k_{obsd} = k_f[NH_3] + k_r \qquad (13)$$

If the experiments were indeed carried out under reversible conditions, the higher observed formation rate constants could be explained by considering the reversibility and Equation 7, making it unnecessary to invoke bromination by OBr^-.

Because the [NH$_3$] for the most basic data is not given by Wajon and Morris,[4] it is not possible to reanalyze the data based on reversibility. However, a qualitative determination of the ratio of NH$_2$Br to total uncombined bromine ([Br]$_{Total}$ = [HOBr] + [OBr$^-$]) can be made if the 2 x 10^{-4} M to 5 × 10^{-4} M range of ammonia concentrations used by Wajon and Morris at pH 7.1 and 10.4 is typical of the ammonia concentrations used at pH 12.5.

The overall equilibrium constant expression for Equation 12 can be expressed:

$$K[NH_3] = \frac{[NH_2Br]}{[HOBr]} \qquad (14)$$

At pH 12.5, using pK_a^{HOBr} = 8.50 (Reference 9) and letting [HOBr] = [Br]$_{Total}Q$, where $Q = [H^+]/(1 + K_a^{HOBr}[H^+])$, the value of Q is 10^{-4}. Therefore,

$$(10^{-4})K[NH_3] = \frac{[NH_2Br]}{[Br]_{Total}} \qquad (15)$$

and the ratio of combined-to-free bromine is 2.6 × 10^4[NH$_3$]. Using the 2 × 10^{-4} to 5 × 10^{-4} M range of ammonia concentrations, the ratio of combined-to-free bromine lies between 5.2 and 13, indicating that the formation rate constants at high base may have been measured under reversible conditions.

CONCLUSIONS

Under the condition of excess ammonia, the decomposition reactions of bromamines in aqueous solution can be studied using integrated rate expressions of simple order in bromamine, analyzing at least 90% of the observed reaction. Where NH$_2$Br and NHBr$_2$ predominate, the reactions follow second-order-equal kinetics when observed via any of the bromamine absorbance bands. Where NBr$_3$ predominates, the reaction follows first-order kinetics. When NBr$_3$ and NHBr$_2$ are both present in significant concentration, the reaction is complex, presumably a mixture of both pathways. At pH 12.7, the dependence of [NH$_3$] on the observed rate constant indicates two pathways, one dependent on [NH$_3$]$^{-1}$ and one dependent on [NH$_3$]$^{-2}$. The pH dependence above pH 5 is given, and the mechanism is still under scrutiny.

The disproportionation of NH$_2$Br to NHBr$_2$ has been investigated. The reaction is first order under the condition of excess ammonia and the reaction appears to exhibit both general and specific-acid catalysis.

REFERENCES

1. Johnson, J. D., and R. Overby. "Bromine and Bromamine Disinfection Chemistry," *J. Sanit. Eng. Div. Amer. Soc. Civ. Eng.* 97:617 (1971).
2. Inman, G. W., Jr., T. L. LaPointe, and J. D. Johnson. "Kinetics of Nitrogen Tribromide Decmposition in Aqueous Solutions," *Inorg. Chem.* 15:3037–3042 (1976).
3. Cromer, J. L., G. W. Inman, and J. D. Johnson. "Dibromamine Decomposition Kinetics," in *Chemistry of Wastewater Technology*, A. J. Rubin, Ed. (Ann Arbor, MI: Ann Arbor Science Publishers, Inc., 1978), pp. 213–225.
4. Wajon, J. E., and Morris, J. C. "Rates of Formation of N-Bromo Amines in Aqueous Solution," *Inorg. Chem.* 21:4258–4263 (1982).
5. Galal-Gorchev, H., and J. C. Morris. "Formation and Stability of Bromamide, Bromimide, and Nitrogen Tribromide in Aqueous Solution," *Inorg. Chem.* 71:899–905 (1965).
6. Gray, E. T., Jr., and H. J. Workman. "Simultaneous Kinetic Analysis of Chlorine and Chloramines in Aqueous Solution at Micromolar Concentrations," in *Water Chlorination: Environmental Impact and Health Effects, Vol. 4*, R. L. Jolley, W. A. Brungs, J. A. Cotruvo, R. B. Cumming, J. S. Mattice, and V. A. Jacobs, Eds. (Ann Arbor, MI: Ann Arbor Science Publishers, Inc., 1983), pp. 723–731.
7. Margerum, D. W., E. T. Gray, Jr., and R. P. Huffman. "Chlorination and the Formations of N-Chloro Compounds in Water Treatment," in *Organometals and Metaloids: Occurrence and Fate in the Environment*, F. E. Brinkman and J. M. Bellama, Eds. (Washington, DC: American Chemical Society, 1978), pp. 278–291.
8. Inman, G. W., and J. D. Johnson. "Kinetics of Monobromamine-Dibromamine Formation in Aqueous Ammonia Solutions," *Environ. Sci. Technol.* 18:219–224 (1984).
9. Anbar, M., and H. Taube. "The Exchange of Hypochlorite and of Hypobromite Ions with Water," *J. Am. Chem. Soc.* 80:1073–1077 (1958).
10. Gray, E. T., Jr. *Kinetics and Equilibria of the Formation and Interconversion of Chloramines, Alkylchloramines, and Chloramino Acids in Aqueous Solution*, PhD Thesis, (Lafayette, IN: Purdue University, 1977).
11. Latimer, W. M., *The Oxidation States of the Elements and Their Potentials in Aqueous Solutions*, 2d ed. (New York: Prentice Hall, Inc., 1952).

CHAPTER 61

Influence of Sodium, Potassium, and Lithium on Hypochlorite Solution Equilibria

Charles N. Haas and Dolores M. Brncich

In 1972, Scarpino et al.[1] reported that free chlorine was a more effective disinfectant of viruses at high pH than at low pH, in contradiction to numerous previous reports.[2,3] This was subsequently attributed to the presence of high concentrations of alkaline cations in the high-pH buffer used.[4] Further work indicated that the presence of sodium or potassium ions in solutions of free chlorine increased viral inactivation at high pH.[5-7] In addition, it has been found that sodium ions enhance the rate of inactivation of *Escherichia coli* by free chlorine.[8] The explanation proposed for this phenomenon has been that alkali cations form ion pairs with the hypochlorite anion of greater intrinsic biocidal potency than free hypochlorite anions.[8]

In prior work, the data of Sugam and Helz[9] and Jensen et al.[7] have been reanalyzed by Haas[10] and found to be consistent with the existence of the NaOCl ion pair having a dissociation constant between 0.9 and 2.3 M. However, to our knowledge, no previous attempts have been made to measure the equilibrium constants for alkali cation hypochlorite ion-pair formation. Therefore, the goal of this study was to measure the equilibrium constants for the information of the ion pairs of sodium, potassium, and lithium with OCl$^-$.

MATERIALS AND METHODS

All experiments were conducted using chlorine demand-free water prepared according to Engelbrecht et al.[11] The chlorine used was high-purity gas, which was sparged into demand-free deionized water.

Experiments were conducted with no added salt and with 0.33, 0.67, 1.0 and 2.0 M of either $NaNO_3$, KNO_3, or $LiNO_3$. Chlorine and salt solutions were mixed, and the pH was adjusted to approximately 10.0 by the addition of sodium hydroxide (in all cases, the sodium ion concentration was negligible with respect to the salt concentration added). Acid titration using standardized nitric acid was then conducted using constant volume increments. The pH was measured using a Horizon Model 5998-10 digital pH meter and an Orion combination electrode. The solution was continuously agitated using a magnetic stirring apparatus. Chlorine residuals were determined using the forward amperometric procedure.[12]

It was necessary to correct for alkali cation interference in the pH measurement. This was accomplished by adding 0.33, 0.67, 1.0, and 2.0 M of the appropriate salts to aliquots of standard pH 4, 7, and 10 buffer solutions and measuring the pH. Calibration curves were thus developed between the measured pH and the pH of the standard buffer solution at each ionic strength. These plots were then used to correct the pH measured in the titration experiments to that of the apparent zero ionic strength value before the data were further reduced. This procedure does not account for the ionic strength effect on proton activity coefficients and, hence, this procedure does not produce accurate measurements of species activity coefficients. However, by extrapolation to zero ionic strength, reasonable estimates of the infinite dilution equilibrium constants may be obtained.

DATA ANALYSIS

Considering the formation of a 1:1 cation hypochlorite ion pair, and a formal dissociation constant given by Equation (1), for the case K_D larger than the hypochlorite ion concentration, the equation for the titration curve can be shown to be that of Equation (2).

$$K_D = \frac{[Na^+][OCl^-]}{[NaOCl]} B \qquad (1)$$

where

$$B = (\gamma_{Na^+} \gamma_{OCl^-})/(\gamma_{NaOCl})$$

$$C_A = \frac{\{H^+\}}{\gamma_{H^+}} + C_{Na} - N - \frac{K_W}{\{H^+\}\gamma_{OH^-}}$$

$$- \frac{C_{Cl}\left(1 + \frac{C_{Na}B}{K_D}\right)}{1 + \frac{\{H^+\}A}{K_A} + \frac{BC_{Na}}{K_D}} \qquad (2)$$

In Equation (2), A is the ratio of the activity coefficients of hypochlorite anion to hypochlorous acid, N is the initial concentration of nitrate, C_{Cl} is the free chlorine concentration, and it is understood that for C_{Na}, as appropriate, Na^+ may be replaced by either K^+ or Li^+.

Conventionally, the buffer intensity is defined as $-dC_A/dpH$. As long as the concentration of chlorine is in excess of approximately $10^{-5}\ M$, the contribution of the water terms to the buffer intensity, β, will be negligible, and it may be shown that

$$\frac{\beta}{2.303} = \{H^+\}\left(1 + \frac{C_{Na}B}{K_D}\right)(AC_{Cl}/K_A) \bigg/ \left(1 + \frac{\{H^+\}A}{K_A} + \frac{BC_{Na}}{K_D}\right)^2 \quad (3)$$

By differentiation of Equation (3), it can be determined that the hydrogen ion concentration for which the buffer intensity is a maximum, denoted by x, is

$$x = (K_A/A)(1 + BC_{Na}/K_D) \quad (4)$$

Equation (4) can be substituted into Equation (3) to determine the maximum value of the buffer intensity. Subsequently, Equation (3) can be solved to determine the hydrogen ion concentrations for which the buffer intensity is half its maximum value. The difference between these two hydrogen ion concentrations, the width at half height of the curve of buffer intensity vs hydrogen ion concentration, is denoted by y, and is given by

$$y = \frac{8K_A}{A}\left(1 + \frac{BC_{Na}}{K_D}\right)^3 \left[1 - \frac{1}{2\left(1 + \frac{BC_{Na}}{K_D}\right)^2}\right]^{1/2} \quad (5)$$

Equations (4) and (5) can be used to provide explicit relationships for the equilibrium constants as a function of observable parameters of the titration curve (x and y), cation concentration, and activity coefficient ratio (A and B) as follows:

$$\frac{A}{K_A} = \frac{1}{2x}\left[1 + \left(1 + \frac{y^2}{4x^2}\right)^{1/2}\right]^{1/2} \quad (6)$$

$$\frac{BC_{NA}}{K_D} = \frac{1}{2}\left[1 + \left(1 + \frac{16y^2}{x^2}\right)^{1/2}\right]^{1/2} - 1 \quad (7)$$

Equations (6) and (7) can be used to estimate the values for K_A/A and K_D/B from titration curves. By extrapolation of these latter values to zero ionic strength, where the activation coefficient ratios (A and B) approach unity, the infinite dilution equilibrium constants may be obtained. Use of this method does not require a precise determination of the free chlorine concentration, merely the assumption that this concentration is sufficiently high to swamp the buffer strength of water and low with respect to the ion-pair dissociation constant.

To use this method, it was necessary to determine the pH values at which the buffer intensity was maximum and half maximum. Because of the inherent noise from numerical differentiation necessary in the calculation of buffer intensity, a data smoothing technique was used. This technique used five-point running average smoothing, followed by a five-point estimation of the derivatives.[13,14] Using the derivatives (i.e., estimates for β), polynomials of second through sixth order in pH were fitted by least squares. The highest-order polynomial for which a statistically significant improvement in fit was obtained was used to calculate the values of x and y for use in Equations (6) and (7).

RESULTS AND DISCUSSION

Figure 1 is a typical plot of the computed values for buffer intensity and the best-fit polynomial. A complete description of the data sets has been presented elsewhere.[15] At each concentration, experiments were repeated in duplicate or triplicate. Table I summarizes the computed values of the equilibrium constant/activity ratios for each experiment. As can be seen by comparing the results from replicate experiments, computed values for $\log(K_A/A)$ and $\log(K_D/B)$ generally agree within 0.1 to 0.2 log units between experiments.

It is also evident that the effect of salt concentration on the computed values of these constants is significant. This was anticipated because of the incorporation of activity coefficient terms in the variables A and B. From the information in Table I, linear regressions between either $\log(K_A/A)$ or $\log(K_D/B)$ and salt concentration were computed and extrapolated to zero ionic strength to produce an estimate for the thermodynamic equilibrium constants. Table II presents a summary of the computed infinite dilution constants and includes a comparison to the value for K_A reported by Morris.[16]

Examination of Table II indicates that the lithium ion pair exhibits the strongest association [i.e., the smallest value of $\log(K_D)$], whereas the potassium ion pair exhibits the weakest association. This is in agreement with the order of atomic weights. The computed value for the dissociation constant of the sodium ion pair (0.1808 M) is somewhat less than that computed by Haas[10] from data of Sugam and Helz.[9]

Given the values for the apparent infinite dilution dissociation constants, the significance of these may be determined by comparison with implied interionic separation distances, which would occur if only electrostatic interactions were of significance. The theory of ion association has been discussed by Bjerrum[17] and others. Considering electrostatic interactions between monovalent ions, Fuoss[18] developed the following expression for K_D:

$$K_D = \frac{4\Pi N a^3}{3000} \exp\left(\frac{E^2}{akTD}\right) \tag{8}$$

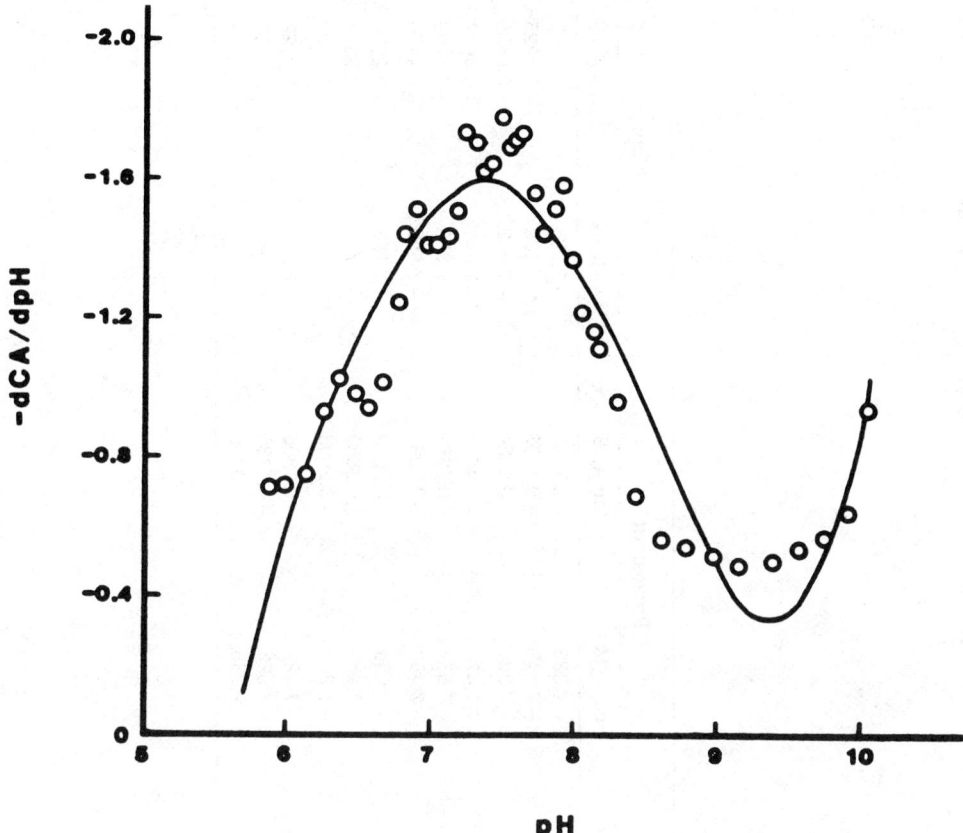

Figure 1. Buffer intensity as a function of pH for the titration of 2.0 M KNO$_3$ and free chlorine. Conditions: 0.004 M free chlorine; 2.0 M KNO$_3$; 0.1 M HNO$_3$; 0.0045 M NaOH; 25.5°C; log K_A/A = −7.525, log K_D/B = −0.1769.

where N is Avogadro's number, k is the Boltzman constant, T is absolute temperature, D is the coulomb force constant in water (product of coulomb force constant in vacuo and the dielectric constant), E is the charge on an electron, and a is the interionic separation distance. Given the values of K_D from Table II, all values in Equation (8) are known except for a. For lithium, sodium, and potassium, respectively, Equation (8) is solved to yield separation distances of 14.0 Å, 12.9 Å, and 12.3 Å. These distances are considerably greater than the ordinary ionic radii for the cations or the hypochlorite anion,[19,20] indicating that the cation-hypochlorite pairing is weaker than would be expected purely on the basis of electrostatic interaction. It is hypothesized that (1) some solvation with water molecules may occur, which results in failure to approach smaller separation distances; or (2) the repulsive forces, perhaps between the electropositive chlorine atom and the approaching cation, intervene to reduce the strength of the ionic interaction.

Table I. Apparent Equilibrium Constants Observed

Salt Concentration (M)	Sodium		Potassium		Lithium	
	$\log K_A/A$	$\log K_D/B$	$\log K_A/A$	$\log K_D/B$	$\log K_A/A$	$\log K_D/B$
0.33	−7.521	−0.7873	−7.333	−0.7398	−7.663	−0.6683
	−7.490	−0.6863	−7.247	−0.4893	−7.639	−0.8223
	−7.542	−0.6284	−7.365	−0.5734	−7.632	−0.8435
0.67	−7.451	−0.3559	−7.478	−0.4548	−7.841	−0.4647
	−7.484	−0.3143	−7.532	−0.5351	−7.873	−0.5703
	−7.865	−0.5163	−7.468	−0.5284	−7.859	−0.4170
1.00	−7.602	−0.2681	−7.199	−0.3202	−7.757	−0.2979
	−7.728	−0.2809	−7.269	−0.3559	−7.777	−0.3341
	−7.636	−0.1708	−7.287	−0.2595		
2.00	−7.802	0.0318	−7.324	−0.1305	−8.307	−0.0127
	−7.729	0.1011	−7.525	−0.1768	−8.081	−0.0406
	−7.949	−0.0192	−7.286	−0.0388		

Table II. Infinite Dilution Equilibrium Constants

System	log (K_A)[a]	log (K_D)
Sodium	-7.468 ± 0.145	-0.743 ± 0.1395
Lithium	-7.610 ± 0.098	-0.827 ± 0.1365
Potassium	-7.356 ± 0.208	-0.675 ± 0.1128
Morris[b]	-7.537	

[a]Range indicated is based on 95% confidence limit of regression intercept.
[b]See Reference 16.

perhaps between the electropositive chlorine atom and the approaching cation, intervene to reduce the strength of the ionic interaction.

SUMMARY AND CONCLUSIONS

This study has determined that measurable association exists between the cations sodium, potassium, lithium, and the hypochlorite anion. The extrapolated zero ionic strength dissociation constants ranged from 0.15 to 0.2 M, which are somewhat less than the values (uncorrected for ionic strength) calculated from data of Sugam and Helz[9] and Haas.[10]

Comparison of the measured dissociation constants with estimates from the theory of electrostatic interaction indicates that the observed ion pairing is somewhat weaker than would be expected. We suggest that this is a result of either solvation by water molecules or hinderance of ion pairing by intra-ionic interactions involving the chlorine and oxygen atoms in hypochlorite.

ACKNOWLEDGMENT

Charles N. Haas is Associate Professor in the Pritzker Department of Environmental Engineering at the Illinois Institute of Technology. At the time this research was conducted, Dolores M. Brncich was a graduate student in this department, and this work forms a portion of her M.S. thesis in environmental engineering.

REFERENCES

1. Scarpino, P. V., G. Berg, S. L. Chang, D. Dahlings, and M. Lucas. "A Comparative Study of the Inactivation of Viruses in Water By Chlorine," *Water Res.* 6:959–965 (1972).
2. Fair, G. M. "The Behavior of Chlorine as a Water Disinfectant," *J. Am. Water Works Assoc.* 40:1051–1061 (1948).

3. Weidenkopf, S. J. "Inactivation of Type I Poliomyelitis Virus with Chlorine," *Virology* 5:56–67 (1958).
4. Hoff, J. C., and E. E. Geldreich. "Comparison of the Biocidal Efficiency of Alternate Disinfectants," *J. Am. Water Works Assoc.* 73:40–44 (1981).
5. Engelbrecht, R. S., M. J. Weber, B. L. Salter, and C. A. Schmidt. "Comparative Inactivation of Viruses by Chlorine," *Appl. Environ. Microbiol.* 40:249–256 (1980).
6. Sharp, D. G., D. C. Young, R. Floyd, and J. D. Johnson. "Effect of Ionic Environment on the Inactivation of Poliovirus in Water by Chlorine," *Appl. Environ. Microbiol.* 39:530–534 (1980).
7. Jensen, H., K. Thomas, and D. G. Sharp. "Inactivation of Coxsackieviruses B3 and B5 in Water by Chlorine," *Appl. Environ. Microbiol.* 40:633–40 (1980).
8. Haas, C. N., and M. A. Zapkin. "Enhancement of Chlorine Inactivation of *Escherichia coli* by Sodium Ions," in *Water Chlorination: Environmental Impact and Health Effects, Vol. 4*, R. L. Jolley, W. A. Brungs, J. A. Cotruvo, R. B. Cumming, J. S. Mattice, and V. A. Jacobs, Eds. (Ann Arbor MI: Ann Arbor Science Publishers, Inc., 1983), pp. 1087–96.
9. Sugam, R., and G. R. Helz. "Apparent Ionization Constant of Hypochlorous Acid in Seawater," *Environ. Sci. Technol.* 10:384–386 (1976).
10. Haas, C. N. "Sodium Alterations of Chlorine Equilibrium: Quantitative Description," *Environ. Sci. Technol.* 15:1243–1248 (1981).
11. Engelbrecht, R. S., B. F. Severin, M. T. Masarik, S. Farooq, S. H. Lee, C. N. Haas, and A. Lalchandani. "New Microbial Indicators of Disinfection Efficiency," EPA 600/2-77-052 (Cincinnati: U.S. Environmental Protection Agency, 1977).
12. *Standard Methods for the Examination of Water and Wastewater*, 15th ed. (Washington: DC: American Public Health Association, 1980).
13. Hildebrand, F. B. *Introduction to Numerical Analysis* (New York: McGraw Hill, 1956).
14. Pollard, J. H. *Numerical and Statistical Techniques* (London: Cambridge University Press, 1977).
15. Brncich, D. M. *The Determination of Stability Constants for Sodium, Lithium and Potassium Ion Pairs with Hypochlorite*, MS Thesis (Chicago: Illinois Institute of Technology, 1984).
16. Morris, J. C. "The Acid Ionization Constant of HOCl from 5 to 35°C," *J. Phys. Chem.* 70:3798 (1966).
17. Bjerrum, N. "Ionic Association. I. Influence of Ionic Association on the Activity of Ions at Moderate Degrees of Association," *Mat.-Fys. Medd.-K. Dan. Vidensk. Selsk.* 7(9):1–48 (1926).
18. Fuoss, R. M. "Ionic Association III. The Equilibrium Between Ion Pairs and Free Ions," *J. Am. Chem. Soc.* 80:5059 (1958).
19. Robinson, R. A., and Stokes, R. H. *Electrolyte Solutions* (New York: Academic Press, 1959).
20. Harned, H., and Owen, B. *The Physical Chemistry of Electrolyte Solutions* (New York: Reinhold Publishing Company, 1958).

CHAPTER 62

The Chemistry of Oxo-Chlorine Compounds Relevant to Chlorine Dioxide Generation

E. Marco Aieta and Paul V. Roberts

Within the last decade, halogenated organic compounds formed in drinking water by disinfection with chlorine have caused considerable concern throughout the drinking water industry.[1] The U.S. Environmental Protection Agency (EPA) has regulated the concentration of THMs to a maximum of 100 μg/L total THMs in drinking water as a first step in an effort to control the chlorinated organic compounds formed during the disinfection process. Three strategies for meeting this regulation are removal of the organic precursors; removal of the THMs that are formed; or substitution of an alternative disinfectant that does not form THMs.

One of the alternative disinfectants that has received considerable attention is chlorine dioxide (ClO_2), which offers several attributes that make it an acceptable substitute for chlorine.[2] It does not form THMs. It does, however, produce some chlorinated organic compounds in aqueous solution, but under conditions encountered in drinking water treatment the concentrations of these compounds are 10 to 100 times less than the concentration produced by chlorine under the same conditions.[3] Chlorine dioxide is an effective disinfectant over a broad pH range and in some cases significantly more effective than chlorine.[4] Chlorine dioxide can be measured easily at the point of use of the disinfected water, permitting residual disinfectant measurements to be used as an indicator of the microbiological safety of the delivered water.

Chlorine dioxide is an unstable explosive gas at concentrations greater than about 10% in air.[5] The gas, either pure or in mixtures, cannot be compressed and stored; hence, ClO_2 must be generated on-site for immediate use. For drinking water treatment, ClO_2 is most commonly generated by the oxidation of sodium chlorite by chlorine.

This chapter reviews the chemistry of oxo-chlorine compounds so that the reactions used to generate ClO_2 for water treatment may be more fully understood. This chapter also includes results from our research on the kinetics of the oxidation of sodium chlorite by chlorine. This reaction is the preferred ClO_2 generation reaction for water treatment, because of the high yields that can be achieved and because there are no waste streams that pose disposal problems.

AQUEOUS CHEMISTRY OF CHLORINE AND OXO-CHLORINE COMPOUNDS

Table I lists the major chlorine compounds of the oxo-chlorine family and their oxidation states.

Chlorine Hydrolysis Reaction

Chlorine gas is moderately soluble in water. In addition to solvated halogen molecules, hypochlorous acid and chloride ion are present in aqueous solution due to the hydrolysis/disproportionation reaction of the solvated molecular chlorine. These reactions are

$$Cl_2(g) = Cl_2(aq) \tag{1}$$

$$Cl_2(aq) + H_2O = HOCl + Cl^- + H^+ \tag{2}$$

The chlorine hydrolysis constant is given by

$$K_H = \frac{[HOCl][H^+][Cl^-]}{[Cl_2(aq)]} \tag{3}$$

The rate of chlorine hydrolysis has been widely studied. The interpretation of the results from early studies was complicated because of confusion about the mechanism of the chlorine hydrolysis reaction. Shilov and Solodushenkov[6] assumed the reaction to be as shown by Equation (2). Their results, however, indicated that the rate constant decreased as the reaction proceeded. Morris[7] recalculated the results of Shilov and Solodushenkov,[6] assuming the reaction involved the hydroxide ion

$$Cl_2(aq) + OH^- = HOCl + Cl^- \tag{4}$$

Table I. Chlorine and Oxo-Chlorine Species

Oxidation State	Oxidation No.	Compounds	Name
Chlorine (−I)	−1	HCl, Cl$^-$	Hydrochloric acid, chloride
Chlorine (0)	0	Cl$_2$	Molecular chlorine
Chlorine (I)	1	HOCl, OCl$^-$	Hypochlorous acid, hypochlorite ion
Chlorine (III)	3	HClO$_2$, ClO$^-_2$	Chlorous acid, chlorite ion
Chlorine (IV)	4	ClO$_2$	Chlorine dioxide
Chlorine (V)	5	HClO$_3$, ClO$^-_3$	Chloric acid, chlorate ion
Chlorine (VII)	7	HClO$_4$, ClO$^-_4$	Perchloric acid, perchlorate ion

and found much better constancy in the calculated second-order rate constant. The value of the second-order rate constant determined by Morris was 5×10^{14} M^{-1} s^{-1}, an extremely high value. The Morris[7] calculations also indicated zero activation energy for the chlorine hydrolysis reaction. A reinvestigation of the chlorine hydrolysis reaction by Shilov and Solodushenkov[8] did not reproduce the decreasing rate constant observed in their first study, so that Morris' argument did not seem to apply. Lifshitz and Perlmutter-Hayman[9] tried to establish which of Equations (2) or (4) represented the hydrolysis in pure water. Their findings supported the original assumption of Shilov and Solodushenkov[6] that Equation (2) correctly represented the chlorine hydrolysis reaction in pure water. Furthermore, Lifshitz and Perlmutter-Hayman[9] pointed out that the maximum second-order rate constant to be expected in solution corresponded to the diffusion-controlled limit of 1×10^{10} M^{-1} s^{-1}, a much lower value than that calculated by Morris,[7] and that the rate of formation of hydroxide ions was not rapid enough in acid solution for Equation (4) to contribute significantly to the rate of chlorine hydrolysis.

Morris'[7] ideas relating to the mechanism of chlorine hydrolysis were not totally in error, however, as later demonstrated by Spalding.[10] The equilibrium pH of a chlorine solution in pure, unbuffered water at a given temperature is controlled by the total chlorine concentration of the system, as given by the chlorine hydrolysis equilibrium constant, Equation (3). In the studies of Shilov and Solodushenkov[6,8] and of Lifshitz and Perlmutter-Hayman,[9] chlorine gas was dissolved in pure, unbuffered water so that the pH of the reaction solution was near pH 2 and may have been somewhat lower at the lower temperatures of these studies. Eigen and Kustin[11] proposed a general mechanism for the hydrolysis of halogens, including chlorine. Both Equations (2) and (4) are included in the general mechanisms. Eigen and Kustin's[11] results indicated that at pH 2.2, the highest pH they investigated, the hydrolysis mechanism represented by Equation (4) did not contribute to the rate of chlorine hydrolysis. Spalding,[10] however, absorbed chlorine gas into water at initial pH values between 3 and 10.2. His results indicated that below an initial pH of 10.2, the primary reaction mechanism for the hydrolysis of chlorine was that given in Equation (2); however, above pH 12.5, the reaction mechanism was that proposed by Morris,[7] which is given in Equation (4). It should be emphasized that the actual pH value reported by Spalding,[10] at which Equation (2) became dominant, was the initial pH of the absorbing fluid and that the contact time was very short. Under different experimental conditions this pH value might be shifted. Spalding[10] estimated the first-order chlorine forward rate constant for Equation (2) to be 20.9 s^{-1} at 25°C. His estimate for the second-order rate constant of Equation (4) was 1×10^6 M^{-1} s^{-1} at 25°C. More recent work by Sandal et al.,[12] who studied the absorption of chlorine gas in strong sodium hydroxide solution, gave a value for the second-order rate constant of Equation (4) as 2.7×10^7 M^{-1} s^{-1} at 0°C.

Figure 1 shows a comparison of the first-order forward rate constants for the chlorine hydrolysis reaction determined in our laboratory with the results

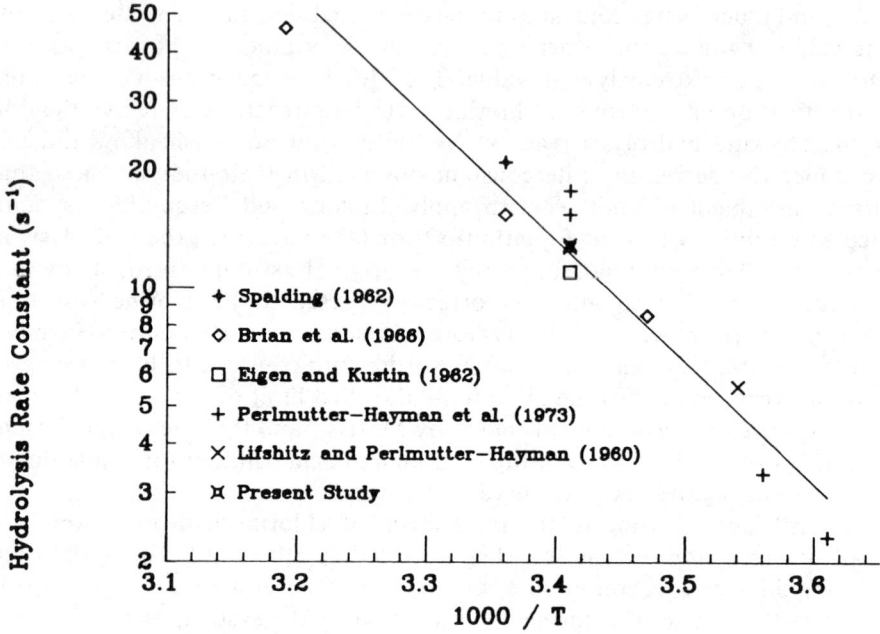

Figure 1. Chlorine hydrolysis rate constant as a function of temperature.

from the work of others. A rather wide range of values for the chlorine hydrolysis rate constant is reported at 20°C. The kinetic methods used include temperature jump relaxation[11,13] and stopped flow techniques,[9,13] as well as gas-liquid mass transfer techniques.[10,14,15] If the data in Figure 1 are analyzed according to the Arrhenius rate law by the regression of ln(k) vs 1/T, the activation energy is found to be 60.3 kJ/mol, with a 95% confidence interval of 49.4 to 71.2 kJ/mol. The preexponential factor is 6.76 × 10¹¹ and the coefficient of determination, r^2, is 0.94. The value of the chlorine hydrolysis first-order forward rate constant predicted from all the data is 12.2 s⁻¹ at 20°C, with a 95% confidence interval of 10.7 to 14.0 s⁻¹.

Aqueous Reactions of Hypochlorite, Chlorate, and Chlorite

Hypochlorite ion tends to disproportionate to form chlorate ion and chloride ion,

$$3\ OCl^- = ClO_3^- + 2\ Cl^- \tag{5}$$

This reaction is slow at or below room temperature, but in hot (e.g., 80°C) solutions, the reaction is rapid and produces high yields of chlorate ion. Equation (5) forms the basis of the industrial production of chlorate. Chlorine dioxide is produced for the pulp and paper industry by the reduction of chlorate in strong acid medium. Either hydrochloric or sulfuric acid may be used. The most common reducing agents are chloride, sulfur dioxide, and methanol. The reduction of chlorate by chloride in acid medium is given by

$$2H^+ + ClO_3^- + Cl^- = ClO_2 + \tfrac{1}{2} Cl_2 + H_2O \tag{6}$$

There is a competing parallel reaction that produces only chlorine,

$$6H^+ + ClO_3^- + 5Cl^- = 3Cl_2 + 3H_2O \tag{7}$$

Commercially, chlorites are produced from the selective reduction of ClO_2. The ClO_2 used in the manufacture of chlorites is produced as described above and is reduced in alkaline solution in the presence of a weak reducing agent to form the metal chlorite.

Chlorites are used commercially as bleaching agents and for small-scale ClO_2 production for water treatment. In alkaline solution, chlorite ion is very stable, even at 100°C. In acid solution, chlorous acid rapidly disproportionates to form ClO_2, chlorate ion, and chloride ion,

$$4 HClO_2 = 2 ClO_2 + ClO_3^- + Cl^- + 2 H^+ + H_2O \tag{8}$$

In the presence of appreciable amounts of chloride ion, only small amounts of chlorate are formed and the stoichiometry is approximated by

$$5HClO_2 = 4ClO_2 + Cl^- + 2H_2O + H^+ \tag{9}$$

Equation (9) forms the basis for some commercial, small-scale ClO_2 generation systems. These systems find only limited application in water treatment systems because of the availability of more efficient ClO_2 generation systems that use the oxidation of chlorites by chlorine.[16]

Chlorine – Chlorite Reaction and Mechanism

The pioneering work of Taube and Dodgen[17] using radioactive chlorine provided significant insight into the reaction mechanisms of the oxidation of chlorite by chlorine (or hypochlorous acid), the reduction of chlorate by chloride ion (and the reverse reaction), and the disproportionation of chlorous acid. The observations of Taube and Dodgen[17] led them to postulate an unsymmetrical activated complex that was common to all three reaction mechanisms. Subsequent work by Emmenegger and Gordon[18] has amplified and refined the mechanism proposed by Taube and Dodgen,[17] but their original findings are still the foundation for many of the subsequent studies of the kinetics and mechanisms of aqueous chlorine chemistry.

In the oxidation of aqueous chlorite by chlorine or hypochlorous acid, both ClO_2 and chlorate appear as products. In acid solution, where the chlorine is present mainly as dissolved molecular chlorine, the stoichiometries of the two reactions are

$$Cl_2 + 2ClO_2^- = 2ClO_2 + 2Cl^- \tag{10}$$

and

$$Cl_2 + ClO_2^- + H_2O = ClO_3^- + 2Cl^- + 2H^+ \tag{11}$$

In solutions near neutral pH, where chlorine is present largely as hypochlorous acid, the stoichiometries are

$$HOCl + 2ClO_2^- = 2ClO_2 + Cl^- + OH^- \tag{12}$$

and

$$HOCl + ClO_2^- + OH^- = ClO_3^- + Cl^- + H_2O \tag{13}$$

In alkaline solutions in which the chlorine is present as hypochlorite ion, the reaction is very slow and the only product formed is chlorate ion:

$$OCl^- + ClO_2^- = ClO_3^- + Cl^- \tag{14}$$

The product ratio of ClO_2 to chlorate has been observed to vary as a function of the experimental conditions such that:

1. The reaction between chlorite ion (or chlorous acid) and chlorine (in acid or neutral solution) is second order overall, being first order in both chlorine and chlorite.
2. Acidic pH values favor the formation of ClO_2, whereas at neutral and alkaline pH, chlorate is the primary reaction product.
3. For a given pH value, an increase in chlorite concentration results in relatively more ClO_2 production.
4. Higher chloride ion concentrations favor the formation of ClO_2 relative to chlorate, especially at acidic pH.
5. Proportional increases in all reactant concentrations favor the formation of ClO_2 relative to chlorate.

The mechanism for the chlorine-chlorite reaction proposed by Taube and Dodgen[17] as consistent with experimental observations is presented below. The starred (*) chlorine atoms trace the fate of the chlorine originally present as molecular chlorine or hypochlorous acid. Hypothetical, metastable, intermediate species are enclosed in brackets { }. For the reaction of chlorite with molecular chlorine,

$$*Cl_2 + ClO_2^- = \{*Cl\text{-}Cl{<}^O_O\} + *Cl^- \qquad (15)$$

and with hypochlorous acid

$$H^+ + HO*Cl + ClO_2^- = \{*Cl\text{-}Cl{<}^O_O\} + H_2O \qquad (16)$$

For the production of ClO_2,

$$2\{*Cl\text{-}Cl{<}^O_O\} = 2ClO_2 + *Cl_2 \qquad (17)$$

For the production of chlorate,

$$\{*Cl\text{-}Cl{<}^O_O\} + H_2O = ClO_3^- + *Cl^- + 2H^+ \qquad (18)$$

This mechanism satisfies the experimental observations discussed above according to the following rationale. The metastable intermediate, $\{Cl_2O_2\}$, can decompose by either a first-order process, Equation (18), to give chlorate ion or by a second-order process, Equation (17), to give ClO_2. The formation of the intermediate, $\{Cl_2O_2\}$, is, in either case, the rate-limiting step in the mechanism as determined from the observed rate law.[18] Reaction conditions that promote the formation of higher concentrations of the intermediate will favor the product ClO_2 via the second-order route over the product chlorate ion via the first-order route. Emmenegger and Gordon[18] have shown that the rate of oxidation of chlorite by molecular chlorine, Equation (15), is considerably faster than the oxidation by hypochlorous acid, Equation (16). Therefore, reaction conditions that favor molecular chlorine as opposed to hypochlorous acid will result in higher intermediate concentrations and relatively more ClO_2. In aqueous systems, higher hydrogen ion concentrations and chloride ion concentrations shift the chlorine hydrolysis equilibrium, Equation (3), toward increased molecular chlorine concentration. These reaction conditions are, in fact, those that are observed to produce relatively more ClO_2 both experimentally and in practice. An excess of chlorite also promotes higher concentrations of the intermediate, as does increasing the absolute concentrations of all reactants, while maintaining the same reactant ratios, which results in proportionally more ClO_2.

In our research, chlorine gas was contacted with aqueous sodium chlorite solution in a gas-liquid system. As shown by Emmenegger and Gordon,[18] the rate of reaction of the dissolved molecular chlorine gas with the chlorite ion is much faster than the rate of chlorine hydrolysis. Hence, the chlorine molecule reacted preferentially with the chlorite ion, and the reaction solution pH did not influence the rate of reaction in this gas-liquid system as it had been shown to do in a liquid-liquid system.

The details of this kinetic apparatus and the methodology have been presented elsewhere.[15] The results of these experiments are presented in Table II.

Table II. Summary of Experimentally Determined Second-Order Rate Constants of the Chlorine–Chlorite Reaction

Experiment[a]	N[b]	I (M)	T (°C)	Rate Constant, × 10^{-4} (M^{-1} s^{-1})	Standard Deviation, × 10^{-4} (M^{-1} s^{-1})	Standard Error, × 10^{-4} (M^{-1} s^{-1})	95% CI[c] × 10^{-4} (M^{-1} s^{-1})	CV[d] (%)
1	28	1.42	20	1.31	0.22	0.04	1.23 – 1.40	16.5
2	6	0.12	20	1.49	0.30	0.12	1.19 – 1.79	20.2
3	16	0.17	20	1.69	0.36	0.09	1.50 – 1.88	21.5
4	25	1.39	10	0.55	0.15	0.03	0.49 – 0.61	26.6
5	12	4.26	20	1.16	0.28	0.08	0.98 – 1.33	24.3
6	19	1.42	30	1.67	0.25	0.06	1.55 – 1.79	14.9

[a]Experiment = experiment identification number.
[b]N = Number of runs.
[c]CI = Confidence interval.
[d]CV = Coefficient of variation = standard deviation/mean.

END–H/1308

Effect of Ionic Strength

As discussed by Moore and Pearson[19] for ionic strengths $>0.01\ M$, the logarithm of the reaction rate constant should be a linear function of the first power of ionic strength for reactions involving a neutral molecule and an ion. The chlorine-chlorite reaction falls into this category.

Experiments 1, 3, and 5 reported in Table II were all performed at 20°C, but the absorbing liquids were of different ionic strengths. A regression analysis of the natural logarithms of the second-order rate constants, k_2, of the chlorine-chlorite reaction from these experiments vs ionic strength yielded the relationship:

$$\ln\left(\frac{k_2}{k_{2,0}}\right) = -0.0851 \tag{19}$$

where $k_{2,0}$ is the second-order rate constant determined by extrapolation to zero ionic strength and is $1.62 \times 10^4\ M^{-1}s^{-1}$. The 95% confidence interval for the zero ionic strength second-order rate constant is 1.49×10^4 to 1.76×10^4 $M^{-1}s^{-1}$. The 95% confidence interval for the slope in Equation (19) is -0.046 to -0.124. This relationship is shown in Figure 2 together with the means of the experimental data and the respective standard deviations.

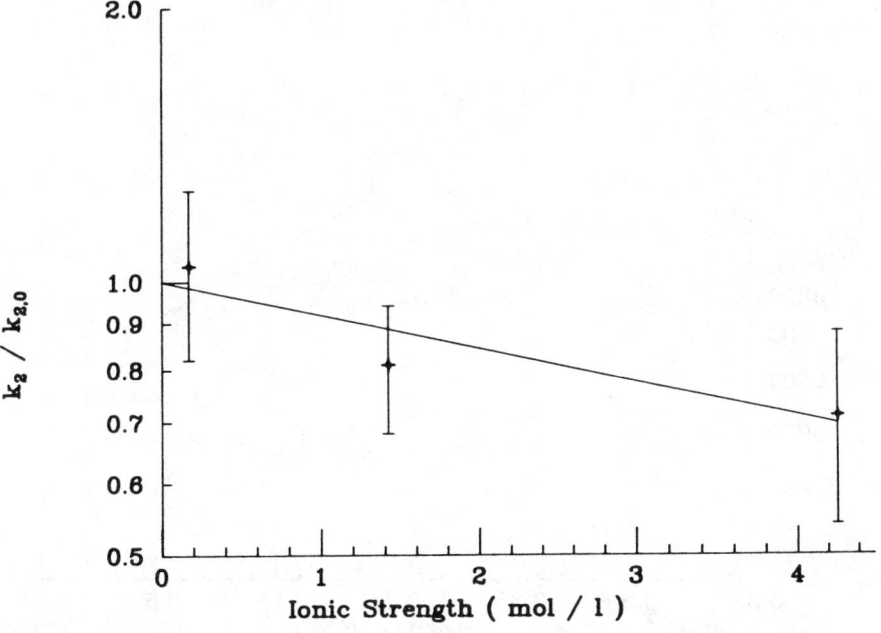

Figure 2. The effect of ionic strength on the second-order rate constant of the chlorine-chlorite reaction.

Effect of Temperature

The rate of the chlorine-chlorite reaction increases with increasing temperature. The temperature dependence can be described by the Arrhenius relationship.[19] The data from Experiments 1, 4, and 6 shown in Table II for the second-order rate constant of the chlorine-chlorite reaction at 20, 10, and 30°C, respectively, were used to determine the parameters of the Arrhenius expression. Correction of the second-order rate constant determined from Experiment 4 at ionic strength of 1.39 to the ionic strength of Experiments 1 and 6 resulted in a decrease in the value given in Table II of only 0.03%. A regression analysis of ln (rate) vs 1/T yielded

$$\ln k_2 = 25.6 - 4766 \, (1/T) \tag{20}$$

or

$$k_2 = 1.31 \times 10^{11} \cdot \exp\left\{\frac{-39.9 \text{ kJ} \cdot \text{mol}^{-1}}{RT}\right\} \tag{21}$$

Equation (21) is plotted in Figure 3 along with the means of the experimental data points and their respective standard deviations. The 95% confidence interval for the activation energy is 35.6 to 45.2 kJ mol.

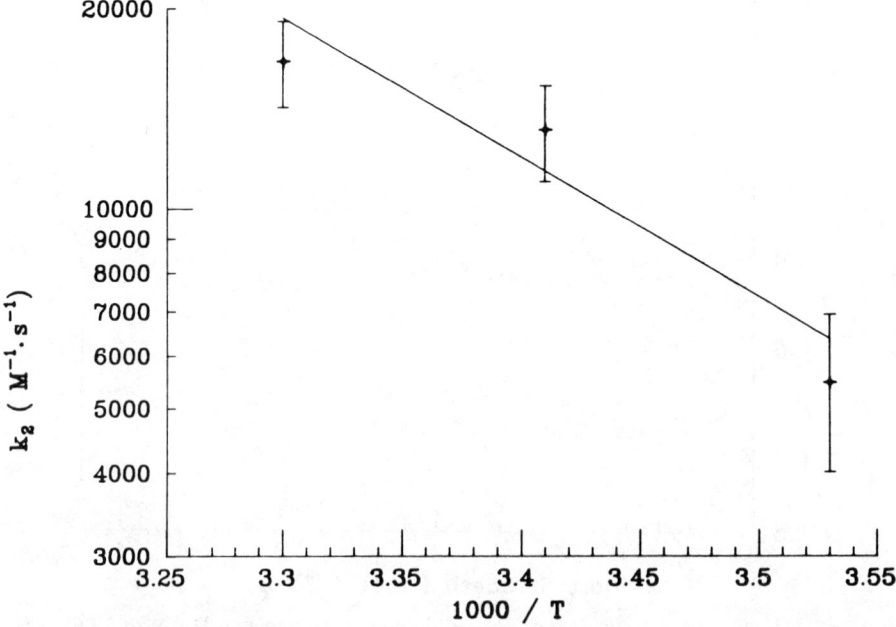

Figure 3. The effect of temperature on the second-order rate constant of the chlorine-chlorite reaction.

SUMMARY

A review of oxo-chlorine chemistry and results from recent experimental investigations have been presented. Relevant to ClO_2 generation for water treatment is the observation that ClO_2 yield can be optimized by selecting reaction conditions such that high concentrations of molecular chlorine are in contact with high concentrations of sodium chlorite.

ACKNOWLEDGMENT

This work was funded by the U.S. Environmental Protection Agency under Research Grant R-808686. This paper has not been subjected to the Agency's required administrative review, and therefore does not necessarily reflect the views of the Agency, nor should any official endorsement be inferred.

REFERENCES

1. Bellar, T. A., J. J. Lichtenberg, and R. D. Kroner. "The Occurrence of Organohalides in Chlorinated Drinking Water," *J. Am. Water Works Assoc.* 66:703-706 (1974).
2. Aieta, E. M., P. V. Roberts, and M. Hernandez. "Determination of Chlorine Dioxide, Chlorine, Chlorite, and Chlorate in Water," *Am. Water Works Assoc.* 76(1):64-70 (1984).
3. Chow, B. M., and P. V. Roberts. "Halogenated By-product Formation by Chlorine Dioxide and Chlorine," *J. Env. Eng. Div. Am. Soc. Civ. Eng.* 107(4):609 (1981).
4. Berg, J. D., E. M. Aieta, P. V. Roberts, and R. C. Cooper. "Effectiveness of Chlorine Dioxide as a Wastewater Disinfectant," in *Progress in Wastewater Disinfection Technology*, A. Venosa, Ed., EPA-600/9-79-018 (Cincinnati: U.S. Environmental Protection Agency, 1979).
5. Kirk-Othmer. *Encyclopedia of Chemical Technology*, Vol. 5, 3rd ed., (New York: John Wiley & Sons, 1978), pp. 580-647.
6. Shilov, E. A., and S. N. Solodushenkov. "The Velocity of Hydrolysis of Chlorine," *Compt. Rend. Acad. Sci. USSR 3:15-19 (1936)*.
7. Morris, J. C. "The Mechanism of the Hydrolysis of Chlorine," *J. Am. Chem. Soc.* 68(9):1692-1694 (1946).
8. Shilov, E. A., and S. N. Solodushenkov. "The Mechanism of Hydrolysis of Chlorine," *J. Phys. Chem. (USSR)* 21:1159 (1947).
9. Lifshitz, A., and B. Perlmutter-Hayman. "The Kinetics of the Hydrolysis of Chlorine. I. Reinvestigation of the Hydrolysis in Pure Water," *J. Phys. Chem.* 64:1663-1665 (1960).
10. Spalding, C. W. "Reaction Kinetics in the Absorption of Chlorine into Aqueous Media," *AIChE J.* 8(5):685-689 (1962).
11. Eigen, M., and K. Kustin. "The Kinetics of Halogen Hydrolysis," *J. Am. Chem. Soc.* 84:1355-1361 (1962).
12. Sandal, O. C., et al. "Solubility and Rate of Hydrolysis of Chlorine in Aqueous Sodium Hydroxide at 273 K," *AIChE J.* 27(5):856-859 (1981).
13. Perlmutter-Hayman, B., H. Wieder, and M. H. Wolff. "The Role of Cations in Some Inorganic Hydrolysis Reactions," *Isr. J. Chem.* 11(1):27-36 (1973).

14. Brian, P. L. T., J. E. Vivian, and C. Piazza. "The Effect of Temperature on the Rate of Absorption of Chlorine into Water," *Chem. Eng. Sci.* 21:551–558 (1966).
15. Aieta, E. M., and P. V. Roberts, "Applications of Mass Transfer Theory to the Kinetics of Fast Gas-Liquid Reaction," (submitted for publication).
16. Aieta, E. M., and P. V. Roberts. "Chlorine Dioxide Chemistry: Generation and Residual Analysis." in *Chemistry and Chemical Analysis of Waste/Wastewater Intended for Reuse*, W. J. Cooper, Ed., (Ann Arbor, MI: Ann Arbor Science Publishers, Inc., 1981), pp. 429–452.
17. Taube, H., and H. Dodgen. "Applications of Radioactive Chlorine to the Study of the Mechanisms of Reactions Involving Changes in the Oxidation State of Chlorine," *J. Am. Chem. Soc.*, 71:3330–3336 (1949).
18. Emmenegger, F., and G. Gordon. "The Rapid Interaction between Sodium Chlorite and Dissolved Chlorine," *Inorg. Chem.* 6(3):633–635 (1967).
19. Moore, J. W., and R. G. Pearson. *Kinetics and Mechanism*, 3rd ed. (New York: John Wiley & Sons, 1981).

SECTION X

Chlorine Demand Reactions: Proteins and Other Organics

Oxidizing chemicals introduced into water during chlorination first disappear rapidly, and then more slowly. This process is usually described by stating that natural waters possess a chlorine demand.

George R. Helz, Douglas A. Dotson, and **Anne C. Sigleo**
Chlorine Demand: Studies Concerning Its Chemical Basis, 1981

"Curiouser and curiouser!" cried Alice.
Lewis Carroll, 1832–1898
Alice's Adventures in Wonderland

CHAPTER 63

Chlorination of the Peptide Nitrogen

R. C. Ayotte and E. T. Gray, Jr.

The widely accepted tenet that the proton of the amide nitrogen present in the peptide linkage (peptide nitrogen) cannot be exchanged for the oxidizing chlorine in HOCl or Cl_2 is based on various observations that are the results of experiments designed for other purposes. For example, when the simplest peptide, glycylglycine, is placed in solution at pH 7 and subjected to increasing quantities of chlorine (HOCl), two molar equivalents of chlorine rapidly exchange for the protons on the amine nitrogen. Any excess chlorine appears to remain as free chlorine, but the beginning of the breakpoint process masks any slower reactions that may simultaneously occur at the peptide nitrogen. Although observations of this type have led to the assumption that the peptide nitrogen has no chlorine demand, no definitive experiment has been reported in the literature designed specifically to prove that the peptide nitrogen cannot be chlorinated and that it has no chlorine demand. Because recent reports show that most of the nitrogen in potable and wastewaters is present as amino acids and peptides,[1] chlorination of the peptide nitrogen and the implication of a chlorine demand would have an impact on the models currently used for water chlorination, as well as on the interpretation of current analytical methods, which purport to separate classes of free and combined chlorine, such as the DPD method.

This chapter discusses the design of a definitive experiment to test the hypothesis that the peptide nitrogen can be chlorinated, reports the successful execution of that experiment, and presents the preliminary results of ongoing efforts to characterize the proposed chlorinated peptide nitrogen moiety.

EXPERIMENTAL DESIGN

Three points must be considered in designing a definitive experiment to maximize the possibility of observing the chlorination of the peptide nitrogen, either directly or through some effect that can be unambiguously attributed to the chlorination of the peptide nitrogen. First, the solution conditions should afford the largest possible rate constant, so that if the amide is chlorinated, the chlorination reaction will occur as rapidly as possible. Second, the experiment should be executed under conditions that are consistent with a favorable formation equilibrium constant. Last, the choice of the peptide species should be made to minimize or, if possible, to eliminate all possible side reactions that might mask the reaction in question.

Solution Conditions for Rapid Kinetics

Figure 1 shows the relationship between the basicity of the nitrogen moiety and the second-order rate constant for the addition of chlorine from HOCl to amines and chloramines.[2] This plot suggests that chlorination by HOCl is controlled by the basicity of the incoming nucleophile (the amine or chloramine) and that, except for ammonia, the rate constant for the chlorination of primary and secondary amines of known basicity can be determined from this plot. For the purposes of experimental design, the basicity of the peptide nitrogen was estimated from similar model compounds such as formamide and acetamide. This approximate value for log (K_H) was chosen to be −1, indicative of an extremely weak base.

$$RC(=O)NHR' + H^+ \rightleftarrows RC(=O)\overset{+}{N}H_2R' \qquad K_H = 0.10 \qquad (1)$$

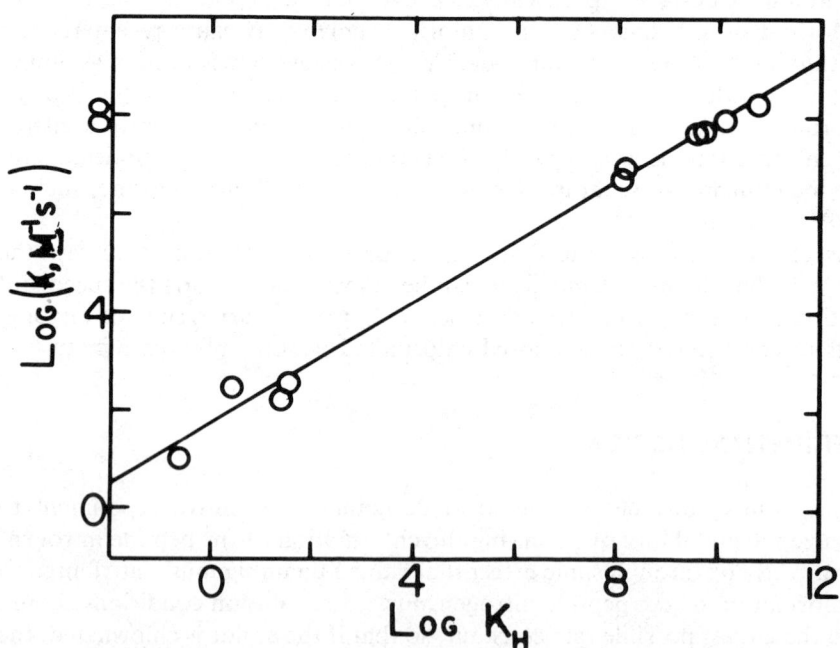

Figure 1. Dependence of the rate constant for the HOCl chlorination of amines and chloramines on the protonation constant of the amine or chloramine (ionic strength = 0.50 M NaClO$_4$; 25.0°C; from Reference 2).

Using this value for the equilibrium constant with Equation (2), which is the equation for the solid line in Figure 1,

$$\log(k) = 0.61 \, [\log(K_H)] + 1.7 \qquad (2)$$

the calculated second-order rate constant is 13 $M^{-1}s^{-1}$. Consequently, a 0.01 M solution of peptide would be predicted to have an observed pseudo-first-order rate constant of 0.13 s^{-1} if the peptide concentration were in strong excess over the total chlorine concentration. This is not very slow at all, and if Cl_2 were used as the chlorinating agent, the reaction should be even faster.[2] Therefore, the solution conditions necessary to promote a rapid chlorination reaction would presumably be at a low enough pH and high enough [Cl⁻] to ensure full conversion of uncombined chlorine to Cl_2 from HOCl.

Predicted Equilibrium Constant

Data in the literature for equilibrium constants of chloramine formation reactions are limited. Table I gives the equilibrium constants, K_{HOCl}, for three amines forming chloramines with HOCl, along with the log (K_H) values for the parent amines.[2] Although the sample of values is small, the trend implies a lower K_{HOCl} with a lower K_H. To design the experiment, if the relationship was 1:1, the K_{HOCl} associated with a $K_H = 0.1$ M^{-1} would be approximately 100 M^{-1}. Furthermore, the equilibrium constant with Cl_2 as the chlorinating agent would be expected to be higher.[1] This supports the design of low pH, high [Cl⁻] solution conditions for the experiment and suggests the possibility of chlorinating the peptide moiety.

Table I. Summary of the Equilibrium Constants for Three Amines for the Formation of Chloramines with HOCl, K_{HOCl}, vs the Protonation Constants K_H for the Parent Amines (Ionic Strength = 0.50 M NaClO$_4$, 25.0°C).[a]

log K_{HOCl}	log K_H
11.18	9.43
13.58	10.15
14.08	10.80

[a]Taken from Reference 2.

Choice of Peptide

To eliminate any side reactions of the amine group of the peptide, the amine functional group must be blocked or removed. The model compound acetylglycine

$$\begin{array}{c} \text{H} \quad \text{O} \quad \text{O} \\ | \quad \parallel \quad \parallel \\ \text{H-C-C-N-C-OH} \\ | \quad \quad | \\ \text{H} \quad \quad \text{H} \end{array}$$

in which the amine group of glycylglycine is replaced by a proton, was chosen for the experiment both because it is the analog of the simplest peptide and because it is easily prepared[3-5] via a one-step condensation of glycine with acetic anhydride.

EXPERIMENTAL

Preparation of Acetylglycine

The procedure of Bandrowski[3] and Baeyer,[4] as modified by Ruggli,[5] was used, except that only the first cut of product was taken, and the model peptide was subsequently recrystallized twice from water. The compound was confirmed by melting point, IR, and NMR.

Reagents

All solutions were Baker reagent grade. Sodium hypochlorite was standardized spectrophotometrically[6] at 292 nm, using a molar absorptivity of 350 $M^{-1}\text{cm}^{-1}$. Solutions of sodium hydroxide were prepared from electrolytic pellets (low carbonate) and standardized against KHP. Sodium perchlorate was prepared by neutralization of sodium bicarbonate with $HClO_4$. Sodium iodide, sodium chloride, and sodium nitrate were used without further purification.

Apparatus

All spectrophotometric measurements were made with a Cary 219 interfaced to a Z-80 microprocessor (Tandy). Kinetics experiments on the Cary 219 were initiated with a hand-driven rapid-mixing unit and a glass Y-mixer mounted upstream of the 1.00-cm flow cell. All pH measurements were made with an Orion model 701 meter, a Beckman pH 0–14 glass electrode, and a commercial calomel reference electrode. The reference electrode was bridged to the test solution from a saturated KCl solution through an agar-agar salt bridge (3 M $NaNO_3$). The meter was standardized using standard pH 6.86 and pH 4.01 buffers (Fisher), followed by $HClO_4$/NaOH titrations held at an ionic strength of 0.5 M $NaClO_4$. This standardization resulted in the relationship $-\log (H^+)$ = 1.02 pH - 0.12.

The stopped-flow instrument used for these experiments was constructed in our laboratory. It employs the optical path and power supply of a Beckman DU-2 for the deuterium lamp with electronic modifications to reduce noise. The visible lamp and the photomultiplier have separate power supplies. Solutions are pneumatically driven through an Aminco mixer and flow cell (path length, 1.183 cm). The interface and computer used to collect the data are triggered upon impact of the stopping syringe. Kinetic data are analyzed by TRS-80 microcomputers using a combination of assembled code and a compiled-BASIC (Microsoft) main program for linear data analyses, graphic display (Integral Data Systems 440 Printer), and nonlinear data analysis. All kinetic measurements were carried out at 25.0 ± 0.1°C, and 0.5 M ionic strength was maintained with $NaClO_4$.

Solution Preparation

All chloramine solutions were prepared through a two-jet or a double two-jet mixer to ensure rapid mixing.

RESULTS AND DISCUSSION

Spectral Observation of Mixtures of Acetylglycine and Chlorine in Acid

The spectral results of mixing chlorine with an excess of acetylglycine in strongly acidic solution are presented in Figure 2. The spectrum of the reactants was obtained by placing the acetylglycine solution and the chlorine solution on opposite sides of a split, 1.00-cm cell. The product spectrum was obtained by mixing equal volumes of the reagent solutions and then placing the mixed solution in both cell compartments. Sample preparation plus one spectral scan takes approximately 10 min. The spectrum of the product mixture was scanned repeatedly for 1 h and then scanned again after 48 h. No change in the spectrum was observed. Because, as shown in Figure 2, the Cl_2 absorption band is not present in the product spectrum, it can be concluded that either the chlorine was destroyed or that the peptide nitrogen was chlorinated, had no absorption maximum above 200 nm, and was stable over the 48-h period.

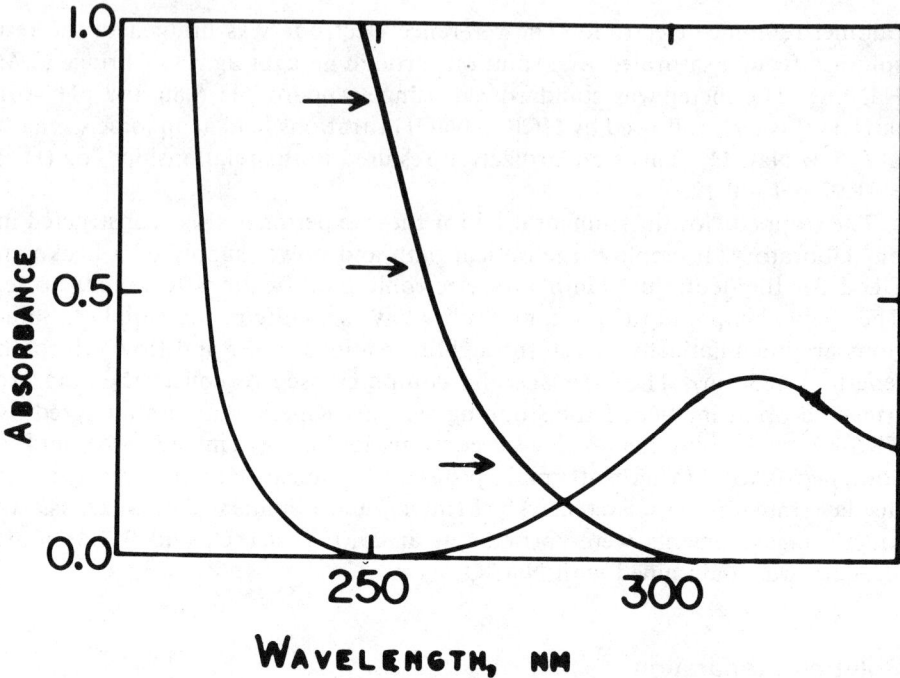

Figure 2. Spectra of acetylglycine and Cl_2 before mixing (solid line with asorbance maximum at 323 nm) and after mixing (arrows). [Acetylglycine] = 0.0502 M; [Cl_2] = 0.00507 M; [Cl^-] = 0.01 M; [H^+] = 0.10 M $HClO_4$; 1.00-cm cell, 25.0°C.

The experiment was reproduced (pH 1, 0.0502 M acetylglycine, and 0.00507 M chlorine), but this time two solutions were prepared, one with acetylglycine and one without, as a control. The chlorine content was the same in both solutions. Immediately after mixing, 1.00 mL of each solution was introduced into separate flasks, each containing 50 mL of a 0.05 M KI solution (pH 1). The spectra of the resulting I_3^- solutions were identical within experimental error. The same analysis was carried out after 1 h had elapsed and gave the same results. Although the spectrum of the Cl_2 was not observed in the presence of an excess of acetylglycine, the oxidizing power of the solution was retained.

Variation of pH

A solution containing 0.0125 M acetylglycine and 0.00126 M chlorine was prepared at pH 2. The spectrum is shown in Figure 3. The solution was

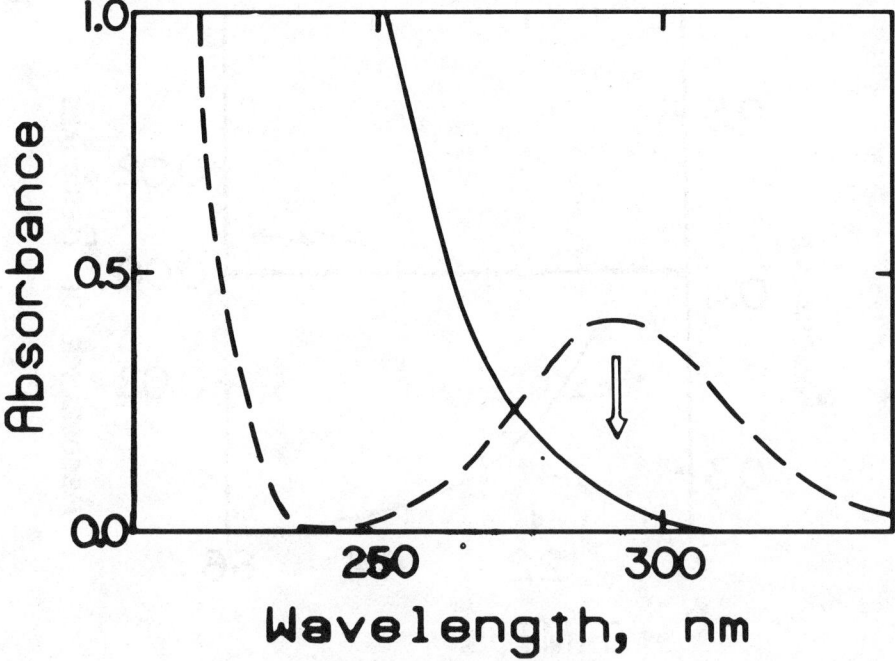

Figure 3. Spectrum of a mixture of acetylglycine and chlorine at pH 1 (solid line) and pH 11 (dashed line). [Acetylglycine] = 0.0125 M; [Cl_2] = 0.00126 M; [Cl^-] = 0.01 M; 1.00-cm cell, 25.0°C.

increased to pH 11 by adding drops of sodium hydroxide. This spectrum is also shown in Figure 3. The pH 11 spectrum clearly shows the presence of the OCl^- band centered about 292 nm. This solution was cycled to acid and back to base; the OCl^- band was lost in acid and reappeared in base. The reaction occurred within the time necessary to change pH, rinse and fill the cell, and take the spectrum.

One change was observed. While the solution was basic, repetitive scans showed a slow decrease of the OCl^- band with time. This decay was arrested when the sample was made acidic but continued when the solution was returned to pH 11; that is, the spectrum taken just before the solution was made acidic was the same as the spectrum taken after that acidic solution was again made basic. The basic solution was tested with an acidic iodide solution after 12 h. No evidence of oxidizing power remained. Therefore, although an acidic solution is stable for 2 d, a solution of chlorine and the model peptide at pH 11 shows a clear chlorine demand.

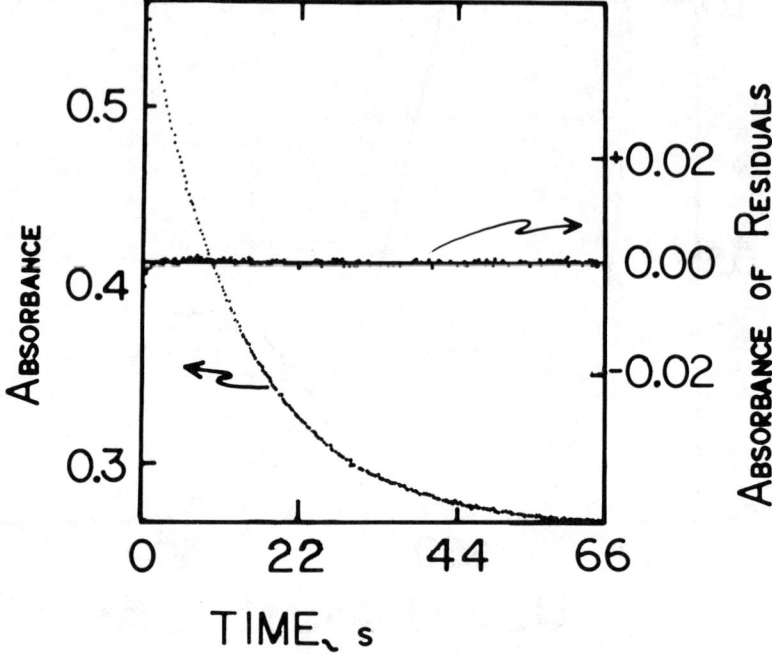

Figure 4. Plots of the loss of absorbance at 323 nm following the reaction of Cl_2 with acetylglycine and of the residual line (horizontal line) indicating the reaction follows pseudo-first-order kinetics. k_{obsd} = 0.066 s^{-1}; [acetylglycine] = 0.098 M; [Cl_2] = 0.00419 M; [Cl^-] = 0.05 M; [$NaClO_4$] = 0.50 M; pH 1.10, 25.0°C.

Preliminary Kinetic Observations

Figure 4 is an example of the chlorination reaction of the model peptide in very acidic solution. The reaction is first order with respect to chlorine over the entire reaction. When this reaction is observed at higher pH, some non-first-order character is observed in the first 10% of the reaction. This observation is concurrent with the change in the [Cl_2]/[HOCl] ratio from a predominance of Cl_2 to a predominance of HOCl. The irregularity may be caused by a mechanistic change in the formation of the chloramide, by a small residual impurity of glycine from incomplete purification of the acetylglycine, or by hydrolysis of the acetylglycine before mixing with HOCl. However, the remaining 90% of the observed reaction is first order with respect to HOCl, and the rate constant is about a factor of 10 slower than that with Cl_2. It is noteworthy that a

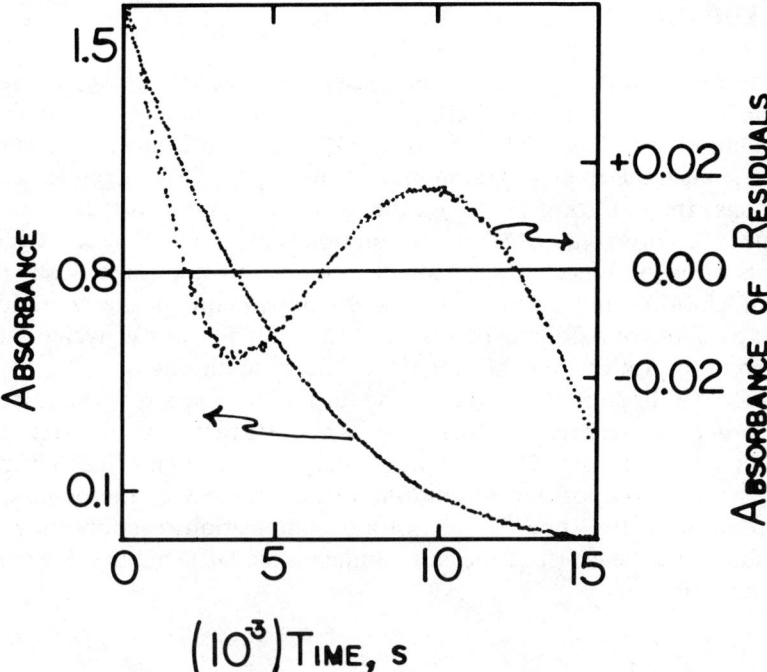

Figure 5. The loss of absorbance at 292 nm following the decomposition of OCl$^-$ in the presence of acetylglycine at pH 11.6, and the residual line (wavy line) indicating that the reaction does not follow pseudo-first-order kinetics. [Acetylglycine] = 0.0427 M; [OCl$^-$] = 0.00423 M; [Cl$^-$] = 0.05 M; [NaClO$_4$] = 0.50 M; 25.0°C.

reaction is seen when either Cl$_2$ or HOCl is the predominant form of free chlorine, but not when OCl$^-$ predominates. This observation is currently being exploited in an effort to extract an equilibrium constant for the formation of the chloramide.

Figure 5 is a plot of the decrease of the absorbance of OCl$^-$ as a function of time in the presence of excess acetylglycine. The nonlinear residual line indicates that the data do not correspond to the integrated first-order rate expression. In fact, the absorbance vs time function can be explained by no single reaction order. Although the mechanism for this decomposition is not yet understood, the reaction does occur. The peptide nitrogen does have a chlorine demand. The full mechanism of this chlorine demand is currently under investigation.

CONCLUSIONS

It has been shown that the model compound acetylglycine can take up chlorine in acidic solution; however, it is not a strong enough base to compete for the chlorine in basic solution where OCl⁻ is the prevalent species of free chlorine. When the chlorine is combined with the peptide nitrogen under acidic conditions, the oxidizing power of the solution is stable for days, and free chlorine is rapidly obtained when the solution is made basic. When significant amounts of uncombined chlorine are observed spectrophotometrically in mixtures of chlorine and acetylglycine, a decomposition of the free chlorine, evident by a loss of oxidizing power, is observed. Thus, acetylglycine catalyzes the decomposition of free chlorine under these conditions.

When Cl_2 is the source of chlorine, the formation reaction of the chloramide is first order with respect to chlorine and is about a factor of 10 faster than the analogous reaction when HOCl is the primary source of chlorine. The mechanism for the decomposition of chlorine in the presence of the peptide under basic conditions, the apparent lack of a decomposition reaction under acidic conditions, and the effect of the chloramide on the DPD method are currently under investigation.

REFERENCES

1. Ram, N. M., and J. C. Morris. "Selective Passage of Hydrophilic Nitrogenous Organic Materials Through Macroreticular Resins," *Environ. Sci. Technol.* 16(3):170–174 (1982).
2. Margerum, D. W., E. T. Gray, Jr., and P. R. Huffman. "Chlorination and the Formation of N-Chloro Compounds in Water Treatment," in *Organometals and Organometalloids, Occurrence and Fate in the Environment*, F. E. Brinckman and J. M. Belloma, Eds. (Washington, DC: American Chemical Society, 1979), pp. 278–291.
3. Bandrowski, E. "Ueber Acetylendicarbonsaure," *Ber. Dtsch. Chem. Ges.* 10:838–842 (1877).
4. Baeyer, A. "Ueber Polyacetylenverbindungen," *Ber. Dtsch. Chem. Ges.* 18:674–681 (1885).
5. Ruggli, P. "Ueber Versuche zur Darstellung von Derivaten des Diamido-acetylens," *Helv. Chim. Acta.* 3:559–572 (1920).
6. Galal-Gorchev, H., and J. C. Morris. "Formation and Stability of Bromamide, Bromimide, and Nitrogen Tribromide in Aqueous Solutions," *Inorg. Chem.* 4(6):899–905 (1965).

CHAPTER 64

Contribution of Proteins to the Formation of Trihalomethanes on Chlorination of Natural Waters

Frank E. Scully, Jr., Robert Kravitz, G. Dean Howell,
Mark A. Speed, and Richard P. Arber

Organic nitrogen compounds are abundant and essential components of nature. In several studies of natural waters,[1,2] dissolved organic nitrogen compounds comprised as much as one-third of the total dissolved organic material. Organic nitrogen levels in natural waters can vary widely with water source and season, but levels of 0.5 to 1.0 mg/L are not uncommon.[2-6] For instance, Speed has followed the yearly fluctuations of organic nitrogen concentrations in shallow gravel lakes, which serve as the main drinking water supply for Thornton, Colorado. Levels in June have measured 3.8 mg/L, whereas in February, levels of 1.2 mg/L are more typical.[3] Ram and Morris have found one freshwater source with as much as 21.7 mg/L organic nitrogen and correlated this with the bloom of a blue-green algae.[5] Several studies have attempted to subdivide the organic nitrogen fraction into amino and non-amino nitrogen compounds. Gardner and Stephens[7] found that 20% of the dissolved organic nitrogen in a marine water sample was amino nitrogen, whereas in a freshwater system Gocke found as much as one-third was organic amino nitrogen compounds.[1] All studies agree that the major portion of the amino nitrogen present in natural waters is tied up in proteins and polypeptides rather than as free amino acids.

The presence of high concentrations of proteins and amino acids in natural waters seems to be related to algal activity. For instance, Gardner and Lee[8] have found substantially higher concentrations of amino acids during a period of rapid algal decomposition. Tuschall and Brezonik[2] have found that 71% of the dissolved organic nitrogen in the filtrate of a culture of *Anabaena* sp. was proteinaceous. The full implications of these observations for the treatment of natural water affected by algae growth have not yet been fully realized, but Williams[9] has described problems of taste, odor, disinfection requirements, and cost of treatment associated with the presence of proteins in drinking water supplies.

Oliver and Shindler[10] and Hoehn et al.[11,12] have recently reported that algae, both their biomass and their extracellular products, can be important sources of trihalomethane (THM) precursors. Because proteins are important intercellular components of all cells as well as extracellular by-products of algae cultures, algae growth and seasonal variations of THM production from chlorination seem to be related to the presence of proteins in some natural waters. The question of whether this relationship is direct or coincidental has been part of the rationale for the study described here.

Because Morris et al.[13] have demonstrated that several of the amino acids found in most proteins can produce chloroform in chlorinated water, a study was undertaken to evaluate the significance of proteins in the production of THMs in both natural and model water systems and to assess the effectiveness of current treatment methods for their removal.

EXPERIMENTAL

General

Model proteins used in this work were obtained from Sigma Chemical Company. To determine the yield of chloroform from chlorination of proteins, stock solutions of the proteins were prepared in chlorine-demand-free water and dialyzed overnight against deionized water using dialysis tubing with a molecular weight cut-off of 8000. These dialyzed stock solutions were quantitated spectrophotometrically before they were diluted to the desired concentration with chlorine-demand-free water. Control experiments showed that in all but one case (rennin), dialysis did not reduce the amount of chloroform produced.

Trihalomethane Analysis

Solutions were chlorinated with standard hypochlorite (1000 ± 10 mg/L) prepared as described in EPA Method 510.1. Trihalomethanes were isolated and analyzed by the liquid-liquid extraction method prescribed by EPA protocol.[14]

Samples of raw water were obtained in 1-L glass bottles and sealed without headspace using Teflon®-lined caps. The samples were shipped overnight to Norfolk where they were refrigerated at 4°C. They were analyzed within 1 week for protein concentration. Dissolved hydrolyzable amino acid nitrogen levels (protein concentrations) were measured by a variation of the method of Gardner and Stephens.[7] To eight vacuum hydrolysis tubes were added 0.5-mL aliquots of a sample of water to be analyzed. Standard additions of glycine were made to the tubes, followed by 0.5 mL of a 50% solution of distilled propionic acid and ultrapure concentrated HCl.[15] The samples were hydrolyzed for 30 min at 145°C, cooled, and lyophilized before being derivatized with fluorescamine and analyzed for fluorescence emission at 480 nm (excitation at 390 nm).

RESULTS AND DISCUSSION

Yields of THMs from Proteins

Four commercially available proteins were chosen as models of proteins present in nature on the basis of their different molecular weights, which span the range found by Tuschall and Brezonik.[2] Bovine serum albumin (BSA) is a readily available compound frequently used in model studies because its purity, secondary and tertiary chemical structures, activity, and function have been well characterized. It has the highest molecular weight of the compounds examined. Pepsin and rennin are digestive enzymes found in the stomachs of many mammals. Cytochrome c, which contains an iron-porphyrin group, is a pigmented protein used by aerobic cells to transfer electrons to oxygen. Although the structure of cytochrome c isolated from different cells varies slightly, a cytochrome of the c-type has been isolated from blue-green algae.[16]

Terminal (5 d) THM concentrations were measured using solutions of each protein with a carbon content of 2 to 3 mg/L in 0.025 M potassium phosphate at pH 7.0, chlorinated to 20 mg/L Cl_2, and incubated in the dark at 20°C. The results are recorded in Table I. Humic acid obtained from Aldrich Chemical Company was used for comparison. Surprisingly, the yields of chloroform produced by proteins of widely varying structure and molecular weight are quite similar. Although they seem low, they are about one-half that of humic acid, which also gives low yields.

Table I. Yields of Trihalomethanes Formed 5 d After Chlorination of Solutions of Proteins

Chlorination Substrate	Mol wt	% Carbon	THM	
			µg/L	% Yield
Bovine serum albumin	66,000	45	97	0.44
Pepsin	30,000	46	117	0.51
Rennin[a]	30,000	54	117	0.49
Rennin	30,000	54	47	0.20
Cytochrome c	12,500	45	93	0.42
Humic acid	Unknown	35	135	0.78

[a]This solution of rennin was not dialyzed.

Modeling of Proteins vs Humic Acid

Because the yield of chloroform from BSA appeared to be representative of that of other proteins, BSA was chosen as a model for studying the production from proteins. Figure 1 is a plot of the chloroform produced with time when a solution of either BSA (5 mg/L) or humic acid (HA)(5 mg/L) is chlorinated to an initial concentration of 20 mg/L. Qualitatively, the curve obtained for the formation of chloroform from BSA is similar to that obtained from the chlorination of HA. However, there is one noteworthy difference. The THM formation from BSA is much slower than that from HA at incubation times less than 24 h. It is probable that before chloroform can be produced from the reactive sites on the protein, the protein must be oxidized sufficiently to allow the protein to unfold enough to expose these sites.

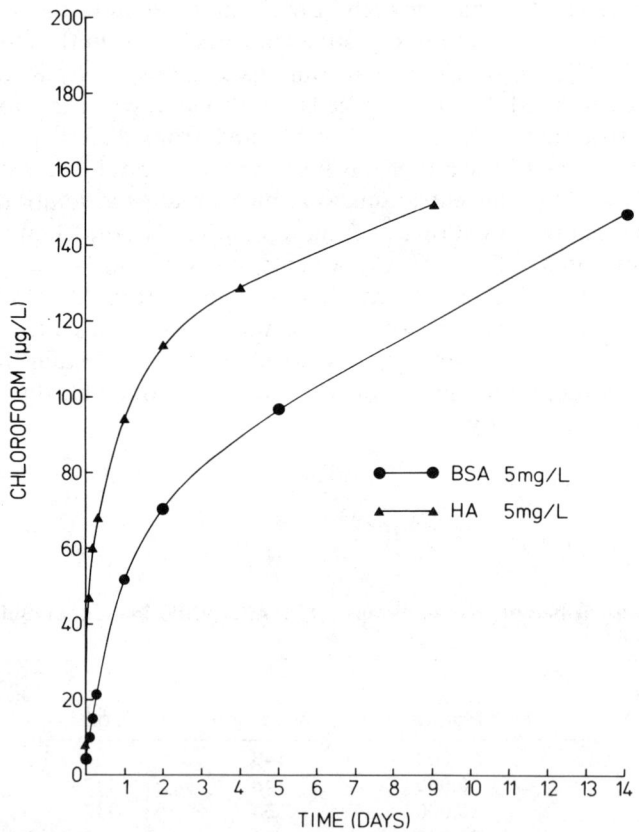

Figure 1. The amount of chloroform produced with time after solutions of 5 mg/L BSA (2.2 mg/L as carbon) or 5 mg/L humic acid (1.7 mg/L as carbon), both in 0.025 M aqueous potassium phosphate at pH 7.0, were chlorinated to 20 mg/L Cl_2 and incubated in the dark at 20°C (±0.5°C).

The effect of increasing the initial Cl_2 dosage on the chloroform produced from a constant concentration of protein over 5 d is plotted in Figure 2. Because 5 mg of BSA contains 2.2 mg total organic carbon (TOC), Figure 2 shows that up to a Cl_2:TOC weight ratio of 1, increasing the chlorine concentration has no significant effect on the chloroform production. Between a Cl_2:TOC ratio of 1 and 5, the chloroform production increases, but above a ratio of about 5:1, increasing the chlorine concentration has little effect. These results are very similar to those of Engerholm and Amy[17] who studied HA. However, one additional fact should be noted. On the same graph are plotted the total residual chlorine levels measured shortly before the solutions were analyzed for chloroform. No residual chlorine remained after 5 d to produce more chloroform when solutions were chlorinated to a Cl_2:TOC ratio below 5:1. This is presented somewhat differently in Figure 3, which shows the effect of increasing the Cl_2:TOC ratio on the THM formation over a period of 5 d. It appears that doubling the chlorine concentration from 10 to 20 mg/L doubles the amount of chloroform produced. However, it should be pointed out again that no free residual chlorine remains by 72 h after initially dosing solutions to 15 mg/L or less. Consequently, the apparent chlorine dosage dependency observed in Figures 2 and 3 results primarily because the free chlorine is completely consumed over 5 d when the Cl_2:TOC ratio is less than 5.

Trussel and Umphres[18] noted the same phenomenon with HA and divided a curve like Figure 2 into three regions. The first region occurs at low chlorine

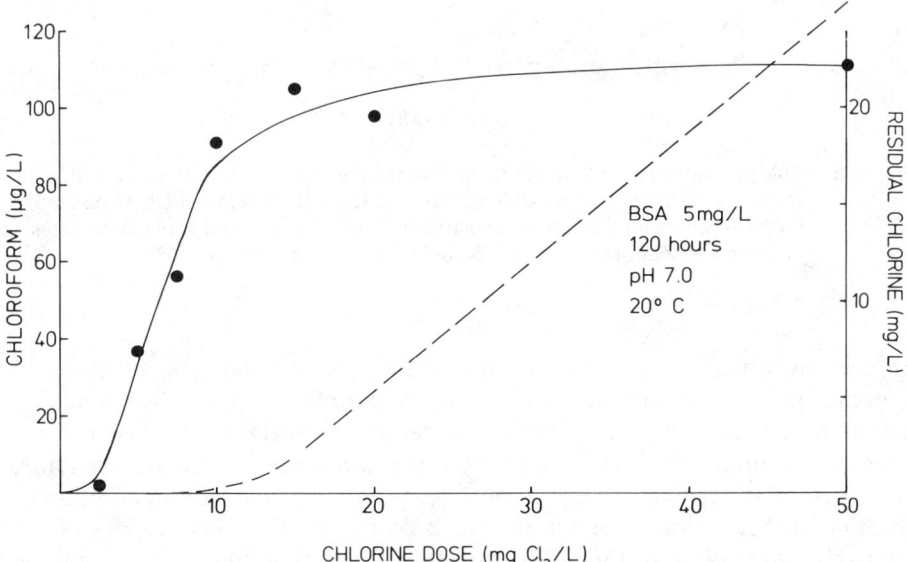

Figure 2. Chloroform produced and residual chlorine remaining 120 h after a solution of 5 mg/L BSA (2.2 mg/L as carbon) in 0.025 M aqueous potassium phosphate (pH 7.0) was chlorinated to various levels and stored in the dark in a constant-temperature bath at 20°C (±0.5°).

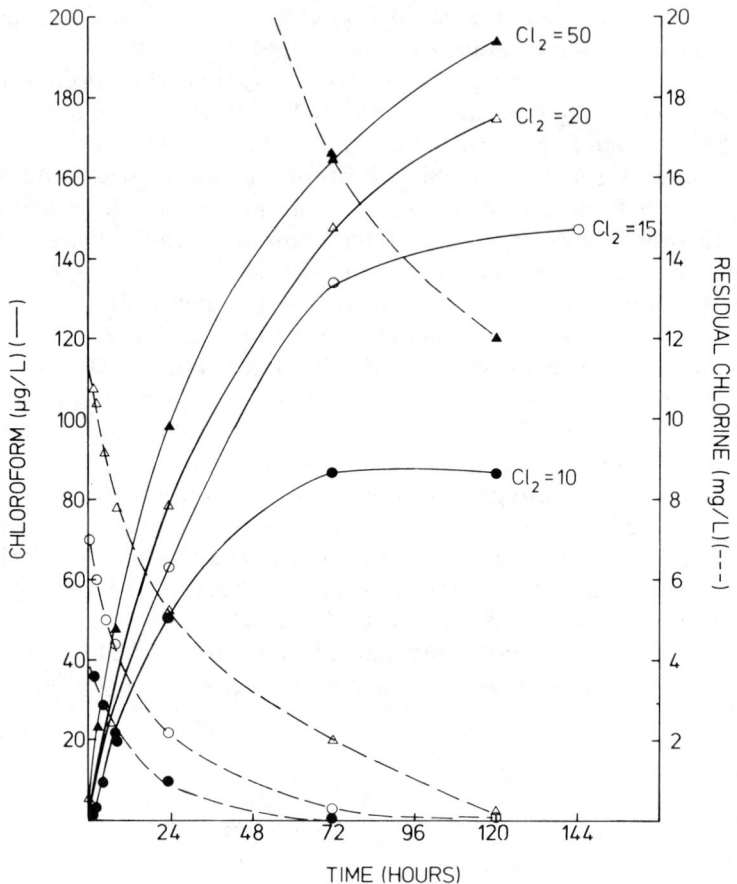

Figure 3. The amount of chloroform produced and the corresponding total residual chlorine remaining at various times after solutions of 10 mg/L BSA (4.4 mg/L as carbon) in 0.025 M aqueous potassium phosphate at pH 7.0 were chlorinated to varying chlorine concentrations and incubated in the dark at 20°C (±0.5°C).

dosages in which immediate chlorine demand is exerted. As Trussel and Umphres point out, in natural water samples this effect is typically exerted by inorganic species such as ammonia, sulfide, and iron(II). In the case of proteins, labile amino acid residues such as lysine, arginine, cysteine, hystidine, and tryptophan are likely to be responsible. Whatever the reactant, addition of chlorine in this region results in limited production of THMs because of the relatively slow rate of the THM formation reaction. The second region described by Trussel and Umphres occurs after the immediate chlorine demand is satisfied. In this region chloroform production is directly proportional to the amount of chlorine added. The amount of chloroform produced is limited

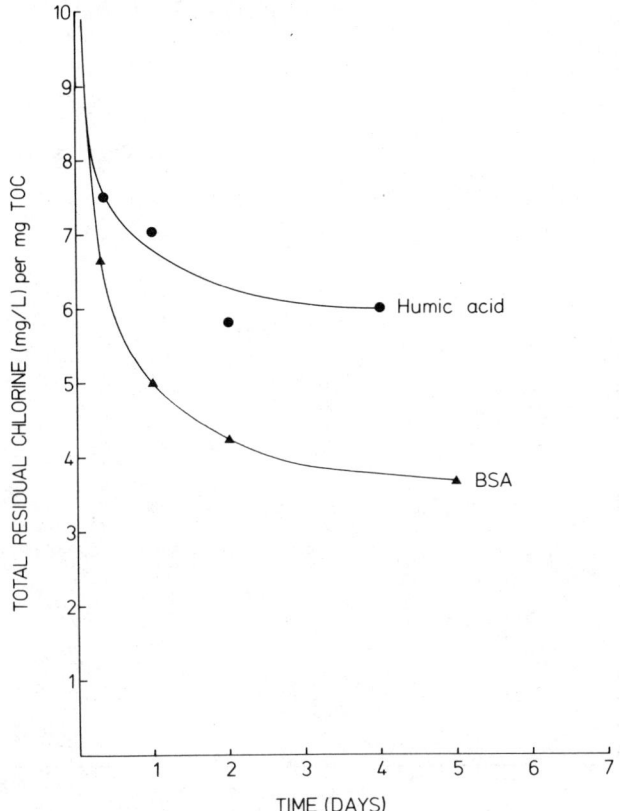

Figure 4. The residual chlorine remaining at varying times after solutions of humic acid (1 mg/L as carbon) or BSA (1 mg/L as carbon), both in 0.025 M potassium phosphate buffer at pH 7.0, were chlorinated to 11.9 mg/L Cl_2 and stored in the dark at 20°C ($\pm 0.5°$) before being analyzed.

because there is not enough chlorine present to fully oxidize the organics to chloroform. At a Cl_2:TOC ratio of 5:1, sufficient chlorine is present to completely meet the demand needed to produce chloroform and comparatively little effect is observed with the addition of more chlorine.

Proteins appear to exert a greater chlorine demand per milligram of TOC than HA. This is demonstrated in Figure 4 where the total residual chlorine per milligram of TOC is plotted as a function of time for solutions of HA and BSA, both of which were initially dosed to 11.9-mg/L chlorine. There is a much greater demand by the BSA than by the HA in the first few hours after dosing. After the first 24 h, the chlorine dissipation rate levels off to more similar rates for each solution.

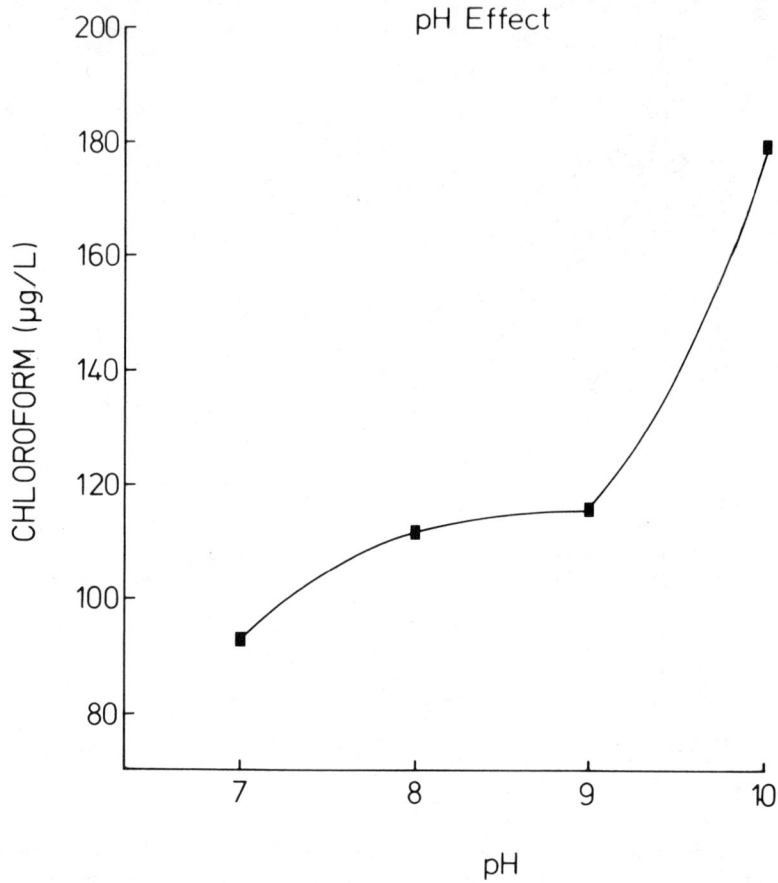

Figure 5. The yield of chloroform obtained 5 d after solutions of 5 mg/L BSA in 0.025 M aqueous potassium phosphate (pH 7.0 and 8.0) or 0.025 M aqueous sodium borate (pH 9.0 and 10.0) were chlorinated to 20 mg/L Cl_2 and incubated in the dark at 20°C (±0.5°).

It has been reported that increasing the pH of a solution of HA dramatically increases the yields of THMs formed for a specific Cl_2 dose and TOC concentration. BSA also shows a marked increase in yield of chloroform as the pH is raised, especially above pH 9.0 (Figure 5). At pH 10.0, the yield from BSA is almost twice that at pH 7.0.

One of the goals of our work was to develop a mathematical model for predicting the yield of THMs from chlorination of solutions containing known concentrations of proteins. Combined with measured concentrations of proteins in natural waters, this model could be used to predict the contribution proteins make to the THM formation potential of a given raw water.

The results of time-dependent THM formation curves for three different levels of BSA at four different initial chlorine dosages were used as parameters in the algorithm developed by Engerholm and Amy:[17]

$$\ln (CHCl_3) = \ln K + x \ln (TOC) + y \ln (Cl_2/TOC) + z \ln(t) \qquad (1)$$

In this equation, K is a reaction constant that is a function of temperature and pH; TOC is the initial total organic carbon content of the solution in milligrams per liter of precursor; Cl_2 is the initial chlorine concentration in milligrams per liter; and t is the time in hours after dosing. The terms x, y, and z are constants characteristic of the particular trihalomethane precursor. In this particular study, the reaction temperature was kept constant at 20°C and the pH was maintained at 7.0, using a 0.025 M phosphate buffer.

The multiregression analysis yielded values of K (0.35), x (1.17), y (0.467), and z (0.855) with a correlation coefficient, R^2, of 0.92. The regression was then repeated using a reduced data base, excluding data produced from samples that did not have any residual chlorine at the time of the analysis or that had large standard residuals from the predicted values. These last values corresponded to the data obtained in the first 2 h after dosing, when the protein is believed to be unfolding. In these first hours, the THM levels are low and the relative error in the measurement is largest. With removal of these points, the multiregression analysis yields values of K (1.35), x (0.97), y (0.27), and z (0.64) with a correlation coefficient, R^2, of 0.96.

Engerholm and Amy[17] obtained values for a natural HA of K (16.8), x (0.95), y (0.28), and z (0.22) under the same conditions of temperature and pH ($R^2 = 0.98$). The values Amy obtained for x and y are identical to those we obtained with BSA, whereas those he calculated for K and z differ dramatically. The similarity of the x values in the two studies is understandable. For limiting concentrations of TOC, one might expect the amount of chloroform to double when the concentration of carbon substrate is doubled. The similarity in the value of y obtained for the two substrates was not expected.

The major difference between proteins and HA is the rate of the THM formation reaction. This is supported by the relative importance of the time term in the algorithm for the chloroform production from BSA. The value of z is 0.64 in this expression compared with the value of 0.22 in the equation for the HA, which suggests that THM formation from proteins is much slower than from HA. On the other hand, because proteins produce THMs in lower yield than HA, they would require a longer incubation time to produce the same THM levels as a similarly concentrated solution of humic acid. Nevertheless, even when this is taken into account, the rate of chloroform formation from proteins is still slower than from HA. This can be illustrated in the following manner. Consider a solution of a humic acid that obeys the model developed by Amy. If this solution contains 1.28 mg (as TOC) HA and is chlorinated to a Cl_2:TOC level of 6, then after 120 h (5 d), 100 µg of chloroform will be produced. Because of the lower yield of chloroform per gram of

TOC, it requires 2.21 mg protein to produce 100 μg of chloroform if a solution of BSA that follows our model is chlorinated to a Cl_2:TOC level of 6 and allowed to sit for 5 d. Even though a higher Cl_2 dose is required to achieve a Cl_2:TOC level of 6 in the solution of the protein, the two solutions approach the 100 μg maximum at different rates. This difference is illustrated graphically in Figure 6, where it can be seen that the humic acid solution produces far more chloroform in the first 24 h than the protein solution.

The objective of developing the model presented here was to measure the relative importance of the contribution of proteins to the overall production of THMs in a natural water. The model is only an approximation of how proteins may produce THMs on chlorination. A greater understanding of the distribution of proteins in natural waters is required before the full significance of their role in the formation of THMs can be determined.

Proteins in Natural Waters

The main source of raw water for the city of Thornton, Colorado, is a series of shallow gravel lakes adjacent to the South Platte River, 1 mile downstream from the discharge point of the Denver Metro Wastewater Treatment Plant. Besides periodic groundwater intrusion from the river, the lakes are affected regularly by algal blooms. Throughout the year they are found to be high in ammonia and dissolved organic nitrogen.

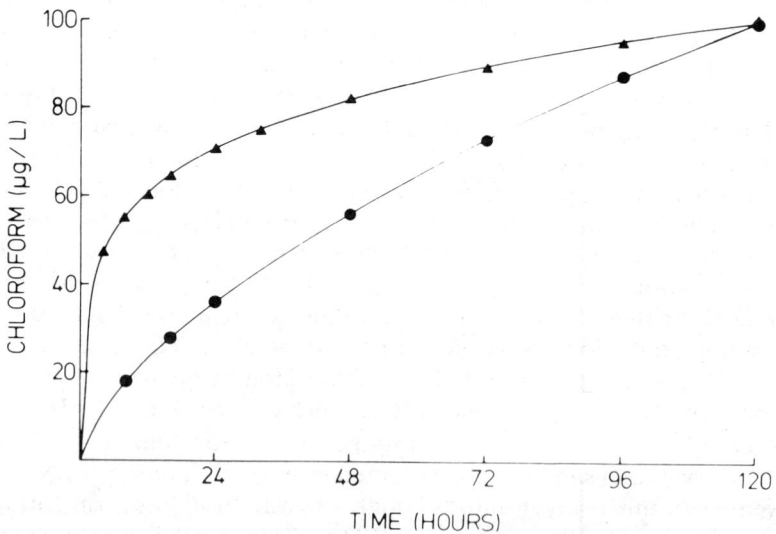

Figure 6. The predicted yield of chloroform with time from solutions of humic acid (▲ – – – – – – ▲, 1.28 mg/L as carbon) and BSA (● – – – – –●, 2.21 mg/L as carbon), both in 0.025 M aqueous potassium phosphate at pH 7.0 assuming an initial Cl_2:TOC level of 6. The algorithm of Engerholm and Amy[17] was used to generate the curve for humic acid, and the algorithm developed in the current study was used to generate the curve for BSA.

The concentration of proteins in these lakes were measured regularly for 6 months. In general, levels were relatively low, as might be expected during winter months. There was a slight seasonal variation for one of the lakes, South Gravel Lake, to which an algicide is not added. Protein concentrations have varied from 0.17 ($\pm$ 0.08) mg/L on February 2, 1984, to 0.39 ($\pm$ 0.04) mg/L on February 22 and 0.31 ($\pm$ 0.05) on March 21. During an algal bloom, the protein concentration in another of the lakes jumped from an average of about 0.1 mg/L to 0.92 ($\pm$ 0.16) mg/L. Because these levels are measured in milligrams per liter as glycine (mol wt = 75), and because the molecular weight of the average amino acid in a protein is 120, the approximate weight of protein as BSA can be calculated by multiplying the concentration in milligrams per liter as glycine by 1.6 (i.e., 120/75). Because BSA is 45% carbon by TOC measurement, the weight of proteinaceous carbon is (milligrams per liter as glycine) $\times$ 1.6 $\times$ 45% = 0.72 $\times$ (milligrams per liter as glycine). Therefore, when there was about 0.39 ($\pm$ 0.04) mg/L protein (as glycine) in South Gravel Lake on February 22, there was 0.28 mg/L proteinaceous carbon in South Gravel Lake. A total of 5.1 mg/L TOC was measured in the lake water sampled on that day. If the assumption is made that no unusual interaction occurs among proteins and other organic compounds in the water, then, if the water were chlorinated to a free residual chlorine level of 10 mg/L (Cl_2:TOC = 2), after 5 d (120 h) the protein would contribute

$$CHCl_3 \text{ (micrograms)} = K \text{ (protein as C)}^{0.97} (Cl_2:TOC)^{0.27} (120)^{0.64}, \quad (2)$$

or about 10 μg/L.

The city of Thornton is bounded on one side by Lower Clear Creek and on the other by the South Platte River. Lower Clear Creek contains elevated concentrations of nitrogen and TOC because of industrial and municipal wastewater discharge upstream. On February 22, it contained 4.6 mg/L TOC and 0.75 mg/L protein as glycine (0.54 mg/L protein as carbon). In 5 d with the same chlorine dose as South Gravel Lake, Lower Clear Creek would produce 20 μg/L of chloroform. In fact, chlorination of this Lower Clear Creek sample to a 1-h free residual chlorine concentration of 2 ppm produced, after 5 d, 54 μg/L of chloroform, of which a significant percentage was likely to have been produced by chlorination of proteins.

On February 28, a study was conducted to determine the effect of different chlorine dosages along a breakpoint curve on the destruction of proteins and the concomitant formation of THMs over 5 d (Table II). The sample was collected from one of the gravel lakes, which was found to contain about 0.38 ($\pm$ 0.06) mg/L protein as glycine, (0.27 mg/L protein as carbon), and had a 1-h chlorine demand of about 4 mg/L. The concentrations of THMs began to increase dramatically at dosage levels of 5 mg/L and above. However, the protein levels remained relatively constant in samples that contained no free chlorine after 5 d. When free chlorine was measured in a sample, the protein concentrations began to decrease. This would seem to suggest that HA precursors are oxidized to THMs before proteins.

In one study, water was obtained from a reservoir used as the main water supply for a community in southeastern Pennsylvania. This reservoir is affected by heavy agricultural runoff and was suspected of containing high concentrations of proteins. Water was obtained from several points in the local treatment plant to determine the effect of the different phases of water treatment on the removal of proteins. The utility adds powdered activated carbon to its raw water followed within 5 min by a rapid mix with chlorine and alum, and sometimes lime. The water is flocculated for 20 min, settled for 4 to 5 h, and filtered through a dual medium of sand and anthracite coal. Ammonia is added after filtration, followed by the addition of lime and chlorine before the water is held in the clear well. A sample of raw water was obtained along with samples of filter effluent and finished water. The samples were dechlorinated with sodium metabisulfite and sealed in glass bottles (Teflon®-lined caps) with no headspace. These samples were shipped to Norfolk and analyzed for protein concentrations. The raw water contained 0.96 ± 0.16 mg/L protein as glycine, the filter effluent contained 0.37 ± 0.06 mg/L protein as glycine, and the finished water contained 0.26 ± 0.06 mg/L. It appears that flocculation, sedimentation, and filtration removed slightly more than one-half of the proteins but did not remove them completely.

CONCLUSION

Although the levels of THMs produced by proteins in these samples are relatively low, their contribution may be quite significant during summer months of high algal growth. In one study, it was found that conventional methods of water treatment removed about one-half of the protein from a raw water sample. Trihalomethanes are only one example of organohalogen compounds that may be produced by chlorination of proteins. More work is needed to investigate the other organohalogen compounds that are produced.

Table II. Protein Concentrations and 5-d Terminal Trihalomethane Concentrations Produced from Chlorination of Raw Water from Gravel Lakes at Well 28, Thornton, Colorado.

Chlorine Dose (mg/L)	Residual Chlorine (mg/L)	5-d TTHM (μg/L)	Protein (mg/L as glycine)
0	0		0.38 ± 0.06
2	0	26	0.42 ± 0.05
5	0	169	0.32 ± 0.04
7	0	217	0.41 ± 0.07
10	0.6	274	0.29 ± 0.05
12	1.8	286	0.19 ± 0.02

ACKNOWLEDGMENTS

This work was funded by the U.S. Environmental Protection Agency under assistance agreement CR-809333 with the agency's Metropolitan Environmental Research Laboratory, Cincinnati, Ohio (Mr. Ben Lykins, Project Officer).

REFERENCES

1. Gocke, K. "Untersuchungen uber Abgabe und Aufname von Aminosauren und Polypeptiden durch Plankton-organismen." *Arch. Hydrobiol.* 67:285-367 (1970).
2. Tuschall, J. R., Jr., and P. L. Brezonik. "Characterization of Organic Nitrogen in Natural Waters: Its Molecular Size, Protein Content and Interactions with Heavy Metals," *Limnol. Oceanogr.* 25(3):495-504 (1980).
3. Speed, M. Unpublished results, (Thornton, CO: City of Thornton, Utilities Department, 1983).
4. Stanbro, W. D. "The Chemistry of Amino Acids and Peptides in Power Plant Cooling Towers," *Chesapeake Sci.* 18(1):126-128 (1977).
5. Ram, N. M., and J. C. Morris. "Environmental Significance of Nitrogenous Organic Compounds in Aquatic Sources," *Environ. Int.* 4:397-405 (1980).
6. Manny, B. A. "Seasonal Changes in Dissolved Organic Nitrogen in Six Michigan Lakes," *Verh. Int. Verein. Limnol.* 18:147-156 (1972).
7. Gardner, W. S., and J. A. Stephens. "Stability and Composition of Terrestrially Derived Dissolved Organic Nitrogen in Continental Shelf Surface Waters," *Mar. Chem.* 6:335-342 (1978).
8. Gardner, W. S., and G. F. Lee. "The Role of Amino Acids in the Nitrogen Cycle of Lake Mendota," *Limnol. Oceanogr.* 20(3):379-388 (1975).
9. Williams, D. B. "The Organic Nitrogen Problem," *J. Am. Water Works Assoc.* 43:837-846 (1951).
10. Oliver, B. G., and D. B. Shindler. "Trihalomethanes from the Chlorination of Aquatic Algae," *Environ. Sci. Technol.* 14:1502-1505 (1980).
11. Hoehn, R. C., D. B. Barnes, B. C. Thompson, C. W. Randall, T. J. Grizzard, and P. T. B. Shaffer. "Algae as Sources of Trihalomethane Precursors," *J. Am. Water Works Assoc.* 72:344-350 (1980).
12. Hoehn, R. C., K. L. Dixon, J. K. Malone, J. T. Novak, and C. W. Randall. "Biologically Induced Variations in the Nature and Removability of THM Precursors by Alum Treatment," *J. Am. Water Works Assoc.* 76:134-141 (1984).
13. Morris, J. C., N. M. Ram, B. Baum, and E. Wajon. *Formation and Significance of N-Chloro Compounds in Water Supplies*, EPA 600/2-80-031 (Cincinnati: U.S. Environmental Protection Agency, 1980).
14. "Analysis of Trihalomethanes in Drinking Water. Appendix C-Part II," *Fed. Regist.* 44(231):68683-68690 (1979).
15. Westall, F., and H. Hesser. "Fifteen-Minute Acid Hydrolysis of Peptides," *Anal. Biochem.* 61:610-613 (1974).

16. Fogg, G. E., W. D. P. Stewart, P. Fay, and A. E. Walsby. *The Blue-Green Algae* (London: Academic Press, Inc., 1973), pp. 49–51.
17. Engerholm, B. A., and G. L. Amy. "A Predictive Model for Chloroform Formation from Humic Acid," *J. Am. Water Works Assoc.* 75:418–423 (1983).
18. Trussel, R. R., and M. D. Umphres. "The Formation of Trihalomethanes," *J. Am. Water Works Assoc.* 70:604–612 (1978).

CHAPTER **65**

Evolution of Amino Acids in Water Treatment Plants and the Effect of Chlorination on Amino Acids

C. Le Cloirec and G. Martin

Numerous nitrogenous organic molecules may be found in surface waters used for drinking water treatment, including naturally occurring substances (e.g., humic substances, chlorophylls, proteins, peptides, amino acids), agricultural products (e.g., herbicides, pesticides), and industrial products (e.g., amines, nitrophenols).[1,2]

In an extensive 1981 study carried out during production of drinking water in western France, we observed the production of various chemical families, especially amino acids.[3] These small biologically important molecules, which are released in the environment by microorganisms and bacteria, are very water soluble and are found in concentrations of 2 to 400 μg/L.[4]

This chapter discusses the general findings of that study[3] and specifically describes the methods and results for the case of alanine production and reactions.

EVOLUTION OF AMINO ACIDS DURING WATER TREATMENT

In the initial study, the following sequence of treatments was monitored in ten drinking water treatment plants:[3] prechlorination (chlorine or sodium hypochlorite), flocculation-sedimentation, sand filtration, and ozonation followed by post chlorination.

Quantity and nature of the free amino acids were followed at five stages: raw water, prechlorinated water, filter water, ozonated water, and treated water.[5] Examples of data are given in Figures 1 and 2.

1. We found that during prechlorination, the amino acids present are oxidized and produce aldehydes and nitriles. These amino acids appear to be responsible for a great part of the surface water chlorine demand.

2. An increasing concentration of amino acids developed during prechlorination. These apparently came from partial hydrolysis or depolymerization of proteins and peptides found in untreated water (example in Figure 3).

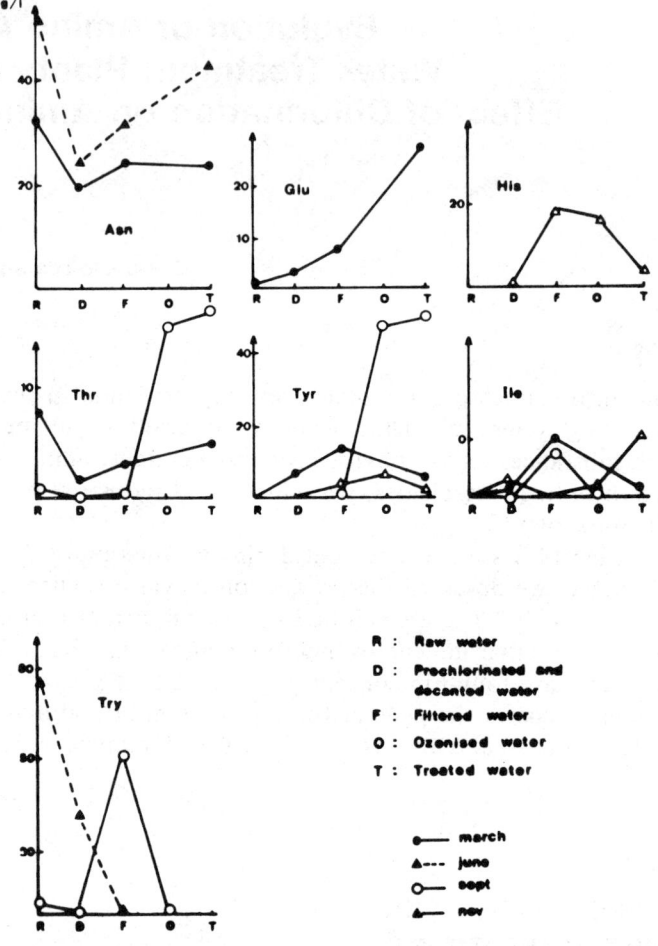

Figure 1. Examples of amino acid production in the Arzal water treatment plant. Amino acid designations: aspargine, asn; glutamic acid, glu; histidine, his; isoleucine, ile; threonine, thr; tyrosine, tyr.

3. The flocculation-decantation itself does not affect free amino acids except by some adsorption on flocs (example in Figure 4) or on activated carbon powder when injected.

4. In some cases, the release of amino acids is noticed after filtration, perhaps due to a sampling too close to the washing period or from microorganisms on the filters. This release is often significant after an activated carbon biological filter.[6-7]

5. Ozone may induce an increasing concentration of amino acids by the breakdown of the peptide linkages of the small proteins not eliminated by the previous treatments.[8]

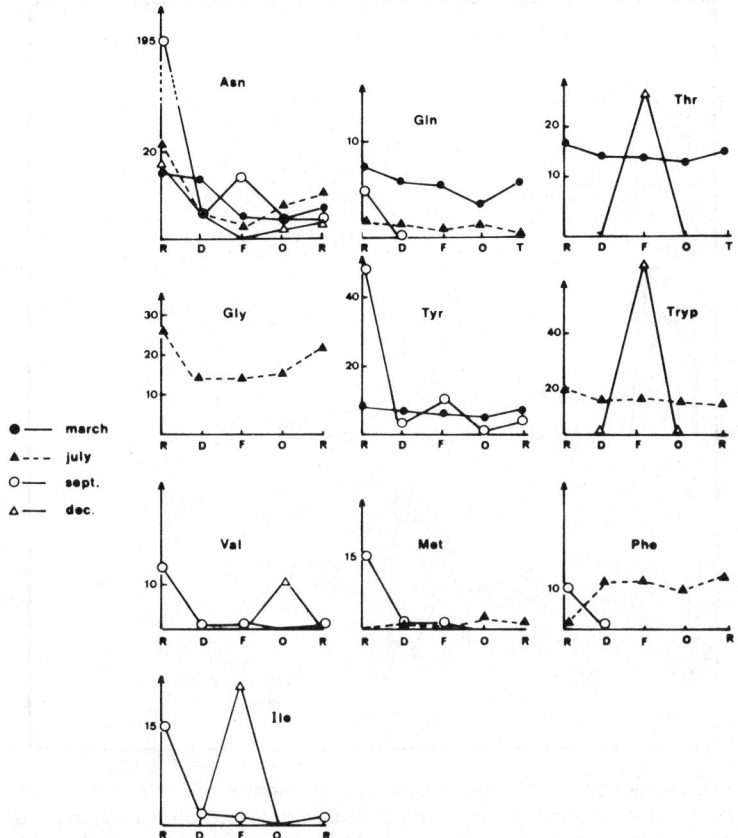

Figure 2. Examples of amino acid production in the Nantes water treatment plant. Amino acid designations: aspargine, asn; glutamine, gln; glycine, gly: isoleucine, ile; methionine, met; phenylalanine, phe; threonine, thr; tyrosine, tyr; tryptophan, tryp; valine, val.

These observations show that amino acid concentrations are reduced slightly by the first steps of the water treatment, and they are changed by the oxidation with ozone and chlorine. Consequently, a more complete study was made of the effect of chlorination on selected α-amino acids, alanine, phenylalanine, and tyrosine.

ALANINE CHLORINATION

Although chlorination of amino acids has been investigated previously, researchers are divided about the chemical nature of the final products.[9,10] For instance, Kantouch and Abdel-Fattah, studying the reaction of sodium hypochlorite with amino acid, reported that the products depended on pH and

824 WATER CHLORINATION

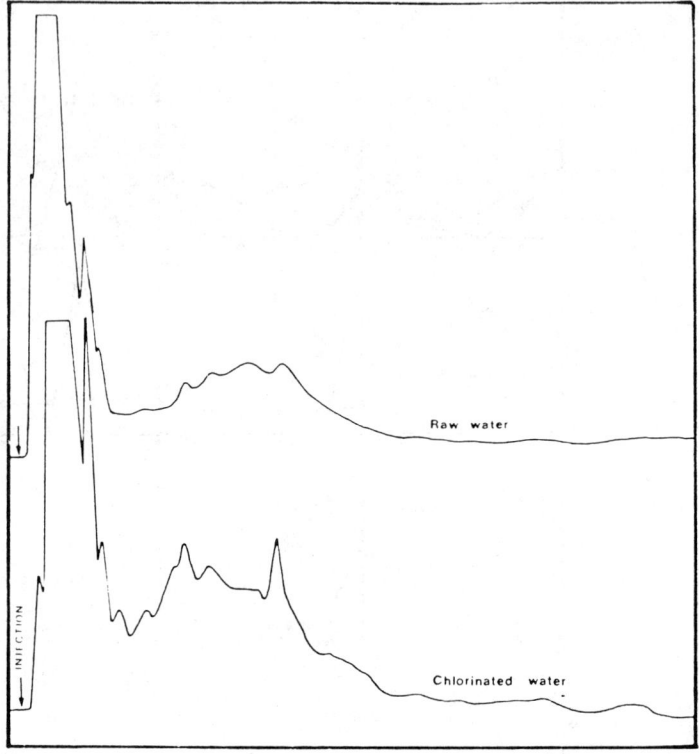

Figure 3. Chromatograms comparing amino acid concentrations in a river water sample before and after chlorination. The increased amino acid concentrations apparent after chlorination are probably caused by breakdown of the peptide linkages.

temperature, and always included aldehydes.[11] Other authors reported nitrile formation, this property being used for amino acid identification in gas chromatography.[10,12]

The purpose of our study was to clarify the knowledge of the mechanism(s) of chlorination of amino acids in dilute aqueous medium, and to understand what occurs during breakpoint chlorination.

METHODS

Our experiments were conducted at well-defined pH (under conditions very close to those that occur naturally).

The first tested amino acid was alanine, chlorinated at 18°C in darkness by concentrated sodium hypochlorite, in buffered aqueous solutions (pH = 5, 7,

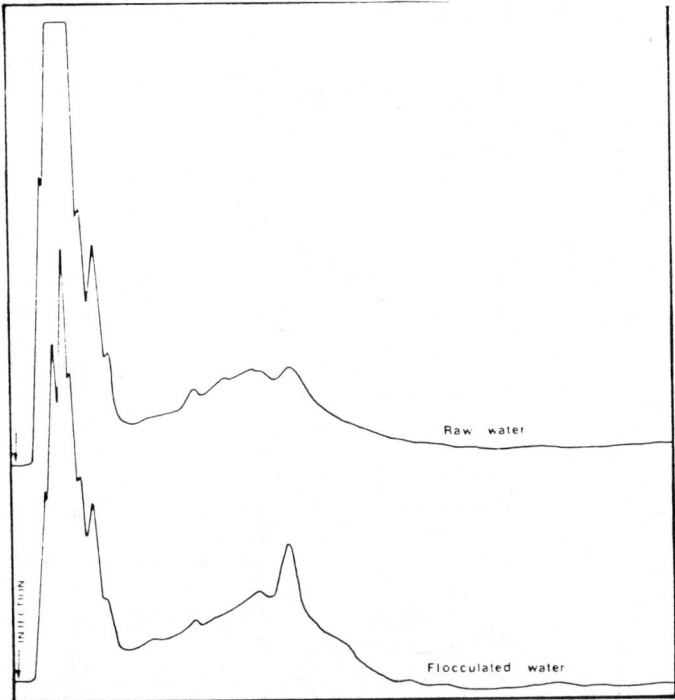

Figure 4. Chromatograms comparing amino acid concentrations in a river water sample before and after flocculation. The decreased amino acid concentrations apparent in the first 5-min retention time for the flocculated water indicate slight adsorption on the floc.

or 9). Residual amino acid was detected by fluorescence after derivatization and HPLC analysis.[13,14] The products of chlorination were analyzed directly in the aqueous samples by gas chromatography on a PORAPAK Q column (specially conditioned for aqueous injections).

RESULTS AND DISCUSSION

N-Chloramine Formation

The plots of residual chlorine vs introduced chlorine for reaction of chlorine with α-amino acids (alanine, phenylalanine, and tyrosine) at various pH values, show breakpoints similar to that for ammonia (Figure 5). The breakpoints correspond to chloramine removal in the presence of excess chlorine.[15] These data on the formation of N-chloramines (mono-, di-, and trichloramines)

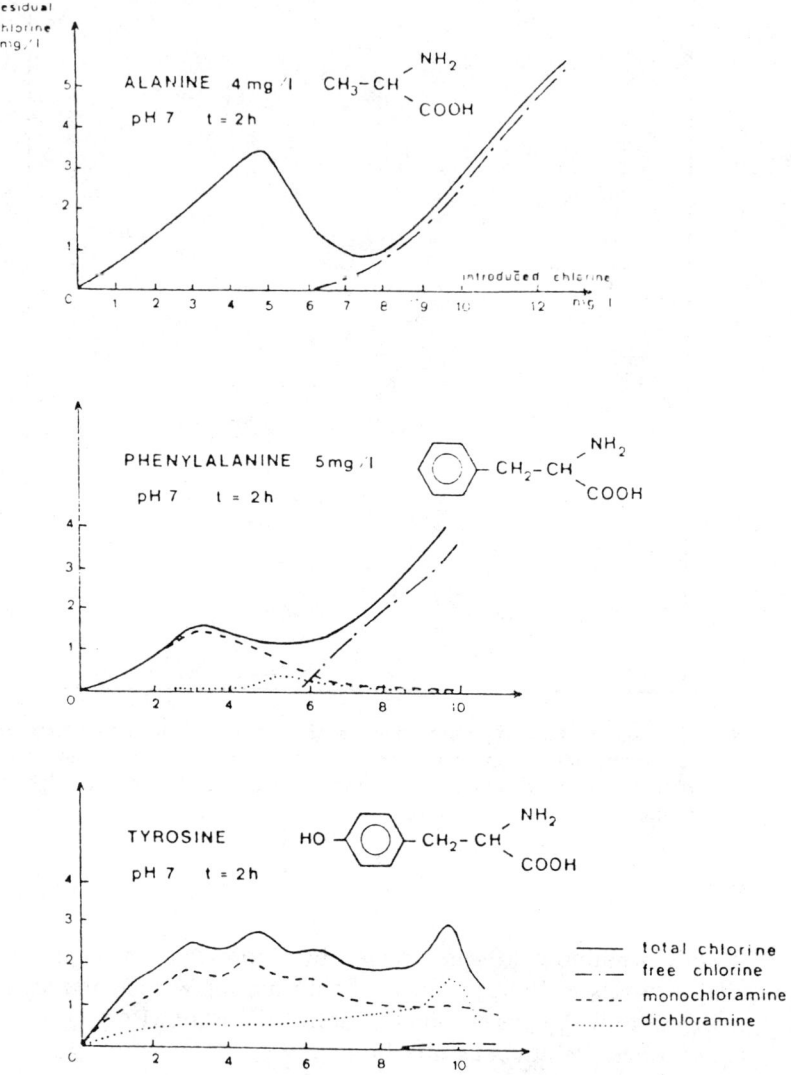

Figure 5. Comparison of the chlorine demand of three α-amino acids, alanine, phenylalanine, and tyrosine.

demonstrate that during chlorination the α-amino acids first give a monochloramine, followed by an unstable dichloramine, which disappears with chlorine excess (see phenylalanine and tyrosine in Figure 5). The dichloramines are unstable with time and are formed in lower concentration that those coming from β-amino acids (e.g., β-alanine), whose behavior is similar to the amines (Figure 6). These observations suggest the formation of other oxidation products.

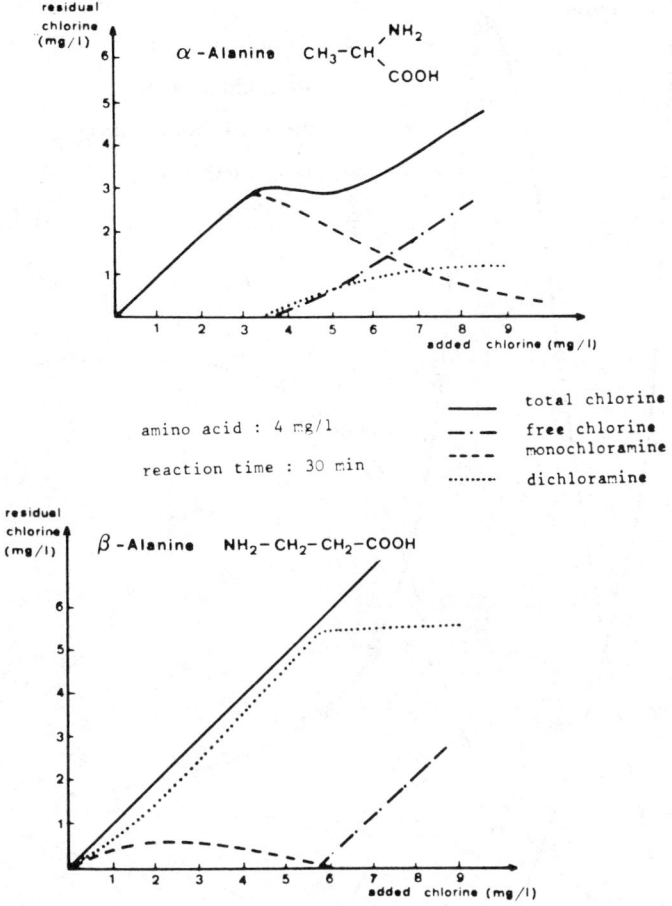

Figure 6. Comparison of α- and β-alanine for chlorine demand and mono- and dichloramine production.

Aldehyde or Nitrile Formation

To demonstrate aldehyde or nitrile formation during chlorination, a 0.250 mM aqueous solution of alanine was chlorinated at increasing amounts of sodium hypochlorite (x is the molar ratio of active chlorine to amino acid), for 24 h in darkness. From the data shown in Figure 7, it was concluded that both acetaldehyde and acetonitrile are formed and that alanine has completely disappeared for x ≥ 2.5. After 24 h at pH 5, there is no more free chlorine or monochloramine in the solutions, but for pH 7 and 9 only free chlorine has disappeared for low x values. Dichloramine appears for x > 2.5, and acetalde-

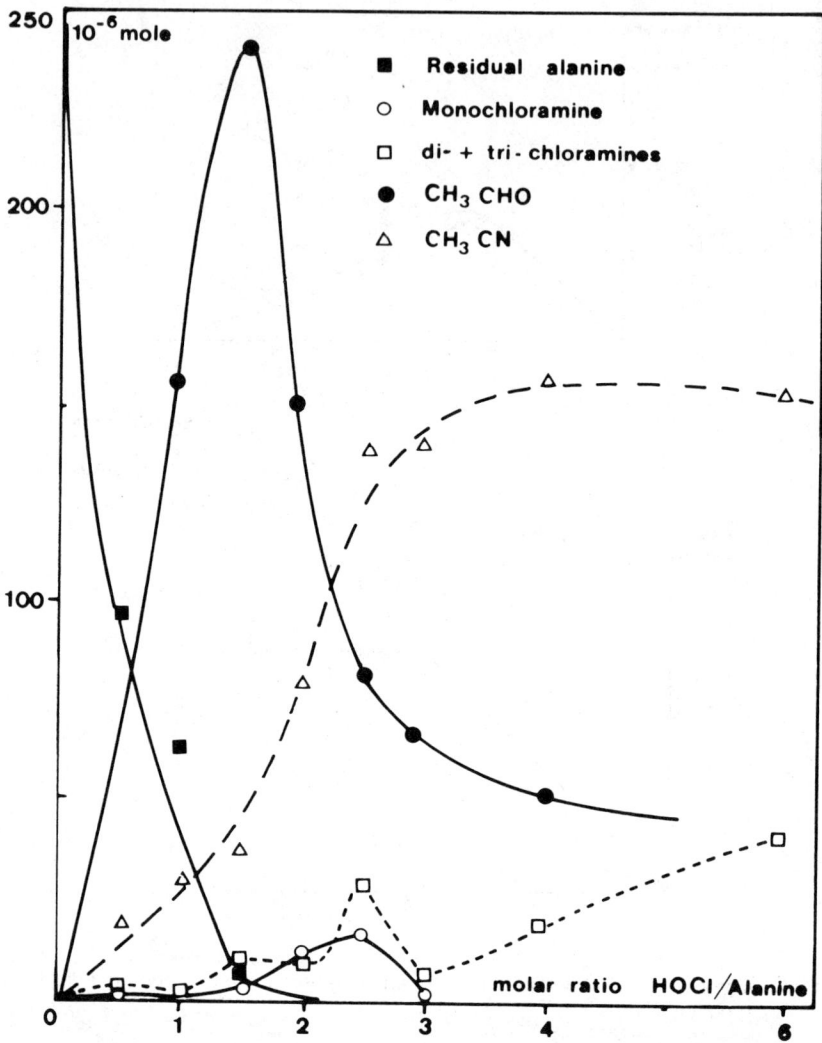

Figure 7. α-Alanine chlorination at pH 7. Products were analyzed after 24 h for increasing x values.

hyde and acetonitrile are simultaneously present in nearly all samples (i.e., at every pH between 5 and 9, and at every x value).

The last observation is inconsistent with the literature data, which state that either the aldehyde or the nitrile is present, depending on the reaction pH. However, the acetonitrile concentration is higher at an acidic pH than at alkaline pH values, as shown below:

$$[CH_3CN]_{max} = \left.\begin{array}{l} 0.190 \text{ m}M \text{ at pH} = 5 \\ 0.155 \text{ m}M \text{ at pH} = 7 \\ 0.120 \text{ m}M \text{ at pH} = 9 \end{array}\right\} \text{ for } (3 \leq x \leq 6)$$

Furthermore, acetaldehyde is found in higher concentrations when x is low. Therefore, for an increasing x, CH_3CHO concentration decreases and the CH_3CN concentration is higher. This fact suggests a direct relation between aldehyde and nitrile, which has been shown by the following experiments with acetaldehyde as the starting reactant.

Acetaldehyde and Acetonitrile Formation

The species likely to react with acetaldehyde (CH_3CHO) in the reaction medium may be mono- or dichloramines. To test this hypothesis, chloramines formed by chlorination of ammonium ion (from NH_4OH or NH_4Cl) have been treated with acetaldehyde under the same conditions as previously used (concentration, pH, temperature, darkness, and stoppered flasks). In this series of experiments, acetaldehyde disappearance is observed simultaneously with the formation of acetonitrile as confirmed by gas chromatography (Figure 8).

Acetonitrile is produced for low x values, that is, in the presence of monochloramine rather than dichloramine. The concentration of CH_3CN formed coincides with the CH_3CHO lost for all stages of the test (Figure 8). The overall following reaction can be written:

$$CH_3CHO + NH_2Cl \rightarrow CH_3CN + HCl + H_2O$$

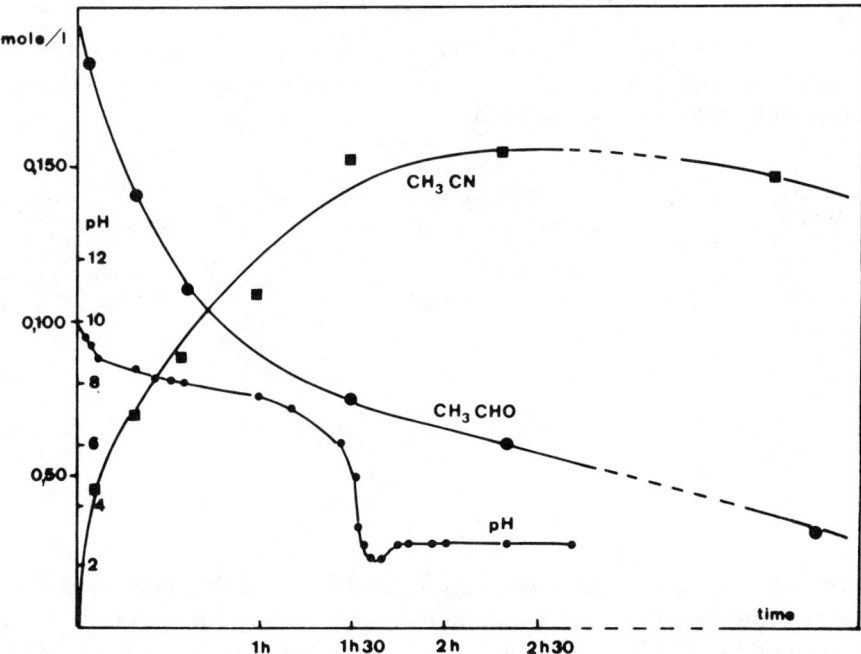

Figure 8. Acetonitrile production from acetaldehyde showing pH and concentration change with reaction time. Initial concentrations of CH_3CHO and NH_2Cl were equimolar.

WATER CHLORINATION

which is characterized by a solution acidification (see Figure 8). The pH of the unbuffered reaction mixture decreases to 2 during the course of the reaction.

The first reaction step is an amine formation. After the reaction between gaseous chlorine and ammonia on concentrated acetaldehyde, acetaldamine has been isolated and characterized by NMR and IR spectra. This amine leads to the corresponding nitrile but regenerates the aldehyde with the addition of water.

Reaction Scheme

A general scheme of chlorination of α-alanine is therefore proposed:

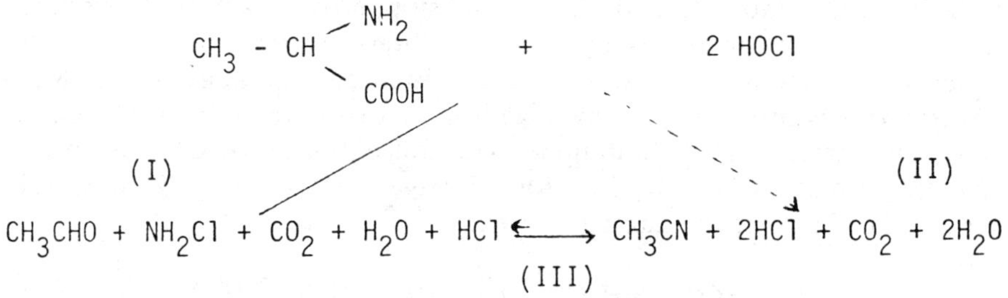

$$CH_3-CH(NH_2)(COOH) + 2\,HOCl$$

(I)
$$CH_3CHO + NH_2Cl + CO_2 + H_2O + HCl \;\longleftrightarrow\; CH_3CN + 2HCl + CO_2 + 2H_2O$$
(III)
(II)

Reaction I. Analysis of the chlorination products of α-alanine has shown the rapid formation of acetaldehyde and monochloramine. The following reaction sequence can be considered:

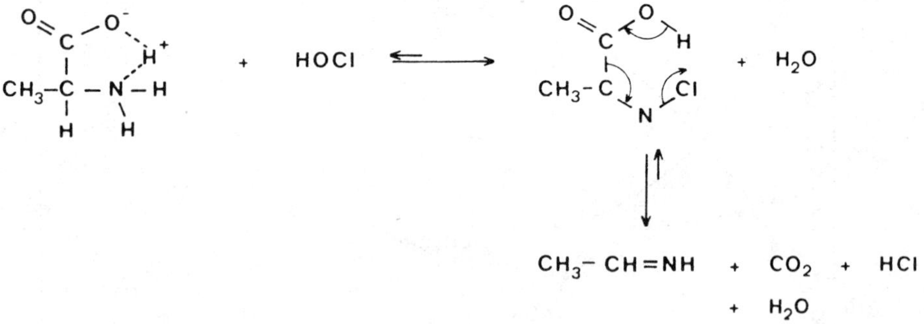

$$\underset{H\;\;H}{\underset{|\;\;\;\;\;|}{CH_3-\underset{|}{C}-\overset{}{N}-H}}\overset{O^-}{\underset{}{\overset{\|}{\underset{}{C}}}}\!\!\!\overset{O^-}{}\cdot H^+ \;+\; HOCl \;\longleftrightarrow\; CH_3-C\overset{O}{\underset{N}{\overset{\|}{}}}\!\!\!\overset{O}{}\!\!H,\;Cl \;+\; H_2O$$

$$\updownarrow$$

$$CH_3-CH=NH \;+\; CO_2 \;+\; HCl \;+\; H_2O$$

After substitution of a chlorine atom to produce an N-chloramine, an electronic rearrangement of the N-chloramine leads to the formation of three separate molecules: HCl, CO_2, and the corresponding imine. This unstable imine reacts very quickly in the presence of a new hypochlorous acid molecule.

$$CH_3-CH=NH \quad + \quad HOCl \quad \longrightarrow \quad CH_3-\underset{H}{\overset{O-H}{\underset{|}{C}}}-N\overset{H}{\underset{Cl}{\diagdown}}$$

$$\downarrow$$

$$CH_3-C\overset{O}{\underset{H}{\diagdown}} \quad + \quad NH_2Cl$$

Aldehyde and mineral monochloramine products have been identified and quantified. The amine has been "perceived" in a few samples, but we cannot confirm this amine is that of reaction I or III.

This sequence is based on the first assumptions of Fox and Bullock[16] for their synthesis of indoleactic acid from glutamic acid (see also Stanbro and Smith[13] and Maierski[17]).

Reaction II. We have previously stated that acetonitrile appeared in higher quantities at acidic pH (pH 5) than alkaline pH (pH 9), analogous to dichloramines which are formed in highest yield in acidic solutions. Therefore, we propose the hypothetical nitrile formation from the N,N-dichloramines, according to a similar mechanism, that is, an electronic transfer in the 6-membered ring of the reaction intermediate.

$$\underset{CH_3-\underset{H}{\overset{|}{C}}-NH_2}{\overset{O=C-O^- \cdots H^+}{}} \quad + \quad HOCl \quad \longrightarrow \quad \underset{CH_3-\underset{H}{\overset{|}{C}}-N\overset{H}{\underset{Cl}{\diagdown}}}{\overset{O=C-O^- \cdots H^+}{}} \quad + \quad H_2O$$

$$\downarrow HOCl$$

$$\underset{+2H_2O \; + \; HCl \; + \; CO_2}{\overset{CH_3}{\underset{H}{\diagdown}}C=N-Cl} \quad \longleftarrow \quad \underset{CH_3-\underset{H}{\overset{|}{C}}\overset{}{\underset{N}{\diagdown}}\overset{}{\underset{Cl}{\diagdown}}}{\overset{O=C-O-H}{\overset{|}{\underset{Cl}{}}}} \quad + \quad 2H_2O$$

$$\downarrow$$

$$CH_3-C\equiv N \quad + \quad 2HCl \quad + \quad 2H_2O \quad + \quad CO_2$$

Reaction III. We have shown that it is also possible to obtain acetonitrile from acetaldehyde in aqueous solution by reaction with monochloramine. The mechanism can be defined as follows:

$$CH_3-C\underset{H}{\overset{O}{\diagdown}} + NH_2Cl \rightleftharpoons CH_3-\underset{H}{\overset{O^-}{\underset{|}{C}}}-\underset{Cl}{\overset{H}{N-H}}$$

$$CH_3-\underset{H}{\overset{CH_3}{C=N-Cl}} + H_3O^+ \longleftarrow CH_3-\underset{H}{\overset{OH}{\underset{|}{C}}}\underset{Cl}{\overset{H}{N}}$$

$$CH_3-C\equiv N + H_2O + HCl$$

Finally, in our experimental conditions, the chlorination products of acetaldehyde (for example, chloral) or acetonitrile (chloroacetonitirle) have never been detected in the presence of excess chlorine, confirming the observations of Trehy and Bieber.[18-20]

Other Amino Acids Oxidation

In addition to the chlorine demand of about 20 amino acids whose reaction curves generally present a characteristic breakpoint, we chlorinated phenylalanine and tyrosine. Our data obtained with phenylalanine show similar results: chlorine demand curves with a breakpoint; chloramine degradation (especially monochloramine) with time or chlorine excess; and rapid aldehyde formation (phenylacetaldehyde, with a characteristic jasmine odor) identified by HPLC in the first minutes of reaction.

The ozonation of these amino acids (alanine and phenylalanine) has been thoroughly studied. The oxidation products are the corresponding aldehydes and acids. In addition, condensation products (macromolecules) were formed with phenylalanine.[8]

Tyrosine behaves differently when chlorinated. It appears that chlorine substitution reactions on the aromatic ring also occur. Kinetics studies are being conducted to advance our understanding of these phenomena.

CONCLUSION

Contrary to the literature, where most of the experiments were conducted in concentrated solutions and at extreme pH, we have obtained aldehyde *and* nitrile formation from a chlorination of α-alanine in dilute aqueous medium and at pH between 5 and 9 (i.e., under conditions very close to reality). The nitrile is the product of the reaction of previously formed N-chloramines and acetaldehyde. This observation explains the N-chloramine degradation and the opposing results of early workers which were obtained under more drastic conditions that promote nitrile formation or stopping the reaction at the aldehyde step. The direct production of acetonitrile is not excluded.

REFERENCES

1. Elmghari-Tabib, M. *Analyse de la Micropollution Azotée des Eaux en Cours de Potabilisation. Action de l'Ozone sur Quelques Constituants*, Thesis, (Rennes: Université de Rennes, 1981).
2. Le Cloirec, C., P. Le Cloirec, J. Morvan, and G. Martin. "Forms of Organic Nitrogen in Surface Waters: Crude Waters or in Drinking Water Treatment Steps," *Rev. Fr. Sci. Eau* 2:25-39 (1983).
3. Le Cloirec, C., M. Elmghari-Tabib, J. Morvan, and G. Martin. *Suivi de la Qualité des Eaux de Dix Unités de Potabilisation: Evolution des Composés Azotés*, (Bretagne: Ministère de la Santé—DRASS, 1982).
4. Le Cloirec-Renaud, C. *Analyse et Evolution de la Micropollution Organique Azotée dans les Stations d'Eau Potable. Effet de la Chloration sur des Acides Aminés*, Thesis (Rennes: Université de Rennes, 1984).
5. Le Cloirec, C., P. Le Cloirec, J. Morvan, and G. Martin. "Evolution of Nitrogenous Organic Products (P.O.A.) in Different Water Plants," *J. Fr. Hydrol.* 14(1,No.40):59-74 (1983).
6. Le Cloirec, P. *Elimination de Polluants Organiques de l'Eau au Moyen de Filtres Biologiques à Charbon Actif*, Thesis, (Rennes: Université de Rennes, 1983).
7. Le Cloirec, P., C. Le Cloirec, G. Martin, and M. M. Bourbigot. "Evolution of Nitrogenous Organic Products on Biological Activated Carbon Filters," *J. Fr. Hydrol.* 14(1,No.40):75-87 (1983).
8. Le Cloirec, C., Y. Wei, A. Laplanche, and J. Poncin. "Action Comparée du Chlore et de l'Ozone sur des Acides Aminés en Solution," in *Symposium on Ozone and Biology*, Rennes, April 1984, (Cleveland: International Ozone Association, 1984).
9. Langheld, K. "Uber das Verhalten von α-Aminosäuren gegen Natriumhypochlorit," *Ber. Dtsch. Chem. Ges.* 42(392):2360-2374 (1909).
10. Dakin, H. D. "The Oxidation of Amino Acids to Cyanides," *Biochem. J.* 10:319-323 (1916).
11. Kantouch, A., and S. H. Abdel-Fattah. "Action of Sodium Hypochlorite on α-Amino Acids," *Chem. Zvesti.* 25:222-230 (1971).
12. Dardenne, G. A., M. Severin, and N. M. Arlier. "Chromatographie en Phase Gazeuse de Nitriles. III. Application à l'Identification d'Acides α-Aminés Isomeres," *J. Chromatogr.* 47:182-185 (1970).
13. Stanbro, W. D., and W. D. Smith. "Kinetics and Mechanism of the Decomposition of N-Chloroalanine in Aqueous Solution," *Environ. Sci. Technol.* 13(4):446-451 (1979).

14. Le Cloirec, C., P. Le Cloirec, M. Elmghari-Tabib, J. Morvan, and G. Martin. "Concentration and Analysis of Numerous Nitrogenous Organic Substances in Natural Waters." *Int. J. Environ. Anal. Chem.* 14:127–145 (1983).
15. Poncin, J., C. Le Cloirec, and G. Martin. "Kinetic Studies on the Chlorination of Methylamine by Sodium Hypochlorite in Dilute Aqueous Medium," *Environ. Technol. Let.* 5:263–274 (1984).
16. Fox, S. W., and M. W. Bullock. "Synthesis of Indoleacetic Acid from Glutamic Acid and a Proposed Mechanism for the Conversion," *J. Am. Chem Soc.* 73:2754–2755 (1951).
17. Maierski, H. *Vorkommen, Bildung und Verminderung Chlororganischer Verbindungen bei der Aufbereitung von Schwimmbad Wasser*, Thesis, (Munchen: Technische Universitat Munchen, 1983).
18. Trehy, M. L., and T. I. Bieber. "Effects of Commonly Used Water Treatment Processes on the Formation of THMs and DHANs," in *Proceeding of the Annual Meeting of the American Water Works Association,* Denver, 1980, pp. 125–138.
19. Bieber, T. I., and M. L. Trehy. "Dihaloacetonitriles in Chlorinated Natural Waters," in *Water Chlorination: Environmental Impact and Health Effects, Vol. 4*, R. L. Jolley, W. A. Brungs, J. A. Cotruvo, R. B. Cumming, J. S. Mattice, and V. A. Jacobs, Eds. (Ann Arbor, MI: Ann Arbor Science Publishers, Inc., 1983), pp. 85–96.
20. Trehy, M. L., and T. I. Bieber. "Detection, Identification and Quantitative Analysis of Dihaloacetonitriles in Chlorinated Natural Waters," in *Advances in the Identification and Analysis of Organic Pollutants in Water*, L. H. Keith, Ed. (Ann Arbor, MI: Ann Arbor Science Publishers, Inc., 1981), pp. 941–978.

CHAPTER **66**

Characterization of the Products from the Reaction of Hydroxybenzoic and Hydroxycinnamic Acids with Aqueous Solutions of Chlorine, Chlorine Dioxide, and Chloramine

Robert M. Carlson and Sechoing Lin

Phenolic acids are ubiquitous in nature, present notably in plant materials[1-4] and as the degradation products of humic substances.[5] Consequently, they represent common components of the organic material in water undergoing disinfection.[6-8]

Since the initial identification of chloroform from disinfected drinking water,[9] much effort has gone into identification of chloroform precursors.[10-16] For example, it was observed that 2,4-dihydroxy-, 2,6-dihydroxy-, 3,5-dihydroxy-, and 2,4,6-trihydroxybenzoic acids were particularly well disposed to chloroform production; higher conversions occurred at increased pH and chlorine concentration. However, because the chlorine demand remains high for all the phenolic acids, it is not surprising that chloroproducts other than chloroform have been observed.[6-8,13,15,17] It is well recognized that only a small portion of the chlorinated organic content has been identified, and efforts therefore continue to better evaluate the overall chemical and biological consequences of chemical disinfection.

The current study provides additional insight into the disinfection chemistry of hydroxybenzoic and hydroxycinnamic acids by comparing the reactions of chlorine with two commonly suggested disinfectants: chlorine dioxide and chloramine.

EXPERIMENTAL

Instruments and Apparatus

A Hewlett-Packard 5995C GC-MS was equipped with a split/splitless interface and a cross-linked methyl silicone-fused silica column (Hewlett-Packard, 15 m by 0.2 mm ID); GC temperatures were programmed from 50 to 280°C. The high-performance liquid chromatography (HPLC) instrumentation consisted of a Beckman 110A pump, a Rhodyne 7126 injector, a resin column (10-μm particle size, Hamilton, 18 cm by 4.6 mm ID, PRP-1), a Shoeffel SF

770 variable wavelength UV detector, and a Hewlett-Packard 3390A integrator. Chromatographic conditions were varied to achieve adequate retention for the organic substrates under investigation. The mobile phase was 3 to 30% acetonitrile in 0.01 M KH_2PO_4, brought to pH 2.0 with H_3PO_4, at a flow rate of 1.0 mL/min. The pH was measured by using a Graphics Controls PHM 7900 pH meter having a Fisher combination electrode.

Chemicals

All the phenolic acids were purchased from Aldrich Chemical Company and used without further purification. The monobasic potassium phosphate, phosphoric acid, sodium thiosulfate, sodium hypochlorite, and acetonitrile (HPLC grade) were obtained from Fisher Chemical Company. The ammonium hydroxide was purchased from DuPont. Ether was obtained from Columbus Chemical Industries. The water was purified by passage through a Continental deionizer and a Milli-Q (Millipore Corporation) system.

Procedure

The sodium hypochlorite and chlorine dioxide reagent solutions were prepared by dilution of concentrated stock solutions and were titrated (iodometric) prior to use. Sodium hypochlorite (5%, Fisher) was used as a stock solution. The concentrated chlorine dioxide was generated as described in the literature.[18] Chloramine solutions were freshly prepared by mixing the appropriate amount of ammonium hydroxide solution with hypochlorite to achieve an ammonia-to-hypochlorite ratio of 3:1.[19]

Solutions (1.0 × 10^{-4} M) of the substrates were prepared by stirring a carefully weighed amount of the organic into 200 mL of phosphate-buffered water (pH 7.0). Each reaction was initiated at 24 ± 2°C by addition, with stirring, of a predetermined volume of the oxidizing reagent to the bottle containing the organic solution. The initial molar ratio of the oxidizing reagent to the organic was 3.0. The pH and organic content (HPLC analysis) were monitored throughout the reaction period, and the reactions were terminated using sodium thiosulfate. The contact time for the reaction was 20 min unless otherwise specified.

Sample solutions were acidified to pH 1 with hydrochloric acid and concentrated (three times) by liquid-liquid extraction (ether). The ether was then dried with anhydrous sodium sulfate and concentrated to 0.5 to 1 mL using a rotary evaporator; additional concentration was achieved under a stream of nitrogen. A fraction of the sample was directly analyzed by GC-MS. The remaining fraction was treated with diazomethane and analyzed again.

Table I. Phenolic Acids, Chlorinating Species, and Products

Phenolic Acid	Chlorinating Species	Products[a,b]	Reference
4-Hydroxybenzoic acid	Cl_2	Mono- and dichloro addition products (M), phenol (T), mono-, di- and trichlorophenols (M)	8,15
	ClO_2	1,4-Benzoquinone (M), phenol (T), 2,2-dichloro-4-cyclopentene-1,3-dione (T)	
	$ClNH_2$[c]	Mono- and dichloro addition products (T), monochlorophenol (T)	
2,4-Dihydroxybenzoic acid	Cl_2	Mono- and dichloro addition products (T), 2,4-dihydroxybenzene (T), mono-, di-, and trichloro-2,4-dihydroxybenzenes (T), 5,5-dichloro-3-chlorocyclopentene-1,2-dione (T), 3,5,5-trichloro-3-cyclopentene-1,2-dione (M), chloroform (M)	17
	ClO_2	2-Hydroxy-1,4-benzoquinone (T), 2,2-dichloro-4-cyclopentene-1,3-dione (M), 2,2-dihydroxy-3,3-dichloro-1,4-benzoquinone (M)	21
	$ClNH_2$[d]	Mono- and dichloro addition products (M), 2,4-dihydroxybenzene (T), mono-, di-, and trichloro-2,4-dihydroxybenzenes (T)	
2,4,6-Trihydroxybenzoic acid	Cl_2	Mono- and dichloro addition products (T), 1,3,5-trihydroxybenzene (T), mono- and dichloro-1,3,5-trihydroxybenzenes (T), and unknown polar products (M).	
	ClO_2	Unknown products	
	$ClNH_2$	Mono- and dichloro addition products (M), 1,3,5-trihydroxybenzene (T), mono-, di-, and trichloro-1,3,5-trihydroxybenzenes (T)	
4-Hydroxy-3-methoxybenzoic acid	Cl_2	Monochloro addition product (M) mono- and dichloro-3-methoxyphenols (T)	13,15
	ClO_2	2-Methoxy-1,4-benzoquinone (M)[e]	
	$ClNH_2$[c]	Monochloro-4-hydroxy-3-methoxybenzoic acid (T), mono- and dichloro-3-methoxyphenol (T)	

Table I. Phenolic Acids, Chlorinating Species, and Products

Phenolic Acid	Chlorinating Species	Products[a,b]	Reference
4-Hydroxy-3,5-dimethoxybenzoic acid	Cl_2 ClO_2	2-Chloro-4-hydroxy-3,5-dimethoxybenzoic acid (T) 2,6-Dimethoxy-1,4-benzoquinone (M)	8
4-Hydroxy-3-methoxycinnamic acid	Cl_2	1-(4-Hydroxy-3-methoxyphenyl)-ethylene (T), 1-(monochloro-4-hydroxy-3-methoxyphenyl)-ethylene (M), 1-(dichloro-4-hydroxy-3-methoxyphenyl)-ethylene (M), 1-(4-hydroxy-3-methoxyphenyl)-2-chloroethylene (M), 1-(monochloro-4-hydroxy-3-methoxyphenyl)-2-chloroethylene (M), mono- and dichloro-4-hydroxy-3-methoxycinnamic acid (T), 1-(5-chloro-4-hydroxy-3-methoxyphenyl)-2,2-dichloroethanol (T), 2-methoxyphenol (T), monochloro-2-methoxyphenol (T)	13
	ClO_2 $ClNH_2$[d]	Unknown products 1-(4-Hydroxy-3-methoxyphenyl)-ethylene (T), 1-(monochloro-4-hydroxy-3-methoxyphenyl)-ethylene (T), 1-(4-hydroxy-3-methoxyphenyl)-2-chloroethylene (M), monochloro-4-hydroxy-3-methoxycinnamic acid (M)	

[a]M, major product; T, trace amount observed.
[b]Detected less than 1% of chloroform in all cases except in the reaction of 2,4-dihydroxybenzoic acid with chlorine.
[c]Recovered most of phenolic acid after a contact time of 24 h.
[d]Contact time of 24 h.
[e]Unstable product.

RESULTS AND DISCUSSION

General and very efficient conversion of *p*-hydroxybenzoic acids to *p*-benzoquinones was observed. Table I summarizes the results. This is of some environmental concern because of the toxicity that has been attributed to quinones.[20] Moreover, hydroxyquinones represent a β-dicarbonyl equivalent that facilitates chlorine incorporation and enhanced chloroform production in the presence of chlorine/chlorine dioxide mixtures.[21]

$$\text{HO-C}_6\text{H}_3(\text{R})\text{-CO}_2\text{H} \xrightarrow{\text{ClO}_2} \text{O=C}_6\text{H}_2(\text{R})\text{=O} \quad (R = H, OH, OCH_3)$$

The significant reaction of chloramine was observed with only the most reactive of the phenolic acids. Moreover, a comparison of the reaction products from chlorine addition with chloramine to those formed in the presence of chlorine only shows that the presence of chloramine results in the formation of chlorohydroxybenzoic acids, whereas chlorine alone leads to further oxidation. However, in reactions with chloramines, both the chloramine and chlorine are believed to be potential chlorine donors.[22,23] This interesting result continues to occur when low ratios of chlorine/substrate (i.e., 1:10) are used. Therefore, although the initial aromatic substitution appears to be the same, the mediating effect of the chloramine mitigates further chlorination processes that would ultimately lead to chloroform production.

ACKNOWLEDGMENT

The research described in this chapter has been funded wholly by the U.S. Environmental Protection Agency under assistance agreement numbers R807455 and R809695 to the University of Minnesota.

REFERENCES

1. Roston, D. A., and P. T. Kissinger. "Liquid Chromatographic Determination of Phenolic Acids of Vegetable Origin," *J. Liq. Chromatogr.* 5:75–103 (1982).
2. Schwarzenbach, R. "High-Performance Liquid Chromatography of Carboxylic Acids," *J. Chromatogr.* 251:339–358 (1982).
3. Casteele, K. V., H. Geiger, and C. F. Van Sumere. "Separation of Phenolics (Benzoic Acids, Cinnamic Acids, Phenylacetic Acids, Quinic Acid Esters, Benzaldehydes and Acetophenones, Miscellaneous Phenolics) and Coumarins by Reversed-

Phase High-Performance Liquid Chromatography," *J. Chromatogr.* 258:111–124 (1983).
4. Andersen, J. M., and W. B. Pedersen. "Analysis of Plant Phenolics by High-Performance Liquid Chromatography," *J. Chromatogr.* 259:131–139 (1983).
5. Stevenson, F. J. *Humus Chemistry — Genesis, Composition, Reactions* (New York: John Wiley & Sons, Inc., 1982).
6. Shimizu, Y., and R. Y. Hsu. "Interaction of Chlorine and Selected Plant Phenols in Water," *Chem. Pharm. Bull.* 23(9):2179–2181 (1975).
7. Hsu, R. Y., and Y. Shimizu, "Phenylpropanoids in Chlorination," *Chesapeake Sci.* 18(1):129 (1977).
8. McCreary, J. J., V. L. Snoeyink, and R. A. Larson. "Comparison of the Reaction of Aqueous Free Chlorine with Phenolic Acids in Solution and Adsorbed on Granular Activated Carbon," *Environ. Sci. Technol.* 16(6):339–344 (1982).
9. Rook, J. J. "Formation of Haloforms During Chlorination of Natural Waters," *Water Treat. Exam.* 23:234–243 (1974).
10. Christman, R. F., D. L. Norwood, D. S. Millington, J. D. Johnson, and A. A. Stevens. "Identity and Yields of Major Halogenated Products of Aquatic Fulvic Acid Chlorination," *Environ. Sci. Technol.* 17(10):625–628 (1983).
11. Rook, J. J. "Chlorination Reactions of Fulvic Acids in Natural Waters," *Environ. Sci. Technol.* 11(5):478–482 (1977).
12. Rook, J. J. "Possible Pathways for the Formation of Chlorinated Degradation Products During Chlorination of Humic Acids and Resorcinol,"in *Water Chlorination: Environmental Impact and Health Effects, Vol. 3,* R. L. Jolley, W. A. Brungs, and R. B. Cumming, Eds. (Ann Arbor, MI: Ann Arbor Science Publishers, Inc., 1980), pp. 85–98.
13. Norwood, D. L., J. D. Johnson, R. F. Christman, J. R. Hass, and M. J. Bobenrieth. "Reactions of Chlorine with Selected Aromatic Models of Aquatic Humic Material," *Environ. Sci. Technol.* 14(2):187–190 (1980).
14. Rockwell, A. L., and R. A. Larson. "Aqueous Chlorination of Some Phenolic Acids," in *Water Chlorination: Environmental Impact and Health Effects, Vol. 2*, R. L. Jolley, H. Gorchev, and D. H. Hamilton, Jr., Eds. (Ann Arbor, MI: Ann Arbor Science Publishers, Inc., 1978), pp. 67–74.
15. Larson, R. A., and A. L. Rockwell. "Chloroform and Chlorophenol Production by Decarboxylation of Natural Acids during Aqueous Chlorination," *Environ. Sci. Technol.* 13(3):325–329 (1979).
16. Boyce, S. D., and J. F. Hornig. "Formation of Chloroform from the Chlorination of Diketones and Polyhydroxybenzenes in Dilute Aqueous Solution," in *Water Chlorination: Environmental Impact and Health Effects, Vol. 3*, R. L. Jolley, W. A. Brungs, and R. B. Cumming, Eds. (Ann Arbor, MI: Ann Arbor Science Publishers, Inc., 1980), pp. 131–140.
17. Boyce, S. D., and J. F. Hornig. "Reaction Pathways of Trihalomethane Formation from the Halogenation of Dihydroxyaromatic Model Compounds for Humic Acid," *Environ. Sci. Technol.* 17(4):202–211 (1983).
18. Granstrom, M. L., and G. F. Lee. "Generation and Use of Chlorine Dioxide in Water Treatment," *J. Am. Water Works Assoc.* 50:1453–1466 (1958).
19. Sisler, H. H. "Chloramine," *J. Chem. Educ.* 60(11):1002–1004 (1983).
20. Ames, B. N. "Dietary Carcinogens and Anticarcinogens," *Science* 221:1256–1264 (1983).
21. Lin, S., R. J. Liukkonen, R. E. Thom, J. G. Bastian, M. T. Lukasewycz, and

R. M. Carlson. "Increased Chloroform Production From Model Components of Aquatic Humus and Mixtures of Chlorine Dioxide/Chlorine," *Environ. Sci. Technol.,* 18(12):932–935 (1984).
22. Isaac, R. A., and J. C. Morris. "Transfer of Active Chlorine from Chloramine to Nitrogenous Organic Compounds. 1. Kinetics," *Environ. Sci. Technol.* 17(12):738–742 (1983).
23. Lin, S., and R. M. Carlson. "Susceptibility of Environmentally Important Heterocycles to Chemical Disinfection: Reactions with Aqueous Chlorine, Chlorine Dioxide, and Chloramine," *Environ. Sci. Technol.,* 18(10):743–748 (1984).

CHAPTER 67

Formation of Aryl-Chlorinated Aromatic Acids and Precursors for Chloroform in Chlorination of Humic Acid

Ed W. B. de Leer, Jaap S. Sinninghe Damsté, and Leo de Galan

The formation of chloroform when humic substances are chlorinated is well known.[1] Other chlorinated products that may be formed are chloral, di- and trichloroacetic acid, chlorinated C-4 diacids, and α-chlorinated aliphatic acids.[2-5] Several of these compounds are formed in molar yields comparable to chloroform.[5-7]

The mechanism for the formation of these products is still largely unknown, due to the complex structure of humic material. Humic materials are geopolymers formed from lignin, carbohydrates, proteins, and fatty acids by microbial degradation and enzymatic or autooxidative coupling reactions. Although humic materials are not well defined organic compounds, several structures have been proposed[8] which contain moieties that may be converted into chloroform by chlorine in aqueous medium.

For example, 1,3-dihydroxybenzenes,[2,9,10] 1,3-diketo compounds,[11] natural acids such as citric acid,[12] and compounds with activated C-H bonds such as indoles[13] or methylketones[14] can form chloroform in high yield on chlorination in aqueous medium. For humic substances, 1,3-dihydroxybenzenes appear to be likely chloroform precursor candidates as suggested by the fact that 3,5-dihydroxybenzoic acid is formed in the degradation of humic material with $CuSO_4$–NaOH at 175 to 180°C.[15] However, the products of KOH fusion may be of no diagnostic value for the structure of humic substances.[16] In recent $KMnO_4$ degradation studies of humic and fulvic acids no 1,3-dihydroxybenzene structures were detected,[17,18] probably because of complete oxidation of these structures.[19]

Although the possibility of 1,3-dihydroxybenzene structures as the precursor fragments for chloroform remains to be proven, Rook proposed a mechanism based on the chemistry of the reaction between chlorine and resorcinol.[2,20] The identification of reaction intermediates is necessary to achieve a better understanding of this mechanism and to assist in the identification of the structural fragments in humic material that are converted into chloroform and chlorinated acids.[21]

In this study we describe (1) the identification and structural assignment of such intermediates in the reaction between terrestrial humic acid and chlorine in aqueous medium at pH 7.2, and (2) the attempts to demonstrate the presence of 1,3-dihydroxybenzene structures in humic acids by means of Curie-point pyrolysis/gas chromatography/mass spectrometry (Py/GC/MS) and nuclear magnetic resonance (NMR) before and after the chlorination reaction.

EXPERIMENTAL

Isolation and Chlorination of Humic Acid

The humic acid (HA) used in this research was extracted from a peat soil in the Liesselse Peel. The elementary composition of the final HA was C, 50.7%; H, 5.1%; N, 1.5%; O, 39.4%; ash, 1.5%. The HA was chlorinated under the following experimental conditions: HA, 0.83 g/L; pH, 7.2 (0.5 M HPO_4^{2-}/$H_2PO_4^-$ buffer); Cl_2/C molar ratio, 0.39 to 3.35; total volume, 300 mL; reaction time, 24 h.

After completion of the reaction, a possible excess of chlorine was removed by adding solid sodium arsenite and the pH was lowered to 1 with concentrated HCl. The solution was extracted with 50 mL freshly distilled diethyl ether and 50 mL freshly distilled ethyl acetate. Both extracts were dried with anhydrous sodium sulfate, concentrated in a Kuderna-Danish evaporator to about 5 mL, methylated by passing diazomethane gas through the solution, and finally concentrated to 100 μL by a gentle stream of nitrogen.

Isolation of Humic Acid after Chlorination

In the experiment with an initial Cl_2/C molar ratio of 0.39, a precipitate was formed on acidification to pH 1. This product was removed by filtration, redissolved in 0.1 M NaOH, and reprecipitated with concentrated HCl. After dissolution in water, the product (yield, 140 mg) was freeze-dried. This product will be referred to as chlorinated humic acid (CHA).

Instrumental Analysis

Gas Chromatography

For a fingerprint analysis of the reaction products, a Varian 3700 capillary gas chromatograph equipped with a flame ionization detector (FID) and a 63 Ni electron capture detector (ECD) was used. The GC conditions were; column, fused silica CP-SIL-5 (Chrompack, Middelburg), 0.23 mm by 25 m; injector, 280°C; detector, 300°C; oven temp., 35°C (5 min) programmed to 300°C (15 min) with a program rate 6°/min; carrier gas, N_2, flow rate, 0.8 mL/min.

Gas Chromatography/Mass Spectrometry

The gas chromatograph/mass spectrometer (GC/MS) was a Varian MAT 44, capillary GC-quadrupole MS, with a computerized data system of our own design. The capillary column was the same type as above and was connected to the MS by an open atmospheric split. GC conditions were as above, except that helium was used as the carrier gas. Electron impact (EI) spectra were obtained at 80 eV and chemical ionization (CI) spectra at 160 eV with isobutane as the reagent gas. The ionization current was 700 and 200 µA, respectively. Cyclic scanning from m/z = 50 to 500 was used with a cycle time of 2 s.

Curie-Point Pyrolysis—Gas Chromatography/Mass Spectrometry

Py/GC/MS was carried out with a Curie-point pyrolysis system as described by Meuzelaar et al.[22] and modified according to van de Meent et al.[23] HA and CHA samples (100–200 µg) were brought onto a ferromagnetic Ni/Fe wire (Curie temperature 610°) in the form of a suspension (20 mg/mL) in methanol. After evaporation of the methanol, the wire was heated to the Curie-temperature in 0.15 s. The pyrolysis products were separated and identified in the capillary GC/MS system as described above.

Nuclear Magnetic Resonance Spectroscopy

About 100 mg HA or CHA was dissolved in 1-mL DMSO-d6. ^{1}H-NMR spectra were measured at 200 MHz with a Nicolet NT 200 WB spectrometer using the FT-technique.

RESULTS AND DISCUSSION

Chlorination Products

Figure 1 shows the chromatograms of the GC/FID analysis of the methylated ethyl acetate extract of HA chlorinated at a Cl_2/C molar ratio of 3.35. Ethyl acetate was shown to be more effective in extracting polar chlorination products than diethyl ether. Notably, the aromatic polycarboxylic acids and the cyano-substituted alkanoic acids were found mainly in the ethyl acetate extract.

The chlorine dose used strongly influenced the composition of the product mixture. With a high chlorine dose more products were found which appeared early in the chromatogram, whereas at low chlorine dose most products were found to elute late in the chromatogram. Structures were assigned to more than 100 different reaction products by the combined use of GC/MS with EI and CI. The principal products for the different classes of organic compounds are given in Table I.

Table I. Principal Reaction Products for Different Classes of Organic Compounds in the Chlorination of Terrestrial Humic Acid

Compounds Class	Compounds Identified (No.)	Principal Compound
Nonchlorinated products		
Aliphatic monobasic acids	25	Hexacosanoic acid
Aliphatic dibasic acids	8	Butanedioic acid
Cyano-substituted acids	2	3-Cyanopropanoic acid
Aromatic carboxylic acids	13	1,2,4-Benzenetricarboxylic acid
Heterocyclic acids	2	Methylfuranedicarboxylic acid
Miscellaneous	6	Indole
Chlorinated Products		
Aliphatic monobasic acids		
α-Monochlorinated	6	2-Chloropentanoic acid
α,α-Dichlorinated	6	Dichloroethanoic acid
Other substitution	9	Trichloroethanoic acid
Unsaturated	7	2,3-Dichloropropenoic acid
Aliphatic dibasic acids		
α-Monochlorinated	4	Chlorobutanedioic acid
α,α-Dichlorinated	5	2,2-Dichlorobutanedioic acid
Other substitution	5	Tetrachlorohexanedioic acid
Unsaturated	10	Dichlorobutenedioic acid
Aromatic carboxylic acids	6	2-Chlorophenylacetic acid
Chloroform precursors	11	See Table II
Miscellaneous	6	Chloral

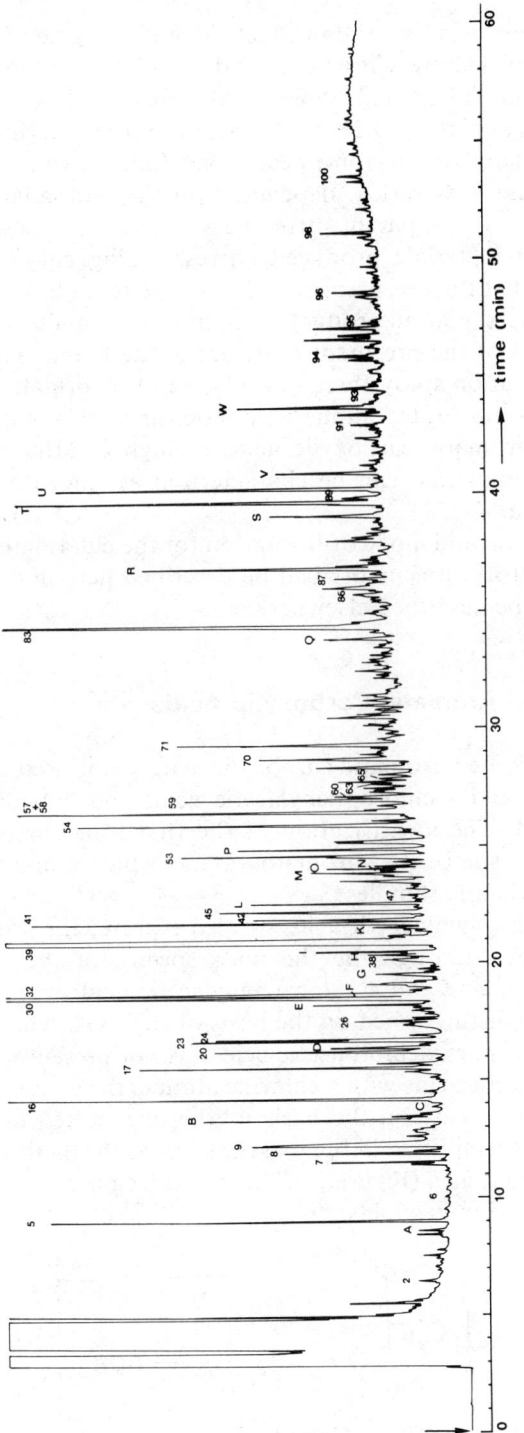

Figure 1. Capillary gaschromatogram of HA chlorination products. Methylated ethyl acetate extract. Chlorine dose 3.35 mol Cl_2 per mol C. The numbers refer to a full list of identified products (see Reference 26). Letters denote products that were found in the ethyl acetate extract and not in the preceding diethyl ether extract.

At the high chlorine dose, the products identified agreed very well with the compounds identified by Christman and co-workers after chlorination of aquatic humic and fulvic acid.[4,5] We used terrestrial HA with a H:C molar ratio of 1.20 and an O:C of 0.58, which placed our HA in the normal position in the van Krevelen diagram for a peat HA.[24] Lake or river fulvic acids show different H:C and O:C ratios, depending on the molecular weight fraction isolated.[25] Despite the apparent differences in overall structure, aquatic and terrestrial humic materials produced corresponding chlorination products, which indicates that the precursor structures in both materials must be similar and that differences (e.g., in product yield) must be explained by differences in the concentration of the precursor structure in the humic material.

In the identification study three new classes of chlorination products were found: (1) cyano substituted aliphatic monobasic acids at high chlorine dose, (2) chlorinated aromatic carboxylic acids at high chlorine doses, and (3) a group of compounds that can be characterized as chloroform precursors at low chlorine dose.

The identification and mode of formation for the chlorinated aromatic acids and some chloroform precursors will be described here in detail. Most other compounds will be described elsewhere.[26]

Aryl-Chlorinated Aromatic Carboxylic Acids

The aryl chlorinated aromatic carboxylic acids comprised 2- and 4- chlorobenzoic acid, 2- and 4-chlorophenylacetic acid, and 2,4- and 2,6-dichlorophenylacetic acid. The identification of the first four compounds was confirmed by comparison of the chromatographic behavior and the mass spectra with those of authentic samples.

In the case of the phenylacetic acids we also expected the presence of phenylchloroacetic acid. Although the the mass spectra of the methyl esters of phenylchloroacetic acid and 4-chlorophenylacetic acid showed much similarity, they could be distinguished on the basis of their GC retention. Therefore, we concluded that phenylchloroacetic acid was not present.

Chlorophenylacetic acids with a chlorine atom in the ortho position could be characterized quite easily on the basis of the unexpected loss of the ortho-chlorine atom as exemplified in the mass spectra of the methyl esters of 2- and 4-chlorophenylacetic acid (Figure 2). The major fragments can be explained as simple primary fission products:

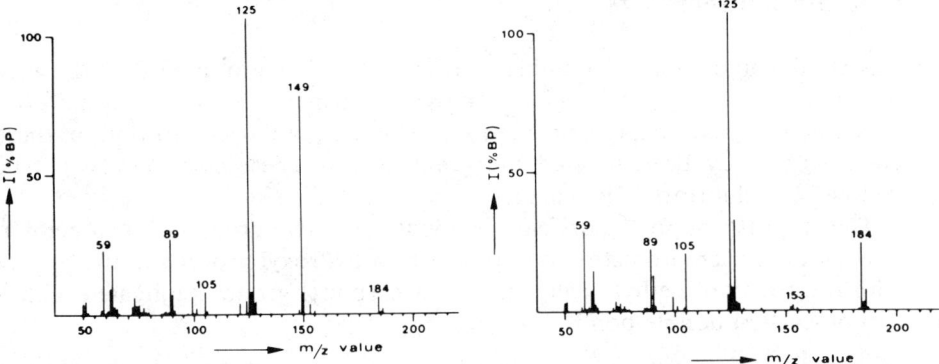

Figure 2. EI mass spectra for 2-chlorophenylacetic acid (left) and 4-chlorophenylacetic acid (right). Note the intense fragment at m/z = 149 corresponding with the loss of a chlorine atom from the molecular ion in the spectrum of 2-chlorophenylacetic acid.

The formation of the aryl-chlorinated phenylacetic acid derivatives seems possible through the direct chlorination of phenylacetic acid, which was found as one of the products after chlorination of HA.[12,26] The ortho/para substitution pattern is in accordance with this assumption. For benzoic acid, which was also found after chlorination of HA,[26] direct chlorination would yield 3-chlorobenzoic acid and not the 2- and 4-chlorobenzoic acid found in this study. This direct chlorination is also very improbable because of the deactivating effect of the carboxyl group.

An explanation for the occurrence of both chlorinated benzoic acids and phenylacetic acids may be given by the assumption that first an alkenyl substituted aromatic ring in the humic acid core is chlorinated, followed by an oxidative breakdown of the alkenyl side chain as shown in Scheme I.

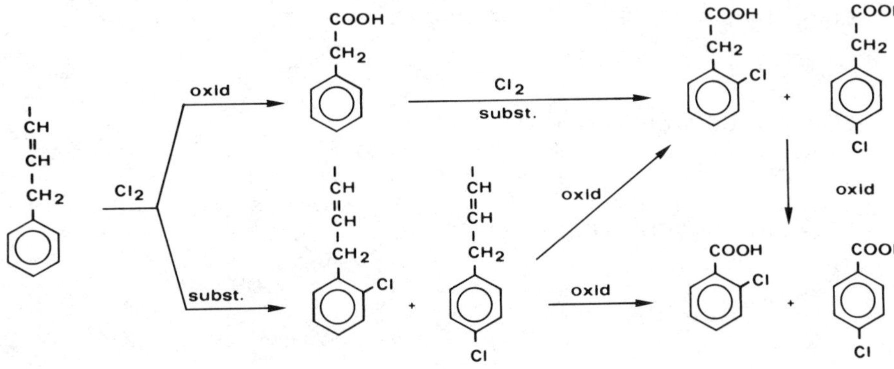

Scheme I. The formation of aryl-substituted benzoic and phenylacetic acid derivatives.

Chloroform Precursors

Several compounds containing a trichloromethyl group (Table II) were detected as chlorination products, most of them in the low chlorine dose experiment. Upon further chlorination, oxidation, and decarboxylation, these compounds may be converted into chloroform. These compounds will be denoted as "chloroform precursors."

The chloroform precursors can be divided into two groups (shown below), one with the trichloromethyl group next to a hydroxyl group and the second with the trichloromethyl group next to a carbonyl group conjugated with a carbon-carbon double bond.

$$-\underset{\underset{H}{|}}{\overset{\overset{OH}{|}}{C}}-CCl_3 \qquad -CH=CH-\overset{\overset{O}{\|}}{C}-CCl_3$$

The first group may be converted into chloroform by oxidation of the hydroxyl group to a keto group followed by nucleophilic substitution accord-

Table II. Chloroform Precursors

Hydroxyl Type	Trichloroacetyl Type
$CCl_3-\underset{\underset{OH}{|}}{CH}-COOH$	$CCl_3-\overset{\overset{O}{\|}}{C}-CH=CH-COOH$
$CCl_3-\underset{\underset{OH}{|}}{CCH_3}-COOH$	$CCl_3-\overset{\overset{O}{\|}}{C}-CH=CCH_3-COOH$ + isomer
$CCl_3-\underset{\underset{OH}{|}}{CH}-CH_2-COOH$	$CCl_3-\overset{\overset{O}{\|}}{C}\underset{HOOC}{\diagdown}C=C\underset{COOH}{\diagup}^{Cl}$
$CCl_3-\underset{\underset{OH}{|}}{CH}-CCl_2-CH_2-COOH$	
$CCl_3-\underset{\underset{OH}{|}}{CH}-CCl_2-CHCl-COOH$	$CCl_3-\overset{\overset{O}{\|}}{C}\underset{Cl}{\diagdown}C=C\underset{COOH}{\diagup}^{COOH}$
	$CCl_3-\overset{\overset{O}{\|}}{C}-CCl=CCl-CCl_2-COOH$

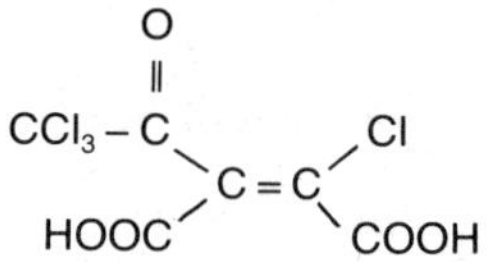

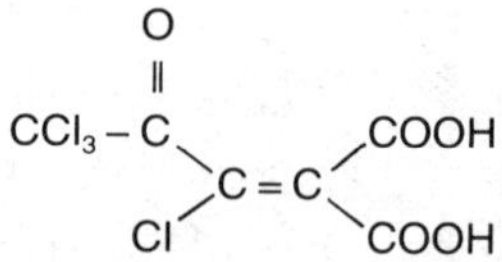

CHLORINE DEMAND REACTIONS: PROTEINS AND OTHER ORGANICS

ing to the classical haloform reaction. The first step is expected to be slow and the second step may need further activation if it is to be achieved at neutral pH and normal temperatures.[14]

$$\text{Oxidation}: -\underset{\underset{\text{OH}}{|}}{\text{CH}}-\text{CCl}_3 \xrightarrow{\text{Cl}_2} -\underset{\underset{\text{O}}{\|}}{\text{C}}-\text{CCl}_3 \quad \text{(slow)}$$

$$\text{Hydrolysis}: -\underset{\underset{\text{O}}{\|}}{\text{C}}-\text{CCl}_3 \xrightarrow[\text{H}_2\text{O}]{\text{OH}^{\ominus}} -\text{COO}^{\ominus} + \text{CHCl}_3 \quad \text{(fast)}$$

The second group of chloroform precursors already contains a trichloroacetyl group, but in these cases hydrolysis to chloroform is hampered by resonance stabilization of the carbonyl group with a carbon-carbon double bond.

$$-\underset{\underset{}{}}{\overset{\overset{\text{O}}{\|}}{\text{C}}}-\underset{|}{\text{C}}=\underset{|}{\text{C}}- \quad \longleftrightarrow \quad -\underset{}{\overset{\overset{\text{O}^{\ominus}}{|}}{\text{C}}}=\underset{|}{\text{C}}-\underset{|}{\overset{\oplus}{\text{C}}}-$$

The structural assignments for all chloroform precursors are based on a priori interpretation of the EI and CI spectra. The application of CI mass spectrometry was essential, because the molecular ion was missing in all cases, and the presence of a trichloromethyl group (CCl_3) could be deduced from the combined use of both techniques only.

For example, two isomers were found with CI-mass spectrometry at M + 1 = 221 (Figure 3). From the isotope distribution pattern of the quasi molecular ion the presence of three chlorine atoms could be deduced. From the EI spectra, the presence of the CCl_3 group could not be deduced directly from a peak at m/z = 117 but only indirectly from the loss of a CCl_3 group (-117) from the molecule, which gives the fragment at m/z = 103. The presence of a COOCH$_3$ group (m/z = 59) leaves a mass of 44 or a –C$_2$H$_4$O-group for the rest of the molecule.

Several isomers seem possible, but in view of the stability of the m/z = 103 ion (α-cleavage next to a hydroxyl group), a CCl$_3$-$\overset{|}{\underset{|}{C}}$-OH group is to be prefer-

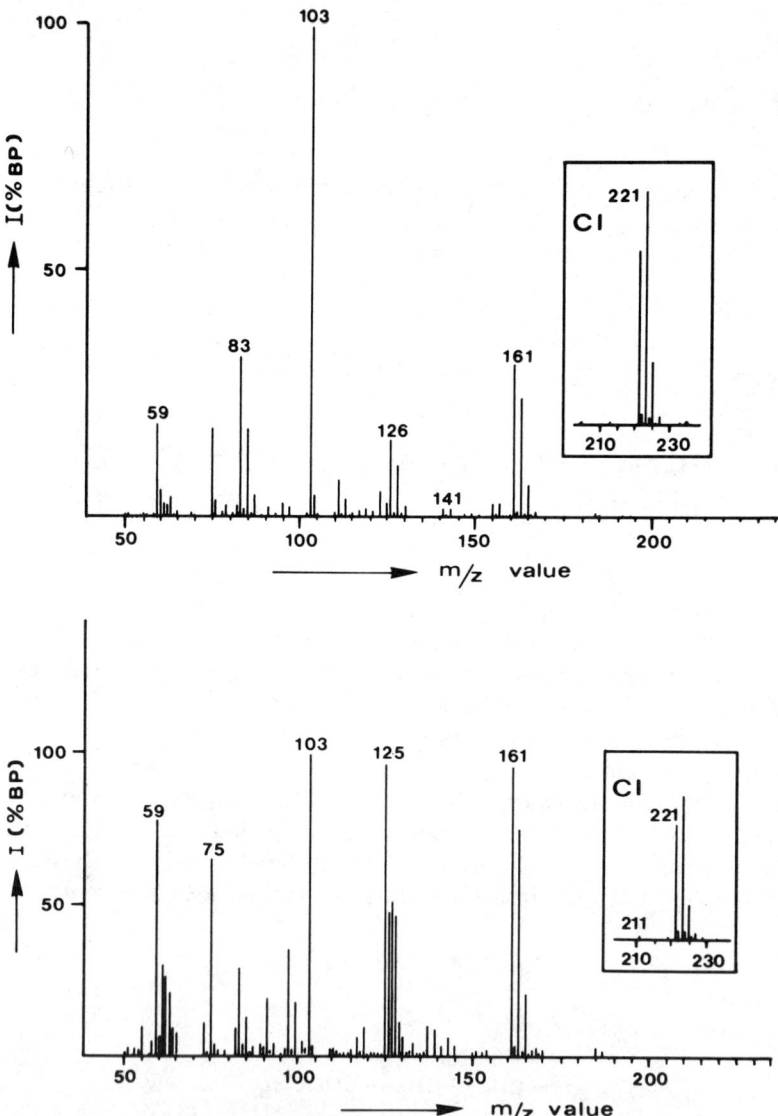

Figure 3. EI and CI mass spectra of two C$_5$H$_7$O$_3$Cl$_3$ isomers. The assigned structures are 3,3,3-trichloro-2-hydroxy-2-methylpropanoic acid methyl ester (lower) and 4,4,4-trichloro-3-hydroxybutanoic acid methyl ester (upper).

red. The final decision between 3,3,3-trichloro-2-hydroxy-2-methylpropanoic acid methyl ester (Figure 3, left) and 4,4,4-trichloro-3-hydroxy-butanoic acid methyl ester (Figure 3, right) is based on the intensity of the m/z = 103 ion (tertiary or secondary C) and the loss of Cl or HCl from the m/z = 161 ion.

The structure of many chloroform precursors was confirmed independently from the isolation of identical products in the chlorination of substituted resorcinol derivatives.[26] The formation of those compounds can be explained on the ground of the mechanism proposed by Rook[2] and extended by Boyce and Hornig[10] and points to the important role of 1,3-dihydroxybenzene structures in the formation of chloroform in the chlorination of HA.

The formation of chloroform precursors also explains the difference in chloroform yield as found by extraction with pentane followed by GC-analysis and the direct aqueous injection (DAI) method.[10] Trichloromethyl-substituted products will be decomposed into chloroform in the heated injection port on DAI, whereas these products are not extracted from the reaction mixture with pentane.

With the HA used in this study 1.3% of the carbon content could be converted into chloroform. When we consider that resorcinol gives a yield of 14.2% mol per mol $CHCl_3$ per mol C according to Rook,[2] this yield is quite high. Indeed, if only resorcinol structures are responsible for the chloroform production in the chlorination of HA, about 10% of the organic carbon must be present as free or fused 1,3-dihydroxybenzene structures.

Pyrolysis – GC/MS and NMR Investigation of HA and CHA

We tried to detect the presence of 1,3-dihydroxybenzene structure in HA by applying Curie-point pyrolysis–GC/MS and 200 MHz NMR. After Py/GC/MS about 160 pyrolysis products could be identified, which are summarized in Table III.

Py/GC/MS showed the presence of proteins, carbohydrates, lignin, and lipids which give characteristic pyrolysis products.[28] There was no indication of the presence of 1,3-dihydroxybenzene structures. This could be due to the destruction of these structures during pyrolysis at 610°C. Indeed in the pyrolysis of several model compounds like hesperitin, florhizin, and rutin, only one 1,3-dihydroxybenzene product could be identified. However, because no structure could be assigned for many compounds, no definitive answers can be given.

Comparison of the pyrolysis products of HA and CHA showed that in CHA the number of aromatic structures was drastically reduced. For example, the intensity ratio between 4-vinylguaiacol, an aromatic lignin pyrolysis product, and 1-tridecene, the decarboxylation product of tetradecanoic acid, changed from 10:1 for HA to 1:5 for CHA.

Table III. Typical Curie-Point Pyrolysis Products of Humic Acid

Carbohydrate Origin	Lipid Origin
Acetic acid	Alkanes $CH_3-(CH_2)_n-CH$
2-Methylfurane	$n = 4-29$
Furfural	Alkenes $CH_3-(CH_2)_n-CH=CH_2$
	$n = 4-26$
Levoglucosan	1-Pristene

(structure of levoglucosan shown)

Lignin Origin	Protein Origin
	Toluene, C-2-benzenes
	Phenol
(substituted phenol structure)	Benzonitrile
	Phenylacetonitrile
$Y = H, -CH_3, -CH_2-CH_3,$	Pyrrole
$-CH=CH_2, -CHO$	
$R = H$ or OCH_3	

Py/GC/MS of CHA showed chloroform as the only chlorinated product. Chloroform may be formed during pyrolysis from trichloromethyl substituted products that are still bound to the HA frame in an analogous way as the formation of chloroform in the heated injection port of a gas chromatograph on DAI.

The disappearance of aromatic structure from HA on chlorination was also confirmed from the 200 MHz NMR spectra of HA and CHA, which show a strong reduction of the aromatic signal from 6.5 to 8.5 ppm as compared with the aliphatic part of the spectrum for CHA (Figure 4).

CONCLUSION

Chlorination of terrestrial HA, at a chlorine-to-carbon ratio commonly used in the production of drinking water, produces several compounds that can be regarded as precursors for chloroform. Whether these precursors form the

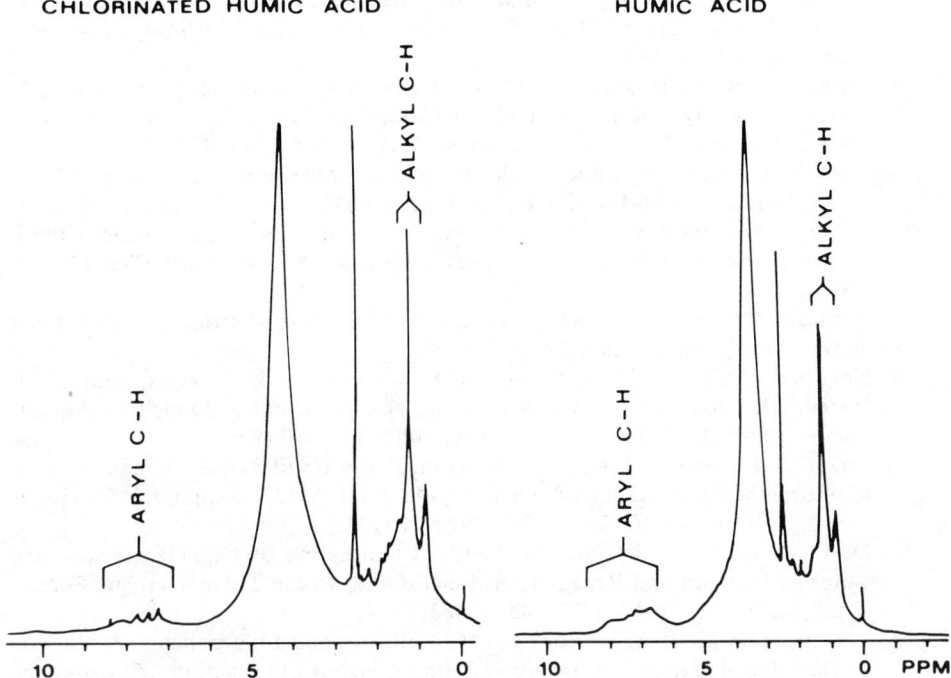

Figure 4. 200-MHz NMR spectra of CHA and HA.

only source of chloroform cannot be concluded, but they point strongly to the important role of 1,3-dihydroxybenzene structures as the primary source. However, it remains to be demonstrated that these structures form a significant part of the HA frame.

It was demonstrated by Py/GC/MS and 200-MHz NMR that chlorination destroys the aromatic part of the HA-structure.

REFERENCES

1. Rook, J. J. "Formation of Haloforms During Chlorination of Natural Waters," *Water Treat. Exam.* 23(2):234–243 (1974).
2. Rook, J. J. "Chlorination of Fulvic Acids in Natural Waters," *Environ. Sci. Technol.* 11(5):478–482 (1977).
3. Quimby, B. D., M. F. Delaney, P. C. Uden, and R. M. Barnes. "Determination of the Aqueous Chlorination Products of Humic Substances by Gas Chromatography with Microwave Emission Detection," *Anal. Chem.* 52(2):259–263 (1980).

4. Johnson, J. D., R. F. Christman, D. L. Norwood, and D. S. Millington. "Reaction Products of Aquatic Humic Substances with Chlorine," *Environ. Health Perspect.* 46(1):63-71 (1982).
5. Christman, R. F., D. L. Norwood, D. S. Millington, J. D. Johnson, and A. A. Stevens. "Identity and Yields of Major Halogenated Products of Aquatic Fulvic Acid Chlorination," *Environ. Sci. Technol.* 17(10):625-628 (1983).
6. Uden, P. C., and J. W. Miller."Chlorinated Acids and Chloral in Drinking Water," *J. Am. Water Works Assoc.* 75(10):524-527 (1983).
7. Miller, J. W., and P. C. Uden. "Characterization of Nonvolatile Aqueous Chlorination Products of Humic Substances," *Environ. Sci. Technol.* 17(3):150-157 (1983).
8. Schnitzer, M., and S. U. Khan. *Humic Substances in the Environment*, (New York: Marcel Dekker, Inc., 1972).
9. Norwood, D. L., J. D. Johnson, R. F. Christman, J. R. Hass, and M. J. Bobenrieth. "Reactions of Chlorine with Selected Aromatic Models of Aquatic Humic Material," *Environ. Sci. Technol.* 14(2):187-190 (1980).
10. Boyce, S. D., and J. F. Hornig. "Reaction Pathways of Trihalomethane Formation from the Halogenation of Dihydroxy-Aromatic Model Compounds for Humic Acid," *Environ. Sci. Technol.* 17(4):202-211 (1983).
11. De Laat, J., N. Merlet, and M. Dore. "Chlorination of Organic Compounds: Chlorine Demand and Reactivity in Relationship to the Trihalomethane Formation," *Water Res.* 16(10):1437-1450 (1982).
12. Larson, R. A., and A. L. Rockwell."Chloroform and Chlorophenol Production by Decarboxylation of Natural Acids during Aqueous Chlorination," *Environ. Sci. Technol.* 13(3):325-329 (1979).
13. Morris, J. C., and B. Baum. "Precursors and Mechanisms of Haloform Formation in the Chlorination of Water Supplies," in *Water Chlorination: Environmental Impact and Health Effects, Vol. 2*, R. L. Jolley, H. Gorchev, and D. H. Hamilton, Eds. (Ann Arbor, MI: Ann Arbor Science Publishers, Inc., 1978), pp. 29-48.
14. Gurol, M. D., A. Wowk, S. Myers, and I. H. Suffet. "Kinetics and Mechanism of Haloform Formation: Chloroform Formation from Trichloroacetone," in *Water Chlorination: Environmental Impact and Health Effects, Vol. 4*, R. L. Jolley, W. A. Brungs, J. A. Cotruvo, R. B. Cumming, J. S. Mattice, and V. A. Jacobs, Eds. (Ann Arbor, MI: Ann Arbor Science Publishers, Inc., 1983), pp. 269-284.
15. Christman, R. F., and M. Ghassemi. "Chemical Nature of Organic Color in Water," *J. Am. Water Works Assoc.* 58(6):723-741 (1966).
16. Cheshire, M. V., P. A. Cranwell, and R. D. Haworth. "Humic Acid-III," *Tetrahedron* 24(14):5155-5167 (1968).
17. Liao, W., R. F. Christman, J. D. Johnson, D. S. Millington, and J. R. Hass. "Structural Characterization of Aquatic Humic Material," *Environ. Sci. Technol.* 16(7):403-410 (1982).
18. Reuter, J. H., M. Ghosal, E. S. K. Chian, and M. Giabbi. "Oxidative Degradation Studies on Aquatic Humic Substances," in *Aquatic and Terrestrial Humic Materials*, R. F. Christman and E. Gjessing, Eds. (Ann Arbor, MI: Ann Arbor Science Publishers, Inc., 1983), pp. 107-125.
19. Randall, R. B., M. Benger, and C. M. Groocock. "The Alkaline Permanganate Oxidation of Organic Substances Selected for Their Bearing Upon the Chemical Constitution of Coal," *Proc. Roy. Soc.* 165(A1):432-452 (1938).

20. Rook, J. J. "Possible Pathways for the Formation of Chlorinated Degradation Products During Chlorination of Humic Acids and Resorcinol," in *Water Chlorination: Environmental Impact and Health Effects, Vol. 3*, R. L. Jolley, W. A. Brungs, and R. B. Cumming, Eds. (Ann Arbor, MI: Ann Arbor Science Publishers, Inc., 1980), pp. 85-98.
21. Christman, R. F., and E. Gjessing. "Priorities in Humic Research," in *Aquatic and Terrestrial Humic Materials*, R. F. Christman, and E. Gjessing, Eds. (Ann Arbor, MI: Ann Arbor Science Publishers, Inc., 1983), pp. 517-528.
22. Meuzelaar, H. L. C., H. G. Ficke, and H. C. den Harrinck. "Fully Automated Curie-Point Pyrolysis Gas Liquid Chromatography," *J. Chromatogr. Sci.* 13(1):12-17 (1975).
23. Meent, D. van de, S. C. Brown, R. P. Philip, and B. R. T. Simoneit. "Pyrolysis-High Resolution Gas Chromatography and Pyrolysis Gas Chromatography-Mass Spectrometry of Kerogens and Kerogen Precursors," *Geochim. Cosmochim. Acta* 44(7):999-1013 (1980).
24. Van Krevelen, D. W. "Graphical-Statistical Method for the Study of Structure and Reaction Processes of Coal," *Fuel* 29(12):269-284 (1950).
25. Visser, S. A. "Application of Van Krevelen's Graphical Statistical Method for the Study of Aquatic Humic Material," *Environ. Sci. Technol.* 17(7):412-417 (1983).
26. De Leer, E. W. B., J. S. Sinninghe Damsté, C. Erkelens, and L. de Galan. "The Identification of Intermediates Leading to Chloroform and C-4 Diacids in the Chlorination of Humic Acid," *Environ. Sci. Technol.* (in press).
27. Peters, C. J., R. J. Young, and R. Perry. "Factors Influencing the Formation of Haloforms in the Chlorination of Humic Materials," *Environ. Sci. Technol.* 14(11):1391-1395 (1980).
28. Meent, D. van de, J. W. de Leeuw, and P. A. Schenck. "Chemical Characterization of Non-Volatile Organics in Suspended Matter and Sediments of the River Rhine Delta," *J. Anal. Appl. Pyr.* 2:249-263 (1980).

CHAPTER 68

Formation of Acidic Trace Organic By-Products from Chlorination of Humic Acids

Dennis R. Seeger, Leown A. Moore, and Alan A. Stevens

The discovery that trihalomethanes (THMs) were formed in drinking water during chlorination for disinfection,[1,2] follow-up studies demonstrating the ubiquitous occurrence of THMs in chlorinated water,[3,4] and the extrapolation of results from animal studies[5] to suggest that ingestion of THMs posed a risk to human health led to the promulgation of federal regulations in 1979 limiting the concentration of total THMs in drinking water to 0.10 mg/L.[6] Concurrent research indicating that naturally occurring humic substances were likely precursors to THM formation[1,7,8] has led to studies investigating other possible by-products from the chlorination of water containing humic materials.

In chlorinated drinking waters, THMs have been found to account for only 22 to 67% (depending on the source water) of total organic halogen (TOX),[9] and, in studies using a concentrated solution of an isolated aquatic fulvic acid, trichloroacetic acid was the major chlorination product, accounting for 32.1% of the final TOX, compared with 17.3% for chloroform.[10] Another study by Norwood et al.[11] has identified numerous individual by-products from the treatment of concentrated solutions of aquatic humic and fulvic acids with chlorine.

The work described in this chapter attempts to examine by-products resulting from the chlorination of humic substances at concentrations likely to be found in drinking water with the following goals in mind: (1) to develop methods suitable for isolating and identifying specific compounds from large volumes of water; (2) to qualitatively and quantitatively compare compounds found in chlorinated humic acid solutions to the corresponding nonchlorinated solutions; (3) to qualitatively and quantitatively compare compounds formed from the chlorination of humic substances from different sources; and (4) to qualitatively compare the results of chlorinating humic acids at low concentration to those of other investigators using highly concentrated humic acid solutions.

EXPERIMENTAL

Humic substances from four different sources were used: two were humic acids from commercial suppliers, Aldrich Chemical Company and Fluka

Chemical Corporation; a third was an aquatic humic acid isolated from Black Lake, North Carolina;[12] and the fourth was an unconcentrated highly colored lake water from southern Florida that is used as a drinking water source. The humic acid solutions from the first three sources were prepared by dissolving the solid material in a basic solution, neutralizing with acid, buffering with phosphate to pH 7, filtering, and diluting to 40 L to yield a solution of 5–6 mg/L total organic carbon (TOC). The Florida lake water (28 mg/L TOC) was only buffered for pH control. The samples were then split into two 20-L portions; one was chlorinated to 20 mg/L, and the other was maintained as an unchlorinated control. The Florida lake water required a larger chlorine dose (40 mg/L) to maintain a free residual, presumably to satisfy the chlorine demand of its higher initial TOC. After 3 d at ambient temperature, each chlorinated sample was checked to ensure that a free chlorine residual remained and was then dechlorinated with sodium sulfite. Both the chlorinated sample and the control were acidified to pH 2 to prepare for the adsorption step. Ten to twenty liters of each of the samples was pumped at about 8 mL/min through separate 1.3-cm-diam glass columns containing a 25 cm depth of XAD-8 resin.

The organics were desorbed from the resin by a continuous flow of ethyl ether condensate for 3 h. This was accomplished by attaching a side arm to the adsorption column, which allowed vapors from 75 mL of peroxide-free ethyl ether in a heated collection flask connected to the base of the column to pass to a condenser at the top of the column. The condensed ether then drained through the adsorption media and returned to the collection flask. The aqueous layer (from residual water on the resin column) was further acidified to pH 1, separated from the ether layer, and extracted with an additional 25 mL of ethyl ether. The combined ethyl ether fractions were dried with sodium sulfate and concentrated to 5 mL using Kuderna-Danish equipment. The samples were methylated with diazomethane gas following the procedure described by Schlenk and Gellerman,[13] with 5% methanol added to aid methylation and to minimize precipitation in the sample, and held at room temperature for at least 16 h. The samples were remethylated by the same procedure just prior to analysis to ensure thorough methylation because the yellow color of the extracts prevented visual examination for the presence of diazomethane.

An internal standard, 1-chlorododecane, was added to a 0.1-mL aliquot of the sample just prior to analysis by splitless injection onto a 50-m by 0.20-mm-ID fused silica capillary column coated with a methyl silicone fluid. A Finnigan model 9500 gas chromatograph was used, with the column exiting directly into the ion source of a Finnigan model 3300 quadrupole mass spectrometer operated at 70 eV in the electron impact mode. The data were collected and processed using the Finnigan INCOS model 2300 data system.

The identification and integration of the individual chromatographic peaks were greatly facilitated by the ability to automate the peak searching and quantitation procedures on the INCOS data system. The first set of samples, however, required extensive interaction by the analyst. First, the mass spectra of all chromatographic peaks, minus background, were compared to the libraries of 38,728 spectra from the EPA/NIH Mass Spectral Data Base and of 105 spectra identified by researchers at the University of North Carolina from products of humic acid chlorination using a gas chromatograph/mass spectrometer (GC/MS) capable of accurate mass measurements.[14] If the analyst concluded that a library spectrum matched that of the sample peak, its identity, mass fragment intensities, retention time relative to the internal standard, and a principal fragment ion for future quantitation purposes were entered into a special sample library. If the analyst concluded that the sample peak spectrum was valid (i.e., that it was not overly influenced by background or other interfering peaks) but did not match any library spectra, it was entered into the sample library as a numbered unknown, again including mass fragment intensities, relative retention time (RRT), and a principal fragment ion.

In subsequent samples, much of the analyst's interaction was eliminated by using the automated procedures of the data system to find and integrate the peak areas of compounds that had been entered into the sample library from previous samples. These procedures used retention time and spectral data from each sample library entry to search for that compound in the new sample data file. If a matching peak was found, the area of the principal ion from the library entry was integrated and reported with the compound name from the sample library. To find and identify peaks that had not been previously entered into the sample library, a modified Biemann-Biller peak-finding algorithm[15] was used to list peaks that had not matched any of the sample library entries. These peaks were then examined by the analyst and added to the sample library either with identities obtained from the standard libraries or as unknowns.

RESULTS AND DISCUSSION

Because few of the named compounds in this work have been confirmed by the analysis of authentic standards or by any means other than electron impact MS, all identifications listed in this chapter are considered tentative unless specifically noted otherwise. Also, in many cases, a library match or specific interpretation could not be made, but the spectra exhibited unique characteristics such as clusters of fragment ion peaks or repetitive ion peak patterns. These characteristics allowed classification of the spectra with similar compounds. For example, several compounds characterized as "aliphatic acid, methyl ester" in the sample library exhibited the characteristically strong fragment ions at m/z 74 and 87, but weak responses in the higher mass range precluded any further identification or molecular weight assignment. For other compounds that were characterized only as chlorine-containing compounds, clusters of ions resulting from the natural isotopic ratio of ^{35}Cl to ^{37}Cl were observed, but further characterization based on library matches or spectral interpretation could not be made. Characterization of compounds in this way provided a useful tool to monitor the classes of compounds found in the different sample runs. Because capillary column retention time as well as spectral matching was used, a level of confidence was added in determining the recurrence of individual compounds in succeeding samples, even when compounds of nearly identical spectra, such as isomers or members of a homologous series, were present.

The computer-reconstructed total ion chromatograms of nonchlorinated and chlorinated Fluka humic acid (Figure 1) visually illustrate the increases in both the numbers and intensities of the peaks produced by chlorination. The largest peak in each chromatogram is the added internal standard, which was approximately equivalent to a concentration of 10 μg/L in the original humic acid solutions.

Four broad groups of compounds are shown in Tables I and II, and the individual peaks that have been tentatively identified or characterized are listed within these groups with a numerical value corresponding to area counts of the peak rounded to the nearest 1000 after normalization on the internal standard. Because response factors were not calculated from known standards, the intensities cannot be related directly to concentration, and the intensity numbers can only be compared with other intensity numbers for the same compound (i.e., within horizontal rows of the table). Even then, the large quantitative variability (discussed below) must be considered. Forty-one additional compounds were found in more than one of the chlorinated humic samples. They were characterized only as to whether they contain chlorine, and lack of further specific spectral characteristics precluded their assignment to the groups in Tables I and II. All 26 of the chlorine-containing compounds and 13 of the 15 remaining compounds not containing chlorine were nearly absent in the corresponding nonchlorinated humic acid controls.

The monobasic and dibasic aliphatic acids, mono- and polycarboxylic aromatic acids in Table I, and various chloroaliphatic acids listed in Table II generally agree with compounds identified by other investigators where the reaction conditions were more harsh.[10,11] Ring-chlorinated aromatic acids, however, were not reported in those studies using similar substrates. One explanation is that the ring-chlorinated aromatic acids may survive intact at the relatively low concentration of chlorine and carbonaceous substrates used in this work, but the aromatic ring structures are further oxidized, or other mechanisms prevail, at the higher concentrations used by other researchers. Conversely, another explanation may be that the extraction and concentration method used in this study selectively enhanced the concentration of the ring-chlorinated aromatic acids relative to the other chlorination by-products.

Chlorophenols[16] and chloromethoxybenzenes[17] have been reported in chlorinated humic acid samples and highly chlorinated reservoir water, respectively, and 2-chlorobenzoic acid methyl ester, has been reported (tentatively identified by GC retention time only) by one investigator.[18] These are the only previously reported specific identifications of ring-halogenated compounds resulting from the chlorine treatment of humic materials. Super chlorination

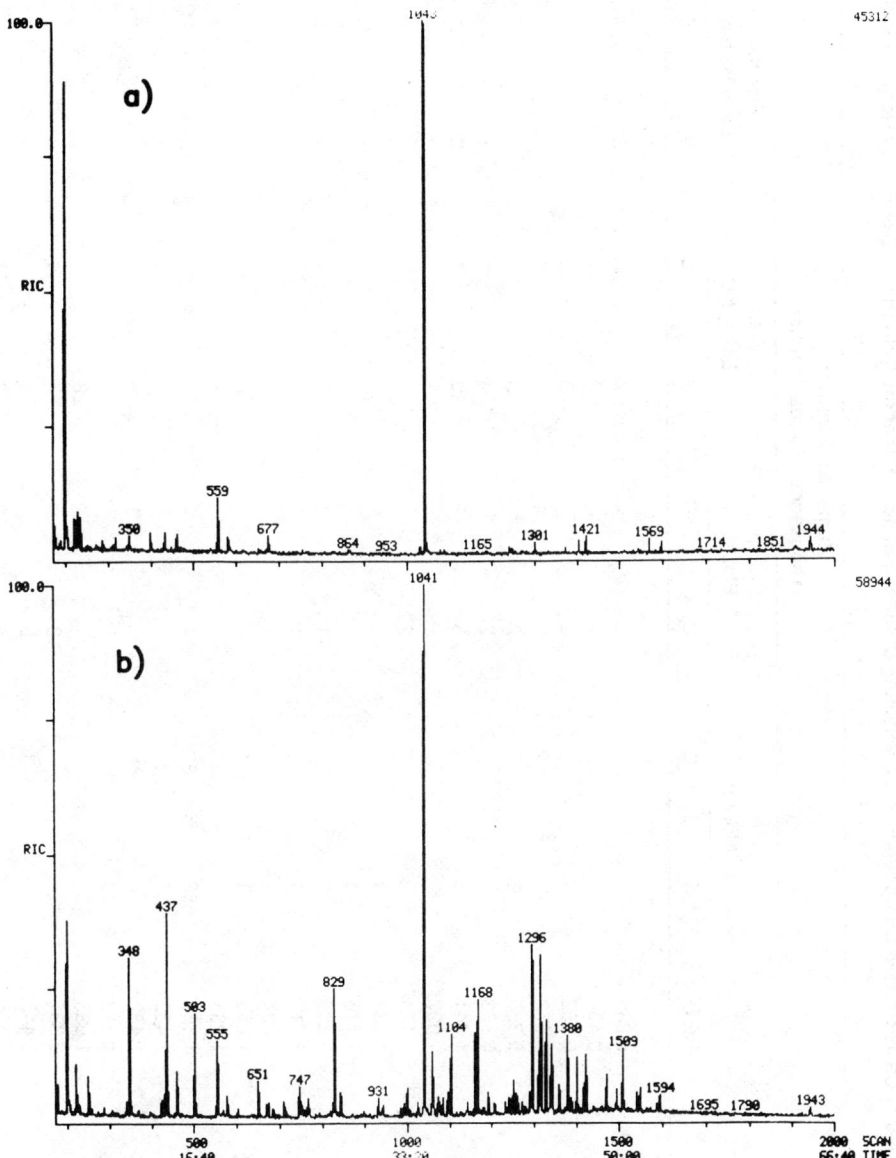

Figure 1. Computer-reconstructed chromatograms of methylated concentrated extracts from (a) nonchlorinated and (b) chlorinated Fluka humic acid solution. GC conditions: 2.0 μL splitless injection at 35°C; programmed at 4°C/min to 240°C after 5 min. Mass range scanned from 35 to 435 amu every 2 s.

of municipal wastewaters[19] and pulp bleaching operations[20,21] also yield some ring-chlorinated compounds, and aqueous chlorination of simple phenols to form odorous ring-chlorinated derivatives is also well known in the drinking water industry. The work described in this chapter, however, illustrates that

Table I. Tentatively Identified or Characterized Compounds Not Containing Chlorine Found in Either the Chlorinated Samples or Nonchlorinated Controls

		\multicolumn{8}{c}{Humic Acid Sources (Thousands of Area Counts)}							
		Aldrich		Fluka		Black Lake, NC		Florida Lake Water	
Compound	RRT[a]	S[b]	C[c]	S	C	S	C	S	C
Aliphatic acid, methyl esters									
2-Hydroxypropanoic	0.249					39	41	3	4
Hexanoic	0.436	1	1	1		3	1	1	
Heptanoic	0.552	2		2		3	1	1	
Octanoic	0.662	4	2	3	1	6	1	1	1
Nonanoic	0.765	3	3	2		3	1	1	
Decanoic	0.863	4		2		4	1	1	
Dodecanoic	1.041	5	4	6	2	10	1	2	1
Compound 265	1.097					1			
Compound 123	1.124	2		2					
Compound 126	1.134	1		1					
Compound 132	1.175	1							
Tetradecanoic	1.203	10	7	11	1	17	4	4	2
Compound 139	1.213	1		1					
Compound 227	1.232						1		
Compound 145	1.258	1	1			4		2	
Compound 204	1.285					5	2	2	
Hexadecanoic	1.355	52	32	14	4	70	28	15	1
Compound 230	1.432						1		
Compound 377	1.482				1				
Octadecanoic	1.513	10				28	21		
Compound 206	1.533								
Ethanedioic	0.331			13		7	6	6	3
Propanedioic	0.445	2	1	2		1	1	1	1
Butanedioic[d]	0.556			14	12	2	1	1	

Octanedioic	0.965	1						
Nonanedioic	1.052	7		2		1		
Aromatic acid, methyl esters								
Benzoic[d]	0.628	11		18		10	28	
Benzenedicarboxylic	0.982	20	5	9	4	10	15	16
Benzenedicarboxylic	1.018	9	2	7	7	3	6	3
Benzenedicarboxylic	1.029	8		9		4	17	1
Benzenetricarboxylic	1.294	44	5	25		18	10	
Benzenetricarboxylic	1.340	10		6		4	5	
Benzenetricarboxylic	1.454	2				1		
Benzenetetracarboxylic	1.534			6				
3-Methoxybenzoic	0.909	3		1		2	2	
Methylbenzenedicarboxylic	1.078	4		1			6	
Methylbenzenedicarboxylic	1.103	1		1				
Methylfurancarboxylic	0.601	2		1		2	1	
Methylfurandicarboxylic	0.959	32		14		9	11	

[a] Retention time relative to internal standard.
[b] Sample.
[c] Control.
[d] Identification confirmed by mass spectrum and retention time of an authentic standard.

Table II. Tentatively Identified or Characterized Chlorine Containing Compounds Found in Either the Chlorinated Samples or Nonchlorinated Controls

		Humic Acid Sources (Thousands of Area Counts)							
		Aldrich		Fluka		Black Lake, NC		Florida Lake Water	
Compound	RRT[a]	S[b]	C[c]	S	C	S	C	S	C
Aliphatic acid, methyl esters									
Chloroacetic[d]	0.256	1		1		6		2	
Dichloroacetic[d]	0.343	18		49		83	5	46	
2,2-Dichloropropanoic	0.364	2		2		2		7	
Trichloroacetic[d]	0.428	32		64				64	
Bromochloroacetic	0.434							1	
3,3-Dichloropropenoic	0.451	13						11	
Dichloroacetic acid, ethyl ester	0.420	1							
Trichloroacetic acid, ethyl ester	0.504	1							
Trichloroacetic acid	0.859					1		20	
Aromatic Acid, methyl esters									
5-Chloro-2-methoxybenzoic[d]	1.022	8		2		1			
Dichloromethoxybenzoic	1.060	20		11		19		3	
Dichloromethoxybenzoic	1.070	1		1					
Chloromethoxybenzenedicarboxylic	1.231	9				2		1	
Chloromethoxybenzenedicarboxylic	1.240	4		1					
Chloromethoxybenzenedicarboxylic	1.267	63		21		8		11	
Dichloromethoxybenzenedicarboxylic	1.291	13		2		7			
Chloromethoxybenzenedicarboxylic	1.314	17		5					
Dichloromethoxybenzenedicarboxylic	1.331	41		7		11		2	
Chloromethoxybenzenetricarboxylic	1.466	24		1		6			
Chloromethoxybenzenetricarboxylic	1.534	16		4		4			

[a] Retention time relative to internal standard.
[b] Sample.
[c] Control.
[d] Identification confirmed by mass spectrum and retention time of an authentic standard.

ring-chlorinated aromatic acids may constitute a more important class of by-products from the chlorination of humic materials under drinking water treatment conditions than previously recognized.

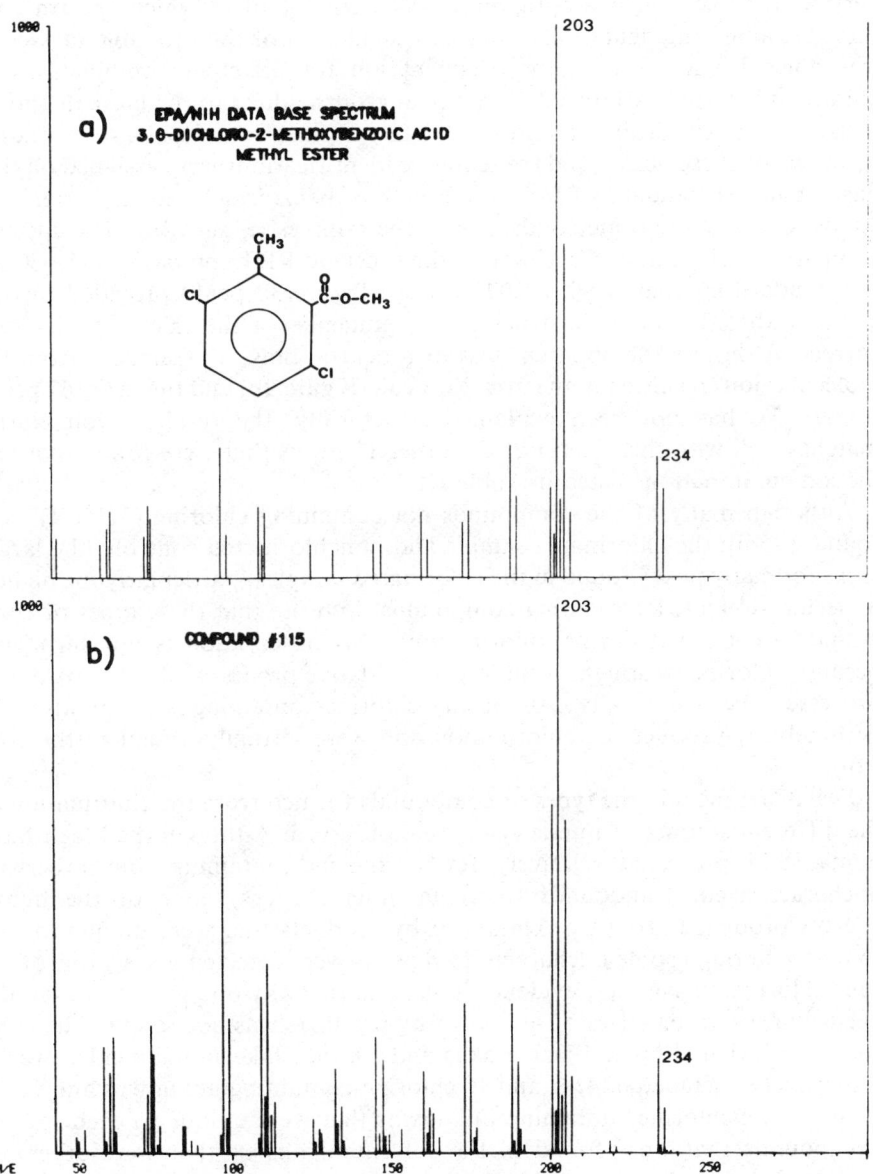

Figure 2. Mass spectra of (a) 3,6-dichloro-2-methoxy-benzoic acid, methyl ester, from the EPA/NIH Mass Spectral Data Base; and (b) compound 115 from the methylated extract of a chlorinated Aldrich humic acid solution.

The tentative identification of the peak having the spectrum shown in Figure 2b was initially based solely on its match with the standard spectrum found in the EPA/NIH Mass Spectral Data Base (Figure 2a). The general fragmentation pattern and the major m/z peaks, except for the m/z 201 ion, matched very well. Other chromatographic peaks were found in which spectra contained similar fragmentation patterns. Calculation of the probable masses of the molecular ion and major fragment ion for different combinations of chloro and methyl carboxylate functional groups added to the basic methoxybenzene structure (Table III) provided likely explanations for the spectral data for many of these peaks, and the tentative identifications were assigned on that basis. Later, a standard of 5-chloro-2-methoxybenzoic acid, methyl ester, was analyzed under the same conditions as the samples (Figure 3a). The spectral comparison with Figure 3b as well as the excellent RRT comparison (1.020 for the standard compared with 1.022 for the Figure 3b peak) provided further evidence that the tentative structural assignments for the sample peaks were correct. Although the apparent loss of a neutral mass 33 fragment from the molecular ion, resulting in the m/z 201 peak (Figure 2b) and the m/z 167 peaks (Figure 3), has not been explained structurally, the resulting ion cluster matches well with the spectra of the other 11 peaks that were found that had the ion combinations listed in Table III.

Although many of the compounds not containing chlorine (Table I) were found in both the chlorinated sample and nonchlorinated controls, the larger areas for many peaks found in the chlorinated samples, particularly the higher-molecular-weight, later-eluting compounds, indicate that these types of compounds are formed during chlorination. This observation is not surprising because chlorine treatment is mainly an oxidative process and other oxidative processes occur naturally. All of the chlorine-containing compounds were definitely by-products of chlorination and were virtually absent in the controls.

Few differences in the types of compounds formed from the chlorination of the different sources of humic acids were observed. Although the Black Lake humic acid produced relatively fewer chlorine-containing, but otherwise uncharacterized, compounds than the other sources, none of the humic sources produced groups or classes of by-products that were unique to that source. Although some uncharacterized peaks were detected in only one of the four chlorinated samples, evidence indicating that any one source produced a particular group or series of unknown by-products was not found. The chlorinated Aldrich, Fluka, Black Lake, and Florida lake humic acid extracts, respectively, included 6, 4, 6, and 10 chlorine-containing unknowns and 12, 8, 6, and 4 nonchlorine-containing unknowns that were unique to each source. Compounds that were found in the solvent blanks or that were otherwise suspected of being method artifacts have not been included in the tables. Many compounds that were found at about equal concentrations in both chlorinated and nonchlorinated samples were also omitted from the tables. These included alkanes, alkyl-substituted benzenes, phthalic acid esters other than methyl esters, and eight unknowns.

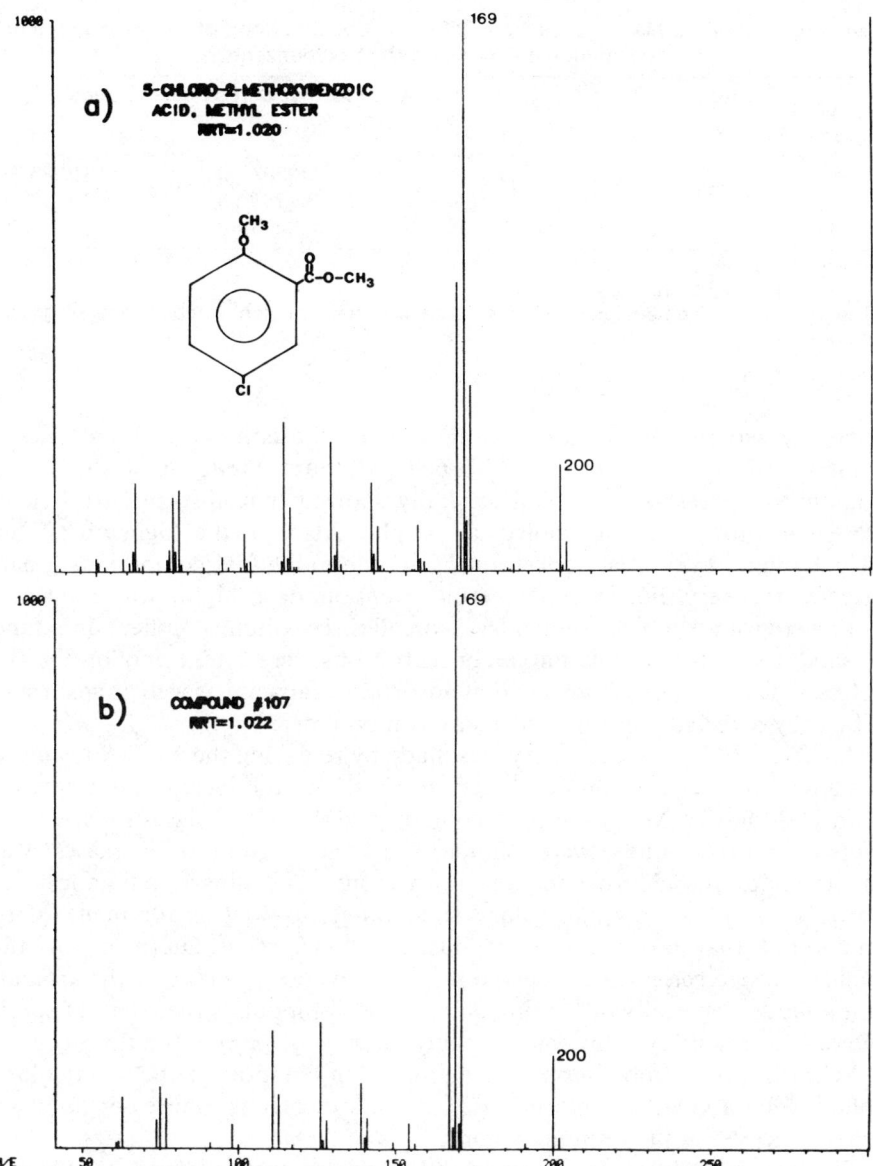

Figure 3. Mass spectra of (a) 5-chloro-2-methoxybenzoic acid, methyl ester standard run under the same GC/MS conditions as the samples; and (b) compound 107 from the methylated extract of a chlorinated Aldrich humic acid solution.

Several additional experiments were included in this study to determine the recovery and reproducibility of the adsorption/extraction method and to investigate alternative procedures. Because breakthrough studies showed poor removal of TOX and THMs, a chlorinated Aldrich humic acid solution was passed through three XAD-8 resin columns in series in an attempt to deter-

Table III. Expected Mass Peaks from Different Combinations of Chlorine and Methyl Carboxylate Functional Groups on Methoxybenzene

No. of Cl Atoms	Number of Methyl Carboxylate Groups		
	1	2	3
1	200[a],169[b](1)[c]	258[a],227[b](5)[c]	316[a],285[b](2)[c]
2	234,203 (3)	292,261 (2)	350,329 (0)
3	268,237 (0)	326,295 (0)	

[a] Molecular ion, m/z.
[b] Principal fragment ion, m/z.
[c] The number of isomers actually found that matched the ion combinations is given in parentheses.

mine, by separate extraction and analysis of the columns, which types of compounds were more readily adsorbed. Although hindered by the lack of quantitative precision, the data generally showed that adsorption efficiency increased with increasing molecular weight. Also, in the higher molecular weight range, the initial concentration of the individual compounds greatly affected recovery. For example, certain compounds at high initial concentrations were found at nearly equal levels on all three columns. Other compounds of similar structure at low initial concentrations were found only on the first column. These findings indicate that adsorption capacity as well as adsorption efficiency affected recoveries of many compounds.

An effort to increase recovery was made by replacing the XAD-8 resin with granular activated carbon (GAC). Figure 4 shows the increased removals of both TOX and THMs by the GAC column, and the peak areas for many of the individual compounds were similarly affected. Notable increases were observed for the chloro and nonchloro mono- and dibasic aliphatic acids, particularly in the low-molecular-weight range, as well as for many of the compounds that have not been attributed to chlorination. On the other hand, significant decreases were observed in the recoveries of many of the aromatic compounds, especially in the ring-chlorinated compounds that were of special interest in this study. This was probably because of poor extraction from the GAC rather than poor initial adsorption. Combination of the resin column followed by a GAC column in series appears to offer promise for improved overall recoveries in continuing work.

Finally, to evaluate the reproducibility of the methodology, the Aldrich humic acid run was repeated under conditions duplicating those of the first run. For each of 50 compounds that were found in both concentrated extracts of the chlorinated samples, 54% of the area count pairs were reproducible within a factor of 2 and 82% within a factor of 3, after adjusting for the internal standard response. This level of precision was not sufficient to make all the desired quantitative evaluations of the data. The low and variable recoveries from the resin by the adsorption and extraction procedures were probably the chief causes for the poor reproducibility.

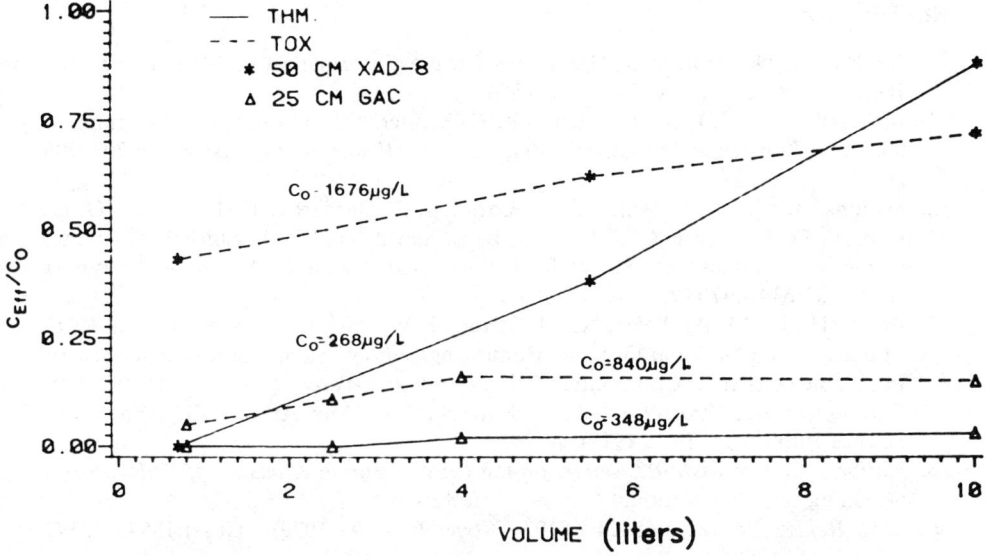

Figure 4. Adsorption of total organic halogen and trihalomethane from chlorinated Aldrich humic acid solution on 50 cm of XAD-8 resin compared to 25 cm of granular activated carbon.

SUMMARY

Most of the compounds and compound groups tentatively identified in this work on chlorination of humic substances at low concentrations have been found previously by others in work that was conducted at high concentrations. This supports the validity of extrapolating high concentration work to predict by-products in drinking water treatment applications. Further research will be required to determine the significance of the tentative identification of numerous ring-chlorinated aromatic acid methyl esters.

No major differences were found in the types of compounds produced with the different sources of humic substances used. This consistency is important because the composition of humic materials in a water supply can vary greatly from one geographic area to another, or even from one season to another. Research in this field would be much more complicated if different by-products were produced from different humic sources. This consistency also lends credibility to the health effects studies and other research that has been conducted using commercial sources of humic acid as models.

Finally, this study demonstrated the practicality of using the described method for concentrating and identifying specific compounds from water samples at low concentrations. Improvements, particularly in the areas of recovery and quantitation, are needed so that specific conclusions can be reached about the amount and types of by-products formed when comparing different chlorine treatment practices.

REFERENCES

1. Rook, J. J. "Formation of Haloforms During Chlorination of Natural Waters," *Water Treat. Exam.* 23:234–243 (1974).
2. Bellar, T. A., J. J. Lichtenberg, and R. C. Kroner. "The Occurrence of Organohalides in Chlorinated Drinking Water," *J. Am. Water Works Assoc.* 66:703–706 (1974).
3. Symons, J. M., T. A. Bellar, J. K. Carswell, J. DeMarco, K. L. Kropp, G. G. Robeck, D. R. Seeger, C. J. Slocum, B. L. Smith, and A. A. Stevens. "National Organics Reconnaissance Survey for Halogenated Organics," *J. Am. Water Works Assoc.* 67:634–647 (1975).
4. Brass, H. J., M. A. Feige, T. Halloran, J. W. Mello, D. Munch, and R. F. Thomas. "The National Organic Monitoring Survey: Sampling and Analysis for Purgeable Organic Compounds," in *Drinking Water Quality Enhancement Through Source Protection*, R. B. Pojasek, Ed. (Ann Arbor, MI: Ann Arbor Science Publishers, Inc., 1977), p. 393.
5. *National Cancer Institute Report on the Carcinogenesis Bioassay of Chloroform*, (Washington, DC: National Cancer Institute, 1976).
6. *Fed. Regist.* 44 (231):68624–68707 (November 29, 1979); 45(49):15542–15547 (March 11, 1980).
7. Stevens, A. A., C. J. Slocum, D. R. Seeger, and G. G. Robeck. "Chlorination of Organics in Drinking Water," in *Water Chlorination: Environmental Impact and Health Effects, Vol. 1*, R. L. Jolley, Ed. (Ann Arbor, MI: Ann Arbor Science Publishers, Inc., 1978), pp. 77–101.
8. Rook, J. J. "Haloforms in Drinking Water," *J. Am. Water Works Assoc.* 68:168 (1976).
9. Symons, J. M., A. A. Stevens, R. M. Clark, E. E. Geldreich, O. T. Love, Jr., and J. DeMarco. *Treatment Techniques for Controlling Trihalomethanes in Drinking Water* (Denver, CO: American Water Works Association, 1982), p. 186.
10. Christman, R. F., D. L. Norwood, D. S. Millington, J. D. Johnson, and A. A. Stevens. "Identity and Yields of Major Halogenated Products of Aquatic Fulvic Acid Chlorination," *Environ. Sci. Technol.* 17:625–628 (1983).
11. Norwood, D. L., J. D. Johnson, and R. F. Christman. "Chlorinated Products from Aquatic Humic Material at Neutral pH," in *Water Chlorination: Environmental Impact and Heatlh Effects, Vol. 4*, R. L. Jolley, W. A. Brungs, J. A. Cotruvo, R. B. Cumming, J. S. Mattice, and V. A. Jacobs, Eds. (Ann Arbor, MI: Ann Arbor Science Publishers, Inc., 1983), pp. 191–200.
12. Colclough, C. A., J. D. Johnson, D. S. Millington, and R. F. Christman. "Organic Reaction Products of Chlorine Dioxide and Natural Aquatic Fulvic Acids," in *Water Chlorination: Environmental Impact and Health Effects, Vol. 4*, R. L. Jolley, W. A. Brungs, J. A. Cotruvo, R. B. Cumming, J. S. Mattice, and V. A. Jacobs, Eds. (Ann Arbor, MI: Ann Arbor Science Publishers, Inc., 1983), pp. 219–229.
13. Schlenk, H., and J. L. Gellerman. "Esterification of Fatty Acids with Diazomethane on a Small Scale," *Anal. Chem.* 32(11):1412–1414 (1960).
14. Christman, R. F., W. T. Liao, D. S. Millington, and J. D. Johnson. "Oxidative Degradation of Aquatic Humic Material," in *Advances in the Identification and Analysis of Organic Pollutants in Water*, Vol. 2, L. H. Keith, Ed. (Ann Arbor, MI: Ann Arbor Science Publishers, Inc., 1981), pp. 979–999.

15. *MSDS System Reference Manual*, (Sunnyvale, CA: Finnigan Instrument Corp., 1978).
16. McCreary, J. J., and V. L. Snoeyink. "Reaction of Free Chlorine with Humic Substances Before and After Adsorption on Activated Carbon," *Environ. Sci. Technol.* 15:193–197 (1981).
17. Rook, J. J. "Chlorination Reactions of Fulvic Acids in Natural Waters," *Environ. Sci. Technol.* 11:478–482 (1977).
18. Quimby, B. D., M. F. Delaney, P. C. Uden, and R. M. Barnes. "Determination of the Aqueous Chlorination Products of Humic Substances by Gas Chromatography with Microwave Emission Detection," *Anal. Chem.* 52:259–263 (1980).
19. Glaze, W. H., J. E. Henderson, IV, and G. Smith, "Analysis of New Chlorinated Compounds Formed by Chlorination of Municipal Wastewater," in *Water Chlorination: Environmental Impact and Health Effects, Vol. 1*, R. L. Jolley, Ed. (Ann Arbor, MI: Ann Arbor Science Publishers, Inc., 1978), pp. 139–159.
20. Leach, J. M. "Loadings and Effects of Chlorinated Organics from Bleached Pulp Mills," in *Water Chlorination: Environmental Impact and Health Effects, Vol. 3*, R. L. Jolley, W. A. Brungs, and R. B. Cumming, Eds. (Ann Arbor, MI: Ann Arbor Science Publishers, Inc., 1980), pp. 325–334.
21. Claeys, R. R., L. E. LaFleur, and D. L. Borton. "Chlorinated Organics in Bleach Plant Effluents of Pulp and Paper Mills," in *Water Chlorination: Environmental Impact and Health Effects, Vol. 3*, R. L. Jolley, W. A. Brungs, and R. B. Cumming, Eds. (Ann Arbor, MI: Ann Arbor Science Publishers, Inc., 1980), pp. 335–345.

CHAPTER **69**

Nonpurgeable Organohalide Formation on Chlorination of Algal Extracellular Material

Jan K. Wachter and Julian B. Andelman

In 1974, chloroform ($CHCl_3$), an animal carcinogen, was identified by Rook[1] and Bellar et al.[2] as a major by-product of chlorination reactions involving naturally occurring organic compounds in water supplies. Subsequently, research has intensified to isolate and identify the precursors and by-products associated with chlorination reactions in water and wastewater systems. It has been shown that algal biomass and extracellular products (ECP) can also generate chloroform upon chlorination.[3-5] However, these studies have been limited to the detection of purgeable organic halide (POX), including chloroform, from algal biomass and ECP and not the characterization and quantification of nonpurgeable organohalide (NPOX) compounds. This NPOX fraction is important because studies have shown that upon chlorination of organics in water supplies (e.g., humic acids), the amount of nonpurgeable organic-bound chlorine generally exceeds the volatile fraction produced.[6-8] In addition, recent studies have shown that nonpurgeable organics in various water samples after chlorination have greater mutagenic potential than the organics present before chlorination.[9,10]

The purpose of this investigation is to assess the NPOX-forming potential of algal ECP and to characterize the NPOX fraction formed under a variety of environmental conditions. The chlorine demand associated with the generation of this NPOX will also be determined.

Two green algae, *Chlorella vulgaris* and *Chlorella pyrenoidosa*, and a blue-green alga, *Anabaena flos-aquae*, were chosen in this study to generate ECP because of their widespread occurrence in water supplies. Because chlorophyll-a is a typical extracellular product of algae, it was included in the study as a model ECP.

MATERIALS AND METHODS

Chlorella vulgaris (UTEX No. 262), *C. pyrenoidosa* (UTEX No. 26), and *A. flos-aquae* (Utex No. 1444) cultures were obtained from the Texas Culture Collection, Austin, Texas. The algal cultures were grown at 22.7 ± 0.7°C in aerated, autoclaved, organic-free media using a diurnal cycle with fluorescent lights. Four to six large Pyrex bottles containing 15-L media were used in the culturing of each alga. For each algal culture and 5 mg/L chlorophyll solution, there was a corresponding control consisting of only algal media exposed to the same laboratory environment as the algal cultures. The algal cultures were harvested after attaining the stationary phase of growth, following which algal biomass was separated from the media containing algal ECP by passage through a cream separator and two glass microfibre filters. The detailed materials and methods used to generate the ECP in this experiment are described elsewhere.[11]

Following pH adjustment, aliquots of the algal and chlorophyll filtrates and their controls were chlorinated in 160-mL vials sealed without headspace using Teflon®-faced septa. A standardized stock solution of sodium hypochlorite (NaOCl) was used to chlorinate by injection through the Teflon septum at an initial concentration of 20 mg/L Cl_2, producing a free chlorine residual under all reaction conditions. The standard chlorination conditions used in this experiment were pH 7, 24°C, a contact time of 24 h, and exposure to light to simulate a typical environment of natural water systems and drinking water reservoirs undergoing chlorination. Parameters were also individually varied as follows while maintaining the others at standard conditions: pH, temperature, contact time, light conditions, and chlorine dose. At the conclusion of each experiment, the free available chlorine concentration remaining was determined in one of the bottles to assess the chlorine demand, using the DPD ferrous titrimetric technique.[12] All the remaining bottles were then quenched with excess sodium sulfite to eliminate the chlorine, and the samples were stored in the dark at 4 to 7°C prior to analysis.

Nonpurgeable organic halide (NPOX) and total organic carbon (TOC) levels were calculated and measured in each group of chlorinated algal filtrate and chlorinated control samples. Results from the chlorinated controls were subtracted from those of chlorinated supernatant samples to obtain the final readings. NPOX values were obtained by calculating the difference between total organic halide (TOX) and purgeable organic halide (POX) concentrations measured using the DX-20 Total Organic Halide Analyzer (Dohrmann-Envirotech), according to the proposed ASTM Standard Practice for TOX analysis.[13] Total organic carbon (TOC) measurements were made using the Oceanography International Corporation Total Organic Carbon Analyzer, Model 524 (100-μL injections were performed) and Dohrmann-Envirotech Organic Analyzer and Ultra-Low Organics Model, DC-54 (10-mL samples used). On the average, NPOX analyses were repeated three times and TOC analyses twice per set of environmental conditions of each algal system investigated.

The molecular size distribution of NPOX formed upon chlorination of algal and model ECP was determined using size-exclusion high-performance liquid chromatography (HPLC), followed by NPOX detection of the eluted fractions. Three liters of *A. flos-aquae* and *C. pyrenoidosa* filtrates, as well as their controls, were adjusted to pH 7 and chlorinated at 20 mg/L Cl_2 at 24°C for 1 d. The filtrates and their controls were then concentrated to between 30 and 45 mL using rotary vacuum evaporation. Between 2 and 3 mL of each concentrate was injected via multiple runs onto a HP liquid chromatograph equipped with a size-exclusion column. Four eluted fractions were collected for each concentrate, corresponding to >10,000; 10,000 to 1,000; 1,000 to 100, and <100 daltons. NPOX analyses were performed on these fractions. NPOX results are reported as if 1 L of unconcentrated filtrate had been injected into the HPLC and its fractions analyzed for NPOX content. Control NPOX values were subtracted from NPOX values measured in ECP filtrate fractions to arrive at values presented in this chapter.

In conjunction with the above investigation, the organohalide formation potentials of molecular weight fractions of unchlorinated ECP filtrates were determined. This was accomplished by concentrating 3 to 9 L of unchlorinated algal ECP filtrates (and their controls) by rotary evaporation to 30 to 45 mL. Aliquots of 4 to 6 mL of these concentrates were injected onto the HPLC column via repeated injections. Eluted fractions were collected from the column corresponding to the four aforementioned molecular size ranges. These fractions were then chlorinated at 24°C, pH 7, and 1-d contact time. The total amount of chlorine added to a set of fractions was based on the amount required to arrive at a concentration of 20 mg/L Cl_2 in the volume of the original sample having theoretically passed through the column [i.e., if 4 L of ECP filtrate was concentrated to 40 mL and 4 mL of concentrate was injected, then 8 mg chlorine would have to be added to one-tenth (400 mL) of the original sample, which theoretically passed through the column to arrive at a chlorine concentration of 20 mg/L]. This total chlorine amount was distributed among the four fractions based on volume proportions. Organohalide analyses were performed on these fractions after quenching.

A Perkin-Elmer Series 3B liquid chromatograph equipped with a size-exclusion column (25 cm by 0.8 cm I.D. SHODEX S-803/S, having an exclusion limit of 50,000 daltons) was used to separate by size the NPOX fraction generated from algal ECP. The column was calibrated by detection at ~192 nm with dextrans and sugars of various molecular sizes. The LC conditions used in experimentation were as follows; flow, 1 mL/min water; detection, TOX analyzer.

To identify the specific ECP compounds present in the filtrates and their halogen-containing derivatives, gas chromatography/mass spectrometry (GC/MS) analyses were performed on chlorinated (under standard conditions) and unchlorinated algal supernatant samples, as well as their corresponding control solutions. Generally, 6 L of a particular filtrate were divided and adjusted to pH 2 and 7; serially extracted using methylene chloride; concentrated using a Kuderna-Danish evaporator apparatus; reduced to 50 μL in volume by a nitrogen gas stream; methylated (acid extracts only) using ethereal diazomethane; and then subjected to GC/MS analyses. More detailed descriptions of the analytical methods used are described elsewhere.[14] The typical conditions in the GC/MS analyses of methylene chloride extracts, using a Hewlett Packard 5985 instrument, were as follows: column—10 ft × 0.25 in. I.D. glass packed with 3% SP-2250 on 80/100 mesh Supelcoport; temperature programming—100°C for 3 min, 8°C/min to 260°C, hold at 260°C for 15 min; gas flow—helium at ~28 mL/min; injection volume—2- to 5-μL extract; electron multiplier voltage—2600 eV; mass to charge ratios detected—40–600 m/e. Presumptive identification of compounds was made by comparing m/e ratios of known compounds to those for compounds in the methylene chloride extracts. Available standards were obtained for the presumptively identified compounds and confirmation was made by (1) standard addition techniques and (2) matching retention times of the standards with the compounds contained in the sample extracts.

RESULTS AND DISCUSSION

Effect of Various Parameters on NPOX Formation

Table I illustrates the effect of various parameters on NPOX formation on chlorination of algal ECP. Because a previous study by the authors[11] has determined that the algae strains used in this experiment did not naturally excrete organically bound halide into their media, these measurements represent reactions between chlorine and algal ECP in the filtrate samples.

Increases in both reaction time and temperature generally caused increases in NPOX formation. The effect of reaction time on NPOX formation in *A. flos-aquae* filtrates and chlorophyll solutions was typical of that for organic substrates in water[15–19] in that the rate of NPOX generation was rapid during the first hour of chlorination, eventually decreasing and approaching zero after 1 week of reaction. This decrease in rate of NPOX generation could be the result of the chlorine saturation of ECP addition/substitution sites with time; however, the decrease of TOC content of chlorinated ECP filtrates with contact time suggests the destruction of potential sites for chlorine incorporation with time. The rate of NPOX formation in *C. vulgaris* filtrates increased with time (24 to 168 h rate vs 1 to 24 h rate), typical of chlorination reactions involving algal biomass, described as being a relatively slow process.[3,4,20] The increases in NPOX formation with temperature were linear only for the *C. pyrenoidosa* and *A. flos-aquae* systems. These results indicate generally that if ECPs were present in water supplies during chlorine disinfection, the rate of NPOX formation would be greatest during the first hour after chlorination and would increase with water temperature.

The formation of NPOX did not appear to correlate with either the presence or absence of light. Given the instrumental imprecision of the TOX measurement and the variability of ECP content within sample sets,[14] there were no differences in the NPOX concentrations generated under light and dark conditions.

As pH decreased, NPOX generation increased in all the systems. At pH 2, NPOX accounted for 97% of the TOX produced, 74% at pH 7, and 26% at

Table I. Effect of Various Parameters on Formation of NPOX (in micrograms of Cl⁻ per liter) on Chlorination of Algal ECP

	Algal Species							
	Chlorella vulgaris		Chlorella pyrenoidosa		Anabaena flos-aquae		Chlorophyll	
Parameter	NPOX	% of TOX	NPOX	% of TOX	NPOX	% of TOX	NPOX	% of TOX
Reaction time, h								
1	76	87	0	0	162	83	77	85
24	81	83	15	54	380	83	243	77
168	241	91	48	43	439	71	309	62
pH								
2	168	97	65	100	613	97	494	93
7	81	83	15	54	380	83	243	77
12	7	18	6	5	161	33	79	48
Light conditions								
light	81	83	15	54	380	83	243	77
dark	92	82	24	75	322	79	241	81
Temperature, °C								
7	75	88	0	0	259	85	181	87
24	81	83	15	54	380	83	243	77
35	125	85	20	53	453	75	210	60

pH 12. Oliver[6] has shown a similar effect of pH on NPOX formation from naturally occurring organics (fulvic acids) in water supplies. These results indicate that an acid-catalyzed promoted or dependent mechanism may be responsible for NPOX formation. Results of more detailed experiments investigating the effect of pH on NPOX generation from algal ECP follow.

Excluding the data generated at high pH, the results in Table I indicate that the majority of organic halide generated on chlorination of algal ECP was nonpurgeable. These findings are in agreement with those of many other researchers who have indicated a similar propensity for naturally occurring organic compounds in water supplies to incorporate the majority of chlorine into nonvolatile (rather than purgeable) derivatives, under both field and laboratory conditions.[6,7,8,21-24] In addition, similar experiments conducted at the same standard chlorinating conditions used in this experiment have shown that the percentages of organic halide as NPOX derived from algal biomass were greater (91-99%) than those reported herein for the corresponding ECP.[25] Generally, 90% of the purgeable organic halide (POX) generated upon chlorination of algal ECP in these experiments was accounted for by chloroform.[11]

There was an obvious effect of algal species on the amount of NPOX generated, as shown in Table I. At a given set of chlorinating conditions, the filtrates of *A. flos-aquae*, a blue-green alga, generated the most NPOX. The filtrates of the two green algae, *C. vulgaris* and *C. pyrenoidosa*, generally generated the next highest and lowest NPOX concentrations, respectively.

NPOX Yields Based on Organic Carbon, Chlorine Concentration and Chlorine Demand

Because the observed effects of algal species on NPOX formation could have been the result of differences in TOC content (i.e., ECP excretion rates) of the algal filtrates, NPOX concentrations were normalized to organic carbon concentrations by calculating the molar ratio of NPOX as Cl^- to organic carbon as C. This ratio is an indication of the degree (percentage) of halide incorporation into nonpurgeable ECP carbon and was calculated using TOC concentrations remaining after the prescribed contact time, because TOC levels changed during chlorination. Typically, after chlorination they ranged from 5 to 10 mg/L TOC. As shown in Table II, the NPOX yields based on TOC concentrations obtained at standard chlorinating conditions ranged from 0.14 to 1.9% (14-fold difference), in contrast to the 25-fold difference in NPOX concentrations used to calculate these ratios (15 to 380 μg/L Cl^-). Thus, it appears that only half of the observed differences among algae species in forming NPOX could be attributed to differences in the respective amounts of ECP excreted into solution. A wide range of NPOX/TOC yields per algal system occurred at 1-week contact time and the lowest yield corresponded to chlorination at pH 12 or at 1 h. Since chlorine addition to a compound generally increases its toxicity,[26] these NPOX/TOC ratios are of special interest

Table II. Range of Percentage NPOX Yields Generated by Chlorinating Algal ECP Under Various Conditions, Based on Initial Chlorine Concentration, Chlorine Demand, and Total Organic Carbon

Species	NPOX Yield Based on Initial Chlorine Concentration[a] (%)		NPOX Yield Based on Chlorine Demand[b] (%)		NPOX Yield Based on Organic Carbon Concentration (Halogen Substitution of ECP Carbon from NPOX Formation)[c] (%)	
	Standard Chlorinating Conditions[d]	Range	Standard Chlorinating Conditions[d]	Range	Standard Chlorinating Conditions[d]	Range
Chlorella vulgaris	0.81	0.07 – 2.4	7.7	0.74 – 100	0.92	0.08 – 3.6
Chlorella pyrenoidosa	0.15	0 – 0.65	0.52	0 – 1.9	0.14	0 – 0.54
Anabaena flos-aquae	3.8	1.6 – 6.1	13	3.5 – 19	1.6	0.39 – 3.0
Chlorophyll	2.4	0.77 – 4.9	12	3.4 – 100	1.9	0.33 – 2.6

[a] % NPOX Yield = $\left(\dfrac{\text{NPOX, mg/L Cl}^-}{\text{initial reactive chlorine, mg/L Cl}^+}\right) 100.$

[b] % NPOX Yield = $\left(\dfrac{\text{NPOX, mg/L Cl}^-}{\text{chlorine demand, mg/L Cl}^+}\right) 100.$

[c] % NPOX Yield = $\left(\text{NPOX, mg/L Cl}^- \times \dfrac{\text{mmol Cl}}{35 \text{ mg Cl}} \times \dfrac{1}{\text{mg/L TOC}} \times \dfrac{12 \text{ mg TOC}}{\text{mmol TOC}}\right) 100.$

[d] Chlorination at pH 7, 24°C, 24-h contact time, light conditions.

because they are indicative of the degree of halide substitution on ECP carbon. On the average under standard chlorinating conditions, only 1% of ECP carbon combined with chlorine. However, this percentage increased to nearly 4% under some chlorinating conditions (1-week reaction time).

Jolley et al.[27] determined that 0.5 to 3.1% of the chlorine dosage used in the chlorination of cooling waters from electric power generating plants and in effluents from domestic sanitary sewage treatment plants was incorporated into organic compounds. Using this range as a basis for comparison, the percentages of initial chlorine concentration incorporated into algal ECP as NPOX were determined for the four filtrates under standard conditions. These yields were calculated by dividing the NPOX generated (μg/L Cl$^-$) by the initial reactive chlorine concentration (μg/L Cl$^+$; i.e., chlorine in the oxidation state of +1). As shown in Table II, the range of percentage incorporation under standard chlorinating conditions was 0.15 to 3.8%, comparable to that reported by Jolley et al.[27] Under varying chlorinating conditions, the percentages were greater than 4% (*A. flos-aquae* using 1-week contact time, 35°C, and pH 2, and chlorophyll at pH 2). The percentages were lowest at pH 12 and 1-h reaction time.

Although the above NPOX yields are of interest, a better indication of the tendency for compounds to incorporate chlorine in solutions having a free chlorine residual is the percentage of chlorine incorporated into organic compounds (assessed via NPOX measurement) based on the amount of reactive chlorine consumed (chlorine demand) in all of its aqueous reactions. Using this percentage, one could assess the magnitude of NPOX-forming reactions to other (i.e., oxidation) reactions. Jolley et al.[27] and other researchers have estimated that oxidation reactions account for 99% of the chlorine consumed in water treatment. However, the data in Table II show that under standard chlorinating conditions, NPOX-forming reactions accounted for sizeable percentages (0.52 to 13%; average, 8.3%) of the chlorine demand in these algal filtrates. Furthermore, given the appropriate environmental conditions (*C. vulgaris* and chlorophyll filtrates at 7°C) all the chlorine demand was incorporated into organically bound chlorine. These results indicate that NPOX-forming reactions may account for the chlorine demand of water supplies to a larger degree than previously estimated, at least for systems containing algae.

Effect of pH on NPOX Formation

To investigate more thoroughly the role of pH on NPOX formation, *A. flos-aquae* and chlorophyll solutions were chlorinated at pH 2, 4, 6, 8, 10, and 12 (40 mg/L Cl$_2$, 1-d contact time, 24°C, and in light). As observed previously, NPOX concentrations increased with decreases in pH. The plots of log$_{10}$ NPOX concentrations vs pH for the two algal filtrates are shown in Figure 1. The empirical equation describing the relationship in Figure 1, at a

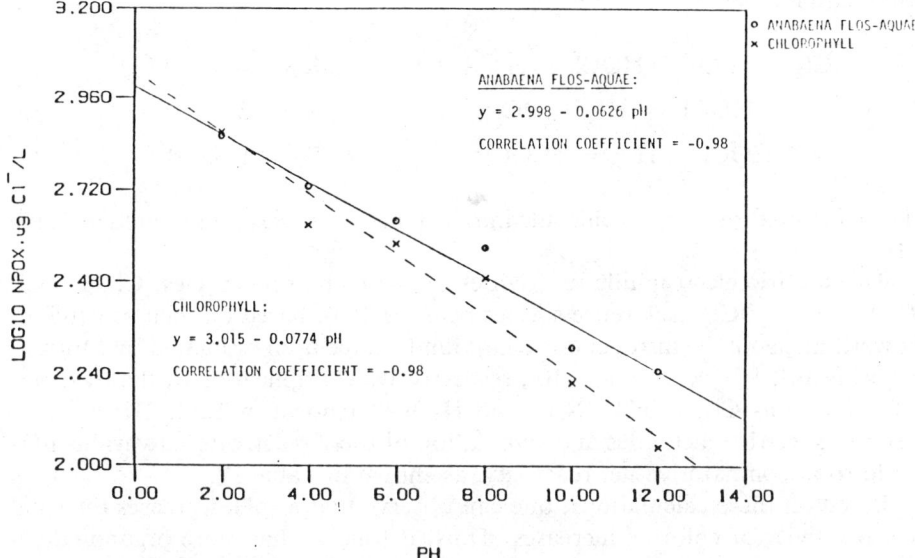

Figure 1. The effect of pH on log NPOX concentration formed on chlorination of *A. flos-aquae* and chlorophyll filtrates (40 mg/L Cl_2; reaction time, 24 h; temp, 24°C; light conditions).

$HOCl_{TOT}$ concentration sufficient to maintain a free available chlorine concentration after time t, was found to be

$$[NPOX]_t = a[H^+]^n \text{ or } \log[NPOX]_t = a' - n(pH) \quad (1)$$

For both *Anabaena* and chlorophyll, linear regression analyses resulted in a' values of about 3, and the n values were 0.063 and 0.077, respectively. For each regression, the correlation coefficient was –0.98, significant at $p < 0.005$.

One explanation for the decrease in NPOX formation with pH could be the direct dependence of chlorine speciation, and thus chlorine electrophilicity, on hydrogen ion concentration. For instance, the concentration of H_2OCl^+, ClOCl, and Cl_2, which are better electrophiles than OCl^- and HOCl, would generally increase with decreased pH. Because most of the possible organohalide-forming reactions involve electrophilic attack of chlorine on organic molecules, this explanation is supported by various studies of NPOX-forming reactions, indicating that the more electrophilic species, such as Cl_2O and H_2OCl^+, were the responsible chlorinating agents.[28-30]

The fractions of total reactive chlorine as Cl_2, HOCl, OCl^-, and H_2OCl^+ were calculated at various pH levels under the conditions used in this experiment (i.e., [Cl⁻] ≅ 35 mg/L), using the following chlorine speciation equilibrium equations:

$$Cl_2 + H_2O = HOCl + H^+ + Cl^- \quad K_h = 4 \times 10^{-4} \quad (2)$$

$$HOCl = H^+ + OCl^- \quad K_a = 3 \times 10^{-8} \quad (3)$$

$$HOCl + H^+ = H_2OCl^+ \quad K_f = 1 \times 10^{-3} \quad (4)$$

The fractionation of total chlorine into its reactive species is presented in Table III.

The specific electrophilic reactivities of these chlorine species, Cl_2, HOCl, OCl⁻, and H_2OCl^+ (reference species being HOCl), based on their reactivities toward nucleophilic nitrogenous compounds, have been estimated by Morris[31] as being 10^3, 1, 1×10^{-4}, and 10^5, respectively. Multiplication of the fractions of chlorine as Cl_2, HOCl, OCl⁻, and H_2OCl^+ (shown in Table III) by their relative specific reactivities and summation of these reactivities at various pHs yield total comparative net reactivity, as shown in Table IV.

Based on these calculations, one can observe that as pH decreases the total net reactivity of chlorine increases. Thus, it follows that more organohalide-forming reactions would occur as pH decreases. It should be noted that chlorine monoxide, Cl_2O, was not included in the above analyses primarily because its contribution to net chlorine reactivity would be negligible, based on its estimated reactivity and low formation constant.

Figure 2 is a plot of log 10 net chlorine reactivity vs pH. The empirical equation, which describes the relationship shown in Figure 2 between net chlorine reactivity and pH, is

or

$$\text{net chlorine reactivity} = b[H^+]^m$$
$$\log[\text{net chlorine reactivity}] = b' - m(pH) \quad (5)$$

Linear regression analyses resulted in a b' value of ~2.6, an m value of 0.5, and a correlation coefficient of -0.96.

As can be seen in Equations (1) and (5), the empirical equations generated to describe the effect of pH on net chlorine reactivity and NPOX formation are identical in form, which tends to support the hypothesis that increases in NPOX formation with decreases in pH may be the result of increases in chlorine electrophilicity. However, the slopes for Equations (1) and (5) differ by a factor of 7, which indicates that further research into the kinetics of organohalide formation in complex mixtures and its competing (oxidation) reactions is warranted.

Table III. Fractions of Total Reactive Chlorine as a Function of pH

Chlorine Species	Fraction of Total Reactive Chlorine					
	pH 2	pH 4	pH 6	pH 8	pH 10	pH 12
Cl_2	0.025	3×10^{-4}	2×10^{-6}	7×10^{-9}	1×10^{-12}	1×10^{-16}
HOCl	0.975	0.999+	0.974	0.270	0.004	4×10^{-5}
OCl^-	3×10^{-6}	3×10^{-4}	0.026	0.730	0.996	0.999+
H_2OCl^+	1×10^{-5}	1×10^{-7}	1×10^{-9}	3×10^{-12}	4×10^{-16}	4×10^{-20}

Table IV. Relative Reactivities of Chlorine Species as a Function of pH

Chlorine Species	Relative Reactivity					
	pH 2	pH 4	pH 6	pH 8	pH 10	pH 12
Cl_2	25	0.25	2×10^{-3}	7×10^{-6}	1×10^{-9}	1×10^{-13}
HOCl	0.975	0.999	0.974	0.270	0.004	4×10^{-5}
OCl^-	3×10^{-10}	3×10^{-8}	2.6×10^{-6}	7.3×10^{-5}	1.0×10^{-4}	1.0×10^{-4}
H_2OCl^+	1.0	1×10^{-2}	1×10^{-4}	3×10^{-7}	4×10^{-11}	4×10^{-15}
Total Net Reactivity	27.0	1.26	0.977	0.270	0.004	1×10^{-4}

Identification of Compounds in Chlorinated and Unchlorinated Algal Filtrates

Specific ECP structures and their by-products generated on chlorination were identified using gas chromatography/mass spectrometry (GC/MS) techniques. A list of compounds identified in the chlorinated and unchlorinated extracts is presented in Table V. Halogenated compounds that could account for NPOX concentrations observed in the previous tables were not identified in the chlorinated samples. This finding supports the research results by McCreary and Snoeyink,[32] who concluded that <0.1% of the total organic halide generated on chlorination of humic acid was gas-chromatographable, and those by Nolle,[33] who also did not identify gas-chromatographable organohalides on chlorination of *C. vulgaris* and *C. pyrenoidosa* biomass.

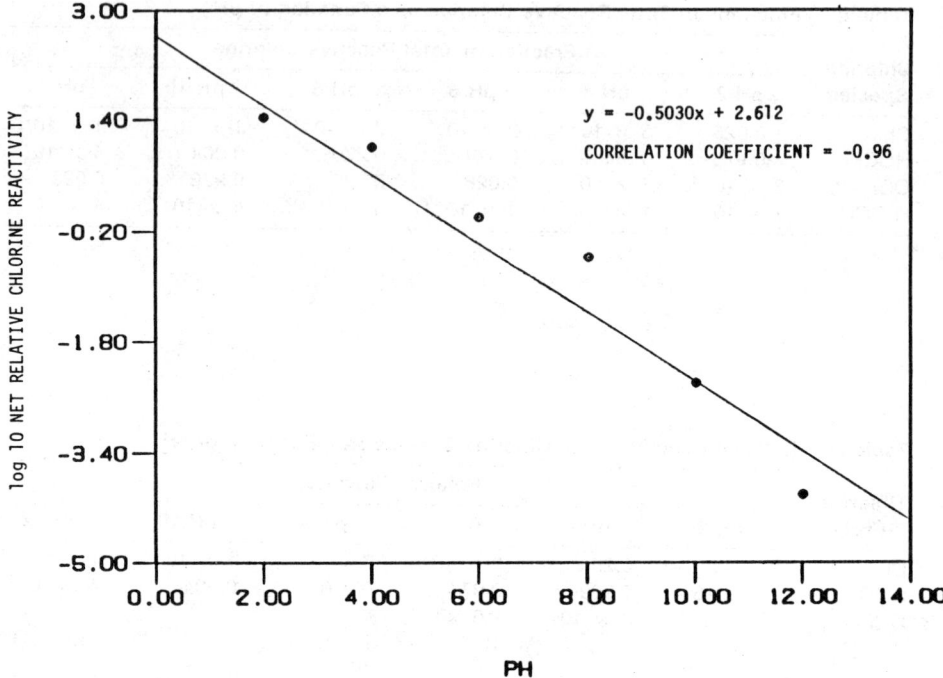

Figure 2. The effect of pH on log net relative chlorine reactivity.

Table V indicates that the two main classes of compounds in the acid extracts were substituted phthalic acids and saturated fatty acids (methylated). These fatty acids were exclusively even numbered (10, 12, 14, 16, 18), indicating a natural origin of these compounds. Because these fatty acids were not observed in the control samples, they are presumed to be excreted by the algae. Nolle[33] also determined that the major constituents of acid extracts from chlorinated algal biomass were even numbered, saturated fatty acids. The most commonly occurring constituents present in the neutral fractions from these experiments were phthalates and tributyl phosphate; however, the former class of compound was also identified in control samples.

Molecular Size Distribution of NPOX

Molecular size distributions of NPOX generated in chlorinated algal filtrates and the organohalide-forming potentials of various molecular size fractions of unchlorinated algal ECP were determined. The efficiencies of the concentration/elution method used in this experiment were first calculated.

Table V. Identification of Compounds in Neutral and Acid (Methylated) Methylene Chloride Extracts of Chlorinated and Unchlorinated Algal Filtrates Using Gas Chromatography/Mass Spectrometry Techniques

	Unchlorinated Filtrate		Chlorinated Extract	
Sample	Acid Extract (methylated)	Neutral Extract	Acid Extract (methylated)	Neutral Extract
C. vulgaris	bis-(2-Ethoxyethyl ether)[a] di-2-Ethylhexyl phthalate di-Nor-Octylphthalate Dioctyl adipate Isoamyl butyrate Methyl alpha-phenyllactate Methyl decanoate N-butylbenzenesulphonamide	Methyl alpha-phenyllactate N-Butylbenzenesulphonamide Oleamide Phthalate[b] Phthalate[b]	bis-(2-Ethoxyethyl) ether[a] Butylbenzenesulphonamide di-2-Ethylhexylphthalate Isoamyl butyrate Methyl decanoate Methyl 4-formylbenzoate Methyl undecanoate Nor-butylisobutyrate Oleamide 2-Phenyl-2-propanol	Methyl alphaphenyllactate Phthalate[b]
C. pyrenoidosa	Methyl dehydroabietate Methyl dodecanoate Methyl ethyl terephthalate Methyl hexadecanoate Methyl isotetradecanoate Methyl octadecanoate Tributyl phosphate	di-2-Ethylhexylphthalate diethylphthalate Glycerol triacetate	di-2-Ethylhexyl phthalate di-Nor-Butyl phosphate Methyl butyl phthalate Methyl dehydroabietate Methyl ethylterephthalate Methyl hexadecanoate Tributyl phosphate	di-n-Butyl phthalate di-2-Ethylhexyl phthalate Diethylphthalate Tributylphosphate
A. flos-aquae	Diethyl phthalate Methyl dehydroabietate Methylethyl terephthalate Methyl hexanoate Methyl isotetradecanoate Methyl octadecanoate Tributyl phosphate 1,1,3-Trimethyl-3-phenyl-indane	di-2-Ethylhexyl phthalate Phthalate[b] Tributyl phosphate	Diethyl phthalate di-2-Ethylhexyl phthalate Methyl benzoate Methyl decanoate Methyl dehydroabietate Methyl dodecanoate Methyl octadecanoate Methyl phenyllactate	Benzyl cyanide Diethylphthalate Tributyl phosphate

Table V, continued

Sample	Unchlorinated Filtrate		Chlorinated Extract	
	Acid Extract (methylated)	Neutral Extract	Acid Extract (methylated)	Neutral Extract
Combined Controls	di-Nor-Butyl phthalate Dioctyl adipate Methyl alpha-phenyllactate Oleamide	Ethyl phthalate Methyl alpha-phenyllactate N-(nor-butyl)-benzenesulphonamide Oleamide 2-Phenyl-3-methyl butanol-2 phthalate[b]	di-Nor-Octyl phthalate Methyl alpha-phenyllactate N-(nor-butyl)-benzene sulphonamide Oleamide Phthalate[b]	di-2-Ethylhexyl phthalate Iso-butyl norbutyrate Methyl alpha-phenyllactate N-(nor-butyl)benzenesulphonamide Oleamide Phthalate[b]

[a]Used in methylation process.
[b]Substitution on phthalate molecule could not be discerned by mass spectral data.

Table V, continued

Sample	Unchlorinated Filtrate		Chlorinated Extract	
	Acid Extract (methylated)	Neutral Extract	Acid Extract (methylated)	Neutral Extract

Table VI. Characterization of NPOX Generated Upon Chlorination of Algal ECP by Molecular Size Using Size-Exclusion HPLC[a]

Molecular Size Fraction (Daltons)	Chlorella pyrenoidosa				Anabaena flos-aquae			
	Filtrates chlorinated; then fractionated		Filtrates fractionated; then chlorinated[b]		Filtrates Chlorinated then fractionated		Filtrates fractionated; then chlorinated	
	µg Cl⁻	% of Total	µg Cl⁻	% of Total	µg Cl⁻	% of Total	µg Cl⁻	% of Total
>10,000	2.5	16	45	25	98	31	209	32
10,000 – 1,000	9.5	59	88	49	133	42	308	47
1,000 – 100	4.1	25	32	18	59	19	99	15
<100	0.0	0	14	8	26	8	35	5
Total	16.1	100	179	100	316	100	651	99

[a]Chromatography results based on elution of 1-L samples; chlorination occurred at pH 7, 20 mg/L Cl₂, 24-h contact time, light conditions. Actual chlorine dose higher than 20 mg/L in the fractions.
[b]Results may include measurement of purgeable organic halide.

The NPOX concentration in unconcentrated chlorinated *A. flos-aquae* filtrate was 380 µg/L Cl⁻, whereas that in the combined four HPLC effluent fractions (after subtraction of NPOX content in chlorinated control and normalizing results to an injection volume of 1-L sample) was 315 µg/L Cl⁻. Thus the efficiency for this procedure in this case was 83%. Similarly, the efficiency for the *C. pyrenoidosa* system was 103%.

Table VI shows that 75 and 73% of NPOX produced via chlorination of *C. pyrenoidosa* and *A. flos-aquae* filtrates, respectively, were associated with a molecular size fraction exceeding 1000 daltons. Thus, a reasonable explanation for the lack of GC/MS identification of compounds contributing to NPOX concentrations in this study is that these NPOX-containing compounds were not amenable to GC/MS analyses because of their high molecular weight (insufficient volatilities).

When algal ECP compounds were fractionated by size first and then chlorinated, the TOX distribution by size (on chlorination of these fractions, POX as well as NPOX formed) followed that of their organohalide-containing ECP derivatives. This indicated that the ECP precursors in NPOX-generating reactions generally were greater than 1000 daltons in molecular size, and on chlorination the molecular size of their NPOX by-products did not change greatly from their precursor molecules. These results are in agreement with those of McCahill et al.,[34] who showed that wastewater chlorination led to significant accumulations of high molecular size (>1000 daltons) organically bound chlorine and Glaze et al.,[17] who characterized the NPOX-forming potentials of high-molecular-weight organics in lake water. Lastly, it is not surprising that algal ECP precursors in NPOX-forming reactions were of high molecular weight, because phytoplankton are known to release high-molecular-weight compounds, especially during the stationary phase of growth.[35]

ACKNOWLEDGMENTS

The authors gratefully acknowledge the support for this research by the U.S. Environmental Protection Agency (EPA), Health Effects Research Laboratory, Cincinnati, Ohio, cooperative agreement No. CR-807365-03-3. W. Emile Coleman was the EPA Project Officer and Herbert Pahren, also of the EPA, provided advice and encouragement. Some additional EPA support was provided under cooperative agreement No. CR-810543 between the EPA Office of Research and Development, Washington, DC, and the University of Pittsburgh Center for Environmental Epidemiology.

REFERENCES

1. Rook, J. J. "Formation of Haloforms during Chlorination of Natural Waters," *Water Treatment Exam.* 23:234–243 (1974).

2. Bellar, T. A., J. J. Lichtenberg, and R. C. Kroner. "The Occurrence of Organohalides in Chlorinated Drinking Water," *J. Am. Water Works Assoc.* 66(12):703–706 (1974).
3. Briley, K. F., R. F. Williams, K. E. Longley, and C. A. Sorber. "Trihalomethane Production from Algal Precursors," in *Water Chlorination: Environmental Impact and Health Effects, Vol. 3*, R. L. Jolley, W. A. Brungs, and R. B. Cumming, Eds. (Ann Arbor, MI: Ann Arbor Science Publishers, Inc., 1980), pp. 117–129.
4. Hoehn, R. C., D. B. Barnes, B. C. Thompson, C. W. Randall, T. J. Grizzard, and T. B. Shaffer. "Algae as Sources of Trihalomethane Precursors," *J. Am. Water Works Assoc.* 72(6):344–350 (1980).
5. Oliver, B. G., and D. B. Shindler. "Trihalomethanes from the Chlorination of Aquatic Algae," *Environ. Sci. Technol.* 14(12):1502–1505 (1980).
6. Oliver, B. G. "Chlorinated Non-volatile Organics Produced by the Reaction of Chlorine with Humic Materials," *Can. Res.* 11(6):21–22 (1978).
7. Glaze, W. H., G. R. Peyton, F. Y. Saleh, and F. Y. Huang. "Analysis of Disinfection By-products in Water and Wastewater," *Int. J. Environ. Anal. Chem.* 7:143–160 (1979).
8. Rook, J. J. "Possible Pathways for the Formation of Chlorinated Degradation Products During Chlorination of Humic Acids and Resorcinol," in *Water Chlorination: Environmental Impact and Health Effects, Vol. 3*, R. L. Jolley, W. A. Brungs, and R. B. Cumming, Eds. (Ann Arbor, MI: Ann Arbor Science Publishers, Inc., 1980), pp. 85–98.
9. Maruoka, S., and S. Yamanaka. "Production of Mutagenic Substances by Chlorination of Waters," *Mutat. Research* 79:381–386 (1980).
10. Fallon, R. D., and C. B. Fliermans. *Chemosphere* 9:385–391 (1980).
11. Wachter, J. K., and J. B. Andelman. "Organohalide Formation on Chlorination of Algal Extracellular Products," *Environ. Sci. Technol.* 18(11):811–817 (1984).
12. *Standard Methods for the Examination of Water and Wastewater*, 15th ed. (Washington, DC: American Public Health Association, 1980), pp. 289–291.
13. *Proposed Method of Test for Organic Halides in Water by Carbon Adsorption-Microcoulometric Detection.* ASTM Committee D-19 on Water. R. J. Joyce, Task Group Chairman (1981).
14. Wachter, J. K. *Characterization of Organohalide Formation Upon Chlorination of Algal Extracellular Matter*, Sc.D. Dissertation, (Pittsburgh, PA: University of Pittsburgh, 1982).
15. Stevens, A. A., C. J. Slocum, D. R. Seeger, and G. G. Robeck. "Chlorination of Organics in Drinking Water," in *Water Chlorination: Environmental Impact and Health Effects, Vol. 1*, R. L. Jolley, Ed. (Ann Arbor, MI: Ann Arbor Science Publishers, Inc., 1978), pp. 77–104.
16. Morris, J. C., and B. Baum. "Precursors and Mechanisms of Haloform Formation in the Chlorination of Water Supplies," in *Water Chlorination: Environmental Impact and Health Effects, Vol. 2*, R. L. Jolley, H. Gorchev, and D. H. Hamilton, Jr., Eds. (Ann Arbor, MI: Ann Arbor Science Publishers, Inc., 1978), pp. 29–48.
17. Glaze, W. H., F. Y. Saleh, and W. Kinstley. "Characterization of Nonvolatile Halogenated Compounds Formed during Water Chlorination," in *Water Chlorination: Environmental Impact and Health Effects, Vol. 3*, R. L. Jolley, W. A. Brungs, and R. B. Cumming, Eds. (Ann Arbor, MI: Ann Arbor Science Publishers, Inc., 1980), pp. 99–108.

18. Oliver, B. G. "Effect of Temperature, pH and Bromide Concentration on Trihalomethane Reaction of Chlorine with Aquatic Humic Material," in *Water Chlorination: Environmental Impact and Health Effects, Vol. 3*, R. L. Jolley, W. A. Brungs, and R. B. Cumming, Eds. (Ann Arbor, MI: Ann Arbor Science Publishers, Inc., 1980), pp. 141–149.
19. Minear, R. A., and J. C. Bird. "Trihalomethanes: Impact of Bromide Ion Concentration on Yield, Species Distribution, Rate of Formation and Influence of Other Variables," in *Water Chlorination: Environmental Impact and Health Effects, Vol. 3*, R. L. Jolley, W. A. Brungs, and R. B. Cumming, Eds. (Ann Arbor, MI: Ann Arbor Science Publishers, Inc., 1980), pp. 151–160.
20. Echelberger, W. F., Jr., J. L. Pavoni, P. C. Singer, and M. W. Tenney. "Disinfection of Algal Laden Waters," *Am. Soc. Civil. Eng. J. Sanit. Eng. Div.* 172(SA5):721 (1971).
21. Jekel, M. R., and P. Roberts. "Total Organic Halogen as a Parameter for the Characterization of Reclaimed Waters: Measurement, Occurrence, Formation and Removal," *Environ. Sci. Technol.* 14(8):970–975 (1980).
22. Wachter, J. K., J. B. Andelman, J. M. Beck, and S. Nolle. "Organic Chemicals and Other Factors in Water Reuse at a Poultry Processing Plant," in *Proceedings of Water Reuse Symposium II, Vol. 2* (Washington, DC: American Waterworks Association Research Foundation, 1981), pp. 862–880.
23. Sontheimer, H., E. Heilker, M. Jekel, H. Nolte, and F. H. Vollmer. "The Mülheim Process," *J. Am. Water Works Assoc.* 70(7):393–396 (1978).
24. Sanders, R., W. Kuhn, and H. Sontheimer. "Studies on the Reaction of Chlorine with Humic Substances," *Z. Wasser Abwasser Forsch.* 10(5):155–160 (1977).
25. Johnson, L. C. *Chloroorganics Formation from the Chlorination of Chlorella vulgaris.* Master of Science (Hyg.) thesis. (Pittsburgh, PA: University of Pittsburgh, 1982).
26. Fairhall, L. T. *Industrial Toxicology.* (New York, New York: Hafner Publishing Company, 1969).
27. Jolley, R. L., G. Jones, W. W. Pitt, and J. E. Thompson. "Chlorination of Organics in Cooling Waters and Process Effluents," in *Water Chlorination: Environmental Impact and Health Effects, Vol. 1*, R. L. Jolley, Ed. (Ann Arbor, MI: Ann Arbor Science Publishers, Inc., 1978), pp. 105–138.
28. Swain, C. G., and D. R. Christ. "Mechanisms of Chlorination by Hypochlorous Acid. The Last of the Chlorinium Ion, Cl^+," *J. Am. Chem. Soc.* 94(9):3195–3200 (1972).
29. Reinhard, M., and W. Stumm. "Kinetics of Chlorination of p-Xylene in Aqueous Solution," in *Water Chlorination: Environmental Impact and Health Effects, Vol. 3*, R. L. Jolley, W. A. Brungs and R. B. Cumming, Eds. (Ann Arbor, MI: Ann Arbor Science Publishers, Inc., 1980), pp. 210–218.
30. Snider, E. H., and F. C. Alley. "Kinetics of Biphenyl Chlorination in Aqueous Systems in the Neutral and Alkaline pH Ranges," in *Water Chlorination: Environmental Impact and Health Effects, Vol. 3*, R. L. Jolley, W. A. Brungs and R. B. Cumming, Eds. (Ann Arbor, MI: Ann Arbor Science Publishers, Inc., 1980), pp. 219–225.
31. Morris, J. C. "The Chemistry of Aqueous Chlorine in Relation to Water Chlorination," in *Water Chlorination: Environmental Impact and Health Effects, Vol. 1*, R. L. Jolley, Ed. (Ann Arbor, MI: Ann Arbor Science Publishers, Inc., 1978), pp. 21–35.

32. McCreary, J. J., and V. L. Snoeyink. "Reaction of Free Chlorine with Humic Substances Before and After Adsorption on Activated Carbon," *Environ. Sci. Technol.* 15(2):193-197 (1981).
33. Nolle, S. *Characterization of Organic Compounds Generated upon Chlorination of Algal Biomass*, Master of Public Health thesis, (Pittsburgh, PA: University of Pittsburgh, 1982).
34. McCahill, M. P., L. E. Conroy, and W. J. Maier. "Determination of Organically Combined Chlorine in High Molecular Weight Aquatic Organics," *Environ. Sci. Technol.* 14(2):201-203 (1980).
35. Dunstall, T. G., and C. Nalewajko. "Extracellular Release in Planktonic Bacteria," *Verh. Internat. Verein. Limnol.* 19:2643-2649 (1975).

CHAPTER **70**

Novel Precursor of Trihalomethanes

William J. Cooper and Delia M. Kaganowicz

It is well known that the use of chlorine during water treatment produces trihalomethanes (THMs) in the presence of natural aquatic humic substances.[1] It is also known that there is considerable variability in the chemical nature of humic material and this variability can give rise to different THM formation potentials.[2,3] To gain a better understanding of the chemical reactions that occur when chlorine is added to waters containing natural aquatic organics, some research has focused on the use of model organic compounds (i.e., model precursors) with structures similar to those in natural aquatic humic substances such as aromatic polyhydroxy compounds[4-12] and other compounds.[11-14]

There are probably numerous precursor molecules having diverse functional groups of natural origin.[11] This observation is supported in part by the relatively elusive structure of naturally occurring humic and fulvic acids, the apparent long reaction times observed in natural waters (assuming an excess chlorine), and experimental work with model compounds of diverse structure. Although several mechanisms could lead to the formation of THMs, the relative significance of the mechanisms is not known. It is possible that all are operative to a greater or lesser extent in the formation of THMs in natural waters.

Our recent studies indicate that α-methylbenzylamine can act as a precursor for THMs. The overall mechanism can be divided into two series, the first producing acetophenone and the second, the classical haloform reaction, forming THMs and benzoic acid (Figure 1). The first series of reactions, resulting in the formation of acetophenone, was reported using phase transfer catalysis.[15] We have studied the formation of THMs when aqueous solutions of α-methylbenzylamine (α-MBA) are chlorinated.

This chapter discusses the formation of the trihalomethanes as a function of the variables, bromide and pH at a chlorine (Cl_2) to α-MBA mole ratio of 1.0. These results serve as a model for THM formation in natural waters and may contribute to a better understanding of the effect of bromide in natural waters and the formation and distribution of trihalomethanes.

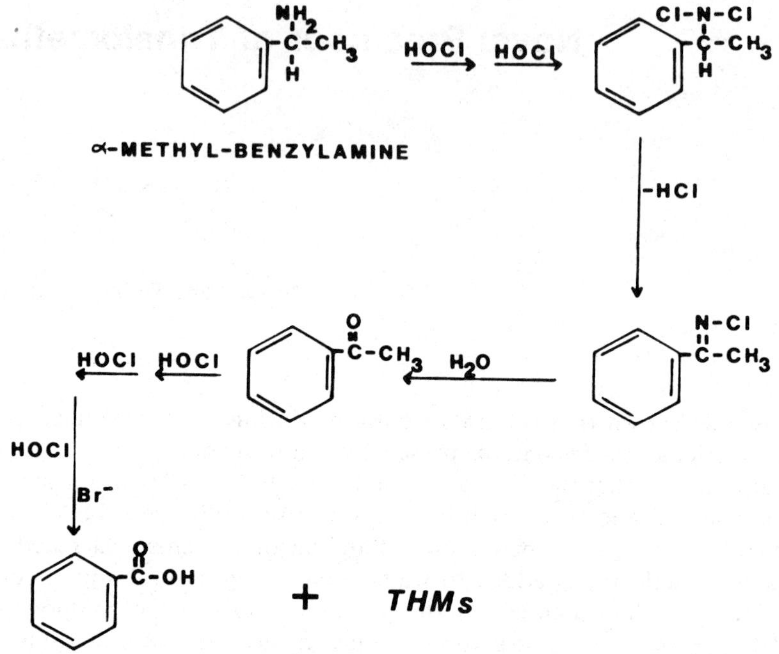

Figure 1. Reaction pathway of α-methylbenzylamine and HOCl.

EXPERIMENTAL

The formation and distribution of the four chlorinated and brominated THMs were studied at a precursor (α-MBA) concentration of 10^{-4} M. The concentration of Cl_2 was also held constant at 10^{-4} M, while Br^- and pH were varied.

All experiments were conducted in chlorine-demand-free water.[16] Sufficient bromide was added as KBr (infrared grade) to give final concentrations of 1×10^{-4} M (8.0 mg/L Br^-), 1×10^{-5} M (0.8 mg/L Br^-), and 1×10^{-6} M (0.08 mg/L Br^-). Chlorine solutions were prepared from a NaOCl solution (Fisher Reagent Grade) and standardized using iodometric titration.[17] Chlorine concentrations during the reactions were determined using a continuous total chlorine analyzer (CHLORTECT®, EPCO, Woodburry, Connecticut).[18] After chlorine was added to the samples, the reaction was allowed to proceed for 2, 6, 24, 72, and 144 h in the dark. For any one reaction series, sodium thiosulfate (0.1 g/60-mL sample) was added to stop any further reaction at the predesignated times. The samples were then stored in the refrigerator (4°C) in 60-mL serum vials with Teflon®-lined silicon septa secured with aluminum crimp-topped seals until THM analysis. (In no case were the samples stored for more than more than 10 d). In separate experiments we have shown that the THMs are invariant in concentration for at least 60 d when stored under the conditions used in this experiment.) To adjust the pH of the samples at 5.0, 7.0, 9.0, and 11.0, we used KH_2PO_4, K_2HPO_4 + KH_2PO_4, $Na_2B_4O_7$, and Na_2CO_3 at 0.01 M, respectively. THMs were determined by liquid-liquid extraction[19] followed by gas chromatography using a Hall electrolytic conductivity detector (Tracor Model 700).

RESULTS AND DISCUSSION

The results of the formation of the THMs at an α-MBA/Cl_2 mole ratio of 1 and Br^- concentration of 10^{-6} M are shown in Figure 2. The total chlorine concentration was measured at each reaction time and is also presented. It was observed that at every pH except 5, $CHCl_3$ was the predominant THM formed and increased in concentration with time. Bromoform was either below the detection limit or detected only at trace levels at pH 5 and 7. At pH 9 and 11, $CHBr_3$ was observed but only at low concentrations. The concentration of the total THMs (THMs as carbon, THM-C) increased with time and pH. The increased pH appeared to favor the formation of $CHCl_3$. The THM-C data are summarized in Figure 3. From this it appears that at a Br^- concentration of 10^{-6} M, the formation of the THMs, as moles per liter of carbon, is approximately linear on a log-log plot of THM-carbon concentration vs OH^- concentration, at all reaction times.

The results of the formation of THMs at a Br^- concentration of 10^{-5} M are presented in Figure 4. At pH 5 and 7 the predominant THM formed was $CHCl_2Br$, whereas at pH 9 the $CHClBr_2$ predominates slightly over the $CHCl_2Br$. At pH 11, initially the more brominated THMs, $CHClBr_2$ and $CHBr_3$, predominated. The formation of $CHBr_3$ was essentially complete within 6 h, whereas the formation of $CHClBr_2$ reached a plateau (minimal additional increase in concentration) after 24 h. At this time, the Br^- had all been converted into organic bromine. With excess chlorine the chlorinated THMs, $CHCl_2Br$ and $CHCl_3$, began to increase significantly until at 144 h the $CHCl_3$ was the predominant THM formed, representing 56% of the total THM-C.

At pH 5, $CHBr_3$ was observed as only a trace ($<10 \times 10^{-9}$ M). As the pH of the reaction was increased, $CHBr_3$ was formed; at pH 7, it reached concentrations that could be quantitated only after 72 and 144 h. At pH 9, $CHBr_3$ formed rapidly and in a higher concentration than $CHCl_3$, but less than the mixed halogen species. The formation of $CHBr_3$ at pH 11 was extremely rapid and appeared to reach maximum concentration after 6 h reaction time.

Figure 5 summarizes the formation of the THMs at Br^- concentration of 10^{-5} M. The formation of the THM-C is again approximately linearly related to OH^- concentration (when plotted as log-log).

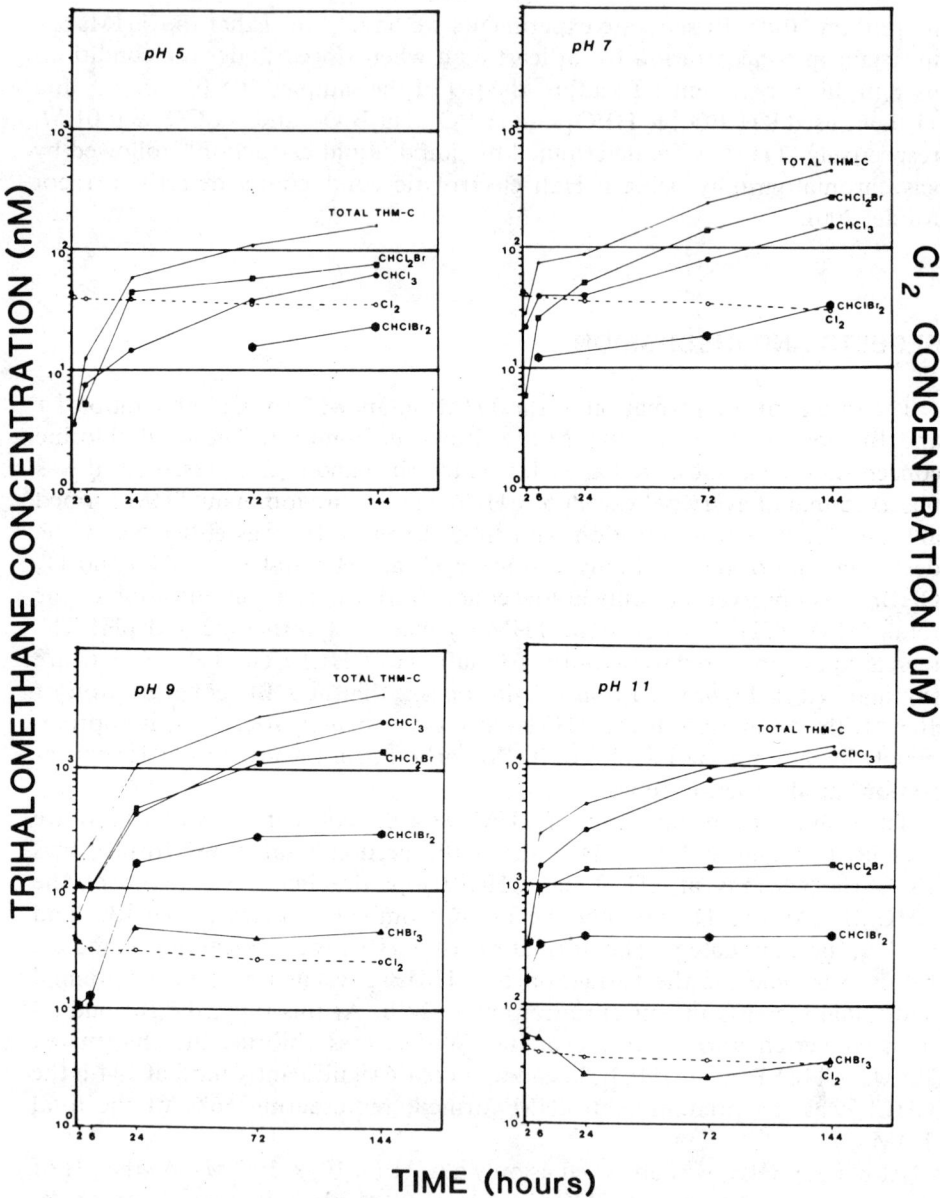

Figure 2. Trihalomethane formation at an α-methylbenzylamine/Cl$_2$ mole ratio of 1 and Br$^-$ = 10^{-6} M.

The results of the formation of the THMs at Br$^-$ concentration of 10^{-4} M are presented in Figure 6. In this experiment the α-MBA, Cl$_2$, and Br$^-$ concentrations were all 10^{-4} M. At almost every pH and reaction time, CHBr$_3$ is the predominant THM. At pH 5, CHCl$_3$ is formed at concentrations greater than

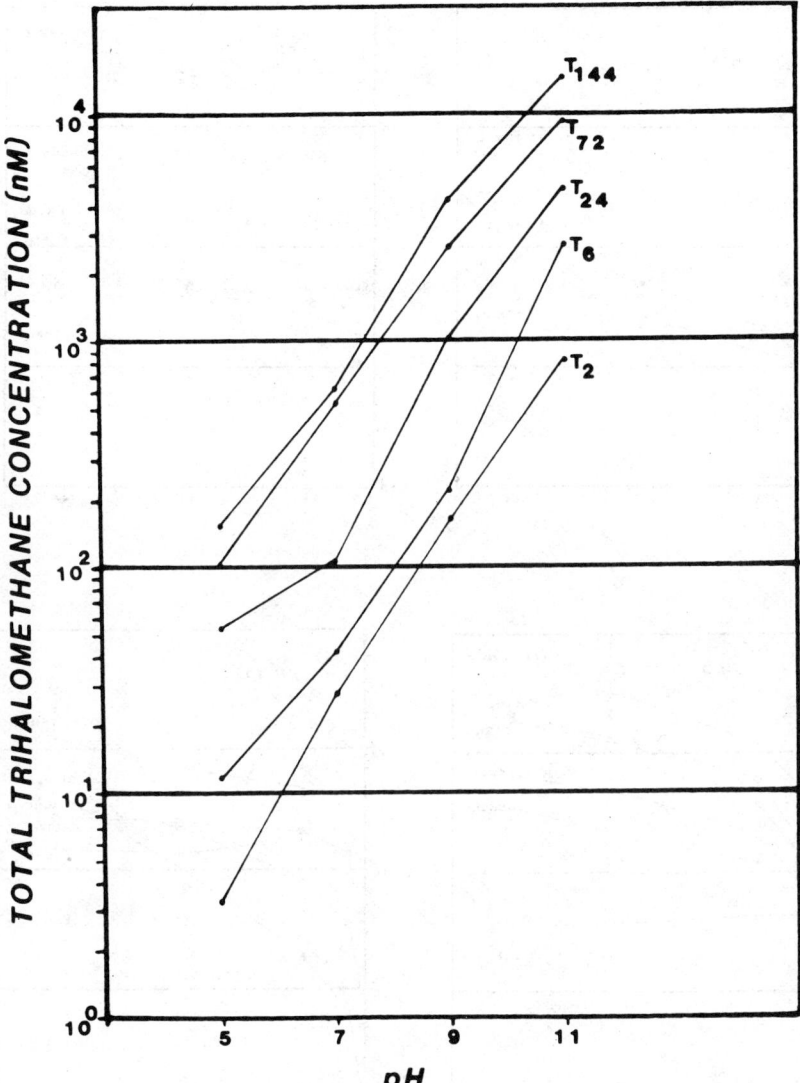

Figure 3. Total trihalomethane formation, $Br^- = 10^{-6}$ M.

the mixed halogen THMs. The increase in pH favored the formation of the $CHClBr_2$ and $CHBr_3$. The concentrations of $CHBr_3$ and $CHClBr_2$ both increased with time and pH, as did the total THM concentration as carbon. Figure 7 shows the result of the THM-C concentration at different reaction times and pH. The THM-C concentration appeared to decrease at pH 7 when compared with that at both pH 5 and 9. This effect is not understood as yet but indicates that, at this concentration of Br^- (10^{-4} M), an acid/base catalyzed step was rate limiting (i.e., slowed down THM formation).

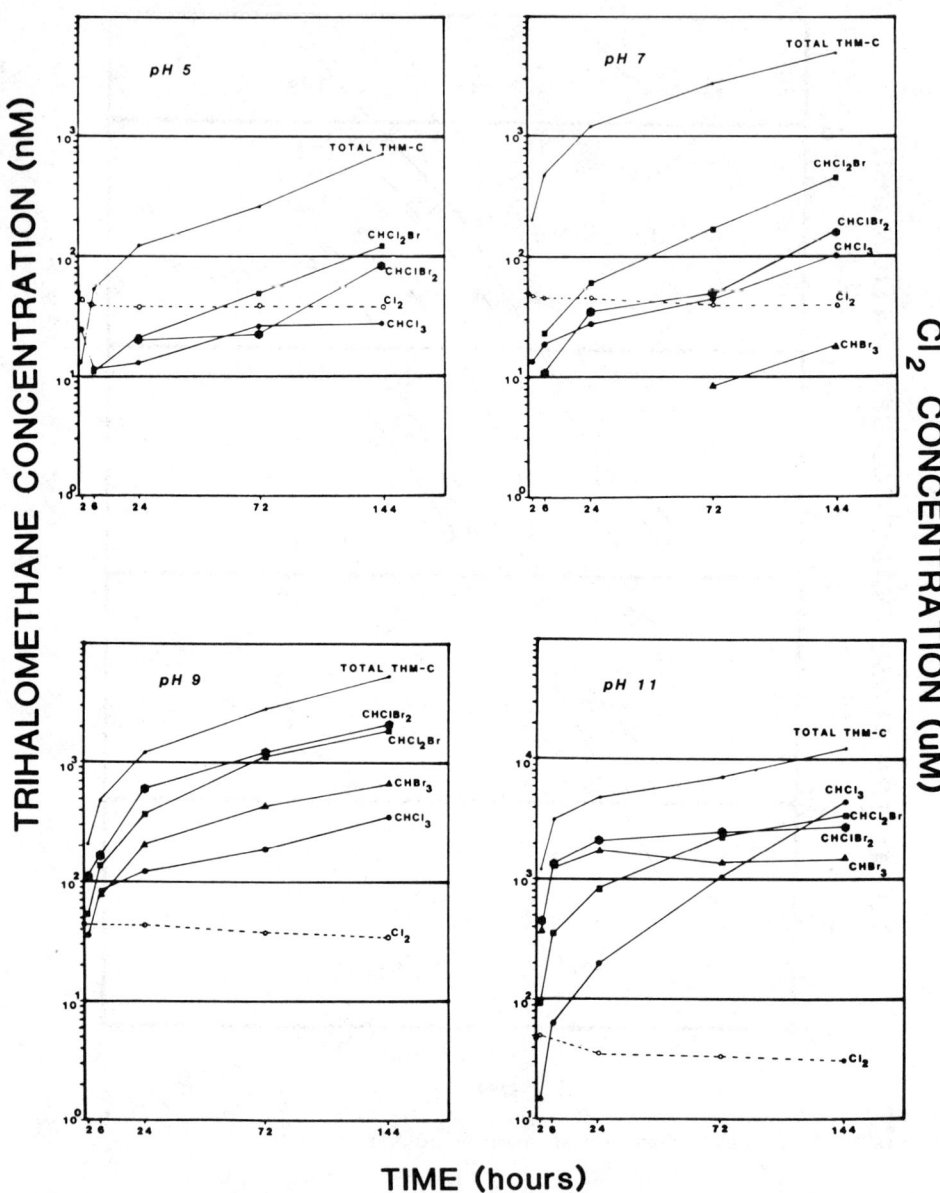

Figure 4. Trihalomethane formation at an α-methylbenzylamine/Cl$_2$ mole ratio of 1 and Br$^-$ = 10^{-5} M.

It has been reported that, in general, when Br$^-$ is present, the molar concentration of the THMs increases.[4,6,7] By comparing the results of the total THM-C curves in the above figures, it can be seen that in this study, the addition of Br$^-$ did not increase the total THM-C concentration. At pH 5, 10^{-5} M Br$^-$ increased the total THM-C relative to the samples containing 10^{-6} and 10^{-4} M

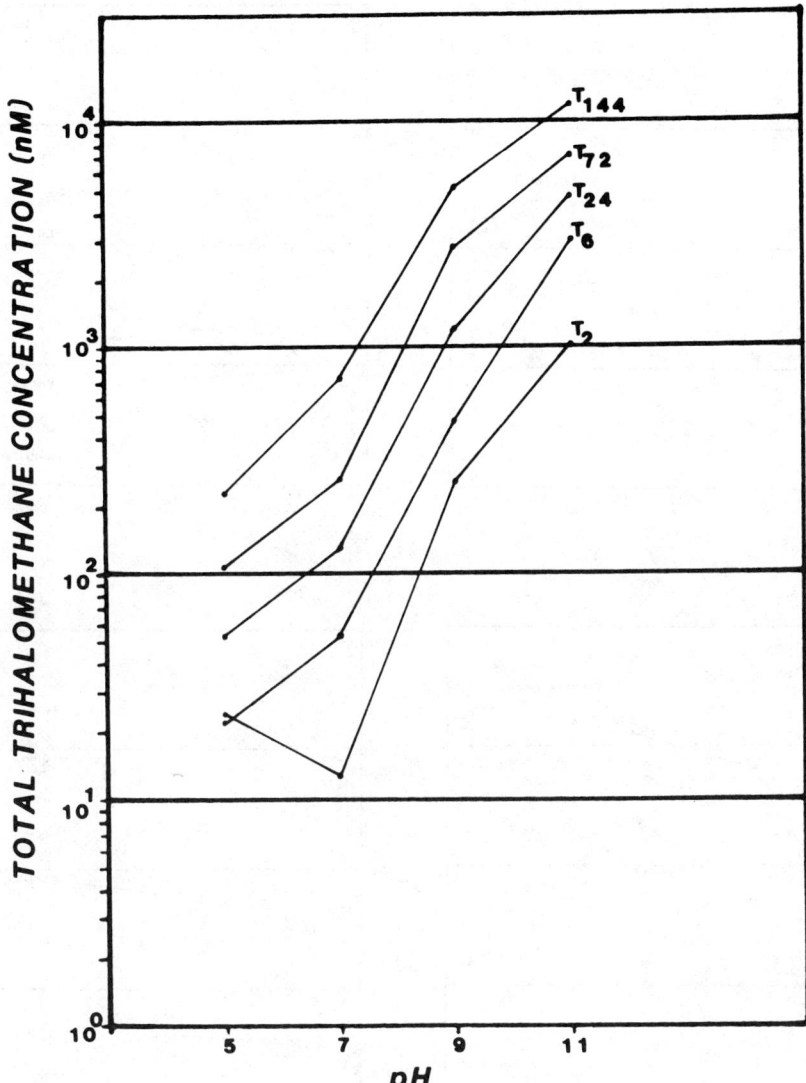

Figure 5. Total trihalomethane formation, $Br^- = 10^{-5}\ M$.

Br^-. At pH 9, the total THM-C concentration is similar at the two lower Br^- concentrations; however, when Br^- was added ($10^{-4}\ M$), the concentration of total THM-C decreased by approximately 50% at 144 h reaction time. The Br^- concentration had little effect on the formation of total THM-C at pH 11.

Amy et al.[20] studied the effect of Br^- on the formation and distribution of THMs in natural waters. They found that as the pH increased the relative conversion of Br^- to THM-Br also increased. We found that at a reaction time of 144 h and pH 5, 7, and 11 the relative conversion of Br^- to THM-Br

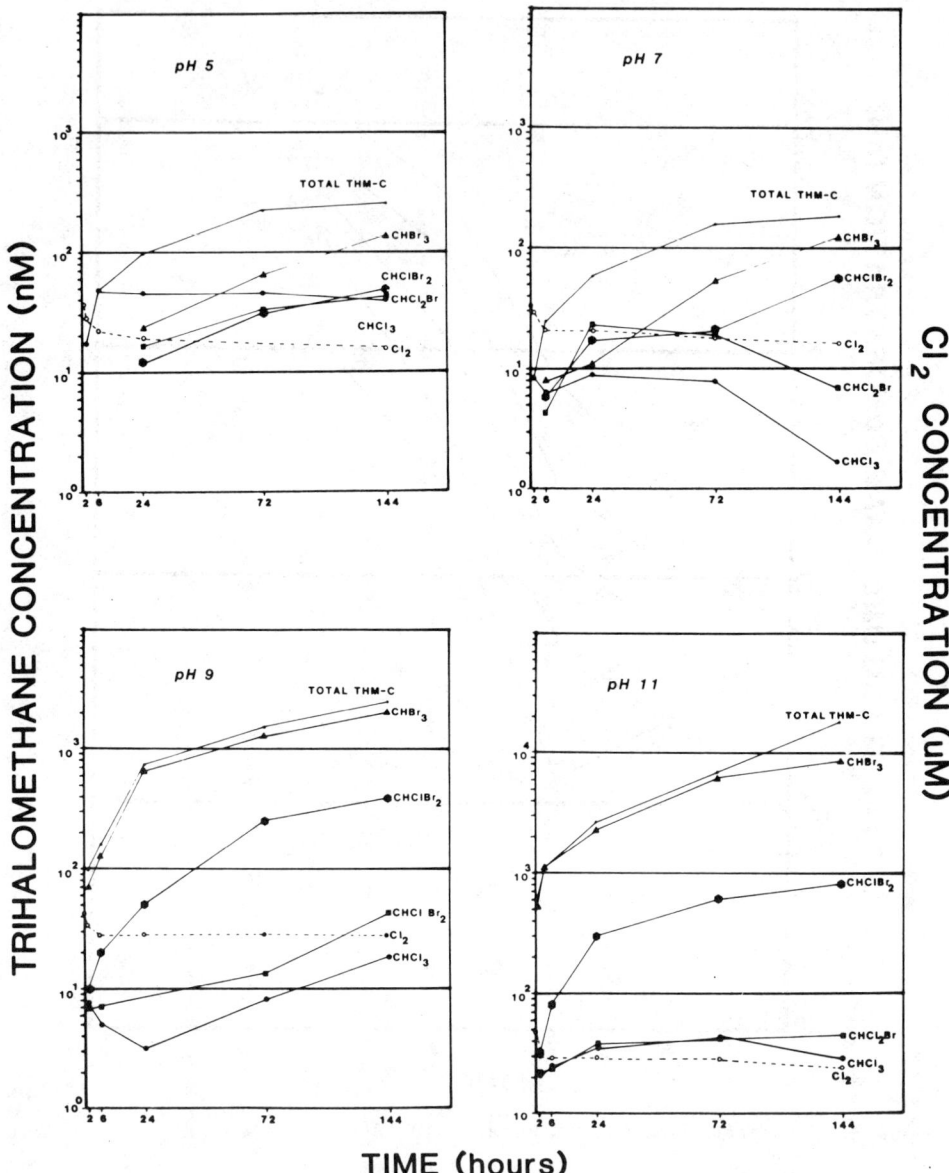

Figure 6. Trihalomethane formation at an α-methylbenzylamine/Cl$_2$ mole ratio of 1 and Br$^-$ = 10^{-4} M.

increased with decreasing bromide concentration (Figure 8). From this it appears that the conversion of Br$^-$ (inorganic) to THM-Br (organic bromine) increases both as a function of decreasing bromide and increasing pH. The apparent >100% conversion observed at pH 9 and 11 probably reflects the summation of errors in the THM measurements. Slight contamination by Br$^-$,

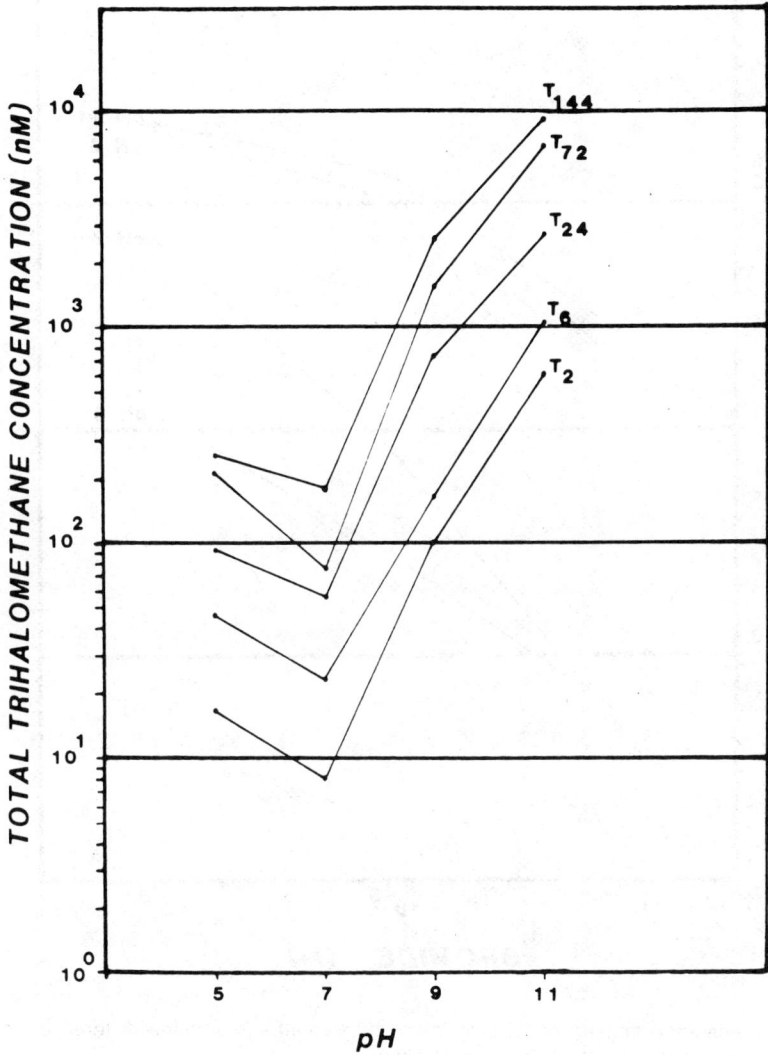

Figure 7. Total trihalomethane formation, $Br^- = 10^{-4}$ M.

either in the chlorine-demand-free water or in the buffer reagents, is not the cause of the increase because when no bromide was added, only $CHCl_3$ was observed.

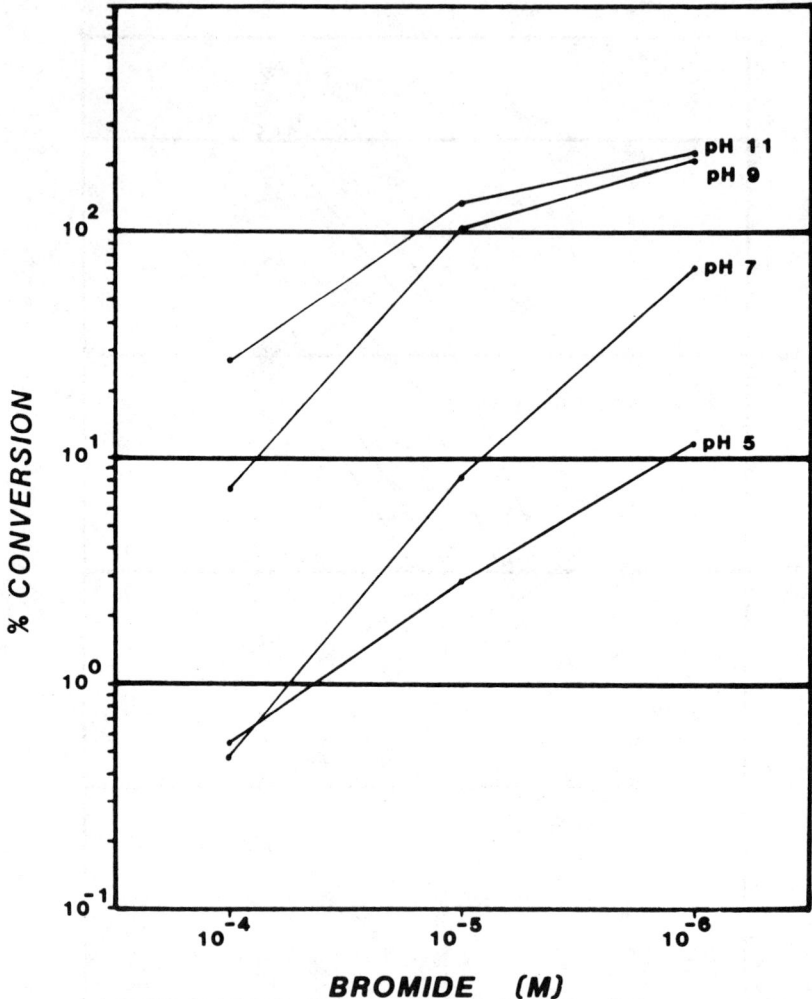

Figure 8. Percentage conversion of bromide to trihalomethane-bromine at three bromide concentrations, after 144-h reaction time.

SUMMARY AND CONCLUSIONS

A better understanding of the mechanisms and the factors affecting the formation and distribution of trihalomethanes in aqueous solution is needed if water treatment processes for controlling their formation are to be optimized. We have shown that α-methylbenzylamine is a good model compound for studying the formation and distribution of THMs. Advantages of using model compounds are more exact control of the concentration of the starting compounds and more detailed studies of by-products. Although this study reports only on the formation and distribution of THMs, further studies are in progress to determine other halogenated and nonhalogenated by-products.

THMs are formed when α-methylbenzylamine is chlorinated and, in the presence of Br⁻, the mixed haloforms and $CHBr_3$ are formed as predicted. The formation of the individual THMs is dependent on the concentration of Br⁻, pH, and reaction time. The formation curves obtained for the total THM as carbon closely follow those of natural waters, indicating that α-MBA is a good candidate for studying THM formation rates.

ACKNOWLEDGMENTS

We gratefully acknowledge the technical help of Mehrzad F. Mehran, Linda M. Meyer, Alessandra Vanzella, Duane Baur, Rose Ann Slifker, and Rafael Diaz and discussions with Daisey Macias and Arthur Herriott. Financial support for the research was provided by the U.S. Environmental Protection Agency Cooperative Agreement CR-810277-01 and the Drinking Water Research Center, Florida International University.

REFERENCES

1. Symons, J. J., A. A. Stevens, R. M. Clark, E. E. Geldreich, O. T. Love, Jr., and J. DeMarco. *Treatment Techniques for Controlling Trihalomethanes in Drinking Water.* EPA-600/2-81-156, (Cincinnati, OH: U.S. Environmental Protection Agency, 1981).
2. Oliver, B. G., and J. Lawrence. "Haloforms in Drinking Water: A Study of Precursors and Precursor Removal," *J. Am. Water Works Assoc.* 71:161-163 (1979).
3. Oliver, B. G., and E. M. Thurman. "Influence of Aquatic Humic Substance Properties on Trihalomethane Potential," in *Water Chlorination: Environmental Impact and Health Effects, Vol. 4*, R. L. Jolley, W. A. Brungs, J. A. Cotruvo, R. B. Cumming, J. S. Mattice, and V. A. Jacobs, Eds. (Ann Arbor, MI: Ann Arbor Science Publishers, Inc., 1983), pp. 231-241.
4. Boyce, S., A. C. Barefoot, D. R. Britton, and J. F. Hornig. "The Formation of Trihalomethanes from Halogenation of 1,3-Dihydroxy-Benzenes in Dilute Aqueous Solution. The Synthesis of 2-¹³C-Resorcinol and its Reaction with Chlorine and Bromide," in *Water Chlorination: Environmental Impact and Health Effects, Vol. 4,* R. L. Jolley, W. A. Brungs, J. A. Cotruvo, R. B. Cumming, J. S. Mattice, and V. A. Jacobs, Eds. (Ann Arbor, MI: Ann Arbor Science Publishers, Inc., 1983), pp. 253-267.
5. Christman, R. F., J. D. Johnson, J. R. Hoss, F. K. Pfaender, W. T. Liao, D. L. Norwood, and H. J. Alexander. "Natural and Model Aquatic Humics: Raction with Chlorine," in *Water Chlorination: Environmental Impact and Health Effects, Vol. 1*, R. L. Jolley, Ed. (Ann Arbor, MI: Ann Arbor Science Publishers, Inc., 1978), pp. 15-28.
6. Norwood, D. L., J. D. Johnson, R. F. Christman, J. R. Hoss, and M. J.

Bobenrieth. "Reactions of Chlorine with Selected Aromatic Models of Aquatic Humic Material," *Environ. Sci. Technol.* 14:187-189 (1980).
7. Rook, J. J. "Possible Pathways for the Formation of Chlorinated Degradation Products During Chlorination of Humic Acids and Resorcinol," in *Water Chlorination: Environmental Impact and Health Effects, Vol. 3*, R. L. Jolley, W. A. Brungs, and R. B. Cumming, Eds. (Ann Arbor, MI: Ann Arbor Science Publishers, Inc., 1980) pp. 85-98.
8. Onodera, S., M. Tabata, S. Suzuki, and S. Ishikura. "Gas Chromatographic Identification and Determination of Chlorinated Quinones Formed During Chlorination of Dihydric Phenols with Hypochlorite in Dilute Aqueous Solution," *J. Chromatogr.* 200:137 (1980).
9. Stevens, A. A. "Formation of Non-Polar Organo-Chlorine Compounds as Byproducts of Chlorination," in *Proceedings—Oxidation Techniques in Drinking Water Treatment, Sept. 11-13, 1978. Karlsruhe, F.R.G.*, EPA-570/9-79020 (Washington, DC: U.S. Environmental Protection Agency, 1979), pp. 145-160.
10. Boyce, S. D., and J. F. Hornig. "Reaction Pathways of Trihalomethane Formation from the Halogenation of Dihydroxyaromatic Model Compounds for Humic Acid," *Environ. Sci. Technol.* 17:202-211 (1983).
11. Boyce, S. D., and J. F. Hornig. "Formation of Chloroform from the Chlorination of Diketones and Polyhydroxybenzenes in Dilute Aqueous Solution," in *Water Chlorination: Environmental Impact and Health Effects, Vol. 3*, R. L. Jolley, W. A. Brungs, and R. B. Cumming, Eds. (Ann Arbor, MI: Ann Arbor Science Publishers, Inc., 1980), pp. 131-140.
12. Boyce, S. D., and J. F. Hornig. "Reaction Processes Affecting the Analysis of Chloroform by Direct Aqueous Injection Gas Chromatography," *Water Res.* 17:685-697(1983).
13. Rook, J. J. "Chlorination Reactions of Fulvic Acids in Natural Waters," *Environ. Sci. Technol.* 11:478-482 (1978).
14. Morris, J. C., and B. Baum. "Precursors and Mechanisms of Haloform Formation in the Chlorination of Water Supplies," in *Water Chlorination: Environmental Import and Health Effects, Vol. 2*, R. L. Jolley, H. Gorchev, and D. H. Hamilton, Jr., Eds. (Ann Arbor, MI: Ann Arbor Science Publishers, Inc., 1978), pp. 29-48.
15. Lee, G. A., and H. H. Freedman. "Phase Transfer Catalyzed Oxidations of Alcohols and Amines by Aqueous Hypochlorite," *Tetrahedron Lett.* pp. 641-44.
16. Guter, K. J., W. J. Cooper, and C. A. Sorber. "Evaluation of Existing Field Test Kits for Determining Free Chlorine Residuals in Aqueous Solutions," *J. Am. Water Works Assoc.* 66:38-43 (1974).
17. *Standard Methods for the Analysis of Water and Wastewater*, 15th ed. (Washington DC: American Public Health Association, 1980).
18. Cooper, W. J., M. F. Mehran, R. A. Slifker, D. A. Smith, J. T. Villate, and P. H. Gibbs. "Comparison of Several Instrumental Methods for Determining Chlorine Residuals in Drinking Water," *J. Am. Water Works Assoc.* 74:546-552 (1982).
19. Mehran, M. F., R. A. Slifker, and W. J. Cooper. "A Simplified Liquid-Liquid Extraction Method for Analysis of Trihalomethanes in Drinking Water," *J. Chromatogr. Sci.* 22:241-243 (1984).
20. Amy, G. L., P. A. Chadik, Z. K. Chowdhury, P. H. King, and W. J. Cooper. "Factors Affecting the Incorporation of Bromide into Brominated Trihalomethanes During Chlorination," Chapter 71, this volume.

CHAPTER 71

Factors Affecting Incorporation of Bromide into Brominated Trihalomethanes During Chlorination

Gary L. Amy, Paul A. Chadik, Zaid K. Chowdhury,
Paul H. King, and William J. Cooper

Trihalomethanes (THMs) are formed during the chlorination of waters containing humic substances and other precursor compounds. Although chloroform is most often the predominant THM species, brominated haloforms can occur in waters containing bromide ion through the oxidation of bromide to hypobromous acid (HOBr) by hypochlorous acid (HOCl).[1] Thus, when waters containing bromide are chlorinated, chlorinated and brominated THMs are formed. An important concern in THM formation is the relative distribution of chloroform and brominated haloforms. Individual THM species vary in their amenability to removal by various treatment technologies as well as their suspected health hazards. Chloroform is the most volatile THM species and is removed effectively by air stripping, whereas bromoform is the species most effectively adsorbed by activated carbon.[2]

It is well known that bromide ion concentration affects both the rate of formation and the species distribution of THMs.[3,4] Although chlorine preferentially acts as an oxidant, bromine is more effective as a halogen-substituting agent.[5] Moreover, precursors from different sources may vary in their susceptibility toward either chlorination or bromination reactions. Stevens[6] noted that bromine substitution is favored over chlorine even when chlorine is present in large excess compared with the bromide concentration. If HOBr acts as an oxidant, it will be reduced to bromide ion, which may then be reoxidized by chlorine. These factors are responsible for the higher percentage of bromine vs chlorine incorporated into THM species.

Several researchers[3,4,6] have observed trends in species distribution as a function of initial bromide concentration when all other parameters are held constant. Bromoform increases while chloroform decreases as a nonlinear function of initial bromide. Bromodichloromethane increases with bromide to a maximum concentration of bromodichloromethane and thereafter decreases. Dibromochloromethane behaves in a similar manner except that the maximum concentration occurs at a higher initial bromide level. Based on work with humic acid, Stevens[6] found the total molar yield of trihalomethanes to increase with increasing bromide. A similar trend was noted by Oliver[4] in analyzing an aquatic fulvic acid. Minear and Morrow,[7] in studying a natural water spiked with humic acid, also observed an increase in THM yield as bromide increased, both on a weight and a molar basis. They noted that the increase in THM formation caused by bromide appeared to level off at high bromide concentrations, ostensibly because of the limiting amount of nonvolatile total organic carbon (NVTOC) available for reaction.

A percentage of the ambient bromide present in a water is transformed and incorporated into THM species. Cooper et al.[8] suggested that some bromide not recovered in THMs may have become associated with higher-molecular-weight organobromine compounds or, in the presence of ammonia, may appear as bromamines.

Cooper et al.[8] studied bromide incorporation during the chlorination of a highly colored groundwater. For a reaction time of 144 h, the percentage of bromide incorporated into THM species increased as a function of bromide concentration to a maximum and thereafter decreased. Given the ambient bromide level of the water[9] and the incremental addition of further bromide, this maximum occurred at a Br^- concentration of approximately 330 μg/L. Luong et al.[10] studied the effects of spiking a diluted natural water with bromide. Based on bromide additions of 50, 500, and 2000 μg/L, the percentages of bromide incorporated into THMs were determined to be 15, 28, and 11%, respectively. Considering the low ambient level of bromide (<30 μg/L), a maximum occurred at about 500 μg/L. Minear and Morrow[7] studied the effects of adding bromide to a river water spiked with humic acid. Based on data derived from a 96-h reaction time, the percentage of the bromide incorporated into brominated haloforms increased through a maximum of 40% at very low bromide levels (approximately 50 to 60 μg/L) and then declined steadily with increasing bromide until reaching a minimum of 5% at a bromide concentration of 8 mg/L.

Variables that affect THM formation and speciation include temperature, pH, chlorine dose, bromide ion concentration, precursor source and concentration, and reaction time.[11] The same variables may also influence the extent of conversion of inorganic bromide to organically bound bromine incorporated into THMs (THM-Br). Our objective in this chapter is to define the effects of each of the variables mentioned on the formation of THM-Br and the related conversion of inorganic bromide to THM-Br, expressed as "percentage conversion" [(THM-Br/Initial Br^-) × 100]. It should be noted that bromine may be incorporated into organic compounds other than THMs; however, these other forms of organic bromine are not included in THM-Br as defined herein.

EXPERIMENTAL

To study bromide incorporation, nine natural waters were subjected to experimental evaluation: the Edisto River (South Carolina), the Scioto River

(Ohio), the Verde River (Arizona), the Biscayne Aquifer (Florida), the Ilwaco Reservoir (Washington), the Kaw Reservoir (Oklahoma), the Grasse River (New York), the Pearl River (Mississippi), and the James River (South Dakota).

As indicated in Tables I and II, these waters embrace a broad range of raw water quality characteristics: nonvolatile total organic carbon (NVTOC), 3.0 to 13.8 mg/L; ambient pH, 6.1 to 8.3; ambient Br^-, 10 to 245 µg/L; UV absorbance (at 254 nm and pH 7), 0.063 to 0.489; and color, 5 to 93 color units. Each water was studied in a series of experiments encompassing the following parameters and ranges of conditions:

Temperature = 10, 20, 30°C
pH = ambient, ambient + 1.5 units, ambient − 1.5 units (Waters were held constant at the respective pH values with 0.1 M buffers.)
Br^- = ambient, ambient + 0.25 mg/L, ambient + 0.50 mg/L, ambient + 1.0 mg/L
Cl_2/NVTOC = 0.5, 1.0, 3.0, 5.0
Reaction time = 0.1, 0.5, 1, 2, 4, 8, 24, 48, 96, 168 h

Kinetic experiments (i.e., THM formation as a function of reaction time) were run for various combinations of experimental conditions. A set of "baseline" conditions was defined as follows: 20°C, ambient pH, ambient Br^-, and Cl_2/NVTOC ratio of 3.0. This set of conditions constitutes a "baseline" experiment, which represents the first kinetic experiment run for each water. Next, a related set of experiments was run in which each parameter was varied separately while other parameters were held constant at baseline levels. Thus, considering the above parameter ranges, two additional experiments were run to evaluate the effects of temperature and pH, respectively. Similarly, three additional experiments were run to study the effects of chlorine-to-NVTOC ratio and bromide. Subsequently, two more experiments were conducted to bracket the effects of the specified parameters: a "high extreme" experiment involving parameter levels most conducive to THM formation (i.e., 30°C, ambient pH + 1.5 units, ambient Br^- + 1.0 mg/L, and Cl_2/NVTOC = 5) and a "low extreme" experiment based on parameter levels least conducive to THM formation (i.e., 10°C, ambient pH − 1.5 units, ambient bromide, and Cl_2/NVTOC = 0.5).

A Dohrman DC-80 Carbon Analyzer was used for NVTOC determinations and a Perkin-Elmer UV-Visible Spectrophotometer (Model 200) was used for UV absorbance measurements. Bromide analyses were accomplished by two methods: a spectrophotometric technique[12] applicable to levels ranging from 1 to 100 µg/L (with higher levels measured by dilution) and a single-column ion chromatograph (Dionex Model 10) having a lower practical detection limit of about 100 µg/L. Trihalomethane formation potential (THMFP) was measured by introducing a buffered sample (phosphate or borate buffer) into a headspace-free septa-sealed serum vial. Sodium hypochlorite was then introduced via syringe, the vial was placed in a constant-temperature module, and the THM reaction was allowed to proceed for a designated reaction time. The

Table I. Summary of Raw Water Characteristics

Water	NVTOC (mg/L)	UV Absorbance[a]	THMFP[b] µg/L	µmol/L	Ambient pH
Edisto	11.3	0.489	1083	9.00	7.30
Scioto	6.25	0.152	336	2.73	7.56
Verde	3.00	0.063	97	0.75	8.31
Biscayne	6.50	0.251	296	2.33	7.31
Ilwaco	6.00	0.329	405	3.27	6.10
Kaw	5.22	0.153	267	2.01	7.72
Grasse	6.56	0.288	490	4.09	6.83
Pearl	5.62	0.136	303	2.49	6.55
James	13.8	0.296	694	5.56	8.00

[a] 254 nm, pH 7.0.
[b] 168 h, pH 7.0, 20°C, Cl_2/NVTOC = 3:1.

Table II. Summary of Bromide Analyses

Water	Ambient Bromide Concentration (µg/L) Spectrophotometric Method	Ion Chromatographic Method	Assumed Value[a]
Edisto	74	<100	74
Scioto	96	100	98
Verde	109	112	111
Biscayne	161	140	151
Ilwaco	129	100	115
Kaw	162	142	152
Grasse	10	[b]	10
Pearl	51	[b]	51
James	245		245

[a] Based on average or spectrophotometric method if level below sensitivity of IC method.
[b] Not detected.

reaction was terminated with excess thiosulfate, and the resultant THM species were measured using liquid/liquid extraction and a Hewlett Packard 5794 gas chromatograph equipped with an electron capture (EC) detector.

RESULTS AND DISCUSSION

Data for the various reaction times were examined to study the kinetics of inorganic bromide conversion and the corresponding formation of THM-Br. To determine the effects of various factors on the percentage incorporation of bromide, a series of linear regression analyses was performed using that portion of the data base that would isolate the effects of a given variable (i.e., data in which a given variable was varied while all other variables were held constant). Linear regressions were separately performed for both long-term data (arbitrarily defined as a reaction time of 96 h) and short-term data (arbitrarily defined as a reaction time of 1 h). These regressions evaluated percentage conversion vs either pH, temperature, applied chlorine concentration, or initial bromide concentration. Precursor-related parameters (NVTOC, UV absorbance) were also evaluated, but little correlation was observed.

Effects of Reaction Time

The formation of THM-Br as a function of reaction time for selected waters is shown in Figure 1. The data shown in this figure correspond to the baseline data previously discussed. For the waters shown, levels of THM-Br ranged from <10 to >50μg/L. These THM-Br formation curves mimic the general shape of THM formation curves in which an initial phase of rapid formation is followed by a subsequent phase of slower formation and, eventually, a leveling off as any reactant approaches depletion. The data in Figure 1 suggest that THM-Br levels off at different times for different water samples following an initial rapid phase formation and a subsequent slower phase of formation. We also observed that although THM-Br levels off at different times further chloroform formation continued for these same waters. This may be consistent with the results of other researchers showing that the formation kinetics for each THM species differ.[13]

An alternative representation of data appears in Figure 2, portraying percent of conversion as a function of reaction time. This approach normalizes the data shown in Figure 1. It is apparent that a positive nonlinear trend exists between percent of conversion and reaction time. Linearization of the data could be accomplished by data transformation in which percent of conversion was regressed against the log of reaction time.

From Figure 2 it appears that the slope of the curves, percent of conversion vs reaction time, is generally the same for all four waters. However, there does not appear to be a simple relationship between the percent of conversion and NVTOC, Br$^-$, or chlorine concentration. For example, it can be seen that two waters having significantly different bromide levels, the Kaw Reservoir and Grasse River, exhibit approximately the same percent of conversion. On the other hand, the water obtained from the Biscayne Aquifer and the Edisto River, although markedly different in Br$^-$ and NVTOC, gave similar percentage conversion curves as a function of time. One variable not included in this analysis that could account for some of these variations is ammonia, because chloramines (and, to a lesser extent, bromamines) could reduce the THM-Br. The formation of monochloramine would inhibit the oxidation of Br$^-$, and the formation of bromamines might inhibit the incorporation of organically bound bromine in THMs.

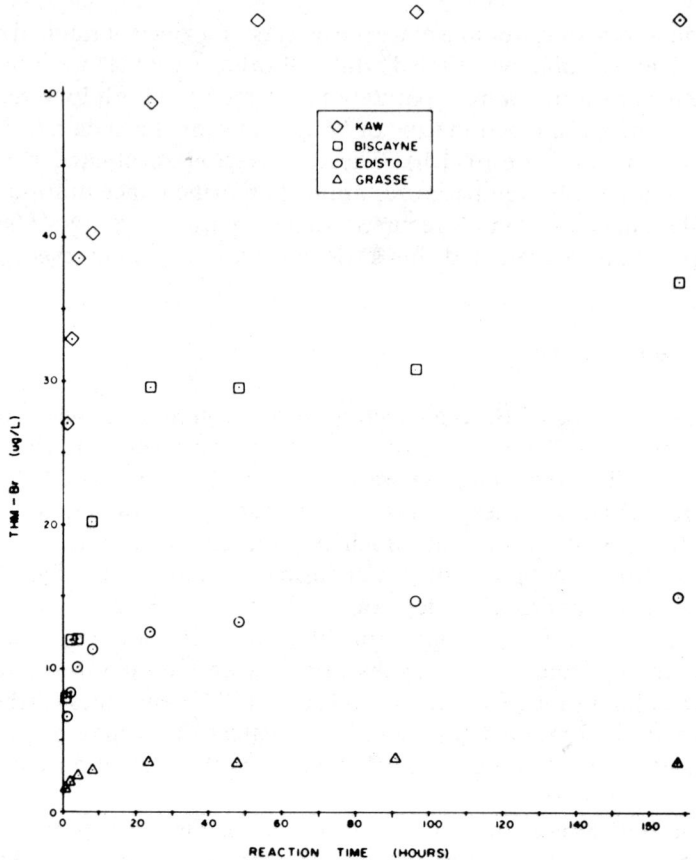

Figure 1. Formation of THM-Br as a function of reaction time for selected waters.

Effects of pH

In general, it was observed that TTHM yield (in micromoles per liter and in micrograms per liter) increased with increasing pH. All other variables being constant, a strong positive correlation was observed between percentage conversion and pH for each water. Linear regressions were performed for both long-term (96 h) and short-term (1 h) data for each water, using portions of the data base that isolate the effects of pH. Values of the coefficient of determination, r^2, based on long-term data were >0.9 for seven of the nine waters, whereas corresponding r^2 values for short-term data were >0.9 for five of the nine waters.

A graphical representative of percentage conversion vs pH is presented in Figure 3 for selected waters. Generally, slopes for long-term data were greater

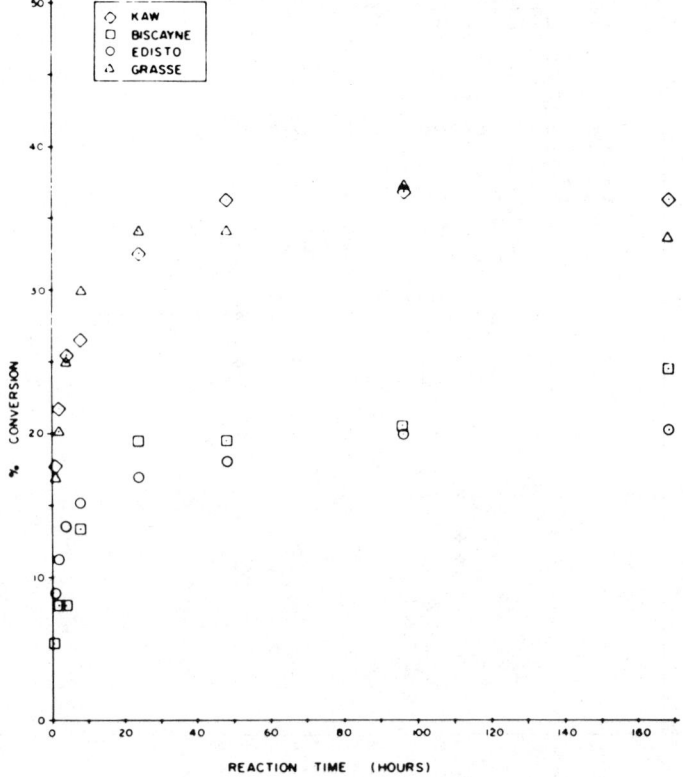

Figure 2. Percentage conversion as a function of reaction time for selected waters.

than corresponding slopes for short-term data, suggesting that differences between short-term and long-term bromide incorporation are greater at high pH levels. It is evident that the effect of pH on percentage conversion is a function of the nature of the particular water under evaluation. Based on the diversity in statistically derived estimates of slope and intercept, a universal response to pH is not exhibited by all waters. Incremental increases of about 2 to 7% per unit increase in pH were found for long-term reaction times, whereas for short-term reaction times, incremental increases ranged from 1 to 4%.

Effects of Temperature

An analysis of the data isolating the effects of temperature indicated that TTHM yield increased (on both molar and weight bases) as a function of

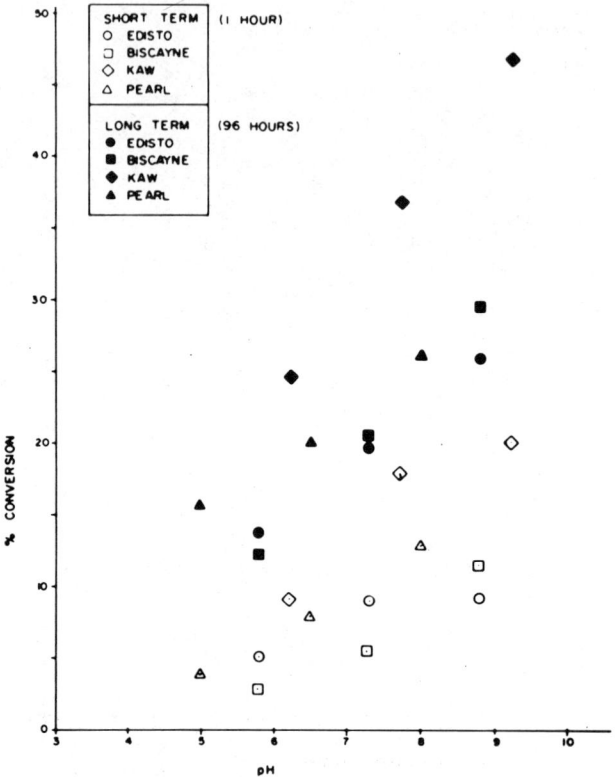

Figure 3. Percentage conversion as a function of pH for selected waters.

temperature. As with pH, a strong positive correlation was observed between percentage conversion and temperature in the range under consideration. A graphical representation of temperature effects is presented in Figure 4 for selected waters. As noted with respect to pH, each water responded uniquely to changes in temperature. Slopes associated with long-term data were generally steeper than corresponding slopes for short-term data, indicating more pronounced differences at higher temperatures.

It should be noted that in this evaluation of temperature (as well as subsequent evaluations of chlorine, bromide, and precursor) pH was held constant at the ambient level for a given water. Ambient pH levels for the various waters ranged from 6.1 to 8.3. Therefore, although most of the variation in response from water to water may result from changes in temperature (or another variable as discussed in subsequent discussions), some variation may also be attributed to differences in ambient pH.

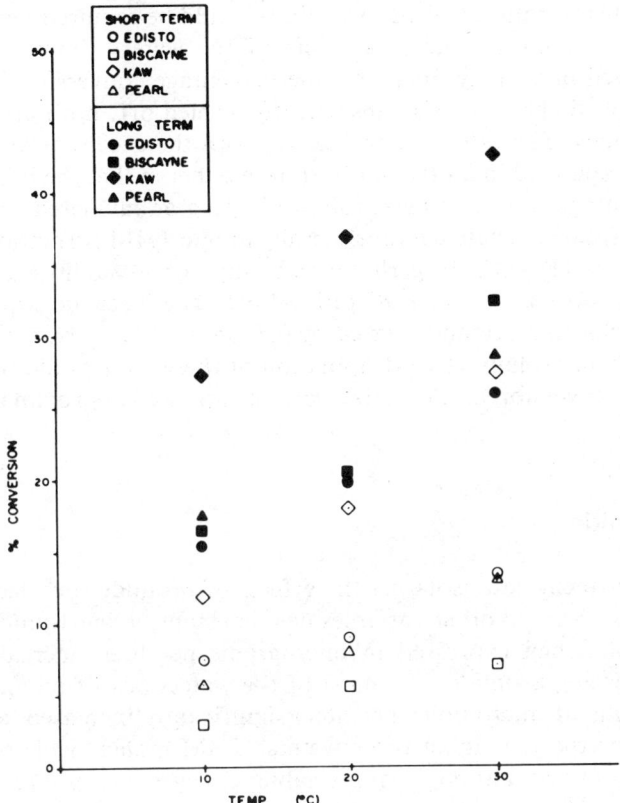

Figure 4. Percentage conversion as a function of temperature for selected waters.

Linear regression analysis of either long-term or short-term data gave values of r^2 that were greater than 0.9 for seven of the nine waters, indicating general linear trends. Considering statistically derived estimates of the slopes of regression equations, it can be expected that about a 0.5 to 0.9 % increase in percentage conversion will occur per degree centigrade increase for a reaction time of 96 h, whereas for the short-term reaction time, a corresponding increase of about 0.2 to 0.8 % can be expected.

Effects of Applied Chlorine Concentration

It was observed that TTHM increased as a function of applied chlorine. A positive, although nonlinear, correlation was observed between percentage conversion and applied chlorine. A graphical representative of the effects of chlorine on the incorporation of bromide appears in Figure 5 for selected waters. Data transformation (for linearization) would be required before analyzing these data by linear regression. One of the waters shown, the Kaw Reservoir, behaved in a unique manner: the percentage conversion increased sharply with applied chlorine and subsequently leveled off, both at long and short reaction times. The other waters shown, with the exception of the Biscayne Aquifer, responded in a slightly different manner in that the initial rapid increase in percentage conversion was followed by a gradual increase at higher chlorine concentrations. These curves generally mimic THM formation curves as a function of Cl_2/NVTOC. Engerholm and Amy[14] observed little additional THM formation above a Cl_2/NVTOC ratio of 6:1. The Biscayne Aquifer was found to have a chlorine demand exerted by 0.8 mg/L of NH_3-N, a characteristic that would help explain why extrapolation of these data would indicate a zero percentage conversion at a chlorine concentration of approximately 6 to 7 mg/L Cl_2.

Effects of Bromide

Analysis of pertinent data isolating the effects of bromide revealed several interesting trends. With all other variables held constant, it was found that, in all cases, TTHM (when expressed in micrograms per liter) increased as a function of increasing bromide. For most of the waters (six of nine), TTHM (expressed in terms of micromoles per liter) significantly increased with bromide, whereas for the remaining three waters, THM molar yields remained approximately constant although appreciable changes in speciation were observed.

The percentage conversion trends as a function of bromide for all waters varied. As bromide levels increased in the range of the experimental matrix, percentage conversion decreased for five waters, increased for two waters, and increased through a maximum before decreasing for two waters. Trends for both short- and long-term data were similar. Previously run experiments using larger bromide concentration ranges helped explain this phenomenon. The results of these experiments (168-h reaction time) are portrayed in Figure 6, which shows percentage conversion as a function of bromide level for three selected waters. For the Biscayne Aquifer, the decreasing percentage conversion is demonstrated throughout the bromide range. For the Verde River, an increase is shown up to a bromide level of 0.361 mg/L and then a decrease is shown at higher bromide levels. The Scioto River curve is similar to that of the Verde, but the maximum is located at a higher bromide level. Similar trends of increasing followed by decreasing percentage conversion as a function of bromide ion concentration were observed for other waters, but the bromide level for maximum percentage conversion was different in all cases. For two of the waters, the apparent maximum occurred at a bromide concentration of greater than the highest level studied during the kinetic experiments (ambient + 1.0 mg/L). This explains why analysis of the long-term and short-term data indi-

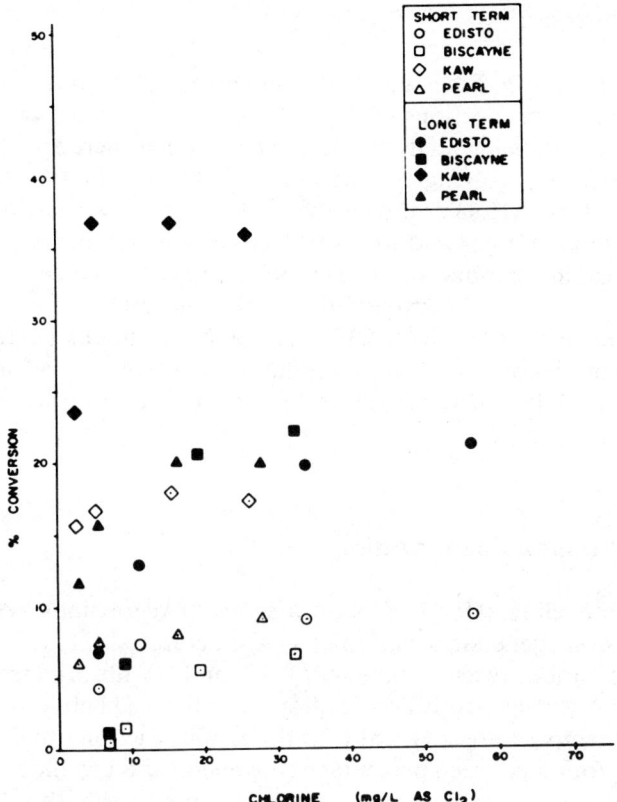

Figure 5. Percentage conversion as a function of chlorine concentration for selected waters.

cated a positive correlation between percentage conversion and bromide for these particular waters.

Effects of Precursor Source

Based on the previous discussion, it is apparent that each water exhibited a unique response to changes in experimental variables, which suggests that the precursor source and concentration affect bromide incorporation. However, simple linear regressions between percentage conversion and two precursor-related parameters, NVTOC and UV absorbance, yielded r^2 of <0.1 when either short-term or long-term data associated with the baseline experiment were analyzed. This finding suggests that precursor effects are more complex than indicated by the above analysis. The role of precursor may be related to multiplicative effects with other factors such as ammonia, pH, and temperature.

Multiplicative Effects

Each of the previously discussed effects on percentage conversion of Br^- to THM-Br were based on portions of the data base that isolated the effects of the particular variable being analyzed. It is expected that there are also various multiplicative effects involving two or more of these variables that are not simply additive. Table III summarizes the percentage conversion observed in the previously defined high- and low-THM extreme experiments corresponding to conditions most conducive to THM formation. The percentage conversion ranged from 0.5% for 1-h experiments in the low THM extreme to 64.6% for 96-h experiments in the high THM extreme. It should be recognized, however, that not all conditions most conducive to THM formation are most conducive to conversion of Br^- to THM-Br (i.e., Br^- concentration).

Modeling of Bromide Incorporation

Preliminary modeling efforts involved a series of regressions between percentage conversion (dependent variable) vs temperature, pH, chlorine dose, bromide concentration, reaction time, NVTOC, and UV absorbance (independent variables). A strong positive correlation was observed between percentage conversion and temperature, pH, chlorine dose, and reaction time. An inverse correlation was found between percentage conversion and bromide for most of the waters, whereas a very poor correlation was noted for both NVTOC and UV absorbance. Plots of percentage conversion vs each of the independent variables (x) indicated some departures from linearity. In many cases, data transformation (ln x, x^2, $\sqrt{x}$, 1/x) improved linearity and facilitated the use of multiple linear regression techniques in formulating a model.

A computerized multiple linear regression program[15] was used to analyze the data. A series of stepwise multiple linear regressions was performed to evaluate various forms and combinations of variables. Separate regressions were conducted for each water in addition to an overall regression of data for all waters. Not all of the data base was used for these regressions. Data derived from experiments involving incremental bromide additions, as well as data associated with the high and low extreme experiments, were not included because of the complexity of modeling the nonlinear effects of bromide. All other data were used (nine waters, four Cl_2/NVTOC ratios, three pH levels, three temperatures, and ten reaction times).

The regressions involved percentage conversion vs arithmetic as well as transformed values of each independent variable. Based on analysis of partial correlation coefficients, the stepwise multiple linear regression technique, when analyzing all data, entered the following variables into the equation during the first five steps in the indicated sequence; (1) ln (reaction time),

Table III. Percentage Conversion Observed in Low- and High-THM Extreme Experiments

Water	Short-Term (1 h)		Long-Term (96 h)	
	Low-THM Extreme	High-THM Extreme	Low-THM Extreme	High-THM Extreme
Edisto	2.8	23.5	10.1	52.4
Scioto	3.4	20.7	10.7	52.5
Kaw	5.5	18.1	16.5	64.3
Ilwaco	0.7	15.0	2.5	34.1
Pearl	1.3	23.9	6.3	50.3
James	0.5	26.5	3.7	64.6
Grasse	2.7	22.3	15.1	51.8

(2) (temperature)^2, (3) (1/bromide), (4) (pH)^2, and (5) (1/chlorine). The multiple coefficient of determination (R^2) was 0.66 for the following equation:

$$\text{percentage conversion} = -1.73 + 2.50 \ln(\text{reaction time}) + 0.0118 (\text{temperature})^2 + 0.104 (1/\text{bromide}) + 0.129 (\text{pH})^2 - 17.1 (1/\text{chlorine})$$

It can be seen that the regression coefficient for the chlorine-related term, inverse chlorine, is negative. This, of course, implies that chlorine concentration will have a positive effect on percentage conversion. This particular form of the chlorine variable was selected before other forms by the stepwise regression techniques, based on a comparison of partial correlation coefficients for the variables remaining after the previous step. Also, during the stepwise analysis, a precursor-related parameter was not entered into the equation until the eighth step, in which (NVTOC)^2 was entered; however, this entry into the equation provided an incremental change in R^2 of only 0.005 over the previous step. This suggests that the relationship between bromide incorporation and precursor is more complex than the above type of analysis assumes.

Next, the same stepwise technique was applied to data for each of the waters individually. Bromide-related variables were not considered, because the bromide level was constant for each water. In most, but not all cases, the same variables in the previous equation (except for 1/bromide) were entered into the equations before other transformed variables although the sequence of entered variables differed somewhat for each of the waters. Multiple R^2 values after four steps were >0.84 for seven of the waters and >0.90 for five of the waters. Finally, as an alternative analysis to the stepwise approach, data for

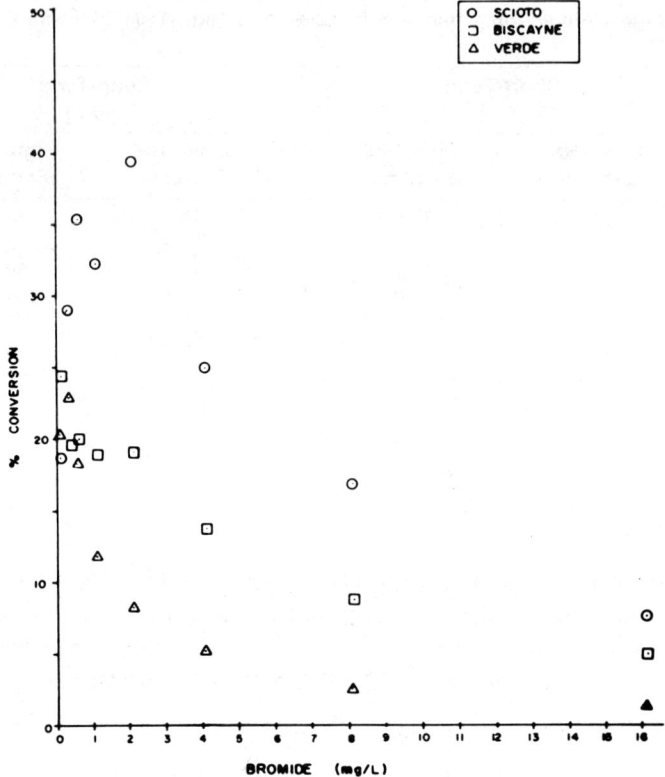

Figure 6. Percentage conversion as a function of bromide concentration for selected waters (168-h reaction time).

each water were regressed against the specific variables determined for the general model shown above (except 1/bromide). Multiple R^2 values were >0.82 for seven of the waters and >0.92 for four of the waters. The higher R^2 values observed for the individual water regressions compared to the overall regression suggests that, although the various waters respond in a somewhat similar manner to the variables, the magnitude of the responses varies from water to water.

CONCLUSIONS

In THM formation a portion of the inorganic bromide ion concentration is converted to organically bound bromine in the three common brominated THM species. This percentage conversion (1) increases rapidly in the first stages of THM formation and approaches a plateau at longer reaction times; (2) increases with increasing pH and increasing temperature; (3) increases with chlorine concentration but approaches a plateau at high concentrations; (4)

initially increases with Br⁻ concentration to a maximum percentage conversion and subsequently decreases at higher Br⁻ concentrations.

It was difficult to define clear trends for the variation of percentage conversion with THM precursor source and concentration presumably because of multiplicative effects of variables as well as parameters such as ammonia concentration that were not examined in this study.

ACKNOWLEDGMENTS

Financial support for the research was provided by the U.S. Environmental Protection Agency under Contract R809935-01 (G. L. Amy) and Cooperative Agreement CR810277-01 (W. J. Cooper).

REFERENCES

1. Rook, J. J., et al. "Bromide Oxidation and Organic Substitution in Water Treatment," *J. Environ. Sci. Health*, 13:91-116 (1978).
2. Symons, J. M., et al. "Treatment Techniques for Controlling Trihalomethanes in Drinking Water," EPA-600/2-81-156, (Cincinnati, OH: U.S. Environmental Protection Agency, 1981).
3. Minear, R. A., and J. C. Bird. "Trihalomethanes: Impact of Bromide Ion Concentration on Yield, Species Distribution, Rate of Formation, and Influence of Other Variables," in *Water Chlorination: Environmental Impact and Health Effects, Vol. 3*, R. L. Jolley, W. A. Brungs, and R. B. Cumming, Eds. (Ann Arbor, MI: Ann Arbor Science Publishers, Inc., 1978), pp. 151-160.
4. Oliver, G. B. "Effect of Temperature, pH, and Bromide Concentration on the Trihalomethane Reaction of Chlorine with Aquatic Humic Material," in *Water Chlorination: Environmental Impact and Health Effects, Vol. 3*, R. L. Jolley, W. A. Brung, and R. B. Cumming, Eds. (Ann Arbor, MI: Ann Arbor Science Publishers, Inc., 1978), pp. 141-149.
5. Rook, J. J. "Possible Pathways for the Formation of Chlorinated Degradation Products During Chlorination of Humic Acids and Resorcinol," in *Water Chlorination: Environmental Impact and Health Effects, Vol. 3*, R. L. Jolley, W. A. Brung, and R. B. Cumming, Eds. (Ann Arbor, MI: Ann Arbor Science Publishers, Inc., 1978), pp. 85-98.
6. Stevens, A. "Formation of Non-Polar Organo-Chloro Compounds as Byproducts of Chlorination," in *Oxidative Techniques in Drinking Water Treatment*, W. Kuhn and H. Sontheimer, Eds., EPA-570/9-79-020, (Cincinnati, OH: U.S. Environmental Protection Agency, 1979), pp. 145-160.
7. Minear, R. A., and C. M. Morrow. "Raw Water Bromide: Levels and Relationship to Distribution of Trihalomethanes in Finished Drinking Water," Research Report No. 91, (Knoxville, TN: Water Resources Research Center, University of Tennessee, 1983).
8. Cooper, W. J., L. M. Meyer, C. C. Bofill, and E. Cordal. "Quantitative Effects of Bromine on the Formation and Distribution of Trihalomethanes in Groundwater with a High Organic Content," in *Water Chlorination: Environmental Impact and Health Effects, Vol. 4*, R. L. Jolley, W. A. Brungs, J. A. Cotruvo, R. B. Cum-

ming, J. S. Mattice, and V. A. Jacobs, Eds. (Ann Arbor, MI: Ann Arbor Science Publishers, Inc., 1983), pp. 285–296.
9. Amy, G. L., P. A. Chadik, P. H. King, and W. J. Cooper. "Chlorine Utilization During Trihalomethane Formation in the Presence of Ammonia and Bromide," *Environ. Sci. Technol.* 18:781–786 (1984).
10. Luong, T. V., J. P. Peters, and R. Perry. "Influence of Bromide and Ammonia Upon the Formation of Trihalomethanes Under Water Treatment Conditions," *Environ. Sci. Technol.* 16:473–479 (1982).
11. Trussell, R. R., and M. D. Umphres. "The Formation of Trihalomethanes," *J. Am. Water Works Assoc.* 70:604–612 (1978).
12. Fishman, M. J., and M. W. Skougstad. "Indirect Spectrophotometric Determination of Traces of Bromide in Water," *Anal. Chem.* 35:146–149 (1963).
13. Cooper, W. J., J. T. Villate, E. M. Ott, R. A. Slifker, and F. Z. Pearsons. "Formation of Organohalogen Compounds in Chlorinated Secondary Wastewater Effluent," in *Water Chlorination: Environmental Impact and Health Effects, Vol. 4*, R. L. Jolley, W. A. Brungs, J. A. Cotruvo, R. B. Cumming, J. S. Mattice, and V. A. Jacobs, Eds. (Ann Arbor, MI: Ann Arbor Science Publishers, Inc., 1983), pp. 483–497.
14. Engerholm, B. A. and G. L. Amy. "A Predictive Model for Chloroform Formation From Humic Acid," *J. Am. Water Works Assoc* 75:418–422 (1983).
15. Nie, N. H., et al. *Statistical Package for the Social Sciences*, 2nd ed. (New York: McGraw-Hill, 1975).

CHAPTER 72

Formation of Iodinated Trihalomethanes

Joseph P. Gould, Maurizio Giabbai, and Jong-Soo Kim

Although much information is available on the formation of the four chlorine- and bromine-containing trihalomethanes (THMs)[1-3], very little research has been performed on the formation and occurrence of iodinated THMs.[4,5] That these compounds might be quite uncommon is reasonable in the context of the very low concentrations of iodide in surface waters and most groundwaters as determined in previous studies[6-9] (Table I). There are exceptions to this, however, such as oil-field brines and waters in which iodine has been added as a disinfectant. In both cases the potential for formation of iodine-containing THMs exists. The research reported herein represents a preliminary investigation into formation of these compounds under controlled laboratory conditions.

MATERIALS AND METHODS

Experimental Design

A series of 37 runs was conducted. These studies were designed to cover a reasonable range of chlorine, bromide, and iodide concentrations and to explore a few extreme cases as well. The precursor chosen was 2,4,6-trihydroxyacetophenone (2,4,6-THA) at a concentration of 210 μM in all cases. All runs were conducted at a pH of 8.3, and 0.1 M NaHCO$_3$ was used for pH control. Chlorine concentrations of 8 and 20 mM were provided by use of distilled HOCl. Bromide and iodide concentrations ranging from 4 to

Table I. Typical Iodide Levels in Natural Waters

Water	Iodide	Reference
Seawater	50 μg/L	6
Missouri surface water	5 μg/L	7
Missouri groundwater	50 μg/L	7
Oilfield brines	40 mg/L	8
Oklahoma brines	>100 mg/L	9

800 μM were used for most studies, although one series using concentrations as high as 8 μM was also run. Several controls lacking either or both of the halides were also run. Table II provides details of the composition of all runs.

The samples were run in 50-mL glass vials capped with silicone rubber septum caps. The reactions were run for 4.5 h in the dark at an ambient temperature of 23°C. This time was chosen on the basis of previous studies in our laboratories which indicated that formation of THMs from 2,4,6-THA

Table II. Design of Experiments and Halogen Incorporation Coefficients[a]

Run	$[HOCl]_o$ (mM)	$[Br^{-1}]_o$ (μM)	$[I]_o$ (μM)	ηBr	ηI
1	8	4	4	0.01	0.00
2	8	4	80	0.01	0.02
3	8	4	800	0.01	0.07
4	8	80	4	0.29	0.00
5	8	80	80	0.34	0.02
6	8	80	800	0.56	0.07
7	8	800	4	2.01	0.00
8	8	800	80	1.54	0.01
9	8	800	800	1.88	0.02
10	20	4	4	0.04	0.00
11	20	4	80	0.02	0.01
12	20	4	800	0.01	0.07
13	20	80	4	0.17	0.00
14	20	80	80	0.20	0.02
15	20	80	800	0.17	0.06
16	20	800	4	1.44	0.00
17	20	800	80	1.21	0.01
18	20	800	800	2.14	0.03
19	8	0	4	0.00	0.00
20	8	0	80	0.00	0.03
21	8	0	800	0.00	0.10
22	20	0	4	0.00	0.00
23	20	0	80	0.00	0.03
24	20	0	800	0.00	0.06
25	8	8000	0	2.74	0.00
26	8	6000	2000	2.31	0.01
27	8	4000	4000	0.21	0.39
28	8	2000	6000	0.02	0.48
29	8	0	8000	0.00	0.58
30	8	0	0	0.00	0.00
31	20	0	0	0.00	0.00
32	8	4	0	0.02	0.00
33	8	80	0	0.32	0.00
34	8	800	0	2.11	0.00
35	20	4	0	0.03	0.00
36	20	80	0	0.16	0.00
37	20	800	0	1.53	0.00

[a] $[2,4,6\text{-THA}]_o = 210$ μM; pH = 8.3; $[HCO_3^-] = 0.1$ M.

was usually complete after about 3 h.[10,11] The samples were quenched by use of about 250 mg of sodium thiosulfate per 30-mL sample.

Analysis

A liquid-liquid extraction method (LLE) was used rather than the purge and trap method. This choice was based on the probable low volatility and high affinity for water of the iodine-rich THMs such as iodoform.

An examination of the literature relating to iodoform[12] indicated that its solubility in the typical LLE extracts, pentane and cyclohexane, was very low, suggesting an unfavorable partition between the solvent and water, and consequently poor extraction efficiencies. To circumvent this problem, all extractions were conducted using a mixture of 5.0 mL of *n*-pentane and 3.0 mL of benzene for each sample. Each sample/solvent mixture was agitated vigorously by hand for 2 min and the phases allowed to separate.

GC-MS Analysis

A Hewlett-Packard 5830-A gas chromatograph equipped with a split-splitless capillary injection system and interfaced to a Finnigan 4023 mass spectrometer was used for the analysis of the samples. A glass capillary column (30 m by 0.30 mm ID) deactivated by the persilylation method[13] and coated by the static method[14] with SE-54 silicone gum phase was used for the separation of the organic mixture. Fused silica tubing served as sample transfer line between the column effluent and the ionization source.[15] The GC conditions were as follows: injection volume, 1 µL; injection mode, splitless; oven temperature, 40°C (3 min), then increased to 280°C at 15°C/min. Hexamethylbenzene was used as an internal standard for the evaluation of both relative retention time and quantitative data. The MS conditions were as follows: ionization mode, electron impact; electron multiplier, 1500 V; electron energy, 70 eV; emission current, 0.5 mA; mass range, 40 to 450 amu.; and scan rate, 0.95 s/decade. The mass spectrometer was tuned with perfluorotributylamine, and a solution of decafluorotriphenylphosphine was subsequently injected onto the GC to verify the tuning thus obtained. The identification of the halogenated organic constituents was carried out by automatic search against the library stored in the Incos Data System capable of 25,000 spectra, by comparison with the mass spectra compendia,[16,17] and by considering the basic principles of fragmentation in MS. Figures 1 through 3 show a typical chromatogram and two mass spectra from these studies.

Quantification

Although reference standards are readily available for the four chlorobromo-THMs and iodoform, there are no commercial sources for

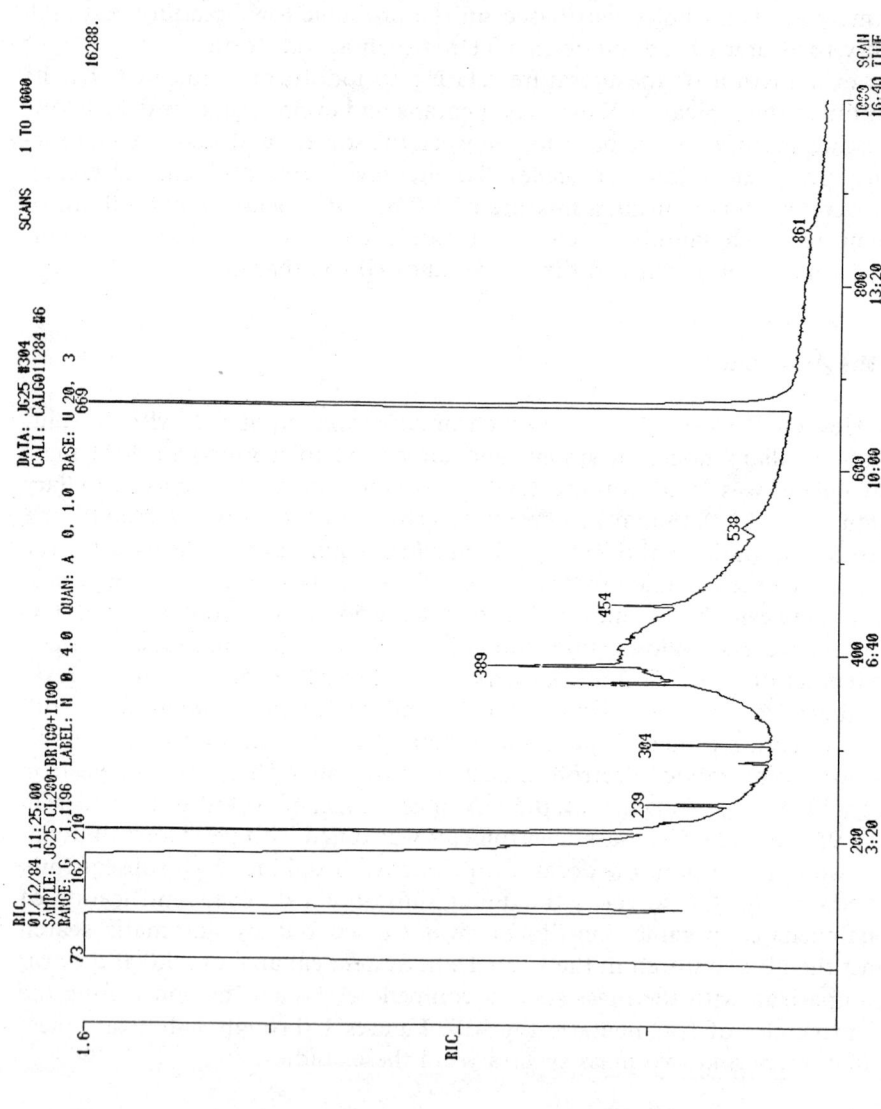

Figure 1. Reconstructed chromatogram for Run No. 27.

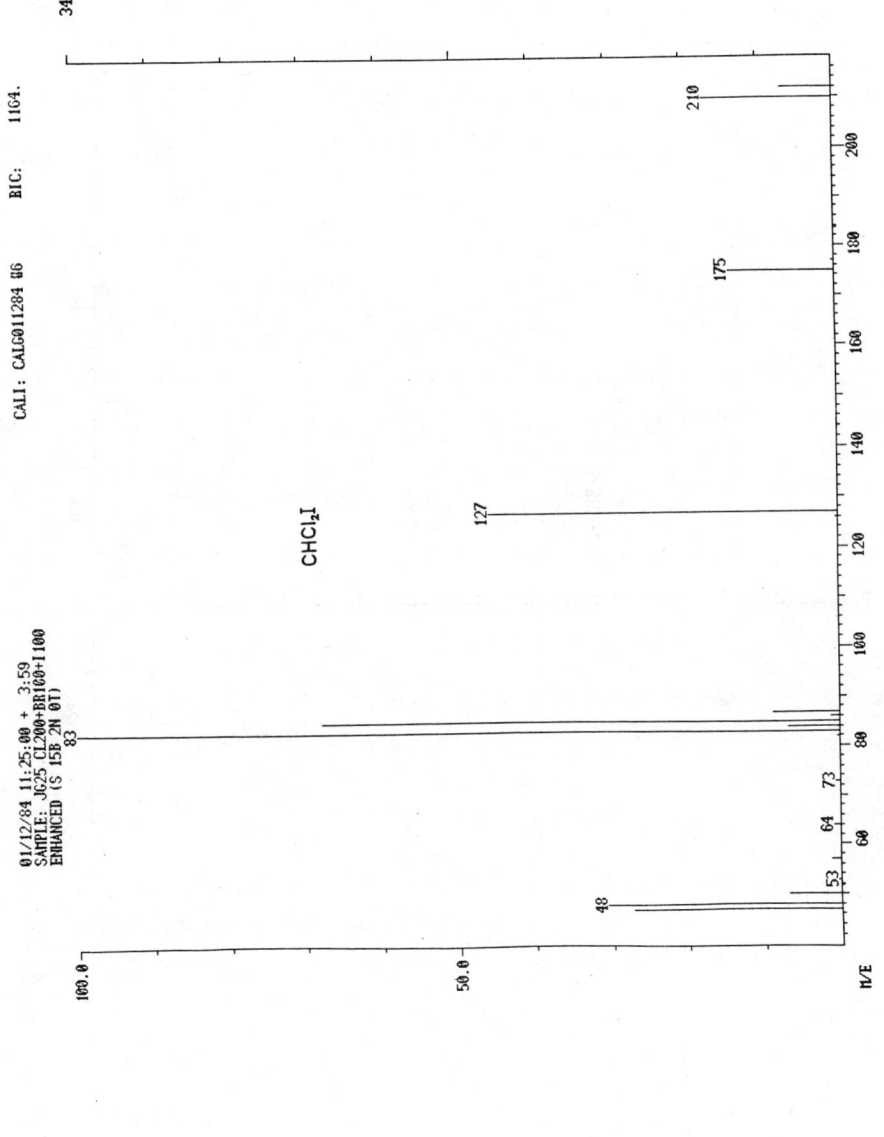

Figure 2. Mass spectrum of $CHCl_2I$.

928 WATER CHLORINATION

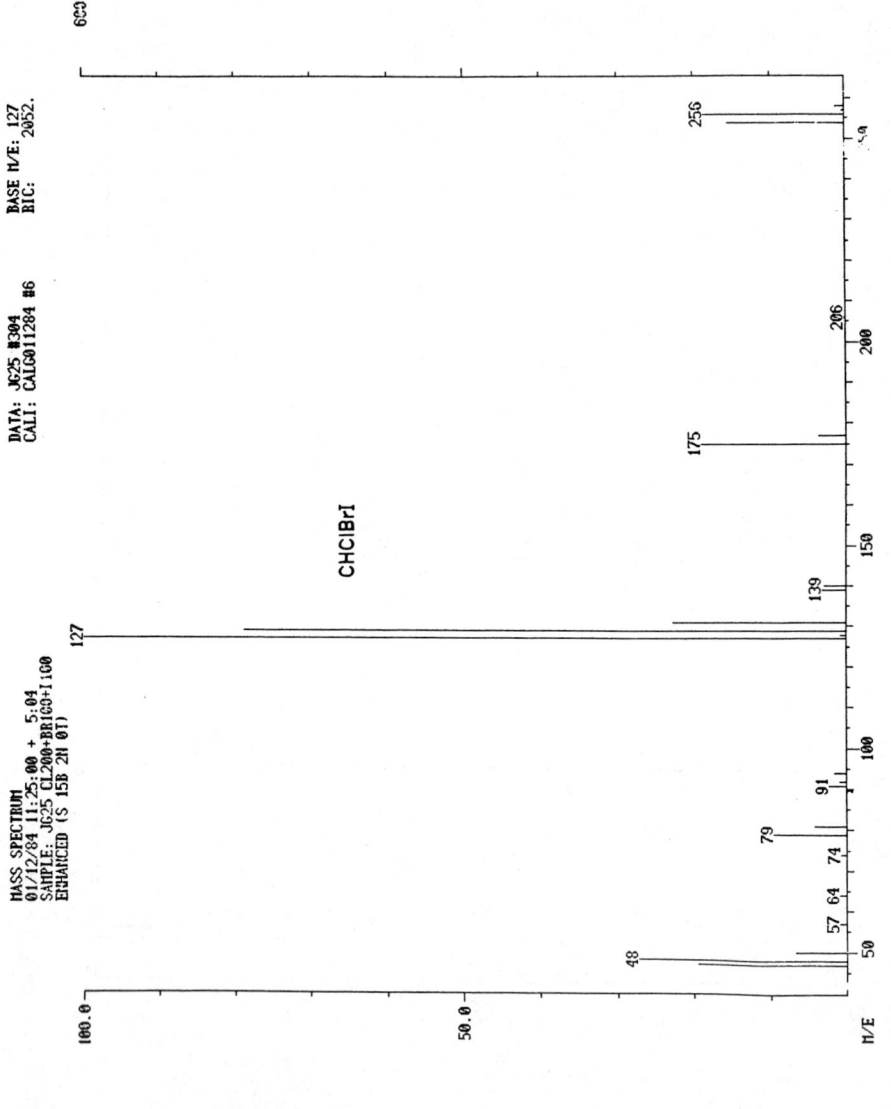

Figure 3. Mass spectrum of CHClBrI.

$CHCl_2I$, $CHClI_2$, $CHBr_2I$, $CHBrI_2$, and CHClBrI. These compounds are difficult to synthesize in a pure form and are quite unstable, losing iodine readily on exposure to air and light. Thus, although efforts to prepare reference standards of the various iodo-THMs are ongoing, reference standards are, at present, unavailable. To obtain usable concentration data of sufficient quality to permit a preliminary analysis of the behavior of these systems, the data were analyzed by comparison with an internal standard (chlorobromomethane) added in known concentration to each solvent layer immediately prior to GC-MS analysis. The ratio of the integrated response area of a given THM to the area of the internal standard was multiplied by the molar concentration of the standard to give an approximate molar concentration of the THM in the solvent. The uncertainties in this approach are obvious and, clearly, the resulting concentrations are only approximate and must be considered as such. However, for the purpose of providing some initial understanding of the general process of iodinated THM formation, these "concentration" data have proved entirely adequate. Although the application of rigorous analytical methodology will certainly serve to refine details of the systems, the general trends are so clear-cut that no substantial modifications are likely.

Analysis of Data

To simplify the manipulation and analysis of data sets that can include as many as ten THM concentrations per sample, incorporation coefficients [Equation (1)] discussed elsewhere will be used.[10] Application of these

$$n_x = \frac{\Sigma \, a \, [CHX_aY_bZ_c]}{\Sigma \, [CHX_aY_bZ_c]} \quad (1)$$

coefficients permits the computation of a composite or average THM for each THM mixture. Thus, a THM mixture containing iodinated THMs would have a composite composition characterized by the formula $CHCl_{\eta Cl}Br_{\eta Br}I_{\eta I}$, where ηCl, ηBr, and ηI are incorporation coefficients for chlorine, bromine, and iodine, respectively, and should add up to 3.00.

All THM concentrations will be expressed and analyzed in micromolar units.

RESULTS AND DISCUSSION

The primary object of this study was to evaluate the extent to which iodine participated in the formation of THMs. Included in Table II are the values of the bromine and iodine incorporation coefficients obtained in each study. A brief examination of these data indicates that even under the most extreme conditions the extent of iodine incorporation into THMs was slight. Thus, even Run 29, which included 8 mM of iodide and no bromide, generated a ηI value of about 0.6, or 20 mol % of the total halogens in the THM mixture. More striking is the fact that those systems containing between 4 and 800 μM of iodide never exceeded ηI values of 0.1, or about 3 mol % of TTHM halogens. The initial discussion will focus on the 4- to 800-μM halide systems.

Iodide was, on the whole, unable to compete successfully with bromide with respect to incorporation into THMs. In all systems that had equal initial concentrations of bromide and iodide, the ηBr was significantly larger than the ηI. Furthermore, the value of ηBr/ηI increased from about 14 at 80 μM of each halide to about 80 at 800 μM. It was not until the total concentrations of each halide reached about 4 mM that iodide began to compete effectively with bromide.

All systems containing 4 μM of initial iodide were devoid of iodinated THMs even when no competing bromide was present.

These general observations indicate strongly that iodine will, at usual environmental levels, not participate strongly in THM formation. There are two probable explanations for this fact. The first is that carbon-iodine bonds are far more easily broken than carbon-bromine or carbon-chlorine bonds. Thus, as explained earlier, aliphatic iodocompounds lose iodine readily. For example, solutions of iodoform in nonpolar solvents will rapidly turn violet because of liberation of elemental iodine from the CHI_3. A similar reaction has been reported[5] in the case of $CHCl_2I$ and is probably characteristic of all other iodinated THMs. Thus, it is probable that a complex of reactions exists which will result in the decomposition of iodinated THMs at a rate that will comprise a significant fraction of their formation rate, a factor of considerably lesser importance in the case of chloro- and bromo-THMs.

The second factor is the steric hindrance associated with the iodine atom, one of the bulkiest of common atoms. This steric effect will strongly retard the formation of all iodinated THMs and, in particular, those containing more than a single iodine atom. This conclusion is supported by the observation that an average of more than 90% of the iodine incorporated into THMs was found in three species, $CHCl_2I$, $CHBr_2I$, and $CHClBrI$. Less than 10% was found in the polyiodomethanes. Shown in Figure 4 is the relationship among the average mole percent of each iodo-THM obtained in the 18 low-halide studies and the molecular weight. The strong negative correlation of abundance to molecular size suggests that steric factors were highly significant in determining the level of formation of these compounds.

Effect on Halogen Incorporation

The comparative impacts of iodine and bromine on the degree of chlorine and bromine incorporation are shown in Figures 5 and 6. In these figures, the

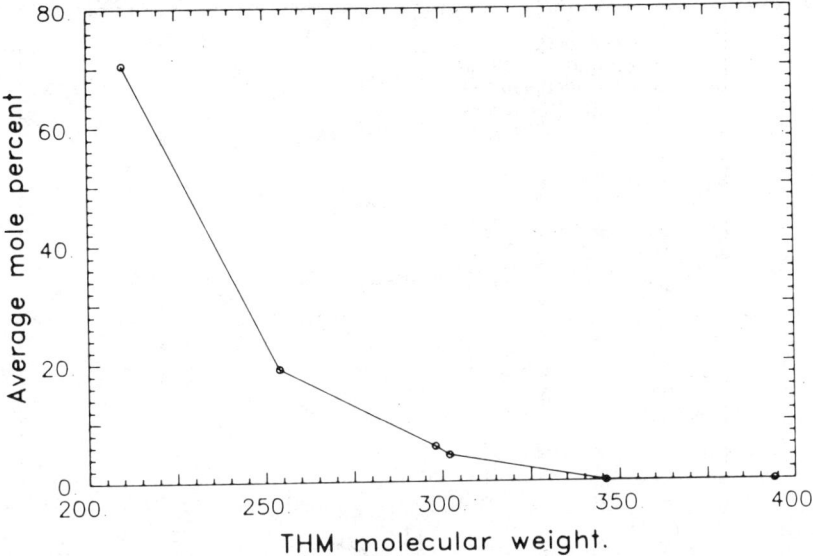

Figure 4. Mol percent of individual iodinated THMs as a function of the THM molecular weight.

chlorine incorporation terms are plotted as a function of initial iodide concentration at a fixed bromide level (Figure 5) and of initial bromide concentration with the iodide level held constant (Figure 6). Chlorine and bromine incorporation were not systematically related to iodide levels, whereas a clear trend of decreasing chlorine η values with increasing bromide concentration was evident. This probably reflects the significantly greater ability of bromide to compete with chlorine in THM formation than that of iodine. Thus, bromine was far more able to replace chlorine in these reactions than was iodine.

Formation of TTHMs

Of the three factors studied, only one, the bromide concentration, seemed to influence TTHM levels. Although the initial concentrations of chlorine and iodide had relatively little impact on the levels of TTHMs, there was a clear trend toward decreasing molar levels of TTHMs with increasing bromide concentrations (Figure 7). Only at the lowest initial bromide concentrations was there evidence that higher initial chlorine levels would yield substantially increased concentrations of TTHMs. The discrepancy decreased rapidly as the initial bromide level increased, becoming insignificant at 800 μM bromide. The impact of iodide conversely was negligible (Figure 8) and the response to initial chlorine levels erratic.

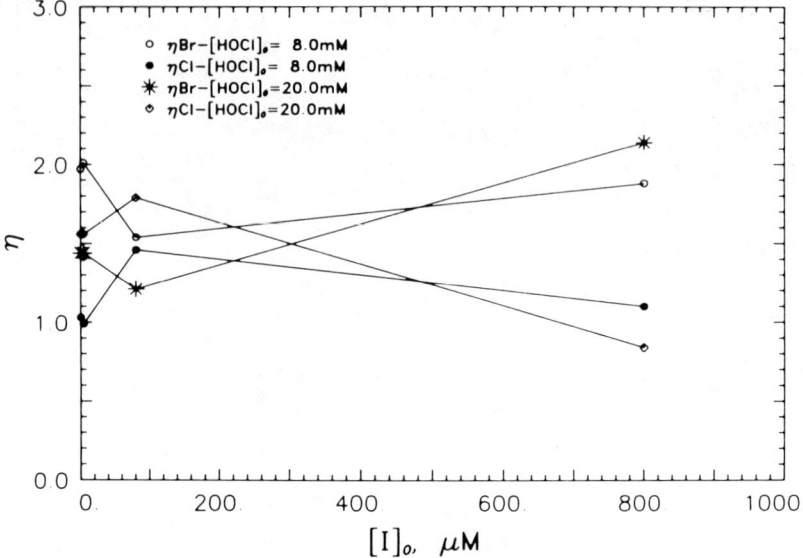

Figure 5. Bromine and chlorine incorporation coefficients as a function of the initial iodide concentration. $[Br]_0 = 800\ \mu M$.

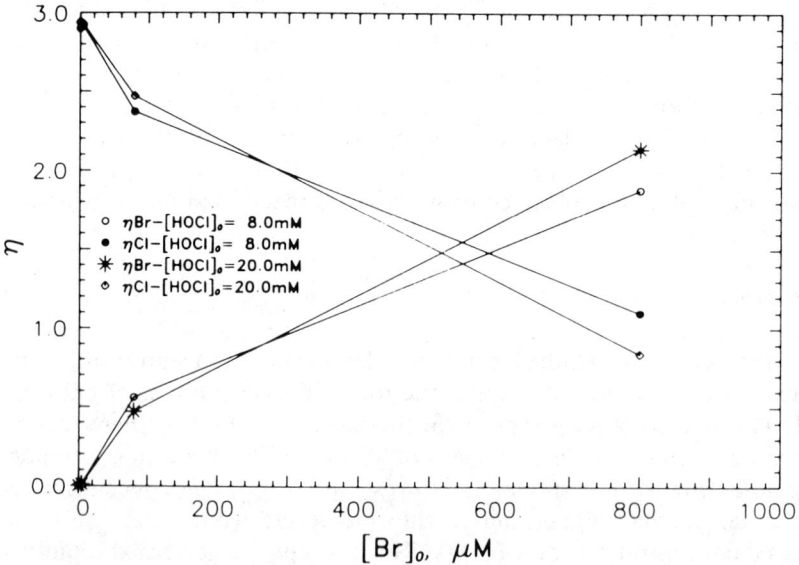

Figure 6. Bromine and chlorine incorporation coefficients as a function of the initial bromide concentration. $[I]_0 = 800\ \mu M$

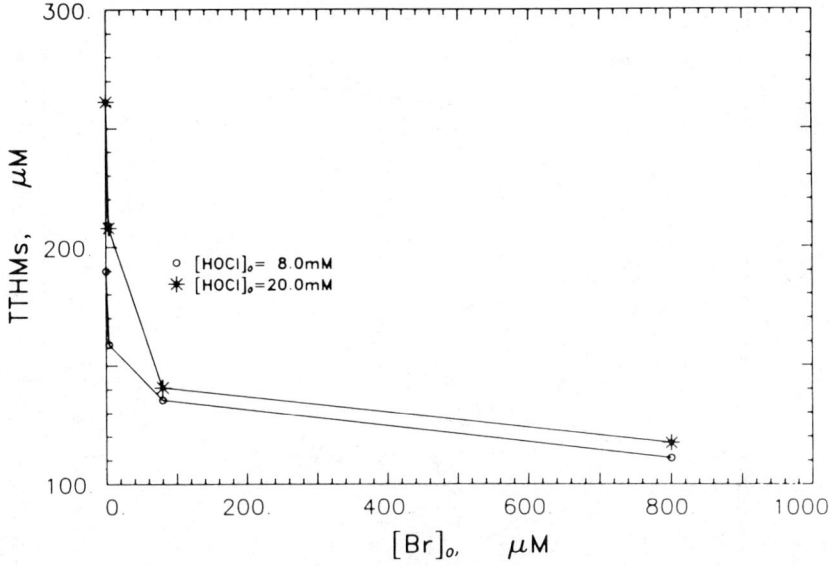

Figure 7. Total THM generation as a function of the initial concentration of bromide.

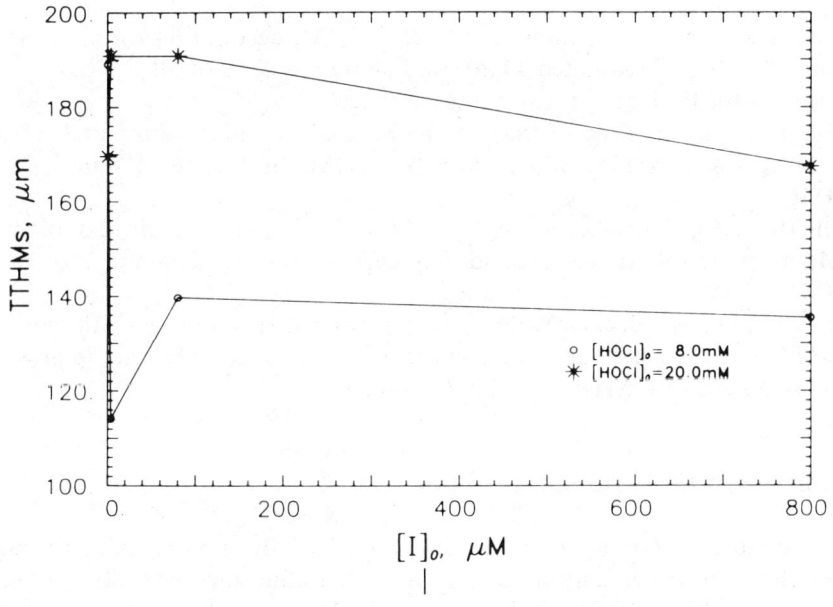

Figure 8. Total THM generation as a function of the initial concentration of iodide.

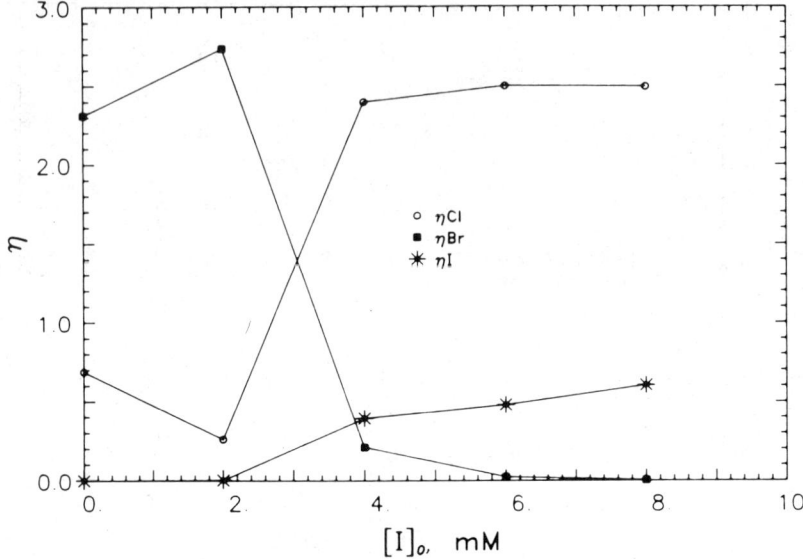

Figure 9. Halogen incorporation coefficients as a function of the initial iodide level in systems where $[Br]_0 + [I]_0 = 8$ mM.

Specific Iodinated THMS

$CHCl_2I$ was the most common iodinated THM, accounting for an average 70.5 mol % of total iodinated THMs and about 6 mol % of all THMs. This is consistent with findings of other researchers.[5]

CHClBrI was the second most common and abundant iodinated THM, accounting for 19.1 mol % of iodinated THMs and about 1.7 mol % of TTHMs.

$CHClI_2$ and $CHBr_2I$ accounted for 4.4 and 5.9 mol % each of iodinated THMs, respectively. In no case did they account for as much as 1 mol % of TTHMs.

$CHBrI_2$ and CHI_3 occurred so infrequently and at such low levels in the 4- to 800-μM halide runs as to be of no significance at all. Their presence in any but the most unusual of systems is highly unlikely.

High-Concentration Halide Systems

The systems containing 8-mM total halides initially showed more complex results than the lower concentration systems. Iodine incorporation in these systems was substantial at iodide levels about 2.0 mM, whereas the impact of bromide on chlorine incorporation is obvious (Figure 9). It is also notable that for the four samples of this group containing bromide, the average generation of TTHMs was 143 μM, whereas the bromide-free (iodide = 8 mM) sample produced 257 μM of TTHMs. This was consistent with the comparative impact of bromide and iodide on TTHM formation in the lower halide concentration mixtures. These high-bromide and high-iodide systems, which might be characteristic of certain subsurface brines such as oil field brines, were the only ones in which formation of iodinated THMs was an important process.

CONCLUSIONS

Although iodine will react to form iodinated THMs, it is by far the least active of the three significant halogens in this regard. Indeed, at levels similar to those to be expected in normal groundwaters and surface waters, iodide will neither lead to significant levels of iodinated THMs nor modify the overall THM reactions. In fact, only $CHCl_2I$ was produced with reasonable consistency in these systems, a finding in accord with the results of other studies. Only in unusual circumstances—oil field brines, iodine disinfection, high-iodide industrial wastes—might we expect iodine to be an important factor in THM formation. The very strong negative correlation between the molecular weight of the iodo-THMs and their average abundance in these systems suggests that the factor of greatest significance in controlling and retarding the formation of these compounds was the steric hindrance associated with the bulky iodine atom.

REFERENCES

1. Rook, J. J. "Chlorination of Fulvic Acids in Natural Waters," *Environ. Sci. Technol.* 11:478 (1977).
2. Cooper, W. J., L. M. Meyer, C. C. Bofill, and E. Cordal. "Quantitative Effects of Bromine on the Formation and Distribution of a Groundwater with a High Organic Content," in *Water Chlorination: Environmental Impact and Health Effects, Vol. 4,* R. L. Jolley, W. A. Brungs, J. A. Cotruvo, R. B. Cumming, J. S. Mattice and V. A. Jacobs, Eds. (Ann Arbor, MI: Ann Arbor Science Publishers, Inc., 1983), pp. 285-296.
3. Minear, R. A., and J. C. Bird. "Trihalomethanes: Impact of Bromide Ion Concentration on Yield, Species Distribution, Rate of Formation and Influence of other Variables." *Water Chlorination: Environmental Impact and Health Effects, Vol. 3,* R. L. Jolley, W. A. Brungs, and R. B. Cumming, Eds., (Ann Arbor, MI: Ann Arbor Science Publishers, Inc., 1980), pp. 151-160.
4. Bunn, W. W., B. B. Haas, E. R. Deane, and R. D. Kloepfer. "Formation of Trihalomethanes by Chlorination of Surface Water," *Environ. Lett.* 10:205 (1975).
5. Thomas, R. F., M. J. Weisner, and H. J. Brass. "The Fifth Trihalomethane: Dichloroiodomethane, Its Stability and Occurrences in Chlorinated Drinking Water," in *Water Chlorination: Environmental Impact and Health Effects, Vol. 3,*

R. L. Jolley, W. A. Brungs, and R. B. Cumming, Eds. (Ann Arbor, MI: Ann Arbor Science Publishers, Inc., 1980), pp. 161–168.
6. *Handbook of Chemistry and Physics*, 53rd ed. (Cleveland: Chemical Rubber Company, 1973).
7. Feder, G. L. "Geochemical Survey of Missouri: Geochemical Survey of Waters of Missouri," Professional Paper 954-E (Washington, DC: U.S. Geological Survey, 1979).
8. Collins, A. G., W. P. Zelinski, and C. A. Pearson. "Bromide and Iodide in Oilfield Brines in Some Tertiary and Cretaceous Formations in Mississippi and Alabama," U.S. Bureau of Mines Report of Investigations No. 6959 (Washington, DC: U.S. Dept. of Interior, 1967).
9. Arndt, R. H., and C. J. Mankin. "Minerals in the Economy of Oklahoma," State Mineral Profiles of the U.S. Bureau of Mines (Washington, DC: U.S. Dept. of Interior, 1979).
10. Gould, J. P., L. E. Fitchhorn, and E. Urheim. "Formation of Brominated Trihalomethanes: Extent and Kinetics," in *Water Chlorination: Environmental Impact and Health Effects, Vol. 4*, R. L. Jolley, W. A. Brungs, R. B. Cumming, J. S. Mattice, and W. A. Jacobs, Eds. (Ann Arbor, MI: Ann Arbor Science Publishers, Inc., 1983), pp. 297–310.
11. Gould, J. P., and L. E. Fitchhorn. "Kinetics of Formation of TTHMs from a Low Molecular Weight Precursor," (submitted for publication).
12. *Dictionary of Organic Compounds*. 5th ed. (New York: Chapman and Hall, 1982).
13. Grob, K. "Persilylation of Glass Capillary Columns," *J. High Resolut. Chromatogr. Chromatogr. Commun.* 3(10):493–496 (1980).
14. Giabbai, M., M. Shoults, and W. Bertsch. "Static Coating of Glass Capillary Columns, Some Practical Observations," *J. High Resolut. Chromatogr. Chromatogr. Commun.* 1:277(1978).
15. Giabbai, M., L. Roland, and E. S. K. Chian. In *Chromatography in Biochemistry, Medicine and Environmental Research*, A. Frigerio, Ed., (Amsterdam: Elsevier, 1983), p. 41.
16. *Registry of Mass Spectral Data*, Vols. 1–4, E. Stenhagen, S. Abrahamsson, and F. McLafferty, Eds., (New York: John Wiley & Sons, 1974).
17. *Eight Peak Index of Mass Spectra*, Vols. I and II, (Aldermaston, Reading, UK: Mass Spectrometry Data Centre, AWRE, 1974).

SECTION XI

Chemistry of Chloramination

Virtually all of the active chlorine discharged in nonnitrified effluents from wastewater treatment plants is in the form of various chloramines.

Russell A. Isaac and **J. Carrell Morris**
Rate of Transfer of Active Chlorine Between Nitrogenous Substrates, 1979

Of all the possible halamines, only chloramines and bromamines have thus far been recognized as resulting from water chlorination. . . . The purpose of this work is to demonstrate unequivocally the existence of N-bromo-N-chloramines and to characterize their physical properties so that they may be identified in studies of chlorination effects . . .

Werner R. Haag
Formation of N-Bromo-N-Chloramines in Chlorinated Seawater, 1980

CHAPTER 73

Characterization of the Reaction Between Monochloramine and Isolated Aquatic Fulvic Acid

James N. Jensen, Jessica J. St. Aubin, Russell F. Christman, and J. Donald Johnson

Humic and fulvic substances in natural waters were first suggested as trihalomethane precursors by Rook[1] and Bellar et al.[2] After these discoveries, research in the field assumed one of two directions. First, some investigators began looking at the nonvolatile organic chlorine fraction of chlorinated natural organics. Second, other researchers attempted to find ways of reducing trihalomethane (THM) concentrations in treated waters. Approaches have included reducing precursor concentrations, removing THMs after formation, and changing disinfectants. With regard to this last approach, Stevens and his colleagues[3] showed that the addition of ammonia drastically reduced THM production by the chlorination process. This observation regenerated interest in monochloramine as an alternative to chlorine in water treatment.

This work combines the previously mentioned two research directions by examining the interactions between monochloramine and naturally occurring organics, and by conducting characterization studies of the chloramine-produced, nonpurgeable total organic halide (NPTOX).

TRIHALOMETHANE AND TOTAL ORGANIC HALIDE PRECURSORS

Several classes of organic compounds in natural waters have been shown to be THM and total organic halide (TOX) precursors. Among these are fulvic and humic acids and their degradation products,[4,5] plant pigments,[6,7] algal biomass,[7,8] amino acids and pyrimidines,[6] and industrial effluents such as phenol.[9,10] Of these classes, fulvic acid is generally found in the highest concentrations, accounting for 45% of the dissolved organic carbon in surface waters.[11] Thus, fulvic acid is an attractive substrate for modeling disinfectant-precursor interactions in natural waters.

COMPARISON OF CHLORINE AND CHLORAMINE REACTIVITY

Chlorine-organic and chloramine-organic interactions may be compared in two ways. First, one may choose to compare overall parameters such as THM or TOX formation potential. Chloramines are known to produce only very small amounts of THMs ($\leq 3\%$ of the chloroform produced by chlorine at oxidant levels of ≤ 20 mg/L).[12] However, chloramines produce significant quantities of NPTOX relative to chlorine. Table I demonstrates that chloramines produce from 9 to 49% as much NPTOX as chlorine under the same reaction conditions.[12,13] A second way in which chlorine and chloramines can be compared is by examining reactions with model compounds. Three examples are shown in Table II.[14-16] In all three cases, the power of chlorine as an oxidizing agent may be seen. Chloramines, on the other hand, appear much more likely to form substitution products (resulting in TOX formation) with these model compounds.

From this discussion one can conclude that (1) fulvic acid apparently is a reasonable model of organics in natural waters, and (2) chloramines have the capability of forming significant quantities of NPTOX through interactions with fulvic and humic materials.

EXPERIMENTAL METHODS

Fulvic acid was extracted from Bay Tree Lake (formerly Black Lake), a surface water in southeastern North Carolina. The extraction procedure has been previously described.[17] Monochloramine was prepared by mixing equal volumes of buffered sodium hypochlorite and ammonium chloride (both Fisher Scientific Company, Pittsburgh, PA), with the latter in a threefold molar excess. Reaction conditions were as follows: pH 9 (held with chlorine demand-free borate buffer), molar chlorine-to-carbon ratio (Cl:C) = 10, organic carbon equal to 21 mg/L, and reaction time equal to 24 h. The high pH value was necessary to ensure stable and pure monochloramine solutions.

Three types of experiments were performed. First, the demand exerted by the fulvic material on monochloramine was determined by measuring the amount of oxidant consumed (oxidant-loss studies). Second, attempts were made to identify products of the monochloramine-fulvic acid reaction. The analytical scheme for the product identification studies is very similar to that used by other investigators to study the reactions between fulvic acid and chlorine,[18] chlorine dioxide,[19] and ozone.[20] Third, monochloramine-produced TOX and its characteristics were investigated. Demand measurements were corrected for blanks that consisted of buffered monochloramine solutions. Buffered fulvic acid blanks were also run in parallel with the demand, product, and TOX experiments.

Table I. Comparison of Quantities of Chlorine- and Chloramine-Produced NPTOX[a]

Substrate	Conditions	Chloramine-Produced NPTOX		Reference
		μg Cl$^-$/mg OC[b]	Cl$_2$ NPTOX (%)	
Peat fulvic acid	pH 7, TOC[c] = 1–5 mg/L; T[d] = 100 h; Cl$_2$ dose = 20 mg/L	27–31	12–13	12
Humic acid	pH 7, TOC = 3 mg/L; T = 100 h; Cl$_2$ dose = 20 mg/L	12–52	20	12
Peat fulvic			13	
Groundwater			9	
Groundwater fulvic			17	
Secondary effluent			37	
Rhein River humic	pH 6.9, TOC = 0.8–8 mg/L; T = 0.5 h; Cl$_2$ dose = 15 mg/L	15–58	33–49	13
	pH 9.2	7–30	28	

[a] Nonpurgeable total organic halide.
[b] Organic carbon.
[c] Total organic carbon.
[d] Reaction time.

Table II. Comparison of the Reactions of Chlorine and Chloramines with Model Compounds

Phenol[14]
  Chlorine—oxidation and substitution; all oxidation end products after 3 h
  Chloramine—primarily substitution; phenol + NH_2Cl → chlorophenols
Amino acids
  Chlorine[15]—oxidation; alanine + HOCl → pyruvic acid
  Chloramine[16]—substitution; alanine + NH_2Cl → N-chloroalanine
Protein[15]
  Chlorine—oxidation
    Glycylglycylglycine + HOCl → hydrolysis and deamination
  Chloramine—substitution
    Glycylglycylglycine + NH_2Cl → terminal N-chloro derivative

For TOX extraction experiments, 100 mL of aqueous solution was batch extracted with 10 + 5 + 5 mL of diethyl ether. The sample and system control were purged for 1 h with helium. Total organic halide column chromatography experiments were conducted with Amberlite XAD-8 resin (Rohm and Haas Chemical Company, Philadelphia). Chromatographic conditions were similar to those of Liao et al.[21] Columns were eluted with 0.01 N NaOH. No attempt was made to trap or measure volatile products in any of this work.

Residual chlorine or monochloramine measurements were made by amperometric titration[22] using a recording polarograph (Sargent-Welch Scientific Company, Skokie, IL). Gas chromatography was accomplished with a DB-1 15-m capillary column (J & W Scientific, Inc., Rancho Cordova, CA) and a Varian Model 3700 gas chromatograph (Varian Associates, Inc., Palo Alto, CA) using conditions described previously.[23] Total organic carbon (TOC) values were determined from an elemental analysis of the solid fulvic acid.[21] The TOC values derived by this method were within 18% of those measured by a TOC analyzer (Model 915B, Beckman Instruments, Fullerton, CA). Total organic halide was determined by the granular activated carbon (GAC) (Filtrasorb 400, Calgon Corp., Pittsburgh, PA) procedure[24] using a commercially available adsorption/microcoulometry system (Dohrmann AD-2 Adsorption Module and MCTS-20 Microcoulometry System, Dohrmann/Xertex, Santa Clara, CA).

RESULTS AND DISCUSSION

Oxidant Loss

Oxidant loss data are shown in Tables III and IV. Preliminary work with the fulvic acid model compound resorcinol (1,3-dihydroxybenzene) showed that the ultimate demand exerted by this compound for chlorine and mono-

Table III. Demand Exerted by Resorcinol

Oxidant	Demand[a]	Conditions	Reference
Chlorine	1.10 (40 min)	pH 7 Cl:C = 1.93	9
Monochloramine	0.57 ± 0.18 (1 h) 1.10 ± 0.15 (24 h)	pH 9 Cl:C = 2	This work

[a]In units of mol oxidant per mol carbon.

Talbe IV. Demand Exerted by Black Lake Fulvic Acid

Oxidant	Demand[a]	Conditions	Reference
Monochloramine	0.06 ± 0.17[b] (24 h)	pH 9 Cl:C = 2	This work
Chlorine	0.27 ± 0.17 (24 h)	pH 8.9 Cl:C = 2	This work
Chlorine dioxide[c]	0.3	pH 7.8 Cl:C = 1.65	19
Ozone[d]	0.6	pH 7 O_3:C = 1	20

[a]In units of mol oxidant per mol carbon.
[b]Plus or minus one standard deviation.
[c]Colclough et al.[19] define demand as the quantity of ClO_2 needed to produce a finite residual after 24 h.
[d]Data are uncorrected for blank values. Anderson[20] labels the data "consumption" rather than "demand."

chloramine is the same (see Table III). This demand value (1.10 mol as chlorine dioxide per mol carbon, or 6.60 mol as chlorine per mol resorcinol) is similar to the predictions of Boyce and Hornig.[10] They suggest that 5 to 6 mol of chlorine would be consumed per mol resorcinol.

Table IV compares the demand exerted by Bay Tree Lake fulvic acid for several common disinfectant chemicals. Chlorine, chlorine dioxide, and ozone are all consumed to a much greater extent by fulvic acid than is monochloramine. Note also that monochloramine is not consumed to a statistically significant degree. Fleischacker and Randtke[12] provide data (not corrected for the autodecomposition of monochloramine) which suggest that the demand exerted by groundwater fulvic acid on monochloramine is very large. They state, and our study verifies, that monochloramine decomposition is significant at the higher monochloramine concentrations.

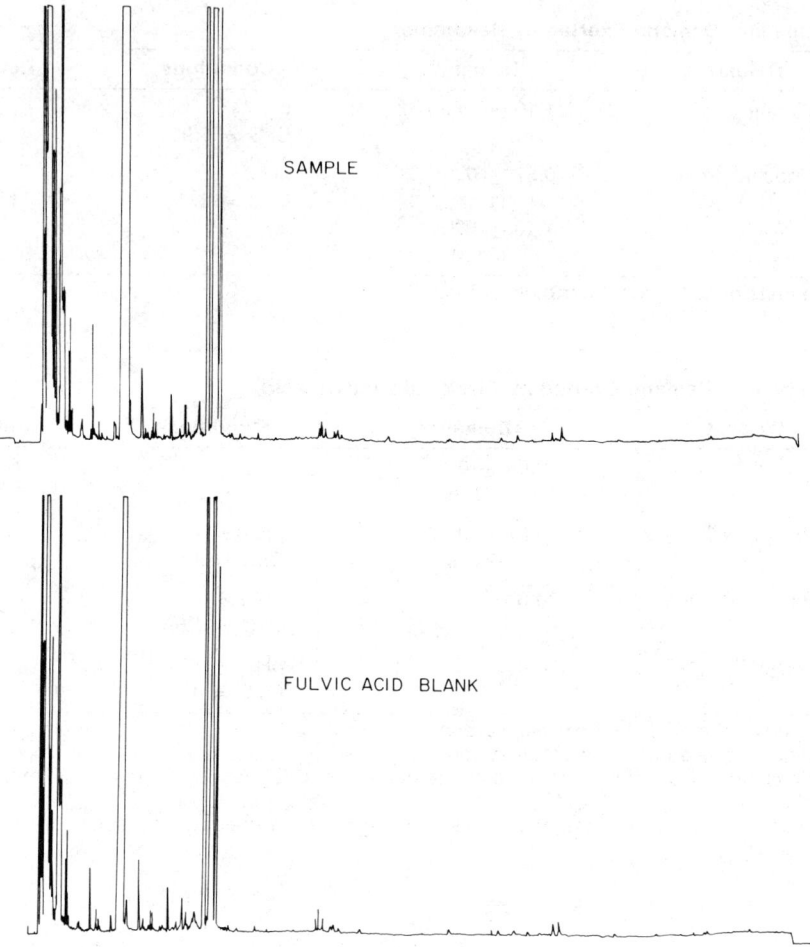

Figure 1. Sample and fulvic acid-containing blank gas chromatograms.

Product Identification

Typical chromatograms from the product identification experiments are shown in Figure 1. The large numbers of peaks are a result of the concentration of chromatographable fulvic acid components in the ether extracts. Differences between the sample (top of Figure 1) and the fulvic acid-containing blank (bottom of Figure 1) are minimal. This suggests that no compounds, ether extractable from an acidified aqueous solution and detectable by gas chromatography/flame ionization detection (GC/FID), are produced by the reaction between monochloramine and fulvic acid at pH 9. The significance of this conclusion is emphasized by comparing this result with other oxidant-fulvic acid systems. Using techniques similar to those described here (but using mass spectrometric detection), Johnson and co-workers[18] were able to identify over 100 compounds from the chlorination of fulvic acid; Colclough et al.[19] identified over 60 compounds from the chlorine dioxide-fulvic acid reaction. Compared with chlorine, monochloramine is relatively unreactive toward fulvic acid.

Total Organic Halide Characterization

The following characteristics of monochloramine-produced TOX were investigated: quantity, adsorbability, ether extractability, and XAD-8 chromatographic distribution. Total organic halide produced by monochloramine and chlorine is quantified in Table V.[25] Note that the yield of monochloramine-produced TOX in Table V is in the lower range of the literature values presented in Table I. This may be because of the pH 9 value used in this study.

In terms of adsorption onto GAC, monochloramine-produced TOX breaks through to a greater extent than does chlorine-produced TOX. Of the TOX produced by monochloramine (pH 9, Cl:C = 10), 15% appears on the second of two serial GAC columns. With chlorine (pH 9, Cl:C = 0.5 or 5), only about 5% breaks through to the second column. Since experimental conditions allow for the adsorption of over 99% of the organic carbon (calculated from the work of Snoeyink and co-workers),[26] competition between chlorinated and unchlorinated organic carbon for GAC sites is probably not important. The remaining hypothesis is that monochloramine-produced TOX is more hydrophilic than chlorine-produced TOX.

Table V. Comparison of Monochloramine- and Chlorine-Produced Total Organic Halide (TOX)

Property	Monochloramine	Chlorine
Quantity of TOX in Cl$^-$/mg OC (μg)	17 (Cl:C = 10)	124 (Cl:C = 0.5)
		205 (Cl:C = 4.9)
Oxidant consumed, %	5 (Cl:C = 10)	10 (Cl:C = 0.5)
		5 (Cl:C = 4.9)
GAC adsorption of total TOX in 1st of 2 GAC columns, %	85 (Cl:C = 10)	95 (Cl:C = 0.5)
		96 (Cl:C = 4.9)
Extraction		
TOX extractable, %	20 ± 20 (Cl:C = 10)	>53[a]
XAD-8 chromatography		
TOX in 1st peak, %	85 (Cl:C = 8.5)	0 (Cl:C = 1)
TOX in 2nd peak, %	15	100

[a]The work of Christman et al.[25] at Cl:C = 4 shows that 53% of the TOX (43% of nonvolatile TOX) is chromatographable; therefore, at least 53% is extractable.

Batch extraction of the chloraminated solution shows that only 20% (with a standard deviation also of 20%, so the confidence interval includes zero) of the TOX is extractable with ether. Degradation with chlorine[25] shows that at least 43% of the chlorine-produced NPTOX is extractable under similar extraction conditions. That only a small quantity of the monochloramine-produced TOX is extractable is consistent with the observation that apparently no halogenated products are gas chromatographable under the conditions used in this study. Also, the extraction results are consistent with the hypothesis that monochloramine-produced TOX is more hydrophilic than chorine-produced TOX.

Earlier work[21] showed that undegraded fulvic acid yields two peaks within the exclusion and inclusion limits of the column when chromatographed on XAD-8 resin at pH 11. Liao et al.[21] suggested that the first peak represents a larger molecular size or lower polarity fraction, and the second peak represents a smaller molecular size or higher polarity fraction. Our XAD-8 chromatography results are presented in Table V. Note that most of the monochloramine-produced TOX and none of the chlorine-produced TOX is present in the first peak. This implies that monochloramine-produced TOX is larger in molecular size or less polar (i.e., less hydrophilic) than chlorine-produced TOX. Taken with the other TOX characterization experiments, these data are more consistent with the hypothesis that monochloramine-produced TOX is more hydrophilic and larger in molecular size than chlorine-produced TOX.

CONCLUSIONS

Relative to chlorine, monochloramine is less reactive to aquatic fulvic acid based on demand values, product identification results, and TOX production. The monochloramine-fulvic acid reaction generates no products that are ether extractable from an acidified solution and detectable by GC/FID techniques. Total organic halide studies show monochloramine produces TOX that is more hydrophilic and larger than chlorine-produced TOX.

APPLICATIONS TO WATER TREATMENT

The results from this work have direct application to the treatment of drinking water. First, the product identification experiments suggest that production of small chlorinated acids by monochloramine is minimal. These chloroacids (particularly trichloroacetic acid and dichloroacetic acid) have been identified in large concentrations in chlorinated humic and fulvic preparations[18,25,27-29] and in treated drinking water.[30] We note that the amount of trichloroacetic acid produced at pH 9 by chlorine reacting with fulvic acid is only about one-half the maximum amount (produced at pH 5).[29]

Second, the finding of increased hydrophilicity of monochloramine-produced TOX relative to chlorine-produced TOX suggests that the monochloramine-produced TOX may be more difficult to remove on GAC or other hydrophobic materials.

ACKNOWLEDGMENTS

We express our thanks to Theodore Walters for preparation of the fulvic acid samples. We would also like to thank Linda Anderson (currently with the U.S. Environmental Protection Agency, Atlanta), David Millington (currently at Duke University), Daniel Norwood, and Gavin Thompson (both at the University of North Carolina) for their helpful comments throughout the course of this work. This research was supported by Cooperative Agreement CR 810532-01 to the University of North Carolina (A. A. Stevens, Project Officer).

REFERENCES

1. Rook, J. J. "Formation of Haloforms During Chlorination of Natural Waters," *Water Treat. Exam.* 23(2):234-243 (1974).
2. Bellar, T. A., J. J. Lichtenberg, and R. C. Kroner. "The Occurrence of Organohalides in Finished Drinking Water," *J. Am. Water Works Assoc.* 66(12):703-706 (1974).
3. Stevens, A. A., C. J. Slocum, D. R. Seeger, and G. G. Robeck. "Chlorination of Organics in Drinking Water," *J. Am. Water Works Assoc.* 63(11):615-620 (1976).
4. Rook, J. J. "Chlorination of Reactions of Fulvic Acids in Natural Waters," *Environ. Sci. Technol.* 11(5):478-482 (1977).
5. Christman, R. F., J. D. Johnson, J. R. Hass, F. K. Pfaender, W. T. Liao, D. L. Norwood, and H. J. Alexander. "Natural and Model Aquatic Humics: Reactions with Chlorine," in *Water Chlorination: Environmental Impact and Health Effects, Vol. 2*, R. L. Jolley, H. Gorchev, and D. H. Hamilton, Jr., Eds. (Ann Arbor, MI: Ann Arbor Science Publishers, Inc., 1978), pp. 15-28.
6. Morris, J. C., and B. Baum."Precursors and Mechanisms of Haloform Formation in Chlorination of Water Supplies," in *Water Chlorination: Environmental Impact and Health Effects, Vol. 2*, R. L. Jolley, H. Gorchev, and D. H. Hamilton, Jr., Eds. (Ann Arbor, MI: Ann Arbor Science Publishers, Inc., 1978), pp. 29-48.
7. Wachter, J. K. "Characterization of Organohalide Formation upon Chlorination of Algal Extracellular Matter," ScD. Dissertation, (Pittsburgh, PA: University of Pittsburgh, 1982).
8. Hoehn, R. C., D. B. Barnes, B. C. Thompson, C. W. Randall, T. J. Grizzard, and P. T. B. Shaffer. "Algae as Sources of Trihalomethane Precursors," *J. Am. Water Works Assoc.* 72(6):334-350 (1980).
9. Norwood, D. L., J. D. Johnson, R. F. Christman, J. R. Hass, and M. J. Bobenrieth. "Reactions of Chlorine with Selected Aromatic Model of Aquatic Humic Material," *Environ. Sci. Technol.* 14(2):187-190 (1980).

10. Boyce, S. D., and J. F. Hornig. "Reaction Pathways of Trihalomethane Formation from the Halogenation of Dihydroxyaromatic Model Compounds for Humic Acid," *Environ. Sci. Technol.* 17(4):202–211 (1983).
11. Malcolm, R. L. "Geochemistry of Humic Substances in the Dissolved and Sediment Phases of Streams and Rivers," presented at the First International Meeting of the International Humic Substances Society, Estes Park, CO, August 1983.
12. Fleischacker, S. J., and S. J. Randtke. "Formation of Organic Chlorine in Public Water Supplies," *J. Am. Water Works Assoc.* 75(3):132–138 (1983).
13. Sander, R. "Untersuchungen zum Optimalen Einsatz von Chlor bei der Aufbereitung von Oberflaschenwassern," Dr.-Ing. Dissertation, (Karlsruhe, FRG: Universitat Fridericiana Karlsruhe, 1981).
14. Burttschell, R. H., A. A. Rosen, F. M. Middleton, and M. B. Ettinger. "Chlorine Derivatives of Phenol Causing Taste and Odor," *J. Am. Water Works Assoc.* 51(2):205–214 (1959).
15. Ingols, R. S., H. A. Wyckoff, T. W. Kethley, H. W. Hodgden, J. C. Fincher, J. C. Hildebrand, and J. E. Mandel. "Bactericidal Studies of Chlorine," *Ind. Eng. Chem.* 45(5)966–1000 (1953).
16. Isaac, R. A., and J. C. Morris. "Transfer of Active Chlorine from Chloramine to Nitrogenous Organic Compounds. 1. Kinetics," *Environ. Sci. Technol.* 17(12):738–742 (1983).
17. Christman, R. F., W. T. Liao, D. S. Millington, and J. D. Johnson. "Oxidative Degradation of Aquatic Humic Material," in *Advances in the Identification and Analysis of Organic Pollutants in Water, Vol. 2*, L. H. Keith, Ed. (Ann Arbor, MI: Ann Arbor Science Publishers, Inc., 1981), pp. 979–1000.
18. Johnson, J. D., R. F. Christman, D. L. Norwood, and D. S. Millington. "Reaction Products of Aquatic Fulvic Acid Chlorination," *Environ. Health Perspect.* 46:63–71 (1982).
19. Colclough, C. A., J. D. Johnson, R. F. Christman, and D. S. Millington. "Organic Reaction Products of Chlorine Dioxide and Natural Aquatic Fulvic Acids," in *Water Chlorination: Environmental Impact and Health Effects, Vol. 4*, R. L. Jolley, W. A. Brungs, J. A. Cotruvo, R. B. Cumming, J. S. Mattice, and V. A. Jacobs, Eds. (Ann Arbor, MI: Ann Arbor Science Publishers, Inc., 1983), pp. 219–229.
20. Anderson, L. J. "The Reaction of Ozone with Isolated Fulvic Acid," MSPH Technical Report, (Chapel Hill, NC: University of North Carolina, 1983).
21. Liao, W., R. F. Christman, J. D. Johnson, D. S. Millington, and J. R. Hass. "Structural Characterization of Aquatic Humic Material," *Environ. Sci. Technol.* 16:403–410 (1982).
22. *Standard Methods for the Examination of Water and Wastewater*, 15th ed. (Washington, DC: American Public Health Association, 1980).
23. Jensen, J. N. "Characterization of the Reaction Between Monochloramine and Isolated Aquatic Fulvic Acid," MSPH Technical Report, (Chapel Hill: University of North Carolina, 1983).
24. *Total Organic Halide: Method 450.1*, Interim. (Cincinnati, OH: U.S. Environmental Protection Agency, 1980).
25. Christman, R. F., D. L. Norwood, D. S. Millington, J. D. Johnson, and A. A. Stevens. "Identity and Yields of Major Halogenated Products of Aquatic Fulvic Acid Chlorination," *Environ. Sci. Technol.* 17(10):625–628 (1983).

26. Snoeyink, V. L., J. J. McCreary, and C. J. Murin. *Activated Carbon Adsorption of Trace Compounds*, EPA 600/2-77-223 (Cincinnati, OH: U.S. Environmental Protection Agency, 1977).
27. Miller, J. W., and P. C. Uden. "Characteristics of Nonvolatile Aqueous Chlorination Products of Humic Substances," *Environ. Sci. Technol.* 17(3):150–157 (1983).
28. Coleman, W. E., J. W. Munch, W. H. Kaylor, H. P. Ringhand, and J. R. Meier. "GC/MS Analysis of Mutagenic Extracts of Aqueous Chlorinated Humic Acids — A Comparison of the By-Products to Drinking Water Contaminants," Extended Abstracts, American Chemical Society, Division of Environmental Chemistry, 23(1):53–55 (1983).
29. Reckhow, D. A., and P. C. Singer. "Removal of Organic Halide Precursors by Pre-Ozonation and Alum Coagulation," *J. Am. Water Works Assoc.* 76(4):151–157 (1984).
30. Uden, P. C., and J. W. Miller. "Chlorinated Acids and Chloral in Drinking Water," *J. Am. Water Works Assoc.* 75(10):524–527 (1983).

CHAPTER 74

Significant Findings Related to Formation of Chlorinated Organics in the Presence of Chloramines

Richard Arber, Mark A. Speed, and Frank Scully

The primary objective of the drinking water industry is to produce safe, potable water for its consumers at the lowest possible cost. As the definition of acceptable quality changes via federal regulations, many communities are experiencing difficulty in meeting this objective. Of particular concern for some communities is the ability to meet the trihalomethane (THM) standard. For those communities faced with poor raw water quality, the alternatives for treatment may be very costly.

Previous research indicates that the use of combined chlorine as an alternative to free chlorine disinfection is one way that a utility can meet the THM standard. Unfortunately, there are other quality considerations that must also be addressed when evaluating treatment alternatives. Currently, the U.S. Environmental Protection Agency (EPA) is concerned about the practice of chloramination as a means of disinfection because of its lower bactericidal control characteristics, as well as the possible formation of unidentified byproducts. Recent research performed at Thornton, Colorado, has indicated that these concerns are warranted and that the quality characteristics of a water disinfected with chloramines varies considerably with the method of formation of combined chlorine.

BACKGROUND

The City of Thornton, Colorado, is located just north and downstream of Denver along the South Platte River. The city-owned water system provides drinking water to approximately 75,000 people. Thornton's primary raw water source (shown in Figure 1) is a series of small gravel-mined lakes adjacent to the South Platte River.

Because of the location of Thornton's raw water supply relative to the urbanized area, the quality of raw water available to Thornton suffers from contamination resulting from urban runoff and point wastewater discharges. In 1978, nitrite (NO_2) concentrations approaching 2 mg/L as nitrogen were discovered within the distribution system. Because of the acute toxicity of

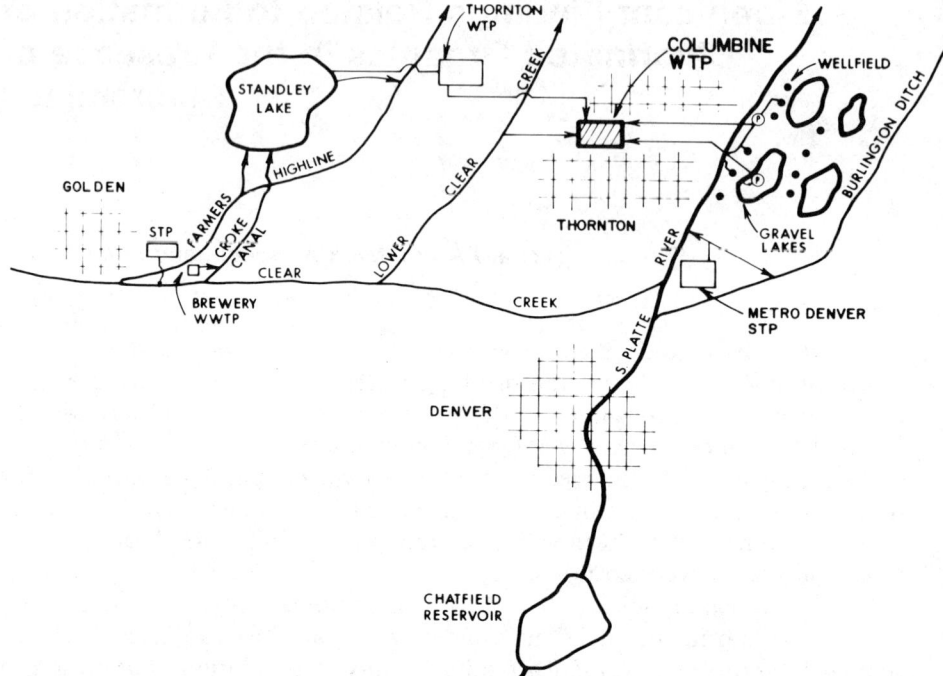

Figure 1. Water system for Thornton, Colorado.

nitrite, emergency measures were taken to solve the severe water quality problem.

Analytical testing of the raw water showed that sources from the South Platte River contained concentrations of ammonia as high as 18 mg/L. As seen in Figure 1, the Metropolitan Denver Sewage Disposal District No. 1 secondary treatment plant discharges ~170 mgd (7.45 m³/s) to the river less than 2 miles (12,180 m) upstream from the alluvial lakes. The influence of this discharge on Thornton's gravel lakes resulted in ammonia concentrations of 4 mg/L as nitrogen, with an average concentration of ~1.5 mg/L as nitrogen.

Breakpoint chlorination was selected as the most practical means for removing the ammonia from raw water at the Columbine Plant because of its relative simplicity. Although no data were available at the time, it was speculated that relatively high concentrations of organic material were present in the raw water. Because of concerns for the formation of THMs and other chlorinated organics during the breakpoint chlorination process, it was concluded that modifications to the treatment process should be considered. A process scheme consisting of conventional water treatment followed by granular activated carbon (GAC) was proposed. However, prior to implementation of a full-scale process modification, pilot testing was recommended, along with implementation of a comprehensive analytical program to fully identify the raw water characteristics. The Thornton Pilot Program formally began in October 1981, jointly funded by the EPA and the City of Thornton.[1]

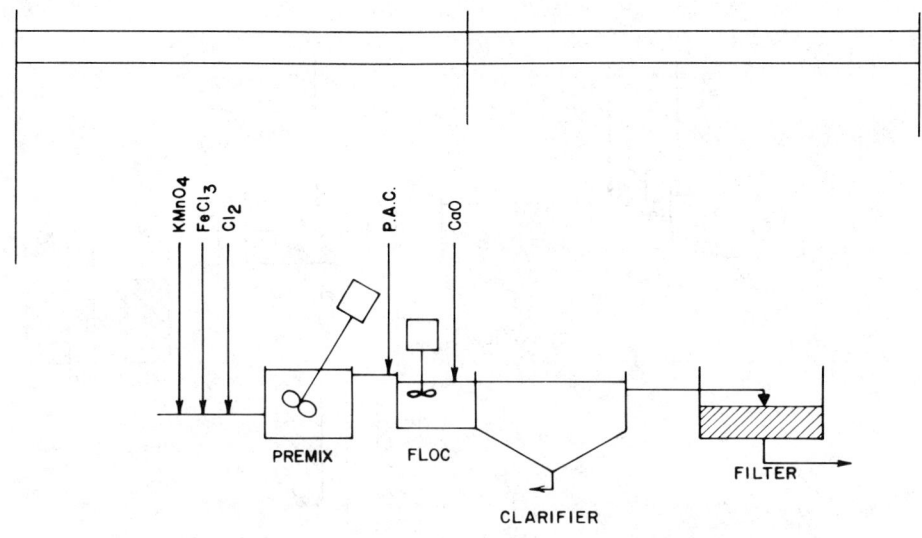

Figure 2. Columbine Water Treatment Plant.

PILOT PLANT PROGRAM

The overall goal of the Thornton Pilot Plant Program was to identify the most appropriate treatment process for meeting both current and future water quality standards. A comprehensive water quality monitoring program was established to characterize the quality of raw water from various sources available to Thornton, as well as the performance of individual processes. As part of this program, a thorough quality assurance (QA) program was established. This QA program ensured the integrity of the data collected during the program.

The process flow diagram for the 20-mgd (8.76-m^3/s) Columbine Water Treatment Plant, which is the city's primary treatment facility, is shown in Figure 2. The process consists of adding potassium permanganate (0.4 to 0.6 mg/L) for iron and manganese removal, ferric chloride (5 to 20 mg/L) as the primary coagulant, and chlorine for predisinfection. Following the addition of chemicals, the water enters a basin for ~30-min detention time, and then is vigorously mixed to promote the chemical reactions. Powdered activated carbon is added to the premix basin effluent at dosages of 2 to 4 mg/L for control of taste and odors.

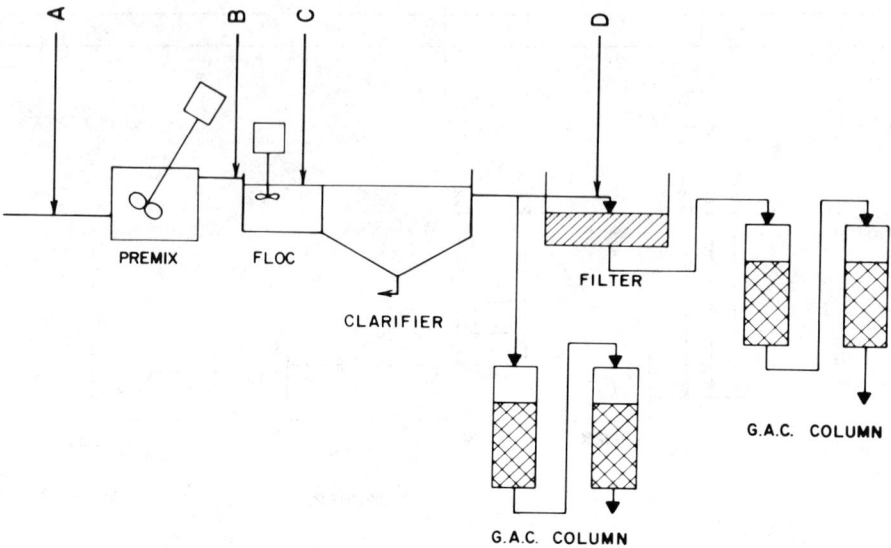

Figure 3. Thornton pilot plant.

Flocculation and sedimentation occur in center-feed circular basins; total detention time is ~135 min. Lime may be added to the clarifier at doses up to 20 mg/L for pH control. Sludge is withdrawn from the clarifiers and sent to holding basins for dewatering and disposal. The decant water is returned to the head of the plant for reuse.

The settled water is filtered with dual-media filters loaded at ~5 gal/min (0.0034 m/s). Chlorine is added to the filter influent for postdisinfection. High-pressure finished water pumps deliver water to the distribution system, which has a maximum travel time of about 5 d.

Figure 3 and Table I show the pilot plant process configuration and chemical addition points for the various modes of operation. The pilot plant consists of

Table I. Pilot Plant Chemical Addition Points for Respective Modes of Operation (see Figure 3)

Mode	$KMnO_4$	$FeCl_3$	Cl_2	PAC[a]	CaO	NH_2Cl	$(NH_4)_2SO_4$	ClO_2
1	A	A	A	B	C			
2	A	A	D	B	C			
3	A	A	D		C			
4	A	A	A	B	C		B	
5	A	A		B	C	A		
6	A	A		B	C	A		A

[a]Powdered activated carbon.

essentially the same processes as contained in the full-scale plant, with the addition of GAC for removal of organics. The flow rate for the pilot plant was established at 6 gal/min (3.79 × 10^{-4} m^3/s), a condition that most closely matched the full-scale facility detention time and resulted in similar process performance. The carbon columns used for the pilot program consist of four 6-in. (0.15-m)-diam by 8-ft (2.44-m)-tall downflow glass columns operated in series. Each column has 5.5 ft (1.68 m) of coal-base GAC as media, with six sample taps for monitoring organic breakthrough. The carbon was selected from bench-scale laboratory tests on several different samples of commercially available carbon. The columns can be operated in series or in parallel, with each having flow rate and retention times ranging from 1 to 4 gal $min^{-1}ft^{-2}$ (6.79 × 10^{-4} to 2.72 × 10^{-3} m/s). Two pairs of columns are provided as part of the pilot plant. One pair treats clarifier effluent, and the other treats filter effluent. The columns treating the clarifier effluent were intended to evaluate the viability of retrofitting the existing sand filters with GAC.

In addition to the analysis normally performed, many short-term side studies were conducted to evaluate unexpected results from the program.[1]

TREATMENT MODES

To date, the pilot plant has been operated in five different modes, and a sixth mode is being evaluated. Each mode represents a variation or refinement of a previous mode. The six modes are described briefly.

Mode 1

Mode 1 is identical to the full-scale facility, with the exception of the addition of GAC columns. This mode was to ensure that the performance of the pilot plant resembled that of the full-scale facility. Performance results indicated that the pilot plant in this mode matched full-scale treatment plant effectiveness at similar chemical dosages and detention times.

Mode 2

Mode 2 varies from Mode 1 in that the location for breakpoint chlorination was moved downstream of the flocculation/sedimentation processes. This modification was to maximize the removal of organic precursor material prior to the addition of chlorine, thus reducing the formation of chlorinated organics. Early results showed that it was impractical to treat the raw water without some form of disinfection at the head of the plant. Excessive biological slime growths occurred almost immediately and needed to be controlled. Thus, small amounts of chlorine were added to the raw water in quantities sufficient to form chloramines. This mode of operation prevented the formation of slime while allowing some precursor material to be removed. However, results showed that the relocation of the breakpoint chlorination process did not materially alter the amount of THMs formed in the finished water.

Mode 3

This mode was the same as Mode 2 with the exception that addition of powdered activated carbon (PAC) was eliminated from the process. This mode was to evaluate the effectiveness of PAC on the removal of organic precursor material. A comparison of Modes 2 and 3 indicated that PAC, at the dosages used at Thornton (4 to 6 mg/L), was generally ineffective in the removal of organic precursor material.

Mode 4

This mode consisted of breakpoint chlorination in the premix basin, followed immediately by the addition of ammonium sulfate to form chloramines. It was believed that breakpoint chlorination followed by rapid conversion to chloramine could reduce the time available for the formation of THMs by the free chlorine. The results of Mode 4 showed that the time for initiation of reactions between chlorine and organic precursor material is very rapid at Thornton. Large increases from instantaneous to 5-d terminal trihalomethanes (TTHM) were observed as these reactions proceeded across the plant.

Mode 5

Since Mode 4 suggested a competitive reaction between the chlorine and organic precursor material vs the chlorine and ammonia, preparation of the chloramine solution prior to introduction into the process stream was considered as an alternative. This approach was used for Mode 5 and indicated that, indeed, the formation of THMs is minimized with "premanufactured" chloramines. However, this mode of operation resulted in much lower bactericidal effectiveness, since no free chlorine was ever in contact with the raw water.

Mode 6

In an effort to avoid the bacteriological problems associated with Mode 5 (while avoiding formation of chlorinated organics) a dual disinfection mode is being evaluated. This mode of operation consists of the use of both chloramines and chlorine dioxide fed to the raw water.

RESULTS AND DISCUSSION

Several significant observations have been made during the first phase of the Thornton Pilot Program. Below are brief descriptions of the major findings:

1. Chloramines have been formed at Thornton under four separate conditions: (1) far to the left of breakpoint, primarily where monochloramines predominate; (2) just to the left of breakpoint where dichloramines predominate; (3) following breakpoint, with subsequent addition of ammonia; and (4) during preparation of chloramine solution prior to its addition to the main process water stream. Each condition shows an identical total residual chlorine concentration as measured by the DPD method; however, each has widely varying quality characteristics relative to THM, TOX, organic chloramines (OCA), inorganic chloramine, dichloroacetonitrile, and bacteriological quality. Relative comparisons of the quality of the various points along the breakpoint curve can be seen in Figure 4. The relative value of these parameters may change somewhat with source and season.

2. It appears that THMs and TOX are being formed under conditions where chloramines are the predominant products of chlorination reactions. This observation has been made during many experiments conducted at Thornton (to various degrees) on five separate water sources.

Under the Mode 4 configuration of the pilot plant, raw water is coagulated using $FeCl_3$, iron and manganese are oxidized using $KMnO_4$, and chlorine is added to accomplish breakpoint.

Immediately following breakpoint chlorination, ammonium sulfate is added to react with the free chlorine to form chloramines. The water is subsequently flocculated, settled, filtered, and passed through GAC columns. No additional chlorine is added to the T-THM samples prior to storage.

Measurements for residual chlorine and THMs are both made instantaneously after a 5-d holding period. The data consistently show the virtual absence of free chlorine after ammoniation. The instantaneous THM values are also quite low (generally less than 20 $\mu g/L$). However, after the 5-d holding period at ambient water temperature, the THM values are at least double the instantaneous values, without the addition of more chlorine. The magnitude of the final THM concentration during the critical summer months at Thornton has approached 100 $\mu g/L$ when using chloramines formed by the Mode 4 method of treatment.

Figure 5 shows the results of several other tests that have been performed using other sources of raw water. Study 23 was conducted using commercially available ultrapure water, whereas other waters are from various surface sources in the Denver area. The waters that did not contain natural ammonia were spiked with ammonia. Breakpoint curves were developed by adding chlorine to the sample and measuring the residual chlorine. At each chlorine dosage point, THM samples were collected. The results consistently show the formation of THM to the left of breakpoint. It appears that a competitive

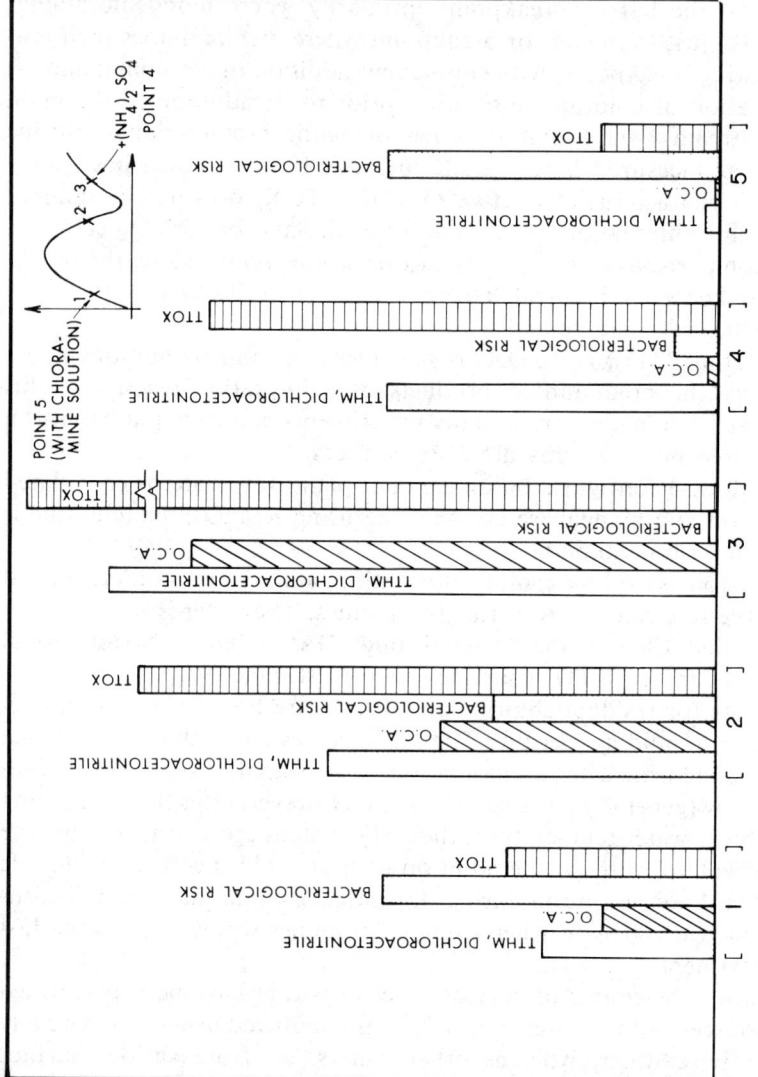

Figure 4. Water quality spectrum as a function of location on the breakpoint curve.

CHEMISTRY OF CHLORAMINATION 959

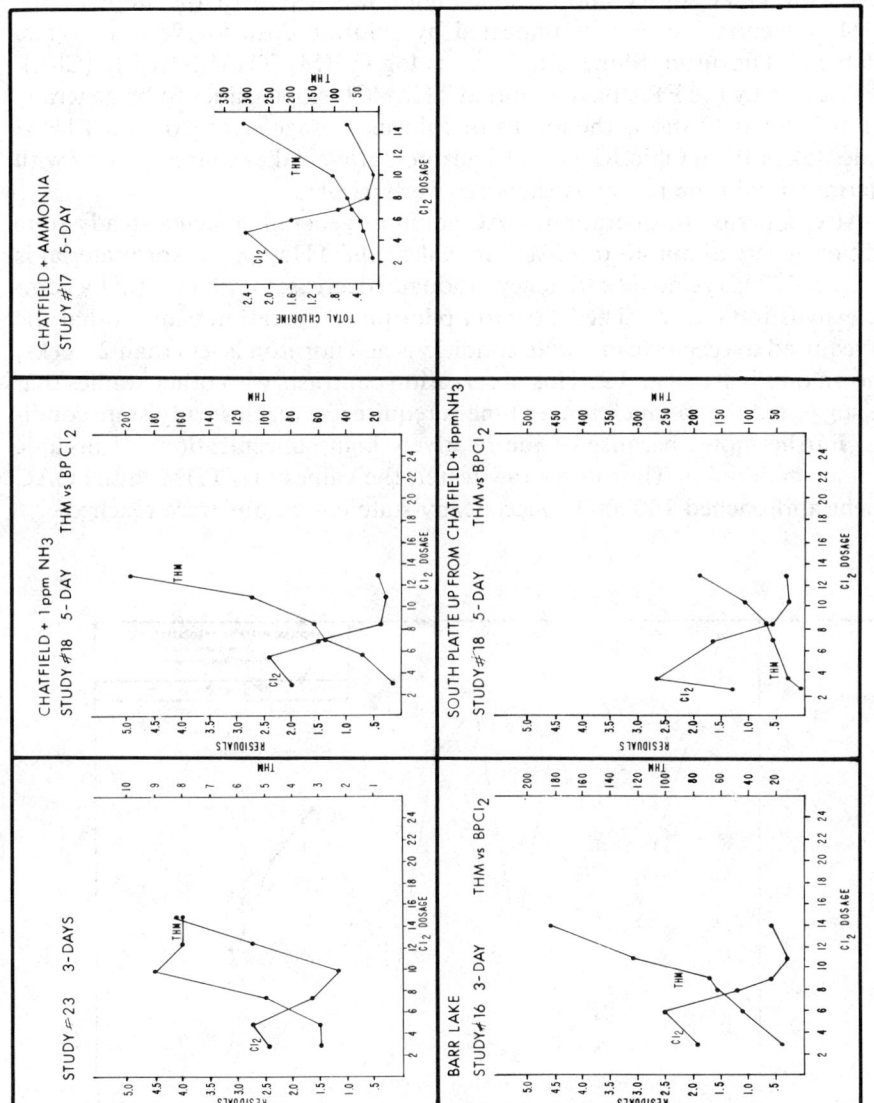

Figure 5. Test results using other sources of raw water.

reaction takes place between the formation of chloramines and THMs with the available chlorine.

3. The significant effect of chlorine dose on TTHM has been observed at Thornton. The extent to which chlorine dose affects THM formation varies widely with water source and season. Figure 6 is a plot of the slope of the TTHM concentration as it is impacted by chlorine dose for various waters available to Thornton. Slope is defined as $\log (THM_2/THM_1)/[(Cl_2)_2/(Cl_2)_1]$. Other studies by the EPA have reported THM/Cl_2 slope values to be generally about 0.3. Figure 7 shows the results of chlorine dosage effects on 3-d TTHM samples taken from Ohio River and Thornton gravel lakes sources, along with the terminal chlorine residuals shown as vertical bars.

4. After a period of operation, GAC columns generally reach a steady-state condition where about 40 to 60% removal of the THM precursor material is experienced. This removal efficiency gradually decreases until eventually complete exhaustion occurs. The Thornton pilot plant operation showed that the time required to reach steady-state conditions at Thornton is less than 2 weeks, as shown on Figures 8 and 9. This observation contrasts with other studies that have suggested that a much longer time is required to reach steady-state conditions. Furthermore, because of the relatively high concentrations of organic precursor material in Thornton's raw water, the values of TTHM in the GAC effluent approached 100 µg/L once steady-state conditions were reached.

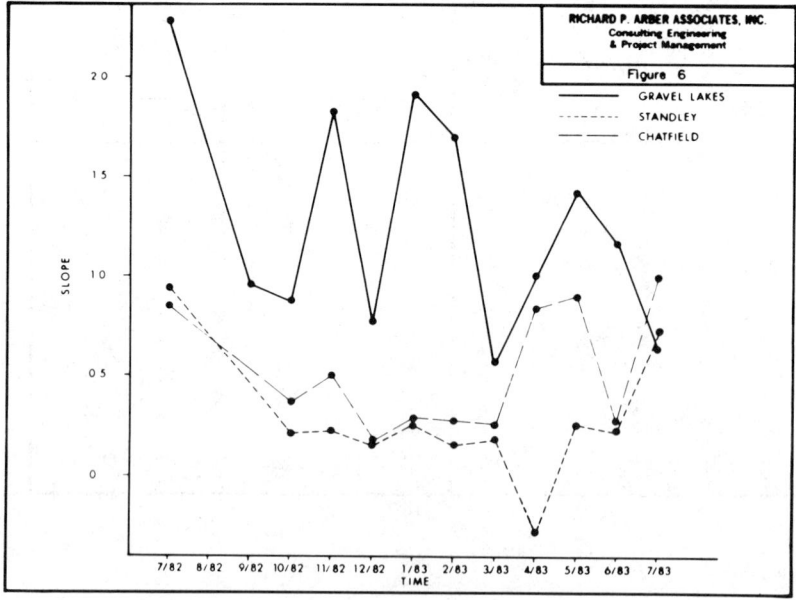

Figure 6. Plot of the slope of terminal trihalomethane concentration as it is impacted by chlorine dose for various waters available to Thornton.

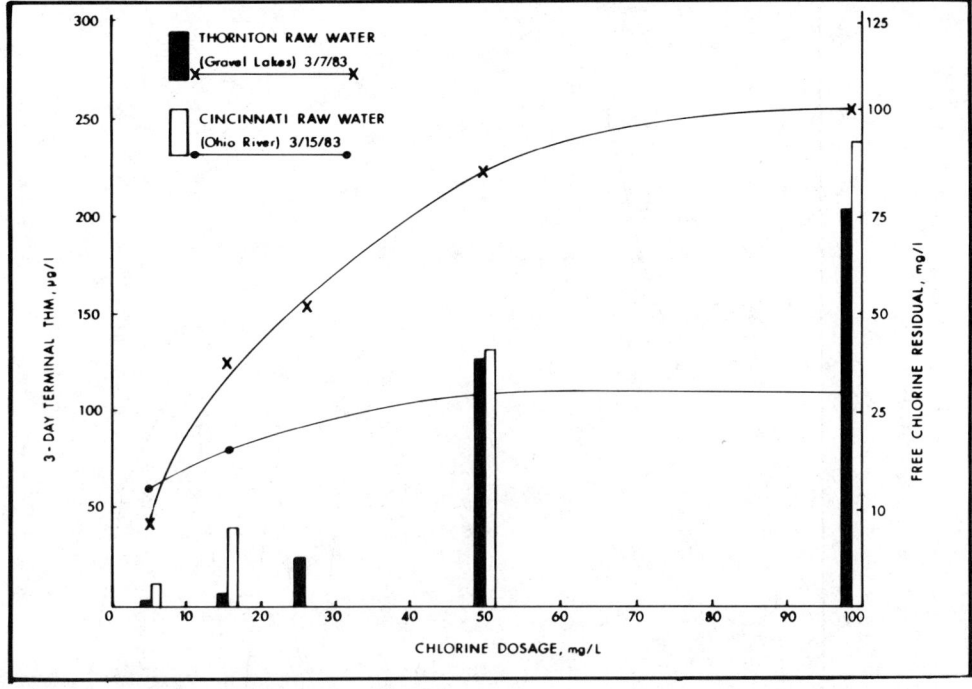

Figure 7. Effect of chlorine dose on THM concentrations for surface waters.

SUMMARY

The results of work being done at Thornton indicate that alternative methods of chlorination may result in widely varying water quality. The byproducts of disinfection using monochloramine, dichloramine, and free chlorination include varying amounts of THMs, TOX, organic chloramines, and dichloroacetonitrile.

Although it appears that chloramines themselves do not form significant amounts of THMs, the competitive reaction between the chlorine and ammonia or organic precursors suggests that the technique used for chloramination be scrutinized. Off-line chloramine formation followed by subsequent introduction of the solution to the main process steam has minimized the THM formation at Thornton; however, concerns about bactericidal effectiveness remain.

The effect of chlorine dose on the concentration of TTHMs must also be considered, especially when interpreting data. The experience at Thornton suggests that chlorine residuals should be reported with TTHMs, so that the response of THM formation with varying chlorine dose can be determined.

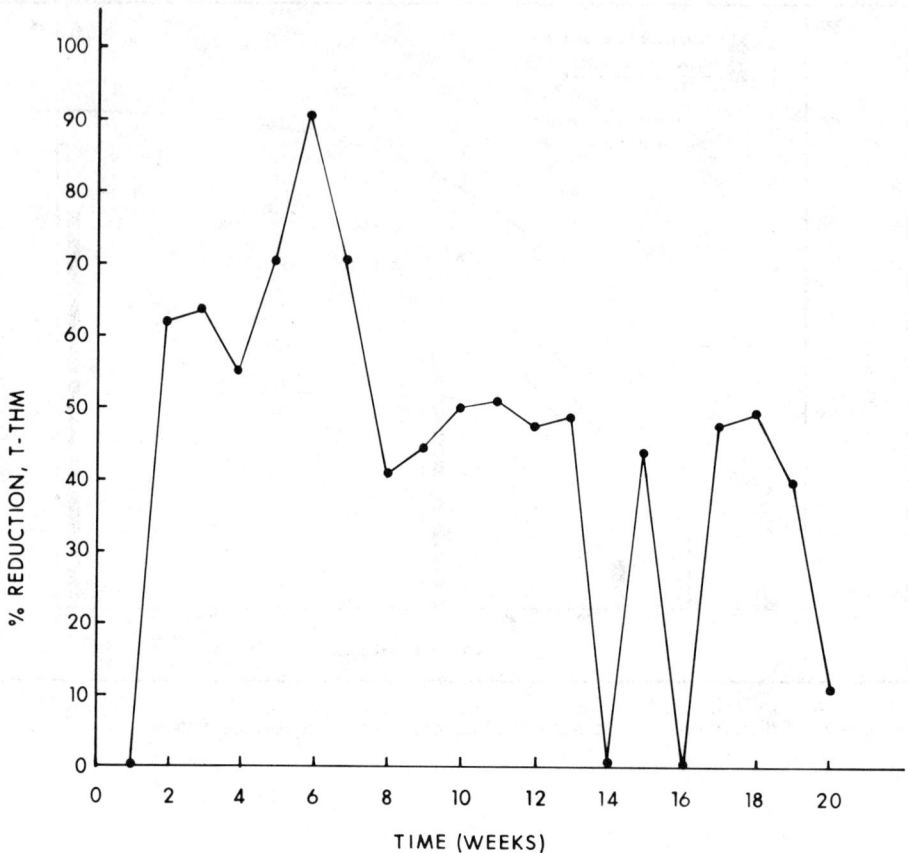

Figure 8. Mode 1 GAC performance.

This procedure should be used at various times throughout the year since dosage dependency may change.

It is important for water utilities to recognize the quality effects associated with practicing different types of disinfection of drinking water supplies. Although a specific quality problem may be avoided, other possibly less desirable by-products may result.

ACKNOWLEDGMENT

The information in this document has been funded in part by the U.S. Environmental Protection Agency (EPA) under Cooperative Agreement No. CR-809333-01 to the City of Thornton, Colorado.

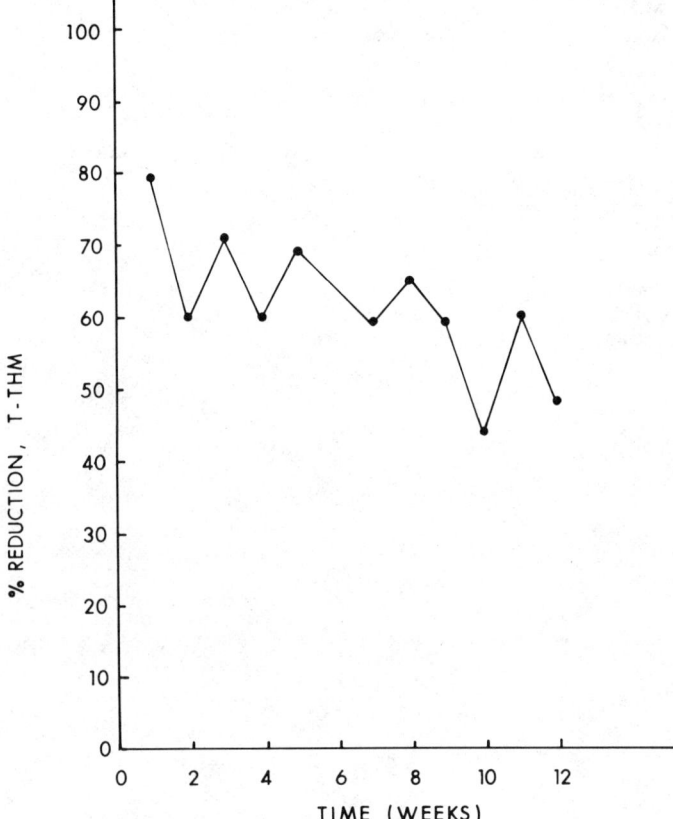

Figure 9. Mode 5 GAC performance.

REFERENCE

1. All analyses were performed according to accepted standard procedures. *Standard Methods for Examination of Water and Wastewater*, 15th ed. (Washington, DC: American Public Health Association, 1980).

CHAPTER 75

Analysis and Formation Mechanisms of Mixed N-Halogenated Methylamines

Terrence J. Kearney and Francis J. Sansone

Studies showing good correlations between the reduction of chlorine-produced oxidants (CPOs) and the production of trihalomethanes (THMs) from organic carbon have led to the conclusion that organic carbon plays an important role in the fate of CPOs.[1-3] Nevertheless, it has been suggested[4] that the amount of organic matter in seawater is insufficient to account for the reduction of more than a fraction of added hypochlorite (OCl^-). Recently, it has been shown that chlorinated oligotrophic seawater produces extremely low amounts of THM, presumably because of the very low levels of organic carbon in these waters.[5]

Although substitution reactions giving rise to THM production in temperate waters have been observed to reach completion in <20 min, Helz and Hsu[1] noted that these reaction rates are very slow in comparison to the oxidation of bromide (Br^-) by OCl^- or the halogenation of amines by either OCl^- or OBr^-. Helz and Hsu[1] concluded that fast reactions involving the redistribution of the oxidizing capacity of the initial OCl^- dose among species containing Cl, Br, and N will occur within 2 s.

Aqueous methylamine (MA), $CH_3NH_3^+$, was chosen as a model compound for the investigation of the reaction mechanisms of amine halogenation in seawater. This amine was chosen primarily because its molecular structure is the most analogous of all organic amines to the ammonium ion (the single most abundant amine in seawater, and the one on which most data exist).

EXPERIMENTAL

Time-course experiments were conducted using 40-mL amber glass vials with Teflon®-lined silicon septum caps (Pierce Chemical, Rockford, IL). Samples were transferred to the vials, chlorinated with 10 μL of dilute OCl^- solution, and then reacted in the dark at 0°C.

The CPOs were measured by differential pulse polarography to determine the loss of excess 0.00564 N phenylarsine oxide (Fisher Scientific) that was added to terminate individual experiments.[6-8] Samples were measured at pH 3.5 to 4.5, with 50 mg of potassium iodide added per 10-mL sample. This method gave results comparable to those from amperometric titrations for combined residual oxidants.[9]

Nitrogen-halogenated methylamines (N-HMAs) were extracted from 40-mL samples with 4 mL of methyl-t-butyl ether.[8] The extracts were analyzed by ambient-temperature gas chromatography (Varian 3700) using a 30-m DB-5 (SE-54 equivalent) fused-silica capillary column with a thick (1.0 μm) bonded stationary phase (J&W Scientific, Rancho Cordova, CA); a model 700A Hall electrolytic conductivity detector (Tracor, Austin, TX) was used in either the halogen- or nitrogen-selective mode. Compound identifications were made using a Finnigan-MAT 1020B gas chromatograph/mass spectrometer (GC/MS). In both cases a split-mode injector was used with a split ratio of <50:1.

Ammonia analyses were made with a Technicon Autoanalyzer II using standard spectrophotometric techniques.[10] Artificial seawater was prepared according to the method of Kester et al.[11]

RESULTS AND DISCUSSION

Analysis of N-HMAs

Ambient-temperature gas chromatography (GC) with a bonded-phase fused-silica capillary column was found to be a quick and precise means of analyzing N-HMAs. The chromatograms obtained showed good separation of these compounds, without the severe peak tailing frequently seen with basic compounds (Figure 1). We attribute the success of this technique largely to the inert nature of fused silica, which limits tailing to that resulting from the use of a Hall detector.

The assignment of GC peaks was made by GC/MS (Figure 2); the resultant spectra were easily interpreted by the identification of the M^+, $(M + 2)^+$, $(M - 1)^+$, and $(M - X_n)^+$ ions.[8] Spectra corresponding to CH_3NHBr were never observed with our samples.

The partition coefficients for CH_3NHCl and CH_3NCl_2 between methyl-t-butyl ether and water were measured to be 42:1 and 101:1, respectively.[8] The average recovery of a chlorinated MA solution not containing bromide (Figure 1B), as measured by summing both CH_3HNCl and CH_3NCl_2 peaks and comparing with the amount of OCl^- added, was 90% with a coefficient of variability of 11%, when using a GC split ratio of <30:1.[8]

Distribution of N-HMAs

Figure 3 illustrates the production of N-HMAs in artificial seawater (Br^- constant at 0.8 mM) as functions of the molar ratios of additions of OCl^- and methylamine. The most direct relationship observed in these data is the uniform decrease of all N-HMAs as the molar OCl^-:MA ratio drops from 3:1 to 1:1 to 1:3.

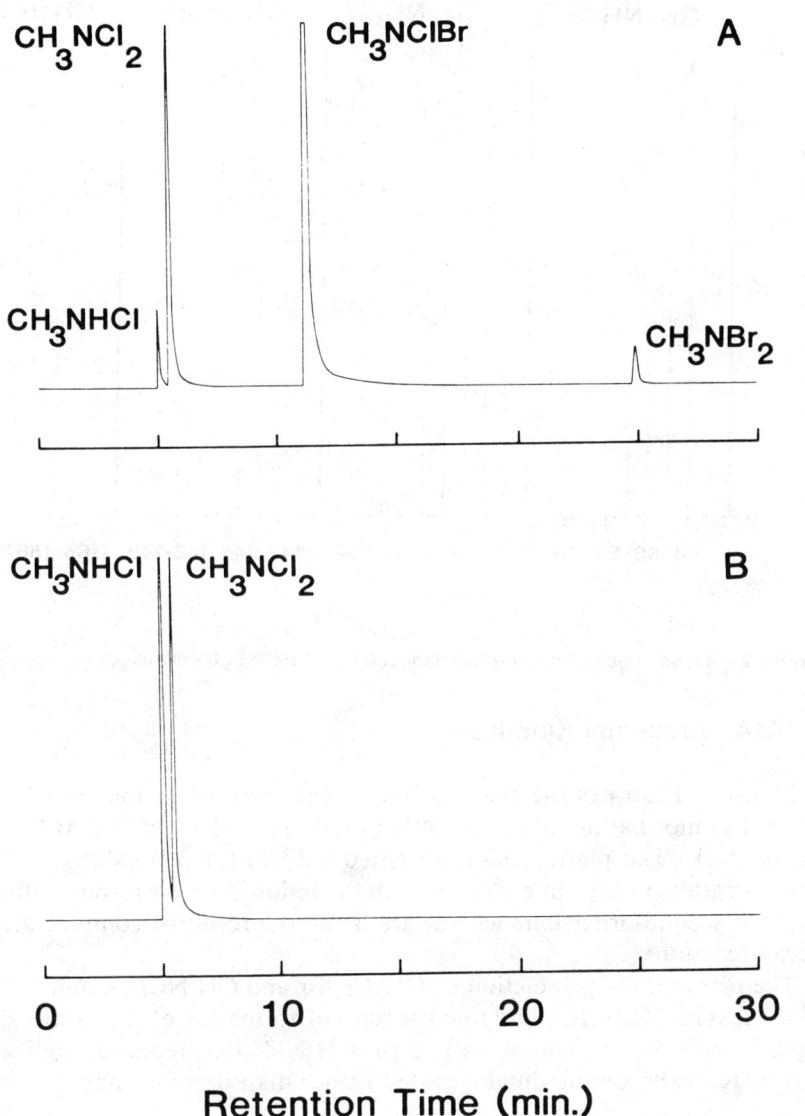

Figure 1. Representative chromatograms of N-HMAs: (A) 8 mM MA, 12 mM OCl$^-$, 30 mM Br$^-$; (B) 16 mM MA, 12 mM OCl$^-$, no Br$^-$. A Hall detector in the nitrogen-selective mode was used for these analyses.

N-HMA production resulting from different initial concentrations of Br$^-$, MA, and OCl$^-$ (Figure 3) indicates that the production of any given N-HMA is dependent either directly or indirectly on these initial concentrations. In a sense this conclusion is a unification of observations by Haag,[12] who observed trends in the distribution of N-HMA production with respect to the Br$^-$:MA ratio, and Peron and Courtot-Coupez,[13] who studied the distribution of inorganic halamines in response to changing OCl$^-$:NH$_4^+$ ratios.

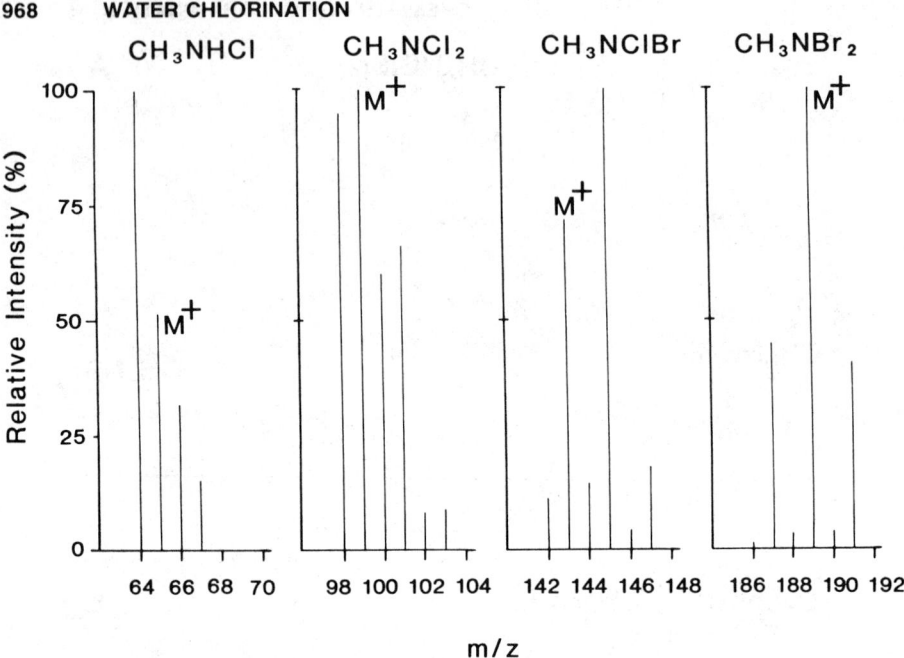

Figure 2. Mass spectra showing the molecular ions (M⁺) of N-HMAs.

N-HMA Production Kinetics

Figure 4 illustrates the changes in concentration of various N-HMAs produced by chlorinating solutions containing 17 mM MA and 30 mM KBr at 0°C in the dark. The plotted data are fitted with linear regressions of N-HMA concentration vs log time. None of the reaction data correspond directly to first- or second-order kinetics but are likely the result of complex growth or decay reactions.

The fact that the production of CH_3NClBr and CH_3NCl_2 accounts for 93% of the loss in CH_3NHCl, and that the rates of formation of the former decrease significantly as the concentration of CH_3NHCl is reduced, indicate that CH_3NHCl is becoming dihalogenated rather than decomposing.

Proposed Mechanism of N-HMA Production

The production of CH_3NClBr observed in Figure 4 may be the result of four different possible processes:[6,12] (1) chlorination (with HOCl/OCl⁻) of a bromamine, (2) halogen exchange between halogenated amines, (3) bromination (with HOBr/OBr⁻) of a chloramine, and (4) slow reaction of Br⁻ with monochloramine. Mechanisms (1) and (2) are not consistent with the data presented here, because no N-HMA other than CH_3NHCl shows a decrease in concentration concurrent with the production of other N-HMAs. Unless different mech-

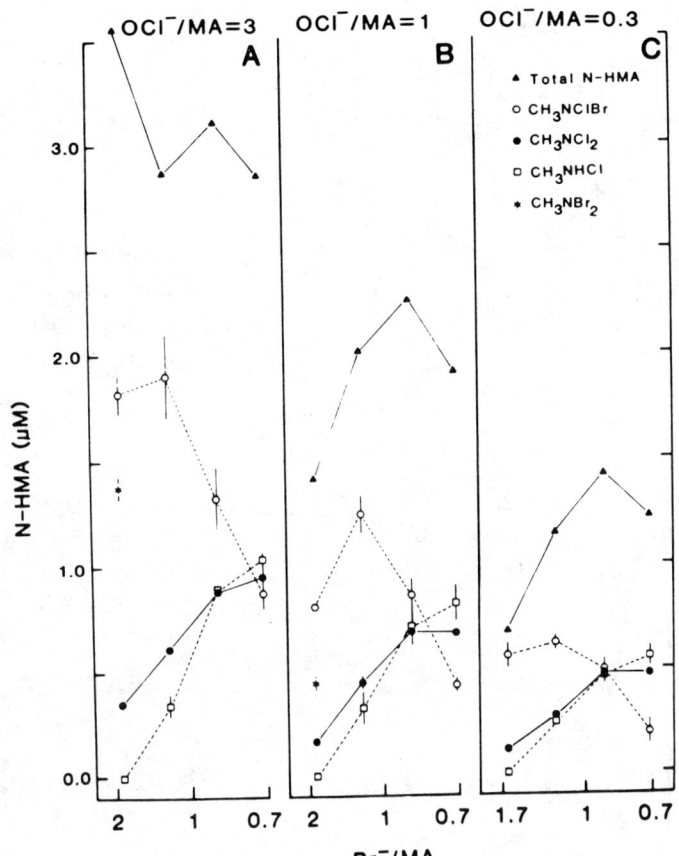

Figure 3. Distribution of N-HMAs at various OCl^-:Br^-:MA molar ratios at 0°C after 5 to 90 min chlorination in the dark. Data are averages of four chromatograms; vertical bars are standard deviations of replicates. Points without error bars have standard deviations smaller than the point symbols.

anisms control the production of CH_3NClBr as the halogenations proceed, the faster of mechanisms (3) and (4) must be responsible for the observation that the CH_3NClBr concentration equals twice the concentration of CH_3NCl_2 at the first time point (Figure 4).

This assumption of a single fast mechanism is supported by the observation that HOBr is more reactive by a factor of 2.5 to 20 toward amines than is HOCl.[14] The fact that CH_3NHCl is observed at all in the presence of such rapid bromination is explained by the competition between the initial reactions of OCl^- to form CH_3NHCl and HOBr. Inman and Johnson[15] observed that in seawater with a Br^-:NH_4^+ ratio of 2.6×10^4:1, the reaction rates of the formation of both HOBr and NH_2Cl are equal. A reduction of this ratio by four orders of magnitude, as in these experiments, would presumably greatly favor the production of NH_2Cl.

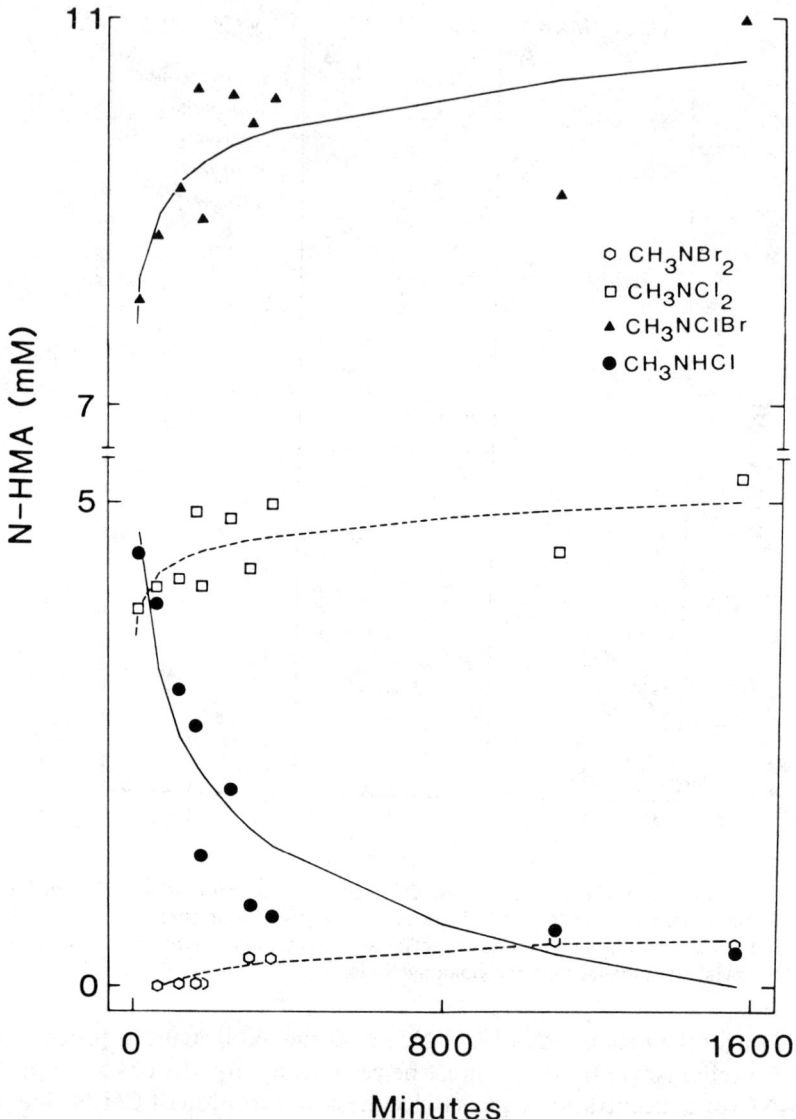

Figure 4. Time-course production and loss of N-HMAs in chlorinated aqueous solutions of MA and KBr at 0°C in the dark. Data are fit by linear regressions of N-HMA = $k \log t + b$. Initial MA = 17 mM; initial Br$^-$ = 30 mM.

Thus, the initial production of CH_3NHCl should be favored early in the reaction. However, because HOBr is being formed concurrently, and because the latter has a greater reactivity than HOCl to amines, there is an initial presence of both CH_3NHCl and CH_3NClBr. During the continued formation of CH_3NClBr, the concentration of CH_3NHCl apparently decreases sufficiently to slow the production of CH_3NClBr.

While 64% of the loss of CH_3NHCl (Figure 4) is accounted for by the production of CH_3NClBr, 29% may also be the result of CH_3NCl_2 production. The fact that the initial concentration of CH_3NCl_2 is less than CH_3NClBr may also reflect the relative reactivities of HOBr vs HOCl, as noted previously (in this case, toward CH_3NHCl).

The initial partitioning reaction between CH_3NHCl and HOBr production, and the subsequent production of CH_3NClBr and CH_3NCl_2 from the initial rapid formation of CH_3NHCl, may also be responsible for (1) the absence of detectable CH_3NHBr, (2) the kinetically slow production of CH_3NBr_2 (Figure 4), and (3) the limitation of production of CH_3NBr_2 to Br^-:MA ratios >2:1 and OCl^-:MA ratios >1:1 (Figure 3). The possibility of HOBr exchange with chlorine in CH_3NHCl is unlikely considering the reactivity of HOBr toward the nitrogenous hydrogen, which promotes the production of CH_3NClBr. Thus, fast production of CH_3NBr_2 would be limited to high Br^-:MA and OCl^-:MA ratios, which would permit the initial bromination of MA by HOBr to compete with the production of CH_3NHCl (Figure 5). This conclusion is supported by the rate-constant data reported for the reaction of OCl^- with MA and Br^- in saline solutions.[16,17]

Artificial Seawater Kinetic Experiments

Ammonia and methylamine were added to artificial seawater (ASW) to observe their effects on the reduction of reactive CPO in a series of time-course experiments. The results obtained with ASW blanks and with 25 μM solutions of ammonia and methylamine are illustrated in Figure 6; the data are fitted with linear regressions of CPO concentration vs log time. A logarithmic relationship fits the data well, as was true in the time-course experiments of N-HMA production in Br^- solutions (Figure 4).

The observed decrease in CPO reduction rates in chlorinated ASW resulting from the addition of MA contrasts with the observed increase in rates of CPO reduction in chlorinated ASW with added ammonia (Figure 6). The fact that inorganic halamines are less stable than their organically substituted analogues[18] may explain these trends. Thus, the increased loss of reactive CPOs we observed in chlorinated seawater with added ammonia is apparently caused by the decomposition of, or halogen transfer from, the pool of operationally defined reactive (total) CPOs. This conclusion is substantiated by the work of Peron and Courtot-Coupez.[13]

In contrast, the observed stability of N-HMAs and their apparent oxidative strength presumably result in a stabilization of the concentration of the analytically reactive CPOs by their inclusion into this operationally defined pool. Since there are no published data on the oxidative power of either organic or inorganic halamines, nor on what range of oxidizing power a compound must have to be measured as a reactive CPO by an electrochemical method, this evaluation of the trends illustrated in Figure 6 is an area of speculation.

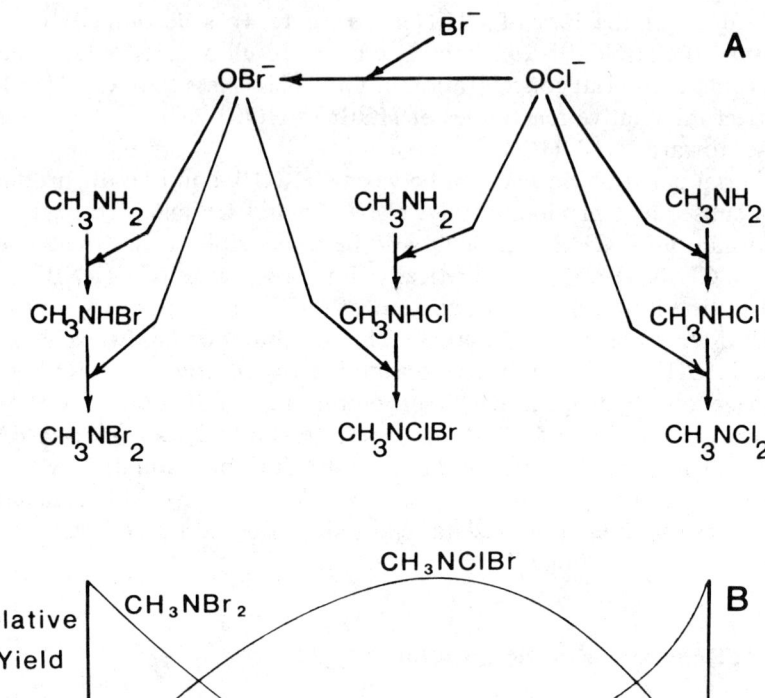

Figure 5. (A) Proposed mechanism for the formation of mixed N-HMAs; (B) proposed distribution of N-HMAs in chlorinated methylamine solutions containing varying amounts of initial Br^- (initial OCl^-:MA = 1).

Nevertheless, work by Isaac and Morris[19] on the transfer of positive halogens between nitrogenous substrates, and the analysis of halamines by electrochemical methods,[20] indicate that these compounds could be considered functionally reactive CPOs.

ACKNOWLEDGMENTS

We thank S. Kansako and R. Goo of the Sand Island (Honolulu) Wastewater Treatment Plant for access to their GC/MS.

This work was supported by the U.S. Department of Energy and State of Hawaii funds administered through the Hawaii Energy Institute (DOE grant

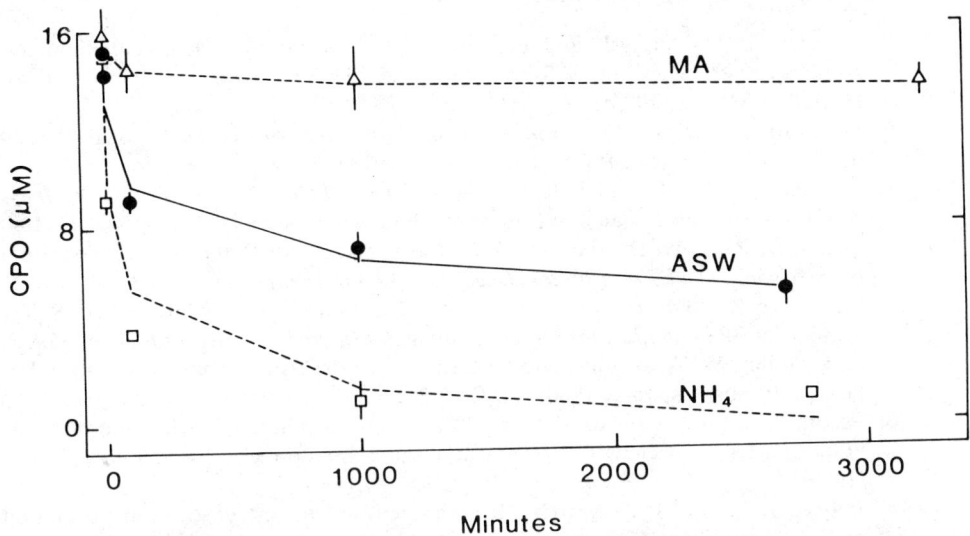

Figure 6. Time-course reduction of CPOs at 0°C in the dark in artificial seawater with and without 25-μM additions of nitrogenous species; data are means of 12 replicates. Data are fit by linear regressions of CPO = $-m \log t + b$. (MA, $r = -0.91$; ammonium, $r = -0.97$; ASW, $r = -0.98$).

DE-FG03-81ER10250). Hawaii Institute of Geophysics Contribution No. 1616.

REFERENCES

1. Helz, G. R., and R. Y. Hsu. "Volatile Chloro- and Bromocarbon in Coastal Waters," *Limnol. Oceanogr.* 23:858 (1978).
2. Morris, J. C., and B. Baum. "Kinetics of Reactions Between Aqueous Chlorine and Nitrogen Compounds," in *Water Chlorination: Environmental Impact and Health Effects, Vol. 2*, R. L. Jolley, H. Gorchev, and D. H. Hamilton, Jr., Eds. (Ann Arbor, MI: Ann Arbor Science Publishers, Inc., 1978), pp. 29-48.
3. Carpenter, J. H., and C. A. Smith. "Reactions in Chlorinated Seawater," in *Water Chlorination: Environmental Impact and Health Effects, Vol. 2*, R. L. Jolley, H. Gorchev, and D. H. Hamilton, Jr., Eds. (Ann Arbor, MI: Ann Arbor Science Publishers, Inc., 1978), pp. 195-308.
4. Wong, G. T., and J. A. Davidson. "The Role of Bromide in the Dissipation of Chlorine in Seawater," *Water Res.* 16:335-343 (1977).
5. Sansone, F. J., and T. J. Kearney. "Chlorination Kinetics of Surface and Deep Tropical Seawater," Chapter 59, this volume.
6. Trofe, T. W., G. W. Inman, and J. D. Johnson. "The Kinetics of Monochloramine Decomposition in the Presence of Bromine," *Environ. Sci. Technol.* 14:544 (1980).
7. Smart, R. B., J. H. Lowry, and K. H. Mancy. "Analysis of Ozone and Residual Chlorine by Differential Pulse Polarography of Phenylarsine Oxide," *Environ. Sci. Technol.* 13:89 (1979).

8. Kearney, T. J. "Analysis and Formation Mechanisms of N-Halomethylamines: Application to Seawater Chlorination," M.S. Thesis, (Honolulu: University of Hawaii at Manoa, 1983).
9. *Standard Methods for the Examination of Water and Wastewater*, 14th ed. (Washington, DC: American Public Health Association, 1979).
10. Strickland, J. D. H., and T. R. Parson. *A Practical Handbook of Seawater Analysis*, 2nd ed. (Ottawa, Canada: Publishing Supply and Services, 1977), pp. 133–134.
11. Kester, D. R., I. W. Duedall, D. N. Connors, and R. M. Pytkowicz. "Preparation of Artificial Seawater," *Limnol. Oceanogr.* 12:176 (1967).
12. Haag, W. R. "Formation of N-Bromo-N-Chloramines in Chlorinated Saline Waters," in *Water Chlorination: Environmental Impact and Health Effects, Vol. 3*, R. L. Jolley, W. A. Brungs, and R. Cumming, Eds. (Ann Arbor, MI: Ann Arbor Science Publishers, Inc., 1980), pp. 193–201.
13. Peron, A., and J. Courtot-Coupez. "Etude Physiochimique de la Chloration de l'Eau de Mer Artificielle Contenant de l'Azote Ammoniacal," *Water Res.* 14:883 (1980).
14. Wajon, J. E., and J. C. Morris. "Bromamination Chemistry: Rates of Formation of NH_2Br and Some N-Bromamino Acids," in *Water Chlorination: Environmental Impact and Health Effects, Vol. 3*, R. L. Jolley, W. A. Brungs, and R. B. Cumming, Eds. (Ann Arbor, MI: Ann Arbor Science Publishers, Inc., 1980), pp. 171–181.
15. Inman, G. W., and J. D. Johnson. "The Effect of Ammonia Concentration on the Chemistry of Chlorinated Seawater," in *Water Chlorination: Environmental Impact and Health Effects, Vol. 2*, R. L. Jolley, H. Gorchev, and D. H. Hamilton, Jr., Eds. (Ann Arbor, MI: Ann Arbor Science Publishers, Inc., 1978), pp. 235–252.
16. Weil, I., and J. C. Morris. "Kinetic Studies on the Chloramines. I. The Rates of Formation of Monochloramine, N-Chlormethylamine and N-Chlordimethylamine," *J. Am. Chem. Soc.* 71:1664 (1949).
17. Johnson, J. D., G. W. Inman, and T. W. Trofe. "Cooling Water Chlorination: The Kinetics of Chlorine, Bromine, and Ammonia in Seawater," NUREG/CR-1552 (Washington, DC: U.S. Nuclear Regulatory Commission, 1982).
18. Jander, J., and V. Engelhardt. "Nitrogen Compounds of Chlorine, Bromine, and Iodine," in *Developments in Inorganic Nitrogen Chemistry*, Vol. 2, C. B. Colburn, Ed. (Amsterdam: Elsevier Scientific Publishing Co., 1973), pp. 70–203.
19. Isaac, R. A., and J. C. Morris. "Rates of Transfer of Active Chlorine Between Nitrogen Substrates," in *Water Chlorination: Environmental Impact and Health Effects, Vol. 3*, R. L. Jolley, W. A. Brungs, and R. B. Cumming, Eds. (Ann Arbor, MI: Ann Arbor Science Publishers, Inc., 1980), pp. 183–191.
20. H. C. Marks, D. B. Williams, and G. V. Glasgow. "Determination of Residual Chlorine Compounds," *J. Am. Water Works Assoc.* 43:201 (1951).

CHAPTER **76**

Disappearance of Monochloramine in the Presence of Nitrite

Richard L. Valentine

Nitrite (NO_2^-) may play a fundamental role in the chemistry of aqueous chlorine-ammonia systems characterized by the chemistry of the ammoniacal chloramines. The oxidation of nitrite by hypochlorous acid (HOCl) is well known. Chloramines are also oxidizing agents that would be expected to oxidize NO_2^- through an indirect reaction involving hydrolysis to HOCl. However, a direct reaction with a chloramine itself cannot be ruled out. Additionally, nonoxidative reactions involving NO_2^- and chloramines could occur. The potential for a reaction between NO_2^- and one or more chloramines exists in several environmentally important systems.

The treatment of domestic wastewater frequently involves biological nitrification to convert ammonia to nitrate. This process occurs through the formation of nitrite and may result in an effluent containing relatively high concentrations of nitrite in addition to ammonia. Chlorination of this effluent would presumably yield a mixture containing both nitrite and chloramines. Nitrite could, therefore, potentially represent a source of chlorine demand via direct and indirect reactions with chloramines.

Nitrite, in addition to nitrogen gas and nitrate, may be produced as the result of ammonia nitrogen oxidation in chlorinated waters.[1-3] The formation of nitrate under breakpoint conditions has been hypothesized to occur via the formation and subsequent oxidation of nitrite by free chlorine, although chloramines are also present.[4,5] Leao[6] also suggested that nitrite or nitrate may be produced at significant concentrations because of the relatively slow redox reactions occurring in the presence of excess ammonia when only chloramines are measurable oxidants.

Reactions between chloramines and nitrite are believed to be very slow;[7] however, the kinetics have not been fully investigated. Without a better understanding of these kinetics, potential reactions cannot be automatically discounted under all environmentally important conditions. This chapter presents the results of a study aimed at partly filling this knowledge gap and reports on the disappearance of NH_2Cl in the presence of NO_2^-.

EXPERIMENTAL

Reagents

All reagent solutions were prepared with chlorine demand-free water (CDFW) obtained by chlorinating distilled deionized water and then dechlorinating with sunlight. Stock solutions of reagent-grade Na_2SO_4, Na_2HPO_4, and NaH_2PO_4 were passed through a 0.45-μm millipore filter, chlorinated, and then dechlorinated with sunlight. Stock solutions of $NaNO_2$ and $(NH_4)_2SO_4$ were also filtered prior to use. Sodium hypochlorite was prepared by sparging ultrapure chlorine gas through a solution of sodium hydroxide. A 75 μM solution of NH_2Cl (total nitrogen-to-chlorine mol ratio of 2.5:1.0) was obtained by the slow addition of sodium hypochlorite to a vigorously stirred solution of excess ammonium sulfate buffered at pH 9.1.

Procedures

Batch experiments were conducted in a light-proof flask maintained at 25°C by immersion in a thermostated water bath. The experiments were initiated by adding an aliquot of monochloramine to a temperature-equilibrated flask containing 0.1 M phosphate and 0.17 M Na_2SO_4 reaction media prepared with CDFW and appropriate additions of stock $NaNO_2$ and $(NH_4)_2SO_4$. Initial monochloramine concentrations were measured using both the DPD-FAS[8] titrimetric method and by UV spectrophotometry at 245 nm. Both methods compared within ± 2%.

Nitrite and ammonia nitrogen were kept in excess of monochloramine. The concentration of one species was varied while others were held constant. Initial reactant concentrations ranged from 0.039 to 1.37 mM NH_2Cl, 4.7 to 47 mM NO_2^-, 4.3 to 74.4 mM excess ammonia nitrogen ($[N_T]_x$), and pH from 7.03 to 7.88.

A procedure based on the DPD-FAS[8] titrimetric method was used to periodically measure total oxidant in the presence of nitrite. A modification involved the use of a more concentrated buffer. Nitrite was not found to interfere at concentrations used in this study. Experimental conditions were chosen that minimized the potential formation of dichloramine, and, therefore, it was assumed that the total oxidant concentration was comprised essentially of NH_2Cl.

Data Analysis

Data analysis was based on a simple single-term generalized-rate expression,

$$\frac{d[NH_2Cl]}{dt} = -k\,[NH_2Cl]^a[NO_2^-]^b[H^+]^c[N_T]_x^d \tag{1}$$

which was anticipated as being applicable if NH_2Cl hydrolysis governed. Integration under pseudo first-order conditions yields the approximation

$$-\ln \frac{[NH_2Cl]}{[NH_2Cl]_0} \simeq k_{ob} t \qquad (2)$$

where

$$k_{ob} = [NH_2Cl]^{a-1} [NO_2^-]^b [H^+]^c [N_T]_x^d \qquad (3)$$

and is exact if $a = 1$.

Pseudo first-order rate constants (k_{ob} values) were evaluated by linear regression plots of Equation (2) for ordinate values lying between 0 and 1.0, and by log k_{ob} – log species concentration plots used to obtain reaction dependencies.

RESULTS

Figure 1 shows a typical plot of $-\ln[NH_2Cl]/[NH_2Cl]_0$ vs time used to obtain k_{ob} values. Linear results were consistently observed to approximately a 64% disappearance in NH_2Cl concentration, with a noticeable slowing of the rate thereafter. The k_{ob} values appear to have only a very weak dependence, if any, on NH_2Cl concentration, as indicated by Figure 2.

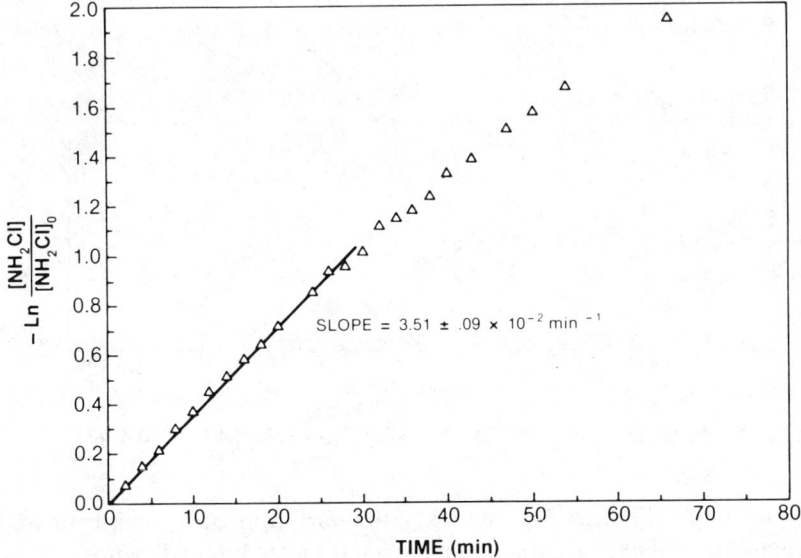

Figure 1. First-order NH_2Cl disappearance. $[NH_2Cl]_0 = 0.58$ mM, $[NO_2^-] = 18.8$ mM, pH 7.52, $[N_T]_x = 9.42$ mM.

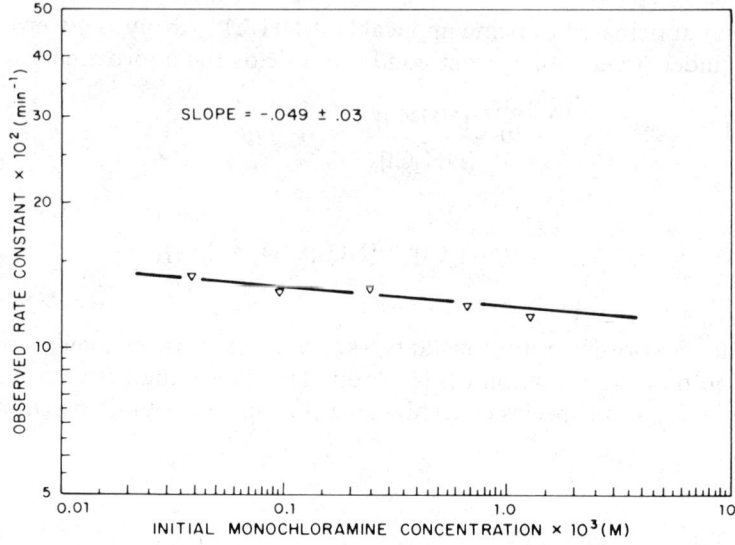

Figure 2. Initial monochloramine dependence. $[NO_2^-] = 18.8$ mM, pH 7.03, $[N_T]_x = 19.4$ mM.

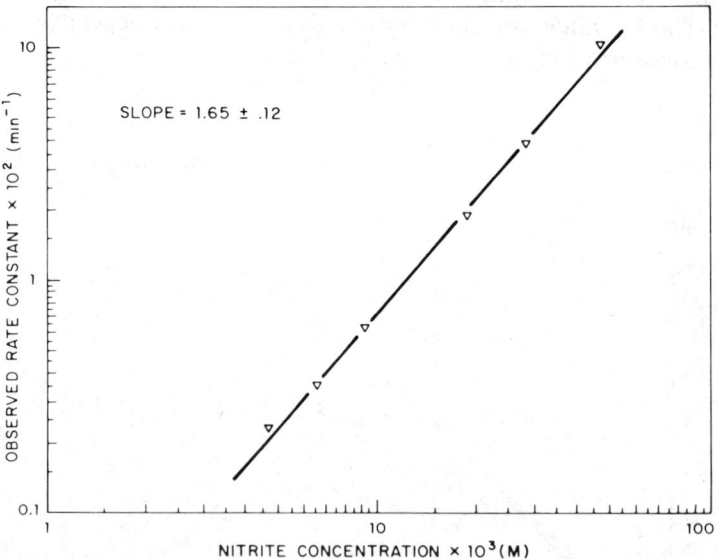

Figure 3. Nitrite dependence. $[NH_2Cl]_0 = 0.60$ mM, pH 7.48, $[N_T]_x = 19.4$ mM.

Log-log plots (Figures 3–5) showing the variation of k_{ob} with nitrite, the hydrogen ion, and excess ammonia nitrogen indicate linear dependencies ($r^2 > 0.99$) having slopes equal to reaction orders b, c, and d, respectively. Measured reaction orders are summarized in Table I.

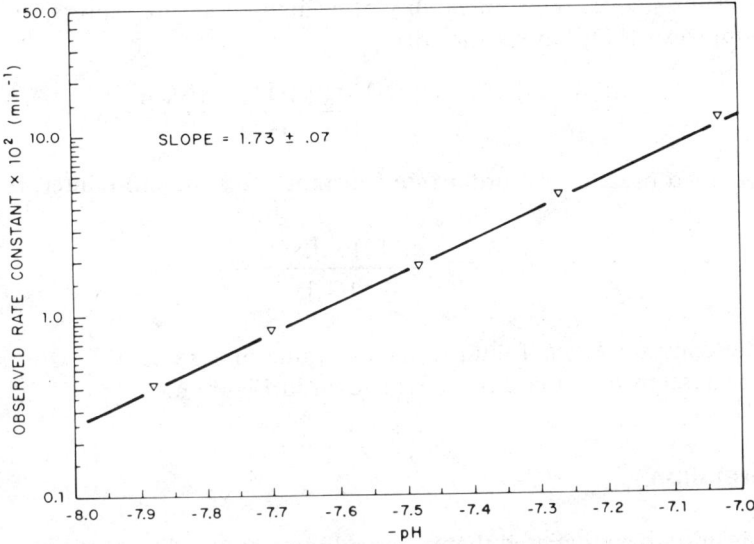

Figure 4. Hydrogen ion dependence. [NH_2Cl] = 0.06 mM, [NO_2^-] = 18.8 mM, [N_T]$_x$ = 19.4 mM.

Table I. Measured Reaction Orders (±95% confidence level indicated)

Monochloramine	0.95 ± 0.03
Nitrite	1.65 ± 0.12
Hydrogen Ion	1.73 ± 0.07
Excess Ammonia	0.90 ± 0.05

DISCUSSION

Reaction Dependencies

A first-order dependence on NH_2Cl concentration is strongly suggested. The apparent small dependence on NH_2Cl concentration may be caused by the formation of some $NHCl_2$ and increased free excess ammonia produced during the course of the reaction. An inverse dependency on excess ammonia nitrogen is also suggested.

Most striking is the apparent similar noninteger dependencies of H^+ and NO_2^-, which are clearly representative of some value between 1 and 2. A value of 1.7 has been assigned.

The results suggest the applicability of a single-term rate expression, at least as an empirical tool, having the form

$$\frac{d[NH_2Cl]}{dt} = -k \frac{[NH_2Cl][H^+]^{1.7}[NO_2^-]^{1.7}}{[N_T]_x} \quad (4)$$

The observed pseudo first-order rate constant k_{ob} is therefore interpretable as

$$k_{ob} = \frac{k[H^+]^{1.7}[NO_2^-]^{1.7}}{[N_T]_x} \quad (5)$$

The rate constant k was found to have a value of $1.58 \times 10^{12}\ M^{1.7}\ min^{-1}$ by linear regression of Equation (5), as shown in Figure 6.

Interpretation

It was initially anticipated that observed rates of NH_2Cl loss could be attributed to a simple reaction between NO_2^- and $HOCl$ produced by the hydrolysis of NH_2Cl according to the mechanism

$$NH_2Cl + H_2O \underset{}{\overset{K_h}{\rightleftarrows}} HOCl + NH_3 \quad (6)$$

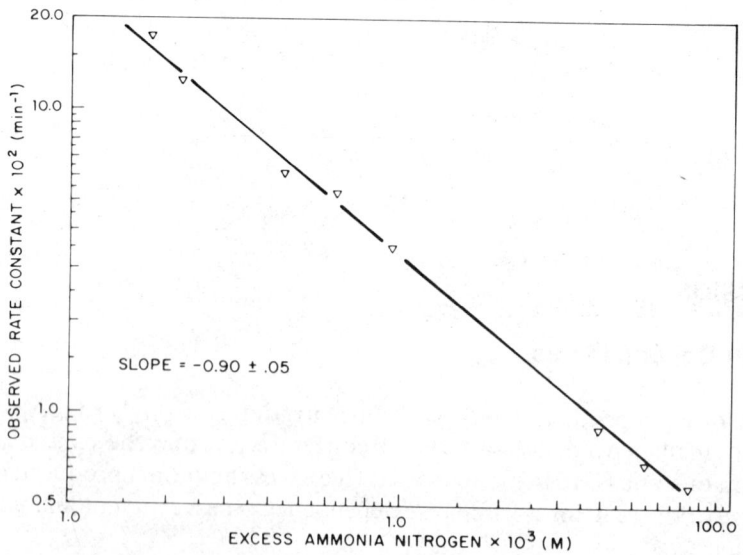

Figure 5. Excess ammonia nitrogen dependence. $[NH_2Cl]_0 = 0.60$ mM, $[NO_2^-] = 18.8$ mM, pH 7.52.

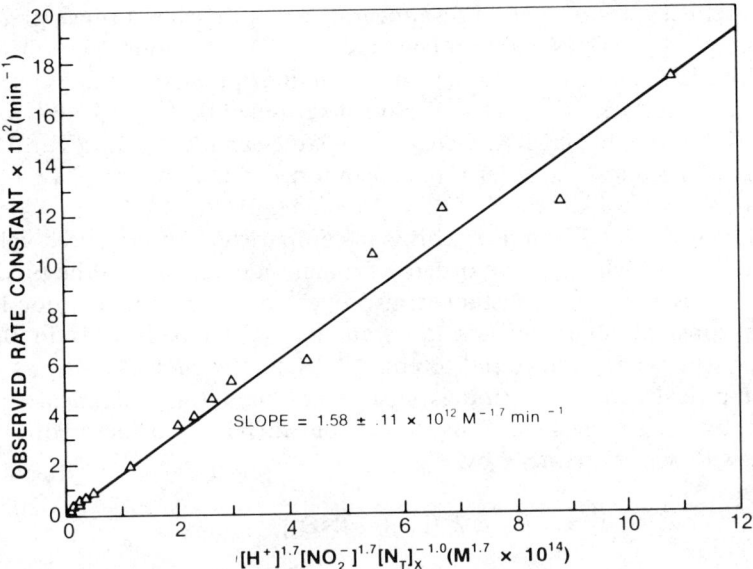

Figure 6. Determination of empirical single-term reaction-rate constant.

$$NH_4^+ \underset{}{\overset{K_a}{\rightleftharpoons}} NH_3 + H^+ \qquad (7)$$

$$HOCl + NO_2^- \underset{fast}{\overset{k_N}{\rightarrow}} NO_3^- + H^+ + Cl^- \qquad (8)$$

This mechanism results in a single-term rate expression that can be simplified at pH values less than about 8.0 to yield

$$-\frac{d[NH_2Cl]}{dt} = \frac{k_N K_h [H^+] [NH_2Cl] [NO_2^-]}{K_a [N_T]_x} \qquad (9)$$

Integration under conditions of constant pH, nitrite, and excess ammonia results in

$$-\ln \frac{[NH_2Cl]}{[NH_2Cl]_0} = \bar{k} t \qquad (10)$$

where

$$\bar{k} = \frac{k_n K_h [H^+] [NO_2^-]}{K_a [N_T]_x} \qquad (11)$$

which may be used to calculate a predicted pseudo first-order rate constant, $\bar{k}$, for purposes of comparison.

The observed rates of NH_2Cl disappearance greatly exceed those predicted by the simple hydrolysis model. The rate of NO_2^- oxidation by HOCl was studied by Lister and Rosenblum,[9] whose findings predict that $k_N = 4.29 \times 10^2$ M^{-1} min^{-1} at 25°C. Gray et al.[10] have determined that $K_h = 6.67 \times 10^{-12}$ M^{-1} at 25°C in 0.5 M NaClO$_4$. Using these values and $K_a = 5.01 \times 10^{-10}$ M, the predicted pseudo first-order rate constant describing loss of NH_2Cl under conditions shown in Figure 1 ([NO_2^-] = 18.8 mM, pH 7.52, [N_T]$_x$ = 9.42 mM) has a value of 3.44×10^{-7} min^{-1}. This is in comparison to a measured value of 3.51×10^{-2} min^{-1} which is five orders of magnitude larger than the predicted value. The observed rate constant is also over an order of magnitude larger than the constant characterizing the rate of NH_2Cl hydrolysis in 0.5 M NaClO$_4$,[11] which has been found to equal 1.14×10^{-3} min^{-1}.

The empirical rate expression is clearly not of a form identical to that predicted by the simple hydrolysis model, but differs by a factor of $[H^+]^{0.7}$ $[NO_2^-]^{0.7}$, with k_{ob} related to $\bar{k}$ by,

$$k_{ob} \propto \bar{k} \, [H^+]^{0.7} \, [NO_2^-]^{0.7} \qquad (12)$$

assuming that a single-term rate expression is indeed appropriate. The reader should be reminded that fractional dependencies are approximate, and since these dependencies may relate to ratios of integers, several such ratios (e.g., 1:2, 2:3) may be equally capable of being rationalized.

The chloramine disappearance presumably involves the oxidation of nitrite. Unfortunately, no products were identified to resolve the fate of NH_2Cl and NO_2^-. Several products, such as N_2O_4 and NO_3^-, are possible. Additionally, the observed loss could also involve the oxidation of nitrogen in the –III oxidation state.

The observed first-order dependence on NH_2Cl suggests that the reaction involves some oxidant species in direct proportion to the concentration of NH_2Cl. The inverse dependence on excess ammonia suggests an equilibrium with the rate-determining oxidant that produces some species chemically similar to NH_2Cl. These two dependencies are consistent with a mechanism in which the rate-determining oxidant behaves similarly to HOCl. The first-order dependence on NH_2Cl is also consistent with a reaction producing NO_3^-. However, a first-order dependence on NO_2^- would also be expected, since a one-to-one stoichiometry is involved.

The non-first-order and apparently identical reaction dependencies on H^+ and NO_2^- could reflect the catalysis of NH_2Cl hydrolysis by some species whose formation involves both H^+ and NO_2^-. The presence of nitrite has been shown to increase the rate of oxidation of bromide by NH_2Cl in a catalytic manner.[12] A catalytic process involving phosphate or sulfate could also be involved in the rapid formation of an oxidizing species that behaves similarly to HOCl in the presence of ammonia and is an intermediate in the reaction leading to NH_2Cl loss. Buffer salts have been found to catalyze reactions of chloramines[13] and bromamines.[14] Lastly, a direct but complex reaction between NH_2Cl and NO_2^- cannot be ruled out.

SUMMARY

Clearly, the loss of NH_2Cl is the result of some complex process. Several important observations have been made within the scope of this rather exploratory study.

The rate of NH_2Cl disappearance in the presence of NO_2^- was found to greatly exceed that anticipated, based on a simple NH_2Cl hydrolysis model. Furthermore, an empirically derived rate expression describing NH_2Cl disappearance was not identical to that derived from consideration of a simple NH_2Cl hydrolysis model. An unexpected dependence on H^+ and NO_2^- was observed, which precludes the existence of a simple mechanism. Catalytic processes or a direct reaction of NH_2Cl with NO_2^- may explain the enhanced rates.

The greatly increased rates of NH_2Cl disappearance above that predicted by simple NH_2Cl hydrolysis and the inverse dependence on excess ammonia suggest that the observed phenomena could be important under conditions of low ammonia concentration, such as during the rapid breakpoint reaction, particularly if the active oxidant is not HOCl. Additionally, reaction at relatively high ammonia levels may out-compete oxidation of NO_2^- by hydrolytically produced HOCl. The findings also suggest that the inorganic environment may have an important role in observed chloramine reaction kinetics and that extrapolation of laboratory results to environmental conditions should be done with extreme care.

ACKNOWLEDGMENTS

This work was supported in part by the U.S. Department of Energy under contract No. W-7405-eng-48 and The University of Iowa.

REFERENCES

1. Snoeyink, V. L., and D. Jenkins. *Water Chemistry* (New York: John Wiley and Sons, 1980), pp. 392, 399.
2. Palin, A. T. "A Study of the Chloro Derivatives of Ammonia and Related Compounds with Special Reference to Their Formation in the Chlorination of Natural and Polluted Waters," *Water Water Eng.* 54:151 (1950).
3. Pressley, T. A., D. F. Bishop, and S. G. Roan. "Ammonia-Nitrogen Removal by Breakpoint Chlorination," *Environ. Sci. Technol.* 6(7):622–628 (1972).
4. Wei, I. W., and J. C. Morris. "Dynamics of Breakpoint Chlorination," in *Chemistry of Water Supply Treatment and Distribution*, A. J. Rubin, Ed. (Ann Arbor, MI: Ann Arbor Science Publishers, Inc., 1974), pp. 297–332.
5. Saunier, B., and R. Selleck. "Kinetics of Breakpoint Chlorination and of Disinfection," Sanitary Engineering Research Laboratory Report No. 76-2 (Berkeley, CA: University of California, 1976).
6. Leao, S. F. "Kinetics of Combined Chlorine: Reactions of Substitution and Redox", Ph.D. Thesis, (Berkeley, CA: University of California, 1981).

7. White, G. C. *Handbook of Chlorination*, (New York: Van Nostrand Reinhold Co., 1972), p. 201.
8. *Standard Methods for the Examination of Water and Wastewater*, 14th ed. (Washington, DC: American Public Health Association, 1976).
9. Lister, M. W., and P. Rosenblum. "The Oxidation of Nitrite and Iodate Ions by Hypochlorite Ions," *Can. J. Chem.* 39:1645-1651 (1961).
10. Gray, E. T., Jr., D. W. Margerum, and R. P. Huffman. "Chloramine Equilibria and the Kinetics of Disproportionation in Aqueous Solution," in *Organometals and Organometaloids, Occurrence and Fate in the Environment*, F. E. Brinkman and J. M. Bellama, Eds. (Washington, DC: American Chemical Society, 1978).
11. Margerum, D. W., E. T. Gray, Jr., and R. P. Huffman. "Chlorination and the Formation of N-Chloro Compounds in Water Treatment," in *Organometals and Organometaloids, Occurrence and Fate in the Environment*, F. E. Brinkman and J. M. Bellama, Eds. (Washington, DC: American Chemical Society, 1978).
12. Valentine, R. L. "The Disappearance of Chloramines in the Presence of Bromide and Nitrite," DOE/NBM-1056 (DE38003460), (Washington, DC: U.S. Department of Energy, 1983).
13. Granstrom, L. M. "The Disproportionation of Monochloramine," Ph.D. Thesis, (Cambridge, MA: Harvard University, 1954).
14. Inman, G. W., Jr., and J. D. Johnson. "Kinetics of Monobromamine Disproportionation—Dibromamine Formation in Aqueous Solutions," *Environ. Sci. Technol.* 18(4):219-224 (1984).

CHAPTER 77

Subbreakpoint Modeling of the HOBr-NH$_3$-Org-N Reactions

Russell A. Isaac, Johannes Edmund Wajon, and J. Carrell Morris

The disinfection of wastewater has been practiced on a routine basis since the beginning of the twentieth century. The mainstay of this process, since the invention of a reliable gas feed system by Wallace and Tiernan, has been chlorine. The primary reason for disinfection remains the protection of public health, but like nearly everything else, trade-offs are involved. First, active chlorine in nearly all forms is acutely toxic to aquatic organisms at the concentrations used for wastewater disinfection. Second, minute, but nevertheless undesirable, concentrations of carbon-chlorinated compounds are formed during chlorination. The concentrations of these compounds not only may be toxic to aquatic organisms on chronic exposure, but they also are of concern because some bioaccumulate. Third, while chlorination is generally efficacious, the concentrations and forms of chlorine present in wastewater effluents do not control all of the infectious agents that may be present.

The most likely alternatives to chlorine for disinfection of wastewater are (1) ozone, (2) UV radiation, (3) bromine, (4) no disinfection.

Pilot plant studies on bromine have been or are being conducted and several full-scale applications of ozone and UV radiation are being implemented.[1] The proposal to eliminate disinfection, while anathema to many, is receiving consideration. Although some pilot-scale applications of various bromine compounds have been conducted, the chemistry of the HOBr-nitrogen compound system can be examined through the use of a mathematical model. This chapter discusses various aspects of bromine chemistry and its implications for wastewater disinfection.

Bromine has several apparent advantages over chlorine: first, bromamines generally are better disinfectants than their chloro counterparts; second, bromine residuals in wastewater effluents are less stable than those of chlorine, and this instability contributes to lower toxicity to aquatic life;[2] third, at pH values normally encountered in domestic wastewater effluents, a greater fraction of HOBr than HOCl is present. This provides better disinfection for the short period of time (~0.5 s) that the unreacted hypohalous acid is present.

To compare some of the chemical aspects of HOBr and HOCl, two major efforts have been undertaken: (1) a review of relevant aspects of bromine chemistry to select the most appropriate reactions to consider and the best estimate of the associated reaction rate constants; (2) the formulation of a simple model to compare results from this model with a similar one for chlorination.

The reasons for these efforts are to (1) organize and summarize the existing data; (2) identify the most important reactions for the time frame involved (maximum of 15 min); (3) provide as sound a comparison as possible between predictions for reactions involving chlorine, which is used extensively for wastewater disinfection, and bromine, which has been suggested as an alternative.

FORMATION OF INORGANIC BROMAMINES

Formation of NH_2Br

$$HOBr + NH_3 \rightarrow NH_2Br + H_2O \quad (1)$$

Although the products of $HOBr-NH_3$ reactions have been identified, the rates at which they form generally remain unquantified other than being rapid. The most extensive studies of $HOBr-NH_3$ kinetics have been done by Wajon[3] and Wajon and Morris[4,5] who focused on the formation of NH_2Br. They also investigated the rates of formation of several organic N-bromamines. The formation of NH_2Br varies with pH in a pattern similar to that for the formation of NH_2Cl; however, the maximum observed rate constant occurs at about pH 9 for NH_2Br as opposed to pH 8.3 for NH_2Cl. Wajon and Morris[5] reported that both HOBr and OBr^- will react with NH_3 to form NH_2Br, but the OBr^- reaction becomes significant only at high pH values (>pH 11). The reaction was determined to be second order, first order in each reactant. The temperature dependence was determined from data at 5 and 20°C, with the resulting Arrhenius expression of

$$k = 4.7 \times 10^{10} \exp(-15.7/RT) \; M^{-1} \; s^{-1} \quad (2)$$

with the activation energy in kilojoules per mol. This equation predicts a value for the rate constant at 25°C within a factor of 2 of that reported recently by Inman and Johnson.[6]

Formation of $NHBr_2$

$$HOBr + NH_2Br \rightarrow NHBr_2 + H_2O \quad (3)$$

No direct measurements of this reaction have been reported. Haag and Lietzke[7] estimated a reaction rate coefficient of about $7 \times 10^5 \; M^{-1} \; s^{-1}$ at 25°C. This value has been used for reactions at 20°C in the present model. This rate constant is adjusted for the portion of the total HOBr present as unionized HOBr at each pH.

$$NH_2Br + NH_2Br \rightarrow NHBr_2 + NH_3 \qquad (4)$$

The disproportionation of NH_2Br to form $NHBr_2$ and the back reaction have been investigated by Inman and Johnson.[6] Their recent report provides the following rate constants at 25°C for

$$2NH_2Br + H^+ \rightarrow NHBr_2 + NH_4^+ \quad k = 2.4 \times 10^8 \, M^{-2} \, s^{-1} \qquad (5)$$

$$NH_4^+ + NHBr_2 \rightarrow 2NH_2Br + H^+ \qquad (6)$$

The equilibrium constant $K_{eq} = 2.0 \times 10^{-10} \, M$. Because NH_3 provides a free pair of electrons for an exchange site and NH_4^+ does not, the reaction might be better described as

$$NH_3 + NHBr_2 \rightarrow 2 \, NH_2Br \qquad (7)$$

While these rate constants were determined at 25°C, other available data are limited to 20°C. Therefore, the rate constants for the forward and back reactions were adjusted to 20°C on the assumption of a twofold change in their values for a 10°C change in temperature.

Formation of NBr$_3$

Because the NH_3–N:HOBr ratio in a nonnitrified effluent is likely to be high (e.g., 15 mg/L ($1.07 \times 10^{-3} \, M$) NH_3–N, 1 mg/L ($6.26 \times 10^{-6} \, M$) Br_2, NH_3–N:HOBr molar ratio of 171), the predominant forms of inorganic bromamines will be NH_2Br and $NHBr_2$ once the initial reactions are completed. In addition, the presence of organic nitrogenous compounds that can react to form organic bromamines tends to limit the formation of di- and tribromamines. For these reasons, the formation of NBr_3 is assumed to be insignificant under these conditions and is not considered in the model.

FORMATION OF N-BROMO-ORGANIC COMPOUNDS

As is the case with HOCl, hypobromous acid will react with amines, amino acids, and other organic nitrogenous substrates to form N-bromo-organic compounds. The bromamines can form very rapidly and will constitute part of the total residual bromine in any brominated waters containing the nitrogenous substrates. The rates of formation of N-bromodimethylamine, bromoglycine, and bromoglutamate were determined by Wajon[3] and Wajon and Morris[5] using stopped-flow spectrophotometry. Experiments were conducted under pseudo first-order conditions, and the reactions were determined to be second-order—first-order in each reactant. The specific rate constants for the reaction of HOBr and nitrogenous substrate to form N-bromoglycine and N-bromoglutamate were determined at 20°C to be $3.8 \times 10^8 \, M^{-1}s^{-1}$ and $3.5 \times 10^8 \, M^{-1}s^{-1}$, respectively, but that for N-bromodimethylamine was too fast to be determined under the conditions and the equipment used. This implies that the specific rate constant for the formation of this latter product is about $3 \times 10^9 \, M^{-1}s^{-1}$; that is, the reaction appears to be diffusion limited. Specific rate constants of reactions between OBr$^-$ and the nitrogenous substrates are three to four orders of magnitude slower than those involving HOBr. Thus, for pH values normally encountered in environmental work, the reaction involving HOBr predominates.

The specific rate constants for the formation of N-halo compounds by the reaction of HOCl, NH$_2$Cl, or HOBr with organic nitrogenous compounds increase with increasing basicity of the nitrogen reactant.[3,5,8-10] For HOBr, the reaction apparently becomes diffusion limited for nitrogenous compounds with pK_b values between that for glutamate (pK_b = 4.17 at 20°C) and dimethylamine (pK_b = 3.44 at 20°C). Most amino acids are less basic than dimethylamine; therefore, formation of N-bromoamino acids is probably chemically controlled at the temperatures and pH values normally encountered in the environment.

DISAPPEARANCE OF TOTAL RESIDUAL BROMINE

The decrease of total residual bromine or any other oxidant in natural waters or wastewaters can be ascribed to at least two types of processes. One is the reduction of the oxidant through the reaction with another compound, a process generally referred to as "demand." The second is the decomposition of the active bromine autonomously through various pathways, including disproportionation. Few data are available on the demand reactions, for which the rates probably vary significantly. They will be considered no further at this time, and the focus will be on the direct (or autonomous) pathway.

Disappearance of Inorganic Bromamines

Some studies have investigated the decomposition of inorganic bromamines, but the process is not as well understood as it is for chloramines. It is tempting to speculate that NHBr$_2$ plays as much of a key role as NHCl$_2$ does in the disappearance of inorganic chloramines. The data available appear to present a more complicated picture; hence, this topic needs to be pursued. Wajon[3] has summarized qualitatively the effects of pH and NH$_3$:Br$_2$ ratios on the rate of decomposition of bromamines, as presented in Table I.

Monobromamine (NH$_2$Br) decomposes in apparently a first-order reaction. Wajon reported rate constants at 25°C that varied with pH. The constants ranged from 0.1×10^{-4} s^{-1} at pH 9 to 1.4×10^{-4} s^{-1} at pH 12, as summarized in Table II.[11,12] At pH 9, the $t_{1/2}$ is estimated to be 170 min, well beyond the 100-s time horizon used in the model. Therefore, this reaction has not been included

Table I. Effect of pH and Ammonia:Br$_2$ Ratio on Rate of Decomposition of Bromamines[a]

Compound	Effect on Decomposition Rate of Increasing	
	pH	NH$_3$:Br$_2$
NH$_2$Br	Increase	Decrease
NHBr$_2$	Decrease	Decrease
NBr$_3$	Increase	Increase

[a]From Reference 3.

Table II. First-Order Decay of NH$_2$Br

pH	N:Br	Temp (°C)	k$_1$ (h^{-1})	k$_1$ (s^{-1})	Ref.
12	≥ 100	25	5.1 × 10^{-1}	1.4 × 10^{-4}	11
10.5	≥ 100	25	2.0 × 10^{-1}	0.6 × 10^{-4}	11
9	450	25	0.5 × 10^{-1}	0.1 × 10^{-4}	11
	46	25	3.6 × 10^{-1}	1.0 × 10^{-4}	
	20	25	11 × 10^{-1}	3.1 × 10^{-4}	
	5	25	17 × 10^{-1}	4.7 × 10^{-4}	
Alkaline		0	3.4 × 10^{-1}		12
		15	21 × 10^{-1}		
		24	60 × 10^{-1}		

in the present model but will be incorporated into a revised model that will consider longer time frames.

Although all of the bromamines are unstable, dibromamine is the least stable, with more than 90% of the available bromine being lost in 30 min. The average rate of NHBr$_2$ disappearance has been characterized by Cromer et al.[13] and reported by Wajon[3] as

$$-\frac{d[NHBr_2]}{dt} = 3.3\ M^{-1}s^{-1}\ \frac{[NHBr_2]^{2.5}}{[NH_3]^{0.5}} + 13\ M^{-1}s^{-1}\ [NHBr_2]^2 \quad (8)$$

at 20°C.

The reaction for tribromamine (NBr$_3$) decomposition is second order with respect to OH$^-$ but varies inversely with HOBr concentration. Since very little if any NBr$_3$ would be formed under the conditions modeled, this reaction has not been incorporated into the present model.

Disappearance of Organic Bromamines

There are few studies on the decomposition of organic bromamines. Wajon investigated the decomposition kinetics of N-bromoglycine and N-bromoglutamate.[3] The results of the investigation revealed first-order rate constants that varied from $\sim 1 \times 10^{-4}$ s^{-1} to 3×10^{-3} s^{-1} (Table III) for N-bromoglycine over the pH range of 8.4 to 11.5 at 20°C. Wajon considered these estimates preliminary for many reasons, including the lack of verifying results from UV spectrophotometry with measurements of loss of titre. The rate of decomposition of N-bromoglutamate was also estimated from the loss of absorbance in the UV spectrum. The estimated value of the first-order rate constant was 1.2×10^{-3} s^{-1} at 20°C and pH 9.5.

DISINFECTION

The principal reason for treating an effluent with chlorine is to provide disinfection. Total and fecal coliform concentrations generally are the measures used to monitor the effectiveness of disinfection, and, although the appropriateness of these indicators is often questioned, they remain the standard by which effluents normally are judged. Therefore, the impact of alternative disinfection techniques on the indicators, as well as on target organisms, is of keen interest. The situation is complicated because the neutral and ionic forms of the free halogens have different effects on various organisms, and not always in a consistent relationship. While some free bromine or chlorine will exist during the initial mixing of the disinfectant in the wastewater effluent, and this transient (generally < 1 s) situation needs to be accounted for, most of the contact time will involve the exposure of organisms to halamines. Monochloramine (NH$_2$Cl) does have some disinfecting properties, but it is much less effective than free chlorine. Organic chloramines encountered in wastewater effluents appear to have little or no disinfectant capability. In contrast, inorganic bromamines appear to be nearly as effective disinfectants as free bromine. Although HOCl seems to be more effective than HOBr against enteric bacteria, HOBr is more effective than OCl$^-$, and OBr$^-$ appears to be more effective than OCl$^-$. The general relationship for enteric bacteria, based mainly on Wajon's review,[3] is

$$\text{HOCl} > \text{HOBr} = \text{NH}_2\text{Br} > \text{NHBr}_2 > \text{OBr}^- = \text{OCl}^- = \text{NHCl}_2 > \text{NH}_2\text{Cl} > \text{NBG} \tag{9}$$

Table III. Rate of Decomposition of N-Bromoglycine at 20°C[a]

pH	N:Br	k_1 (s^{-1})	$t_{1/2}$ (min)
11.5	500	2.9×10^{-3}	4
10.5	500	1.1×10^{-3}	10
9.5	250	2.2×10^{-4}	55
8.4	500	1.1×10^{-4}	105

[a]From Ref. 3.

(OBr⁻ is positioned on the basis of experiments with bacterial spores instead of with enteric bacteria, and $NHCl_2$ is positioned on the basis of judgment.)

Also, the organic bromamines may have some disinfecting power, as illustrated by the presence of N-bromoglycine (NBG) in Equation (9).

The exposure of organisms to disinfectant is calculated in terms of HOCl equivalents integrated over time. The relationships are:

Chlorine model

$$Exp = (HOCl + 0.01 * OCl^- + 0.0025 * NH_2Cl + 0.01 * NHCl_2) * \Delta t$$

Bromine model

$$Exp = (0.25 * HOBr + 0.0025 * OBr^- + 0.25 * NH_2Br + 0.125 * NHBr_2 + 6.25 \times 10^{-4} * NBG) * \Delta t$$

MODELING THE HOBr – NH₃ – OrgN SYSTEM

Given the limited but useful information previously summarized, a model of the active bromine in the HOBr–NH₃–OrgN system was constructed.

The model selected, previously discussed elsewhere,[14] describes substitution and autonomous decomposition reactions but not oxidation–reduction reactions involving carbonaceous compounds. Apparently, one cannot yet generalize about these demand reactions other than to state that they are highly variable and can be considered only on a case-by-case basis. The reactions and rate constants were chosen on what is judged to be the best information available that reflects conditions most likely to be encountered in nonnitrified wastewater effluents. Our objective was to compare the predictions for the HOBr–NH₃ system with those for the HOCl–NH₃ system under similar conditions. In each case, the distribution of the active halogen was assessed after a simulation of 100 s. This arbitrary time was chosen to ensure that all of the free halogen would be reacted and early patterns in the reactions would be revealed. The reactions considered for each system are presented in Tables IV and V.[15] These are considered to be the important reactions for each system under the conditions and constraints noted. In each case, the distribution of 1 mg/L of active halogen was assessed. Because of the lack of observed temperature dependence for many of the reactions in the HOBr system, predictions were restricted to 20°C.

Distributions of active halogen for both systems were calculated using first glycine (Gly) and then glycine ethyl ester (GEE) as the organic nitrogenous substrates. The rate constant for the HOBr-glycine ethyl ester reaction was estimated by extrapolation based on an assumed linear relationship between log k and pK_b of the nitrogenous compounds. The relationship was defined by using data from Wajon[3] and Wajon and Morris[5] on glycine and glutamate reactions with HOBr.

Table IV. Reactions and Rate Constants[a] for HOBr-NH$_3$-OrgN Model

Reactions modeled		
HOBr + NH$_3$	→ NH$_2$Br + H$_2$O	k$_1$
HOBr + NH$_2$Br	→ NHBr$_2$ + H$_2$O	k$_2$
NH$_2$Br + NH$_2$Br	→ NHBr$_2$ + NH$_3$	k$_3$
NHBr$_2$ + NH$_3$	→ NH$_2$Br + NH$_2$Br	k$_4$
NHBr$_2$	→ Decomposition[b]	k$_5$, k$_6$
HOBr + AMN	→ AMNBr + H$_2$O	k$_7$

Rate constants[a] (20°C)	Reference
k$_1$ = 7.44 × 10^7 M^{-1} s^{-1}	5
k$_2$ = 7.0 × 10^5 M^{-1} s^{-1}	6
k$_3$ = 2.4 × 10^8 M^{-2} s^{-1}	7
k$_4$ = 8.67 M^{-1} s^{-1}	7c
k$_5$ = 3.3 M^{-1} s^{-1}	11 as reported in 3
k$_6$ = 13 M^{-1} s^{-1}	11 as reported in 3
k$_7$ = 3.8 × 10^8 M^{-1} s^{-1} (Glycine)	3,5
8.6 × 10^6 M^{-1} s^{-1} (Glycine ethyl ester)	Estimate

[a]Fundamental rate constant.

[b] $$\frac{d\,(NHBr_2)}{dt} = \frac{-k_5\,(NHBr_2)^{2.5}}{(NH_3)^{0.5}} - k_6\,(NHBr_2)^2$$

[c]Calculated from information in this reference.

Table V. Reactions and Rate Constants[a] for HOCl-NH$_3$-OrgN Model

Reactions modeled		
HOCl + NH$_3$	→ NH$_2$Cl + H$_2$O	k$_1$
HOCl + NH$_2$Cl	→ NHCl$_2$ + H$_2$O	k$_2$
NHCl$_2$ + ½ H$_2$O	→ ½ N$_2$ + ½ HOCl + 3/2 H$^+$ + 3/2 Cl$^-$	k$_3$
AMN + HOCl	→ AMN−Cl + H$_2$O	k$_4$

Rate constants[a] (20°C)	Reference
k$_1$ = 3.81 × 10^6 M^{-1} s^{-1}	15
k$_2$[b] = 3.15 × 10^2 M^{-1} s^{-1}	15
k$_3$ = 8.56 × 10^4 s^{-1}	9
k$_4$ = 7.66 × 10^7 M^{-1}s^{-1} (Glycine)	9
6.71 × 10^6 M^{-1} s^{-1} (Glycine ethyl ester)	

[a]Fundamental rate constants.
[b]3.44 × 10^2 M^{-1} s^{-1} at pH 6, which reflects impact of acid (i.e., H$^+$) catalysis.

RESULTS

Under the conditions of the simulation, and with the time horizon of 100 s, monobromamine (NH$_2$Br) is the dominant species present at both pH 6 and 8. Significant concentration of dibromamine (NHBr$_2$) is predicted to be present at pH 6 but not at pH 8 (Figures 1 and 2). Organic bromamine is present at

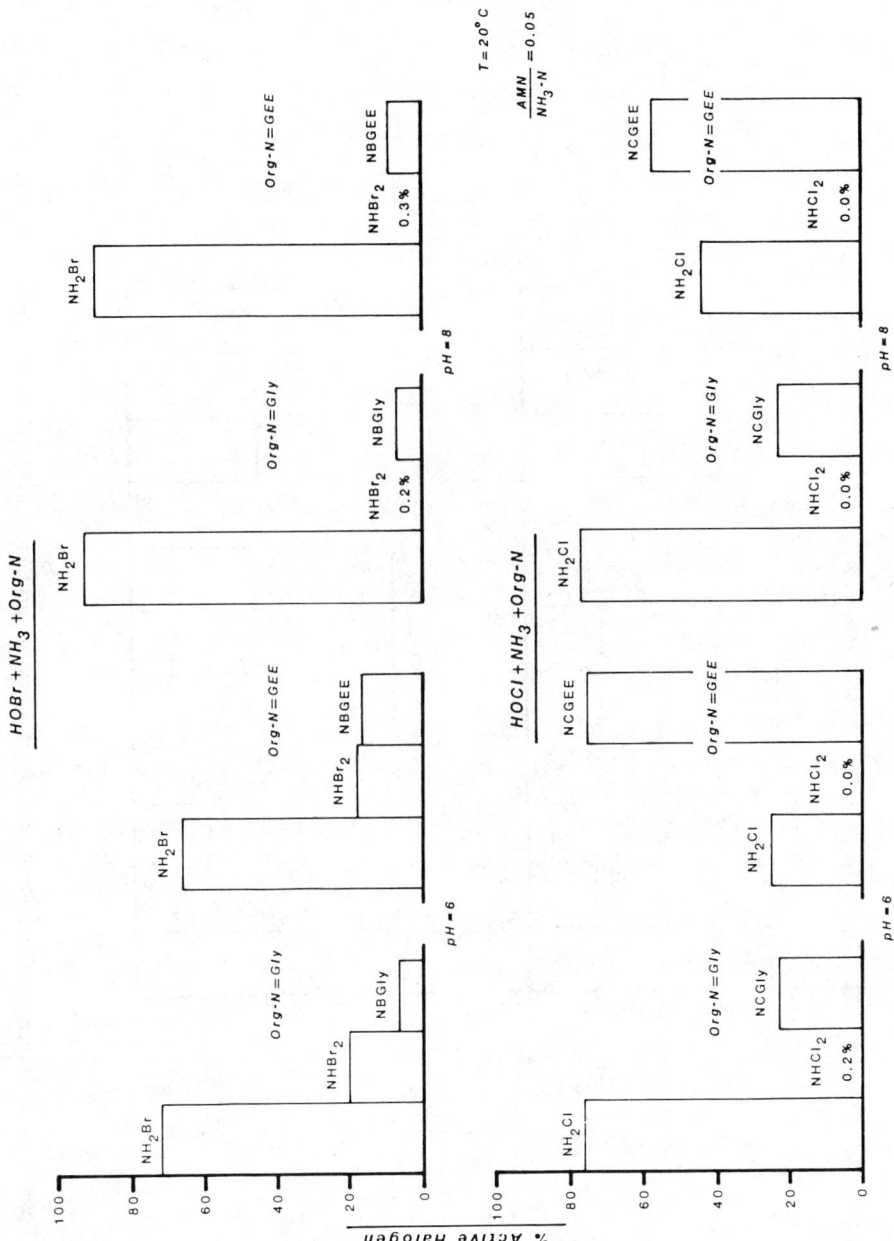

Figure 1. Model results: influence of Org-N species. Total active halogen, 1 mg/L; simulation time, 100 s.

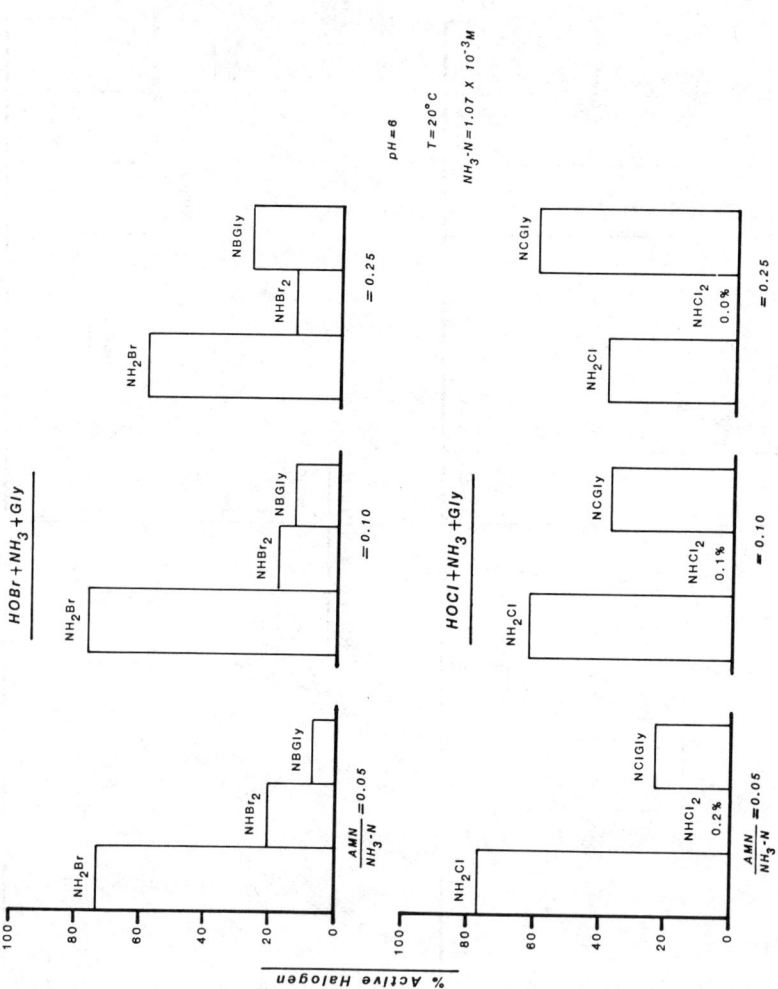

Figure 2. Model results: influence of Org-N:NH_3-N ratio on active halogen distribution at pH 6, T = 20°C. Total active halogen, 1 mg/L; simulation time, 100 s.

both pH values. The concentration of the organic bromamine not only varies directly with the relative concentration of the organic nitrogenous substrate, as depicted in Figure 3, but also varies in a more complicated way with the nature of the organic nitrogenous compound. For the conditions used, there is a higher concentration of organic bromamine predicted when glycine ethyl ester is used as the model organic compound than when glycine is used (Figures 1 and 2). This is despite the fact that the rate constant for the HOBr-Gly reaction is over an order of magnitude larger than that for the HOBr-GEE reaction, as presented in Table IV. This apparent discrepancy is derived from the difference in rate constants being more than offset by the higher concentration of the reactive form of GEE being present in this pH range compared with that of Gly. The observed rates are represented by

$$R_1 = k_1 (HOBr)(Gly) = k_1obs (HOBr)_T (Gly)_T \qquad (10)$$

$$R_2 = k_2 (HOBr)(GEE) = k_2obs (HOBr)_T (GEE)_T \qquad (11)$$

where (HOBr), (Gly), and (GEE) are the actual concentrations of the reacting species (neutral HOBr and anions of Gly and GEE), whereas $(HOBr)_T$, $(Gly)_T$, and $(GEE)_T$ are the total concentrations.

The major difference between the $HOCl-NH_3$-Org-N system and the HOBr system is the much greater predicted amount of $NHBr_2$, as compared with $NHCl_2$ at pH 6. Also, the HOCl system is impacted to a greater degree by the concentration and type of organic compound than is the HOBr system, as illustrated in Figures 1 through 3. In each system, the total active halogen remained 1 mg/L for the reactions and time (100 s) modeled.

Some runs were extended to 15 min, a time commonly used for wastewater disinfection, to assess the exposure to disinfectant in each system. The comparison is based on equating each species of disinfectant to an equivalent concentration of HOCl and then integrating with time.

Four observations can be made:

1. Disinfection is predicted to be not affected between pH 6 and 8 for the same halogen, but it is about 40 times greater in the HOBr system than in the HOCl system using the relationships discussed previously. This observation is made without considering the fact that total residual bromine disappears more rapidly in effluents than does total residual chlorine.

2. The decomposition of $NHBr_2$ is not significant at the concentrations predicted and in the time frame (15 min) used. In fact, the decomposition of NH_2Br, with a first-order rate constant of 10^{-4} s^{-1}, would produce a decrease of about 10% of the monobromamine in 15 min. In contrast to second-order reactions, the $t_{1/2}$ for a first-order reaction is not affected by the concentration of the substance. Since this reaction was not considered in the present model, a revised formulation is being developed to include it along with decomposition of the organic bromamine.

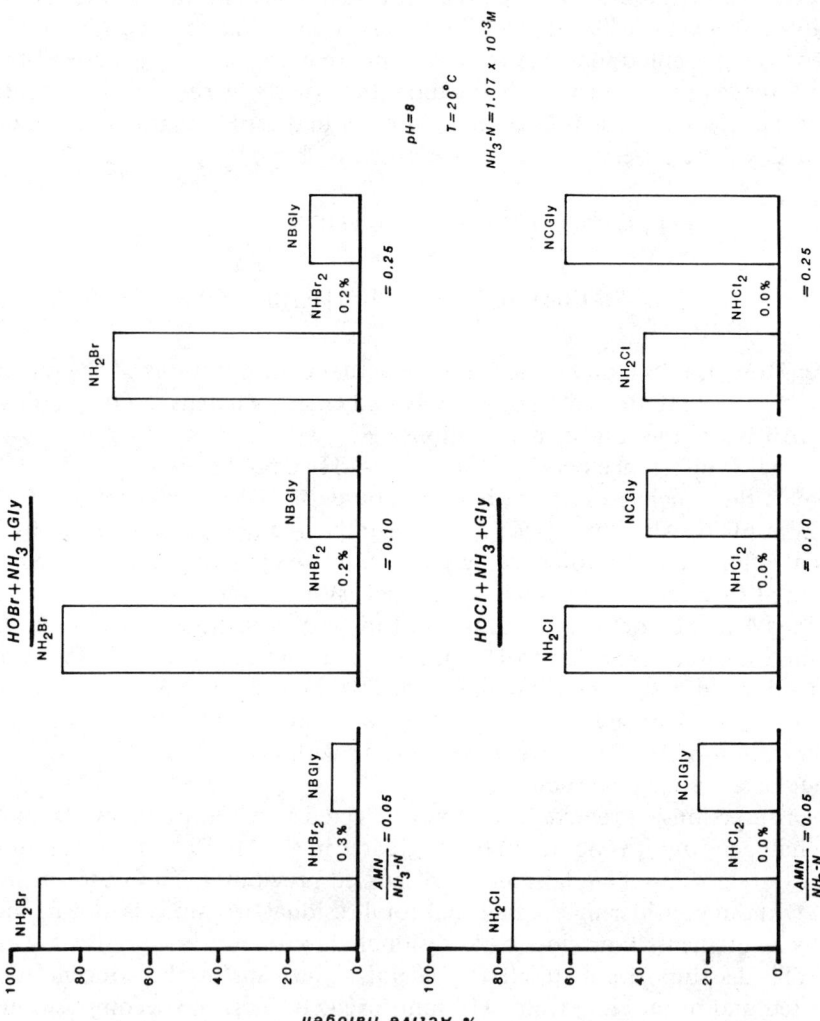

Figure 3. Model results: influence of Org-N:NH$_3$-N ratio on active halogen distribution at pH 8, T = 20°C. Total active halogen, 1 mg/L; simulation time, 100 s.

3. Since the so-called autonomous decomposition of active bromine would produce only about a 10% loss in 15 min, the relatively fast disappearance of total residual bromamine (and consequently acute toxicity) in pilot studies involving wastewater effluents apparently may be due primarily to demand reactions (or, probably less likely, to catalysis of the autonomous decomposition).

4. Total residual halogen is conserved at 1 mg/L in the present models for the 15-min times as well as in runs for 100 s.

Based on the results of the present bromine model, at least four areas require better definition:

1. Rate constant for the reaction of HOBr with NH_2Br.
2. Rate constants for demand reactions, especially for NH_2Br and $NHBr_2$ that form in effluents, but also for those involving HOBr that would affect the dose required to achieve 1 mg/L total residual bromine in an effluent. Some of the demand reactions involving HOBr or HOCl lead to the formation of carbon-chlorinated organic compounds, which are undesirable. The amount of compound formed may decrease with increasing concentrations of NH_3-N and Org-N because of the competitive nature of the two types of reactions. N-halamines presumably do not react to form carbon-halogenated compounds.
3. Transfer of active bromine from NH_2Br to organic nitrogenous compounds. Transfer is known to occur when NH_2Cl is involved, but the rate is relatively slow.[9,10,16]
4. Autonomous decomposition reactions of bromamines, especially organic forms.

ACKNOWLEDGMENTS

A portion of this work was supported by the U.S. Army Medical Research and Development Command Contract DAMD 17-77-C-7051, with David Rosenblatt as project manager. The authors thank Robert J. Kerrigan for the graphic illustrations and Theresa A. Vigneault for preparing the manuscript.

REFERENCES

1. Venosa, A. D. "Current State of the Art of Wastewater Disinfection," *J. Water Pollut. Control Fed.* 55(5):457–466 (1983).
2. Ward, R. W., R. D. Griffin, and G. M. DeGrave. *Disinfection Efficiency and Residual Toxicity of Several Wastewater Disinfectants VII*, EPA-600/2-77-203, (Washington, DC: U.S. Environmental Protection Agency, 1977).
3. Wajon, J. E. *Kinetics of N-Bromination Reactions*, Ph.D. Dissertation, (Cambridge, MA: Harvard University, 1980).
4. Wajon, J. E., and J. C. Morris. "Bromination Chemistry: Rates of Formation of NH_2Br and Some N-Bromamino Acids," in *Water Chlorination: Environmental Impact and Health Effects, Vol. 3*, R. L. Jolley, W. A. Brungs, and R. B. Cumming, Eds., (Ann Arbor, MI: Ann Arbor Science Publishers, Inc., 1980), pp. 171–181.

5. Wajon, J. E., and J. C. Morris. "Rates of Formation of N-Bromo Amines in Aqueous Solution," *Inorg. Chem.* 21:4258-4264 (1982).
6. Inman, G. W., Jr., and J. D. Johnson. "Kinetics of Monobromamine Disproportionation−Dibromamine Formation in Aqueous Ammonia Solutions," *Environ. Sci. Technol. 18* (4):219-224 (1984).
7. Haag, W. R., and M. H. Lietzke. "A Kinetic Model for Predicting the Concentrations of Active Halogen Species in Chlorinated Saline Cooling Waters," in *Water Chlorination: Environmental Impact and Health Effects, Vol. 3*, R. L. Jolley, W. A. Brungs, and R. B. Cumming, Eds. (Ann Arbor, MI: Ann Arbor Science Publishers, Inc., 1980), pp. 415-426.
8. Friend, A. G. *Rates of N-Chlorination of Amino Acids,* Ph.D. Dissertation, (Cambridge, MA: Harvard University, 1956).
9. Isaac, R. A. *Transfer of Active Chlorine from NH_2Cl to Organic Nitrogenous Compounds*, Ph.D. Dissertation, (Cambridge, MA: Harvard University, 1981).
10. Snyder, M. P., and D. W. Margerum. "Kinetics of Chlorine Transfer from Chloramines to Amines, Amino Acids, and Peptides," *Inorg. Chem.* 21:2545-2550 (1982).
11. Galal-Gorchev, H., and J. C. Morris. "Formation and Stability of Bromoamide, Bromoimide and Nitrogen Tribromide in Aqueous Solution," *Inorg. Chem.*, 4:899 (1965).
12. Moldenhauer, W., and M. Burger. "Monobromamine," *Chem. Ber.* 62:1615 (1929).
13. Cromer, J. L., G. W. Inman, Jr., and J. D. Johnson. "Dibromamine Decomposition Kinetics," in *Chemistry of Wastewater Technology*, A. J. Rubin, Ed. (Ann Arbor, MI: Ann Arbor Science Publishers, Inc., 1979), pp. 213-225.
14. Wei, I. W., and J. C. Morris. "Chlorine Ammonia Breakpoint Reactions: Model Mechanisms and Computer Simulation," presented to the 157th National Meeting of the American Chemical Society, April 13-18, 1969.
15. Morris, J. C., and R. A. Isaac. "A Critical Review of Kinetic and Thermodynamic Constants for the Aqueous Chlorine-Ammonia System," in *Water Chlorination: Environmental Impact and Health Effects, Vol. 4.*, R. L. Jolley, W. A. Brungs, J. A. Cotruvo, R. B. Cumming, J. S. Mattice, and V. A. Jacobs, Eds., (Ann Arbor, MI: Ann Arbor Science Publishers, 1983), pp. 49-62.
16. Isaac, R. A., and J. C. Morris. "Transfer of Active Chlorine from Chloramine to Nitrogenous Organic Compounds 1. Kinetics," *Environ. Sci. Technol.* 17:738-742 (1983).

CHAPTER 78

Reversibility in the Reactions of Chloramines with Bromide: Dimethylchloramine Reaction

Werner R. Haag

Residual mutagenic organic oxidants,[1] possibly chloramines,[2,3] are found in natural waters (to 9 μM),[2] wastewaters (to 25 μM),[2] and drinking waters (to 1 μM).[3] Stable chloramines can be produced from the chlorination of amines or their breakdown products[4] or by transfer of chlorine from monochloramine to organic amines.[5] Stable chloramines may go undetected because they sometimes respond in the free chlorine fractions of conventional analytical methods for combined chlorine,[6] although some may even survive dechlorination by sulfur dioxide.[7]

A possible fate of organic chloramines in cooling waters or in wastewaters discharged into saline waters is the reaction with bromide. For example, Helz et al.[8] believe that nearly all of the residual oxidants they found in a chlorinated estuarine cooling-water discharge were in the form of organic bromamines or chloramines, the former of which could be produced from the latter by reaction with the Br$^-$ that is present. Trofe et al.[9] demonstrated that monochloramine oxidizes Br$^-$ slowly at neutral pH and that the rate increases with decreasing pH. Antelo et al.[10] found no effect of 0.9 M Br$^-$ on the rapid decomposition of N-chloro-2-(methylamino)ethanol at pH 12.4; however, this does not preclude a reaction with Br$^-$ at neutral pH where the decomposition is much slower (days) and the Br$^-$ reaction should be much faster.

This study is an initial attempt to characterize the reactions of bromide with organic chloramines. Dimethylchloramine (Me$_2$NCl) was chosen as a substrate because it is relatively stable and only forms one halamine product (dimethylbromamine, Me$_2$NBr), thus greatly simplifying the interpretation of results.

EXPERIMENTAL

Chemicals

Chemicals were generally reagent grade used without further purification. Hypochlorous acid solutions were obtained from commercial chlorine bleach by adding enough AgNO$_3$ to precipitate the Cl$^-$, adjusting to pH 7.5 with

HNO_3, and vacuum distilling on a rotary evaporator at 30 to 35°C into a flask containing 5 drops of concentrated $HClO_4$ per 100 mL of solution. The preparation had a pH of 2 and was stable for several months in the dark at 4°C. Bromide-free hypobromous acid solutions were obtained by mixing HOCl and NaBr in an equal final concentration at pH 3. Me_2NCl and Me_2NBr were prepared by mixing buffered dimethylamine with HOCl or HOBr in an equal final concentration. Phosphate buffers (50 mM in final solution) were used to control the pH during kinetic and equilibrium measurements.

Analyses

Concentrations of hypohalites and halamines were determined by UV spectrophotometry and checked by iodometric titration (see Ref. 11 for absorbance data).

Kinetics and Equilibrium

Reactions were generally followed at 22 ± 1°C in 1- to 10-cm UV cuvettes by recording the increase or decrease in absorbance at 300 nm (the λ_{max} for Me_2NBr[11]). The spectra obtained before and after kinetic runs in the absence of Cl⁻ yielded molar absorptivities in agreement with complete conversion of Me_2NCl to Me_2NBr. The equilibrium constant was determined in the presence of Cl⁻ and Br⁻ in at least tenfold excess of the halamines (with catalytic quantities of HOBr) from the absorbances at 300 nm using the equation,

$$K_1 = \left(\frac{A - A_{Cl}}{A_{Br} - A}\right)\frac{[Cl^-]}{[Br^-]} \qquad (1)$$

where A is absorbance of the equilibrium mixture, and A_{Br} and A_{Cl} are absorbances of the corresponding pure solutions of Me_2NBr and Me_2NCl, respectively.

RESULTS AND DISCUSSION

Induction Period

Initial experiments over several hours and under conditions similar to those of seawater (pH 8, 1 mM Br⁻, 0.1 mM Me_2NCl) showed little or no reaction between Me_2NCl and Br⁻. Attempts to speed up the reaction to a measurable rate by lowering the pH and increasing the Br⁻ concentration resulted in curves typified by Figure 1. The observance of an induction period suggested that the direct reaction

$$Me_2NCl + Br^- \rightarrow Me_2NBr + Cl^- \qquad (2)$$

is very slow even at pH 5, but it is catalyzed by one of the reaction products. Addition of Cl^- did little to shorten the induction period, whereas addition of Me_2NBr greatly shortened the period but did not eliminate it.

At this point it was noted that the reacted solutions at low pH and high Br^- concentration always had an unexpected absorbance increase at 265 nm, and that this increase became even greater when Me_2NBr was added (Me_2NBr has λ_{max} at 300 nm[11]). It was therefore hypothesized that the increase in absorbance was because of Br_3^- (λ_{max} 265 nm, ϵ_{max} 38,000 $M^{-1}cm^{-1}$),[11] resulting from a partial breakdown of Me_2NBr (footnote 6 in Ref. 11 points out that Br_3^- does not brominate NH_4^+),

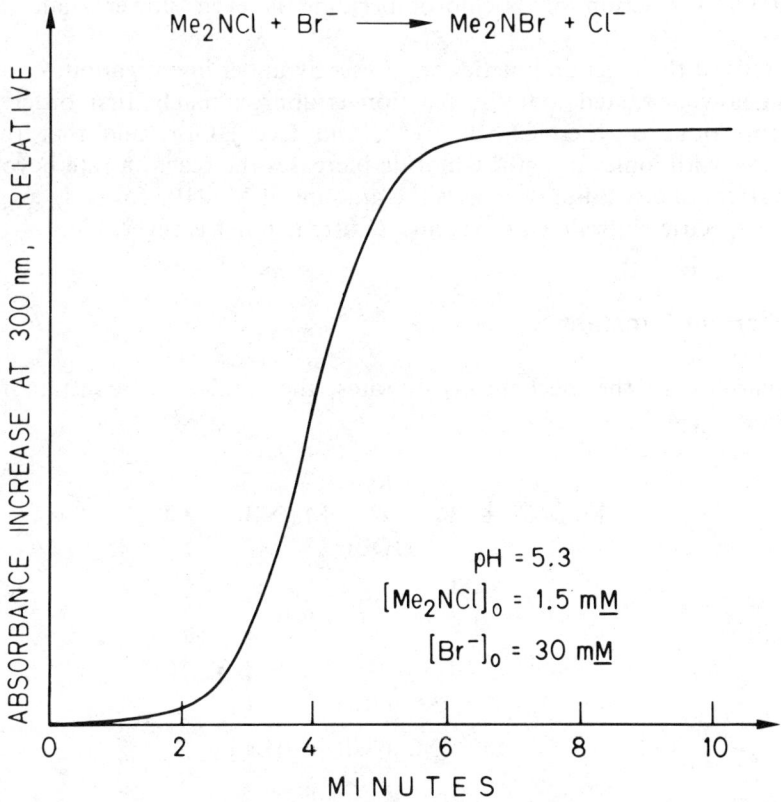

Figure 1. Concentration-time curve for the formation of Me_2NBr from Me_2NCl and Br^- in the absence of added HOBr.

$$Me_2NBr \underset{H^+}{\overset{}{\rightleftarrows}} Me_2\overset{+}{N}HBr \overset{Br^-}{\rightleftarrows} \begin{array}{c} Br_2 \overset{Br^-}{\rightleftarrows} Br_3^- \\ + \\ Me_2NH \underset{H^+}{\overset{}{\rightleftarrows}} Me_2\overset{+}{N}H_2 \end{array} \quad (3)$$

and that one of the active forms of bromine (e.g., $Me_2\overset{+}{N}HBr$, Br_2, HOBr) was responsible for the catalysis of the Me_2NCl + Br^- reaction. Indeed, the addition of small amounts of HOBr to mixtures of Br^- and Me_2NCl (without excess amine) greatly increased the reaction rate and completely eliminated the induction period. Thus, the curve in Figure 1 is explained as the result of a slow initial reaction (k < 0.001 $M^{-1}s^{-1}$) followed by a steady rate increase as the catalyst (product) concentration increases, then continuing until the limiting reactant (Me_2NCl) is consumed. Similar results were obtained using N-chloropiperidine, except that the induction period was longer, indicating that the uncatalyzed reaction of N-chloropiperidine is even slower than that of Me_2NCl.

Details of the reaction kinetics are presently under investigation;[12] however, it can now be stated that the reaction is approximately first order in the concentrations of Me_2NCl, Br^-, H^+, and free HOBr, and that the rate increases with ionic strength. Chloride increases the reaction rate beyond the ionic strength effect but decreases the amount of Me_2NBr formed, suggesting both a specific chloride catalysis and that reaction 1 is reversible.

Equilibrium Constant

Regardless of the mechanisms or rates, the equilibrium constant for the reaction

$$Me_2NCl + Br^- \underset{HOBr}{\overset{K_1}{\rightleftarrows}} Me_2NBr + Cl^- \quad (4)$$

where

$$K_1 = \frac{[Me_2NBr]}{[Me_2NCl]} \cdot \frac{[Cl^-]}{[Br^-]} \quad (5)$$

could be quantified when catalytic amounts of HOBr were added to mixtures of Me_2NCl, Br^-, and Cl^-, or when Cl^- was added to mixtures of Me_2NBr and Br^- containing some HOBr. Equilibrium was attained within several minutes at

HOBr concentrations (0.1 mM) too low to affect the absorbance spectra significantly, and K_1 was independent of the HOBr concentration. Representative spectra of equilibrium mixtures are given in Figure 2. The observance of an isosbestic point indicates smooth conversion of Me_2NBr to Me_2NCl without decomposition. The average of 14 measurements at pH 7 and 22°C gave K_1 = 42 ± 2 (dimensionless; error is 95% confidence limit).

The occurrence of Equation (4) is significant because it shows that Cl^- cannot be considered inert, as is often presumed; therefore, the equilibrium ratio [Me_2NBr]:[Me_2NCl] will depend on the [Cl^-]:[Br^-] ratio of the medium [see Equation (5)]. In seawater and estuarine water, [Cl^-]:[Br^-] equals 660, which yields an [Me_2NBr]:[Me_2NCl] ratio of 0.06, corresponding to a mixture of 94% Me_2NCl and 6% Me_2NBr. Since kinetic considerations suggest that the HOBr concentrations (0.1 mM) too low to affect the absorbance spectra significantly, and K_1 was independent of the HOBr concentration. Representative spectra of equilibrium mixtures are given in Figure 2. The observance of an

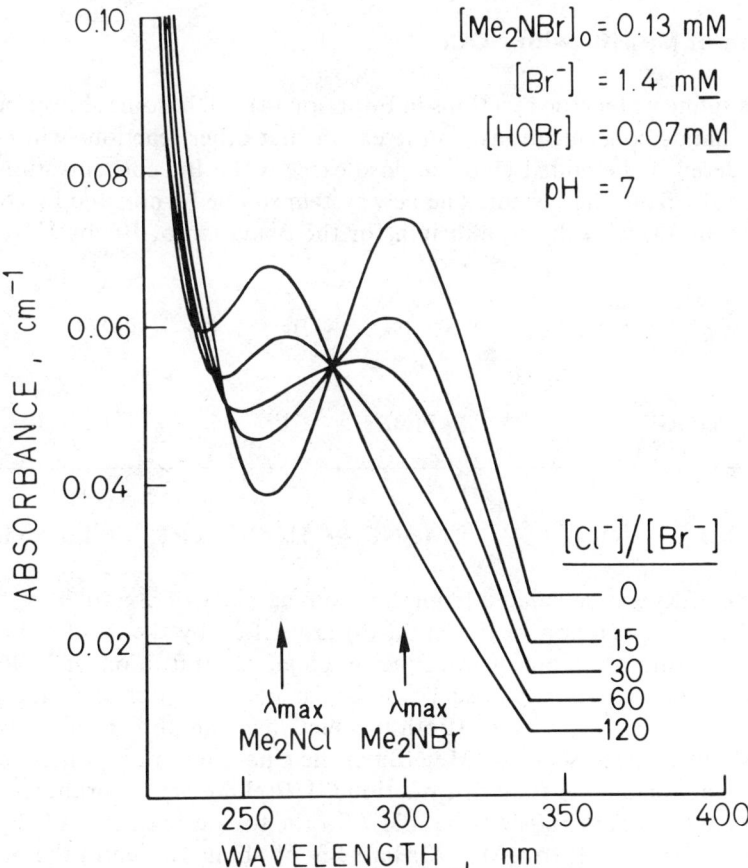

Figure 2. Spectra of equilibrium dimethylhalamine mixtures obtained several minutes after addition of Cl^- to solutions of Me_2NBr, Br^-, and HOBr.

isosbestic point indicates smooth conversion of Me_2NBr to Me_2NCl without decomposition. The average of 14 measurements at pH 7 and 22°C gave $K_1 = 42 \pm 2$ (dimensionless; error is 95% confidence limit).

The occurrence of Equation (4) is significant because it shows that Cl^- cannot be considered inert, as is often presumed; therefore, the equilibrium ratio $[Me_2NBr]:[Me_2NCl]$ will depend on the $[Cl^-]:[Br^-]$ ratio of the medium [see Equation (5)]. In seawater and estuarine water, $[Cl^-]:[Br^-]$ equals 660, which yields an $[Me_2NBr]:[Me_2NCl]$ ratio of 0.06, corresponding to a mixture of 94% Me_2NCl and 6% Me_2NBr. Since kinetic considerations suggest that the initial products of chlorination should contain a much higher percentage of the brominated amine, the main subsequent reaction in seawater would be the oxidation of chloride by dimethylbromamine rather than the reverse reaction. Note that this may be expected to occur at a low but significant rate,[12] because an excess of HOBr over amines and ammonia is expected to be present upon chlorination of a moderately clean seawater[13] and the reaction will be catalyzed.

Reaction of Me₂NBr with HOCl

In less saline waters the reactions in Equation (4) will become slower because of lower halide concentrations, to the extent that other reactions will need to be considered if the added chlorine dose exceeds the Br^- concentration, thus removing Br^- from the system. The new system can be formulated by combining Equation (4) with the equilibrium for the oxidation of Br^- by HOCl:

$$Me_2NBr + Cl^- \underset{}{\overset{1/K_1}{\rightleftarrows}} Me_2NCl + Br^- \quad 1/K_1 = 0.024 \quad (7)$$

$$HOCl + Br^- \underset{}{\overset{K_2}{\rightleftarrows}} HOBr + Cl^- \quad K_2 = 1.7 \times 10^5 \quad (8)$$

$$Me_2NBr + HOCl \underset{}{\overset{K_2/K_1}{\rightleftarrows}} Me_2NCl + HOBr \quad K_2/K_1 = 4.0 \times 10^3 \quad (9)$$

The value of K_2 was calculated from thermodynamic data presented by Sugam and Helz.[14] The reaction in Equation (9) is verified by the results shown in Figure 3. Within a minute of injecting an equal concentration of HOCl to a solution of Me_2NBr in the absence of Br^- at pH 7, complete formation of Me_2NCl + HOBr is observed. (Both products have an absorption maximum near 260 nm.) The absence of Me_2NBr in the final mixture is consistent with the high equilibrium constant of Equation (9). Preliminary experiments yield a rate constant of about 100 $M^{-1}s^{-1}$ at pH 7 for the forward reaction of Equation (9), which corresponds to a Me_2NBr half-life of about 10 min in the presence of 1 mg/L (14 μM) of free HOCl. Thus, even if Me_2NBr is formed initially as a kinetic product in the chlorination of waters containing traces of Br^- and Me_2NH, it will be unstable with respect to conversion to Me_2NCl if free HOCl is present.

GENERALIZATION OF RESULTS: REVERSIBILITY OF INORGANIC HALAMINE REACTIONS

One might speculate that Br^- oxidation by inorganic chloramines is also reversible. Based on thermodynamic data,[14] the reaction

$$NH_2Cl + Br^- \underset{}{\overset{K_3}{\rightleftharpoons}} NH_2Br + Cl^- \qquad (10)$$

has an equilibrium constant

$$K_3 = \frac{[NH_2Br]}{[NH_2Cl]} \cdot \frac{[Cl^-]}{[Br^-]} = 1 \times 10^3 \qquad (11)$$

at room temperature, thus giving a $[NH_2Br]:[NH_2Cl]$ ratio of only 1.5 at the $[Cl^-]:[Br^-]$ ratio of seawater. Experiments at pH 7 with active halogen to total

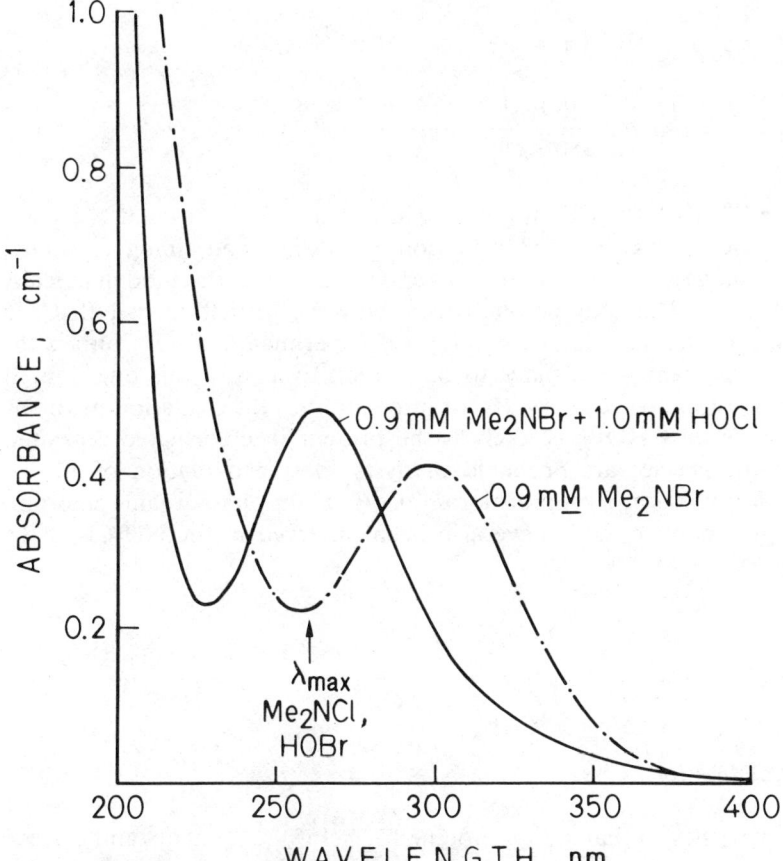

Figure 3. Spectra showing the formation of Me_2NCl + HOBr within 1 min of adding HOCl to Me_2NBr at pH 7 in the absence of Br^-.

ammonia ratios of 1:1000 (where NH_2Br does not disproportionate significantly to $NHBr_2$ and NH_4^+[15]) indeed have shown that NH_2Cl is consumed by Br^- at a $[Cl^-]:[Br^-]$ ratio of 200; however, NH_2Cl is produced (absorbance increase at 245 nm) from NH_2Br at a ratio of 7000. Thus, the equilibrium constant is roughly verified, which indicates that at higher halogen-to-ammonia ratios where disproportionation (i.e., multiple halogenation) occurs,[9]

$$NH_2Cl + NH_2Br + H^+ \rightarrow NHBrCl + NH_4^+ \qquad (12)$$

the oxidation of Br^- by NH_2Cl in saline waters goes to completion not because the reaction in Equation (10) is highly favorable, but because a product (NH_2Br) is continually removed.

Similar calculations, assuming a free energy of formation for NBr_2Cl of 71.0 kcal/mol [a reasonable estimate since the values[14] for NBr_3 (70.7) and NCl_3 (71.5) are very close], yield:

$$NBr_2Cl + Br^- \underset{}{\overset{K_4}{\rightleftarrows}} NBr_3 + Cl^-$$

$$K_4 = \frac{[NBr_3]}{[NBr_2Cl]} \cdot \frac{[Cl^-]}{[Br^-]} = 1 \times 10^5$$

This gives an $[NBr_3]:[NBr_2Cl]$ ratio of 170 in seawater. Thus, even if NBr_2Cl is an initial kinetic product in the chlorination of seawater containing ammonia, as has been suggested,[13,16] it does not appear to be a thermodynamically favored halamine. This may be one reason why the formation of NBr_2Cl in chlorinated seawater has not been observed experimentally. Of course, the kinetics of the system would have to be studied to answer this question. It would be interesting to see if free HOBr also catalyzes the oxidation of Br^- by NBr_2Cl, since excess HOBr is likely to be present in chlorinated seawater. Active bromine species are probable catalysts for the oxidation of Br^- by inorganic chloramines and bromochloramines (in addition to organic chloramines), since induction periods have also been observed in the NH_2Cl + Br^- system.[17]

ACKNOWLEDGMENT

The author wishes to thank Jürg Hoigné for helpful comments and discussion and for his support of this study.

REFERENCES

1. Cheh, A. M., J. Skochdopole, P. Koski, and L. Cole. "Nonvolatile Mutagens in Drinking Water: Production by Chlorination and Destruction by Sulfite," *Science* 207:90-92 (1980).
2. Larson, R. A., K. Smykowski, and L. L. Hunt. "Occurrence and Determination of Organic Oxidants in Rivers and Wastewaters," *Chemosphere* 10(11/12):1335-1338 (1981).
3. Cooper, W. J., M. F. Mehran, R. A. Slifker, D. A. Smith, J. T. Villate, and P. H. Gibbs. "Comparison of Several Instrumental Methods for Determining Chlorine Residuals in Drinking Water," *J. Am. Water Works Assoc.* 74(10:546-552 (1982).
4. Gould, J. P., and T. R. Hay. "The Nature of the Reactions Between Chlorine and Purine and Pyrimidine Bases: Products and Kinetics," *Water Sci. Technol.* 14:629-640 (1982).
5. Isaac, R. A., and J. C. Morris. "Rates of Transfer of Active Chlorine Between Nitrogenous Substrates," in *Water Chlorination: Environmental Impact and Health Effects, Vol. 3,* R. L. Jolley, W. A. Brungs, and R. B. Cumming, Eds. (Ann Arbor, MI: Ann Arbor Science Publishers, Inc., 1982), pp. 183-191.
6. Wajon, J. E., and J. C. Morris. "The Analysis of Free Chlorine in the Presence of Nitrogenous Organic Compounds," *Environ. Int.* 3:41-47 (1980).
7. Stanbro, W. D., and M. J. Lenkevich. "Slowly Dechlorinated Organic Chloramines," *Science* 215:967-977 (1982).
8. Helz, G. R., R. Sugam, and A. C. Sigleo. "Chemical Modifications of Estuarine Water by a Power Plant Using Continuous Chlorination," *Environ. Sci. Technol.* 18(3):192-199 (1984).
9. Trofe, T. W., G. W. Inman, and J. D. Johnson. "Kinetics of Monochloramine Decomposition in the Presence of Bromide," *Environ. Sci. Technol.* 14(5):544-549 (1980).
10. Antelo, J. M., F. Arce, J. Casado, R. Castro, M. E. Sanchez, and A. Varela. "Kinetics and Mechanism of the Decomposition of N-Chloro-2-(methylamino)ethanol in Aqueous Solution," *Environ. Sci. Technol.* 18(2):97-100 (1984).
11. Galal-Gorchev, H., and J. C. Morris. "Formation and Stability of Bromamide, Bromimide, and Nitrogen Tribromide in Aqueous Solution," *Inorg. Chem.* 4(6):899-905 (1965).
12. Haag, W. R. "Kinetics of the Oxidation of Bromide Ion by Dimethylchloramine," (Submitted for publication).
13. Haag, W. R., and M. H. Lietzke. "A Kinetic Model for Predicting the Concentrations of Active Halogen Species in Chlorinated Saline Cooling Waters," ORNL/TM-7942 (Oak Ridge, TN: Oak Ridge National Laboratory, 1981).
14. Sugam, R., and G. R. Helz. "Chlorine Speciation in Seawater; a Metastable Equilibrium Model for Cl^I and Br^I Species," *Chemosphere* 10(1):41-57 (1981).
15. Johnson, J. D., and R. Overby. "Bromine and Bromamine Disinfection Chemistry," *J. Sanit. Eng. Div., Am. Soc. Civ. Eng.* 97:617-628 (1971).
16. Haag, W. R., and R. L. Jolley. "Ultraviolet Absorption Spectra of Bromochloramine and Dibromochloramine in Ether," in *Water Chlorination: Environmental Impact and Health Effects, Vol. 4,* R. L. Jolley, W. A. Brungs, J. A. Cotruvo, R. B. Cumming, J. S. Mattice, and V. A. Jacobs, Eds. (Ann Arbor, MI: Ann Arbor Science Publishers, Inc., 1983), pp. 77-83.

17. Valentine, R. L., and R. E. Selleck. "Effect of Bromide and Nitrite on the Degradation of Monochloramine," in *Water Chlorination: Environmental Impact and Health Effects, Vol. 4,* R. L. Jolley, W. A. Brungs, J. A. Cotruvo, R. B. Cumming, J. S. Mattice and V. A. Jacobs, Eds. (Ann Arbor, MI: Ann Arbor Science Publishers, Inc., 1983), pp. 125–139.

SECTION XII

Photochemistry of Oxidants

The fact that light produces chemical reactions has been known since time immemorial.

     H. A. Olander
     Presentation of Nobel Prize for Chemistry,
     1967

But what is light really? Is it a wave or a shower of photons?

     Albert Einstein, 1879–1955, and
     Leopold Infeld, 1898–
     The Evolution of Physics, 1938

SECTION XII

Photochemistry of Oxidants

CHAPTER **79**

Degradation of Compounds in Water By Singlet Oxygen

Werner R. Haag and Jürg Hoigné

Singlet oxygen ($^1O_2$) is an electronically excited state of oxygen that is produced when wastewaters or natural waters are exposed to sunlight.[1-8] It oxidizes organic and inorganic compounds much more rapidly than does ground-state oxygen, and its products are generally more highly oxygenated and, therefore, less toxic and more readily biodegradable[9] (although the intermediate formation of unstable toxic peroxides is possible).[10] In addition, $^1O_2$ is a good algicide and disinfectant when present in a high enough concentration.[11,12] This chapter presents a summary of how singlet oxygen is formed in surface waters, what concentrations might be expected to be found, and the significance of such singlet oxygen concentrations for the removal of anthropogenic compounds.

Singlet oxygen has been studied extensively but primarily in nonaqueous systems. For general reviews see Kearns[9] or Wilkinson and Brummer;[13] for overviews concerning natural waters, see Zepp and Baughman[14] or Zepp et al.[3,4]

SINGLET OXYGEN FORMATION

Since oxygen does not absorb sunlight directly, a sensitizer (S) is required to absorb the radiation (see Figure 1).[5] Excited sensitizer is formed in its singlet state (^1_1S), which rapidly either deactivates or converts to the triplet state (^3_1S). Sensitizer triplets are generally so long-lived ($\geq$ milliseconds) that in toxic waters, the bulk of them are quenched by ground-state oxygen ($^3O_2$) to produce singlet oxygen.[14] The $^1O_2$ itself is deactivated by water within 4 μs,[15] resulting in low steady-state concentrations that are proportional to the production rates. The production rates, in turn, are proportional to the light intensity and the amount of sensitizer present (when absorption is low).[4,8] Evidence for this is given in Figure 2.

The steady-state singlet oxygen concentrations ($[^1O_2]_{SS}$) also depend on the absorption coefficients and quantum efficiencies (Φ^1O_2) of the sensitizers.[4] In natural waters and wastewaters, the sensitizers are the dissolved organic materials. Absorption of sunlight below 700 nm by dissolved organics varies from <5% to >90%, depending on the water source and the depth of the mixed

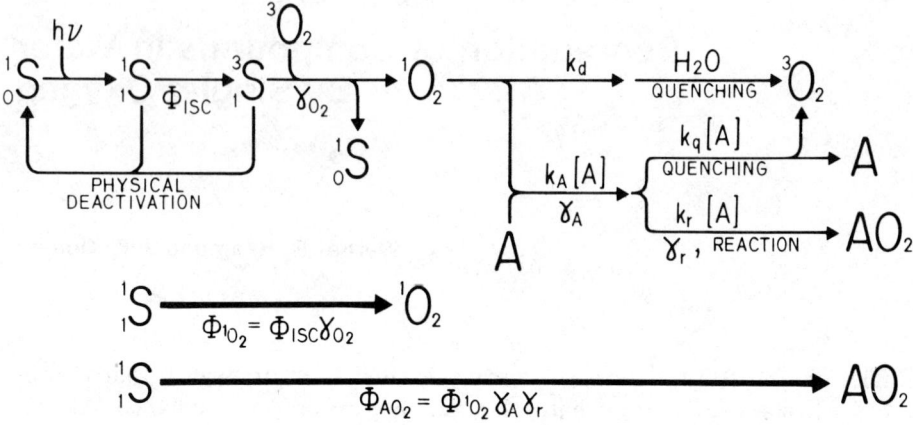

Figure 1. Photochemical pathways for singlet oxygen formation and consumption in water.[5] γ_x is defined as the fraction of precursor molecules following the pathway x, and k_x is the associated rate constant; S is any sensitizer (rose bengal, DOC, etc.); A is any $^1O_2$ acceptor (FFA, DMF, a micropollutant, etc.); and AO_2 is any product formed by consumption of both A and O_2.

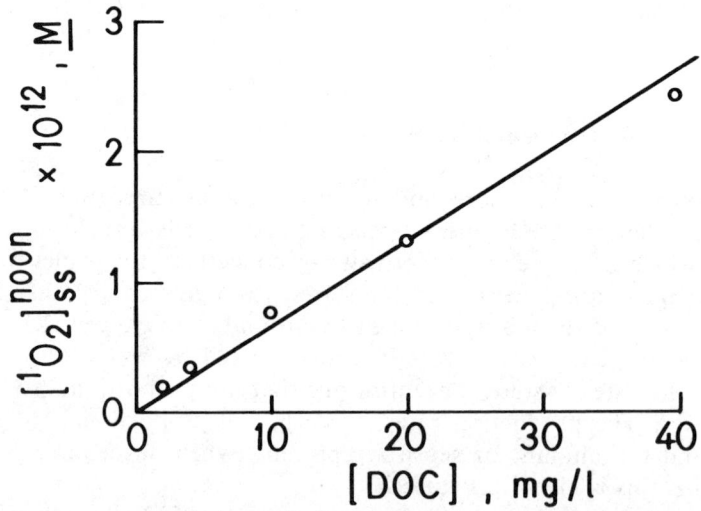

Figure 2. Steady-state singlet oxygen concentrations as a function of humic acid concentration, measured in test tubes under noon, summer, Dübendorf sunlight (47.5°N).[8] HA = humic acid, isolated from Black Lake, North Carolina.

layer considered. Quantum yields above 300 nm are typically low and decrease with increasing wavelength,[6,14,16] as shown in Figure 3. This results in steady-state concentrations of 10^{-13} to 10^{-12} M[1,6] during midday sunlight in the surface layer of the water. Sensitizers with high absorption coefficients and quantum yields such as riboflavin, rose bengal, or methylene blue may be added, which increases the concentrations of $^1O_2$ to about 10^{-10} M. This approach has been proposed by Acher and co-workers[11,12,17] for the disinfection and removal of organics from wastewaters. However, the method suffers from the difficulty of removing the sensitizer after the water has been treated. Other interesting types of sensitizers are oils as liquid films on water, which result from spillage.[10,18] In such cases, the aromatic hydrocarbons act both as sensitizers and oxidizable substrates, and the extremely high concentrations help maximize the efficiency of $^1O_2$ formation. However, in this case, the $^1O_2$ is produced at the oil and air interface and has little chance of entering the aqueous phase to oxidize dissolved organic compounds. We should point out that in all cases described above, other mechanisms such as radical reactions may also contribute to the oxidation of organics.

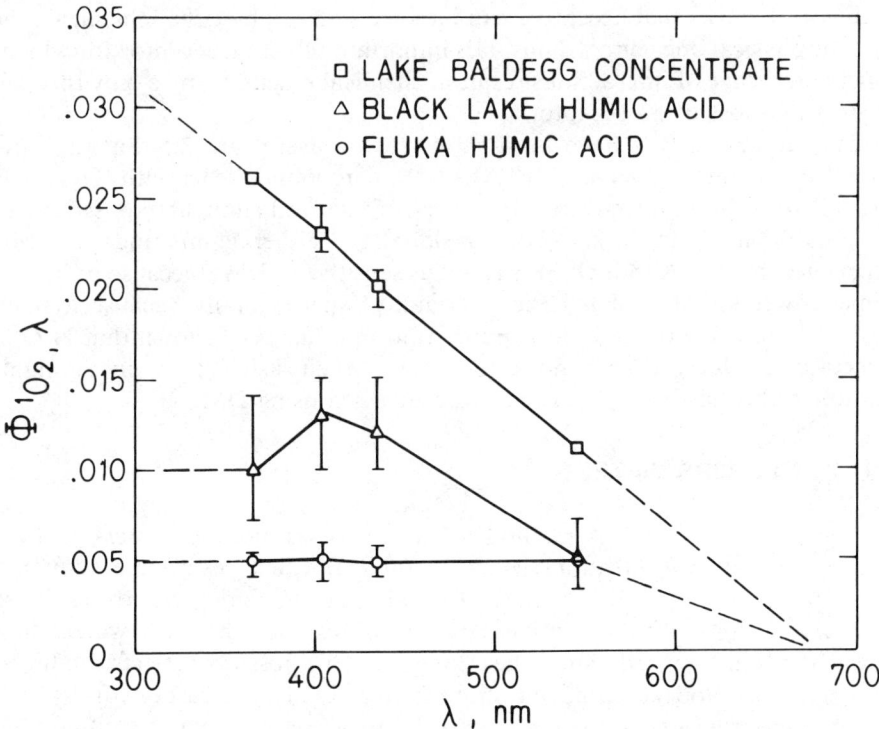

Figure 3. Quantum yields of singlet oxygen formation ($\Phi_{^1O_2,\lambda}$) as a function of wavelength.[6] Lake Baldegg is in Switzerland, Black Lake in North Carolina, and Fluka humic acid is a soil-derived commercial product.

MEASUREMENT OF SINGLET OXYGEN CONCENTRATIONS

The most convenient way to measure $[^1O_2]_{SS}$ is by adding a $^1O_2$-selective acceptor (A) to the water sample in a quartz tube and measuring its rate of disappearance when exposed to sunlight.[1,5] The acceptor loss should be first order, provided it is added in low enough concentration so as not to reduce the lifetime of the $^1O_2$ (normally $< 10^{-4}$ M)[4]. The $[^1O_2]_{SS}$ is then easily calculated from the slope of ln[A] vs time plots using the known value of k_r for the acceptor:

$$-\frac{d[A]}{dt} = k_r[^1O_2]_{SS}[A] = k_{meas}[A] \quad (1)$$

$$\ln[A]/[A]_0 = -k_{meas} t \quad (1a)$$

$$[^1O_2]_{SS} = \frac{k_{meas}}{k_r} \quad (2)$$

Figure 1 shows that acceptors can interact with $^1O_2$ by either chemical reaction or physical quenching. Thus, it is important to use an acceptor for which a high percentage of interactions result in chemical reaction, since only this path results in a loss of an acceptor.

Two acceptors that have been shown to be useful are 2,5-dimethylfuran (DMF)[1] and furfuryl alcohol (FFA).[5] Both compounds react with $^1O_2$ rapidly ($k_r > 10^8$ $M^{-1}s^{-1}$) and specifically (at least in natural sunlight-exposed waters containing no halogen or ozone residuals). Neither compound physically quenches the $^1O_2$. Although FFA is not as sensitive as DMF because of its five-times lower k_r value, it is easier to handle experimentally because it is not volatile from aqueous solution. Furthermore, it has been shown that H_2O_2, a reaction product, causes no interference when using FFA in sunlight,[5] although this has been suggested to occur when using DMF.[2]

$[^1O_2]_{SS}$ IN SWISS WATERS

Application of the FFA method to various types of Swiss waters gave the singlet oxygen concentrations listed in Table I.[7,8] The results are similar to those found in surface waters in the United States.[1,6] Notice that rivers, lakes, and wastewaters all give similar concentrations, but the wastewaters have lower DOC-normalized values (last column). This possibly reflects the higher percentage of biodegradable but photochemically inactive carbohydrate material in the wastewater compared to natural surface waters, which contain more refractory, polymeric, photoactive, humic materials. Supportive of this is the increase in the DOC-normalized value for the Werdhölzli wastewater after activated sludge treatment.

Table I. Steady-State Singlet Oxygen Concentrations[a] in Swiss Waters[b] Under Noon, Summer Sunlight

DOC Source	DOC (mg/L)	$t_{1/2}$ of FFA (h)	$[^1O_2]_{SS} \times 10^{14}$ (M)	$[^1O_{2SS}] \times 10^{14}$ (M per mg/L DOC)
Rivers				
Rhein	3.2	27	5.9	1.8
Kleine Emme	3.2	16	10	3.2
Glatt	4.1	15	11	2.6
Lakes				
Türlersee	8.3	24	6.7	0.8
Greifensee	3.5	20	8.0	2.3
Lützelsee	7.9	12	13	1.7
Etang de Gruère	13	6	25	1.9
Communal wastewaters				
Glatt, Secondary Effluent	8.6	11	14	1.1
Werdhölzli, Influent	31	15	11	0.3
Secondary Effluent	15	14	11	0.8
Zünikon Waste Stabilization Pond				
Inflow	20	11	15	0.7
Outflow	14	13	12	0.9
Added dyes				
5 mg/L rose bengal or methylene blue in pH 8 buffered distilled water	~2	0.02	7000	3500

[a]Measured in 1.8-cm OD quartz tubes and divided by 1.5 to normalize to flat water bodies.[7,8]
[b]Filtered through 0.45-μm pore size membrane.

The sample from Zünikon was taken in an attempt to see if $^1O_2$ could be an important oxidant for the removal of organics in waste stabilization ponds. Zünikon, a farm town of about 150 people, performs no wastewater treatment other than allowing it to biodegrade in the pond. The pond is eutrophic, and its water has a residence time of about 2 months. We found $^1O_2$ concentrations similar to those of other small lakes in both the effluent and the water near the influent, suggesting that the organic load (during the winter when the samples were taken) was not high enough to contribute significantly to the $^1O_2$ formation potential of the pond. Thus, even here, natural sensitizers probably play the dominant role in producing $^1O_2$.

IMPORTANCE OF $^1O_2$ FOR THE DEGRADATION OF COMPOUNDS IN SUNLIGHT

At the low concentrations formed in natural waters or wastewaters (Table I), $^1O_2$ cannot be important for the mineralization of the DOC compared to microbiological processes. (DOC removal is only observed when a sensitizer dye is added.[17]) However, it can be significant for the degradation of specific micropollutants. One can calculate the half-lives of micropollutants for which k_r is known by using the integrated form of Equation (1) and the measured value of $[^1O_2]_{ss}$. Figure 4 presents rate constants for various compounds[13] and their half-lives at a $^1O_2$ concentration typical of Swiss waters (2×10^{-13} M). (The chemical symbols were purposely made so large because the accuracy by which k_r values are known is generally worse than a factor of 2.[13])

Figure 4 shows that singlet oxygen is a highly selective, electrophilic oxidant. It reacts predominantly with compounds containing specific structural reaction centers (Diels-Alder type reactions), electron-rich double bonds, or those that are otherwise easily oxidizable. Any factors that tend to reduce the electron density at the active site also serve to slow the rate. Thus, protonation of the imidazole ring of histidine causes a reduction in rate constant from 10^8 $M^{-1}s^{-1}$ to an immeasurably low value ($< 10^6$ $M^{-1}s^{-1}$).[19] Similarly, chlorination of trimethylstyrene reduces the rate, but the effect is much less. Substituent effects are seen more clearly in Table II.[13,20] Here again the chlorinated derivative reacts slower than the parent compound, suggesting that water chlorination can reduce its activity towards $^1O_2$. For example, 8-chloroxanthine, a possible mutagen found in chlorinated effluents,[21] is likely to be much less reactive to $^1O_2$ than xanthine itself. Another chlorinated uracil derivative, terbacil, was shown to be degraded to less toxic products, but only after the addition of a sensitizer.[22] Chloroform cannot be destroyed by $^1O_2$, nor can Br^- be oxidized;[13] since NH_3 does not react significantly,[23] NH_2Cl should not either. Interaction with HOCl or HOBr must be very slow, because $^1O_2$ is produced in high concentration by the reaction of H_2O_2 with NaOCl or NaOBr.[13] Thus, $^1O_2$ does not contribute to the photochemical decomposition of oxidants, except indirectly, in that it may be involved in the sunlight-

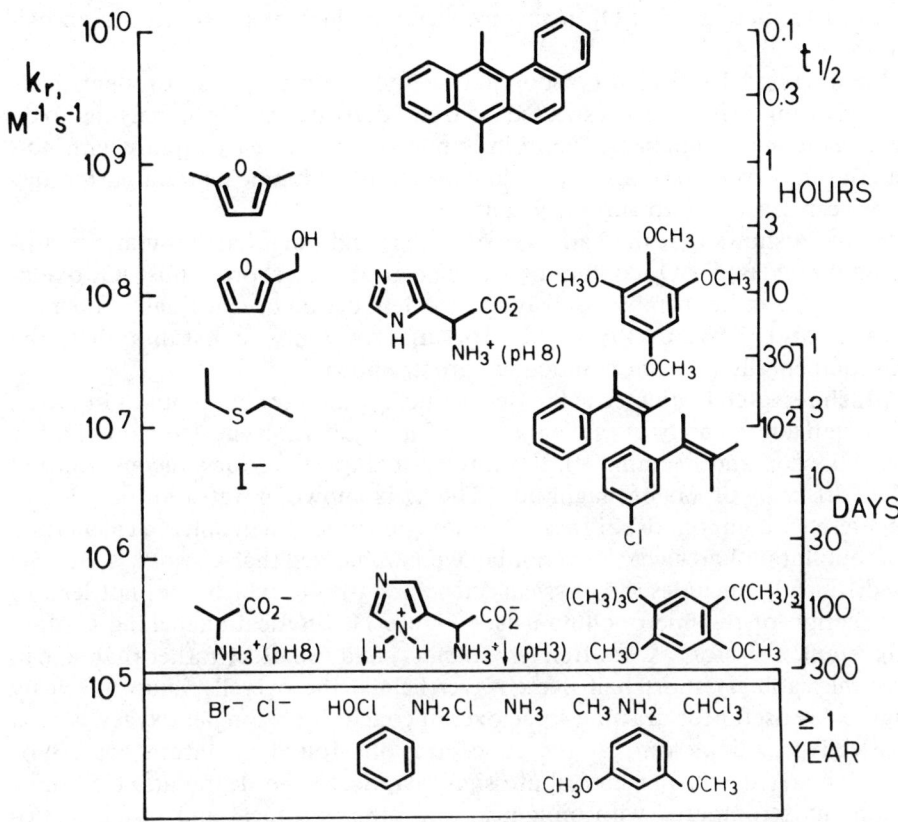

Figure 4. Rate constants for reaction of $^1O_2$ with various compounds[13,20] and their half-lives, assuming a steady-state concentration of 2×10^{-13} M. Rate constants[13] are either for aqueous solution or estimated as twice the value obtained in methanol or D_2O.

Table II. Rate Constants for Reaction of Singlet Oxygen with Substituted N,N-Dimethylanilines in MeOH[13]

Substituent	k_A $(M^{-1}s^{-1})$	Hammet Substituent Constant[20] (σ)
p-CN	2×10^6	+ 1.00[a]
p-CHO	2×10^6	+ 0.99[a]
m-Cl	$\sim 1 \times 10^7$	+ 0.35
p-Br	$\sim 2 \times 10^7$	+ 0.27
m-OCH$_3$	9×10^7	+ 0.14
-H	1×10^8	0.00
p-CH$_3$	2×10^8	− 0.17
p-OCH$_3$	3×10^8	− 0.32

[a]Values for the dissociation of protonated anilines; others are for benzoic acid dissociation.

induced formation of H_2O_2 in surface waters, which generally causes an oxidant demand.[24]

Figure 4 and Table II also show that methyl and methoxy substituents have an activating effect on substituted benzene derivatives. Thus, polymethoxylated moieties, as might be found in humic acids, can react rapidly with $^1O_2$, and this may be a cause for the gradual bleaching of humic materials after they have been leached into surface waters.[1,25]

Figure 4 shows that the half-lives of furans and polycyclic aromatic hydrocarbons (PAHs) in $^1O_2$-containing waters can be very short. Thus, $^1O_2$ oxidation may be an important pathway for the destruction of chlorinated dibenzofurans and certain PAHs. Unfortunately, rate constants for the environmentally relevant compounds are unknown.

Much research is still needed to determine k_r values for compounds in water. Although a large body of rate data exists for organic solvents (see compilation by Wilkinson and Brummer[13]), the corresponding values for aqueous solution often differ by orders of magnitude. The k_A is known in water for only about 30 reference compounds; of these, k_r is known for less than half. In calculating micropollutant half-lives, it cannot be overemphasized that k_r, not k_A, must be used, since k_A includes the physical quenching of $^1O_2$, which does not lead to destruction of the micropollutant (see Figure 1). Physical quenching is often important; therefore, k_r is often less than k_A, and use of k_A rather than k_r can give misleadingly short half-lives. Nevertheless, the overall picture given by Figure 4 is useful for drawing some overall conclusions: Singlet oxygen is such a selective oxidant that, at the concentrations found in natural waters or wastewaters, it will not contribute significantly to the degradation of most compounds compared with biological and other chemical and physical processes. Only when organic dyes are added do $^1O_2$ concentrations become high enough so that many types of compounds will be oxidized and the chemical oxygen demand will be reduced.

ACKNOWLEDGMENTS

We wish to thank A. M. Braun and E. Gassman for their valuable discussions, and W. Stumm for his continued support of these studies.

REFERENCES

1. Zepp, R. G., N. L. Wolfe, G. L. Baughman, and R. C. Hollis. "Singlet Oxygen in Natural Waters," *Nature* 267(5610):421-423 (1977).
2. Wolff, C. J. M., M. T. H. Malmans, and H. B. van der Heijde. "The Formation of Singlet Oxygen in Surface Waters," *Chemosphere* 10:59-62 (1981).
3. Zepp, R. G., G. L. Baughman, and P. F. Schlotzhauer. "Comparison of Photochemical Behavior of Various Humic Substances in Water: I. Sunlight Induced

Reactions of Aquatic Pollutants Photosensitized by Humic Substances," *Chemosphere* 10:109–117 (1981).
4. Zepp, R. G., G. L. Baughman, and P. F. Schlotzhauer. "Comparison of Photochemical Behavior of Various Humic Substances in Water: II. Photosensitized Oxygenations," *Chemosphere* 10:119–126 (1981).
5. Haag, W. R., J. Hoigné. E. Gassman, and A. M. Braun. "Singlet Oxygen in Surface Waters—Part I: Furfuryl Alcohol as a Trapping Agent," *Chemosphere*, 13(5/6):631–640 (1984).
6. Haag, W. R., J. Hoigné, E. Gassman, and A. M. Braun. "Singlet Oxygen in Surface Waters—Part II: Quantum Yields of its Production by Some Natural Humic Materials as a Function of Wavelength," *Chemosphere*, 13(5/6):641–650 (1984).
7. Haag, W. R., and J. Hoigné. "Singlet Oxygen in Surface Waters—Part III: Steady State Concentrations in Swiss Waters," submitted for publication.
8. Haag, W. R., J. Hoigné, E. Gassman, and A. M. Braun. "Steady State Singlet Oxygen Concentrations in Natural Waters Measured Using Furfuryl Alcohol," extended abstract, presented at the Conference on Gas-Liquid Chemistry of Natural Waters, Brookhaven National Laboratory, NY, April 1984.
9. Kearns, D. R. "Physical and Chemical Properties of Singlet Molecular Oxygen," *Chem. Rev.* 71(4):395–427 (1971).
10. Larson, R. A., L. L. Hunt, and D. W. Blankenship. "Formation of Toxic Products from a No. 2 Fuel Oil Photooxidation," *Environ. Sci. Technol.* 11(5):492–496 (1977).
11. Acher, A. J., and A. Elgavish. "The Effect of Photochemical Treatment of Water on Algae Growth," *Water Res.* 14(5):539–543 (1980).
12. Acher, A. J., and B. I. Juven. "Destruction of Fecal Coliform in Sewage Water by Dye-sensitized Photooxidation," *J. Appl. Environ. Microbiol.* 33:1019–1023 (1977).
13. Wilkinson, F., and J. G. Brummer. "Rate Constants for the Decay and Reactions of the Lowest Excited Singlet State of Molecular Oxygen in Solution," *J. Phys. Chem. Ref. Data* 10(4):809–999 (1981).
14. Zepp, R. G., and G. L. Baughman. "Prediction of Photochemical Transformation of Pollutants in the Aquatic Environment," in *Aquatic Pollutants: Transformation and Biological Effects*, O. Hutzinger, I. H. van Lelyveld, and B. C. J. Zoeteman, Eds. (Oxford: Pergamon Press, 1978), pp. 237–263.
15. Rodgers, M. A. J., and P. T. Snowden. "Lifetime of O_2 ($^1\Delta_g$) in Liquid Water as Determined by Time-Resolved Infrared Luminescence Measurements," *J. Am. Chem. Soc.* 104(20):5541–5543 (1982).
16. Zepp, R. G., P. F. Schlotzhauer, and R. M. Sink. "Photosensitized Transformations Involving Electronic Energy Transfer in Natural Waters: Role of Humic Substances," *Environ. Sci. Technol.* 19(1):74–81 (1985).
17. Acher, A. J., and I. Rosenthal. "Dye-sensitized Photooxidation—a New Approach to the Treatment of Organic Matter in Sewage Effluents," *Water Res.* 11:559–562 (1977).
18. Aksnes, G., and A. Iversen. "Photooxidation of Diphenylmethane and 1,2,3,4-Tetrahydronaphthalene as Liquid Film on Water," *Chemosphere* 12(3):385–396 (1983).
19. Sluyterman, L. A. "Photooxidation, Sensitized by Proflavine, of Furfuryl Alcohol, N-Allyl-Thiourea and Histidine," *Recueil Trav. Chim. Pay-Bas* 80:989–1002 (1961).

20. Dean, J. A., Ed. *Lange's Handbook of Chemistry, 12th Ed.* (New York: McGraw-Hill Book Co., 1979).
21. Jolley, R. L. "Chlorine-Containing Organic Constituents in Sewage Effluents," *J. Water Pollut. Control Fed.* 47:601–618 (1975).
22. Saltzman, S., A. J. Acher, N. Brates, M. Horowitz, and A. Gevelberg. "Removal of Phytotoxicity of Uracil Herbicides by Photodecomposition," *Pestic. Sci.* 13:211–217 (1982).
23. Joussot-Dubien, J., and A. Kadiri. "Photosensitized Oxidation of Ammonia by Singlet Oxygen in Aqueous Solution and in Seawater," *Nature* 227:700–701 (1970).
24. Helz, G. R., and R. Kieber. "Hydrogen Peroxide: A Minor but Significant Contributor to Chlorine Demand in Estuarine Waters," Chapter 81, this volume.
25. Gjessing, E. T. *Physical and Chemical Characteristics of Aquatic Humus* (Ann Arbor, MI: Ann Arbor Science Publishers, Inc., 1976), pp. 11–14.

CHAPTER **80**

Utilization of Molecular Oxygen and Sunlight in the Oxidative Purification of Water

William Cherry and Brian Jessen

Acher and Rosenthal[1] have proposed the use of dye-sensitized photooxidation for the treatment of wastewater as an alternative to biological treatment, chlorination, or ozonation. Photosensitized oxidation has been studied extensively on a laboratory scale, and much work has been done to discover the mechanisms of the process.[2] Chief among these is the singlet oxygen mechanism, by which an electronically excited dye molecule passes its energy on to an oxygen molecule, raising the latter from its triplet ground state to a more energetic singlet excited state. The more energetic oxygen molecule can enter into reactions that ground-state oxygen cannot, leading to oxidation of such substrates as olefins, amines, phenols, divalent sulfur compounds, and many heterocyclic compounds.

Several other mechanisms, not involving singlet oxygen, can readily occur under the proper circumstances. Dyes of many classes sensitize reactions initiated by visible light; among the more extensively studied dye classes are the thiazines, xanthenes, porphyrins, and flavines. The biological counterpart to photosensitize oxidation is traditionally called photodynamic action, by which organisms ranging from bacteria to cattle are killed or injured by light in the presence of sensitizing agents and air.[3]

Direct photolytic oxidation of organic substances in water by irradiation with the UV component of mercury lamp light has been successful.[4] However, the far UV radiation of sunlight is almost entirely absorbed by the ozone layer of the atmosphere. The role of sunlight in modifying organic compounds in the environment has been recognized,[5] and a few reports of sensitized oxidations, using the visible component of sunlight, have appeared. Sargent and Sanks reported the oxidation of phenols by methylene blue in solution.[6] Acher and Rosenthal have demonstrated reductions in the coliform count and the chemical oxygen demand of aerated sewage treated with methylene blue and exposed to sunlight.[1] Acher and Juven,[7] Gerba et al.[8] and Hobbs et al.[9] have observed photodynamic inactivation of coliforms and virus in sewage and other waters by methylene blue and solar or artificial illumination.

An obvious objection to the use of dye-sensitized photooxidation for water treatment is that the dye must be removed from the water after treatment or, if the dye is degraded during the process, its degradation products may themselves be toxic. To avoid this problem, the dye can be bound to polymer beads so that oxidation takes place in a heterogeneous system. The dye is then easily removed from the purified water by simple filtration. A further advantage is that binding to a support stabilizes many dyes to photolysis, thus reducing the amount of decomposition of the dye.

To determine the usefulness of polymer-bound (PB) dyes in water purification, the oxidation of chlorinated phenols has been examined. These were selected for study because they are common pollutants that are difficult to remove. Initially, oxidation in homogeneous solution was examined so that the oxidation products could be characterized. In particular, the fate of the chlorine atom and the aromatic ring was determined. The next step was to determine the effect of binding the dye onto the polymer. It was possible that the binding would lower the efficiency of oxidation and render the process useless on a commercial scale.

EXPERIMENTAL

The chlorophenols were obtained from Eastman Kodak Company and were purified by vacuum distillation. Pentachlorophenol was obtained from Aldrich Chemical Company (Gold Label, >99% purity) and used as received. The PB rose bengal was obtained from Hydron Labs and purified according to literature procedures.[10]

All ^{1}H NMR spectra were taken on a Varian EM-360 spectrometer and are reported in parts per million (ppm) relative to TMS. The ^{13}C spectra were taken on a Varian CFT-20 spectrometer and reported in ppm relative to TMS. In both cases, the solvent was $CDCl_3$ with 1% TMS. Infrared spectra were recorded on a Beckman IR-8 that was calibrated using the 1601 cm^{-1} absorption of polystyrene. Mass spectra were recorded on a Finnigan GC/MS spectrometer. Elemental analyses were carried out by Galbraith Labs, Inc., Knoxville, Tennessee.

Initially, the photolysis was performed in an immersion-type apparatus. In a typical experiment, 0.5 g chlorophenol, 0.22 g KOH (1 equiv), 40 mg rose bengal, and 300 mL ethanol were added to the reaction vessel. The lamp was a Sylvania FFJ 600W tungsten-halogen, which was run at 50 V. During photolysis, oxygen was slowly bubbled through the solution to ensure efficient formation of singlet oxygen.

When the PB rose bengal was used as the sensitizer, the photolysis was carried out in a stirred tube to prevent the polymer from settling. The tube was placed in front of the cooled lamp.

RESULTS AND DISCUSSION

Products of Photolysis

When *p*-chlorophenol is oxidized under the reaction conditions discussed above, inorganic chloride was produced and the aromatic ring was broken. Several major products resulted, two of which were purified by column chromatography (silica gel) followed by thin-layer chromatography (2-mm silica gel).

The first product had a molecular formula of $C_{12}H_{18}O_6$ as revealed by elemental analysis (55.05% C and 7.07% H) and mass spectrosocpy (mol wt = 258). The ^{1}H NMR showed signals due to the incorporation of three solvent molecules as well as several other singlets (NMR:7.15 δ, singlet, 2H; 4.50 δ singlet, 1H; 4.5 to 3.68, multiplet, 6H; 1.5 to 0.8 δ, multiplet, 9H). The H-decoupled ^{13}C NMR spectrum had the following lines: 129.0, 104.9, 68.4, 62.7, 60.5, 60.2, 36.1, 15.1, 14.0, and 13.8 δ (relative to TMS). The infrared spectrum showed C-H stretches (3050, 2940, and 2920 cm^{-1}), two carbonyl stretches (1745 and 1725 cm^{-1}) and an olefin stretch (1615 cm^{-1}). The above physical data can be accommodated by the diester (I).

(I)

The two different ethyl groups clearly show in the ^{1}H NMR. The other three protons comprise the remaining singlets. The ^{13}C NMR shows the expected olefin carbons (129 and 104 δ) as well as the methylene and methyl signals for three slightly different ethyl groups.

The second product had a molecular weight of 302 amu and a molecular formula $C_{14}H_{22}O_7$. The ^{1}H NMR consisted of three multiplets centered at 5.4 δ, 4.0 δ, and 1.2 δ, which integrated in the ratio of 1:8:12. The ^{13}C NMR consisted of lines at 129.1, 120.2, 104.0, 68.3, 62.1, 60.1, 60.1, 57.8, 36.2, 24.6, 15.1, 13.9, and 13.7 δ. The IR spectrum for this compound was virtually identical to that of the previous product except for the "fingerprint" region (600 to 1200 cm^{-1}). The above spectral data can be accommodated by the ethyl vinyl ether (II).

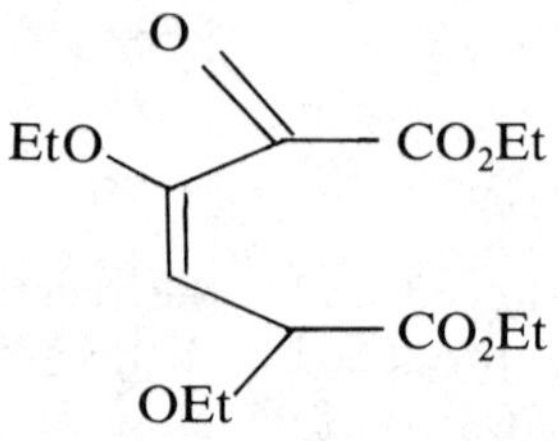

(II)

Several mechanisms for the production of I and II are reasonable. For example, the phenol could be oxidized to benzoquinone, which reacts further to yield the products. However, we attempted the oxidation of benzoquinone under the same reaction conditions and found that it was inert. Hence, it cannot be an intermediate. Another possible mechanism begins by electron transfer from the phenolate ion to singlet oxygen yielding the phenoxyl radical and the superoxide anion. The radical is then attacked in a nucleophilic manner by O_2^-. To test this hypothesis, we examined the oxidation of chlorophenol by O_2^- in acetonitrile and found no reaction. This indicates that O_2^- is not the reactive species.

A plausible scheme for the oxidation of chlorophenol is shown in Figure 1. The initial step is oxidative addition of $^1O_2$ to the phenolate anion, producing the dioxetane. Cleavage of the dioxetane results in formation of a dialdehyde, which subsequently undergoes nucleophilic addition and oxidation to the diester.

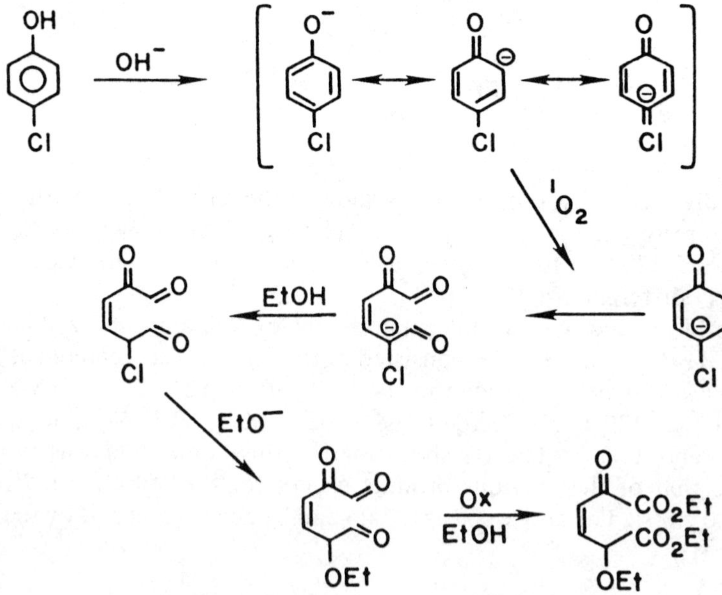

Figure 1. Plausible scheme for the oxidation of p-chlorophenol by $^1O_2$.

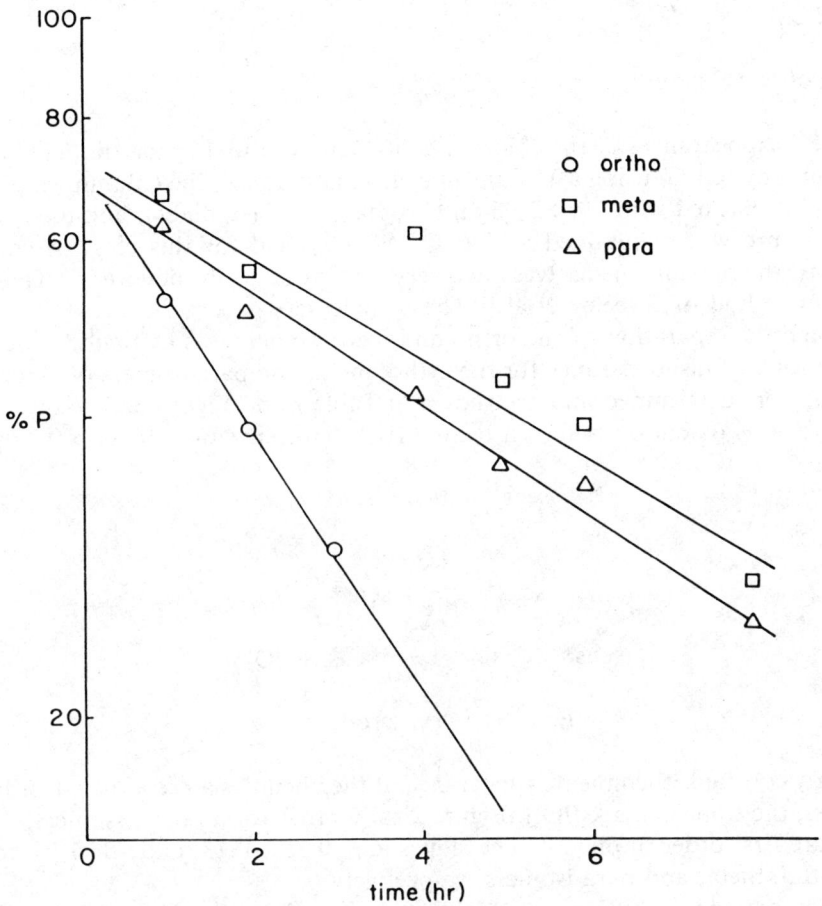

Figure 2. Disappearance of o-, m-, and p-chlorophenol in methanol.

Table I. Disappearance of o-, m, and p-Chlorophenol in Methanol

Time (h)	Relative Amount Chlorophenol		
	Ortho	Meta	Para
0	1.0	1.0	1.0
1	0.52	0.67	0.62
2	0.38	0.56	0.51
3	0.29		
4	0.08	0.61	0.42
5		0.43	0.36
6		0.39	0.34
8		0.28	0.24

Kinetics

In Alcoholic Solvents

The disappearances of the chlorophenols were followed by gas/liquid chromatography (GLC) using a Varian 1300 chromatograph. The column was 6 ft by 1/8 in., packed with 5% SF-96 on chromasorb W (Analabs), and the oven temperature was maintained at 140°C. As is typical for this class of compounds, the phenols themselves gave very poor results, so they were silylated by the method of Sweeley et al.[11] These silylethers gave good peak shapes, although the separation of the ortho and meta isomers was difficult.

The rates of disappearance for the ortho, meta, and para isomers of chlorophenol were determined and are shown in Table I and Figure 2. The mechanism for this oxidation is shown below. If the concentration of $^1O_2$

$$\text{sens} + h\nu \rightarrow \text{sens}^{*1}$$

$$\text{sens}^{*1} \xrightarrow{\text{isc}} \text{sens}^{*3}$$

$$\text{sens}^{*3} + {}^3O_2 \xrightarrow{k_q} \text{sens} + {}^1O_2$$

$$P + {}^1O_2 \xrightarrow{k_2} \text{products}$$

is nearly constant, then the disappearance of the phenol is a pseudo-first-order process, the slope being $k_2[^1O_2]$. Figure 2 shows that for all three isomers, the rates are first order in phenol. The slopes are -0.25, -0.13, and -0.17 h^{-1} for the ortho, meta, and para isomers, respectively.

In contrast to the above example, the oxidation of the phenols in the absence of base did not result in the disappearance of phenol. Apparently, the only reactive form is the phenolate anion. This is in agreement with previous examples.[12-14]

In Water

The oxidation of phenol in water gave results very similar to those in methanol. Again, the disappearance was followed by GLC. At specific times during the reaction, 3-mL aliquots were taken and the water removed in a vacuum evaporator. The phenols were then silylated and analyzed.

The results for the ortho, meta, and para isomers are shown in Table II and Figure 3. The rates for the ortho and para isomers are approximately the same as those in methanol. However, the ortho isomer is not as reactive in this solvent.

Table II. Disappearance of o-, m, and p-Chlorophenol in Water

Time (h)	Relative Amount Chlorophenol		
	Ortho	Meta	Para
0	1.0	1.0	1.0
1	0.62	0.87	0.79
2	0.48		0.65
3	0.38	0.52	0.58
4	0.08	0.32	0.46
5	0.10	0.29	0.29

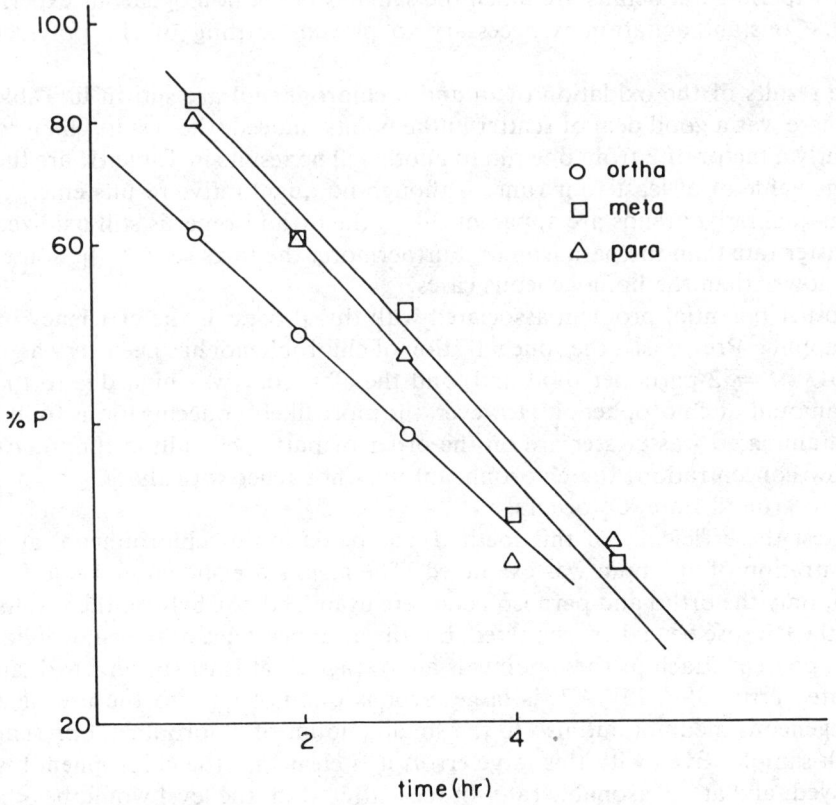

Figure 3. Disappearance of o-, m-, and p-chlorophenol in water.

Again, there was no reaction if base was not added. Furthermore, the solution pH gradually decreased as the reaction progressed, demonstrating that acid was being produced, probably hydrochloric acid, which was produced from the nucleophilic substitution of the chlorine.

Oxidation Using PB Dye

The greatest problem with this method of water purification is the requirement that a dye be present to sensitize the production of $^1O_2$. This obviously contaminates the water and may result in no net purification, but rather substitution of one impurity for another. A method of overcoming this problem is to immobilize the dye on a polymer support. This renders the dye insoluble and, hence, easily removable. On the other hand, PB dye is likely to be a less effective catalyst. To determine if PB rose bengal is an effective sensitizer, we have examined the kinetics of chlorophenol oxidation using rose bengal bound to hydrophilic polymers.[10]

The experimental details are much the same as in the homogeneous experiments. Constant agitation is necessary to prevent settling of the polymer beads.

The results of the oxidation of o- and p-chlorophenol are shown in Table III. There was a good deal of scatter in the points; indeed, the results changed by nearly a factor of 2 from one run to another. The results in Table III are the average value of at least four runs. Although no quantitative results emerge, certain qualitative trends are apparent. First, the ortho isomer is still oxidized at a faster rate than the para isomer. Furthermore, the rates seem to be somewhat slower than the homogeneous cases.

Another potential problem associated with this process is the efficiency of $^1O_2$ trapping. Previously, the concentration of chlorophenol has been very high ($\sim 0.013\ M$ = 2 parts per thousand) and the efficiency was high due to the large amount of chlorophenol. However, the more likely concentrations found in contaminated wastewater are on the order of parts per million (ppm). At this low concentration, the chlorophenol may not react with the $^1O_2$ before decay to ground state $^3O_2$ occurs.

To test the efficiency of this method, the oxidation of chlorophenol at a concentration of 100 ppm was examined. The results are shown in Table IV. Again, only the ortho and para isomers were examined. As before, the results using the PB rose bengal are scattered, but the disappearance of the chlorophenol is apparent. Each of these points is an average of at least six runs with an estimated error of $\pm 15\%$. This large error is due not only to the use of a heterogeneous medium but also to the small amount of chlorophenol present in each sample. Even with this large error, it is clear that the chlorophenol is destroyed, and at a reasonable rate. In fact, after 10 h, the level would be ≤ 5 ppm; hence, this method is sufficient to oxidize chlorophenol at any concentration.

Table III. Disappearance of o- and p-Chlorophenol Using Polymer-Bound Rose Bengal

Time (h)	Relative Amount Chlorophenol	
	Ortho	Para
0	1.0	1.0
1	0.63	0.83
2	0.63	0.82
3	0.48	0.79
4	0.38	0.69

Table IV. Disappearance of o- and p-Chlorophenol in Water at 100 ppm Using Polymer-Bound Rose Bengal

Time (h)	Relative Amount Chlorophenol	
	Ortho	Para
0	1.0	1.0
0.25	0.74	0.81
0.50	0.62	0.71
0.75	0.56	0.67
1.00	0.63	0.72
1.25	0.49	0.63
1.50	0.42	0.55
2.00	0.39	0.57
2.50	0.28	0.46
3.00	0.30	0.55
4.00	0.22	0.52
5.00	0.20	0.38

Oxidation of Pentachlorophenol

The pentachlorophenol was oxidized in the same manner as the monochlorophenols. Only the reaction using the PB rose bengal in water was studied, and the results are listed in Table V. The disappearance of the pentachlorophenol is much faster than that of the monochlorophenol. In fact, the pentachlorophenol is completely decomposed after only 3 h. Consequently, this would be an extremely efficient method for removing pentachlorophenol from wastewater.

CONCLUSION

Singlet oxygen is very effective in oxidizing chlorinated phenols. The aromatic ring is destroyed, and the chlorine becomes an inorganic chloride ion.

Table V. Disappearance of Pentachlorophenol (PCP) Using Polymer-Bound Rose Bengal in Water

Time (h)	Relative Amount PCP
0	1.00
0.5	0.81
1.0	0.60
1.5	0.41
2.0	0.32
3.0	0.13
3.5	0.02

The kinetics of the oxidation are pseudo first order, and the rates are reasonable with respect to the potential use of $^1O_2$ for the purification of wastewater. The oxidation is feasible even when the dye is bound to a solid support such as a polymer bead. Even at low concentrations of chlorinated phenol, the oxidation is efficient. Consequently, this seems to be a reasonable method for the oxidative purification of wastewater.

Although our results clearly indicate the potential of this process, it is still only in the developmental stages. Further work on the oxidation of the other common contaminants is required. Also, a large-scale experiment using actual wastewater is necessary. It is possible that trace contaminants in the wastewater may deactivate the $^1O_2$ before any oxidation occurs.

REFERENCES

1. Acher, A. J., and I. Rosenthal. "Dye Sensitized Photo-oxidation—A New Approach to the Treatment of Organic Matter in Sewage Effluents," *Water Res.* 11:557 (1977).
2. Foote, C. S. "Mechanism of Photosensitized Oxidation," *Science* 162:963 (1968).
3. Spikes, J. D. "Photodynamic Action," in *Photophysiology*, Vol. III, A. C. Giese, Ed., (New York: Academic Press, 1968).
4. Schorr, V., B. Boval, V. Hancil, and J. M. Smith. "Photo-oxidation Kinetics of Organic Pollutants in Municipal Wastewater," *Ind. Eng. Chem. Process Des. Dev.* 10:509 (1971).
5. Zepp, R. G. and D. M. Cline. "Rates of Direct Photolysis in Aquatic Environment." *Environ. Sci. Technol.* 11:359 (1977).
6. Sargent, J. W., and R. L. Sanks. "Light-Energized Oxidation of Organic Wastes," *J. Water Pollut. Control Fed.* 46:2547 (1974).
7. Acher, A. J., and B. J. Juven. "Destruction of Coliforms in Water and Sewage Water by Dye-Sensitized Photo-oxidation," *Appl. Environ. Microbiol.* 33:1019 (1977).
8. Gerba, C. P., C. Wallis, and J. J. Melnick. "Disinfection of Wastewater by Photodynamic Oxidation," *J. Water Pollut. Control Fed.* 49:575 (1977).

9. Hobbs, M. E., et al, "Photodynamic Inactivation of Infectious Agents," *J. Environ. Eng. Div. ASCE*, 103:459 (1977).
10. Schapp, A. P., A. L. Thayer, E. C. Blossey, and D. C. Nekers, "Polymer-Based Sensitizers for Photo-oxidations," *J. Am. Chem. Soc.* 97:3741 (1975).
11. Sweely, C. C., R. Bentley, M. Makita, and W. W. Wells. "Gas-Liquid Chromatography of Trimethylsilyl Derivatives of Sugars and Related Substances," *J. Am. Chem. Soc.* 85:2497 (1963).
12. Seely, G. R., and R. L. Hart. "Preparation of Stained Beads for Photosensitized Oxidation of Organic Pollutants," *Environ. Sci. Technol.* 11:623 (1977).
13. Seely, G. R., and R. L. Hart. "Photosensitized Oxidation by Stained Alginate Beads," *Photochem. Photobiol.* 26:655 (1977).
14. Seely, G. R., and R. L. Hart. "The Photosensitized Oxidation of Tyrosine Derivatives in the Presence of Alginate-I: Reaction under Homogeneous Conditions," *Photochem. Photobiol.* 23:1 (1976).

CHAPTER 81

Hydrogen Peroxide in Estuarine Waters: A Minor But Significant Contributor to Chlorine Demand

George R. Helz and Robert J. Kieber

At present, the exact chemical mechanisms that account for chlorine decay in natural waters are not very well known. Particularly poorly understood are the very fast processes that eliminate (within a few minutes) a major fraction of the chlorine used in fouling control at power plants.[1,2] Although much of this fast chlorine decay is probably caused by reactions with organic components, the role of inorganic components has not yet been thoroughly evaluated. One potentially significant component is hydrogen peroxide (H_2O_2).

We know that H_2O_2 reacts rapidly with free chlorine[3-5] to produce singlet molecular oxygen and chloride ions; thus, H_2O_2 has been proposed as a dechlorinating agent.[6] However, its relatively slow reaction with combined chlorine limits its usefulness.

Relatively little is known about H_2O_2 concentrations in natural waters. More than a decade ago, Van Baalen and Marler[7] reported concentrations of 0.05 to 0.7 μM in surface waters of the Gulf of Mexico; Sinel'nikov[8] reported concentrations in the range 1 to 3 μM in river water. More recently, it has been recognized that rainwater can contain more than 10^{-5} M H_2O_2.[9,10] Zika and Zelmer[11] reported typical concentrations near 0.1 μM in surface ocean water. Quite recently, Cooper and Zika[12] produced 10^{-5} M H_2O_2 in photolyzed groundwaters, and Drapper and Crosby[13] produced similar levels in photolyzed wastewater and surface water. These results suggest that H_2O_2 might be sufficiently abundant to provide significant amounts of chlorine demand.

The first extensive measurements of H_2O_2 in estuarine surface waters are reported in this chapter. Since H_2O_2 is thought to be produced in natural waters by a photochemical reaction involving organic matter,[11,12] estuaries supposedly might contain higher concentrations than the ocean because of their higher organic carbon content and because of their intense density stratification, which limits the depths to which surface-produced species can be mixed. About one-third of the cooling water used by the power industry in the United States is obtained from estuarine or coastal sources.[1]

METHODS

Surface samples were collected using a polyethylene container attached to the end of a pole. All samples were analyzed immediately after collection. The H_2O_2 concentration was measured using a fluorescence decay technique[14] similar to that of Perschke and Broda.[15] The method involves recording the decrease in fluorescence of a pH 7 phosphate-buffered sample on addition of a horseradish peroxidase (HRP) catalyst. All measurements were done on a Turner Designs Model 10 fluorometer with excitation and emission wavelengths of 365 and 490 nm.

The H_2O_2 concentration in a sample was determined by adding 100 μL of a pH 7 phosphate buffer and 40 μL of scopoletin (6-methyl-7-hydroxyl-1,2-benzopyrene) to 20 mL of the water of interest. The solution was mixed constantly by means of a magnetic stir bar at the bottom of the cell. The decrease in fluorescence upon addition of the HRP catalyst was recorded, and the concentration was read from a H_2O_2 vs change-in-fluorescence standard curve. The standard curve was prepared by taking another 20 mL of the same water and adding the buffer, HRP, and scopoletin in the same proportions as before. Aliquots of a stock solution of H_2O_2 were added to the mixture, and the change in fluorescence after each addition was recorded and plotted vs H_2O_2 added.

RESULTS

Figure 1 shows surface water data obtained over 24 h at Benedict Bridge on the Patuxent estuary, a tributary to Chesapeake Bay. When these samples were collected, in early November 1983, a distinct diurnal variation was observed, such as might be expected for a photochemically produced compound. The concentration rose steadily during the daylight hours and then began to drop after dark. Initially, the drop was rapid, but after an hour or so the decay began to follow a slower, first-order pattern with a half-life of about 9 h.

The classic diurnal pattern in Figure 1 was observed on only one occasion. The weather at this time was extremely calm. On other occasions, the 24-h pattern was completely erratic, with H_2O_2 concentrations fluctuating by a factor of 2 or 3. These fluctuations did not correlate with tides or light intensity. Currently, we believe that the windy conditions most often prevailing on the Patuxent cause mixing of surface and deep waters in an irregular way that obscures the diurnal pattern. For a similar reason, concentrations of H_2O_2 measured along the length of the estuary, covering a salinity range from 0 to about 10, varied somewhat erratically by a factor or 2 or 3.

Despite the irregular pattern of the analytical results obtained over 24 h at one station or at a series of stations along the estuary, the seasonal variations were of greater magnitude and more regular. This is shown in Figure 2. The points in this figure represent the mean value at Benedict, and the vertical bars enclose the entire range of data obtained on the day that samples were collected.

The primary control on the seasonal pattern is the variation in light intensity, which is assumed to control the production of H_2O_2. However, the intensity of summer-winter oscillations in concentration is probably enhanced by hydrodynamic factors. In winter, density stratification is weaker than in summer, so it is possible for compounds produced in surface waters to be mixed to greater depth, resulting in correspondingly greater dilution. It is also possible that the seasonal fluctuations in concentration are influenced by variations in the H_2O_2 decay rate or in the quantum efficiency of H_2O_2 production.

During the period July 26-28, 1983, surface samples were collected and analyzed over the entire 300-km length of Chesapeake Bay. The H_2O_2 concentrations displayed a patchy variability and did not correlate systematically with salinity, with geographic coordinates, or with time of day. Therefore, these data are presented in Figure 3 simply in the form of a histogram. The distribution is approximately log normal, with a mean near 2×10^{-6} M. There is a cluster of points at about 3×10^{-7} M; according to the field notes, algal

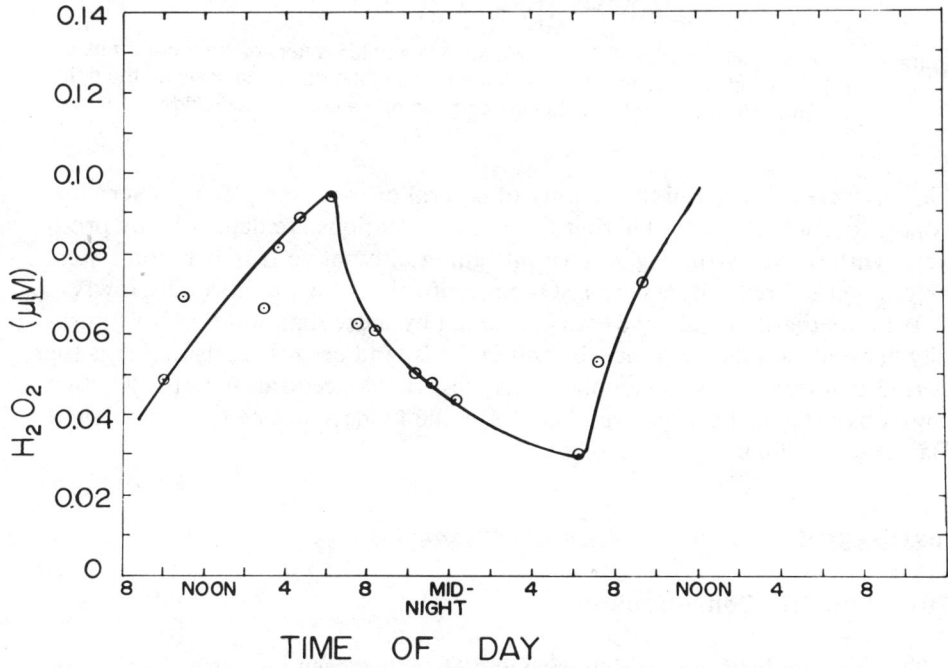

Figure 1. Diurnal variation in H_2O_2 concentration in surface waters of the Patuxent estuary at Benedict, Maryland. Samples collected in early November 1983.

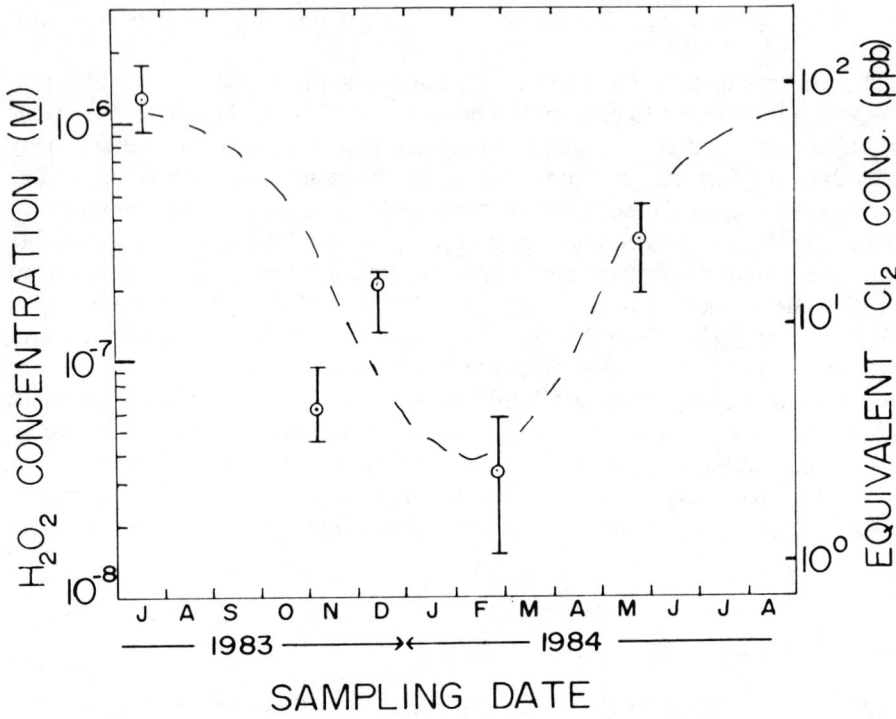

Figure 2. Seasonal variations in H_2O_2 concentration in surface waters of the Patuxent estuary. The vertical bars enclose the entire range of the data. The scale on the right indicates the stoichiometric chlorine demand of the observed peroxide.

blooms were observed in the vicinity of several of these samples with very low concentrations. It is possible that H_2O_2 concentrations are depressed by products synthesized by the algae. (An intriguing alternative is that blooms were able to get started only where H_2O_2 concentrations were abnormally low.)

We note that late July 1983 was preceded by more than a month of unusually dry and cloudless weather. Because of this, and because midsummer is the period of most intense solar photon fluxes, the concentrations of H_2O_2 that were observed at this time are likely to be the highest attained in Chesapeake Bay under ordinary circumstances.

DISCUSSION

Stoichiometric Considerations

The reaction between free chlorine and H_2O_2 has been established as a two-electron transfer reaction, rather than a free-radical process,[4,5] and can be represented as follows:

$$HOCl + HO_2^- \rightarrow Cl^- + H_2O + O_2(\text{singlet})$$

One mol HOCl destroys 1 mol H_2O_2 and, under neutral-to-alkaline conditions, 1 mol of singlet oxygen is produced. Some of the chemical properties of singlet oxygen are discussed elsewhere in this volume.

In both Figures 2 and 3, the amount of free chlorine that would be reduced by complete reaction with H_2O_2 has been indicated. During the maximum photic period in July, for example, Chesapeake Bay waters contained enough H_2O_2 to reduce about 10% of a typical estuarine power plant chlorine dose. In the colder months of the year, the H_2O_2 concentrations were far smaller than in July. However, in the temperate latitudes to which our results apply, chlorination is used very moderately during cold weather because fouling problems are minimal. Thus, the seasonal pattern of H_2O_2 abundance crudely mimics chlorine usage.

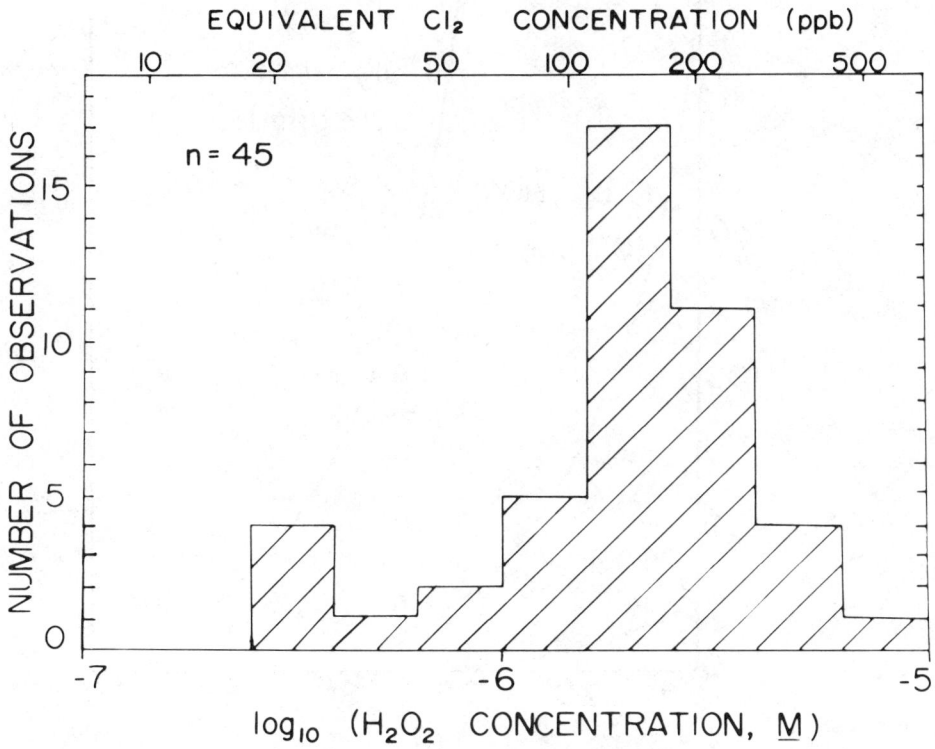

Figure 3. Histogram of H_2O_2 concentrations plotted on a logarithmic scale in surface waters of Chesapeake Bay during July 26–28, 1983, showing plots of 45 analyses. The scale at the top of the figure indicates the stoichiometric chlorine demand of the peroxide.

Kinetic Considerations

The reaction rate of free chlorine with H_2O_2 has been studied in some detail.[3-5] At least three parallel mechanisms exist, although under the neutral-to-alkaline conditions prevailing in natural waters, the so-called alkaline pathway, represented by the reaction previously given, will predominate. Under conditions of excess free chlorine, the second-order reaction becomes pseudo-first order, and the half-life of H_2O_2 can be computed at a given temperature and free chlorine concentration. This has been done in Figure 4 at 25°C using the data of Held et al.[4] The numbers in this figure indicate the free chlorine concentrations assumed in calculating each curve. It is apparent that for pH values greater than 7 and free chlorine concentrations greater than 10^{-5} M, the half-life of H_2O_2 will be less than 1 min. At the pH of seawater (8.3), the half-life in the presence of more than 10^{-5} M (0.7 ppm) free chlorine will be a few seconds or less.

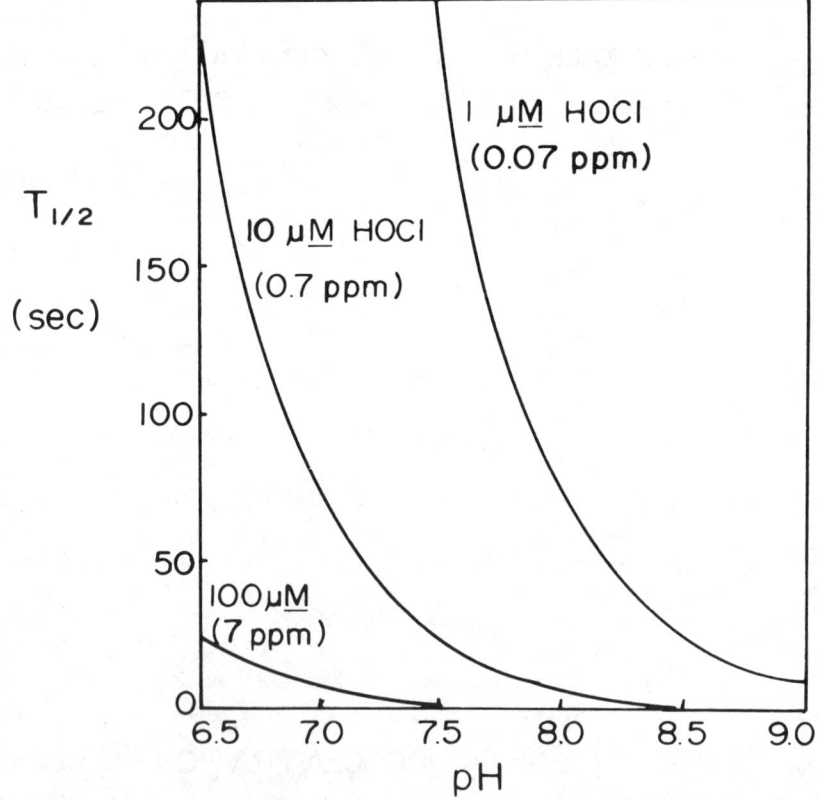

Figure 4. Calculated half-life at 25°C of hydrogen peroxide in the presence of excess free chlorine. The numbers associated with the curves designate amounts of free chlorine.

Unfortunately, the actual situation in chlorinated seawater is more complicated because of the rapid oxidation of Br^- to $HOBr$.[16] The reaction rate of $HOBr$ with H_2O_2 has been studied only in acidic solutions.[17] However, we know from general experience that the rate of attack by $HOBr$ on nucleophiles such as HO_2^- is usually faster than attack by $HOCl$, so it is likely that in estuarine and marine waters, H_2O_2 is destroyed even faster than implied by Figure 4.

A critical consideration with regard to the importance of H_2O_2 as a contributor to chlorine demand is the competitive rate at which $HOCl$ will react with NH_3 and organic amines in the presence of H_2O_2. Once chloramines have formed, the H_2O_2 demand can be exerted only slowly, if at all. Based on the data of Held et al.[4] and Weil and Morris,[18] the competitive rates for reaction of $HOCl$ with H_2O_2 and NH_3 can be computed at neutral-to-alkaline pH and 25°C as follows:

$$\frac{d(H_2O_2)_T}{d(NH_3)_T} = 7.2 \frac{(H_2O_2)_T (1 + H^+/K_{NH_4^+})}{(NH_3)_T (1 + H^+/K_{H_2O_2})}$$

Here, $(H_2O_2)_T$ and $(NH_3)_T$ represent the total analytical concentrations of H_2O_2 and ammonia, and the K's represent the acid ionization constants of the subscripted species. Below pH 8,

$$\frac{d(H_2O_2)_T}{d(NH_3)_T} = 0.028 \frac{(H_2O_2)_T}{(NH_3)_T}$$

Since total NH_3 will almost always exceed total H_2O_2 in nature, it is apparent that the formation of monochloramine will occur much more rapidly than oxidation of H_2O_2. Inasmuch as organic amines tend to react more rapidly than ammonia, and considering that chloramines react only slowly with H_2O_2,[6,19] the potential chlorine demand represented by H_2O_2 in natural waters will be exerted only if free chlorine remains after chloramine formation is essentially complete.

In summary, we have found H_2O_2 to be a ubiquitous trace constituent of estuarine waters. The concentrations vary seasonally, but reach sufficiently high levels in summer so that they could exert a chlorine demand equivalent to about 10% of chlorine doses typically used for fouling control at power plants. The reaction rate of H_2O_2 is quite fast but not fast enough to compete with monochloramine formation.

ACKNOWLEDGMENTS

We gratefully acknowledge financial support from the Maryland Power Plant Siting Program and from the Maryland Sea Grant Program. Rod Zika and Robert Petasne generously shared with us their knowledge concerning the analysis of H_2O_2 and trained one of us (R. Kieber) in the method. James

Sanders let us share ship time on one of his cruises. Werner Haag called our attention to Reference 17,

REFERENCES

1. Helz, G. R., R. Sugam, and A. C. Sigleo. "Chemical Modifications of Estuarine Water by a Power Plant Using Continuous Chlorination," *Environ. Sci. Technol.* 18:192-199 (1984).
2. Jaworske, D. A., and G. R. Helz. "Rapid Oxidant Demand: Methods For Study," Chapter 86, this volume.
3. Connick, R. E. "The Interaction of Hydrogen Peroxide and Hypochlorous Acid in Acidic Solutions Containing Chloride Ion," *J. Am. Chem. Soc.* 69:1509-1514 (1947).
4. Held, A. M., D. J. Halko, and J. K. Hurst. "Mechanisms of Chlorine Oxidation of Hydrogen Peroxide," *J. Am. Chem. Soc.* 100:5732-5740 (1978).
5. Kozlov, Yu. N., A. P. Parmal, and A. K. Uskov. "The Two-Electron Mechanism of the Oxidation of Hydrogen Peroxide by Hypochlorous Acid," *Russian J. Phys. Chem.* 54:992-994 (1980).
6. Harrison, J. R., L. B. Fournier, and C. H. Lemke. "The Use of Hydrogen Peroxide In Dechlorination Of Industrial Effluents," *AIChE Symp. Ser.* 71:64-69 (1975).
7. van Baalen, C., and J. E. Marler. "Occurrence of Hydrogen Peroxide in Sea Water," *Nature* 211:951 (1966).
8. Sinel'nikov, V. Ye. "Hydrogen Peroxide Content of River Water and a Method For Determining It," *Hydrobiol. J.* 7:96-99 (1971).
9. Kok, G. L. "Measurements of Hydrogen Peroxide in Rainwater," *Atmos. Environ.* 14:653-655 (1980).
10. Zika, R., E. Saltzman, W. L. Chameides, and D. D. Davis. "H_2O_2 Levels in Rainwater Collected in South Florida and the Bahama Islands," *J. Geophys. Res.* 87:5015-5017 (1982).
11. Zika, R. G., and P. P. Zelmer. "Photochemical Generation and Decay of Hydrogen Peroxide in Seawater," *Trans. Am. Geophys. U.* 61:1010 (1980).
12. Cooper, W. J., and R. G. Zika. "Photochemical Formation of Hydrogen Peroxide in Surface and Ground Waters Exposed to Sunlight," *Science* 220:711 (1983).
13. Drapper, W. M., and D. G. Crosby. "The Photochemical Generation of Hydrogen Peroxide in Natural Waters," *Arch. Environ. Contam. Toxicol.* 12:121-126 (1983).
14. Zika, R. G. University of Florida, unpublished data.
15. Perschke, H., and E. Broda. "Determination of Very Small Amounts of Hydrogen Peroxide," *Nature* 190:257-259 (1961).
16. Helz, G. R., and R. Y. Hsu. "Volatile Chloro- and Bromocarbons in Coastal Waters," *Limnol. Oceanogr.* 25:858-869 (1978).
17. Bray, W. C., and R. S. Livingston. "The Rate of Oxidation of Hydrogen Peroxide by Bromine and Its Relation to the Catalytic Decomposition of Hydrogen Peroxide in a Bromine-Bromide Solution," *J. Am. Chem. Soc.* 50:1654-1665 (1928).
18. Weil, I., and J. C. Morris. "Kinetic Studies of the Chloramines. I. The Rates of Formation of Monochloramine, N-Chloromethylamine, and N-Chlorodimethylamine," *J. Am. Chem. Soc.* 71:1664-1671 (1949).
19. Hurst, J. K., P. A. G. Carr, F. E. Hovis, and R. J. Richardson. "Hydrogen Peroxide Oxidation by Chlorine Compounds. Reaction Dynamics and Singlet Oxygen Formation," *Inorg. Chem.* 20:2435-2438 (1981).

CHAPTER 82

Sunlight-Induced Photodecomposition of Chlorine Dioxide

Rod G. Zika, Cynthia A. Moore, Louis T. Gidel, and William J. Cooper

Many water treatment facilities, at some point after or immediately before addition of the oxidant, expose the water being treated to direct sunlight. Because of the significant sunlight absorption by some of the oxidants, the chemical pathway leading to their decomposition can be dramatically altered. Potentially, this can have a pronounced effect on the nature of the inorganic by-products of the oxidant decomposition and on the halogenated organic materials produced via interaction with the oxidant photoproducts. In sunlight, direct photolytic decomposition will only be significant for those oxidants that exhibit appreciable light absorption at wavelengths longer than 300 nm. Of all the water treatment oxidants, ClO_2 is most prone to photochemical decomposition during sunlight exposure. This is because of its high light absorption in the near-UV and visible region of the spectrum and high quantum yields, reported to be 1.0 mol/Einstein at 405 nm and lower wavelengths.

The gas-phase photochemistry of ClO_2 has been studied extensively.[1-4] However, far less attention has been paid to its photolysis in aqueous solution.[5-8] In the gas phase, the primary photochemcial reaction is the homolytic fission of the chlorine oxygen bond:

$$ClO_2 \rightarrow ClO + O \qquad (1)$$

The products of this primary reaction then go on to generate secondary products such as chlorine peroxide (ClOO), excited state oxygen (O_2^*), chlorine (Cl_2), chlorine trioxide (Cl_2O_3), chlorine hexoxide (Cl_2O_6), and the recently proposed chlorine perchlorate ($ClOClO_3$).[9]

In aqueous solution, the primary photoreaction product ClO is not observed, and O appears to be a minor product since little O_3 is found in oxygenated aqueous solutions.[8] It is postulated that the reaction proceeds, as in the gas phase, to give ClO and O, but these products then undergo solvent cage recombination to generate chlorine atoms and molecular oxygen. In addition, a rapid bulk solution reaction between ClO and ClO_2 produces Cl_2O_3, and the reaction between Cl (or Cl_2^-) and ClO_2 leads to Cl_2O_2. Rapid hydrolysis of Cl_2O_3 and Cl_2O_2 leads to the formation of the observed products ClO_3^-, ClO^- and Cl^-. The overall reaction is as follows:

$$ClO_2 \rightarrow (ClO + O) \text{ cage} \qquad (2)$$
$$(ClO + O) \text{ cage} \rightarrow Cl + O_2 \qquad (3)$$
$$ClO + ClO_2 \rightarrow Cl_2O_3 \qquad (4)$$
$$Cl_2O_3 + H_2O \rightarrow ClO^- + ClO_3^- + 2H^+ \qquad (5)$$
$$Cl + Cl^- \rightarrow Cl_2^- \qquad (6)$$
$$Cl \text{ (or } Cl_2^-\text{)} + ClO_2 \rightarrow Cl_2O_2 \qquad (7)$$
$$Cl_2O_2 + H_2O \rightarrow Cl^- + ClO_3^- + 2H^+ \qquad (8)$$
$$O + O_2 \rightarrow O_3 \qquad (9)$$

Interestingly, although ClO_2 has the distinct advantage in water treatment of not having the capacity to oxidize Br^-, many of its transient photolysis products do. The appearance of transitory species during the flash photolysis of ClO_2 solutions, which contained low levels of Br^-, have been reported.[8] The transients are believed to be Br_2^- or $BrCl^-$. The information on the reactions of these radicals with organic compounds is scarce; however, it has been shown that they do react at appreciable rates with alcohols.[9] By analogy with the Cl_2^- radical, bromine radical anions should undergo addition to alkenes and aromatic compounds.

In this chapter, the light-induced decomposition of ClO_2 is examined in an attempt to characterize inorganic products and trihalomethane (THM) formation, and to describe the extent to which sunlight-induced photodecomposition might occur during water treatment.

EXPERIMENTAL

Chlorine dioxide used in this study was prepared from sodium chlorite (80% purity) obtained from Alfa Products and potassium persulfate (reagent grade) obtained from Mallinckrodt. A standard procedure was used to prepare and handle the ClO_2.[10] Stock solutions thus prepared were stored refrigerated and protected from light. They were standardized immediately before being used in experiments.

Chlorine dioxide solutions for sunlight reaction studies were prepared by adding the concentrated stock solution to organic-demand-free water in 10-cm quartz spectrophotometer cells. The cells were sealed so that no headspace existed. The initial concentration was calculated from the measured absorbance at 360 nm using a molar extinction coefficient, Σ, of 1035 L mol^{-1} cm^{-1}. Sunlight exposure experiments were conducted on the roof of the laboratory. The integrated solar flux during the experiments was determined with an Eppley Laboratory Ultraviolet Radiometer, Model TUVR. Changes in the ClO_2 and light absorbing product concentrations during the course of the experiments were determined from the UV-vis spectra. Controls were run simultaneously by storing sample cells in the dark at the same temperature as the light-exposed samples. Experiments involving fluorescent room lighting were conducted in a similar manner. The samples were exposed to standard laboratory fluorescent lights on a bench ~7 ft below overhead light fixtures.

Quantum yield studies were conducted using a Kratos/Schoeffel 1000-W mercury-xenon monochromatic irradiation system. Quantum yields were determined at the mercury lines 296.7, 313.0, 334.1, 366, 404.5, and 435.8 nm, with a bandwidth of ~7 nm. The intensity at each line was determined using ferrioxalate actinometry.[11] The measurements were performed in 10-cm quartz spectrophotometry cells in the same way as described for the sunlight experiments. Time intervals for light exposure varied (i.e., 10 to 120 s) and were optimized to examine the reaction during the first 10% of the decay curve.

Natural water samples were placed in 60-mL reaction vials, spiked with ClO_2 at ~1.5 × $10^{-4}M$ (10 mg/L), topped with additional water, and capped with Teflon-lined septa. The vials were then either placed in the dark as controls or exposed to sunlight on the roof. The sunlight was monitored with the Eppley Radiometer, and the samples were exposed for a period in excess of twice the half-life for ClO_2^- (a product of ClO_2 photodecomposition) photolysis. This interval was usually about 4 h at midday. The dark- and light-exposed samples were then quenched by adding a slight excess of sodium thiosulfate. The capped samples were refrigerated and analyzed by gas chromatography for THMs.[12] Samples for anion analysis were prepared by purging residual ClO_2 with N_2 gas to quench the reaction. Analyses were conducted using ion chromatography on a Dionex Model 10 chromatograph with a "fast anion" column and 0.0015 M Na_2CO_3 as eluant. The column flow rate was 3.0 mL/min. With these conditions, the approximate retention times in minutes for anions were ClO_2^- at 2.40; BrO_3^- at 3.10; Cl^- at 3.45; Br^- at 8.10; and ClO_3^- at 9.25.

RESULTS AND DISCUSSION

Sunlight Studies

Aqueous solutions of ClO_2 in sealed 10-cm spectrophotometer cells were exposed to summer midday solar radiation. The resultant photochemical decay of ClO_2 is shown in Figure 1. The results indicate that ClO_2 should have a lifetime of <0.5 min in water exposed to bright sunlight. To evaluate the wavelength dependence of the reaction, quantum yields for the decomposition of ClO_2 were measured at various wavelengths. The quantum yields were compared with those determined earlier (Table I) by Bowen and Cheung.[5] The reasons for the differences in quantum yields determined in the two studies are not apparent but could be the result of the differences in methodologies. There is, for instance, no information in the Bowen and Cheung report on the pH of the reaction solutions. Although differences between pH 7.0 and 9.0 appear to be negligible, it is possible that the spread could be significantly larger at more-

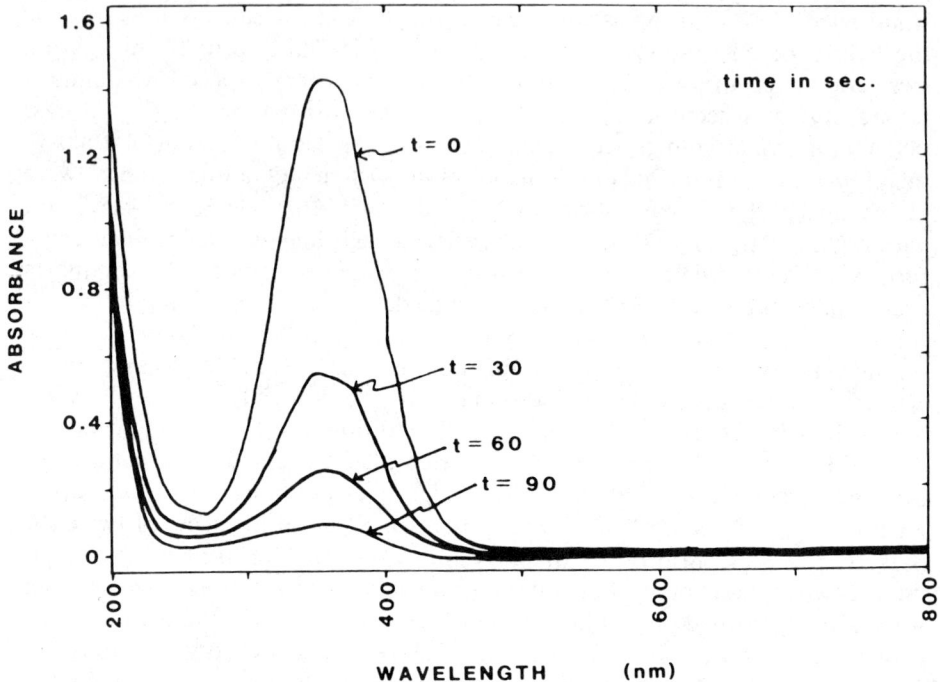

Figure 1. Photochemical decomposition of chlorine dioxide aqueous solution in sunlight.

Table I. Decomposition Quantum Yields (Φ) of ClO_2

Wavelength (nm)	From Bowen and Cheung[5]	This Work at pH 7.0	This Work at pH 9.0
296.7		1.40 ± 0.03	1.47 ± 0.09
300	1.00		
313.0		0.98 ± 0.02	0.86 ± 0.01
334.1		0.90 ± 0.02	0.90 ± 0.02
366.0	0.76	0.46 ± 0.02	0.46 ± 0.02
404.5	0.51	0.13 ± 0.02	
435.8	0.21		

acidic pH values. In water treatment applications, however, pH values in the 7 to 9 range are more likely to be encountered in ClO_2 treatment applications.

Another complication in comparing the results is caused by the possible differences in the bandwidth of the incident radiation. In the work presented here, a high-energy monochromator with a bandwidth of less than 7 nm was used to isolate mercury lines. It is clear from the Bowen and Cheung report that for the 300-nm value of $\Phi = 1.00$, a broad spectral region of 316 to 270 nm was used. The rapidly changing values of Φ and ClO_2 absorbance in this region make it unlikely that the quantum yields from the two studies would agree closely. This same problem could contribute to the lack of agreement at other wavelengths. Unfortunately, the filter combinations used in the Bowen and Cheung report to isolate other mercury lines could not be reproduced in our laboratory because the glass filters used are apparently no longer available.

Reaction Products

When ClO_2 is used as an oxidant in water treatment applications, the resultant thermal and photochemical decomposition pathways become considerably more complex than they are in pure water solutions of ClO_2. Compositional variations in the particular water can have a pronounced effect on the mechanism and the products generated. The occurrence of oxidizable substrates in the water will contribute to the amount of ClO_2^- produced from ClO_2 via dark oxidation reactions. The photochemical decomposition of ClO_2 will then also involve ClO_2^-, which has a moderately strong electronic absorption spectra in the near-UV region of sunlight. The complex nature of the mechanism can be observed from the changing electronic absorption spectra of ClO_2^- solutions during sunlight exposure (Figure 2). It is obvious that ClO_2^- has a lifetime of < 10 min in full sunlight and that its reaction products are ClO_2 and OCl^-. The photochemistry of ClO_2^- has been extensively covered elsewhere[13,14] and will not be further discussed here. It is worth noting, however, that the photochemistry of ClO_2 in disinfection application will to some extent, depending on the characteristics of the water and the light-spectral distribution, involve the oxidants ClO_2, ClO_2^- and OCl^-.

The major stable end products of ClO_2 photolysis in aqueous solution are Cl^-, ClO_3^- and O_2.[5] The Cl^- and ClO_3^- are produced through hydrolysis of Cl_2O_3 and Cl_2O_2, with ClO^- and Cl also involved in the reaction sequence.[2-8] An advantage of ClO_2 in disinfection is its inability to oxidize Br^-; however, in solution exposed to light, these intermediate oxidants could initiate reactions such as

$$OCl^- + Br^- \rightarrow OBr^- + Cl^- \qquad (10)$$

$$Cl + Br^- \rightarrow ClBr^- \qquad (11)$$

In solutions of $1.5 \times 10^{-4} M$ ClO_2, which were exposed to direct sunlight, ion chromatographic analysis revealed that the only major ionic products were Cl^- and ClO_3^-. When 1×10^{-4} M Br^- was added to the solution, BrO_3^- was also observed as a product (Figure 3). Although the ClO_2 was photolyzed during the first few minutes of the reaction, the BrO_3^- gradually increased on contin-

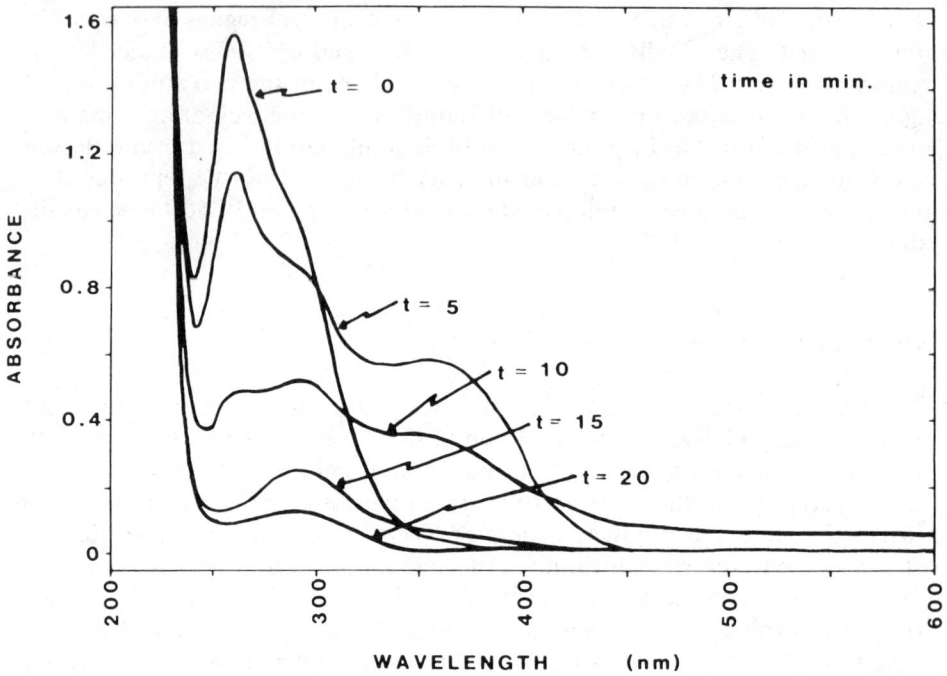

Figure 2. Photochemical decomposition of chlorite aqueous solution in sunlight.

ued exposure to sunlight. The amount of Br⁻ concentration decreased initially to 50% of that added and then increased as the reaction proceeded. This is interpreted as an initial rapid oxidation of Br⁻ to OBr⁻, which then reacts photochemically[15]

$$3 \text{ OBr}^- \rightarrow \text{BrO}_3^- + 2 \text{ Br}^- \tag{12}$$

to generate the observed product distributions. The OBr⁻ would not be detected using the ion chromatographic determination as described previously.

In water containing significant amounts of labile organic materials, the organic product yields might vary because of available organic scavengers. When ClO$_2$ was added to groundwater and exposed to sunlight, THM analysis revealed a linear increase in bromoform formation with increasing Br⁻ concentration (Figure 4). The onset of significant CHBr$_3$ formation in this water appears at about 1 mg/L Br⁻. A low yield of CHCl$_3$ was also observed which, within the limits of analytical precision, appeared to be independent of Br⁻ concentration.

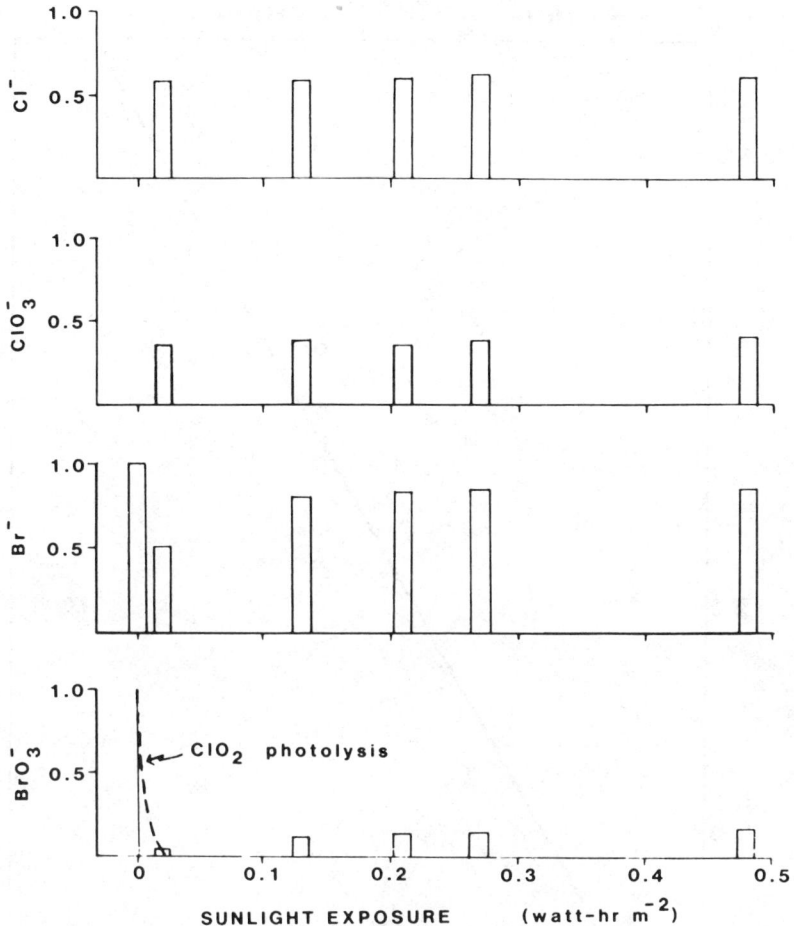

Figure 3. Major anionic products of ClO_2 photolysis in solutions containing 1×10^{-4} M Br^- and 1×10^{-4} M ClO_2 at pH 7.0. Vertical axes represent relative molar concentration.

To examine the effect of natural water composition on $CHBr_3$ formation, water from three different wells in the South Florida area was enriched with Br^-, dosed with $1.5 \times 10^{-4} M$ ClO_2, and exposed to sunlight. The yields of $CHBr_3$ are shown in Table II for pH 5, 7, and 9. The highest yields were observed for all water samples at pH 5, which is not consistent with the usual trend of increasing THM production with increasing pH. This decreasing yield could be due to an increased photolysis decomposition rate of the active bromine species at elevated pH.[16] There is the added complication that water composition also affects the $CHBr_3$ production, as is indicated for the Orr well 2 sample, which displayed less pH dependence than the other two wellwater

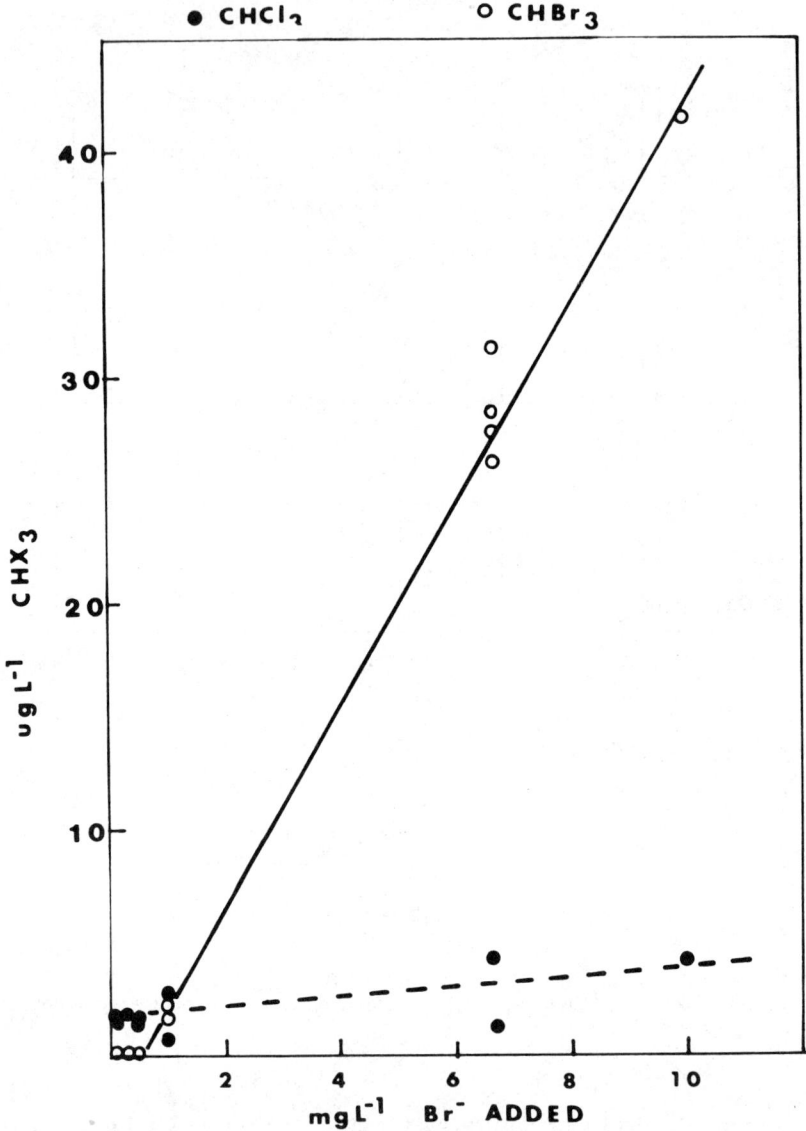

Figure 4. Effect of Br⁻ concentration on THM formation when groundwater samples, to which 1×10^{-4} M ClO_2 has been added, are exposed to sunlight.

samples. There is no direct correlation to organic carbon content, since Preston well 5 and Northwest well 3 had approximately the same level of total organic carbon (i.e., 6.3 mg/L) but very different concentrations of $CHBr_3$. The Orr well was comparable to the Preston well in $CHBr_3$ production at pH 5, but had only 2.2 mg/L TOC. The extent to which halogen radicals (i.e., Cl_2^-,

Table II. Bromoform Formation from ClO_2 (µg/L)

Source of Water	pH 5		pH 7		pH 9	
	Light	Dark	Light	Dark	Light	Dark
Preston well 5	50	ND[a]	7	ND	0	ND
Orr well 2	44	ND	21	ND	17	ND
Northwest well 3	13	ND	ND	ND	ND	ND

[a]Not determined.

Br_2^-, $ClBr^-$) are involved in THM formation is unknown. If, however, they are involved, then composition factors such as the concentration of inorganic anions may play an important role. High levels of HCO_3^- or CO_3^{2-} would rapidly quench active halogen radicals, thereby limiting $CHBr_3$ production.

Another interesting facet of Br^- involvement in ClO_2 photolysis is shown in Figures 5 and 6. In these experiments, $1.5 \times 10^{-4} M$ ClO_2 solutions containing $1 \times 10^{-4} M$ Br^- were exposed to alternating 12-h light and dark periods. Irradiation during the light periods was supplied by overhead fluorescent room lights. It is apparent from Figure 5 that without added Br^-, no significant ClO_2 decay occurred during the dark periods. When Br^- was present, however, the initial light exposure period apparently generated photoproducts that enhanced the ClO_2 decay reaction during subsequent dark and light exposure periods.

In summary, the nature of the compositional variations that affect ClO_2 decomposition and $CHBr_3$ formation are still obscure, and the reaction mechanisms are complex.

PHOTOCHEMICAL DECOMPOSITION MODELING

The photolysis rate and half-life of ClO_2 were calculated for waters with organic contents of 0.53, 6.20, and 17.6 mg/L TOC using a modified version of the photochemical model of Zepp and Cline.[17] The photolysis rate is calculated based on the assumption that the loss occurs as a result of primary processes. Any secondary reactions that may lead to a loss of ClO_2 would lead to a faster rate. This model calculates the attenuation of light in the water column resulting from absorption and Rayleigh scattering within the water column itself and from the layer average amount of light absorbed between the surface and a depth D. Solar fluxes at ground level are specified as a function of time of day, season, total ozone column amounts, and the latitude based on the regression analysis of Benner.[18] Other critical input parameters are water absorbance, molar extinction coefficient, and quantum yield, all as a function of wavelength. Except for surface photolysis rates, all other calculated rates represent averages over the depth of shallow holding ponds with weak thermal stratification and sufficiently strong wind. Thermally or mechanically driven mixing cycles the water from the bottom to the surface on a time scale faster

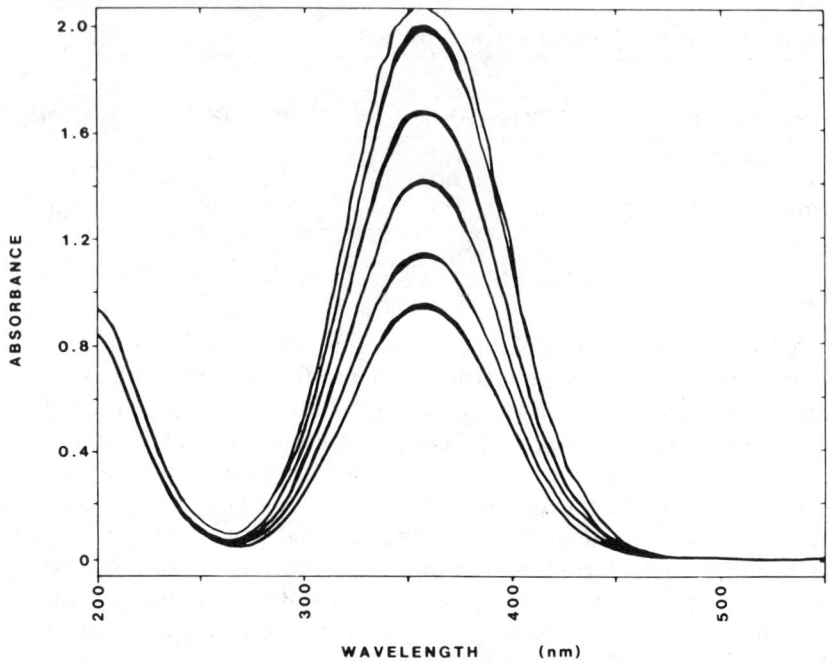

Figure 5. Alternating 12-h periods of exposure to room fluorescent light levels (white areas) and to the dark (shaded areas). Initial concentration of ClO_2, 1×10^{-4} M.

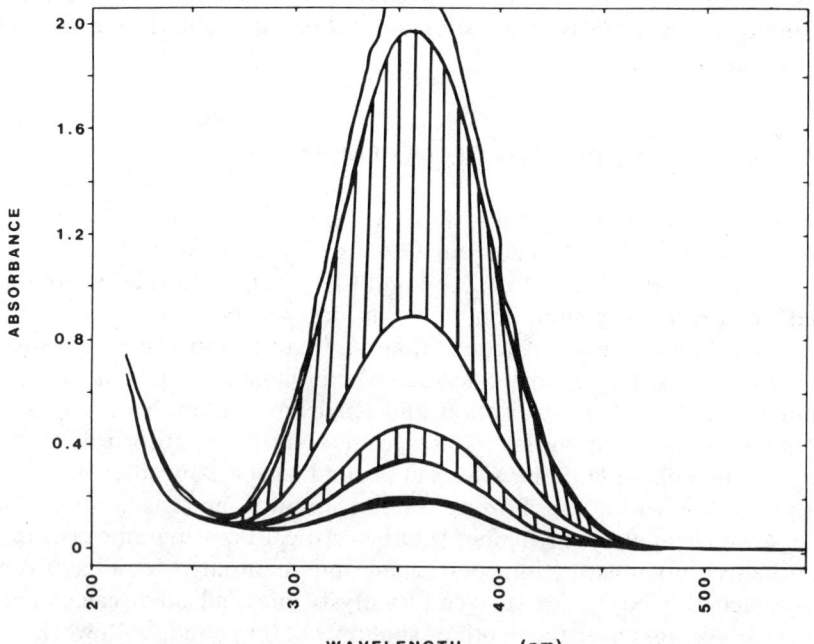

Figure 6. Alternating 12-h periods of exposure to room fluorescent light levels (white areas) and to the dark (shaded areas). Initial concentration of ClO_2, 1×10^{-4} M with 1×10^{-4} M of Br^- present.

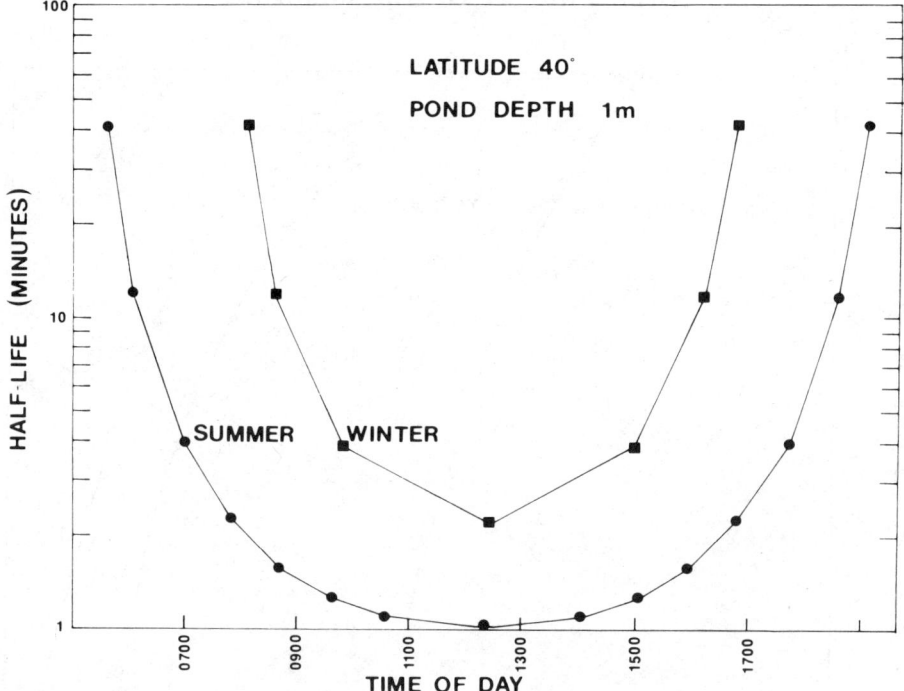

Figure 7. Time-of-day variation in half-life of ClO_2 for sunlight exposure for midsummer and midwinter in contact pond water with a depth of 1 m.

than the half-life because of photolysis, so that photolysis rather than mixing is the rate-limiting step in the destruction of ClO_2. As the photolysis rate increases, the assumption is no longer valid; the calculated half-life is then only a lower limit, with mixing in the pond becoming the rate-limiting step in the loss process.

The time-of-day variation in the half-life for a 1-m-deep pond at 40° latitude in winter and summer is shown in Figure 7. At midday in summer, the lifetime is about 1 min and is always <10 min during the daylight hours. In winter, photolysis is only important during the middle 6 h of the day. Figure 8 shows the variation of the half-life with the depth of the holding pond for waters of different clarities ranging from clear to dark-colored (i.e., 17.6 mg/L). For clear water (i.e., 0.53 mg/L), the half-life is nowhere as sensitive to the depth of the pond as for colored water. In 17.6 mg/L, most of the light is absorbed in the upper 10 cm; therefore, the loss rate will be dependent on the strength of the mixing.

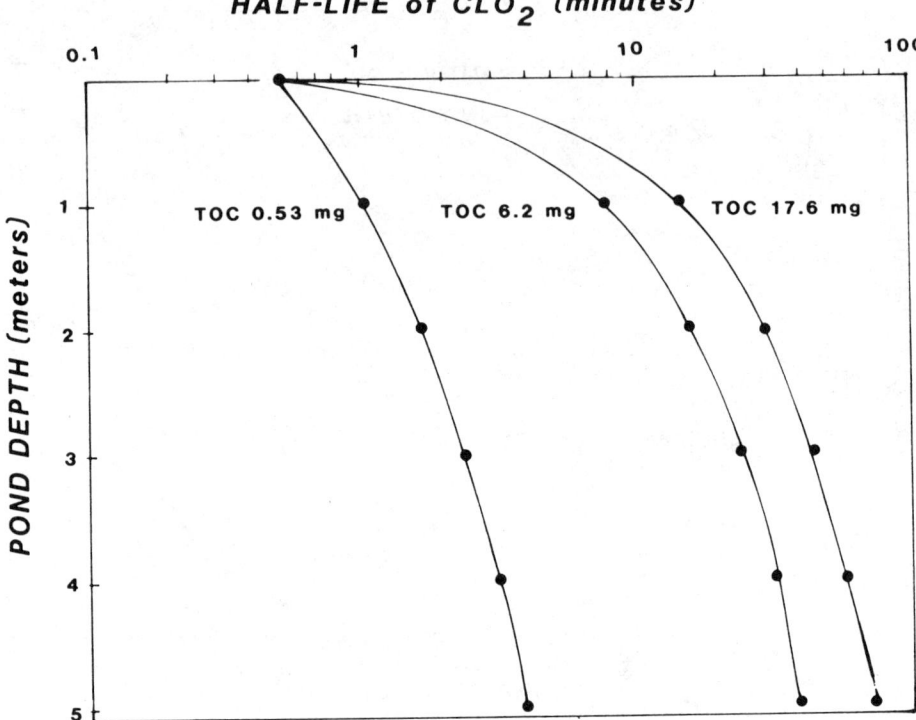

Figure 8. Variations of half-life of ClO_2 in ponds of different depths and with different total organic carbon concentrations when exposed to sunlight at 40° latitude in midsummer.

CONCLUSIONS

Both ClO_2 and ClO_2^- are readily decomposed by sunlight and fluorescent lights. This can lead to significant losses during water treatment. The characteristics of the water play an important role in the nature of products resulting from light-initiated reactions. Significant levels of ClO_3^-, BrO_3^-, and bromoform can be produced. The bromide ion was found to play a particularly important role in THM formation and in initiating light reactions that accelerated the decomposition of ClO_2 in the dark.

ACKNOWLEDGMENTS

Financial support for this research was provided by the U.S. Environmental Protection Agency, Cooperative Agreement CR-810277-01, and the Drinking Water Research Center, Florida International University.

REFERENCES

1. Spinks, J. W. T., and J. M. Porter. "Photodecomposition of Chlorine Dioxide," *J. Am. Chem. Soc.* 56:264-270 (1934).
2. Basco, N., and S. K. Dogra. "Some New Features in the Flash Photolysis of Chlorine Dioxide," *Chem. Comm.* 18:1071-1073 (1968).
3. Basco, N., and R. D. Morse. "Reactions of Halogen Oxides Studied by Flash Photolysis IV. Vacuum Ultraviolet Kinetic Spectroscopy Studies on Chlorine Dioxide," *Proc. R. Soc. Lond. A.* 336:495-505 (1974).
4. Gole, J. L. "Photochemical Isomerization of ClO_2 and the Low-Lying Electronic State of the Asymmetric Isomer. Possible Implications for Matrix Isolation Spectroscopy," *J. Phys. Chem.* 84:1333-1340 (1980).
5. Bowen, E. J., and W. M. Cheung. "The Photodecomposition of Chlorine Dioxide Solutions," *J. Chem. Soc.* 1200-1207 (1932).
6. Taube, H. "Photochemical Reactions of Ozone in Solution," *Trans. Faraday Soc.* 53:656-665 (1957).
7. Dardelet, S., and A. Robert. "Stability of Aqueous Solutions of Chlorine Dioxide," *Assoc. Tech. Ind. Papet. Bull.* 22:123-128 (1968).
8. Mialocq, J. C., F. Barat, L. Gilles, B. Hickel, and B. Lesigne. "Flash Photolysis of Chlorine Dioxide in Aqueous Solution," *J. Phys. Chem.* 77:742-749 (1973).
9. Schell-Sorokin, A. J., O. S. Bethune, J. R. Lankard, M. M. T. Loy, and P. P. Sorokin. "Chlorine Perchlorate. A Major Photolysis Product of Chlorine Dioxide," *J. Phys. Chem.* 86:4653-4655 (1982).
10. *Standard Methods for the Examination of Water and Wastewater*, 15th ed. (Washington, DC: American Public Health Association, 1981), pp. 304-310.
11. Hatchard, C. G., and C. A. Parker. "A New Sensitive Chemical Actinometer II. Potassium Ferrioxalate as a Standard Chemical Actinometer," *Proc. R. Soc. London, Ser. A*: 235:518-536 (1956).
12. Mehran, M. F., R. A. Slifker, and W. J. Cooper. "A Simplified Liquid-Liquid Extraction Method for Analysis of Trihalomethanes in Drinking Water," *J. Chromatogr. Sci.* 22:241-243 (1984).
13. Buxton, G. V., and M. S. Subhani. "Radiation Chemistry and Photochemistry of Oxychlorine Ions. Part I. Radiolysis of Aqueous Solutions of Hypochlorite and Chlorite Ions," *J. Chem. Soc. Faraday Trans. I.* 68:947-957 (1972).
14. Buxton, G. V., and M. S. Subhani. "Radiation Chemistry and Photochemistry of Oxychlorine Ions. Part 3. Photodecomposition of Aqueous Solutions of Chlorite Ions," *J. Chem. Soc. Faraday Trans. I.* 68:970-977 (1972).
15. Lewin, M., and M. Avrahami. "The Decomposition of Hypochlorite-Hypobromite Mixtures in the pH Range 7-10," *J. Chem. Soc.* 77:4491 (1965).
16. Zika, R. G., R. G. Petasne, and W. J. Cooper. "Sunlight-induced Photodecomposition of $HOCl/OCl^-$, $HOBr/OBr^-$ and NH_2Cl During Water Treatment," presented at the Fifth Conference on Water Chlorination: Environmental Impact and Health Effects, Williamsburg, Virginia, June 3-8, 1985.
17. Zepp, R. C., and D. M. Cline. "Rates of Direct Photolysis in Aquatic Environments," *Environ. Sci. Technol.* 11:359-366 (1977).
18. Benner, P. *Approximate Values of Intensity of Natural Ultraviolet Radiation for Different Amounts of Atmospheric Ozone,* U.S. Army Report DAJA 37-68-C-1017, Davos Platz, Switzerland (1972).

CHAPTER 83

Photodegradation of Water Pollutants in Chlorinated Water

Lisa H. Nowell and Donald G. Crosby

Chlorinated organic compounds are ubiquitous in "purified" drinking water and chlorinated wastewater. Aromatic compounds with electron-donating groups are known to react readily with hypochlorite to form halomethanes and other chlorinated products, but aromatics with electron-withdrawing substituents have been found to be stable to hypochlorite under the conditions of water treatment.[1-4] However, irradiation at sunlight wavelengths may cause even "deactivated" compounds to react with hypochlorite.[5]

To determine the effect of substitution patterns on photochemical reactivity, model compounds having electron-withdrawing substituents, including several important water pollutants, were chlorinated under standardized conditions and their relative reactivities were determined in both dark and light. The loss of residual chlorine also was monitored in both dark and irradiated samples.

EXPERIMENTAL METHODS

Model Compounds

Aromatic model compounds were selected according to two criteria: the presence of an electron-withdrawing substituent that would deactivate the ring toward thermal chlorination, and significance as a priority water pollutant. Model compounds screened included benzoic acid, 3-hydroxybenzoic acid, nitrobenzene, and 2-, 3-, and 4-nitrophenol.

Reaction Conditions

Standardized screening conditions were selected to approximate those used in water treatment: substrate concentration was 7×10^{-5} M (8–12 ppm) and NaOCl concentration 7×10^{-4} M, to give a molar chlorine-to-substrate ratio of 10:1 in distilled water buffered at pH 6 or 8. Model compounds were added directly (without a carrier solvent) to deionized distilled water buffered with 0.01 M phosphate and dissolved by sonication and stirring. Hypochlorite solution was Clorox bleach (Clorox Company, Oakland, California), nominally 5.25% NaOCl, verified using an Orion residual chlorine electrode.

Dark reactions were allowed to proceed at ambient temperature (20–25°C) in Pyrex flasks covered with aluminum foil to exclude light. Light solutions, also in Pyrex flasks, were irradiated in a photoreactor consisting of a 14-in. by 5-ft cylindrical housing lined with reflective foil and illuminated with six F40BL fluorescent lamps.[6] Experiments with a *p*-nitroanisole-pyridine actinometer[7] gave a half-life for *p*-nitroanisole of 2 to 3 h, which corresponds to that obtained in spring sunlight.

In some experiments, substrate solutions were treated with a lower dose of hypochlorite to permit detection of aromatic intermediates. 3-Hydroxybenzoic acid solutions, buffered at pH 6, 8 and 10, were treated at a molar chlorine-to-substrate ratio of 1.0, 0.46, and 0.16 so that the presence of thermal chlorination products could be investigated. Nitrobenzene solutions buffered at pH 4, 6, 8, and 10 were treated at a chlorine-to-substrate ratio of 0.25 so that the presence of photochemical reaction products could be investigated.

Sample Preparation

Samples (100 mL) were immediately treated with excess sodium bisulfite to destroy the unreacted chlorine (verified using the Orion residual chlorine electrode) and then acidified with HCl. Nitrobenzene and nitrophenol samples were saturated with NaCl, passed through cyclohexyl bonded-phase extraction columns and eluted with methanol according to published methods for phenols[8]; methanol extracts were analyzed by reverse-phase HPLC. Recoveries for standards of nitrobenzene, 2-, 3-, and 4-nitrophenol, and 2-, 3-, and 4-chloronitrobenzene were all greater than 91%.

Nitrophenol, benzoic acid, and 3-hydroxybenzoic acid samples to be analyzed by gas chromatography (GC) were extracted with ethyl acetate; extracts were concentrated under vacuum and methylated with diazomethane. Recoveries exceeded 86% for nitrophenol isomers and 93% for the acids.

Analysis

Hypochlorite concentrations were determined using the Orion Model 97–70 residual chlorine electrode, which measures free chlorine, hypochlorites, and chlorine bound to nitrogenous materials (as a simplified alternative to the iodometric ASTM methods, with the same interferences).

Model compounds and their derivatives were analyzed by reverse-phase HPLC with a UV absorbance detector (254 nm) or by GC with an FID, EC, or AFID detector. Products were confirmed by GC-MS.

RESULTS AND DISCUSSION

Thermal Chlorination

Both benzoic acid and nitrobenzene were stable in the dark to hypochlorite under screening conditions (pH 6 and 8; chlorine-to-substrate ratio, 10). Hydroxylated aromatics, despite the presence of an electron-withdrawing group (-COOH or -NO_2), reacted thermally with hypochlorite at both pH 6 and 8.

When 3-hydroxybenzoic acid solutions were treated at a molar chlorine-to-substrate ratio of 10, no parent compound remained in methylated sample extracts at the earliest sampling time (t ~20 s) and no aromatic products were detectable by GC-FID. At a lower hypochlorite dose (chlorine-to-substrate ratio of 1.0), some parent compound remained, and mono-, di-, and trichlorinated derivatives of 3-hydroxybenzoic acid were detected (Figure 1). Our results are not completely consistent with those of Larson and Rockwell,[9] who reported that, although 2- and 4-hydroxybenzoic acid were thermally chlorinated under their reaction conditions (unbuffered, chlorine-to-substrate ratio of 1.0), 3-hydroxybenzoic acid was quantitatively recovered from chlorinated solution, even after heating to 90°C. We performed experiments at still lower hypochlorite doses (chlorine-to-substrate ratio of 0.46 and 0.16) and observed both loss of parent compound and appearance of a monochlorinated derivative in samples taken at t <1 min.

All three nitrophenol isomers appeared to be more persistent than 3-hydroxybenzoic acid in (dark) aqueous hypochlorite solutions, probably because -NO_2 is an even stronger electron-withdrawing group than -COOH. In experiments with each of the three isomers at a chlorine-to-substrate ratio of 10, the parent compound remained in the earliest samples taken (t <1 min) and several ring-chlorinated products were detected (Figure 1). Successive ring chlorination occurred, as well as displacement of the -NO_2 by a -Cl to form 2,4,6-trichlorophenol in 2- and 4-nitrophenol solutions and 2,3,4,6-tetrachlorophenol in 3-nitrophenol solutions. The displacement of a -NO_2 by a -Cl in chlorinated aqueous 4-nitrophenol solution was previously reported by Smith et al.,[3] although they used different reaction conditions: very high concentrations of reactants (2000 ppm 4-nitrophenol and 5000 ppm HOCl), pH 6, and a reaction time of 24 h.

Photolysis of Hypochlorite

When aqueous hypochlorite solutions, buffered with 0.01 M phosphate at pH values between 5.7 and 10, were irradiated, the loss of total residual chlorine followed first-order kinetics through at least 1 to 2 half-lives. Between pH 5.7 and 8, the rate of loss of total residual chlorine appeared to be pH-

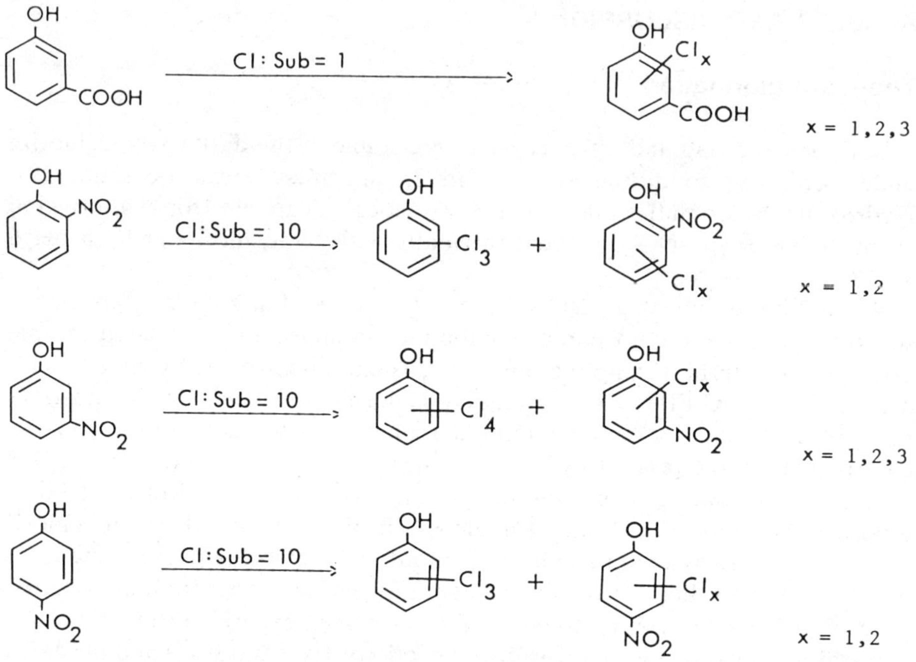

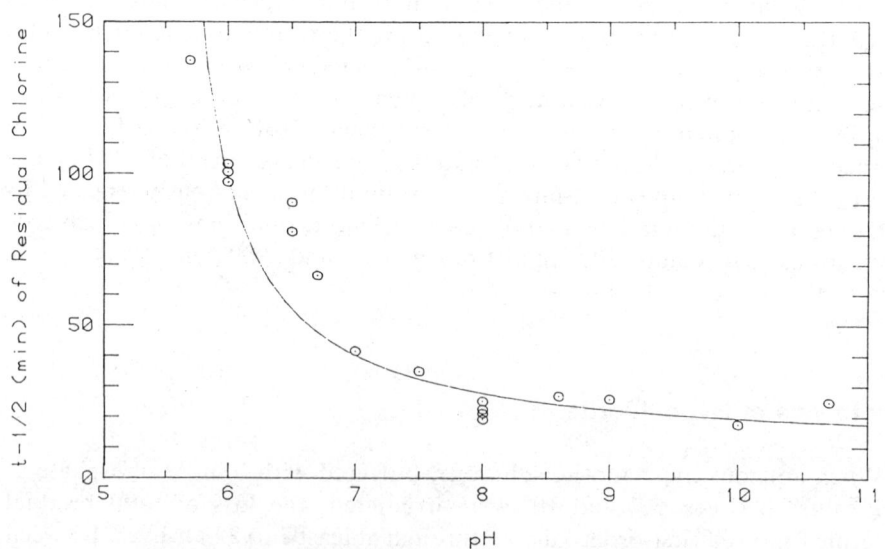

Figure 1. Products formed in dark or thermal chlorination experiments with hydroxylated model compounds. [Substrate] = 70 μM; [NaOCl] = 70 or 700 μM; pH 6; molar chlorine-to-substrate ratio (Cl:Sub) = 1 or 10.

Figure 2. Photolysis of HOCl/OCl$^-$: Half-life of residual chlorine (in minutes) as a function of pH (Initial [NaOCl] = 700 μM).

dependent, so that the loss was slower at low pH. Above pH 8, the $t_{1/2}$ remained approximately constant (Figure 2).

When 0.1% methanol (0.025 M) was added to the hypochlorite solution, loss of residual chlorine was markedly accelerated, especially at lower pH. Its half-life decreased from 21.0 ± 1.7 to 6.11 min when 0.1% methanol was added at pH 8, and from 100.3 ± 3.0 to 0.78 min at pH 6. The loss of residual chlorine on irradiation was also accelerated by the presence of deactivated substrates such as nitrobenzene or benzoic acid, although to a lesser extent.

HOCl and OCl⁻, which exist in equilibrium together (pK_a = 7.5), have different photodegradation pathways. According to Oliver and Carey,[5] the primary photochemical process of HOCl at 350 nm is

a. $HOCl \rightarrow OH + Cl$

Buxton and Subhani[10] have shown that the primary photoprocesses for OCl⁻ are wavelength-dependent:

	Photoprocess	Primary Quantum Yields 365 nm	313 nm
b.	$OCl^- \rightarrow Cl^- + O(^3P)$	0.28 ± 0.03	0.075 ± 0.015
c.	$OCl^- \rightarrow Cl + O^-$	0.08 ± 0.02	0.127 ± 0.014
d.	$OCl^- \rightarrow Cl^- + O(^1D)$	0	0.020 ± 0.015
e.	$O(^1D) + H_2O \rightarrow 2\ OH$		
f.	$O^- + H_2O \rightarrow OH + OH^-$		

Thus, the primary photoprocesses of OCl⁻ in sunlight are (b) and (c), although (d) will occur to a lesser extent. There should be hydroxyl radicals generated even at high pH via reactions (e) and (f).

Methanol is a good free-radical scavenger and will be attacked by the powerful oxidant hydroxyl radical. In the absence of methanol, radicals produced on irradiation may recombine with each other or react with neutral species. According to Buxton and Subhani,[10] both generation of nonreactive oxychlorine species (such as ClO_2^- and ClO_3^-) and regeneration of reactive chlorine species (OCl⁻ and Cl_2O) occur. In the presence of excess methanol, free radicals produced on irradiation would instead be scavenged by reaction with methanol. Regeneration of reactive chlorine would not occur, thus accelerating the loss of residual chlorine on irradiation.

Light-Induced Reactions

Nitrobenzene and benzoic acid, both stable to hypochlorite at pH 6 and 8, were degraded when solutions were chlorinated at a chlorine-to-substrate ratio of 10 and irradiated. No significant amounts of aromatic products were detected by GC-FID for benzoic acid or by HPLC for nitrobenzene. However, on treatment of nitrobenzene solutions at a lower dose of hypochlorite (chlorine-to-substrate ratio of 0.25), 2-, 3-, and 4-nitrophenol were detected within 5 min of irradiation. Between pH 4 and 10, the total yield of nitrophenols observed after 60 min of irradiation was inversely related to pH, with the highest yield at pH 4 (Table I). There was no loss of nitrobenzene or formation of nitrophenols in dark chlorinated controls. Oliver and Carey[5] also observed light-induced degradation of benzoic acid in hypochlorite solution (at pH 7, chlorine-to-substrate ratio of 10, with irradiation at 350 nm). They measured the weight percent organic chlorine of reaction products as a function of irradiation time. They detected salicylic acid at short irradiation times and concluded that thermal chlorination of phenols generated photochemically was probably responsible for the large increase in weight percent chlorination observed after the initial 15-min period of irradiation.

In the presence of 0.1% methanol (0.025 M), we found that both benzoic acid and nitrobenzene were stable to hypochlorite, even on irradiation; parent substrate was quantitatively recovered after 4 h to several days. This was the case even when the chlorine-to-substrate ratio was increased to 25 or when solutions were irradiated outdoors in spring sunlight (Davis, California). Inhibition of light-induced degradation by methanol, an excellent free-radical scavenger, indicates that the light-induced reactions are probably free radical in nature rather than photochemical reactions of an excited-state substrate with hypochlorite.

The distribution of nitrophenol isomers obtained in nitrobenzene solution (chlorine-to-substrate ratio of 10) after 60 min of irradiation are shown in Table I. At all pH values, 2-nitrophenol was the predominant isomer detected.

Table I. Distribution and Yield of Nitrophenol Isomers[a] On Irradiation of Chlorinated Aqueous Nitrobenzene Solutions (chlorine-to-substrate ratio = 0.25)

pH	Total Yield Nitrophenols[b] (%)		Distribution of Nitrophenols in Light[c] (%)		
	Dark	Light	Ortho	Meta	Para
4.00	0	3.19	56.6	22.6	20.7
6.00	0	2.09	54.0	25.2	20.8
8.00	0	1.52	55.4	14.6	30.0
10.00	0	0.920	55.3	6.47	38.2
Unbuffered[d]	0	2.21	51.0	24.2	24.8
7[e]	0	1.97	50.0	29.5	20.5

[a][Nitrobenzene] = 70 μM; [NaOCl] = 17.6 μM; contact or irradiation time = 60 min.
[b]Molar concentration of total nitrophenols as a percentage of initial nitrobenzene concentration.
[c]Concentration of each isomer as a percentage of the concentration of total nitrophenols.
[d]Initial pH = 6.10.
[e]For irradiation of 500-μM nitrobenzene and 100-μM H_2O_2 for 18 h. From Ref. 11.

Statistically, we would expect the ratio of free-radical attack at the ortho, meta, and para positions to be 2:2:1. Because it is not, we may be seeing an inductive effect as well. The isomer distributions obtained at pH 4 and 6 correspond remarkably well with those obtained by Draper and Crosby[11] on irradiation of nitrobenzene in the presence of hydrogen peroxide at pH 7 (Table I). This suggests that hydroxyl radical may be the oxidant involved.

From the work of Buxton and Subhani[10] discussed earlier, it appears that more hydroxyl radical is generated on irradiation of HOCl/OCl- at sunlight wavelengths at lower pH. This may explain the pH-dependence of the yield of nitrophenols observed after 60 min of irradiation. Higher yield may be a function of higher rate of hydroxyl radical generation, which corresponds to a higher concentration of HOCl relative to OCl-.

CONCLUSIONS

In certain common environmental situations, such as waters receiving chlorinated effluents, holding ponds, and swimming pools, chlorinated water is exposed to sunlight. Both primary and secondary photochemical reactions can occur, as well as subsequent thermal chlorination of any resulting activated reaction products. At least for aromatic compounds, free-radical oxidants formed from the photodegradation of HOCl/OCl⁻ appear to hydroxylate even deactivated rings, generating phenols that will react thermally with residual chlorine. It therefore may be important to consider the role of sunlight in both the fate of reactive chlorine and the reactivity of organic compounds in chlorinated waters in our environment.

ACKNOWLEDGMENTS

We gratefully acknowledge the financial support of the N.I.H. Training Grant No. 2T32 ES 07059-06 (to L.H.N.) and the technical advice and assistance of Basil Bowers, Richard Higashi, Kei Miyano, and Clayton Reece.

REFERENCES

1. Carlson, R. M., R. E. Carlson, H. L. Kopperman, and R. Caple. "Facile Incorporation of Chlorine Into Aromatic Systems During Aqueous Chlorination Processes," *Environ. Sci. Technol.* 9(7):674–675 (1975).
2. Rockwell, A. L., and R. A. Larson. "Aqueous Chlorination of Some Phenolic Acids," in *Water Chlorination: Environmental Impact and Health Effects, Vol. 2*, R. L. Jolley, H. Gorchev, and D. H. Hamilton, Jr., Eds. (Ann Arbor, MI: Ann Arbor Science Publishers, Inc., 1978), p. 67.

3. Smith, J. G., S. Lee, and A. Netzer. "Model Studies in Aqueous Chlorination: The Chlorination of Phenols in Dilute Aqueous Solutions," *Water Res.* 10:985–990 (1976).
4. Norwood, D. L., J. D. Johnson, R. F. Christman, J. R. Hass, and M. J. Bobenrieth. "Reactions of Chlorine with Selected Aromatic Models of Aquatic Humic Material," *Environ. Sci. Technol.* 14(2):187–190 (1980).
5. Oliver, B. G., and J. H. Carey. "Photochemical Production of Chlorinated Organics in Aqueous Solutions Containing Chlorine," *Environ. Sci. Technol.* 11(19):893–895 (1977).
6. Crosby, D. G., and A. S. Wong."Photodecomposition of 2,4,5-Trichlorophenoxyacetic Acid (2,4,5-T) in Water," *J. Agric. Food Chem.* 21(6):1052–1054 (1973).
7. Dulin, D., and T. Mill. "Development and Evaluation of Sunlight Actinometers," *Environ. Sci. Technol.* 16(11):815–820 (1982).
8. Dimson, P. "Isolation of Phenol and Substituted Phenols Using a Cyclohexyl Bonded-Phase Extraction Column with HPLC Analysis," *Liq. Chromatogr.* 1(4):236–237 (1983).
9. Larson, R. A., and A. L. Rockwell. "Chloroform and Chlorophenol Production by Decarboxylation of Natural Acids During Aqueous Chlorination," *Environ. Sci. Technol.* 13(3):325–329 (1979).
10. Buxton, G. V., and M. S. Subhani. "Radiation Chemistry and Photochemistry of Oxychlorine Ions, Part 2—Photodecomposition of Aqueous Solutions of Hypochlorite Ions," *Trans. Faraday Soc. I.* 68:958 (1972).
11. Draper, W. M., and D. G. Crosby. "Solar Photooxidation of Pesticides in Dilute Hydrogen Peroxide," *J. Agric. Food Chem.* 32(2):231–237 (1984).

SECTION XIII

Chemical Methods

Every fragment derived from original structure and containing more than one structure element offers a piece of information about the structure if we know the size and composition of the fragment.
> A. Tiselius
> *Some General Relations Between the Composition of Fragments and the Structure from which they Originate*, 1959

What science strives for is an utmost acuteness and clarity of concepts as regards their mutual relation and their correspondence to sensory data. . . . The connections between concepts and statements on the one hand and the sensory data on the other hand is established through acts of counting and measuring whose performance is sufficiently well determined.
> **Albert Einstein**, 1879–1955
> *The Common Language of Science*, 1941

CHAPTER **84**

Analytical Determination of Inorganic Chlorination Products in Water Treatment by an HPLC Method

M. Dreux, M. Lafosse, M. Gibert, and A. Blaison

The diversity of anions (mono- and polyatomic anions, mono- and polycharged species, mineral and organic anions) normally requires different methods for their determination. A one-step method for analysis of complex anion mixtures involving liquid chromatography would be useful. Recently much research has been reported on ion chromatography using conductivity measurements for detection. However, two problems remain to be solved: adequate separation and sensitive determination. These are mutually dependent needs, because a good separation does not always allow sensitive detection or vice versa.

GOALS AND RATIONALE OF THIS WORK

Optimum conditions were sought for separation of complex anion mixtures on two chromatographic substrates, Zorbax C18 and Zorbax C8. Separations were performed with either ion exchange chromatography or ion pair chromatography, and the separated constituents were detected by either conductivity or UV spectrophotometry. If solutes have a chromophore, a direct determination may be made; if not, an indirect determination may be obtained providing the mobile phase is a UV-absorbing eluent.[1] Conductimetric detection requires low-conductance (Λ) eluents because of the relative measure:

$$\frac{\Delta \Lambda}{\Lambda} = \frac{\Lambda_{elute} \pm \Lambda_{solute}}{\Lambda_{elute}} \qquad (1)$$

The minimum detectable conductance change is 1/3000.[2] Thus, in ion exchange chromatography two methods can be explored: use of eluents with low ionic strength (low conductance), and use of eluents with high salt concentrations (high conductance). In the first case, ion exchangers with low capacity, such as the one proposed by Gjerde[3] (not commercially available), must be used. In the second case, the conductance of eluents must be reduced to low values before detection. This method has been proposed by Small et al.[4] and it was commercialized by the Dionex Society. Its advantages and disadvantages have been previously discussed.[5]

In ion pair chromatography, the chromatographic parameters influencing the separation are more numerous than in ion exchange. We will show that they permit the use of mobile phases with low conductance and that the detectability is good with either conductimetry or UV spectrophotometry. With UV spectrophotometry, good detectability is directly obtained if the ions that have a chromophore are detected on a wavelength of their absorptions maxima. When the ions have no chromophore, Denkert et al. have developed a method of indirect photometry.[6] The sensitivity of the detection depends on the spectral characteristics of the mobile phase, which results in all anions being determined at the same wavelength. This principle can be used with any type of chromatography. Small used it for ion exchange and Denkert et al. for ion pair chromatography.[7,8]

In this study, ion pair chromatography is used for the analysis of the chlorinated anions produced during water treatment with chlorine dioxide. This method could be usefully applied during chlorine dioxide manufacture and for analysis of treated waters.

EXPERIMENTAL

The high-performance liquid chromatographic (HPLC) equipment included the following components: a Gilson model 302 pump with a 5.S model pump head, and a Rheodyne model 7125 injection valve. For UV-absorbing compounds, a Schoeffel model GM 770 spectrophotometer was used; for conductimetry a Vydac model 6000 CD conductivity meter was used. Signals from the two detectors were recorded by a two-channel Vitatron model 2001 recorder. Columns were slurry packed in the laboratory with 7-μm Zorbax C8 and C18 (DuPont, Inc.) at 6000 psi using carbon tetrachloride followed by methanol. The column, with 3-μm Zorbax C18, was purchased from DuPont (Golden series 6.2 mm ID by 80 mm).

The eluents were mixtures of deionized water from a Milli Q purification system (Millipore Corp.) and ammonium salts. The pH was adjusted with the addition of the counter ion (acid) to the calculated quantity of tetrabutylammonium hydroxide. With the long chain length amine, the calculated quantity of acid was solubilized in water and the amine was added to provide the desired pH. Amines (octylamine and nonylamine) and tetrabutylammonium hydroxyde were obtained from Eastman Kodak and Merck, Inc. Acids (p-toluene sulfonic acid, 3-nitrobenzene sulfonic acid, phthalic acid, and benzoic acid) were purchased from Prolabo (Paris) and Aldrich.

RESULTS AND DISCUSSION

In earlier works, we described the use of the ion pair chromatography for determination of inorganic anions in aqueous medium and discussed retention, selectivity, and the factors influencing the dynamic equilibrium of these systems.[9,10]

Optimal conditions for the chromatographic separations of the chlorinated inorganic compounds such as chloride, chlorite, chlorate, and perchlorate (this last was not a necessity of real determination) were developed with two different apolar stationary phases (octyl and octadecyl bonded silica) and with pure aqueous eluents using cationic (ammonium) surfactants. The counter ions had an aromatic group for the indirect photometric determination of solutes.

A simple reaction scheme is

$$RN^+SO_3^-Ar \rightleftharpoons RN^+ \text{ (pairing reagent)} + ArSO_3^- \text{(counter ion)} \quad (2)$$

$$RN^+ + X^- \text{ (solute)} \rightleftharpoons [RN^+ \ X^-] \text{ (ion pair complex)} \quad (3)$$

Ion Pair Chromatography with an Octylbonded Silica (Zorbax C8)

Two different ammonium surfactants were chosen. One has a long linear hydrophobic chain (i.e., octylammonium or OA). Longer linear chains lead to irreversible adsorptions.[11] The second was the most used pairing-ion reagent (i.e., tetrabutylammonium or TBA) in which the chain length is short.

Table I shows the results obtained with two different counter ions: the p-toluene sulfonic acid (PTSA) and the phthalic acid (PA).[12] Eluents were used at pH 6 and were not buffered for a low conductivity value or to stabilize the chlorine dioxide. The surfactant concentration was 1 mM.

Above pH 6, chlorine dioxide may change into chlorite and chlorate in the presence of buffers.[13] Because hypochlorous acid is not highly dissociated at pH 6, determination of hypochlorous acid with conductimetric measure or indirect photometry is impossible. Hypochlorous acid does not react with our chromatographic eluents. Consequently, no false determination occurs for the other anions. Hypochlorite determination with an electrochemical method was described by Dionex.[14]

From the results of the first four experiments (Table I), a choice of the eluent for the fifth experiment was made to optimize the separation of anions with the sensitivity of detection.

Table I. Ion Pair Chromatography with Octyl-Bonded Silica.[a] Conditions: Column 240 mm by 4 mm ID; stationary phase Zorbax C8, 7 μm; eluents 1 mM, pH 6, flow rate 1.5 mL/min.

Experiment	Ion pairing surfactant[b]	Conductivity (μS)	Retention volume (V_R) (mL)					Selectivity	V_R (mL)		Selectivity
			Cl^-	ClO_2^-	ClO_3^-	ClO_4^-		$\alpha_{ClO_2^-/Cl^-}$	NO_3^-	SO_4^{2-}	$\alpha_{SO_4^{2-}/ClO_3^-}$
1	TBA-PTSA	55	5.5	5.75	14	15.2		1.06	10	26.7	2.02
2	TBA-PA	105	7	8.5	29			1.27	17	26.8	0.92
3	OA-PTSA	55	5.1	6	10.2	16.5		1.26	7.4	36	4.00
4	OA-PA	115	3.7	4.3	8.4	15.5		1.29	5.7	14.4	1.83
5[c]	OA-PTSA	28	3.7	5.7	7.5	9		1.54	4.8	21.0	3.29

[a] From Ref. 12.
[b] Definitions: TBA, tetrabutylammonium; PTSA, p-toluene sulfonic acid; PA, phthalic acid; OA, octylamine.
[c] Eluent 0.5 mM.

Table II. Ion Pair Chromatography with Octyl-Bonded Silica.[a] Conditions: Column 150mm by 4.7 mm ID; stationary phase Zorbax C18, 7 μm; eluent ≤ 0.5 mM.

Experiment	Column length (mm)	Ion pairing surfactant[a]	Concentration (mM)	pH	Retention volume (V_R) (mL)						Selectivity	
					Cl^-	ClO_2^-	ClO_3^-	NO_3^-	SO_4^-		$\alpha_{ClO_2^-/Cl^-}$	$\alpha_{ClO_3^-/NO_3^-}$
6	150	OA-PTSA	0.5	5.6	3.8	4.8	7.1	5.4	22.4		1.18	1.43
7	150	OA-NBSA	0.5	5.6	4	5	7.4	5.6	26		1.17	1.44
8	150	NA-PTSA	0.5	5.6	5.7	7	12	9	LR[b]		1.38	1.4
9	150	NA-NBSA	0.5	5.6	6.3	8.4	12.9	10	LR[b]		1.43	1.34
10	150	NA-NBSA	0.3	5.6	7.2	9.5	14.7	10.5	48		1.40	1.46
11	250	NA-BA	0.1	6.2	12	14.2	24	17.9	LR[b]		1.22	1.40
12[b]	80	NA-BA	0.1	6	6.1	7.5	10.6	7.8	LR[b]		1.27	1.41

[a] OA, Octylamine; PTSA, p-toluene sulfonic acid; NA, nonylamine; NBSA, 3-nitrobenzene sulfonic acid; PA, phthalic acid; BA, benzoic acid.
[b] LR, large retention and broad peak.
[c] Diameter of particles, 3 μm; column, 6.2 mm ID.

Separation of Anions

Selectivity is the ability to separate anions. The selectivity value (α) is defined as the retention value of one species divided by the retention value of the second species:

$$\alpha_{ClO_2^-/Cl^-} = V_R \text{ for } ClO_2^-/V_R \text{ for } Cl^- \qquad (4)$$

A low selectivity value with chlorite-chloride was obtained in the first experiment. A large value was desirable because of the concentration difference between chloride and chlorite. Chlorine dioxide is manufactured with chlorine and sodium chlorite and chloride is also produced. Furthermore, water may contain high concentrations of chloride. Similar values of chlorite-chloride selectivity were obtained with experiments 2, 3, and 4. Experiment 2 involved a long analysis time, and a low selectivity value for the chlorate-sulfate mixture was obtained (Table I). Because of the relatively poor sulfate-chlorate separation, and because sulfate is often found in waters, the experimental conditions of experiment 2 were not investigated further.

The selectivity in experiments 3 and 4 appears adequate. A greater intensity of the "system peak" occurred with the tetrabutylammonium surfactant. The system peak is a group of superfluous peaks that is produced by the displacement of the dynamic ion pair equilibrium; a more appropriate name is "induced peaks."[15] We also noted a lower intensity of the induced peaks with pure aqueous eluents.[10]

In Table I, we have shown the retentions of nitrates because they are present in water; we did not discuss them because their retentions were different enough from the retention of the other ions. Carbonates, phosphates, and silicates are also present in water; carbonates are in the monoanionic form at the pH of eluents, and they have a shorter retention than the chlorides. Phosphates have poor detectability both in conductimetry and indirect photometry. Silicates, which are not ionized at eluent pH, are not detectable.

Sensitivity of Detection

With conductivity detection, the detectability is better with eluents that have low conductance, so p-toluene sulfonic acid (PTSA) is better than phthalic acid (PA). Phthalic acid is a dianion at pH 6, and, at an identical ammonium concentration, it will lead to eluents with almost double conductance value. In indirect photometry, with the carboxylic acid (CA), the unbuffered eluents show a bad baseline with high sensitivity of detection. This problem occurs with conductivity detection also. The reason is a slightly variable ionization of the carboxylic acid at the pH of eluents. For these reasons, the PTSA counter ion is more attractive and, consequently, was used in the fifth experiment.

Use of PTSA leads to the following results: better separation, shorter analysis time, and better detectabilities with both UV and conductimetry. For example, with a 100-μL injection, the detection limit for the chlorate is 0.5 ppm by conductimetry and 1 ppm in indirect photometry at 254 nm (instead of about 2 ppm for the third experiment).

Ion Pair Chromatography with an Octadecyl-Bonded Silica (Zorbax ODS)

We have previously shown that the octyl-bonded silica required a prewetting of the column before the equilibration with the chromatographic eluent.[10] Other studies have shown the utility of the octadecyl-bonded silica.[15] Consequently, we investigated the use of an octadecyl phase commonly used by chromatographers (Table II). Because the interactions between the apolar chain of the surfactant and the apolar chain of the silica is greater with octadecyl than octyl, we can use eluents with lower concentrations of the surfactant and shorter columns. (Table I: eluent 1 mM, column length 240 mm; Table II: eluent $\leq$ 0.5 mM, column 150 mm.)

The sixth experiment in Table II shows the transposition C8-C18 stationary phase with the OA-PTSA 0.5 mM eluent. The selectivity of the system and the analysis times are still good.

The seventh experiment (Table II) used a counter ion that has a greater UV absorptivity than the PTSA, that is, 3-nitrobenzenesulfonic acid (NBSA). Also NBSA improves the detection sensitivity according to the principles developed by Schill.[8] With good chromatographic separation, the NBSA counter ion permits a limit of detection of 0.25 ppm (at 238 nm) for chlorate using a 100-μL sample injection. The conductimetric sensitivity is the same as before (0.50 ppm).

A decrease of the concentration of the surfactant often leads to a decrease of retention. Conversely, an increase of the ammonium chain length leads to an increase of the solute retention. The eighth and ninth experiments (Table II) illustrated the influence of a longer chain length of the ammonium (nonylammonium, NA). A better selectivity for the chlorite-chloride ions was observed in these experiments. The tenth experiment showed the influence of a decrease of the surfactant concentration (Table II). We noted that the retention was very important for sulfate; consequently the sulfate peak became very large. Often at the concentration found in water, the sulfate peak is in the baseline drift. In this tenth experiment, we used an eluent with low conductance to increase the limit of the detection in conductimetry. For a 100-μL injection, the chlorate detectability was about 0.3 ppm by conductimetry and about 0.2 ppm by indirect photometry at 238 nm.

To further improve analysis time, we examined the effect of eluent conductivity. This required the choice of another counter ion. Use of benzoic acid (BA) as counter ion in the eleventh experiment (Table II) provided an eluent with a very weak conductance and a reasonable analysis time using a column 250 mm long. We noted poorer selectivity for the chlorite-chloride pair. For this surfactant concentration, the baseline of the detection at high sensitivities remains good. The detectability with conductimetry is equivalent to that obtained in the tenth experiment.

The last experiment (experiment 12) involved a repeat of the eleventh experiment but with a different column (Table II). The Golden series, commercialized by DuPont, uses a C18 bonded silica of 3 μm. The poor selectivity for the nitrate-chlorite pair prevents the use of this system. If there is no chlorite, this system has the advantage of a rapid analysis. The detection is almost unchanged by conductimetry.

Detection Sensitivity

Two methods were explored to improve the detection sensitivity:

1. Narrow-bore columns (2-mm ID) permitted much reduced flow-rates of the mobile phase. The baseline noise was reduced and the sensitivity improved using conductivity for detection.
2. Superior detectors improved detectability. With eluents of low conductance, we found better results with the Tacussel conductivity meter, model CD 810 (Solea-Tacussel, Villeurbanne, France), equipped with a 2-μL cell for HPLC. New UV spectrometers, such as the Kratos model 773 (Kratos Analytical Instruments, Ramsey, New Jersey) permitted improved detectability.

CONCLUSIONS

The detectable limits observed in these ion pair chromatography experiments are inadequate for a direct determination of the chlorinated inorganic anions in common drinking water. However, detection sensitivity may be improved by using narrow-base columns and superior detectors.

REFERENCES

1. Eksborg, S., and G. Schill. "Ion-Pair Partition Chromatography of Organic Ammonium Compounds," *Anal. Chem.* 45:2092-2100 (1973).
2. The Separation Group, Inc. *Vydac Conductivity Detector Model 600 CD Instruction Manual.*
3. Gjerde, D. T., J. S. Fritz, and G. Schmuckler. "Anion Chromatography with Low Conductivity Eluents," *J. Chromatogr.* 186:509-519 (1979).
4. Small, H., T. S. Stevens, and W. C. Bauman. "Novel Ion Exchange Chromatographic Methods Using Conductimetric Detection," *Anal. Chem.* 57:1801-1809 (1975).

5. Girard, J. E., and J. A. Glatz. "Ion Chromatography with Conventional HPLC Instrumentation," *Int. Lab.* 1981 (November-December):62-68.
6. Denkert, M., L. Hackzell, G. Schill, and E. Sjogren. "Reversed-Phase Ion Pair Chromatography with UV-Absorbing Ions in the Mobile Phase," *J. Chromatogr.* 218:41-43 (1981).
7. Small, H., and T. H. Miller, Jr. "Indirect Photometric Chromatography," *Anal. Chem.* 54:462-469 (1982).
8. Schill, G. "Detector in Reversed-Phase Liquid Chromatography by Use of Ion-Pairing Probes," *Anal. Proc.* 20:359-362 (1983).
9. Dreux, M., M. Lafosse, and M. Pequignot. "Separation of Inorganic Anions by Ion-Pair, Reverse Phase Liquid Chromatography Monitored by Indirect Photometry," *Chromatographia* 15:653-656 (1982).
10. Dreux, M., M. Lafosse, and P. Agbo-Hazoume. "Ion-Pair Chromatography: A New Approach to Dynamic Equilibrium," *Chromatographia* 18:15-18 (1984).
11. Knox, J. H., and R. A. Hartwick. "Mechanism of Ion-Pair Liquid Chromatography of Amines, Neutrals, Zwitterions, and Acids Using Anionic Hetaerons," *J. Chromatogr.* 204:3-21 (1981).
12. Agbo-Hazoume, P. Unpublished results (Orléans: Université d'Orléans).
13. Medir, M., and F. Giralt. "Stability of Chlorine Dioxide in Aqueous Solution," *Water Res.* 16:1379-1382 (1982).
14. Dionex Corporation. "Analysis of Hypochlorite and Chlorate in Bleach Using Combined Conductivity and Electrochemical Detection," Dionex Application Note 29, (Sunnyvale, CA: Dionex Corporation).
15. Bidlingmeyer, B. A., and F. W. Warren, Jr. "Effect of Ionic Strength on Retention and Detector Response in Reversed-Phase Ion-Pair Liquid Chromatography with Ultraviolet-Absorbing Reagents," *Anal. Chem.* 54:2351-2356 (1982).

CHAPTER 85

Instrument for Total Chlorine Amperometric Back Titration Using Coulometric Iodine Generation

Daniel H. Raab, and Calvin O. Huber

The currently accepted standard method for determining total residual chlorine in wastewaters containing high concentrations of dissolved organic carbon (DOC) is the back titration method using an amperometric endpoint detector.[1-3] This method is superior to others because of both the wet chemical and endpoint detection methods used. The chemistry does not allow iodine to react with solution constituents (DOC, etc.) because of its greater rate of reaction with excess phenyl arsineoxide (PAO).[2]

The amperometric endpoint is the standard for comparison because of its greater inherent accuracy as well as freedom from color and turbidity interferences. This freedom from interference makes amperometric back titration the method of choice for determining total residual chlorine (TRC) in wastewater. However, for accurate and precise work at trace levels of TRC, the method requires significant amounts of operator skill because of the small volumes of dilute iodine titrant used. Thus, a method was developed that eliminates the volumetric titration.

The new instrumental method described here differs from the standard amperometric back titration in two ways: (1) the iodine titrant is added in situ electrochemically rather than as a solution, thus eliminating the need for preparation, maintainence, and manipulation of relatively dilute standard iodine solutions; and (2) all phases of data collection, data processing, equivalence point determination, calculation of TRC concentration, and operator prompting are computer controlled. Thus, the advantages of the amperometric back titration are retained, and the disadvantages are removed.

MATERIALS AND METHODS

Apparatus

The apparatus consists of three major components: (1) A constant current source for in situ coulometric generation of iodine. The constant current source uses an operational amplifier having the electrochemical cell in the feedback loop. This configuration has shown excellent stability over both the short and long term. (2) An amperometric detector consisting of a standard three-electrode potentiostat and current follower. (3) A microcomputer that completely controls data aquisition, data analysis, equivalence point determination, calculation of TRC concentration, and operator prompting. The microcomputer used is a Timex Sinclair 1000 outfitted with a Memotech 16K RAM, a Microdevelopments IO expansion board, and a Microdevelopments AD/DA expansion board. Platinum wire electrodes were used for both the iodine generator and amperometric detector circuits. As in the standard amperometric back titration procedure, the additional apparatus consists of a 400 mL beaker, a magnetic stir bar, and stirring motor.

Chemicals

Chemicals used in this procedure are (1) ACS certified potassium iodide (KI) (MCB Reagents); (2) standard 0.00564 N phenyl arsineoxide solution (PAO) (Anderson Labs and Ricca Chemical); and (3) pH 4 acetate buffer, as per *Standard Methods*.[2]

Experimental

Field experiments were conducted at the Milwaukee Metropolitan Sewage District, Jones Island facility from August 5 to December 10, 1983. Fieldwork was performed outdoors under a tarp or inside a covered nonlaboratory area of the facility. Grab samples of about 3 L were obtained within 10 s after chlorination. These samples were gently stirred in an uncovered 4 L beaker. For determination of TRC, 20 mL samples from the beaker were diluted to 200 mL, so that the resultant TRC concentrations were in the range of 0 to 1 ppm. TRC determination was performed in a 400 mL beaker using a 35-mm-long Teflon® stir bar spinning at ~700 rpm. Iodine was generated by passing a constant current of 5.00 mA at a current density of 3.51 mA/cm². When the standard methods procedure was performed, a 0.01-mL-graduated, 1.00 mL buret was used for the addition of standardized iodine solution. Operator responsibilities are limited to initiation of the computer program and to sample preparation; complete operator prompting and data analysis are then provided by the instrument.

RESULTS AND DISCUSSION

To apply Faraday's Law for the quantitative electrochemical generation of iodine, evaluation of current efficiency for the electrochemical in situ generation of iodine was necessary. The only constraints demanded by the electrochemical method are that (1) current efficiency be both reproducible and known and (2) the time for which current is passed is accurately known.

Ideally, an invariant 100% efficiency is most desirable. Background electrolyte was prepared by adding 4 mL of pH 4 acetate buffer to 200 mL of distilled water. Current efficiency for electrochemical generation of iodine was determined in distilled water using two methods. In the first method, iodine was generated in an open beaker and then titrated with PAO. At a current density of 5.04 mA/cm^2, the current efficiency was observed to be 99.2 ± 0.7%. Deviations from 100% are postulated to be the result of volatilization of iodine, a 1% deviation corresponding to the volatilization of only 28 nmol of iodine. The second method was that of Marinenko and Taylor.[1] Current efficiencies were determined for current densities of 0.5 to 15.0 mA/cm^2 at various KI concentrations. The minimum efficiency observed was >99.99% at a current density of 0.53 mA/cm^2 and a KI concentration of 0.0147 M (0.4989 g KI in 200 mL distilled water and 4 mL buffer). These results are similar to those of Marinenko and Taylor.[1]

In chlorinated samples, other methods for determination of current efficiency were devised. In chlorinated effluent and tap water, current efficiencies of between 99.9 ± 0.2% and 100.0 ± 0.5% were observed, based on a current efficiency of 100.0% in distilled water. The large estimated standard deviations may be an artifact of the method, which used two volumetric 1.00-mL additions of PAO from a pipet.

PRECISION AND ACCURACY OF THE INSTRUMENTAL METHOD

TRC in chlorinated effluent samples was determined at five different times using the standard amperometric back titration. TRC concentration decreases with time because of numerous biological and chemical factors.[4]

Our observations additionally show that most of this decrease is caused by volatilization of chlorine species. Because of this time dependence, the data points for TRC vs time for the standard method were fit using a quadratic least-squares function. A typical plot is shown in Figure 1. TRC in the same sample was also determined five times using the new instrument. These data points were compared to the quadratic curve to test for bias. One example of such data analysis is shown in Table I.[5] The deviations when comparing the two methods were examined using the t test at the 95% confidence level. Data from three samples were analyzed in this manner. In all cases the calculated t values were smaller than the tabulated t values. Thus, the new method has no bias at the 95% confidence level, with respect to the standard amperometric back titration.

Precision for field data was evaluated by determining the standard error of the estimate for the fit of the data to the quadratic least-squares function. The mean standard error of the estimate for 37 TRC vs time curves was 13.5 ppb TRC under field test conditions, using chlorinated effluent samples. In the laboratory, samples of chlorinated distilled water were kept in a closed con-

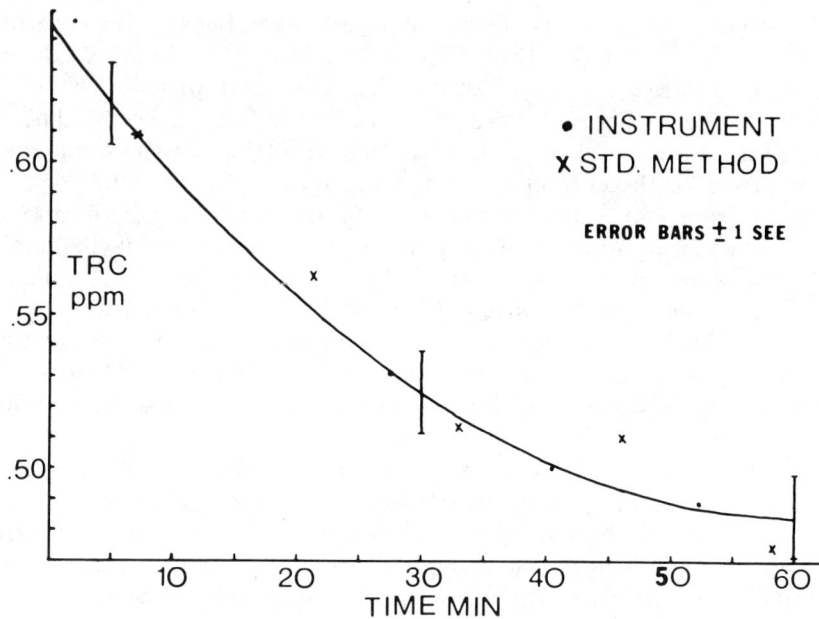

Figure 1. A typical TRC vs time plot.

tainer to prevent volatilization of chlorine species. Under these conditions, TRC concentration vs time was observed to be a constant. The standard deviation for the determination of TRC for six samples, each determined either four or five times, was 7.0 ppb. The difference between laboratory and field precision value may be the result of (1) the water used in the laboratory (i.e., chlorinated distilled water allows for a more reproducible sample than chlorinated effluent), (2) the differences between the laboratory and field samples and their handling, and (3) approximations inherent in the curve-fitting technique.

Field data were additionally analyzed using the f and t statistical parameters to determine the effect of chlorine dosage and TRC concentration on precision and accuracy of the new instrumental method. At the 95% confidence level, precision and accuracy were shown to be statistically independent of chlorine dosage in the range of 0.1 to 7.5 ppm, as well as resultant TRC concentrations in the range of 0.01 to 7.0 ppm. Field data were analyzed to determine the effect of operator familiarity with the instrument on precision. Data obtained by several nonchemists showed precision that was statistically indistinguishable from that obtained by the chemist-developer of the instrument.

The software developed for determination of the equivalence point, and subsequent calculation of the final value for TRC found, ultimately depend on response of the amperometric detector electrode to background electrolyte and

Table I. Multiple Samples t Test at the 95% Confidence Level[a]

Time (min)	ppm TRC		Difference, D (ppm)
	Standard Method	New Instrument	
2.0	0.646	0.629	−0.017
15.2	0.550	0.579	0.029
27.5	0.533	0.541	0.008
40.4	0.501	0.509	0.008
52.2	0.489	0.487	−0.002

[a] $s_d = 0.017$, $s_d = \left[\dfrac{\Sigma(D_i - \bar{D})^2}{N-1}\right]^{1/2}$, $\bar{D} = 0.005$

$t_{calc} = 0.694$, $t_{calc} = \dfrac{\bar{D}}{s_d}[N]^{1/2}$, $t_{table} = 2.776$

If t_{calc} is less than t_{table} at the 95% confidence level, then the new method does not exhibit bias as compared to the standard method.

to iodine. Any factors that affect this response could, therefore, alter the final value of parts per million TRC reported. The effects of stirring rate and KI concentration were investigated experimentally. TRC in a sample of chlorinated distilled water, sealed to the atmosphere, was determined five times each with three different stirring bars (26, 28, and 35 mm in length, each spinning at 700 rpm). Another set of five measurements at each of three different iodide concentrations (0.5, 1.0, and 2.0 g KI in 200 mL sample and 4 mL buffer) was also performed. In each of these sets, no statistically significant difference in precision or accuracy for determination of TRC concentration was found at the 95% confidence level. Errors caused by electrooxidation of phenyl arsineoxide, drift of the constant current source from the nominal value, and deviations in total sample solution volume in the beaker were calculated. As can be seen in Table II, the effects of these variables on accuracy and precision of the new instrumental technique are minimal.

Significantly erroneous readings can be caused by improper electrode maintainence and improper sampling. The improper electrode maintenance most likely to occur involves allowing the electrolyte to dry on the electrodes. This has been observed to cause extreme deviations in electrode behavior. Inappropriate samples include aged chlorinated samples (12 h old) and samples having excessive particulate matter. Such samples typically result in TRC values that are too large, probably because of the oxidation of iodide to iodine by oxidized DOC and/or particulate matter. Typical field samples possessing particulates, for which no difficulties were encountered, exhibited turbidities of 7 to 15 NTU, as determined by surface scatter turbidimetry.[2]

Table II. Experimental Parameters and Effects

Parameter	Maximum Effect on Reported TRC Concentration	
	Accuracy	Precision
Stirring	No effect	No effect
26-, 28-, 35-mm stir bars at 700 rpm		
KI added to sample	No effect	No effect
0.5, 1.0, 2.0 g		
Electrooxidation of PAO[a]	+0.006 ppb	2.6×10^{-6} ppb
Constant current	±2.0 ppb[b]	±0.8 ppb[c]
Source drift[a,d]		
Sample dilution	−0.4 ppb	
(for dilution to 225 mL rather than 200 mL)		

[a]For a titration time of 100 s.
[b]Maximum observed long-term drift.
[c]Maximum observed short-term drift.
[d]Attributed primarily to measuring device.

Instrumental failure occurred in only two cases: (1) for ambient temperatures above 93° F and (2) humidity in excess of 85% combined with temperatures less than ~45° F. Failures of this type represent the inability of electronic components to operate under these unusually difficult conditions.

CONCLUSIONS

The new instrumentation described here for determination of TRC in chlorinated effluent has achieved several goals. The new instrumentation uses the proven chemistry of the standard methods amperometric back titration. The new technique is accurate with respect to the standard method at the 95% confidence level. The new instrumentation allows for routine determinations of parts per billion TRC in wastewater samples with estimated standard deviations of 14 ppb or less. Precision and accuracy are independent of chlorine dosage in the range of 0.1 to 7.5 ppm, as well as resultant TRC concentrations in the range of 0.01 to 7.0 ppm. No dependence on operator familiarity with the instrument was observed. The new instrumentation decreases demands on operator skill by eliminating the use of iodine solutions and the associated amperometric back titration procedure. The instrument is dependable under a wide range of environmental conditions.

Future work will examine the dependence of method precision on PAO addition to the sample, as well as methods for improving this "weak link" in the procedure.

ACKNOWLEDGMENTS

This work was supported in part by the Electric Power Research Institute contract No. 1435 and by the Milwaukee Metropolitan Sewage District.

REFERENCES

1. Marinenko, G., and J. K. Taylor. "High-Precision Coulometric Iodimetry," *Anal. Chem.* 39(13):1568–1571 (1967).
2. *Standard Methods For the Examination of Water and Wastewater.*, 14th ed., Sect. 409.B.4.a. (Washington, DC: American Public Health Association, 1975), pp. 318–321.
3. *Methods for Analysis of Water and Wastes.* EPA Method No. 330.2, EPA-600/4-79-020 (Washington, DC: U.S. Environmental Protection Agency, 1979).
4. Helz, G. R., R. Sugam, and A. C. Sigleo. "Chemical Modifications of Estuarine Water by a Power Plant Using Continuous Chlorination," *Environ. Sci. Technol.* 18:192–199 (1984).
5. Christian, G. D. *Analytical Chemistry.*, 3rd ed. (New York: John Wiley & Sons, 1980), p. 77.

CHAPTER **86**

Rapid Oxidant Demand: Methods for Study

Donald A. Jaworske and George R. Helz

The toxicity of chlorine-produced oxidants to aquatic organisms is a function of exposure time as well as oxidant concentration and speciation.[1] Therefore, interest exists in trying to develop kinetic models to predict the rates of oxidant decay or, stated inversely, to predict the rates of oxidant demand exertion. To improve existing kinetic models, particular attention needs to be directed toward understanding the earliest and fastest part of the decay process, when in some cases a major fraction of the total demand is exerted.[2-4]

This chapter discusses three approaches to the study of the rapid initial oxidant decay processes. The first approach involves tracing the decay process through an operating power plant cooling system. We will show that limited access to sampling ports and inhomogeneities in the cooling water severely limit the usefulness of this approach. The second approach involves continuous electrochemical monitoring of oxidant concentrations during manual mixing experiments in the laboratory. This approach is also shown to be of limited use because much of the decay occurs more rapidly than reagents can be mixed manually. The third approach involves the use of a rotating ring disc electrode to generate and detect bromine in situ. This superior approach has shown that significant amounts of oxidant demand are exerted on a millisecond time scale.

METHODS

Measurement of Residual Oxidant

Residual oxidant measurements were made at the Mercer Generating Station on July 10, 1981. Mercer is located in the freshwater region of the Delaware River (hence, chlorine oxidants rather than bromine oxidants were of interest). The methods used to measure residual oxidant in the samples collected from the power plant plumbing have been described in detail elsewhere.[5-6] Special care was taken in the order in which the reagents were added to the sample.[7] The samples were analyzed within minutes of collection via amperometric titration at pH 4. This type of titration yields total residual oxidant values that not only include HOCl and HOBr, but also the various chloramines and bromamines. The transit times through the power plant plumbing were estimated from dye studies made previously by the plant personnel. The initial chlorine concentration was calculated from chlorination rate and water flow data provided by the plant operators.

Measurement of Oxidant Demand with a Rotating Disc Electrode

A platinum (Pt) rotating disc electrode (RDE), purchased from the Pine Instrument Company, was used to monitor the concentration of HOBr in 26 mL of buffered electrolyte solution. The electrolyte was 0.1 M in potassium chloride, buffered to pH 8 with 0.02 M phosphate buffer. The initial concentration of HOBr was 5.5×10^{-5} M. A PAR-174 polarograph was used to polarize the disc to 0.0 V vs SCE. The disc was rotated at 400 rpm, which served to stir the solution. After the base-line current from HOBr was established (usually 5 min), a 1.00-mL aliquot of a natural water sample was injected into the electrolyte. The blank was obtained in a similar fashion, except that 1.00 mL of the 0.1 M KCl electrolyte was injected rather than the natural water sample. The disc current was monitored for several additional minutes before terminating the experiment. Amperometric titrations, performed at the beginning and end of the experiment, used the standard methods already mentioned.

Measurement of Oxidant Demand with a RRDE

The oxidant demand of estuarine Patuxent River water was measured with a rotating ring disc electrode (RRDE). All samples were analyzed at ambient pH. Enough $NaBr_{(s)}$ was added to each sample to yield a solution 0.1 M in Br^-. Bromine was produced electrochemically at the central disc electrode[8] and detected at the annular ring electrode. The coulombic current efficiency for the Br^-/Br_2 couple is very near 100%.[9] The disc circuit and ring circuit were completely independent. This was accomplished by using a battery-powered variable dc power supply in the disc circuit and a Metrohm model E585 polarizer in the ring circuit. The variable dc power supply was attached to the Pt disc and a Pt counter electrode. A Tektronix model 502 differential amplifier measured the voltage drop across a series resistor in the disc circuit, yielding a measure of the disc current. The ring circuit consisted of the Pt ring electrode vs a saturated calomel reference electrode. Care was taken to replace the spent calomel reference electrode with a fresh calomel electrode as needed. All ring current measurements were made at +0.4 V vs SCE.

The theory of the rotating ring disc electrode has been described in detail elsewhere.[10-11] Briefly, the spinning action of the disc causes the solution adjacent to the disc to flow across the surface centrifugally. The highly polished surface of the disc ensures a laminar flow. At the same time, fresh solution from the bulk is brought up axially as a replacement. Nascent bromine, generated at the disc, is swept across to the annular ring. The important parameter is the angular velocity of the disc, because the angular velocity controls the flow rate across the surface of the disc, and hence controls the transit time from disc to ring.

The ring current vs disc current titration curves gathered from the Patuxent River water samples exhibited second-order behavior, that is, a curvilinear relationship between ring current and disc current similar to the one shown in Figure 1. Extrapolation of the linear portion of the curve to zero ring current gave the endpoint of the diffusion layer titration, according to Equation (1).

$$C_b = \frac{(7.36 \times 10^{-6}) I_{do}}{w^{1/2}} \qquad (1)$$

The concentration of bromine consuming substrate (C_b) is a function of the disc current at zero ring current (I_{do}) and the angular velocity of the electrode (w). The constant in Equation (1) represents a collection of terms including various hydrodynamic and geometric parameters.[12]

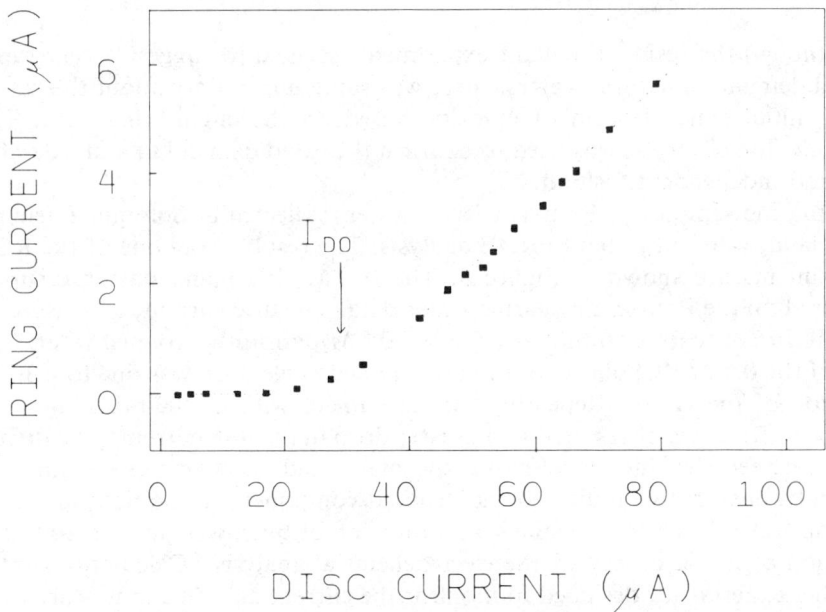

Figure 1. Second-order ring current vs disc current titration curve for a Patuxent River water sample, with the disc current at zero ring current (I_{do}) highlighted. Experimental conditions: 0.1 M Br$^-$, pH 7.7, and 3600 rpm.

RESULTS AND DISCUSSION

Results of the Residual Oxidant Measurements

The results of the residual oxidant measurements, from samples collected along the plumbing of the Mercer Generating Station, are summarized in Figure 2. In every case, the total residual oxidant was substantially less than the calculated applied dose, even at the sampling port closest to the point of chlorine application (~3 s downstream from the chlorinators). If complete mixing of the chlorine occurs at the chlorinators, then a certain fraction of the total oxidant demand must be very prompt indeed. The residual oxidant measurement at the circulating pump seems erroneously low. This may result from the location of the sampling port. The valve was located on the perimeter of the pump housing, so perhaps the sample was representative of only a thin layer of fluid rather than the bulk of solution.

The rapid oxidant demand, observed at this freshwater site, was similar to the oxidant demand observed by Helz et al.[2] The power plant they studied was located at an estuarine site where bromine-based oxidants are important. In their experiments, the head of the cooling canal was the sampling station closest to the point of application (~5 min downstream of the chlorinators).

Results of the RDE Experiments

Although the residual oxidant experiments seemed to suggest a very rapid initial demand in natural waters, there was some uncertainty about the calculated initial concentration of chlorine based on the engineering data. The rotating disc electrode was used to confirm the rapid demand in a more rigorous and independent fashion.

Estuarine samples of Patuxent River water, collected at Solomon's Island, Maryland, were subjected to RDE analysis. The results from one of the RDE experiments are shown in Figure 3. The 0.1 M KCl blank data are superimposed on the Patuxent estuarine water data. The disc current, a measure of the HOBr concentration (initially 5.5×10^{-5} M), abruptly dropped when 1.00 mL of the 0.1 M KCl blank solution was added. This drop was due to a 3.7% dilution of the HOBr. Repeating the experiment with a 1.00-mL aliquot of Patuxent River water resulted in a greater drop in the disc current. The difference between the line representing the blank and the line representing the natural water sample indicated the demand component in the natural water. The downward slopes in the lines, as a function of time, were attributed to the attrition of HOBr caused by the electrochemical analysis. (Calculations from another experiment, designed to monitor the current and time more carefully, confirmed the electrochemical contribution.) The subtle changes in disc current were observed only because the gain on the amplifier of the polarograph was set rather high. Because the disc current is a function of pH, the electrolyte

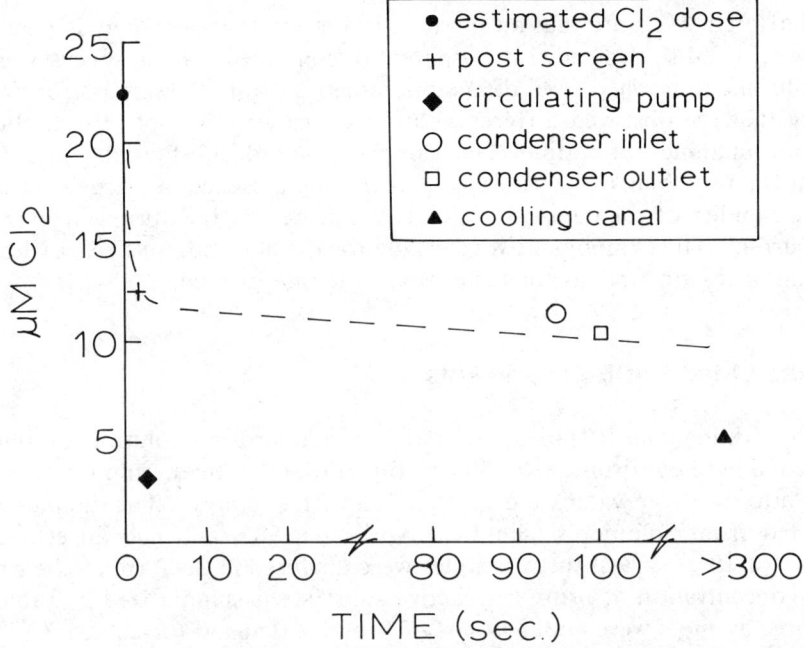

Figure 2. Total oxidant, measured via amperometric titration, in samples collected from the plumbing of the Mercer Generating Station. Cumulative transit times, from the point of Cl_2 injection, were estimated from dye studies done by plant operators.

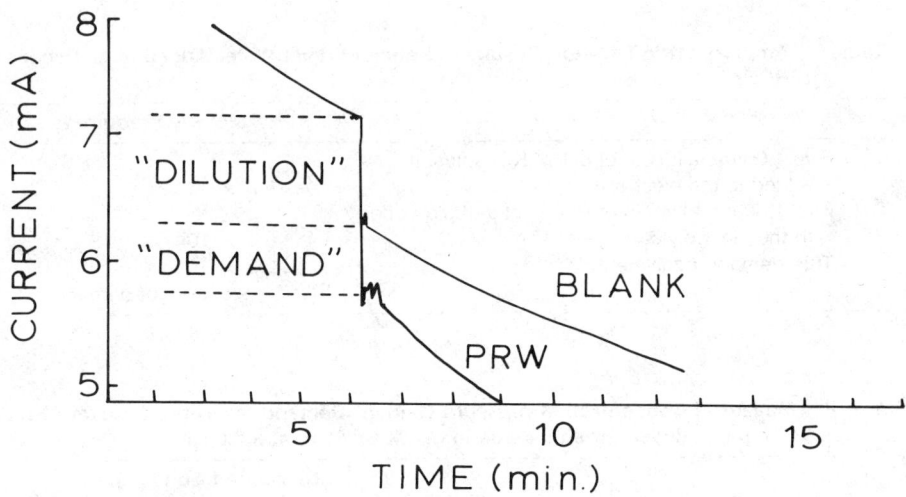

Figure 3. Current resulting from 50 μM HOBr in 26 mL of 0.1 M KCl electrolyte monitored with an RDE as a function of time. At t = 6 min, either 1.00 mL of Patuxent River water (10 g/kg) or 1.00 mL of 0.1 M KCl was injected into the electrolyte.

was buffered to ensure that an increase in pH was not mistaken as demand. Likewise, the disc electrode is transparent to combined oxidant species such as monobromamine; therefore, dual amperometric titrations were performed to ensure that the observed differences in disc current were not attributable to monobromamine formation. The sample required substantially less PAO titrant than the blank (see Table I), proving the presence of a true demand.

The rapidity of the demand is clearly illustrated by the abrupt drop in the disc current. The components responsible for the demand must react with the oxidant at least as fast as (or faster than) the rate of mixing.

Results of the RRDE Experiments

After testing our RRDE system with a standardized solution of phenyl arsine oxide to confirm the validity of Equation (1),[12] three samples of estuarine Patuxent River water were subjected to RRDE analysis. The samples were collected from Solomon's Island on April 26, 1983, and had an estimated salinity of 10 g/kg. All three samples were analyzed at 3600 rpm. The calculated concentration of bromine-reactive substrate is summarized in Table II. All three aliquots were about 14 μM of bromine demand (or 28 μeq/L). This value is in excellent agreement with Helz et al.[2]

Table I. Amperometric Titration Results of Patuxent River Water Added to an Excess of HOBr

	Mol PAO required
Five 1.00-mL aliquots of 0.1 M KCl solution added to the electrolyte	8.9×10^{-7}
Five 1.00-mL aliquots of Patuxent sample added to the electrolyte	6.6×10^{-7}
True demand equivalent	2.3×10^{-7} (92 μeq/L of demand)

Table II. Measured Concentration of Rapid Oxidant Demand for Three Patuxent River Samples. Measurements Made in 0.1 M Br$^-$ and at 3600 rpm.

Sample	pH	Calculated Conc (μM)
I	7.7	13.4
II	7.8	14.2
III	7.9	13.8

At 3600 rpm, the transit time from disc to ring was estimated to be 0.037 s.[13] Based on this transit time information, one can estimate the magnitude of the second-order rate constant for the reaction of bromine and substrate. The sharp curvilinear shape of the ring current vs disc current titration curve implies that the reaction between bromine and substrate reaches completion in transit from disc to ring. If five reaction half-lives constitute essentially complete reaction, then the half-life for the reaction of bromine and substrate must be equal to or less than 0.0074 s. Given that the concentration of substrate is approximately 14 μM, then the second-order rate constant must be equal to or greater than 10^7 M^{-1} s^{-1}.

Although this chapter only presents the results from a single natural water sample, the RRDE technique has potential for further laboratory and field applications. It has been used successfully to study the kinetics of several model compounds,[14] and it has also been used on board ship to make rapid oxidant demand measurements in the field.[15] Stopped-flow equipment is capable of measuring reaction kinetics on the same time scale as the RRDE, but the delicate nature of stopped-flow equipment may prohibit its use in some field applications.

CONCLUSIONS

Three different experimental approaches have been used to progressively isolate the time scale of rapid oxidant demand reactions in natural waters. In the first case, total residual oxidant measurements were obtained at several points along the plumbing of a once-through power plant cooling system. The oxidant concentration at the sampling port closest to the chlorinators was significantly less than the applied dose. Judging from the flow rate through the system, the rapid oxidant demand must have occurred within seconds of contact with the natural water. In the second case, a rotating disc electrode was used to monitor the oxidant concentration in a well-mixed solution. Rapid injection of an aliquot of natural water decreased the oxidant concentration beyond that caused by dilution. These results isolated the rapid oxidant demand reaction to the time scale of mixing, perhaps less than a second. Finally, a rotating ring disc electrode was used to carry out a diffusion layer titration between electrochemically generated bromine and estuarine Patuxent River water. The titration results showed that the titer of the rapid oxidant demand was approximately 14 μM. Furthermore, the sharpness of the titration curve indicated that the demand reaction had gone to completion during transit from disc to ring, isolating the rapid oxidant demand reaction to a time scale of less than 37 ms. Based on transit time calculations, the second-order rate constant of the rapid demand reaction was estimated to be about 10^7 M^{-1} s^{-1}.

ACKNOWLEDGMENTS

We gratefully acknowledge the Maryland Power Plant Siting Program and the Electric Power Research Institute for their financial support, the Public Service Electric and Gas Company for their cooperation during the power plant sampling experiments, and Rick Sugam and Calvin Huber for their critical review of the manuscript.

REFERENCES

1. Mattice, J. S., and H. E. Zittel. "Site-Specific Evaluation of Power Plant Chlorination," *J. Water Pollut. Control Fed.* 48(10):2284–2308 (1976).
2. Helz, G. R., R. Sugam, and A. C. Sigleo. "Chemical Modifications of Estuarine Water by a Power Plant Using Continuous Chlorination," *Environ. Sci. Technol.* 18(3):192–199 (1984).
3. Eppley, R. W., E. H. Renger, and P. M. Williams. "Chlorine Reactions with Seawater Constituents and the Inhibition of Photosynthesis of Natural Marine Photoplankton," *Estuarine Coastal Mar. Sci.* 4:147–161 (1976).
4. Horstgaard-Jensen, P., J. Klitgaard, and K. M. Pederson. "Chlorine Decay in Cooling Water and Discharge into Seawater," *J. Water Pollut. Control Fed.* 49:1832–1841 (1977).
5. *Standard Methods for the Examination of Water and Wastewater*, 14th ed., (Washington, DC: American Public Health Association, 1976).
6. Sugam, R., and G. R. Helz. "The Chemistry of Chlorine in Estuarine Waters," PPRP-26 (Annapolis, MD: Maryland Power Plant Siting Program, 1977).
7. Helz, G. R., D. A. Jaworske, and L. Kosak-Channing. "Experience with Amperometric Titrations for Total Chlorine in the Microgram-per-liter Range: Limitations to Accuracy," in *Water Chlorination: Environmental Impact and Health Effects, Vol. 4*, R. L. Jolley, W. A. Brungs, J. A. Cotruvo, R. B. Cumming, J. S. Mattice, and V. A. Jacobs, Eds. (Ann Arbor, MI: Ann Arbor Science Publishers, Inc., 1983), pp. 667–680.
8. Rubinstein, I. "Electrode Kinetics of the Br_2/Br^- Couple," *J. Phys. Chem.* 85:1899–1906 (1981).
9. Nagy, G., Z. Feher, K. Toth, and E. Pungor. "A Novel Titration Technique for the Analysis of Streamed Samples—the Triangle-Programmed Titration Technique. Part 3. Titrations with Electrically Generated Bromine," *Anal. Chim. Acta* 100:181–191 (1978).
10. Albery, W. J., and M. L. Hitchman. *Ring-Disc Electrodes* (London: Oxford University Press, 1971), pp. 1–170.
11. Bard, A. J., and L. J. Falkner. *Electrochemical Methods: Fundamentals and Applications* (New York: John Wiley and Sons, Inc., 1980), pp. 280–313.
12. Jaworske, D. A. "The Kinetics of Rapid Oxidant Consumption: Measurements with a Rotating Ring Disc Electrode," Ph.D. Thesis, (College Park, MD: University of Maryland, 1983).

13. Bruckenstein S., and G. A. Feldman. "Radial Transport Times at Rotating Ring-Disk Electrodes. Limitations on the Detection of Electrode Intermediates Undergoing Homogeneous Chemical Reactions," *J. Electroanal. Chem. 9*:395–399 (1965).
14. Jaworske, D. A., and G. R. Helz. "Use of a Rotating Ring Disc Electrode to Study Fast Bromine Demand Reactions," *Int. J. Environ. Anal. Chem.* 19:189–202 (1985).
15. Jaworske, D. A., and G. R. Helz. "Rapid Consumption of Bromine Oxidants in Estuarine Waters" (submitted for publication).

CHAPTER **87**

Anodic Voltammetric Determination of Monochloramine in Water

Debra A. Davies and Calvin O. Huber

Chlorine in the form of chlorine gas or hypochlorite, commonly used for disinfection or to inhibit biofouling, will react quickly with ammonia to form the corresponding chloramine. The free chlorine species and the chloramines are distinctively toxic to different forms of aquatic life. Consequently, the capability to monitor levels of specific species is important.

Although several methods exist for the determination of free chlorine or the combined concentration of free chlorine and the chloramines, a technique to provide a direct, rapid, selective measure of the concentration of monochloramine is needed. The new method presented here for that purpose is based on the oxidation of monochloramine at a platinum anode, in contrast to previous techniques based on nonselective cathodic reactions or iodometry.[1,2] A prototype instrument using the flow injection technique and an inexpensive microcomputer has been designed and constructed. The instrument minimizes dependence on operator skill.

EXPERIMENTAL

The flow injection technique was used. The flow-through electrochemical detector consists of a platinum-wire working electrode (0.5 by 10 mm), a platinum counter electrode, and a saturated calomel reference electrode (SCE). All reported potentials are vs SCE. Potential control was maintained using op-amp–based conventional potentiostat circuitry. The potentiostat is supplemented with a timing circuit to automatically control the electrochemical pretreatment and sample injection timing. Samples were introduced onstream with either a manual rotary injection valve or a solenoid-actuated, four-way stream exchange valve.[3]

The signal resulting from the injection of monochloramine is a current peak, the height of which is proportional to concentration. Peak heights were determined manually, using either a potentiometric strip chart recorder or peak detection; measurement, calibration, and readout were obtained via an interfaced microcomputer. The computer was a Timex Sinclair 1000 with a Microsystems Development I/O expansion board and ADDA S-3 board. With this

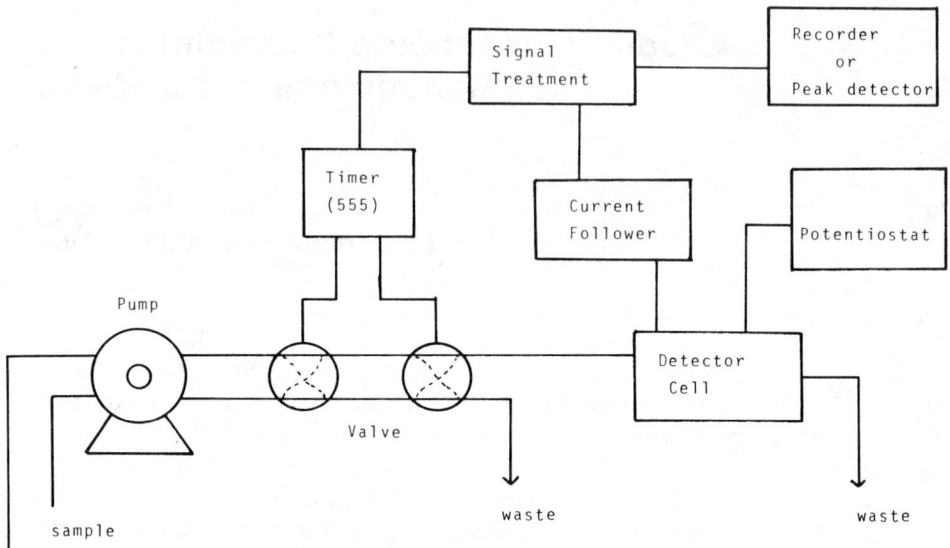

Figure 1. Schematic representation of prototype instrumentation.

instrumentation, the operator is responsible only for placement of standards or samples and initiation of the measurement sequence. Electrode pretreatment, sample injection, operator prompting, calibration, and concentration readout are automated. A schematic diagram is shown in Figure 1.

All samples of monochloramine were prepared by mixing the appropriate solutions of sodium hypochlorite with at least a twofold molar excess of ammonium nitrate, or with ammonium sulfate solutions at pH 7 and 8 using 0.1 M phosphate or 1 mM carbonate buffers.

RESULTS AND DISCUSSION

Initial potential scan experiments for anodic processes with monochloramine indicated that oxidation occurs in the range of 0.7 to 0.9 V vs SCE.

The response to monochloramine was investigated and was found to be linear with concentration. The response decreased (10–20%) during preparation of successive calibration curves. After 1 h, the electrode activity had decreased by about 80%. The loss of activity is apparently the result of accumulation of a passive oxide surface layer on the working electrode.

To control electrode surface activity, it was necessary to use a pretreatment regime. This was done by stepping to a potential of −1.0 V for 10 s, during which the platinum surface oxides were reduced. The potential was then stepped to +0.80 V, the analytical potential. The anodic background current was allowed to decay for a specified time before injection of the sample.

Figure 2 shows that monochloramine is oxidized at potentials more positive than 0.5 V. At potentials more positive than 0.7 V, this current decreases due to the formation of passivating platinum oxide. Also, at potentials more positive than 0.7 V, hypochlorite yields a small anodic current. Ammonia, not shown in Figure 2, also yields a very small anodic current in this potential range. Table I shows the selectivity for monochloramine over hypochlorite and ammonia. Based on these data, 0.80 V was selected for optimum response and selectivity.

During the injection time (the interval between stepping to the analytical voltage and injecting the sample), the background current is decaying so that the analytical signal is a peak anodic current on a decreasing anodic current background. The injection time can be prescribed by time or by background current magnitude. Experiments showed that precisions for either were statistically equivalent. A time designation was chosen for experimental convenience and instrumental design simplicity. The exponential decrease in background current is attributed to charging current and increasing surface oxide coverage. This decay is observable for several minutes. It causes a corresponding decrease of electrode sensitivity with time (i.e., increasing electron transfer overvoltage with oxide coverage). For shorter injection times, however, the rapidly decreasing background current tends to attenuate the anodic peak signals. These opposing effects resulted in selection of injection times of 60 or 90 s for most samples, although for samples with concentrations of monochloramine greater than 1 μg/L Cl, a 30-s injection time with attendant speed and sensitivitiy advantages could be used.

The effect of pH on analytical signal is summarized in Table II. The increase in electrode activity from pH 8 to 7 can be attributed to a decrease in the surface oxidation of the electrode. There was little difference in sensitivity between pH 6 and 7. A pH of 7 was chosen because of the instability of monochloramine at lower pH.[4]

Analytical signal dependence on buffer concentration shows a maximum at about 0.25 M (Figure 3). Below 0.25 M, the buffer concentration is apparently too low to maintain the pH in the reaction layer. The decrease in current at concentrations >0.25 M is probably the result of electrode surface adsorption of HPO_4^{2-} and $H_2PO_4^-$ that inhibits the oxidation of monochloramine. Both of these anions are known to be strong adsorbers on platinum.[5] Table III shows that even sulfate in this concentration range apparently adsorbs and inhibits the signal.

To avoid the need to buffer samples individually, the flow injection apparatus was designed so that on-stream dispersion yielded 0.25 M buffer capacity for injected samples. The flow injection system contained a 45 cm length of coiled tubing between the injection valve and the detector to yield a twofold dispersion of injected samples. A corresponding buffer concentration of 0.5 M was used for the background carrier solution.

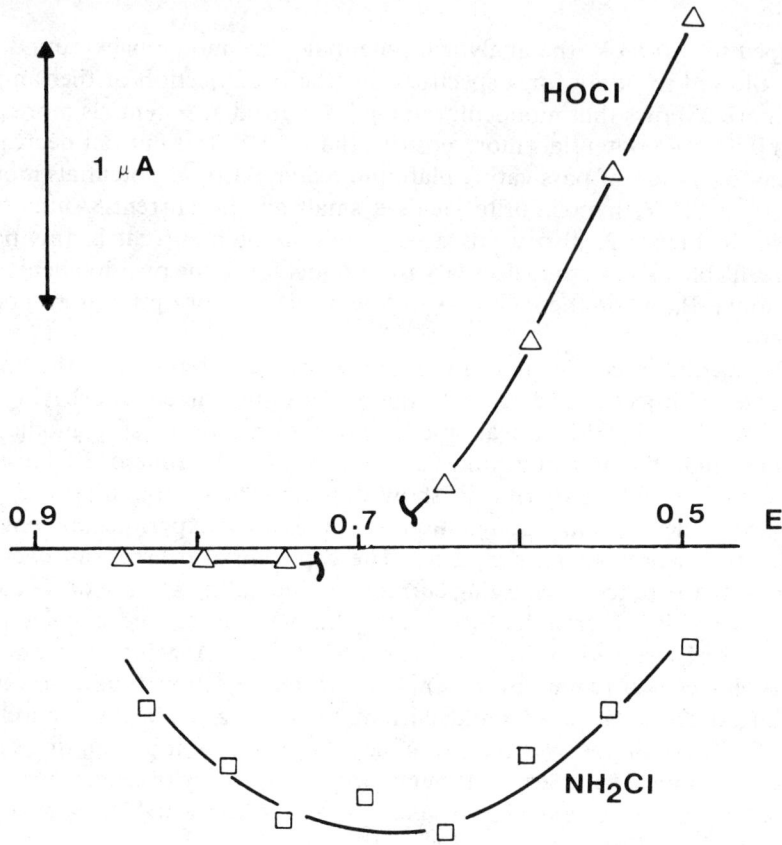

Figure 2. Peak current vs applied potential for 140 M monochloramine and 140 M hypochlorite. Background electrolyte is 0.25 M, pH 8.0 phosphate buffer. Each point is one sample injection.

Table I. Selectivity for Monochloramine over Hypochlorite and Ammonia

E_{app}, V	$\dfrac{iNH_2Cl}{iHOCl}$	$\dfrac{iNH_2Cl}{iNH_3}$
0.50	0.20	24
0.55	0.40	43
0.60	1.0	80
0.65	4.5	120
0.70		115
0.75	27.0	210
0.80	26.0	210
0.85	31.0	160

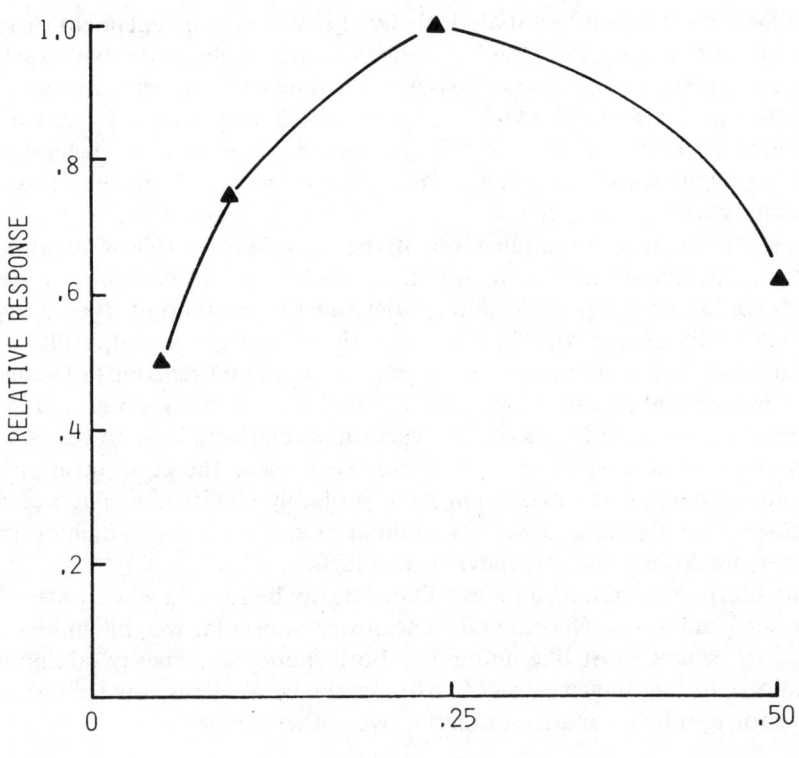

Figure 3. Variation of observed signal with buffer concentration (pH 8, phosphate).

Table II. Effect of pH on Sensitivity

pH	Slope (nA/μM)
6	9.6
7	10.8
8	5.2

Table III. Dependence of Response for 60 μM Monochloramine on Added Sulfate in a pH 7.4, 0.10 M Phosphate Buffer

Conc (M) SO_4^{2-}	Peak Current (nA)
0	480
0.1	460
0.3	380
0.6	320

Figure 4 indicates that the analytical signal is virtually independent of temperature over the range examined. Increasing rates of electrode oxide surface coverage apparently offset the increases of current with temperature expected for electron or mass-transfer kinetics. The positive temperature effect for the background currents reflects the expected positive temperature dependence. The low or negligible peak current temperature effect is, of course, an advantage in analytical applications.

Figure 5 illustrates the application of the technique to follow breakpoint chemistry. The initial increase in current with additions of hypochlorite is due to the formation of monochloramine. Beyond the breakpoint, where hypochlorite molarity exceeds that of ammonia, the decrease in signal results from the decomposition of monochloramine via "breakpoint" reactions. The usual increase in signal at excess chlorine additions (i.e., "superchlorination") is not registered here because the electrode reaction is relatively inert to active chorine species. Because short contact times were used, the concentrations of dichloramine beyond the maximum were probably significant. The negative slope observed in the data suggests a minimal, if any, response to dichloramine and other breakpoint intermediates or products.

Monochlorinated organic amines will ordinarily be formed when water containing such amines is chlorinated. The lower-molecular-weight amines are expected to behave most like ammonia. Both monochloromethyl amine and monochloroglycine, however, yield virtually no signal. Thus, the technique is selective for monochlorinated ammonia over other amines.

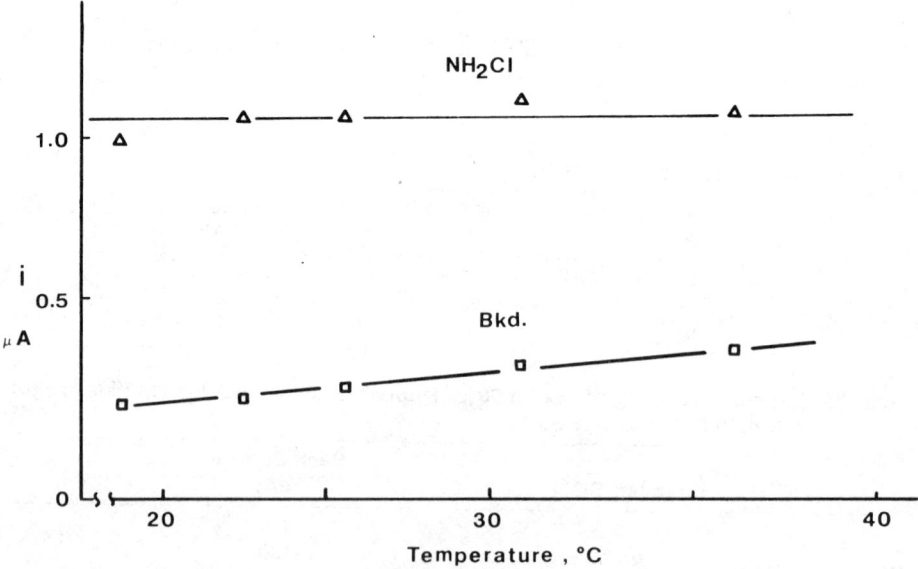

Figure 4. Dependence of analytical and background current on temperature.

The analytical characteristics for the instrumental technique are summarized as follows:

Sensitivity, m = 7.2 nA/μM (0.22 A mg^{-1} L^{-1} Cl)
Std deviation (1 mg/L Cl, n = 16), ~2%
Std error of estimate, s' = 1.9 nA
Lower detection limit, x_{min} = 0.81 μM (0.027 mg/L Cl)
Linear range maximum, x_{max} = 150 μM (5.0 mg/L Cl)
Selectivity, relative signals: HOCl, 0.038; NH$_3$, 0.0048; NH$_2$CH$_3$COOH, <0.001; NHClCH$_3$COOH, 0.0059; NH$_2$CH$_3$ and NHClCH$_3$, <0.02.

The instrument was designed to minimize operator skill requirements. Table IV shows the precision obtained for chlorinated municipal tap water by four different operators with varied chemical knowledge and experience. Each operator was given about a half-page set of operator instructions and only a few minutes of hands-on training with the first sample. Examination of the precision obtained by the various operators using the F parameter indicated no distinctions at the 95% confidence level.

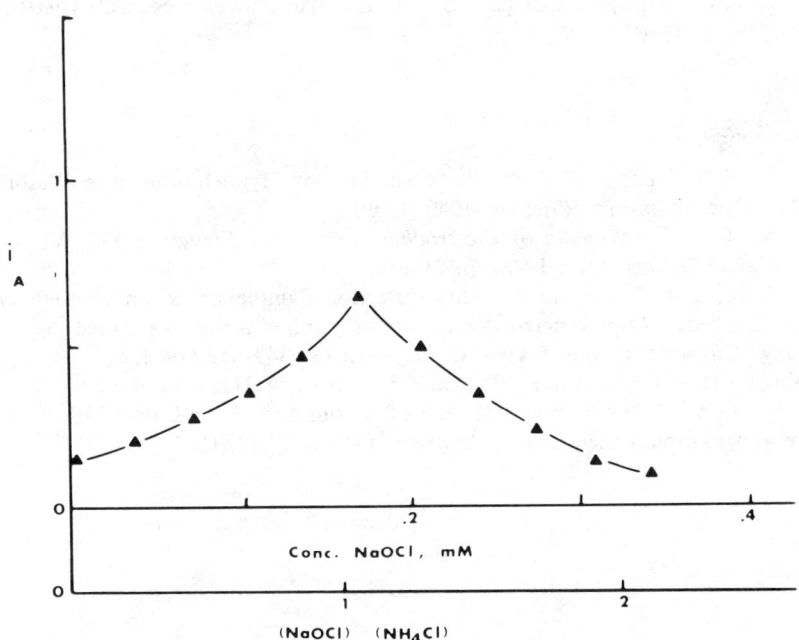

Figure 5. Breakpoint curve (2-min intervals between hypochlorite additions, with stirring; lake water, pH 8).

Table IV. Dependence of Precision on Operator Skill

Operator Description	Sample	Observed Measurement (mg/L Cl)	Std dev, n = 6 (mg/L Cl)
Designer of Instrument	A	0.48	0.020
Chemist	B	0.66	0.023
High School Chemistry Student	C	0.73	0.024
Housewife	C	0.71	0.027

CONCLUSIONS

Selective anodic amperometric measurement of monochloramine at a platinum electrode can be performed with little operator training. The instrument components are relatively inexpensive. The various analytical characteristics of the new technique suggest its usefulness for chlorination chemistry and field applications.

ACKNOWLEDGMENT

This work was supported in part by the Electric Power Research Institute, Contract No. 1435-01.

REFERENCES

1. Filippov, T. S., and Y. V. Dobrov. "Oxidation of Hypochlorite at a Platinum Anode," *Electrokhimiya* 5(9):1046–1049 (1969).
2. Bard, A., Ed. *Encyclopedia of Electrochemistry of the Elements*, Vol. VI (New York: Marcel Dekker, Inc., 1976), p. 210.
3. Kafil, J. B., and C. O. Huber. "Flow Injection Sample Processing with Nickel Oxide Electrode Amperometric Detection of Amino Acids Separated by Ion-Exchange Chromatography," *Anal. Chim. Acta* 139:347–352 (1982).
4. Chapin, R. M. "Dichloramine," *J. Am. Chem. Soc.* 51:2112–2117 (1929).
5. Snell, K. D., and A. G. Keenan. "Effects of Anions and pH on Ethanol Electrooxidation at a Platinum Electrode," *Electrochim. Acta* 27(12):1683–1696 (1982).

CHAPTER 88

Broad-Spectrum Analysis of Organics in Drinking Water Using Macroreticular Resins— A Quality Assurance Evaluation

J. Gibs, B. Najar, and I. H. Suffet

Broad-spectrum GC capillary analysis of trace organic chemicals from water can be defined as a method that simultaneously isolates for analysis the largest possible number of chemicals contained in a sample. Samples to which broad-spectrum organic analysis is best applied are those (1) requiring minimum pretreatment and (2) containing a large molecular weight range of organic components. The usefulness of broad-spectrum analysis stems from the capacity to interpret large quantities of chromatographic data at one time.

The typical applications of broad-spectrum environmental analysis are to observe the concentration differences of many compounds simultaneously. For example, one might wish to compare sample analyses (1) before and after a unit process, (2) at one location over a time interval (a temporal sampling program), or (3) at different locations at the same time (a spatial sampling program).

Broad-spectrum chromatographic analysis has been used as a method to determine order of magnitude, semiquantitative changes between samples by (1) evaluating peak-by-peak changes in gas chromatography (GC) response,[1,2] (2) visual impressions of the overall relative numbers of peaks and their areas,[3] (3) counting peaks that have different concentration values,[4] and (4) changes in the response in parts of different chromatograms.[1,3,4]

Macroreticular resin (MRR) samplers are widely used to collect nanogram-per-liter quantities of a wide range of nonpolar trace organics from large volumes of water. Thus, this sampling technique is well matched for analysis by broad-spectrum chromatography. However, little information is currently available in the literature about the interaction between the sampling parameters and the final analytical results. Therefore, sample and instrument quality assurance becomes critical for reliable comparison of chromatographic data.

A good quality assurance program is made up of two components:[5] (1) quality control to maintain an analysis within known error limits (e.g., within a 10% coefficient of variation for reliable quantification by gas chromatography) and (2) quality assessment to determine the quality of the measurement process and its results (i.e., to define the acceptable limits of the technique).

In the current study, an actual field sampling method is assessed. The reproducibility, but not the accuracy of the method, is determined. The accuracy cannot be determined by the use of environmental samples that can vary temporally with concentration. The accuracy of the macroreticular resin-isolation technique has been defined in terms of percentage recovery of specific compounds during numerous laboratory investigations.[6-8] This chapter will discuss resin artifacts, reproducibility of resin samplers, and the breakthough of resin samplers when compositing large volumes of water. This study presents the quality assessment in terms of precision of the sampling and analysis techniques.

RESIN CLEANING ARTIFACTS

The ability to use XAD macroreticular resins as an isolation technique is hindered by the presence of organic contaminants in the eluants that originated from the resins. Table I shows the impurities from XAD-2 and XAD-8 after sequential Soxhlet extraction by methanol, acetonitrile, and diethyl ether and characterization using GC/MS. The compounds in Table I are possible contaminants in the resin eluants because they are primarily monomers and polymer fractions of the resin.

In this study, an added step was taken to determine the cleanliness of the resin. Columns of 100 mL of resin were eluted with 200 mL of ether, which was then concentrated to 1 mL and analyzed by GC with a flame ionization detector. This analysis concluded that an additional cleanup step was needed. This step was to alternately elute the resin for 10 min each with successive bed volumes of ether, methanol, and ether. The last diethyl ether elution was concentrated to 1 mL for GC analysis. This was repeated until the final diethyl ether elution had a reproducible minimum number and level of contaminant peaks that could be practically achieved. The total process of cleaning and assuring cleanliness took 2 weeks per resin batch and consumed large quantities of high-purity solvents. Resin artifacts are observed to reappear when the resin sampler is reused. The primary artifacts that reappear are noted in Table I by a superscript f. These peaks could be identified by GC/MS.

EXPERIMENTAL

Delaware River Water from a pilot plant at the City of Philadelphia Baxter (Torresdale) plant was used for this study.[9] The Delaware River was subjected to conventional water treatment processes excluding chlorination (i.e., sedimentation, flocculation, coagulation, and rapid sand filtration). The water was then ozonated. Samples were collected before ozonation.

Table I. XAD Resin Cleanup By Soxhlet Extractions—GC/MS Tentative Identifications By Computer Search[a,b]

Compounds	XAD-8 CH$_3$OH	XAD-8 CH$_3$C≡N	XAD-8 Ether[c]	XAD-2 CH$_3$OH	XAD-2 CH$_3$C≡N	XAD-2 Ether[c]
Toluene	X[d]	X				
C6-Alcohol (e.g., 4-methyl-2-pentanol)	2[e]					
C2-Benzene	3	3				
Benzaldehyde	X					
C4-benzene	2	2				
C2-styrene	4	2				
Divinyl benzene	X					
Naphthalene	X			X	X	
1-Undecanol	X					
Diethyl phthalate	X					
C-17 HC, branched	X					
Nonamide	X					
Dichlorobenzophenone	X					
N-(p-chlorophenyl)-4-chlorobenzamide	X					
Methyl benzoate[f]				X		
C2-Benzaldehyde[f]				X		
C2-Acetophenone[f]				X		
Methyl naphthalene				2		
Biphenyl				X		
C2-Naphthalene				X		
Methyl styrene						3
Methacrylic acid	X					
Unknown[f,g]	2					

[a]Method: 900 mL solvent/700 mL resin successively soxhlet extracted for 24 h with CH$_3$OH, CH$_3$C≡N, and ether. A sample was injected directly into the GC/MS.
[b]Tentative identification. No retention time has yet been correlated with the RRT of known standards. Identification was completed by MS only.
[c]Ethyl acetate is a solvent impurity of ether.
[d]Compound present.
[e]Number of isomers present.
[f]Primary resin artifacts are always observed in resin samples. C-2 methyl benzoate is occasionally found as an artifact in XAD-8 samples, although it was not found in these extracts.
[g]Unknown identified by computer search as nitrooctane. Chemical ionization MS does not confirm this.

Reproducibility and Breakthrough of Resin Samplers Collecting Large Volumes of Water

A reproducibility study was performed in which five MRR samplers received the same influent water (>130 L). A blank elution and recleaning of the sampler were completed before a 3-d composite sample was collected. The sample was then eluted and concentrated for capillary GC analysis. Three resin samplers of the same design were placed after three of the MRR samplers to study breakthrough.

The resin sampler was a mixed bed of 100 mL XAD-2 and 100 mL XAD-8 (Rohm and Haas Co., Philadelphia). For the first sampling run, a ¼-in. Teflon® line was drawn from the water source. An ⅛-in. Teflon line was run into the ¼-in. line from a 40-L reagent feed bottle, which contained 10% H_3PO_4 and 10% Na_2SO_3 (by weight). The pH of the sample was adjusted to fall between 2.9 and 3.5. The ¼-in. influent line was split into ⅛-in Teflon lines. Five of these lines were fed into the resin samplers. The resin samplers had been previously used in the monitoring studies at the plant. Each of the five samplers consisted of a glass column (750 mm by 50 mm ID) with a glass frit and a Teflon stopcock at the bottom. The ⅛-in. influent lines fit into a Teflon flow controller at the top of each sampler. This design allowed the samplers to be closed and pressurized to ~20 psi.

Breakthrough samplers were connected to the bottom of three of the samplers. Each breakthough sampler consisted of three Teflon column sections, giving overall column dimensions of about 600 mm by 45 mm ID. A ¼-in. ID PVC line ran from a Cole-Palmer Masterflex pump to the effluent end of each sampling column. The pump was adjusted to draw ~35 mL/min through each sampler.

A second run was carried out under somewhat different conditions. The glass covers, with Teflon flow controllers, were removed from the samplers. A glass tube section was added with an arm piece to drain off overflow water. This design change left the system open and, therefore, at atmospheric pressure. Reagent feed water was added for the entire 70-h run. All other run conditions were identical to run No. 1. Tables II and III show the water quality and flow data for runs 1 and 2.

Ether Elution and Concentration

After a 70-h sampling run, the water was drained off the columns. The resin was then extracted with ether in three elutions of 150, 100, and 100 mL, respectively. The three aliquots of ether were combined and stored (below 0°C to freeze out water) until the concentration step. All of the ether was passed through about 30 mL of sodium sulfate in a 1.5-cm diam. column to remove residual water. The sodium sulfate column was carefully prerinsed with solvent. The samples were then evaporated to 10 mL using a Kuderna-Danish (K-D) apparatus. A micro K-D system was subsequently used to concentrate the sample under a nitrogen stream to 1 mL.

After Kuderna-Danish concentration, quality assurance samples were spiked with n-alkane hydrocarbon standards. The n-alkane range was C_8 through C_{26}, except for C_{24}. The alkane concentration in the 1-mL concentrated sample was 20 μg/mL, except for C_{21}, C_{23}, C_{25}, and C_{26}, which was 10 μg/mL. The reasons for using internal standards in this project were (1) to normalize quantification; (2) to determine the column resolution efficiency, as seen by the peak shape of the standards; and (3) to normalize retention time into Kovats Indices. The C_{14} hydrocarbon peak response was used to normalize quantification.

Gas Chromatography

The gas chromatograph used was a Carlo Erba 2150 with a split/splitless injector and a flame ionization detector. Conditions were as follows:

Temperature of injector and detector: 275°C
Makeup gas for flame: 0.3 kp/cm² H_2 (prepurified) and 0.9 kp/cm² air (zero grade)
Septum purge on injector: 1–2 mL/min

A 0.9 to 1.5 μL sample was pulled into a 5 μL syringe. The needle was inserted into the injector with the splitter closed and held. The 5-s hot-needle technique was used with a cool oven and the oven door open. After 5 s, the syringe was withdrawn from the injector. At 20 s, the oven door was closed and the temperature program started. After 45 s, the splitter was opened. The temperature program used was 50–250°C at 3.5°C/min, with a 3-min hold at 50°C and a 15-min hold at 250°C.

Table II. Water Quality Parameters for Runs 1 and 2

Parameter	Run 1 Value	Run 1 Date	Run 2 Value	Run 2 Date
pH, adjusted	3.2	10/1	3.3	3/3
	8.0	10/3[a]	3.4	3/4
			3.25	3/6
TOC, ppm	1.91	9/29	1.83	3/3
	1.51	10/6	1.96	3/9
			1.79	3/10

[a] 20 h into Run 1, the reagent feed water (10% H_3PO_4 and 10% Na_2SO_3) was no longer added to the inlet line. The pH for the remainder of the run was ~8. The run lasted 70 h.

Table III. Sampler Flow Rates and Total Volumes for Runs 1 and 2

	Run 1 Flow Rate (mL/min)			Run 1 Total Volume (L)	Run 2 Flow Rate (mL/min)			Run 2 Total Volume (L)
Sampler	10/1	10/2	10/3		3/3	3/4	3/6	
16/1	36	33	37	152	36	36	34	152
15/2[a]	37	35	34	156	38	34	34	152
19/3	37	36	35	154	37	35	35	152
21/4	36	34	35	150	36	34	35	150
14/5	32	33	33	131	36	34	35	150

[a] Sampler 15/2 ran for 1 h extra at 70 mL/min.

Chromatography columns differed for the two runs. Run No. 1 used a 30-m by 0.33-mm-ID SE-30 glass capillary column. A 60-m by 0.33-mm-ID SE-30 fused silica capillary column was used for the second run. The carrier gas for all gas chromatography was hydrogen. The linear velocity was 0.5 m/s for both sampling runs, as determined by methane gas retention time.

RESULTS

Qualitative Reproducibility of GC Peaks

Figures 1 and 2 are chromatographic plots of the first third of chromatograms containing over 300 peaks from the two reproducibility studies. Figure 3 is a plot of the standard deviation of the retention time vs the carbon number of the internal standards for run No. 1. The standard deviations are low overall for standards in solvent alone. The higher standard deviation for the sample compared to the blank is attributed to sample components interfering with the GC analysis technique.

GC capillary column performance was evaluated by running a Grob test mixture on the column.[10] Analysis of column testing showed very little degradation for the 30-m SE-30 glass column used in run 1. Run 2 used a 60-m SE-30 fused silica column. Two-thirds of the way through the analysis of run 2 samples, "tailing" was observed in the three methyl esters in the Grob test mixture. Breaking off a small length of column at the injector end and then flushing one column with solvent decreased the tailing satisfactorily. Comparison of n-alkane internal standard peak areas showed no significant change in column performance.

Quantitative Reproducibility of Peaks

Sample analysis contains several steps that can affect the precision of the concentration measurement of a sample.

E_1 Sampling differences; f(flowrate, organic loading)

E_2 Ether elution of resin bed; f(technique, volume, elution time)

E_3 Kuderna-Danish concentration; f(evaporation rate, contamination, temperature, N_2 flowrate)

E_4 Addition of internal standard; f(amount added)

E_5 Gas chromatography f(conditions, injection size, column phase integrity).

where E is the variance (σ^2).

To develop a quality assessment of the measurement process, we have assumed that these steps are additive in an analytical precision model. The hypothesis made is that each step contributed a portion of the total variability.

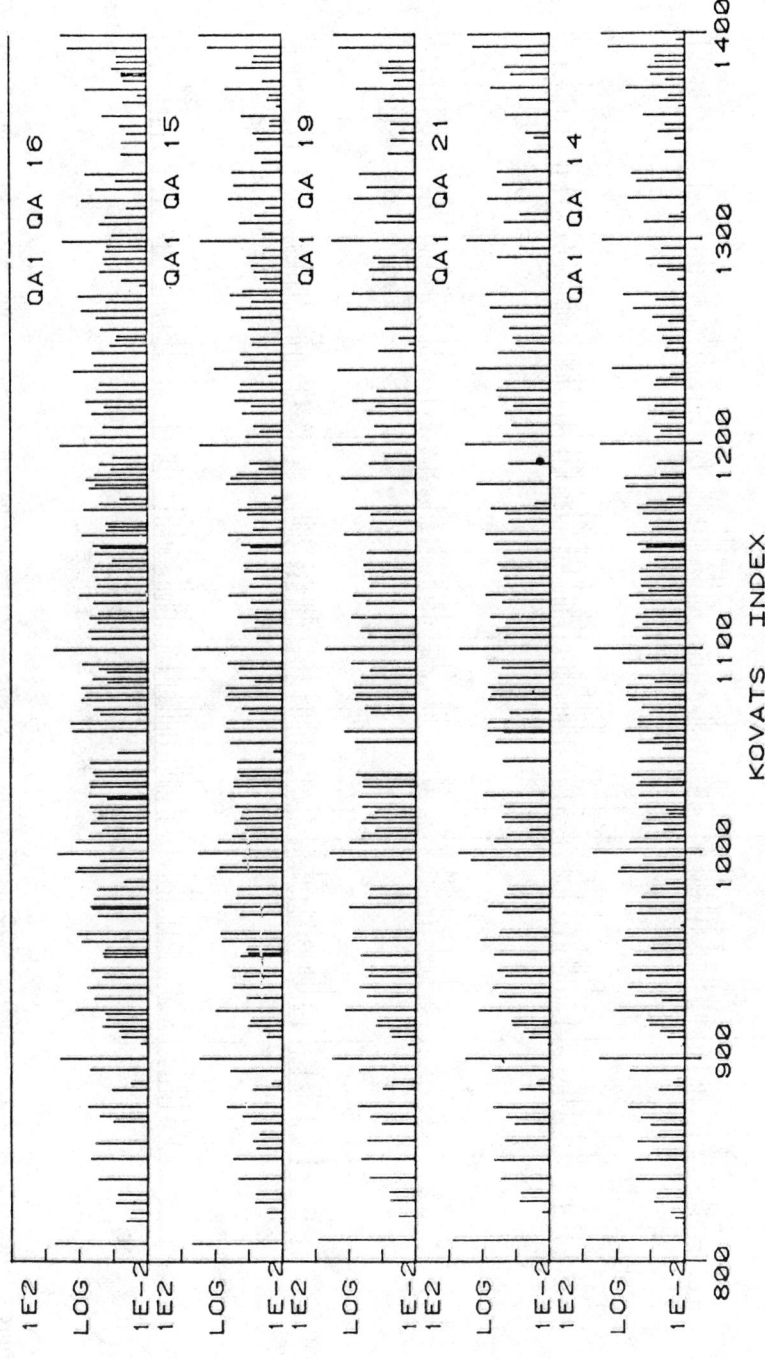

Figure 1. Reproducibility study—run No. 1 GC capillary chromatograms of five resin extracts collected from the same location at the same time (first third of chromatogram). Assume 100% breakthrough. Log 1E−O = 0.06 ppb

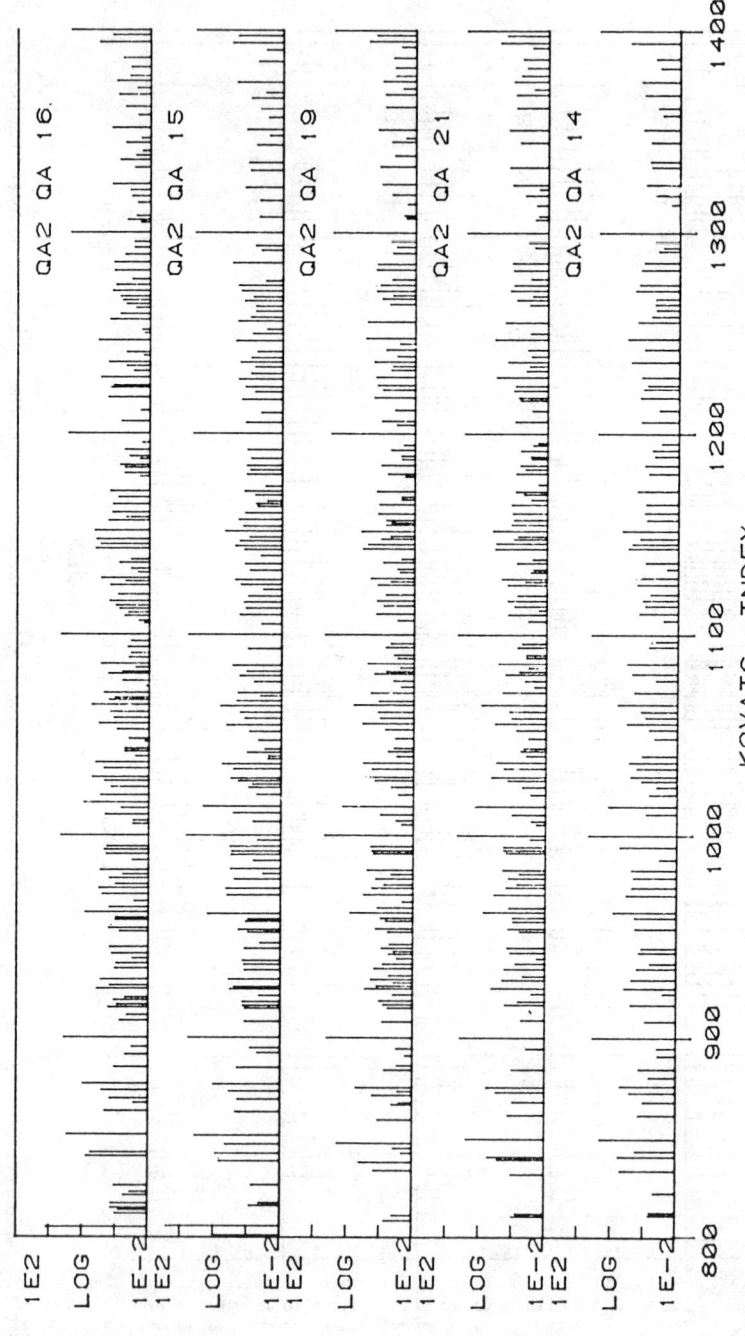

Figure 2. Reproducibility study—run No. 2 GC capillary chromatograms of five resin extracts collected from the same location at the same time (first third of chromatogram). Assume 100% breakthrough. Log 1E−0 = 0.06 ppb

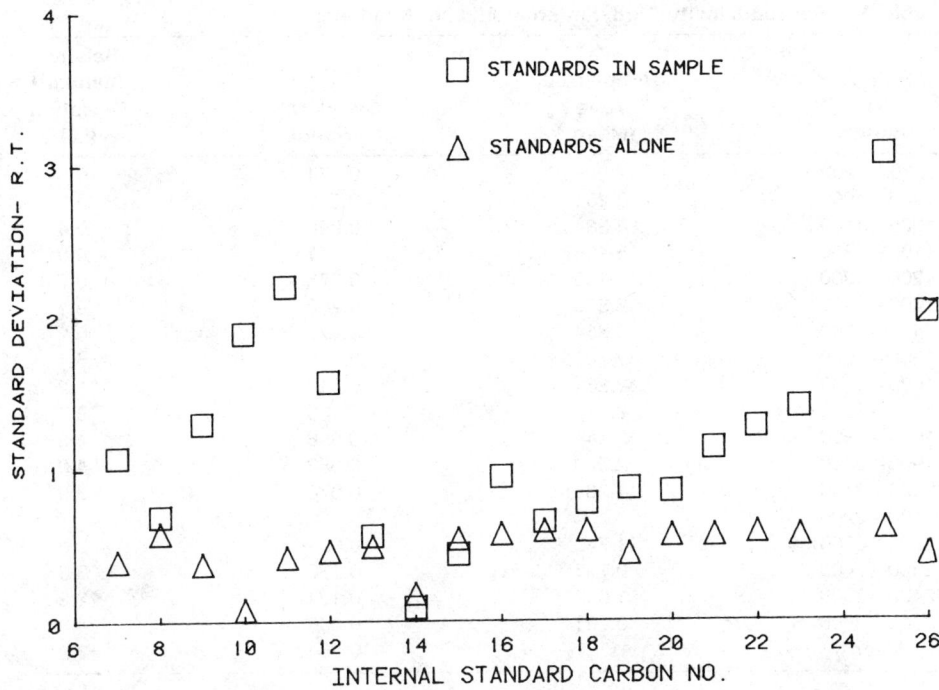

Figure 3. Sample matrix effect on reproducibility of internal standard retention time (in minutes).

$$E_T = E_1 + E_2 + E_3 + E_4 + E_5$$

Each step should be considered when attempting to measure precision of analysis between replicate samples and thus define the acceptable limits of the technique.

Table IV represents the precision (reproducibility) when only sampling steps E_4 and E_5 are included. Table V represents the precision of the whole analysis (E_T). GC chromatographic peaks were arbitrarily selected by a Kovats number ± 1 Kovats unit from each Kovats window for five replicate samples in Table V. Peak areas were normalized to an internal standard (C-21). The mean ($\bar{x}$), standard deviation (σ), and relative standard deviation ($\sigma/\bar{x}$) were computed and compared.

Of the two tests, the relative standard deviation for detector response is more important than the standard deviation because a change in peak response is a change in detector sensitivity, rather than a lower limit of detection. The standard deviation represents the y-intercept of a standard curve (lower limit of detection). The slope of the curve or detector response is represented by the

Table IV. Reproducibility Study: Internal Alkane Standards

KI Window	Normalized Area (Mean)	Standard Deviation	Relative Standard Deviation (%)
800 – 900	2.909	0.274	9.4
900 – 1000	3.261	0.296	9.1
1000 – 1100	4.384	0.896	20.4
1100 – 1200	5.166	0.341	6.6
1200 – 1300	3.155	0.275	8.7
1300 – 1400	2.868	0.260	9.1
1400 – 1500	2.854	0.177	6.2
1500 – 1600	2.707	0.138	5.1
1600 – 1700	2.831	0.317	11.2
1700 – 1800	2.348	0.205	8.7
1800 – 1900	2.141	0.096	4.5
1900 – 2000	2.023	0.082	4.0
2000 – 2100	1.965	0.076	3.9
2100 – 2200			
2200 – 2300	1.743	0.052	3.0
2300 – 2400	0.778	0.230	3.0
2400 – 2500	1.009	0.124	12.3
2500 – 2600	0.761	0.036	4.8
Mean		0.204	7.2

Table V. Reproducibility Study: Arbitrarily Selected Sample Peaks from Kovats Indices Windows from Run No. 1

KI Window	Normalized Area (Mean)	Standard Deviation	Relative Standard Deviation (%)
800 – 900	0.251	0.293	11.6
900 – 1000	0.133	0.028	21.1
1000 – 1100	0.164	0.061	37.4
1100 – 1200	0.275	0.099	36.0
1200 – 1300	0.703	0.215	30.6
1300 – 1400	0.176	0.020	11.4
1400 – 1500	0.078	0.022	27.6
1500 – 1600	0.049	0.024	48.3
1600 – 1700	0.051	0.007	13.7
1700 – 1800	0.678	0.327	48.2
1800 – 1900	0.026	0.003	33.5
1900 – 2000	0.173	0.047	27.2
2000 – 2100	0.741	0.044	6.0
2100 – 2200	0.057	0.029	50.8
2200 – 2300	0.047	0.015	31.4
2300 – 2400	0.008	0.007	21.4
2400 – 2500	0.041	0.005	11.4
2500 – 2600	0.035	0.007	20.8
Mean		0.054	27.1

relative standard deviation. A hierarchical ANOVA was conducted to see if the data in Tables IV and V fit the hypothesized model. The ANOVA showed that the data fit the model at the 95% confidence limit.

The precision data show the largest errors (variance) in the sampling, ether elution, and concentration steps (E_1, E_2, and E_3), not in the addition of internal standards and gas chromatography (E_4 and E_5). The average of all the relative standard deviations calculated was $E_T = 27.1\%$ (Table V), where E_T is an estimate of the total sample variability. This includes the relative standard deviation associated with the addition of internal standards and gas chromatography ($E_4 + E_5$), shown in Table IV to be 7.2%, which is the mean relative standard deviation for all the internal standards. The value of the square root of $E_4 + E_5$ was independently estimated by Coyle et al.[11] to be 7.5% for a parallel study in the same laboratory.

The large relative standard deviation of 27.1% indicates that small differences in chromatographic peak height are probably the result of system noise. Therefore, chromatograms are plotted on a log-scale y-axis. The log scale minimizes small differences in peak size and highlights major differences between peak sizes in a broad-spectrum chromatographic analysis.

Breakthrough of Organics by Resin Accumulators

Breakthrough of organics from the MRR accumulators was suspected, although the degree to which this occurred and the classes of compounds breaking through was unknown. Breakthrough is a function of the number of bed volumes of water processed by the resin. The combined XAD-2/XAD-8 resin beds in the samplers have a volume of ~200 mL. Water flowed through the bed at about 35 mL/min for 3 d, for a total number of 700 bed volumes. The design and operational conditions of the breakthrough columns was identical to the samplers (Table III). The use of breakthrough columns to check the samplers is a necessity in a good quality assurance study.

Figure 4 shows a chromatographic plot of a typical sample, breakthrough sample, and blank. Linear velocity through the bed has been shown to be an important parameter affecting organic breakthrough on XAD-2 resin.[6] Using the equation $L = F/r$, where the flow (F) = 35 mL/min, and the radius of the sampler (r) = 2.5 cm, the approximate linear velocity (L) for this study is 1.8 cm/min. Harris et al.[6] indicate that if the linear velocity is below 30 cm/min, the breakthrough curve for resin is independent of flowrate. Their study considered only one compound at a time; therefore, no competition was present to affect volumetric capacity of the resin. In this study, competition appears to be an important consideration. It is suggested that competition of a temporally varying influent lowers the value of linear velocity below which breakthrough is independent of flowrate in multicomponent sampling.

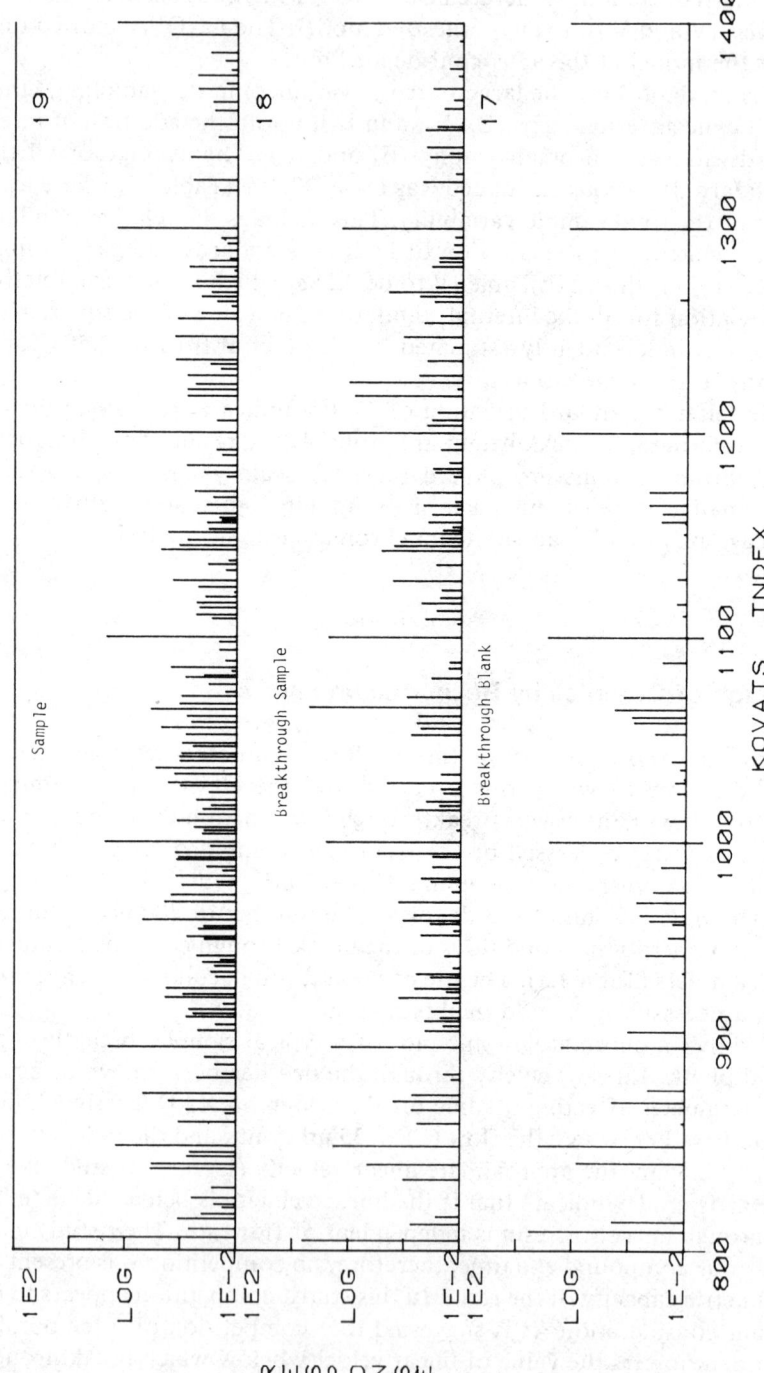

Figure 4. Breakthrough of organics adsorbed on XAD resin bed sampler after 700 bed volumes. GC capillary chromatogram of sample, breakthrough sample, and breakthrough blank (first third of chromatogram).

Breakthrough analysis of runs 1 and 2 included using chi-square contingency tables[12] to compare peak populations for normal and breakthrough samples. Observed and expected data were split into six peak-response ranges and four Kovats Index ranges (Table VI). The null hypothesis for the chi-square test of both runs was that the peak populations from the normal and breakthrough samples were homogeneous (i.e., did complete breakthrough occur for the run?). Table VI presents the results of the chi-square contingency table results for run 2. The test for run 2 shows the null hypothesis to be true; the peak populations are homogeneous, so complete breakthrough seems to have occurred. Chi-square comparison for run 1 rejects the null hypothesis that the two samples are homogeneous. Further chi-square analysis of run 1 determined that the major deviations between the normal and breakthrough samples occurred when the peak response range 0.03 to 0.1 (represents 0.9 to 6 ppt equivalent concentration) and over the Kovats Index range 800 to 1100.

The data from this study indicate that the samplers with pH <3.5 (run 2) were operated such that the columns in series with the original samplers were not different at $p = 0.05$ by chi-square contingency table analysis (Table VI). Thus, the recovery of organics from the resin accumulator is very low for ~ 750 bed volumes of sampled water, 150 L (Table III). This is caused by the complete breakthrough that was observed. A conservative estimate of the percentage breakthrough would be $>50\%$, assuming the the breakthrough column did not lose any organics. The breakthrough run 1 (partially at pH 7.5) revealed that the amount of breakthrough is substantially less than the samplers at the lower pH. This is verified by chi-square contingency table analysis, which determined that the breakthrough and the sampler chromatograms are not homogeneous at $p = 0.05$.

From the breakthough results, it can be concluded that the number of bed volumes (700) sampled was much too high. Decreasing this ratio would reduce breakthrough. Additionally, the resin could be used at a pH of ~ 8. The higher pH prevents the adsorption of fulvic acid,[13] which is a major fraction of the aqueous total organic carbon (TOC). This would result in a reduction in the TOC adsorbed that can displace the nonpolar compounds of interest. This proposed change in the sampled water pH is supported by comparison of the breakthrough data at pH 3.0 and 8.0. The purpose of the low pH in the sampling protocol is to hinder biodegradation of the adsorbed organic compounds on the resin and to allow acids such as phenol to be adsorbed. The use of some other means of hindering biodegradation, for example, a heavy metal such as silver or mercury, will permit the use of a higher pH in the sampled water. A lower volume of water and/or a higher volume of resin is the practical alternative to reduce breakthrough.

The quality assessment program that was completed in our study has shown that the considerable amount of breakthough of organics from the XAD resin was reproducible. It was not anticipated, and it highlights the problems of estimating the amount of resin, the flowrate, and sample volume needed to sample organics whose concentrations vary with compositing time. Further

Table VI. Chi-Square Analysis of Breakthrough Peak Distribution from Run 2[a,b]

Kovats Range		0.01[c] to 0.3	0.3[c] to 0.1	0.1[c] to 0.31	Peak Response 0.31[c] to 1.0	1.0[c] to 3.16	3.16[c] to 31.6	No. of Peaks
800–1100								
Sample	Observed	6.2	25	28	10.4	2.4	3	75
	Expected	13.38	27.4	20.6	5.8	3.94	3.78	
Breakthrough	Observed	7.5	15	12	1.5	2.5	6	44.5
	Expected	7.94	16.25	12.2	3.47	2.35	2.24	
1100–1400								
Sample	Observed	13	31.4	19.6	3.6	1.2	1.8	70.6
	Expected	12.60	25.79	19.4	5.5	3.73	3.55	
Breakthrough	Observed	11	11.5	11.5	1.5	2	3.5	41
	Expected	7.31	14.98	11.3	3.2	2.16	2.06	
1400–2000								
Sample	Observed	17.6	30.8	20.8	6.8	3.2	1.0	80.2
	Expected	14.31	29.8	22.07	6.25	4.23	4.04	
Breakthrough	Observed	8.5	15	15	4	3	4.04	50
	Expected	8.92	18.25	13.76	3.9	2.64	2.52	
2000–2600								
Sample	Observed	7.4	15.2	4.8	2.4	4.0	0.5[d]	34.3
	Expected	6.12	12.53	9.44	2.57	1.8	1.73	
Breakthough	Observed	2.5	7	2	2	3.5	0.5[d]	17.5
	Expected	3.12	6.39	4.82	1.36	0.92	0.88	
No. of Peaks		73.7	150.7	113.7	32.2	21.8	20.8	413.1

[a]Degrees of freedom 7(5) = 35; critical value of $\chi^2(0.05)$ = 49.88; calculated χ^2 value = 45.96.
[b]Ho: The two populations are homogeneous; accept Ho.
[c]Relative response to C_{21}.
[d]True observed value was zero but 0.5 arbitrarily substituted.

work is needed to define a method to accurately predict these variables before field sampling. Many of the existing laboratory recovery and breakthrough studies at constant concentration are inadequate, since they do not investigate competitive adsorption effects of TOC on the trace organic elements that the resin sampler was designed to collect.

CONCLUSIONS

The following limitations of the XAD resin samplers were observed in this quality assessment study.

1. The method is reproducible for the water sampled, as multiple samples are shown to be reproducible.
2. Artifacts are observed in each sample collected, but the artifacts are consistent within the experimental error.
3. Reproducibility is acceptable for order-of-magnitude estimates using log-scale response to indicate large differences. The error of the analysis was 27% (relative standard deviation). Therefore, the method is considered semiquantitative as it is currently used.
4. Laboratory evaluations of the accuracy of the macrorecticular resin isolation technique, defined in terms of percentage recovery and breakthrough of a specific compound in distilled water, did not present sufficient guidance to design a broad-spectrum method for field sampling with samples that contain TOC and vary temporally in type of compound(s) present and their relative concentrations.

ACKNOWLEDGMENTS

The research work was supported by the Philadelphia Water Department, W. Marrazo, Water Commissioner, and Drinking Water Research Division of the Municipal Environmental Research Laboratory of U.S. EPA, J. Keith Carswell, Project Officer (EPA Contract # 806256-02).

REFERENCES

1. Suffet, I. H., L. Brenner, J. T. Coyle, and P. R. Cairo. "Evaluation of the Capability of Granular Activated Carbons and XAD-2 Resin to Remove Trace Organics from Treated Drinking Water," *Environ. Sci. Technol.* 12:1315–1322 (1978).
2. Yohe, T. L., I. H. Suffet, and J. T. Coyle. "Monitoring and Analysis of Aqueous Chlorine Effects on GAC Pilot Contactors," in *Activated Carbon Adsorption of Organics from the Aqueous Phase*, Vol. 2, M. J. McGuire and I. H. Suffet, Eds. (Ann Arbor, MI: Ann Arbor Science Publishers, Inc., 1980), p. 37.
3. Stevens, A. A., D. R. Seeger, C. J. Slocum, and M. M. Domino. "Gas Chromatographic Techniques for Controlling Organics Removal Processes," *J. Am. Water Works Assoc.* 73:548–554 (1981).

4. Van Rensberg, J. F. J., A. Hassett, S. Theron, and S. C. Wiecher. "The Fate of Organic Micropollutants Through an Integrated Waste Water/Water Reclamation System," *Prog. Water Technol.* 12:537–552 (1980).
5. Keith, L. H., et al. "Principles of Environmental Analysis," *Anal. Chem.* 55:2210–2218 (1983).
6. Harris, J. C., et al. *Evaluation of Solid Sorbents for Water Sampling*, Report to U.S. EPA 1 ERC, C-82480-42 (New York, A. D. Little, Inc.,1981).
7. Junk, G. A., et al. "Use of Macroreticular Resins in the Analysis of Water for Trace Organic Contaminants," *J. Chromatogr.* 99:745–762 (1974).
8. Junk, G. A. "Synthetic Polymers for Accumulating Organic Compounds from Water," Extended Abstract, *Div. Environ. Chem., Am. Chem. Soc.* 24(2):249–251 (1984).
9. Neukrug, H. M., et al. *Removing Organics from Drinking Water by Combined Ozonation and Adsorption*, EPA 600/52-83-048 (Cincinnati: U.S. Environmental Protection Agency, 1983).
10. Grob, K., Jr., G. Grob, and K. Grob. "Comprehensive, Standardized Quality Test for Capillary Glass Columns," *J. Chromatogr.* 156:1–20 (1978).
11. Coyle, J. T., J. Gibs, S. W. Maloney, and I. H. Suffet. "Broad-Spectrum Analysis of the Removal of Trace Organics in an Ozone-Granular Activated Carbon Potable-Water Pilot Plant Study," in *Water Chlorination: Environmental Impact and Health Effects, Vol. 4,* R. L. Jolley, W. A. Brungs, J. A. Cotruvo, R. B. Cumming, J. S. Mattice, and V. A. Jacobs, Eds. (Ann Arbor, MI: Ann Arbor Science Publishers, Inc., 1983), pp. 421–443.
12. Gibs, J. *Broad Spectrum Analysis of Trace Organics in Drinking Water,* Ph.D. Thesis, (Philadelphia: Drexel University, 1983).
13. Thurman, E., and R. L. Malcolm. "Preparative Isolation of Aquatic Humic Substances," *Environ. Sci. Technol.* 15:463–466 (1981).

CHAPTER 89

Monitoring Trichloroacetic Acid in Municipal Drinking Water

Daniel L. Norwood, Gavin P. Thompson, J. D. Johnson, and R. F. Christman

Scientific interest has recently focused on the hydrophobic halogenated by-products of drinking water disinfection with chlorine, principally chloroform and its sister trihalomethanes (THMs).[1] Natural product organic materials such as humic and fulvic acids are the likely precursors for these volatile substances. There is, however, a growing body of literature indicating that THM formation cannot account for most of the total organic halide (TOX) produced from the aqueous chlorination of these precursor materials.[2-5] Furthermore, it has been demonstrated that the mutagenic activity of chlorinated drinking water is more closely associated with the nonvolatile fraction of TOX.[6,7] Efforts have thus been made to identify and quantify the individual components of nonvolatile TOX in drinking water.

Varieties of small aliphatic chlorinated hydrophilic compounds were determined in our laboratory to result from the chlorination of an extracted aquatic humic material under various reaction conditions.[8-10] The dominant components found in each case were dichloroacetic acid (DCAA) and trichloroacetic acid (TCAA). Other workers have confirmed and extended these results using a variety of precursor humic materials.[11-13] In all these studies, it was demonstrated that TCAA was produced in greater quantities than chloroform. Concurrently, it was shown by Rook that DCAA and TCAA were the principal constituents of a methylene chloride extract of Rotterdam finished drinking water subjected to breakpoint chlorination.[14] More recently, Uden and Miller have shown chlorinated acids to be major components in two tap water samples, again with DCAA and TCAA being the dominant hydrophilic structures.[15]

In our laboratory we have undertaken to extend our work on chlorination by-products to raw waters and actual finished drinking water samples.[16] We report on a limited quantitative survey of finished drinking water samples for TCAA using an isotope diluton gas chromatography/mass spectrometry (GC/MS) method[17] for determining this substance at the microgram per liter level. Comparison is made with the TOX found and, in selected instances, with the corresponding chloroform value.

EXPERIMENTAL

Finished water samples for organic halide screening were collected from selected points in the respective distribution systems in 1-L glass bottles (washed with chromic acid) and stored at room temperature in a head-space-free condition until workup. The procedure for spiking and workup of aqueous samples for GC/MS analysis is similar to that described previously.[16,17] After removal of residual chlorine by adding pre-Soxhlet extracted sodium arsenite, measured 300-mL aliquots of each reaction mixture and blank were spiked with precise amounts (~50 μg/L) of the compound $^{13}C_1$-trichloroacetic acid (KOR, Inc., Cambridge, MA).

Stock solutions of this isotopically labeled internal standard were made up in deionized/distilled water. After allowing at least 2 h for the internal standard to equilibrate with the sample matrix, each aqueous solution was acidified to pH 0.7 with concentrated hydrochloric acid and extracted with three 100-mL aliquots of distilled ether. The resulting ether extracts were dried by liquid nitrogen freezing and equilibration with pre-Soxhlet extracted magnesium sulfate and then concentrated to ~10 mL in a Kuderna-Danish apparatus. A 5-mL aliquot of this concentrate was methylated to excess with ethereal diazomethane produced by the reaction of sodium hydroxide with N-methyl-N-nitro-N-nitrosoguanidine.[18] Any further concentration of samples was achieved by evaporating the ether with a gentle stream of dry nitrogen. Note that it is not necessary to use special procedures to obtain quantitative recovery of the TCAA in this method.[16]

Calibration curves were prepared by spiking variable amounts of unlabeled TCAA along with a constant amount of the labeled analogue into 300-mL aliquots of a natural water matrix free of endogenous TCAA. The TCAA used was of the highest purity available. Standard solutions were subjected to the same workup procedure as the samples.

The GC/MS system used in our experiments consisted of a VG Micromass 7070F double-focusing sector mass spectrometer of Nier-Johnson geometry (VG Analytical, Altrincham, Cheshire, UK) interfaced to an HP5710A capillary gas chromatograph (Hewlett-Packard, Inc., Avondale, PA). The GC was equipped with an SGE OCI-2 capillary on-column injector (Scientific Glass and Engineering, Austin, TX) and a 30-m DB-1 fused silica capillary column (J&W Scientific, Rancho Cordova, CA). Data acquisition and manipulation were controlled by a VG 2035 data system. For these studies the mass spectrometer was operated in the voltage switching, selected-ion-monitoring (SIM) mode under the control of a VG digital scan unit. Each concentrated, methylated ether extract was analyzed by an isotope dilution method to quantitate the TCAA. The method incorporates procedures similar to those outlined in EPA Method 1625[19] and has been described previously.[16]

The mass spectrometer was operated in the electron ionization (EI) mode with 70-eV electrons, a 200-μA trap current, and 4-kV accelerating potential at a mass resolution of 1000 (10% valley definition). In the SIM mode, only three ions were monitored by rapid accelerating voltage switching. Two of the ions were those from the analyte (m/z 141, methyl trichloroacetate) and the labeled analogue (m/z 142, ^{13}C-methyltrichloroacetate), respectively. The third was a reference ion (m/z 136, methyl benzoate) that was continuously leaked into the ion source from a heated reservoir. The dwell time in the reference mass channel was 50 ms, with a channel delay of 50 ms. Corresponding parameters for the analyte channels were 100 and 50 ms, respectively. The ion area ratio m/z 141:142 was calculated for all standards and samples. In all GC/MS runs, 1 to 2 µL of sample was injected on-column at room temperature. After the solvent front eluted, the column temperature was rapidly raised to 60°C, held there for 2 min, and then programmed at 6°C per min until methyltrichloroacetate and its labeled analogue coeluted.

Measured ion area ratios were converted to relative responses (RR) and corrected for potential isotope overlap using a formula described by Colby et al.[20]

$$RR = \frac{(R_y - R_m)(R_x + 1)}{(R_m - R_x)(R_y + 1)}$$

where R_m = ratio (m/z 141:142) for any standards or sample
R_x = ratio for the pure unlabeled compound
R_y = ratio for the pure labeled internal standard

This equation was generally used only when

$$2 R_y < R_m < 0.5 R_x$$

to minimize propagated errors. In a few cases where a ratio violated the above criteria to a slight degree, it was still used and the error was accepted.

Samples for total organic halide analysis were taken from the sample bottle just after residual chlorine removal and placed in acid washed, 60-mL septum-capped (Teflon®) vials in a head-space-free condition. These samples were refrigerated prior to analysis. The measurements were made using EPA Interim Method 450.1[21] with a Dohrman MCTS-20 TOX instrument (Dohrman, Inc., Santa Clara, CA).

RESULTS AND DISCUSSION

The selection of municipal finished drinking waters for organic halide analysis included samples from three major types of raw water sources. These are listed in Table I. The first group represented rivers and included Pittsboro, North Carolina (Haw River), Sanford and Fayetteville, North Carolina (Cape Fear River), Washington, D.C. (Potomac River), and Richmond, Virginia (James River). The second group was representative of impoundments that are used as drinking water supplies and included Chapel Hill (University Lake), Durham (Lake Michie), and Raleigh (Lake Wheeler), North Carolina. The third group represented chlorinated groundwater sources and included New Bern, (an inland groundwater from Cove City) and Emerald Isle (a coastal groundwater from Bogue Banks), North Carolina. These sources represent a reasonable cross section of North Carolina waters, along with two other major metropolitan areas for comparison.

Table I. Sources of Finished Drinking Waters Sampled in this Study

Source Type	Area Served	Source
River	Pittsboro, NC[a,b]	Haw River
	Sanford, NC[a,b]	Cape Fear River
	Fayetteville, NC[a,b]	Cape Fear River
	Washington, DC[c]	Potomac River
	Richmond, VA[c]	James River
Reservoir	Chapel Hill, NC[a,b]	University Lake
	Durham, NC[a,b]	Lake Michie
	Raleigh, NC[a,b]	Lake Wheeler
Groundwater	New Bern, NC[c]	Cove City, NC, inland groundwater
	Emerald Isle, NC[c]	Bogue Banks, NC, coastal groundwater

[a]Samples taken in October 1983.
[b]Chlorinated raw water previously analyzed.[17]
[c]Samples taken in April 1984.

Analysis of these samples was divided into two parts, as indicated in Table I, with each group of unknown samples having its own set of spiked calibration standards and standard curve. The standard curve used for the October 1983 group of samples is shown as a log-log plot in Figure 1. Note that the trace is linear over approximately three orders of magnitude in TCAA concentration (5 to 1000 μg/L), with a slope of 1 (0.95) and a y-intercept near the origin (0.033). The points shown are actual spiked standard values, and the line is derived from the logarithmically weighted linear regression on these points. The bars near the regression at each point represent the 95% prediction interval for an unknown at that particular point on the line. These prediction intervals were used to estimate an error range for each unknown value. The standard curve used for the April 1984 group of samples is very similar in slope, intercept, and 95% prediction interval width to that in Figure 1. In our experience, these are typical in TCAA analysis by the isotope dilution GC/MS method.

The results of these analyses are given in Table II. Note that TCAA was detected in each sample, with levels varying from a maximum of 53.8 μg/L (Emerald Isle) to a minimum of 4.23 μg/L (New Bern). From these data it is attractive to hypothesize that TCAA, like chloroform, is ubiquitous in drinking waters derived from chlorinated raw waters. These data also compare favorably with those of Uden and Miller,[15] who used a different method to determine TCAA in two Amherst, Massachusetts, tap water samples. These data showed TCAA values of 33.6 μg/L in a tap water sample from Atkins Reservoir and 161 μg/L in a sample from Pelham Reservoir. The TCAA concentration of the Atkins sample increased to 72.8 μg/L after a 24-h holding time, with the Pelham sample holding constant at 160 μg/L.

Uden and Miller found zero-time chloroform value to be approximately the

Table II. Results of Municipal Drinking Water Survey for Trichloroacetic Acid (TCAA)

Source	TCAA (µg/L)	Range	TOX[a] (µg Cl/L)	TOX Accounted for by TCAA (%)
Pittsboro, NC	52.0	46.5 – 56.0	501	6.76
Sanford, NC	20.5	18.1 – 22.1	325	4.11
Fayetteville, NC	16.2	14.4 – 17.8	248	4.26
Washington, DC	23.4	18.3 – 30.0	146	10.4
Richmond, VA	36.1	28.5 – 46.5	310	7.59
Chapel Hill, NC	40.3	35.9 – 43.7	403	6.52
Durham, NC	44.0	39.0 – 47.0	378	7.58
Raleigh, NC	24.5	21.5 – 26.5	382	4.18
New Bern, NC	4.23	3.20 – 5.20	156	1.77
Emerald Isle, NC	53.8	42.0 – 67.0	566	6.19

[a] Total organic halide.

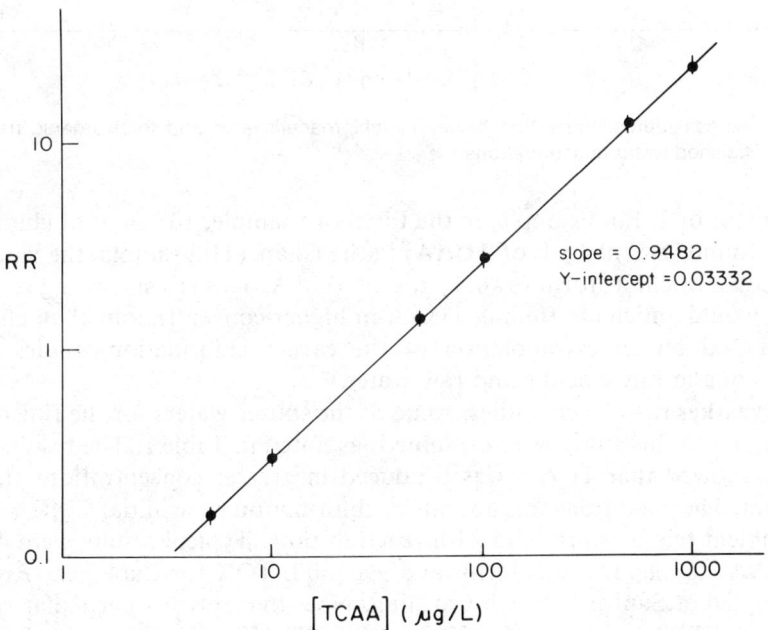

Figure 1. Standard curve for the isotope dilution GC/MS analysis of trichloroacetic acid in the October 1983 samples.

same as the TCAA value (39.6 µg/L for Atkins, 190 µg/L for Pelham). Several of the samples were also analyzed for chloroform using the isotope dilution GC/MS method with ^{13}C-chloroform as an internal standard.[17] In each case the measured chloroform concentration exceeded that of TCAA, usually by

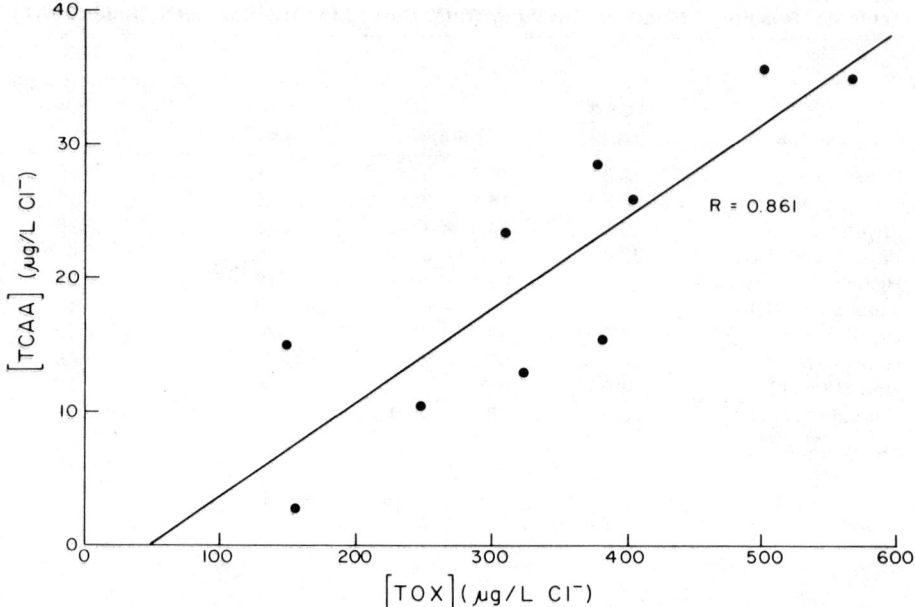

Figure 2. Across-source correlation beween trichloroacetic acid and total organic halide finished water concentrations.

about a factor of 2. For example, in the Pittsboro sample, 107 μg/L of chloroform was found vs 52.0 μg/L of TCAA; in the Chapel Hill sample, the values were 118 μg/L of chloroform vs 40.3 μg/L of TCAA. This is a surprising result since one would anticipate finding TCAA in higher concentrations than chloroform, based on an extrapolation of the earlier chlorination studies on extracted aquatic fulvic acid[16] and raw water.[17]

In these earlier raw water studies, some of the source waters for the finished water samples in this study were examined, as noted in Table I. The results in each case showed that TCAA was produced in greater concentrations than chloroform. The conditions for laboratory chlorination were initial $Cl_2:C = 4$, pH 7, ambient temperature, and 24-h reaction time. Typical results were 410 μg/L TCAA, 281 μg/L chloroform, and 990 μg/L TOX for Cape Fear River water sampled at Sanford, North Carolina. Note that for this particular raw water sample, TCAA accounts for 27.0% of the TOX produced, also a typical result. In the finished water samples, however, TCAA accounts for a much smaller TOX fraction, as indicated in Table II. In each case, less than 11% of the TOX can be accounted for by TCAA; for the corresponding Sanford sample, the value is only 4.11%.

It is instructive to examine the correlation between finished water TCAA and TOX concentrations shown in Figure 2 (R = 0.861). The chlorinated raw waters showed correlations between TCAA and TOX, as well as chloroform and TOX, with greatly reduced residuals and somewhat better correlation coefficients (R = 0.989 and 0.954, respectively).[17] However, even with the somewhat poorer across-source correlation for the finished waters, one would anticipate that it might be possible to use a specific regulation on finished-water TCAA concentration to regulate TOX within a certain error margin. This sort of dual regulating capacity has been envisioned for THM regulation.[22]

The differences between finished water values and those derived from raw water chlorination are probably the result of differences in reaction conditions between water treatment plants and laboratory (lower Cl_2:C in water plants, temperature differences, variable reaction times, prechlorination treatment practices, etc.). The improved specific compound/group parameter correlations observed in the raw water study can be explained by the constancy of conditions in the laboratory vs the variability between water treatment plants. Taking this into consideration, the similarities in these results appear more striking than the differences.

CONCLUSIONS

The results reported here demonstrate that trichloroacetic acid is an important and quite possibly ubiquitous by-product of drinking water disinfection with chlorine. The suspected importance of hydrophilic organic compounds in drinking water is thus confirmed. Further studies are needed to determine what other hydrophilic structures (such as dichloroacetic acid)[15] are present and what fraction of the total organic halide each accounts for under a given set of treatment conditions. The interrelationships in the formation of these compounds should also be examined so that improved alternative water treatment practices can be instituted to minimize all of them.

REFERENCES

1. Rook, J. J. "Formation of Haloforms During Chlorination of Natural Waters," *Water Treat. Exam.* 23(2):234–243 (1974).
2. Oliver, B. G. "Chlorinated Non-Volatile Organics Produced by the Reaction of Chlorine with Humic Materials," *Can. J. Res.* 11(6):21–22 (1978).
3. Glaze, W. H., and G. R. Peyton. "Soluble Organic Constituents of Natural Waters and Wastewaters Before and After Chlorination," in *Water Chlorination: Environmental Impact and Health Effects, Vol. 2*, R. L. Jolley, H. Gorchev, and D. H. Hamilton, Jr., Eds (Ann Arbor, MI: Ann Arbor Science Publishers, Inc., 1978), p. 3–14.
4. Glaze, W. H., F. Y. Saleh, and W. Kintsley. "Characterization of Nonvolatile Halogenated Compounds Formed During Water Chlorination," in *Water Chlorination: Environmental Impact and Health Effects, Vol. 3*, R. L. Jolley, W. A. Brungs, and R. B. Cumming, Eds. (Ann Arbor, MI: Ann Arbor Science Publishers, Inc., 1980), p. 99–108.
5. Watts, C. D., B. Crathorne, M. Fielding, and S. D. Killops. "Nonvolatile Organic Compounds in Treated Waters," *Environ. Health Perspect.* 46:87–99 (1982).
6. Bull, R. J. "Health Effects of Drinking Water Disinfectants and Disinfectant By-Products," *Environ. Sci. Technol.* 16(10):554A–559A (1982).

7. Kool, H. J., C. F. van Kreije, E. de Greef, and H. J. van Kronen. "Presence, Introduction, and Removal of Mutagenic Activity During Preparation of Drinking Water in the Netherlands," *Environ. Health Perspect.* 46:207-211 (1982).
8. Christman, R. F., J. D. Johnson, F. K. Pfaender, D. L. Norwood, M. R. Webb, J. R. Hass, and M. J. Bobenrieth. "Chemical Identification of Aquatic Humic Chlorination Products," in *Water Chlorination: Environmental Impact and Health Effects, Vol. 3*, R. L. Jolley, W. A. Brungs, and R. B. Cumming, Eds. (Ann Arbor, MI: Ann Arbor Science Publishers, Inc., 1980), p. 75-83.
9. Johnson, J. D., R. F. Christman, D. L. Norwood, and D. S. Millington. "Reaction Products of Aquatic Humic Substances with Chlorine," *Environ. Health Perspect.* 46:63-71 (1982).
10. Norwood, D. L., J. D. Johnson, D. S. Millington, and R. F. Christman. "Chlorinated Products from Aquatic Humic Material at Neutral pH," in *Water Chlorination: Environmental Impact and Health Effects, Vol. 4*, R. L. Jolley, W. A. Brungs, J. A. Cotruvo, R. B. Cumming, J. S. Mattice, and V. A. Jacobs, Eds. (Ann Arbor, MI: Ann Arbor Science Publishers, Inc., 1983), p. 191-200.
11. Quimby, B. D., M. F. Delaney, P. C. Uden, and R. M. Barnes. "Determination of the Aqueous Chlorination Products of Humic Substances by Gas Chromatography with Microwave Emission Detection," *Anal. Chem.* 52:259-263 (1980).
12. Coleman, W. E., J. W. Munch, W. H. Kaylor, H. P. Ringhand, and J. R. Meier. "GC/MS Analysis of Mutagenic Extracts of Aqueous Chlorinated Humic Acids – A Comparison of the By-Products to Drinking Water Contaminants," Extended Abstract, *Am. Chem. Soc. Div. Environ. Chem.* 23(1):53-55 (1983).
13. Miller, J. W., and P. C. Uden. "Characterization of Nonvolatile Chlorination Products of Humic Substances," *Environ. Sci. Technol.* 17(3):150-157 (1983).
14. Rook, J. J. "Possible Pathways for the Formation of Chlorinated Degradation Products during Chlorination of Humic Acids and Resorcinol," in *Water Chlorination: Environmental Impact and Health Effects, Vol. 3*, R. L. Jolley, W. A. Brungs, and R. B. Cumming, Eds. (Ann Arbor, MI: Ann Arbor Science Publishers, Inc., 1980), p. 85-98.
15. Uden, P. C., and J. W. Miller. "Chlorinated Acids and Chloral in Drinking Water," *J. Am. Water Works Assoc.* 75(10):524-527 (1983).
16. Christman, R. F., D. L. Norwood, D. S. Millington, J. D. Johnson, and A. A. Stevens. "Identity and Yields of Major Halogenated Products of Aquatic Fulvic Acid Chlorination," *Environ. Sci. Technol.* 17(10):625-628 (1983).
17. Norwood, D. L., G. P. Thompson, J. J. St. Aubin, D. S. Millington, and R. F. Christman. "By-Products of Chlorination: Specific Compounds and Their Relationship to Total Organic Halogen," in press (1984).
18. Fales, H. M., T. M. Jaorine, and J. F. Babashek. "Simple Device for Preparing Ethereal Diazomethane without Resorting to Codistillation," *Anal. Chem.* 45(3):2301 (1973).
19. "Semivolatile Organic Compounds by Isotope Dilution GC/MS," (Washington, DC: U.S. Environmental Protection Agency).
20. Colby, B. N., A. E. Rosecrance, and M. E. Colby. "Measurement Parameter Selection for Quantitative Isotope Dilution Gas Chromatography/Mass Spectrometry," *Anal. Chem.* 53:1907-1911 (1981).
21. Billets, S., and J. J. Lichtenberg. *Total Organic Halide: Method 450.1 Interim*, EPA 600/4-81-056, (Cincinnati: U.S. Environmental Protection Agency, 1980).
22. Cotruvo, J. "THM's in Drinking Water," *Environ. Sci. Technol.* 15(3):268-274 (1981).

CHAPTER 90

Monitoring for Volatile Organohalides Using Purgeable and Total Organic Halide as Surrogates

R. K. Sorrell, E. Daly, L. Boyer, and H. J. Brass

Recent surveys of groundwater supplies have provided enough data on the presence of volatile organic chemicals (VOCs) to justify consideration of a comprehensive monitoring program.[1] The VOCs that occur most frequently at highest concentration are trichloroethylene, 1,1,1-trichloroethane, tetrachloroethylene, and cis/trans-dichloroethylene.[2] The analytical methods likely to be used in such a program—purge and trap (P&T) or liquid-liquid extraction (LLE) followed by gas chromatography—require the determination of individual VOCs. However, other methodology may be suitable. Two such methods that could be appropriate for monitoring these compounds are total organic halide (TOX) and purgeable organic halide (POX).

The expected ability of TOX and POX to detect and quantitate the halogen content of a wide variety of organic chemicals has led to their use as surrogates in unit process control.[3] TOX has been incorporated into federal regulations for monitoring groundwaters near hazardous waste sites.[4] Its use as a surrogate has also been suggested for monitoring individual halogen-containing VOCs in waters intended for drinking.[1]

This chapter presents information concerning the availability, cost, and suitability of TOX and POX analysis as compared with individual compound analysis for volatile organohalides.

EXPERIMENTAL

Apparatus

A Dohrmann DX-20 total organic halide analyzer system, composed of an AD-2 adsorption module and an MC-1 microcoulometer module, was used to determine TOX and POX. The total chlorine residual was analyzed with the HACH CN-66 total chlorine test kit, which uses the DPD method of analysis.[5] The VOCs were determined using two Tekmar LSC-2 liquid sample concentrators interfaced to a Hewlett-Packard 5710A gas chromatograph (GC) with dual 6-ft glass columns containing 1% SP-1000 on Carbopack B. The two columns were interfaced to a Tracor 700 Hall electrolytic conductivity detector (E1CD) and an HNU photoionization detector (PID), respectively. Thus, halogenated and unsaturated VOCs could be determined simultaneously. A Thelco 32 MR incubator was used for controlled sample storage. Inorganic chlorine was determined using an Orion chloride electrode, Model 96-17.

Procedure

The analysis for TOX (EPA Method 450.1) makes use of the carbon adsorption/combustion process and microcoulometric titration.[6] 4,5-Dichlorophthalic acid was substituted for trichlorophenol as a standard.

Determination of VOCs was performed using purge and trap/GC/ElCD and PID methods. These methods have been described previously.[7,8] The ElCD was the primary detector used for quantitation.

Analysis for POX was accomplished via direct purging of the VOCs into the furnace, followed by microcoulometric titration. This procedure is described in the Dohrmann DX-20 manual.[9]

Laboratory samples for VOC recovery studies were prepared in a calibrated 2000-mL separatory funnel, which was filled with either VOC-free groundwater or reagent water. The stoppered funnel was placed on a magnetic stirrer for mixing. Depending on the intended concentration, between 4 and 12 μL of a VOC stock solution in methanol was injected through the stopcock while stirring. After stirring the dosed water for ~1 min, the separatory funnel was placed on a ring stand for ~5 min. The stopper was then removed and 400 mL of dosed water was drained through the stopcock and discarded. The sample bottles (60- and 250-mL amber screw-top bottles with Teflon®-lined caps) were filled, and the remaining 500 mL of dosed water was discarded. The first and last bottles were analyzed by gas chromatography to determine any sampling bias. Sample analysis was completed within 24 h.

Raw Groundwater

These waters were collected in 60- and 250-mL amber screw-top bottles with Teflon-lined caps. The samples were shipped in ice by overnight express. Upon receipt, they were immediately analyzed or stored at 4°C for no longer than 3 weeks.

Quality Assurance

The analytical procedure was evaluated with a quality assurance (QA) program that included monitoring of the activated carbon to ensure a consistently low chlorine background for the TOX analysis. The microcoulometer was evaluated with sodium chloride and responded linearly at concentrations between 1 and 2000 µg/L. The performance of the microcoulometer was monitored by injecting 10 µg Cl prior to analysis each day. During the course of the project, the observed values averaged 10.1 µg Cl ± 0.8% for the period July to October 1983, and 9.98 µg Cl⁻ ± 0.7% for the period December 1983 to May 1984.

An additional part of the QA program for the DX-20 was the analysis of QA standards that were internally generated from 4,5-dichlorophthalic acid concentrates sealed in glass ampules. Aliquots (at 31 µg/L as Cl⁻) had an average accuracy of 95 ± 8% for the project. At 94 µg/L as Cl⁻, the QA standard accuracy was 95 ± 3%.

For P&T/GC/ElCD, standards were evaluated against externally generated quality control standards available from the U.S. Environmental Protection Agency (EPA). In addition, gas chromatography was used to examine raw water prior to dosing to ensure that it was free of VOCs. After dosing, the waters were again submitted to GC analysis for two reasons. The first was quantitative, to confirm the concentration of the spiked VOC. The second was qualitative, to ensure that no other halo-VOCs were present (i.e., contamination), which would lead to inaccurate recoveries as measured by TOX or POX.

RESULTS AND DISCUSSION

Availability and Cost

Because of the Resource Conservation and Recovery Act (RCRA) requirements[4] and general interest in TOX as a surrogate, the use of TOX has become more widespread. At present, it is estimated there are between 50 and 75 commercial laboratories and universities that perform TOX analysis using the Dohrmann DX-20 system.[10]

A survey of 33 laboratories was conducted to determine their capabilities. This is about half of the laboratories that perform TOX analyses using the DX-20 system. It was determined that only about half of all laboratories performing TOX analyses also analyze for POX.

These same laboratories were asked to provide price quotes for the measurement of TOX, POX, and trihalomethanes (THMs) by purge-and-trap (P&T) and liquid-liquid extraction techniques. Also requested were quotes for P&T analyses of volatile organic chemicals containing halogenated VOCs. TOX, by Method 450.1, requires duplicate analysis for each sample, and this was also applied to POX. The prices for THM and VOC analyses are based on a single measurement, because a duplicate is not required according to their respective methods. No information was obtained on the quality of analytical data generated by these laboratories.

The average costs of TOX and POX analyses were $99 and $77, respectively, for the 33 and 18 responding laboratories. The ranges of cost for sample analysis varied by approximately an order of magnitude. The THM measurements by P&T (15 laboratories) had an average price of $86 per sample, which was more costly than measurement by liquid-liquid extraction (8 laboratories), which averaged $59 per sample. Cost data show there is no savings in using TOX or POX as surrogate measures of the presence of THMs. Analysis of halogen-containing VOCs had average prices of $114 (5 laboratories) and $207 (18 laboratories), using ElCD and mass spectrometers, respectively.

General caveats must be placed on these cost comparisons. The basis for contacting laboratories was their ability to perform TOX measurements using the Dohrmann DX-20 system. THM and VOC cost data from these laboratories may not accurately reflect analytical costs for a wide range of laboratories. The costs appear reasonable, however, based on our direct experience in dealing with laboratories specializing in THM and VOC measurement.

Suitability

To properly interpret TOX or POX results, it is necessary to understand the ability of these analytical techniques to reflect the concentration of specific halogen-containing chemicals, that is, the appropriate mass balances. One aspect to bear in mind is that the response of TOX or POX varies from compound to compound and is dependent on the halogen species as well as the proportional amount of halogen in specific compounds. For example, equal concentrations (on a mass basis) of chlorobenzene and carbon tetrachloride afford only one-third of the halide for chlorobenzene when compared with carbon tetrachloride. Because TOX and POX are reported as chloride, bromoform provides only one-half the response of chloroform for an equal mass concentration.

Previously reported investigations provide an initial mass balance data base for TOX and POX accuracy. TOX accuracy data for volatile organohalides ranged from 73 to 110% (Table I)[11,12] and included such compounds as chloroform, bromoform, and bromobenzene in reagent water at concentrations of 98 to 443 µg/L. POX data (Table I) from the same authors[13,14] indicated a 98% recovery for bromoform. An 80% recovery for THM was obtained with a vitrified insert tube in the measurement system. The tube was believed to have caused the reduced recovery, since more nearly complete recovery of chloroform was reported when the tube was replaced.

Additional POX recovery data were obtained for a wider variety of volatiles at higher concentrations by Riggin et al. (Table II).[15] These recoveries were also from reagent waters dosed at 1000 µg/L. Riggin et al. reported compound recoveries from 47 to 106%, which were from 11 (chloroform) to 51% (bromodichloromethane) lower than those previously reported.[13,14] Recoveries were also determined at lower concentrations for five selected VOCs (Table III). With the exception of chloroform, recoveries were lower at lower-dosed concentrations.

Table I. Accuracy of TOX and POX for VOCs in Reagent Water

Model Compounds	Concentration μg/L	Concentration μg/L as Cl⁻	Average Recovery (%)	Reference
TOX				
Chloroform	98	87	89	12
	112	99.7	94	11
Bromodichloromethane	160	104	98	12
Dibromochloromethane	155	79.0	86	12
	374	191	73	12
Bromoform	160	67.2	110	12
	238	100	100	11
Bromobenzene	443	100	95	11
POX				
Bromoform	100		98	13
THMs[a]	140		80[b]	14
Chloroform			100	14

[a]THMs are the sum concentration of chloroform, bromodichloromethane, dibromochloromethane, and bromoform.
[b]Vitrified combustion tube insert.

Table II. Recovery for Various Purgeable Organic Halide (POX) Compounds Spiked at 1000 μg/L into Reagent Water – Dohrmann DX-20 System[a]

Compound	Recovery[b] (%)
Methylene chloride	87 ± 4
Chloroform	81 ± 3
Trans-1,2-Dichloroethylene	106 ± 0.2
1,1-Dichloroethane	84 ± 3
1,1-Dichloroethylene	78 ± 2
1,1,2,2-Tetrachloroethane	88 ± 6
Tetrachloroethylene	86 ± 0.6
Carbon tetrachloride	86 ± 1
1,1,2-Trichloroethane	76 ± 8
1,2-Dichloropropane	76 ± 2
Trichlorofluoromethane	79 ± 7
Trichloroethylene	76 ± 2
1,1,1-Trichloroethane	68 ± 10
1,2-Dichloroethane	70 ± 4
1,3-Dichloropropene	60 ± 4
Chlorobenzene	48 ± 7
1,2-Dichlorobenzene	65 ± 5
1,3-Dichlorobenzene	51 ± 7
1,4-Dichlorobenzene	61 ± 7
Bromodichloromethane	47 ± 9
Bromoform	64 ± 8

[a]From Ref. 15.
[b]With relative standard deviation.

Table III. Average Recovery from 7 POX Analyses of Selected Compounds[a]

Compound	Spike Level		Amount of POX Found (μg/L as Cl$^-$)		Recovery (%)
	μg/L	μg/L as Cl$^-$	Average	SD[b]	
Chloroform	12.5	11	11	1.4	100
Trichloroethene	14	10	6	0.7	60
Tetrachloroethene	14	10	5	1.0	50
Chlorobenzene	25	8	3	0.6	38
Bromoform[c]	30	13	ND[d]		<10

[a]From Ref. 15.
[b]Standard deviation.
[c]Recoveries for bromoform at 300 μg/L (130 μg/L as Cl$^-$) were 48 and 49% for duplicate analyses.
[d]Not detected.

In-House Recovery Study

To add to this data base, eight compounds were evaluated by TOX and POX. The VOCs chosen were bromoform, carbon tetrachloride, chlorobenzene, chloroform, p-dichlorobenzene, t-dichloroethylene, tetrachloroethylene, and trichloroethylene. Each compound was dosed into a separate groundwater or reagent water. Three concentrations of each compound (10, 30, and 100 μg/L) were examined, since these values were generally within the range of those reported in contaminated groundwater.[2] The dosed waters were submitted to TOX, POX, and P&T/GC/ElCD analysis. The results of the analyses are given in Table IV.

The VOCs at 10 μg/L concentrations gave a recovery range of 64 to 140% for TOX and 25 to 88% for POX. The precision of these measurements was, in most cases, greater than 10% relative standard deviation (RSD). This was not unexpected since the detection limit of the method is 2 to 5 μg/L as Cl$^-$. Therefore, at 10 μg/L of a VOC, where most of the variability was observed, organic halide concentration was 3 to 9 μg/L of Cl$^-$. As the concentration of VOCs increased, the percent RSD decreased so that only 1 of 16 sets at 100 μg/L afforded a percent RSD > 10.

At 30 μg/L of VOC, the recoveries varied from 63 to 100% for TOX and 42 to 76% for POX analyses. Precision at this concentration improved with recoveries for only three compounds >10% (RSD), all via POX. A comparison of TOX and POX revealed that TOX achieved a more efficient recovery. This trend was also observed at 100 μg/L concentrations. Recoveries of the eight chlorinated compounds at 100 μg/L were 60 to 120% using TOX and 59 to 91% using POX. The precision again improved, with only one compound >10% RSD.

Table IV. Average Accuracy of TOX, POX, and Purge-and-Trap Analyses for VOC-Dosed Groundwater and Reagent Water Samples[a]

Compound[b]	Calculated Concentration		TOX Recovery (%)	POX Recovery (%)	P&T Confirmation Analysis, Percent Recovery
	µg/L	µg/L as Cl⁻			
Bromoform, GW	10.0	4.2	140[c]	62[c]	120
	30.5	12.6	93	66	100
	100	42	120[c]	91	110
Carbon tetrachloride, RW	10.0	9.2	64[c]	25[c]	120
	30.1	27.7	72	43	98
	101	92.9	72	65	110
Chlorobenzene, RW	10.0	3.10	110	50[c]	95
	29.9	9.27	100	46[c]	100
	101	31.3	94	67	100
Chloroform,[d] GW	10.0	8.90	79[c]	25[c]	94
	29.9	26.6	76	43[c]	110
	100	89.0	81	76	96
p-Dichlorobenzene, RW	10.0	4.8	68	71	110
	30.0	14.4	82	42	110
	101	48.5	82	64	99
t-Dichloroethylene, GW	10.1	7.37	84[c]	53	97
	30.1	22.0	63	55	90
	98.4	71.8	60	59	99
Tetrachloroethylene, GW	9.83	8.40	79[c]	88[c]	91
	30.2	25.8	75	70[c]	100
	101	86.4	78	70	110
Trichloroethylene, RW	9.96	8.07	69[c]	60[c]	110
	30.3	24.5	81	76[c]	100
	100	81.0	75	75	99

[a] Based on triplicate analyses for TOX, POX, and duplicate for P&T.
[b] GW denotes groundwater sample; RW denotes reagent water sample.
[c] Denotes precision was >10%; the ranges of these percent relative standard deviations were 15 to 61% at 10 µg/L; 22 to 25% at 30 µg/L; 27% at 100 µg/L.
[d] A repeat experiment at 29.8 µg/L of $CHCl_3$ gave a TOX recovery of 84% and a POX recovery of 61%.

Table V. Interlaboratory VOC Recovery Study

Compound Spike (μg/L)	Recovery[a] (%)			
	TOX		POX	
	TSD[b]	DWRD[c]	TSD	DWRD
$CHCl_3$ (50)	81 (3.7)	90 (4.9)	77 (5.1)	56 (6.2)
$CHCl_3$ (150)	77 (1.0)	73 (9.7)	70 (1.6)	66 (4.2)
$CHBr_3$ (150)	85 (1.3)	87 (14)	60 (1.5)	58 (11)
t-DCE[d] (150)	76 (0.1)	91 (17)	73 (3.1)	61 (8.2)

[a]Relative standard deviation for triplicate measurements is given in parentheses.
[b]Technical Support Division (of EPA).
[c]Drinking Water Research Division (of EPA).
[d]trans-Dichloroethylene.

Within the variability of these data, there did not appear to be any relationship between TOX efficiency and concentration. However, for POX, there were some indications of a concentration dependency, with an average recovery of 55% at 30 μg/L and 71% at 100 μg/L. The mean recoveries for all of the VOCs at 30 and 100 μg/L concentrations were 79% (TOX) and 64% (POX). Analysis by P&T/GC/ElCD was also performed on the dosed samples. The samples were generally shown to be within 10% of their intended value.

While P&T/GC/ElCD can provide data to demonstrate the integrity of the dosed samples, a question still remains: Were these TOX and POX results unique to one laboratory? To answer this question an interlaboratory study was conducted. Dosed waters were split between the Technical Support Division of this laboratory and the EPA Drinking Water Research Division (DWRD). Reagent water was spiked with $CHCl_3$, $CHBr_3$, and t-DCE. Results from this study gave similar efficiencies (Table V). The TOX and POX recoveries ranged from 73 to 91% and 56 to 77%, respectively, for the three halo-VOCs. The results obtained from both laboratories generally agreed within 10% of each other. Again, there seemed to be a slight recovery bias in favor of TOX over POX.

Raw Groundwater

The VOC data presented thus far have predominantly been from spiked reagent water or a single source of groundwater. The ability of TOX and POX to measure the presence of VOCs in a variety of groundwaters, and the procedure for doing this, have not been fully investigated. To gain additional information, a limited survey was begun. Fifteen cities surveyed used a groundwater source. The TOX concentration of these raw waters ranged from <5 to >2000-μg Cl per liter (Table VI). The POX results ranged from <2 to 2300-μg

Table VI. Comparison of TOX, POX, and Purge-and-Trap VOC Concentrations (μg/L as Cl$^-$) for Selected Groundwaters

City	TOX[a]	POX[a]	VOCs via P&T
66951	21 ($\pm$21)	<2	ND[b]
66989	10 ($\pm$56)	<2	0.3 ($\pm$4)
66952	18 ($\pm$46)	<2	0.6[c]
66990	27 ($\pm$10)	21 ($\pm$5)	20 ($\pm$7)
66954	30 ($\pm$15)	<2	ND
66991	20 ($\pm$12)	<2	ND
66993	25 ($\pm$27)	<2	ND
66998	9 ($\pm$50)	<2	ND
66999	10 ($\pm$5)	2 ($\pm$50)	6 ($\pm$6)
67000	<5	<2	ND
840242	300 ($\pm$2)	300 ($\pm$2)	330[c]
840317	>2000	2300 ($\pm$4.9)	1800 ($\pm$5)
840318	80 ($\pm$50)	110 ($\pm$11)	75 ($\pm$5)
840319	37 ($\pm$4)	28 ($\pm$4)	52 ($\pm$4)
840320	9 ($\pm$56)	<2	ND

[a]Based on triplicate analyses for TOX and POX; duplicate for P&T; the percent relative standard deviation (or percent difference for P&T) is given in parentheses.
[b]Not detected.
[c]Denotes single analysis only.

Cl per liter. P&T/GC/ElCD analysis again was used as a reference method to compare the TOX and POX concentrations with those of individual VOCs. This comparison indicated that POX gave a positive result in 6 of the 15 samples only when VOCs were present. The two contaminated waters not identified by POX (66989 and 66952) contained volatile organohalide concentrations below the POX detection limit of 2 μg Cl per liter. TOX analysis, on the other hand, gave positive results in all but 1 of 15 samples. Six of the positive TOX samples did not contain any detectable VOCs. The average TOX value obtained for the six false positives was 19 μg Cl$^-$ per liter.

Information collected in a previous EPA survey also indicated that a non-VOC background could interfere with the use of TOX as a surrogate for low concentrations of VOCs.[16] The survey found an average TOX of 19 μg/L as Cl$^-$ in raw groundwater, with a range of <5 to 85 μg/L as Cl$^-$. Less than 1 to less than 40% of the TOX was accounted for by the presence of VOCs. A high TOX, 85 μg/L as Cl$^-$, did not indicate a high VOC concentration, since <1% could be accounted for by halogenated VOC in this sample. Surface waters also seem to possess a variable TOX background, ranging from <5 to 49 μg/L as Cl$^-$. Less than 1% of the TOX was accounted for by VOCs in all samples. In both groundwater and surface waters, TOX analysis would have to detect VOC concentrations above background interference, the nature and variability of which is yet undefined.

One possible explanation for background interference could be the presence of inorganic chloride from water that was not removed by the nitrate wash used in the analytical method. To test this possibility, seven groundwater samples (selected from the cities in Table VI) were analyzed for inorganic chloride. The concentration of the inorganic halide was then compared with the nonvolatile TOX background. The values of the nonvolatile TOX were estimated by subtracting the POX concentration in the sample from the TOX concentration. When a least-squares regression was performed, the resulting correlation coefficient ($r^2 = 0.18$) demonstrated that no relationship existed between the two parameters.

CONCLUSIONS

From the data generated in this study, we make the following conclusions:
1. The average costs of TOX and POX analysis (in duplicate) were $99 and $77, respectively. Although these costs are competitive with chromatographic analytical techniques, the relative lack of qualitative and quantitative information regarding compounds and their specific concentrations may limit the usefulness of TOX and POX, for example, to unit process control or screening.
2. The average accuracy of TOX for analysis of halo-VOCs (30 and 100 µg/L) was 79%, whereas POX gave an average recovery of 64%.
3. Based on the limited data available, POX appears to be a more reliable technique in screening uncharacterized groundwaters for halo-VOCs at low concentrations. POX gave no false positives. In turn, TOX gave false positives for six of seven groundwaters, with no quantifiable VOCs. The precise nature of the TOX background has yet to be determined; however, inorganic chloride does not appear to be a dominant factor.

ACKNOWLEDGMENTS

We would like to express our appreciation to Mr. Ronald Dressman and Mr. Gilbert Contner of the EPA Drinking Water Research Division, Cincinnati, for their participation in the interlaboratory study. We would also like to thank Dorothy Davis and Cynthia Bultman for the preparation of this manuscript.

REFERENCES

1. USEPA. "National Revised Primary Drinking Water Regulations, Volatile Synthetic Organic Chemicals in Drinking Water; Advance Notice of Proposed Rulemaking," *Fed. Reg.* 47:43:9352 (1982).
2. Westrick, J. J., J. W. Mello, and R. F. Thomas, "The Groundwater Supply Survey," *J. Am. Water Works Assoc.* 76(5):52–59 (1984).

3. Kuehn, W., and H. Sontheimer, Eds., *Oxidationsverfahren in der Trinkwaseraufbereitung (Oxidation Procedures in Drinking Water Treatment)*, (Frankfurt: ZfGW Publishers, 1979).
4. USEPA. "Hazardous Waste Management System, Part VII; Standards and Interim Status Standards for Owners and Operators of Hazardous Waste Treatment, Storage, and Disposal Facilities," *Fed. Reg.*, 45:98:33239 (1980).
5. *Standard Methods for the Examination of Water and Wastewater*, 15th ed. (Washington, DC: American Public Health Association, 1980), p. 292.
6. Billets, S., and J. J. Lichtenberg. *Total Organic Halide, Method 450.1 — Interim*, EPA 600/4-81-056, (Cincinnati: U.S. Environmental Protection Agency, 1981).
7. Bellar, T. A., and J. J. Lichtenberg. *The Determination of Halogenated Chemicals by the Purge and Trap Method — Method 502.1*, EPA 600/4-81-059 (Cincinnati: U.S. Environmental Protection Agency, 1981).
8. Bellar, T. A., and J. J. Lichtenberg, *The Analysis of Aromatic Chemicals in Water by the Purge and Trap Method*, EPA 600/4-81-057 (Cincinnati: U.S. Environmental Protection Agency, 1981).
9. *Dohrmann DX-20 Total Organic Halide Analyzer System — Equipment Manual*, (Santa Clara, CA: Xertex Corp., 1979).
10. Dohrmann Division, Xertex Corporation, Santa Clara, CA; personal communication.
11. Takahashi, Y., and R. Moore, "Measurement of Total Organic Halides (TOX) in Water by Carbon Adsorption/Microcoulometric Determination," presented at the 177th National Meeting of the American Chemical Society, Div. Environ. Chem., Honolulu (1979).
12. Dressman, R., B. Najar, and R. Redzikowski, "The Analysis of Organohalides (OX) in Water as a Group Parameter," Proceedings of the Seventh American Water Works Association, Water Quality Technology Conference, Philadelphia (1979).
13. Takahashi, Y., R. Moore, and R. Joyce, "Measurement of Total Organic Specific Concentrations May Limit the Usefulness of TOX and POX Halides in Water Using Carbon Adsorption and Microcoulometric Determination," Presented at the 179th Meeting of the American Chemical Society, Div. of Envir. Chem., Houston (1980).
14. Dressman, R. and A. Stevens, "The Analysis of Organohalides in Water — An Evaluation Update," *J. Am. Water Works Assoc.* 75(8):431 (1983).
15. Riggin, R., et al. *Development and Evaluation of Methods for Total Organic Halide and Purgeable Organic Halide in Wastewater*, EPA-600/54-84-008 (Cincinnati: U.S. Environmental Protection Agency, 1984).
16. Preliminary Report — "Community Water Supply Survey — Resample" (Cincinnati: Drinking Water Quality Assessment Branch, TSD, USEPA, 1981).

SECTION XIV

Drinking Water Treatment

Probably no other public health issue affects a larger proportion of the U.S. population than that of drinking water disinfection.

Richard J. Bull
Health Risks of Drinking Water Disinfectants and Disinfection By-Products,
1981

The objective of disinfection treatment is attainment of the maximum control of biological contaminant while introducing the minimum possible amounts of potentially toxic chemical by-products.

Joseph A. Cotruvo
Regulatory Aspects of Disinfection, 1984

CHAPTER 91

Influence of Water Treatment Processes on the Formation of Organic Halogens and Mutagenic Activity by Postchlorination

J. C. Kruithof, A. Noordsij, L. M. Puijker, and M. A. van der Gaag

In the years following the first discovery of trihalomethane (THM) production during drinking water chlorination,[1,2] the policy of the Dutch Water Works has been to reduce the side effects of chlorination as much as possible. Initially much attention was paid to minimizing the formation of THM. It was thought that a reduction of the THM content should be obtained by lowering the dose of chlorine, especially in the first stages of the purification, and by removal of THM precursors prior to chlorination. Postchlorination should be maintained as a final hygienic barrier.[3]

Along these lines the Netherlands Water Works Testing and Research Institute, KIWA Ltd., together with the Dutch Water Works, carried out many investigations. These showed that under Dutch drinking water practice, removal of THM precursors prior to postchlorination did not give a substantial reduction of the THM content. Chlorination of carbon filter effluents gave a shift to the production of brominated THM, especially at short running times.[4,5] Relatively high carbon-adsorbable organohalogen concentrations were found after postchlorination with a low chlorine dose. Further study was therefore focussed on the postchlorination under Dutch drinking water practice.[6]

Recently, KIWA has developed an integrated methodology for the assessment of water quality, with special reference to health effects. The basic approach is that analytical chemical measurements (GC-MS, surrogates) and toxicological bioassays (short-term assays for mutagenicity) are performed on the same isolate of a water sample.[7-10] An increase of XAD-adsorbable organic halogens, which cannot be identified by GC-MS (molecular weight >350), was observed after chlorination together with increased mutagenic activity.[8,9] A rapid breakthrough of GAC filters by organic material leading to the formation of mutagenic organohalogens by chlorination was also observed.

Based on these preliminary results, KIWA set up a systematic pilot plant investigation to determine the influence of preceding treatment processes on the side effects of a postchlorination. In this chapter the results of the first year are presented. The relationship among chemical and biological parameters will be discussed, focussing on their usefulness for the assessment of water quality. Preliminary conclusions will be formulated about the influence of a post-chlorination with respect to previous purification steps.

EXPERIMENTAL

The pilot plant investigation started in February 1983. In the first year about 85 samples were isolated and analyzed for THM, surrogates (purgeable, extractable, XAD-adsorbable and carbon-adsorbable organohalogen, nitrogen, phosphorus, and sulfur), and the *Salmonella*/microsome assay (Ames test). The experiments were still in progress in 1984 and will continue to focus principally on the adsorption process.

Purification System

At Nieuwegein, water is taken from the river Rhine and treated according to the scheme presented in Figure 1.

Coagulation is carried out with iron(III) chloride (7 mg/L Fe) at pH 7.8 to

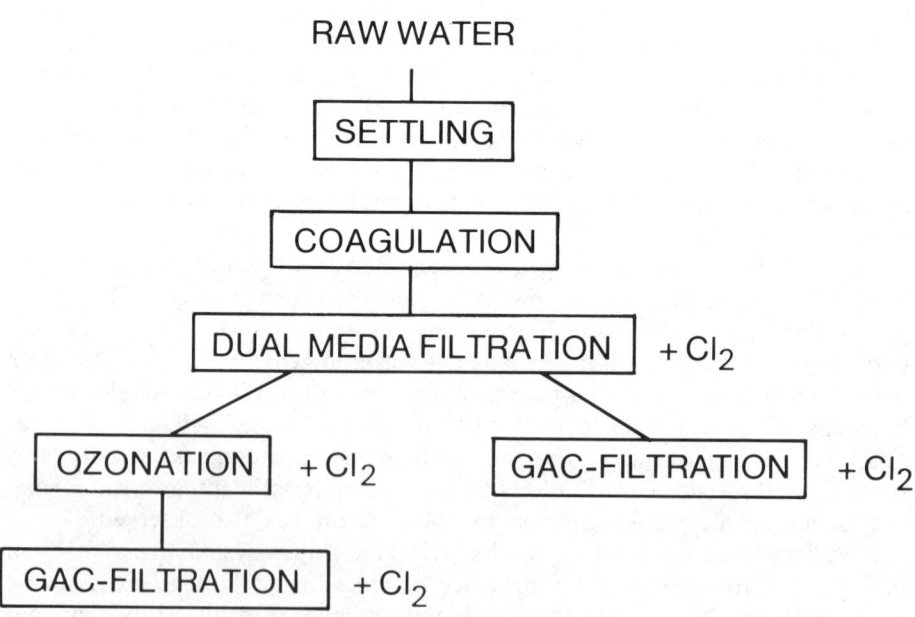

Figure 1. Schematic design of the purification system.

8.0 with a flocculation time of about 20 min. The flocs are removed in a plate separator with a surface load of 1.0 m/h and a residence time of 3 min. Rapid filtration is carried out in a dual-media filter consisting of 0.45-m anthracite and 0.55-m sand with a filtration rate of 5.2 m/h. After rapid filtration, part of the water is ozonated (ozone dose, 2.7 mg/L) with an initial contact time in a bubble column of 7.0 min, followed by a residence time of 1.3 h in a contact chamber.

The rapid filtrate as well as the ozonated water are carbon filtered (Norit ROW 0.8 S). The bed length of the carbon is 1.0 m and the empty bed contact time is 20 min. Finally, the rapid filtrate, the ozonated water, and both GAC filtrates are postchlorinated according to a criterion of 0.2 mg/L free chlorine after a contact time of 20 min.

Isolation and Sample Treatment

The determination of surrogates (organohalogen, OX; organophosphorus, OP; organonitrogen, ON; and organosulfur, OS), GC/MS investigations, and the Ames test cannot be carried out directly in the water because the concentration of organics is generally too low. Noordsij et al. have developed an isolation and concentration technique that allows chemical and biological measurements to be carried out in the same concentrate.[7] The organic material is adsorbed sequentially on two XAD columns at pH 7 and 2. Each fraction is eluted and concentrated in ethanol. Organohalogen (XOX), –phosphorus (XOP), –nitrogen (XON), and –sulfur (XOS) contents are measured in the eluate.

The following techniques were also used to isolate organic micropollutants from the same water samples.

Adsorption on Tenax After Purging

Water samples (100 mL) are purged with nitrogen at 40 mL/min for 15 min at 95°C. After passing a condenser, the purged material is adsorbed on a Tenax column and used for organohalide measurements (POX).

Extraction with Petroleum Ether

Water samples (1 L) are shaken with two proportions of 100-mL petroleum ether. The combined extracts are dried with sodium sulfate and concentrated to 1 mL in a Kuderna-Danish apparatus and finally with a gentle stream of nitrogen. Organohalogen (EOX), –phosphorus (EOP), –nitrogen (EON), and –sulfur (EOS) are measured in the extracts.

Activated Carbon

Water samples (50 mL) are purged with nitrogen to remove volatile compounds and thoroughly mixed with 5 mL of a solution containing 10 mg carbon, sodium nitrate, and nitric acid. Subsequently, the sample is filtered over a polycarbonate filter. Carbon and filter are washed with a sodium nitrate solution and used for organohalide measurements (AOX).

Principles of Measurements

Organohalide is measured with a microcoulometric titration system after pyrolysis of the sample at 850 to 1000°C in an oxygen atmosphere. For POX determination the Tenax column is heated to 250°C, then desorbed and led into the oven with a stream of argon. The organic solvents for the EOX determination are injected directly in a hot zone (500°C). The X_7OX, X_2OX and AOX samples are manually introduced into the oven tube using a boat.

Organophosphorus and organosulfur (EOP, EOS, X_7 + X_2OP, and X_7 + X_2OS) are measured after conversion of the organic compounds into phosphine and hydrogen sulfide in a hydrogen atmosphere at 1100°C. The reaction products are collected in a cold trap, separated, and detected with an on-line flame photometer.

Organonitrogen (EON and X_7ON) is measured after conversion into ammonia in a hydrogen atmosphere at 900°C using nickel as the catalyst. Ammonia is detected on-line in a microcoulometric set-point titration system. Total organic carbon (POC and DOC) is measured after complete photooxidation with UV irradiation. The CO_2 thus formed is reduced to CH_4 and detected with a flame ionization detector.

The total trihalomethanes (TTHM) in aqueous samples are measured with a head-space/GC/ECD technique using a capillary column (WSCOT, 25 m, phase CP-sil 5). Ultraviolet absorption spectrophotometry (254 nm) is used to monitor the organic matter content of the water.

The Ames test with strains TA98 and TA100 is carried out according to Maron and Ames[11] with slight modifications.[12] The liver homogenate (S9) is prepared by the CIVO-TNO Institute (Zeist, The Netherlands) from Aroclor-induced Sprague Dawley rats. In the assays with S9 mix, 0.5 mL S9 mix containing 0.075 mL S9 is added to the top agar. The XAD samples are tested at six dose levels, ranging from 10- to 140-μL ethanol per plate, depending on the type of water tested. In all samples, organic substances from 1 L of water are concentrated to 25 μL ethanol.

RESULTS

Dual-Media Filtrate

The measurements carried out in dual-media filtrate are summarized in Tables I and II. In the untreated rapid filtrate, the organic halide content was rather low. The highest values have been found for AOX (25 µg/L). The mutagenic activity of the filtrate was almost identical to that of Rhine water. It was highest in strain TA 98.

As expected, chlorination caused an increase of all organohalide parameters. The average TTHM production was 24 µg/L with about 6 µg/L for each THM. During the summer period a shift toward more brominated THM was observed (Figure 2). POX followed the TTHM content. The average rise in EOX-content was less than 2.5 µg/L. Relatively high values were found for AOX and XOX. Chlorination did not change the values for organonitrogen, -sulfur, and -phosphorus. Chlorination caused an increase of the number of revertants for TA98 and TA100. The mutagenic activity in TA100 without S9 was eight to nine times higher for both pH fractions. In contrast to the unchlorinated dual-media filtrate, the mutagenic effect in TA100 was reduced by S9 in both fractions.

Ozonate

Measurements carried out in the ozonate are given in Tables III and IV. Ozonation gave only a small reduction of the organic halide content. The average mutagenic activity was reduced, especially in the presence of S9.

Once again chlorination caused a rise in the organic halide content. The average TTHM formation was 12 µg/L with the highest concentration of $CHBr_3$ (7 µg/L). The EOX formation was very moderate. The AOX and XOX production were surprisingly low. Chlorination did not change the organonitrogen, -sulfur, and -phosphorus content.

After chlorination of the ozonate, the number of revertants in TA100 was two- to threefold higher than before. Chlorination had little or no effect on the mutagenicity in TA98.

GAC-Effluents

The most important results for both GAC-effluents are graphically presented as a function of the running time of the filters for the organohalides as well as the mutagenicity assays (Figures 3 through 11). The results for organonitrogen, -sulfur, and -phosphorus are given in Table V.

For the GAC filtrate not preceded by ozonation, THM formation started after about 2800 bed volumes (bv) and increased sharply to about 17,000 bv.

Table I. Organohalide and Ames Test Water Quality Data for Unchlorinated and Chlorinated Dual-Media Filtrate.

Parameter	Dual-media filtrate			Chlorinated dual-media filtrate			Δ Ave
	Min	Max	Ave	Min	Max	Ave	
UV, m^{-1}	5.6	9.2	6.7	4.6	7.4	5.7	−1.0
TOC, mg/L	2.3	3.1	2.8	2.1	3.3	2.8	0
TTHM, µg/L	0.2	1.0	0.6	12.8	55.8	24.7	24.1
CHCl$_3$, µg/L	0.2	1.0	0.6	2.2	12.0	6.9	6.3
CHBrCl$_2$, µg/L	<0.1	<0.1	<0.1	2.1	11.0	6.1	6.1
CHBr$_2$Cl, µg/L	<0.1	<0.1	<0.1	1.2	17.8	6.3	6.3
CHBr$_3$, µg/L	<0.5	<0.5	<0.5	<0.5	22.0	5.6	5.6
POX, µg Cl/L	1.1	4.3	2.9	9.8	23.0	15.1	12.2
EOX, µg Cl/L	0.5	4.8	1.8	2.4	7.3	4.3	2.5
AOX, µg Cl/L	16	39	25	25	104	68	43
X$_7$OX, µg Cl/L	3.5	11.0	6.9	16	61	32	25
X$_2$OX, µg Cl/L	2.9	9.8	6.5	29.5	52	37	31
TA98[a]							
pH 7 − S9	170	460	290	395	800	550	+[b]
pH 2 − S9	45	190	100	130	225	190	+[b]
pH 7 + S9	390	2000	900	440	1500	790	−[b]
pH 2 + S9	85	205	130	65	140	100	−/0[b]
TA100[a]							
pH 7 − S9	75	205	170	670	1800	1240	++[b]
pH 2 − S9	40	130	90	570	1300	880	++[b]
pH 7 + S9	110	505	270	240	905	460	+[b]
pH 2 + S9	25	145	80	115	295	210	+[b]

[a] Induced revertants per 1.6 L equivalent.
[b] Increase, +; strong increase, ++; decrease, −; slight decrease to no effect, −/0.

Table II. Organonitrogen, Sulfur, and Phosphorus Water Quality Data for Unchlorinated and Chlorinated Dual-Media Filtrate

Parameter (μg/L)	Dual-media filtrate				Chlorinated dual-media filtrate			
	No.[a]	Min	Max	Ave	No.[a]	Min	Max	Ave
EON	2	1.1	1.7	1.4	1	0.8	0.8	0.8
X_7ON	5	18.5	25.0	19	2	19.5	26.5	23
X_2ON	0				0			
EOS	1	<1.0	<1.0	<1.0	1	<1.0	<1.0	<1.0
X_7OS	4	9	19	14	3	12	18	15
X_2OS	3	17	18	17	3	15	21	18
EOP	1	<0.1	<0.1	<0.1	1	<0.1	<0.1	<0.1
X_7OP	4	0.2	0.6	0.4	3	0.2	0.5	0.4
X_2OP	3	0.2	0.5	0.3	3	0.3	0.5	0.4

[a]Number of measurements.

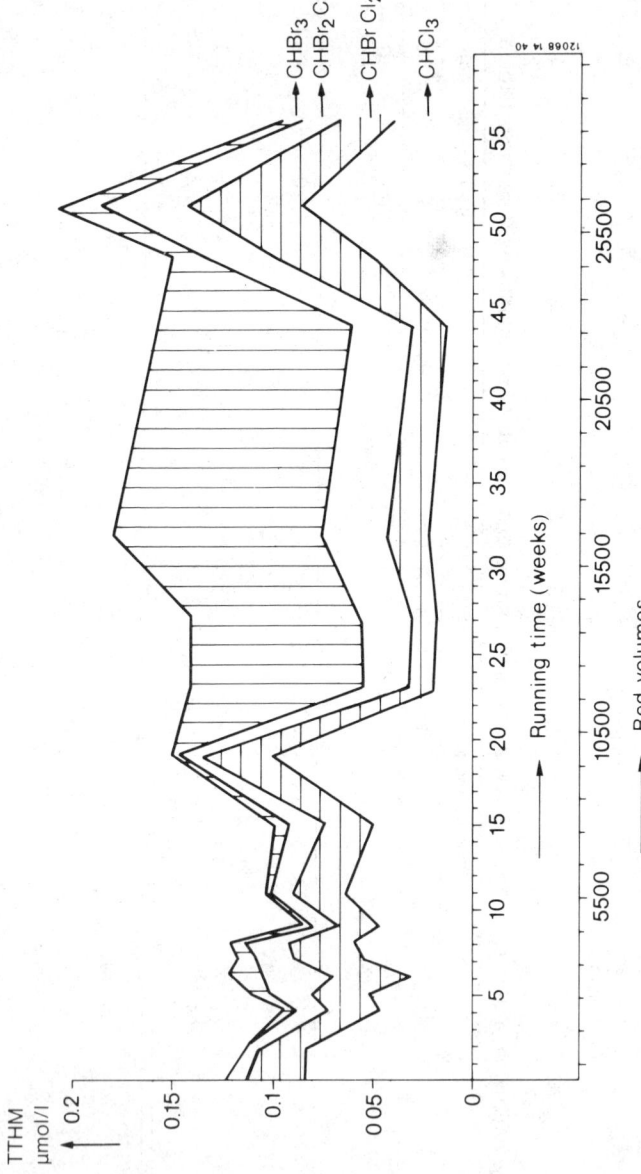

Figure 2. Formation of THM by chlorination of dual-media filtrate.

Table III. Organohalide and Ames Test Water Quality Data for Unchlorinated and Chlorinated Ozonate

Parameter	Ozonate			Chlorinated ozonate			
	Min	Max	Ave	Min	Max	Ave	△ Ave
UV, m^{-1}	2.2	7.3	3.1	2.2	3.5	2.6	-0.5
TOC, mg/L	2.3	3.0	2.7	2.1	2.9	2.6	-0.1
TTHM, μg/L	0.3	0.6	0.4	3.1	34.6	11.7	11.3
CHCl$_3$, μg/L	0.3	0.6	0.4	0.8	2.0	1.4	1.0
CHBrCl$_2$, μg/L	<0.1	<0.1	<0.1	0.2	1.9	0.9	0.9
CHBr$_2$Cl, μg/L	<0.1	<0.1	<0.1	0.6	7.7	2.7	2.7
CHBr$_3$, μg/L	<0.5	<0.5	<0.5	0.8	23	6.8	6.8
POX, μg Cl/L	0.9	2.1	1.3	2.9	5.7	4.7	3.4
EOX, μg Cl/L	0.3	1.5	1.0	0.7	3.2	1.8	0.8
AOX, μg Cl/L	10	18	14	8	24	16	2
X$_7$OX, μg Cl/L	2.0	6.3	3.7	4.8	9.7	7.4	3.7
X$_2$OX, μg Cl/L	2.6	8.5	4.8	6.8	12.5	9.5	4.7
TA98[a]							
pH 7 − S9	50	260	160	95	245	160	0[b]
pH 2 − S9	20	55	30	25	50	35	0[b]
pH 7 + S9	30	90	50	25	95	60	0[b]
pH 2 + S9	nm[c]	25	15	nm[c]	20	10	0[b]
TA100[a]							
pH 7 − S9	nm[c]	205	90	85	270	190	+[b]
pH 2 − S9	nm[c]	175	85	110	275	170	+[b]
pH 7 + S9	40	175	105	45	160	110	0[b]
pH 2 + S9	nm[c]	135	70	15	110	70	0[b]

[a] Induced revertants per 1.6 L equivalent.
[b] No change, 0; increase, +.
[c] Not measurable.

Table IV. Organonitrogen, Sulfur, and Phosphorus Water Quality Data for Unchlorinated and Chlorinated Ozonate

Parameter (μg/L)	Ozonate				Chlorinated ozonate			
	No.[a]	Min	Max	Ave	No.[a]	Min	Max	Ave
EON	2	<0.5	<0.9	<0.9	2	<0.5	<0.9	<0.9
X_7ON	2	11.5	12.0	12	1	10.5	10.5	10.5
X_2ON	0				0			
EOS	0				0			
X_7OS	2	2	2	2	1	2	2	2
X_2OS	2	5	5	5	1	5	5	5
EOP	0				0			
X_7OP	2	<0.1	0.1	<0.1	1	0.1	0.1	0.1
X_2OP	2	0.3	0.2	0.3	1	0.2	0.2	0.2

[a]Number of measurements.

Table V. Organonitrogen, Sulfur, and Phosphorus Water Quality Data for Unchlorinated and Chlorinated Granular Activated Carbon (GAC) Filtrates

Parameter (μg/L)	GAC filtrate			Chlorinated GAC filtrate			Ozonated/ GAC filtrate			Chlorinated–Ozonated/ GAC filtrate		
	No.[a]	Min	Max	No.[a]	Min	Max	No.[a]	Min	Max	No.[a]	Min	Max
EON	3	<9.0	<0.9	2	<0.9	<0.9	1		<0.5	1		<0.5
X$_1$ON	6	0.6	10.0	4	1.5	9.9	3	0.5	4.1	2	0.6	5.4
X$_2$ON	0			0			0			0		
EOS	1		<1.0	1		<1.0	1		<1.0	1		<1.0
X$_1$OS	5	<0.5	6	2	3	9	2	<0.5	1	3	<0.5	2
X$_2$OS	5	3	20	2	7	19	2	0.5	6	3	0.5	6
EOP	1		<0.1	1		<0.1	1		<0.1	1		<0.1
X$_1$OP	5	0.1	0.4	2	0.1	0.3	2	0.1	0.1	3	0.1	0.2
X$_2$OP	5	0.1	0.5	2	0.3	0.3	2	0.1	0.3	3	0.1	0.3

[a]Number of measurements.

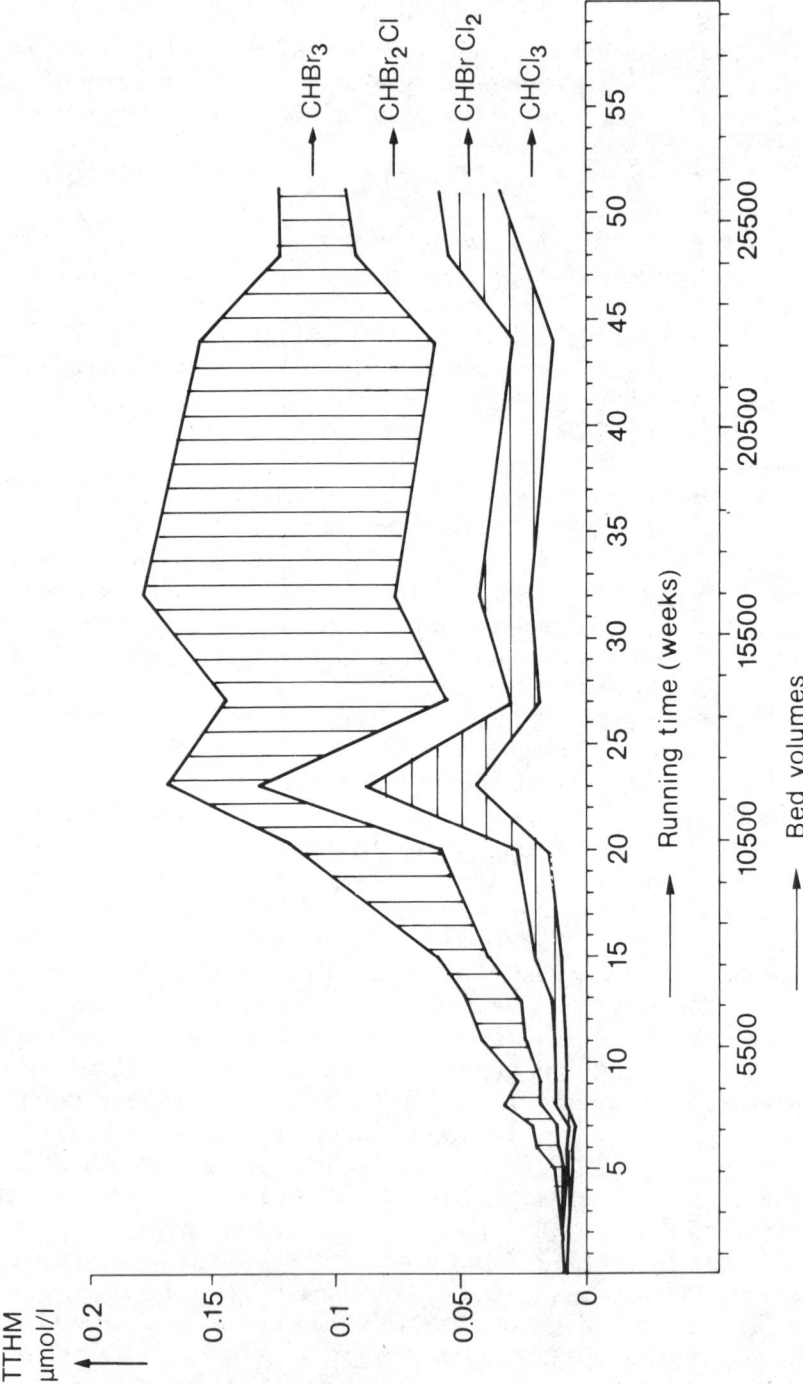

Figure 3. Formation of THM by chlorination of GAC filtrate (carbon filter influent: dual-media filtrate).

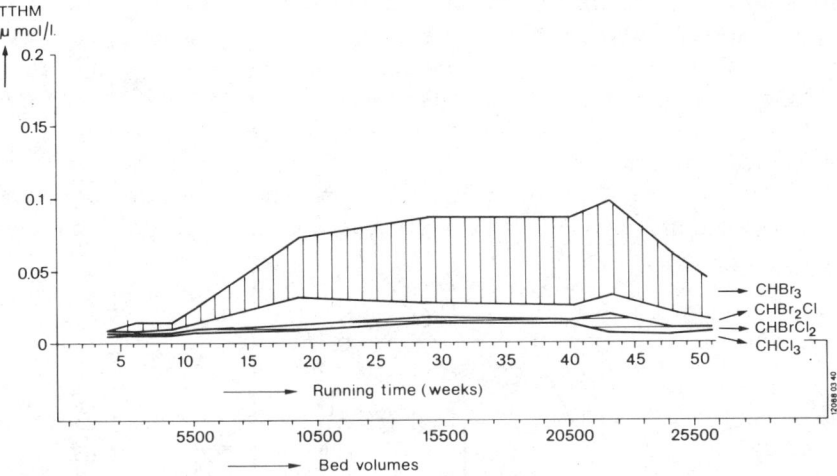

Figure 4. Formation of THM by chlorination of GAC filtrate (carbon filter influent: ozonate).

POX followed the TTHM concentration. Breakthrough of EOX was not very clear. Formation of EOX started after about 5000 bv and remained rather low. Breakthrough of AOX started after 10,000 bv, and formation upon chlorination started after about 3800 bv. XOX breakthrough started at about the same time as the AOX breakthrough. Directly after the start of the filter run, XOX formation upon chlorination took place, and the amount produced rose gradually.

No mutagenic activity was detected in the effluent of the GAC filter up to 23,000 bv. However, formation of mutagenic activity by chlorination was detected in an early phase of the filter run. Chlorination caused an increasing mutagenic activity in TA100 without S9 mix for the pH 2 fraction after 3000 bv and for the pH 7 fraction after 4000 bv. S9 inactivates the mutagenic effect of the pH 2 fraction more than that of the pH 7 fraction. For the effluents of the GAC filter fed with ozonated water, once again there was only a small breakthrough of $CHCl_3$. THM production upon chlorination started after 2300 bv and increased up to 9000 bv. The total production was lower than for GAC filtration only. POX followed the TTHM concentration. There was very little EOX breakthrough. EOX formation started after 6000 bv and remained constant at 1 μg/L. Up to about 10,000 bv no AOX was formed by chlorination. Even at longer filter runs the AOX formation was very moderate. This was also the case for the formation of XOX. For the combination of ozonation and GAC filtration no mutagenic activity was detected for filter runs up to 23,000 bv. Chlorination caused mutagenicity for strain TA100 without S9 mix. In the pH 2 fraction, this effect stabilized at a level of 200 induced revertants per 6 L equivalent after 6000 bv. In the pH 7 fraction the first mutagenicity

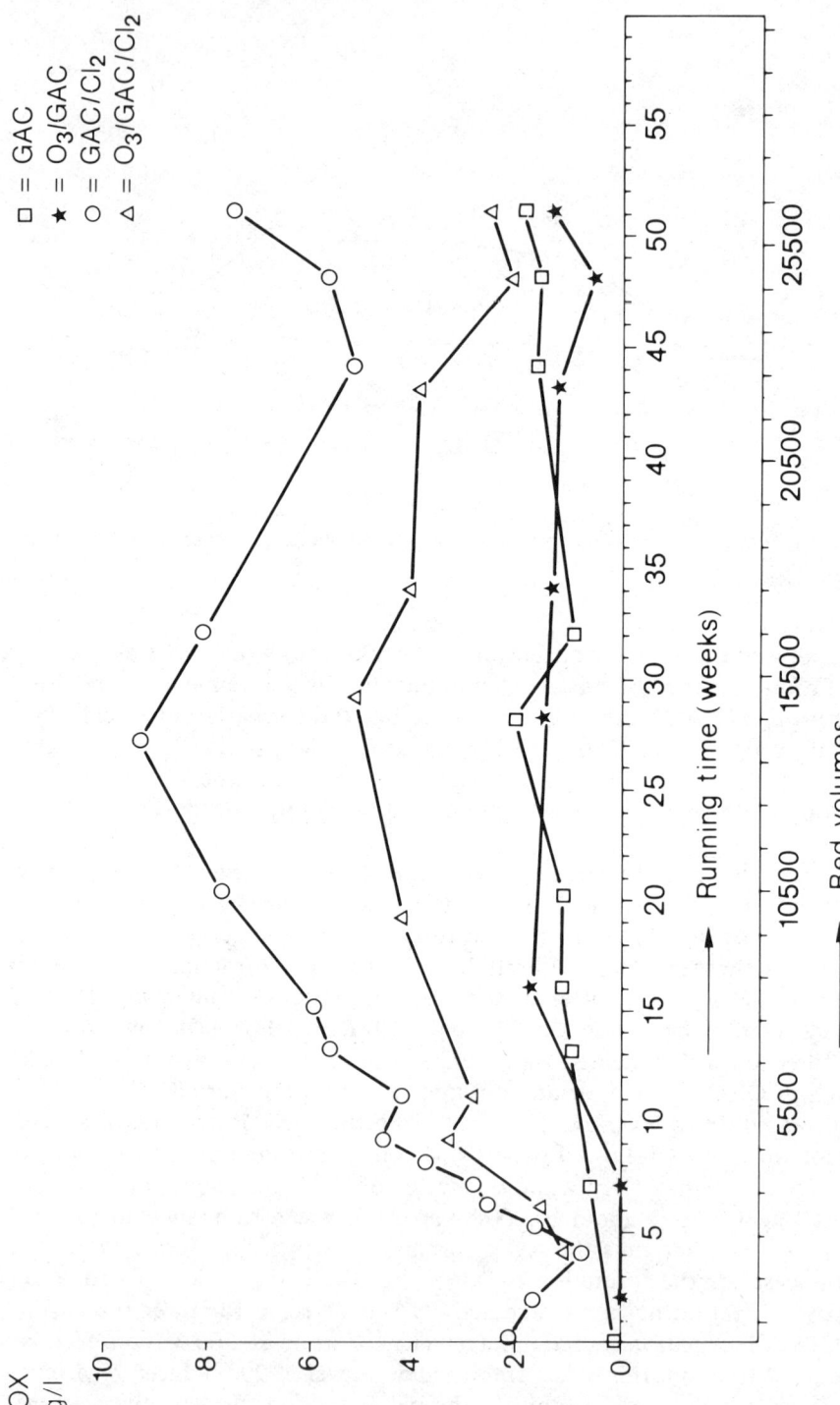

Figure 5. POX content of unchlorinated and chlorinated GAC effluents.

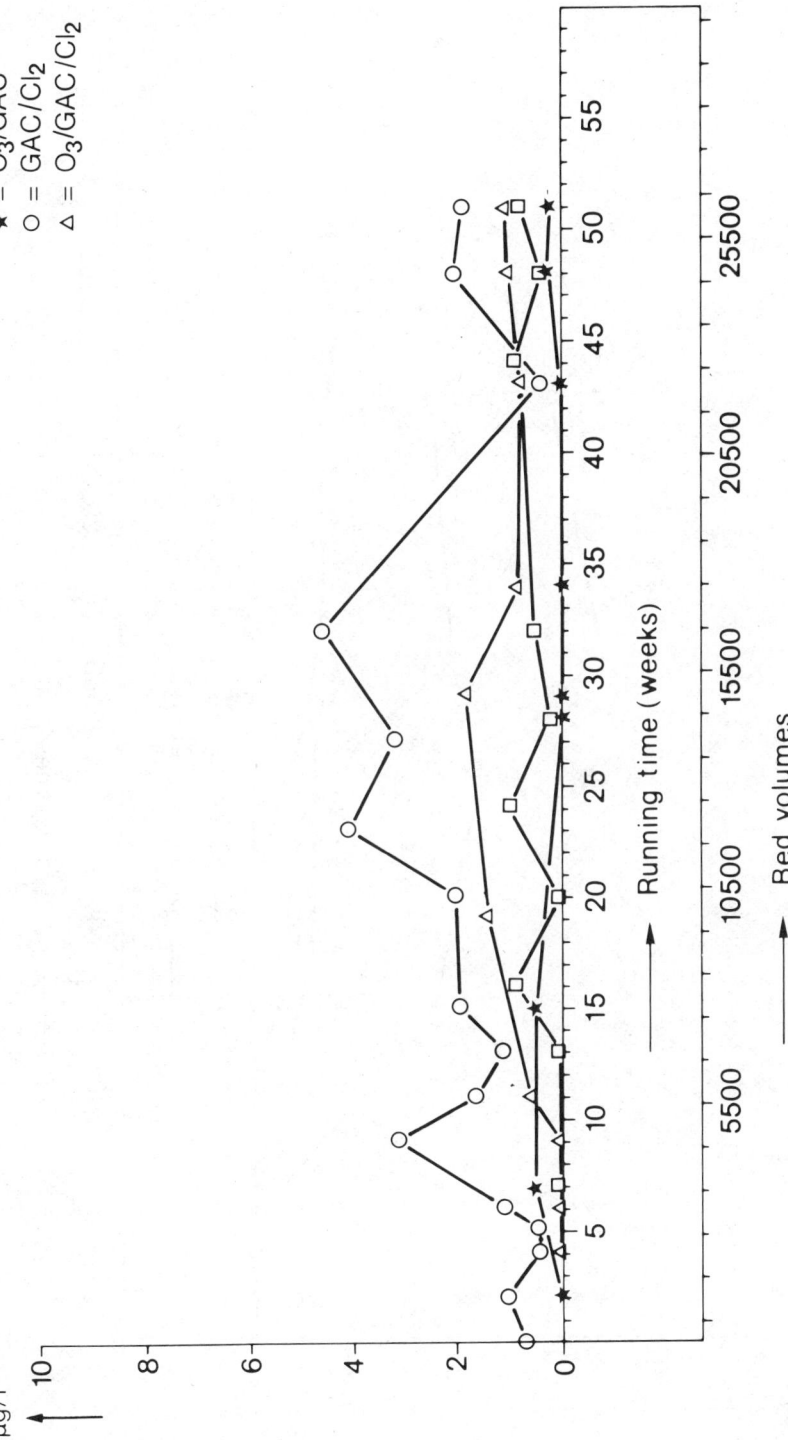

Figure 6. EOX content of unchlorinated and chlorinated GAC effluents.

1152 **WATER CHLORINATION**

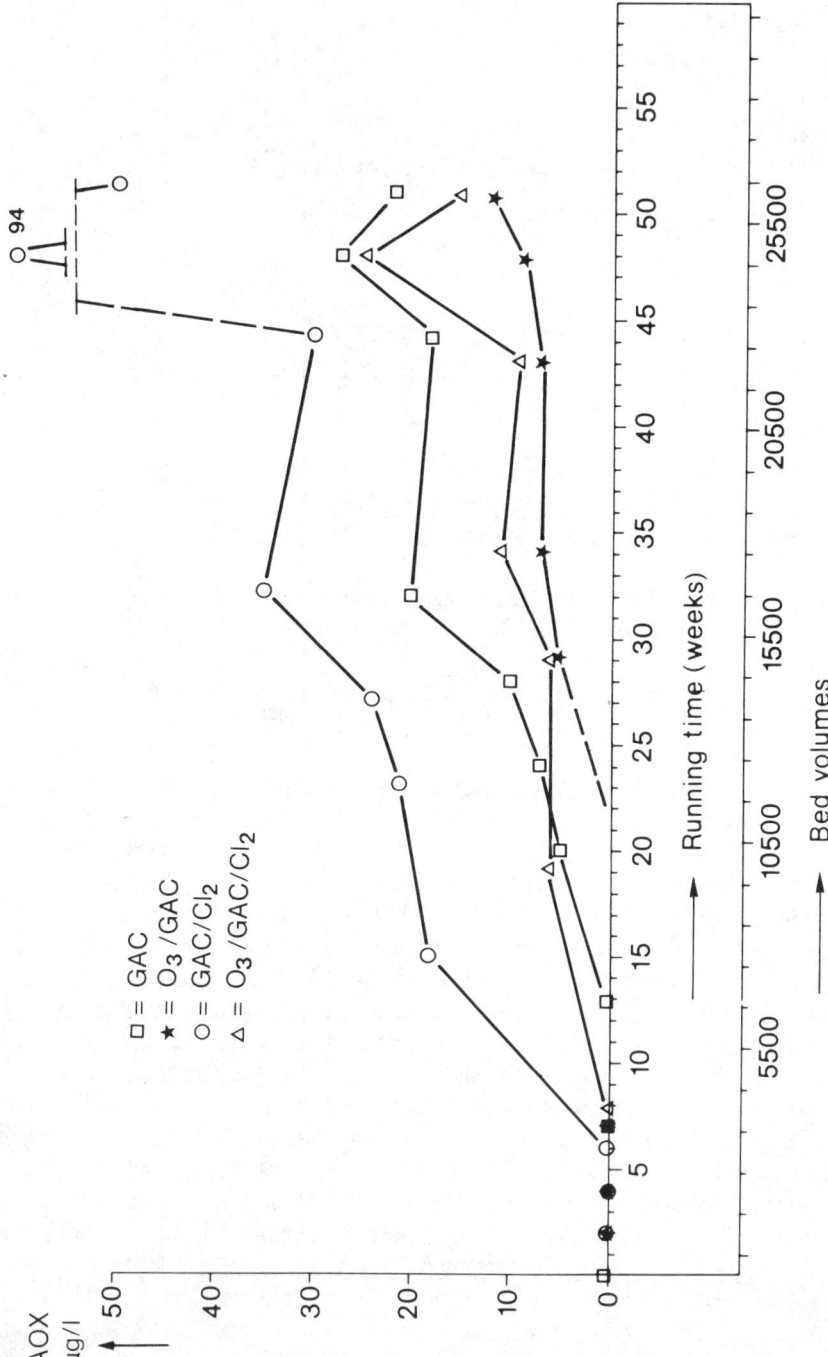

Figure 7. AOX content of unchlorinated and chlorinated GAC effluents.

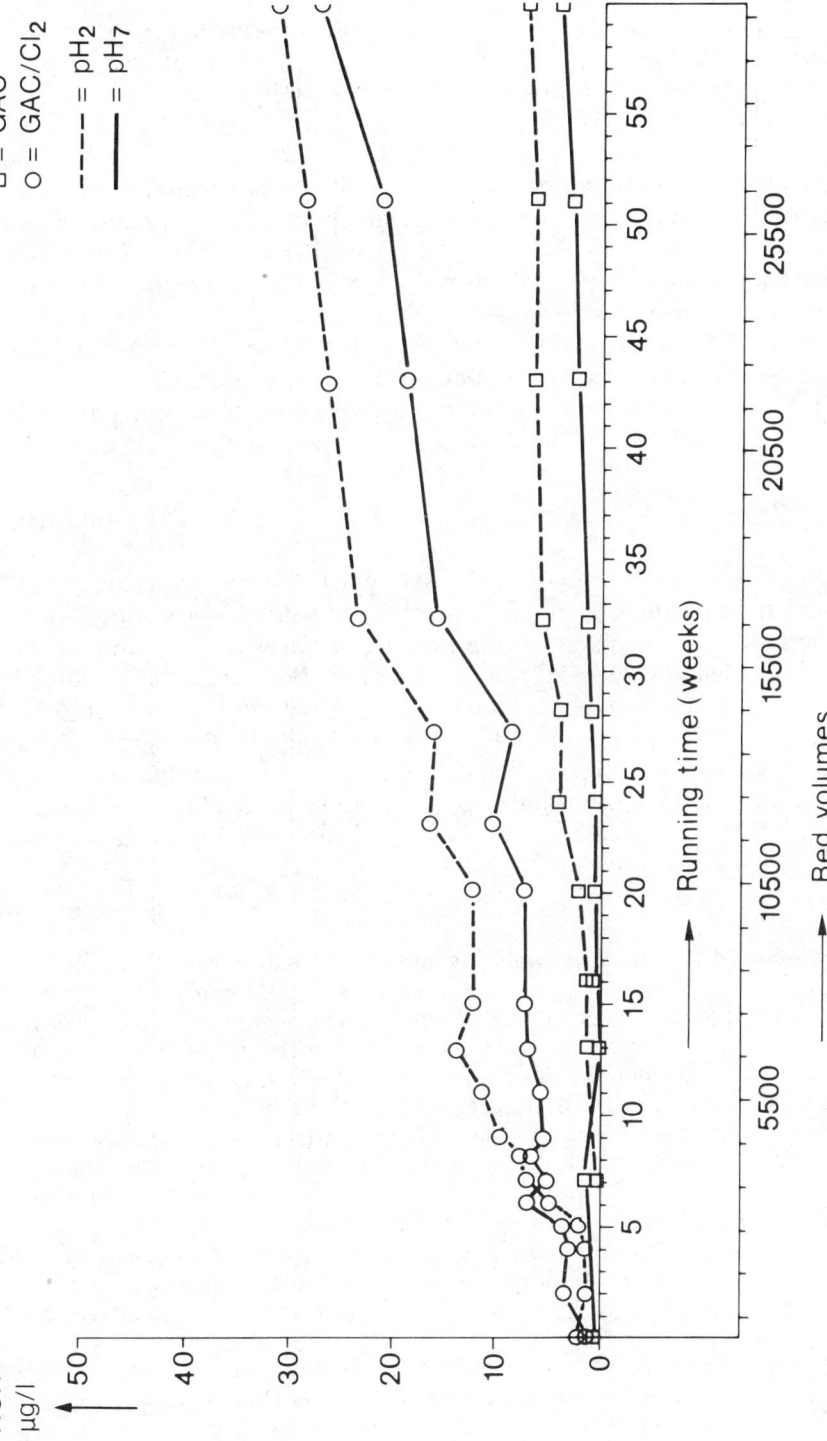

Figure 8. XOX content of unchlorinated and chlorinated GAC filtrate (carbon filter influent: dual-media filtrate).

1154 WATER CHLORINATION

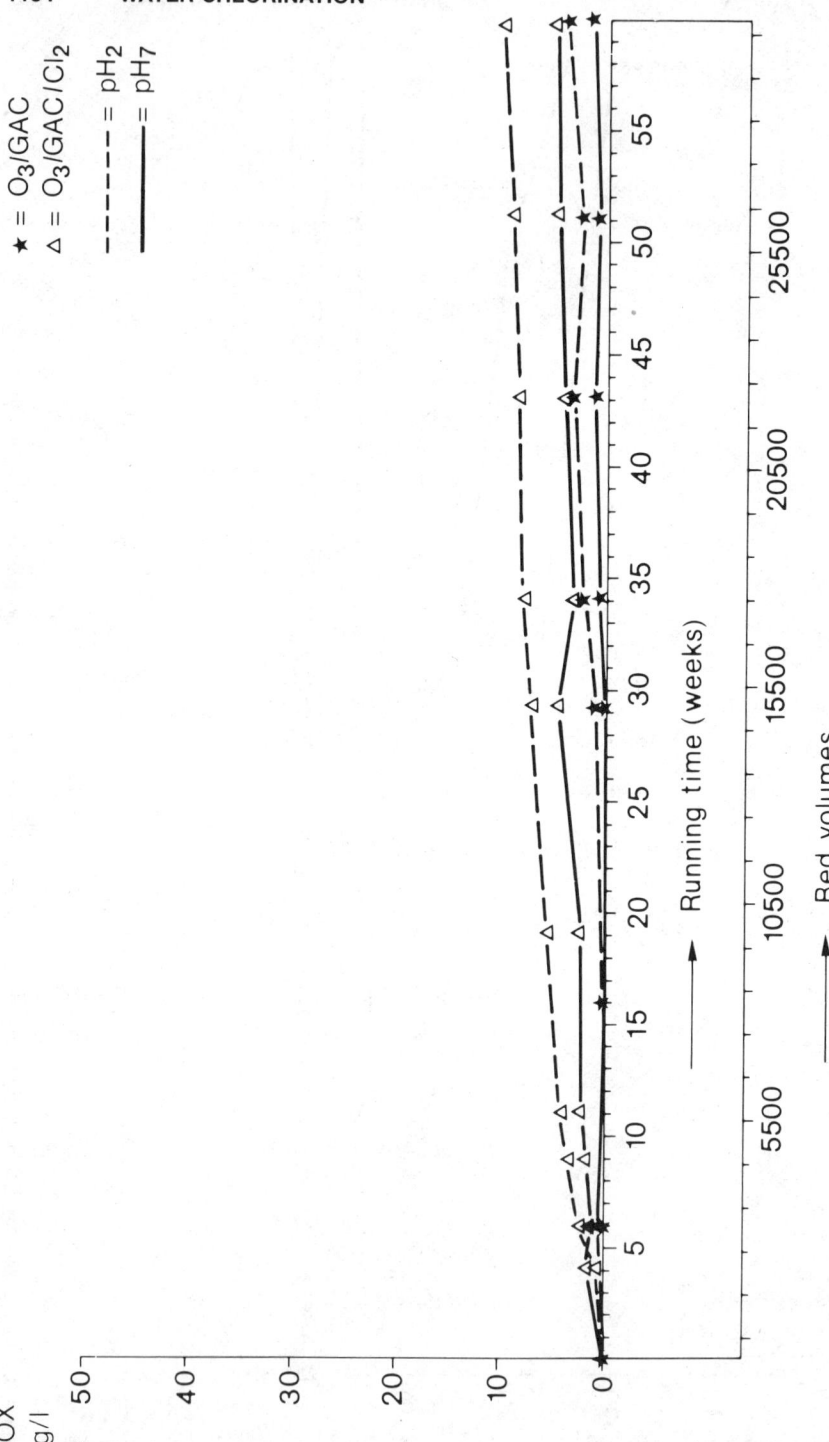

Figure 9. XOX content of unchlorinated and chlorinated GAC filtrate (carbon filter influent: ozonate).

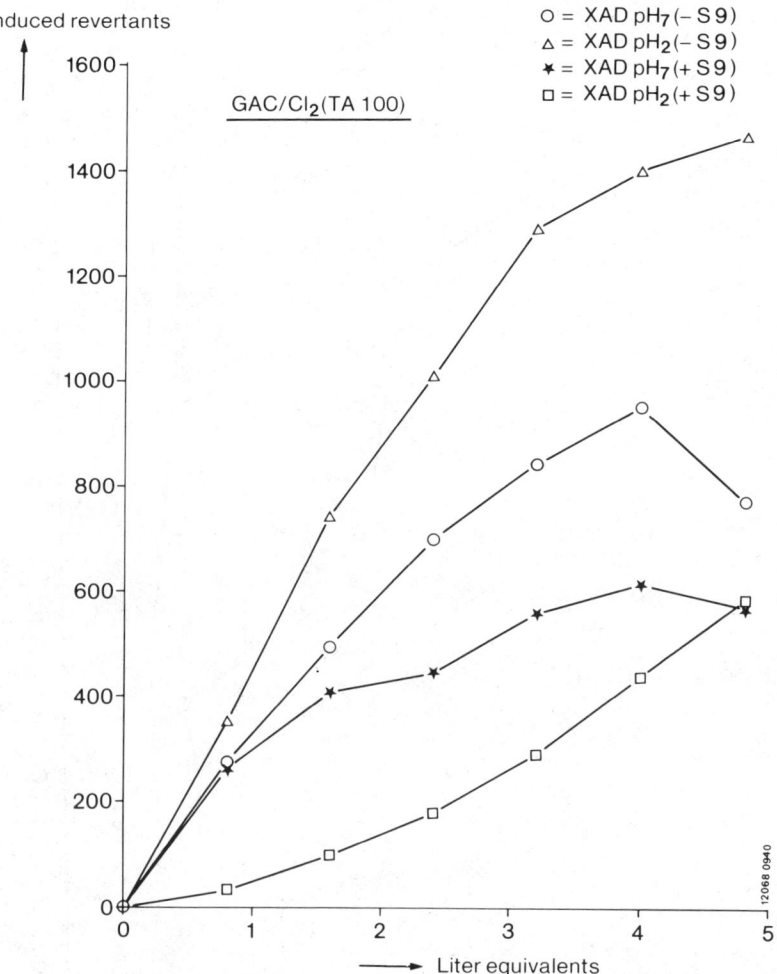

Figure 10. Induced revertants as a function of the number of liter equivalents of GAC effluent (carbon filter influent: dual-media filtrate). (Spontaneous revertants: 162 without S9-mix; 160 with S9.)

was detected after 18,000 bv reaching about 200 induced revertants per 6 L equivalent after 22,000 bv. The mutagenic effect for TA100 was inactivated by S9.

DISCUSSION

The results presented in this chapter are part of a project that includes the determination of carbon life with technological, chemical, analytical, toxicological, and microbiological parameters. Only aspects directly related to chlorination are discussed in this context.

1156 WATER CHLORINATION

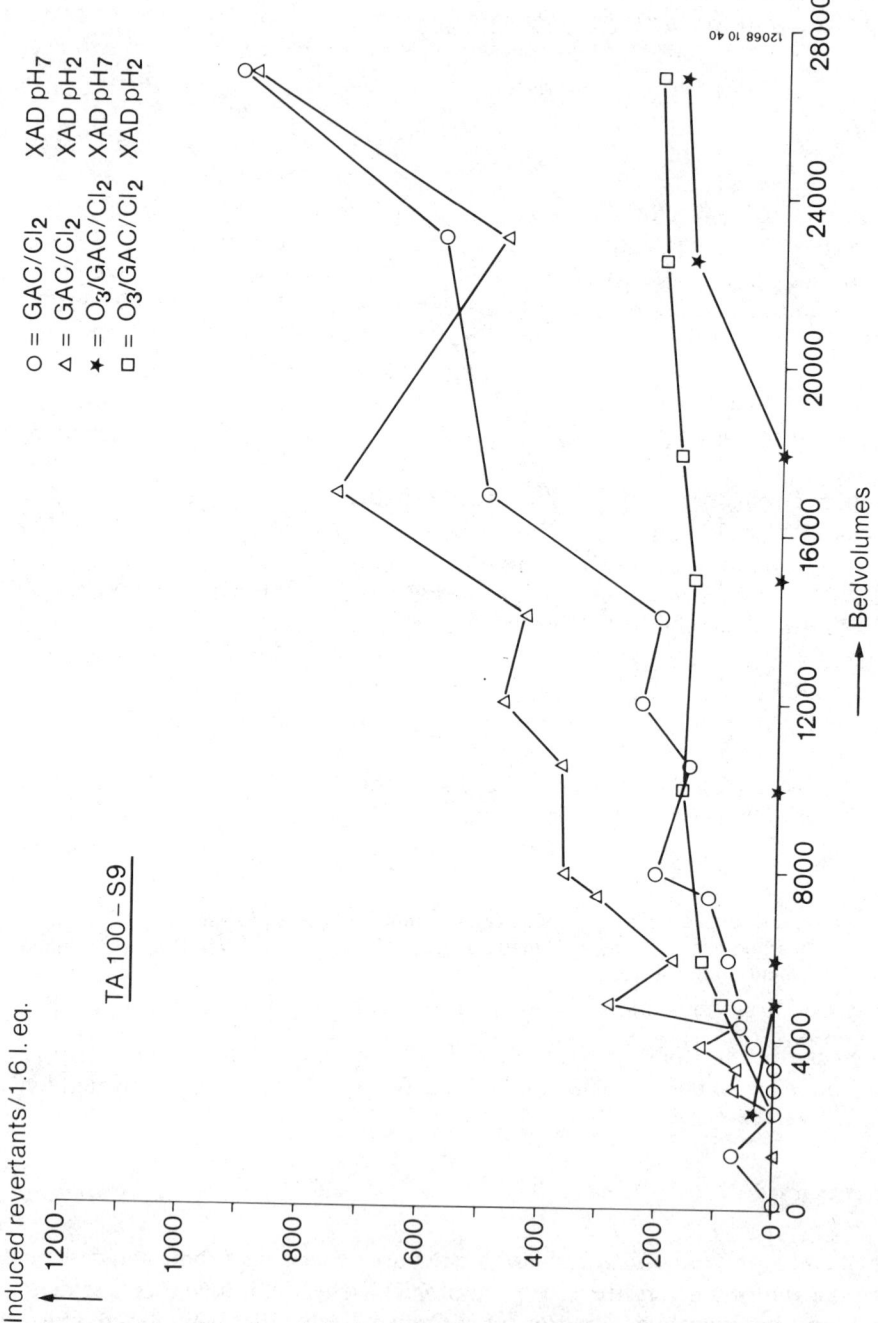

Figure 11. Induced revertants in chlorinated GAC effluents (Spontaneous revertants: 135 – 183 without S9 and 132 – 172 with S9.)

Monitoring of Water Quality Changes Due to Chlorination

Among the different analytical chemical parameters available, THM and organohalogen surrogates are most appropriate for screening the effects of chlorination. Extractable and XAD-adsorbable organonitrogen, -sulfur and -phosphorus are not influenced by chlorination (Tables II, IV, and V). For that reason these parameters are not useful to follow the effects of this process.

Initially, THMs have been the focus of attention. Apart from THM few or no purgeable organohalogens are formed during chlorination. This is illustrated by converting the THM content (in $\mu g/L$) to the organohalide content (in mmol/L) expressed as organochlorine, assuming that organobromide is determined microcoulometrically with a recovery of 60%. The graph of calculated vs measured POX shows a linear relationship with a slope of about one, indicating that besides THM little purgeable organohalide is produced (Figure 12). Thus, the determination of the POX content before and after chlorination does not provide additional information and can be omitted.

During chlorination, only a few lipophilic organohalogens are formed, as can be seen by the relatively small increase of EOX. The large increase of carbon- and XAD-adsorbable organohalogens demonstrates the importance of good preconcentration techniques to isolate organics from water. The amount of organohalogens after both isolations is many times higher than in extraction samples. Both the $X_7 + X_2OX$ and AOX isolates contain approximately the same amount of material.

In the Ames test, strain TA100 without S9 is most sensitive for mutagens formed during chlorination. Increased mutagenic activity in TA100 always occurred with a higher $X_7 + X_2OX$ content (Figure 13). When using chemical and·biological parameters to assess water quality changes, one of the goals is to determine adverse effects on health. The Ames mutagenicity assay is, however, only a first step in predicting mutagenic or even carcinogenic effects in human beings, and the isolation of organic material from water is not complete.

Although other sources have already demonstrated possible effects of nonvolatile chlorination products in higher organisms, additional testing will be needed to confirm reactivity in eukaryotic systems and to allow further risk estimation.

Influence of Treatment Processes on Side Effects of Postchlorination

Before discussing the results, we must mention that all investigations have been carried out under Dutch postdisinfection conditions. The average chlorine dose varies between 0.09 to 0.80 mg/L; the residual concentration is 0.02

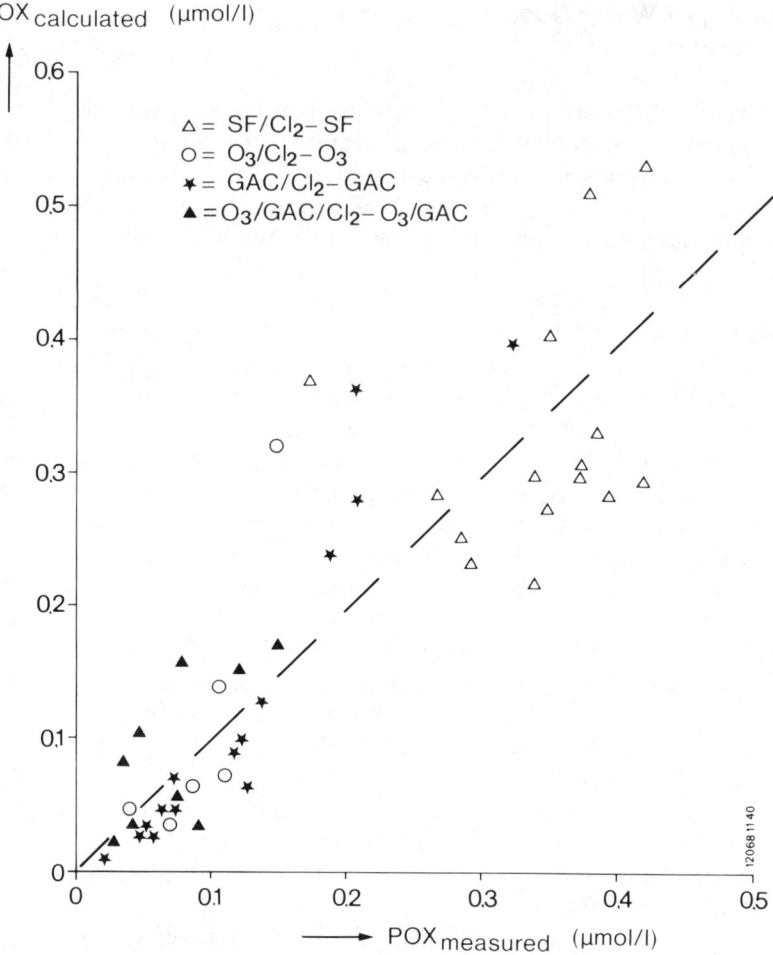

Figure 12. POX calculated from THM production as a function of the measured POX content formed during chlorination.

to 0.49 mg/L after a residence time of 2 min to 48 h.[6] In our investigations we used the most common criterion (i.e., a rest concentration of 0.2 mg/L free chlorine after a contact time of 20 min).

Trihalomethanes

The composition of THM in chlorinated dual-media filtrate fluctuated with time. The $CHCl_3$ concentrations were highest in the winter period (first 23 and final 12 weeks). Due to extremely low water flow of the Rhine in this particular summer, bromide concentrations were probably high, leading to formation of

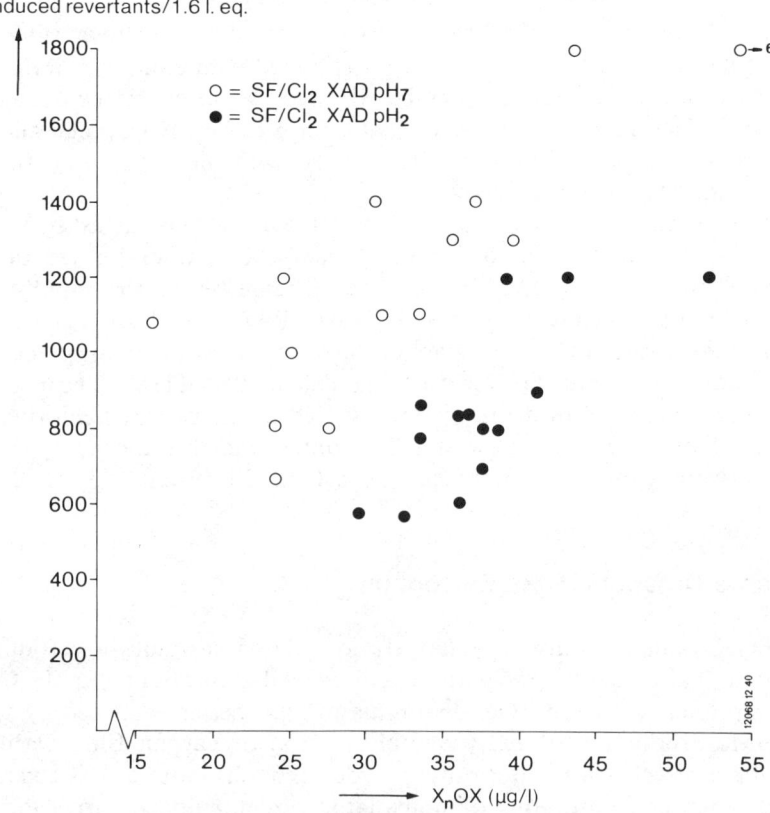

Figure 13. Induced revertants as a function of the XOX content for chlorinated rapid filtrate. (Spontaneous revertants: 135–183 without S9-mix and 132–172 with S9.)

mainly $CHBr_3$ (Figure 2). No THMs were formed using ozonation only. Chlorination of the ozonated water caused about 60% less THM than chlorination of the dual-media filtrate. Therefore, in this situation ozonation gave a relatively high decomposition of THM precursors. During the testing period there was a shift to the production of more brominated THM.

During the first 2800 bv no THMs were formed by chlorination of the GAC effluent. THM formation started after 2800 bv, reaching the highest concentration after about 17,000 bv. This indicates a gradual breakthrough of THM precursors during this filter run. Compared with ozonated water, an even stronger shift to the production of brominated THM took place during the chlorination of the GAC effluent. Contrary to earlier results[4] this shift was found not only for relatively short filter runs, but generally throughout the complete running time. The interpretation of these results is complicated because of the unknown variability of the bromide content.

In the GAC effluent of the ozonated water no THMs were formed during the first 2300 bv, then THM production started, rising up to a filter run corresponding to 9000 bv. This point corresponded with complete THM precursor breakthrough, so that the combined effect is caused by the ozonation only. Thus, THM precursors after ozonation are rather hydrophilic and are therefore poorly adsorbed by GAC. Compared with the ozonate no further shift to brominated THM occurred.

The shift to the formation of more brominated THM was caused by a drop in the THM precursor/bromide ratio, so that the first major reaction is between chlorine and bromide. The bromine product reacts very rapidly with THM precursors, causing brominated THM. When the THM precursor/bromide ratio rises, an initial reaction between chlorine and precursors becomes more favorable, leading to more chlorinated THM. The reaction mechanism is explained in detail by Kruithof et al.[5] During postchlorination THMs are always produced. Lowest THM content may be reached by chlorination following ozonation combined with a GAC filtration.

Nonvolatile Organohalogen Surrogates

The EOX values are always relatively low. Highest values are found in chlorinated dual-media filtrate with an average value of about 4 μg/L. Compounds represented by the EOX determination are better adsorbed by GAC than THM. Production of EOX by chlorination of carbon filter effluents starts after relatively short filter runs and soon remains more or less constant. Relatively high concentrations of nonvolatile organohalogen surrogates are found in the carbon-adsorbable fraction. By chlorination of the dual-media filtrate the average AOX content rises from 25 to 68 μg/L. Ozonation reduces the AOX content from 25 to 14 μg/L. Note that chlorination causes only a small rise in the AOX content of the ozonated water (only 2 μg/L). As expected, carbon-adsorbable compounds are adsorbed very well by GAC. Formation of AOX starts after about 3800 bv and increases to about 25 μg/L. In the GAC effluent fed with ozonated water, chlorination does not lead to a high production of AOX. Even after 30,000 bv only 3 μg/L AOX is formed.

The most favorable results are obtained by either ozonation or a combination of ozonation and GAC filtration.

XAD Investigations

In this phase of the study we will only discuss the influence of treatment processes on X_7OX and X_2OX and the Ames test. The average total XOX content of dual-media filtrate is 13 μg/L. Chlorination adds an additional 56 μg/L. Ozonation reduces XOX, especially in the neutral fraction. Once again, the low rise of the XOX caused by chlorination of the ozonated water is notable.

A gradual breakthrough of XOX occurs for the GAC filtration of dual-media filtrate, as well as ozonated water, especially in the pH-2 fraction. Chlorination of the GAC filtrate fed with dual-media filtrate leads to an immediate formation of XOX, which rises gradually. Highest XOX values are found in the acid fraction. The GAC filter fed with ozonated water shows a different picture. After about 6000 bv, 7 μg/L XOX is formed upon chlorination, and this quantity does not rise for very long filter runs.

The mutagenic activity in TA100 without S9 follows the same trends. Chlorination of rapid filtrate causes a high increase, which is greatly lowered by ozonation following rapid filtration. GAC filtration of both water types removes the mutagenic effect completely for more than 23,000 bv. Chlorination of the GAC effluents fed with dual-media filtrate causes a gradual rise in the mutagenic activity. After 26,000 bv the mutagenic activity is of the same magnitude as for chlorinated dual-media filtrate. Again the GAC filter fed with ozonated water gives much better results. In the pH 2 fraction there is a rapid production of mutagenicity upon chlorination, but the number of induced revertants remains low and stabilizes after 6000 bv at the same level as with the chlorinated ozonate. Production of mutagens in the pH 7 fraction only starts after about 18,000 bv.

CONCLUSIONS

From this research we concluded that the application of a postchlorination always leads to a number of secondary effects. This can be seen from the rise of THM, POX, EOX, AOX, X_7OX, and X_2OX content, and finally from the results of the Ames assay. As long as the risks of these side effects have not been evaluated for humans, postchlorination should be omitted when the biological quality of the water is sufficient and the distribution system does not need protection.

Based on these principles, the Dutch Water Works is studying the following disinfection philosophy:

1. No chemical disinfection is applied when sufficient physical, mechanical, and biological barriers (slow sand filtration, UV disinfection) are present.

2. When chemical disinfection is needed, use of alternative disinfectants (chloramine, chlorine dioxide) can be considered when the side effects have been evaluated and compared with those of chlorine.

3. During the period of investigation and evaluation, when disinfection is needed a limited dose of chlorine may be applied.

Currently, the first two points are under study. The results of those studies, together with the results presented in this chapter, will make a decision about the disinfection method to be followed possible.

As long as chlorine is still used, the side effects of chlorination must be reduced as much as possible. Initially, GAC filtration gives satisfactory results, but only for relatively short running times of the filter. Ozonation gives a better reduction of the side effects of a following chlorination. By far the best results are obtained by GAC filtration of ozonated water. This combination gives a sharp reduction of all side effects for very long filter runs. However the effect of ozonation does not seem to be quite consistent.[13] Therefore, on-site investigations must be carried out before making final decisions about purification systems.

ACKNOWLEDGMENTS

This project is financed by the Netherlands Waterworks Association (VEWIN). We wish to thank all the staff of the Procestechnological, Analytical-Chemical and Biological Departments of KIWA who participated in this project.

REFERENCES

1. Rook, J. J. "Formation of Haloforms during Chlorination of Natural Water," *Water Treat. Exam.* 23:234–245 (1974).
2. Bellar, T. A., J. J. Lichtenberg, and R. C. Kroner. "The Occurrence of Organohalides in Chlorinated Drinking Water," *J. Am. Water Works Assoc.* 66:703–706 (1974).
3. Sybrandi, J. C., A. P. Meyers, A. Graveland, C. L. M. Poels, J. J. Rook, and G. J. Piet. *Problems Concerning Haloforms*, KIWA Communication No. 57 (Rijswijk, The Netherlands: KIWA, 1978).
4. Graveland, A., J. C. Kruithof, and P. A. N. M. Nuhn. "Production of Volatile Halogenated Compounds by Chlorination After Carbon Filtration," presented at the 181st National Meeting of the American Chemical Society, Atlanta, March 1981.
5. Kruithof, J. C., P. A. N. M. Nuhn, and J. A. M. van Paassen. "Removal of Trihalomethanes and Precursors for Trihalomethanes by Granular Activated Carbon Filtration," *H$_2$O* 15:277–284 (1982).
6. Kruithof, J. C. *Side Effects of Chlorination*, KIWA Communication No. 74 (Nieuwegein, The Netherlands: KIWA, 1984).
7. Noordsij, A., J. van Beveren, and A. Brandt. "Isolation of Organic Compounds from Water for Chemical Analysis and Toxicological Testing," *Int. J. Environ. Anal. Chem.* 13:205–217 (1983).
8. Van der Gaag, M. A., A. Noordsij, C. L. M. Poels, and J. C. Schippers. "Preliminary Investigations with Analytical Chemical and Genotoxicological Methods of Water Treatment Processes," *H$_2$O* 15:539–546 (1982).
9. Van der Gaag, M. A., A. Noordsij, and J. P. Oranje. "Presence of Mutagens in Dutch Surface Water and Effects of Water Treatment Processes for Drinking

Water Preparation," in *Mutagens in Our Environment*, M. Sorsa and H. Vaïnio, Eds. (New York: Alan R. Liss, 1982), pp. 277-286.
10. Puijker, L. M., H. M. Janssen, and H. G. A. Kampert. "Developments of Methods for Measurement of Organohalide in Water," H_2O 17:318-322 (1984).
11. Maron, D. M., and B. N. Ames. "Revised Methods for the Salmonella Mutagenicity Test," *Mutat. Res.* 113:173-215 (1983).
12. Van der Gaag, M. A., and J. P. Oranje. "The Use of the Ames Test (*Salmonella*/ Microsomal Mutagenicity-test) for Genotoxicity Studies of Water Samples," H_2O 17:257-261 (1984).
13. Janssens, J. G., F. van Hoof, and J. Diricks. "Ozonation and Activated Carbon Filtration, A Critical Evaluation," *Aqua* 2:102-107 (1984).

CHAPTER **92**

Characterization of Total Halogenated Compounds During Various Water Treatment Processes

A. Bruchet, Y. Tsutsumi, J. P. Duguet, and J. Mallevialle

In 1974 Rook detected trihalomethanes (THM) in potable waters and triggered numerous worldwide studies on their toxicity, the mechanism of their formation, and methods for their removal.[1] It was soon recognized that other halogenated organic compounds are also present. Volatile chlorinated compounds, such as dihaloacetonitriles (DHAN),[2,3] trichloroacetone,[4-8] and chloropicrin,[9-11] that potentially have a high toxicity have been detected. However, mass balance measurements indicated that volatile organics represent only a minor fraction of the total chlorinated by-products (total organic halogen, TOX).[12,13] Furthermore, studies of chlorination by-products of humic and fulvic acids suggested that chlorinated carboxylic acids such as dichloroacetic acid (DCAA) and trichloroacetic acid (TCAA) could account for a high percentage of the nonvolatile TOX formed in chlorinated potable waters.[12-15] These potentially very important water constituents are difficult to identify and require sophisticated analytical techniques, including mass spectrometry. This chapter presents some research on the characterization of these substances and changes that occur in two different potable water treatment processes.

MATERIAL AND METHODS

Volatile and semivolatile compounds were tentatively identified by mass spectrometry using different extraction techniques, while nonvolatile compounds were characterized by pyrolysis–gas chromatography/mass spectrometry.

TOX Measurements

Measurements of TOX were based on activated carbon filtration at pH 2 and were performed following a technique previously described.[16,17]

Gel Permeation Chromatography

The molecular weight distribution was determined by gel permeation chromatography (GPC) on Sephadex G25 (Sephadex Pharmacia, Uppsala, Sweden). Conditions were as follows: column size, 0.25 by 90 cm; eluent, water; eluent flowrate, 100 mL/h. Water samples were first concentrated with a rotary evaporator to obtain a total organic carbon (TOC) concentration between 100 and 200 mg/L. Ten mL of this concentrate was then injected into the Sephadex column. UV absorbance, TOC (Dohrman DC80, Dohrman, Envirotech Corp., Santa Clara, CA), and TOX were measured on the 10 mL-fractions collected.

Macroreticular Resin Extraction

Extractions of trace organics were performed with automatic systems (SERES, Seres, Aix en Provence, France) using macroreticular resins (100 mL XAD8/100 mL XAD2) at pH 2. After passing about 100 L water, elution from the XAD resins is effected successively with 200 mL dichloromethane and 200 mL methanol. The use of dichloromethane for extraction permits the elution of more substances for gas chromatographic analysis than extraction with methanol alone. Methanol elutes the more polar substances of higher molecular weight.[18]

High-Performance Liquid Chromatography Analysis

For TCAA determination, the high-performance liquid chromatography (HPLC) method of Skelly[19] was slightly modified. Separation was carried out on a Du Pont ODS 25-cm column (Du Pont Company, Wilmington, DE). The initial mobile phase, pumped at 2 mL/min, was 0.01 M octylamine adjusted to pH 6 with phosphoric acid. Elution was completed using 25% acetonitrile at 2 mL/min. A 500-μL aliquot of a 25-fold water concentrate was injected. Eluted constituents were detected by UV absorbance at 200 nm.

Gas Chromatography (GC)

Trihalomethanes and chlorinated solvents were measured by head-space injection into a Carlo Erba 2150 gas chromatograph (Carlo Erba, Strumentazione, Milan) on a capillary fused silica 30-m SPB5 column (bath temperature of 40°C). Detection was by electron capture (EC).

Gas Chromatography—Mass Spectrometry (GC/MS)

Samples from water treatment lines were collected into 1-L glass bottles, then stripped in a closed-loop stripping apparatus (CLSA) according to the Grob method.[20] Extracts in 20-μL carbon disulfide were injected into a Carlo Erba 4160 GC interfaced with a Ribermag R10–10C mass spectrometer (Ribermag, SWG Nermag, Rueil Malmaison, France). The conditions were as follows: 50-m fused silica capillary OV1701 column, film thickness 0.25 μm, 0.32 mm ID; column injection at 25°C; programmed from 25 to 180°C at 3°C/min; full-spectrum electron impact at 70 eV (source temperature, 200°C) and mass fragmentography were used for identification and quantitation.

Mass Spectrometry Analysis

The methanol extracts of macroreticular resins were introduced by a desorption chemical ionization probe into a VG ZAB HT mass spectrometer (VG Laboratory Systems, Stanford, CA) using electron impact at 70eV. The mass range was 100 to 900 amu with a cycle time of 3 s.

Pyrolysis–GC/MS Analysis

Flash pyrolysis was performed using a pyroprobe 100 (VG Laboratory Systems, Stanford, CA) temperature control system. Samples were pyrolyzed from 150 to 900°C with a temperature program of 20°C/ms with final hold for 20 s. After pyrolysis, the fragments were separated on a 25-m CP WAX 57 CB fused silica capillary column (programmed from 25 to 220°C at 3°C/min) connected with the R10–10 C mass spectrometer working at 70 eV and scanned from m/e 20 to 400.

TREATMENT PLANTS

The Plant at Vigneux

The pilot unit at Vigneux treats 4 m^3/h of the Seine river upstream from Paris (Figure 1). The water has a rather low organic load (TOC between 2 and 3 mg/L). The treatment process uses an upflow floc blanket clarifier (Pulsator by Degremont, Rueil Malmaison, France) without prechlorination, followed by rapid sand filtration. The sand-filtered water (SFW) is then fed to two treatment lines to evaluate the efficiency of ozone and/or granular activated carbon (GAC) filtration. Final disinfection with chlorine is performed with doses allowing a 0.2 mg/L residual chlorine after 1 h.

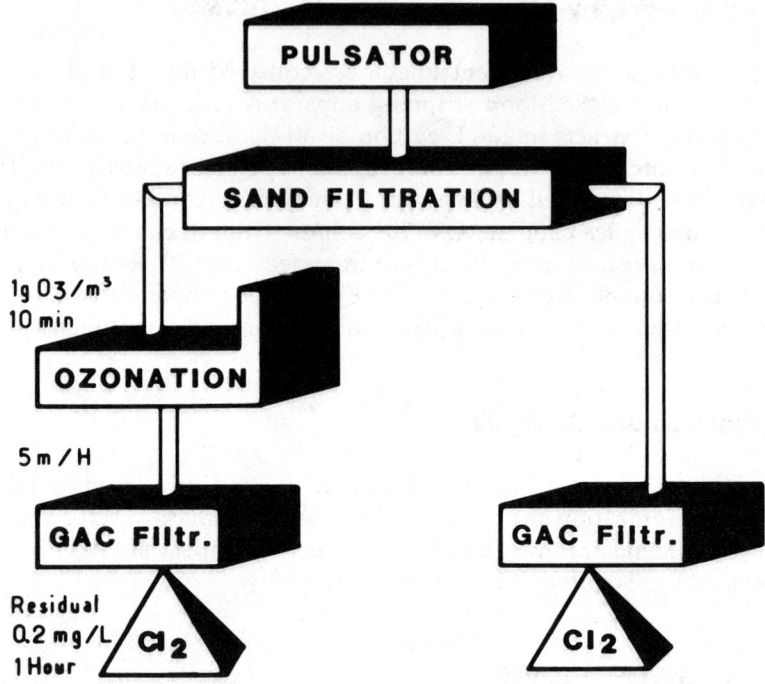

Figure 1. Schematic diagram of the Vigneux pilot plant.

The Plant at Cholet

The Cholet plant treats 1500 m^3/h of a reservoir water highly loaded with natural organic matter (between 8 and 12 mg/L of TOC). Raw water is prechlorinated above the breakpoint and then distributed over three treatment lines operating in parallel. Each line includes coagulation, sedimentation, rapid sand filtration, ozonation, and final chlorination.

RESULTS AND DISCUSSION

Fate of TOX Along the Treatment Lines at Vigneux Water Treatment Plant

Figures 2 and 3, respectively, show the evolution of TOX on lines 1 and 2 of the pilot plant during 1983.

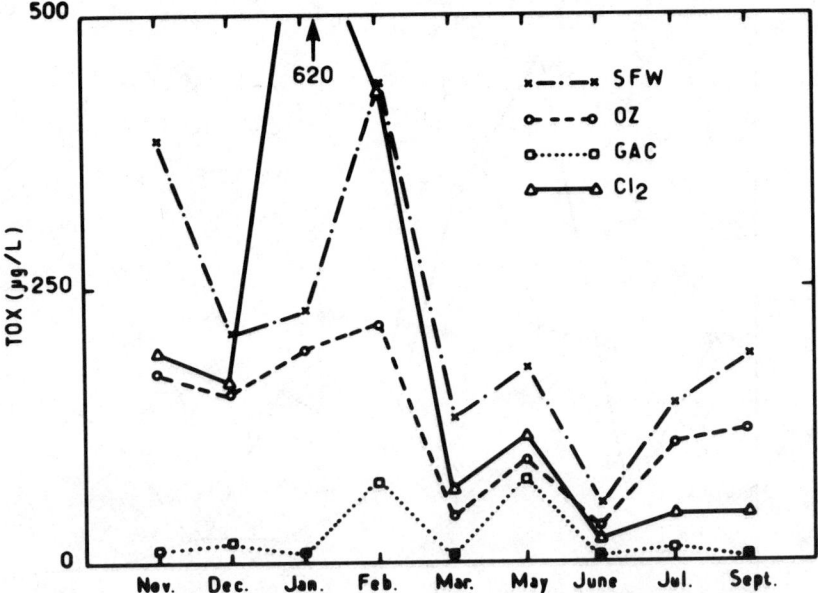

Figure 2. Production of TOX along line I (Vigneux pilot plant).

The level of TOX in the SFW varied from 50 to 440 µg/L during the experiment. Ozonation appeared to be effective because 30 to 65% of TOX was removed. The remaining TOX was significantly adsorbed by the GAC filter. For example, there was > 80% adsorption even after TOC breakthrough, which occurred after 3 months of operation.

Although TOX values after GAC were generally below 20 µg/L, an important increase could be noticed after chlorine disinfection. The concentrations observed in the treated water and in the SFW water were of the same order of magnitude.

Molecular Weight Distribution of TOX

Samples were submitted to GPC to separate the organics according to their molecular weight. An example of Sephadex chromatograms of line 2 is shown in Figures 4–6. Three major fractions were detected, corresponding respectively to apparent molecular weights greater than 5000 (fraction G1), between 5000 and 1000 (fraction G3), and below 1000 (fraction G5). Table I shows the

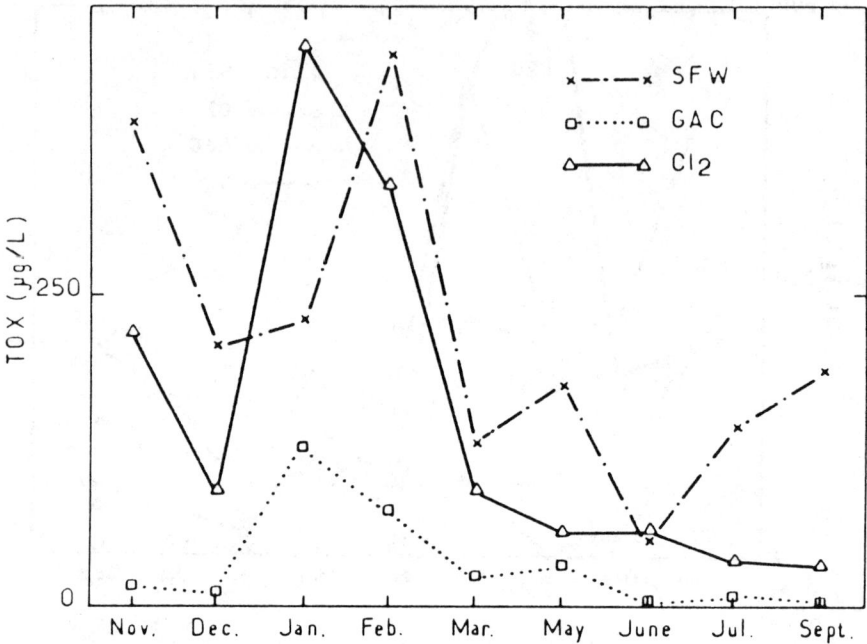

Figure 3. Production of TOX along line II (Vigneux pilot plant).

percentage of TOC and TOX in each fraction for SFW and chlorinated water of line 2.

According to TOC measurements half of the organics had an apparent molecular weight < 1000 and only 10% were included in the heaviest fraction (> 5000). Most of the TOX was found in the lowest fraction (70%) and the percentage found in high-molecular-weight compounds (HMW) was extremely low (2%). The principal effect of GAC filtration was the removal of a significant amount of fraction G5. Chlorination did not appear to affect the TOC distribution.

Surprisingly, the percentage of TOX found after chlorination on HMW (G1) was still very low (2%). Since TOC and TOX/TOC values for this fraction were unchanged, it was concluded that HMW has low reactivity with chlorine. Despite the TOC reduction of G5 due to GAC filtration, the majority of TOX formed was still included in this low-molecular-weight fraction. Both G3 and G5 show an increase of TOX/TOC ratios.

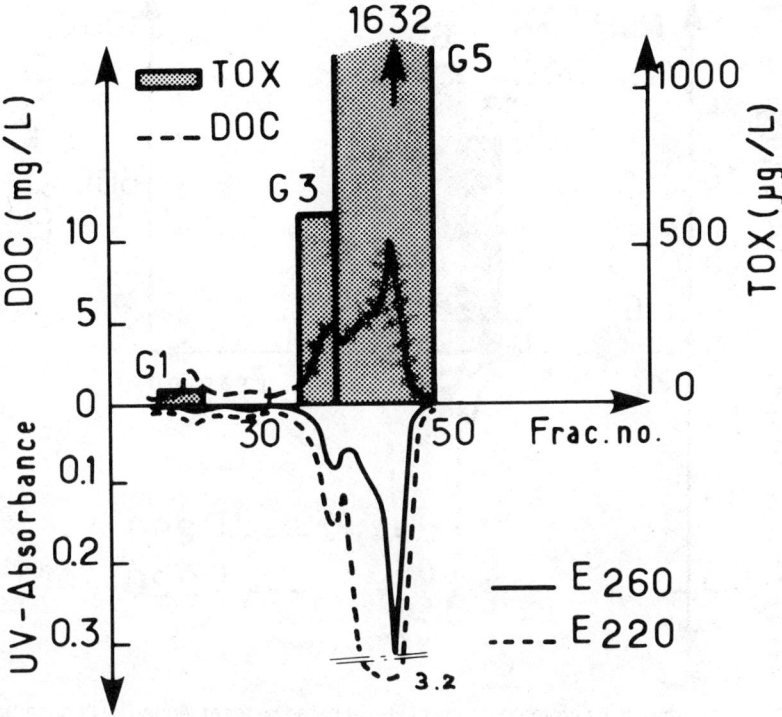

Figure 4. Gel permeation chromatography of sand-filtered water (Vigneux); UV absorbance, dissolved organic carbon (DOC), and TOX distribution.

Volatile and Semivolatile Chlorinated Compounds

It was anticipated from the GPC results that a significant part of the low-molecular-weight chlorinated molecules should be detectable with conventional GC or GC/MS. Total chlorinated solvents measured by head-space analysis were generally about micrograms per liter (mainly trichloroethylene) and were practically unaffected by the treatment, whereas THM levels after chlorination appeared to be below the sensitivity limits of the technique (< 1 μg/L). The total amount of other chlorinated products found in the 20 μL carbon disulfide CLSA extracts or 1-mL dichloromethane composite MRR extracts (from 100 to 150 L of water) were always below 1 μg/L. Identified compounds included, for example, chlorinated benzenes (di-, tri-, and tetra-chlorobenzenes), polychlorinated biphenyls (PCBs), atrazine, simazine, trichlorophenol, 3-chloro-3-methylbenzene for SFW, and THMs for treated waters. All these compounds varied widely with time in a concentration range of 1 to 100 ng/L.

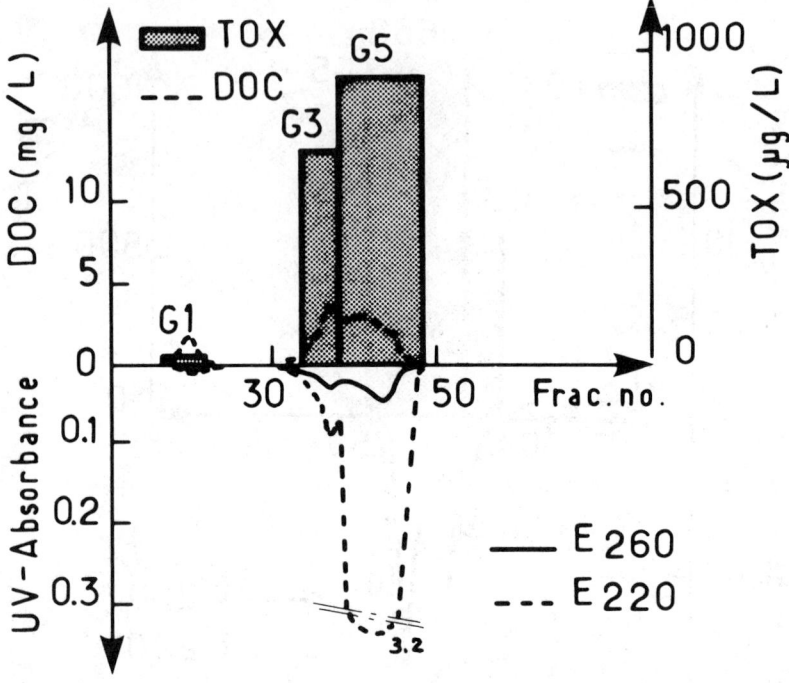

Figure 5. Gel permeation chromatography of GAC-filtered water at Vigneux; UV absorbance, DOC, and TOX distribution.

Direct introduction of MRR methanol extracts (preseparated with preparative HPLC) in a mass spectrometer, permitted the detection of only rare chlorinated compounds. The spectrum shown in Figure 7 presents a typical chlorinated isotope pattern that corresponds to a molecular weight ≥ 570. Definitive identification was not achieved in this case because the amount of sample was too low. We found that the great bulk of chlorinated organics present in water, despite their low molecular weight, cannot be analyzed by GC. They probably correspond to molecules that are too polar (and consequently nonvolatile) to be extracted by conventional procedures.

Nonvolatile Compounds

Pyrolysis GC/MS was used to get more information on the high-molecular-weight nonreactive fraction. This technique produces typical fragmentation of the macromolecules present in the samples: polyhydroxyaromatic compounds produce molecules of the phenol or methoxyphenol type; sugars produce furan derivatives; and proteins produce derivatives of pyrrole, indole, and nitriles.[21]

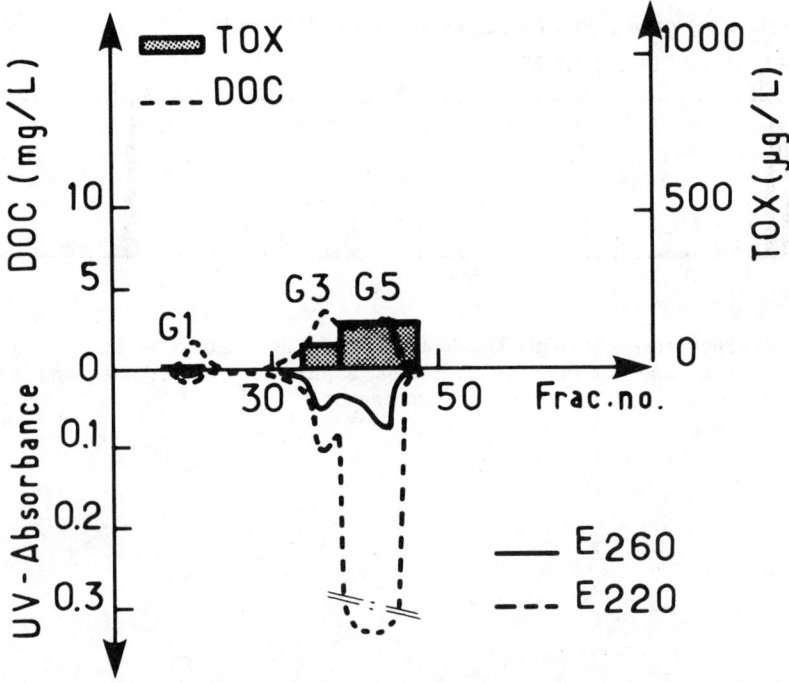

Figure 6. Gel permeation chromatography of chlorinated water (after ozonation, GAC filtration, and chlorination) at Vigneux; UV absorbance, DOC, and TOX distribution.

Table I. Molecular Weight Distribution of TOC and TOX at Vigneux Plant

Mol wt fraction	% TOC	TOX/TOC (μg/mg)	TOX[a] (μg/L)	TOX (%)
Sand-filtered water				
5000 (G1)	16	10–15	46	2
1000–5000 (G3)	32	100	610	27
1000 (G5)	52	170	1632	71
Chlorinated Water				
5000 (G1)	21	10–15	28	2
1000–5000 (G3)	41	180	684	42
1000 (G5)	38	260	910	56

[a]Concentrations directly measured on the Sephadex fractions.

The pyrochromatogram of the line-1 treated water (fraction G1) shows mainly pyrrole, indole, and furan derivatives, the peak of phenol being very low (Figure 8). This suggests the presence of significant concentrations of sugars and proteins rather than polyhydroxyaromatics. This observation is in agreement with the low UV absorbance and could explain the very low yield of TOX formed; that is, proteins and sugars show a lower potential of TOX formation than humic or fulvic acids, as illustrated in Table II.

1174   WATER CHLORINATION

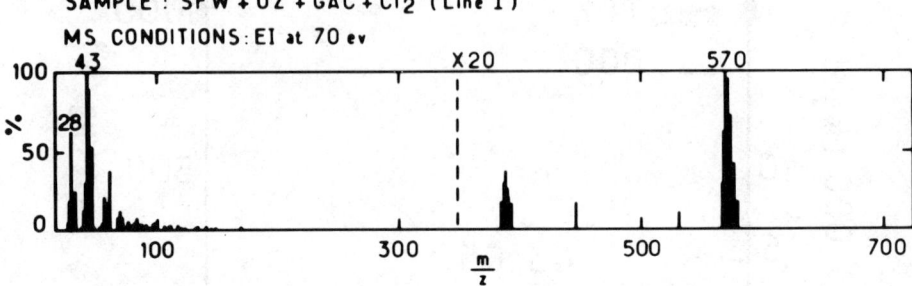

Figure 7. High-molecular-weight chlorinated compounds detected by direct introduction mass spectrometry of HPLC fraction. Sample was collected after sand filtration, ozonation, GAC filtration, and chlorination.

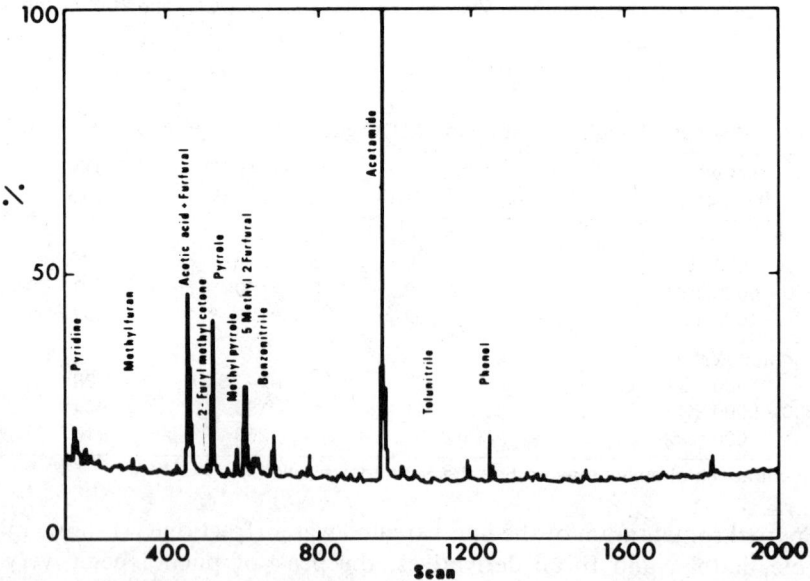

Figure 8. Pyrolysis GC/MS analysis of fraction G1 (mol wt > 5000) of treated water (Vigneux pilot plant).

Table II. Comparison of TOX Formation Potentials for Different Types of Compounds[a]

	TOX/TOC (μg/mg)
Dextran, mol wt: 40,000	34
Protein, BSA	60
Humic acid, Fluka	86
Fulvic acid, Contech	118

[a]Conditions: chlorine dose 1 mg/mg product; contact time 24 h.

Table III. Example of TOX Evolution Along the Cholet Plant, April 24, 1984

	TOX (μg/L)
Raw water	50
Prechlorinated water	560
Sand-filtered water	265
Treated water, O_3 + Cl_2	550

Fate of TOX Along the Treatment Processes at Cholet Water Treatment Plant

Table III gives an example of TOX production along one of the treatment lines at Cholet. The TOX of this natural water is highly increased by the prechlorination, which involves significant chlorine doses (5 mg/L chlorine). The positive effects of clarification and ozonation appear to be balanced by the effect of final chlorination, and the final supplied water often shows high TOX levels.

Molecular Weight Distribution of TOX

Because of the high values of TOX formed during the Cholet prechlorination, it was easier to determine the mass balance for TOX than in the Vigneux plant. The GPC chromatogram of the prechlorinated water is shown in Figure 9. During the sample preparation (filtration and vacuum rotary evaporation), 48% of TOX was lost, including 38% identified as THM and 0.7% as various volatile compounds. The total amount of TOX recovered after gel permeation represents 52% of the original TOX, and the major portion is found in fraction G5 (Table IV). The total yield of low-molecular-weight chlorinated compounds, therefore, represents 78.8% of TOX (40.1% of nonvolatile compounds and 38.7% of volatile).

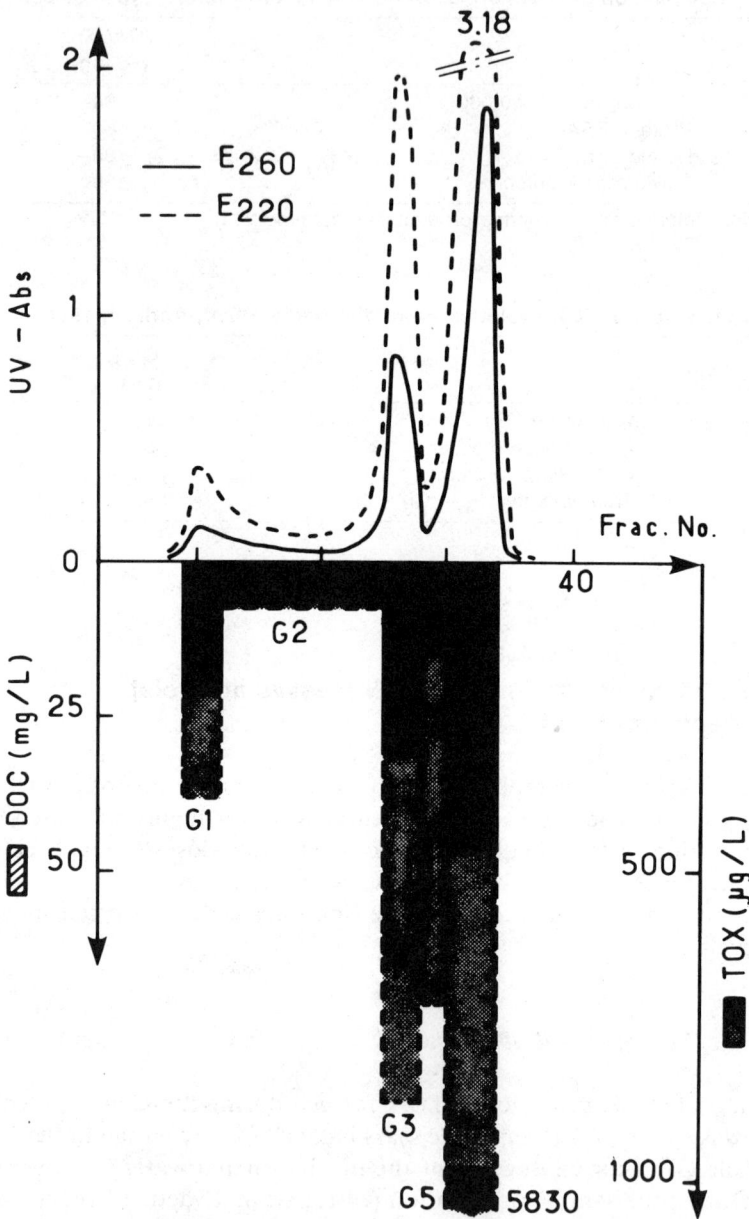

Figure 9. Gel permeation chromatography of the Cholet prechlorinated water; UV absorbance, DOC, and TOX distribution.

Table IV. Molecular Weight Distribution of TOX and TOC for the Cholet Prechlorinated Water[a]

Mol wt fractions		TOX (μg)	TOX/DOC (μg/mg)	Total TOX (%)
5000	G1	11.7	23	2
	G2	10	9	1.7
1000–5000	G3	33	34	5.7
	G4	14	53	2.4
1000	G5	283	128	40.3

[a] Conditions: Sample preparation

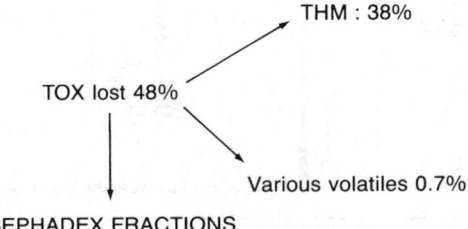

SEPHADEX FRACTIONS

Volatile Compounds

In this case many volatile compounds were identified by head-space and CLSA analysis. The CLSA chromatogram of the prechlorinated water is presented in Figure 10. In addition to THMs, DHANs, and chloropicrin, 1, 1, 1-trichloroacetone and numerous compounds tentatively identified as chlorinated ketones (i.e., with base peak at m/e 43 or 57 and low-chlorinated-isotope patterns) were present (Table V).

Nonvolatile Compounds

The three main fractions (G1, G3, G5) obtained by GPC of prechlorinated water were examined by pyrolysis-GC/MS to get structural information about the chlorinated compounds formed and, in particular, to answer the question, "Is TOX mostly in an aromatic or an aliphatic form?"

The pyrochromatograms presented in Figures 11–14 show the usual fragments of peptides, polyhydroxyaromatics, and sugars. Each chromatogram shows an important peak of chloroform at its beginning. All other peaks were carefully examined and other chlorinated fragments were not detected, except for G5 where traces of dichloroacetonitrile and an unidentified chlorinated aliphatic were present (Figure 14). Because chloroform itself and other volatiles present in the chlorinated water were completely lost during the sample preparation, the presence of chloroform as the main chlorinated decomposi-

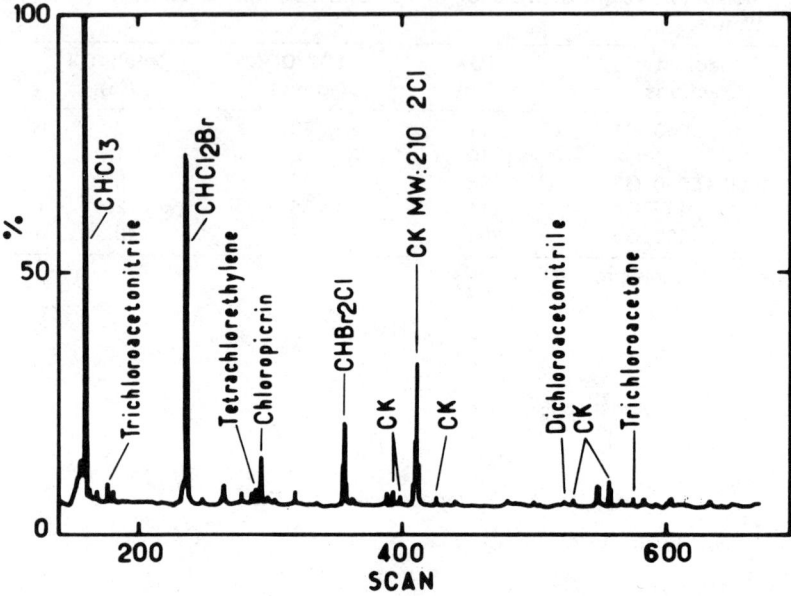

Figure 10. Analysis of chlorinated volatile organics by closed loop stripping of the Cholet prechlorinated water (CK, chlorinated ketone).

Table V. Volatile Compounds Concentration in the Cholet Prechlorinated Water, January 17, 1984

Compound	Concentration	Total TOX (%)
THMs	131 µg/L	38
Chloropicrin	776 ng/L	
Trichloroacetonitrile	112 ng/L	0.2
Dichloroacetonitrile	141 ng/L	
Chlorinated ketones	4400 ng/L	0.5
TOTAL		38.7

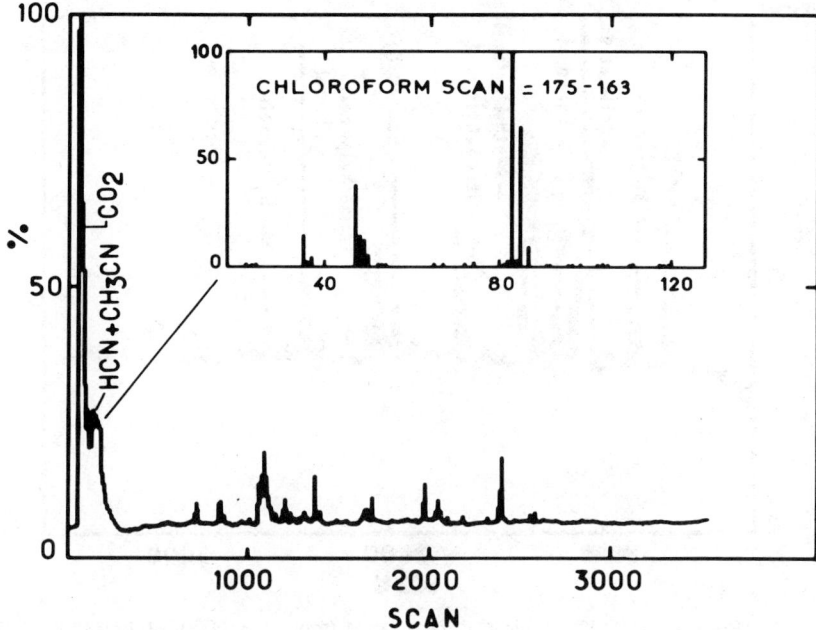

Figure 11. Pyrolysis GC-MS analysis of fraction G1 (mol wt > 5000) of the Cholet prechlorinated water.

tion fragment suggests that TOX is mainly in the form of aliphatic molecules; for example, carboxylic acids such as trichloroacetic acid break down into chloroform in the pyrolysis unit.

In all cases no chlorinated aromatic fragments were identified. Under the pyrolysis conditions used here many chlorinated aromatics (e.g., chlorophenols, dichlorophenols, chlorinated benzenes) are stable. This is illustrated in Figure 14, which represents the pyrochromatogram of a chlorinated humic acid model. Therefore, we conclude that the major portion of chlorine is not fixed on aromatic rings.

Trichloroacetic acid (TCAA) is described by Johnson and co-workers,[12,14] Reckhow and Singer,[13] and Onodera et al.[15] as the principal nonvolatile chlorination by-product of phenols and humic acids. We found TCAA concentrations of 25 μg/L, which only represents 6% of the nonvolatile TOX. Because the chlorine dose per mass unit of organic carbon (0.46 mg Cl/mg TOC) and reaction times (1 h) used at Cholet are much lower than in the experiments described by the above authors,[12-15] the reactions in the Cholet plant should be less advanced. This may explain why about 30% of TOX does not consist of

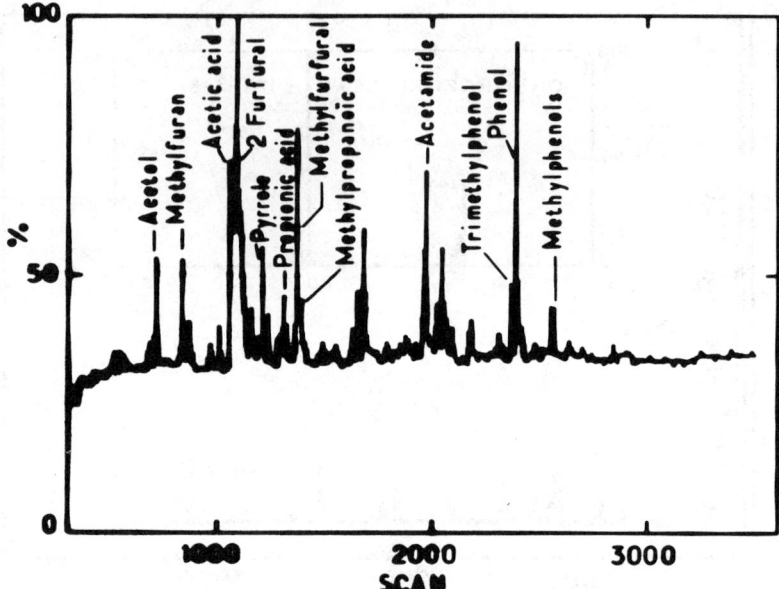

Figure 12. Pyrolysis GC-MS analysis of fraction G1 (mol wt > 5000) of the prechlorinated water.

simple carboxylic acids such as DCAA or TCAA but probably corresponds to more complex acids or to organic chloramines (concentrations of free and bound amino acids in the raw water are 30 and 300 µg/L, respectively). These more complex compounds also could give chloroform as the main chlorinated fragment.

CONCLUSIONS
ch92

In waters with low organic content, the low concentrations of chlorinated compounds make identification difficult. Because prechlorination was not used in the Vigneux pilot plant experiments, only a small concentration of TOX, corresponding to that present in the Seine river, was detected. Almost all of this TOX was removed by the combined treatments of ozone and GAC, but final disinfection with chlorine produced significant concentrations of chlorinated compounds.

In waters with higher organic content such as at the Cholet site, it is easier to determine mass balances. At Cholet the identified chlorinated compounds represent 45% of the TOX. Approximately 38.7% of the TOX consists of THMs and various volatile products, and 6% of the TOX consists of TCAA.

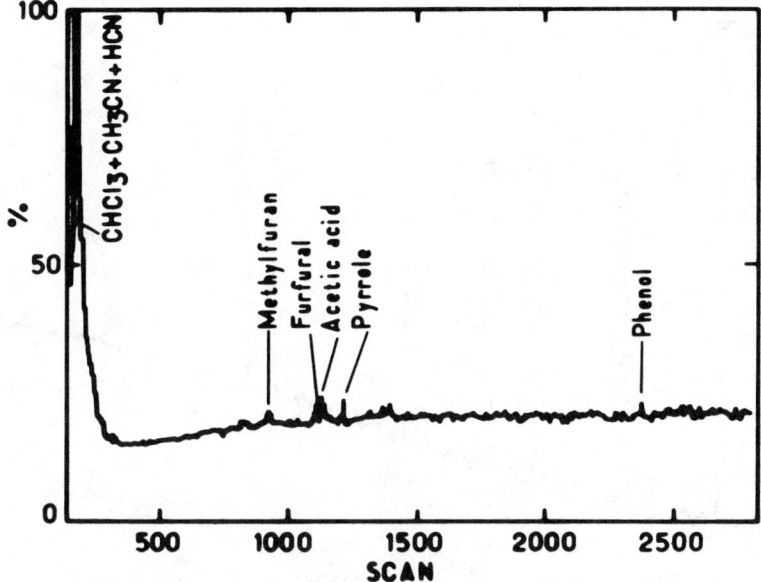

Figure 13. Pyrolysis GC/MS analysis of fraction G3 (mol wt 1000–5000) of the Cholet prechlorinated water.

In addition, in the overall study several chlorinated ketones and chloropicrin were identified in the volatile fractions.

The main fraction of TOX in surface or chlorinated waters corresponds to low-molecular-weight organics (< 1000). At Cholet, for example, they represent 80% of the total. Most of these compounds cannot be analyzed by GC and are difficult to extract by conventional techniques. This behavior is typical for polar molecules. The pyrolysis GC/MS investigations gave valuable information and, in particular, indicated that chlorine is not fixed directly on aromatic rings but on aliphatic compounds or alkyl chains of more complex molecules. The relatively low yield of TCAA found at Cholet, compared with those reported during batch chlorination of polyhydroxyaromatic extracts or models, can be explained by the higher diversity of organics present in this surface water and the less arduous chlorination conditions used (lower ratio of chlorine added/TOC and less reaction time). Proteinaceous material and sugars have shown a certain potential of TOX formation, and part of the unidentified TOX may be due to their degradation products (e.g., organic chloramines). The fact that only 10% of the nonvolatile TOX was identified as TCAA in the Cholet study indicates the need for more adequate methods for the measurement in water of the numerous carboxylic degradation products.

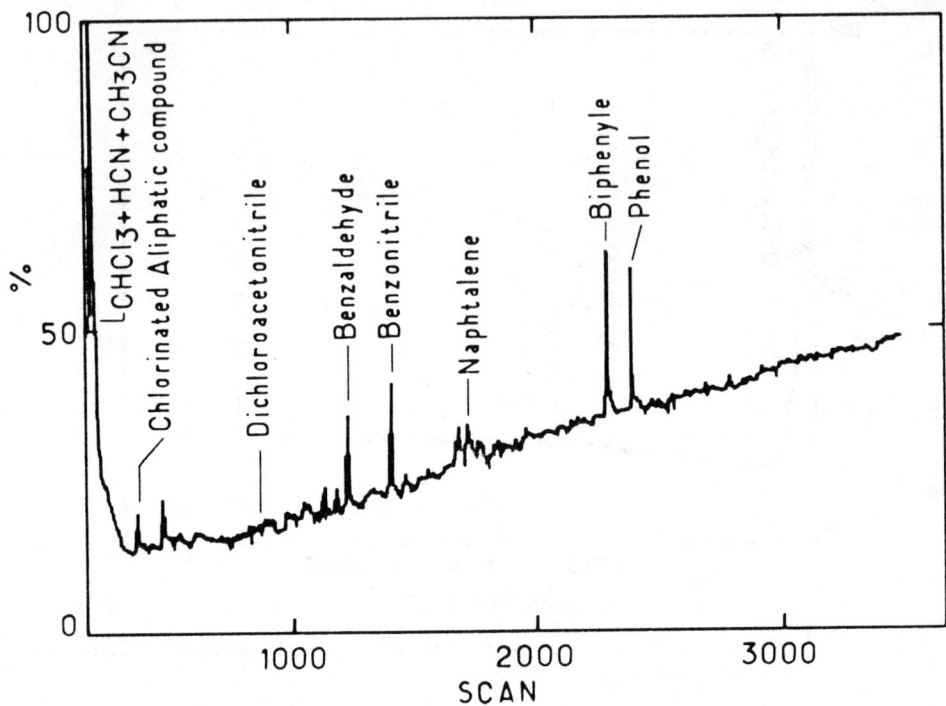

Figure 14. Pyrolysis GC-MS analysis of fraction G5 (mol wt < 1000) of the Cholet pre-chlorinated water.

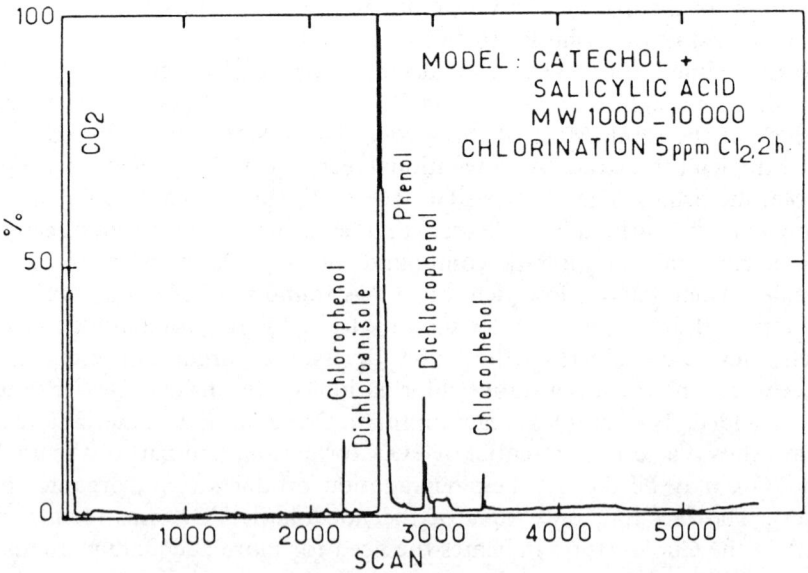

Figure 15. Pyrolysis GC/MS analysis of chlorinated Model Humic Acid.

REFERENCES

1. Rook, J. J. "Formation of Haloforms during Chlorination of Natural Waters," *Water Treat. Exam.* 23:234–243 (1974).
2. Trehy, M. L., and T. I. Bieber. "Certain Amino Acids and Probable Precursors of Dihaloacetonitriles in Chlorinated Natural Waters," American Chemical Society, Division of Environmental Chemistry, 20(2):447–450 (1980).
3. Oliver, B. G. "Dihaloacetonitriles in Drinking Water: Algae and Fulvic Acid as Precursors," *Environ. Sci. Technol.* 17(2):80–83 (1983).
4. Suffet, I. H., L. Brenner, and B. Silver. "Identification of 1,1,1-Trichloroacetone (1,1,1-Trichloropropanone) in Two Drinking Waters: A Known Precursor in Haloform Reaction," *Environ. Sci. Technol.* 10(13):1273–1275 (1976).
5. Giger, W., M. Reinhard, C. Schaffer, and F. Zurcher. "Analyses of Organic Constituents in Water by High-Resolution Gas Chromatography in Combination with Specific Detection and Computer-Assisted Mass Spectrometry," in *Identification and Analysis of Organic Pollutants in Water*, L. H. Keith, Ed. (Ann Arbor, MI: Ann Arbor Science Publishers, Inc., 1976).
6. Rook, J. J. Rotterdam Waterworks, personal communication.
7. Mallevialle, J. Société Lyonnaise des Eaux et de l'Eclairage, personal communication.
8. Gurol, M. D., A. Wowk, and S. Myers. "Kinetics and Mechanism of Haloform Formation: Chloroform Formation from Trichloroacetone," in *Water Chlorination: Environmental Impact and Health Effects, Vol. 4*, R. L. Jolley, W. A. Brungs, J. A. Cotruvo, R. B. Cumming, J. S. Mattice, and V. A. Jacobs, Eds. (Ann Arbor, MI: Ann Arbor Science Publishers, Inc., 1983), pp. 269–284.
9. Sayato, Y., K. Nakamuro, S. Matsui. "Studies on Mechanism of Volatile Chlorinated Organic Compound Formation (III): Mechanism of Formation of Chloroform and Chloropicrin by Chlorination of Humic Acid," *Suishitu Odaku Kenkyu* 1982:127–134.
10. Maier, D., and C. Becke. "Bildungsmoglichkeiten fur Nitrierte Organische Wasserinhaltsstoffe bei der Trinkwasseraufbereitung," in Proceedings of the International Ozone Association Symposium on Ozonization, Environmental Impact and Benefit, Brussels, Sept. 12–13, 1983.
11. Duguet, J. P., Y. Tsutsumi, J. Mallevialle, and F. Fiessinger. "Chloropicrin in the Water: Conditions of its Formation and Evolution in Different Treatment Plants," Chapter 94, this volume.
12. Johnson, J. D., R. F. Christman, D. L. Norwood, and D. S. Millington. "Reaction Products of Aquatic Humic Substances with Chlorine," *Environ. Health. Perspect.* 46:63–71 (1982).
13. Reckhow, D. A., and P. C. Singer. "Removal of Organic Halide Precursors by Preozonation and Alum Coagulation," presented at the Am. Water Works Assoc. TTHM Seminar, Las Vegas, NV, June 5, 1983.
14. Johnson, J. D., "THM and TOX Formation Routes, Rates and Precursors," presented at the Am. Water Works Assoc. TTHM Seminar, Las Vegas, NV, June 5, 1983.
15. Onodera, S., T. Udagawa, M. Tabata, S. Ishikura, and S. Suzuki. "Isotachophoretic Determination of Chlorinated Carboxylic Acids formed during Chlorination of Phenol with Hypochlorite in Dilute Aqueous Solution," *J. Chromatogr.* 287:176–182 (1984).

16. Montiel, A., M. Bigoit, J. Mallevialle, R. Lheritier, and N. Houel. "Mise au Point du Dosage du Chlore Organique Total par Pyrohydrolyse et Microcoulométrie," présenté aux Journées Electrochimie, Brussels, May 1981.
17. Mallevialle, J., F. Fiessinger, S. W. Maloney, and P. Charles. "Activated Carbon Surface Evaluation; A Non-Specific Measure of Carbon Exhaustion," in *Practical Aspects of GAC Adsorption*, M. J. McGuire and I. H. Suffet, Eds. (Washington, DC: American Chemical Society), in press.
18. Mallevialle, J., and A. Bruchet. "Oxidation of Organic Traces in Ground Water Treatment," presented at the First Atlantic Workshop, Nashville, TN, Dec. 9-10, 1983.
19. Skelly, N. E. "Separation of Inorganic and Organic Anions on Reversed-Phase Liquid Chromatography Columns," *Anal. Chem.* 54:712-715 (1982).
20. Grob, K. "Organic Substances in Potable Water and in Its Precursors. Part. I. Method for Their Determination by Gas-Liquid Chromatography," *J. Chromatogr.* 84:255-273 (1973).
21. Mallevialle, J., A. Bruchet, and E. Schmitt. "Nitrogenous Organic Compounds: Identification and Significance in Several French Water Treatment Plants," presented at Am. Water Works Assoc. Conf., Norfolk, VA, Dec. 4-7, 1983.

CHAPTER 93

Trihalomethane Formation and Control Through a Direct Filtration Water Treatment System

John N. Veenstra and Parweiz A. Khan

Several investigators[1,2] of the direct filtration process have concluded that it offers both capital and operating cost savings when compared with a conventional coagulation-sedimentation plant. However, several questions about the direct filtration process need to be addressed, such as what is the maximum amount of turbidity this process can handle? Turbidity values reported in the literature,[2-11] as illustrated in Table I, cover a wide range.

Another concern of the direct filtration technology is the regulation promulgated by the U.S. Environmental Protection Agency (EPA)[12] on November 29, 1979, that limits trihalomethanes (THMs) to 100 μg/L in water supplies serving over 10,000 people. The THMs are derived when water is chlorinated for disinfection. The chlorine reacts with humic and fulvic acids already present in the water to form THMs.[13] In general, the natural organics are derived from the presence of humic substances (i.e., humic and fulvic acids). Of these organics, humic acids have the highest molecular weights and fulvic acids the lowest, about a few thousand. The major fraction of organic matter in natural water is fulvic acids. Other questions then arise: How well does the direct filtration process remove THM precursors, and, if modifications to the disinfection procedure of a direct filtration plant must be made to control THMs, what effect does this have on the bacterial quality of the treated water?

The objectives of this work were to examine the water quality produced by the THMs formed in a direct filtration pilot plant and to evaluate methods of controlling the THM formation. To ensure that quality was not compromised, the bacteriologic quality of the water was monitored during the attempts to modify the system to control THMs.

EXPERIMENTAL METHODS

This work was conducted on a direct filtration pilot plant located at the Oklahoma State University Water Treatment Plant (OSU Plant), Stillwater, Oklahoma. A schematic of the plant is shown in Figure 1. Water samples for analysis were taken from the flow just after the chlorination tank, rapid-mix unit, flocculation tank, and the filters. A separate chlorination tank was included in the pilot plant to compare the pilot plant with the OSU Plant that

Table I. Recommended Raw-Water Quality Maximums for Direct Filtration

Investigator	Ref. No.	Maximum Turbidity (NTU)	Maximum Color (units)	Conditions
Culp	2	25	25	Low color
		200		Low turbidity
Amy	3	10		
Kawamura	4	10	100	
Tredgett	5	50		
Walder et al.	6	25–30		
Baumann	7	50–60		
Hutchinson	8	20–25		Using alum
		150		Using cationic polymers
Am. Water Works Assoc. Filtration Committee	9	15	30–40	
McCormick and King	10	25		Low color and algae
Khan	11	60		

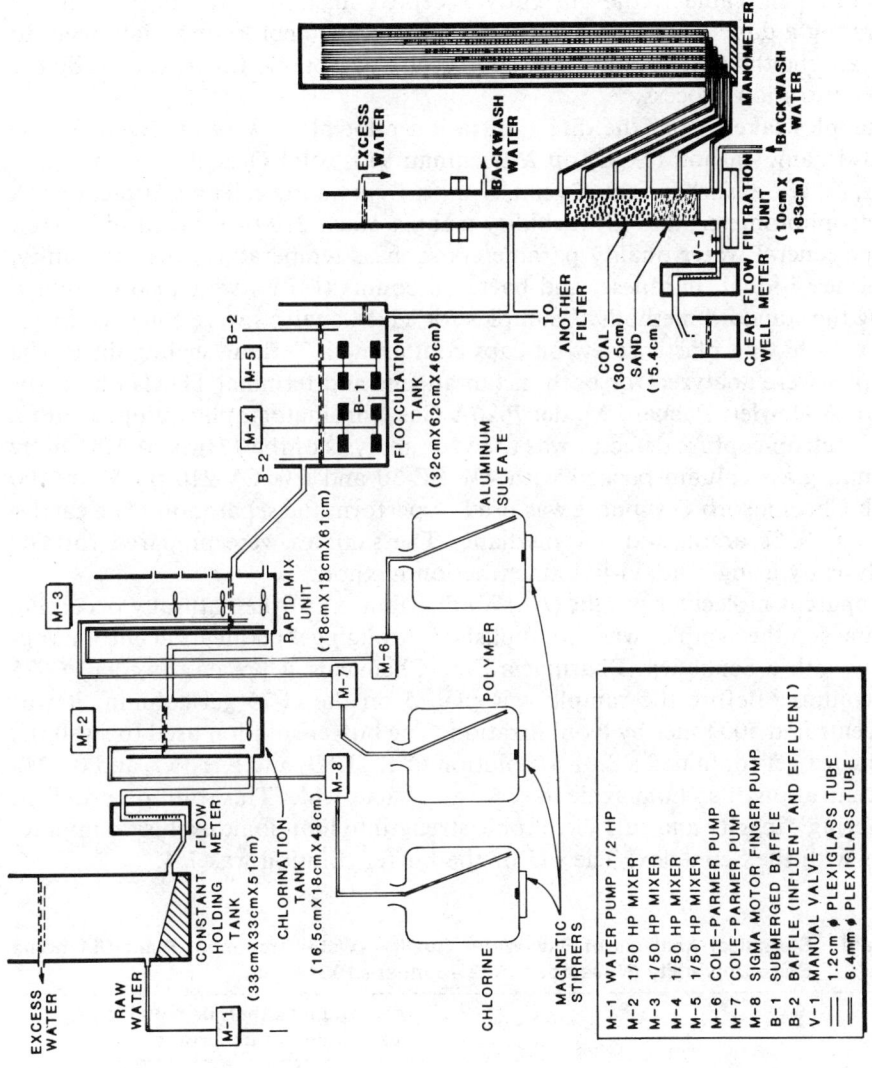

Figure 1. Schematic of pilot-scale direct filtration system.

uses conventional coagulation, flocculation, and sedimentation prior to filtration. The OSU Plant uses an old cascade aeration basin to chlorinate the influent water before it reaches the rapid mix basin. Lakes Carl Blackwell and McMurtry are the supply sources for the plant. A statistical summary[14] of the quality of the influent to the OSU Plant is given in Table II. Although, as illustrated in Table I, the currently accepted maximum turbidity limit for operating a direct filtration plant is subject to much conjecture, data seem to indicate the the OSU Plant water sources are acceptable for treatment by the direct filtration process.

Samples taken from the direct filtration pilot plant were analyzed for (1) total organic carbon (TOC) on a Beckman 915 Total Organic Carbon Analyzer, (2) UV absorbance at 254 nm, using a 1-cm quartz cell on a Hitachi 800A spectrophotometer; and (3) turbidity using a Hach 2100A ratio turbidimeter. Other general water quality parameters such as temperature, pH, alkalinity, chlorine residual, hardness, and bacterial counts (MPN) were also examined using the standard methods.[15] Samples for THM analysis were collected in 40-mL vials having plastic screw-on caps containing a Teflon® sealing liner. The samples were analyzed for both instantaneous and terminal THMs (20°C for 72 h). A Hewlett Packard Model 7626A Gas Chromatograph equipped with a ^{63}Ni electron-capture detector was used for analysis of the THMs. A 1.83-m by 6.4-mm glass column packed with 4% SE-30 and 6% OV-210 on 80 to 100 mesh Chromosorb Q support was used to perform the separation. The carrier gas was 95% argon and 5% methane. The samples were prepared for GC analysis by using a liquid-liquid extraction method.[16]

Apparent molecular weight (AMW) distribution of the naturally occurring organics in the samples was accomplished through gel permeation chromatography with a Sephadex (Pharmacia Fine Chemicals, Uppsala, Sweden) G-75 gel column. Before the sample was placed on the G-75 gel column, it was concentrated 500 times by lyophilization. The buffer solution used to swell the Sephadex gel contained a 0.01 M solution of K_2HPO_4 and KH_2PO_4 and 0.02% concentration of sodium azide to serve as a bactericide. This solution provided buffering capacity and sufficient ionic strength to limit ionic exclusion interactions with the Sephadex. The pH of the buffer solution was 7.4.

Table II. Statistical Analysis of Raw-Water Quality—Water Treatment Plant Oklahoma State University (November 1979 – January 1984)[a]

Parameter	Percentage less than or equal to probability of occurrence	
	50%	95%
Turbidity, NTU	24	46
Total hardness, mg/L	156	192
Total alkalinity, mg/L	108	142
pH	7.9	8.3

[a]From Reference 14.

The Sephadex column, 100 cm by 1.5 cm ID, was operated in the downflow mode with a constant head of 150 cm of H_2O. This head resulted in a flowrate of about 30 mL/h. The eluant was collected in eighty 2.5-mL fractions in 13- by 100-mm test tubes by a Gilson FC-80 Micro-Fractionator operating in a drop-counting mode. The column was previously calibrated using compounds of known molecular weights.

RESULTS AND DISCUSSION

Physical and Chemical Variables

Prior to data collection on the formation of THMs, both the physical and chemical variables of the direct filtration pilot plant were optimized.[11] This involved the initial screening by jar tests of 42 different polymers alone or in combination with aluminum sulfate (alum); 14 different polymers plus alum were eventually tested in the pilot plant. During this study, more than 300 filtration runs were conducted to optimize the chemical and physical variables. Table III lists the optimized chemical variables. The optimized variables were chosen on the basis of turbidity removal and allowable filter run length. Filter runs were terminated because of excessive headloss or turbidity breakthrough (1 NTU).

Table III. Optimum Coagulant Dosages as Determined by Jar Tests and Pilot Plant Studies

Coagulant	Dosage (mg/L)
Alum	12 (summer)
	16 (winter)
Polymer, magnifloc 572 C	2.0
Alum + polymer	8.0 + 0.3 (summer)
	12.0 + 0.4 (winter)

Various Coagulant Regimes

During this phase, the pilot plant was operated using both prechlorination ahead of rapid mix tank and postchlorination of the filter effluent to match the operating mode of the OSU Plant. The pilot plant THM production at coagulant dosages resulting in the longest filter runs is shown in Figure 2. As seen in the figure, none of the coagulants was able to drop the terminal THM value below the 100-μg/L limit. During the 4-month period of comparison, of the three coagulant regimes, the raw-water terminal THMs averaged 215 μg/L, whereas the raw-water TOC averaged 11.0 mg/L. The average percentage

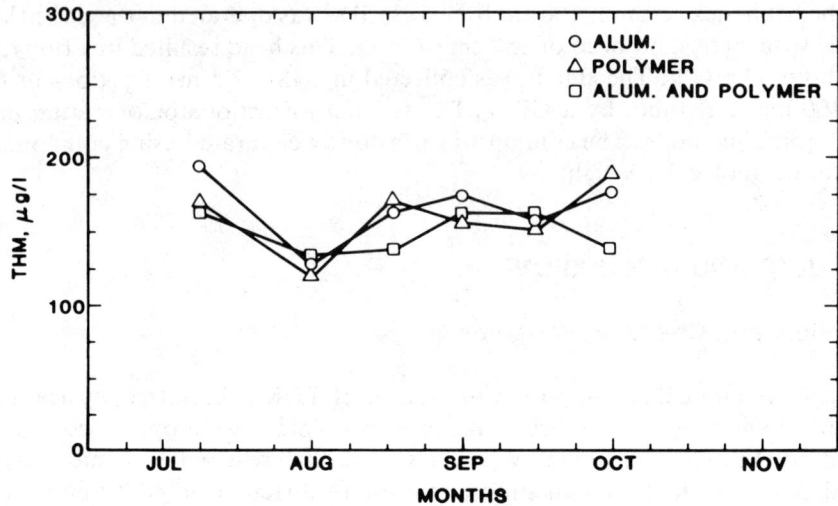

Figure 2. Comparison of terminal THM concentrations produced by pilot plant direct filtration process using alum, polymer, and alum plus polymer.

Table IV. Average Percentage Organic Removal by Direct Filtration Using Various Coagulants

Treatment and location	TOC	UV (at 254 nm)	THMFP
Pilot plant effluent			
Polymer	27 ± 10	90 ± 3	51 ± 16
Alum	21 ± 8	90 ± 6	49 ± 13
Alum + polymer	24 ± 6	92 ± 5	50 ± 9

reductions of TOC, UV absorbance, and THM forming potential (THMFP) are shown in Table IV. The results show that all three coagulant regimes removed the organics that act as THM precursors better than TOC.

Babcock and Singer[17] have previously shown that alum removes chloroform-precursor organics better than TOC. Alum, acting as the sole coagulant, showed the lowest removal for TOC and THMFP. The polymer alone showed the highest degree of removal for both the TOC and THMFP. The UV-absorbance removal by alum and the polymer were equal. The terminal THMs formed by chlorinating a solution of the polymer were minimal, < 10 μg/L. The combination of alum plus the polymer produced removals for

TOC and THMFP in the range between the values obtained for the two coagulants alone. However, the combined coagulant regime did display the highest UV absorbance removal. It should be noted that the removals obtained for TOC, THMFP, and UV absorbance were all very close and none of the coagulant regimes established itself as better than the others. At the $\alpha = 0.05$ level, no significant difference was observed between the treatment means based on TOC, UV and THMFP. Previous direct filtration studies[18,19] conducted in upstate New York found alum to be more effective in removing TOC and THMFP when compared with polymers. The difference in the results of this study compared with the earlier two is reflected in the different raw water sources used for the studies.

Apparent Molecular Weight Distribution

To more closely monitor where the THM precursor removal was taking place, effluent samples from each unit were subjected to gel permeation chromatography on a Sephadex column to determine changes in the distribution of the apparent molecular weight of the naturally occurring organics. Samples of raw, flocculated, and filtered water were collected from (1) the direct filtration process, using either alum, polymer, or alum plus polymer; and (2) the OSU Plant. Before they were lyophilized, samples were centrifuged at 2500 rpm to remove suspended materials. Figures 3 and 4 illustrate the UV absorbance of the eluant from the Sephadex G-75 column. Figure 3 illustrates the UV absorbance trace of a raw water sample. The profile shown in this figure indicates the predominance of one large peak containing organics in the apparent molecular weight range of 1000 to 9000. A second smaller peak containing organics with an apparent molecular weight of greater than 50,000 was also observed in the raw water. This second peak was observed infrequently and only during August and September of both 1981 and 1982. Figure 4 includes samples from the direct filtration pilot plant using alum, polymer, or polymer plus alum as the coagulant, as well as samples from the OSU Plant, which uses alum as coagulant. A UV tracing of the raw water is included in the figure. Also shown in the figure are samples taken after each unit process in the direct filtration pilot plant, along with samples taken after flocculation and filtration in the OSU Plant. A complete removal of the high molecular weight peak ($>50,000$) was observed by the time the raw water passed through the prechlorination unit. Both the direct filtration pilot plant and the OSU Plant were operated with a chlorine dosage that gave a 1.0 mg/L free residual. Apparently, chlorine may fragment the organics into smaller molecular weight fractions as measured by UV absorbance. These findings were similar for the OSU Plant. On the other hand, a partial reduction of the second dominant peak, representing apparent molecular weights of 1000 to 9000, was observed as the water passed through the flocculation and filtration processes.

In comparing the molecular weight removal data from the three coagulant

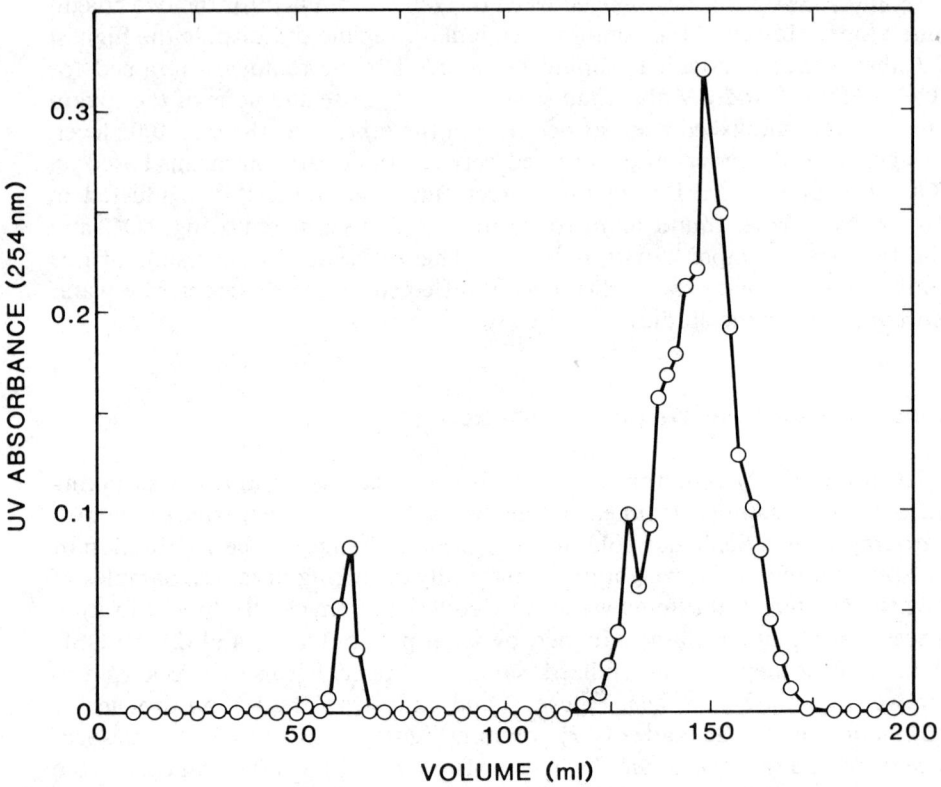

Figure 3. Ultraviolet absorbance chromatogram showing the fractionation of the raw water sample for September 1, 1982, by a Sephadex G-75 Column.

regimes used in the direct filtration pilot plant, no significant ($\alpha = 0.05$) difference was found between the removal when using alum, polymer, or polymer plus alum. Agreement among the regimes was also shown in the THMFP removals. Very little difference was observed in the amount of organics in each molecular weight group removed by the direct filtration pilot plant and the OSU Plant when both used alum.

Moderate TOC removal efficiency, <30%, was found in this study, probably because the direct filtration plant was unable to remove the fulvic acids completely. This study confirmed that the low-molecular-weight organics, fulvic acids, present in the raw water could not be removed completely by the conventional or the direct filtration processes. Successful removals of humic acids by coagulation and flocculation of synthetic raw waters have been reported by other authors.[17,20]

From Figure 4, the location in the treatment train where these naturally occurring organics are removed can be approximated. In the pilot plant, a

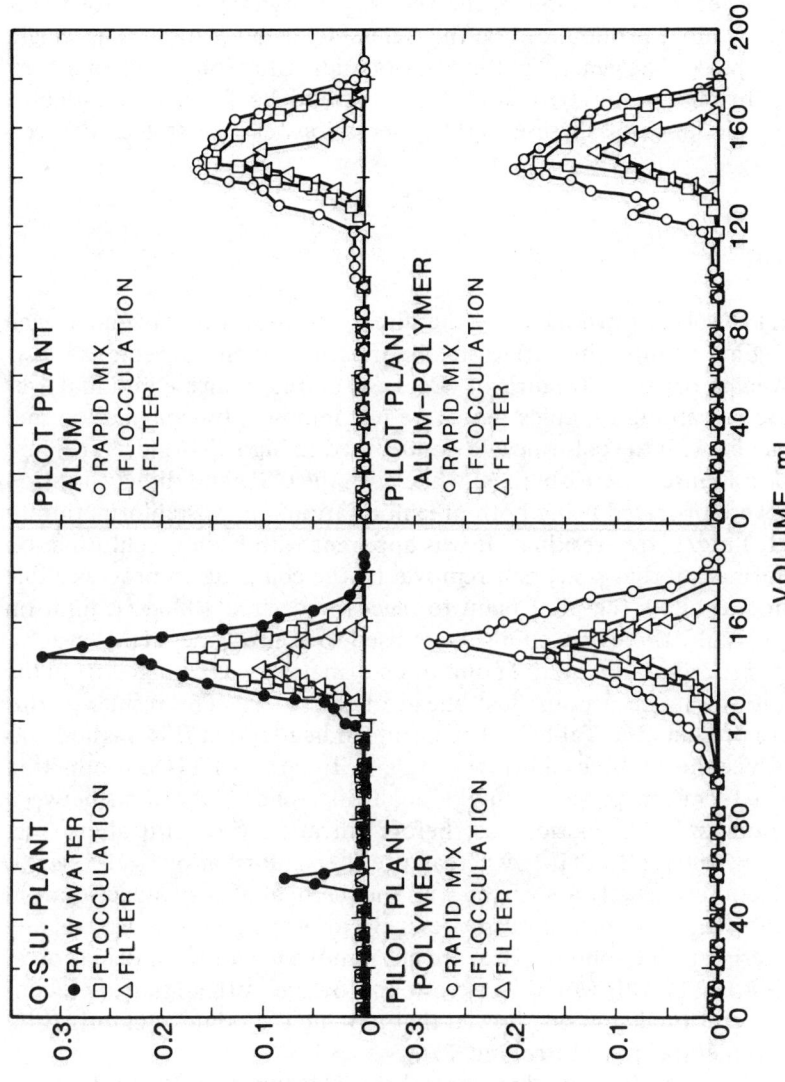

Figure 4. Ultraviolet absorbance chromatograms from a G-75 column tracing the removal of various molecular weight organics through the different unit processes in a conventional water treatment plant and direct filtration pilot plant.

decrease in the width of the eluting profile occurred between the rapid mix and the flocculator effluent. This decrease took place on the left side or the larger molecular weight portion of the peak. This indicates a reduction in larger molecular weight organics as the water passes through the flocculation tank. In addition, a general reduction in the remaining organics occurs following filtration that is more pronounced on the right side or lower molecular weight portion of the peak. The general pattern of organics reduction based on molecular weight through water treatment plants should be further investigated before the significance of these general observations can be stated with certainty.

THM Control

The high THM levels produced by the direct filtration process under each coagulant regime resulted in part from the presence of the unremoved low-molecular-weight organics (Figure 4). Because a free chlorine residual was available, the remaining organics that were not removed by coagulation and filtration reacted with the chlorine and contributed to high THM levels (Figure 5). The data in Figure 5 were obtained while both the OSU and direct filtration pilot plants were operated using both prechlorination and postchlorination to maintain a 1.0 mg/L free residual. It was apparent with both prechlorination and postchlorination that precursor removal by the coagulation process alone was not going to allow the pilot plant to meet the desired 100 μg/L limit on THMs. As a result, three procedures were used to control the THM level.

In the first procedure, the initial point of chlorination was changed from the prechlorination point to a point just ahead of the filters. The results of this technique are presented in Table V. The results indicated that this method was unable to reduce the THM level below 100 μg/L. In terms of THM production for the direct filtration process, only a small difference was found between prechlorination and chlorination just before filtration (i.e., with alum, prechlorination = 180 μg/L THM, whereas prefilter chlorination = 153 μg/L THM). Reduction is usually associated with the removal of suspended solids in the coagulation and sedimentation process. Because the solids as well as any organics adhering to the solid surface, are not removed prior to chlorination in this system, little THMFP would be expected. However, other studies[21,22] conducted on conventional systems showed that moving the point of chlorination is an efficient method for controlling THMs.

The other two techniques used to control the THM level involved the addition of ammonia to the filter effluent. In one technique, ammonia was added to the water that was prechlorinated to a 1.0 mg/L free residual in the chlorine contact chamber. Using this technique, the water possessed a free chlorine residual for about 35 min before ammonia addition to the filter effluent. In the other technique, ammonia was added to water that was chlorinated directly ahead of the filters. In this case, the water possessed a 1.0 mg/L free residual

DRINKING WATER TREATMENT 1195

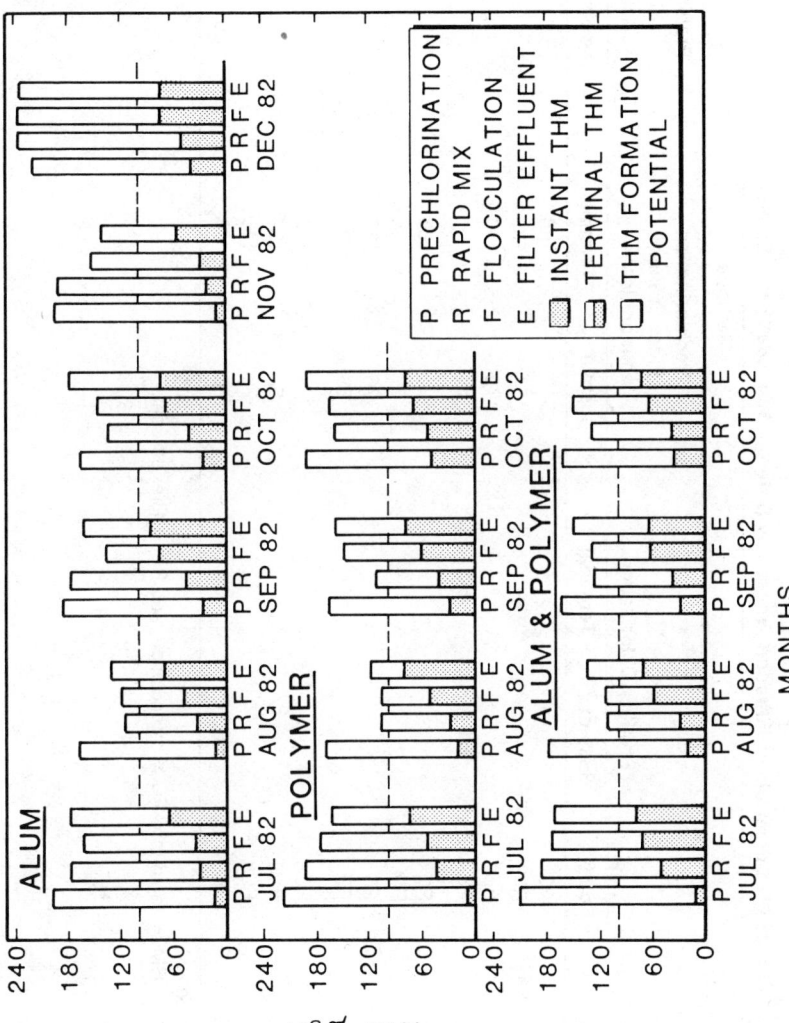

Figure 5. Average monthly instantaneous and terminal THM concentrations at different treatment units for the pilot plant direct filtration process.

Table V. Reduction of THMs for Pilot Plant Direct Filtration Process by Different Methods (5-month averages)[a]

Types of Treatment	Sampling Point[b]	Methods					
		Prefilter Chlorination		Prechlorination Filter Effluent Ammonification		Prefilter Chlorination Filter Effluent Ammonification	
		Instant	Terminal	Instant	Terminal	Instant	Terminal
Direct filtration, alum = 16.0 mg/L	PC			54	160		
	RM			70	133		
	FC			66	140	73	72
	FE	64	153		67		
Direct filtration, polymer = 2.0 mg/L	PC			67	167		
	RM			76	135		
	FC			68	133	70	73
	FE	70	158		70		
Direct Filtration, alum = 12.0 mg/L polymer = 0.4 mg/L	PC			52	170		
	RM			70	137		
	FC			68	145	60	63
	FE	68	167		70		

[a]THM concentrations are μg/L.
[b]PC = prechlorination tank, RM = rapid mix, FC = flocculation, and FE = filter effluent.

for about 5 min, until the water passed through the filter and was quenched by the addition of ammonia to the filter effluent. The ammonia was added as ammonium sulfate at a weight ratio of 1:4 ($NH_4:Cl_2$). Chloramines are formed by adding ammonia during chlorination to consume free chlorine, thus reducing its availability for THM formation. During the use of ammonia, an excellent reduction of THM level was observed in the direct filtration process. The effluent THM level averaged about 60 to 70 µg/L, or about 30 to 40% below the EPA regulation of 100 µg/L. Comparing the effluent terminal THM concentration of 72 µg/L from the pilot plant with the average raw water THM concentration of 180 µg/L indicates that, indeed, ammonia addition is an effective method for controlling THMs. The results of the ammonification studies also showed the average THM reduction to be about 60% regardless of which coagulant was used: alum, polymer, or alum plus polymer. Results of a statistical analysis of the data showed there was no significant difference among the THM levels produced using the different coagulants while ammonia was being added.

Bacterial Quality

To evaluate the effect of the THM control procedures on the bacterial quality of the finished water, bacterial samples were taken before and during the implementation of the three procedures. The results of the bacterial tests (MPN) during these runs showed no coliform bacteria in the filter effluent except for three isolated samples taken while the raw water turbidity was below 60 NTU. Duplicates of these three samples showed no coliforms. On this basis, these three isolated samples are believed to have been contaminated during laboratory handling. Thus a >98.8% bacterial removal was observed using alum, polymer, and alum plus polymer for the direct filtration process under the conditions of these runs.

The excellent bacterial removal found in this study was mainly the result of very low turbidity (0.25 NTU) in the effluent water. The low-turbidity effluent was accomplished through a very effective coagulation and filtration process. This was brought to light by the results of pilot plant runs that were attempted when the raw water turbidity exceeded 60 NTUs. Above 60 NTUs turbidity in the raw water, the direct filtration pilot plant was unable to meet the 1 NTU effluent turbidity limit and was also unable to meet the bacterial standards. This held true for all pilot plant runs on water with turbidity in excess of 60 NTU. In addition to this, the short chlorine contact time in direct filtration had no adverse effect on bacterial removal as compared with the OSU Plant. Similar bacterial removal efficiencies have also been reported[9] for the direct filtration process.

SUMMARY

The data collected on a direct filtration pilot plant located at the OSU Water Treatment Plant showed that the finished water quality, including turbidity and bacterial removal, for the conventional OSU Plant and the direct filtration pilot plant were comparable. However, when the raw water exceeded 60 NTUs, the direct-filtration pilot plant failed to meet either the turbidity or bacterial standards.

During this study, the terminal THM concentration in the finished water from the direct-filtration plant (using the optimum dosages of alum, polymer, or alum plus polymer) and the finished water from the OSU Water Treatment Plant were higher than the 100 μg/L limit. The high THM levels observed in both processes were probably the result of unremoved natural organics of low apparent molecular weight that were present in the treated water. At the $\alpha = 0.05$ level there was no significant difference among the treatment means of the three coagulant regimes, based on TOC, UV, and THMFP. The molecular weight measurements obtained by gel permeation chromatography were found to be useful in monitoring changes in the distribution of the apparent molecular weight of the naturally occurring organics as they pass through the individual units of the treatment plants.

The apparent molecular weight distribution of the organics in the raw water was in two molecular weight ranges. The first of these was the group range greater than 50,000, and the second and predominant group was in the 1000 to 9000 range. A complete removal of the first weight group organics ($>50,000$) was observed by the time the raw water passed through the prechlorination units in both the pilot plant and the conventional OSU Plant. A partial reduction of the dominant peak, representing the second apparent molecular weight group organics (1000 to 9000), was observed as the water passed through flocculation and filtration.

Because high THM levels were showing up in the pilot plant effluent, three methods were considered for reducing the THM concentrations. Moving the point of initial chlorination to immediately ahead of the filters was not effective in reducing the terminal THMs to less than 100 μg/L. The two procedures using ammonification, at a ratio of 1:4 ($NH_4:Cl_2$) by weight, were effective in limiting the terminal THM concentration to less than 100 μg/L. In both cases, ammonia was added to the filter effluent, the major difference being the points of initial chlorination. In the case of prechlorinated water, chlorine was added at the chlorine contact chamber immediately ahead of the rapid mix unit, whereas in prefilter chlorination, chlorine was added to the water immediately ahead of the filters. The results of all the bacterial tests showed that the chloramines formed by ammonification did not lower the bacterial quality of the finished water from the direct filtration pilot plant.

ACKNOWLEDGMENTS

This work was supported by the Water Research Center of Oklahoma State University, Stillwater, Oklahoma.

REFERENCES

1. Tate, C. H., J. S. Land, and H. L. Hutchinson. "Pilot Plant Tests of Direct Filtration," *J. Am. Water Works Assoc.* 69(7):379-384 (1977).
2. Culp, R. L. "Direct Filtration," *J. Am. Water Works Assoc.* 69(7):375-378 (1977).
3. Amy, W. T. *A Laboratory Study of Factors Affecting the Applicability of Direct Filtration Water Treatment*, M.S. Thesis (Blacksburg, VA: Virginia Polytechnic Institute and State University, 1978).
4. Kawamura, S. "Design and Operation of High Rate Filters—Part 1," *J. Am. Water Works Assoc.* 67(10):535-544 (1975).
5. Tredgett, R. G. "Direct Filtration Studies for Metropolitan Toronto," *J. Am. Water Works Assoc.* 66(2):103-109 (1974).
6. Walder, E. J., W. A. Hays, and P. W. Prendiville. "World's Largest Direct Filtration Facility, The Proposed Prospect Water Treatment Plant," *J. Am. Water Works Assoc.* 67(7):353-359 (1975).
7. Baumann, R. E. "Granular-Media Deep Bed Filtration," Presented at the Water Treatment Plant Design, Fifth Environmental Engineering Conference, Montana State University, Bozeman (1976).
8. Hutchinson, W. R. "High Rate Direct Filtration," *J. Am. Water Works Assoc.* 68(6):292-298 (1976).
9. American Water Works Association Filtration Committee. "The Status of Direct Filtration," *J. Am. Water Works Assoc.*, 72(7):405-411 (1980).
10. McCormick, R. F., and P. H. King. "Direct Filtration of Virginia Surface Waters: Feasibility and Costs," Bulletin 129 (Blacksburg, Virginia: Virginia Water Resources Research Center, 1980).
11. Khan, P. A. *Interaction of Polymers, Natural Organics, and Chlorine in a Direct Filtration System*, Ph.D. Thesis (Stillwater, OK: Oklahoma State University, 1983).
12. "National Interim Primary Drinking Water Regulations: Control of Trihalomethanes in Drinking Water; Final Rule," *Fed. Reg.* November 29, 1979.
13. Rook, J. J. "Formation of Haloforms During Chlorination of Natural Waters," *Water Treat. Exam* 23:234-243 (1974).
14. Williams Brothers Engineering Company. "Water Treatment Plant Study," Prepared for Stillwater Utilities Authority (Stillwater, OK: March 1980).
15. *Standard Methods for the Examination of Water and Wastewater*, 15th ed. (Washington, DC: American Public Health Association, 1980).
16. Richards, J. J., and G. A. Junk. "Liquid Extraction for the Rapid Determination of Halomethanes in Water," *J. Am. Water Works Assoc.* 69(1):62-64 (1977).
17. Babcock, D. B., and P. C. Singer. "Chlorination and Coagulation of Humic and Fulvic Acids," *J. Am. Water Works Assoc.* 71(3):149-153 (1979).

18. Wattier, K. L. *Removal of Trihalomethane Precursors From the Glenmore Reservoir by Direct Filtration and Conventional Water Treatment Methods*, M.S. Thesis, (Potsdam, NY: Clarkson College of Technology, 1982).
19. Tambini, S. J. *Direct Filtration Studies of the Grasse River*, M.S. Thesis (Potsdam, NY: Clarkson College of Technology, 1982).
20. Edzwald, J. K. "A Preliminary Feasibility Study of the Removal of Trihalomethane Precursors by Direct Filtration" (Cincinnati: U.S. Environmental Protection Agency, 1979).
21. Duke, D. T., J. W. Siria, B. R. Burton and D. W. Amundsen, Jr. "Control of Trihalomethanes in Drinking Water," *J. Am. Water Works Assoc.* 72(8):470–476 (1980).
22. Harms, L. L., and R. W. Looyenga. "Chlorination Adjustment to Reduce Chloroform Formation," *J. Am. Water Works Assoc.* 69(6):258–263 (1977).

CHAPTER **94**

Chloropicrin in Potable Water: Conditions of Formation and Production During Treatment Processes

J. P. Duguet, Y. Tsutsumi, A. Bruchet, and J. Mallevialle

Chloropicrin has been found in the prechlorinated water of the drinking water plant at Dunkerque and Cholet, in the north and west of France, respectively.[1] This compound has a structure similar to that of chloroform and is, in fact, produced industrially by reacting nitric acid with chloroform.[2] It is a highly toxic substance that was first used as a poison gas during World War I. In France it is now used for the control of small mammals, particularly foxes, which represent potential spreading agents of rabies.

Chloropicrin as a chlorination by-product has been reported only rarely. Maier and Becke postulated that chloropicrin could be formed either by chlorination of previously ozonated waters and reaction with the nitrogen oxides (NOx) produced in the ozonator tubes.[3] Sayato et al. have shown that chloropicrin can be formed by the action of chlorine on humic acids, amino acids, and nitrophenols and that the presence of nitrite seems to increase the reaction yield.[4]

Our research had two main objectives: (1) to obtain a better understanding of the conditions for chloropicrin formation from various precursors, and (2) to determine changes in chloropicrin concentration resulting from the various processes in a water treatment line.

MATERIALS AND METHODS

Chloropicrin Analysis

Preparation of Standard Solutions

Chloropicrin (>98%) was obtained from Fluka AG, Buchs, Switzerland. Ten microliters chloropicrin was dissolved in 10 mL methanol (solution 1). Twenty microliters solution 1 was then dissolved in 10 mL of methanol (solution 2). Aliquots of solution 2 were added to 7 mL of bottled water (Evian-France) contained in 10-mL vials closed with septa and aluminum screw caps.

Gas Chromatography-Mass Spectrometry (GC-MS) Analysis

Samples from water treatment lines were collected in 1-L glass bottles, then stripped in a closed-loop stripping apparatus (CLSA), according to the Grob method.[5] Extracts in 20 μL carbon disulfide were injected into a Carlo Erba 4160 gas chromatograph (Carlo Erba, Strumentazione, Milan, Italy) interfaced with a Ribermag R 10-10C mass spectrometer (Ribermag, SWG Nermag, Reuil Malmaison, France). The conditions were as follows: 50-m fused silica capillary OV1701 column, film thickness 0.25 μm, 0.32 mm ID; cold on column injection at 25°C, programmed from 25 to 180°C at 3°C/min; full spectrum electron impact at 70 eV (source temperature, 200°C); and mass fragmentography with ions at m/e 30, 117, 119, 121 were used for identification and quantitation.

Gas Chromatographic Analysis

Head-space injection into a Carlo Erba 2150 gas chromatograph (bath temperature of 40°C) on a capillary fused silica 30-m SPB5 column, at room temperature with EC detection, permitted sensitivity limits of 0.5 μg/L. This technique was used for batch experiments with standard compounds.

Gel Permeation Chromatography

Size exclusion chromatography on Sephadex G25 (Sephadex Pharmacia, Uppsala, Sweden) was used to separate the organics. The conditions were as follows: column size, 0.25 × 90 cm; eluent flowrate, 100 mL/h with Millipore water (Millipore, Bedford, MA) as eluent.

Water samples were first concentrated with a rotary evaporator to obtain a TOC concentration between 100 and 200 mg/L. A 10-mL aliquot of this concentrate was then injected into the Sephadex column. UV absorbance, fluorescence, and TOC (Dohrman DC80, Dohrman, Envirotech Corp., Santa Clara, CA) were measured on the 10-mL fractions collected.

Removal of Residual Chlorine in Samples

To control the reaction times of chlorine with the organic matter, a reducing agent such as sodium thiosulfate was generally added to destroy the residual chlorine and thus stop the reaction. An excess of reducing agent decreases the concentration of the chloropicrin measured. This effect is probably the reason why chloropicrin has not often been identified in chlorinated waters. The effect of three reducing agents (sodium sulfite, sodium thiosulfate, and ascorbic acid) on chloropicrin is shown in Figure 1. Experiments with standard solutions of chloropicrin showed that the following reaction is involved:

$$CCl_3NO_2 \xrightarrow{\text{reductor}} CHCl_3$$

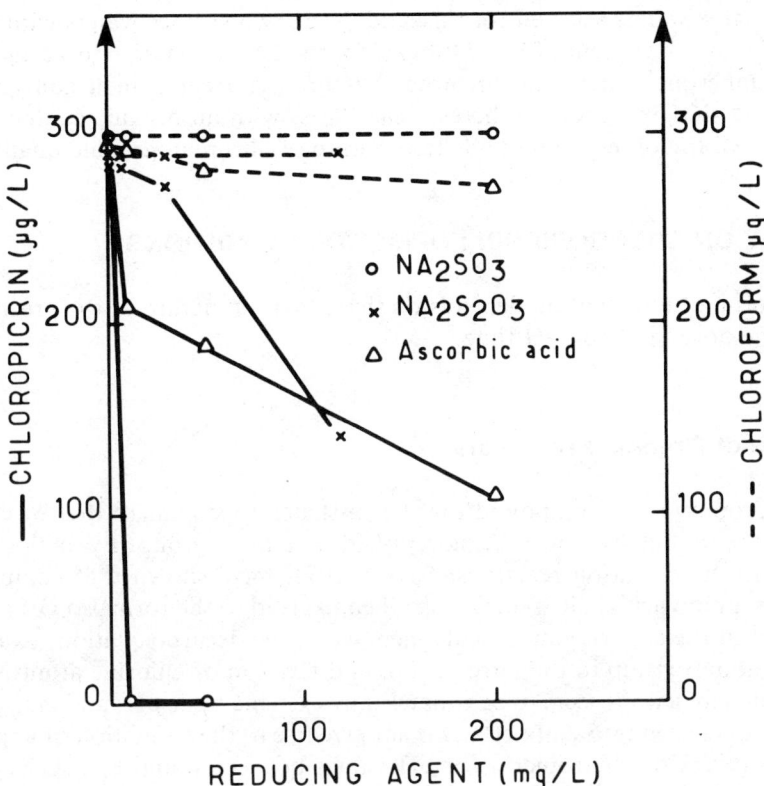

Figure 1. Effect of reducing agent and dose on chloropicrin and chloroform reduction.

In practice, it would be preferable to use thiosulfate in an exact stoichiometric quantity sufficient to only neutralize the residual chlorine in the water samples.

Standard Compounds and Investigated Waters

In the investigation of possible chloropicrin precursors, two types of polyhydroxyaromatic compounds, fulvic and humic acids, were examined. The fulvic acid (obtained from Contech, ETC Ltd., Ottawa, Canada) had the following elemental composition: carbon, 50.9%; hydrogen, 0.75%; nitrogen, 0.25%. The humic acid (obtained from Fluka AG, Buchs, Switzerland) had the following: carbon, 52.4%; hydrogen, 5.13%; nitrogen, 0.4%.

The three waters selected for this study were two surface waters with a high organic carbon content of 8 to 14 mg/L for the Cholet water, 6 to 12 mg/L for the Dunkerque water, and an urban wastewater treated in a conventional activated sludge process (Achères, near Paris) with an organic content of 12 mg/L. All the other compounds tested were of chromatographic quality.

STUDY ON CHLOROPICRIN FORMATION CONDITIONS

Chloropicrin formation depends on three factors: nature of the precursors, chlorine dose, and contact time.

Nature of Organic Precursors

Chloropicrin is a compound that has a structure similar to that of chloroform, the mobile hydrogen being replaced by a nitro group. By analogy with chloroform formation reactions, Sayato et al. have shown that compounds such as amino acids, nitro- or nitrosophenols, lead to the formation of chloropicrin.[4] In the case of amino acids there would be decarboxylation, oxidation of the amino group to the nitro group, and fixation of chlorine atoms. However, these reactions seem to be enhanced by extreme basic pH, and most of the precursors tested by Sayato et al. do not give rise to the formation of appreciable quantities of chloropicrin for pH values between 6 and 8, that is, at the customary conditions of water treatment.

Moreover, numerous tests run on other nitrogen organic compounds at pH 7 (chlorophyll, dichloroacetonitrile, tetramethyl ammonium chloride, N-butylbenzenesulfonamide, Fluka humic acids) gave similar low results (i.e., <5 µg/L chloropicrin for 100 mg/L of standard in the presence of 100 mg/L chlorine for 2 h). In fact, even for humic acids that are good precursors at pH 11 (because of adequate carbon and nitrogen in their structure), the formation of chloropicrin at pH 7 seems to require an appropriate nitrogen mineral form (Figure 2).

Influence of Mineral Forms of Nitrogen

It was equally important to determine which of the mineral sources of nitrogen could participate in chloropicrin formation. As expected, we found that nitrate does not play a prominent role whereas nitrite appears to do so (Table I). Figure 2 shows that the chloropicrin yield rapidly increases with nitrite concentration as long as residual chlorine is present. Batch experiments with fulvic acids (Figure 3) or humic acids indicate that chloropicrin formation can be increased by a factor of 10 in the presence of nitrite.

Table I. Effect of Nitrite and Nitrate on Chloroform and Chloropicrin Formation During Chlorination of Fluka Humic Acid[a]

Constituent	Experiment		
	1	2	3
Nitrites, mg/L	0	0	50
Nitrates, mg/L	0	50	0
Chloropicrin, µg/L	5.8	6	200
Chloroform, µg/L	1000	1000	1000

[a]Conditions: Fluka humic acid, 100 mg/L; chlorine dose, 400 mg/L; pH 7; contact time 24 h.

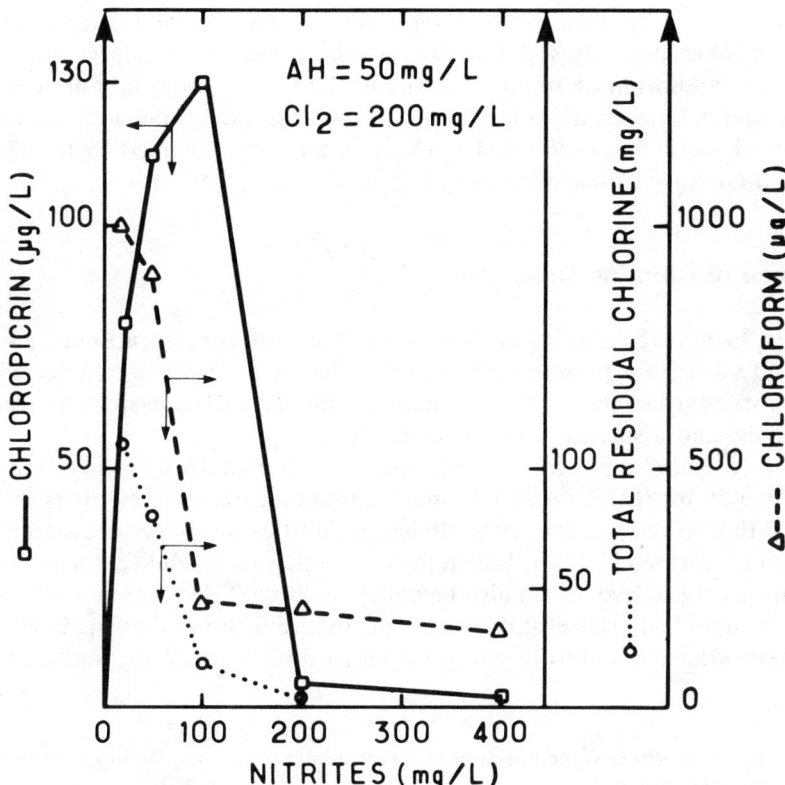

Figure 2. Effect of nitrite concentration on chloropicrin and chloroform formation during chlorination of Fluka humic acid (AH) at pH 7.

Because the conditions of these batch experiments were extreme as compared with industrial water treatment conditions, the following experiment was run. A 20 mg/L solution of resorcinol (TOC, 13 mg/L) was chlorinated with a chlorine dose of 5 mg/L for 2 h. After neutralization, samples were analyzed by the CLSA method. The results indicate that the presence of nitrite promotes the formation of chloropicrin (Table II), and that for customary TOC levels, low nitrite concentrations are sufficient to explain the levels of chloropicrin actually found in full-scale water treatment plants.

Influence of Contact Time

Chlorination tests performed on Contech fulvic acid solutions (Figure 3) as well as on the Cholet water (Figure 4) show that, in the presence of an excess of chlorine, half of the chloropicrin is formed in less than 1 h, whether or not 20 mg/L nitrite has been added to the solution. For contact times greater than 1 h, the reaction appears to be slower. The comparison of Figures 3 and 4 reveals another interesting point. Despite the presence of a large excess of chlorine, the addition of 20 mg/L nitrite to the Cholet water (about 20 mg/L organic matter) has clearly a much lower effect than the same addition to the solution of fulvic acids (50 mg/L). These results are explained by the difference in the nature of organic content.

Influence of Chlorine Dose

The influence of the chlorine dose depends on the concentration of ammonium and nitrite ions present in the solution. Three examples were selected for study: a standard solution of Fluka humic acids, an urban wastewater treated at Achères, and a surface water at Dunkerque.

Figures 5 and 6 represent chlorination of humic acid solutions (50 mg/L) containing 50 mg/L nitrite and 45 mg/L ammonia, respectively. It is seen in Figure 5 that in the presence of sufficient quantities of nitrite, the amount of chloropicrin formed is closely linked to the quantity of residual chlorine. From the shape of this curve, it can also be concluded that THM formation kinetics are more rapid than those of chloropicrin. Figure 6 shows that the beginning of the appearance of chloropicrin coincides with the start of the breakdown of

Table II. Effect of Nitrite Concentration on Chloropicrin Formation During Chlorination of Resorcinol[a]

Nitrite concentration (mg/L)	Chloropicrin concentration (μg/L)
0	0
0.05	1
0.10	2.1
0.20	2.4

[a]Conditions: resorcinol concentration, 20 mg/L; chlorine dose, 5 mg/L; pH 7; contact time, 2 h.

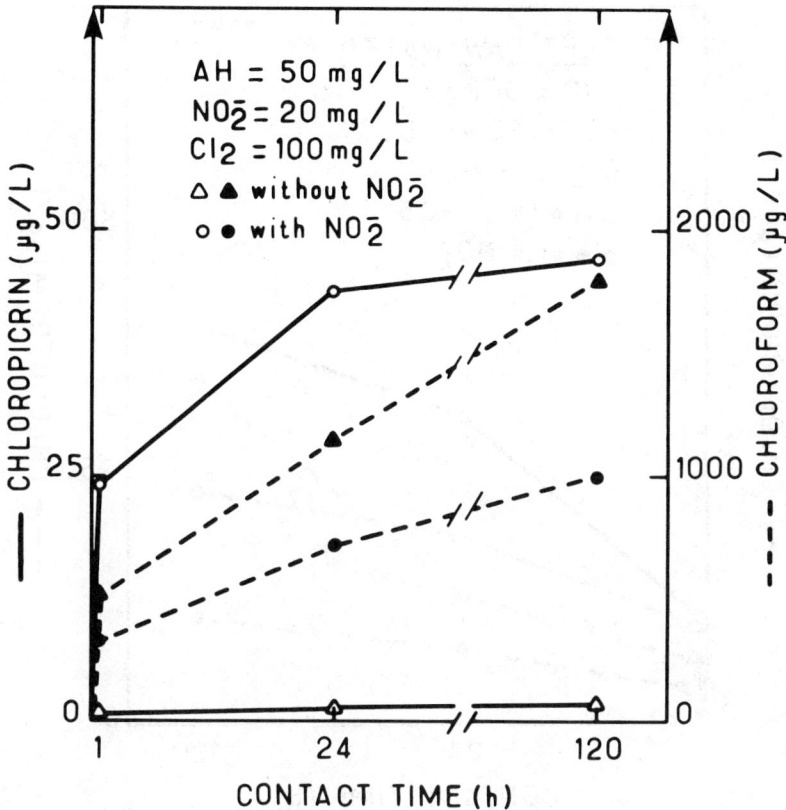

Figure 3. Effect of contact time on chloropicrin and chloroform formation during chlorination of Contech fulvic acid (AH) with and without nitrite present.

monochloramine. The fact that the chloropicrin content remains low and tends to diminish after the breakpoint is due to the lack of nitrogen present in a usable form (i.e., nitrite).

For comparison, we chlorinated two waters with organic carbon and nitrite content of the same order of magnitude but with very different ammonia concentrations, 25 mg/L for the Achères water (Figure 7) and 0.8 mg/L for the Dunkerque water (Figure 8). Although the curves showing chloropicrin formation vs Cl:N ratio have similar shapes (i.e., slight formation before the breakpoint and change in the slope above the breakpoint), the absolute chloropicrin values obtained were much greater in the Achères case. This difference is probably due to the greater concentration of the oxidant introduced, a hypothesis confirmed by the similar differences observed in the chloroform contents.

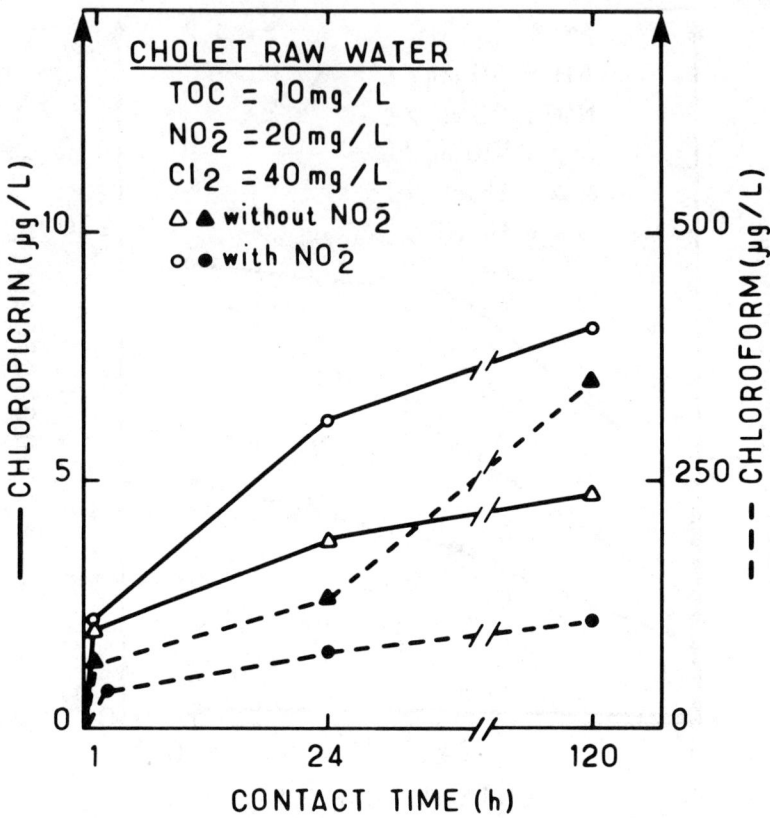

Figure 4. Effect of contact time on chloropicrin and chloroform formation during chlorination of the Cholet raw water.

FATE OF CHLOROPICRIN ALONG VARIOUS TREATMENT LINES

Two particularly significant examples, the Cholet and Dunkerque plants, were studied to determine the effects of treatment processes on chloropicrin content.

Cholet Plant

This plant treats 1500 m³/h of a reservoir water highly loaded with natural organic matter (11 mg C/L at time of experiment), which also contains traces of ammonia and nitrite (0.1 mg/L each).

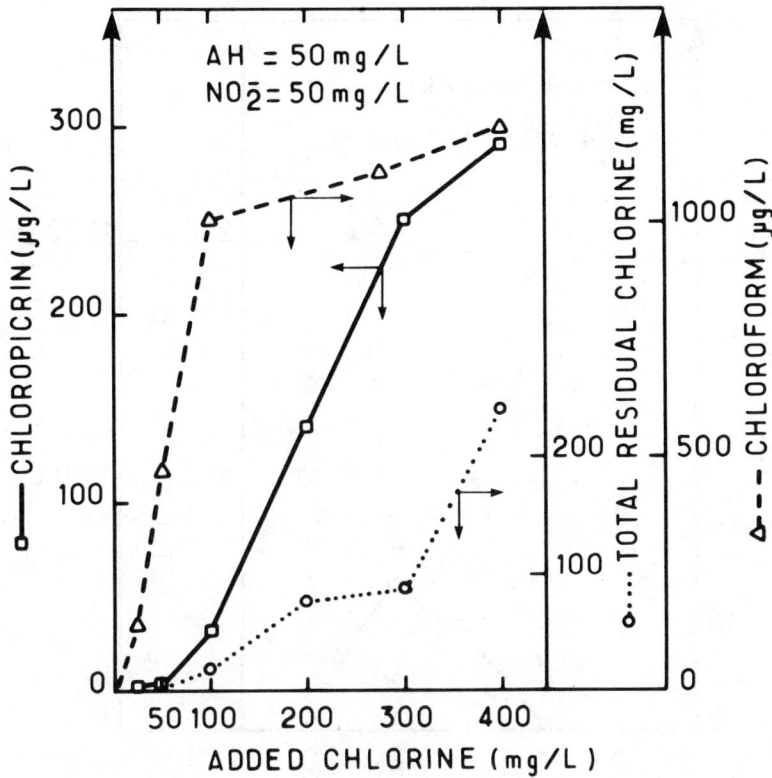

Figure 5. Effect of chlorine dose on chloropicrin and chloroform formation during chlorination of Fluka humic acid (AH) with nitrite present.

The raw water is prechlorinated above the breakpoint and then distributed over three lines operating in parallel. Each line includes coagulation, settling, sand filtration, ozonation, and postchlorination. Table III gives a typical example of the results obtained on one of these lines. The value given for tap water corresponds in fact to the mixture of the three treated waters. The chloropicrin concentrations were always about $\mu g/L$.

Possible sources of nitrogen were investigated. The total organic nitrogen concentration was less than 0.4 mg/L, and determinations of free and combined amino acids gave values of 30 and 370 $\mu g/L$, respectively. Because of their low reactivity at pH 7, these concentrations cannot explain the formation of chloropicrin.

Because of the humic content of this water and traces of nitrite in this plant, and because of the experiment with resorcinol where TOC, nitrite concentration, and chlorination conditions were similar to those of the Cholet plant, we concluded that humic acid and nitrite are most probably the most important factors in chloropicrin formation, even if other chloropicrin precursors cannot

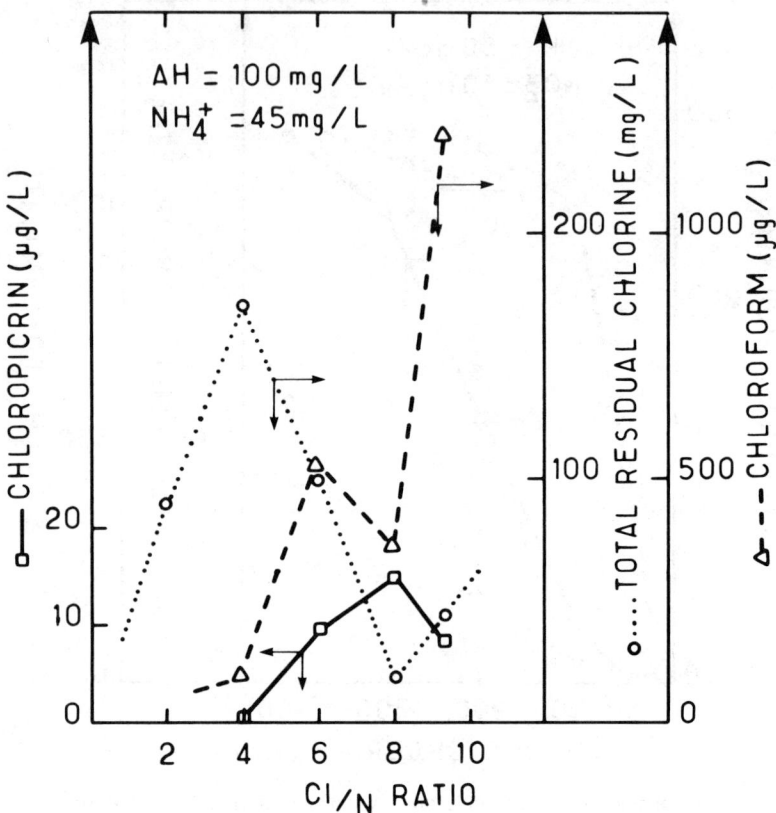

Figure 6. Effect of chlorine dose on the chloropicrin and chloroform formation during chlorination of Fluka humic acid (AH) with ammonia present.

be completely excluded. Separation of organic matter from the raw water by exclusion chromatography (Sephadex G25) into three molar weight ranges (greater than 5000, between 1000 and 5000, and less than 1000) followed by determination of chloropicrin formation potential, showed that the organic precursors were almost homogeneous within the various molar weight ranges.

Dunkerque Plant

This plant treats 1000 m³/h of a surface water that is rich in organic matter (TOC, 8 mg/L) and contains 2 mg/L ammonia and 0.35 mg/L nitrite.

The plant treatment consists of breakpoint prechlorination, coagulation with iron chlorosulfate followed by flotation, and filtration on granular activated carbon, before the water passes into the recharge basins.

DRINKING WATER TREATMENT

Table III. Fate of Chloropicrin in the Cholet Plant

Process	Chloropicirin concentration (ng/L)
Raw water	< 10
Sand filtration	2400
Ozonation	1700
Tap water	1000

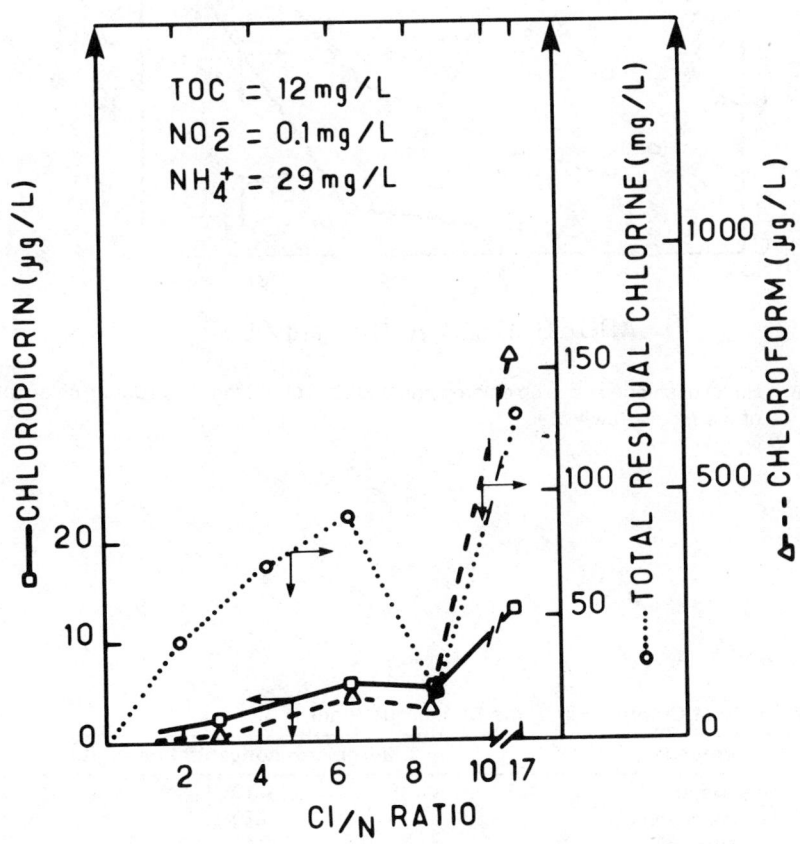

Figure 7. Effect of chlorine dose on chloropicrin and chloroform formation during chlorination of the Achères water.

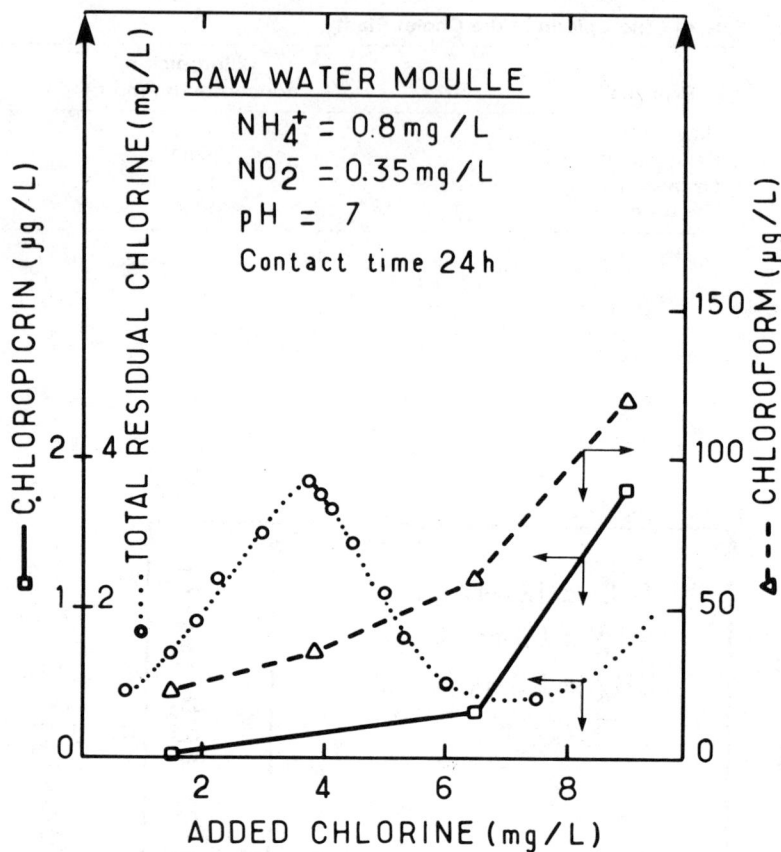

Figure 8. Effect of chlorine dose on chloropicrin and chloroform formation during chlorination of the Moulle raw water.

Table IV. Fate of Chloropicrin in the Dunkerque Plant

Process	Chloropicrin concentration (ng/L)
Raw water	<10
Prechlorination	270
Air flotation	110
GAC filtration	20
Groundwater	<10

Table IV gives an example of the results generally obtained in this plant. The quantity of chloropicrin formed during prechlorination is partially reduced at the flotation stage, probably by a stripping effect. Adsorption on activated carbon then removes practically all the chloropicrin, which has never been found in the groundwater.

CONCLUSIONS

Chloropicrin may be found in prechlorinated surface waters. Its presence can be detected only when residual chlorine is stoichiometrically removed with thiosulfate. Chloropicrin is present in the microgram-per-liter range in surface water with an organic content of about 10 mg/L TOC. Because of its toxicity, this fact raises some concern about the mechanism of its formation and its fate along the drinking water treatment processes. Precursors to its formation may include a variety of organic nitrogen compounds, but nitrite and residual chlorine, as well as organic compounds analogous to the trihalomethane precursors (humic acids), are most important.

Chloropicrin is effectively removed by granular activated carbon filtration, but the simplest control is the abandonment of prechlorination.

REFERENCES

1. Mallevialle, J., A. Bruchet, and E. Schmitt. "Nitrogenous Organic Compounds: Identification and Significance in Several French Water Treatment Plants," presented at the American Water Works Association Congress, Norfolk, VA, Dec. 4–7, 1983.
2. Grignard, V. *Précis de Chimie Organique*, (Paris: Masson, 1958), p. 248.
3. Maier, D., and C. Becke. "Bildungsmoglichkeiten fur Nitrierte Organische Wasserinhaltstoffe bei der Trinkwasseraufbereitung," presented at the International Ozone Association Congress on Environmental Impact and Benefit of Ozone, Brussels, Sept. 12–13, 1983.
4. Sayato, Y., K. Nakamuro, and S. Matsui. "Studies on Mechanism of Volatile Chlorinated Organic Compound Formation (III). Mechanism of Formation of Chloroform and Chloropicrin by Chlorination of Humic Acid," *Suishitsu Odaku Kenkyu* 1982:127–134.
5. Grob, K. "Organic Substances in Potable Water and in its Precursors. Part I. Methods for their Determination by Gas-Liquid Chromatography," *J. Chromatogr.* 84:255–273 (1973).

CHAPTER **95**

Preozonation in Drinking Water Treatment: Nondisinfection Applications of Ozone

Rip G. Rice

The application of ozone to the treatment of drinking water has been practiced continuously since the beginning of this century. The city of Nice, France, has been using ozone since 1906 to disinfect its drinking water. When used as a disinfectant, ozone is applied as the last or second-to-last treatment step conducted in the plant.

During an Environmental Protection Agency (EPA)-sponsored survey of European uses for ozone and chlorine dioxide in drinking water treatment, Miller et al.[1] found more than 1000 water treatment plants using ozone. About half of these plants were located in France, and most were using ozone as the primary disinfectant. This means that the bulk of the disinfection work is performed by ozone, after which a small quantity of chlorine, chlorine dioxide, or chloramine is added to maintain a residual in the distributed water.

On the other hand, the other 50% of the plants found to be using ozone in 1977 were exploiting the oxidation benefits of ozone for purposes other than disinfection. These applications include taste and odor control, color removal, iron and manganese oxidation, and others that will be discussed in more detail.

When used as a chemical oxidant, ozone is applied in the initial stages of water treatment, and always before a filtration step. Thus, when ozone is used as a chemical oxidant, the process is rapidly becoming known as "preozonation."

There are currently ten known major preozonation applications in U.S., European, and Canadian water treatment plants,[2] none of which are related directly to disinfection:

1. control of algae
2. removal of tastes and odors
3. color removal
4. removal of iron and manganese
5. microflocculation (oxidation of soluble micropollutants to produce insoluble, but readily filterable, materials)
6. removal of turbidity (by oxidative flocculation)
7. organics removal (by oxidation of phenols, detergents, some pesticides, etc.)
8. partial oxidation of dissolved organics for subsequent biological oxidation of total organic carbon and conversion of ammonia to nitrate

9. control of halogenated organic compounds
10. control of trihalomethane (THM) formation potential

With the recent realization that excessive use of chlorination during water treatment produces halogenated by-products of organic materials that may increase public health risks to consumers, water treatment professionals around the world have been seeking alternative treatment approaches to lower the levels of halogenated and other organic materials in treated water supplies. The judicious use of the chemical oxidant ozone in many water treatment plants is lowering these levels and providing high-quality drinking water in a cost-effective manner. How this is being accomplished with preozonation is the subject of this chapter.

MAJOR APPLICATIONS FOR OZONE

To fully understand the potential benefits of preozonation processes, it is important to differentiate clearly between the two major applications of ozone in water treatment—disinfection and chemical oxidation. Applications for ozone discussed under the topic "Preozonation" are those in which ozone is applied as a chemical oxidant.

Disinfection

Conditions for using ozone as a disinfectant are well established[3,4] and have been adopted as standards by public health officials in France and by the World Health Organization. When a concentration of 0.4 mg/L dissolved ozone is maintained for a minimum of 4 min, 99.9% inactivation of poliovirus types I, II, and III is obtained.

In many water treatment plants, the required 0.4 mg/L residual ozone concentration is customarily developed in the first of two or more ozone contact chambers, then is maintained for 6 to 10 min in the second contact chamber. Ozone demand of the water being treated is satisfied in the first chamber and maintained in the second.

Preozonation

In 1977 only two U.S. water treatment plants were using ozone—Whiting, Indiana, for taste and odor control, and Strasburg, Pennsylvania, for disinfection. By early 1985, however, 24 U.S. plants were operating with ozonation, 17 of which use preozonation for a variety of purposes. The number of Canadian plants using ozone has also grown, from 24 in 1977 to at least 55 in early 1984.[5] The number of European plants using ozone also continues to grow, but no official tabulation has been made since 1977.

Table I lists the 20 water treatment plants in the United States known to be using ozone as of early 1984, including five additional plants under construction, two more being designed, and two sites known to be conducting pilot plant testing of ozonation. Most of the full-scale preozonation applications are for taste and odor removal; however, two plants remove iron and manganese, seven remove color, four oxidize THM precursors, two use ozone to improve flocculation, and one controls algae. Some of these plants use ozone for more than one purpose.

Multiple-Stage Ozonation

In a number of European water treatment plants, in which ozonation was installed many years ago for disinfection at the end of the process, preozonation has been added as an economical second step with the following beneficial effects: (1) lowering disinfectant demand, (2) lowering dissolved organic contents, (3) eliminating the need for prechlorination and lowering the amount of other chemicals required in former pretreatment steps, and (4) producing higher-quality water for distribution.

Two articles describe the German "Mülheim process" installed at the Dohne plant in 1977, in which preozonation replaced breakpoint chlorination, and a second ozonation step now precedes granular activated carbon (GAC) adsorption.[6,7] The costs of installing two-stage ozonation in Mülheim, doubling the GAC adsorber bed depths and operating the new process, were offset by the savings obtained in chemicals and labor costs.

The Rheinisch-Westfälischen Water Works Company installed this same two-stage process in two other nearby treatment plants during the period 1980–82.

Three large, French water treatment plants in the suburbs of Paris have installed still a third stage of ozonation within the past 3 years, between the preozonation and ozone disinfection steps, to take advantage of promoting biological activity and lowering the levels of organic contamination by GAC adsorption without increasing the concentrations of undesired chemicals.[8]

Significantly, annual costs for the new process, which includes two preozonation steps, GAC adsorbers, and ozone disinfection, are only 6% higher than the costs for the older process, which had no GAC adsorption but included ozone disinfection. The older process required an average of 4 mg/L ozone dosage for disinfection only. The new process requires a total dosage of 4.5 mg/L ozone for all three stages of ozonation.

Three U.S. plants have recently installed two-stage ozonation processes, and a fourth is under design (see Table I).

Table I. U.S. Potable Water Treatment Plants Using Ozone as of May 1984

Location	Primary Purpose of Ozone	Startup Date	Av Flow Rate mgd	Av Flow Rate m³/d
In Operation				
Whiting, IN	T & O[a]	1940	4	15,142
Strasburg, PA	Disinfection	1973	0.1	379
Grandin, ND	Fe & Mn[b]	1978	0.05	189
Saratoga, WY	T & O	1978	3.5	13,249
Bay City, MI	T & O	1978	40	162,487
Monroe, MI	T & O	1979	18	73,119
Newport, DE	Disinfection	1979	0.25	1,016
Newport, RI	THM precursors; T&O; color	1980	5	20,311
Tarrytown, NY	T & O	1980	1.2	4,875
Kennewick, WA	Color; T&O	1980	3	12,187
Elizabeth City, NC	Color	1981	5	20,311
Casper, WY	Disinfection	1982	5	20,311
Ephrata Borough, PA	T & O	1982	0.145	549
New Ulm, MN	Fe & Mn	1982	2.6	10,562
South Bay, FL	Color	1982	2.2	8,937
Rockwood, TN[c]	Flocculation; THM precursors	1982	6	22,712
Potsdam, NY	Color	1984	1	3,786
Beria, OH[c]	THM precursors; T&O	1984	3.6	14,348
Belle Glade, FL[c]	Color, THM precursors; algae	1984	6	22,172
Stillwater, OK	Color	1984	5	18,927

DRINKING WATER TREATMENT

Under Construction			
Los Angeles, CA	Microflocculation & organics	1986	2,195,531
New York, NY	Organics	1986	11,358
Hackensack, NJ	Color, Fe & Mn; THM precursors	1986 (bid)	378,540
Ormand Beach, FL	THM precursors; color	1985 (bid)	15,142
Costa Mesa, CA	H_2S; color	1985	18,927
Under Design			
Myrtle Beach, SC[c]	Color; THM precursors	1987	121,865
Pilot Plant Studies			
Rocky Mount, NC	Color; THM precursors		580
Celina, OH	T&O; THM precursors		3 (pilot)
			100
			4
			5
			30

[a] Taste and odor removal.
[b] Iron and manganese removal.
[c] Two-stage ozonation plants.

PREOZONATION APPLICATIONS

Control of Algae

At the Dohne plant in Mülheim, Federal Republic of Germany, where preozonation dosages of about 1 mg/L are used to aid in microflocculation, an additional 1 mg/L dosage of ozone during periods of algal growth controls the problem effectively.[9] Because algae generally occur seasonally, it is cost-effective to couple this application of preozonation with a second application, which requires more frequent use of ozone.

Removal of Tastes and Odors

Phenols and other compounds that cause organoleptic problems are often readily oxidized to inoffensive products by small (<3 mg/L) dosages of ozone. Ozonation for this purpose should be conducted before or concurrently with chemical treatment and before filtration.

Color Removal

Ozone is particularly reactive toward chromophoric groups, cleaving the carbon-carbon double bonds to produce ketones, aldehydes, or carboxylic acids, depending on other substituents on the carbon atoms affected, the amount of ozone applied, the ozone contact time involved, and the temperature. As soon as the conjugation has been disrupted by ozone, the color disappears.

Polar groupings formed on ozonation can now combine with polyvalent cations such as aluminum or ferric ions added as flocculating agents, and thus can be more easily precipitated from the treated water. Therefore, when using ozone for color removal, it should be applied before chemical addition and be followed by filtration.

Removal of Iron and Manganese

Ferrous iron is rapidly oxidized by ozone to ferric ions, which then hydrolyze, coagulate, and precipitate as the insoluble ferric hydroxide.

Similarly, divalent manganous ions are easily oxidized to the tetravalent manganic ions, which then hydrolyze and rapidly form the insoluble manganese dioxide. However, excessive ozonation of manganese compounds will continue the oxidation process and produce the soluble septavalent permanganate ion, which is pink in color.

Microflocculation

During ozonation of some clear waters, such as when ozone disinfection is conducted on waters containing relatively large quantities of dissolved organics, the turbidity of the ozonized water sometimes will increase, because the polar moieties can now combine with polyvalent cations also present to form higher-molecular-weight structures that will precipitate. This phenomenon of turbidity formation after ozonation has been termed "microflocculation" by Maier.[10] This is another reason to follow ozonation with filtration.

Removal of Turbidity

When turbidity is caused by colloidal-sized suspended solids, ozonation sometimes will change the chemical nature of the particle surfaces, allowing the particles to coagulate and settle. In such instances, ozonation decreases the level of turbidity. The process of removing turbidity is aided by the addition of coagulating agents.

In 1986, the City of Los Angeles, California, will start up what will be the largest drinking water treatment plant in the world using ozone. Some 3311 kg (7300 lb)/d ozone will be generated from oxygen to treat 2.2 million m^3/d (580 Mgd) to lower turbidity to <0.3 unit. The ozonized and chemically treated (with iron salts and cationic polymer) raw water will be coagulated, flocculated, then filtered through 6 ft of anthracite coal at a rate of 13.5 gal min^{-1} ft^{-2}. This high filtration rate is made possible by using ozone as a microflocculant, without which the filtration rates of the coagulated and flocculated water could not exceed 9 gal min^{-1} ft^{-2}.[11]

Ozone Pretreatment for Biological Processing

Under ozonation conditions normally applied in treating drinking water, dissolved organic compounds are partially oxidized, and the dissolved oxygen content of the water is increased, thereby improving conditions for growth of aerobic microorganisms. Such organisms can convert dissolved organic impurities to CO_2 and water, but can also convert ammonia to nitrate.

The biological treatment of drinking water following preozonation is conducted in a variety of manners:

1. Storage of preozonized raw water several days in a reservoir.
2. Ozonation of plant intake waters in conjunction with chemical addition, then filtration. Nitrification also occurs in sand filters, along with some removal of dissolved organic carbon.
3. GAC filters/adsorbers placed after sand filter beds. These also contain considerable aerobic bioactivity, which degrades dissolved organic carbon.

Because of the increased biodegradation that occurs in the GAC adsorbers, the breakthrough time for the GAC is postponed. In European water treatment plants that use this biological activated carbon process, it has been shown that the useful lifetime of the GAC can be extended by factors up to six times, simply by incorporating ozone oxidation prior to filtration through sand and anthracite.[12]

Extension of the GAC operating life results in cost savings that, in some instances, have been shown to pay for the costs of installing preozonation.

Ozone Treatment of THMs or THM Precursors

It is axiomatic that once THMs are produced, chlorine has been wasted and the THMs are costly to remove. Thus, the discerning water treatment engineer should modify the treatment process to minimize the concentrations of THM precursors before the addition of chlorine.

Preozonation of raw waters, followed by chemical addition, flocculation, sedimentation, and filtration, appears to be one effective approach to lowering the concentrations of THM precursors, and, therefore, the concentration of THMs ultimately produced. This means that prechlorination should be abandoned in favor of precursor minimization techniques.

MULTIPLE-STAGE OZONATION

Belle Glade, Florida

One of the first two-stage ozonation water treatment plants in the United States recently began operating at Belle Glade, Florida.[13] Raw water for this plant is taken from Lake Okeechobee, which has very high contents of humic color and other naturally occurring organic materials, turbidity, and biological growths. Normally the lake waters exhibit color values of about 100 platinum-cobalt units, chemical oxygen demand levels in excess of 60 mg/L, and total organic carbon (TOC) levels about 30 mg/L. After storms, these levels have risen to as high as 100 and 75 mg/L, respectively, with color levels peaking at 500 color units.

Historically, this raw water has been prechlorinated for algae control and partial color removal, followed by lime softening and treatment with alum, synthetic polymers, and recycled lime sludge, then recarbonation prior to filtration and chlorine disinfection. By this process, the Belle Glade plant normally produces total THM levels of 400 to 1600 μg/L (ppb) on a seasonal basis.

Moving the point of chlorination was not effective in that serious algae blooms were then encountered within the treatment plant. The respiratory products of the algae and the saprophytic bacteria were not removed by the conventional treatment process without prechlorination; postchlorination caused taste, odor, and additional THM problems.[14]

Figure 1 shows a schematic of the new two-stage ozonation process that was developed for treating the Lake Okeechobee water at Belle Glade. The primary objective of the process was to remove THM precursors via ozonation while maintaining free-residual chlorination.

Raw water is preozonized to modify the structure of the THM precursors, making them more susceptible to physical removal by means of added lime and polyelectrolytes after flash mixing. After clarification, the water is recarbonated for pH control, postozonized to polish remaining organics, filtered to

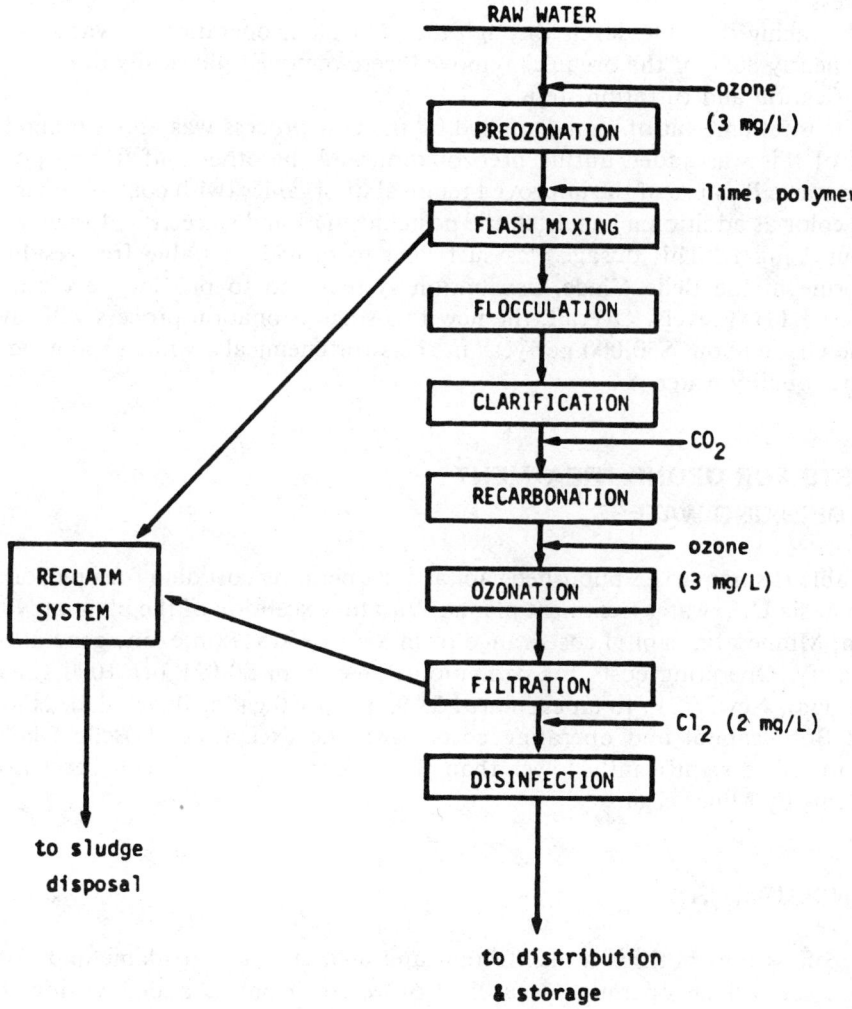

Figure 1. Two-stage ozone treatment system[14] at Belle Glade, Florida.

remove any remaining apparent color and filterable organic material, and postchlorinated to develop a free residual of chlorine for the distribution system.

This process was demonstrated over a 12-week period in a 5-gal/min pilot plant unit. During this period, terminal THM concentrations were consistently maintained below 100 μg/L (ppb). Using free chlorine disinfection, the THM content could be consistently maintained as low as 25 ppb. With combined chlorine residual, the THM content could be adjusted to even lower levels. During the same time period, the conventional process generated 600 to 1600 ppb of THMs in the main plant. Thus, an 80% reduction in total THM formation was obtained during the test period using the two-stage ozonation process.

By tracing the TOC levels through the pilot plant operation, it was shown that nearly 80% of the organics removed were removed physically during the clarification and filtration steps.

The total amount of ozone applied by the new process was about 6 mg/L; half of this was added during preozonation and the other half during postozonation. Because of the improved removal of organics (with control of algae and color as additional benefits), the postchlorination disinfection dosage was about 2 mg/L. This dosage was sufficient to provide a stable free residual chlorine in the Belle Glade distribution system and to provide very much lowered THM levels. Overall, the new two-stage ozonation process will save Belle Glade about $50,000 per year in costs for chemicals, while producing a higher-quality water.[16]

COSTS FOR OZONE TREATMENT OF DRINKING WATERS

Table II summarizes published capital and operating cost data for preozonation at six U.S. water treatment plants. With the exception of the plant at New Ulm, Minnesota, capital costs range from $661 to $1517/lb ozone generation capacity. Operating costs for ozonation range from $0.021 per 1000 gal at Potsdam, New York, to an estimated $3.95 per 1000 gal at Belle Glade, Florida. Both capital and operating costs (with the exception of Belle Glade) reported are significantly lower than those found for European ozonation systems by Miller et al.[1]

CONCLUSIONS

Ozone serves both as a disinfectant and as a chemical oxidant, and both processes will be operative regardless of where ozone is added during the treatment process. When used as a disinfectant, ozone is applied at or near the end of the treatment process; when used as an oxidant, ozone normally is applied at an early stage.

Table II. Summary of Ozone Capital and Operating Costs at Operational U.S. Water Treatment Plants

Plant	Year Started	Size (m³/d)	Size (Mgd)	Amount O₃ Generated (kg/d)	Amount O₃ Generated (lb/d)	Capital Cost ($/lb capacity)	Operating Cost (¢/1000 gal treated)
Monroe, MI[15]	1979	68,137	18	113	225	1,050	0.636
Bay City, MI[16]	1978	151,416	40	454	1,000	661	0.448
New Ulm, MN[17]	1982	9,811	2.6	341	75	6,400	Not given
Potsdam, NY[18]	1983	9,842	2.6	68	150	1,517	0.021
Los Angeles, CA[11]	1984	2,200,000	580	3,311	7,300	716	0.295
Belle Glade, FL[13]	1984	22,712	6	227	500	1,340[a]	3.95
						1,060	

[a]Includes costs of two-stage contacting plus housing.

The ability of preozonation to partially oxidize organic materials in raw waters leads to their more efficient removal by subsequent flocculation and filtration. In many cases, preozonation saves chemical costs. At the new Los Angeles plant, preozonation with chemical treatment allows a 50% faster filtration rate. At Belle Glade, two-stage ozonation will save $50,000/year in chemicals while producing a higher-quality drinking water.

Preozonation has replaced breakpoint prechlorination at many European water treatment plants and at Belle Glade, thus moving the point of chlorine addition until much later in the treatment process after most of the organics have been removed. This results in much lower quantities of halogenated organics being produced.

The most modern French water treatment plants are now using triple-stage ozonation (preozonation prior to reservoir storage, a second preozonation prior to GAC adsorption, and ozone disinfection) to produce high-quality drinking waters. This is being accomplished with < 30% more ozone than that originally required for disinfection alone.

Although the capital costs for installing ozonation equipment are high compared with other oxidants, operating costs are low. Capital costs for ozone generation in some of the newer U.S. water treatment plants range from $661 to $1500/lb of daily ozone generating capacity. Operating costs range from $0.02 to $3.95 per 1000 gal treated.

Most organic materials cannot be oxidized completely to carbon dioxide and water under ozonation conditions normally used in water treatment plants. Partial oxidation products generally are more biodegradable than the original compounds. Thus, ozonation can be used to promote biological activity, which, when properly designed and operated within the water treatment plants, can lower the total organic carbon and ammonia contents of the treated water.

Whenever any chemical oxidant is added to waters containing high concentrations of dissolved organic materials, oxidation products will form. Although each chemical oxidant produces organic oxidation products specific to its oxidation chemistry, many of the same products are formed by all chemical oxidants. Minimizing the formation of organic oxidation products requires maximizing the removal of organics by physical and chemical techniques before oxidants are added.

REFERENCES

1. Miller, G. W., R. G. Rice, C. M. Robson, R. L. Scullin, W. Kühn, and H. Wolf. "An Assessment of Ozone and Chlorine Dioxide Technologies for Treatment of Municipal Water Supplies," EPA Report No. 600/2-78-147 (Washington, DC: U.S. Environmental Protection Agency, 1978).
2. Rice, R. G., C. M. Robson, G. W. Miller, and A. G. Hill. "Uses of Ozone in Drinking Water Treatment," *J. Am. Water Works Assoc.* 73(1):44–57 (1981).

3. Coin, L., C. Hannoun, and C. Gomella. "Inactivation of Poliomyelitis Virus by Ozone in the Presence of Water," *Presse Med.* 72(37):2153 (1964).
4. Coin, L., C. Gomella, C. Hannoun, and J. C. Trimoreau. "Inactivation of Poliomyelitis Virus in Water," *Presse Med.* 75(38):1883 (1967).
5. Larocque, R. L., Hankin Environmental Systems, Toronto, private communication (1984).
6. Sontheimer, H., E. Heilker, M. Jekel, H. Nolte, and F.-H. Vollmer. "The Mülheim Process," *J. Am. Water Works Assoc.* 70(7):393–396 (1978).
7. Heilker, E. "The Mülheim Process for Treating Ruhr River Water," *J. Am. Water Works Assoc.* 71(11):623–627 (1979).
8. Rapinat, M. "Recent Developments in Water Treatment in France," *J. Am. Water Works Assoc.* 74(12):610–617 (1982).
9. Vollmer, F.-H. Mülheim Water Works, Mülheim, Federal Republic of Germany, private communication (1979).
10. Maier, D. "Microflocculation With Ozone," in *Oxidation Techniques in Drinking Water Treatment*, W. Kühn and H. Sontheimer, Eds., EPA 570/9-79-020 (Washington, DC: U.S. Environmental Protection Agency, 1979), pp. 394–417.
11. Stolarik, G. "Ozonation—Direct Filtration of Los Angeles Drinking Water," presented at the Sixth Ozone World Congress of the International Ozone Association, Washington, DC, May 1983.
12. Rice, R. G., and C. M. Robson. *Biological Activated Carbon* (Stoneham, MA: Ann Arbor Science Publishers, Inc., 1983).
13. Elefritz, R. A., D. W. Porter, and S. F. Morris. "The Application of Ozone in Softening Processes for Cost-Effective THM Control: Two Case Histories," presented at a Seminar on Strategies for the Control of Trihalomethanes, American Water Works Association Southeast Annual Conference, Jekyll Island, GA (April 1984).
14. Wagner, R., and R. A. Elefritz. "Ozonation for Effective THM Control," *Public Works* 114(4):46–48 (1983).
15. LePage, W. L. "The Anatomy of an Ozone Plant," *J. Am. Water Works Assoc.* 73(2):105–111 (1981).
16. Croy, R. S. "Bay City Plant is Newest, Largest in U.S. Using Ozone for Drinking Water Treatment," Part I: *OZONEws* 8(6) (1980), Part II: *OZONEws* 8(7) (1980), International Ozone Association, Norwalk, CT.
17. Kirk, J. T. "Ozonation Reduces Manganese Concentrations," *Public Works* 115(1):40–42 (1984).
18. Simcoe, W. D. "Potsdam, New York Water Treatment Plant: A Case History," *Ozone Sci. Eng.* 5(1):51–60 (1983).

CHAPTER 96

Mechanisms of Organic Halide Formation During Fulvic Acid Chlorination and Implications with Respect to Preozonation

David A. Reckhow and Philip C. Singer

The chlorination of raw drinking waters and humic substances has been extensively investigated for the past decade in an effort to better understand the formation of trihalomethanes (THMs) and other potentially harmful chlorination by-products.[1-5] While several empirical models have been developed for the formation of THMs,[6-8] there is much progress yet to be made in understanding the chemical transformations that occur during the formation of THMs and other organic halides from natural organic material. It is only by understanding some of the chemical mechanisms involved in organic halide formation that the impact of control strategies and alternative treatment techniques (e.g., coagulation, ozonation) on organic halide formation can be predicted.

Aside from THMs, many other organic halides comprising the TOX (total organic halide) content have been found in chlorinated humic solutions and drinking waters. Two of the major nonvolatile products are trichloroacetic acid (TCAA) and dichloroacetic acid (DCAA).[5,9,10] These two halogenated acids plus the THMs have been found to account for as much as 50% of the TOX in chlorinated fulvic acid solutions.[4,11] A large number of other halogenated acids, alkyl nitriles, and ketones have also been identified.[9,12]

Three volatile halogenated products that have been studied to a greater extent are dichloroacetonitrile (DCAN), 1,1,1-trichloroacetone (TCAC), and chloral hydrate. These compounds can all undergo base-catalyzed hydrolysis and are therefore considered to be long-lived intermediates rather than end products in the chlorination of natural organics.

Dichloroacetonitrile hydrolyzes slowly at pH 7. By analogy to its brominated analogue, the final product may be DCAA.[13] The decomposition of DCAN is greatly accelerated by the presence of aqueous chlorine.[14] Although the products formed from the chlorination of DCAN are not well known, trichloroacetonitrile and TCAA are distinct possibilities. The concentration of DCAN in finished drinking waters has been found to average about 10% of the THM concentration.[14]

Gurol et al.[15] studied chloroform production from the base-catalyzed hydrolysis of TCAC in the absence of chlorine. They noted that in the presence of aqueous chlorine, an additional pathway exists that greatly increases the rate of chloroform production. Gurol and co-workers also developed kinetic expressions for the two TCAC degradation pathways at 25°C in the presence of a high-ionic-strength phosphate buffer (I = 0.200, $[PO_4]_T$ = 0.090). However, their kinetic expressions probably do not apply at other buffer concentrations because such reactions (whose rate-limiting step is a proton transfer) are particularly sensitive to general base catalysis.[16,17] Thus, the kinetics of the latter reaction require further study.

In addition to TCAC, chloral hydrate is also known to yield chloroform as the principal hydrolysis product.[18] Miller and Uden[5] found about 40 µg/L of chloral hydrate in a fulvic acid solution (5 mg/L TOC) that was chlorinated at pH 7. Chloral hydrate is especially interesting, because it has also been shown to undergo oxidation to trichloroacetic acid at pH 7 in the presence of calcium hypochlorite.[19] Thus, it may be a precursor common to the two major trichloromethyl chlorination products of humic materials and raw drinking waters, $CHCl_3$ and TCAA. Although the kinetics of hydrolysis of this intermediate at neutral pH to form chloroform are probably too slow to be of significance,[20] the importance of chloral hydrate as a TCAA precursor at pH 7 merits investigation.

This chapter explores some possible mechanisms in the formation of chloroform, TCAA, and DCAA from humic materials by (1) studying the yields of these organic halides from an aquatic fulvic acid and from model organic compounds representing a variety of chemical structures, and (2) evaluating the role and importance of some intermediates in the formation of organic halides from the fulvic acid and model compounds. The chemical insights gained from these studies are combined into a generalized model for organic halide formation. This model is then used to explain the effects of preozonation on organic halide formation potentials observed with an aquatic fulvic acid.[11]

EXPERIMENTAL

Fulvic acid was extracted from Black Lake, near Elizabethtown, North Carolina, according to the procedure of Thurman and Malcolm.[21] The fulvic acid from this lake has been well characterized.[9,22-25]

Model organics (reagent grade or better) were obtained commercially. They were used as received without further purification. 1,1,1-Trichloroacetone was prepared according to the procedure of Pouwels.[26] The product, which was purified by fractional distillation, contained small amounts of 1,1-dichloroacetone. The identity of the TCAC was confirmed using capillary gas chromatography-mass spectrometry.

All chlorinations in the model compound study were carried out under the following conditions:

pH 7 (or 12) with phosphate buffer (I = 0.028); 20 mg/L chlorine dose; 72-h reaction time, quenched with sodium arsenite; and 20°C.

Residual chlorine was measured by the DPD-titrimetric procedure.[27] A commercial carbon adsorption-pyrolysis-microcoulometric detection analyzer (Dohrmann Instruments) was used to measure the TOX concentration. Chloroform, dichloroacetonitrile, and 1,1,1-trichloroacetone were measured from neutral pentane extracts using a gas chromatograph equipped with an electron capture detector. A more detailed description of these analytical methods has been presented elsewhere.[11,28]

Base-hydrolyzable chloroform (Base-CHCl$_3$) was measured using a procedure similar to that of Morris and Baum.[29] Immediately after quenching the residual chlorine with arsenite, the sample was divided into two aliquots. One was immediately analyzed for CHCl$_3$; the other was spiked with NaOH to raise the pH to 12 and was analyzed 3 h later. Base-CHCl$_3$ was calculated as the difference between the two values.

RESULTS

Model Compound Chlorinations

Twenty-five simple organic compounds were chlorinated and analyzed for CHCl$_3$, TCAA, DCAA, and TCAC. In addition to the specific organic halides, TOX, chlorine consumption, and Base CHCl$_3$ were also measured. The results of these analyses are presented in Tables I through IV. In each case, a "percent incorp" value was calculated. This value—an estimate of the percent of chlorine consumed that became covalently bound to the model organic compound—was calculated as follows:

$$\text{Percent incorp} = \frac{\text{TOX}^{\text{corr}}/35{,}500}{\text{chlorine consumption}/71} \times 100 \quad (1)$$

where

$$\text{TOX}^{\text{corr}} = \text{TOX} + \frac{3(35.5)}{119.36}(1 - 0.85)\,\text{CHCl}_3 \quad (2)$$

Because the TOX recovery for chloroform was low (85% or 0.85), it was believed that a better estimate for the percent of chlorine incorporated could be made using the measured TOX value corrected for the loss resulting from chloroform [i.e., TOX^{corr}, in accordance with Equation (2)].

Tables I through IV also show the organic halide yields expressed as a percent of TOX^{corr}. An important point to note from these data is that chloroform, TCAA, and DCAA are all common chlorination products of the aro-

Table I. Chlorination of Selected Aromatics ⟨⟩-R

R	pH	Initial Conc (μM)	HOCl Cons[a] (M/M)	TOX	Yield (M/M)					Percent Incorp	Percent of TOX^{corr}				
					CHCl$_3$	TCAA	DCAA	TCAC	Base CHCl$_3$		CHCl$_3$	Base CHCl$_3$	TCAA	DCAA	TCAC
-COOH	7	238	<0.01	<0.005	<0.0005	<0.0005	<0.0005	<0.0005	<0.0005		<1	<1		<1	
-CHO	7	714	0.025	0.0044	<0.0001	0.0004	<0.0004	<0.0001	<0.0001	17.6	<1	<0.2	29	<1	<0.1
-COCH$_3$	7	143	0.266	0.191	-0.0740	0.00085	0.0002	<0.0001	<0.0005	84.3	98.8	2.9	1.3	0.1	<0.1
-OH	7	7.29	9.76	1.49	0.030	0.355	0.018	<0.001	0.014	15.4	5.2	0.8	71.4	2.4	<0.1
-OCH$_3$	7	71.4	0.61	0.344	0.0009	0.0075	0.0006	<0.0001	0.0009	56.5	0.7	5.5	6.6	0.3	<0.1
-CH(CH$_3$)$_2$	7	714	0.093	0.0122	0.00086	0.00020	0.00010	0.00003	0.00026	17.3	18.0		4.9	1.6	0.6

[a] 72-h HOCl consumption in mol HOCl per mol substrate.

Table II. Chlorination of Dihydroxybenzenes

	pH	Initial Conc (μM)	HOCl Cons (M/M)	TOX	Yield (M/M)				Percent Incorp	Percent of TOX^{corr}				
					CHCl$_3$	TCAA	DCAA	TCAC	Base CHCl$_3$		CHCl$_3$	Base CHCl$_3$	TCAA	DCAA	TCAC
meta	7	9.92	7.35	3.10	0.888	0.054	0.003	<0.001	0.045	47.6	73.1	4.4	5.3	0.2	<0.1
ortho	7	21.6	3.98	0.10	0.0024	0.0014	<0.0005	<0.0001	0.0016	2.5	6.2	4.1	4.1	<0.5	<0.1
para	7	21.6	3.73	0.21	0.0079	0.0098	0.0011	<0.0001	0.0048	5.7	9.8	6.0	14.3	1.0	<0.1

Table III. Chlorination of Two Model Humic Monomers and Black Lake Fulvic Acid

	pH	Initial Conc (μM)	HOCl Cons (M/M)	Yield (M/M)						Percent Incorp	Percent of TOX^corr				
				TOX	CHCl$_3$	TCAA	DCAA	TCAC	Base CHCl$_3$		CHCl$_3$	Base CHCl$_3$	TCAA	DCAA	TCAC
COOH / OCH$_3$ / OH	7	12.2	8.54	1.55	0.030	0.268	0.019	<0.001	0.013	18.3	5.0	2.5	52.1	2.4	<0.2
CHO / H$_3$CO–OCH$_3$ / OH	7	11.9	8.26	1.80	0.003	0.532	0.030	<0.001	0.018	21.8	0.4	3.0	88.5	3.3	<0.01
	12	11.9	3.97	0.935	0.192	0.001	<0.001	<0.0001		25.7	52.4		0.4	<0.02	<0.02
Black Lake fulvic acid[a]	7	46	2.25	0.63	0.044	0.033	0.015	0.00010	0.009	28.9	20.8	4.3	15.7	4.7	0.5
	12	46	1.94	0.36	0.060	0.0004	0.017	<0.00001		20.1	49.3	—	0.3	9.5	<0.1

[a] Values expressed as mol per 9 mol of carbon

Table IV. Chlorination of Some Simple Aliphatic Compounds

Compound	pH	Initial Conc (μM)	HOCl Cons (M/M)	Yield (M/M)						Percent Incorp	Percent of TOX^{corr}					
				TOX	$CHCl_3$	TCAA	DCAA	TCAC	Base $CHCl_3$		$CHCl_3$	Base $CHCl_3$	TCAA	DCAA	TCAC	
$(CH_3)_2CHCH_2COOH$	7	714	<0.002	<0.002	<0.0001	<0.0001	<0.0001	<0.0001	<0.0001	86.0	<0.2	46.7	<0.2	13.1	<0.2
H H \ / C=C / \ HOOC COOH	7	833	0.030	0.0258	<0.00002	<0.00002	0.00169	<0.00002	0.00472	<2	<0.2				
HOOC–COOH	7	238	0.237	<0.005	<0.0002	<0.0002	<0.0002	<0.0002	<0.0002	74.2	<0.1	<0.1	<0.1	0.9	<0.1
$HOOC–CH_2–COOH$	7	39.7	1.63	1.21	<0.0005	<0.0005	0.0056	<0.0005	<0.0005						
$HOOC–CH_2CH_2–COOH$	7	737	<0.005	<0.001	<0.0001	<0.0001	0.0002	<0.0001	0.0003						
O ‖ $CH_3C–COOH$	7	71.4	1.01	0.051	0.0015	0.0118	0.0008	0.0011	0.0015	5.1	7.4	8.7	70.1	3.2	6.6
	12	71.4	2.33	1.32	0.497	0.0012	0.531	<0.0001		66.3	96.1	0.3	8.1	<0.1
O ‖ $CH_3CCH_2C–O–Et$	7	35.7	1.74	1.62	0.0019	0.0017	0.760	0.0007	0.0011	93.2	0.3	0.2	0.3	94.0	0.1
OHC–CHO	7	64.9	1.26	<0.05	<0.001	<0.001	<0.001	<0.001	<0.001	28.4	0.5	38.3	<0.5	<0.5	<0.5
CH_3CHO	7	100	0.074	0.021	<0.0001	<0.0001	<0.0001	<0.0001	0.0031	70.6	1.1	0.3	<0.1	<0.1	0.4
$CH_3(CH_2)_3CHO$	7	357	0.301	0.212	0.0009	<0.00005	<0.00005	0.0003	0.0002	72.2					
$CH_3(CH_2)_6CHO$	7	200	0.247	0.178	0.0008	<0.0001	<0.0001	0.0001	0.0004		1.1	0.6	<0.1	<0.1	0.1
O ‖ CH_3CCH_3	7	714	0.165	0.127	0.0234	0.00008	0.0150	<0.0100	0.0153	83.4	46.8	30.7	0.2	23.6	<20
O ‖ $CH_3CH_2CCH_3$	7	714	0.061	0.0433	0.0106	0.00021	0.00024	0.00004	0.0007	78.8	62.3	4.5	1.5	1.1	0.3
OH | $CH_3CH_2CHCH_3$	7	714	0.004	0.003	0.00074	0.00005	<0.00001	0.00006	0.00022	80	59	20	5	1	6

matic and aliphatic compounds examined. Also, the two compounds chlorinated at both pH 7 and 12, syringaldehyde and pyruvic acid (see Tables III and IV), showed a shift in TOX speciation from TCAA to $CHCl_3$ as pH was raised.

Fulvic Acid Chlorinations

Black Lake fulvic acid solutions were prepared at a concentration of 4.1 mg/L TOC and chlorinated under a variety of conditions. Figure 1 shows the concentrations of TOX (uncorrected), $CHCl_3$, TCAA, DCAA, TCAC, and DCAN as a function of time at pH 7. Note that both TCAC and DCAN peaked within the first few hours and then declined, whereas all other organic halides showed a monotonic increase with time. Figures 2 and 3 show the effect of pH on these same analytes. As anticipated from the literature,[14,15] TCAC and DCAN showed reduced concentrations at high pH. The effect of chlorine dose on the yields of TOX, $CHCl_3$, TCAA, DCAA, TCAC, and DCAN is shown in Figure 4. The reduction in TCAC and DCAN formation at

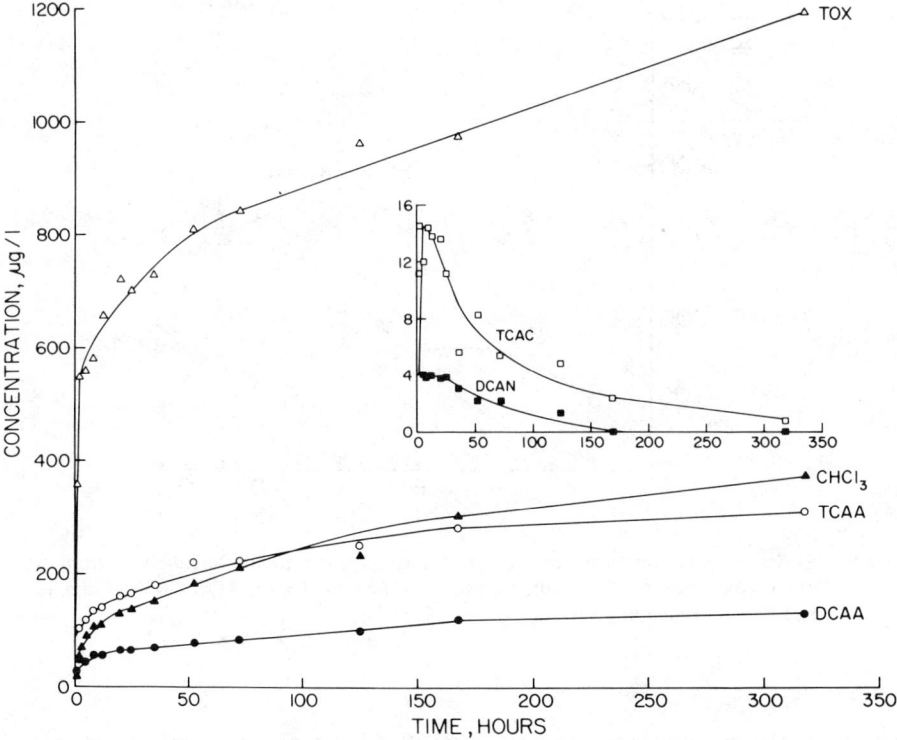

Figure 1. Formation of organic halides from Black Lake fulvic acid as a function of chlorine contact time. Conditions: 4.1 mg/L TOC, pH 7.0, 20 mg/L applied HOCl.

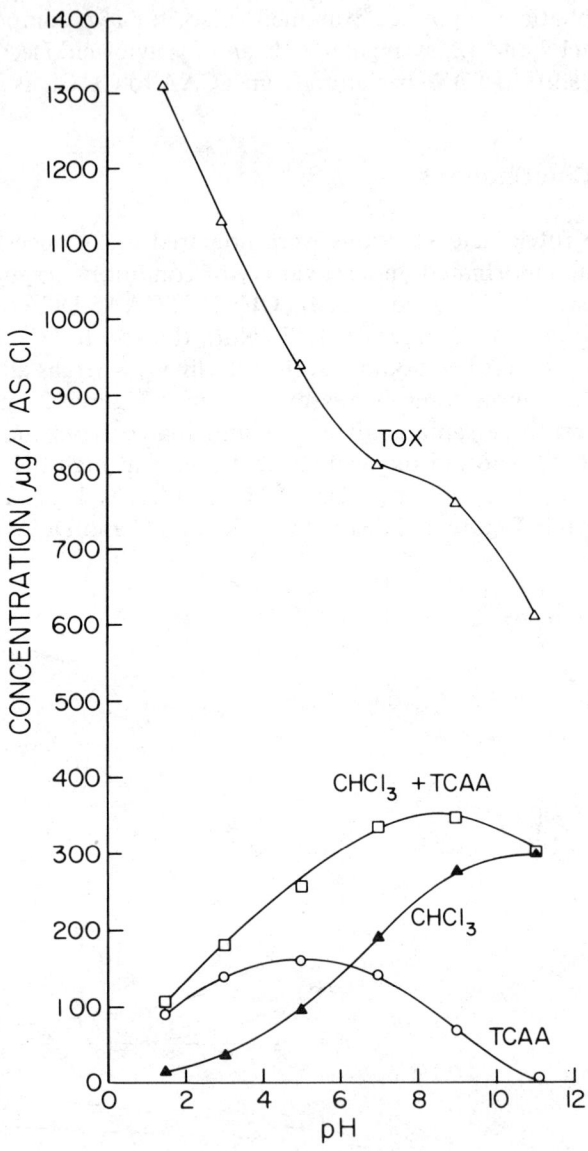

Figure 2. Effect of pH on the formation of TOX, chloroform, and trichloroacetic acid from the chlorination of Black Lake fulvic acid. Conditions: 4.1 mg/L TOC, 20 mg/L applied HOCl, 72-h reaction time.

elevated chlorine doses was perhaps a result of the direct reaction between chlorine and these two relatively long-lived intermediates. Figure 5 shows the formation of $CHCl_3$, TCAA, DCAA, TCAC, and DCAN as a function of

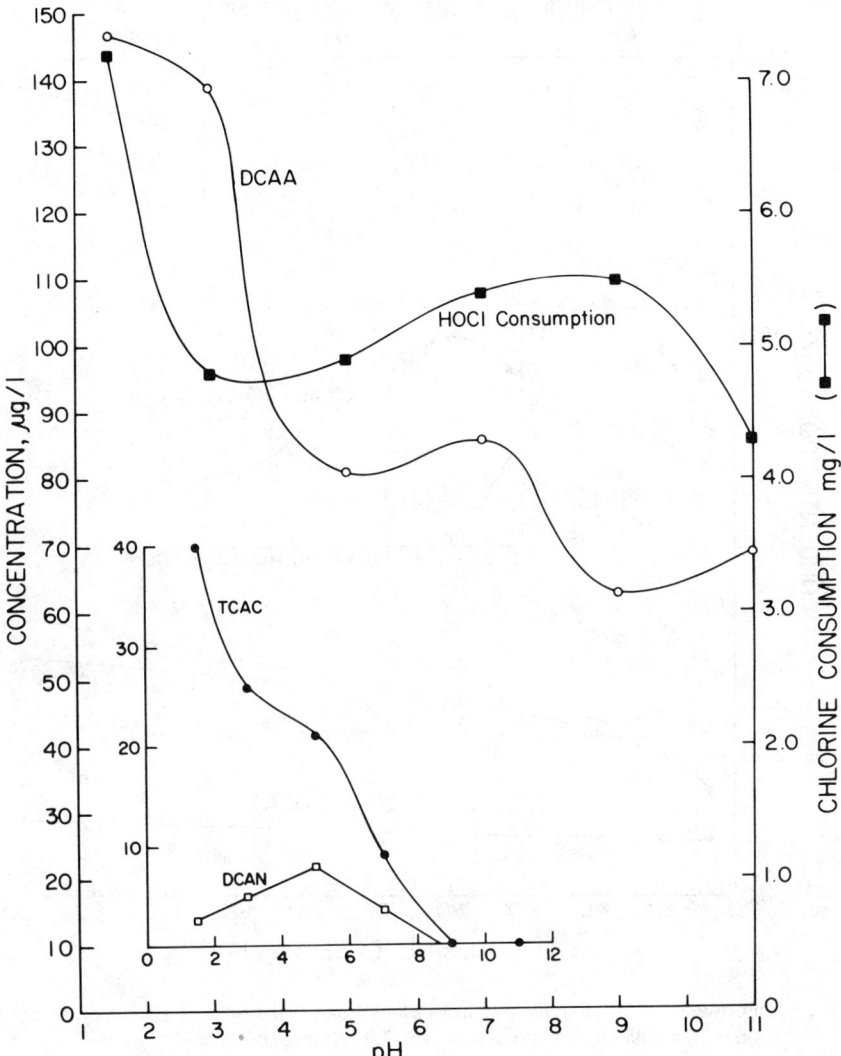

Figure 3. Effect of pH on chlorine consumption and the formation of dichloroacetic acid, trichloroacetone, and dichloroacetonitrile from the chlorination of Black Lake fulvic acid. Conditions: 4.1 mg/L TOC, 20 mg/L applied HOCl, 72-h reaction time.

phosphate buffer concentration. Trichloroacetic acid showed a slight increase; TCAC showed a significant decline with increasing phosphate concentration and ionic strength (pH drift was negligible throughout these experiments).

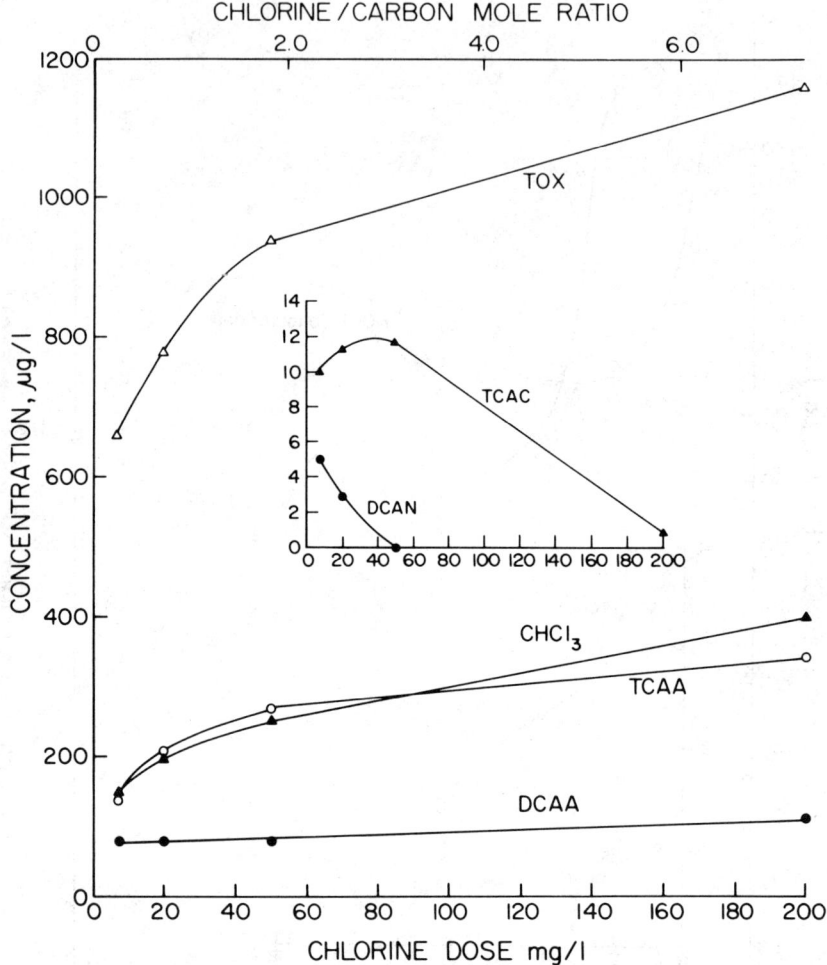

Figure 4. Formation of organic halides from Black Lake fulvic acid as a function of chlorine dose. Conditions: 4.1 mg/L TOC, pH 7.0, 72-h reaction time.

Kinetics of Product Formation from Intermediates

Trichloroacetone

To determine the importance of TCAC as a THM precursor in Black Lake fulvic acid, the kinetics of $CHCl_3$ formation from TCAC under the conditions used in this study (pH 7, 20°C, phosphate buffer with $[PO_4]_T = 0.0145\ M$) had to be established. A series of kinetic experiments were performed using low

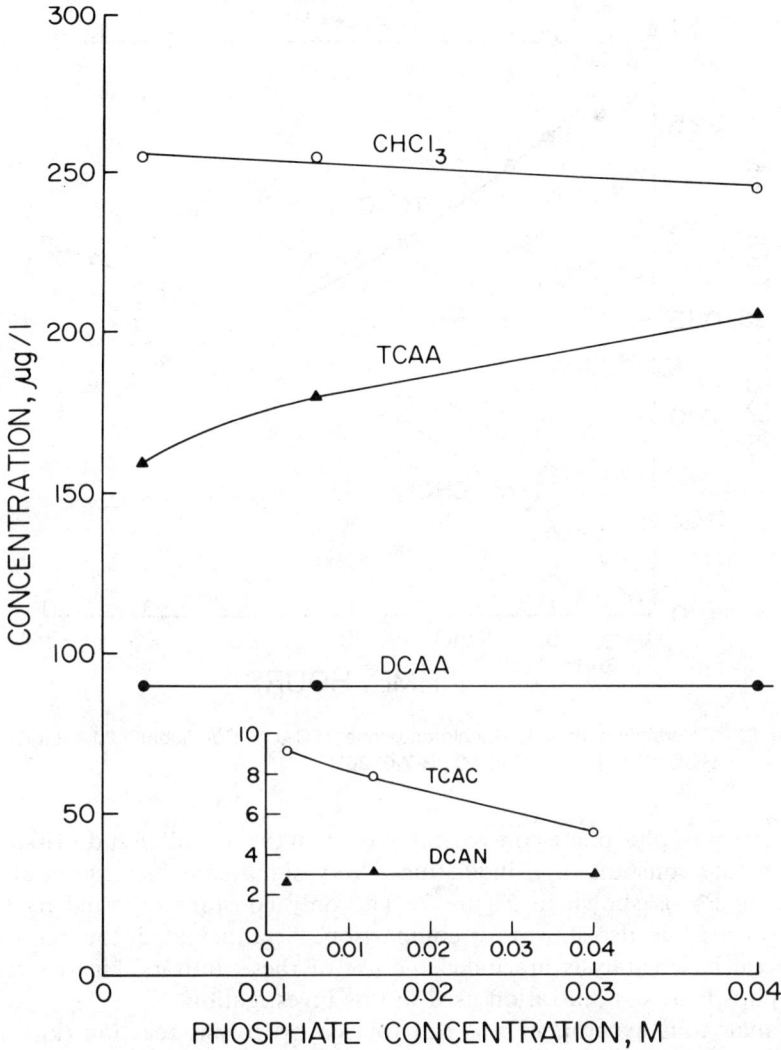

Figure 5. Effect of phosphate buffer strength on the formation of organic halides from Black Lake fulvic acid. Conditions: 4.1 mg/L TOC, pH 7.0, 20 mg/L applied HOCl, 72-h reaction time.

concentrations of TCAC (<100 μg/L) in the presence of milligram-per-liter levels of chlorine. As a result, the chlorine concentration remained essentially constant and pseudo-first-order kinetics were observed throughout the experiments. Figure 6 shows TCAC and $CHCl_3$ concentrations as a function of time for a typical run.

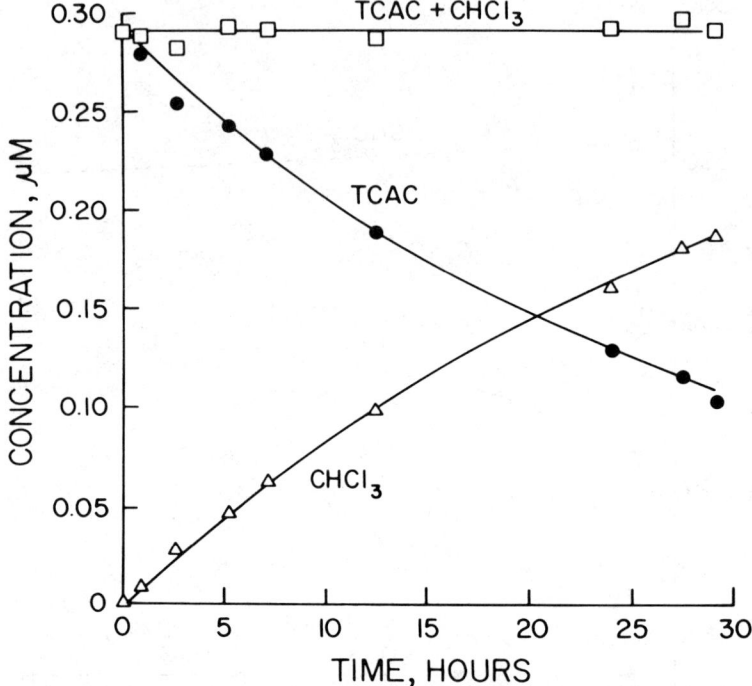

Figure 6. Chlorination of 1,1,1-trichloroacetone (TCAC). Conditions: 20.4 mg/L applied HOCl, $[PO_4]_T$ = 0.0145 M, pH 7.0, 20°C.

The effect of phosphate concentration (P_T) on the overall pseudo-first-order reaction rate constant (k_T), indicating a very significant increase in k_T with increasing P_T, is shown in Figure 7. The only constant reported by Gurol et al.[15] at pH 7 is also shown for comparison. As anticipated, the importance of general base catalysis precluded the use of these authors' kinetics for the lower phosphate concentration used in this investigation.

An analogous experiment was run for the hydrolysis reaction (k_H) in the absence of chlorine. Figure 8 shows that only very small changes in k_H occurred with changing buffer strength. The hydrolysis reaction rate constant k_H (in h^{-1}) at pH 7 was found to be

$$\ln k_H = -4.81 - 1.4 \sqrt{I} \qquad (3)$$

This gives a rate constant of 0.0064 h^{-1} for the buffer system used in the fulvic acid chlorination (I = 0.028).

The effect of chlorine concentration on the overall reaction rate is shown in Figure 9. The dotted line at the bottom of the figure represents the decay constant in the absence of chlorine. A roughly linear relationship was seen between k_T and $[HOCl]_T$ in accordance with the following expression

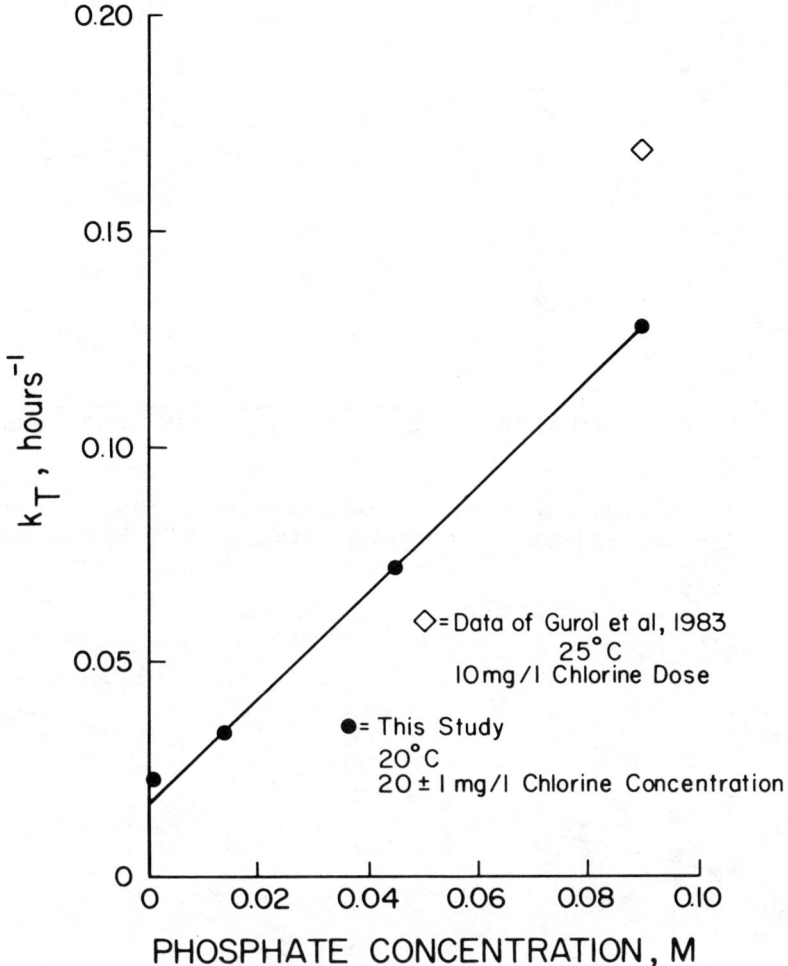

Figure 7. Pseudo first-order reaction rate constants for the chlorination of 1,1,1-trichloroacetone as a function of phosphate concentration at pH 7.0.

$$k_T = 0.024 + 32\,[\text{HOCl}]_T \quad (4)$$

where $[\text{HOCl}]_T$ is in units of mol per liter of residual chlorine and k_T is in reciprocal hours. Note that the apparent intercept (0.024) is significantly greater than that which one would expect, based on the hydrolysis data in the absence of chlorine (0.0064). Similar data (not shown) at a higher phosphate concentration gave an even larger apparent intercept.

Using this kinetic expression along with TCAC and chlorine concentration data from the experiments with Black Lake fulvic acid, the rate of chloroform

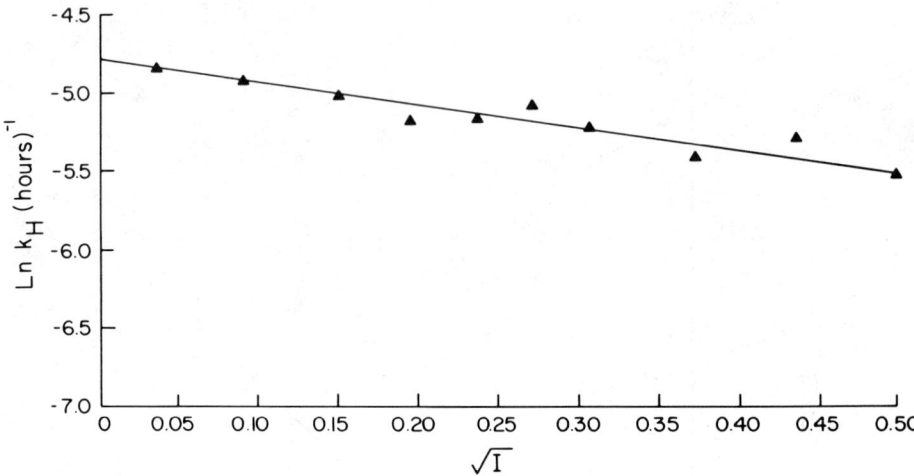

Figure 8. Pseudo first-order reaction rate constants for the hydrolysis of 1,1,1-trichloroacetone as a function of ionic strength. Conditions: pH 7.0, 20°C, no added chlorine.

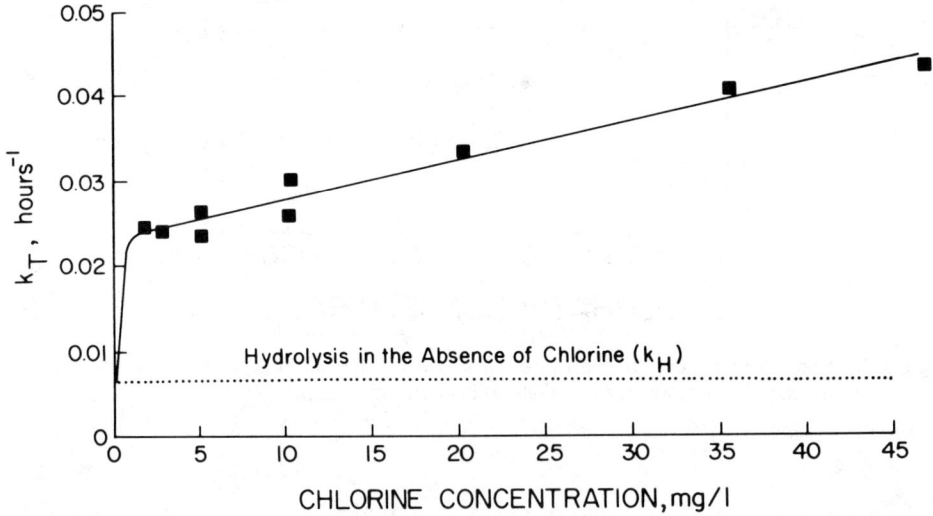

Figure 9. Pseudo first-order reaction rate constants for the disappearance of 1,1,1-trichloroacetone as a function of chlorine dose. Conditions: pH 7.0, 20°C, $[PO_4]_T$ = 0.0145 M.

formation from TCAC was calculated for a reaction time of 0 to 85 h. These rates were then integrated over time, and the total chloroform formation attributable to the TCAC intermediate was calculated as a function of total reaction time. Figure 10 shows these results along with the measured TCAC

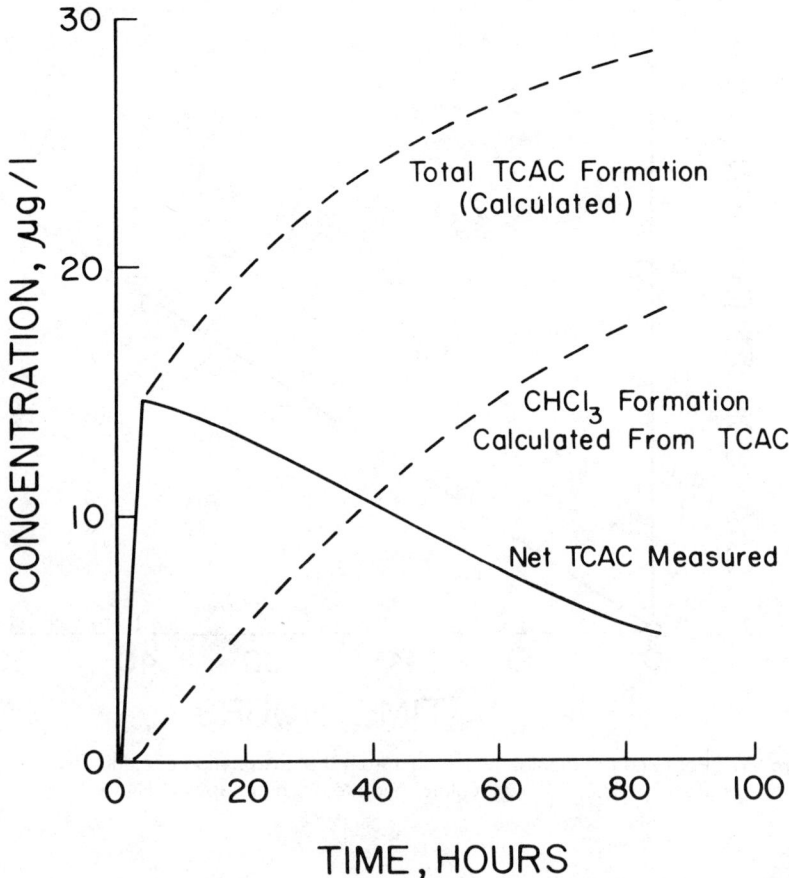

Figure 10. Estimated contribution of 1,1,1-trichloroacetone to the total chloroform yield from Black Lake fulvic acid. Conditions: 4.1 mg/L TOC, pH 7.0, 20 mg/L applied HOCl.

concentration (from Figure 1) for comparison. Also shown is the total TCAC produced during the reaction, which is calculated as the sum of TCAC measured plus TCAC decomposed (equivalent on a molar basis to the $CHCl_3$ formed).

In comparing chloroform formation curves in Figures 10 and 1, it is evident that TCAC cannot account for any of the fast chloroform formation (reaction time < 1 h). A significant fraction of the slow $CHCl_3$ formation, however, does pass through TCAC. The fraction of chloroform formation attributable to this intermediate increased with time such that at 72 h, it was approximately equal to 7.5% of the total $CHCl_3$ produced.

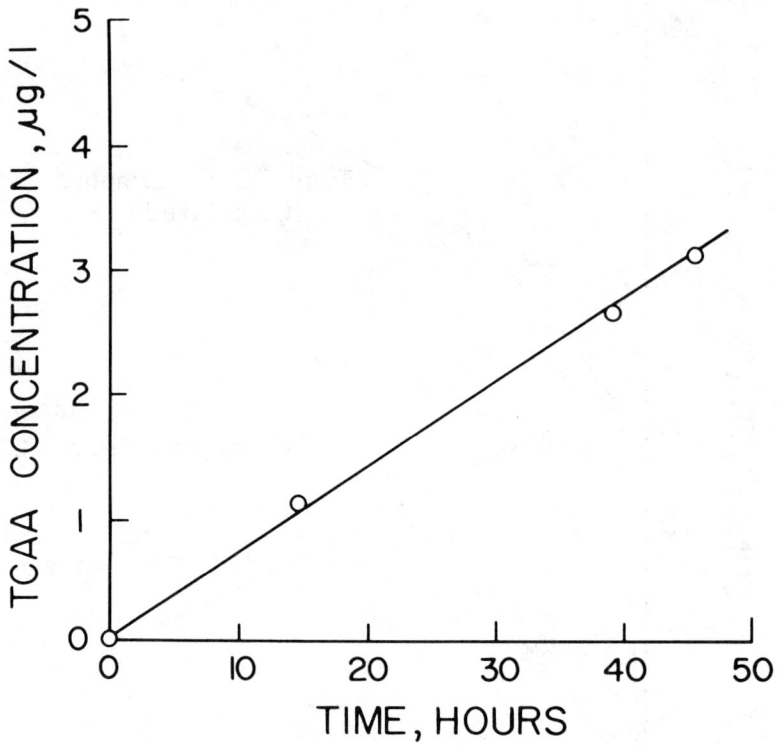

Figure 11. Formation of trichloroacetic acid from the chlorination of chloral hydrate. Conditions: pH 7.0, 200 µg/L chloral hydrate, 20 mg/L applied HOCl.

Chloral Hydrate

To determine the importance of chloral hydrate as a TCAA precursor, a relatively concentrated solution of $CCl_3CH(OH)_2$ (200 µg/L) was chlorinated under standard conditions, and samples were withdrawn at several times for TCAA analysis. Figure 11 shows the formation of TCAA in this experiment as a function of time. Since the reaction was quite slow, the concentrations of chloral hydrate and aqueous chlorine did not change appreciably over the course of the experiment. As a result, a constant rate of about 0.007 µg/L-h TCAA formation was observed. If this value is compared with the high yield of TCAA (~700 µg/L) and the low yield of chloral hydrate (~40 µg/L) observed by Miller and Uden,[5] it is apparent that this intermediate cannot account for a significant fraction of the TCAA formation. Furthermore, chloral hydrate concentrations measured over the course of this study (not shown) were generally lower than those observed by Miller and Uden.[5]

DISCUSSION

Intermediates

One can rationalize the results shown in Figures 6 through 9 for TCAC decomposition using a three-path mechanism (see Figure 12). In accordance with the observations of Gurol et al.,[15] base hydrolysis (k_4) would occur in the absence of chlorine. In the presence of small amounts of chlorine, a fast reaction that relies on a preequilibrium strongly dependent on phosphate concentration would predominate (k_1 and k_2). The enolate formation might be such a preequilibrium. As chlorine dose is increased, this preequilibrium might become the rate-limiting step, and the overall rate for this pathway would shift from first-order to zero-order in chlorine. At this point, a third and much slower pathway (k_3) that does not require preequilibrium and is not necessarily dependent on phosphate concentration might become important. This pathway could possibly involve the nucleophilic attack of OCl^- on the carbonyl carbon catalyzing the hydrolysis of TCAC much like OH^-. One can calculate that under the conditions used in the fulvic acid and model compound chlorinations ($[PO_4]_T = 0.0145\ M$, average chlorine concentration = 10–20 mg/L), roughly 60% of the $CHCl_3$ produced from TCAC would pass through the second pathway (k_2) and thereby also form DCAA (compare Figures 9 and 12). It may be significant that the ratio of DCAA to $CHCl_3$ formed from the chlorination of acetone (see Table IV) is also roughly 60%. One might carry this a step further by proposing that TCAC could account for a portion of the DCAA formation from fulvic acid equivalent to 60% of the $CHCl_3$ resulting from this intermediate. Using the data from Figures 1 and 10, one can calculate that 11 μg/L of DCAA, or about 15% of its total production, may be attributable to TCAC after 72 h of reaction.

Just as the decay of TCAC is catalyzed by phosphate, so might the formation of TCAC be subject to general base catalysis. If this were the case, the importance of TCAC as an organic halide intermediate might be an artifact of the use of phosphate buffers in such experiments. However, the evidence at hand suggests that this is not true. Because TCAC was formed so rapidly (see Figure 1), it is likely that the precursors to TCAC are far more acidic than acetone (and TCAC itself) and, therefore, not as subject to general base catalysis. Note from Figure 5 that TCAC decreased with increasing phosphate concentration. This is what one might expect from the phosphate catalysis of TCAC decomposition in the absence of significant catalysis of TCAC formation. Also note that the overall formation of TCAC degradation products ($CHCl_3$ and DCAA) was unaffected by phosphate concentration (see Figure 5).

1246 WATER CHLORINATION

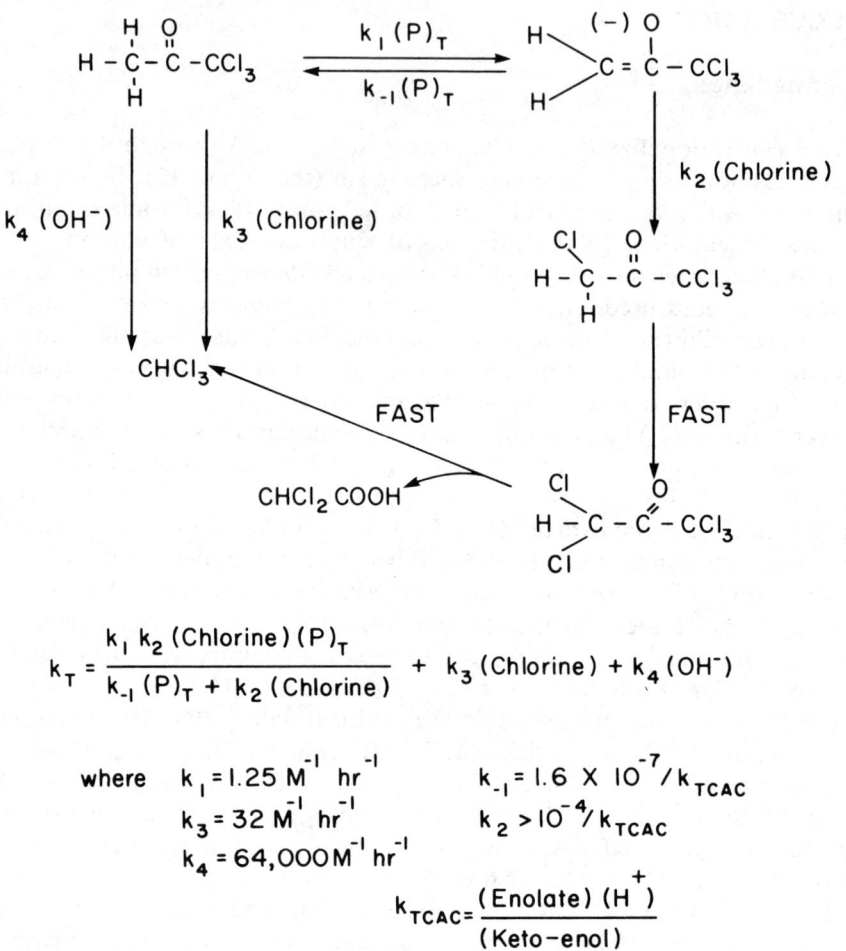

Figure 12. Hypothetical mechanism for the decomposition of trichloroacetone to chloroform in the presence of intermediate concentrations of phosphate at pH 7.

General Trichloromethyl Product Precursors

Although chloral hydrate was found to be insufficiently reactive to play a substantial role in the formation of TCAA and chloroform, the existence of other TCAA/$CHCl_3$ common precursors of the type

$$R - \overset{O}{\underset{\|}{C}} - CCl_3$$

cannot be ruled out. In support of a common link between the two major trichloromethyl species, Figure 2 shows the formation of TCAA, $CHCl_3$ and the sum of the two in micrograms per liter as chlorine as a function of pH. Note that the maximum yield of TCAA plus $CHCl_3$ is close to the pK_a of HOCl. The data of Miller and Uden[5] show a similar behavior. Also, Peters et al.[30] and Boyce[20] reported that the total potential chloroform from the chlorination of aquatic humic materials reached a maximum between pH 7 and 8. The total potential chloroform probably represents the sum of $CHCl_3$, TCAA, and Base-$CHCl_3$. Note that local maxima near pH 7.6 were also observed for DCAA formation and chlorine consumption (see Figure 3 and Reckhow et al.[31]). Even the TOX formation showed a plateau in this region (see Figure 2 and Fleischacker et al.[32]). This pH dependence resembles the behavior one would expect with an HOCl-carbanion reaction of the type observed for acidic ketones. Thus, the data suggest that a single class of reactions may be responsible for most of the trichloromethyl product formation in humic substances. The speciation within this group might then be a function of various reaction conditions (e.g., pH).

The link between TCAA and chloroform has been directly observed with resorcinol. Boyce and Hornig[33] reported that when ^{13}C-labeled resorcinol (the No. 2 carbon was labeled) was chlorinated, the chloroform and a minor unquantified product, TCAA, both contained the label. The data on resorcinol from Table II show that the molar sum of the yield of trichloromethyl groups ($CHCl_3$ + TCAA + Base-$CHCl_3$) is 98.7%. This suggests that the No. 2 carbon in resorcinol may be completely converted into a mixture of trichloromethyl products upon chlorination, with little or no $CHCl_3$ and TCAA coming from other parts of the molecule. The relatively minor yields of trichloromethyl species observed from the chlorination of isomers not containing the doubly activated No. 2 carbon (catechol and hydroquinone, see Table II) support this hypothesis.

Because of the extremely reactive nature of resorcinol and other activated aromatics with respect to chlorine, it is difficult to develop a reaction pathway and thereby propose a structure for the immediate TCAA/$CHCl_3$ precursor in these model organics. Aliphatic compounds, however, seem to react in a more predictable fashion (see Table IV). The pyruvic acid case is particularly instructive, because this compound shows the same predominance of trichloromethyl products, with chloroform dominating at high pH. In this case, the choice for a common TCAA/$CHCl_3$ precursor (trichloropyruvate) is obvious (see Figure 13). At low pH, this intermediate could undergo oxidative decarboxylation to form TCAA. At high pH, base-catalyzed hydrolysis would prevail, thus giving chloroform. The oxidative decarboxylations proposed in Figure 13 are analogous to the chlorine-induced oxidative decarboxylation seen for α-keto pentanoic acid.[34]

1248 WATER CHLORINATION

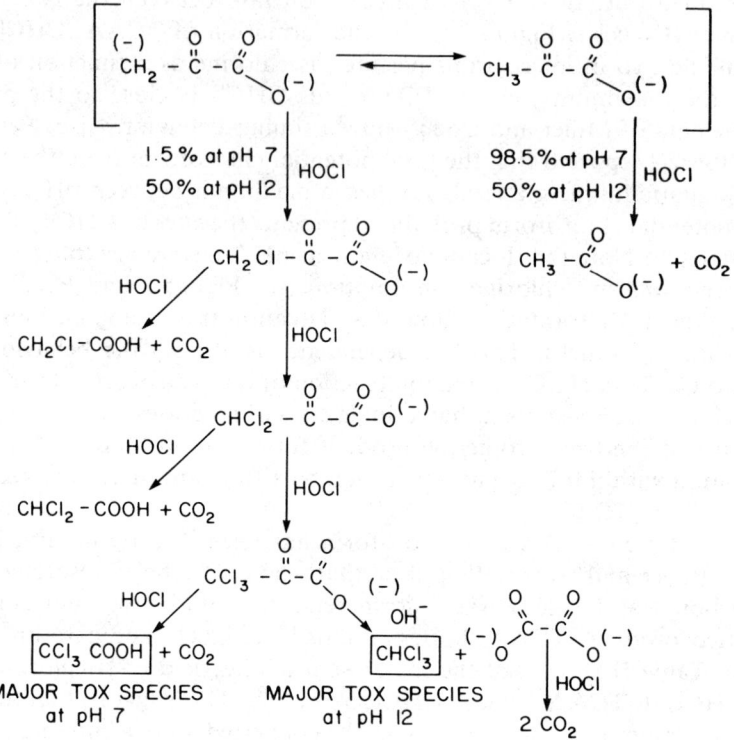

Figure 13. Proposed pathway for the chlorination of pyruvate.

Generalized Conceptual Model

Some of the observations discussed in this chapter can be applied to the general case of a polyfunctional unsaturated organic molecule with both aliphatic and aromatic components such as fulvic acid. For this purpose, a highly simplified conceptual model was developed.

In the absence of carbon-carbon triple bonds, chlorine addition reactions cannot by themselves account for the formation of dichloromethyl and trichloromethyl products. This implies that either radical reactions or activated ionic substitution reactions play an important role in the formation of $CHCl_3$, TCAA, and DCAA from humic substances. Model compound studies have shown that only highly activated structures produce chloroform at a rate equivalent to that of fulvic acid.[1] Typically, these highly activated compounds contain β-diketone moieties or structures that can be readily oxidized by chlorine to give β-diketone moieties.

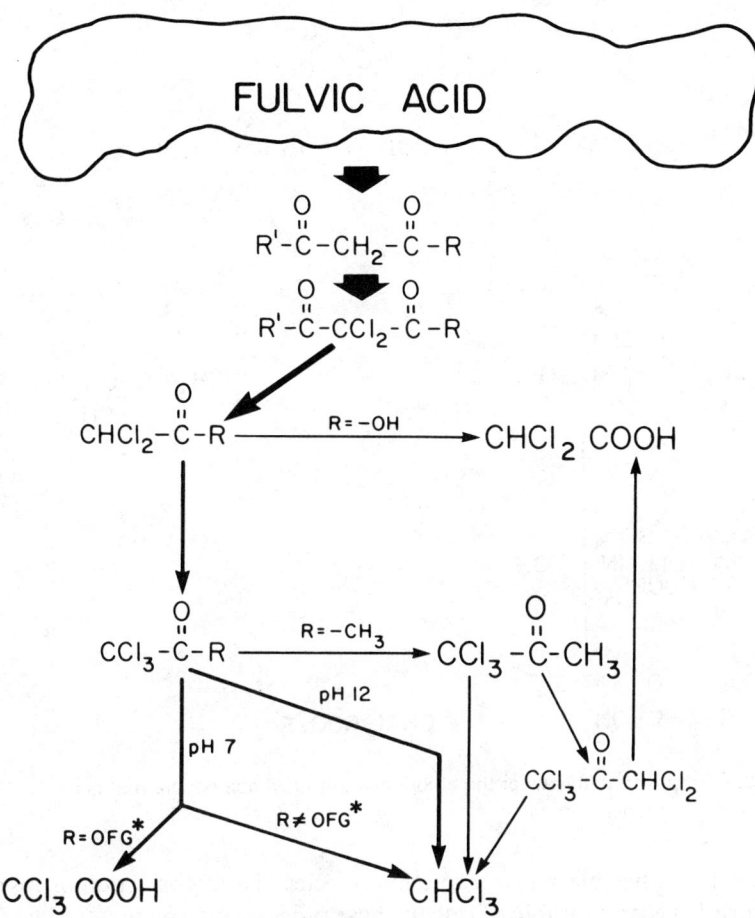

Figure 14. Simplified conceptual model for the formation of major organic halide products from fulvic acid (OFG = oxidizable functional group).

Most of the reactions between chlorine and humic materials result in oxidation rather than chlorine incorporation.[28,35] Since these oxidation reactions would be expected to lead to an increase in the number of oxygenated functional groups, the probability of forming transient β-diketone groups may be significant. With the formation of a β-diketone-type moiety, the activated carbon would quickly become fully substituted with chlorine (see Figure 14). Hydrolysis would then occur rapidly in a fashion analogous to the hydrolysis of dichloroacetylacetone.[36,37] If the remaining R group is –OH (or –OR), then the reaction would stop at that point, giving DCAA. This is illustrated by the chlorination pathway for ethyl acetoacetate hypothesized from the data in Table IV (see Figure 15). Otherwise, the structure would be further chlorinated

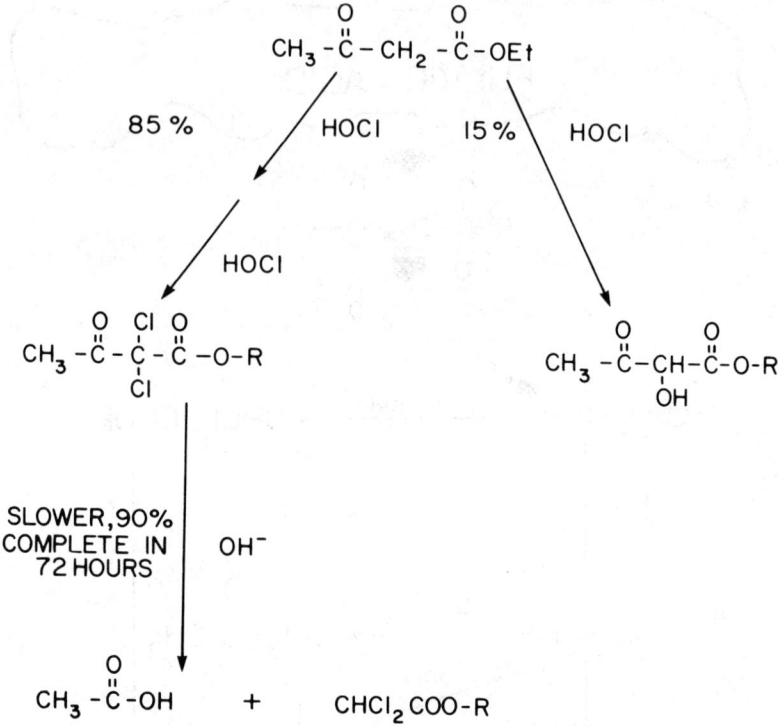

Figure 15. Proposed pathway for the chlorination of ethyl acetoacetate at pH 7 (R = −H or −CH₂CH₃).

to a base-hydrolyzable trichloromethyl species. Note that trichloropyruvate (see Figure 13) is an example of one of these types of intermediates. The fate of this precursor at pH 7 would perhaps depend on the nature of the "R" group. If it is capable of readily donating the electron pair joining it to the rest of the molecule, as is the case with trichloropyruvate, TCAA might be the major product. In the absence of such an oxidative cleavage, hydrolysis could prevail, giving chloroform.

One can speculate on the types of readily oxidizable functional groups that might encourage the formation of TCAA. The most common group of this type one might expect to find in chlorinated humic materials would have either α-hydroxy functions or conjugated hydroxy functions that could readily donate their electrons to the α-carbon (see Figure 16). Note that according to this scheme, the formation of TCAA would either be preceded by a proton transfer (and resulting donation of electrons) or occur simultaneously with one. This might explain the apparent general base catalysis seen in the formation of TCAA (see Figure 5). As discussed earlier, when the precursor is TCAC (i.e., R = −CH₃) the unsubstituted methyl group can become slowly chlorinated to give CHCl₃ and some DCAA.

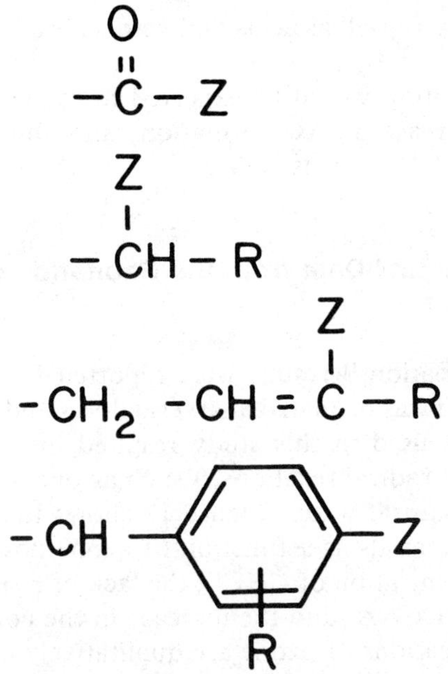

$$-\overset{\overset{O}{\|}}{C}-Z$$

$$-\overset{\overset{Z}{|}}{CH}-R$$

$$-CH_2-CH=\overset{\overset{Z}{|}}{C}-R$$

$$-CH-\underset{R}{\langle\!\langle\pm\rangle\!\rangle}-Z$$

Figure 16. Examples of readily oxidizable functional groups [Z = an electron-donating substituent (e.g., $-OH$, $-NH_2$)].

Implications with Respect to Preozonation

The implications of this scheme with respect to preozonation are numerous. Like chlorine, ozone might produce β-diketone structures and subsequently react with them, thereby precluding their reaction with chlorine. This could lead to a net reduction in subsequent TOX formation. Also, ozone would oxidize other parts of the molecule, perhaps increasing the chances that "–R" from Figure 14 is a hydroxyl group. This would have the effect of increasing DCAA formation at the expense of $CHCl_3$ and TCAA formation.

The effect of ozonation on $TCAA/CHCl_3$ speciation would be mixed. The formation of acids would favor TCAA, whereas the destruction of conjugated olefinic bonds and the formation of aldehydes and ketones might favor $CHCl_3$. (Note: Ozone is known to readily cleave double bonds via direct molecular O_3 attack, yielding acids, aldehydes, and ketones.[38])

The formation of ketones might also increase the TCAC yield along with DCAA formation. As shown in Figure 14, TCAC would be expected to be formed from a doubly activated methyl ketone (R = CH_3). It is unlikely, however, that chlorine oxidation would lead to a significant increase in the concentration of methyl ketones. Like many oxidants, chlorine tends to produce acids on reaction with fulvic acid; thus, one would expect R to be –COOH or R'''–COOH.[9] In contrast, ozone is known to produce simple ketones and aldehydes as well as acids and ketoacids after reaction with unsaturated organics.[38]

In summary, ozonation would be expected to reduce the concentration of TOX formation, increase TCAC formation, and shift the TOX speciation from $CHCl_3$ and TCAA towards DCAA.

Comparison with Actual Data from the Ozonation of Fulvic Acid

In an earlier publication,[11] results were reported for an experiment where Black Lake fulvic acid was ozonated at several doses and then chlorinated. The ozonation conditions used in this study resulted in a mixture of direct O_3 reactions and indirect radical reactions,[31] such as one might encounter during the ozonation of a natural water. Figure 17 shows the $CHCl_3$, DCAA, and TCAC formation potentials as a function of ozone dose. Note that the large reduction in the concentration of $CHCl_3$, the lack of a significant reduction in the concentration of DCAA, and the increase in the concentration of TCAC with increasing applications of ozone are qualitatively in accordance with the conceptual model developed here.

When the relative concentrations of the organic halide precursors are plotted on the same set of axes in a normalized fashion as a function of ozone dose (not shown), one observes the following order of increasing precursor destruction for nearly all of the doses investigated:

$$TCAC < DCAA < TOX \leq CHCl_3 \leq TCAA < DCAN.$$

The highly efficient destruction of DCAN precursors might be expected from the high degree of reactivity between ozone and nitrogenous organic compounds.[38,39] The greater removal of $CHCl_3$ and TCAA precursors over the TOX precursors was expected from the model. Although the differences among these three are small, additional experiments using organics from different sources have shown the same trends.[31] As previously mentioned, the very poor removal of DCAA precursors and the enhancement of TCAC precursors were expected.

The large increase in TCAC formation and the significant decline in $CHCl_3$ formation suggest a major shift in the nature of the $CHCl_3$ precursors as a result of preozonation. Earlier, it was shown that 7.5% of the $CHCl_3$ precursors from unozonated Black Lake fulvic acid passed through the TCAC intermediate. If one assumes that preozonation increases the yield of TCAC without changing the shape of the TCAC formation vs time curve, the $CHCl_3$ produced via this pathway increases with ozonation in direct proportion to the increase in TCAC yield. Thus, from a threefold increase in the production of TCAC (see Figure 17), one might propose a corresponding threefold increase in TCAC-derived chloroform. Coupling this with the greatly reduced overall

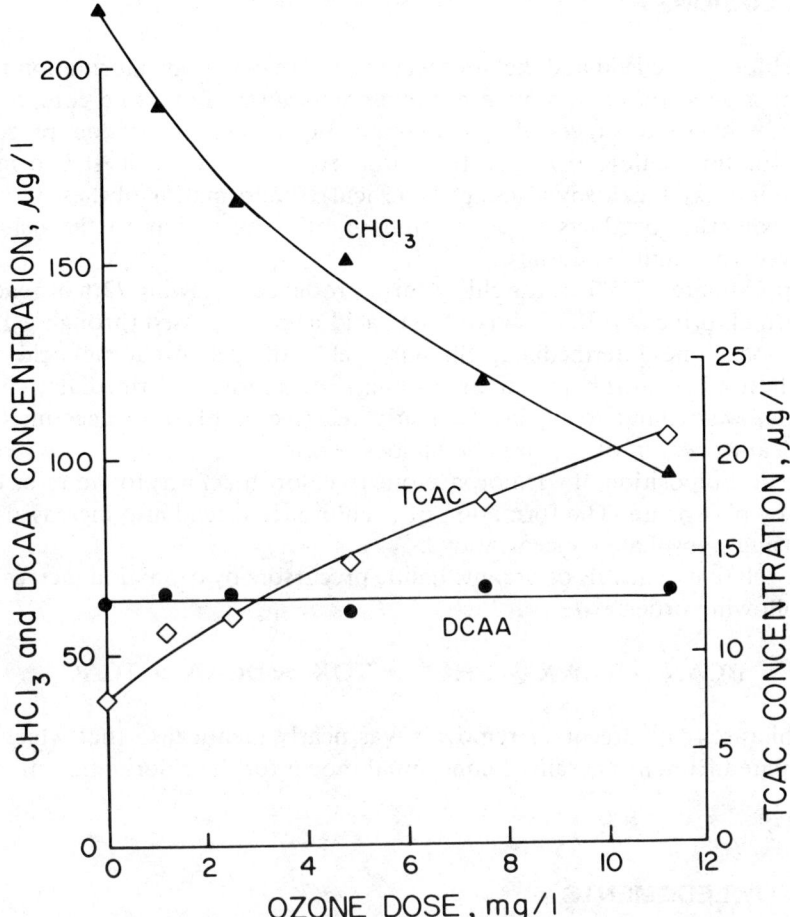

Figure 17. Effect of ozone dose on the formation of chloroform, dichloroacetic acid, and trichloroacetone following chlorination. Conditions: 4.1 mg/L TOC, pH 7.0, 20 mg/L applied HOCl, 72-h reaction time.

$CHCl_3$ formation, one can calculate that at the highest ozone dose investigated (0.74 M/M), 60% of the $CHCl_3$ passed through the TCAC intermediate. Similarly, the importance of TCAC in the formation of DCAA might also be magnified by preozonation.

Since chloroform formation from TCAC is rather slow, one would expect the chloroform vs chlorine contact time curve for ozonated fulvic acid to show a reduced fast formation and a more pronounced long-term formation. This is, in fact, what Riley[40] observed when a commercial terrestrial fulvic acid was ozonated and chlorinated at pH 7.

CONCLUSIONS

Trichloroacetic acid and dichloroacetic acid are common chlorination products of a wide range of simple organic compounds. For some compounds, there is evidence to suggest that a common trichloromethyl product precursor exists for both chloroform and trichloroacetic acid. For at least two model compounds (syringaldehyde and pyruvic acid), the formation of these terminal trichloromethyl products as a function of pH was similar to the behavior observed for humic materials.

Approximately 7.5% of the chloroform produced following 72 h of reaction between chlorine and Black Lake fulvic acid at pH 7 passed through a 1,1,1-trichloroacetone intermediate. Similar yields of dichloroacetic acid were hypothesized from this precursor. In contrast, another intermediate, chloral hydrate, was found to be insufficiently reactive at pH 7 to account for a significant fraction of the organic halides formed.

The decomposition of trichloroacetone to chloroform was found to be catalyzed by phosphate. The formation of trichloroacetic acid also increased with increasing phosphate concentrations.

The relative removals of organic halide precursors by ozonation increased in the following order:

$$DCAN > TCAA \geq CHCl_3 \geq TOX > DCAA > TCAC.$$

This hierarchy of precursor removals was nearly identical to that which was anticipated from a generalized conceptual model for the chlorination of fulvic acid.

ACKNOWLEDGMENTS

This research was supported by the U.S. Environmental Protection Agency under research grant R-810235. D. A. Reckhow thanks the U.S. Public Health Service for providing him with a traineeship. The authors thank Russell F. Christman, J. Donald Johnson, Bernard Legube, Andrea Dietrich, Daniel Cutugno, and Dan Norwood for their advice and technical assistance.

REFERENCES

1. Rook, J. J. "Haloforms in Drinking Water," *J. Am. Water Works Assoc.* 68(4):168 (1976).
2. Riley, T. L., K. H. Mancy, and E. O. Boettner. "The Effect of Preozonation on Chloroform Production in the Chlorine Disinfection Process," in *Water Chlorination: Environmental Impact and Health Effects, Vol. 2,* R. L. Jolley, H. Gorchev, and D. H. Hamilton, Jr., Eds. (Ann Arbor, MI: Ann Arbor Science Publishers, Inc., 1978), p. 593.

3. Kuhn, W., and R. Sander. "Formation and Behavior of Polar Organic Chlorine Compounds," *Oxidation Techniques in Drinking Water Treatment*, EPA-570/9-79-020 (Washington, DC: U.S. Environmental Protection Agency, 1979).
4. Christman, R. F., D. L. Norwood, D. S. Millington, J. D. Johnson, and A. A. Stevens. "Identity and Yields of Major Halogenated Products of Aquatic Fulvic Acid Chlorination," *Environ. Sci. Technol.* 17(10):625-628 (1983).
5. Miller, J. W., and P. C. Uden. "Characterization of Nonvolatile Aqueous Chlorination Products of Humic Substances," *Environ. Sci. Technol.* 17(3):150-157 (1983).
6. Trussell, R. R., and M. D. Umphries. "The Formation of Trihalomethanes," *J. Am. Water Works Assoc.* 70(11):604-612 (1978).
7. Moore, G. S., R. W. Tuthill, and D. W. Polakoff. "A Statistical Model for Predicting Chloroform Levels in Chlorinated Surface Water Supplies," *J. Am. Water Works Assoc.* 71(1):3-39 (1979).
8. Engerholm, B. A., and G. L. Amy. "An Empirical Model for Predicting Chloroform from Humic and Fulvic Acids," in *Water Chlorination: Environmental Impact and Health Effects, Vol. 4,* R. L. Jolley, W. A. Brungs, J. A. Cotruvo, R. B. Cumming, J. S. Mattice, and V. A. Jacobs, Eds. (Ann Arbor, MI: Ann Arbor Science Publishers, Inc., 1983), pp. 243-252.
9. Norwood, D. L., G. P. Thompson, J. J. St. Aubin, D. S. Millington, R. F. Christman, and J. D. Johnson. "By-products of Chlorination: Specific Compounds and Their Relationship to Total Organic Halogen," presented at the meeting of the American Chemical Society Division of Environmental Chemistry, Washington, DC, 1983).
10. Norwood, D. L., G. P. Thompson, J. D. Johnson, and R. F. Christman. "Monitoring Trichloroacetic Acid in Municipal Drinking Water," Chapter 89, this volume.
11. Reckhow, D. A., and P. C. Singer. "The Removal of Organic Halide Precursors by Pre-ozonation and Alum Coagulation," *J. Am. Water Works Assoc.* 76(4):151-157 (1984).
12. Coleman, W. E., J. W. Munch, W. H. Kaylor, H. P. Ringhand, and J. R. Meier. "GC/MS Analysis of Mutagenic Extracts of Aqueous Chlorinated Humic Acids: A Comparison of the By-products of Drinking Water Contaminants," *Environ. Sci. Technol.* 18(9):674-681 (1984).
13. Exner, J. H., G. A. Burk, and D. Kyriacou. "Rates and Products of Decomposition of 2,2-Dibromo-3-nitrilopropionamide," *J. Agr. Food Chem.* 21(5):838-842 (1973).
14. Oliver, B. G. "Dihaloacetonitriles in Drinking Water: Algae and Fulvic Acid as Precursors," *Environ. Sci. Technol.* 17(2):80-83 (1983).
15. Gurol, M. D., A. Wowk, S. Myers, and I. H. Suffet. "Kinetics and Mechanism of Haloform Formation: Chloroform Formation from Trichloroacetone," in *Water Chloroform: Environmental Impact and Health Effects, Vol. 4,* R. L. Jolley, W. A. Brungs, J. A. Cotruvo, R. B. Cumming, J. S. Mattice, and V. A. Jacobs, Eds. (Ann Arbor, MI: Ann Arbor Science Publishers, Inc., 1983), pp. 269-284.
16. Jencks, W. P. *Catalysis in Chemistry and Enzymology*, (New York: McGraw-Hill Book Co., 1969).
17. Bell, R. P., and K. Yates. "Kinetics of the Chlorination of Acetone in Aqueous Solution," *J. Chem. Soc.* 1927-1933 (1962).
18. Merck Index. *The Merck Index*, 9th ed. (Rahway, NJ: Merck and Company, Inc., 1976).

19. Plump, R. E. "Preparation of Trichloro-Aliphatic Carboxylic Acids," *Chem. Abstr.* 42:7320e (1948).
20. Boyce, S. D. "The Formation of Chloroform and Related Products from the Chlorination of Polyhydroxybenzenes and Diketones in Dilute Aqueous Solution: Possible Models for Aquatic Humic Material," PhD Dissertation, (Hanover, NH: Dartmouth College, 1980).
21. Thurman, E. M., and R. L. Malcolm. "Preparative Isolation of Aquatic Humic Substances," *Environ. Sci. Technol.* 15(4):463–466 (1981).
22. Liao, W., R. F. Christman, J. D. Johnson, D. S. Millington, and J. R. Hass. "Structural Characterization of Aquatic Humic Material," *Environ. Sci. Technol.* 16(7):403–410 (1982).
23. Colclough, C. A., J. D. Johnson, R. F. Christman, and D. S. Millington. "Organic Reaction Products of Chlorine Dioxide and Natural Aquatic Fulvic Acids," in *Water Chlorination: Environmental Impact and Health Effects, Vol. 4*, R. L. Jolley, W. A. Brungs, J. A. Cotruvo, R. B. Cumming, J. S. Mattice, and V. A. Jacobs, Eds. (Ann Arbor, MI: Ann Arbor Science Publishers, Inc., 1983), pp. 219–229.
24. Anderson, L. J. "The Reaction of Ozone with Isolated Aquatic Fulvic Acid," M.S. Thesis, (Chapel Hill, NC: University of North Carolina, 1984).
25. Jensen, J. N., J. J. St. Aubin, R. F. Christman, and J. D. Johnson. "Characterization of the Reaction Between Monochloramine and Isolated Aquatic Fulvic Acid," Chapter 73, this volume.
26. Pouwels, H. "Preparation of 1,1,1-Trihalo 2-Ketones," *Chem. Abstr.* 63:2900e (1965).
27. *Standard Methods for the Examination of Water and Wastewater*, 15th ed. (Washington, DC: American Public Health Association, 1980).
28. Reckhow, D. A. "Organic Halide Formation and the Use of Pre-ozonation and Alum Coagulation to Control Organic Halide Precursors," PhD Dissertation (Chapel Hill, NC: University of North Carolina, 1984).
29. Morris, J. C., and B. Baum. "Precursors and Mechanisms of Haloformation in the Chlorination of Water Supplies," in *Water Chlorination: Environmental Impact and Health Effects, Vol. 2*, R. L. Jolley, H. Gorchev, and D. H. Hamilton, Jr., Eds. (Ann Arbor, MI: Ann Arbor Science Publishers, Inc., 1978), pp. 24–48.
30. Peters, C. J., R. J. Young, and R. Perry. "Factors Influencing the Formation of Haloforms in the Chlorination of Humic Materials," *Environ. Sci. Technol.* 14(11):1391–1395 (1980).
31. Reckhow, D. A., B. Legube, and P. C. Singer. "The Ozonation of Organic Halide Precursors: Effect of Bicarbonate," submitted for publication, *Water Res.* (1984).
32. Fleischacker, S. J., D. E. Johnson, and S. J. Randtke. "Formation and Control of Organic Chlorine in Public Water Supplies," *J. Am. Water Works Assoc.* 75(3):132–138 (1983).
33. Boyce, S. D., and J. F. Hornig. "Formation of Chloroform from the Chlorination of Diketones and Polyhydroxybenzenes in Dilute Aqueous Solution," in *Water Chlorination: Environmental Impact and Health Effects, Vol. 3*, R. L. Jolley, W. A. Brungs, and R. B. Cumming, Eds. (Ann Arbor, MI: Ann Arbor Science Publishers, Inc., 1980), pp. 131–140.
34. Nwaukwa, S. O., and P. M. Keehn. "Oxidative Cleavage of α-Diols, α-Diones, α-Hydroxyketones, and α-Hydroxy- and α-Keto Acids With Calcium Hypochlorite," *Tetrahedron Lett.* 23(31):3135–3138 (1982).

35. Morris, J. C. *Formation of Halogenated Organics by Chlorination of Water Supplies*, EPA 600/1-75-002 (Washington, DC: U.S. Environmental Protection Agency, 1975).
36. DeLaat, J. "Contribution a l'Etude du Mecanisme de Formation des Trihalomethanes," Dissertation, (Poitiers, France: University of Poitiers, 1981).
37. DeLaat, J., N. Merlet, and M. Dore. "Chloration de Composes Organiques: Demande en Chlore et Reactivite Vis-a-vis de la Formation des Trihalomethanes," *Water Res.* 16(10):1437–1450 (1982).
38. Bailey, P. S. "Organic Groupings Reactive Toward Ozone, Mechanisms in Aqueous Media," in *Ozone in Water and Wastewater Treatment*, F. L. Evans III, Ed. (Ann Arbor, MI: Ann Arbor Science Publishers, Inc., 1972), pp. 29–59.
39. Hoigne, J. "Mechanisms, Rates, and Selectivities of Oxidations of Organic Compounds Initiated by Ozonation of Water," in *Handbook of Ozone Technology and Its Applications, Vol. 1,* R. G. Rice and A. Netzer, Eds. (Ann Arbor, MI: Ann Arbor Science Publishers, Inc., 1982).
40. Riley, T. L. "The Effect of Preozonation on Chloroform Production in the Chlorine Disinfection Process," PhD Dissertation, (Ann Arbor, MI: University of Michigan, 1978).

CHAPTER 97

Effect of Cyanuric Acid, A Chlorine Stabilizer, on Trihalomethane Formation

Caren M. Feldstein, Janet Rickabaugh, and Richard J. Miltner

Cyanuric acid (Figure 1) has been widely used as a stabilizer of free chlorine in swimming pool disinfection since the mid-1950s.[1] When chlorine and cyanuric acid are mixed in solution, mono-, di-, and trichloroisocyanurate are formed, depending on pH and concentration. Cyanuric acid acts as a reservoir for free chlorine in solution; as free chlorine is consumed, more free chlorine is released from chlorinated isocyanurates.[2] The equilibrium is

$$H_3Cy + xHOCl \rightleftharpoons Cl_xH_{3-x}Cy + xH_2O \qquad (1)$$

The interactions of cyanuric acid and its chlorinated derivatives are very complex and include twelve equilibria expressions in addition to the equilibria of aqueous chlorine.[3] The time required to reach equilibrium between free chlorine and cyanuric acid is <1 min.[2] Studies of swimming pools indicate that chlorinated isocyanurates are at least as effective as chlorine in bactericidal efficiency.[1,4,5]

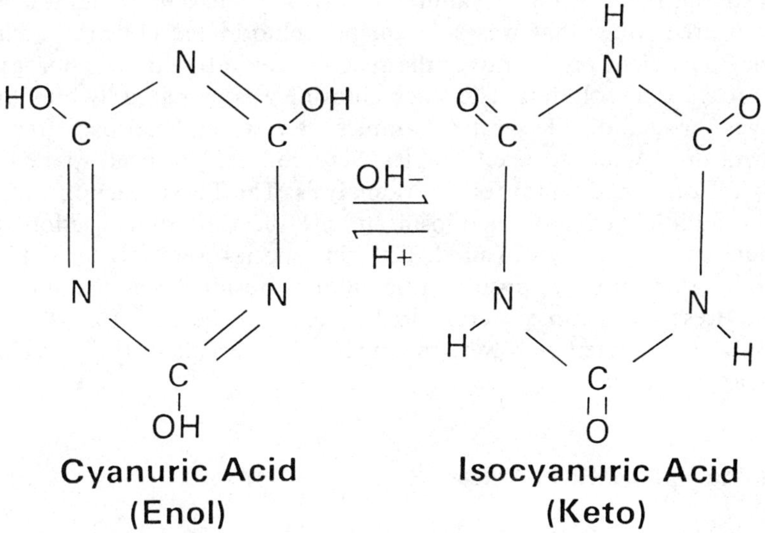

Figure 1. Tautomeric forms of cyanuric acid.

Cyanuric Acid (Enol)   Isocyanuric Acid (Keto)

The objective of this study was to investigate the effect of cyanuric acid on trihalomethane (THM) formation in potable water. Based on the classic haloform reaction[6] as a mechanism of THM formation, no reduction in THM concentration would be expected, because the rate-determining step is the conversion of the methyl ketone from the keto to enol form. However, if the mechanism of THM formation were chlorine-dose dependent, it would be expected that the addition of cyanuric acid would inhibit THM formation because of the reservoir effect of cyanuric acid with chlorine [Equation (1)].

EXPERIMENTAL

Ohio River water was studied under two conditions. First, the effect of cyanuric acid on THM formation in river water was examined. Second, the effect of cyanuric acid on THM formation in bromide-spiked river water was studied.

River Water

Ohio River water was filtered through anthracite/sand filters using the pilot plant facility of the Environmental Protection Agency (EPA) Drinking Water Research Division in Cincinnati. Typically, this water has a turbidity below 1.0 NTU and a total organic carbon concentration of 2 to 3 mg/L. The filtered river water was buffered at the appropriate pH with a phosphate buffer for a final strength of 0.004 M. Cyanuric acid (Kodak reagent grade) was added to the buffered water that was subsequently chlorinated (13 mg/L, chosen to provide sufficient residual over the reaction period) with an appropriate aliquot of chlorine solution. The stock chlorine was prepared by bubbling chlorine gas through distilled water. Samples were stored headspace free at room temperature. After storage, samples were collected in head-space free vials (with Teflon®-lined septa) for THM analysis. The THM reaction was stopped by the addition of sodium thiosulfate. At the time of collection, pH and residual chlorine were measured. Chlorine species were measured by a DPD titration procedure.[7] A positive free chlorine residual was maintained at all times. Preserved samples were stored at 20°C in the dark for ≤7 d prior to analysis. Trihalomethanes were measured by a purge and trap gas chromatographic procedure.[8]

Bromide-Spiked River Water

After filtered river water was spiked with bromide (chlorine: bromine = 22:1 molar ratio), the procedures described above were carried out.

RESULTS AND DISCUSSION

Trihalomethane Formation in River Water

Unchlorinated river water had a total trihalomethane (TTHM) concentration below the detection level. Chlorinated river water produced a characteristic TTHM formation curve[9] reaching 0.756 μmol/L at 7 d (Figure 2). The mean distribution of THM species was 76, 21, 4, and 0.7% for chloroform, bromodichloromethane, dibromochloromethane, and bromoform, respectively.

The addition of cyanuric acid (175 mg/L) in a 7.5:1 molar ratio (defined as the ratio of applied cyanuric acid concentration and applied chlorine concentration) to chlorine inhibited TTHM formation by a mean 29% each day over a 13-d reaction period (Figure 2). However, increasing the concentration of cyanuric acid (15:1 molar ratio) showed no further reduction in TTHM concentration. The distribution of THM compounds was the same with or without cyanuric acid.

Because of the reservoir effect and the rapidity with which chlorinated isocyanurates liberate free chlorine as free chlorine is consumed, chlorinated isocyanurates appear as false-positive free chlorine in common analytical procedures.[10,11] Thus, in a cyanuric acid-aqueous chlorine system, these procedures measure both free chlorine and chlorinated isocyanurates. For purposes

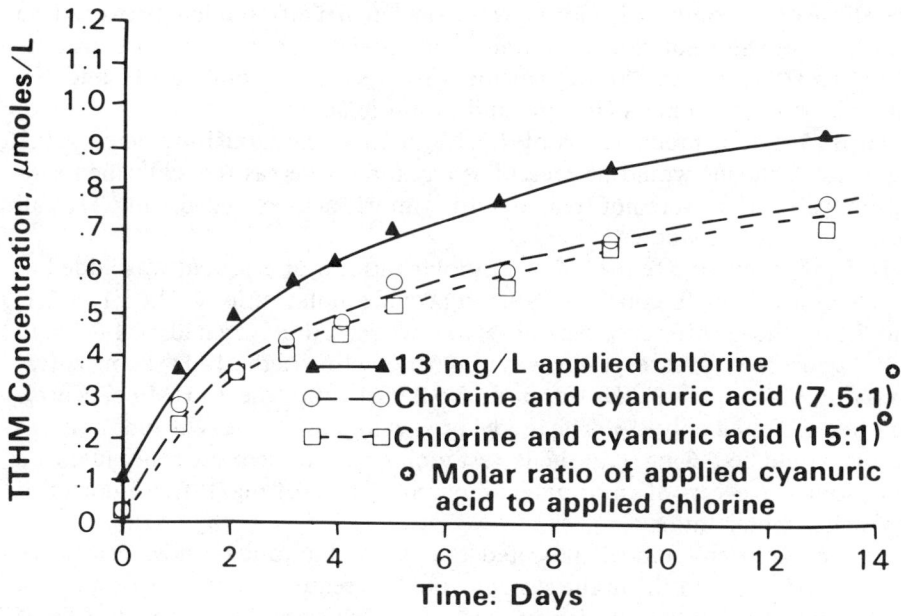

Figure 2. Effect of cyanuric acid on TTHM formation in river water.

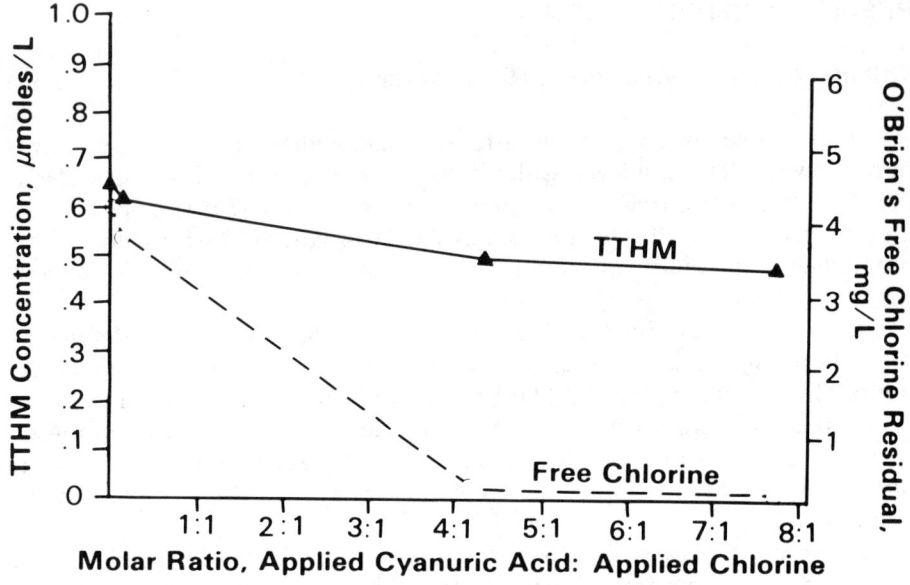

Figure 3. Effect of molar ratio on TTHM formation in river water after 7 d at pH 7.

of this study, the sum of these species was termed total chlorine concentration since little or no combined chlorine was detected. To differentiate between free chlorine and chlorinated isocyanurates, the model and equilibrium constants of O'Brien[3] were used. Model calculations were based on the pH and the concentrations of applied chlorine and cyanuric acid.

At pH 7 and at molar ratios of 7.5:1 and 15:1, the model predicted <0.1 mg/L free chlorine would be present at any time, whereas free chlorine residuals in waters that were not treated with cyanuric acid exceeded 6 mg/L at all times during these studies.

To further investigate the effect of molar ratio, cyanuric acid was added at 2, 100, and 175 mg/L concentrations to provide molar ratios of 0.08:1, 4.3:1, and 7.5:1. Little difference was observed between those treated at 4.3:1 and 7.5:1 ratios. In both cases, there was a 24% reduction in the TTHM concentration after 7 d (Figure 3). However, at the 0.08:1 ratio, the TTHM concentration was inhibited only 7%. The model predicted that the free chlorine concentration would be 5.6 mg/L at the lowest molar ratio (as expected, because of a stoichiometric excess of chlorine), compared with <0.1 mg/L free chlorine at the higher molar ratios.

At 7 d, O'Brien's model indicated that at molar ratios below 1.1:1, the amounts of free chlorine in the system would increase dramatically (Figure 3). With most of the total chlorine present as free chlorine, the levels of TTHM were comparable to those with no cyanuric acid present.

The effect of pH was also studied. THM concentrations were measured after a 7-d reaction time at pH 6, 7, 8, and 9. River water treated with 13 mg/L chlorine showed an increasing TTHM concentration with increasing pH (Figure 4), in agreement with the literature.[12] The distribution of the THM species was similar to that discussed earlier. With the addition of cyanuric acid at a molar ratio of 7.5:1, TTHM concentrations were, respectively, 24, 23, 22, and 6% less than the TTHM concentrations in the system containing no cyanuric acid. Based on O'Brien's model, the free chlorine was <0.2 mg/L at pH 6, 7, 8, and increased to 0.75 mg/L at pH 9.

The large difference in the concentration of chlorine present (6 mg/L in water not treated with cyanuric acid compared with <0.1 mg/L calculated for cyanuric acid-treated water), coupled with only a 29% difference in TTHM concentration, can be explained by the classic haloform reaction for the formation of THMs. Since the rate-determining step in this reaction is the conversion of the methyl ketone substrate from the keto to the enol form, the overall reaction rate for chloroform formation is independent of chlorine concentration. However, not all precursor material in natural waters are methyl ketones, which may react by a mechanism other than the classic haloform reaction where the rate-determining step may be chlorine-dose dependent, thus accounting for some observed decrease in TTHM concentration when cyanuric acid was present.

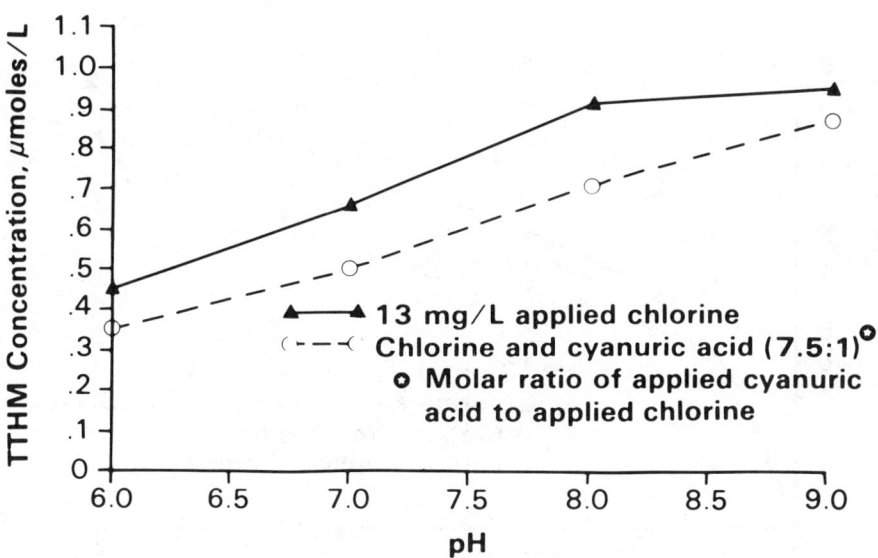

Figure 4. Effect of pH on TTHM formation in river water in the presence of cyanuric acid after 7 d.

Trihalomethane Formation in Bromide-Spiked River Water

River water spiked with 0.67-mg bromide per liter and treated with chlorine reached a TTHM concentration of 1.17 μmol/L after 7 d reaction time, with larger concentrations of brominated THMs caused by bromide addition. The application of cyanuric acid in a 7.5:1 molar ratio with applied chlorine in the bromide-spiked water resulted in a TTHM concentration of 1.06 μmol/L after 7 d (Figure 5). This shows only a 9% reduction in TTHM concentration with the addition of cyanuric acid.

In bromide-spiked river water treated with chlorine only, the mean distribution of species was 8% chloroform, 23% bromodichloromethane, 46% dibromochloromethane, and 20% bromoform. However, with the application of cyanuric acid (molar ratio 7.5:1) the mean distribution changed to 1, 10, 35, and 53%, respectively. (In river water containing naturally occurring levels of bromide, no such change in species distribution was observed with the addition of cyanuric acid. This might be explained, however, by the low levels of brominated species formed and the analytical imprecision in that range.)

O'Brien's model does not account for chlorine's oxidation of bromide to bromine. No attempt was made to modify the model. Although the relative amounts of free chlorine and free bromine were not known, increasing amounts of cyanuric acid favored an increase of brominated species. When cyanuric acid was added at concentrations of 0, 2, 25, 100, and 175 mg/L to

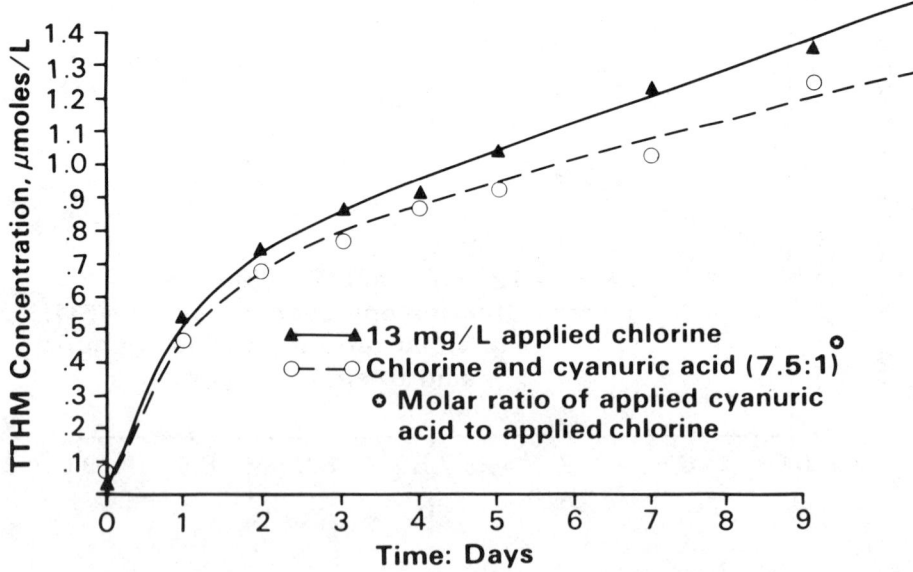

Figure 5. Effect of cyanuric acid on TTHM formation in bromide-spiked river water.

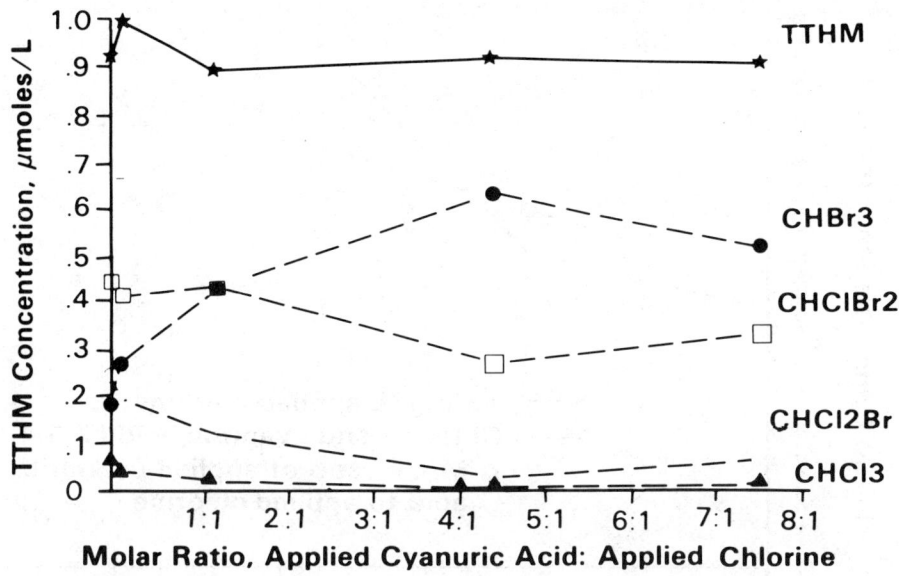

Figure 6. Effect of molar ratio on TTHM formation in bromide-spiked river water after 7 d at pH 7.

bromide-spiked river water and was subsequently chlorinated, giving molar ratios of 0:1, 0.08:1, 1.1:1, 4.3:1, and 7.5:1, the 7-d TTHM concentrations were 0.929, 1.02, 0.990, 0.936, and 0.924 μmol/L, respectively (Figures 6). The distribution of species at 0:1 and 0.08:1 found dibromochloromethane the predominant species—similar to the system that received no cyanuric acid. At 1.1:1, 4.3:1, and 7.5:1, the predominant species was bromoform.

The TTHM concentrations of bromide-spiked river water were measured after 7 d at varying pH (Figure 7). The mean distribution of the THM species was 6%, chloroform, 25% bromodichloromethane, 46% dibromodichloromethane, and 24% bromoform. After treatment with cyanuric acid, the mean distribution was 1, 12, 36, and 55%, respectively. The TTHM concentrations were higher in the cyanuric acid-treated system at pH 6 and 7, but they decreased in the same system at pH 8 and 9 (Figure 7).

The data for bromide-spiked water can also be explained by the classic haloform reaction as a mechanism of THM formation. Chlorine and bromine were present at all times and were available for THM formation once the enol form of the methyl ketone was present. Therefore, very little difference in total THM concentration was observed with the addition of cyanuric acid. In the cyanuric acid treated water, the ratio of HOBr to HOCl is higher than in the noncyanuric acid treated water, because the HOCl concentration is reduced as a result of its equilibrium with chlorinated isocyanurates. In the other system, HOCl is in stoichiometric excess of HOBr. Although the overall reaction rate is not affected by the halogen concentration, the product distribution is affected

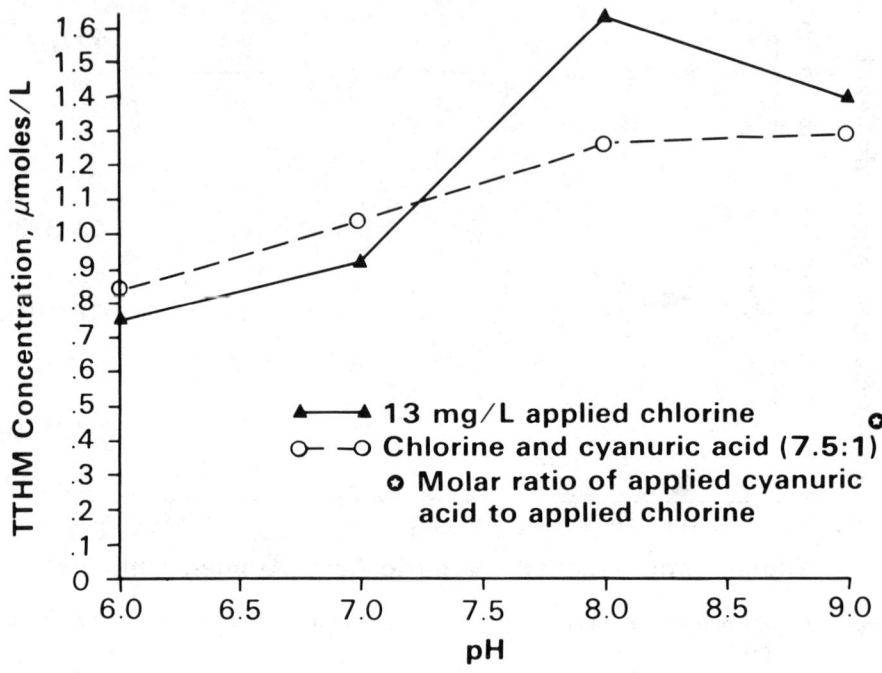

Figure 7. Effect of pH on TTHM formation in the presence of cyanuric acid in bromide-spiked river water after 7 d.

according to the relative amounts and rates for HOCl and HOBr. Therefore, in the bromide-spiked water, the addition of cyanuric acid would be expected to favor formation of more heavily brominated THM species to a greater extent than the noncyanuric acid treated water, and this was observed. In this attempt to explain brominated THM formation, a comparison of the kinetics of bromide oxidation by HOCl with the kinetics of cyanuric acid complexation of free chlorine was not made because of twelve pH-dependent equilibria expressions that define the latter.

CONCLUSIONS

These studies show that cyanuric acid, when used in combination with 13 mg/L chlorine at an applied molar ratio greater than 1.1:1, or at pH <9, inhibited total THM formation by a mean 29% in filtered river water; at lower ratios, cyanuric acid had little or no effect. In bromide-laden waters, cyanuric acid demonstrated only a 9% decrease in total THMs, but favored the formation of brominated over chlorinated THM species. These data can be explained in part by the classical haloform reaction as a mechanism for THM formation.

It is not known whether a 9 to 29% reduction in total THM formation would occur at the concentrations at which chlorine is typically applied during water treatment (i.e., below 13 mg/L) or what the complicating effects of other treatment processes would be. Cyanuric acid probably holds little promise for THM control in drinking water treatment. Before cyanuric acid could be applied to potable water, it would be necessary to undertake an extensive study of its potential health effects and to ensure that finished water microbial standards would be met.

ACKNOWLEDGMENTS

This research was funded by the EPA Drinking Water Research Division, Cincinnati, and by a University of Cincinnati Summer Research Fellowship. At EPA, we thank Bradford Smith for analytical support, Kenneth Kropp for laboratory support, Alan Stevens, O. Thomas Love and Leown Moore for their review of the manuscript, and Pat Pierson and Maura Lilly for typing the manuscript.

REFERENCES

1. Morgan, G. B., F. W. Gilcreas, and P. P. Bubbins. "Cyanuric Acid—an Evaluation," *Swimming Pool Age*, 31-38 (May 1966).
2. Hu, H. C., "Chronoamperometric Determination of Free Chlorine by Using a Wax-Impregnated Carbon Electrode," *J. Am. Water Works Assoc.* 73(3):150-153 (1981).
3. O'Brien, J. E., "Hydrolytic and Ionization Equilibria of Chlorinated Isocyanurate in Water," Ph.D. Dissertation, (Cambridge, MA: Harvard University, 1972).
4. Linda, F. W., and R. C. Hollenback. "The Bactericidal Efficacy of Cyanurates—A Review," *J. Environ. Health* 40(6):308-314 (1978).
5. Kowalski, X., and T. B. Hilton. "Comparison of Chlorinated Cyanurates with other Chlorine Disinfectants," *Public Health Rep.* 81(3):282-288 (1966).
6. Morris, J. C., and B. Baum. "Precursors and Mechanisms of Haloform Formation in the Chlorination of Water Supplies," in *Water Chlorination; Environmental Impact and Health Effects, Vol. 2*, R. L. Jolley, H. Gorchev, and D. H. Hamilton, Jr., Eds. (Ann Arbor, MI: Ann Arbor Science Publishers, Inc., 1977), pp. 29-48.
7. *Standard Methods for the Examination of Water and Wastewater*, 15th ed. (Washington, DC: American Public Health Association, 1980).
8. *Methods for Organic Chemical Analysis for Municipal and Industrial Water and Wastewater, Procedure 601, Purgeable Halocarbons*, (Cincinnati, OH: U.S. Environmental Protection Agency, 1982).
9. Bellar, T. A., J. J. Lichtenberg, and R. C. Kroner. "The Occurrence of Organohalides in Drinking Water," *J. Am. Water Works Assoc.* 66:703-706 (1974).
10. Wajon, J. E., and J. C. Morris. "The Analysis of Free Chlorine in the Presence of Nitrogenous Organic Compounds," *Environ. Int.* 3(1):41-47 (1980).

11. Whittle, G. P. "Recent Advances in Determining Free Chlorine," presented at the American Society of Civil Engineers Conference on Disinfection, University of Massachusetts, Amherst (1970).
12. Symons, J. M., et al. "Treatment Techniques for Controlling Trihalomethanes in Drinking Water," EPA-600/2-81-156, (Cincinnati: U.S. Environmental Protection Agency, 1981).

CHAPTER 98

Potential New Water Disinfectants

S. D. Worley, D. E. Williams, H. D. Burkett,
S. B. Barnela, and L. J. Swango

For the past several years the development of a new water disinfectant has been our primary goal. The disinfectant should satisfy several desirable requirements such as being a nontoxic solid that is stable when dissolved in water or stored for extensive periods in the dry form. The primary disinfectants in use today (chlorine, ozone, chlorine dioxide, and inorganic chloramines) are toxic gases that are not stable for extensive time periods dissolved in water and which can be hazardous to handle in treatment facilities.

The ideal disinfectant should not react with organic contaminants in water to produce toxic substances such as trihalomethanes (THMs).[1] Organic chloramines are less reactive with such contaminants than is free chlorine.[2,3] The ideal disinfectant must be a rapid and efficient bactericide. Chloramines tested in water treatment thus far are weaker disinfectants than are free chlorine, ozone, and chlorine dioxide.[3] Finally, the ideal disinfectant should not impart color, odor, or taste to the water, and it must not be toxic to humans or animals drinking the water.

A considerable amount of work has been done in our laboratories concerning the organic N-chloramine, 3-chloro-4,4-dimethyl-2-oxazolidinone (agent I) (Figure 1). Agent I was first prepared and shown to be bactericidal by Kaminski and co-workers.[4,5] In a laboratory-scale water treatment plant we demonstrated that agent I kills all bacteria in raw lake water when used with the chemicals (other than chlorine) used by the City of Auburn municipal treatment plant and appropriate sand filtration.[6] Agent I is stable indefinitely

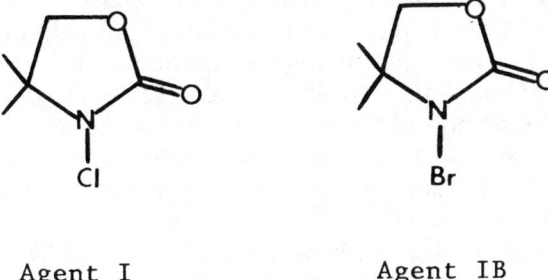

Agent I   Agent IB

Figure 1. The chemical structures of agent I and agent IB.

in nonbuffered, nondemand-free distilled water at pH 5.1 and ambient temperatures, whereas for all other disinfectants tested the total chlorine content degraded rapidly under these conditions.[7] In a series of experiments with a large number of chickens drinking water containing agent I at high concentration (200 ppm), we detected no toxic effects relative to control groups.[8]

We previously presented a general summary of the properties of agent I[9] and the possible cellular mechanisms of action of agent I in inhibiting bacterial DNA, RNA, and protein synthesis.[10] Agent I was compared to several commercial disinfectants in its action against 11 species of bacteria at ambient temperatures.[11] However, in the latter study the distilled water samples were not buffered or rendered chlorine demand free.

This chapter describes further work concerning agent I. Agent I is compared to free chlorine (calcium hypochlorite) regarding activity against various microorganisms; stability as a function of temperature, pH, and water quality; action on materials; and tendency to produce trihalomethanes under carefully controlled conditions. Preliminary studies on the brominated analog of agent I (3-bromo-4,4-dimethyl-2-oxazolidinone, agent IB; Figure 1) are also discussed.

EXPERIMENTAL

Materials

Agents I and IB, which are both crystalline solids, were synthesized and purified by established procedures.[4,5,12] Final purities of the two compounds were greater than 99% (NMR analyses and amperometric titration). Calcium hypochlorite (HTH from Olin Chemical Company: 65% calcium hypochlorite, 35% inert ingredients) was used without further purification. Stock solutions of the three compounds were checked for total halogen content immediately before use by either amperometric or iodometric titration.[13] Buffer salts were reagent grade chemicals.

Most of the studies used buffered solutions of the three compounds in demand-free water. The buffered solutions were rendered demand free by the addition of 1 to 2 mg/L total chlorine (sodium hypochlorite) followed by standing for 24 h and dechlorination by exposure to direct sunlight for 8 to 14 h. The buffered solutions were autoclaved at 121°C for 15 min and checked for residual chlorine and pH immediately before use. Glassware was made demand free by soaking in a 3 to 5 mg/L total chlorine solution (sodium hypochlorite) for 24 h, followed by rinsing in demand-free, distilled, deionized water and drying in sunlight.

For some experiments a synthetic demand water (SDW)[14] was used that contained calcium chloride, magnesium chloride, potassium chloride, and sodium chloride (375 mg/L each); 50 mg/L bentonite clay; 30 mg/L humic acid; 0.01% final concentration heat-treated horse serum; and 5×10^5 cells/

mL heat-killed *Saccharomyces cerevisiae*. The inorganic salts, bentonite clay, and humic acid were dissolved (suspended) in pH 9.5 buffer and autoclaved. The dead yeast cells and horse serum were added to the treatment vessel at the time of the experiments.

Microorganisms

A broad variety of microorganisms was tested including bacteria (*Staphylococcus aureus* ATCC 25923, *Pseudomonas aeruginosa* ATCC 27853, and *Shigella boydii* ATCC E9207), poliovirus type 1 (ATCC VR-192 attenuated Chat strain), protozoa (*Entamoeba invadens* IP-1 strain and *Giardia lamblia* Portland strain), and fungi (*Candida albicans* ATCC 44506 and *Rhodotorula rubra* ATCC 16639). Solutions containing disinfectant and about 10^6 cfu/mL final concentration of each microorganism were prepared. Aliquots were withdrawn periodically, quenched with 0.02 N sodium thiosulfate, and examined for survival of each microorganism. Variables in the study included pH, temperature, and water quality (demand free or SDW). Agent I and calcium hypochlorite were tested extensively against all of the microorganisms; agent IB was evaluated for two species of bacteria. Experimental details are presented elsewhere.[15]

Stability Studies

Agent I and calcium hypochlorite were tested for stability in water and dry storage. A 10 mg/L total chlorine solution of each agent was prepared in a 1-L flask, which was lightly stoppered with a sterile porous cotton plug. Aliquots were withdrawn periodically and titrated amperometrically for total chlorine. Variables in the study included pH, temperature, and water quality. Solid samples of the two compounds were also stored in vials containing porous cotton plugs. The total chlorine contents of these samples were monitored at the beginning of the study and following a 111-d period of standing at ambient temperature. A preliminary stability study for agent IB at pH 7.0, 22°C in demand-free water was also conducted using an initial 22.5 mg/L total bromine content (the molar equivalent to 10 mg/L total chlorine for the other two compounds).

Action on Materials

Solutions of agent I and calcium hypochlorite were exposed to various materials used by the military in field water treatment. In one experiment, water containing agent I at 10 mg/L total chlorine was pumped through a membrane test cell unit similar to that used in a ROWPU field water treatment unit. Deterioration of the membrane filters was monitored as a function of time. In a second set of experiments, samples of two bladder materials used for military water storage were exposed to agent I and calcium hypochlorite at 100 mg/L total chlorine content for 28 d. At the end of the test period, the samples were examined for deterioration by electron microscopy and tested for mechanical tensile strength. The solutions were analyzed by gas chromatography for the presence of any organic matter which might have leached out of the materials.

Production of Trihalomethanes

Demand-free solutions of agent I and calcium hypochlorite buffered at pH 7.0 and containing 10 mg/L total chlorine concentration were prepared. Two test solutions for each compound were doped with 3.8% by weight spectroscopic grade acetone, a precursor for chloroform in the haloform reaction. Two other solutions for each agent were doped with organic load in the form of SDW, except at pH 7.0. Appropriate control samples for each compound buffered at pH 7.0 also were prepared. One solution of each agent for each class (control, acetone, SDW) was exposed to sunlight for 10 h at 25°C. A second solution of each agent and a control for the solutions doped with acetone were kept in darkness for 7 h. At the end of the reaction periods all solutions were quenched by 0.056 N sodium thiosulfate, extracted with isooctane, and analyzed by gas chromatography.[15]

RESULTS

Microorganisms

Typical graphs showing the efficacies of agent I, agent IB, and calcium hypochlorite against *Staphylococcus aureus* in demand-free water and in SDW and against *Shigella boydii* in demand-free water are shown in Figures 2-4. In demand-free water calcium hypochlorite (2.5 to 5 mg/L as chlorine) killed both of the organisms very rapidly so that no organisms survived after 1 min; thus curves for calcium hypochlorite are not shown in Figures 2 and 4. Agent IB completely disinfected *S. aureus* in less than 15 s at 11.25 mg/L total bromine (the molar equivalent to 5 mg/L total chlorine), while agent I required 60 min (Figure 2). Figure 3 shows that agent IB was a more rapid disinfectant than calcium hypochlorite in SDW. The actions of all three compounds were inhibited significantly by the presence of the organic demand and lower temperature (e.g., contact time needed for complete disinfection by agent I was extended to 4 h). We concluded that organic demand is a more significant inhibitor of disinfectant action than is lowering temperature.[15] Furthermore, if the total chlorine concentration is lowered to 5 mg/L, agent I still

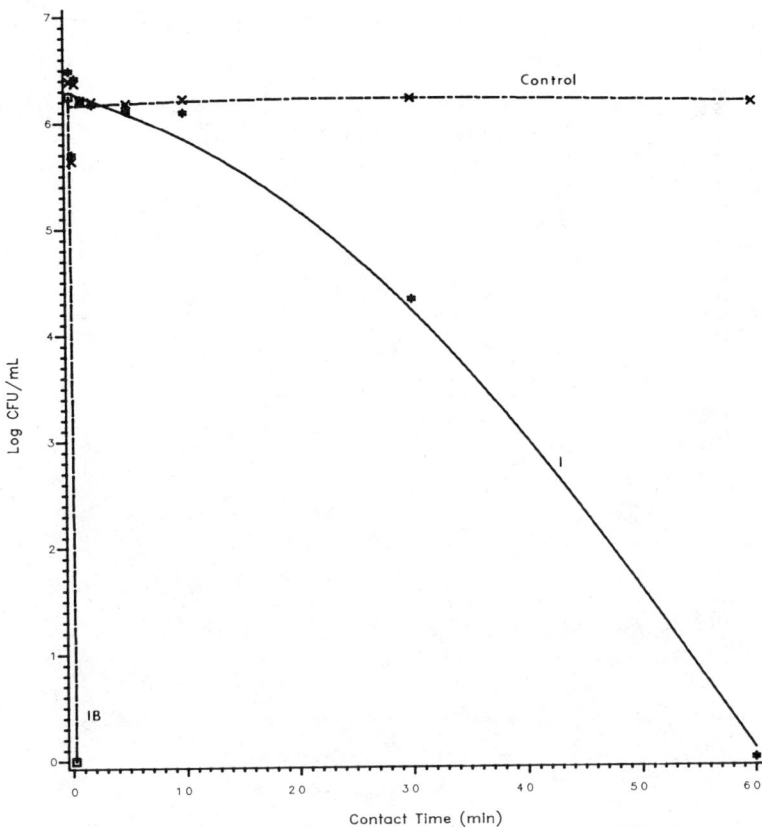

Figure 2. The actions of agent I and agent IB (5 mg/L total chlorine; 11.25 mg/L total bromine) against *Staphylococcus aureus* in demand-free water at pH 7.0, 22°C.

kills *S. aureus* in SDW, but calcium hypochlorite is no longer able to effect disinfection (i.e., chlorine from the calcium hypochlorite reacts with the organic demand in SDW before disinfection can occur). A 10 mg/L total chlorine solution of calcium hypochlorite is able to cause disinfection before the free chlorine is completely neutralized. Agent IB has not yet been studied at the molar equivalent to 5 mg/L total chlorine in SDW. Agent I was a significantly better disinfectant against *Shigella boydii* than against *Staphylococcus aureus* (Figure 4), although it was not as rapid as agent IB or calcium hypochlorite in demand-free water.

Table I presents a comparison of the actions of agent I and calcium hypochlorite against all the microorganisms included in this investigation under

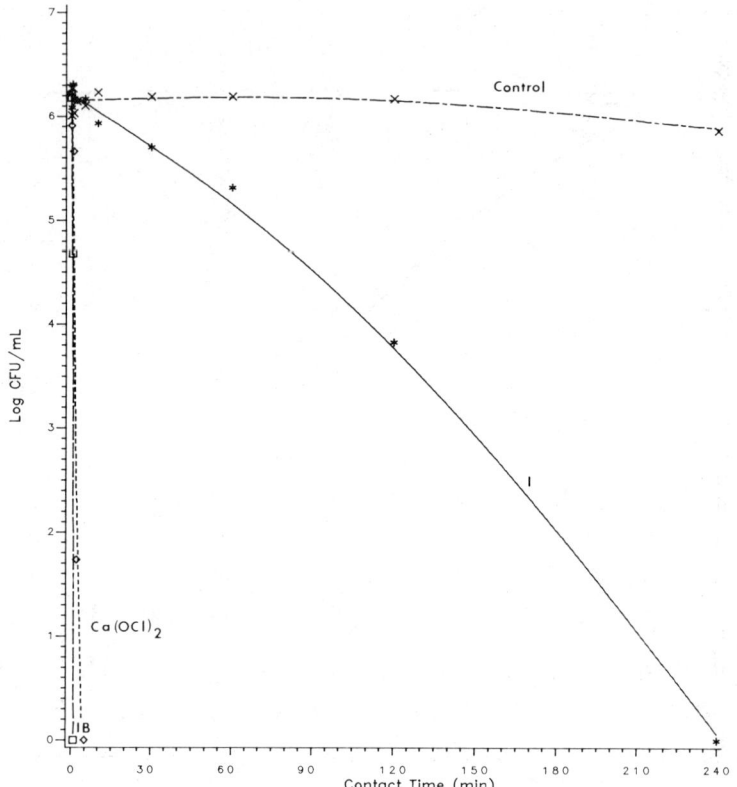

Figure 3. The actions of agent I, agent IB, and calcium hypochlorite (10 mg/L total chlorine; 22.5 mg/L total bromine) against *Staphylococcus aureus* in synthetic demand water at pH 9.5, 4°C.

various laboratory conditions. Agent IB was not included in the table because it was not evaluated for most of the microorganisms.

Calcium hypochlorite was a better disinfectant than agent I in demand-free water against all of the microorganisms studied except for the two protozoa species. However, agent I was the better disinfectant for the solutions containing organic demand (SDW). Also, agent I killed all microorganisms given sufficient contact time at all concentrations and conditions investigated.[15]

Stability Studies

Figure 5 indicates the stability of agent I, agent IB, and calcium hypochlorite in demand-free water at pH 7.0, 22°C as a function of time. The order of

DRINKING WATER TREATMENT

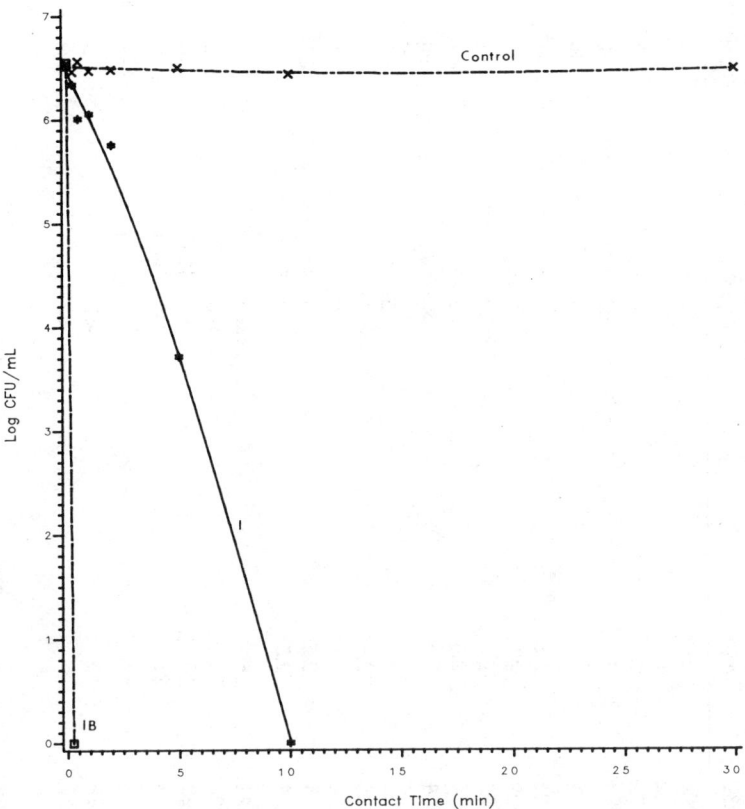

Figure 4. The actions of agent I and agent IB (2.5 mg/L total chlorine; 5.6 mg/L total bromine) against *Shigella boydii* in demand-free water at pH 7.0, 22°C.

stability is agent I > agent IB > calcium hypochlorite. Figure 6 shows the stabilities of agent I and calcium hypochlorite in SDW at 4°C as a function of time. The calcium hypochlorite solution lost over one-half of its titratable total chlorine content in less than 1 h; the half-life of agent I in SDW under these conditions was greater than 100 h. The stability of agent IB in SDW was not evaluated.

The half-lives of agent I and calcium hypochlorite in demand-free water at pH 9.5, 22°C are 43.8 and 33.2 d, respectively; at pH 4.5, 22°C they are both more stable than at pH 7.0, 22°C.[15]

Calcium hypochlorite in the solid state lost 90.71% of its total chlorine content, whereas solid-state agent I lost only 2.46% of its total chlorine over 111 d. These losses would have been much less significant if the compounds had been stored in sealed containers.

Table I. Disinfection Time (min) for Agent I and Calcium Hypochlorite (Ca(OCl)$_2$) Against Various Microorganisms (10^6 cfu/mL) Under Varied Laboratory Conditions

Microorganism	Laboratory Conditions			Disinfection Time[a] (min)	
	pH	°C	mg/L	Agent I	Ca(OCl)$_2$
Staphylococcus aureus	7.0	22	5[b]	30–60	<1
	4.5	22	5	30–60	<1
	9.5	22	5	10–20	<1
	9.5	4	5	60–120	2–5
	SDW[c]	4	5	120–240	>240
Pseudomonas aeruginosa	7.0	22	2.5	5–10	<1
	4.5	22	2.5	5–10	<1
	9.5	22	2.5	2–5	<1
	9.5	4	2.5	2–5	<1
	SDW[c]	4	2.5	30–60	>60
Shigella boydii	7.0	22	2.5	5–10	<1
	4.5	22	2.5	2–5	<1
	9.5	22	2.5	2–5	<1
	7.0	4	2.5	5–10	<1
	SDW[c]	4	2.5	2–5	<1
Poliovirus Type I	7.0	22	25[b]	>1440	<10
	4.5	22	25	>1440	<10
	9.5	22	25	60–120	<10
	9.5	4	25	240–1440	<10
	SDW[c]	4	25	>1440	120–240
Entamoeba invadens[d]	7.0	22	5	<2	5–10
	9.5	22	5	<2	2–5
	9.5	4	5	<2	5–10
Giardia lamblia	7.0	22	1	<2	>10
	9.5	22	1	<2	2–5
	9.5	4	1	<2	>10

Candida albicans				
7.0	22	25	20–30	<1
4.5	22	25	60–1440	<1
9.5	22	25	10–20	1–5
9.5	4	25	60–1440	10–20
SDW[c]	4	25	60–1440	60–1440
Rhodotorula rubra				
7.0	22	10	60–1440	<1
4.5	22	10	60–1440	<1
9.5	22	10	30–60	10–20
9.5	4	10	60–1440	30–60
SDW[c]	4	10	60–1440	60–1440

[a]100% kill.
[b]Concentration in mg/L total ionizable chlorine.
[c]Synthetic demand water.
[d]Agent I will kill this organism at 1.0 mg/L total chlorine concentration in less than 2 min.

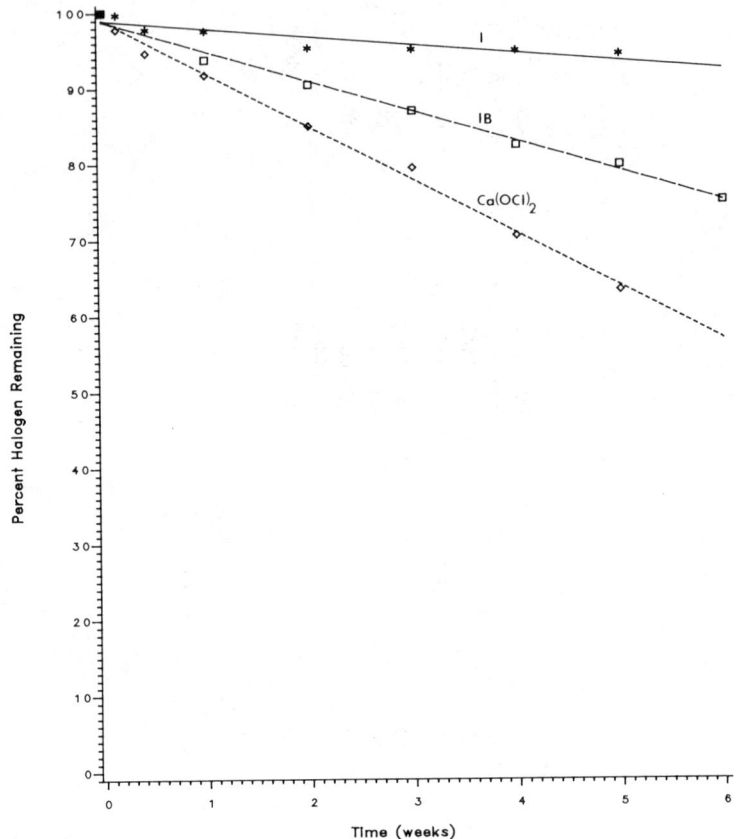

Figure 5. The stabilities of agent I, agent IB, and calcium hypochlorite in demand-free water at pH 7.0, 22°C.

Action on Materials

Even at the 10 mg/L total chlorine concentration level, agent I caused no detectable deterioration of membrane filters during 330 h of flow through a membrane test cell unit. Deterioration was noted only after the chlorine content was quenched by sodium bisulfite and after an additional 170 h of flow. It was concluded that although agent I may cause gradual deterioration of the membrane filters, the deterioration would be significantly lower than that caused by 10 mg/L total chlorine from hypochlorite and that a ROWPU field unit should function satisfactorily for at least 1 month in the presence of 10 mg/L total chlorine supplied by agent I.[16]

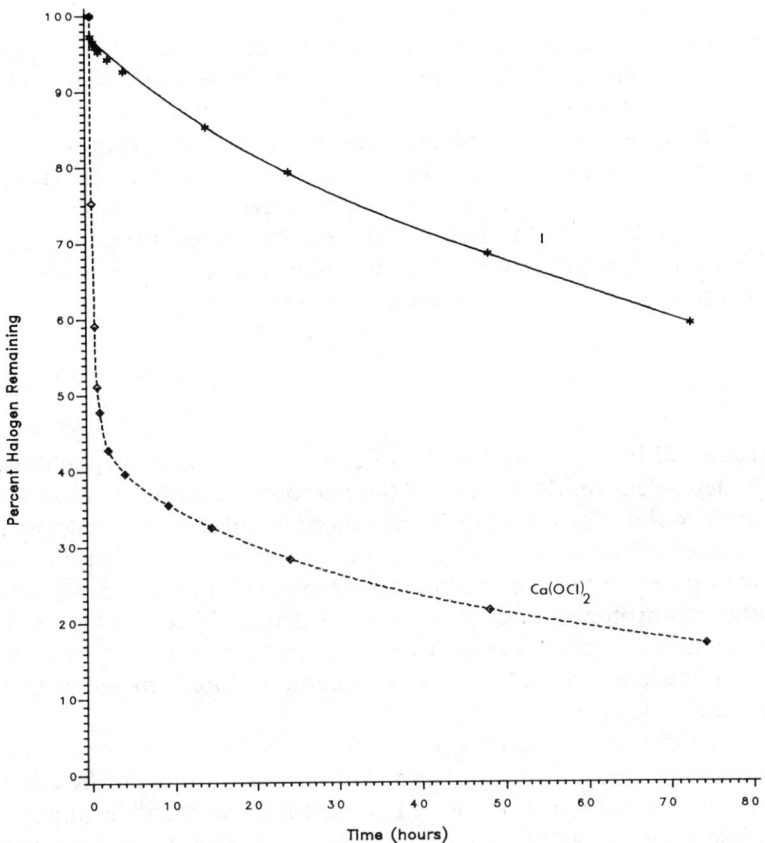

Figure 6. The stabilities of agent I and calcium hypochlorite in synthetic demand water at pH 9.5, 4°C.

An electron microscopic examination of two water bladder materials exposed to water solutions of the two compounds revealed some surface damage caused by both, but considerably more caused by calcium hypochlorite. Tensile strength testing led to the same conclusion. Gas chromatography studies of the two solutions exposed to the materials revealed no detectable organic impurities. However, the particular instrument used for these studies was only capable of detecting concentrations in excess of 100 mg/L volatile organics such as chloroform. Therefore, it cannot be concluded that low levels of organic impurities did not leach from the materials into the solutions of the two compounds.

Production of Trihalomethanes

The results of the experiments concerning production of THMs have been summarized in Table II. Chloroform was the only THM detected, although the procedure used would have detected other THMs as well. Some chloroform was detected in the samples containing agent I which were exposed to direct sunlight, but none was observed for similar samples held in darkness. The calcium hypochlorite samples contained significantly greater amounts of chloroform than did the agent-I samples, with sunlight accentuating the formation of the THM. The control samples contained low concentration levels of chloroform caused by the organic chemical additives.

DISCUSSION

The relative disinfection efficiencies of agent I and calcium hypochlorite are primarily dependent on the nature of the microorganism in question and the quality of the water. The free chlorine produced by calcium hypochlorite under demand-free conditions was more effective against six of the eight microorganisms investigated. Agent I containing combined chlorine was more effective against the two protozoa species, even under demand-free conditions. In the presence of heavy organic load, however, agent I becomes at least equally effective to calcium hypochlorite at reasonable disinfection levels (5 mg/L total chlorine or lower).

Agent I causes less deterioration in materials and lower production of THMs than does calcium hypochlorite. These observations can be attributed directly to the great stability of agent I. Calcium hypochlorite is almost completely dissociated in water to free chlorine, while agent I is less than 1%

Table II. Concentration of Chloroform Detected in Solutions Containing Agent I or Calcium Hypochlorite and Organic Load

Test conditions[a]	Chloroform (μg/L)	Chloroform[b] (μg/L)
Control + acetone + sunlight	5[c]	
Agent I + acetone + sunlight	9	4
Ca(OCl)$_2$ + acetone + sunlight	56	51
Control + SDW + sunlight	4	
Agent I + SDW + sunlight	14	10
Ca(OCl)$_2$ + SDW + sunlight	58	54
Control + acetone + darkness	11[c]	
Agent I + acetone + darkness	11	0
Ca(OCl)$_2$ + acetone + darkness	29	18

[a]See Experimental Section.
[b]Corrected for control.
[c]Two different lots of acetone were used in these two sets of experiments, which explains the difference noted in background chloroform content in the controls.

dissociated in water at ambient temperatures and under neutral, acidic, or mildly basic pH conditions.[6,7] Thus it is evident that free chlorine is more reactive than combined chlorine with organic materials in water. For this reason, agent I may be a better disinfectant than calcium hypochlorite for use in long-term disinfection applications, particularly for water containing appreciable organic load, even though free chlorine may provide more rapid disinfection for most microorganisms. An obvious experimental disinfectant would be one that contains a small amount of calcium hypochlorite for rapid disinfection and a larger amount of agent I for long-term disinfection. We are currently evaluating such mixtures of the two compounds.

The reason for the greater stability of agent I relative to other organic N-chloramines is that the molecule contains two electron-donating methyl groups in the 4-position of the oxazolidinone ring adjacent to the N-Cl moiety. These methyl groups can destabilize any developing negative charge on nitrogen that occurs upon loss of Cl^+ and thus serve to stabilize the N-Cl bond.

The stability studies presented in this work showed that agent IB is intermediate between agent I and calcium hypochlorite in stability in demand-free water at neutral pH. Also agent IB appears to be as effective as calcium hypochlorite as a disinfectant for those microorganisms examined thus far. Obviously much testing is yet to be done for agent IB, including toxicity evaluations; however, agent IB now appears to be an exciting candidate as a new all-purpose disinfectant for water.

CONCLUSIONS

Agent I is considerably more stable than calcium hypochlorite in water or in dry storage. The great stability of agent I renders it less reactive with organic materials than is free chlorine. Thus, few problems with degradation of water purification unit materials or with production of toxic trihalomethanes should result from use of agent I as a disinfectant. On the other hand, calcium hypochlorite is a more rapid disinfectant against most microorganisms than is agent I, exceptions being the two protozoa examined in this work and several bacteria studied earlier.[11] Agent I should be an excellent disinfectant for long-term disinfection applications. Finally, preliminary investigations of agent IB have shown that the brominated analog of agent I may have considerable potential as an all purpose disinfectant.

ACKNOWLEDGMENTS

We deeply appreciate the research contributions of W. B. Wheatley, H. H. Kohl, U. Barnela, and N. Bryson of the Department of Chemistry, Auburn University, and of M. H. Attleberger, C. M. Hendrix, D. Geiger, and T. Clark of the School of Veterinary Medicine, Auburn University.

We gratefully acknowledge the generous support of the U.S. Army Medical Research and Development Command at Ft. Detrick and the U.S. Air Force Engineering and Services Center at Tyndall Air Force Base through Contract DAMD 17-82-C-2257. We also thank the Water Resources Research Institute at Auburn University for the administration of the project.

We also gratefully acknowledge the help of of J. F. Price of the Alabama Department of Environmental Management Laboratory (GC detection of trihalomethanes), A. Riedinger of UOP, Inc. (action of agent I on membrane filters), E. C. Mora of Auburn University (electron microscopy studies), and B. Z. Jang of Auburn University (tensile strength studies).

REFERENCES

1. Brodtmann, N. V., and P. J. Russo. "The Use of Chloroamines for Reduction of Trihalomethanes and Disinfection of Drinking Water," *J. Am. Water Works Assoc.* 71(1):40-42 (1979).
2. Vogt, C., and S. Regli. "Controlling Trihalomethanes While Attaining Disinfection," *J. Am. Water Works Assoc.* 73(1):33-40 (1981).
3. Hoff, J. C., and E. E. Geldreich. "Comparison of the Biocidal Efficiency of Alternative Disinfectants," *J. Am. Water Works Assoc.* 73(1):40-44 (1981).
4. Kosugi, M., J. J. Kaminski, S. H. Selk, I. H. Pitman, N. Bodor, and T. Higuchi. "N-Halo Derivatives VI: Microbiological and Chemical Evaluations of 3-Chloro-2-Oxazolidinones," *J. Pharm. Sci.* 65(12):1743-1746 (1976).
5. Kaminski, J. J., M. H. Huycke, S. H. Selk, N. Bodor, and T. Higuchi. "N-Halo Derivatives V: Comparative Antimicrobial Activity of Soft N-Chloramine Systems," *J. Pharm. Sci.* 65(12):1737-1742 (1976).
6. Burkett, H. D., J. H. Faison, H. H. Kohl, W. B. Wheatley, S. D. Worley, and N. Bodor. "A Novel Chlorine Compound for Water Disinfection," *Water Res. Bull.* 17(5):874-879 (1981).
7. Worley, S. D., W. B. Wheatley, H. H. Kohl, J. A. Van Hoose, H. D. Burkett, and N. Bodor. "The Stability in Water of a New Chloramine Disinfectant," *Water Res. Bull.* 19(1):97-100 (1983).
8. Mora, E. C., H. H. Kohl, W. B. Wheatley, S. D. Worley, J. H. Faison, H. D. Burkett, and N. Bodor. "Properties of a New Chloramine Disinfectant and Detoxicant," *Poul. Sci.* 61:1968-1971 (1982).
9. Worley, S. D., W. B. Wheatley, H. H. Kohl, H. D. Burkett, J. H. Faison, J. A. Van Hoose, and N. Bodor. "A Novel Bactericidal Agent for Treatment of Water," in *Water Chlorination: Environmental Impact and Health Effects, Vol. 4*, R. L. Jolley, W. A. Brungs, J. A. Cotruvo, R. B. Cumming, J. S. Mattice, and V. A. Jacobs, Eds. (Ann Arbor: Ann Arbor Science Publishers, 1983), pp. 1105-1113.
10. Kohl, H. H., W. B. Wheatley, S. D. Worley, and N. Bodor. "Antimicrobial Activity of N-chloramine Compounds," *J. Pharm. Sci.* 69(11):1292-1295 (1980).
11. Worley, S. D., W. B. Wheatley, H. H. Kohl, H. D. Burkett, J. A. Van Hoose, and N. Bodor. "A New Water Disinfectant: A Comparative Study," *Ind. Eng. Chem. Prod. Res. Dev.* 22(4):716-718 (1983).
12. Kaminski, J. J., and N. Bodor. "3-Bromo-4,4-Dimethyl-2-Oxazolidinone," *Tetrahedron* 32:1097-1099 (1976).

13. *Standard Methods for the Examination of Water and Wastewater*, 15th ed. (Washington, DC: American Public Health Association, 1980), pp 280-289.
14. Kenyon, K. F. "Free Available Chlorine Disinfection Criteria for Fixed Army Installation Primary Drinking Water," Technical Report 8108 (Ft. Detrick, MD: U.S. Army Medical Bioengineering Research and Development Laboratory, 1981).
15. Worley, S. D., et al. (manuscript in preparation).
16. Riedinger, A. Private communication (San Diego, CA: UOP, Inc.).

CHAPTER 99

Properties of Ferrate(VI) in Aqueous Solution: An Alternate Oxidant in Wastewater Treatment

James D. Carr, Paul B. Kelter, Alireza Tabatabai, David Splichal, John Erickson, and C. William McLaughlin

The ferrate(VI) ion is an extremely strong oxidizing agent, with a standard reduction potential of 2.2 V in acid solution.[1] The salt, potassium ferrate(VI), K_2FeO_4, is rapidly and highly soluble in water, where it forms the ferrate ion which acts to oxidize dissolved or suspended materials or to oxidize the water itself. The iron-containing product of this reaction is iron(III), usually in the form of hydrated ferric oxide which precipitates from solution. This precipitate is very effective at removing heavy metals or suspended clays from water by coprecipitation.[2-4] Ferrate treatment has also been shown to be effective in destroying many forms of waterborne microorganisms.[4] This chapter describes some details of the reactions of the ferrate ion in aqueous solution.

The oxidation of water by the ferrate(VI) ion has been the subject of several studies.[5-7] Molecular oxygen is formed by the reaction

$$2 \text{ FeO}_4^{2-} + 5 \text{ H}_2\text{O} \longrightarrow 2 \text{ Fe(III)} + \tfrac{3}{2} \text{ O}_2 + 10 \text{ OH}^- \qquad (1)$$

The most detailed examination of the kinetics of this reaction did not distinguish among a three-halves order dependence on ferrate(VI) and parallel first- and second-order reactions.[8] The ferrate ion has been shown to exist in aqueous solution as a tetrahedral ion with all four oxygen atoms equivalent and slow to exchange with solvent water oxygen atoms.[8] The exchange rate is accelerated by higher hydrogen ion concentration.

EXPERIMENTAL

Chemicals

Potassium ferrate was prepared by the method of Thompson et al.[9] and assayed both by the chromite method[10] and by visible spectrophotometry. Potassium ferrate used in these studies ranged from 70 to 97% pure, although a few experiments with material of 50% purity showed kinetic behavior the same as the purest material. The principal impurity in K_2FeO_4 is ferric oxide, but water and KOH are also present in small amounts.

The buffer choice for these reactions is critical because of the additional requirements that the buffer not be oxidizable and that it complex the Fe(III) product to prevent ferric oxide precipitation. Such precipitation would obscure spectrophotometric determinations and has been shown to catalyze the reduction of ferrate(VI).[8] The buffer used in most experiments was 0.2 M orthophosphate, although borate, carbonate, and pyrophosphate buffers were also used either alone or in combination with phosphate. Adjustments of pH were made with nitric or sulfuric acid or sodium hydroxide. All solutions were prepared from deionized water that was distilled in an all-glass still prior to use. The method of Scholder et al.[11] was used to prepare Na_4FeO_4.

Procedures

The disappearance of ferrate(VI) was monitored spectrophotometrically on a Beckman 26, Cary 14, or Hewlett-Packard 8450 UV-visible spectrophotometer, or on a Durrum-Gibson stopped-flow spectrophotometer. Conventional speed experiments were initiated by adding known volumes of appropriate thermostatted buffer to weighed portions of solid K_2FeO_4. The resulting solution was then transferred to a thermostatted cuvette, and the absorbance was monitored at 505 nm. Stopped-flow experiments were with solutions of K_2FeO_4 in water, which were reacted with appropriate buffers or buffered solutes. An unbuffered K_2FeO_4 solution rapidly attains a pH of about 10.5, at which the decomposition rate is quite slow and solutions are fairly stable. Nonetheless, fresh solutions are made up about once an hour.

Spectrophotometric data were evaluated on a Horizon Northstar microcomputer. The stopped-flow spectrophotometer was interfaced directly to the computer, and conventional-speed data were entered manually. Data were evaluated as described earlier.[12,13]

Several series of reactions were carried out in the presence of 1,10-phenanthroline, which was useful in determining the presence of iron(II) intermediates in the reaction. To give better measurement of the time dependence of the formation of ferrous-*tris*-1,10-phenanthroline (ferroin), a double (or reactive blank) experiment was conducted. In such an experiment, two solutions were prepared that were equal in pH and ionic strength; one solution, however, included a known amount of 1,10-phenanthroline and the other had none. Equal amounts of potassium ferrate were mixed simultaneously with the two solutions. The solution containing phenanthroline was put into the sample side of the HP 8450 spectrophotometer, and the solution without phenanthroline was put into the reference side. The ferrate(VI) disappeared at the same rate in the two solutions; therefore, the observed absorbance difference was due to the formation of ferroin in the sample cell.

Oxygen evolution experiments were monitored by a Clark-type voltammetric electrode on a Yellow Springs ARC 54A dissolved oxygen meter with an attached recorder. Selective ion electrode and pH measurements were made on an Orion 701 pH meter.

RESULTS

Kinetics

Kinetic absorbance data were evaluated both by initial and instantaneous rates to determine the reaction order, as well as by the computer-aided evaluation described earlier.[12,13]

$$\text{Rate} = \frac{-dC_{Fe(VI)}}{dt} = \frac{-1}{\Delta \epsilon b} \frac{dA}{dt} \qquad (2)$$

where dC/dt is the time rate of change of iron(VI) concentration
dA/dt is the time rate of absorbance change
$\Delta \epsilon$ is the difference in molar absorptivities of products and reactants
b is the spectrophotometer cell path length.

The best agreement of the rate data is to a rate law incorporating a mixed first- and second-order dependence of iron (VI) in parallel reactions [Equations (3) and (4)], where $k_{1(obs)}$ and $k_{2(obs)}$ are first- and second-order rate constants, respectively, and A and A_e are absorbances at time t and at equilibrium, respectively.

$$\text{Rate} = k_{1(obs)}C_{Fe(VI)} + k_{2(obs)}C^2Fe(VI) \qquad (3)$$

$$\frac{-dA/dt}{(A-A_e)} = k_{1(obs)} + \frac{k_{2(obs)}}{\Delta \epsilon b} \qquad (4)$$

Plots of Equation (4) using either initial rates from several reactions of constant pH but at different ferrate concentrations, or from several points on the absorbance-time curve for a single reaction, are usually linear and yield values of k_1 and k_2 while verifying the rate law (Figure 1). The occasional curvature of these plots is taken as evidence for the rapid interconversion of FeO_4^{2-} and $Fe_2O_7^{2-}$, as will be described later.

Rate constants based on Equation (3), as determined by the mixed-order computer method,[12] are presented in Table I and Figures 2 and 3 and are derived from reactions designed to test for specific alkali metal ion effects, pH effects, buffer composition effects, and ionic strength effects. The kinetic evaluation program also allows extrapolation of absorbance to time zero; therefore, the absorbance, and hence the molar absorptivity of ferrate(VI), can be calculated before any reduction has occurred. The results of these measurements are shown in Figure 4.

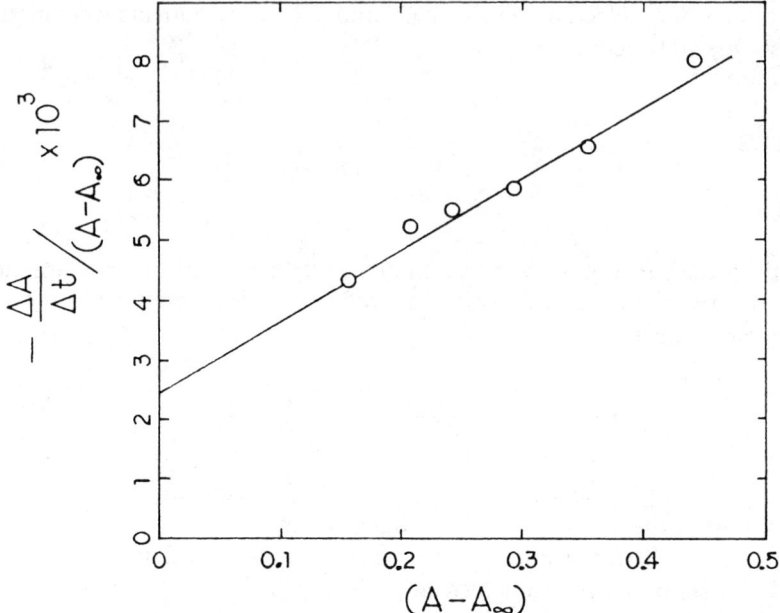

Figure 1. Dependence of reaction rate on ferrate concentration. Conditions: pH 8.25, $C°_{FC(VI)} = 4.57 \times 10^{-4}$; plot of Equation (4).

DISCUSSION

First-Order Kinetics

The pH dependence of the reaction of ferrate(VI) with water is interpreted in terms of the different protonated forms of ferrate(VI) present in solution, and in terms of general acid catalysis by the phosphate ion. The molar absorptivities in Figure 4 are interpreted in terms of three equilibria of H_2FeO_4 acting as diprotic acid. The values of $pK_{a1} = 3.5$ and $pK_{a2} = 7.8$, as well as

$$\epsilon_{H_2FeO_4} = 30 \text{ L/mol-cm}, \quad \epsilon_{HFeO_4^-} = 450,$$

and

$$\epsilon_{FeO_4^{2-}} = 1030 \text{ at 505 nm} \tag{5}$$

are resolved from the best fit to Figure 4. These equilibrium constants and molar absorptivities have not been measured before because of the rapid reduction of iron(VI) in water at pH values where appreciable amounts of protonated materials can exist.

Table I. Rate Constants for Decomposition of Ferrate(VI) in Aqueous Buffer (T = 25°C; phosphate buffer, 0.2 M unless otherwise noted)

pH	$k_{1(obs)}(s^{-1})$	$k_{2(obs)}(M^{-1}s^{-1})$	C_{PO_4} (M)	μ
2.53	18.0	3.71×10^5		
2.75	15.1	2.60×10^5		
2.95	14.4	1.44×10^5		
3.57	9.50	7.04×10^4		
3.68	8.71			
3.98	3.02	3.95×10^4		
4.25	1.89	1.23×10^4		
4.58	0.952	2.35×10^4		
4.92	0.831	1.38×10^4		
4.97	0.764	1.44×10^4		
5.34	0.442	7.88×10^3		
5.45	0.916	1.02×10^4		
5.50	0.699	1.48×10^4		
5.95	1.18	7.35×10^3		
5.96	0.342	4.79×10^3		
5.99	1.24	5.12×10^3		
6.01	1.25	8.94×10^3		
6.39	0.278	2.94×10^3		
6.45	0.659	5.46×10^3		
6.73	0.151	2.25×10^3		
6.75	0.434	3.24×10^3		
6.92	0.271	1.10×10^3		
6.95	0.270	3.31×10^3		
6.96	0.272	1.22×10^3		
7.03	0.309	6.68×10^2		
7.13	0.041	2.76×10^2		
7.15	0.194	1.97×10^3		
7.16	0.034	9.37×10^2		
7.32	0.034	2.16×10^3		
7.35	0.110	1.88×10^3		
7.36	4.8×10^{-3}	3.86×10^2		
7.41	6.5×10^{-3}	9.45×10^1		
7.68	2.09×10^{-3}	3.89×10^1		
7.74	1.68×10^{-3}	4.73×10^0		
7.76	8.12×10^{-4}	5.93×10^1		
7.78	6.7×10^{-3}	1.64×10^2		
7.81	2.44×10^{-2}	2.42×10^2		
7.90	1.37×10^{-3}	2.77×10^1		
8.03	6.80×10^{-4}	1.63×10^1		
8.08	6.0×10^{-4}	1.36×10^1		
8.25	3.7×10^{-4}	2.99×10^1		
8.42	1.7×10^{-4}	4.61		
8.50[a]	3.0×10^{-4}	5.0		
8.71		5.99		
8.91	7.2×10^{-5}			
9.11	2.4×10^{-5}	1.56		
9.12[a]	1.0×10^{-5}	2.0		
9.31	2.0×10^{-6}	0.19		

Table I, continued

pH	$k_{1(obs)}(s^{-1})$	$k_{2(obs)}(M^{-1}s^{-1})$	C_{PO_4} (M)	μ
5.13	0.275		0.05	0.50
5.13	0.346		0.10	0.50
5.13	0.361		0.25	0.50
5.13	0.669		0.35	0.50
5.13	1.019		0.49	0.50
5.03	0.265		0.05	0.50
5.03	0.238		0.10	0.50
5.08	0.264		0.25	0.50
5.05	0.375		0.35	0.50
5.06	0.822		0.49	0.50
5.50	0.225		0.15	0.156
5.50	0.225		0.15	0.206
5.50	0.251		0.15	0.256
5.50	0.287		0.15	0.306
5.50	0.408		0.15	0.356
5.50	0.577		0.15	0.406

[a] Borate buffer.

Additionally, if K_{a1} and K_{a2} are treated as variables to be fitted to a proposed rate law, estimates of their values can be obtained while the rate constants are also being determined. The values of the conditional rate constants $k_{1(obs)}$ shown in Table I are fitted to a detailed rate law:

$$k_{1(obs)} = k_{2H}\alpha_{H_2FeO_4} + k_{1Hu}\alpha_{HFeO_4^-} + k_{1Hc}\alpha_{HFeO_4^-}\alpha_{H_2PO_4}\cdot C_{PO_4} + k_{OH}\alpha_{FeO_4^{2-}} \quad (6)$$

where α is the fraction of ferrate or phosphate in the designated form, and k_{2H}, k_{1Hu}, k_{1Hc}, and k_{OH} are rate constants for the reaction of H_2FeO_4, $HFeO_4^-$, $HFeO_4^-$ (as catalyzed by H_2PO_4), and FeO_4^{2-}. Iterative manual calculations using acidity constants of phosphoric acid of 7.5×10^{-3} and 6.2×10^{-8}, respectively, and of H_2FeO_4 of 2.5×10^{-4} and 5.0×10^{-8}, respectively, lead to the best fit to the $k_{1(obs)}$ data. The theoretical line through these data is shown in Figure 2, and the values of the rate constants are $k_{2H} = 20\ s^{-1}$, $k_{1Hu} = <0.1\ s^{-1}$, $k_{1Hc} = 1.7\ s^{-1}$, and $k_{OH} < 1 \times 10^{-5}$. In the presence of phosphate buffer, the general acid catalyzed term, k_{1Hc}, so dominates the uncatalyzed term, k_{1Hu}, that the entire pH profile of $k_{1(obs)}$ can be described by only the two terms involving the reaction of H_2FeO_4 and the phosphate catalyzed term k_{1Hc}. Monoprotonated phosphate is not effective at catalyzing this reaction, as evidenced by the lack of phosphate dependence of the rate constant at pH 8.8. Details of the phosphate dependence of the rate constant at pH 5.5 are shown in Figure 5.

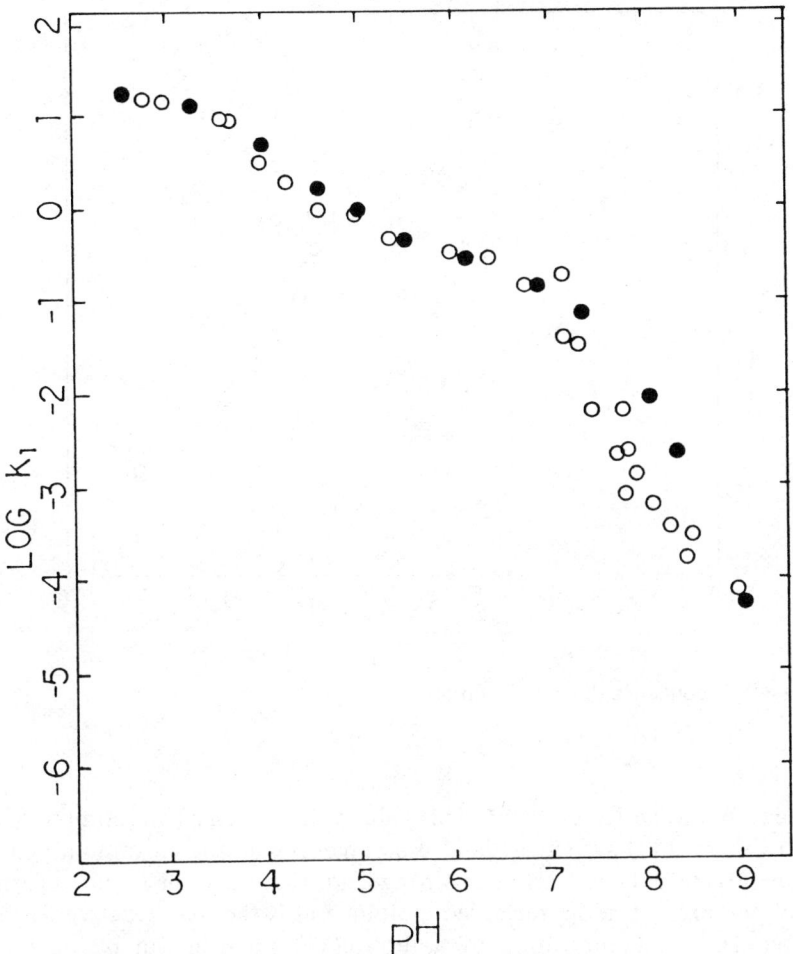

Figure 2. Dependence of log $k_{1(obs)}$ on pH; ○, observed points; ●, theoretical points from Equation 5 and best-fit values.

Second-Order Kinetics

The rate law term involving a second-order reaction in ferrate(VI) is thought to be caused by the reaction of an equilibrium amount of diferrate ($Fe_2O_7^{2-}$) formed:

$$2\ FeO_4^{2-} + 2H^+ \quad \underset{}{\overset{K_D}{\rightleftharpoons}} \quad Fe_2O_7^{2-} + H_2O \qquad (7)$$

The value of the dimerization equilibrium constant K_D has not been measured in the iron system; however, in the analogous chromium(VI) system, K_D has

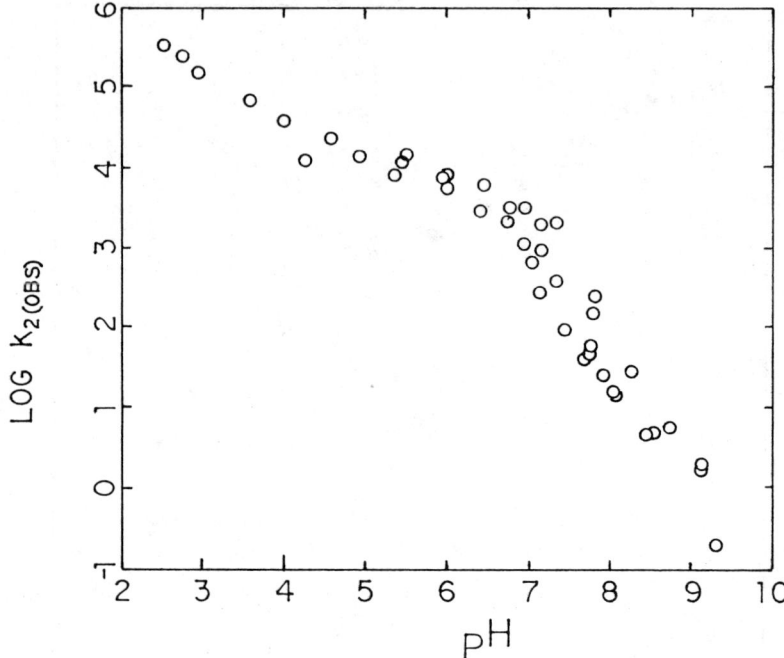

Figure 3. Dependence of log $k_{2(obs)}$ on pH.

been measured to be 1.0×10^{14}.[14] This value will be used as an estimate of the value in the iron(VI) system. It is also assumed, as in the chromium system, that $H_2Fe_2O_7$ and $HFe_2O_7^-$ are such strong acids that the unprotonated form is the only one present in appreciable amounts and is the only reactive form of diferrate(VI). The composition of a ferrate(VI) solution can be calculated from the analytical concentration of ferrate(VI) and K_{a1}, K_{a2}, and K_D in the following manner.

$$C_{Fe(VI)} = [H_2FeO_4] + [HFeO_4^-] + [FeO_4^-] + 2[Fe_2O_7^{2-}] \quad (8)$$

Appropriate rearrangement of Equation (8) leads to the following quadratic equation:

$$[FeO_4^{2-}]^2 (2K_D[H^+]^2) + [FeO_4^{2-}]\left(1 + \frac{[H^+]}{K_{a2}} + \frac{[H^+]^2}{K_{a1}K_{a2}}\right) - C_{Fe(VI)} = 0 \quad (9)$$

Implicit in this derivation is the assumption that all of these equilibria are attained rapidly. The solution of this quadratic equation yields directly the concentration of the ferrate dianion at any pH and analytical ferrate(VI)

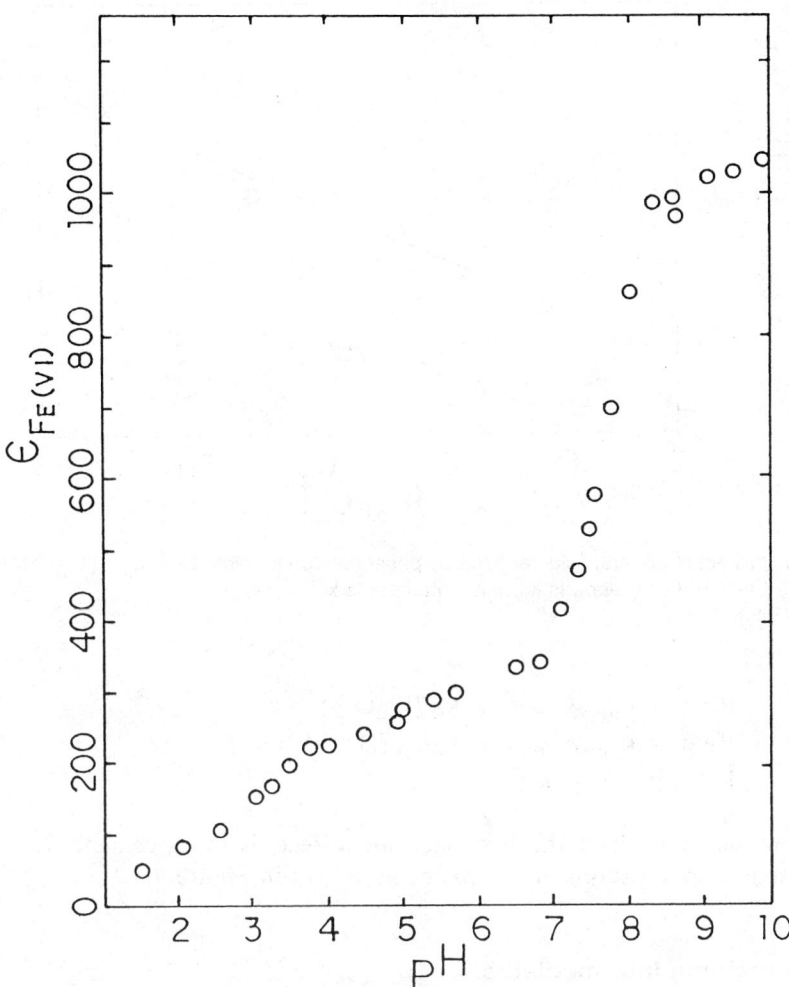

Figure 4. Molar absorptivity of ferrate(VI) vs pH.

concentration. This value and the appropriate equilibrium constants then allow the calculation of the concentrations of other species. Under conditions of our reaction, the fraction of ferrate(VI) present as the diferrate dianion never exceeds 1.5×10^{-4}. As $C_{Fe(VI)}$ decreases during the course of a given reaction, Equation (9) predicts and allows the calculation of the decrease in the fraction of diferrate in solution. Since diferrate is thought to be responsible for the second-order term in the rate law, a decrease in its concentration relative to that of the monoferrate species predicts that the value of $k_{2(obs)}$ should decrease during a given reaction. The curvature of plots, as in Figure 1, is evidence of this effect. The rate law then has the following form when written to emphasize the reactivity of diferrate.

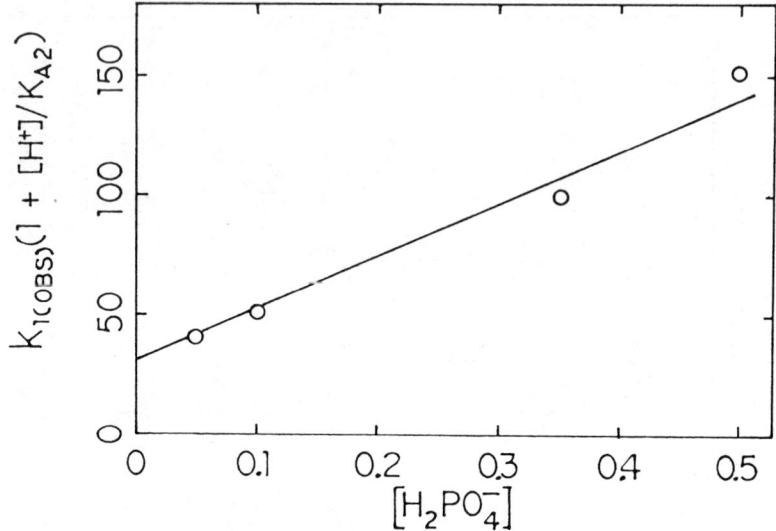

Figure 5. Effect of general acid catalysis by phosphate on k_1; plot of log $k_{1(obs)}[1 + (H^+)/K_{a2}]$ vs $(H_2PO_4^-)$; slope is k_{1Hc} and intercept is k_{1Hu}.

$$\text{Rate} = k_{1(obs)}C_{Fe(VI)} + k_D [Fe_2O_7^{2-}] \qquad (10)$$

$$\text{Rate} = k_{1(obs)}C_{Fe(VI)} + k_D K_D [H^+]^2 [FeO_4^{2-}]^2 \qquad (11)$$

$$k_D K_D [H^+]^2 = k_{2(obs)} \qquad (12)$$

This model, especially the hydrogen ion effect, is in agreement with the experimental observations above pH 5, as shown in Figure 3.

Iron-Containing Intermediates

The experiments with 1,10-phenanthroline (phen) show that iron(II), a more reduced form of iron than the final product, iron(III), is formed as an intermediate in the reduction of ferrate(VI). Ferroin is formed in a step slower than the disappearance of ferrate(VI), leading to a minimum in the absorbance-time curve (Figure 6). Ferroin is not formed when phen is added to an equilibrium solution after ferrate decomposition is complete. Subsequent addition of hydroxylamine to such a solution, however, gives instantaneous color formation of an amount of ferroin expected from the initial ferrate(VI) concentration. The addition of phen to the phosphate-buffered iron(II) solution at pH 8 makes formation of the ferroin color too rapid for measurement on the stopped-flow spectrophotometer.

These observations of iron(II) intermediates make sense only if the reduction of ferrate(VI) occurs by two-electron steps, as shown in Equations (13)

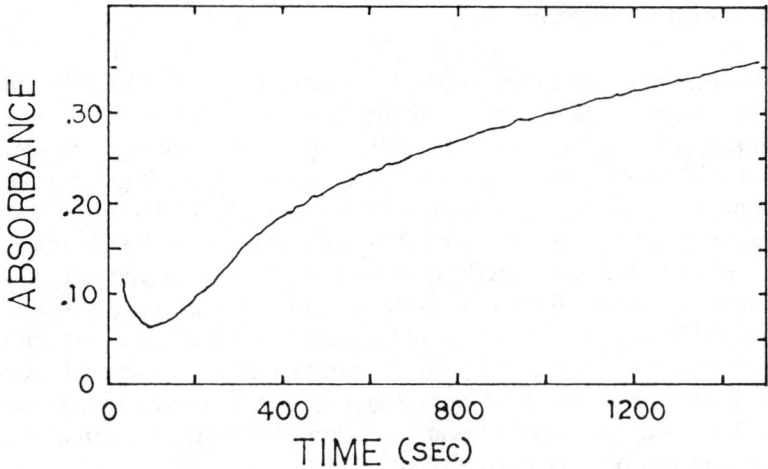

Figure 6. Absorbance-time curve of ferrate decomposition in the presence of methanol (0.2 M) and 1,10-phenanthroline (5 × 10^{-3} M), pH = 8.75, 506 nm.

and (14), where Sol signifies water or a solute that is oxidized to some form designated Sol_o.

$$Fe(VI) + Sol \longrightarrow Fe(IV) + Sol_o \qquad (13)$$

$$Fe(IV) + Sol \longrightarrow Fe(II) + Sol_o \qquad (14)$$

The fate of iron(II) depends on solution conditions. If phen is present, Fe(II) will react preferentially to form ferroin, $Fe(phen)_3$, as in Equation (15). If no ligand is present that can stabilize Fe(II), some oxidizing agent can oxidize the Fe(II) to form Fe(III). Available oxidants include Fe(VI), Fe(IV), and O_2. For simplicity, Equation (16) is shown using Fe(IV) as the oxidant. Additionally, it is known that Fe(IV) will disproportionate rapidly in water to give Fe(VI) and Fe(III).[11] This is shown in Equation (17).

$$Fe(II) + 3\ phen \longrightarrow Fe(phen)_3^{2+} \qquad (15)$$

$$Fe(II) + Fe(IV) \longrightarrow 2\ Fe(III) \qquad (16)$$

$$3\ Fe(IV) \longrightarrow Fe(VI) + 2\ Fe(III) \qquad (17)$$

The presence of a high concentration of reactive solute enhances the rate of reaction (14) and thus favors the production of Fe(II) and, hence, ferroin. In the absence of such a solute, the disproportionation reaction is favored, and less ferroin is formed. These statements were tested and verified by the use of methanol as solute.

Oxygen Evolution Studies

Experiments were carried out in which the rate of evolution and the amount of evolved oxygen were measured. When a reactive solute is present, only small amounts of oxygen are formed. When no reactive solute is present, the amount of evolved oxygen was less than that expected from Equation (1). The observed amounts of evolved oxygen are typically 65 to 70% of theoretical. Also, the evolution of oxygen occurred at the same rate as the disappearance of the absorbance due to ferrate(VI); however, little or no oxygen was evolved at the slower rate characteristic of Equation (14). It appears, therefore, that the product of iron(IV) reduction is, in large part, not molecular oxygen. This oxidizing power not accounted for by the product oxygen is termed "residual oxidant." Experiments designed to clarify this feature of the reaction are in progress. The residual oxidant is completely gone by the time the final concentration of iron(III) has been established.

Oxidation of Solutes

The kinetics and products of ferrate(VI) oxidation of a large number of solutes have been studied,[15-25] which have shown that ferrate(VI) is a strong but selective oxidant. Rate constants for the second-order oxidation of aqueous solutes by ferrate(VI) are strongly pH dependent, and, at a given pH, vary over several orders of magnitude among the compounds studied. Representative values for several compounds at pH 8 are shown in Table II. Also included is the concentration of this solute at which the rate of solution oxidation is equal to the first-order oxidation of water by ferrate(VI) at pH 8.

Usefulness of Ferrate(VI) in Wastewater Treatment

Potassium ferrate(VI) will be most useful in wastewater treatment for discharge into natural water systems or for water reuse if the original wastewater contains highly objectionable materials in the absence of high biological oxygen demand (BOD). In the case of phenol and cyanide, the oxidation reaction by ferrate(VI) is so fast that it could compete successfully with the slower oxidation of most components of BOD and with the oxidation of water. However, most other solutes are oxidized more slowly; therefore, the competing reactions would also have to be proportionally slower to allow selective oxidation to occur.

A single reagent that can serve as a destructive oxidant, a biocide, a flocculant for heavy metals and suspended material, and one that leaves no residual in the treated water other than molecular oxygen should be seriously considered as an alternate to chlorine in wastewater treatment.

Table II. Oxidation Rate Constants of Select Aqueous Solutes by Potassium Ferrate(VI) (T = 25°C; Phosphate Buffer, 0.1 M; pH = 8)

Compound	$k_S(M^{-2}s^{-1})$	$[S]_{50\%}$[a]
Methyl alcohol	0.03	2×10^{-2}
Ethyl alcohol	0.08	1×10^{-2}
Isopropyl alcohol	0.06	1×10^{-2}
Neopentyl alcohol	0.1	7×10^{-3}
Formaldehyde	0.5	1×10^{-3}
Acetaldehyde	0.4	2×10^{-3}
Trimethylaldehyde	2.3	3×10^{-4}
Chloral	6	1×10^{-4}
Formic acid	0.4	2×10^{-3}
Nitrilotriacetic acid	2	3×10^{-4}
N-methyliminodiacetic acid	2	3×10^{-4}
Dimethylglycine	2.5	3×10^{-4}
Iminodiacetic acid	100	7×10^{-4}
Sarcosine	120	6×10^{-6}
Glycine	100	7×10^{-6}
Ethylene glycol	0.04	2×10^{-2}
Glycolaldehyde	3	2×10^{-4}
Glycolic acid	0.4	2×10^{-3}
Glyoxal	300	2×10^{-7}
Glyoxylic acid	700	1×10^{-6}
Oxalic acid	0.1	7×10^{-3}
Methylamine	40	2×10^{-5}
Dimethylamine	200	3×10^{-6}
Diethylamine	0.7	1×10^{-3}
Diethylsulfide	100	1×10^{-4}
Thiodiethanol	100	1×10^{-6}
Dimethylsulfoxide	1	1×10^{-4}
Ammonia	0.1	1×10^{-3}
Nitrite	2	3×10^{-4}
Cyanide	$>10^4$	1×10^{-7}
Phenol	$>10^4$	1×10^{-7}

[a] $[S]_{50\%} = \dfrac{k_1}{k_S}$; $k_1 = 7 \times 10^{-4}$ at pH 8.

ACKNOWLEDGMENTS

We gratefully acknowledge the contributions of former graduate and undergraduate research students in the Chemistry Department of the University of Nebraska-Lincoln: Mark G. Cherwin, Ronald L. Bartzatt, O. Femi Kotoye, Alfred T. Erickson III, Charles Butler, Kola E. Alabi, William H. Smith, and Donna Latwaitis.

REFERENCES

1. Wood, R. H. "Heat, Free Energy, and Entropy of Ferrate(VI) Ion," *J. Am. Chem. Soc.* 80:2038 (1958).
2. Murmann, R. K., and P. K. Robinson. "Experiment Utilizing FeO^{2-} for Purifying Water," *Water Res.* 8:543 (1974).
3. Deyrup, A. J., and J. R. Mills. "Water Purification Process," U.S. Patent 2,758,084, August 7, 1956.
4. Waite, T. D. "Management of Wastewater Residuals with Iron(VI) Ferrate," Final Report, NSF-Rann Grant #ENV76-83897 (Washington, DC: National Science Foundation, 1979).
5. Veprek-Siska, J., and V. Ettel. "Reactions of Very Pure Substances: Decomposition of Mn(VII), Fe(VI), and Ru(VII) Oxyanions in Alkaline Solution," *Chem. Ind. (London)*, 548 (1967).
6. Wagner, W. F., J. R. Gump, and E. H. Hart. "Factors Affecting the Stability of Aqueous Potassium Ferrate(VI) Solutions," *Anal. Chem.* 24:1497 (1952).
7. Schreyer, J. M., and L. T. Ockerman. "Stability of the Ferrate(VI) Ion in Aqueous Solution," *Anal. Chem.* 23:1312 (1951).
8. Goff, H., and R. K. Murmann. "Studies on the Mechanism of Isotopic Oxygen Exchange and Reduction of Ferrate(VI) Ion," *J. Am. Chem. Soc.* 93:6058 (1971).
9. Thompson, G. W., L. T. Ockerman, and J. M. Schreyer. "Preparation and Purification of Potassium Ferrate(VI)," *J. Am. Chem. Soc.* 73:1279 (1951).
10. Schreyer, J. M., G. W. Thompson, and L. T. Ockerman. "Oxidation of Cr(III) with K_2FeO_4," *Anal. Chem.* 22:1426 (1950).
11. Von Scholder, R., H. V. Bunsen, and W. Zeiss. "Uber Orthoferrate(IV)," *Z. Anorg. und Allg. Chem.* 283:330 (1956).
12. Kelter, P. B., and J. D. Carr. "A Noniterative Method for Computer Evaluation of Second-Order Kinetic Data," *Anal. Chem.* 51:1825 (1979).
13. Kelter, P. B., and J. D. Carr. "A Microcomputer Compatible Method of Resolving Rate Constants in Mixed First- and Second-Order Kinetic Rate Laws," *Anal. Chem.* 51:1828 (1979) and *Anal. Chem.* 52:1552 (1980).
14. Martell, A. E., Ed. "Stability Constants," Special Publication 17, (London: The Chemical Society).
15. Carr, J. D., and P. B. Kelter. "Ferrate(VI) Oxidation of Nitrilotriacetic Acid," *Environ. Sci. Technol.* 15:184 (1981).
16. Kelter, P. B., Ph.D. Thesis, (Lincoln, NE: University of Nebraska, 1980).
17. Splichal, D., M.S. Thesis, (Lincoln, NE: University of Nebraska, 1983).
18. Buller, C. K., M.S. Thesis, (Lincoln, NE: University of Nebraska, 1983).
19. Tabatabai, A., Ph.D. Thesis, (Lincoln, NE: University of Nebraska, 1980).
20. Cherwin, M. G., Ph.D. Thesis, (Lincoln, NE: University of Nebraska, 1977).
21. Kotoye, O. F., M.S. Thesis, (Lincoln, NE: University of Nebraska, 1982).
22. Ericson, A. T. M.S. Thesis, (Lincoln, NE: University of Nebraska, 1975).
23. Bartzatt, R. L., M.S. Thesis, (Lincoln, NE: University of Nebraska, 1980).
24. Bartzatt, R. L., Ph.D. Thesis, (Lincoln, NE: University of Nebraska, 1982).
25. Alabi, K. E., M.S. Thesis, (Lincoln, NE: University of Nebraska, 1981).

CHAPTER **100**

Effect of a Spill Event on an Ozone – Granular Activated Carbon Treatment Plant

Howard M. Neukrug, Matthew G. Smith, Stephen W. Maloney, and Irwin H. Suffet

Extensive research has been conducted in the past decade to determine the effectiveness of granular activated carbon (GAC) in reducing organic levels in drinking water.[1-7] The results of these studies have indicated that GAC is an excellent adsorbent of a broad spectrum of organics, although its bed life varies considerably, depending on the water quality goal(s) desired. If bed life is short, as it is for many of the volatile halogenated organics (VHOs) of potential health concern, the costs associated with GAC application are high.[8] Figure 1 shows the exponential relationship that exists between bed life and operation and maintenance (O&M) costs. The examples of water quality criteria and their associated bed lives and costs (shown in Figure 1) demonstrate how stringent water quality goals for GAC treatment can often lead to a severe economic burden on the water utility.

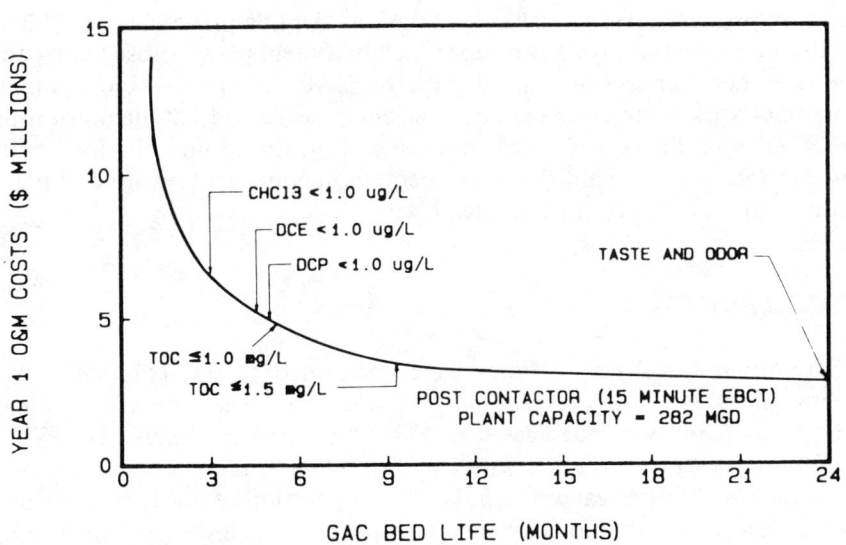

Figure 1. Relationship between GAC bed life and first-year operation and maintenance costs.

An alternate philosophy for implementing GAC treatment is to extend its design service life well past the point of volatile organic halogen breakthrough to water quality criteria based on total organic carbon (TOC) reductions or taste and odor control. TOC removal has been associated with reductions in trihalomethane (THM) precursors[9] as well as general removal of organic matter. Taste and odor are chronic problems in Philadelphia drinking water that were shown to be effectively controlled through GAC treatment.[1] Using either TOC or taste and odor as a criterion, GAC may be useful only as a "protective barrier" for volatile halogenated organics, protecting the water supply from sudden increases in concentration. This chapter discusses the effect of extended GAC bed life on specific organic concentrations for compounds that have already reached 100% breakthrough. In particular, the response of these "exhausted" GAC systems to a spill event involving chloroform is examined.

Spill events can be defined as any pollution episode where the influent concentrations of a specific compound significantly exceed typical raw water levels by fourfold to tenfold. These levels need not signify dangerous or regulated levels but do allow us to observe the effectiveness of the GAC during a spill event.

This chapter also outlines the effectiveness of ozone pretreatment on GAC performance for the removal of VHOs. Ozone pretreatment has been proposed as a potential mechanism for extending GAC bed life, which in the past has been the exponentially rising portion of the cost curve.[10] Preozonation (1) oxidizes some organics to form readily biodegradable compounds; (2) completely oxidizes other organics to form carbon dioxide; (3) strips organics from the process stream by aeration; and (4) does not form THMs or other chlorinated by-products. However, an evaluation of pilot plant data at Philadelphia[11] has shown that, for a prechlorinated treatment plant, subsequent ozonation does not increase the capacity of the GAC for the less biodegradable, more poorly oxidizable organics such as chloroform and 1,2-dichloropropane (DCP) (Figure 2). This chapter concentrates on the results obtained from a nonchlorinated pilot plant that was operated in conjunction with and parallel to the chlorinated plant in Philadelphia.[11]

PLANT FACILITIES

Ozone and GAC pilot studies were performed at the Torresdale Water Treatment Plant (WTP) located on the Delaware River Estuary in Philadelphia. (This plant was rededicated in 1983 as the Samuel S. Baxter WTP.) A 30,000 gpd advanced water treatment pilot plant (Figure 3), situated in the Torresdale WTP filter gallery, simulated the operation of the Torresdale plant, with the exception that no chlorine was applied through the pilot plant. Operating conditions of the postfiltration-GAC and ozone-GAC pilot systems are outlined in Tables I and II.

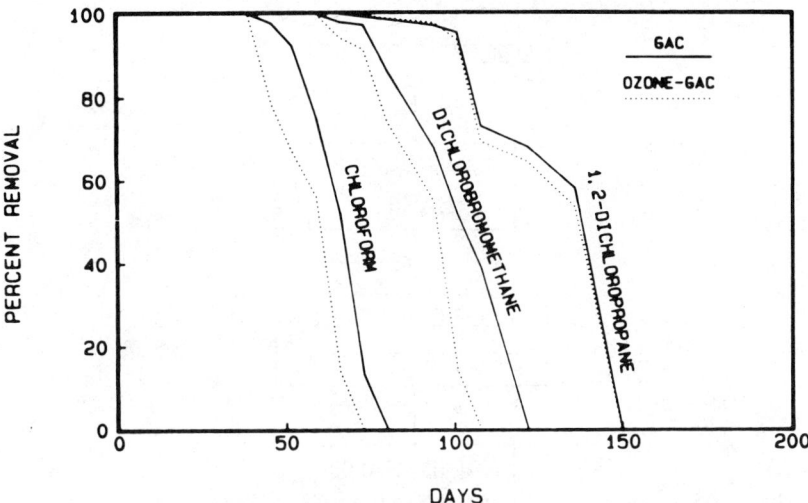

Figure 2. Comparison of GAC and ozone-GAC breakthrough characteristics for three VHOs in prechlorinated water.

EXPERIMENTAL PROCEDURES

Samples for VHO analysis were collected once a week as 30-min composites, sequenced to account for the theoretical detention times of the unit processes, and stored at 4°C in 40-mL clear screw-cap vials. Samples were analyzed by a Tracor Model 222 gas chromatograph (GC) with a Hewlett-Packard Model 3352C GC-data system. A Hall detector was used according to U.S. Environmental Protection Agency method 601,[12] with two minor modifications: a 3-m (10-ft) glass column was used instead of a 2.4-m (8-ft) column, and the temperature program was 60°C for 4 min and then 60°C/min to 215°C.

Total organic carbon was analyzed by a Dohrmann DC-54 low-level analyzer. Prior to analysis the samples were pressure filtered through a 0.2-μm surfactant-free polycarbonate membrane to remove carbon fines; thus, all results expressed as TOC are levels of dissolved organic carbon.

RESULTS

Ozonation

The ozone unit process can reduce organic concentrations in the nonchlorinated filter effluent by two removal mechanisms: oxidation and volatilization. Figure 4 shows the cumulative distribution function (CDF) curves for

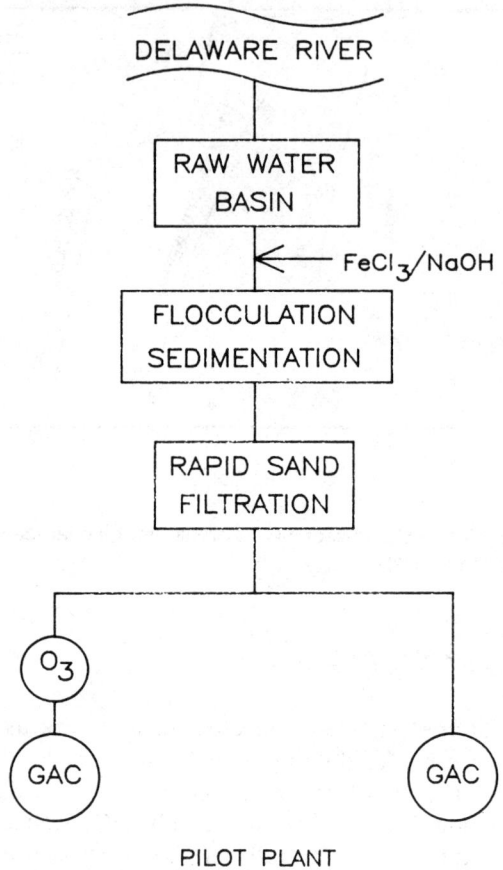

Figure 3. Pilot plant treatment process schematic.

Table I. GAC System Operating Parameters

Parameter	Value
Material	Glass, PFTE, 316 SS
Bed Diam, in.	6
Bed Depth, ft	4
Media	GAC (F-400)
Surface Loading Rate, gpm/ft^2	2.0
EBCT, min	15

Table II. Ozone System Operating Parameters

Parameter	Value
Contact time	
Contactor, min	22
Retention tank, min	25
Flow rate	
Gas, scfh	60 – 80
Water, gpm	2.0
Gas-to-water ratio	4:1 to 5:1
Ozone concentration	
Applied dosage, mg/L air	1.2
Transfer efficiency, %	25
Residual in water, mg/L water	0.3

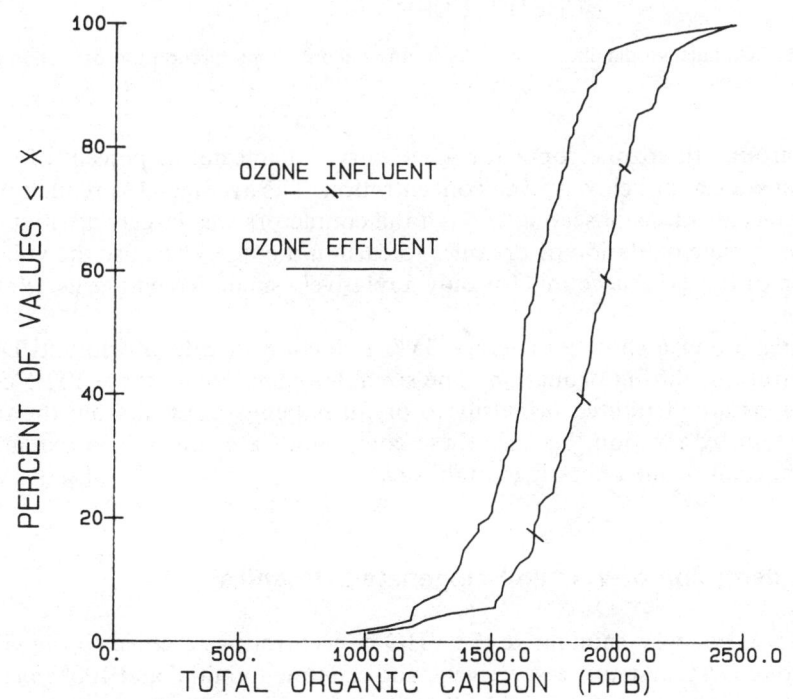

Figure 4. Cumulative distribution of TOC concentrations through the ozonation process.

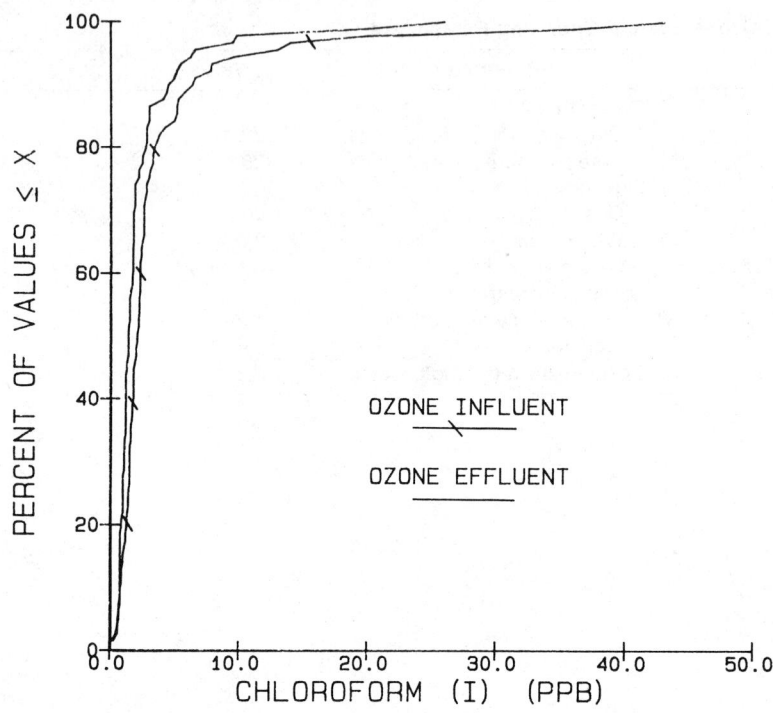

Figure 5. Cumulative distribution of chloroform concentrations through the ozonation process.

TOC through the ozone contactor. CDF curves delineate the percent of samples that were at or below a given concentration. The average 11% reduction in TOC concentrations observed through the contactors was largely attributable to the complete oxidation of organics to carbon dioxide,[1] because the volatile fraction of the TOC accounts for only a relatively small percentage (<1%) of the TOC.

Figures 5 and 6 show an average 35% reduction in chloroform and DCP concentrations during ozonation. The sizeable reductions in these VHO concentrations are attributed primarily to organics being stripped from the process stream by aeration, because these compounds are resistant to oxidation under normal ozone operating conditions.

GAC Adsorption of Volatile Halogenated Organics

Tables III and IV summarize the VHO concentrations applied to the GAC and ozone-GAC systems, respectively, and the time to initial and 100% breakthrough. As observed for the chlorinated plant[1] (Figure 2), the time to breakthrough was not significantly affected by preozonation, even though VHO and

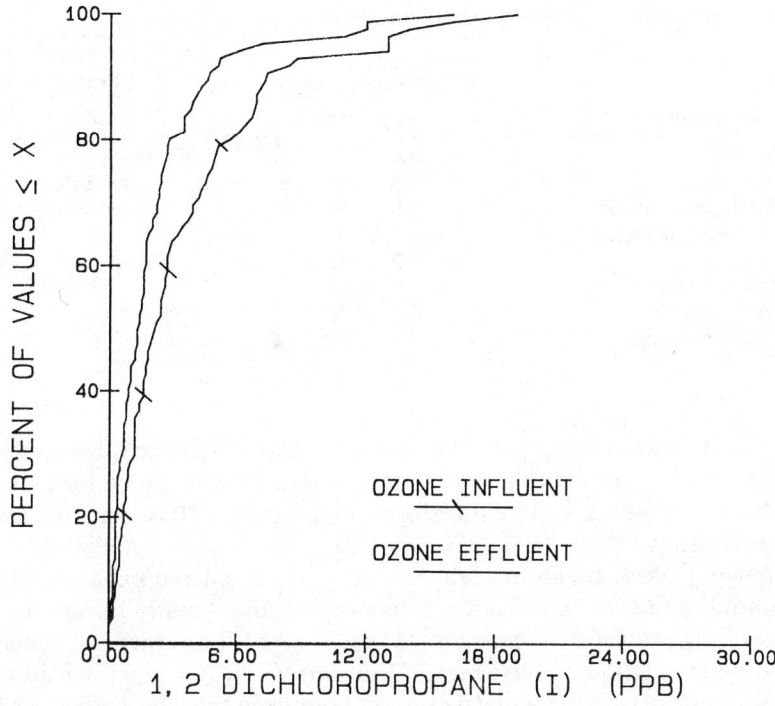

Figure 6. Cumulative distribution of DCP concentrations through the ozonation process.

Table III. GAC Breakthrough Characteristics

Compound	Concentration, µg/L Avg (Max)	Weeks to Breakthrough Initial	100%
Total THMs	5.6 (53)	Immediate	11
Chloroform	4.3 (43)	Immediate	11
Dichlorobromomethane	1.0 (11)	2	25
Dibromochloromethane	0.3 (2.7)	25	34
Bromoform	BD (0.1)	Never	Never
1,2-Dichloropropane	4.5 (19)	17	22
1,2-Dichloroethane	1.2 (9.0)	14	22
1,1,1-Trichloroethane	0.2 (0.8)	19	25

Table IV. Ozone-GAC Breakthrough Characteristics

Compound	Concentration, µg/L Avg (Max)	Weeks to Breakthrough Initial	100%
Total THMs	3.8 (33)	Immediate	11
Chloroform	2.8 (26)	Immediate	11
Dichlorobromomethane	0.8 (7.9)	Immediate	27
Dibromochloromethane	0.2 (2.3)	27	34
Bromoform	BD (0.1)	Never	Never
1,2-Dichloropropane	3.0 (16)	14	22
1,2-Dichloroethane	1.0 (7.3)	14	16
1,1,1-Trichloroethane	0.1 (0.4)	34	34

TOC loadings were lower onto the ozone-GAC bed. One possible explanation for this phenomenon is that lower-molecular-weight organics formed during ozonation competed with chloroform and other VHOs for the available adsorption sites.[1,11]

Following 100% breakthrough, the GAC is considered exhausted for that compound, and GAC adsorption behavior becomes greatly influenced by (1) changes in the adsorptive capacity of the GAC with changes in influent concentration (solid–liquid phase reequilibration), and (2) competitive adsorption (or displacement) among a matrix of compounds in the nanograms-per-liter to micrograms-per-liter range and TOC in the milligrams-per-liter range. During this exhausted phase, effluent concentrations were observed to exceed influent concentrations (the chromatographic effect) a significant portion of the time (up to 61%).

Chloroform Spill Event

Seven weeks after GAC exhaustion for chloroform at the 1- to 3-µg/L range, a spill event occurred where influent chloroform concentrations ranged between 6.6 and 43.0 µg/L (Figure 7). These concentrations far exceeded the levels for the previous 100 d. The GAC, which had been exhibiting the chromatographic effect over 50% of the time during exhaustion, began acting as an effective adsorber when the equilibrium surface concentration responded to the increased aqueous-phase concentration. Thus, although the influent levels increased from the 1- to 3-µg/L range to 40-µg/L, the effluent levels (which had also reached the 1- to 3-µg/L level) did not reach 10 µg/L. Immediately after the end of the spill event, the GAC contactor underwent a long period of chloroform desorption in which the effluent remained greater than the influent 100% of the time. After the desorption period, the GAC contactor again entered a phase in which the influent and effluent levels closely followed each other, similar to characteristics observed after 100% breakthrough of chloroform.

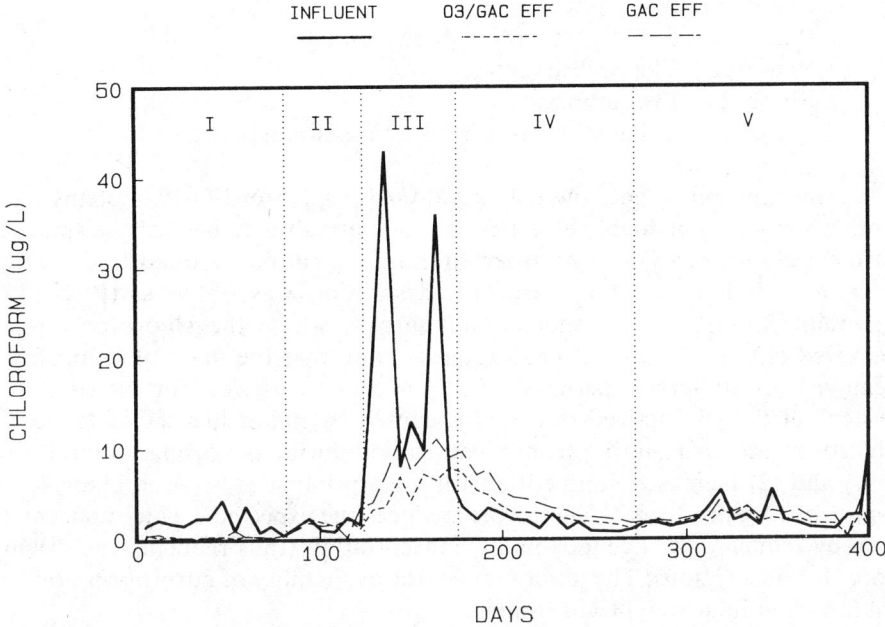

Figure 7. Chloroform concentrations through the pilot plant.

DISCUSSION

The use of GAC to remove undesirable organic compounds has been well established.[13] Similarly, competitive displacement[14,15] and the chromatographic effect[14] have been consistently observed in GAC applications. As previously discussed, the bed life of a GAC filter (and the associated cost) is very dependent on the water quality criteria that must be met (Figure 1). An important question that remains is: How will the GAC affect water quality for compounds not chosen as water quality criteria? Regardless of the cost or criteria selected, it is essential to avoid reducing the quality of the water during a treatment process.

One must be careful when determining whether water quality has been compromised. For example, whenever a chromatographic effect occurs, it could be reasonably argued that water quality has been reduced. Using chloroform as an example, we have attempted to clarify the action of the GAC after exhaustion has been reached and during a period in which chromatographic effects predominated. GAC adsorption of chloroform has been followed well past exhaustion and may best be described by the following five phases as shown on the breakthrough curve:

Phase   I. Adsorption
Phase II. Exhaustion, 100% breakthrough
Phase III. The Spill Event
Phase IV. Reequilibration
Phase   V. Exhaustion, return to Phase II conditions

During the spill event, the exhausted-GAC and ozone-GAC systems were both observed to be highly effective at moderating the impact of the spill. As influent chloroform levels returned to more typical background levels, however, a period of steady chloroform desorption was observed (Phase IV. Reequilibration). This is shown in Figure 8, where the chloroform mass removed is plotted against mass applied. Note that the mass of chloroform removed (in milligrams-per-gram GAC) is consistently less for the ozonated system. It is hypothesized that this is caused by the influences of (1) lower chloroform levels resulting from volatilization during the ozonation unit process, and (2) increased competition for adsorption sites between chloroform and ozone by-products.[12] The former reduces the expected surface concentration by reducing the aqeuous-phase concentration (thus reducing the driving force for adsorption). The latter reduces the availability of adsorption sites for all low-molecular-weight adsorbates.

The ratio of the actual chloroform adsorbed per gram GAC to its theoretical adsorptive capacity is shown in Figure 9. The actual chloroform surface concentration is calculated from mass balances, as shown in Figure 8. The theoretical adsorptive capacity is determined by calculating the surface concentration based on the influent. This value allows the effects of reduced chloroform concentrations in the ozone-GAC system to be removed from the analysis. Because of the variable influent, some deviation from isotherm predictions is inevitable. The GAC system, however, has consistently higher ratios than the ozone-GAC system after Phase I. Both systems have <100% capacity used (i.e., less than predicted equilibrium concentrations) during the spill event (phase III). During Phase IV, both systems exhibited >100% of capacity used. Thus, both systems were capable of releasing chloroform, and, as shown in Figure 7, a large amount of previously adsorbed chloroform was desorbing.

The smaller percentage of adsorptive capacity used by the ozone-GAC system (as compared to GAC) is indicative of the greater degree of chloroform displacement observed in this study when the influent was preozonated.

CONCLUSIONS

GAC treatment has been shown to be effective in reducing VHO concentrations; however, bed life is short and costs are high. Methods to increase GAC bed life for VHO removal (and thus reduce costs) have been investigated in this study through the ozonation of a nonchlorinated water prior to GAC treatment. However, preozonation was shown to have little impact on the time to

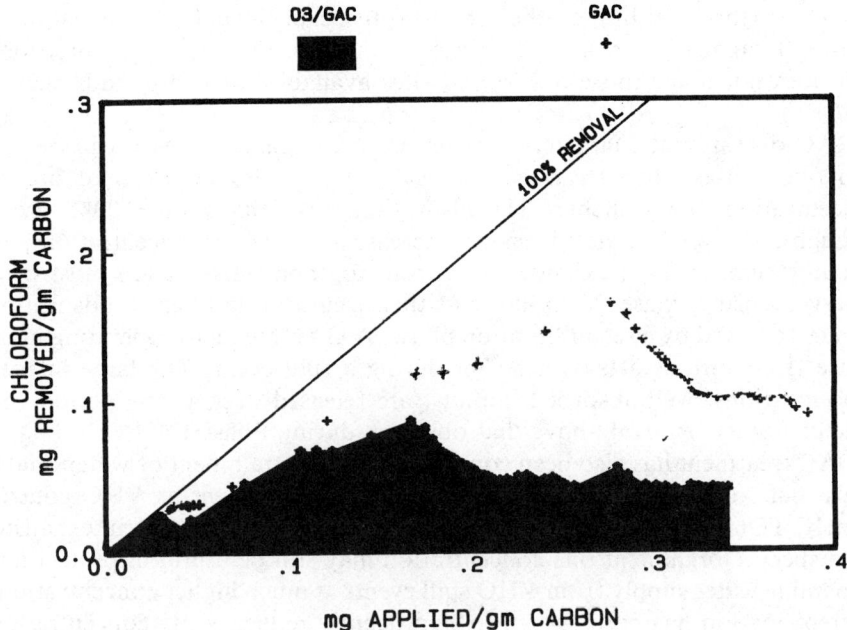

Figure 8. Cumulative mass removal of chloroform.

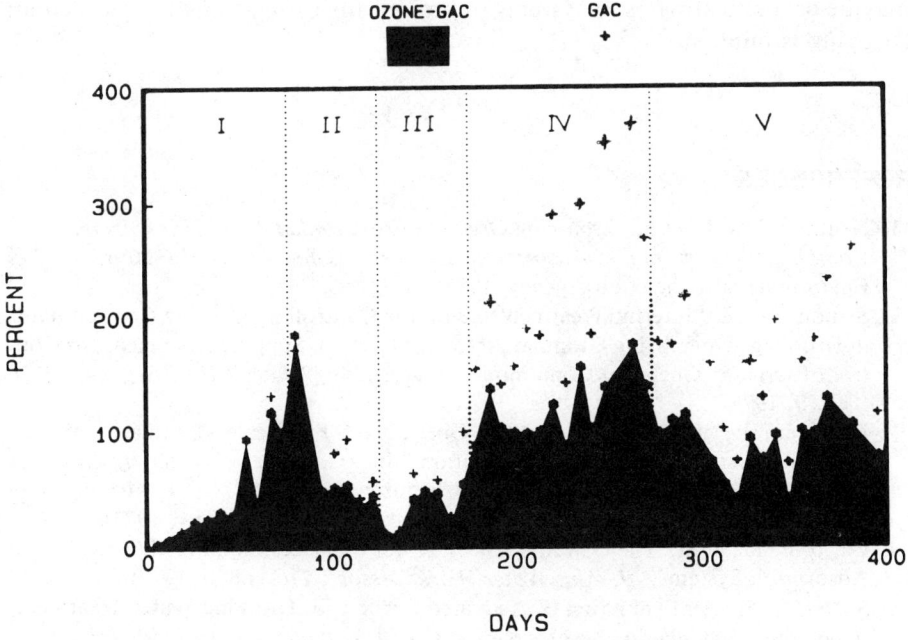

Figure 9. Predicted GAC adsorptive capacity used for chloroform.

VHO breakthrough. In fact, the net adsorptive capacity and the percent of the predicted capacity used both decreased with preozonation. Thus, preozonation does not make more adsorption sites available for compounds such as VHOs.

GAC operational characteristics for specific organic removal can be split into five phases when treating water subject to wide variations of organic concentration. They include (1) adsorption, (2) exhaustion (100% breakthrough), (3) spill event (dramatic increase in organic concentration), (4) reequilibrium, and (5) exhaustion. Chromatographic effects are most often observed during Phase IV. Spill events that occur during Phase I, adsorption, are not followed by a reequilibration phase. A GAC contactor operating under Phase II conditions acts as a buffer during a spill event. The large influent concentrations, well absorbed initially, are released over a long period with little or no net removal above that observed during Phase I.

GAC treatment has also been considered for the attainment of water quality goals that would not require as frequent bed replacement as VHO control, namely, TOC removal or taste and odor control. The GAC system exhausted for a specific organic at one concentration may still be useful in protecting a vulnerable water supply from VHO spill events at much higher concentrations.

Preozonation has been shown to significantly reduce VHO concentrations through volatilization. This reduction translates into lower VHO effluent concentrations in an ozone-GAC system, as compared with a GAC-only system. During a spill event, this volatilization effect is most pronounced. Removal may be dramatically less in systems optimized for ozone transfer in which air stripping is minimal.

REFERENCES

1. Neukrug, H. M., et al. *Removing Organics From Philadelphia Drinking Water by Combined Ozonation and Adsorption*, EPA-600/52-83-048 (Washington, DC: US Environmental Protection Agency, 1983).
2. Symons, J. M. "Interim Treatment Guide for Controlling Organic Contaminants in Drinking Water Using Granular Activated Carbon. Support Document for Control of Organic Chemical Contaminants in Drinking Water," *Fed. Reg.*, 43: 5756 (Feb. 9, 1978).
3. DeMarco, J., et al. "Experiences in Operating a Full-Scale Granular Activated Carbon System with On-Site Reactivation," in *Treatment of Water by Granular Activated Carbon*, ACS Advances in Chemistry Series 202, M. J. McGuire and I. H. Suffet, Eds. (Washington, DC: American Chemical Society, 1983).
4. Cairo, P. R., et al."The US Experiences—Pilot Plant Testing of Activated Carbon Adsorption Systems," *J. Am. Water Works Assoc.*, 71(11):660 (1979).
5. Suffet, I. H. "An Evaluation of Activated Carbon for Drinking Water Treatment: A National Academy of Science Report," *J. Am. Water Works Assoc.*, 72 (1):41 (1980).

6. Miller, R., and D. J. Hartman. *Feasibility Study of Granular Activated Carbon and On-Site Regeneration*, EPA-600/2-82-087, (Cincinnati: U.S. Environmental Protection Agency, 1982).
7. Brodtmann, N. V., Jr., et al. "Critical Study of Large-Scale Granular Activated Carbon Filter Units for the Removal of Organic Substances from Drinking Water," in *Activated Carbon Adsorption of Organics from Aqueous Phase*, Vol. 2, M. J. McGuire and I. H. Suffet, Eds. (Ann Arbor, MI: Ann Arbor Science Publishers Inc., 1980).
8. Neukrug, H. M. et al. "Biological Activated Carbon—At What Cost?" *J. Am. Water Works Assoc.*, 76(4):158 (1984).
9. Singer, P. C., et al. "Trihalomethane Formation in North Carolina Drinking Waters," *J. Am. Water Works Assoc.* 73(8):392 (1981).
10. Rice, R. G., et al. *Biological Processes in the Treatment of Municipal Water Supplies*, EPA-600/2-82-020, (Cincinnati: U.S. Environmental Protection Agency, 1982).
11. Maloney, S. W., et al. "Ozone-GAC for Drinking Water Treatment," in Fundamentals of Carbon Adsorption, A. Myers and G. Belhart, Eds. (New York: The Engineering Foundation, 1984), p. 325.
12. "Guidelines Establishing Test Procedures for the Analysis of Volatile Halogenated Organic Compounds," *Fed. Reg.,* 44:233 (Dec. 3, 1979).
13. Weber, W. J., et al. "Effectiveness of Activated Carbon for Removal of Volatile Halogenated Hydrocarbons from Drinking Water," in *Viruses and Trace Contaminants in Water and Wastewater*, J. A. Borchardt, et al., Eds. (Ann Arbor, MI: Ann Arbor Science Publishers, Inc., 1977).
14. McGuire, M. J., I. H. Suffet, and J. V. Radziul. "Assessment of Unit Processes for the Removal of Trace Organic Compounds from Drinking Water," *J. Am. Water Works Assoc.*, 70(10):565 (1978).
15. Thacker, W. E., V. L. Snoeyink, and J. C. Crittenden. "Desorption of Compounds During Operation of GAC Adsorption Systems," *J. Am. Water Works Assoc.*, 75(3):144 (1983).

CHAPTER 101

Activated Carbon: An Oxidant Producing Hydroxylated PCBs

Evangelos A. Voudrias, Richard A. Larson, Vernon L. Snoeyink, and A. S.-C. Chen

Activated carbon, in powdered and granular forms (PAC and GAC, respectively), is widely used in drinking water and wastewater treatment for the removal of hydrophobic organic compounds. In use, activated carbon frequently contacts disinfectants such as free chlorine, often in a prechlorination step. For example, the drinking water treatment plant in Hopewell, Virginia, uses prechlorination and backwashing of the GAC bed with up to 50-ppm combined chlorine.[1]

Activated carbon is known to react with many types of oxidizing agents. During this reaction, acidic surface oxides (carboxyl, phenolic hydroxyl, etc.) accumulate on the surface. In the case of chlorine, the concentration of oxides increases until 15 mmol of free chlorine has reacted per gram of carbon.[2] Puri et al.[3] showed that a large fraction of the surface oxygen formed on charcoal by reaction with chlorine water, when heated under vacuum, is evolved from the carbon surface as CO_2.

Oxidations of organic compounds at activated carbon surfaces have occasionally been noted. For example, partial oxidative conversion of n-butyl thiol to butyl disulfide was observed.[4] Activated carbon also promotes many inorganic oxidations. For example, sodium nitrite, potassium arsenite, sodium sulfite, and potassium ferrocyanide were oxidized in aqueous solutions in the presence of sugar charcoal.[5] Activated carbon is an effective catalyst for the decomposition of hydrogen peroxide; the rate of decomposition has been used for comparing carbons for their catalytic activity.[6]

There is a variety of surface functional groups on all carbons; the surface groups most often suggested are carboxyl, phenolic-OH, and quinone carbonyl groups. Less commonly named are ether, peroxide, lactone, anhydride, and cyclic peroxide groups. Extensive reviews of the subject have been made by Snoeyink and Weber[7] and Boehm.[8]

The existence of free radicals in carbon has been established by several techniques. For example, some time ago Donnet and Henrich determined the concentration of free radicals on carbon black surfaces by the fixation of externally generated radicals. The number of radicals bound by the surface coincided satisfactorily with the number of unpaired electrons determined by electron spin resonance (ESR).[9,10] Free radicals on the carbon surface play an important role in its ability to catalyze some reactions. The oxidation of cyclohexene to cyclohexenone and cyclohexenol in the presence of activated carbon and carbon black is a free radical reaction; the action of the carbon in initiating the oxidation is analogous to the generally accepted scheme (involving one-electron transfers) for the action of metal ion catalysis.[11]

In water treatment processes involving activated carbon, some of the added chlorine is destroyed by reactions with dissolved organic molecules. Trihalomethanes, other chlorinated organic compounds, and oxidation products are known to form. Some chlorine will reach the activated carbon and will react both with activated carbon and with compounds adsorbed on it. Recently, we[12] have shown that phenolic acids which readily react with HOCl in dilute aqueous solutions also reacted when adsorbed, but different and unusual reaction products were formed, thus indicating a catalytic or promoting effect of the GAC surface. In addition to the expected substitution products (chlorinated phenolic acids), compounds with introduced hydroxyl groups, quinones, and decarboxylation and demethylation products were observed. Also, when the products of a reaction between HOCl and adsorbed humic substances were compared with those of a similar reaction between HOCl and humic substances in solution, several dihydroxybenzenes and chlorinated dihydroxybenzenes were found in the presence of carbon.[13]

Chlorine dioxide[14,15] and chlorite[16] likewise produced very different compounds with GAC-adsorbed organic molecules than with the same substrates in aqueous solution.

We report that simple phenolic compounds undergo a variety of oxidative transformations on GAC treated with HOCl. The reactions observed include hydroxyl substitution, carboxylation, and dimer formation. The principal dimers are hydroxylated biphenyls; when chlorophenols are applied to the GAC surface, chlorinated biphenyls having hydroxyl substitution are major products.

EXPERIMENTAL

Chlorine

Chlorine solutions were prepared by bubbling high-purity chlorine gas (Linde, New York) into distilled-deionized water made alkaline with NaOH. This stock solution was added to distilled-deionized water to achieve the required concentration (8–10 mg/L as Cl_2 unless indicated otherwise). Chlorine solutions were buffered to the required pH with 0.001-M phosphate salts. The chlorine concentration was determined by the DPD method.[17]

Carbon

The carbon used in all experiments was F-400 GAC (Calgon, Pittsburgh), 20 × 30 or 40 × 60 mesh. It was prepared by grinding the particles, sieving to the desired mesh size, and washing with distilled-deionized water to remove fine particles. The GAC was then baked at 175°C for 1 week to remove volatile impurities, and was subsequently kept at 105°C.

For some experiments, the carbon surface was modified by oxidation, using an HOCl solution at room temperature. Usually, a 1 × 5-cm carbon column received 20 L of 10 to 18 mg/L (as Cl_2) of HOCl at 10 mL/min. This oxidized carbon, after prolonged washing with distilled-deionized water, was used in subsequent experiments.

Chlorine-GAC-Organic Compound Experiments

Batch reactor tests were used to determine the products of the free chlorine–organic compound reactions without GAC. The appropriate amount of organic compound (phenol, p-chlorophenol, or 2,4-dichlorophenol) was dissolved in 50 mL H_2O or 15 mL CH_3OH and added to 10 L of buffered free chlorine solution. After 3-min reaction time, the excess chlorine was destroyed with Na_2SO_3 after acidification to pH 3 with H_2SO_4.

Several column experiments were conducted. In all cases, carbon was secured between Pyrex glass wool plugs in Pyrex glass columns. Teflon® tubing was used to transfer the solution from the container to the column. In one experiment, a phenol solution (3 mg/L, 40 L) and a chlorine solution (8–10 mg/L, 40 L) were pumped separately through metering pumps into a mixing chamber at the bottom of a GAC column.[13] After a 3-min reaction time, the mixture was pumped onto the carbon. In other experiments, the organic compounds (25–50 mg/L) were adsorbed individually from solution onto GAC columns (1.5-g F-400 GAC in columns of 1 cm diam and 5–8 cm long). One column receiving 2,4-dichlorophenol was preloaded with 17.6-mg peat fulvic acid per gram GAC. A chlorine solution was subsequently passed through each of the columns.

For preoxidized GAC studies, a 1 × 5-cm column containing GAC oxidized with HOCl (as described) was treated with 8 L of a 25 mg/L solution of the organic compound at 10 mL/min.

Workup and Product Analysis

The 10-L batch reaction solutions were acidified and passed through a 1 × 8-cm column of XAD-2 macroreticular resin (Rohm and Haas, Philadelphia) previously cleaned by sequential Soxhlet extractions with ether, acetonitrile, and methanol. After adsorption, the bed was rinsed with 2 mL methanol, followed by 75 mL methylene chloride (Burdick and Jackson, distilled in glass, Muskegon, MI). F-400 GAC was also used as the column material in a similar concentration procedure.

Carbon samples were removed from the columns, centrifuged, and Soxhlet extracted with methanol-methylene chloride for 24 h. The organic extracts were dried and concentrated to 1 mL in a Kuderna-Danish evaporative concentrator.

The concentrated organic extracts were methylated with diazomethane generated from N-methyl-N-nitroso-p-toluenesulfonamide (Diazald®). A 2-μL injection was made into a SP-2100 20-m glass or DB-1 30-m fused silica capillary column in the splitless mode [with flame ionization detection (FID)]. Under identical conditions, the samples were also injected into a Hewlett-Packard 5985 GC/MS system. Most of the samples were also analyzed by GC/MS before they were methylated with diazomethane. Compounds were tentatively identified by comparing their mass spectra with those in standard reference collections, those from reference compounds of known structure, or by comparison to or extrapolation from literature data.

RESULTS AND DISCUSSION

2,4-Dichlorophenol

Free chlorine and 2,4-dichlorophenol reacted in aqueous solution to give 2,4,6-trichlorophenol.[18] When they were allowed to react in the presence of GAC, however, many additional products were formed. Compounds tentatively identified from such a chlorination reaction are listed in Table I; a corresponding total ion current chromatogram of the Soxhlet extract (experiment 1 in Table I) is shown in Figure 1.

One type of product formed only on GAC was the group of chlorinated or hydroxylated derivatives of benzoquinone (compounds 1, 2, 3, and 8 in Table I). Chlorinated and hydroxylated benzoic acids (12 and 17, Table I) were also identified in the same carbon extract. These compounds are derived by carboxylation of the phenol ring system at the GAC surface; similar compounds have been observed in several previous studies of phenolic compound–GAC–disinfectant reactions.[16]

Another type of product is a group of dimers (25–27, 29 and 30, Table I). The mass spectra of compounds 25–27 are almost identical. One such spectrum (27) is depicted in Figure 2. This fragmentation pattern corresponds to trichlorodihydroxybiphenyls and is very similar to that of trichloromonohydroxybiphenyls.[19] Since fragments and fragment intensities in the mass spectra of compounds 25–27 are about the same, the spectra cannot be used to identify specific isomers.

Table I. Reaction Products from HOCl – 2,4-Dichlorophenol – GAC Reactions

Compound	Mol. Wt	I[a]	II[b]	III[c]	IV[d]
1. Chloro-*p*-benzoquinone	142	+			
2. Chloro-*o*-benzoquinone	142	+			
3. Chlorohydroxy-*p*-benzoquinone	158	+			
4. 2,4-Dichlorophenol[e]	162	+	+	+	+
5. Unknown[f]	176	+			
6. Unknown[f]	198	+			
7. 2,4,6-Trichlorophenol	196	+	+	+	+
8. Chloromethoxy-*p*-benzoquinone	172	+			
9. Unknown[f]	194	+			
10. Unknown[f]	220	+	+	+	
11. S_6[g]	192	+			+
12. Methyl 2-hydroxy-3,5-dichlorobenzoate	220	+	+	+	+
13. Unknown[f]	256	+			
14. Unknown[f]	270	+			
15. Unknown[f]	248	+	+		
16. S_7[g]	224	+			
17. Methyl trichlorohydroxybenzoate	254	+			
18. Unknown[f]	244	+			
19. Unknown[f]	280	+			
20. Unknown[f]	258	+	+	+	
21. Unknown[f]	290	+			
22. Unknown[f]	314	+			
23. S_8[g]	256	+	+	+	
24. Unknown[f]	268	+	+	+	+
25. Trichlorodihydroxybiphenyl	288	+	+	+	+
26. Trichlorodihydroxybiphenyl	284	+	+	+	+
27. 3,5,5'-Trichloro-2,4'-dihydroxybiphenyl	288	+	+	+	+
28. Unknown[f]	302	+			
29. 3,3',5,5'-Tetrachloro-2,2'-dihydroxybiphenyl	322	+	+	+	+
30. Tetrachlorodihydroxybiphenyl	322	+	+	+	+
31. Unknown[f]	298	+			

[a]Preadsorbed 2,4-dichlorophenol reacted with 10 mg/L HOCl as Cl_2.
[b]Preadsorbed 2,4-dichlorophenol reacted with 1.5 mg/L HOCl as Cl_2.
[c]Preadsorbed 2,4-dichlorophenol reacted with 1.5 mg/L HOCl as Cl_2 in the presence of peat fulvic acid.
[d]2,4-Dichlorophenol reacted with chlorine-preoxidized F-400 GAC.
[e]Starting material.
[f]Chloroderivatives.
[g]Impurity in water.

GC/MS analysis of the methylated extract revealed three trichlorodihydroxybiphenyls. The mass spectrum of one of these compounds shows a strong molecular ion (m/z 302) and an even stronger M-50 (M-CH$_3$Cl) parent peak. The high intensity of the M-50 fragment indicates that the –OCH$_3$ group is in the 2-position. Loss of CH$_3$Cl would then result in a stable dibenzofuran structure.[20] The mass spectrum of compound 27 after methylation shows a

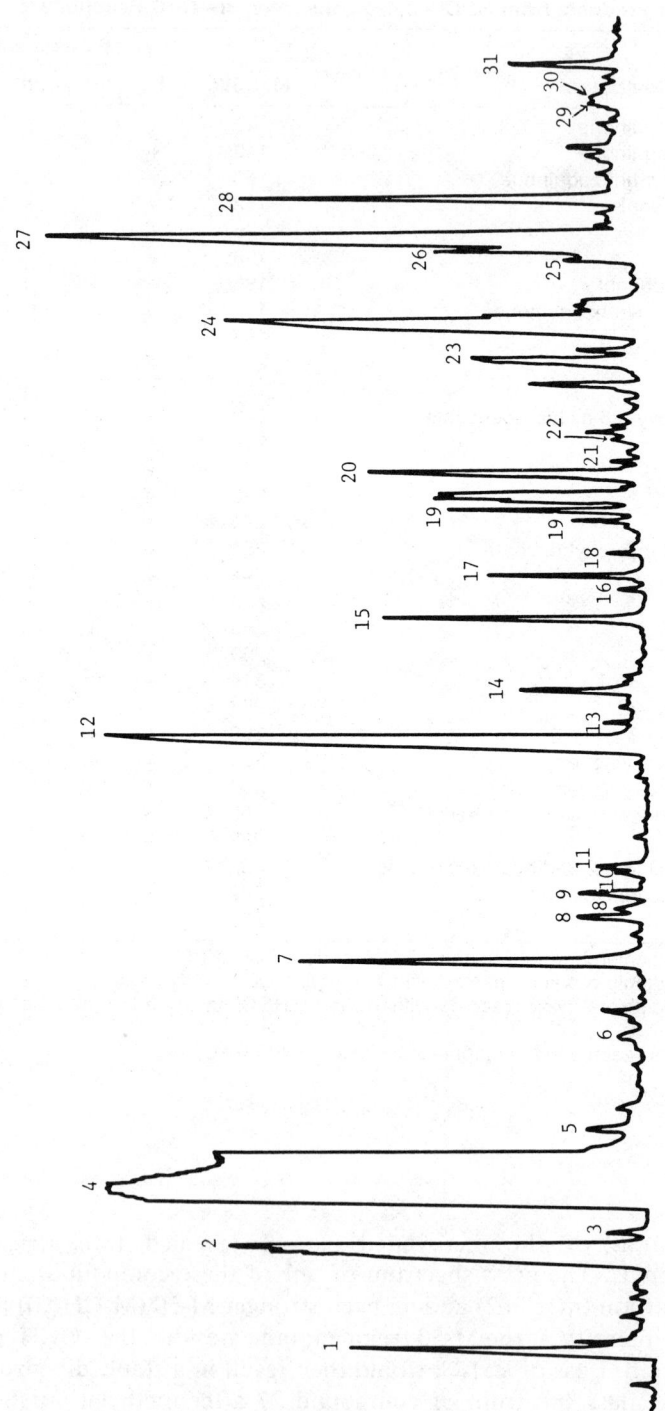

Figure 1. Total ion current chromatogram of products from the 2,4-dichlorophenol-GAC-HOCl (10 mg/L as Cl_2) reaction.

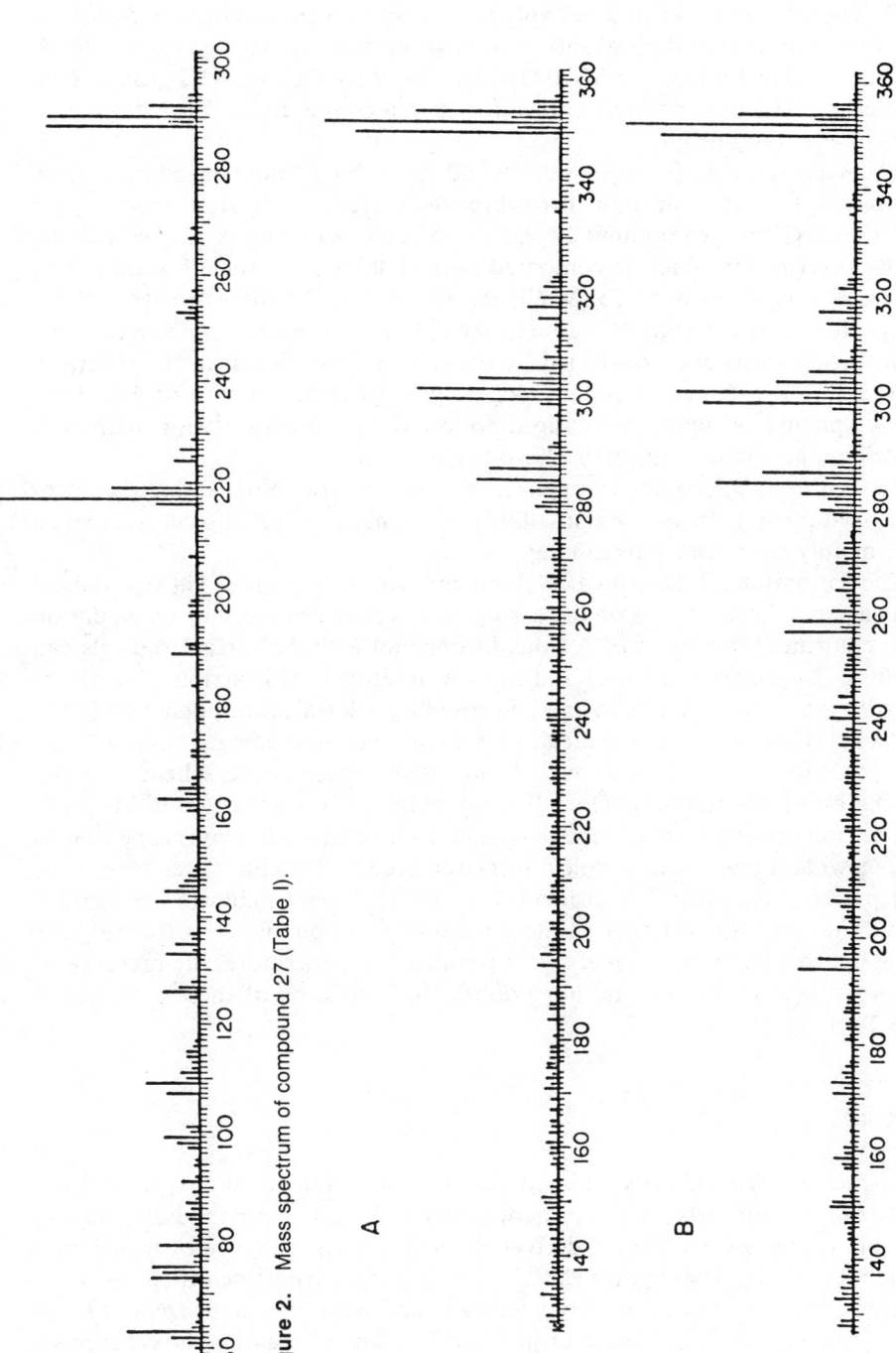

Figure 2. Mass spectrum of compound 27 (Table I).

Figure 3. Mass spectra of 3,3',5,5'-tetrachloro-2,2'-dimethoxybiphenyl: (A) derived from 2,4-dichlorophenol-GAC reaction, (B) authentic standard.

strong M-15 (M-CH$_3$) and small fragments at M-43 (M-CH$_3$CO) and M-70 (M-2Cl). The strong M-15 could be explained by the formation of a stable quinoid ion in which charge delocalization is possible (an -OCH$_3$ group in the 4-position).[20] The presence of M-70 (M-2Cl) might indicate an -OH group in the 2-position. The only possible choice for such a compound is 3,5,5'-trichloro-2,4'-dihydroxybiphenyl.

The mass spectra of compounds 29 and 30 (Table I) show the characteristic fragments for tetrachlorodihydroxybiphenyls reported by Lotjonen et al.[19] After methylation, compound 29 was converted to a dimethoxy derivative, the mass spectrum of which is compared with that of an authentic standard in Figure 3. Fragments of M-35 (M-Cl), M-50 (M-CH$_3$Cl), and M-65 (M-CH$_3$Cl-CH$_3$) are consistent with 2,2'-dimethoxy-3,3',5,5'-tetrachlorobiphenyl.[21] This identification was also confirmed by the satisfactory agreement of spectra in Figure 3. The reference standard was prepared by chlorination of 2,2'-dihydroxybiphenyl in aqueous solution, followed by extraction with methylene chloride, and methylation with diazomethane.

Tri- and tetrachlorodihydroxybiphenyls are isomeric to tri- and tetrachlorohydroxydiphenyl ethers. The possibility of diphenyl ethers, however, is ruled out, mainly for reasons given later.

The formation of chlorohydroxybiphenyls can be explained via free radical mechanisms. Free radicals on the carbon surface are produced by its oxidation with chlorine. Interaction of 2,4-dichlorophenol with the surface radicals can produce 2,4-dichlorophenoxy radicals by hydrogen abstraction. Trichlorodihydroxybiphenyls can be formed by coupling a 2,4-dichlorophenol molecule and a 2,4-dichlorophenoxy radical, with displacement of halogen; this reaction has precedent in the literature.[22] A postulated mechanistic scheme for the formation of tri- and tetrachlorodihydroxybiphenyls is presented in Figure 4.

The concentration of chlorine (8-10 mg/L) used in the previous experiments is somewhat higher than is typically encountered in drinking water treatment. A separate experiment at considerably lower (1.5 mg/L) chlorine concentration, however, showed that, although fewer compounds were formed, the same hydroxylated PCB dimers were produced. Furthermore, the presence of adsorbed peat fulvic acid did not prevent the formation of these compounds (see Table I).

Phenol

In the presence of GAC, phenol gave, in addition to the expected ring-chlorination products, a mixture of other oxidized compounds, including quinones, polyhydroxyphenyl derivatives, and biphenyls, some of which contained chlorine. The hydroxylated phenol derivatives (identified as their methyl ethers) included dimethoxybenzenes and tetramethoxybenzenes (derivatives of compounds 2, 3, and 11 in Table II). One of the dimethoxybenzenes corresponds to *o*- and the other to *p*-dimethoxybenzene (matching retention

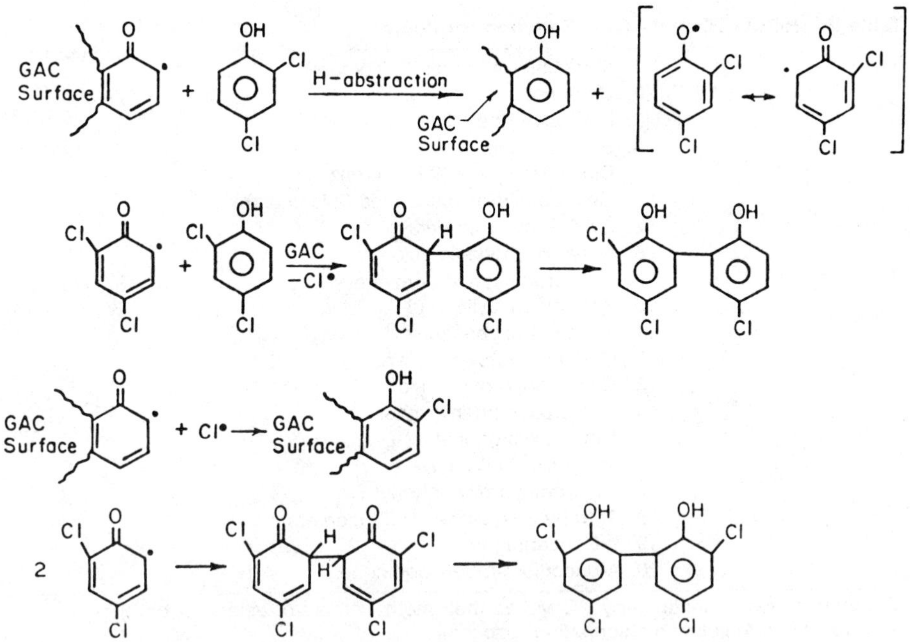

Figure 4. Postulated mechanism for formation of hydroxylated PCBs.

time and mass spectra with authentic standards). The peak assigned as tetramethoxybenzene has a mass spectrum typical of this class of compounds,[23] that is, molecular ion at m/z 198 with fragment ions of 183, 155, 140, and 125. Products 6, 9, and 12 (Table II) may be formed by chlorination of the corresponding hydroxylated derivatives of phenol.

Hydroxylated biphenyls were the predominant products. Hydroxydiphenyl ethers are theoretically possible, and one such compound (14, Table II) was detected. The dimers may be formed by coupling phenoxy radicals[24] that are formed on GAC. The polyhydroxylated and chlorinated biphenyls (hydroxylated PCBs) are possibly formed by reactions analogous to those forming hydroxy and chlorohydroxy derivatives of phenol.

Mass spectra and retention times of compounds isomeric to hydroxybiphenyls, such as hydroxydiphenyl ethers, show significant differences. Table III depicts differences in the mass spectra of some hydroxydiphenyl or methoxydiphenyl ethers and isomeric hydroxybiphenyls or methoxybiphenyls. The literature also indicates that hydroxybiphenyls generally predominate as products of phenol coupling, as compared to hydroxydiphenyl ethers.[25]

p-Chlorophenol

Compounds tentatively identified in methylated and nonmethylated extracts from the chlorination reaction of GAC-adsorbed p-chlorophenol are shown in

Table II. HOCl – Phenol – GAC Reaction Products[a]

1. Monochlorophenols (2 isomers)[b]
2. Catechol
3. Hydroquinone
4. Salicylic acid
5. Dichlorophenols (2 isomers)[b]
6. Chlorodihydroxybenzenes (2 isomers)
7. 2,4,6-Trichlorophenol[b]
8. p-Hydroxybenzoic acid
9. Dichlorodihydroxybenzene
10. 3-Chlorosalicylic acid
11. Tetrahydroxybenzene
12. Chlorotrihydroxybenzene
13. Dichlorohydroxybenzoic acid
14. 4-Hydroxydiphenyl ether
15. Dihydroxybiphenyl
16. Trihydroxybiphenyl
17. Chlorodihydroxybiphenyl
18. Tetrahydroxybiphenyls (3 isomers)
19. p-Benzoquinone
20. Monochloro-p-benzoquinone

[a]Compounds were identified by GC/MS as their methyl ethers or esters (or both).
[b]Also identified in solution chlorination reactions.

Table III. Mass Spectra of Some Isomeric Hydroxymethoxybiphenyls and Hydroxymethoxydiphenyl Ethers

Compound	Mass Spectra (%)
2,2'-Dihydroxybiphenyl	186(100), 171(3), 169(5), 168(4), 157(30) 158(28), 139(12), 131(23), 129(10), 128(15)
2-Hydroxydiphenyl ether	186(100), 169(2), 157(5), 131(3), 129(5), 128(4), 80(20), 78(22), 77(30)
4-Hydroxy-4'-methoxybiphenyl	200(100), 199(25), 186(3), 185(25), 184(7), 171(5), 170(8), 169(52), 168(18), 167(15), 157(23), 141(18), 139(25), 129(24), 128(35), 127(21)
4-Methoxydiphenyl ether	200(100), 186(4), 185(35), 170(2), 157(2), 129(12), 128(2), 123(8)
Dimethoxyhydroxybiphenyl	230(100), 215(47), 214(85), 200(20), 199(70), 198(15), 197(25), 187(11), 186(14), 185(14), 184(14), 171(40)
2,2'-Dimethoxydiphenyl ether	230(100), 215(4), 185(7), 184(49), 128(5), 121(10), 92(24), 91(27)

Figure 5. Di- and trichlorophenol (3 and 4, Figure 5) are the expected chlorination products in aqueous solution. In the carbon extract, a chlorinated analog of methyl salicylate (5, a carboxylation product, Figure 5) was identified from its mass spectrum. A group of unidentified isomers having molecular weight 310 were present in both methylated and nonmethylated extracts. This group represents rather major reaction products (about 12% of original *p*-chlorophenol).

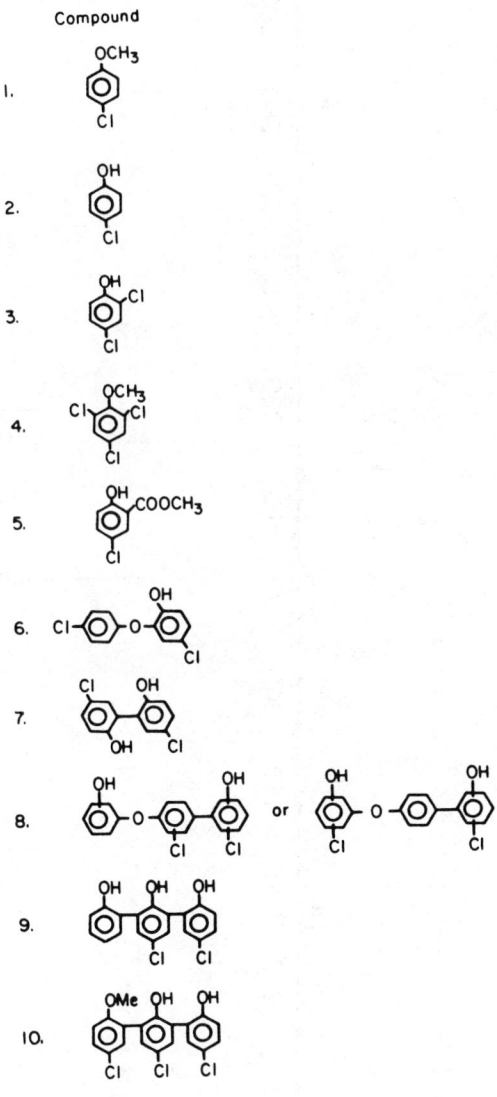

Figure 5. Products from HOCl-*p*-chlorophenol-GAC reaction.

1324 WATER CHLORINATION

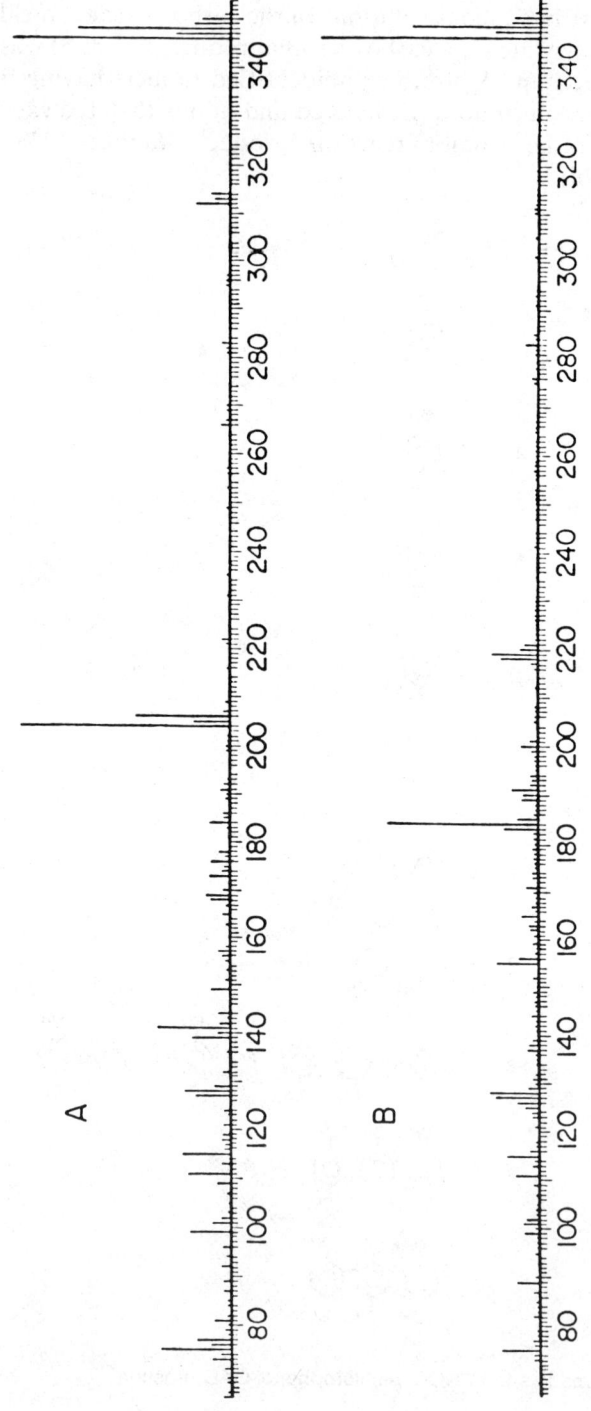

Figure 6. Mass spectra of (A) compound 8 and (B) compound 9 (Figure 5).

Table IV. Reaction Products Between Phenol and Chlorine-Preoxidized F-400 GAC

Compound
1. p-Benzoquinone[a]
2. Methoxybenzene
3. Phenol
4. Chloro-p-benzoquinone[a]
5. Chlorodimethoxybenzene
6. Dimethoxybiphenyl
7. Dimethoxybiphenyl
8. Dimethoxybiphenyl
9. 4-Methoxydiphenyl ether
10. MW 216
11. Trihydroxytrimethoxybiphenyl

[a]Detected in nonmethylated extracts

Both chlorohydroxybiphenyl (hydroxylated PCBs) and chlorohydroxydiphenyl ether-type dimers and trimers (6–10, Figure 5) were found. Information on the mass spectrometry of compound 7 (Figure 5) was found in the literature.[19] The remainder of these compounds were tentatively identified from their mass spectra only. Compounds 8 and 9 (Figure 6) are isomers, but they can be distinguished from their mass spectra. Compounds of the same type have been produced by the oxidation of p-cresol with one-electron oxidizing agents.[24] The results suggest that these compounds were formed by coupling chlorophenoxy radicals produced on the GAC surface. The yield of dimers corresponded to ~20% of the original p-chlorophenol and that of trimers to ~25%.

Preoxidized GAC plus Phenols

Free chlorine was allowed to react with a column of F-400 GAC, and residual chlorine was removed by extensive washing of the column with deionized water. Phenol was then applied. As expected, the GC-MS analysis of the methylated carbon extract revealed the formation of products previously observed from the reaction of chlorine with adsorbed phenol. No aqueous chlorination products (i.e., chlorophenols) were detected. The identified products are listed in Table IV.

In the next experiment, phenol was substituted by a 2,4-dichlorophenol solution with pH 8.5. No buffer was used in this case. The other experimental parameters and conditions remained unchanged. A total ion chromatogram of the carbon extract is shown in Figure 7. Numbered peaks are identified in Table I. A blank experiment was also conducted in which nonoxidized F-400 received a 2,4-dichlorophenol solution at pH 8.5. The total ion chromatogram showed only one peak, 2,4-dichlorophenol, which suggests that the com-

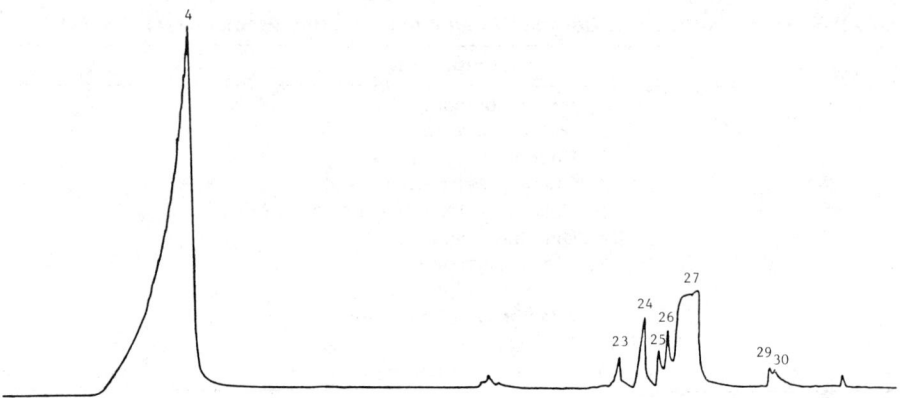

Figure 7. Total ion current chromatogram of products from the reaction of 2,4-dichlorophenol with preoxidized GAC.

pounds in Figure 7 were formed because of pretreatment of the GAC surface with chlorine.

SUMMARY

Activated carbon oxidized with chlorine became very reactive toward phenols. Phenol, 2,4-dichlorophenol, and p-chlorophenol, when allowed to react with chlorine at a GAC surface rather than in solution, gave a variety of oxidation and dimerization products in addition to the expected substitution products. The dimers were principally biphenyl rather than diphenyl ether types. Dimers from p-chlorophenol and 2,4-dichlorophenol were hydroxylated chlorobiphenyls. p-Chlorophenol also gave chlorinated trimers. Preoxidized carbon treated with phenol produced compounds previously observed from the reaction of chlorine with adsorbed phenol: hydroxyphenols, hydroxybiphenyls, and quinones. Preoxidized carbon loaded with 2,4-dichlorophenol produced tri- and tetrachlorodihydroxybiphenyls (hydroxylated PCBs), even without chlorine being present. One trichlorodihydroxybiphenyl (27, Table I) was produced in significant amounts: approximately 18% of the original dichlorophenol was converted to compound 27. A free radical mechanism has been postulated to explain the formation of hydroxylated PCBs.

Since prechlorination in water treatment results in contact of GAC with chlorine, hydroxylated PCBs may be formed under drinking water treatment conditions. These compounds are in vivo metabolites of PCBs, and their acute toxicity can be higher than their parent compounds.[26] In addition, some of these products do not adsorb very well on GAC and can be displaced in the effluent by more strongly adsorbed materials.[27] Therefore, processes that result in contact of chlorine with GAC having adsorbed organic compounds may need to be reevaluated.

ACKNOWLEDGMENT

We thank the National Science Foundation for financial support (Grant No. CEE-81-10024).

REFERENCES

1. Culp, R. L., J. A. Faisst, and C. E. Smith. "Granular Activated Carbon Installations," EPA Report PB82-102492, (Cincinnati: U.S. Environmental Protection Agency, 1981).
2. Snoeyink, V. L., H. T. Lai, J. H. Johnson, and J. F. Yang. "Active Carbon: Dechlorination and the Adsorption of Organic Compounds," in *Chemistry of Water Supply and Distribution*, A. J. Rubin, Ed. (Ann Arbor, MI: Ann Arbor Science Publishers, Inc., 1974), p. 233.
3. Puri, B. R., D. D. Singh, J. Chander, and L. R. Sharma. "Interaction of Charcoal with Chlorine Water," *J. Indian Chem. Soc.* 35:181 (1958).
4. Ishizaki, C., and J. T. Cookson, Jr. "Influence of Surface Oxides on Adsorption and Catalysis with Activated Carbon," in *Chemistry of Water Supply and Distribution*, A. J. Rubin, Ed. (Ann Arbor, MI: Ann Arbor Science Publishers, Inc., 1974), p. 201.
5. Puri, B. R., L. R. Sharma, and D. D. Singh. "Studies in Catalytic Reactions of Charcoal. I. Catalytic Oxidation and Decomposition of Salt Solutions," *J. Indian Chem. Soc.* 35:765 (1958).
6. Puri, B. R., L. R. Sharma, and D. D. Singh. "Studies in Catalytic Reactions of Charcoal. II. Catalytic Decomposition of Hydrogen Peroxide," *J. Indian Chem. Soc.* 35:770 (1958).
7. Snoeyink, V. L., and W. J. Weber, Jr. "Surface Functional Groups on Carbon and Silica," in *Progress in Surface and Membrane Science*, Vol. 5, J. F. Danielli, Ed. (New York: Academic Press, 1972), p. 63.
8. Boehm, H. P. "Chemical Identification of Surface Groups," *Adv. Catal.* 16:179 (1966).
9. Donnet, J. B., and G. Henrich. "Sur le Caractere Radicalaire du Noir de Carbone," *C. R. Hebd. Acad. Sci.* 246:3230 (1958).
10. Donnet, J. B., and G. Henrich. "Reactions Radicalaires et Chimie Superficielle du Noir de Carbone," *Bull. Soc. Chim. Fr.* 1609 (1960).
11. Tomita, A., S. Mori, and Y. Tamai. "Catalytic Behavior of Carbon on the Oxidation of Cyclohexene," *Carbon* 9:224 (1971).
12. McCreary, J. J., V. L. Snoeyink, and R. A. Larson. "A Comparison of the Reaction of Aqueous Free Chlorine with Phenolic Acids in Solution and Adsorbed on Granular Activated Carbon," *Environ. Sci. Technol.* 16:339 (1982).
13. McCreary, J. J., and V. L. Snoeyink. "Reaction of Free Chlorine with Humic Substances Before and After Adsorption on Activated Carbon," *Environ. Sci. Technol.* 15:193 (1981).
14. Chen, A. S.-C., R. A. Larson, and V. L. Snoeyink. "Reactions of Chlorine Dioxide with Hydrocarbons: Effects of Activated Carbon," *Environ. Sci. Technol.* 16:268 (1982).

15. Chen, A. S.-C., R. A. Larson, and V. L. Snoeyink. "Importance of Surface Free Radicals in an Aqueous Chlorination Reaction (Indan - ClO_2) Promoted by Granular Activated Carbon," *Carbon* 22:63 (1984).
16. Voudrias, E. A., L. M. J. Dielmann III, V. L. Snoeyink, R. A. Larson, J. J. McCreary, and A. S.-C. Chen. "Reactions of Chlorite with Activated Carbon and with Vanillic Acid and Indan Adsorbed on Activated Carbon," *Water Res.* 17:1107 (1983).
17. *Standard Methods for the Examination of Water and Wastewater*, 15th ed. (Washington, DC: American Public Health Association, 1980).
18. Burttschell, R. H., A. A. Rosen, F. M. Middleton, and M. B. Ettinger. "Chlorine Derivatives of Phenol Causing Taste and Odor," *J. Am. Water Works Assoc.* 51:205 (1959).
19. Lotjonen, S., P. Ayras, and H. Pyysalo. "Glass Capillary Gas Chromatographic Detection and Mass Spectral Analysis of Some Hydroxyl Derivatives of Polychlorinated Biphenyls (PCBs)," *Finn. Chem. Lett.* 57 (1979).
20. Janson, B., and A. Sundstrom. "Mass Spectrometry of the Methyl Ethers of Isomeric Hydroxybiphenyls—Potential Metabolites of Chlorobiphenyls," *Biomed. Mass Spectrom.* 1:386 (1974).
21. Tulp, M. T. M., K. Olie, and O. Hutzinger. "Identification of Hydroxyhalobiphenyls as Their Methyl Ethers by Gas Chromatography-Mass Spectrometry," *Biomed. Mass Spectrom.* 4:310 (1977).
22. Blanchard, H. S., H. L. Finkbeiner, and G. A. Russell. "Polymerization by Oxidative Coupling. IV. Polymerization of 4-Bromo and 4-Chloro-2,6-dimethylphenol and Preparation and Decomposition of the Silver and Copper Salts of Certain Other Phenols," *J. Polym. Sci.* 58:469 (1962).
23. McLafferty, F. W. *Interpretation of Mass Spectra*, 2nd ed. (Reading, MA: Benjamin-Cummings, 1973).
24. Nonhebel, D. C., and J. C. Walton. *Free Radical Chemistry*. (Cambridge: University Press, 1974).
25. Taylor, W. I., and A. R. Battersby. *Oxidative Coupling of Phenols* (New York: Marcel Dekker, 1967).
26. Yoshimura, H., H. Yamamoto, and H. Kinoshita. "Metabolic Fate of PCBs and Their Toxicological Evaluation," in *New Methods in Environmental Chemistry and Toxicology*, F. Coulston, F. Korte, and M. Goto, Eds. (Tokyo: Academic Printing Co., 1973), p. 291.
27. Voudrias, E. A., V. L. Snoeyink, and R. A. Larson, unpublished data.

CHAPTER 102

Mutagenic Residues Recovered from Granular Activated Carbon After Use in Drinking Water Treatment

John C. Loper, M. Wilson Tabor, and Laura Rosenblum

In addition to its use in the removal of compounds that cause taste, odor, or color problems in drinking water, granular activated carbon (GAC) has received increasing attention as a treatment for the removal of compounds known or suspected to be of toxicological significance.[1-3] An experimental full-scale GAC system for such water treatment has been developed at the Cincinnati Water Works.[4,5] We have recently described the results of a study that examined this system throughout the adsorptive life of the GAC, monitoring trends both in the production of nonvolatile mutagens by chlorine disinfection and in the removal of such bioactive compounds from the water.[6] Mutagenicity values were determined with *Salmonella* tester strains in assays of water residue organics extracted using XAD-2. None of the samples of settled unchlorinated water was mutagenically active in that study; however, samples were consistently active following chlorination. Mutagenic activity in chlorinated water was effectively removed by GAC treatment throughout the study, even after the removal of total organic carbon (TOC) had reached a steady-state level of about 65% of that contained in the water influent to the column.

We then demonstrated that residues extracted from the used GAC were mutagenic. Differences in that activity were shown for residues recovered from the top, middle, or bottom of the used GAC adsorber. The data presented in Figure 1 are reproduced from our earlier publication of that work.[6]

This chapter presents further analyses of GAC extracted organics. The mutagens can be separated into two populations, one of which contains relatively nonpolar compounds and shows mutagenic activity similar to that seen among other residues from the drinking water source, whether extracted by us using XAD resins or by others using reverse osmosis. In the other population, more-polar compounds predominate (including mutagenic compounds), which are poorly recovered using our XAD-2/XAD-7 resin procedure.

MATERIALS AND METHODS

Solvents were nanograde (Mallinckrodt), high-performance liquid chromatography (HPLC) grade (Fisher Scientific), or distilled-in-glass (Burdick and

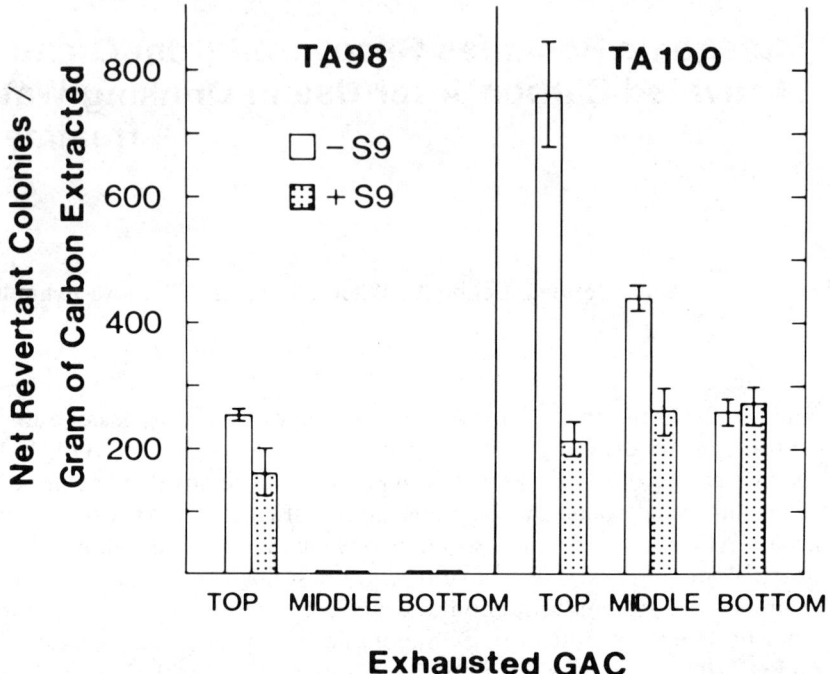

Figure 1. Mutagenic activity recovered from carbon taken from different levels of granular activated carbon following its use after 35 weeks of water treatment. Open bars: assayed in the absence of microsomal activation, −S9; stippled bars, +S9. Reproduced from Ref. 6.

Jackson). Water of type-1 quality was prepared by using a water conditioning system (Continental).[7] Other chemicals were purchased as reagent grade from commercial sources and were used as obtained. Reagent XAD-2 and XAD-7 resins prepared according to U.S. Environmental Protection Agency (EPA) criteria[8] were obtained from the Applied Science Division of the Milton-Roy Company. Resins were stored until use in amber bottles under methanol and columns were formed by gravity from slurries as previously described.[6] Based on a method of LeBel et al.,[9] columns were eluted using four column volumes of an 85:15 (vol:vol) mixture of hexane-acetone.

HPLC Separation

Separations by HPLC were performed with a Waters Associates model ALC/GPC 204. The unit was fitted with a Waters Associates radial compression module column system with 8-mm by 10-cm columns. For reverse-phase HPLC separations, a stationary phase of 10-μm silica particles bonded with octadecylsilane was used. A stationary phase of 10-μm silica was used for normal phase separations. Operating conditions for the system are detailed in a previous study.[7]

Analytical-scale (microgram level) reverse-phase separations were accomplished using a linear solvent gradient from water to acetonitrile.[7] Preparative-scale (milligram level) normal-phase HPLC separations were accomplished by a series of isocratic and linear gradient elutions as follows: 100% hexane for 2 min, then a 2-min linear gradient to 80% hexane:20% methylene chloride, fraction 1; isocratic for 5 min, then a 3-min linear gradient to 30% hexane:70% methylene chloride, fraction 2; isocratic for 2 min, then a 2-min linear gradient to 100% methylene chloride, fraction 3; isocratic for 5 min, then a 5-min linear gradient to 100% methanol, fraction 4; isocratic for 5 min or until all 254-nm absorbing components eluted, fraction 5. Each HPLC fraction was gently evaporated at 35°C under a stream of dry nitrogen; the residue was then dissolved in dimethyl sulfoxide for mutagenesis testing.

Mutagenicity Assay

The *Salmonella* mammalian microsome mutagenicity assay using strains TA98 and TA100 was based on the method described by Ames et al.[10] Strain verification tests and minor assay modifications are described by Loper et al.[6]

GAC Used in Water Treatment

Figure 2 is a schematic diagram of the California plant of the Cincinnati Water Works and of the GAC treatment procedure. Specific features of the adsorber are summarized in Table I (further details of water treatment procedures are available in Ref. 6). One GAC adsorber filled with freshly regenerated carbon was used to treat sand-filtered chlorinated water at 1 Mgd for 9 months.[6]

Extraction of GAC

Samples of carbon were taken from the adsorber prior to its use and at the end of 9 months. Aliquots of damp carbon samples were Soxhlet extracted with methylene chloride for 24 h. The extracts were then passed through anhydrous sodium sulfate to remove any water; residues were concentrated by rotary evaporation. Some studies have examined residues prepared in the same manner from carbon removed from a separate GAC adsorber used several months earlier. (See Ref. 6.)

WATER CHLORINATION

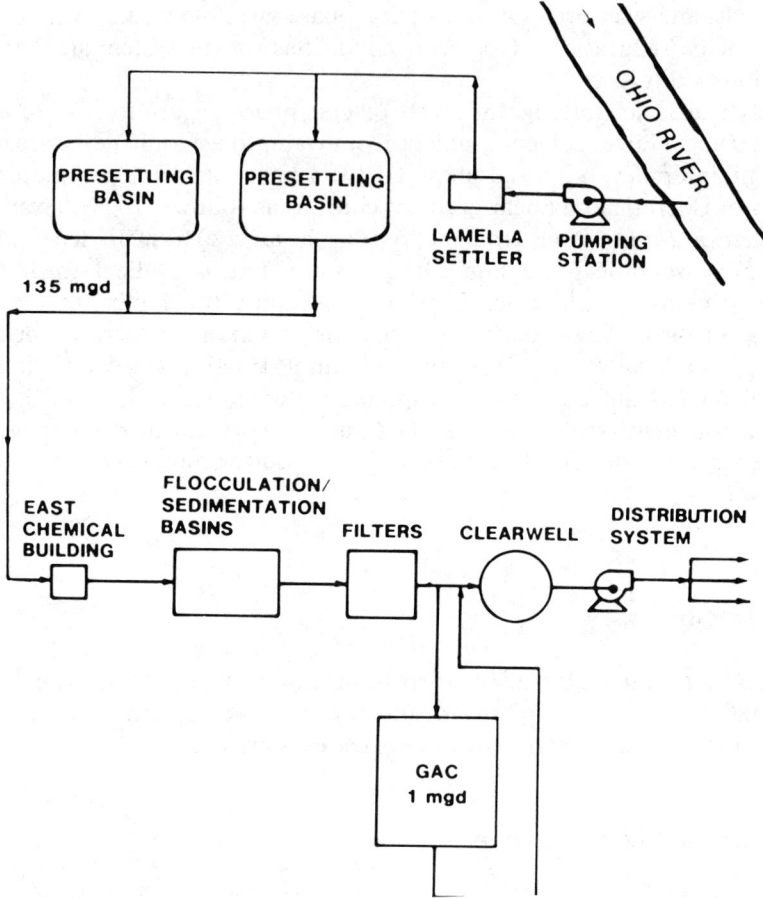

Figure 2. Schematic of Cincinnati Water Plant. All mutagenic activity detected in water influent to the granular activated carbon (GAC) arose following chlorination at East Chemical Building. (See Ref. 6.)

Table I. Characteristics of Granular Activated Carbon (GAC) Adsorber

GAC Data		Hydraulic Data	
Manufacturer and type	Westvaco, WV-G	Diam, ft	11
Mesh size	12 × 40	Bed height, ft	14.8
Total BET surface area, m^2/g	1,100 (minimum)	Surface loading, gpm/ft^2	6.94
Mean particle diameter, mm	0.90 – 1.20	Flow per day, Mgd	1
Carbon weight, lb	42,000	Residence time, min	15

RESULTS

Figure 1 summarizes the diversity of mutagenic properties that we observed earlier for residues extracted from used GAC.[6] We were interested in further investigating the strain TA100 mutagens. Unlike the case with mutagenic activity for TA98, residues mutagenic for TA100 were recovered from carbon from all three levels of the adsorber. Thus, mutagens for this strain were the most likely to break through the GAC adsorber into the water supply.

The first experiments were designed to test how well our XAD adsorption procedure could extract these compounds. Residues extracted from used carbon obtained from the top and bottom of the GAC adsorber were dissolved in small volumes of acetone and then diluted into separate volumes of type I water to form solutions of 2.5 mg/L. The water samples were processed using separate sets of columns of XAD-2 and XAD-7 in sequence. The residues were examined by reverse-phase (RP) HPLC and by mutagenicity assay using TA100.

Figure 3A shows 254-nm absorption chromatographic elution profiles for two control solutions: a solvent blank and the residue extracted from 350 mg of carbon removed from the top of the regenerated GAC adsorber at the start of the 9-month experiment. Few peaks were observed for either of these controls. By contrast, the chromatograms in Figure 3B show the results when the same RP-HPLC procedure was applied to residues extracted from carbon removed from the top or bottom of the used GAC adsorber. The top and bottom chromatographic elution patterns are for residues from 35 and 23 mg of carbon, respectively. Numerous 254-nm absorbing components were separated over a wide range of polarity for each sample; however, relative to the top carbon sample, the bulk of the absorbing components for the bottom sample were eluted earlier by more-polar portions of the gradient.

This relatively polar material was poorly recovered from reconstituted aqueous samples using XAD-2 and XAD-7. The RP-HPLC chromatograms in Figure 3C are for residues recovered by XAD-2. Although some loss of 254-nm absorbing material was apparent for the top sample, the major losses occurred for the polar portions of the bottom sample, as may be seen by comparing chromatograms in Figures 3B and 3C. Extracts from the XAD-7 columns yielded negligible amounts of residue, which revealed only trace levels of 254-nm absorbing components when examined by RP-HPLC (data not shown).

A qualitatively similar effect was seen when these residues recovered by XAD were examined for mutagenic activity. As summarized in Figure 4, appreciable activity was lost from each residue by this procedure, but the loss was most pronounced for the residue sample originally obtained from the bottom of the GAC. The XAD-2 column recovered 22% of this mutagenic activity vs 37% of that contained in the residue from the top of the GAC. No additional mutagenic activity was recovered by the XAD-7 column used in

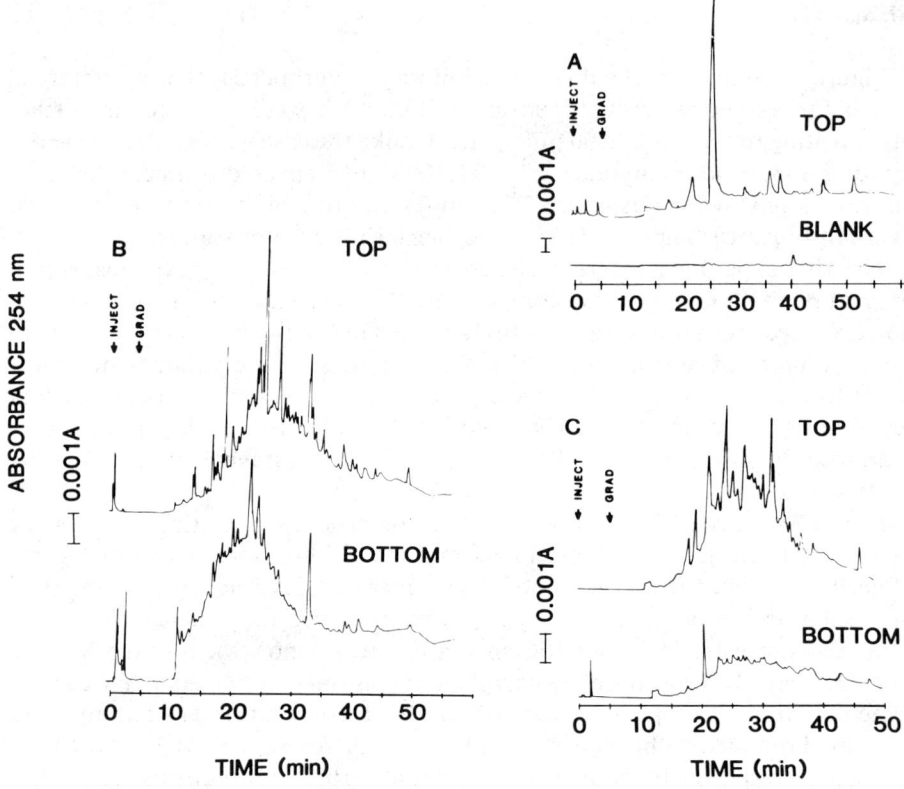

Figure 3. Adsorption profiles (by reverse-phase HPLC) of residues extracted from carbon samples removed from the same GAC absorber used for Figure 1. A: Residue from 350 mg of virgin carbon, solvent blank. B: Residues from 35 and 23 mg of carbon of the used adsorber taken from the top and bottom, respectively. C: Residues of B were dissolved in water and reextracted with XAD-2.

tandem (Figure 4), although this resin was included on the prospect that it might extract compounds more polar than those absorbed by XAD-2.

In a parallel study of drinking water organics from the Cincinnati Water Works, we have examined similar residues for the mutagenic properties of subfractions obtained by using preparative-scale HPLC (Tabor et al., unpublished data). Residues were extracted from carbon of another GAC adsorber that earlier had been used extensively for water treatment. Figure 5 shows a 254-nm absorption chromatographic elution profile for a sample of parent residue separated by normal-phase HPLC. Mutagenic activity was detected in subfractions containing nonpolar compounds that were eluted by hexane near the beginning of the HPLC separation. Similar activity was also detected in subfractions containing polar compounds that were eluted by methanol at the

DRINKING WATER TREATMENT 1335

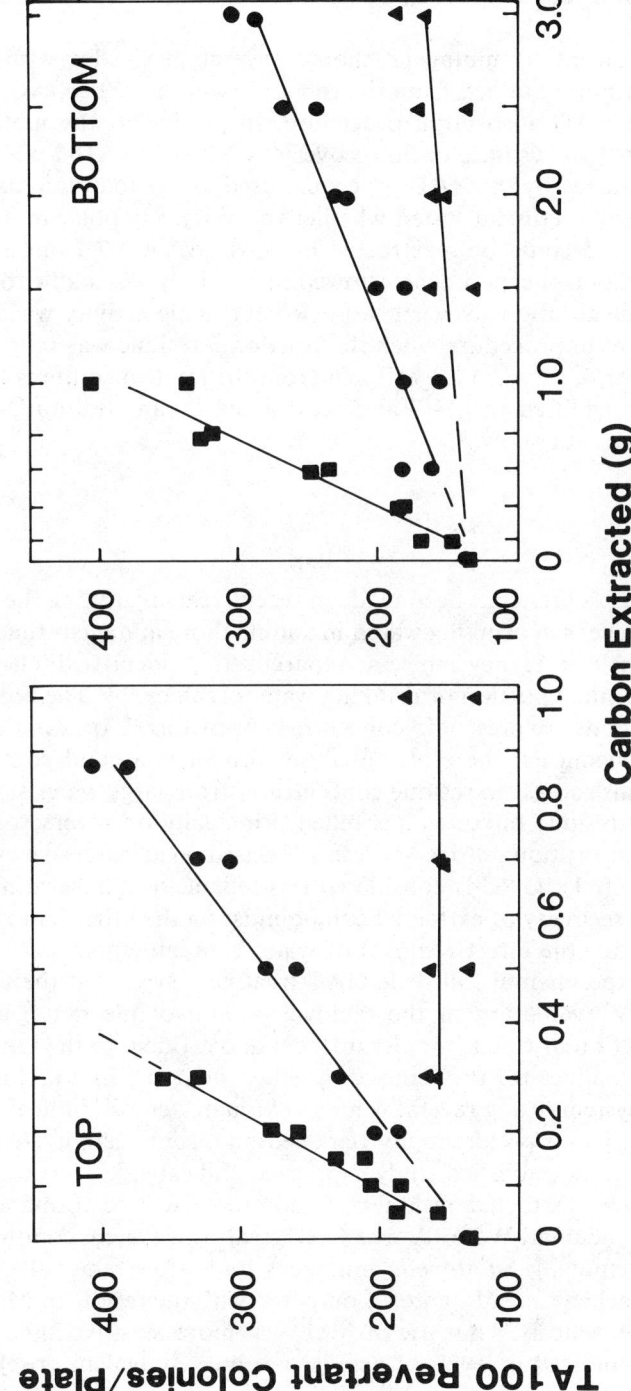

Figure 4. Recovery (by XAD) of mutagenic residues originally isolated from the top and bottom of the used GAC adsorber. Activity of parent residue from top or bottom carbon is indicated by -■-. Activities from water solutions of these residues reextracted by XAD are XAD-2, -●-; and XAD-7, -▲-.

end of the separation. These mutagenic fractions were combined into fractions 1 and 5 (Figure 5).

The two populations of mutagens shared several properties with those recovered from carbon obtained from the top and bottom of the GAC adsorber used with the XAD adsorption procedure. In particular, the mutagenic activity of fraction 5 shows little or no recovery by XAD-2/XAD-7 when that fraction was reextracted by the XAD procedure used earlier to obtain data for Figure 4. A further experiment tested whether this relatively polar mutagenic material in fraction 5 could be reextracted by XAD-2/XAD-7 from a water solution at pH 2. Control experiments showed this activity was stable to acidification and reneutralization. Nevertheless, no mutagenic activity was recovered by the XAD resin procedure when the fraction 5 residue was redissolved in water at pH 2, or 6, or when the effluent from the tandem columns for the pH 6 sample was acidified to pH 2 and reextracted by an additional set of XAD columns (data not shown).

DISCUSSION

Although major progress has been made in recent years regarding the origin and nature of mutagens in drinking water, including those mutagens that result from chlorination,[11] it is nevertheless apparent that identified chemicals account for only a small portion of drinking water organics.[11,12] The unknown mutagenic compounds are present in concentrations of 1 μg/L or less among a variety of other chemicals; therefore, analyses for bioassay and compound identification require access to residue concentrates from large water samples. Recovery of such residue mixtures has relied principally on reverse osmosis techniques or on adsorption onto XAD resins.[13] Based on an extensive analysis (summarized in Ref. 1), GAC is considered less reliable as a general method for representative recovery of extracted compounds; on the other hand, it is a preferred method for the direct removal of water contaminants.

Thus, for the experimental full-scale GAC treatment system at the Cincinnati Water Works, evaluation of the removal of nonvolatile mutagens has involved the collection of water samples influent and effluent to the GAC and the extraction of residues for the samples by other methods. In a preliminary test of this GAC system, using reverse osmosis, Monarca et al.[14] indicated that the GAC efficiently and preferentially removed mutagenic activity from the water. Our more recent published study supported and extended those conclusions. Using XAD-2 extracted residues to monitor GAC performance, we showed that the Cincinnati Water Works GAC treatment system continued to be effective for removing *Salmonella* mutagens long after removal of total organic carbon reached a steady state.[6] Compared with the results of Monarca et al. using reverse osmosis,[14] our use of XAD was more sensitive in detecting mutagen levels (expressed in levels of activity per liter equivalent dose).

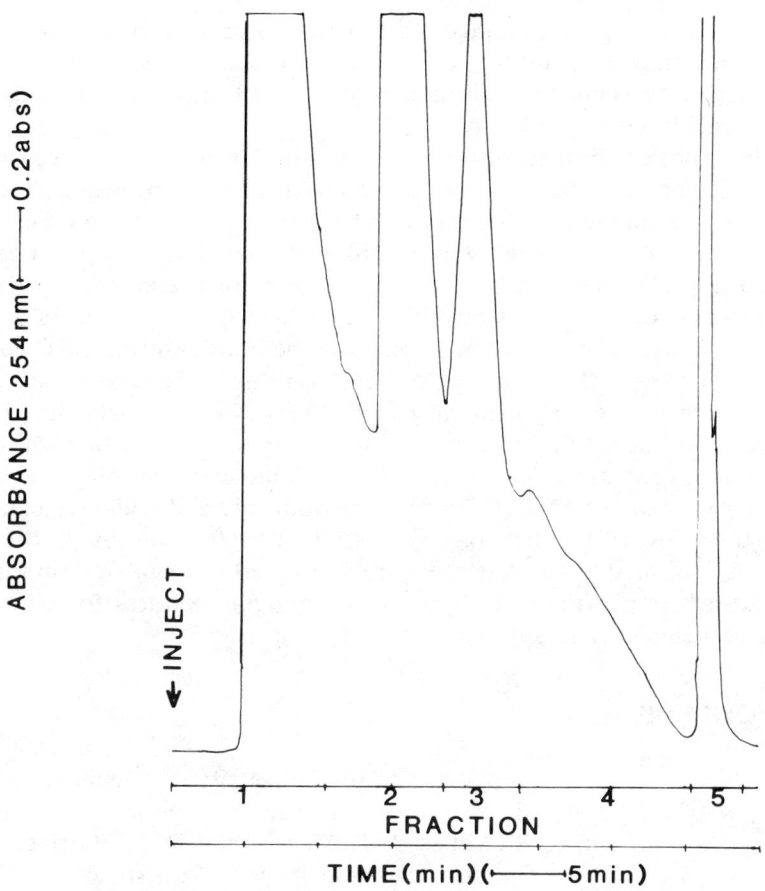

Figure 5. Normal-phase HPLC separation of residue organics extracted from used GAC. The separation was accomplished via a series of isocratic and linear gradient elutions using hexane, methylene chloride, and methanol mobile phases. Mutagenic activity was recovered in fractions 1 and 5.

However, based on the studies reported in this chapter, there is one class of relatively polar mutagens recoverable from GAC used in water treatment that is less capable of being absorbed and recovered using XAD-2/XAD-7. These mutagens are active for strain TA100 assayed -S9. They were recovered from the bottom carbon of a used GAC adsorber and probably are also represented in the mutagens among the relatively polar compounds of fraction 5, that is, a fraction obtained with normal-phase HPLC separation of residues extracted from a random aliquot of carbon taken from a holding tank containing previously used GAC. The mutagenic activity in fraction 5 was not reextracted from reconstituted water solutions by XAD-2/XAD-7, whether these solutions were at pH 6 or 2. Thus, these mutagens differ from a previously reported specific class of drinking water mutagens that does not adsorb on XAD resins from normal drinking water but does adsorb on XAD resins from drinking water solutions acidified to pH 2.[3]

We have not established whether this class of TA100 mutagens was present in the chlorinated water influent to the GAC, or was formed on the GAC during the several months that the adsorber was in use. Although the influent water contained an average 2 ppm chlorine, low levels of bacteria were detected in the effluent from this system prior to rechlorination;[15] therefore, some biotransformation was possible. Nevertheless, since this class of mutagens was recovered from carbon taken from the bottom of the GAC adsorber at a level just above the effluent, these mutagens might be expected eventually to be desorbed into the drinking water supply. In that case, the effective adsorptive life of a GAC treatment column for waterborne mutagens could best be monitored using GAC itself. Modest volumes of carbon at the bottom of the GAC adsorber could be removed periodically and then extracted using methylene chloride[6] for the bioassay of strain TA100 mutagens. In the meantime, as is indicated by the data in Figure 5, residues extractable from full-scale GAC adsorbers provide a ready source of organic mixtures for analyses of different chemical classes of waterborne mutagens.

ACKNOWLEDGMENTS

Part of this study was supported at the University of Cincinnati by EPA grant CR808603, J. R. Meier, project officer. Our special appreciation is expressed to him and to R. Miller, J. DeMarco, and D. J. Hartman of the Cincinnati Water Works for their encouragement and assistance.

REFERENCES

1. National Academy of Sciences-National Research Council. "An Evaluation for Activated Carbon," in *Drinking Water and Health*, Vol. 2 (Washington, DC: National Academy of Science, 1980), pp. 251–380.
2. Rook, J. J. "Comparison of Removal of Halogenated and Other Organic Compounds by Six Types of Carbon in Pilot Filters," in *Treatment of Water by Granular Activated Carbon*, M. J. McGuire and I. H. Suffet, Eds., Advances in Chemistry Series, Vol. 202, (Washington, DC: American Chemical Society, 1983), pp. 455–479.
3. Kool, H. J., and C. F. Van Kreijl. "Formation and Removal of Mutagenic Activity During Drinking Water Preparation," *Water Res.* 18:1011–1016 (1984).
4. DeMarco, J., A. A. Stevens, and D. J. Hartman. "Application of Organic Analysis for Evaluation of Granular Activated Carbon Performance in Drinking Water Treatment," in *Advances in the Identification and Analysis of Organic Pollutants in Water*, L. H. Keith, Ed. (Ann Arbor, MI: Ann Arbor Science Publishers, Inc., 1981), pp. 907–940.

5. DeMarco, J., R. Miller, D. Davis, and C. Cole. "Experiences in Operating a Full Scale Granular Activated-Carbon System with On-site Reactivation," in *Treatment of Water by Granular Activated Carbon*, M. J. McGuire and I. H. Suffet, Eds., Advances in Chemistry Series, Vol. 202 (Washington, DC: American Chemical Society, 1983), pp. 525-563.
6. Loper, J. C., M. W. Tabor, L. Rosenblum, and J. DeMarco. "Continuous Removal of Both Mutagens and Mutagen Forming Potential by an Experimental Full Scale Granular Activated Carbon Treatment System," *Environ. Sci. Technol.* 19:333-339 (1985).
7. Tabor, M. W., and J. C. Loper. "Separation of Mutagens from Drinking Water Using Coupled Bioassay/Analytical Fractionation," *Int. J. Environ. Anal. Chem.* 8:197-215 (1980).
8. U.S. Environmental Protection Agency, Office of Research and Development. *IERL-RTP Procedures Manual: Level I Environmental Assessment* 2nd ed., USEPA-600/7-78-201 PB-293795. (Washington, DC: 1980).
9. LeBel, G. L., D. T. Williams, G. Griffith, and F. M. Benoit. "Isolation and Concentration of Organophosphorus Pesticides from Drinking Water at the ng/L Level Using Macroreticular Resin," *J. Assoc. Off. Anal. Chem.* 62:241-249 (1979).
10. Ames, B. N., J. McCann, and E. Yamasaki. "Methods for Detecting Carcinogens and Mutagens with the *Salmonella*/Mammalian-Microsome Mutagenicity Test," *Mutat. Res.* 31:347-364 (1975).
11. Kool, H. J., C. F. Van Kreijl, and B. C. J. Zoeteman. "Toxicology Assessment of Organic Compounds in Drinking Water, *CRC Crit. Rev. Environ. Control* 12:307-359 (1982).
12. Coleman, W. E., J. W. Munch, W. H. Kaylor, R. P. Steicher, H. P. Ringhand, and J. R. Meier. "Gas Chromatography/Mass Spectroscopy Analysis of Mutagenic Extracts of Aqueous Chlorinated Humic Acid. A Comparison of the By-products to Drinking Water Contaminants," *Environ. Sci. Technol.* 18:674-681 (1984).
13. Jolley, R. L. "Concentrating Organics in Water for Biological Testing," *Environ. Sci. Technol.* 15:874-880 (1981).
14. Monarca, S., J. R. Meier, and R. J. Bull. "Removal of Mutagens from Drinking Water by Granular Activated Carbon," *Water Res.* 17:1015-1026 (1983).
15. DeMarco, J. Personal Communication (Cincinnati: U.S. Environmental Protection Agency).

CHAPTER **103**

Effect of Dechlorinating Agents on the Mutagenic Activity of Chlorinated Water Samples

Philip Wilcox and Susan Denny

Concentrated extracts of drinking water derived from surface sources have been consistently shown to possess mutagenic activity in bacterial test systems.[1-3] One of the sources of mutagenic chemicals in drinking water may be naturally occurring organic compounds that react with chlorine and result in the generation of electrophiles capable of reacting with DNA.[4,5] At the Water Research Centre, we are engaged in a program of research to investigate the mutagenicity and carcinogenicity of water samples.[6] Although the ultimate aim of the program is to identify the mutagenic agents and assess the potential hazard to humans from these waterborne mutagens, it would be shortsighted to await the outcome of these studies before investigations are undertaken on possible ways to reduce the level of mutagenic chemicals in treated drinking water.

We have conducted a number of experiments to investigate the effect of various water treatment practices on the mutagenic activity of chlorinated water samples. Part of this work has involved a study of the effect of different dechlorinating agents on mutagenicity. Partial dechlorination is frequently used in the United Kingdom as the final stage of water treatment to reduce residual chlorine levels so that the water is aesthetically more acceptable to the consumer while retaining some disinfection capacity throughout the distribution system. The most commonly used dechlorinating agent in the water industry is sulfur dioxide (SO_2), although other sulfur IV compounds such as sulfite and metabisulfite can be used.

The ability of dechlorinating agents such as sulfite to reduce the mutagenic activity of drinking water samples was first noted by Cheh et al.[7] They reported that total dechlorination with sulfite caused a 50 to 80% reduction in the mutagenic activity of concentrated XAD-4 extracts prepared from chlorinated water samples. Other nucleophiles have been reported to reduce mutagenicity when added directly to the concentrated extracts.[8] Dechlorination has also been found to reduce the acute toxicity of chlorinated waters to certain aquatic species.[9-12]

Our initial interest in dechlorinating agents stemmed from the development of a standardized laboratory chlorination technique to survey raw waters collected from different sites. It was intended that dechlorination with thiosulfate would be included as part of the standard method for two reasons: (1) chlorination reactions could be terminated after a fixed contact time; and (2) if the samples still contained free chlorine, it seemed possible that atypical chlorination reactions might occur during the subsequent freeze-drying process used in the laboratory for sample concentration. However, preliminary results indicated that samples dechlorinated with thiosulfate were substantially less mutagenic than corresponding samples that had not been dechlorinated. It was decided, therefore, to instigate a more extensive study on the effects of different dechlorinating agents, with a view to evaluating the effectiveness of dechlorination as a means of reducing the mutagenic activity of chlorinated drinking water.

MATERIALS AND METHODS

Laboratory Chlorination and Dechlorination Procedures

For some experiments, raw water samples were collected from particular sites in stainless steel containers. At the laboratory, the samples were filtered through Whatman GF/C filters and the pH adjusted to 7.0 using HCl or NaOH as appropriate. In initial studies, samples were chlorinated at a fixed dose of 6 mg/L chlorine by the addition of an appropriate amount of a standardized sodium hypochlorite solution. The sample was then left at room temperature on a magnetic stirrer for a contact period of 2 h. In later studies the chlorine dose was reduced to 5 mg/L, and the contact period was shortened to 1 h. Chlorine residuals were measured at the beginning (after 5 min) and at the end of the contact period using the DPD-ferrous ammonium sulfate titrimetric method.[13] In other experiments, chlorinated water samples were collected directly from the chlorination tank of a water treatment plant.

In experiments requiring dechlorination, concentrated solutions of dechlorinating agents were prepared immediately before use. The agents used in this study were sodium thiosulfate, sodium metabisulfate, sodium sulfite, sulfur dioxide, and biotin. The volume required to achieve the necessary degree of dechlorination was calculated (based on the stoichiometric reactions with free chlorine), added to the water sample, and mixed thoroughly. The samples were left at room temperature for 5 min, and chlorine residuals were then redetermined. It was usually necessary to add a second small dose of the dechlorinating agent to achieve the required final chlorine levels. Water samples were placed in trays of the freeze dryer for the initial freezing phase, which takes about 3 h in our machine. Reactions between the dechlorinating agents and organic compounds could theoretically still proceed during this time.

Concentrated Extracts

Concentrated extracts of water samples were prepared by freeze-drying in an Edwards Minifast freeze dryer. This machine can freeze-dry 8 L of water in ~26 h. The residual solids were extracted three times with methanol, and the solvent extracts were concentrated by rotary evaporation at 30°C and finally under a stream of nitrogen. In early experiments the concentrated methanol extracts were tested directly in the fluctuation assay. Although these samples gave satisfactory results in the absence of S9, they were often toxic to the bacteria in the presence of the activation system. We have demonstrated that under the conditions used in our experiments, S9 enzymes can metabolize methanol to formaldehyde. When formaldehyde was added directly to a bacterial suspension at the level detected with the methanol/S9 mix, toxicity was observed. In later experiments the methanol extracts were exchanged into dimethyl sulfoxide (DMSO) prior to testing in an attempt to alleviate this problem.

Bacterial Mutagenicity Assay

Concentrated extracts were tested for mutagenic activity in two-step bacterial fluctuation assays[14,15] using *Salmonella typhimurium* TA98 and TA100. Details of the precise method have been given in a previous publication from our laboratory.[16] The extracts were routinely tested at dose levels equivalent to 0.02, 0.05, and 0.1 L of original water per milliliter of test medium. Each experiment contained appropriate negative (no additions), solvent, and positive controls (tested at a single dose level), and each assay was repeated at least once on a different day. Initial studies indicated that the presence of aroclor-induced rat liver S9 fraction reduced the activity of the DMSO extracts; consequently, the use of S9 was omitted in many of the experiments.

The results of the fluctuation assays were analyzed using the GLIM (generalized linear interactive modeling) statistical package.[17] Using the results from the two repeat experiments, GLIM will determine whether the variability exhibited by the observed number of wells containing revertants is significantly greater than would be expected on the basis of binomial variation. A probability value is calculated that the observed results could have arisen by chance in the absence of a genuine dose effect; p-values at or below 0.05 (after adjustment for multiple comparisons) were taken to indicate significant mutagenicity. The data are plotted as estimated number of revertants per well vs dose. This term can be calculated from the zero term of the Poisson distribution.[18] This parameter, rather than the number of positive wells, is often linearly related to dose.[19] The GLIM program also indicates the position of the best-fitted straight line through the data and the slope value of this line with a 95% confidence limit. Comparisons of slope values before and after dechlorination were made using Welch's t-test; p-values of 0.05 or below indicated a significant difference in mutagenic response.

RESULTS

Slope values from the dose response plots (obtained with various samples before and after dechlorination) and the level of significance with respect to mutagenic activity are listed in Table I. The results obtained in each assay with positive control mutagens are not shown; however, on all occasions they were as expected, indicating that the assay was functioning satisfactorily.

Sodium Thiosulfate

The graphs in Figure 1 provide an example of the dose-response plots from which the slope values were determined. In this experiment a chlorinated water sample taken directly from the chlorine contact tank of a water treatment plant was tested for mutagenic activity before and after total dechlorination with sodium thiosulfate. The chlorine residuals of the sample were 1.26 mg/L total chlorine and 0.84 mg/L free chlorine, which was totally dechlorinated by the addition of 1.3 mL/L of a 3.5-mg/mL sodium thiosulfate solution. The extract prepared from the sample that was not dechlorinated gave a dose-related response in *Salmonella typhimurium* TA100, which detects primarily base-pair substitution mutations, and weaker activity in the frameshift-detecting strain TA98. Treatment with thiosulfate produced a significant reduction in activity observed in strain TA100, indicating that thiosulfate not only reacts with free chlorine but also with mutagenic organic compounds produced by chlorination. Activity in strain TA98 was marginally reduced by thiosulfate treatment, but the difference in slope values failed to achieve the necessary level of significance (5%). Thiosulfate also reduced mutagenic activity when added directly to the final methanol extract and tested without exchanging into DMSO (data not shown).

Sodium Metabisulfite

The slope values obtained in a similar experiment, using metabisulfite as the dechlorinating agent, are given in Table I. On this occasion the water from the chlorine contact tank had a total chlorine residual of 1.30 mg/L (0.80 mg/L free chlorine) when measured in the laboratory. The water was totally dechlorinated by the addition of 1.43 mL/L of a 2.68-mg/mL sodium metabisulfite solution. Total dechlorination was found to significantly reduce the mutagenic activity of the sample in both TA98 and TA100.

Table I. Mutagenic Activity of Concentrated Water Samples After Treatment with Various Dechlorinating Agents (Slope Values Taken from Linear Dose Response Plots).

Source of Water Sample	Dechlorinating Agent	Extent of Dechlorination	Slope Values (with approximate standard errors)[a]	
			TA100	TA98
Chlorine contact tank	None		20.43 (2.49)[b]	8.53 (1.84)[b]
	Thiosulfate	Total	3.26 (0.92)[c]	2.56 (2.34)[c]
Chlorine contact tank	None		20.19 (2.99)[b]	5.11 (1.07)[b]
	Metabisulfite	Total	8.02 (2.46)[b]	1.91 (1.03)[d]
Raw water, laboratory chlorinated	None		29.50 (1.27)[b]	NT[e]
	Biotin	Total	5.46 (1.92)[b]	NT
Raw water, laboratory chlorinated	None	0.8 mg/L Free chlorine	25.06 (1.59)[b]	NT
	Sulfite	Total	21.25 (2.03)[b]	NT
		Excess	3.52 (2.44)[f]	NT
			1.09 (2.86)[f]	NT
Chlorine contact tank	None	0.5 mg/L Free chlorine	13.15 (1.26)[b]	NT
	Sulfite	0.35 mg/L Free chlorine	9.97 (0.83)[b]	NT
		Total	5.01 (0.70)[b]	NT
			3.63 (0.81)[b]	NT
Chlorine contact tank	None	Partial (0.5 mg/L total)	16.81 (1.26)[b]	7.16 (1.45)[b]
	Sulfur dioxide		4.04 (0.58)[b]	3.12 (1.10)[c]
Chlorine contact tank	None		11.78 (1.82)[b]	5.89 (0.51)[b]
	Sulfur dioxide	Total	2.71 (0.74)[b]	0.95 (0.71)[f]
Chlorine contact tank	Sulfur dioxide	Total, then rechlorinated to 0.5 mg/L total	2.14 (0.76)[b]	0.99 (1.36)[f]
			6.86 (1.15)[b]	1.64 (1.53)[f]
Chlorine contact tank	None	Total	14.44 (0.87)[b]	NT
	Sulfite	Total, then chloraminated to 0.44 mg/L total	4.00 (0.76)[b]	NT
			4.12 (0.68)[b]	NT

[a] A significance of response in fluctuation assay.
[b] Significant at 0.1% level.
[c] Significant at 1% level.
[d] Significant at 5% level.
[e] Not tested.
[f] Not significant.

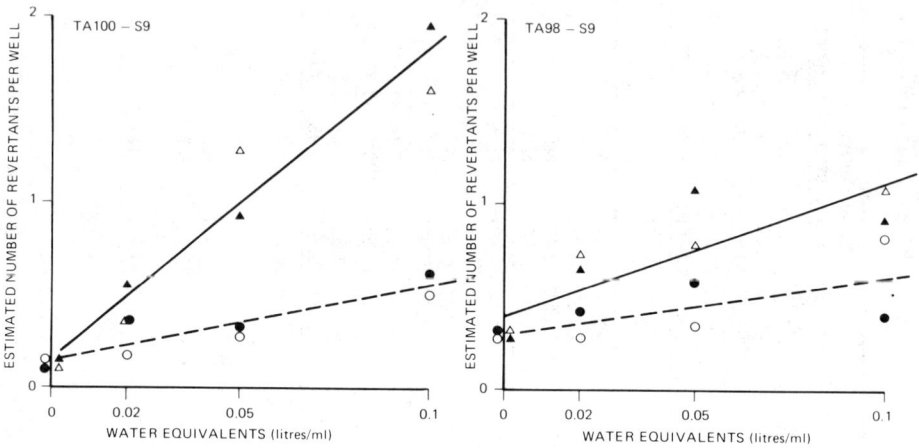

Figure 1. Effect of total dechlorination (with sodium thiosulfate) on the mutagenic activity of a chlorinated water sample. ▲ and △, water from chlorine contact tank (experiments 1 and 2); ● and ○, dechlorinated with sodium thiosulfate (experiments 1 and 2).

Biotin

In early work at the Water Research Centre on mutagenicity testing of water samples, experiments were carried out with unconcentrated drinking water.[20] As a control for this work, tests were conducted to check that components of the bacterial growth medium did not react with residual chlorine in the water samples and produce mutagenic compounds. One of the findings of these experiments was that biotin was capable of acting as a dechlorinating agent. Since biotin was already an essential component of the mutagenicity test, if it could also serve as a dechlorinating agent, this would eliminate the need to add another extraneous compound that might interfere with the performance of the assay. Preliminary experiments to examine the effectiveness of biotin to dechlorinate a sodium hypochlorite solution indicated that the addition of 4 mg to each liter of water was required to quench 1 mg/L free chlorine. It was also shown that the chlorination of biotin did not result in the generation of mutagenic compounds. However, when biotin was used to dechlorinate a raw water sample chlorinated in the laboratory at a fixed dose of 6 mg/L, a marked reduction in mutagenic activity (see Table I for slope values) was also observed. In contrast to thiosulfate, however, biotin did not reduce activity when added to the final methanol extract (data not shown).

Sodium Sulfite

Results from previous experiments showed that total dechlorination of chlorinated water samples markedly reduced the mutagenic activity of the sample.

In practice, however, most water treatment works only partially dechlorinate the final water to maintain a residual in the distribution system. It was decided, therefore, to investigate the effect of dechlorinating different chlorine residuals on mutagenic activity. In the first experiment, a raw water sample was chlorinated in the laboratory with a single dose of 6 mg/L and left for 2 h. The chlorine residuals were then determined to be 1.85 mg/L free chlorine and 3.97 mg/L total chlorine. Different amounts of sodium sulfite were added to water samples to partially dechlorinate (leaving a free residual of 0.8 mg/L and a total residual of 2.19 mg/L), totally dechlorinate (no residual chlorine), or chlorinate to excess (two times the amount of sulfite required for total dechlorination). The samples were then concentrated by freeze-drying–methanol extraction and tested in the fluctuation assay against TA100 in the absence of S9. The results of this experiment (see Table I) indicated that partial dechlorination did not significantly affect the mutagenic activity of the sample, whereas total dechlorination and to excess abolished the activity.

In the second experiment, chlorinated water was taken from the contact tank of a water treatment plant; samples were dechlorinated in the laboratory with sulfite to leave a free residual of 0.5 mg/L, 0.35 mg/L, or total dechlorination. The slope values obtained with these samples in strain TA100 were proportional to the degree of dechlorination (see Table I). On this occasion, however, mutagenic activity was not completely abolished in the totally dechlorinated sample. Tests for statistical significance between the various treatments indicated that the difference in response between no dechlorination and dechlorinating to 0.5 mg/L free chlorine, and between 0.35 mg/L free chlorine and total dechlorination, was not significant at the 5% level. All other comparisons gave statistically significant differences. It appeared, therefore, that unless the water was dechlorinated beyond 0.5 mg/L free chlorine, there was little benefit in terms of reducing the mutagenic activity of the sample.

Sulfur Dioxide

In all the studies discussed thus far, dechlorination was carried out under laboratory conditions. For the study on sulfur dioxide, therefore, experiments were designed to investigate whether similar effects could be demonstrated in the field. This was achieved by taking samples from appropriate points at a water treatment plant that dechlorinated with sulfur dioxide as part of treatment. The normal practice at the plant involves direct abstraction from a lowland river, prechlorination, alum sedimentation, rapid gravity filtration, and secondary chlorination in the chlorine contact tank. Immediately before leaving the plant, the water is usually dechlorinated with sulfur dioxide to leave a total residual of about 0.5 mg/L (about 0.2 mg/L free chlorine). We were fortunate to be able to arrange the study when the final water was being totally dechlorinated with sulfur dioxide prior to treating the distribution system with the insecticide Permethrin. Samples from the contact tank and of the final water after partial dechlorination were taken before the period of total dechlorination to provide control data. Corresponding samples were taken the following week when the water was begin totally dechlorinated. Freeze-dried–methanol extracts were prepared from the four samples and tested in the fluctuation assay against strains TA98 and TA100. The results, given in Table I and Figures 2 and 3, indicated that even under normal practices, the application of sulfur dioxide produced a significant reduction in the mutagenic activity of the samples in both TA98 and TA100. In samples that were totally dechlorinated, activity was further reduced in TA100, but not totally abolished. In strain TA98, however, the totally dechlorinated sample did not show a statistically significant increase in mutagenic activity over the negative control.

Dechlorination-Rechlorination

As a follow-up to this work, we decided to investigate the effect of total dechlorination and then rechlorinating to a low level suitable for maintaining a residual throughout distribution. We wanted to know whether mutagens destroyed by dechlorination would regenerate if chlorine were applied a second time. For these experiments, final water samples were taken from the treatment plant during the period of total dechlorination. The samples were then rechlorinated in the laboratory with hypochlorite to give a total residual of 0.5 mg/L (i.e., the same as the normal final water) after a contact time of

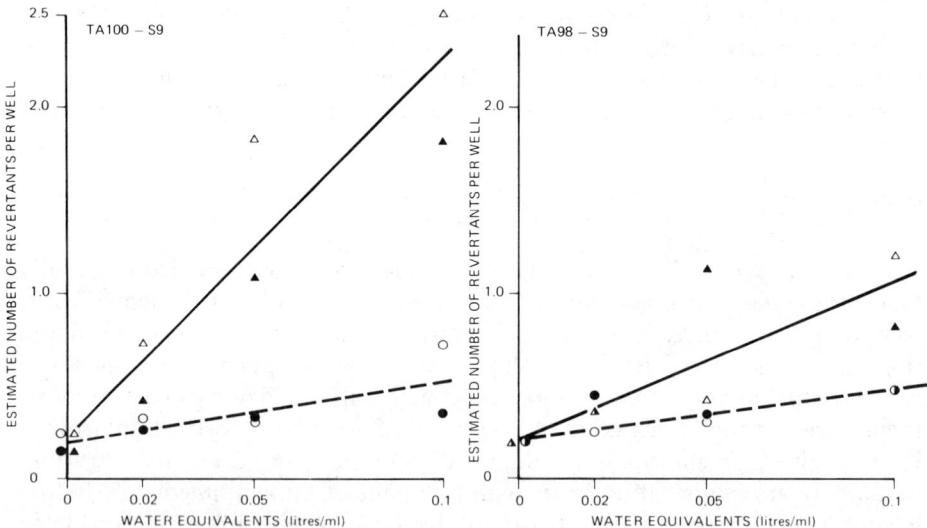

Figure 2. Effect of partial dechlorination (with sulfur dioxide) on the mutagenic activity of a chlorinated water sample. ▲ and △, water from chlorine contact tank (experiments 1 and 2); ● and ○, normal final water dechlorinated to 0.5 mg/L total (experiments 1 and 3).

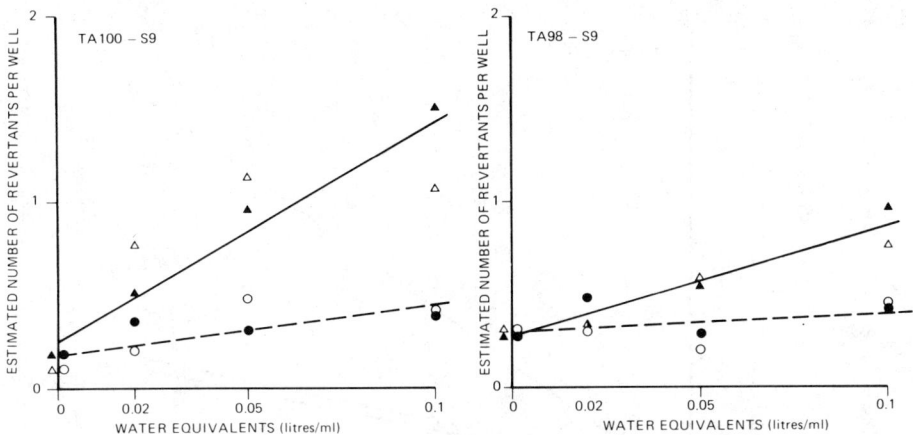

Figure 3. Effect of total dechlorination (with sulfur dioxide) on the mutagenic activity of a chlorinated water sample. ▲ and △, water from chlorine contact tank (experiments 1 and 2); ● and ○, dechlorinated totally with sulfur dioxide (experiments 1 and 2).

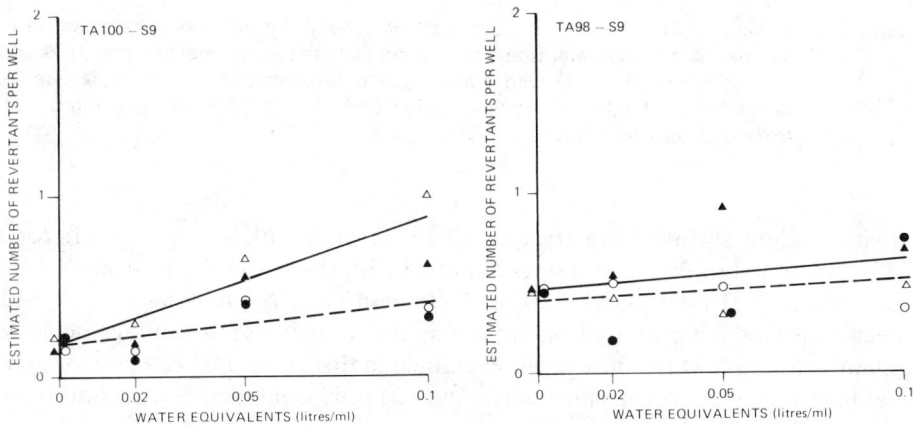

Figure 4. Effect of rechlorination on the mutagenic activity of a totally dechlorinated water sample. ● and ○, contact tank water dechlorinated totally with sulfur dioxide (experiments 1 and 2); ▲ and △, rechlorinated to 0.5 mg/L total (experiments 1 and 2).

2 h. In strain TA100, rechlorination appeared to result in a restoration of mutagenic activity to the original level (see Table I and Figure 4). In strain TA98, however, neither the totally dechlorinated nor the dechlorinated and rechlorinated samples showed significant levels of mutagenicity at the dose levels examined.

In a second experiment of this type, samples of totally dechlorinated water

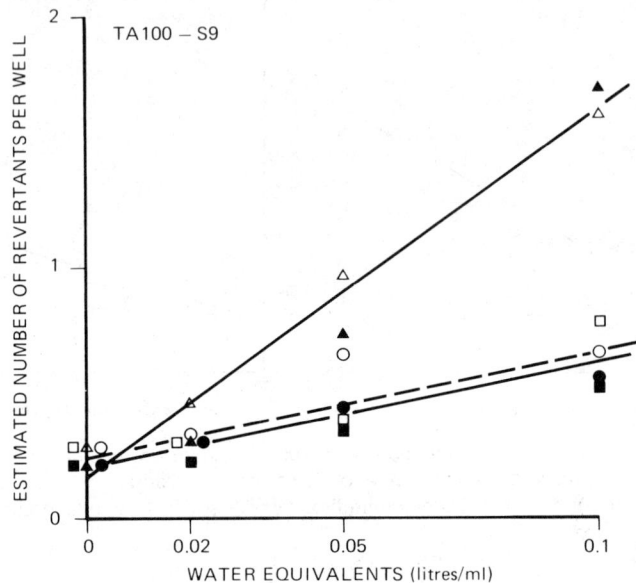

Figure 5. Effect of chloramination on the mutagenic activity of a totally dechlorinated water sample. ▲ and △, water from chlorine contact tank (experiments 1 and 2); ● and ○, dechlorinated totally with sulfur dioxide (experiments 1 and 2); ■ and □, dechlorinated totally with sulfur dioxide and chloraminated at 0.44 mg/L total (experiments 1 and 2).

(with sodium sulfite) were treated with chloramine (0.44 mg/L total) for a contact period of 2 h. The results obtained with these samples in strain TA100 are given in Table I and Figure 5. In contrast to previous results with rechlorination, chloramination did not appear to increase the mutagenic activity of the sample at the dose levels examined in this assay. In TA98, the original samples that were not dechlorinated showed only weak activity, and all treatments gave a similar response in this strain (data not shown).

DISCUSSION AND CONCLUSIONS

The most widely used dechlorinating agent in the water industry is sulfur dioxide (SO_2). In small installations that are not equipped to handle gaseous SO_2, dechlorination can be achieved by adding sodium sulfite or sodium metabisulfite in their solid form. Sodium thiosulfate has often been used for laboratory dechlorination studies, but it is not recommended for use at water treatment plants. This is because reactions with chlorine appear to proceed in steps, thus creating a time factor for completion of the reaction. It has also been known to contribute to taste and odor problems.[21]

The specific reactions of sulfur dechlorinating agents with chlorine species have not been well documented in the literature. In aqueous solution, sodium sulfite and sodium metabisulfite ionize to the corresponding ions, which are oxidized by free chlorine to give sulfate (SO_4^{2-}) and chloride (Cl^-) ions. Sulfur dioxide introduced into water rapidly hydrates to form $SO_2 \cdot xH_2O$,[22] which then ionizes to an equilibrium mixture of bisulfite and sulfite ions.[23] These ions react with free chlorine as indicated.

Sulfur dechlorinating agents have also been reported to react with monochloramine and organic chloramines according to the following reactions:[23]

$$SO_3^{2-} + NH_2Cl + H_2O \longrightarrow NH_4^+ + SO_4^{2-} + Cl^-$$
$$SO_3^{2-} + R\text{-}NHCl + H_2O \longrightarrow RNH_2 + SO_4^{2-} + Cl^- + H^+$$

It is frequently stated that sulfur IV dechlorinating agents react rapidly with all residual chlorine fractions.[21] Stanbro and Lenkevich,[24] however, have shown that at least some organic chloramines react relatively slowly with aqueous SO_3^{2-} under typical environmental conditions.[24]

The ability of nucleophiles such as sulfite to react with chlorination-derived mutagens in water samples has been reported by Cheh and co-workers.[7,8] This was demonstrated primarily by the addition of nucleophiles to concentrated XAD extracts derived from chlorinated waters, but also by dechlorination of unconcentrated drinking water samples. A method was developed for titrating unknown, direct-acting mutagenic agents with nucleophiles to produce estimates of their minimal potencies and maximum concentration in the samples.[8]

We have extended the work of Cheh's group by studying the effect of a number of different dechlorinating agents on the mutagenic activity of chlorinated water samples. Throughout these studies we have attempted to use chlorination and dechlorination conditions that are representative of those used in water treatment plants. The conclusions of our work to date are summarized as follows:

1. Total dechlorination of chlorinated drinking waters with a variety of dechlorinating agents can substantially reduce the mutagenic activity of the samples both in strains TA100 and TA98. This has been demonstrated both in laboratory studies and by taking appropriate samples from a water treatment plant.
2. Partial dechlorination had little effect on mutagenic activity unless the free residual was reduced to below 0.5 mg/L.
3. Low-level rechlorination (i.e., to 0.5 mg/L total) of totally dechlorinated samples restored activity in TA100 but not in TA98.
4. Chloramination of totally dechlorinated samples resulted in lower activity than rechlorination.

Further studies are in progress to confirm the result with chloramine and to look at the effects of using chlorine dioxide in a similar rechlorination scheme. Clark and Johnston[25] have reported that chloramination of drinking water can result in the production of S9-dependent mutagenicity. It will be necessary, therefore, to confirm the benefits of chloramination in the presence of an in vitro S9 activation system.

The data presented in this chapter add support to the proposal made by Cheh et al.[7,8] that dechlorination could provide a relatively cheap means by which the mutagenic activity of drinking water could be reduced. Unfortunately, only rarely did dechlorination result in the abolition of activity at the dose levels examined. This could mean that the reactions between the dechlorinating agent and the organic mutagens do not proceed to completion under the conditions used in these studies. If this is the case, it may be possible to optimize dechlorination conditions for the destruction of mutagenic compounds, for example, by using a longer contact time. Alternatively, the results may indicate that there are several different mutagenic species in chlorinated water, at least one of which is resistant to inactivation by dechlorinating agents.

ACKNOWLEDGMENTS

This work was funded by the Department of the Environment, to whom we are grateful for permission to publish.

REFERENCES

1. Nestmann, E. R., G. L. Le Bel, D. T. Williams, and D. J. Kowbel. "Mutagenicity of Organic Extracts from Canadian Drinking Water in the *Salmonella*/Mammalian Microsome Assay," *Environ. Mutagen* 1:337–345 (1979).
2. Loper, J. C. "Mutagenic Effects of Organic Compounds in Drinking Water," *Mutat. Res.* 76:241–268 (1980).
3. Kool, H. J., C. F. Van Kreijl, and B. C. J. Zoeteman. "Toxicological Assessment of Organic Compounds in Drinking Water," *CRC Crit. Rev. Environ. Control* 12(4):307–357 (1983).
4. Meier, J. R., R. D. Lingg, and R. J. Bull. "Formation of Mutagens Following Chlorination of Humic Acids," *Mutat. Res.* 118:25–41 (1983).
5. Kringstad, K. P., P. O. Ljungquist, F. de Sousa, and L. M. Stromberg. "On the Formation of Mutagens in the Chlorination of Humic Acid," *Environ. Sci. Technol.* 17:553–555 (1983).
6. Forster, R. "Tracking the Micropollutant," *Water* 3:7–9 (1980).
7. Cheh, A. M., J. Skochdopole, P. M. Koski, and L. Cole. "Nonvolatile Mutagens in Drinking Water: Production by Chlorination and Destruction by Sulfite," *Science* 20:90–92 (1980).
8. Cheh, A. M., J. Skochdopole, C. Heilig, P. M. Koski, and L. Cole. "Destruction of Direct-Acting Mutagens in Drinking Water by Nucleophiles. Implications for Mutagen Identification and Mutagen Elimination from Drinking Water," in *Water Chlorination: Environmental Impact and Health Effects, Vol. 3,* R. L. Jolley, W. A. Brungs, and R. B. Cumming, Eds. (Ann Arbor, MI: Ann Arbor Science Publishers, Inc., 1980), pp. 803–817.
9. Hall, L. W., D. T. Burton, W. C. Graves, and S. L. Margrey. "Effects of Dechlorination on Early Life Stages of Striped Bass (*Morone saxatilis*)," *Environ. Sci. Technol.* 15:573–578 (1981).

10. Ward, R. W., and G. M. De Graeve. "Residual Toxicity of Several Disinfectants in Domestic Wastewater," *J. Water Pollut. Control Fed.* 50:46–60 (1978).
11. Ward, R. W., and G. M. De Graeve. "Residual Toxicity of Several Disinfectants in Domestic and Industrial Wastewaters," *J. Water Pollut. Control Fed.* 50:2703–2722 (1978).
12. Arthur, J. W., R. W. Andrew, V. R. Mattson, D. T. Olson, and G. E. Glass. "Comparative Toxicity of Sewage-Effluent Disinfection to Freshwater Aquatic Life," EPA/600/3-75-012 (Washington, DC: U.S. Environmental Protection Agency, 1975).
13. "Chemical Disinfecting Agents in Water and Effluents and Chlorine Demand," (London: Her Majesty's Stationery Office, 1980).
14. Green, M. H. L., W. J. Muriel, and B. A. Bridges. "Use of a Simplified Fluctuation Test to Detect Low Levels of Mutagens," *Mutat. Res.* 38:33–42 (1976).
15. Green, M. H. L., B. A. Bridges, A. M. Rogers, G. Horspool, W. J. Muriel, J. W. Bridges, and J. R. Fry. "Mutagen Screening by a Simplified Bacterial Fluctuation Test: Use of Microsomal Preparations and Whole Liver Cells for Metabolic Activation," *Mutat. Res.* 48:287–294 (1977).
16. Forster, R., and I. Wilson. "The Application of Mutagenicity Testing to Drinking Water," *J. Inst. Water Eng.* 35:259–274 (1981).
17. Baker, R. J., and J. A. Nelder. *The GLIM System: Release 3-Generalized Linear Interactive Modelling*, (London: Royal Statistical Society, 1978).
18. Venitt, S. "UKEMS Collaborative Genotoxicity Trial: Bacterial Mutation Tests of 4-Chloromethylbiphenyl, 4-Hydroxymethylbiphenyl, and Benzyl Chloride: Analysis of Data from Seventeen Laboratories," *Mutat. Res.* 100:91–109 (1982).
19. Venitt, S., R. Forster, and E. Longstaff. "Bacterial Mutagenicity Assays," in *Report of the UKEMS Sub-committee on Guidelines for Mutagenicity Testing*, Part 1, B. Dean, Ed. (UKEMS Publication, 1983), pp. 5–40.
20. Forster, R., M. H. L. Green, R. D. Gwilliam, A. Priestley, and B. A. Bridges. "Use of the Fluctuation Test to Detect Mutagenic Activity in Unconcentrated Samples of Drinking Waters in the UK," in *Water Chlorination: Environmental Impact and Health Effects, Vol. 4*, R. L. Jolley, W. A. Brungs, J. A. Cotruvo, R. B. Cumming, J. S. Mattice, and V. A. Jacobs, Eds. (Ann Arbor, MI: Ann Arbor Science Publishers, Inc., 1983), pp. 189–1197.
21. White, G. C. In *Handbook of Chlorination* (New York: Van Nostrand Reinhold Company, 1972), pp. 346–347.
22. Cotton, F. A., and G. Wilkinson. *Advanced Inorganic Chemistry. A Comprehensive Text*, 4th ed. (New York: John Wiley and Sons, 1980), pp. 532–533.
23. Helz, G. R., and L. Kosak-Channing. "Dechlorination of Wastewater and Cooling Water," *Environ. Sci. Technol.* 18(2):48–55 (1984).
24. Stanbro, W. D, and M. L. Lenkevich. "Slowly Dechlorinating Organic Chloramines," *Science* 215:967–968 (1982).
25. Clark, R. R., and J. B. Johnston. "Mutagens Associated with the Distribution System of a Public Water Supply," presented at a U.S. Environmental Protection Agency Symposium on Health Effects of Drinking Water Disinfectants and Disinfectant By-products (April 1981).

SECTION XV

Cooling Water Treatment

Biofouling is a general phenomenon which is an important natural process . . . In a larger sense it represents a problem for society because failure to mitigate the effects of biofouling can have a measurable effect on our national energy problems.

        Roger M. Jorden
        Remarks to EPRI Condenser Biofouling Control Symposium, 1979

An ethic to supplement and guide the economic relation to (water) presupposes the existence of some mental image of (water) as a biotic mechanism. We can be ethical only in relation to something we can see, feel, understand, love, or otherwise have faith in.

        Aldo Leopold
        A Sand County Almanac, 1949

We have an unknown distance yet to run; an unknown river yet to explore.

        John Wesley Powell, 1834–1902
        Exploration of the Colorado River of the West and its Tributaries (1875),
        Log of August 13, 1869

CHAPTER 104

Analysis of Sediment Matter for Halogenated Products from Chlorination of Power Plant Cooling Water

Roger M. Bean, Berta L. Thomas, and Duane A. Neitzel

The chlorination of cooling water to control biofouling at electrical power stations is known to cause formation of low concentrations of stable halogenated organic products.[1,2] Many of these halogenated products, including the trihalomethanes (THMs) and halogenated phenols, are considered lipophilic because they strongly adsorb to macroeticular organic materials such as XAD resins.[3] Because of the lipophilic character of these chlorination by-products, there is the potential for concentration in environmental compartments by adsorption on suspended matter or in biological tissues. Sediment organic matter has been correlated with the capacity of sediments for adsorption of organics,[4] and the bioconcentration potential of certain halogenated organics is well documented.[5] It is therefore appropriate to investigate the environments affected by the cooling water discharges of power plants to determine if chlorination products are being concentrated. We have examined the adsorption of halogenation products on suspended matter in the discharge of a coastal power station. We have also studied the concentrations of halogenated organics in sediments exposed to the discharge plumes at three freshwater-cooled plants.

SAMPLING AT POWER STATIONS

Suspended matter from the discharge of the Redondo Generating Station (oil-fueled units No. 7 and 8, 950 MW, at Redondo Beach, California) was collected by filtering large volumes (about 1,500 L) of discharge through a high-flow filtration device consisting of eight 30-cm filters in parallel.[6] A subsample of the filtered water (210 L) was collected in a polyvinyl fluoride-lined barrel. The suspended-matter-free water was sampled by pumping it over 2.5 by 23-cm stainless steel columns filled with 80 mL XAD resin.[2] XAD sampling was in duplicate. In addition, triplicate 1-L samples of filtered water were extracted twice with 50 mL methylene chloride for phenol analysis. Sampling of the Redondo discharge was conducted during two chlorination cycles [normal chlorination, 0.15 mg/L total residual oxidant (TRO); high chlorination, 0.48 mg/L TRO] and during a period of no chlorination. In addition, a sample of seawater ~2 km away from the discharge was taken as a control.

Surface sediments (1-L samples from the top 2 cm) and 5 by 15-cm sediment cores were taken from water bodies where there was direct contact with the discharge plumes from three nuclear power stations. The three plants studied were Beaver Valley Power Station (885 MW), Shippingport, Pennsylvania, on the Ohio River; Trojan Nuclear Plant (1130 MW), Rainier, Oregon, on the Columbia River; and Palisades Nuclear Generating Plant (715 MW), South Haven, Michigan, on Lake Michigan. In addition to sediments from the discharge plumes, sediment samples were also taken from locations not affected by the cooling water plumes, that is, upstream, or in the case of the Palisades site, 10 km from the plant along the Lake Michigan shoreline.

ANALYTICAL PROCEDURES

Extraction of Suspended Matter and Sediments and Separation of Extracts into Acid and Base-Neutral Fractions

Suspended matter on filters and 100-g samples of sediment were Soxhlet-extracted with a benzene-methanol azeotrope. Hexane and water were added to the Soxhlet extract to form two layers; the organic layer was freed from methanol and enriched in organic extract by appropriate backwashings.[7] The organic phase was extracted with 0.1 N NaOH to remove the chlorinated phenols, leaving the base-neutrals to be analyzed for total organic halogen (TOX) by microcoulometry.[2]

Extraction of XAD Columns and Analysis of Extracts for Trihalomethanes and Organic Halogen

Extraction and analysis of XAD columns for organohalogen compounds has been reported.[2,3] Ethyl ether extracts of the columns are directly analyzed for THMs using electron capture gas chromatography. Phenols are extracted from the ether extracts with 0.1 N NaOH. TOX is determined microcoulometrically on the ether extracts prior to phenol extraction.

Analysis of Halogenated Phenols in Extracts of Sediment, Suspended Matter, and Water

Methylene chloride extracts of discharge water were treated with 0.1 N NaOH to remove phenols. This base extract and other 0.1 N NaOH extracts (obtained from previously described procedures) were treated with acetic anhydride to form phenol acetates that were analyzed by capillary glass chromatography after extraction into hexane. Procedural details and quantitation methods are given in Bean et al.[2] Recoveries of phenols from the methylene chloride extraction of water approached 90% and were not recovery corrected; however, recoveries of phenols from suspended matter and sediments ranged from 24 to 74%, and the reported data were corrected for losses.

Determination of Volatile Halocarbons in Sediment Samples

Previous analyses of discharge waters indicated volatile chlorocarbons were present at the 1-μg/L level.[2] Assessments of the relative volatility and water solubility of these compounds suggested that they would not be found in sediments in concentrations much higher than in water. Excessive sample manipulation or evaporation of sediment solvent extracts would result in unacceptable losses of volatiles. Thus, a purge-and-trap method[8] was used to analyze the volatiles in sediments. Purging of the sediment with ultrapure nitrogen was conducted in a modified purge-and-trap apparatus (Figure 1). Analysis of volatiles after adsorption onto a Tenax® trap was accomplished using electron-capture gas chromatography to achieve the required sensitivity. Sediment samples were allowed to thaw at 4°C prior to analysis, and the supernatant water was removed by pipette. A 10-g sediment sample was removed by coring with a 13-mm ID glass tube and dropping the core into a preweighed purge apparatus that was immediately capped and weighed again to determine sample weight. An internal standard (1,3-dibromopropane) was then added in 2 μL of methanol. The sparger, sintered on the sides of the frit so that nitrogen gas only came from the bottom (Figure 1), was fitted to the purge apparatus, which was placed in a 50°C ultrasonic bath, and the trap was allowed to collect the purged volatiles for 20 min. The trap was then introduced to the electron-capture gas chromatograph for thermal desorption and analysis.

Chromatography was performed on a 2-mm by 1.8-m glass column (0.2% carbowax 1500 on Carbopack®). Conditions were as follows: 0°C for 5.5 min (sample introduction), 30°C/min for 1.5 min, then 8°C/min to 160°C. Figure 2 shows a chromatogram of a 10-g sediment sample spiked with 1 to 2 ng of THMs together with a procedural blank. Acceptable blanks could only be obtained by prolonged sparging of boiled water prior to sample introduction. Component retention times and detection limits are listed in Table I along with data obtained for surface sediments.

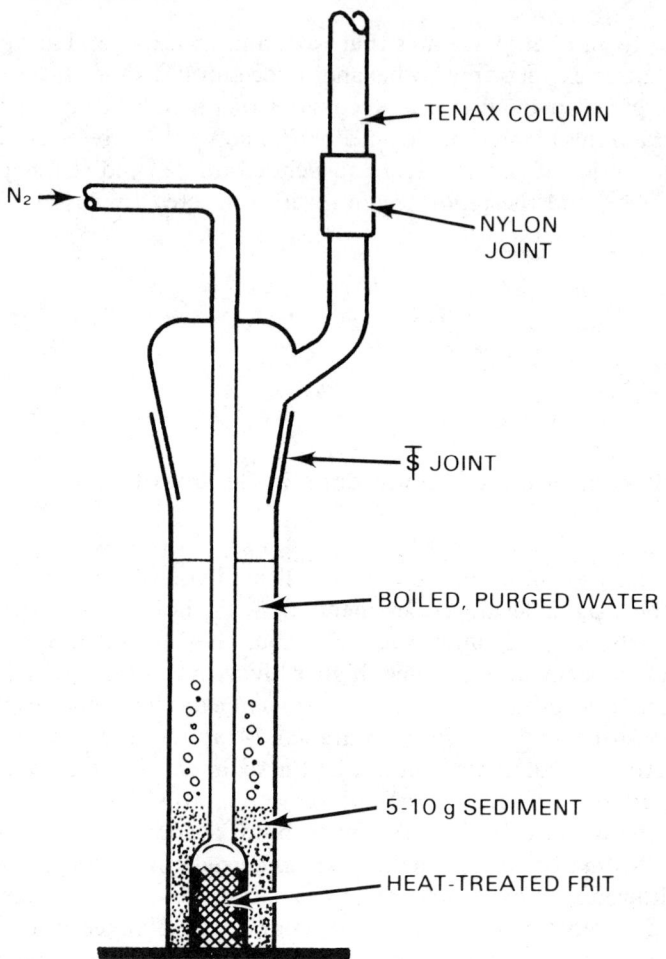

Figure 1. Modified purge-and-trap apparatus to determine volatiles in sediments.

REDONDO BEACH STATION: PARTITIONING OF CHLORINATION PRODUCTS BETWEEN WATER AND SUSPENDED MATTER

The total organic bromine in filtered discharge water and in collected suspended material is presented in Table II. Also included in the table is the concentration of organobromine in filtered water, which is accounted for as bromoform, a major product of seawater chlorination.[9] Analysis of bromoform in suspended matter was not practical under the sampling and analytical conditions used. Bromine in suspended matter is reported in Table II as micro-

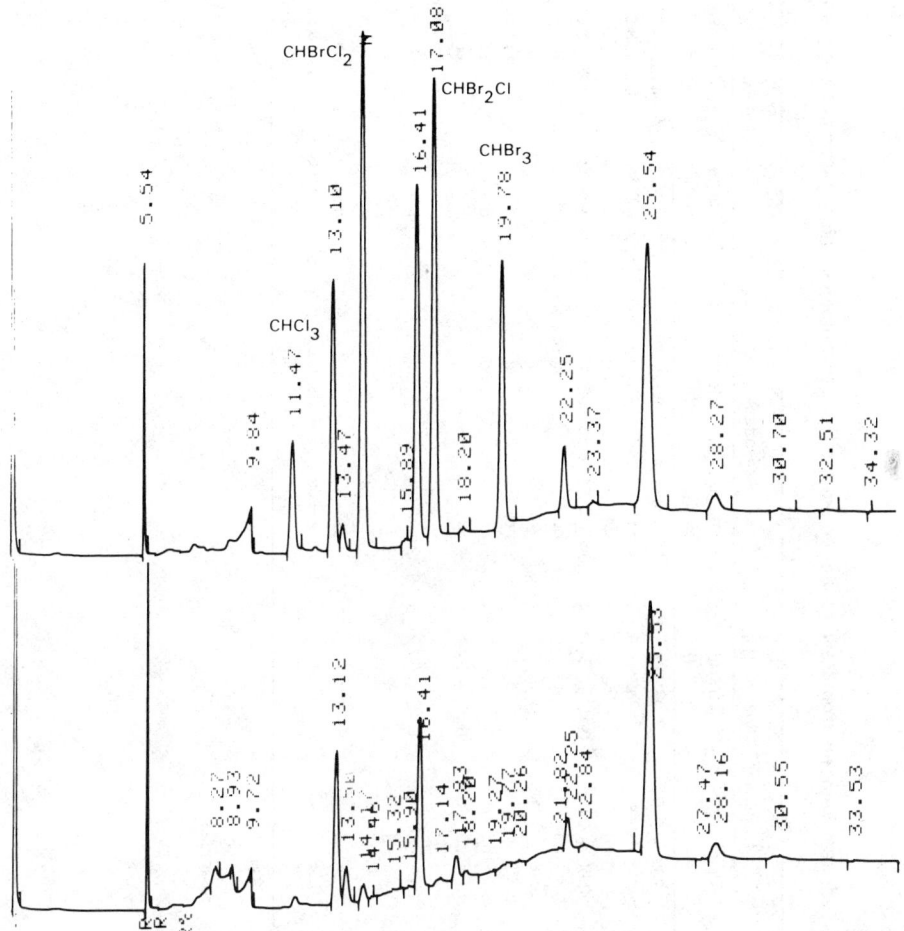

Figure 2. Chromatograms of haloform standards and a procedural blank. Top: Haloform standards. $CHCl_3$, 1.20 ng; $CHBrCl_2$, 1.58 ng; $CHBr_2Cl$, 1.96 ng; and $CHBr_3$, 2.312 ng. Coefficient of variation for n = 5, $CHCl_3$ = 4.7%, $CHBrCl_2$ = 7.1%, $CHBr_2Cl$ = 9.5%, and $CHBr_3$ = 18.2%. Recovery from sediment at these concentrations was 85%. Bottom: Procedural blank.

grams per liter water sampled to allow comparison with water column concentrations, and as micrograms per gram suspended matter collected on the filter. About 50% of the bromine collected from chlorinated seawater on XAD samplers was accounted for as bromoform; this is consistent with results obtained at another marine power plant.[2] However, a rather high value (16.8 µg/L) for total bromine was obtained when nonchlorinated discharge was

Table I. Analysis of Volatile Halocarbons in Bulk Sediment Samples from Power Plant Locations (concentrations in micrograms per kilogram of dry sediment)

Component	Retention Time	Minimum Detection Level (μg/kg)	Beaver Valley Discharge (n = 4)	Beaver Valley Control (n = 3)	Palisades Discharge (n = 4)	Palisades Control (n = 3)	Trojan Discharge (n = 4)	Trojan Control (n = 2)
Methylene chloride	6.6	20	230 ± 210	890[a]	120 ± 45	80 ± 28	220 ± 200	440 ± 40
Trichlorofluoromethane	8.3	0.1	1.0–0.7[b]	d[c]	nd[d]	nd	1.7 ± 0.1[b]	3.2 ± 3.3
Trans-1,2-dichloroethylene	10.8	3.4	nd	nd	nd	nd	d	nd
Chloroform	11.5	0.04	0.08 ± 0.02	0.15 ± 0.11[b]	0.04 ± 0.01	0.03 ± 0.01	0.04 ± 0.02	0.09 ± 0.01
1,2-Dichloroethane	12.1	1.2	nd	nd	nd	nd	nd	d
1,1,1-Trichloroethane	13.1	0.04	0.16 ± 0.16[b]	d	0.13 ± 0.08[b]	nd	d	d
Carbon tetrachloride	13.5	0.01	nd	nd	nd	nd	d	nd
Bromodichloromethane	14.4	0.01	nd	nd	d	nd	d	nd
1,2-Dichloropropane trans-1,3-Dichloropropene	15.8	0.4	nd	0.90 ± 0.41	d[e]	d[e]	d[e]	d[e]
Trichloroethylene	16.5	0.01	0.02 ± 0.01[b]	0.13 ± 0.12	0.02 ± 0.01[b]	d	nd	d
Dibromochloromethane cis-1,3-Dichloropropene	17.1	0.01	nd	d	0.03 ± 0.03[b]	d	d	nd
Bromoform	19.7	0.02	nd	nd	d	nd	d	0.04[b]
Tetrachloroethylene	22.3	0.01	0.08 ± 0.03	0.13[a]	0.09 ± 0.09	0.03 ± 0.01	0.04 ± 0.03	0.04 ± 0.02
1,1,2,2-tetrachloroethane								

[a] Variance too high to compute standard deviation.
[b] Not present in all subsamples.
[c] Component present in trace quantities in some (but not all) subsamples.
[d] Not detected.
[e] Detected at the minimum detectable level (0.4 ng/g).

Table II. Bromine in Redondo Discharge: Water vs Suspended Matter[a]

Sample	Water		Suspended Matter		Calculated Concentration Factor (C_{SM}/C_{water})
	Bromine as $CHBr_3$ (μg/L)	Total Bromine (μg/L)	Total Bromine (μg/L, water)	Total Bromine (μg/g SM)	
Normal Cl_2 (0.15 mg/L TRO)	4.9 ±1.5	8.5 ±1.9	0.22 ±0.09	30.7 ±14.5	3,600
High Cl_2 (0.48 mg/L TRO)	12.1 ±0.3	22.5 ±1.6	1.59 ±0.12	302 ±23.8	13,400
Not chlorinated	<0.01	16.8 ±0.8	0.16 ±0.04	23.3 ± 5.5	1,400
Open ocean	0.04 ±0.03	1.7 ±0.1	0.35 ±0.06	52.1 ±8.9	30,000

[a]Suspended matter (SM) concentration in water samples: 6.4 ± 1.4 mg/L, n = 7).

sampled; the absence of bromoform demonstrated that the high value was not a result of inadvertent chlorination. Total bromine levels for nonchlorinated ocean samples are normally 1 to 2 μg/L,[2,3] as determined for an "open ocean" sample taken outside the harbor that enclosed the discharge (Table II). The source of the high nonchlorinated value has not been determined, but was probably not a phenomenon related to water biocidal chlorination.

Concentrations of bromine determined in suspended matter were 3 to 4 orders of magnitude higher than in the water column, indicating that there was substantial adsorption of halogenated material on the particulate fraction of the water column. A comparison of micrograms per liter bromine in water with that measured identically in suspended matter (Table II) shows that the bromine absorbed on particulates is, in general, a small percentage of the total organic bromine found; however, the concentration factors (Table II) are substantial. This suggests that suspended matter may play a significant role in the environmental concentration of chlorination products.

Although the data based on microcoulometric halogen analyses indicate significant concentrations of chlorination products in suspended matter, the analytical problems associated with this method (e.g., interference by high concentrations of sulfur)[3] cast some doubt on the result. Analysis of water and suspended matter for halogenated phenols, presented in Tables III and IV, respectively, helps to validate the result obtained from the total bromine analyses. The relatively high concentration obtained for monobromophenols in water during normal chlorination sampling (Table II) was a probable result of contamination of the water samples by milligram-per-liter concentrations of

Table III. Brominated Phenols from Solvent Extraction of Filtered Redondo Discharge Water (concentrations in nanograms per liter)

Sample	Monobromo[a]	2,4-Dibromo	2,6-Dibromo	2,4,6-Tribromo
Normal chlorination, 0.15 mg/L TRO	300	1	7	100
High chlorination, 0.48 mg/L TRO	nd[b]	nd	2	150
No chlorination	nd	nd	nd	20
Open ocean	nd	12	27	30

[a] p-Bromophenol identified from standards; another monobromophenol identified by mass spectral analysis.
[b] Not detected.

bisphenol A. This contaminant was not present in other samples, but does raise a question as to the reliability of this sample analysis. The source of contamination is undetermined. There were three separate sampling operations being conducted at the time we sampled.

An analysis of suspended matter for bromophenols (Table IV) indicates a considerable concentration of phenols in particulates. The quantities of monobromophenols and dibromophenols present were only slightly above detection levels (excepting the questionable monobromo compound); however, the levels of tribromophenol during chlorination permitted the computation of concentration factors that are in the same range as those obtained from the total bromine data in Table II. Since tribromophenol was found in both control samples (nonchlorinated and open ocean), concentration factors are also presented in Table IV; however, the levels present were not sufficiently high to give a reliable result.

The data obtained for both total bromine and brominated phenols are evidence that chlorination products can be concentrated at the parts-per-million level in suspended matter, although the quantity adsorbed constitutes only a minor amount of the total organohalogen released during chlorination. The environmental concentration of halogenated materials has the potential for eliciting deleterious environmental effects. Hence, we examined environmental samples at other power plants that chlorinate cooling water to determine whether there is environmental accumulation of halogenated materials in sedimentary material.

HALOGENATED PRODUCTS AND TOTAL ORGANIC HALOGEN IN SEDIMENTS FROM POWER PLANT LOCATIONS

Although volatile halocarbons were detected in surface sediments (Table I), generally at levels below 1 μg/kg, no indication can be obtained from the data

Table IV. Brominated Phenols on Suspended Matter in Redondo Cooling Water Discharge[d]

Sample	Monobromo		Dibromo		2,4,6-Tribromo		Concentration Factor for Tribromophenol (C_{SM}/C_{water})
	(ng/L)	(ng/mg)	(ng/L)	(ng/mg)	(ng/L)	(ng/mg)	
Normal chlorination, 0.15 mg/L TRO	2.0	0.28	<0.01	<<0.01	6.0	0.86	8,600
High chlorination, 0.48 mg/L TRO	0.09	0.02	1.1	0.20	9.6	1.8	12,000
No chlorination	nd[b]	nd	nd	nd	0.05	0.01	500
Open ocean	nd	nd	nd	nd	0.05	0.01	300

[a]Suspended matter (SM) concentration in water samples: 6.4 ± 1.4 mg/L (n = 7).
[b]Not detected.

that the sediments from the chlorination plumes contained increased concentrations of volatile halocarbons and, in particular, increased concentrations of THMs. All samples were found to contain chromatographic peaks corresponding to tetrachloroethylene and methylene chloride, neither of which is a product of chlorination of natural waters.[3] None of the reported compounds, identified by retention time, were confirmed by mass spectrometry because of the minute quantities present in the samples. The presence of a trichlorofluoromethane peak in the Beaver Valley and Trojan samples may have been the result of contamination from packing the sediment samples with dry ice for shipment to the laboratory. Core sediments were subsampled for volatiles analysis at points several centimeters below the surface. These samples invariably were free from most volatiles with the exception of chloroform, tetrachloroethylene, and methylene chloride. This finding suggests that much of the volatile material in the surface sediments may have been associated with residue from the water column rather than with the sediment inorganic or organic matter.

The variation of halophenol concentrations (Table V) from sample to sample was high, as might be expected from a heterogeneous sediment matrix. Concentrations of phenols approached detection limits, and specific verification of the presence of individual phenol types was only possible by conducting reconstructed single-ion mass spectrometer scans on extracts from 300-g sediment samples. In some cases (see analyses of Palisades samples, Table V) phenols detected by mass spectrometry were not found as isolated peaks in the electron-capture capillary chromatograms; therefore, they are listed as not detected in the quantitative analyses.

For both Beaver Valley and Trojan sediments, there were significant contributions of trichlorophenols other than the 2,4,6,-trichloro isomer. This indicates that not all halogenated phenols found in sediments originated from power plant or other water chlorination sources, since the 2,4,6 isomer is the predominant trichloro isomer observed from low-level freshwater chlorination.[2] Beaver Valley and Trojan plants are situated in the vicinity of considerable industrial activity, which may contribute substantially to the chemical input to sediments.

The phenol concentrations given in Table V for Beaver Valley discharge samples, while highly variable, are consistently higher than those found in the control sediments, an apparent indication that the chlorinated plume affected sediment composition. When sediment organic extracts were analyzed for total organic halogen, the discharge sediments from both Beaver Valley and Trojan locations were found to be higher in TOX than the corresponding controls (Table VI) when expressed as micrograms-per-gram dry sediment—again, an apparent indication of influence from chlorination. However, the heterogeneous nature of the sediments must be taken into account in making comparisons. For example, the Beaver Valley discharge sediment was a thick mud, whereas the corresponding control sediment contained a substantial proportion of coarse gravel. To compensate for these differences in composition,

Table V. Analysis of Halogenated Phenols in Sediments Exposed to Power Plant Discharge Plume vs Control Sediments (concentrations in micrograms-per-kilogram Dry Sediment)

Phenols Identified by GC/MS in Discharge Sediment	RRT[a]	Phenols Quantitated by Electron-Capture Gas Chromatography					
		Discharge			Control		
Beaver Valley							
Dichloro	0.880	5.4	2.2	6.1	1.7	0.2	1.5
	0.951[b]	2.3	nd[c]	4.1	0.8	nd	<0.3
Trichloro	1.000	1.4	2.3	1.1	0.6	nd	0.3
	1.065	1.4	0.6	0.6	0.3	0.3	<0.3
	1.071[b]	2.8	0.8	1.4	0.3	0.6	<0.3
Dibromochloro	1.206	0.5	0.4	0.0	0.1	nd	<0.1
	1.214[b]	0.8	<0.1	0.5	0.1	nd	<0.1
Tribromo	1.316	0.8	0.5	0.3	0.3	<0.1	<0.1
Trojan							
Dichloro	0.873	1.0	nd	2.0	1.0	nd	0.5
	0.945	nd	nd	nd	nd	nd	nd
Bromochloro	0.976	<0.2	nd	nd	<0.2	nd	nd
	0.985	1.1	<0.2	nd	<0.2	nd	nd
Trichloro	1.000	1.4	<0.3	0.3	1.1	<0.3	0.6
	1.065	<0.3	0.3	nd	nd	<0.3	nd
Palisades							
Dichloro	0.864	nd	nd	nd	nd	nd	nd
Bromochloro	0.970	<0.2	0.7	<0.2	<0.2	0.7	0.2
	0.967	nd	nd	nd	nd	nd	nd
	0.980	nd	nd	nd	nd	nd	nd

[a]Retention time relative to 2,4,6-trichlorophenol.
[b]Peaks corresponding to these RRTs found in blank and subtracted from analytical chromatograms.
[c]Not detected

concentrations of TOX and halogenated phenols have been calculated on the basis of the organic carbon content of the sediments (Table VI).

The rationale for correlating concentrations of chloroorganics with organic carbon was based on the strong correlation between sediment organic content and adsorption of organic matter.[4] When haloorganic content is based on organic carbon, the differences observed between discharge and control become less pronounced for Beaver Valley sediments. For Palisade sediments, a higher carbon content in the discharge sediments resulted in a higher calculated concentration of organohalogen per gram carbon in the control sediment than in the discharge sediment. The results for halogenated phenols are similar (Table VI). When based on organic carbon content, an apparent difference between Beaver Valley discharge sediments and controls becomes less pronounced. Considering the variability of the phenol concentrations in sediments (Table V), there is little evidence to suggest a difference between discharge and

Table VI. Total Organic Halogen and Halogenated Phenols in Sediment Samples from Power Stations

Sample	Percent Organic Carbon in Sediment (dry basis)	Organic Halogen, n = 3		Halogenated Phenols in Dry Sediment (μg/kg)	Halogenated Phenols in Organic Carbon (μg/g)
		Dry Sediment (μg Cl/g)	Organic Carbon (μg Cl/g)		
Beaver Valley					
Control	3.5	0.39 ± 0.01	11.1 ± 0.3	2.7	0.08
Discharge	9.9	2.33 ± 0.11	23.6 ± 1.1	12.3	0.12
Trojan					
Control	1.6	0.32 ± 0.06	20.3 ± 4.0	1.1	0.07
Discharge	1.7	0.47 ± 0.05	27.8 ± 2.8	2.1	0.12
Palisades					
Control	0.18	0.03 ± 0.00	15 ± 3	0.3	0.17
Discharge	0.51	0.02 ± 0.00	3.0 ± 0.3	0.3	0.06

control locations in terms of halogenated phenol content. With the possible exception of the Beaver Valley samples, evidence for differences in total organic halogen between discharge and control sediments is not compelling when based on carbon content. It is of some interest to note that concentrations of TOX and halophenols, when based on organic carbon content, are in the same ranges as those found adsorbed on suspended matter during chlorination at the Redondo power station.

CONCLUSIONS

Analysis of water and suspended matter at the Redondo Generating Station demonstrated that while most of the chlorination products are in the filterable water column, a small portion remains adsorbed on suspended matter in concentrations in the micrograms-per-gram range. This represents a considerable environmental magnification from the micrograms-per-liter concentrations normally reported for cooling water chlorination by-products. Examination of sediments taken from both discharge and control locations at operating power plants did not produce evidence for the accumulation of chlorination by-products in sediments exposed to chlorinated cooling water discharges, although concentrations of halogenated organics, when expressed as micrograms-per-gram sediment organic carbon, were of the same order of magnitude as those determined on suspended matter at the Redondo facility.

ACKNOWLEDGMENTS

We are pleased to acknowledge the support of the Nuclear Regulatory Commission (NRC), Silver Spring, Maryland, and the Southern California Electric Company, Rosemead, for this work.

REFERENCES

1. Jolley, R. L., W. W. Pitt, Jr., F. G. Taylor, Jr., S. J. Hartmann, G. Jones, Jr., and J. E. Thompson. "An Experimental Assessment of Halogenated Organics in Waters from Cooling Towers and Once-Through Systems," in *Water Chlorination: Environmental and Health Effects, Vol. 2*, R. L. Jolley, H. Gorchev, and D. H. Hamilton, Eds. (Ann Arbor, MI: Ann Arbor Science Publishers, Inc., 1978), pp. 695-706.
2. Bean, R. M., D. C. Mann, and D. A. Neitzel. "Organohalogens in Chlorinated Cooling Waters Discharged from Nuclear Power Stations," in *Water Chlorination: Environmental Impact and Health Effects, Vol. 4*, R. L. Jolley, W. A. Brungs, J. A. Cotruvo, R. B. Cumming, J. S. Mattice, and V. A. Jacobs, Eds. (Ann Arbor, MI: Ann Arbor Science Publishers, Inc., 1983), pp. 383-390.
3. Bean, R. M, D. C. Mann, and R. G. Riley. "Analysis of Organohalogen Products from Chlorination of Natural Waters Under Simulated Biofouling Control Conditions," NUREG/CR-1301, Battelle Pacific Northwest Laboratories for the U.S. Nuclear Regulatory Commission (1980).

4. Karickhoff, S. W., D. S. Brown, and T. A. Scott. "Sorption of Hydrophobic Pollutants on Natural Sediments," *Water Res.* 13:241-248 (1979).
5. Veith, G. D., D. W. Kuehl, F. A. Puglisi, G. E. Glass, and J. G. Eaton. "Residues of PCBs and DDT in the Western Lake Superior Ecosystem," *Arch. Environ. Contam. Toxicol.* 5:487-499 (1977).
6. Silker, W.B., R. W. Perkins, and H. G. Rieck. "A Sampler for Concentrating Radionuclides from Large Volume Samples," *Ocean Eng.* 2:49-50 (1971).
7. Bean, R. M., R. L. Wilson, D. A. Neitzel, and M. A. O'Malley. "Analysis of Organohalogen and Other Organic Chemicals in Cooling Waters Discharged from Redondo Generating Station," (Richland, WA: Battelle Pacific Northwest Laboratories, 1982).
8. Bellar, T. A., and J. J. Lichtenberg. "Determining Volatile Organics at Microgram-Per-Liter Levels by Gas Chromatography," *J. Am. Water Works Assoc.* 66:739-744 (1980).
9. Carpenter, J. H., and C. A. Smith, "Reactions in Chlorinated Seawater," in *Water Chlorination: Environmental Impact and Health Effects, Vol. 2*, R. L. Jolley, H. Gorchev, D. H. Hamilton, Jr., Eds. (Ann Arbor, MI: Ann Arbor Science Publishers, Inc., 1978), pp. 195-208.

CHAPTER 105

Halogenated Compounds Discharged from a Coastal Power Plant

Robert S. Grove, Edward J. Faeder, Jean Ospital, and Roger M. Bean

This study was initiated to analyze organic chemicals in aquatic environments adjacent to a Southern California Edison Company (SCE) electric generating facility and to assess their potential health effects. Sodium hypochlorite is periodically added to the cooling water to prevent microbial fouling of the electric facility's condenser tubes. The principal source and sink of cooling water at SCE's fossil-fueled and nuclear facilities is the Pacific Ocean. Since 1963, SCE has conducted research to determine the effects of coastal electric stations (including chlorination practices)[1-4] on the marine environment. The present study focuses on organohalogen products formed in the cooling water discharge from the reactions of chlorine.

The potential for the formation of haloorganics from the chlorination of cooling water has been noted for years.[5,6] However, the published literature is limited concerning the formation of halogenated organic components in chlorinated cooling waters of coastal electric generating stations.

Helz et al.[7] reported evidence of the formation of brominated macromolecular material during the chlorination of estuarine cooling water. Carpenter and Smith[6] also detected bromoform as a major product from the low-level chlorination of salt water. However, only one marine electric facility (Millstone Nuclear Power Station, located on Long island Sound in Connecticut) is known to have been sampled[8] using state-of-the-art analytical techniques to assess the organic components in cooling waters.

In this study, SCE investigated the organic chemistry of cooling water discharged from units 7 and 8 of the Redondo Generating Station located at King Harbor in the city of Redondo Beach, California. Sampling was done over 3 d in June 1981. Potential first-order impacts on the aquatic life of the organic components of chlorination were estimated.

METHODS

Field Sample Collection

Samples were collected from the discharge of Redondo units 7 and 8 during chlorination and under varied chlorine dosage conditions. Samples were also collected from the discharge while the station was not chlorinating and from a control area 1 km away (outside of King Harbor) at a depth of 1 m.

Standard amperometric titration was used to measure total residual oxidants (TRO). Chlorination occurs twice a day at the Redondo station, and each chlorination procedure consists of two 20-min cycles.[1]

Samples were obtained with a hose attached to the discharge pipe opening (Figure 1). They were passed either through a suspended-matter high-volume sampler containing eight filters, each 30 cm in diameter with 0.5-μm pore-size openings arranged in parallel, or through a bypass line. Suspended matter from 1500- to 1700-L samples was also collected on filters having 0.5-μm pore-size openings. Aliquots of unfiltered water were collected for volatile analysis by purge and trap. Sodium sulfite (Na_2SO_3) was added to the purge-and-trap samples. Filtered water (210 L) was collected in polyvinyl fluoride-lined barrels (to preclude iron-chlorine interaction). Samples were acidified, treated with Na_2SO_3, and pumped through two 2.5 by 22.9-cm stainless steel columns containing a macroreticular resin (80-ml XAD-2) that adsorbs organic halogen. The columns were placed in series so that adsorption efficiency could be measured. Purge and trap samples were also taken from the barrel samples. One-liter samples of filtered water were extracted with methylene chloride for phenol analysis.

Sample Analysis Procedures

The analytical procedures were developed during a study of chlorinated natural waters for the Nuclear Regulatory Commission by Battelle Pacific Northwest Laboratories.[8] The specific procedures for this study included (1)

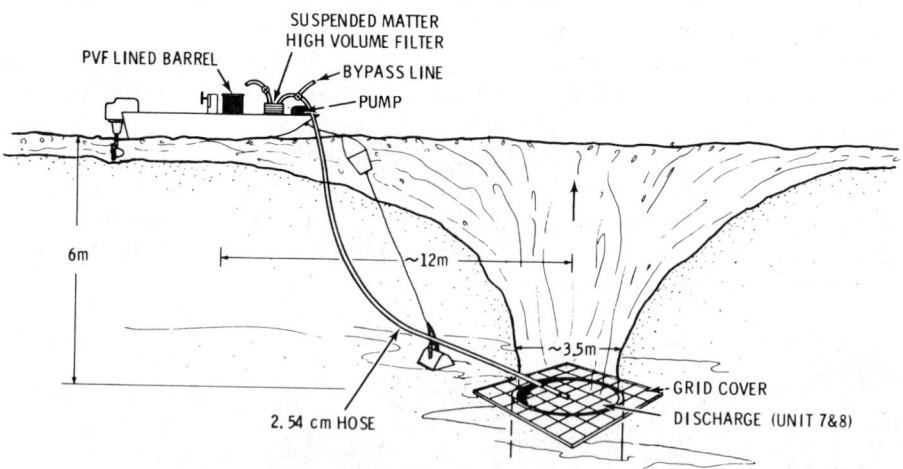

Figure 1. Position of boat and sample gear over the submerged discharge outlet of Redondo Generating Station units 7 and 8.

analyses of purge-and-trap samples for volatile organics; (2) extraction and analysis of XAD-2 columns, including derivatization and separation of phenols, analysis of extracts from organic halogen, and the determination of total organic bromine through microcoulometric titration; (3) extraction and analysis of suspended matter, including phenols, total organic halogen, and base-neutral components; and (4) analysis of methylene chloride extracts for phenols. The details of these procedures are described in another report.[9]

RESULTS AND DISCUSSION

Field Sample Findings

Total Residual Oxidants

Chlorination at the Redondo station was varied on the two consecutive sample days for comparative purposes. On June 23, 1981, the maximum value of total residual oxidants (TRO) measured was 0.15 mg/L; on June 24, the maximum value for TRO was 0.5 mg/L. Values were determined at the head of the generating station's discharge pipe. The discharge volume on June 23 and 24 during sampling was 29,521 L/s. Station operating levels were 850 MW (89% capacity) on June 23 and 880 MW (92% capacity) on June 24.

Volatile Organics

Several purge-and-trap samples were analyzed using gas chromatography/mass spectrometry (GC/MS). Results of the analyses of discharge samples and open-ocean samples are summarized in Table I. Bromoform was the principal volatile component formed in the cooling water because of chlorination. Concentrations at the discharge pipe opening ranged from 3 to 34 µg/L. Other

Table I. Volatile Organics: Summary of Purge-and-Trap Analyses of Redondo Discharge Samples and Open-Ocean Samples (values reported as micrograms per liter).

	Discharge			Open Ocean
	Chlorinated 6-23-81 (0.15 mg/L TRO)	Nonchlorinated 6-24-81	Chlorinated 6-24-81 (0.5 mg/L TRO)	6-25-81
Chloroform	0.4	0.1	0.5	nm[a]
Bromodichloromethane	<0.1	<0.1	<0.1	<0.1
Dibromochloromethane	0.3	<0.1	1.3	<0.1
Bromoform	6.7	<0.1	34.0	0.1
Trichloroethylene	0.1	<0.1	0.1	<0.1
Tetrachloroethylene	0.8	0.1	0.3	<0.1
Methylene chloride	0.5	0.5	0.5	0.6
1,1,1-Trichloroethane	0.1	0.2	0.7	<0.1

[a]Not measured.

haloforms were found in considerably smaller concentrations (Table I). Overall concentrations were in the same range as those determined by others.[8] Other volatile halocarbons found in the chlorinated discharge were trichloroethylene, tetrachloroethylene, trichloroethane, and methylene chloride. These compounds were also measured in nonchlorinated discharge and open-ocean samples. Our measurements demonstrate that volatile organics are not produced in chlorinated cooling water in significant amounts, which is consistent with previous findings.[8,10] Lower TRO levels (0.15 mg/L) produced less bormoform than higher (0.5 mg/L) levels. Additionally, sample analyses suggested that the potential for increased bromoform production with time is much greater at the higher oxidant level.[9] Samples taken at higher TRO levels appeared to continue to produce bromoform over several hours.

Halogenated Phenols

Levels of brominated phenols produced by chlorination in filtered water (Table II) were about 0.1 to 0.3 µg/L—somewhat lower than the 0.5 µg/L found at Millstone Nuclear Power Station in Connecticut.[10] Only 10% of the halogenated phenols were associated with suspended matter. Bisphenol A and monobromobisphenol A were found in the chlorinated discharge of June 23. These compounds were not the result of power station operation but may have been the result of the sampling procedure.

Base-Neutral Components

Basic and neutral U.S. Environmental Protection Agency (EPA) priority pollutants in the discharge of Redondo units 7 and 8 were determined using resin-filtered samples. Analysis of open-ocean samples was also required to distinguish between organic components being discharged because of plant operations and those components that are existing contaminants of either ocean water in the vicinity of the plant or of the sampling resin.

Table III lists priority pollutants found in the Redondo samples. Only one component was found in the discharge samples at a level approaching 0.1 µg/L that was not found in the ocean control. Butylbenzylphthalate was found at 0.06 µg/L in both chlorinated and nonchlorinated discharge samples. Although we do not know why this compound appears in King Harbor, we believe it to be a constituent added by other human activities.

Toxicity

The volatile halogenated and other organic chemicals detected in the Redondo station's receiving waters are summarized in Table IV. The values in Table IV are compared with EPA standard concentrations[11] for acute toxicity for saltwater organisms. The comparison does not consider the initial dilution

Table II. Phenols from Solvent Extraction of Filtered Water (concentrations in nanograms per liter)

Sample	Monobromo[a]	2,4-Dibromo	2,6-Dibromo	2,4,6-Tribromo	M = 390 Phenol[b]
Chlorinated 6/23/81	300	1.0	7.0	101	~1000
Nonchlorinated 6/24/81	nd[c]	nd	nd	23	
Chlorinated 6/24/81	nd	nd	2.0	153	
Ocean 6/25/81	nd	12.0	27.0	32	

[a] p-Bromophenol identified from standards; another monobromophenol was identified by mass spectral analysis.
[b] A monobrominated diphenol with mass 390, 392 was identified by mass spectrometry to be bromobisphenol A.
[c] Not detected.

Table III. Compounds Identified in XAD Extracts and Suspended Matter (component concentration expressed in micrograms per liter)

Compound	Retention Time (min)	Chlorinated 6-23-81		Chlorinated 6-24-81		Nonchlorinated 6-24-81		Ocean (Control) 6-25-81		Procedural Blank[b]	
		XAD-2	SM[a]	XAD-2	SM[a]	XAD-2	SM[a]	XAD-2	SM[a]	XAD-2	SM[a]
1,3-Dichlorobenzene	15.9			0.01							
1,4-Dichlorobenzene	16.0			0.04	<0.01	0.10		0.01			
1,2-Dichlorobenzene	17.0			0.19		0.68		0.04			
1,2,4-Trichlorobenzene	22.8			<0.01		0.01					
Naphthalene	22.9	<0.01		0.11	<0.01	0.01		0.01			
Dimethylphthalate	32.4	<0.01	<0.01	0.10	<0.01	<0.01	<0.01	0.02			
Diethylphthalate	37.0			0.10		0.01		0.06			
Phenanthrene	43.0	<0.01		0.02	<0.01	<0.01		<0.01			
Anthracene	43.9				<0.01						
Di-N-Butylphthalate	47.3	0.06	<0.01	0.50	0.02	0.03	<0.01	0.01	<0.01	0.03	0.10
Fluoranthrene	50.6			0.01	<0.01						
Pyrene	51.9			<0.01		0.03	<0.01				
Butylbenzylphthalate	57.8			0.06		0.06					
Benzo(a)anthracene	63.0				<0.01						

[a] Suspended matter.
[b] Reported on the basis of a 100-L water sample.

Table IV. Summary of Volatile Halogenated and Other Organic Chemicals in the Redondo Generating Station Receiving Waters, and a Comparison with Acute Toxicity Concentrations and Bioconcentration Factors (values in micrograms per liter).[a,b]

Compound	Background Concentration in Open Ocean	Background Concentration in King Harbor	Maximum Concentration in Discharge During Chlorination[a]	Saltwater Acute Aquatic Toxicity Concentrations (LC_{50})[a,b]	Weighted Average Bioconcentration Factor[b,c]
Bromoform	0.1	0.1	34.0	24,400	ND[c]
Tetrachloroethylene	0.1	0.1	0.8	10,200	30.6
Dibromochloromethane	0.1	0.1	0.1	2,000	10.6
Bromodichloromethane	0.1	0.1	1.3	ND	ND
Chloroform	ND	0.1	0.5	81,500	3.75
Dichlorobenzene	0.04	0.68	0.19	1,970	55.6
Phthalates	None	None	0.50	2,944	130.0
2,4,6-Tribromophenol	0.003	0.002	0.015	970	1.4

[a]EPA criteria concentrations are 24-h/d exposure values, whereas Redondo chlorination cycle values are peak values for one of four 20-min cycles per day.
[b]Based on data in Reference 11.
[c]Weighted average biocentration factors is for the edible portion of all aquatic organisms consumed by Americans.

factor as the cooling water effluent disperses from the discharge pipe outlet; nor does it consider that the peak discharge sample values represent a maximum value that would only be seen for a few minutes per day at most. Chlorination is not continuous and the four 20-min-per-day chlorination cycles do not result in a constant 20-min high output of chlorine.[1]

Risk of harmful environmental effects from halogenated and other organic compounds released or found in seawater from the chlorination of cooling water (as evidenced in Table IV) appears to be very low. Maximum concentration peaks are on average < 4 µg/L. Bromoform is the major organic product of chlorination, having a peak value of 34 µg/L. Comparisons with LC_{50} numbers for ocean organisms show that the acute toxicity to fish from these concentrations is insignificant. Human risk from chlorination products at a generating station such as Redondo would be attributed primarily to fish consumption. The only other pathways for human exposure would be from direct ingestion of seawater during the chlorination cycle, which is not a plausible assumption, and exposure to volatile compounds by transmission through the atmosphere. The latter pathway assumes that a boater or fisherman would be in close proximity to the submerged discharge structure during the few minutes of the daily chlorination cycle (a possibility that would lead to a brief and negligible exposure at the Redondo station). Any substance that enters the air would be quickly dispersed and diluted, and the potential for exposure to human populations would be insignificant.

Fish uptake and bioaccumulation of the Redondo chlorination products are low at the concentrations observed. For example, the bioaccumulation factors for chloroform and 1,2-dichlorobenzene are 3.75 and 55.6, respectively.[11] Bioconcentration factors, as presented in Table IV, range from 1.4 to 130. Table V lists Redondo Generating Station chlorination product concentrations

Table V. Comparison of Selected Chlorination Product Concentrations from Redondo Generating Station with EPA Human Health Ambient Water Quality Criteria Based on Consumption of Fish and Shellfish

Compounds	Maximum Concentration in Discharge During Chlorination (µg/L)	Ambient Water Quality Criteria for the Protection of Human Health
Bromoform	34.0	15.7[a] µg/L
Tetrachloroethylene	0.8	8.85 µg/L
Trichloroethylene	0.1	>27.0 µg/L
Chloroform	0.5	15.7 µg/L
Dichlorobenzene	0.19	2.6 mg/L
Phthalates		
Dimethyl	0.10	2.9 g/L
Diethyl	0.10	1.8 g/L
Dibutyl	0.50	154.0 mg/L
Diethylhexyl	0.10	50.0 mg/L

[a]Based on chloroform toxicity.

in comparison with EPA human health ambient water quality criteria, based on consumption of fish and shellfish only.[11] These data represent potential harm to humans exposed to these compounds generated by a Redondo chlorination cycle based on the assumptions that the generating station is discharging continuously at these levels with no further dilution in the ocean discharge zone, and that the fish are constantly exposed to these levels. These assumptions are not correct; therefore, the values are very conservative. The only high concentration of potential concern found at Redondo, according to Table V, is bromoform. There is no direct EPA bromoform criteria value with which to make this comparison—only the chloroform value. Given the conservative assumptions used, however, it is clear that the potential for exposure to halogenated organics from the consumption of seafood in the Redondo area is low, and human health risk is probably negligible.

CONCLUSIONS

A detailed analysis of the organic components in the discharge revealed no halogenated products other than haloforms and halogenated phenols from the addition of chlorine to the cooling water. Phthalates and chlorobenzenes were the only EPA priority pollutants found in any significant concentrations, and these are chemicals that are found in almost all natural water samples, including the open-ocean control. The results of these investigations indicate that the nonchlorinated cooling water discharged from one coastal power plant studied is little different from the surrounding water in terms of organic content; however, concentrations of bromoform and brominated phenols in the parts-per-billion range can be produced from the chlorination.

Redondo chlorination by-product concentrations were compared with EPA water quality criteria, assessing acute and chronic toxicity values for organic compounds found in discharge waters during the chlorination cycle. A comparison of these values with LC_{50} numbers for aquatic organisms shows that acute toxicity to fish is insignificant. Further, the risk to human health appears to be negligible, based on EPA water quality criteria for the consumption of fish and shellfish and the intermittent periods of chlorination (without consideration of the initial dilution factors in the receiving waters).

ACKNOWLEDGMENTS

We thank Dr. Philip Dorn and Colleen P. Doyle, who arranged the analysis program and coordinated the fieldwork at the Redondo Generating Station; the staff and students of Occidental College who helped in the field and at the station during the sampling; and Richard L. Wilson, Duane A. Neitzel, and Marty L. O'Malley, who represented the Battelle research staff through the completion of the analysis work.

REFERENCES

1. Grove, R. S. "Dispersion of Chlorine at Seven Southern California Coastal Generating Stations," in *Water Chlorination: Environmental Impact and Health Effects, Vol. 4*, R L. Jolley, W. A. Brungs, J. A. Cotruvo, R. B. Cumming, J. S. Mattice, and V. A. Jacobs, Eds. (Ann Arbor, MI: Ann Arbor Science Publishers, Inc., 1983), pp. 333-346.
2. "San Onofre Nuclear Generating Station, Unit 1, Environmental Technical Specifications, Annual Operating Report, Oceanographic Data Analysis—1977," Vol. 1 (Rosemead, CA: Southern California Edison Company, March 1978).
3. Hose, J. E., T. D. King, K. E. Zerba, R. J. Stoffel, J. S. Stephens, Jr., and J. A. Dickinson. "Does Avoidance of Chlorinated Seawater Protect Fish Against Toxicity? Laboratory and Field Observations," in *Water Chlorination: Environmental Impact and Health Effects, Vol. 4*, R. L. Jolley, W. A. Brungs, J. A. Cotruvo, R. B. Cumming, J. S. Mattice, and V. A. Jacobs, Eds. (Ann Arbor, MI: Ann Arbor Science Publishers, Inc., 1983), pp. 967-982.
4. Hose, J. E., R. J. Stoffel, and K. E. Zerba. "Behavioral Responses of Selected Marine Fishes to Chlorinated Seawater," *Mar. Environ. Res.* 9:37-59 (1983).
5. Jolley, R. L. "Identification of Organic Halogen Products," *Chesapeake Sci.* 18:122-125 (1977).
6. Carpenter, J. H., and C. A. Smith. "Reactions in Chlorinated Seawater," in *Water Chlorination: Environmental Impact and Health Effects, Vol. 2*, R. L. Jolley, H. Gorchev, and D. H. Hamilton, Jr., Eds. (Ann Arbor, MI: Ann Arbor Science Publishers, Inc., 1978), pp. 195-208.
7. Helz, G. R., R. Sugan, and R. Y. Hsu. "Chlorine Degradation and Halocarbon Production in Estuarine Water," in *Water Chlorination: Environmental Impact and Health Effects, Vol. 2*, R. L. Jolley, H. Gorchev, and D. H. Hamilton, Jr., Eds. (Ann Arbor MI: Ann Arbor Science Publishers, Inc., 1978), pp. 209-222.
8. Bean, R. M., D. C. Mann, and R. G. Riley. "Analysis of Organohalogen Products, from Chlorination of Natural Waters Under Simulated Biofouling Conditions," NUREG/CR-1301, (Washington, DC: U.S. Nuclear Regulatory Commission, 1980).
9. Bean, R. M., R. L. Wilson, D. A. Neitzel, and M. A. O'Malley. "Analysis of Organohalogen and Other Organic Chemicals in Cooling Waters Discharged from Redondo Generating Stations," a research report prepared for Southern California Edison Company (Richland, WA: Battelle Pacific Northwest Laboratories, 1982).
10. Bean, R. M., D. C. Mann, and D. A. Neitzel. "Organohalogens in Chlorinated Cooling Waters Discharged from Nuclear Power Stations," in *Water Chlorination: Environmental Impact and Health Effects, Vol. 4*, R. L. Jolley, W. A. Brungs, J. A. Cotruvo, R. B. Cumming, J. S. Mattice, and V. A. Jacobs, Eds. (Ann Arbor, MI: Ann Arbor Science Publishers, Inc., 1983), pp. 383-390.
11. U.S. Environmental Protection Agency. "Ambient Water Quality Criteria Documents for: Halomethanes, Chlorinated Phenols, Phenol, Trichloroethylene, Tetrachloroethylene, Chloroform, Chlorinated Benzenes, Dichlorobenzenes, and Phthalates" (Washington, DC: National Technical Information Service, 1980).

CHAPTER **106**

Results of Analyzing Simulated Cooling Tower Blowdown for Organic Priority Pollutants

James Rios

A laboratory study was carried out to assess the effects of discharging chlorine-bearing cooling tower blowdown water into the Susquehanna River from two steam electric units. To assess these effects, it was necessary to determine the residual concentrations of chloroorganic compounds in the cooling tower blowdown that would result from chlorination of the condenser cooling water. Of particular interest were trihalomethane (THMs) concentrations, because the EPA National Interim Primary Drinking Water Regulations 40 CFR 141, amended by 44 FR 68641 (29 November 1979), limited the concentration of total THMs to 100 μg/L in public water systems.

Bechtel Power Corporation, consulting to the utility, recommended that laboratory tests be conducted on concentrated Susquehanna River water to simulate the cooling water chlorination and subsequent aeration that takes place in the cooling tower. Because the results were needed for a scheduled public hearing, sufficient time was not available to design and manufacture a recirculating cooling system with a miniature cooling tower. Therefore, laboratory apparatus was used to simulate actual plant operating conditions as closely as possible.

The Battelle Columbus Laboratories were selected to concentrate, chlorinate, and analyze the samples according to Bechtel's instructions. Susquehanna River water samples were taken by Icthyological Associates according to Battelle's directions and analyzed by gas chromatograph — mass spectrometer (GC-MS) for volatile (purgeable) and semivolatile organic compounds. Strict quality assurance procedures were observed in taking, transporting, handling and analyzing samples, because even minute quantities of contaminant could render the results meaningless.

Battelle's GC-MS was computerized to search semiautomatically for 111 organic compounds designated as priority pollutants by the EPA, and all of the chloroorganic compounds of concern are included.

This chapter includes a description of the laboratory apparatus used to simulate plant conditions, analytical methods, quality assurance, analytical results, discussion of results, and conclusions.

EXPERIMENTAL PROCEDURE

Approach

Samples were taken monthly for 5 months from a location adjacent to the power station intake structure on the North Branch of the Susquehanna River. The samples were shipped to Battelle's laboratory, where sample aliquots were concentrated ~3.8 times and aliquots of the concentrate were chlorinated to produce a 0.3 mg/L free chlorine residual (FCR). This FCR had to be measurable 2 min after chlorination (chlorine demand).

Concentrated chlorinated river water samples were aerated for 5 and 30 min. Analyses for volatile and semivolatile priority pollutants were carried out on the samples with and without aeration. Aliquots of the samples were dechlorinated immediately after the 5- and 30-min aeration period to determine the effect on formation of chloroorganic compounds.

To prevent precipitation of calcium carbonate while concentrating the river water, the alkalinity of the samples was reduced to the range of 15 to 25 mg/L by sulfuric acid addition. This is somewhat different from actual plant operation where the sulfuric acid is injected into the concentrated circulating water. The calcium carbonate alkalinity introduced with the makeup water is diluted with circulating water and converted to the more soluble compound of calcium sulfate by reaction with the acid, thus minimizing the scaling potential on the condenser tubes.

Aliquots were taken of each monthly sample to form a set of samples that was analyzed for 111 priority organic pollutants. Each of the five sample sets, referred to here as sets A through E, consisted of:

- Unconcentrated, untreated
- Unconcentrated, chlorinated
- Concentrated, untreated
- Concentrated, chlorinated
- Concentrated, chlorinated, aerated 5 min
- Concentrated, chlorinated, aerated 30 min
- Concentrated, chlorinated, aerated 5 min, dechlorinated
- Concentrated, chlorinated, aerated 30 min, dechlorinated

Sampling and Quality Assurance

Battelle Columbus Laboratories prepared sample containers by washing with soap and water, rinsing with distilled water, and heating to 450°C for 8 h to remove any organic contamination. Sample containers consisted of one 5-gal glass jug, four 40-mL vials, and a 1-gal jug. The cap for the 5-gal jug was Teflon®-lined and rinsed with water, methanol, and methylene chloride prior to shipment. The Teflon-lined septa lids of the 40-mL screw cap vials were rinsed only with purged distilled water to prevent solvent contamination of the samples for purgeable organic analysis. The containers were shipped to Ichthyological Associates for collection of samples.

A 5-gal sample taken adjacent to the station intake structure was used for semivolatile organic compound analysis, with aliquots concentrated prior to analysis. The two 40-mL vials for analysis of volatile organics were filled to overflowing, capped tightly, and packed in ice in the inverted position. No preservatives were added to the samples at the time of collection. For quality assurance, two separate 40-mL vials containing reagent water were opened to the atmosphere during sampling at the site to serve as a field blank for volatile organics. Similarly, a 1-gal jug containing reagent water was opened to serve as a blank for semivolatile organic analysis.

The collected samples were packed with leak-proof ice packs and packing material and shipped by bus to the laboratory in an insulated Battelle shipping container.

Laboratory Concentrating-Aerating Apparatus

The laboratory apparatus used to concentrate river water samples is shown schematically in Figure 1. Samples were evaporated in a 1-gal jug into which four extra-coarse, sintered glass sparging tubes (Pyrex No. 12 EC) were inserted to introduce air in the bottom for bubble dispersion. The evaporation jug was placed in a water bath set to be maintained at 98.6 to 100.4°F.

The intent was to concentrate the sample similarly to plant condenser cooling water, which would evaporate while flowing countercurrently to air in the cooling tower. The warm condenser cooling water was expected to be at approximately 100°F and the countercurrent air flow would average 5 cfm/gal. Because the air flow required to simulate station cooling tower operation was much larger than available in the laboratory, it was necessary to concentrate the river water samples in batches, beginning with 3-L aliquots. Water volume was measured in graduated cylinders before and after concentrating to keep a running record of the approximate cycles of concentration. Each 3-L batch was evaporated to approximately 750 mL as visualized by a calibrated mark on the evaporation jug.

The concentrated sample batches were progressively composited to form a single ~1-gal concentrated sample, which was then stored at ~4°C. Unconcentrated river water was also stored at 4°C. The actual cycles of concentration attained were determined by analyses of chloride ion concentration before and after evaporation.

To prevent organic contamination of the samples, laboratory compressed air was passed through a molecular-sieve trap after first passing through granular activated charcoal and Drierite. The two Drierite (calcium sulfate) cartridges removed water vapor that could clog the cold molecular sieve. Two granular activated charcoal columns were used in series to remove organics from the air. The charcoal was reactivated at 400°C for 8 h before reuse.

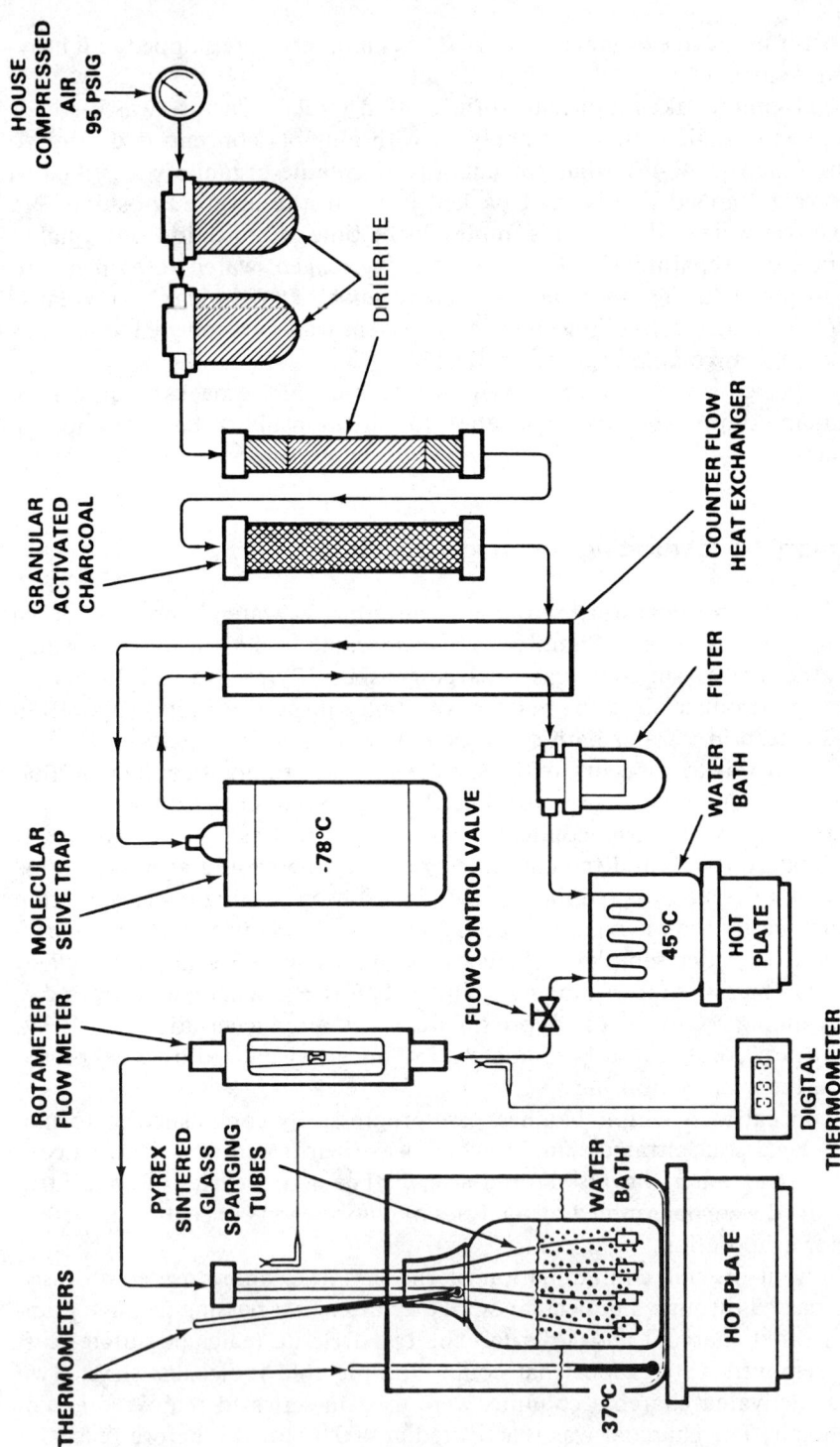

Figure 1. Schematic diagram of the concentration-aeration apparatus.

Removal of the last traces of organic contaminants from the air was achieved in a 22-cm-long molecular sieve column having an inside diameter of 2.2 cm and containing 800 g of 4 and 5 A molecular sieve (Linde, spherical 4 to 8 mesh). The sieve temperature was held at $-78°C$ by inserting the column into a Dewar flask containing dry ice pellets in methanol. To minimize dry ice consumption and reduce the reheating requirement of the exiting air from the cold molecular sieve, a counterflow heat exchanger was installed upstream of the molecular sieve.

To protect the rotameter flow gage and prevent clogging of the sintered glass sparging tubes, a sintered bronze filter element was located after the heat exchanger to remove any particulate material originating from the Drierite, activated charcoal, or molecular sieve.

Because the molecular sieve was held at $-78°C$, the temperature of the air passing through was reduced, which in turn excessively lowered the sample temperature.

To obtain better control of the temperature of the sample being evaporated, it was necessary to heat the organic-free air by means of a coil immersed in a warm water bath. The water bath temperature was set to provide a filtered air temperature of 98.6 to 102.2°F as measured by a chromel-alumel thermocouple located at the sparging tube inlet. Because of the cooling effect of evaporation on the sample, the spent sparging air temperature was lowered to a range of 95.0 to 98.6°F, as monitored at the mouth of the evaporation vessel. A flow control needle valve and rotameter were located downstream of the filtered-air water bath. Air flow was set to a constant of 2.2 cfm.

Water Analysis

Alkalinity, chloride ion concentration, pH, conductivity, and 2-min chlorine demand were measured for all unconcentrated and concentrated river water samples. Chlorine demand was defined as the difference between the amount of FCR after 2 min and the amount of standardized chlorine solution added. All chlorine demand determinations were performed in duplicate.

GC-MS Analysis of Volatile (Purgeable) Organics and Semivolatile (Extractable) Organics

The purgeable organic compounds were analyzed by the purge-trap desorption (PTD) method using a Tekmar Model LCS-1 liquid-sample concentrator fitted with a 5-mL sample container and a 30-cm adsorbent trap containing 15-cm Tenax and 8-cm silica gel; a gas chromatograph equipped with a 2-m by 2-mm-ID glass column packed with 60/80 mesh Carbopak B coated with 1% SP-1000; and a Finnigan Model 4000 quadrupole mass spectrometer controlled by an INCOS data system. The procedure used for analysis was essentially the same as that prescribed by EPA for purgeable pollutants.[1]

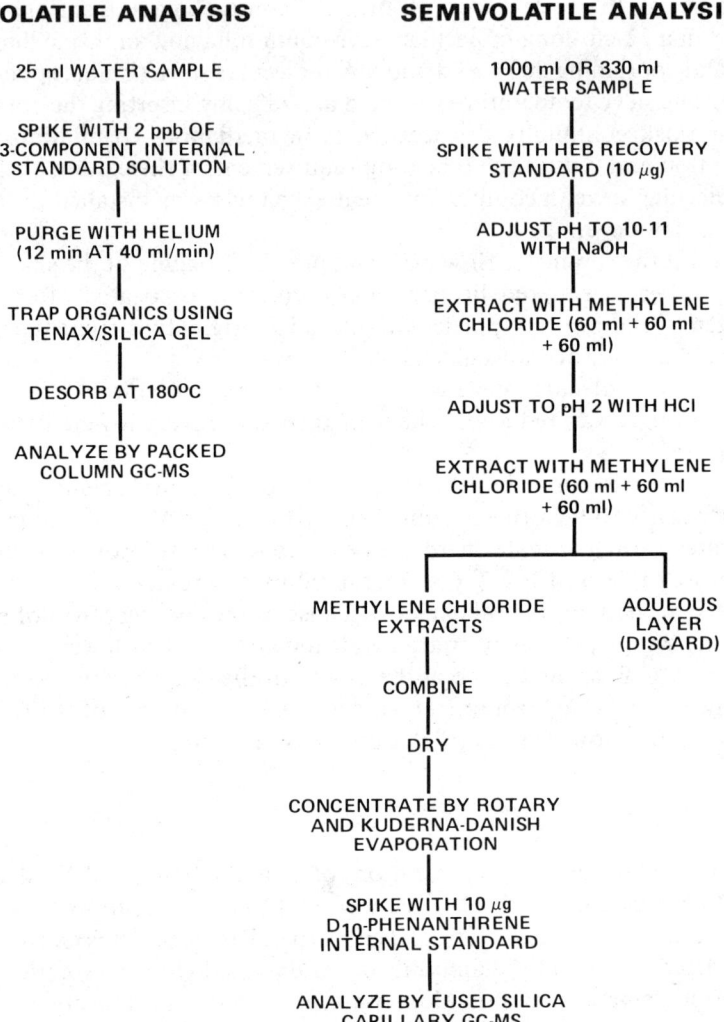

Figure 2. Scheme for analysis of water samples.

The sample extracts containing the semivolatile organic compounds were analyzed using a Finnigan Model 9610 gas chromatograph equipped with a 30-m × 0.2-mm ID fused silica capillary column coated with SE-52 methyl silicone; a Finnigan Model 4000 quadrupole mass spectrometer; and a Hewlett-Packard Model 7672A automatic sample injection system. The extraction and semivolatile analysis method used was that prescribed by EPA for base/neutrals, acids, and pesticides.[2]

A scheme for the analysis of the water samples for priority organic pollutants is shown in Figure 2.

Quality Assurance

Quality assurance began with preparation of the sampling containers as described previously. The cooler containing the 5-gal water sample, purgeable sample vials, and field blanks kept all samples cool during shipment and receipt at the laboratory. In Battelle's laboratory the samples were kept cold until analysis. The quality control plan involved performing the following analyses in addition to those performed on the actual samples:

- Method blanks for purgeables
- Fixed blanks for purgeables
- Field blanks for purgeables
- Spiked river water for purgeables
- Field blanks for semivolatiles
- Spiked reagent water for semivolatiles
- Calibration standards, analyzed every fourth sample to ensure accurate quantification of sample constitutents

For volatile organics analysis, the mass spectrometer was tuned using EPA Method 624[1] and for semivolatile organics it was tuned using EPA Method 625.[2]

DISCUSSION OF RESULTS

Water Analysis

The results of water analyses for chlorine demand, chloride, alkalinity, pH, and conductivity of unconcentrated and concentrated river water sample sets A through E are presented in Table I. There is bound to be some effect on the parameters being measured because of the time delay between sampling and analysis.

Apparently, the pH of the river water sample was affected, possibly by pickup of carbon dioxide from the air, because the river water pH normally was above 6.8 and the pH for sample sets B through E was $\leq$ 6.2 (Table I). The samples were cooled in transit to minimize the possible effect of the delay time on the organic analysis results.

The chlorine demand tabulated in Table I is the difference between the FCR concentration 2 min after chlorine addition and the original concentration of chlorine added to the sample. Two minutes was selected because in the plant design it would take approximately that for the condenser cooling water to travel from the point of chlorine injection to the outlet of the condenser, which is the location at which a measurable FCR is desired. The river water chlorine demand range was between 0.57 and 0.94 mg/L and, when concentrated approximately 3.8 times, the range was between 1.08 and 2.63 mg/L.

Table I. Results of Water Analysis of All Five Sample Sets

	Set A[a]		Set B		Set C		Set D		Set E	
	Unconc.	Conc.	Unconc.	Conc.	Unconc.	Conc.	Unconc.	Conc.	Unconc.	Conc.
Chlorine demand,[b] mg/L	0.57	1.08	0.66	2.30	0.92	2.63	0.94	2.61	0.57	2.06
Alkalinity, mg/L	51	153	48	65	49	74	57	35	58	5
Chloride ion conc., mg/L	18	66	12	44	12	46	16	65	24	82
pH	6.9	7.7	6.2	6.4	6.2	6.4	6.1	5.6	6.0	5.1
Conductivity, μmho/cm	161	728	171	590	183	730	252	915	240	904

[a]Unconcentrated is North Branch Susquehanna River water and concentrated is river water concentrated approximately 3.8 times by heating and aerating.
[b]Chlorine demand is the difference between the FCR concentration measured 2 min after chlorine addition and the original concentration of chlorine added.

The chlorine demand of the concentrated samples in sets B and E increased in approximately direct proportion to the number of times the water was concentrated, 3.6 and 3.4 cycles, respectively. For samples A, C, and D, the chlorine demand increased to 50 to 70% of the product of cycles of concentration and river water demand, which indicates that some of the chlorine demand may have been caused by volatile organics that were stripped off during the concentrating process. These results indicate that it is possible to estimate chlorine injection dosages required to satisfy chlorine demand in recirculating cooling water systems incorporating cooling towers if the chlorine demand of the cooling tower makeup water is known. A conservative design estimate can be made by taking the product of the design cycles of concentration and the makeup water chlorine demand. Additional chlorine demand can develop in cooling systems that are not chlorinated with sufficient frequency to restrict biolife growth.

To simulate the plant condenser tube scale control system design by acid injection, the alkalinity of samples B through D was reduced to a concentration range of 15 to 25 mg/L prior to concentrating. This explains why the alkalinity and pH of the unconcentrated and concentrated samples are fairly close. Acid was overfed to sample E before concentrating and the effects show up as lowered alkalinity and pH. No acid was fed to sample A before concentrating; therefore, the alkalinity increased almost in direct proportion to the cycles of concentration.

The only reason for analyzing for chloride ion concentration was to determine the actual cycles of concentration attained. Analyses for chloride were carried out on the concentrated samples before chlorine was added. Conductivity measurements confirm that they can be used as an approximation of cycles of concentration. On samples B through E, the increase in conductivity was within $\pm 10\%$ of the product of river water conductivity and the cycles of concentration as determined by chloride ion concentration. On sample A, the conductivity measurements, if used to determine cycles of concentration, would have led to an 18% higher value than that determined by chloride.

Purgeable (Volatile) Organics

It would have been desirable to have extended the monthly tests over a 1-year period to determine the effects of monthly and seasonal variations. However, this was not possible because the sampling and analysis had to be completed before scheduled public hearings.

The results of the analysis for 28 volatile priority pollutants in sample set B are presented in Table II. The results of seven of these sample analyses, identified at the bottom of the table, are reported in columns 1 through 7. The analysis results for the eighth sample, concentrated river water, chlorinated, aerated for 30 min, and dechlorinated, were not included in the table to allow

Table II. Results of GC-MS Analysis of Simulated Cooling Tower Blowdown Samples for Purgeable Priority Pollutants (Sample Set B)

Volatile Compounds	Concentration, μg/L (ppb) Sample No.[a]			
	1	2	3	4
Acrolein	BDL[b]	BDL	BDL	BDL
Acrylonitrile	BDL	BDL	BDL	BDL
Benzene	BDL	0.1	0.05	0.08
Bromoform	BDL	BDL	BDL	BDL
Carbon tetrachloride	BDL	BDL	BDL	BDL
Chlorobenzene	BDL	BDL	BDL	BDL
Chlorodibromomethane	BDL	0.2	BDL	1.0
Chloroethane	BDL	BDL	BDL	BDL
2-Chloroethylvinyl ether	BDL	BDL	BDL	BDL
Chloroform	BDL	17.9	2.7	41.5
Dichlorobromomethane	BDL	2.6	BDL	10.4
1,1-Dichloroethane	BDL	BDL	BDL	BDL
1,2-Dichloroethane	BDL	BDL	BDL	BDL
1,1-Dichloroethylene	BDL	BDL	BDL	BDL
1,2-Dichloropropane	BDL	BDL	BDL	BDL
1,3-Dichloropropylene	BDL	BDL	BDL	BDL
Ethylbenzene	BDL	BDL	BDL	BDL
Methyl bromide	BDL	BDL	BDL	BDL
Methyl chloride	BDL	BDL	BDL	BDL
Methylene chloride	BDL	6.2	1.1	1.2
1,1,2,2-Tetrachloroethane	BDL	BDL	BDL	BDL
Tetrachloroethylene	BDL	0.1	BDL	BDL
Toluene	BDL	0.1	BDL	BDL
1,2-Trans-Dichloroethylene	BDL	BDL	BDL	BDL
1,1,1-Trichloroethane	0.3	1.8	1.6	1.1
1,1,2-Trichloroethane	BDL	BDL	BDL	BDL
Trichloroethylene	BDL	0.1	0.07	0.2
Vinyl chloride	BDL	BDL	BDL	BDL

room for Column 8, which contains the estimated detection limits established for this set of samples. However, a separate table is included to provide the results of dechlorinating the concentrated, chlorinated samples that were aerated for 5 and 30 min for sets A through E.

The results of the analysis for the 28 volatile priority pollutants for untreated river water are presented in Column 1 of Table II and are fairly typical of the results obtained for river water on sample sets A through E. The North Branch of the Susquehanna River appears to be fairly free of volatile priority pollutants, with only 1,1,1-trichloroethane appearing slightly above the detectable limit in sample 1 of set B. Chloroform was barely detectable in two other river water samples, with concentrations of 0.1 μg/L and 0.6 μg/L.

Table II(Cont'd). Results of GC-MS Analysis of Simulated Cooling Tower Blowdown Samples for Purgeable Priority Pollutants (Sample Set B)

Volatile Compounds	Concentration, µg/L (ppb) Sample No.[a]			Estimated Detection Limit[c]
	5	6	7	
Acrolein	BDL[b]	BDL	BDL	6
Acrylonitrile	BDL	BDL	BDL	0.6
Benzene	BDL	BDL	0.06	0.05
Bromoform	BDL	BDL	BDL	0.2
Carbon tetrachloride	BDL	BDL	BDL	0.2
Chlorobenzene	BDL	BDL	BDL	0.06
Chlorodibromomethane	0.2	0.1	0.1	0.05
Chloroethane	BDL	BDL	BDL	0.2
2-Chloroethylvinyl ether	BDL	BDL	BDL	0.2
Chloroform	16.7	7.2	12.8	0.02
Dichlorobromomethane	2.0	0.8	0.7	0.2
1,1-Dichloroethane	BDL	BDL	BDL	0.05
1,2-Dichloroethane	BDL	BDL	BDL	0.5
1,1-Dichloroethylene	BDL	BDL	BDL	0.06
1,2-Dichloropropane	BDL	BDL	BDL	4
1,3-Dichloropropylene	BDL	BDL	BDL	0.1
Ethylbenzene	BDL	BDL	BDL	0.1
Methyl bromide	BDL	BDL	BDL	0.2
Methyl chloride	BDL	BDL	BDL	6
Methylene chloride	BDL	BDL	BDL	0.05
1,1,2,2-Tetrachloroethane	BDL	BDL	BDL	0.1
Tetrachloroethylene	BDL	BDL	BDL	0.1
Toluene	BDL	BDL	BDL	0.1
1,2-Trans-Dichloroethylene	BDL	BDL	BDL	0.05
1,1,1-Trichlorethane	BDL	BDL	BDL	0.1
1,1,2-Trichloroethane	BDL	BDL	BDL	0.2
Trichloroethylene	BDL	BDL	BDL	0.05
Vinyl chloride	BDL	BDL	BDL	5

[a]1. Unconcentrated, untreated river water (as received)
2. Unconcentrated, chlorinated river water
3. Concentrated river water
4. Concentrated river water, chlorinated.
5. Concentrated river water, chlorinated, 5-min aeration
6. Concentrated river water, chlorinated, 30-min aeration
7. Concentrated river water, chlorinated, 5-min aeration, dechlorinated
[b]BDL = Below detection limit.
[c]The estimated detection limits are determined from analysis of calibration standards.

Table III contains the results of the analyses for all sample sets for the five volatile priority pollutants (chloroorganic compounds) that were detected in the concentrated river water before and after chlorination.

Chloroform, methylene chloride, and 1,1,1-trichloroethane were detected in the concentrated river water samples of all sets, which indicates that they may have been present just at the detectable limit so they would appear upon

Table III. Results of Analysis for Volatile Chloroorganics in Sample Sets A – E

Volatile Compound	Concentration, μg/L (ppb) Sample Set				
	A	B	C	D	E
Conc. river water					
Chloroform	C[a]	2.7	1.8	2.6	0.1
Chlorodibromomethane	BDL[b]	BDL	BDL	BDL	BDL
Dichlorobromomethane	BDL	BDL	BDL	BDL	BDL
Methylene Chloride	1.5	1.1	1.7	1.4	0.3
1,1,1-Trichloroethane	0.7	1.6	0.9	2.2	0.2
Chlorinated, conc. river water					
Chloroform	17.0	41.5	18.0	35.0	4.0
Chlorodibromomethane	0.4	1.0	0.2	1.0	1.0
Dichlorobromomethane	3.4	10.4	3.3	9.4	4.0
Methylene Chloride	2.4	1.2	3.0	0.7	2.0
1,1,1-Trichloroethane	0.4	1.1	1.3	2.5	4.0
Chlorinated, conc., aereated 5 min					
Chloroform	4.8	16.7	8.8	13.2	3.0
Chlorodibromomethane	BDL	0.2	BDL	0.2	BDL
Dichlorobromomethane	BDL	2.0	0.7	1.8	BDL
Methylene Chloride	0.5	BDL	BDL	BDL	0.2
1,1,1-Trichloroethane	0.1	BDL	BDL	BDL	0.2
Chlorinated, conc., aerated 30 min					
Chloroform	2.0	7.2	4.8	6.4	3.0
Chlorodibromomethane	BDL	0.2	BDL	0.2	BDL
Dichlorobromomethane	BDL	0.8	0.1	0.7	BDL
Methylene Chloride	0.7	BDL	BDL	BDL	0.7
1,1,1-Trichloroethane	BDL	BDL	BDL	BDL	0.3

[a]C = Contaminated accidentally.
[b]BDL = Below detectable limit; for detection limits, see Table II.

concentration (see Table III). The remaining volatile priority pollutants were not present or were below detectable limits in the concentrated river water samples.

The volatile organic compounds formed upon chlorination of the unconcentrated and concentrated river water samples were chloroform, dichlorobromomethane, chlorodibromomethane (all three are THMs), methylene chloride, and 1,1,1-trichloroethane. Table III shows the concentrations found in sample sets A through E. Constituent concentration ranges in the concentrated chlorinated samples were 4 through 41.5 μg/L for chloroform, 3.3 through 10.4 μg/L for dichlorobromomethane, and 0.2 through 1.0 μg/L for chlorodibromomethane. Mixed results were obtained on concentrations of methylene chloride and 1,1,1-trichloroethane. Results indicate that these compounds were formed when the unconcentrated river water was chlorinated, but the concentrating technique (evaporation and aeration) may have removed some organic precursors to these compounds before chlorination.

Five min of aeration removed from 56 to 77% of the THMs formed during chlorination of samples A through E (Table III). Thirty min of aeration on samples A through D (concentrated, chlorinated river water) resulted in an overall removal of 73 to 86% of the THMs. Sample E of Table III started out with the lowest THM concentration (9 μg/L) and 66% of this was stripped away in 5 min of aeration, but the chloroform portion of the THMs was barely affected (25% removal). An additional 25 min of aeration did not remove any additional chloroform from sample E. Since all samples contained chloroform even after 30 min of aeration, it appears that the other two THMs, chlorodibromomethane and dichlorobromomethane, are more volatile than chloroform and that it becomes more difficult to remove the last traces of chloroform. The aeration time may have to be extended substantially to reduce the chloroform residual below 3 μg/L. Methylene chloride and 1,1,1-trichloroethane were easily stripped away with 5 min of aeration.

Table IV. Results of Dechlorination on Volatile Chloroorganics in Sample Sets A – E

Volatile Compound	Concentration, μg/L (ppb) Sample Set				
	A	B	C	D	E
Chlorinated, conc., aerated 5 min					
Chloroform	4.8	16.7	8.8	13.2	3.0
Chlorodibromomethane	BDL[a]	0.2	BDL	0.2	BDL
Dichlorobromomethane	BDL	2.0	0.7	1.8	BDL
Chlorinated, conc., aerated 5 min, dechlorinated					
Chloroform	4.3	12.8	NA[b]	4.2	3.0
Chlorodibromomethane	BDL	0.1	NA	0.1	BDL
Dichlorobromomethane	BDL	0.7	NA	0.4	BDL
Chlorinated, conc., aerated 30 min					
Chloroform	2.0	7.2	4.8	6.4	3.0
Chlorodibromomethane	BDL	0.2	BDL	0.2	BDL
Dichlorobromomethane	BDL	0.8	0.1	0.7	BDL
Chlorinated, conc., aerated 30 min, dechlorinated					
Chloroform	1.6	6.7	3.0	NA	3.0
Chlorodibromomethane	BDL	0.07	BDL	NA	BDL
Dichlorobromomethane	BDL	0.4	BDL	NA	BDL

[a]BDL = Below detectable limit.
[b]NA = Not analyzed.

Samples were dechlorinated by the addition of a stoichiometric amount of freshly prepared sodium bisulfite solution sufficient for the removal of a chlorine residual of 0.3 mg/L immediately following the 5- and 30-min aeration of the concentrated chlorination samples. Table IV presents the results of dechlorination on the concentration of THMs for sample sets A through E.

There was a measurable reduction in the concentration of chloroform and other THMs when dechlorinating immediately after the 5- and 30-min aeration of concentrated chlorinated river water samples. The reduction was substantially larger for those samples where higher concentrations of chloroform were formed during chlorination. In column B of Table IV, the chloroform reduction by dechlorination of the 5-min aerated sample was approximately 23% and for sample D it was 68%. This indicates that formation of chloroform and other THMs continues between the time the aeration is discontinued and when the samples are finally analyzed.

Semivolatile Priority Pollutants

Table V lists the 83 semivolatile priority pollutants looked for in all samples of sets A through E. No semivolatile priority pollutants were found in the river water samples either before or after concentrating or after chlorinating any of the samples. The estimated detection limits for these semivolatile priority pollutants, as established for sample set B, are presented in Table V. Estimated detection limits vary with each sample set, depending on variables such as sample complexity and sensitivity of the mass spectrometer, but the limits presented in Table V were fairly typical of all five sample sets with minor variations.

Comparison of Laboratory Results with an Operating Cooling Tower

A comparison was made of data obtained in this study with data obtained from analyses of samples taken of the Oak Ridge High Flux Isotope Reactor (HFIR) cooling tower basin water.[3] Recirculating cooling water system design data for both the HFIR and the Boiling Water Reactor plant for which the laboratory studies were made are presented in Table VI. Results of the analyses for chloroform in samples from the HFIR cooling water and concentrated, chlorinated, aerated Susquehanna River water are also presented.

To allow a comparison of laboratory and HFIR cooling system data, the following assumptions were made:

1. The 38 µg/L of chloroform measured in the HFIR cooling tower basin immediately after chlorination would not change in recirculating to the cooling tower inlet distributor. Therefore, this value could be taken as the initial chloroform concentration prior to aeration.

Table V. Semivolatile Priority Pollutants Analyzed (GC-MS) with Estimated Detection Limits (Sample Set B)

Semivolatile Compounds	Estimated Detection Limit, μg/L (ppb)	Semivolatile Compounds	Estimated Detection Limit, μg/L (ppb)
Acid Compounds		**Base/Neutral Compounds**	
2-Chlorophenol	0.8	Acenaphthene	0.1
2,4-Dichlorophenol	1.0	Acenaphthylene	0.1
2,4-Dimethylphenol	0.5	Anthracene	0.1
4,6-Dinitro-o-cresol	2.0	Benzidine	20
2,4-Dinitrophenol	4.0	Benzo(a)anthracene	0.1
2-Nitrophenol	2.0	Benzo(a)pyrene	0.1
4-Nitrophenol	4.0	3,4-Benzofluoranthene	0.1
p-Chloro-m-Cresol	1.0	Benzo(g,h,i)perylene	0.2
Pentachlorophenol	1.5	Benzo(k)fluoranthene	0.1
Phenol	1.0	Bis(2-chloroethoxy)methane	0.2
2,4,6-Trichlorophenol	0.9	Bis(2-chloroethyl)ether	0.2
		Bis(2-chloroisopropyl)ether	0.2
Pesticides		Bis(2-ethylhexyl)phthalate	0.2
		4-Bromophenyl Phenyl Ether	0.5
Aldrin	1	Butyl Benzyl Phthalate	0.2
α-BHC	1	2-Chloronaphthalene	0.1
β-BHC	1	4-Chlorophenyl Phenyl Ether	1.0
γ-BHC	1	Chrysene	0.1
δ-BHC	1	Dibenzo(a,h)anthracene	0.2
Chlordane	10	1,2-Dichlorobenzene	0.2
4,4'-DDT	0.5	1,3-Dichlorobenzene	0.2
4,4'-DDE	0.5	1,4-Dichlorobenzene	0.2
4,4'-DDD	0.8	3,3-Dichlorobenzidine	0.5
Dieldrin	1.2	Diethyl Phthalate	0.1
α-Endosulfan	5	Dimethyl Phthalate	0.1
β-Endosulfan	5	Di-n-Butyl Phthalate	0.1
Endosulfan Sulfate	10	2,4-Dinitrotoluene	0.5
Endrin	5	2,6-Dinitrotoluene	0.5
Heptachlor	1.5	Di-n-Octyl Phthalate	0.1
Heptachlor Epoxide	2	1,2-Diphenylhydrazine[a]	—
PCB-1242	1	Fluoranthene	0.1
PCB-1254	1	Fluorene	0.1
PCB-1221	1	Hexachlorobenzene	0.5
PCB-1232	1	Hexachlorobutadiene	0.5
PCB-1248	1	Hexachlorocyclopentadiene	2.0
PCB-1260	1	Hexachloroethane	0.5
PCB-1016	1	Indeno(1,2,3-cd)pyrene	0.2
Toxaphene	10	Isophorone	2.0
Tetrachlorobiphenyl	10	Naphthalene	0.1
Hexachlorobiphenyl	10	Nitrobenzene	0.5
		N-Nitrosodimethylamine	20
		N-Nitrosodi-N-Propylamine	0.5
		N-Nitrosodiphenylamine	0.2
		Phenanthrene	0.1
		Pyrene	0.1
		1,2,4-Trichlorobenzene	0.2

[a] Compound is not stable in water.

Table VI. Comparison of Cooling Water System Design Data and Chloroform Concentrations for HFIR and BWR Plants

	HFIR[a]	BWR Plant
Cooling water		
System volume, gal	600,000	8,000,000
Recirculating rate, gpm	26,000	478,000
Time for system volume turnover, min	23	17
Cycles of concentration	4.7	3.8
Makeup, gpm	700	19,000
Blowdown, gpm	150	5,000
Tower basin water		
Chloroform content—initial, µg/L	38[b]	
Chloroform content after 120 min, µg/L	6.2	
Estimated air-water contact time, min	1.3[c]	
Volume turnovers in 120 min	5.2	
Laboratory concentrated water sample from set D		
Chloroform content—initial, µg/L		35.0
Chloroform content—5-min aeration, µg/L		13.2
Chloroform content—30-min aeration, µg/L		6.4

[a]High Flux Isotope Reactor at Oak Ridge.
[b]Initial content was taken as chloroform concentration of sample collected from the tower basin immediately after chlorine addition.
[c]Based on an estimated travel time for cooling water of 15 sec from top of cooling tower packing to tower basin and 23 min for 1 volume turnover.

2. The time of transit for the water from the cooling tower water distributor to the cooling tower basin is approximately 15 s for most towers. This allows calculating the actual time that water and air are in intimate contact during each full-volume turnover of the cooling water.

The reduction of chloroform concentration from 38 to 6.2 µg/L in the HFIR cooling tower took 2 h. This is equivalent to 5.2 cooling system volume turnovers, resulting in 1.3 min of contact time between air and water. This is based on 15 s of contact time per volume turnover. It took 30 min of aeration (air sparging through a warm sample) to obtain a similar reduction in chloroform concentration in the laboratory (35 to 6.4 µg/L) (Table VI). The cooling tower was much more efficient in reducing chloroform concentration, taking only approximately 1.3 min of actual air-water contact time as compared with 30 min in the laboratory to accomplish about the same removal. This is not surprising, because the cooling tower acts as a packed tower in a stripping operation, bringing large surfaces of water in contact with air, something that cannot be achieved with finely dispersed bubbles of air in water.

The power station's cooling water volume turnover is only 17 min as compared with 23 min for the HFIR, which should result in achieving the same level of chloroform reduction in approximately 1.5 h rather than 2 h. A rough approximation of chloroform reduction per pass through the HFIR cooling tower was 30%. A higher reduction could be expected in the first pass because of the higher concentration of chloroform (greater driving force), and the percentage reduction would decrease with each succeeding pass through the cooling tower.

Impact of Discharging Cooling Tower Blowdown to the Susquehanna River

The effect of discharging blowdown from the two cooling towers to the Susquehanna River would be negligible. Operating procedures call for chlorinating one volume of cooling water every 8 h and minimizing chlorine injection to obtain ≤ 0.3 μg/L of FCR. The maximum concentration of THMs formed based on the laboratory study was 52.9 μg/L (sample set B). With an estimated 30% reduction by stripping in the cooling tower, the maximum concentration of THMs in the cooling tower blowdown would be 37 μg/L, of which most would be chloroform. This is well below the EPA National Interim Primary Drinking Water Regulation limit of 100 μg/L total THMs, even before dilution takes place in the river.

The average river flow would provide a dilution factor of 550; therefore, the concentration of THMs contributed to the river would be below the detectable limit of analysis. The duration of chlorination for each of the two condenser cooling water systems is approximately 17 min and is carried out every 8 h. Dechlorination will be carried out to reduce chlorine residuals to undetectable limits. This, in turn, will prevent further production of THMs in the discharged cooling tower blowdown.

The only other two priority pollutants detected during the laboratory tests, methylene chloride and 1,1,1-trichloroethane, were produced in such low concentrations and are stripped away by air so easily that they would be below detection limits in the cooling tower blowdown.

CONCLUSIONS

Based on our laboratory study the only priority pollutants formed by chlorinating the recirculating cooling water at the two cooling towers using North Branch Susquehanna River water makeup were THMs, methylene chloride, and 1,1,1-trichloroethane. The concentration of THMs formed was below the EPA drinking water limit of 100 μg/L, and the concentrations of the other two volatile compounds found were very close to their detectable limits. All of these compounds would be stripped away by air flowing countercurrently to the water in the cooling tower, so in 1.5 h they would be almost undetectable. Further, these compounds would be below their detectable limits after diluting in the river.

Because of their volatility, THMs will be stripped away by air in the cooling towers within several hours. Chloroform appears to be less volatile than chlorodibromomethane and dichlorobromomethane.

In the report on assessment of halogenated organics in waters from cooling towers,[3] it was concluded that the formation of chloroform continues in the discharged effluent. The results of this study seem to confirm this and indicate that dechlorination can prevent or minimize the further formation of chloroform in cooling tower blowdown.

To minimize the formation of THMs in condenser cooling water systems containing cooling towers, the dechlorination chemical would be best injected into the effluent pipeline of the condenser. The disadvantages are increased consumption of dechlorination chemical, increased equipment cost, and buildup of biolife on the cooling tower fill. Another means of reducing the quantity of THMs discharged in cooling tower blowdown where intermittent chlorination of short duration (20 to 60 min) is practiced is to close off the blowdown valve for several hours. With cooling water systems containing large volumes of water such as those presented in Table VI, the increase in the total dissolved solids due to stopping blowdown for several hours will be negligible.

REFERENCES

1. U.S. Environmental Protection Agency. "Guidelines Establishing Test Procedure for the Analysis of Pollutants, Purgeables — Method 624," *Fed. Regist.* 44(23):69532 (1979).
2. U.S. Environmental Protection Agency. "Guidelines Establishing Test Procedures for the Analysis of Pollutants, Base/Neutrals, Acids and Pesticides — Method 625," *Fed. Regist.* 44(23):69540 (1979).
3. Jolley, R. L., W. W. Pitt, Jr., F. G. Taylor, Jr., S. J. Hartmann, G. Jones, Jr., and J. E. Thompson. "An Experimental Assessment of Halogenated Organics in Waters From Cooling Towers and Once-Through Systems," in *Water Chlorination: Environmental Impact and Health Effects, Vol. 2*, R. L. Jolley, H. Gorchev, and D. H. Hamilton, Jr., Eds. (Ann Arbor, MI: Ann Arbor Science Publishers, Inc., 1978), pp. 695–705.

CHAPTER 107

Chlorination of Coal Slurry Transport Waters

John W. Davis, M. Carrington Reid, Roger A. Minear, and Gary S. Sayler

The bulk transportation of coal as a 50:50 (w/v%) water-prepared slurry represents a cost efficient and reliable alternative process for delivering coal to power plants. However, regional water resource limitations, especially in the western United States, may require substantial reuse of large volumes of transport wastewater. One potential reuse application is for in-plant cooling water needs.

Such water reuse application and eventual environmental release are dependent on the water quality characteristics of the transport wastewater. Recently, we have reported on changes in transport water quality resulting from the leaching of organic materials during the transport process.[1] These organic compounds have been characterized as polar, water soluble, aromatic polymers of moderate but variable molecular weight (<5000) most likely originating from humic and fulvic acids present in the parent coal. Since dissolved organic carbon (DOC) concentrations can reach as high as 370 mg/L, depending on the coal source, potential water reuse applications and eventual disposal of the wastewater may be limited by these elevated carbon levels.

One potential area of concern is whether these organic materials serve as precursors for the formation of trihalomethanes (THM) during chlorination of cooling waters. Numerous investigators[2-6] have implicated humic and fulvic acids as direct precursors of THM during chlorination. In addition, from studies of chlorinated drinking water throughout the United States, it has been concluded that there is a strong relationship between the concentration of nonvolatile organic carbon and the concentration of THM produced during chlorination.[7]

The specific objective of this investigation was to determine the potential of slurry wastewater organics for serving as precursors for the formation of THM when transport waters are used as cooling waters and thus subject to chlorination. Secondary objectives were to determine the qualitative fraction of the organic pool responsible for THM production and the influence of bromide on the THM formation process.

METHODS

The procedures used in this study are summarized in Table 1.[1,8-10] Simulated coal slurry wastewaters were generated from three subbituminous western coals: Montana Rosebud, Black Mesa, and Wyodak. A fourth wastewater sample was obtained from the Mohave Power Generation Station in Nevada; the station receives coal via the Black Mesa pipeline. After coal fines were removed, the dissolved organics contained in the wastewaters were separated according to the humic acid isolation procedure of Thurman and Malcolm.[8] The desalting step prior to lyophilization was omitted during sample preparation. The humic/fulvic acid fractions of the Wyodak and Black Mesa pipeline wastewaters were further fractionated, according to molecular weight (MW), using membrane ultrafiltration.[1] This procedure resulted in two distinct fractions: >1000 MW and <1000 MW.

Wastewaters (and subfractions) were diluted to ~10-mg carbon per liter with either distilled water or phosphate buffer (pH 7.0) and chlorinated by sodium hypochlorite treatment at a concentration ratio of 2:1 (chlorine:carbon). At designated time intervals the reaction was quenched with sodium thiosulfate. The effects of bromide on THM production were conducted in an identical format except for the addition of 0.1 or 0.5 mg/L bromide to the desired sample.

Table I. Summary of Methods

Procedure	Description	Reference
Slurry wastewater preparation	Pulverized coal and water (distilled or lake) mixed in a 1:1 ratio (w/v%) Coal fines removed after 3 d Three subbituminous western coals: (1) Black Mesa; (2) Wyodak; (3) Montana Rosebud Fourth wastewater: Black Mesa pipeline water	1
Fractionation of slurry wastewater	Isolation of humic/fulvic acids	8
	Molecular weight fractionation of humic/fulvic fraction using ultrafiltration	1
Chlorination	Sample dilute to ≤10 ppm DOC Chlorinate with sodium hypochlorite at a concentration of 2:1 (chlorine:carbon) Bromide added to selected samples at a concentration of 0.1 or 0.5 mg/L pH 7 samples were buffered; others achieved final pH as a result of initial adjustment of chlorine solution followed by reactions	10
Trihalomethane detection	Purge and trap procedure	9
	Hall electrolytic detector	10

The purge-and-trap procedure was utilized to quantitate the product THM. Operating and column conditions for the gas chromatograph were those of Minear and Morrow.[10]

RESULTS

THM Detection in Raw Slurry Wastewater

Slurry wastewaters generated from different coal types exhibit wide variability in their DOC levels.[1] The DOC in the wastewater in this investigation ranged from 15 to 180 ppm, depending on the coal type (Table II). Prior to chlorination, all wastewaters were diluted to a DOC concentration of ~5 to 10 ppm, and the THM production for each sample was normalized on a per-milligram-carbon basis. This procedure provides a basis for comparing THM production between samples regardless of their initial DOC content.

The THM formation potential (THMFP) of coal slurry wastewaters generated in three chlorinated western subbituminous samples is presented in Table II as micromoles chloroform per milligram organic carbon (μmol $CHCl_3$/mg carbon). Time series data are presented for 1-, 2-, and 4-d reaction times. Typically, 90 to 100% of maximum value is obtained in 96 h. Chloroform was the only THM detected in the chlorinated wastewater.

Table II. Trihalomethane (THM) Production Generated in Three Western Subbituminous Coal Slurry Wastewaters

Sample	Undiluted DOC (mg/L)	Reaction pH	THM, μmol/mg carbon		
			24 h	48 h	96 h
Wyodak[a]					
Slurry 1	180	11.4	0.25	0.33	0.40
Slurry 2	130	7.5	0.20		0.29
Slurry 3	153	6.9	0.15	0.22	0.24
Montana Rosebud[a]					
Slurry 1	160	10.4	0.12	0.12	0.16
Slurry 2	144	7.2	0.13	0.13	0.14
Slurry 3	119	6.5	0.07	0.08	0.12
Black Mesa[a]					
Slurry 1	130	9.4	0.09	0.16	0.14
Slurry 2	15	6.9	0.25	0.35	0.45
Slurry 3	33	7.1	0.26	0.27	0.31
Black Mesa[b]					
Pipeline centrate	48	7.0			0.50

[a]All laboratory-prepared slurry systems were from the same coal batch except Black Mesa, where slurry 1 was prepared from a coal sample 1 year older than coal used for slurry 2 and slurry 3.
[b]Real pipeline field sample taken after coal separation via centrifugation.

Little difference in THMFP between Wyodak and Black Mesa wastewaters was observed. However, both coal types yielded on average twice the amount of chloroform per unit carbon as compared with Montana Rosebud wastewaters. The influence of reaction pH is shown for the Wyodak and Montana Rosebud coal slurries, that is, an increase in THM formation with increasing pH. This effect is not clear with the Black Mesa slurry. The apparent reversal of the expected pH effect, however, is likely explained by the difference in coal used for Black Mesa slurry 1 and that used for Black Mesa slurries 2 and 3. The latter two were from the same coal sample, but slurry 1 was prepared from a separate coal batch that had been obtained nearly 1 year earlier.

At first observation, it would appear that the THMFP for the Wyodak and Black Mesa slurries were similar and jointly about twice the value for the Montana Rosebud slurry. However, when Black Mesa slurry 2 and 3 values are compared with Black Mesa slurry 1 values, we see that the difference is as great as that between Montana Rosebud values and Black Mesa slurry 2 and 3 values. On the other hand, a Black Mesa pipeline centrate sample yielded a value comparable to the laboratory sample value obtained for Black Mesa slurry 2. Thus, the variation in THMFP within a coal type from batch to batch may be as great as the variation between coal types.

These results indicate that DOC contained in the slurry wastewater will serve as precursors in the formation of THM. Coal slurry wastewaters containing elevated DOC concentrations and those chlorinated prior to disposal may represent a significant source of THM. For example, if undiluted Wyodak, Black Mesa, and Montana Rosebud slurry wastewaters were chlorinated under conditions similar to those used in this investigation, approximately 5600-, 2100-, and 2400-μg $CHCl_3$ per liter would be formed.

Fulvic Acid Content of Slurry Wastewater

Although initial DOC levels were approximately the same at the time of chlorination (≤ 10 ppm), the THMFP for individual slurry wastewaters exhibited considerable variation. In part, the variations in pH would be expected to yield different values. However, the behavior of Black Mesa slurries made from different Black Mesa coal batches, and the differences between other coal-type slurries, likely reflect qualitative differences in the class of organic compounds that leached from the coal into transport waters during the slurry process. One specific difference in slurry wastewater DOC was reported by Reid and co-workers.[1] They demonstrated that the molecular weight distribution of the DOC varied among wastewaters generated from different coal types.

Since previous reports[11] have shown that humic and fulvic acids contribute to the DOC of slurry wastewater, the acid content of the waters was assayed. Isolated fractions were then chlorinated to determine whether a relationship existed between THM production and humic or fulvic content. The fractionation was performed on Black Mesa and Wyodak wastewaters only. The relative composition of humic and fulvic acids in the two wastewaters is given in Table III. Based on a DOC analysis, greater than 98% of the organic compounds contained in the humic and fulvic acid fraction were soluble at a pH 2. Thus, this fraction was considered to be essentially all fulvic acids. Overall, there was a 2.7-fold higher level of DOC in Wyodak wastewaters as compared with Black Mesa wastewaters. Fulvic acids comprised 56 and 66% of the Wyodak and Black Mesa wastewater's DOC, respectively. Although the fulvic acid components comprised approximately the same fraction of DOC in the two wastewaters, greater differences were noted when the fulvics were further subdivided according to their molecular weight (Table III). The Wyodak slurry water contained a greater proportion of fulvic acid with MW < 1000 (82%) than was observed for Black Mesa wastewaters (30%).

Chlorination results on the fractions are presented in Figure 1 and Table IV. Only chloroform was detected, thus substantiating the absence of bromide in the slurry water. For the Black Mesa sample, all THMFP was represented by the fulvic acid component. However, the whole-sample chlorination yielded 0.50 μmol $CHCl_3$/mg carbon; the fulvic acid fraction yielded only a value of 0.59 μmol $CHCl_3$/mg carbon. A somewhat higher yield would be expected from the fulvic acid fraction as it comprises only 66% of the nonvolatile total organic carbon (NVTOC); the nonfulvic fraction did not contribute to the THMFP. From Figure 1, the Wyodak results show a similar dominance of the THMFP by the fulvic material, but the difference is a factor of 3 over the nonfulvic contribution (0.33 vs 0.12 μmol $CHCl_3$/mg carbon at 96 h).

THMFP data on molecular weight size fractions of the fulvic acid material were obtained for Black Mesa samples. Yields for the lower-molecular-weight fraction were much higher than for the higher-molecular-weight fraction (1.32 vs 0.55 μmol $CHCl_3$/mg carbon). These results would indicate much greater contribution to THMFP from low-molecular-weight components. However, the Wyodak slurry is comprised of a much greater fraction of low-molecular-weight fulvic acids than the Black Mesa sample (82% vs 30%), yet has a much lower yield from the fulvics in general (0.33 vs 0.59 μmol $CHCl_3$/mg carbon).

Table III. Mass Carbon Balance on the Fractionated Black Mesa Pipeline and Wyodak Coal Wastewater[a]

	Wyodak	Black Mesa Pipeline Wastewater
Original slurry wastewater	130	48
Amount recovered after fractionation	118 (91%)	41 (85%)
Fulvic acid	66 (56%)	27 (66%)
> 1000 MW	12 (18%)	19 (70%)
< 1000 MW	54 (82%)	8 (30%)
Nonfulvic fraction	52 (44%)	14 (34%)

[a] All values in mg carbon per liter unless otherwise noted.

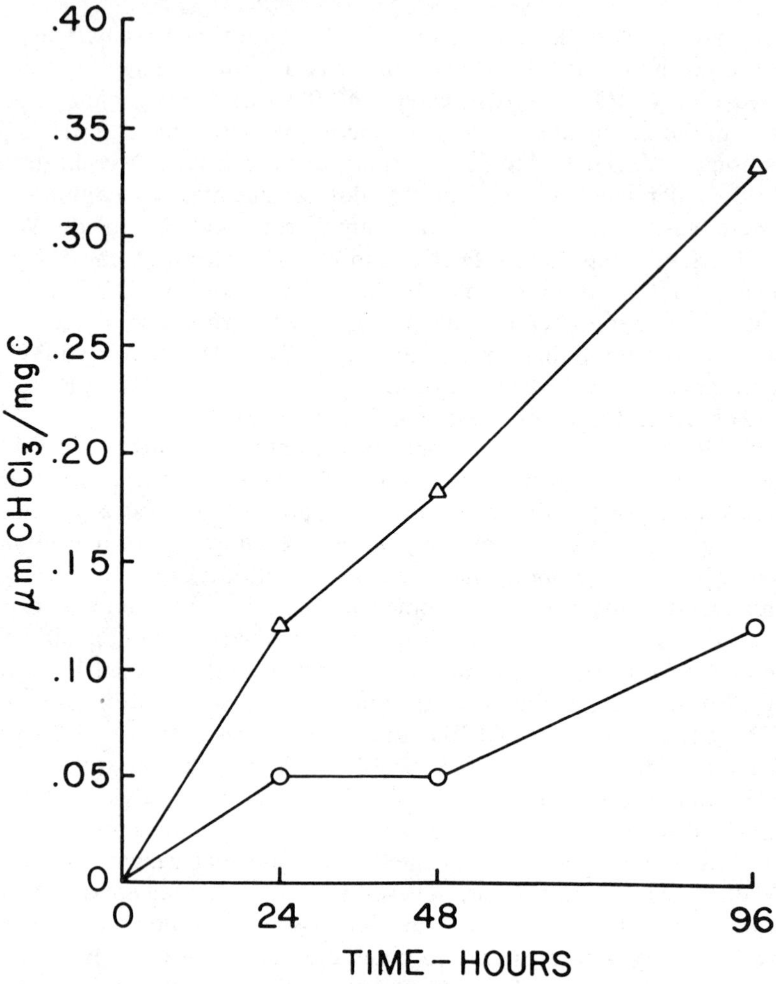

Figure 1. The production of chloroform in fractionated Wyodak coal slurry wastewaters. ○, Wyodak nonfulvic fraction; △, Wyodak fulvic fraction.

These contradictory results make generalization about molecular weight influence difficult without a more extensive data base.

What is clear is that fulvic materials represent the major contribution to the THMFP. These results indicate that the fulvic acid fractions are primarily, if not exclusively, responsible for the THMs produced in the slurry wastewater after chlorination. The extent of THM formation for these wastewaters is comparable to those yields reported for other aqueous humic/fulvic systems (Table V),[3,10,12-15] further demonstrating the role of fulvic acids as THM precursors.

Table IV. Production of THM in Fractionated and Nonfractionated Black Mesa Pipeline Wastewaters

Sample	Reaction Time (h)	μmol $CHCl_3$/mg carbon	DOC (mg/L)	pH
Black Mesa pipeline wastewater (BMPW)	96	0.50	5.2	7.0
BMPW fulvic acids	96	0.59	4.5	7.0
BMPW nonfulvic acids	96	ND[a]	5.0	7.0
BMPW fulvic acids < 1000 MW	96	1.32	4.0	7.0
BMPW fulvic acids > 1000 MW	96	0.55	7.4	7.0

[a]None detected.

Effects of Bromide Addition

Previous work has clearly defined a strong dependence of THM formation on the presence of relatively low bromide ion concentrations.[10,12,16-19] Although laboratory-generated slurry waters and Black Mesa pipeline waters did not contain an appreciable bromide ion concentration (noted by the absence of brominated THMs), it cannot be presupposed that other slurry wastewater systems or the pipeline terminus water systems will not contain bromide. Consequently, chlorination experiments were repeated on the Wyodak fulvic and nonfulvic fractions with the addition of bromide ion.

The results in Table VI show that little effect is observed with the nonfulvic fractions at 0.1 mg/L Br$^-$. The apparent decrease is more likely experimental variation. On the other hand, an increase of 33% in THM was obtained for the fulvic fraction at 0.1 mg/L Br$^-$. At 0.5 mg/L Br$^-$, an increase of 245% was observed. Bird's work[13] found increases of 44% at 0.1 mg/L Br$^-$ and 72% at 0.4 mg/L Br$^-$ with humic acid solutions.

In comparison with Bird[13] and Minear and Morrow,[10] the change in distribution of the THMs produced is modest. At 0.1 mg/L Br$^-$, the product THMs are still 98% chloroform. This distribution shifts to 67% at 0.5 mg/L Br$^-$, with the balance represented only by the monobromo species.

The efficiency of bromine incorporation has been shown to be strongly related to the initial bromide concentration.[18,20] Peak incorporation typically occurs at very low Br$^-$ ($\sim$0.01 to 0.02 mg/L) and reaches 25 to 30%. For the fractionated Wyodak slurry waters, percent bromine incorporation was $\sim$4% for both the fulvic acid and nonfulvic acid fractions at an initial bromide concentration of 0.1 mg/L. These data are consistent with earlier observations that the incorporation trails off at higher bromide concentrations.[20] However, the bromine incorporation was 32% for the 0.5 mg/L Br$^-$/fulvic acid fraction system. This result was unexpectedly high and the reverse of what would be expected.

Table V. Trihalomethane Production of Selected Humic/Fulvic Sources

Source	Reference	μmol CHCl$_3$/mg carbon	Cl$_2$ Dose (mg/L)	pH	Reaction Time (h)	Reaction Temp (°C)
Aquatic fulvic acid	13	0.05	8	7	48	10
Humic acid	12	0.70	10	6.7	96	25
Wyodak coal fulvic acid		0.33	10	7	96	20
Aquatic humic acid	3	0.11	8	8	48	15
Black Mesa coal fulvic acid		0.59	9	7	96	20
Commercial humic acid	13	0.39 – 0.50	5	7	96	20
River water with commercial humic acids	10	0.25 – 0.33	10	7	96	20
Peat humic acid	14	1.14 – 1.59	10	6.5	96	20
Commercial humic acid	15	0.73	10 – 20	7	96	20
Florida water[a]	15	0.3 – 0.5	24 – 36	7	96	20
Florida water[b]	15	0.3 – 0.5	19.5 – 32.5	7	96	20

[a] 0.25 mg/L Br$^-$ in water.
[b] 0.16 mg/L Br$^-$ in water.

Table VI. Trihalomethane (THM) Formation of Slurry Wastewaters in the Presence of Added Bromide[a]

	Bromide (ppm)	$CHCl_3$ in Unbrominated Wastewater	$CHCl_3$	$CHCl_2Br$	$CHClBr_2$	Total THM
Wyodak fulvic fraction	0.1	0.33	0.43	0.01	ND[b]	0.44
	0.5		0.76	0.38	ND	1.14
Wyodak nonfulvic fraction	0.1	0.12	0.08	0.01	ND	0.09

[a] Measured in μmol THM/mg carbon; all reaction times were 96 h at pH 7.0.
[b] None detected.

Predictive Modeling

Morrow and Minear[21] presented a predictive model that incorporated the major variables in THM formation. For a series of drinking waters, very good predictive capability was obtained. Of the three successful multilinear convergence techniques used, that using the SAS DUD routine (statistical analysis system–multivariate secant or false position method), a nonlinear regression technique, is given:

$$THM = -3.94 + [Br^-]^{0.19} + 0.35 \log [Cl_2 \text{ dose}] + 0.24 \text{ pH} + 10^{0.009t} + 0.27 [NVTOC]$$

Application of the model to the data obtained for chlorinating coal slurry waters and subfractions thereof is presented in Table VII. The only reasonable agreement with the model results is seen for the Wyodak slurry. The remainder of the whole samples and the fractionated Wyodak samples do not fit the model well. This is not necessarily surprising, because the model was developed from data on a single water source (Tennessee River water), with NVTOC adjustment accomplished by the addition of commercial humic acid.

CONCLUSIONS

The results of this investigation demonstrate that chlorination of coal slurry wastewaters will result in the formation of THM. Fulvic acids leached during the slurry process were found to be the predominant THM precursors in chlorinated Wyodak slurry wastewaters and exclusive THM precursors in chlorinated Black Mesa transport pipeline wastewaters. Thus, the extent to which THM formation occurs will be in part dependent on the relative humic and fulvic acid content of the leached organic materials. Variations in THMFP

Table VII. Comparison of Model-Predicted Trihalomethane (THM) Values with Measured THM Concentrations for Coal Slurry Waters and Functions

Slurry	Predicted THM (μmol/L)	Actual THM (μmol/L)	Percent Difference
Unfractionated			
Wyodak 1	3.21	3.59	11
Wyodak 2	1.66	1.97	16
Wyodak 3	1.76	1.83	4
Black Mesa 1	2.01	0.94	−114
Black Mesa 2	−0.17	0.67	125
Black Mesa 3	1.48	2.05	28
Black Mesa pipeline	1.04	2.6	60
Montana Rosebud 1	2.80	1.28	−119
Montana Rosebud 2	1.67	1.02	−64
Montana Rosebud 3	1.13	0.71	−59
Fractionated			
Wyodak fulvic	1.04	1.72	40
Wyodak fulvic + 0.1-ppm Br$^-$	1.69	2.29	26
Wyodak fulvic + 0.5-ppm Br$^-$	1.94	5.90	67
Wyodak nonfulvic	1.03	0.63	−63
Wyodak nonfulvic + 0.1-ppm Br$^-$	1.68	0.47	−257

were noted both within and among coal types. These differences likely reflect the broad spectrum of THM precursors present in coal matrices that may leach into waters during slurry transport. In addition, this process is complicated by the fact that progressive oxidation of coal during mining, preparation, and storage processes results in increased DOC loads in slurry transport waters as well as likely changes in the qualitative nature of the leachable organics.[22,23] These phenomena make generalizations concerning THM formation vs relative DOC content in slurry wastewaters at best a difficult endeavor.

Results from the MW fractionation study suggest that the MW of leached organic materials may not be a good general predictive parameter in terms of assessing THMFP for various coal types.

The chlorination of fractionated coal slurry wastewater organics in the presence of the bromide ion resulted in an increase in THM yield. However, percent incorporation behavior differed from previous research. Further studies are warranted to better define this phenomenon.

With the exception of unfractionated Wyodak slurry samples, the predictive model was of little use unless specifically calibrated for individual slurry systems. The fact that good results were obtained with one coal type and not others underscores the variability that is present among slurry systems. Additional experimentation in the area of qualitative identification of THM precursors would greatly enhance our understanding of this subject area.

REFERENCES

1. Reid, M. C., J. W. Davis, G. S. Sayler, and R. A. Minear. "Characterization and Biological Treatability of Coal Slurry Wastewaters," *Water Res.*, in press.
2. Oliver, B. G., and S. A. Visser. "Chloroform Production from the Chlorination of Aquatic Humic Material: The Effect of Molecular Weight, Environment, and Season," *Water Res.* 14:1137-1141 (1980).
3. Peters, C. J., R. J. Young, and R. Perry. "Factors Influencing the Formation of Haloforms in the Chlorination of Humic Materials," *Environ. Sci. Technol.* 14(1):1391-1395 (1980).
4. Quimby, B. D., M. F. Delaney, P. C. Uden, and R. M. Barnes. "Determination of the Aqueous Chlorination Products of Humic Substances by Gas Chromatography with Microwave Emission Detection," *Anal. Chem.* 52:259-263 (1980).
5. Trussel, R. R., and M. D. Umphres. "The Formation of Trihalomethanes," *J. Am. Water Works Assoc.* 70(11):604-611 (1978).
6. Oliver, B. G., and E. M. Thurman. "Influence of Aquatic Humic Substance Properties on Trihalomethane Potential," in *Water Chlorination: Environmental Impact and Health Effects, Vol. 4*, R. L. Jolley, W. A. Brungs, J. A. Cotruvo, R. B. Cumming, J. S. Mattice, and V. A. Jacobs, Eds. (Ann Arbor, MI, Ann Arbor Science Publishers, Inc., 1983), pp. 234-241.
7. Symons, J. M., T. A. Bellar, J. K. Carswell, J. DeMarco, K. L. Kropp, G. G. Robeck, D. R. Seeger, C. J. Slocum, B. L. Smith, and A. A. Stevens. "National Organic Reconnaissance Survey for Halogenated Organics," *J. Am. Water Works Assoc.* 67:634-64 (1975).
8. Thurman, E. M., and R. L. Malcolm. "Preparative Isolation of Aquatic Humic Substances," *Environ. Sci. Technol.* 15(4):463-466 (1981).
9. Bellar, T. A., and J. J. Lichtenberg. "Determining Volatile Organics of Microgram-per-Liter Levels by Gas Chromatography," *J. Am. Water Works Assoc.* 66:739-744 (1974).
10. Minear, R. A., and C. M. Morrow. *Raw Water Bromide: Levels and Relationship to Distribution of Trihalomethanes in Finished Drinking Water*, Research Report 91 (Knoxville: Tennessee Water Resources Research Center, University of Tennessee, 1983).
11. Godwin, J., and S. E. Manahan. "Interchange of Metals and Organic Matter Between Water and Sub-bituminous Coal or Lignite Under Simulated Coal Slurry Pipeline Conditions," *Environ. Sci. Technol.* 13:1100-1104 (1979).
12. Stevens, A. A., C. J. Slocum, D. R. Seeger, and G. G. Robeck. "Chlorination of Organics in Drinking Water," *J. Am. Water Works Assoc.* 68:615-620 (1976).
13. Bird, J. C. *The Effect of Bromide on Trihalomethane Formation*, MS Thesis, (Knoxville, TN: University of Tennessee, 1979).
14. Babcock, D. B., and P. C. Singer. "Chlorination and Coagulation of Humic/Fulvic Acids," presented at the 97th Annual American Water Works Conference, Anaheim, CA, May 8-13, 1977.
15. Amy, G. L., P. A. Chadik, P. H. King, and W. J. Cooper. "Chlorine Utilization During Trihalomethane Formation in the Presence of Ammonia and Bromide," *Environ. Sci. Technol.*, in press (1985).
16. Cooper, W. J., L. M. Meyer, C. C. Bofill, and E. Cordal. "Quantitative Effects of Bromine on the Formation and Distribution of Trihalomethanes in Groundwater with a High Organic Content," in *Water Chlorination: Environmental Impact and*

Health Effects, Vol. 4, R. L. Jolley, W. A. Brungs, J. A. Cotruvo, R. B. Cumming, J. S. Mattice, and V. A. Jacobs, Eds. (Ann Arbor, MI: Ann Arbor Science Publishers,Inc., 1983), pp. 285-296.
17. Oliver, B. G. "Effect of Temperature, pH, and Bromide Concentrations on the Trihalomethane Reaction of Chlorine with Aquatic Humic Material," in *Water Chlorination: Environmental Impact and Health Effects, Vol. 3*, R. L. Jolley, W. A. Brungs, and R. B. Cumming, Eds. (Ann Arbor, MI: Ann Arbor Science Publishers, Inc., 1980), pp. 141-149.
18. Luong, T. V., C. J. Peters, and R. Perry. "Influence of Bromide upon THM Formation During Water Chlorination," in *Water Industry '81. International Conference*, (Brighton, UK: CEP Consultants Ltd., 1981).
19. Minear, R. A., and J. C. Bird. "Trihalomethanes: Impact of Bromide Ion Concentration on Yield, Species Distribution, Rate of Formation, and Influence of Other Variables," in *Water Chlorination: Environmental Impact and Health Effects, Vol. 3,* R. L. Jolley, W. A. Brungs, and R. B. Cumming, Eds. (Ann Arbor, MI: Ann Arbor Science Publishers, Inc., 1980), pp. 151-160.
20. Minear, R. A. *Production, Fate, and Removal of Trihalomethanes in Municipal Drinking Water Systems*, Tennessee Water Resources Research Report 78 (Knoxville: University of Tennessee, 1980).
21. Morrow, C. M., and R. A. Minear. "A Nonlinear Regression Model Linking Raw Water Characteristics to Trihalomethane Concentrations in Drinking Water," Reprints of Papers, American Chemical Society, Division of Environmental Chemistry, 23(1):184-187 (1983).
22. Sayler, G. S., R. A. Minear, M. C. Reid, and J. W. Davis. *Enhanced Reuse Potential of Coal Slurry Transport Water: Toxic Organics Assessment and Removal*, RU-84/4, (Washington, DC: U.S. Department of Interior, Office of Water Resources Research, 1983).
23. Plummer, A. H., B. L. Jordan, R. H. Derammelaere, and A. S. Sandau. "The ETSI Coal Evaluation Plant Wastewater Characterization Study," presented at the Eighth International Technical Conference on Slurry Transportation, San Francisco, 1983.

CHAPTER 108

Effects of Chlorination on the Levels of Mutagens in Contaminated Estuarine Sediments

Carol B. Daniels, Sandra M. Baksi, Allen D. Uhler, and Jay C. Means

The widespread use of chlorine as a primary agent for the disinfection of water has prompted concern over the potential for related health effects. Recent studies have demonstrated the formation of mutagens resulting from the reaction of chlorine with the organic constituents present in drinking water and sewage effluent.[1-3] However, although the production or enhancement of mutagenicity by treatment of organic matter with chlorine is quite plausible, the action of chlorine actually may be twofold. By producing hypochlorite ions in solution, chlorine may serve to decrease the mutagenicity associated with organic constituents by acting as a strong oxidizing agent. On the other hand, by acting as an electrophile (i.e., by giving up chlorine) chlorine may serve to enhance mutagenicity.

The presence of Br$^-$ ions in estuarine systems also gives rise to highly reactive species (OBr$^-$) when exchange reactions occur with chlorine, thereby increasing the likelihood of chemical reactivity with organic compounds. The addition of the halogen moiety to organic compounds would enhance compound hydrophobicity, thereby increasing the bioconcentration potential and the potential toxicity of the newly formed product.

Sediments contain a complex mixture of natural and anthropogenic compounds with varying potential for mutagenicity or toxicity. The possibility that the mutagenic potential of this complex mixture may be altered by reaction with chlorine and its aqueous decay products has not been fully studied. Large quantities of water containing suspended sediments (10–100 mg/L) are circulated through cooling towers by power plants. This process, which often involves the use of chlorine as an antibiofouling agent, may serve as a significant opportunity for the reaction of chlorine with particulate organic matter. The fact that sediments that may contain 1000-fold more natural organic matter than the water column are continually being mixed with chlorinated water due to tidal resuspension in estuaries provides ample opportunity for the formation of a complex mixture of sorbed reaction products. Depending on the nature of the complex mixture formed, these newly formed products may desorb from the sediment or may remain sorbed to the sediment surface. Ultimately this affects their bioavailability and potential toxicity. Similarly, sorption reactions would also play a significant role in determining the fate of mutagens produced or altered by chlorination reactions in the solution phase.

For the present study, we have assessed the mutagenic potential of estuarine sediments with and without chlorine treatment. Chlorinated and unchlorinated sediment extract sets were used in conjunction with a microbial reversion assay to provide information on the influence of chlorination on the levels and types of mutagens associated with these sediments.

METHODS

Samples of sediment were collected at three sites in the Patapsco River (Figure 1). Bulk samples of surficial sediment (top 2–5 cm) were collected using a Van Veen grab and stored in the dark in large pails until assayed.

Sediments were divided into three 100-g aliquots with fractions 2 and 3 receiving four successive treatments, each of 1-h duration, with 10 mg HOCl/kg sediment and 50 mg HOCl/kg sediment, respectively. The chlorine-treated and unchlorinated control sediments were then extracted exhaustively for 24 h in a Soxhlet apparatus with 250 mL of acetonitrile. These extracts were subsequently back-extracted with three successive volumes of hexane (50 mL per wash). The hexane fractions were pooled, dried over sodium sulfate, and reduced to a volume of 5 mL via rotary evaporation at 55°C. The concentrated extract was then transferred to 50 mL methanol. Thirty milliliters of the resulting extract was dried under nitrogen and taken up in 6 mL of dimethylsulfoxide.

The mutagenic activity associated with each of the sediment extracts was assessed using the standard *Salmonella*/microsome mutagenicity assay[4] and a modified *Salmonella* reversion assay.[5] Dilutions of the sediment extracts (0.1 mL) ranging from 10^0 to 10^7 were applied to the top agar in petri dishes in

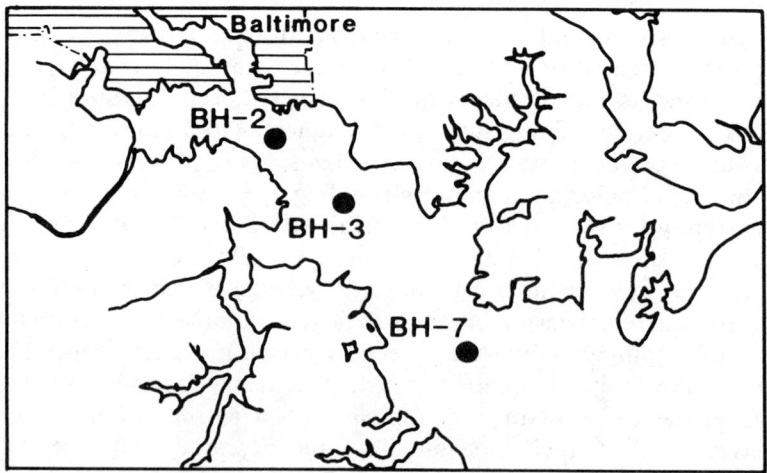

Figure 1. Sampling sites along the Patapsco River.

conjunction with 0.1 mL of each of the two tester strains TA98 and TA100. To assess the requirement for metabolic activation of promutagens, 0.5 mL of rat liver microsomal fraction (S-9) was allowed to preincubate at 37°C in a sterile tube with 0.1 mL of the sediment extract and bacteria for 20 min (preincubation assay) or was added to the top agar (standard assay). Each dilution was then plated three times, with each complete assay being duplicated. Extract dilutions that produced a twofold increase in the number of histidine revertants above spontaneous controls were initially considered to be mutagenic. A dose-response of revertants per plate vs extract concentration expressed as the equivalent microgram sediment per plate was also required to confirm the mutagenicity. Those tests showing less than two times the activity associated with the control (i.e., spontaneous revertants) were considered to be nonmutagenic.

Since most mutagens are toxic to some extent,[5] a relative toxicity rating scale was developed to assess the extent of toxicity associated with the sediment extracts (Table I). Toxicity to the tester strains was denoted by the disruption

Table I. Toxicity Rating Scale[a,b]

ppm	100 µg sediment/ plate	500 µg sediment/ plate	1,000 µg sediment/ plate	5,000 µg sediment/ plate	10,000 µg sediment/ plate
	Sediment Extract 3 Without Activation				
	TA98				
0	+	+	+ +	+ + +	+ + +
10	+ +	+ + +	+ +	+ + +	+ + +
50	+ + +	+ + +	+ +	+ +	+ + +
	TA100				
0	−	−	−	+	+ +
10	+	+	+ +	+	+ +
50	+	+	+ +	+ +	+ +
	Sediment Extract 2 With Activation				
	TA98				
0	+	−	−	−	+
10	−	−	−	+	+
50	−	−	+	+	+
	TA100				
0	−	−	−	−	+
10	−	−	−	+	+
50	+ + +	−	−	−	+

[a]Data shown are from treatment of bacterial strains, TA98 and TA100, with sediment extracts both in the presence and absence of metabolic activation in the standard plate incorporation assay.
[b]− No toxicity; + to + + + indicates increasing toxicity.

or, in extreme cases, the lack of the background "lawn" of bacterial cells. In addition, toxicity was reflected in cases where the background lawn remained relatively intact but a reverse in the slope of the dose-response curve was noted.

RESULTS

Standard Assay

Data from the standard plate incorporation assay indicated a general lack of mutagenicity associated with both the control (untreated) and chlorine-treated (10- and 50-ppm Cl_2) sediment extracts in the absence of S-9 for both TA98 and TA100 (data not shown). In all cases, the revertants per plate were at or below the spontaneous reversion frequency for the strain exposed. Expression of mutagenic activity may have been obscured by the toxic nature of both the chlorinated and unchlorinated sediment extracts without metabolic activation (Table I). The highest concentrations of sediment extracts were toxic in all treatment groups. This was often expressed as the disruption or lack of a background lawn of bacterial growth. For both strains TA98 and TA100, as the level of chlorine treatment to sediment 3 increased, the toxicity increased, with even the lowest dose levels of the 50-ppm treatment resulting in a toxic response (Table I). In general, unactivated sediment extracts were more toxic than the S-9 activated sediment extracts. Increases in the level of chlorination tended to increase the toxicity observed at higher dose levels, but the disruption of the background lawn rarely occurred in the S-9 activated treatment groups.

Treatment of strain TA98 with sediment extract 2 in the presence of S-9 resulted in a toxic effect occurring at lower concentrations of extract as the chlorine dose was increased from 0 to 50 ppm. This effect was less evident for strain TA100 under the same conditions (Table I).

A significant increase in mutagenicity was noted with the addition of S-9 to the sediment extracts from stations 2 and 3. Both base-pair substitution and frameshift mutagens were present in the sediment extracts, since mutagenicity was observed with both strains TA100 and TA98 (Figures 2 and 3). The level of mutagenic activity was found to be both strain- and sediment-dependent. For example, strain TA100 exhibited a maximum of 1659 rev/500 μg sediment/plate with the unchlorinated extract of sediment 2 in the presence of S-9 (Figure 2). On the other hand, the highest mutagenic activity with TA98, 1684 rev/1000 μg sediment/plate, was observed in response to an S-9 treated control (unchlorinated) extract of sediment 3 (Figure 3).

For both strains TA98 and TA100, treatment with extracts from sediments 2 and 3 in the presence of S-9 resulted in positive dose-response curves (Figures 2 and 3). There was an initial rise in mutagenicity in most cases, followed by a decrease in the number of revertants per plate. The reverse in the slope of the

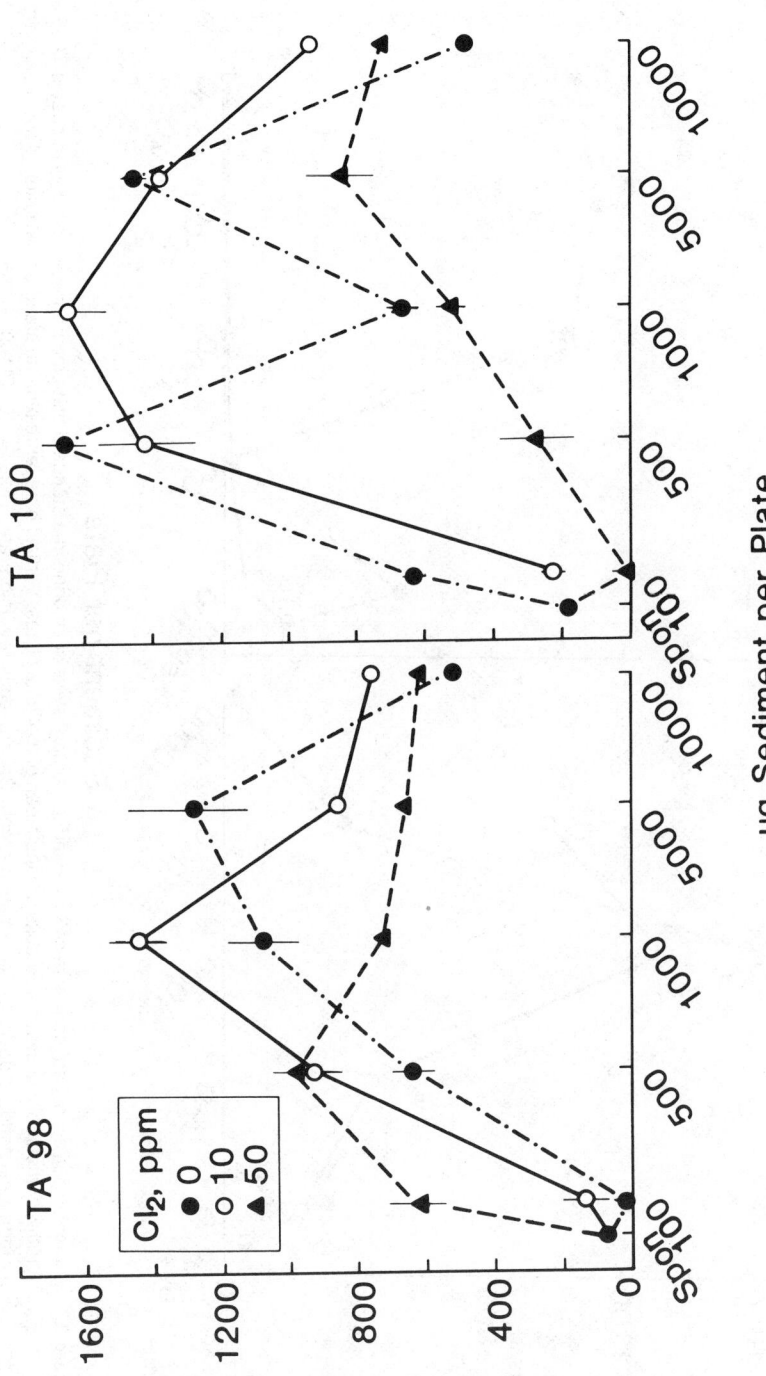

Figure 2. Mutagenic activity of sediment extract 2 exposed to varying levels of chlorine in a standard plate incorporation assay with strains TA98 and TA100 with metabolic activation. Results are the mean of triplicate plates. Vertical bars represent the standard deviation. Absence of vertical bar indicates standard deviation too small to graph. Spontaneous reversion rate: TA98 = 29, TA100 = 195.

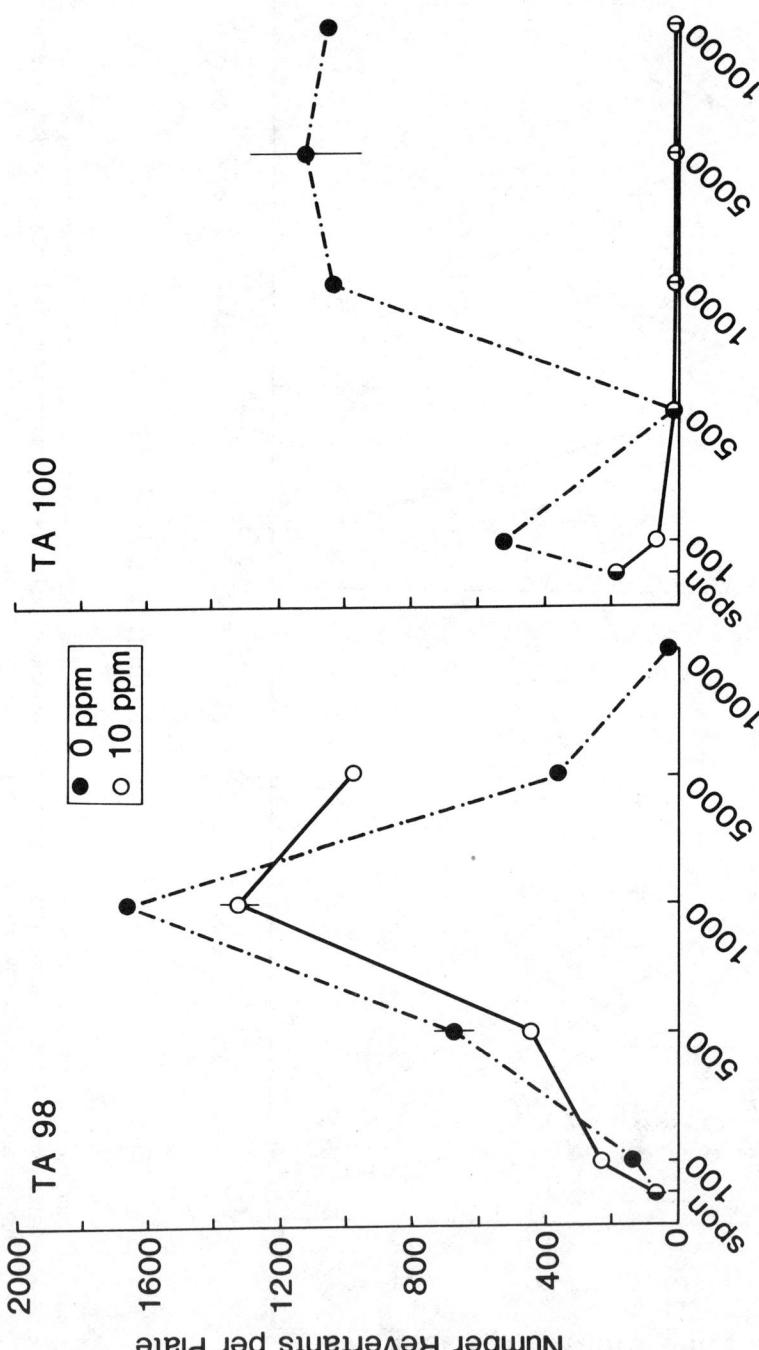

Figure 3. Mutagenic activity of sediment extract 3 exposed to varying levels of chlorine in a standard plate incorporation assay with strains TA98 and TA100 with metabolic activation. Results are the mean of triplicate plates. Vertical bars represent the standard deviation. Absence of vertical bar indicates standard deviation too small to graph. Spontaneous reversion rate: TA98 = 29, TA100 = 195.

dose-response curve was probably the result of toxicity occurring at the higher concentrations.

At the lower concentrations, there was an increase in mutagenicity with an increase in chlorine dose levels for strain TA98 in the presence of sediment 2 extracts and S-9 (Figure 2). As the concentrations of extracts increased, there was a reverse in the slope of the dose-response curves. This reversal occurred at lower concentrations as chlorine dose increased and may be indicative of higher toxicity associated with a higher chlorine dose. When strain TA98 was exposed to sediment 3 extracts in the presence of S-9, the 10-ppm chlorine-treated sediment extract was more mutagenic than the control sediment extract at the lowest concentration tested (100 μg sediment per plate) (Figure 3). However, at higher concentrations, more mutagenicity was associated with the unchlorinated extracts. This may be the result of the destruction of mutagenic components of the sediment by their reactions with chlorine. The unchlorinated extract of sediment 3 was much more toxic at higher dose levels than the unchlorinated sediment 2 extract (Figures 2 and 3).

The unchlorinated sediment 2 extract with the addition of S-9 was more mutagenic at a lower concentration to strain TA100 than to TA98 (Figure 2). The addition of 10 ppm chlorine tended to slightly depress the mutagenic activity of this extract to TA100. A large decrease in mutagenicity was observed for the 50-ppm treated sediment extract. This response was difficult to interpret because of the competing effects of toxicity with mutagenicity. When strain TA100 was exposed to sediment 3 in the presence of S-9, the unchlorinated extract resulted in a lower mutagenic response than was observed for TA100 and unchlorinated sediment 2 extract (Figures 2 and 3). The 10-ppm chlorinated sediment 3 extract was toxic at all dose levels tested, with the number of revertants per plate much lower than the spontaneous reversion frequency (Figure 3).

The assay plates for sediment 7 were contaminated, and therefore no trends could be reported for this sediment extract.

In looking at the effect of chlorination on the mutagenic activity associated with these sediments, we found with strain TA98 that as the level of chlorine increased, the mutagenic activity also increased (Figures 2 and 3). Because of the competing effects of toxicity and mutagenic response, this phenomenon is best observed when one holds the level of dilution constant at 10^4 (equivalent to 100 μg sediment per plate), since at higher concentrations (lower dilutions) the toxic nature of the extracts tends to mask mutagenic activity (Figure 4). For example, the relative levels of mutagenicity observed in sediment 2 extracts at this dilution in TA98 with S-9 were 30 rev/plate, 115 rev/plate, and 605 rev/plate for unchlorinated, 10 mg/kg and 50 mg/kg chlorine-treated sediment, respectively, and in sediment 3 extracts 143 rev/plate and 234 rev/plate for the unchlorinated and 10 mg/kg chlorine-treated sediment, respectively. The effect of chlorine on mutagenicity may be further illustrated by the shifting of the reversion maxima for TA98 from lower to higher dilutions as the level of chlorine dose is increased from 0 to 50 ppm (Figure 2). In sediment 2 extracts,

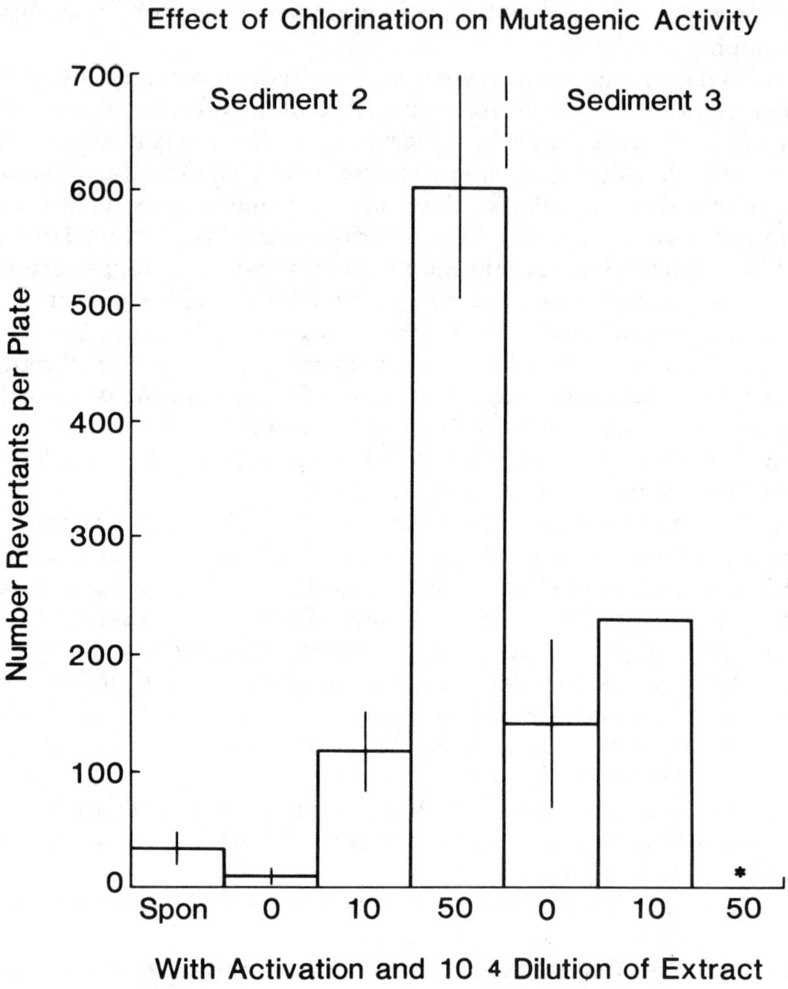

Figure 4. Effect of chlorination on mutagenicity of nonpreincubated sediment extracts diluted to 10^4 (equivalent to 100 μg sediment/plate) for strain TA98. Vertical bars represent the standard deviation. Absence of vertical bar indicates standard deviation too small to graph. Data from the 50-ppm chlorine treatment of sediment 3 are not presented because of contamination of the assay plates.

in TA98 with S-9 activation, the maximum mutagenic responses were observed at the 5000-μg sediment/plate dose for the unchlorinated sediment and at 1000 and 500-μg sediment/plate dilutions for the 10- and 50-ppm chlorinated sediments, respectively. This shift in maxima was not observed for TA98 with sediment 3 extracts. The levels of mutagens from sediment 3 were similar to those in sediment 2.

Preincubation Assay

Preincubation of strain TA100 with sediment extracts and a buffered medium (i.e., no metabolic activation) revealed sediments 2, 3, and 7 to be nonmutagenic (Figure 5, for example), which indicates that these sediments do not contain detectable levels of direct-acting mutagens. For reasons not known at this time, much less toxicity was observed with this assay technique than was observed in the standard assay. The addition of the microsomal activation system to the reaction vessel of the preincubated extracts exposed to strain TA100 did little to alter the level of mutagenicity observed without activation. Although a slightly larger number of revertants was noted for the activated samples as compared with the unactivated samples, a doubling of the number of revertants above the spontaneous reversion frequency was never observed. Therefore, the overall response of TA100 should be described as nonmutagenic for all three sediments (Figure 5, for example).

Preincubation of the three sediment extracts with TA98 in the absence of metabolic activation showed all three to be nonmutagenic, again suggesting the absence of direct-acting mutagens (Figures 6 and 7). Toxicity was again

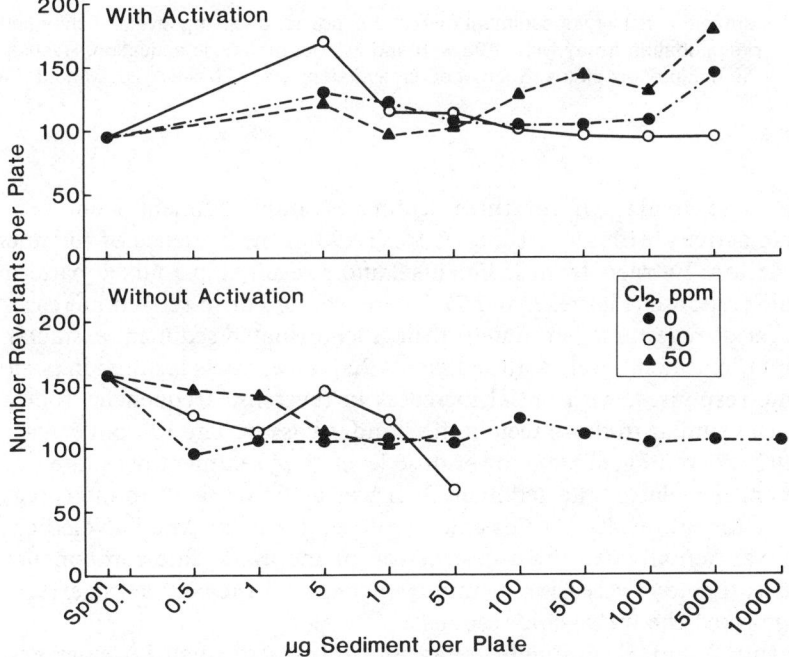

Figure 5. Mutagenic activity of sediment extract 7 exposed to varying levels of chlorine in a preincubation assay with TA100 with and without metabolic activation. Results are the mean of six plates. Spontaneous reversion rate: with S-9 = 97, without S-9 = 156.

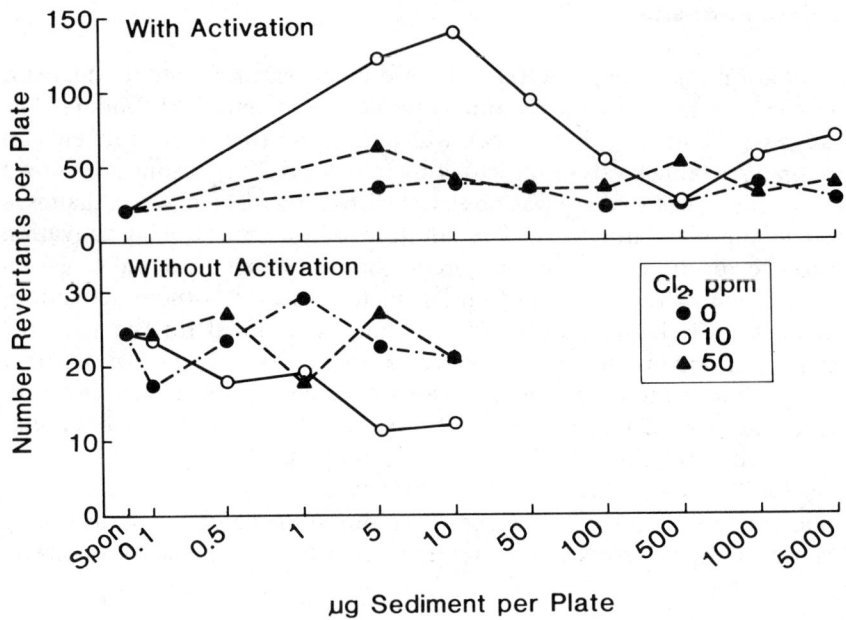

Figure 6. Mutagenic activity of sediment extract 2 exposed to varying levels of chlorine in a preincubation assay with TA98 with and without metabolic activation. Results are the mean of six plates. Spontaneous reversion rate: with S-9 = 20, without S-9 = 24.

lower in these assays. Of the three sediments, only sediment 7 showed no mutagenic activity with any treatment for TA98 in the presence of the microsomal fraction. Extracts from sediments 2 and 3 resulted in a mixed pattern of mutagenic response (Figures 6 and 7). There were no dose-related increases in the number of revertants per plate for the unchlorinated sediment extracts. At the 10-ppm treatment level, both sediments 2 and 3 extracts resulted in positive mutagenic responses, with initial increases in reversion frequencies followed by decreases similar to those seen in the standard assay. The 10-ppm extract of sediment 2 was mutagenic at a lower dose level, 5 μg sediment per plate vs 500 μg sediment per plate, than sediment 3. It was again difficult to interpret the 50-ppm treatment results. At this chlorine dose, it was not known whether the chlorine was actually causing a destruction of the mutagenic components of the extract, thereby decreasing its mutagenicity, or if the competing effects of toxicity masked the mutagenic response.

Sediments 2 and 3, although mutagenic, displayed much lower reversion frequencies (highest level of mutagenicity observed being 136 rev/10 μg sediment/plate for sediment 2, treated with 10 ppm Cl_2 with S-9 activation) than were observed in the standard assay (as many as 1450 rev/1000 μg sediment/plate were noted for sediment 2, 10 ppm Cl_2 with metabolic activa-

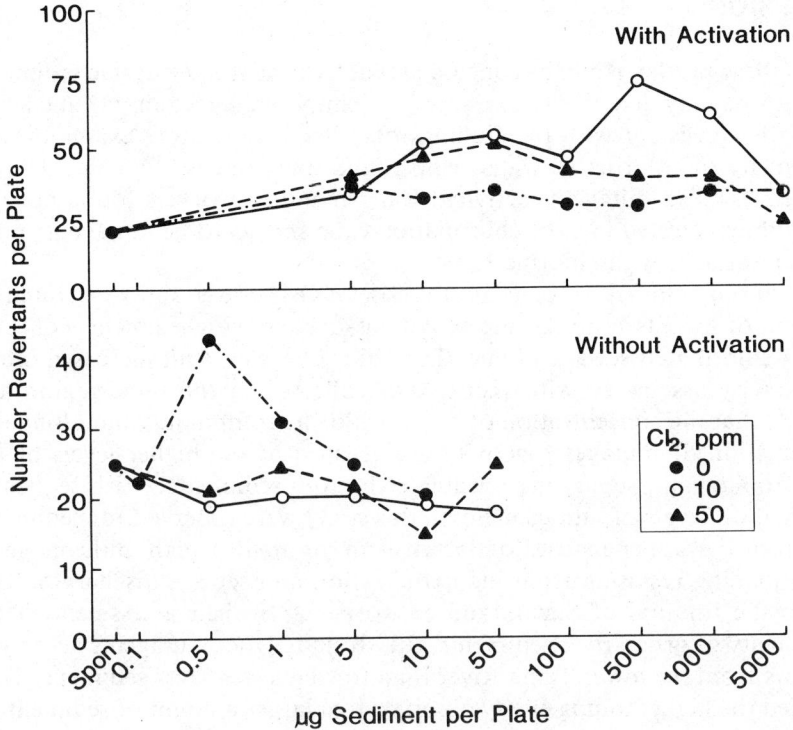

Figure 7. Mutagenic activity of sediment extract 3 exposed to varying levels of chlorine in a preincubation assay with TA98 with and without metabolic activation. Results are the mean of six plates. Spontaneous reversion rate: with S-9 = 20, without S-9 = 24.

tion) (Figures 2 and 6). Although the reversion frequencies were lower in the preincubation assay, a positive mutagenic response was observed at a lower extract concentration than in the standard assay. For example, the lowest extract concentration that resulted in a doubling of the number of revertants over the spontaneous reversion frequency for strain TA98, treated with the sediment 2, 10-ppm chlorine extract with metabolic activation, was 500 µg sediment per plate in the standard assay, compared to 5 µg sediment per plate in the preincubation assay (Figures 2 and 6).

When compared with data from the standard plate incorporation assay, the preincubated extracts tended to display less mutagenicity and less toxicity than observed in the standard assay plates. This may be the result of the degradation of some toxicants and direct-acting mutagens by the microsomal enzyme system during the prolonged exposure time. However, a similar trend was also observed in the preincubation assays without S-9, suggesting that other mechanisms of metabolism and detoxification may be active.

DISCUSSION

From these results, it can be concluded that treatment of estuarine sediments with chlorine may give rise to extracts that exhibit enhanced mutagenic activity. This is consistent with the findings of other investigators examining the mutagenicity of chlorinated water, who found that increases in chlorine dose tended to increase mutagenic activity.[6] Rapson and co-workers found none of the compounds tested in their chlorination experiments to be mutagenic prior to their treatment with chlorine.[7]

Some of our findings are contrary to those of Osborne et al.,[8] who found no indication of increased mutagenic activity associated with sediments collected within a chlorinated sewage plume. They did, however, find increased mutagenic activity associated with fish tissues collected in the same region and suggested that the concentration of organochlorine compounds and ultimately the expression of mutagenic activity are greatest at the higher levels of the aquatic food chain, suggesting a water → bottom sediment → fish sequence.

The highest levels of mutagenicity in this study were observed for sediments taken from the upper reaches of the river in an area of high anthropogenic organic loading resulting from industrialization and sewage discharges. This parallels the findings of Suzuki and co-workers.[9] In their assessment of the mutagenicity of urban river sediments, they found higher mutagenic activity in sediments from the lower Tama River than for the upper river sediment. They attributed the higher mutagenicity observed to a larger amount of sedimentary organic matter associated with the lower portion of the river.

High toxic organic loading in the upper river could also explain the high toxicity of sediments 2 and 3 in the standard assay without microsomal activation. The fact that metabolic activation was necessary for the expression of mutagenicity was expected, because other investigators found concentrated sediment extracts and suspended particulate matter to be extremely toxic, thus supressing the mutagenic response.[10-12] A study of the mutagenicity of Chesapeake Bay sediments and water conducted by Voll and co-workers[12] revealed the sediment extracts to be highly toxic but proved inconclusive in determining the mutagenicity associated with the sediments. The disparity between our findings can probably be attributed to the fact that their investigations utilized the Ames "spot test," a less sensitive procedure than either the plate incorporation or preincubation techniques used in this study.

CONCLUSIONS

Patapsco River sediments contain both frameshift and base-pair substitution mutagens. Higher levels of mutagenic activity are associated with sediments collected from a highly industrialized portion of the river. Microsomal activation increased the mutagenic activity and decreased the cytotoxic effects of the sediment extracts. In the absence of metabolic activation, cytotoxicity may mask the effects of direct-acting mutagens. At lower extract concentrations, chlorine treatment caused an increase in mutagenic activity in some sediments. This relationship was masked at higher concentrations by cytotoxicity.

The data presented here indicate the potential threat to aquatic organisms posed by the chlorination of sediment-sorbed organic matter. Further study of the mechanisms of reactions of chlorine with these organic constituents should generate valuable information concerning the mutagenic properties of chlorination products. Further research could include the fractionation of the crude extracts to separate the toxic components from those that are mutagenic. In addition, the role of sorption/desorption phenomena in determining the fate of mutagens associated with sediments needs to be studied.

ACKNOWLEDGMENTS

We would like to thank Dr. B. N. Ames for supplying us with the bacterial tester strains, Ms. Fran Younger for the graphic illustrations, and Ms. Gail Canaday for the typing of this manuscript. This research was sponsored in part by the Maryland Sea Grant Program (Grant NA84AA-D-0014) and the U.S. Department of Interior (Grant CT 372203). Contribution No. 1578, Center for Environmental and Estuarine Studies of the University of Maryland.

REFERENCES

1. Cumming, R. B. "Potential for Increased Mutagenic Risk to the Human Population due to Products of Water Chlorination," in *Water Chlorination: Environmental Impact and Health Effects, Vol. 1*, R. L. Jolley, Ed. (Ann Arbor, MI: Ann Arbor Science Publishers, Inc., 1976), pp. 229-241.
2. Meier, J. R., R. D. Lingg, and R. J. Bull. "Formation of Mutagens Following Chlorination of Humic Acid: A Model for Mutagen Formation During Drinking Water Treatment," *Mutat. Res.* 118:25-41 (1983).
3. Loper, J. C. "Mutagenic Effects of Organic Compounds in Drinking Water," *Mutat. Res.* 76:241-268 (1980).
4. Ames, B. N., J. McCann, and E. Yamasaki. "Methods for Detecting Carcinogens and Mutagens with the *Salmonella*/Mammalian-Microsome Mutagenicity Test," *Mutat. Res.* 31:347-363 (1975).
5. Maron, D. M., and B. N. Ames. "Revised Methods for the *Salmonella* Mutagenicity Test," *Mutat. Res.* 113:173-215 (1983).
6. Stover, E. L., R. B. Cumming, N. E. Lee, and R. L. Jolley. "Chlorine vs Ozone at Marlborough, Massachusetts: Disinfection and Mutagenic Activity Screening," in *Water Chlorination: Environmental Impact and Health Effects, Vol. 4*, R. L. Jolley, W. A. Brungs, J. A. Cotruvo, R. B. Cumming, J. S. Mattice, and V. A. Jacobs, Eds. (Ann Arbor, MI: Ann Arbor Science Publishers, Inc., 1983), pp. 1249-1260.
7. Rapson, W. H., M. A. Nazar, and V. V. Butsky. "Mutagenicity Produced by Aqueous Chlorination of Organic Compounds," *Bull. Environ. Contam. Toxicol.* 24:590-596 (1980).

8. Osborne, L. L., R. W. Davies, K. R. Dixon, and R. L. Moore. "Mutagenicity of Fish and Sediments in the Sheep River, Alberta," *Water Res.* 16:899–902 (1982).
9. Suzuki, J., T. Sadamasu, and S. Suzuki. "Mutagenic Activity of Organic Matter in an Urban Sediment," *Environ. Pollut. Ser. A* 29(2):91–99 (1982).
10. Dexter, R. N., S. P. Pavlou, and R. M. Kocan. "Mutagenicity of Puget Sound Sediment Extract: Feasibility Study," PB80-152770 (Boulder, CO: National Oceanic Atmospheric Administration, 1979).
11. Moore, R. L., L. L. Osborne, and R. W. Davies. "The Mutagenic Activity in a Section of the Sheep River, Alberta Receiving a Chlorinated Sewage Effluent," *Water Res.* 14:917–920 (1980).
12. Voll, M. J., J. Isbister, L. Isaki, M. McCommas, and R. R. Colwell. "Effects of Microbial Activity on Aquatic Pollutants," *Ann. N.Y. Acad. Sci.* 298:104–110 (1977).

CHAPTER 109

Chlorine Minimization in Macrofouling Control in The Netherlands

Henk A. Jenner

In the Netherlands, macrofouling continues to be a nuisance to power plants and other industries using marine or fresh water. Besides the well-known problems with *Mytilus edulis* at coastal plants,[1-3] problems also arise with the freshwater "zebra" mussel, *Dreissena polymorpha*,[4-8] whose increased abundance can probably be traced to amelioration of water quality.

Most plants have no way to combat macrofouling other than by chlorination. Although heat treatment and mechanical cleaning are far better options, the use of heat is often restricted to new plants designed for this method. A yearly mechanical cleaning often is not compatible with the operation schedule. This problem is further complicated by the tightening of regulations for the treatment of water with chlorine. The question now is how to reduce chlorine consumption while maintaining an acceptable suppression of macrofouling.

Probably the most important abiotic regulatory mechanism that determines the size of the mussel population is the availability of solid substrate for settlement of mussel spat. Unfortunately, each intake structure forms an excellent substrate for byssal attachment by mussels. The lack of predators and the continuous flow of water with food and oxygen from the biotic factor allow rapid growth of the mussel population.

Many antifouling methods have been investigated over the years. Of these, the best options from an environmental point of view have been heat treatment and mechanical cleaning.[9,10] This chapter surveys the possibilities of reducing chlorine consumption in marine and fresh water with the aid of a newly designed mussel monitor.[11] The monitor enables plant operators to obtain insight to the occurrence and growth of fouling and the effectiveness of the chlorination scheme.

DESIGN AND OPERATION OF THE MONITOR

The monitor (Figure 1) is designed as a bypass of the cooling water conduit to follow and observe mussel settlement and growth and the effect of chlorination. The monitor is a closed polyvinyl chloride (PVC) container in which two larger and four smaller PVC plates are placed. Water flows from top to

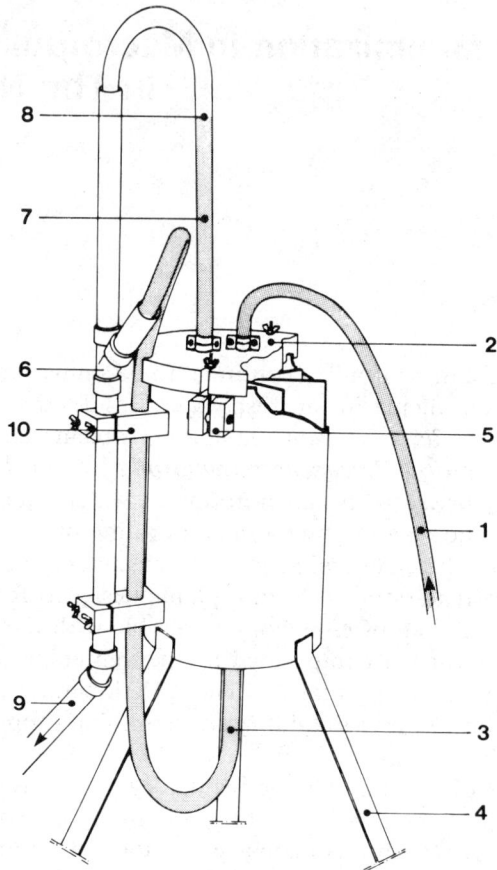

Figure 1. Design mussel monitor: (1) water supply; (2) detachable cover; (3) drain hose; (4) stand; (5) mounting clamps for cover; (6) tee, with the end of the drain hose; (7) overflow hose; (8) water level during operation; (9) outlet end of PVC tube; (10) holder for PVC tube and drain hose.

bottom. A PVC tube is fitted vertically against the container. A 3/4-in. drain hose permits water to flow freely in the tube on a level with the container cover. This avoids siphoning the water. With a flowrate of ~40 L/min. the average flow near the drain hole and in the drain hose is high enough ($\pm$ 3 m/s) to prevent silting of the container and settlement of mussels in the drain. If the water supply stops, the container level does not fall below the highest point in the drain hose. A 3/4-in. overflow hose runs from the container cover to the upper part of the PVC tube. The top of the tube is 1.5 to 2.5 m above the cover. Should the drain at the container bottom be obstructed, water would drain through the overflow, thus avoiding flooding. Normally, the effective pressure of a 1.5 to 2.5-m water column (visible in the overflow hose) is sufficient to maintain a high flow in the monitor and prevent silting.

To achieve an optimal environment for mussel growth in the monitor, the design and location of the withdrawal point must be chosen with care to ensure that spat concentrations in the monitor correspond to those in the cooling water flow. The velocity in the conduit should preferably be high enough (> 2 m/s) to prevent sedimentation of the spat. If the velocity is lower, the monitor should be located directly behind the pumps, especially when long conduits are to be monitored.

SEAWATER CHLORINATION

Conventionally, seawater chlorination begins when water temperatures rise above 10°C; at this temperature the first veligers are found in the zooplankton.

Depending on the experience of plant chemists, a continuous or discontinuous chlorination scheme [with concentrations of 0.1 to 0.5 mg/L free available oxidant (FAO)] was chosen.

The first step in this study was the introduction of the mussel monitor. The entrance of mussel spat was easy to observe by sampling the PVC plates of the monitor. It became apparent that spat settles about 4 to 6 weeks later than at the 10°C level. Mussel fouling begins with the settlement of spat and not with the appearance of veligers in zooplankton samples.[12] The spat in monitor samples must have a length of >750 μm, the so-called second settlement spat, which can settle at relatively high water velocities. Because of the low flow in the monitor (±1 cm/s), small first settlement spat are also found (350–750 μm). Furthermore, it was found that chlorination can stop when the water temperature drops below 12°C, which can lead to a considerable saving of chlorine. At the power plant where most of the research was done, a saving of more than 30% on hypochloride was realized.

The second step was to investigate the effectiveness of continuous and discontinuous (4 h, on and off) chlorination at low level. Continuous and discontinuous low-level chlorination at 0.1 to 0.2 mg/L (FAO) diminishes settling; however, settlement will occur. The numbers of spat are reduced by about a factor of 10 (Figure 2). The sampled spat are divided into two groups (< and > 750 μm). This distinction is made to differentiate between spat of the first and second settlements. The mussel problem is created when the second-settlement group seeks a more final place for settling, which shows that settlement behavior dominates the toxic action of chlorine more than expected.

With discontinuous chlorination, growth is reduced but not inhibited; the reduction is about 50% (Figure 3). The mean size of mussels at discontinuous chlorination is only 3.8 mm, as compared with mussels of 7.2 mm that grow on ropes in the intake area. More information about growth retardation is given by Khalanski and Bordet.[13] It is noteworthy that settled spat are not very active (in crawling) and do not migrate, but rather prefer to be immobile during chlorination. This is contrary to the earlier observations of James[14] of so-called exomotive chlorination at low levels.

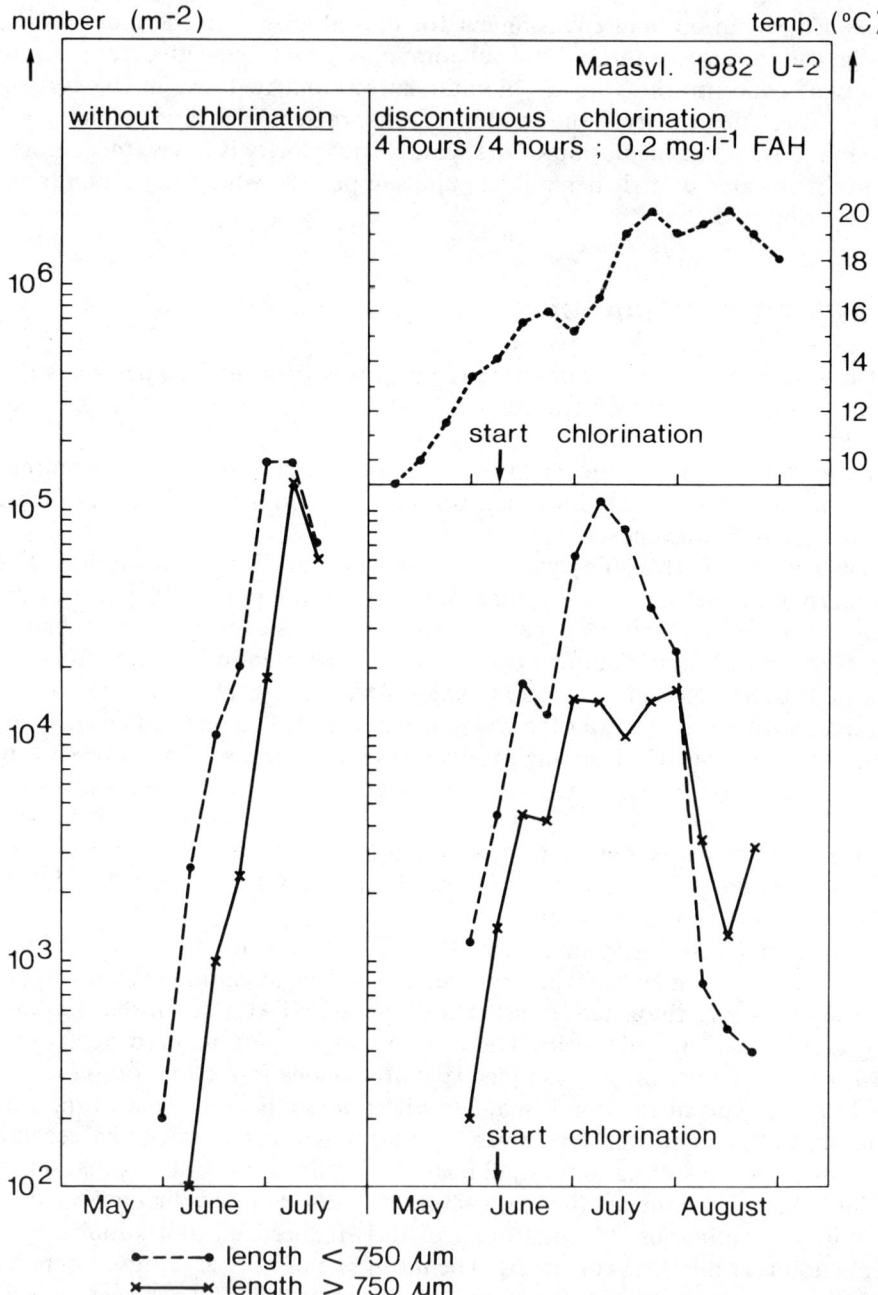

Figure 2. Sequence of mussel spat density (m^{-2}) at continuous exposition, sampled (monitors) at weekly intervals, together with the inlet water temperatures; power station Maasvlakte (two units of ±540 MW).

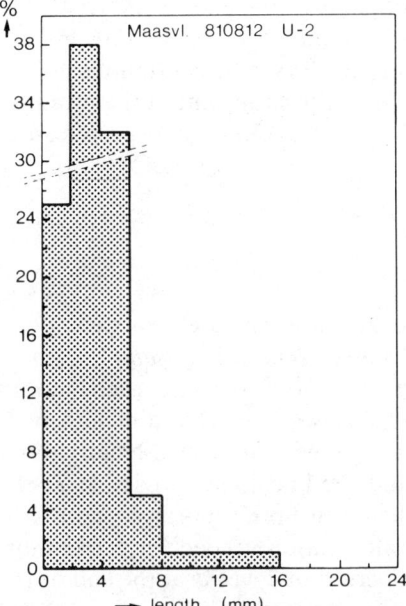 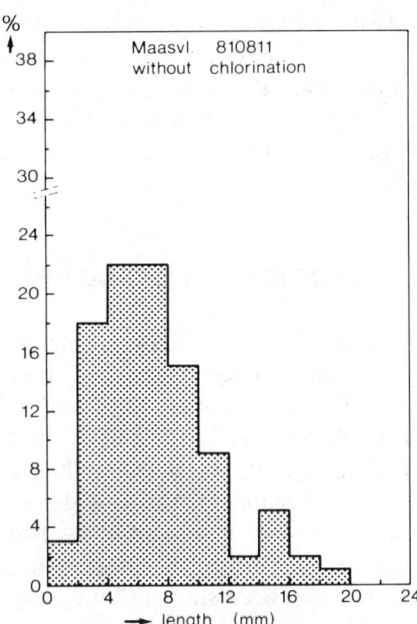

Figure 3. Length frequency distribution of mussels sampled in a conduit subjected to discontinuous chlorination (left) and on a rope just before the inlet without chlorination (right). Macrofouling periods were from September 1980 to August 1981 (left) and from May 1981 to August 1981 (right). Discontinuous chlorination began June 16, 1981.

Mortality occurs after extended periods of continuous chlorination. According to Khalanski and Bordet,[13] mortality occurred over a period of 15 to 135 d at residual chlorine concentrations of 0.2 to 1.0 mg/L. The resistance of mussels to chlorination was found to be related to conditions, food, and genetic factors.

It can be concluded that low-level continuous chlorination will eventually kill all macrofouling. The necessary amount of chlorine will be considerable, as will the environmental effects. A lower chlorine consumption can be realized by taking the following actions:

1. Monitoring spat settlement, and, after the appearance of mussel spat >750 μm (and not earlier), beginning continuous chlorination, especially during peak settlement (2–3 weeks);

2. Applying discontinuous chlorination during the summer months with the purpose of reducing growth rather than killing. If a second settlement period occurs (as observed in the monitor), continuous chlorination for about 2 weeks can be applied. If a second settlement does not occur, the discontinuous scheme can be continued until the water temperature reaches about 12°C.

1430   WATER CHLORINATION

If manual cleaning of the system is not possible at the end of the first year, then the second year must begin with discontinuous chlorination (for growth retardation) at a water temperature of about 12°C. Again, continuous chlorination can be applied after the beginning of mussel settlement in the monitor. Manual cleaning is necessary in the second year to avoid an unexpected shutdown.

FRESHWATER CHLORINATION

In recent years, fouling problems with the freshwater mussel *Dreissena polymorpha* have been increasing in rivers and lakes. Where *Dreissena* became a problem, continuous chlorination was applied for 1 week each month (June–October). At the end of the first year, *Dreissena* reaches a mean length of about 8 mm, which is small as compared with *Mytilus* (mean length about 20 mm). Because of this low growth rate and the known sensitivity of *Dreissena* for chlorine, research has been conducted in both laboratory and field experiments to find the shortest period of chlorination needed for 100% mortality. It has been shown that a chlorination period of 2 weeks at the end of the growing season will kill more than 95% of the mussels.

Figure 4 shows mortality rates and the influence of water temperature in two chlorination experiments, both at 0.5 mg/L total residual oxidant (TRO) at a large petrochemical plant. Mussel monitors were installed to observe spatfall

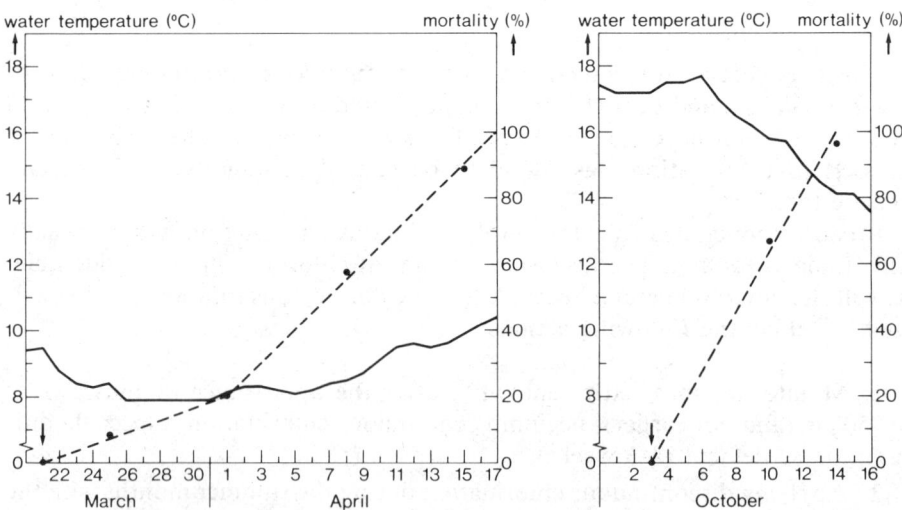

Figure 4. Mortality rate of *Dreissena* at continuous chlorination (0.4 mg/L TRO) in spring and autumn at a large petrochemical plant to show dependence on water temperature. (– – –) percent mortality; (———) water temperature (°C) at beginning of chlorination.

and growth during the summer months and during chlorination. Additional caged mussels were placed in the monitors to determine mortality rates. In the spring, with low water temperatures, it takes more than 3 weeks for 100% mortality; however, at the end of the summer, with water temperatures of about 15°C, only 2 weeks are required. The killed mussels drop from the walls of the conduits and pass through the system if they are not too large. Figure 5 shows the results of laboratory experiments at three different chlorine concentrations. All experiments were carried out in a once-through system (2 × 50 mussels per experiment). The results of laboratory experiments are in good agreement with those in the field at 0.5 mg/L TRO. Doubling the chlorine concentration (1 mg/L TRO) has less effect on the mortality rate than expected. On the other hand, half the concentration (0.25 mg/L TRO) would be enough to kill the mussels in about 3 weeks.

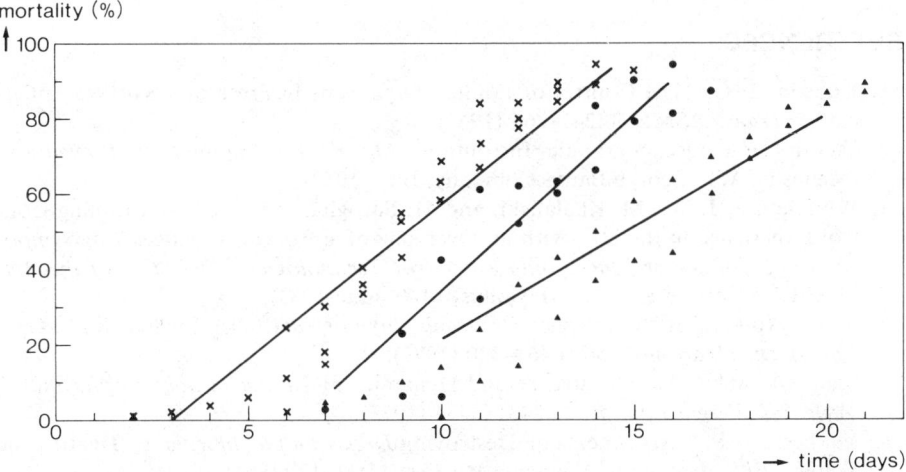

Figure 5. Mortality rate of *Dreissena* in laboratory experiments at three chlorine concentrations. Lines represent linear correlation. Water temperature was 12 to 15°C. x = 1.0 mg/L TRO, 3 experiments in duplicate, r = 0.972; ● = 0.5 mg/L TRO, 2 experiments in duplicate, r = 0.988; ▲ = 0.25 mg/L TRO, 2 experiments in duplicate, r = 0.956.

CONCLUSIONS

With most macrofouling problems, there is no solution other than chlorination. However, from an environmental point of view, far better antifouling methods exist, such as regular mechanical cleaning or heat treatment.

Nevertheless, chlorine consumption can be reduced by better-fitted chlorination regimes based on the use of mussel monitors. With mussel monitors, plant operators can observe the occurrence and frequency of macrofouling in the inlet system. The beginning of chlorination and its subsequent effectiveness can be determined accurately. With seawater, the regime can change from a continuous to an alternating continuous/discontinuous scheme at low concentrations (0.1 to 0.2 mg/L TRO), with the purpose of retarding settlement density and growth. With fresh water, continuous chlorination for 2 weeks after the breeding season would be necessary for a complete kill of all macrofouling.

The present situation, at least for The Netherlands, can be summarized as follows: Macrofouling problems increase with the amelioration of water quality which in turn, at first instance, can increase the yearly consumption of chlorine by industry; however, with the application of more-optimized antifouling methods, a considerable decrease in the amount of chlorine used is possible.

REFERENCES

1. Dobson, J. G. "The Control of Fouling Organisms in Fresh and Salt Water Circuits," *Trans. ASME* 68:247-265 (1945).
2. Woods Hole Oceanographic Institution. *Marine Fouling and its Prevention*, (Menasha, WI: George Banta Company, Inc., 1952).
3. Whitehouse, J. W., M. Khalanski, and M. Saroglia. "Marine Macrofouling Control Experience in the UK, with an Overview of European Practices," in *Symposium On Condenser Macrofouling Control Technologies—The State-of-the-Art*, I. A. Diaz-Tous et al., Eds. (Hyannis, MA: June 1983).
4. Stanczykowska, A. "Ecology of *Dreissena polymorpha* (Pall.) (Bivalvia) in Lakes," *Pol. Arch. Hydrobiol.* 24(4):461-530 (1977).
5. Stanczykowska, A. "Occurrence and Dynamics of *Dreissena polymorpha* Pall.," *Verh. Int. Verein Limnol.* 20:2431-2434 (1978).
6. Mikheer, V. P. "Experiments on Destroying *Dreissena polymorpha* by Heating the Water," *Bjul. Inst. Biol. Vodochr. Moskwa* 11:11-12 (1961).
7. Kirpichenko, M. Ya., V. P. Mikheer, and E. P. Stern. "Battling Overgrowths of *Dreissena* at Hydroelectric Power Plants," translation from *The Russian Elektricheskie Stantsii* 5:30-32, (1962), ORNL-tr-4705, (Oak Ridge, TN: Oak Ridge National Laboratory).
8. Mattice, J. S. "Freshwater Macrofouling and Control with Emphasis on Corbicula," in *Symposium on Condenser Macrofouling Control Technologies—The State-of-the-Art*, I. A. Diaz-Tous et al., Eds. (Hyannis, MA: June 1983).
9. Burton, D. T., and L. H. Liden. "Biofouling Control Alternatives to Chlorine for Power Plant Cooling Water Systems: An Overview," in *Water Chlorination; Environmental Impact and Health Effects, Vol. 2*, R. L. Jolley, W. A. Brungs, and R. B. Cumming, Eds. (Ann Arbor, MI: Ann Arbor Science Publishers, Inc., 1978) pp. 717-734.
10. Jenner, H. A. "Control of Mussel Fouling in the Netherlands: Experimental and Existing Methods," in *Symposium On Condenser Macrofouling Control Technologies—The State-of-the-Art*, I. A. Diaz-Tous, et al., Eds. (Hyannis, MA: June 1983).

11. Jenner, H. A. "A Microcosm Monitoring Mussel Fouling", in *Symposium On Condensor Macrofouling Control Technologies—The State-of-the-Art*, I. A. Diaz-Tous et al. Eds. (Hyannis, MA: June 1983).
12. Bayne, B. L. "Primary and Secondary Settlement in *Mytilus edulis* L. (Mollusca)", *J. Anim. Ecol.* 33:513–523 (1964).
13. Khalanski, M., and F. Bordet. "Effects of Chlorination on Marine Mussels," in *Water Chlorination: Environmental Impact and Health Effects, Vol. 3*, R. L. Jolley, W. A. Brungs, and R. B. Cumming, Eds. (Ann Arbor,MI: Ann Arbor Science Publishers, Inc., 1980), pp. 557–568.
14. James, W. G. "Mussel Fouling and Use of Exomotive Chlorination," *Chem. Ind.* London 1967:994–996.

CHAPTER **110**

A Predictive Model for Destruction of Biofilms with Chlorine

Meletios Platon and Thomas D. Waite

The formation of a microbiological film on solid surfaces creates a costly problem for many industries. Although the mechanisms by which microorganisms become attached to solid surfaces are not well understood, they have created a great deal of interest, especially as related to condenser fouling. The fouling of condenser systems in the electric power industry is a major problem, because it results in the excessive use of fuels during the power generation process. Traditionally, the common method for controlling slime growth in electric generating plant cooling systems has been with intermittent or continuous chlorine treatment. Chlorine has been proven effective for controlling the growth of biofilms. In closed cooling systems, a residual of chlorine in the circulating cooling water can be maintained for biological control. In open systems using natural waters, the chlorine concentration must be limited to produce a minimum residual in the discharge.[1] The U.S. Environmental Protection Agency has recently limited chlorine concentrations in cooling water to 0.2 ppm for 2 h/d.

It has been observed that biofouling of condensers is a seasonal phenomenon that increases in the summer because of higher cooling water temperatures. Additionally, the demand for electrical power is higher during the summer months, thus creating an even bigger problem for utilities. It has also been shown that fuel savings of $100 million per year can be realized by reducing turbine backpressure by only 0.1 in. mercury absolute.[2] Considering that common backpressure reductions are in the range of 0.2 to 0.5 in. mercury, we see that a need exists for keeping condensers clean, especially as fuel prices remain high.[1-3]

Many utilities are now faced with the dilemma of determining what minimum dose of chlorine can be applied to maintain condenser cleanliness. This usually requires a chlorine minimization study, and many of these studies require model condensers to be fabricated, instrumented, and monitored over long periods of time at the utility site. During a study, different combinations of chlorine concentration and contact time must be monitored to generate a matrix of data for determining the most effective combination. Considering that the fouling process may take at least a month before reliable data can be obtained for each concentration and contact period, experiments often become burdensome tasks.

There is a clear need for a predictive model to estimate the effects of chlorine on biofilm accumulation in condensers. Although several models have been developed predicting biofilm accumulation rates, none have effectively coupled chlorination effects to a biofilm accumulation rate model. Equation (1) shows a biofilm accumulation rate expression as proposed by Characklis.[4]

$$\frac{dx}{dt} = af_1(x) + \delta f_2(x) - \zeta f_3(x) \tag{1}$$

In this relationship, x = biofilm thickness; a, δ, and ζ are all constants relating different growth phenomena, and $f_1(x)$, $f_2(x)$, and $f_3(x)$ are different functions of x. These processes were reduced by Characklis to the simple expression:

$$\frac{dx}{dt} = \beta(x) \tag{2}$$

where β = the overall growth constant, representing effects resulting from biomass growth rate, biofilm accumulation rate, and Reynolds number.

Perhaps the most intensive modeling effort of biofilm growth is that reported by Kirkpatrick et al.[5] They have developed a model of heat and mass transfer for a fluid flowing in a hollow cylinder coated with an axially symmetrical biofilm. For a typical heat exchanger, it was found that the presence of a biofilm significantly decreases heat transfer. The biofilm thickness was nearly independent of axial distance in the tube for isothermal turbulent flow. When combined heat and mass transfer was considered, the biofilm thickness varied substantially with fluid temperature.

The Kirkpatrick et al.[5] model is somewhat difficult to use because it requires that many rate constants be determined. The values of these constants will probably be unique to individual utilities and therefore not usable on a general scale. In addition, only biofilm accumulation with time and distance can be predicted by this model.

There is a clear need for a predictive model to estimate the effect of chlorine on biofilm accumulation in power plant condensers. Although several models have been developed predicting biofilm accumulation rates, few have attempted to couple chlorination effects to biofilm accumulation. This chapter describes such a model; once calibrated, it can be used to predict biofilm destruction by chlorine.

Many studies of the disinfection kinetics of suspended bacterial systems have been undertaken. There is a plethora of data in the literature relating bacterial survival vs time when in contact with chlorine. Power plants, however, have been practicing chlorination for years to minimize biofouling, but very little information about biofilm destruction is available. It is therefore assumed that the kinetics of biofilm destruction would be similar to a typical suspended system; therefore, Equation (2) could be modified to include a decay term caused by chlorination.

$$\frac{dx}{dt} = \beta(x) - \gamma(x) \tag{3}$$

where γ is a decay constant related to the disinfection constant in a suspended bacteria system by a factor ϕ. That is, $\gamma = \phi \cdot \kappa$, where κ is the decay constant from a monomolecular rate equation of bacterial inactivation in the presence of chlorine.

The solution to Equation (3) is

$$x_t = x_o e^{(\beta - \gamma) \Delta t} \tag{4}$$

Because different time intervals are used for chlorination, Equation (4) was modified to

$$x_t = x_o e^{\beta t} - x_o e^{\gamma t^*} \tag{5}$$

which is valid for small values of x_o.

Equation (5) is a model representing the destruction of biofilm as a function of chlorine concentration and time of contact. The initial biofilm thickness is x_o, β is the biofilm growth rate constant, γ is the chlorine destruction constant, and t, t* are the time of biofilm growth and time of chlorination, respectively; ϕ is a calibration factor that will be defined later.

To effectively use Equation (5), β, γ, and ϕ must be established as follows:

Biofilm accumulation rates (β) were determined from actual field data. Figure 1 shows typical friction factor data resulting from biofilm accumulation in a condenser tube that was not treated with chlorine. Note that a substantial biofilm formation accumulated over a 40-d period. Several days were required for an initial accumulation, and growth then appeared to follow exponential kinetics. A regression analysis of the data showed that an exponential growth model did indeed fit best with a growth rate of $\beta = 0.084625$ per day and a correlation coefficient of $R = 0.9267$. This value of β was used in all calculations.

Assuming that the kinetics of biofilm destruction by chlorine can be related to disinfection kinetics of suspended systems, the model can be calibrated to predict biofilm destruction. Figure 2 shows typical disinfection kinetics for chlorine in a suspended bacterial system. The figure represents bacterial die-off for chlorine concentration of 0.5, 1, and 1.5 ppm.[6] Rate constants for the three chlorine concentrations were estimated for the raw data as 1.0 h^{-1} for Cl_2 = 0.5 ppm, 2.4 h^{-1} for Cl_2 = 1.0 ppm, and 3.4 h^{-1} for Cl_2 = 1.5 ppm. The rate constants were plotted vs chlorine concentration to extrapolate values for lower concentrations. The correlation coefficient for this line was found to be $R = 0.9954$. These data are used only as an example to show how the model is to be calibrated. Actual in situ disinfection experiments would normally be undertaken to develop these data.

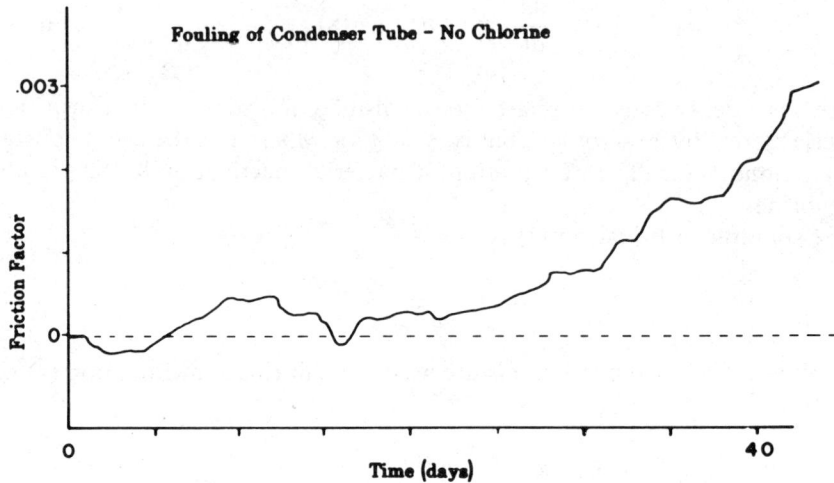

Figure 1. Friction factor changes in an operating condenser with no biofilm control.

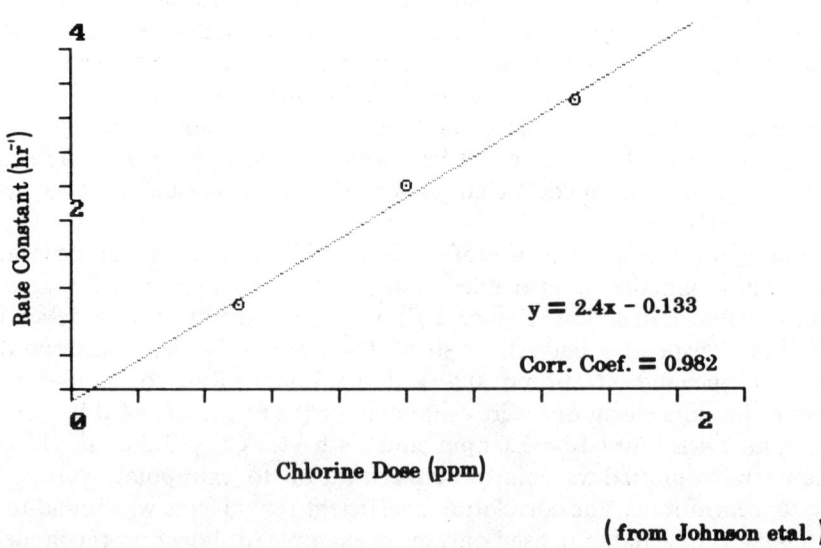

Figure 2. Effects of chlorine concentration on disinfection rate constants.

The values of rate constants for different chlorine doses were calibrated to be used in our model. To do this, we used fouling data from a power plant being treated with chlorine (see Figure 3). In this case, 0.1 ppm chlorine was introduced for 1 h/d, and the γ value was determined to be 0.053778/h by a curve-fitting program using least-squares regression. This value was then used to determine ϕ, since both γ and κ represented disinfection performance for the same concentration of chlorine. The correlation factor ϕ was calculated to be 0.52. It is understood that many factors influence disinfection processes, and it is probable that the value of ϕ reflects a combination of all these factors. However, it is interesting to note that the 0.52 value is close to the 0.50 value, which represents the ratio of the visible surface area of spheres when placed next to each other on a surface. If it is assumed that bacterial cells have an approximate spherical shape, it may be that the cell surface available for reaction is an important parameter in the disinfection process.

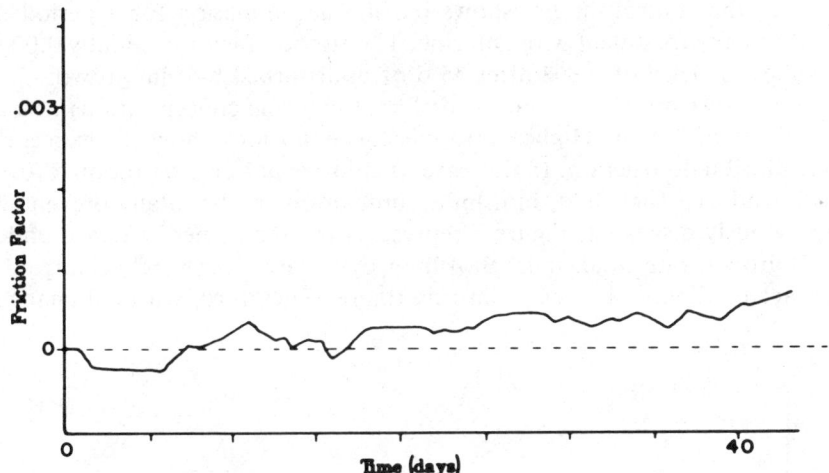

Figure 3. Friction factor changes in an operating condenser with chlorine dosed at 0.1 ppm for 1 h/d.

Assuming that the same factor applies for all chlorine concentrations, rate constants were determined for different chlorine concentrations (Table I).

A computer program was created to calculate biofilm growth at any time, as well as biofilm destruction by chlorine introduced at any concentration for any time interval. By simply introducing the initial biofilm friction factor, the biogrowth constant β, decay constant γ, and the desired time of chlorination, the biofilm friction factor can be predicted at any moment, and the effect of chlorine dosage can be determined.

The program's flexibility will permit any utility to use ambient fouling rates (e.g., different β values for different seasons or different γ values for any chlorine concentration) to monitor and predict biofilm accumulation.

Table I. Rate Constants for Varied Chlorine Concentrations

Chlorine Conc (ppm)	γ Values (h^{-1})
0.1	0.053778
0.2	0.168421
0.15	0.118421
0.07	0.0131578
0.05	0.003681

A series of runs using the calibrated model was performed, and the results are shown in Figures 4 through 11. Most of the runs were made for a period of 35 d as a comparison with actual data obtained from Figures 1 and 3. However, Figure 4 shows a friction factor for only 7 d to show the biofilm destruction during the input of chlorine on a more expanded scale.

Figure 5, the control model, shows biofilm accumulation for a period of 35 d without any treatment with chlorine. The friction factor is initially 0.0002 and reaches the level of 0.003 after 35 d of undisturbed biofilm growth.

Figure 6 shows a series of runs at different chlorine concentrations and at contact times of 1 h/d. Higher chlorine concentrations show, as expected, stronger biofilm destruction. In this case, if chlorine at a concentration of 0.15 ppm is introduced for 1 h/d, biofilm accumulation can be totally prevented.

As previously discussed, Figure 7 represents the model performance under different growth-rate conditions. Biofilm growth rate can severely change the biofilm accumulation, as seen from this figure. Therefore, seasonal changes

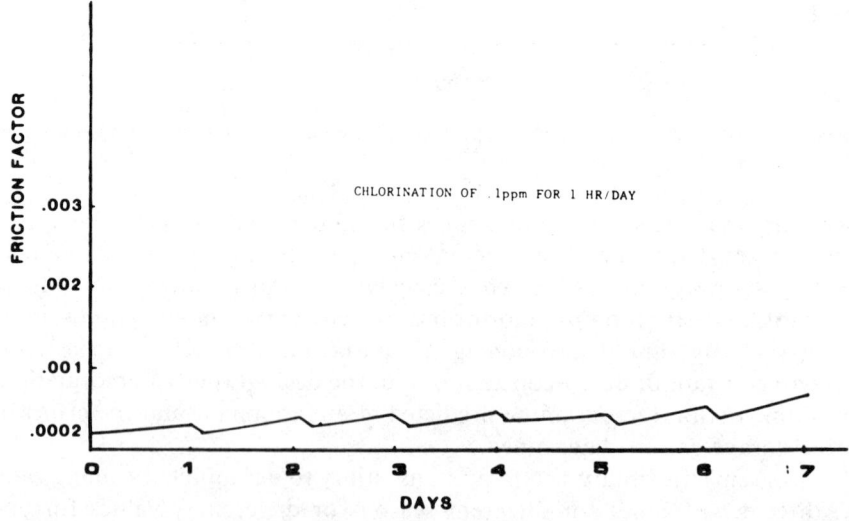

Figure 4. Predicted friction factor changes for 7 d with chlorination at 0.1 ppm for 1 h/d.

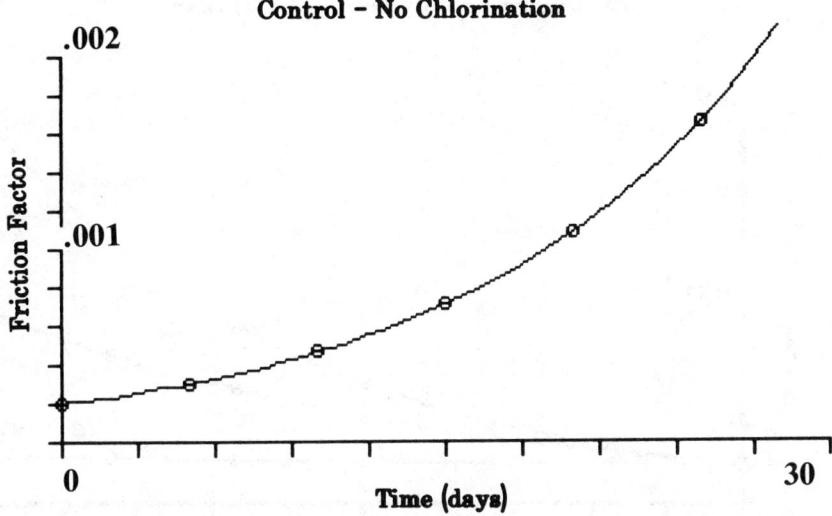

Figure 5. Predicted friction factor change with time and no chlorination.

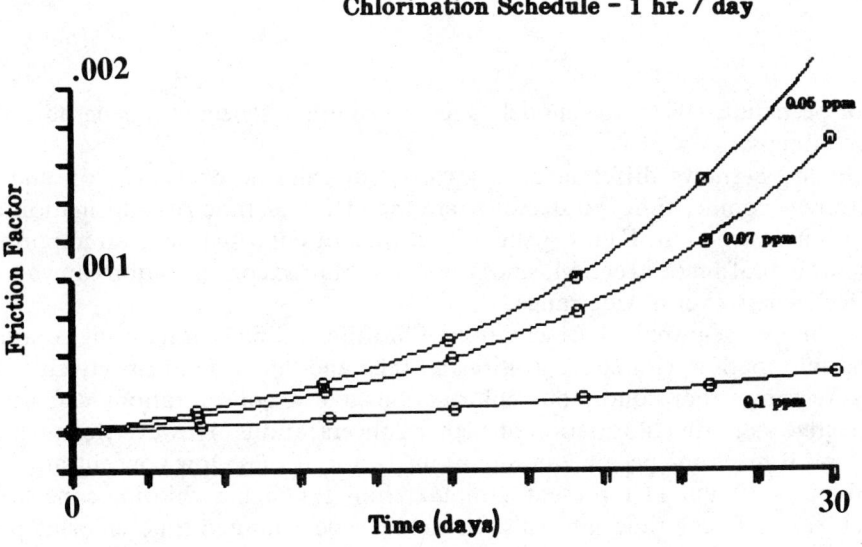

Figure 6. Predicted friction factor change with time for various chlorine doses at 1 h/d.

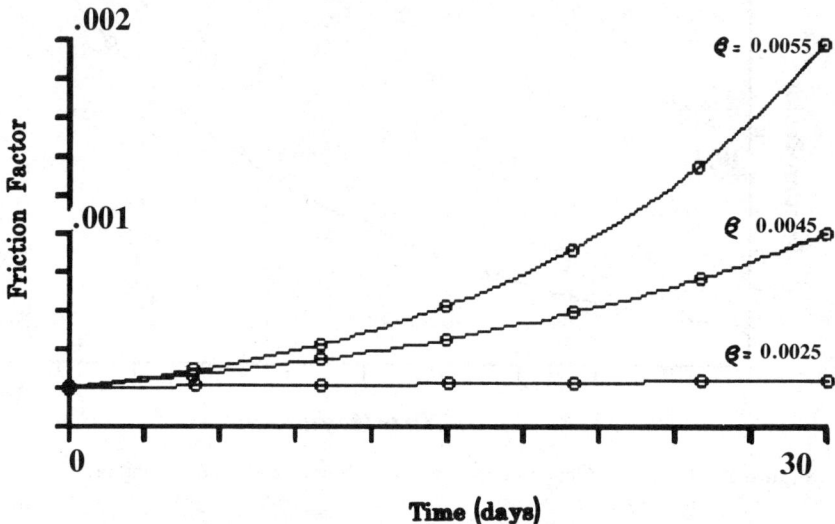

Figure 7. Predicted friction factor change with time for different biofilm growth rates at a chlorine dose of 0.1 ppm for 1 h/d.

can be handled with the model, and chlorination schemes can be adjusted accordingly.

Figure 8 shows different runs at 0.1-ppm chlorine concentration and at different contact times to demonstrate the effect of time of chlorination on biofilm destruction. Figure 8 shows that time of chlorination is an important factor in biofilm destruction, since the same chlorine concentration can vary in effectiveness over a wide range.

Figure 9 is a graphic representation of biofilm accumulation during continuous chlorination with concentrations of 0.05 and 0.07 ppm, respectively. The figure shows that continuous chlorination at low concentrations can be as effective as pulse chlorination at higher concentrations. However, competing chemical reactions become an important factor at these low concentrations.

Figures 10 and 11 represent computer runs for higher chlorine concentrations at different time intervals. Some extremely limited-time chlorinations were used (e.g., 0.2 ppm at 6 min/d) to study the model performance in such a condition. We observed that higher chlorine concentrations of 6 min/d can be as effective as lower concentrations (e.g., 0.05 ppm) for 1 h/d.

COOLING WATER TREATMENT 1443

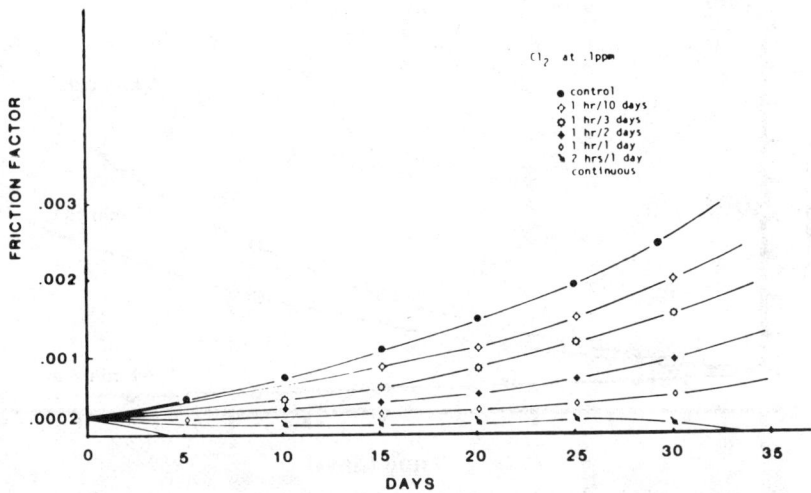

Figure 8. Predicted friction factor change with time for various dosing schedules at a chlorine concentration of 0.1 ppm.

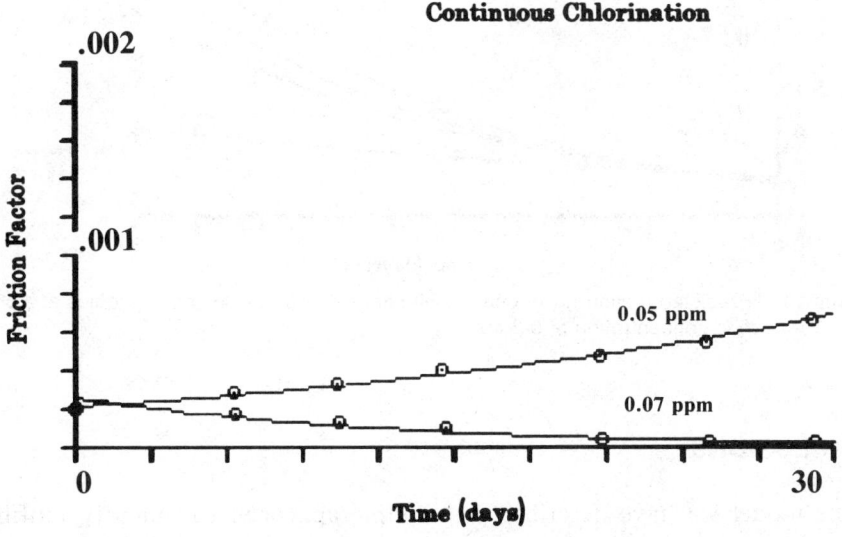

Figure 9. Predicted friction factor change with time under continuous chlorination.

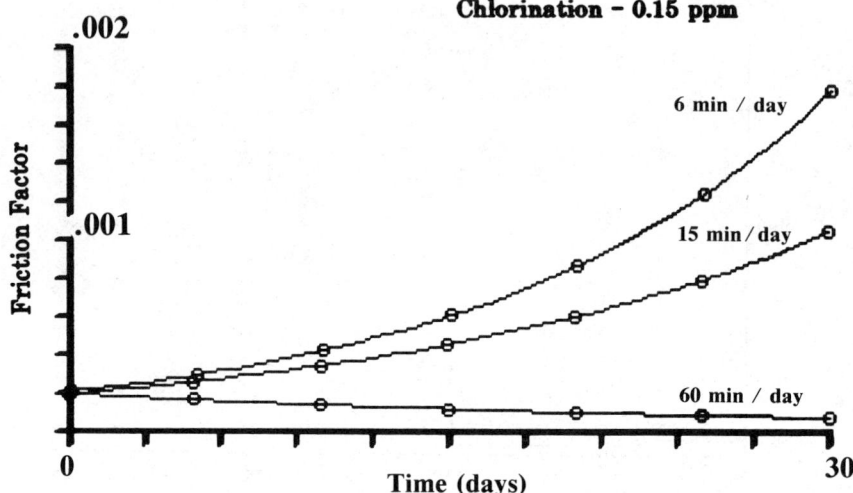

Figure 10. Predicted friction factor change with time for various dosing schedules at a chlorine concentration of 0.15 ppm.

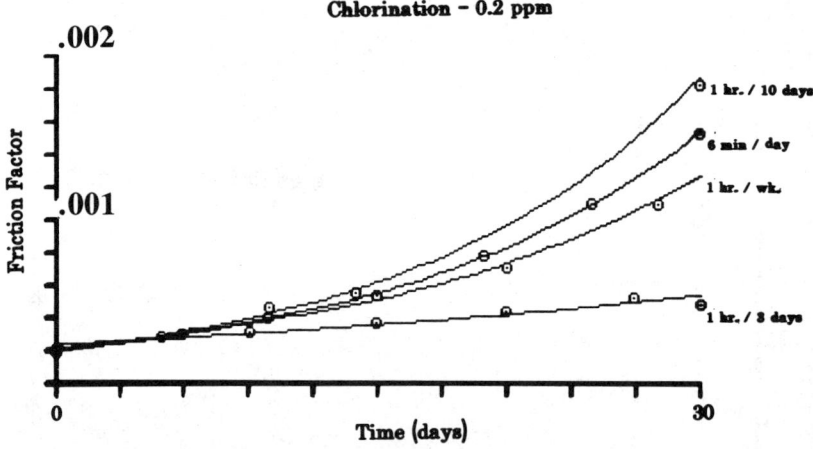

Figure 11. Predicted friction factor change with time for various dosing schedules at a chlorine concentration of 0.2 ppm.

CONCLUSIONS

The model we have described is a simple approach to quantify biofilm formation and destruction in a power plant condenser system. It is possible to compare the performance of chlorine under variable conditions. As an example, continuous chlorination vs pulsing chlorination can be compared easily from the output. For example, a chlorine concentration of 0.2 ppm dosed 1 h per 3 d is as effective as 0.07 ppm dosed for 4 h/d, or 0.1 ppm dosed at 1 h/d. The most economical dosage can be determined by comparing the three values for the same period of time (e.g., 35 d):

1. 0.2 ppm at 1 h per 3 d gives chlorine = 0.2 mg/L × (35 d) × (1/3 h) = 2.33 mg h^{-1} L^{-1}
2. 0.07 ppm at 4 h/d gives chlorine = 0.07 mg/L × (35 d) × (4 h) = 9.8 mg h^{-1} L^{-1}
3. 0.1 ppm at 1h/d gives chlorine = 0.1 mg/L × (35 d) × (1 h) = 3.5 mg h^{-1} L^{-1}

We see that 0.2 ppm for 1 h over 3 d is the most economical scheme, since less chlorine is required over 35 d to maintain the same level of cleanliness in the condenser tube. These data indicate that the total chlorine dose applied to biofilms is not necessarily the determining factor for regulating treatment efficiency. This implies that alterations of chlorine application schemes may be effective ways of optimizing condenser treatments.

The proposed modeling approach can be a useful tool for electric power utilities. Since the model can easily accommodate different input data according to individual needs, it can readily be used to maximize chlorination efficiency. Most utilities have enough operational data for β value calibration, and simple disinfection data can be obtained on site for any disinfecting chemicals under consideration. The model is then capable of providing information relating to dosage optimization and maximum treatment efficiency.

REFERENCES

1. Helz, G. R., and L. Kosak-Channing. "Dechlorination of Wastewater and Cooling Waters," *Environ. Sci. Technol.* 18(2):48–55 (1984).
2. Garey, J., R. M. Jorden, A. H. Aitken, D. T. Burton, and R. H. Gray. *Condenser Biofouling Control*, (Ann Arbor, MI: Ann Arbor Science Publishers, Inc., 1980).
3. Garey, J. "Possible Alternatives to Chlorination for Controlling Fouling in Power Station Cooling Water Systems: Final Report," (Sandwich, MA: Marine Research, Inc., 1976).
4. Characklis, W. G. "Fouling Biofilm Development: A Process Analysis," *Biotech. Bioeng.* 23:1923–1960 (1981).
5. Kirkpatrick, J. P., L. V. McIntire, and W. G. Characklis. "Mass and Heat Transfer in a Circular Tube with Biofouling," *Water Res.* 14:117–127 (1980).
6. Johnson, B. A., J. H. Reynolds, J. L. Wight, and E. J. Middlebrooks. "Disinfection of Stabilization Pond Effluents," Utah Water Research Laboratory, Utah State University, Logan (1978).

CHAPTER 111

Targeted Chlorination: Design and Field Studies

Robert D. Moss, Stephen P. Gautney, and Patrick A. March

Because the overall operating theme for the 1980s is concerned with system efficiencies, utilities have embarked on various methods to make existing systems more fuel efficient, require less unscheduled maintenance, and achieve longer operating lives. Complex computerized systems have been installed to monitor all phases of the steam cycle.

One area, however, that has received relatively little attention is the condenser—the largest heat sink in the system. By means of the transfer of heat by cooling water flowing through many tubes in the condenser, steam leaving the turbine is condensed and forms a vacuum on the turbine. This simple process saves millions of dollars annually in fuel costs. However, a reduction in the heat transfer will severely reduce the vacuum at the turbine outlet. One of the primary reasons for reduced vacuum is fouling of the heat transfer surfaces. Although there are many types of foulants, one of the most significant is slime growth on the cooling water side of the condenser tubes.

For many years, slime growth on condenser tubes has been alleviated by feeding chlorine. Chlorine has been a cheap, relatively safe, and effective biocide. However, because there are some environmental concerns regarding chlorination, the U.S. Environmental Protection Agency (EPA) began regulating chlorine to minimize any potential hazards to other inhabitants of the ecosystem near the plant.

In 1976, EPA limited the discharge of free residual chlorine (FRC) to an average of 0.2 mg/L. The Tennessee Valley Authority (TVA) conducted the first comprehensive chlorine minimization study in the industry and found that a residual of 0.2 mg/L FRC at the outlet of the condenser appeared to maintain condenser efficiency only marginally at the test plant.[1] It was also noted, after conducting minimization studies at four more plants, that chlorination requirements were very site specific.[2-5]

The EPA has now issued even lower guidelines for chlorine. The new limits are 0.2 mg/L total residual chlorine (TRC). On a once-through system with high ammonia concentration in the water, such low chlorine levels may be difficult to meet without serious adverse effects on condenser efficiency.

There are several alternatives that can be used to meet the new EPA guidelines, but all have serious deficiencies. Alternative biocides only postpone meeting potential regulations that may seriously affect their use. Severe cost penalties are associated with manual condenser cleanings, whether they be for manpower and replacement power costs or for equipment in the case of an abrasive ball cleaning system. Dechlorination only treats the symptoms and not the problem, and it introduces additional chemicals into the receiving water.

TARGETED CHLORINATION: CONCEPT AND DESIGN

The most encouraging technology being tested for dealing effectively with environmental concerns and concerns for operating efficiencies and economy is called targeted chlorination (TC). The basic idea behind TC is to chlorinate only a few condenser tubes at a time with sufficient chlorine to maintain biofouling control. At the same time, the cooling water flow from the unchlorinated tubes will mix with the chlorinated water to yield concentrations of chlorine at the end of the discharge pipe below the detection limit of the amperometric titration method. This process may provide adequate fouling control that would improve the unit's heat rate and still meet the need for a clean environment, without the addition of other chemicals (as with dechlorination) and without the high additional costs associated with retrofit mechanical cleaning systems.

TVA's project has the goal of conceptualizing, designing, and testing a TC system at a TVA power plant to develop and verify an advanced methodology for maintaining condenser efficiency while meeting more stringent National Pollutant Discharge Elimination System (NPDES) permit requirements for chlorinated water effluent quality.

The objectives of the project are to (1) determine if TC is a viable chlorination regime for TVA; (2) determine the most effective type of TC system; (3) evaluate a TC system at an operating plant; (4) determine if TC is cost beneficial; (5) provide a methodology for applying TC so that condenser performance is maintained and NPDES permit limitations are met; and (6) compare costs of TC to the costs of dechlorination.

The project scope is to (1) assess various designs to target chlorine in a condenser and (2) perform generic mathematical and physical modeling to develop designs for a variety of condenser water-box configurations.

The project approach was divided into three phases:

PHASE I — Project Planning and System Design
- Conceptualize and evaluate chlorination systems
- Decide on basic system
- Design system

PHASE II — Physical Modeling (concurrent with Phase I)
- Build condenser model
- Test design for chlorine concentrations, effluent limits, efficiency, and operability
- Rank alternative TC methods

- Design outlet waterbox sampling system

PHASE III — Field Evaluation
- Perform full-scale mock-up tests
- Install system at plant
- Test system
- Final report

The basic design of a TC system had to meet the following criteria:

1. The system must feed chlorine to only a small percent of the condenser tubes at a time.
2. The system must feed chlorine to at least 98% of the condenser tubes during a chlorination cycle.
3. The system must be able to effect flow reduction in targeted tubes during chlorination to allow longer contact between chlorine and the fouling mass, and to allow shear stress of the normal cooling water flow to cause the dead biofilm layer to slough off.[6]
4. The system must be constructed of materials to withstand forces in the condenser water box and must not be subject to excessive corrosion.
5. The system must be reliable and require only minimal operator attention.

To fulfill these criteria, a manifold approach was required. However, if a manifold was to be used, it must operate in harmony with the shape of the tube sheet and the water box. The unit 7 condenser at Kingston (Tennessee) Steam Plant was chosen for the TC system tests; therefore, a thorough study of the tube-sheet configuration and system operations was conducted.

The Kingston unit 7 steam generator is rated at 200 MW capacity. The heat rejection system is a Westinghouse single-pass surface condenser. The condenser water box is divided in half, with each side containing 5871 7/8-in.-OD tubes. The total number of tubes is, therefore, 11,742, which will provide about 80,000 ft^2 of surface area to handle 887,000 lb of steam per hour at 2 in. mercury back pressure. The cooling water flowrate through the condenser is 120,000 gpm at a tube velocity of 6.9 fps. The tube sheet configuration is circular. Each inlet water box is semicircular, with an 8-ft 1.5-in. radius from the condenser centerline and a depth of 6-ft 8-in. from the water box face to the tube sheet.

It was decided that a rotary manifold similar to a windshield wiper would satisfy the operations criteria (Figures 1–5). The manifold is 6 in. wide and constructed of carbon steel. Holes are drilled at specific intervals and sized so that the flux of chlorinated water through the manifold is constant.

A seal surrounds the entire manifold. The seal, when contacting the tube sheet, provides a significant flow reduction (2000 to 200 gpm). It is wedge-shaped so that all the tubes (300 tubes) are chlorinated for the same length of time (3 min). The manifold on one side of the water box feeds chlorine to the tubes as it slowly makes a 180° arc in 2 h. At that time the manifold on the

1450   WATER CHLORINATION

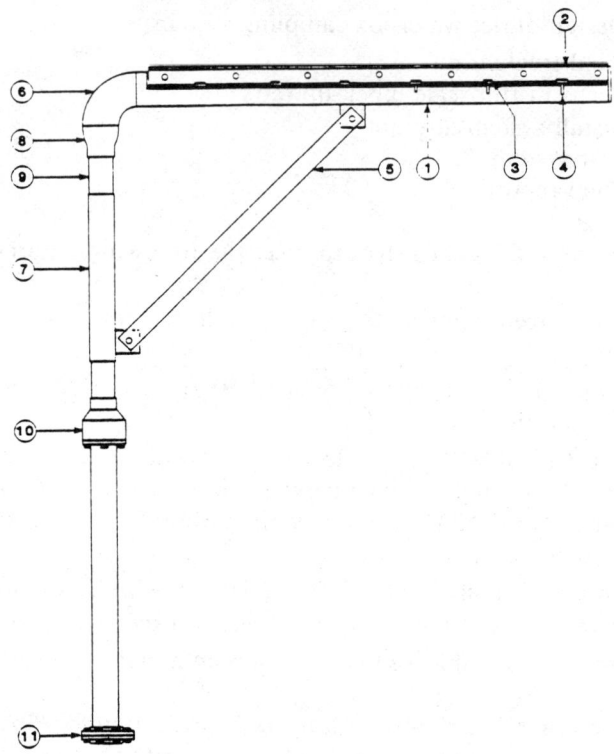

Figure 1. Overhead view of manifold and drive shaft. Components are designated as follows: 1. Manifold; 2. Seal; 3. Seal adjustment; 4. Support gusset; 5. Support bar; 6. 90° ell; 7. Drive shaft; 8. 4″ to 6″ expansion; 9. Turn down for bearings; 10. Drive shaft coupling; 11. Water box wall seal.

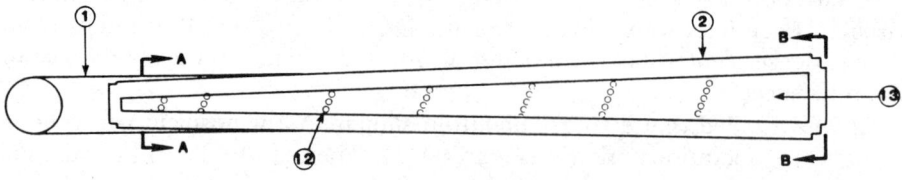

Figure 2. Frontal view of manifold. Components are designated as follows: 1. Manifold; 2. Seal; 12. Diffuser holes; 13. Diffuser plate.

COOLING WATER TREATMENT 1451

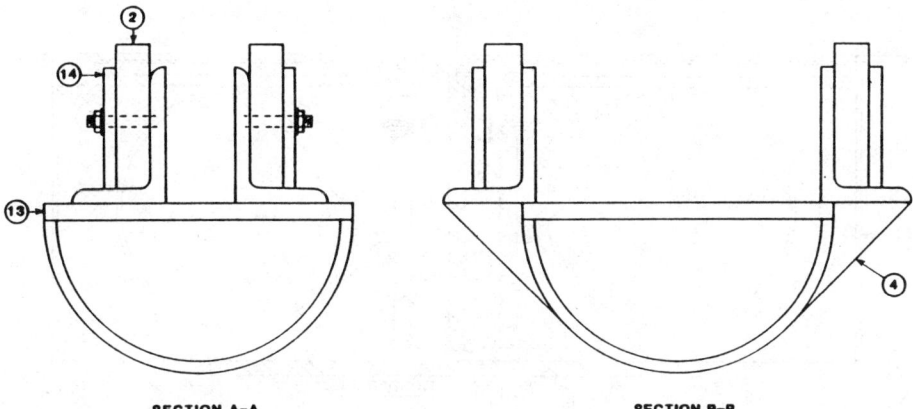

SECTION A-A **SECTION B-B**

Figure 3. Seal configuration on manifold. The locations of Sections A-A and B-B are shown in Figure 2. Components are designated as follows: 2. Seal; 4. Support gusset; 13. Diffuser plate; 14. Seal support.

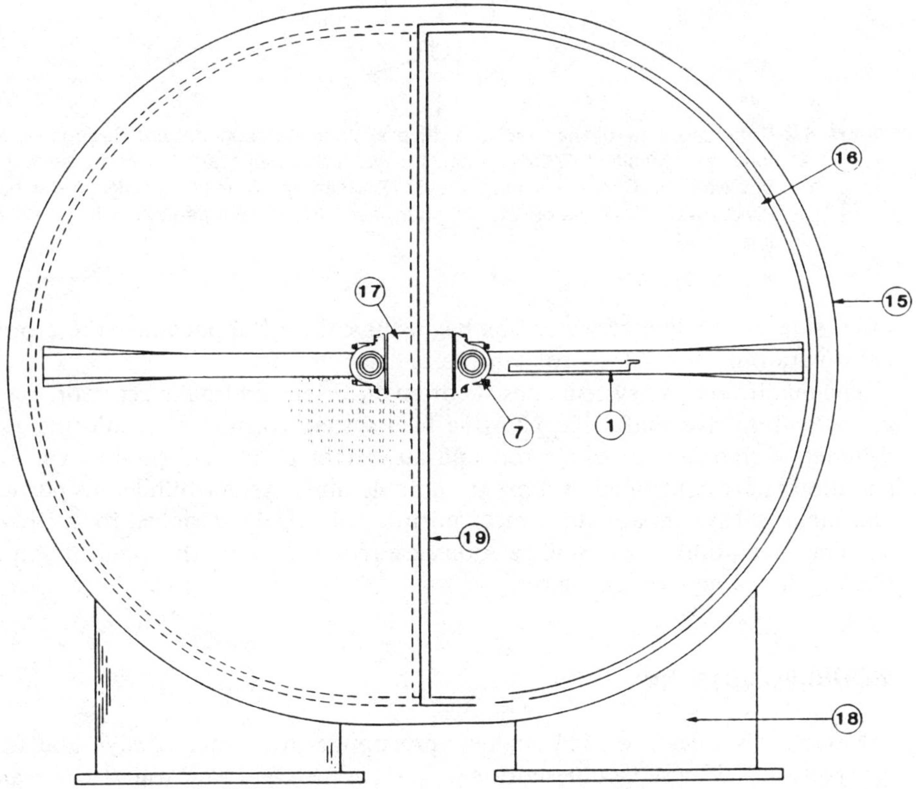

Figure 4. Locations of manifolds in water box. Components are designated as follows: 1. Manifold; 7. Drive shaft; 15. Condenser water box wall; 16. Condenser tube sheet; 17. Bearing; 18. Inlet pipe; 19. Water box divider wall.

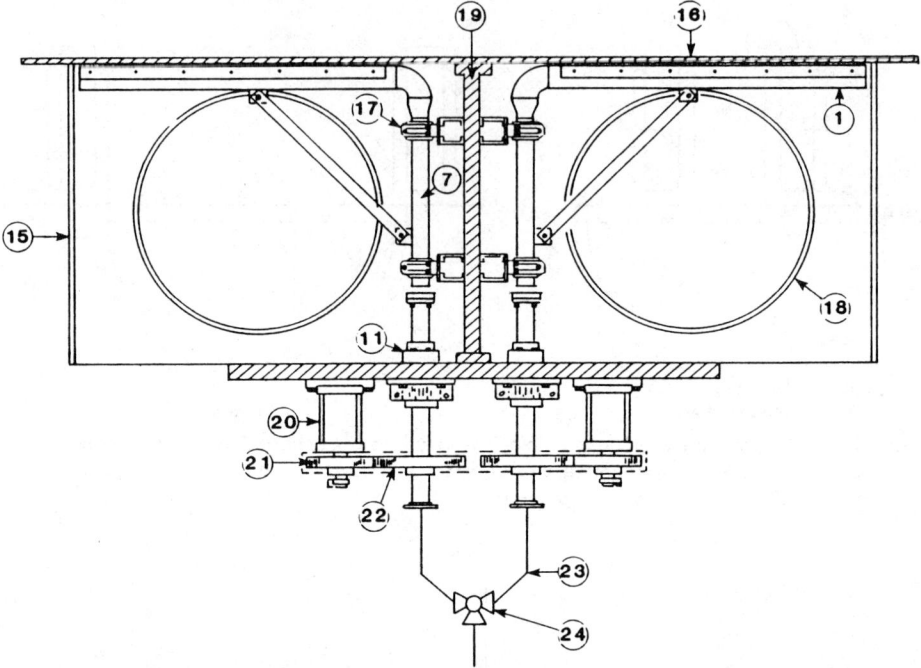

Figure 5. Overhead view of manifold and drive system. Components are designated as follows: 1. Manifold; 7. Drive shaft; 11. Water box wall seal; 15. Condenser water box wall; 16. Condenser tube sheet; 17. Bearing; 18. Inlet pipe; 19. Water box divider wall; 20. Actuator; 21. 16″ spur gear; 22. 24″ spur gear; 23. Chlorine feed line; 24. 3-way valve.

other side of the divided water box begins its chlorine application; therefore, chlorination of unit 7 lasts for 4 h.

The manifold drive system consists of an air-driven hydraulic actuator, spur gears, and a drive shaft. The chlorine feed system consists of a chlorine gas cylinder, a chlorinator, an ejector, and an ejector pump. All pumps, valves, and motors are controlled by a programmable timer system. While this particular manifold system and drive mechanism is specifically designed for a circular tube sheet, only slight modifications are required to use the manifold on a rectangular or square tube sheet.

MODELING STUDIES

Modeling studies provided engineering support for system design and for subsequent field-test troubleshooting. Tests were conducted on a 1:6 scale model of the Kingston unit 7 condenser to provide velocity data, flow distributional data, and concentrations data regarding three types of TC systems: manifold against tube sheet, nozzles in inlet pipe, and nozzles in inlet water box. The tests also provided information such that, if a condenser water box did not lend itself to a manifold-type system, alternative systems would be available. (We have noted that alternative systems must be used if some other TVA plants are to be retrofitted with a TC system.)

Tests on the seal device and on mixing chlorinated water with unchlorinated water in the condenser outlet water box and in the outlet culvert were also conducted.

The results of the modeling tests were:

Seal Tests

- Average friction coefficient between seal and simulated tube sheet was 0.85 under dry conditions and 0.78 under wet conditions.
- Shear tests on potential tube sheet obstructions were conducted to size the full-scale system.
- Seal wear rate—40% wear in first 100 d, almost no wear for next 500 d.
- Seal leakage was determined to be ~44 gpm at 13-ft head pressure differential.

Dilution Tests

- Using a 1:20 scale outlet culvert model, total flow was demonstrated to be well mixed both vertically and horizontally within three outlet culvert diameters (24 ft) downstream of the outlet water box.

Physical Model Tests

- Velocity reduction via the manifold design criteria was achieved; tests showed a drop from 6.9 to 1.2 fps when the manifold was over a section of tubes.
- Relatively little mixing of chlorinated and unchlorinated water occurred in the outlet water box, but substantial mixing occurred in the outlet pipe.
- The nozzle injection system in the inlet pipe provided poor targeting of condenser tubes.
- The nozzle injection system in the inlet water box near the tube sheet would provide a viable alternative to the manifold design if flow reduction is not deemed an important criterion for operation.

FIELD TESTS

Full-scale mock-up tests demonstrated that all mechanical parts functioned as designed. Equipment was installed at the plant during the unit's scheduled outage from May 15 to July 10, 1983. A 3-in. line was run from the chlorination building into the plant where it directly feeds unit 7 (~2000 ft). The drive system was mounted on the outside face of the condenser water box, and the manifold was installed inside the water box. The manifold and drive shaft were designed so that installation could be accomplished without removing the doors on the condenser water box.

Field tests began on July 11, 1983, and continued through August 31, 1984. The data collected each day include:

- Four measurements of inlet chlorine concentration
- Eight measurements of tube exit chlorine concentrations
- Eight measurements of discharge chlorine concentrations with a Lawrence Livermore polarographic analyzer
- One water sample each week analyzed for ammonia, total suspended solids, pH, temperature, total organic carbon, organic nitrogen, copper, iron, alkalinity, and 5-, 10-, and 30-min chlorine demands; and
- Complete condenser efficiency tests each day on unit 7, and once per week on each of units 5, 6, 8, and 9.

RESULTS AND CONCLUSIONS FROM FIELD TESTS

Results to date are preliminary and should not be construed as final. The seal around the manifold has provided a significant reduction in flow velocity (from 6.9 to 1.75 fps, see Figure 6).

Chlorine concentrations in the system were (a) chlorine feed line, 22 mg/L FRC; (b) at condenser tube inlet, 15 mg/L FRC; and (c) at end of discharge pipe, undetectable TRC.

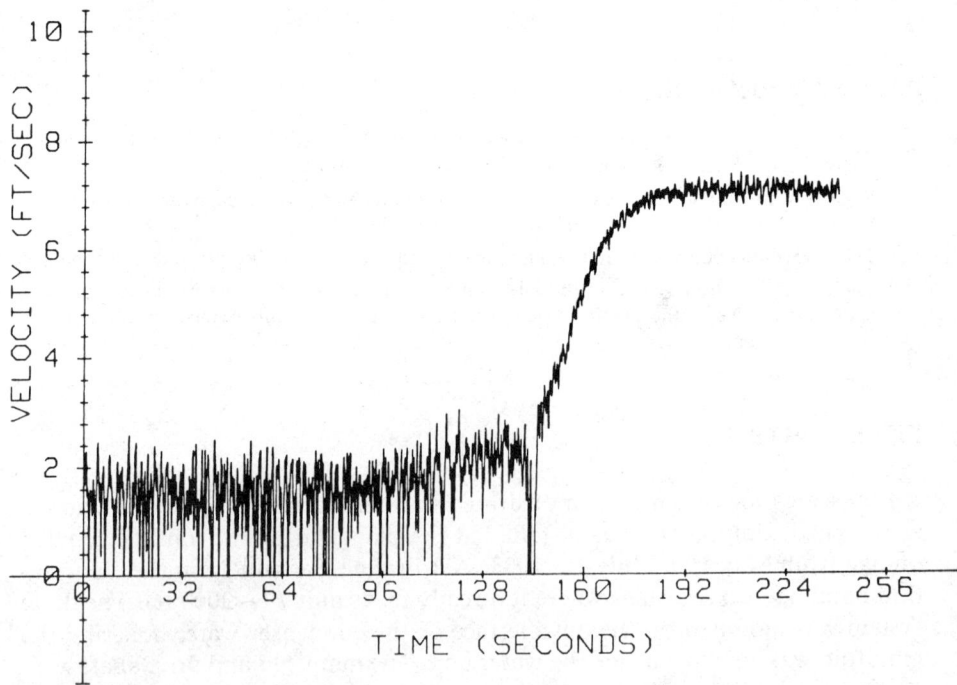

Figure 6. Velocity during and after targeted chlorination.

Performance data on unit 7 had much less variability than other units. Data of special concern include turbine back pressure, differential water box pressure, an apparent cleanliness factor, and chlorine concentrations of the system.

There are insufficient data to state that TC has increased the condenser's performance; however, there are sufficient data to conclude that the system has not made the condenser operations any worse than other units or the unit 7 history. The overall prognosis is that TC is effective for controlling biological fouling and should prove to be beneficial to the overall plant operating efficiency.

Discharge concentrations were below 0.02 mg/L TRC. The Lawrence Livermore analyzer demonstrated that accurate measurements below 0.02 mg/L in river water are not possible. In many cases, background concentrations before and after chlorination were higher than measurements made during chlorination, as shown in Table I.

Total chlorine usage on a daily basis was reduced by 75%, whereas higher concentrations than ever before were fed to individual condenser tubes.

A preliminary cost assessment of equipment and manpower was made. The fabrication of the manifold, installation of the system, and costs of shakedown and troubleshooting were:

Equipment and materials	$ 60,000
Construction manpower	68,000
Engineering manpower	25,000
Troubleshooting manpower	20,000
Total	$173,000

If, for instance, additional systems are required for the other eight units at Kingston, the costs will be less for equipment and materials.

It should be noted that this type of system is probably the most expensive for plant and equipment. If other types of systems such as nozzles are used, costs will be much less than the reasonably priced system now being tested. A very generalized cost-benefit analysis was conducted. It was determined that if targeted chlorination caused a 0.3-in. mercury turbine back pressure reduction over a yearly operation, a savings of approximately $200,000 per unit per year in fuel costs alone could be realized. Savings in manpower and replacement power costs would be in addition to fuel savings. Therefore, a 9-unit plant such as Kingston could save $1.8 million/year in fuel costs if all assumptions were met.

REFERENCES

1. Moss, R. D., et al. *Chlorine Minimization/Optimization for Condenser Biofouling Control: Final Report*, EPA 600/7-80-143, (Washington, DC: U.S. Environmental Protecton Agency, 1980).

Table I. Chlorine Concentrations at Discharge using a Lawrence Livermore Analyzer

1983	Background (μg/L)	During (μg/L)	Difference[a]	1984	Background (μg/L)	During (μg/L)	Difference[a]
07/19	18.55	11.20	−7.35	01/09	3.42	6.85	3.43
10/11	21.75	23.80	2.05	01/11	1.43	0.00	−1.43
10/12	21.00	15.30	−5.70	01/25	2.85	4.20	1.35
10/13	13.20	15.20	2.00	01/27	0.15	5.30	5.15
10/14	14.75	15.50	0.75	02/06	2.55	6.40	3.85
10/17	25.80	19.50	−6.30	02/09	0.30	5.30	5.00
10/18	19.40	17.60	−1.80	02/13	0.00	3.60	3.60
10/19	9.60	10.00	0.40	02/14	0.00	1.10	1.10
10/20	12.80	10.50	−2.30	02/16	1.75	2.60	0.85
10/25	7.15	6.90	−0.25	02/17	2.20	4.10	1.90
10/26	7.35	10.70	3.35	02/21	0.00	2.00	2.00
10/28	11.75	14.10	2.35	02/22	1.95	3.42	1.47
11/01	7.55	8.40	0.85	02/23	0.00	5.14	5.14
11/02	9.90	10.50	0.60	02/24	4.45	9.10	4.65
11/03	7.65	8.00	0.35	02/29	1.95	1.90	−0.05
11/04	7.20	8.20	1.00	03/01	2.35	3.80	1.45
11/07	7.80	6.30	−1.50	03/05	5.40	2.70	−2.70
11/08	7.40	6.80	−0.60	03/06	2.00	1.30	−0.70
11/09	8.40	9.40	1.00	03/07	0.80	4.80	4.00
11/10	6.15	5.80	−0.35	03/08	0.85	1.90	1.05
11/14	4.80	5.00	0.20	03/12	3.40	9.10	5.70
11/15	6.45	9.60	3.15	03/13	1.45	5.00	3.55
11/16	5.15	8.20	3.05	03/14	2.50	2.70	0.20
11/17	3.70	7.00	3.30				
11/21	5.00	7.50	2.50				
11/22	4.40	8.20	3.80				
11/23	6.55	7.50	0.95				
11/25	0.65	4.30	3.65				

[a]Average difference is 1.14 μg/L; standard deviation is 2.5 μg/L.

2. Moss, R. D. *The Minimization of Chlorine at the Paradise Steam Plant*, EDT-132, (Chattanooga, TN: Tennessee Valley Authority, 1980).
3. Moss, R. D., and R. A. Hiltunen. *Chlorine Minimization for Condenser Biofouling Control at the Thomas H. Allen Steam Plant*, EDT-133, (Chattanooga, TN: Tennessee Valley Authority, 1980).
4. Moss, R. D., and R. A. Hiltunen. *Chlorine Minimization for Condenser Biofouling Control at the Shawnee Steam Plant*, EDT-134, (Chattanooga, TN: Tennessee Valley Authority, 1980).
5. Moss, R. D., and R. A. Hiltunen. *A Chlorine Minimization Study at the Kingston Steam Plant*, EDT-135, (Chattanooga, TN: Tennessee Valley Authority, 1980).
6. Characklis, W. G. *Biofilm Development and Destruction*, RP-902-1/CS-1554, (Palo Alto, CA: Electric Power Research Institute, 1980).

CHAPTER 112

Concentrations of Chlorine around Marine Cooling Water Outfalls: Validation of a Model

Jack Coughlan and Martin H. Davis

Chlorine is widely used to control fouling in cooling circuits. For example, a continuous dose of 2 mg/L is specified to control macrofouling for a 1000 MWe direct-cooled (30 m^{3}s^{-1}) power station; this represents between 1000 to 2500 kg (2 to 5 tonnes) of chlorine each day.[1] Some 5 to 15 min after chlorination this water is discharged. The outfall terminal is usually designed and sited to maximize the rate of dilution of effluent and thereby minimize environmental levels of chlorine. Until recently there has been little interest in the fate of chlorine at sea; it was simply accepted that after discharge it soon disappeared. Several recent investigators have proposed kinetic models for the chlorination of seawater (e.g., see Reference 2).

Physical models cannot easily handle nonconservative pollutants such as chlorine since they require a tracer having a similar die-off profile but with a half-life scaled in accordance with the time-scale of water movements in the model. Thus, decay is seldom modelled and the dispersion of a conservative tracer is merely factored by an appropriate decay rate. With wastewater, for example, the organic material (BOD) is degraded at about 10% and the microbial populations at about 90% per tidal cycle.[3] Wide errors can be accepted in these factors, since in most cases dispersion will not be very sensitive to decay.[4] However, it will be shown below that decay is an extremely significant factor in the dissipation of chlorine in the marine environment.

Decay may more readily be incorporated into a numerical model but often there is little realistic information as to rates. For this reason, and the aforementioned lack of sensitivity, it is frequently omitted. Chlorine is a case in point: this nonspecific biocide has been discharged for many years, but there have been virtually no measurements of environmental concentrations and relatively few theoretical estimates. These latter have assumed that the chlorine concentrations decline in parallel with the dilution of the heated effluent and completely ignore a fairly rapid rate of decay. Consequently, the areal extent of the chlorine plume and the concentration at any point within it were both overestimated.

This chapter describes the first field validation of our nonconservative model of chlorine dissipation. The use of in situ decay rates and a novel term, demand-enhanced dilution, generate values for chlorine concentration that correspond closely with the prototype measurements. This is particularly apparent in poorly mixed regions of the plume where conservative (dilution only) models grossly overestimate concentrations.

MODEL

The procedures for determining the rate constants for chlorine decay and the development of the model itself have been described previously.[5] The exponential basis for the model came from the observation that decay depends only on time and that decay rate was not proportional to initial concentration, which indicate first-order (or a close approximation to first-order) kinetics. The model embraces the following terms and assumptions: after chlorine addition there is an instantaneous demand and the remaining chlorine (C_{in}-C_{id}) decays at rate k. Decay continues after the chlorinated effluent leaves the outfall and begins to become diluted by the receiving water at a rate D. This water introduces an additional demand for chlorine fD, which serves to enhance the effect of dilution. The dimensionless factor, f, is numerically equivalent to the rate constant for chlorine decay in the receiving water. Then C_t, the concentration of chlorine in the plume at time t min from discharge, is given by:

$$C_t = \frac{(C_{in} - C_{id})e^{-kt}}{(D + fD)^t} \quad (1)$$

The above model differs from that published previously[5] only in the substitution of f, a dimensionless factor, in place of a decay constant k. This change was made to achieve dimensional balance (i.e., mathematical nicety) and in no way changes the construction or practical application of the model.

Because plume temperature decreases with dilution, the constants k and f should be varied continuously; in practice this could be ignored, since the temperature differentials were only about 6 to 10 °C and a mean value for k and f sufficed.

VALIDATION METHOD

Thermal discharges, with few exceptions, rise to form a relatively shallow layer at the surface. There is some loss of heat by evaporative cooling but the greatest drop in temperature in the near field is due to dilution. Thus heat may be treated as a conservative pollutant, with temperature at any position along the plume signifying the extent of dilution, or D. Time t from discharge can be calculated from the local current velocity and the distance between the first and the nth sample. The best result was obtained by allowing the sampling vessel to drift in the plume. Where this was not feasible because of nearby obstructions or of wind driving the vessel, a "dan-buoy" (subsurface float with vertical spar) was launched into the plume and visited for sampling. The position of the vessel at the time of sampling was determined by electronic trilateralization ("Trisponder"). Discrete samples of water were collected from the surface by bucket and temperature and chlorine concentration were measured immediately.

Residual chlorine (TRO) was measured by the Lovibond colorimetric DPD procedure.[6] This was the only chlorine analytical method sufficiently rapid for this application. Moreover it is sufficiently robust and cheap to be used on the deck of small vessels. Initially, the water samples were split and a subsample was passed to a technician using a proprietary chlorine electrode. However, this method could process samples at only half the rate of the colorimetric comparator and produced increasingly erratic values with time, requiring more frequent interspersion of standards and hence fewer samples processed. Eventually the electronics of the chlorine electrode instrument failed, presumably due to the humid, salt-laden atmosphere, despite the protected conditions in which it was being used. The rate constants k and f were determined on board by the simulation technique.[5] The initial chlorine concentration was kept below what we have loosely termed the breakpoint, because this gives rise to discontinuities during decay.[5] It was not possible to obtain a water sample from the actual discharge, which was submerged and surmounted by a large structure. Instead a sample was collected as close to the outfall as possible and this value C replaces (C_{in}-C_{id}) in the calculation.

In September and October 1981, chlorine concentration and water temperature were measured in surface water samples collected by a vessel tracking the thermal discharge plume from Sizewell 480 MWe nuclear power station. This provided both a synoptic picture of environmental chlorine concentrations and data for model validation. Sampling runs were made with routine levels of gas chlorination and with elevated levels. The latter enabled us to track the plume over a greater distance and so provided a range of data for the validation.

This part of the Suffolk coastline is relatively straight and aligned north/south (Figure 1). The cooling water (CW) outfall is sited about 100 m offshore and consists of vertical shafts terminating just above seabed level. The depth of water here is from 2 to 5 m, depending on tidal state. A shallow bank lying just offshore from the outfall confines the plume laterally. On the flood tide there is an elongated plume lying to the south of the outfall. On the ebb, the plume is wider and lies to the north of the outfall. At slack water the effluent ponds around the outfall and this creates a broad pulse of warm water as the tide starts to run. Soon afterwards the last portion of the aged plume from the previous tide is set back across the outfall. We made one or two runs in the pond but chose to avoid the complex situation as the tide began to move; most sampling runs were made in the developed plume to the north or south of the outfall.

The sampling vessel was engaged primarily in a water temperature profiling survey, following a preplanned course which frequently took the ship out of the plume. This resulted in a gap in our coverage (e.g., Figure 2). Some at-sea

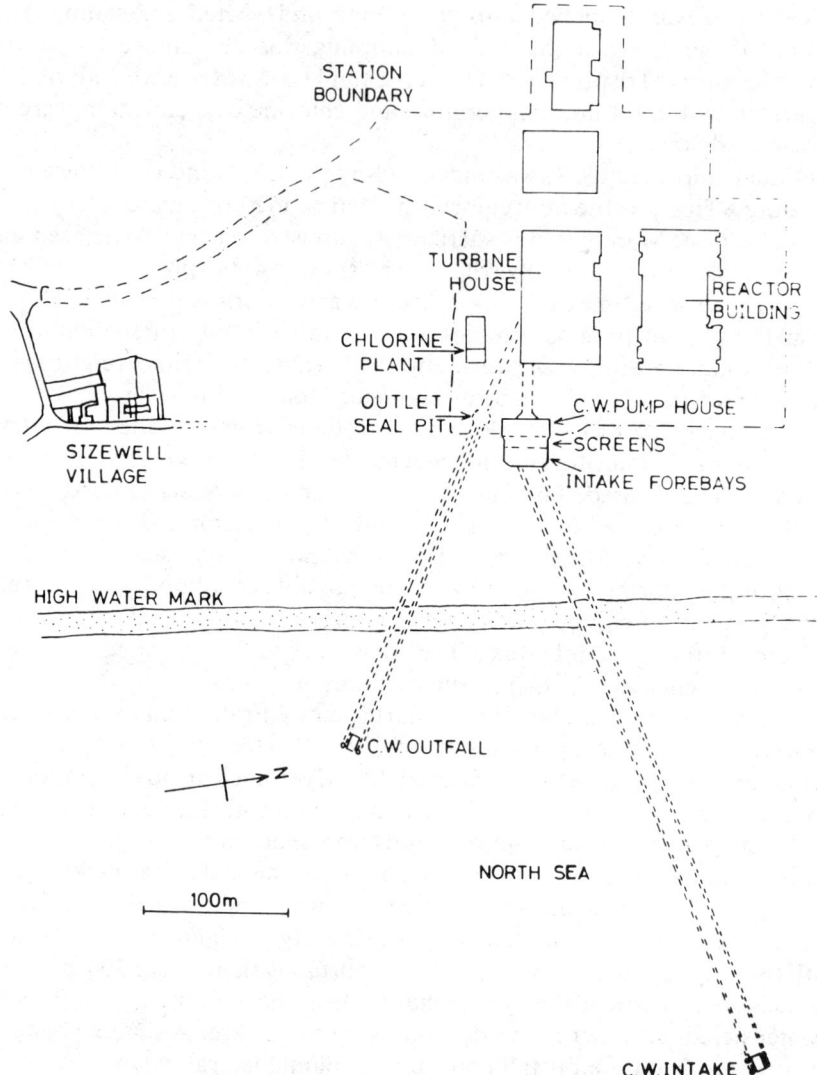

Figure 1. Plan of Sizewell Power Station showing coastline and positions of outfall and intake.

time was lost through bad weather, but 18 chlorine runs were successfully completed.

RESULTS

The Trisponder ranges for the sampling positions were converted to UK Ordinance Survey coordinates and plotted with time, chlorine concentration, and temperature. Figures 2 to 7 all follow the same convention and can be

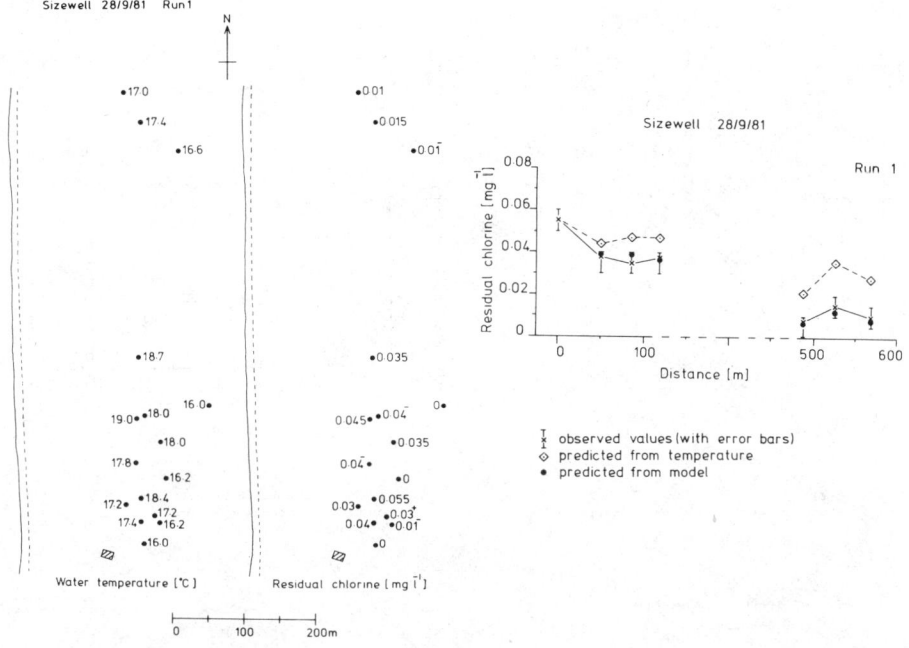

Figure 2. (Left) Surface water temperatures and chlorine concentrations measured around Sizewell outfall, September 28, 1981. The tide is ebbing so that the plume lies to the north of the outfall. The gap in the readings shows where the sampling vessel left and later reentered the plume.
(Right) Comparison of actual and predicted chlorine concentrations.

orientated by reference to Figure 1; note the position of the shoreline and outfall. The left side of each of these figures shows a duplicate plot of water temperature (far left) and chlorine concentration (center left). The modeling result is summarized in the graphs to the right of these figures. This shows the actual chlorine concentrations measured during the survey (with error bars), the concentrations predicted from the present model, and, for comparison, the concentrations that would have been predicted at the same sampling positions by a dilution-only estimate of chlorine concentration. The x-axis is the distance down plume from the initial sample C. The divergence between the conservative and nonconservative approach is most marked in regions of the plume where there is little mixing and, hence, virtually no drop in temperature (Figures 3–5).

The gap in the data (Figure 2) shows where the survey vessel left the plume. When it later reentered and chlorine measurements could be resumed, the model predictions were still valid, showing that close-interval sampling is not essential.

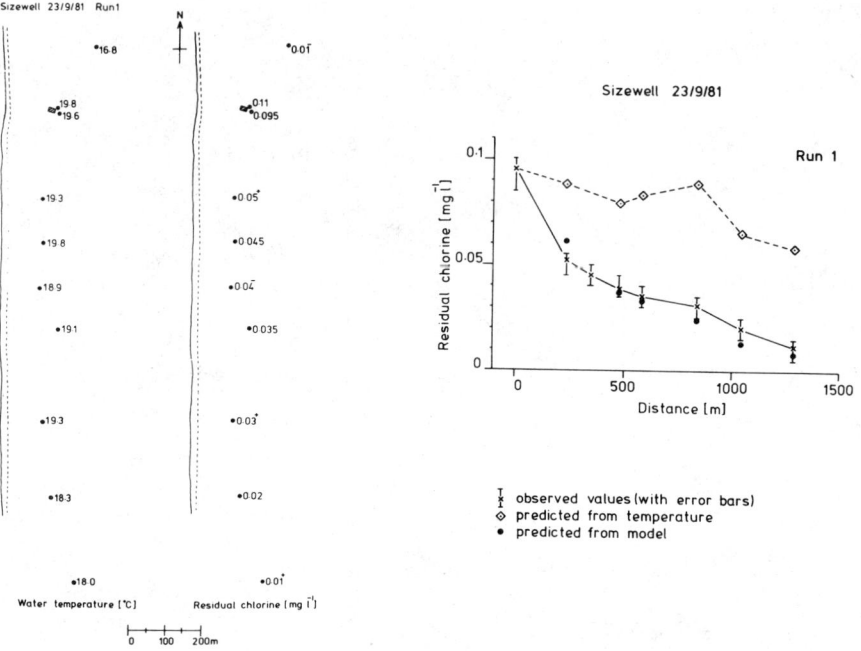

Figure 3. (Left) Surface water temperatures and chlorine concentrations measured around Sizewell outfall, September 23, 1981. The tide is flooding so that the plume lies to the south of the outfall.
(Right) Comparison of actual and predicted chlorine concentrations. There was little mixing and consequently little change in temperature. Note how the present model, which includes decay, predicts the observed value more closely than the dilution-only model.

DISCUSSION

This model differs from most of its predecessors in three respects. First, it supposes exponential decay. Second, it accommodates discrete rate constants for the effluent and for the receiving water to allow for the considerable horizontal or vertical separation of intake and outfall at some sites. Third, it introduces the concept of demand-enhanced dilution, although we cannot attempt any explanation of how additional chlorine demand exerts itself at a late stage in the reactions. In short, we have a model that works, but there are gaps in our underlying theory.

A recent freshwater model[7] apportioned chlorine loss between dilution, decay, phototransformation, and volatilization. In seawater, volatilization is said to be negligible[8] despite the inevitable smell of chlorine around outfalls, whereas the evidence for phototransformation is conflicting.[5]

COOLING WATER TREATMENT

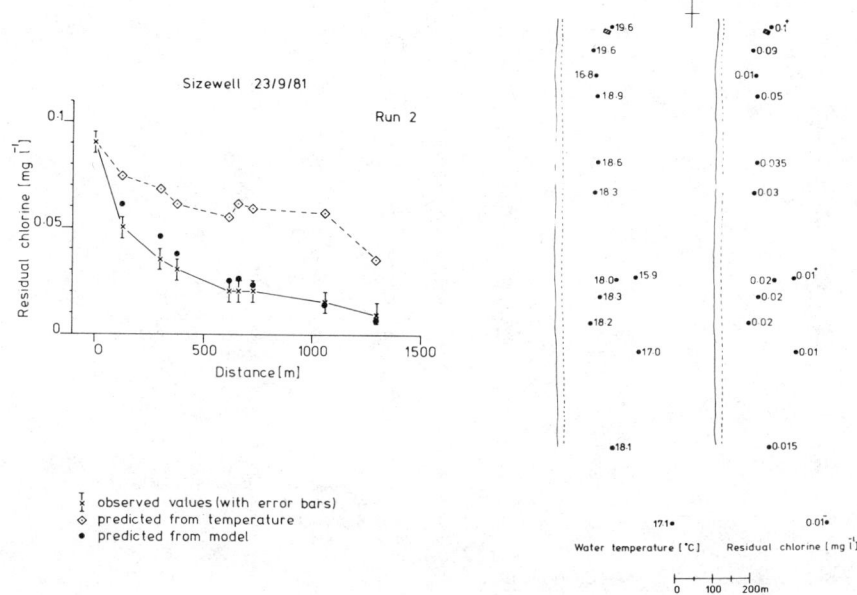

Figure 4. (Left) Surface water temperatures and chlorine concentrations measured around Sizewell outfall, September 23, 1981. As with Figure 3, this is an ebb tide and measurements extended for more than 1 km.
(Right) Comparison of actual and predicted chlorine concentrations.

Model validation was confined to the axis of the plume since, at Sizewell, this is the maximum dimension of the chlorine field. Plume width was confirmed on occasion by tracking the sampling vessel at right angles to the axis, and the good agreement between model and prototype held at the periphery. However, our priority was to establish the elapsed time and dilution components for the validation.

There exist adequate mathematical descriptions of the spread of buoyant plumes in flowing water from which elapsed time at any position may be calculated together with the extent of dilution. Similar information is available from hydraulic (physical) models and to some extent from aerial infrared surveys. The latter requires ground-truth information on plume velocity to determine elapsed time. All three sources require in situ simulation of chlorine decay to generate the decay constants for use in the model.

The surveys have shown that the chlorine field around Sizewell outfall is more restricted in area, with lower concentrations, than would have been

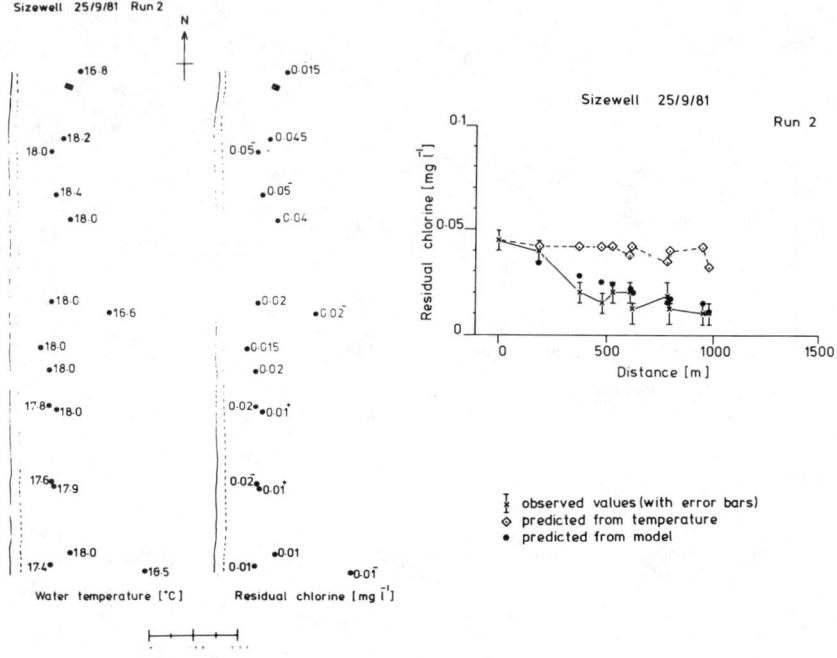

Figure 5. (Left) Surface water temperatures and chlorine concentrations measured around Sizewell outfall, September 25, 1981. The narrowness of the plume is shown by the 16.5° and 16.6°C values.
(Right) Comparison of actual and predicted chlorine concentrations.

estimated from thermal data. Moreover, we have a model that faithfully reproduces the prototype. This permits first-order calculations to be made as to the scale of acute effects of the discharged total residual oxidant on organisms in the receiving water. Obviously, it takes us no further toward resolving the question of chronic or cumulative effects that might arise from the nonoxidant components of the chlorinated discharge.

We consider that this empirically developed model is an adequate tool to describe the environmental concentrations of chlorine at existing and planned outfalls. Furthermore, the approach should be applicable to other nonconservative pollutants, although neutral or negatively buoyant plumes will necessitate vertical sampling; the buoyant plume in this study greatly simplified sampling. If the plume being studied has no convenient conservative tracer, it will be necessary to inject Rhodamine dye or a similar readily detectable tracer by which to estimate dilution.

COOLING WATER TREATMENT 1467

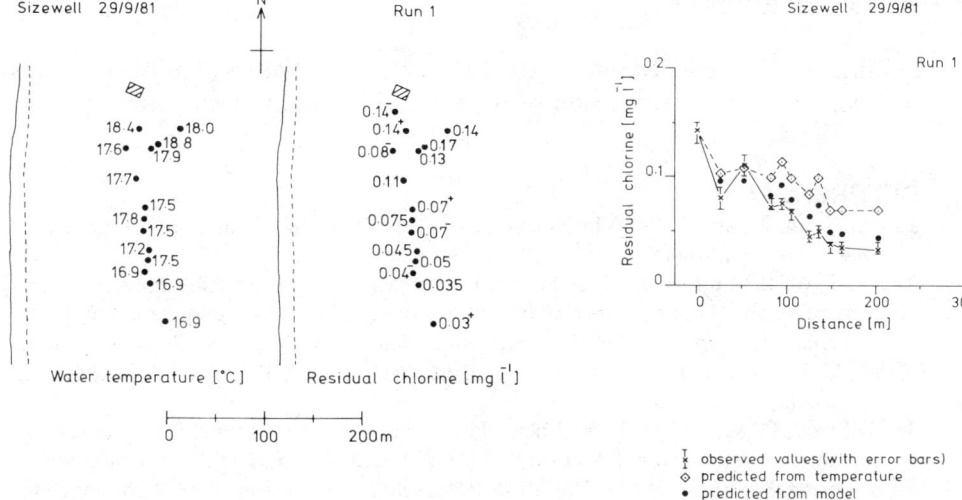

Figure 6. (Left) Surface water temperatures and chlorine concentrations measured around Sizewell outfall, September 29, 1981. The rate of chlorine injection has been increased slightly above routine levels.
(Right) Comparison of actual and predicted chlorine concentrations. In contrast to the preceding figures there has been significant mixing and hence dilution in the plume.

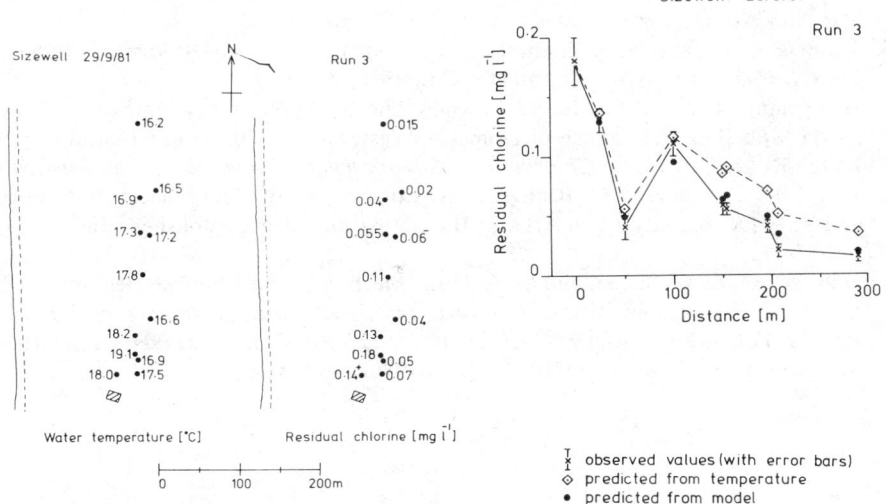

Figure 7. (Left) Surface water temperatures and chlorine concentrations measured around Sizewel outfall, September 29, 1981. The rate of chlorine injection has been increased slightly above routine levels.
(Right) Comparison of actual and predicted chlorine concentrations.

ACKNOWLEDGMENT

This work was carried out for the Central Electricity Research Laboratories and is published by permission of the Central Electricity Generating Board.

REFERENCES
1. Coughlan, J., and J. W. Whitehouse. "Aspects of Chlorine Utilization in the U.K.," *Chesapeake Sci.* 18(1):102–111 (1977).
2. Haag, W. R., and M. H. Lietzke. "A Kinetic Model for Predicting the Concentrations of Active Halogen Species in Chlorinated Saline Cooling Waters," in *Water Chlorination: Environmental Impact and Health Effects, Vol. 3*, R. L. Jolley, W. A. Brungs, and R. B. Cumming, Eds. (Ann Arbor, MI: Ann Arbor Science Publishers, Inc., 1980), pp. 415–426.
3. Helliwell, P. R., and N. B. Webber. "The Use of the Solent Model for an Investigation into Sewer Outfall Location," in *Mathematical and Hydraulic Modelling of Estuarine Pollution,* Water Pollution Research Technical Paper 13, (London, Her Majesty's Stationery Office, 1973), pp. 211–220.
4. Ackers, P., and L. J. Jaffrey "Applicability of Hydraulic Models to Pollution Studies," in *Mathematical and Hydraulic Modelling of Estuarine Pollution*, Water Pollution Research Technical Paper 13, (London, Her Majesty's Stationery Office, 1973), pp. 173–190.
5. Davis, M. H., and J. Coughlan. "A Model for Predicting Chlorine Concentrations within Marine Cooling Circuits and its Dissipation at Outfalls," in *Water Chlorination: Environmental Impact and Health Effects, Vol. 4*, R. L. Jolley, W. A Brungs, J. A. Cotruvo, R. B. Cumming, J. S. Mattice, and V. A Jacobs, Eds. (Ann Arbor MI: Ann Arbor Science Publishers Inc., 1983), pp. 347–357.
6. Thomas, L. C., and G. J. Chamberlin. *Colorimetric Chemical Analytical Methods, 9th Ed.* (Salisbury, U.K.: Tintometer Co., 1980) p. 626.
7. Heinemann, T. J., G. F. Lee, R. A. Jones, and B. W. Newbry. "Summary of Studies on Modelling Persistence of Domestic Wastewater Chlorine in Colorado Front Range Rivers," in *Water Chlorination: Environmental Impact and Health Effects, Vol. 4*, R. L. Jolley, W. A Brungs, J. A. Cotruvo, R. B. Cumming, J. S. Mattice, and V. A Jacobs, Eds. (Ann Arbor, MI: Ann Arbor Science Publishers, Inc., 1983), pp. 97–112.
7. Helz, G. R., A. C. Sigleo, and C. A. Hill. "Mechanisms of Chlorine Degradation in Estuarine Waters," in *Water Chlorination: Environmental Impact and Health Effects, Vol. 3*, R. L. Jolley, W. A. Brungs, and R. B. Cumming, Eds. (Ann Arbor, MI: Ann Arbor Science Publishers, Inc., 1980), pp. 387–394.

CHAPTER **113**

Predicting Chlorine Compounds in Power Plant Cooling Tower Systems

Vito L. Punzi and Rutton D. Patel

Chlorine is used intermittently as a cooling system defouling agent in the electric utility industry, because it destroys bacterial slimes that form on heat exchanger surfaces. When ammonia is present, the reactions of chlorine in natural waters lead to the formation and discharge of numerous compounds, principally chloramines, which are known as combined residual chlorine. Although the effects of these compounds are well documented,[1,2] there are currently few models[3-5] that quantitatively predict the presence of these compounds in power plant cooling tower systems and discharges.

Although previous models are each increasingly more sophisticated, it is necessary that a model be developed that incorporates the valuable features of previous models while also addressing aspects not previously considered.

This chapter describes one such model, which can be used to predict free and combined residual chlorine concentrations and specific species found in cooling tower systems and discharges, and apply the data to a specific power plant cooling tower system. A detailed description of the analyses summarized here appears in a previous publication.[6]

DESCRIPTION OF THE STUDY

Overview

The model can be used to study several aspects that were not included in previous investigations and a number of phenomena that were previously examined together:

1. Identification of the important chemical species, their chemistry, and the rates of the reactions that occur.
2. The effect of slime presence on residual chlorine concentrations and vice versa.
3. A description of mass transfer phenomena and chemical reactor considerations involved in the cooling system of a power plant, along with the mathematical techniques needed to describe the phenomena.

Each of these items is discussed in the following sections.

Table I. Equilibrium and Rate-Dependent Reactions Considered in Power Plant Chlorination Model Development

Equilibrium reactions (hydrolysis/ionization)

Chlorine	$Cl_2 + H_2O \rightleftarrows H^+ + Cl^- + HOCl$
	$HOCl \rightleftarrows H^+ + OCl^-$
Ammonia	$NH_3 + H_2O \rightleftarrows NH_4^+ + OH^-$
Typical organic compound (X)	$X + H_2O \rightleftarrows XH^+ + OH^-$

Instantaneous conversion to chloride ion: immediate chlorine demand (typical)

Nitrates (NO_2^-)	$NO_2^- + HOCl \rightarrow NO_3^- + HCl$
Ferrous iron (Fe^{2+})	$Fe^{2+} + HOCl + 2HCl \rightarrow FeCl_2 + HCl + H_2O$
Manganese (Mn^{2+})	$Mn^{2+} + Cl_2 \rightarrow Mn^{4+} + 2Cl^-$
Hydrogen sulfide (H_2S)	$H_2S + HOCl \rightarrow S + HCl + H_2O$ or
	$H_2S + 4HOCl \rightarrow H_2SO_4 + 8HCl$
Cyanide (CN^-)	$CN^- + HOCl \rightarrow CNCl + OH^-$

Homogeneous rate-dependent reactions

Reactions between chlorine and ammonia compounds	$NH_3 + HOCl \rightleftarrows NH_2Cl + H_2O$
	$NH_2Cl + HOCl \rightleftarrows NHCl_2 + H_2O$
	$2NH_2Cl \rightleftarrows NHCl_2 + NH_3$
Reactions between chlorine and a typical organic compound (X)	$X + HOCl \rightleftarrows X'Cl + H_2O$
	$X'Cl + HOCl \rightleftarrows X''Cl_2 + H_2O$

Heterogeneous rate-dependent reactions (slime destruction)

Reactions between slime (S_1) and chlorine compounds	$3HOCl + S_1 \rightarrow$ Products
	$NH_2Cl + 2S_1 \rightarrow$ Products

Chemical Species/Chlorination Chemistry

Several types of chemical reactions occur when freshwater cooling tower systems are chlorinated (as shown in Table I).

1. The nearly instantaneous hydrolysis and ionization reactions that occur when chlorine is added to water, along with the hydrolysis products of ammonia (NH_3) and typical nitrogen-containing organic compounds (X).

2. The instantaneous reactions between certain inorganic species and hypochlorous acid (HOCl) that convert HOCl to the nondisinfecting chloride ion form ("immediate" chlorine demand).[6]

3. The homogeneous, rate-dependent reactions that result in chloramine formation.

Table II. Chlorination Chemistry: Equilibrium Considerations

Reaction	Equilibrium Constant		
		Relationship	Typical Value[a]
$Cl_2 + H_2O \rightleftarrows H^+ + Cl^- + HOCl$	K_1	$= \dfrac{(H^+)(Cl^-)(HOCl)}{(Cl_2)}$	4.5×10^{-4} mol^2/L$^{2\,[b]}$
$HOCl \rightleftarrows H^+ + OCl^-$	K_G	$= \dfrac{(H^+)(OCl^-)}{(HOCl)}$	3.7×10^{-8} mol/L
$NH_3 + H_2O \rightleftarrows NH_4^+ + OH^-$	K_H	$= \dfrac{(NH_4^+)(OH^-)}{(NH_3)}$ or	1.81×10^{-5} mol/L
	K_H	$= \dfrac{K_W(NH_4^+)}{(NH_3)(H^+)}$	
where	K_W	$= (H^+)(OH^-)$	1.03×10^{-14} mol^2/L^2
$X + H_2O \rightleftarrows XH^+ + OH^-$	K_I	$= \dfrac{(XH^+)(OH^-)}{(X)} = \dfrac{K_W(XH^+)}{(X)(H^+)}$	5.0×10^{-4} mol/L

[a]Equilibrium constants are temperature dependent; a typical value at 25°C (77°F) is shown, although values at other temperatures are available.[5]
[b]The large value of K_1 indicates nearly complete conversion to HOCl; therefore, presence of elemental chlorine can be neglected. In addition, this implies that the initial molar concentration of HOCl equals the initial molar concentration of elemental chlorine.

4. The heterogeneous, rate-dependent reactions between hypochlorous acid and slime and between monochloramine (NH_2Cl) and slime that actually occur in a power plant condenser.

Equilibrium or rate constant data (or both) were obtained from the literature[6,7] and are summarized in Tables II and III.

Slime Destruction

Conditions suitable for microbial growth lead to the formation of bacterial slimes that adhere to heat exchanger surfaces. A mechanism that models slime destruction via chlorination has been developed. It is based on mechanisms[2,8-12] and simplifying assumptions that permit it to be modeled using chemical kinetics.[6]

Table III. Kinetic Data for N-Chlorination of Ammonia and Organic Nitrogen Compounds

Reaction	Rate Constants[a]	
	Theoretical	Observed[b]
Monochloramine formation[c]		Forward
$NH_3 + HOCl \rightleftarrows NH_2Cl + H_2O$	$k_{AT} = 9.7 \times 10^8 \exp(-3000/RT)$	$k_A = \dfrac{k_{AT}}{1 + \dfrac{K_G K_H}{K_W} + \dfrac{K_G}{(H^+)} + \dfrac{K_H(H^+)}{K_W}}$
		Reverse
		$k_B = 8.7 \times 10^7 \exp(-17000/RT)$
Dichloramine formation[d]		$k_C = \dfrac{k_{CT} K_W}{K_H(H^+)}$
$NH_2Cl + HOCl \rightleftarrows NHCl_2 + H_2O$	$k_{CT} = 7.6 \times 10^7 \exp(-7300/RT)$	
Monochloramine disproportionation		$k_D = 80 \exp(-4300/RT)$
$2NH_2Cl \rightarrow NHCl_2 + NH_3$		
Organic monochloramine formation		$k_E = \dfrac{k_{ET}}{1 + \dfrac{K_G K_I}{K_W} + \dfrac{K_G}{(H^+)} + \dfrac{K_I(H^+)}{K_W}}$
$X + HOCl \rightarrow X'Cl + H_2O$	$k_{ET} = 3.0 \times 10^8$ L/(mol·s)	
Organic dichloramine formation		$k_F = 1.0 \times 10^3$ L/(mol·s)
$X'Cl + HOCl \rightarrow X''Cl_2 + H_2O$		

[a] $R = 1.987$ cal/(mol·degree K); temperature (T) units are (K).
[b] K_G, K_H, K_I, and K_W are shown in Table II.
[c] Forward reaction, second order; reverse reaction, first order.
[d] Second-order overall; negligible reverse reaction.

The literature indicates that the sulfhydryl (–SH) groups on the amino acid cysteine can be oxidized according to

$$CHNH_2COOHCH_2SH + 3HOCl \longrightarrow CHNH_2COOHCH_2SO_3H + 3HCl$$

$$2CHNH_2COOHCH_2SH + NH_2Cl \longrightarrow$$

$$CHNH_2COOHCH_2S-SCH_2CHNH_2COOH + NH_3 + HCl$$

When pure reagents are involved, these reactions occur instantaneously. However, the presence of the microbial cell wall slows the apparent rate of these reactions. Microbe survival data[12] are used to obtain[6] the apparent rate constant and reaction order; therefore, the overall rate of slime destruction R_S (mol per liter-second) is expressed in terms of the concentration of hypochlorous acid, C_{HOCl} (mol per liter), monochloramine, C_{NH_2Cl} (mol per liter), and slime C_{Sl} (mol cysteine per liter condenser volume):

$$R_S = -K_{S1}C_{HOCl}C_{Sl} - K_{S2}C_{NH_2Cl}^2 C_{Sl}$$

The constants K_{S1} and K_{S2} are 621.5 L mol^{-1}s^{-1} and 7.8633 × 10^6 L^2 mol^{-2}s^{-1}, respectively.

Modeling the Circulating Water System

A schematic showing the major components of the circulating water system at a power plant that uses a cooling tower is shown in Figure 1. Although modeling the system as a series of chemical reactors or physical mass exchangers permits the concentrations of all chlorine-containing compounds and slime to be determined at any point within the system, emphasis is placed on the condenser exit and the cooling tower blowdown pipe, the sampling locations used by Draley.[4]

The circulating water pipelines and the condenser are modeled as plug-flow reactors (PFR); the general equation that applies is:

$$\frac{\partial C_i}{\partial t} + u \frac{\partial C_i}{\partial Z} = R_i$$

where C_i is the molar concentration of the ith constituent (mol per liter), u is the average velocity (meters per second), Z is the axial position (m), and R_i represents the rate of generation of the ith constituent (mol per liter-second) in the PFR. This approach yields a family of differential equations that are solved numerically.[6] Expressions for the reaction rate (R_i) terms are presented in Table IV.

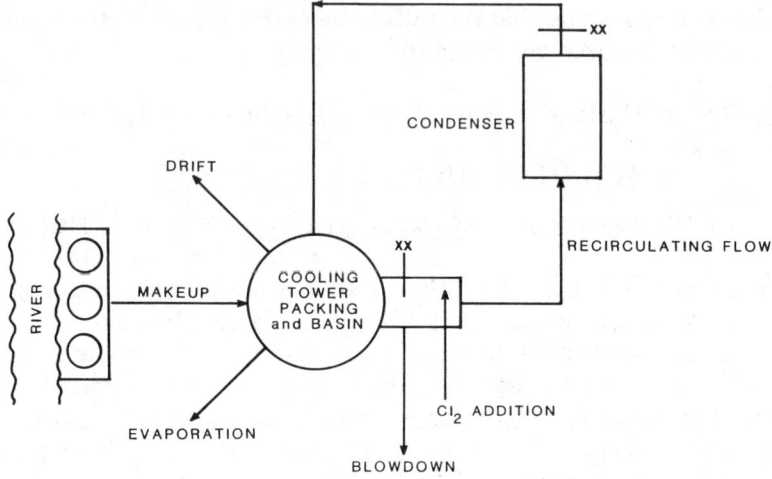

Figure 1. Plant flow sheet. Sampling points used by Draley are indicated by xx (Ref. 4).

In modeling the packing section of the tower, it is necessary to develop algebraic expressions that relate the concentration of each species in the liquid and vapor phases. Since only NH_3 and NH_2Cl exert a significant vapor pressure, a modified version of Raoult's law[13] is used to describe the physical transport from the liquid to the vapor. Specifically, the concentration of compounds in the exiting air stream is assumed to be a fraction (f) of the concentration that would have been in equilibrium with the incoming liquid if an infinite packing height (i.e., f = 1) had been used. In this study, this empirical factor was estimated to be 0.15 and is not related to the factor F developed by Nelson[3] and Draley.[4]

The cooling tower basin is modeled as a perfectly mixed continuous-stirred tank reactor (CSTR) so that the concentrations of each constituent in the basin, blowdown, and the recirculating stream leaving the basin are identical. The general equation that applies to such a reactor is:

$$V \frac{dC_i}{dt} = R'C_{RE_i} + MC_{M_i} - RC_i - BC_i - R_iV$$

where C_i and R_i are as previously defined, and R' and C_{RE_i} represent the flowrate (liters per minute) and the concentration (mol per liter), respectively, of the ith constituent in the liquid entering the basin. The terms M and C_{M_i} are similarly defined for the makeup water stream, and V, R, and B represent the tower basin volume (L^3), recirculating water flowrate (liters per minute), and

Table IV. Reaction Rates Used in the Model Analysis

Constituent	Concentration Nomenclature	Reaction Rate Nomenclature
HOCl	C_A	R_A
NH_3	C_B	R_B
NH_2Cl	C_C	R_C
$NHCl_2$	C_D	R_D
X	C_E	R_E
X'Cl	C_F	R_F
X"Cl_2	C_G	R_G
Slime	C_S	R_S

Reaction rate expressions (entire cooling tower system)

$$R_B = -k_A C_B C_A + k_B C_C + \tfrac{1}{2} k_D C_C^2$$
$$R_D = k_C C_C C_A + \tfrac{1}{2} k_D C_C^2$$
$$R_E = -k_E C_E C_A$$
$$R_F = k_E C_E C_A - k_F C_F C_A$$
$$R_G = k_F C_F C_A$$

Reaction rate expressions (entire cooling tower system except the condenser)

$$R_A = -k_A C_B C_A + k_B C_C - k_C C_C C_A - k_E C_E C_A - k_F C_F C_A$$
$$R_C = k_A C_B C_A - k_B C_C - k_C C_C C_A - k_D C_C^2$$
$$R_S = 0$$

Reaction rate expressions (condenser only)

$$R_A = -k_A C_B C_A + k_B C_C - k_C C_C C_A - k_E C_E C_A - k_F C_F C_A - 3k_{S1} C_A C_S$$
$$R_C = k_A C_B C_A - k_B C_C - k_C C_C C_A - k_D C_C^2 - \tfrac{1}{2} k_{S2} C_C^2 C_S$$
$$R_S = -k_{S1} C_A C_S + k_{S2} C_C^2 C_S$$

blowdown flowrate (liters per minute), respectively. Since C_{M_i} is usually zero, the above equation reduces to:

$$\frac{dC_i}{dt} = \left(\frac{R'}{V}\right) C_{RE_i} - \left(\frac{B+R}{V}\right) C_i + R_i$$

This yields a family of differential equations that are also solved numerically.[6] Expressions for the reaction rate (R_i) terms are presented in Table IV.

Specification of Physical/Chemical Parameters

Table V lists the parameters required to run the model and the values used in the simulation runs. The data were either collected by Draley[4] or obtained from the utility involved in the plant design.[14-19] In some cases, the model used data (obtained from the utility) that conflicted with the data reported by Draley.[4]

The model required data for four parameters that had to be estimated since no data were available for the plant used in the simulation. The parameters included the immediate chlorine demand, the initial ammonia-nitrogen concentration, the value of factor f (previously identified), and the residence (or lag) time of the system. The assumed values for these parameters are shown in Table VI.

RESULTS OF ANALYSIS

Summary of Previous Experiments

Draley[4] conducted two runs at the Appalachian Power Company's Amos plant during which chlorine was added to the plant's circulating water system. Samples were taken at the condenser exit and the cooling tower basin discharge and analyzed for free and/or total residual chlorine. Details of the experiments and the results have been described previously;[4,6] the most significant data are presented in Table VII.

Simulation of Amos Plant Chlorination Experiments: Chlorination Period

The two runs conducted by Draley[4] are simulated using the data presented in Tables V and VI; the model predictions for chlorine addition periods are shown in Figures 2 and 3.

Free residual chlorine (FRC) predictions for run 1 (not shown) indicate that less FRC was present at the condenser exit than was sampled by Draley. The FRC prediction for run 2 (not shown) indicates that the model predicts the same trend as Draley's data, although quantitative agreement was poor, which is in contrast to the behavior observed during run 1, where predictions and data were inexplicably different.

Combined residual chlorine (CRC) predictions for runs 1 and 2 (not shown) indicate rapid and nearly complete ammonia conversion to chloramines, compared with the gradual trend noted by Draley,[4] probably indicating that actual reaction rates are different than those observed in the laboratory.

The model predictions of total residual chlorine (TRC) at the condenser exit for runs 1 and 2, shown in Figures 2 and 3, respectively, indicated that the model was reasonably accurate, even though the distribution between FRC

Table V. Physical and Chemical Parameters Used in Model Simulation Runs[a]

Parameter	Draley[b]		Model Values[c]	
	Run 1	Run 2[b]	Run 1	Run 2[d]
Average system pH	7.5	*	7.5	*
Average system temperature, °C(°F)	37.8(100)	31.7(89)	32.2(90)	*
Chlorine feed rate, kg/min (lb/min)	1.23(2.1)	1.51(3.33)	1.23(2.71)	1.51(3.33)
Chlorine feed concentration, mg/L	1.31	1.61	1.31	1.61
Chlorination period, min	30	*	30	*
Ammonia-nitrogen concentration, mg/L	0		0.135	0.180
Organic compound concentration, µg/L	0	0	1.0	
Immediate chlorine demand, mg/L as HOCl	0.45–0.6 (assumed)	*	1.25 (Initial) 0.70 (Level)	1.65 (Initial) 0.75 (Level)
Initial slime concentration, mol/L			1.87322×10^{-4}	*
Cooling tower basin volume, L	2.08×10^7	*	1.25×10^7	*
gal	5.5×10^6	*	3.3×10^6	*
Recirculating flowrate, L/min	938,800	*	938,800	*
gal/min	248,000	*	248,000	*
Blowdown flowrate, L/min	15,140	0	2,910	0
gal/min	4,000	0	770	0
Blowdown flow frequency	1.5 h/d		Continuous	*
Drift rate, L/min			1,900	*
gal/min			500	*
Evaporation rate, L/min	26,500	*	20,820	*
gal/min	7,000	*	5,500	*
Makeup flowrate, L/min	26,500	*	25,630	*
gal/min	7,000	*	6,770	*
Air inflow rate, kg/min			6.36×10^{5e}	*
lb/min			1.4×10^{6e}	*
Average pipeline water velocity, m/s (ft/s)			2.13(7.0)	*
Average condenser water velocity, m/s (ft/s)			2.29(7.5)	*

Table V, continued

Parameter	Draley[b]		Model Values[c]	
	Run 1	Run 2[b]	Run 1	Run 2[d]
Pipeline length (cooling tower basin to condenser), m (ft)			103.6(340)	*
Pipeline length (condenser to cooling tower basin), m (ft)			152.4(500)	*
Condenser length, m (ft)			18.3(60)	*
System residence time, min	22	*	1–2	*
f, Mass transfer factor[f]			0.15	*

[a]From References 4, 14–19.
[b]See Reference 4.
[c]In those cases where model values are different than the values presented by Draley,[4] the model values reflect information provided by American Electric Power Service Corp.[15–19]
[d]Values of parameters indicated by (*) are identical to those in run 1.
[e]Calculated from typical natural draft cooling tower design specifications.[14]
[f]The factor f is a factor that affects liquid-to-vapor-phase mass transfer of species in the evaporation stream.

Table VI. Parameter Estimates Based on Study Model

Parameter	Run 1	Run 2
Factor f	0.15	0.15
Ammonia nitrogen	0.135 mg/L	0.180 mg/L
Chlorine demand[a]	1.25 mg/L (Initial)	1.65 mg/L (Initial)
	0.70 mg/L (Level)	0.75 mg/L (Level)
System time lag		
Condenser	1 min[b]	Same as run 1
Cooling tower basin	2 min[b]	Same as run 1

[a]An analysis of chlorine demand behavior determined that a model using a continuously decreasing demand until a specified level is achieved, at which time the chlorine demand remains constant for the remainder of the chlorination cycle, provides the best fit of the data taken by Draley.[4]
[b]An additional 1-min lag time between the condenser and the cooling tower basin occurs because of the presence of a connecting pipeline between these two points.

and CRC was not predicted accurately. The difference between TRC predictions and Draley's data averaged 20% in run 1 and 23% in run 2.

The predictions of TRC in the cooling tower blowdown for run 1 (shown in Figure 2) not only indicated the same form as the data obtained by Draley, but predicted values that were in close agreement with Draley's data. The differences averaged 15%, with a number of predictions within 10%. Since cooling tower blowdown is the only stream that leaves the plant, favorable TRC predictions for this stream are probably the most critical. Further, the results seemed to indicate that TRC predictions for cooling tower blowdown were less sensitive to inaccuracies than the predictions at the condenser exit.

The TRC predictions for cooling tower blowdown during run 2 (shown in Figure 3) also indicated the same behavior as Draley's data. However, while the predictions deviated considerably from actual data, it appeared that shifting model predictions by 9 min would yield excellent agreement. That is, a comparison of Draley's data and unadjusted model predictions indicated an average difference of 54%. However, a comparison of Draley's data and the model predictions shifted or delayed by 9 min reduced the average difference to 35%. This phenomenon cannot be explained theoretically, although Draley observed TRC in the cooling tower blowdown 12 min after the onset of chlorination during run 1 and 22.5 min after the onset of chlorination during run 2. The difference of 10.5 min is greater than would be expected for two runs conducted under similar conditions. This is very close to the 9-min "displacement" that would make the run-2 TRC predictions in cooling tower blowdown agree very well with the data.

Table VII. Summary of Data Obtained by Draley[a]

	Chlorine Addition Data			Condenser Outlet Data[b]			Cooling Tower Blowdown Data[b]		
Run	Mass (kg)	Time (min)	Concentration (mg/L)	FRC (mg/L)	CRC (mg/L)	TRC (mg/L)	FRC (mg/L)	CRC (mg/L)	TRC (mg/L)
1	36.8	30	1.31	0.10[c]	0.56[c]	0.62[c]	0.0	0.32[d]	0.32[d]
2	45.5	30	1.61	0.26[e]	f	0.90[c]	0.14[e]	f	0.51[c]

[a]See Reference 4.
[b]FRC = free residual chlorine, CRC = combined residual chlorine, TRC = total residual chlorine.
[c]Maximum concentration measured during the run.
[d]Maximum concentration measured during the run; TRC declined to 0.1 mg/L in about 50 min. An additional 90 min was required before concentrations fell to below detectable levels.
[e]Maximum concentration measured during the run, although unaccountable fluctuations were observed.
[f]In run 2, FRC or TRC was determined analytically; therefore, CRC is estimated from one measured and one interpolated value.

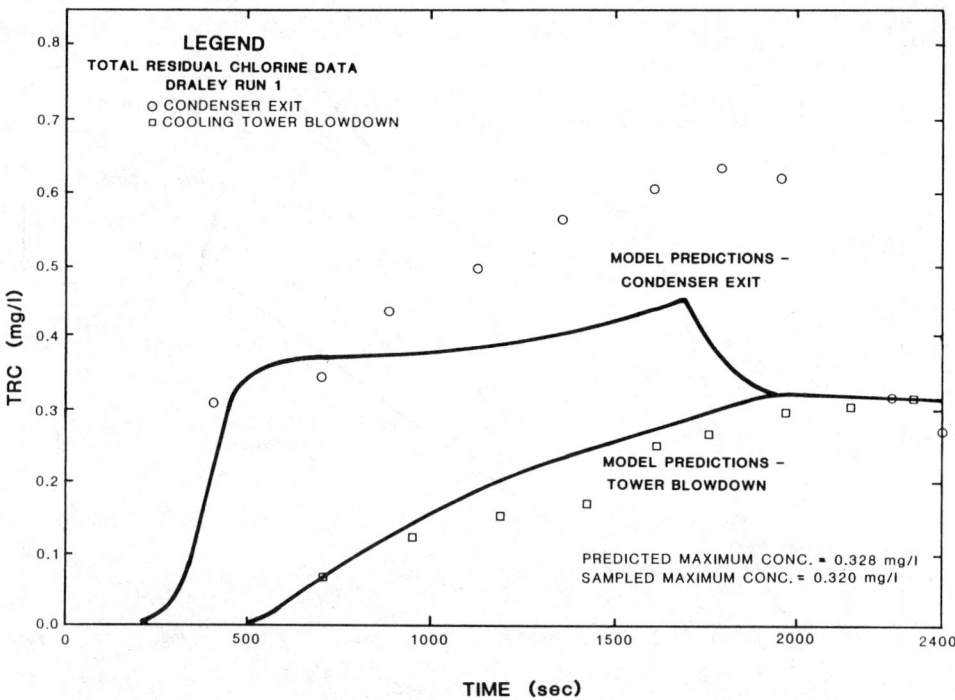

Figure 2. Total residual chlorine at the condenser exit and in tower blowdown; run 1 data and model predictions for chlorine feed period.

Simulation of Amos Plant Chlorination Experiments: Postchlorination Period

Experimental data and model predictions for runs 1 and 2 indicated that any FRC present at the condenser exit at the end of the chlorine feed period was dissipated rapidly. In run 1, Draley's data indicated that no FRC would remain 3 min after the end of the chlorine feed period; the model predicted no FRC 1 min after the end of the chlorine feed period. Comparable values for run 2 are 8 min (Draley) and 5 to 15 min (predicted). Neither the data nor the model predictions indicated that any FRC would be present in the cooling tower blowdown at any time during run 1.

The data and model predictions for run 2 indicated that FRC was present in the cooling tower blowdown. After chlorine addition ends, the data indicated that FRC in the cooling tower blowdown was dissipated in 22 to 30 min; the model predicted 26 to 28 min. However, this agreement does not extend to TRC predictions both at the condenser exit and in the cooling tower blowdown, in both runs 1 and 2.

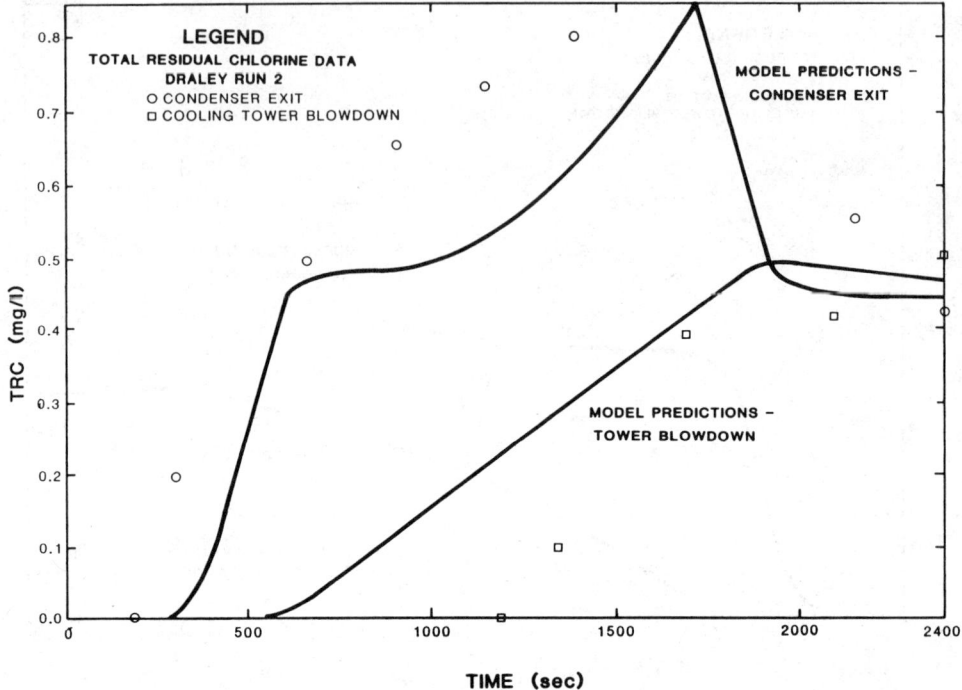

Figure 3. Total residual chlorine at the condenser exit and in tower blowdown; run 2 data and model predictions for chlorine feed period.

Figures 4 and 5 present data and model predictions for TRC at the condenser exit and in the cooling tower basin, respectively, during run 1. Predicted TRC levels were higher than measured and decreased at a slower rate than the data. However, the decay rate (apparent from the data) beyond an elapsed time of 50 min was closely parallelled by the model predictions beyond ~150 min. These results indicated that a mechanism not included in this model operates throughout the entire chlorination cycle. This mechanism, which dominates after chlorine addition ends but is relatively insignificant during chlorine addition, resulted in CRC dissipation at a higher rate than predicted by the model. Possible mechanisms are discussed later.

The model can be used to estimate the dimensionless factor F defined by Nelson[3] and Draley.[4] The factor F is dependent on the ratio of the water flowrate evaporated to the water flow returning to the tower from the condenser, which is usually quite small. Thus, a value of 0.5[3,4] greatly overestimates the amount of chlorine-containing compounds lost via evaporation. Using the above definition, F has been estimated to be 0.01 to 0.02.[6]

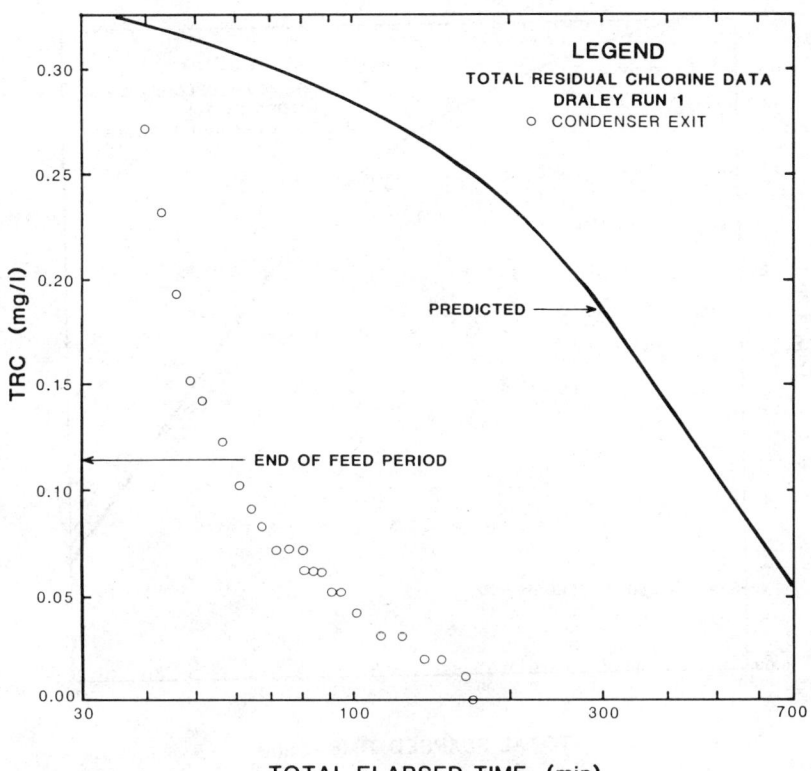

Figure 4. Total residual chlorine at the condenser exit; run 1 data and model predictions for postchlorination period.

Slime Destruction

Figure 6 is a plot of the fraction of slime present at three locations in the condenser as a function of time after the onset of chlorination. Figure 6 indicates that the greatest slime destruction occurs near the inlet to the condenser where the highest disinfectant concentrations are present, and that an average of 70% of the slime is destroyed during the chlorine feed period.

DISCUSSION OF RESULTS

Chlorination Period

The model predictions presented show a general agreement with field data, particularly the cooling tower blowdown TRC predictions, although a complete set of data would significantly improve the quality of the predictions.

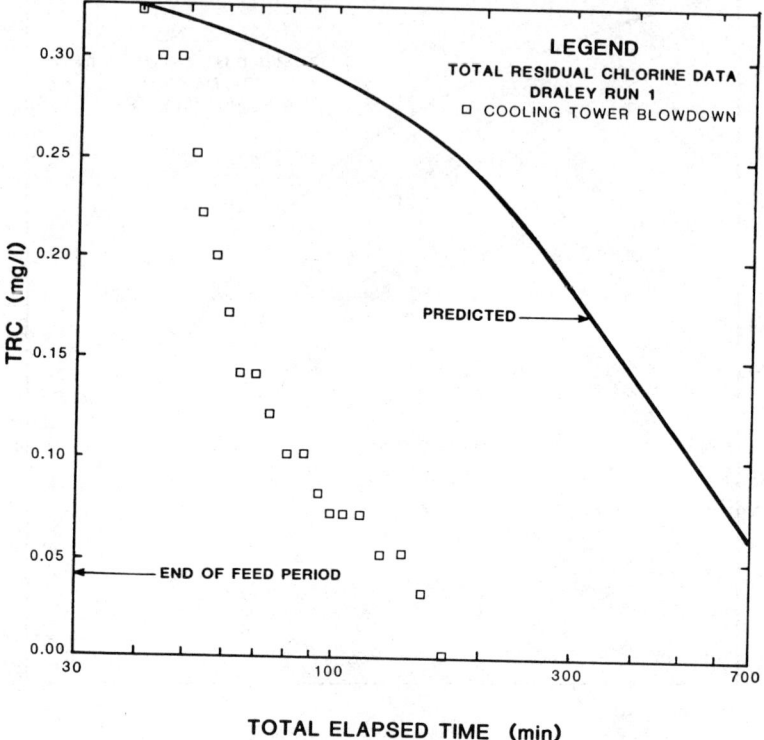

Figure 5. Total residual chlorine in tower blowdown; run 1 data and model predictions for postchlorination period.

Other process-related variables such as tower basin lag time require additional study. A more accurate model for the cooling tower basin that would account for stagnant regions is needed. A study of the appropriate reactions under cooling tower conditions is also needed to determine the effect of chemical constituents (presently believed to be extraneous) on the reaction kinetics.

Postchlorination Period

Model predictions for the postchlorination period do not adequately represent the behavior observed.[4] One possible mechanism that would explain the difference is the potential reactions between sulfur-containing compounds (e.g., SO_2) and HOCl or NH_2Cl:

$$SO_2 + H_2O \longrightarrow H_2SO_3$$

$$H_2SO_3 + HOCl \longrightarrow H_2SO_4 + HCl$$

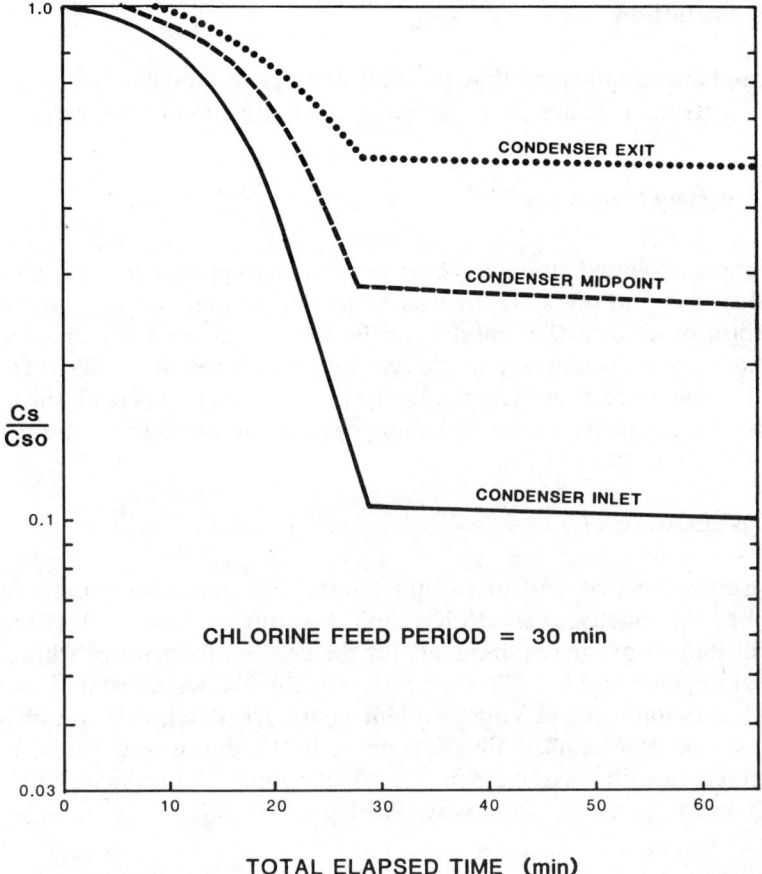

Figure 6. Fraction of slime remaining at three condenser locations after onset of chlorination.

$$H_2SO_3 + NH_2Cl + H_2O \longrightarrow NH_4HSO_4 + HCl$$

At the Amos plant, the cooling tower basin outlet is equipped with adjacent chlorine and sulfuric acid feed facilities that use similar diffuser pipes to introduce the chemicals into the circulating water flow. Further, the basin sampling point used by Draley[4] is located immediately upstream of these facilities. Since sulfuric acid typically contains 0.015% SO_2 (>250 mg/L) as a contaminant,[20] the potential exists for backmixing and reaction, thereby resulting in the rapid loss of chloramines observed by Draley. Further study of this possible mechanism is needed.

Slime Destruction

The predictions indicated that the first attempt at modeling slime destruction yields results that are consistent with qualitative field observations.

CONCLUSIONS

A theoretically based model has been developed to predict concentrations of chemical species and slime presence as a function of time for any intermittent chlorination procedure. The model can be used to determine a chlorination policy that minimizes chlorine discharges and maximizes slime destruction.

Methods that more accurately model the postchlorination period, the behavior of the cooling tower basin, and slime destruction phenomena are needed.

ACKNOWLEDGMENTS

The results presented here are taken from the dissertation submitted by Vito L. Punzi to the faculty of the Polytechnic Institute of New York (PINY) in partial fulfillment of the requirements for the degree of Doctor of Philosophy (Chemical Engineering) in 1979. Currently, Dr. Punzi is an Assistant Professor of Chemical Engineering at Villanova University. Dr. Patel, who was an Associate Professor of Chemical Engineering at PINY during this study, is currently an engineering associate in the Technology Department of Exxon Research and Engineering Company, Florham Park, NJ.

REFERENCES

1. Brungs, W. A. "Effects of Residual Chlorine on Aquatic Life," *J. Water Pollut. Control Fed.* 45(10):2180–2193 (1973).
2. Mattice, J. S., and H. E. Zittel. "Site-Specific Evaluation of Power Plant Chlorination," *J. Water Pollut. Control Fed.* 48(10)2284–2308 (1976).
3. Nelson, G. R. "Predicting and Controlling Total Residual Chlorine in Cooling Tower Blowdown," EPA-R2-73-273, (Washington, DC: U.S. Environmental Protection Agency, 1973).
4. Draley, J. E. "Chlorination Experiments at the John E. Amos Plant of the Appalachian Power Company," (Argonne, IL: Argonne National Laboratory, 1973).
5. Lietzke, M. H. "A Kinetic Model for Predicting the Composition of Chlorinated Water Discharged from Power Plant Cooling Systems," ORNL/NUREG-13 (Oak Ridge, TN: Oak Ridge National Laboratory) 1977.
6. Punzi, V. L. "Modelling and Predicting Free and Combined Residual Chlorine Concentrations in Cooling Tower Systems," Ph.D. Dissertation, (Brooklyn: Polytechnic Institute of New York, 1979).

7. Lietzke, M. H. "A Kinetic Model for Predicting the Composition of Chlorinated Water Discharged from Power Plant Cooling Systems," in *Water Chlorination: Environmental Impact and Health Effects, Vol. 1*, R. L. Jolley, Ed. (Ann Arbor, MI: Ann Arbor Science Publishers, Inc., 1978), pp. 367–378.
8. Ingolls, R. S., H. A. Wyckoff, T. W. Kithley, H. W. Hodgden, E. L. Fincher, J. C. Hildebrand, and J. E. Mandel. "Bacterial Studies of Chlorine," *Ind. Eng. Chem.* 45(5):996–1000 (1953).
9. Pelczar, M. J., and R. D. Reid. *Microbiology*, 2nd ed., (New York: McGraw-Hill, Inc., 1965).
10. Green, D. E., and P. K. Stumpf. "The Mode of Action of Chlorine," *J. Am. Water Works Assoc.* 38(11):1301–1305 (1946).
11. Wyss, O. "Disinfection by Chlorine: Theoretical Aspects," *Water Sewage Works* 109(12):R155–R158 (1962).
12. Hom, L. W. "Kinetics of Chlorine Disinfection in an Ecosystem," in *Proceedings of the National Specialty Conference on Disinfection*, American Society of Chemical Engineers (1970), pp. 515–537.
13. King, C. J. *Separation Processes* (New York: McGraw-Hill, Inc., 1971).
14. Kennedy, J. F. "Wet Cooling Towers," in *Engineering Aspects of Heat Disposal from Power Generation*, Massachusetts Institute of Technology Short Course Document (June 1972), Chapter 13.
15. American Electric Power Service Corporation. "Amos Plant Units 1 and 2, Circulating Water Chlorination System," Descriptive Article A-1-3-9 (New York: AEP Service Corporation, August 1970).
16. American Electric Power Service Corporation. "Amos Plant Units 1 and 2, Circulating Water System," Descriptive Article A-1-3-4 (New York: AEP Service Corporation, March 1972).
17. American Electric Power Service Corporation. "Amos Plant Units 1 and 2, Main and F-P Turbine Condensers," Descriptive Article A-1-3-2 (New York: AEP Service Corporation, January 1971).
18. American Electric Power Service Corporation. "Amos Plant Units 1 and 2, Cooling Tower," Descriptive Article A-1-3-15 (New York: AEP Service Corporation, August 1970).
19. Miskimen, T., J. Balletto, and S. Sosnowski. American Electric Power Service Corporation, personal communication.
20. Kirk, R. E., and D. F. Othmer. "Sulfuric Acid and Sulfur Trioxide," Vol. 19, *Kirk-Othmer Encyclopedia of Chemical Technology*, 2nd ed. (New York: John Wiley and Sons, 1969), pp. 441–482.

CHAPTER **114**

Prediction of Total Residual Chlorine in Power Plant Discharges and Receiving Waters: Application to Effluent Limitations and Water Quality Standards

John P. Lawler, Thomas B. Vanderbeek, and Peter M. Cumbie

This chapter summarizes the results of a project[1] funded by the Utility Water Act Group (UWAG). The objective of the project was to provide procedures suitable for predicting total residual chlorine (TRC) in the discharges of once-through cooling water systems in steam electric generating stations located on freshwater rivers and lakes.

The procedures include (1) predictive mathematical models to delineate dilution and decay of TRC after condenser passage, and (2) field and laboratory procedures to permit evaluation of model parameters and to provide an alternative empirical means of estimating TRC dilution and decay. This chapter also discusses the use of chlorine predictions and measurements and presents an actual application. Chlorine chemistry and model development are discussed in more detail elsewhere.[2]

PURPOSE

The original purpose of the predictive procedures, as envisioned by UWAG, was to assist in the preparation of a 301(g) waiver application, which is a request for modifications of the Best Available Technology Economically Achievable requirements for nonconventional pollutants under Section 301(g) of the Clean Water Act. The U.S. Environmental Protection Agency (EPA) draft technical guidance was developed in 1982.[3] However, no formally proposed substantive 301(g) variance regulations existed as of April 1984.

UWAG's chemical committee had recognized that, regardless of the eventual details of expected 301(g) rulemaking, and regardless of mixing zone considerations, a means of predicting TRC concentrations at any point in time and space in the receiving watercourse would be an important tool for demonstrating in a 301(g) waiver application that water quality standards or criteria would be met outside the mixing zone.

The original purpose of the project remained the primary purpose, although two additional reasons for developing TRC predictive techniques emerged during early discussions of the project between UWAG and Lawler, Matusky & Skelly Engineers (LMS).

Predictive models and measurement procedures could also be used in the discharge channel itself to determine the actual TRC at the point of discharge, when direct measurement would be impossible or impractical, for example, a deep offshore submerged discharge. If this measurement could be shown to meet the effluent limitation, application could be made for an NPDES or SPDES permit without any need for a 301(g) waiver application.

The procedures could also be used to demonstrate compliance with regulations for water quality limited streams [Section 303(d)] by either showing that water quality standards or criteria would be met for a given proposed discharge, or by determining (by back calculation) the maximum loading for which standards or criteria would still be met.

SCOPE

The project required that TRC predictive models and measurement procedures be developed for once-through cooling systems discharging to freshwater rivers and lakes. Although saltwater systems and/or cooling tower systems (closed or otherwise) were not formally included in the project's scope, the literature review included material that addressed both topics; such material provided useful insights. Furthermore, most of the models and measurement procedures can also be used in saltwater and cooling tower applications, albeit with some modification or at least with care, caution, and recognition of the differences.

Although none of the dilution models are tidal, and the chlorine chemistry of salt water is materially different from that of freshwater, the conceptual approach set forth here is certainly similar to what would be used for saltwater systems.

In the case of recirculating cooling towers (or lakes or ponds), the in-plant (flume) procedures would have to be modified to recognize buildup, as is done in certain of the available models (e.g., the TVA model of Zielke and Moss.[4] For discharge of blowdown to a receiving freshwater river or lake, the procedures for chlorine prediction in the receiving water body are applicable, provided that calculations begin in the blowdown stream.

RELEVANT REGULATORY FRAMEWORK

Final Effluent Limitations Guidelines

EPA issued final best available technology (BAT) effluent limitations guidelines, pretreatment standards, and new source performance standards for the steam electric power generating point source category on Friday, November 19, 1982 (*Fed. Regis.*:47 52290-52309).

For chlorine, BAT and NSPS (new source performance standards for new plants) are defined. A daily maximum limitation for TRC is based on a final discharge point concentration of 0.2 mg/L, for not more than 2 h/d per generating unit.

Section 301(g) Rulemaking

Draft 301(g) procedural regulations were developed by EPA in 1982;[3] however, there are no proposed substantive regulations at this time. The language of the act itself, however, requires that any BAT effluent limitation modified under a 301(g) variance "not interfere with the attainment of maintenance of that water quality which shall assure protection of public water supplies. . . . " Presumably, this regulation requires that water quality standards be met at the edge of a prescribed mixing zone.

Although the new effluent limitations specify TRC only, eventual 301(g) rulemaking could conceivably require meeting TRC, combined residual chlorine (CRC), and/or free available chlorine (FAC) standards or criteria, as appropriate. In 47 FR 52300, EPA recognizes the toxicity of all three, but indicates that its present water quality criteria for chlorine are based on TRC. However, recognizing that state or federal standards or criteria can change, the models developed for TRC prediction permit specification to CRC and FAC.

Discussions with EPA personnel responsible for developing 301(g) guidelines suggested that:

1. EPA recommends the use of several specific dilution models, all of which reportedly are available from EPA.
2. EPA recognizes the significance of decay of chlorine within the mixing zone and accordingly recognizes the appropriateness of modifying the recommended dilution models to reflect chlorine decay.

Numerous calls to various EPA offices during the project were unsuccessful in locating fully documented, working dilution model programs. LMS, therefore, developed documentation for the working programs from program listings obtained from EPA.

Mixing Zone Considerations

Strictly speaking, the procedures developed are independent of mixing zone considerations. The UWAG Request for Proposal (RFP) stated it well:

"Disregarding the legal and policy questions of how much of a mixing zone should be allowed and how they are defined in the various state Water Quality Standards, UWAG believes that technical guidance is needed on how to calculate chlorine concentrations in the chlorine plume resulting from both cyclical and continuous discharges of TRC in power plant chlorinated once-through cooling water."

Assuming that mixing zones are allowed by the 301(g) applicant's state, the application will have to determine whether these calculated chlorine concentrations indicate compliance with chlorine criteria outside the prescribed mixing zone for the type of receiving water body and the state involved. Alternatively, the discharger and the permitting authority will have to agree that a proposed mixing zone that shows compliance at its boundary is reasonable.

Toward these ends, two facts need to be recognized. First, in at least some cases, with a source water possessing relatively high ammonia or organic amines coupled with a station circulating a significant percentage of the source water flow, consistent compliance with chlorine criteria outside a fixed mixing zone may not be possible, or at least may be very difficult to demonstrate. Second, most chlorination is intermittent, whereas most prescribed mixing zones are not time variable.

APPROACH TO THE STUDY

Project Perspectives

In essence, predicting the distribution of TRC in a receiving water body requires a knowledge of how TRC decays after chlorine injection into plant intake water, and how much dilution is available along the path traveled by the decaying TRC. Application of this knowledge to current plant chlorination practice will determine whether effluent limitation guidelines and/or water quality standards are being or can be met.

Though formally outside the scope of the project, dosage reduction (chlorine minimization, in federal terms) was clearly an integral part of the exercise. Decay and dilution calculations that demonstrate the difficulty in meeting limitations or standards can also be used to determine the dosage and/or discharge water-box (DWB) residual that will achieve compliance. For cases of minimal available dilution and combined residuals, chlorine minimization may be a real necessity.

Operators are encouraged to institute test programs to determine real chlorine requirements. Although probably less saddled with ammonia-laden or organic-rich source waters and warm temperatures, many utilities in the northeast have either totally eliminated condenser chlorination or have reduced it sharply over the past 10 to 15 years. Many practitioners believe they can maintain condensers satisfactorily without maintaining any free chlorine residual at the DWB, much less the common present practice of holding to 0.5 to 1.0 mg/L FAC.

Some overall concepts, in addition to those already presented, are:

1. Dilution is always ultimately limited by the size of the receiving water body, regardless of the assumptions inherent in some dilution models. For rivers, this is the river flow during the period of chlorination. If chlorination is year round, seasonal low flow, corresponding to the season or seasons of

highest effluent TRC loading and/or persistence, will probably control. For lakes, maximum dilution will be related to total lake volume and annual throughput.

2. Residual chlorine decay beyond the discharge point may be rather slow, particularly if ammonia and/or organic amines are present in the water. However, inclusion of decay kinetics will always result in lower predicted TRC values in both the discharge channel and receiving water plume. Numerous field studies have confirmed this fact.[5,6]

PROJECT SUMMARY

The report produced from this study[1] consists of six chapters and multiple appendices. A brief description of Chapters 2 through 6 follows.

Chapter 2.0—Chlorine Chemistry

This chapter provides an understanding of the chlorine decay curve, its characteristics, and the factors that affect it. Initial rapid demand, continuing slow demand, organic and inorganic demand, ammonia and organic amines, free and combined residual, reaction kinetics, reaction order, breakpoint, temperature, and pH are among the topics presented. The entire discussion is designed to promote enough of an understanding of the factors influencing decay to assist a permit or 301(g) waiver applicant to make accurate estimates of TRC decay in the discharge and/or plume. Simple desk-top procedures to calculate decay as a function of travel time are given.

Chapter 3.0—Model Review

Five receiving water body dilution models suggested by EPA as appropriate for use in 301(g) waiver applications are discussed. In addition, two other models are presented. Steps taken by LMS to determine the applicability and correctness of each model are given. Available computer program listings were key-punched, run, debugged, and recommended as acceptable or not acceptable for utility use. Procedures for obtaining working tapes are given.

Several available, relatively complex chlorine decay models applicable to cooling water discharges are presented. The EPA dilution models were modified by LMS to include decay kinetics of the receiving water body. Receiving-water demand and TRC are included. An essentially instantaneous decay is used for the reaction of the FAC fraction with receiving water demand. First-order kinetics are used to decay any remaining TRC. Procedures for using nonlinear-order kinetics are also given.

Chapter 4.0—Model Use

Typical model runs, with typical data, are provided. Required parameters and the variety of means to estimate each are identified. Model accuracy and possible pitfalls are discussed. Procedures for estimating the costs of using these models for various permit applications are also given.

Chapter 5.0—Data Collection, Analysis, and Costs

Procedures are given for measuring chlorine residual decay demand as a function of dosage and rate coefficients. Field and laboratory alternatives are presented, and the advantages and disadvantages of each are discussed. A brief review of available analytical techniques is provided, along with procedures for estimating the cost of an experimental program.

Chapter 6.0—Chlorine Predictions Using Existing Data

Triaxial temperature distributions have been made by many utilities in discharge plumes and surrounding vicinity in conjunction with permit applications, 316(a) waiver applications, and/or permit monitoring requirements. Since the discharged TRC becomes an integral part of the thermal plume, undergoing the same hydrodynamic dilution phenomena, measured temperature distributions can be used to describe dilution as a function of space and/or travel time. Procedures are given for superimposing chlorine decay kinetics on these measurements, including calculation of travel time to any given point in space.

Appendices

The appendices in the report constitute the users manuals for the supplied models. They include program listings, sample input and output, and instructions on how to obtain copies of the models from LMS.

Chlorine Literature Reviewed

Water treatment, sewage treatment, industrial waste treatment, and power plant water-use literature are replete with reports on virtually all aspects of chlorination. Several observations regarding the use of this literature in the project and the report are as follows:

1. The relevant literature was integrated, as appropriate, throughout the report. The primary use is in Chapters 2 and 3.

2. Numerous advances in chlorine chemistry have been made since 1970. In general, the greatest attention has been paid to reports published after 1975, although many earlier studies have provided good data on decay rates.

3. Of particular note as source references for chlorine chemistry are the preceding volumes in this series of proceedings,[7-10] the Electric Power Research Institute's (EPRI) *Power Plant Chlorination—A Biological and Chemical Assessment*,[11] and the *Handbook of Chlorination*.[12]

4. With regard to EPA's suggested dilution models, the original reports on each were found and reviewed. Additionally, two comprehensive studies of hydrodynamic and water quality models were reviewed and cited as excellent source references for this area of the project. These are MIT's *An Assessment of Techniques for Hydrothermal Prediction*[13] and the Argonne National Laboratory report on *Surface Thermal Plumes: Evaluation of Models for the Near and Complete Field*.[14]

5. The literature has not been used to aquaint the reader with the details of chlorine chemistry per se, but rather to mold a practical approach to the calculation of TRC concentrations in cooling water discharges and receiving water bodies.

Models Presented in the Report

DKHPLM.

This is a near-field multiple-port submerged diffuser model based on work by Hirst[15] and Kannberg and Davis.[16] This model is capable of analyzing a positively buoyant plume in an arbitrarily stratified environment as a progression of three zones: zone of flow establishment (ZFE), zone of established flow (ZEF), and zone of merging plumes. Documentation on the ZFE and ZEF was found,[15] but the zone of merging plumes was not referenced in the EPA manual[3] and its discussion is relatively limited.

MOBEN.

This is a near-field integral model that assumes a rectangular discharge canal. Developed by Motz and Benedict,[17] it does not account for shoreline or bottom effects. It assumes that the depth of the jet remains constant; hence, the model should be viewed with caution if used in other than a shallow water body.

PDS.

This is another integral model developed by Prych[18] and modified extensively by Shirazi and Davis.[19] It is a three-dimensional surface plume program that assumes a rectangular discharge and no plume contact with the bottom or shoreline. It includes the effects of buoyancy and the lateral spreading associated with it.

PDSM.

This is a modification of PDS that was determined to have limited use.

PSY.

Although explained somewhat vaguely in the EPA manual,[3] a review of the original publication[20] showed the PSY to be a very useful far-field model, with a concentration-specified shoreline discharge. The model is fairly simple to use and has the ability to break up the river into reaches with differing geometry and dispersive characteristics.

LMS3D.

This model was originally developed for slug releases into estuaries. It is the analytical solution to the unsteady-state convection-diffusion equation that describes the concentration distribution of a substance released as an instantaneous source. The solution differs from most conventional distributions in that it incorporates finite boundaries in the lateral and vertical directions. This solution was originally developed by Cleary and Adrian[21] for constant-velocity rivers. It was modified by LMS to include first-order decay and velocity as any function of time.

TVA.

This is one of the most recently developed chlorine kinetic models for freshwater cooling systems prior to discharge to a mixing zone. It incorporates a system of eight rate equations into a nonsite-specific model. The model was developed by Zielke et al.[22] and Zielke and Moss[4] and validated for both once-through and closed-cycle cooling systems. Zielke and Moss first reviewed existing kinetic models, isolated the deficiencies, and developed a model "that would duplicate a cooling system's characteristics, such that the chlorination point and the makeup point can be anywhere in the system, require minimal input data, simulate the loss of chlorine across the cooling tower, simulate breakpoint chlorination, and [be] flexible enough to identify the cooling system."[4]

Model Modifications for Chlorine

Near-field model modifications were designed to compute the FAC and CRC in the discharge plume, with the following points in mind:

1. All dilution models predict maximum dilution, average dilution, and travel time.
2. The plant discharge can contain FAC and/or CRC, but no demand.
3. The receiving water body can contain two types of FAC demand. The first is the nitrogenous demand, which converts FAC to CRC, and the second is the sum of the other rapid demands that remove FAC from TRC. All reactions are assumed to be instantaneous and to compete equally.
4. The receiving water body can contain two types of CRC demand. The first is a very rapid (almost instantaneous) consumption, and the second is a slower decay of any order. The instantaneous CRC demand of receiving waters can be a major contributor to the overall disappearance of CRC.
5. The receiving water body can have a background level of CRC but not FAC; however, if there is an instantaneous CRC demand, there can be no background CRC.
6. Reactions do not occur until the zone of established flow is reached (a conservative assumption).
7. Centerline dilution values and times are used for model computation (another conservative assumption).

The subroutine for doing the computations has been programmed so that it can be used in virtually any dilution model (listed in Appendix C of the report). All that is required are initial values of plant discharge CRC and FAC, river CRC, and demand characteristics. The current time and dilution factors are passed into the subroutine, which then returns the new values of CRC and FAC. At present, first-order decay of CRC is hard-coded in the program, but this can be easily modified for other site-specific orders. These are documented in the report.

Since reactions of FAC occur quickly, and CRC decay (after any initial demand) is exerted much more slowly, the near field is the only area where both are of concern. For a far-field model, the only constituent that needs to be considered is the CRC component of TRC.

Most far-field models can have first-order decay added to the solution without any problem; this was done with PSY and LMS3D. Models with numerical solutions will be able to handle other orders of decay. Analytical models can have other orders of decay incorporated only if the receiving water body is completely mixed, with no lateral or vertical variations.

If substantial CRC demand exists in the receiving water body, there would be no need to run a far-field model; all the FAC and CRC would be consumed in the near field.

Model Use and Misuse

This section might be more aptly named, "What comes out can be no better than what goes in." When using any model, one must be keenly aware of the limitations and pitfalls inherent in any method of approximation. There are two basic areas where a model could be misused. First, one could try to apply it to a situation for which it was not designed. Second, although it might be appropriate for the situation at hand, the output may not be interpreted correctly. A common example of the latter is when a near-field model predicts more dilution than is possible in a receiving water body (usually caused by the assumption of an ambient receiving water that is infinite in extent).

Several general and specific uses and potential problems when using the UWAG models are discussed in this report. The reader is cautioned concerning some of the modeling pitfalls.

APPLICATION

The methodology developed in this project was used recently at a southeastern nuclear facility to answer questions raised regarding the impact of condenser and cooling tower chlorination on the receiving cooling water lake.

The plant and the cooling lake were analyzed separately so that the results of one model could be used independently of those of the other. The plant model was divided into three sections; the cooling tower basin, the circulating water piping, and the cooling tower itself. For a given chlorination schedule, the model computed total chlorine concentrations at the three sections and the quantity of chlorine that was released through the blowdown. The model accounted for the immediate inorganic demand and other losses of chlorine via first-order decay and evaporative removal by the cooling tower. The removal coefficient, which is the least definitive parameter, was determined by calibration to observed data.

The lake model used the total chlorine output from the plant blowdown (lb/d) and computed an average completely mixed concentration in a specified volume. Chlorine in the blowdown from the plant cooling tower sump was decayed over the 2.4-h travel time to the discharge point (NPDES) in the lake. A 200-acre mixing zone was chosen as conservatively representative of the mixing volume. The choice of a completely mixed system is overly conservative; a plug flow representation would result in concentrations of effectively zero because of the long retention time in the mixing zone. If the total volume of the lake were used, chlorine concentrations would also be zero, even with the completely mixed assumption.

Using a background chlorine demand of 1.3 mg/L and a constant dose of 3.0 mg/L over 30 min, the model computed an average 2-h TRC level in the blowdown stream at the sump of 0.4 mg/L, which was composed totally of CRC (no FAC remained). About 75% of the discharge occurred in the first hour after initiation of chlorination.

Current regulations[23] on cooling tower blowdown limit the discharge of FAC to a daily average of 0.2 mg/L for no more than 2-h/d per unit. Since all the FAC was consumed or converted to CRC, this regulation was met.

The plant discharge to the lake occurs through a 48-in.-diam pipe about 3.5 miles long. Since the discharge port is the point where NPDES permit limitations are applied, the decay during the travel through the pipe was accounted for. With the discharge of 27 cfs, the travel time is 2.4 h. Then, using a conservative first-order decay rate of 4/d, the discharge to the lake was reduced to 67% of the levels at the plant. This resulted in a release to the lake of an average 2-h CRC level of 0.27 mg/L (corresponding to the 30-min chlorination period).

Assuming (conservatively) no immediate demand on chlorine released to the lake, the lake was modeled as a completely mixed system. This requires knowing the amount of chlorine discharged (from the plant model), the lake outflow, the decay rate of CRC, and the volume of the lake available for mixing.

The average lake outflow is 43 cfs. The entire volume of the lake—72,000 acre-ft—could be used for mixing. However, this would be guaranteed only if the plant discharge were at the upstream end of the lake (which it is not). The whole lake may take part but, in general, the mixing is only partial and some short circuiting may occur. A very conservative assumption would be to use the volume of the mixing zone. This not only used substantially less volume than is actually available but also enabled the prediction of CRC levels at the edge of the zone. Assuming a depth of 40 ft in the 200-acre mixing zone, the computer CRC level at the edge of the mixing zone was 0.07 ppb (using a decay rate of 4/d).

Most of the large decrease is the result of residence time and decay. As a sensitivity analysis, the decay coefficient was decreased by a factor of 10. The resulting concentration was 0.7 ppb. Then, when the mixing zone area was cut to 5 acres, the resulting concentration was 14 pbb. Using the original decay rate and a 5-acre zone, the concentration was 2.6 ppb.

The above results were considered conservative for several reasons. No sedimentation has been incorporated; several chlorinated by-products have an affinity for particulates that will settle out of the water column and eventually be buried and removed from active participation in the lake ecosystem. No immediate demand in the lake that could be exerted on CRC has been considered. The assumption of complete mixing is very conservative, particularly since steady state is also assumed.

The alternative of plug flow was also considered. This is not an unreasonable assumption, given the location and direction of the discharge port and the use of a 200-acre mixing zone vs the whole lake. Since a small, localized area of the lake is being used as a mixing zone, plug flow behavior is possible.

For a 200-acre mixing zone, a flow of 43 cfs, and a decay rate of 4/d, the resulting CRC level was effectively zero. A very low decay rate of 0.4/d and a blowdown concentration of 0.27 mg/L computed to a discharge concentration of 1×10^{-17} mg/L.

ACKNOWLEDGMENT

The "Chlorine Plume Modeling Study" was performed by Lawler, Matusky & Skelly Engineers. The assistance of members of the Chemical Committee of the Utility Water Act Group; UWAG's legal counsel, Hunton & Williams, Richmond, Virginia; and Mr. James K. Rice, Olney, Maryland, is gratefully acknowledged.

REFERENCES

1. *Chlorine Plume Modeling Study*, LMSE-83/0091&356/003, prepared for the Utility Water Act Group (Pearl River, NY: Lawler, Matusky & Skelly Engineers, 1983).
2. Lawler, J. P., T. B. Vanderbeek, and P. Cumbie. "Predicton of Total Residual Chlorine in Power Plant Discharges and Receiving Waters: Chlorine Kinetics and Modeling," presented at the Fifth Conference on Water Chlorination: Environmental Impact and Health Effects, Williamsburg, VA, June 3-8, 1984.
3. "Technical Guidance Manual for the Regulation Promulgated Pursuant to Section 301(g) of the Clean Water Act of 1977, 40 CFR part 125 (Subpart F)," draft report, (Washington, DC: U.S. Environmental Protection Agency, 1982).
4. Zielke, R. L., and R. D. Moss. "Validation of a Kinetic Model for Predicting Chlorine Residual Concentrations in a Closed-Cycle Cooling System," EDT-108, staff reference document (Chattanooga, TN: Tennessee Valley Authority, 1980).
5. Høstgaard-Jensen, P., J. Klitgaard, and K. M. Peterson. "Chlorine Decay in Cooling Water and Discharge into Sea Water," *J. Water Pollut. Control Fed.* 49(8):1832-1841 (1977).
6. Lee, G. F. "Persistence of Chlorine in Cooling Water from Electric Generating Station," *Am. Soc. Civil Eng., J. Environ. Eng. Div.* 105 (EEA):757-773 (1979).
7. Jolley, R. L., Ed. *Water Chlorination: Environmental Impact and Health Effects, Vol. 1* (Ann Arbor, MI: Ann Arbor Science Publishers, Inc., 1978).
8. Jolley, R. L., H. Gorchev, and D. H. Hamilton, Jr., Eds. *Water Chlorination: Environmental Impact and Health Effects, Vol. 2* (Ann Arbor, MI: Ann Arbor Science Publishers, Inc., 1978).
9. Jolley, R. L., W. A. Brungs, and R. B. Cumming, Eds. *Water Chlorination: Environmental Impact and Health Effects, Vol. 3* (Ann Arbor, MI: Ann Arbor Science Publishers, Inc., 1980).
10. Jolley, R. L., W. A. Brungs, J. A. Cotruvo, R. B. Cumming, J. S. Mattice, and V. A. Jacobs, Eds. *Water Chlorination: Environmental Impact and Health Effects, Vol. 4* (Ann Arbor, MI: Ann Arbor Science Publishers, Inc., 1983).
11. Hall, L. W., G. R. Helz, and D. T. Burton. *Power Plant Chlorination: A Biological and Chemical Assessment*, (Ann Arbor, MI: Ann Arbor Science Publishers, Inc., 1981).
12. White, C. *Handbook of Chlorination* (New York: Van Nostrand Reinhold, 1972).
13. Jirka, G. H., G. Abraham, and D. R. F. Harleman. *An Assessment of Techniques for Hydrothermal Prediction*, Ralph M. Parsons Laboratory for Water Resources and Hydrodynamics Rep. 203 (Cambridge, MA: Massachusetts Institute of Technology, 1975).

14. Dunn, W. E., A. J. Policastro, and R. A. Paddock. *Surface Thermal Plumes: Evaluation of Mathematical Models for the Near and Complete Field*, ANL/WR-75-3, (Argonne, IL: Argonne National Laboratory, 1975).
15. Hirst, E. *Analysis of Buoyant Jets Within the Zone of Flow Establishment*, ORNL-TM-3470 (Oak Ridge, TN: Oak Ridge National Laboratory, 1971).
16. Kannberg, L. D., and L. R. Davis. *An Experimental/Analytical Investigation of Deep Submerged Multiple Buoyant Jets*, EPA-600/3-76-001 (Corvallis, OR: U.S. Environmental Protection Agency, 1976).
17. Motz, L. H., and B. A. Benedict. "Surface Jet Model for Heated Discharges," *Am. Soc. Civil Eng., J. Hydraul. Div.* 98(HY1):181–200 (1972).
18. Prych, E. *A Warm Water Effluent Analyzed as a Buoyant Surface Jet*, Hydraulic Research Report 21 (Swedish Meteorological and Hydrological Institute, 1972).
19. Shirazi, M. A., and L. R. Davis. *Workbook of Thermal Plume Prediction, Vol. 2: Surface Discharge*, EPA-R2-72-005b (Washington, DC: U.S. Environmental Protection Agency, 1974).
20. Paily, P. P., and J. F. Kennedy. *A Computational Model for Predicting the Thermal Regimes of Rivers*, Iowa Institute of Hydraulic Research Report 169 (Iowa City: University of Iowa, 1974).
21. Cleary, R. W., and D. D. Adrian. "New Analytical Solutions for Dye Diffusion Equations," *Am. Soc. Civil Eng., J. Environ. Eng. Div.* 99(EE):213–227 (1973).
22. Zielke, R. L., H. B. Flora, II, and S. K. Macey. "Validation of a Kinetic Model to Predict Total Residual Chlorine in Fresh Water," in *Water Chlorination: Environmental Impact and Health Effects, Vol. 3*, R. L. Jolley, W. A. Brungs, and R. B. Cumming, Eds. (Ann Arbor, MI: Ann Arbor Science Publishers, Inc., 1980), pp. 445–452.
23. *Steam Electric Power Generating Point Source Category; Effluent Limitations Guidelines, Pretreatment Standards, and New Source Performance Standards*, 47FR52290 (Washington, DC: U.S. Environmental Protection Agency, 1982).

SECTION XVI

Wastewater Treatment

Around 1900 a debate was in progress over how the country should save its freshwater from increasing pollution. . . . (some) proponents claimed the nation had enough water to serve indefinitely, both as water supply and to dilute municipal sewage discharged into streams and other natural bodies of water. . . . Arguing to the contrary . . . were other proponents, who advocated obeying nature's inviolable law of return by sending our used water back to the natural cleansing system of soil, plants, air and sunshine for reclamation and reuse, over and over again.

John R. Sheaffer and **Leonard A. Stevens**
Future Water, 1983

The goal of life is living in agreement with nature.
Zeno, 335–263 B.C.
Diogenes Laertius

SECTION XVI

Wastewater Treatment

CHAPTER 115

Discharge of Halogenated Octylphenol Polyethoxylate Residues in a Chlorinated Secondary Effluent

Harold A. Ball and Martin Reinhard

Chlorination of wastewater prior to effluent discharge is a standard practice in the United States and many parts of the world. It has long been recognized that chlorinated organics are formed as a result of this disinfection procedure.[1,2] Most of these products appear to originate from uncharacterized precursors such as humic and fulvic acids.[3] Recently, chlorinated wastewater has been found to contain a range of brominated alkylphenol polyethoxy carboxylates (APECs).[4] The brominated APECs were shown to be formed from APECs during chlorination with 6 to 10 mg/L of chlorine. APECs are the biological metabolites of alkylphenol polyethoxylate (APE) decomposition.

APEs are nonionic surfactants commonly used for industrial and agricultural purposes. In 1982, more than 115,000 metric tons of APEs were produced in the United States,[5] generally for use as nonionic detergents. APE compounds consist of alkylphenol (branched nonyl or octyl) attached to a hydrophilic polyethyleneglycol chain having from 4 to 30 ethoxy (EO) units. The primary aerobic biodegradation pathway for branched APE is through shortening of the EO chain with the possible formation of a carboxylic acid intermediate.[6-8] Ultimate biodegradation of quaternary branched APEs has been postulated[9] but has not been proven under wastewater treatment plant conditions. In this chapter, we denote the multiple APE homologues with a number following the acronym to indicate the number of EO units attached to the benzene ring. For APEC compounds, the number indicates the sum of unaltered EO units plus the terminal $-O-CH_2-COOH$ group. Compounds with a branched octyl side chain as the alkyl substituent are referred to as OPE and OPEC.

This study summarizes quantitative analyses of APE residues in chlorinated wastewater from the Palo Alto secondary wastewater treatment facility. Significant concentrations of nonhalogenated, brominated, and chlorinated OPE metabolites were found in these secondary effluents. Experimental protocol to develop synthetic analytical standards is presented.

EXPERIMENTAL

Palo Alto Regional Water Quality Control Plant

The Palo Alto Regional Water Quality Control Plant (RWQCP), Palo Alto, California, is a secondary wastewater treatment facility that treats an average flow of 1.4 m³/s (Figure 1). Following mechanical screening, the influent flows through primary sedimentation tanks, fixed film reactors, activated sludge aeration tanks, secondary clarifiers, dual media filters, and a chlorine plug flow conduit and contact basin prior to discharge to the San Francisco Bay. The wastewater is chlorinated at three points: prior to the dual media filters, prior to the plug flow conduit, and prior to the chlorine contact basin. During the sampling period, the measured average residual chlorine concentrations were 6.5 mg/L (range, 1.2 to 7.8 mg/L) after the dual media filters, and 6.4 mg/L (range, 3.8 to 8.7 mg/L) after the chlorine contact basin. The total contact time through the filters, conduit, and contact basin is about 2 h. The wastewater effluent is dechlorinated with sulfur dioxide gas prior to discharge into San Francisco Bay.

Sampling

Grab samples (1 L) of the dechlorinated effluent were collected at 9 a.m., 2 p.m., and 7 p.m. for five continuous days in July 1984. Samples were promptly refrigerated and stored for up to 2 d until analyzed. Flow-weighted 1-L composite samples were mixed for each day based on the wastewater flow at the plant at the time of collection.

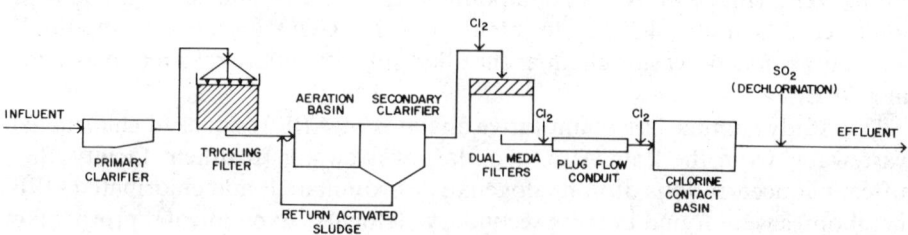

Figure 1. Schematic of Palo Alto Regional Water Quality Control Plant.

Extraction and Methylation

The surrogate standards p-bromophenylacetic acid and 3,6-dimethylphenanthrene were added to the composite samples at a concentration of 10 μg/L. Prior to extraction, samples were acidified to pH 1.5 with concentrated nitric acid. The organic fraction was extracted three times with 100-mL portions of dichloromethane (CH_2Cl_2). Approximately 30 g sodium chloride was added to the sample before extraction to decrease the formation of water/solvent emulsions. The CH_2Cl_2 extract was dried over anhydrous sodium sulfate and filtered. The combined extracts were then concentrated to approximately 5 mL under reduced pressure in a 30°C water bath, followed by evaporation to dryness under a purified nitrogen stream. The extract was redissolved in 1 mL of a 1:1 ether-methanol solution and methylated with diazomethane prior to GC/MS analysis.[4] The final volume of the extract was 1 mL.

Preparation of Standards

Tert-octylphenol polyethoxylate (3 mol of ethylene oxide units/molecule) was purchased from Chem Service, West Chester, Pennsylvania. This technical grade OPE was analyzed with GC/MS and found to contain molecules with 1 to 6 EO units. Brominated products were formed in purified water (MQ system, Millipore Corp., Bedford, Massachusetts) containing 1.2 g/L Br_2. Approximately 400 mg OPE dissolved in methanol was added to 200 mL of the aqueous bromine solution. After 1 h, excess bromine was destroyed by adding 0.8 N sodium sulfite solution. Chlorination was performed in MQ water containing approximately 0.6 g/L Cl_2. Approximately 400 mg OPE dissolved in 5 mL methanol was added to 200 mL of the aqueous chlorine solution. After 10 min, excess chlorine was destroyed by adding 0.8 N sodium sulfite solution. Each reaction mixture was extracted in a 0.5-L separatory funnel using two 25-mL portions of ether. The extracts were dried over sodium sulfate, filtered, and evaporated under reduced pressure in a warm water bath. Approximately 65% halogenated product was recovered. For analysis, a known quantity (approximately 800 μg) of the products was redissolved in 1 mL methanol to which 110 μg of the internal standard 3,6-dimethylphenanthrene was added.

Some OPE was converted from the alcohol to the corresponding carboxylic acid by oxidation with Jones reagent.[4,10] The resultant OPEC was extracted two times with 50-mL portions of ether, dried over sodium sulfate, filtered, and evaporated under reduced pressure. The OPEC solution was purified using silica gel chromatography,[4] in which the acid fraction was eluted with methanol. Approximately 10% carboxylated product was ultimately recovered. Portions of the purified OPEC were brominated and chlorinated according to the procedures described above. For analysis, a known quantity (approximately 1 mg) of the products was redissolved in 1 mL of a 1:1 ether--methanol solution to which 140 µg of the internal standard p-bromophenylacetic acid was added. The solutions were then methylated with diazomethane and concentrated to a volume of 1 mL. It was observed that during silica gel chromatography a significant fraction of the free acids became methylated. Hence, in subsequent studies[11] the liquid-extraction method was used to purify the OPEC standards.

Inorganic Analyses

Separate portions of each grab sample were set aside for inorganic anion analysis. A high-pressure liquid chromatograph was equipped with an anion column (Model 269–001, Wescan Instruments, Santa Clara, California). The ion chromatograph was calibrated for the determination of bromide and chloride.

Chlorine residual was measured after the dual media filters and in the chlorine contact basin by Palo Alto RWQCP personnel using the amperometric titration method.[12] A glass electrode was used to measure pH.

GC/FID Analysis

A Carlo Erba gas chromatograph (GC) (Model 2100) with a flame ionization detector (FID) and a Spectra-Physics integrator (Model 4020) was used for quantitative analysis of the synthesized standards. The GC was equipped with a DB-5 fused silica capillary column (15 m, 0.25 mm ID, J&W Scientific Inc., Rancho Cordova, California). The temperature program used was 70°C isothermal for 1 min, increasing to 310°C at 5°C/min. Helium was used as a carrier gas with an inlet pressure of 0.5 atm. The percent composition of each standard sample was determined by proportioning the total mass of the sample among individual peaks according to relative area calculated by the GC/FID integrator, assuming equal response factors for the OPE and OPEC homologues on a mass basis.

GC/MS Analysis

Gas chromatograph/mass spectrometry (GC/MS) was done on a Finnigan 4000 with an INCOS data system (Finnigan, Sunnyvale, California). A DB-5 fused silica capillary column (30 m, 0.25 mm ID) was connected directly to the ion source. The temperature program used was 70°C isothermal for 1 min, increasing to 310°C at 3°C/min. Helium was used as a carrier gas with an inlet pressure of 0.5 atm. A 1-µL sample of extract was injected splitless for 30 s. Mass spectra were acquired at an electron energy of 70 eV, and emission current of 0.4 mA. A mass range of 41 to 550 amu was scanned at a rate of one scan/2 s. The prepared standards were analyzed by GC/MS. Similar chromatography on the GC/MS and GC/FID allowed the direct comparison of the chromatograms from the two instruments. On this basis, GC/MS response factors for characteristic ions in the synthesized sample and internal standards were calculated. This method assumes equal recovery of the standards and OPE residues in the extraction procedure. A GC/MS database of the individual compounds and varying polyethoxy chain length homologues was generated containing the quantitation information. Mass spectral characteristics and the ions used to quantify these compounds are shown in Table I.

RESULTS

Selected mass spectra of the OPE residues are shown in Figure 2. The loss of C_5H_{11} (71 amu) from the molecular ion is characteristic of these compounds due to the favored breaking of the benzylic bond in the octyl side chain. Additional detail regarding spectra identification of OPE residues appears elsewhere.[4,11,13] Mass spectra of nonylphenol polyethoxylates are similar but lose C_6H_{13} (85 amu) from the molecular ion.[14,15]

Results for the OPE residue analyses, as well as inorganic data, are presented in Table II. The major residue in all samples was OPEC-2 followed by OPEC-1 and OPEC-3 in approximately equal concentrations. OPE-2 was the most prevalent of the neutral species. Nonylphenol polyethoxylates were also detected but were not quantified. Their concentrations appear to be lower than those of the OPE residues.

The extent of bromination varied widely, but BrOPEC-2 was the dominant halogenated species in all wastewater extracts analyzed. Other brominated acid homologues found were BrOPEC-1, BrOPEC-3, and BrOPEC-4. Measurable quantities of the chlorinated diethoxy carboxylate (C1OPEC-2) were found in four of the five wastewater extracts analyzed. However, the typical concentration of this species was much lower than that of BrOPEC-2. Halogenated OPEs were found in low concentrations in two of the five wastewater extracts analyzed.

DISCUSSION

OPECs, particularly the diethoxy OPEC, were found to be present in the wastewater extracts at the highest concentration and averaged approximately 64 µg/L. Chlorination of the wastewater effluent produced a range of halogenated derivatives. The concentration of halogenated products formed due to chlorination could not be positively correlated with any of the inorganic characteristics measured (Cl^-, Br^-, pH, or chlorine residual). The residual chlorine data of dual media filter effluent were measured only twice a day and, therefore, may not reflect a true average.

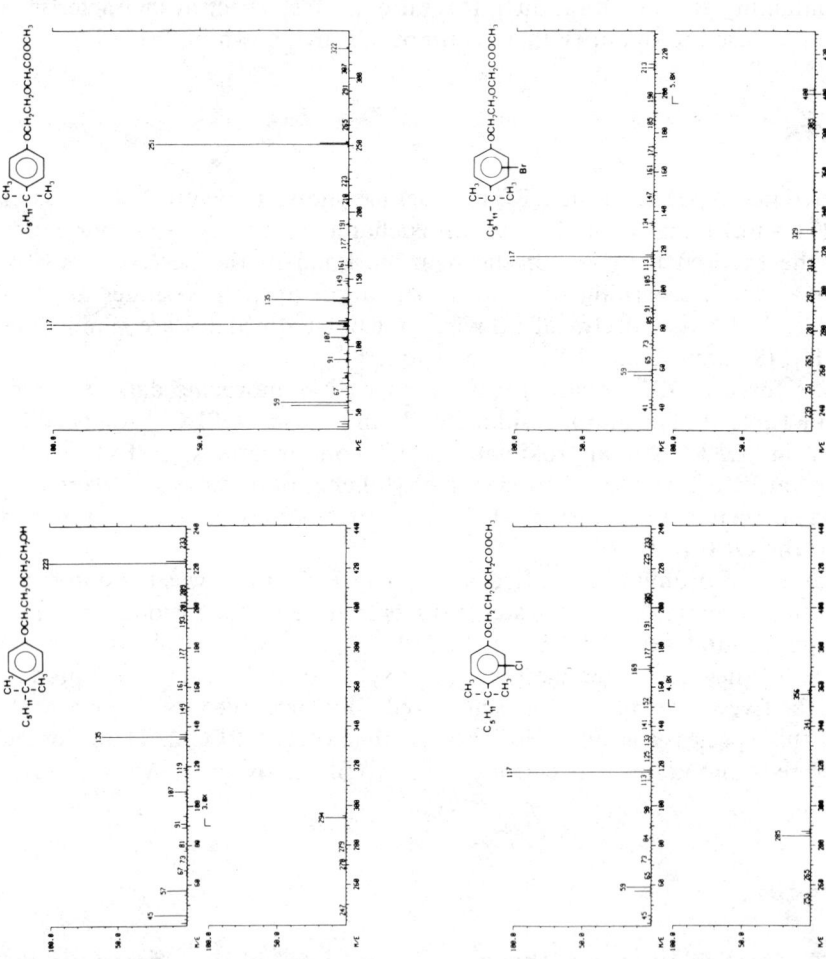

Figure 2. Selected electron impact (70 eV) mass spectra of octylphenol polyethoxylate residues.

Table I. Characteristic Ions used for Identification and Quantification of Octylphenol Polyethoxylates

Compound	Molecular Wt	MW-71	Base Peak	Quantitation Ion
OPE-1	250	179	179	179
OPE-2	294	223	223	223
OPE-3	338	267	267	267
ClOPE-2	328	257	257	257
BrOPE-1	328	257	257	213
BrOPE-2	372	301	301	213
OPEC-1	278	207	207	207
OPEC-2	322	251	251	117
OPEC-3	366	295	295	161
OPEC-4	410	339	339	117
OPEC-5	454	383	117	117
ClOPEC-2	356	285	117	169
ClOPEC-3	400	329	329	329
ClOPEC-4	444	373	117	117
BrOPEC-1	356	285	135	285
BrOPEC-2	400	329	117	329
BrOPEC-3	444	373	117	117
BrOPEC-4	488	417	117	117

In contrast to the investigation at WF21,[4] chlorinated APE residues were detected along with brominated APE residues. However, the brominated APE residues were present at significantly higher concentrations, indicating that aromatic bromination reactions occur at a faster rate. Similarly, in a laboratory chlorination study using 2,4,6-trihydroxyacetophenone as a model for humic substances, significantly faster formation of brominated products than chlorinated products was noted.[16] The identity of the highly reactive brominating agent during wastewater chlorination is yet unknown but merits further investigation.

Stefanou and Giger[14] analyzed extracts from six activated sludge, secondary wastewater treatment plants in Switzerland for nonylphenol polyethoxylate (NPE) compounds. They found high (total) concentrations of NPE with one to three EO groups in three of the plants ranging from 36 to 202 µg/L, but no data on the carboxylates were given. The plants where these compounds were detected "generally received wastewaters with greater contributions from industrial sources." The wastewater extracts were also found to contain up to 35 µg/L of the persistent and toxic 4-nonylphenol. APE residues apparently degrade to the alkylphenol during stabilization of sewage sludge under aerobic and anaerobic conditions as suggested by findings of Giger et al.[17]

No halogenated or nonhalogenated alkylphenols were detected in the current study. Our findings agree with reports of laboratory studies that biodegradation under wastewater treatment plant conditions produces relatively

Table II. Water Quality Characteristics of Palo Alto RWQCP Samples

Constituent	Concentration (μg/L)[a]				
	Day 1	Day 2	Day 3	Day 4	Day 5
OPE-1	NQ[b]	0.8	NQ	NQ	0.9
OPE-2	3.0	6.0	6.0	2.6	1.5
ClOPE-2				0.5	
BrOPE-1			0.9	NQ	
BrOPE-2			1.5		
OPEC-1	4.9	9.9	11	5.6	6.0
OPEC-2	24	57	84	36	42
OPEC-3	3.6	9.6	8.1	7.2	5.7
OPEC-4	NQ	2.2	NQ	NQ	1.9
OPEC-5		0.5		NQ	NQ
ClOPEC-2		2.4	3.0	8.7	2.0
ClOPEC-3		NQ		6.1	
ClOPEC-4				3.9	
BrOPEC-1	NQ	NQ	NQ	NQ	NQ
BrOPEC-2	0.5	0.8	19	17	3.2
BrOPEC-3			6.4	15	NQ
BrOPEC-4				NQ	
Cl_2[c,d]	6.7	5.3	7.2	7.2	6.2
Cl_2[c,e]	6.1	6.0	7.1	6.0	6.9
pH	6.5	6.4	6.3	6.5	6.4
Cl^-[c]	300.	275.	285.	295.	285.
Br^-[c]	0.5	0.6	0.5	0.5	0.7

[a]Detection limit: 0.5 μg/L.
[b]Detected but not quantifiable.
[c]Concentrations in mg/L.
[d]Average chlorine residual after dual media filtration (measured twice daily).
[e]Average chlorine residual in contact basin prior to dechlorination (measured every 2 h).

recalcitrant APE metabolites with two ethoxy units remaining on the aromatic ring.[8,18,19] In the laboratory, microorganisms have been found which carry out further degradation of this metabolite to the phenol under aerobic conditions.[7] Aerobic biodegradation of the common herbicide, 2,4-dichlorophenoxyacetic acid (2,4-D), which is structurally similar to the chlorinated APEC-1, begins with transformation into the phenol metabolite, 2,4-dichlorophenol.[20] This suggests that the halogenated APE metabolites in chlorinated wastewater effluents may be biotransformed in the environment to halogenated phenols. Because ultimate biodegradation of highly branched octylphenol has, to our knowledge, not yet been demonstrated, steady accumulation of these products in sediments and possibly organisms is suspected. Clearly, further studies are indicated to investigate both environmental fate and toxicology of these compounds.

In summary, a wide range of halogenated APE residues were detected in a chlorinated secondary effluent. A brominated OPE carboxylate was the predominant halogenated species, with lesser amounts of chlorinated products present. Of an estimated 8 kg/d OPE residues discharged, 1 to 30% were brominated and 0 to 20% were chlorinated. The factors influencing the extent of halogenation are as yet not understood.

ACKNOWLEDGMENT

We thank Jim Graydon for his assistance with the GC/MS analysis.
The work presented in this chapter was supported by a grant from the National Science Foundation (CEE-81-17561).

REFERENCES

1. Jolley, R. L., R. B. Cumming, N. E. Lee, L. R. Lewis, J. E. Thompson, and C. I. Mashni. "Nonvolatile Organics in Disinfected Wastewater Effluents: Chemical Characterization," in *Water Chlorination: Environmental Impact and Health Effects, Vol. 4,* R. L. Jolley, W. A Brungs, J. A. Cotruvo, R. B. Cumming, J. S. Mattice, and V. A. Jacobs, Eds. (Ann Arbor, MI: Ann Arbor Science Publishers, Inc., 1983), pp. 499-523.
2. Glaze, W. H., and J. E. Henderson IV. "Formation of Organochlorine Compounds from the Chlorination of a Municipal Secondary Effluent," *J. Water Pollut. Control Fed.* 47:2511-2515(1975).
3. Christman, R. F., D. L. Norwood, D. S. Millington, and J. D. Johnson. "Identity and Yields of Major Halogenated Products of Aquatic Fulvic Acid Chlorination," *Environ. Sci. Technol.* 17:625-628 (1983).
4. Reinhard, M., N. Goodman, and K. E. Mortelmans. "Occurrence of Brominated Alkylphenol Polyethoxy Carboxylates in Mutagenic Wastewater Concentrates," *Environ. Sci. Technol.* 16:351-362 (1982).
5. *Synthetic Organic Chemicals, United States Production and Sales, 1982*, Publication #1422 (Washington, DC: U.S. International Trade Commission, 1983), p. 187.
6. Swisher, R. D. *Surfactant Biodegradation*, Surfactant Science Series, Vol. 3 (New York: Marcel Dekker Inc., 1970).
7. Baggi, G., L. Baretta, E. Galli, C. Scolastico, and V. Trecanni. "Biodegradation of Alkylphenol Polyethoxylates," in *The Oil Industry and Microbial Ecosystems*, K. W. A. Chater and H. J. Somerville, Eds. (London: Institute of Petroleum, 1978), pp. 129-136.
8. Osburn, G. W., and J. H. Benedict. "Polyethoxylated Alkylphenols; Relationship of Structure to Biodegradation Mechanism," *J. Am. Oil Chem. Soc.,* 43:141-146 (1966).
9. Cain, R. B. "Microbial Degradation of Surfactants and 'Builder' Components," in *Microbial Degradation of Xenobiotics and Recalcitrant Compounds*, T. Leisinger et al., Eds. (London: Academic Press, 1981), pp. 325-370.

10. Lee, D. G. "Oxidation with Transition Metal Compounds," in *Oxidation*, Vol. I, R. L. Augustine, Ed. (New York: Marcel Dekker Inc., 1969), pp. 53–118.
11. Stephanou, E., M. Reinhard, H. A. Ball, and P. L. McCarty. "Identification of Halogenated and Nonhalogenated Alkylphenol Polyethoxylate Residues by GC/MS Using Electron-Impact and Chemical Ionization," (in preparation).
12. *Standard Methods for the Examination of Water and Wastewater*, 15th ed. (Washington, DC: American Public Health Association, 1980), pp. 286–289.
13. Stefanou, E. "Identification of Nonionic Detergents by GC/CI-MS: 1. A Complementary Method or an Attractive Alternative to GC/EI-MS and Other Methods," *Chemosphere* 13:43–51 (1984).
14. Stefanou, E., and W. Giger. "Persistent Organic Chemicals in Sewage Effluents: 2. Quantitative Determinations of Nonylphenols and Nonylphenol Ethoxylates by Glass Capillary Gas Chromatography," *Environ. Sci. Technol.* 16:800–805 (1982).
15. Giger, W., E. Stefanou, and C. Schaffner. "Persistent Organic Chemicals in Sewage Effluents: 1. Identifications of Nonylphenols and Nonylphenolpolyethoxylates by Glass Capillary Gas Chromatography/Mass Spectrometry," *Chemosphere*, 10:1253–1263 (1981).
16. Gould, J. P., L. E. Fitchhorn, and E. Urheim. "Formation of Brominated Trihalomethanes: Extent and Kinetics," in *Water Chlorination: Environmental Impact and Health Effects, Vol. 4*, R. L. Jolley, W. A. Brungs, J. A. Cotruvo, R. B. Cumming, J. S. Mattice, and V. A. Jacobs, Eds. (Ann Arbor, MI: Ann Arbor Science Publishers, Inc., 1983), pp. 297–310.
17. Giger, W., H. P. Brunner, and C. Schaffner. "4-Nonylphenol in Sewage Sludge: Accumulation of Toxic Metabolites from Nonionic Surfactants," *Science* 225:623–625 (1984).
18. Rudling, L. and P. Solyom. "Investigation of Biodegradability of Branched Nonyl Phenol Ethoxylates," *Water Res.* 8:115–119 (1974).
19. Kravetz, L., K. F. Guin, W. T. Shebs, L. S. Smith, and H. Stupel. "Ultimate Biodegradation of an Alcohol Ethoxylate and a Nonylphenol Ethoxylate under Realistic Conditions," *Soap Cosmet. Chem. Spec.* 58(4):34–42, 102B (1982).
20. Loos, M. A., R. N. Roberts, and M. Alexander. "Phenols as Intermediates in the Decomposition of Phenoxyacetates by an Arthrobacter Species," *Can. J. Microbiol.* 13:679–699 (1967).

CHAPTER **116**

Effect of Ozonation and Chlorination on Environmental Protection Agency Priority Pollutants

Yun-Shen Lee and Joseph V. Hunter

In 1893, chlorine was first used in the United States to disinfect wastewater at Brewster, New York, for the protection of the Croton Watershed, which is part of the New York City water supply. Since this time, and in particular as the adoption of the National Pollution Discharge Elimination Systems (NPDES) in 1977 required that most municipal wastewater treatment plants meet effluent bacterial standards, the use of chlorine to disinfect wastewater treatment plant effluents has become prevalent.[1]

As well as acting as a disinfectant, chlorine also can either oxidize organic compounds or substitute chlorine or bromine atoms for hydrogen in organic compounds. Because certain halogenated organics are suspected carcinogens, there has been considerable concern about the use of chlorine in potable water disinfection, and the literature abounds in references on this subject.

Chlorinated wastewater treatment plant effluents frequently are discharged into waters eventually used as potable water supplies; this concern has also been evidenced in the wastewater treatment area, because a large number of halogenated organics have been detected in wastewater effluents.[2-4]

Since chlorine can act as an oxidant as well as a substitution agent and many suspected toxic or carcinogenic organics are not halogenated substances, chlorine also has the potential to destroy hazardous organics as well as form them.

In June 1976, a Consent Decree was issued by the U.S. District Court in Washington, D.C., mandating that effluent limitations be established for 65 specific families of toxic substances by December 31, 1979. The Environmental Protection Agency redefined these families to an initial list of 129 compounds, 114 of which were organic, and which are usually referred to as EPA Priority Pollutants. These organics are usually classified by what amounts to analytical groups: PCB/pesticides; purgeables; base/neutrals; and acids (phenols). Of these, the last three groups are determined by gas chromatography-mass spectrometry (GC-MS) analysis and the first group by GC analysis.

The purpose of the research described in this chapter was to observe the effect of chlorination on the EPA Priority Pollutants, and to compare this effect with a disinfectant/oxidant that would not be expected to rival chlorine in the production of halogenated organics. To simplify the analytical program, only the three groups determined by GC-MS were studied. Of these, only the more commonly available compounds were used (i.e., 75 out of a total of 89).

MATERIALS AND METHODS

Ozonation Procedure

Ozone was generated using a Welsbach Laboratory Ozonator Model T-408, which at a flowrate of 0.6 L/min produced ozone at a rate of 27 mg/min. For the base-neutral and the acid samples, ozone was purged into a reactor containing 1 L of sample for 5 min. The off-gas from this reactor was then transferred to a second reactor containing 1 L of 2% neutral potassium iodide. A standard iodometric titration was used[5] to determine mass of ozone present. This was subtracted from the mass generated during the 5-min period to determine the dosage of ozone to the reaction chamber.

Because this procedure could not be used for the purgeable priority pollutants, ozonized water was prepared by purging ozone into organic free water for 10 min. This was then added to the purgeables sample, and the concentration of ozone was calculated by measuring the ozone concentration in the organic free water[5] and the volumes of sample and ozonized water.

Chlorination Procedure

Sodium hypochlorite solutions were diluted and added to the three priority pollutant mixtures. The dosage was calculated by determining the chlorine concentration of the dilute chlorine sample by the DPD method[5] and the volume of chlorine solution and sample.

Oxidation Conditions

The sample and oxidant concentrations used were as shown in Table I. The chlorine dosages selected are similar to those used in central New Jersey wastewater treatment plants that treat significant industrial wastewaters. The ozone dosages were essentially the maximum levels that could be reasonably achieved with the ozonator.

In the purgeable group, 28 organic compounds were subjected to oxidation; in the base/neutral group there were 37 compounds; and in the acid (phenol) group there were 10 compounds. The concentrations used were determined by extraction efficiency and method sensitivity.

After the oxidant was added to the systems, portions were withdrawn at intervals of 0.25, 0.5, 1.0, 2.0, and in some cases 30 h. The oxidant dosages were sufficient to leave substantial residuals at the end of the oxidation period.

Table I. Sample and Oxidant Concentrations

Group	Compound Conc (mg/L)	Ozone Dosage (mg/L)	Chlorine Dosage (mg/L)
Purgeables	10	7.5	38.5
Base-neutrals	110	21.3	45.5
Phenols	667	28.2	45.5

Analyses

Each sample removed was immediately either purged (for purgeables) or extracted (base/neutrals and acids) and subjected to GC-MS analyses according to the procedure described by the U.S. Environmental Protection Agency (EPA),[6] using a Hewlett Packard model 5895A Gas Chromatograph-Mass Spectrograph.

RESULTS AND DISCUSSION

For ease of comparison among the three groups of compounds studied, the concentrations of the priority pollutants remaining after various contact periods were subtracted from the initial concentrations, and the percent reductions calculated were then rounded off to the nearest percent. Therefore, a reduction of 100% does not imply complete removal, merely that this was the closest whole number to the actual percent reduction.

The percent reductions observed are presented in Tables II through VIII. To facilitate the data interpretation, the discussion will follow the three EPA Priority Pollutant Divisions studied: purgeables, base/neutrals, and acids. Although data were collected over a 2 to 3 h period, comparisons were made on the basis of the removals observed at the longest contact period. No attempt was made to compare rates of removal because of differences in both the oxidant and compound concentrations.

Purgeables

With the exception of bromoform, chlorine had little effect on the saturated halogenated compounds; this was also observed for its effect on the unsaturated purgeables (Tables II and III). Ozone had little effect on the saturated purgeables, but had a variable effect on the unsaturated purgeables. Although ozonation had little effect on benzene, substituted benzenes, and halogenated ethylenes containing three or more chlorine atoms, essentially 100% removal was observed for vinyl chloride, *trans*-dichloroethylene, and dichlorinated

Table II. Reductions of EPA Priority Pollutant Purgeables Through Chlorination and Ozonation: Group 1, Unsaturated Compounds

Reductions (%)

	Chlorine Contact Time (h)					Ozone Contact Time (h)			
	0.25	0.5	1	2	3	0.25	0.5	1	2	3
Vinyl chloride	2	5	14	17	21	100	100	100	100	100
1,1-Dichloroethylene	1	14	15	18	18	65	65	58	56	58
trans-1,2-Dichloroethylene	0	8	11	10	12	100	100	100	100	100
Trichloroethylene	1	7	8	11	12	8	9	5	5	15
Tetrachloroethylene	1	7	9	10	10	0	12	5	5	18
trans-1,3-Dichloropropene	2	4	11	11	13	100	100	100	100	100
cis-1,3-Dichloropropene	1	9	10	12	11	100	100	100	100	100
Benzene	0	6	10	12	16	3	6	8	5	14
Toluene	0	9	15	17	21	2	7	6	8	13
Chlorobenzene	0	7	10	17	22	1	5	6	8	15
Ethylbenzene	0	9	13	15	14	1	7	5	7	15
m-Xylene	1	9	10	17	17	50	49	47	48	50
o- and p-Xylene	1	8	11	18	18	48	52	50	50	54

Table III. Reductions of EPA Priority Pollutant Purgeables Through Chlorination and Ozonation: Group 2, Unsaturated Compounds

Reductions (%)

	Chlorine Contact Time (h)					Ozone Contact Time (h)			
	0.25	0.5	1	2	3	0.25	0.5	1	2	3
Bromomethane	1	15	14	21	28	0	6	5	10	19
Methylene chloride	13	14	17	0	−13	0	0	4	4	4
Trichlorofluoromethane	2	14	16	18	20	0	7	4	7	20
Carbon tetrachloride	3	16	18	20	24	0	12	12	15	21
Chloroform	0	5	7	8	9	1	12	15	16	17
Bromodichloromethane	3	7	19	19	17	7	11	10	10	7
Dibromochloromethane	7	12	15	21	20	0	8	5	5	4
Bromoform	15	28	38	39	41	11	11	27	28	30
1,1-Dichloroethane	0	7	10	11	16	0	4	14	11	17
Chloroethane	4	9	12	12	17	12	11	11	14	24
1,2-Dichloroethane	1	3	11	18	15	0	4	14	11	17
1,1,2-Trichloroethane	8	9	12	20	19	6	13	11	9	9
1,1,2,2-Tetrachloroethane	1	7	8	10	10	0	12	4	4	18
1,1,1-Trichloroethane	1	7	8	12	17	5	13	8	8	16
1,2-Dichloropropane	0	6	14	12	15	14	21	21	20	27

Table IV. Reductions of EPA Priority Pollutant Base/Neutrals Through Chlorination and Ozonation: Group 1, Polynuclear Aromatics

Reductions (%)

	Chlorine Contact Time (h)					Ozone Contact Time (h)			
	0.25	0.5	1	2	3	0.25	0.5	1	2	3
Naphthalene	1	0	22	43	46	99	100	100	100	100
2-Chloronaphthalene	5	16	25	44	51	100	100	100	100	100
Acenaphthalene	20	26	32	50	55	97	97	97	97	98
Acenaphthene	41	46	48	56	56	89	90	91	90	89
Fluorene	44	47	48	49	49	75	79	79	78	76
Phenanthrene, anthracene	18	21	23	26	31	89	93	94	94	94
Fluoranthene	9	8	8	8	11	100	100	100	100	100
Pyrene	44	45	47	47	45	54	73	74	72	72
Chrysene	32	38	38	39	38	68	84	93	93	93
Benzo(a)anthracene	38	44	46	47	47	52	72	76	77	74
Benzo(b)fluoranthene and benzo(k)fluoranthene	6	21	28	39	39	100	100	100	100	100
Benzo(a)pyrene	34	36	39	64	59	100	100	100	100	100

Table V. Reductions of EPA Priority Pollutant Base/Neutrals Through Chlorination and Ozonation: Group 2, Substituted Benzenes

	Reductions (%)								
	Chlorine Contact Time (h)					Ozone Contact Time (h)			
	0.25	0.5	1	2	3	0.25	0.5	1	2	3
1,3-Dichlorobenzene	0	22	36	44	53	34	34	38	39	50
1,4-Dichlorobenzene	38	33	44	57	60	34	36	39	39	41
1,2-Dichlorobenzene	15	16	27	45	48	27	27	28	24	36
1,2,4-Trichlorobenzene	1	11	15	35	30	4	4	12	4	11
Hexachlorobenzene	55	49	49	50	—	22	38	43	43	—
Nitrobenzene	23	32	36	48	57	4	5	10	8	11
2,6-Dinitrotoluene	10	16	17	15	15	8	8	13	11	13
2,4-Dinitrotoluene	34	36	35	35	34	69	60	60	60	60

Table VI. Reductions of EPA Priority Pollutant Base/Neutrals Through Chlorination and Ozonation: Group 3, Phthalate Esters

	Reductions (%)								
	Chlorine Contact Time (h)					Ozone Contact Time (h)			
	0.25	0.5	1	2	3	0.25	0.5	1	2	3
Dimethylphthalate	0	50	63	76	76	3	5	6	—	11
Diethylphthalate	6	13	15	17	19	8	12	10	3	—
di-m-Butylphthalate	15	19	23	23	22	6	7	10	8	15
bis-(2-Ethylhexyl)-phthalate	25	30	34	30	—	25	26	29	32	36
Butylbenzylphthalate	9	17	17	19	23	0	0	0	0	22

Table VII. Reduction of EPA Priority Pollutant Base/Neutrals Through Chlorination and Ozonation: Group 4, Ethers, Amines, Ketones and Polychlorinated Compounds

	Reductions (%)								
	Chlorine Contact Time (h)					Ozone Contact Time (h)			
	0.25	0.5	1	2	3	0.25	0.5	1	2	3
bis-(2-Chloroethyl) ether	22	31	34	46	50	1	2	3	4	8
bis-(2-Chloroisopropyl) ether	15	20	32	37	45	4	3	5	7	8
bis-(2-Chloroethoxy)methane	11	16	19	24	28	2	3	0	1	3
4-Chlorophenylphenyl ether	11	23	27	30	39	49	51	52	52	55
4-Bromophenylphenyl ether	7	16	18	22	30	47	52	55	54	55
Isophorone	20	20	26	27	27	100	100	100	100	100
1,2-Diphenylhydrazine	6	10	13	15	23	33	37	39	35	37
N-Nitrosodiphenylamine	61	88	89	90	92	100	100	100	100	100
Hexachloroethane	17	27	36	55	55	58	57	60	57	66
Hexachlorobutadine	13	18	21	38	33	56	63	57	59	56
Hexachlorocyclopentadiene	8	29	38	53	57	100	100	100	100	100

Table VIII. Reductions of EPA Priority Pollutant Acids (Phenols) Through Chlorination and Ozonation

Reductions (%)

	Chlorine Contact Time (h)			Ozone Contact Time (h)		
	0.25	0.5	1	2	0.25	0.5	1	2
2-Chlorophenol	93	98	99	100	100	100	100	100
2-Nitrophenol	49	71	74	76	100	100	100	100
Phenol	100	100	100	100	100	100	100	100
2,4-Dimethylphenol	100	100	100	100	100	100	100	100
2,4-Dichlorophenol	17	78	83	83	100	100	100	100
2,4-Dinitrophenol	15	13	15	12	100	100	100	100
2,4,6-Trichlorophenol	(−76)	(−113)	(−96)	(−100)	97	97	97	97
Pentachlorophenol	11	12	16	18	100	100	100	100
p-Chloro-m-cresol	98	100	100	100	98	98	98	98
4,6-Dinitro-o-cresol	3	0	0	6	100	100	100	100

END−W/1348

propylenes. Xylenes and 1,1-dichloroethylene exhibited intermediate removals.

Base/Neutrals

Because this division contains the greatest number of compounds of EPA priority pollutants, it was divided into four subgroups to assist interpretation.

Polynuclear Aromatics

Chlorination produces an approximate 50% reduction in polynuclear aromatics, while ozonation results in reductions between 75 and 100% (Table IV). Thus the polynuclear aromatics were quite susceptible to degradation by both chlorine and ozone.

Substituted Benzenes

Both chlorine and ozone produced intermediate removals of substituted benzenes, with chlorine providing somewhat superior removals (Table V).

Phthalate Esters

With the exception of dimethyl phthalate, which indicated a 75% removal on chlorination, neither ozone nor chlorine produced substantial removals of the phthalate esters (Table VI). Thus, along with the saturated halogenated methanes, ethanes, and propanes, the phthalate esters showed the greatest resistance to degradation.

Ethers, Amines, Ketones and Polychlorinated Compounds

Ozone had almost no effect on alkyl ethers, but gave a 50% reduction of aryl ethers. Chlorination resulted in low to intermediate removals of both alkyl and aryl ethers. Diphenylhydrazine was poorly removed by ozonation or chlorination, but N-nitrosodiphenylamine was significantly removed by both. Ozonation and chlorination produced intermediate removals of polychlorinated compounds, with ozonation producing slightly better removals (Table VII).

Table IX. Average Removals at End of Oxidation Period for Various Classes of Organic Compounds

Class of Compounds	Number in Class	Reduction by Oxidizing Agent (%)	
		Chlorine	Ozone
Benzene and alkyl-sbustituted benzenes	6	18	27
Chloro- or nitro-substituted benzenes or alkylbenzenes	9	41	31
Polynuclear aromatic hydrocarbons	12	44	91
Phthalate esters	5	34	17
Phenols	10	66	100
Halogenated methanes, ethanes, and propanes	16	20	20
Halogenated ethylenes, and propylenes	7	14	70
Halogenated dienes	2	45	78
Halogenated ethers	5	38	26
Amines	2	58	69

Acids (Phenols)

Ozonation produced excellent reductions of all phenols studied. Chlorination produced good but highly variable removals, ranging from a 6% reduction for 4,6-dinitro-o-cresol to a 100% reduction for p-chloro-m-cresol (Table VIII). Good reductions were also achieved for 2-chlorophenol and dichlorophenols, which produced an increase in the trichlorophenol concentration. This in turn produced a negative removal of trichlorophenol.

SUMMARY

The effectiveness of ozonation and chlorination for reduction of EPA Priority Pollutants is presented in Table IX, in which the compounds are grouped by chemical structure rather than by analytical procedure.

In general, phenols and amines are most susceptible to attack by both chlorine and ozone, and the phthalate esters and the halogenated methanes, ethanes, and propanes are least susceptible to attack. Chloro- and nitrobenzenes were more susceptible to oxidation by chlorine and ozone than benzene and alkylbenzenes. Polynuclear aromatics were about as susceptible to degradation by chlorine as the chloro- and nitro-benzenes but were very susceptible to degradation by ozone.

Halogenated methanes, ethanes, and propanes were considerably more resistant to degradation by ozone than were halogenated ethylenes and propylenes but exhibited similar degradabilities when chlorinated. The few halogenated dienes were more subject to degradation by chlorine or ozone than were the mono-unsaturated halogenated compounds. Halogenated ethers exhibited degradability similar to that observed with the nitro- and chlorobenzenes and alkylbenzenes.

From Table IX we can see that ozone was more effective than chlorine for six groups, chlorine was more effective than ozone for three groups, and chlorine and ozone were equally effective for one group. On the basis of an individual compound weighted average, ozone produced a 52% reduction in EPA priority pollutants and chlorine a 36% reduction. In addition, ozonation did not produce chlorinated EPA priority pollutant by-products such as trichlorophenol, which was produced in significant concentrations by chlorination. However, although chlorination did form at least two priority pollutants (trichlorophenol and methylene chloride), it also produced a significant reduction in the concentration of most of the other priority pollutants studied.

ACKNOWLEDGMENT

New Jersey Agricultural Experiment Station Publication No. D-07523-1-84 was supported by State Funds from the Agricultural Experiment Station and the New Jersey Department of Environmental Protection.

REFERENCES

1. White, G. C. *Handbook of Chlorination*, (New York: Van Nostrand-Reinhold, 1972).
2. Jolley, R. L. *Chlorination Effects of Organic Constituents in Effluents from Domestic Sanitary Sewage Treatment Plants,* Ph.D. Dissertation, (Knoxville: University of Tennessee, 1972).
3. Carlson, R. M., et al. "Facile Incorporation of Chlorine into Aromatic Systems During Aqueous Chlorination Process," *Environ. Sci. Technol.* 9:674–675 (1975).
4. Morris, J. C. *Formation of Halogenated Organics by Chlorination of Water Supplies*, EPA-600/1-75-002 (Washington, DC: U.S. Environmental Protection Agency, 1975).
5. *Standard Methods for the Examination of Water and Wastewater*, 15th ed. (Washington, DC: American Public Health Association, 1980).
6. *Methods for Organic Chemical Analysis of Municipal and Industrial Wastewater*, EPA-600/4-82-057, (Washington, DC: U.S. Environmental Protection Agency, 1982).

Epilogue

A whole is that which has beginning, middle, and end.

Aristotle, 384–322 B.C.
Poetics

List of Authors

Mohamed S. Abdel-Rahman, University of Medicine and Dentistry of New Jersey, Newark, New Jersey 07103

E. Marco Aieta, Stanford University, Stanford, California, 94305

Elmer W. Akin, U.S. Environmental Protection Agency, Cincinnati, Ohio 45268

Gary Amy, University of Arizona, Tucson, Arizona 85721

Julian B. Andelman, University of Pittsburgh, Pittsburgh, Pennsylvania 15261

Richard P. Arber, Richard P. Arber Associates, Inc., 100 Fillmore, Denver, Colorado 80206

R. C. Ayotte, University of Hartford, West Hartford, Connecticut 06117

M. T. Azure, University of Hartford, West Hartford, Connecticut 06117

David S. Bailey, Environmental Defense Fund, 11 South 12th St., Richmond, Virginia 23219

Alexander E. Bakalian, The Johns Hopkins University, Baltimore, Maryland, 21205

Sandra M. Baksi, University of Maryland, Solomons, Maryland 20688-0038

Harold A. Ball, Stanford University, Stanford, California 94305

S. B. Barnela, University of Auburn, Auburn, Alabama 36849

Douglas S. Baughman, University of South Carolina, Columbia, South Carolina 29208

Roger M. Bean, Battelle Pacific Northwest Laboratories, P.O. Box 999, Richland, Washington 99352

J. Peter Bercz, U.S. Environmental Protection Agency, Cincinnati, Ohio 45268

James D. Berg, Rio Linda Chemical Co., Inc., 2444 Elkhorn Blvd., Rio Linda, California 95673

John Black, Roswell Park Memorial Institute, Buffalo, New York 14263

Penny Black, Roswell Park Memorial Institute, Buffalo, New York 14263

A. Blaison, Centre de Recherche de la Compagnie Generale des Eaux, Chemin de la digue, Maisons Laffitte, France

Walter J. Blogoslawski, National Marine Fisheries Service, 212 Rogers Ave., Milford, Connecticut 06460-6499

Fred Bock, Papanicolaou Cancer Research Institute, 1155 N.W. 14th Street, Miami, Florida 33101

Joseph F. Borzelleca, Medical College of Virginia, Richmond, Virginia 23298

Keith W. Bossung, Indiana American Water Company, Richmond, Indiana 47274

L. Boyer, U.S. Environmental Protection Agency, Cincinnati, Ohio 45268

H. J. Brass, U.S. Environmental Protection Agency, Cincinnati, Ohio 45268

Donald W. Brown, Northwest and Alaska Fisheries Center, 2725 Montlake Blvd. East, Seattle, Washington 98112

Dolores M. Brncich, Illinois Institute of Technology, Chicago, Illinois 60616

A. Bruchet, S.L.E.E. Laboratorie Central, 38 rue du President Wilson, LE PECQ 78230, France

Richard J. Bull, Washington State University, Pullman, Washington 99164

H. D. Burkett, University of Auburn, Auburn, Alabama 36849

Kenneth P. Cantor, National Cancer Institute, Bethesda, Maryland 20205

Robert M. Carlson, University of Minnesota, Duluth, Minnesota 55812

Betsy D. Carlton, Battelle Institute, Columbus, Ohio 43201

James D. Carr, University of Nebraska, Lincoln, Nebraska 68588-0304

Beverly B. Casey, Virginia Institute of Marine Science, Gloucester Point, Virginia 23062

Paul A. Chadik, University of Arizona, Tucson, Arizona 85721

Sin-Lam Chan, National Oceanic and Atmospheric Administration, Northwest and Alaska Fisheries Center, 2725 Montlake Blvd. East, Seattle, Washington 98112

R. C. Chawla, Howard University, Washington, D.C. 20059

A. S.-C. Chen, University of Illinois at Urbana-Champaign, 1005 West Western Avenue, Urbana, Illinois 61801

H. F. Cheng, University of Iowa, Iowa City, Iowa 52242

William Cherry, Louisiana State University, Baton Rouge, Louisiana 70803-1804

Zaid K. Cowdhury, The University of Arizona, Tucson, Arizona 85721

Russell F. Christman, University of North Carolina, Chapel Hill, North Carolina 27514

Paul Chrostowski, EA Engineering, Science, and Technology, Inc., 15 Loveton Circle, Sparks, Maryland 21152

Dean A. Cole, University of Iowa, Iowa City, Iowa 52242

W. E. Coleman, U.S. Environmental Protection Agency, Cincinnati, Ohio 45268

Lyman W. Condie, U.S. Environmental Protection Agency, Cincinnati, Ohio 45268

William J. Cooper, Florida International University, Miami, Florida 33199

Joseph A. Cotruvo, Environmental Protection Agency, 401 M. Street, SW, Washington, D.C. 20460

John A. Couch, U.S. Environmental Protection Agency, Gulf Breeze, Florida 32561

Jack Coughlan, Central Electricity Generating Board, Marine Laboratory, Fawley Power Station, Fawley, Hampshire, United Kingdom

L. Courtney, U.S. Environmental Protection Agency, Gulf Breeze, Florida 32561

Donna L. Cragle, Oak Ridge Associated Universities, Oak Ridge, Tennessee 37830

S. Pasquini Cristina, University of Hartford, West Hartford, Connecticut 06117

Gunther F. Craun, U.S. Environmental Protection Agency, Cincinnati, Ohio 45268

Donald G. Crosby, University of California, Davis, California 95616

Peter M. Cumbie, Duke Power Company, Charlotte, North Carolina 28242

E. Daley, U.S. Environmental Protection Agency, Cincinnati, Ohio 45268

Jaap S. Sinninghe Damste, Delft University of Technology, Jaffalaan 9, 2628 BX Delft, The Netherlands

F. Bernard Daniel, National Institute for Occupational Safety and Health, 4676 Columbia Parkway, Cincinnati, Ohio 45226

Carol B. Daniels, University of Maryland, Solomons, Maryland 20688-0038

Debra A. Davies, The University of Wisconsin-Milwaukee, Milwaukee, Wisconsin 53201

John W. Davis, University of Tennessee, Knoxville, Tennessee 37996-1610

Martin H. Davis, Central Electricity Generating Board, Marine Laboratory, Fawley Power Station, Fawley, Hampshire, United Kingdom

William P. Davis, Environmental Protection Agency, Environmental Research Laboratory–Sabine Island, Gulf Breeze, Florida 32561

Leo de Galan, Delft University of Technology, Jaffalaan 9, 2628 BX Delft, The Netherlands

Ed W. B. de Leer, Delft University of Technology, Jaffalaan 9, 2628 BX Delft, The Netherlands

Susan Denny, Water Research Centre, Henley Road, Medmenham, P.O. Box 16, Marlow, Bucks SL72HD, United Kingdom

Douglas Dotson, University of Maryland, College Park, Maryland 20742

M. Dreux, Laboratorie chimie organique physique et chromatographie, Universite D'Orleans, Orleans, France

Brian Dudley, University of Massachusetts, Amherst, Massachusetts 01003

J. P. Duguet, S.L.E.E. Laboratorie Central, 38 rue du President Wilson, LE PECQ 78230, France

John Erickson, University of Nebraska, Lincoln, Nebraska 68588-0304

Jerry H. Exon, University of Idaho, Moscow, Idaho 83843

Edward J. Faeder, Southern California Edison Company, P.O. Box 800, Rosemead, California 91770

James A. Fava, EA Engineering, Science, and Technology, Inc.,* 15 Loveton Circle, Sparks, Maryland 21152. *Formerly Ecological Analysts, Inc.

Caren M. Feldstein, University of Cincinnati, Cincinnati, Ohio 45221

Joseph S. Ferris, EA Engineering, Science, and Technology, Inc., 15 Loveton Circle, Sparks, Maryland 21152.

Jean-Marie Fiquet, Electricite de France, 93206 Saint Denis, 76013 Rouen, France

Helen Fox, Roswell Park Memorial Institute, Buffalo, New York 14263

Stephen P. Gautney, Tennessee Valley Authority, 1160 Chestnut Street, Tower II, Chattanooga, Tennessee 37401

Maurizio Giabbai, Georgia Institute of Technology, Atlanta, Georgia 30332

J. Gibs, Drexel University, Philadelphia, Pennsylvania 19104

Louis T. Gidel, University of Miami, Miami, Florida 33149

M. Gibert, Centre de Recherche de la Compagnie Generale des Eaux, Chemin de la digue, Maisons Laffitte, France

Joseph P. Gould, Georgia Institute of Technology, Atlanta, Georgia 30332

Margaret L. Gourlay, University of Iowa, Iowa City, Iowa 52242

Edward T. Gray, Jr., University of Hartford, West Hartford, Connecticut 06117

John M. Grizzle, Auburn University, Auburn University, Alabama 36849-4201

Robert S. Grove, Southern California Edison Company, P.O. Box 800, Rosemead, California 91770

Werner R. Haag, SRI International, Menlo Park, California 94025

Charles N. Haas, Illinois Institute of Technology, Chicago, Illinois 60616

James A. Hampton, West Virginia University, Morgantown, West Virginia 26506

M. S. Harakeh, Stanford University, Stanford, California 94305

Patricia Hartge, National Cancer Institute, Bethesda, Maryland 20205

F. S. Hauchman, University of North Carolina, Chapel Hill, North Carolina 27514; The Johns Hopkins University, Baltimore, Maryland 21205

William E. Hawkins, Gulf Coast Research Laboratory, Ocean Springs, Mississippi 39564

Johnnie R. Hayes, Medical College of Virginia, Richmond, Virginia 23298

George R. Helz, University of Maryland, College Park, Maryland 20742

David E. Hinton, West Virginia University, Morgantown, West Virginia 26506

Soon Lin Ho, Virginia Institute of Marine Science, Gloucester Point, Virginia 23062

Harold O. Hodgins, Environmental Conservation Division, Northwest and Alaska Fisheries Center, 2725 Montlake Blvd. East, Seattle, Washington 98112

John C. Hoff, U.S. Environmental Protection Agency, Cincinnati, Ohio 45268

Jurg Hoigne, Swiss Federal Institute for Water Resources and Water Pollution Control, CH-8600 Dubendorf, Switzerland

G. Holdsworth, Oak Ridge Research Institute, 113 Union Valley Road, Oak Ridge, Tennessee 37830

Robert Hoover, National Cancer Institute, Bethesda, Maryland 20205

G. Dean Howell, Old Dominion University, Norfolk, Virginia 23508

J. Hrubec, National Institute for Water Supply, P.O. Box 150, 2260 AD Leidschendam, The Netherlands

Calvin O. Huber, The University of Wisconsin–Milwaukee, Milwaukee, Wisconsin 53201

Joseph V. Hunter, Cook College, Rutgers University, New Brunswick, New Jersey 08903

Russell A. Isaac, Massachusetts Division of Water Pollution Control, Westborough, Massachusetts 01581

Bonnie Isacovics, University of Toronto, Toronto, Ontario, Canada M5S 2W7

J. C. Jacangelo, The Johns Hopkins University, Baltimore, Maryland 21205

Donald A. Jaworske, NASA–Lewis Research Center, Cleveland, Ohio 44135

Henk A. Jenner, N.V. KEMA, P.O. Box 9035, 6800 ET Arnhem, The Netherlands

James N. Jensen, The University of North Carolina, Chapel Hill, North Carolina 27514

Brian Jessen, Louisiana State University, Baton Rouge, Louisiana 70803-1804

J. Donald Johnson, The University of North Carolina, Chapel Hill, North Carolina 27514

C. Ian Johnson, University of Toronto, Toronto, Ontario, Canada M5S 2W7

Delia M. Kaganowicz, Florida International University, Miami, Florida 33199

Terrence J. Kearney, University of Hawaii at Manoa, 1000 Pope Road, Honolulu, Hawaii 96822

Paul B. Kelter, University of Nebraska, Lincoln, Nebraska 68588-0304

Parweiz A. Khan, Oklahoma State University, Stillwater, Oklahoma 74078

Kamel M. Khater, Illinois Institute of Technology, Chicago, Illinois 60616

Robert J. Kieber, University of Maryland, College Park, Maryland 20742

Joog-Soo Kim, Georgia Institute of Technology, Atlanta, Georgia 30332

Paul H. King, University of Arizona, Tucson, Arizona 85721

Loren D. Koller, University of Idaho, Moscow, Idaho 83843

H. J. Kool, National Institute for Water Supply, P.O. Box 150, 2260 AD Leidschendam, The Netherlands

F. C. Kopfler, U.S. Environmental Protection Agency, Cincinnati, Ohio 45268

Yehuda Kott, Environmental and Water Resources Engineering, Technion—Israel Institute of Technology, Haifa, Israel

Margaret M. Krahn, National Oceanic and Atmospheric Administration, Northwest and Alaska Fisheries Center, 2725 Montlake Blvd. East, Seattle, Washington 98112

Robert Kravitz, Old Dominion University, Norfolk, Virginia 23508

Dr. Herman F. Kraybill, National Cancer Institute (Retired), 14112 Chelmsford Rd., Rockville, Maryland 20853

C. F. van Kreijl, National Institute for Water Supply, P.O. Box 150, 2260 AD Leischendam, The Netherlands

J. C. Kruithof, The Netherlands Waterworks Testing and Research Institute, KIWA N.C., P.O. Box 1072, 3430 BB Nieuwegein, The Netherlands

Susan M. Laborde, Tennessee Valley Authority, Muscle Shoals, Alabama 35660

Richard A. Larson, University of Illinois at Urbana-Champaign, 1005 West Western Avenue, Urbana, Illinois 61801

John P. Lawler, Lawler, Matusky and Skelly Engineers, One Blue Hill Plaza, Pearl River, New York 10968

C. LeCloirec, Laboratorie CNGE, ENSCR, Rennes, Paris, France

M. Lafosse, Laboratorie chimie organique physique et chromatographie, Universite d'Orleans, Orleans, France

Yun-Shen Lee, Department of Environmental Protection, State of New Jersey, Trenton, New Jersey 08628

Lynn I. Levin, National Cancer Institute, Bethesda, Maryland 20205

Edith L. C. Lin, U.S. Environmental Protection Agency, Cincinnati, Ohio 45268

Sechoing Lin, University of Minnesota, Duluth, Minnesota 55812

Paul A. Lindholm, University of Iowa, Iowa City, Iowa 52242

Paul T. Liu, University of Iowa, Iowa City, Iowa 52242

John C. Loper, University of Cincinnati, 231 Bethesda Avenue, Cincinnati, Ohio 45267

Ernest D. Lowther, Indiana American Water Company, Richmond, Indiana 47274

Donald C. Malins, National Oceanic and Atmospheric Administration, Northwest and Alaska Fisheries Center, 2725 Montlake Blvd. East, Seattle, Washington 98112

James P. Malley, University of Massachusetts, Amherst, Massachusetts 01003

J. Mallevialle, S.L.E.E. Laboratorie Central, 38 rue du President Wilson, LE PECQ 78230, France

Stephen W. Maloney, Drexel University, Philadelphia, Pennsylvania 19104

C. Steve Manning, Gulf Coast Research Laboratory, Ocean Springs, Mississippi 39564

Patrick A. March, Tennessee Valley Authority, 1160 Chestnut Street, Tower II, Chattanooga, Tennessee 37401

G. Martin, Laboratoire CNGE, ENSCR, Rennes, Paris, France

Thomas J. Mason, National Cancer Institute, Bethesda, Maryland 20205

Abdul Matin, Stanford University, Stanford, California 94305

Jack S. Mattice, Electric Power Research Institute, P.O. Box 10412, Palo Alto, California 94303

Kathryn Mazina, Old Dominion University, Norfolk, Virginia

L. J. McCabe, U.S. Environmental Protection Agency, Cincinnati, Ohio 45268

Bruce B. McCain, National Oceanic and Atmospheric Administration, Northwest and Alaska Fisheries Center, 2725 Montlake Blvd. East, Seattle, Washington 98112

P. McCauley, U.S. Environmental Protection Agency, Cincinnati, Ohio 45268

Patricia A. McCuskey, West Virginia University, Morgantown, West Virginia 26506

C. William McLaughlin, University of Nebraska, Lincoln, Nebraska 68588-0304

Jay C. Means, University of Maryland, Solomons, Maryland 20688-0038

John R. Meier, U.S. Environmental Protection Agency, Cincinnati, Ohio 45268

John W. Meldrim, Harza Engineering Company, 150 South Wacker Drive, Chicago, Illinois 60606

Christopher D. Metcalfe, McMaster University, Hamilton, Ontario, Canada L8S 4K1

Douglas P. Middaugh, U.S. Environmental Protection Agency, Gulf Breeze Environmental Research Laboratory, Sabine Island, Gulf Breeze, Florida 32561

Richard J. Miltner, U.S. Environmental Protection Agency, Cincinnati, Ohio 45268

Roger A. Minear, University of Illinois at Urbana, 408 South Goodwin Avenue, Urbana, Illinois 61801

Thomas A. Miskimen, American Electric Power Service Corporation, Canton, Ohio 44701

Cynthia A. Moore, University of Miami, Miami, Florida 33149

Leown A. Moore, U.S. Environmental Protection Agency, Cincinnati, Ohio 45268

J. Carrell Morris, International Institute for Hydraulic and Environmental Engineering, P.O. Box 3015, 2601 DA, Delft, The Netherlands

Robert D. Moss, Tennessee Valley Authority, 1160 Chestnut Street, Tower II, Chattanooga, Tennessee 37401

Sylvia A. Murray, Tennessee Valley Authority, Muscle Shoals, Alabama 35660

Mark S. Myers, National Oceanic and Atmospheric Administration, Northwest and Alaska Fisheries Center, 2725 Montlake Blvd. East, Seattle, Washington 98112

B. Najar, Drexel University, Philadelphia, Pennsylvania 19104

Duane A. Neitzel, Battelle Pacific Northwest Laboratories, P.O. Box 999, Richland, Washington 99352

Howard M. Neukrug, Philadelphia Water Department, Philadelphia, Pennsylvania 19102

A. Noordsij, The Netherlands Waterworks Testing and Research Institute, KIWA N.C., P.O. Box 1072, 3430 BB Nieuwegein, The Netherlands

Daniel L. Norwood, University of North Carolina, Chapel Hill, North Carolina 27514

C. I. Noss, University of South Florida, 13301 North 30th Street, Tampa, Florida 33612

Lisa H. Nowell, University of California, Davis, California 95616

H. Okrend, Howard University, Washington, D.C. 20059

Vincent P. Olivieri, The Johns Hopkins University, Division of Environmental Health Engineering, Baltimore, Maryland 21205

Betty H. Olson, University of California, Irvine, California 92717

Lewis L. Osborne, University of Illinois at Urbana-Champaign, Urbana, Illinois 61801-3882

T. R. Osborne, Oak Ridge Research Institute, 113 Union Valley Road, Oak Ridge, Tennessee 37830

Jean Ospital, Southern California Edison Company, P.O. Box 800, Rosemead, California 91770

Edward O. Oswald, University of South Carolina, Columbia, South Carolina 29208

Robin M. Overstreet, Gulf Coast Research Laboratory, Ocean Springs, Mississippi 39564

Cynthia A. Parks, University of Massachusetts, Amherst, Massachusetts 01003

Rutton D. Patel, Polytechnic Institute of New York, Brooklyn, New York 11201

Michael A. Pereira, U.S. Environmental Protection Agency, Cincinnati, Ohio 45268

Hans Plugge, EA Engineering, Science, and Technology, Inc., 15 Loveton Circle, Sparks, Maryland 21152

Ronald J. Pfohl, Miami University, Oxford, Ohio 45056

Meletios Platon, University of Miami, Coral Gables, Florida 33124

L. M. Puijker, The Netherlands Waterworks Testing and Research Institute, KIWA N.C. P.O. Box 1072, 3430 BB Nieuwegein, The Netherlands

Vito L. Punzi, Villanova University, Villanova, Pennsylvania 19085

Robert G. Qualls, University of North Carolina, Chapel Hill, North Carolina 27514

Daniel H. Raab, The University of Wisconsin-Milwaukee, Milwaukee, Wisconsin 53201

Neil M. Ram, GCA Corporation, Technology Division, 213 Burlington, Bedford, Massachusetts 01730

W. Howard Rapson, University of Toronto, Toronto, Ontario, Canada M5S 2W7

David A. Reckhow, University of North Carolina, Chapel Hill, North Carolina 27514

M. Carrington Reid, University of Tennessee, Knoxville, Tennessee 37996-1610

Martin Reinhard, Stanford University, Stanford, California 94305

Nathaniel W. Revis, Oak Ridge Research Institute, 113 Union Valley Road, Oak Ridge, Tennessee 37830

James K. Rice, Consultant, Onley, Maryland 20832

Rip G. Rice, 1331 Patuxent Drive, Ashton, Maryland 20861

Janet Rickabaugh, University of Cincinnati, Cincinnati, Ohio 45221

H. P. Ringhand, U.S. Environmental Protection Agency, Cincinnati, Ohio 45268

James Rios, Bechtel Power Corporation, P.O. Box 3965, San Francisco, California 94119

Morris H. Roberts, Jr., Virginia Institute of Marine Science, Gloucester Point, Virginia 23062

Paul V. Roberts, Stanford University, Stanford, California 94305

Merrell Robinson, U.S. Environmental Protection Agency, Cincinnati, Ohio 45268

Laura Rosenblum, University of Cincinnati, 231 Bethesda Avenue, Cincinnati, Ohio 45267

William T. Roubal, National Oceanic and Atmospheric Administration, Northwest and Alaska Fisheries Center, 2725 Montlake Blvd. East, Seattle, Washington 98112

William J. Rue, EA Engineering, Science, and Technology, Inc., 15 Loveton Circle, Sparks, Maryland 21152

Jessica J. St. Aubin, University of North Carolina, Chapel Hill, North Carolina 27514

Tommy I. Sammons, University of South Carolina, Columbia, South Carolina 29208

Francis J. Sansone, University of Hawaii at Manoa, 1000 Pope Road, Honolulu, Hawaii 96822

Gary S. Sayler, University of Tennessee, Knoxville, Tennessee 37996-1610

Geoffrey I. Scott, University of South Carolina, Columbia, South Carolina 29208

Frank E. Scully, Jr., Old Dominion University, Norfolk, Virginia 23508

Dennis R. Seeger, U.S. Environmental Protection Agency, Cincinnati, Ohio 45268

Carl M. Shy, University of North Carolina, Chapel Hill, North Carolina 27514

Edward J. Siff, University of North Carolina, Chapel Hill, North Carolina 27514

Debra T. Silverman, National Cancer Institute, Bethesda, Maryland 20205

Philip C. Singer, University of North Carolina, Chapel Hill, North Carolina 27514

Maryrose K. Smith, U.S. Environmental Protection Agency, Cincinnati, Ohio 45268

Matthew G. Smith, Philadelphia Water Department, Philadelphia, Pennsylvania 19102

Vernon L. Snoeyink, University of Illinois at Urbana-Champaign, 1005 West Western Avenue, Urbana, Illinois 61801

Daniel E. Sonenshine, Old Dominion University, Norfolk, Virginia 23508

Ronald A. Sonstegard, McMaster University, Hamilton, Ontario, Canada L8S 4K1

R. Kent Sorrell, U.S. Environmental Protection Agency, Cincinnati, Ohio 45268

Mark A. Speed, City of Thornton Utilities Department, Thornton, Colorado 80229

David Splichal, University of Nebraska, Lincoln, Nebraska 68588-0304

John E. Stein, National Oceanic and Atmospheric Administration, Northwest and Alaska Fisheries Center, 2725 Montlake Blvd. East, Seattle, Washington 98112

Alan A. Stevens, U.S. Environmental Protection Agency, Cincinnati, Ohio 45268

Reggie H. Stevens, University of Iowa, Iowa City, Iowa 52242

Mary Elizabeth Stewart, Hospital of Saint Raphael, 1450 Chapel Street, New Haven, Connecticut 06511

Robert J. Struba, Medical College of Georgia, Augusta, Georgia 30912

Irwin H. Suffet, Drexel University, Philadelphia, Pennsylvania 19104

L. J. Swango, University of Auburn, Auburn, Alabama 36849

Alireza Tabatabai, University of Nebraska, Lincoln, Nebraska 68588-0304

M. Wilson Tabor, University of Cincinnati, 231 Bethesda Avenue, Cincinnati, Ohio 45267

Douglas H. Taylor, Miami University, Oxford, Ohio 45056

Kenneth J. Tennessen, Tennessee Valley Authority, Muscle Shoals, Alabama 35660

Berta L. Thomas, Battelle Pacific Northwest Laboratories, P.O. Box 999, Richland, Washington 99351

Gavin P. Thompson, University of North Carolina, Chapel Hill, North Carolina 27514

G. Torrence, Howard University, Washington, D.C. 20059

Y. Tsutsumi, S.L.E.E. Laboratorie Central, 38 rue du President Wilson, LEPECQ 78230, France

Allen D. Uhler, University of Maryland, Solomons, Maryland 20688-0038

Richard L. Valentine, The University of Iowa, Iowa City, Iowa 55242

Thomas B. Vanderbeek, Lawler, Matusky and Skelly Engineers, One Blue Hill Plaza, Pearl River, New York 10968

M. A. Van der Gaag, The Netherlands Waterworks Testing and Research Institute, KIWA N.C., P.O. Box 1072, 3430 BB Nieuwegen, The Netherlands

Usha Varanasi, National Oceanic and Atmospheric Administration, Northwest and Alaska Fisheries Center, 2725 Montlake Blvd. East, Seattle, Washington 98112

M. M. Varma, Howard University, Washington, D.C. 20059

R. Vasl, Reschov Esriel 15/7, Givat Shaul, Jerusalem, Israel

John N. Veenstra, Oklahoma State University, Stillwater, Oklahoma 74078

Craig D. Vogt, U.S. Environmental Protection Agency, Washington, D.C. 20460

Evangelos A. Voudrias, University of Illinois at Urbana-Champaign, 1005 West Western Avenue, Urbana, Illinois 61801

Jan K. Wachter, University of Pittsburgh, Pittsburgh, Pennsylvania 15261

Thomas D. Waite, University of Miami, Coral Gables, Florida 33124

Johannes Edmund Wajon, Western Australian Institute of Technology, Perth, Australia

William W. Walker, Gulf Coast Research Laboratory, Ocean Springs, Mississippi 39564

Philip Wilcox, Water Research Centre, Henley Road, Medmenham, P.O. Box 16, Marlow, Bucks SL72HD, United Kingdom

D. E. Williams, University of Auburn, Auburn, Alabama 36849

Allen T. Wojtas, Illinois Institute of Technology, Chicago, Illinois 60616

Roy L. Wolfe, University of California, Irvine, California 92717

H. J. Workman, University of Hartford, West Hartford, Connecticut 06117

S. D. Worley, Auburn University, Auburn, Alabama 36849

Rodney G. Zika, University of Miami, 4600 Rickenbacker Causeway, Miami, Florida 33149

INDEX

Abdel-Rahman, Mohamed S. 281
absorption of sunlight 1011
absorptivity 1293
Acanthamoeba castellani 7
accidents 123
accuracy 1075,1099
acenaphthalene 1520
acenaphthene 1395,1520
acenaphthylene 1395
acetaldehyde 281,827,1297
acetic acid 854
acetone 281,704
acetonitrile 828,1100,1331,1412
acetophenone 1101
acetovanillone 237
acetylacetone 281,704
acetylglycine 800
acid catalysis 770,1294
acids 94,166
Acinetobacter 563
acrolein 1390
acrylonitrile 1390
activated carbon 25,907,953,1140,1313
 See also granular-activated carbon
activated sludge treatment 27,268,1014
acute toxicity 493,500
 oral 333
adenine 230,578,591
adenocarcinoma 254
adenosine 578
adenoviruses 6
adipose tissue 307
adsorbable organic chlorine 190
adsorbed organohalogen (AOX) 1140
adsorption, Tenax 1139
advance notice of proposed rule
 making 91
Aeromonas 563
aflatoxin B1 235,397
aflatoxin B2 439
African green monkey (*Cercopitheous
 aethiops*) 346,355
agricultural runoff 818
 use 691
Aieta, E. Marco 783
air stripping 907
Akin, Elmer W. 99
alanine 570,583,619,823

Alcaligenes 563
alcohols 94,281,1101
aldehydes 94,166,821
 formation 827
aldrin 1395
algae 7,42,1186
 blue-green 807
 control 1215
 cultures 876
 extracellular material 875
algal biomass 875,939
algicide 1011
aliphatic compounds 1234
 dibasic acids 846
 monobasic acids 846
alkalinity 1188
alkanes 854,868
alkenes 854,1042
alkylating agents 193
alkyl nitriles 1229
alkyl substituted benzenes 868
altered foci induction 262
alternative disinfectant 105,295
 by-products 218
 ferrate 1285
alternatives 77,282
 See also bromine chloride;
  chloramines; chlorine dioxide;
  ferrate
alum 1186
 coagulation 25
aluminum sulfate 268
ambient water quality criteria 1377
American Public Power Association 65
Ames test 161,187,207,221,237,252,
 265,452,457,1138,1331,1343,1412
 "spot test" 1422
 See also assay; bioassay;
  mutagenicity
Amherst, Massachusetts 1120
amines 521,714,798,821,965,1021,1524
amino acids 169,238,478,521,570,575,
 723,807,821,939,942,1180,1201
 consumption of monochloramine
 580
 N-chlorination 721
 oxidation 832
amino nitrogen compounds 807

aminoanthracene 212
aminoazotoluene 439
ammonia 24,513,521,543,714,723,737,
  756,812,952,975,1016,1039,1093,
  1197,1297,1447,1459,1469
  breakpoint reactions 723
ammonium ions, stabilizing effect 748
amoebic dysentery 100
AMP 578
amperometric back titration 1073
  instrument 1073
  See also oxidant analysis
amperometric titration 22,494,513,522,
  542,556,575,652,667,725,775,942,
  1073,1372
  See also oxidant analysis
amperometric titrator 494
Amy, Gary L. 907
Anabaena flos-aqua 7
Anabaena sp. 807
anadromous fish 30
analytical instrument 1073
analytical methods 797,1073,1123,1358,
  1372
  See also oxidant analysis
ancylostoma, duodenal 7
Andelman, Julian B. 875
anhydride 1313
animal studies 296
anion mixtures 1065
anions, separation 1069
anodic voltammetric determination
  1091
anthracene 419,1375,1395,1520
anthracenecarboxyaldehyde 406
anthracite filtration 1139
anthraquinone 406
antithyroid effect 114,345
antitumor immunity 251
Aphanizomenon flos-aqua 7
Apicomplexan disease 509
Appoquinimink River 494
aquatic models 375
aquatic organisms 53,68,73
  potential threat 1423
aquatic toxicology 81
Arber, Richard P. 807,951
arginine 583,812
Arno River 192
Aroclor 1254 208,267,407
aromatic acid 843,846,865
aromatic amines 193
aromatic compounds 1042,1232
arsenic 145

arteriosclerotic heart disease 124
artifacts 113
artificial seawater 966
aryl-chlorinated aromatic acids 843,
ascariasis 7
Ascaris lumbricoides 7
ascorbic acid 260,1202
Asiatic clam 45,48
asparagine 583,619,823
aspartic acid 583,591,619
aspermatogenesis 113,121
assay 1414
  enzyme-linked immunosorbent 309
  genotoxicity 267
  immunologic 309
  inactivation 556
  liver carcinogen 377
  mouse skin 190
  See also Ames test; bioassay;
    mutagenicity; Sencar mouse
assessment input 505
  methodology 75
association reactions 705
atherosclerosis 365
Atlanta 146
Atlantic silverside 493
Atlantic croaker 503
Atlantic menhaden 503
atrazine 1171
Australia 14
autoimmune disorders 307
available chlorine, See chlorine
avoidance response chemical effect 493
avoidance studies 496
Ayotte, R. C. 797
Azure, M. T. 763

bacillary dysentery 6
bacilli 556
bacteria 6,87,556,1271
  batch culture 603
  chemostat-grown 615
  chlorine uptake 677
  continuous culture 603
  disinfectant-tolerant 561
  indicator 99
  influence of subculturing 557
  recovery techniques 565
  resistance to chlorine 558
  standard plate count 652
  See also coliform; fecal coliform
bacterial assays, See Ames test, assay,
  bioassay, mutagenicity
  counts 555,691,1188

INDEX 1545

disease (*Aeromonas* sp.) 418
kill 691
mutagenicity assay 1343
quality 1197
response 615
slimes 1469
virus, See also disinfection
bactericidal efficiency 1259
See also disinfection
bacteriophage 584,681,683
Bailey, David S. 85
Bakalian, A. E. 651
Baksi, Sandra M. 1411
balantidiasis 7
Balantidium coli 7
Ball, Harold A. 1505
barnacles 42,48,505
Barnela, S. B. 1269
base-catalyzed hydrolysis 1229
base-neutral components 1374,1517
BAT, See best available treatment technology
batch-culture-grown cells, sensitivity 617
Baughman, Douglas S. 463
Bay Anchovy 503
Bay Tree Lake 940
Bean, Roger M. 1357,1371
behavioral parameters 337,356
benthic organisms 505
bentonite clay 1270
benzaldehyde 1101,1182
benzanthracene 402,419,1375,1395, 1520
benzene 27,92,145,188,1101,1390,1517
dicarboxylic acid 865
substituted 1524
benzenetetracarboxylic acid 865
benzenetricarboxylic acid 846,865
benzidine 1395
benzofluoranthrene 188,419,1395,1520
benzofluorene 419
benzoic acid 849,865,1055,1066,1070
benzonitrile 854,1182
benzoperylene 419,1395
benzopyrene 188,212,416,419,1395, 1520
benzoquinone 837,1024,1322,1325
benzyl cyanide 887
Bercz, J. Peter 345
Berg, James D. 603
best available treatment technology 64
economically achievable 1489
effluent guidelines 64

BGM, See Buffalo green monkey
BHC 1395
bicarbonate 1049
bioaccumulation 30,78,407
bioassay 261,1137,1414
breast cancer induction 261
carcinogen 251,309,441
liver foci alteration 261
long-term 226
models 429
promoter 441
pulmonary tumor induction 261
skin tumor induction 261
tier scheme 252
tumor promotion 261
See also Ames test; assay; Chinese hamster; mutagenicity; sister chromatid exchange
bioavailability 404
biocide 64,1011,1447,1459
See also algicide, disinfection
biodegradation 1505
biodeposition rate measurements 466
biofilm 64,1435
accumulation rate model 1436
destruction 1435
destruction kinetics 1437
formation 44
polysaccharide matrix 47
See also biofouling; macrofouling; macroinvertebrate; microfouling
biofouling 64,73,493,533,541,737,1091, 1357,1435,1459
biofouling control 42,63,73,76,723
alternatives 82
OTEC 760
See also Asiatic clam; bacterial; corbicula; mussels; slime
biological treatment 1021
biotic response 51
biotin 1346
biotransformation 78
biphenyl 1101,1182
Biscayne Aquifer 909
bis(chloroethyoxy)methane 1395
bis(chloroethyl) ether 188,1395
bis(chloroisopropyl) ether 188,1395
bis(2-ethoxyethyl) ether 887
bis(ethylhexyl)phthlate 1395
bivalve mollusks 48,505
black bullhead (*Ictalurus melas*) 451
Black Lake, NC 860,940,1013,1230
Black Mesa pipeline 1400
Black River 415

Black, John 415
Black, Penny 415
bladder cancer, See cancer
Blaison, A. 1065
Blogoslawski, Walter 521
blood plasma, absorption of N-chloropiperidine 182
blood serum 30,286
blowdown 1381
　See also cooling water
Blue Crab 503
Blueback herring 503
Bock, Fred 415
Bogue Banks, NC 118
bone marrow 223,186
　micronuclei 207
borate 1286
Borzelleca, Joseph F. 331
Bossung, K. W. 651
bovine serum albumin 717,809
Boyer, L. 1123
brain growth 355
brain lesions 121
Brass, H. J. 1123
breakpoint reaction 27,190,702,710, 713,739,797,952,975,1097,1115, 1210,1496
　ammonia 723
　dynamics 705
　mechanism 709
　model 724
　See also chlorination
Brncich, Dolores M. 775
broad-spectrum analysis 1099
bromamines 21,23,763,982,985
　decomposition 763,766,988
　formation model 991
　hydrolysis 763
　inorganic 986
　organic 986
　reaction rate constant 992
　See also bromochloramine; dibromamine; halamines; tribromamine
bromate 22,713,738,1043
bromide 22,737,907,999,1016,1039, 1043,1158,1260,1399,1405,1411
　oxidation 907,965,1005
　oxidation catalysis 1006
　reaction with organic chloramines 999
brominated alkylphenol polyethoxy carboxylates 1505

brominated macromolecular material 1371
brominated phenols 1365
brominated trihalomethanes 907,1266
bromine 10,94,985,1081
　decomposition 997
　demand 988
　incorporation coefficient 929
　reaction with ammonia 985
　reaction with nitrogenous compounds 985
　residual disappearance 988
bromine chloride 77,94
bromobenzene 1127
bromochloramine 737
bromochloracetic acid 866
bromochloroacetonitrile 113,222,231
bromochloroiodomethane 928
bromocymene 237
bromodeoxyuridine 267
bromodichloromethane 113,188,200, 237,897,908,1127,1261,1264,1362, 1373,1376,1519
bromodiiodomethane 929
bromodimethyl oxazolidinone 1269
bromodimethylamine 987
bromoform 29,113,255,463,759,897, 907,1046,1127,1261,1264,1305,1360, 1362,1373,1517
　See also trihalomethanes
bromoglutamate 987
bromoglycine 987
bromomethane 1519
bromoorganics, See haloorganics
bromophenol phenyl ether 1395
bromophenyl-acetic acid 1508
bromophenylphenyl ether 1522
brown bullhead (Ictalurus nebulosus) 415,453
Brown, Donald W. 399
Bruchet, A. 1165,1201
Buffalo green monkey (BGM) kidney cells 620
　cell cultures 681
Bull, Richard J. 111,207,221
Burkett, H. D. 1269
butanedioic acid 846,864
butyl disulfide 1313
butyl thiol 1313
butylbenzenesulphonamide 887
butylbenzylphthalate 1375,1395,1521
butylphthalate 1395,1521

by-products 4,81,93,111,112,119,207,
   295,345,1121
   excretion 285
   target organ effects 345
   See also chlorination; disinfection
byssal attachment 1425

caddisflies 490
cadmium 145,403
calcium carbonate 1382
calcium hypochlorite 1270
   See also chlorine; hypochlorite
calcium intake 116
Campylobacter 12
Campylobacter fetus 6
Canada 104
cancer 133,251
   mortality 122,281
   bladder 139,190,252
   breast, induction 261
   colon 138,142,153,252
   liver 139
   pancreas 153,256
   rectal 140,252
   See also carcinogen; carcinogenic;
      carcinogenicity; epidemiology;
      liver; neoplasms; risk; tumors
Candida albicans 1271
Cantor, Kenneth P. 145
Cape Fear River 726,1118
carbazole 406
carbohydrates 717,843,1014
carbon adsorption, See activated
   carbon; granular-activated carbon
carbon dioxide 713
carbon tetrachloride 92,255,1127,1362,
   1390,1519
carbonate 1049,1069,1286
carboxyl 1313
carcinogen 3,145,188,207,251,307,429,
   439,451,875
   colon 259
   human 262
   identification 251
   metabolic activation 261
   monitoring, fish 377
carcinogenic effect 115,415
carcinogenic risk 121,226
carcinogenicity 113,124,187,200,221,
   229,317,1341
   chloroform 119,229
   determination 251
   haloacetonitrile 229
   testing 429

carcinoma 415
   cholangiolar 389,393,401
   hepatocellular 120,121
cardiac hypertrophy 365,366
cardiovascular disease 116
   mortality 122
Carlson, Robert M. 835
Carlton, Betsy D. 295
Carr, James D. 1285
case-comparison mortality studies 140
Casey, Beverly B. 509
catalysis 710,1006
   catalytic role of hypochlorous acid 710
   catalytic role of nitrogen trichloride
      710
catechol 237,1322
catfish (*Ictaluridae*) 424
cationic polymers 1186
cell density 615
cell-mediated immunity 251,309
cell membrane permeability 571
cell viability 668
cell walls 30
Centers for Disease Control 11,93
Cercopitheous aethiops 346,355
cestodes (tapeworms) 7
Chadik, Paul A. 907
Chan, Sin-Lam 399
channel catfish (*Ictalurus punctatus*)
   452
Chapel Hill, NC 1118
charcoal 1313
Chawla, R. C. 635
chemical bonds, binding energies 705
chemical fate 78
chemical oxygen demand 1021
chemostat 603,615
Chen, A. S.-C. 1313
Cheng, H. F. 251
Cherry, William 1021
Chesapeake Bay 89,494,1034
chi-square analysis 1112
Chick-Watson equation 669
Chinese hamster ovary (CHO) 208,221
Chironomidae 486
chloral hydrate 26,843,846,1229,1297,
   1244
chloramide 570
chloramination 951,1351
   by-products 121
chloramines 10,21,78,79,94,95,111,116,
   175,200,281,296,555,570,587,636,
   798,835,942,951,965,975,985,999,
   1091,1161,1197,1215,1269,1351,1469

formation 179,825
formation equilibrium constants 798
interfering nitrogenous compounds 567
metabolism 290
microbicidal effectiveness 567
organic 1180,1269
prereacted 556
reaction with bromide 999
reaction with hydroxybenzoic acid 835
reaction with hydroxycinnamic acid 835
See also bromochloramine; chlorine residual; combined chlorine; dichloramine; halamines; monochloramine; organic chloramines; total residual chlorine
chlorate 95,210,296,345,355,784,1043, 1067
chlordane 1395
chloric acid 784
chloride 281,285,784,1043,1067,1125
metabolism 290
chlorinated aliphatic acids 843
chlorinated alkylbenzenes 27
chlorinated aromatic acids 27
chlorinated benzenes 1171,1179
chlorinated C-4 diacids 843
chlorinated dibenzofurans 1018
chlorinated drinking water
blood pressure effects 367
contractile properties 367
chlorinated effluents 452,481,493,509
chlorinated estuarine waters 493
chlorinated fatty acids 365
chlorinated guaiacols 29
chlorinated humic material, mutagenic activity 168,207
chlorinated hydrocarbons 64,187
chlorinated hydroxyfuranone 169
chlorinated ketones 1178
chlorinated organics 87,951
chlorinated phenolic acids 1314
chlorinated phenols 331
chlorinated phenols
chemical properties 332
physical properties 332
toxic effects 323
transplacental exposure 326
chlorinated seawater 971,1039
chlorinated surface water 142,146
chlorinated syringaldehydes 29

chlorinated tyrosine 240
mutagenicity 237
chlorinated wastewater 451,541,1505, 1515
effluent toxicity 546
fish behavior 481
fish mortality 481
mutagenicity 481
chlorinated water 1338
acute toxicity 1341
mutagenic activity 1341
supplies 125,1055
chlorination
aldehyde formation 827
algal extracellular material 875
amines 999
amino acids 821,827
breakpoint model 724
by-products 119,145,1065
carbon dioxide production 715
chemistry 1470
coal slurry transport waters 1399
conceptual model 1248
continuous 1444
cooling water 724,1357,1371,1399
effectiveness 1525
freshwater 1430
fulvic acid 1229
humic acid 844
intermittent 68,537,1444
issues 3
kinetics 723,755
mutagenicity 265,1411
natural waters 807
nitrite formation
oil refinery effluent 265
peptide, kinetics 804
products 1065
rates 570
reaction dynamics 701
scheme for alanine 830
seawater 1427,1459
targeted 1447
thermal 1057
wood pulp 237
See also alternatives; cooling water; disinfection; drinking water; intermittent chlorination; seawater; superchlorination; wastewater
chlorine 5,21,63,78,111,113,115,187, 207,209,281,463,603,635,667,784, 835,843,885,895,907,923,939,951, 965,975,985,999,1016,1091,1115,

INDEX 1549

1162,1197,1215,1269,1314,1411,
1435,1447,1459,1469,1515
analysis, See oxidant analysis
analyzer 22,896
bacterial inactivation 638
biotoxicity 30
chemistry 1493
continuous exposure 68
decomposition reactions 22
dilemma 85
direct transfer 568
discharges 63,85,89
disinfection 681,1329
efficacy 606
electrophilic substitution 720
exposure criteria 69
gas 79
hydrolysis 784
hydrolysis rate constant 786
metabolism 290
minimization 1425,1435,1447
oxidation reactions 704
plume 1459
predictions 1494
reaction with acetylglycine 797
reaction with alanine 823
reaction with algae 875
reaction with aliphatic compounds 1234
reaction with amines 49
reaction with amino acids 169,721
reaction with ammonia 21,49,705, 737
reaction with aromatics 1232
reaction with bovine serum albumin 809
reaction with bromide 907
reaction with chlorite 787
reaction with chlorophenol 1321
reaction with citrate 286
reaction with cooling waters 28,49
reaction with cytochrome c 809
reaction with dichlorophenol 1316
reaction with dihydroxybenzenes 1232
reaction with food products 29
reaction with fulvic acids 26,702, 939,1235
reaction with glycine 120
reaction with glycylglycine 724
reaction with halogens 49
reaction with humic acids 25,119, 161,723,809,844,859,907
reaction with hydrobenzoic acid 835

reaction with hydroxycinnamic acid 835
reaction with ketoglutaric acid 287
reaction with lignin 242
reaction with macromolecules 94
reaction with methionine 120
reaction with methyl amines 965
reaction with methyl benzylamine 895
reaction with natural organic matter 702
reaction with natural waters 713,723
reaction with nitrite 23
reaction with nitrogenous compounds 587
reaction with organic nitrogen 723
reaction with ozone 23
reaction with paper industry effluents 29
reaction with pepsin 809
reaction with peptides 797
reaction with phenolic acids 26
reaction with phenols 29,1320,1515
reaction with phenylalanine 120,826
reaction with rennin 809
reaction with resorcinol 844
reaction with seawater 22
reaction with stomach contents 169, 175
reaction with tyrosine 27,120,237, 826
reaction with uracil 23
reaction with wastewater 27
reaction with water 21
residual 67,667,1188,1342
sensitivity 667
stabilizer 1259
total residual 494,509,723,885,1073
tracer 49
transfer reactions 570,999
transformation reactions 22
wastewater disinfection 951
water 1313
See also chloramines; combined chlorine; free chlorine; oxidant; total residual chlorine
chlorine decay 67,1460,1491
  chemistry 713
  kinetics 724
  pathway, carbon dioxide-producing process 717
  oxidation 713
  seasonal variation 715

chlorine-chlorite reaction 787
chlorine demand 50,66,678,723,797,
    812,826,942,1033
  fast reactions 66
  fulvic acid 734
  hydrogen peroxide 1033
  kinetics 723
  nitrogen-containing compounds 734
  reactions 66,797,807,821,835
  slow reactions 66
  stomach fluid 177
  See also oxidant demand 1081
chlorine dioxide 10,78,94,95,107,111,
    114,117,187,209,237,281,296,345,
    355,603,615,619,635,783,835,943,
    1041,1066,1161,1215,1270,1314
  antithyroid effect 355
  bacterial inactivation 640
  by-products 121
  efficacy 606
  generation 282,783
  half-life 1051
  iodine binding 348
  metabolism 290
  mode of action 619
  neonatal hypothyroidism 381
  photodecomposition 1041
  photolysis reaction products 1045
  reaction with citrate 287
  reaction with hydroxybenzoic acid
    835
  reaction with hydroxycinnamic acid
    835
  reaction with ketoglutaric acid 297
  reaction with protein 629
  thyroid function 362
  toxicity effects 355
chlorine hexoxide 1041
chlorine perchlorate 1041
chlorine peroxide 1041
chlorine-produced oxidants 20,21,48,
    53,463,466,494,509,521,755,965,
    1081
  analysis 21
  decomposition reactions 22
  reduction kinetics 757
  transformation reactions 22
  See also oxidant
chlorine toxicity 21,87,366
  algae 54
  fish 54
  macroinvertebrate 54
  zooplankton 54
  See also toxicity

chlorine trioxide 1041
chlorite 95,111,114,117,210,281,296,
    345,355,784,1043,1314
chloro-2-(methylamino)ethanol 999
chloroacetaldehyde 237
chloroacetic acids 94,866
chloroacetone 704
chloroacetonitrile 113,166,222
chloroacroleins 169
chloroalanine 942
chlorobenzene 256,331,1127,1378,1390,
    1518
chlorobenzoic acid 848
chlorobenzoquinone 1317,1325
chlorobromomethane 929
chlorobutanedioic acid 846
chlorobutanone 167,169
chlorobutenone 167
chlorocatechol 237
chlorocresol 1395,1523
chlorodibromamine 1006
chlorodibromomethane 897,1390,1397
  See also dibromochloromethane
chlorodichloromethyl hydroxyfuranone
    237
chlorodihydroxybenzene 1322
chlorodihydroxybiphenyl 1322
chlorodiiodomethane 929
chlorodimethyl oxazolidinone 1269
chlorodimethylbenzene 1325
chlorododecane 860
chloroethane 1390,1519
chloroethoxy methane 1522
chloroethyl ether 1522
chloroethylvinyl ether 1390
chloroform 66,78,94,111,113,119,138,
    139,145,164,166,169,188,200,229,
    255,281,703,808,835,843,859,875,
    895,907,923,1115,1127,1201,1229,
    1254,1261,1264,1300,1305,1362,
    1373,1376,1390,1392,1397,1519
  carbon-13 labeled 1120
  formation 810
  formation, predictive model 24
  metabolism 229
  precursors 835,839,850
  spill event 1306
  See also trihalomethanes
chloroguaiacol 237
chlorohydroxybenzoquinone 1317
chloroydroxydimethoxybenzoic acid
    838

chlorohydroxymethoxycinnamic acid
838
chlorohydroxymethoxyphenyl-
chloroethylene 838
chlorohydroxymethoxyphenyl-
chloroethylene 838
chlorohydroxymethoxyphenyldichloro-
ethanol 838
chloroisopropyl ether 1522
chloromercuribenzoate 630
chloromethoxybenzenedicarboxylic acid
866
chloromethoxybenzenes 860
chloromethoxybenzenetricarboxylic
acid 866
chloromethoxybenzoic acid 866
chloromethoxybenzoquinone 1317
chloromethoxyphenols 837
chloromethylbenzene 1171
chloronaphthalene 270,1395,1520
chloronitrobenzene 1056
chloroorganics, See haloorganics
chloropentanoic acid 846
chlorophenols 113,121,237,307,332,
  837,860,942,1023,1179,1182,1315,
  1321,1395,1523
  acute toxicity 229
  immune dysfunction 326
  immune effects 325
  latency 318
  mutagenicity 324
  oral toxicity 337
  oxidation 1027
  postnatal exposure 314
  prenatal exposure 312
  reproductive effects 342
  subchronic toxicity 339
  toxicity 307,331,339
  tumor incidence 318
chlorophenylacetic acid 846,848
chlorophenylphenyl ether 1395,1522
chlorophyll 542,821,876
chloropicrin 1165,1178,1181,1201
  formation 1204
chloropiperidine 1002
chloropropanone 167
chloropropenal 26,166,169,222,237
chlorosalicylic acid 1322
chlorosyringols 29
chlorotrihydroxybenzenes 837,1322
chlorouracil 24,30
chlorouridine 30
chlorous acid 784
chloroxanthine 1016

cholangiocellular carcinoma 401
cholangiolar carcinoma 389,393
cholera 6
cholesterol 365
Chowdhury, Zaid K. 907
Christman, Russell F. 939,1115
chromatography 1043,1099
  broad-spectrum 1099
  See also gas; gel permeation; liquid;
    HPLC; ion exchange
chromic acid 1116
chromite method 1285
chromophore 1066
chromosomal abberation assays 209
  damage 209
chronic heart failure 366
Chrostowski, Paul 73
chrysene 419,1395,1520
cigarette smoking 150
Cincinnati 68
cirrhosis of liver 123
citrate, alcoholic hydrogen 292
citric acid 281,843
Citrobacter fruendii 658
clams 42
clarification 268
Clark-type voltammetric electrode 1286
Clean Water Act 1489
clinical chemistry 349
closed-loop stripping analysis 162
Clostridium perfringens 652
coagulation 25,1100,1138,1185,1189,
  1210,1229
coal 1400
  Black Mesa 1401
  Montana Rosebud 1401
  slurry transport waters, chlorination
    1399
  Wyodak 1401
coastal power plant 1371
cocarcinogens 324,441
Cole, Dean A. 251
Coleman, W. E. 161
coleoptera 486
coliform 555,603,651,691,990,1021
coliphage 692
collagen 368
colon cancer, See cancer
color 1186,1215,1220
Colorado 93
combined bromine 737,743
combined chlorine 52,737,743,951,
  1469,1476

combined residual chlorine, *See* chlorine; combined chlorine; residual chlorine; total residual chlorine
Comptroller General 15
conception rates 299
condenser efficiency 1447
condenser waterbox 1449
Condie, Lyman W. 331,345
conductivity 1065
congenital malformations 123
Connecticut, 146
continuous chlorine exposure 68
Continuous Systems Modeling program 75
cooling systems 40,64,723,1081,1447
cooling tower 40,68,1381,1469
   blowdown 68
cooling water 3,22,23,28,39,63,73,737, 755,999,1033,1357,1371,1459,1411, 1425,1435,1459
   chlorination 1425
   heat treatment 1425
   mechanical cleaning 1425
   system design data 1396
   systems, once-through 1489
   treatment 20
   See also water
Cooper, William J. 895,907,1041
copepods 505
copper sulfate-caustic degradation studies 843
coprecipitation 1285
Corbicula 51
Corbicula fluminea 45
corn oil gavage 119,235
cost-benefit analysis 1455
cost-effective assessment matrix 82
costs of analysis 1125
Cotruvo, Joseph A. 91
Couch, John A. 377
Coughlan, Jack 1459
coulombic current efficiency 1082
coulometric iodine generation 1073
counter ions 1067
Courtney, Lee A. 377
Cove City, NC 1118
coxsackievirus 6,682,683
CPO, *See* chlorine-produced oxidants
Cragle, Donna L. 153
Crassostrea virginica, *See* oyster
Craun, Gunther F. 133
creatine 588
Cristina, S. Pasquini 763

Crosby, Donald G. 1055
Culex mosquito larvae 536
Cumbie, Peter M. 63,1489
Curie-Point pyrolysis-GC/MS 845
cyanate 94
cyanide 94,232,1297,1470
cyanobacteria 7
cyanopropanoic acid 846
cyano-substituted acids 846
cyanuric acid 588,1259
cyclohexene 1314
cyclohexenol 1314
cyclohexenone 1314
cysteine 581,619,630,812
cystine 581,619
cytidine 578
cytochrome 809
cytosine 578
cytosine monophosphate 578

Daly, E. 1123
Damste, J. S. S. 843
dan-bouy 1460
Daniel, F. Bernard 175,229
Daniels, Carol B. 1411
dansyl derivatives 175
Daphnia 51
Davies, Debra A. 1091
Davis, John W. 1399
Davis, Martin H. 1459
Davis, William P. 3,553
DDD 1395
DDE 188,1395
DDT 1395
death 496
decanoic acid 864
decay models 713,1464
dechlorinated sewage 541
   toxicity 548
dechlorinating agent 1033,1341
dechlorination 78,86,512,548,1033, 1341,1348
decision criteria 73,83
decomposition kinetics 763
decomposition reactions 22
defouling agent 1469
degradation of compounds by sunlight 1016
   micropollutants 1016
   singlet oxygen 1011
Delaware River 415,494,713,1081,1100
Delaware River Estuary 1300
delayed effects 533
DEN 439

INDEX 1553

Denny, Susan 1341
deoxyribose 578
detection, sensitivity 1069
detoxification 1421
Detroit 146
di-Nor-octylphthalate 887
diazomethane 860,1116,1316
dibenzanthracene 419,1395
dibromamine 748,765,987
    decomposition 996
    formation 770
dibromoacetonitrile 222
dibromochloromethane 113,237,908,
    1127,1261,1264,1305,1362,1373
dibromoiodomethane 929
dibromomethane 237
dibromopropane 1359
dibromotyrosine 238
dibutylnitrosoamine 235
dibutylphthalate 887,1521
dichloramine 10,52,521,524,706,826,
    831,957
    decomposition 706
    decomposition, monochloramine
        effect 708
    formation 1472
dichloroacetaldehyde 166
dichloroacetic acid 26,113,164,166,169,
    843,866,946,1115,1121,1165,1229,
    1254
dichloroacetone 26,237,288,704
dichloroacetonitrile 113,164,166,169,
    222,229,241,957,1178,1182,1229
dichloroacetylacetone 1249
dichloroanisole 1182
dichlorobenzene 92,200,1127,1375,
    1395,1521
dichlorobenzidine 1395
dichlorobenzophenone 1101
dichlorobromomethane 1305,1392,1397
    See also bromodichloromethane
dichlorobutanedioic acid 846
dichlorobutanone 167
dichlorobutenal 166,169
dichlorocyclopentenedione 837
dichlorodihydroxybenzene 1322
dichloroethane 92,188,255,1127,1305,
    1362,1390,1519
dichloroethanoic acid 846
dichloroethylene 92,1123,1127,1362,
    1390,1517,1518
dichlorohydroxybenzaldehyde 237
dichlorohydroxybenzoic acid 1322
dichloroiodomethane 927
dichloroisocyanurate 1259
dichloromethane 255,1507
dichloromethoxybenzenedicarboxylic
    acid 866
dichloromethoxybenzoic acid 866,867
dichlorophenol 24,113,121,307,332,
    1182,1315,1322,1395,1513,1523
dichlorophenoxyacetic acid (2,4-D)
    1513
dichlorophenylacetic acid 848
dichlorophthalic acid 1124,1125
dichloropropanal 166
dichloropropane 1127,1305,1362,1390,
    1519
dichloropropanenitrile 166
dichloropropanoic acid 866
dichloropropanone 164,167,222
dichloropropenal 166
dichloropropene 1127,1362,1518
dichloropropenenitrile 166
dichloropropenoic acid 846
dichloropropylene 1390
dichlorosuccinic acid 26
dichlorosulfone 29
dichlorotyrosine 27
dichlorouracil 24
dieldrin 1395
Diels-Alder type reactions 1016
diet, polyunsaturated fat 235
diethanolamine 570
diethyl ether 1100
diethyl-p-phenylenediamine method,
    See DPD
diethylamine 570,1297
diethylhexylphthalate 887
diethylnitrosamine 235,439
diethylphthalate 887,1101,1375,1395,
    1521
diethylsulfide 1297
differential pulse polarography 757
diahaloacetonitriles 26,94,229,1165
    See also haloacetonitrile
dihydroconiferyl alcohol 29
dihydroxyaromatic compounds 25
dihydroxybenzene 837,843,1232,1314
dihydroxybenzoic acids 835,843
dihydroxybiphenyl 1322
dihydroxydichlorobenzoquinone 837
diiodotyrosine 238
diketone groups 843,1249
diluter system 511
dilution 66,78
    demand-enhanced 1464
    rate 615

dimethoxybenzoquinone 838
dimethoxybiphenyl 1325
dimethoxydiphenyl ether 1322
dimethyl malonate 288,291
dimethylamine 570,988,1297
dimethylamino-naphthalenesulfinic acid 175
dimethylbenz(a)anthracene 256,402
dimethylbromamine 999
   reaction with hypochlorous acid, 1004
   spectrum 1003
dimethylchloramine 999
   spectrum 1003
dimethylglycine 1297
dimethylhydrazine 259
dimethylnitrosamine 261
dimethyloxyhydroxybiphenyl 1322
dimethylphenol 1395,1523
dimethylphthalate 1375,1395,1521
dimethylsulfate 231
dimethylsulfoxide 1297,1343,1412
dinitrocresol 1395,1523
dinitrophenol 1395,1523
dinitrotoluene 1395,1521
dioctyl adipate 887
diphenylamine 588
diphenylhydrazine 1395,1522
diptera 486
direct filtration 1185
direct-acting mutagens 113
disease 463
   heart 123
   of infancy 123
   transmission 5
disease-free water 3
disinfectants 9,112,281,295,1269,1515
   comparative efficacy 10,1269
   ferrate 1285
   field performance 555
   health risks 122
   mutagenic potential 209
   new 1269
   performance models 555
   regulatory options 112
   stability 1269
   See also alternatives; bromine; bromine chloride; choramines; chlorine; ferrate; iodine; monochloramine; ozone; UV radiation
disinfection 5,85,91,92,96,99,161,541, 555,587,603,667,681,691,1011,1115, 1216,1269,1411

by-products 91,112,119
chlorine dioxide 603
E. coli 587
effect of N-organics 587
efficiency 571,595,648
iodine 935
ozone 691
pH effect 695
potential 635
rate constants 1438
reactor 603
resistance 603
schemes 589
disproportionation 763,1472
   pH effect 769
   phosphate effect 769
dissociation constant 776
dissociation reactions 705
dissolved organic carbon 939,1073
   See also organic carbon; total organic carbon
dissolved organic nitrogen 756
oxygen 514
distribution, molecular weight 1191
dithio-bis(nitrobenzoic acid) 582
diurnal variation 1035
divinyl benzene 1101
DNA 261,578,583,1270
   adducts 230
   binding 410
   damage 229
   repair 410
   strand breaks 231
   synthesis 297,355
   turnover 297
   See also nucleic acids
dodecanoic 864
dose response 108,275
Dotson, Douglas, 713
double-exchange reactions 706
DPD (N,N-diethyl-p-phenylenediamine) methods 21,177, 513,522,533,556,587,635,652,738, 876,976,1314,1461,1516
   ferrous ammonium sulfate titrimetric 1342
   Steadifac 587
   titration 177,513,522,635,738,876, 976,1231
   See also oxidant analysis
Dreux, M. 1065
drinking water 3,5,26,91,99,111,145, 161,187,207,238,251,265,275,281, 295,331,345,355,365,555,575,587,

635,651,783,807,821,859,875,907,
951,999,1055,1099,1115,1130,1137,
1165,1185,1201,1215,1229,1299,1411
 chlorination 229
 chlorination, hypothyroidism 369
 contaminants 91
 disinfection 111
 disinfection epidemiology 133
 microbiological contamination 99
 organic amine compounds 570
 regulation limit 1397
 risk 201
 source 145
 survey, trichloroacetic acid 1119
 treatment 20,1215,1313,1329
 See also water, water treatment plant
Dubuque, IA 13
Dudley, Brian 587
Duguet, J. P. 1165,1201
duodenum 286
Durham, NC 1118
dysentery 7

E. Coli, *See Escherichia coli*
echovirus 6,108,683
ecological niche 19
Edisto River 909
effluent limitations 1489
electric power plants 28,39,63,493,723,
737,1039,1081,1357
 Amos 1476
 Beaver Valley Power Station 1358
 Columbia River 1358
 Dunkirk 737
 Kingston (TN) Steam Plant 1449
 Lake Michigan 1358
 Le Havre 737
 Masvlakte 1428
 Mercer Generating Station 1081
 Millstone Nuclear Power Station 1371
 Mohave Power Generation Station 1400
 Pacific Gas and Electric Co. 65
 Palisades Nuclear Generating Plant 1358
 Paluel Nuclear Plant 737
 Potrero 65
 Rainer, OR 1358
 Redondo Generating Station 1357, 1371
 Shippingport, PA 1358
 Sizewell nuclear power station 1461

 South Haven, MI 1358
 Southern California Edison Company 1371
 Trojan Nuclear Plant 1358
electrochemical detector 1091
electrochemical monitoring 1081
electrode 1078,1081,1093,
 maintenance 1078
 rotating ring disc 1081
 sensitivity 1093
electrolytic conductivity detector 966
electron paramagnetic resonance 410
Emerald Isle, NC 1118
encephalitis 6
endoplasmic reticulum 232
endosulfan sulfate 1395
endpoints, toxicological 30
endrin 1395
energetics 701
energy production 40
English sole (*Parophrys vetulus*) 399
enlargement of the heart 368
Entamoeba histolytica 7,100
Entamoeba invadens 1271
enteric bacteria 5,563,615,658,990
enteric viruses 6,8,104,681,692
enterobacter 563
Enterobacter agglomerans 658
Enterobacter cloacae 658
Enterobius vermicularis 7
enteroviruses 681
environmental carcinogens 251,377,415
environmental hazard 51,63
environmental impact 63,493
environmental impact analyses 66,80
environmental significance 713
enzyme-linked immunosorben assay 309
EPA guidelines 1447
EPA priority pollutants 1378,1381
ephemeroptera 486
epidemiologic associations, validity 133
epidemiologic data, interpretation 138, 140
epidemiologic studies 122,126,133
 association 137
 case-control study 153,145
 confounding bias 134
 ecologic studies 145,153
 observation bias 134
 population characteristics 155
 random misclassification 136
 selection bias 134

selection of subjects 135
types 134
epidermal lesions 415
epilithic periphyton 481
epizootic 375
equilibrium constant 987,1000,1002
equilibrium data analysis 776
equilibrium effects 775
Erickson, John 1285
Escherichia coli 6,99,101,558,587,603, 615,620,635,681,692,775
esters 1524
estuarine organisms 493
estuarine sediments 1411
estuarine waters 1033
Etang de Gruere Lake 1015
ethanedioic acid 864
ether 1313,1524
ethyl acetoacetate 1250
ethyl alcohol 1297
ethyl carbamate 225
ethyl pyruvate 704
ethylaminoacetate 570
ethylbenzene 1390,1518
ethylene blue 1015
ethylene glycol 1297
ethylhexylphthalate 887,1521
ethylnitrosourea 233,311
ethylurea 310
evaluation, conceptual approach 73
evolution of amino acids 821
exchange reactions 706
excited state oxygen 1011,1041
excretion 285
Exon, Jerry H. 307
exploratory behavior 355
exposure 56,73,81
extractable organic chlorine 190
extractable organohalogen (EOX) 1139
extraction 417,1139,1320,1358,1373, 1412,1507

FACTS (free available chlorine test with syringaldazine) 21,587,652
See also oxidant analysis
Faeder, Edward J. 1371
false positive tests 587
far-field model 1497
fast chlorine consumption 731
fast neutron activation analysis 282
fathead minnow (*Pimpehales promelas*) 429
fatty acids 26,843,886

Fava, James A. 73,493,553
Fayetteville, NC 1118
fecal coliform 88,651,691,990
fecal production 466
fecal streptococcus 691
feeding studies, human 107
Feldstein, Caren M. 1259
ferrate 94,1285
ferrate molar absorptivity 1293
ferric oxide 1285
Ferris, Joseph S. 73
ferroin 1286
ferrous iron 1470
fertility 295
fetotoxicity 113
fibrosarcomas 432
field investigations 565
filtration 25,92,101,821,1100,1138, 1188,1210
Fiquet, Jean Marie 737
first-order kinetics 745,764,977,1287
rate constants 977,1287
fish 30,53,375,481,505
bottom-feeding 399,415
carcinogens 415
consumption 1377
cytotoxic response 396
eggs 505
hepatocarcinogenesis assays 377
mobility 411
neoplasia 375
pollution effects 399
tumor induction 429,457
tumors 375,451
flavines 1021
Flavobacterium 557,563
flocculation 101,267,821,954,1100, 1138,1185,1302
flocculation-sedimentation 821
Florida 14,104
Florida water 1406
flotation 1210
flour, bleaching 30
fluorescent lights 757
flow injection technique 1091
flow-through exposure system 430
fluoranthene 419,1375,1395,1520
fluorene 419,1395,1520
fluorenylacetamide 235
fluorescence 1034
fluorescence decay technique 1034
food industry 29
food-chain transfer 404
formaldehyde 1297,1343

formic acid 1297
fouling control 40,1033,1039,1459
    See also biofouling; macrofouling
Fox, Helen 415
France 104
free available chlorine (FAC), See also chlorine
free available oxidant 1427
    See also oxidant
free bromine 737
    See also bromine
free chlorine 40,52,67,494,524,533,555, 587,652,723,882,1091
    reactions 1033
    See also chlorine
free radicals 1313
free residual chlorine 1476
    See also chlorine
freeze dryer 1343
freeze-drying, See lyophilization
freshwater "zebra" mussel, *Dreissena polymorpha* 1425
freshwater 22,1489
    See also natural waters, water
friction factor change 1441
fulvic acid 26,161,281,345,702,724,843, 859,939,1115,1165,1185,1230,1249, 1315,1399,1505
    aquatic 1406
    Black Mesa coal 1406
    ozonation 1252
    Wyodak coal 1406
    See also humic acids
fumaric acid 281
functional groups 1313
fungi 1271
furfural 854
furfuryl alcohol 1014

GAC operational characteristics 1310
    See also granular-activated carbon
Galan, Leo de 843
gall bladder anomalies 121
gas chromatography 845,966,1056, 1099,1100,1103,1123,1166,1188, 1202,1301,1358,1385,1508
    purge and trap 757
gas chromatography/mass spectrometry 161,845,925,966,1100, 1115,1137,1167,1202,1230,1316, 1385,1508,1515
gas-phase photochemistry 1041
gastropoda 486
gastroenteritis 6,103,123
gastrointestinal tract 111
Gautney, Stephen P. 1447
gel electrophoresis 626
gel permeation chromatography 1166, 1188,1202,1507
gelatin 591
general acid catalysis 1294
generalized linear interactive modeling 1343
generally available treatment technologies 92
genetic toxicology 334
genotoxic compounds 265
genotoxicity 113,265
    assays 267
Georgetown, TX 12
germicidal behavior 587
Giabbai, Maurizio 923
Giardia 93,100,104
Giardia lamblia 7,101,108,1271
giardiasis 7
Gibert, M. 1065
Gibs, J. 1099
Gidel, Louis T. 1041
gill development 533
Glatt River 1015
glucose-6-phosphate dehydrogenase activity 297
glutamic acid 238,583,591,831,988,991
glutamine 583,619,823
glutathione 215,355
glycerol triacetate 887
glycine 120,238,567,570,583,591,823, 991,1297
glycine ethyl ester 991
glycogen reserves 477
glycolaldehyde 1297
glycolic acid 1297
glycylglycine 723,797
glycylglycylglycine 570,942
glyoxal 1297
golden shiner (*Notemigonus crysoleucas*) 452
Gould, Joseph P. 923
Gourlay, Margaret L. 251
gram-negative fermentative bacilli 556
gram-negative nonfermentative bacilli 556
gram-positive cocci 556
granular-activated carbon 25,209,870, 951,1138,1167,1210,1213,1217,1299, 1329
    removal of mutagenic activity 1336
    treatment costs 1299

grass shrimp 503
Grasse River 909
Gray, E. T., Jr. 763,797
green algae
   Anabaena flos-aquae 875
      Chlorella pyrenoidosa
      Chlorella vulgaris 875
green sunfish (*Lepomis cyanellus*) 452
Greifensee Lake 1015
Grizzle, John M. 451
groundwater 92,103,125,154,187,281,
   923,941,1118,1123
   chlorination 154
Grove, Robert S. 1371
growth rate 605
   studies 534
   temperature 605
guaiacol 237
guanine 230,578,591
guanosine 578
guanosine monophosphate (GMP) 578
Gulf killifish (*Fundulus grandis*) 429
Gulf of Mexico 1033
guppy (*Poecilia reticulata*) 429

Haag, Werner R. 999,1011
Haas, Charles N. 667,775
halamines 23,50,52
   reactions, reversibility 1004
   See also bromamines; chloramines
hallate formation 22
haloacetone carcinogenic activity 221
haloacetonitrile 120,221,229,295
   carcinogenic activity 221
   metabolism 229
   See also chloracetonitriles;
      bromoacetonitriles
haloacid derivatives 94
halocarbons, production kinetics 759
   See trihalomethanes
haloform reaction 702
haloforms, *See* bromoform;
   chloroform; trihalomethanes;
   specific trihalomethanes
halogen 930
halogen incorporation coefficients 924
halogen radicals 1048
halogenated aldehydes 113,120
halogenated ketones 113,120
halogenated methylamines 965
halogenated octylphenol polyethoxylate
   residues 1505
halogenated phenols 1374
halogenated phenols 28,121,1357,1374

halomethanes, *See* chloroform;
   bromoform; trihalomethanes;
   specific trihalomethanes
haloorganics 51,56
   volatile 48,190
   See also bromoorganics,
      chloroorganics
Hampton, James A. 439
Harakeh, M. S. 615,681
hardness 1188
Hartge, Patricia 145
hatching study 534
   success 537
Hauchman, F. S. 619
Haw River 1118
Hawkins, William E. 429
Hayes, Johnnie R. 331
hazard assessment 75
hazardous waste sites 1123
health effects 73,111,183,295,307,345,
   1137
health risks 8,78,111,122,126,281,
   345122
heart disease 123
heat transfer 1447
heat treatment 1425
heavy metals 1285
helminths 7
Helz, George R. 713,1033,1081
hemangiosarcoma 393
hematocrit 296
hematology 308,321,334,349
hematuria 349
hemoglobin 296
hemolytic anemia 111,113,114,117,345
hepatic lesions 399,407,415
hepatic neoplasms 399,402,415,430
hepatitis 6,103
hepatocarcinogenesis, inhibitor 233
hepatocellular carcinoma 386,393,401
hepatotoxic 113
hepatotoxins 365
heptachlor 188,1395
heptachlor epoxide 1395
heptanoic acid 864
herbicides 821
heterocyclic acids 846
heterocyclic bases 588
heterocyclic ring compounds 588,713,
   846,1021
hexachloroacetone 237
hexachlorobenzene 1395,1521
hexachlorobiphenyl 1395
hexachlorobutadiene 1395,1522

hexachlorocyclopentadiene 167,1395, 1522
hexachloroethane 167,1395,1522
hexachlorophenol 270
hexacosanoic acid 846
hexadecanoic acid 864
hexane 1331
hexanoic acid 864
high-molecular-weight compounds 890
high-performance liquid chromatography, See HPLC
Hinton, David E. 439
histidine 581,619,812
Ho, Soon Lin 541
Hodgins, Harold O. 399
Hoff, John C. 99,603,615
Hogchoker 503
Hoigne, Jurg 1011
holding ponds 1061
Holdsworth, G. 365
Holston River 726
hookworm disease 7
Hoover, Robert 145
horse serum 1270
Howell, G. Dean 807
HPLC 177,270,416,576,876,1066,1166, 1330,1508
Hrubec, J. 187
Huber, Calvin O. 1073,1091
Hudson River 102
human population 303
human risk 1377
human studies 296
humic acid 23,26,119,161,191,207,238, 265,266,273,281,345,702,717,723, 809,821,843,859,895,907,939,1013, 1115,1165,1185,1201,1270,1399, 1406,1505
   chlorination, genotoxic properties 210
   commercial 1406
   elemental analysis 718
   peat 1406
   pyrolysis products 854
   See also fulvic acids
humic materials, See humic acid, fulvic acid
humic substances 1229,1314
   See also humic acid; fulvic acid
humoral immunity 309
Hunter, Joseph V. 1515
hydrazines 193
hydrochloric acid 784
hydrogen peroxide 94,1016,1033,1313

hydrogen sulfide 1470
hydroids 42,48
hydrolysis 770,851
hydroquinone 1322
hydroxybenzoic acid 281,835,1055,1322
hydroxybenzoquinone 837
hydroxycinnamic acids 835
hydroxydimethoxybenzoic acid 838
hydroxydiphenyl ether 1322
hydroxyfluorene 406
hydroxylapatite 231
hydroxylated PCBs 1313
   mechanism for formation 1321
hydroxymethoxybenzoic acid 837
hydroxymethoxybiphenyl 1322
hydroxymethoxycinnamic acid 838
hydroxymethoxyphenyl-chloroethylene 838
hydroxymethoxyphenyl-ethylene 838
hydroxyproline 366
hydroxypropanoic acid 864
Hymenolepsis nana 7
hyperplasia 114,118,399
hyperplasia, foci of cellular alteration 399
hyperplastic lesions 454
hypertension 365
hypobromite 737,1016
   See also bromine
hyprobromous acid 737,771,907,985, 1039
   See also bromine
hypochlorite 10,175,296,603,681,775, 784,885,965,1016,1055,1091
   disproportionating 786
   ion-pair 775
   photolysis 1057
   See also chlorine, hypochlorous acid
hypochlorous acid 10,115,175,681,784, 885,907,975,985,1067
   See also chlorine; hypochlorite
hypohalous acid 985

idiopathic lesions 399
ileum 286
Ilwaco Reservoir 909
iminodiacetic acid 1297
immune response 312,334,337
immunodeficiency disorders 307
immunologic assays 309
inactivation 555
   assay 556
   experiments 682

kinetics 559,571
variability 559
incubation temperature 566
indenopyrene 188,419,1395
India 104
indicator bacteria 555,603,694
indigo red 620
indoleacetic acid 831
indoles 843,846
indoxyl 620
induction period 1001
industrial pollution 265
industrial processes 20
infection density 466
infectious hepatitis, See hepatitis
infectious virions 619
infective dose 107
infertility 295
initiators 223
Inland silverside (*Menidia beryllina*) 429
inorganic amines, See amines
inorganic chlorination products, See chlorination products
instrument 1073
insulin 260
in vivo chlorination of piperidine 181
in vivo formation of N-chloroglycine 181
iodate 22
iodide 923
iodine 10,94
   disinfection 935
   incorporation coefficients 929
   titrant 1073
iodoform 925
iodometric titration 513,1091,1516
   See oxidant analysis
ion association theory 778
ion pairs 775,1065
   chromatography 1065
   dissociation constant 777
ionic strength 791
ionic substitution reactions 1248
ionization reaction 704
Iowa 146
iron 714,812,1215,1220
iron chlorosulfate 1210
Isaac, Russell A. 985
Isacovics, Bonnie 237
isatin 620
isoamyl butyrate 887
Isochrysis galbana 512
isoleucine 583,823

isophorone 1395,1522
isopropyl alcohol 1297
Israel 104

Jacangelo, J. G. 575
James River 510,541,909,1118
Japanese medaka (*Oryzias latipes*) 429, 440
Jaworske, Donald A. 1081
Jenner, Henk A. 1425
Jensen, James N. 939
Jersey City 99
Jessen, Brian 1021
Johnson, C. Ian 237
Johnson, J. Donald 723,939,1115
Jolley, Robert L. 19

Kaganowicz, Delia M. 895
Kaw Reservoir 909
Kearney, Terrence J. 755,965
Kelter, Paul B. 1285
ketoglutaric 281
ketones 167,1229,1524
Khan, Parweiz A. 1185
Khater, K. M. 667
kidney 286
   tumors 229
Kieber, Robert J. 1033
Killen, AL 533
Kim, Jong-Soo 923
kinetic models 1459
kinetic product 1004
kinetics 557,701,744,755,763,798,884, 968,975,1000,1026,1038,1238,1287, 1437,1471
King, Paul H. 907
Klebisiella oxytoca 658
Klebsiella pneumonia 610,615,658
Kleine Emme River 1015
Koller, Loren D. 307
Kool, H. J. 187
Kopfler, F. C. 161
Kott, Yehuda 691
Kovats indices 1102
Krahn, Margaret M. 399
Kravitz, Robert 807
Kraybill, Herman F. 375
Kruithof, J. C. 1137
Kuderna-Danish concentration 860, 877,1102,1116,1139,1316

laboratory models 555
Laborde, Susan M. 533
lactone 1313

Lafosse, M. 1065
lag phase 670
Lake Baldegg 1013
Lake Carl Blackwell 1188
Lake Erie 415
Lake McMurtry 1188
Lake Michie 1118
Lake Okeechobee 1222
Lake Ontario 266,272
Lake Wheeler 1118
Larson, Richard A. 1313
larvae
    survival 523
    growth 509
    stages 505
Lawler, John P. 1489
LC_{50} values 51,216,496
    See also toxicity
Le Cloirec, C. 821
Leadenwah Creek 464
Lee, Yun-She 1515
Leer, Ed W. B. de 843
Legionella pneumophila 6,557,603,605
    sensitivity to chlorine dioxide 605
Legionellosis 5
leiomyosarcomas 432
lethal concentration or dose (median), See LC_{50}
lethal factor 648
leucine 570,581,619
leuco crystal violet method 620
    See also oxidant analysis
leukemia 121
Levin, Lynn I. 145
levoglucosan 854
light-induced reactions 1059
lignin 237,843
Lin, Edith L. C. 229
Lin, Sechoing 835
lindane 188
Lindholm, Paul A. 251
lipids 717
lipophilicity 30
liquid chromatography 1065
    See also HPLC
liquid-liquid extraction (LLE) 162,808, 925,1123
lithium 775
lithium hypochlorite, dissociation 775
litter size 312
Little Patuxent River 713
Liu, Paul T. 251
liver 286,439
    carcinogen assays 377

homogenate (S9) 208,1140, See also S9
lesions 395,443
lesions, chemically induced 395
microsomal activities 334
microsomal fraction (S9) 1413
neoplastic development 377
tumor 234
tumor biology 447
tumor model 439
locomotor activity 355,359
Loper, John C. 1329
Louisiana 140
Lower Clear Creek 817
Lowther, E. D. 651
lung 286
    adenomas 225
lymphomas 121
lyophilization 1400
lysine 583,812

macrofouling 42,47,64,1425,1459
    See also biofouling
macroinvertebrates 481,505
    communities 481
    fouling, See biofouling
macrophage function 309
macroreticular resin, See XAD
malic acid 281
malignant neoplasms 123
Malins, Donald C. 399
Mallevialle, J. 1165,1201
Malley, James P., Jr. 587
Maloney, Stephen W. 1299
malonic acid 288,292
malonic ester 704
MAM 439
mammalian test systems, See mutagenicity
management 3,73
manganese 714,1215,1220,1470
Manning, C. Steven 429
March, Patrick A. 1447
marine organisms 463
Martin, G. 821
Mason, Thomas J. 145
mass spectrometry 860,1316,1366, 1510
    See also gas chromatography/mass spectrometry
Matin, Abdul 603,615
Mattice, Jack S. 39
maximum contaminant levels 91,112

mayfly (Hexagenia bilineata) 533
    hatching success 533
    life stages 533
Mazina, Kathryn E. 175
McCabe, L. J. 111
McCain, Bruce B. 399
McCauley P. 365
McCuskey, Patricia A. 439
McLaughlin, C. William 1285
Means, Jay C. 1411
mechanical cleaning 1425
medaka fish, Japanese 395,439
    tumor model 439
median survival times 496
megalocytic hepatosis 399
Meier, John R. 161,207
melanomas 415
Meldrim, John W. 493
membrane electrode 587
membrane filtration technique 691
Menchville, Virginia 542
meningitis 6
metabolic activation 261
metathetical reactions 706
Metcalfe, Christopher D. 265
methacrylic acid 1101
methanol 1100,1331
methemoglobin 299
methemoglobinemia 117,345
methionine 120,581,591,823
methoxybenzene 1325
methoxybenzoic acid 865
methoxydiphenyl ether 1322
methoxyphenol 838
methyl alcohol 1297
methyl alpha-phenyllactate 887
methyl benzoate 887,1101,1117
methyl bromide 1390
methyl chloride 1390
methyl decanoate 887
methyl dehydroabietate 887
methyl dodecanoate 887
methyl formylbenzoate 887
methyl hexadecanoate 887
methyl hexanoate 887
methylhydroxydichlorobenzoate 1317
methyl isotetradecanoate 887
methyl octadecanoate 887
methyl trichloroacetate 1117
methyl trichlorohydroxybenzoate 1317
methyl undecanoate 887
methyl-t-butyl ether 966
methylamine 570,724,965,1297
methylanthracene 419

methylazoxymethanol acetate 429,439
methylbenzenedicarboxylic acid 865
methylbenzylamine 895
methylchrysene 402
methylene blue 1021
methylene chloride 1115,1127,1331,
    1362,1373,1390,1519
methylethyl terephthalate 887
methylfurancarboxylic acid 865
methylfurandicarboxylic acid 846,865
methylfurane 854
methylketones 843
methylmethane sulfonate 231
methylnaphthalene 1101
methylphenanthrene 419
methylstyrene 1101
Meuse River 188
Mexico 104
mice 395
    A/J 225,297
    B6C3F1 116,121,208,221,229,235
    C57L/J 297
    CD-1 208,221,233
    Sencar 221,229
    Swiss IcR 417
microbial contaminants 91,611
microbial inactivation 100
microbiology 92
microcomputer 1091
microcoulometric titration 1124,1140
microcoulometry 942
Microcystis aeruginosa 7
microflagellates 512
microflocculation 1215
microfouling 42
    control 46,47
    See also biofouling
microorganisms 1272
micropollutants 1016,1139
    degradation 1016
microsomal assay, See mutagenicity
Middaugh, Douglas P. 463
Minear, Roger A. 1399
minimal infectious dose 8
Minnesota 88
Miskimen, Thomas A. 63
Mississippi River 13,102,140,209,282
mixed function oxidase 232,324,451
mixing zone 1491
mobile organisms 505
mobility, species 505
model
    biofilm accumulation rate 1436
    chlorine dissipation 1459

circulating water system 1473
dilution 1491
DKHPLM 1495
LMS3D 1496
MOBEN 1495
modeling 66,919,991,1049,1407, 1435,1452,1469,1493
PDS 1495
PDSM 1496
photochemical decomposition 1049
predictive 1435,1490
PSY 1496
simulation runs 1479
TVA 1490,1496
validation 1459
molecular weight fractionation 162
monitor
   biofouling 1425
   mussel 1426
monitoring 1123
monobromamine 763,765,1006
   disproportionation 763,768,987
   hydrolyis 770
monobromophenols 1363
monochloramine 10,64,65,70,113,207, 296,345,521,524,575,667,706,759, 826,831,939,957,975,993,999,1091
   anodic voltammetric determination 1091
   demand 668
   disappearance 975–983
   formation 1472
   hydrolysis 568,980
   mode of action 575
   reaction with bromide 999
   reaction with cysteine 583
   reaction with cystine 583
   reaction with fulvic acid 939
   reaction with histidine 583
   reaction with leucine 583
   reaction with methionine 583
   reaction with sulfhydryl group 581
   reaction with threonine 583
   reaction with tryptophan 583
   reaction with tyrosine 583
   See also chloramines
monochloroacetone 288
monochlorobenzoquinone 1322
monochloroglycine 1096
monochloroisocyanurate 1259
monochloromaleic acid 25
monochloromethyl amine 1096
monochlorophenol 1322
monochlorotyrosine 27

Moore, Cynthia A. 1041
Moore, Leown A. 859
Moraxella 563
morpholine 570
Morris, J. Carrell 701,985
mortality 466,537
   rate 1431
Moss, Robert D. 1447
most-probable-number method 691
motile sperm 300
mouse skin initiation/promotion test 190,199
MSX epidemic 509
mucochloric acid 237
Mülheim process 1217
Mummichog 503
municipal water systems 555
Murray, Sylvia A. 533
mussel 42,48
   fouling, *See* biofouling
   mortality rate 1431
   Mytilus edulis 1425
   spat 1425
mutagenesis 113,115
mutagenic by-products 124
mutagenic compounds 3,161,188,207
mutagenic, *See also* mutagens
mutagenicity 79,111,115,119,187,207, 210,223,237,265,1137,1329,1341, 1411
   assay 189,1331
   chlorinated humic material 168,207
   chlorine dioxide effect 193
   chlorine effect 190
   ozone effect 193
   UV effect 193
mutagens 251,365,457,1329,1411
   nucleophile reaction 1351
   production by disinfection with chlorine 247
Mycobacteria 557
Mycobacterium tuberculosis 6
Myers, Mark S. 399
myocardial hypertrophy 113
myocardial structure 365
Mytilus edulis 478

N,N-dichloroglycylglycine 724
   decay 726
N,N-diethyl-p-phenylenediamine (DPD), *See* DPD
N-(chlorophenyl)chlorobenzamide 1101
N-acetylimidazole 632
N-chloramines, *See* chloramines

N-chloroamino acids 587
N-chloromethylamine 570
N-chloroglycine 52,175
N-chloroglycylglycine 724
N-chloropiperidine 175
N-halamines 94
   See also bromamines; chloramines; halamines
N-hydroxy-acetylaminofluorene 324
N-methyl-N'-nitro-N-nitrosoguanidine 429
N-methyliminodiacetic acid 1297
N-nitrosodiethylamine 377,397
N-organic compounds 587
Naegleria fowleri 7
Najar, B. 1099
naphthalene 407,1101,1182,1375,1395,1520
National Pollutant Discharge Elimination Systems (NPDES) 39,86,1490,1515
   permit requirements 144
National Primary Drinking Water Regulations 91
National Toxicology Program 116
   Bioassay 95
natural populations, sensitivity 617
natural waters 713,723,939,1011,1016,1033,1043
   hydrogen peroxide 1033
   iodide levels 923
   organic nitrogen 807
near-field model 1497
Necator americanus 7
necropsy 298,334
Neitzel, Duane A. 1357
nematodes (roundworms) 7
neonatal hypothyroidism 361
neopentyl alcohol 1297
neoplasms 399,451,453
   cholangiocellular 399
   hepatic 399,402,407,415,430
   hepatocellular 399
   mesenchymal 399
   See also cancer
neoplastic disease 415
nephropathy 349
Netherlands 28,187,1425
Neukrug, Howard M. 1299
neurobehavioral development 355
neurotoxic 113
New Bern, NC 1118
New Haven Harbor 521
New Haven, CT 521

New Jersey 14,146
New Mexico 146
New Orleans 146
New York City 16
Niagara River 399,415
Nice, France 1215
Nieuwegein 1138
nitrate 94,513,543,756,975,1470
nitric acid 1201,1286
nitrification 975
nitriles 166,821
   formation 827
nitrilotriacetic acid 1297
nitrile 24,94,310,513,543,714,756,951,975,1201,1297
   oxidation 975
nitro aromatics 193
nitrobenzene 1055,1395,1521
   sulfonic acid 1066,1070
nitrobenzyl-pyridine 229
nitrofluorene 212
nitrogen gas 975
nitrogen oxides 1201
nitrogen tribromide, See tribromamine
nitrogen trichloride, See trichloramine
nitrogenous organic compounds 567,587
nitrophenol 821,1055,1201,1395,1523
nitroso compounds 193
nitrosodimethylamine 1395
nitrosodiphenylamine 1395,1522,1524
nitrosodipropylamine 1395
nonamide 1101
nonanedioic acid 865
nonanoic acid 864
nonionic detergents 1505
nonpurgeable organohalide 875,939
nonvolatile hydrophilic haloorganics 48
nonvolatile lipophilic haloorganics 48
nonvolatile mutagens 1336
nonvolatile total organic carbon (NVTOC) 908,1403
nonvolatile total organic halide (NVTOX) 1115,1165
nonylamine 1066
Noordsij, A. 1137
nor-butylisobutyrate 887
North Carolina 153
North Edisto River Estuary 464
Norwalk viruses 6
Norwood, Daniel L. 1115
Noss, C. I. 619
Nowell, Lisa H. 1055

NPDES, *See* National Pollutant Discharge Elimination System
nuclear magnetic resonance spectroscopy 845
nuclear power plants 66
   See also electric power plant
nuclear reactors, *See* electric power plant; Oak Ridge High Flux Isotope Reactor
nucleic acids 575,579,619,688
   consumption of monochloramine 576
   See also DNA, RNA
nucleophilic carbanion 702
nucleophilic carbon centers 704
nucleophilic terminal amino groups 723
nutrient concentrations 545
nutrient-limited growth 603
nymphs
   growth 533
   survival 534

Oak Ridge High Flux Isotope Reactor 1394
ocean thermal energy conversion (OTEC) plants 755
ocean
   as energy source 755
   water 1033
ocean water, *See also* seawater; saline water
octadecanoic acid 864
octane/water partition coefficient 52
octanedioic acid 865
octanoic acid 864
octylamine 1066
octylphenol polyethoxylate 1507
octylphenol polyethoxylate residues, mass spectra 1510
octylphthalate 1395
ocular lesions 113,121
Ohio River 68,200,218,1260
oil refinery effluents 265
oilfield brines 923
Oklahoma brines 923
Okrend, H. 635
oleamide 888
olefins 1021
oligochaeta 486
oligotrophic seawater 965
Olivieri, Vincent P. 5,575,619,651
Olson, Betty H. 555
once-through chlorination 724
oncogenesis 307

oncogenicity 79
operation and maintenance costs 1299
operator skill 1098
Orange County, California 555
organic amines 713,723,1039,1096
   See also amines
organic bromamines 991
   decomposition 996
   See also bromamines
organic carbon 713,1048
   See also dissolved organic carbon; total organic carbon
organic chloramines 113,725,957,999, 1269
   formation 1472
   See also chloramines
organic concentrates 189,209
organic halides 27,876,1229
   formation 1229
   See also TOX
organic halogen, total 876
organic loading, toxic 1422
organic matter 713
organic nitrogen 723,807
organic precursors 96,783
   removal 96
   See also precursors
organic priority pollutants 1381
   See also EPA
organic sulfur 713
organics
broad spectrum analysis 1099
   oxidation 1011
organohalide 1137
organohalogen 1139
   surrogates 1156
   See also halogenated; organic halogen
organoleptic properties 331
organonitrogen 1139
organophosphorus 1139
organosulfur 1139
oriniter 441
orthophosphate 1286
Osborne, Lewis L. 481
Osborne, T. R. 365
Osborne-Mendel rat 119
Ospital, Jean 1371
Oswald, Edward O. 463
OTEC 755
Overstreet, Robin M. 429
Oxalic acid 1297
oxazolone 309
oxidant 1285

analysis, SNORT 587
  starch-iodine titrimetric method 22
  See also amperometric; anodic; DPD; FACTS; iodometric; leuco-crystal violet; SNORT; starch-iodine; syringaldazine
decomposition 1041,1081
demand 1081
  See also chlorine demand
monitoring, rotating disc electrode 1087
photochemistry 1041
  See also bromamines; bromine chloride; chloramines; chlorine; chlorine dioxide; hydrogen peroxide; ozone; singlet oxygen
oxidation 30,187,713,851,975,1023, 1030
  kinetics 1030
  of solutes 1297
  organics 1013
  polymer-bound dye 1028
  singlet oxygen 1011
  treatments 187
  water 1285
oxidative decarboxylation 1247
oxidative purification of water 1021
oxidizing agent 1285,1411
oxo-chlorine compounds 783
oxygen
evolution 1286
  excited state 1011
oyster, American eastern (*Crassostrea virginica*) 463,521
  chlorination exposure 463
  chlorine-produced oxidant exposed 478
  host-parasite interactions 478
  larvae 510,521
  metamorphosis 509
  set 526
  spatfall 509
  spawn 526
  survival 463
ozonated waters 1201
ozonation 122,191,691,821,932,1021, 1100,1138,1162,1168,1211,1229, 1301,1515
  by-products 122
  costs 1226
  effectiveness 1525
  fulvic acid 1252
  multiple stage 1217

ozone 10,24,77,78,89,95,107,111,114, 118,187,209,282,296,943,985,1215, 1269,1300,1516
  pretreatment 1300

Pacific Ocean 1371
Pacific staghorn sculpin (*Leptocottus armatus*) 399
packed cells 286
Palo Alto, CA 1506
pancreas cancer, See cancer
paper industry effluents 23,29
papillomas 199,415,451
parasitism 463
paratyphoid fever 6
Paris 199
Parks, Cynthia A. 587
partitioning 78
Patapsco River 1412
Patel, Rutton D. 1469
pathogens 5,87,100,555
  chlorine resistant 4
Patuxent estuary 1034
Patuxent River 714,1034,1082
Pavlova lutheri 512
PCB, See hydroxylated PCBs; polychlorinated biphenyls
peak-finding algorithm 860
Pearl River 909
peat, See fulvic acid; humic acid
pediveliger larvae 512
Pennsylvania 103,818
pentachlorobutenone 167
pentachlorophenol 307,1029,1395,1523
  oxidation 1029
pentachloropropanone 167,222
pentachloropropene 167,237
pepsin 809
peptides 570,723,797,821
  nitrogen 723
  nitrogen chlorination 797
perchlorate 784,1067
perchloric acid 784
perchloroethylene 145
Pereira, Michael A. 229
perinatal survival 299
Perkinsus marinus 463
permanganate, See potassium permanganate
peroxidase 1034
peroxide 1313
perylene 419
pesticides 821
petroleum ether 1139

Pfohl, Ronald J. 355
pharmacokinetics 281
    dose-response modeling 79
    parameters 290
phenanthrene 419,1375,1395,1520
phenanthroline 1286
phenol 237,282,837,854,939,943,1021,
    1026,1182,1220,1297,1313,1315,
    1320,1325,1395,1513,1517,1523
    oxidation 1026
phenolic acids 26,1314
phenyl propanol 887
phenylacetonitrile 854
phenylalanine 120,238,583,591,619,823
phenylarsine oxide 757,965,1073
phenylmethyl butanol 888
phenylphenol 615
pheromonic communications 30
Philadelphia 415,1300
phloroglucinol 24
phosphate 578,1069
phosphorous, total dissolved 756
photochemical reactivity 1055
photochemistry 1011,1033
    oxidant 1041
    quantum efficiencies 1011
    quantum yield studies 1043
photodecomposition, sunlight-induced
    1041
photodegradation 1055
photolysis 1057
photolyzed groundwaters 1033
photolyzed surface water 1033
photolyzed wastewater 1033
photooxidation 1021,1140
photoprocess 1059
photosynthesis
    formula of Strickland 544
    inhibition 541,546
    phytoplankton 541
phototransformation 1464
phthalates (phthalic acids) 26,868,886,
    887,1066,1069,1376,1378,1524
    See also specific phthalates
phytoplankton communities 541
    photosynthesis 541
pigeons, white Carneau 365
pilot plant 691
pinworms 7
piperidine 175
Pittsboro, NC 1118
plankton counting 523
planktonic bacteria 481
    organisms 505

plant pigments 939
plasma 286
Platon, Meletios 1435
plecoptera 486
Plesimonas shigelloides 14
Plugge, Hans 73
polarograph 942
policy 73
poliomyelitis 6
poliovirus 6,101,108,619,683,692,1271
    inactivation 619,685
pollution indicators 451
polychlorinated acetone 27
polychlorinated biphenyls 191,402,
    1171,1395
    See also hydroxylated PCBs
polychlorinated compounds 1524
polycyclic aromatic hydrocarbons 191,
    275,402,416,1018,1524
polyethoxylate residues 1505
polymer-bound dyes 1022
polyneuropathy 120
polynuclear aromatic hydrocarbons,
    See polycyclic aromatic
    hydrocarbons
polypeptides 807
    See also peptides; macromolecules
population dynamics 557
porphyrins 1021
postchlorination 821,1137
    See also chlorination
postdisinfection 101
    See also disinfection
postnatal exposure 314
potable water 1165,1201
    See also drinking water, tap water
potassium 775
potassium arsenite 1313
potassium chloride 1270
potassium ferrate 1285
potassium ferrocyanide 1313
potassium hydroxide 1285
potassium hypochlorite, dissociation
    775
potassium pemanganate 94,953
    degradation studies 843
potassium persulfate 1042
Potomac River 1118
poultry 30
power plant 1425
    chlorination 541
    chlorination model 1470
    chlorination, *See also* chlorination;
        cooling waters

cooling system 1081
cooling towers 1411
See also electric power plant
preammoniation 556
prechlorination 101,821,1213,1313
See also chlorination
precision 1075,1098
precursors 3,24,161,281,843,850,875, 907,939,1115,1137,1185,1204,1246, 1300,1399
See also trihalomethanes, organic precursors
prenatal exposure 312
preozonation 1215,1229,1251,1300
prestraight hinge larvae 516
prezoosporangia 468
priority pollutants
chlorination 1515
destruction 1515
ozonation 1515
removal 1515
See also EPA; organic priority pollutants
pristene 854
proline 583
promoter 441
propanedioic acid 864
proteins 238,717,723,807,816,821,843, 942,1270
in natural waters 816
synthesis repression 626
protistan parasite 463
prototype instrument 1091
protozoa 7,667,1271
Pseudomonas 563
Pseudomonas aeruginosa 13,557,1271
public health 3,99,145,611,1216
See also health
Puget Sound 399,403
Puijker, L. M. 1137
pulmonary illness 6
pulmonary tumor induction 261
pulp bleaching processes 29
pulse-dose tests 430
Punzi, Vito L. 1469
purge and trap method 1123,1360,1385
See also gas chromatography
purgeable (volatile) organics 1389,1517
purgeable organic halide (POX) 875, 1123,1139
See also volatile
purine 591
pyelonephritis 349
pyrene 419,1395,1520

pyrimidines 939
pyrolysis-gas chromatography/mass spectrometry 1165
pyrophosphate 1286
pyrrole 591,854
pyruvic acid 281,704,942,1254

quality assessment 1099
assurance 1099,1124,1387
control 1099
of life 19
Qualls, Robert G. 723
quantum efficiencies 1011
yield studies 1043
quinone 237
carbonyl groups 1313

Raab, Daniel H. 1073
rabbits, New Zealand white 365
radiometer 1043
radionuclides 91
rainbow trout (Salmo gairdneri) 395, 410,452
rainwater 1033
Raleigh, NC 1118
Ram, Neil M. 587
rapid oxidant demand 1081
See also chlorine demand; oxidant demand
rapid sand filtration 1100,1302
Rapson, W. Howard 237
rate-determining step 704,709
rats 355,395
CD 333
Long-Evans 296
Osborne-Mendel 229
Sprague-Dawley 283,297,308,346, 356
reactions
bromide 999
chains 703
dynamics 701
four-center 706
mechanisms 701,713
nitrate formation 710
orders 701
products 48,944
rate constants 987,1017,1241
rates 1475
reversibility 999
three-center 706
two-center 705
with amino acids 575
with nucleic acids 575

See also chlorination; chlorine;
  monochloramine; ozone
rechlorination 1358
Reckhow, David A. 1229
recovery
  study 1128
  techniques 565,571
recreational waters 12
rectal cancer, *See* cancer
recurrent coliforms 651
red blood cell count 296
Redondo Beach, CA 1357
reduction potentials, bromine systems
    772
  chlorine systems 772
refinery effluents 266
regrowth 667
regulation 39,86
regulatory aspects 91
regulatory decision 126
regulatory options 112
Reid, M. Carrington 1399
Reinhard, Martin 1505
relative risk 146
renal toxicity 113
rennin 809
reoviruses 6
reproducibility of peaks 1104
reproduction 308,312
reproductive effects 115,295
reproductive toxicology 334
residual chlorine 39,48,63,94,452,483,
    713,1202,1231,1461
  decay 66
  effects 485
  limits 40
  See also chlorine; chloramines;
    combined chlorine, oxidant;
    residual oxidant
residual oxidant 48,743,1081
  measurement 1081
  See also chlorine; chlorine-produced
    oxidant; oxidant
resin
  accumulators 1109
  artifacts 1100
  samples 1100
  See also XAD
resistance 615
  to chlorine, physiological basis 558
resorcinol 24,843,942,1209
Resource Conservation and Recovery
  Act 1125

respiration 466
  rates 467
retinal lesions 436
retinoblastomas 436
reverse osmosis 199
reverse-phase HPLC, *See also* HPLC
Revis, N. W. 365
rhabdomyosarcomas 432
Rhine River 188,941,1015,1138,1141
rhodamine dye 1466
Rhodotorula rubra 1271
ribonucleic acids, *See* RNA
ribose 578
Rice, James K. 63
Rice, Rip G. 1215
Richmond, VA 1118
Rickabaugh, Janet 1259
Ringhand, H. P. 161
Rios, James 1381
risk 99,133,145,201,252,295
  assessment 4,73,80,111,201
  calculation 79
  chemical-specific 79
  estimation 80,201
  infectious disease 108
  waterborne disease 107
river sediment 415,717
  extracts 415
  See also sediment
river water 1406
  See also specific rivers
rivulus (*Rivulus marmoratus*) 429
RNA 578,583,620,688,1270
RNA, *See also* nucleic acids,
Roberts, Morris H., Jr. 3,509,541
Roberts, Paul V. 603,783
Robinson, Merrell 221
rock sole (*Lepidopsetta bilineata*) 399
Romania 104
Rome, NY 16
Rose bengal 1015,1028
Rosenblum, Laura 1329
rotary manifold 1449
rotating ring disc electrode 1081
rotaviruses 6,681,683
Rotterdam 1115
Roubal, William T. 399
Rue, William J. 73

S9 192,261,267,1343,1412
  activation, *See also* mutagenicity
  reversion assay 412
  See also liver homogenate
Saccharomyces cerevisiae 1271

Safe Drinking Water Act 91
salicylic acid 1322
saline waters 49,1004
    See also seawater; ocean water
salinity 514
Salmonella 5,1329,1343
Salmonella kanton 693
Salmonella parathyphi 6
Salmonella reversion assay, See Ames test; assay; bioassay
Salmonella Shigella 691
Salmonella typhimurium 189,208,222, 229,246,251,691,1343
    See also Ames test; assay; bioassay; TA98; TA100; TA1538
Salmonella typhosa 6,99,673
Salmonellosis 6
Sammons, Tommy I. 463
sample analysis, reliability 1364
sampling 1078
San Francisco 146
San Francisco Bay 1506
San Joaquin Reservoir 560
sand filtration 821,1139,1168
Sand Shrimp 503
Sanford, NC 1118
Sansone, Francis J. 755,965
sarcomas 432
sarcosine 588,1297
Sayler, Gary S. 1399
scavengers 1046
Schuylkill river 415
Scioto River 909
scopoletin 1034
Scott, Geoffrey I. 463
Scully, Frank E. Jr. 175,807,951
seasonal variation 1036
Seattle 146,402
seawater 22,588,755,923,965
    bromination 737,741
    bromination kinetics 750
    chemistry 755
    chlorination 737,755,1360,1427
    chlorination kinetics 744
    See also saline waters; ocean water
second-order reaction 745
    kinetics 764,1291
    rate constants 790,1287
secondary effluents 691,941
    See also wastewater
Sedgwick-Rafter cell 512
sedimentation 25,101,954,1100,1185, 1302
sediments 404,1357

associated chemicals 404
estuarine 1411
mutagenic potential 1412
    See also river sediment
Seeger, Dennis R. 859
selected-ion monitoring 1116
selectivity value 1069
semivolatile analysis 1386
Sencar mouse skin bioassay 161
sensitivity, species 506
    to disinfection 607,615
sensitizer 1011
Sephadex resin 1188
serine 570,583,591,619
serum chemistry 334
    cholesterol 113,116
serum glucose 121
serum lactic acid 121
serum thyroxine 118,355,366
sessile bacteria 481,505
sewage effluent 1411
    See also wastewater
shear tests 1453
Sheep River, Alberta 481
sheepshead minnow (*Cyprinodon variegatus*) 377,429
Shelford-Allee apparatus 494
shellfish 8,375
Shigella 5,6
Shigella boydii 1271
Shigella dysenteriae 108
Shigella flexneri 108
Shoal Creek Embayment 533
Shy, Carl M. 153
Siff, Edward J. 153
silica gel column chromatography 266
    See also liquid chromatography
silicates 1069
silver carp (*Hypopththalmichthys molitrix*) 452
Silverman, Debra T. 145
simazine 1171
simian rotavirus 683
Singer, Philip C. 1229
singlet oxygen 1011,1021,1033
    measurement 1014
    photochemical pathways for formation 1012
    reaction rate constants 1017
    reaction with substituted dimethylanilines 1017
sister chromatid exchange 120,207,221, 265,337
    See also assay; bioassay

site-specific limits 40
skeletal anomalies 297
skin tumors 116
    induction 261
slime 42,489,1471,1483
    destruction 1471,1483
slurry wastewater 1400
Smith, Maryrose K. 295
Smith, Matthew G. 1299
Snoeyink, Vernon L. 1313
SNORT 587
    See also oxidant analysis
sodium 775
sodium azide 212
sodium bisulfite 548
sodium chloride 1270
sodium chlorite 1042
sodium hydroxide 1286
sodium hypochlorite 116,463,1342,1516
    dissociation 775
    See also chlorine; hypochlorite
sodium metabisulfite 1344
sodium nitrite 1313
sodium sulfite 1202,1313,1346,1372
sodium thiosulfate 78,512,1202,1344
Solomon's Island, MD 1084
Sonenshine, Daniel E. 175
Sonstegard, Ronald A. 265
Sorrell, R. K. 1123
South Gravel Lake 817
South Platte River 816,951
Soxhlet extraction 1116,1412
species selection 80
specific dilution models 1491
spectrophotometry 575,579,1140,1285
Speed, Mark A. 807,951
sperm
    count 303
    drive range 301
    evaluation 295
    morphology 300
    mortality 301
    motility 114
    spermhead abnormalities 113,124,
        207,223
spill event, chloroform 1299
spleen 286
Splichal, David 1285
sponges 48
spongiosis hepatis 391,447
Spot 503
Sprague-Dawley rats 161,176,208,1140
    See also rats
St. Aubin, Jessica J. 939

Staphylococcus aureus 1271
starch-iodine titrimetric method 22
    See also oxidant analysis
steam distillation 163
steatosis/hemosiderosis 399
Stein, John E. 399
steric effect 930
sterigmatocystin 439
Stevens, Alan A. 859
Stevens, Reggie H. 251
Stewart, Mary Elizabeth 521
stomach 286
    contents 169,175
    fluid, amino nitrogen 178
stoneflies 490
Stone River 464
stopped-flow
    experiments 764,801,987,1286
    instrument 764,801
    spectrophotometry 987
Strongloides stercoralis 7
Struba, Robert J. 153
styrene 1101
subcultured bacteria 571
subculturing 557
substitution reactions 706
sucrose 260
Suffet, Irwin H. 1099,1299
sulfhydryl groups 630,688
sulfide 812
sulfur compounds 1021
sulfur dioxide 1341,1347
sulfuric acid 1286
sunlight 757,1011
    diurnal variation 1051
    studies 1043
superchlorination 860,1096
    See also chlorination
surface waters 125,187,272,281,923,
    961,1014,1033
    chlorination 154
    hydrogen peroxide 1033
survival ratio 615,670
survival studies 534
Susquehanna River 1381
Swango, L. J. 1270
swimmers itch 13
swimming pools 1061
    disinfection 1259
Swiss waters 1014
synergistic interaction 235
synthetic demand water 1269
synthetic organic chemicals 91

syringaldazine colorimetric method 575,587
See also oxidant analysis
syringaldehyde 1254

TA100 222,190,208,238,267,452,1138, 1331,1343,1413
TA1538 192
TA98 190,208,267,452,1140,1331,1343, 1413
    See also Ames test; *Salmonella typhimurium*
Tabatabai, Alireza 1285
Tabor, M. Wilson 1329
Taenia saginata 7
Taenia solium 7
taeniasis 7
Tama River 1422
tannic acid 266,273
targeted chlorination 1447
taste and odor 1215,1220
taurine 591
Taylor, Douglas H. 355
temperature effects 485,615
Tenax 1139
Tennessee River 726,1407
Tennessen, Kenneth J. 533
teratogenicity 79,297
terbacil 1016
testes 286
tetrabutylammonium hydroxyde 1066
tetrachloroacetone 237
tetrachlorobiphenyl 270,1395
tetrachlorobutanedioic acid 846
tetrachlorobutenone 167
tetrachlorocyclopropene 167
tetrachlorodihydroxybiphenyl 1317
tetrachlorodimethyoxybiphenyl 1319
tetrachloroethane 200,1127,1362,1390, 1519
tetrachloroethylene 92,237,1123,1127, 1362,1373,1376,1390,1518
tetrachloromethane 188,200
tetrachloropropanone 167,237
tetrachloropropene 237
tetrachlorosulfone 29
tetrachlorothiophene 167
tetradecanoic acid 854,864
tetrahydroxybiphenyls 1322
tetrahydroxybenzene 1322
Tetrahymena pyriformis 667
Texas 88,103,104
theophylline 260
thermal chlorination 1057

thiazines 1021
thiocyanate 232
thiodiethanol 1297
THMs, See trihalomethanes
Thomas, Berta L. 1357
Thompson, Gavin P. 1115
Thornton, CO 807,816,951
threonine 581,823
thymidine 578
thymine 578,591
thyroid 286,346
    iodine uptake 346
    metabolism 117
    studies 347
thyroxine 238
Tidewater Silverside 503
TMP 578
    See also nucleic acids
toluene 854,1101,1390,1518
    sulfonic acid 1067,1069
Torch Lake, MI 399
Torrence, G. 635
total bacterial counts 555,691
total chlorine, See chlorine; residual chlorine; total residual chlorine
total dissolved phosphorus 756
total organic carbon 190,714,756,860, 876,1188,1300,1329
    See also organic carbon; dissolved organic carbon
total organic halogen 26,65,704,859, 876,1115,1123,1165,1358
    analyzer 876
    See also nonvolatile TOX
total residual bromine, disappearance 988
total residual chlorine 509,723,1073, 1447,1489
    See also chlorine; combined chlorine; residual chlorine
total residual oxidant 1372,1430
total trihalomethanes, See THM; trihalomethanes
toxaphene 1395
toxicant-free water 3
toxicity 30,51,63,307,493,498
toxicological effects 112
toxicological endpoints 80
toxicology 115
    aquatic 81
transformation reactions 22,78
transitory species 1042
transport exposure analysis 78

treatment plant, granular-activated-carbon 1299
tribromamine 741,772,987,1006
    decomposition 989
tribromophenol 1364,1376
tributyl phosphate 887
trichloramine 706,1006
trichloroacetaldehyde 166
trichloroacetic acid 25,26,27,113,164, 166,169,843,859,866,946,1115,1121, 1165,1179,1229,1254
    See also chloroacetic acid
trichloroacetone 288,1165,1229,1238, 1254
trichloroacetonitrile 113,264,166,222, 231,1178,1229
    See also haloacetonitrile; chloroacetonitrile
trichlorobenzenes 200,1375,1395,1521
trichlorobutanal 166
trichlorobutanone 167
trichlorocyclopentenedione 167
trichlorodihydroxybiphenol 1317
trichloroethane 92,1123,1127,1305, 1362,1373,1390,1392,1519
trichloroethanoic acid 846
trichloroethylene 92,145,200,237,430, 1123,1127,1362,1373,1390,1518
trichlorofluoromethane 1127,1362, 1366,1519
trichlorohydroxybutanoic acid 853
trichlorohydroxymethylpropanoic acid 853
trichloroisocyanurate 1259
trichloromethane formation 1185
trichloromethyl product precursors 1246
trichloropentenone 237
trichlorophenol 78,13,167,307,332, 1124,1317,1322,1395,1523
    See also chlorophenol
trichloropropanal 166
trichloropropanone 164,167,222
trichloropropenal 166,222
trichloropropenenitrile 166
trichloropropionitrile 27
trichloropyruvate 1250
trichlorosulfone 29
trichoptera 486
trichuriasis 7
Trichuris trichiura 7
tridecene 854
trihalomethanes 23,28,49,79,91,93,105, 111,119,124,138,145,161,190,229, 281,295,345,355,575,635,701,783, 807,843,859,875,895,907,1055,1125, 1127,1137,1157,1165,1178,1185, 1229,1269,1272,1300,1397
    control 25,1194
    formation 24,895,907,911,913,916, 919,923,1042,1049,1216,1259, 1399
    formation potential 911,1217
    iodinated 923
    mechanism for formation 843
    precursors 807,895
    precursors, See also precursors
    regulations 575
    See also bromodiiodomethane; chlorobromoiodomethane; bromoform; bromodichloromethane; chlorodibromomethane; chlorodiiodomethane; chloroform; THM
trihydroxybenzoic acids 835
trihydroxybiphenyl 1322
trihydroxytrimethoxybiphenyl 1325
triiodothyroxine 366
trimethylaldehyde 1297
trimethylphenylindane 887
trimethylstyrene 1016
tropical seawater 755
tryptophan 238,581,591,619,631,812, 823
Tsutsumi, Y. 1165,1201
tuberculosis 5,6
tumor 439
    histogenesis 442
    induction 375,429
    initiators 223
    promotors 113,115,190,262,324
    See also cancer; neoplasms
tumor-inducing substance 261
tumorigenesis 375,443
    dose-response studies 442
    temperature effects 446
tumorigenic response 120
tumorigenicity 190,307
turbidity 1186,1215
Turlersee Lake 1015
Tuskegee, Alabama 452
typhoid fever 6,99
tyrosine 237,581,591,619,630,821

Uhler, Allen D. 1411
ulcer of stomach and duodenum 123
ultrafiltration 1400
ultraviolet, See UV

undecanol 1101
United Kingdom 12,41,664,1341
University Lake 1118
uracil 24,578,1016
uridine 578
uridine monophosphate (UMP) 578
urinalysis 349
urinary excretion 232
Utah 146
UV radiation 89,187,757,985,1140
UV spectrophotometry 976,1065

Valentine, Richard L. 975
valine 583,619,823
van der Gaag, M. A. 1137
van Kreijl, C. F. 187
Vanderbeek, Thomas B. 1489
Varanasi, Usha 399
Varma, M. M. 635
vascular lesions 123
Vasl, R. 619
Veenstra, John N. 1185
vegetative protozoa 667
Verde River 909
Vibrio cholerae 6
Vibrio parahaemolyticus 14
vinyl chloride 92,1390,1517
vinylguaiacol 854
vinylidene chloride 430
viral adsorption 687
Virginia 85,104
virion structural integrity 623
virus 93
    attachment function 623
    chlorine uptake 677
    inactivation 100,619,691
Vogt, Craig D. 91
volatile halocarbons 1359
volatile halogenated organics 48,190, 1123,1299
volatile organic chemicals (VOCs) 1123
volatile organics 1373
volatile organohalides, *See* halogenated organics
volatile synthetic organic chemicals 91, 92
volatiles analysis 1386
volatilization 78,1464
Voudrias, Evangelos A. 1313

Wachter, Jan K. 875
Waite, Thomas D. 1435
Wajon, Johannes Edmund 985
Walker, William W. 429

Warwick River 519
Washington County, Maryland 139
Washington State 12
Washington, DC 1118
wastewater 23,27,73,951,999,1011, 1021,1055,1073,1399,1411,1459,1505
    chlorination 27,667
    disinfection 575,985
    effluent 452,681,691
    oxidation pond 451
    treatment 20,451,1285,1313
    Werdholzli, secondary effluent 1015
    See also secondary effluents; water
wastewater treatment plant 65,1515
    East Shore 521
    Haifa 692
    Hampton Roads Sanitation District 509
    Henrico County 89
    James River 511,541
    Nansemond 509
    Palo Alto Regional Water Quality Control Plant 1506
    Turner Valley 481
    Tuskegee 452
    Werdholzli 1015
    Zunikon 1015
water
    analysis 1385,1387
    disinfectants 1269
    distribution systems 651
    mutagenicity 1341
    quality 1137
    reuse 1399
    sample, carcinogenicity 1341
    standards 68,89,1489
waterborne diseases 92,101,102,112
    transmission 99
    transmission barrier 103
    See also drinking water; cooling water; wastewater
water treatment
    photo oxidation 1021
    polymer-bound dyes 1021
    production of amino acids 821
water treatment plant 96,555,821
    Baxter (Torresdale), Philadelphia 1100
    Bay City, MI 1218,1225
    Belle Glade, FL 1218,1222,1225
    Beria, OH 1218
    Casper, WY 1218
    Celina, OH 1219
    Cholet 1168,1201

Cincinnati 1329
City of Auburn 1269
Columbine 953
Costa Mesa, CA 1218
Dohne 1217
Dunkerque 1201
Dutch Water Works 1137
Elizabeth City, NC 1218
Ephrata Borough, PA 1218
Grandin, ND 1218
Hackensack, NJ 1218
Hopewell, VA 1313
Kennewick, WA 1218
Los Angeles, CA 1218,1225
Monroe, MI 1218,1225
Myrtle Beach, SC 1219
New Ulm, MN 1218,1225
New York, NY 1218
Newport, DE 1218
Newport, RI 1218
Nieuwegein 1138
Oklahoma State University 1185
Ormand Beach, FL 1218
Potomac 1118
Potsdam, NY 1218,1225
Rheinisch-Westfalischen 1217
Rockwood, RN 1218
Rocky Mount, NC 1219
Samuel S. Baxter 1300
Saratoga, WY 1218
South Bay, FL 1218
Stillwater, OK 1218

Strasburg, PA 1218
Tarrytown, NY 1218
Torresdale 1300
Vigneux 1167
Whiting, IN 1218
Weakfish 503
wellwater 1047
White perch 493
whole animal test 252
Wilcox, Philip 1341
Williams, D. E. 1269
Wisconsin 88
Wojtas, A. T. 667
Wolfe, Roy L. 555
Woodward, Herbert N. 553
Workman, H. J. 763
Worley, S. D. 1269

X-rays 260
XAD resin 189,209,266,270,860,942,
   1099,1100,1137,1139,1166,1315,
   1329,1358,1372
   See also resin
xanthenes 1021
xanthine 1016
xylene 1518

yellow bullhead (*Ictalurus natalis*) 453
Yersinia enterocolitica 6,610,615
York River 511

Zika, Rod G. 1041